Helping Teachers
and Students
Succeed Together

WILEY

Principles of

HUMAN ANATOMY

13th Edition

Gerard J. Tortora
Bergen Community College

Mark T. Nielsen
University of Utah

WILEY

PUBLISHER	Kaye Pace	ILLUSTRATION EDITOR	Claudia Volano
ASSOCIATE PUBLISHER	Kevin Witt	SENIOR PHOTO EDITOR	MaryAnn Price
EXECUTIVE EDITOR	Bonnie Roesch	SENIOR PRODUCT DESIGNER	Linda Muriello
ASSOCIATE CONTENT EDITOR	Lauren Elfers	MEDIA SPECIALIST	Svetlana Barskaya
DEVELOPMENTAL EDITOR	Karen Trost	DESIGN DIRECTOR	Harry Nolan
ASST. CONTENT EDITOR	Brittany Cheetham	SENIOR DESIGNER/	
MARKETING MANAGER	Maria Guarascio	COVER DESIGNER	Madelyn Lesure
PRODUCTION MANAGER	Juanita Thompson	TEXT DESIGNER	Brian Salisbury
PRODUCTION EDITOR	Barbara Russiello	COVER PHOTO	Mark Nielsen
EDITORIAL ASSISTANT	Grace Bagley		

This book was set in 10/12 Janson Text LT Std by Aptara®, Inc. Printed and bound by RR Donnelley/Von Hoffmann. This book is printed on acid-free paper. ∞

Founded in 1807, John Wiley & Sons, Inc. has been a valued source of knowledge and understanding for more than 200 years, helping people around the world meet their needs and fulfill their aspirations. Our company is built on a foundation of principles that include responsibility to the communities we serve and where we live and work. In 2008, we launched a Corporate Citizenship Initiative, a global effort to address the environmental, social, economic, and ethical challenges we face in our business. Among the issues we are addressing are carbon impact, paper specifications and procurement, ethical conduct within our business and among our vendors, and community and charitable support. For more information, please visit our website: www.wiley.com/go/citizenship.

ISBN-13   978-1-118-34499-6
BRV ISBN 978-1-118-34441-5

Printed in the United States of America

10 9 8 7 6

ABOUT THE AUTHORS

Jerry Tortora is Professor of Biology and former Biology Coordinator at Bergen Community College in Paramus, New Jersey, where he teaches human anatomy and physiology as well as microbiology. He received his bachelor's degree in biology from Fairleigh Dickinson University and his master's degree in science education from Montclair State College. He is a member of many professional organizations, including the Human Anatomy and Physiology Society (HAPS), the American Society of Microbiology (ASM), American Association for the Advancement of Science (AAAS), National Education Association (NEA), and the Metropolitan Association of College and University Biologists (MACUB).

Above all, Jerry is devoted to his students and their aspirations. In recognition of this commitment, Jerry was the recipient of MACUB's 1992 President's Memorial Award. In 1996, he received a National Institute for Staff and Organizational Development (NISOD) excellence award from the University of Texas and was selected to represent Bergen Community College in a campaign to increase awareness of the contributions of community colleges to higher education.

Jerry is the author of several best-selling science textbooks and laboratory manuals, a calling that often requires an additional 40 hours per week beyond his teaching responsibilities. Nevertheless, he still makes time for four or five weekly aerobic workouts that include biking and running. He also enjoys attending college basketball and professional hockey games and performances at the Metropolitan Opera House.

To Reverend Dr. James F. Tortora, my brother, my friend, and my role model.
His life of dedication has inspired me in so many ways, both personally and professionally,
and I honor him and pay tribute to him with this dedication. **G.J.T.**

Mark Nielsen is a Professor in the Department of Biology at the University of Utah. For the past twenty-eight years he has taught anatomy, neuroanatomy, embryology, human dissection, comparative anatomy, and an anatomy teaching course to over 23,000 students. He developed the anatomy course for the physician assistant program at the University of Utah School of Medicine, where he taught for five years, and taught in the cadaver lab at the University of Utah School of Medicine. He developed the anatomy and physiology program for the Utah College of Massage Therapy, and his course materials are used by massage schools throughout the country. His graduate training is in comparative anatomy, and his anatomy expertise has a strong basis in dissection. He has prepared and participated in hundreds of dissections of both humans and other vertebrate animals. All his courses incorporate a cadaver-based component to the training with an outstanding exposure to cadaver anatomy. He is a member of the American Association of Anatomists (AAA), the Human Anatomy and Physiology Society (HAPS), and the Anatomical Society of Great Britain and Ireland (ASGBI).

Mark has a passion for teaching anatomy and sharing his knowledge with his students. In addition to the many students to whom he has taught anatomy, he has trained and served as a mentor for over 1,000 students who have worked in his anatomy laboratory as teaching assistants. His concern for students and his teaching excellence have been acknowledged through numerous awards. He received the prestigious Presidential Teaching Scholar Award at the University of Utah for excellence in teaching and was an initial recipient of the Beacons of Excellence Award for developing exceptional programs for student mentoring. He is a five-time recipient of the University of Utah Student Choice Award for Outstanding Teacher and Mentor, a two-time winner of the Outstanding Teacher in the Physician Assistant Program, recipient of the American Massage Therapy Association Jerome Perlinski Teacher of the Year Award, and a two-time recipient of Who's Who Among America's Teachers.

He enjoys sports, photography, good food, traveling, and exploring with his lovely wife and playing with his grandchildren.

To my father and mother, the best mentors a son could ever have.
Thank you for your neverending support and love, and for teaching me the value of hard work. **M.T.N.**

PREFACE

Principles of Human Anatomy, thirteenth edition, is designed for introductory courses in human anatomy. The highly successful approach of previous editions—to provide students with an accurate, clearly written, and expertly illustrated presentation of the structure of the human body, to offer insights into the connections between structure and function, and to explore the practical and relevant applications of anatomical knowledge to everyday life and career development—has been enhanced in this edition by innovations designed to increase student motivation and success.

An anatomy course can be the gateway to a satisfying career in a host of health-related professions. It can also be incredibly challenging. We have designed the organization and flow of content within these pages based on our deep experience teaching anatomy and interacting with students over many years. We are cognizant of the fact that the teaching and learning environment has changed significantly to rely more heavily on the ability to access the rich content in this printed text in a variety of digital ways, anytime and anywhere. We are pleased that this 13th edition meets these changing standards and offers dynamic and engaging choices to make your experience in this course more rewarding and fruitful.

New for This Edition

The thirteenth edition of **Principles of Human Anatomy** has been updated throughout, paying careful attention to include the most current terminology in use and including an enhanced glossary. The design has been refreshed to ensure that the content is clearly presented and easy to access. Clinical Connections that help students understand the relevance of anatomical structures and the functions they support by considering what happens when they don't work the way they should have been updated throughout and in some cases are now placed alongside related illustrations to strengthen these connections for students. The all-important illustrations that support this most visual of sciences have been scrutinized and revised as needed throughout. Nearly every chapter of the text has either a new or revised illustration or photograph.

We are most excited about the enhanced digital experience now available with the thirteenth edition of this text. **WileyPLUS** now includes a powerful new adaptive learning component called **ORION**. **WileyPLUS** with **ORION** allows students to take charge of their study time in ways they have not previously experienced and prepares them for more meaningful classroom and laboratory interactions. **Real Anatomy**, so popular with professors and students alike for its outstanding photography and deep possibilities for exploration of a human cadaver, has been updated and is now fully web-based and integrated into *WileyPLUS*. **WileyPLUS** itself has been refreshed with a new design that allows easier discoverability and access to the rich resources included, in addition to updated assessment questions throughout. New for the thirteenth edition is a digital alternative called **All Access Pack** for **Principles of Human Anatomy**, 13e. This choice offers you a full etext to download and to keep, full access to **WileyPLUS** with **ORION**, and a Study Resource Guide to use as a basis for taking notes in class and studying later. It provides you with everything you need for your course, anytime, anywhere, on any device.

TAKE A TOUR

The challenges of learning anatomy can be complex and time consuming. This textbook and *WileyPLUS* with **ORION** have been specifically designed to maximize your time studying by simplifying the choices you make in deciding what to study, how to study it, and in assessing your understanding of the content along the way.

Chapter Beginnings and Ends

Each chapter is effectively bookended with stunning chapter introductions designed to grab students' interest and engage them in the topic at hand, and chapter summaries that not only highlight the important concepts of the chapter but point students to the media resources that will support greater understanding of those concepts.

23 THE RESPIRATORY SYSTEM

INTRODUCTION Have you ever swallowed something and had it go down the wrong tube, making you cough or choke uncontrollably? This uncomfortable (and sometimes embarrassing) situation occurs because the respiratory and digestive systems both arise from the embryonic gut tube and share the nose, mouth, and throat as a common initial pathway. While the majority of the gut tube gives rise to the digestive system (Chapter 24), the tube that will become the respiratory system forms a highly branched network of respiratory airways that terminate in the lungs. The respiratory tubes have the basic design fea-

...sounds, and rids the body of small amounts of water and heat in exhaled air.

The branch of medicine that deals with the diagnosis and treatment of disease of the ears, nose, and throat (ENT) is called **otorhinolaryngology** (o′-tō-rī-nō-lar-in-GOL-ō-jē; *oto-*=ear; *rhino-*=nose; *laryngo-*=voice box, *-logy*=study of). A **pulmonologist** (pul-mō-NOL-ō-gist; *pulmo-*=lung) is a specialist in the diagnosis and treatment of diseases of the lungs. •

Did you ever wonder how smoking affects the respiratory system?

CHAPTER REVIEW AND RESOURCE SUMMARY

WileyPLUS

Review

23.1 Respiratory System Anatomy

1. The respiratory system consists of the nose, pharynx, larynx, trachea, bronchi, and lungs. The respiratory system acts with the cardiovascular system to supply oxygen (O_2) to and remove carbon dioxide (CO_2) from the blood.
2. The external nose is made of cartilage and skin and is lined with a mucous membrane. Openings to the exterior are the external nares. The internal nose communicates with the paranasal sinuses and nasopharynx through the internal nares. The nasal cavity is divided by a nasal septum. The anterior portion of the cavity is called the nasal vestibule. The nose warms, moistens, and filters air and functions in olfaction and speech.
3. The pharynx (throat) is a muscular tube lined by a mucous membrane. The anatomical regions of the pharynx are the nasopharynx, oropharynx, and laryngopharynx. The nasopharynx functions in respiration. The oropharynx and laryngopharynx function both in respiration and digestion.
4. The larynx (voice box) is a passageway that connects the pharynx with the trachea. It contains the thyroid cartilage (Adam's apple); the epiglottis, which prevents food from entering the larynx; the cricoid cartilage, which connects the larynx and trachea; and the paired arytenoid, corniculate, and cuneiform cartilages. The vocal folds of the larynx produce sound as they vibrate; taut folds produce high pitches, and relaxed ones produce low pitches.
5. The trachea (windpipe) extends from the larynx to the main bronchi. It is composed of C-shaped rings of cartilage and smooth muscle and is lined with pseudostratified ciliated columnar epithelium.
6. The bronchial tree consists of the trachea, main bronchi, lobar bronchi, segmental bronchi, bronchioles, and terminal bronchioles. Walls of bronchi contain rings of cartilage; walls of bronchioles contain increasingly smaller plates of cartilage and increasing amounts of smooth muscle.
7. Lungs are paired organs in the thoracic cavity enclosed by the pleural membrane. The parietal pleura is the superficial layer that lines the thoracic cavity; the visceral pleura is the deep layer that covers the lungs. The right lung has three lobes separated by two fissures; the left lung has two lobes separated by one fissure, along with a depression, the cardiac notch.

Resource

Anatomy Overview - Overview of Respiratory Organs
Anatomy Overview - Nasal Cavity and Pharynx
Anatomy Overview - Respiratory Tissues
Anatomy Overview - Respiratory Epithelium
Figure 23.3 - Larynx
Figure 23.7 - Relationship of the Pleural Membrane to the Lung
Figure 23.11 - Structure of an Alveolus
Exercise - Build an Airway
Exercise - Concentrate on Respiratory Structures
Concepts and Connections - Functional Anatomy of the Respiratory System

727

Anatomy Is a Visual Science

Studying the figures in this book is as important as reading the narrative. The tools described here will help you understand the concepts being presented in any figure and assure you get the most out of the visuals.

❶ LEGEND Read this first. It explains what the figure is about.

❷ KEY CONCEPT STATEMENT Indicated by a "key" icon, this reveals a basic idea portrayed in the figure.

❸ ORIENTATION DIAGRAM Added to many figures, this small diagram helps you understand the perspective from which you are viewing a particular piece of anatomical art.

❹ FIGURE QUESTIONS Found at the bottom of each figure and accompanied by a "question mark" icon, these serve as a self-check to help you understand the material as you go along.

❺ FUNCTIONS BOXES Included with selected figures, these provide brief summaries of the functions of the anatomical structure or system depicted.

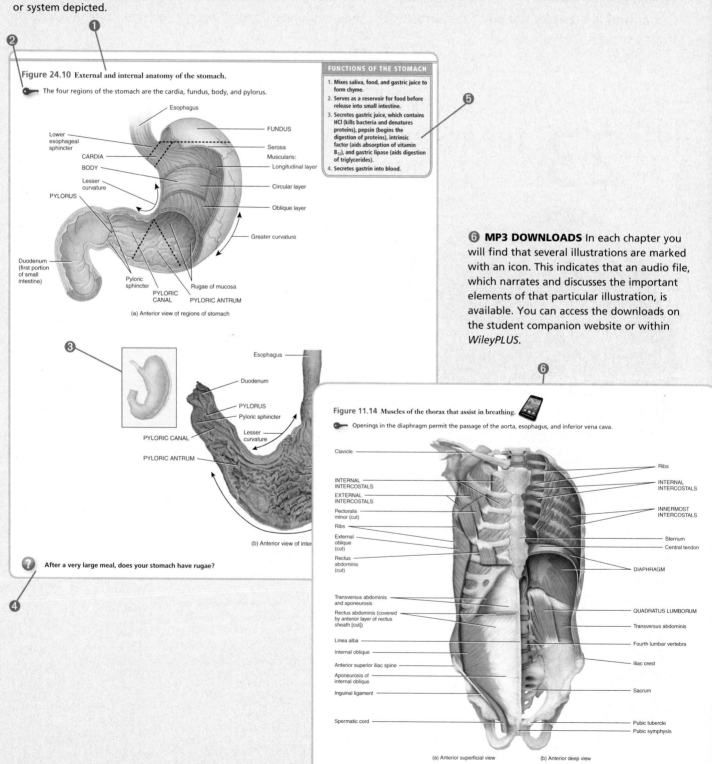

Figure 24.10 External and internal anatomy of the stomach.

The four regions of the stomach are the cardia, fundus, body, and pylorus.

FUNCTIONS OF THE STOMACH

1. Mixes saliva, food, and gastric juice to form chyme.
2. Serves as a reservoir for food before release into small intestine.
3. Secretes gastric juice, which contains HCl (kills bacteria and denatures proteins), pepsin (begins the digestion of proteins), intrinsic factor (aids absorption of vitamin B$_{12}$), and gastric lipase (aids digestion of triglycerides).
4. Secretes gastrin into blood.

(a) Anterior view of regions of stomach

(b) Anterior view of inter

After a very large meal, does your stomach have rugae?

❻ MP3 DOWNLOADS In each chapter you will find that several illustrations are marked with an icon. This indicates that an audio file, which narrates and discusses the important elements of that particular illustration, is available. You can access the downloads on the student companion website or within *WileyPLUS*.

Figure 11.14 Muscles of the thorax that assist in breathing.

Openings in the diaphragm permit the passage of the aorta, esophagus, and inferior vena cava.

(a) Anterior superficial view   (b) Anterior deep view

vi

Clinical Connections

It is easier to understand the relevance of anatomical structures and the functions they support by considering what happens when they don't work the way they should. The Clinical Connections, which appear throughout the text, present a variety of clinical perspectives related to the text discussion.

CLINICAL CONNECTION | Rheumatism and Arthritis

Rheumatism (ROO-ma-tizm) is any painful disorder of the supporting structures of the body—bones, ligaments, tendons, or muscles—that is not caused by infection or injury. **Arthritis** is a form of rheumatism in which the joints are swollen, stiff, and painful. It afflicts about 45 million people in the United States, and is the leading cause of physical disability among adults over age 65. **Osteoarthritis (OA)** (os'-tē-ō-ar-THRĪ-tis) is a degenerative joint disease in which joint cartilage is gradually lost. It results from a combination of aging, obesity, irritation of the joints, muscle weakness, and wear and abrasion. Commonly known as "wear-and-tear" arthritis, osteoarthritis is the most common type of arthritis and the most common reason for hip- and knee-replacement surgery. In **gout**, a form of arthritis, sodium urate crystals are deposited in the soft tissues of the joints. Gouty arthritis most often affects the joints of the feet, especially at the base of the big toe. The crystals irritate and erode the cartilage, causing inflammation, swelling, and acute pain.

Rheumatoid arthritis (RA) is an autoimmune disease in which the immune system of the body attacks its own tissues—in this case, its own cartilage and joint linings. RA is characterized by inflammation of the joint, which causes swelling, pain, and loss of function (Figures A and B). As noted above, this form of arthritis usually occurs bilaterally: If one wrist is affected, the other is also likely to be affected, although they are often not affected to the same degree.

The primary symptom of RA is inflammation of the synovial membrane. If untreated, the membrane thickens, and synovial fluid accumulates. The resulting pressure causes pain and tenderness. The membrane then produces an abnormal granulation tissue, called *pannus*, that adheres to the surface of the articular cartilage and sometimes erodes the cartilage completely. When the cartilage is destroyed, fibrous tissue joins the exposed bone ends. The fibrous tissue ossifies and fuses the joint so that it becomes immovable—the ultimate crippling effect of rheumatoid arthritis. The growth of the granulation tissue causes the distortion of the fingers that characterizes hands of RA sufferers. •

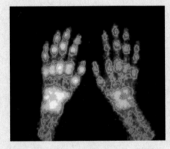

(A) Gamma ray photograph of swollen joints (bright spots) due to RA

(B) Photograph of an individual with severe RA

WileyPLUS offers you opportunities to take Clinical Connections even further with animated and interactive case studies that relate specifically to one body system or another. Look for these under additional chapter resources as an interesting and engaging break from traditional study routines.

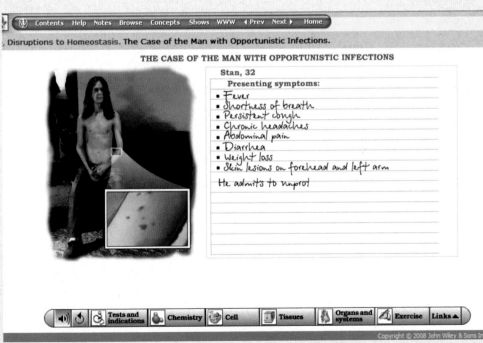

Exhibits Organize Complex Anatomy into Manageable Modules

Many topics in this text have been organized to bring together all the anatomical information into a simple-to-navigate content module. You will find Exhibits for bones, joints, skeletal muscles, nerves, blood vessels, and surface anatomy.

❶ Objective to focus your study

❷ Overview narrative of structure(s)

❸ Table summarizing key features of structure(s)

❹ Illustrations and photographs

❺ Checkpoint question assesses your understanding

❻ Clinical Connection provides relevance for learning the details

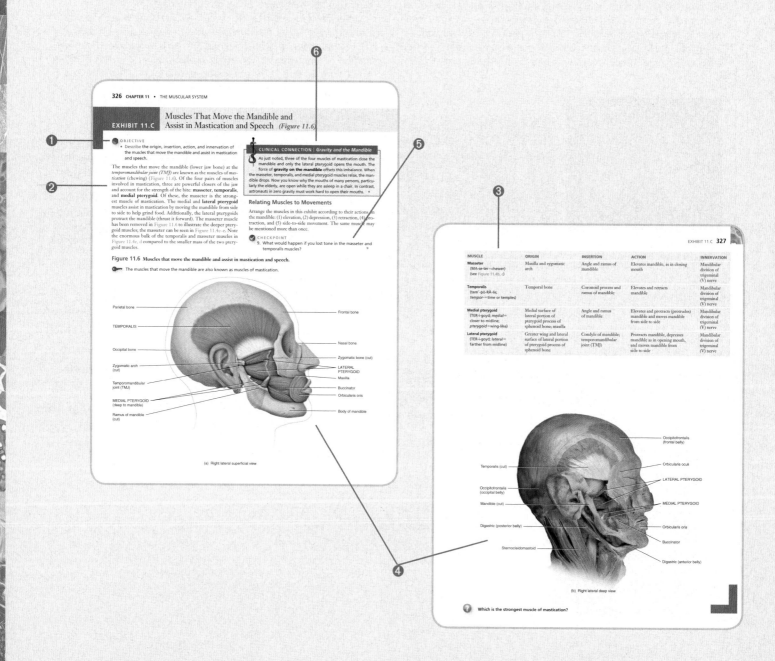

Chapter Resources Help You Focus and Review

Your book has a variety of special features that will make your time studying anatomy a more rewarding experience. These have been developed based on feedback from students—like you—who have used previous editions of the text. Their effectiveness is even further enhanced within *WileyPLUS*.

Objectives at the start of each section help you focus on what is important as you read. All of the content within *WileyPLUS* with ORION is tagged to these specific learning objectives so that you can organize your study or review what is still not clear in simple, more meaningful ways.

Checkpoint questions at the end of each section help you assess if you have absorbed what you have read.

Mnemonics are a memory aid that can be particularly helpful when learning specific anatomical features. Mnemonics are included throughout the text, some displayed in figures, tables, or Exhibits, and some included within the text discussion. We encourage you not only to use the mnemonics provided but also to create your own to help you learn the multitude of terms involved in your study of human anatomy.

Key Medical Terms at the end of chapters include selected terms dealing with both normal and pathological conditions.

Critical Thinking Questions are word problems that allow you to apply the concepts you have studied in the chapter to specific situations.

Mastering the Language of Anatomy

The terminology in this edition is based on *Terminologia Anatomica*. Throughout the text we have included **Pronunciations** and, sometimes, **Word Roots** for many terms that may be new to you. These appear in parentheses immediately following the new words, and the pronunciations are repeated in the glossary at the back of the book. Look at the words carefully and say them out loud several times. Learning to pronounce a new word will help you remember it and make it a useful part of your medical vocabulary. Take a few minutes to read the pronunciation key, found at the beginning of the

Glossary at the end of this text (page G-1), so it will be familiar as you encounter new words.

To provide more assistance in learning the language of anatomy, a full **Glossary** of terms with phonetic pronunciations appears at the end of the book. The basic building blocks of medical terminology—**Combining Forms, Word Roots, Prefixes, and Suffixes**—are listed inside the back cover, as is a listing of **Eponyms**, traditional terms that include reference to a person's name, along with the current terminology.

WileyPLUS *with* ORION

WileyPLUS with ORION helps students learn by learning about them.

ORION is a new addition to *WileyPLUS* that provides students with a personal, adaptive learning experience to help them build their proficiency on topics and use study time most efficiently.

WileyPLUS with ORION is great as:

- an adaptive **pre-lecture tool** that assesses your students' conceptual knowledge so they come to class better prepared,
- a **personalized study guide** that helps students understand both strengths and area where they need to invest more time, especially in preparation for quizzes and exams.

BEGIN

Unique to ORION, students **begin** by taking a quick **diagnostic** for any chapter. This will determine each student's baseline proficiency on each topic in the chapter. Students see their individual diagnostic report to help them decide what to do next with the help of ORION's recommendations.

PRACTICE

For each topic, students can either Study or Practice. **Study** directs the student to the specific topic they choose in *WileyPLUS*, where they can read from the e-textbook or use the variety of relevant resources available there. Students can also **practice**, using questions and feedback powered by ORION's adaptive learning engine. Based on the results of their diagnostic and ongoing practice, ORION will present students with questions appropriate for their current level of understanding and will continuously adapt to each student, helping them build their proficiency.

MAINTAIN

ORION includes a number of reports and ongoing recommendations for students to help them **maintain** their proficiency over time for each topic. Students can easily access ORION from multiple places within *WileyPLUS*. It does not require any additional registration, and there will not be any additional charge for students using this adaptive learning system.

ADDITIONAL RESOURCES

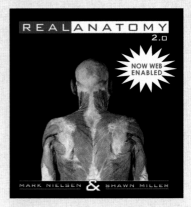

Real Anatomy 2.0
Mark Nielsen and Shawn Miller, University of Utah

Real Anatomy is 3-D imaging software that allows you to dissect through multiple layers of a three-dimensional real human body to study and learn the anatomical structures of all body systems.

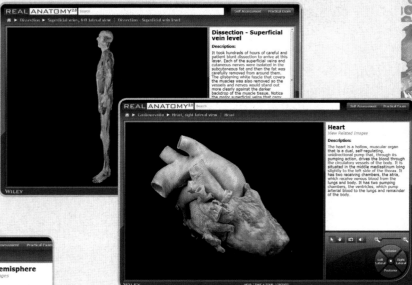

NEW to Real Anatomy 2.0

• Now available on the Web, accessible by iPad and Android tablets.

• All possible highlight structures on an image are now accessible via a drop-down list as well as being searchable.

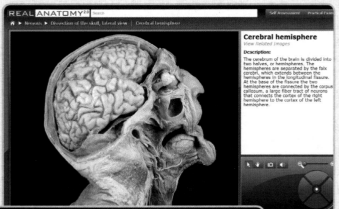

• New crumb trail navigation shows context of system, image, structure.

• Fully integrated into *WileyPLUS* with ORION.

• Dissect through up to 40 layers of the body and discover the relationships of the structures to the whole.

• Rotate the body as well as major organs to view the image from multiple perspectives.

• Use a built-in zoom feature to get a closer look at detail.

• A unique approach to highlighting and labeling structures does not obscure the real anatomy on view.

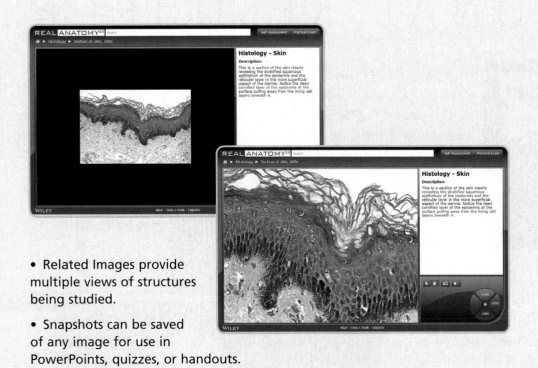

- Snapshots can be saved of any image for use in PowerPoints, quizzes, or hand-outs

- Audio pronunciation of all labeled structures is readily available

- Related Images provide multiple views of structures being studied.

- Snapshots can be saved of any image for use in PowerPoints, quizzes, or handouts.

- View histology micrographs at varied levels of magnification with the virtual microscope.

- Audio pronunciation of all labeled structures is readily available.

REAL ANATOMY

Photographic Atlas of Human Anatomy, First Edition

Mark Nielsen and Shawn Miller, University of Utah

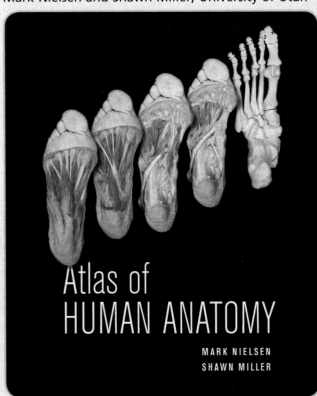

This beautiful atlas filled with outstanding photographs of meticulously executed dissections of the human body is a strong teaching and learning solution, not just a catalog of photographs. Organized around body systems, each chapter includes a narrative overview of the body system followed by detailed photographs that accurately and realistically represent the anatomical structures. Histology is included. *Photographic Atlas of Human Anatomy* will work well in your laboratories, as a study companion to your textbook, and as a print companion to the *Real Anatomy* 2.0.

ACKNOWLEDGMENTS

We wish to thank especially several academic colleagues for their helpful contributions to this edition. We are very grateful to our colleagues who have reviewed the manuscript, participated in focus groups and meetings, or offered suggestions for improvement. Most importantly, we thank those who have contributed to the creation and integration of this text and media, particularly *WileyPLUS* with ORION. The improvements and enhancements for this edition are possible in large part because of the expertise and input of the following group of people:

Kathleen Anderson, *University of Iowa*

Frank Baker, *Golden West Community College*

Celina Bellanceau, *University of South Florida*

Evelyn Biluk, *Lake Superior College*

Lois Borek, *Georgia State University*

Betsy Brantley, *Valencia College*

Stephen Burnett, *Clayton State University*

Jeanne D'Brant, *SUNY Farmingdale*

Kash Dutta, *University of New England*

Heather Dy, *Long Beach Community College*

Linda Flores, *Delaware Community College*

Lynn Gargan, *Tarrant County Community College*

Harold Grau, *Christopher Newport University*

Wanda Hargroder, *Louisiana State University*

Jane Horlings, *Saddleback College*

Cynthia Kincer, *Wytheville Community College*

Thomas Lancraft, *St. Petersburg College*

Shawn Miller, *University of Utah*

Erin Morrey, *Georgia Perimeter College*

Gloria Nusse, *San Francisco State University*

Izak Paul, *Mount Royal University*

Julie Porterfield, *Tulsa Community College*

Benjamin Predmore, *University of South Florida*

Amanda Rosenzweig, *Delgado Community College*

Jay Zimmer, *South Florida Community College*

Finally, our hats are off to everyone at Wiley. We enjoy collaborating with this enthusiastic, dedicated, and talented team of publishing professionals. Our thanks to the entire team: Bonnie Roesch, Executive Editor; Karen Trost, Developmental Editor; Lauren Elfers, Associate Content Editor; Grace Bagley, Editorial Assistant; MaryAnn Price, Senior Photo Editor; Claudia Volano, Illustration Editor; Madelyn Lesure, Senior Designer; Linda Muriello, Senior Product Designer; Maria Guarascio, Marketing Manager; Barbara Russiello, Production Editor; Shelley Flannery, copy editor; Joe Ford, proofreader; and WordCo Indexing Services.

GERARD J. TORTORA

Department of Science and Health, S229
Bergen Community College
400 Paramus Road
Paramus, NJ 07675
gjtauthor01@optonline.com

MARK NIELSEN

Department of Biology
University of Utah
257 South 1400 East
Salt Lake City, UT 84112
marknielsen@biology.utah.edu

BRIEF CONTENTS

CONTENTS

9 JOINTS 248

10 MUSCULAR TISSUE 284

11 THE MUSCULAR SYSTEM 310

12 THE CARDIOVASCULAR SYSTEM: BLOOD 414

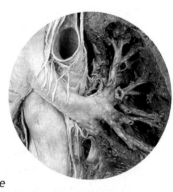

CLINICAL CONNECTIONS

1 AN INTRODUCTION TO THE HUMAN BODY

INTRODUCTION You are about to begin a study of the human body to learn how it is organized and how it functions. In order to understand what happens when the body is injured, diseased, or placed under stress, you must know how it is put together and how its different parts work. Just as an auto mechanic must be familiar with the details of the structure and function of a car, health-care professionals and others who work in human performance and care professions must have intimate knowledge of the structures and functions of the human body. This knowledge can be one of your most effective tools. Much of what you study in this chapter will help you understand how anatomists visualize the body, and the basic anatomical vocabulary presented here will help you describe the body in a language common to both scientists and professionals. •

 Did you ever wonder why an autopsy is performed?

1.1 ANATOMY DEFINED

 OBJECTIVE

• Define anatomy and physiology, and name several branches of anatomy.

Anatomy (a-NAT-ō-mē; *ana-*=up; *=-tomy*=process of cutting) is primarily the study of *structure* and the relationships among structures. It was first studied by **dissection** (dis-SEK-shun; *dis-*= apart; *-section*=act of cutting), the careful cutting apart of body structures to study their relationships. Today, a variety of imaging techniques also contribute to the advancement of anatomical knowledge. We will describe and compare some common imaging techniques in Table 1.3, which appears later in this chapter (see Section 1.8). The anatomy of the human body can be studied at various levels of structural organization, ranging from microscopic (visible only with the aid of a microscope) to macroscopic (visible without the use of a microscope). These levels and the different methods used to study them provide the basis for the branches of anatomy, several of which are described in Table 1.1.

Anatomy deals mostly with structures of the body. A related discipline, **physiology** (fiz′-e-OL-o-je; *physio-*=nature; *-logy*= study of), deals with *functions* of body parts—that is, how they work. Because function cannot be separated completely from

structure, you will learn how the structure of the body often reflects its functions. Some of the structure–function relationships are visibly obvious, such as the tight connections between the bones of the skull, which protect the brain. In contrast, the bones of the fingers are more loosely joined to permit movements such as playing an instrument, grasping a baseball bat, or retrieving a small object from the floor. The shape of the external ear assists in the collection and localization of sound waves, which facilitates hearing. Other relationships are not as visibly obvious; for example, the passageways that carry air into the lungs branch extensively when they reach the lungs. Tiny air sacs—about 300 million—cluster at the ends of the large number of airway branches. Similarly, the vessels carrying blood into the lungs branch extensively to form tiny tubes that surround the small air sacs. Because of these anatomical features, the total surface area within the lungs is about the size of a handball court. This large surface area is the key to the primary function of the lungs: the efficient exchange of oxygen and carbon dioxide between the air and the blood.

 CHECKPOINT

1. Which branches of anatomy would be used when dissecting a cadaver?
2. Give several examples of connections between structure and function in the human body.

TABLE 1.1

Selected Branches of Anatomy

BRANCH	STUDY OF
Embryology (em′-brē-OL-ō-jē; *embry-*=embryo; *-logy*=study of)	In humans, the first eight weeks of development after fertilization of the egg
Developmental biology	The complete developmental history of an individual from fertilization to death
Cell biology	Cellular structure and function
Histology (his′-TOL-ō-jē-; *hist-*=tissue)	Microscopic structure of tissues
Sectional anatomy	Internal structure and relationships of the body through the use of sections
Gross anatomy	Structures that can be examined without using a microscope
Systemic anatomy	Structure of specific systems of the body such as the nervous or respiratory systems
Regional anatomy	Specific regions of the body such as the head or chest
Surface anatomy	Surface markings of the body to understand the relationships of deep or internal anatomy through visualization and palpation (gentle touch)
Imaging anatomy	Body structures that can be visualized with x-rays, CT scans, MRI, and so on
Pathological anatomy (path′-ō-LOJ-i-kal; *path-*=disease)	Structural changes (from gross to microscopic) associated with disease

CLINICAL CONNECTION | *Noninvasive Diagnostic Techniques*

Several noninvasive diagnostic techniques are commonly used by health-care professionals and students to assess certain aspects of body structure and function. A **noninvasive diagnostic technique** is one that does not involve insertion of an instrument or device through the skin or into a body opening. In **inspection**, the first noninvasive diagnostic technique, the examiner observes the body for any changes that deviate from normal (Figure A). For example, a physician may examine the mouth cavity for evidence of disease. In **palpation** (pal-PĀ-shun; *palpa-*=to touch) the examiner feels body surfaces with the hands (Figure B). An example is palpating the neck to detect enlarged or tender lymph nodes. In **auscultation** (aus'-cul-TĀ-shun; *ausculta-*=to listen to) the examiner listens to body sounds to evaluate the functioning of certain organs, often using a stethoscope

to amplify the sounds (Figure C). An example is auscultation of the lungs during breathing to check for crackling sounds associated with abnormal fluid accumulation in the air spaces of the lungs. In **percussion** (pur-KUSH-un; *percus-*=to beat) the examiner taps on the body surface with the fingertips and listens to the resulting sound. Hollow cavities or spaces produce a different sound than solid organs do (Figure D). For example, percussion may reveal the abnormal presence of fluid in the lungs or air in the intestines. It is also used to reveal the size, consistency, and position of an underlying structure. An understanding of anatomy is important for the effective application of most of these techniques. Also, clinicians use these terms and others covered in this chapter to annotate their findings following a clinical examination. •

(A) Inspection of oral (mouth) cavity

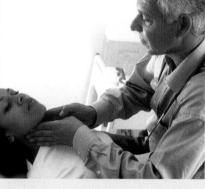
(B) Palpation of lymph nodes in neck

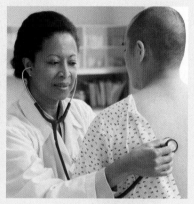
(C) Auscultation of lungs

(D) Percussion of lungs

1.2 LEVELS OF BODY ORGANIZATION AND BODY SYSTEMS

● OBJECTIVES

• Describe the levels of structural organization that make up the human body.
• Outline the 11 systems of the human body, list the organs present in each, and explain their general functions.

The levels of organization of a language—letters of the alphabet, words, sentences, paragraphs, and so on—can be compared to the levels of organization of the human body. Your exploration of the human body will extend from some of the smallest body structures and their functions to the largest structure—an entire person. Organized from smallest to largest, six levels of organization will help you to understand anatomy: the chemical, cellular, tissue, organ, system, and organismal levels of organization (Figure 1.1).

❶ The **chemical level**, which can be compared to the *letters of the alphabet*, includes **atoms**, the smallest units of matter that participate in chemical reactions, and **molecules**, two or more atoms joined together. Certain atoms, such as carbon (C), hydrogen (H), oxygen (O), nitrogen (N), phosphorus (P), and calcium (Ca), are essential for life. Two familiar molecules found in the body are deoxyribonucleic acid (DNA), the genetic material passed from one generation to the next, and glucose, commonly known as blood sugar.

❷ At the **cellular level**, molecules combine to form cells, which can be compared to assembling letters into words. **Cells** are structures composed of chemicals and are the basic structural and functional units of an organism. Just as *words* are the

smallest building blocks of language, cells are the smallest living units in the human body. Among the many kinds of cells in your body are muscle cells, nerve cells, and blood cells. Figure 1.1 shows a smooth muscle cell, one of three types of muscle cells in the body. The cellular level of organization is the focus of Chapter 2.

❸ The next level of structural organization is the **tissue level. Tissues** are groups of cells and the materials surrounding them that work together to perform a particular function, similar to the way words are put together to form *sentences*. There are just four basic types of tissue in your body: epithelial tissue, connective tissue, muscular tissue, and nervous tissue. *Epithelial tissue* covers body surfaces, lines hollow organs and cavities, and forms glands. *Connective tissue* connects, supports, and protects body organs while distributing blood vessels to other tissues. *Muscular tissue* contracts (shortens) to make body parts move and generates heat. *Nervous tissue* carries information from one part of the body to another. Chapter 3 describes the tissue level of organization in greater detail. Shown in Figure 1.1 is smooth muscle tissue, which consists of tightly packed smooth muscle cells.

❹ At the **organ level,** different types of tissues are joined together. Similar to the relationship between sentences and *paragraphs*, **organs** are structures that are composed of two or more different types of tissues; they have specific functions and usually have recognizable shapes. Examples of organs are the stomach, heart, liver, lungs, and brain. Figure 1.1 shows how several tissues make up the stomach. The stomach's outer covering is a layer of epithelial and connective tissues that reduces friction when the stomach moves and rubs against other organs. Underneath these layers is a type of muscular tissue called *smooth muscle tissue*, which contracts to churn and mix food and push it on to the next digestive organ, the small intestine. The innermost lining, the epithelial tissue layer, produces fluid and chemicals responsible for digestion in the stomach.

⑤ The next level of structural organization in the body is the **system level,** also called the **organ-system level. A system** (or *chapter* in our language analogy) consists of related organs (paragraphs) with a common function. An example is the digestive system, which breaks down and absorbs food. Its organs include the mouth, salivary glands, pharynx (throat), esophagus (tube that carries food from the throat to the stomach), stomach, small intestine, large intestine, liver, gallbladder, and pancreas. Sometimes an organ is part of more than one system. For example, the pancreas, which has multiple functions, is included in the digestive and endocrine systems.

⑥ The largest organizational level is the **organismal level.** An **organism** (OR-ga-nizm), any living individual, can be compared to a *book* in our analogy. All the parts of the human body functioning together constitute the total organism.

In the following chapters, you will study the anatomy and some physiology of the body systems. Table 1.2 introduces the components and functions of these systems in the order they are discussed in the book.

 CHECKPOINT
3. Define the following terms: atom, molecule, cell, tissue, organ, system, and organism.
4. Which body systems help eliminate wastes? (*Hint:* Refer to Table 1.2.)

Figure 1.1 Levels of structural organization in the human body.

🔑 The levels of structural organization are chemical, cellular, tissue, organ, system, and organismal.

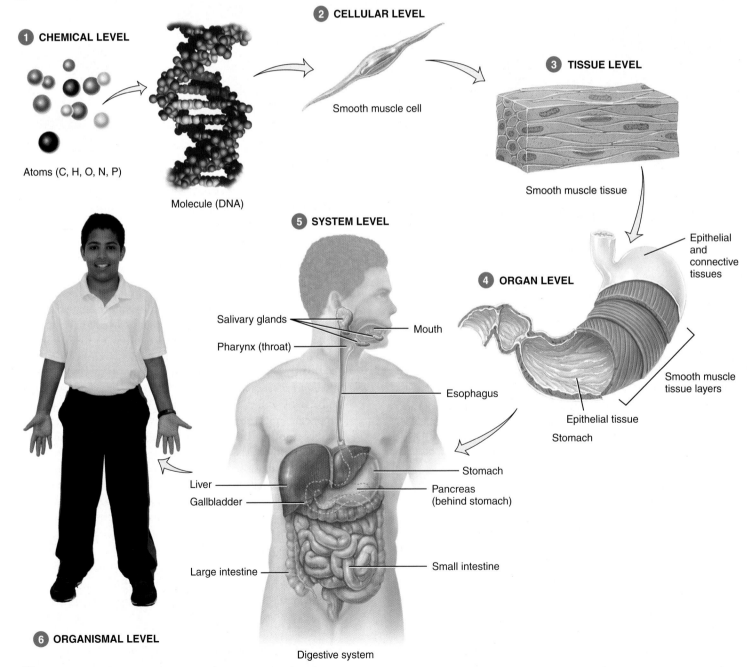

① CHEMICAL LEVEL

Atoms (C, H, O, N, P)

Molecule (DNA)

② CELLULAR LEVEL

Smooth muscle cell

③ TISSUE LEVEL

Smooth muscle tissue

Epithelial and connective tissues

④ ORGAN LEVEL

Smooth muscle tissue layers

Epithelial tissue

Stomach

⑤ SYSTEM LEVEL

Salivary glands
Pharynx (throat)
Mouth
Esophagus
Stomach
Pancreas (behind stomach)
Liver
Gallbladder
Large intestine
Small intestine

⑥ ORGANISMAL LEVEL

Digestive system

 Which level of structural organization is composed of two or more different types of tissues that work together to perform a specific function?

TABLE 1.2

The Eleven Systems of the Human Body

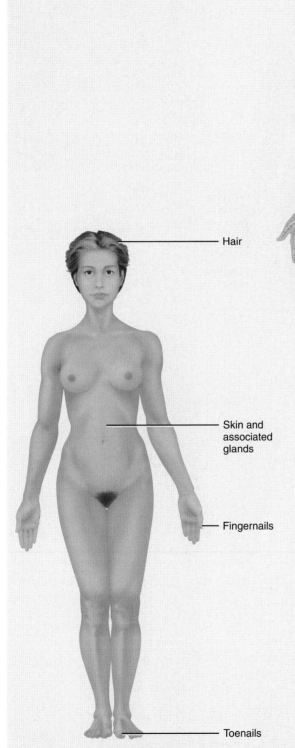

INTEGUMENTARY SYSTEM (CHAPTER 5)

Components: Skin, and structures associated with it, such as hair, fingernails and toenails, sweat glands, and oil glands and the subcutaneous layer.

Functions: Protects the body; helps regulate body temperature; eliminates some wastes; helps make vitamin D; and detects sensations such as touch, pain, warmth, and cold; stores fat and provides insulation.

SKELETAL SYSTEM (CHAPTERS 6–9)

Components: Bones and joints of the body and their associated cartilages.

Functions: Supports and protects the body; provides a surface area for muscle attachments; aids body movements; houses cells that produce blood cells; stores minerals and lipids (fats).

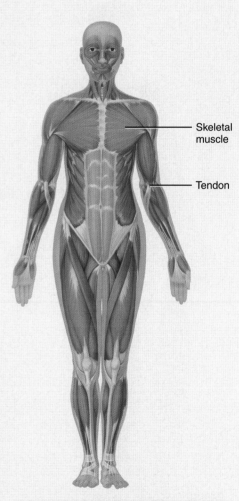

MUSCULAR SYSTEM (CHAPTERS 10, 11)

Components: Specifically refers to skeletal muscle tissue, which is muscle usually attached to bones (other muscle tissues include smooth and cardiac).

Functions: Participates in bringing about body movements, such as walking, maintains posture, and produces heat.

TABLE 1.2 CONTINUES

TABLE 1.2 CONTINUED

The Eleven Systems of the Human Body

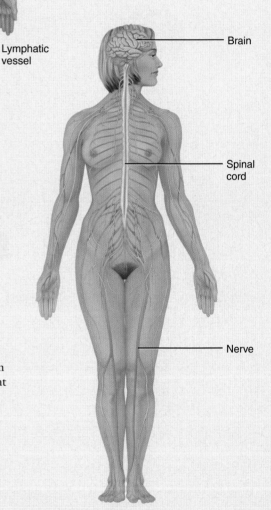

Pharyngeal tonsil

Palatine tonsil

Lingual tonsil

Thymus

Thoracic duct

Spleen

Red bone marrow

Lymph node

Lymphatic vessel

LYMPHATIC SYSTEM AND IMMUNITY (CHAPTER 15)

Components: Lymphatic fluid, lymphatic vessels, spleen, thymus, lymph nodes, and tonsils; cells that carry out immune responses (B cells, T cells, and others).

Functions: Returns proteins and fluid to blood; carries lipids from gastrointestinal tract to blood; contains sites of maturation and proliferation of B cells and T cells that protect against disease-causing microbes.

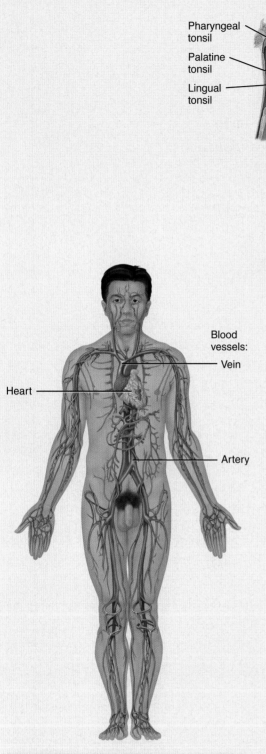

Blood vessels:

Vein

Heart

Artery

CARDIOVASCULAR SYSTEM (CHAPTERS 12–14)

Components: Blood, heart, and blood vessels.

Functions: Heart pumps blood through blood vessels; blood carries oxygen and nutrients to cells and carbon dioxide and wastes away from cells and helps regulate acid–base balance, temperature, and water content of body fluids; blood components help defend against disease and repair damaged blood vessels.

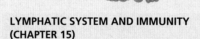

Brain

Spinal cord

Nerve

NERVOUS SYSTEM (CHAPTERS 16–21)

Components: Brain, spinal cord, nerves, and special sense organs, such as the eyes and ears.

Functions: Generates action potentials (nerve impulses) to regulate body activities; detects changes in the body's internal and external environments, interprets the changes, and responds by causing muscular contractions or glandular secretions.

RESPIRATORY SYSTEM (CHAPTER 23)

Components: Lungs and air passageways such as the pharynx (throat), larynx (voice box), trachea (windpipe), and bronchial tubes within the lungs.

Functions: Transfers oxygen from inhaled air to blood and carbon dioxide from blood to exhaled air; helps regulate acid–base balance of body fluids; air flowing out of lungs through vocal cords produces sounds.

ENDOCRINE SYSTEM (CHAPTER 22)

Components: Hormone-producing glands (pineal gland, hypothalamus, pituitary gland, thymus, thyroid gland, parathyroid glands, adrenal glands, pancreas, ovaries, and testes) and hormone-producing cells in several other organs.

Functions: Regulates body activities by releasing hormones, which are chemical messengers transported in blood from an endocrine gland or tissue to a target organ.

DIGESTIVE SYSTEM (CHAPTER 24)

Components: Organs of gastrointestinal tract—a long tube that includes the mouth, pharynx (throat), esophagus, stomach, small and large intestines, and anus; also includes accessory organs that assist in digestive processes, such as the salivary glands, liver, gallbladder, and pancreas.

Functions: Achieves physical and chemical breakdown of food; absorbs nutrients; eliminates solid wastes.

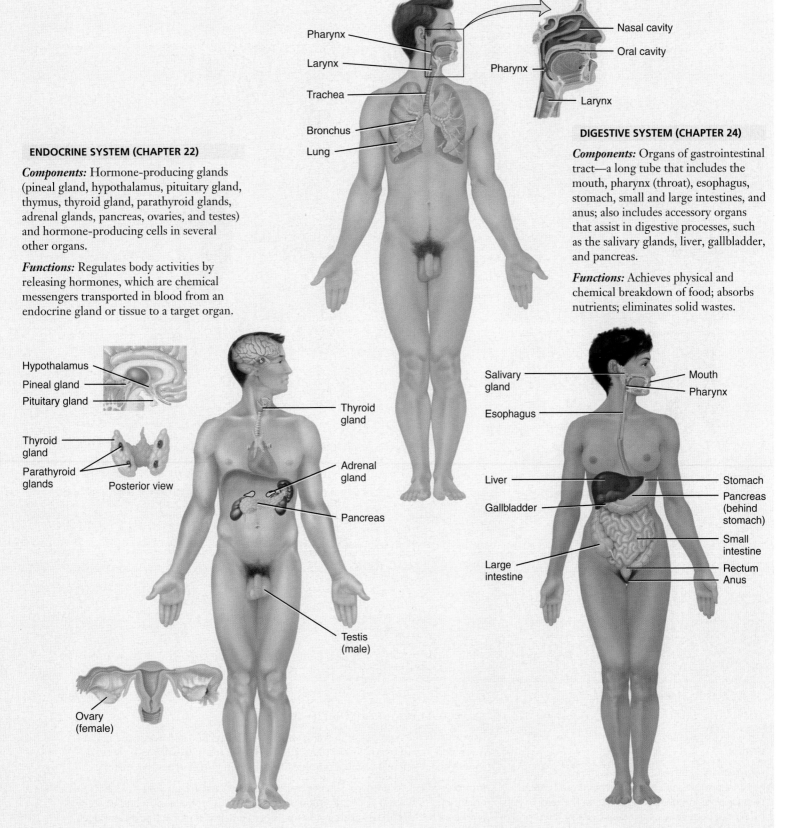

TABLE 1.2 CONTINUES

TABLE 1.2 CONTINUED

The Eleven Systems of the Human Body

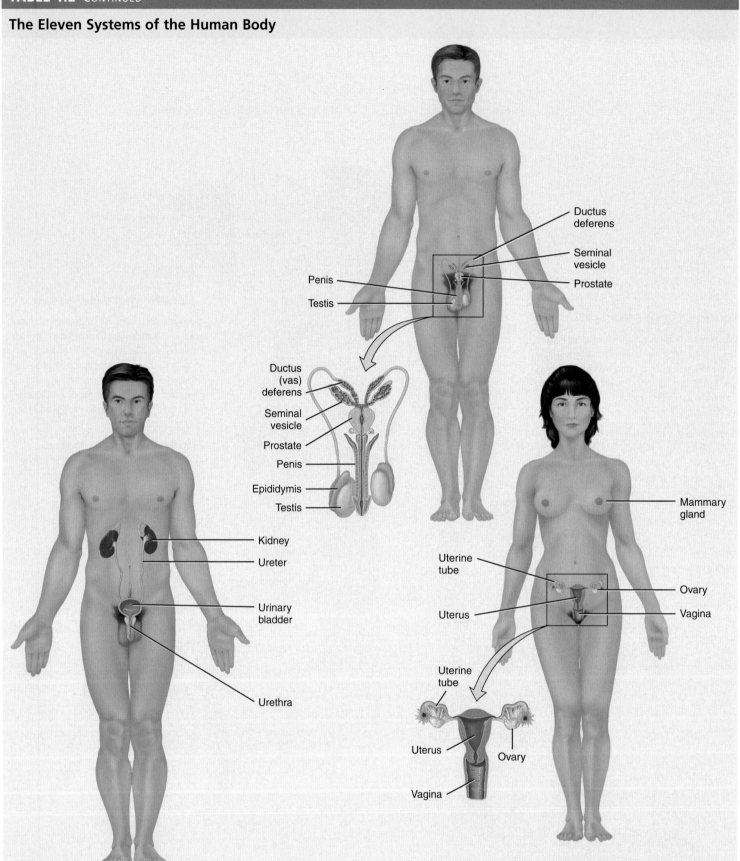

URINARY SYSTEM (CHAPTER 25)

Components: Kidneys, ureters, urinary bladder, and urethra.

Functions: Produces, stores, and eliminates urine; eliminates wastes and regulates volume and chemical composition of blood; helps maintain the acid–base balance of body fluids; maintains body's mineral balance; helps regulate production of red blood cells.

REPRODUCTIVE SYSTEMS (CHAPTER 26)

Components: Gonads (testes in males and ovaries in females) and associated organs (such as the uterine or fallopian tubes, uterus, and vagina in females and epididymides, seminal vesicles, prostate, ductus deferenses, and penis in males).

Functions: Gonads produce gametes (sperm or oocytes) that unite to form a new organism; gonads also release hormones that regulate reproduction and other body processes; associated organs transport and store gametes; mammary glands produce milk.

1.3 LIFE PROCESSES

 OBJECTIVE

• Define the important life processes of humans.

All living organisms have certain characteristics that set them apart from nonliving things. The following are six important life processes of humans:

1. **Metabolism** (me-TAB-ō-lizm) is the sum of all the chemical processes that occur in the body. It includes the breakdown of large, complex molecules into smaller, simpler ones (catabolism) and the building up of complex molecules from smaller, simpler ones (anabolism). For example, food proteins are broken down into amino acids, building blocks that can then be used to build new proteins that make up muscles and bones.

2. **Responsiveness** is the body's ability to detect and respond to changes in its internal (inside the body) or external (outside the body) environment. Different cells in the body detect different sorts of changes and respond in characteristic ways. Nerve cells respond to changes in the environment by generating electrical signals, known as nerve impulses. Muscle cells respond to nerve impulses by contracting, which generates force to move body parts.

3. **Movement** includes motion of the whole body, individual organs, single cells, and even structures inside cells. For example, the coordinated action of several muscles and bones enables you to move your body from one place to another by walking or running. After you eat a meal that contains fats, your gallbladder (an organ) contracts and releases bile into the gastrointestinal tract to help digest them. When a body tissue is damaged or infected, certain white blood cells move from the bloodstream into the affected tissue to help clean up and repair the area. And inside individual cells, various cell structures move from one position to another to carry out their functions.

4. **Growth** is an increase in body size. It may be due to an increase in (1) the size of existing cells, (2) the number of cells, or (3) the amount of material surrounding cells.

5. **Differentiation** (dif′-er-en-shē-Ā-shun) is the process unspecialized cells go through to become specialized cells. Such precursor cells, which can divide and give rise to cells that undergo differentiation, are called *stem cells*. Specialized cells differ in structure and function from the unspecialized cells that gave rise to them. For example, specialized red blood cells and several types of white blood cells differentiate from the same unspecialized cells in red bone marrow. Similarly, a single fertilized human egg cell undergoes tremendous differentiation to develop into a unique individual who is similar to, yet quite different from, either of his or her parents.

6. **Reproduction** (rē-prō-DUK-shun) refers to the formation of new cells through cell division. The production of a new individual occurs through the fertilization of an ovum by a sperm cell to form a zygote, followed by repeated cell divisions and the differentiation of these cells.

Although not all of these processes occur in cells throughout the body all of the time, when any one of them ceases to occur properly, cell death may occur. When cell death is extensive and leads to organ failure, the result is death of the organism.

 CHECKPOINT

5. What types of movement can occur in the human body?

1.4 BASIC ANATOMICAL TERMINOLOGY

 OBJECTIVES

• Describe the orientation of the human body in the anatomical position.
• Relate the common names to the corresponding anatomical descriptive terms for various regions of the human body.
• Define the anatomical planes, the anatomical sections, and the directional terms used to describe the human body.

Scientists and health-care professionals use a common language of special terms when referring to body structures and their functions. The language of anatomy has precisely defined meanings that allow us to communicate clearly and unambiguously. For example, take the statement "The wrist is above the fingers." This might be true if your upper limbs (described shortly) are at your sides. But if you held your hands up above your head, your fingers would be above your wrists. To prevent this kind of confusion, anatomists use a standard anatomical position and a special vocabulary for relating body parts to one another.

Anatomical Position

In anatomy, the **anatomical position** (an′-a-TOM-i-kal) is the standard position of reference for the description of anatomical structures. In the anatomical position, the subject stands erect facing the observer, with the head level and the eyes facing directly forward. The lower limbs are parallel and the feet are flat on the floor and directed forward. The upper limbs are at the sides with the palms facing forward (Figure 1.2). With the body in the anatomical position, it is easier to visualize and understand its organization into various regions and describe relationships of various structures.

As just described, in the anatomical position, the body is upright. There are two terms used to describe a reclining body. If the body is lying face down, it is in the **prone** position. If the body is lying face up, it is in the **supine** position.

Regional Names

The human body is divided into several major regions that can be identified externally. These are the head, neck, trunk, upper limbs, and lower limbs. The **head** consists of the skull and face. The *skull* encloses and protects the brain, while the *face* is the front portion of the head that includes the eyes, nose, mouth, forehead, cheeks, and chin. The **neck**, a modified portion of the trunk, supports the head and attaches it to the remainder of the trunk. The **trunk** consists of the *neck, thorax, abdomen,* and *pelvis.* Each **upper limb (extremity)** is attached to the trunk and consists of the *shoulder, armpit, arm* (portion of the limb from the shoulder to the elbow), *forearm* (portion of the limb from the elbow to the wrist), *wrist,* and *hand.* Each **lower limb (extremity)** is also attached to the trunk and consists of the *buttock, thigh* (portion of the limb from the buttock to the knee), *leg* (portion of the limb from the knee to the ankle), *ankle,* and *foot.* The **groin** is the area on the front surface of the body marked by a crease on each side, where the trunk attaches to the thighs. Understanding the precise meaning of arm and forearm in the upper limb and thigh and leg in the lower limb is very important when reading or describing a clinical assessment.

Figure 1.2 shows the anatomical and common names of major parts of the body. The anatomical term appears first followed by the corresponding common name (in parentheses). For example, if you receive a tetanus shot in your *gluteal region*, it is in the *buttock*. Why is the anatomical term for a body part different from its common name? The anatomical term is based on a Greek or Latin word or "root." For example, the Latin word *axilla* (ak-SIL-a) is the armpit region. Thus, the axillary nerve is one of the nerves passing within the armpit region. Understanding the word roots of anatomical terms can help you learn the terms more easily. The word roots will become more familiar as you read this book, so by the time you finish the course you'll be able to tell your roommate with confidence that the funnybone she just hit on the door jamb is the olecranon region (elbow) of her brachium (arm) (not that it will help much with the pain).

Planes and Sections

As you have just seen, referencing various body regions enables you to study the surface anatomy of the body. It is also possible to study the internal structure of the body by slicing the body in different ways and examining the sections. The terms that follow describe the different planes and sections you will encounter in your anatomical studies; they are also used in many medical procedures. **Planes** are imaginary flat surfaces that pass through the body (Figure 1.3). A **sagittal plane** (SAJ-i-tal; *sagitta-*=arrow) is a vertical plane that divides the body or organ into right and left sides. More specifically, when such a plane passes through the midline of the body and divides it into *equal* right and left sides, it is called a **midsagittal plane**, or a **median plane**. The *midline* is an imaginary vertical line that

Figure 1.2 The anatomical position. The anatomical names and corresponding common names (in parentheses) are indicated for specific body regions. For example, the cephalic region is the head.

In the anatomical position, the subject stands erect facing the observer with the head level and the eyes facing forward. The lower limbs are parallel and the feet are flat on the floor and directed forward, and the upper limbs are at the sides with the palms facing forward.

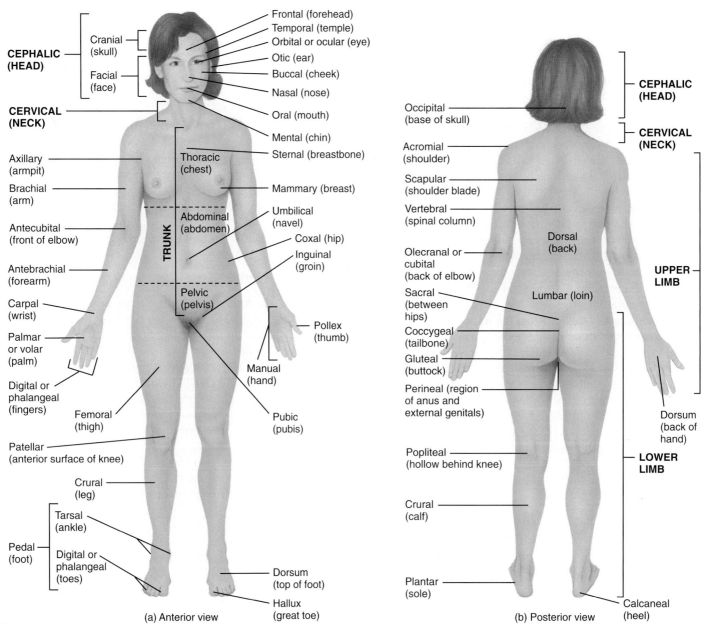

(a) Anterior view

(b) Posterior view

? **Why is it important to define one standard anatomical position?**

divides the body into equal left and right sides. If the sagittal plane does not pass through the midline but instead divides the body into *unequal* right and left sides, it is called a **parasagittal plane** (*para-*=beside, near). A **frontal**, or **coronal**, **plane** (kō-RŌ-nal; *corona*=crown) divides the body or an organ into front and back portions. A **transverse plane** divides the body or an organ into upper and lower portions. A transverse plane may also be termed a *cross-sectional plane* or *horizontal plane*. Sagittal, frontal, and transverse planes are all at right angles to one another. An **oblique plane** (ō-BLĒK), by contrast, passes through the body or organ at an oblique angle (any angle other than a 90° angle).

When you study a body region, you often view it in section. **Sections** are cuts of the body or one of its organs made along one of the planes just described. A section produces a flat two-dimensional surface of the original three-dimensional structure. It is important to know the plane of the section so you can understand the anatomical relationship of one part to another. Figure 1.4 indicates how three different sections—a *transverse (cross) section*, a *frontal section*, and a *midsagittal section*—provide different views of the brain.

Figure 1.4 Planes and sections through different parts of the brain. The diagrams (left) show the planes, and the photographs (right) show the resulting sections. **Note:** The arrows in the diagrams indicate the direction from which each section is viewed. This aid is used throughout the book to indicate viewing perspective.

 Planes divide the body in various ways and the resulting cuts made along a plane are called sections.

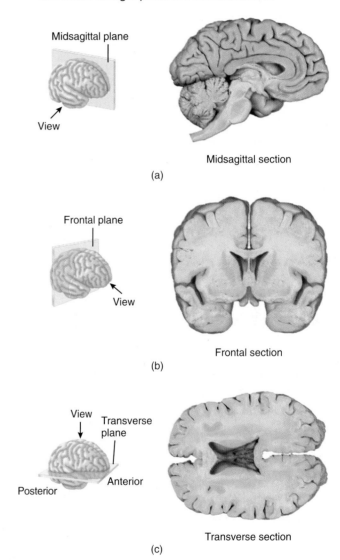

Figure 1.3 Planes through the human body.

 Midsagittal, parasagittal, frontal, transverse, and oblique planes divide the body in specific ways.

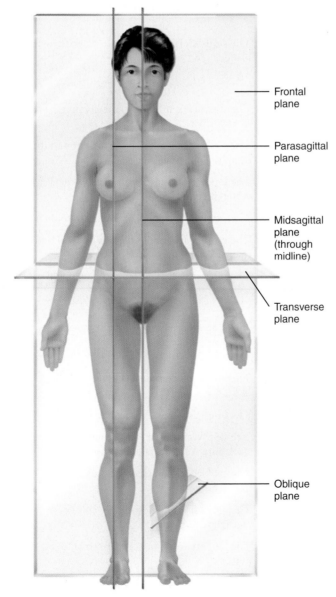

Anterior view

 Which plane divides the heart into anterior and posterior portions?

 Which plane divides the brain into unequal right and left portions?

✓ CHECKPOINT
6. Describe the anatomical position and explain why it is used.
7. Locate each region shown in Figure 1.2 on your own body, and then identify it by its anatomical descriptive form and corresponding common name.
8. Which of the planes that divide the body are vertical?
9. What is the difference between a plane and a section?

EXHIBIT 1.A   Directional Terms (*Figure 1.5*)

 OBJECTIVE

• Define each directional term used to describe the human body.

Overview

To help improve communication when discussing the basic parts of the body and the relationships those parts have to one another, anatomists and health-care professionals use specific **directional terms**, words that describe the position of one body part relative to another. Most of the directional terms used to describe the relationship of one part of the body to another can be grouped into pairs that have opposite meanings. For example, *superior* means toward the upper part of the body, and *inferior* means toward the lower part of the body. It is important to understand that directional terms have relative meanings; they make sense only when used to describe the position of one structure relative to another. For example, your knee is superior to your ankle, even though both are located in the inferior half of the body. Study the directional terms below and the example of how each is used. As you read the examples, look at Figure 1.5 to see the location of each structure.

 CHECKPOINT

10. Which directional terms can be used to specify the relationships between (1) the elbow and the shoulder, (2) the left and right shoulders, (3) the sternum and the humerus, and (4) the heart and the diaphragm?

DIRECTIONAL TERM	DEFINITION	EXAMPLE OF USE
Superior (soo′-PĒR-ē-or)	Above or higher in position; toward the head. (Not used in reference to the limbs.)	The heart is superior to the liver.
Cranial (KRĀN-ē-al) or **cephalic** (se-FAL-ik)	Relating to the skull or head; toward the head. (This is a more flexible term than superior because it can be applied to all animals, whether they stand upright on two limbs or on all four limbs.)	The stomach is more cranial than the urinary bladder.
Inferior (in′-FĒR-ē-or)	Below or lower in position; toward the feet. (Not used in reference to the limbs.)	The stomach is inferior to the lungs.
Rostral (ROS′-tral)	Relating to the nose and mouth region; toward the face.	The frontal lobe of the brain is rostral to the occipital lobe (see Figure 18.12b).
Caudal (KAWD-al)	Relating to the tail; at or near the tail or posterior part of the body.	The lumbar vertebrae are caudal to the cervical vertebrae (see Figure 7.15a).
Anterior (an-TER-ē-or)	Nearer to or at the front of the body.	The sternum (breastbone) is anterior to the heart.
Posterior (pos-TER-ē-or)	Nearer to or at the back of the body.	The esophagus (food tube) is posterior to the trachea (windpipe).
Ventral (VEN-tral)	Relating to the belly side of the body; toward the belly. (Used synonymously with anterior in human anatomy.)	The intestines are ventral to the vertebral column.
Dorsal (DORS-al)	Relating to the back side of the body; toward the back. (Used synonymously with posterior in human anatomy.)	The kidneys are dorsal to the stomach.
Medial (MĒ-dē-al)	Nearer to the midline (an imaginary vertical line that divides the body into equal right and left sides).	The ulna is medial to the radius.
Lateral (LAT-er-al)	Farther from the midline.	The lungs are lateral to the heart.
Intermediate (in′-ter-MĒ-dē-at)	Between two structures.	The transverse colon is intermediate to the ascending colon and descending colon.
Ipsilateral (ip-si-LAT-er-al)	On the same side of the body's midline as another structure.	The gallbladder and ascending colon are ipsilateral.
Contralateral (CON-tra-lat-er-al)	On the opposite side of the body's midline from another structure.	The ascending and descending colons are contralateral.
Proximal (PROK-si-mal)	Nearer to the attachment of a limb to the trunk; nearer to the origination of a structure.	The humerus (arm bone) is proximal to the radius.
Distal (DIS-tal)	Farther from the attachment of a limb to the trunk; farther from the origination of a structure.	The phalanges (finger bones) are distal to the carpals (wrist bones).
Superficial (soo′-per-FISH-al)	Toward or on the surface of the body.	The ribs are superficial to the lungs.
Deep	Away from the surface of the body.	The ribs are deep to the skin of the chest and back.

DIRECTIONAL TERM	DEFINITION	EXAMPLE OF USE
External (ex-STERN-al)	Toward the outside of a structure. (Is typically used when describing relationships of individual organs.)	The visceral pleura is on the external surface of the lungs (see Figure 1.7a).
Internal (in-TERN-al)	Toward the inside of a structure. (Is typically used when describing relationships of individual organs.)	The mucosa forms the internal lining of the stomach (see Figure 24.11a).

Figure 1.5 Directional terms.

🔑 Directional terms precisely locate various parts of the body relative to one another.

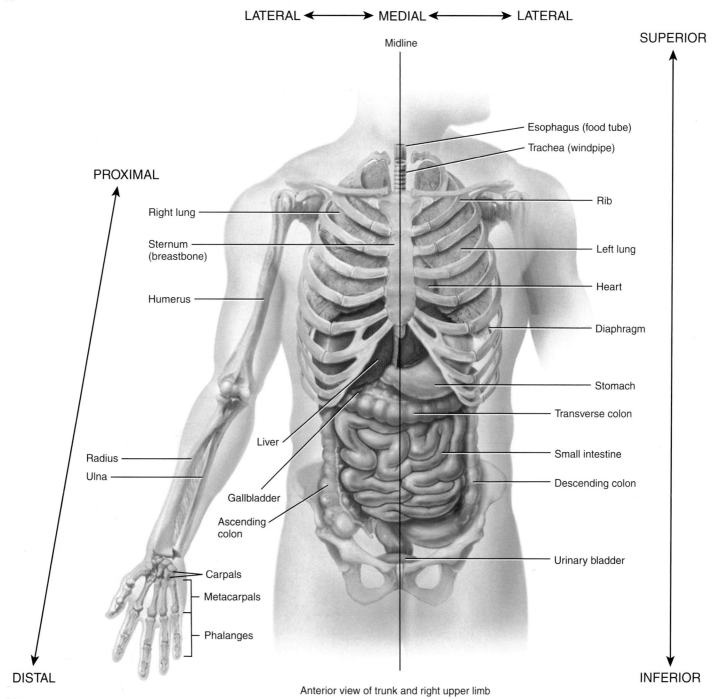

Anterior view of trunk and right upper limb

 Is the radius proximal to the humerus? Is the esophagus anterior to the trachea? Are the ribs superficial to the lungs? Is the urinary bladder medial to the ascending colon? Is the sternum lateral to the descending colon?

1.5 BODY CAVITIES

 OBJECTIVE

• Describe the major body cavities, the organs they contain, and their associated linings.

Body cavities are spaces within the body that house internal organs. Bones, muscles, and ligaments separate the various body cavities from one another. Here we discuss several body cavities (Figure 1.6).

The cranial bones form a hollow space of the head called the **cranial cavity** (KRĀ-nē-al), which contains the brain. The bones of the vertebral column (backbone) form the **vertebral (spinal) canal** (VER-te-bral), which contains the spinal cord and the beginnings of the spinal nerves. The cranial cavity and vertebral canal are continuous with one another. Three layers of protective tissue, the **meninges** (me-NIN-jēz), and a shock-absorbing fluid surround the brain and spinal cord.

The major body cavities of the trunk are the thoracic and abdominopelvic cavities. The **thoracic cavity** (thor-AS-ik; *thorac-*=chest), or chest cavity (Figures 1.6 and 1.7), is formed by the ribs, the muscles of the chest, the sternum (breastbone), and the thoracic (chest) portion of the vertebral column. Within the thoracic cavity are two **pleurae (serous sacs)** (PLOOR-ē; *pleur-*=rib or side; singular is *pleura* [PLOO-ra]). Each double-layered pleura surrounds one lung and also contains a small amount of lubricating fluid in a potential space between the layers. The pleurae will be described in more detail shortly. The central portion of the thoracic cavity is an anatomic region called the **mediastinum** (mē-dē-as-TĪ-num; *media-*=middle; *-stinum*=partition). The mediastinum is between the medial walls of the two pleurae and extends from the sternum to the vertebral column, and from the first rib to the diaphragm (Figure 1.7a, b). The mediastinum contains all thoracic organs except the lungs themselves. Among the structures in the mediastinum are the heart, esophagus, trachea, thymus, and several large blood vessels that enter and leave the heart. The **pericardium** (serous sac) (per-i-KAR-dē-um; *peri-*=around; *-cardial*=heart) surrounds the heart and contains a small amount of lubricating fluid. The **diaphragm** (DĪ-a-fram= partition or wall) is a dome-shaped muscle that separates the thoracic cavity from the abdominopelvic cavity.

The **abdominopelvic cavity** (ab-dom-i-nō-PEL-vik; see Figure 1.6) extends from the diaphragm to the groin and is encircled by the abdominal muscle wall and the bones and muscles of the pelvis. As the name suggests, the abdominopelvic cavity is divided into two portions, even though no wall separates them. The superior portion, the **abdominal cavity** (ab-DOM-i-nal; *abdomin-*=belly), contains the kidneys, adrenal glands, stomach, spleen, liver, gallbladder, pancreas, small intestine, and most of the large intestine. The inferior portion, the **pelvic cavity** (PEL-vik; *pelv-*=basin), contains the urinary bladder, portions of the large intestine, and internal organs of the reproductive

Figure 1.6 Body cavities. The black dashed lines in (a) and (b) indicate the border between the abdominal and pelvic cavities.

🔑 The major body cavities of the trunk are the thoracic and abdominopelvic cavities.

CAVITY	COMMENTS
Cranial cavity	Formed by cranial bones and contains brain.
Vertebral canal	Formed by vertebral column and contains spinal cord and the beginnings of spinal nerves.
Thoracic cavity*	Chest cavity; contains pleural and pericardial sacs and the mediastinum.
Pleural cavity	A potential space between the layers of the pleura that surrounds a lung.
Pericardial cavity	A potential space between the layers of the pericardium that surrounds the heart.
Mediastinum	Central portion of thoracic cavity between the lungs; extends from sternum to vertebral column and from first rib to diaphragm; contains heart, thymus, esophagus, trachea, and several large blood vessels.
Abdominopelvic cavity	Subdivided into abdominal and pelvic cavities.
Abdominal cavity	Contains stomach, spleen, liver, gallbladder, small intestine, and most of large intestine.
Pelvic cavity	Contains urinary bladder, portions of large intestine, and internal organs of reproduction.

(a) Right lateral view (b) Anterior view

CRANIAL CAVITY

VERTEBRAL CANAL

THORACIC CAVITY

Diaphragm

ABDOMINOPELVIC CAVITY:

ABDOMINAL CAVITY

PELVIC CAVITY

* See Figure 1.7 for details of the thoracic cavity.

system. Organs inside the thoracic and abdominopelvic cavities are termed the **viscera** (VIS-er-a).

Thoracic and Abdominal Cavity Membranes

A **membrane** is a thin pliable tissue that covers, lines, partitions, or connects structures. One example is a slippery double-layered membrane associated with body cavities that does not open directly to the exterior called a **serous membrane.** It covers the viscera within the thoracic and abdominal cavities and also lines the walls of the thorax and abdomen. The parts of a serous membrane are (1) the *parietal layer* (pa-RĪ-e-tal=wall of a cavity), a thin epithelium that lines the walls of the body cavities, and (2) the *visceral layer* (VIS-er-al), a thin epithelium that covers and adheres to the viscera within the body cavities. Because the parietal and visceral membranes are continuous with one another, they form a serous sac. The organs of the body cavity push into this serous sac, similar to pushing your hand into a balloon (Figure 1.7e). Between the parietal and visceral layers is a potential called the *serous cavity*. It contains a small amount of lubricating fluid called *serous fluid* that reduces friction between the two layers, allowing the viscera to slide freely during movements such as the pumping of the heart or the inflation and deflation of your lungs when you breathe in and out.

The serous membrane associated with the lungs is called the **pleura** (Figure 1.7a, c, d). The *visceral pleura* clings to the surface of the lungs (the part of the balloon touching your fist); the *parietal pleura* lines the chest wall and covers the superior surface of the diaphragm. In between is the serous cavity called the *pleural cavity* (analogous to the inside of the balloon), filled with a small volume of lubricating serous fluid. The serous membrane of the heart is the **pericardium** (per'-i-KAR-dē-um) (see Figure 1.7a, c, d). The *visceral pericardium* covers the surface of the heart, and the *parietal pericardium* lines the fibrous pericardium that surrounds the heart. Between them is the serous cavity called the *pericardial cavity*, which contains a small amount of lubricating serous fluid. The **peritoneum** (per'-i-to-NE-um) is the serous membrane of the abdominal cavity (see Figure 24.3a). The *visceral peritoneum* covers the abdominal viscera, and the *parietal peritoneum* lines the abdominal wall and covers the inferior surface of the diaphragm. Between them is the serous cavity called the *peritoneal cavity*, which contains a small amount of lubricating serous fluid. Most abdominal organs are surrounded by the peritoneum and are referred to as *intraperitoneal* (in'-tra-per'-i-tō-NĒ-al; *intra-*=within). These include the stomach, spleen, liver, gallbladder, jejunum and ileum of the small intestine, and the cecum, appendix, and tranverse colon of the large intestine. However, some are not surrounded by the peritoneum or are only partially covered by peritoneum and lie behind the peritoneum. Such organs are said to be *retroperitoneal* (re-trō-per'-i-tō-NĒ-al; *retro*=behind). The kidneys, adrenal glands, pancreas, duodenum of the small intestine, ascending and descending colons of the large intestine, and the abdominal aorta and inferior vena cava are retroperitoneal organs.

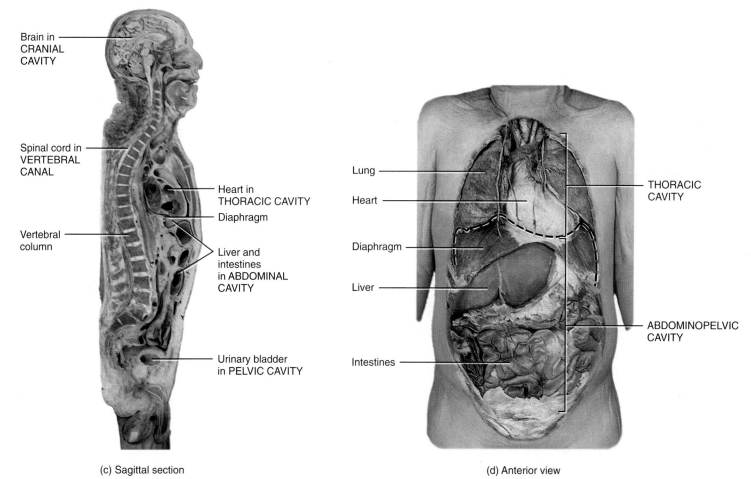

(c) Sagittal section

(d) Anterior view

 In which cavities are the following organs located: urinary bladder, stomach, heart, small intestine, lungs, internal female reproductive organs, thymus, spleen, liver? Use the following symbols for your responses: T = thoracic cavity, A = abdominal cavity, or P = pelvic cavity.

In addition to the major body cavities just described, you will learn about other regional cavities in later chapters. These include the *oral (mouth) cavity*, which contains the tongue and teeth (see Figure 24.4); the *nasal cavity* in the nose (see Figure 23.1a); the *orbital cavities (orbits)*, which contain the eyeballs (see Figure 7.2a); the *middle ear cavities* (middle ears), which contain small bones and muscles in the middle ear (Figure 21.11a); and *synovial cavities*, which are found in freely movable joints and contain synovial fluid (see Figure 9.3a). A summary of the major body cavities and their membranes is presented in the table in Figure 1.6.

✔ CHECKPOINT

11. What structures separate the various body cavities from one another?
12. Describe the contents of the mediastinum.

Figure 1.7 The thoracic cavity. The black dashed lines in (a) and (b) indicate the borders of the mediastinum. Notice that the pericardium surrounds the heart, and that the pleurae surround the lungs. **Note:** When transverse sections such as those shown in (b) and (c) are viewed inferiorly (from below), the anterior aspect of the body appears on the top of the illustration and the left side of the body appears on the right side of the illustration.

🔑 The mediastinum is an anatomical region that is between the lungs and extends from the sternum to the vertebral column and from the first rib to the diaphragm.

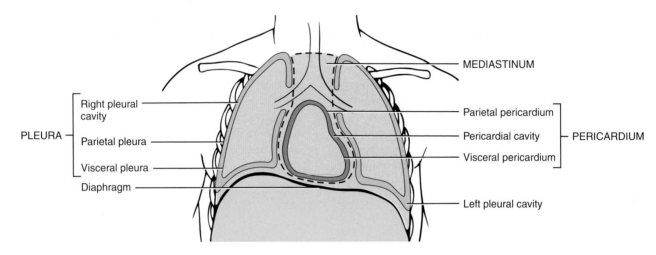

(a) Anterior view of thoracic cavity

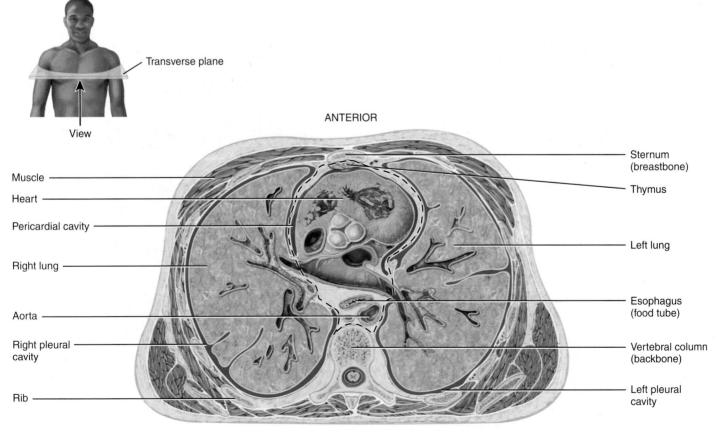

(b) Inferior view of transverse section of thoracic cavity

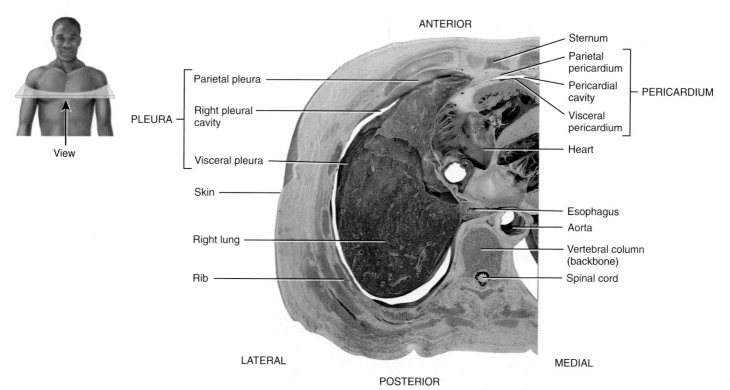

ANTERIOR

Sternum

Parietal pleura

Right pleural cavity

PLEURA

Visceral pleura

Parietal pericardium

Pericardial cavity

Visceral pericardium

PERICARDIUM

Heart

View

Skin

Esophagus

Aorta

Right lung

Vertebral column (backbone)

Rib

Spinal cord

LATERAL

MEDIAL

POSTERIOR

(c) Inferior view of transverse section of thoracic cavity

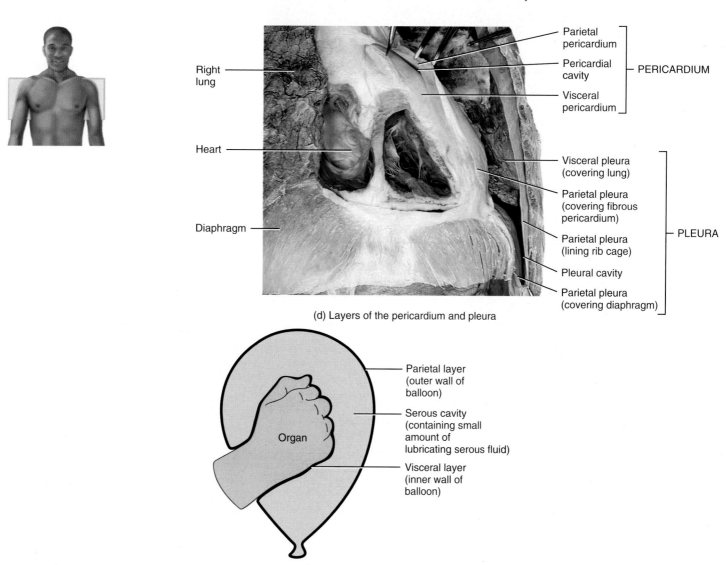

Right lung

Heart

Diaphragm

Parietal pericardium

Pericardial cavity

Visceral pericardium

PERICARDIUM

Visceral pleura (covering lung)

Parietal pleura (covering fibrous pericardium)

Parietal pleura (lining rib cage)

Pleural cavity

Parietal pleura (covering diaphragm)

PLEURA

(d) Layers of the pericardium and pleura

Parietal layer (outer wall of balloon)

Serous cavity (containing small amount of lubricating serous fluid)

Organ

Visceral layer (inner wall of balloon)

(e) The concept of a serous sac

Which of the following structures are contained in the mediastinum: right lung, heart, esophagus, spinal cord, trachea, rib, thymus, left pleural cavity?

1.6 ABDOMINOPELVIC REGIONS AND QUADRANTS

OBJECTIVE

• Name and describe the abdominopelvic regions and the abdominopelvic quadrants.

To describe the location of the many abdominal and pelvic organs more easily, anatomists and clinicians use two methods of dividing the abdominopelvic cavity into smaller areas. In the first method, two transverse and two vertical lines, aligned like a tick-tack-toe grid, partition this cavity into nine **abdominopelvic regions** (Figure 1.8). The top horizontal line, the *subcostal line* (*sub-*=below; *costal*=rib), is drawn just inferior to the right

Figure 1.8 Abdominopelvic regions. (a) The nine regions in surface view. (b) The nine regions with the greater omentum removed. The greater omentum is a double fold of the serous membrane that contains fatty tissue and covers some of the abdominal organs (see Figure 24.3). (c) Organs or parts of organs in the nine regions. The internal reproductive organs in the pelvic cavity are shown in Figures 26.1 and 26.11.

🔑 The nine-region designation is used for anatomical studies.

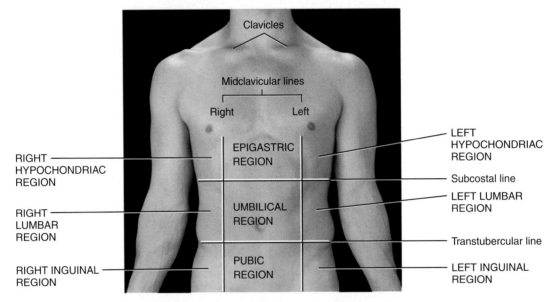

(a) Anterior view showing abdominopelvic regions

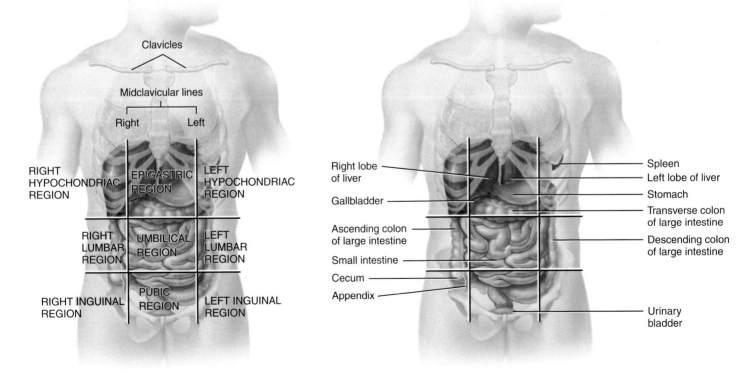

(b) Anterior view showing location of abdominopelvic regions

(c) Anterior superficial view of organs in abdominopelvic regions

 In which abdominopelvic region is each of the following found: most of the liver, transverse colon, urinary bladder, spleen?

and left lateral margins of the ribs; the bottom horizontal line, the *transtubercular line* (trans-too-BER-kū-lar), intersects the iliac tubercles, landmarks near the top of the right and left hip bones. Two vertical lines, the left and right *midclavicular lines* (mid-kla-VIK-ū-lar), are drawn through the midpoints of the clavicles (collar bones), just medial to the nipples. The four lines divide the abdominopelvic cavity into a larger middle section and smaller left and right sections. The names of the nine abdominopelvic regions are **right hypochondriac** (hī′-pō-KON-drē-ak), **epigastric** (ep-i-GAS-trik), **left hypochondriac, right lumbar, umbilical** (um-BIL-i-kal), **left lumbar, right inguinal (iliac)** (IN-gwi-nal), **hypogastric (pubic)**, and **left inguinal (iliac)**. Note which organs and parts of organs are in the different regions by carefully examining Figure 1.8c. The organs of the abdominopelvic cavity will be discussed in detail in later chapters.

The second method is simpler and divides the abdominopelvic cavity into **quadrants** (KWOD-rantz; *quad-*=one-fourth), as shown in Figure 1.9. In this method, a transverse line, the *transumbilical line*, and a midsagittal line, the *median line*, are passed through the **umbilicus** (um-BIL-i-kus or um-bi-LĪ-kus; *umbilic-*=navel) or *belly button*. The names of the abdominopelvic quadrants are **right upper quadrant (RUQ), left upper quadrant (LUQ), right lower quadrant (RLQ)**, and **left lower quadrant (LLQ)**. The nine-region division is more widely used for anatomical studies to determine organ location; quadrants are more commonly used by clinicians for describing the site of abdominopelvic pain, tumor, injury, or other abnormality.

Figure 1.9 Quadrants of the abdominopelvic cavity. The two lines intersect at right angles at the umbilicus (navel).

 The quadrant designation is used to locate the site of pain, a tumor, or some other abnormality.

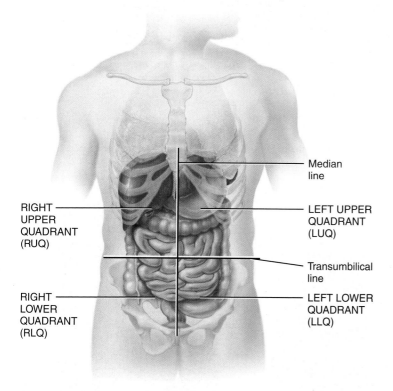

Median line

LEFT UPPER QUADRANT (LUQ)

Transumbilical line

LEFT LOWER QUADRANT (LLQ)

RIGHT UPPER QUADRANT (RUQ)

RIGHT LOWER QUADRANT (RLQ)

Anterior view

 In which quadrant would the pain from appendicitis (inflammation of the appendix) be felt?

 CHECKPOINT

13. Locate the abdominopelvic regions and the abdominopelvic quadrants on yourself. Which one(s) contain(s) your spleen? Your colon? Your urinary bladder?

1.7 THE HUMAN BODY AND DISEASE

OBJECTIVE
• Distinguish between a symptom and a sign of disease.

A **disorder** is any abnormality of structure and/or function. **Disease** is a more specific term for an illness characterized by a recognizable set of symptoms and signs in which body structures and functions are altered in characteristic ways. A person with a disease may experience **symptoms**, *subjective* changes in body functions that are not apparent to an observer. Examples of symptoms are headache, nausea, and anxiety. *Objective* changes that a clinician can observe and measure are called **signs**. Signs of disease can be either anatomical or physiological. An anatomical sign of disease is referred to as a **lesion** (organ or tissue damage resulting from injury or disease), such as swelling, a rash, an ulcer, a wound, or a tumor. Physiological signs of disease include fever, high blood pressure, and paralysis. A *local disease* (such as a sinus infection) affects one part or a limited region of the body; a *systemic disease* (for example, influenza) affects either the entire body or several parts of it.

The science that deals with why, when, and where diseases occur and how they are transmitted among individuals in a community is known as **epidemiology** (ep′-i-dē-mē-OL-ō-jē; *epi*=upon; *-demi*=people). **Pharmacology** (far′-ma-KOL-ō-jē; *pharmac*=drug) is the science that deals with the uses and effects of drugs in the treatment of disease.

Diagnosis (dī′-ag-NŌ-sis; *dia*=through; *-gnosis*=knowledge) is the science and skill of distinguishing one disorder or disease from another. The patient's symptoms and signs, his or her medical history, a physical exam, and laboratory tests provide the basis for making a diagnosis. Taking a *medical history* consists of collecting information about events that might be related to a patient's illness. These include the chief complaint (primary reason for seeking medical attention), history of present illness, past medical problems, family medical problems, social history, and review of symptoms and signs. A *physical examination* is an orderly evaluation of the body and its functions. This process includes the noninvasive techniques of inspection, palpation, auscultation, and percussion that you learned about earlier in the chapter, along with measurement of vital signs (temperature, pulse, respiratory rate, and blood pressure), and sometimes laboratory tests.

 CHECKPOINT

14. Classify each of the following as a sign or symptom: high blood pressure, fever, headache, rapid pulse.

1.8 MEDICAL IMAGING

🔴 OBJECTIVE
• Describe the principles of medical imaging procedures and their importance in the evaluation of organ functions and the diagnosis of disease.

Medical imaging refers to techniques and processes used to create images of the human body. Various types of medical imaging allow visualization of structures inside our bodies and are being used

---

TABLE 1.3

Common Medical Imaging Procedures

RADIOGRAPHY

Procedure: A single barrage of x-rays passes through the body, producing an image of interior structures on x-ray-sensitive film. The resulting two-dimensional image is a *radiograph* (RĀ-dē-o-graf′), commonly called an *x-ray*.

Comments: Radiographs are relatively inexpensive, quick, and simple to perform, and usually provide sufficient information for diagnosis. X-rays do not easily pass through dense structures so bones appear white. Hollow structures, such as the lungs, appear black. Structures of intermediate density, such as skin, fat, and muscle, appear as

varying shades of gray. At low doses, x-rays are useful for examining soft tissues such as the breast (**mammography**) and for determining bone density (**bone densitometry**).

It is necessary to use a substance called a contrast medium to make hollow or fluid-filled structures visible in radiographs. X-rays make structures that contain contrast media appear white. The medium may be introduced by injection, orally, or rectally, depending on the structure to be imaged.

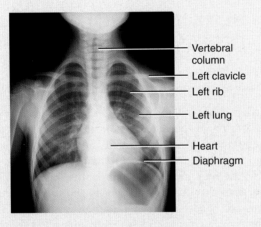

Radiograph of thorax in anterior view

Vertebral column
Left clavicle
Left rib
Left lung
Heart
Diaphragm

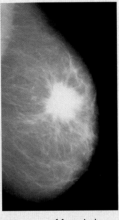

Mammogram of female breast showing cancerous tumor (white mass with uneven border)

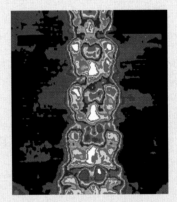

Bone densitometry scan of lumbar spine in anterior view

Contrast x-rays are used to image blood vessels (**angiography**), the urinary system (**intravenous urography**), and the gastrointestinal tract (**barium contrast x-ray**).

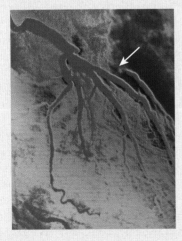

Angiogram of adult human heart showing blockage in coronary artery (arrow)

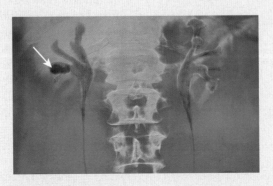

Intravenous urogram showing kidney stone (arrow) in right kidney

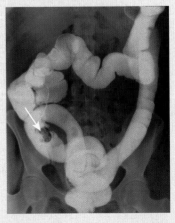

Barium contrast x-ray showing cancer of the ascending colon (arrow)

more and more to increase the precision of diagnosis of a wide range of anatomical and physiological disorders. The grandparent of all medical imaging techniques is conventional radiography (x-rays), in medical use since the late 1940s and still in widespread use today. Newer imaging technologies not only improve diagnostic capabilities, but also have advanced our understanding of normal anatomy and physiology. Table 1.3 describes and illustrates the images generated by some commonly used medical imaging techniques.

 CHECKPOINT

15. Which forms of medical imaging would be used to show a blockage in an artery of the heart?
16. Which one of the medical imaging techniques outlined in Table 1.3 best reveals the physiology of a structure?
17. Which medical imaging technique would you use to determine whether a bone was broken?

MAGNETIC RESONANCE IMAGING (MRI)

Procedure: The body is exposed to a high-energy magnetic field, which causes protons (small positive particles within atoms, such as hydrogen) in body fluids and tissues to arrange themselves in relation to the field. Then a pulse of radio waves "reads" these ion patterns, and a color-coded image is assembled on a video monitor. The result is a two- or three-dimensional blueprint of cellular chemistry.

Comments: Relatively safe, but can't be used on patients with metal in their bodies. Shows fine details for soft tissues but not for bones. Most useful for differentiating between normal and abnormal tissues. Used to detect tumors and artery-clogging fatty plaques, reveal brain abnormalities, measure blood flow, and detect a variety of musculo-skeletal, liver, and kidney disorders.

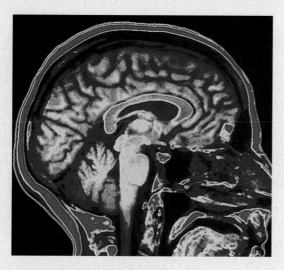

Magnetic resonance image of brain in sagittal section

CORONARY (CARDIAC) COMPUTED TOMOGRAPHY ANGIOGRAPHY (CCTA)

Procedure: Computer-assisted radiography in which an iodine-containing contrast medium is injected into a vein and a beta-blocker is given to decrease heart rate. Then, numerous x-ray beams trace an arc around the heart and a scanner detects the x-ray beams and transmits them to a computer, which transforms the information into a three-dimensional image of the coronary blood vessels on a monitor. The image procured is called a *CCTA scan* and can be generated in less than 20 seconds.

Comments: Used primarily to determine whether there are any coronary artery blockages (for example, atherosclerotic plaque or calcium) that may require an intervention such as angioplasty or stent. The CCTA can be rotated, enlarged, and moved at any angle. Since the procedure can take thousands of images of the heart within the time of a single heartbeat, it provides a great amount of detail about the heart's structure and function.

COMPUTED TOMOGRAPHY (CT) [formerly called computerized axial tomography (CAT) scanning]

Procedure: Computer-assisted radiography in which an x-ray beam traces an arc at multiple angles around a section of the body. The resulting transverse section of the body, called a *CT scan*, is shown on a video monitor.

Comments: Visualizes soft tissues and organs with much more detail than conventional radiographs. Differing tissue densities show up as various shades of gray. Multiple scans can be assembled to build three-dimensional views of structures (described next). Whole-body CT scanning is also used. Typically, such scans actually target the torso. Whole-body CT scanning appears to provide the most benefit in screening for lung cancers, coronary artery disease, and kidney cancers.

ANTERIOR

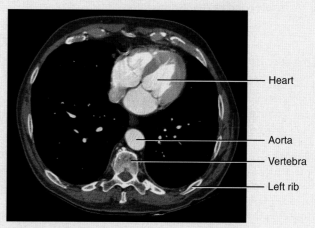

— Heart

— Aorta

— Vertebra

— Left rib

POSTERIOR

Computed tomography scan of thorax in inferior view

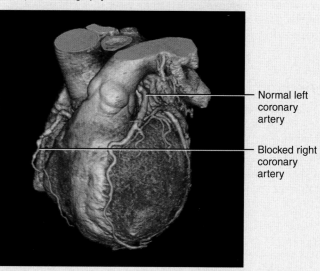

— Normal left coronary artery

— Blocked right coronary artery

CCTA scan of coronary arteries

TABLE 1.3 CONTINUES

TABLE 1.3 CONTINUED

Common Medical Imaging Procedures

ULTRASOUND SCANNING

Procedure: High-frequency sound waves produced by a handheld wand reflect off body tissues and are detected by the same instrument. The image, which may be still or moving, is called a *sonogram* (SON-ō-gram) and is shown on a video monitor.

Comments: Safe, noninvasive, painless, and uses no dyes. Most commonly used to visualize the fetus during pregnancy. Also used to observe the size, location, and actions of organs and blood flow through blood vessels (**doppler ultrasound**).

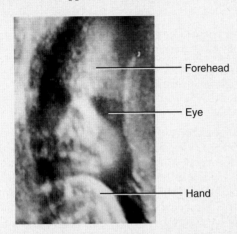

— Forehead

— Eye

— Hand

Sonogram of fetus (Courtesy of Andrew Joseph Tortora and Damaris Soler)

RADIONUCLIDE SCANNING

Procedure: A *radionuclide* (radioactive substance) is introduced intravenously into the body and carried by the blood to the tissue to be imaged. Gamma rays emitted by the radionuclide are detected by a gamma camera outside the subject and fed into a computer. The computer constructs a *radionuclide scan* and displays it in color on a video monitor. Areas of intense color take up a lot of the radionuclide and represent high tissue activity; areas of less intense color take up smaller amounts of the radionuclide and represent low tissue activity. **Single-photon-emission computed tomography (SPECT) scanning** is a specialized type of radionuclide scanning that is especially useful for studying the brain, heart, lungs, and liver.

Comments: Used to study activity of a tissue or organ, such as searching for malignant tumors in body tissues or scars that may interfere with heart muscle function.

POSITRON EMISSION TOMOGRAPHY (PET)

Procedure: A substance that emits positrons (positively charged particles) is injected into the body, where it is taken up by tissues. The collision of positrons with negatively charged electrons in body tissues produces gamma rays (similar to x-rays) that are detected by gamma cameras positioned around the subject. A computer receives signals from the gamma cameras and constructs a *PET scan* image, displayed in color on a video monitor. The PET scan shows where the injected substance is being used in the body. In the PET scan image shown here, the black and blue colors indicate minimal activity; the red, orange, yellow, and white colors indicate areas of increasingly greater activity.

Comments: Used to study the physiology of body structures, such as metabolism in the brain or heart.

ANTERIOR

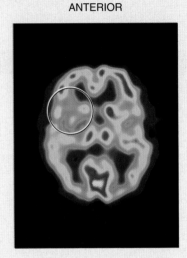

POSTERIOR

Positron emission tomography scan of transverse section of brain (circled area at upper left indicates where a stroke has occurred)

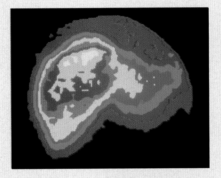

Radionuclide (nuclear) scan of normal human liver

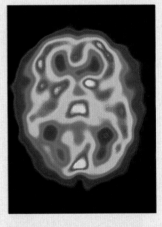

Single-photon-emission computerized tomography (SPECT) scan of transverse section of brain (green area at lower left indicates a migraine attack)

ENDOSCOPY

Procedure: The visual examination of the inside of body organs or cavities using a lighted instrument with lenses called an *endoscope.* The image is viewed through an eyepiece on the endoscope or projected onto a monitor.

Comments: Examples of endoscopy include colonoscopy, laparoscopy, and arthroscopy. *Colonoscopy* is used to examine the interior of the colon, which is part of the large intestine. *Laparoscopy* is used to examine the organs within the abdominopelvic cavity. *Arthroscopy* is used to examine the interior of a joint, usually the knee.

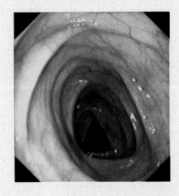

Interior view of colon as shown by colonoscopy

1.9 MEASURING THE HUMAN BODY

 OBJECTIVE

- Explain the importance of measurements in the evaluation of the human body.

In order to describe the body and understand how it works, you need to use *measurement*—determination of the dimensions of an organ, its weight, and the length of time it takes for a physiological event to occur. Measurements also have clinical importance, such as determining the dose of a particular medication. As you will see, measurements involving time, weight, temperature, size, and volume are a routine part of your studies in a medical science program.

Whenever you come across a measurement in this text, it will be given in metric units. The metric system is the standard used in the sciences because it is universal and is based on units of ten (10). In some cases, to help you compare the metric unit to a familiar unit, the approximate U.S. equivalent may also be given in parentheses directly after the metric unit. For example, in Chapter 4 you will learn that a fetus is 7.5 cm (3 in.) long at the beginning of the fetal period (see Section 4.2). To help you understand the correlation between the metric system and the U.S. system of measurement, see the tables in Appendix A.

 CHECKPOINT

18. After stepping off the scale, your roommate complains that she gained 453.6 grams since the beginning of the semester. How much weight did she gain in pounds?

CHAPTER REVIEW AND RESOURCE SUMMARY   **WileyPLUS**

Review

Resource

1.1 Anatomy Defined

1. Anatomy is the science of body structures and the relationships among structures; physiology is the science of body functions.
2. Branches of anatomy include embryology, developmental biology, cell biology, histology, surface anatomy, sectional anatomy, gross anatomy, systemic anatomy, regional anatomy, radiographic anatomy, and pathological anatomy (see Table 1.1).

1.2 Levels of Body Organization and Body Systems

1. The human body consists of six levels of structural organization: chemical, cellular, tissue, organ, system, and organismal.
2. Cells are the basic structural and functional living units of an organism and the smallest living units in the human body.
3. Tissues are groups of cells and the materials surrounding them that work together to perform a particular function.
4. Organs are composed of two or more different types of tissues; they have specific functions and usually have recognizable shapes.
5. Systems consist of related organs that have a common function.
6. An organism is any living individual.
7. Table 1.2 introduces the 11 systems of the human organism: the integumentary, skeletal, muscular, nervous, endocrine, cardiovascular, lymphatic, respiratory, digestive, urinary, and reproductive systems.

Anatomy Overviews:
  The Integumentary System
  The Skeletal System
  The Muscular System
  The Nervous System
  The Endocrine System
  The Cardiovascular System
  The Lymphatic and Immune Systems
  The Respiratory System
  The Digestive System
  The Urinary System
  The Reproductive Systems
Exercise - Find the System Outsiders
Exercise - Concentrate on Systemic
  Functions

1.3 Life Processes

1. All living organisms have certain characteristics that set them apart from nonliving things.
2. Among the life processes in humans are metabolism, responsiveness, movement, growth, differentiation, and reproduction.

Animation - Communication,
  Regulation, and Homeostasis
Animation - Homeostatic
  Relationships

1.4 Basic Anatomical Terminology

1. Descriptions of any region of the body assume the body is in the anatomical position, in which the subject stands erect facing the observer, with the head level and the eyes facing directly forward. The lower limbs are parallel and the feet are flat on the floor and directed forward. The upper limbs are at the sides, with the palms turned forward.
2. A body lying face down is prone, and a body lying face up is supine.
3. Regional names are terms given to specific regions of the body. The principal regions are the head, neck, trunk, upper limbs, and lower limbs.
4. Within the regions, specific body parts have anatomical names and correponding common names. Examples are thoracic (chest), nasal (nose), and carpal (wrist).
5. Planes are imaginary flat surfaces that are used to divide the body or organs into definite areas. A sagittal plane divides the body or organ into right and left sides. A midsagittal plane divides the body into *equal* right and left sides; a parasagittal plane divides the body into *unequal* right and left sides; a frontal plane divides the body or organ into anterior and posterior portions; a transverse plane divides the body or organ into superior and inferior portions; and an oblique plane passes through the body or an organ at an angle.
6. Sections are cuts of the body or one of its organs made along a plane. Sections are named according to the plane along which the cut is made, and include transverse, frontal, and sagittal.
7. Directional terms indicate the relationship of one part of the body to another.
8. Exhibit 1.A summarizes commonly used directional terms.

Figure 1.2 -
  The Anatomical Position
Figure 1.5 -
  Directional Terms

Concept

Resource

Anatomy Overview - Serous
  Membrane
Figure 1.6 -
  Body Cavities
Figure 1.7 -
  The Thoracic Cavity

1.5 Body Cavities

1. Spaces in the body that house and support internal organs are called body cavities.
2. The cranial cavity contains the brain, and the vertebral canal contains the spinal cord. The meninges are protective tissues that line the cranial cavity and vertebral canal.
3. The diaphragm separates the thoracic cavity from the abdominopelvic cavity. Viscera are organs within the thoracic and abdominopelvic cavities. The membrane of a serous sac lines the wall of the cavity and covers the outside of the viscera.
4. Within the thoracic cavity are two pleurae, each of which surrounds one lung.
5. The central portion of the thoracic cavity is an anatomical region called the mediastinum. It is located between the medial wall of each pleural cavity and extends from the sternum to the vertebral column and from the first rib to the diaphragm. It contains all thoracic viscera except the lungs. The pericardium surrounds the heart.
6. The abdominopelvic cavity is divided into a superior abdominal cavity and an inferior pelvic cavity.
7. Viscera of the abdominal cavity include the kidneys, adrenal glands, stomach, spleen, liver, gallbladder, pancreas, small intestine, and most of the large intestine.
8. Viscera of the pelvic cavity include the urinary bladder, portions of the large intestine, and internal organs of the reproductive system.
9. Sac-like serous membranes line the walls of the thoracic and abdominal cavities and cover the organs within them. They include the pleurae, associated with the lungs; the pericardium, associated with the heart; and the peritoneum, associated with the abdominal cavity.
10. Figure 1.6 summarizes the body cavities and their membranes.

1.6 Abdominopelvic Regions and Quadrants

1. To describe the location of organs easily, the abdominopelvic cavity may be divided into nine regions by drawing four imaginary lines (left midclavicular, right midclavicular, subcostal, and transtubercular).
2. The names of the abdominopelvic regions are right hypochondriac, epigastric, left hypochondriac, right lumbar, umbilical, left lumbar, right iliac (inguinal), hypogastric (pubic), and left iliac (inguinal).
3. To locate the site of an abdominopelvic abnormality in clinical studies, the abdominopelvic cavity may be divided into quadrants by passing an imaginary transverse line (transumbilical) and a midsagittal line (median line) through the umbilicus.
4. The names of the abdominopelvic quadrants are right upper quadrant (RUQ), left upper quadrant (LUQ), right lower quadrant (RLQ), and left lower quadrant (LLQ).

1.7 The Human Body and Disease

1. A disorder is a general term for any abnormality of structure and/or function. A disease is an illness with a definite set of symptoms and signs.
2. Symptoms are subjective changes in body functions that are not apparent to an observer. Signs are objective changes that can be observed and/or measured.

1.8 Medical Imaging

1. Medical imaging refers to techniques and processes used to create images of the human body. They allow visualization of internal structures to diagnose abnormal anatomy and deviations from normal physiology.
2. Table 1.3 describes and illustrates several medical imaging techniques.

1.9 Measuring the Human Body

1. Measurements involving time, weight, temperature, size, and volume are used in clinical situations.
2. Measurements in this book are given in metric units; in many cases these are followed by U.S. equivalents in parentheses.

CRITICAL THINKING QUESTIONS

1. Eight-year-old Taylor was going for the record for the longest upside-down hang from the monkey bars. She didn't make it and she may have done extensive damage to her entire upper limb. The emergency room technician would like an x-ray film of her arm in the anatomical position. Use the proper anatomical terms to describe the position of Taylor's arm, forearm, and hand in the x-ray.

2. An alien landed in your backyard, abducted your cat, and flew off. Being an observant student of anatomy, you later described the alien's appearance to the FBI as follows: "It had two caudal extensions, six bilateral extremities, four axillae, and one oral orifice in place of an umbilicus." What did the alien look like?

3. Your anatomy professor displays an MRI scan that shows a parasagittal section of the torso taken through the midpoint of the left mammary gland. Name five organs you would expect to see in this image. (You may consult a general photo of the human body.)

4. Mikhail has been diagnosed with a ruptured appendix, which has allowed bacteria from his intestinal tract to infect his peritoneum. The doctors are very concerned. Why do they consider this condition (peritonitis) to be so dangerous?

? ANSWERS TO FIGURE QUESTIONS

1.1 Organs are composed of two or more different types of tissues that work together to perform a specific function.

1.2 Having one standard anatomical position allows directional terms to be clearly defined, so that any body part can be described in relation to any other part.

1.3 The frontal plane divides the heart into anterior and posterior portions.

1.4 The parasagittal plane divides the brain into unequal right and left portions.

1.5 No, the radius is distal to the humerus; No, the esophagus is posterior to the trachea; Yes, the ribs are superficial to the lungs; Yes, the urinary bladder is medial to the ascending colon; No, the sternum is medial to the descending colon.

1.6 Urinary bladder = P, stomach = A, heart = T, small intestine = A, lungs = T, internal female reproductive organs = P, thymus = T, spleen = A, liver = A.

1.7 Structures in the mediastinum include the heart, esophagus, trachea, and thymus.

1.8 The liver is mostly in the epigastric region; the transverse colon is in the umbilical region; the urinary bladder is in the hypogastric region; the spleen is in the left hypochondriac region.

1.9 The pain associated with appendicitis would be felt in the right lower quadrant (RLQ).

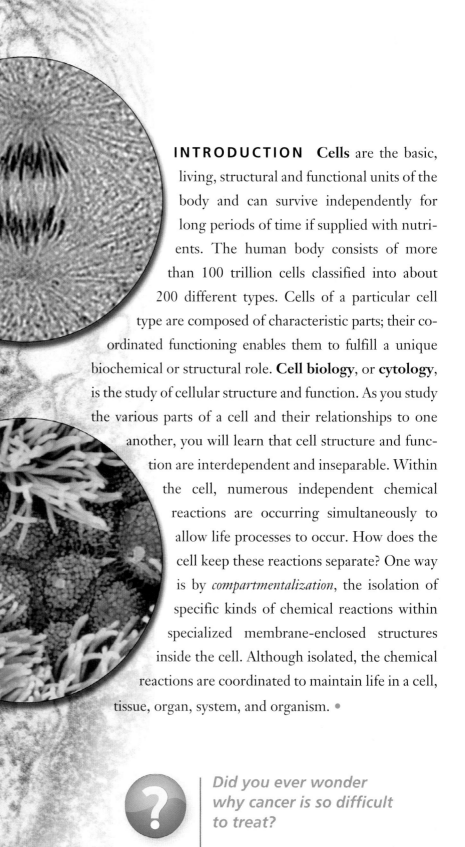

2 | CELLS

INTRODUCTION **Cells** are the basic, living, structural and functional units of the body and can survive independently for long periods of time if supplied with nutrients. The human body consists of more than 100 trillion cells classified into about 200 different types. Cells of a particular cell type are composed of characteristic parts; their coordinated functioning enables them to fulfill a unique biochemical or structural role. **Cell biology**, or **cytology**, is the study of cellular structure and function. As you study the various parts of a cell and their relationships to one another, you will learn that cell structure and function are interdependent and inseparable. Within the cell, numerous independent chemical reactions are occurring simultaneously to allow life processes to occur. How does the cell keep these reactions separate? One way is by *compartmentalization*, the isolation of specific kinds of chemical reactions within specialized membrane-enclosed structures inside the cell. Although isolated, the chemical reactions are coordinated to maintain life in a cell, tissue, organ, system, and organism. •

Did you ever wonder why cancer is so difficult to treat?

2.1 A GENERALIZED CELL

 OBJECTIVE
- Name and describe the three principal parts of a cell.

As you just learned, there are about 200 different types of cells in the body. All cells share many features, but each cell type is unique in some way. Figure 2.1 is a composite of many different cell types. Most cells have many of the features shown in this diagram. For ease of study, a cell can be divided into three principal parts: plasma membrane, cytoplasm, and nucleus.

1. The **plasma membrane** forms the cell's flexible outer surface; it separates the cell's *internal environment* (everything inside the cell) from its *external environment* (everything outside the cell). This selective barrier regulates the flow of materials into and out of a cell, helping to establish and maintain the appropriate environment for normal cellular activities. The plasma membrane also plays a key role in communication among cells and between cells and their external environment. The structure–function relationships of the cell membrane are responsible for many of the functions of the human body that you will encounter in your studies.

Figure 2.1 The cell.

 The cell is the basic, living, structural and functional unit of the body.

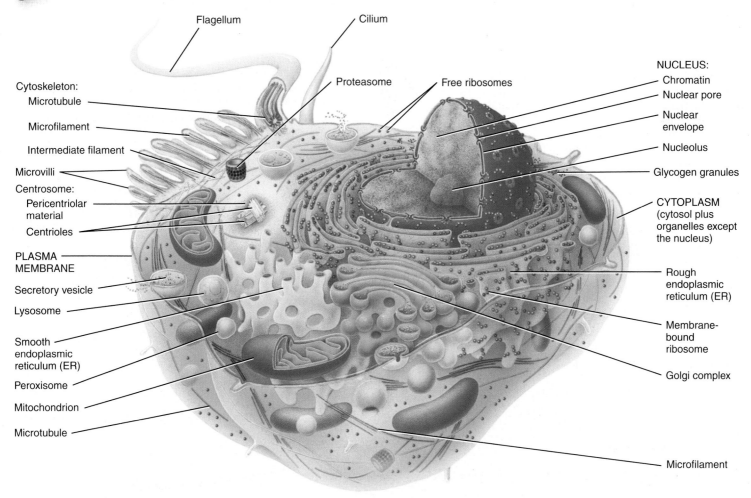

Sectional view

 What are the three main parts of a cell?

2. The **cytoplasm** (SĪ-tō-plazm; *-plasm*=formed or molded) is all of the cellular contents between the plasma membrane and the nucleus. This compartment has two components: the cytosol and organelles. **Cytosol** (SĪ-tō-sol), the fluid portion of cytoplasm, also called **intracellular fluid**, contains water, dissolved solutes, and suspended particles. Within the cytosol are several different types of **organelles** (or-gan-ELS=little organs). Each organelle has a characteristic shape and specific functions. Examples include ribosomes, endoplasmic reticulum, Golgi complex, lysosomes, peroxisomes, and mitochondria.

3. The **nucleus** (NOO-klē-us=nut or kernel) is a large organelle that houses most of a cell's DNA. Within the nucleus, each **chromosome** (KRŌ-mō-sōm; *chromo-*=colored), a single molecule of DNA associated with several proteins, contains thousands of hereditary units called **genes** that control most aspects of cellular structure and function.

 CHECKPOINT

1. Describe the general features of the three main parts of a cell.

2.2 THE PLASMA MEMBRANE

 OBJECTIVES

• Describe the structure and functions of the plasma membrane.
• Outline the processes that transport substances across the plasma membrane.

The plasma membrane is a flexible yet sturdy barrier that surrounds and contains the cytoplasm of a cell. Its structure is described using the **fluid mosaic model**. According to this model, the molecular arrangement of the plasma membrane resembles a continually moving sea of lipids that contains a "mosaic" of many different proteins (Figure 2.2). The proteins may float freely (like icebergs) or may be anchored at specific locations (like islands). The lipids act as a barrier to the passage of various substances into and out of the cell, while some of the proteins in the plasma membrane act as "gatekeepers," regulating the travels of other molecules and ions (charged particles) between the cell's external and internal environments.

Structure of the Membrane

The basic structural framework of the plasma membrane is the **lipid bilayer**, two back-to-back layers made up of three types of lipid molecules—phospholipids, cholesterol, and glycolipids (Figure 2.2). About 75 percent of the membrane lipids are *phospholipids*, lipids that contain phosphorus. About 20 percent of plasma membrane lipids are *cholesterol* molecules, which are interspersed among the other lipids in both layers of the membrane. *Glycolipids*, which are lipids attached to carbohydrates (*glyco-*= carbohydrate), account for the other 5 percent.

Membrane proteins are divided into two categories—integral and peripheral—according to whether they are firmly embedded in the membrane (Figure 2.2). **Integral proteins** extend into or through the lipid bilayer and are firmly embedded in it. Most integral proteins are **transmembrane proteins**, which means that they span the entire lipid bilayer and protrude into both the cytosol and extracellular fluid. **Peripheral proteins** (pe-RIF-er-al), by contrast, are not as firmly embedded in the membrane and are attached to membrane lipids or integral proteins at the inner or outer surface of the membrane.

Figure 2.2 Structure of the plasma membrane.

 The basic framework of the plasma membrane is the lipid bilayer.

FUNCTIONS OF THE PLASMA MEMBRANE

1. Acts as a barrier separating inside and outside of the cell.
2. Controls the flow of substances into and out of the cell.
3. Helps identify the cell to other cells (e.g., immune cells).
4. Participates in intercellular signalling.

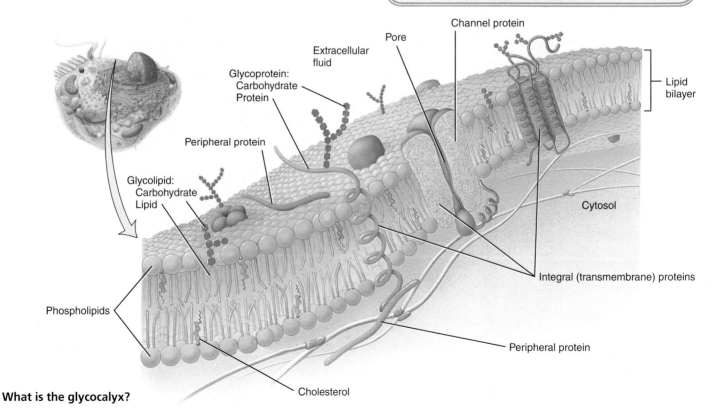

 What is the glycocalyx?

Many integral membrane proteins are **glycoproteins**, proteins with carbohydrate groups attached to the ends that protrude into the extracellular fluid. The carbohydrate portions of glycolipids and glycoproteins form an extensive sugary coat called the **glycocalyx** (glī-kō-KĀL-iks), which has a number of important functions. The composition of the glycocalyx acts like a molecular "signature" that enables cells to recognize one another. For example, a white blood cell's ability to detect a "foreign" glycocalyx is one basis of the immune response that helps your body destroy invading organisms. In addition, the glycocalyx enables cells to adhere to one another in some tissues, and it protects cells from being digested by enzymes in the extracellular fluid. Enzymes are proteins that speed up a chemical reaction, such as digestion. The chemical properties of the glycocalyx attract a film of fluid to the surface of many cells. This action makes red blood cells slippery as they flow through narrow blood vessels and protects cells that line the airways and the gastrointestinal tract from drying out.

Functions of Membrane Proteins

Generally, the types of lipids in cellular membranes vary only slightly from one type of membrane to another. It is the remarkable assortment of proteins in the membranes of different cells and in various intracellular organelles that determine many of the membrane's functions.

- Some integral membrane proteins form **ion channels**, *pores* or holes through which specific ions, such as potassium ions (K^+), can flow through to gain entry or leave the cell.
- Other integral proteins act as **carriers** or **transporters**, selectively moving a polar substance (one having two opposite poles) or ion from one side of the membrane to the other.
- Integral proteins called **receptors** serve as cellular recognition sites. Each type of receptor recognizes and binds a specific type of molecule. For instance, insulin receptors bind the hormone insulin. A specific molecule that binds to a receptor is called a **ligand** (LĪ-gand; *liga-*=tied) of that receptor.
- Some integral proteins are **enzymes** that catalyze specific chemical reactions at the inside or outside surface of the cell.
- Integral proteins may also serve as **linkers**, proteins that anchor the plasma membranes of neighboring cells to one another or to protein filaments inside and outside the cell. Peripheral proteins also serve as linkers.
- Membrane glycoproteins and glycolipids often serve as **cell-identity markers**, enabling a cell to (1) recognize other cells of the same kind during tissue formation or (2) recognize and respond to potentially dangerous foreign cells. The ABO blood-type markers are one example of cell-identity markers. When you receive a blood transfusion, the blood type must be compatible with your own, or red blood cells may clump together.

In addition, peripheral proteins help support the plasma membrane, anchor integral proteins, and participate in mechanical activities such as moving materials and organelles within cells, changing cell shape in dividing cells and muscle cells, and attaching cells to one another.

Membrane Permeability

The term *permeable* means that a structure permits the passage of substances through it, while *impermeable* means that a structure does not permit the passage of substances. Although plasma membranes are not completely permeable to any substance, they do permit some substances to pass more readily than others. This property of membranes is called **selective permeability** (per′-mē-a-BIL-i-tē).

The lipid bilayer portion of the membrane is permeable to molecules such as oxygen, carbon dioxide, and steroids, but is impermeable to ions and molecules such as glucose. It is also permeable to water. Transmembrane proteins that act as channels and transporters increase the plasma membrane's permeability to a variety of small- and medium-sized charged substances (including ions) that cannot cross the lipid bilayer without help. These proteins are very selective—each one helps only a specific molecule or ion to cross the membrane. Macromolecules, such as proteins, cannot pass through the plasma membrane except by the processes of endocytosis and exocytosis (discussed later in this chapter).

 CHECKPOINT

2. What is the composition of the lipid bilayer?
3. Distinguish integral proteins from peripheral proteins.
4. What are the major functions of membrane proteins?

Transport Across the Plasma Membrane

Before discussing how materials move into and out of cells, you need to understand the locations of the various fluids through which the substances move (Figure 2.3). Fluid within cells is called **intracellular fluid (ICF)** (*intra-*=within, inside). Fluid outside body cells, called **extracellular fluid (ECF)** (*extra-*=outside), is found in several locations: (1) The ECF filling the microscopic spaces between the cells of tissues is called

Figure 2.3 Body fluids. Intracellular fluid (ICF) is the fluid within cells. Extracellular fluid (ECF) is found outside cells: in blood vessels as plasma, in lymphatic vessels as lymph, and between tissue cells as interstitial fluid.

 Plasma membranes regulate fluid movements from one compartment to another.

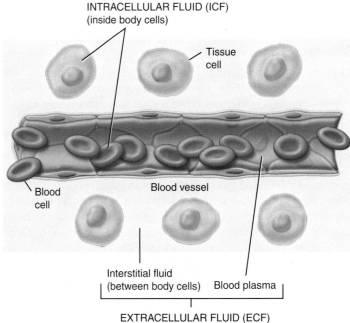

INTRACELLULAR FLUID (ICF)
(inside body cells)

Tissue cell

Blood cell

Blood vessel

Interstitial fluid
(between body cells)   Blood plasma

EXTRACELLULAR FLUID (ECF)
(outside body cells)

 What is another name for intracellular fluid? (*Hint*: *intra-* means "within.")

interstitial fluid (in-ter-STISH-al) (*inter-*=between; *stit*=to place or set), or *intercellular fluid*. (2) The ECF in blood vessels is termed **plasma**; in lymphatic vessels it is called **lymph**. Among the substances in extracellular fluid are gas molecules, nutrient molecules, and ions—all needed for the maintenance of life.

Extracellular fluid circulates through the blood vessels (circulatory vessels), returns through lymphatic vessels (accessory drainage vessels), and moves into and out of the spaces between the tissue cells and capillaries of these two categories of vessels. Thus, it is in constant motion throughout the body. Essentially, all body cells are surrounded by the same fluid environment. The movement of substances across a plasma membrane and across membranes within cells is essential to the life of the cell and the organism. Certain substances—oxygen, for example—must move into the cell to support life, and waste materials or harmful substances must be moved out. Plasma membranes regulate the movements of such materials.

Substances generally move across cellular membranes via transport processes that can be classified as passive or active, depending on whether they require cellular energy. In **passive processes**, a substance moves down its concentration gradient or electrical gradient to cross the membrane using only its own kinetic energy. There is no input of energy from the cell. In **active processes**, cellular energy is used to drive the substance "uphill" against its concentration or electrical gradient. The cellular energy used is usually in the form of **adenosine triphosphate (ATP)**.

Materials may cross plasma membranes by using (1) kinetic energy, (2) transporter proteins, or (3) vesicles. Some materials get across simply by moving through the lipid bilayer or membrane channels using their own *kinetic energy* (energy of motion). Kinetic energy is intrinsic to the particles that are moving. Processes that rely on kinetic energy to pass through the plasma membrane include diffusion and osmosis. Other substances must bind to specific transporter proteins to piggyback across the membrane, as in facilitated diffusion and active transport. Still other substances pass through cellular membranes within small, spherical sacs called *vesicles* that bud off from an existing membrane. Examples include endocytosis, in which vesicles detach from the plasma membrane while bringing materials into a cell, and exocytosis, the merging of vesicles with the plasma membrane to release materials from a cell.

Kinetic Energy Transport

DIFFUSION Diffusion (di-FŪ-zhun; *diffus-*=spreading) is a passive process in which the *net* movement of a substance is from a region of higher concentration to a region of lower concentration—that is, the substance moves from an area where there is more of it to an area where there is less of it. The substance moves because of its kinetic energy; diffusion continues until *equilibrium* (ē′-kwi-LIB-rē-um) is reached, that is, the substance becomes evenly distributed. A good example of diffusion in the body occurs in the lungs. As we inhale, air with a high concentration of oxygen enters the lungs. The lungs receive blood from the heart that has a high concentration of carbon dioxide. Within the lungs the oxygen moves from its site of high concentration in the air spaces in the lungs into the blood and the carbon dioxide moves from its site of high concentration in the blood into the air spaces. This movement brings oxygen into the blood and removes carbon dioxide, which we exhale.

OSMOSIS Another passive process is **osmosis** (oz-MŌ-sis; *osmo-*=a pushing), the net movement of water molecules through a selectively permeable membrane from an area of higher water concentration (lower concentration of *solutes*, dissolved substances) to an area of lower water concentration (higher solute concentration). Due to their kinetic energy, water molecules pass through aquaporins (a-kwa-POR-ins), pores (holes) made of integral proteins, and between neighboring phospholipid molecules in the membrane, and movement continues until equilibrium is reached. Water moves between various compartments of the body by osmosis.

Transport by Transporter Proteins

FACILITATED DIFFUSION **Facilitated diffusion** is a passive process that is accomplished with the assistance of transmembrane proteins functioning as carriers. This process allows some molecules that are too large to fit through the protein pores and others that are insoluble in lipids to pass through the plasma membrane. Among these are various sugars, especially glucose. Glucose is the body's preferred energy source for making ATP. In facilitated diffusion, glucose binds to a specific carrier protein on one side of the plasma membrane, the carrier changes shape, and glucose is released on the opposite side.

ACTIVE TRANSPORT The process by which substances are transported across plasma membranes with the expenditure of energy by the cell, typically from an area of lower concentration to an area of higher concentration, is called **active transport**. In active transport the substance being moved, usually an ion, makes contact with a specific site on a transporter protein. Then ATP splits, and the energy from its breakdown causes a change in the shape of the transporter protein that expels the substance on the opposite side of the membrane. Active transport is considered an active process because energy is required for transporter proteins to move substances across the membrane against a concentration gradient. Active transport is vitally important in maintaining ion concentrations in both body cells and extracellular fluids. For example, before a nerve cell can conduct a nerve impulse, the concentration of potassium ions (K^+) must be considerably higher inside the nerve cell than outside, and the concentration of sodium ions (Na^+) must be higher outside, than inside. Active transport makes this balancing act possible.

Transport in Vesicles

A **vesicle** (VES-i-kul; *vescula*=little blister or bladder) is a small, spherical, membranous sac formed by budding off from an existing membrane. Vesicles transport substances from one structure to another within cells, take in substances from extracellular fluid, or release substances into extracellular fluid. In **endocytosis** (en′-dō-sī-TŌ-sis; *endo-*=within), materials move into a cell in a vesicle formed from the plasma membrane. In **exocytosis** (ek-sō-sī-TŌ-sis; *exo-*=out), materials move out of a cell by the fusion of vesicles formed inside a cell with the plasma membrane. Both endocytosis and exocytosis require cellular energy supplied by the breakdown of ATP. Thus transport in vesicles is an active process.

ENDOCYTOSIS Here we consider three types of endocytosis: receptor-mediated endocytosis, phagocytosis, and bulk-phase endocytosis.

In **receptor-mediated endocytosis**, which is highly selective, cells take up specific ligands. (Recall that *ligands* are molecules that bind to specific receptors.) A vesicle forms after a receptor protein in the plasma membrane recognizes and binds to a particular particle in the extracellular fluid. For instance, cells take up cholesterol contained in low-density lipoproteins (LDLs), transferrin (an iron-transporting protein in the blood), some vitamins, antibodies, and certain hormones by receptor-mediated endocytosis. Receptor-mediated endocytosis of LDLs (and other ligands) occurs as follows (Figure 2.4):

❶ *Binding.* On the extracellular side of the plasma membrane, an LDL particle that contains cholesterol binds to a specific receptor in the plasma membrane to form a receptor–LDL complex. The receptors are integral membrane proteins that are concentrated in regions of the plasma membrane called *clathrin-coated pits* (KLATH-rin). Here, a protein called *clathrin* attaches to the membrane on its cytoplasmic side. Many clathrin molecules come together, forming a basketlike structure around the receptor–LDL complexes that causes the membrane to *invaginate* (fold inward).

❷ *Vesicle formation.* The invaginated edges of the membrane around the clathrin-coated pit fuse, and a small piece of the membrane pinches off. The resulting vesicle, known as a *clathrin-coated vesicle*, contains the receptor-LDL complexes.

❸ *Uncoating.* Almost immediately after it is formed, the clathrin-coated vesicle loses its clathrin coat to become an *uncoated vesicle.* Clathrin molecules either return to the inner surface of the plasma membrane or help form coats on other vesicles inside the cell.

❹ *Fusion with endosome.* The uncoated vesicle quickly fuses with another vesicle known as an *endosome.* Within an endosome, the LDL particles separate from their receptors.

❺ *Recycling of receptors to plasma membrane.* Most of the receptors accumulate in elongated protrusions of the endosome (the arms of the cross-shaped vesicle at the center of the figure). These pinch off, forming transport vesicles that return the receptors to the plasma membrane. An LDL receptor is returned to the plasma membrane about 10 minutes after it enters a cell.

❻ *Degradation in lysosomes.* Other transport vesicles, which contain the LDL particles, bud off the endosome and soon fuse with a *lysosome.* Lysosomes contain many digestive enzymes. Certain enzymes break down the large protein and lipid molecules of the LDL particle into amino acids, fatty acids, and cholesterol. These smaller molecules then leave the lysosome. The cell uses cholesterol for rebuilding its membranes and for synthesis of steroids, such as estrogen. Fatty acids and amino acids can be used for ATP production or to build other molecules needed by the cell.

Phagocytosis (fag-ō-sī-TŌ-sis; *phago-*=to eat) is a form of endocytosis in which the cell engulfs large solid particles, such as worn-out cells, whole bacteria, or viruses (Figure 2.5). Only a few body cells, termed **phagocytes** (FAG-ō-sīts), are able to carry out phagocytosis. Two main types of phagocytes are *macrophages*, located in many body tissues, and *neutrophils*, a type of white blood cell. Phagocytosis begins when the particle binds to a plasma membrane receptor on the phagocyte, causing it to extend **pseudopods** (SOO-dō-pods; *pseudo-*=false; *-pods*=feet), projections of its plasma membrane and cytoplasm. Pseudopods surround the particle outside the cell, and the membranes fuse to form a vesicle called a *phagosome*, which enters the cytoplasm. The phagosome fuses with one or more lysosomes, and lysosomal enzymes break

Figure 2.4 Receptor-mediated endocytosis of a low-density lipoprotein (LDL) particle.

🔑 Receptor-mediated endocytosis imports materials that are needed by cells.

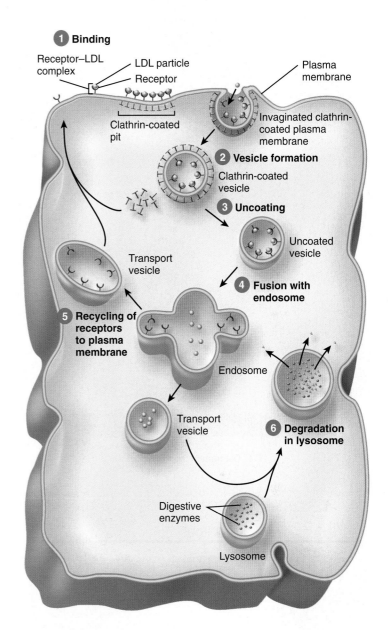

CLINICAL CONNECTION | *Viruses and Receptor-mediated Endocytosis*

Although receptor-mediated endocytosis normally imports needed materials, some viruses are able to use this mechanism to enter and infect body cells. For example, the human immunodeficiency virus (HIV), which causes acquired immunodeficiency syndrome (AIDS), can attach to a receptor called CD4. This receptor is present in the plasma membrane of white blood cells called helper T cells. After binding to CD4, HIV enters the helper T cell via receptor-mediated endocytosis. •

❓ **What are several other examples of ligands that can undergo receptor-mediated endocytosis?**

Figure 2.5 Phagocytosis. Pseudopods surround a particle and the membranes fuse to form a phagosome.

🔑 Phagocytosis is a vital defense mechanism that helps protect the body from disease.

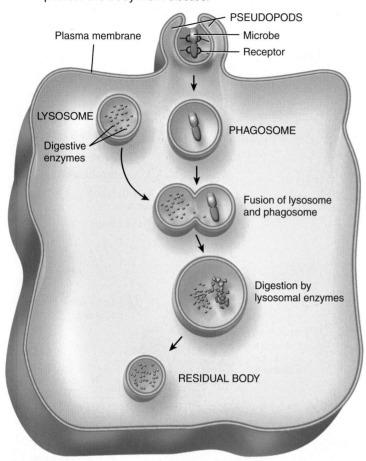

(a) Diagram of the process

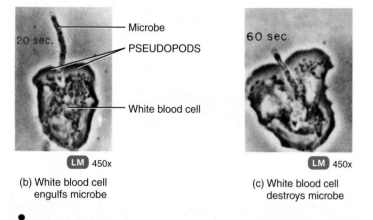

(b) White blood cell engulfs microbe

(c) White blood cell destroys microbe

CLINICAL CONNECTION | *Phagocytosis and Microbes*

The process of phagocytosis is a vital defense mechanism that helps protect the body from disease. Through phagocytosis, macrophages dispose of invading microbes and billions of aged, worn-out red blood cells every day; neutrophils also help rid the body of invading microbes. **Pus** is a mixture of dead neutrophils, macrophages, tissue cells, and fluid in an infected wound. •

❓ **What triggers pseudopod formation?**

down the ingested material. In most cases, any undigested materials in the phagosome remain indefinitely in a vesicle called a *residual body*. The residual bodies are then either secreted by the cell via exocytosis, or they remain stored indefinitely in the cell as lipofuscin granules.

Most body cells carry out **bulk-phase endocytosis**, also called **pinocytosis** (pī-nō-sī-TŌ-sis; *pino-*=to drink), a form of endocytosis in which tiny droplets of extracellular fluid are taken into the cell. No receptor proteins are involved; all solutes dissolved in the extracellular fluid are brought into the cell. During bulk-phase endocytosis, the plasma membrane folds inward and forms a vesicle containing a droplet of extracellular fluid. The vesicle detaches or "pinches off" from the plasma membrane and enters the cytosol. Within the cell, the vesicle fuses with a lysosome, where enzymes degrade the engulfed solutes. The resulting smaller molecules, such as amino acids and fatty acids, leave the lysosome to be used elsewhere in the cell. Bulk-phase endocytosis occurs in most cells, especially absorptive cells in the intestines and kidneys.

EXOCYTOSIS In contrast with endocytosis, which brings materials into a cell, **exocytosis** releases materials from a cell. Just remember that *endo* means "in," and *exo* means "out." All cells carry out exocytosis, but it is especially important in two types of cells: (1) secretory cells that liberate digestive enzymes, hormones, mucus, or other secretions; and (2) nerve cells that release substances called *neurotransmitters*. In some cases, wastes are also released by exocytosis. During exocytosis, membrane-enclosed vesicles called *secretory vesicles* form inside the cell, fuse with the plasma membrane, and release their contents into the extracellular fluid (see Figure 2.1).

Segments of the plasma membrane lost through endocytosis are recovered or recycled by exocytosis. The balance between endocytosis and exocytosis keeps the surface area of a cell's plasma membrane relatively constant. Membrane exchange is quite extensive in certain cells. In your pancreas, for example, the cells that secrete digestive enzymes can recycle an amount of plasma membrane equal to the cell's entire surface area in 90 minutes.

TRANSCYTOSIS Transport in vesicles may also be used to successively move a substance into, across, and out of a cell. In this active process, called **transcytosis** (tranz′-sī-TŌ-sis), vesicles undergo endocytosis on one side of a cell, move across the cell, and then undergo exocytosis on the opposite side. As the vesicles fuse with the plasma membrane, the vesicular contents are released into the extracellular fluid. Transcytosis occurs most often across the epithelial cells that line blood vessels and is a means for materials to move between blood plasma and interstitial fluid. For instance, when a woman is pregnant, some of her antibodies cross the placenta into the fetal circulation via transcytosis.

Table 2.1 summarizes the processes by which materials move into and out of cells.

✓ CHECKPOINT

5. How are passive and active processes similar? How do they differ?
6. What are the roles of simple diffusion, facilitated diffusion, osmosis, and active transport in the homeostasis of the human body?
7. Describe each type of transport in vesicles and explain its importance to the body.

TABLE 2.1

Transport of Materials Into and Out of Cells

TRANSPORT PROCESS	DESCRIPTION
Kinetic energy transport	
Diffusion	A passive process in which a substance moves from an area of higher to lower concentration until equilibrium is reached.
Osmosis	A passive process that involves the movement of water molecules across a selectively permeable membrane from an area of higher water concentration to an area of lower water concentration until equilibrium is reached.
Transport by transporter proteins	
Facilitated diffusion	Passive movement of a substance down its concentration gradient via transmembrane proteins that act as transporters.
Active transport	An active process in which cell expends energy to move a substance across the membrane against its concentration gradient through transmembrane proteins that act as transporters.
Transport in vesicles	An active process that involves the movement of substances into or out of a cell in vesicles that bud from the plasma membrane.
Endocytosis	Movement of substances into a cell in vesicles.
Receptor-mediated endocytosis	Ligand–receptor complexes trigger infolding of a clathrin-coated pit that forms a vesicle containing ligands.
Phagocytosis	"Cell eating"; movement of a solid particle into a cell after pseudopods engulf it to form a phagosome.
Bulk-phase endocytosis	"Cell drinking"; movement of extracellular fluid into a cell by infolding of plasma membrane to form a vesicle.
Exocytosis	Movement of substances out of a cell in secretory vesicles that fuse with the plasma membrane and release their contents into the extracellular fluid.
Transcytosis	Movement of a substance through a cell as a result of endocytosis on one side and exocytosis on the opposite side.

2.3 CYTOPLASM

 OBJECTIVE

• Describe the structure and function of cytoplasm, cytosol, and organelles.

As you have already learned, cytoplasm consists of all the cellular contents inside the plasma membrane except for the nucleus. It has two components: (1) cytosol and (2) organelles, tiny structures that perform different functions in the cell.

Cytosol

The *cytosol (intracellular fluid)*, the fluid portion of the cytoplasm that surrounds organelles (see Figure 2.1), constitutes about 55 percent of total cell volume. Although it varies in composition and consistency from one part of a cell to another, cytosol is 75–90 percent water plus various dissolved and suspended components. Among these are different types of ions, glucose, amino acids, fatty acids, proteins, lipids, ATP, and waste products. Also present in some cells are various organic molecules that aggregate into masses for storage. These aggregations may appear and disappear at different times in the life of a cell. Examples include *lipid droplets* that contain triglycerides, and clusters of glycogen molecules called *glycogen granules* (see Figure 2.1). Triglycerides are fats and oils and are the body's most concentrated source of energy. Glycogen is a stored form of glucose (blood sugar).

The cytosol is the site of many chemical reactions required for a cell's existence. For example, enzymes in cytosol catalyze numerous chemical reactions that release and capture energy to drive cellular activities. In addition, some of these reactions provide the building blocks for maintaining cell structure, function, and growth.

The **cytoskeleton** is a network of protein filaments that extends throughout the cytosol (see Figure 2.1). Three types of filamentous proteins contribute to the structure of the cytoskeleton and other organelles. In the order of their increasing diameter, these structures are microfilaments, intermediate filaments, and microtubules.

Microfilaments (mī-krō-FIL-a-ments), the thinnest elements of the cytoskeleton, are concentrated at the periphery (near the plasma membrane) of a cell (Figure 2.6a). They are composed of the proteins *actin* and *myosin* and have two general functions: movement and mechanical support. With respect to movement, microfilaments are involved in muscle contraction, cell division, and cell locomotion. Cell locomotion occurs during the migration of embryonic cells during development, the invasion of tissues by white blood cells to fight infection, and the migration of skin cells during wound healing.

Microfilaments provide much of the mechanical support that is responsible for the basic strength and shapes of cells. They anchor the cytoskeleton to integral proteins in the plasma membrane. Microfilaments also provide mechanical support for nonmotile, microscopic fingerlike projections of the plasma membrane called **microvilli** (mī-krō-VIL-ī; *micro-*=small; *-villi*=tufts of hair; singular

Figure 2.6 Cytoskeleton.

 The cytoskeleton is a network of three kinds of protein filaments that extend throughout the cytosol: microfilaments, intermediate filaments, and microtubules.

FUNCTIONS OF THE CYTOSKELETON

1. Aids movement of organelles within the cell, of chromosomes during cell division, and of whole cells such as phagocytes.

2. Serves as a scaffold that helps to determine a cell's shape and to organize the cellular contents.

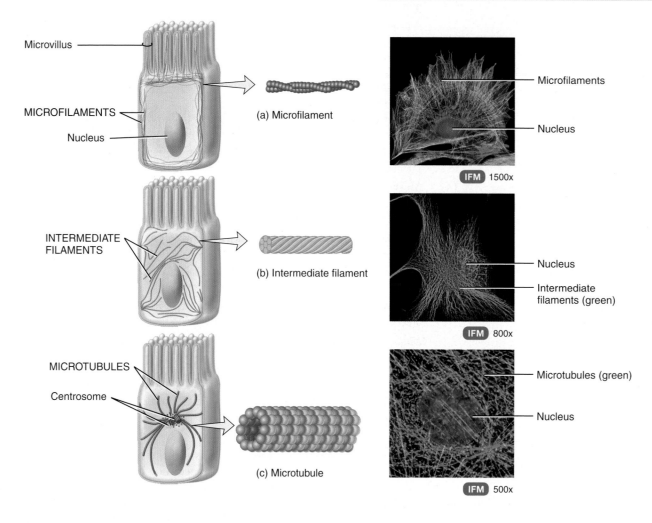

(a) Microfilament

(b) Intermediate filament

(c) Microtubule

Which cytoskeletal component helps form the structure of centrioles, cilia, and flagella?

is *microvillus*). A core of parallel microfilaments within a microvillus supports it and attaches it to other parts of the cytoskeleton (Figure 2.6a). Microvilli increase the surface area of the plasma membrane and are abundant on the surfaces of cells involved in absorption, such as the epithelial cells that line the small intestine. Some microfilaments extend beyond the plasma membrane and help cells attach to one another or to extracellular materials.

As their name suggests, **intermediate filaments** are thicker than microfilaments but thinner than microtubules (Figure 2.6b). Several different proteins can compose intermediate filaments, which are exceptionally strong. Found in parts of cells subject to mechanical stress, they help anchor organelles such as the nucleus and attach cells to one another.

The largest of the cytoskeletal components, **microtubules** (mī-krō-TOO-būls) are long, unbranched hollow tubes composed mainly of a protein called *tubulin*. The centrosome (discussed shortly) serves as the initiation site for the assembly of microtubules. The microtubules grow outward from the centrosome toward the periphery of the cell (Figure 2.6c). Microtubules

help determine cell shape and function in the intracellular transport of organelles, such as secretory vesicles, and the migration of chromosomes during cell division. They also participate in the movement of specialized cell projections such as cilia and flagella.

Organelles

As noted earlier in the chapter, *organelles* are specialized structures within the cell that have characteristic shapes and perform specific functions in cellular growth, maintenance, and reproduction. Despite the many chemical reactions going on in a cell at any given time, there is little interference because the reactions are confined to different organelles. Each type of organelle has its own set of enzymes that carry out specific reactions, and serves as a functional compartment for specific biochemical processes. The numbers and types of organelles vary in different cells, depending on the cell's function. Although the nucleus is technically an organelle, it is discussed in a separate section because of its special importance in directing the life of a cell.

Centrosome

The **centrosome** (SEN-trō-sōm), located near the nucleus, consists of two components: a pair of centrioles and pericentriolar material (Figure 2.7a). The two **centrioles** (SEN-trē-ōls) are cylindrical structures, each composed of nine clusters of three microtubules (triplets) arranged in a circular pattern (Figure 2.7b). The long axis of one centriole is at a right angle to the long axis of the other (Figure 2.7a). Surrounding the centrioles is **pericentriolar material** (per′-ē-sen′-trē-Ō-lar), which contains hundreds of ring-shaped complexes composed of the protein *tubulin*. These tubulin complexes are the organizing centers for growth of the mitotic spindle, which plays a critical role in cell division, and for microtubule formation in nondividing cells. During cell division, centrosomes replicate so that succeeding generations of cells have the capacity for cell division.

Cilia and Flagella

Microtubules are the dominant components of cilia and flagella, which are motile projections of the cell surface (see Figure 2.1). **Cilia** (SIL-ē-a=eyelashes; singular is *cilium*) are numerous, short, hairlike projections that extend from the surface of the cell (see Figure 2.1). Each cilium contains a core of 20 microtubules surrounded by plasma membrane (Figure 2.8a). These 20 microtubules are arranged with one pair in the center surrounded by

nine clusters of two fused microtubules (doublets). Each cilium is anchored to a *basal body* just below the surface of the plasma membrane. A basal body is similar in structure to a centriole and functions in initiating the assembly of cilia and flagella.

The coordinated movement of many cilia on the surface of a cell causes the steady movement of fluid along the cell's surface. Many cells of the respiratory tract, for example, have hundreds of cilia that help sweep foreign particles trapped in mucus away from the lungs. The extremely thick mucus secretions that are produced in people who suffer from cystic fibrosis interfere with ciliary action and the normal functions of the respiratory tract.

Flagella (fla-JEL-a=whip; singular is *flagellum*) are similar in structure to cilia but are typically much longer (see Figure 2.1). Flagella usually move an entire cell. The only example of a flagellum in the human body is a sperm cell's tail, which propels the sperm toward the oocyte in the uterine tube.

Ribosomes

Ribosomes (RĪ-bō-sōms; *-somes*=bodies) are sites of protein synthesis. These tiny structures are packages of **ribosomal RNA (rRNA)** and many ribosomal proteins. Ribosomes are so named because of their high content of *ribonucleic* acid. Structurally, a ribosome consists of two subunits, one about half the size of the

---

Figure 2.7 Centrosome.

 Located near the nucleus, the centrosome consists of a pair of centrioles and pericentriolar material.

FUNCTIONS OF CENTROSOMES
1. The pericentriolar material of the centrosome contains tubulins that build microtubules in nondividing cells.   2. The pericentriolar material forms the mitotic spindle during cell division.

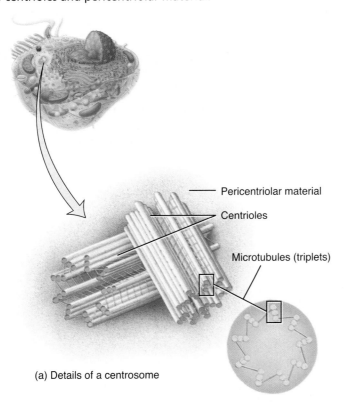

(a) Details of a centrosome

Pericentriolar material

Centrioles

Microtubules (triplets)

(b) Arrangement of microtubules in centrosome

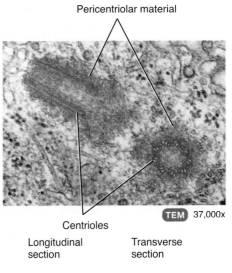

Pericentriolar material

TEM 37,000x

Centrioles

Longitudinal section   Transverse section

(c) Centrioles

 If you observed that a cell did not have a centrosome, what could you predict about its capacity for cell division?

Figure 2.8 Cilia and flagella.

 A cilium contains a core of microtubules with one pair in the center surrounded by nine clusters of doublet microtubules.

FUNCTIONS OF CILIA AND FLAGELLA

1. Cilia move fluids along a cell's surface.

2. A flagellum moves an entire cell.

CLINICAL CONNECTION | *Cilia and Smoking*

The movement of cilia is also paralyzed by nicotine in cigarette smoke. For this reason, smokers cough often to remove foreign particles from their airways. Because cells that line the uterine (fallopian) tubes also have cilia that sweep oocytes (egg cells) toward the uterus, females who smoke have an increased risk of ectopic (outside the uterus) pregnancy. •

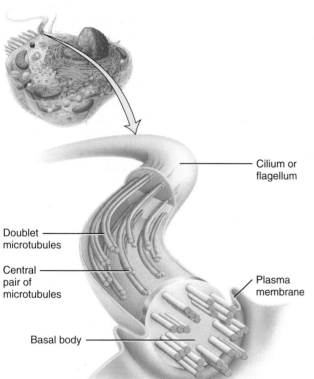

Cilium or flagellum

Doublet microtubules

Central pair of microtubules

Plasma membrane

Basal body

(a) Arrangement of microtubules in a cilium or flagellum

Cilia

SEM 3000x

(b) Cilia lining the trachea

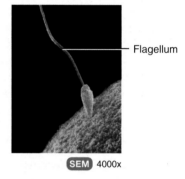

Flagellum

SEM 4000x

(c) Flagellum of a sperm cell

? **What is the functional difference between cilia and flagella?**

Figure 2.9 Ribosomes.

 Ribosomes are the sites of protein synthesis.

FUNCTIONS OF RIBOSOMES

1. Ribosomes associated with endoplasmic reticulum synthesize proteins destined for insertion in the plasma membrane or secretion from the cell.

2. Free ribosomes synthesize proteins used in the cytosol.

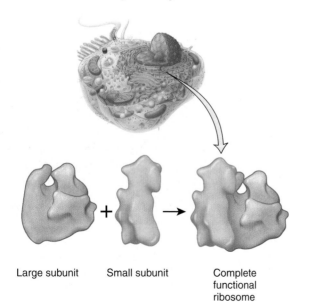

Large subunit   Small subunit   Complete functional ribosome

Details of ribosomal subunits

? **Where are subunits of ribosomes synthesized and assembled?**

other (Figure 2.9). The two subunits are made separately in the nucleolus, a spherical body inside the nucleus (see Section 2.4). Once produced, they exit the nucleus and join together in the cytosol, where they become functional.

Some ribosomes, called *free ribosomes*, are unattached to any structure in the cytoplasm. Free ribosomes primarily synthesize proteins used *inside* the cell. Other ribosomes, called *membrane-bound ribosomes*, attach to the nuclear membrane and to an extensively folded membrane called the endoplasmic reticulum (see Figure 2.10). These ribosomes synthesize proteins destined for specific organelles, for insertion in the plasma membrane, or for export from the cell. Ribosomes are also located within mitochondria, where they synthesize mitochondrial proteins.

Endoplasmic Reticulum (ER)

The **endoplasmic reticulum** (en-dō-PLAS-mik re-TIK-ū-lum; *-plasmic*=cytoplasm; *reticulum*=network) is a network of membranes in the form of flattened sacs or tubules (Figure 2.10). The ER extends from the nuclear envelope (membrane around the nucleus), to which it is connected, and projects throughout the cytoplasm. The ER is so extensive that it constitutes more than half of the membranous surfaces within the cytoplasm of most cells.

Cells contain two distinct forms of ER, which differ in structure and function. **Rough ER** is continuous with the nuclear membrane and usually is folded into a series of flattened sacs. The outer surface of rough ER is studded with ribosomes, the sites of protein synthesis. Proteins synthesized by ribosomes attached to the rough ER enter spaces within the ER for processing and sorting. In some cases, enzymes attach the proteins to carbohydrates

to form glycoproteins. In other cases, enzymes attach the proteins to phospholipids, also synthesized by rough ER. These molecules (glycoproteins and phospholipids) may be incorporated into the membranes of organelles, inserted into the plasma membrane, or secreted via exocytosis. Thus rough ER produces secretory proteins, membrane proteins, and many organellar proteins.

Smooth ER extends from the rough ER to form a network of membrane tubules (Figure 2.10). Unlike rough ER, smooth ER does not have ribosomes on the outer surfaces of its membrane. However, smooth ER contains unique enzymes that make it functionally more diverse than rough ER. Because it lacks ribosomes, smooth ER does not synthesize proteins, but it does synthesize fatty acids and steroids, such as estrogens and testosterone. In liver cells, enzymes of the smooth ER help release glucose into the bloodstream and inactivate or detoxify lipid-soluble drugs or potentially harmful substances, such as alcohol, pesticides, and carcinogens (cancer-causing agents). In liver, kidney, and intestinal cells a smooth ER enzyme removes the phosphate group from glucose-6-phosphate, which allows the "free" glucose to enter the bloodstream. In muscle cells, the calcium ions (Ca^{2+}) that trigger contraction are released from the sarcoplasmic reticulum, a form of smooth ER.

Golgi Complex

Most of the proteins synthesized by ribosomes attached to rough ER are ultimately transported to other regions of the cell. The first step in the transport pathway is through an organelle called the **Golgi complex** (GOL-jē). It consists of 3 to 20 **cisternae** (sis-TER-nē-=cavities; singular is *cisterna*), small, flattened membranous sacs with bulging edges that resemble a stack of

---

Figure 2.10 Endoplasmic reticulum.

The endoplasmic reticulum is a network of membrane-enclosed sacs or tubules that extend throughout the cytoplasm and connect to the nuclear envelope.

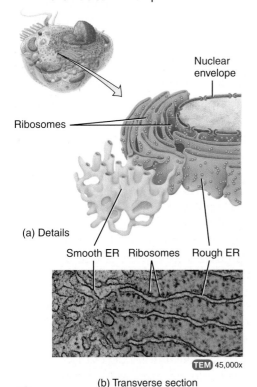

Ribosomes

Nuclear envelope

(a) Details

Smooth ER Ribosomes Rough ER

TEM 45,000x

(b) Transverse section

FUNCTIONS OF ENDOPLASMIC RETICULUM

1. Rough ER synthesizes glycoproteins and phospholipids that are transferred into cellular organelles, inserted into the plasma membrane, or secreted during exocytosis.

2. Smooth ER synthesizes fatty acids and steroids, such as estrogens and testosterone; inactivates or detoxifies drugs and other potentially harmful substances; removes the phosphate group from glucose-6-phosphate; and stores and releases calcium ions that trigger contraction in muscle cells.

CLINICAL CONNECTION | *Smooth ER and Drug Tolerance*

One of the functions of smooth ER, as noted earlier, is to detoxify certain drugs. Individuals who repeatedly take such drugs, such as the sedative phenobarbital, develop changes in the smooth ER in their liver cells. Prolonged administration of phenobarbital results in increased tolerance to the drug; the same dose no longer produces the same degree of sedation. With repeated exposure to the drug, the amount of smooth ER and its enzymes increases to protect the cell from its toxic effects. As the amount of smooth ER increases, higher and higher dosages of the drug are needed to achieve the original effect. This could result in an increased possibility of overdose and increased drug dependence. •

 What are the structural and functional differences between rough and smooth ER?

Figure 2.11 Golgi complex.

🔑 The opposite faces of a Golgi complex differ in size, shape, content, and enzymatic activity.

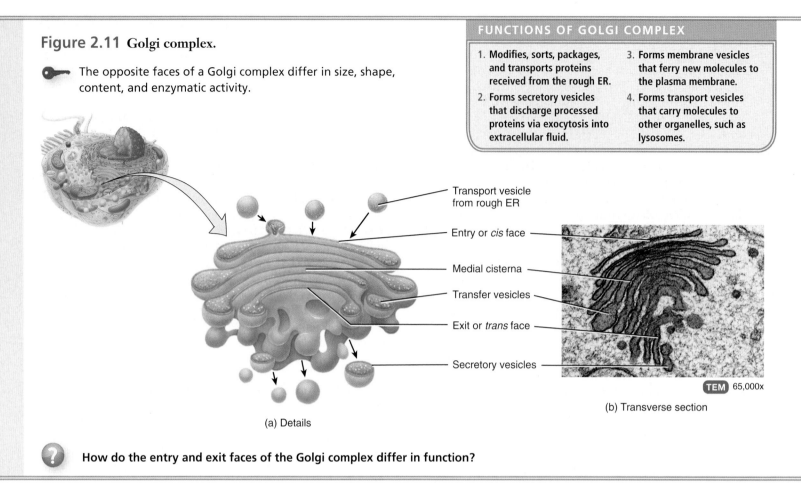

FUNCTIONS OF GOLGI COMPLEX

1. Modifies, sorts, packages, and transports proteins received from the rough ER.
2. Forms secretory vesicles that discharge processed proteins via exocytosis into extracellular fluid.
3. Forms membrane vesicles that ferry new molecules to the plasma membrane.
4. Forms transport vesicles that carry molecules to other organelles, such as lysosomes.

Transport vesicle from rough ER
Entry or *cis* face
Medial cisterna
Transfer vesicles
Exit or *trans* face
Secretory vesicles

TEM 65,000x

(b) Transverse section

(a) Details

❓ **How do the entry and exit faces of the Golgi complex differ in function?**

pita bread (Figure 2.11). The cisternae are often curved, giving the Golgi complex a cuplike shape. Most cells have several Golgi complexes, and Golgi complexes are more extensive in protein-secreting cells (a clue to the organelle's role in the cell).

The cisternae at the opposite ends of a Golgi complex differ from each other in size, shape, and enzymatic activity. The convex **entry** or *cis* **face** is a cisterna that faces the rough ER. The concave **exit** or ***trans* face** is a cisterna that faces the plasma membrane. Cisternae between the entry and exit faces are called **medial cisternae**. Transport vesicles (described shortly) from the ER merge to form the entry face. From the entry face, the cisternae are thought to mature, in turn becoming medial and then exit cisternae.

Different enzymes in the entry, medial, and exit cisternae of the Golgi complex permit each of these areas to modify, sort, and package proteins into vesicles for transport to different destinations. The entry face receives and modifies proteins produced by the rough ER. The medial cisternae add carbohydrates to proteins to form glycoproteins, and add lipids to proteins to form lipoproteins. The exit face modifies the molecules further and then sorts and packages them for transport to their destinations.

Proteins arriving at, passing through, and exiting the Golgi complex do so through maturation of the cisternae and exchanges that occur via transfer vesicles (Figure 2.12):

❶ Proteins synthesized by ribosomes on the rough ER are surrounded by a piece of the ER membrane, which eventually buds from the membrane surface to form a **transport vesicle**.

❷ Transport vesicles move toward the entry face of the Golgi complex.

❸ Fusion of several transport vesicles creates the entry face of the Golgi complex and releases proteins into its lumen (space).

❹ The proteins move from the entry face into one or more medial cisternae. Enzymes in the medial cisternae modify the proteins to form glycoproteins, glycolipids, and lipoproteins. **Transfer vesicles** that bud from the edges of the cisternae move specific enzymes back toward the entry face and move some partially modified proteins toward the exit face.

❺ The products of the medial cisternae move into the lumen of the exit face.

❻ Within the exit face cisterna, the products are further modified, sorted, and packaged.

❼ Some of the processed proteins leave the exit face and are stored in **secretory vesicles**. These vesicles deliver the proteins to the plasma membrane, where they are discharged by exocytosis into the extracellular fluid. For example, certain pancreatic cells release the hormone insulin in this way.

❽ Other processed proteins leave the exit face in **membrane vesicles** that deliver their contents to the plasma membrane for incorporation into the membrane. In doing so, the Golgi complex adds new segments of plasma membrane as existing segments are lost and modifies the number and distribution of membrane molecules.

❾ Finally, some processed proteins leave the exit face in transport vesicles that will carry the proteins to another cellular destination. For instance, transport vesicles carry digestive enzymes to lysosomes; the structure and functions of these important organelles are discussed next.

Figure 2.12 Processing and packaging of synthesized proteins by the Golgi complex.

All proteins exported from the cell are processed in the Golgi complex.

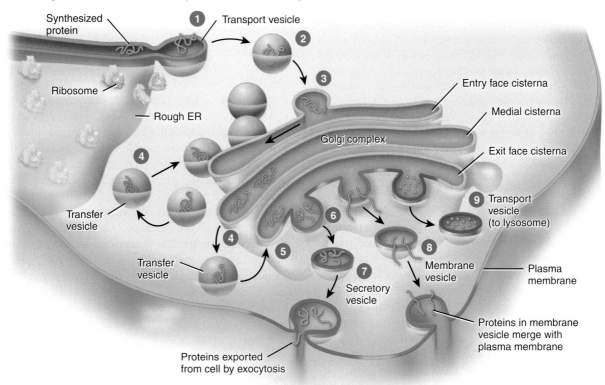

What are the three general destinations for proteins that leave the Golgi complex?

Lysosomes

Lysosomes (LĪ-sō-sōms; *lyso-*=dissolving; *-somes*=bodies) are membrane-enclosed vesicles that form from the Golgi complex (Figure 2.13). They can contain as many as 60 kinds of powerful digestive enzymes capable of breaking down a wide variety of molecules. After lysosomes fuse with vesicles formed during endocytosis, the lysosomal enzymes break down the contents of the vesicles. Proteins in the lysosomal membrane allow the final products of digestion, such as sugars, fatty acids, and amino acids, to be transported into the cytosol for use by the cell. In a similar way, lysosomes in phagocytes can break down and destroy microbes, such as bacteria and viruses.

Figure 2.13 Lysosomes.

Lysosomes contain several kinds of powerful digestive enzymes.

FUNCTIONS OF LYSOSOMES

1. Digest substances that enter a cell via endocytosis and transport final products of digestion into cytosol.
2. Carry out autophagy, the digestion of worn-out organelles.
3. Implement autolysis, the digestion of the entire cell.
4. Accomplish extracellular digestion.

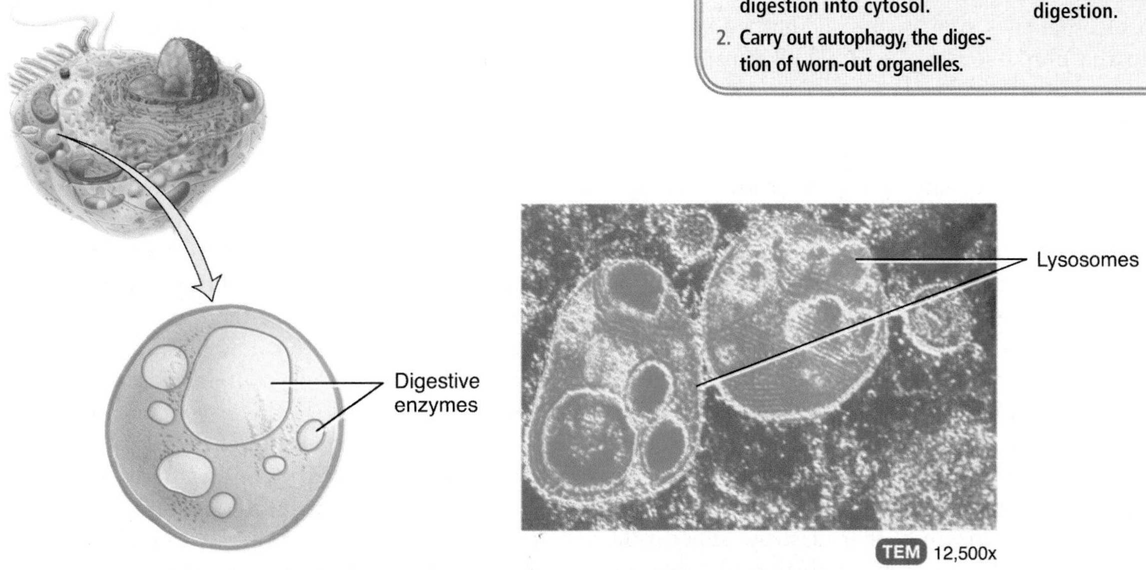

(a) Lysosome

(b) Several lysosomes

TEM 12,500x

What is the name of the process by which worn-out organelles are digested by lysosomes?

Some disorders are caused by faulty or absent lysosomal enzymes. For instance, **Tay-Sachs disease** (TĀ-SAKS), which most often affects children of Ashkenazi (eastern European Jewish) descent, is an inherited condition characterized by the absence of a single lysosomal enzyme called Hex A. This enzyme normally breaks down a membrane glycolipid called ganglioside G_{M2} that is especially prevalent in nerve cells. As the excess ganglioside G_{M2} accumulates, the nerve cells function less efficiently. Children with Tay-Sachs disease typically experience seizures and muscle rigidity. They gradually become blind, demented, and uncoordinated and usually die before the age of 5. Tests can now reveal whether an adult is a carrier of the defective gene. •

Lysosomal enzymes also recycle the cell's own structures. A lysosome can engulf another organelle, digest it, and return the digested components to the cytosol for reuse. In this way, old organelles are continually replaced. The process by which entire worn-out organelles are digested is called **autophagy** (aw-TOF-a-jē; *auto-*=self; *-phagy*=eating). In autophagy, the organelle to be digested is enclosed by a membrane derived from the ER to create a vesicle called an **autophagosome** (aw-tō-FA-gō-sōm); the vesicle then fuses with a lysosome. In this way, cells such as human liver cells recycle about half their cytoplasmic contents every week. Autophagy also contributes to cellular differentiation, control of growth, tissue remodeling, adaptation to adverse environments, and cell defense. Lysosomal enzymes may also destroy the entire cell that contains them, a process known as **autolysis** (aw-TOL-i-sis). Autolysis occurs in some pathological conditions and is responsible for the tissue deterioration that occurs immediately after death.

Although most of the digestive processes involving lysosomal enzymes occur within a cell, in some cases the enzymes operate in extracellular digestion. One example is the release of lysosomal enzymes during fertilization. Enzymes from lysosomes in the head of a sperm cell help the sperm cell penetrate the surface of the ovum.

Peroxisomes

Another group of organelles similar in structure to lysosomes, but smaller, are the **peroxisomes** (pe-ROKS-i-sōms; *peroxi-*=peroxide; *-somes*=bodies; see Figure 2.1). Peroxisomes, also called *microbodies*, contain several *oxidases*, enzymes that can oxidize (remove hydrogen atoms from) various organic substances. For instance, amino acids and fatty acids are oxidized in peroxisomes as part of normal metabolism. In addition, enzymes in peroxisomes oxidize toxic substances, such as alcohol. Thus, peroxisomes are very abundant in the liver, where detoxification of alcohol and other damaging substances occurs. A byproduct of the oxidation reactions is hydrogen peroxide (H_2O_2), a potentially toxic compound, and associated free radicals such as superoxide. However, peroxisomes also contain an enzyme called *catalase*, which decomposes H_2O_2. Because production and degradation of H_2O_2 occur within the same organelle, peroxisomes protect other parts of the cell from the toxic effects of H_2O_2. Peroxisomes also contain enzymes that destroy superoxide. Without peroxisomes, byproducts of metabolism could accumulate inside a cell and result in cellular death. Peroxisomes can self-replicate. New peroxisomes may form from preexisting ones by enlarging and dividing. They may also form by a process in which components accumulate at a given site in the cell and then assemble into a peroxisome.

Proteasomes

As you have just learned, lysosomes degrade proteins delivered to them in vesicles. Cytosolic proteins also require disposal at certain times in the life of a cell. Continuous destruction of unneeded, damaged, or faulty proteins is the function of **proteasomes** (PRŌ-tē-a-sōms=protein bodies), tiny barrel-shaped structures consisting of four stacked rings of proteins around a central core. For example, proteins that are part of metabolic pathways need to be degraded after they have accomplished their function. Such protein destruction plays a part in negative feedback by halting a pathway once the appropriate response has been achieved. A typical body cell contains many thousands of proteasomes, in both the cytosol and the nucleus. Discovered only recently because they are far too small to discern under the light microscope and do not show up well in electron micrographs, proteasomes were so named because they contain myriad *proteases* (PRŌ-tē-ās-es), enzymes that cut proteins into small peptides. Once the enzymes of a proteasome have split a protein into peptides, other enzymes break down the peptides into amino acids, which can be recycled into new proteins.

Some diseases could result from failure of proteasomes to degrade abnormal proteins. For example, clumps of misfolded proteins accumulate in brain cells of people with Parkinson's disease and Alzheimer's disease. Discovering why the proteasomes fail to clear these abnormal proteins is a goal of ongoing research. •

Mitochondria

Because they generate most of the ATP through aerobic (oxygen-requiring) respiration, **mitochondria** (mī-tō-KON-drē-a; *mito-*=thread; *-chondria*=granules; singular is *mitochondrion*) are referred to as the "powerhouses" of the cell. A cell may have as few as a hundred or as many as several thousand mitochondria, depending on the activity of the cell. Active cells that use ATP at a high rate, such as those found in the muscles, liver, and kidneys, have a large number of mitochondria. For example, regular exercise can lead to an increase of mitochondria in muscle cells. This allows the muscle cells to function more efficiently. Mitochondria are usually located where oxygen enters the cell or where the ATP is used, for example, among the contractile proteins in muscle cells.

A mitochondrion consists of an **outer mitochondrial membrane** and an **inner mitochondrial membrane** with a small fluid-filled space between them (Figure 2.14). Both membranes are similar in structure to the plasma membrane. The inner mitochondrial membrane contains a series of folds called **mitochondrial cristae** (KRIS-tē=ridges). The central fluid-filled cavity of a mitochondrion, enclosed by the inner mitochondrial membrane, is the **mitochondrial matrix**. The elaborate folds of the cristae provide an enormous surface area for the chemical reactions that are part of the aerobic phase of *cellular respiration*, the reactions that produce most of a cell's ATP. The enzymes that catalyze these reactions are located on the cristae and in the matrix of the mitochondria.

Figure 2.14 Mitochondria.

 Within mitochondria, chemical reactions of aerobic cellular respiration generate ATP.

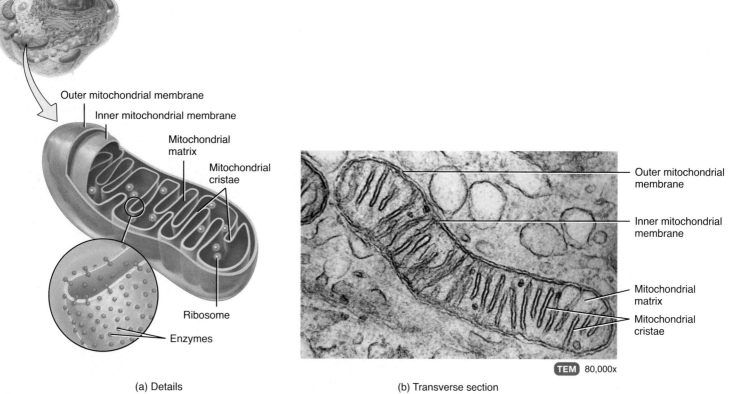

(a) Details

(b) Transverse section

TEM 80,000x

 How do the cristae of a mitochondrion contribute to its ATP-producing function?

Mitochondria also play an important and early role in **apoptosis** (ap′-op-TŌ-sis or ap′-ō-TŌ-sis=a falling off), the orderly, genetically programmed death of a cell. In response to stimuli such as large numbers of destructive free radicals, DNA damage, growth factor deprivation, or lack of oxygen and nutrients, certain chemicals are released from mitochondria following the formation of a pore in the outer mitochondrial membrane. One of the chemicals released into the cytosol of the cell is cytochrome *c*, which while inside the mitochondria is involved in aerobic cellular respiration. In the cytosol, however, cytochrome *c* and other substances initiate a cascade of activation of protein-digesting enzymes that bring about apoptosis.

Like peroxisomes, mitochondria self-replicate, a process that occurs during times of increased cellular energy demand or before cell division. Synthesis of some of the proteins needed for mitochondrial functions occurs on the ribosomes in the mitochondrial matrix. Mitochondria even have their own DNA, in the form of multiple copies of a circular DNA molecule that contains 37 genes. These mitochondrial genes control the synthesis of 2 ribosomal RNAs, 22 transfer RNAs, and 13 proteins that build mitochondrial components.

Although the nucleus of each somatic cell contains genes from both your mother and your father, mitochondrial genes are inherited only from your mother. This is due to the fact that all mitochondria in a cell are descendants of those that were present in the oocyte (egg) during the fertilization process. The head of a sperm (the part that penetrates and fertilizes an oocyte) normally lacks most organelles, such as mitochondria, ribosomes, endoplasmic reticulum, and the Golgi complex, and any sperm mitochondria that do enter the oocyte are soon destroyed. Since all mitochondrial genes are inherited from the maternal parent, mitochondrial DNA can be used to trace maternal lineage (to determine whether two or more individuals are related through their mother's side of the family).

✓ CHECKPOINT

8. What does cytoplasm have that cytosol lacks?
9. Which organelles are surrounded by a membrane and which are not?
10. Name the organelles that contribute to synthesizing protein hormones and package them into secretory vesicles.
11. What happens on the cristae and in the matrix of mitochondria?

2.4 NUCLEUS

OBJECTIVE

• Describe the structure and functions of the nucleus.

The **nucleus** is a spherical or oval-shaped structure that usually is the most prominent feature of a cell (Figure 2.15). Most cells have a single nucleus, although some, such as mature red blood cells, have none. In contrast, skeletal muscle cells and a few other types of cells have multiple nuclei. A double membrane called the **nuclear envelope** separates the nucleus from the cytoplasm. Both layers of the nuclear envelope are lipid bilayers similar to the plasma membrane. The outer membrane of the nuclear envelope is continuous with rough ER and resembles it in structure. Many openings called **nuclear pores** extend through the nuclear envelope. Each nuclear pore consists of a circular arrangement of proteins surrounding a large central opening that is about 10 times wider than the pore of a channel protein in the plasma membrane.

Nuclear pores control the movement of substances between the nucleus and the cytoplasm. Small molecules and ions move through the pores passively by diffusion. Most large molecules, such as RNAs and proteins, cannot pass through the nuclear pores by diffusion. Instead, their passage involves an active transport process in which the molecules are recognized and selectively transported through the nuclear pore into or out of the nucleus. For example, proteins needed for nuclear functions move from the cytosol into the nucleus; newly formed RNA molecules move from the nucleus into the cytosol in this manner.

Inside the nucleus are one or more spherical bodies called **nucleoli** (noo-KLĒ-ō-lī; singular is *nucleolus*) that function in producing ribosomes. Each nucleolus is simply a cluster of protein, DNA, and RNA that is not enclosed by a membrane. Nucleoli

Figure 2.15 Nucleus.

The nucleus contains most of a cell's genes, which are located on chromosomes.

FUNCTIONS OF NUCLEI	
1. Control cellular structure.	3. Produce ribosomes in nucleoli.
2. Direct cellular activities.	

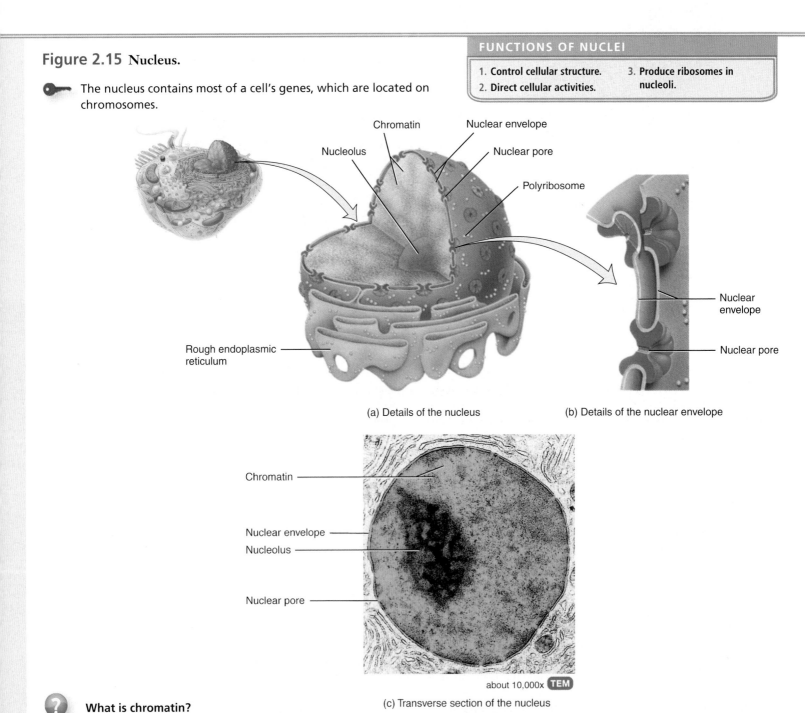

(a) Details of the nucleus

(b) Details of the nuclear envelope

about 10,000x **TEM**

(c) Transverse section of the nucleus

What is chromatin?

Figure 2.16 Packing of DNA into a chromosome in a dividing cell. When packing is complete, two identical DNAs and their histones form a pair of chromatids, held together by a centromere.

 A chromosome is a highly coiled and folded DNA molecule that is combined with protein molecules.

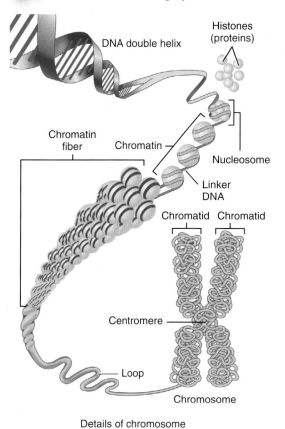

DNA double helix

Histones (proteins)

Chromatin fiber

Chromatin

Nucleosome

Linker DNA

Chromatid   Chromatid

Centromere

Loop

Chromosome

Details of chromosome

? **What are the components of a nucleosome?**

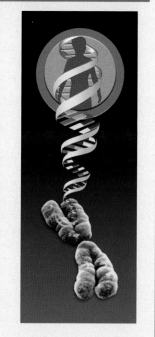
are the sites of rRNA synthesis and the assembly of rRNA and proteins into ribosomal subunits. Nucleoli are quite prominent in cells that synthesize large amounts of protein, such as muscle and liver cells. Nucleoli disperse and disappear during cell division and reorganize once new cells are formed.

Within the nucleus are most of the cell's hereditary units, called **genes**, which control cellular structure and direct cellular activities. Genes are arranged along **chromosomes** (KRŌ-mō-sōms; *chromo*=colored). Human somatic (body) cells have 46 chromosomes, 23 inherited from each parent. Each chromosome is a long molecule of DNA that is coiled together with several proteins (Figure 2.16). This complex of DNA, proteins, and some RNA is called **chromatin** (KRŌ-ma-tin). The total genetic information carried in a cell or an organism is its **genome** (JĒ-nōm).

In cells that are not dividing, the chromatin appears as a diffuse, granular mass. Electron micrographs reveal that chromatin has a beads-on-a-string structure. Each bead is a **nucleosome** (NOO-klē-ō-sōm) that consists of double-stranded DNA wrapped twice around a core of eight proteins called **histones**, which help organize the coiling and folding of DNA. The string between the beads, called **linker DNA**, holds adjacent nucleosomes together. In cells that are not dividing, another histone promotes coiling of nucleosomes into a larger-diameter **chromatin fiber**, which then folds into large loops. Just before cell division takes place, the DNA replicates (duplicates) and the loops condense even more, forming a pair of **chromatids** (KRŌ-ma-tids). As you will see shortly, during cell division each pair of chromatids constitutes a chromosome.

The main parts of a cell, their descriptions, and their functions are summarized in Table 2.2.

✓ CHECKPOINT

12. How is DNA packed into the nucleus?
13. What is the importance of nuclear pores?
14. How do chromosomes and chromatids differ?

2.5 CELL DIVISION

● OBJECTIVES

• Discuss the stages, events, and significance of somatic cell division.
• Describe the stages, events, and significance of reproductive cell division.

Most cells of the human body undergo **cell division**, the process by which cells reproduce themselves. The two types of cell division—somatic cell division and reproductive cell division—accomplish different goals for the organism.

A **somatic cell** (sō-MAT-ik; *soma*=body) is any cell of the body other than a germ cell. A **germ cell** is a gamete (sperm or oocyte) or any precursor cell destined to become a gamete. In **somatic cell division**, a cell undergoes a nuclear division called **mitosis** (mī-TŌ-sis; *mitos*=thread) and a cytoplasmic division called **cytokinesis** (sī-tō-ki-NĒ-sis; *cyto*=cell; *kinesis*=movement) to produce two genetically identical cells, each with the same number and kind of chromosomes as the original cell. Somatic cell division replaces dead or injured cells and adds new ones during tissue growth.

Reproductive cell division is the mechanism that produces gametes, the cells needed to form the next generation of sexually

TABLE 2.2

Cell Parts and Their Functions

PART	DESCRIPTION	FUNCTIONS
Plasma membrane	Fluid-mosaic lipid bilayer (phospholipids, cholesterol, and glycolipids) studded with proteins; surrounds cytoplasm.	Protects cellular contents; makes contact with other cells; contains channels, transporters, receptors, enzymes, cell-identity markers, and linker proteins; mediates the entry and exit of substances.
Cytoplasm	Cellular contents between the plasma membrane and nucleus—cytosol and organelles.	Site of all intracellular activities except those occurring in the nucleus.
Cytosol	Composed of water, solutes, suspended particles, lipid droplets, and glycogen granules. Contains cytoskeleton, a network of three types of protein filaments: microfilaments, intermediate filaments, and microtubules.	Fluid in which many of cell's metabolic reactions occur. Cytoskeleton maintains shape and general organization of cellular contents; responsible for cellular movements.
Organelles	Specialized structures with characteristic shapes.	Each organelle has specific functions.
Centrosome	A pair of centrioles plus pericentriolar material.	The pericentriolar material contains tubulins, which are used for growth of the mitotic spindle and microtubule formation.
Cilia and flagella	Motile cell surface projections that contain 20 microtubules and a basal body.	Cilia move fluids over a cell's surface; flagella move an entire cell.
Ribosome	Composed of two subunits containing ribosomal RNA and proteins; may be free in cytosol or attached to rough ER.	Protein synthesis.
Endoplasmic reticulum (ER)	Membranous network of flattened sacs or tubules. Rough ER is covered by ribosomes and is attached to the nuclear envelope; smooth ER lacks ribosomes.	Rough ER synthesizes glycoproteins and phospholipids that are transferred to cellular organelles, inserted into the plasma membrane, or secreted during exocytosis. Smooth ER synthesizes fatty acids and steroids; inactivates or detoxifies drugs; removes the phosphate group from glucose-6-phosphate; and stores and releases calcium ions in muscle cells.
Golgi complex	Consists of 3–20 flattened membranous sacs called cisternae; structurally and functionally divided into entry (*cis*) face, medial cisternae, and exit (*trans*) face.	Entry (*cis*) face accepts proteins from rough ER; medial cisternae form glycoproteins, glycolipids, and lipoproteins; exit (*trans*) face modifies the molecules further, then sorts and packages them for transport to their destinations.
Lysosome	Vesicle formed from Golgi complex; contains digestive enzymes.	Fuses with and digests contents of endosomes, phagosomes, and vesicles formed during bulk-phase endocytosis and transports final products of digestion into cytosol; digests worn-out organelles (autophagy), entire cells (autolysis), and extracellular materials.
Peroxisome	Vesicle containing oxidases (oxidative enzymes) and catalase (decomposes hydrogen peroxide); new peroxisomes bud from preexisting ones.	Oxidizes amino acids and fatty acids; detoxifies harmful substances, such as hydrogen peroxide and associated free radicals.
Proteasome	Tiny barrel-shaped structure that contains proteases (proteolytic enzymes).	Degrades unneeded, damaged, or faulty proteins by cutting them into small peptides.
Mitochondrion	Consists of an outer and an inner mitochondrial membrane, cristae, and matrix; new mitochondria form from preexisting ones.	Site of aerobic cellular respiration reactions that produce most of a cell's ATP and play an important early role in apoptosis.
Nucleus	Consists of a nuclear envelope with pores, nucleoli, and chromosomes, which exist as a tangled mass of chromatin in interphase cells.	Nuclear pores control the movement of substances between the nucleus and cytoplasm, nucleoli produce ribosomes, and chromosomes consist of genes that control cellular structure and direct cellular functions.

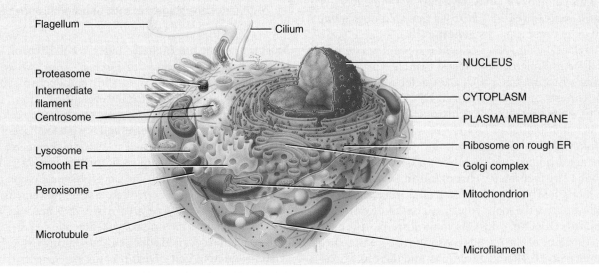

reproducing organisms. This process consists of a special two-step division called **meiosis** (mī-Ō-sis), in which the number of chromosomes in the nucleus is reduced by half.

Somatic Cell Division

The **cell cycle** is an orderly sequence of events in which a somatic cell duplicates its contents and divides in two. Some cells divide more frequently than others. Human cells, such as those in the brain, stomach, and kidneys, contain 23 pairs of chromosomes, for a total of 46. One member of each pair is inherited from each parent. The two chromosomes that make up each pair are called **homologous chromosomes** (hō-MOL-o-gus; *homo*=same) or **homologs**; they contain similar genes arranged in the same (or almost the same) order. When examined under a light microscope, homologous chromosomes generally look very similar. The exception to this rule is one pair of chromosomes called the **sex chromosomes**, designated X and Y. In females the homologous pair of sex chromosomes consists of two large X chromosomes; in males the pair consists of an X and a much smaller Y chromosome. Because somatic cells contain two sets of chromosomes, they are called **diploid cells** (DIP-loyd; *dipl-*= double; *-oid*=form), symbolized **2n**.

When a cell reproduces, it must replicate (duplicate) all its chromosomes to pass its genes to the next generation of cells. The cell cycle consists of two major periods: interphase, when a cell is not dividing, and the mitotic (M) phase, when a cell is dividing (Figure 2.17).

Interphase

During **interphase** (IN-ter-fāz), the cell replicates its DNA through a process that will be described shortly. It also produces additional organelles and cytosolic components in anticipation of cell division. Interphase is a state of high metabolic activity; it is during this time that the cell does most of its growing. Interphase consists of three phases: G_1, S, and G_2 (Figure 2.17). The phase designated S stands for *synthesis* of DNA. Because the G phases are periods when there is no activity related to DNA duplication, they are thought of as *gaps* or interruptions in DNA duplication.

The **G_1 phase** is the interval between the mitotic phase and the S phase. During G_1 the cell is metabolically active; it replicates most of its organelles and cytosolic components but not its DNA. Replication of centrosomes also begins in the G_1 phase. Virtually all the cellular activities described in this chapter happen during G_1. For a cell with a total cell cycle time of 24 hours, G_1 lasts 8 to 10 hours. However, the duration of this phase is quite variable. It is very short in many embryonic cells or cancer cells. Cells that remain in G_1 for a very long time, perhaps destined never to divide again, are said to be in the **G_0 phase**. Most nerve cells are in the G_0 phase. Once a cell leaves the G_1 phase and enters the S phase, it is committed to go through the rest of the cell cycle.

The **S phase**, the interval between G_1 and G_2, lasts about 8 hours. During the S phase, DNA replication occurs. As a result of DNA replication, the two identical cells formed during cell division later in the cycle will have the same genetic material. The **G_2 phase** is the interval between the S phase and the mitotic phase. It lasts 4 to 6 hours. During G_2 cell growth continues, enzymes and other proteins are synthesized in preparation for cell division, and replication of centrosomes is completed. When

DNA replicates during the S phase, its helical structure partially uncoils, and the two strands separate at the points where hydrogen bonds connect base pairs. Each exposed base of the old DNA strand then pairs with the complementary base of a newly synthesized nucleotide. A new DNA strand takes shape as chemical bonds form between neighboring nucleotides. The uncoiling and complementary base pairing continues until each of the two original DNA strands is joined with a newly formed complementary DNA strand. The original DNA molecule has become two identical DNA molecules.

A microscopic view of a cell during interphase shows a clearly defined nuclear envelope, a nucleolus, and a tangled mass of chromatin (Figure 2.18a). Once a cell completes its activities during the G_1, S, and G_2 phases of interphase, the mitotic phase begins.

Mitotic Phase

The **mitotic (M) phase** of the cell cycle, which results in the formation of two identical cells, consists of a nuclear division (mitosis) and a cytoplasmic division (cytokinesis). The events that occur during mitosis and cytokinesis are plainly visible under a microscope because chromatin condenses into discrete chromosomes.

NUCLEAR DIVISION: MITOSIS *Mitosis*, as noted earlier, is the distribution of two sets of chromosomes into two separate nuclei. The process results in the *exact* partitioning of genetic information. For convenience, biologists divide the process into four stages: prophase, metaphase, anaphase, and telophase. However, mitosis is a continuous process; one stage merges seamlessly into the next.

Figure 2.17   The cell cycle. Not illustrated is cytokinesis, division of the cytoplasm, which usually occurs during late anaphase of the mitotic phase.

 In a complete cell cycle, a cell duplicates its contents and divides into two identical cells.

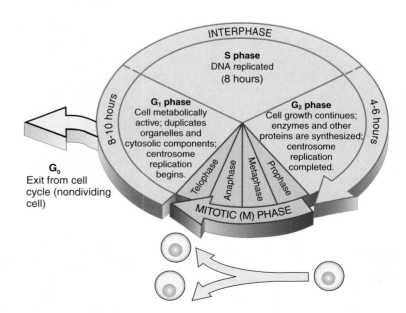

During which phase of the cell cycle does DNA replication occur?

Figure 2.18 Cell division: mitosis and cytokinesis. Begin the sequence at the top of the figure and read clockwise until you complete the process.

In somatic cell division, a single diploid cell divides to produce two identical diploid cells.

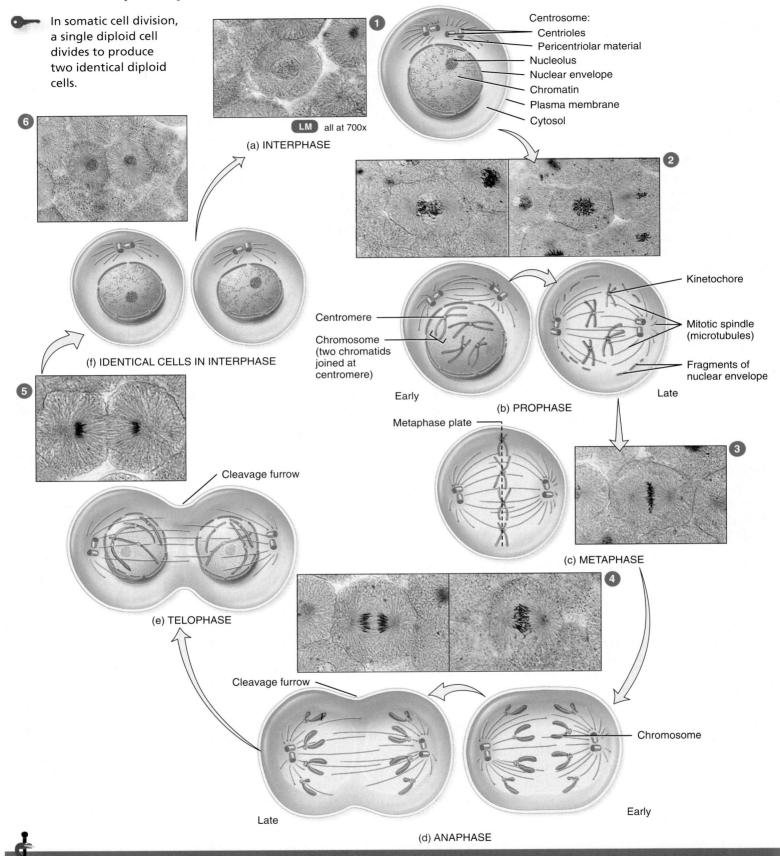

(a) INTERPHASE

Centrosome:
- Centrioles
- Pericentriolar material
- Nucleolus
- Nuclear envelope
- Chromatin
- Plasma membrane
- Cytosol

LM all at 700x

Kinetochore

Centromere

Chromosome (two chromatids joined at centromere)

Mitotic spindle (microtubules)

Fragments of nuclear envelope

Early

Late

(b) PROPHASE

(f) IDENTICAL CELLS IN INTERPHASE

Metaphase plate

Cleavage furrow

(c) METAPHASE

(e) TELOPHASE

Cleavage furrow

Chromosome

Late

Early

(d) ANAPHASE

During which phase of mitosis does cytokinesis begin?

1. **Prophase** (PRŌ-fāz). During early prophase, the chromatin fibers condense and shorten into chromosomes that are visible under the light microscope (Figure 2.18b). The condensation process may prevent entangling of the long DNA strands as they move during mitosis. Because longitudinal DNA replication took place during the S phase of interphase, each prophase chromosome consists of a pair of identical strands called **chromatids**. A constricted region called a **centromere** (SEN-trō-mēr) holds the chromatid pair together. At the outside of each centromere is a protein complex known as the **kinetochore** (ki-NET-ō-kor). Later in prophase, tubulins in the pericentriolar material of the centrosomes start to form the **mitotic spindle**, a football-shaped assembly of microtubules that attach to the kinetochore (Figure 2.18b). As the microtubules lengthen, they push the centrosomes to the poles (opposite ends) of the cell so that the spindle extends from pole to pole. The mitotic spindle is responsible for the separation of chromatids to opposite poles of the cell. Then, the nucleolus disappears and the nuclear envelope breaks down.

2. **Metaphase** (MET-a-fāz). During metaphase, the microtubules of the mitotic spindle align the centromeres of the chromatid pairs at the exact center of the mitotic spindle (Figure 2.18c). This midpoint region is called the **metaphase plate.**

3. **Anaphase** (AN-a-fāz). During anaphase, the centromeres split, separating the two members of each chromatid pair, which move toward opposite poles of the cell (Figure 2.18d). Once separated, the chromatids are termed *chromosomes*. As the chromosomes are pulled by the microtubules of the mitotic spindle during anaphase, they appear V-shaped because the centromeres lead the way, dragging the trailing arms of the chromosomes toward the pole.

4. **Telophase** (TEL-ō-fāz). The final stage of mitosis, telophase, begins after chromosomal movement stops (Figure 2.18e). The identical sets of chromosomes, now at opposite poles of the cell, uncoil and revert to the threadlike chromatin form. A nuclear envelope forms around each chromatin mass, nucleoli reappear in the identical nuclei, and the mitotic spindle breaks up.

CYTOPLASMIC DIVISION: CYTOKINESIS As noted earlier, division of a cell's cytoplasm and organelles into two identical cells is called *cytokinesis* (sī-tō-ki-NĒ-sis; *-kinesi*=motion). This process usually begins in late anaphase with the formation of a **cleavage furrow**, a slight indentation of the plasma membrane, and is completed after telophase. The cleavage furrow usually appears midway between the centrosomes and extends around the periphery of the cell (Figures 2.18d and e). Actin microfilaments that lie just inside the plasma membrane form a *contractile ring* that pulls the plasma membrane progressively inward. The ring constricts the center of the cell, like tightening a belt around the waist, and ultimately pinches it in two. Because the plane of the cleavage furrow is always perpendicular to the mitotic spindle, the two sets of chromosomes end up in separate cells. When cytokinesis is complete, interphase begins (Figure 2.18f).

The sequence of events can be summarized as

$$G_1 \text{ phase} \rightarrow S \text{ phase} \rightarrow G_2 \text{ phase} \rightarrow \text{mitosis} \rightarrow \text{cytokinesis}$$

Table 2.3 summarizes the events of the cell cycle in somatic cells.

Control of Cell Destiny

A cell has three possible destinies: (1) to remain alive and functioning without dividing, (2) to grow and divide, or (3) to die. Homeostasis is maintained when there is a balance between cell proliferation and cell death. The signals that tell a cell when to exist in the G_0 phase, when to divide, and when to die have been the subjects of intense and fruitful research in recent years.

Within a cell, enzymes called **cyclin-dependent protein kinases (Cdks)** can transfer a phosphate group from ATP to a protein to activate the protein; other enzymes can remove the phosphate group from the protein to deactivate it. The activation and deactivation of Cdks at the appropriate time is crucial in the initiation and regulation of DNA replication, mitosis, and cytokinesis.

Switching the Cdks on and off is the responsibility of cellular proteins called **cyclins** (SĪK-lins), so named because their levels rise and fall during the cell cycle. The joining of a specific cyclin and Cdk molecule triggers various events that control cell division.

TABLE 2.3

Events of the Somatic Cell Cycle

PHASE	ACTIVITY
Interphase	Period between cell divisions; chromosomes not visible under light microscope.
G₁ phase	Metabolically active cell duplicates most of its organelles and cytosolic components; replication of chromosomes begins. (Cells that remain in the G₁ phase for a very long time, and possibly never divide again, are said to be in the G₀ state.)
S phase	Replication of DNA and centrosomes.
G₂ phase	Cell growth, enzyme and protein synthesis continues; replication of centrosomes complete.
Mitotic phase	Parent cell produces identical cells with identical chromosomes; chromosomes visible under light microscope.
Mitosis	Nuclear division; distribution of two sets of chromosomes into separate nuclei.
Prophase	Chromatin fibers condense into paired chromatids; nucleolus and nuclear envelope disappear; each centrosome moves to an opposite pole of the cell.
Metaphase	Centromeres of chromatid pairs line up at metaphase plate.
Anaphase	Centromeres split; identical sets of chromosomes move to opposite poles of cell.
Telophase	Nuclear envelopes and nucleoli reappear; chromosomes resume chromatin form; mitotic spindle disappears.
Cytokinesis	Cytoplasmic division; contractile ring forms cleavage furrow around center of cell, dividing cytoplasm into separate and equal portions.

The activation of specific cyclin–Cdk complexes is responsible for progression of a cell from G_1 to S to G_2 to mitosis in a specific order. If any step in the sequence is delayed, all subsequent steps are delayed in order to maintain the normal sequence. The levels of cyclins in the cell are very important in determining the timing and sequence of events in cell division. For example, the level of the cyclin that helps drive a cell from G_2 to mitosis rises throughout the G_1, S, and G_2 phases and into mitosis. The high level triggers mitosis, but toward the end of mitosis, the level declines rapidly and mitosis ends. Destruction of this cyclin, as well as others in the cell, is by proteasomes.

Cellular death is also regulated. Throughout the lifetime of an organism, certain cells undergo apoptosis, an orderly, genetically programmed death. In **apoptosis**, a triggering agent from either outside or inside the cell causes "cell-suicide" genes to produce enzymes that damage the cell in several ways, including disruption of its cytoskeleton and nucleus (see Section 2.3, which describes the role of mitochondria in apoptosis). As a result, the cell shrinks and pulls away from neighboring cells. Although the plasma membrane remains intact, the DNA within the nucleus fragments and the cytoplasm shrinks. Nearby phagocytes then ingest the dying cell via a complex process that involves the binding of a receptor protein in the phagocyte plasma membrane to a lipid in the dying cell's plasma membrane. Apoptosis removes unneeded cells during fetal development, such as the webbing between digits. It continues to occur after birth to regulate the number of cells in a tissue and eliminate potentially dangerous cells such as cancer cells.

Apoptosis is a normal type of cell death; in contrast, **necrosis** (ne-KRŌ-sis=death) is a pathological type of cell death that results from tissue injury. In necrosis, many adjacent cells swell, burst, and spill their cytoplasm into the interstitial fluid. The cellular debris usually stimulates an inflammatory response by the immune system, a process that does not occur in apoptosis.

Reproductive Cell Division

In this process, also called *sexual reproduction*, each new organism is the result of the union of two different gametes (fertilization), one produced by each parent. If gametes had the same number of chromosomes as somatic cells, the number of chromosomes would double at fertilization. Meiosis (mī-Ō-sis; *mei-*=lessening; *-osis*=condition of), the reproductive cell division that occurs in the gonads (ovaries and testes), produces gametes with half the number of chromosomes. As a result, gametes contain a single set of 23 chromosomes and thus are **haploid (n) cells** (HAP-loyd; *hapl-*=single). Fertilization restores the diploid number of chromosomes.

Meiosis

Unlike mitosis, which is complete after a single round, meiosis occurs in two successive stages: **meiosis I** and **meiosis II**. During the interphase that precedes meiosis I, the chromosomes of the diploid cell start to replicate. As a result of replication, each chromosome consists of two sister (genetically identical) chromatids, which are attached at their centromeres. This replication of chromosomes is similar to the one that precedes mitosis in somatic cell division.

MEIOSIS I Meiosis I, which begins once chromosomal replication is complete, consists of four phases: prophase I, metaphase I, anaphase I, and telophase I (Figure 2.19a). Prophase I is an extended phase in which the chromosomes shorten and thicken, the nuclear envelope and nucleoli disappear, and the mitotic spindle forms. Two events that are not seen in mitotic prophase occur during prophase I of meiosis. First, the two sister chromatids of each pair of homologous chromosomes pair off, an event called **synapsis** (sin-AP-sis) (Figure 2.19b). The resulting four chromatids form a structure called a **tetrad** (TE-trad; *tetra*=four). Second, parts of the chromatids of two homologous chromosomes may be exchanged with one another. This exchange between parts of nonsister (genetically different) chromatids, called **crossing-over** (Figure 2.19b), permits an exchange of genes between the chromatids. Crossing-over results in *genetic recombination*—the formation of new combinations of genes—and accounts in part for the great genetic variation among humans and other organisms that form gametes via meiosis.

In metaphase I, the tetrads formed by the homologous pairs of chromosomes line up along the metaphase plate of the cell, with homologous chromosomes side by side (Figure 2.19a). During anaphase I, the members of each homologous pair of chromosomes separate as they are pulled to opposite poles of the cell by the microtubules attached to the centromeres. The paired chromatids, held by a centromere, remain together. (Recall that during mitotic anaphase, the centromeres split and the sister chromatids separate.) Telophase I and cytokinesis of meiosis are similar to telophase and cytokinesis of mitosis. The net effect of meiosis I is that each resulting cell contains the haploid number of chromosomes because it contains only one member of each pair of the homologous chromosomes present in the starting cell.

MEIOSIS II The second stage of meiosis, meiosis II, also consists of four phases: prophase II, metaphase II, anaphase II, and telophase II (Figure 2.19a). These phases are similar to those that occur during mitosis; the centromeres split, and the sister chromatids separate and move toward opposite poles of the cell.

In summary, meiosis I begins with a diploid starting cell and ends with two cells, each with the haploid number of chromosomes. During meiosis II, each of the two haploid cells formed during meiosis I divides; the net result is four haploid gametes that are genetically different from the original diploid starting cell.

Figure 2.20 and Table 2.4 compare the events of mitosis and meiosis.

 CHECKPOINT

15. Distinguish between somatic and reproductive cell division and explain the importance of each.
16. Define interphase. When during interphase does DNA replicate?
17. Outline the major events of each stage of the mitotic phase.
18. How are apoptosis and necrosis similar? How do they differ?
19. What is the difference between haploid (*n*) and diploid (2*n*) cells?
20. What are homologous chromosomes?

Figure 2.19 **Meiosis, reproductive cell division.** Details of each of the stages shown are discussed in the text.

In reproductive cell division, a single diploid starting cell undergoes meiosis I and meiosis II to produce four haploid gametes that are genetically different from the starting cell that produced them.

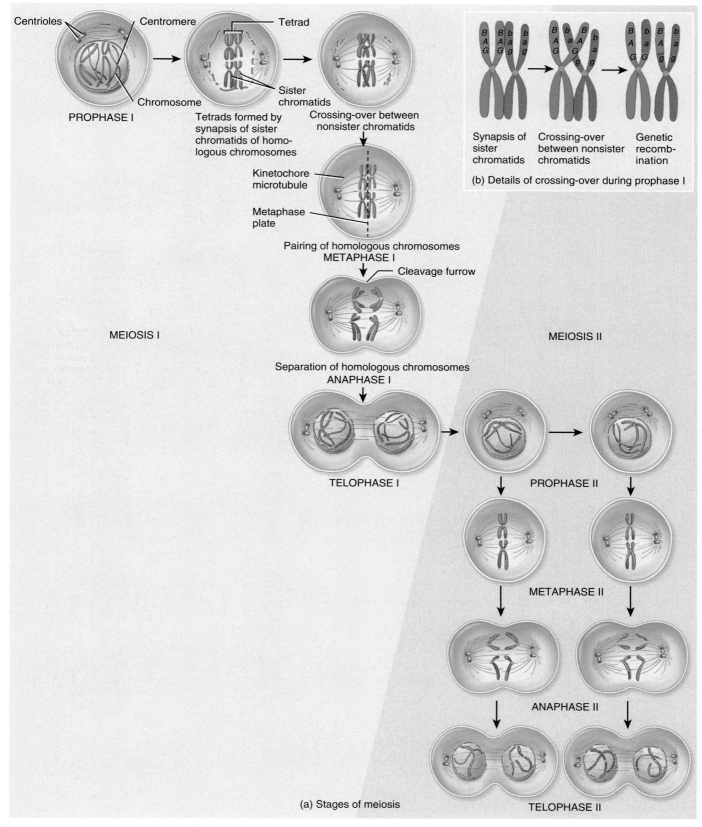

(b) Details of crossing-over during prophase I

(a) Stages of meiosis

 How does crossing-over affect the genetic content of the four haploid gametes?

Figure 2.20 **Mitosis and meiosis.** Comparison between mitosis (left) and meiosis (right) in which the parent cell has two pairs of homologous chromosomes.

The phases of meiosis II and mitosis are similar.

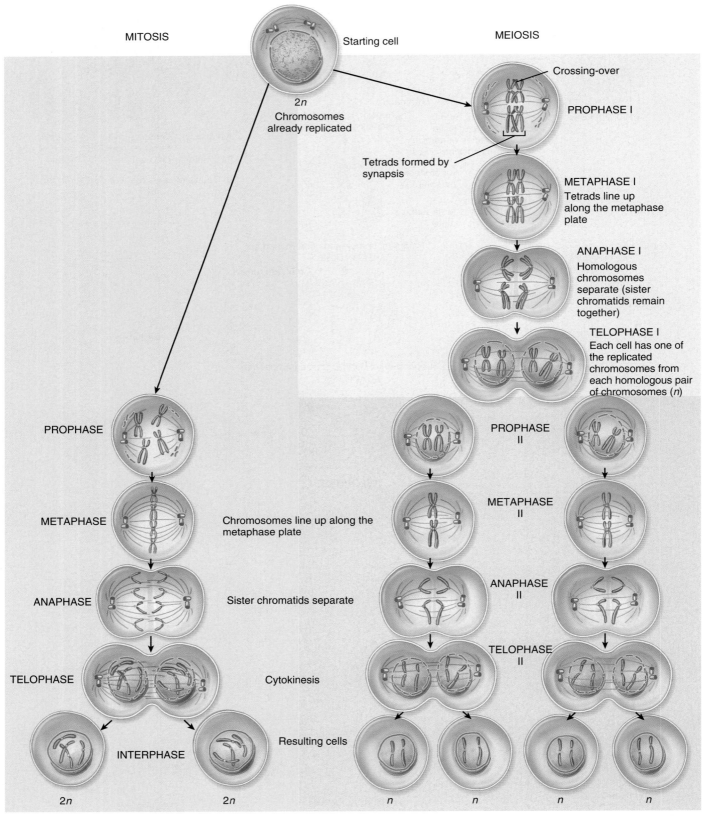

Somatic cells with diploid number
of chromosomes (not replicated)

Gametes with haploid number of
chromosomes (not replicated)

 How does anaphase I of meiosis differ from anaphase of mitosis?

TABLE 2.4

Comparison Between Mitosis and Meiosis

POINT OF COMPARISON	MITOSIS	MEIOSIS
Cell type	Somatic	Gamete
Number of divisions	1	2
Stages	Interphase	Interphase I only
	Prophase	Prophase I and II
	Metaphase	Metaphase I and II
	Anaphase	Anaphase I and II
	Telophase	Telophase I and II
Copy DNA?	Yes, interphase	Yes, interphase I; No, interphase II
Tetrads?	No	Yes
Number of cells	2	4
Number of chromosomes per cell	46, or two sets of 23; this makeup, called diploid, is identical to the chromosomes in the starting cell	One set of 23; this makeup, called haploid, represents half of the chromosomes in the starting cell

2.6 CELLULAR DIVERSITY

 OBJECTIVE
- Describe cellular differences in size and shape.

As noted earlier, the body of an average human adult is composed of nearly 100 trillion cells. All of these cells can be classified into about 200 different cell types. Cells vary considerably in size. The sizes of cells are measured in units called *micrometers* (mī-KROM-i-ters). One micrometer (μm) is equal to 1 one-millionth of a meter, or 10^{-6} (1/25,000th of an inch). High-powered microscopes are needed to see the smallest cells of the body. The largest cell, a single oocyte, has a diameter of about 140 μm and is barely visible to the unaided eye. A red blood cell has a diameter of 10 μm. To better visualize this, an average hair from the top of your head is approximately 100 μm in diameter.

The shapes of cells also vary considerably (Figure 2.21). They may be round, oval, flat, cube-shaped, column-shaped, elongated, star-shaped, cylindrical, or disc-shaped. A cell's shape is related to its function in the body. For example, a sperm cell has a long whiplike tail (flagellum) that it uses for locomotion. The disc shape of a red blood cell gives it a large surface area that enhances its ability to pass oxygen to other cells. The long, spindle shape of a relaxed smooth muscle cell shortens as it contracts. This change in shape allows groups of smooth muscle cells to narrow or widen the passage for blood flowing through blood vessels. In this way, they regulate blood flow through various tissues. Some cells contain microvilli, which greatly increase their surface area. Microvilli are common in the epithelial cells that line the small intestine, where the large surface area speeds the absorption of digested food. Nerve cells have long extensions that permit them to conduct nerve impulses over great distances. As you will see in the next several chapters, cellular diversity also permits organization of cells into more complex tissues and organs.

 CHECKPOINT
21. How is the shape of a cell related to its function?

Figure 2.21 Diverse shapes and sizes of human cells. The relative difference in size between the smallest and largest cells is actually much greater than shown here.

 The nearly 100 trillion cells in an average adult can be classified into about 200 different cell types.

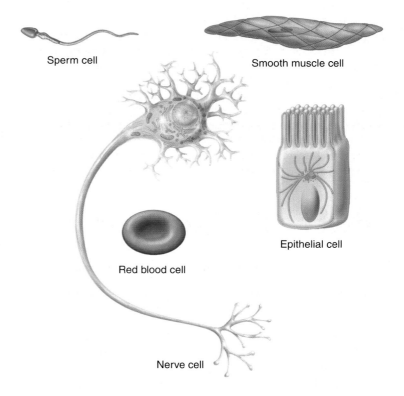

Sperm cell

Smooth muscle cell

Epithelial cell

Red blood cell

Nerve cell

 Why are sperm the only body cells that need to have a flagellum?

CLINICAL CONNECTION | *Cancer*

Cancer is a group of diseases characterized by uncontrolled or abnormal cell division. When cells in a part of the body divide without control, the excess tissue that develops is called a **tumor** or **neoplasm** (NĒ-ō-plazm; *neo-*=new). The study of tumors is called **oncology** (on-KOL-o-jē; *onco-*=swelling or mass). Tumors may be cancerous and often fatal, or they may be harmless. A cancerous neoplasm is called a **malignant tumor** (ma-LIG-nant) or **malignancy**. One property of most malignant tumors is their ability to undergo **metastasis** (me-TAS-ta-sis), the spread of cancerous cells to other parts of the body.

A **benign tumor** (bē-NĪN) is a neoplasm that does not metastasize. An example is a wart. Most benign tumors may be removed surgically if they interfere with normal body function or become disfiguring. Some benign tumors can be inoperable (due to tumor size or location) and perhaps fatal.

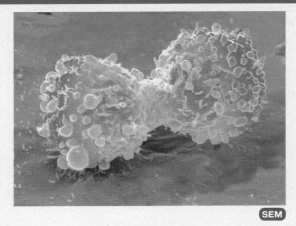

SEM

Lung cancer cell dividing

Growth and Spread of Cancer

Cells of malignant tumors duplicate rapidly and continuously. As malignant cells invade surrounding tissues, they often trigger **angiogenesis** (an'-jē-ō-JEN-e-sis), the growth of new networks of blood vessels. Proteins that stimulate angiogenesis in tumors are called **tumor angiogenesis factors (TAFs)**. The formation of new blood vessels can occur either by overproduction of TAFs or by the lack of naturally occurring angiogenesis inhibitors. As the cancer grows, it begins to compete with normal tissues for space and nutrients. Eventually, the normal tissue decreases in size and dies. Some malignant cells may detach from the initial (primary) tumor and invade a body cavity or enter the blood or lymph, then circulate to and invade other body tissues, establishing secondary tumors. Malignant cells resist the antitumor defenses of the body. The pain associated with cancer develops when the tumor presses on nerves or blocks a passageway in an organ so that secretions build up pressure, or as a result of dying tissue or organs.

Causes of Cancer

Several factors may trigger a normal cell to lose control and become cancerous. One cause is environmental agents: substances in the air we breathe, the water we drink, and the food we eat. A chemical agent or radiation that produces cancer is called a **carcinogen** (car-SIN-ō-jen). Carcinogens induce **mutations** (mū-TĀ-shuns), permanent changes in the DNA base sequence of a gene. The World Health Organization estimates that carcinogens are associated with 60–90 percent of all human cancers. Examples of carcinogens are hydrocarbons found in cigarette tar, radon gas from the earth, and ultraviolet (UV) radiation in sunlight.

Intensive research efforts are now directed toward studying cancer-causing genes, or **oncogenes** (ON-ko-jēnz). When inappropriately activated, these genes have the ability to transform a normal cell into a cancerous cell. Most oncogenes are derived from normal genes called **proto-oncogenes** that regulate growth and development. The proto-oncogene undergoes some change that causes it (1) to be expressed inappropriately, (2) to make its products in excessive amounts, or (3) to make its products at the wrong time. Some oncogenes cause excessive production of growth factors, chemicals that stimulate cell growth. Others may trigger changes in a cell-surface receptor, causing it to send signals as though it were being activated by a growth factor. As a result, the growth pattern of the cell becomes abnormal.

Proto-oncogenes in every cell carry out normal cellular functions until a malignant change occurs. It appears that some proto-oncogenes are activated to oncogenes by mutations in which the DNA of the proto-oncogene is altered. Other proto-oncogenes are activated by a rearrangement of the chromosomes so that segments of DNA are exchanged. Rearrangement activates proto-oncogenes by placing them near genes that enhance their activity.

Damage to genes called **tumor-suppressor genes**, which produce proteins that normally inhibit cell division, causes some types of cancer. Loss or alteration of a tumor-suppressor gene called *p53* on chromosome 17 is the most common genetic change leading to a wide variety of tumors, including breast and colon cancers. The normal p53 protein arrests cells in the G_1 phase, which prevents cell division. Normal p53 protein also assists in repair of damaged DNA and induces apoptosis in the cells where DNA repair was not successful. For this reason, the p53 gene is nicknamed "the guardian angel of the genome."

2.7 AGING AND CELLS

OBJECTIVE

• Describe the cellular changes that occur with aging.

Aging is a normal process accompanied by a progressive alteration of the body's homeostatic adaptive responses to maintain normal conditions. It produces observable changes in structure and function and increases vulnerability to environmental stress and disease. The specialized branch of medicine that deals with the medical problems and care of elderly persons is **geriatrics** (jer-ē-AT-riks; *ger-*=old age; *-iatrics*=medicine). **Gerontology** (jer-on-TOL-ō-jē) is the scientific study of the process and problems associated with aging.

Although many millions of new cells normally are produced each minute, several kinds of cells in the body—including skeletal muscle cells and nerve cells—do not divide because they are arrested permanently in the G_0 phase (see Section 2.5). Experiments have shown that many other cell types have only a limited capability to divide. Normal cells grown outside the body divide only a certain number of times and then stop. These observations suggest that cessation of mitosis is a normal, genetically programmed event. According to this view, "aging genes" are part of the genetic blueprint at birth. These genes have an important function in normal cells but their activities slow over time. They bring about aging by slowing down or halting processes vital to life.

Another aspect of aging involves **telomeres** (TE-lō-merz), specific DNA sequences found only at the tips of each chromosome. These pieces of DNA protect the tips of chromosomes from erosion and from sticking to one another.

Some cancers have a viral origin. Viruses are tiny packages of nucleic acids, either RNA or DNA, that can reproduce only while inside the cells they infect. Some viruses, termed **oncogenic viruses**, cause cancer by stimulating abnormal proliferation of cells. For instance, the *human papillomavirus (HPV)* causes virtually all cervical cancers in women. The virus produces a protein that causes proteasomes to destroy p53, a protein that normally suppresses unregulated cell division. In the absence of this suppressor protein, cells proliferate uncontrollably.

Recent studies suggest that certain cancers may be linked to an abnormal number of chromosomes. As a result, the cell could potentially have extra copies of oncogenes or too few copies of tumor-suppressor genes; in either case, uncontrolled cell proliferation could result. There is also some evidence suggesting that cancer may be caused by normal stem cells that develop into cancerous stem cells capable of forming malignant tumors.

Inflammation is a defensive response to tissue damage. It appears that inflammation contributes to various steps in the development of cancer. Some evidence suggests that chronic inflammation stimulates the proliferation of mutated cells and enhances their survival, promotes angiogenesis, and contributes to invasion and metastasis of cancer cells. There is a clear relationship between certain chronic inflammatory conditions and the transformation of inflamed tissue into a malignant tissue. For example, chronic gastritis (inflammation of the stomach lining) and peptic ulcers may be causative factors in 60–90 percent of stomach cancers. Chronic hepatitis (inflammation of the liver) and cirrhosis of the liver are believed to be responsible for about 80 percent of liver cancers. Colorectal cancer is 10 times more likely to occur in patients with chronic inflammatory diseases of the colon, such as ulcerative colitis and Crohn's disease. And the relationship between asbestosis and silicosis (two chronic lung inflammatory conditions) and lung cancer has long been recognized. Chronic inflammation is also an underlying contributor to rheumatoid arthritis, Alzheimer's disease, depression, schizophrenia, cardiovascular disease, and diabetes.

Carcinogenesis: A Multistep Process

Carcinogenesis (kar′-si-nō-JEN-e-sis) is a multistep process of cancer development in which as many as 10 distinct mutations may have to accumulate in a cell before it becomes cancerous. The progression of genetic changes leading to cancer is best understood for colon (colorectal) cancer. Such cancers, as well as lung and breast cancer, take years or decades to develop. In colon cancer, the tumor begins as an area of increased cell proliferation that results from a single mutation. This growth then progresses to abnormal, but noncancerous, growths called *adenomas*. After two or three additional mutations, a mutation of the tumor-suppressor gene p53 occurs and a carcinoma develops. The fact that so many mutations are needed for a cancer to develop indicates that cell growth is normally controlled with many sets of checks and balances. Thus it is not surprising that a compromised immune system contributes significantly to carcinogenesis.

Treatment of Cancer

Many cancers are removed surgically. However, cancer that is widely distributed throughout the body or exists in organs with essential functions such as the brain, which might be greatly harmed by surgery, may be treated with chemotherapy and radiation therapy instead. Sometimes surgery, chemotherapy, and radiation therapy are used in combination. Chemotherapy involves administering drugs that cause death of cancerous cells. Radiation therapy breaks chromosomes, thus blocking cell division. Because cancerous cells divide rapidly, they are more vulnerable to the destructive effects of chemotherapy and radiation therapy than are normal cells. Unfortunately for the patients, hair follicle cells, red bone marrow cells, and cells lining the gastrointestinal tract also are rapidly dividing. Hence, the side effects of chemotherapy and radiation therapy include hair loss due to death of hair follicle cells, vomiting and nausea due to death of cells lining the stomach and intestines, and susceptibility to infection due to slowed production of white blood cells in red bone marrow.

Treating cancer is difficult because it is not a single disease and because the cells in a single tumor population rarely behave all in the same way. Although most cancers are thought to derive from a single abnormal cell, by the time a tumor reaches a clinically detectable size, it may contain a diverse population of abnormal cells. For example, some cancerous cells metastasize readily and others do not. Some are sensitive to chemotherapy drugs and some are drug-resistant. Because of differences in drug resistance, a single chemotherapeutic agent may destroy susceptible cells but permit resistant cells to proliferate.

Another potential treatment for cancer that is currently under development is *virotherapy*, the use of viruses to kill cancer cells. The viruses employed in this strategy are designed so that they specifically target cancer cells without affecting the healthy cells of the body. For example, proteins (such as antibodies) that specifically bind to receptors found only in cancer cells are attached to viruses. Once inside the body, the viruses bind to cancer cells and then infect them. The cancer cells are eventually killed once the viruses cause cellular lysis.

Researchers also are investigating the role of *metastasis regulatory genes* that control the ability of cancer cells to undergo metastasis. Scientists hope to develop therapeutic drugs that can manipulate these genes and, therefore, block metastasis of cancer cells. •

However, in most normal body cells each cycle of cell division shortens the telomeres. Eventually, after many cycles of cell division, the telomeres can be completely gone and even some of the functional chromosomal material may be lost. These observations suggest that erosion of DNA from the tips of our chromosomes contributes greatly to aging and death of cells. It has recently been learned that individuals who experience high levels of stress have significantly shorter telomere length.

Glucose, the most abundant sugar in the body, plays a role in the aging process. It is haphazardly added to proteins inside and outside cells, forming irreversible cross-links between adjacent protein molecules. With advancing age, more cross-links form, which contributes to the stiffening and loss of elasticity that occur in aging tissues.

Some theories of aging explain the process at the cellular level, while others concentrate on regulatory mechanisms operating within the entire organism. For example, the immune system may start to attack the body's own cells. This *autoimmune response* might be caused by changes in cell-identity markers at the surface of cells (certain plasma membrane glycoproteins and glycolipids) that cause antibodies to attach to and mark the cell for destruction. As changes in the proteins on the plasma membrane of cells increase, the autoimmune response intensifies, producing the well-known signs of aging. In the chapters that follow, we will discuss the effects of aging on each body system in sections similar to this one.

 CHECKPOINT

22. Why do some tissues become stiffer as they age?

CLINICAL CONNECTION | *Free Radicals*

Free radicals are atoms or groups of atoms that are *oxidized* (have an extra electron); they damage lipids, proteins, or nucleic acids by "stealing" an electron to accompany their unpaired electrons. Some effects are wrinkled skin, stiff joints, and hardened arteries. Normal cellular metabolism—for example, aerobic cellular respiration in mitochondria—produces some free radicals. Others are present in air pollution, radiation, and certain foods we eat. Naturally occurring enzymes in peroxisomes and in the cytosol normally dispose of free radicals. Certain dietary substances, such as vitamin E, vitamin C, beta carotene, zinc, and selenium, are referred to as *antioxidants* because they inhibit the formation of free radicals. •

KEY MEDICAL TERMS ASSOCIATED WITH CELLS

Note to the Student *Most chapters in this text are followed by a glossary of key medical terms that include both normal and pathological conditions. You should familiarize yourself with the terms because they will play an essential role in your medical vocabulary.*

Anaplasia (an'-a-PLĀ-zē-a; *an-*=not; *-plasia*=to shape) The loss of tissue differentiation and function that is characteristic of most malignancies.

Atrophy (AT-rō-fē; *a-*=without; *-trophy*=nourishment) A decrease in the size of cells, with a subsequent decrease in the size of the affected tissue or organ; wasting away.

Biopsy (BĪ-op-sē; *bio-*=life; *-opsy*=viewing) The removal and microscopic examination of tissue from the living body for diagnosis.

Dysplasia (dis-PLĀ-zē-a; *dys-*=abnormal) Alteration in the size, shape, and organization of cells due to chronic irritation or inflammation; may progress to neoplasia (tumor formation, usually malignant) or revert to normal if the irritation is removed.

Hyperplasia (hī-per-PLĀ-zē-a; *hyper-*=over) Increase in the number of cells of a tissue due to an increase in the frequency of cell division.

Hypertrophy (hī-PER-trō-fē) Increase in the size of cells without cell division.

Metaplasia (met-a-PLĀ-zē-a; *meta-*=change) The transformation of one type of cell into another.

Necrosis (ne-KRŌ-sis; =death) A pathological type of cell death, resulting from tissue injury, in which many adjacent cells swell, burst, and spill their cytoplasm into the interstitial fluid; the cellular debris usually stimulates an inflammatory response, which does not occur in apoptosis.

Progeny (PROJ-e-nē; *pro-*=forward; *-geny*=production) Offspring or descendants.

Progeria (pro-JER-ē-a) A disease characterized by normal development in the first year of life followed by rapid aging. It is caused by a genetic defect in which telomeres are considerably shorter than normal. Symptoms include dry and wrinkled skin, total baldness, and birdlike facial features. Death usually occurs around age 13.

Proteomics (prō'-tē-Ō-miks; *proteo-*=protein) The study of the proteome (all of an organism's proteins) in order to identify all the proteins produced; it involves determining how the proteins interact and ascertaining the three-dimensional structure of proteins so that drugs can be designed to alter protein activity to help in the treatment and diagnosis of disease.

Tumor marker A substance introduced into circulation by tumor tissue that indicates the presence of a tumor as well as the specific type. Tumor markers may be used to screen, diagnose, assess prognosis, evaluate a response to treatment, and monitor for recurrence of cancer.

Werner syndrome (VER-ner) A rare, inherited disease that causes a rapid acceleration of aging, usually while the person is only in his or her twenties. It is characterized by wrinkling of the skin, graying of the hair and baldness, cataracts, muscular atrophy, and a tendency to develop diabetes mellitus, cancer, and cardiovascular disease. Most afflicted individuals die before age 50. Recently, the gene that causes Werner syndrome has been identified. Researchers hope to use this information to gain insight into the mechanisms of aging, as well as to help those suffering from the disorder.

CHAPTER REVIEW AND RESOURCE SUMMARY

WileyPLUS

Review   **Resource**

Introduction

1. A cell is the basic, living, structural and functional unit of the body.
2. Cell biology is the scientific study of cellular structure and function.

2.1 A Generalized Cell

1. Figure 2.1 provides an overview of the typical structures in body cells.
2. The principal parts of a cell are (1) the plasma membrane; (2) the cytoplasm, the cellular contents between the plasma membrane and nucleus; and (3) the nucleus.

Anatomy Overview - Cell Structure and Function
Figure 2.1 - The Cell

2.2 The Plasma Membrane

1. The plasma membrane, which surrounds and contains the cytoplasm of a cell, is composed of proteins and lipids that are held together by forces other than chemical bonds. According to the fluid mosaic model, the membrane is a mosaic of proteins floating like icebergs in a lipid bilayer sea. The lipid bilayer consists of two back-to-back layers of phospholipids, cholesterol, and glycolipids.
2. Integral proteins extend into or through the lipid bilayer; peripheral proteins associate with membrane lipids or integral proteins at the inner or outer surface of the membrane. Many integral proteins are glycoproteins, with sugar groups attached to the ends that face the extracellular fluid. Together with glycolipids, the glycoproteins form a glycocalyx on the extracellular surface of cells.
3. Membrane proteins have a variety of functions. Integral proteins are channels and transporters that help specific solutes cross the membrane; receptors that serve as cellular recognition sites; enzymes

Anatomy Overview - Plasma Membrane Structure
Animation - Membrane Functions
Animation - Transport Across the Plasma Membrane
Figure 2.2 - Structure of the Plasma Membrane
Exercise - Paint a Cell Membrane
Exercise - Animation Membrane Function Matchup

Review

that catalyze specific chemical reactions; and linkers that anchor proteins in the plasma membranes to protein filaments inside and outside the cell. Peripheral proteins serve as enzymes and linkers; support the plasma membrane; anchor integral proteins; and participate in mechanical activities. Membrane glycoproteins function as cell-identity markers.

4. The membrane's selective permeability permits some substances to pass more readily than others. The lipid bilayer is permeable to most nonpolar, uncharged molecules. It is impermeable to ions and charged or polar molecules other than water and urea. Channels and transporters increase the plasma membrane's permeability to small- and medium-sized polar and charged substances, including ions, that cannot cross the lipid bilayer. Substances cross plasma membranes by using kinetic energy, by binding to specific transporter proteins, and by utilizing vesicles.

5. In passive processes, including diffusion and osmosis, a substance moves down its concentration gradient across the membrane using its own kinetic energy of motion. In diffusion, molecules or ions move from an area of higher concentration to an area of lower concentration until equilibrium is reached. Osmosis is the net movement of water through a selectively permeable membrane from an area of higher water concentration to an area of lower water concentration.

6. In active processes, including facilitated diffusion, active transport, and transport in vesicles, cellular energy is used to drive the substance "uphill" against its concentration gradient. In facilitated diffusion, a solute such as glucose binds to a specific transporter on one side of the membrane and is released on the other side after the transporter undergoes a change in shape. Active transport is the movement of a substance (usually ions) across a cell membrane from lower to higher concentration using energy derived from ATP and a carrier protein.

7. Transport in vesicles includes both endocytosis and exocytosis. Receptor-mediated endocytosis is the selective uptake of large molecules and particles (ligands) that bind to specific receptors in membrane areas called clathrin-coated pits. A second type of endocytosis, phagocytosis, is the ingestion of solid particles; it is an important process used by some white blood cells to destroy bacteria that enter the body. Bulk-phase endocytosis is the ingestion of extracellular fluid. In this process, the fluid becomes surrounded by a pinocytic vesicle.

8. Exocytosis involves movement of secretory or waste products out of a cell by fusion of vesicles with the plasma membrane.

2.3 Cytoplasm

1. Cytoplasm, all the cellular contents between the plasma membrane and the nucleus, consists of cytosol and organelles. Cytosol, the fluid portion of cytoplasm, contains mostly water, plus ions, glucose, amino acids, fatty acids, proteins, lipids, ATP, and waste products. Cytosol is the site of many chemical reactions required for a cell's existence. The cytoskeleton, a network of several kinds of protein filaments that extend throughout the cytosol, provides a structural framework for the cell and is responsible for cell movements; its components include microfilaments, intermediate filaments, and microtubules.

2. Organelles are specialized structures with characteristic shapes and specific functions. The centrosome is an organelle that consists of pericentriolar material and a pair of centrioles. The pericentriolar material organizes microtubules in nondividing cells and the mitotic spindle in dividing cells.

3. Cilia and flagella, motile projections of the cell surface, are formed by basal bodies. Cilia move fluid along the cell surface; flagella move an entire cell.

4. Ribosomes consist of two subunits made in the nucleus that are composed of ribosomal RNA and ribosomal proteins. They serve as sites of protein synthesis.

5. Endoplasmic reticulum (ER) is a network of membranes that form flattened sacs or tubules; it extends from the nuclear envelope throughout the cytoplasm. Rough ER is studded with ribosomes that synthesize proteins; the proteins then enter the space within the ER for processing and sorting. Rough ER produces secretory proteins, membrane proteins, and organelle proteins; forms glycoproteins; synthesizes phospholipids; and attaches proteins to phospholipids. Smooth ER lacks ribosomes. It synthesizes fatty acids and steroids; inactivates or detoxifies drugs and other potentially harmful substances; removes phosphate from glucose-6-phosphate; and stores and releases calcium ions that trigger contraction in muscle cells.

6. The Golgi complex consists of flattened sacs called cisternae. The entry, medial, and exit regions of the Golgi complex contain different enzymes that permit each to modify, sort, and package proteins for transport in secretory vesicles, membrane vesicles, or transport vesicles to different cellular destinations.

7. Lysosomes are membrane-enclosed vesicles that contain digestive enzymes. Endosomes, phagosomes, and vesicles formed during bulk-phase endocytosis deliver materials to lysosomes for degradation. Lysosomes function in digestion of worn-out organelles (autophagy), digestion of a host cell (autolysis), and extracellular digestion.

8. Peroxisomes contain oxidases that oxidize amino acids, fatty acids, and toxic substances; the hydrogen peroxide produced in the process is destroyed by catalase.

9. The proteases contained in proteasomes continually degrade unneeded, damaged, or faulty proteins by cutting them into small peptides.

10. Mitochondria consist of a smooth outer membrane, an inner membrane containing mitochondrial cristae, and a fluid-filled cavity called the mitochondrial matrix. These so-called "powerhouses" of the cell produce most of a cell's ATP and play an important early role in apoptosis.

Exercise - Create a Transport Condition
Exercise - Osmosis—Move It!
Concepts and Connections - Membrane Functions

Animation - Cell Respiration
Animation - Protein Synthesis
Exercise - Concentrate on Cellular Functions
Exercise - Target Practice
Exercise - Protein-producing Processes
Exercise - Synthesize and Transport a Protein
Concepts and Connections - Human Cell
Concepts and Connections - Membranous Organelles

Review	Resource

2.4 Nucleus

1. The nucleus consists of a double nuclear envelope; nuclear pores, which control the movement of substances between the nucleus and cytoplasm; nucleoli, which produce ribosomes; and genes arranged on chromosomes, which control cellular structure and direct cellular activities.
2. Human somatic cells have 46 chromosomes, 23 inherited from each parent. The total genetic information carried in a cell or an organism is its genome.

Anatomy Overview - The Nucleus

2.5 Cell Division

1. Cell division is the process by which cells reproduce themselves. It consists of a nuclear division (mitosis or meiosis) and a cytoplasmic division (cytokinesis). Cell division that replaces cells or adds new ones is called somatic cell division and involves mitosis and cytokinesis. Cell division that results in the production of gametes (sperm and ova) is called reproductive cell division and consists of meiosis and cytokinesis.
2. The cell cycle, an orderly sequence of events in which a somatic cell duplicates its contents and divides in two, consists of interphase and a mitotic phase. Before the mitotic phase, the DNA molecules, or chromosomes, replicate themselves so that identical sets of chromosomes can be passed on to the next generation of cells. A cell between divisions that is carrying on every life process except division is said to be in interphase, which consists of three phases: G_1, S, and G_2.
3. During the G_1 phase, the cell replicates its organelles and cytosolic components, and centrosome replication begins; during the S phase, DNA replication occurs; during the G_2 phase, enzymes and other proteins are synthesized and centrosome replication is completed.
4. Mitosis is the splitting of the chromosomes and the distribution of two identical sets of chromosomes into separate and equal nuclei; it consists of prophase, metaphase, anaphase, and telophase.
5. In cytokinesis, which usually begins in late anaphase and ends once telophase is complete, a cleavage furrow forms at the cell's metaphase plate and progresses inward, pinching in through the cell to form two separate portions of cytoplasm.
6. A cell can either remain alive and functioning without dividing, grow and divide, or die. The control of cell division depends on specific cyclin-dependent protein kinases and cyclins. Apoptosis is normal, programmed cell death. It first occurs during embryological development and continues throughout the lifetime of an organism. Certain genes regulate both cell division and apoptosis. Abnormalities in these genes are associated with a wide variety of diseases and disorders.
7. In sexual reproduction, each new organism is the result of the union of two different gametes, one from each parent. Human somatic cells contain 23 pairs of homologous chromosomes and are thus diploid ($2n$). Gametes contain a single set of chromosomes (23) and thus are haploid (n). Meiosis is the process that produces haploid gametes; it consists of two successive nuclear divisions called meiosis I and meiosis II.
8. During meiosis I, homologous chromosomes undergo synapsis (pairing) and crossing-over; the net result is two haploid cells that are genetically unlike each other and unlike the starting diploid parent cell that produced them.
9. During meiosis II, two haploid cells divide to form four haploid cells.

Animation - The Cell Cycle and Division Processes
Figure 2.20 - Mitosis and Meiosis
Exercise - Mitosis Matchup
Exercise - Meiosis Matchup
Exercise - Cell Cycle Quiz

2.6 Cellular Diversity

1. The almost 200 different types of cells in the body vary considerably in size and shape.
2. The sizes of cells are measured in micrometers. One micrometer (μm) equals 10^{-6} meter (1/25,000th of an inch). Cells in the body range from 8 μm to 140 μm in size.
3. A cell's shape is related to its function.

2.7 Aging and Cells

1. Aging is a normal process accompanied by progressive alteration of the body's homeostatic adaptive responses.
2. Many theories of aging have been proposed, including genetically programmed cessation of cell division, buildup of free radicals, and an intensified autoimmune response.

CRITICAL THINKING QUESTIONS

1. You may inherit your brown eyes from either your mother or your father but some traits can be passed down only from mother to child. "Maternal inheritance" is due to genetic material that is not located in the nucleus. Can you suggest an explanation?

2. In the "old days," intravenous solutions did not come "prepackaged" and had to be mixed at the hospital. Maureen wasn't very good at arithmetic and misplaced her decimal point when calculating how much glucose to add to the intravenous solution for her patient. Instead of making a 0.9 percent saline solution, she made a 9.0 percent saline solution. Using your knowledge of osmosis, predict what would happen to her patient's blood cells if she injected this solution into her patient's bloodstream.

3. A child was brought to the emergency room after eating rat poison containing arsenic. Arsenic kills rats by blocking the function of the mitochondria. What effect would the poison have on the child's body functions?

4. Imagine that researchers have discovered a new chemotherapy agent that disrupts microtubules in cancerous cells but leaves normal cells unaffected. What effect would this agent have on the cancer cells?

5. "Gene therapy" relies on delivering healthy genes into cells containing defective genes. One of the current processes undergoing research is to attach healthy genes to a virus that can "carry" the gene into the cells. With your knowledge of transport processes, provide an explanation of how this can work, and speculate on potential obstacles.

? ANSWERS TO FIGURE QUESTIONS

2.1 The three main parts of the cell are the plasma membrane, cytoplasm, and nucleus.

2.2 The glycocalyx is a coat on the extracellular surface of the plasma membrane that is composed of the carbohydrate portions of membrane glycolipids and glycoproteins.

2.3 Another name for intracellular fluid is cytosol.

2.4 Transferrin, vitamins, and hormones are examples of ligands that can undergo receptor-mediated endocytosis.

2.5 The binding of particles to a plasma membrane receptor triggers pseudopod formation.

2.6 Microtubules help form the structure of centrioles, cilia, and flagella.

2.7 A cell without a centrosome probably would not be able to undergo cell division.

2.8 Cilia move fluids across cell surfaces; flagella move entire cells.

2.9 Large and small ribosomal subunits are synthesized in the nucleolus of the nucleus and assembled in the cytoplasm.

2.10 Rough ER has attached ribosomes; smooth ER does not. Rough ER synthesizes proteins that will be exported from the cell; smooth ER is associated with lipid synthesis and other metabolic reactions.

2.11 The entry face receives and modifies proteins from rough ER. The exit face modifies, sorts, packages, and transports molecules to other destinations.

2.12 Some proteins are discharged from the cell by exocytosis, some are incorporated into the plasma membrane, and some occupy storage vesicles that become lysosomes.

2.13 Digestion of worn-out organelles by lysosomes is called autophagy.

2.14 Mitochondrial cristae increase the surface area available for chemical reactions and contain the enzymes needed for ATP production.

2.15 Chromatin is a complex of DNA, proteins, and some RNA.

2.16 A nucleosome is a double-stranded molecule of DNA wrapped twice around a core of eight histones (proteins).

2.17 DNA replication occurs during the S phase of interphase in the cell cycle.

2.18 Cytokinesis usually starts in late anaphase.

2.19 The result of crossing-over is that the four haploid gametes are genetically unlike each other and genetically unlike the starting cell that produced them.

2.20 During anaphase I of meiosis, the paired chromatids are held together by a centromere and do not separate. During mitotic anaphase the centromeres split and sister chromatids separate.

2.21 Sperm, which use the flagella for locomotion, are the only body cells required to move considerable distances.

3 TISSUES

INTRODUCTION As you learned in Chapter 2, a cell is a complex collection of compartments, each of which carries out a host of biochemical reactions that make life possible. However, a cell seldom functions as an isolated unit. Instead, cells usually work together in groups called tissues. A **tissue** is a group of cells that usually have a common origin in an embryo and function together to carry out specialized activities. The structure and properties of a specific tissue are influenced by factors such as the nature of the extracellular material that surrounds the tissue cells and the connections between the cells that compose the tissue. Tissues may be hard, semisolid, or even liquid in their consistency, a range exemplified by bone, fat, and blood. In addition, tissues vary tremendously with respect to the kinds of cells present, how the cells are arranged, and the types of fibers present, if any.

Think about taking a shower. As you wash your body, your skin feels soft and pliable, but the bones beneath your skin are hard and inflexible. What makes these structures so different? Why doesn't the water that splashes on your body penetrate through its surface? The water you drink is absorbed from your digestive system into your blood. What features of the skin and digestive system lining are responsible for this difference in water permeability? Occasionally, people break bones. While these breaks can be very painful, they are usually repaired relatively quickly, leaving little or no evidence of the once painful damage. On the other hand, cartilage damage in a joint can lead to pain and misery that lasts for years. What accounts for these differences?

In order to answer these questions you must understand the nature of different types of tissues. Tissues form an important level in the hierarchy of body design introduced in Chapter 1. Cells form tissues, and tissues make up the organs that work together to maintain stability in the internal fluid environment. As you study and understand the structure of tissues in this chapter, you will gain more insight into their functional roles in maintaining normal body conditions.

Histology (his′-TOL-ō-jē; *histo-*=tissue; *-logy*=study of) is the science that deals with the study of tissues. A **pathologist** (pa-THOL-ō-gist; *patho-*=disease) is a physician who examines cells and tissues to help other physicians make accurate diagnoses. One of the principal functions of a pathologist is to examine tissues for any changes that might indicate disease. •

Did you ever wonder whether the complications of liposuction outweigh the benefits?

3.1 TYPES OF TISSUES

 OBJECTIVE

• Name the four basic types of tissues that make up the human body, and state the characteristics of each.

Body tissues can be classified into four basic types according to their structure and function (Figure 3.1):

1. **Epithelial tissue** covers body surfaces, and lines hollow organs, body cavities, and ducts; it also forms glands. This tissue allows the body to interact with both its internal and external environments.
2. **Connective tissue** protects and supports the body and its organs. Various types of connective tissue bind organs together, store energy reserves as fat, and help provide the body with immunity to disease-causing organisms.
3. **Muscular tissue** is composed of cells specialized for contraction and generation of force. In the process, muscular tissue generates heat that warms the body.
4. **Nervous tissue** detects changes in a variety of conditions inside and outside the body and responds by generating electrical signals called nerve action potentials (nerve impulses) that activate muscular contractions and glandular secretions.

CLINICAL CONNECTION | Biopsy

A **biopsy** (BĪ-op-sē; bio-=life; -opsy=to view) is the removal of a sample of living tissue for microscopic examination. This procedure is used to help diagnose many disorders, especially cancer, and to discover the cause of unexplained infections and inflammations. Both normal and potentially diseased tissues are removed for purposes of comparison. Once the tissue samples are removed, either surgically or through a needle and syringe, they may be preserved, stained to highlight special properties, or cut into thin sections for microscopic observation. Sometimes a biopsy is conducted while a patient is anesthetized during surgery to help a physician determine the most appropriate treatment. For example, if a biopsy of thyroid tissue reveals malignant cells, the surgeon can proceed immediately with the most appropriate procedure. •

Epithelial tissue and most types of connective tissue, except cartilage, bone, and blood, are more general in nature and have a wide distribution in the body. These tissues are components of most body organs and have a wide range of structure and function. We will look at epithelial tissue and connective tissue in some detail in this chapter. The general features of bone tissue and blood will be introduced here, but their detailed discussion is presented in Chapters 6 and 12, respectively. Similarly, the structure and

Figure 3.1 Types of tissues.

 Each of the four types of tissues has different cells that vary in shape, structure, function, and distribution.

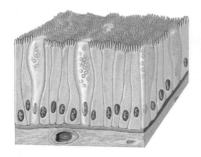 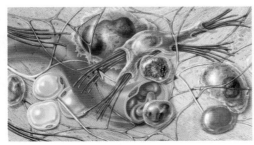  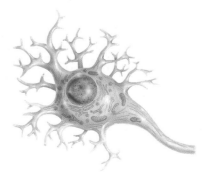

(a) Epithelial tissue

(b) Connective tissue

(c) Muscular tissue

(d) Nervous tissue

 What are some key differences among the four tissue types?

function of muscular tissue and nervous tissue are introduced here and examined in detail in Chapters 10 and 16, respectively.

CHECKPOINT

1. Define a tissue.
2. What are the basic types of human tissues?

3.2 CELL JUNCTIONS

OBJECTIVE

• Describe the structure and functions of the five main types of cell junctions.

Before looking more specifically at the types of tissues, we will first examine how cells are held together to form tissues. Most epithelial cells and some muscle and nerve cells are tightly joined into functional units. **Cell junctions** are contact points between the plasma membranes of tissue cells. Here we consider the five most important types of cell junctions: tight junctions, adherens junctions, desmosomes, hemidesmosomes, and gap junctions (Figure 3.2).

Tight Junctions

Tight junctions consist of weblike strands of transmembrane proteins that fuse together the outer surfaces of adjacent plasma membranes to seal off passageways between adjacent cells (Figure 3.2a). Cells of epithelial tissue that line the stomach, intestines, and urinary bladder have many tight junctions. They inhibit the passage of substances between cells and prevent the contents of these organs from leaking into the blood or surrounding tissues.

Adherens Junctions

Adherens junctions (ad-HER-ens) contain *plaque* (PLAK), a dense layer of proteins on the inside of the plasma membrane that

Figure 3.2 Cell junctions.

Most epithelial cells and some muscle and nerve cells contain cell junctions.

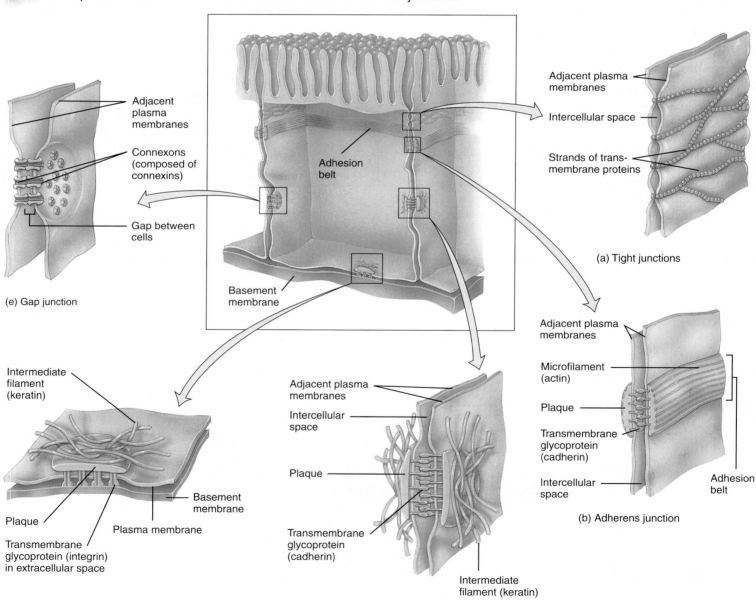

(a) Tight junctions

(b) Adherens junction

(c) Desmosome

(d) Hemidesmosome

(e) Gap junction

 Which type of cell junction functions in communication between adjacent cells?

attaches both to membrane proteins and to microfilaments of the cytoskeleton (Figure 3.2b). Transmembrane glycoproteins called **cadherins** join the cells. Each cadherin inserts into the plaque from the opposite side of the plasma membrane, partially crosses the intercellular space (the space between the cells), and connects to a cadherin of an adjacent cell. In epithelial cells, adherens junctions often form extensive zones called *adhesion belts* because they encircle the cell similar to the way a belt encircles your waist. Adherens junctions help epithelial surfaces resist separation during various contractile activities, as when food moves through the intestines.

Desmosomes

Like adherens junctions, **desmosomes** (DEZ-mō-sōms; *desmo-* =band) contain plaque and have transmembrane glycoproteins (cadherins) that extend into the intercellular space between adjacent cell membranes and attach cells to one another (Figure 3.2c). However, unlike adherens junctions, the plaque of desmosomes does not attach to microfilaments. Instead, a desmosome plaque attaches to elements of the cytoskeleton known as intermediate filaments, which consist of the protein keratin. The intermediate filaments extend from desmosomes on one side of the cell across the cytosol to desmosomes on the opposite side of the cell. This structural arrangement contributes to the stability of the cells and tissue. These spot-weld-like junctions are common among the cells that make up the epidermis (the outermost layer of the skin) and among cardiac muscle cells in the heart. Desmosomes prevent epidermal cells from separating under tension and cardiac muscle cells from pulling apart during contraction.

Hemidesmosomes

Hemidesmosomes (*hemi-*=half) resemble desmosomes but they do not link adjacent cells. The name arises from the fact that they look like half of a desmosome (Figure 3.2d). However, the transmembrane glycoproteins in hemidesmosomes are **integrins** rather than cadherins. On the inside of the plasma membrane, integrins attach to intermediate filaments made of the protein keratin. On the outside of the plasma membrane, the integrins attach to the protein *laminin*, which is present in the basement membrane (discussed shortly). Thus, hemidesmosomes anchor cells not to each other but to the basement membrane.

Gap Junctions

At **gap junctions**, membrane proteins called **connexins** form tiny fluid-filled tunnels called *connexons* that connect neighboring cells (Figure 3.2e). Unlike the fused plasma membranes of tight junctions, the plasma membranes of gap junctions are separated by a very narrow intercellular gap (space). Through the connexons, ions and small molecules can diffuse from the cytosol of one cell to another, but the passage of large molecules such as vital intracellular proteins is prevented. The transfer of nutrients, and perhaps wastes, takes place through gap junctions in avascular tissues such as the lens and cornea of the eye. Gap junctions allow the cells in a tissue to communicate with one another. In a developing embryo, some of the chemical and electrical signals that regulate growth and cell differentiation travel via gap junctions. Gap junctions also enable nerve or muscle impulses to spread rapidly among cells, a process that is crucial for the normal operation of some parts of the nervous system and for the contraction of muscle in the heart, gastrointestinal tract, and uterus.

 CHECKPOINT

3. Which type of cell junction prevents the contents of organs from leaking into surrounding tissues?
4. Which types of cell junctions are found in epithelial tissue?

3.3 COMPARISON BETWEEN EPITHELIAL AND CONNECTIVE TISSUES

 OBJECTIVE

• Outline the main differences between epithelial and connective tissues.

Before examining epithelial tissue and connective tissue in more detail, let's compare these two widely distributed tissues. Major structural differences between an epithelial tissue and a connective tissue are immediately obvious under a light microscope (Figure 3.3). The first obvious difference is the number of cells in relation to the extracellular matrix (the substance between cells). In an epithelial tissue, many cells are tightly packed together with little or no extracellular matrix, while in a connective tissue a large amount of extracellular material separates cells that are usually widely scattered. The second obvious difference is that an epithelial tissue has no blood vessels, while most connective tissue has significant networks of blood vessels. Another key difference is that epithelial tissue always forms surface layers and is not covered by another tissue, except within blood vessels where the blood constantly passes over the epithelial tissue lining the vessels. While these key structural distinctions account for some of the major functional differences between these tissue types, they also lead to a common bond. Because epithelial tissue lacks blood vessels and forms surfaces, it is always found immediately adjacent to blood-vessel-rich connective tissue, which enables it to make the exchanges with blood necessary for the delivery of oxygen and nutrients and the removal of wastes that are critical processes for its survival and function.

 CHECKPOINT

5. In what sense is epithelial tissue dependent on connective tissue?

Figure 3.3 Comparison between epithelial tissue and connective tissue.

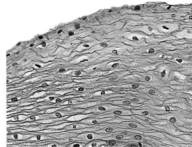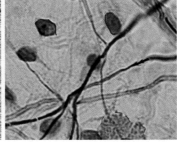

The ratio of cells to extracellular matrix is a major difference between epithelial tissue and connective tissue.

(a) Epithelial tissue with many cells tightly packed together and little to no extracellular matrix

(b) Connective tissue with a few scattered cells surrounded by large amounts of extracellular matrix

 What relationship between epithelial tissue and connective tissue is important for the survival and function of epithelial tissue?

3.4 EPITHELIAL TISSUE

 OBJECTIVES

- Describe the general features of epithelial tissue.
- List the location, structure, and function of each different type of epithelial tissue.

An **epithelial tissue** (ep-i-THĒ-lē-al) or **epithelium** (plural is *epithelia*) consists of cells arranged in continuous sheets, in either single or multiple layers. Because the cells are closely packed and are held tightly together by many cell junctions, there is little intercellular space between adjacent plasma membranes. Epithelial tissue forms coverings and linings throughout the body. It is not covered by another tissue, so it always has a free surface. Epithelial tissue has three major functions: It (1) serves as a selective barrier to limit or aid the transfer of substances into and out of the body; (2) releases products produced by the cells onto its free surfaces; and (3) protects against the abrasive influences of the environment.

The various surfaces of epithelial cells often differ in structure and have specialized functions. The **apical (free) surface** of an epithelial cell faces the body surface, a body cavity, the lumen (interior space) of an internal organ, or a tubular duct that receives cell secretions (Figure 3.4). Apical surfaces may contain cilia or microvilli. The **lateral surfaces** of an epithelial cell, which face the adjacent cells on either side, may contain tight junctions, adherens junctions,

Figure 3.4 Surfaces of epithelial cells and the structure and location of the basement membrane.

 The basement membrane is found between an epithelial tissue and a connective tissue.

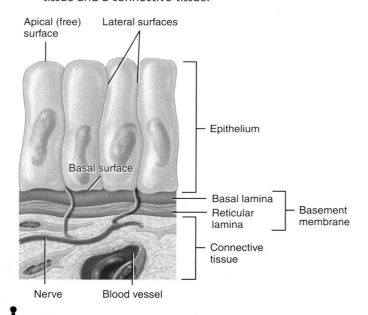

CLINICAL CONNECTION | *Basement Membranes and Disease*

Under certain conditions, basement membranes become markedly thickened, due to increased production of collagen and laminin. In untreated cases of diabetes mellitus, the basement membrane of small blood vessels (capillaries) thickens, especially in the eyes and kidneys. Because of this, the blood vessels cannot function properly and blindness and kidney failure may result. •

? **What are the functions of the basement membrane?**

desmosomes, and/or gap junctions. The **basal surface** of an epithelial cell is opposite the apical surface. The basal surfaces of the deepest layer of epithelial cells adhere to extracellular materials such as the basement membrane. Hemidesmosomes in the basal surfaces of the deepest layer of epithelial cells anchor the epithelium to the basement membrane (described next). In discussing epithelia with multiple layers the term *apical layer* refers to the most superficial layer of cells, and the *basal layer* is the deepest layer of cells.

The **basement membrane** is a thin extracellular layer that commonly consists of two layers, the basal lamina and reticular lamina. The *basal lamina* (*lamina*=thin layer) is closer to—and secreted by—the epithelial cells. It contains proteins such as laminin and collagen (described shortly), as well as glycoproteins and proteoglycans (also described shortly). As you have already learned, the laminin molecules in the basal lamina adhere to integrins in hemidesmosomes and thus attach epithelial cells to the basement membrane (see Figure 3.2d). The *reticular lamina* is closer to the underlying connective tissue and contains proteins such as collagen produced by connective tissue cells called *fibroblasts* (see Figure 3.8). Basement membranes have other functions in addition to attaching to and supporting the overlying epithelial tissue. They form a surface along which epithelial cells migrate during growth or wound healing, restrict passage of larger molecules between epithelium and connective tissue, and participate in filtration of blood in the kidneys.

Epithelial tissue has its own nerve supply, but as mentioned previously is **avascular** (*a-*=without; *vascular*=vessel), relying on the blood vessels of the adjacent connective tissue to bring nutrients and remove wastes. Exchange of substances between an epithelial tissue and connective tissue occurs by diffusion.

Because epithelial tissue forms boundaries between the body's organs, or between the body and the external environment, it is repeatedly subjected to physical stress and injury. A high rate of cell division allows epithelial tissue to constantly renew and repair itself by sloughing off dead or injured cells and replacing them with new ones. Epithelial tissue has many different roles in the body; the most important are protection, filtration, secretion, absorption, and excretion. In addition, epithelial tissue combines with nervous tissue to form special organs for smell, hearing, vision, and touch.

Epithelial tissue may be divided into two types. (1) **Covering and lining epithelium** forms the *outer* covering of the skin and some internal organs. It also forms the *inner* lining of blood vessels, ducts, and body cavities, and the interior of the respiratory, digestive, urinary, and reproductive systems. (2) **Glandular epithelium** makes up the secreting portion of glands such as the thyroid gland, adrenal glands, and sweat glands.

Classification of Epithelial Tissue

Types of epithelial tissue are classified according to two characteristics: the arrangement of cells into layers and the shapes of the cells.

1. *Arrangement of cells in layers* (Figure 3.5). The cells are arranged in one or more layers depending on function:
   a. *Simple (unilaminar) epithelium* is a single layer of cells that functions in diffusion, osmosis, filtration, secretion, or absorption. **Secretion** is the production and release of substances such as mucus, sweat, or enzymes. **Absorption** is the intake of fluids or other substances such as digested food from the intestinal tract.
   b. *Pseudostratified epithelium* (*pseudo-*=false) appears to have multiple layers of cells because the cell nuclei lie at different

Figure 3.5 Cell shapes and arrangement of layers for covering and lining epithelium.

🔑 Cell shapes and arrangement of layers are the bases for classifying covering and lining epithelium.

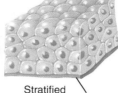

Arrangement of layers

Simple   Pseudostratified   Stratified

Basement membrane

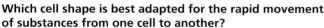

Cell shape

Squamous   Cuboidal   Columnar

Basement membrane

❓ **Which cell shape is best adapted for the rapid movement of substances from one cell to another?**

levels and not all cells reach the apical surface, but it is actually a simple epithelium because all its cells rest on the basement membrane. Cells that do extend to the apical surface may contain cilia; others (goblet cells) secrete mucus.

  c. *Stratified (multilaminar) epithelium* (*stratum*=layer) consists of two or more layers of cells that protect underlying tissues in locations where there is considerable wear and tear.

2. *Cell shapes* (Figure 3.5). Epithelial cells vary in shape depending on their function:

  a. *Squamous* cells (SKWĀ-mus=flat) are thin, which allows for the rapid passage of substances through them.

  b. *Cuboidal* cells are as tall as they are wide and are shaped like cubes or hexagons. They may have microvilli at their apical surface and function in either secretion or absorption.

  c. *Columnar* cells are much taller than they are wide, like columns, and protect underlying tissues. Their apical surfaces may have cilia or microvilli, and they often are specialized for secretion and absorption.

  d. *Transitional* cells change shape, from squamous to cuboidal and back, as organs such as the urinary bladder stretch (distend) to a larger size and then collapse to a smaller size.

If you combine the two characteristics (arrangements of layers and cell shapes), you get the following types of epithelia:

  I. Simple epithelium
    A. Simple squamous epithelium
    B. Simple cuboidal epithelium
    C. Simple columnar epithelium (nonciliated and ciliated)
    D. Pseudostratified columnar epithelium (nonciliated and ciliated)

  II. Stratified epithelium
    A. Stratified squamous epithelium (keratinized, when surface cells are dead and become hardened, and nonkeratinized, when surface cells remain alive)*
    B. Stratified cuboidal epithelium*
    C. Stratified columnar epithelium*
    D. Transitional epithelium

We will now examine the important features of covering and lining epithelium.

*This classification is based on the shape of the cells at the apical surface.

A **Papanicolaou test** (pa-pa-NI-kō-lō), also called a *Pap test* or *Pap smear*, involves collection and microscopic examination of epithelial cells that have been scraped off the apical layer of a tissue. A very common type of Pap test involves examining the cells from the nonkeratinized stratified squamous epithelium of the vagina and cervix (inferior portion) of the uterus. This type of Pap test is performed mainly to detect early changes in the cells. The cells are scraped from the tissue and are then smeared on a microscope slide. The slides are sent to a laboratory for analysis. It is recommended that Pap tests should be performed every three years beginning at age 21. It is further recommended that females aged 30 to 65 should have Pap testing and HPV (human pappilomavirus) testing (cotesting) every five years or a Pap test alone every three years. Females with certain high risk factors may need more frequent screening or even continue screening beyond age 65. •

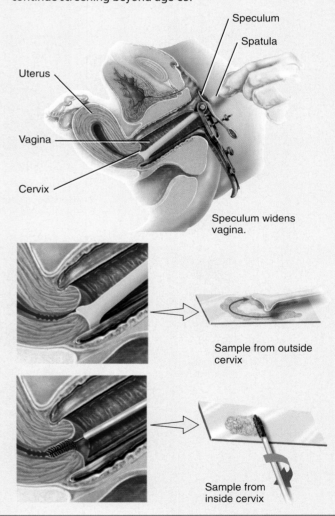

Speculum

Spatula

Uterus

Vagina

Cervix

Speculum widens vagina.

Sample from outside cervix

Sample from inside cervix

Covering and Lining Epithelium

As noted earlier, covering and lining epithelium forms the outer covering of the skin and some internal organs. It also forms the inner lining of blood vessels, ducts, and body cavities, and the interior of the respiratory, digestive, urinary, and reproductive systems. Table 3.1 describes covering and lining epithelium in more detail. The discussion of each type consists of a photomicrograph, a corresponding diagram, and an inset that identifies a major location of the tissue in the body. Descriptions, locations, and functions of the tissues accompany each illustration.

TABLE 3.1

Epithelial Tissue: Covering and Lining Epithelium

A. SIMPLE SQUAMOUS EPITHELIUM

Description: Single layer of flat cells that resembles a tiled floor when viewed from the apical surface; centrally located nucleus that is flattened and oval or spherical in shape.

Location: Occurs in two common locations in the body. The simple squamous epithelium that lines the cardiovascular and lymphatic system, that is, the heart, blood vessels, and lymphatic vessel linings, is known as **endothelium** (en′-dō-THĒ-lē-um; *endo-*=within; *-thelium*=covering). The simple squamous epithelium that forms the **epithelial** layer of serous membranes, such as the peritoneum, pleura, or pericardium, is called **mesothelium** (mez′-ō-THĒ-lē-um; *meso-*=middle). It is also found in the air sacs of lungs, glomerular (Bowman's) capsule of kidneys, and inner surface of the tympanic membrane (eardrum).

Function: Present at sites where the processes of filtration (such as blood filtration in the kidneys), diffusion (such as diffusion of oxygen into blood vessels of the lungs), and secretion (in serous membrane) occur. It is not found in body areas that are subject to mechanical stress (wear and tear).

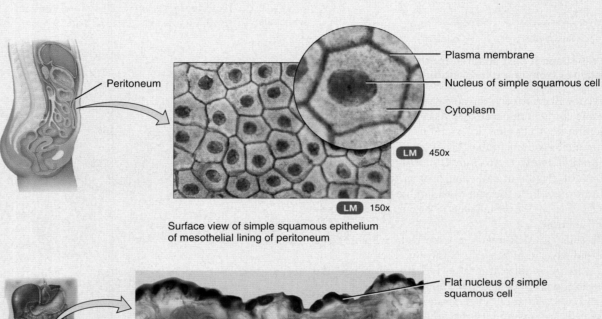

Surface view of simple squamous epithelium of mesothelial lining of peritoneum

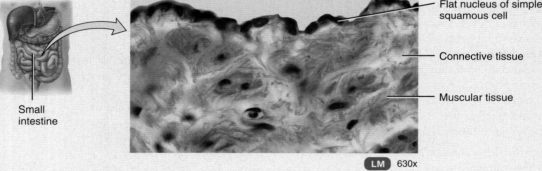

Sectional view of simple squamous epithelium (mesothelium) of peritoneum of small intestine

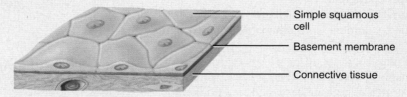

Simple squamous epithelium

B. SIMPLE CUBOIDAL EPITHELIUM

Description: Single layer of cube-shaped cells; round, centrally located nucleus. The cuboidal shape of the cells in simple cuboidal epithelium is obvious when the tissue is sectioned and viewed from the side. Note that cells that are strictly cuboidal could not form small tubes; such cuboidal cells are more pie-shaped but they are still nearly as high as they are wide (at the base).

Location: Covers surface of ovary, lines anterior surface of capsule of the lens of the eye, forms the pigmented epithelium at the posterior surface of the retina of the eye, lines kidney tubules and smaller ducts of many glands, and makes up the secretory portion of some glands such as the thyroid gland and the ducts of some glands such as the pancreas.

Function: Secretion and absorption.

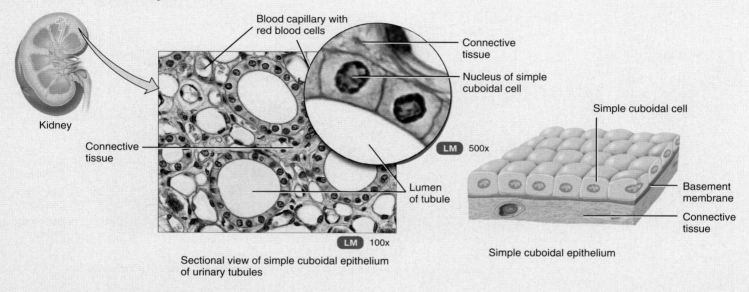

Sectional view of simple cuboidal epithelium of urinary tubules

Simple cuboidal epithelium

C. NONCILIATED SIMPLE COLUMNAR EPITHELIUM

Description: Single layer of nonciliated column-like cells with oval nuclei near base of cells; contains two types of cells—columnar epithelial cells with microvilli at their apical surface, and goblet cells. **Microvilli**, fingerlike cytoplasmic projections, increase the surface area of the plasma membrane (see Figure 2.1), thus increasing the rate of absorption by the cell. **Goblet cells** are modified columnar epithelial cells that secrete mucus, a slightly sticky fluid, at their apical surfaces. Before it is released, mucus accumulates in the upper portion of the cell, causing it to bulge out and making the whole cell resemble a goblet or wine glass.

Location: Lines the gastrointestinal tract (from the stomach to the anus), ducts of many glands, and gallbladder.

Function: Secretion and absorption, but the larger columnar cells contain more organelles and are therefore capable of a higher level of secretion and absorption than are cuboidal cell. Secreted mucus serves as a lubricant for the linings of the digestive, respiratory, and reproductive tracts, and most of the urinary tract. Mucus also helps prevent destruction of the stomach lining by acidic gastric juice secreted by the stomach.

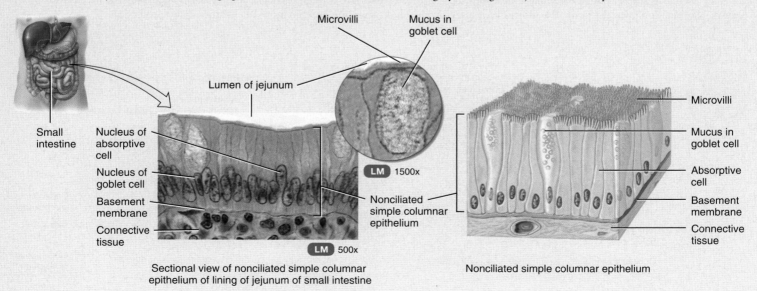

Sectional view of nonciliated simple columnar epithelium of lining of jejunum of small intestine

Nonciliated simple columnar epithelium

TABLE 3.1 CONTINUES

TABLE 3.1 CONTINUED

Epithelial Tissue: Covering and Lining Epithelium

D. CILIATED SIMPLE COLUMNAR EPITHELIUM

Description: Single layer of ciliated column-like cells with oval nuclei near base of cells. In certain parts of the smaller airways of the upper respiratory tract, goblet cells are interspersed among ciliated columnar epithelia.

Location: Lines some bronchioles (small tubes) of respiratory tract, uterine (fallopian) tubes, uterus, some paranasal sinuses, central canal of spinal cord, and ventricles of the brain.

Function: The cilia beat in unison, moving the mucus and any foreign particles toward the throat, where they can be coughed up and swallowed or spit out. Coughing and sneezing speed up the movement of cilia and mucus. Cilia also help move oocytes expelled from the ovaries through the uterine (fallopian) tubes into the uterus.

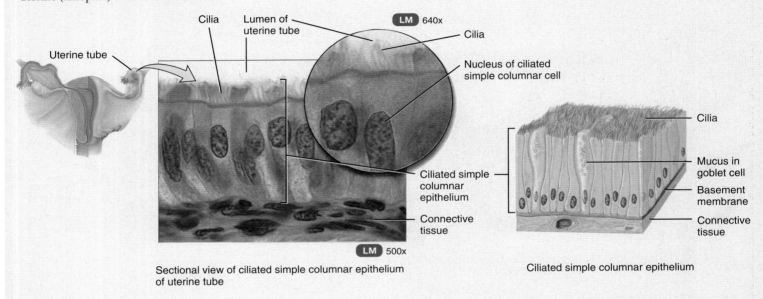

Sectional view of ciliated simple columnar epithelium of uterine tube

Ciliated simple columnar epithelium

E. PSEUDOSTRATIFIED COLUMNAR EPITHELIUM

Description: Appears to have several layers because the nuclei of the cells are at various levels. Even though all the cells are attached to the basement membrane in a single layer, some cells do not extend to the apical surface. When viewed from the side, these features give the false impression of a multilayered tissue—thus the name pseudostratified epithelium (*pseudo-* = false). *Pseudostratified ciliated columnar epithelium* contains cells that extend to the surface and either secrete mucus (goblet cells) or bear cilia. *Pseudostratified nonciliated columnar epithelium* contains cells without cilia and also lacks goblet cells.

Location: Pseudostratified ciliated columnar epithelium lines the airways of most of the upper respiratory tract; pseudostratified nonciliated columnar epithelium lines larger ducts of many glands, epididymis, and part of male urethra.

Function: Pseudostratified ciliated columnar epithelium secretes mucus that traps foreign particles, and the cilia sweep away the mucus for eventual elimination from the body. Pseudostratified nonciliated columnar epithelium functions in absorption and protection.

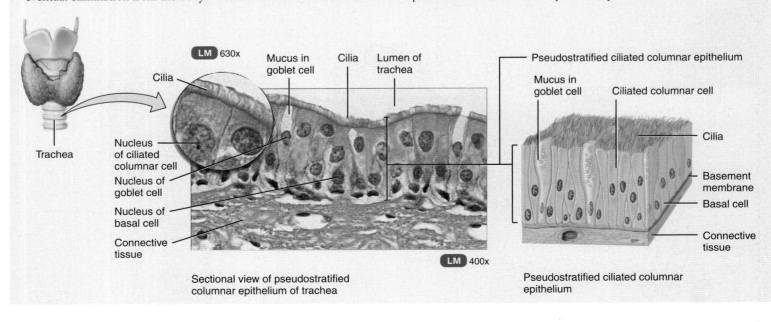

Sectional view of pseudostratified columnar epithelium of trachea

Pseudostratified ciliated columnar epithelium

F. STRATIFIED SQUAMOUS EPITHELIUM

Description: Consists of two or more layers of cells; cells in the apical layer and several layers deep to it are squamous; cells in the deeper layers vary from cuboidal to columnar. As basal cells divide, cells arising from the cell divisions push upward toward the apical layer. As they move toward the surface and away from their blood supply in the underlying connective tissue, they become dehydrated and less metabolically active as they shrink in size. As their cytoplasm is reduced, the percentage of tough proteins predominate within the cell and they become tough, hard structures that eventually die. At the apical layer, after the dead cells lose their cell junctions they are sloughed off, but they are replaced continuously as new cells emerge from the basal cells. Stratified squamous epithelium exists in both keratinized and nonkeratinized forms. *Keratinized stratified squamous epithelium* develops a tough layer of keratin in the apical layer of cells and several cell layers deep to it (see Figure 5.3). **Keratin** is a tough, fibrous intracellular protein that helps protect the skin and underlying tissues from heat, microbes, and chemicals. The relative amount of keratin increases in the cells as they move away from their nutritive blood supply and the organelles die. *Nonkeratinized stratified squamous epithelium* does not contain large amounts of keratin in the apical layer and several layers deep to it and is constantly moistened by mucus secreted from salivary and mucous glands. The organelles are not replaced in this epithelium.

Location: Keratinized variety forms superficial layer of skin; nonkeratinized variety lines wet surfaces, such as lining of the mouth, esophagus, part of epiglottis, part of pharynx, and vagina, and covers the tongue.

Function: Protection against abrasion, water loss, ultraviolet radiation, and foreign invasion. Both types of stratified squamous epithelium form the first line of defense against microbes.

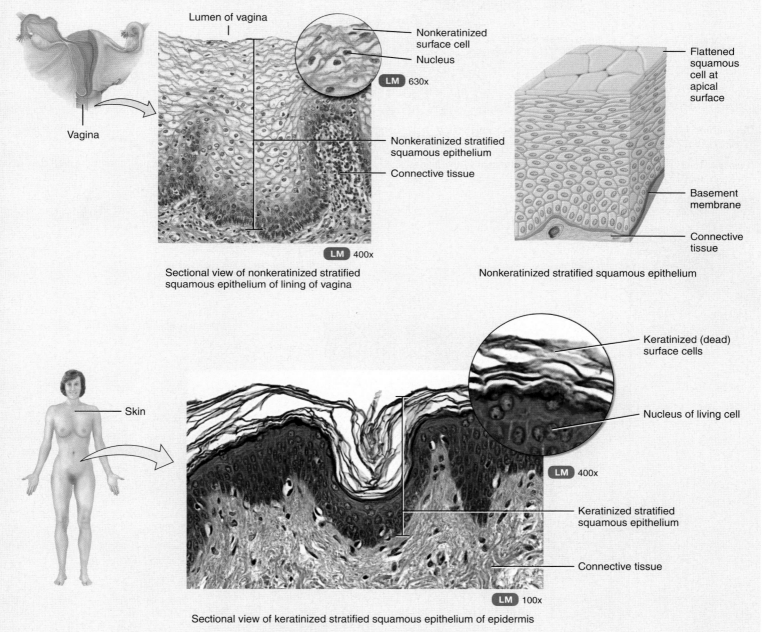

Sectional view of nonkeratinized stratified squamous epithelium of lining of vagina

Nonkeratinized stratified squamous epithelium

Sectional view of keratinized stratified squamous epithelium of epidermis

TABLE 3.1 CONTINUES

TABLE 3.1 CONTINUED

Epithelial Tissue: Covering and Lining Epithelium

G. STRATIFIED CUBOIDAL EPITHELIUM

Description: Two or more layers of cells in which the cells in the apical layer are cube-shaped. This is a fairly rare type of epithelium.

Location: Ducts of adult sweat glands and esophageal glands and part of male urethra.

Function: Protection and limited secretion and absorption.

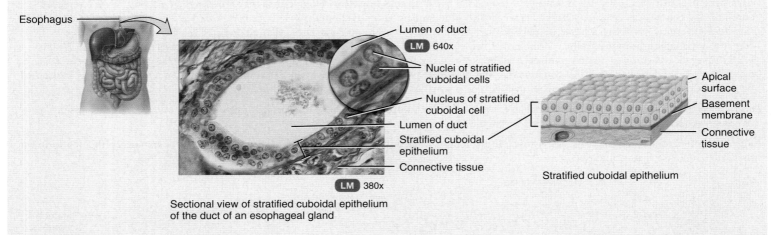

Esophagus

Lumen of duct
LM 640x
Nuclei of stratified cuboidal cells
Nucleus of stratified cuboidal cell
Lumen of duct
Stratified cuboidal epithelium
Connective tissue

LM 380x

Sectional view of stratified cuboidal epithelium of the duct of an esophageal gland

Apical surface
Basement membrane
Connective tissue

Stratified cuboidal epithelium

H. STRATIFIED COLUMNAR EPITHELIUM

Description: Usually the basal layers consist of shortened, irregularly shaped cells; only the apical layer has cells that are columnar in shape. Like stratified cuboidal epithelium, stratified columnar epithelium also is uncommon.

Location: Lines part of urethra, large excretory ducts of some glands, such as esophageal glands, small areas in anal mucous membrane, and part of the conjunctiva of the eye.

Function: Protection and secretion.

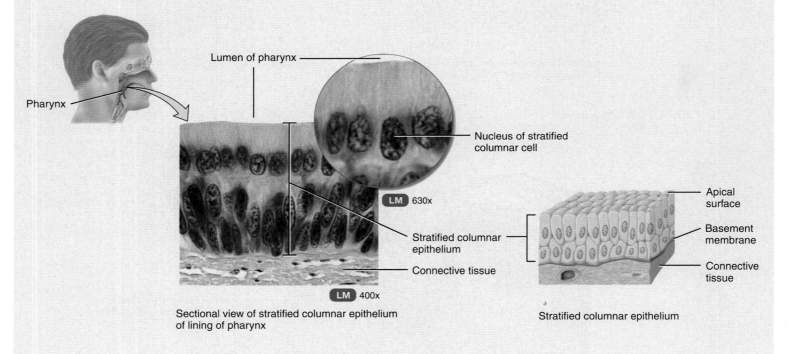

Pharynx

Lumen of pharynx
Nucleus of stratified columnar cell
LM 630x
Stratified columnar epithelium
Connective tissue

LM 400x

Sectional view of stratified columnar epithelium of lining of pharynx

Apical surface
Basement membrane
Connective tissue

Stratified columnar epithelium

TABLE 3.1 CONTINUED

Epithelial Tissue: Covering and Lining Epithelium

I. TRANSITIONAL EPITHELIUM

Description: Appearance is variable (transitional). In its relaxed or unstretched state, transitional epithelium looks like stratified cuboidal epithelium, except that the cells in the apical layer tend to be large and rounded. As the tissue is stretched, its cells become flatter, giving the appearance of stratified squamous epithelium. Because of its multiple layers and elasticity, it is ideal for lining hollow structures (such as the urinary bladder) that are subject to expansion from within.

Location: Lines urinary bladder and portions of ureters and urethra.

Function: It allows the urinary organs to stretch to hold a variable amount of fluid without rupturing, while still serving as a protective lining.

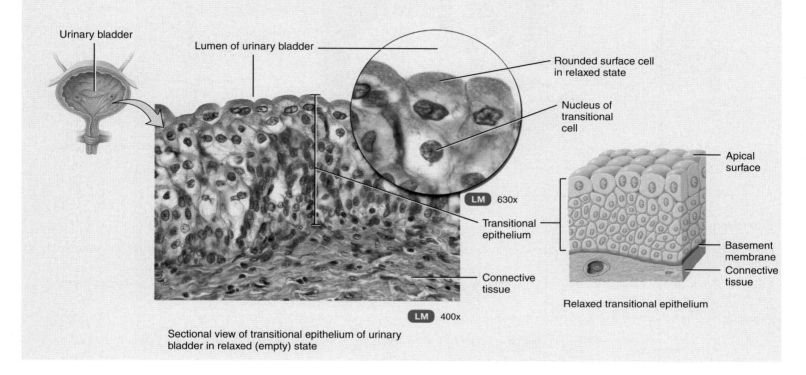

Sectional view of transitional epithelium of urinary bladder in relaxed (empty) state

Relaxed transitional epithelium

Glandular Epithelium

The function of glandular epithelium—secretion—is accomplished by glandular cells that often lie in clusters deep to the covering and lining epithelium. A **gland** may consist of a single cell or a group of cells that secrete substances via ducts (tubes) onto a surface, or into the blood in the absence of ducts. All glands of the body are classified as either endocrine or exocrine.

The secretions of **endocrine glands** (EN-dō-krin; *endo-*=*inside*, *-crine*=secretion; Table 3.2) enter the interstitial fluid and then diffuse directly into the bloodstream without flowing through a duct. These secretions, called *hormones*, regulate many metabolic and physiological activities to maintain homeostasis. The pituitary, thyroid, and adrenal glands are examples of endocrine glands. Endocrine glands will be described in detail in Chapter 22. Endocrine secretions have far-reaching effects because they are distributed throughout the body by the bloodstream.

Exocrine glands (EX-ō-krin; *exo-*=outside; *-crine*=secretion; Table 3.2) secrete their products into ducts that empty onto the surface of a covering or lining epithelium such as the skin surface or the lumen of a hollow organ. The secretions of exocrine glands have limited effects and some of them would be harmful if they entered the bloodstream. Exocrine secretions include digestive enzymes, mucus, sweat, oil, earwax, and saliva. Examples of exocrine glands include sudoriferous (sweat) glands, which

produce sweat to help lower body temperature, and salivary glands, which secrete saliva. Saliva contains mucus and digestive enzymes among other substances. As you will learn, some glands of the body, such as the pancreas, ovaries, and testes, are mixed glands that contain both endocrine and exocrine tissue.

Structural Classification of Exocrine Glands

Exocrine glands are classified as unicellular or multicellular. As the name implies, **unicellular glands** are single-celled glands. Goblet cells are important unicellular exocrine glands that secrete mucus directly onto the apical surface of a lining epithelium. Most exocrine glands are **multicellular glands**, composed of many cells that form a distinctive microscopic structure or macroscopic organ. Examples include sudoriferous, sebaceous (oil), and salivary glands.

Multicellular glands are categorized according to two criteria: (1) whether their ducts are branched or unbranched and (2) the shape of the secretory portions of the gland (Figure 3.6). If the duct of the gland does not branch, it is a **simple gland**. If the duct branches, it is a **compound gland**. Glands with tubular secretory parts are **tubular glands**; those with rounded secretory portions are **acinar glands** (AS-i-nar; *acin-*=berry), also called *alveolar glands*. **Tubuloacinar glands** have both tubular and more rounded secretory parts.

TABLE 3.2

Epithelial Tissue: Glandular Epithelium

A. ENDOCRINE GLANDS

Description: The secretions, called *hormones*, made by endocrine glands enter the interstitial fluid and then diffuse directly into the bloodstream without flowing through a duct. Endocrine glands will be described in detail in Chapter 22.

Location: Examples include pituitary gland at base of brain, pineal gland in brain, thyroid and parathyroid glands near larynx (voice box), adrenal glands superior to kidneys, pancreas near stomach, ovaries in pelvic cavity, testes in scrotum, and thymus in thoracic cavity.

Function: Hormones regulate many metabolic and physiological activities to maintain homeostasis.

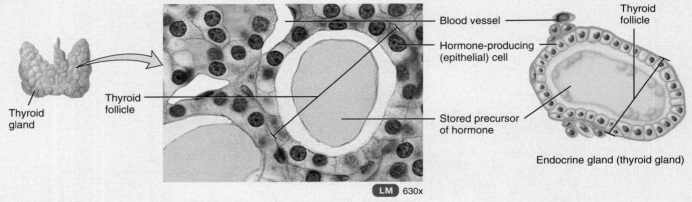

Sectional view of endocrine gland (thyroid gland)

LM 630x

Endocrine gland (thyroid gland)

B. EXOCRINE GLANDS

Description: Secretory products released into ducts that empty onto the surface of a covering and lining epithelium such as the skin surface or the lumen of a hollow organ.

Location: Sweat, oil, and earwax glands of the skin; digestive glands such as salivary glands, which secrete into mouth cavity, and pancreas, which secretes into the small intestine.

Function: Produce substances such as sweat (to help lower body temperature), oil, earwax, saliva, or digestive enzymes.

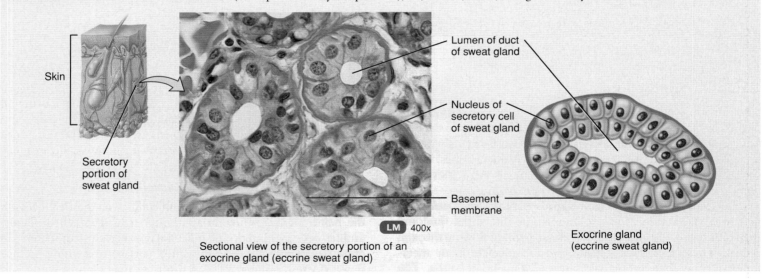

Sectional view of the secretory portion of an exocrine gland (eccrine sweat gland)

LM 400x

Exocrine gland (eccrine sweat gland)

Combinations of these features are the criteria for the following structural classification scheme for multicellular exocrine glands:

I. **Simple glands**
   A. **Simple tubular.** Tubular secretory part is straight and attaches to a single unbranched duct. Example: glands in the large intestine.
   B. **Simple branched tubular.** Tubular secretory part is branched and attaches to a single unbranched duct. Example: gastric glands.
   C. **Simple coiled tubular.** Tubular secretory part is coiled and attaches to a single unbranched duct. Example: sweat glands
   D. **Simple acinar.** Secretory portion is rounded and attaches to a single unbranched duct. Example: glands of the penile urethra.
   E. **Simple branched acinar.** Rounded secretory part is branched and attaches to a single unbranched duct. Example: sebaceous glands.

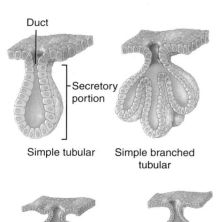

Duct

Secretory portion

Simple tubular   Simple branched tubular   Simple coiled tubular   Simple acinar   Simple branched acinar

Compound tubular   Compound acinar   Compound tubuloacinar

Figure 3.6 Multicellular exocrine glands. Pink represents the secretory portion; lavender represents the duct.

 Structural classification of multicellular exocrine glands is based on the branching pattern of the duct and the shape of the secreting portion.

 How do simple multicellular glands differ from compound ones?

II. Compound glands

A. **Compound tubular.** Secretory portion is tubular and attaches to a branched duct. Example: bulbourethral (Cowper's) glands.

B. **Compound acinar.** Secretory portion is rounded and attaches to a branched duct. Example: mammary glands.

C. **Compound tubuloacinar.** Secretory portion is both tubular and rounded and attaches to a branched duct. Example: acinar glands of the pancreas.

Functional Classification of Exocrine Glands

The functional classification of exocrine glands is based on how their secretions are released. Each of these secretory processes begins with the endoplasmic reticulum and Golgi complex working together to form intracellular secretory vesicles that contain the secretory product. Secretions of **merocrine glands**

(MER-ō-krin; *mero-*=a part), also known as **eccrine glands** (EK-rin), are synthesized on ribosomes attached to rough ER; processed, sorted, and packaged by the Golgi complex; and released from the cell in secretory vesicles via exocytosis (Figure 3.7a). Most exocrine glands of the body are merocrine glands. Examples include the salivary glands and pancreas. **Apocrine glands** (AP-ō-krin; *apo-*=from) accumulate their secretory product at the apical surface of the secreting cell. Then, that portion of the cell pinches off from the rest of the cell to release the secretion (Figure 3.7b). The remaining part of the cell repairs itself and repeats the process. Electron microscopy has recently confirmed that this is the mechanism of secretion of milk fats in the mammary glands. Recent evidence reveals that the sweat glands of the skin, named apocrine sweat glands after this mode of secretion, actually undergo merocrine secretion. The cells of **holocrine glands** (HŌ-lō-krin; *holo-*=entire) accumulate a secretory product in their cytosol. As the secretory cell matures, it ruptures

Figure 3.7 Functional classification of multicellular exocrine glands.

 The functional classification of exocrine glands is based on whether a secretion is a product of a cell or consists of an entire or a partial glandular cell.

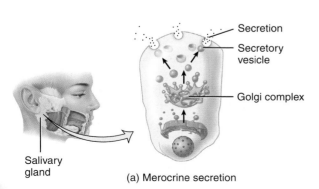

Salivary gland

Secretion

Secretory vesicle

Golgi complex

(a) Merocrine secretion

 What class of glands are sebaceous (oil) glands? Salivary glands?

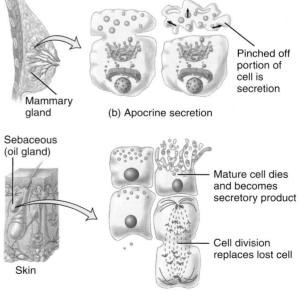

Mammary gland

(b) Apocrine secretion

Pinched off portion of cell is secretion

Sebaceous (oil gland)

Skin

Mature cell dies and becomes secretory product

Cell division replaces lost cell

(c) Holocrine secretion

and becomes the secretory product (Figure 3.7c). Because the cell ruptures in this mode of secretion, the secretion consists of a considerable amount of lipids from the plasma membrane and intracellular membranes. The sloughed-off cell is replaced by a new cell. One example of a holocrine gland is a sebaceous gland of the skin.

 CHECKPOINT

6. Describe the various layering arrangements and cell shapes of epithelium.
7. What characteristics are common to all epithelial tissue?
8. How is the structure of the following kinds of epithelium related to their functions: simple squamous, simple cuboidal, simple columnar (nonciliated and ciliated), pseudostratified columnar (ciliated and nonciliated), stratified squamous (keratinized and nonkeratinized), stratified cuboidal, stratified columnar, and transitional?
9. What is the difference between endocrine glands and exocrine glands?
10. Name and give examples of the three functional classes of exocrine glands.

3.5 CONNECTIVE TISSUE

 OBJECTIVES

• Describe the general features of connective tissue.
• Outline the structure, location, and function of the various types of connective tissue.

Connective tissue is one of the most abundant and widely distributed tissues in the body. In its many forms, connective tissue has a variety of functions:

• It binds together, supports, and strengthens other body tissues.
• It protects and insulates internal organs.
• It compartmentalizes structures such as skeletal muscles.
• Blood, a fluid connective tissue, serves as the major transport system within the body.
• Adipose (fat) tissue is the primary location of stored energy reserves.
• It is the main source of immune responses.

General Features of Connective Tissue

Connective tissue consists of two basic elements: extracellular matrix and cells. A connective tissue's **extracellular matrix** (MĀ-triks) is the material located between its widely spaced cells. The extracellular matrix consists of *protein fibers* and *ground substance*, the material between the cells and the fibers. The extracellular fibers are secreted by the connective tissue cells and account for many of the functional properties of the tissue in addition to controlling the surrounding watery environment via specific proteoglycan molecules (described shortly). The structure of the extracellular matrix determines much of the tissue's qualities. For instance, in cartilage, the extracellular matrix is firm but pliable. The extracellular matrix of bone, by contrast, is hard and inflexible.

Recall that, in contrast to epithelium, connective tissue does not usually occur on body surfaces. Also, unlike epithelium, connective tissue usually is highly vascular; that is, it has a rich blood supply. Exceptions include cartilage, which is avascular, and tendons, with a scant blood supply. Except for cartilage, connective tissue, like epithelium, is supplied with nerves.

Connective Tissue Cells

Embryonic cells called mesenchymal cells give rise to the cells of connective tissue. Each major type of connective tissue contains an immature class of cells with a name ending in *-blast*, which means "to bud or sprout." These immature cells are called *fibroblasts* in loose and dense connective tissue (described shortly), *chondroblasts* in cartilage, and *osteoblasts* in bone. Blast cells retain the capacity for cell division and secrete the extracellular matrix that is characteristic of the tissue. In cartilage and bone, once the extracellular matrix is produced, the immature cells differentiate into mature cells with names ending in *-cyte*, namely fibrocytes, chondrocytes, and osteocytes. Mature cells have reduced capacities for cell division and extracellular matrix formation and are mostly involved in monitoring and maintaining the extracellular matrix.

The types of connective tissue cells vary according to the type of tissue and include the following (Figure 3.8):

1. **Fibroblasts** (FĪ-brō-blasts; *fibro-*=fibers) are large, flat cells with branching processes. They are present in all the general connective tissue, and usually are the most numerous. Fibroblasts migrate through the connective tissue, secreting the fibers and certain components of the ground substance of the extracellular matrix.

2. **Macrophages** (MAK-rō-fā-jez; *macro-*=large; *-phages*=eaters) develop from *monocytes*, a type of white blood cell. Macrophages have an irregular shape with short branching projections and are capable of engulfing bacteria and cellular debris by phagocytosis. *Fixed macrophages* reside in a particular tissue; examples include alveolar macrophages in the lungs or splenic macrophages in the spleen. *Wandering macrophages* have the ability to move throughout the tissue and gather at sites of infection or inflammation to carry out phagocytosis.

3. **Plasma cells** are small cells that develop from a type of white blood cell called a *B lymphocyte*. Plasma cells secrete antibodies, proteins that attack or neutralize foreign substances in the body. Thus, plasma cells are an important part of the body's immune response. Although they are found in many places in the body, most plasma cells reside in connective tissue, especially in the gastrointestinal and respiratory tracts. They are also abundant in the salivary glands, lymph nodes, spleen, and red bone marrow.

4. **Mast cells** are abundant alongside the blood vessels that supply connective tissue. They produce histamine, a chemical that dilates small blood vessels as part of the inflammatory response, the body's reaction to injury or infection. In addition, researchers have recently discovered that mast cells can bind to, ingest, and kill bacteria.

5. **Adipocytes** (A-di-pō-sīts), also called fat cells or *adipose* cells, are connective tissue cells that store triglycerides (fats). They are found deep to the skin and around organs such as the heart and kidneys.

Figure 3.8 **Representative cells and fibers present in connective tissue.**

Fibroblasts are usually the most numerous connective tissue cells.

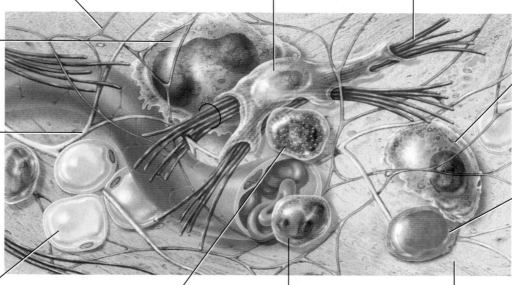

RETICULAR FIBERS
are made of collagen and glycoproteins. They provide support in blood vessel walls and form branching networks around various cells (fat, smooth muscle, nerve).

FIBROBLASTS
are large flat cells that move through connective tissue and secrete fibers and ground substance.

COLLAGEN FIBERS
are strong, flexible bundles of the protein collagen, the most abundant protein in your body.

MACROPHAGES
develop from monocytes and destroy bacteria and cell debris by phagocytosis.

MAST CELLS
are abundant along blood vessels. They produce histamine, which dilates small blood vessels during inflammation and kills bacteria.

ELASTIC FIBERS
are stretchable but strong fibers made of proteins, elastin, and fibrillin. They are found in skin, blood vessels, and lung tissue.

PLASMA CELLS
develop from B lymphocytes. They secrete antibodies that attack and neutralize foreign substances.

ADIPOCYTES
or fat cells store fats. They are found below the skin and around organs (heart, kidney).

EOSINOPHILS
are white blood cells that migrate to sites of parasitic infection and allergic responses.

NEUTROPHILS
are white blood cells that migrate to sites of infection that destroy microbes by phagocytosis.

GROUND SUBSTANCE
is the material between cells and fibers. It is made of water and organic molecules (hyaluronic acid, chondroitin sulfate, glucosamine). It supports cells and fibers, binds them together, and provides a medium for exchanging substances between blood and cells.

 What is the function of fibroblasts?

6. **White blood cells** are not found in significant numbers in normal connective tissue. However, in response to certain conditions they migrate from blood into connective tissue. For example, *neutrophils* gather at sites of infection, and *eosinophils* migrate to sites of parasitic invasions and allergic responses.

Connective Tissue Extracellular Matrix

Each type of connective tissue has unique properties, based on the specific extracellular materials between the cells. The extracellular matrix consists of two major components: (1) the ground substance, and (2) the fibers.

Ground Substance

As noted earlier, the **ground substance** is the component of a connective tissue between the cells and fibers. The ground substance may be fluid, semifluid, gelatinous, or calcified. It supports cells, binds them together, stores water, and provides a medium for exchange of substances between the blood and cells. It plays an active role in how tissues develop, migrate, proliferate, and change shape, and in how they carry out their metabolic functions.

Ground substance contains water and an assortment of large organic molecules, many of which are complex combinations of polysaccharides and proteins. The polysaccharides include hyaluronic acid, chondroitin sulfate, dermatan sulfate, and keratan sulfate. Collectively, they are referred to as **glycosaminoglycans (GAGs)** (glī-kos-a-mē′-nō-GLĪ-kans). Except for hyaluronic acid, the GAGs are associated with proteins called **proteoglycans** (prō-tē-ō-GLĪ-kans). The proteoglycans form a core protein and the GAGs project from the protein like the bristles of a brush. One of the most important properties of GAGs is that they trap water, making the ground substance more jellylike.

Hyaluronic acid (hī′-a-loo-RON-ik) is a viscous, slippery substance that binds cells together, lubricates joints, and helps maintain the shape of the eyeballs. White blood cells, sperm cells, and some bacteria produce *hyaluronidase*, an enzyme that breaks apart hyaluronic acid, thus causing the ground substance of connective tissue to become more liquid. The ability to produce hyaluronidase helps white blood cells move more easily through connective tissue to reach sites of infection and aids penetration of an oocyte by a sperm cell during fertilization. It also accounts for the rapid spread of bacteria through connective tissue. **Chondroitin sulfate** (kon-DROY-tin) provides support and adhesiveness in cartilage, bone, skin, and blood vessels. The skin, tendons, blood vessels, and heart valves contain **dermatan sulfate**; bone, cartilage, and the cornea of the eye contain **keratan sulfate**. Also present in the ground substance are **adhesion proteins**, which are responsible for linking components of the ground substance to one another and to the surfaces of cells. The main adhesion protein of connective tissue is **fibronectin**, which binds to both collagen fibers (discussed shortly) and ground substance, linking them together. Fibronectin also attaches cells to the ground substance.

Fibers

Three types of **fibers** are embedded in the extracellular matrix between the cells: collagen fibers, elastic fibers, and reticular fibers (Figure 3.8). They function to strengthen and support connective tissue.

Collagen fibers (KOL-a-jen; *colla*=glue) are very strong and resist pulling forces (tension), but they are not stiff, which allows tissue flexibility. The properties of different types of collagen fibers vary from tissue to tissue. For example, the collagen fibers found in cartilage and bone form different associations with surrounding molecules. As a result of these associations, the collagen fibers in cartilage are surrounded by more water molecules than those in bone, which gives cartilage a more cushioning effect. Collagen fibers often occur in parallel bundles (see Table 3.5, dense regular connective tissue). The bundle arrangement adds great tensile strength to the tissue. Chemically, collagen fibers consist of the protein *collagen*, which is the most abundant protein in your body, representing about 25 percent of the total. Collagen fibers are found in most types of connective tissue, especially bone, cartilage, tendons (which attach muscle to bone), and ligaments (which attach bone to bone).

Elastic fibers, which are smaller in diameter than collagen fibers, branch and join together to form a fibrous network within a tissue. An elastic fiber consists of molecules of the protein *elastin* surrounded by a glycoprotein named *fibrillin*, which adds strength and stability. Because of their unique molecular structure, elastic fibers are strong but can be stretched up to 150 percent of their relaxed length without breaking. Equally important, elastic fibers have the ability to return to their original shape after being stretched, a property called *elasticity*. Elastic fibers are plentiful in skin, blood vessel walls, and lung tissue.

Reticular fibers (*reticul-*=net), consisting of *collagen* arranged in fine bundles with a coating of glycoprotein, provide support in the walls of blood vessels and form a network around the cells in some tissues, such as areolar connective tissue, adipose tissue, nerve fibers, and smooth muscle tissue. Produced by fibroblasts, reticular fibers are much thinner than collagen fibers and form branching networks. Like collagen fibers, reticular fibers provide support and strength. Reticular fibers are plentiful in reticular connective tissue, which forms the **stroma** (STRŌ-ma=bed or covering) or supporting framework of many soft organs, such as the spleen and lymph nodes. These fibers also help form the basement membrane.

Classification of Connective Tissue

Because of the diversity of cells and extracellular matrix and the differences in their relative proportions, the classification of connective tissue is not always clear-cut and several classifications exist. We offer the following classification scheme:

I. Embryonic connective tissue
   A. Mesenchyme
   B. Mucous connective tissue
II. Mature connective tissue
   A. Loose connective tissue
      1. Areolar connective tissue
      2. Adipose tissue
      3. Reticular connective tissue
   B. Dense connective tissue
      1. Dense regular connective tissue
      2. Dense irregular connective tissue
      3. Elastic connective tissue
   C. Cartilage
      1. Hyaline cartilage
      2. Fibrocartilage
      3. Elastic cartilage
   D. Bone tissue
   E. Liquid connective tissues
      1. Blood
      2. Lymph

Embryonic Connective Tissue

Note that our classification scheme has two major subclasses of connective tissue: embryonic and mature. **Embryonic connective tissue** is of two types: *mesenchyme* and *mucous connective tissue*. Mesenchyme is present primarily in the *embryo*, the developing human from fertilization through the first two months of pregnancy. Mucous connective tissue is found in the *fetus*, the developing human from the third month of pregnancy to birth (Table 3.3).

Mature Connective Tissue

The second major subclass of connective tissue, **mature connective tissue**, is present in the newborn. Its cells arise primarily from mesenchyme. In this section we explore the numerous types of mature connective tissue. The five types of mature connective tissue are (1) *loose connective tissue*, (2) *dense connective tissue*, (3) *cartilage*, (4) *bone tissue*, and (5) *liquid connective tissues* (*blood and lymph*). We now examine each in detail.

Loose Connective Tissue

The fibers of **loose connective tissue** (Table 3.4) are *loosely* arranged between cells. The types of loose connective tissue are areolar connective tissue, adipose tissue, and reticular connective tissue.

TABLE 3.3

Embryonic Connective Tissue

A. MESENCHYME

Description: Consists of irregularly shaped mesenchymal cells embedded in a semifluid ground substance that contains delicate reticular fibers.

Location: Found almost exclusively in the embryo, under skin and along developing bones of embryo; some mesenchymal cells are found in adult connective tissue, especially along blood vessels.

Function: Forms almost all other types of connective tissue.

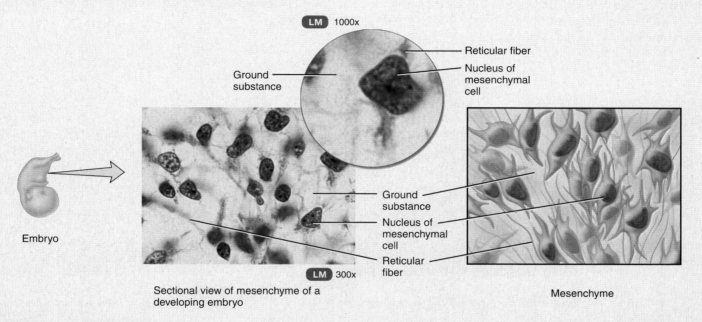

Sectional view of mesenchyme of a developing embryo

Mesenchyme

B. MUCOUS CONNECTIVE TISSUE

Description: Consists of widely scattered fibroblasts embedded in a viscous, jellylike ground substance that contains fine collagen fibers.

Location: Umbilical cord of fetus.

Function: Support.

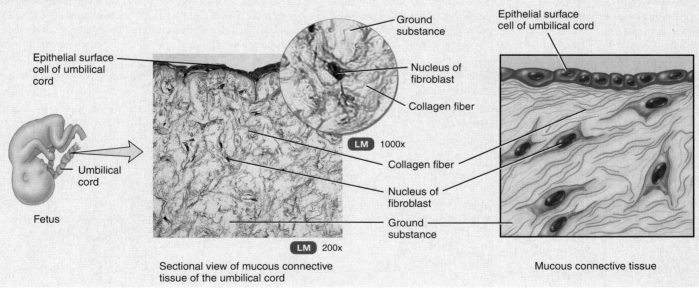

Sectional view of mucous connective tissue of the umbilical cord

Mucous connective tissue

TABLE 3.4

Mature Connective Tissue: Loose Connective Tissue

A. AREOLAR CONNECTIVE TISSUE (a-RĒ-ō-lar; *areol-*=small space)

Description: Consists of fibers (collagen, elastic, and reticular) arranged randomly and several kinds of cells (fibroblasts, macrophages, plasma cells, adipocytes, mast cells, and a few white blood cells) embedded in a semifluid ground substance (hyaluronic acid, chondroitin sulfate, dermatan sulfate, and keratan sulfate).

Location: One of the most widely distributed connective tissues in the body, it is found in the subcutaneous layer deep to skin; papillary (superficial) region of dermis of skin; lamina propria of mucous membranes; and around blood vessels, nerves, and body organs. Called the "packing material" of the body since it is found in and around nearly every structure of the body.

Function: Strength, elasticity, and support.

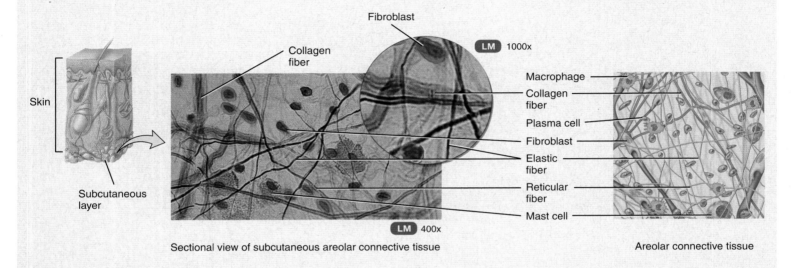

Sectional view of subcutaneous areolar connective tissue

Areolar connective tissue

B. ADIPOSE TISSUE

Description: The cells of adipose tissue, called *adipocytes* (*adipo-*=fat), are specialized for storage of triglycerides (fats) as a large centrally located droplet. Because the cell fills up with a single, large triglyceride droplet, the cytoplasm and nucleus are pushed to the periphery of the cell. As a person gains weight, the amount of adipose tissue increases and new blood vessels form. Thus, an obese person has many more blood vessels than does a lean person, a situation that can cause high blood pressure, since the heart has to work harder. Most adipose tissue in adults is *white adipose tissue*, the type just described. Another type, called *brown adipose tissue* (BAT), obtains its darker color from a very rich blood supply, along with numerous pigmented mitochondria that participate in aerobic cellular respiration. Although BAT is widespread in the fetus and infant, in adults only small amounts are present.

Location: Found wherever areolar connective tissue is located. Subcutaneous layer deep to skin, around heart and kidneys, yellow bone marrow, and padding around joints and behind eyeball in eye socket.

Function: Reduces heat loss through skin, serves as an energy reserve, and supports and protects organs. In newborns, brown adipose tissue generates considerable heat that helps maintain proper body temperature. The heat generated by the many mitochondria is carried away to other body tissues by the extensive blood supply.

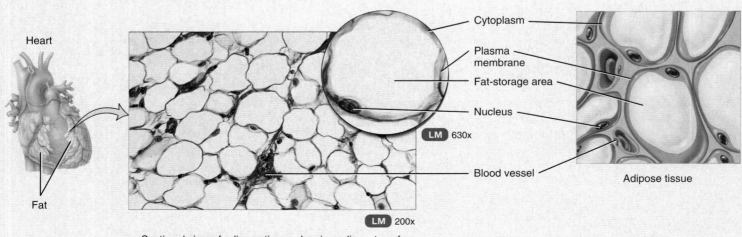

Sectional view of adipose tissue showing adipocytes of white fat and details of an adipocyte

Adipose tissue

C. RETICULAR CONNECTIVE TISSUE

Description: Consists of fine interlacing network of reticular fibers (a thin form of collagen fiber) and reticular cells.

Location: Stroma (supporting framework) of liver, spleen, lymph nodes; red bone marrow, which gives rise to blood cells; reticular lamina of the basement membrane; and around blood vessels and muscles.

Function: Forms stroma of organs; binds together smooth muscle tissue cells; filters and removes worn-out blood cells in the spleen and microbes in lymph nodes.

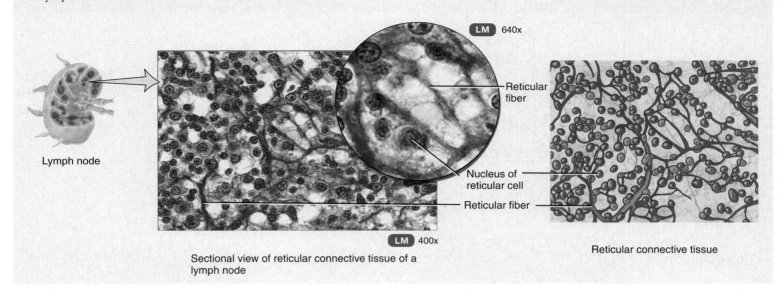

Lymph node

LM 640x

Reticular fiber

Nucleus of reticular cell

Reticular fiber

LM 400x

Sectional view of reticular connective tissue of a lymph node

Reticular connective tissue

CLINICAL CONNECTION | *Liposuction and Cryolipolysis*

A surgical procedure called **liposuction** (LIP-ō-suk'-shun; *lip-* =fat) or *suction lipectomy* (*-ectomy*=to cut out) involves suctioning out small amounts of adipose tissue from various areas of the body. In one type of liposuction, an incision is made in the skin, the fat is removed through a stainless steel tube, called a cannula, with the assistance of a powerful vacuum pressure unit that suctions out the fat. Ultrasound and laser can also be used to liquify fat for removal. The technique can be used as a body-contouring procedure in regions such as the thighs, buttocks, arms, breasts, and abdomen, and to transfer fat to another area of the body. Postsurgical complications that may develop include fat that may enter blood vessels broken during the procedure and obstruct blood flow, infection, loss of feeling in the area, fluid depletion, injury to internal structures, and severe postoperative pain.

Cryolipolysis (*cryo*=cold) or **CoolSculpting** refers to the destruction of fat cells by the external application of controlled cooling. Since fat crystallizes faster than cells surrounding adipose tissue, the cold temperature kills the fat cells while sparing damage to nerve cells, blood vessels, and other structures. Within a few days of the procedure, apoptosis (genetically programmed death) begins, and within several months, the fat cells are removed. •

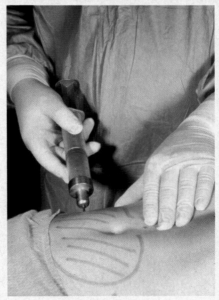

Abdominal liposuction

Dense Connective Tissue

Dense connective tissue contains more fibers, which are thicker and more *densely* packed, but has considerably fewer cells than loose connective tissue. There are three types: **dense regular** connective tissue, dense irregular connective tissue, and elastic connective tissue (Table 3.5).

TABLE 3.5

Mature Connective Tissue: Dense Connective Tissue

A. DENSE REGULAR CONNECTIVE TISSUE

Description: Extracellular matrix looks shiny white. Consists mainly of collagen fibers *regularly* arranged in bundles with fibroblasts present in rows between the bundles. The collagen fibers are not living (they are protein structures secreted by the fibroblasts); since tendons and ligaments have very few blood vessels, they are slow to heal following damage.

Location: Forms tendons (attach muscle to bone), most ligaments (attach bone to bone), and aponeuroses (sheetlike tendons that attach muscle to muscle or muscle to bone).

Function: Provides strong attachment between various structures. The tissue structure withstands pulling (tension) along the long axis of the fibers.

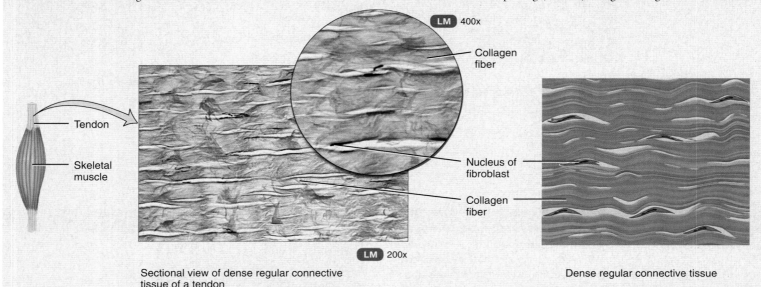

Sectional view of dense regular connective tissue of a tendon

Dense regular connective tissue

B. DENSE IRREGULAR CONNECTIVE TISSUE

Description: Contains tightly packed, woven meshes of collagen fibers that are usually *irregularly* arranged.

Location: This tissue often occurs in sheets, such as fasciae (tissue beneath skin and around muscles and other organs), reticular (deeper) region of dermis of skin, fibrous pericardium of heart, periosteum of bone, perichondrium of cartilage, joint capsules, membrane capsules around various organs (kidneys, liver, testes, lymph nodes). Also found in heart valves.

Function: Provides tensile (pulling) strength in many directions.

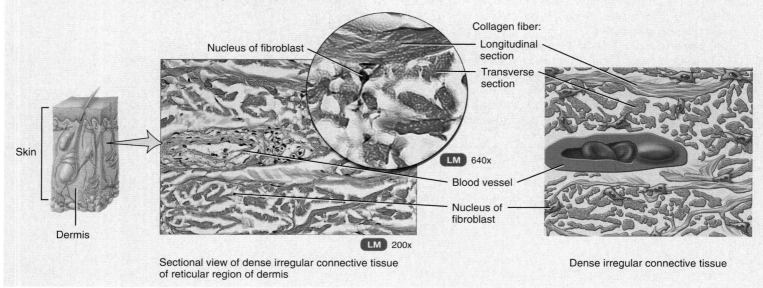

Sectional view of dense irregular connective tissue of reticular region of dermis

Dense irregular connective tissue

C. ELASTIC CONNECTIVE TISSUE

Description: Consists predominantly of elastic fibers with fibroblasts in spaces between the fibers. The unstained tissue has a yellowish color.

Location: Lung tissue, walls of elastic arteries, trachea, bronchial tubes, true vocal cords, suspensory ligaments of penis, and some ligaments between vertebrae.

Function: Allows stretching of various organs. Elastic connective tissue is quite strong and can recoil to its original shape after being stretched. Elasticity is important to the normal functioning of lung tissue, which recoils as you exhale, and elastic arteries, which recoil between heartbeats to help maintain blood flow.

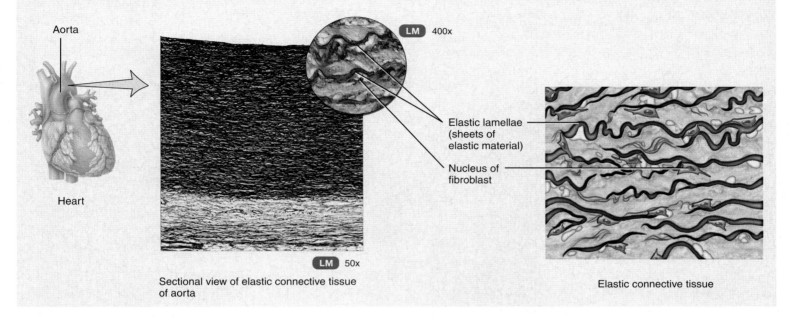

Sectional view of elastic connective tissue of aorta

Elastic connective tissue

Cartilage

Cartilage (KAR-ti-lij) consists of a dense network of collagen fibers and elastic fibers firmly embedded in chondroitin sulfate, a gel-like component of the ground substance. Cartilage can endure considerably more stress than loose and dense connective tissue. The tensile strength of cartilage is due to its collagen fibers. *Tensile strength* is the maximum strength that it can withstand while being stretched or pulled. The *resilience* of cartilage (ability to assume its original shape after deformation) and compression strength are due to chondroitin sulfate. *Compression strength* is the maximum strength that it can withstand while being crushed or squashed.

Like other connective tissue, cartilage has few cells and large quantities of extracellular matrix. It differs from other connective tissue, however, in not having nerves or blood vessels in its extracellular matrix. Interestingly, cartilage does not have a blood supply because it secretes an *antiangiogenesis factor* (an'-tē-an'-jē-ō-JEN-e-sis; *anti-*=against; *angio-*=vessel; *genesis*=production), a substance that prevents blood vessel growth. Because of this property, antiangiogenesis factor is being studied as a possible cancer treatment: If cancer cells can be stopped from promoting new blood vessel growth, their rapid rate of cell division and expansion could be slowed or even halted.

The cells of mature cartilage, called **chondrocytes** (KON-drō-sīts; *chondro-*=cartilage), occur singly or in groups within spaces called *lacunae* (la-KOO-nē=little lakes; singular is *lacuna*, pronounced la-KOO-na) in the extracellular matrix. A covering of dense irregular connective tissue called the **perichondrium** (per'-i-KON-drē-um; *peri-*=around) surrounds the surface of most cartilage, contains blood vessels and nerves, and is the source of new cartilage cells. Since cartilage has no blood supply, it heals poorly following an injury.

The cells and collagen-embedded extracellular matrix of cartilage form a strong, firm material that resists tension (stretching), compression (squeezing), and shear (pushing in opposite directions). The chondroitin sulfate in the extracellular matrix is largely responsible for cartilage resilience and compression strength. Because of these properties, cartilage plays an important role as a support tissue in the body. It is also a precursor to bone, and forms almost the entire embryonic skeleton. Though bone gradually replaces cartilage during further development, cartilage persists after birth as growth plates (epiphyseal plates) that allow bones to increase in length during the growing years (see Section 6.7). Cartilage also persists throughout life as the lubricated articular surfaces of most joints (see Chapter 9).

There are three types of cartilage: **hyaline cartilage**, **fibrocartilage**, and **elastic cartilage** (Table 3.6).

TABLE 3.6

Mature Connective Tissue: Cartilage

A. HYALINE CARTILAGE

Description: Hyaline (*hyalinos*=glassy) cartilage contains a resilient gel as its ground substance and appears in the body as a bluish-white, shiny substance (can stain pink or purple when prepared for microscopic examination). The fine collagen fibers are not visible with ordinary staining techniques, and prominent chondrocytes are found in lacunae. Most hyaline cartilage is surrounded by a perichondrium. The exceptions are the articular cartilage in joints and the cartilage of epiphyseal plates, the regions where bones lengthen as a person grows.

Location: Most abundant cartilage in the body, it is found at the ends of long bones, anterior ends of ribs, nose, parts of larynx, trachea, bronchi, bronchial tubes, and embryonic and fetal skeleton.

Function: Provides smooth surfaces for movement at joints, as well as flexibility and support. It is the weakest of the three types of cartilage and can be fractured.

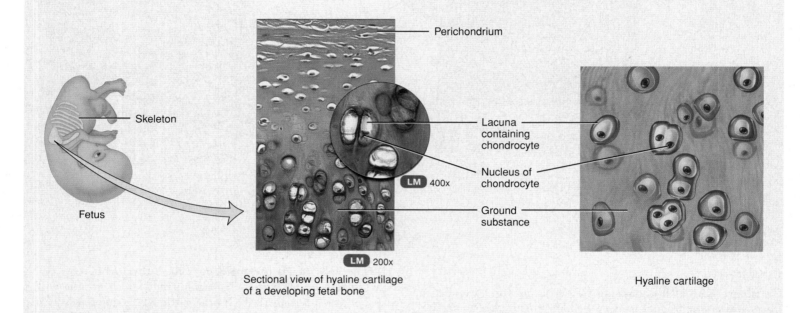

Sectional view of hyaline cartilage of a developing fetal bone

Hyaline cartilage

B. FIBROCARTILAGE

Description: Chondrocytes are scattered among clearly visible thick bundles of collagen fibers within the extracellular matrix of fibrocartilage. Fibrocartilage lacks a perichondrium.

Location: Pubic symphysis (point where hip bones join anteriorly), intervertebral discs (discs between vertebrae), menisci (cartilage pads) of knee, and portions of tendons that insert into cartilage.

Function: Support and joining structures together. With a combination of strength and rigidity, this tissue is the strongest of the three types of cartilage.

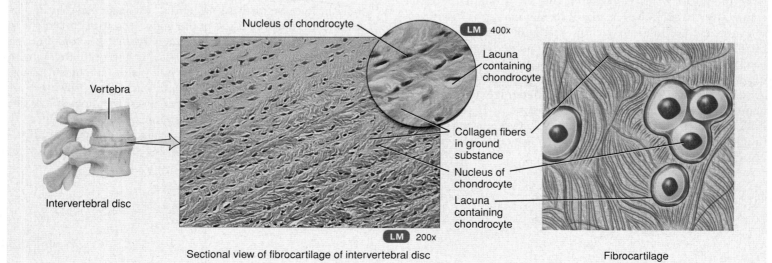

Sectional view of fibrocartilage of intervertebral disc

Fibrocartilage

C. ELASTIC CARTILAGE

Description: Consists of chondrocytes located in a threadlike network of elastic fibers within the extracellular matrix. A perichondrium is present.

Location: Lid on top of larynx (epiglottis), part of external ear (auricle), and auditory (eustachian) tubes.

Function: Provides strength and elasticity and maintains shape of certain structures.

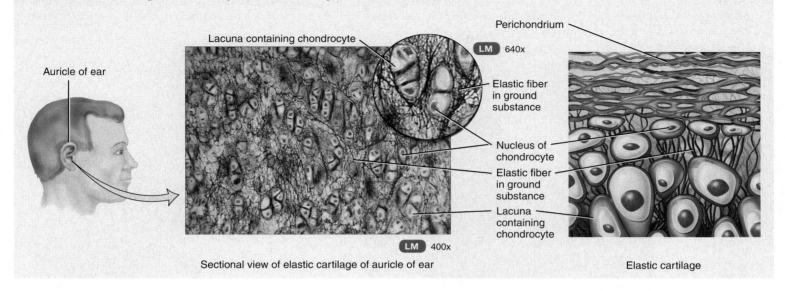

Sectional view of elastic cartilage of auricle of ear

Elastic cartilage

Bone Tissue

Cartilage, joints, and bones make up the skeletal system. The skeletal system supports soft tissues, protects delicate structures, and works with skeletal muscles to generate movement. Bones (Table 3.7) store calcium and phosphorus; house red bone marrow, which produces blood cells; and contain yellow bone marrow, a storage site for triglycerides. Bones are organs composed of several different connective tissues, including **bone tissue** or **osseous tissue** (OS-ē-us), the periosteum, red and yellow bone marrow, and the endosteum (a membrane that lines a space within bone that stores yellow bone marrow). Bone tissue is classified as either compact or spongy, depending on how its extracellular matrix and cells are organized.

The basic unit of **compact bone** is an **osteon** or **haversian system** (Table 3.7). Each osteon has four parts:

1. The **lamellae** (la-MEL-lē=little plates; singular is *lamella*) are concentric rings of extracellular matrix that consist of mineral salts (mostly calcium and phosphates), which give bone its hardness and compressive strength, and collagen fibers, which give bone its tensile strength. The lamellae are responsible for the compact nature of this type of bone tissue.
2. **Lacunae** (la-KOO-nē; singular is *lacuna*, prounced la-KOO-na) are small spaces between lamellae that contain mature bone cells called **osteocytes**.
3. Projecting from the lacunae are **canaliculi** (kan-a-LIK-ū-lī=little canals), networks of minute canals containing the processes of osteocytes. Canaliculi provide routes for nutrients to reach osteocytes and for wastes to leave them.
4. A **central (haversian) canal** contains blood vessels and nerves.

Spongy bone lacks osteons. Rather, it consists of slender columns of bone called **trabeculae** (tra-BEK-ū-lī=little beams), which contain lamellae, osteocytes, lacunae, and canaliculi. Spaces between trabeculae are filled with red bone marrow. Chapter 6 presents bone tissue histology in more detail.

CLINICAL CONNECTION | *Tissue Engineering*

The technology of **tissue engineering**, which combines synthetic material with cells, has allowed scientists to grow new tissues in the laboratory to replace damaged tissues in the body. Tissue engineers have already developed laboratory-grown versions of skin and cartilage using scaffolding beds of biodegradable synthetic materials or collagen as substrates that permit body cells to be cultured. As the cells divide and assemble, the scaffolding degrades; the new, permanent tissue is then implanted in the patient. Other structures currently under development include bones, tendons, heart valves, bone marrow, and intestines. Work is also under way to develop insulin-producing cells for diabetics, dopamine-producing cells for Parkinson's disease patients, and even entire livers and kidneys. •

Liquid Connective Tissue

BLOOD A **liquid connective tissue** has a liquid as its extracellular matrix. **Blood**, one of the liquid connective tissues, has a liquid extracellular matrix and formed elements. The extracellular matrix is called blood plasma. The **blood plasma** is a pale yellow fluid that consists mostly of water with a wide variety of dissolved substances—nutrients, wastes, enzymes, plasma proteins, hormones, respiratory gases, and ions. Suspended in the blood plasma are **formed elements**—red blood cells (erythrocytes), white blood cells (leukocytes), and platelets (thrombocytes) (Table 3.8). **Red blood cells** transport oxygen to body cells and remove some carbon dioxide from them. **White blood cells** are involved in phagocytosis, immunity, and allergic reactions. **Platelets** (PLĀT-lets) participate in blood clotting. You will learn more about blood in Chapter 12.

TABLE 3.7

Mature Connective Tissue: Bone Tissue

COMPACT BONE

Description: The basic unit of compact bone tissue is an osteon (haversian system) that contains lamellae, lacunae, osteocytes, canaliculi, and central (haversian) canals (see below). By contrast, spongy bone tissue (see Figure 6.4a, b) consists of thin columns called trabeculae; spaces between trabeculae are filled with red bone marrow.

Location: Both compact and spongy bone tissue make up the various parts of bones of the body.

Function: Support, protection, storage; houses blood-forming tissue; serves as levers that act with muscle tissue to enable movement.

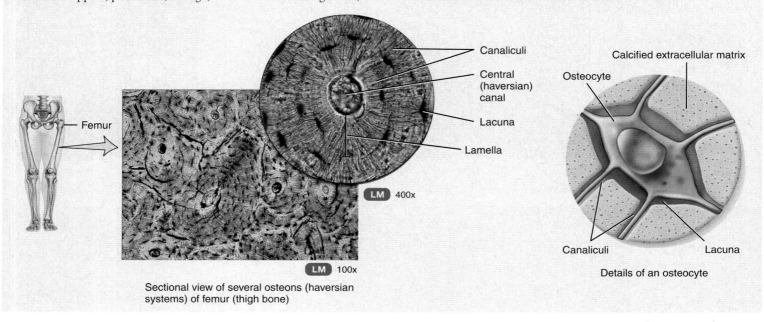

Femur

Canaliculi

Central (haversian) canal

Lacuna

Lamella

LM 400x

LM 100x

Sectional view of several osteons (haversian systems) of femur (thigh bone)

Calcified extracellular matrix

Osteocyte

Canaliculi

Lacuna

Details of an osteocyte

TABLE 3.8

Mature Connective Tissue: Liquid Connective Tissue

BLOOD

Description: Consists of blood plasma and formed elements: red blood cells (erythrocytes), white blood cells (leukocytes), and platelets (thrombocytes).

Location: Within blood vessels (arteries, arterioles, capillaries, venules, and veins) and within the chambers of the heart.

Function: Red blood cells transport oxygen and some carbon dioxide; white blood cells carry on phagocytosis and are involved in allergic reactions and immune system responses; platelets are essential for the clotting of blood.

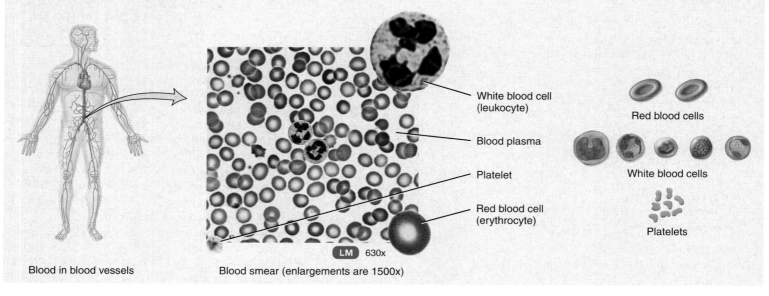

White blood cell (leukocyte)

Blood plasma

Platelet

Red blood cell (erythrocyte)

Red blood cells

White blood cells

Platelets

LM 630x

Blood in blood vessels

Blood smear (enlargements are 1500x)

LYMPH **Lymph** is the extracellular fluid that flows in lymphatic vessels. This connective tissue consists of several types of cells in a clear liquid extracellular matrix that is similar to blood plasma but with much less protein. The composition of lymph varies from one part of the body to another. For example, lymph leaving lymph nodes includes many lymphocytes, a type of white blood cell, in contrast to lymph from the small intestine, which has a high content of newly absorbed dietary lipids. The details of lymph are considered in Chapter 15.

 CHECKPOINT

11. How does connective tissue differ from epithelial tissue?
12. What are the features of the cells, ground substance, and fibers that make up connective tissue?
13. Explain the classification of connective tissue and list the various types.
14. Describe how the structures of the following connective tissue relate to their functions: areolar connective tissue, adipose tissue, reticular connective tissue, dense regular connective tissue, dense irregular connective tissue, elastic connective tissue, hyaline cartilage, fibrocartilage, elastic cartilage, bone tissue, blood tissue, and lymph.

3.6 MEMBRANES

OBJECTIVES

• Define a membrane.
• Describe the classification of membranes.

Membranes are flat sheets of pliable tissue that cover or line a part of the body. The majority of membranes consist of an epithelial layer and an underlying connective tissue layer and are called **epithelial membranes**. The principal epithelial membranes of the body are mucous membranes, serous membranes, and the cutaneous membrane, or skin. Another type of membrane, a **synovial membrane**, lines joints and contains connective tissue but no epithelium.

Epithelial Membranes

Mucous Membranes

A **mucous membrane** or **mucosa** (mū-KŌ-sa) lines a body cavity that opens directly to the exterior. Mucous membranes line the entire digestive, respiratory, and reproductive tracts, and much of the urinary tract. They consist of a lining layer of epithelium and an underlying layer of connective tissue (Figure 3.9a).

The epithelial layer of a mucous membrane is an important feature of the body's defense mechanisms because it is a barrier that microbes and other pathogens have difficulty penetrating. Usually, tight junctions connect the cells, so materials cannot leak in between them. Goblet cells and other cells of the epithelial layer of a mucous membrane secrete mucus, and this slippery fluid prevents the cavities from drying out. It also traps particles in the respiratory passageways and lubricates food as it moves through the gastrointestinal tract. In addition, the epithelial layer secretes some of the enzymes needed for digestion and is the site of food and fluid absorption in the gastrointestinal tract. The epithelia of mucous membranes vary greatly in different parts of the body. For example, the mucous membrane of the small intestine is nonciliated simple columnar epithelium (see Table 3.1), and the large airways to the lungs consist of pseudostratified ciliated columnar epithelium (see Table 3.1).

The connective tissue layer of a mucous membrane, called the **lamina propria** (LAM-ī-na PRŌ-prē-a; *propria*=one's own), is areolar connective tissue. It is so named because it belongs to (is owned by) the mucous membrane. The lamina propria supports the epithelium, binds it to the underlying structures, allows some flexibility of the membrane, and affords some protection for underlying structures. It also holds blood vessels in place and is the vascular source for the overlying epithelium. Oxygen and nutrients diffuse from the lamina propria to the covering epithelium; carbon dioxide and wastes diffuse in the opposite direction.

Serous Membranes

A **serous membrane** (SĒR-us; *serous*=watery) or **serosa** lines a body cavity that does not open directly to the exterior (thoracic or abdominal cavities), and it covers the organs that are within the cavity. Serous membranes consist of areolar connective tissue covered by mesothelium (simple squamous epithelium) (Figure 3.9b). You will recall from Chapter 1 that serous membranes have two continuous layers: The layer attached to and lining the cavity wall is called the **parietal layer** (pa-RĪ-e-tal; *pariet-*=wall); the layer that covers and adheres to the organs within the cavity is the **visceral layer** (*viscer-*=body organ) (see Figure 1.7a). The mesothelium of a serous membrane secretes **serous fluid**, a watery lubricant that allows organs to glide easily over one another or to slide against the walls of cavities.

Recall from Chapter 1 that the serous membrane lining the thoracic cavity and covering the lungs is the **pleura**. The serous membrane lining the heart cavity and covering the heart is the **pericardium**. The serous membrane lining the abdominal cavity and covering the abdominal organs is the **peritoneum**.

Cutaneous Membrane

The **cutaneous membrane** (kū-TĀ-nē-us) or **skin** covers the entire surface of the body and consists of a superficial portion called the *epidermis* and a deeper portion called the *dermis* (Figure 3.9c). The epidermis consists of keratinized stratified squamous epithelium, which protects underlying tissues. The dermis consists of dense irregular connective tissue and areolar connective tissue. Details of the cutaneous membrane are presented in Chapter 5.

Synovial Membranes

Synovial membranes (sin-Ō-vē-al; *syn-*=together, referring here to a place where bones come together; *ova*=egg, because of their resemblance to the slimy egg white of an uncooked egg) line the cavities (joint cavities) of freely movable joints. Like serous membranes, synovial membranes line structures that do not open to the exterior. Unlike mucous, serous, and cutaneous membranes, they lack an epithelium and are therefore not epithelial membranes. Synovial membranes are composed of a discontinuous layer of cells called **synoviocytes** (si-NŌ-vē-ō-sīts), which are closer to the synovial cavity (space between the bones), along with a layer of connective tissue (areolar and adipose) deep to the synoviocytes (Figure 3.9d). Synoviocytes secrete some of the components of synovial fluid. **Synovial fluid** lubricates and nourishes the cartilage covering the bones at movable joints and contains macrophages that remove microbes and debris from the joint cavity.

 CHECKPOINT

15. Define the following types of membranes: mucous, serous, cutaneous, and synovial. How do they differ from one another?
16. Where is each type of membrane located in the body? What are their functions?

Figure 3.9 Membranes.

A membrane is a flat sheet of pliable tissues that covers or lines a part of the body.

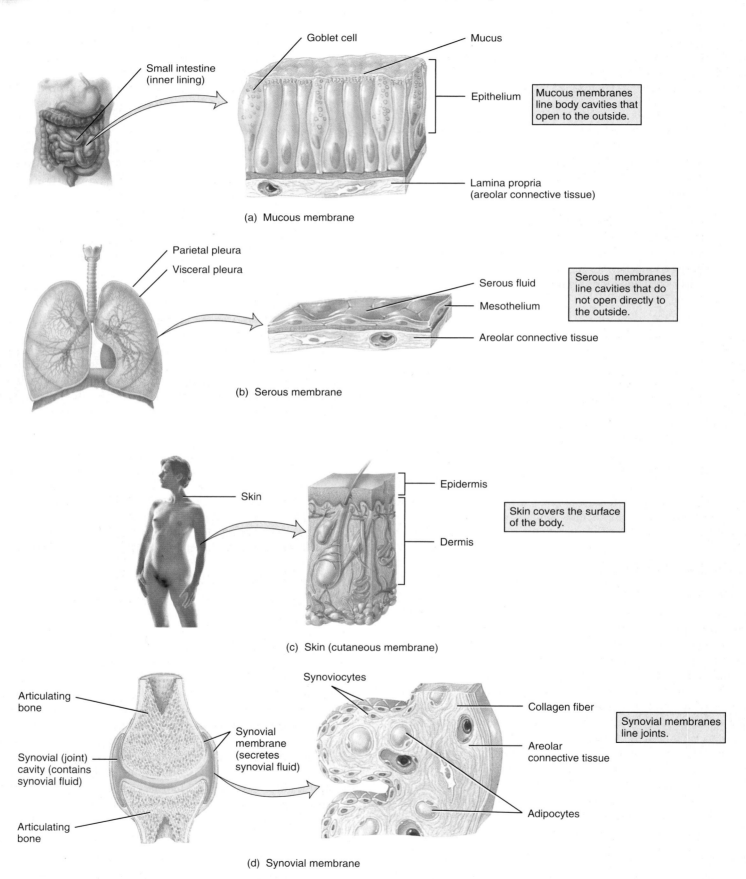

Goblet cell

Mucus

Small intestine (inner lining)

Epithelium

Mucous membranes line body cavities that open to the outside.

Lamina propria (areolar connective tissue)

(a) Mucous membrane

Parietal pleura

Visceral pleura

Serous fluid

Mesothelium

Serous membranes line cavities that do not open directly to the outside.

Areolar connective tissue

(b) Serous membrane

Skin

Epidermis

Skin covers the surface of the body.

Dermis

(c) Skin (cutaneous membrane)

Synoviocytes

Articulating bone

Collagen fiber

Synovial membrane (secretes synovial fluid)

Synovial membranes line joints.

Synovial (joint) cavity (contains synovial fluid)

Areolar connective tissue

Articulating bone

Adipocytes

(d) Synovial membrane

What is an epithelial membrane?

3.7 MUSCULAR TISSUE

 OBJECTIVES

• **Describe** the general features of muscular tissue.
• **Compare** the structure, location, and mode of control of skeletal, cardiac, and smooth muscle tissue.

Muscular tissue consists of elongated cells called *muscle fibers* or *myocytes* that can use ATP to generate force. As a result, muscular tissue produces body movements, maintains posture, and generates heat. It also provides protection. Based on its location and certain structural and functional features, muscular tissue is classified into three types: *skeletal*, *cardiac*, and *smooth* (Table 3.9). Chapter 10 provides a more detailed discussion of muscular tissue.

✓ CHECKPOINT

17. Which types of muscular tissue are striated? Which are smooth?
18. What are gap junctions, and which types of muscular tissue have them?

TABLE 3.9

Muscular Tissue

A. SKELETAL MUSCLE TISSUE

Description: Consists of long, cylindrical, striated fibers (alternating light and dark bands within the fibers called *striations* that are visible under a light microscope). Skeletal muscle fibers can vary greatly in length, ranging from a few centimeters in short muscles to up to 30–40 cm (about 12–16 in.) in your longest muscles. Roughly cylindrical in shape, a muscle fiber is a multinucleated cell with the nuclei located at the cell's periphery. Skeletal muscle is considered *voluntary* because it can be made to contract or relax by conscious control.

Location: Usually attached to bones by tendons.

Function: Motion, posture, heat production, and protection.

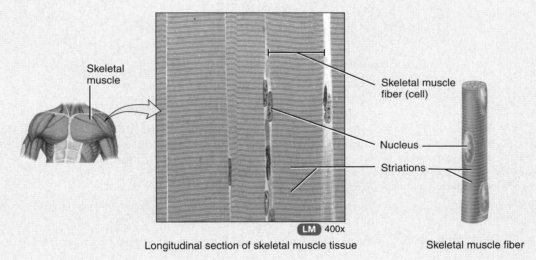

Skeletal muscle

Skeletal muscle fiber (cell)

Nucleus

Striations

LM 400x

Longitudinal section of skeletal muscle tissue

Skeletal muscle fiber

TABLE 3.9 CONTINUES

TABLE 3.9 CONTINUED

Muscular Tissue

B. CARDIAC MUSCLE TISSUE

Description: Cardiac muscle fibers are branched and usually have only one centrally located nucleus; an occasional cell has two nuclei. They attach end to end by transverse thickenings of the plasma membrane called *intercalated discs* (in-TER-ka-lāt-ed; *intercalate*=to insert between), which contain both desmosomes and gap junctions. Intercalated discs are unique to cardiac muscle tissue. The desmosomes strengthen the tissue and hold the fibers together during their vigorous contractions. The gap junctions provide a route for quick conduction of electrical signals called muscle action potentials throughout the heart. *Involuntary* control, meaning it is not consciously controlled.

Location: Heart wall.

Function: Pumps blood to all parts of the body.

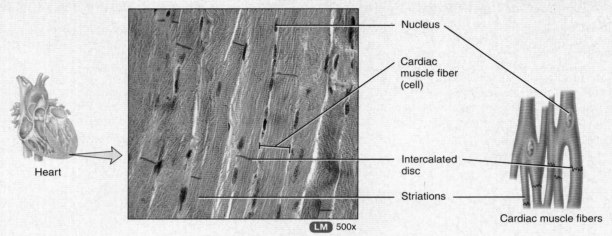

Heart

Nucleus

Cardiac muscle fiber (cell)

Intercalated disc

Striations

LM 500x

Longitudinal section of cardiac muscle tissue

Cardiac muscle fibers

C. SMOOTH MUSCLE TISSUE

Description: Smooth muscle fibers are usually *involuntary*, and they are nonstriated (lack striations), hence the term *smooth*. A smooth muscle fiber is a small spindle-shaped cell that is thickest in the middle and tapers at each end. It contains a single, centrally located nucleus. Gap junctions connect many individual fibers in some smooth muscle tissues, for example, in the wall of the intestines. Such muscle tissues can produce powerful contractions as many muscle fibers contract in unison. In other locations, such as the iris of the eye, smooth muscle fibers contract individually, like skeletal muscle fibers, because gap junctions are absent.

Location: Iris of the eyes, walls of hollow internal structures such as blood vessels, airways to the lungs, stomach, intestines, gallbladder, urinary bladder, and uterus.

Function: Motion (constriction of blood vessels and airways, propulsion of foods through gastrointestinal tract, contraction of urinary bladder and gallbladder).

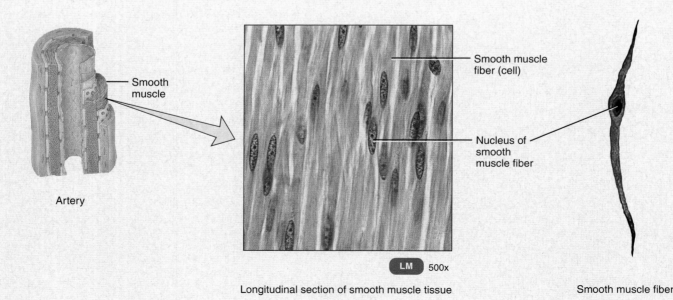

Smooth muscle

Artery

Smooth muscle fiber (cell)

Nucleus of smooth muscle fiber

LM 500x

Longitudinal section of smooth muscle tissue

Smooth muscle fiber

3.8 NERVOUS TISSUE

OBJECTIVE

• Describe the structural features and functions of nervous tissue.

Despite the tremendous complexity of the nervous system, the **nervous tissue** that makes up this system is a highly specialized cellular tissue consisting of only two principal types of cells: neurons and neuroglia (glial cells). Little extracellular material is present. **Neurons** (NOO-rons; *neuro-*=nerve) comprise a broad category of unusually shaped cell types with one common structural feature: Long cytoplasmic processes (extensions) known as dendrites and axons project from the main **cell body** where the nucleus and other typical organelles are located (Table 3.10). These processes can vary in number (from two to thousands) and length (from fractions of a millimeter to a meter). Neurons form complex networks within the central nervous system (the brain and spinal cord), somewhat comparable to the circuit boards of sophisticated computers. About 98 percent of nervous tissue is located in the central nervous system. Outside the central nervous system, long neuronal processes are bundled together to form peripheral nerves that travel between the central nervous system and the various tissues of the body.

Functionally, neurons are highly excitable cells with the ability to initiate and propagate electrical signals called **nerve action potentials (nerve impulses)** when excited by physical or chemical agents in their environment. **Dendrites** (*dendr-*=tree) are tapering, highly branched, and usually short cell processes (extensions). The dendrites of many neurons receive information from other sources, most commonly from other neurons, and electrically transmit it toward the cell body. The **axon** (*axo-*=axis) is a single, thin, cylindrical process that can be very long, and in most neurons carries action potentials away from the cell body to other cells. When an action potential reaches the end of the axon, the neuron releases a **neurotransmitter**, a locally acting chemical messenger that influences another cell (or cells), which can be another neuron, a muscle cell, or a gland cell. Because of this unique combination of structure and function, nervous tissue plays important roles in communicating information from one part of the body to another and in regulating muscular and glandular tissues.

Neuroglia (noo-RŌG-lē-a; *-glia*=glue) do not generate or conduct nerve impulses. Instead, they support neurons both physically and metabolically. Without neuroglia, the neurons could not survive and function. Neuroglia, which outnumber the neurons approximately five to one, come in a variety of shapes, each specialized to accomplish specific functions. The detailed structure and function of neurons and neuroglia are considered in Chapter 16.

CHECKPOINT

19. Name the major parts of a neuron, and describe the functions of each.

 # 3.9 AGING AND TISSUES

OBJECTIVE

• Describe the effects of aging on tissues.

In later chapters, the effects of aging on specific body systems will be considered. With respect to tissues, epithelial tissue gets progressively thinner and connective tissue becomes more fragile with aging. This is evidenced by an increased incidence of skin and mucous membrane disorders, wrinkles, more susceptibility to bruises, increased loss of bone density, higher rates of bone fractures, and increased episodes of joint pain and disorders. There is also an effect of aging on muscle tissue as evidenced by loss of skeletal muscle mass and strength, decline in the efficiency of pumping action of the heart, and decreased activity of smooth muscle–containing organs such as the organs of the gastrointestinal tract.

Generally, tissues heal faster and leave less obvious scars in the young than in the aged. In fact, surgery performed on fetuses leaves no scars at all. The younger body is generally in a better nutritional state, its tissues have a better blood supply, and its

---

TABLE 3.10

Nervous Tissue

Description: Consists of neurons (nerve cells) and neuroglia. Neurons consist of a cell body and processes extending from the cell body (one to multiple dendrites and a single axon). Neuroglia do not generate or conduct nerve impulses but have other important supporting functions.

Location: Nervous system.

Function: Exhibits sensitivity to various types of stimuli, converts stimuli into nerve impulses (action potentials), and conducts nerve impulses to other neurons, muscle fibers, or glands.

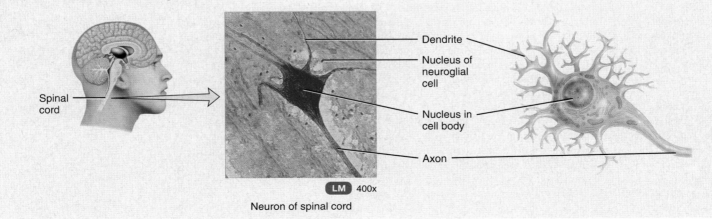

LM 400x

Neuron of spinal cord

cells have a higher metabolic rate. Thus, its cells can synthesize needed materials and divide more quickly. The extracellular components of tissues also change with age. Glucose, the most abundant sugar in the body, plays a role in the aging process. As the body ages, glucose is haphazardly added to proteins inside and outside cells, forming irreversible cross-links between adjacent protein molecules. With advancing age, more cross-links form, which contributes to the stiffening and loss of elasticity that occur in aging tissues. Collagen fibers, responsible for the strength of tendons, increase in number and change in quality with aging. Changes in the collagen of arterial walls affect the flexibility of arteries as much as the fatty deposits associated with atherosclerosis (see Section 13.7). Elastin, another extracellular component, is responsible for the elasticity of blood vessels and skin. It thickens, fragments, and acquires a greater affinity for calcium with age—changes that may also be associated with the development of atherosclerosis, the deposition of fatty materials in arterial walls.

 CHECKPOINT

20. What common changes occur in epithelial and connective tissues with aging?

KEY MEDICAL TERMS ASSOCIATED WITH TISSUES

Adhesions (ad-HĒ-zins; *adhaero*=to stick to) An abnormal joining of tissues. Adhesions commonly form in the abdomen around a site of previous inflammation such as an inflamed appendix, and they can develop after surgery. Although adhesions do not always cause problems, they can decrease tissue flexibility, cause obstruction (such as in the intestine), and make a subsequent operation, such as a c-section, more difficult. In rare cases, adhesions can result in infertility. An *adhesiotomy*, the surgical release of adhesions, may be required.

Atrophy (AT-rō-fē; *a-*=without; *-trophy*=nourishment) A decrease in the size of a tissue or organ because of a decrease in the size of its cells.

Hypertrophy (hī-PER-trō-fē; *hyper-*=above or excessive) Increase in the size of a tissue because its cells enlarge without undergoing cell division.

Tissue rejection An immune response of the body directed at foreign proteins in a transplanted tissue or organ. Immunosuppressive drugs, such as cyclosporine, have largely overcome tissue rejection in heart-, kidney-, and liver-transplant patients.

Tissue transplantation The replacement of a diseased or injured tissue or organ. The most successful transplants involve use of a person's own tissues or those from an identical twin.

Xenotransplantation (zen′-ō-trans-plan-TĀ-shun; *xeno-*=strange, foreign) The replacement of a diseased or injured tissue or organ with cells or tissues from an animal. Porcine (from pigs) and bovine (from cows) heart valves are used for some heart-valve replacement surgeries.

CHAPTER REVIEW AND RESOURCE SUMMARY

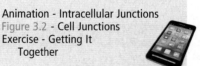

Review

Resource

3.1 Types of Tissues

1. A tissue is a group of cells, usually of similar embryological origin, that is specialized for a particular function.
2. The various tissues of the body are classified into four basic types: epithelial, connective, muscular, and nervous.

3.2 Cell Junctions

1. Cell junctions are points of contact between adjacent plasma membranes.
2. Tight junctions form fluid-tight seals between cells; adherens junctions, desmosomes, and hemidesmosomes anchor cells to one another or to the basement membrane; and gap junctions permit electrical and chemical signals to pass between cells.

Animation - Intracellular Junctions
Figure 3.2 - Cell Junctions
Exercise - Getting It Together

3.3 Comparison Between Epithelial and Connective Tissue

1. Epithelial tissue has many cells tightly packed together and are avascular.
2. Connective tissue has relatively few cells with lots of extracellular material.

Anatomy Overview - Epithelial Tissue
Anatomy Overview - Connective Tissue

3.4 Epithelial Tissue

1. Epithelial tissue consists mostly of cells with little extracellular material between adjacent plasma membranes. The apical, lateral, and basal surfaces of epithelial cells are modified in various ways to carry out specific functions. Epithelial tissue is arranged in sheets and attached to a basement membrane. Although it is avascular, it has a nerve supply. Epithelial tissue has a high capacity for renewal.
2. Epithelial layers can be simple (one layer), pseudostratified (appears to have several layers but is actually a single layer), or stratified (several layers). The cell shapes may be squamous (flat), cuboidal (cubelike), columnar (rectangular), or transitional (variable). The subtypes of epithelial tissue include covering and lining epithelium and glandular epithelium.
3. Simple squamous epithelium, which consists of a single layer of flat cells (Table 3.1A), is found in parts of the body where filtration or diffusion are priority processes. One type, called endothelium, lines the heart and blood vessels. Another type, called mesothelium, forms the serous membranes that line the thoracic and abdominopelvic cavities and cover the organs within them.
4. Simple cuboidal epithelium consists of a single layer of cube-shaped cells that function in secretion and absorption (Table 3.1B). It is found covering the ovaries, in the kidneys and eyes, and lining some glandular ducts.

Anatomy Overview - Epithelial Tissue
Figure 3.5 - Cell Shapes and Arrangements of Layers for Covering and Lining Epithelium
Exercise - Concentrate on Tissue Function
Exercise - Name That Tissue
Exercise - Tissue Identification
Concepts and Connections - Human Tissue

Review	Resource

5. Nonciliated simple columnar epithelium, a single layer of nonciliated rectangular cells (Table 3.1C), lines most of the gastrointestinal tract and contains specialized cells that perform absorption and secrete mucus. Ciliated simple columnar epithelium, which consists of a single layer of ciliated rectangular cells (Table 3.1D), is found in a few portions of the upper respiratory tract, where it moves foreign particles trapped in mucus out of the respiratory tract. A ciliated variety of pseudostratified columnar epithelium (Table 3.1E) contains goblet cells and lines most of the upper respiratory tract; a nonciliated variety has no goblet cells and lines ducts of many glands, the epididymis, and part of the male urethra. The ciliated variety moves mucus in the respiratory tract. The nonciliated variety functions in absorption and protection.

6. Stratified epithelium consists of several layers of cells. Cells of the apical layer of stratified squamous epithelium and several layers deep to it are flat (Table 3.1F); a nonkeratinized variety lines the mouth, and a keratinized variety forms the epidermis, the most superficial layer of the skin. Cells at the apical layer of stratified cuboidal epithelium are cube-shaped (Table 3.1G); found in adult sweat glands and a portion of the male urethra, it protects and provides limited secretion and absorption. Cells of the apical layer of stratified columnar epithelium have a columnar shape (Table 3.1H); this type of stratified epithelium is found in a portion of the male urethra and large excretory ducts of some glands, and functions in protection and secretion.

7. Transitional epithelium consists of several layers of cells whose appearance varies with the degree of stretching (Table 3.1I). It lines the urinary bladder.

8. A gland is a single cell or a group of epithelial cells adapted for secretion. There are two types of glands: endocrine and exocrine. Endocrine glands secrete hormones into interstitial fluid and then into the blood (Table 3.2A). Exocrine glands (mucous, sweat, oil, and digestive glands) secrete into ducts or directly onto a free surface (Table 3.2B).

9. The structural classification of exocrine glands includes unicellular and multicellular glands. The functional classification of exocrine glands includes merocrine, apocrine, and holocrine glands.

3.5 Connective Tissue

1. Connective tissue, one of the most abundant body tissues, consists of relatively few cells and an abundant extracellular matrix of ground substance and fibers. It does not usually occur on free surfaces, has a nerve supply (except for cartilage), and is highly vascular (except for cartilage, tendons, and ligaments).

2. Cells in connective tissue, which are derived from mesenchymal cells, include fibroblasts (secrete extracellular matrix), macrophages (perform phagocytosis), plasma cells (secrete antibodies), mast cells (produce histamine), adipocytes (store fat), and white blood cells (migrate from blood in response to infections).

3. The ground substance and fibers make up the extracellular matrix; it supports and binds cells together, provides a medium for the exchange of materials, stores water, and is active in influencing cell functions. Substances found in the ground substance include water and polysaccharides such as hyaluronic acid, chondroitin sulfate, dermatan sulfate, and keratan sulfate (glycosaminoglycans). Proteoglycans and adhesion proteins are also present.

4. The fibers in the extracellular matrix, which provide strength and support, are of three types: (a) Collagen fibers (composed of collagen) are found in large amounts in bone, tendons, and ligaments. (b) Elastic fibers (composed of elastin, fibrillin, and other glycoproteins) are found in skin, blood vessel walls, and lungs. (c) Reticular fibers (composed of collagen and glycoprotein) are found around fat cells, nerve fibers, and skeletal and smooth muscle cells.

5. The two major subclasses of connective tissue are embryonic connective tissue (found in the embryo and fetus) and mature connective tissue (present in the newborn). The embryonic connective tissues are mesenchyme, which forms almost all other connective tissues (Table 3.3A), and mucous connective tissue, found in the umbilical cord of the fetus, where it gives support (Table 3.3B). Mature connective tissue differentiates from mesenchyme, and is subdivided into several types: loose or dense connective tissue, cartilage, bone tissue, and liquid connective tissue.

6. Loose connective tissue includes areolar connective tissue, adipose tissue, and reticular connective tissue. Areolar connective tissue consists of the three types of fibers (collagen, elastic, and reticular), several types of cells, and a semifluid ground substance (Table 3.4A); it is found in the subcutaneous layer, in mucous membranes, and around blood vessels, nerves, and body organs. Adipose tissue consists of adipocytes, which store triglycerides (Table 3.4B); it is found in the subcutaneous layer, around organs, and in yellow bone marrow. Brown adipose tissue (BAT) generates heat. Reticular connective tissue consists of reticular fibers and reticular cells and is found in the liver, spleen, and lymph nodes (Table 3.4C).

7. Dense connective tissue includes dense regular connective tissue, dense irregular connective tissue, and elastic connective tissue. Dense regular connective tissue consists of parallel bundles of collagen fibers and fibroblasts (Table 3.5A); it forms tendons, most ligaments, and aponeuroses. Dense irregular connective tissue usually consists of randomly arranged collagen fibers and a few fibroblasts (Table 3.5B); it is found in fasciae, the dermis of skin, and membrane capsules around organs. Elastic connective tissue consists of branching elastic fibers and fibroblasts (Table 3.5C) and is found in the walls of large arteries, lungs, trachea, and bronchial tubes.

8. Cartilage contains chondrocytes and has a rubbery extracellular matrix (chondroitin sulfate) containing collagen and elastic fibers. Hyaline cartilage, which consists of a gel-like ground substance and appears bluish white in the body, is found in the embryonic skeleton, at the ends of bones, in the nose, and in

Resource

Anatomy Overview - Connective Tissue
Figure 3.7 - Functional Classification of Multicellular Exocrine Glands
Figure 3.8 - Representative Cells and Fibers Present in Connective Tissue
Exercise - Concentrate on Tissue Function
Exercise - Name That Tissue
Exercise - Tissue Identification
Concepts and Connections - Human Tissue

Review	Resource

respiratory structures (Table 3.6A); it is flexible, allows movement, and provides support, and is usually surrounded by a perichondrium. Fibrocartilage is found in the pubic symphysis, intervertebral discs, and menisci (cartilage pads) of the knee joint (Table 3.6B); it contains chondrocytes scattered among clearly visible bundles of collagen fibers. Elastic cartilage maintains the shape of organs such as the epiglottis of the larynx, auditory (eustachian) tubes, and external ear (Table 3.6C); its chondrocytes are located within a threadlike network of elastic fibers, and it has a perichondrium.

9. Bone or osseous tissue consists of an extracellular matrix of mineral salts and collagen fibers that contribute to the hardness of bone, and osteocytes that are located in lacunae (Table 3.7). It supports and protects the body, provides a surface area for muscle attachment, helps the body move, stores minerals, and houses blood-forming tissue.

10. There are two types of liquid connective tissue: blood and lymph. Blood consists of blood plasma and formed elements—red blood cells, white blood cells, and platelets (Table 3.8); its cells transport oxygen and carbon dioxide, carry on phagocytosis, participate in allergic reactions, provide immunity, and bring about blood clotting. Lymph, the extracellular fluid that flows in lymphatic vessels, is a clear fluid similar to blood plasma but with less protein.

3.6 Membranes

1. An epithelial membrane consists of an epithelial layer overlying a connective tissue layer. Types include mucous, serous, and cutaneous membranes.
2. Mucous membranes line cavities that open to the exterior, such as the gastrointestinal tract.
3. Serous membranes line closed cavities (pleura, pericardium, peritoneum) and cover the organs in the cavities. These membranes consist of parietal and visceral layers.
4. The cutaneous membrane is the skin. It covers the entire body and consists of a superficial epidermis (epithelium) and a deep dermis (connective tissue).
5. Synovial membranes line joint cavities and consist of areolar connective tissue; and do not have an epithelial layer.

Anatomy Overview - Epithelial Membranes
Exercise - Concentrate on Tissue Function
Exercise - Name That Tissue
Exercise - Tissue Identification
Concepts and Connections - Human Tissue

3.7 Muscular Tissue

1. Muscular tissue consists of cells called muscle fibers or myocytes that are specialized for contraction. It provides motion, maintenance of posture, heat production, and protection.
2. Skeletal muscle tissue is attached to bones and is striated and voluntary (Table 3.9A).
3. The action of cardiac muscle tissue, which forms most of the heart wall and is striated, is involuntary (Table 3.9B).
4. Smooth muscle tissue is found in the walls of hollow internal structures (blood vessels and viscera) and is nonstriated and involuntary (Table 3.9C).

Anatomy Overview - Muscular Tissue
Exercise - Concentrate on Tissue Function
Exercise - Name That Tissue
Exercise - Tissue Identification
Concepts and Connections - Human Tissue

3.8 Nervous Tissue

1. The nervous system is composed of neurons (nerve cells) and neuroglia (protective and supporting cells) (Table 3.10).
2. Neurons respond to stimuli by converting the stimuli into electrical signals called nerve action potentials (nerve impulses), and conducting nerve impulses to other cells.
3. Most neurons consist of a cell body and two types of processes, dendrites and axons.

Anatomy Overview - Nervous Tissue
Exercise - Concentrate on Tissue Function
Exercise - Name That Tissue
Exercise - Tissue Identification
Concepts and Connections - Human Tissue

3.9 Aging and Tissues

1. Tissues heal faster and leave less obvious scars in the young than in the aged; surgery performed on fetuses leaves no scars.
2. The extracellular components of tissues, such as collagen and elastic fibers, also change with age.

CRITICAL THINKING QUESTIONS

1. Mike has had a series of respiratory tract infections this winter. His doctor has just prescribed a mucus-thinning drug. Using your knowledge of the structure of the mucous membrane lining the respiratory tract, how do you think this type of drug will help Mike get better?

2. As part of a routine pelvic exam for all women, a biopsy of cervical and vaginal tissue is taken (*Pap smear*). Describe the kinds of cells that you would see in the normal smear. How would the cells appear in *cervical dysplasia*?

3. Janelle has been an anorexic for several years. As a result of her chronically low daily caloric intake, her adipocytes are storing little or no triglycerides. What structural problems might Janelle suffer as a result?

4. In order to show-off to his fifth-grade classmates, Jonathan stuck a common pin and sewing needle into his fingertip. His classmates were amazed that there was no blood. What type of tissue did Jonathan pierce, and why was there no visible bleeding?

5. Sometime during your daily activities you come in contact with a colony of bacteria that gets onto your skin. You have no cuts in the area, and your skin is in good condition. However, these microbes manage to penetrate your tissues and get into your blood. What specific structural barriers would the bacteria have to overcome to accomplish this?

3.1 Epithelial tissue covers the body, lines various structures, and forms glands. Connective tissue protects, supports, binds organs together, stores energy, and helps provide immunity. Muscular tissue contracts and generates force and heat. Nervous tissue detects changes in the environment and generates nerve impulses that activate muscular contraction and glandular secretion.

3.2 Gap junctions allow cellular communication via passage of electrical and chemical signals between adjacent cells.

3.3 Since epithelial tissue is avascular, it depends on the blood vessels in connective tissue for oxygen and nutrients and waste disposal.

3.4 The basement membrane provides physical support for the epithelial tissue, forms a surface along which epithelial cells migrate during growth or wound healing, restricts passage of larger molecules between epithelium and connective tissue, and participates in filtration of blood in the kidneys.

3.5 Because they are so thin, substances would move most rapidly through squamous cells.

3.6 Simple multicellular exocrine glands have a nonbranched duct; compound multicellular exocrine glands have a branched duct.

3.7 Sebaceous (oil) glands are holocrine glands, and salivary glands are merocrine glands.

3.8 Fibroblasts secrete the fibers and ground substance of the extracellular matrix.

3.9 An epithelial membrane is a membrane that consists of an epithelial layer and an underlying layer of connective tissue.

4 | DEVELOPMENT

INTRODUCTION Think for a moment about a complex machine designed and built by humans. A computer might come to mind, or—better yet—what about the space shuttle? Regardless of the complexity of the machine that crosses your mind, its design and manufacture pales in comparison to the developmental processes that transform a single cell into the approximately 100 trillion cells of the human body. Before we examine the first body system in Chapter 5 (The Integumentary System), let's take a look at how body systems develop. Knowledge of the origins of the different systems of the human body will enhance your understanding of their structures and how they work. Later in the text you will learn more about development in the context of the various body systems.

As you learned in Chapter 2, **sexual reproduction** is the process by which organisms produce offspring by making sex cells called **gametes** (GAM-ēts=spouses). Male gametes are called **sperm (spermatozoa)** and female gametes are called **secondary oocytes**. The organs that produce gametes are called **gonads**; these are the testes in the male and the ovaries in the female. The details of sperm formation in the testes and secondary oocyte formation in the ovaries are discussed in Chapter 26. Once sperm have been deposited in the female reproductive tract and a secondary oocyte has been released from the ovary, fertilization can occur. This process initiates a cascade of developmental events that, when completed properly, produces a healthy newborn baby.

Pregnancy is a sequence of events that begins with fertilization, proceeds to implantation, embryonic development, and fetal development, and ideally ends with birth about 38 weeks later, or 40 weeks after the mother's last menstrual period.

Developmental biology is the study of the sequence of events from the fertilization of a secondary oocyte by a sperm cell to the formation of an adult organism. From fertilization through the eighth week of development, the **embryonic period**, the developing human is called an **embryo** (*em-*=into; *-bryo*=grow). **Embryology** (em-brē-OL-ō-jē) is the study of development from the fertilized egg through the eighth week. The **fetal period** begins at week nine and continues until birth. During this time, the developing human is called a **fetus** (FĒ-tus=offspring).

Did you ever wonder why the heart, blood vessels, and blood begin to form so early in the developmental process?

Prenatal development (prē-NĀ-tal; *pre-* = before; *natal* = birth) is the time from fertilization to birth and includes both the embryonic and fetal periods. Prenatal development is divided into periods of three calendar months each, called **trimesters**.

1. During the **first trimester**, the most critical stage of development, all of the major organ-systems begin to form. Because of the extensive, widespread activity, it is also the period when the developing organism is most vulnerable to the effects of drugs, radiation, and microbes.

2. The **second trimester** is characterized by the nearly complete development of organ systems. By the end of this stage, the fetus assumes distinctively human features.

3. The **third trimester** represents a period of rapid fetal growth in which the weight of the fetus doubles. During the early stages of this period, most of the organ systems become fully functional.

The female reproductive organs involved in the major events of the embryonic and fetal periods are the ovaries, uterine tubes, uterus, and vagina. The *ovaries* are paired organs in the superior portion of the pelvic cavity on either side of the uterus (Figure 4.1). During the reproductive years of a female, they produce secondary oocytes and discharge them into the peritoneal cavity each month, a process called *ovulation*. Each of the two *uterine (fallopian) tubes*, which extend laterally from the uterus, provides an exit route from the peritoneal cavity for the secondary oocyte. In the uterine tubes, a secondary oocyte can come in contact with sperm, and the tubes can transport a fertilized (or unfertilized) oocyte to the uterus. The *uterus*, an inverted pear-shaped organ in the pelvic and lower abdominal cavity, consists of a superior portion (fundus), middle portion (body), and inferior portion (cervix). The interior of the body of the uterus is called the *uterine cavity*, and the interior of the cervix is known as the *cervical canal*. After puberty, in preparation for implantation of the fertilized ovum, the uterus forms a vascular, glandular lining called the *endometrium* during each

menstrual cycle. Part of the endometrium is shed during menstruation. Deep to the endometrium is a muscular layer called the *myometrium*. The largest mass of smooth muscle in the human body, the myometrium is responsible for the strong contractions that expel the endometrium during menstruation and help push the fetus out of the uterus during labor. The uterus serves as the site of implantation of a fertilized ovum and development of the fetus during pregnancy. The cervix of the uterus connects to the *vagina*, a multipurpose canal that opens to the exterior. The vagina serves as the receptacle for the penis during sexual intercourse, the outlet for menstrual flow, and the passageway for childbirth. •

Figure 4.1 Uterus and associated structures. The left side of the uterine tube, uterus, and vagina have been sectioned to show the interior of these structures.

The ovaries produce female gametes called secondary oocytes.

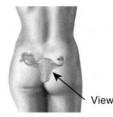

View

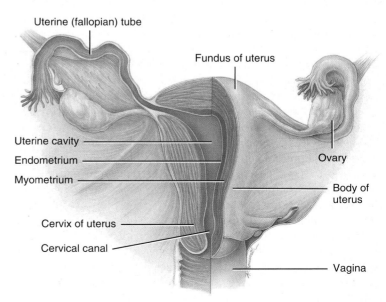

Uterine (fallopian) tube

Fundus of uterus

Uterine cavity

Endometrium

Myometrium

Ovary

Body of uterus

Cervix of uterus

Cervical canal

Vagina

Posterior view

 What are the functions of the uterine tubes?

4.1 EMBRYONIC PERIOD

OBJECTIVE

- Explain the major developmental events that occur during the embryonic period.

First Week of Development

The first week of development is characterized by several significant events including fertilization, cleavage of the zygote, blastocyst formation, and implantation.

Fertilization

During **fertilization** (fer-til-i-ZĀ-shun; *fertil-*=fruitful) the genetic material from a haploid sperm cell and a haploid secondary oocyte merges into a single diploid nucleus. Of the approximately 200 million sperm introduced into the vagina, fewer than 2 million (1 percent) reach the cervix and only about 200 (0.0001 percent) reach the secondary oocyte. Fertilization normally occurs in the uterine (fallopian) tube within 12 to 24 hours after ovulation. Sperm can remain viable for about 48 hours after deposition in the vagina, although a secondary oocyte is viable for only about 24 hours after release from the ovary. Thus, pregnancy is *most likely* to occur if intercourse takes place during a 3-day window—from 2 days before ovulation to 1 day after ovulation.

Sperm swim from the vagina into the cervical canal propelled by the whiplike movements of their tails (flagella). The passage of sperm through the rest of the uterus and into the uterine tube results mainly from contractions of the walls of these organs. Sperm that reach the vicinity of the oocyte within minutes after ejaculation *are not capable* of fertilizing it until about seven hours later. During this time in the female reproductive tract, mostly in the uterine tube, sperm undergo **capacitation** (ka-pas′-i-TĀ-shun; *capacit-*=capable of), a series of functional changes that cause the sperm's tail to beat even more vigorously and prepare its plasma membrane to fuse with the oocyte's plasma membrane. During capacitation, secretions in the female reproductive tract remove cholesterol, glycoproteins, and proteins from the plasma membrane around the head of the sperm cell. Only capacitated sperm are capable of being attracted by and responding to chemical factors produced by the surrounding cells of the ovulated oocyte.

For fertilization to occur, a sperm cell first must penetrate the **corona radiata** (ko-RŌ-na=crown; rā-dē-A-ta=to shine), the cells that surround the secondary oocyte, and the **zona pellucida** (ZŌ-na=zone; pe-LOO-si-da=allowing passage of light), the clear glycoprotein layer between the corona radiata and the oocyte's plasma membrane (Figure 4.2a). The **acrosome** (AK-rō-sōm), a helmetlike structure that covers the head of a sperm, contains several enzymes (see Figure 26.6). Enzymes and strong tail movements by the sperm help it penetrate the cells of the corona radiata and come in contact with the zona pellucida. One of the glycoproteins in the zona pellucida acts as a sperm receptor, binding to specific membrane proteins on the acrosome to trigger the **acrosomal reaction**, the release of the contents of the acrosome. The acrosomal enzymes digest a path through the zona pellucida as the lashing sperm tail pushes the sperm cell onward. Although many sperm undergo acrosomal reactions, only the first sperm cell to penetrate the entire zona pellucida and reach the oocyte's plasma membrane undergoes **syngamy** (*syn-*=coming together; *-gamy*=marriage), fusion with the oocyte. This results in a reaction within the oocyte that makes the zona pellucida impermeable, blocking **polyspermy** (POL-ē-sper′-mē), fertilization by more than one sperm cell.

Figure 4.2 Selected structures and events in fertilization.

 During fertilization, genetic material from a sperm cell and a secondary oocyte merge to form a single diploid nucleus.

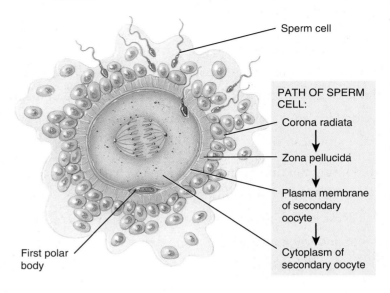

PATH OF SPERM CELL:

Corona radiata ↓ Zona pellucida ↓ Plasma membrane of secondary oocyte ↓ Cytoplasm of secondary oocyte

(a) Sperm cell penetrating a secondary oocyte

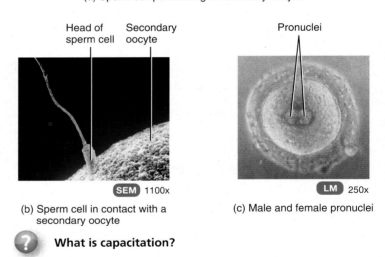

SEM 1100x

(b) Sperm cell in contact with a secondary oocyte

LM 250x

(c) Male and female pronuclei

? **What is capacitation?**

Once a sperm cell enters a secondary oocyte, the oocyte completes meiosis II (see Figure 2.20). It divides into a larger ovum (mature egg) and a smaller second polar body, which contains DNA but no cytoplasm. The polar body eventually fragments and disintegrates (see Figure 26.15). The nucleus in the head of the sperm develops into the **male pronucleus**, and the nucleus in the fertilized ovum develops into the **female pronucleus** (Figure 4.2c). After the pronuclei form, they fuse, producing a single nucleus with 23 chromosomes from each pronucleus. Fusion of the two haploid (n) pronuclei restores the diploid number ($2n$) of 46 chromosomes in the fertilized ovum, now called a **zygote** (ZI-gōt; *zygon*=yolk). Fertilization takes approximately 24 hours.

Dizygotic (fraternal) twins are produced from the independent release of two secondary oocytes and the subsequent fertilization of each of them by different sperm. They are the same age and in the uterus at the same time, but genetically they are as dissimilar as any other siblings. Dizygotic twins may or may not be the same sex. Because **monozygotic (identical) twins** develop from a single fertilized ovum, they contain exactly the same genetic material and are always the same sex, but the expression of that genetic information may differ. Monozygotic twins arise from separation of the developing zygote into two embryos; 99 percent of the time this occurs within 8 days of fertilization. Separations that occur later than 8 days are likely to produce **conjoined twins**, a situation in which the twins are joined together and share some body structures. The surgical separation of conjoined twins is a very delicate and risky procedure. Mortality rates for the procedure depend on the type of connection and the organs shared.

CLINICAL CONNECTION | *Ectopic Pregnancy*

Ectopic pregnancy (ek-TOP-ik; *ec-*=out of; *-topic*=place), the development of an embryo or fetus outside the uterine cavity, usually occurs when movement of the fertilized ovum through the uterine tube is impaired by scarring due to prior tubal infection, decreased movement of uterine tube smooth muscle, or abnormal tubal anatomy. The most common site is the uterine tube, but they may also occur in the ovary, abdominal cavity, or uterine cervix. Women who smoke are twice as likely to have an ectopic pregnancy because nicotine in cigarette smoke paralyzes the cilia in the lining of the uterine tube. Scars from pelvic inflammatory disease, previous uterine tube surgery, and previous ectopic pregnancy may also hinder movement of the fertilized ovum.

The signs and symptoms of ectopic pregnancy include one or two missed menstrual cycles followed by bleeding and acute abdominal and pelvic pain. Unless removed, the developing embryo can rupture the uterine tube, often resulting in death of the mother. Treatment options include surgery or the use of a cancer drug called methotrexate, which causes embryonic cells to stop dividing and eventually disappear. •

Cleavage of the Zygote

After fertilization, the zygote undergoes mitotic cell divisions called **cleavage** (KLĒV-ij) that initially increase the number of cells without increasing the overall size of the cell mass (Figure 4.3). The first division of the zygote begins about 24 hours after fertilization and is completed about 6 hours later. Each succeeding division takes slightly less time. By the second day after fertilization, the second cleavage is completed and there are four cells (Figure 4.3b). By the end of the third day, there are 16 cells. The progressively smaller cells produced by cleavage are called **blastomeres** (BLAS-tō-merz; *blasto-*=germ or sprout; *-meres*=parts). Successive cleavages eventually become more rapid and produce a solid sphere of cells called the **morula** (MOR-ū-la; *morula*=mulberry). The morula is still surrounded by the zona pellucida and is about the same size as the original zygote (Figure 4.3c).

Blastocyst Formation

By the end of the fourth day, the number of cells in the morula increases as it continues to move through the uterine tube toward the uterine cavity. When the morula enters the uterine cavity on day 4 or 5, a glycogen-rich secretion from the glands of the endometrium passes into the uterine cavity and enters the morula through the zona pellucida. This secretion, called **uterine milk**, along with nutrients stored in the cytoplasm of the blastomeres of the morula, provides nourishment for the developing morula. At the 32-cell stage, the fluid enters the morula, collects between the blastomeres, and reorganizes them around a large fluid-filled cavity called the **blastocyst cavity** (BLAS-tō-sist; *blasto-*=germ or sprout; *-cyst*=bag) or *blastocele* (BLAS-tō-sēl) (Figure 4.3e). Once the blastocyst cavity is formed, the developing mass is referred to

as the **blastocyst**. Though it now has hundreds of cells, the blastocyst is still about the same size as the original zygote.

During the formation of the blastocyst, two distinct cell populations arise: the embryoblast and trophoblast (Figure 4.3e). The **embryoblast** (EM-brē-ō-blast′), or **inner cell mass**, is located internally and eventually develops into the embryo and some of the extraembryonic membranes. The **trophoblast** (TRŌF-o-blast; *tropho-*=develop or nourish) is the outer superficial layer of cells that forms the sphere-like wall of the blastocyst. It will ultimately develop into the outer chorionic sac that surrounds the fetus and the fetal portion of the *placenta*, the site of exchange of nutrients and wastes between the mother and fetus. On about the fifth day after fertilization, the blastocyst is released from the zona pellucida by digesting a hole in it with an enzyme, and then squeezing through the hole. This shedding of the zona pellucida is necessary in order to permit the next step, implantation (attachment) into the vascular, glandular endometrial lining of the uterus.

Figure 4.3 Cleavage and the formation of the morula and blastocyst.

🔑 Cleavage refers to the early, rapid mitotic divisions of a zygote.

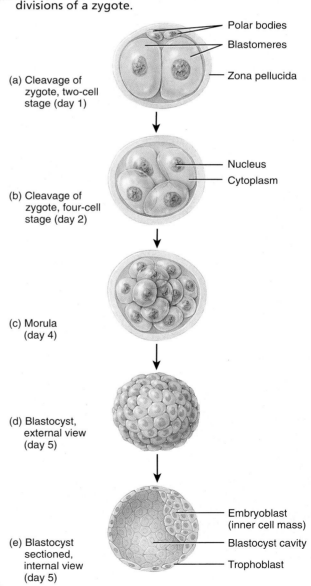

(a) Cleavage of zygote, two-cell stage (day 1)
— Polar bodies
— Blastomeres
— Zona pellucida

(b) Cleavage of zygote, four-cell stage (day 2)
— Nucleus
— Cytoplasm

(c) Morula (day 4)

(d) Blastocyst, external view (day 5)

(e) Blastocyst sectioned, internal view (day 5)
— Embryoblast (inner cell mass)
— Blastocyst cavity
— Trophoblast

 What is the histological difference between a morula and a blastocyst?

Implantation

The blastocyst remains free within the uterine cavity for about 2 days before it attaches to the uterine wall. During this period it is nourished by secretions from the endometrium. About 6 days after fertilization, the blastocyst loosely attaches to the endometrium in a process called **implantation** (im′-plan-TĀ-shun) (Figure 4.4). The blastocyst usually implants either in the posterior portion of the fundus or in the body of the uterus, orienting itself with the embryoblast toward the endometrium (Figure 4.4b). About seven days after fertilization, the blastocyst attaches to the endometrium more firmly, endometrial glands in the vicinity enlarge, and the endometrium becomes more vascular by forming new blood vessels.

Following implantation, the functional layer of the endometrium is known as the **decidua** (dē-SID-ū-a=falling off). The decidua separates from the endometrium after the fetus is delivered, much as the superficial layer of the endometrium does in normal menstruation. Different regions of the decidua have different names based on their positions relative to the site of the implanted blastocyst (Figure 4.4c). The **decidua basalis** is the portion of the endometrium beneath the implanting embryo; it provides large amounts of glycogen and lipids for the developing embryo and fetus and later becomes the maternal part of the placenta. The **decidua capsularis** is the portion of the endometrium that will cover the embryo after it implants in the endometrium. The **decidua parietalis** (par-rī-e-TAL-is) is the remaining modified endometrium that lines the noninvolved areas of the rest of the uterus. As the embryo and later the fetus enlarges and pushes into the uterine cavity, the decidua capsularis becomes thin and eventually disappears as the enlarged fetus fills the uterine cavity and pushes against the surrounding decidua parietalis. By about 27 weeks, the decidua capsularis degenerates and disappears.

The major events associated with the first week of development are summarized in Figure 4.5.

Figure 4.4 Relation of a blastocyst to the endometrium of the uterus at the time of implantation.

Implantation, the attachment of a blastocyst to the endometrium, occurs about 6 days after fertilization.

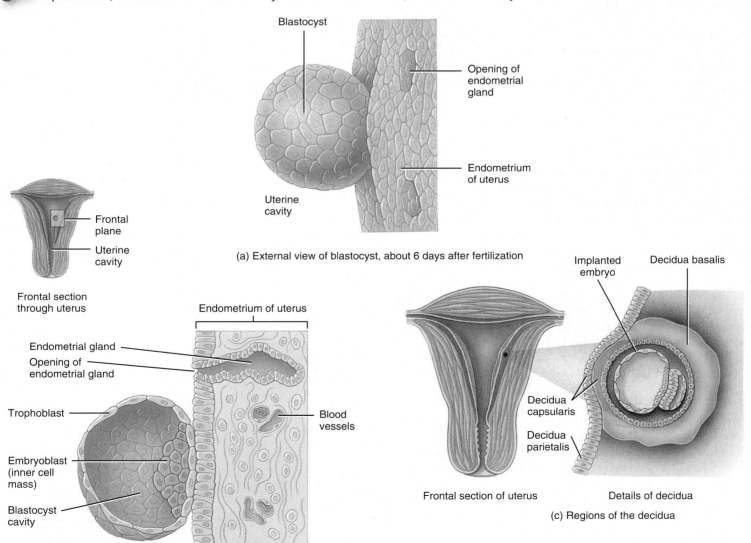

(a) External view of blastocyst, about 6 days after fertilization

(b) Frontal section through endometrium of uterus and blastocyst, about 6 days after fertilization

(c) Regions of the decidua

 How does the blastocyst merge with and burrow into the endometrium?

CLINICAL CONNECTION | *Stem Cell Research*

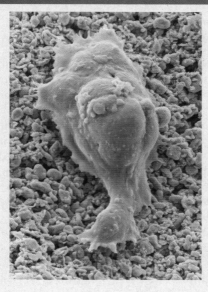

Human embryonic stem cell

Stem cells are unspecialized cells that have the ability to divide for indefinite periods and give rise to specialized cells. A stem cell with the potential to form an entire organism is known as a *totipotent stem cell* (tō-TIP-ō-tent; *totus-*= whole; *-potentia*=power). The cells of a zygote and an embryo through the morula stage of development are totipotent stem cells. Embryoblast cells of a blastocyst, by contrast, can give rise to many (but not all) different types of cells. Such stem cells are called *pluripotent stem cells* (ploo-RIP-ō-tent; *plur-*=several). Later, pluripotent stem cells can undergo further specialization into cells that have a specific function and give rise to a closely related family of cells. The primary role of these cells, called *multipotent stem cells* (mul-TIP-ō-tent), is to maintain or repair a tissue. *Oligopotent stem cells* (ō-LIG-op-ō-tent) give rise to a few different cell types, such as myeloid and lymphoid stem cells that develop into the different types of blood cells. *Unipotent stem cells* (ū-NIP-ō-tent) produce only one cell type, but can renew themselves, which distinguishes them from non-stem cells. Examples include stem keratinocytes in the skin or satellite cells in the muscles. Pluripotent stem cells currently used in research are derived from (1) the embryoblast of embryos in the blastocyst stage that were destined to be used for infertility treatments but were not needed and from (2) nonliving fetuses terminated during the first three months of pregnancy.

Scientists are also investigating the potential clinical applications of *adult oligopotent and unipotent stem cells*—stem cells that remain in the body throughout adulthood. Recent experiments suggest that the ovaries of adult mice contain stem cells that can develop into new ova (eggs). If these same types of stem cells are found in the ovaries of adult women, scientists could potentially harvest some of them from a woman about to undergo a sterilizing medical treatment (such as chemotherapy), store them, and then return the stem cells to her ovaries after the medical treatment is completed in order to restore fertility. Studies have also suggested that stem cells in human adult red bone marrow have the ability to differentiate into cells of the liver, kidney, heart, lung, skeletal muscle, skin, and organs of the gastrointestinal tract. In theory, adult stem cells from red bone marrow could be harvested from a patient and then used to repair other tissues and organs in that patients's body without having to use stem cells from embryos. •

Figure 4.5 Summary of events associated with the first week of development.

 Fertilization usually occurs in the uterine tube.

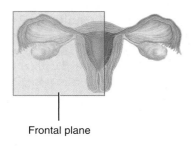

Frontal plane

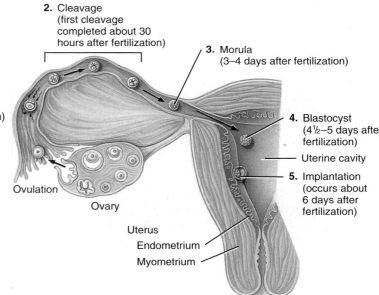

1. Fertilization (occurs within uterine tube 12–24 hours after ovulation)

2. Cleavage (first cleavage completed about 30 hours after fertilization)

3. Morula (3–4 days after fertilization)

4. Blastocyst (4½–5 days after fertilization)

Uterine cavity

5. Implantation (occurs about 6 days after fertilization)

Ovulation

Ovary

Uterus

Endometrium

Myometrium

Frontal section through uterus, uterine tube, and ovary

 In which portion of the uterus does implantation usually occur?

CHECKPOINT

1. Where does fertilization normally occur?
2. What is a morula, and how is it formed?
3. Describe the layers of a blastocyst and their eventual fates.
4. When, where, and how does implantation occur?
5. On what basis are the three regions of the decidua named?

Second Week of Development

Development of the Trophoblast

About 8 days after fertilization, the trophoblast develops into two layers in the region of contact between the blastocyst and endometrium. These layers are a **syncytiotrophoblast** (sin-sīt-ē-ō-TRŌF-ō-blast), an area without distinct cell boundaries, and a **cytotrophoblast** (sī-tō-TRŌF-ō-blast), an area between the embryoblast and syncytiotrophoblast that has distinct cell boundaries (Figure 4.6a). The two layers of trophoblast become part of the chorion (one of the fetal membranes) as they undergo further growth (see Figure 4.11a). During implantation, the syncytiotrophoblast secretes enzymes that digest and liquefy the endometrial cells so the blastocyst can penetrate the uterine lining. Eventually, the blastocyst becomes buried in the endometrium and inner one-third of the myometrium (muscle layer of the uterus). Another secretion of the trophoblast is human chorionic gonadotropin (hCG), the key hormone in the maintenance of pregnancy and the hormone that is detected by a home pregnancy test. Its presence in the maternal urine or blood is an indication that an implanted embryo is present in the uterus. Human chorionic gonadotropin mimics a maternal hormone, luteinizing hormone (LH), which maintains the endocrine secretions of progesterone and estrogens from the ovary. In the typical menstrual cycle, LH production declines, resulting in uterine contractions and the expulsion of the endometrium from the uterus. If this were to happen during pregnancy the embryo would be aborted. By producing hCG, the embryo ensures the continuation of pregnancy.

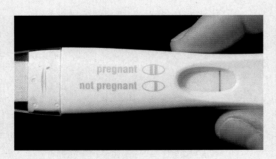

CLINICAL CONNECTION | *Early Pregnancy Tests*

Early pregnancy tests detect the tiny amounts of human chorionic gonadotropin (hCG) in the urine that begin to be excreted about 8 days after fertilization. The test kits can detect pregnancy as early as the first day of a missed menstrual period—that is, at about 14 days after fertilization. Chemicals in the kits produce a color change if a reaction occurs between hCG in the urine and hCG antibodies included in the kit.

Several of the test kits available at pharmacies are as

Home pregnancy test

sensitive and accurate as test methods used in many hospitals. Still, false-negative and false-positive results can occur. A false-negative result (the test is negative, but the woman is pregnant) may be due to testing too soon or to an ectopic pregnancy. A false-positive result (the test is positive, but the woman is not pregnant) may be due to excess protein or blood in the urine or to hCG produced from a rare type of uterine cancer. Thiazide diuretics, hormones, steroids, and thyroid drugs may also affect the outcome of an early pregnancy test. •

Peak secretion of hCG occurs about the ninth week of pregnancy, at which time the placenta is well developed and can now produce the progesterone and estrogens on its own.

Development of the Bilaminar Embryonic Disc

Cells of the embryoblast also differentiate into two layers around 8 days after fertilization: a **hypoblast (primitive endoderm)** and **epiblast (primitive ectoderm)** (Figure 4.6a). Cells of the hypoblast and epiblast together form a flat disc referred to as the **bilaminar embryonic disc** (bi-LAM-in-ar=two-layered). During further development, the cells of the hypoblast continue as a single squamous (flat) layer of cells, while the epiblast cells become stratified. Soon a small cavity appears within the layers of epiblast cells and eventually enlarges to form the **amniotic cavity** (am-nē-OT-ik; *amnio-*=lamb).

Development of the Amnion

As the amniotic cavity enlarges, a single layer of squamous cells forms a dome-like roof above the epiblast cells called the **amnion** (AM-nē-on) (Figure 4.6a). Thus, the amnion forms the roof of the amniotic cavity and the epiblast forms the floor. Initially, the amnion overlies only the bilaminar embryonic disc. However, as the embryonic disc increases in size and begins to fold, the amnion eventually surrounds the entire embryo (see Figure 4.11a, inset), creating the amniotic cavity that becomes filled with amniotic fluid. Most amniotic fluid is initially derived from maternal blood. Later, the fetus contributes to the fluid by excreting urine into the amniotic cavity. Amniotic fluid serves as a shock absorber for the fetus, helps regulate fetal body temperature, helps prevent the fetus from drying out, and prevents adhesions between the skin of the fetus and surrounding tissues. The amnion usually ruptures just before birth; it and its fluid constitute the "bag of waters." Embryonic cells are normally sloughed off into amniotic fluid. They can be examined in a procedure called *amniocentesis*, which involves withdrawing some of the amniotic fluid that bathes the developing fetus and analyzing the fetal cells and dissolved substances (see Key Medical Terms Associated with Development at the end of this chapter).

Development of the Yolk Sac

Also on the eighth day after fertilization, cells of the hypoblast migrate and cover the inner surface of the blastocyst wall (Figure 4.6a). The migrating cells form a thin membrane called the **exocoelomic membrane** (ek-sō-sē-LŌ-mik; *exo-*=outside; *-koilos*=space). Combined with the hypoblast, the exocoelomic membrane forms the wall of the **yolk sac**, the former blastocyst cavity (Figure 4.6b). As a result, the bilaminar embryonic disc is now positioned between the amniotic cavity and yolk sac.

Since human embryos receive their nutrients from the endometrium, the yolk sac is relatively empty, small, and decreases in size as development progresses (see Figure 4.11a, inset). Nevertheless, the yolk sac has several important functions in humans. It supplies nutrients to the embryo during the second and third weeks of development, is the source of blood cells from the third through sixth weeks, contains the first cells (*primordial germ cells*) that will eventually migrate into the developing gonads and differentiate into gametes, forms part of the gut (gastrointestinal tract), and helps prevent drying out of the embryo.

Figure 4.6 Principal events of the second week of development.

 About 8 days after fertilization, the trophoblast develops into a syncytiotrophoblast and a cytotrophoblast; the embryoblast (inner cell mass) develops into a hypoblast and epiblast (bilaminar embryonic disc).

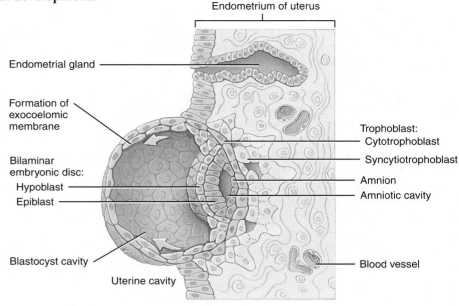

Endometrium of uterus

Endometrial gland

Formation of exocoelomic membrane

Bilaminar embryonic disc:
Hypoblast
Epiblast

Blastocyst cavity

Uterine cavity

Trophoblast:
Cytotrophoblast
Syncytiotrophoblast

Amnion
Amniotic cavity

Blood vessel

(a) Frontal section through endometrium of uterus showing blastocyst, about 8 days after fertilization

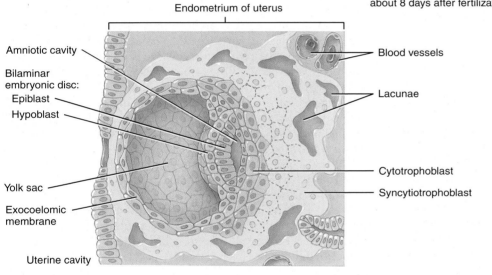

Endometrium of uterus

Amniotic cavity

Bilaminar embryonic disc:
Epiblast
Hypoblast

Yolk sac

Exocoelomic membrane

Uterine cavity

Blood vessels

Lacunae

Cytotrophoblast

Syncytiotrophoblast

(b) Frontal section through endometrium of uterus showing blastocyst, about 9 days after fertilization

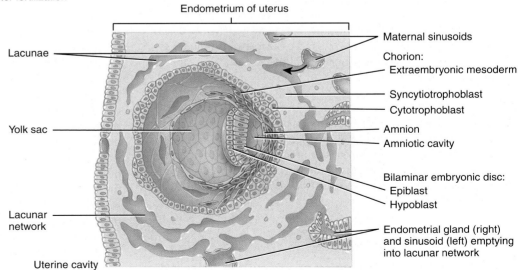

Endometrium of uterus

Lacunae

Yolk sac

Lacunar network

Uterine cavity

Maternal sinusoids

Chorion:
Extraembryonic mesoderm

Syncytiotrophoblast
Cytotrophoblast

Amnion
Amniotic cavity

Bilaminar embryonic disc:
Epiblast
Hypoblast

Endometrial gland (right) and sinusoid (left) emptying into lacunar network

(c) Frontal section through endometrium of uterus showing blastocyst, about 12 days after fertilization

How is the bilaminar embryonic disc connected to the trophoblast?

Development of Sinusoids

On the ninth day after fertilization, the blastocyst becomes completely embedded in the endometrium. As the syncytiotrophoblast expands into the endometrium and around the yolk sac, small spaces called **lacunae** (la-KOO-nē=little lakes) develop within it (Figure 4.6b).

By the twelfth day of development, the lacunae fuse to form larger, interconnecting spaces called **lacunar networks** (Figure 4.6c). Endometrial capillaries (microscopic maternal blood vessels) around the developing embryo expand and are referred to as **maternal sinusoids** (SĪ -ne-soids). As the syncytiotrophoblast erodes some of the sinusoids and endometrial glands, maternal blood and glandular secretions enter the lacunar networks, which serve as both a rich source of materials for embryonic nutrition and a disposal site for the embryo's wastes.

Development of the Extraembryonic Coelom

About the twelfth day after fertilization, the **extraembryonic mesoderm** develops. These mesodermal cells are derived from the yolk sac and form a connective tissue (mesenchyme) around the amnion and yolk sac (Figure 4.6c). Soon, numerous large cavities develop in the extraembryonic mesoderm, which then fuse to form a single, even larger cavity called the **extraembryonic coelom** (SĒ-lom).

Development of the Chorion

The extraembryonic mesoderm and the two layers of the trophoblast (the cytotrophoblast and the syncytiotrophoblast) together form the **chorion** (KOR-ē-on=membrane) (Figure 4.6c). The chorion surrounds the embryo and, later, the fetus (see Figure 4.11a, inset). Eventually the chorion becomes the principal embryonic part of the placenta, the structure for exchange of materials between mother and fetus. The chorion also protects the embryo and fetus from the immune responses of the mother in two ways: (1) It secretes proteins that block antibody production by the mother; and (2) it promotes the production of T lymphocytes that suppress the normal immune response in the uterus. Finally, the chorion produces human chorionic gonadotropin (hCG), an important hormone of pregnancy.

The inner layer of the chorion eventually fuses with the amnion. With the development of the chorion, the extraembryonic coelom is now referred to as the **chorionic cavity**. By the end of the second week of development, the bilaminar embryonic disc is connected to the trophoblast by a band of extraembryonic mesoderm called the **connecting (body) stalk** (see Figure 4.7, inset), the future umbilical cord.

 CHECKPOINT

6. What are the functions of the trophoblast?
7. How is the bilaminar embryonic disc formed?
8. Describe the formation of the amnion, yolk sac, and chorion and explain their functions.
9. Why are sinusoids important during embryonic development?

Third Week of Development

The third embryonic week begins a six-week period of rapid development and differentiation. During the third week, the three primary germ layers are established, which lays the groundwork for organ development in weeks four through eight.

Gastrulation

The first major event of the third week of development, **gastrulation** (gas-troo-LĀ-shun), occurs about 15 days after fertilization. In this process, the bilaminar (two-layered) embryonic disc, consisting of epiblast and hypoblast, is transformed into a **trilaminar** (three-layered) **embryonic disc** consisting of three primary germ layers: the ectoderm, mesoderm, and endoderm. The **primary germ layers** are the major embryonic tissues from which the various tissues and organs of the body develop.

Gastrulation involves the well-coordinated and important rearrangement and migration of cells from the epiblast to set the stage for the important interactions of the newly positioned cells. The first evidence of gastrulation is the formation of the **primitive streak**, a faint groove on the dorsal surface of the epiblast that elongates from the posterior to the anterior part of the embryo (Figure 4.7a). The primitive streak clearly establishes the head and tail ends of the embryo, as well as its right and left sides. At the head end of the primitive streak a small group of epiblast cells forms a rounded structure called the **primitive node**.

Following formation of the primitive streak, cells of the epiblast move inward below the primitive streak and detach from the epiblast (Figure 4.7b). Once the cells begin to migrate beneath the overlying epiblast, some of them displace the original hypoblast by pushing it laterally from beneath the epiblast and completely replace it with a new layer of cells. This new layer of cells forming the roof of the yolk sac is the **endoderm** (*endo-*=inside; *-derm*=skin). Other migrating cells remain between the epiblast and newly formed endoderm to form the **mesoderm** (*meso-*=middle). Cell placement and differentiation of cells within this layer are crucial to inducing further development of the final body plan. Cells remaining in the epiblast then form the **ectoderm** (*ecto-*=outside).

Table 4.1 provides details about the fates of these primary germ layers; coverage is also included in later chapters in the context of the various body systems.

About 16 days after fertilization, mesodermal cells from the primitive node migrate toward the head end of the embryo and form a hollow tube of cells in the midline called the **notochordal process** (nō-tō-KOR-dal) (Figure 4.8). By days 22–24, the notochordal process becomes a solid cylinder of cells called the **notochord** (nō-tō-KORD; *noto-*=back; *-chord-*=cord) (see Figure 4.9a). This structure plays an extremely important role in **induction** (in-DUK-shun), a process in which one tissue (*inducing tissue*) stimulates the specialization of an adjacent tissue (*responding tissue*). An inducing tissue usually produces a chemical substance that influences the responding tissue. The notochord induces certain neighboring mesodermal cells to develop into parts of vertebrae (back bones), induces the overlying ectoderm to fold inward to form the nervous system, and contributes to the formation of intervertebral discs between vertebrae (see Figure 7.15).

Also during the third week of development, two faint depressions appear on the dorsal surface of the embryo where the ectoderm and endoderm make contact but lack mesoderm between them. The structure closer to the head end is called the **oropharyngeal membrane** (or-ō-fa-RIN-jē-al; *oro-*=mouth; *-pharyngeal*=pertaining to the pharynx) (Figure 4.8a, b). It breaks down during the fourth week to connect the mouth cavity to the pharynx (throat) and the remainder of the gastrointestinal tract. The structure closer to the tail end, called the **cloacal membrane** (klō-Ā-kul=sewer), degenerates in the seventh week to form the openings of the anus and urinary and reproductive tracts.

Figure 4.7 Gastrulation. The most important event during the third week of development is gastrulation.

 Gastrulation involves the rearrangement and migration of cells from the epiblast.

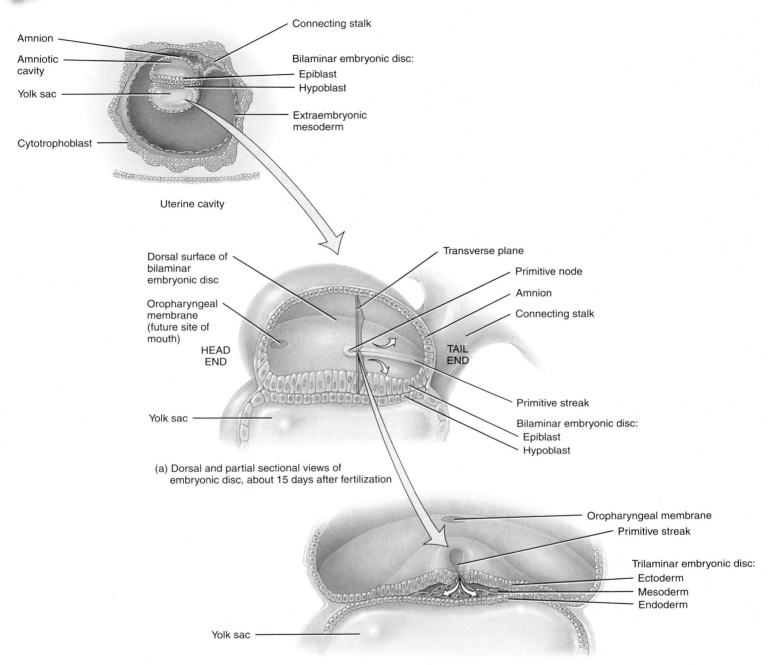

(a) Dorsal and partial sectional views of embryonic disc, about 15 days after fertilization

(b) Transverse section of trilaminar embryonic disc, about 16 days after fertilization

? **What are the events of gastrulation?**

TABLE 4.1

Structures Produced by the Three Primary Germ Layers

ENDODERM	MESODERM	ECTODERM
Epithelial lining of gastrointestinal tract (except the oral cavity and anal canal) and the epithelium of its glands	All skeletal and cardiac muscle tissue and most smooth muscle tissue	All nervous tissue
Epithelial lining of urinary bladder, gallbladder, and liver	Most cartilage, bone, and other connective tissues	Epidermis of skin
Epithelial lining of pharynx, auditory (eustachian) tubes, tonsils, tympanic (middle ear) cavity, larynx, trachea, bronchi, and lungs	Blood, red bone marrow, and lymphatic tissue	Hair follicles, arrector pili muscles, nails, epithelium of skin glands (sebaceous and sudoriferous), and mammary glands
Epithelium of thyroid gland, parathyroid glands, pancreas, and thymus	Blood vessels and lymphatic vessels	Lens, cornea, and internal eye muscles
	Dermis of skin	Internal and external ear
Epithelial lining of prostate and bulbourethral (Cowper's) glands, vagina, vestibule, urethra, and associated glands such as the greater vestibular (Bartholin's) and lesser vestibular glands	Fibrous tunic and vascular tunic of eye	Neuroepithelium of sense organs
	Mesothelium of thoracic, abdominal, and pelvic cavities	Epithelium of oral cavity, nasal cavity, paranasal sinuses, salivary glands, and anal canal
	Kidneys and ureters	Epithelium of pineal gland, pituitary gland, and adrenal medullae
	Adrenal cortex	Melanocytes (pigment cells)
	Gonads and genital ducts (except germ cells)	Almost all skeletal and connective tissue components of the head
Gametes (sperm and oocytes)	Dura mater	Arachnoid mater and pia mater

When the cloacal membrane appears, the wall of the yolk sac forms a small vascularized outpouching called the **allantois** (a-LAN-tō-is; *allant-*=sausage) that extends into the connecting stalk (Figure 4.8b). In nonmammalian organisms enclosed in an amnion (reptiles and birds), the allantois is used for gas exchange and waste removal. Because the mammalian placenta performs these functions, the allantois is not a prominent structure in most mammals (see Figure 4.11a, inset). Nevertheless, it does function in early formation of blood and blood vessels and it is associated with the development of the urinary bladder.

Neurulation

As mentioned previously, in addition to inducing mesodermal cells to develop into parts of vertebrae, the notochord also in-duces ectodermal cells above it to form the **neural plate** (Figure 4.9a). By the end of the third week the neural plate begins the process of **invagination** (in-vaj-in-NĀ-shun), an infolding of the ectoderm neural plate cells into the underlying mesoderm. During this process, the lateral edges of the neural plate become more elevated and form the **neural folds** (Figure 4.9b). The depressed region or groove between the neural folds is called the **neural groove** (Figure 4.9c). As the fold deepens, the neural folds approach each other and fuse, closing the neural groove and converting the neural plate into a **neural tube** that is pushed beneath the surface ectoderm into the underlying mesoderm (Figure 4.9d). Neural tube cells then develop into the brain and spinal cord. The process of formation of the neural plate, neural folds, and neural tube is called **neurulation** (noor-oo-LĀ-shun).

Figure 4.8 Development of the notochordal process.

 The notochordal process develops from the primitive node and later becomes the notochord.

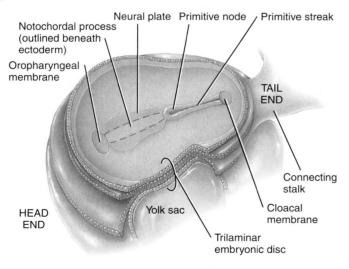
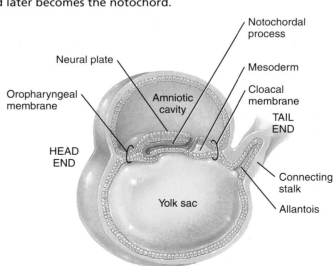

(a) Dorsal and partial sectional views of trilaminar embryonic disc, about 16 days after fertilization

(b) Sagittal section of trilaminar embryonic disc, about 16 days after fertilization

 What is the significance of the notochord?

As the neural tube forms, some of the ectodermal cells migrate dorsolaterally to form several layers of cells called the **neural crest**. This population of cells is more extensive in the head end of the embryo where it forms a large mass of cells dorsolateral to the neural tube. Neural crest cells form all the sensory neurons and postganglionic motor neurons of the peripheral nerves, adrenal medullae, melanocytes (pigment cells) of the skin, the arachnoid mater and pia mater of the brain and spinal cord, and almost all of the skeletal and connective tissue components of the head.

Approximately four weeks after fertilization, the head end of the neural tube develops into three enlarged areas called **primary brain vesicles** (see Figure 18.1). The parts of the brain that develop from the various brain vesicles are described in Section 18.1.

Development of Somites

By about the seventeenth day after fertilization, the mesoderm adjacent to the notochord and neural tube forms paired longitudinal columns of **paraxial mesoderm** (par-AK-sē-al; *para-*=near) (Figure 4.9b). The mesoderm lateral to the paraxial mesoderm forms paired cylindrical masses called **intermediate mesoderm**. The mesoderm lateral to the intermediate mesoderm consists of a pair of flattened sheets called **lateral plate mesoderm**. The paraxial mesoderm soon segments into a series of paired, cube-shaped structures called **somites** (SŌ-mīts=little bodies) (Figure 4.9c, d). By the end of the fifth week, 42–44 pairs of somites are present. The number of somites that develop over a given period can be correlated to the approximate age of the embryo.

Figure 4.9 Neurulation and the development of somites.

🔑 Neurulation is the process by which the neural plate, neural folds, and neural tube form.

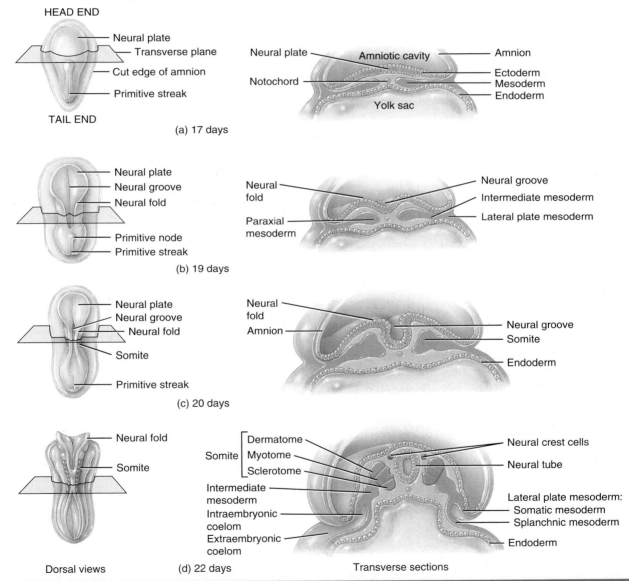

Dorsal views   (d) 22 days   Transverse sections

CLINICAL CONNECTION | *Neural Tube Defects*

Neural tube defects (NTDs) are caused by arrest of the normal development and closure of the neural tube. These include spina bifida (discussed in Exhibit 7.L) and anencephaly. In **anencephaly** (an'-en-SEPH-a-lē; *an-*=without; *encephal*=brain), the cranial bones fail to develop and certain parts of the brain remain in contact with amniotic fluid and degenerate. Usually, a part of the brain that controls vital functions such as breathing and regulation of the heart is also affected. Infants with anencephaly are stillborn or die within a few days after birth. The condition occurs about once in every 1000 births and is 2 to 4 times more common in female infants than males. •

❓ **Which structures develop from the neural tube and somites?**

Each somite differentiates into three regions: a **myotome** (MĪ-o-tōm), a **dermatome** (DERM-a-tōm), and a **sclerotome** (SKLE-ro-tōm) (see Figure 10.11b, c). The myotomes develop into the skeletal muscles of the neck, trunk, and limbs; the dermatomes form connective tissue, including the dermis of the skin; and the sclerotomes give rise to the vertebrae and ribs.

Development of the Intraembryonic Coelom

In the third week of development, small spaces appear in the lateral plate mesoderm. These spaces soon merge to form a larger cavity called the **intraembryonic coelom** (SĒ-lōm=cavity). This cavity splits the lateral plate mesoderm into two parts called the splanchnic mesoderm and somatic mesoderm (Figure 4.9d). **Splanchnic mesoderm** (SPLANGK-nik=visceral), which is adjacent to the endoderm and yolk sac, forms the heart and the visceral layer of the serous pericardium, blood vessels, the smooth muscle and connective tissues of the respiratory and digestive organs, and the visceral layer of the serous membrane of the pleurae and peritoneum. **Somatic mesoderm** (sō-MAT-ik; *soma*-=body), which is adjacent to the ectoderm and amnion, gives rise to the bones, ligaments, blood vessels, and connective tissue of the limbs and the parietal layer of the serous membrane of the pericardium, pleurae, and peritoneum. During the second month of development, the intraembryonic coelom is partitioned into the pericardial, pleural, and peritoneal cavities.

Development of the Cardiovascular System

At the beginning of the third week, **angiogenesis** (an-jē-ō-JEN-e-sis; *angio*-=vessel; -*genesis*=production), the formation of blood vessels, begins in the extraembryonic mesoderm of the yolk sac, connecting stalk, and chorion. This early development is necessary because there is insufficient yolk in the yolk sac and ovum to provide adequate nutrition for the rapidly developing embryo. Angiogenesis is initiated when mesodermal cells differentiate into **hemangioblasts** (hē-MAN-jē-ō-blasts). These then develop into cells called **angioblasts**, which aggregate to form isolated masses of cells referred to as **blood islands** (see Figure 14.18). As the blood islands throughout the embryonic mesoderm grow they fuse together, forming an extensive system of blood vessels within the embryo.

About 3 weeks after fertilization, blood cells and blood plasma begin to develop *outside* the embryo—in the walls of the yolk sac, allantois, and chorion—from hemangioblasts in blood vessels. These then develop into pluripotent stem cells that form blood cells. Blood formation begins *within* the embryo at about the fifth week in the liver and around the twelfth week in the spleen, red bone marrow, and thymus.

The heart forms from splanchnic mesoderm in the head end of the embryo on day 18 or 19 after fertilization. This region of mesodermal cells is called the **cardiogenic area** (kar-dē-ō-JEN-ik; *cardio*-=heart; -*genic*=producing). In response to induction signals from the underlying endoderm, these mesodermal cells ultimately form a pair of **endocardial tubes** (see Figure 13.12b, c). The tubes then fuse to form a single **primitive heart tube**. By the end of the third week, the primitive heart tube bends on itself, becomes S-shaped, and begins to beat. It then joins blood vessels in other parts of the embryo, connecting stalk, chorion, and yolk sac to form a primitive cardiovascular system.

Development of the Chorionic Villi and Placenta

As the embryonic tissue invades the uterine wall, maternal uterine vessels are eroded and maternal blood fills spaces, called **lacunae** (la-KOO-nē), within the invading tissue. By the

end of the second week of development, **chorionic villi** (ko-rē-ON-ik VIL-ī) begin to develop. These fingerlike projections consist of chorion (syncytiotrophoblast surrounded by cytotrophoblast) that project into the endometrial wall of the uterus (Figure 4.10a). By the end of the third week, blood capillaries develop in the chorionic villi (Figure 4.10b). Blood vessels in

Figure 4.10 Development of chorionic villi.

 Blood vessels in chorionic villi connect to the embryonic heart via the umbilical arteries and umbilical vein.

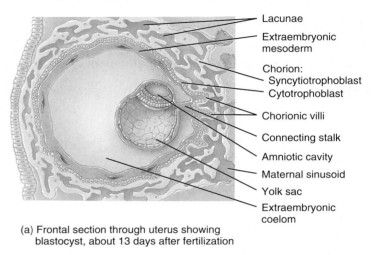

(a) Frontal section through uterus showing blastocyst, about 13 days after fertilization

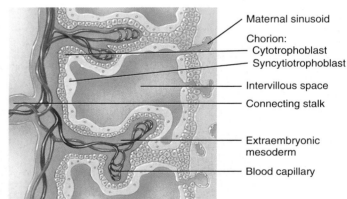

(b) Details of two chorionic villi, about 21 days after fertilization

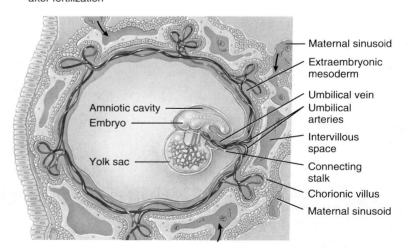

(c) Frontal section through uterus showing an embryo and its vascular supply, about 21 days after fertilization

? **Why is development of chorionic villi important?**

the chorionic villi connect to the embryonic heart by way of the umbilical arteries and umbilical vein through the connecting (body) stalk, which will eventually become the umbilical cord (Figure 4.10c). The fetal blood capillaries within the chorionic villi project into the lacunae, which unite to form the **intervillous spaces** (in′-ter-VIL-us) where the chorionic villi (and the fetal blood vessels within them) are bathed in maternal blood. However maternal blood and fetal blood do not mix directly. Instead, oxygen and nutrients in the mother's blood diffuse from the intervillous spaces across the plasma membranes of the chorion and the capillaries of the villi. Waste products such as carbon dioxide diffuse in the opposite direction.

Placentation (plas-en-TĀ-shun) is the process of forming the **placenta** (pla-SEN-ta=flat cake), the site of exchange of nutrients and wastes between the mother and fetus (Figure 4.11). The placenta also produces hormones needed to sustain the pregnancy. The placenta is unique because it develops from the tissues of two separate individuals, the mother and the fetus.

By the beginning of the twelfth week, the placenta has two distinct parts: (1) the fetal portion formed by the chorionic villi of the chorion and (2) the maternal portion formed by the maternal blood in the intervillous spaces and the decidua basalis of the endometrium (Figure 4.11a). When fully developed, the placenta is a rounded, disc-like structure (Figure 4.11b).

Functionally, the placenta allows oxygen and nutrients to diffuse from maternal blood into fetal blood while carbon dioxide and wastes diffuse from fetal blood into maternal blood. The placenta also serves as a protective barrier, because most microorganisms cannot pass through it. However, certain viruses, such as those that cause AIDS, German measles, chickenpox, measles, encephalitis, and poliomyelitis, can cross the placenta, as can many drugs, alcohol, and some substances that can cause birth defects. The placenta stores nutrients such as carbohydrates, proteins, calcium, and iron, which are released into fetal circulation as required.

The actual connection between the placenta and embryo, and later the fetus, is through the **umbilical cord** (um-BIL-i-kul=navel),

Figure 4.11 Placenta and umbilical cord.

🔑 The placenta is formed by the chorionic villi of the embryo and the decidua basalis of the endometrium of the mother.

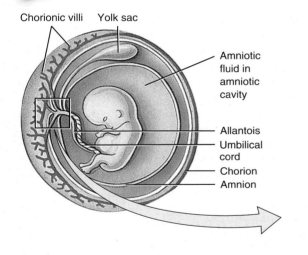

Labels: Chorionic villi, Yolk sac, Amniotic fluid in amniotic cavity, Allantois, Umbilical cord, Chorion, Amnion

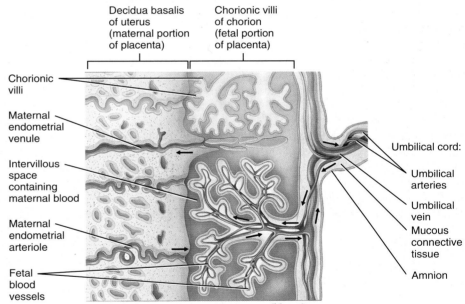

Labels: Decidua basalis of uterus (maternal portion of placenta), Chorionic villi of chorion (fetal portion of placenta), Chorionic villi, Maternal endometrial venule, Intervillous space containing maternal blood, Maternal endometrial arteriole, Fetal blood vessels, Umbilical cord:, Umbilical arteries, Umbilical vein, Mucous connective tissue, Amnion

K. Somerville

(a) Details of placenta and umbilical cord

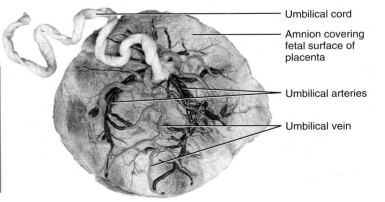

Labels: Umbilical cord, Amnion covering fetal surface of placenta, Umbilical arteries, Umbilical vein

(b) Fetal surface of placenta

 What is the function of the placenta?

which develops from the connecting stalk and is usually about 2 cm (1 in.) wide and about 50–60 cm (20–24 in.) in length. The umbilical cord consists of two umbilical arteries that carry deoxygenated fetal blood to the placenta, one umbilical vein that carries oxygen and nutrients acquired from the mother's intervillous spaces into the fetus, and supporting mucous connective tissue called Wharton's jelly (WOR-tons) derived from the allantois. A layer of amnion surrounds the entire umbilical cord and gives it a shiny appearance (Figure 4.11b). In some cases, the umbilical vein is used to transfuse blood into a fetus or to introduce drugs for various medical treatments.

In about 1 in 200 newborns, only one of the two umbilical arteries is present in the umbilical cord. It may be due to failure of the artery to develop or degeneration of the vessel early in development. Nearly 20 percent of infants with this condition develop cardiovascular defects.

After the birth of the baby, the placenta detaches from the uterus and is then called the **afterbirth**. At this time, the umbilical cord is tied off and then severed. The small portion (about an inch) of the cord that remains attached to the infant begins to wither and falls off, usually within 12 to 15 days after birth. The area where the cord was attached becomes covered by a thin layer of skin, and scar tissue forms. The scar is the **umbilicus** (um-BIL-i-kus) or navel.

 CHECKPOINT

10. What is the significance of gastrulation?
11. How do the three primary germ layers form? Why are they important?
12. What is the function of the allantois?
13. Describe how neurulation occurs. Why is it important?
14. What are the functions of somites?
15. How do blood vessels and the heart develop?
16. Outline the process of placenta formation and explain the importance of this structure.

Fourth Week of Development

The fourth through eighth weeks are very significant in embryonic development because all major organs appear during this time. The term **organogenesis** (or′-ga-nō-JEN-e-sis) refers to the formation of body organs and systems. By the end of the eighth week, all major body systems have begun to develop,

although their functions for the most part are minimal. Organogenesis requires the presence of blood vessels to supply developing organs with oxygen and other nutrients. However, recent studies suggest that blood vessels play a significant role in organogenesis even before blood begins to flow within them. The endothelial cells of blood vessels apparently provide some type of developmental signal, either a secreted substance or a direct cell-to-cell interaction, that is necessary for organogenesis.

During the fourth week of development, the embryo undergoes dramatic changes in shape and nearly triples its size. It is essentially converted from a flat, two-dimensional trilaminar embryonic disc to a three-dimensional cylinder, via a process called **embryonic folding** (Figure 4.12). The cylinder consists of endoderm in the center (gut lining), ectoderm primarily on the outside (epidermis), and mesoderm in between. The cylinder does contain some internal ectoderm from neurulation (the nervous system and most of the structures of the head). The main force responsible for embryonic folding is the different rates of growth of various parts of the embryo, especially the rapid longitudinal growth of the nervous system (neural tube). Folding in the median plane produces a **head fold** and a **tail fold**, while folding in the horizontal plane results in the two **lateral folds**. As a result of the folding, the embryo curves into a C-shape.

The head fold brings the developing heart and mouth into their eventual adult positions. The tail fold brings the developing anus into its eventual adult position. The lateral folds form as the lateral margins of the trilaminar embryonic disc bend ventrally, curving toward the open part of the C. As they move toward the midline, the lateral folds incorporate the dorsal part of the yolk sac into the embryo as the **primitive gut**, the forerunner of the gastrointestinal tract (Figure 4.12b). The primitive gut differentiates into an anterior **foregut**, an intermediate **midgut**, and a posterior **hindgut** (Figure 4.12c). The fates of the foregut, midgut, and hindgut are described in Section 24.12.

Recall that the oropharyngeal membrane is located in the head end of the embryo (see Figure 4.8). It separates the future pharyngeal (throat) region of the foregut from the **stomodeum** (stō-mō-DĒ-um; *stomo-*=mouth), the future oral (mouth) cavity. Because of head folding, the oropharyngeal membrane moves downward and the foregut and stomodeum move closer to their final positions. When the oropharyngeal membrane ruptures during the fourth week, the pharyngeal region of the pharynx and stomodeum are brought into contact with each other.

In a developing embryo, the last part of the hindgut expands into a cavity called the **cloaca** (klō-Ā-ka) (see Figure 25.13a). On the outside of an embryo is a small cavity in the tail region called the **proctodeum** (prok′-tō-DĒ-um; *procto-*=anus) (Figure 4.12c). The **cloacal membrane** separates the cloaca from the proctodeum (see Figure 4.8). During embryonic development, the cloaca divides into a ventral urogenital sinus and a dorsal anorectal canal. As a result of tail folding, the cloacal membrane moves downward and the urogenital sinus, anorectal canal, and proctodeum move closer to their final positions. When the cloacal membrane ruptures during the seventh week of development, the urogenital and anal openings are created.

Figure 4.12 Embryonic folding.

 Embryonic folding converts the two-dimensional trilaminar embryonic disc into a three-dimensional cylinder.

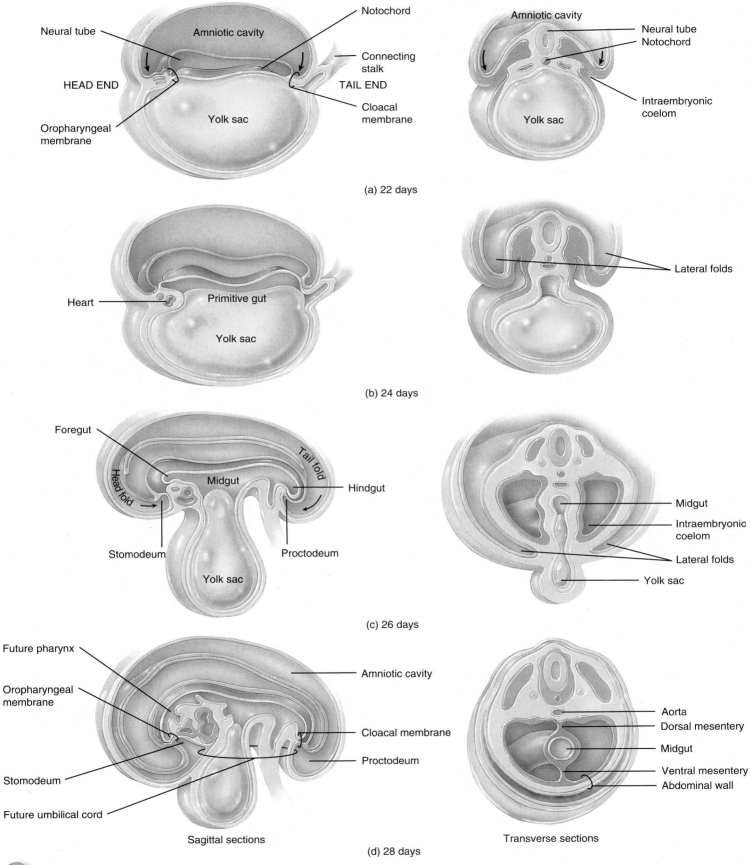

(a) 22 days

(b) 24 days

(c) 26 days

Sagittal sections

Transverse sections

(d) 28 days

What are the results of embryonic folding?

In addition to embryonic folding, development of somites, and development of the neural tube, five pairs of **pharyngeal arches** (fa-RIN-jē-al) or **branchial arches** (BRAN-kē-al; *branch*=gill) begin to develop on each side of the future head and neck regions during the fourth week (Figure 4.13). These five paired structures begin their development on the twenty-second day after fertilization and form swellings on the surface of the embryo. Each pharyngeal arch consists of an outer covering of ectoderm and an inner covering of endoderm, with mesoderm in between. Within each pharyngeal arch there is an artery, a cranial nerve, skeletal cartilaginous rods that support the arch, and skeletal muscle tissue that attaches to and moves the cartilage rods. On the ectodermal surface of the pharyngeal region, each pharyngeal arch is separated by a groove called a **pharyngeal cleft** (Figure 4.13a). The pharyngeal clefts meet corresponding balloonlike outgrowths of the endodermal pharyngeal lining called **pharyngeal (branchial) pouches**. Where the pharyngeal cleft and pouch meet to separate the arches, the outer ectoderm of the cleft contacts the inner endoderm of the pouch and there is no mesoderm between (Figure 4.13b).

Just as the somite gives rise to specified structures in the body wall, each pharyngeal arch, cleft, and pouch gives rise to specified structures in the head and neck. Each pharyngeal arch is a developmental unit and includes a skeletal component, muscles, a distinct cranial nerve, and blood vessels.

In the human embryo, there are four obvious pharyngeal arches and two less distinct arches. Each of these arches develops into a specific and unique component of the head and neck region. For example, the first pharyngeal arch is often called the *mandibular arch* because it forms the jaws (the *mandible* is the lower jawbone).

The first sign of a developing ear is a thickened area of ectoderm, the **otic placode** (PLAK-ōd) (future internal ear), which can be distinguished about 22 days after fertilization (see Figure 4.13a). A thickened area of ectoderm called the **lens placode** (see Figure 4.13a), which will become the eye, also appears at this time.

By the middle of the fourth week, the upper limbs begin their development as ectoderm-covered outgrowths called **upper limb buds** (see Figure 8.13a). By the end of the fourth week, the **lower limb buds** develop. The heart also forms a distinct projection on the ventral surface of the embryo called the **heart prominence**, just caudal to the pharyngeal arches (see Figure 8.13b), and a **tail** becomes visible (see Figure 8.13a).

Fifth Through Eighth Weeks of Development

During the fifth week of development, growth of the head is considerable because of very rapid development of the brain. By the end of the sixth week, the head has grown even larger relative to the trunk, and the limbs show substantial development (see Figure 8.13b). In addition, the neck and trunk begin to straighten, and the heart is now four-chambered. By the seventh week, the various regions of the limbs become distinct and the beginnings of digits appear (see Figure 8.13c). At the start of the eighth week, the final week of the embryonic period, the digits of the hands are short and webbed and the tail is still visible, but shorter. In addition, the eyes are open and the auricles of the ears are visible (see Figure 8.13b). By the end of the eighth week, all regions of limbs are apparent; the digits are distinct and no longer webbed due to removal of cells via apoptosis (see Section 2.5). Also, the eyelids come together and may fuse, the tail shortens and becomes imperceptible, and the external genitals begin to differentiate. The embryo now has clearly human characteristics.

Figure 4.13 Development of pharyngeal arches, pharyngeal clefts, and pharyngeal pouches.

 The five pairs of pharyngeal pouches consist of ectoderm, mesoderm, and endoderm and contain blood vessels, cranial nerves, cartilage, and muscular tissue.

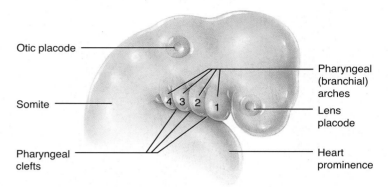

(a) External view, about 28-day embryo

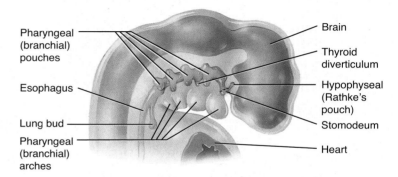

(b) Sagittal section, about 28-day embryo

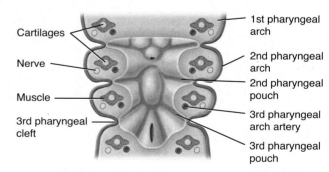

(c) Transverse section of the pharynx, about 28-day embryo

 What is the importance of pharyngeal arches, clefts, and pouches?

CHECKPOINT

17. How does embryonic folding occur?
18. How does the primitive gut form and what is its significance?
19. What is the origin of the structures of the head and neck?
20. What are limb buds?
21. What changes occur in the limbs during the second half of the embryonic period?

4.2 FETAL PERIOD

OBJECTIVE

• **Describe** the major events of the fetal period.

During the fetal period, tissues and organs that developed during the embryonic period grow and differentiate. Very few new structures appear during the fetal period, but the rate of body growth is remarkable, especially during the second half of intrauterine life. For example, during the last two-and-one-half months of intrauterine life, the weight of the fetus doubles. At the beginning of the fetal period, the head is half the length of the body. By the end of the fetal period, the head size is only one-quarter the length of the body. During the same period, the limbs also increase in size from one-eighth to one-half the fetal length. During the fetal period, the fetus is less vulnerable to the damaging effects of drugs, radiation, and microbes than it was as an embryo.

A summary of the major developmental events of the embryonic and fetal period is illustrated in Figure 4.14 and presented in Table 4.2.

Figure 4.14 Summary of representative developmental events of the embryonic and fetal periods. The embryos and fetuses are not shown at their actual sizes.

Development during the fetal period is mostly concerned with the growth and differentiation of tissues and organs formed during the embryonic period.

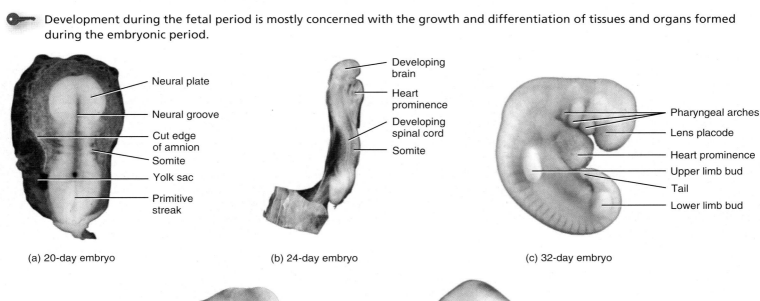

(a) 20-day embryo

Neural plate
Neural groove
Cut edge of amnion
Somite
Yolk sac
Primitive streak

(b) 24-day embryo

Developing brain
Heart prominence
Developing spinal cord
Somite

(c) 32-day embryo

Pharyngeal arches
Lens placode
Heart prominence
Upper limb bud
Tail
Lower limb bud

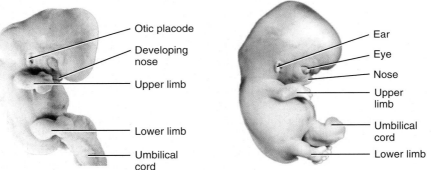

(d) 44-day embryo

Otic placode
Developing nose
Upper limb
Lower limb
Umbilical cord

(e) 52-day embryo

Ear
Eye
Nose
Upper limb
Umbilical cord
Lower limb

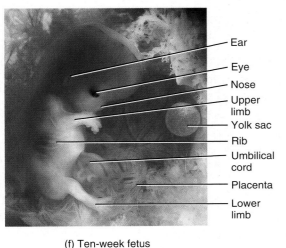

(f) Ten-week fetus

Ear
Eye
Nose
Upper limb
Yolk sac
Rib
Umbilical cord
Placenta
Lower limb

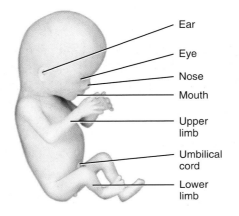

(g) Thirteen-week fetus

Ear
Eye
Nose
Mouth
Upper limb
Umbilical cord
Lower limb

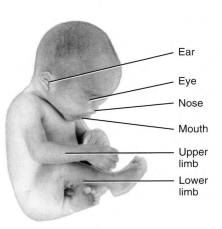

(h) Twenty-six-week fetus

Ear
Eye
Nose
Mouth
Upper limb
Lower limb

How much weight is gained by a fetus during the last two-and-a-half months of intrauterine life?

TABLE 4.2

Summary of Changes During Embryonic and Fetal Development

TIME	APPROXIMATE SIZE AND WEIGHT	REPRESENTATIVE CHANGES
EMBRYONIC PERIOD		
1–4 weeks	0.6 cm (3/16 in.)	Primary germ layers and notochord develop. Neurulation occurs. Primary brain vesicles, somites, and intraembryonic coelom develop. Blood vessel formation begins and blood forms in yolk sac, allantois, and chorion. Heart forms and begins to beat. Chorionic villi develop and placental formation begins. The embryo folds. The primitive gut, pharyngeal arches, and limb buds develop. Eyes and ears begin to develop, tail forms, and body systems begin to form.
5–8 weeks	3 cm (1.25 in.) 1 g (1/30 oz)	Limbs become distinct and digits appear. Heart becomes four-chambered. Eyes are far apart and eyelids are fused. Nose develops and is flat. Face is more humanlike. Bone formation begins. Blood cells start to form in liver. External genitals begin to differentiate. Tail disappears. Major blood vessels form. Many internal organs continue to develop.
FETAL PERIOD		
9–12 weeks	7.5 cm (3 in.) 30 g (1 oz)	Head constitutes about half the length of the fetal body, and fetal length nearly doubles. Brain continues to enlarge. Face is broad, with eyes fully developed, closed, and widely separated. Nose develops a bridge. External ears develop and are low set. Bone formation continues. Upper limbs almost reach final relative length but lower limbs are not quite as well developed. Heartbeat can be detected. Gender is distinguishable from external genitals. Urine secreted by fetus is added to amniotic fluid. Red bone marrow, thymus, and spleen participate in blood cell formation. Fetus begins to move, but its movements cannot be felt yet by the mother. Body systems continue to develop.
13–16 weeks	18 cm (6.5–7 in.) 100 g (4 oz)	Head is relatively smaller than rest of body. Eyes move medially to their final positions, and ears move to their final positions on the sides of the head. Lower limbs lengthen. Fetus appears even more humanlike. Rapid development of body systems occurs.
17–20 weeks	25–30 cm (10–12 in.) 200–450 g (0.5–1 lb)	Head is more proportionate to rest of body. Eyebrows and head hair are visible. Growth slows but lower limbs continue to lengthen. Vernix caseosa (fatty secretions of oil glands and dead epithelial cells) and lanugo (delicate fetal hair) cover fetus. Brown fat forms and is the site of heat production. Fetal movements are commonly felt by mother (quickening).
21–25 weeks	27–35 cm (11–14 in.) 550–800 g (1.25–1.5 lb)	Head becomes even more proportionate to rest of body. Weight gain is substantial, and skin is pink and wrinkled. Fetuses 24 weeks and older usually survive if born prematurely.
26–29 weeks	32–42 cm (13–17 in.) 1100–1350 g (2.5–3 lb)	Head and body are more proportionate and eyes are open. Toenails are visible. Body fat is 3.5 percent of total body mass and additional subcutaneous fat smoothes out some wrinkles. Testes begin to descend toward scrotum at 28 to 32 weeks. Red bone marrow is major site of blood cell production. Many fetuses born prematurely during this period survive if given intensive care because lungs can provide adequate ventilation, and central nervous system is developed enough to control breathing and body temperature.
30–34 weeks	41–45 cm (16.5–18 in.) 2000–2300 g (4.5–5 lb)	Skin is pink and smooth. Fetus assumes upside-down position. Body fat is 8 percent of total body mass.
35–38 weeks	50 cm (20 in.) 3200–3400 g (7–7.5 lb)	By 38 weeks, circumference of fetal abdomen is greater than that of head. Skin is usually bluish-pink, and growth slows as birth approaches. Body fat is 16 percent of total body mass. Testes are usually in scrotum in full-term male infants. Even after birth, an infant is not completely developed; an additional year is required, especially for complete development of the nervous system.

Throughout the text we will discuss the developmental biology of the various body systems in their respective chapters. The following list of these sections is presented for your reference.

Integumentary System (Section 5.6)

Skeletal System (Section 8.6)

Muscular System (Section 10.7)

Heart (Section 13.8)

Blood and Blood Vessels (Section 14.3)

Lymphatic System (Section 15.3)

Nervous System (Section 18.1)

Eyes and Ears (Section 21.5)

Endocrine System (Section 22.10)

Respiratory System (Section 23.5)

Digestive System (Section 24.12)

Urinary System (Section 25.5)

Reproductive Systems (Section 26.5)

 CHECKPOINT

22. What are the general developmental trends during the fetal period?
23. Using Table 4.2 as a guide, select any one body structure in weeks 9 through 12 and trace its development through the remainder of the fetal period.

CLINICAL CONNECTION | *Premature Infant*

Delivery of a physiologically immature baby carries certain risks. A **premature infant** or "preemie" is generally considered a baby who weighs less than 2500 g (5.5 lb) at birth. Poor prenatal care, drug abuse, history of a previous premature delivery, and mother's age below 16 or above 35 increase the chance of premature delivery. The body of a premature infant is not yet ready to sustain some critical functions, and thus its survival is uncertain without medical intervention. The major problem after delivery of an infant under 36 weeks gestation is respiratory distress syndrome (RDS) of the newborn due to insufficient *surfactant* (a mixture of phospholipids and lipoproteins produced by lung cells that decreases surface tension and reduces the tendency of air sacs to collapse). RDS can be eased by use of artificial surfactant and a ventilator that delivers oxygen until the lungs can operate on their own. •

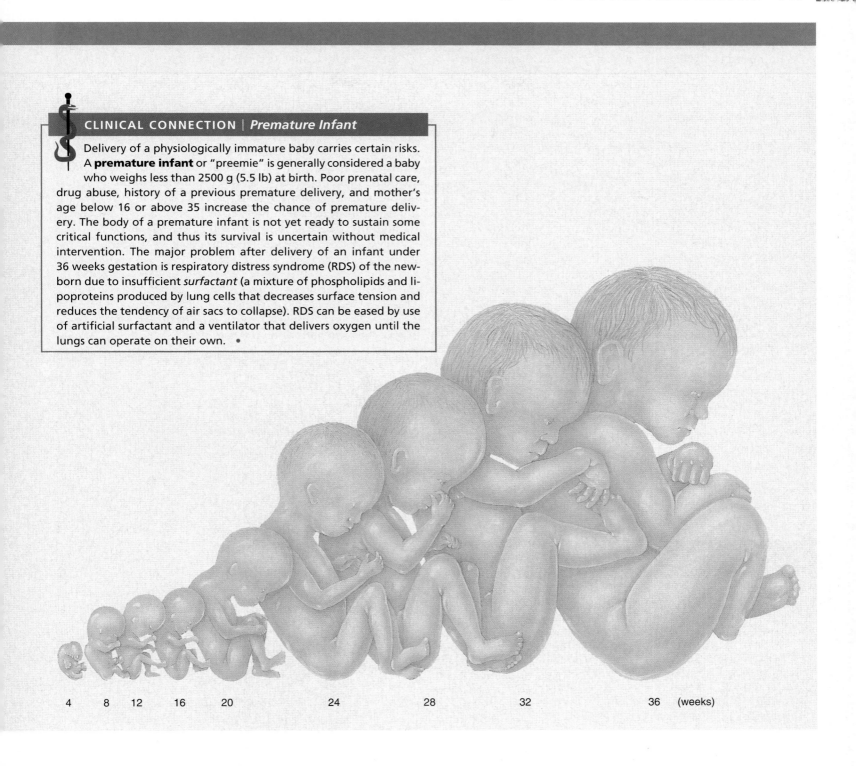

4   8   12   16   20   24   28   32   36   (weeks)

4.3 MATERNAL CHANGES DURING PREGNANCY

OBJECTIVE

• Describe the effects of pregnancy on various body systems.

Near the end of the third month of pregnancy, the uterus occupies most of the pelvic cavity. As the fetus continues to grow, the uterus extends higher into the abdominal cavity. Toward the end of a full-term pregnancy, the uterus fills almost the entire abdominal cavity. It pushes the maternal intestines, liver, and stomach superiorly, elevates the diaphragm, and widens the thoracic cavity. In the pelvic cavity, compression of the ureters and urinary bladder occurs, as well as pressure on the veins returning blood from the pelvis and lower limbs.

Besides the anatomical changes associated with pregnancy, pregnancy-induced physiological changes also occur, including weight gain due to the fetus, amniotic fluid, the placenta, uterine enlargement, and increased total body water; increased storage of nutrients; marked breast enlargement in preparation for milk secretion and ejection; and lower back pain due to *lordosis* (hollow back or sway back; see Section 7.4).

Several changes occur in the maternal cardiovascular system. For example, heart rate increases and blood volume increases, mostly during the second half of pregnancy, to meet the additional demands of the fetus for nutrients and oxygen. The total volume of air inhaled and exhaled per minute also increases to meet the added oxygen demands.

With regard to the digestive system, pregnant women experience an increase in appetite. A general decrease in gastrointestinal tract movement and the pressure of the enlarged uterus on the rectum can cause constipation. Nausea and vomiting are caused by elevated hCG levels, which are highest early in the pregnancy. Pressure on the stomach may force the stomach contents superiorly into the esophagus, resulting in heartburn. Heartburn can also result from hormonally induced changes in stomach motility and smooth muscle tone.

Pressure on the urinary bladder by the enlarging uterus can produce urinary symptoms, such as increased frequency and urgency of urination, and incontinence (inability to retain urine). Pressure on pelvic veins leads to increased edema (fluid retention) in tissues of the lower limbs, which can result in swollen ankles and knees. An increase in blood flow through the kidneys allows faster elimination of the extra wastes produced by the fetus. The kidneys, ureters, and urethra also lengthen.

Changes in the skin during pregnancy are more apparent in some women than in others, and can include increased pigmentation around the eyes and cheekbones in a masklike pattern called *chloasma* (klō-AZ-ma) and in the circular area around the nipples of the breasts. Some expectant mothers also exhibit a dark, vertical line along the lower abdomen called the *linea nigra*. Stretch marks over the abdomen can occur as the uterus enlarges, and hair loss increases.

Changes in the reproductive system include swelling of and increased blood supply to the external genitals, and enhanced pliability of the vagina in preparation for delivery. The uterus increases from its nonpregnant mass of 60–80 g to 900–1200 g at term due to growth of muscle fibers in the myometrium.

| CLINICAL CONNECTION | *Pregnancy-Induced Hypertension* |

About 10–15 percent of all pregnant women in the United States experience **pregnancy-induced hypertension (PIH)**, an elevation in blood pressure that is associated with pregnancy. The major cause is **preeclampsia** (prē-ē-KLAMP-sē-a), an abnormal condition of pregnancy characterized by sudden hypertension, large amounts of protein in the urine, and generalized edema that typically appears after the twentieth week of pregnancy. Other signs and symptoms include generalized edema, blurred vision, and headaches. Preeclampsia might be related to an autoimmune or allergic reaction resulting from the presence of a fetus. Treatment involves bed rest and various drugs. When the condition is also associated with convulsions and coma, it is a life-threatening condition termed **eclampsia**. •

✔ CHECKPOINT

24. What structural and functional changes does the mother experience during pregnancy?

4.4 LABOR

● OBJECTIVE

• Explain the events associated with the three stages of labor.

Obstetrics (ob-STET-riks; *obstetrix*=midwife) is the branch of medicine that deals with the management of pregnancy, labor, and the **neonatal period**, the first 28 days after birth. **Labor** is the process by which the fetus is expelled from the uterus through the vagina. **Parturition** (par-toor-ISH-un; *parturit-*=childbirth) also means giving birth.

Uterine contractions occur in waves that start at the top of the uterus and move downward, eventually expelling the fetus. **True labor** begins when uterine contractions occur at regular intervals, usually producing pain. As the interval between contractions shortens, the contractions intensify. Another symptom of true labor in some women is localization of pain in the back that is intensified by walking. The reliable indicator of true labor is dilation (expansion) of the cervix and the "show," a discharge of a blood-containing mucus that appears in the cervical canal during labor. In **false labor**, pain is felt in the abdomen at irregular intervals, but it does not intensify and walking does not alter it significantly. There is no "show" and no cervical dilation.

True labor can be divided into three stages (Figure 4.15):

Figure 4.15 Stages of true labor.

🔑 The term parturition refers to birth.

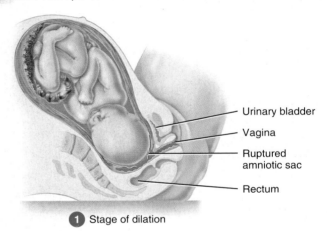

Urinary bladder

Vagina

Ruptured amniotic sac

Rectum

1 Stage of dilation

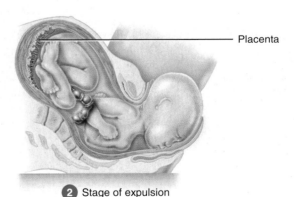

Placenta

2 Stage of expulsion

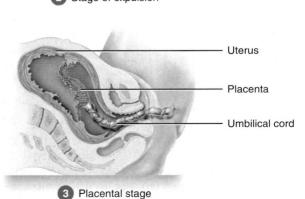

Uterus

Placenta

Umbilical cord

3 Placental stage

 What event marks the beginning of the stage of expulsion?

❶ *Stage of dilation.* The time from the onset of labor to the complete dilation of the cervix is the *stage of dilation*. This stage, which typically lasts 6–12 hours, features regular contractions of the uterus, usually a rupturing of the amniotic sac, and complete dilation (to 10 cm) of the cervix. If the amniotic sac does not rupture spontaneously, it is ruptured intentionally.

❷ *Stage of expulsion.* The time (10 minutes to several hours) from complete cervical dilation to delivery of the baby is the *stage of expulsion.*

❸ *Placental stage.* The time (5–30 minutes or more) after delivery until the placenta or "afterbirth" is expelled by powerful uterine contractions is the *placental stage.* These contractions also constrict blood vessels that were torn during delivery, thereby reducing the likelihood of hemorrhage.

As a rule, labor lasts longer with first babies, typically about 14 hours. For women who have previously given birth, the average duration of labor is about 8 hours—although the time varies enormously among births.

About 7 percent of pregnant women do not deliver by 2 weeks after their due date. Such cases carry an increased risk of brain damage to the fetus, and even fetal death, due to inadequate supplies of oxygen and nutrients from an aging placenta. Post-term deliveries may be facilitated by inducing labor, initiated by administration of oxytocin (Pitocin®), or by surgical delivery (cesarean section).

CLINICAL CONNECTION | *Dystocia*

Dystocia (dis-TO-se-a; *dys-*=painful or difficult; *toc-*=birth), or difficult labor, may result either from an abnormal position (presentation) of the fetus or a birth canal of inadequate size to permit vaginal delivery. In a **breech presentation**, for example, the fetal buttocks or lower limbs, rather than the head, enter the birth canal first; this occurs most often in premature births. If fetal or maternal distress prevents a vaginal birth, the baby may be delivered surgically through an abdominal incision. A low, horizontal cut is made through the abdominal wall and lower portion of the uterus, through which the baby and placenta are removed. Even though it is popularly associated with the birth of Julius Caesar, the true reason this procedure is termed a **cesarean section (C-section)** is because it was described in Roman Law, *lex cesarea*, about 600 years before Julius Caesar was born. Even a history of multiple C-sections need not exclude a pregnant woman from attempting a vaginal delivery. •

✔ CHECKPOINT

25. What is the difference between false labor and true labor?
26. What happens during each of the stages of true labor?

KEY MEDICAL TERMS ASSOCIATED WITH DEVELOPMENT

Amniocentesis (am′-nē-ō-sen-TĒ-sis; *amnio-*=amnion; *-centesis*=puncture to remove fluid) A prenatal diagnostic test in which some of the amniotic fluid that bathes the developing fetus is withdrawn and the fetal cells and dissolved substances are analyzed.

Chorionic villi sampling (CVS) (ko-rē-ON-ik VIL-ī) A prenatal diagnostic test in which a catheter is guided through the vagina and cervix of the uterus and then advanced to the chorionic villi, and about 30 milligrams of tissue are suctioned out and prepared for chromosomal analysis.

Conceptus (kon-SEP-tus) Includes all structures that develop from a zygote and includes an embryo plus the embryonic part of the placenta and associated membranes (chorion, amnion, yolk sac, and allantois).

Cryopreserved embryo (krī-ō-PRĒ-servd; *cryo-*=cold) An early embryo produced by in vitro fertilization (fertilization of a secondary oocyte in a laboratory dish) that is preserved for a long period by freezing it. After thawing, the early embryo is implanted into the uterine cavity. Also called a **frozen embryo**.

Deformation (de-for-MĀ-shun; *de-*=without; *-forma*=form) A developmental abnormality due to mechanical forces that mold a part of the fetus over a prolonged period of time. Deformations usually involve the skeletal and/or muscular system and may be corrected after birth. An example is clubfeet, which can be corrected by manipulation and casting.

Emesis gravidarum (EM-e-sis gra-VID-ar-um; *emeo*=to vomit; *gravida*=a pregnant woman) Episodes of nausea and possibly vomiting that are most likely to occur in the morning during the early weeks of pregnancy; also called **morning sickness**. Its cause is unknown, but the high levels of human chorionic gonadotropin (hCG) secreted by the placenta, and of progesterone secreted by the ovaries, have been implicated. If the severity of these symptoms requires hospitalization for intravenous feeding, the condition is known as **hyperemesis gravidarum**.

Epigenesis (ep-i-GEN-e-sis; *epi-*=upon; *-genesis*=creation) The development of an organism from an undifferentiated cell.

Fertilization age Two weeks less than the gestational age since a secondary oocyte is not fertilized until about two weeks after the last normal menstrual period (LNMP).

Fetal alcohol syndrome (FAS) A specific pattern of fetal malformation due to intrauterine exposure to alcohol. FAS is one of the most common causes of mental retardation and the most common preventable cause of birth defects in the United States.

Fetal surgery A surgical procedure performed on a fetus; in some cases the uterus is opened and the fetus is operated on directly. Fetal surgery has been used to repair diaphragmatic hernias and remove lesions in the lungs.

Gestational age (jes-TĀ-shun-al; *gestatus*=to bear) The age of an embryo or fetus calculated from the presumed first day of the last normal menstrual period (LNMP).

Karyotype (KAR-ē-ō-tīp; *karyo-*=nucleus) The chromosomal characteristics of an individual presented as a systematic arrangement by size and according to the position of the centromere; useful in judging whether chromosomes are normal in number and structure.

Klinefelter syndrome A sex chromosome *aneuploidy* (any deviation from the human diploid number of 46) of males, due to an extra X sex chromosome (XXY) that occurs once in every 500 births. Such individuals are somewhat mentally disadvantaged, sterile males with undeveloped testes, scant body hair, and enlarged breasts.

Lethal gene (LĒ-thal JĒN; *lethum*=death) A gene that, when expressed, results in death either in the embryonic state or shortly after birth.

Maternal alpha-fetoprotein (AFP) test A prenatal diagnostic test in which the mother's blood is analyzed for the presence of AFP, a protein synthesized in the fetus that passes into the maternal circulation.

Metafemale syndrome A sex chromosome aneuploidy of females characterized by an extra X sex chromosome (XXX) that occurs about once in every 700 births. These females have underdeveloped genital organs and limited fertility, and most are mentally retarded.

Mutation (mū-TĀ-shun) Any change in the sequence of bases in a DNA molecule resulting in a permanent alteration of an inherited trait.

Primordium (prī-MOR-dē-um; *primus-*=first; *-ordior*=to begin) The beginning or first discernible indication of the development of an organ or structure.

Puerperal fever (pū-ER-per-al; *puer*=child) An infectious disease of childbirth, also called puerperal sepsis and childbed fever. The disease, which results from an infection originating in the birth canal, affects the maternal endometrium. It may spread to other pelvic structures and lead to septicemia.

Teratogen (TER-a-tō-jen; *terato-*=monster; *-gen*=creating) Any agent or influence that causes developmental defects in the embryo.

Examples include alcohol, pesticides, industrial chemicals, antibiotics, thalidomide, LSD, and cocaine.

Turner syndrome A sex chromosome aneuploidy in females caused by the presence of a single X sex chromosome (designated XO); occurring about once in every 5000 births, it produces a sterile female with virtually no ovaries and limited development of secondary sex characteristics. Other features include short stature, webbed neck, underdeveloped breasts, and widely spaced nipples. Intelligence usually is normal.

CHAPTER REVIEW AND RESOURCE SUMMARY

WileyPLUS

Review | **Resource**

4.1 Embryonic Period

1. Pregnancy is a sequence of events that begins with fertilization, and proceeds to implantation, embryonic development, and fetal development. It normally ends in birth. During fertilization a sperm cell penetrates a secondary oocyte and their pronuclei unite. Penetration of the zona pellucida is facilitated by enzymes in the sperm's acrosome. The resulting cell is a zygote. Normally, only one sperm cell fertilizes a secondary oocyte.

2. Early rapid cell division of a zygote is called cleavage, and the cells produced by cleavage are called blastomeres. The solid sphere of cells produced by cleavage is a morula. The morula develops into a blastocyst, a hollow ball of cells differentiated into a trophoblast and an embryoblast (inner cell mass) that attaches to the endometrium in a process called implantation. The trophoblast develops into the syncytiotrophoblast and cytotrophoblast, both of which become part of the chorion. The embryoblast differentiates into hypoblast and epiblast, the bilaminar (two-layered) embryonic disc. The amnion is a thin protective membrane that develops from the cytotrophoblast.

3. During the second week the exocoelomic membrane and hypoblast form the yolk sac, which transfers nutrients to the embryo, forms blood cells, produces primordial germ cells, and forms part of the gut. Erosion of sinusoids and endometrial glands provides blood and secretions, which enter lacunar networks to supply nutrition to and remove wastes from the embryo. The extraembryonic coelom forms within extraembryonic mesoderm. The extraembryonic mesoderm and trophoblast form the chorion, the principal embryonic portion of the placenta.

4. The third week of development is characterized by gastrulation, the conversion of the bilaminar disc into a trilaminar (three-layered) embryonic disc consisting of ectoderm, mesoderm, and endoderm. The first evidence of gastrulation is formation of the primitive streak and then the primitive node, notochordal process, and notochord. The three primary germ layers form all tissues and organs of the developing organism (see Table 4.1). Also during the third week, the oropharyngeal and cloacal membranes form. The wall of the yolk sac forms a small vascularized outpouching called the allantois, which functions in blood formation and development of the urinary bladder.

5. The process by which the neural plate, neural folds, and neural tube form is called neurulation. The brain and spinal cord develop from the neural tube.

6. Paraxial mesoderm segments to form somites from which skeletal muscles of the neck, trunk, and limbs develop. Somites also form connective tissues and vertebrae.

7. Blood vessel formation, called angiogenesis, begins in mesodermal cells called angioblasts. The heart develops from mesodermal cells called the cardiogenic area. By the end of the third week, the primitive heart beats and circulates blood. Chorionic villi, projections of the chorion, connect to the embryonic heart so that maternal and fetal blood vessels are brought into close proximity; this allows nutrients and wastes to be exchanged between maternal blood and fetal blood.

8. Placentation refers to formation of the placenta, the site of exchange of nutrients and wastes between the mother and fetus. The placenta also functions as a protective barrier, stores nutrients, and produces several hormones to maintain pregnancy. The actual connection between the placenta and embryo (and later the fetus) is the umbilical cord.

9. Organogenesis, the formation of body organs and systems, occurs during the fourth week of development. Conversion of the flat, two-dimensional trilaminar embryonic disc to a three-dimensional cylinder occurs by a process called embryonic folding. Embryonic folding brings various organs into their final adult positions and helps form the gastrointestinal tract. Pharyngeal arches, clefts, and pouches give rise to the structures of the head and neck. By the end of the fourth week, upper and lower limb buds develop, and by the end of the eighth week the embryo has clearly human features.

Resource column:

Animation - Fertilization
Anatomy Overview - Developmental Stages
Animation - Embryonic and Fetal Development
Figure 4.3 - Cleavage and the Formation of the Morula and Blastocyst
Figure 4.5 - Summary of Events Associated with the First Week of Development
Figure 4.7 - Gastrulation

4.2 Fetal Period

1. The fetal period is primarily concerned with the growth and differentiation of tissues and organs that develop during the embryonic period.

2. The rate of body growth is remarkable, especially during the ninth and sixteenth weeks.

3. The principal changes associated with embryonic and fetal growth are summarized in Table 4.2.

Resource column:

Anatomy Overview - Developmental Stages
Animation - Embryonic and Fetal Development

Review	Resource

4.3 Maternal Changes During Pregnancy

1. During pregnancy, numerous anatomical and physiological changes occur in the mother.
2. The uterus nearly fills the abdominal cavity toward the end of full-term pregnancy, pushing the viscera out of their normal positions.
3. Physiological changes include weight gain, increased skin pigmentation in certain areas, and various alterations in the cardiovascular, respiratory, digestive, urinary, and reproductive systems.

Animation - Hormonal Regulation of Pregnancy

4.4 Labor

1. Labor is the process by which the fetus is expelled from the uterus through the vagina to the outside.
2. True labor involves dilation of the cervix, expulsion of the fetus, and delivery of the placenta.

Anatomy Overview - Hypothalamic Reproductive Hormones
Animation - Regulation of Labor and Birth
Animation - Positive Feedback Control of Labor
Figure 4.15 - Stages of True Labor
Exercise - Pregnancy, Birth, and Lactation

CRITICAL THINKING QUESTIONS

1. Fetal alcohol syndrome (FAS) is one of the most common causes of mental retardation and the most common preventable cause of birth defects in the United States. How is the alcohol that the woman drinks able to affect her developing fetus?

2. Some disorders of the nervous system may cause particular skin problems. For example, a person with a type of nervous system tumor may show coffee-colored spots on the skin. How are the structures of the nervous system and the skin related?

3. Josefina, in the last 2 weeks of her first pregnancy, anxiously called her doctor to ask if she should leave for the hospital. She was experiencing irregular "labor" pains that were thankfully eased by walking. She had no other signs to report. The doctor told Josefina to stay home—it wasn't time yet. Why did the doctor tell Josefina to stay home? How did he know it "wasn't time"?

4. Infection of the mother by certain viruses is known to result in severe birth defects in the child. If there is no mixing of maternal and fetal blood in the placenta, how can a viral infection in the mother cause problems in the child? At what point during the pregnancy do you think the greatest risk to the child would occur?

5. Larry's wife Elena is eight months pregnant. He has been fussing at her lately about the increase in their weekly grocery bill "even though there are still only two of us," and he has complained that they have been "going through toilet paper and antacids like crazy lately." When he wakes up in the morning, he discovers that Elena has stolen his extra pillow to prop herself up higher. What are some of the specific anatomical and physiological changes Elena has been experiencing that might explain these lifestyle changes?

ANSWERS TO FIGURE QUESTIONS

4.1 The uterine tubes provide an exit route from the peritoneal cavity for the secondary oocyte, a route for sperm to reach a secondary oocyte, and help transport a fertilized (or unfertilized) oocyte to the uterus.

4.2 Capacitation is the series of changes in sperm after they have been deposited in the female reproductive tract that enable them to fertilize a secondary oocyte.

4.3 A morula is a solid ball of cells; a blastocyst consists of a rim of cells (trophoblast) surrounding a cavity (blastocyst cavity) and an embryoblast (inner cell mass).

4.4 The blastocyst secretes digestive enzymes that eat away the endometrial lining at the site of implantation.

4.5 Implantation usually occurs in the posterior portion of the fundus or body of the uterus.

4.6 The bilaminar embryonic disc is attached to the trophoblast by the connecting stalk.

4.7 Gastrulation converts a bilaminar (two-layered) embryonic disc into a trilaminar (three-layered) embryonic disc. This results in the formation of the primary germ layers from which all tissues of the body develop.

4.8 The notochord induces mesodermal cells to develop into parts of vertebrae and forms a portion of the intervertebral discs.

4.9 The neural tube forms the brain and spinal cord; somites develop into skeletal muscles, connective tissue, and the vertebrae.

4.10 Chorionic villi help to bring the fetal and maternal blood vessels into close proximity, allowing for the more efficient movement of oxygen and nutrients from maternal blood into fetal blood and movement of carbon dioxide and wastes from fetal blood into maternal blood.

4.11 The placenta exchanges materials between the fetus and the mother, protects the fetus against many microbes, and stores nutrients.

4.12 Because of embryonic folding, the embryo curves into a C-shape, various organs are brought into their eventual adult positions, and the primitive gut is formed.

4.13 Pharyngeal arches, clefts, and pouches give rise to structures of the head and neck.

4.14 During the last two-and-a-half months of development, fetal weight doubles.

4.15 Complete dilation of the cervix marks the onset of the stage of expulsion.

5 | THE INTEGUMENTARY SYSTEM

INTRODUCTION   When you meet someone for the first time, your first impression is based largely on the most visible part of the body, the skin and its associated structures. For example, the distribution, color, length, and condition of hair give clues about the health and age of your new acquaintance. You might also note freckles, moles, and other pigmentation differences. When you are cold, you develop goose bumps and your hairs stand on end. If you are hot or nervous, you sweat. Most of us are so conscious of the importance of first impressions that we expend considerable time, money, and effort in our daily grooming rituals to "put our best face forward."

Why are there different colors and textures of skin? What is a freckle, a blister, a callus? Why is the skin thicker in some areas of the body than in others? What are fingerprints? How does hair grow, and what makes it curly or straight? Why does it turn gray? These are just a few of the questions you will be able to answer as you learn about the structure and functions of this important body system.

The **integumentary system** (in-teg-ū-MEN-tar-ē; *inte-*=whole; *-gument*=body covering) is composed of organs such as the skin, hair, oil and sweat glands, nails, and sensory receptors. It helps maintain a constant body temperature, protects the body, and provides sensory information about the surrounding environment. Of all the body's organs, none is more easily inspected or more exposed to infection, disease, and injury. Although its location makes it vulnerable to trauma, sunlight, microbes, and environmental pollutants, the skin has protective features that ward off most such damage. Changes in skin color may indicate homeostatic imbalances. Abnormal skin eruptions or rashes may reveal systemic infections or diseases of internal organs. **Dermatology** (der′-ma-TOL-ō-jē; *dermato-*=skin; *-logy*=study of) is the medical specialty that deals with the structure, function, and disorders of the integumentary system. •

Did you ever wonder why it is so difficult to save the life of someone with extensive third-degree burns?

5.1 STRUCTURE OF THE SKIN

 OBJECTIVES

• Describe the layers of the epidermis and the cells that compose them.
• Compare the composition of the papillary and reticular regions of the dermis.
• Explain the anatomical basis of differences in skin color.

The **skin**, also referred to as the **cutaneous membrane** (kū -TĀ-nē-us), covers the external surface of the body and is the largest organ of the body in weight. In adults, the skin covers an area of about 2 square meters (22 square feet) and weighs 4.5–5 kg (10–11 lb), about 7 percent of total body weight. It ranges in thickness from 0.5 mm (.02 in.) on the eyelids to 4.0 mm (.16 in.) on the heels. However, over most of the body it is 1–2 mm (.04–.08 in.) thick. Structurally, the skin consists of two main parts (Figure 5.1). The superficial, thinner portion, which is composed of *epithelial tissue*, is the avascular **epidermis** (ep-i-DERM-is; *epi-*=above; *dermis*=skin). For this reason, if you scratch the epidermis there is no bleeding. The deeper, thicker, *connective tissue* portion is the **dermis**. It is vascular and a cut that penetrates to the dermis produces bleeding. Deep to the dermis, but not part of the skin, is the **subcutaneous (subQ) layer** (*sub*=below; *cutis*=skin), or **hypodermis** (*hypo-*=below), which consists of areolar and adipose tissues. The skin and subcutaneous layer form the **integument**.

Figure 5.1 Components of the integumentary system.
The skin consists of a superficial, thin epidermis and a deep, thicker dermis. Deep to the skin is the subcutaneous layer, which attaches the dermis to the fascia of the body.

The integumentary system includes organs such as the skin and hairs and its accessory structures—nails and skin glands—along with associated muscles and nerves.

FUNCTIONS OF THE INTEGUMENTARY SYSTEM
1. Regulates body temperature.
2. Stores blood.
3. Protects body from external environment.
4. Detects cutaneous sensations.
5. Excretes and absorbs substances.
6. Synthesizes vitamin D.

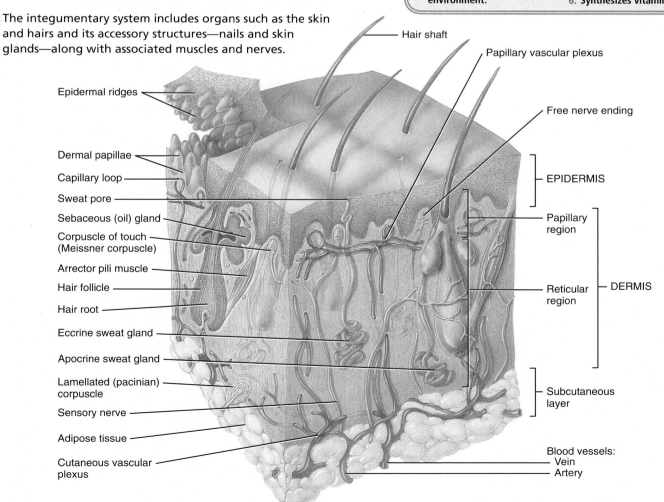

(a) Sectional view of skin and subcutaneous layer

FIGURE 5.1 CONTINUES ▶

● FIGURE **5.1** CONTINUED ▶

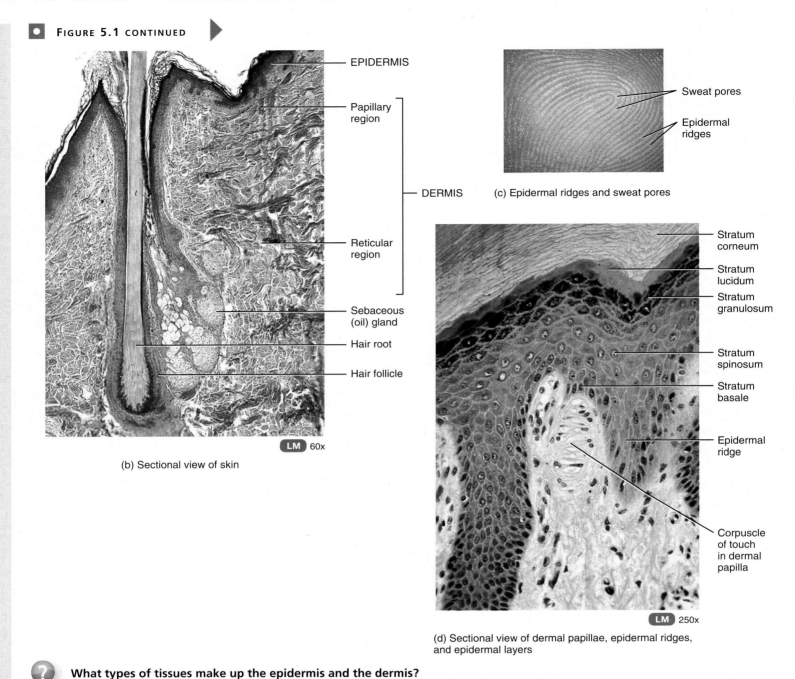

EPIDERMIS

Papillary region

DERMIS

Reticular region

Sebaceous (oil) gland

Hair root

Hair follicle

LM 60x

(b) Sectional view of skin

Sweat pores

Epidermal ridges

(c) Epidermal ridges and sweat pores

Stratum corneum

Stratum lucidum

Stratum granulosum

Stratum spinosum

Stratum basale

Epidermal ridge

Corpuscle of touch in dermal papilla

LM 250x

(d) Sectional view of dermal papillae, epidermal ridges, and epidermal layers

❓ **What types of tissues make up the epidermis and the dermis?**

Epidermis

The **epidermis** is composed of keratinized stratified squamous epithelium (see Section 3.4). It contains four principal types of cells: keratinocytes, melanocytes, intraepidermal macrophage cells, and tactile epithelial cells (Figure 5.2). About 90 percent of epidermal cells are **keratinocytes** (ker-a-TIN-ō-sīts; *keratino-*=hornlike, after the toughness of animal horns; *-cytes*=cells), which are arranged in four or five layers and produce the protein **keratin** (KER-a-tin) (Figure 5.2a). Recall from Chapter 3 that keratin is a tough, fibrous intracellular protein that helps protect the skin and underlying tissues from abrasions, heat, microbes, and chemicals. Keratinocytes also produce lamellar granules, which release a water-repellent sealant that decreases water entry and water loss and inhibits the passage of foreign materials.

About 8 percent of the epidermal cells are **melanocytes** (MEL-a-nō-sīts; *melano*=black), which develop from the neural crest of a

developing embryo and produce the pigment melanin (Figure 5.2b). Melanocytes have a cuboidal cell body with long armlike processes that project between the neighboring keratinocytes. By means of these processes, one melanocyte can contact approximately 30 neighboring keratinocytes. Their long, slender projections transfer melanin granules to the neighboring keratinocytes. **Melanin** (MEL-a-nin) is a yellow-red or brown-black pigment that contributes to skin color and absorbs damaging ultraviolet (UV) light. Once inside keratinocytes, the melanin granules cluster to form a protective veil over the nucleus, on the side toward the skin surface. In this way, they shield the nuclear DNA from damage by UV light. Although their melanin granules effectively protect keratinocytes, melanocytes themselves are particularly susceptible to damage by UV light.

Intraepidermal macrophage cells, or **Langerhans cells** (LANG-er-hans), arise from red bone marrow and migrate to the epidermis (Figure 5.2c), where they constitute a small fraction of the epidermal cells. Like melanocytes, intraepidermal macrophage

Figure 5.2 Types of cells in the epidermis. (See also Figure 5.1d.) Besides keratinocytes, the epidermis contains melanocytes, which produce the pigment melanin; intraepidermal macrophage cells, which participate in immune responses; and tactile epithelial cells, which function in the sensation of touch.

 Most of the epidermis consists of keratinocytes, which produce the protein keratin (that protects underlying tissues) and lamellar granules (that contain a waterproof sealant).

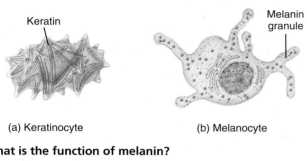

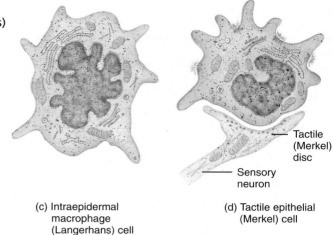

(a) Keratinocyte (b) Melanocyte

(c) Intraepidermal macrophage (Langerhans) cell

(d) Tactile epithelial (Merkel) cell

 What is the function of melanin?

cells have long armlike processes and situate themselves among many surrounding keratinocytes. They participate in immune responses mounted against microbes that invade the skin. Intraepidermal macrophage cells and other cells of the immune system recognize a foreign microbe or substance so that it can be destroyed. Intraepidermal macrophage cells are easily damaged by UV light.

Tactile epithelial cells, or *Merkel cells* (MER-kel), are the least numerous of the epidermal cells. Tactile epithelial cells are located in the deepest layer of the epidermis, where they contact the flattened process of a sensory neuron (nerve cell), a structure called a **tactile (Merkel) disc** (Figure 5.2d). Tactile epithelial cells and tactile discs detect touch sensations.

Several distinct layers of keratinocytes in various stages of development form the epidermis (Figure 5.3). In most regions of the body the epidermis has four strata or layers—stratum basale, stratum spinosum, stratum granulosum, and a thin stratum corneum. This is called **thin skin.** Where exposure to friction is greatest, such as in the fingertips, palms, and soles, the epidermis has five layers—stratum basale, stratum spinosum, stratum granulosum, stratum lucidum, and a thick stratum corneum. This is called **thick skin.** The details of thin and thick skin are discussed later in the chapter (see Section 5.3).

Stratum Basale

The deepest layer of the epidermis, the **stratum basale** (ba-SA-lē; *basal*=base), is composed of a single row of cuboidal or columnar keratinocytes, some of which are *stem cells* that undergo cell division to continually produce new keratinocytes. The nuclei of these keratinocytes are large, and their cytoplasm contains many ribosomes, a small Golgi complex, a few mitochondria, and some rough endoplasmic reticulum. The cytoskeleton within keratinocytes of the stratum basale includes scattered intermediate filaments called *keratin intermediate filaments* (*tonofilaments* of electron microscopy). The keratin intermediate filaments form the tough protein keratin in the more superficial epidermal layer. Keratin protects the deeper layers from injury. Keratin intermediate filaments attach to desmosomes, which bind cells of the stratum basale to each other, to the cells of the adjacent stratum spinosum, and to hemidesmosomes, which bind the keratinocytes to the basement membrane positioned between the epidermis

and the dermis. Melanocytes and tactile epithelial cells (with their associated tactile discs) are scattered among the keratinocytes of the basal layer. The stratum basale is sometimes referred to as the **stratum germinativum** (jer-mi-na-TĒ-vum; *germ*=sprout) to indicate its role in forming new cells.

Figure 5.3 Layers of the epidermis.

 The epidermis consists of keratinized stratified squamous epithelium.

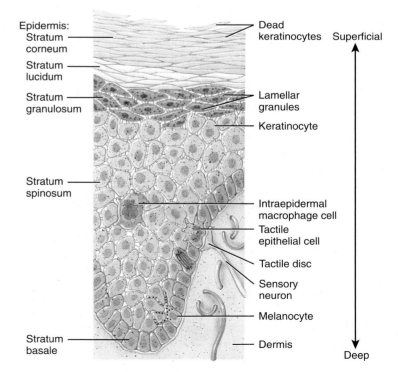

Location of four principal cell types in epidermis of thick skin

 Which epidermal layer includes stem cells that continually undergo cell division?

Stratum Spinosum

Superficial to the stratum basale is the **stratum spinosum** (spi-NŌ-sum; *spinos*=thornlike). Other than an occasional tactile epithelial cell, this stratum consists of numerous keratinocytes produced by the stem cells in the basal layer, arranged in 8–10 layers. Cells in the more superficial layers become somewhat flattened. The keratinocytes in the stratum spinosum have the same organelles as cells of the stratum basale, and some retain their ability to divide. The keratinocytes of this region produce coarser bundles of keratin intermediate filaments than those of the basal layer. Although they are rounded and larger in living tissue, cells of the stratum spinosum shrink and pull apart when prepared for microscopic examination, so they appear to be covered with thornlike spines (thus the name) (Figure 5.3a). At each spinelike projection, bundles of keratin intermediate filaments insert into desmosomes, which tightly join the cells to one another. This arrangement provides both strength and flexibility to the skin. Intraepidermal macrophage cells and projections of melanocytes are also present in the stratum spinosum.

Stratum Granulosum

At about the middle of the epidermis, the **stratum granulosum** (gran-ū-LŌ-sum; *granulos*=little grains) consists of three to five layers of flattened keratinocytes that are undergoing apoptosis. (Recall from Chapter 2 that *apoptosis* is an orderly, genetically programmed cell death in which the nucleus fragments before the cells die.) The nuclei and other organelles of these cells begin to degenerate as they have moved farther from their source of nutrition (the dermal blood vessels). Even though keratin intermediate filaments are no longer being produced by these cells, they become more apparent because the organelles in the cells are regressing. A distinctive feature of cells in the stratum granulosum is the presence of darkly staining protein granules; this protein, called **keratohyalin** (ker'-a-tō-HĪ-a-lin), is involved in assembling the keratin intermediate filaments into keratin. Also present in the keratinocytes are membrane-enclosed **lamellar granules** (la-MEL-lar) which fuse with the plasma membrane and release a lipid-rich secretion. This secretion is deposited in the spaces between cells of three epidermal layers: the stratum granulosum, stratum lucidum, and stratum corneum.

The lipid-rich secretion acts as a water-repellent sealant, retarding loss of body fluids and entry of foreign materials. As their nuclei break down during apoptosis, the keratinocytes of the stratum granulosum can no longer carry on vital metabolic reactions, and they die. Thus, the stratum granulosum marks the transition between the deeper, metabolically active strata and the dead cells of the more superficial strata.

Stratum Lucidum

The **stratum lucidum** (LOO-si-dum; *lucid*=clear) is present in the thick skin of areas such as the fingertips, palms, and soles. It consists of four to six layers of clear, flat, dead keratinocytes that contain large amounts of keratin and thickened plasma membranes. The keratin is more regularly arranged parallel to the skin surface. This probably provides an additional level of toughness in this region of thick skin.

Stratum Corneum

The **stratum corneum** (KOR-nē-um; *corne*=horn or horny) consists on average of 25 to 30 layers of flattened dead keratinocytes, but can range in thickness from a few cells in thin skin to 50 or more cell layers in thick skin. The cells, which are extremely thin, flat, plasma-membrane-enclosed packages of keratin, are called **corneocytes** or **squames** (SKWĀMS, *squame*=scale). Corneocytes no longer contain nuclei or any internal organelles. They are the final product of the differentiation process of the keratinocytes. The corneocytes within each layer overlap one another like the scales on the skin of a snake. Neighboring layers of corneocytes also form strong connections with one another. The plasma membranes of adjacent corneocytes are arranged in complex, wavy folds that fit together like pieces of a jigsaw puzzle to hold the layers together. In this outer stratum of the epidermis, often referred to as the *cornified layer*, cells are continuously shed and replaced by cells from the deeper strata. Its multiple layers of dead cells help the stratum corneum to protect deeper layers from injury and microbial invasion. Constant exposure of skin to friction stimulates increased cell production and keratin production, and that results in the formation of a **callus**, an abnormal thickening of the stratum corneum.

Keratinization and Growth of the Epidermis

Newly formed cells in the stratum basale are pushed slowly through the various layers to the skin's surface. As the cells move from one epidermal layer to the next, they accumulate more keratin in a process called **keratinization** (ker'-a-tin-i-ZĀ-shun). Then they undergo apoptosis. Eventually the keratinized cells slough off and are replaced by underlying cells that in turn become keratinized. This process accounts for the changes in characteristics of the keratinocytes as they mature into corneocytes. The whole process by which cells form in the stratum basale, rise to the surface, become keratinized, and slough off takes about four to six weeks in an average epidermis of 0.1 mm (.04 in.) thickness. The rate of cell division in the stratum basale increases when the outer layers of the epidermis are stripped away, as in abrasions and burns. The mechanisms that regulate this remarkable growth are not well understood, but hormonelike proteins such as **epidermal growth factor (EGF)** play a role. An excessive amount of keratinized cells shed from the skin of the scalp is called **dandruff**.

Table 5.1 presents a summary of the distinctive features of the epidermal strata.

TABLE 5.1

Summary of Epidermal Strata (See also Figure 5.3)

STRATUM	DESCRIPTION
Basale	Deepest layer, composed of a single row of cuboidal or columnar keratinocytes that contain scattered keratin intermediate filaments (tonofilaments); stem cells undergo cell division to produce new keratinocytes; melanocytes and tactile epithelial cells associated with tactile discs are scattered among the keratinocytes.
Spinosum	Eight to ten rows of many-sided keratinocytes with bundles of keratin intermediate filaments; contains armlike projections of melanocytes and intraepidermal macrophage cells.
Granulosum	Three to five rows of flattened keratinocytes, in which organelles are beginning to degenerate; cells contain the protein keratohyalin, which converts keratin filaments into keratin, and lamellar granules, which release a lipid-rich, water-repellent secretion.
Lucidum	Present only in skin of fingertips, palms, and soles; consists of four to six rows of clear, flat, dead keratinocytes with large amounts of orderly arranged keratin.
Corneum	Twenty-five to thirty rows of dead, flat keratinocytes that contain mostly keratin.

 CHECKPOINT

1. Describe the types of cells found in the epidermis and explain their functions.
2. List the distinctive features of the epidermal layers from deepest to most superficial.

Dermis

The second, deeper part of the skin, the **dermis**, is composed of a strong dense irregular connective tissue containing collagen and elastic fibers. It is much thicker than the epidermis and this thickness varies from region to region in the body, reaching its greatest thickness on the palms and soles. Because the dermis is typically thinner in women than in men, many women have the appearance of dimples in the skin referred to as *cellulite*. The dermis has great *tensile strength* (resistance to pulling or stretching forces). It also has the ability to stretch and recoil easily. Leather, used for belts, shoes, baseball gloves, and basketballs, is dried and treated animal dermis. As is typical of all general connective tissue, the cells present in the dermis are scattered and include fixed cells and wandering cells. The predominant fixed cells are fibroblasts; the wandering cells include macrophages, mast cells, eosinophils, neutrophils, and dermal interstitial dendritic cells (immune surveillance cells). Blood vessels and nerves, along with glands and hair follicles (two epithelial invaginations of the epidermis), are embedded in the dermal layer. The dermis is essential to the survival of the epidermis, and these adjacent layers form many important structural and functional relations. Based on its tissue structure, the dermis can be divided by an indistinct boundary into a thin, superficial papillary region and a thick, deeper reticular region.

The **papillary region** makes up about one-fifth of the thickness of the total layer (see Figure 5.1). It contains thin collagen fibers and fine elastic fibers. Its surface area is greatly increased by small, fingerlike structures that project into the undersurface of the epidermis called **dermal papillae** (pa-PIL-ē=nipples); these greatly increase the surface contact between the papillary region and the epidermis. The dermal papillae can vary greatly in size and number throughout different parts of the dermis; they are taller and more numerous in sensitive regions of skin that experience more mechanical stress. In the thin skin covering most of the body, the dermal papillae are relatively few in number, small, and irregularly scattered. By contrast, in the thick skin of the palmar surfaces of the hands and plantar surfaces of the feet, the papillae are relatively numerous, tall, and arranged in patterned rows. All dermal papillae contain **capillary loops** (blood capillaries). Some contain encapsulated touch receptors called **corpuscles of touch**, or **Meissner corpuscles** (MĪS-ner). As their name implies, corpuscles of touch are sensitive nerve endings that detect touch. Still other dermal papillae contain **free nerve endings**, dendrites that lack any apparent structural specialization. Different free nerve endings initiate signals that produce sensations of warmth, coolness, pain, tickling, and itching (see Section 20.2).

The **reticular region** (*reticul*=netlike), which is attached to the subcutaneous layer, contains bundles of thick collagen fibers, scattered fibroblasts, various wandering cells (such as macrophages), and some coarse elastic fibers (see Figure 5.1). In addition, some adipose cells can be present in the deepest part of the layer. The collagen fibers in the reticular region are arranged in a netlike manner and in a more regular formation than those in the papillary region. The more regular orientation of the large collagen fibers aligns with the local tensile forces to help the skin resist stretching. Blood vessels, nerves, hair follicles, sebaceous (oil) glands, and sudoriferous (sweat) glands occupy the spaces between fibers.

In certain regions of the body, collagen fibers within the reticular region of the dermis tend to orient more in one direction than another because of natural tension resulting from bony projections, orientation of muscles, and movements at joints. **Tension lines (lines of cleavage)** in the skin indicate the predominant direction of underlying collagen fibers (Figure 5.4). Knowledge

Figure 5.4 Tension lines.

 Tension lines in the skin indicate the predominant direction of underlying collagen fibers in the reticular region.

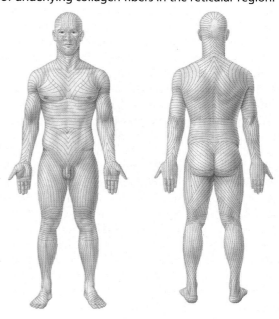

(a) Anterior view   (b) Posterior view

 Why are tension lines clinically important?

of tension lines is especially important to plastic surgeons. For example, a surgical incision running parallel to the collagen fibers will heal with only a fine scar. A surgical incision made across the rows of fibers disrupts the collagen, and the wound tends to gape open and heal in a broad, thick scar.

Both of the layers of the dermis contain dense horizontal networks of small blood vessels. These vascular networks arise from vessels in the underlying skeletal muscles that send branches into the subcutaneous layer and then to the dermis (see Figure 5.1). These dermal vascular networks feed the extensive capillary loops in the dermal papillae. The blood supply of the skin is discussed in detail in Section 5.5.

The combination of collagen and elastic fibers in the reticular region provides the skin with strength, **extensibility** (ek-sten′-si-BIL-i-tē), the ability to stretch, and **elasticity** (e-las-TISS-i-tē), the ability to return to original shape after stretching. The extensibility of skin can be seen readily around joints and in pregnancy and obesity.

CLINICAL CONNECTION | *Stretch Marks*

Because of the collagenous, vascular structure of the dermis, **striae** (STRI-ē =streaks), or **stretch marks**, a form of intrinsic scarring, can result from the internal damage to this layer. When the skin is overstretched, the mechanical demands placed on this layer exceed the strength of the collagen bonding and the stretch available from both the elastic fibers and the creases in the skin. As a result, the lateral bonding of adjacent collagen fibers is disrupted and small dermal blood vessels rupture. Stretch marks initially appear as reddish streaks at these sites. Later, after a poorly vascularized scar tissue forms at these sites of dermal breakdown, the stretch marks remain visible as silvery white streaks. Stretch marks often occur in the abdominal skin during pregnancy, on the skin of weight-lifters where the skin is stretched by a rapid increase in muscle mass, and in the stretched skin accompanying gross obesity. •

The surfaces of the palms, fingers, soles, and toes are marked by series of ridges and grooves. They appear either as straight lines or as a pattern of loops and whorls, as on the tips of the digits. These **epidermal ridges** are produced during the third month of fetal development as the epidermis projects downward into the dermis between the dermal papillae of the papillary region (see Figure 5.1a, c). The epidermal ridges serve multiple functions: (1) They increase the surface area of the epidermis and thus function to increase the grip of the hand or foot by increasing friction; (2) the interdigitating pattern between epidermal ridges and dermal papillae creates a stronger bond between the epidermis and dermis in regions of high mechanical stress; and (3) they greatly increase the surface area, which increases the number of corpuscles of touch and thus increases tactile sensitivity. Because the ducts of sweat glands open on the tops of the epidermal ridges as sweat pores, the sweat and ridges form **fingerprints** (or **footprints**) when a smooth object is touched. The epidermal ridge pattern is in part genetically determined, but even identical twins have different patterns. Normally, the ridge pattern does not change

during life, except to enlarge, and thus can serve as a permanent basis for identification. The study of the pattern of epidermal ridges that is concerned with the identification and classification of fingerprints is referred to as **dermatoglyphics** (der-ma-tō-GLIF-iks; *glyphe*=carved work).

In addition to forming epidermal ridges, the complex papillary surface of the dermis has other functional properties. The dermal papillae greatly increase the contact area between the dermis and epidermis. This increased dermal contact area, with its extensive network of small blood vessels, serves as an important source of nutrition for the overlying epidermis. Molecules diffuse from the small blood capillaries in the dermal papilla to the cells of the stratum basale, allowing the basal epithelial stem cells to divide and the keratinocytes to grow and develop. As keratinocytes push toward the surface and away from the dermal blood source, they are no longer able to obtain the nutrition they require, leading to the eventual breakdown of their organelles.

The dermal papillae fit together with the complementary epidermal ridge to form an extremely strong junction between the two layers. This jigsaw puzzle–like connection strengthens the skin against forces that attempt to separate the epidermis from the dermis.

Table 5.2 summarizes the structural features of the papillary and reticular regions of the dermis.

CHECKPOINT

3. Compare the structure and functions of the epidermis and dermis.
4. Compare the composition of the papillary and reticular regions of the dermis.
5. How are epidermal ridges formed?

TABLE 5.2

Summary of Papillary and Reticular Regions of the Dermis (See also Figure 5.1b)

REGION	DESCRIPTION
Papillary region	The superficial portion of the dermis (about one-fifth); consists of areolar connective tissue with thin collagen fibers and fine elastic fibers; contains dermal papillae that house blood capillaries, corpuscles of touch, and free nerve endings.
Reticular region	The deeper portion of the dermis (about four-fifths); consists of dense irregular connective tissue with bundles of thick collagen and some coarse elastic fibers. Spaces between fibers contain some adipose cells, hair follicles, nerves, sebaceous glands, and sudoriferous glands.

The Structural Basis of Skin Color

Melanin, hemoglobin, and carotene are three pigments that impart a wide variety of colors to skin. The amount of **melanin** causes the skin's color to vary from pale yellow to reddish-brown to black. The difference between the two forms of melanin, *pheomelanin* (fē-ō-MEL-a-nin) (yellow to red) and **eumelanin** (ū-MEL-a-nin) (brown to black), is most apparent in the hair. Melanocytes, the melanin-producing cells, are most plentiful in the epidermis of the penis, nipples and areolae (area just around the nipples), face, and limbs. They are also present in mucous membranes. Because the *number* of melanocytes is about the same in all people, differences in skin color are due mainly to the *amount of pigment* the melanocytes produce and transfer to keratinocytes. In some people, melanin accumulates in patches called **freckles**. There are two types of freckles: **ephelides** (ef-Ē-li-dēz), light freckles, and **lentigines** (len-TIJ-i-nēz), dark or sunburn freckles. Light freckles are most often found on light-skinned individuals and appear to be an inherited trait. Dark, or sunburn, freckles are darker than common (light) freckles and arise because of exposure to the sun. As a person gets older, **age (liver) spots** may develop. These flat blemishes, which are a form of dark freckles, look like freckles and range in color from light brown to black. Like freckles, age spots are accumulations of melanin. A round, flat, or raised area that usually develops in childhood or adolescence and represents a benign localized overgrowth of melanocytes is called a **nevus** (NĒ-vus), or a **mole**.

Melanocytes synthesize melanin from the amino acid *tyrosine* in the presence of an enzyme called *tyrosinase*. Synthesis occurs in an organelle called a **melanosome** (MEL-an-ō-sōm). Exposure to ultraviolet (UV) light increases the enzymatic activity within melanosomes and thus increases melanin production. This activity results from DNA damage from the UV light. The DNA damage stimulates melanin production to protect against further damage. Both the amount and darkness of melanin increase upon UV exposure, which gives the skin a tanned appearance and helps protect the body against further UV radiation. A tan is lost when the melanin-containing keratinocytes are shed from the stratum corneum. Melanin absorbs UV radiation, prevents damage to DNA in epidermal cells, and neutralizes free radicals that form in the skin following damage by UV radiation. Thus, within limits, melanin serves a protective function. As you will see later in this section, however, repeatedly exposing the skin to UV light may cause skin cancer.

CLINICAL CONNECTION | Albinism and Vitiligo

Albinism (AL-bin-izm; *albin*=white) is the inherited inability of an individual to produce melanin. Most **albinos** (al-BĪ-nōs), people affected by albinism, have melanocytes that are unable to synthesize tyrosinase. Melanin is missing from their hair, eyes, and skin. This results in problems with vision and a tendency of the skin to burn easily from overexposure to sunlight.

In another condition, called **vitiligo** (vit-i-LĪ-gō), the partial or complete loss of melanocytes from patches of skin produces irregular white spots. The loss of melanocytes may be related to an immune system malfunction in which antibodies attack the melanocytes. •

Dark-skinned individuals have large amounts of melanin in the epidermis, so their skin color ranges from yellow to reddish-brown to black. The skin of light-skinned individuals, which has little melanin in the epidermis, appears translucent and its color ranges from pink to red depending on the oxygen content of the blood moving through capillaries in the dermis. The red color is

due to **hemoglobin** (hē-mō-GLŌ-bin), the oxygen-carrying pigment in red blood cells.

Carotene (KAR-o-tēn; *carot*=carrot) is a yellow-orange pigment that gives egg yolk and carrots their color. This precursor of vitamin A, used to synthesize pigments needed for vision, is stored in the stratum corneum and fatty areas of the dermis and subcutaneous layer in response to excessive dietary intake. Too much carotene may be deposited in the skin after eating large amounts of carotene-rich foods so that the skin actually turns orange, which is especially apparent in light-skinned individuals. Decreasing carotene intake alleviates the problem.

CLINICAL CONNECTION | Skin Color as a Diagnostic Clue

The color of skin and mucous membranes can provide clues for diagnosing certain conditions. When blood is not picking up an adequate amount of oxygen from the lungs, as in someone who has stopped breathing, the mucous membranes, nail beds, and skin appear bluish or **cyanotic** (sī-a-NOT-ik; *cyan-*=blue). **Jaundice** (JON-dis; *jaund-*=yellow) is due to a buildup of the yellow pigment bilirubin in the skin. This condition gives a yellowish appearance to the skin and the whites of the eyes, and usually indicates liver disease. **Erythema** (er-e-THĒ-ma; *eryth-*=red), redness of the skin, is caused by engorgement of capillaries in the dermis with blood due to skin injury, exposure to heat, infection, inflammation, or allergic reactions. This reddish tint also appears in facial blushing, when the relatively thin skin of the face fills with blood during a moment of embarrassment. **Pallor** (PAL-or), or paleness of the skin, may occur in conditions such as shock and anemia. Certain regions of skin, such as the lips, have such a thin stratum corneum that the blood-filled capillaries in the underlying dermis show through, giving the skin a permanent reddish tint. All skin color changes are observed most readily in people with lighter-colored skin and may be more difficult to discern in people with darker skin. However, examination of the nail beds and gums can provide some information about circulation in individuals with darker skin. •

Tattooing and Body Piercing

Tattooing is a permanent coloration of the skin in which a foreign pigment is deposited with a needle into the dermis. It is believed that the practice originated in Ancient Egypt between 4000 and 2000 B.C. Today, tattooing is performed in one form or another by nearly all peoples of the world, and it is estimated that about one in three U.S. college students has one or more. Tattoos are created by injecting ink with a needle that punctures the epidermis and moves between 50 and 3000 times per minute as the ink is deposited in the dermis. Since the dermis is stable (unlike the epidermis, which is shed about every four to six weeks), tattoos are permanent. However, they can fade over time due to exposure to sunlight, improper healing, picking scabs, and flushing away of ink particles by the lymphatic system. Tattoos can be removed by lasers in a series of treatments that use concentrated beams of light. In the procedure, the tattoo inks and pigments selectively absorb the high-intensity laser light without destroying normal surrounding skin tissue. The laser causes the tattoo to dissolve into small ink particles that are eventually removed by the immune system. Laser removal of tattoos involves a considerable investment of time and money and can be quite painful.

Body piercing, the insertion of jewelry through an artificial opening, is also an ancient practice that was employed by Egyptian pharaohs and Roman soldiers, and is a current tradition among many Americans. Today it is estimated that about one in two U.S. college students has had a body piercing. For most piercing locations, the practitioner cleans the skin with an antiseptic, retracts the skin with forceps, and pushes a needle through the skin. Then the jewelry is connected to the needle and pushed through the skin. Total healing can take up to a year. Among the sites that are pierced are the ears, nose, eyebrows, lips, tongue, nipples, navel, and genitals. Potential complications of body piercing are infections, allergic reactions, and anatomical damage (such as nerve damage or cartilage deformation). In addition, body-piercing jewelry may interfere with certain medical procedures such as masks used for resuscitation, airway management procedures, urinary catheterization, radiographs, and delivery of a baby.

 CHECKPOINT

6. What are the three pigments in the skin, and how do they contribute to skin color?
7. How do albinism and vitiligo differ?

Subcutaneous Layer or Hypodermis

The **subcutaneous** (*under the skin*) **layer**, which is deep to the dermis, is also referred to as the **hypodermis** (*hypo*=below) (see Figure 5.1). This connective tissue layer, which is not part of the skin, differs from region to region in the body. In some areas it is a thin layer comprised of the loose connective tissue called *areolar tissue*, while in other regions it is a thick, tough layer of fibrous

bands of collagen accompanied by adipose tissue. Fibers that extend from the dermis anchor the skin to the subcutaneous layer, which in turn attaches to the underlying *fascia*, the connective tissue that surrounds muscles and bones. The subcutaneous layer serves as a storage depot for fat and contains large blood vessels that supply and drain the capillaries of the skin. The amount of fat deposited in subcutaneous regions of adipose tissue varies greatly among different individuals. A lean person could have a very thin layer with minimal fat deposits, while an obese individual could have a layer of fat four to six inches thick. Like some areas of the dermis, the subcutaneous layer also contains encapsulated nerve endings called **lamellated (pacinian) corpuscles** (pa-SIN-ē-an) that are sensitive to pressure (see Figure 5.1).

The subcutaneous layer has multiple functions. It functions as a loose binding tissue that unites the upper layers of the skin to deeper structures, while at the same time allowing the skin to move freely over these deeper structures. In certain regions, such as the skin of the soles and palms, the subcutaneous layer forms tough fat pads, composed of fibrous bands of collagen and adipose tissue, which absorb shock and protect underlying muscle and bone. As noted previously, the subcutaneous layer is also the principal site of energy storage in the body. Adipose tissue is an active metabolic tissue with numerous nerve endings and rich vascular networks that help regulate and mobilize the energy stores. Finally, the subcutaneous layer serves as a layer of insulation that helps retard heat loss from the body.

 CHECKPOINT

8. How does the structure of the subcutaneous layer vary from one body region to another?
9. State three functions of the subcutaneous layer.

CLINICAL CONNECTION | *Skin Cancer*

Excessive exposure to the sun causes virtually all of the 1 million cases of **skin cancer** diagnosed in the United States annually. There are three common forms of skin cancer. **Basal cell carcinomas** account for about 78 percent of all skin cancers. The tumors arise from cells in the stratum basale of the epidermis and rarely metastasize. **Squamous cell carcinomas**, which account for about 20 percent of all skin cancers, arise from the stratum spinosum of the epidermis, and they have a variable tendency to metastasize. Most arise from preexisting lesions of damaged tissue in sun-exposed skin. Basal and squamous cell carcinomas are together known as *non-melanoma skin cancer* and are 50 percent more common in males than in females. **Malignant melanomas** arise from melanocytes and account for about 2 percent of all skin cancers. The American Academy of Dermatology estimates that the lifetime risk of developing melanoma is currently 1 in 75, double the risk only 15 years ago. In part, this increase is due to depletion of the ozone layer, which absorbs some UV light high in the atmosphere. However, the main reason for the increase is that more people are spending more time in the sun. Malignant melanomas metastasize rapidly and can kill a person within months of diagnosis.

The key to successful treatment of malignant melanoma is early detection. The early warning signs of malignant melanoma are identified by the acronym ABCD (compare Figures A and B). A is for asymmetry; malignant melanomas tend to lack symmetry. B is for border; malignant melanomas have notched, indented, scalloped, or indistinct borders. C is for color; malignant melanomas have uneven coloration and may contain several different colors. D is for diameter; ordinary moles tend to be smaller than 6 mm (0.25 in.), about the size

 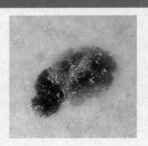

(A) Normal nevus (mole)   (B) Malignant melanoma

of a pencil eraser. Once a malignant melanoma has the characteristics of A, B, and C, it is usually larger than 6 mm.

Among the risk factors for skin cancer are the following:

1. *Skin type*. Individuals with light-colored skin who never tan but always burn are at high risk.
2. *Sun exposure*. People who live in areas with many days of sunlight per year and at high altitudes (where ultraviolet light is more intense) have a higher risk of developing skin cancer. Likewise, people who engage in outdoor occupations and those who have suffered three or more severe sunburns have a higher risk.
3. *Family history*. Skin cancer rates are higher in some families than in others.
4. *Age*. Older people are more prone to skin cancer owing to longer total exposure to sunlight.
5. *Immunological status*. Individuals who are immunosuppressed have a higher incidence of skin cancer. •

5.2 ACCESSORY STRUCTURES OF THE SKIN

● OBJECTIVE

• Compare the structure, distribution, and functions of hair, skin glands, and nails.

Accessory structures of the skin—hair, skin glands, and nails—develop from the embryonic epidermis. Among their important functions are protection of the body by hair and nails, and regulation of body temperature by sweat glands.

Hair

The human body has a dense covering of **hairs**, or *pili* (PĪ-lī) (Figure 5.5), but they are not nearly as numerous as the hairs on the skin of some other mammals. The distribution of hairs varies from approximately 600 per cm^2 (3900 per $in.^2$) on facial skin to 60 per cm^2 (390 per $in.^2$) on the skin over the rest of the body. When you look at a square centimeter of your skin, it is difficult to notice 600 or even 60 hairs on this small patch of skin, because hairs can range in length from a fraction of a millimeter (.04 in.) to over a meter (1.09 yards), with widths from 0.005 mm (the

Figure 5.5 Hair.

🔑 Hairs are growths of dead, keratinized epidermal cells.

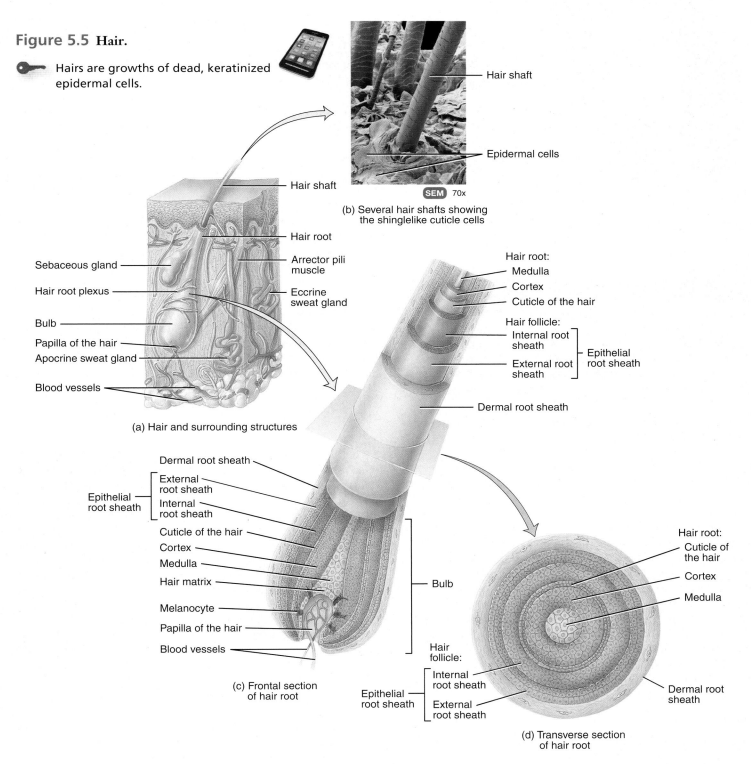

(b) Several hair shafts showing the shinglelike cuticle cells

(a) Hair and surrounding structures

(c) Frontal section of hair root

(d) Transverse section of hair root

FIGURE 5.5 CONTINUES ▶

◼ FIGURE 5.5 CONTINUED ▶

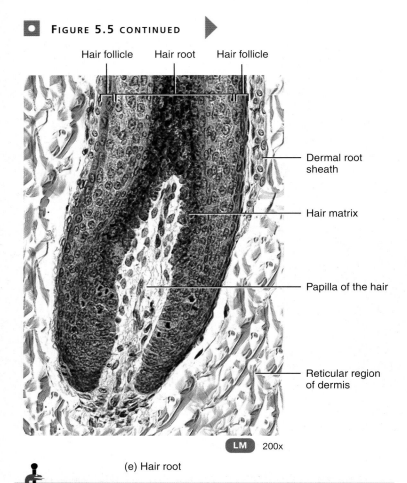

Hair follicle   Hair root   Hair follicle

Dermal root sheath

Hair matrix

Papilla of the hair

Reticular region of dermis

LM   200x

(e) Hair root

CLINICAL CONNECTION | Hair Removal

A substance that removes hair is called a **depilatory** (de-PIL-a-tō-rē). It dissolves the protein in the hair shaft, turning it into a gelationous mass that can be wiped away. Because the hair root is not affected, regrowth of the hair occurs. In **electrolysis**, an electric current is used to destroy the hair matrix so the hair cannot regrow. **Laser treatments** may also be used to remove hair. •

? **Why does it hurt when you pluck a hair but not when you have a haircut?**

diameter of some of the smallest cells in the body) to 0.6 mm. So, while some hairs are very obvious to the unaided eye, others are visible only with the use of a microscope.

Hairs are absent from the palms, palmar surfaces of the fingers, soles, and plantar surfaces of the toes. In adults, hair is usually most heavily distributed across the scalp, over the brows of the eyes, in the armpits, and around the external genitalia. Genetic and hormonal influences largely determine the thickness and pattern of hair distribution.

Although the protection it offers is limited, hair on the head guards the scalp from injury and the sun's rays. It also decreases heat loss from the scalp. Eyebrows and eyelashes protect the eyes from foreign particles, similar to the way that hair in the nostrils and in the external ear canal defend those structures. Touch receptors associated with hair follicles (hair root plexuses) are activated by the slightest movement of a hair.

Anatomy of a Hair

Each hair is composed of columns of dead, keratinized epidermal cells bonded together by extracellular proteins. The **shaft** is the superficial portion of the hair, most of which projects from the

surface of the skin (Figure 5.5a). The shape of the shafts of hairs varies with different ethnic groups. Straight hair has round shafts; wavy hair has oval shafts; and curly hair has kidney-shaped shafts.

The **root** is the portion of the hair deep to the shaft that penetrates into the dermis, and sometimes into the subcutaneous layer. The shaft and root both consist of three concentric layers (Figure 5.5c–e). The inner *medulla* is composed of two or three rows of irregularly shaped cells containing pigment granules and air spaces. The middle *cortex* forms the major part of the shaft and consists of elongated cells that contain pigment granules in dark hair but mostly air spaces between the cells in gray hair and the absence of pigmentation in white hair. The *cuticle* of the hair, the outermost layer, consists of a single layer of thin, flat cells that are the most heavily keratinized. Cuticle cells are arranged like shingles on the roof of a house, with their free edges pointing toward the free end of the hair (Figure 5.5b).

Surrounding the root of the hair is the **hair follicle** (FOL-li-kul), which is made up of an external root sheath and an internal root sheath (Figure 5.5c–e). The *external root sheath* is a downward continuation of the epidermis. Near the surface of the skin, it contains all the epidermal layers. At the base of the hair follicle, the external root sheath contains only the stratum basale. The *internal root sheath* is produced by the hair matrix (described shortly) and forms a cellular tubular sheath of epithelium between the external root sheath and the hair. Together, the external and internal root sheath are referred to as the **epithelial root sheath**. The dense dermis surrounding the hair follicle is called the **dermal root sheath**.

The base of each hair follicle and its surrounding dermal root sheath is an onion-shaped structure, the **bulb** (Figure 5.5c). This structure houses a nipple-shaped indentation, the **papilla of the hair**, which contains areolar connective tissue and many blood vessels that nourish the growing hair follicle. The bulb also contains a germinal layer of cells called the **hair matrix**. Hair matrix cells arise from the stratum basale, the site of cell division. Hence, hair matrix cells are responsible for the growth of existing hairs, and they produce new hairs when old hairs are shed. This replacement process occurs within the same follicle. Hair matrix cells also give rise to the cells of the internal root sheath.

CLINICAL CONNECTION | Chemotherapy and Hair Loss

Chemotherapy is the treatment of disease, usually cancer, by means of chemical substances or drugs. Chemotherapeutic agents interrupt the life cycle of rapidly dividing cancer cells. Unfortunately, the drugs also affect other rapidly dividing cells in the body, such as the hair matrix cells. It is for this reason that individuals undergoing chemotherapy experience hair loss. Since about 15 percent of the hair matrix cells of scalp hairs are in the resting stage, these cells are not affected by chemotherapy. Once chemotherapy is stopped, the hair matrix cells replace lost hair follicles and hair growth resumes. Other side effects of chemotherapy include red bone marrow suppression (which can result in infections, bleeding problems, and anemia), nausea and vomiting, appetite and weight changes, diarrhea or constipation, fatigue, nervous system disorders, reproductive disorders, and liver and kidney damage. •

Sebaceous (oil) glands (discussed shortly) and a bundle of smooth muscle cells are also associated with hairs (Figure 5.5a). The smooth muscle, called **arrector pili** (a-REK-tor PĪ-lī; *arrect*=to raise), extends from the superficial dermis of the skin to the connective tissue sheath around the hair follicle. In its normal position, hair emerges at a less than 90° angle to the surface of the skin. Under physiological or emotional stress, such as cold, fright, or anger, autonomic nerve endings stimulate the arrector pili muscles to contract, which pulls the hair shafts perpendicular to the skin surface. This action, which results in a slight elevation around each shaft of hair, causes "goose bumps" or "goose flesh." Why does this happen? It is difficult to see the significance of this function in humans. However, in animals with a lot of hair, this mechanism can be very important for survival. By erecting the hairs on their bodies, most mammals can trap dead (unmoving) air between the hairs, creating dead air space, the most effective form of insulation known. Similar to the principle behind double-pane insulated glass windows, this thin layer of dead air space helps reduce heat loss when the body is exposed to cold. Hair erection can also make mammals look bigger and more formidable, which might help them intimidate predators. Humans continue to exhibit hair erection, but it has lost its effectiveness in reducing heat loss or intimidating enemies because our hair has become relatively sparse.

Dendrites of neurons surround each hair follicle; these form a **hair root plexus** (PLEK-sus), which is sensitive to touch (Figure 5.5a). The hair root plexus initiates nerve impulses if the hair shaft is moved, as when an insect bumps into it when crawling across your skin.

Hair Growth

Normal hair loss in an adult scalp is about 70–100 hairs per day. Both the rate of growth and the replacement cycle can be altered by illness, diet, high fever, surgery, blood loss, severe emotional stress, and gender. Rapid weight-loss diets that severely restrict the intake of calories or protein increase hair loss. An increase in the rate of hair shedding can also occur with certain drugs, after radiation therapy for cancer, and for 3–4 months after childbirth. **Alopecia** (al-ō-PĒ-shē-a), the partial or complete lack of hair, may result from genetic factors, aging, endocrine disorders, chemotherapy for cancer, or skin disease.

CLINICAL CONNECTION | Hair and Hormones

At puberty, when the testes begin secreting significant quantities of androgens (masculinizing sex hormones), males develop the typical male pattern of hair growth, including a beard and a hairy chest. In females at puberty, the ovaries and the adrenal glands produce small quantities of androgens, which promote the growth of coarse hairs in the axillae and pubic region. Occasionally, a tumor of the adrenal glands, testes, or ovaries produces an excessive amount of androgens. The result in females or prepubertal males is **hirsutism** (HER-soo-tizm; *hirsut-*=shaggy), excessive body hair, or hairiness in areas that are usually not hairy. •

Each hair follicle goes through a growth cycle; the cycle consists of a growth stage, a regression stage, and a resting stage. During the **growth stage**, cells of the hair matrix divide. As new cells from the hair matrix are added to the base of the hair root, existing cells of the hair root are pushed upward and the hair grows longer. While the cells of the hair are being pushed upward, they become keratinized and die. Following the growth stage is the **regression stage**, when cells of the hair matrix stop dividing, the hair follicle atrophies (shrinks), and the hair stops growing. After the regression stage, the hair follicle enters a **resting stage**. Following the resting stage, a new growth cycle begins. The old hair root falls out or is pushed out of the hair follicle, and a new hair begins to grow in its place. Scalp hair is in the growth stage for 2 to 6 years, the regression stage for 2 to 3 weeks, and the resting stage for about 3 months. At any time, about 85 percent of scalp hairs are in the growth stage. Visible hair is dead, which is why, despite the fears of many children, haircuts don't hurt a bit. However, until the hair is pushed out of its follicle by a new hair, portions of its root within the scalp are alive and surrounded by nerve endings, which is the reason for the verbal explosions that result if those same children yank on the hair of their siblings.

CLINICAL CONNECTION | Baldness

Surprisingly, androgens also must be present for occurrence of the most common form of baldness, **androgenic alopecia** (an'-drō-JEN-ik al'-ō-PĒ-shē-a) or **male-pattern baldness**. In genetically predisposed adults, androgens inhibit hair growth. In men, hair loss usually begins with a receding hairline followed by hair loss in the temples and crown. Women are more likely to have thinning of hair on top of the head. The first drug approved for enhancing scalp hair growth was minoxidil (Rogaine®). It causes vasodilation (widening of blood vessels), thus increasing circulation; direct stimulation of hair follicle cells to pass into growth-stage follicles; and inhibition of androgens. In about a third of the people who try it, minoxidil improves hair growth, causing scalp follicles to enlarge and lengthening the growth cycle. For many, however, the hair growth is meager. Minoxidil does not help people who already are bald. •

Types of Hairs

Hair follicles develop at about 12 weeks after fertilization as downgrowths of the stratum basale of the epidermis into the dermis (see Figure 5.8e). Usually by the fifth month of development, the follicles produce very fine, nonpigmented hairs called **lanugo** (la-NOO-gō=wool or down) that cover the body of the fetus. This hair is shed before birth, except from the scalp, eyebrows, and eyelashes. A few months after birth, slightly thicker hairs replace these hairs. Over the remainder of the body of an infant, a new growth of short, fine hair occurs. These hairs, known as **vellus hairs** (VEL-us=fleece), are commonly called "peach fuzz." The hairs that develop at puberty, together with those of the head, eyebrows, and eyelashes, are called **terminal hairs**. About 95 percent of body hair on males is terminal hair (the other 5 percent is vellus hair). Only about 35 percent of body hair on females is terminal hair; the other 65 percent is vellus hair.

Hair Color

Similar to its activity in the epidermis, hair color is the result of melanocyte activity in the stratum basale at the base of the hair follicle. During the active phase of the growth cycle, large melanocytes in the central portion of the bulb distribute melanosomes to the keratinocytes of the medulla and cortex of the developing hair (Figure 5.5c). These melanocytes form melanosomes containing melanin pigment as either pheomelanin (yellow to red) or eumelanin (brown to black). The resulting hair color arises from the ratio of the different pigment-containing melanosomes produced. With age, the melanocyte number and activity level at the base of the follicle declines, which leads to a decrease in the pigmentation of the hair. White hair results from lack of pigment for the medulla of the hair. In addition, the number of air spaces within the medulla of the hair increases. The air spaces change the way the hair reflects light, resulting in gray hair.

Hair coloring is a process that adds or removes pigment. *Temporary hair dyes* coat the surface of a hair shaft and usually wash out within 2 or 3 shampoos. *Semipermanent hair dyes* penetrate the hair shaft moderately and do fade and wash out of hair after about 5 to 10 shampoos. *Permanent hair dyes* penetrate deeply into the hair shaft and do not wash out, but are eventually lost as the hair grows out.

Skin Glands

Recall from Chapter 3 that *glands* are single or groups of epithelial cells that secrete a substance. Several kinds of exocrine glands are associated with the skin: sebaceous (oil) glands, sudoriferous (sweat) glands, ceruminous glands, and mammary glands. Mammary glands, which are specialized sudoriferous glands that secrete milk, are discussed in Chapter 26 along with the female reproductive system.

Sebaceous Glands

Sebaceous glands (se-BĀ-shus; *sebace*=greasy), or **oil glands**, are simple, branched acinar (rounded) glands. With few exceptions, they are connected to hair follicles (Figure 5.6a; see also Figures 5.1 and 5.5a). The secreting portion of a sebaceous gland lies in the dermis, typically situated in the angle the arrector pili

muscle forms with the outer wall of a hair follicle, and usually opens into the neck of the hair follicle. In other locations, such as the lips, glans penis, and labia minora, sebaceous glands open directly onto the surface of the skin. Sebaceous glands, which vary in size and shape, are found in the skin over all regions of the body except the palms and soles. Because of their relationship to hairs, sebaceous glands are most numerous where hairs are most numerous. They are small in most areas of the trunk and limbs, but large in the skin of the breasts, face, neck, and upper chest.

The sac-like base of a sebaceous gland has a lining of cuboidal cells that resemble basal epithelial cells. Like the basal epithelial cells, these cells divide and produce cells that are pushed away from the lining as newer generations of cells are produced. As multiple generations of cells pile on top of each other, the lumen gradually fills with cells. These cells differentiate by developing large lipid-filled vesicles in their cytoplasm, and eventually become so distended that they rupture and fill the duct of the gland with an oily secretion called **sebum** (SĒ-bum; *sebum*=oily substance), which is derived from the released lipid-filled vesicles and cellular debris. Sebum is a mixture of fats, cholesterol, proteins, and inorganic salts. Sebum coats the surface of hairs and helps keep them from drying and becoming brittle. Sebum also prevents excessive evaporation of water from the skin, keeps the skin soft and pliable, and inhibits the growth of certain bacteria.

Figure 5.6 Histology of skin glands.

🔑 Sebaceous glands are simple branched acinar glands; eccrine and apocrine sweat glands are simple coiled tubular glands.

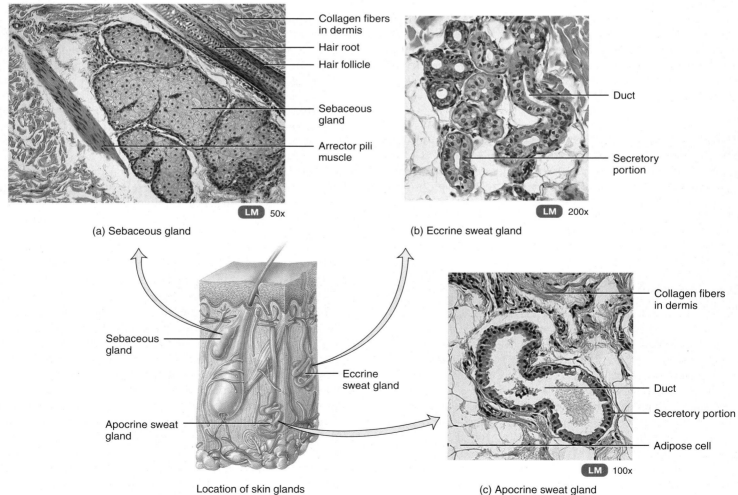

(a) Sebaceous gland

(b) Eccrine sweat gland

Location of skin glands

(c) Apocrine sweat gland

❓ **What is the main function of eccrine sweat glands?**

CLINICAL CONNECTION | *Acne*

During childhood, sebaceous glands are relatively small and inactive. At puberty, androgens from the testes, ovaries, and adrenal glands stimulate sebaceous glands to grow in size and increase their production of sebum. **Acne** is an inflammation of sebaceous glands that usually begins at puberty, when the sebaceous glands are stimulated by androgens. Acne occurs predominantly in sebaceous follicles that have been colonized by bacteria, some of which thrive in the lipid-rich sebum. The infection may cause a cyst or sac of connective tissue cells to form, which can destroy and displace epidermal cells. This condition, called **cystic acne**, can permanently scar the epidermis. •

Sudoriferous Glands

There are three to four million **sweat glands**, or **sudoriferous glands** (soo'-dor-IF-er-us; *sudor*=sweat; *-ferous*=bearing), in the body. The cells of sweat glands release their secretions by exocytosis and empty them onto the skin surface through pores or into hair follicles. Depending on their structure and type of secretion, the glands are classified as either eccrine or apocrine.

Eccrine sweat glands (EK-krin; *eccrine*=secreting outwardly), also known as **merocrine sweat glands**, are simple, coiled tubular glands (Figure 5.6b) that are much more common than apocrine sweat glands. They are distributed throughout the skin of most parts of the body, except for the margins of the lips, nail beds of the fingers and toes, glans penis, glans clitoris, labia minora, and eardrums. Eccrine sweat glands are most numerous in the skin of the forehead, palms, and soles; their density can be as high as 450 per square centimeter (3000 per square inch) in the palms. The secretory portion of eccrine sweat glands is located mostly in the deep dermis (sometimes in the upper subcutaneous layer). The excretory duct projects through the dermis and epidermis and ends as a pore at the surface of the epidermis (see also Figures 5.1 and 5.4c). On the palms and soles, they open all along the apex of the epidermal ridges. The secretions of eccrine sweat glands form the fingerprints that allow your lab partner to pinpoint who "borrowed" their dissection kit and didn't clean it up afterwards.

The sweat produced by eccrine sweat glands (about 600 mL per day) consists primarily of water, with small amounts of ions (mostly Na^+ and Cl^-), urea, uric acid, ammonia, amino acids, glucose, and lactic acid. The main function of eccrine gland sweat is to help regulate body temperature through evaporation. As sweat evaporates, large quantities of heat energy leave the body surface. The homeostatic regulation of body temperature is known as **thermoregulation**. The role of eccrine sweat glands in helping the body to achieve thermoregulation is known as **thermoregulatory sweating**. During thermoregulatory sweating, sweat first forms on the forehead and scalp and then extends to the rest of the body, forming last on the palms and soles. Sweat that evaporates from the skin before it is perceived as moisture is termed **insensible perspiration** (*in*=not). Sweat that is excreted in larger amounts and is seen as moisture on the skin is called **sensible perspiration**.

The sweat produced by eccrine sweat glands also plays a small role in eliminating wastes such as urea, uric acid, and ammonia. However, as you will learn in Chapter 25, the kidneys play a much greater role in the excretion of these waste products from the body.

Eccrine sweat glands also release sweat in response to an emotional stress such as fear or embarrassment. This type of sweating is referred to as **emotional sweating** or a *cold sweat*. In contrast to thermoregulatory sweating, emotional sweating first occurs on the palms, soles, and axillae and then spreads to other areas of the body.

Another type of sweat gland, called the apocrine sweat gland, is also active during emotional sweating. **Apocrine sweat glands** (AP-ō-krin; *apo*=separated from), like eccrine sweat glands, are simple, coiled tubular glands but they have larger ducts and lumens than the eccrine sweat glands (see Figures 5.1 and 5.4a). They are found mainly in the skin of the axilla (armpit), groin, areolae (pigmented areas around the nipples) of the breasts, and bearded regions of the face in adult males. These glands were once thought to release their secretions in an apocrine manner (see Section 3.4 and Figure 3.5b)—by pinching off a portion of the cell. We now know that their secretion is released by exocytosis, which is characteristic of the release of secretions by merocrine glands (see Figure 3.5a). Nevertheless, the term *apocrine* is still used. The secretory portion of these sweat glands is located mostly in the subcutaneous layer, and the excretory duct opens into hair follicles (see also Figure 5.1).

The secretory portion of an apocrine gland has a simple cuboidal lining of secretory cells whose free surfaces project into the lumen of the gland (Figure 5.6c). Surrounding the secretory portion of the gland is a vascular connective tissue containing numerous nerve endings. Surrounding the upper portions of the coils and the duct of the gland is a thin coat of **myoepitheliocytes** (mī'-ō-ep'-i-THĒ-lē-ō-sīts; *myo*=muscle). These muscle-like contractile epithelial cells help squeeze out the secretions from the gland into the hair follicle. Their secretory product is slightly viscous compared to eccrine secretions and contains the same components as eccrine sweat plus lipids and proteins. At first this substance is odorless, but as it spreads onto the hairs, bacteria decompose the proteins, creating a strong musky odor that is called *body odor*. In women, cells of apocrine sweat glands enlarge at about the time of ovulation and shrink during menstruation. Eccrine sweat glands start to function soon after birth, but apocrine sweat glands do not begin to function until puberty. Apocrine sweat glands are stimulated during emotional stress and sexual excitement; these secretions are commonly known as a "cold sweat." In contrast to eccrine sweat glands, apocrine sweat glands are not active during thermoregulatory sweating.

Ceruminous Glands

Modified sweat glands in the external ear, called **ceruminous glands** (se-ROO-mi-nus; *cer*=wax), produce a yellowish waxy lubricating secretion. The secretory portions of ceruminous glands lie in the subcutaneous layer, deep to sebaceous glands. Their excretory ducts open either directly onto the surface of the external auditory canal (ear canal) or into ducts of sebaceous glands. The combined secretion of the ceruminous and sebaceous glands is called **cerumen** (se-ROO-men), or earwax. Cerumen in the external auditory canal provides a sticky barrier that prevents the entrance of foreign bodies, such as insects. Cerumen also waterproofs the canal and keeps bacteria and fungi from entering cells.

Some people produce an abnormally large amount of cerumen in the external auditory canal. If it accumulates until it becomes impacted (firmly wedged), sound waves may be prevented from reaching the eardrum. Treatments for **impacted cerumen** include periodic ear irrigation with enzymes to dissolve the wax and removal of wax with a blunt instrument by trained medical personnel.

Table 5.3 presents a summary of skin glands.

TABLE 5.3

Summary of Skin Glands (See also Figures 5.1 and 5.4a)

FEATURE	SEBACEOUS (OIL) GLANDS	ECCRINE SWEAT GLANDS	APOCRINE SWEAT GLANDS	CERUMINOUS GLANDS
Distribution	Large in lips, glans penis, labia minora, and tarsal glands; small in trunk and limbs; absent in palms and soles	Throughout skin of most regions of the body, especially in skin of forehead, palms, and soles	Skin of axilla, groin, areolae, bearded regions of face, clitoris, and labia minora	External auditory canal
Location of secretory portion	Dermis	Mostly in deep dermis	Mostly in subcutaneous layer	Subcutaneous layer
Termination of excretory duct	Mostly connected to hair follicles	Surface of epidermis	Hair follicle	Surface of external auditory canal or into ducts of sebaceous glands
Secretion	Sebum (mixture of triglycerides, cholesterol, proteins), and inorganic salts	Less viscous; consists of water, ions (Na^+, Cl^-), urea, uric acid, ammonia, amino acids, glucose, and lactic acid	More viscous; consists of the same components as eccrine sweat glands plus lipids and proteins	Cerumen, a waxy material
Functions	Prevent hairs from drying out, prevent water loss from skin, keep skin soft, and inhibit growth of some bacteria	Regulation of body temperature, waste removal, and stimulation during emotional stress	Stimulation during emotional stress and sexual excitement	Impedes entrance of foreign bodies and insects into external ear canal, waterproofs canal, and prevents microbes from entering cells
Onset of function	Relatively inactive during childhood; activated during puberty	Soon after birth	Puberty	Soon after birth

Nails

Nails are plates of tightly packed, hard, dead keratinized epidermal cells. The cells form a clear, solid covering over the dorsal surfaces of the distal portions of the digits. Each nail consists of a nail body, a free edge, and a nail root (Figure 5.7). The **nail body (plate)** is the visible portion of the nail. It is comparable to the stratum corneum of the epidermis of the skin, but its flattened, keratinized cells fill with a harder type of keratin and the cells are not shed. Below the nail body is a layer of epithelium and a deeper layer of dermis. Most of the nail body appears pink because of blood flowing through underlying capillaries. The **free edge** is the part of the nail body that may extend past the distal end of the digit. The free edge is white because there are no underlying capillaries. The **nail root** is the part of the nail that is buried in a fold of skin. The whitish, crescent-shaped area of the proximal end of the nail body is called the **lunula** (LOO-noo-la=little moon). It appears whitish because the thickened stratum basale in this area does not allow the vascular tissue underneath to show through. Beneath the free edge is a thickened region of stratum corneum called the **hyponychium** (hī-pō-NIK-ē-um; *hypo*=below; *-onych*=nail), which secures the nail to the fingertip. The **nail bed** is the skin below the nail plate that extends from the lunula to the hyponychium. The epidermis of the nail bed lacks a stratum granulosum. The **eponychium** (ep-ō-NIK-ē-um;

ep=above) or **cuticle** is a narrow band of epidermis that extends from and adheres to the margin (base and lateral border) of the nail wall. It occupies the proximal border of the nail and consists of stratum corneum.

The portion of the epithelium proximal to the nail root is known as the **nail matrix**. The superficial nail matrix cells divide mitotically to produce new nail cells. In the process, the harder outer layer is pushed forward over the stratum basale. The growth rate of nails is determined by the rate at which nail matrix cells divide, which is influenced by factors such as your age, health, and nutritional status. Nail growth also varies according to the season, the time of day, and environmental temperature. The average growth in the length of fingernails is about 1 mm (.04 in.) per week. The growth rate is somewhat slower in toenails. The longer the digit, the faster the nail grows.

Nails have a variety of functions:

1. They protect the distal end of the digits.
2. They provide support and counterpressure to the palmar surface of the fingers to enhance touch perception and manipulation.
3. They allow us to grasp and manipulate small objects, and they can be used to scratch and groom the body in various ways.

Figure 5.7 **Nails.** Shown is a fingernail.

Nail cells arise by transformation of superficial cells of the nail matrix.

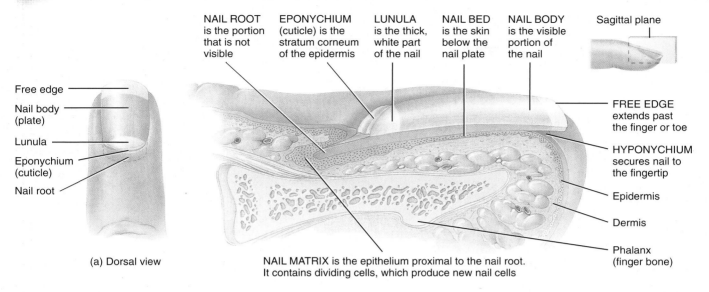

NAIL ROOT is the portion that is not visible

EPONYCHIUM (cuticle) is the stratum corneum of the epidermis

LUNULA is the thick, white part of the nail

NAIL BED is the skin below the nail plate

NAIL BODY is the visible portion of the nail

Sagittal plane

Free edge
Nail body (plate)
Lunula
Eponychium (cuticle)
Nail root

FREE EDGE extends past the finger or toe

HYPONYCHIUM secures nail to the fingertip

Epidermis

Dermis

Phalanx (finger bone)

(a) Dorsal view

NAIL MATRIX is the epithelium proximal to the nail root. It contains dividing cells, which produce new nail cells

(b) Sagittal section showing internal detail

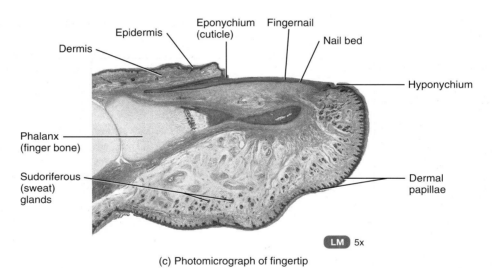

Epidermis
Dermis
Eponychium (cuticle)
Fingernail
Nail bed
Hyponychium
Phalanx (finger bone)
Sudoriferous (sweat) glands
Dermal papillae

LM 5x

(c) Photomicrograph of fingertip

 Why are nails so hard?

 CHECKPOINT

10. Describe the structure of a hair.
11. What produces "goosebumps"?
12. Contrast the locations and functions of sebaceous (oil) glands, sudoriferous (sweat) glands, and ceruminous glands.
13. Describe the principal parts of a nail.

5.3 TYPES OF SKIN

 OBJECTIVE

• Compare structural and functional differences in thin and thick skin.

Although the skin over the entire body is similar in structure, there are quite a few local variations related to thickness of the epidermis, strength, flexibility, degree of keratinization, distribution and type of hair, density and types of glands, pigmentation, vascularity (blood supply), and innervation (nerve supply). Two major types of skin are recognized on the basis of certain structural and functional properties: **thin (hairy) skin** and **thick (hairless) skin** (see also Section 5.1). The greatest contributor to epidermal thickness is the increased number of layers in the stratum corneum. This arises in response to the greater mechanical stress in regions of thick skin.

Table 5.4 presents a comparison of the features of thin and thick skin.

 CHECKPOINT

14. What criteria are used to distinguish thin and thick skin?

TABLE 5.4

Comparison of Thin and Thick Skin

FEATURE	THIN SKIN	THICK SKIN
Distribution	All parts of the body except areas such as palms and palmar surface of digits, and soles	Areas such as the palms, palmar surface of digits, and soles
Epidermal thickness	0.10–0.15 mm (0.004–0.006 in.)	0.6–4.5 mm (0.024–0.18 in.), due mostly to a thicker stratum corneum
Epidermal strata	Stratum lucidum essentially lacking; thinner strata spinosum and corneum	Strata lucidum present; thicker stratum spinosum and corneum
Epidermal ridges	Lacking due to poorly developed and fewer and less-well-organized dermal papillae	Present due to well-developed and more numerous dermal papillae organized in parallel rows
Hair follicles and arrector pili muscles	Present	Absent
Sebaceous glands	Present	Absent
Sudoriferous glands	Fewer	More numerous
Sensory receptors	Sparser	Denser

5.4 FUNCTIONS OF THE SKIN

 OBJECTIVE

- Describe how the skin contributes to the regulation of body temperature, storage of blood, protection, sensation, excretion and absorption, and synthesis of vitamin D.

Now that you have a basic understanding of the structure of the skin, you can better appreciate its many functions. Among the numerous functions of the integumentary system (mainly the skin) are the following:

1. **Thermoregulation.** The skin contributes to *thermoregulation*, the homeostatic regulation of body temperature, in two ways: by liberating sweat at its surface and by adjusting the flow of blood in the dermis. In response to high environmental temperature or heat produced by exercise, sweat production from eccrine sweat glands increases; the evaporation of sweat from the skin surface helps lower body temperature. In addition, blood vessels in the dermis of the skin dilate (become wider); consequently, more blood flows through the dermis, which increases the amount of heat loss from the body. In response to low environmental temperature, production of sweat from eccrine sweat glands is decreased, which helps conserve heat. Also, the blood vessels in the dermis of the skin constrict (become narrow), which decreases blood flow through the skin and reduces heat loss from the body. In addition, skeletal muscle contractions generate body heat.

2. **Blood reservoir.** The dermis houses an extensive network of blood vessels that carry 8–10 percent of the total blood flow in a resting adult. For this reason, the skin is considered a *blood reservoir*.

3. **Protection.** The skin provides *protection* to the body in various ways. Keratin protects underlying tissues from microbes, abrasion, heat, and chemicals; the tightly interlocked keratinocytes resist invasion by microbes. Lipids released by lamellar granules inhibit evaporation of water from the skin surface, thus protecting against dehydration; they also retard entry of water across the skin surface during showers

and swims. The oily sebum from the sebaceous glands protects skin and hairs from drying out and contains *bactericidal chemicals* (substances that kill bacteria). The acidic pH of perspiration retards the growth of some microbes. The pigment melanin helps shield against the damaging effects of ultraviolet light. Two types of skin cells carry out protective functions that are immunological in nature. Intraepidermal macrophage cells alert the immune system to the presence of potentially harmful microbial invaders by recognizing and processing them, and macrophages in the dermis phagocytize bacteria and viruses that manage to bypass the intraepidermal macrophage of the epidermis.

4. **Cutaneous sensations.** *Cutaneous sensations* are sensations that arise in the skin, including tactile sensations—touch, pressure, vibration, and tickling—as well as thermal sensations such as warmth and coolness. Another cutaneous sensation, pain, usually is an indication of impending or actual tissue damage. There is a wide variety of nerve endings and receptors distributed throughout the skin; you have already read about the tactile discs of the epidermis, the corpuscles of touch in the dermis, and hair root plexuses around each hair follicle. Chapter 21 provides more details on the topic of cutaneous sensations.

5. **Excretion and absorption.** The skin normally has a small role in *excretion*, the elimination of substances from the body, and *absorption*, the passage of materials from the external environment into body cells. Despite the almost waterproof nature of the stratum corneum, about 400 mL of water evaporates through it daily. A sedentary person loses an additional 200 mL per day as sweat; a physically active person loses much more. Besides removing water and heat from the body, sweat also is the vehicle for excretion of small amounts of salts, carbon dioxide, and two organic molecules that result from the breakdown of proteins—ammonia and urea.

The absorption of water-soluble substances through the skin is negligible, but certain lipid-soluble materials do penetrate the skin. These include fat-soluble vitamins (A, D, E, and K), certain drugs, and oxygen and carbon dioxide gases. Toxic materials that can be absorbed through the

CLINICAL CONNECTION | *Burns*

A **burn** is tissue damage caused by excessive heat, electricity, radioactivity, or corrosive chemicals that denature (destroy) the proteins in the skin cells. Burns destroy some of the skin's important contributions to homeostasis—protection against microbial invasion and dehydration, and regulation of body temperature.

Burns are graded according to their severity. A *first-degree burn* involves only the epidermis (Figure A). It is characterized by mild pain and erythema (redness) but no blisters. Skin functions remain intact. The pain and damage caused by a first-degree burn may be lessened by immediately flushing it with cold water. Generally, a first-degree burn will heal in about 3–6 days and may be accompanied by flaking or peeling. One example of a first-degree burn is a mild sunburn.

A *second-degree burn* destroys a portion of the epidermis and possibly parts of the dermis (Figure B). Some skin functions are lost. In a second-degree burn, redness, blister formation, edema, and pain result. (Blister formation is caused by separation of the epidermis from the dermis due to the accumulation of tissue fluid between the layers.) Associated structures, such as hair follicles, sebaceous glands, and sweat glands, usually are not injured. If there is no infection, second-degree burns heal without skin grafting in about 3–4 weeks, but scarring may result. First- and second-degree burns are collectively referred to as *partial-thickness burns*.

A *third-degree burn*, or *full-thickness burn*, destroys a portion of the epidermis, the underlying dermis, and associated structures (Figure C). Most skin functions are lost. Such burns vary in appearance from marble-white to mahogany colored to charred, dry wounds. There is marked edema, and the burned region is numb because sensory nerve endings have been destroyed. Regeneration occurs slowly, and much granulation tissue forms before being covered by epithelium. Skin grafting may be required to promote healing and to minimize scarring.

The injury to the skin tissues directly in contact with the damaging agent is the *local effect* of a burn. Generally, however, the *systemic*

effects of a major burn are a greater threat to life. The systemic effects of a burn may include (1) a large loss of water, plasma, and plasma proteins, which causes shock; (2) bacterial infection; (3) reduced circulation of blood; (4) decreased production of urine; and (5) diminished immune responses.

The seriousness of a burn is determined by its depth and extent of area involved, as well as the person's age and general health. According to the American Burn Association's classification of burn injury, a major burn includes third-degree burns over 10 percent of body surface area; second-degree burns over 25 percent of body surface area; or any third-degree burns on the face, hands, feet, or *perineum* (per'-i-NĒ-um, which includes the anal and urogenital regions). When the burn area exceeds 70 percent, more than half the victims die.

A quick means for estimating the surface area affected by a burn in an adult is the **rule of nines** (Figure D).

1. Count 9 percent if both the anterior and posterior surfaces of the head and neck are affected.
2. Count 9 percent for both the anterior and posterior surfaces of each upper limb (total of 18 percent for both upper limbs).
3. Count four times nine or 36 percent for both the anterior and posterior surfaces of the trunk, including the buttocks.
4. Count 9 percent for the anterior and 9 percent for the posterior surfaces of each lower limb as far up as the buttocks (total of 36 percent for both lower limbs).
5. Count 1 percent for the perineum.

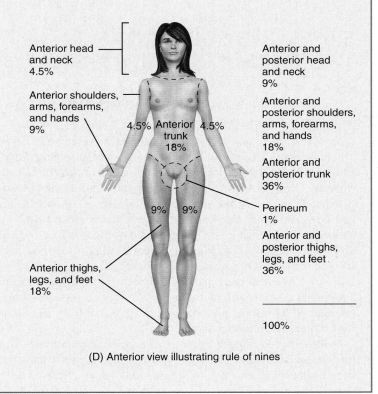

(A) First-degree burn (sunburn)

(B) Second-degree burn (note the blisters in the photograph)

(C) Third-degree burn

(D) Anterior view illustrating rule of nines

skin include organic solvents such as acetone (in some nail polish removers) and carbon tetrachloride (dry-cleaning fluid); salts of heavy metals such as lead, mercury, and arsenic; and the toxins in poison ivy and poison oak. Since topical (applied to the skin) steroids, such as cortisone, are lipid-soluble, they move easily into the papillary region of the dermis. Here, they exert their anti-inflammatory properties

by inhibiting the production of histamines by mast cells (recall that histamine contributes to inflammation; see Section 3.5). Certain drugs that are absorbed through the skin may be administered by applying adhesive patches to the skin. These are discussed in the listing for *transdermal drug administration* in the Key Medical Terms section at the end of the chapter.

6. **Synthesis of vitamin D.** Synthesis of vitamin D requires activation of a precursor molecule in the skin by ultraviolet (UV) rays in sunlight. Enzymes in the liver and kidneys then modify the activated molecule, finally producing *calcitriol*, the most active form of vitamin D. Calcitriol is a hormone that aids in the absorption of calcium from foods in the gastrointestinal tract into the blood. Only a small amount of exposure to UV light (about 10 to 15 minutes at least twice a week) is required for vitamin D synthesis. People who avoid sun exposure and individuals who live in colder, northern climates may require vitamin D supplements to avoid vitamin D deficiency. Most cells of the immune system have vitamin D receptors and the cells activate vitamin D in response to an infection, especially a respiratory infection, such as influenza. Vitamin D is believed to enhance phagocytic activity, increase the production of antimicrobial substances by phagocytes, regulate immune functions, and help reduce inflammation.

 CHECKPOINT

15. In what two ways does the skin help regulate body temperature?
16. How does the skin serve as a protective barrier?
17. What sensations arise from stimulation of neurons in the skin?
18. What types of molecules can penetrate the stratum corneum?

5.5 BLOOD SUPPLY OF THE INTEGUMENTARY SYSTEM

 OBJECTIVE

• Describe the blood supply of the integumentary system.

Although the epidermis is avascular, the dermis is well supplied with blood (see Figure 5.1). The arteries supplying the dermis are generally derived from branches of arteries supplying skeletal muscles in a particular region, but some arteries supply the skin directly. One plexus (network) of arteries, the **cutaneous arterial plexus**, is located at the junction of the dermis and subcutaneous layer and sends branches that supply the sebaceous (oil) and sudoriferous (sweat) glands, the deep portions of hair follicles, and adipose tissue. The **papillary arterial plexus**, formed at the level of the papillary region, sends branches that supply the capillary loops in the dermal papillae, sebaceous (oil) glands, and the superficial portion of hair follicles. The arterial plexuses are accompanied by venous plexuses that drain blood from the dermis into larger subcutaneous veins.

Nutrients and oxygen diffuse to the avascular epidermis from blood vessels in the dermal papillae. The epidermal cells of the stratum basale, which are closest to these blood vessels, receive most of the nutrients and oxygen. These cells are the most active metabolically and continuously undergo cell division to produce new keratinocytes. As the new keratinocytes are pushed further from the blood supply by continuing cell division, the epidermal strata above the stratum basale receive fewer nutrients and the cells become less active and eventually die.

 CHECKPOINT

19. How do the distributions of the cutaneous plexus and the papillary plexus differ?

5.6 DEVELOPMENT OF THE INTEGUMENTARY SYSTEM

 OBJECTIVE

• Describe the development of the epidermis, its accessory structures, and the dermis.

As you learned in Chapter 4, the *epidermis* is derived from the **ectoderm**, which covers the surface of the embryo. Initially, at about the fourth week after fertilization, the epidermis consists of only a single layer of ectodermal cells (Figure 5.8a). At the beginning of the seventh week the single layer, called the **basal layer**, divides and forms a superficial protective layer of flattened cells called the **periderm** (Figure 5.8b). The peridermal cells are continuously sloughed off and, by the fifth month of development, secretions from sebaceous glands mix with them and with hairs to form a fatty substance called **vernix caseosa** (VER-niks KĀ-sē-ō-sa; *vernix*=varnish, *caseosa*=cheese). This substance covers and protects the skin of the fetus from the constant exposure to the amniotic fluid in which it is bathed. In addition, the slippery vernix caseosa facilitates the birth of the baby and protects the skin from being damaged by the nails.

By about 11 weeks, the basal layer forms an **intermediate layer** of cells (Figure 5.8c). Proliferation of the basal cells eventually forms all layers of the epidermis, which are present at birth (Figure 5.8d). *Epidermal ridges* form along with the epidermal layers (Figure 5.8c). By about the eleventh week, cells from the neural crest (see Figure 19.26b) migrate into the dermis and differentiate into **melanoblasts** (Figure 5.8c). These cells soon enter the epidermis and differentiate into **melanocytes**. Later in the first trimester of pregnancy, *intraepidermal macrophages*, which arise from red bone marrow, invade the epidermis. *Tactile epithelial cells*, also derived from the neural crest cells, appear in the epidermis between 8 and 12 weeks.

The *dermis* arises from **mesoderm** from the **lateral plate mesoderm** (see Figure 4.9b) and the **dermatome of somites** (see Figure 10.10). The mesoderm from these two sources is located deep to the surface ectoderm. The mesoderm gives rise to a loosely organized embryonic connective tissue called **mesenchyme** (MEZ-en-kīm; see Figure 5.8a). By 11 weeks, the mesenchymal cells differentiate into fibroblasts and begin to form collagen and elastic fibers. As the epidermal ridges form, parts of the superficial dermis project into the epidermis and develop into the *dermal papillae*, which contain capillary loops, corpuscles of touch, and free nerve endings (Figure 5.8c).

Hair follicles develop at about 12 weeks as downgrowths of the basal layer of the epidermis into the deeper dermis called **hair buds** (Figure 5.8d). As the hair buds penetrate deeper into the dermis, their distal ends become club-shaped and are called *hair bulbs* (Figure 5.8e). Invaginations of the hair bulbs, called *papillae of the hair*, fill with mesoderm, in which blood vessels and nerve endings develop (Figure 5.8f). Cells in the center of a hair bulb develop into the *hair matrix*, which forms the *hair*. The peripheral cells of the hair bulb form the *epithelial root sheath* (Figure 5.8g). Mesenchyme in the surrounding dermis develops into the *dermal root sheath* and *arrector pili muscle* (Figure 5.8g). By the fifth month, the hair follicles produce lanugo (delicate fetal hair; see Section 5.2). It is produced first on the head and then on other parts of the body, and is usually shed prior to birth.

Most *sebaceous (oil) glands* develop as outgrowths from the sides of hair follicles at about four months and remain connected to the follicles (Figure 5.8e). Most *sudoriferous (sweat)*

Figure 5.8 Development of the integumentary system.

The epidermis develops from ectoderm and the dermis develops from mesoderm.

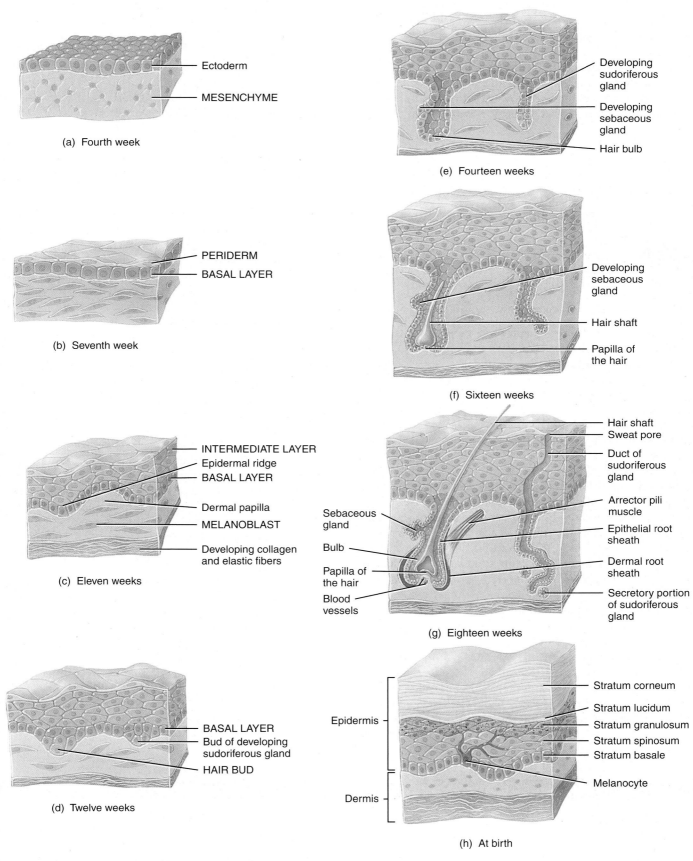

(a) Fourth week

Ectoderm

MESENCHYME

(b) Seventh week

PERIDERM

BASAL LAYER

(c) Eleven weeks

INTERMEDIATE LAYER

Epidermal ridge

BASAL LAYER

Dermal papilla

MELANOBLAST

Developing collagen
and elastic fibers

(d) Twelve weeks

BASAL LAYER

Bud of developing
sudoriferous gland

HAIR BUD

(e) Fourteen weeks

Developing
sudoriferous
gland

Developing
sebaceous
gland

Hair bulb

(f) Sixteen weeks

Developing
sebaceous
gland

Hair shaft

Papilla of
the hair

(g) Eighteen weeks

Sebaceous
gland

Bulb

Papilla of
the hair

Blood
vessels

Hair shaft

Sweat pore

Duct of
sudoriferous
gland

Arrector pili
muscle

Epithelial root
sheath

Dermal root
sheath

Secretory portion
of sudoriferous
gland

(h) At birth

Epidermis

Dermis

Stratum corneum

Stratum lucidum

Stratum granulosum

Stratum spinosum

Stratum basale

Melanocyte

What is the composition of vernix caseosa?

glands are derived from downgrowths (buds) of the stratum basale of the epidermis into the dermis (Figure 5.8d). As the buds penetrate into the dermis, the proximal portion forms the duct of the sweat gland and the distal portion coils and forms the secretory portion of the gland (Figure 5.8g). Sweat glands appear at about five months on the palms and soles and a little later in other regions.

Nails are developed at about 10 weeks. Initially they consist of a thick layer of epithelium called the **primary nail field**. The nail itself is keratinized epithelium and grows distally from its base. It is not until the ninth month that the nails actually reach the tips of the digits.

 CHECKPOINT

20. Which structures develop as downgrowths of the basal layer of epidermal cells?

5.7 AGING AND THE INTEGUMENTARY SYSTEM

● OBJECTIVE

• Describe the effects of aging on the integumentary system.

Most of the age-related changes begin at about age 40 and occur in the proteins in the dermis. Collagen fibers in the dermis begin to decrease in number, stiffen, break apart, and disorganize into a shapeless, matted tangle. Elastic fibers lose some of their elasticity, thicken into clumps, and fray, an effect that is greatly accelerated in the skin of smokers. Fibroblasts, which produce both collagen and elastic fibers, decrease in number. As a result, the skin forms the characteristic crevices and furrows known as *wrinkles*.

The pronounced effects of skin aging do not become noticeable until people reach their late forties. Intraepidermal macrophage cells dwindle in number and become less-efficient phagocytes, thus decreasing the skin's immune responsiveness. Decreased size of sebaceous glands leads to dry and broken skin that is more susceptible to infection. The production of sweat diminishes, which probably contributes to the increased incidence of heat stroke in the elderly. There is a decrease in the number of functioning melanocytes, resulting in gray hair and atypical skin pigmentation. Hair loss increases with aging as hair follicles stop producing hairs. About 25 percent of males begin to show signs of hair loss by age 30 and about two-thirds have significant hair loss by age 60. Both males and females develop pattern baldness. An increase in the size of some melanocytes produces pigmented blotching (age spots). Walls of blood vessels in the dermis become thicker and less permeable, and subcutaneous adipose tissue is lost. Aged skin

(especially the dermis) is thinner than young skin, and the migration of cells from the basal layer to the epidermal surface slows considerably. With the onset of old age, skin heals poorly and becomes more susceptible to pathological conditions such as skin cancer and pressure ulcers. **Rosacea** (ro-ZĀ-shē-a=rosy) is a skin condition that affects mostly fair-skinned adults between the ages of 30 and 60. It is characterized by redness, tiny pimples, and noticeable blood vessels, usually in the central area of the face.

Growth of nails and hair slows during the second and third decades of life. The nails also may become more brittle with age, often due to dehydration or repeated use of cuticle remover or nail polish.

Several cosmetic anti-aging treatments are available to diminish the effects of aging or sun-damaged skin. These include the following:

• *Topical products* that bleach the skin to tone down blotches and blemishes (hydroquinone) or decrease fine wrinkles and roughness (retinoic acid)

• *Microdermabrasion* (mī-krō-DER-ma-brā-zhun; *mikros*= small; *derm*=skin; *-abrasio*=to wear away), a process that uses tiny crystals under pressure to remove and vacuum the skin's surface cells to improve skin texture and reduce blemishes

• *Chemical peel*, the application of a mild acid (such as glycolic acid) to the skin to remove surface cells to improve skin texture and reduce blemishes

• *Laser resurfacing*, use of a laser to clear up blood vessels near the skin surface, even out blotches and blemishes, and decrease fine wrinkles; an example is the IPL Photofacial®

• *Dermal fillers*, injections of human collagen (Cosmoderm)®, hyaluronic acid (Restylane® and Juvaderm®), calcium hydroxylapatite (Radiesse®), or polylactic acid (Sculptra®) that plumps up the skin to smooth out wrinkles and fill in furrows, such as those around the nose and mouth and between the eyebrows

• *Fat transplantation*, a process in which fat from one part of the body is injected into another location, such as around the eyes

• *Botulinum toxin* or *Botox*®, a diluted version of the toxin that inhibits nerve impulse transmission to skeletal muscles, which is injected into the skin to paralyze muscles that cause the skin to wrinkle

• *Radio frequency nonsurgical facelift*, a procedure that uses radio frequency emissions to tighten the deeper layers of the skin of the jowls, neck, and sagging eyebrows and eyelids

• *Facelift, browlift,* or *necklift*, invasive surgery in which loose skin and fat are removed surgically and the underlying connective tissue and muscle are tightened

 CHECKPOINT

21. What factors contribute to the susceptibility of aging skin to infection?

KEY MEDICAL TERMS ASSOCIATED WITH THE INTEGUMENTARY SYSTEM

Abrasion (a-BRĀ-shun; *ab*=away; *-rasion*=scraped) An area where skin has been scraped away.

Blister A collection of serous fluid within the epidermis or between the epidermis and dermis, due to short-term but severe friction. The term *bulla* (BUL-a) refers to a large blister.

Callus (KAL-lus=hard skin) An area of hardened and thickened skin that is usually seen in palms and soles and is due to persistent pressure and friction.

Cold sore A lesion, usually in the oral mucous membrane, caused by Type I herpes simplex virus (HSV) transmitted by oral or respiratory routes. The virus remains dormant until triggered by factors such as ultraviolet light, hormonal changes, and emotional stress. Also called a *fever blister.*

Contusion (kon-TOO-shun; *contundere*=to bruise) Condition in which tissue deep to the skin is damaged, but the epidermis is not broken.

Corn A painful conical thickening of the stratum corneum of the epidermis found principally over toe joints and between the toes; often caused by friction or pressure.

Cyst (SIST=sac containing fluid) A sac with a distinct connective tissue wall, containing a fluid or other material.

Eczema (EK-ze-ma, *ekzeo*=to boil over) An inflammation of the skin characterized by patches of red, blistering, dry, extremely itchy skin. It occurs mostly in skin creases in the wrists, backs of the knees, and front of the elbows. It typically begins in infancy, and many children outgrow the condition. The cause is unknown but is linked to genetics and allergies.

Erysipelas (er′-i-SIP-e-las) A streptococcal infection of the skin that may, if not properly treated, become systemic and involve the lymphatic and cardiovascular systems. A cardinal sign of erysipelas is a very sharp margin between the red and tender involved skin and the uninvolved skin.

Frostbite Local destruction of skin and subcutaneous tissue on exposed surfaces as a result of extreme cold. In mild cases, the skin is blue and swollen and there is slight pain. In severe cases there is considerable swelling, some bleeding, no pain, and blistering. If untreated, gangrene may develop. Frostbite is treated by rapid rewarming.

Hemangioma (hē-man′-jē-Ō-ma; *hem*=blood; *-angi*=blood vessel; *-oma*=tumor) Commonly called a *birthmark*, a hemangioma is a localized benign tumor of the skin and subcutaneous layer that results from an abnormal increase in the number of blood vessels. One type is a **port-wine-stain**, a flat, pink, red, or purple lesion present at birth, usually at the nape of the neck. Certain hemangiomas last a lifetime; others gradually fade and may disappear.

Hives Condition of the skin marked by reddened elevated patches that are often itchy. Most commonly caused by infections, physical trauma, medications, emotional stress, food additives, and certain food allergies. Also called *urticaria* (ūr-ti-KAR-ē -a).

Keloid (KĒ-loid; *kelis*=tumor) An elevated, irregular darkened area of excess scar tissue caused by collagen formation during healing. It extends beyond the original injury and is tender and frequently painful. It occurs in the dermis and underlying subcutaneous tissue, usually after trauma, surgery, a burn, or severe acne; more common in people of African descent.

Keratosis (ker-a-TŌ-sis; *kera*=horn) Formation of a hardened growth of epidermal tissue, such as a *solar keratosis*, a relatively common premalignant lesion of the sun-exposed skin of the face and hands.

Laceration (las-er-Ā-shun; *lacer*=torn) An irregular tear of the skin.

Lice Contagious arthropods that include two basic forms. **Head lice** are tiny, jumping arthropods that suck blood from the scalp. They lay eggs, called nits, and their saliva causes itching that may lead to complications. **Pubic lice** are tiny, non-jumping arthropods that look like miniature crabs.

Papule (PAP-ūl; *papula*=pimple) A small, round skin elevation less than 1 cm (about 1/2 inch) in diameter. One example is a pimple.

Pressure ulcers A lesion through the skin or mucous membrane caused by a constant deficency of blood flow to tissues. Typically the affected tissue overlies a bony projection that has been subjected to prolonged pressure against an object such as a bed, cast, or splint. Also known as *decubitus ulcers* (de-KŪ-bi-tus) or *bedsores*.

Pruritus (proo-RĪ-tus; *pruri*=to itch) Itching, one of the most common dermatological disorders. It may be caused by skin disorders (infections), systemic disorders (cancer, kidney failure), psychogenic factors (emotional stress), or allergic reactions.

Psoriasis (sō-RĪ-a-sis) A common and chronic skin disorder in which keratinocytes divide and move more quickly than normal from the stratum basale to the stratum corneum. As a result, the surface cells never get a chance to cycle into the later keratinizing stages and are shed immaturely; when it occurs on the scalp it is referred to as *dandruff*.

Self-tanning lotions (sunless tanners) Topically applied substances that contain a color additive (dihydroxyacetone) that produces a tanned appearance by interacting with proteins in the skin.

Sunscreens Topically applied preparations that contain various chemical agents (such as benzophenone or one of its derivatives), which absorb UVB rays but let most of the less-harmful UVA rays pass through.

Sunblocks Topically applied preparations that contain substances such as zinc oxide, which reflect and scatter both UVB and UVA rays.

Tinea corporis A fungal infection characterized by scaling, itching, and sometimes painful lesions that may appear on any part of the body; also known as **ringworm**. Fungi thrive in warm, moist places such as skin folds of the groin, where it is known as **tinea cruris (jock itch)** or between the toes, where it is called **tinea pedis (athlete's foot)**.

Topical In reference to a medication, applied to the skin surface rather than ingested or injected.

Transdermal (transcutaneous) drug administration The administration of a drug contained within an adhesive skin patch that passes across the epidermis and into the blood vessels of the dermis. The drug is released continuously at a controlled rate over a period of one to several days. A growing number of drugs are available for transdermal administration, including nitroglycerine, for prevention of angina pectoris, which is chest pain associated with heart disease (nitroglycerine can also be given under the tongue and intravenously); scopolamine, for motion sickness; estradiol, used for estrogen-replacement therapy during menopause; ethinyl estradiol and norelgestromin in contraceptive patches; nicotine, used to help people stop smoking; and fentanyl, used to relieve severe pain in cancer patients.

Wart Mass produced by uncontrolled growth of epithelial skin cells as a result of infection by a papilloma virus. Most warts are noncancerous.

CHAPTER REVIEW AND RESOURCE SUMMARY

WileyPLUS

Review

Resource

5.1 Structure of the Skin

1. The integumentary system consists of organs such as the skin and hair, and other structures such as nails.
2. The skin is the largest organ of the body in surface area and weight. The principal parts of the skin are the epidermis (superficial) and dermis (deep).
3. The subcutaneous layer (hypodermis) is deep to the dermis and not part of the skin. It anchors the dermis to underlying tissues and organs, and it contains lamellated corpuscles.
4. The types of cells in the epidermis are keratinocytes, melanocytes, intraepidermal macrophage cells, and tactile epithelial cells.
5. The epidermal layers, from deepest to most superficial, are the stratum basale (undergoes cell division and produces all other layers), stratum spinosum (provides strength and flexibility), stratum granulosum (contains keratin and lamellar granules), stratum lucidum (present only in palms and soles), and stratum corneum (sloughs off dead skin) (see Table 5.1). Stem cells in the stratum basale undergo continuous cell division, producing keratinocytes for the other layers.
6. The dermis consists of dense irregular connective tissue containing collagen and elastic fibers. It is divided into papillary and reticular regions. The papillary region contains thin collagen and fine elastic

Anatomy Overview - The Skin
Figure 5.2 - Types of Cells in the Epidermis
Figure 5.3 - Layers of the Epidermis

Review   **Resource**

fibers, dermal papillae, and corpuscles of touch. The reticular region contains bundles of thick collagen and some coarse elastic fibers, adipose tissue, hair follicles, nerves, sebaceous (oil) glands, and ducts of sudoriferous (sweat) glands. See Table 5.2.

7. Epidermal ridges provide the basis for fingerprints and footprints.

8. The color of skin is due to melanin, carotene, and hemoglobin.

9. In tattooing, a pigment is deposited with a needle in the dermis. Body piercing is the insertion of jewelry through an artificial opening.

5.2 Accessory Structures of the Skin

Anatomy Overview - Hair
Anatomy Overview - Nails
Figure 5.5 - Hair

1. Accessory structures of the skin—hair, skin glands, and nails—develop from the embryonic epidermis.

2. A hair consists of a shaft, most of which is superficial to the surface, a root that penetrates the dermis and sometimes the subcutaneous layer, and a hair follicle. Associated with each hair follicle is a sebaceous (oil) gland, an arrector pili muscle, and a hair root plexus.

3. New hairs develop from division of hair matrix cells in the bulb; hair replacement and growth occur in a cyclic pattern consisting of growth, regression, and resting stages.

4. Hairs offer a limited amount of protection—from the sun, heat loss, and entry of foreign particles into the eyes, nose, and ears. They also function in sensing light touch.

5. Lanugo hairs of the fetus are shed before birth. Most body hair on males is terminal (coarse, pigmented); most body hair on females is vellus (fine).

6. Sebaceous (oil) glands are usually connected to hair follicles; they are absent from the palms and soles. Sebaceous glands produce sebum, which moistens hairs and waterproofs the skin. Clogged sebaceous glands may produce acne.

7. There are two types of sudoriferous (sweat) glands: eccrine and apocrine. Eccrine sweat glands have an extensive distribution; their ducts terminate at pores at the surface of the epidermis. Eccrine sweat glands are involved in thermoregulation and waste removal and are stimulated during emotional stress. Apocrine sweat glands are limited to the skin of the axillae, groin, and areolae; their ducts open into hair follicles. Apocrine sweat glands are stimulated during emotional stress and sexual excitement. See Table 5.3.

8. Ceruminous glands are modified sudoriferous glands that secrete cerumen (ear wax). They are found in the external auditory canal (ear canal).

9. Nails are hard, dead, keratinized epidermal cells over the dorsal surfaces of the distal portions of the digits. The principal parts of a nail are the nail body, free edge, nail root, lunula, nail bed, hyponychium, eponychium, and nail matrix. Cell division of the matrix cells produces new nails.

5.3 Types of Skin

1. Thin skin covers all parts of the body except for the palms, palmar surfaces of the digits, and the soles.

2. Thick skin covers the palms, palmar surfaces of the digits, and soles.

3. See Table 5.4.

5.4 Functions of the Skin

Anatomy Overview - The Integument and Disease Resistance
Animation - Nonspecific Disease Resistance
Concepts and Connections - The Role of the Integument in Disease Resistance

1. Skin functions include body temperature regulation (thermoregulation), blood storage, protection, sensation, excretion and absorption, and synthesis of vitamin D.

2. The skin participates in thermoregulation by liberating sweat at its surface and functions as a blood reservoir.

3. The skin provides physical, chemical, and biological barriers that help protect the body.

4. Cutaneous sensations include touch, hot and cold, and pain.

5.5 Blood Supply of the Integumentary System

1. The epidermis is avascular.

2. The dermis is supplied by the cutaneous and papillary arterial plexuses.

5.6 Development of the Integumentary System

1. The epidermis develops from the embryonic ectoderm, and the accessory structures of the skin (hair, nails, and skin glands) are epidermal derivatives.

2. The dermis is derived from mesodermal cells.

5.7 Aging and the Integumentary System

1. Most effects of aging begin to occur when people reach their late forties.

2. Among the effects of aging are wrinkling, loss of subcutaneous fat, atrophy of sebaceous glands, and decrease in the number of melanocytes and intraepidermal macrophage cells.

CRITICAL THINKING QUESTIONS

1. Your 65-year-old aunt has spent every sunny day at the beach for as long as you can remember. Her skin looks a lot like a comfortable lounge chair—brown and wrinkled. Recently her dermatologist removed a suspicious growth from the skin of her face. What would you suspect is the problem and its likely cause?

2. Lindsay is partying with his friends on a cold winter night. The outside temperature is below freezing, but they are not wearing their coats when they step outside to smoke. They say they don't need their coats because they are drinking alcoholic beverages, which they say keeps them warm enough. You have learned that alcohol dilates the blood vessels in the skin. Do you think Lindsay and his friends have chosen an effective way to control their body temperature? Why or why not?

3. Courtney and her college roommate made travel plans to go to Cancun for spring break. Neither one of them ever bothered to wear sunscreen, thinking that they would get a better suntan

without it. When the women reached their forties, they were both diagnosed with malignant melanoma. Why has the risk of developing melanoma doubled over the last 15 years? What is the prognosis for someone diagnosed with malignant melanoma?

4. Your nephew has been learning about cells in science class and now refuses to take a bath. He asks, "If all cells have a semipermeable membrane, and my skin is made out of cells, then won't I swell up and pop when I take a bath?" Explain this dilemma to your nephew before he starts attracting flies.

5. People always say, "It's not the heat; it's the humidity," when complaining about summer weather. With regard to the integumentary system, why do you think people feel hotter when it's 95 degrees Fahrenheit and 95 percent humidity than when it's 95 degrees and 30 percent humidity? (Humidity refers to the amount of water vapor in the air.)

? ANSWERS TO FIGURE QUESTIONS

5.1 The epidermis is composed of epithelial tissue, and the dermis is made up of connective tissue.

5.2 Melanin protects the DNA of the nucleus of keratinocytes from being damaged by UV light.

5.3 The stratum basale is the layer of the epidermis with stem cells that continually undergo cell division.

5.4 Tension lines are clinically important because surgical incisions parallel to them leave only fine scars, but a surgical incision made across several layers of tension lines will tend to heal in a broad, thick scar.

5.5 Plucking a hair stimulates hair root plexuses in the dermis, some of which are sensitive to pain. Because the cells of a hair shaft are already dead and the hair shaft lacks nerves, cutting hair is not painful.

5.6 The main function of eccrine sweat glands is to help regulate body temperature through evaporation.

5.7 Nails are hard because they are composed of tightly packed, dead, keratinized epidermal cells that do not shed and are filled with a harder type of keratin than in the epidermis.

5.8 Vernix caseosa consists of secretions from sebaceous glands, sloughed off peridermal cells, and hairs.

6 BONE TISSUE

INTRODUCTION A **bone** is an organ made up of several different tissues working together: bone (osseous) tissue, cartilage, dense connective tissue, epithelium, adipose tissue, and nervous tissue. The entire framework of bones and their cartilages constitute the **skeletal system**. Bone tissue is a complex and dynamic living tissue. It continually engages in a process called *remodeling*—the building of new bone tissue and breaking down of old bone tissue. In the early days of space exploration, young, healthy men in prime physical shape returned from their space flights only to alarm their physicians. Physical examinations of the astronauts revealed that they had lost up to 20 percent of their total bone density during their extended stay in space. The zero-gravity (weightless) environment of space, coupled with the fact that the astronauts traveled in small capsules that greatly limited their movement for extended periods of time, placed minimal strain on their bones. In contrast, athletes subject their bones to great forces, which place significant strain on the bone tissue. Accomplished athletes show an increase in overall bone density. How is bone capable of changing in response to the different mechanical demands placed on it? Why do high activity levels that strain bone tissue greatly improve bone health? This chapter surveys the various components of bones to help you understand how bones form, how they age, and how exercise affects their density and strength. The study of bone structure and the treatment of bone disorders is referred to as **osteology** (os-tē-OL-o-jē; *osteo-*=bone; *-logy*=study of). •

Did you ever wonder why more females than males are affected by osteoporosis?

6.1 FUNCTIONS OF BONE AND THE SKELETAL SYSTEM

OBJECTIVE

• Describe the six main functions of the skeletal system.

The skeletal system performs several basic functions:

1. ***Support.*** The skeleton serves as the structural framework for the body by supporting soft tissues and providing attachment points for the tendons of most skeletal muscles.
2. ***Protection.*** The skeleton protects the most important internal organs from injury. For example, cranial bones protect the brain, vertebrae (backbones) protect the spinal cord, and the rib cage protects the heart and lungs.
3. ***Assistance in movement.*** Most skeletal muscles attach to bones; when they contract, they pull on bone to produce movement.
4. ***Mineral storage and release.*** Bone tissue makes up about 18 percent of the weight of the human body and stores several minerals, especially calcium and phosphorus, which contribute to the strength of bone. Bone tissue stores about 99 percent of total body calcium. On demand, bone releases minerals into the blood to maintain critical mineral balances and to distribute the minerals to other parts of the body.
5. ***Blood cell production.*** Within certain bones, a connective tissue called **red bone marrow** produces red blood cells, white blood cells, and platelets in a process called **hemopoiesis** (hēm-ō-poy-Ē-sis; *hemo-*=blood; *poiesis-*=making) or **hematopoiesis** (hē-ma-tō-poy-Ē-sis). Red bone marrow consists of developing blood cells, adipocytes, fibroblasts, and macrophages within a network of reticular fibers. It is present in developing bones of the fetus and in some adult bones, such as the hip bones (pelvic bones), ribs, sternum (breastbone), vertebrae (backbones), skull, and ends of the humerus (arm bone) and femur (thigh bone). In a newborn, all bone marrow is red and is involved in hemopoiesis. With increasing age, much of the bone marrow changes from red to yellow.
6. ***Triglyceride storage.*** **Yellow bone marrow** consists mainly of adipose cells, which store triglycerides. The stored triglycerides are a potential chemical energy reserve.

CHECKPOINT

1. How does the skeletal system function in support, protection, and storage of minerals?
2. Describe the role of bones in blood cell production.

6.2 TYPES OF BONES

OBJECTIVE

• Classify bones on the basis of shape and location.

Bones can serve as a library of information about the human body. They serve as an enduring record of an individual's life because bones do not deteriorate following death as quickly as soft tissues do. For example, age, size, stature, gender, health, and race can all be determined by examining the skeleton. Such information can be useful to forensic specialists seeking to identify the skeletal remains of a victim or anthropologists who study ancient skeletal remains. The shape of a given bone also reveals a great deal of information about its functional role in the body, such as its physical strength and the type of forces it experienced as it was moved by muscles. Each little bump, groove, hole, projection, and ridge on a bone has a story to tell.

The 206 individual bones in an adult skeleton come in a variety of sizes and shapes. They range in size from the small ear bones in the skull to the nearly two-foot-long femur (thigh bone) of an adult. The various shapes of the bones are designed to accomplish particular functions. Anatomists recognize five types of bones in the skeleton on the basis of shape: long, short, flat, irregular, and sesamoid (Figure 6.1).

Long bones have greater length than width and consist of a *diaphysis* (shaft) and a variable number of *epiphyses* or extremities (ends). They are slightly curved for strength. A curved bone absorbs the stress of the body's weight at several different points so that it is evenly distributed. If such bones were straight, the weight of the body would be unevenly distributed and the bone would fracture easily. Long bones consist mostly of *compact bone tissue*, which is dense and has smaller spaces, but they also contain considerable amounts of *spongy bone tissue*, which has larger spaces (see Figure 6.4). Long bones include the humerus (arm bone), ulna and radius (forearm bones), femur (thigh bone), tibia and fibula (leg bones), metacarpals (hand bones), metatarsals (foot bones), and phalanges (finger and toe bones).

Short bones are somewhat cube-shaped and nearly equal in length, width, and depth. They consist of spongy bone except at the surface, where there is a thin layer of compact bone. Examples of short bones are most carpal (wrist) bones and most tarsal (ankle) bones.

Flat bones are generally thin and composed of two nearly parallel plates of compact bone enclosing a layer of spongy bone. The layers of compact bone are called external and internal *tables*. In cranial bones, the spongy bone is referred to as **diploë** (DIP-lō-ē)

Figure 6.1 Types of bones based on shape. The bones are not drawn to scale.

 The shapes of bones largely determine their functions.

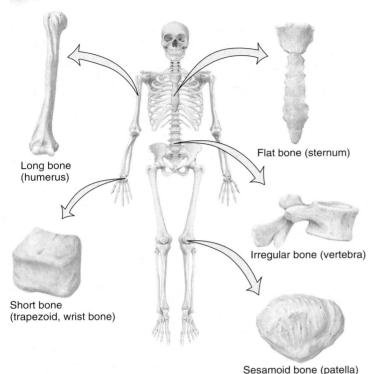

Long bone (humerus)

Flat bone (sternum)

Irregular bone (vertebra)

Short bone (trapezoid, wrist bone)

Sesamoid bone (patella)

Which type of bone primarily protects and provides a large surface area for muscle attachment?

(see Figure 6.6). Flat bones afford considerable protection and provide extensive areas for muscle attachment. They include the cranial (skull) bones, which protect the brain; the sternum (breastbone) and ribs, which protect organs in the thorax; and the scapulae (shoulder blades).

Irregular bones have complex shapes and cannot be grouped into any of the three categories just described. They also vary in the amounts of spongy and compact bone they contain. Such bones include the vertebrae (backbones), certain facial bones, and the calcaneus (heel bone).

Sesamoid bones (SES-a-moyd=shaped like a sesame seed) develop in certain tendons where there is considerable friction, compression, and physical stress. They are not always completely ossified and measure only a few millimeters to centimeters in diameter except for the two patellae (kneecaps), the largest of the sesamoid bones. Sesamoid bones vary in number from person to person except for the patellae, which are located in the quadriceps femoris tendon (see Figure 11.24a, b) and are normally present in all individuals. Functionally, sesamoid bones protect tendons from excessive wear and tear, and they can alter the direction of pull of a tendon, which improves the mechanical advantage at a joint.

In the upper limbs, sesamoid bones usually occur only in the joints of the palmar surface of the hands. Two frequently encountered sesamoid bones are in the tendons of the adductor pollicis and flexor pollicis brevis muscles at the metacarpophalangeal joint of the thumb (see Figure 8.6a). In the lower limbs, there are two constant sesamoid bones in addition to the patellae; these occur on the plantar surface of each foot in the tendons of the flexor hallucis brevis muscle at the metatarsophalangeal joint of the great (big) toe (see Figure 8.12b).

An additional type of bone is not classified by shape, but rather by location. **Sutural bones** (SOO-chur-al; *sutura*=seam) or *wormian bones* (named after a Danish anatomist, O. Worm, who lived from 1588 to 1654) are small bones located within the sutures (joints) of certain cranial bones (see Figure 7.5a). The number of sutural bones varies greatly from person to person.

✓ CHECKPOINT

3. Give examples of long, short, flat, irregular, and sesamoid bones and explain the function of each type of bone.

6.3 ANATOMY OF A BONE

● OBJECTIVE

• Describe the parts of a long bone.

Let's examine the structure of a bone using the humerus as an example. A typical long bone consists of the following parts (Figure 6.2):

1. The **diaphysis** (dī-AF-i-sis=growing between) is the bone's shaft, or body—the long, cylindrical, main portion of the bone.
2. The **epiphyses** (e-PIF-i-sēz=growing own; singular is *epiphysis*) or *extremities* are the proximal and distal ends of the bone.
3. The **metaphyses** (me-TAF-i-sēz; *meta-*=between; singular is *metaphysis*) are the regions between the diaphysis and the epiphyses. In a *growing bone*, each metaphysis contains an **epiphyseal (growth) plate** (ep-i-FIZ-ē-al), a layer of hyaline cartilage that allows the diaphysis of the bone to grow in length (a process described later in this chapter). When bone growth in length stops somewhere between the ages of 14 and 24, the cartilage in the epiphyseal plate is replaced by bone and the resulting bony structure is known as the **epiphyseal line**.
4. The **articular cartilage** is a thin layer of hyaline cartilage covering the part of the epiphysis where the bone forms an articulation (joint) with another bone. Articular cartilage reduces friction and absorbs shock at freely movable joints. Because articular cartilage lacks a perichondrium and lacks blood vessels, repair of damage is limited.
5. The **periosteum** (per-ē-OS-tē-um; *peri-*=around) is a tough connective tissue sheath and its associated blood supply that surrounds the bone surface wherever it is not covered by articular cartilage. It is composed of an *outer fibrous layer* of dense irregular connective tissue and an *inner osteogenic layer* that consists of cells. Some of the cells enable bone to grow in thickness, but not in length. The periosteum also protects the bone, assists in fracture repair, helps nourish bone tissue, and serves as an attachment point for ligaments and tendons. The periosteum is attached to the underlying bone by **perforating (Sharpey's) fibers**, thick bundles of collagen that extend from the periosteum into the bone extracellular matrix.
6. The **medullary cavity** (MED-ū-lar-ē; *medulla-*=marrow, pith), or **marrow cavity**, is a hollow, cylindrical space within the diaphysis that contains fatty yellow bone marrow and numerous blood vessels in adults. This cavity minimizes the weight of the bone by reducing the dense bony material where it is least needed. The long bones' tubular design provides maximum strength with minimum weight.
7. The **endosteum** (end-OS-tē-um; *endo-*=within) is a thin membrane that lines the medullary cavity. It contains a single layer of bone-forming cells and a small amount of connective tissue.

Figure 6.2 Parts of a long bone. The spongy bone tissue of the epiphysis and metaphysis contains red bone marrow, and the medullary cavity of the adult diaphysis contains yellow bone marrow.

A long bone is covered by articular cartilage at the articular surfaces of its proximal and distal epiphyses and by periosteum around all other parts of the bone.

FUNCTIONS OF BONE TISSUE

1. Supports soft tissue and provides attachment for skeletal muscles.
2. Protects internal organs.
3. Assists in movement along with skeletal muscles.
4. Stores and releases minerals.
5. Contains red bone marrow, which produces blood cells.
6. Contains yellow bone marrow, which stores triglycerides (fats), a potential chemical energy reserve.

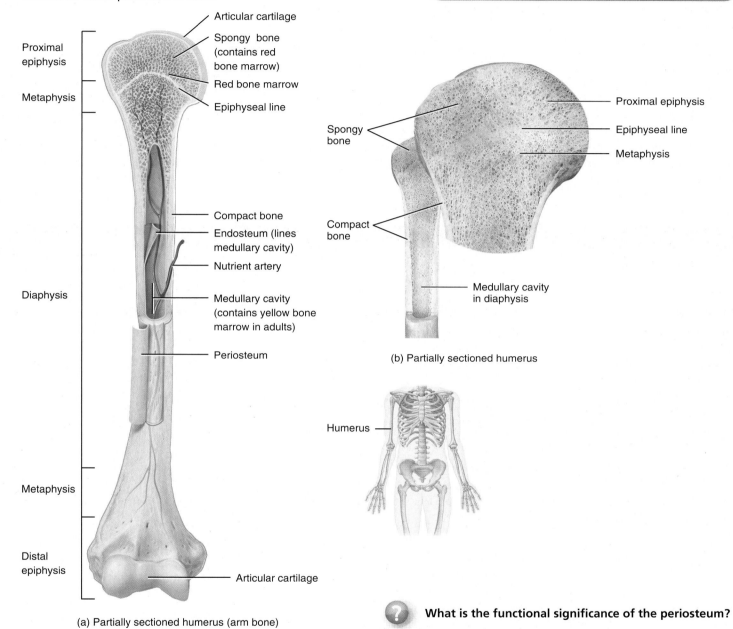

(a) Partially sectioned humerus (arm bone)

(b) Partially sectioned humerus

 What is the functional significance of the periosteum?

Using the design of the long bone as a model, the structure of the other bone types is not difficult to understand. Each of the other bone types—short, flat, irregular, and sesamoid—have designs that closely resemble the epiphysis of a long bone. In fact, long bones are the only bones with a diaphysis and a medullary cavity. The other bone types have outer plates of compact bone covering an inner core of spongy bone.

 CHECKPOINT

4. Diagram the parts of a long bone, and explain the functions of each part.

6.4 BONE SURFACE MARKINGS

● OBJECTIVE

• Describe the principal surface markings on bones and the functions of each.

In addition to having a variety of shapes, bones have different surface textures and contours. A closer look at the surface of a bone reveals a great deal of information. The surface of a bone is marked by a variety of bumps, grooves, indentations, projections, and holes. These characteristic features of bones are called **surface markings** or **osseous landmarks**, and they are structural features adapted for specific functions. Most are not present at birth but develop later in response to certain forces; they are most prominent during adult life. Various bumps, ridges, or rough areas indicate where soft tissues, such as tendons and ligaments, attach to the bone. (*Tendons* typically attach skeletal muscle to bone; *ligaments* typically attach one bone to another bone.) Groove-like impressions or holes indicate locations where nerves and blood vessels pass through the bones. Smooth surfaces indicate areas of movement between neighboring bones. In general, a smooth-surfaced depression on one bone accommodates a smooth-surfaced projection from another bone to form the joint surfaces so that bones can fit together. In response to tension on a bone surface where tendons, ligaments, aponeuroses, and fasciae pull on the periosteum of bone, new bone is deposited, resulting in raised or roughened areas.

There are two major types of surface markings: (1) *depressions and openings*, which form joints or allow the passage of soft tissues (such as blood vessels and nerves), and (2) *processes*, projections or outgrowths that either (a) help form joints or (b) serve as attachment points for connective tissue (such as ligaments and tendons). Table 6.1 describes the various surface markings and provides references to illustrations of each, which appear in the next two chapters of this text in the descriptions of the axial skeleton (Chapter 7) and the appendicular skeleton (Chapter 8).

 CHECKPOINT

5. List and describe several bone surface markings, and give an example of each. Check your list against Table 6.1.

6.5 HISTOLOGY OF BONE TISSUE

 OBJECTIVES

- Explain why bone tissue is classified as a connective tissue.
- Describe the cellular composition of bone tissue and the functions of each component.
- Outline the structural and functional differences between compact and spongy bone tissue.
- Describe the histological features of bone tissue.

We will now examine the structure of bone at the microscopic level. Like other connective tissues, **bone**, or **osseous tissue** (OS-ē-us), contains an abundant extracellular matrix that surrounds widely separated cells. The extracellular matrix is about

TABLE 6.1

Bone Surface Markings

MARKING	DESCRIPTION	EXAMPLE
DEPRESSIONS AND OPENINGS: Sites allowing the passage of soft tissue (nerves, blood vessels, ligaments, tendons) or formation of joints		
Fissure (FISH-ur)	Narrow slit between adjacent parts of bones through which blood vessels or nerves pass	Superior orbital fissure of the sphenoid bone (Figure 7.6f and 7.10b)
Foramen (fō-RĀ-men=hole; plural is *foramina*)	Opening through which blood vessels, nerves, or ligaments pass	Optic foramen (canal) of the sphenoid bone (Figure 7.6f and 7.10b)
Fossa (FOS-a=trench; plural is fossae, FOS-ē)	Shallow depression (*fossa*=trench)	Coronoid fossa of the humerus (Figure 8.4a)
Sulcus (SUL-kus=groove; plural is *sulci*, SUL-sī)	Furrow along a bone surface that accommodates a blood vessel, nerve, or tendon	Intertubercular sulcus (groove) of the humerus (Figure 8.4a)
Meatus (mē-Ā-tus=passageway; plural is *meati*, me-Ā-tī)	Tubelike opening	External and internal auditory meati of the temporal bone (Figure 7.3c and 7.4a)
PROCESSES: Projections or outgrowths on bone that form joints or attachment points for connective tissue, such as ligaments and tendons		
Processes that form joints:		
Condyle (KON-dīl; *condylus*=knuckle)	Large, round protuberance with a smooth articular surface at the end of a bone	Lateral condyle of the femur (Figure 8.11a)
Facet (FAS-et or fa-SET)	Smooth, flat, slightly concave or convex articular surface	Superior articular facet of a vertebra (Figure 7.16a)
Head	Usually rounded articular projection supported on the neck (constricted portion) of a bone	Head of the femur (Figure 8.11a)
Processes that form attachment points for connective tissue:		
Crest	Prominent ridge or elongated projection	Iliac crest of the hip bone (Figure 8.9b)
Epicondyle (*epi-*=above)	Typically roughened projection above a condyle	Medial epicondyle of the femur (Figure 8.11a)
Line	Long, narrow ridge or border (less prominent than a crest)	Linea aspera of the femur (Figure 8.11b)
Spinous process	Sharp, slender projection	Spinous process of a vertebra (Figure 7.16a)
Trochanter (trō-KAN-ter)	Very large projection	Greater trochanter of the femur (Figure 8.11a)
Tubercle (TOO-ber-kul; *tuber*=knob)	Variable sized rounded projection	Greater tubercle of the humerus (Figure 8.4a)
Tuberosity	Variable sized projection that has a rough, bumpy surface	Ischial tuberosity of the hip bone (Figure 8.9b)

15 percent water, 30 percent collagen fibers, and 55 percent crystallized mineral salts. [Dry bones (the nonliving bones that are studied in the laboratory) are 60 percent inorganic minerals and 40 percent organic substances by weight.] The most abundant mineral salt is calcium phosphate $[Ca_3(PO_4)_2]$. It combines with another mineral salt, calcium hydroxide $[Ca(OH)_2]$, to form crystals of **hydroxyapatite** $[Ca_{10}(PO_4)_6(OH)_2]$ (hī-drok′-sē-AP-a-tīt). As the crystals form, they combine with still other mineral salts, such as calcium carbonate $(CaCO_3)$, and ions such as magnesium, fluoride, potassium, and sulfate. As these mineral salts are deposited in the framework formed by the collagen fibers of the extracellular matrix, they crystallize and the tissue hardens. This process, called **calcification** (kal-si-fi-KĀ-shun), is initiated by bone-building cells called *osteoblasts*, and will be described shortly.

It was once thought that calcification simply occurred when enough mineral salts were present to form crystals. However, we now know that the process occurs only in the presence of collagen fibers. Mineral salts begin to crystallize in the microscopic spaces between collagen fibers. After the spaces are filled, mineral crystals accumulate around the collagen fibers. The combination of crystallized salts and collagen fibers is responsible for the characteristics of bone.

Although a bone's *hardness* depends on the crystallized inorganic mineral salts, a bone's *flexibility* depends on its collagen fibers. Like reinforcing metal rods in concrete, collagen fibers and other organic molecules provide *tensile strength*, resistance to being stretched or torn apart. Soaking a bone in an acidic solution, such as vinegar, dissolves its mineral salts, causing the bone to become rubbery and flexible. As you will see shortly, bone cells called *osteoclasts* secrete enzymes and acids that break down the extracellular matrix of bone.

Four types of cells are present in bone tissue: osteoprogenitor cells, osteoblasts, osteocytes, and osteoclasts (Figure 6.3).

1. **Osteoprogenitor cells** (os-tē-ō-prō-JEN-i-tor; -*genic*=producing) are unspecialized bone stem cells derived from mesenchyme, the tissue from which almost all connective tissues are formed. They are the only bone cells to undergo cell division; the resulting cells develop into osteoblasts. Osteoprogenitor cells are found along the inner portion of the periosteum, in the endosteum, and in the canals within bone that contain blood vessels.

2. **Osteoblasts** (OS-tē-ō-blasts; -*blasts*=buds or sprouts) are bone-building cells. They synthesize and secrete collagen fibers and other organic components needed to build the extracellular matrix of bone tissue, and they initiate calcification. As osteoblasts surround themselves with extracellular matrix, they become trapped in their secretions and become osteocytes. (*Note:* Cells with the suffix *blast* in bone or any other connective tissue secrete extracellular matrix.)

3. **Osteocytes** (OS-tē-ō-sīts; -*cytes*=cells), mature bone cells, are the main cells in bone tissue and maintain its daily metabolism, such as the exchange of nutrients and wastes with the blood. Like osteoblasts, osteocytes do not undergo cell division. (*Note:* Cells with the suffix *cyte* in bone or any other tissue maintain and monitor the tissue.)

4. **Osteoclasts** (OS-tē-ō-klasts; -*clast*=break), huge cells derived from the fusion of as many as 50 monocytes (a type of white blood cell), are concentrated in the endosteum. The plasma membrane of an osteoclast is deeply folded into a *ruffled border* on the side of the cell that faces the bone surface. Here the cell releases powerful lysosomal enzymes and acids that digest the protein and mineral components of the underlying

Figure 6.3 Types of cells in bone tissue.

🔑 Osteoprogenitor cells undergo cell division and develop into osteoblasts, which secrete bone extracellular matrix.

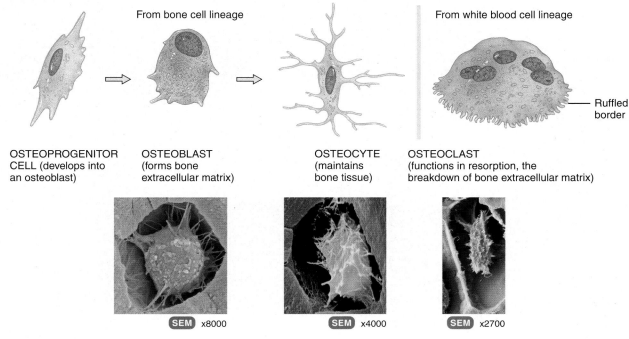

From bone cell lineage

From white blood cell lineage

Ruffled border

OSTEOPROGENITOR CELL (develops into an osteoblast)

OSTEOBLAST (forms bone extracellular matrix)

OSTEOCYTE (maintains bone tissue)

OSTEOCLAST (functions in resorption, the breakdown of bone extracellular matrix)

SEM x8000   SEM x4000   SEM x2700

 Why is bone resorption important?

extracellular matrix of bone. This breakdown of the extracellular matrix of bone, termed **resorption** (re-SORP-shun), is part of the normal development, growth, maintenance, and repair of bone. (*Note:* Cells with the suffix *clast* in bone break down extracellular matrix.) As you will see later, osteoclasts help regulate blood calcium level in response to certain hormones. They are also the target cells for drug therapy used to treat osteoporosis.

A mnemonic that will help you to remember the difference between the functions of osteoblasts and osteoclasts is as follows: osteo*B*lasts *B*uild bone, while osteo*C*lasts *C*arve out bone.

Compact Bone Tissue

Compact bone, also referred to as *cortical* or *dense bone*, is the type of bone tissue observed at the surface of a bone, but it also can extend deeper into the bone tissue. It makes up the bulk of the diaphyses of long bones (see Figure 6.2). To the unaided eye,

compact bone looks like a dense, solid material. However, when viewed under a microscope, compact bone is quite porous, with an abundance of microscopic spaces and canals. It provides protection and support and resists the stresses produced by weight and movement.

Compact bone tissue is composed of repeating structural units called **osteons**, or **haversian systems** (ha-VER-shan). Each osteon consists of concentric lamellae arranged around a **central (haversian) canal**. Resembling the growth rings of a tree, the **concentric lamellae** (LA-mel-ē) are circular plates of mineralized extracellular matrix of increasing diameter, surrounding a small network of blood vessels and nerves located in the central canal (Figure 6.4a). These tube-like units of bone generally form a series of parallel cylinders that, in long bones, tend to run parallel to the long axis of the bone. Between the concentric lamellae are small spaces called **lacunae** (la-KOO-nē=little lakes; singular is *lacuna*), which contain osteocytes. Radiating in all directions from the lacunae are tiny **canaliculi** (kan-a-LIK-ū-lī=small channels), which are filled with extracellular fluid. Inside the canaliculi are

Figure 6.4 Histology of compact and spongy bone. (a) Sections through the diaphysis of a long bone, from the surrounding periosteum on the right, to compact bone in the middle, to spongy bone and the medullary cavity on the left. The inset at the upper right shows an osteocyte in a lacuna. (b and c) Details of spongy bone. See part (d) for a photomicrograph of compact bone tissue and the Clinical Connection on osteoporosis in Section 6.7 for a scanning electron micrograph of spongy bone tissue.

Bone tissue is arranged in concentric lamellae around a central (haversian) canal in compact bone, and in irregularly arranged lamellae in the trabeculae of spongy bone.

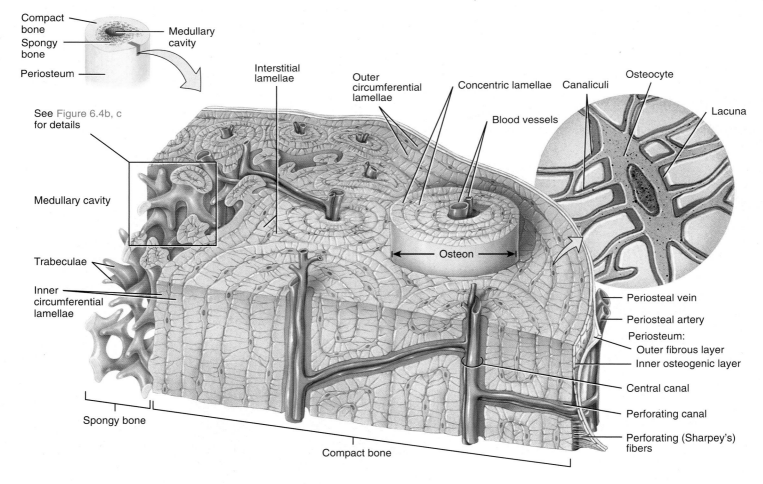

(a) Osteons (haversian systems) in compact bone and trabeculae in spongy bone

slender fingerlike processes of osteocytes (see inset at right of Figure 6.4a). Neighboring osteocytes communicate via gap junctions (see Section 3.2). The canaliculi connect lacunae with one another and with the central canals, forming an intricate, miniature system of interconnected canals throughout the bone. This system provides many routes for nutrients and oxygen to reach the osteocytes and for the removal of wastes.

Osteons in compact bone tissue are aligned in the same direction and are parallel to the length of the diaphysis. As a result, the shaft of a long bone resists bending or fracturing even when considerable force is applied from either end. Compact bone tissue tends to be thickest in those parts of a bone where stresses are applied in relatively few directions. The lines of stress in a bone are not static. They change as a person learns to walk and in response to repeated strenuous physical activity, such as weight training. The lines of stress in a bone also can change because of fractures or physical deformity. Thus, the organization of osteons is not static but changes over time in response to the physical demands placed on the skeleton.

The areas between neighboring osteons contain lamellae called **interstitial lamellae** (in′-ter-STISH-al), which also have lacunae with osteocytes and canaliculi. Interstitial lamellae are fragments of older osteons that have been partially destroyed during bone rebuilding or growth.

Blood vessels and nerves from the periosteum penetrate the compact bone through transverse **perforating canals** or *Volkmann canals* (FOLK-man). The vessels and nerves of the perforating canals connect with those of the medullary cavity, periosteum, and central canals.

Arranged around the entire outer and inner circumference of the shaft of a long bone are lamellae called **circumferential lamellae** (ser′-kum-fer-EN-shē-al). They develop during initial bone formation. The circumferential lamellae directly deep to the periosteum are called *outer circumferential lamellae*. They are connected to the periosteum by **perforating fibers** or **Sharpey's fibers.** The circumferential lamellae that line the medullary cavity are called *inner circumferential lamellae* (Figure 6.4a).

Spongy Bone Tissue

In contrast to compact bone tissue, **spongy bone tissue**, also referred to as *trabecular* or *cancellous bone tissue*, does not contain osteons (Figure 6.4b, c). Spongy bone tissue is always located in the *interior* of a bone, protected by a covering of compact bone. It consists of lamellae that are arranged in an irregular pattern of thin columns called **trabeculae** (tra-BEK-ū-lē=little beams; singular is *trabecula*). Between the trabeculae are spaces that are visible to the unaided eye. These macroscopic spaces are filled with red bone

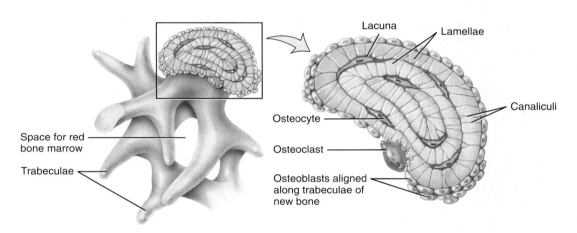

(b) Enlarged aspect of spongy bone trabeculae

(c) Details of a section of a trabecula

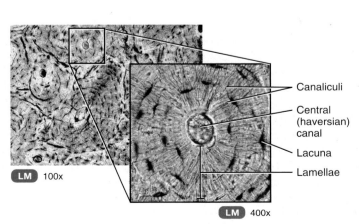

(d) Sectional view of several osteons (haversian systems) of femur (thigh bone) and details of one osteon

 As people age, some central (haversian) canals may become blocked. What effect would this have on the osteocytes?

marrow in bones that produce blood cells, and yellow bone marrow (adipose tissue) in other bones. Both types of bone marrow contain numerous small blood vessels that provide nourishment to the osteocytes. Each trabecula consists of concentric lamellae, osteocytes that lie in lacunae, and canaliculi that radiate outward from the lacunae.

Spongy bone tissue makes up most of the interior bone tissue of short, flat, sesamoid, and irregularly shaped bones. In long bones it forms the core of the epiphyses beneath the paper-thin layer of compact bone, and forms a variable narrow rim bordering the medullary cavity of the diaphysis. Spongy bone is always covered by a layer of compact bone for protection.

At first glance, the trabeculae of spongy bone tissue may appear to be less organized than the osteons of compact bone tissue. However, they are precisely oriented along lines of stress, a characteristic that helps bones resist stresses and transfer force without breaking. Spongy bone tissue tends to be located where bones are not heavily stressed or where stresses are applied from many directions. The trabeculae do not achieve their final arrangement until locomotion is completely learned. In fact, the arrangement can even be altered as lines of stress change due to a poorly healed fracture or a deformity.

Spongy bone tissue is different from compact bone tissue in two respects. First, spongy bone tissue is light, which reduces the overall weight of a bone. This reduction in weight allows the bone to move more readily when pulled by a skeletal muscle. Second, the trabeculae of spongy bone tissue support and protect the red bone marrow. Spongy bone in the hip bones, ribs, sternum (breastbone), vertebrae, and the proximal ends of the humerus and femur is the only site where red bone marrow is stored and, thus, the site where hemopoiesis (blood cell production) occurs in adults.

 CHECKPOINT

6. Why is bone considered a connective tissue?
7. Explain the functions of the four types of cells in bone tissue.
8. What is the composition of the extracellular matrix of bone tissue?
9. Distinguish the microscopic appearance, location, and function of compact and spongy bone tissue.

6.6 BLOOD AND NERVE SUPPLY OF BONE

 OBJECTIVE

• Describe the blood and nerve supply of bone.

Bone is richly supplied with blood. Blood vessels, which are especially abundant in portions of bone containing red bone marrow, pass into bones from the periosteum. Let's consider the blood supply of a long bone using the mature tibia (shin bone) shown in Figure 6.5.

Periosteal arteries (per'-ē-OS-tē-al), small arteries accompanied by nerves, enter the diaphysis through numerous perforating (Volkmann) canals and supply the periosteum and outer part of the compact bone (see Figure 6.4a). Near the center of the diaphysis, a large **nutrient artery** enters the compact bone at an oblique angle through a hole called the **nutrient foramen** (Figure 6.5). The path of the artery through the bone is always away from the dominant growth end of the bone. (This is true of all the long bones of the limbs. To help you learn and remember this, use the following learning device: nutrient canals always "go to the elbow and flee the knee," indicating their orientation away from the knee in the femur, tibia, and fibula and toward the elbow in the humerus, ulna, and radius.) On entering the medullary cavity, the nutrient artery divides into proximal and distal branches that course toward each end of the bone. These branches supply both the inner part of compact bone tissue of the diaphysis and the spongy bone tissue and red bone marrow as far as the epiphyseal plates (or lines). Most bones, like the tibia, have only one nutrient artery entering the diaphysis; others, like the femur (thigh bone), have several. The ends of long bones are supplied by the metaphyseal and epiphyseal arteries, which arise from arteries that supply the associated joint. The **metaphyseal arteries** (met-a-FIZ-ē-al) enter the metaphyses of a long bone and, together with

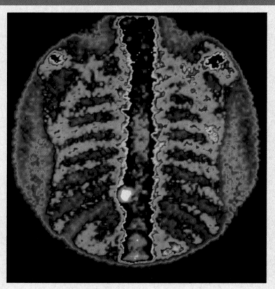

Figure 6.5 Blood supply of a mature long bone, the tibia (shinbone).

 Bone is richly supplied with blood vessels.

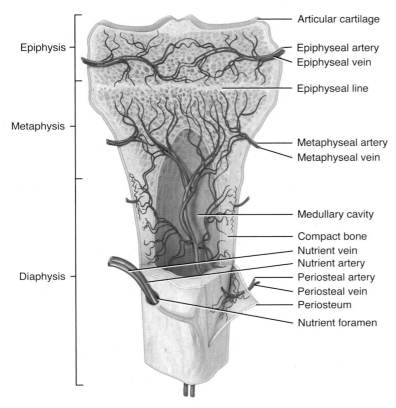

Partially sectioned tibia (shin bone)

 Where do periosteal arteries enter bone tissue?

the nutrient artery, supply the red bone marrow and bone tissue of the metaphyses. The **epiphyseal arteries** (ep-i-FIZ-ē-al) enter the epiphyses of a long bone and supply the red bone marrow and bone tissue of the epiphyses.

Veins that carry blood away from long bones are evident in three places: (1) One or two **nutrient veins** accompany the nutrient artery and exit through the diaphysis; (2) numerous **epiphyseal veins** and **metaphyseal veins** accompany their respective arteries and exit through the epiphyses; and (3) many small **periosteal veins** accompany their respective arteries and exit through the periosteum. The periosteum surrounding the bone has numerous lymphatic capillaries and lymph vessels, but there is no evidence of any lymphatic vessels within the bone tissue.

Nerves accompany the blood vessels that supply bones. The periosteum is rich in sensory nerves, some of which carry pain sensations. These nerves are especially sensitive to tearing or tension, which explains the severe pain resulting from a fracture or a bone tumor. For the same reason there is some pain associated with a bone marrow needle biopsy. In this procedure, a needle is inserted into the middle of the bone to withdraw a sample of red bone marrow for microscopic examination; conditions such as leukemias, metastatic neoplasms, lymphoma, Hodgkin disease, and aplastic anemia are often diagnosed

through the use of a bone marrow needle biopsy. As the needle penetrates the periosteum, pain is felt. Once it passes through, there is little pain.

 CHECKPOINT

10. Explain the location and roles of the nutrient arteries, nutrient foramina, epiphyseal arteries, and periosteal arteries.

6.7 BONE FORMATION

OBJECTIVES

• **Distinguish** between intramembranous ossification and endochondral ossification.
• **Explain** the importance of the different types of bone formation during different phases of a person's lifetime.
• **Describe** the process of bone remodeling.

The process by which bone forms is called **ossification** (os'-i-fi-KĀ-shun; *ossi-*=bone; *-fication*=making). Bone formation occurs in four principal situations: (1) the initial formation of bones in an embryo and fetus, (2) the growth of bones during infancy, childhood, and adolescence until their adult sizes are reached, (3) the remodeling of bone (replacement of old bone by new bone tissue throughout life), and (4) the repair of fractures (breaks in bones) throughout life. We will discuss the first three situations in this section; fractures will be explained in Section 6.8.

Initial Bone Formation in an Embryo and Fetus

We will first consider the initial formation of bone in an embryo and fetus. The embryonic "skeleton" is at first composed of mesenchyme in the general shape of bones. These become the sites where subsequent cartilage formation and then ossification occurs. (Recall that *mesenchyme* is a connective tissue found mostly in an embryo and is the tissue from which most other connective tissues develop.) This begins during the sixth week of embryonic development and follows one of two patterns.

The two methods of bone formation, which both involve the replacement of a preexisting connective tissue with bone, do not lead to differences in the structure of mature bones, but are simply different methods of bone development. In the first type of ossification, called **intramembranous ossification** (in'-tra-MEM-bra-nus; *intra-*=within; *membran*=membrane), bone forms directly within condensed mesenchyme, which is arranged in sheetlike layers that resemble membranes. In the second type, **endochondral ossification** (en'-dō-KON-dral; *endo-*=within; *chondral*=cartilage), bone forms within hyaline cartilage that develops from mesenchyme.

Intramembranous Ossification

Intramembranous ossification is the simpler of the two methods of bone formation. The flat bones of the skull, most of the facial bones, mandible (lower jawbone), and the medial part of the clavicle (collar bone) are formed in this way. The remaining bones form by endochondral ossification, which will be described later in this section. Also, the "soft spots" that help the fetal skull pass through the birth canal later harden as they undergo intramembranous ossification, which occurs as follows (Figure 6.6):

❶ *Development of the ossification center.* At the site where the bone will develop, specific chemical messages cause the mesenchymal cells to cluster together and differentiate, first into osteoprogenitor cells and then into osteoblasts. The site of such a cluster is called an **ossification center**. Osteoblasts secrete the organic extracellular matrix of bone until they are surrounded by it.

❷ *Calcification.* Next, the secretion of extracellular matrix stops and the cells, now called osteocytes, lie in lacunae and extend their narrow cytoplasmic processes into canaliculi that radiate in all directions. Within a few days, calcium and other mineral salts are deposited and the extracellular matrix hardens or calcifies (calcification).

❸ *Formation of trabeculae.* As the bone extracellular matrix forms, it develops into trabeculae that fuse with one another to form spongy bone around the network of blood vessels in the tissue. Connective tissue that is associated with the blood vessels in the trabeculae differentiates into red bone marrow.

❹ *Development of the periosteum.* In conjunction with the formation of trabeculae, the mesenchyme at the periphery of the bone condenses and develops into the periosteum. Eventually, a thin layer of compact bone replaces the surface layers of the spongy bone, but spongy bone remains in the center. Much of the newly formed bone is remodeled (destroyed and reformed) as the bone is transformed into its adult size and shape.

Figure 6.6 Intramembranous ossification. Illustrations ❶ and ❷ show a smaller field of vision at higher magnification than illustrations ❸ and ❹.

🔑 Intramembranous ossification involves the formation of bone within mesenchyme arranged in sheetlike layers that resemble membranes.

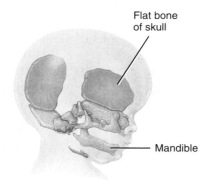

Flat bone of skull

Mandible

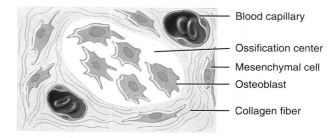

Blood capillary
Ossification center
Mesenchymal cell
Osteoblast
Collagen fiber

1 Development of ossification center: osteoblasts secrete organic extracellular matrix

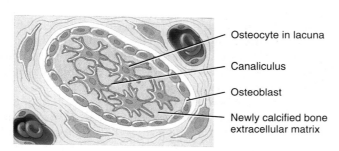

Osteocyte in lacuna
Canaliculus
Osteoblast
Newly calcified bone extracellular matrix

2 Calcification: calcium and other mineral salts are deposited and extracellular matrix calcifies (hardens)

Mesenchyme condenses
Blood vessel
Spongy bone trabeculae
Osteoblast

3 Formation of trabeculae: extracellular matrix develops into trabeculae that fuse to form spongy bone

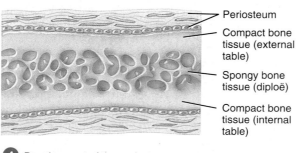

Periosteum
Compact bone tissue (external table)
Spongy bone tissue (diploë)
Compact bone tissue (internal table)

4 Development of the periosteum: mesenchyme at the periphery of the bone develops into the periosteum

 Which bones of the body develop by intramembranous ossification?

Endochondral Ossification

Although most bones of the body are formed via endochondral ossification, the process is best observed in a long bone. It proceeds as follows (Figure 6.7):

❶ **Development of the cartilage model.** At the site where the bone is going to form, specific chemical messages cause the cells in mesenchyme to crowd together in the general shape of the future bone, and then develop into chondroblasts. The chondroblasts secrete cartilage extracellular matrix, producing a **cartilage model** consisting of hyaline cartilage. A mesenchymal covering called the **perichondrium** (per-i-KON-drē-um) develops around the cartilage model.

Figure 6.7 Endochondral ossification.

During endochondral ossification, bone gradually replaces a cartilage model.

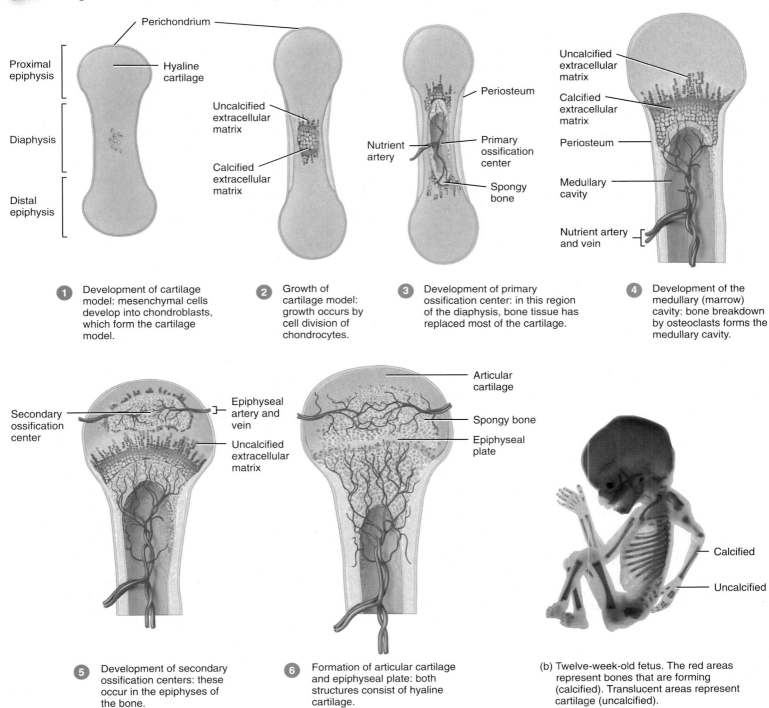

❶ Development of cartilage model: mesenchymal cells develop into chondroblasts, which form the cartilage model.

❷ Growth of cartilage model: growth occurs by cell division of chondrocytes.

❸ Development of primary ossification center: in this region of the diaphysis, bone tissue has replaced most of the cartilage.

❹ Development of the medullary (marrow) cavity: bone breakdown by osteoclasts forms the medullary cavity.

❺ Development of secondary ossification centers: these occur in the epiphyses of the bone.

❻ Formation of articular cartilage and epiphyseal plate: both structures consist of hyaline cartilage.

(b) Twelve-week-old fetus. The red areas represent bones that are forming (calcified). Translucent areas represent cartilage (uncalcified).

(a) Sequence of events

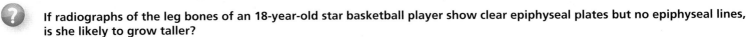

If radiographs of the leg bones of an 18-year-old star basketball player show clear epiphyseal plates but no epiphyseal lines, is she likely to grow taller?

❷ *Growth of the cartilage model.* Once chondroblasts become deeply buried in the cartilage extracellular matrix, they are called chondrocytes. The cartilage model grows in length by continual cell division of chondrocytes, accompanied by further secretion of the cartilage extracellular matrix. This type of cartilaginous growth, called **interstitial (endogenous) growth** (growth from within), results in an increase in length. In contrast, growth of the cartilage in thickness is due mainly to the deposition of extracellular matrix material on the cartilage surface of the model by new chondroblasts that develop from the perichondrium in a process called **appositional (exogenous) growth** (a-pō-ZISH-i-nal), meaning growth of the outer surface (described shortly).

As the cartilage model continues to grow, chondrocytes in its midregion hypertrophy (increase in size), and the surrounding cartilage extracellular matrix begins to calcify. Other chondrocytes within the calcifying cartilage die because nutrients can no longer diffuse quickly enough through the extracellular matrix. As these chondrocytes die, the spaces left behind by the dead chondrocytes merge into small cavities called lacunae.

❸ *Development of the primary ossification center.* Primary ossification proceeds *inward* from the external surface of the bone. A nutrient artery penetrates the perichondrium and the calcifying cartilage model through a nutrient foramen in the midregion of the cartilage model, stimulating osteoprogenitor cells in the perichondrium to differentiate into osteoblasts. Once the perichondrium starts to form bone, it is known as the **periosteum.** Near the middle of the model, periosteal capillaries grow into the disintegrating calcified cartilage, inducing growth of a **primary ossification center**, a region where bone tissue will replace most of the cartilage. Osteoblasts then begin to deposit bone extracellular matrix over the remnants of calcified cartilage, forming spongy bone trabeculae. Primary ossification spreads from this central location toward both ends of the cartilage model.

❹ *Development of the medullary (marrow) cavity.* As the primary ossification center grows toward the ends of the bone, osteoclasts break down some of the newly formed spongy bone trabeculae. This activity leaves a cavity, the medullary (marrow) cavity, in the diaphysis (shaft). Eventually, most of the wall of the diaphysis is replaced by compact bone.

❺ *Development of the secondary ossification centers.* When branches of the epiphyseal artery enter the epiphyses, **secondary ossification centers** develop, usually around the time of birth. Bone formation is similar to what occurs in primary ossification centers. However, in the secondary ossification centers spongy bone remains in the interior of the epiphyses (no medullary cavities are formed here). In contrast to primary ossification, secondary ossification proceeds *outward* from the center of the epiphysis toward the outer surface of the bone.

❻ *Formation of articular cartilage and the epiphyseal (growth) plate.* The hyaline cartilage that covers the epiphyses becomes the articular cartilage. Prior to adulthood, hyaline cartilage remains between the diaphysis and epiphysis as the epiphyseal (growth) plate, the region responsible for

the lengthwise growth of long bones that you will learn about next.

 CHECKPOINT

11. What are the major events of intramembranous ossification and endochondral ossification? How are they similar? How do they differ?

Bone Growth During Infancy, Childhood, and Adolescence

During infancy, childhood, and adolescence, bones throughout the body grow in thickness by appositional growth, and long bones lengthen by interstitial growth.

Growth in Length

The growth in length of a long bone involves (1) interstitial growth of cartilage on the epiphyseal side of the epiphyseal plate and (2) replacement of cartilage with bone by endochondral ossification on the diaphyseal side of the epiphyseal plate.

To understand how a bone grows in length, you need to know some of the details of the structure of the epiphyseal plate (Figure 6.8). The **epiphyseal (growth) plate** (ep′-i-FIZ-ē-al) is a layer of hyaline cartilage in the metaphysis of a growing bone that consists of four zones (Figure 6.8b):

1. *Zone of resting cartilage.* This layer is nearest the epiphysis and consists of small, scattered chondrocytes. The term *resting* is used because the cells do not function in bone growth. Rather, they anchor the epiphyseal plate to the epiphysis of the bone.
2. *Zone of proliferating cartilage.* Slightly larger chondrocytes arranged like stacks of coins undergo interstitial growth as they divide and secrete extracellular matrix. The chondrocytes in this zone divide to replace those that die at the diaphyseal side of the epiphyseal plate.
3. *Zone of hypertrophic cartilage* (hī-per-TRŌ-fik). This layer consists of large, maturing chondrocytes arranged in columns.
4. *Zone of calcified cartilage.* The final zone of the epiphyseal plate is only a few cells thick and consists mostly of chondrocytes that are dead because the extracellular matrix around them has calcified. Osteoclasts dissolve the calcified cartilage, and osteoblasts and capillaries from the diaphysis invade the area. The osteoblasts lay down bone extracellular matrix, replacing the calcified cartilage by the process of endochondral ossification. As a result, the zone of calcified cartilage becomes "new diaphysis" that is firmly cemented to the rest of the diaphysis of the bone.

The activity of the epiphyseal plate is the only way that the diaphysis can increase in length. As a bone grows, chondrocytes proliferate on the epiphyseal side of the plate. New chondrocytes replace older ones, which are destroyed by calcification. Thus, the cartilage is replaced by bone on the diaphyseal side of the plate. In this way the thickness of the epiphyseal plate remains relatively constant, but the bone on the diaphyseal side increases in length. If a bone fracture damages the epiphyseal plate, the fractured bone may be shorter than normal once adult stature is reached. This is because damage to cartilage, which is avascular, accelerates closure of the epiphyseal plate due to the cessation of

Figure 6.8 The epiphyseal (growth) plate is a layer of hyaline cartilage in the metaphysis of a growing bone. The epiphyseal plate appears as a dark band between whiter calcified areas in the radiograph (x-ray) shown in part (a).

 The epiphyseal plate allows the diaphysis of a bone to increase in length.

(a) Radiograph showing the epiphyseal plate of the femur of a 3-year-old

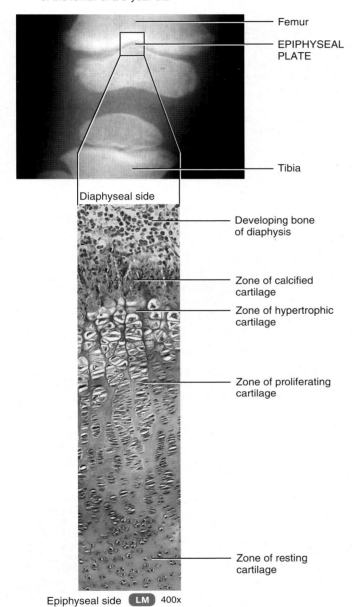

Epiphyseal side **LM** 400x

(b) Histology of the epiphyseal plate

 What activities of the epiphyseal plate account for the lengthwise growth of the diaphysis?

cartilage cell division, thus inhibiting lengthwise growth of the bone.

When adolescence comes to an end (at about age 18 in females and age 21 in males), the epiphyseal plates close; that is, the epiphyseal cartilage cells stop dividing and bone replaces all remaining cartilage. The epiphyseal plate fades, leaving a bony structure called the **epiphyseal line.** With the appearance of the epiphyseal line, bone growth in length stops completely.

Closure of the epiphyseal plate is a gradual process and the degree to which it occurs is useful in determining bone age, to predict adult height, and to establish age at death from skeletal remains, especially in infants, children, and adolescents. For example, an open epiphyseal plate indicates a younger person, while a partially closed epiphyseal plate or a completely closed one indicates an older person. It should also be kept in mind that closure of the epiphyseal plate, on average, takes place 1–2 years earlier in females.

In the long bones of the limbs, growth in length does not occur equally at both ends of the bones; one end is always the dominant growing end. The dominant growing end is always directed away from the orientation angle of the nutrient foramen in the diaphysis. Therefore, the ends of the femur, tibia, and fibula toward the knee are the dominant growing epiphyseal plates and the ends of the humerus, ulna, and radius at the ends opposite the elbow are the dominant growing epiphyseal plates (see Figure 6.2).

Growth in Thickness

Like cartilage, bone can grow in thickness (diameter) only by appositional growth (Figure 6.9a):

❶ At the bone surface, periosteal cells differentiate into osteoblasts, which secrete the collagen fibers and other organic molecules that form bone extracellular matrix. The osteoblasts become surrounded by extracellular matrix and develop into osteocytes. This process forms bone ridges on either side of a periosteal blood vessel. The ridges slowly enlarge and create a groove for the periosteal blood vessel.

❷ Eventually, the ridges fold together and fuse, and the groove becomes a tunnel that encloses the blood vessel. The former periosteum now becomes the endosteum that lines the tunnel.

❸ Osteoblasts in the endosteum deposit bone extracellular matrix, forming new concentric lamellae. The formation of additional concentric lamellae proceeds inward toward the periosteal blood vessel. In this way, the tunnel fills in, and a new osteon is created.

❹ As an osteon is forming, osteoblasts under the periosteum deposit new circumferential lamellae, further increasing the thickness of the bone. As additional periosteal blood vessels become enclosed as in step ❶, the growth process continues.

Recall that as new bone tissue is being deposited on the outer surface of bone, the bone tissue lining the medullary cavity is destroyed by osteoclasts in the endosteum. In this way, the medullary cavity enlarges as the bone increases in thickness (Figure 6.9b).

✓ CHECKPOINT

12. Describe each of the zones of the epiphyseal plate and their functions.
13. What are the differences between interstitial growth and appositional growth?
14. What is the significance of the epiphyseal line?

Remodeling of Bone

Like skin, bone forms before birth but continually renews itself thereafter. **Bone remodeling** is the ongoing replacement of old bone tissue by new bone tissue. It involves **bone resorption**, the removal of minerals and collagen fibers from bone by osteoclasts,

Figure 6.9 Bone growth in thickness.

🔑 As new bone tissue is deposited on the outer surface of bone by osteoblasts, the bone tissue lining the medullary cavity is destroyed by osteoclasts in the endosteum.

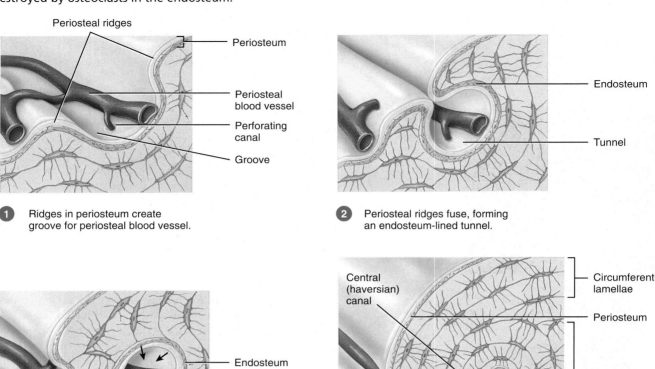

① Ridges in periosteum create groove for periosteal blood vessel.

② Periosteal ridges fuse, forming an endosteum-lined tunnel.

③ Osteoblasts in endosteum build new concentric lamellae inward toward center of tunnel, forming a new osteon.

④ Bone grows outward as osteoblasts in periosteum build new circumferential lamellae. Osteon formation repeats as new periosteal ridges fold over blood vessels.

(a) Microscopic details

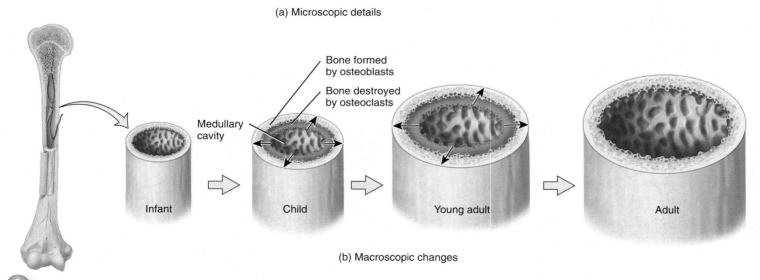

(b) Macroscopic changes

❓ **How does the medullary cavity enlarge during growth in thickness?**

and **bone deposition,** the addition of minerals and collagen fibers to bone by osteoblasts. Thus, bone resorption results in the breakdown of bone extracellular matrix, and bone deposition results in its formation. At any given time, about 5 percent of the total bone mass in the body is being remodeled. The renewal rate for compact bone tissue is about 4 percent per year and for spongy

bone tissue it is about 20 percent per year. Remodeling also takes place at different rates in different regions of the body. The distal portion of the femur is replaced about every four months. By contrast, bone in certain areas of the shaft of the femur will not be replaced completely during an individual's life. Even after bones have reached their adult shapes and sizes, old bone is continually destroyed and new bone is formed in its place. Remodeling also removes injured bone, replacing it with new bone tissue. Remodeling may be triggered by factors such as exercise, lifestyle modifications, and changes in diet.

Remodeling has several other benefits. Since the strength of bone is related to the degree to which it is strained, if newly formed bone is subjected to heavy loads, it will grow thicker and therefore be stronger than the old bone. Also, the shape of a bone can be altered for proper support based on the strain patterns experienced during the remodeling process. Finally, new bone is more resistant to fracture than old bone.

Orthodontics (or-thō-DON-tiks) is the branch of dentistry concerned with the prevention and correction of poorly aligned teeth. The movement of teeth by braces places a stress on the bone that forms the sockets that anchor the teeth. In response to this artificial stress, osteoclasts and osteoblasts remodel the sockets so that the teeth align properly.

 CHECKPOINT

15. What is bone remodeling? Why is it important?

CLINICAL CONNECTION | *Paget's Disease*

A delicate balance exists between the actions of osteoclasts and osteoblasts. Should too much new tissue be formed, the bones become abnormally thick and heavy. If too much mineral material is deposited in the bone, the surplus may form thick bumps, called *spurs,* on the bone that interfere with movement at joints. Excessive loss of calcium or tissue weakens the bones, and they may break, as occurs in osteoporosis, or they may become too flexible, as in rickets and osteomalacia. In **Paget's disease,** there is an excessive proliferation of osteoclasts so that bone resorption occurs faster than bone deposition. In response, osteoblasts attempt to compensate, but the new bone is weaker because it has a higher proportion of spongy to compact bone, mineralization is decreased, and the newly synthesized extracellular matrix contains abnormal proteins. The newly formed bone, especially that of the pelvis, limbs, lower vertebrae, and skull, becomes enlarged, hard, and brittle and fractures easily. •

CLINICAL CONNECTION | *Osteoporosis*

Osteoporosis (os'-tē-ō-pō-RŌ-sis; *por*=passageway; *-osis*= condition), literally a condition of porous bones, affects 10 million people a year in the United States. In addition, 18 million people have low bone mass (*osteopenia*), which puts them at risk for osteoporosis. The basic problem is that bone resorption (breakdown) outpaces bone deposition (formation). In large part this is due to depletion of calcium from the body—more calcium is lost in urine, feces, and sweat than is absorbed from the diet. Bone mass becomes so depleted that bones fracture, often spontaneously, under the mechanical stresses of everyday living. For example, a hip fracture might result from simply sitting down too quickly. In the United States, osteoporosis causes more than one and a half million fractures a year, mainly in the hips, wrists, and vertebrae. Osteoporosis afflicts the entire skeletal system. In addition to fractures, osteoporosis causes shrinkage of vertebrae, height loss, hunched backs, and bone pain.

Osteoporosis primarily affects middle-aged and elderly people, 80 percent of them women. Older women suffer from osteoporosis more often than men for two reasons: (1) Women's bones are less massive than men's bones, and (2) production of estrogens in women declines dramatically at menopause, while production of the main androgen, testosterone, in older men wanes gradually and only slightly. Estrogens and testosterone stimulate osteoblast activity and synthesis of bone matrix. Besides gender, risk factors for developing osteoporosis include a family history of the disease, European or Asian ancestry, thin or small body build, an inactive lifestyle, cigarette smoking, a diet low in calcium and vitamin D, more than two alcoholic drinks a day, and the use of certain medications.

Osteoporosis is diagnosed by taking a family history and undergoing a *bone mineral density* (BMD) test. Performed like x-rays, BMD tests measure bone density. They can also be used to confirm a diagnosis of osteoporosis, determine the rate of bone loss, and monitor the effects of treatment. There is also a relatively new tool called *FRAX®* that incorporates risk factors besides bone mineral density to accurately estimate fracture risk. Patients fill out an online survey of risk factors such as age, gender, height, weight, ethnicity, prior fracture history, parental history of hip fracture, use of glucocorticoids (for example, cortisone), smoking, alcohol intake, and rheumatoid arthritis. Using the data, FRAX® provides an estimate of the probability that a person will suffer a fracture of the hip or other major bone in the spine, shoulder, or forearm due to osteoporosis within ten years.

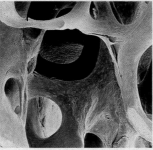
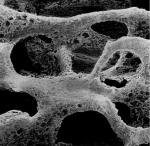

SEM 30x   SEM 30x

(A) Normal bone   (B) Osteoporotic bone

Comparison of spongy bone tissue from (A) a normal young adult and (B) a person with osteoporosis.

Treatment options for osteoporosis are varied. With regard to nutrition, a diet high in calcium is important to reduce the risk of fractures. Vitamin D is necessary for the body to utilize calcium. In terms of exercise, regular performance of weight-bearing exercises has been shown to maintain and build bone mass. These exercises include walking, jogging, hiking, climbing stairs, playing tennis, and dancing. Resistance exercises, such as weight lifting, also build bone strength and muscle mass.

Medications used to treat osteoporosis are generally of two types: (1) **antireabsorptive drugs** slow down the progression of bone loss, and (2) **bone-building drugs** promote increasing bone mass. Among the antireabsorptive drugs are (1) *bisphosphonates,* which inhibit osteoclasts (Fosamax®, Actonel®, Boniva®, and calcitonin); (2) *selective estrogen receptor modulators,* which mimic the effects of estrogens without unwanted side effects (Raloxifene®, Evista®); and (3) estrogen replacement therapy (ERT), which replaces estrogens lost during and after menopause (Premarin®), and hormone replacement therapy (HRT), which replaces estrogens and progesterone lost during and after menopause (Prempro®). ERT helps maintain and increase bone mass after menopause. Women on ERT have a slight increased risk of stroke and blood clots. HRT also helps maintain and increase bone mass. Women on HRT have increased risks of heart disease, breast cancer, stroke, blood clots, and dementia.

Among the bone-building drugs is parathyroid hormone (PTH), which stimulates osteoblasts to produce new bone (Forteo®). Others are under development. •

6.8 FRACTURES

 OBJECTIVE

• **Describe** the process of fracture repair.

A **fracture** is any break in a bone. Fractures are named according to their severity, the shape or position of the fracture line, or even the physician who first described them. Some of the common types of fractures are shown and described in Table 6.2.

In some cases, a bone may fracture without visibly breaking. A **stress fracture** is a series of microscopic fissures in bone that forms without any evidence of injury to other tissues. In healthy adults, stress fractures result from repeated, strenuous activities such as running, jumping, or aerobic dancing. Stress fractures are quite painful and also result from disease processes that disrupt normal bone calcification, such as osteoporosis. About 25 percent of stress fractures involve the tibia. Although standard x-ray images often fail to reveal the presence of stress fractures, they show up clearly in a bone scan.

TABLE 6.2

Some Common Fractures

FRACTURE	DESCRIPTION	ILLUSTRATION	RADIOGRAPH
Open (Compound)	The broken ends of the bone protrude through the skin. Conversely, a **closed (simple) fracture** does not break the skin.		Humerus Radius Ulna
Comminuted (KOM-i-noo-ted; *com-*=together; *-minuted*=crumbled)	The bone is splintered, crushed, or broken into pieces at the site of impact, and smaller bone fragments lie between the two main fragments.		Humerus
Greenstick	A partial fracture in which one side of the bone is broken and the other side bends; similar to the way a green twig breaks on one side while the other side stays whole, but bends; occurs only in children, whose bones are not fully ossified and contain more organic material than inorganic material.		Ulna Radius Wrist bones

Treatments for fractures vary according to age, type of fracture, and the bone involved. The ultimate goals of fracture treatment are realignment of the bone fragments, immobilization to maintain realignment, and restoration of function. For bones to unite properly, the fractured ends must be brought into alignment. This process, called **reduction**, is commonly referred to as setting a fracture. In **closed reduction**, the fractured ends of a bone are brought into alignment by manual manipulation, and the skin remains intact. In **open reduction**, the fractured ends of a bone are brought into alignment by a surgical procedure using internal fixation devices such as screws, plates, pins, rods, and wires. Following reduction, a fractured bone may be kept immobilized by a cast, sling, splint, elastic bandage, external fixation device, or a combination of these devices. •

FRACTURE	DESCRIPTION	ILLUSTRATION	RADIOGRAPH
Impacted	One end of the fractured bone is forcefully driven into the interior of the other.		Humerus
Pott	Fracture of the distal end of the lateral leg bone (fibula), with serious injury of the distal tibial articulation.		Tibia Fibula Ankle bones
Colles (KOL-ēz)	Fracture of the distal end of the lateral forearm bone (radius) in which the distal fragment is displaced posteriorly.		Radius Ulna Wrist bones

Figure 6.10 Steps involved in repair of a bone fracture.

 Bone heals more rapidly than cartilage because its blood supply is more plentiful.

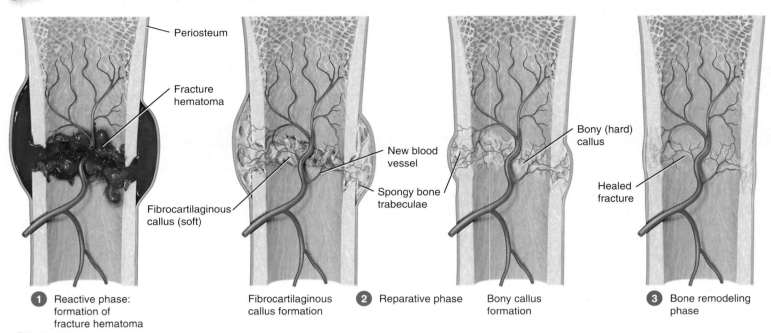

① Reactive phase: formation of fracture hematoma

Fibrocartilaginous callus formation

② Reparative phase

Bony callus formation

③ Bone remodeling phase

 Why does it sometimes take months for a fracture to heal?

The repair of a bone fracture involves the following phases (Figure 6.10):

① **Reactive phase.** Blood vessels crossing the fracture line are broken. As blood leaks from the torn ends of the vessels, it forms a mass of blood (usually clotted) around the site of the fracture. This clot, called a **fracture hematoma** (hē-ma-TŌ-ma; *hemat-*=blood; *-oma*=tumor), usually forms 6 to 8 hours after the injury. Because the circulation of blood stops at the site where the fracture hematoma forms, nearby bone cells die. Swelling and inflammation occur in response to dead bone cells, producing additional cellular debris. Phagocytes (neutrophils and macrophages) and osteoclasts begin to remove the dead or damaged tissue in and around the fracture hematoma. This stage may last up to several weeks.

② **Reparative phase.** This phase is characterized by the formation of a fibrocartilaginous callus and a bony callus.

Blood vessels grow into the fracture hematoma and phagocytes begin to clean up dead bone cells. Fibroblasts from the periosteum invade the fracture site and produce collagen fibers. In addition, cells from the periosteum develop into chondroblasts and begin to produce fibrocartilage in this region. These events lead to the development of a **fibrocartilaginous (soft) callus** (fi-brō-kar-ti-LAJ-i-nus), a mass of repair tissue consisting of collagen fibers and cartilage that bridges the broken ends of the bone. Formation of the fibrocartilaginous callus takes about three weeks.

In areas closer to well-vascularized healthy bone tissue, osteoprogenitor cells develop into osteoblasts, which begin to produce spongy bone trabeculae. The trabeculae join living and dead portions of the original bone fragments. In time, the fibrocartilage is converted to spongy bone, and the callus is then referred to as a **bony (hard) callus.** The bony callus lasts about three to four months.

③ **Bone remodeling phase.** The final phase of fracture repair is **bone remodeling** of the callus. Dead portions of the original fragments of broken bone are gradually resorbed by osteoclasts. Compact bone replaces spongy bone around the periphery of the fracture. Sometimes, the repair process is so thorough that the fracture line is undetectable, even in a radiograph (x-ray). However, a thickened area on the surface of the bone remains as evidence of a healed fracture.

Although bone has a generous blood supply, healing sometimes takes months. The calcium and phosphorus needed to strengthen and harden new bone are deposited only gradually, and the bone cells themselves generally grow and reproduce slowly.

✓ CHECKPOINT
16. Define a fracture and outline the steps involved in fracture repair.

6.9 EXERCISE AND BONE TISSUE

OBJECTIVE
• Describe how exercise and mechanical stress affect bone tissue.

Within limits, bone tissue has the ability to alter its strength in response to changes in the strain it experiences. When placed under stress, bone tissue responds to the strain it experiences and becomes stronger through increased deposition of mineral salts and production of collagen fibers by osteoblasts. Without mechanical stress, bone does not remodel normally because bone resorption occurs more quickly than bone formation. Research has shown that high-impact intermittent strains more strongly influence bone deposition than lower-impact constant strains. Therefore, running and jumping stimulate bone remodeling more dramatically than walking.

The main mechanical stresses on bone are those that result from the pull of skeletal muscles and the pull of gravity. If a person is bedridden or has a fractured bone in a cast, the strength of the unstressed bones diminishes because of the loss of bone minerals and decreased numbers of collagen fibers. Astronauts subjected to the microgravity of space also lose bone mass. In any of these cases, bone loss can be dramatic—as much as 1 percent per week. In contrast, the bones of athletes, which are repetitively and highly stressed, become notably thicker and stronger than those of astronauts or nonathletes. Weight-bearing activities, such as walking or moderate weight lifting, help build and retain bone mass. Adolescents and young adults should engage in regular weight-bearing exercise prior to the closure of the epiphyseal plates to help build total mass prior to its inevitable reduction with aging. However, people of all ages can and should strengthen their bones by engaging in weight-bearing exercise.

CHECKPOINT
17. How do mechanical stresses strengthen bone tissue?

6.10 AGING AND BONE TISSUE

OBJECTIVE
• Describe the effects of aging on bone tissue.

From birth through adolescence, more bone tissue is produced than is lost during bone remodeling. In young adults the rates of bone deposition and resorption are about the same. As the level of sex steroids diminishes during middle age, especially in women after menopause, a decrease in bone mass occurs because bone resorption outpaces bone deposition. In old age, loss of bone through resorption by osteoclasts occurs more rapidly than

bone gain by osteoblasts. Because women's bones generally are smaller and less massive than men's bones to begin with, loss of bone mass in old age typically has a greater adverse effect in women. These factors contribute to a higher incidence of osteoporosis in females.

There are two principal effects of aging on bone tissue: loss of bone mass and brittleness. Loss of bone mass results from **demineralization** (dē-min′-er-al-i-ZĀ-shun), the loss of calcium and other minerals from bone extracellular matrix. This loss usually begins after age 30 in females, accelerates greatly around age 45 as levels of estrogens decrease, and continues until as much as 30 percent of the calcium in bones is lost by age 70. Once bone loss begins in females, about 8 percent of bone mass is lost every 10 years. In males, calcium loss typically does not begin until after age 60, and about 3 percent of bone mass is lost every 10 years. The loss of calcium from bones is one of the contributing factors in osteoporosis.

The second principal effect of aging on the skeletal system, brittleness, results from a decreased rate of protein synthesis. Recall that the organic part of bone matrix, mainly collagen fibers, gives bone its tensile strength. The loss of tensile strength causes the bones to become very brittle and susceptible to fractures. In some elderly people, collagen fiber synthesis slows, in part, due to diminished production of human growth hormone. In addition to increasing the susceptibility to fractures, loss of bone mass also leads to deformity, pain, loss of height, and loss of teeth.

CHECKPOINT
18. What is demineralization, and how does it affect the functioning of bone?
19. What changes occur in the organic part of bone extracellular matrix with aging?

6.11 FACTORS AFFECTING BONE GROWTH

OBJECTIVE
• Explain why minerals, vitamins, and hormones are important in bone growth.

Bone growth in the young, bone remodeling in the adult, and the repair of fractured bone depend on several factors. Among these factors are minerals (such as calcium and phosphorus), vitamins (A, C, D, K, B₁₂), hormones (such as human growth hormone, thyroid hormones, and sex hormones), exercise, and aging. These are described in Table 6.3.

CHECKPOINT
20. Outline the factors that affect bone growth.

TABLE 6.3

Summary of Factors That Affect Bone Growth

FACTOR	COMMENT
MINERALS	
Calcium and phosphorus	Make bone extracellular matrix hard.
Magnesium	Helps form bone extracellular matrix.
Fluoride	Helps strengthen bone extracellular matrix.
Manganese	Activates enzymes involved in synthesis of bone extracellular matrix.
VITAMINS	
Vitamin A	Needed for the activity of osteoblasts during remodeling of bone; deficiency stunts bone growth; toxic in high doses.
Vitamin C	Needed for synthesis of collagen, the main bone protein; deficiency leads to decreased collagen production, which slows down bone growth and delays repair of broken bones.
Vitamin D	Active form (calcitriol) is produced by the kidneys; helps build bone by increasing absorption of calcium from gastrointestinal tract into blood; deficiency causes faulty calcification and slows down bone growth; may reduce the risk of osteoporosis but is toxic if taken in high doses. People who use sunscreens, have minimal exposure to ultraviolet rays, or do not take vitamin D supplements may not have sufficient vitamin D to absorb calcium. This interferes with calcium metabolism.
Vitamins K and B$_{12}$	Needed for synthesis of bone proteins; deficiency leads to abnormal protein production in bone extracellular matrix and decreased bone density.
HORMONES	
Human growth hormone (hGH)	Secreted by the anterior lobe of the pituitary gland; promotes general growth of all body tissues, including bone, mainly by stimulating production of insulinlike growth factors.
Insulinlike growth factors (IGFs)	Secreted by the liver, bones, and other tissues upon stimulation by human growth hormone; promotes normal bone growth by stimulating osteoblasts and by increasing the synthesis of proteins needed to build new bone.
Thyroid hormones (thyroxine and triiodothyronine)	Secreted by thyroid gland; promote normal bone growth by stimulating osteoblasts.
Insulin	Secreted by the pancreas; promotes normal bone growth by increasing the synthesis of bone proteins.
Sex hormones (estrogens and testosterone)	Secreted by the ovaries in women (estrogens) and by the testes in men (testosterone); stimulate osteoblasts and promote the sudden "growth spurt" that occurs during the teenage years; shut down growth at the epiphyseal plates around age 18–21, causing lengthwise growth of bone to end; contribute to bone remodeling during adulthood by slowing bone resorption by osteoclasts and promoting bone deposition by osteoblasts.
Parathyroid hormone (PTH)	Secreted by the parathyroid glands; promotes bone resorption by osteoclasts; enhances recovery of calcium ions from urine; promotes formation of the active form of vitamin D (calcitriol).
Calcitonin (CT)	Secreted by the thyroid gland; inhibits bone resorption by osteoclasts.
EXERCISE	
	Weight-bearing activities stimulate osteoblasts and, consequently, help build thicker, stronger bones and retard loss of bone mass that occurs as people age.
AGING	
	As the level of sex hormones diminishes during middle age to older adulthood, especially in women after menopause, bone resorption by osteoclasts outpaces bone deposition by osteoblasts, which leads to a decrease in bone mass and an increased risk of osteoporosis.

CLINICAL CONNECTION | *Hormonal Abnormalities That Affect Height*

Excessive or deficient secretion of hormones that normally control bone growth can cause a person to be abnormally tall or short. Oversecretion of hGH during childhood produces **giantism**, in which a person becomes much taller and heavier than normal. Undersecretion of hGH produces **pituitary dwarfism**, in which a person has short stature. (The usual adult height of a *dwarf* is under 4 feet 10 inches.) Although the head, trunk, and limbs of a pituitary dwarf are smaller than normal, they are proportionate. The condition can be treated medically with hGH until epiphyseal plate closure. Oversecretion of hGH during adulthood is called **acromegaly** (ak′-rō-MEG-a-lē). Although hGH cannot produce further lengthening of the long bones because the epiphyseal plates are already closed, the bones of the hands, feet, and jaws thicken and other tissues enlarge. In addition, the eyelids, lips, tongue, and nose enlarge, and the skin thickens and develops furrows, especially on the forehead and soles.

Achondroplasia (a-kon-drō-PLĀ-zē-a; *a*=without; *chondro*=cartilage; *-plasia*=to mold) is an inherited condition in which the conversion of cartilage to bone is abnormal. It results in the most common type of dwarfism, called **achondroplastic dwarfism**. These individuals are typically about four feet tall as adults. They have an average-size trunk, short limbs, and a slightly enlarged head with a prominent forehead and flattened nose at the bridge. The condition is essentially untreatable, although some individuals opt for limb-lengthening surgery. •

KEY MEDICAL TERMS ASSOCIATED WITH BONE TISSUE

Osteoarthritis (os'-tē-ō-ar-THRĪ -tis; *arthr*=joint) The degeneration of articular cartilage so that the bony ends touch; the resulting friction of bone against bone worsens the condition. Usually occurs in older individuals.

Osteosarcoma (os'-tē-ō-sar-KŌ-ma; *sarcoma*=connective tissue tumor) Bone cancer that primarily affects osteoblasts and occurs most often in teenagers during their growth spurt; the most common sites are the metaphyses of the thighbone (femur), shinbone (tibia), and arm bone (humerus). Metastases occur most often in lungs; treatment consists of multidrug chemotherapy and removal of the malignant growth, or amputation of the limb.

Osteomyelitis (os'-tē-ō-mī-el-Ī -tis) An infection of bone characterized by high fever, sweating, chills, pain, and nausea; pus formation, edema, and warmth over the affected bone; and rigid overlying muscles. It is often caused by bacteria, usually *Staphylococcus aureus*. The bacteria may reach the bone from outside the body (through open fractures, penetrating wounds, or orthopedic surgical procedures); from other sites of infection in the body (abscessed teeth, burn infections, urinary tract infections, or upper respiratory infections) via the blood; and from adjacent soft tissue infections (as occurs in diabetes mellitus).

Osteopenia (os'-tē-ō-PĒ-nē-a; *-penia*=poverty) Reduced bone mass due to a decrease in the rate of bone synthesis to a level insufficient to compensate for normal bone resorption; any decrease in bone mass below normal. Osteoporosis is an example of severe osteopenia.

CHAPTER REVIEW AND RESOURCE SUMMARY

WileyPLUS

Review	Resource

Introduction

1. A bone is made up of several different tissues: bone (osseous tissue), cartilage, dense connective tissues, epithelium, various blood-forming tissues, adipose tissue, and nervous tissue.
2. The entire framework of bones and their cartilages constitutes the skeletal system.

6.1 Functions of Bone and the Skeletal System

1. The skeletal system functions in support, protection, movement, mineral storage and release, blood cell production, and triglyceride storage.

Anatomy Overview - Bone Structure and Tissues

6.2 Types of Bones

1. On the basis of shape, bones are classified as long, short, flat, irregular, or sesamoid. Sesamoid bones develop in tendons or ligaments.
2. Sutural bones are found within the sutures of certain cranial bones.

6.3 Anatomy of a Bone

1. Parts of a typical long bone are the diaphysis (shaft), proximal and distal epiphyses (ends), metaphyses, articular cartilage, periosteum, medullary (marrow) cavity, and endosteum.

Exercise - Growing Long Bone

6.4 Bone Surface Markings

1. Surface markings are structural features visible on the surfaces of bones.
2. Each marking—whether a depression, an opening, or a process—is structured for a specific function, such as joint formation, muscle attachment, or passage of nerves and blood vessels (see Table 6.1).

6.5 Histology of Bone Tissue

1. Bone tissue consists of widely separated cells surrounded by large amounts of extracellular matrix.
2. The four principal types of cells in bone tissue are osteoprogenitor cells (bone stem cells), osteoblasts (bone-building cells), osteocytes (maintain daily activity of bone), and osteoclasts (bone-destroying cells).
3. The extracellular matrix of bone contains abundant mineral salts (mostly hydroxyapatite) and collagen fibers.
4. Compact bone tissue consists of osteons (haversian systems) with little space between them.
5. Compact bone tissue lies over spongy bone tissue in the epiphyses and makes up most of the bone tissue of the diaphysis. Functionally, compact bone tissue protects, supports, and resists stress.
6. Spongy bone tissue does not contain osteons. It consists of trabeculae surrounding spaces filled with red bone marrow.
7. Spongy bone tissue forms most of the structure of short, flat, and irregular bones, and the epiphyses of long bones. Functionally, spongy bone tissue trabeculae offer resistance along lines of stress; support and protect red bone marrow; and make bones lighter for easier movement.

Anatomy Overview - Compact Bone
Anatomy Overview - Spongy Bone
Figure 6.3 - Types of Cells in Bone Tissue
Figure 6.4 - Histology of Compact and Spongy Bone

6.6 Blood and Nerve Supply of Bone

1. Long bones are supplied by periosteal, nutrient, metaphyseal, and epiphyseal arteries; veins accompany the arteries.
2. Nerves accompany blood vessels in bone; the periosteum is rich in sensory neurons.

Review	Resource

6.7 Bone Formation

1. The process by which bone forms, called ossification, occurs in four principal situations: (1) the initial formation of bones in an embryo and fetus, (2) the growth of bones during infancy, childhood, and adolescence until their adult sizes are reached, (3) the remodeling of bone (replacement of old bone by new bone tissue throughout life), and (4) the repair of fractures (breaks in bones) throughout life.

2. Bone development begins during the sixth or seventh week of embryonic development. The two types of ossification, intramembranous and endochondral, involve the replacement of a preexisting connective tissue with bone.

3. Intramembranous ossification refers to bone formation directly within mesenchyme arranged in sheet-like layers that resemble membranes.

4. Endochondral ossification refers to bone formation within hyaline cartilage that develops from mesenchyme. The primary ossification center of a long bone is in the diaphysis. Cartilage degenerates, leaving cavities that merge to form the medullary cavity. Osteoblasts lay down bone. Next, ossification occurs in the epiphyses, where bone replaces cartilage, except for the epiphyseal (growth) plate.

5. The epiphyseal plate consists of four zones: resting cartilage, proliferating cartilage, hypertrophic cartilage, and calcified cartilage.

6. Because of the activity at the epiphyseal plate, the diaphysis of a bone increases in length.

7. Bone grows in width as a result of the addition of new bone tissue by periosteal osteoblasts around the outer surface of the bone (appositional growth).

8. Bone remodeling is the replacement of old bone tissue by new bone tissue.

9. Old bone tissue is constantly destroyed by osteoclasts; new bone is constructed by osteoblasts.

6.8 Fractures

1. A fracture is any break in a bone.

2. Types of fractures include closed (simple), open (compound), comminuted, greenstick, impacted, stress, Pott, and Colles (see Table 6.2).

3. Fracture repair involves formation of a fracture hematoma during the reactive phase, fibrocartilaginous callus and bony callus formation during the reparative phase, followed by a bone remodeling phase.

6.9 Exercise and Bone Tissue

1. Mechanical stress increases bone strength by increasing deposition of mineral salts and production of collagen fibers.

2. Removal of mechanical stress weakens bone through demineralization and collagen fiber reduction.

6.10 Aging and Bone Tissue

1. The principal effect of aging is a loss of calcium from bones, which may result in osteoporosis.

2. Another effect is decreased production of extracellular matrix proteins (mostly collagen fibers), which makes bones more brittle and thus more susceptible to fracture.

6.11 Factors Affecting Bone Growth

1. Normal growth depends on minerals (calcium, phosphorus, magnesium fluoride, and manganese), vitamins (A, C, D, K, and B_{12}), hormones (human growth hormone, insulinlike growth factors, insulin, thyroid hormones, sex hormones, parathyroid hormone, and calcitonin), and weight-bearing exercise.

Resource

Animation - Bone Formation
Animation - Bone Elongation and Bone Widening
Animation - Bone Remodeling
Animation - Regulation of Blood Calcium
Animation - Bone Dynamics and Tissue
Animation - Regulation of Bone Growth
Exercise - Observe the Ossification
Exercise - Bone Growth Sequencing
Exercise - Concentrate on Bone Tissue
Exercise - Regulation Sequences
Figure 6.6 - Intramembranous Ossification

Figure 6.10 - Steps Involved in Repair of a Bone Fracture

CRITICAL THINKING QUESTIONS

1. Lynda, a petite 55-year-old couch potato who smokes heavily, wants to lose 50 pounds before her next class reunion. Her diet consists mostly of diet soda and crackers. Explain the effects of her age and lifestyle on her bone composition.

2. Hannah and her anthropology classmates were studying some bones from their museum collection. Scott noticed some thick bumps on some of the bones, and noted that these were *spurs*. In terms of bone remodeling, can you offer an explanation as to how the spurs were formed?

3. Aunt Edith is 95 years old today. She comments that she's been getting shorter every year and soon she'll fade out altogether. What's happening to Aunt Edith?

4. Astronaut John Glenn exercised every day while in space, yet he and the other astronauts experienced weakness upon their return to earth. Why?

5. Chantal was concerned that her new baby was a "cone head" when she saw him for the first time. Later on, after carefully following the advice to always lay the baby on his back for sleeping, she became concerned that the back of his head was getting flat. If bone is hard, why does the baby's head keep changing shape?

ANSWERS TO FIGURE QUESTIONS

6.1 Flat bones protect and provide a large surface area for muscle attachment.

6.2 The periosteum is essential for growth in bone thickness, bone repair, and bone nutrition. It also serves as a point of attachment for ligaments and tendons.

6.3 Bone resorption is necessary for the development, maintenance, and repair of bone.

6.4 The central (haversian) canals are the main blood supply to the osteocytes of an osteon (haversian system), so their blockage would lead to death of the osteocytes.

6.5 Periosteal arteries enter bone tissue through perforating (Volkmann) canals.

6.6 Flat bones of the skull, most of the facial bones, mandible (lower jawbone), and medial part of the clavicle develop by intramembranous ossification.

6.7 Yes, she probably will grow taller. The absence of epiphyseal lines, indications of growth zones that have ceased to function, indicates that her bones are still lengthening.

6.8 The lengthwise growth of the diaphysis is caused by cell divisions in the zone of proliferating cartilage and replacement of the zone of calcified cartilage with bone (new diaphysis).

6.9 The medullary cavity enlarges by activity of the osteoclasts in the endosteum.

6.10 Healing of bone fractures can take months because calcium and phosphorus deposition is a slow process, and bone cells generally grow and reproduce slowly.

7 THE SKELETAL SYSTEM: THE AXIAL SKELETON

INTRODUCTION Without bones, you would be unable to perform movements such as walking or grasping. The slightest blow to your head or chest could cause fatal damage to your brain or heart. It would even be impossible for you to do something as simple as chewing food. Every day your skeleton performs a wide range of activities. Think about all of the movements and forces your skeleton experiences from the moment you get out of bed in the morning. As you walk, run, climb stairs, or lift things as light as a test booklet or as heavy as a box of anatomy textbooks, your skeleton must support the weight of the body and the various loads that it carries. In addition to providing support, the skeleton must also allow you to move freely from place to place. The wonderful capacities of the skeleton become even more impressive when you consider the loads and strains it encounters during various exercise and sports activities, such as gymnastics meets, figure-skating competitions, and basketball games.

Movements such as throwing a ball, running, and jumping require interactions between bones and muscles. To understand how muscles produce movements from high-fives to three-point shots, you will need to learn where the muscles attach to individual bones and what types of joints are involved. Together, the bones, muscles, and joints form an integrated system called the **musculoskeletal system**. The branch of medical science concerned with the prevention or correction of disorders of the musculoskeletal system is called **orthopedics** (or′-th-ō-PĒ-diks; *ortho-*=correct; *pedi*=child).

Because the skeletal system forms the framework of the body, a familiarity with the names, shapes, and positions of individual bones will help you locate other organ systems. For example, the radial artery, the site where pulse is usually taken, is named for its proximity to the radius, the lateral bone of the forearm, and the frontal lobe of the brain lies deep to the frontal (forehead) bone. Parts of certain bones also outline the lungs, heart, and abdominal and pelvic organs. •

Did you ever wonder what causes many people to become measurably shorter as they age?

7.1 DIVISIONS OF THE SKELETAL SYSTEM

● OBJECTIVE
- Describe how the skeleton is organized into axial and appendicular divisions.

The adult human skeleton consists of 206 named bones; most of them are paired, with one member of each pair on the right side of the body and the other on the left. The skeletons of infants and children have more than 206 bones because some of their bones fuse later in life. Examples are the hip bones and some bones (such as the sacrum and coccyx) of the vertebral column (also known as the backbone).

Bones of the adult skeleton are grouped into two principal divisions: the **axial skeleton** and the **appendicular skeleton** (*appendic-*=to hang onto). Table 7.1 presents the 80 bones of the axial skeleton and the 126 bones of the appendicular skeleton.

TABLE 7.1

The Bones of the Adult Skeletal System

DIVISION OF THE SKELETON	STRUCTURE	NUMBER OF BONES	DIVISION OF THE SKELETON	STRUCTURE	NUMBER OF BONES
Axial skeleton	**Skull**		**Appendicular skeleton**	**Upper limbs**	
	Cranium	8		**Pectoral (shoulder) girdles**	
	Face	14		Clavicle	2
	Hyoid bone	1		Scapula	2
	Auditory ossicles	6		**Free upper limbs**	
	Vertebral column	26		Humerus	2
	Thorax			Ulna	2
	Sternum	1		Radius	2
	Ribs	24		Carpals	16
	Number of bones = 80			Metacarpals	10
				Phalanges	28
				Lower limb	
				Pelvic (hip) girdle	
				Hip, pelvic, or coxal bone	2
				Free lower limbs	
				Femur	2
				Patella	2
				Fibula	2
				Tibia	2
				Tarsals	14
				Metatarsals	10
				Phalanges	28
				Number of bones = 126	
				Total in adult skeleton = 206	

Figure 7.1 Divisions of the skeletal system. The axial skeleton is indicated in blue. Note the position of the hyoid bone in Figure 7.4a.

 The adult human skeleton consists of 206 bones grouped into two divisions: the axial skeleton and the appendicular skeleton.

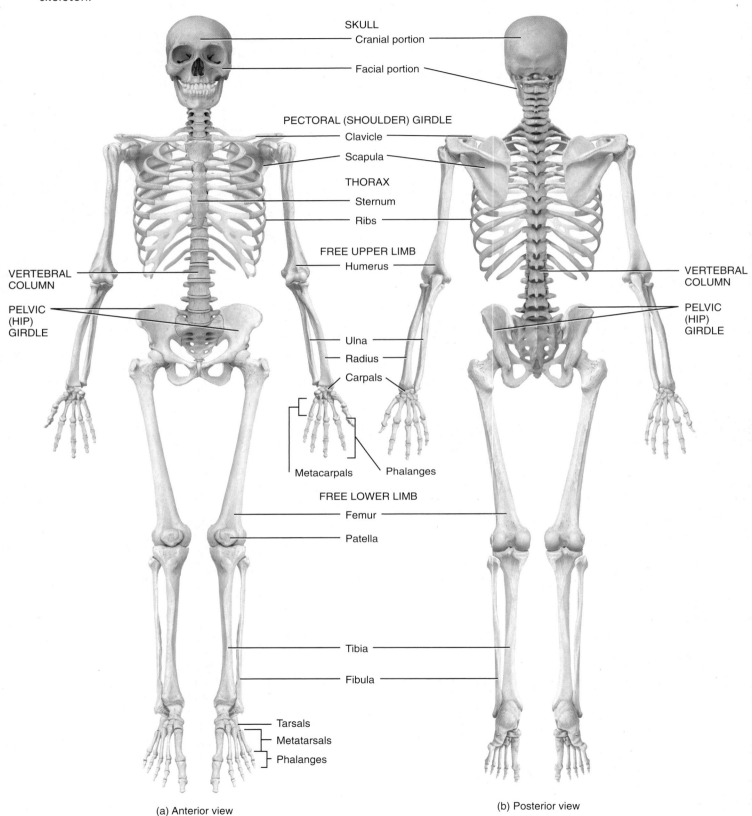

(a) Anterior view

(b) Posterior view

 Which of the following structures are part of the axial skeleton, and which are part of the appendicular skeleton? Skull, clavicle, vertebral column, shoulder girdle, humerus, pelvic girdle, and femur.

Both divisions join to form the complete skeleton shown in Figure 7.1 (the bones of the axial skeleton are shown in blue). You can remember the names of the divisions if you think of the axial skeleton as consisting of the bones that lie around the longitudinal *axis* of the human body, an imaginary vertical line that runs through the body's center of gravity from the head to the space between the feet: skull bones, auditory ossicles (ear bones), hyoid bone (see Figure 7.4a), ribs, sternum (breastbone), and bones of the vertebral column. The appendicular skeleton consists of the bones of the **upper** and **lower limbs**, or *appendages*. The limbs are made up of bones that form the **limb girdles** and bones that form the **free limbs**.

We will organize our study of the skeletal system around the two divisions of the skeleton, with emphasis on how the many bones of the body are interrelated. In this chapter we focus on the axial skeleton, looking first at the skull and then at the bones of the vertebral column and the chest. We then take a closer look at the appendicular skeleton in Chapter 8.

 CHECKPOINT

1. On what basis is the skeleton grouped into axial and appendicular divisions?

7.2 SKULL

 OBJECTIVES

- **Name** the cranial bones and facial bones and indicate the number of each.
- **Describe** the following unique features of the skull: sutures, paranasal sinuses, and fontanels.
- **Outline** the age-related changes and sexual differences in the skull.

The **skull** is the bony framework of the head and contains 22 bones, not counting the bones of the middle ears. It rests on the superior end of the vertebral column and includes two sets of bones: cranial bones and facial bones (Table 7.2). The **cranial bones** (*crani-*=skull) form the cranial cavity, which encloses and protects the brain. The eight cranial bones are the frontal bone, two parietal bones, two temporal bones, the occipital bone, the sphenoid bone, and the ethmoid bone. Fourteen **facial bones** form the face: two nasal bones, two maxillae (or maxillas), two zygomatic bones, the mandible, two lacrimal bones, two palatine bones, two inferior nasal conchae, and the vomer. Figures 7.2 through 7.10 in Exhibits 7.A through 7.I illustrate the bones of the skull from different views.

General Features and Functions

In addition to forming the large cranial cavity that houses the brain, the skull also forms several smaller cavities, including the *nasal cavity* and *orbits* (eye sockets), which open to the exterior. Certain skull bones also contain cavities called *paranasal sinuses* that are lined with mucous membranes and open into the nasal cavity. Also within the skull are small *middle* and *inner ear cavities* in the temporal bones that house the structures that are involved in hearing and equilibrium (balance).

The auditory ossicles, occipital bone, and mandible are the only bones of the skull that are movable. Most of the skull bones are held together by sutures, joints that attach the bones and that are especially noticeable on the outer surface of the skull (see Figure 7.3).

The skull has numerous openings through which blood vessels and nerves pass. You will learn the names of important surface markings as the various bones are described.

The cranial bones have other functions in addition to protecting the brain. Their inner surfaces attach to membranes (meninges) that stabilize the positions of the brain, blood vessels, and nerves. The outer surfaces of cranial bones provide large areas of attachment for muscles that move various parts of the head. The bones also provide attachment for some muscles that are involved in producing facial expressions. Besides forming the framework of the face, the facial bones protect and provide support for the entrances to the digestive and respiratory systems. Together, the cranial and facial bones protect and support the delicate special sense organs for vision, taste, smell, hearing, and equilibrium. The auditory ossicles of the middle ear help amplify sound waves to make hearing possible. The inner ear contains sensory structures important for hearing and others that monitor the position and movement of the head, factors important in the sense of balance.

TABLE 7.2
Summary of Bones of the Adult Skull

CRANIAL BONES	FACIAL BONES
Frontal (1)	Nasal (2)
Parietal (2)	Maxillae (2)
Temporal (2)	Zygomatic (2)
Occipital (1)	Mandible (1)
Sphenoid (1)	Lacrimal (2)
Ethmoid (1)	Palatine (2)
	Inferior nasal conchae (2)
	Vomer (1)

The numbers in parentheses indicate how many of each bone are present. The small bones of the middle ear are not included in this summary.

EXHIBIT 7.A   Cranial Bones—Frontal Bone *(Figure 7.2)*

 OBJECTIVE

• Identify the location and surface features of the frontal bone.

Description

The **frontal bone** forms the forehead (the anterior part of the cranium), the roofs of the *orbits* (eye sockets), and most of the anterior part of the cranial floor (Figure 7.2). In most individuals it is an unpaired bone. Soon after birth the left and right sides of the frontal bone are united by a suture called the *metopic suture* (me-TŌ-pik; *metophoron*=forehead), which usually disappears between the ages of 6 and 8.

Surface Features

If you examine the anterior view of the skull in Figure 7.2, you will note the *frontal squama*, a thick, scalelike plate of bone that forms the forehead. It gradually slopes inferiorly from the coronal suture (its joint with the paired parietal bones) on top of the skull (see Figure 7.3b), then angles abruptly and becomes almost vertical above the orbits. At the superior border of the orbits the frontal bone thickens, forming the *supraorbital margin* (supra-=above; *orbital*=circular). From this margin the frontal bone extends posteriorly as a horizontal plate of bone to form the roof of the orbit and part of the floor of the cranial cavity. Within the supraorbital margin, slightly medial to its midpoint, is a hole called the *supraorbital foramen* through which the supraorbital nerve and artery pass. Sometimes this foramen is incomplete and is called the *supraorbital notch*. Near the midline, within the vertical portion of the frontal squama, the bone is hollow. These hollow spaces are the paranasal sinuses called the *frontal sinuses*. Paranasal sinuses, mucous membrane–lined cavities within certain skull bones, will be discussed later in the chapter.

 CHECKPOINT

2. What structures pass through the supraorbital foramen?

Figure 7.2 Anterior view of the skull.

 The skull consists of cranial bones and facial bones.

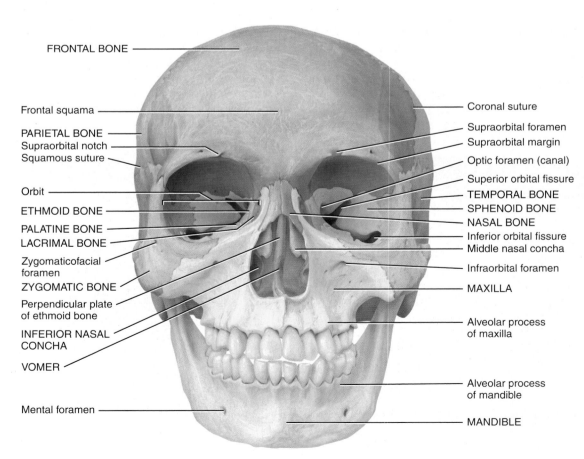

FRONTAL BONE

Frontal squama
PARIETAL BONE
Supraorbital notch
Squamous suture
Orbit
ETHMOID BONE
PALATINE BONE
LACRIMAL BONE
Zygomaticofacial foramen
ZYGOMATIC BONE
Perpendicular plate of ethmoid bone
INFERIOR NASAL CONCHA
VOMER
Mental foramen

Coronal suture
Supraorbital foramen
Supraorbital margin
Optic foramen (canal)
Superior orbital fissure
TEMPORAL BONE
SPHENOID BONE
NASAL BONE
Inferior orbital fissure
Middle nasal concha
Infraorbital foramen
MAXILLA
Alveolar process of maxilla
Alveolar process of mandible
MANDIBLE

(a) Anterior view

EXHIBIT 7.A   **169**

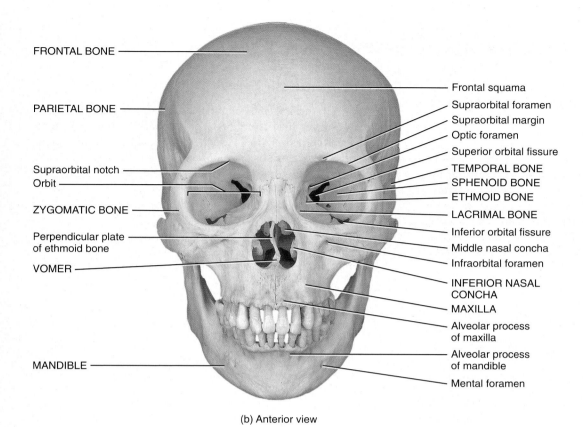

FRONTAL BONE

PARIETAL BONE

Supraorbital notch
Orbit

ZYGOMATIC BONE

Perpendicular plate
of ethmoid bone

VOMER

MANDIBLE

Frontal squama
Supraorbital foramen
Supraorbital margin
Optic foramen
Superior orbital fissure
TEMPORAL BONE
SPHENOID BONE
ETHMOID BONE
LACRIMAL BONE
Inferior orbital fissure
Middle nasal concha
Infraorbital foramen
INFERIOR NASAL
CONCHA
MAXILLA
Alveolar process
of maxilla
Alveolar process
of mandible
Mental foramen

(b) Anterior view

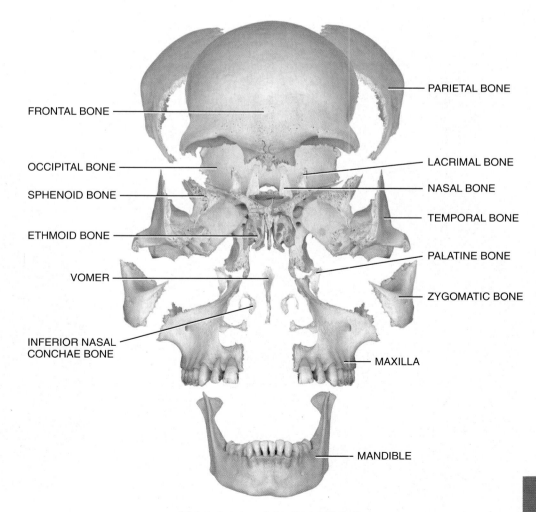

FRONTAL BONE

OCCIPITAL BONE

SPHENOID BONE

ETHMOID BONE

VOMER

INFERIOR NASAL
CONCHAE BONE

PARIETAL BONE

LACRIMAL BONE

NASAL BONE

TEMPORAL BONE

PALATINE BONE

ZYGOMATIC BONE

MAXILLA

MANDIBLE

(c) Anterior view of disarticulated skull

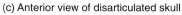   **Which of the bones shown here are cranial bones?**

EXHIBIT 7.B Cranial Bones—Parietal Bones *(Figure 7.3)*

  OBJECTIVE

• Identify the location and surface features of the parietal bones.

Description

The two **parietal bones** (pa-RĪ-e-tal; *pariet-*=wall) are large, quadrilateral (four-sided) bones that form the greater portion of the sides and roof of the cranial cavity (Figure 7.3). Each bone articulates with five other bones. The inferior border forms a beveled articular surface, while the anterior, posterior, and superior borders form deeply denticulate (toothlike) articular surfaces.

Surface Features

The external surface of each of these bones is slightly *convex* (curved outward, like the outside of a sphere), while the internal surface is concave. The internal surfaces of the parietal bones contain many protrusions and depressions that accommodate the blood vessels supplying the dura mater, the superficial membrane (meninx) covering the brain.

  CHECKPOINT

3. How do the parietal bones relate to the cranial cavity?

Figure 7.3 Superior and right lateral view of the skull.

  The zygomatic arch is formed by the zygomatic process of the temporal bone and the temporal process of the zygomatic bone.

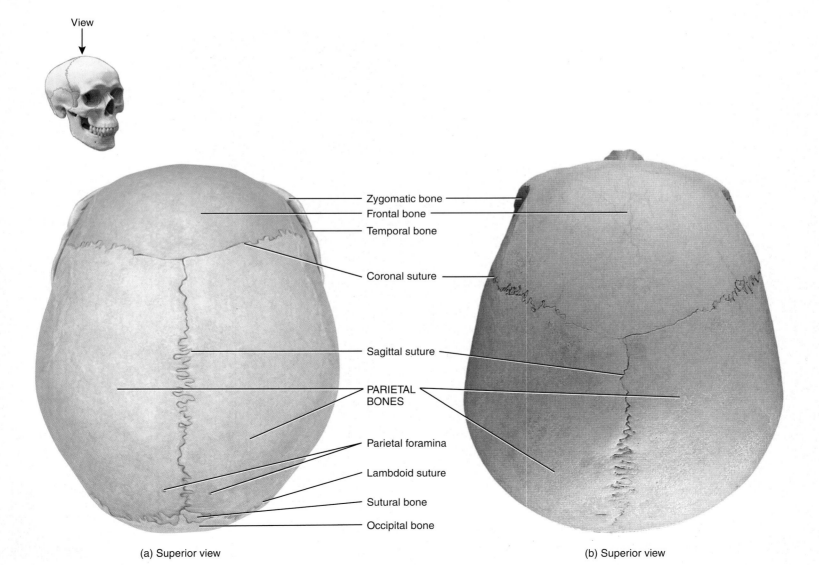

(a) Superior view

(b) Superior view

EXHIBIT 7.B   **171**

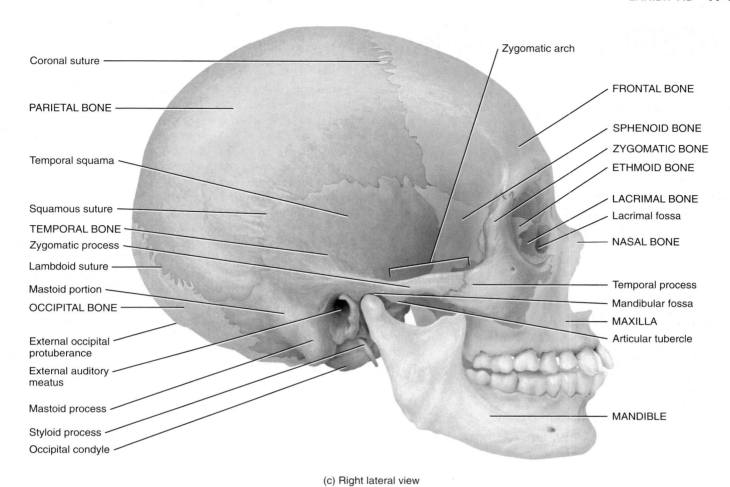

Coronal suture

PARIETAL BONE

Temporal squama

Squamous suture

TEMPORAL BONE

Zygomatic process

Lambdoid suture

Mastoid portion

OCCIPITAL BONE

External occipital protuberance

External auditory meatus

Mastoid process

Styloid process

Occipital condyle

Zygomatic arch

FRONTAL BONE

SPHENOID BONE

ZYGOMATIC BONE

ETHMOID BONE

LACRIMAL BONE

Lacrimal fossa

NASAL BONE

Temporal process

Mandibular fossa

MAXILLA

Articular tubercle

MANDIBLE

(c) Right lateral view

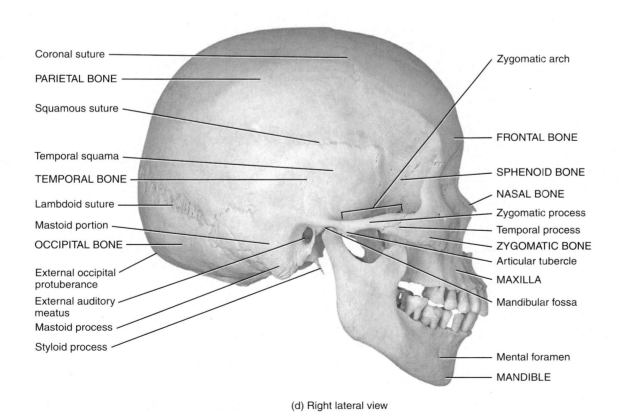

Coronal suture

PARIETAL BONE

Squamous suture

Temporal squama

TEMPORAL BONE

Lambdoid suture

Mastoid portion

OCCIPITAL BONE

External occipital protuberance

External auditory meatus

Mastoid process

Styloid process

Zygomatic arch

FRONTAL BONE

SPHENOID BONE

NASAL BONE

Zygomatic process

Temporal process

ZYGOMATIC BONE

Articular tubercle

MAXILLA

Mandibular fossa

Mental foramen

MANDIBLE

(d) Right lateral view

What are the major bones on either side of the squamous suture, the lambdoid suture, and the coronal suture?

EXHIBIT 7.C Cranial Bones—Temporal Bones *(Figure 7.4)*

OBJECTIVE

• Identify the location and surface features of the temporal bones.

Description

The two **temporal bones** (*tempor-*=temple) form the inferior lateral aspects of the cranium and part of the cranial floor. The terms *temporal* and *temple* are derived from the Latin word *tempus,* meaning "time," in reference to the graying of hair in the temple area, a sign of time's passing.

Surface Features

In the lateral view of the skull (see Figure 7.3c, d), note the *temporal squama,* the thin, flat portion of the temporal bone that forms the anterior and superior part of the *temple* (the region of the cranium around the ear). Projecting anteriorly from the inferior portion of the temporal squama is the *zygomatic process,* which articulates (forms a joint) with the temporal process of the zygomatic (cheek) bone. Together, the zygomatic process of the temporal bone and the temporal process of the zygomatic bone form the *zygomatic arch.* You can easily palpate this horizontal arch of bone immediately anterior to the ear.

On the inferoposterior surface of the zygomatic process of the temporal bone is a socket called the *mandibular fossa.* Anterior to the mandibular fossa is a rounded elevation, the *articular tubercle* (see Figure 7.3c, d). The mandibular fossa and articular tubercle articulate with the mandible (lower jawbone) to form the *temporomandibular joint (TMJ).*

The *mastoid portion* (*mastoid*=breast-shaped) (see Figure 7.3c, d) is located posterior and inferior to the *external auditory meatus* (*meatus*=passageway), or ear canal, which directs sound waves into the ear. In the adult, this portion of the bone contains several *mastoid air cells* that communicate with the hollow space of the middle ear (tympanic cavity). These tiny air-filled

Figure 7.4 Medial view of sagittal section of the skull. Although the hyoid bone is not part of the skull, it is included in the illustration for reference.

 The cranial bones are the frontal, parietal, temporal, occipital, sphenoid, and ethmoid bones. The facial bones are the nasal bone, maxillae, zygomatic bones, lacrimal bones, palatine bones, mandible, and vomer.

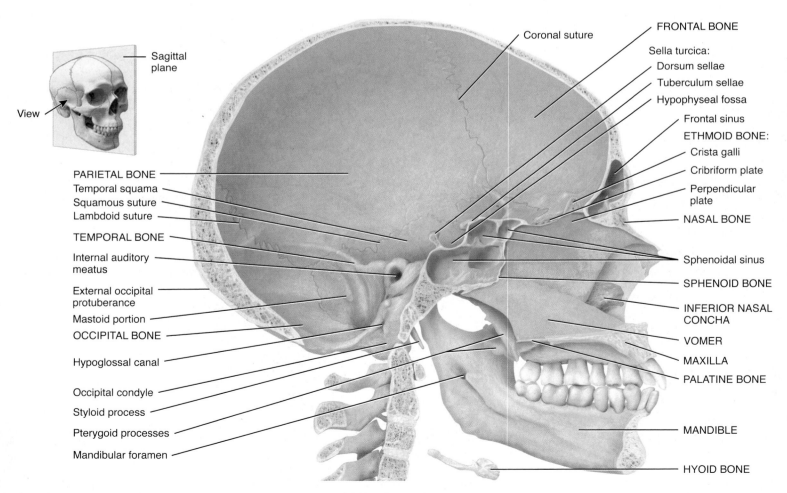

(a) Medial view of sagittal section

EXHIBIT 7.C **173**

compartments are separated from the brain by thin bony partitions. Middle ear infections that go untreated can spread into the mastoid air cells, causing a painful inflammation referred to as **mastoiditis** (mas′-toy-DĪ-tis).

The *mastoid process* is a rounded projection of the mastoid portion of the temporal bone posterior and inferior to the external auditory meatus that serves as a point of attachment for several neck muscles (see Figure 7.3c, d). The *internal auditory meatus* (Figure 7.4) is the opening through which the facial (VII) and vestibulocochlear (VIII) cranial nerves pass. The *styloid process* (*styl-*=stake or pole) projects inferiorly from the inferior surface of the temporal bone and serves as a point of attachment for muscles and ligaments of the tongue and neck (see Figure 7.3c, d). Between the styloid process and the mastoid process is the *stylomastoid foramen*, through which the facial (VII) nerve and stylomastoid artery pass (see Figure 7.6c, d).

At the floor of the cranial cavity (see Figure 7.6a, b) is the *petrous portion* (*petrous*=rock) of the temporal bone. This portion is pyramidal (having the shape of a pyramid) and located at the base of the skull between the sphenoid and occipital bones. The petrous portion houses the internal ear and the middle ear, structures involved in hearing and equilibrium. The middle ear contains three small *auditory ossicles*, the malleus, incus, and stapes. It also contains the *carotid foramen*, through which the carotid artery passes (see Figure 7.6a–d). Posterior to the carotid foramen and anterior to the occipital bone is the *jugular foramen*, a passageway for the jugular vein, formed by adjacent notches in the temporal and occipital bones (see Figure 7.6c).

✓ CHECKPOINT

4. What structures form the zygomatic arch?

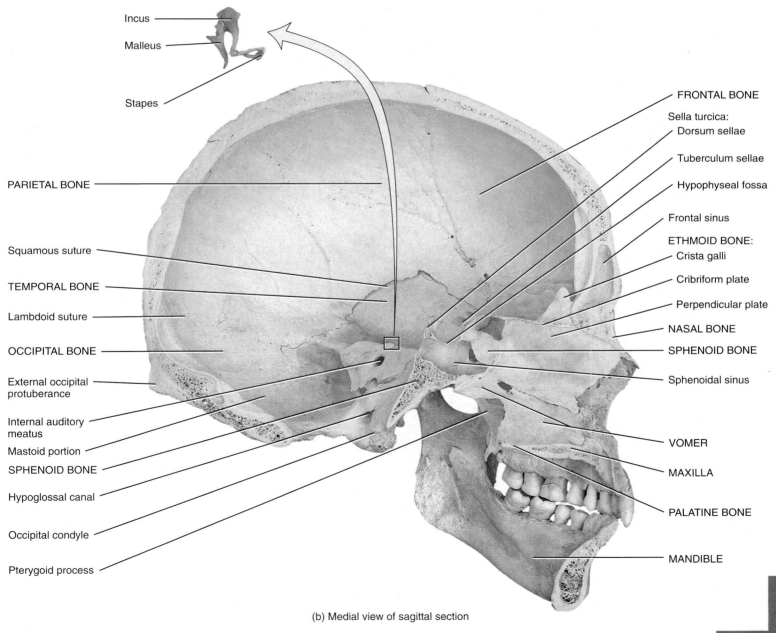

(b) Medial view of sagittal section

❓ With which bones does the temporal bone articulate?

EXHIBIT 7.D Cranial Bones—Occipital Bone *(Figure 7.5)*

 OBJECTIVE
• Identify the location and surface features of the occipital bone.

Description

The **occipital bone** (ok-SIP-i-tal; *occipit-*=back of head) forms the posterior part and most of the base of the cranium. When viewed from behind, it appears as a platelike bone with a somewhat triangular shape. Its inferior portion is a thick, blocklike region that surrounds the junction of the brain and spinal cord (Figure 7.5; also see Figure 7.3c, d).

Surface Features

The *foramen magnum* (=large hole) is in the inferior part of the bone. Within this foramen, the medulla oblongata (inferior part of the brain) connects with the spinal cord. The vertebral arteries, spinal arteries, and accessory (XI) nerve also pass through this foramen. The *occipital condyles* are two oval processes with convex surfaces, one on either side of the foramen magnum (see Figure 7.6a–d). They articulate with depressions on the first cervical vertebra (atlas) to form the *atlanto-occipital joints*. Superior to each occipital condyle on the inferior surface of the skull is the *hypoglossal canal* (*hypo-*=under; *-glossal*=tongue), through which

Figure 7.5 Posterior view of the skull. The sutures are exaggerated for emphasis.

 The occipital bone forms most of the posterior and inferior portions of the cranium.

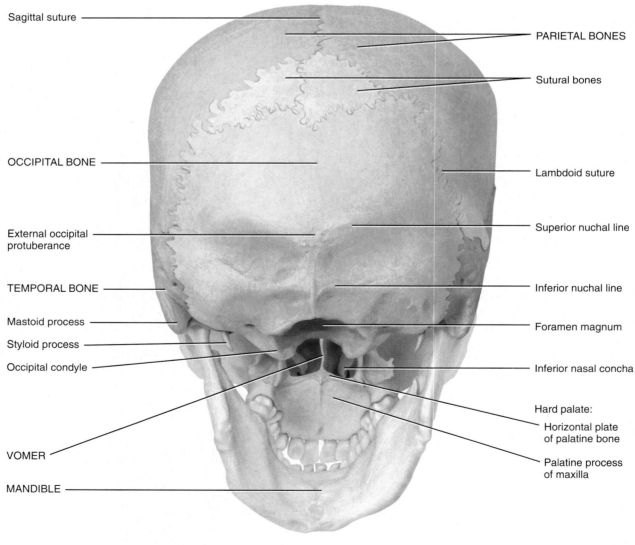

(a) Posteroinferior view

EXHIBIT 7.D **175**

the hypoglossal (XII) nerve and a branch of the ascending pharyngeal artery pass (see Figure 7.4).

The *external occipital protuberance* is the most prominent midline projection on the posterior surface of the bone just superior to the foramen magnum. You may be able to feel this structure as a definite bump, the most prominent protrusion on the back of your head, just above your neck (Figure 7.5b). A large fibrous, elastic ligament, the *ligamentum nuchae* (*nucha-*=nape of neck), which helps support the head, extends from the external

occipital protuberance to the seventh cervical vertebra. Extending laterally from the protuberance are two curved lines, the *superior nuchal lines*, and below these are two *inferior nuchal lines*, which are areas of muscle attachment (Figure 7.5b). It is possible to view the parts of the occipital bone, as well as surrounding structures, in the inferior view of the skull in Figure 7.6c, d.

 CHECKPOINT
5. What structures pass through the hypoglossal canal?

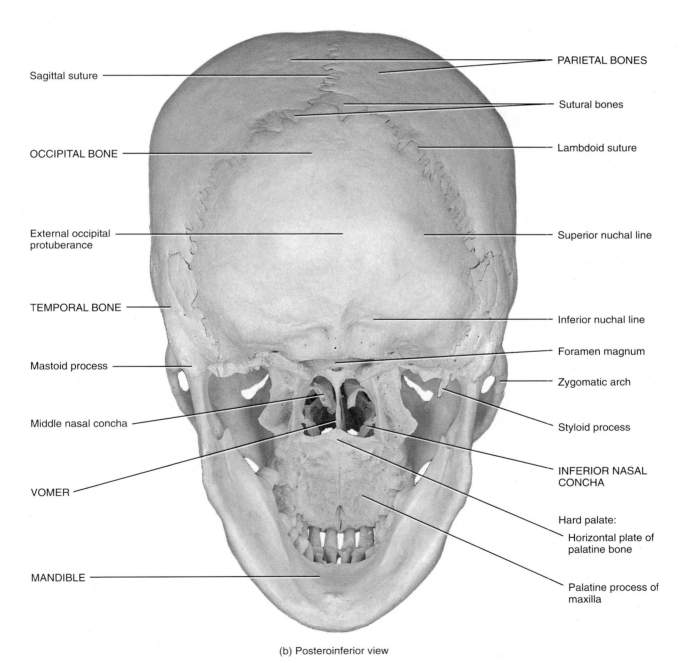

(b) Posteroinferior view

 Which bones form the posterior, lateral portion of the cranium?

EXHIBIT 7.E Cranial Bones—Sphenoid Bone *(Figure 7.6)*

● OBJECTIVE

• Identify the location and surface features of the sphenoid bone.

Description

The **sphenoid bone** (SFĒ-noyd=wedge-shaped) lies at the middle part of the base of the skull (Figure 7.6a–d). This bone is called the keystone of the cranial floor because it articulates with all the other cranial bones, holding them together. When you view the floor of the cranium superiorly (Figure 7.6a, b), note the sphenoid articulations: anteriorly with the frontal and ethmoid bones, laterally with the temporal and parietal bones, anterolaterally with the parietal bones, and posteriorly with the occipital

bone. The sphenoid bone lies posterior and slightly superior to the nasal cavity and forms part of the floor, sidewalls, and rear wall of the orbit (see Figure 7.12).

Surface Features

The shape of the sphenoid resembles a butterfly with outstretched wings (Figure 7.6e, f). The *body* of the sphenoid is the hollowed, cubelike medial portion between the ethmoid and occipital bones. The hollow of the body is the *sphenoidal sinus*, which drains via a narrow opening into the superior aspect of the nasal cavity (see Figure 7.13c). The *sella turcica* (SEL-a TUR-si-ka; *sella*=saddle; *turcica*=Turkish) is a bony, saddle-shaped structure on the superior surface of the body of the sphenoid (Figure 7.6a, b).

Figure 7.6 Sphenoid bone.

⚷ The sphenoid bone is called the keystone of the cranial floor because it articulates with all other cranial bones, holding them together.

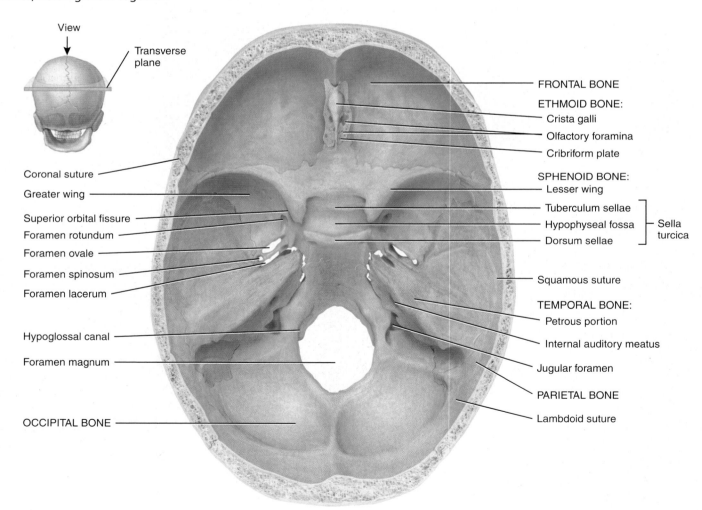

(a) Superior view of sphenoid bone in floor of cranium

EXHIBIT 7.E **177**

The anterior part of the sella turcica, which forms the horn of the saddle, is a ridge called the *tuberculum sellae*. The seat of the saddle is a depression, the *hypophyseal fossa* (hī-po-FIZ-ē-al), which contains the pituitary gland. The posterior part of the sella turcica, which forms the back of the saddle, is another ridge called the *dorsum sellae*.

The *greater wings* of the sphenoid project laterally from the body, forming the anterolateral floor of the cranium (Figure 7.6a, b, e, f). The greater wings also form part of the lateral wall of the skull just anterior to the temporal bone and can be viewed externally. The *lesser wings*, which are smaller than the greater wings, form a ridge of bone anterior and superior to the greater wings. They form part of the floor of the cranium and the posterior part of the orbit of the eye.

Between the body and lesser wing just anterior to the sella turcica is the *optic foramen* or *canal* (*optic*=eye), through which the optic (II) nerve and ophthalmic artery pass into the orbit (Figure 7.6e, f). Lateral to the body between the greater and lesser wings is an elongated, triangular slit called the *superior orbital fissure*. This fissure may also be seen in the anterior view of the orbit in Figure 7.12. Blood vessels and cranial nerves pass through this fissure.

The *pterygoid processes* (TER-i-goyd=winglike) extend from the inferior part of the sphenoid bone (Figure 7.6c–f). These structures project inferiorly from the points where the body and greater wings unite and they form the lateral posterior region of the nasal cavity. Some of the muscles that move the mandible attach to the pterygoid processes. At the base of the lateral pterygoid process in the greater wing is the *foramen ovale* (=oval),

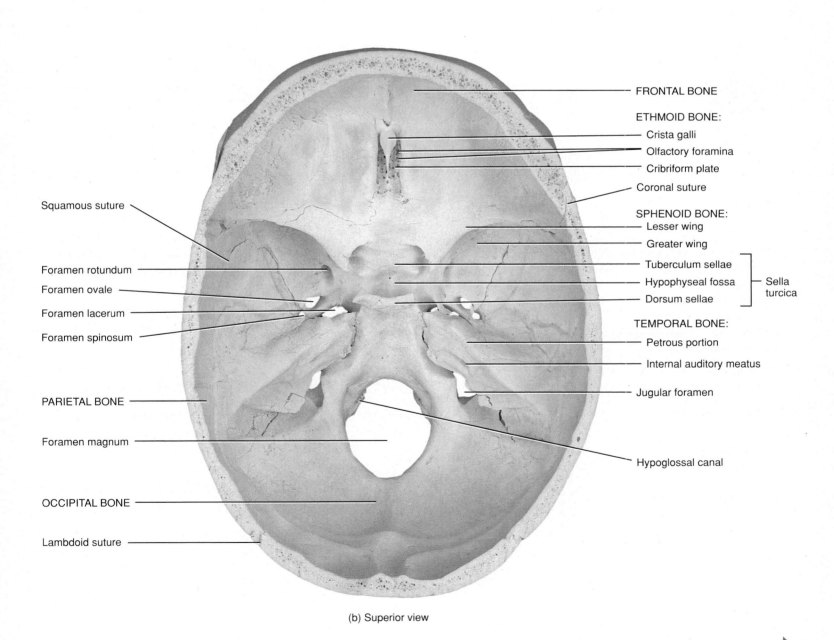

Squamous suture

Foramen rotundum

Foramen ovale

Foramen lacerum

Foramen spinosum

PARIETAL BONE

Foramen magnum

OCCIPITAL BONE

Lambdoid suture

FRONTAL BONE

ETHMOID BONE:
Crista galli
Olfactory foramina
Cribriform plate
Coronal suture

SPHENOID BONE:
Lesser wing
Greater wing
Tuberculum sellae
Hypophyseal fossa — Sella turcica
Dorsum sellae

TEMPORAL BONE:
Petrous portion
Internal auditory meatus
Jugular foramen

Hypoglossal canal

(b) Superior view

FIGURE 7.6 CONTINUES ▶

EXHIBIT 7.E Cranial Bones—Sphenoid Bone *(Figure 7.6)* CONTINUED

an opening for the mandibular branch of the trigeminal (V) nerve (Figure 7.6a–d). Another foramen, the *foramen spinosum* (=resembling a spine, because of its proximity to the sharp spine of the sphenoid), lies at the posterior angle of the sphenoid and transmits the middle meningeal blood vessels. The *foramen lacerum* (=lacerated) is bounded anteriorly by the sphenoid bone and posteriorly by the temporal and occipital bones. This foramen is covered in part by a layer of fibrocartilage in living subjects and is a joint uniting the three bones. It transmits a branch of the

ascending pharyngeal artery and numerous emissary veins. Another foramen associated with the sphenoid bone is the *foramen rotundum* (=round) located at the junction of the anterior and medial parts of the sphenoid bone. The maxillary branch of the trigeminal (V) nerve passes through the foramen rotundum.

✓ CHECKPOINT

6. Why is the sphenoid bone called the keystone of the cranial floor?

⬛ **FIGURE 7.6** CONTINUED ▶

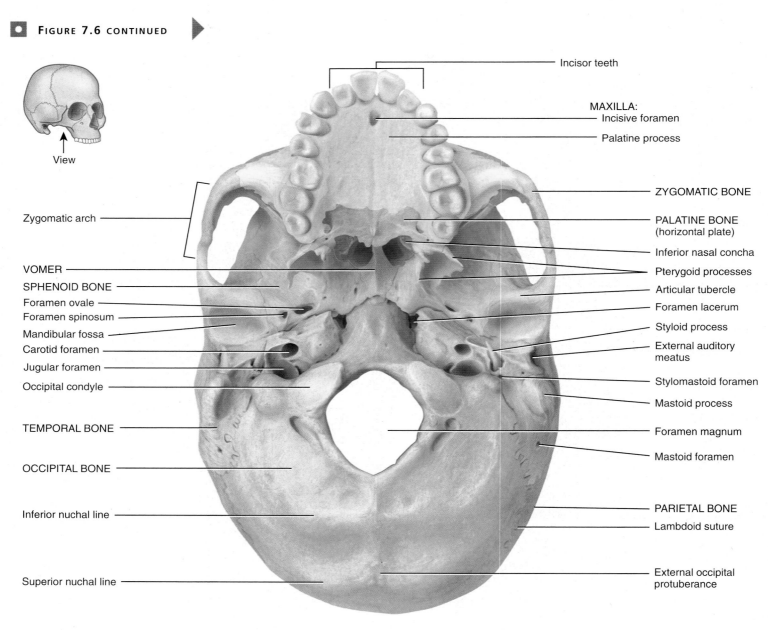

(c) Inferior view

EXHIBIT 7.E **179**

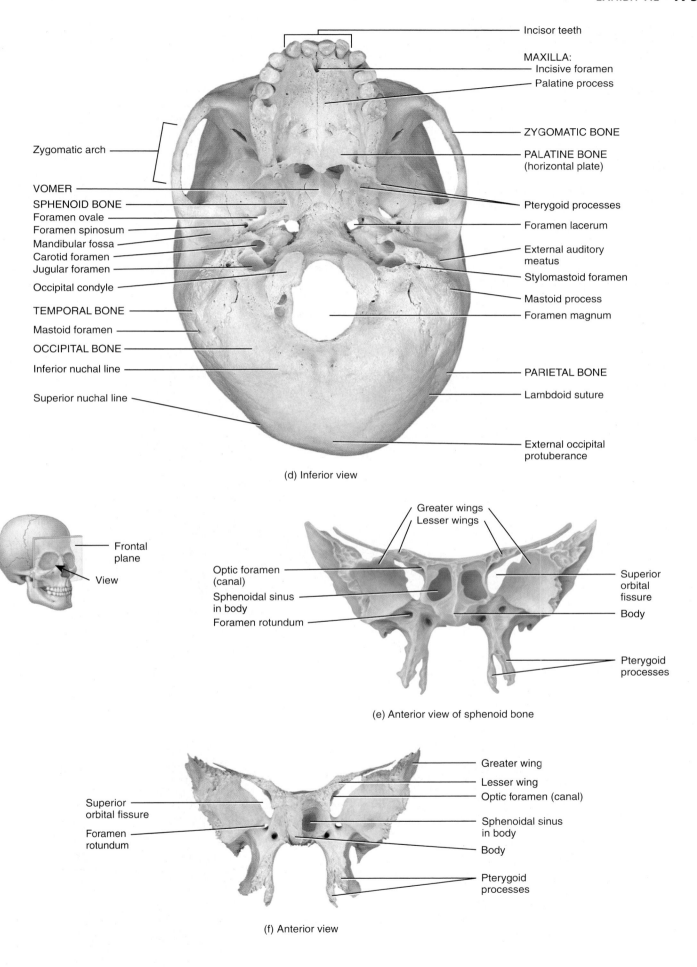

Incisor teeth

MAXILLA:
Incisive foramen
Palatine process

ZYGOMATIC BONE

PALATINE BONE
(horizontal plate)

Zygomatic arch

VOMER

SPHENOID BONE
Foramen ovale
Foramen spinosum
Mandibular fossa
Carotid foramen
Jugular foramen

Occipital condyle

TEMPORAL BONE

Mastoid foramen

OCCIPITAL BONE

Inferior nuchal line

Superior nuchal line

Pterygoid processes

Foramen lacerum

External auditory
meatus
Stylomastoid foramen

Mastoid process
Foramen magnum

PARIETAL BONE

Lambdoid suture

External occipital
protuberance

(d) Inferior view

Frontal
plane

View

Greater wings
Lesser wings

Optic foramen
(canal)
Sphenoidal sinus
in body
Foramen rotundum

Superior
orbital
fissure

Body

Pterygoid
processes

(e) Anterior view of sphenoid bone

Superior
orbital fissure

Foramen
rotundum

Greater wing

Lesser wing

Optic foramen (canal)

Sphenoidal sinus
in body

Body

Pterygoid
processes

(f) Anterior view

Name the bones that articulate with the sphenoid bone, starting at the crista galli of the ethmoid bone and going in a clockwise direction.

| EXHIBIT 7.F | Cranial Bones—Ethmoid Bone *(Figure 7.7)* |

OBJECTIVE

• Identify the location and surface features of the ethmoid bone.

the two orbits and is spongelike in appearance (Figure 7.7). It is anterior to the sphenoid bone and posterior to the nasal bones. The ethmoid bone forms (1) part of the anterior portion of the cranial floor; (2) the thin, medial wall of the orbits; (3) the superior portion of the nasal septum, a partition that divides the nasal cavity into right and left sides; and (4) most of the superior sidewalls of the nasal cavity. The ethmoid bone is a major superior supporting structure of the nasal cavity and forms extensive surface area in the nasal cavity.

Description

The unpaired **ethmoid bone** (ETH-moyd=like a sieve) is a delicate bone located in the anterior part of the cranial floor between

Figure 7.7 Ethmoid bone.

 The ethmoid bone forms part of the anterior portion of the cranial floor, the medial wall of the orbits, the superior portions of the nasal septum, and most of the sidewalls of the nasal cavity.

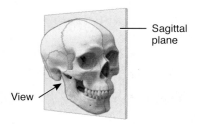

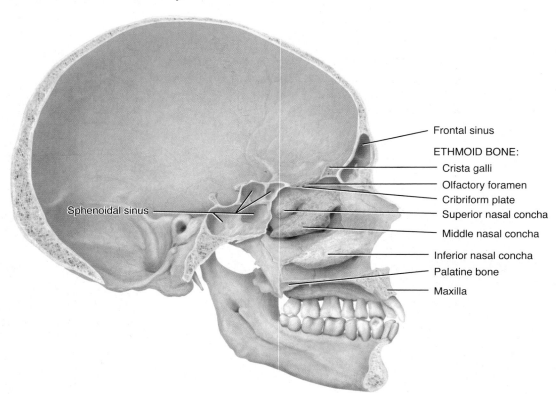

(a) Medial view of sagittal section

POSTERIOR

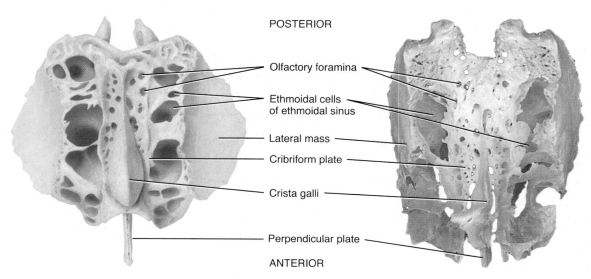

ANTERIOR

(b) Superior view

EXHIBIT 7.F **181**

Surface Features

The *lateral masses* of the ethmoid bone compose most of the wall between the nasal cavity and the orbits. They contain 3 to 18 air spaces called *ethmoidal cells*. The ethmoidal cells together form the *ethmoidal sinuses* (see Figure 7.13a, b). The midline *perpendicular plate* forms the superior portion of the nasal septum (see Figure 7.11). The horizontal *cribriform plate* (*cribri-*=sieve) lies in the anterior floor of the cranium and forms the roof of the nasal cavity. The cribriform plate contains the *olfactory foramina* (*olfact-*=smell), through which the olfactory (I) nerves pass. Projecting superiorly from the cribriform plate is a triangular process called the *crista galli* (*crista*=crest; *galli*=cock), which serves as a point of attachment for the falx cerebri, the membrane that separates the two hemispheres (sides) of the brain.

The lateral masses of the ethmoid bone contain two thin, scroll-shaped projections lateral to the nasal septum. These are called the *superior nasal concha* (KONG-ka=shell) or *superior nasal turbinate* and the *middle nasal concha (middle nasal turbinate)*. The plural form is *conchae* (KONG-kē). A third pair of conchae, the inferior nasal conchae, are separate bones (discussed shortly). The conchae increase the vascular and mucous membrane surface area in the nasal cavity, which warms, moistens, and humidifies inhaled air before it passes into the lungs. The conchae also cause inhaled air to swirl; the result is that many inhaled particles become trapped in the mucus that lines the nasal cavity. This action of the conchae helps cleanse inhaled air before it passes into the rest of the respiratory passageways. The superior nasal conchae are near the olfactory foramina of the cribriform plate where the sensory receptors for olfaction (smell) terminate in the mucous membrane of the superior nasal conchae. Thus, they increase the surface area for the sense of smell.

 CHECKPOINT

7. The ethmoid bone forms which other cranial structures?

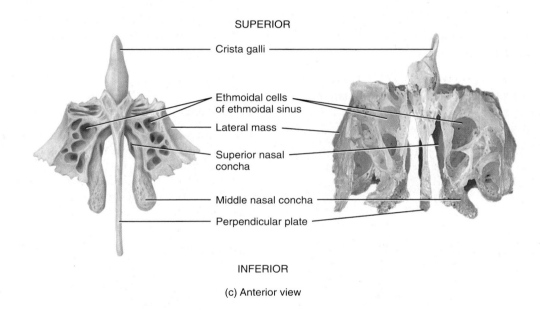

SUPERIOR

Crista galli

Ethmoidal cells of ethmoidal sinus

Lateral mass

Superior nasal concha

Middle nasal concha

Perpendicular plate

INFERIOR

(c) Anterior view

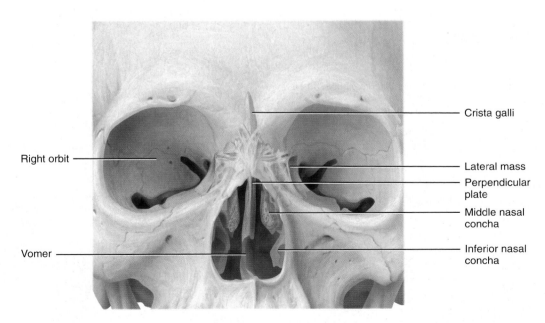

Crista galli

Right orbit

Lateral mass

Perpendicular plate

Middle nasal concha

Vomer

Inferior nasal concha

(d) Anterior view of position of ethmoid bone in skull
(projected to the surface)

 What part of the ethmoid bone forms the superior part of the nasal septum? The medial walls of the orbits?

EXHIBIT 7.G

Facial Bones—Nasal, Lacrimal, Palatine, Inferior Nasal Conchae, and Vomer *(Figure 7.8)*

 OBJECTIVE

- Identify the location of the nasal, lacrimal, and palatine bones, inferior nasal conchae, and vomer.

NASAL BONES

Description

The paired **nasal bones** are small, flattened, rectangular-shaped bones that form the bridge of the nose (see Figures 7.2 and 7.8a). These small bones protect the upper entry to the nasal cavity and provide attachment for a couple of thin muscles of facial expression. For those of you who wear glasses, they are the bones that form the resting place for the bridge of the glasses. The major structural portion of the nose consists of cartilage.

Figure 7.8 **Nasal, lacrimal, and palatine bones, inferior nasal conchae, and vomer.**

 All of the bones in this series contribute to the walls of the nasal cavity, as well as the orbit (lacrimal and palatine) and oral cavity (palatine).

LACRIMAL BONES

Description

The paired **lacrimal bones** (LAK-ri-mal; *lacrim-*=teardrops) are thin and roughly resemble a fingernail in size and shape (see Figures 7.2, 7.3, 7.8b, and 7.12). These bones, the smallest bones of the face, are posterior and lateral to the nasal bones and form a part of the medial wall of each orbit.

Surface Features

The lacrimal bones each contain a *lacrimal fossa*, a vertical tunnel formed with the maxilla, that houses the lacrimal sac, a structure that gathers tears and passes them into the nasal cavity (see Figures 7.8b and 7.12). The medial surface of the bones is covered with mucous membrane and forms part of the upper wall of the nasal cavity.

PALATINE BONES

Description

The two L-shaped **palatine bones** (PAL-a-tīn) form the posterior portion of the hard palate. In addition to the roof of the mouth they form the posterior portion of the floor and walls of the nasal cavity. The superior aspect of their perpendicular plate contributes to a small portion of the floor of the orbit bordering the optic foramen (canal) (see Figures 7.6c, d and 7.8d).

Surface Features

The posterior portion of the hard palate, which separates the nasal cavity from the oral cavity, is formed by the *horizontal plates* of the palatine bones (see Figure 7.6c, d and 7.8d). The perpendicular plates of the palatine bones contribute to the walls of the nasal cavity and orbits.

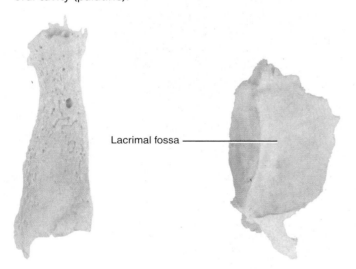

Lacrimal fossa

(a) Right nasal bone, anterior view

(b) Right lacrimal bone, anterior view

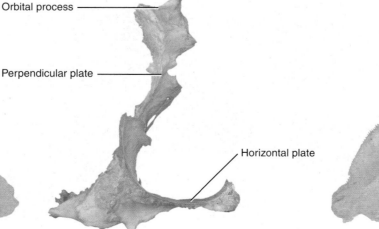

Orbital process

Perpendicular plate

Horizontal plate

(c) Right inferior nasal conchae, medial view

(d) Right palatine bone, anterior view

(e) Vomer, lateral view

EXHIBIT 7.G **183**

INFERIOR NASAL CONCHAE

Description

The two **inferior nasal conchae** are inferior to the middle nasal conchae of the ethmoid bone (see Figures 7.2, 7.7a, and 7.8c). These scroll-like bones form a part of the inferior lateral wall of the nasal cavity and project into the nasal cavity. The inferior nasal conchae are separate bones; they are not part of the ethmoid bone. All three pairs of nasal conchae (superior, middle, and inferior) increase the mucosal-covered surface of the nasal cavity and help swirl and filter air before it passes into the lungs. However, only the superior nasal conchae of the ethmoid bone are involved in the sense of smell.

VOMER

Description

The unpaired **vomer** (VŌ-mer=plowshare) is a roughly triangular bone on the floor of the nasal cavity that articulates superiorly with the perpendicular plate of the ethmoid bone and the inferior surface of the body of the sphenoid bone. Inferiorly, the vomer articulates with both the maxillae and palatine bones along the midline (see Figures 7.2, 7.4, 7.5, and 7.8e, f). It forms the inferior portion of the bony nasal septum, the partition that divides the nasal cavity into right and left sides.

✔ CHECKPOINT

8. Which bones form the hard palate?

> **CLINICAL CONNECTION | *Black Eye***
>
> A **black eye** is a bruising around the eye, commonly due to an injury to the face, rather than an eye injury. In response to trauma, blood and other fluids accumulate in the space around the eye, causing the swelling and dark discoloration. One cause might be a blow to the sharp ridge just superior to the supraorbital margin that fractures the frontal bone, resulting in bleeding. Another is a blow to the nose. Certain surgical procedures (face lift, eyelid surgery, jaw surgery, or nasal surgery) can also result in black eyes. •

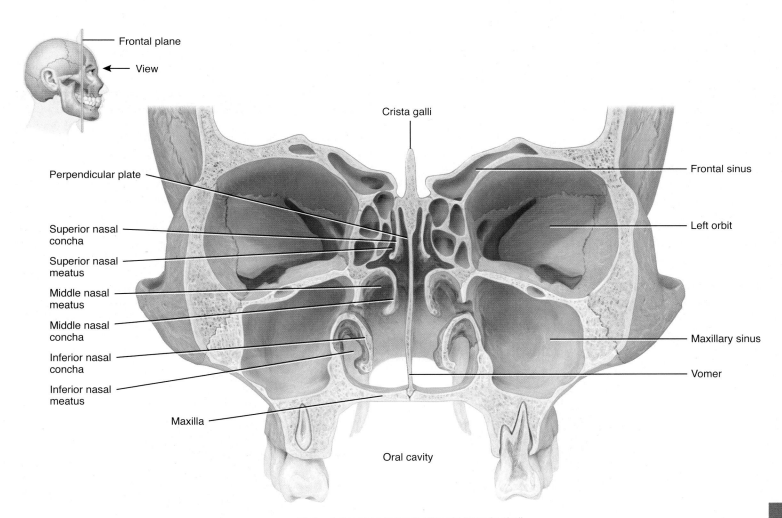

(f) Frontal section through ethmoid bone in skull

 What is the role of the nasal conchae during inhalation?

EXHIBIT 7.H Facial Bones—Maxillae and Zygomatic Bones *(Figure 7.9)*

OBJECTIVE
• Identify the location and surface features of the maxillae and zygomatic bones.

MAXILLAE

Description

The paired **maxillae** (mak-SIL-ē =jawbones; singular is *maxilla*) unite to form the upper jawbone. They articulate with every bone of the face except the mandible (lower jawbone) (see Figures 7.2, 7.3, and 7.9a). The maxillae form part of the floors of the orbits, part of the lateral walls and floor of the nasal cavity, and most of the hard palate. The *hard palate* is the bony roof of the mouth, and is formed by the palatine processes of the maxillae and horizontal plates of the palatine bones. The hard palate separates the nasal cavity from the oral cavity.

Surface Features

Each maxilla has a central body, which is hollowed out internally to form the mucosal-lined *maxillary sinus*. Like the previously mentioned paranasal sinuses, the mucosa of the maxillary sinus is continuous with the mucosa of the nasal cavity (see Figure 7.13). The *alveolar process* (al-VĒ-ō-lar; *alveol-*=small cavity) is the ridge-like arch that contains the *alveoli* (sockets) for the maxillary (upper) teeth (see Figure 7.9a). The *palatine process* is a horizontal projection of the maxilla that forms the anterior three-quarters of the hard palate (see Figure 7.6c, d). The union and fusion of the maxillary bones is normally completed before birth.

The *infraorbital foramen* (*infra-*=below; *orbital*=orbit), which can be seen in the anterior view of the skull in Figures 7.2 and 7.9a, is an opening in the maxilla inferior to the orbit. Through it passes the infraorbital blood vessels and nerve, a branch of the maxillary division of the trigeminal (V) nerve. Another prominent foramen in the maxilla is the *incisive foramen* (=incisor teeth) just posterior to the incisor teeth (see Figure 7.6c, d). It transmits branches of the greater palatine blood vessels and nasopalatine nerve. A final structure associated with the maxilla and sphenoid bone is the *inferior orbital fissure*, which is located between the greater wing of the sphenoid and the posterior aspect of the maxilla (see Figure 7.12).

ZYGOMATIC BONES

Description

The two **zygomatic bones** (*zygo-*=yokelike), commonly called cheekbones, form the prominences of the cheeks and part of the lateral wall and floor of each orbit (see Figures 7.9b and 7.12). They articulate with the frontal, maxilla, sphenoid, and temporal bones.

Surface Features

The *temporal process* of the zygomatic bone projects posteriorly and articulates with the zygomatic process of the temporal bone to form the *zygomatic arch* (see Figures 7.3c, d and 7.9b). Posteriorly, the bone forms a concave temporal surface that contributes to the temporal region of the skull and houses the tendon of the strong jaw-closing muscle called the temporalis. A foramen called the *zygomaticofacial foramen*, located near the center of the zygomatic bone (Figure 7.9b), transmits the zygomaticofacial nerve and vessels.

CHECKPOINT
9. Which structures pass through the zygomaticofacial foramen?

 CLINICAL CONNECTION | *Cleft Palate and Cleft Lip*

Usually the palatine processes of the maxillary bones unite during weeks 10 to 12 of embryonic development. Failure can result in **cleft palate**. The condition may also involve incomplete fusion of the horizontal plates of the palatine bones (see Figure 7.6c, d). Another condition, **cleft lip**, involves a split in the upper lip. Cleft lip and cleft palate often occur together. Depending on extent and position, speech and swallowing may be affected. In addition, children with cleft palate tend to have many ear infections, which can lead to hearing loss. Closure of cleft lip during the first few weeks following birth is recommended. Repair of cleft palate typically is done between 12 and 18 months of age, ideally before the child begins to talk. Speech and orthodontic therapy may be needed. Recent research strongly suggests that supplementation with folic acid during early pregnancy decreases the incidence of cleft palate and cleft lip.   •

Figure 7.9 Maxilla and zygomatic bone.

The maxilla houses the upper teeth, and the zygomatic bone is the cheekbone.

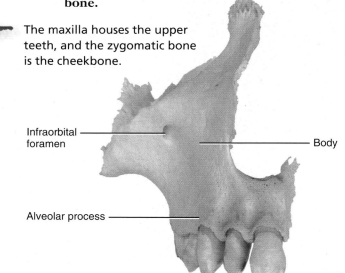

Infraorbital foramen
Body
Alveolar process

(a) Right maxilla, anterior view

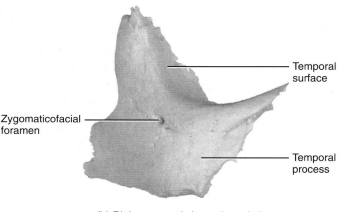

Temporal surface
Zygomaticofacial foramen
Temporal process

(b) Right zygomatic bone, lateral view

 Which part of the maxillae separate the nasal and oral cavities?

EXHIBIT 7.I **185**

EXHIBIT 7.I Facial Bones—Mandible *(Figure 7.10)*

OBJECTIVE
• Identify the location and surface features of the mandible.

Description

The **mandible** (*mand-*=to chew), or lower jawbone, is the largest, strongest facial bone (Figure 7.10). This large, arch-shaped bone is the only movable skull bone other than the auditory ossicles, the small bones of the ear and occipital bone.

Surface Features

In the lateral views shown in Figure 7.10, you can see that the mandible consists of a curved, horizontal portion, the *body*, and two perpendicular portions, the *rami* (RĀ-mī=branches). The *angle* of the mandible is the area where each *ramus* (singular form) meets the body. Each ramus has a posterior *condylar process* (KON-di-lar) that articulates with the mandibular fossa and articular tubercle of the temporal bone (see Figures 7.3c, d and 7.6c, d) to form the *temporomandibular joint (TMJ)*. It also has an anterior *coronoid process* (KOR-ō-noyd) to which the temporalis muscle attaches. The depression between the coronoid and condylar processes is called the *mandibular notch*. The *alveolar process* is the ridge-like arch containing the *alveoli* (sockets) for the mandibular (lower) teeth.

The *mental foramen* (*ment-*=chin) is approximately inferior to the second premolar tooth. It is near this foramen that dentists reach the mental nerve when injecting anesthetics. Another foramen associated with the mandible is the *mandibular foramen* on the medial surface of each ramus, another site often used by dentists to inject anesthetics. The mandibular foramen is the beginning of the *mandibular canal*, which runs obliquely in the ramus and anteriorly to the body. The inferior alveolar nerves and blood vessels pass through the canal and are distributed to the mandibular teeth.

CHECKPOINT
10. What structures form the temporomandibular joint?

CLINICAL CONNECTION | *Temporomandibular Joint Syndrome*

One problem associated with the temporomandibular joint is **temporomandibular joint (TMJ) syndrome.** It is characterized by dull pain around the ear, tenderness of the jaw muscles, a clicking or popping noise when opening or closing the mouth, limited or abnormal opening of the mouth, headache, tooth sensitivity, and abnormal wearing of the teeth. TMJ syndrome can be caused by improperly aligned teeth, grinding or clenching the teeth, trauma to the head and neck, or arthritis. Treatments include application of moist heat or ice, limiting the diet to soft foods, administration of pain relievers such as aspirin, muscle retraining, use of a splint or bite plate to reduce clenching and teeth grinding (especially when worn at night), adjustment or reshaping of the teeth (orthodontic treatment), and surgery. •

Figure 7.10 Mandible.

The mandible is the largest and strongest facial bone.

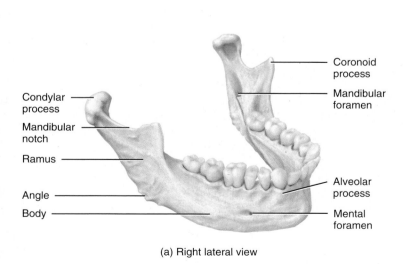

(a) Right lateral view

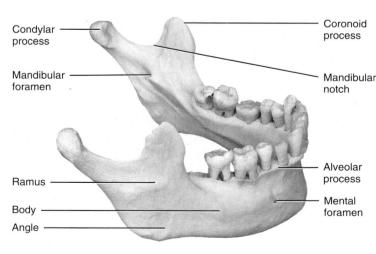

(b) Right lateral view

? What is the functional feature of the mandible that distinguishes it from almost all other skull bones?

Nasal Septum

The nasal cavity is a space inside the skull that is divided into right and left sides by a vertical partition called the **nasal septum**. The three components of the nasal septum are the vomer, the perpendicular plate of the ethmoid bone, and the septal cartilage (Figure 7.11). The inferior border of the perpendicular plate of the ethmoid bone joins the superoposterior border of the vomer to form the more posterior bony part of the septum, referred to as the *bony nasal septum*. The *septal cartilage*, which is hyaline cartilage, articulates with the anterior margins of the two bones to form the anterior portion of the septum. The term "broken nose" usually refers to damage to the septal cartilage rather than the nasal bones.

Orbits

Seven bones of the skull join to form each orbit (eye socket) or *orbital cavity*, which contains the eyeball and associated structures (Figure 7.12). The three cranial bones of the orbit are the frontal, sphenoid, and ethmoid; the four facial bones are the palatine, zygomatic, lacrimal, and maxilla. Each pyramid-shaped orbit has four regions that converge posteriorly:

1. Parts of the frontal and sphenoid bones comprise the *roof* of the orbit.
2. Parts of the zygomatic and sphenoid bones form the *lateral wall* of the orbit.
3. Parts of the maxilla, zygomatic, and palatine bones make up the *floor* of the orbit.
4. Parts of the maxilla, lacrimal, ethmoid, and sphenoid bones form the *medial wall* of the orbit.

Associated with each orbit are five openings:

1. The *optic foramen (canal)* is at the junction of the roof and medial wall.
2. The *superior orbital fissure* is at the superior lateral angle of the apex.
3. The *inferior orbital fissure* is at the junction of the lateral wall and floor.
4. The *supraorbital foramen* is on the medial side of the supraorbital margin of the frontal bone.
5. The *lacrimal fossa* is in the lacrimal bone.

CLINICAL CONNECTION | *Deviated Septum*

A **deviated nasal septum** is one that does not run along the midline of the nasal cavity. It deviates (bends) to one side. A blow to the nose can easily damage, or break, this delicate septum of bone and displace and damage the cartilage. Often, when a broken nasal septum heals, the bones and cartilage deviate to one side or the other. This deviated septum can block airflow into the constricted side of the nose, making it difficult to breathe through that half of the nasal cavity. The deviation usually occurs at the junction of the vomer bone with the septal cartilage. Septal deviations may also occur due to developmental abnormality. If the deviation is severe, it may block the nasal passageway entirely. Even a partial blockage may lead to infection. If inflammation occurs, it may cause nasal congestion, blockage of the paranasal sinus openings, chronic sinusitis, headache, and nosebleeds. The condition usually can be corrected or improved surgically. •

Figure 7.11 Nasal septum.

 The structures that form the nasal septum are the perpendicular plate of the ethmoid bone, the vomer, and septal cartilage.

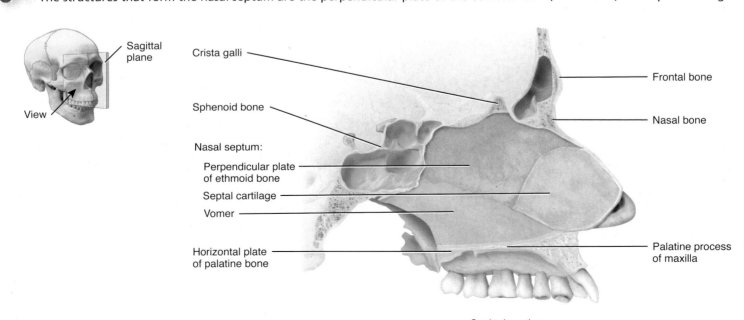

Sagittal section

 What is the function of the nasal septum?

Figure 7.12 Details of the orbit (eye socket).

 The orbit is a pyramid-shaped structure that contains the eyeball and associated structures.

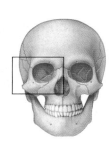

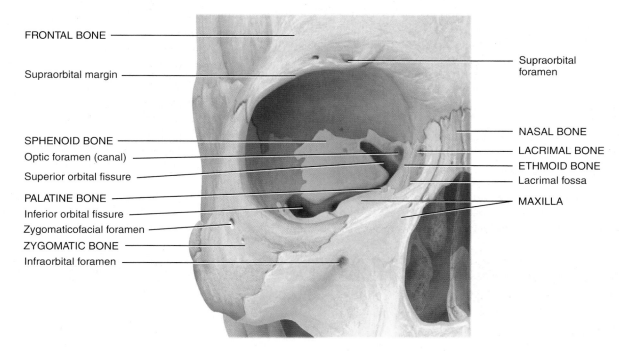

FRONTAL BONE

Supraorbital margin

SPHENOID BONE

Optic foramen (canal)

Superior orbital fissure

PALATINE BONE

Inferior orbital fissure

Zygomaticofacial foramen

ZYGOMATIC BONE

Infraorbital foramen

Supraorbital foramen

NASAL BONE

LACRIMAL BONE

ETHMOID BONE

Lacrimal fossa

MAXILLA

(a) Anterior view showing the bones of the right orbit

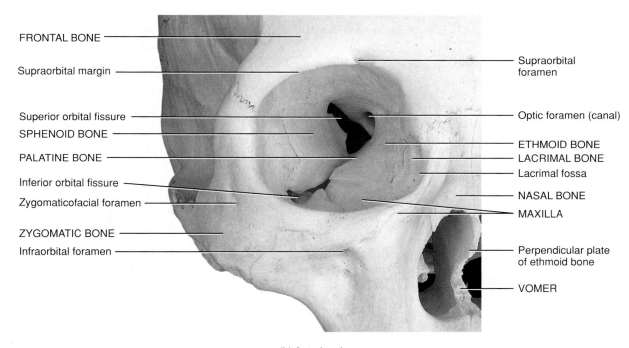

FRONTAL BONE

Supraorbital margin

Superior orbital fissure

SPHENOID BONE

PALATINE BONE

Inferior orbital fissure

Zygomaticofacial foramen

ZYGOMATIC BONE

Infraorbital foramen

Supraorbital foramen

Optic foramen (canal)

ETHMOID BONE

LACRIMAL BONE

Lacrimal fossa

NASAL BONE

MAXILLA

Perpendicular plate of ethmoid bone

VOMER

(b) Anterior view

 Which seven bones form the orbit, and which one is not visible in this figure?

Foramina

We mentioned most of the **foramina** (openings for blood vessels, nerves, or ligaments) of the skull in the descriptions of the cranial and facial bones that they penetrate. As preparation for studying other systems of the body, especially the nervous and cardiovascular systems, these foramina and the structures passing through them are listed in Table 7.3. For your convenience and for future reference, the foramina are listed alphabetically.

Unique Features of the Skull

The skull exhibits several unique features not seen in other bones of the body. These include sutures, paranasal sinuses, and fontanels.

Sutures

A **suture** (SOO-chur=seam) is an immovable joint in most cases in an adult skull that holds most skull bones together. Sutures in the skulls of infants and children, however, often are movable and function as important growth centers in the developing skull. In older individuals, many of the sutures in the skull fuse and become completely immovable. The names of many sutures reflect the bones they unite. For example, the frontozygomatic suture is between the frontal bone and the zygomatic bone. Similarly, the sphenoparietal suture is located between the sphenoid bone and the parietal bone. In other cases, however, the names of sutures are not so obvious. Of the many sutures found in the skull, the following are the most prominent:

- The **coronal suture** (ko-RŌ-nal; *coron-*=relating to the frontal or coronal plane) unites the frontal bone and both parietal bones (see Figure 7.3).
- The **sagittal suture** (SAJ-i-tal; *sagitt-*=arrow) unites the two parietal bones on the superior midline of the skull (see Figure 7.3a, b). The sagittal suture is so named because in the infant, before the bones of the skull are firmly united, the suture and the fontanels (soft spots) associated with it resemble an arrow.

TABLE 7.3

Principal Foramina of the Skull

FORAMEN	LOCATION	STRUCTURES PASSING THROUGH
Carotid (relating to carotid artery in neck)	Petrous portion of temporal bone (Figure 7.6c)	Internal carotid artery and sympathetic nerves for eyes
Hypoglossal canal (*hypo-*=under; *glossus*=tongue)	Superior to base of occipital condyles (Figure 7.6a)	Hypoglossal (XII) nerve and branch of ascending pharyngeal artery
Incisive (=pertaining to incisor teeth)	Posterior to incisor teeth in maxilla (Figure 7.6c)	Branches of greater palatine blood vessels and nasopalatine nerve
Infraorbital (*infra*=below)	Inferior to orbit in maxilla (Figure 7.12a)	Infraorbital nerve and blood vessels and a branch of the maxillary division of trigeminal (V) nerve
Jugular (=throat)	Posterior to carotid canal between petrous portion of temporal bone and occipital bone (Figure 7.6a)	Internal jugular vein, glossopharyngeal (IX), vagus (X), and accessory (XI) nerves
Lacerum (=lacerated)	Bounded anteriorly by sphenoid bone, posteriorly by petrous portion of temporal bone, and medially by the sphenoid bone and occipital bone (Figure 7.6a)	Branch of ascending pharyngeal artery in palatine bones
Magnum (=large)	Occipital bone (Figure 7.6c)	Medulla oblongata and its membranes (meninges), accessory (XI) nerve, and vertebral and spinal arteries
Mandibular (*mand*=to chew)	Medial surface of ramus of mandible (Figure 7.10)	Inferior alveolar nerve and blood vessels
Mastoid (=breast-shaped)	Posterior border of mastoid process of temporal bone (Figure 7.6c)	Emissary vein to transverse sinus and branch of occipital artery to dura mater
Mental (*ment-*=chin)	Inferior to second premolar tooth in mandible (Figure 7.10)	Mental nerve and vessels
Olfactory (*olfact*=to smell)	Cribriform plate of ethmoid bone (Figure 7.6a)	Olfactory (I) nerve
Optic (canal) (=eye)	Between superior and inferior portions of small wing of sphenoid bone (Figure 7.12)	Optic (II) nerve and ophthalmic artery
Ovale (=oval)	Greater wing of sphenoid bone (Figure 7.6a)	Mandibular branch of trigeminal (V) nerve
Rotundum (=round)	Junction of anterior and medial parts of sphenoid bone (Figure 7.6a)	Maxillary branch of trigeminal (V) nerve
Spinosum (=resembling a spine)	Posterior angle of sphenoid bone (Figure 7.6a)	Middle meningeal blood vessels
Stylomastoid (*stylo*=stake or pole)	Between styloid and mastoid processes of temporal bone (Figure 7.6c)	Facial (VII) nerve and stylomastoid artery
Supraorbital (*supra-*=above)	Supraorbital margin of orbit in frontal bone (Figure 7.12a)	Supraorbital nerve and artery

- The **lambdoid suture** (LAM-doyd) unites the two parietal bones to the occipital bone. This suture is so named because of its resemblance to the Greek letter lambda (Λ), as can be seen in Figure 7.5 (with the help of a little imagination). Sutural or wormian bones may occur within the sagittal and lambdoid sutures (see Figure 7.5 and Section 6.2).
- The two **squamous sutures** (SKWĀ-mus; *squam-*=flat, like the flat overlapping scales of a snake) unite the parietal and temporal bones on the lateral aspects of the skull (see Figure 7.3c, d).

Paranasal Sinuses

The **paranasal sinuses** (par′-a-NĀ-zal SĪ-nus-ez; *para-*=beside) are cavities within certain cranial and facial bones near the nasal cavity. They are most evident in a sagittal section of the skull (Figure 7.13). The paranasal sinuses are lined with mucous membranes that are continuous with the lining of the nasal cavity through small openings in the lateral wall of the nasal cavity. Secretions produced by the mucous membrane of the paranasal sinuses drain into the lateral wall of the nasal cavity. Paranasal sinuses are rudimentary or absent at birth and increase in size during two critical periods of facial enlargement—during the eruption of the teeth and at the onset of puberty. They arise as outgrowths of the mucosal lining of the nasal cavity that project into the surrounding bones. Skull bones containing the paranasal sinuses are the frontal, sphenoid, and ethmoid bones, and the maxillae. The paranasal sinuses allow the skull to increase in size without a corresponding change in the mass (weight) of the bone. In addition, the paranasal sinuses serve as resonating (echo) chambers within the skull that intensify and prolong sounds, thereby enhancing the quality of the voice. The influence of the paranasal sinuses on your voice becomes obvious when you have a cold; the passageways through which sound travels into and out of the paranasal sinuses become blocked by excess mucus production, changing the quality of your voice.

CLINICAL CONNECTION | *Sinusitis*

Sinusitis (sīn-ū-SĪ-tis) is an inflammation of the mucous membrane of one or more paranasal sinuses. It may be caused by a microbial infection (virus, bacterium, or fungus), allergic reactions, nasal polyps, or a severely deviated nasal septum. If the inflammation or an obstruction blocks the drainage of mucus into the nasal cavity, fluid pressure builds up in the paranasal sinuses, and a sinus headache may develop. Other symptoms may include nasal congestion, inability to smell, fever, and cough. Treatment options include decongestant sprays or drops, oral decongestants, nasal corticosteroids, antibiotics, analgesics to relieve pain, warm compresses, and surgery. •

Fontanels

The skull of a developing embryo consists of cartilage and mesenchyme arranged in thin plates around the developing brain. Gradually, ossification occurs and bone slowly replaces the cartilage and mesenchyme. At birth, bone formation is incomplete and the mesenchyme-filled spaces become dense connective tissue regions between the incompletely developed cranial bones

---

Figure 7.13 Paranasal sinuses projected to the surface. (a and b) Location of the paranasal sinuses. (c) Openings of the paranasal sinuses into the nasal cavity. (d) Dissection of maxillary sinus showing opening to nasal cavity. Portions of the conchae have been sectioned.

🔑 Paranasal sinuses are mucous membrane–lined spaces in the frontal, sphenoid, and ethmoid bones, and the maxillae that connect to the nasal cavity.

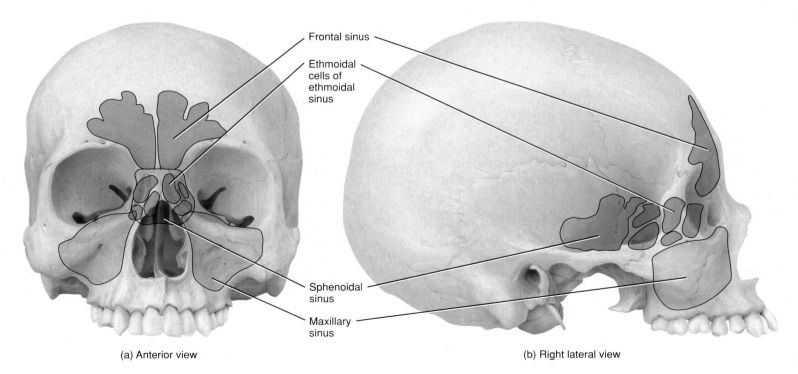

(a) Anterior view

(b) Right lateral view

FIGURE 7.13 CONTINUES ▶

FIGURE 7.13 CONTINUED ▶

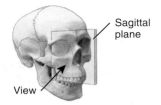

Sagittal plane

View

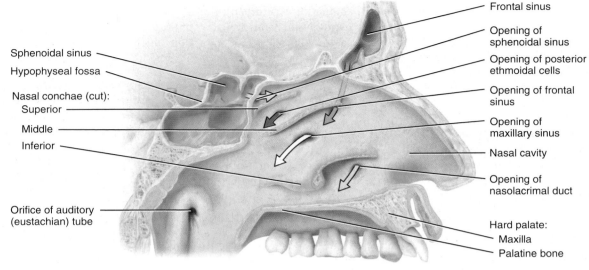

Sphenoidal sinus

Hypophyseal fossa

Nasal conchae (cut):
Superior
Middle
Inferior

Orifice of auditory (eustachian) tube

Frontal sinus

Opening of sphenoidal sinus

Opening of posterior ethmoidal cells

Opening of frontal sinus

Opening of maxillary sinus

Nasal cavity

Opening of nasolacrimal duct

Hard palate:
Maxilla
Palatine bone

(c) Sagittal section

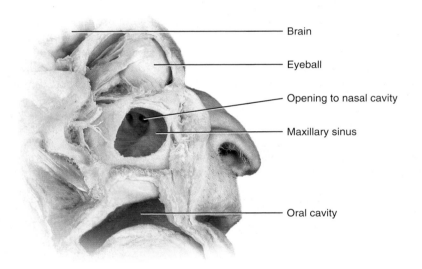

Brain

Eyeball

Opening to nasal cavity

Maxillary sinus

Oral cavity

(d) Dissection into maxillary sinus

? **What are the functions of the paranasal sinuses?**

called **fontanels** (fon-ta-NELZ=little fountains) (Figure 7.14). Commonly called "soft spots," fontanels are the areas where unossified mesenchyme develops into the dense connective tissues of the skull. As bone formation continues after birth, the fontanels are eventually replaced with bone by intramembranous ossification and the thin collagenous connective tissue junctions that remain between neighboring bones become the sutures. Functionally, the fontanels serve as spacers for the growth of neighboring skull bones and provide some flexibility to the fetal skull. They allow the skull to change shape as it passes through the birth canal and permit rapid growth of the brain during infancy. Although an infant may have many fontanels at birth, the form and location of six are fairly constant:

1. The unpaired **anterior fontanel**, located at the midline among the two parietal bones and the frontal bones, is roughly diamond-shaped and is the largest fontanel. It usually closes 18 to 24 months after birth.

2. The unpaired **posterior fontanel** is located at the midline among the two parietal bones and the occipital bone. Because it is much smaller than the anterior fontanel, it generally closes about 2 months after birth.

3. The paired **anterolateral fontanels**, located laterally among the frontal, parietal, temporal, and sphenoid bones, are small and irregular in shape. Normally, they close about 3 months after birth.

4. The paired **posterolateral fontanels**, located laterally among the parietal, occipital, and temporal bones, are irregularly shaped. They begin to close 1 to 2 months after birth, but closure is generally not complete until 12 months.

The amount of closure in fontanels helps a physician gauge the degree of brain development. In addition, the anterior fontanel serves as a landmark for withdrawal of blood for analysis from the superior sagittal sinus (a large midline vein within the covering tissues that surround the brain; see Figure 14.11c).

Figure 7.14 **Fontanels at birth.**

Fontanels are mesenchyme-filled spaces between cranial bones that are present at birth.

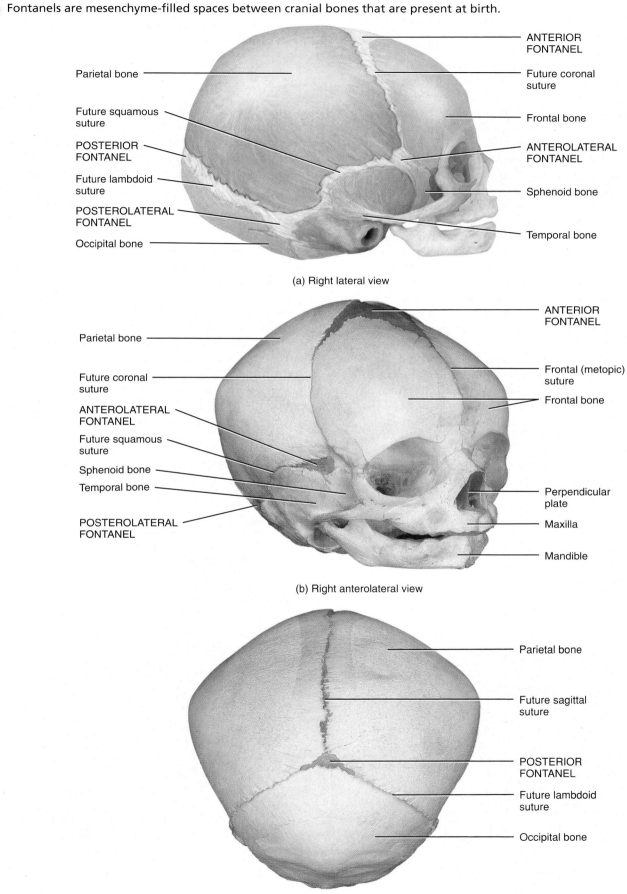

Parietal bone

ANTERIOR FONTANEL

Future coronal suture

Future squamous suture

Frontal bone

POSTERIOR FONTANEL

ANTEROLATERAL FONTANEL

Future lambdoid suture

Sphenoid bone

POSTEROLATERAL FONTANEL

Occipital bone

Temporal bone

(a) Right lateral view

Parietal bone

ANTERIOR FONTANEL

Future coronal suture

Frontal (metopic) suture

ANTEROLATERAL FONTANEL

Frontal bone

Future squamous suture

Sphenoid bone

Temporal bone

Perpendicular plate

POSTEROLATERAL FONTANEL

Maxilla

Mandible

(b) Right anterolateral view

Parietal bone

Future sagittal suture

POSTERIOR FONTANEL

Future lambdoid suture

Occipital bone

(c) Posterior view

Which fontanel is bordered by four different skull bones?

Cranial Fossae

The internal shape and appearance of the cranium resembles the outer surface of the brain. As the cranial bones form around the developing brain, they create a mold that approximates the brain's shape. If the *calvaria*, the roof of the cranium, is removed, the floor of the cranium displays three distinct levels that correspond to the major contours of the inferior surface of the brain. The three distinct regions of the floor of the cranium are called **cranial fossae** (FAWS-ē) (Figure 7.15). Each cranial fossa is at a different level, creating the appearance of stairs that step down from anterior to posterior inside the skull. A specific structural region of the brain fits snugly into each cranial fossa. From anterior to posterior, they are named the anterior cranial fossa, middle cranial fossa, and posterior cranial fossa. The highest level, the *anterior cranial fossa*, is formed largely by the portion of the frontal bone that constitutes the roof of the orbits and nasal cavity, the crista galli and cribriform plate of the ethmoid bone, and the lesser wings and part of the body of the sphenoid bone. This fossa houses the frontal lobes of the cerebral hemispheres of the brain. The rough surface of the frontal bone can lead to tearing of the frontal lobes of the cerebral hemispheres during head trauma. The *middle cranial fossa* is inferior and posterior to the anterior cranial fossa. Like the sphenoid bone that surrounds it, it is shaped like a butterfly, with a small median portion and two expanded lateral portions. The median portion is formed by part of the body of the sphenoid bone, and the lateral portions are formed by the greater wings of the sphenoid bone, the temporal squama, and the parietal bone. The middle cranial fossa cradles the temporal lobes of the cerebral hemispheres and the pituitary gland. The last fossa, forming the most inferior level, is the *posterior cranial fossa*, the largest of the fossae. It is formed mostly by the occipital bone and the petrous and mastoid portions of the temporal bone. It is a very deep fossa that accommodates the cerebellum, pons, and medulla oblongata of the brain.

Age-related Changes in the Skull

From fetus to newborn child to elderly adult, the skull undergoes significant changes as it ages. During fetal life, the skull, like the rest of the skeleton, consists of many more bones than are present later in life. Whether it is the femur in the thigh or the occipital bone at the back of the skull, when a bone first starts to develop and ossify it consists of multiple ossification centers that are each separate bony elements. For example, the occipital bone of the skull is a single bone in adults that typically arises from about a dozen or more ossification centers that fuse during the fetal period into five distinct bones. The five parts of the occipital bone are separated by connective tissue; these connective tissue spacers allow the occipital bone to increase in size during the early months and years of development to accommodate the growing brain. A few years after birth, as the individual bony parts enlarge, they begin to approach one another and soon fuse into a single bone. The final fusion of the five parts of the occipital bone is typically complete by the late teenage years.

Similarly, the individual bones of the skull in the newborn and young skull are separated by connective tissue regions called fontanels, which were described in a previous section of this chapter. As neighboring bones develop and enlarge, the connective tissue

Figure 7.15 Cranial fossae.

 Cranial fossae are levels in the cranial floor that contain depressions for brain convolutions, grooves for blood vessels, and numerous foramina.

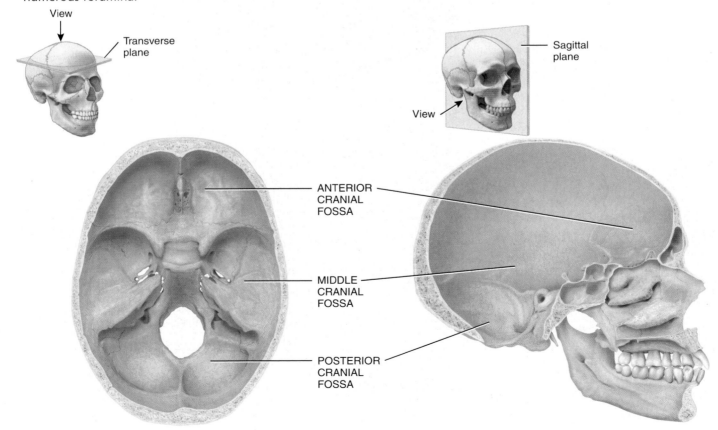

(a) Superior view of floor of cranium

(b) Medial view of sagittal section

Which cranial fossa is the largest?

is replaced and the fontanels decrease in size. Many of the resulting sutures will fuse and even obliterate the boundaries between neighboring bones later in adult life.

Other age-related changes in the skull include the emergence of the paranasal sinuses and the eruption of teeth in the young skull, the wearing down and loss of teeth with advancing age, and the bone loss of the mandible and maxilla (the jaw bones) that accompanies tooth loss in older individuals.

Sexual Differences in the Skull

Sexual dimorphism, the structural differences that exist between males and females, is present in the skull, but skill and awareness of variations are essential when skull characteristics are used to determine the gender of an individual. Determining gender is only possible using the skull of a mature individual. In general, male skulls are always more massive and female skulls more *gracile* (delicate), but there is a zone where gracile males and robust females overlap that makes gender determination difficult. Many television shows and movies would have you believe this gender determination is easy, fast, and foolproof; however, even though the chances of correctly determining the gender of a skull increase with the number of characteristics considered, still only 80 to 90 percent accuracy is standard. Whenever possible, pelvic characteristics should be used in combination with skull characteristics to determine the gender of remains (see Section 8.4). Table 7.4 illustrates and describes some of the key visible characteristic differences in the male and female skulls.

TABLE 7.4

Key Visible Characteristic Gender Differences in the Male and Female Skulls

KEY VISIBLE CHARACTERISTIC	DESCRIPTION
Muscle Attachment Sites ① Temporal lines ② External occipital protuberance ③ Nuchal lines ④ Mastoid process	Muscle attachment sites are typically some of the better distinguishing characteristics in differentiating the gender of skulls. Because of the larger muscle mass of males, these sites of muscle attachment produce more prominent bony landmarks where they apply their forces to the bones. Landmarks ②, ③, and ④ are typically much larger and more developed in males, as these are attachment sites for muscles that stabilize the larger male head during locomotion. Landmark ①, the temporal line, is the attachment site of a jaw closing muscle and typically is not as reliable a diagnostic characteristic.
Bony Prominences ⑤ Angle of the mandible ⑥ Frontal squama (forehead) ⑦ Supraorbital ridge ⑧ Glabella ⑨ Teeth ⑩ Mental eminence (chin)	These bony prominences can be used to further substantiate the sexes; however, they should be used with greater caution. Typically, males have a more pronounced supraorbital ridge ⑦ and a rounded, projecting glabella ⑧, while in the female skull this is flat and does not project forward. The frontal squama of males typically slopes backward toward the top of the skull, while the squama of the female skull is more vertical and rounded. The angle of the mandible ⑤ is typically more square (closer to 90°) in males, while the female angle is more obtuse. Like the glabella, the mental eminence ⑩ of males is typically more prominent and projects forward, creating a more noticeable chin. Teeth ⑨, on average, are larger in the male skull, but this is often a difficult character to judge.
Others (not visible) Paranasal sinuses	On average, the overall volume of the paranasal sinuses is larger in males than in females.

CHECKPOINT

11. Describe the cavities within the skull and the nasal septum.
12. Which area of the orbit does the optic foramen pass through, and what cranial bones are involved?
13. What structures make up the nasal septum?
14. Define the following: foramen, suture, paranasal sinus, and fontanel.
15. Name the cranial fossae from inferior to superior, and indicate which part(s) of the brain each one houses.
16. Using the characteristics listed in Table 7.4, determine whether the skull shown in the table is male or female.

7.3 HYOID BONE

 OBJECTIVE

• Describe the relationship of the hyoid bone to the skull and explain its function.

The single **hyoid bone** (=U-shaped) is a unique component of the axial skeleton because it does not articulate with any other bone. Rather, it is suspended from the styloid processes of the temporal bones by ligaments and muscles (see Figure 11.7a, b). Located in the anterior neck between the mandible and larynx (Figure 7.16a), the hyoid bone supports the tongue, providing attachment sites for some tongue muscles and for muscles of the neck and pharynx. The hyoid bone consists of a horizontal, rectangular *body* and paired projections called the *lesser horns* and the *greater horns* (Figure 7.16b, c). Muscles and ligaments attach to the body and these paired projections.

The hyoid bone, as well as cartilages of the larynx and trachea, are often fractured during strangulation. As a result, they are carefully examined at autopsy when manual strangulation is a suspected cause of death.

 CHECKPOINT

17. What is the relationship between the hyoid bone and the larynx?

Figure 7.16 **Hyoid bone.**

 The hyoid bone supports the tongue, providing attachment sites for muscles of the tongue, neck, and pharynx.

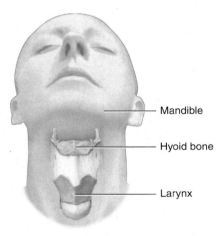

(a) Position of hyoid bone

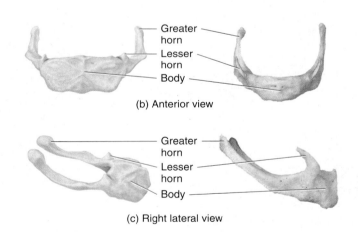

(b) Anterior view

(c) Right lateral view

 In what way is the hyoid bone different from all other bones of the axial skeleton?

7.4 VERTEBRAL COLUMN

OBJECTIVES

* Identify the regions and normal curves of the vertebral column.
* Describe the structural and functional features of the bones in the various regions of the vertebral column.

The **vertebral column**, also called the *spine*, *backbone*, or *spinal column*, makes up about two-fifths of the total height of the body and is composed of a series of bones called **vertebrae** (VER-te-brē; singular is *vertebra*). The vertebral column consists of bone and connective tissue that surrounds and protects the nervous tissue of the spinal cord. The length of the column is about 71 cm (28 in.) in an average adult male and about 61 cm (24 in.) in an average adult female. The vertebral column functions as a strong, flexible rod with elements that can move forward, backward, and sideways, and can rotate. It encloses and protects the spinal cord, supports the head, and serves as a point of attachment for the ribs, pelvic girdle, and muscles of the back and upper limbs.

The total number of vertebrae during early development is 33. Then, several vertebrae in the sacral and coccygeal regions fuse. As a result, the adult vertebral column typically contains 26 vertebrae (Figure 7.17a). These are distributed as follows:

7 **cervical vertebrae** (*cervic-*=neck) in the neck region.

12 **thoracic vertebrae** (*thorax*=chest) posterior to the thoracic cavity.

5 **lumbar vertebrae** (*lumb-*=loin) supporting the lower back.

1 **sacrum** (SĀ-krum=sacred bone) consisting of five fused **sacral vertebrae.**

1 **coccyx** (KOK-siks=cuckoo, because the shape resembles the bill of a cuckoo bird), consisting of four fused **coccygeal vertebrae** (kok-SIJ-ē-al).

The cervical, thoracic, and lumbar vertebrae are movable, but the sacrum and coccyx are not. We will discuss each of these regions in detail shortly.

Normal Curves of the Vertebral Column

When viewed from anterior or posterior, a normal adult vertebral column appears straight (Figure 7.17a). But when viewed from the side, it shows four slight bends called **normal curves** (Figure 7.17b). Relative to the front of the body, the *cervical* and *lumbar curves* are convex (bulging out), and the *thoracic* and *sacral curves* are concave (cupping in). The curves of the vertebral column increase its strength, help maintain balance in the upright position, absorb shocks during walking, and help protect the vertebrae from fracture.

In the fetus, there is only a single anteriorly concave curve throughout the length of the entire vertebral column (Figure 7.17c, left). At about the third month after birth, when an infant begins to hold its head erect, the anteriorly convex cervical curve begins to develop (Figure 7.17c, right). Later, when the child sits up, stands, and walks, the anteriorly convex lumbar curve begins to develop. The thoracic and sacral curves are called *primary curves* because they retain the original curvature of the embryonic vertebral column. The cervical and lumbar curves are known as *secondary curves* because they begin to form later, several months after birth. All curves are fully developed by age 10. However, secondary curves may be progressively lost in old age.

Intervertebral Discs

From the second cervical vertebra to the sacrum, structures called **intervertebral discs** (in′-ter-VER-tē-bral; *inter-*=between) are found between the bodies of adjacent vertebrae (Figure 7.17d). They account for 25 percent of the height of the vertebral column.

CLINICAL CONNECTION | *Abnormal Curves of the Vertebral Column*

Various conditions may exaggerate the normal curves of the vertebral column, or the column may acquire a lateral bend; both of these result in **abnormal curves** of the vertebral column.

Scoliosis (skō-lē-Ō-sis; *scolio-*=crooked) is a lateral bending of the vertebral column, usually in the thoracic region (Figure A). The most common of the abnormal curves, scoliosis may result from congenitally (present at birth) malformed vertebrae, chronic sciatica (pain in the lower back and lower limb), paralysis of muscles on one side of the vertebral column, poor posture, or one leg being shorter than the other.

Kyphosis (kī-FŌ-sis; *kyphos-*=hump) is an increase in the thoracic curve of the vertebral column (Figure B). In tuberculosis of the spine, vertebral bodies may partially collapse, causing an acute angular bending of the vertebral column. In the elderly, degeneration of the intervertebral discs leads to kyphosis. Kyphosis may also be caused by rickets and poor posture. It is also common in females with advanced osteoporosis.

Lordosis (lor-DŌ-sis; *lord-*=bent backward), sometimes called *hollow* or *sway back*, is an increase in the lumbar curve of the vertebral column (Figure C). It may result from increased weight of the abdomen, as in pregnancy, or extreme obesity, poor posture, rickets, osteoporosis, or tuberculosis of the spine. •

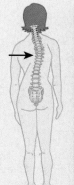 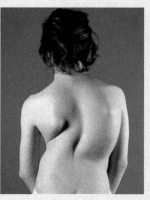 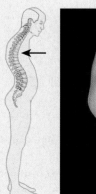 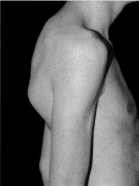 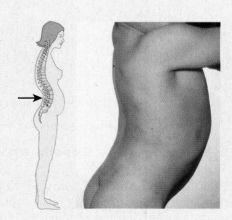

(A) Scoliosis

(B) Kyphosis

(C) Lordosis

Figure 7.17 Vertebral column. The numbers in parentheses in (a) indicate the number of vertebrae in each region. In (d), the relative size of the disc has been enlarged for emphasis.

 The adult vertebral column typically contains 26 vertebrae.

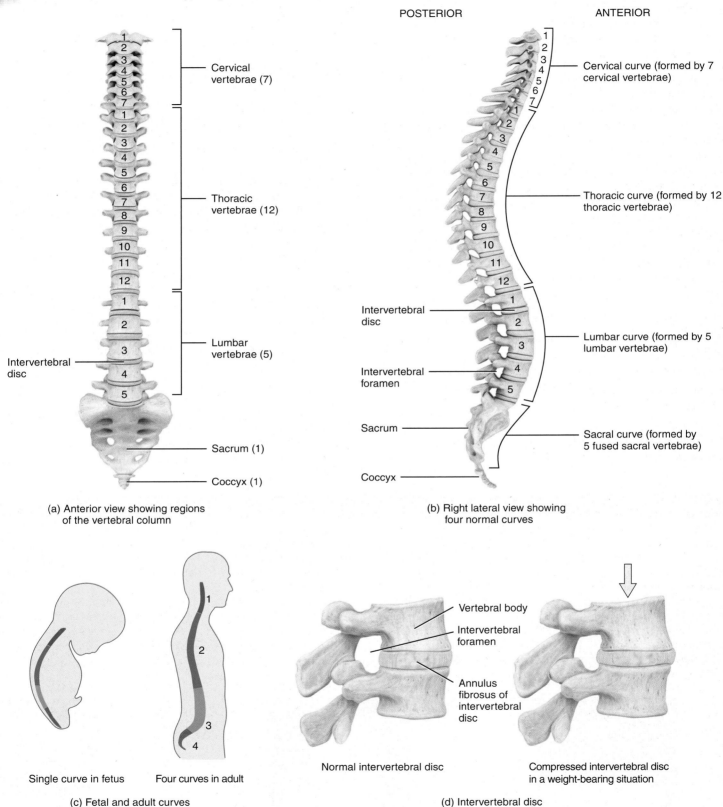

(a) Anterior view showing regions of the vertebral column

(b) Right lateral view showing four normal curves

Single curve in fetus   Four curves in adult

(c) Fetal and adult curves

Normal intervertebral disc

Compressed intervertebral disc in a weight-bearing situation

(d) Intervertebral disc

 Which curves of the adult vertebral column are concave (relative to the anterior side of the body)?

Each disc has an outer fibrous ring consisting of fibrocartilage called the *annulus fibrosus* (*annulus*=ringlike) and an inner soft, pulpy, mucoid substance called the *nucleus pulposus* (*pulposus*= pulplike). With increasing age, this watery mucoid region converts to fibrocartilage as it gradually becomes less hydrated. The superior and inferior surfaces of the disc consist of a thin plate of hyaline cartilage. The discs form strong joints, permit various movements of the vertebral column, and absorb vertical shock.

During the course of a day the discs compress and lose water from their cartilaginous matrix so that we are a bit shorter at night. While we are sleeping there is less compression and the nucleus pulposus rehydrates so that we are taller when we awake in the morning. Recent studies reveal that decrease in vertebral height with age results from a loss of bone in the vertebral bodies and not a decrease in the thickness of the intervertebral discs.

Since intervertebral discs are avascular, the annulus fibrosus and nucleus pulposus rely on blood vessels from the bodies of vertebrae to obtain oxygen and nutrients and remove wastes. Stretching exercises such as yoga decompress discs and increase general blood circulation, both of which speed up the uptake of oxygen and nutrients by discs and the removal of wastes.

Parts of a Typical Vertebra

Even though vertebrae in different regions of the spinal column vary in size, shape, and detail, they are similar enough that we can discuss the structures (and the functions) of a typical vertebra (Figure 7.18). Vertebrae typically consist of a body, a vertebral arch, and several processes.

Vertebral Body

The **vertebral body**, the largest portion of a vertebra, forms the anterior blocklike mass of the bone. All the adjacent vertebral bodies lined up in a row create the columnlike axis of the skeleton. The vertebral bodies serve as the main weight-bearing component of the vertebral column. The cartilaginous intervertebral discs lie between adjacent vertebral bodies. The superior and inferior surfaces of the vertebral bodies are roughened for the attachment of the hyaline cartilage of the intervertebral disc, which helps impede water loss from the disc into the bone of the vertebral body. The anterior and posterior surfaces contain nutrient foramina, openings for blood vessels that deliver nutrients and oxygen and remove carbon dioxide and wastes from bone tissue.

Vertebral Arch

The **vertebral arch** extends posteriorly from the vertebral body of the vertebra and, together with the body of the vertebra, surrounds the spinal cord. The vertebral arch is complex and varies from region to region of the spinal column; it consists of numerous bony projections that provide attachment surfaces and lever arms for muscles. Two short, thick processes, the *pedicles* (PED-i-kuls=little feet), form the base of the vertebral arch. The pedicles project posteriorly from the body to unite with the laminae. The *laminae* (LAM-i-nē=thin layers) are the flat parts that join to form the posterior portion of the vertebral arch. The *vertebral foramen* that lies between the vertebral arch and body contains the spinal cord and its meningeal coverings, adipose tissue, areolar connective tissue, and blood vessels. Collectively, the vertebral foramina of all vertebrae form the *vertebral (spinal) canal*. The pedicles exhibit superior and inferior indentations called *vertebral notches* (see Figure 7.18d and 7.20a). When the vertebral notches are stacked on top of one another, they form an opening between adjoining vertebrae on both sides of the column. Each opening, called an *intervertebral foramen*, permits the passage of a single spinal nerve carrying information to and from the spinal cord (see Figure 7.18b and 7.20a).

Figure 7.18 Structure of a typical vertebra, as illustrated by a thoracic vertebra. In (b), only one spinal nerve has been included, and it has been extended beyond the intervertebral foramen for clarity.

🔑 A vertebra consists of a vertebral body, a vertebral arch, and several processes.

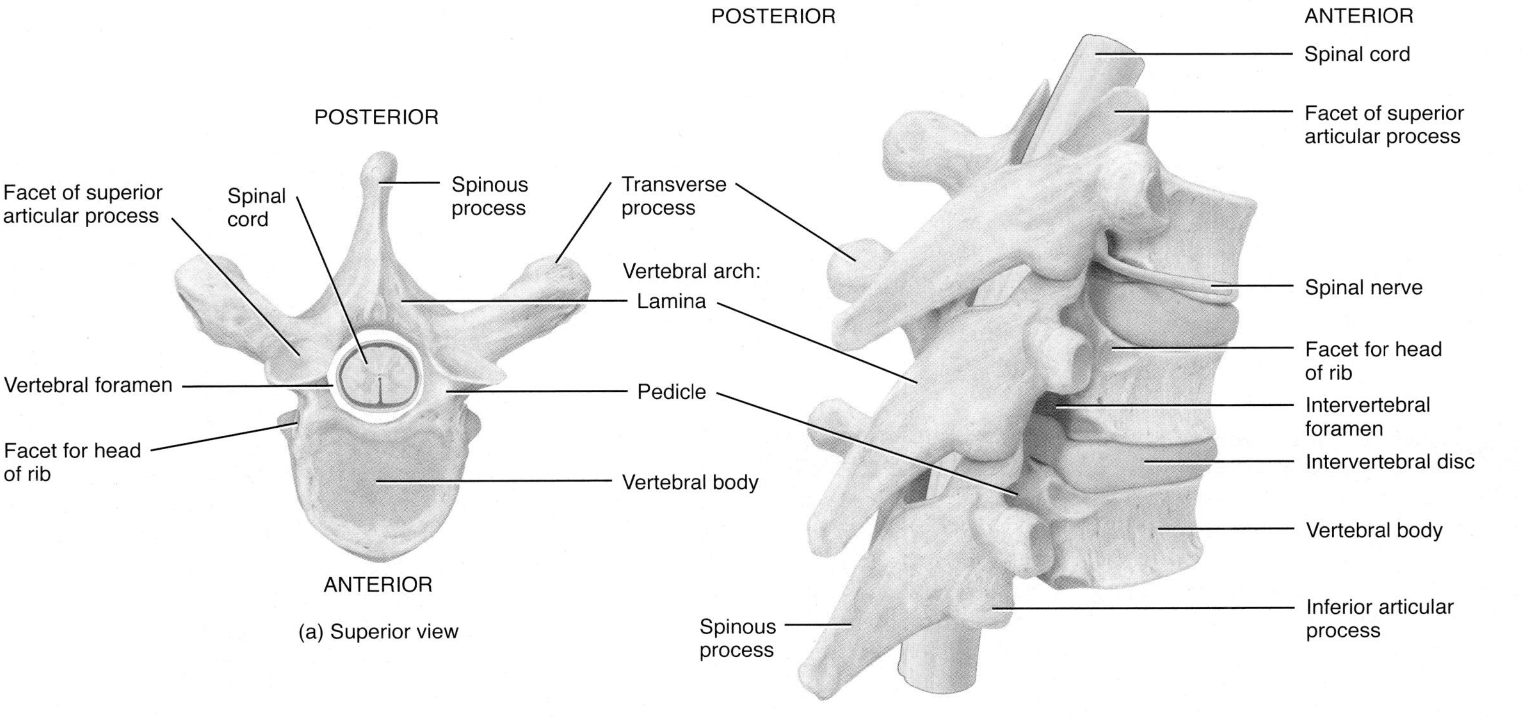

(a) Superior view

(b) Right posterolateral view of articulated vertebrae

FIGURE 7.18 CONTINUES ▶

FIGURE 7.18 CONTINUED

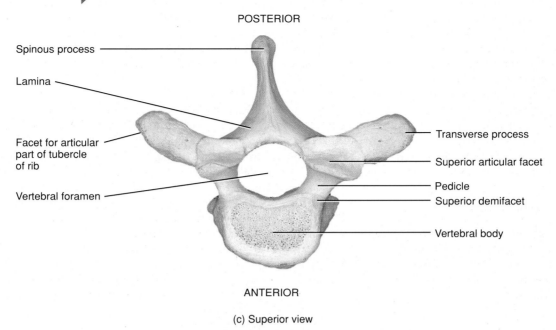

(c) Superior view

(d) Right lateral view

? **What are the functions of the vertebral and intervertebral foramina?**

Processes

Seven **processes** arise from the vertebral arch. At the junction of the lamina and pedicle, *transverse processes* extend posterolaterally, one on each side. A single *spinous process (spine)* projects posteriorly from the junction of the laminae. This process and the two transverse processes serve as lever arms and points of attachment for muscles. The remaining four processes form joints with other vertebrae above or below. The two *superior articular processes* of a vertebra articulate (form joints) with the two inferior articular processes of the vertebra immediately superior to them. In turn, the two *inferior articular processes* of that vertebra articulate with the two superior articular processes of the vertebra immediately inferior to them, and so on. Like the transverse processes, the articular processes typically arise near the junction of the pedicle and lamina.

CLINICAL CONNECTION | *Herniated (Slipped) Disc*

In their function as shock absorbers, intervertebral discs are constantly being compressed. If the anterior and posterior ligaments of the discs become injured or weakened, the pressure developed in the nucleus pulposus may be great enough to rupture the surrounding fibrocartilage (annulus fibrosus). If this occurs, the nucleus pulposus may herniate (protrude) posteriorly or into one of the adjacent vertebral bodies. This condition is called a **herniated (slipped) disc**. Because the lumbar region bears much of the weight of the body, and is the region of the most flexing and bending, herniated discs most often occur in the lumbar area.

Frequently, the nucleus pulposus slips posteriorly toward the spinal cord and spinal nerves. This movement exerts pressure on the spinal nerves, causing local weakness and acute pain. If the roots of the sciatic nerve, which passes from the spinal cord to the foot, are compressed, the pain radiates down the posterior thigh, through the calf, and occasionally into the foot. If pressure is exerted on the spinal cord itself, some of its neurons may be destroyed. Treatment options include bed rest, medications for pain, physical therapy and exercises, and *percutaneous endoscopic discectomy* (removal of disc material using a laser). A person with a herniated disc may also undergo a laminectomy, a procedure in which parts of the laminae of the vertebra and intervertebral disc are removed to relieve pressure on the nerves. •

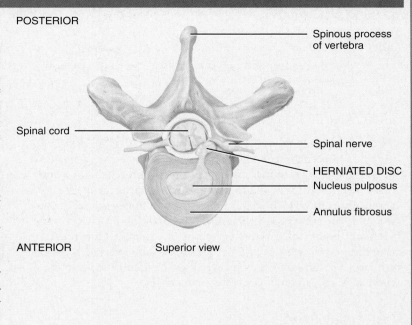

POSTERIOR

Spinal cord

ANTERIOR   Superior view

Spinous process of vertebra

Spinal nerve

HERNIATED DISC
Nucleus pulposus

Annulus fibrosus

The articulating surfaces of the articular processes, referred to as *facets* (FAS-et or fa-SET=little faces), are covered with hyaline cartilage. The articulations formed between the bodies and articular facets of successive vertebrae are called *intervertebral joints*.

Regions of the Vertebral Column

Exhibits 7.J through 7.M (Figures 7.19–7.22) present the five regions of the vertebral column, beginning superiorly and moving inferiorly. The five regions are cervical, thoracic, lumbar, sacral, and coccygeal. Note that vertebrae in each region are numbered in sequence from superior to inferior.

Age-related Changes in the Vertebral Column

With advancing age the vertebral column undergoes changes that are characteristic of the skeletal system in general. These changes include reduction in the mass and density of the bone along with a reduction in the collagen to mineral content within the bone, changes that make the bones more brittle and susceptible to damage. The articular surfaces, those surfaces where neighboring bones move against one another, lose their covering cartilage as they age; in their place, rough bony growths form that lead to arthritic conditions. In the vertebral column, bony growths around the intervertebral discs, called *osteophytes*, can lead to a narrowing (stenosis) of the vertebral canal. This narrowing can lead to compression of spinal nerves and the spinal cord, which can manifest as pain and decreased muscle function in the back and lower limbs.

 CHECKPOINT

18. What are the functions of the vertebral column?
19. When do the secondary vertebral curves develop?

EXHIBIT 7.J Vertebral Regions—Cervical Vertebrae *(Figure 7.19)*

OBJECTIVE

• Identify the location and surface features of the cervical vertebrae.

Description

The **cervical vertebrae** (C1–C7) are the most variable of the vertebrae. This variability is due to the fact that the general structure of the first two vertebrae in this region is modified significantly from the structures of the remaining five cervical vertebrae. The cervical vertebrae form a delicate column of bones that vary considerably in the range of mobility at their joint surfaces. The vertebral bodies of cervical vertebrae are smaller than those of thoracic vertebrae (Figure 7.19a). However, the vertebral arches are larger. A unique characteristic used to identify a vertebra as cervical is a foramen in the transverse process.

Surface Features

All cervical vertebrae have three foramina: one vertebral foramen and two of the transverse foramina mentioned above (Figure 7.19e). The *vertebral foramina* of cervical vertebrae are the largest in the spinal column because they house the cervical enlargement of the spinal cord. As you just learned, each of the two cervical transverse processes contains a *transverse foramen* through which a vertebral artery and its accompanying vein and nerve fibers pass. The spinous processes of C2 through C6 are often *bifid*—that is, at their tips they branch into two small projections (Figure 7.19a, c).

The first two cervical vertebrae differ considerably from the others. Named after the mythological Atlas, who supported the world on his shoulders, the first cervical vertebra (C1), the **atlas**, supports the head (Figure 7.19a, b). The atlas is a ring of bone with *anterior* and *posterior arches* and large *lateral masses*. It lacks a vertebral body and a spinous process. The superior surfaces of the lateral masses, called *superior articular facets*, are concave and articulate with the occipital condyles of the occipital bone to form the paired *atlanto-occipital joints*. These articulations permit the movement seen when moving the head to signify "yes." The inferior surfaces of the lateral masses, the *inferior articular facets*, articulate with the second cervical vertebra. The transverse processes and transverse foramina of the atlas are quite large.

The second cervical vertebra (C2), called the **axis** (Figure 7.19a, d, e), does have a vertebral body. Projecting from its vertebral body

Figure 7.19 Cervical vertebrae.

🔑 The cervical vertebrae are found in the neck region.

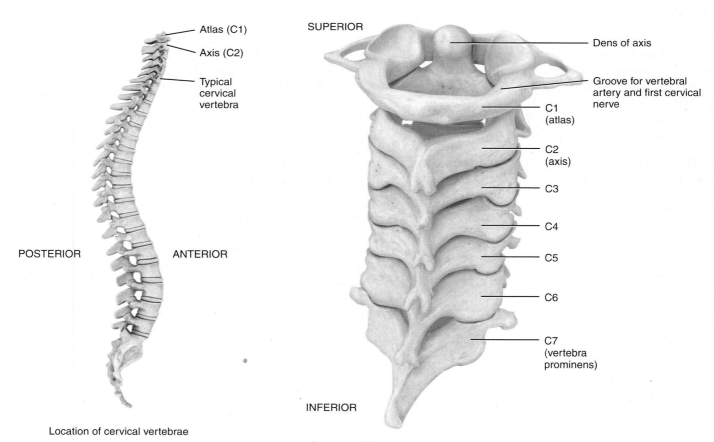

Location of cervical vertebrae

(a) Posterior view of articulated cervical vertebrae

EXHIBIT 7.J **201**

is a peglike process called the *dens* (=tooth) or *odontoid process*. The dens projects superiorly through the anterior portion of the vertebral foramen of the atlas to form a pivot on which the atlas and head rotate. This arrangement permits side-to-side movement of the head, as when you move your head to signify "no." The articulation between the anterior arch of the atlas and dens of the axis, and between their articular facets, which allows you to signify nonverbally that you will not lend your roommate yet another \$20, is called the *atlanto-axial joint*. In some instances of trauma, the dens of the axis may be driven into the medulla oblongata of the brain. This type of injury is the usual cause of death from whiplash injuries.

The third through sixth cervical vertebrae (C3–C6), represented by the vertebra in Figure 7.19c, correspond to the structural pattern of the typical cervical vertebra previously described, with a spinous process, two transverse processes, two superior articular processes, and two inferior articular processes. The seventh cervical vertebra (C7), called the *vertebra prominens*, is somewhat different (Figure 7.19a). It has a non-bifid large spinous process that may be seen and felt at the base of the neck, but otherwise is typical.

The superior and inferior articular facets of the last 5 cervical vertebrae contact each other in the transverse plane. This articular orientation provides the neck region with a varied range of mobility, which includes lateral bending, forward and backward bending, and rotational movements as the articular surfaces slide over one another.

 CHECKPOINT

20. How do the atlas and axis differ from the other cervical vertebrae?

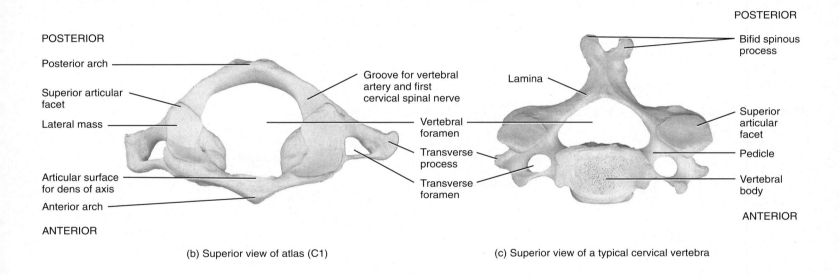

(b) Superior view of atlas (C1)

(c) Superior view of a typical cervical vertebra

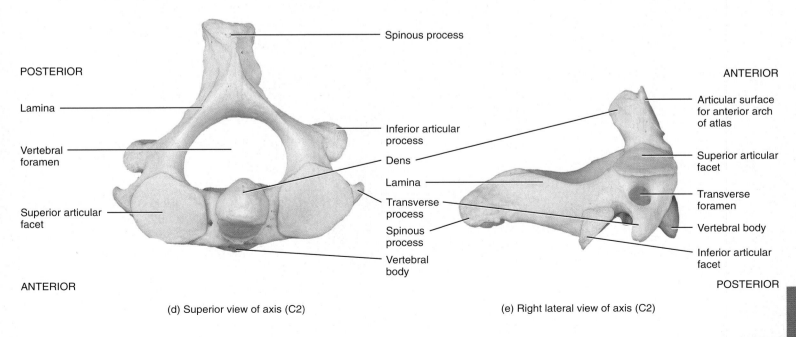

(d) Superior view of axis (C2)

(e) Right lateral view of axis (C2)

? **Which joint permits the movement of the head to signify "no"? Which bones are involved?**

EXHIBIT 7.K Vertebral Regions—Thoracic Vertebrae *(Figure 7.20)*

 OBJECTIVE

• Identify the location and surface features of the thoracic vertebrae.

Description

Thoracic vertebrae (T1–T12; Figure 7.20) are considerably larger and stronger than cervical vertebrae and become progressively larger from superior to inferior. In addition, the spinous processes on T1 through T10 are long, laterally flattened, and directed inferiorly and each one overlaps the process on the vertebra directly inferior to it. In contrast, the spinous processes on T11 and T12 are shorter, broader, and directed more posteriorly. Compared to cervical vertebrae, thoracic vertebrae also have longer and larger transverse processes. They are easily identified by their *costal facets*, which are articular surfaces for the ribs.

Surface Features

The most distinguishing feature of thoracic vertebrae is that they articulate with the ribs. Except for T11 and T12, the transverse processes of thoracic vertebrae have costal facets that form synovial articulations with the *tubercles* of the ribs. Additionally, the bodies of thoracic vertebrae have articular surfaces that form articulations with the *heads* of the ribs. The articular surfaces on the vertebral bodies are called either facets or demifacets. A *facet* is formed when the head of a rib articulates with the body of *one* vertebra. *Demifacets* are formed when the head of a rib articulates with *two* adjacent vertebral bodies. As you can see in Figure 7.20, on each side of the vertebral body T1 has a superior facet for the first rib and an inferior demifacet for the second rib. On each side of the vertebral body of T2–T8 there is a superior demifacet and an inferior demifacet,

Figure 7.20 Thoracic vertebrae.

The thoracic vertebrae are found in the chest region and articulate with the ribs.

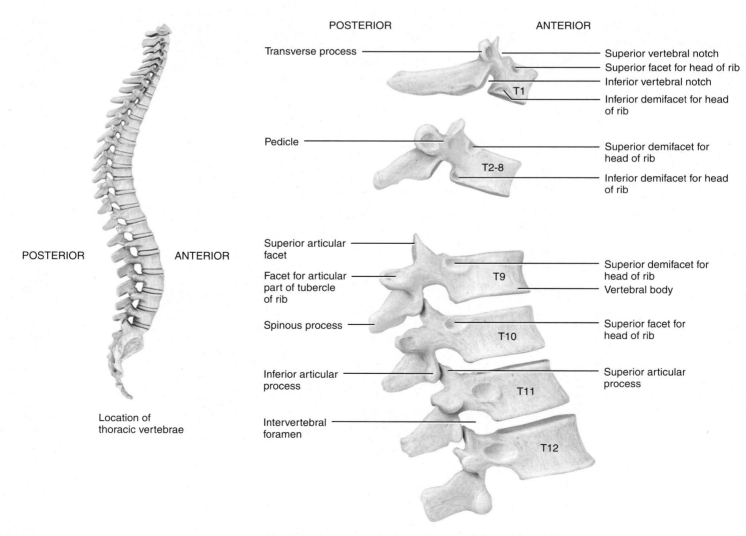

(a) Right lateral view of several articulated thoracic vertebrae

EXHIBIT 7.K **203**

as ribs 2 through 9 articulate with two vertebrae each. T9 has a superior demifacet on each side of the vertebral body, and T10–T12 have a facet on each side of the vertebral body for ribs 10–12. These articulations between the thoracic vertebrae and ribs, called *vertebrocostal joints*, are distinguishing features of thoracic vertebrae. Movements of the thoracic region are limited by thin intervertebral discs and by the attachment of the ribs to the sternum.

The superior and inferior articular processes form flat articular surfaces positioned in the frontal plane. The surface of the superior articular facet faces posteriorly and the surface of the inferior articular facet faces anteriorly. These anterior- and posterior-facing surfaces contact each other and limit forward and backward bending of this region of the vertebral column, while allowing for small amounts of rotational movement and significant lateral bending (as in a side bend).

CLINICAL CONNECTION | *Fractures of the Vertebral Column*

Fractures of the vertebral column often involve C1, C2, C4–T7, and T12–L2. Cervical or lumbar fractures usually result from a flexion–compression type of injury such as might be sustained in landing on the feet or buttocks after a fall or having a weight fall on the shoulders. Cervical vertebrae may be fractured or dislodged by a fall on the head with acute flexion of the neck, as might happen on diving into shallow water, or being thrown from a horse. Spinal cord or spinal nerve damage may occur as a result of fractures of the vertebral column if the fractures compromise the foramina. •

✔ CHECKPOINT

21. Describe several distinguishing features of thoracic vertebrae.

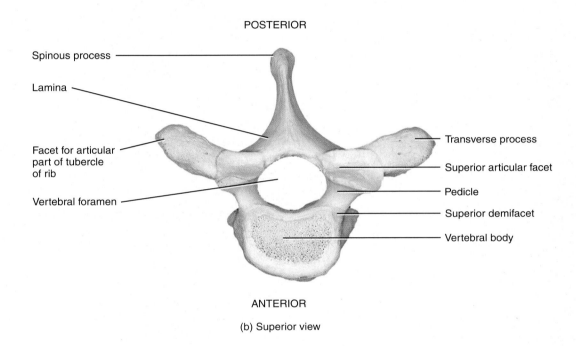

(b) Superior view

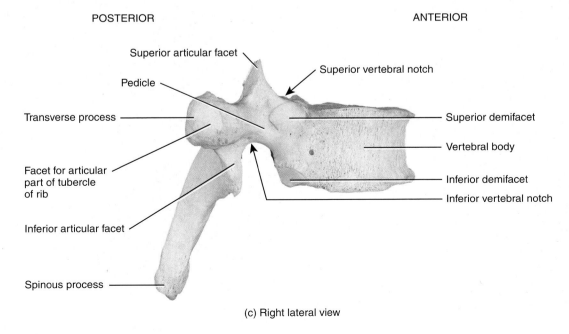

(c) Right lateral view

❓ **Which parts of thoracic vertebrae articulate with the ribs?**

EXHIBIT 7.L Vertebral Regions—Lumbar Vertebrae *(Figure 7.21)*

● **OBJECTIVE**
• Identify the location and surface features of the lumbar vertebrae.

Description

The **lumbar vertebrae** (L1–L5) are the largest and strongest of the unfused vertebrae in the vertebral column (Figure 7.21) because the amount of body weight supported by the vertebrae in-creases toward the inferior end of the backbone. They are readily identifiable by their large size, lack of transverse foramina, and absence of costal articular facets. They have large blocklike bodies with kidney-shaped articular surfaces.

Surface Features

The various projections of the lumbar vertebrae are short and thick. The superior articular processes are directed medially and

Figure 7.21 Lumbar vertebrae.

🔑 Lumbar vertebrae are found in the lower back.

POSTERIOR   ANTERIOR

Location of lumbar vertebrae

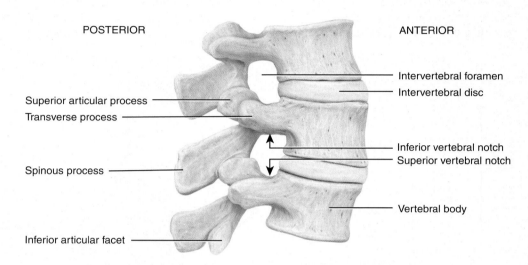

(a) Right lateral view of articulated lumbar vertebrae

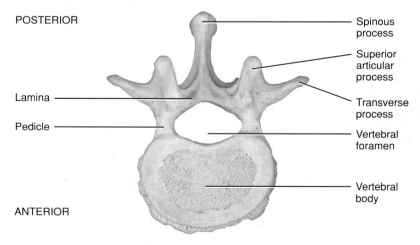

(b) Superior view

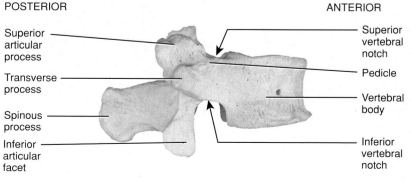

(c) Right lateral view

 Why are the lumbar vertebrae the largest and strongest in the vertebral column?

EXHIBIT 7.L **205**

the inferior articular processes are directed laterally. These surfaces contact each other in the sagittal plane, allowing this region of the vertebral column to bend anteriorly (when you touch your toes) and posteriorly (when the more flexible among you perform a back bend), while limiting rotational movements. The transverse processes are flattened lateral projections that are taller than they are wide, serving as strong lever arms for muscles that bend the vertebral column. The spinous processes are quadrilateral in shape, thick and broad, and project nearly straight posteriorly.

The spinous processes are well-adapted for the attachment of the large back muscles.

A summary of the major structural differences among cervical, thoracic, and lumbar vertebrae is presented in Table 7.5.

● CHECKPOINT
22. What are the distinguishing features of the lumbar vertebrae?

TABLE 7.5

Comparison of Major Structural Features of Cervical, Thoracic, and Lumbar Vertebrae

CHARACTERISTIC	CERVICAL	THORACIC	LUMBAR
Overall structure			
Size	Small	Larger	Largest
Foramina	One vertebral and two transverse	One vertebral	One vertebral
Spinous processes	Slender and often bifid (C2–C6)	Long and fairly thick (most project inferiorly)	Short and blunt (project posteriorly rather than inferiorly)
Transverse processes	Small	Fairly large	Large and blunt
Articular facets for ribs	Absent	Present	Absent
Direction of articular facets			
Superior	Posterosuperior	Posterolateral	Medial
Inferior	Anteroinferior	Anteromedial	Lateral
Size of intervertebral discs	Thick relative to size of vertebral bodies	Thin relative to vertebral bodies	Massive

EXHIBIT 7.M Vertebral Regions—Sacral and Coccygeal Vertebrae *(Figure 7.22)*

OBJECTIVE

• Identify the location and surface features of the sacral and coccygeal vertebrae.

SACRUM

Description

The **sacrum** (SĀ-krum) is a triangular bone formed by the union of five sacral vertebrae (S1–S5) (Figure 7.22a). The **sacral vertebrae** begin to fuse in individuals between 16 and 18 years of age, and the process is usually completed by age 30. The sacrum serves as a strong foundation for the pelvic girdle. It is positioned at the posterior portion of the pelvic cavity where its lateral surfaces, comprised of rib elements, fuse to the two hip bones. The female sacrum is shorter, wider, and more curved between S2 and S3 than the male sacrum (see Table 8.1).

Surface Features

The concave anterior side of the sacrum faces the pelvic cavity. It is smooth and contains four *transverse lines (ridges)* that mark the

joining of the sacral vertebral bodies (Figure 7.22a). At the ends of these lines are four pairs of *anterior sacral foramina*. The lateral portion of the sacrum forms a broad expanded region called the *sacral ala* (ĀL-a=wing; plural is *alae*, ĀL-ē), which is formed by the fused costal (rib) processes of the upper sacral vertebrae.

The convex, posterior surface of the sacrum contains a *median sacral crest*, the fused spinous processes of the upper sacral vertebrae; a *lateral sacral crest*, the fused transverse processes of the sacral vertebrae; and four pairs of *posterior sacral foramina* (Figure 7.22b). These foramina, along with the anterior sacral foramina, connect with the sacral canal to allow passage of nerves and blood vessels. The *sacral canal* is a continuation of the vertebral canal. The laminae of the fifth sacral vertebra, and sometimes the fourth, fail to meet. This leaves an inferior opening in the vertebral canal called the *sacral hiatus* (hī-Ā-tus=opening). On either side of the sacral hiatus is a *sacral cornu* (KOR-noo=horn; plural is *cornua*, KOR-noo-a), an inferior articular process of the fifth sacral vertebra. The sacral cornua are connected by ligaments to the coccyx.

The narrow inferior portion of the sacrum is known as the *apex*. The broad superior portion of the sacrum is called the *base*. The anteriorly projecting border of the base, called the *sacral*

Figure 7.22 Sacrum and coccyx. The sacrum and coccyx are detached in (c) and (d).

🔑 The sacrum is formed by the union of five sacral vertebrae, and the coccyx is formed by the union of usually four coccygeal vertebrae.

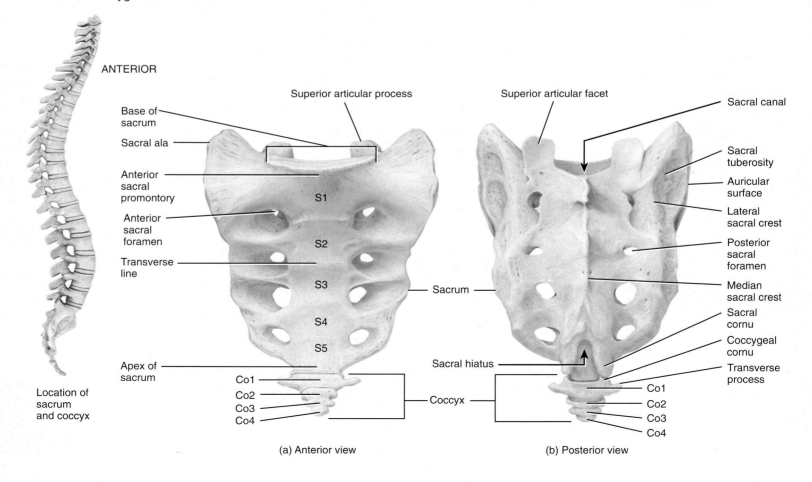

(a) Anterior view   (b) Posterior view

EXHIBIT 7.M **207**

promontory (PROM-on-tō-rē), is one of the points used for measurements of the pelvis. On both lateral surfaces the sacrum has a large, ear-shaped *auricular surface* that articulates with the ilium of each hip bone to form the *sacroiliac joint*. Posterior to the auricular surface is a roughened surface, the *sacral tuberosity*, that contains depressions for the attachment of ligaments. The sacral tuberosity is another surface of the sacrum that unites with the hip bones to form the sacroiliac joints. The *superior articular processes* of the sacrum articulate with the fifth lumbar vertebra, and the base of the sacrum articulates with the body of the fifth lumbar vertebra, to form the *lumbosacral joint*.

CLINICAL CONNECTION | *Caudal Anesthesia*

Anesthetic agents that act on the sacral and coccygeal nerves are sometimes injected through the sacral hiatus, a procedure called **caudal anesthesia**. While this approach is not as common as lumbar epidural block, it is preferred when sacral nerve spread of the anesthetics is preferred over lumbar nerve spread. Because the sacral hiatus is between the sacral cornua, the cornua are important bony landmarks for locating the hiatus. Anesthetic agents also may be injected through the posterior sacral foramina. Since the hiatal and foraminal injection sites are inferior to the lowest portion of the spinal cord, there is little danger of damaging the cord. The lumbar approach is preferred because there is considerable variability in the anatomy of the sacral hiatus, and with advancing age the dorsal ligaments and cornua thicken, making it difficult to identify the hiatal margins. •

COCCYX

Description

The **coccyx** is the terminal end of the vertebral column, indicated in Figure 7.22 as Co1–Co4. In humans it represents a short remnant of longer tail vertebrae found in many other vertebrate species. It is a triangular bone that forms from the fusion of three to five vertebral segments, most commonly representing four fused vertebrae. The first segment is the most characteristic of the vertebral plan, with succeeding segments becoming highly reduced. The **coccygeal vertebrae** fuse when a person is between 20 and 30 years of age.

Surface Features

The dorsal surface of the body of the coccyx contains two long *coccygeal cornua*. They are connected by ligaments to the sacral cornua. The coccygeal cornua are the pedicles and superior articular processes of the first coccygeal vertebra. They are on the lateral surfaces of the coccyx, and are formed by a series of *transverse processes*; the first pair of transverse processes is the largest. The coccyx articulates superiorly with the apex of the sacrum. In females, the coccyx is less curved inferiorly; in males, it is more curved anteriorly (see Table 8.1).

✓ CHECKPOINT

23. How many vertebrae fuse to form the sacrum and the coccyx?

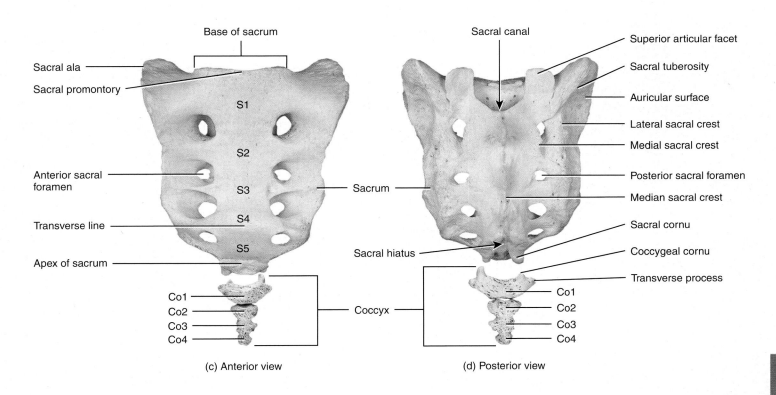

(c) Anterior view   (d) Posterior view

❓ **How many foramina pierce the sacrum, and what is their function?**

7.5 THORAX

OBJECTIVE

• Identify the bones of the thorax and their functions.

The term **thorax** refers to the entire chest region. The skeletal part of the thorax, the **thoracic cage**, is a bony enclosure formed by the sternum, ribs and their costal cartilages, and the bodies of the thoracic vertebrae [see Exhibits 7.N (Figure 7.23)

and 7.O (Figure 7.24)]. The thoracic cage is narrower at its superior end and broader at its inferior end, and is flattened from front to back. It encloses and protects the organs in the thoracic and superior abdominal cavities, provides support for the upper limbs, and plays a role in breathing.

Structures passing between the thoracic cavity and the neck pass through an opening called the **superior thoracic aperture**. Among these structures are the trachea, esophagus, nerves, and blood vessels that supply and drain the head,

---

EXHIBIT 7.N Thoracic Bones—Sternum *(Figure 7.23)*

OBJECTIVE

• Identify the location and surface features of the sternum.

Description

The **sternum**, or breastbone, is a flat, narrow bone located in the center of the anterior thoracic wall that measures about 15 cm (6 in.) in length and consists of three parts (Figure 7.23). The superior part is the **manubrium** (ma-NOO-brē-um=handlelike); the middle and largest part is the **body**, which forms from a se-

ries of smaller fused segments called *sternebrae*; and the inferior, smallest part is the **xiphoid process** (ZĪ-foyd=sword-shaped). The segments of the sternum typically fuse by age 25 and the points of fusion are marked by transverse ridges.

Surface Features

The junction of the manubrium and body forms the *sternal angle*. The manubrium has a depression on its superior surface, the *suprasternal notch*. Lateral to the suprasternal notch are *clavicular*

Figure 7.23 Skeleton of the thorax.

🗝 The bones of the thorax enclose and protect organs in the thoracic cavity and upper abdominal cavity.

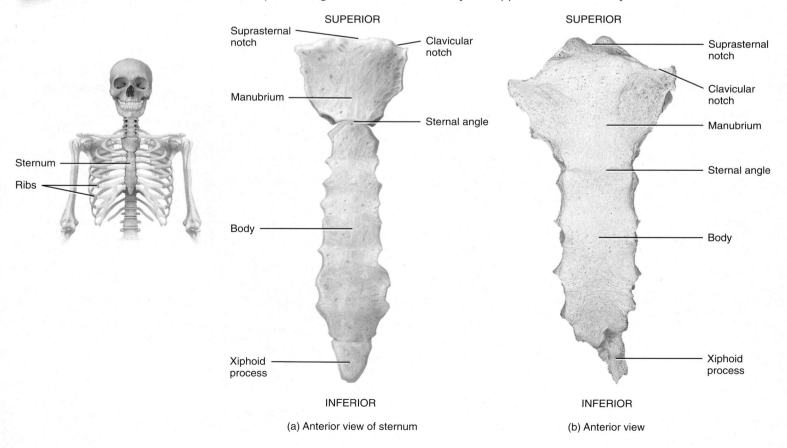

(a) Anterior view of sternum

(b) Anterior view

EXHIBIT 7.N **209**

neck, and upper limbs. The aperture is bordered by the first thoracic vertebra (posteriorly), the first pair of ribs and their cartilages, and the superior border of the manubrium of the sternum. Structures passing between the thoracic cavity and abdominal cavity pass through the **inferior thoracic aperture**. This large opening, which is closed by the diaphragm, allows passage of structures such as the esophagus, nerves, and large blood vessels. The inferior thoracic aperture is bordered

by the twelfth thoracic vertebra (posteriorly), the eleventh and twelfth pair of ribs, the costal cartilages of ribs 7 through 10, and the joint between the body and xiphoid process of the sternum (anteriorly).

 CHECKPOINT

24. What bones form the skeleton of the thorax?
25. What are the functions of the thorax?

notches that articulate with the medial ends of the clavicles to form the *sternoclavicular joints.* The manubrium also articulates with the costal cartilages of the first and part of the second ribs to form the *sternocostal joints.* The body of the sternum articulates directly or indirectly with part of the costal cartilage of the second rib and the costal cartilages of the third through tenth ribs. The xiphoid process consists of hyaline cartilage during infancy and childhood and does not ossify completely until about age 40. No ribs are

attached to it, but the xiphoid process provides attachment for some abdominal muscles. Incorrect positioning of the hands of a rescuer during cardiopulmonary resuscitation (CPR) may fracture the xiphoid process, driving it into internal organs.

 CHECKPOINT

26. What is the clinical significance of the xiphoid process?

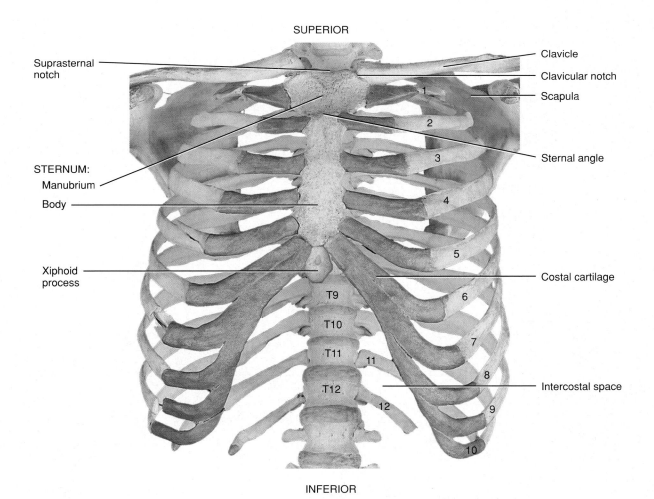

(c) Anterior view of skeleton of thorax

 With which ribs does the body of the sternum articulate?

EXHIBIT 7.O Thoracic Bones—Ribs *(Figure 7.24)*

● OBJECTIVE
- Identify the location and surface features of the ribs.

Description

Twelve pairs of **ribs** numbered 1–12 from superior to inferior give structural support to the sides of the thoracic cavity (see Figure 7.23c). The ribs increase in length from the first through seventh, then decrease in length to the twelfth rib. Each articulates posteriorly with its corresponding thoracic vertebra. Although we describe only the 12 thoracic ribs, there are in reality ribs accompanying each vertebra. The cervical, lumbar, and sacral ribs fuse to their corresponding vertebrae to become the major contributor to what is described as the transverse processes of the cervical and lumbar vertebrae and the alae of the sacrum.

Surface Features

The first through seventh pairs of ribs have a direct anterior attachment to the sternum by a strip of hyaline cartilage called *costal cartilage* (*cost-*=rib). The costal cartilages contribute to the elasticity of the thoracic cage and prevent various blows to the chest from fracturing the sternum and/or ribs. The ribs that have costal cartilages and attach directly to the sternum are called *true (vertebrosternal) ribs*. The remaining five pairs of ribs are termed *false ribs* because their costal cartilages either attach indirectly to the sternum or do not attach to the sternum at all. The cartilages of the eighth, ninth, and tenth pairs of ribs attach to one another and then to the cartilages of the seventh pair of ribs. These false ribs are called *vertebrochondral ribs*. The eleventh and twelfth pairs of ribs are false ribs designated as *floating (vertebral) ribs* because their anterior ends do not attach to the sternum at all. These ribs attach only posteriorly to the thoracic vertebrae. Inflammation of one or more costal cartilages, called *costochondritis*, is characterized by local tenderness and pain in the anterior chest wall that may radiate. The symptoms mimic the chest pain associated with a heart attack (angina pectoris).

Figure 7.24a, c shows the parts of a typical (third through ninth) rib. The *head* is a projection at the posterior end of the rib that contains a pair of articular *facets* (superior and inferior). The facets of the head fit either into a facet on the body of one vertebra or into the demifacets of two adjoining vertebrae to form *vertebrocostal joints*. The *neck* is a constricted portion just lateral to the head. A knoblike structure on the posterior surface where the neck joins the body is called a *tubercle* (TOO-ber-kul). The *nonarticular part* of the tubercle attaches to a ligament (lateral costotransverse

Figure 7.24 The structure of ribs. Each rib has a head, a neck, and a body. The facets and the articular part of the tubercle of a rib articulate with a thoracic vertebra.

🔑 Each rib articulates posteriorly with its corresponding thoracic vertebra.

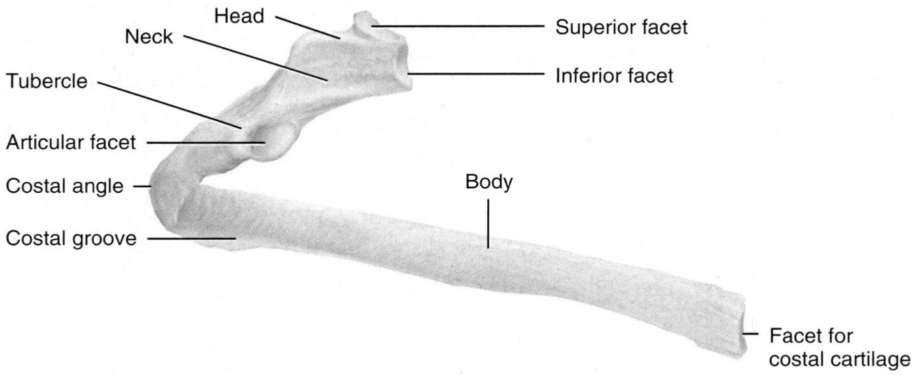

(a) Posterior view of left rib

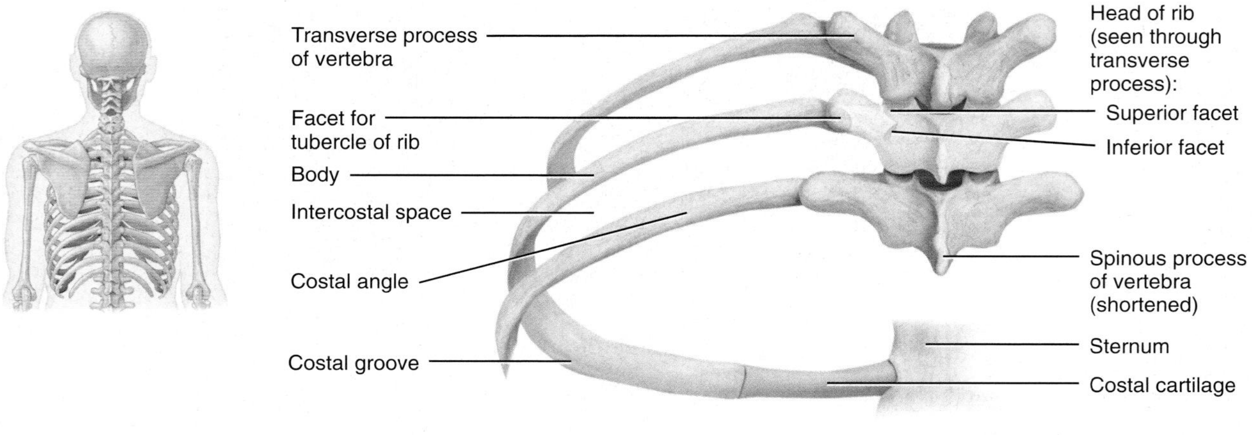

(b) Posterior view of left ribs articulated with thoracic vertebrae and sternum

EXHIBIT 7.0   **211**

ligament), which in turn attaches to the transverse process of a vertebra. The *articular part* of the tubercle articulates with the facet of a transverse process of a vertebra (Figure 7.24d) to form vertebrocostal joints. The *body (shaft)* is the main part of the rib. A short distance beyond the tubercle, an abrupt change in the curvature of the shaft occurs at a point called the *costal angle*. The inner inferior surface of the rib has a *costal groove* that protects the intercostal blood vessels and nerve.

In summary, the posterior portion of the rib is connected to a thoracic vertebra by its head and the articular part of a tubercle. The facet of the head fits either into a facet on the body of one vertebra (T1 only) or into the demifacets of two adjoining vertebrae. The articular part of the tubercle articulates with the facet of the transverse process of the vertebra.

If you examine Figure 7.23c, you will notice that the first rib is the shortest, broadest, and most sharply curved. The first rib is an important landmark because of its close relationship to a number of other structures: the nerves of the *brachial plexus* (the entire nerve supply of the shoulder and upper limb), two ma-

jor blood vessels, the subclavian artery and vein, and two skeletal muscles, the anterior and middle scalene muscles. The superior surface of the first rib has two shallow grooves, one for the subclavian vein and one for the subclavian artery and inferior trunk of the brachial plexus. The second rib is thinner, less curved, and considerably longer than the first. Unlike the paired facets of the typical ribs, the tenth rib has a single articular facet on its head. The eleventh and twelfth ribs also have single articular facets on their heads, but no necks, tubercles, or costal angles.

Spaces between ribs, called *intercostal spaces*, are occupied by intercostal muscles, blood vessels, and nerves. Surgical access to the lungs or other structures in the thoracic cavity is commonly obtained through an intercostal space. Special rib retractors are used to create a wide separation between ribs. The costal cartilages are sufficiently elastic in younger individuals to permit considerable bending without breaking.

 CHECKPOINT

27. How are ribs classified?

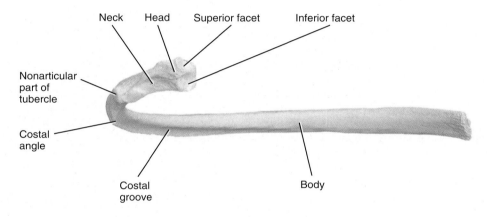

Neck   Head   Superior facet   Inferior facet

Nonarticular part of tubercle

Costal angle

Costal groove

Body

(c) Posterior view

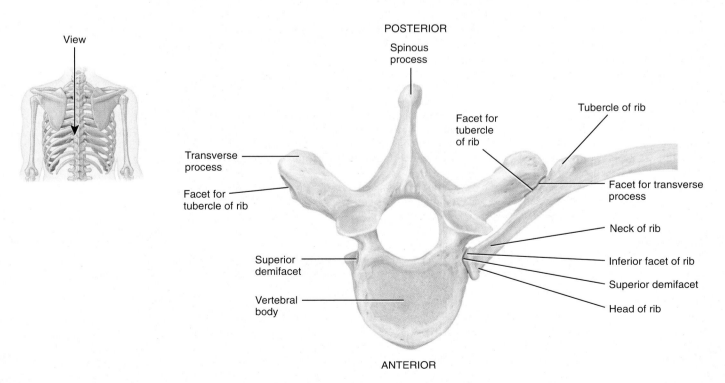

View

POSTERIOR

Spinous process

Tubercle of rib

Facet for tubercle of rib

Transverse process

Facet for transverse process

Facet for tubercle of rib

Neck of rib

Superior demifacet

Inferior facet of rib

Superior demifacet

Vertebral body

Head of rib

ANTERIOR

(d) Superior view of left rib articulated with thoracic vertebra

 How does a rib articulate with a thoracic vertebra?

KEY MEDICAL TERMS ASSOCIATED WITH THE AXIAL SKELETON

Chiropractic (kī-rō-PRAK-tik; *cheir*=hand; *praktikos*=efficient) A holistic health-care discipline that focuses on nerves, muscles, and bones. Treatment involves using the hands to apply specific force to adjust joints of the body (*manual adjustment*), especially the vertebral column.

Craniostenosis (krā′-nē-ō-sten-Ō-sis; *cranio-*=skull; *-stenosis*= narrowing) Premature closure of one or more cranial sutures during the first 18 to 20 months of life, resulting in a distorted skull that can restrict brain growth and development. Premature closure of the sagittal suture produces a long narrow skull; premature closure of the coronal suture results in a broad skull. Surgery is necessary to prevent brain damage.

Craniotomy (krā-nē-OT-ō-mē; *cranio-*=skull; *-tome*=cutting) Surgical procedure in which part of the cranium is removed to extract a blood clot, a brain tumor, or a sample of brain tissue for biopsy.

Laminectomy (lam′-i-NEK-tō-mē; *lamina-*=layer) Surgical procedure involving removal of a vertebral lamina to access the vertebral canal and relieve the symptoms of a herniated disc.

Lumbar spine stenosis (*sten-*=narrowed) Narrowing of the spinal canal in the lumbar part of the vertebral column, due to hypertrophy of surrounding bone or soft tissues. It may be caused by arthritic changes in the intervertebral discs and is a common cause of back and leg pain.

Spinal fusion (FYŪ-zhun) Surgical procedure in which two or more vertebrae of the vertebral column are stabilized with a bone graft or synthetic device to treat a fracture of a vertebra or following removal of a herniated disc.

Whiplash injury Injury to the neck region due to severe hyperextension (backward tilting) of the head followed by severe hyperflexion (forward tilting) of the head, usually associated with a rear-end automobile collision. Symptoms are related to stretching and tearing of ligaments and muscles, vertebral fractures, and herniated vertebral discs.

CHAPTER REVIEW AND RESOURCE SUMMARY   **WileyPLUS**

Review	Resource
Introduction 1. The bones of the skeleton come in a variety of shapes and sizes and must be strong yet relatively lightweight. 2. Bones protect soft body parts and make movement possible; they also serve as landmarks for locating parts of other body systems. 3. The musculoskeletal system is composed of the bones, joints, and muscles working together.	
7.1 Divisions of the Skeletal System 1. The axial skeleton consists of bones arranged along the longitudinal axis. The parts of the axial skeleton are the skull, auditory ossicles (ear bones), hyoid bone, vertebral column, sternum, and ribs. 2. The appendicular skeleton consists of the bones of the upper and lower limbs. The limbs are made up of limb girdles and free limbs.	Anatomy Overview - The Axial Skeleton
7.2 Skull 1. The 22 bones of the skull include cranial bones and facial bones. The eight cranial bones include the frontal, parietal (2), occipital, temporal (2), sphenoid, and ethmoid. The 14 facial bones are the nasal (2), lacrimal (2), palatine (2), inferior nasal conchae (2), vomer, maxillae (2), zygomatic (2), and mandible. 2. The nasal septum consists of the vomer, perpendicular plate of the ethmoid, and septal cartilage. The nasal septum divides the nasal cavity into left and right sides. 3. Seven skull bones form each of the orbits (eye sockets). 4. The foramina of the skull bones provide passages for nerves and blood vessels (Table 7.3). 5. Sutures are immovable joints in adults that connect most bones of the skull. Examples are the coronal, sagittal, lambdoid, and squamous sutures. 6. Paranasal sinuses are cavities in bones of the skull lined with mucous membrane that connect to the nasal cavity. The frontal, sphenoid, and ethmoid bones and the maxillae contain paranasal sinuses. 7. Fontanels are connective tissue-filled spaces between the cranial bones of fetuses and infants. The major fontanels are the anterior, posterior, anterolaterals (2), and posterolaterals (2). After birth, the fontanels fill in with bone and become sutures. 8. The three cranial fossae are regions within the cranium that contain depressions for brain convolutions, grooves for cranial blood vessels, and numerous foramina. 9. Exhibits 7.A through 7.I provide details about the bones of the skull.	Anatomy Overview - Facial Bones Anatomy Overview - Fibrous Joints and Sutures Figure 7.6 - Sphenoid Bone Figure 7.12 - Details of the Orbit (Eye Socket)
7.3 Hyoid Bone 1. The hyoid bone is a U-shaped bone that does not articulate with any other bone; it supports the tongue and provides attachment for some tongue muscles and for some muscles of the throat and neck.	Anatomy Overview - The Hyoid Bone
7.4 Vertebral Column 1. The vertebral column, sternum, and ribs constitute the skeleton of the body's trunk. The 26 bones of the adult vertebral column are the cervical vertebrae (7), the thoracic vertebrae (12), the lumbar vertebrae (5), the sacrum (5 fused sertebrae), and the coccyx (usually 4 fused vertebrae). 2. The adult vertebral column contains four normal curves (cervical, thoracic, lumbar, and sacral) that give strength, support, and balance. 3. Between the bodies of adjacent vertebrae are the cartilaginous intervertebral discs. They form strong joints that permit various movements and absorb vertical shock.	Anatomy Overview - The Vertebral Column Figure 7.17 - Vertebral Column Table 7.5 - Comparison of Major Structural Features of Cervical, Thoracic, and Lumbar Vertebrae Figure 7.22 - Sacrum and Coccyx

Review   **Resource**

4. The vertebrae are similar in structure, each usually consisting of a vertebral body, vertebral arch, and seven processes. Vertebrae in the different regions of the column vary in size, shape, and detail.
5. Exhibits 7.J through 7.M provide details about the bones of the vertebral column.

7.5 Thorax

1. The thoracic skeleton consists of the sternum, ribs and costal cartilages, and thoracic vertebrae. The ribs are classified as true ribs (pairs 1–7) and false ribs (8–12).
2. The thoracic cage protects vital organs in the chest area and upper abdomen.
3. Exhibits 7.N and 7.O provide details about the bones of the thorax.

Anatomy Overview - The Thorax
Exercise - Bones, Bones, Bones
Figure 7.23 - Skeleton of the Thorax

CRITICAL THINKING QUESTIONS

1. Four-year-old Pattee finds a soft spot on her baby brother's skull and announces that the baby needs to go back because "it's not finished yet." Explain the presence of soft spots in the infant.

2. Dave has a tumor on his pituitary gland that requires surgery. The operation will require exceptional surgical skill both to reach the gland and to avoid serious, possibly fatal, complications. Describe the location of the pituitary gland and the associated anatomical features that make such an operation so delicate.

3. The ad reads "New Postureperfect Mattress! Keeps spine perfectly straight—just like when you were born! A straight spine means a great sleep!" Would you buy a mattress from this company?

4. Trina slipped on the ice, fell backward, and hit her head on the ice. Why might a serious blow to the occipital bone be fatal?

5. John had complaints of severe headaches and a yellow-green nasal discharge. His physician ordered x-rays of his head. John was surprised to see several fuzzy-looking holes on the x-ray. Why does John have holes in his head?

? ANSWERS TO FIGURE QUESTIONS

7.1 The skull and vertebral column are part of the axial skeleton. The clavicle, shoulder girdle, humerus, pelvic girdle, and femur are part of the appendicular skeleton.

7.2 The frontal, parietal, sphenoid, ethmoid, and temporal bones are cranial bones.

7.3 The parietal and temporal bones are on either side of the squamous suture. The parietal and occipital bones are on either side of the lambdoid suture. The parietal and frontal bones surround the coronal suture.

7.4 The temporal bone articulates with the parietal, sphenoid, zygomatic, and occipital bones.

7.5 The parietal bones form the posterior, lateral portion of the cranium.

7.6 From the crista galli of the ethmoid bone, the sphenoid articulates with the frontal, parietal, temporal, occipital, temporal, parietal, and frontal bones, ending again at the crista galli.

7.7 The perpendicular plate of the ethmoid bone forms the superior part of the nasal septum, and the lateral masses compose most of the medial walls of the orbits.

7.8 The nasal conchae increase the surface area of the nasal cavity to help swirl and filter air before it passes to the lungs.

7.9 The palatine processes of the maxillae contribute to the hard palate, which separates the nasal and oral cavities.

7.10 The mandible is the only movable skull bone other than the auditory ossicles and occipital bone.

7.11 The nasal septum divides the nasal cavity into right and left sides.

7.12 Bones forming the orbit are the frontal, sphenoid, zygomatic, maxilla, lacrimal, ethmoid, and palatine. The palatine bone is not visible in this figure.

7.13 The paranasal sinuses produce mucus and serve as resonating chambers for vocalization.

7.14 The paired anterolateral fontanels are bordered by four different skull bones: the frontal, parietal, temporal, and sphenoid bones.

7.15 The posterior cranial fossa, which houses the cerebellum, pons, and medulla oblongata, is the largest of the three cranial fossae.

7.16 The hyoid bone is the only bone in the human skeleton that does not articulate with any other bone.

7.17 The thoracic and sacral curves of the vertebral column are concave relative to the anterior of the body.

7.18 The vertebral foramina enclose the spinal cord; the intervertebral foramina provide spaces for spinal nerves to exit the vertebral column.

7.19 The atlas moving on the axis at the atlantoaxial joint permits movement of the head to signify "no."

7.20 The facets and demifacets on the vertebral bodies of the thoracic vertebrae articulate with the heads of the ribs, and the facets on the transverse processes of these vertebrae articulate with the tubercles of the ribs.

7.21 The lumbar vertebrae are the largest and strongest in the body because the amount of weight supported by vertebrae increases toward the inferior end of the vertebral column.

7.22 There are four pairs of sacral foramina, for a total of eight. Each anterior sacral foramen joins a posterior sacral foramen at the intervertebral foramen. Nerves and blood vessels pass through these tunnels in the bone.

7.23 The body of the sternum articulates directly or indirectly with ribs 2–10.

7.24 The facet on the head of a rib fits into a facet on the body of a vertebra, and the articular part of the tubercle of a rib articulates with the facet of the transverse process of a vertebra.

8 THE SKELETAL SYSTEM: THE APPENDICULAR SKELETON

INTRODUCTION As noted in Chapter 7, the two main divisions of the skeletal system are the axial skeleton and the appendicular skeleton. The axial skeleton is the skeletal axis, or core of the body, and helps protect the internal organs. The focus of this chapter is the appendicular skeleton, which consists of the upper limbs and the lower limbs; its primary function is movement. Because they have similar developmental programs, the upper and lower limbs have much in common. Each limb is composed of limb girdles and free limbs. The upper limbs consist of the pectoral girdles and free upper limbs; the lower limbs consist of the pelvic girdles and free lower limbs. Composed of broad, flat bones that form sturdy anchors, the girdles attach the mobile free limbs to the axial skeleton. When you compare the first segments of the free limbs—the arm in the upper limb and the thigh in the lower limb—you will note that there is a single large bone. Proceeding distally to the second segments, the forearm in the upper limb and the leg in the lower limb both have two parallel bones. At the junctions of these second segments with the hand and foot—the wrist and the ankle—there are numerous small bones (8 in the wrist and 7 in the ankle). Finally, the hands and the feet have the same number and arrangement of bones, which form the fingers and toes.

There is one major difference between the upper and lower limbs. The pelvic girdles of the lower limb are firmly anchored to the vertebral column via a strong ligamentous joint, but the pectoral girdles of the upper limb do not form any joints with the vertebral column; they are only weakly joined to the axial skeleton via the junction of the clavicle (collar bone) with the sternum. This primitive feature of all land vertebrates marks the major functional difference in the limbs; the hindlimbs are the locomotor limbs, and the forelimbs are the steering column. When humans took advantage of this primitive difference in limb structure and function and raised the more mobile non-locomotor limb off the ground, we took advantage of its tremendous range of mobility to use the upper limb in many diverse ways. As you progress through this chapter and the two chapters that follow, you will see how the bones of the appendicular skeleton described in this chapter are connected with one another (Chapter 9) and with skeletal muscles (Chapter 10), making possible an array of movements ranging from walking and writing to using a computer, dancing, swimming, and playing a musical instrument. •

Did you ever wonder what causes "runner's knee"?

8.1 SKELETON OF THE UPPER LIMB

● OBJECTIVE

- Identify the bones of the skeleton of the upper limb and explain their functions.

Each upper limb skeleton consists of 32 bones, which form two distinct regions: (1) the pectoral girdle and (2) the free upper limb. **Pectoral** refers to the chest or breast. The **pectoral** (PEK-tō-ral) or **shoulder girdles** attach the bones of the free upper limbs to the axial skeleton (Figure 8.1a, b). Each of the two

Figure 8.1 Skeleton of the right upper limb—pectoral (shoulder) girdle and free upper limb.

Each upper limb skeleton consists of 32 bones.

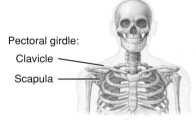

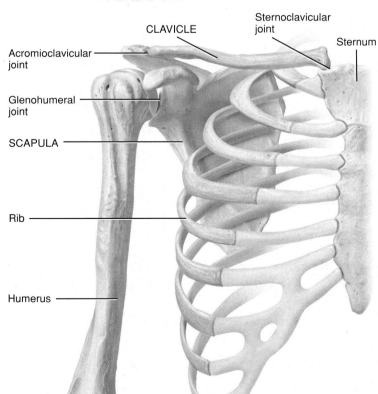

Pectoral girdle:
- Clavicle
- Scapula

Acromioclavicular joint

CLAVICLE

Sternoclavicular joint

Sternum

Glenohumeral joint

SCAPULA

Rib

Humerus

(a) Anterior view of pectoral girdle

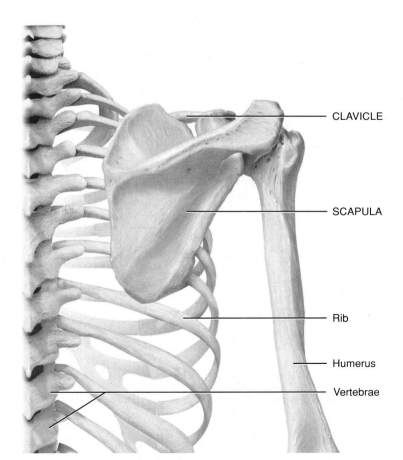

CLAVICLE

SCAPULA

Rib

Humerus

Vertebrae

(b) Posterior view of pectoral girdle

FIGURE 8.1 CONTINUES ▶

⬛ FIGURE **8.1** CONTINUED ▶

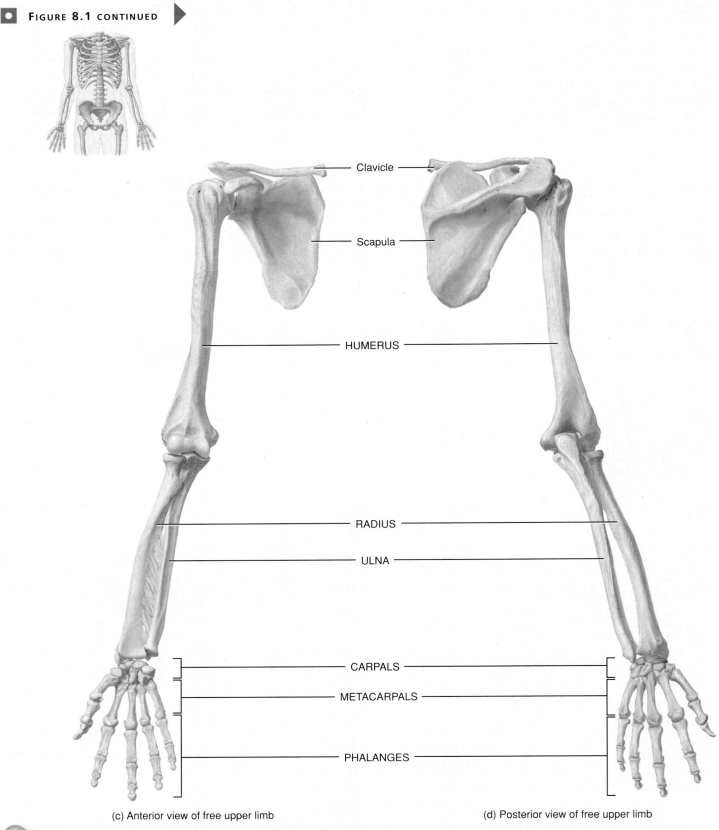

(c) Anterior view of free upper limb

(d) Posterior view of free upper limb

? **Which bones make up the pectoral girdle and free upper limb?**

pectoral girdles consists of a clavicle and a scapula. The clavicle is the anterior bone and the scapula is the posterior bone. As noted in the introduction, the pectoral girdles do not articulate with the vertebral column; they are held in position and supported by a group of large muscles that extend from the vertebral column and rib cage to the scapula. Each **free upper limb (extremity)** has 30

bones in three locations—the arm (humerus); the forearm (ulna and radius); and the hand [8 carpal bones in the carpus (wrist); 5 metacarpal bones in the metacarpus (palm); and 14 phalanges in the digits (fingers and thumb)] (Figure 8.1c).

✔ CHECKPOINT
1. What is the function of the pectoral girdle?

EXHIBIT 8.A | Pectoral Girdle—Clavicle *(Figure 8.2)*

 OBJECTIVE

• Describe the location and surface features of the clavicle.

Description

Each slender, S-shaped **clavicle** (KLAV-i-kul=key), or *collarbone*, lies horizontally across the anterior part of the thorax superior to the first rib (see Figure 8.1a). The medial half of the clavicle is convex anteriorly (curves toward you when viewed in the anatomical position), and the lateral half is concave anteriorly (curves away from you). The junction of the clavicle's two curves is its weakest point. It is typically smoother and straighter in females, and rougher and more curved in males. The clavicle forms the anterior *strut* (a rod or bar that resists compression) of the pectoral girdle that props the shoulder joint away from the rib cage. It is subcutaneous (just under the skin) and easily palpable throughout its length.

Surface Features

The medial end of the clavicle, called the *sternal end* (Figure 8.2), is rounded and articulates with the manubrium of the sternum to form the *sternoclavicular joint* (see Figure 8.1a). The broad, flat, lateral end, the *acromial end* (a-KRŌ-mē-al) (Figure 8.2), articulates with the acromion of the scapula at the *acromioclavicular joint* (see Figure 8.1a).

The *conoid tubercle* (KŌ-noyd=conelike) on the inferior surface of the lateral end of the bone is a point of attachment for the conoid ligament. As its name implies, the *impression for the costoclavicular ligament* on the inferior surface of the sternal end is a point of attachment for the costoclavicular ligament, a ligament that connects the first rib to the clavicle.

✓ **CHECKPOINT**

2. Which joints are formed by the articulation of the clavicle with other bones? Which areas of the clavicle are involved in each joint?

Figure 8.2 Right clavicle.

🔑 The clavicle articulates medially with the manubrium of the sternum and laterally with the acromion of the scapula.

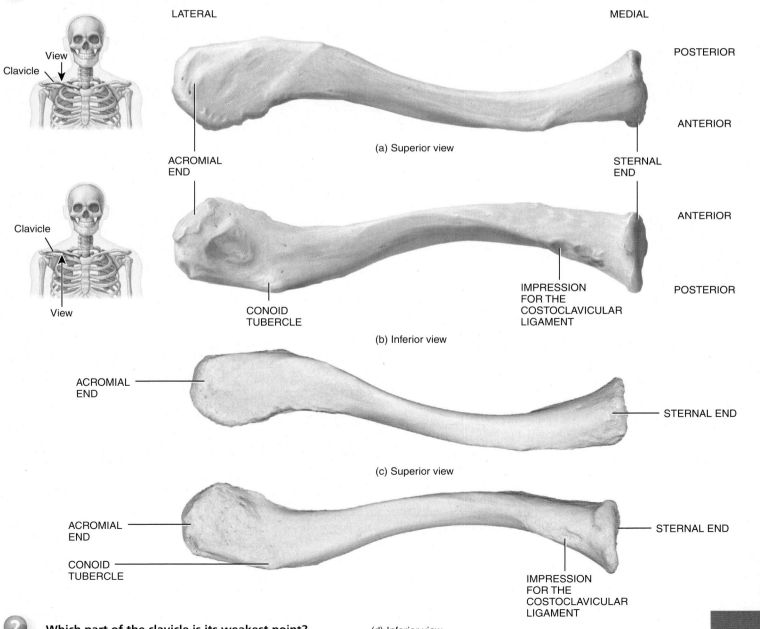

❓ **Which part of the clavicle is its weakest point?**

217

EXHIBIT 8.B Pectoral Girdle—Scapula *(Figure 8.3)*

● OBJECTIVE
• Describe the location and surface features of the scapula.

Description

Each **scapula** (SCAP-ū-la; plural is *scapulae*), or *shoulder blade*, is a large, triangular, flat bone with a ridge on its posterior surface. The scapula occupies the superior part of the posterior thorax between the levels of the second and seventh ribs a few finger breadths lateral to the vertebral column (see Figure 8.1a, b).

Surface Features

A prominent ridge called the *spine* runs diagonally across the posterior surface of the scapula (Figure 8.3c–f). The lateral end of the spine projects as a flattened, expanded process called the *acromion* (a-KRŌ-mē-on; *acro-*=peak; *omos*=shoulder), easily felt as the high point, or peak, of the shoulder. Tailors measure the length of the free upper limb from the acromion. The acromion articulates with the acromial end of the clavicle to form the *acromioclavicular joint*. Inferior to the acromion is a shallow depression, the *glenoid cavity* (Figure 8.3a, d, f), that accepts the head of the humerus (arm bone) to form the *glenohumeral joint*, or shoulder joint (see Figure 8.1a).

Figure 8.3 **Right scapula (shoulder blade).**

🔑 The glenoid cavity of the scapula articulates with the head of the humerus to form the glenohumeral (shoulder) joint.

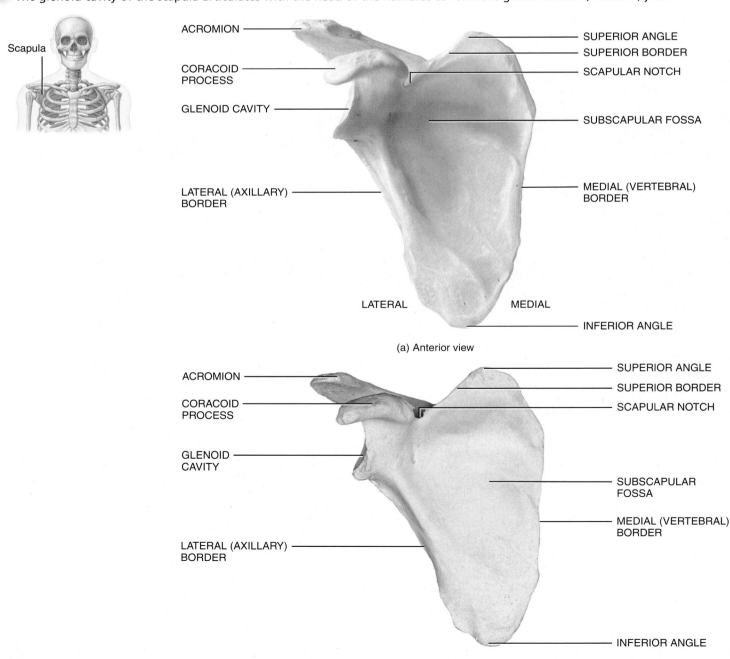

Scapula

ACROMION — SUPERIOR ANGLE
SUPERIOR BORDER
CORACOID PROCESS — SCAPULAR NOTCH
GLENOID CAVITY — SUBSCAPULAR FOSSA
LATERAL (AXILLARY) BORDER — MEDIAL (VERTEBRAL) BORDER
LATERAL   MEDIAL
INFERIOR ANGLE

(a) Anterior view

ACROMION — SUPERIOR ANGLE
SUPERIOR BORDER
CORACOID PROCESS — SCAPULAR NOTCH
GLENOID CAVITY — SUBSCAPULAR FOSSA
MEDIAL (VERTEBRAL) BORDER
LATERAL (AXILLARY) BORDER
INFERIOR ANGLE

(b) Anterior view

EXHIBIT 8.B **219**

The thin edge of the scapula closer to the vertebral column is called the *medial (vertebral) border*. It lies about 5 cm (2 in.) from the vertebral column. The thick edge of the scapula closer to the arm is called the *lateral (axillary) border*. The medial and lateral borders join at the *inferior angle*. The superior edge of the scapula, called the *superior border*, joins the vertebral border at the *superior angle*. The *scapular notch* is a prominent indentation along the superior border through which the suprascapular nerve passes.

At the lateral end of the superior border of the scapula, the tendons of muscles attach to a projection of the anterior surface called the *coracoid process* (KOR-a-koyd=like a crow's beak).

Superior and inferior to the spine are two fossae: the *supraspinous fossa* (sū-pra-SPĪ-nus) and the *infraspinous fossa* (in-fra-SPĪ-nus). The supraspinous fossa and infraspinous fossa serve as surfaces of attachment for the supraspinatus and infraspinatus muscles of the shoulder, respectively. On the anterior surface is a slightly hollowed-out area called the *subscapular fossa*, an attachment surface for the subscapularis muscle.

 CHECKPOINT

3. Which joints are formed by articulations of the scapula with other bones? What are the names of the scapular surfaces that form each joint?

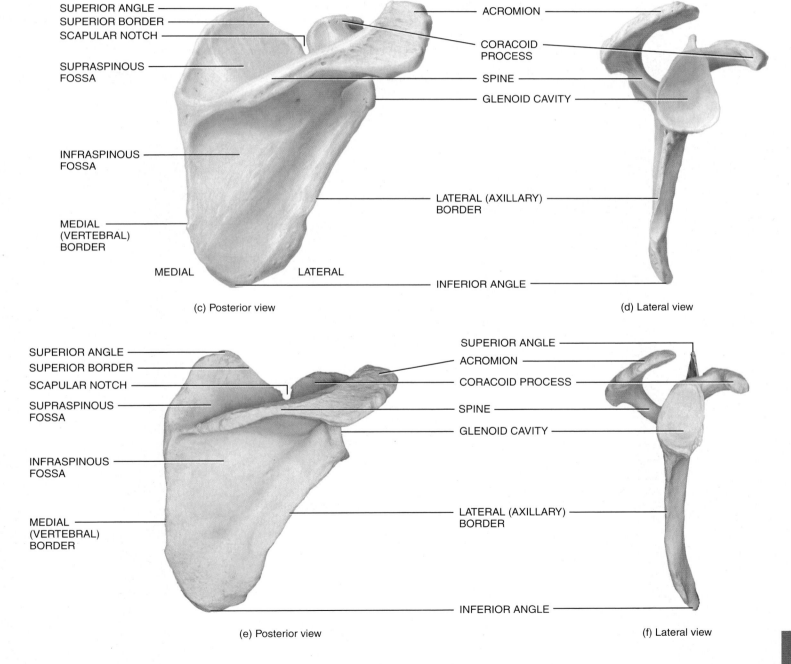

(c) Posterior view

(d) Lateral view

(e) Posterior view

(f) Lateral view

Which part of the scapula forms the high point of the shoulder?

EXHIBIT 8.C Skeleton of the Arm—Humerus *(Figure 8.4)*

OBJECTIVE
• Identify the location and surface landmarks of the humerus.

Description

The **humerus** (HŪ-mer-us), or arm bone, is the longest and largest bone of the free upper limb (Figure 8.4). It has a ball-like proximal end with two prominent projections of bone at the base

Humerus

of the ball, a cylindrical, tubular shaft that makes up the majority of its length, and an expanded flattened distal end. It articulates proximally with the scapula and distally with both the ulna and the radius to form the elbow joint.

Surface Features

The proximal end of the humerus features a rounded *head* that articulates with the glenoid cavity of the scapula to form the glenohumeral (shoulder) joint. Distal to the head is the *anatomical neck*, which is visible as an oblique groove. The anatomical neck is the former site of the epiphyseal (growth) plate in an adult humerus. The *greater tubercle* is a lateral projection distal to the anatomical neck. It is the most laterally palpable bony landmark of the shoulder region and is immediately inferior to the palpable acromion

Figure 8.4 Right humerus in relation to the scapula, ulna, and radius.

🔑 The humerus is the longest and largest bone of the free upper limb.

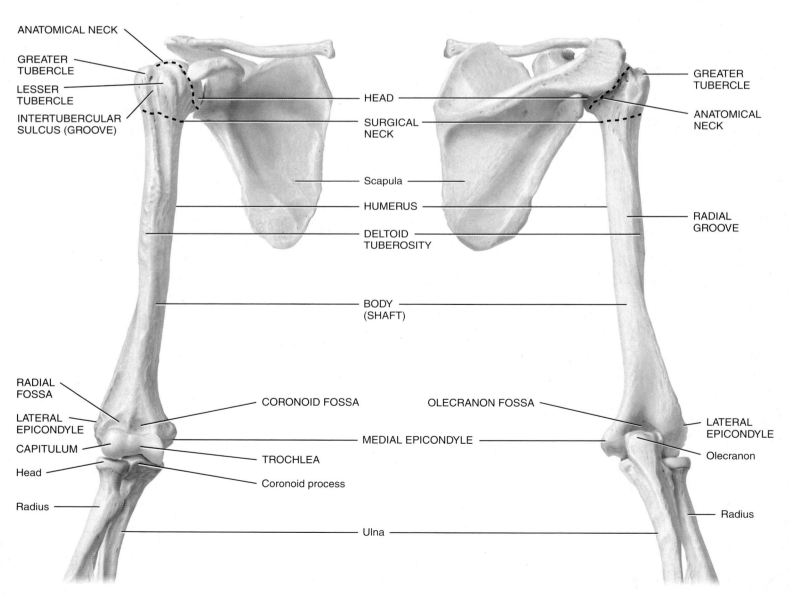

(a) Anterior view   (b) Posterior view

EXHIBIT 8.C **221**

of the scapula mentioned earlier. The *lesser tubercle* projects anteriorly. Between both tubercles runs an *intertubercular sulcus.* The *surgical neck* is a constriction in the humerus just distal to the tubercles, where the head tapers to the shaft; it is so named because fractures often occur here.

The *body (shaft)* of the humerus is roughly cylindrical at its proximal end, but it gradually becomes triangular until it is flattened and broad at its distal end. Laterally, at the middle portion of the shaft, there is a roughened, V-shaped area called the *deltoid tuberosity.* This area serves as a point of attachment for the tendon of the deltoid muscle. The *radial groove* runs along the posterior margin of the deltoid tuberosity on the posterior surface of the humerus. This groove ends at the inferior margin of the deltoid tuberosity and contains the radial nerve.

Several prominent features are evident at the distal end of the humerus. The *capitulum* (ka-PIT-ū-lum; *capit-*=head) is a rounded knob on the lateral aspect of the bone that articulates with the head of the radius. The *radial fossa* is an anterior depression above the capitulum that articulates with the head of the radius when the forearm is flexed (bent). The *trochlea* (TROK-lē-a=pulley), located medial to the capitulum, is a spool-shaped surface that articulates with the ulna. The *coronoid fossa* (KOR-ō-noyd=crown-shaped) is an anterior depression that receives the coronoid process of the ulna when the forearm is flexed. The *olecranon fossa* (ō-LEK-ranon=elbow) is the large posterior depression that receives the olecranon of the ulna when the forearm is extended (straightened). The *medial epicondyle* and *lateral epicondyle* are rough projections on either side of the distal end of the humerus to which the tendons of most muscles of the forearm are attached. The ulnar nerve may be palpated by rolling a finger over the skin surface superficial to the posterior surface of the medial epicondyle. This nerve is the one that makes you feel a very severe pain when you hit your elbow, which for some reason is commonly referred to as the funnybone, even though this event is anything but funny.

 CHECKPOINT

4. Distinguish between the anatomical neck and the surgical neck of the humerus.

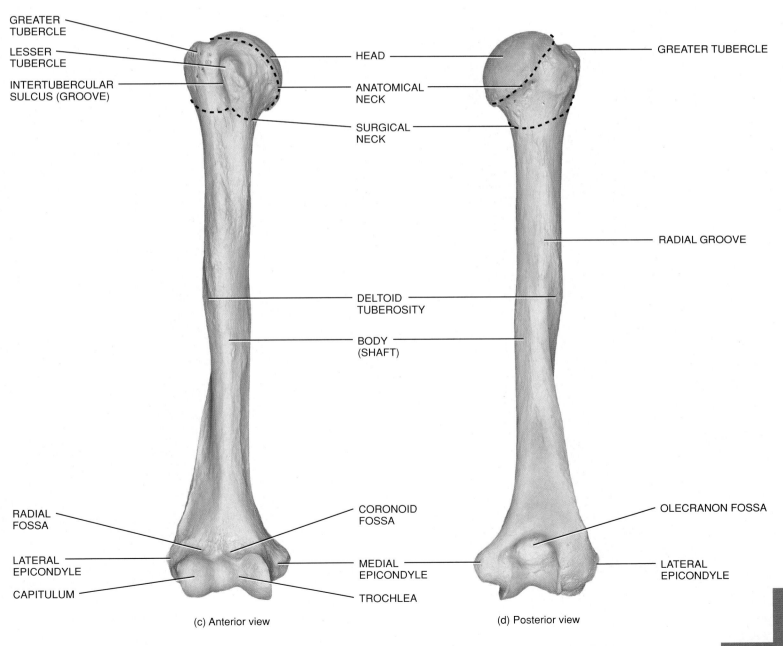

(c) Anterior view

(d) Posterior view

Which parts of the humerus articulate with the radius at the elbow? With the ulna at the elbow?

EXHIBIT 8.D Skeleton of the Forearm—Ulna and Radius *(Figure 8.5)*

● OBJECTIVE
• Identify the location and surface landmarks of the ulna and radius.

ULNA

Description

The **ulna** is located on the medial aspect (the little-finger side) of the forearm and is longer than the radius (Figure 8.5a–d). It is sometimes convenient to use an aid to help you learn and re-

member new information. Such a study aid is called a *mnemonic device* (ne-MON-ik=memory) or simply a *mnemonic*. To help you remember the location of the ulna in relation to the hand, one mnemonic is "p.u." (the *p*inky is on the *u*lna side). The ulna is thick and notched at its proximal end, and its wide triangular shaft tapers to become more narrow and cylindrical distally.

Surface Features

At the proximal end of the ulna (Figure 8.5b, d, f, g) is the *olecranon*, which forms the prominence of the elbow. The *coronoid*

Figure 8.5 **Right ulna and radius.**

In the forearm, the longer ulna is on the medial side, and the shorter radius is on the lateral side.

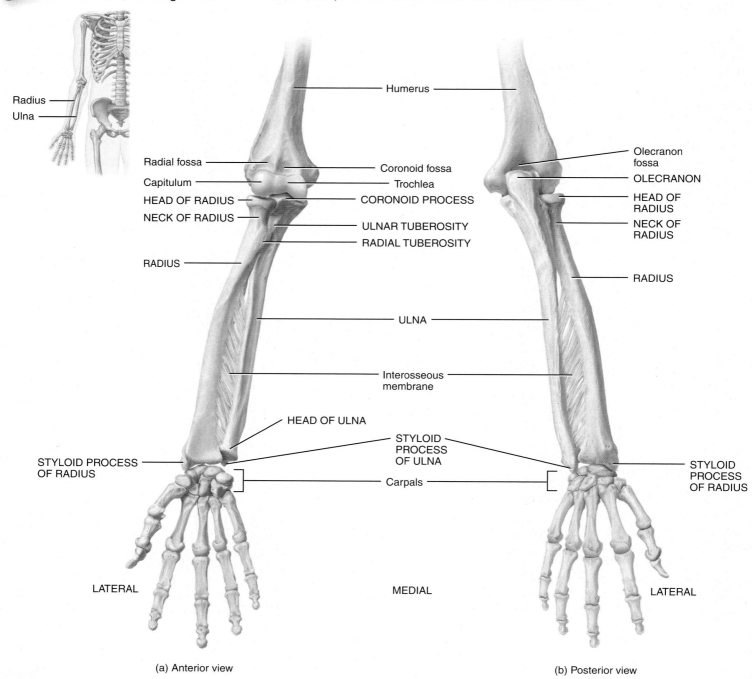

(a) Anterior view   (b) Posterior view

EXHIBIT 8.D **223**

process (*corone*=crow) (Figure 8.5a, c, e, f, g) is an anterior projection distal to a large notch, the *trochlear notch*. This notch, on the anterior side of the olecranon, receives the trochlea of the humerus to form part of the elbow joint (Figure 8.5, c, f, g). The *radial notch* is a depression that is lateral and inferior to the trochlear notch and articulates with the head of the radius. Just inferior to the coronoid process is the *ulnar tuberosity*. The distal end of the ulna consists of a *head* that is separated from the wrist by a disc of fibrocartilage. A *styloid process* is located on the posterior side of the ulna's distal end.

RADIUS

Description

The **radius**, the shorter of the two forearm bones, is located on the lateral aspect (thumb side) of the forearm (Figure 8.5a–d). In contrast to the ulna, the radius is narrow at its proximal end and widens at its distal end.

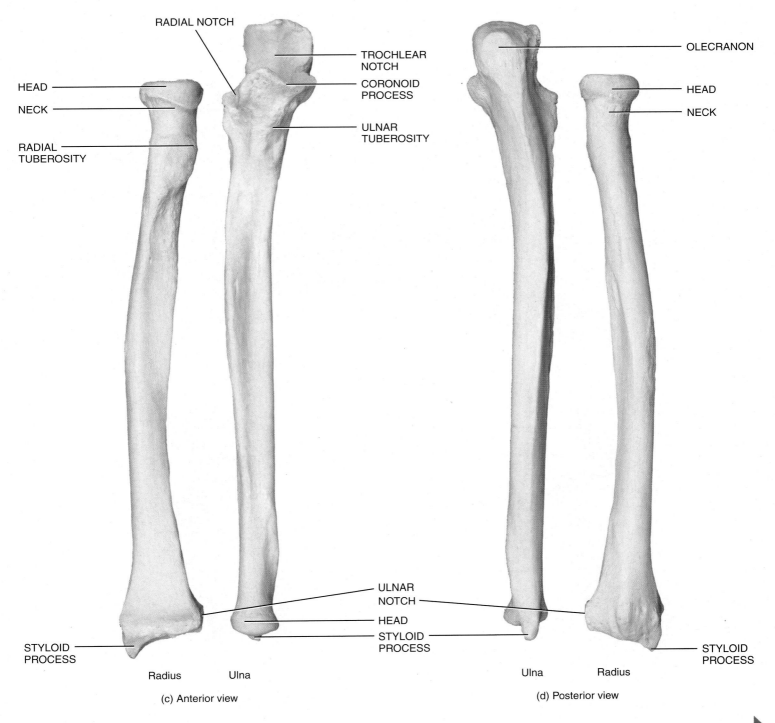

(c) Anterior view

(d) Posterior view

FIGURE 8.5 CONTINUES ▶

EXHIBIT 8.D Skeleton of the Forearm—Ulna and Radius *(Figure 8.5)* CONTINUED

Surface Features

The proximal end of the radius has a disc-shaped *head* that articulates with the capitulum of the humerus and the radial notch of the ulna. Inferior to the head is the constricted *neck*. A roughened area inferior to the neck on the anteromedial side, called the *radial tuberosity*, is a point of attachment for the tendon of the biceps brachii muscle. The shaft of the radius widens distally to form a *styloid process* on the lateral side. Fracture of the distal end of the radius is the most common fracture in adults older than 50 years, typically occurring during a fall.

The ulna and radius articulate with the humerus at the *elbow joint*. The articulation occurs in two places: where the head of

the radius articulates with the capitulum of the humerus (Figure 8.5e), and where the trochlear notch of the ulna receives the trochlea of the humerus (Figure 8.5e, f, g).

The ulna and the radius connect with one another at three sites. First, a broad, flat, fibrous connective tissue called the *interosseous membrane* (in′-ter-OS-ē-us; *inter-*=between, *osse-*= bone) joins the shafts of the two bones (Figure 8.5a, b, e, h). This membrane also provides a site of attachment for some of the deep skeletal muscles of the forearm. The ulna and radius articulate at their proximal and distal ends. Proximally, the head of the radius articulates with the ulna's *radial notch*, a depression that is lateral and inferior to the trochlear notch (Figure 8.5f–g). This articulation is the *proximal radioulnar joint*. Distally, the

○ **FIGURE 8.5 CONTINUED** ▶

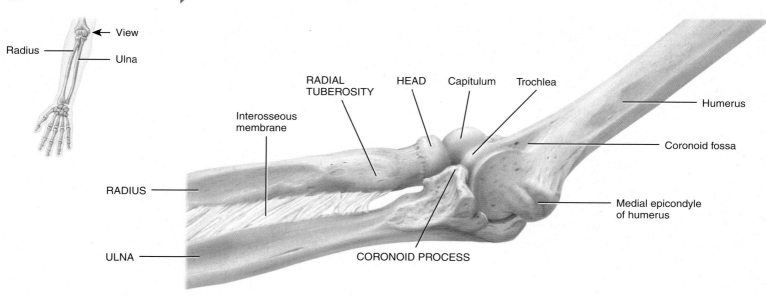

(e) Medial view in relation to humerus

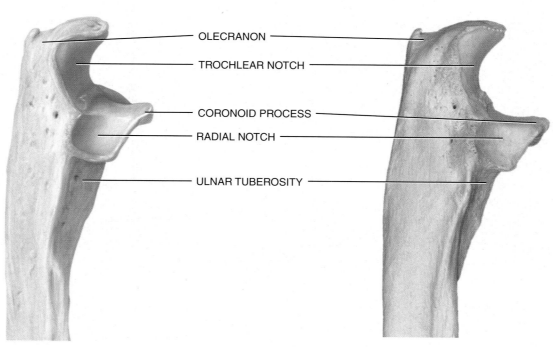

(f) Lateral view of proximal end of ulna (g)

EXHIBIT 8.E **225**

head of the ulna articulates with a narrow concavity, the *ulnar notch* of the radius (Figure 8.5h, i). This articulation is the *distal radioulnar joint*. Finally, the distal end of the radius articulates with two bones of the wrist—the lunate and the scaphoid—to form the *radiocarpal (wrist) joint*.

✓ CHECKPOINT
5. How many joints are formed between the ulna and radius, what are their names, and what surface features are involved?

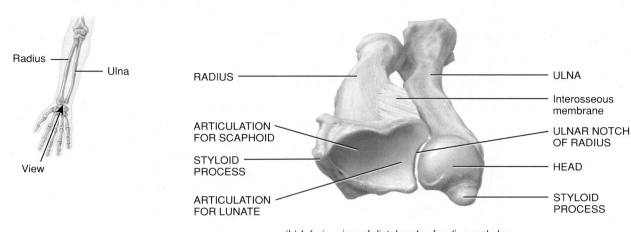

(h) Inferior view of distal ends of radius and ulna

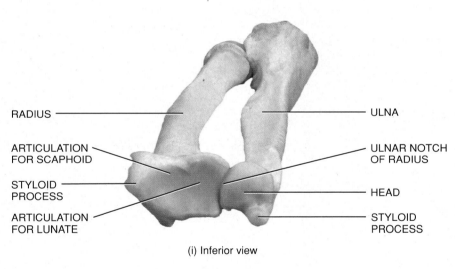

(i) Inferior view

 What part of the ulna is called the "elbow"?

EXHIBIT 8.E   Skeleton of the Hand   *(Figure 8.6)*

● OBJECTIVE
• Identify the location and surface landmarks of the bones of the hand.

CARPALS

Description

The **carpus** (wrist) is the proximal region of the hand and consists of eight small bones, the **carpals,** joined to one another by ligaments (Figure 8.6). Articulations between carpal bones are called *intercarpal joints*. The carpals are arranged in two transverse rows

of four bones each. Their names reflect their shapes; a mnemonic for learning the names of the carpal bones is shown in Figure 8.6.

Surface Features

The carpals in the proximal row, from lateral to medial, are the **scaphoid** (SKAF-oyd=boatlike), **lunate** (LOO-nāt=moon-shaped), **triquetrum** (trī-KWĒ-trum=three-cornered), and **pisiform** (PIS-i-form=pea-shaped). The proximal row of carpal bones articulates with the distal end of the ulna and radius to form the *wrist joint*. The carpals in the distal row, from lateral to medial,

EXHIBIT 8.E Skeleton of the Hand *(Figure 8.6)* CONTINUED

are the **trapezium** (tra-PĒ-zē-um=four-sided figure with no two sides parallel), **trapezoid** (TRAP-e-zoid=four-sided figure with two sides parallel), **capitate** (KAP-i-tāt=head-shaped), and **hamate** (HAM-āt=hooked).

The capitate is the largest carpal bone; its rounded projection, the head, articulates with the lunate. The hamate is named for a large hook-shaped projection on its anterior surface. In about 70 percent of carpal fractures, only the scaphoid is broken. This is because the force of a fall on an outstretched hand is transmit-

ted from the capitate through the scaphoid to the radius. The anterior concave space formed by the pisiform and hamate (on the ulnar side), and the scaphoid and trapezium (on the radial side), with the roof-like covering of the *flexor retinaculum* (a strong fibrous band of connective tissue), is the **carpal tunnel**. The long flexor tendons of the digits and thumb and the median nerve pass through the carpal tunnel. Narrowing of the carpal tunnel may give rise to a condition called *carpal tunnel syndrome* (described in Exhibit 11.R).

Figure 8.6 Right wrist and hand in relation to the ulna and radius.

🔑 The skeleton of the hand consists of the proximal carpals, the intermediate metacarpals, and the distal phalanges.

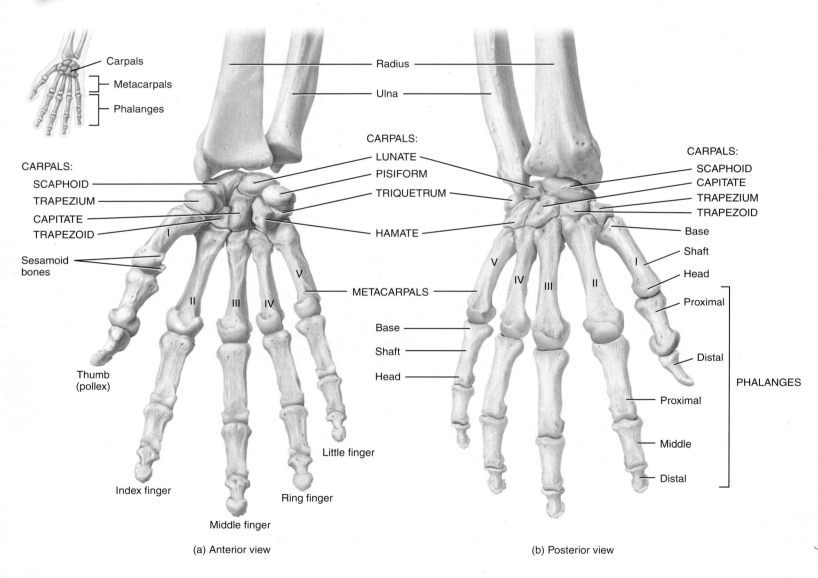

(a) Anterior view   (b) Posterior view

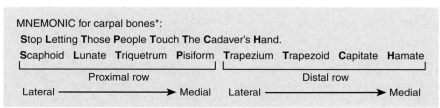

MNEMONIC for carpal bones*:

Stop **L**etting **T**hose **P**eople **T**ouch **T**he **C**adaver's **H**and.

Scaphoid **L**unate **T**riquetrum **P**isiform	**T**rapezium **T**rapezoid **C**apitate **H**amate
Proximal row	Distal row
Lateral ⟶ Medial	Lateral ⟶ Medial

Edward Tanner, University of Alabama, SOM

EXHIBIT 8.E **227**

METACARPALS

Description

The **metacarpus** (*meta-*=beyond), or palm, is the intermediate region of the hand; it consists of five bones called **metacarpals**.

Surface Features

Each metacarpal bone consists of a proximal *base*, an intermediate *shaft*, and a distal *head* (Figure 8.6b). The metacarpal bones are numbered I to V (or 1–5), starting with the thumb, from lateral to medial. The bases articulate with the distal row of carpal bones to form the *carpometacarpal joints*. The heads articulate with the proximal phalanges to form the *metacarpophalangeal (MP) joints*. The heads of the metacarpals, commonly called "knuckles," are readily visible in a clenched fist.

CLINICAL CONNECTION | Boxer's Fracture

A boxer's fracture is a fracture of the fifth metacarpal, usually near the head of the bone. It frequently occurs after a person punches another person or an object, such as a wall. It is characterized by pain, swelling, and tenderness. There may also be a bump on the side of the hand. Treatment is either by casting or surgery and the fracture usually heals in about six weeks. •

PHALANGES

Description

The **phalanges** (fa-LAN-jēz; *phalan-*=a battle line), or bones of the digits, make up the distal region of the hand. There are 14 phalanges in the five digits of each hand. Like the metacarpals, the digits are numbered I to V (or 1–5), beginning with the *thumb (pollex)* (PAWL-lex), from lateral to medial. A single bone of a digit is referred to as a *phalanx* (FĀ-lanks).

Surface Features

Each phalanx consists of a proximal *base*, an intermediate *shaft*, and a distal *head*. The *thumb* (pollex) has two phalanges, called *proximal* and *distal phalanges*. The other four digits have three phalanges called *proximal*, *middle*, and *distal phalanges*. In order from the thumb, these other four digits are commonly referred to as the *index finger*, *middle finger*, *ring finger*, and *little finger*. The proximal phalanges of all digits articulate with the metacarpal bones. The middle phalanges of the fingers (II–V) articulate with their distal phalanges. (The proximal phalanx of the thumb (I) articulates with its distal phalanx.) Joints between phalanges are called *interphalangeal (IP) joints*.

✓ CHECKPOINT

6. Which is more distal, the base or the head of the carpals?

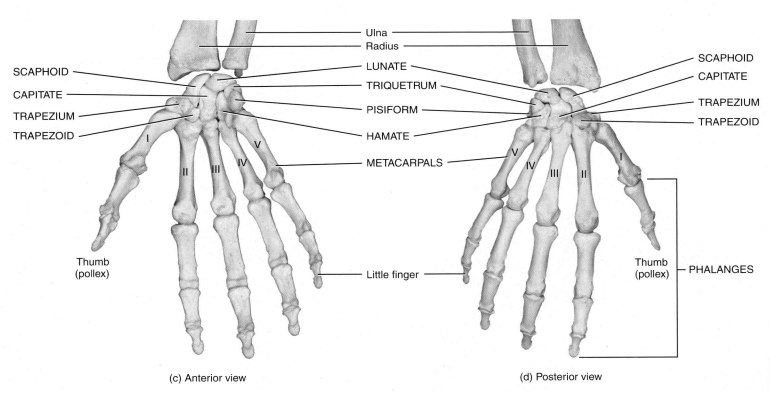

(c) Anterior view   (d) Posterior view

 Which is the most frequently fractured wrist bone?

8.2 SKELETON OF THE LOWER LIMB

OBJECTIVE

• Identify the bones of the skeleton of the lower limb.

Each lower limb skeleton consists of 31 bones, which form two distinct regions: (1) the pelvic girdle and (2) the free lower limb. The **pelvic (hip) girdle** consists of the **hip bone**, also called **coxal (pelvic) bone** (KOK-sal; *cox-*=hip), or **os coxae** (Figure 8.7a, b). The hip bones unite anteriorly at a joint called the **pubic symphysis** (PŪ-bik SIM-fi-sis). They unite posteriorly with the sacrum at the *sacroiliac joints*. The complete ring composed of the hip bones, pubic symphysis, sacrum, and coccyx forms a deep, basinlike structure called the **bony pelvis** (*pelv-*= basin). The plural is *pelves* (PEL-vēz) or *pelvises*. Functionally, the bony pelvis provides a strong and stable support for the vertebral column and pelvic and lower abdominal organs. Each pelvic girdle of the bony pelvis also connects the bones of the free lower limbs to the axial skeleton and transfers forces from the lower limbs to move the entire mass of the body during locomotion.

Figure 8.7 Skeleton of the pelvis and free lower limb.

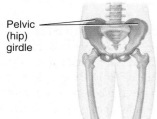
Pelvic (hip) girdle

🔑 Each free lower limb skeleton consists of 30 bones.

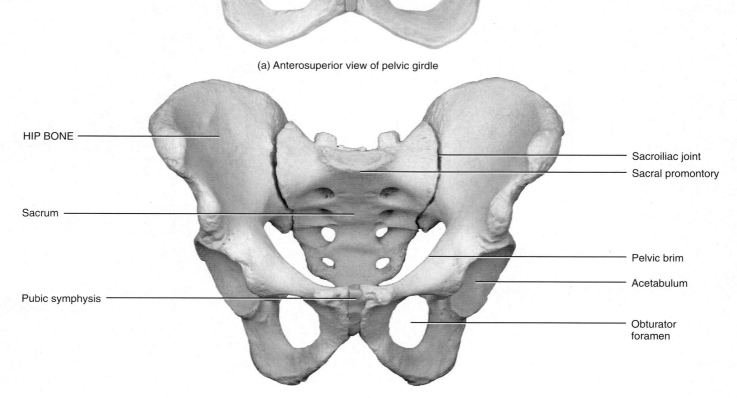

HIP BONE

Sacroiliac joint

Sacral promontory

Sacrum

Pelvic brim

Acetabulum

Coccyx

Pubic symphysis

Obturator foramen

(a) Anterosuperior view of pelvic girdle

HIP BONE

Sacroiliac joint

Sacral promontory

Sacrum

Pelvic brim

Acetabulum

Pubic symphysis

Obturator foramen

(b) Anterosuperior view

Each of the two hip bones of a newborn consists of three bones separated by cartilage: a superior *ilium*, an inferior and anterior *pubis*, and an inferior and posterior *ischium*. By age 23, the three separate bones fuse together (see Figure 8.8a). Although the hip bones function as single bones, anatomists commonly discuss them as though they still consisted of three bones.

Each **free lower limb (extremity)** (Figure 8.7c) has 30 bones in four locations: (1) the femur in the thigh; (2) the patella (knee-cap); (3) the tibia and fibula in the leg; and (4) the 7 tarsal bones in the tarsus (ankle), the 5 metatarsal bones in the metatarsus, and the 14 phalanges (bones of the digits) in the foot.

 CHECKPOINT
7. What is the function of the pelvic girdle?

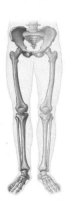

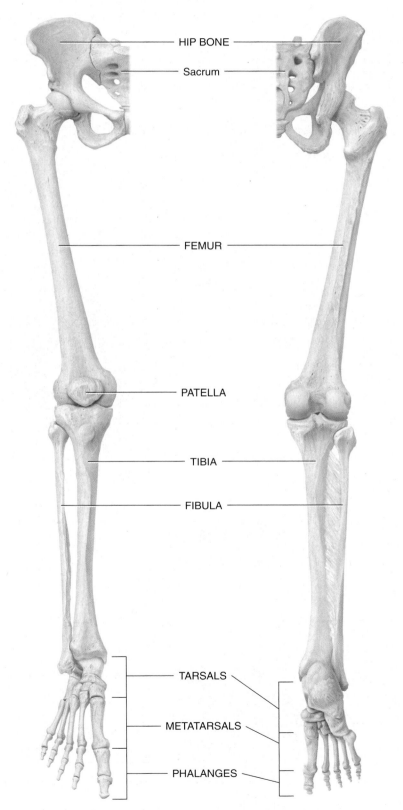

(c) Anterior view of free lower limb   (d) Posterior view of free lower limb

 Which bones make up the pelvic girdle and free part of the lower limb?

EXHIBIT 8.F   Bones of the Pelvic Girdle *(Figure 8.8)*

OBJECTIVE
- Identify the locations and surface features of the three components of the hip bone.

ILIUM

Description

The **ilium** (IL-ē-um=flank) is the largest of the three components of the hip bone (Figure 8.8). It is thick near the hip joint and expands into a large curved plate of bone superiorly.

Figure 8.8 Right hip bone. The lines of fusion of the ilium, ischium, and pubis depicted in (a) are not always visible in an adult.

🔑 The acetabulum is the socket for the head of the femur where the three parts of the hip bone converge and ossify.

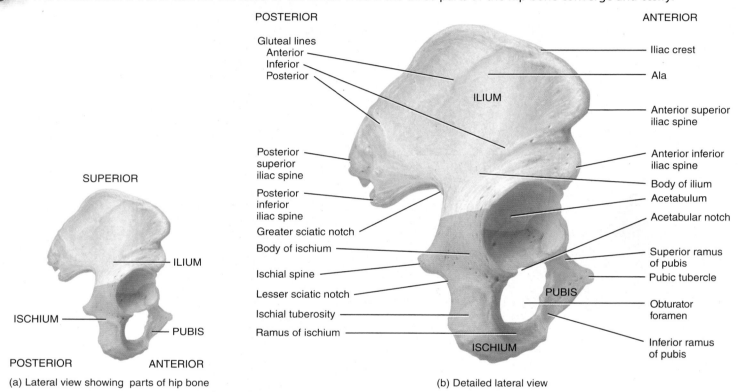

POSTERIOR   ANTERIOR

Gluteal lines
Anterior
Inferior
Posterior

ILIUM

Iliac crest
Ala
Anterior superior iliac spine

Posterior superior iliac spine
Posterior inferior iliac spine
Greater sciatic notch
Body of ischium
Ischial spine
Lesser sciatic notch
Ischial tuberosity
Ramus of ischium

Anterior inferior iliac spine
Body of ilium
Acetabulum
Acetabular notch
Superior ramus of pubis
Pubic tubercle
PUBIS
Obturator foramen
Inferior ramus of pubis

ISCHIUM

SUPERIOR

ILIUM
ISCHIUM
PUBIS

POSTERIOR   ANTERIOR

(a) Lateral view showing parts of hip bone

(b) Detailed lateral view

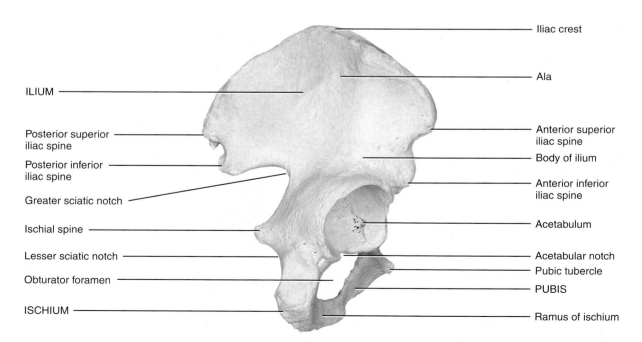

Iliac crest
Ala
ILIUM
Anterior superior iliac spine
Body of ilium
Posterior superior iliac spine
Posterior inferior iliac spine
Greater sciatic notch
Anterior inferior iliac spine
Ischial spine
Acetabulum
Lesser sciatic notch
Acetabular notch
Pubic tubercle
Obturator foramen
PUBIS
ISCHIUM
Ramus of ischium

(c) Lateral view

EXHIBIT 8.F **231**

Surface Features

A superior *ala* (=wing) and an inferior *body* comprise the ilium. The body is one of the components of the *acetabulum*, the socket for the head of the femur. The superior border of the ilium, the *iliac crest*, ends anteriorly in a blunt *anterior superior iliac spine (ASIS)*. Bruising of the anterior superior iliac spine and associated soft tissues, such as occurs in body-contact sports, is called a **hip pointer**. Below this spine is the *anterior inferior iliac spine*

(AIIS). Posteriorly, the iliac crest ends in a sharp *posterior superior iliac spine (PSIS)*. Below this spine is the *posterior inferior iliac spine (PIIS)*. The spines serve as points of attachment for the tendons of the muscles and ligaments of the trunk, hip, and thighs. Below the posterior inferior iliac spine is the *greater sciatic notch* (sī-AT-ik), through which the sciatic nerve passes along with other nerves and muscles; the sciatic nerve is the largest nerve in the body.

The medial surface of the ilium contains the *iliac fossa*, a concavity where the tendon of the iliacus muscle attaches.

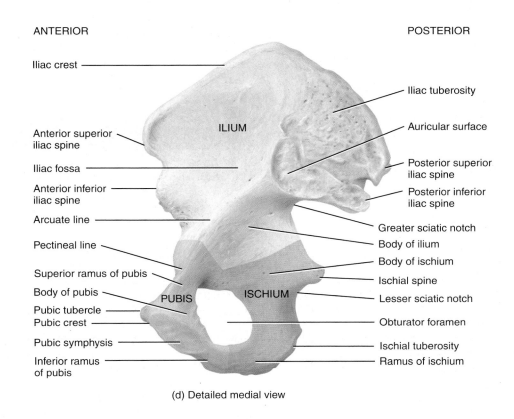

(d) Detailed medial view

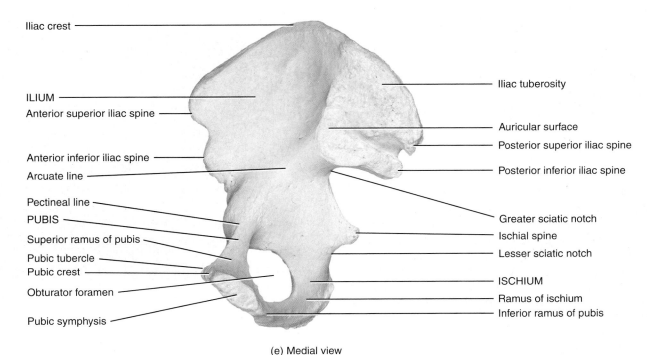

(e) Medial view

 Which part of the hip bone articulates with the femur? With the sacrum?

EXHIBIT 8.F Bones of the Pelvic Girdle *(Figure 8.8)* CONTINUED

Posterior to this fossa are the *iliac tuberosity*, a point of attachment for the sacroiliac ligament, and the *auricular surface* (*auric-* =ear-shaped), which articulates with the sacrum to form the *sacroiliac joint* (see Figure 8.7a, b). Projecting anteriorly and inferiorly from the auricular surface is a ridge called the *arcuate line* (AR-kū-āt; *arc-*=bow).

The other conspicuous markings of the ilium are three arched lines on its lateral surface called the *posterior gluteal line* (*glut-*=buttock), the *anterior gluteal line*, and the *inferior gluteal line*. The gluteal muscles attach to the ilium between these lines.

ISCHIUM

Description

The **ischium** (IS-kē-um=hip) (Figure 8.8), the inferior and posterior portion of the hip bone, is situated between the body of the ilium and the inferior ramus of the pubis. The ischium is a sideways-arched or U-shaped structure, with its concave, notched margin contributing to the posterior two-thirds of the obturator foramen (the large hole on the anterior surface of the hip bone) (Figure 8.8).

Surface Features

The ischium is comprised of a superior *body* and an inferior *ramus* (*ram-*=branch; plural is *rami*). The ramus joins the pubis. Features of the ischium include the prominent *ischial spine*, a *lesser sciatic notch* below the spine, and a rough and thickened *ischial tuberosity*. Because this prominent tuberosity is just deep to the integument, it commonly begins hurting when you sit upright on a hard surface for any length of time, such as the bleachers at a sporting event. Together, the ramus and the pubis surround the *obturator foramen* (OB-too-rā-tōr; *obtur-*=closed up), the largest foramen in the skeleton. The foramen is so-named because, even though blood vessels and nerves pass through it, it is nearly completely closed by the fibrous *obturator membrane*.

PUBIS

Description

The **pubis** (PŪ-bis) is the inferior, anterior portion of the hip bone and, like the ischium, has the form of a sideways arch or a U-shape (Figure 8.8).

Surface Features

A *superior ramus*, an *inferior ramus*, and a *body* between the rami comprise the pubis. The anterior, superior border of the body is the *pubic crest*, and at its lateral end is a projection called the *pubic tubercle*. This tubercle is the beginning of a raised line, the *pectineal line* (pek-TIN-ē-al), which extends superiorly and laterally along the superior ramus to merge with the arcuate line of the ilium. This line, as you will see shortly, is one important landmark for distinguishing the superior (false) and inferior (true) portions of the bony pelvis.

The *pubic symphysis* is the joint between the pubes of the two hip bones (see Figure 8.7a, b). It consists of a disc of fibrocartilage. Inferior to this joint, the inferior rami of the two pubic bones converge to form the *pubic arch*.

In the later stages of pregnancy, the hormone relaxin (produced by the ovaries and placenta) increases the flexibility of the pubic symphysis to ease delivery of the baby. Weakening of the joint, together with an already compromised center of gravity due to an enlarged uterus, also alters the mother's gait during pregnancy.

The *acetabulum* (as-e-TAB-ū-lum=vinegar cup) is a deep fossa formed by the ilium, ischium, and pubis. It functions as the socket that accepts the rounded head of the femur. Together, the acetabulum and the femoral head form the *hip (coxal) joint*. On the inferior side of the acetabulum, a deep indentation, the *acetabular notch*, forms a foramen through which blood vessels and nerves pass. The acetabular notch also serves as a point of attachment for a ligament of the femur called the *ligament of the head of the femur* (see Exhibit 9.D).

CLINICAL CONNECTION | *Hip Fracture*

Although any region of the hip girdle may fracture, the term **hip fracture** most commonly applies to a break in the bones associated with the hip joint—the head, neck, or trochanteric regions of the femur, or the bones that form the acetabulum. In the United States, 300,000–500,000 people sustain hip fractures each year. The incidence of hip fractures is increasing, due in part to longer life spans. Decreases in bone mass due to osteoporosis (which occurs more often in females), along with an increased tendency to fall, predispose elderly people to hip fractures.

Hip fractures often require surgical treatment, the goal of which is to repair and stabilize the fracture, increase mobility, and decrease pain. Sometimes the repair is accomplished by using surgical pins, screws, nails, and plates to secure the head of the femur. In severe hip fractures, the femoral head or the acetabulum of the hip bone may be replaced by *prostheses* (artificial devices). The procedure of replacing either the femoral head or the acetabulum is *hemiarthroplasty* (hem-ē-AR-thrō-plas-tē; *hemi-*=one half; *arthro*=joint; *plasty*=molding). Replacement of both the femoral head and acetabulum is *total hip arthroplasty* (see Clinical Connection on arthroplasty in Section 9.8). •

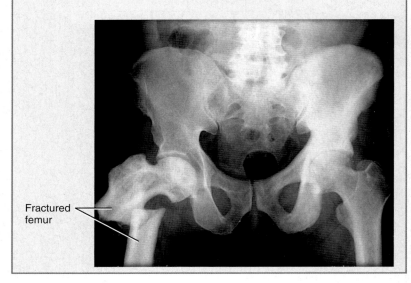

Fractured femur

CHECKPOINT

8. Why is the obturator foramen so named?

8.3 FALSE AND TRUE PELVES

● OBJECTIVE

• Distinguish **between the false and true pelves and explain their clinical importance.**

The bony pelvis is divided into superior and inferior portions by a boundary called the *pelvic brim* that forms the inlet into the pelvic cavity from the abdomen (Figure 8.9a). You can trace the pelvic brim by following the landmarks around parts of the hip bones to form the outline of an oblique plane. Beginning posteriorly at the *sacral promontory*, trace laterally and inferiorly along the *arcuate lines* of the ilium. Continue inferiorly along the *pectineal lines* of the pubis. Finally, trace anteriorly along the *pubic crest* to the superior portion of the *pubic symphysis*. Together, these points form an oblique plane that is higher in the back than in the front. The circumference of this plane is the pelvic brim.

The portion of the bony pelvis superior to the pelvic brim is the **false (greater) pelvis** (Figure 8.9b). It is bordered by the lumbar vertebrae posteriorly, the upper portions of the hip bones laterally, and the abdominal wall anteriorly. The space enclosed by the false pelvis is part of the lower abdomen; it contains the superior portion of the urinary bladder (when it is full) and lower

Figure 8.9 True and false pelves. Shown here is the female pelvis. For simplicity, in part (a) the landmarks of the pelvic brim are shown only on the left side of the body, and the outline of the pelvic brim is shown only on the right side. The entire pelvic brim is shown in Table 8.1.

🔑 The true and false pelves are separated by the pelvic brim.

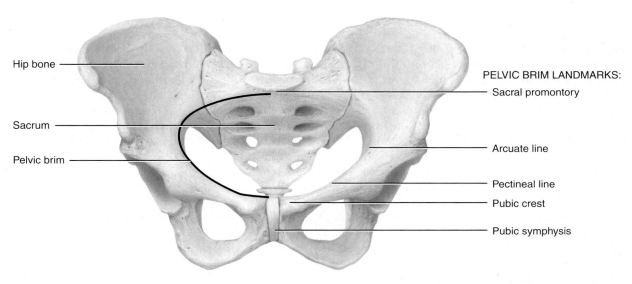

PELVIC BRIM LANDMARKS:

Hip bone

Sacrum

Pelvic brim

Sacral promontory

Arcuate line

Pectineal line

Pubic crest

Pubic symphysis

(a) Anterosuperior view of pelvic girdle

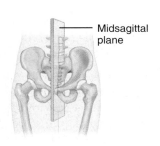

Midsagittal plane

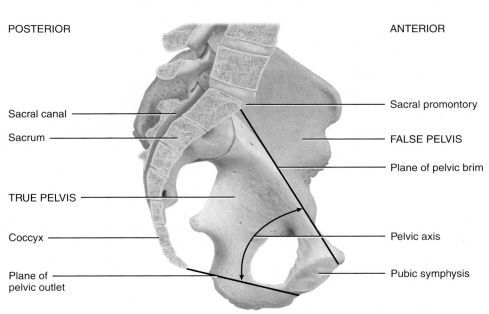

POSTERIOR

ANTERIOR

Sacral canal

Sacrum

TRUE PELVIS

Coccyx

Plane of pelvic outlet

Sacral promontory

FALSE PELVIS

Plane of pelvic brim

Pelvic axis

Pubic symphysis

(b) Midsagittal section indicating locations of true (blue) and false (pink) pelves

FIGURE 8.9 CONTINUES ▶

FIGURE 8.9 CONTINUED

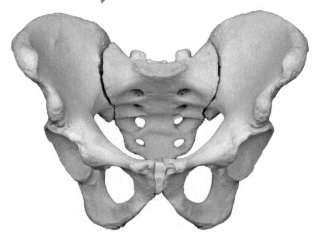 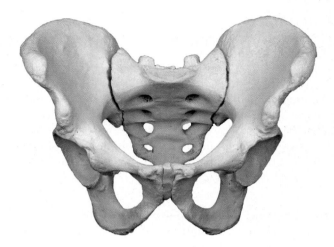

(c) Anterosuperior view of false pelvis (pink)

(d) Anterosuperior view of true pelvis (blue)

 What is the significance of the pelvic axis?

intestines in both genders and the uterus, ovaries, and uterine tubes of the female, which project into this region and are covered by the peritoneal serous membrane. The portion of the bony pelvis inferior to the pelvic brim is the **true (lesser) pelvis** (Figure 8.9b). It is bounded by the sacrum and coccyx posteriorly, inferior portions of the ilium and the ischium laterally, and the pubic bones anteriorly. The true pelvis surrounds the pelvic cavity (see Figure 1.6). It contains the rectum and urinary bladder in both genders, the vagina and cervix of the uterus in females, and the prostate in males. The superior opening of the true pelvis, bordered by the pelvic brim, is called the *pelvic inlet*; the inferior opening of the true pelvis is the bony *pelvic outlet*, which is covered by the muscles at the floor of the pelvis. The *pelvic axis* is an imaginary line that curves through the true pelvis from the central point of the plane of the pelvic inlet to the central point of the plane of the pelvic outlet. During childbirth the pelvic axis is the route taken by the baby's head as it descends through the pelvis.

Pelvimetry is the measurement of the size of the inlet and outlet of the birth canal, which may be done by ultrasonography or physical examination. Measurement of the pelvic cavity in pregnant females is important because the fetus must pass through the narrower opening of the pelvis at birth.

 CHECKPOINT

9. What is the clinical importance of the false pelvis? The true pelvis?

8.4 COMPARISON OF FEMALE AND MALE PELVES

OBJECTIVE

• Describe the principal differences between female and male pelves.

For the purposes of this discussion we assume that the male and female are comparable in age and physical stature. Generally, the bones of a male are larger and heavier than those of a female and have larger surface markings. Sex-related differences in the features of bones can be readily apparent when comparing the adult female and male pelves. Most of the structural differences in the pelves are adaptations to the requirements of pregnancy and childbirth. The female's pelvis is wider and shallower than the male's. Consequently, there is more space in the true pelvis of the female, especially in the pelvic inlet and pelvic outlet, which accommodate the passage of the infant's head at birth. Other significant differences between the pelves of females and males are listed and illustrated in Table 8.1.

 CHECKPOINT

10. Using Table 8.1 as a guide, select three of the easiest ways to distinguish a female from a male pelvis.

8.5 COMPARISON OF PECTORAL AND PELVIC GIRDLES

OBJECTIVE

• Explain the principal differences between the pectoral and pelvic girdles.

Now that you have studied the structures of the pectoral and pelvic girdles, you can understand some of their significant differences. The pectoral girdles do not articulate directly with the vertebral column, but each pelvic girdle does so via the sacroiliac joints. The sockets (glenoid cavities) for the free upper limb in each pectoral girdle are shallow and allow a greater range of motion than the deeper sockets (acetabula) for the free lower limb in each pelvic girdle. Overall, the structure of the pectoral girdles offers more mobility than strength, and that of the pelvic girdles offers more strength than mobility. This difference can be traced back to early evolutionary modifications in the two limbs. The pectoral limb became more mobile because of its modifications as the steering limb, while the pelvic limb became more stable and fixed to the axial skeleton because of its role as the locomotor limb.

 CHECKPOINT

11. Which girdle—pectoral or pelvic—offers more strength? More mobility? Explain your answers.

TABLE 8.1

Comparison of Female and Male Pelves

POINT OF COMPARISON	FEMALE	MALE
General structure	Light and thin	Heavy and thick
False (greater) pelvis	Shallow	Deep
Pelvic brim (inlet)	Wide and more oval	Narrow and heart-shaped
Acetabulum	Small and faces anteriorly	Large and faces laterally
Obturator foramen	Oval	Round
Pubic arch	Greater than 90° angle	Less than 90° angle

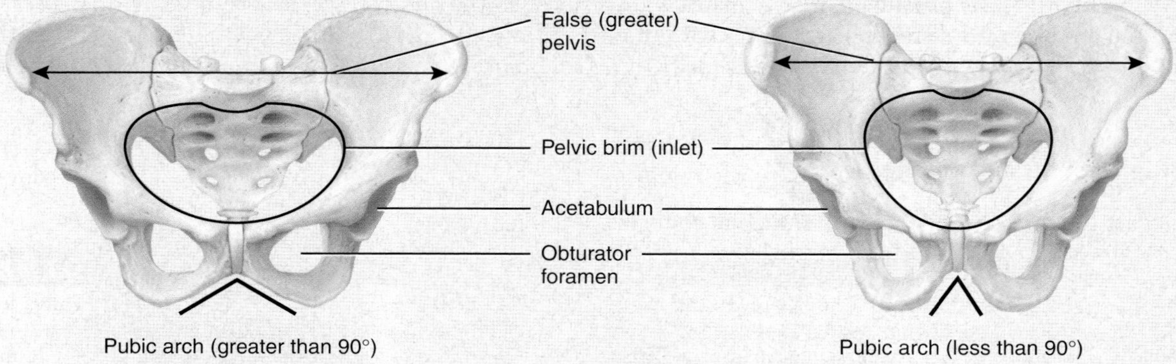

False (greater) pelvis

Pelvic brim (inlet)

Acetabulum

Obturator foramen

Pubic arch (greater than 90°)   Pubic arch (less than 90°)

Anterior views

POINT OF COMPARISON	FEMALE	MALE
Iliac crest	Less curved	More curved
Ilium	Less vertical	More vertical
Greater sciatic notch	Wide (almost 90°)	Narrow (about 70°; inverted V)
Coccyx	More movable and more curved anteriorly	Less movable and less curved anteriorly
Sacrum	Shorter, wider (see anterior views), and less curved anteriorly	Longer, narrower (see anterior views), and more curved anteriorly

Iliac crest

Ilium

Sacrum

Greater sciatic notch

Coccyx

Iliac crest

Ilium

Sacrum

Greater sciatic notch

Coccyx

Right lateral views

POINT OF COMPARISON	FEMALE	MALE
Pelvic outlet	Wider	Narrower
Ischial tuberosity	Shorter, farther apart, and more medially projecting	Longer, closer together, and more laterally projecting

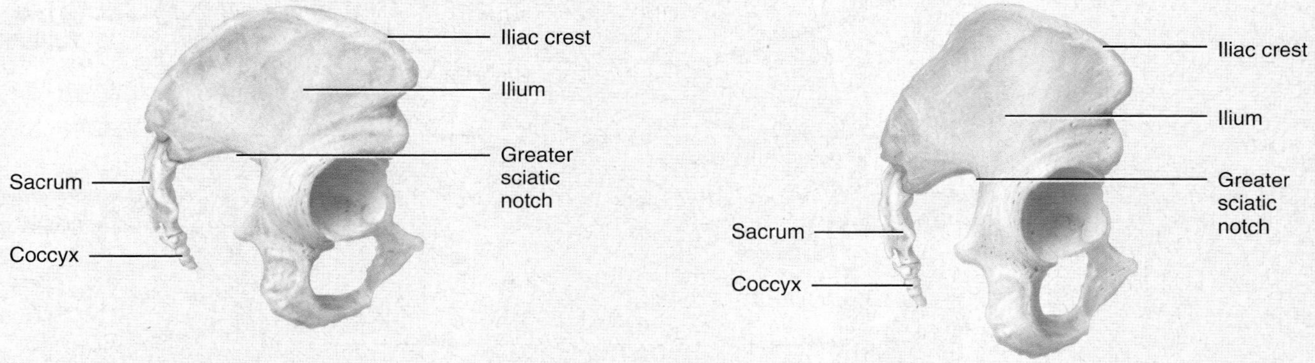

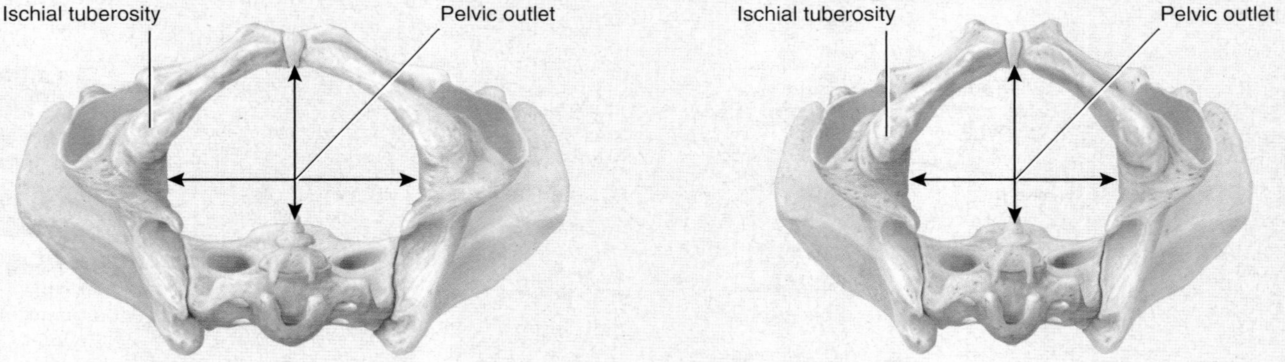

Ischial tuberosity   Pelvic outlet

Ischial tuberosity   Pelvic outlet

Inferior views

EXHIBIT 8.G   Skeleton of the Thigh—Femur and Patella *(Figure 8.10)*

● OBJECTIVE
- Identify the location and surface features of the femur and patella.

FEMUR

Description

The **femur,** or thigh bone, is the longest, heaviest, and strongest bone in the body (Figure 8.10). Its proximal end articulates with the acetabulum of the hip bone. Its distal end articulates with the tibia and patella. The *body (shaft)* of the femur angles medially

and, as a result, the knee joints are closer to the midline than the hip joints. This angle of the femoral shaft (*angle of convergence*) is greater in females because the female pelvis is broader.

Surface Features

The proximal end of the femur consists of a rounded *head* that articulates with the acetabulum of the hip bone to form the *hip (coxal) joint.* The head contains a small, central depression (pit) called the *fovea capitis* (FŌ-vē-a CAP-i-tis; *fovea*=pit; *capitis*=of the head) (Figure 8.10e, f). The ligament of the head of the femur connects the fovea capitis of the femur to the acetabulum

Figure 8.10 Right femur and patella.

🗝 The patella articulates with the lateral and medial condyles of the femur.

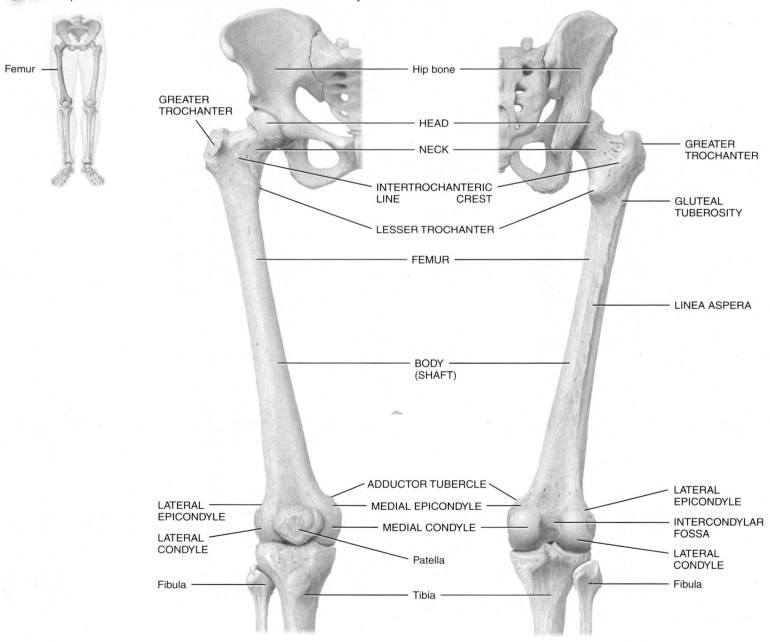

(a) Anterior view   (b) Posterior view

EXHIBIT 8.G **237**

of the hip bone. The *neck* of the femur is a constricted region distal to the head. The *greater trochanter* (trō-KAN-ter) and *lesser trochanter* are projections that serve as points of attachment for the tendons of some of the thigh and buttock muscles. The greater trochanter is the prominence felt and seen anterior to the hollow on the side of the hip. It is a landmark commonly used to locate the site for intramuscular injections into the lateral surface of the thigh. The lesser trochanter is inferior and medial to the greater trochanter. Between the anterior surfaces of the trochanters is a narrow *intertrochanteric line* (Figure 8.10a, c). A ridge called the *intertrochanteric crest* appears between the posterior surfaces of the greater trochanter and lesser trochanter (Figure 8.10b, d, e, f).

Inferior to the intertrochanteric crest on the posterior surface of the body of the femur is a vertical ridge called the *gluteal tuberosity*. It blends into another vertical ridge called the *linea aspera* (LIN-ē-a AS-per-a; *asper*=rough). Both ridges serve as attachment points for the tendons of several thigh muscles.

The expanded distal end of the femur includes the *medial condyle* (condyle=knuckle) and the *lateral condyle*. These articulate with the medial and lateral condyles of the tibia. Superior to the condyles are the *medial epicondyle* and the *lateral epicondyle*, to which ligaments of the knee joint attach. A depressed area between the condyles on the posterior surface is called the *intercondylar fossa* (in-ter-KON-di-lar). The *patellar surface* is located between the condyles on the anterior surface. Just superior to the medial epicondyle is the *adductor tubercle*, a roughened projection that is a site of attachment for the adductor magnus muscle.

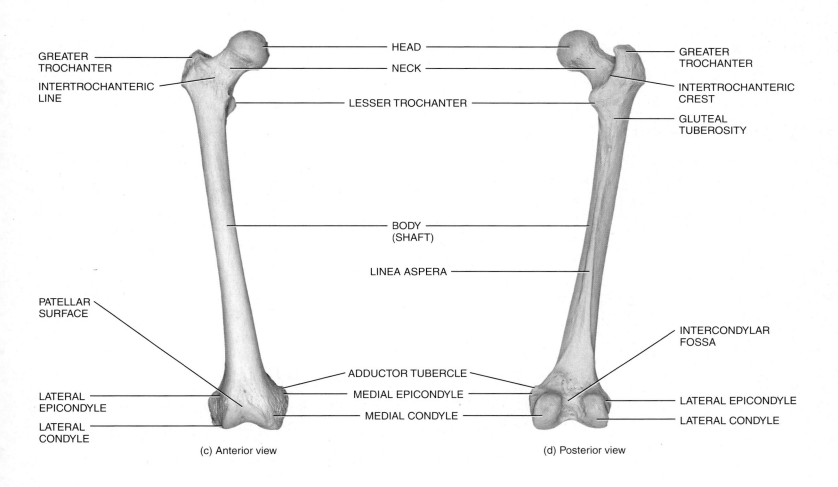

GREATER TROCHANTER
INTERTROCHANTERIC LINE
HEAD
NECK
LESSER TROCHANTER
GREATER TROCHANTER
INTERTROCHANTERIC CREST
GLUTEAL TUBEROSITY
BODY (SHAFT)
LINEA ASPERA
PATELLAR SURFACE
INTERCONDYLAR FOSSA
LATERAL EPICONDYLE
LATERAL CONDYLE
ADDUCTOR TUBERCLE
MEDIAL EPICONDYLE
MEDIAL CONDYLE
LATERAL EPICONDYLE
LATERAL CONDYLE

(c) Anterior view   (d) Posterior view

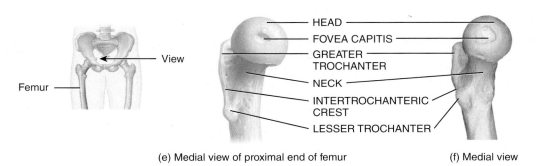

Femur
View
HEAD
FOVEA CAPITIS
GREATER TROCHANTER
NECK
INTERTROCHANTERIC CREST
LESSER TROCHANTER

(e) Medial view of proximal end of femur   (f) Medial view

FIGURE 8.10 CONTINUES ▶

EXHIBIT 8.G Skeleton of the Thigh—Femur and Patella *(Figure 8.10)* **CONTINUED**

PATELLA

Description

The **patella** (=little dish), or kneecap, is a small, triangular bone located anterior to the knee joint (Figure 8.10g–j). It is a sesamoid bone that develops in the tendon of the quadriceps femoris muscle. The functions of the patella are to increase the leverage of the tendon of the quadriceps femoris muscle, to maintain the position of the tendon when the knee is bent (flexed), and to protect the knee joint.

Surface Features

The broad proximal end of the patella is called the *base*. The pointed distal end is the *apex*. The posterior surface contains two *articular facets*, one for the medial condyle of the femur and the other for the lateral condyle of the femur. The patellar ligament attaches the patella to the tibial tuberosity. The *patellofemoral joint*, between the posterior surface of the patella and the patellar surface of the femur, is the intermediate component of the *tibiofemoral (knee) joint*.

CLINICAL CONNECTION | *Patellofemoral Stress Syndrome*

Patellofemoral stress syndrome ("runner's knee") is one of the most common problems runners experience. During normal flexion and extension of the knee, the patella tracks (glides) superiorly and inferiorly in the groove between the femoral condyles. In patellofemoral stress syndrome, normal tracking does not occur; instead, the patella tracks laterally as well as superiorly and inferiorly, and the increased pressure on the joint causes aching or tenderness around or under the patella. The pain typically occurs after a person has been sitting for a while, especially after exercise. It is worsened by squatting or walking down stairs. One cause of runner's knee is constantly walking, running, or jogging on the same side of the road. Other predisposing factors include running on hills, running long distances, and an anatomical deformity called genu valgum or knock-knee (see Key Medical Terms section).

✓ **CHECKPOINT**

12. What is the clinical importance of the greater trochanter?

▣ **FIGURE 8.10 CONTINUED** ▶

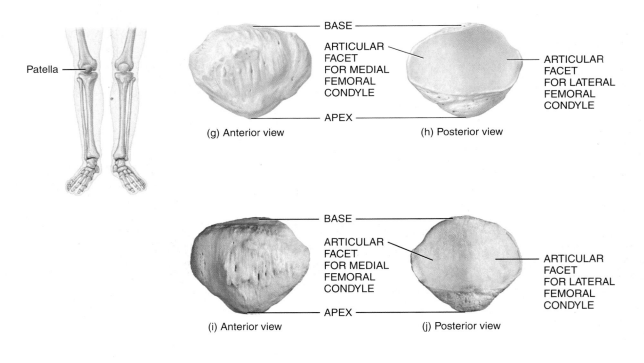

(g) Anterior view   (h) Posterior view

(i) Anterior view   (j) Posterior view

 Why is the angle of convergence of the femurs greater in females than males?

EXHIBIT 8.H **239**

EXHIBIT 8.H | Skeleton of the Leg—Tibia and Fibula *(Figure 8.11)*

OBJECTIVE
- Identify the location and surface features of the tibia and fibula.

TIBIA

Description

The **tibia**, or shin bone, is the larger, medial, weight-bearing bone of the leg (Figure 8.11). The term *tibia* means flute, be-cause in ancient times the tibial bones of birds were used to make musical instruments. The tibia is the second longest bone of the body, exceeded in length only by the femur. The tibia articulates at its proximal end with the femur and fibula and at its distal end with the fibula and the talus bone of the ankle. The tibia and fibula, like the ulna and radius, are connected by an interosseous membrane.

Figure 8.11 Right tibia and fibula.

🔑 The tibia articulates with the femur and fibula proximally, and with the fibula and talus distally.

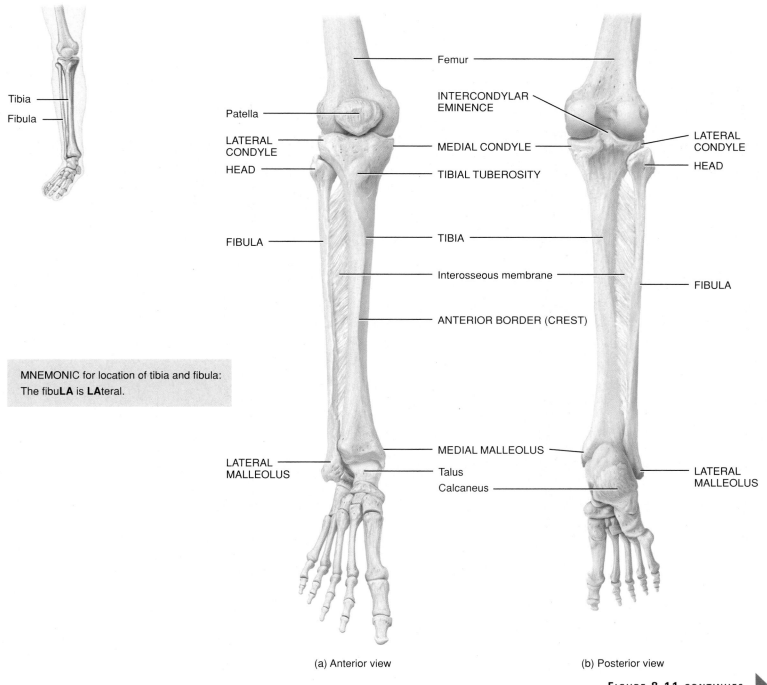

MNEMONIC for location of tibia and fibula:
The fibu**LA** is **LA**teral.

(a) Anterior view   (b) Posterior view

FIGURE 8.11 CONTINUES ▶

EXHIBIT 8.H Skeleton of the Leg—Tibia and Fibula *(Figure 8.11)* CONTINUED

Surface Features

The proximal end of the tibia is expanded into a *lateral condyle* and a *medial condyle* (Figure 8.11a–d). These articulate with the condyles of the femur to form the lateral and medial *tibiofemoral (knee) joints*. The inferior surface of the lateral condyle articulates with the head of the fibula. The slightly concave condyles are separated by an upward projection called the *intercondylar eminence* (Figure 8.11b–d). The *tibial tuberosity* on the anterior surface is a point of attachment for the patellar ligament. Inferior to and continuous with the tibial tuberosity is a sharp ridge that can be felt below the skin and is known as the *anterior border (crest)* or shin. The medial surface of the distal end of the tibia forms the *medial malleolus* (mal-LĒ-ō-lus=hammer). This structure articulates with the talus of the ankle and forms the prominence that can be felt on the medial surface of the ankle. The *fibular notch* (Figure 8.11d–f) articulates with the distal end of the fibula to form the *distal tibiofibular joint*. Of all the long bones of the body, the tibia is the most frequently fractured and is also the most frequent site of an open (compound) fracture.

FIBULA

Description

The **fibula** is a slender, splint-like bone that is slightly expanded at both ends; it is the developmental counterpart of the ulna in the free upper limb. The fibula is parallel and lateral to the tibia, but is considerably smaller. Unlike the tibia, the fibula does not articulate with the femur and is non-weight-bearing, but it does help stabilize the ankle joint.

⬤ **FIGURE 8.11** CONTINUED ▶

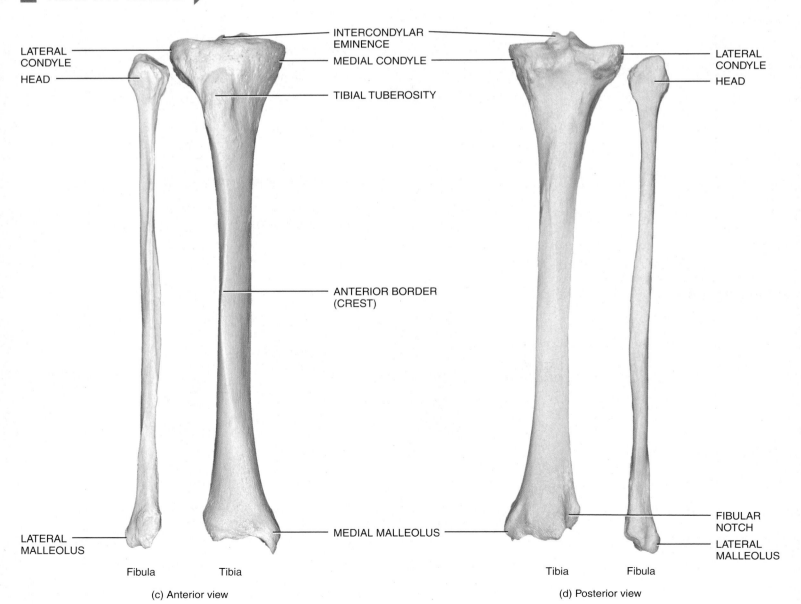

(c) Anterior view

(d) Posterior view

EXHIBIT 8.I **241**

Surface Features

The *head* of the fibula, the proximal end, articulates with the inferior surface of the lateral condyle of the tibia below the level of the knee joint to form the *proximal tibiofibular joint*. The distal end has a projection called the *lateral malleolus* that articulates with the talus of the ankle. This forms the prominence on the lateral surface of the ankle. As noted previously, the fibula also articulates with the tibia at the fibular notch to form the distal tibiofibular joint.

CLINICAL CONNECTION | *Bone Grafting*

Bone grafting generally consists of taking a piece of bone, along with its periosteum and nutrient artery, from one part of the body to replace missing bone in another part of the body. The transplanted bone restores the blood supply to the transplant site and healing occurs as in a fracture. The fibula is a common source of bone for grafting; even after a piece of the fibula has been removed, walking, running, and jumping can all be normal. •

✔ CHECKPOINT

13. Which structures form the medial and lateral prominences of the ankle?

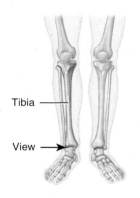

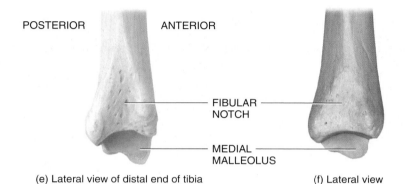

POSTERIOR   ANTERIOR

Tibia

View

FIBULAR NOTCH

MEDIAL MALLEOLUS

(e) Lateral view of distal end of tibia   (f) Lateral view

 Which leg bone bears the weight of the body?

| EXHIBIT 8.I | Skeleton of the Foot *(Figure 8.12)* |

 OBJECTIVE

• Identify the location and surface features of the bones of the foot.

TARSALS

Description

The **tarsus** (ankle) is the proximal region of the foot and consists of seven **tarsal bones** (Figure 8.12a–d). The tarsal bones are much greater in size than the small carpal bones, with the two most proximal bones—the talus and calcaneus—being significantly larger than their more distal counterparts. Joints between tarsal bones are called *intertarsal joints*.

Surface Features

The tarsals include the **talus** (TĀ-lus=ankle bone) and **calcaneus** (kal-KĀ-nē-us=heel), located in the posterior part of the foot. The calcaneus is the largest and strongest tarsal bone. The anterior tarsal bones are the **navicular** (na-VIK-ular=like a little boat), three **cuneiform bones** (KYOO-nē-i-form=wedge-shaped) called the **third (lateral)**, **second (intermediate)**, and **first (medial) cuneiforms**, and the **cuboid** (KŪ-boyd=cube-shaped). The talus, the most superior tarsal bone, is the only bone of the foot that articulates with the fibula and tibia. It articulates on one side with the medial malleolus of the tibia and on the other side with the lateral malleolus of the fibula. These articulations form the *talocrural (ankle) joint*. During walking, the talus transmits about half the weight of the body to the calcaneus. The remainder is transmitted to the other tarsal bones.

METATARSALS

Description

The **metatarsus** is the intermediate region of the foot and consists of five **metatarsal bones** numbered I to V (or 1–5) from medial to lateral (Figure 8.12a–d). They are convex dorsally and concave on their plantar surfaces.

Surface Features

Like the metacarpals of the palm, each metatarsal consists of a proximal *base*, an intermediate *shaft*, and a distal *head*. The metatarsals articulate proximally with the first, second, and third cuneiform bones and with the cuboid to form the *tarsometatarsal joints*. Distally, they articulate with the proximal row of phalanges to form the *metatarsophalangeal joints*. The first metatarsal is thicker than the others because it bears more weight.

EXHIBIT 8.1 Skeleton of the Foot *(Figure 8.12)* **CONTINUED**

Fractures of the metatarsals occur when a heavy object falls on the foot or when a heavy object rolls over the foot. Such fractures are also common among dancers, especially ballet dancers. If a ballet dancer is on the tip of her toes and loses her balance, the full body weight is placed on the metatarsals, causing one or more of them to fracture. •

PHALANGES

Description

The **phalanges** comprise the distal component of the foot and resemble those of the hand both in number and arrangement (Figure 8.12a–d). The toes are numbered I to V (or 1–5) beginning with the great toe (*hallux*), from medial to lateral.

Surface Features

Each *phalanx* (singular) consists of a proximal *base*, an intermediate *shaft*, and a distal *head*. The great or big toe (hallux) has

Figure 8.12 **Right foot and arches of the foot.**

🔑 The skeleton of the foot consists of the proximal tarsals, the intermediate metatarsals, and the distal phalanges.

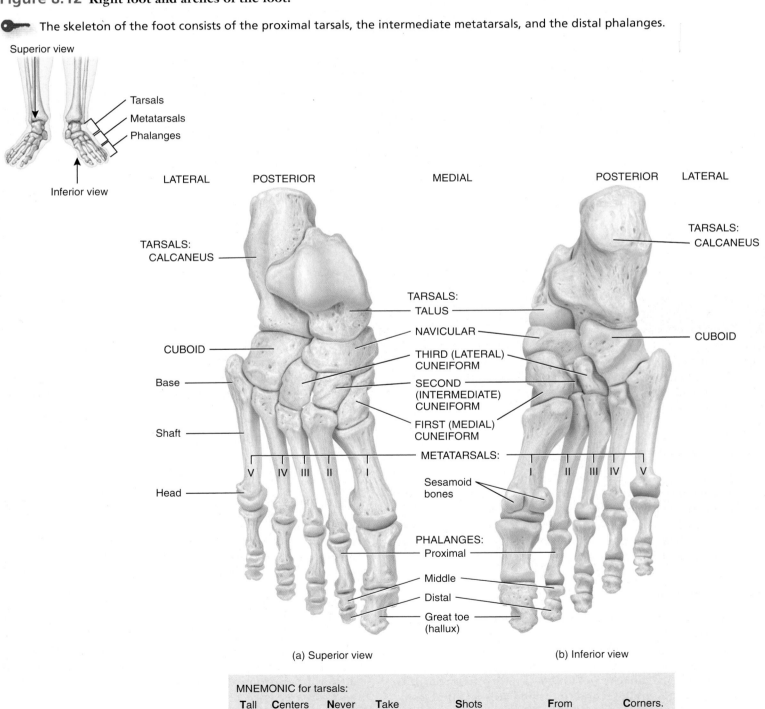

(a) Superior view   (b) Inferior view

MNEMONIC for tarsals:

Tall	**C**enters	**N**ever	**T**ake	**S**hots	**F**rom	**C**orners.
Talus	Calcaneus	Navicular	Third cuneiform	Second cuneiform	First cuneiform	Cuboid

EXHIBIT 8.1 **243**

two large, heavy phalanges called *proximal* and *distal* phalanges. The other four toes each have three phalanges—*proximal*, *middle*, and *distal*. The proximal phalanges of all toes articulate with the metatarsal bones. The middle phalanges of toes II–V articulate with their distal phalanges, while the proximal phalanx of the great toe (I) articulates with its distal phalanx. Joints between phalanges of the foot, like those of the hand, are called *interphalangeal joints*.

ARCHES OF THE FOOT

The bones of the foot are arranged in two **arches** (Figure 8.12e). The arches enable the foot to support the weight of the body, provide an ideal distribution of body weight over the soft and hard tissues of the foot, and provide leverage when walking. The arches are not rigid; they yield as weight is applied and spring back when the weight is lifted, thus helping to absorb

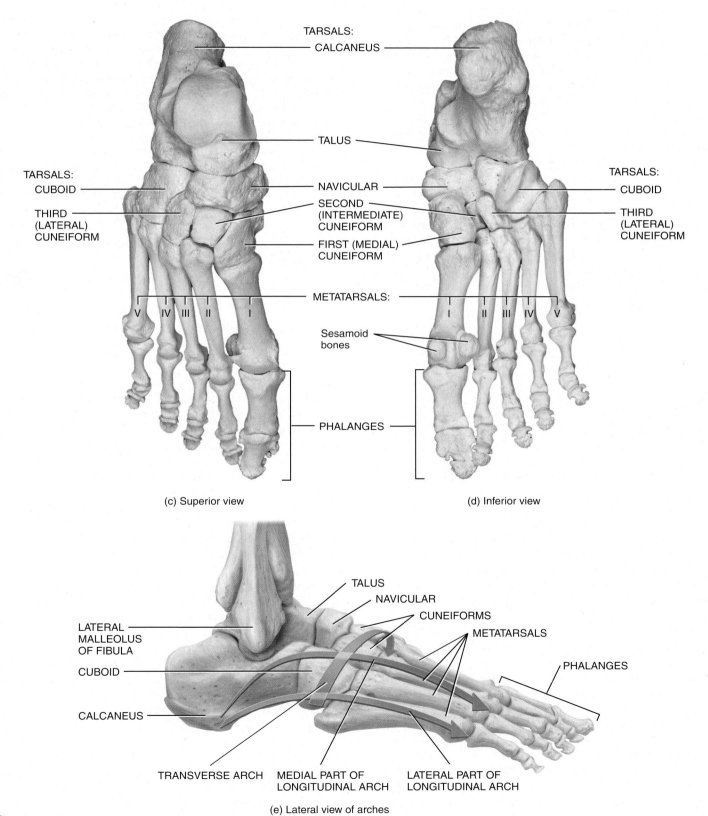

(c) Superior view

(d) Inferior view

(e) Lateral view of arches

Which tarsal bone articulates with the tibia and fibula?

EXHIBIT 8.1 Skeleton of the Foot *(Figure 8.12)* CONTINUED

shocks. Usually, the arches are fully developed by the time children reach age 12 or 13.

The **longitudinal arch** has two parts, both of which consist of tarsal and metatarsal bones arranged to form an arch from the anterior to the posterior part of the foot. The *medial part* of the longitudinal arch, which originates at the calcaneus, rises to the talus and descends through the navicular, the three cuneiforms, and the heads of the three medial metatarsals. The *lateral part* of the longitudinal arch also begins at the calcaneus. It rises at the cuboid and descends to the heads of the two lateral metatarsals. The medial portion of the longitudinal arch is so high that the medial portion of the foot between the ball and heel does not touch the ground when you walk on a hard surface.

The **transverse arch** is found between the medial and lateral aspects of the foot and is formed by the navicular, three cuneiforms, and the bases of the five metatarsals.

As noted earlier, one function of the arches is to distribute body weight over the soft and hard tissues of the foot. Normally, the ball of the foot carries about 40 percent of the weight and the heel carries about 60 percent. The ball of the foot is the padded portion of the sole superficial to the heads of the metatarsals. When a person wears high-heeled shoes, however, the distribution of weight changes so that the ball of the foot may carry up to 80 percent and the heel 20 percent. As a result, the

fat pads at the ball of the foot are damaged, joint pain develops, and structural changes in bones may occur.

CLINICAL CONNECTION | *Flatfoot and Clawfoot*

The bones composing the arches of the foot are held in position by ligaments and tendons. If these ligaments and tendons are weakened, the height of the medial longitudinal arch may decrease or "fall." The result is **flatfoot**, the causes of which include excessive weight, postural abnormalities, weakened supporting tissues, and genetic predisposition. Fallen arches may lead to inflammation of the fascia of the sole (plantar fasciitis), Achilles tendinitis, shinsplints, stress fractures, bunions, and calluses. A custom-designed arch support often is prescribed to treat flatfoot.

Clawfoot is a condition in which the medial longitudinal arch is abnormally elevated. It is often caused by muscle deformities, such as may occur in diabetics whose neurological lesions lead to atrophy of muscles of the foot. •

CHECKPOINT

14. What are the names and functions of the arches of the foot?

8.6 DEVELOPMENT OF THE SKELETAL SYSTEM

OBJECTIVE

• Describe the development of the skeletal system.

Most of the skeletal tissue arises from *mesenchymal cells*, connective tissue cells derived from **mesoderm**. However, much of the skeleton of the skull arises from head mesenchyme, which consists of neural crest cells of ectodermal origin. In either case, the cells condense and form models of bone in areas where the bones themselves will ultimately form. In some cases, the bones form directly within the mesenchyme (intramembranous ossification; see Figure 6.5). In other cases, the mesenchyme first differentiates into hyaline cartilage and then the bones form within the cartilage as they replace it (endochondral ossification; see Figure 6.6).

The *skull* begins development during the fourth week after fertilization. It develops from tissue around the developing brain and consists of two major portions: **neurocranium** (of mesodermal and neural crest origin), which forms the bones of the skull, and **viscerocranium** (of neural crest origin), which forms the bones of the face (Figure 8.13a). The neurocranium is divided into two parts called the **cartilaginous neurocranium** (mesodermal and neural crest in origin) and **membranous neurocranium** (neural crest in origin). The cartilaginous neurocranium consists of hyaline cartilage developed from mesenchyme at the base of the developing skull.

It later undergoes endochondral ossification to form the *bones at the base of the skull*. The membranous neurocranium consists of neural crest mesenchyme and later undergoes intramembranous ossification to form the *flat bones that make up the roof and sides of the skull*. During fetal life and infancy the flat bones are separated by membrane-filled spaces called *fontanels* (see Figure 7.19). The viscerocranium, like the neurocranium, is divided into two parts: **cartilaginous viscerocranium** and **membranous viscerocranium**. The cartilaginous viscerocranium is derived from the cartilage of the first two pharyngeal (branchial) arches (see Figure 4.13). Endochondral ossification of these cartilages forms the *jaw bones*, *ear bones*, and *hyoid bone*. The membranous viscerocranium is derived from neural crest mesenchyme in the frontonasal process and first pharyngeal arch, and, following intramembranous ossification, forms the facial bones.

The *vertebrae* and *ribs* are derived from the sclerotomes of somites (see Figure 10.10). Mesenchymal cells from these regions surround the notochord (see Figure 10.10) at about four weeks after fertilization. The notochord induces the mesenchymal cells to form five distinct centers of bone formation at each segmental level—a *vertebral body*, two *costal (rib) centers*, and two *vertebral arch centers*. Between the vertebral bodies, the notochord induces mesenchymal cells to form the *nucleus pulposus* of an intervertebral disc and induces surrounding mesenchymal cells to form the *annulus fibrosus* of the intervertebral disc. As development continues, the two vertebral arch centers unite with one another and with the vertebral body. The *vertebral arch* then surrounds the spinal cord; failure of

Figure 8.13 **Development of the skeletal system.** Bones that develop from the cartilaginous neurocranium are indicated in light blue; those from the cartilaginous viscerocranium are indicated in dark blue; those from the membranous neurocranium are shown in dark red; and those from the membranous viscerocranium are shaded light red.

After the free limb buds develop, endochondral ossification of the free limb bones begins by the end of the eighth embryonic week.

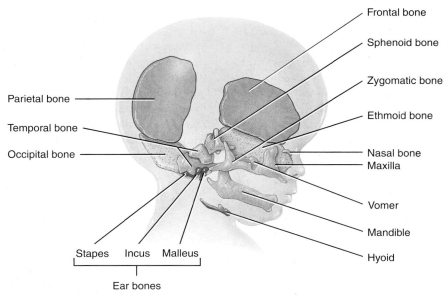

Frontal bone

Sphenoid bone

Zygomatic bone

Parietal bone

Ethmoid bone

Temporal bone

Occipital bone

Nasal bone

Maxilla

Vomer

Mandible

Stapes Incus Malleus

Hyoid

Ear bones

(a) Development of the skull

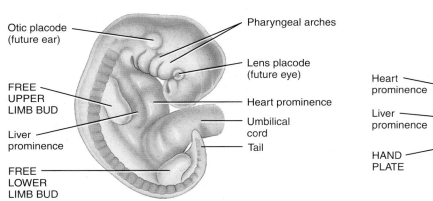

Otic placode (future ear)

Pharyngeal arches

Lens placode (future eye)

FREE UPPER LIMB BUD

Heart prominence

Liver prominence

Umbilical cord

Tail

FREE LOWER LIMB BUD

(b) Four-week embryo showing development of free limb buds

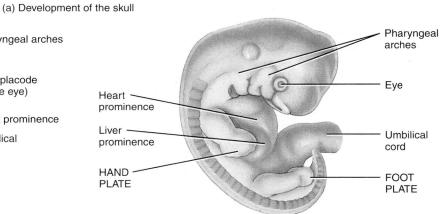

Pharyngeal arches

Eye

Heart prominence

Liver prominence

Umbilical cord

HAND PLATE

FOOT PLATE

(c) Six-week embryo showing development of hand and foot plates

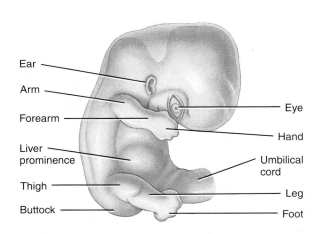

Ear

Arm

Forearm

Eye

Liver prominence

Hand

Thigh

Umbilical cord

Buttock

Leg

Foot

(d) Seven-week embryo showing development of arm, forearm, and hand in free upper limb bud and thigh, leg, and foot in free lower limb bud

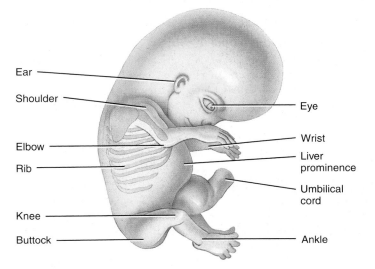

Ear

Shoulder

Eye

Elbow

Wrist

Rib

Liver prominence

Umbilical cord

Knee

Buttock

Ankle

(e) Eight-week embryo in which free limb buds have developed into free upper and lower limbs

Which of the three basic embryonic tissues—ectoderm, mesoderm, and endoderm—gives rise to the skeletal system?

the vertebral arch to develop properly results in a condition called *spina bifida* (see Figure 7.19). In the thoracic region, the costal centers enlarge and grow anteriorly to form the *ribs*. In the cervical, lumbar, and sacral regions the costal centers do not enlarge to the same degree as they do in the thorax, and they also fuse with the body and vertebral arch. The *sternum* develops from mesoderm in the anterior body wall.

The *skeleton of the limb girdles and free limbs* is derived from the somatic region of the lateral plate mesoderm (see Figure 4.9d). During the middle of the fourth week after fertilization, the free upper limbs appear as small elevations at the sides of the trunk called **upper limb buds** (Figure 8.13b). About 2 days later, the **lower limb buds** appear. The limb buds consist of **mesenchyme** covered by **ectoderm**. At this point, a mesenchymal skeleton exists in the limb buds; the masses of mesoderm surrounding the developing bones, which will become the skeletal muscles of the limbs, have migrated into the limbs from the myotome of the somite.

By the sixth week, the limb buds develop a constriction around the middle portion. The constriction produces flattened distal segments of the upper buds called **hand plates** and distal segments of the lower buds called **foot plates** (Figure 8.13c). These plates represent the beginnings of the hands and feet, respectively. A cartilaginous skeleton formed from mesenchyme is present at this stage of limb development. By the seventh week (Figure 8.13d), the *arm*, *forearm*, and *hand* are evident in the upper limb bud, and the *thigh*, *leg*, and *foot* appear in the lower limb bud. By the eighth week (Figure 8.13e), as the shoulder, elbow, and wrist areas become apparent, the free upper limb bud is appropriately called the free upper limb, and the lower limb bud is now the free lower limb.

Endochondral ossification of the limb bones begins by the end of the eighth week after fertilization. By the twelfth week, primary ossification centers are present in most of the limb bones. Most secondary ossification centers appear after birth.

 CHECKPOINT

15. At what stage of development do the free limbs develop? How do they form?

KEY MEDICAL TERMS ASSOCIATED WITH THE APPENDICULAR SKELETON

Clubfoot or **talipes equinovarus** (TAL-i-pēz ek-wīn-ō-VAR-us; *pes-*=foot; *equino-*=horse) An inherited deformity in which the foot is twisted inferiorly and medially, and the angle of the arch is increased; occurs in 1 of every 1000 births.

Fractured clavicle A fracture resulting from excessive mechanical force transmitted from the free upper limb to the trunk (a fall on an outstretched arm), or a blow to the superior part of the anterior thorax (impact following an automobile accident). One of the most frequently broken bones in the body. Usually treated with a figure-of-eight sling to keep the arm from moving outward.

Genu valgum (JĒ-noo VAL-gum; *genu-*=knee; *valgum*=bent outward) A deformity in which the knees are abnormally close together and the space between the ankles is increased due to a lateral angulation of the tibia in relation to the femur. Also called **knock-knee**.

Genu varum (JĒ-noo VAR-um; *varum*=bent toward the midline) A deformity in which the knees are abnormally separated, there is a medial angulation of the tibia in relation to the femur, and the free lower limbs are bowed laterally. Also called **bowleg**.

Hallux valgus (HAL-uks VAL-gus; *hallux*=great toe) Angulation of the great toe away from the midline of the body, typically caused by wearing tightly fitting shoes. When the great toe angles toward the next toe, there is a bony protrusion at the base of the great toe. Also called a **bunion**.

CHAPTER REVIEW AND RESOURCE SUMMARY

WileyPLUS

Review

Resource

8.1 Skeleton of the Upper Limb

1. Each pectoral (shoulder) girdle consists of a clavicle and scapula (see Exhibits 8.A and 8.B).
2. Each pectoral girdle attaches the free upper limb to the axial skeleton.
3. Each of the two free upper limbs (extremities) contains 30 bones.
4. The bones of each of the upper limbs include the humerus, ulna, radius, carpals, metacarpals, and phalanges (see Exhibits 8.C–8.E).

Anatomy Overview - The Appendicular Skeleton: The Pectoral Girdle
Anatomy Overview - The Upper Limb
Figure 8.1 - Skeleton of the Right Upper Limb—Pectoral (Shoulder) Girdle and Free Upper Limb
Figure 8.4 - Right Humerus in Relation to the Scapula, Ulna, and Radius

8.2 Skeleton of the Lower Limb

1. Each pelvic (hip) girdle consists of a hip bone or os coxa.
2. Each hip bone consists of three fused bones: the ilium, pubis, and ischium (see Exhibit 8.F).
3. The hip bones, sacrum, coccyx and pubic symphysis form the bony pelvis. The bony pelvis supports the vertebral column and pelvic viscera and attaches the free lower limbs to the axial skeleton.
4. Each of the two free lower limbs (extremities) contains 30 bones.
5. The bones of each free lower limb include the femur, the patella, the tibia, the fibula, the tarsals, the metatarsals, and the phalanges (see Exhibits 8.G–8.I).
6. The bones of the foot are arranged in two arches, the longitudinal arch and the transverse arch, to provide support and leverage.

Anatomy Overview - The Lower Limb

Review   **Resource**

8.3   False and True Pelves

1. The false pelvis is separated from the true pelvis by the pelvic brim.
2. The true pelvis surrounds the pelvic cavity and houses the rectum and urinary bladder in both genders, the vagina and cervix of the uterus in females, and prostate in males.
3. The false pelvis is the lower portion of the abdomen that is situated superior to the pelvic brim. It contains the superior portion of the urinary bladder (when full) and lower intestines in both genders, and the uterus, uterine tubes, and ovaries of the female.

8.4   Comparison of Female and Male Pelves

1. Bones of males are generally larger and heavier than bones of females, with more prominent markings for muscle attachment.
2. The female pelvis is adapted for pregnancy and childbirth. Gender-related differences in pelvic structure are listed and illustrated in Table 8.1.

8.5   Comparison of Pectoral and Pelvic Girdles

Anatomy Overview - The Pelvic Girdle
Exercise - Bones, Bones, Bones
Figure 8.10 - Right Femur and Patella

1. The pectoral girdle does not articulate directly with the vertebral column; the pelvic girdle does via the sacroiliac joint.
2. The glenoid fossae are shallow and maximize movement; the acetabula are deep and allow less movement.
3. The pectoral girdle affords more mobility than strength; the pelvic girdle offers more strength than mobility.

8.6   Development of the Skeletal System

1. Bone forms from mesoderm by intramembranous or endochondral ossification.
2. Bones of the limbs develop from limb buds, which consist of mesoderm and ectoderm.

CRITICAL THINKING QUESTIONS

1. The *Local News* reported that farmer Bob Ramsey caught his hand in a piece of machinery on Tuesday. He lost the lateral two fingers of his left hand. Science reporter Kent Clark, a high school junior, reports that Farmer Ramsey has 3 remaining phalanges. Is Kent correct or is he anatomy-challenged?

2. Rose had flat feet as a child and was told to take up ballet dancing to correct the condition. Now Rose has hallux valgus and trouble with her talocrural joint, but at least her flat feet are cured. Explain how Rose's problems relate to each other.

3. On the coast of Alaska, archaeologists have unearthed an ancient burial site that contains the skeletal remains of some Native Americans alongside some very large kayaks (a type of paddled boat). The humerus bones show unusually pronounced deltoid tuberosities. Why/how might these projections have grown so large?

4. Amy was visiting her Grandmother Amelia in the hospital. Grandmother said she had a broken hip but when Amy peeked at the chart, it said she had a fractured femur. Explain the discrepancy. Meanwhile, Grandpa Jeremiah is complaining that his "sacroiliac is acting up," making it painful for him to stand or sit for long or to walk very far. Does Grandpa really know what he's talking about? Explain.

5. Derrick, the high school gym teacher, jogs on the same banked high school track at the same time and in the same direction every day. Recently he has been experiencing an ache in one knee, especially after he sits and reads the newspaper following his jog. Nothing else in his routine has changed. What could be the problem with his knee?

? ANSWERS TO FIGURE QUESTIONS

8.1 The pectoral girdle is made up of the clavicle and scapula. Each free upper limb includes the humerus, ulna, radius, carpals, metacarpals, and phalanges.

8.2 The weakest part of the clavicle is its midregion at the junction of the two curves.

8.3 The acromion forms the high point of the shoulder.

8.4 The radius articulates at the elbow with the capitulum and radial fossa of the humerus. The ulna articulates at the elbow with the trochlea, coronoid fossa, and olecranon fossa of the humerus.

8.5 The olecranon is the "elbow" part of the ulna.

8.6 The scaphoid is the most frequently fractured wrist bone.

8.7 The pelvic girdle is made up of the hip bones. The free lower limb includes the femur, patella, tibia, fibula, tarsals, metatarsals, and phalanges.

8.8 The femur articulates with the acetabulum of the hip bone; the sacrum articulates with the auricular surface of the hip bone.

8.9 The pelvic axis is the course taken by a baby's head as it descends through the pelvis during childbirth.

8.10 The angle of convergence of the femurs is greater in females than males because the female pelvis is broader.

8.11 The tibia is the weight-bearing bone of the leg.

8.12 The talus is the only tarsal bone that articulates with the tibia and fibula.

8.13 Most of the skeletal system arises from embryonic mesoderm (skull bones arise from ectoderm).

9 | JOINTS

INTRODUCTION An automotive engine is a complex machine that consists of numerous fixed and moving parts, many of which rub against each other and generate considerable *friction* (resistance between rubbing parts). The various friction-generating parts, which produce the necessary forces required to power the movement of the car, have a limited lifetime because of constant wear and tear. Most people are thrilled if they get 10 to 15 years of use from their car's engine, but our bodies, which endure similar wear and tear, must last a lifetime.

The human skeleton needs to move, but bones are too rigid to bend without being damaged. Fortunately, flexible connective tissues hold bones together while still permitting, in most cases, some degree of movement. A **joint**, also called an **articulation** (ar-tik′-ū-LĀ-shun) or **arthrosis** (ar-THRŌ-sis), is a point of contact between two bones, between bone and cartilage, or between bone and teeth. Think for a moment about the amazing range of motion and the complexity of the coordinated movements that occur as the bones of the body move against one another; movements such as hitting a golf ball or playing a piano are far more complex than those of almost any machine. Many joint actions are repeated daily and produce continuous work from childhood, into adolescence, and throughout our adult lives. How does the structure of a joint make this incredible staying power possible? Why do joints sometimes fail and cause our movements to become painful? How can we prolong the efficient function of our joints? Read on to answer these questions as you learn about the structure and function of the machinery that allows you to go about your everyday activities. The scientific study of joints is termed **arthrology** (ar-THROL-ō-jē; *arthr-*=joint; *ology*=study of). The study of motion of the human body is called **kinesiology** (ki-nē-sē-OL-ō-jē; *kinesi-*=movement). •

Did you ever wonder why pitchers so often require rotator cuff surgery?

248

9.1 JOINT CLASSIFICATIONS

OBJECTIVES
• Describe the structural and functional basis for the classifications of joints.
• Explain the importance of ligaments at joints.

Bones actually do not contact one another directly. Some other connective tissue is always present between the surfaces of the bones that contact one another to make up joints such as your knees, ankles, wrists, and elbows. Joints have a wide variety of structures and just as many functions. The most familiar are junctions between neighboring bones that allow the skeleton to perform actions such as kicking a ball or raising your hand in class, but many joints permit little to no motion. Why do some joints produce a wide range of movements and others allow little to no movement at all? Keep reading and you will soon be able to answer this important question.

Over the years anatomists have used a variety of schemes to classify the various joints of the body; some are based on joint structure, and others use joint movements. Classification schemes based on movement place joints with similar structure in different categories, which is not logical from an anatomical perspective. *Because present-day classification and the current anatomical nomenclature are based on the logic of joint structure, we classify joints by structure in this chapter.* However, we will also make reference to the functional classification of joints.

The functional classification of joints relates to the degree of movement they permit. Functionally, joints are classified as one of the following types:

• **Synarthrosis** (sin′-ar-THRŌ-sis; *syn-*=together; plural is *synarthroses*). An immovable joint.
• **Amphiarthrosis** (am′-fē-ar-THRŌ-sis; *amphi-*=on both sides; plural is *amphiarthroses*). A slightly movable joint.
• **Diarthrosis** (dī-ar-THRŌ-sis=movable joint; plural is *diarthroses*). A freely movable joint; these come in a variety of shapes and permit several different types of movements.

The structure of joints includes the tissues that connect the neighboring bones. At the simplest level, there are two basic ways bones connect with one another to form joints. Bones are either held together by solid masses of connective tissue, or they are joined by a connective tissue capsule that surrounds a lubricated cavity.

Joints that are formed by a solid mass of connective tissue between the neighboring bones include **fibrous joints** (FĪ-brus), which have connective tissue masses of dense irregular connective tissue, and **cartilaginous joints** (kar-ti-LAJ-i-nus), which use some type of cartilage as the connecting tissue between the bones.

Joints that incorporate a lubricated cavity called a synovial cavity are referred to as **synovial joints** (si-NŌ-vē-al; *syn*=together; *ovial*=egg, because the synovial tissues resemble uncooked egg whites). Synovial joints are divided into various types based on the structure of the joint surfaces, but all share the common design of a joint capsule surrounding a lubricated synovial cavity (described shortly).

As a general rule, the more mobile joints are less stable. The highly mobile shoulder joint is the least stable joint in the body; for this reason it is one of the most commonly dislocated joints. The sutures of the skull, which have little to no mobility, are very stable joints. However, because the highly mobile hip joint has the strongest ligaments in the human body, it is more stable than the less mobile elbow, knee, and ankle joints.

The different structural relationships of the binding tissues, the types of connective tissues involved, and the structure of the opposing bone surfaces become the defining factors for further joint classification (see Table 9.1).

Ligaments

Throughout this chapter you will encounter the term *ligament*. **Ligaments** (*ligare*=to bind) refer to dense irregular or dense regular connective tissue structures that bind one bone to another bone (see Figure 9.15). They come in a variety of forms and are integral parts of joints. Ligaments can serve as intrinsic binding structures within the joint itself (such as the sutural ligaments of the skull or the periodontal ligaments of the teeth), or as extrinsic supporting bands that stabilize joints while limiting their range of motion (such as the anterior cruciate ligament of the knee). As you study the various joints of the body in the sections that follow, you will learn about the structures, locations, and functions of a variety of ligaments.

 CHECKPOINT
1. On what basis are joints classified?
2. Describe the importance of ligaments at joints.

9.2 FIBROUS JOINTS

 OBJECTIVE

• Describe the structure and functions of the three types of fibrous joints.

Fibrous joints are joints in which the neighboring bones are joined together by a solid mass of dense irregular connective tissue. The adjoining connective tissue can vary from small fibrous strands of connective tissue, to large thick bands, to extensive membranous sheets. There are three types of fibrous joints: sutures, syndesmoses, and interosseous membranes.

Sutures

A **suture** (SOO-chur; *sutur-*=seam) is a fibrous joint composed of a thin layer of dense irregular connective tissue called the *sutural ligament*. Sutures are found only between bones of the skull. An example is the coronal suture between the frontal and parietal bones (Figure 9.1a). The irregular, interlocking edges of sutures give them added strength and decrease their chance of

fracturing. Sutures are joints that form as the numerous bones of the skull come in contact during development. They are immovable or slightly movable joints. In older individuals, sutures are immovable, but in infants and children they are slightly movable (Figure 9.1b). Sutures play important roles as sites of growth and shock absorption in the skull.

Some sutures, although present during growth of the skull, are replaced by bone in the adult. Such a suture is called a **synostosis** (sīn′-os-TŌ-sis; *os-*=bone), or bony joint—a joint in which there is a complete fusion of two separate bones into one. For example, the frontal bone grows in halves that eventually join together across a suture line. Usually they are completely fused by age 6 and the suture becomes obscure. If the suture persists beyond age 6, it is called a **frontal** or **metopic suture** (me-TŌ-pik; *metopon-*=forehead). A frontal suture is an immovable joint.

Syndesmoses

A **syndesmosis** (sin′-dez-MŌ-sis; *syndesmo-*=band or ligament) is a fibrous joint in which there is a greater distance between

Figure 9.1 Fibrous joints.

🔑 At a fibrous joint the bones are held together by dense irregular connective tissue.

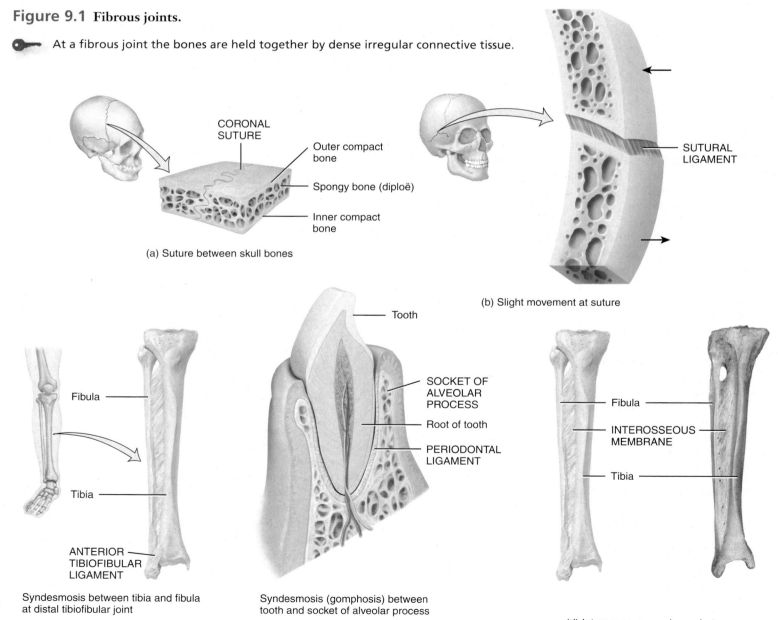

CORONAL SUTURE

Outer compact bone

Spongy bone (diploë)

Inner compact bone

(a) Suture between skull bones

SUTURAL LIGAMENT

(b) Slight movement at suture

Tooth

Fibula

Tibia

ANTERIOR TIBIOFIBULAR LIGAMENT

SOCKET OF ALVEOLAR PROCESS

Root of tooth

PERIODONTAL LIGAMENT

Fibula

INTEROSSEOUS MEMBRANE

Tibia

Syndesmosis between tibia and fibula at distal tibiofibular joint

Syndesmosis (gomphosis) between tooth and socket of alveolar process

(c) Syndesmosis

(d) Interosseous membrane between diaphyses of tibia and fibula

❓ **What are the different types of fibrous joints, and where are they found in the body?**

the articulating surfaces and more dense irregular connective tissue than in a suture. The dense irregular connective tissue is typically arranged as a bundle (ligament) and the joint permits limited movement. One example of a syndesmosis is the distal tibiofibular joint, where the anterior tibiofibular ligament connects the tibia and fibula (Figure 9.1c, left). Another example of a syndesmosis is called a **gomphosis** (gom-FŌ-sis; *gompho-*=bolt or nail) or *dentoalveolar joint*, in which a cone-shaped peg fits into a socket. The only examples of gomphoses in the human body are the articulations between the roots of the teeth and their sockets (alveoli) in the maxillae and mandible (Figure 9.1b, right). The dense irregular connective tissue between a tooth and its socket is the thin periodontal ligament (membrane). A healthy gomphosis permits no movement. Inflammation and degeneration of the gums, periodontal ligament, and bone is called *periodontal disease*.

Interosseous Membranes

The final category of fibrous joints is the interosseous membrane. An **interosseous membrane** (in'-ter-OS-ē-us) is a substantial sheet of dense irregular connective tissue that binds neighboring long bones and permits slight movement. There are two principal interosseous membrane joints in the human body. One occurs between the radius and ulna in the forearm (see Figure 8.5a, b) and the other occurs between the tibia and fibula in the leg (Figure 9.1d). These strong connective tissue sheets not only help hold these adjacent long bones together, they also play an important role in defining the range of motion between the neighboring bones and provide an increased attachment surface for muscles that produce movements of the digits of the hand and foot.

 CHECKPOINT

3. How are the three types of fibrous joints similar? How do they differ?

9.3 CARTILAGINOUS JOINTS

 OBJECTIVE

• Describe the structure and functions of the two types of cartilaginous joints.

In a **cartilaginous joint**, there is solid connective tissue that allows little or no movement. The articulating bones are tightly connected, either by hyaline cartilage or by fibrocartilage (see Table 3.6). The two types of cartilaginous joints are synchondroses and symphyses.

Synchondroses

Not to be confused with the syndesmosis you just read about in the previous section, a **synchondrosis** (sin'-kon-DRŌ-sis; *chondro-*=cartilage; plural is *synchondroses*) is an immovable, cartilaginous joint in which the connecting material is hyaline cartilage. An example of a synchondrosis is the epiphyseal (growth) plate that connects the epiphysis and diaphysis of a growing bone (Figure 9.2a). A photomicrograph of the epiphyseal plate is shown in Figure 6.8b. When bones stop growing in length, bone replaces the hyaline cartilage, and the synchondrosis becomes a synostosis, or bony joint (see Section 9.2).

 CLINICAL CONNECTION | Synchondroses and Bone Growth

In an x-ray of a young person's skeleton, the synchondroses are easily seen as thin dark areas between the white-appearing bone tissue (see Figure 6.8a). This is how a medical professional can view an x-ray and determine that a person still has some growing to do. Breaks in a bone that extend into the epiphyseal plate and damage the cartilage of the synchondrosis can affect further growth of the bone, resulting in abbreviated development in length of the bone, which leads to a bone of shorter length. •

Symphyses

A **symphysis** (SIM-fi-sis=growing together; plural is *symphyses*) is a cartilaginous joint in which the ends of the articulating bones are covered with hyaline cartilage, but the bones are connected by a broad, flat disc of fibrocartilage. All symphyses occur in the midline of the body. The pubic symphysis between the anterior surfaces of the hip bones, which is slightly movable, is one example of a symphysis (Figure 9.2b). This type of joint is also found at the junction of the manubrium and body of the sternum (see Figure 7.21a) and at the intervertebral joints between the bodies of vertebrae (see Figure 7.15a). A portion of the intervertebral disc is made up of fibrocartilage. The structure of the intervertebral disc helps to provide the limited range of motion in the vertebral column while serving as an important shock-absorbing pad between the vertebral bodies.

 CHECKPOINT

4. How do the functions of symphyses and synchondroses differ? Where is each found in the body?

Figure 9.2 Cartilaginous joints.

 At a cartilaginous joint the bones are held together by cartilage.

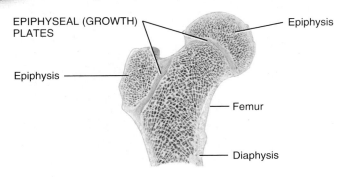

EPIPHYSEAL (GROWTH) PLATES

Epiphysis

Epiphysis

Femur

Diaphysis

(a) Synchondrosis

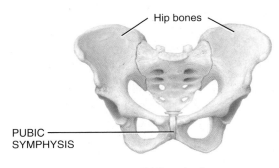

Hip bones

PUBIC SYMPHYSIS

(b) Symphysis

 What is the structural difference between a synchondrosis and a symphysis?

9.4 SYNOVIAL JOINTS

OBJECTIVES

• Describe the structure of synovial joints.
• Outline the structure and function of bursae and tendon sheaths.
• List the six types of synovial joints.

Structure of Synovial Joints

The characteristic of a synovial joint that distinguishes it from other types of joints is the presence of a space called a **synovial (joint) cavity**, which is surrounded by a connective tissue capsule that attaches the articulating bones (Figure 9.3). Synovial joints range from slightly movable to the most mobile joints of the body. For example, the synovial joints between some of the carpal bones have very limited movement, but the shoulder joint can move freely in all directions. The bone surfaces within the capsule of a synovial joint are covered by a layer of hyaline cartilage called **articular cartilage**. The smooth cartilage covers the articulating surfaces of the bones but does not bind them together. The lubricated articular cartilage reduces friction between bones in the joint during movement and helps to absorb shock.

Articular Capsule

Not to be confused with articular cartilage, a sleevelike **articular (joint) capsule** surrounds a synovial joint, encloses the synovial cavity, and unites the articulating bones. The articular capsule is composed of two layers, an outer fibrous membrane and an inner synovial membrane (Figure 9.3a). The **fibrous membrane** usually consists of dense irregular connective tissue (mostly collagen fibers) that attaches to the periosteum of the articulating bones; it is literally a thickened continuation of the periosteum between the two bones. The flexibility of the fibrous membrane permits considerable movement at a joint, and its great tensile strength (resistance to stretching) helps prevent the bones from dislocating, the displacement of a bone from a joint. The fibers of some fibrous membranes are thicker and arranged in parallel bundles that are highly adapted for resisting tensile strains; fiber bundles in such an arrangement are one type of ligament (see Section 9.1). The strength of ligaments is one of the principal mechanical factors that holds bones close together in a synovial joint. The inner layer of the articular capsule, the **synovial membrane**, is composed of a thin arrangement of synovial cells on the surface and areolar connective tissue with elastic fibers beneath. At many synovial joints, the synovial membrane includes accumulations of adipose tissue. The adipose tissue varies in thickness; the thicker regions are called **articular fat pads**. An example is the infrapatellar fat pad in the knee (see Figure 9.15c, d).

A **"double-jointed"** person does not really have extra joints. Individuals who are double-jointed have greater flexibility in their articular capsules and ligaments; the resulting increase in range of motion allows them to entertain fellow partygoers with activities such as touching their thumbs to their wrists and putting their ankles or elbows behind their necks. Unfortunately, such flexible joints are less structurally stable and are more easily dislocated.

Synovial Fluid

The synovial membrane secretes **synovial fluid** (*ov-*=egg), which forms a thin film over the surfaces within the articular capsule. This viscous, clear, or pale-yellow fluid was named for its similarity in appearance and consistency to uncooked egg white. Synovial fluid consists of hyaluronic acid secreted by synovial cells in the synovial membrane and interstitial fluid filtered from blood plasma. Its functions include reducing friction by lubricating the joint, and absorbing shocks. The synovial fluid

Figure 9.3 Structure of a typical synovial joint. Note the two layers of the articular capsule—the fibrous membrane and the synovial membrane. Synovial fluid lubricates the synovial cavity, which is located between the synovial membrane and the articular cartilage.

🔑 The distinguishing feature of a synovial joint is the synovial cavity between the articulating bones.

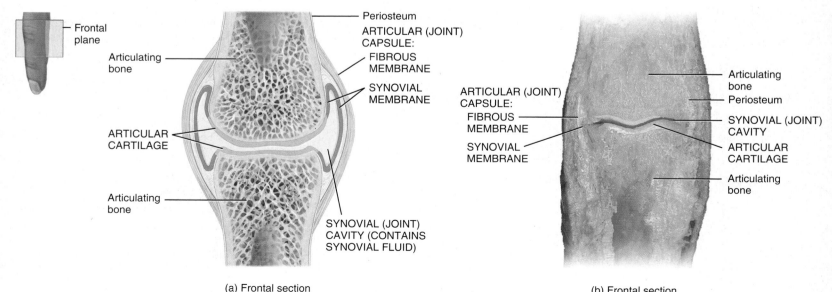

(a) Frontal section

(b) Frontal section

❓ **What structure of synovial joints allows efficient movement between neighboring bones?**

Unlike muscles, ligaments are *noncontractile tissues* (they cannot contract or shorten). Consequently, they resist stretching and are typically damaged when forces are exerted that exceed their strength. Damage to these connective tissue structures, referred to as a **sprain**, is caused by the forcible wrenching or twisting of a joint that stretches or tears its ligaments but does not dislocate the bones. Sprains also may damage surrounding blood vessels, muscles, tendons, or nerves. Severe sprains may be so painful that the joint cannot be moved. There is considerable swelling, which results from hemorrhage of ruptured blood vessels. The ankle is the joint most often sprained; the numerous small joints between the vertebrae of the lower back are also the sites of frequent strains. Ligament injuries are graded on a scale from I to III. *Grade I sprains* involve stretching of a ligament with microscopic tearing of collagen fibers. There is little swelling, little or no loss of function or joint stability, and the person can fully or partially bear weight. *Grade II sprains* involve stretching of a ligament with partial tearing, moderate-to-severe swelling, bruising, moderate functional loss, mild-to-moderate joint instability, and difficulty bearing weight. *Grade III* sprains involve complete rupture of a ligament, immediate and severe swelling, bruising, moderate-to severe joint instability, and inability to bear weight without severe pain.

Initially, sprains should be treated with **PRICE**: protection, rest, ice, compression, and elevation. PRICE therapy may be used on muscle strains (stretched or partially torn muscles), joint inflammation, suspected fractures, and bruises.

The five components of PRICE therapy are:

- **Protection** means protecting the injury from further damage. For example, stop the activity, use padding and protection, and use splints or a sling, or crutches, if necessary.
- **Rest** the injured area to avoid further damage to the tissues. Stop the activity immediately. Avoid exercise or other activities that cause pain or swelling to the injured area. Rest is needed for repair, and exercising before an injury has healed may increase the probability of re-injury.
- **Ice** the injured area as soon as possible. Applying ice slows blood flow to the area, reduces swelling, and relieves pain. Ice works effectively when applied for 20 minutes, off for 40 minutes, back on for 20 minutes, and so on.
- **Compression** by wrap or bandage helps to reduce swelling. Care must be taken to compress the injured area without blocking blood flow.
- **Elevation** of the injured area above the level of the heart, when possible, will reduce potential swelling. •

also supplies oxygen and nutrients to the chondrocytes within articular cartilage, and removes carbon dioxide and metabolic wastes (recall that cartilage is an avascular tissue, so it does not have blood vessels to perform these two functions). Synovial fluid also contains phagocytic cells that remove microbes and the debris that results from normal wear and tear in the joint. When a synovial joint is immobile for a time, the fluid becomes quite viscous (gel-like), but as joint movement increases, the fluid becomes less viscous. One of the benefits of exercise warmups is that they stimulate the production and secretion of synovial fluid; within limits, more fluid means less stress on the joints during exercise.

We are all familiar with the cracking sounds heard as certain joints move, or the popping sounds that arise when a person pulls on his or her fingers to "crack" the knuckles. According to one theory, when the synovial cavity expands, the pressure inside the synovial cavity decreases, creating a partial vacuum. The suction draws carbon dioxide and oxygen out of blood vessels in the synovial membrane, forming bubbles in the fluid. When the joint is flexed (bent), the volume of the cavity decreases and the pressure increases; this bursts the bubbles and creates the cracking or popping sound as the gases are driven back into solution.

Accessory Ligaments, Articular Discs, and Labra

Many synovial joints also contain **accessory ligaments** called extracapsular ligaments and intracapsular ligaments. *Extracapsular ligaments* lie outside the articular capsule, such as the fibular and tibial collateral ligaments of the knee joint (see Figure 9.15a, e). *Intracapsular ligaments* occur within the articular capsule but are excluded from the synovial cavity by folds of the synovial membrane. The anterior and posterior cruciate ligaments of the knee joint are intracapsular ligaments (see Figure 9.15a, h).

A few synovial joints in the body contain **articular discs**, fibrocartilage structures not covered by synovial membrane that divide the synovial cavity into two smaller cavities (see Figure 9.11c).

As you will see later, separate movements can occur in the two cavities, such as those of the temporomandibular joint (TMJ) (see Section 9.7). Peripherally, articular discs attach to the fibrous membrane. In some joints, incomplete discs called **menisci** (me-NIS-sī or me-NIS-kī; singular is *meniscus*) partially divide the joint (see Figure 9.15c, d, g, h). Not to be confused with the menisci you look for in the test tubes in lab, the menisci of joints are crescent-shaped discs that are prominent features of the knee joint. Like the articular discs, during development menisci are sandwiched between the fibrous membrane and synovial membrane and bind strongly to the inside of the fibrous membrane. As the joint becomes functional, the synovial membrane is worn off the surface wherever the meniscus experiences considerable friction. The covering synovial membrane, then, stops at the base of the meniscus and does not continue onto its articular surface. The functions of the menisci are not completely understood, but are known to include the following: (1) absorbing shocks; (2) allowing a better fit between articulating bony surfaces; (3) providing adaptable surfaces for combined movements; (4) distributing weight over a greater contact surface; and (5) distributing synovial lubricant across the articular surfaces of the joint.

A **labrum** (LĀ-brum; the plural is *labra*), prominent in the ball-and-socket joints of the shoulder and hip (see Figures 9.12c, d and 9.14d, e), is the fibrocartilaginous lip that extends from the edge of the joint socket. Triangular in transverse section as it projects from the edge of the joint socket, the labrum helps deepen the joint socket and increases the area of contact between the socket and the ball-like surface of the head of the humerus or the femur.

Nerve and Blood Supply

The nerves that supply a joint are the same as those that supply the skeletal muscles that move the joint. Synovial joints contain many nerve endings that are distributed to the articular capsule and associated ligaments. Some of the nerve endings convey information about pain from the joint to the spinal cord and brain for processing. Others respond to the degree of movement and stretch at a joint, such as when the doctor hits the tendon below your kneecap to test your reflexes. This information is also relayed to the spinal cord and brain, which may respond by sending impulses through different nerves to the muscles to adjust body movements.

Although many of the components of synovial joints are avascular, arteries in the vicinity send out numerous branches that penetrate the ligaments and articular capsule to deliver oxygen and nutrients. Branches from several different arteries typically join together around a synovial joint before penetrating the articular capsule. The chondrocytes in the articular cartilage receive oxygen and nutrients from synovial fluid, which is derived from capillary blood; all other joint tissues are supplied directly by capillaries. Carbon dioxide and wastes pass from chondrocytes of articular cartilage into synovial fluid and then into lymphatic capillaries in the surrounding tissues; carbon dioxide and wastes from all other joint structures pass directly into capillaries that drain into veins.

Bursae and Tendon Sheaths

The various movements of the body create considerable friction between its moving parts. Saclike structures called **bursae** (BER-sē=purses; singular is *bursa*) are strategically situated to alleviate friction around some joints, such as the shoulder and knee joints (see Figures 9.12 a–c and 9.15c–e). Bursae are not strictly parts of synovial joints, but they do resemble joint capsules because their walls consist of an outer fibrous membrane of thin dense connective tissue lined by a synovial membrane. Bursae are filled with a small amount of fluid similar to synovial fluid, and are located between the skin and bone, tendons and bones, muscles and bones, and ligaments and bones. The lubricated bursal sacs reduce friction during the movement of these adjacent body parts.

Structures called tendon sheaths also reduce friction at joints. **Tendon (synovial) sheaths** are tubelike bursae that wrap around tendons experiencing considerable friction on all sides as they pass through *fibro-osseous tunnels* (tunnels formed by connective tissue bands and bone). The tendon sheath protects all sides of the tendon from friction within the tunnel. The tendon of the biceps brachii muscle at the shoulder joint (see Figure 9.12c) and the tendons of wrist, ankle, fingers, and toes are two examples of tendons surrounded by tendon sheaths.

CLINICAL CONNECTION | Bursitis

An acute or chronic inflammation of a bursa, called **bursitis** (bur-SĪ-tis), is usually caused by irritation from repeated, excessive exertion of a joint. The condition may also be caused by trauma, an acute or chronic infection (including syphilis and tuberculosis), or by rheumatoid arthritis. Symptoms include pain, swelling, tenderness, and limited movement. •

✓ CHECKPOINT

5. What are the functions of articular cartilage, synovial fluid, and articular discs?
6. What types of sensations are perceived at joints, and from what sources do joints receive nourishment?
7. In what ways are bursae similar to joint capsules? How do they differ?

9.5 TYPES OF MOVEMENTS AT SYNOVIAL JOINTS

 OBJECTIVE

• Describe the types of movements that can occur at synovial joints.

Anatomists, physical therapists, and kinesiologists (professionals who study the science of human movement and look for ways to improve the efficiency and performance of the human body at work, in sports, and in daily activities) use specific terminology to designate the movements that can occur at synovial joints. These precise terms may indicate the form of motion, the direction of movement, or the relationship of one body part to another during movement. Movements at synovial joints are grouped into four main categories: (1) gliding, (2) angular movements, (3) rotation, and (4) special movements, which occur only at certain joints.

Gliding

Gliding is a simple movement in which nearly flat bone surfaces move from side-to-side and back-and-forth with respect to one another (Figure 9.4). There is no significant alteration of the angle between the bones. Gliding movements are limited in range due to the structure of the articular capsule and associated ligaments and bones; however, these sliding movements can also be combined with rotation. The intercarpal and intertarsal joints are examples of articulations where gliding movements occur.

Angular Movements

In **angular movements**, there is an increase or a decrease in the angle between articulating bones. The major angular movements are flexion, extension, lateral flexion, hyperextension, abduction, adduction, and circumduction. These movements are discussed with respect to the body in the anatomical position (see Figure 1.2).

Figure 9.4 Gliding movements at synovial joints.

 Gliding movements consist of side-to-side and back-and-forth motions.

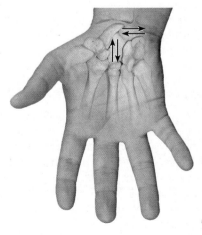

Gliding between carpals (arrows)

 What types of synovial joints permit gliding movements?

Flexion, Extension, Lateral Flexion, and Hyperextension

Flexion and extension are opposite movements. In **flexion** (FLEK-shun; *flex-*=to bend) there is a decrease in the angle between articulating bones; in **extension** (eks-TEN-shun; *exten-*=to stretch out) there is an increase in the angle between articulating bones, often to restore a part of the body to the anatomical position after it has been flexed (Figure 9.5). Both movements usually occur along the sagittal plane. All of the following are examples of flexion (as you have probably already guessed, extension is simply the reverse of these movements):

- Bending the head toward the chest at the atlanto-occipital joints between the atlas (the first vertebra) and the occipital bone of the skull, and at the cervical intervertebral joints between the cervical vertebrae (Figure 9.5a)
- Bending the trunk forward at the intervertebral joints, as in doing a crunch with your abdominal muscles
- Moving the humerus forward at the shoulder joint, as in swinging the arms forward while walking (Figure 9.5b)
- Moving the forearm toward the arm at the elbow joint, between the humerus, ulna, and radius, that is, bending your elbow (Figure 9.5c)
- Moving the palm toward the forearm at the wrist or radiocarpal joint between the radius and carpals, as in the upward movement when doing wrist curls (Figure 9.5d)

- Bending the digits of the hand at the interphalangeal joints between phalanges, as in clenching your fingers to make a fist
- Moving the femur forward at the hip joint between the femur and hip bone, as in walking (Figure 9.5e)
- Moving the heel toward the buttock at the tibiofemoral joint between the tibia, femur, and patella, as occurs when bending the knee (Figure 9.5f)

Although flexion and extension usually occur along the sagittal plane, there are a few exceptions. For example, flexion of the thumb involves movement of the thumb medially across the palm at the carpometacarpal joint between the trapezium and metacarpal of the thumb, as when you touch your thumb to the opposite side of your palm (see Figure 11.21a). Another example is movement of the trunk sideways to the right or left at the waist. This movement, which occurs along the frontal plane and involves the intervertebral joints, is called **lateral flexion** (Figure 9.5g).

Continuation of extension beyond the anatomical position is called **hyperextension** (hī-per-ek-STEN-shun; *hyper-*=beyond or excessive). Examples of hyperextension include:

- Bending the head backward at the atlanto-occipital and cervical intervertebral joints, as in looking up at stars (Figure 9.5a)
- Bending the trunk backward at the intervertebral joints, as in a backbend

Figure 9.5 Angular movements at synovial joints—flexion, extension, hyperextension, and lateral flexion.

🔑 In angular movements, there is an increase or decrease in the angle between articulating bones.

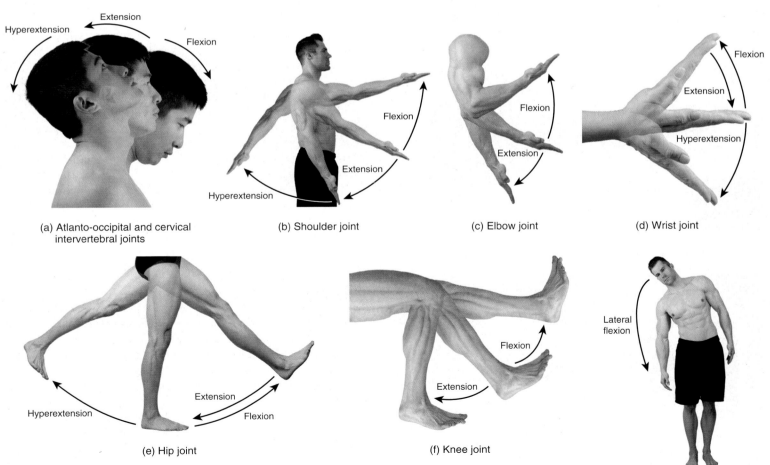

(a) Atlanto-occipital and cervical intervertebral joints

(b) Shoulder joint

(c) Elbow joint

(d) Wrist joint

(e) Hip joint

(f) Knee joint

(g) Intervertebral joints

 What are two examples of flexion that do not occur along the sagittal plane?

- Moving the humerus backward at the shoulder joint, as in swinging the arms backward while walking (Figure 9.5b)
- Moving the palm backward at the wrist joint, as in preparing to shoot a basketball (Figure 9.5d)
- Moving the femur backward at the hip joint, as in walking (Figure 9.5e)

Hyperextension of hinge joints, such as the elbow, interphalangeal, and knee joints, is usually prevented by the arrangement of ligaments and the anatomical alignment of the bones.

Abduction, Adduction, and Circumduction

Abduction (ab-DUK-shun; *ab-*=away; *-duct*=to lead) or *radial deviation* is the movement of a bone away from the midline; **adduction** (ad-DUK-shun; *ad-*=toward) or *ulnar deviation* is the movement of a bone toward the midline. Both movements usually occur along the frontal plane. Examples of abduction include moving the humerus laterally at the shoulder joint, moving the palm laterally at the wrist joint, and moving the femur laterally at the hip joint (Figure 9.6a–c). The movement that returns each of these body parts to the anatomical position is adduction (Figure 9.6a–c).

The midline of the body is *not* used as a point of reference for abduction and adduction of the digits. In abduction of the fingers (but not the thumb), an imaginary line is drawn through the longitudinal axis of the middle (longest) finger, and the fingers

Figure 9.6 Angular movements at synovial joints—abduction and adduction.

 Abduction and adduction usually occur along the frontal plane.

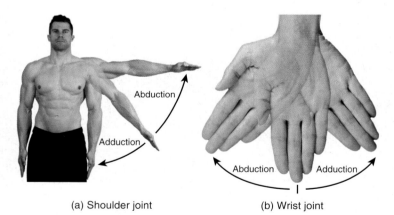

(a) Shoulder joint

(b) Wrist joint

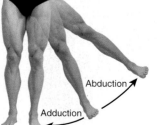

(c) Hip joint

(d) Metacarpophalangeal joints of the fingers (not the thumb)

 In what way is considering adduction as "adding your limb to your trunk" an effective learning device?

Figure 9.7 Angular movements at synovial joints—circumduction.

 Circumduction is the movement of the distal end of a body part in a circle.

(a) Shoulder joint

(b) Hip joint

Which movements in continuous sequence produce circumduction?

move away (spread out) from the middle finger (Figure 9.6d). In abduction of the thumb, the thumb moves away from the palm in the sagittal plane (see Figure 11.21a). Abduction of the toes is relative to an imaginary line drawn through the second toe. Adduction of the fingers and toes returns them to the anatomical position. Adduction of the thumb moves the thumb toward the palm in the sagittal plane (see Figure 11.21a).

Circumduction (ser-kum-DUK-shun; *circ-*=circle) is movement of the distal end of a body part in a circle (Figure 9.7). Circumduction is not an isolated movement by itself but rather a continuous sequence of flexion, abduction, extension, adduction, and rotation of the joint (or in the opposite order). It does not occur along a separate axis or plane of movement. Examples of circumduction are moving the humerus in a circle at the shoulder joint (Figure 9.7a), moving the hand in a circle at the wrist joint, moving the thumb in a circle at the carpometacarpal joint, moving the fingers in a circle at the metacarpophalangeal joints (between the metacarpals and phalanges), and moving the femur in a circle at the hip joint (Figure 9.7b). Both the shoulder and hip joints permit circumduction. Flexion, abduction, extension, and adduction are more limited in the hip joints than in the shoulder joints due to the tension on certain ligaments and muscles and the depth of the acetabulum in the hip joint (see Exhibits 9.B and 9.D).

Rotation

In **rotation** (rō-TĀ-shun; *rota-*=revolve), a bone revolves around its own longitudinal axis. One example is turning the head from side-to-side at the atlanto-axial joint (between the atlas and the axis), as when you shake your head "no" (Figure 9.8a). Another is turning the trunk from side-to-side at the intervertebral joints while keeping the hips and lower limbs in the anatomical position. In the limbs, rotation is defined relative to the midline, and specific qualifying terms are used. If the anterior surface of a bone of the limb is turned toward the midline, the movement is called *medial (internal) rotation*. You can medially rotate the humerus at the shoulder joint as follows: Starting in the anatomical position, flex your elbow and then move your palm across the chest (Figure 9.8b). You can

Figure 9.8 Rotation at synovial joints.

 In rotation, a bone revolves around its own longitudinal axis.

Rotation

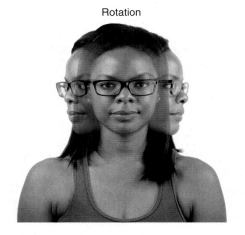

(a) Atlanto-axial joints

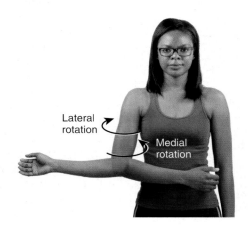

(b) Shoulder joint

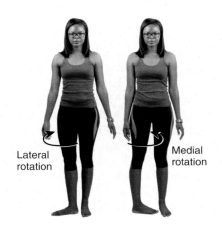

(c) Hip joint

 How do medial and lateral rotation differ?

medially rotate the femur at the hip joint as follows: Lie on your back, bend your knee, and then move your leg and foot laterally from the midline. Although you are moving your leg and foot laterally, the femur is rotating medially (Figure 9.8c). Medial rotation of the leg at the knee joint can be produced by sitting on a chair, bending your knee, raising your lower limb off the floor, and turning your toes medially. If the anterior surface of the bone of a limb is turned away from the midline, the movement is called *lateral (external) rotation* (see Figure 9.8b, c).

Special Movements

Special movements occur only at certain joints. These include elevation, depression, protraction, retraction, inversion, eversion, dorsiflexion, plantar flexion, supination, pronation, and opposition (Figure 9.9):

- **Elevation** (el-e-VĀ-shun = to lift up) is a superior movement of a part of the body, such as closing the mouth at the temporomandibular joint (between the mandible and temporal bone)

Figure 9.9 Special movements at synovial joints.

 Special movements occur only at certain synovial joints.

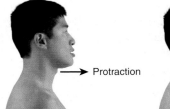

(a)   Temporomandibular joints   (b)

(c)   Temporomandibular joints   (d)

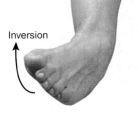

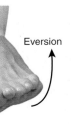

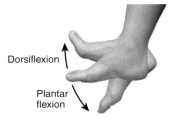

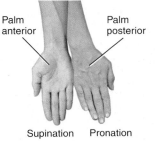

(e)   Intertarsal joints   (f)

(g) Ankle joint

(h) Radioulnar joints

(i) Carpometacarpal joint

 What movement of the shoulder girdle occurs when you bring your arms forward until the elbows touch?

to elevate the mandible (Figure 9.9a) or shrugging the shoulders at the acromioclavicular joint to elevate the scapula and clavicle. Its opposing movement is depression. Other bones that may be elevated (or depressed) include the hyoid and ribs.

- **Depression** (de-PRESH-un = to press down) is an inferior movement of a part of the body, such as opening the mouth to depress the mandible (Figure 9.9b) or returning shrugged shoulders to the anatomical position to depress the scapula and clavicle.
- **Protraction** (prō-TRAK-shun = to draw forth) is a movement of a part of the body anteriorly in the transverse plane. Its opposing movement is retraction. You can protract your mandible at the temporomandibular joint by thrusting it outward (Figure 9.9c) or protract your clavicles at the acromioclavicular and sternoclavicular joints by crossing your arms.
- **Retraction** (rē-TRAK-shun=to draw back) is a movement of a protracted part of the body back to the anatomical position (Figure 9.9d).
- **Inversion** (in-VER-zhun=to turn inward) is movement of the sole medially at the intertarsal joints (between the tarsals) (Figure 9.9e). Its opposing movement is eversion. Physical therapists also refer to inversion combined with plantar flexion of the feet as *supination*.
- **Eversion** (ē-VER-zhun=to turn outward) is a movement of the sole laterally at the intertarsal joints (Figure 9.9f). Physical therapists also refer to eversion combined with dorsiflexion of the feet as *pronation*.
- **Dorsiflexion** (dor-si-FLEK-shun) refers to bending of the foot at the ankle or talocrural joint (between the tibia, fibula, and talus) in the direction of the dorsum (superior surface)

(Figure 9.9g). Dorsiflexion occurs when you stand on your heels. Its opposing movement is plantar flexion.
- **Plantar flexion** (PLAN-tar) involves bending of the foot at the ankle joint in the direction of the plantar or inferior surface (see Figure 9.9g), as when you elevate your body by standing on your toes.
- **Supination** (soo-pi-NĀ-shun) is a movement of the forearm at the proximal and distal radioulnar joints in which the palm is turned anteriorly (Figure 9.9h). This position of the palms is one of the defining features of the anatomical position. Its opposing movement is pronation.
- **Pronation** (prō-NĀ-shun) is a movement of the forearm at the proximal and distal radioulnar joints in which the distal end of the radius crosses over the distal end of the ulna and the palm is turned posteriorly (Figure 9.9h).
- **Opposition** (op-ō-ZISH-un) is the movement of the thumb at the carpometacarpal joint (between the trapezium and metacarpal of the thumb) in which the thumb moves across the palm to touch the tips of the fingers on the same hand (Figure 9.9i). These "opposable thumbs" allow the distinctive digital movement that gives humans and other primates the ability to grasp and manipulate objects very precisely.

A summary of the movements that occur at synovial joints is presented in Table 9.1.

 CHECKPOINT

8. What are the four major categories of movements that occur at synovial joints?
9. On yourself or with a partner, demonstrate each movement listed in Table 9.1.

TABLE 9.1					
Summary of Movements at Synovial Joints					
MOVEMENT	**DESCRIPTION**		**MOVEMENT**	**DESCRIPTION**	
Gliding	Movement of relatively flat bone surfaces back-and-forth and side-to-side over one another; little change in angle between bones		Rotation	Movement of bone around longitudinal axis; in limbs, may be medial (toward midline) or lateral (away from midline)	
Angular	Increase or decrease in angle between bones		Special	Occurs at specific joints	
Flexion	Decrease in angle between articulating bones, usually in sagittal plane		Elevation	Superior movement of body part	
Lateral flexion	Movement of trunk in frontal plane		Depression	Inferior movement of body part	
Extension	Increase in angle between articulating bones, usually in sagittal plane		Protraction	Anterior movement of body part in transverse plane	
Hyperextension	Extension beyond anatomical position		Retraction	Posterior movement of body part in transverse plane	
Abduction	Movement of bone away from midline, usually in frontal plane		Inversion	Medial movement of sole	
Adduction	Movement of bone toward midline, usually in frontal plane		Eversion	Lateral movement of sole	
Circumduction	Flexion, abduction, extension, adduction, and rotation in succession (or in the opposite order); distal end of body part moves in circle		Dorsiflexion	Bending foot in direction of dorsum (superior surface)	
			Plantar flexion	Bending foot in direction of plantar surface (sole)	
			Supination	Movement of forearm that turns palm anteriorly	
			Pronation	Movement of forearm that turns palm posteriorly	
			Opposition	Movement of thumb across palm to touch fingertips on same hand	

The tearing of menisci in the knee, commonly called **torn cartilage**, occurs often among athletes. Such damaged cartilage will begin to wear and may cause arthritis to develop unless the damaged cartilage is treated surgically. Years ago, if a patient had torn cartilage, the entire meniscus was removed by a procedure called a *meniscectomy* (men'-i-SEK-tō-mē). The problem was that over time the articular cartilage was worn away more quickly. Currently, surgeons perform a partial meniscectomy, in which only the torn segment of the meniscus is removed. Surgical repair of the torn cartilage may be assisted by **arthroscopy** (ar-THROS-kō-pē; *scopy*=observation). This minimally invasive procedure involves examination of the interior of a joint, usually the knee, with an arthroscope, a lighted pencil-thin fiber optic camera used for visualizing the nature and extent of damage. Arthroscopy is also used to monitor the progression of disease and the effects of therapy. The insertion of surgical instruments through other incisions also enables a physician to remove torn cartilage and repair damaged cruciate ligaments in

the knee; obtain tissue samples for analysis; and perform surgery on other joints, such as the shoulder, elbow, ankle, and wrist. •

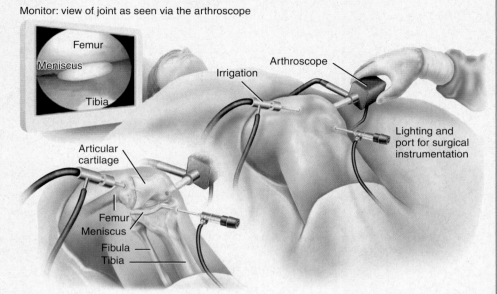

Monitor: view of joint as seen via the arthroscope

9.6 TYPES OF SYNOVIAL JOINTS

● OBJECTIVE

• Describe the six subtypes of synovial joints.

Although all synovial joints share many characteristics in common, the shapes of the articulating surfaces vary; thus, many types of movements are possible. In most joint movements, one bone remains in a fixed position while the other moves around an axis. Synovial joints are divided into six categories based on type of movement: plane, hinge, pivot, condyloid, saddle, and ball-and-socket.

Plane Joints

The articulating surfaces of bones in a **plane joint** (PLĀN), also called a *planar joint* (PLĀ-nar), are flat or slightly curved (Figure 9.10a). Plane joints primarily permit back-and-forth and side-to-side movements between the flat surfaces of bones, but they may also rotate against one another. Many plane joints are *biaxial*, meaning that they permit movement in two axes. An *axis* is a straight line around which a bone rotates (revolves) or slides. If plane joints rotate in addition to sliding, then they are *triaxial (multiaxial)*, permitting movement in three axes. Some examples of plane joints are the intercarpal joints (between carpal bones at the wrist), intertarsal joints (between tarsal bones at the ankle), sternoclavicular joints (between the manubrium of the sternum and the clavicle), acromioclavicular joints (between the acromion of the scapula and the clavicle), sternocostal joints (between the sternum and ends of the costal cartilages at the tips of the second through seventh pairs of ribs), and vertebrocostal joints (between the heads and tubercles of ribs and bodies and transverse processes of thoracic vertebrae).

Hinge Joints

In a **hinge joint**, or *ginglymus joint* (JIN-gli-mus), the convex surface of one bone fits into the concave surface of another bone

(Figure 9.10b). As the name implies, hinge joints produce an angular, opening-and-closing motion like that of a hinged door. Hinge joints are *uniaxial (monaxial)* because they typically allow motion around a single axis. Hinge joints permit only flexion and extension. Examples of hinge joints are the knee (actually a modified hinge joint, which will be described later; see Exhibit 9.E), elbow, ankle, and interphalangeal joints (between the phalanges of the fingers and toes).

Pivot Joints

In a **pivot joint**, or *trochoid joint* (TRŌ-koyd), the rounded or pointed surface of one bone articulates with a ring formed partly by another bone and partly by a ligament (Figure 9.10c). A pivot joint is *uniaxial* because it allows rotation only around its own longitudinal axis. Examples of pivot joints are the atlanto-axial joints, in which the atlas rotates around the axis and permits the head to turn from side to side as when you shake your head "no" (see Figure 9.8a), and the radioulnar joints that enable the palms to turn anteriorly and posteriorly as the head of the radius pivots around its long axis in the radial notch of the ulna (see Figure 9.9h).

Condyloid Joints

In a **condyloid joint** (KON-di-loyd; *condyl-*=knuckle), or *ellipsoidal joint*, the convex oval-shaped projection of one bone fits into the oval-shaped depression of another bone (Figure 9.10d). A condyloid joint is *biaxial* because the movement it permits is around two axes (flexion-extension and abduction–adduction), plus limited circumduction (remember that circumduction is not an isolated movement). Examples of condyloid joints are the radiocarpal (wrist) and metacarpophalangeal joints (between the metacarpals and proximal phalanges) of the second through fifth digits.

Figure 9.10 Types of synovial joints. For each type, a drawing of the actual joint and a simplified diagram are shown.

 Synovial joints are classified into six principal types based on the shapes of the articulating bone surfaces.

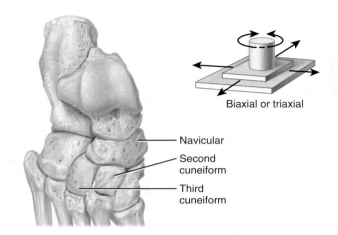

Biaxial or triaxial

Navicular

Second cuneiform

Third cuneiform

(a) Plane joint between navicular and second and third cuneiforms of tarsus in foot

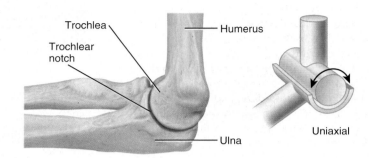

Trochlea

Trochlear notch

Humerus

Ulna

Uniaxial

(b) Hinge joint between trochlea of humerus and trochlear notch of ulna at the elbow

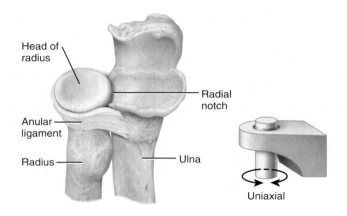

Head of radius

Radial notch

Anular ligament

Radius

Ulna

Uniaxial

(c) Pivot joint between head of radius and radial notch of ulna

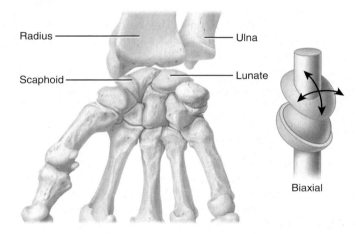

Radius

Ulna

Scaphoid

Lunate

Biaxial

(d) Condyloid joint between radius and scaphoid and lunate bones of carpus (wrist)

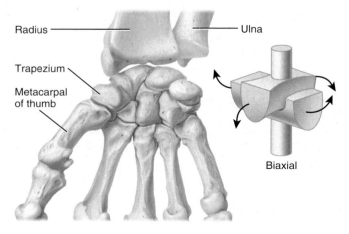

Radius

Ulna

Trapezium

Metacarpal of thumb

Biaxial

(e) Saddle joint between trapezium of carpus (wrist) and metacarpal of thumb

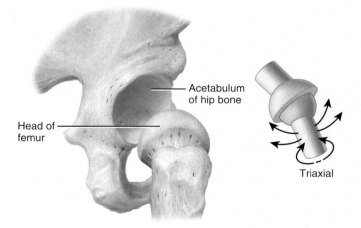

Acetabulum of hip bone

Head of femur

Triaxial

(f) Ball-and-socket joint between head of femur and acetabulum of hip bone

? **What are some examples of pivot joints (other than the one shown in this figure)?**

Saddle Joints

In a **saddle joint** or *sellar joint* (SEL-ar) the articular surface of one bone is saddle-shaped, and the articular surface of the other bone fits into the "saddle" as a sitting rider would sit (Figure 9.10e). The movements at a saddle joint are the same as those at a condyloid joint: *biaxial* (flexion–extension and abduction–adduction) plus limited circumduction. An example of a saddle joint is the carpometacarpal joint between the trapezium of the carpus and metacarpal of the thumb.

Ball-and-Socket Joints

A **ball-and-socket joint** or *spheroid joint* (SFĒ-royd) consists of the ball-like surface of one bone fitting into a cuplike depression of another bone (Figure 9.10f). Such joints are *triaxial (multiaxial),* permitting movements around three axes (flexion–extension, abduction–adduction, and rotation). Examples of ball-and-socket joints are the shoulder and hip joints. At the shoulder joint, the head of the humerus fits into the glenoid cavity of the scapula. At the hip joint, the head of the femur fits into the acetabulum of the hip (coxal) bone.

Table 9.2 summarizes the structural and functional categories of joints.

 CHECKPOINT

10. Which types of joints are uniaxial, biaxial, and triaxial?

TABLE 9.2

Summary of Structural and Functional Classification of Joints

STRUCTURAL CLASSIFICATION	DESCRIPTION	FUNCTIONAL CLASSIFICATION	EXAMPLE
FIBROUS No Synovial Cavity; Articulating Bones Held Together by Fibrous Connective Tissue			
Suture	Articulating bones united by thin layer of dense irregular connective tissue, found between skull bones; with age, some sutures replaced by synostosis (separate cranial bones fused into single bone)	Synarthrosis (immovable) and amphiarthrosis (slightly movable)	Coronal suture
Syndesmosis	Articulating bones united by more dense irregular connective tissue, usually a ligament	Amphiarthrosis (slightly movable)	Distal tibiofibular joint
Interosseous membrane	Articulating bones united by substantial sheet of dense irregular connective tissue	Amphiarthrosis (slightly movable)	Between tibia and fibula
CARTILAGINOUS No Synovial Cavity; Articulating Bones United by Hyaline Cartilage or Fibrocartilage			
Synchondrosis	Connecting material: hyaline cartilage; becomes synostosis when bone elongation ceases	Synarthrosis (immovable)	Epiphyseal plate between diaphysis and epiphysis of long bone
Symphysis	Connecting material: broad, flat disc of fibrocartilage	Amphiarthrosis (slightly movable)	Pubic symphysis and intervertebral joints
SYNOVIAL Characterized by Synovial Cavity, Articular Cartilage, and Articular (Joint) Capsule; May Contain Accessory Ligaments, Articular Discs, and Bursae			
Plane	Articulated surfaces flat or slightly curved	Many biaxial diarthroses (freely movable): back-and-forth and side-to-side movements. Some triaxial diarthroses: back-and-forth, side-to-side, rotation	Intercarpal, intertarsal, sternocostal (between sternum and second to seventh pairs of ribs), and vertebrocostal joints
Hinge	Convex surface fits into concave surface	Uniaxial diarthrosis: flexion–extension	Knee (modified hinge), elbow, ankle, and interphalangeal joints
Pivot	Rounded or pointed surface fits into ring formed partly by bone and partly by ligament	Uniaxial diarthrosis: rotation	Atlanto-axial and radioulnar joints
Condyloid	Oval-shaped projection fits into oval-shaped depression	Biaxial diarthrosis: flexion–extension, abduction–adduction	Radiocarpal and metacarpophalangeal joints
Saddle	Articular surface of one bone is saddle-shaped: articular surface of other bone "sits" in saddle	Biaxial diarthrosis: flexion–extension, abduction–adduction	Carpometacarpal joint between trapezium and metacarpal of thumb
Ball-and-socket	Ball-like surface fits into cuplike depression	Triaxial diarthrosis: flexion–extension, abduction–adduction, rotation	Shoulder and hip joints

9.7 FACTORS AFFECTING CONTACT AND RANGE OF MOTION AT SYNOVIAL JOINTS

 OBJECTIVE

• Describe six factors that influence the type of movement and range of motion possible at a synovial joint.

The articular surfaces of synovial joints contact one another and determine the type and possible range of motion. **Range of motion (ROM)** refers to the range, measured in degrees of a circle, through which the bones of a joint can be moved. The following factors contribute to keeping the articular surfaces in contact and affect range of motion:

1. *Structure or shape of the articulating bones.* The structure or shape of the articulating bones determines how closely they can fit together. The articular surfaces of some bones have a complementary relationship. This spatial relationship is very obvious at the hip joint, where the head of the femur articulates with the acetabulum of the hip bone. An interlocking fit allows rotational movement.

2. *Strength and tension (tautness) of the joint ligaments.* The different components of a fibrous capsule are tense or taut only when the joint is in certain positions. Tense ligaments not only restrict the range of motion but also direct the movement of the articulating bones with respect to each other. In the knee joint, for example, the anterior cruciate ligament is taut and the posterior cruciate ligament is loose when the knee is straightened, and the reverse occurs when the knee is bent.

3. *Arrangement and tension of the muscles.* Muscle tension reinforces the restraint placed on a joint by its ligaments, and thus restricts movement. A good example of the effect of muscle tension on a joint is seen at the hip joint. When the thigh is fixed with the knee extended, the flexion of the hip joint is restricted by the tension of the hamstring muscles on the posterior surface of the thigh, so most of us can't raise a straightened leg more than a 90-degree angle from the floor. But if the knee is also flexed, the tension on the hamstring muscles is lessened, and the thigh can be raised farther, allowing you to raise your thigh to touch your chest (unless, of course, your abdomen gets in the way).

4. *Contact of soft parts.* The point at which one body surface contacts another may limit mobility. For example, if you bend your arm at the elbow, it can move no further once the anterior surface of the forearm meets with and presses against the biceps brachii muscle of the arm. Joint movement may also be restricted by the presence of adipose tissue (such as that contained in the aforementioned abdomen).

5. *Hormones.* Joint flexibility may also be affected by hormones. For example, relaxin, a hormone produced by the placenta and ovaries, increases the flexibility of the fibrocartilage of the pubic symphysis and loosens the ligaments between the sacrum and hip bone toward the end of pregnancy. These changes permit expansion of the pelvic outlet, which assists in delivery of the baby.

6. *Disuse.* Movement at a joint may be restricted if a joint has not been used for an extended period. For example, if an elbow joint is immobilized by a cast, range of motion at the joint may be limited for a time after the cast is removed. Disuse may also result in decreased amounts of synovial fluid, diminished flexibility of ligaments and tendons, and *muscular atrophy*, a reduction in size or wasting of a muscle.

 CHECKPOINT

11. How do the strength and tension of ligaments determine range of motion?

CLINICAL CONNECTION | *Rheumatism and Arthritis*

Rheumatism (ROO-ma-tizm) is any painful disorder of the supporting structures of the body—bones, ligaments, tendons, or muscles—that is not caused by infection or injury. **Arthritis** is a form of rheumatism in which the joints are swollen, stiff, and painful. It afflicts about 45 million people in the United States, and is the leading cause of physical disability among adults over age 65. **Osteoarthritis (OA)** (os′-tē-ō-ar-THRĪ-tis) is a degenerative joint disease in which joint cartilage is gradually lost. It results from a combination of aging, obesity, irritation of the joints, muscle weakness, and wear and abrasion. Commonly known as "wear-and-tear" arthritis, osteoarthritis is the most common type of arthritis and the most common reason for hip- and knee-replacement surgery. In **gout**, a form of arthritis, sodium urate crystals are deposited in the soft tissues of the joints. Gouty arthritis most often affects the joints of the feet, especially at the base of the big toe. The crystals irritate and erode the cartilage, causing inflammation, swelling, and acute pain.

Rheumatoid arthritis (RA) is an autoimmune disease in which the immune system of the body attacks its own tissues—in this case, its own cartilage and joint linings. RA is characterized by inflammation of the joint, which causes swelling, pain, and loss of function (Figures A and B). As noted above, this form of arthritis usually occurs bilaterally: If one wrist is affected, the other is also likely to be affected, although they are often not affected to the same degree.

The primary symptom of RA is inflammation of the synovial membrane. If untreated, the membrane thickens, and synovial fluid accumulates. The resulting pressure causes pain and tenderness. The membrane then produces an abnormal granulation tissue, called *pannus*, that adheres to the surface of the articular cartilage and sometimes erodes the cartilage completely. When the cartilage is destroyed, fibrous tissue joins the exposed bone ends. The fibrous tissue ossifies and fuses the joint so that it becomes immovable—the ultimate crippling effect of rheumatoid arthritis. The growth of the granulation tissue causes the distortion of the fingers that characterizes hands of RA sufferers. •

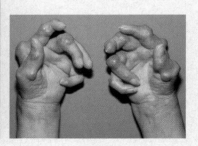

(A) Gamma ray photograph of swollen joints (bright spots) due to RA

(B) Photograph of an individual with severe RA

9.8 SELECTED JOINTS OF THE BODY

OBJECTIVE

* Identify the major joints of the body by location, classification, and movements.

In Chapters 7 and 8 we discussed the major bones and their markings. In this chapter we have examined how joints are classified according to their structure and function, and we have introduced the movements that occur at joints. Table 9.3 (selected joints of the axial skeleton) and Table 9.4 (selected joints of the appendicular skeleton) will help you integrate the information you have learned in all three chapters. These tables list some of the major joints of the body according to their articular components (the bones that enter into their formation), their structural and functional classification, and the type(s) of movement that occur(s) at each joint.

Next, in a series of exhibits we examine six selected joints of the body in detail. Each exhibit considers a specific synovial joint and contains (1) a definition—a description of the type of joint and the bones that form the joint; (2) the anatomical components—a description of the major connecting ligaments, articular disc (if present), articular capsule, and other distinguishing features of the joint; and (3) the joint's possible movements. Clinical Connections are also included, and each exhibit refers you to a figure that illustrates the joint. The joints described are the temporomandibular joint (TMJ), shoulder (humeroscapular or glenohumeral) joint, elbow joint, hip (coxal) joint, knee (tibiofemoral) joint, and ankle (talocrural) joint. Because these joints are described in detail in Exhibits 9.A through 9.F (Figures 9.11 through 9.16), they are not included in Tables 9.3 and 9.4.

CHECKPOINT

12. Using Tables 9.3 and 9.4 as a guide, identify only the cartilaginous joints.

TABLE 9.3

Selected Joints of the Axial Skeleton

JOINT	ARTICULAR COMPONENTS	CLASSIFICATION	MOVEMENTS
Suture	Between skull bones	*Structural:* fibrous *Functional:* slightly movable or immovable	Slight
Atlanto-occipital	Between superior articular facets of atlas and occipital condyles of occipital bone	*Structural:* synovial (condyloid) *Functional:* freely movable	Flexion and extension of head and slight lateral flexion of head to either side
Atlanto-axial	(1) Between dens of axis and anterior arch of atlas and (2) between lateral masses of atlas and axis	*Structural:* synovial (pivot) between dens and anterior arch, and synovial (plane) between lateral masses *Functional:* freely movable	Rotation of head
Intervertebral	(1) Between vertebral bodies and (2) between vertebral arches	*Structural:* cartilaginous (symphysis) between vertebral bodies, and synovial (plane) between vertebral arches *Functional:* slightly movable between vertebral bodies, and freely movable between vertebral arches	Flexion, extension, lateral flexion, and rotation of vertebral column
Vertebrocostal	(1) Between facets of heads of ribs and facets of bodies of adjacent thoracic vertebrae and intervertebral discs between them and (2) between articular part of tubercles of ribs and facets of transverse processes of thoracic vertebrae	*Structural:* synovial (plane) *Functional:* freely movable	Slight gliding
Sternocostal	Between sternum and first seven pairs of ribs	*Structural:* cartilaginous (synchondrosis) between sternum and first pair of ribs, and synovial (plane) between sternum and second through seventh pairs of ribs *Functional:* immovable between sternum and first pair of ribs, and freely movable between sternum and second through seventh pairs of ribs	None between sternum and first pair of ribs; slight gliding between sternum and second through seventh pairs of ribs
Lumbosacral	(1) Between body of fifth lumbar vertebra and base of sacrum and (2) between inferior articular facets of fifth lumbar vertebra and superior articular facets of first vertebra of sacrum	*Structural:* cartilaginous (symphysis) between body and base, and synovial (plane) between articular facets *Functional:* slightly movable between body and base, and freely movable between articular facets	Flexion, extension, lateral flexion, and rotation of vertebral column

TABLE 9.4

Selected Joints of the Appendicular Skeleton

JOINT	ARTICULAR COMPONENTS	CLASSIFICATION	MOVEMENTS
Sternoclavicular	Between sternal end of clavicle, manubrium of sternum, and first costal cartilage	*Structural:* synovial (plane and pivot) *Functional:* freely movable	Gliding, with limited movements in nearly every direction
Acromioclavicular	Between acromion of scapula and acromial end of clavicle	*Structural:* synovial (plane) *Functional:* freely movable	Gliding and rotation of scapula on clavicle
Radioulnar	Proximal radioulnar joint between head of radius and radial notch of ulna; distal radioulnar joint between ulnar notch of radius and head of ulna	*Structural:* synovial (pivot) *Functional:* freely movable	Rotation of forearm
Wrist (radiocarpal)	Between distal end of radius and scaphoid, lunate, and triquetrum of carpus	*Structural:* synovial (condyloid) *Functional:* freely movable	Flexion, extension, abduction, adduction, circumduction, and slight hyperextension of wrist
Intercarpal	Between proximal row of carpal bones, distal row of carpal bones, and between both rows of carpal bones (midcarpal joint)	*Structural:* synovial (plane), except for hamate, scaphoid, and lunate (midcarpal) joint, which is synovial (saddle) *Functional:* freely movable	Gliding plus flexion, extension, abduction, adduction, and slight rotation at midcarpal joint
Carpometacarpal	Carpometacarpal joint of thumb between trapezium of carpus and first metacarpal; carpometacarpal joints of remaining digits formed between carpus and second through fifth metacarpals	*Structural:* synovial (saddle) at thumb and synovial (plane) at remaining digits *Functional:* freely movable	Flexion, extension, abduction, adduction, and circumduction at thumb, and gliding at remaining digits
Metacarpophalangeal and metatarsophalangeal	Between heads of metacarpals (or metatarsals) and bases of proximal phalanges	*Structural:* synovial (condyloid) *Functional:* freely movable	Flexion, extension, abduction, adduction, and circumduction of phalanges
Interphalangeal	Between heads of phalanges and bases of more distal phalanges	*Structural:* synovial (hinge) *Functional:* freely movable	Flexion and extension of phalanges
Sacroiliac	Between auricular surfaces of sacrum and ilia of hip bones	*Structural:* synovial (plane) *Functional:* freely movable	Slight gliding (even more so during pregnancy)
Pubic symphysis	Between anterior surfaces of hip bones	*Structural:* cartilaginous (symphysis) *Functional:* slightly movable	Slight movements (even more so during pregnancy)
Tibiofibular	Proximal tibiofibular joint between lateral condyle of tibia and head of fibula; distal tibiofibular joint between distal end of fibula and fibular notch of tibia	*Structural:* synovial (plane) at proximal joint, and fibrous (syndesmosis) at distal joint *Functional:* freely movable at proximal joint, and slightly movable at distal joint	Slight gliding at proximal joint, and slight rotation of fibula during dorsiflexion of foot
Intertarsal	Subtalar joint between talus and calcaneus of tarsus; talocalcaneo-navicular joint between talus and calcaneus and navicular of tarsus; calcaneocuboid joint between calcaneus and cuboid of tarsus	*Structural:* synovial (plane) at subtalar and calcaneocuboid joints, and synovial (saddle) at talocalcaneonavicular joint *Functional:* freely movable	Inversion and eversion of foot
Tarsometatarsal	Between three cuneiforms of tarsus and bases of five metatarsal bones	*Structural:* synovial (plane) *Functional:* freely movable	Slight gliding

EXHIBIT 9.A **265**

EXHIBIT 9.A Temporomandibular Joint *(Figure 9.11)*

 OBJECTIVE
• Describe the anatomical components of the temporomandibular joint and explain the movements that can occur at this joint.

Description

The **temporomandibular joint (TMJ)** (tem′-pō-rō-man-DIB-ū-lar) (Figure 9.11) is a combined hinge and plane joint formed by the condylar process of the mandible and the mandibular fossa and articular tubercle of the temporal bone. The temporomandibular joint is the only freely movable joint between skull bones (with the exception of the ear ossicles); all other skull joints are sutures and are immovable to slightly movable.

Anatomical Components

1. **Articular disc (meniscus).** Fibrocartilage disc that separates the synovial cavity into superior and inferior compartments, each with a synovial membrane (Figure 9.11c).
2. **Articular capsule.** Thin, fairly loose envelope around the circumference of the joint (Figure 9.11a, b).
3. **Lateral ligament.** Two short bands on the lateral surface of the articular capsule that extend inferiorly and posteriorly from the inferior border and tubercle of the zygomatic process of the temporal bone to the lateral and posterior aspect of the neck of the mandible. The lateral ligament is covered by the parotid gland and helps strengthen the joint laterally and prevent displacement of the mandible (Figure 9.11a).
4. **Sphenomandibular ligament** (sfe-nō-man-DIB-ū-lar). Thin band that extends inferiorly and anteriorly from the spine of the sphenoid bone to the ramus of the mandible (Figure 9.11b). It does not contribute significantly to the strength of the joint.
5. **Stylomandibular ligament** (stī-lō-man-DIB-ū-lar). Thickened band that extends from the styloid process of the temporal bone to the inferior and posterior border of the ramus of the mandible. This ligament separates the parotid gland from the submandibular gland and limits movement of the mandible at the TMJ (Figure 9.11a, b).

Movements

In the temporomandibular joint, only the mandible moves because the temporal bone is firmly anchored to other bones of the skull by sutures. Accordingly, the mandible may function in depression (jaw opening) and elevation (jaw closing), which occur in the inferior compartment, and protraction, retraction, lateral displacement, and slight rotation, which occur in the superior compartment (see Figure 9.9a–d).

 CHECKPOINT
13. What distinguishes the temporomandibular joint from the other joints of the skull?

Figure 9.11 Right temporomandibular joint (TMJ).

 The TMJ is the only movable joint between skull bones.

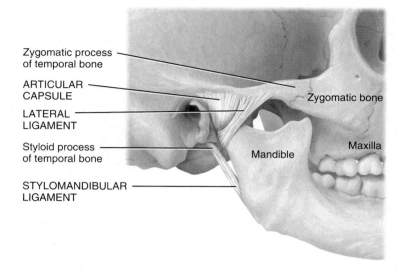

Zygomatic process of temporal bone
ARTICULAR CAPSULE
LATERAL LIGAMENT
Styloid process of temporal bone
STYLOMANDIBULAR LIGAMENT
Zygomatic bone
Maxilla
Mandible

(a) Right lateral view

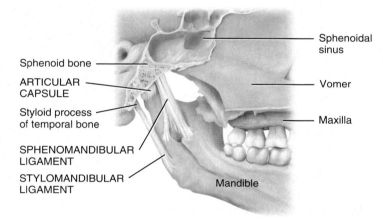

Sphenoid bone
ARTICULAR CAPSULE
Styloid process of temporal bone
SPHENOMANDIBULAR LIGAMENT
STYLOMANDIBULAR LIGAMENT
Sphenoidal sinus
Vomer
Maxilla
Mandible

(b) Left medial view

SYNOVIAL CAVITY
Superior compartment
Inferior compartment
Mandibular fossa of temporal bone
External auditory meatus
Condylar process of mandible
Styloid process of temporal bone
ARTICULAR DISC
Articular tubercle of temporal bone
Mandible

(c) Sagittal section viewed from right

FIGURE 9.11 CONTINUES ▶

EXHIBIT 9.A Temporomandibular Joint *(Figure 9.11)* CONTINUED

CLINICAL CONNECTION | *Dislocated Mandible*

A **dislocation** (dis'-lō-KĀ-shun; *dis-*=apart) or *luxation* (luks-Ā-shun; *luxatio*=dislocation) is the displacement of a bone from a joint with tearing of ligaments, tendons, and articular capsules. It is usually caused by a blow or fall, although unusual physical effort may be a factor. For example, if the condylar processes of the mandible pass anterior to the articular tubercles when you yawn or take a large bite, a **dislocated mandible** (anterior displacement) may occur. When the mandible is displaced in this manner, the mouth remains wide open and the person is unable to close it. This may be corrected by pressing the thumbs downward on the lower molar teeth and pushing the mandible backward. Other causes of a dislocated mandible include a lateral blow to the chin when the mouth is open and a fracture of the mandible. •

FIGURE 9.11 CONTINUED

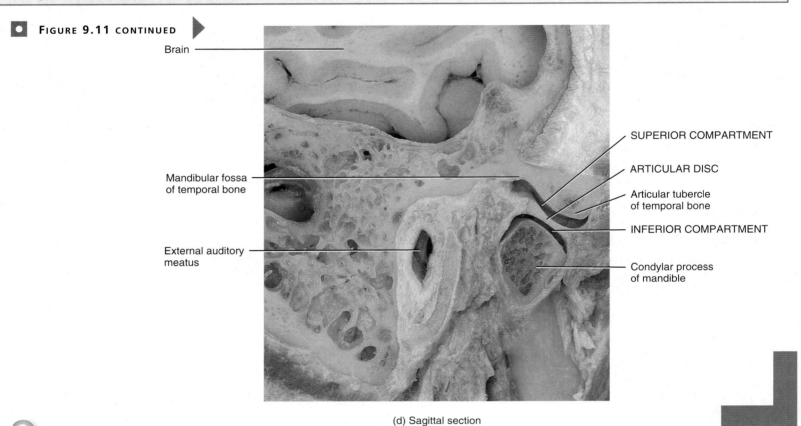

Brain

Mandibular fossa of temporal bone

External auditory meatus

SUPERIOR COMPARTMENT

ARTICULAR DISC

Articular tubercle of temporal bone

INFERIOR COMPARTMENT

Condylar process of mandible

(d) Sagittal section

Which ligament prevents displacement of the mandible?

EXHIBIT 9.B Shoulder Joint *(Figure 9.12)*

OBJECTIVE
• Describe the anatomical components of the shoulder joint and the movements that can occur at this joint.

Description

The **shoulder joint** (Figure 9.12) is a ball-and-socket joint formed by the head of the humerus and the glenoid cavity of the scapula. It also is referred to as the *humeroscapular* or *glenohumeral joint*.

Anatomical Components

1. **Articular capsule.** Thin, loose sac that completely envelops the joint and extends from the glenoid cavity to the anatomical neck of the humerus. The inferior part of the capsule is its weakest area (Figure 9.12a, b).

2. **Coracohumeral ligament** (kōr'-a-kō-HŪ-mer-al). Strong, broad ligament that strengthens the superior part of the articular capsule and extends from the coracoid process of the scapula to the greater tubercle of the humerus (Figure 9.12a, b). The ligament strengthens the superior part of the articular capsule, and reinforces the anterior aspect of the articular capsule.

3. **Glenohumeral ligaments** (gle-nō-HŪ-mer-al). Three thickenings of the articular capsule over the anterior surface of the joint that extend from the glenoid cavity to the lesser tubercle and anatomical neck of the humerus. These ligaments, which are often indistinct or absent and provide only minimal strength (Figure 9.12a, b), play a role in joint stabilization when the humerus approaches or exceeds its limits of motion.

4. **Transverse humeral ligament.** Narrow sheet extending from the greater tubercle to the lesser tubercle of the humerus (Figure 9.12a). The ligament functions as a retinaculum

EXHIBIT 9.B **267**

(retaining band) to hold the long head of the biceps brachii muscle in the intertubercular groove.

5. **Glenoid labrum.** Narrow rim of fibrocartilage around the edge of the glenoid cavity that slightly deepens and enlarges the glenoid cavity (Figure 9.12b–d).

6. **Bursae.** Four bursae (see Section 9.4) are associated with the shoulder joint. They are the *subscapular bursa* (Figure 9.12a), *subdeltoid bursa*, *subacromial bursa* (Figure 9.12a–c), and *subcoracoid bursa*.

Movements

The shoulder joint allows flexion, extension, hyperextension, abduction, adduction, medial rotation, lateral rotation, and circumduction of the arm (see Figures 9.5–9.8). It has more freedom of movement than any other joint of the body because of the looseness of the articular capsule and the shallowness of the glenoid cavity in relation to the large size of the head of the humerus.

Although the ligaments of the shoulder joint strengthen it to some extent, most of the strength results from the muscles that surround the joint, especially *the rotator cuff muscles*. These muscles (supraspinatus, infraspinatus, teres minor, and subscapularis), which you will learn more about in Chapter 11, anchor the humerus to the scapula (see also Figure 11.17). The tendons of the rotator cuff muscles encircle the joint (except for the inferior portion) and intimately surround the articular capsule. The rotator cuff muscles work as a group to hold the head of the humerus in the glenoid cavity.

 CHECKPOINT

14. What is the function of the rotator cuff?

Figure 9.12 Right shoulder (humeroscapular or glenohumeral) joint.

⚬━ Most of the stability of the shoulder joints results from the arrangement of the rotator cuff muscles.

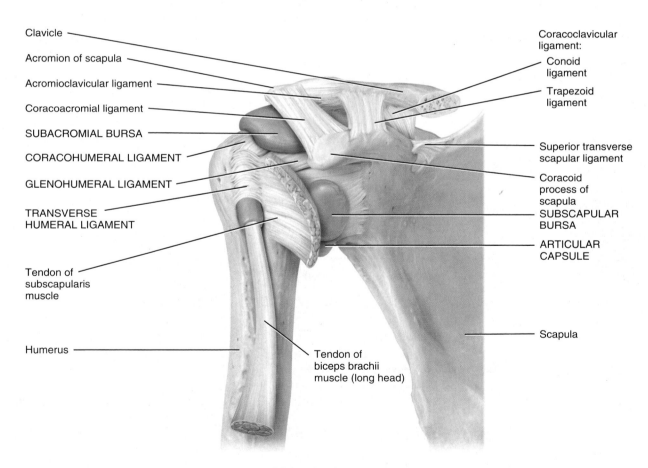

(a) Anterior view

FIGURE 9.12 CONTINUES ▶

EXHIBIT 9.B Shoulder Joint *(Figure 9.12)* CONTINUED

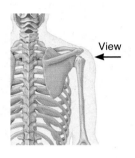

⬤ FIGURE **9.12** CONTINUED ▶

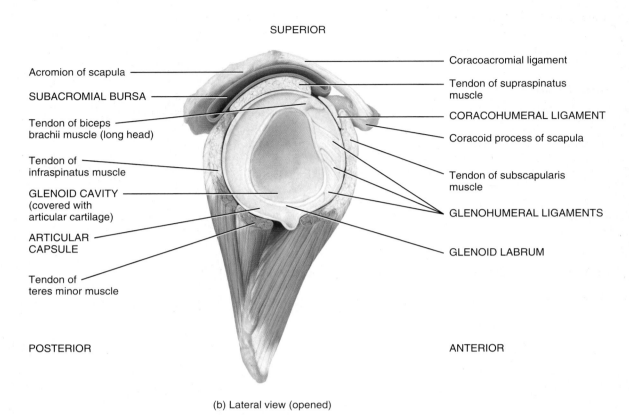

SUPERIOR

Acromion of scapula

SUBACROMIAL BURSA

Tendon of biceps brachii muscle (long head)

Tendon of infraspinatus muscle

GLENOID CAVITY (covered with articular cartilage)

ARTICULAR CAPSULE

Tendon of teres minor muscle

POSTERIOR

Coracoacromial ligament

Tendon of supraspinatus muscle

CORACOHUMERAL LIGAMENT

Coracoid process of scapula

Tendon of subscapularis muscle

GLENOHUMERAL LIGAMENTS

GLENOID LABRUM

ANTERIOR

View

(b) Lateral view (opened)

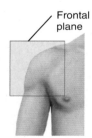

Frontal plane

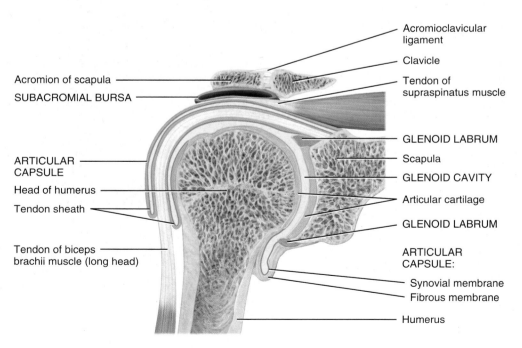

Acromion of scapula

SUBACROMIAL BURSA

ARTICULAR CAPSULE

Head of humerus

Tendon sheath

Tendon of biceps brachii muscle (long head)

Acromioclavicular ligament

Clavicle

Tendon of supraspinatus muscle

GLENOID LABRUM

Scapula

GLENOID CAVITY

Articular cartilage

GLENOID LABRUM

ARTICULAR CAPSULE:

Synovial membrane

Fibrous membrane

Humerus

(c) Frontal section

EXHIBIT 9.B **269**

CLINICAL CONNECTION | *Torn Glenoid Labrum and Dislocated and Separated Shoulder*

In a **torn glenoid labrum**, the fibrocartilaginous labrum may tear away from the glenoid cavity. This causes the joint to "catch" or feel like it's slipping out of place. Indeed, the shoulder may become dislocated as a result of a torn glenoid labrum. The torn labrum is reattached to the glenoid surgically with anchors and sutures. The repaired joint is more stable.

The shoulder joint is the most commonly dislocated joint in adults because its socket is quite shallow and the bones are held together by supporting muscles. Usually, in a **dislocated shoulder** the head of the humerus becomes displaced inferiorly, where the articular capsule is least protected. It is treated by ice, pain relievers, manual manipulation or surgery, followed by use of a sling and physical therapy.

A **separated shoulder** actually refers to an injury not to the shoulder joint itself but to the acromioclavicular joint, a joint formed by the acromion of the scapula and the acromial end of the clavicle. This condition is usually the result of forceful trauma to the joint, as when the shoulder strikes the ground in a fall. Treatment options include rest, ice, pain relievers, physical therapy, use of a sling, or, rarely, surgery. •

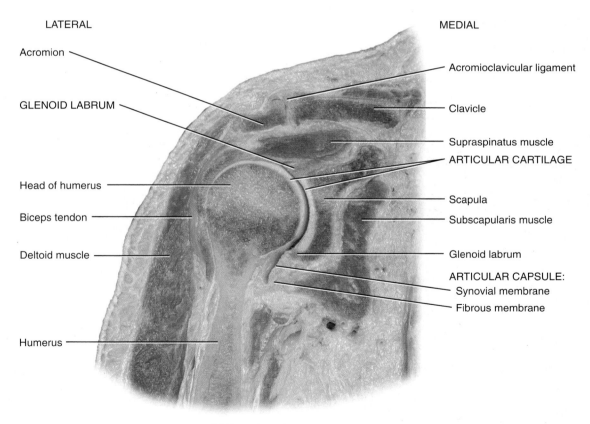

LATERAL

MEDIAL

Acromion

GLENOID LABRUM

Head of humerus

Biceps tendon

Deltoid muscle

Humerus

Acromioclavicular ligament

Clavicle

Supraspinatus muscle

ARTICULAR CARTILAGE

Scapula

Subscapularis muscle

Glenoid labrum

ARTICULAR CAPSULE:
Synovial membrane
Fibrous membrane

(d) Frontal section

Why does the shoulder joint have more freedom of movement than any other joint of the body?

EXHIBIT 9.C | Elbow Joint *(Figure 9.13)*

OBJECTIVE

- Describe the anatomical components of the elbow joint and the movements that can occur at this joint.

Description

The **elbow joint** (Figure 9.13) is a hinge joint formed by the trochlea and capitulum of the humerus, the trochlear notch of the ulna, and the head of the radius.

Anatomical Components

1. **Articular capsule.** The anterior part of the articular capsule covers the anterior part of the elbow joint, from the radial and coronoid fossae of the humerus to the coronoid process of the ulna and the anular ligament of the radius. The posterior part extends from the capitulum, olecranon fossa, and lateral epicondyle of the humerus to the anular ligament of the radius, the olecranon of the ulna, and the ulna posterior to the radial notch (Figure 9.13a, b).

2. **Ulnar collateral ligament.** This thick, triangular ligament extends from the medial epicondyle of the humerus to the coronoid process and olecranon of the ulna (Figure 9.13a). Part of the ulnar collateral ligament deepens the socket for the trochlea of the humerus.

3. **Radial collateral ligament.** This strong, triangular ligament extends from the lateral epicondyle of the humerus to the anular ligament of the radius and the radial notch of the ulna (Figure 9.13b).

4. **Anular ligament of the radius.** This strong band of connective tissue encircles the head of the radius. The anular ligament of the radius holds the head of the radius in the radial notch of the ulna (Figure 9.13a, b).

Movements

The elbow joint allows flexion and extension of the forearm (see Figure 9.5c).

CHECKPOINT

15. At the elbow joint, which ligaments connect (a) the humerus and the ulna, and (b) the humerus and the radius?

Figure 9.13 Right elbow joint.

 The elbow joint is formed by parts of three bones: humerus, ulna, and radius.

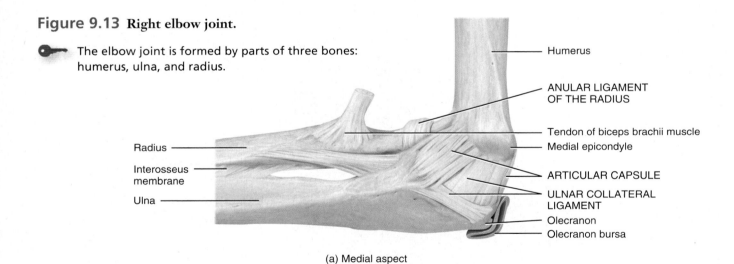

(a) Medial aspect

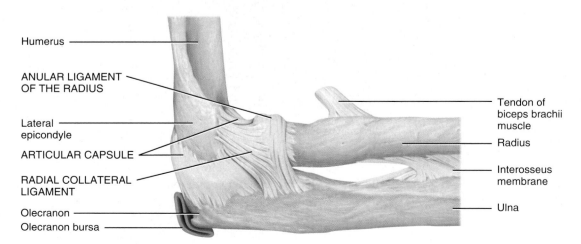

(b) Lateral aspect

EXHIBIT 9.C **271**

CLINICAL CONNECTION | *Tennis Elbow, Little-league Elbow, and Dislocation of the Radial Head*

Tennis elbow most commonly refers to pain at or near the lateral epicondyle of the humerus, usually caused by an improperly executed backhand. The extensor muscles strain or sprain, resulting in pain. **Little-league elbow**, inflammaton of the medial epicondyle, typically develops as a result of a heavy pitching schedule and/or a schedule that involves throwing curve balls, especially among youngsters. In this disorder, the elbow joint may enlarge, fragment, or separate.

A **dislocation of the radial head** is the most common upper limb dislocation in children. In this injury, the head of the radius slides past or ruptures the radial anular ligament, a ligament that forms a collar around the head of the radius at the proximal radioulnar joint. Dislocation is most apt to occur when a strong pull is applied to the forearm while it is extended and supinated, for instance, while swinging a child around with outstretched arms. •

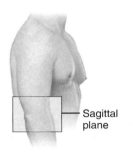

Sagittal plane

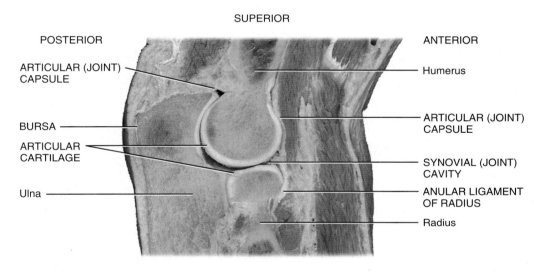

(c) Sagittal section of right elbow joint

 Which movements are possible at a hinge joint?

EXHIBIT 9.D **Hip Joint** *(Figure 9.14)*

● OBJECTIVE
• Describe the anatomical components of the hip joint and the movements that can occur at this joint.

Description

The **hip joint** (*coxal joint*) (Figure 9.14) is a ball-and-socket joint formed by the head of the femur and the acetabulum of the hip bone.

Figure 9.14 Right hip (coxal) joint.

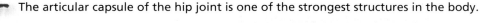 The articular capsule of the hip joint is one of the strongest structures in the body.

Anatomical Components

1. **Articular capsule.** This very dense and strong capsule extends from the rim of the acetabulum to the neck of the femur (Figure 9.14d, e). With its associated accessory ligaments, this is one of the strongest structures of the body. The articular capsule consists of circular and longitudinal fibers. The circular fibers, called the *zona orbicularis*, form a collar around the neck of the femur. Accessory ligaments known as the *iliofemoral*

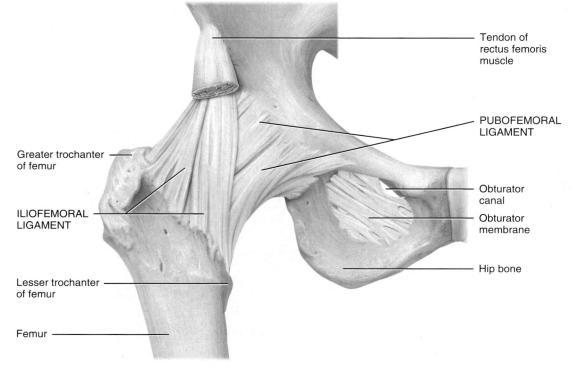

Tendon of rectus femoris muscle

PUBOFEMORAL LIGAMENT

Greater trochanter of femur

Obturator canal

ILIOFEMORAL LIGAMENT

Obturator membrane

Lesser trochanter of femur

Hip bone

Femur

(a) Anterior view

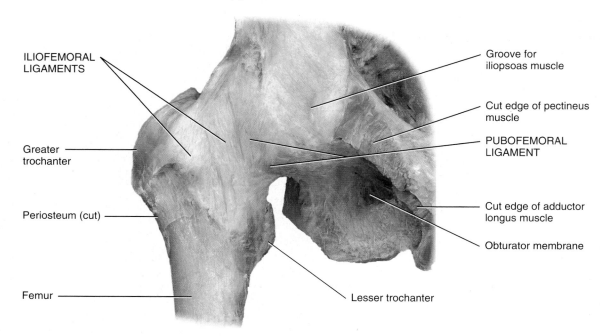

ILIOFEMORAL LIGAMENTS

Groove for iliopsoas muscle

Cut edge of pectineus muscle

Greater trochanter

PUBOFEMORAL LIGAMENT

Periosteum (cut)

Cut edge of adductor longus muscle

Obturator membrane

Femur

Lesser trochanter

(b) Anterior view

EXHIBIT 9.D **273**

ligament, *pubofemoral ligament*, and *ischiofemoral ligament* reinforce the longitudinal fibers of the articular capsule.

2. **Iliofemoral ligament** (il′-ē-ō-FEM-ō-ral). This thickened portion of the articular capsule extends from the anterior inferior iliac spine of the hip bone to the intertrochanteric line of the femur (Figure 9.14a, b, c). The iliofemoral ligament is said to be the body's strongest ligament and prevents hyperextension of the femur at the hip joint.

3. **Pubofemoral ligament** (pū′-bō-FEM-ō-ral). This thickened portion of the articular capsule extends from the pubic part of the rim of the acetabulum to the neck of the femur (Figure 9.14a). The pubofemoral ligament prevents overabduction of the femur at the hip joint and strengthens the articular capsule.

4. **Ischiofemoral ligament** (is′-kē-ō-FEM-ō-ral). A thickened portion of the articular capsule, the ischiofemoral ligament

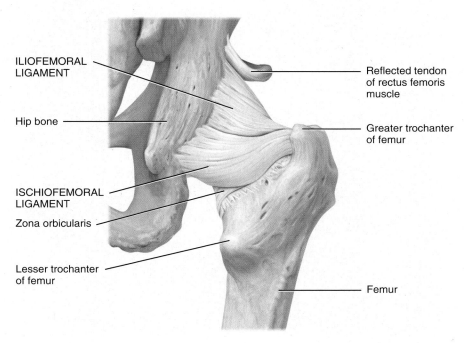

ILIOFEMORAL LIGAMENT

Hip bone

ISCHIOFEMORAL LIGAMENT

Zona orbicularis

Lesser trochanter of femur

Reflected tendon of rectus femoris muscle

Greater trochanter of femur

Femur

(c) Posterior view

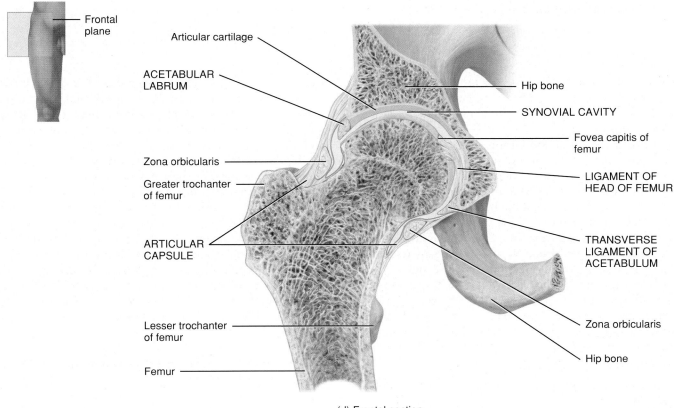

Frontal plane

Articular cartilage

ACETABULAR LABRUM

Zona orbicularis

Greater trochanter of femur

ARTICULAR CAPSULE

Lesser trochanter of femur

Femur

Hip bone

SYNOVIAL CAVITY

Fovea capitis of femur

LIGAMENT OF HEAD OF FEMUR

TRANSVERSE LIGAMENT OF ACETABULUM

Zona orbicularis

Hip bone

(d) Frontal section

Figure 9.14 continues ▶

EXHIBIT 9.D Hip Joint *(Figure 9.14)* CONTINUED

 FIGURE 9.14 CONTINUED

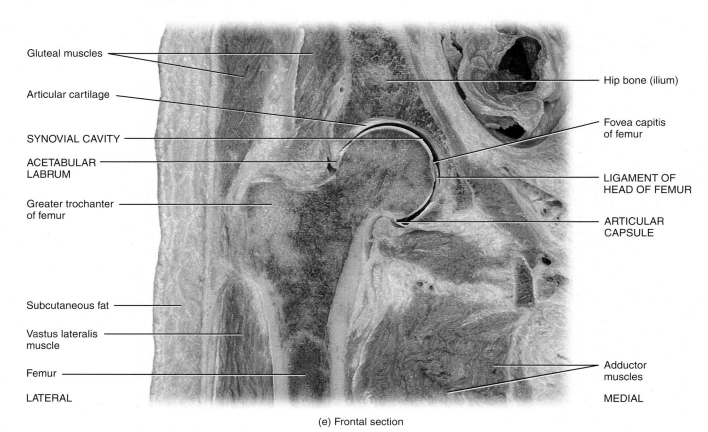

(e) Frontal section

 Which ligaments limit the degree of extension that is possible at the hip joint?

extends from the ischial wall bordering the acetabulum to the neck of the femur (Figure 9.14c). This ligament slackens during adduction and tenses during abduction, and strengthens the articular capsule.

5. **Ligament of the head of the femur.** A flat, triangular band (primarily a synovial fold) that extends from the fossa of the acetabulum to the fovea capitis of the head of the femur (Figure 9.14d, e), the ligament of the head of the femur usually contains a small artery that supplies the head of the femur.

6. **Acetabular labrum** (as-ē-TAB-ū-lar LĀ-brum). This fibrocartilage rim attached to the margin of the acetabulum enhances the depth of the acetabulum. Because the diameter of the acetabular labrum is greater than that of the head of the femur, dislocation of the femur is rare (Figure 9.14d, e).

7. **Transverse ligament of the acetabulum.** A strong ligament that crosses over the acetabular notch, the transverse ligament of the acetabulum supports part of the acetabular labrum and is connected with the ligament of the head of the femur and the articular capsule (Figure 9.14d).

Movements

The hip joint allows flexion, extension, abduction, adduction, lateral rotation, medial rotation, and circumduction of the thigh

(see Figures 9.5–9.8). The extreme stability of the hip joint is related to the very strong articular capsule and its accessory ligaments, the manner in which the femur fits into the acetabulum, and the muscles surrounding the joint. Although the shoulder and hip joints are both ball-and-socket joints, the movements at the hip joints do not have as wide a range of motion. Flexion is limited by the anterior surface of the thigh coming into contact with the anterior abdominal wall when the knee is flexed and by tension of the hamstring muscles when the knee is extended. Extension is limited by tension of the iliofemoral, pubofemoral, and ischiofemoral ligaments. Abduction is limited by the tension of the pubofemoral ligament, and adduction is limited by contact with the opposite limb and tension in the ligament of the head of the femur. Medial rotation is limited by the tension in the ischiofemoral ligament, and lateral rotation is limited by tension in the iliofemoral and pubofemoral ligaments.

✔ CHECKPOINT

16. What factors limit the degree of flexion and abduction at the hip joint?

EXHIBIT 9.E **275**

EXHIBIT 9.E Knee Joint *(Figure 9.15)*

OBJECTIVE
• Describe the main anatomical components of the knee joint and explain the movements that can occur at this joint.

Description

The **knee joint** (*tibiofemoral joint*) is the largest and most complex joint of the body (Figure 9.15). It is a modified hinge joint (because its primary movement is a uniaxial hinge movement) that actually consists of three joints within a single synovial cavity:

1. Laterally is a *tibiofemoral joint*, between the lateral condyle of the femur, lateral meniscus, and lateral condyle of the tibia, which is the weight-bearing bone of the lower leg.
2. Medially is another *tibiofemoral joint*, between the medial condyle of the femur, medial meniscus, and medial condyle of the tibia.
3. An intermediate *patellofemoral joint* is between the patella and the patellar surface of the femur.

Anatomical Components

1. **Articular capsule**. No complete, independent capsule unites the bones of the knee joint. The ligamentous sheath surrounding the joint consists mostly of muscle tendons or their expansions (Figure 9.15e, f, g). There is, however, a thin capsular sheath connecting the articulating bones.
2. **Medial** and **lateral patellar retinacula** (ret′-i-NAK-ū-la). These fused tendons of insertion of the quadriceps femoris muscle and the fascia lata (fascia of thigh) strengthen the anterior surface of the joint (Figure 9.15e).
3. **Patellar ligament**. This continuation of the common tendon of insertion of the quadriceps femoris muscle extends from the patella to the tibial tuberosity. The patellar ligament also strengthens the anterior surface of the joint. The posterior surface of the ligament is separated from the synovial membrane of the joint by an *infrapatellar fat pad* (Figure 9.15c, d, e).
4. **Oblique popliteal ligament** (pop-LIT-ē-al). This broad, flat ligament extends from the intercondylar fossa and lateral condyle of the femur to the head and medial condyle of the tibia (Figure 9.15f, h). The ligament strengthens the posterior surface of the joint.
5. **Arcuate popliteal ligament**. Extending from the lateral condyle of the femur to the styloid process of the head of the fibula, the arcuate popliteal ligament strengthens the inferior lateral part of the posterior surface of the knee joint (Figure 9.15f).
6. **Tibial collateral ligament**. Broad, flat ligament on the medial surface of the joint that extends from the medial condyle of the femur to the medial condyle of the tibia (Figure 9.15a, e, f, g, h). Tendons of the sartorius, gracilis, and semitendinosus muscles, all of which strengthen the medial aspect of the joint, cross the ligament. The tibial collateral ligament is firmly attached to the medial meniscus.
7. **Fibular collateral ligament**. This strong, rounded ligament on the lateral surface of the joint extends from the lateral condyle of the femur to the lateral side of the head of the fibula (Figure 9.15a, e, f, g, h). It strengthens the lateral aspect of the joint. The fibular collateral ligament is covered by the tendon of the biceps femoris muscle, and the tendon of the popliteal muscle is deep to it.
8. **Intracapsular ligaments** (in′-tra-KAP-sū-lar). These two ligaments within the articular capsule connect the tibia and femur. The anterior and posterior **cruciate ligaments** (KROO-shē-āt=like a cross) are named based on their origins relative to the intercondylar area of the tibia. From their origins, they cross on their way to their destinations on the femur.
   a. **Anterior cruciate ligament (ACL)**. The ACL extends posteriorly and laterally from a point anterior to the intercondylar area of the tibia to the posterior part of the medial surface of the lateral condyle of the femur (Figure 9.15a, b, h). The ACL limits hyperextension of the knee (which normally does not occur at this joint) and prevents the anterior sliding of the tibia on the femur. This ligament is stretched or torn in about 70 percent of all serious knee injuries.

      ACL injuries are much more common in females than males, perhaps as much as 3 to 6 times. The reasons are unclear, but may be related to the following factors: (1) there is less space between the femoral condyles in females so that the space for ACL movement is limited; (2) the female pelvis is wider, which creates a greater angle between the femur and tibia and increases the risk for an ACL tear; (3) female hormones allow for greater flexibility of ligaments, muscles, and tendons, but this flexibility doesn't permit them to absorb the stresses put on them so the stresses are transferred to the ACL; and (4) females have less muscle strength so that they rely less on muscles and more on the ACL to hold the knee in place.
   b. **Posterior cruciate ligament (PCL)**. The PCL extends anteriorly and medially from a depression on the *posterior* intercondylar area of the tibia and lateral meniscus to the anterior part of the lateral surface of the medial condyle of the femur (Figure 9.15a, b, h). The PCL prevents the posterior sliding of the tibia (and anterior sliding of the femur) when the knee is flexed, a very important function when you walk down stairs or a steep incline.
9. **Articular discs (menisci)**. Two fibrocartilage discs between the tibial and femoral condyles help compensate for the irregular shapes of the bones and circulate synovial fluid.
   a. **Medial meniscus**. The anterior end of this semicircular, C-shaped piece of fibrocartilage is attached to the anterior intercondylar fossa of the tibia, in front of the ACL. Its posterior end is attached to the posterior intercondylar fossa of the tibia between the attachments of the posterior cruciate ligament and lateral meniscus (Figure 9.15a, b, d, h).
   b. **Lateral meniscus**. This nearly circular piece of fibrocartilage approaches an incomplete O in shape (Figure 9.15a, b, c, h). Its anterior end is attached anteriorly to the intercondylar eminence of the tibia, and laterally and posteriorly to the ACL. Its posterior end is attached posteriorly to the intercondylar eminence of the tibia, and anteriorly to the posterior end of the medial meniscus. The anterior surfaces of the medial and lateral menisci are connected to each other by the *transverse ligament of the knee* (Figure 9.15a) and to the margins of the head of the tibia by the *coronary ligaments* (not illustrated).

EXHIBIT 9.E   Knee Joint *(Figure 9.15)* CONTINUED

10. **Bursae.** The more important bursae of the knee include the following:
    a. **Prepatellar bursa** between the patella and skin (Figure 9.15c, d).
    b. **Infrapatellar bursa** between the superior part of the tibia and the patellar ligament (Figure 9.15c, d, e).
    c. **Suprapatellar bursa** between the inferior part of the femur and the deep surface of the quadriceps femoris muscle (Figure 9.15c, d, e).

Movements

The knee joint allows flexion, extension, slight medial rotation, and lateral rotation of the leg in the flexed position (see Figures 9.5f and 9.8c).

✓ CHECKPOINT
17. What are the opposing functions of the anterior and posterior cruciate ligaments?

Figure 9.15 Right knee (tibiofemoral) joint.

🔑 The knee joint is the largest and most complex joint in the body.

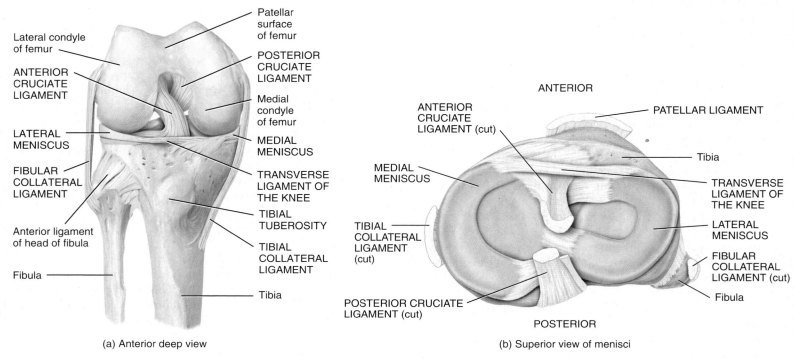

(a) Anterior deep view

(b) Superior view of menisci

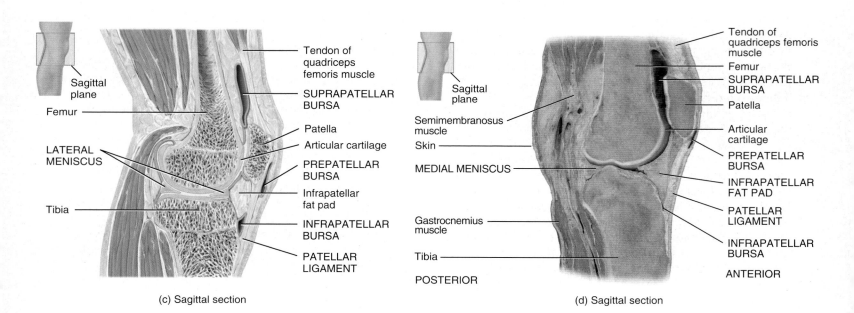

(c) Sagittal section

(d) Sagittal section

EXHIBIT 9.E 277

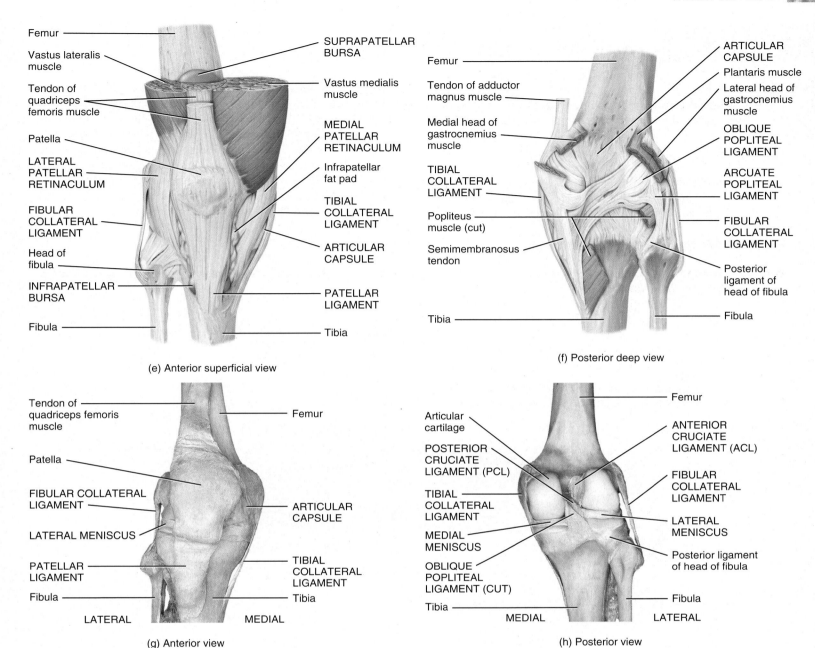

(e) Anterior superficial view

(f) Posterior deep view

(g) Anterior view

(h) Posterior view

What movement occurs at the knee joint when the quadriceps femoris (anterior thigh) muscles contract?

CLINICAL CONNECTION | *Knee Injuries*

The knee joint is the joint most vulnerable to damage because it is a mobile, weight-bearing joint and its stability depends almost entirely on its associated ligaments and muscles. Further, the articular surfaces have only minimal contact throughout the range of motion. Following are several kinds of **knee injuries**.

A **swollen knee** may occur immediately after an injury or many hours later. The initial swelling is due to escape of blood from damaged blood vessels adjacent to areas of injury, including rupture of the anterior cruciate ligament, damage to synovial membranes, torn menisci, fractures, or collateral ligament sprains. Delayed swelling is due to excessive production of synovial fluid, a condition commonly referred to as "water on the knee."

The firm attachment of the tibial collateral ligament to the medial meniscus is clinically significant, because tearing of the ligament also typically results in tearing of the meniscus. Such

an injury may occur in sports such as football and rugby when the knee receives a blow from the lateral side while the foot is fixed on the ground. The force of the blow may also tear the anterior cruciate ligament, which is also connected to the medial meniscus. The term "**unhappy triad**" is applied to a knee injury that involves damage to three components of the knee joint at the same time: the tibial collateral ligament, medial meniscus, and anterior cruciate ligament.

A **dislocated knee** refers to the displacement of the tibia relative to the femur. The most common type is dislocation anteriorly, resulting from hyperextension of the knee. A frequent consequence of a dislocated knee is damage to the popliteal artery.

If no surgery is required, treatment of knee injuries involves PRICE (protection, rest, ice, compression, and elevation) with some strengthening exercises and perhaps physical therapy. •

EXHIBIT 9.F   Ankle Joint *(Figure 9.16)*

OBJECTIVE

• Describe the anatomical components of the ankle joint and explain the movements that can occur at this joint.

Description

The **ankle joint** (*talocrural joint*) (Figure 9.16) is a hinge joint formed by (1) the distal end of the tibia and its medial malleolus with the talus and (2) the lateral malleolus of the fibula with the talus. It is a strong and stable joint due to the shapes of the articulating bones, the strength of its ligaments, and the tendons that surround it.

Anatomically, the ankle is the region that extends from the distal region of the leg to the proximal region of the foot and contains the ankle joint. In this transition region, there is a change in orientation from a vertical orientation of the bones and muscles and associated structures in the leg to a horizontal orientation of the structures in the foot. As a result, there is a turning anteriorly of the tendons, blood vessels, and nerves in the ankle as they enter the foot.

Figure 9.16 **Right ankle (talocrural) joint.**

The strength and stability of the ankle joint are due to the shapes of the articulating bones, the strength of its ligaments, and the tendons that surround it.

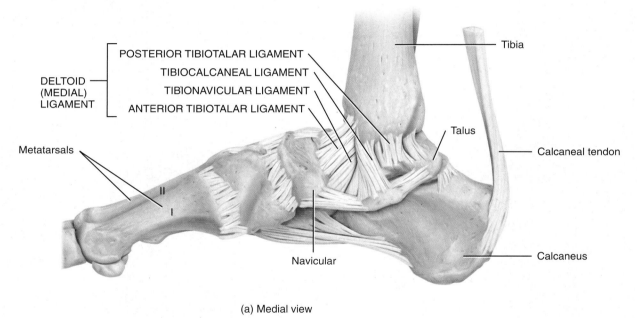

(a) Medial view

(b) Lateral view

How is the ankle defined anatomically?

The structures that pass from the leg into the foot at the ankle are anchored by thickenings of fascia (connective tissue) called **retinacula** (ret-i-NAK-yoo-la; *retineo*=to hold back). Two principal retinacula are the superior and inferior extensor retinacula (see Figure 11.24).

Anatomical Components

1. **Articular capsule.** Completely surrounding the joint, the articular capsule is attached superiorly to the tibia and fibula and inferiorly to the talus. The capsule is thin (and weak) anteriorly and posteriorly to permit dorsiflexion and plantar flexion.
2. **Deltoid (medial) ligament.** This strong, flat, triangular ligament, which extends from the medial malleolus to the talus, navicular, and calcaneus of the tarsus, is divisible into superficial and deep parts (Figure 9.16a). The superficial components from anterior to posterior are the *tibionavicular ligament, tibiocalcaneal ligament,* and *posterior tibiotalar ligament.* The deep component is the *anterior tibiotalar ligament.* The deltoid ligament strengthens the medial aspect of the ankle joint.
3. **Lateral ligament.** Not as strong as the deltoid ligament, the lateral ligament extends from the lateral malleolus to the talus and calcaneus and is divisible into three components: *anterior talofibular ligament, posterior talofibular ligament,* and *calcaneofibular ligament* (Figure 9.16b). The lateral ligament strengthens the lateral aspect of the ankle joint.

CLINICAL CONNECTION | Ankle Sprains

The ankle is the most frequently injured of the major joints of the body. **Ankle sprains**, the most common ankle injuries, often occur in sports that involve running and jumping. Sprains of the lateral ankle occur more frequently than those of the medial ankle and are usually caused by excessive inversion (supination) of the foot with plantar flexion of the ankle. As a result, the weaker lateral ligament is partially torn and there is considerable pain and local swelling. Less common sprains of the medial ankle occur as a result of excessive eversion (pronation). In the process, the deltoid ligament may be torn but the ligament usually does not tear due to its great strength; it may instead break off the tip of the medial malleolus of the tibia. Ankle sprains are treated with PRICE: protection, rest, ice, compression, and elevation. Severe sprains may require cast immobilization or surgery. •

Movements

The ankle joint permits dorsiflexion and plantar flexion (see Figure 9.9g).

 CHECKPOINT
18. What is the function of the deltoid ligament?

 # 9.9 AGING AND JOINTS

 OBJECTIVE
• Explain the effects of aging on joints.

Aging usually results in decreased production of synovial fluid in joints. In addition, the articular cartilage becomes thinner with age, and ligaments shorten and lose some of their flexibility. The effects of aging on joints vary considerably from one person to another and are influenced by genetic factors and wear and tear. Although degenerative changes in joints may begin in people as young as 20 years of age, most changes do not occur until much later. By age 80, almost everyone develops some type of degeneration in the knees, elbows, hips, and shoulders. It is also common for elderly individuals to develop degenerative changes in the vertebral column, resulting in a hunched-over posture and pressure on nerve roots. One type of arthritis, called osteoarthritis (see Clinical Connection in Section 9.7), is at least partially age-related. Nearly everyone over age 70 has evidence of some osteoarthritic changes. Stretching and aerobic exercises that attempt to maintain full range of motion are very important in minimizing the effects of aging. They help to maintain the effective functioning of ligaments, tendons, muscles, synovial fluid, and articular cartilage.

 CHECKPOINT
19. Which joints show evidence of degeneration in nearly all individuals as aging progresses?

CLINICAL CONNECTION | Arthroplasty

Joints that have been severely damaged by diseases such as arthritis, or by injury, may be replaced surgically with artificial joints in a procedure referred to as **arthroplasty** (AR-thrō-plas'-tē; *arthr-*=joint; *plasty*=plastic repair of). Although most joints in the body can be repaired by arthroplasty, the ones most commonly replaced are the hips, knees, and shoulders. About 400,000 hip replacements and 300,000 knee replacements are performed annually in the United States. During the procedure, the ends of the damaged bones are removed and metal, ceramic, or plastic components are fixed in place. The goals of arthroplasty are to relieve pain and increase range of motion.

Partial hip replacements involve only the femur. **Total hip replacements** involve both the acetabulum and head of the femur (Figures A–C). The damaged portions of the acetabulum and the head of the femur are replaced by prefabricated *prostheses* (artificial devices). The acetabulum is shaped to accept the new socket, the head of the femur is removed, and the center of the femur is shaped to fit the femoral component. The acetabular component consists of a plastic such as polyethylene, and the femoral component is composed of a metal such as cobalt-chrome, titanium alloys, or stainless steel. These materials are designed to withstand a high degree of stress and to prevent a response by the immune system. Once the appropriate acetabular and femoral components are selected, they are attached to the healthy portion of bone with acrylic cement, which forms an interlocking mechanical bond.

Knee replacements are actually a resurfacing of cartilage and, like hip replacements, may be partial or total. In a **partial knee replacement (PKR)**, also called a **unicompartmental knee replacement**, only one side of the knee joint is replaced. Once the damaged cartilage is removed from the distal end of the femur, the

CLINICAL CONNECTION | *Arthroplasty* CONTINUED

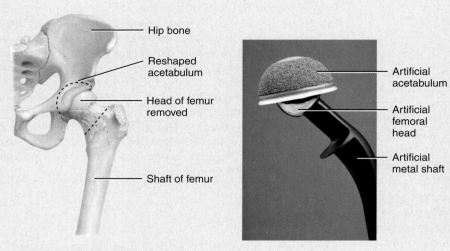

(A) Preparation for total hip replacement

(B) Components of an artificial hip joint prior to implantation

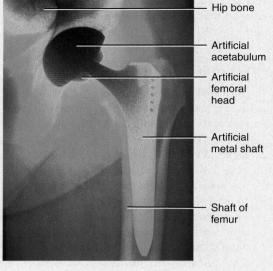

(C) Radiograph of an artificial hip joint

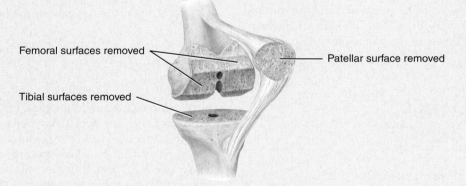

(D) Preparation for total knee replacement

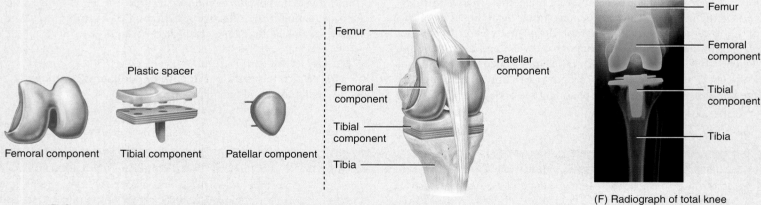

(E) Components of artificial knee joint prior to implantation (left) and implanted (right)

(F) Radiograph of total knee replacement

femur is reshaped and a metal femoral component is cemented in place. Then the damaged cartilage from the proximal end of the tibia is removed, along with the meniscus. The tibia is reshaped and fitted with a plastic tibial component that is cemented into place. If the posterior surface of the patella is badly damaged, the patella is replaced with a plastic patellar component.

In a **total knee replacement (TKR)** (Figures D–F), the entire knee joint is replaced. In the procedure, the damaged cartilage is removed from the distal end of the femur, the proximal end of the tibia, and the back surface of the patella. The femur is reshaped and fitted with

a metal femoral component and cemented in place. The tibia is reshaped and fitted with a plastic tibial component that is cemented in place. If the back surface of the patella is badly damaged, it is replaced with a plastic implant, but if it is not badly damaged, it may be left intact.

Researchers are continually seeking to improve the strength of the cement and devise ways to stimulate bone growth around the implanted area. Potential complications of arthroplasty include infection, blood clots, loosening or dislocation of the replacement components, and nerve injury. •

KEY MEDICAL TERMS ASSOCIATED WITH JOINTS

Arthralgia (ar-THRAL-jē-a; *arthr-*=joint; *algia*=pain) Pain in a joint.

Bursectomy (bur-SEK-to-mē; *ectomy*=removal of) Removal of a bursa.

Chondritis (kon-DRĪ-tis; *chondr-*=cartilage) Inflammation of cartilage.

Subluxation (sub-luks-Ā-shun) A partial or incomplete dislocation.

Synovitis (sin′-ō-VĪ-tis) Inflammation of a synovial membrane in a joint.

CHAPTER REVIEW AND RESOURCE SUMMARY   **WileyPLUS**

Review	Resource

Introduction

1. A joint (articulation or arthrosis) is a point of contact between two bones, between bone and cartilage, or between bone and teeth.
2. The scientific study of joints is termed arthrology and the study of the motion of the human body is called kinesiology.

9.1   Joint Classifications

1. Joints are connective tissue junctions between bones that have varying functions and range from immovable to highly movable junctions.
2. The structural classification of joints is based on the presence or absence of a synovial cavity versus the presence of a solid mass of connective tissue. Solid connective tissue joints can have either cartilage (cartilaginous joints) or dense connective tissue (fibrous joints) as the joining tissue.
3. Functionally, joints are classified as synarthroses (immovable), amphiarthroses (slightly movable), or diarthroses (freely movable).
4. Ligaments are dense irregular or dense regular connective tissue structures that bind one bone to another bone. They stabilize joints and limit the range of motion at a joint.

Anatomy Overview - Joints

9.2   Fibrous Joints

1. The bones of fibrous joints are held together by dense irregular connective tissue.
2. These joints include immovable sutures in adults (found between skull bones), immovable to slightly movable syndesmoses (such as roots of teeth in the sockets in the mandible and maxilla and the distal tibiofibular joint), and slightly movable interosseous membranes (found between the radius and ulna in the forearm and tibia and fibula in the leg).

Anatomy Overview - Fibrous Joints

9.3   Cartilaginous Joints

1. The bones of cartilaginous joints are held together by cartilage.
2. These joints include immovable synchondroses united by hyaline cartilage (epiphyseal plates between diaphyses and epiphyses) and slightly movable symphyses united by fibrocartilage (pubic symphysis).

Anatomy Overview - Cartilaginous Joints

9.4   Synovial Joints

1. Synovial joints (diarthroses) contain a space between bones called the synovial cavity. Other characteristics of synovial joints are the presence of articular cartilage covering the adjacent surfaces of bone and an articular capsule, made up of an outer fibrous membrane and an inner synovial membrane.
2. The synovial membrane secretes synovial fluid, which forms a thin, viscous film over the articular cartilages and other surfaces within the articular capsule. Many synovial joints also contain accessory ligaments (extracapsular and intracapsular) and articular discs or menisci. A labrum is a fibrocartilaginous lip that extends from the edge of the joint socket and helps deepen the socket.
3. Synovial joints contain an extensive nerve and blood supply. The nerves convey information about pain, joint movements, and the degree of stretch at a joint. Blood vessels penetrate the articular capsule and ligaments.
4. Bursae are saclike structures, similar in structure to joint capsules, that alleviate friction as soft tissues such as muscles, tendons, and the skin rub against each other or neighboring bones. They are common around joints such as the shoulder and knee joints. Tendon sheaths are tubelike bursae that wrap around tendons where there is considerable friction, such as in the wrist, fingers, ankles, and toes.

Anatomy Overview - Synovial Joints
Figure 9.3 - Structure of a Typical Synovial Joint
Exercise - Judge That Joint

9.5   Types of Movements at Synovial Joints

1. In a gliding movement, the nearly flat surfaces of bones move back-and-forth and side-to-side.
2. In angular movements, a change in the angle between bones occurs. Examples are flexion–extension, lateral flexion, hyperextension, and abduction–adduction. Circumduction refers to the movement of the distal end of a body part in a circle and involves a continuous sequence of flexion, abduction, extension, adduction, and rotation (or in the opposite order).
3. In rotation, a bone moves around its own longitudinal axis.

Anatomy Overview - Movements Produced at Synovial Joints
Figure 9.5 - Angular Movements at Synovial Joints
Figure 9.6 - Angular Movements at Synovial Joints: Abduction and Adduction

Review	Resource

4. Special movements occur at specific synovial joints. Examples are elevation–depression, protraction–retraction, inversion–eversion, dorsiflexion–plantar flexion, supination–pronation, and opposition. Table 9.1 summarizes the various types of movements at synovial joints.

9.6 Types of Synovial Joints

1. Types of synovial joints are plane, hinge, pivot, condyloid, saddle, and ball-and-socket. In a plane joint the articulating surfaces are flat, and the bones glide back-and-forth and side-to-side (many are biaxial, but some that also permit rotation are triaxial).

2. In a hinge joint, the convex surface of one bone fits into the concave surface of another, and the motion is angular around one axis (uniaxial).

3. In a pivot joint, a round or pointed surface of one bone fits into a ring formed by another bone and a ligament, and movement is rotational (uniaxial).

4. In a condyloid joint, an oval projection of one bone fits into an oval cavity of another, and motion is angular around two axes (biaxial).

5. In a saddle joint, the articular surface of one bone is shaped like a saddle and the other bone fits into the saddle like a sitting rider; motion is angular around three axes (biaxial).

6. In a ball-and-socket joint, the ball-shaped surface of one bone fits into the cuplike depression of another; motion is around three axes (triaxial). Examples include the shoulder and hip joints. Table 9.2 summarizes the categories of joints and their functions (movements).

9.7 Factors Affecting Contact and Range of Motion at Synovial Joints

1. The ways that articular surfaces of synovial joints contact one another determines the type of movement possible.

2. Factors that contribute to keeping the surfaces in contact and affect range of motion are structure or shape of the articulating bones, strength and tension of the ligaments, arrangement and tension of the muscles, contact of soft parts, hormones, and disuse.

9.8 Selected Joints of the Body

1. A summary of selected joints of the body, including articular components, structural and functional classifications, and movements, is presented in Tables 9.3 and 9.4.

2. The temporomandibular joint (TMJ) is between the condyle of the mandible and mandibular fossa and articular tubercle of the temporal bone (Exhibit 9.A).

3. The shoulder (humeroscapular or glenohumeral) joint is between the head of the humerus and glenoid cavity of the scapula (Exhibit 9.B).

4. The elbow joint is between the trochlea of the humerus, the trochlear notch of the ulna, and the head of the radius (Exhibit 9.C).

5. The hip (coxal) joint is between the head of the femur and acetabulum of the hip bone (Exhibit 9.D).

6. The knee (tibiofemoral) joint is between the patella and patellar surface of the femur; the lateral condyle of the femur, the lateral meniscus, and the lateral condyle of the tibia; and the medial condyle of the femur, the medial meniscus, and the medial condyle of the tibia (Exhibit 9.E).

7. The ankle (talocrural) joint is formed by the distal end of the tibia and its medial malleolus with the talus and the lateral malleolus of the fibula (Exhibit 9.F).

9.9 Aging and Joints

1. With aging, a decrease in synovial fluid, thinning of articular cartilage, and decreased flexibility of ligaments occur.

2. Most individuals experience some degeneration in the knee, elbow, hip, and shoulder joints due to the aging process.

CRITICAL THINKING QUESTIONS

1. Burt and Al have been golf partners for 50 years. Burt's golf game improved by 5 points this spring and he credits the hip replacement he had last year. Al's knee has been bothering him for years but when he asked his orthopedist about a new knee joint he was told "it's not that simple." Al told Burt "one joint's like any other joint" and wants a second opinion. What will the new doctor say?

2. After your second Human Anatomy exam, you dropped to one knee, raised your arm over your head with one hand clenched into a fist, pumped your arm up and down, bent your head back, looked straight up, and yelled "*Yes!*" Use the proper terms to describe the movements at the various joints.

3. Lars was just getting the hang of bodysurfing during his first trip to the shore when he got caught in the break of a wave. While he was being rolled in the surf, he felt his shoulder "pop." When Lars finally got back to his towel he was out of breath, in pain, and his arm was hanging at an odd angle. What's the prognosis for the rest of Lars's vacation at the shore?

4. Arthur was in a serious motorcycle accident in which he broke his tibia, fibula, and patella. His knee was twisted sideways in the process. His bones have since healed, but he still experiences considerable pain in the knee joint and says it feels like "there's stuff in there." He also has trouble with stability on the joint while walking. To what specific anatomical problems would you attribute Arthur's joint difficulties?

5. Chuck has gone to the chiropractor because his back hurts. The chiropractor tells him, "You know, your pelvis is out of align-ment." This news came shortly after his orthopedist told him he had slightly bowed legs. On top of all this, he has severe osteoarthritis in his ankle joint. Do you think all of these things could be related? Why or why not?

? ANSWERS TO FIGURE QUESTIONS

9.1 The fibrous joints are sutures found in the skull, syndesmoses found in a variety of places in the skeleton (such as the distal tibiofibular joint), and interosseous membranes found between the radius and ulna and between the tibia and fibula.

9.2 A synchondrosis is held together by hyaline cartilage, and a symphysis is held together by fibrocartilage.

9.3 Synovial joints are capsular joints containing a thin coat of lubricating fluid to reduce friction at the articulating cartilage surfaces.

9.4 Gliding movements occur at plane joints such as the wrist and ankle.

9.5 Two examples of flexion that do not occur in the sagittal plane are flexion of the thumb and lateral flexion of the trunk.

9.6 When you adduct your arm or leg, you bring it closer to the midline of the body, thus "adding" it to the trunk.

9.7 Circumduction involves flexion, abduction, extension, and adduction in continuous sequence.

9.8 The anterior surface of a bone or limb rotates toward the midline in medial rotation, and away from the midline in lateral rotation.

9.9 Bringing the arms forward until the elbows touch is an example of protraction.

9.10 Other examples of pivot joints include the atlanto-axial joints.

9.11 The lateral ligament prevents displacement of the mandible.

9.12 The shoulder joint is the most freely movable joint in the body because of the looseness of its articular capsule and the shallowness of the glenoid cavity in relation to the size of the head of the humerus.

9.13 A hinge joint permits flexion and extension.

9.14 Tension in three ligaments—iliofemoral, pubofemoral, and ischiofemoral—limits the degree of extension at the hip joint.

9.15 Contraction of the quadriceps femoris muscle causes extension at the knee joint.

9.16 Anatomically, the ankle is the region between the distal aspect of the leg and the proximal aspect of the foot.

10 | MUSCULAR TISSUE

INTRODUCTION Machines are part of our daily lives, from simple can openers to complex computers, cars, and copy machines. According to *Webster*'s dictionary, a machine is "an assemblage of parts that transmit forces, motion, and energy one to another in a predetermined manner." So even the most intricate machines are composed of simpler parts such as levers, fulcra, latches, notches, receptors, energy sources, wires, and cables that combine to create the more involved structure we define as a machine. If you think about your body as a machine and its systems and organs as parts, you can more easily comprehend the structure and function of the machinery we call the human body.

One essential component of the machinery of the human body is muscular tissue, which weighs in at 40–50 percent of total body mass (depending on the percentage of body fat, gender, and exercise regimen). Most of the work done by the body, such as pumping blood through the blood vessels, eating, breathing, moving food through the gastrointestinal tract, moving urine out of the urinary bladder, generating heat, speaking, standing up straight, and getting our skeletons to move, is a result of the activity of muscles. In this chapter we will explore muscles from the level of the cell to entire muscles, some of which are quite large, such as the quadriceps muscle that covers most of the anterior surface and sides of your thigh. We will also learn how the simple parts of your muscular machinery work together to produce powerful contractile forces that generate most of the activity of the human body. The scientific study of muscles is known as **myology** (mī-OL-ō-jē; *myo-*= muscle; *-logy*=study of). •

Did you ever wonder what causes rigor mortis?

10.1 OVERVIEW OF MUSCULAR TISSUE

OBJECTIVE

- Compare the three types of muscular tissue with regard to function and special properties.

Types of Muscular Tissue

There are three types of muscular tissue: skeletal, cardiac, and smooth (see Table 3.9). Although the three types of muscular tissue share some properties, they differ from one another in their microscopic anatomy, location, and how they are controlled by the nervous and endocrine systems.

Skeletal muscle tissue is so named because the function of most skeletal muscles is to move the bones of the skeleton. (There are a few that attach to structures other than bone, such as the skin or even other skeletal muscles.) Skeletal muscle tissue is referred to as *striated* because alternating light and dark protein bands (*striations*) are visible when the tissue is examined under a microscope (see Figure 10.4). Skeletal muscle tissue works primarily in a *voluntary* manner; its activity can be consciously (voluntarily) controlled.

Cardiac muscle tissue is found only in the heart, where it forms most of the heart wall. Like skeletal muscle, cardiac muscle is *striated*, but its action is *involuntary*—its alternating contraction and relaxation cannot be consciously controlled. The heart beats because it has a natural pacemaker that initiates each contraction; this built-in (intrinsic) rhythm is called **autorhythmicity**. Several hormones and neurotransmitters adjust heart rate by speeding up or slowing down the pacemaker.

Smooth muscle tissue is located in the walls of hollow internal structures, such as blood vessels, airways, and most organs in the abdominopelvic cavity. It is also attached to hair follicles in the skin. Smooth muscle tissue gets its name from the fact that, under a microscope, it appears *nonstriated* or *smooth*. The action of smooth muscle is usually *involuntary*, and, like cardiac muscle, some smooth muscle tissue has autorhythmicity, such as the muscles that propel food through the gastrointestinal tract. Both cardiac muscle and smooth muscle are regulated by neurons that are part of the autonomic (involuntary) division of the nervous system (see Chapter 20) and by hormones released by endocrine glands.

Functions of Muscular Tissue

Through sustained contraction or alternating contraction and relaxation, muscular tissue has four key functions: producing body movements, stabilizing body positions, storing and moving substances within the body, and generating heat:

1. **Producing body movements.** Total body movements such as walking and running, and localized movements such as grasping a pencil, keyboarding, or raising your hand, rely on the integrated functioning of skeletal muscles, bones, and joints.
2. **Stabilizing body positions.** Skeletal muscle contractions stabilize joints and help maintain body positions, such as standing or sitting. Postural muscles contract continuously when you are awake; for example, sustained contractions in neck muscles hold your head upright when you are listening intently to an anatomy lecture.
3. **Storing and moving substances within the body.** Sustained contractions of ringlike bands of smooth muscles called *sphincters* may prevent outflow of the contents of a hollow organ. Temporary storage of food in the stomach or urine in the urinary bladder is possible because smooth muscle sphincters close off the outlets of these organs. Cardiac muscle contractions pump blood through the body's blood vessels. Contraction and relaxation of smooth muscle in the walls of blood vessels help adjust their diameter and thus regulate the rate of blood flow. Smooth muscle contractions also move food and substances such as bile and enzymes through the gastrointestinal tract, push gametes (sperm and oocytes) through the reproductive systems, and propel urine through the urinary system. Skeletal muscle contractions indirectly promote the flow of lymph throughout the body and aid the return of blood in veins to the heart.
4. **Producing heat.** As muscular tissue contracts, it also produces heat, a process called **thermogenesis** (ther′-mō-JEN-e-sis). Much of the heat released by muscle is used to maintain normal body temperature. Involuntary contractions of skeletal muscles, known as shivering, can dramatically increase the rate of heat production.

Properties of Muscular Tissue

Muscular tissue has four special properties that enable it to perform the functions you just read about and contribute to the homeostasis of the body:

1. **Electrical excitability** (ek-sīt′-a-BIL-i-tē), a property of both muscle and nerve cells, is the ability to respond to certain stimuli by producing electrical signals called **action potentials (impulses)**. Action potentials in muscles are referred to as

muscle action potentials; those in nerves are called *nerve action potentials* or nerve impulses. Action potentials can travel along a cell's plasma membrane due to the presence of specific ion channels. Two main types of stimuli trigger action potentials in muscle cells: electrical and chemical. Autorhythmic *electrical signals* arise in the muscular tissue itself, as in the heart's pacemaker. *Chemical stimuli*, such as neurotransmitters released by neurons, hormones distributed by the blood, or even local changes in pH, can also trigger action potentials in muscle cells.

2. **Contractility** (kon′-trak-TIL-i-tē) is the ability of muscular tissue to contract forcefully when stimulated by an action potential. When a skeletal muscle contracts, it generates tension (force of contraction) while pulling on its attachment points. In some muscle contractions, the muscle develops tension but does not shorten. An example is holding this book in your outstretched hand. In other muscle contractions, the tension generated is greater than resistance, so the muscle shortens and movement occurs. An example is lifting a book off a table.

3. **Extensibility** (ek-sten′-si-BIL-i-tē) is the ability of muscular tissue to stretch, within limits, without being damaged. The connective tissue within the muscle limits the range of extensibility and keeps it within the contractile range of the muscle cells. Normally, smooth muscle is subject to the greatest amount of stretching. For example, each time your stomach fills with food, the muscle in the wall is stretched. Cardiac muscle also is stretched each time the heart fills with blood.

4. **Elasticity** (e-las-TIS-i-tē) is the ability of muscular tissue to return to its original length and shape after contraction or extension.

Skeletal muscle is the focus of much of this chapter. Cardiac muscle and smooth muscle are described more briefly here (see Sections 10.5 and 10.6). The details of cardiac muscle function are discussed in Chapter 13 (the heart), and more about smooth muscle is presented in Chapter 19 (the autonomic nervous system), and in discussions of the various organs containing smooth muscle.

 CHECKPOINT

1. What features distinguish the three types of muscular tissue?
2. Summarize the functions of muscular tissue.
3. Describe the four properties of muscular tissue.

10.2 SKELETAL MUSCLE TISSUE

 OBJECTIVES

• Explain the importance of connective tissue components, blood vessels, and nerves to skeletal muscles.
• Describe the microscopic anatomy of a skeletal muscle fiber.
• Describe the functions of skeletal muscle proteins.
• Outline how a skeletal muscle fiber contracts and relaxes.
• Distinguish between isotonic and isometric contractions.

Each skeletal muscle is a separate organ composed of hundreds to thousands of skeletal muscle cells, also called **muscle fibers** because of their elongated shapes. Connective tissues surround muscle fibers and whole muscles, and carry the blood vessels and nerves that exert their effects on individual muscle fibers (Figure 10.1). To understand how the contraction of a skeletal muscle works, you first need to learn about the gross and microscopic anatomy of its individual fibers.

Structure of a Skeletal Muscle

A typical skeletal muscle consists of a muscle belly connected by tendons to the skeleton (Figure 10.1). The reddish or meatlike appearance that we associate with muscular tissue arises from the large population of well-vascularized muscle cells in the **muscle belly (body)** of the organ. The belly of the muscle can be an elongated, thick, rounded mass, a triangular shape, a thick rectangular mass, or a thin, flat sheet of muscular tissue. In contrast, the **tendons**, tough, glistening white dense regular connective tissue structures that attach the muscle belly to the bones, are minimally vascular, lack muscle cells, and consist primarily of parallel arrangements of collagen fibers. Like the muscle belly, tendons display a great variety of shapes: Some are long, ropelike structures, while others are arranged in flat sheets called **aponeuroses** (ap′-ō-noo-RŌ-sēz; *apo-*=from, *neu-*=a sinew; singular is aponeurosis). An example of an aponeurosis is the epicranial aponeurosis on top of the skull between the occipital and frontal bellies of the occipitofrontalis muscle (see Figure 11.4c). Other tendons are bands or extensions of connective tissue that are so short they make the muscle body appear as if it attaches directly to the bone. Now that you understand the basic parts of a skeletal muscle, let's dig a little deeper and explore the relationship between the body of a muscle and its tendons.

Because muscles generate tremendous forces when they contract, it is important to understand how the muscle belly connects to the tendons and how the tendons connect to the bones. What makes these connections so strong that they do not pull apart when you lift something heavy? Another look at Figure 10.1 will help you find the answer.

CLINICAL CONNECTION | *Tenosynovitis*

Tenosynovitis (ten′-ō-sin-ō-VĪ-tis) is an inflammation of the tendons, tendon sheaths, and synovial membranes surrounding certain joints. The tendons most often affected are at the wrists, shoulders, elbows, finger joints, ankles, and feet. The affected sheaths sometimes become visibly swollen because of fluid accumulation. Tenderness and pain are frequently associated with movement of the body part. The condition often follows trauma, strain, excessive exercise, or other stressors. For example, tenosynovitis of the dorsum of the foot may also be caused by tying shoelaces too tightly. Gymnasts are prone to developing the condition as a result of chronic, repetitive, and maximum hyperextension at the wrists. Other repetitive movements involving activities such as typing, haircutting, carpentry, and assembly line work can also result in tenosynovitis. •

Connective Tissue Coverings

The bulk of the muscle belly consists of striated skeletal muscle fibers. However, a close look at Figure 10.1 will show you that there is more to the muscle belly than just muscle fibers. Surrounding each muscle fiber is a thin wrapping of mostly reticular fibers called the **endomysium** (en′-dō-MĪZ-ē-um; *endo-*=within). This surrounding connective tissue helps to bind the muscle fibers together, yet it is loose enough to allow them to move freely over one another. In addition, the endomysium carries small blood vessels that supply the fibers with nutrients. Next, notice that groups of muscle fibers form bundles wrapped in a thicker layer of connective tissue. This muscle fiber bundle is a **fascicle** (FAS-i-kl=little bundle), also called a *fasciculus*, and its dense irregular connective tissue covering is called the **perimysium** (per′-i-MĪZ-ē-um; *peri-*=around). The perimysium also allows a certain

Figure 10.1 Organization of skeletal muscle and its connective tissue coverings.

🔑 A skeletal muscle consists of individual muscle fibers (cells) bundled into fascicles and surrounded by three connective tissue layers.

FUNCTIONS OF MUSCULAR TISSUE

1. **Produces body motions.**
2. **Stabilizes body positions.**
3. **Stores and moves substances within the body.**
4. **Produces heat.**

Tendon

Transverse plane

Bone

Periosteum

Tendon

Belly of skeletal muscle

Perimysium

Epimysium

Fascicle

Perimysium

Muscle fiber (cell)

Myofibril

Endomysium

Perimysium

Somatic motor neuron

Blood capillary

Endomysium

Nucleus

Muscle fiber

Fascicle

Striations

Transverse sections

Sarcoplasm

Sarcolemma

Myofibril

Filament

❓ **Which connective tissue coat surrounds groups of muscle fibers, separating them into fascicles?**

Components of a skeletal muscle

degree of freedom of motion between neighboring fascicles and transmits blood vessels. Around the periphery of the muscle is a somewhat thicker covering of dense irregular connective tissue called the **epimysium** (ep′-i-MĪZ-ē-um; *epi-*=upon), which binds all the fascicles together to form the muscle belly. So the muscle belly is not just a large mass of individual muscle fibers, but groups of fibers wrapped in connective tissue. Note that, even though we give the connective tissue layers three separate names, they form a continuous interconnected network.

But what does all this have to do with the connections among muscle, tendons, and bones? The muscle fibers in the belly of the muscle eventually taper to round blunt ends, but the connective tissues associated with the fibers continue beyond the blunt ends to become the tendons of the muscle. This continuing mass of collagenous connective tissue takes on a glistening whitish appearance, resulting from highly ordered collagen fibers, reduced numbers of blood vessels, and an absence of muscle cells. So the tendons are a continuous mass of connective tissue that runs through the muscle as the endomysium, perimysium, and epimysium and emerges from the belly of the muscle as the tendons of origin and insertion at either end. This is what makes your muscles so incredibly strong. At its junction with the bone, the surface tissue of the tendon is continuous with the periosteum, while its deeper collagen fibers enter the bone to blend with the collagen of the osseous extracellular matrix. This strong, continuous network of connective tissue is essential to the function of the musculoskeletal system.

The various skeletal muscles of the body are further grouped together and protected by large dense irregular connective tissue sheets, called **fascia** (FASH-ē-a=bandage), which wrap around groups of muscles much like a sock encircles your foot. For example, underneath the skin and subcutaneous layer in the free lower limbs a thin, tough, glistening sheet of dense irregular connective tissue called the *fascia of the free lower limbs* surrounds all the muscles. The free upper limbs have similar sheets of fascia. In the trunk, head, and neck there are multiple fascial layers.

CLINICAL CONNECTION | Fibromyalgia

Fibromyalgia (fī-brō-mī-AL-jē-a; *algia*=painful condition) is a chronic, painful, nonarticular rheumatic disorder that affects the fibrous connective tissue components of muscles, tendons, and ligaments. A striking sign is pain that results from gentle pressure at specific "tender points." Even without pressure, there is pain, tenderness, and stiffness of muscles, tendons, and surrounding soft tissues. In addition to muscle pain, people suffering from fibromyalgia report severe fatigue, poor sleep, headaches, depression, irritable bowel syndrome, and inability to carry out their daily activities. There is no specific identifiable cause. Treatment consists of stress reduction, regular exercise, application of heat, gentle massage, physical therapy, medication for pain, and a low-dose antidepressant to help improve sleep. •

Nerve and Blood Supply

Skeletal muscles are well supplied with nerves and blood vessels (see Figure 11.18c). Most of the skeletal muscles in the body, especially those of the limbs and head, receive one main nerve that carries motor neurons to the muscle and sensory neurons away from the muscle. Other muscles, such as the sheetlike muscles of the body wall, receive multiple nerves that make it possible to independently control different segmental levels within the muscle sheet. Nerves typically enter the muscle along with the main blood vessels of the muscle as a unit called a **neurovascular bundle** (noo-rō-VAS-kū-lar). These neurovascular bundles enter the muscle body near the stable tendon attachment (tendon of origin) and then spread through the muscle via the connective tissue channels formed by perimysium and endomysium as they wrap the muscle cells. The motor fibers initiate the contractile function of muscle cells, while the sensory fibers provide feedback to the nervous system to regulate motor function. The neurons (nerve cells) that stimulate skeletal muscle fibers to contract are called **somatic motor neurons**. A somatic motor neuron has a threadlike extension, called an **axon**, which travels from the neuron cell body in the brain or spinal cord to a group of skeletal muscle fibers in a muscle of the body. You will learn more about the interactions between nerves and muscles later in this section.

Generally, an artery and one or two veins accompany each nerve that penetrates a skeletal muscle. Microscopic blood vessels called *capillaries* are plentiful in muscle tissue; each muscle fiber is in close contact with one or more capillaries (Figure 10.1), which bring oxygen and nutrients to the muscle fibers and remove heat and the waste products of muscle metabolism. Especially during

contraction, a muscle fiber synthesizes and uses considerable ATP (adenosine triphosphate); these reactions require oxygen, glucose, fatty acids, and other substances that are supplied in the blood.

Microscopic Anatomy of a Skeletal Muscle Fiber (Cell)

The most important components of a skeletal muscle are the **muscle fibers** themselves. Mature muscle fibers range from 10 to 100 μm* in diameter. Typical muscle fiber length is about 10 cm (4 in.) in humans, although some are up to 30 cm (12 in.) long, such as those in the free lower limbs. During embryonic development, each skeletal muscle fiber arises from the fusion of a hundred or more small mesodermal cells called *myoblasts* (MĪ-ō-blasts) (Figure 10.2a). Hence, each mature skeletal muscle fiber is a single cell with a hundred or more nuclei. Once fusion has occurred, the muscle fiber loses its ability to undergo cell division. Thus, most skeletal muscle fibers arise before birth, and most of these cells last a lifetime.

The dramatic muscle growth that occurs after birth occurs by enlargement of existing muscle fibers, called **hypertrophy** (hī-PER-trō-fē; *hyper-*=above or excessive; *-trophy*=nourishment), rather than by **hyperplasia** (hī-per-PLĀ-zē-a; *-plasis*=molding), an increase in the number of fibers. Hypertrophy is due to increased production of myofibrils, mitochondria, sarcoplasmic reticulum, and other organelles. It results from very forceful, repetitive muscular activity, such as strength training. Because hypertrophied muscles contain more myofibrils, they are capable of more forceful contractions. During childhood, human growth hormone and other hormones stimulate an increase in the size of skeletal muscle fibers. The hormone testosterone (from the testes in males and in small amounts from other tissues such as the ovaries in females) promotes further enlargement of muscle fibers. A few myoblasts do persist in mature skeletal muscle as *satellite cells* (Figure 10.2a, b). Satellite cells retain the capacity to fuse with one another or with damaged muscle fibers to regenerate functional muscle fibers. However, the number of new skeletal muscle fibers that can be formed by satellite cells is not enough to compensate for significant skeletal muscle damage or degeneration. In such cases, skeletal muscle tissue undergoes **fibrosis**, the replacement of muscle fibers by fibrous scar tissue. Thus, regeneration of skeletal muscle tissue is limited.

Sarcolemma, T Tubules, and Sarcoplasm

The multiple nuclei of a skeletal muscle fiber are located just beneath the **sarcolemma** (sar′-kō-LEM-ma) *sarc-*=flesh; *-lemma*= sheath), the plasma membrane of a muscle fiber (Figure 10.2b, c). Thousands of tiny invaginations of the sarcolemma, called **transverse tubules (T tubules)**, tunnel in from the surface toward the center of each muscle fiber. Because transverse tubules are open to the outside of the fiber, they are filled with interstitial fluid. Muscle action potentials propagate along the sarcolemma and through the transverse tubules, quickly spreading throughout the muscle fiber. This arrangement ensures that all the superficial and deep parts of the muscle fiber become excited by an action potential almost simultaneously.

The sarcolemma surrounds the **sarcoplasm** (SAR-kō-plazm), the cytoplasm of a muscle fiber. The sarcoplasm includes a substantial amount of glycogen, a storage molecule that consists of a chain of linked glucose molecules. When the muscle requires energy and has already depleted its available glucose, glucose molecules from glycogen will be released and utilized for the synthesis of ATP. In addition, the sarcoplasm contains a red-colored

---

*One micrometer (μm) is 10^{-6} meter (1/25,000 in.).

Figure 10.2 **Microscopic organization of skeletal muscle.** (a) During embryonic development, many myoblasts fuse lengthwise to form one skeletal muscle fiber. Once fusion has occurred, a skeletal muscle fiber loses the ability to undergo cell division, but satellite cells retain this ability. (b and c) The sarcolemma of the muscle fiber encloses sarcoplasm and myofibrils, which are striated. Sarcoplasmic reticulum (SR) wraps around each myofibril. Thousands of transverse tubules, filled with interstitial fluid, invaginate from the sarcolemma toward the center of the muscle fiber. A photomicrograph of skeletal muscle tissue is shown in Table 3.9A.

🔑 The contractile elements of muscle fibers are the myofibrils, which contain overlapping thick and thin filaments.

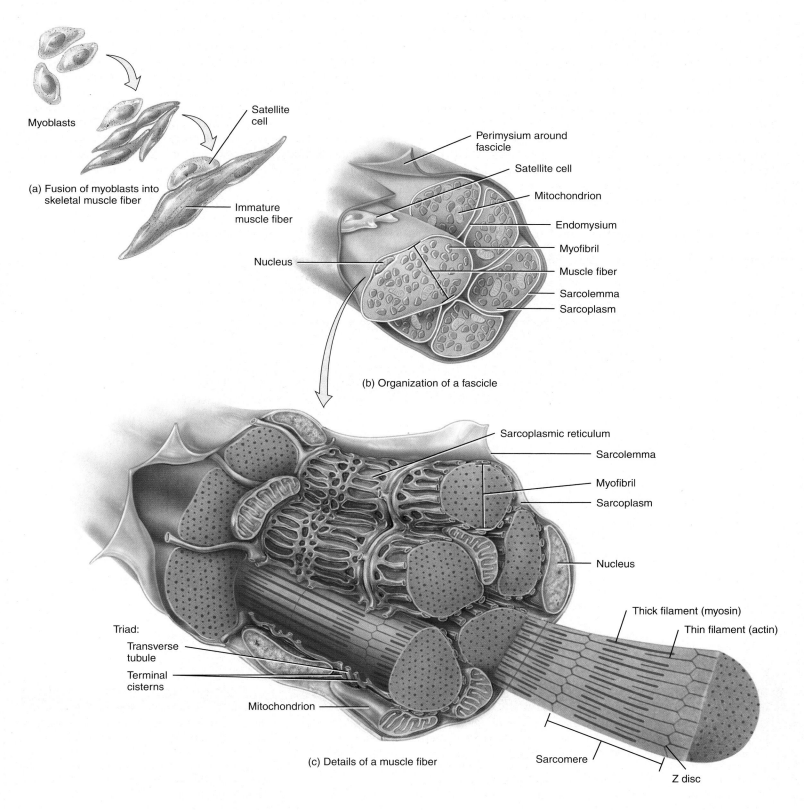

(a) Fusion of myoblasts into skeletal muscle fiber

(b) Organization of a fascicle

(c) Details of a muscle fiber

❓ **Which structure shown here releases calcium ions to trigger muscle contraction?**

protein called **myoglobin** (MĪ-ō-glōb-in). This protein, found only in muscle, binds oxygen molecules that diffuse into muscle fibers from interstitial fluid. Myoglobin releases oxygen when mitochondria need it for ATP production. The mitochondria lie in rows throughout the muscle fiber, strategically close to the contractile muscle proteins that use ATP during contraction so that ATP can be produced quickly as it is needed.

Myofibrils and Sarcoplasmic Reticulum

At high magnification, the sarcoplasm appears stuffed with little threads. These small structures are the contractile elements of skeletal muscle, the **myofibrils** (mī′-ō-FĪ-brils) (Figure 10.2b, c). Myofibrils, which are about $2\mu m$ in diameter and extend the entire length of the muscle fiber, have prominent striations that make the whole muscle fiber look striped (striated).

A fluid-filled system of membranous sacs called the **sarcoplasmic reticulum (SR)** (sar′-kō-PLAZ-mik re-tik′-ū-lum) encircles each myofibril (Figure 10.2c). This elaborate system is similar to smooth endoplasmic reticulum in nonmuscle cells. Dilated end sacs of the sarcoplasmic reticulum called **terminal cisterns** (SISterns=reservoirs) butt up against the transverse tubules from both sides. One transverse tubule and the two terminal cisterns on either

side of it form a **triad** (*tri-*=three) (Figure 10.2c). In a relaxed muscle fiber, the sarcoplasmic reticulum stores calcium ions (Ca^{2+}). When triggered, Ca^{2+} will be released from the terminal cisterns into the sarcoplasm, which triggers muscle contraction.

Filaments and the Sarcomere

Within myofibrils are smaller protein structures called **filaments** or *myofilaments* (Figure 10.2c). *Thin filaments* are about 8 nm[†] in diameter and 1–2 μm long and composed mostly of the protein actin, while *thick filaments* are 16 nm in diameter and 1–2 μm long and composed mostly of the protein myosin. Both thin and thick filaments are directly involved in the contractile process. Overall, there are two thin filaments for every thick filament in the regions of filament overlap. The filaments inside a myofibril do not extend the entire length of a muscle fiber. Instead, they are arranged in compartments called **sarcomeres** (SAR-kōmers; *-mere*=part), the basic functional units of a myofibril (Figure 10.3a). Narrow, plate-shaped regions of dense protein material called **Z discs** separate one sarcomere from the next. Thus, a sarcomere extends from one Z disc to the next Z disc.

[†] One nanometer (nm) is 10^{-9} meter ($0.001\mu m$).

Figure 10.3 The arrangement of filaments within a sarcomere. A sarcomere extends from one Z disc to the next.

Myofibrils contain two types of contractile filaments: thick filaments and thin filaments.

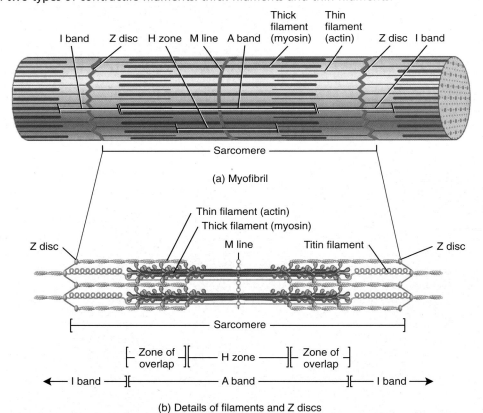

(a) Myofibril

(b) Details of filaments and Z discs

 Among the following, which is smallest: muscle fiber, thick filament, or myofibril? Which is largest?

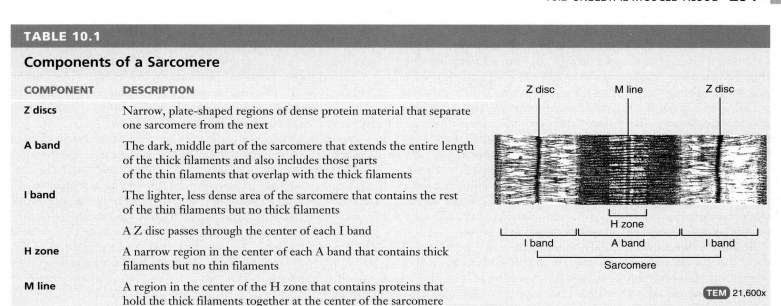

TABLE 10.1

Components of a Sarcomere

COMPONENT	DESCRIPTION
Z discs	Narrow, plate-shaped regions of dense protein material that separate one sarcomere from the next
A band	The dark, middle part of the sarcomere that extends the entire length of the thick filaments and also includes those parts of the thin filaments that overlap with the thick filaments
I band	The lighter, less dense area of the sarcomere that contains the rest of the thin filaments but no thick filaments. A Z disc passes through the center of each I band
H zone	A narrow region in the center of each A band that contains thick filaments but no thin filaments
M line	A region in the center of the H zone that contains proteins that hold the thick filaments together at the center of the sarcomere

TEM 21,600x

The extent of overlap of the thick and thin filaments depends on whether the muscle is contracted, relaxed, or stretched. The pattern of overlap, consisting of a variety of zones and bands (Figure 10.3b), creates the striations that can be seen both in single myofibrils and in whole muscle fibers. The darker middle part of the sarcomere is the **A band,** which extends the entire length of the thick filaments (Figure 10.3b). Toward each end of the A band is a *zone of overlap,* where the thick and thin filaments lie side by side. The **I band** is a lighter, less dense area that contains thin filaments but no thick filaments (Figure 10.3b); a Z disc passes through the center of each I band. A narrow **H zone** in the center of each A band contains thick filaments but no thin filaments. A mnemonic that will help you to remember the composition of the I and H bands is as follows: The letter *I* is thin (contains thin filaments), while the letter *H* is thick (contains thick filaments). Supporting proteins that hold the thick filaments together at the center of the H zone form the **M line**, so named because it is at the *middle* of the sarcomere. Table 10.1 presents a summary of the components of a sarcomere.

Muscle Proteins

Myofibrils are built from three kinds of proteins: (1) contractile proteins, which generate force during contraction; (2) regulatory proteins, which help switch the contraction process on and off; and (3) structural proteins, which keep the thick and thin filaments in the proper alignment, give the myofibril elasticity and extensibility, and link the myofibrils to the sarcolemma and extracellular matrix.

The two *contractile proteins* in muscle, myosin and actin, are components of thick and thin filaments, respectively. Not to be confused with myoglobin, the protein that makes muscle fibers red, **myosin** (MĪ-ō-sin) is the main component of thick filaments and functions as a motor protein in all three types of muscle tissue. By pulling various cellular structures, *motor proteins* convert ATP's chemical energy into the mechanical energy of motion, that is, the production of force. About 300 molecules of myosin form a single thick filament in skeletal muscular tissue. Each myosin molecule is shaped like two golf clubs twisted together (Figure 10.4a). The *myosin tail* (twisted golf club handles) points toward the M line in the center of the sarcomere. Tails of neighboring myosin molecules lie parallel to one another, forming the shaft of the thick filament. The two projections of each myosin molecule (golf club heads) are called *myosin heads*. The heads project outward from the shaft in a spiraling fashion, each extending toward one of the six thin filaments that surround each thick filament.

Figure 10.4 **Structure of thick and thin filaments.** (a) The thick filament contains about 300 myosin molecules, one of which is pictured. The myosin tails form the shaft of the thick filament, and the myosin heads project outward toward the surrounding thin filaments. (b) Thin filaments contain actin, troponin, and tropomyosin.

 Contractile proteins (myosin and actin) generate force during contraction, and regulatory proteins (troponin and tropomyosin) help switch contraction on and off.

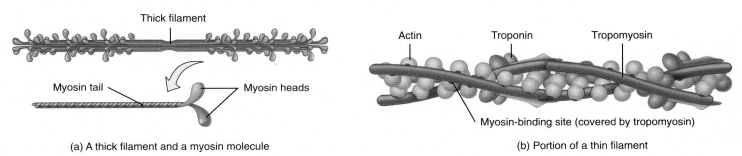

(a) A thick filament and a myosin molecule

(b) Portion of a thin filament

Which proteins connect to the Z disc? Which proteins are present in the A band? In the I band?

Thin filaments extend from anchoring points within the Z discs (see Figure 10.3b). Their main component is the protein **actin** (AK-tin). Individual actin molecules join to form an actin filament that is twisted into a helix (Figure 10.4b). On each actin molecule is a *myosin-binding site*, where a myosin head can attach. Smaller amounts of two *regulatory proteins*—**tropomyosin** (trō-pō-MĪ-ō-sin) and **troponin** (TRŌ-pō-nin)—are also part of the thin filament. In relaxed muscle, myosin is blocked from binding to actin because strands of tropomyosin cover the myosin-binding site on actin. The tropomyosin strand, in turn, is held in place by troponin molecules. You will soon learn that when calcium ions (Ca^{2+}) bind to troponin, it undergoes a change in shape; this change moves tropomyosin away from myosin-binding sites on actin, allowing myosin to bind to actin and muscle contraction to begin.

In addition to contractile and regulatory proteins, muscle contains about a dozen *structural proteins*, which contribute to the alignment, stability, elasticity, and extensibility of myofibrils. Several key structural proteins are titin, myomesin, nebulin, and dystrophin. **Titin** (*titan*=gigantic) is the third most plentiful protein in skeletal muscle (after actin and myosin). This molecule's name reflects its huge size. With a molecular weight of about 3 million daltons, titin is 50 times larger than an average-sized protein. Each titin molecule spans half a sarcomere, from a Z disc to an M line (see Figure 10.3b), a distance of 1 to 1.2 μm in relaxed muscle. Titin anchors a thick filament to both a Z disc and the M line, thereby helping stabilize the position of the thick filament. The part of the titin molecule that extends from the Z disc to the beginning of the thick filament is very elastic. Because it can stretch to at least four times its resting length and then spring back unharmed, titin accounts for much of the elasticity and extensibility of myofibrils. Titin probably helps the sarcom-

eres return to their resting length after a muscle has contracted or been stretched, may help prevent overextension of sarcomeres, and maintains the central location of the A bands.

Molecules of the protein **myomesin** (mī-ō-MĒ-sin) form the M line. The M line proteins bind to titin and connect adjacent thick filaments to one another. Myomesin holds the thick filaments in alignment at the M line.

Nebulin (NEB-ū-lin) helps anchor thin filaments to Z discs and regulates the length of thin filaments during development.

Dystrophin (dis-TRŌ-fin) links thin filaments of the sarcomere to integral membrane proteins of the sarcolemma. In turn, the membrane proteins attach to proteins in the connective tissue extracellular matrix that surrounds muscle fibers. Hence, dystrophin and its associated proteins are thought to reinforce the sarcolemma and help transmit the tension generated by the sarcomeres to the tendons. The relationship of dystrophin to musclar dystrophy is discussed in the Clinical Connection in Section 10.3.

Table 10.2 presents a summary of skeletal muscle fiber proteins and Table 10.3 summarizes the levels of organization within a skeletal muscle.

Contraction and Relaxation of Skeletal Muscle Fibers

Sliding Filament Mechanism

Skeletal muscle shortens during contraction because the thick and thin filaments slide past one another. The model describing this process is known as the **sliding filament mechanism**. Muscle contraction occurs because myosin heads attach to and "walk" along the thin filaments at both ends of a sarcomere, progressively

---

TABLE 10.2

Summary of Skeletal Muscle Fiber Proteins

TYPE OF PROTEIN	DESCRIPTION
Contractile proteins	Proteins that generate force during muscle contractions.
Myosin	A contractile protein that makes up the thick filament. A myosin molecule consists of a tail and two myosin heads, which bind to myosin-binding sites on actin molecules of a thin filament during muscle contraction.
Actin	A contractile protein that is the main component of the thin filament. On each actin molecule is a myosin-binding site where a myosin head of a thick filament binds during muscle contraction.
Regulatory proteins	Proteins that help switch the muscle contraction process on and off.
Tropomyosin	A regulatory protein that is a component of the thin filament. When a skeletal muscle fiber is relaxed, tropomyosin covers the myosin-binding sites on actin molecules, thereby preventing myosin from binding to actin.
Troponin	A regulatory protein that is a component of the thin filament. When calcium ions (Ca^{2+}) bind to troponin, it undergoes a change in shape; this conformational change moves tropomyosin away from myosin-binding sites on actin molecules, and muscle contraction subsequently begins as myosin binds to actin.
Structural proteins	Proteins that keep the thick and thin filaments of the myofibrils in proper alignment, give the myofibrils elasticity and extensibility, and link the myofibrils to the sarcolemma and extracellular matrix.
Titin	A structural protein that connects a Z disc to the M line of the sarcomere, thereby helping to stabilize the position of the thick filament. Because it can stretch and then spring back unharmed, titin accounts for much of the elasticity and extensibility of myofibrils.
Myomesin	A structural protein that forms the M line of the sarcomere; it binds to titin molecules and connects adjacent thick filaments to one another.
Nebulin	A structural protein that wraps around the entire length of each thin filament; it helps anchor the thin filaments to the Z discs and regulates the length of the thin filaments during development.
Dystrophin	A structural protein that links the thin filaments of the sarcomere to integral membrane proteins in the sarcolemma, which are attached in turn to proteins in the connective tissue matrix that surrounds muscle fibers. It is thought that dystrophin helps reinforce the sarcolemma and that it helps transmit tension generated by sarcomeres to tendons.

TABLE 10.3

Levels of Organization within a Skeletal Muscle

LEVEL	DESCRIPTION
Skeletal muscle 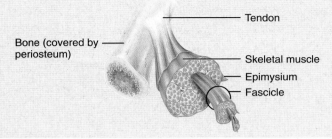 Tendon, Bone (covered by periosteum), Skeletal muscle, Epimysium, Fascicle	A skeletal muscle is an organ made up of fascicles that contain muscle fibers (cells), blood vessels, and nerves. The skeletal muscle is wrapped in epimysium.
Fascicle 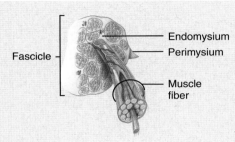 Endomysium, Perimysium, Fascicle, Muscle fiber	A fascicle is a bundle of muscle fibers wrapped in perimysium.
Muscle fiber (cell) 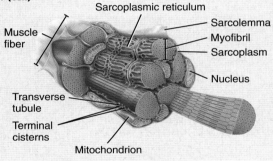 Sarcoplasmic reticulum, Sarcolemma, Myofibril, Sarcoplasm, Muscle fiber, Nucleus, Transverse tubule, Terminal cisterns, Mitochondrion	Long cylindrical cell covered by a vascular endomysium. The cell membrane, the sarcolemma, surrounds the sarcoplasm with its myofibrils, many peripherally located nuclei, mitochondria, transverse tubules, sarcoplasmic reticulum, and terminal cisterns. The fiber has a striated appearance.
Myofibril 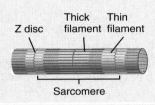 Z disc, Thick filament, Thin filament, Sarcomere	Threadlike contractile elements within the sarcoplasm of a muscle fiber that extend the entire length of the fiber; composed of filaments.
Filaments (myofilaments) 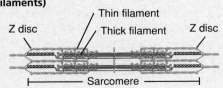 Thin filament, Z disc, Thick filament, Z disc, Sarcomere	Contractile proteins within myofibrils that are of two types: thick filaments composed of myosin and thin filaments composed of actin, tropomyosin, and troponin; the sliding of the thin filaments past the thick filaments produces muscle shortening.

pulling the thin filaments toward the M line (Figure 10.5). As a result, the thin filaments slide inward and meet at the center of a sarcomere. They may even move so far inward that their ends overlap (Figure 10.5c). As the thin filaments slide inward, the Z discs come closer together, and the sarcomere shortens. However, the lengths of the individual thick and thin filaments do not change. Shortening of the sarcomeres causes shortening of the whole muscle fiber, which in turn leads to shortening of the entire muscle.

The Neuromuscular Junction

As noted earlier in the chapter, the neurons that stimulate skeletal muscle fibers to contract are called somatic motor neurons. Each somatic motor neuron has a threadlike axon that extends from the brain or spinal cord to a group of skeletal muscle fibers. A muscle fiber contracts in response to one or more action potentials propagating along its sarcolemma and through its system of T tubules. Muscle action potentials arise at the **neuromuscular junction (NMJ)** (noo-rō-MUS-kū-lar), the synapse between a

Figure 10.5 Sliding filament mechanism of muscle contraction, as it occurs in two adjacent sarcomeres.

During muscle contractions, thin filaments move toward the M line of each sarcomere.

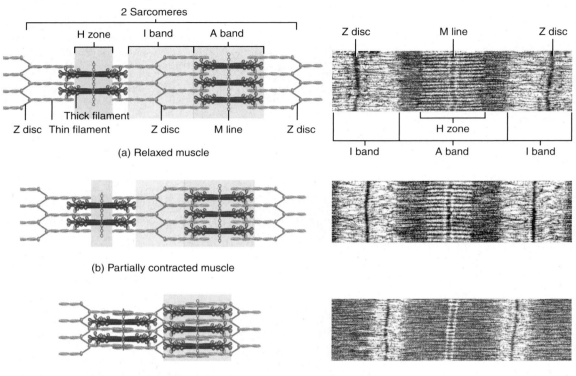

(a) Relaxed muscle

(b) Partially contracted muscle

(c) Maximally contracted muscle

What happens to the I band and H zone during contraction? Do the lengths of the thick and thin filaments change?

somatic motor neuron and a skeletal muscle fiber (Figure 10.6a). A **synapse** is a region where communication occurs between two neurons, or between a neuron and a target cell—in this case, between a somatic motor neuron and a muscle fiber. At most synapses a small gap, called the **synaptic cleft**, separates the two cells. Because the cells do not physically touch, the action potential cannot "jump the gap" from one cell to another. Instead, the first cell communicates with the second by releasing a chemical messenger called a **neurotransmitter**.

At the NMJ, the end of the motor neuron, called the **axon terminal**, divides into a cluster of **synaptic end bulbs** (Figure 10.6a, b) (the neural part of the NMJ). Suspended in the cytosol within each synaptic end bulb are hundreds of membrane-enclosed sacs called **synaptic vesicles**. Inside each synaptic vesicle are thousands of molecules of **acetylcholine (ACh)** (as′-ē-til-KŌ-lēn), the neurotransmitter released at the NMJ.

The region of the sarcolemma opposite the synaptic end bulbs, called the **motor end plate** (Figure 10.6c), is the muscular part of the NMJ. Within each motor end plate are 30 to 40 million **acetylcholine receptors**, integral transmembrane proteins that bind specifically to ACh. These receptors are abundant in **junctional folds**, deep grooves in the motor end plate that provide a large surface area for ACh. As you will see shortly, the ACh receptors are ion channels. A neuromuscular junction thus includes all the synaptic end bulbs of a neuron on one side of the synaptic cleft, plus the motor end plate of the muscle fiber on the other side.

Each skeletal muscle fiber has only a single neuromuscular junction, but the axon of a somatic motor neuron branches out and forms neuromuscular junctions with many different muscle

fibers. A somatic motor neuron plus all the skeletal muscle fibers it stimulates is called a **motor unit**. A single motor neuron makes contact with an average of 150 skeletal muscle fibers, and all muscle fibers in one motor unit contract in unison. Muscles that control precise movements consist of many small motor units. For example, muscles of the larynx (voice box) that control voice production have as few as two or three muscle fibers per motor unit, and muscles controlling eye movements may have 10–20 muscle fibers per motor unit. In contrast, some motor units in skeletal muscles responsible for large-scale and powerful movements, such as the biceps brachii muscle in the arm and the gastrocnemius muscle in the calf of the leg, may have 2000–3000 muscle fibers each. The total strength of a muscle contraction depends in part on how large its motor units are and how many motor units are activated at the same time.

A nerve impulse (or nerve action potential) elicits a muscle action potential in the following way (Figure 10.6c):

❶ Release of acetylcholine. Arrival of the nerve impulse at the synaptic end bulbs stimulates voltage-gated channels to open, and Ca^{2+} enters the synaptic end bulbs. Voltage-gated channels are integral membrane proteins that open in response to a change in membrane potential (voltage). The Ca^{2+} stimulates synaptic vesicles to undergo exocytosis. During exocytosis, the synaptic vesicles fuse with the motor neuron's plasma membrane, liberating ACh into the synaptic cleft. The ACh then diffuses across the synaptic cleft between the motor neuron and the motor end plate.

Figure 10.6 Structure of the neuromuscular junction (NMJ), the synapse between a somatic motor neuron and a skeletal muscle fiber.

Synaptic end bulbs at the tips of axon terminals contain synaptic vesicles filled with acetylcholine.

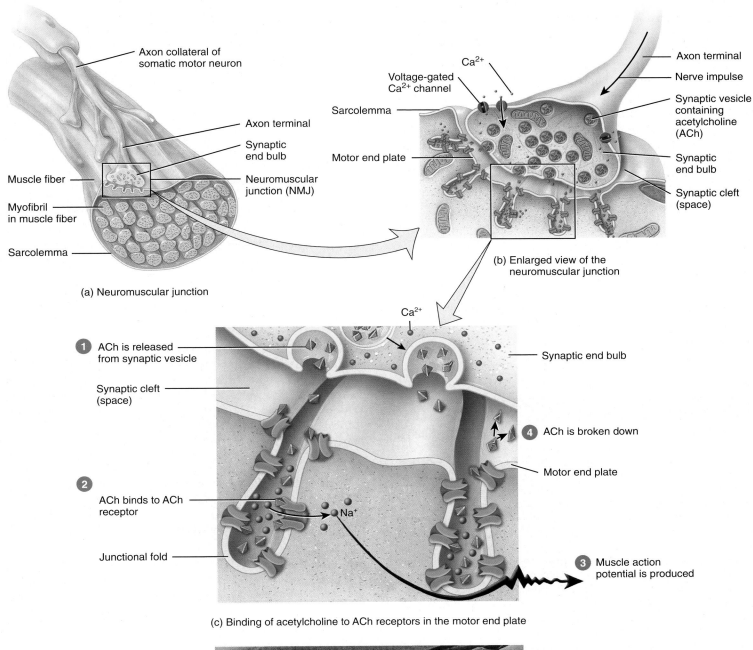

Axon collateral of somatic motor neuron

Axon terminal

Synaptic end bulb

Muscle fiber

Neuromuscular junction (NMJ)

Myofibril in muscle fiber

Sarcolemma

(a) Neuromuscular junction

Ca²⁺

Voltage-gated Ca²⁺ channel

Sarcolemma

Motor end plate

Axon terminal

Nerve impulse

Synaptic vesicle containing acetylcholine (ACh)

Synaptic end bulb

Synaptic cleft (space)

(b) Enlarged view of the neuromuscular junction

Ca²⁺

1 ACh is released from synaptic vesicle

Synaptic cleft (space)

Synaptic end bulb

4 ACh is broken down

Motor end plate

2 ACh binds to ACh receptor

Na⁺

3 Muscle action potential is produced

Junctional fold

(c) Binding of acetylcholine to ACh receptors in the motor end plate

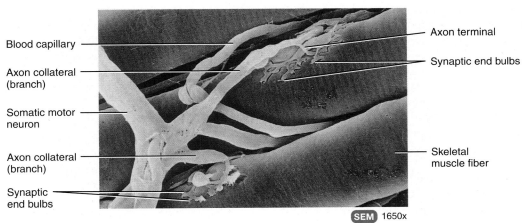

Blood capillary

Axon collateral (branch)

Somatic motor neuron

Axon collateral (branch)

Synaptic end bulbs

Axon terminal

Synaptic end bulbs

Skeletal muscle fiber

SEM 1650x

(d) Neuromuscular junction

 What part of the sarcolemma contains acetylcholine receptors?

❷ *Activation of ACh receptors.* Binding of two molecules of ACh to the receptor on the motor end plate opens an ion channel in the ACh receptor. Once the channel is open, small cations, most importantly Na^+, can flow across the membrane.

❸ *Production of muscle action potential.* The inflow of Na^+ triggers a muscle action potential. Each nerve impulse normally elicits one muscle action potential. The muscle action potential then propagates along the sarcolemma into the system of T tubules. This causes the sarcoplasmic reticulum to release its stored Ca^{2+} into the sarcoplasm and the muscle fiber subsequently contracts.

❹ *Termination of ACh activity.* The effect of ACh binding lasts only briefly because ACh is rapidly broken down by an enzyme called **acetylcholinesterase (AChE)** (as′-ē-til-kō′-lin-ES-ter-ās). This enzyme is attached to collagen fibers in the extracellular matrix of the synaptic cleft. AChE breaks down ACh into acetyl and choline, products that cannot activate the ACh receptor.

If another nerve impulse releases more acetylcholine, steps ❷ and ❸ repeat. When action potentials in the motor neuron cease, ACh is no longer released, and AChE rapidly breaks down the ACh already present in the synaptic cleft. This ends the production of muscle action potentials, and the Ca^{2+} moves from the sarcoplasm of the muscle fiber back into the sarcoplasmic reticulum, and the Ca^{2+} release channels in the sarcoplasmic reticulum membrane close.

CLINICAL CONNECTION | *Myasthenia Gravis*

Myasthenia gravis (mī-as-THĒ-nē-a GRAV-is; *mys-*=muscle; *aisthesis*=sensation) is an autoimmune disease that causes chronic, progressive damage of the neuromuscular junction. The immune system inappropriately produces antibodies that bind to and block some ACh receptors, thereby decreasing the number of functional ACh receptors. As the disease progresses, more ACh receptors are lost. Thus, muscles become increasingly weaker, fatigue more easily, and may eventually cease to function. The muscles of the face and neck are most often affected. Initial symptoms include weakness of the eye muscles, which may produce double vision, and weakness of the throat muscles that may produce difficulty in swallowing. Later, the person has difficulty chewing and talking. Eventually the muscles of the limbs may become involved. Death may result from paralysis of the respiratory muscles, but often the disorder does not progress to this stage.

Anticholinesterase drugs such as pyridostigmine (Mestinon®) or neostigmine, the first line of treatment, act as inhibitors of acetylcholinesterase, the enzyme that breaks down ACh. •

The NMJ usually is near the midpoint of a skeletal muscle fiber. Muscle action potentials that arise at the NMJ propagate toward both ends of the fiber. This arrangement permits nearly simultaneous activation (and thus contraction) of all parts of the muscle fiber.

Several plant products and drugs selectively block certain events at the NMJ. *Botulinum toxin* (bot-ū-LĪN-um), produced by the bacterium *Clostridium botulinum*, blocks exocytosis of synaptic vesicles at the NMJ. As a result, ACh is not released, and muscle contraction does not occur, causing a life-threatening condition known as botulism. The bacteria

proliferate in improperly canned foods, and their toxin is one of the most lethal chemicals known. A tiny amount can cause death by paralyzing skeletal muscles. Breathing stops due to paralysis of respiratory muscles, including the diaphragm. Yet it is also the first bacteria toxin to be used as a medicine (Botox®). Injections of Botox into the affected muscles can help patients who have strabismus (crossed eyes), blepharospasm (uncontrollable blinking), or spasms of the vocal cords that interfere with speech. It is also used to alleviate chronic back pain due to muscle spasms in the lumbar region, and as a cosmetic treatment to relax muscles that cause facial wrinkles.

The plant derivative *curare*, a poison used by South American Indians on arrows and blowgun darts, causes muscle paralysis by binding to and blocking ACh receptors. In the presence of curare, the ion channels do not open. Curare-like drugs are often used during surgery to relax skeletal muscles.

A family of chemicals called *anticholinesterase agents* have the property of slowing the enzymatic activity of acetylcholinesterase, thus slowing removal of ACh from the synaptic cleft. At low doses, these agents can strengthen weak muscle contractions. One example is neostigmine, which is used to treat patients with myasthenia gravis (see Clinical Connection). Neostigmine is also used as an antidote for curare poisoning and to terminate the effects of curare-like drugs after surgery.

The Contraction Cycle

At the onset of contraction, the sarcoplasmic reticulum releases calcium ions (Ca^{2+}) into the sarcoplasm. There, they bind to troponin. Troponin then moves tropomyosin away from the myosin-binding sites on actin. Once the binding sites are "free," the **contraction cycle**—the repeating sequence of events that causes the filaments to slide—begins. The contraction cycle consists of four steps (Figure 10.7):

❶ *ATP hydrolysis.* The myosin head includes an ATP-binding site and an ATPase, an enzyme that breaks down ATP into ADP (adenosine diphosphate) and a phosphate group. This reaction reorients and energizes the myosin head. Notice that the ADP and a phosphate group are still attached to the myosin head.

❷ *Attachment of myosin to actin to form cross-bridges.* The energized myosin head attaches to the myosin-binding site on actin and releases the phosphate group. When the myosin heads attach to actin during contraction, they are referred to as **cross-bridges**.

❸ *Power stroke.* After the cross-bridges form, the power stroke occurs. During the power stroke, the site on the cross-bridge where ADP is still bound opens. As a result, the cross-bridge rotates and releases the ADP. The cross-bridge generates force as it rotates toward the center of the sarcomere, sliding the thin filament past the thick filament toward the M line.

❹ *Detachment of myosin from actin.* At the end of the power stroke, the cross-bridge remains firmly attached to actin until it binds another molecule of ATP. As ATP binds to the ATP-binding site on the myosin head, the myosin head detaches from actin.

The contraction cycle repeats as the myosin ATPase breaks down the newly bound molecule of ATP, and continues as long as ATP is available and the Ca^{2+} level near the thin filament is sufficiently high. The cross-bridges keep rotating back and forth

Figure 10.7 The contraction cycle. Sarcomeres exert force and shorten through repeated cycles during which the myosin heads attach to actin (cross-bridges), rotate, and detach.

 During the power stroke of contraction, cross-bridges rotate and move the thin filaments past the thick filaments toward the center of the sarcomere.

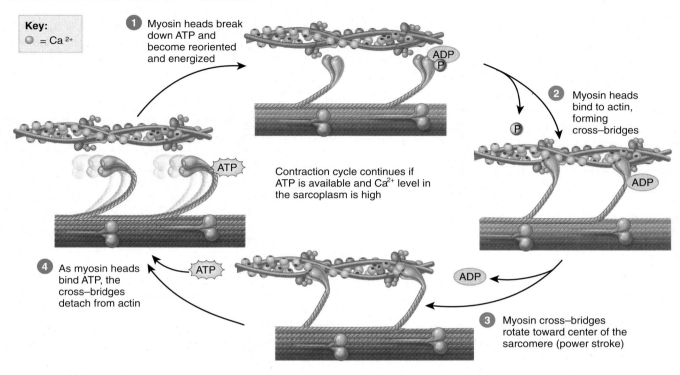

Key:
○ = Ca^{2+}

1 Myosin heads break down ATP and become reoriented and energized

ADP
P

2 Myosin heads bind to actin, forming cross–bridges

P

ATP

Contraction cycle continues if ATP is available and Ca^{2+} level in the sarcoplasm is high

ADP

4 As myosin heads bind ATP, the cross–bridges detach from actin

ATP

ADP

3 Myosin cross–bridges rotate toward center of the sarcomere (power stroke)

? **What would happen if ATP suddenly was not available after the sarcomere had started to shorten?**

CLINICAL CONNECTION | *Electromyography*

Electromyography (EMG) (ē-lek′-trō-mī-OG-ra-fē; *electro-* =electricity; *myo-*=muscle; *-graph*=to write) is a test that measures the electrical activity (muscle action potentials) in resting and contracting muscles. Normally, resting muscle produces no electrical activity; a slight contraction produces some electrical activity; and a more forceful contraction produces increased electrical activity. The electrical activity of the muscle is detected by a recording needle placed on or in a muscle to be tested and displayed as waves on an oscilloscope and heard through a loudspeaker. EMG helps to determine whether muscle weakness or paralysis is due to a malfunction of the muscle itself or the nerves supplying the muscle. EMG is also used to diagnose certain muscle disorders, such as muscular dystrophy, and to understand which muscles function during complex movements. •

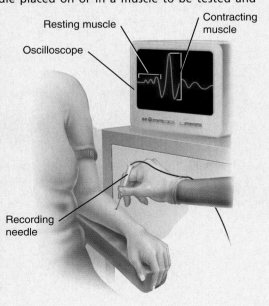

Oscilloscope

Resting muscle

Contracting muscle

Recording needle

with each power stroke, pulling the thin filaments toward the M line. Each of the 600 cross-bridges in one thick filament attaches and detaches about five times per second. The contraction cycle is a continuous process; at any one moment in time, some myosin heads are attached to actin, forming cross-bridges and generating force, and others are detached from actin, getting ready to bind again.

As the contraction cycle continues, movement of cross-bridges applies the force that draws the Z discs toward each other, and the sarcomere shortens. During a maximal muscle contraction, the distance between two Z discs can decrease to half the resting length. The Z discs in turn pull on neighboring sarcomeres, and the whole muscle fiber shortens. As the fibers shorten, they first pull on their connective tissue coverings and tendons. The coverings and tendons stretch and then become taut, and the tension passed through the tendons pulls on the bones to which they are attached, resulting in movement of a part of the body that allows you to perform tasks such as lifting a book with one hand. However, the contraction cycle does not always result in shortening of the muscle fibers and the whole muscle. In some contractions, the cross-bridges rotate and generate tension, but the thin filaments cannot slide inward because the tension they generate is not large enough to move the load on the muscle (such as trying to lift a whole box of books with one hand).

Excitation–Contraction Coupling

An increase in Ca^{2+} concentration in the sarcoplasm starts muscle contraction, and a decrease stops it. When a muscle fiber is relaxed, the concentration of Ca^{2+} in its sarcoplasm is very low. However, a huge amount of Ca^{2+} is stored inside the sarcoplasmic reticulum (Figure 10.8a). As a muscle action potential propagates along the sarcolemma and into the T tubules, it causes **Ca^{2+} release channels** in the SR membrane to open (Figure 10.8b). When these channels open, Ca^{2+} flows out of the SR into the sarcoplasm around the thick and thin filaments, causing the Ca^{2+} concentration in the sarcoplasm to rise tenfold or more. The released calcium ions combine with troponin, causing troponin to change shape. This conformational change moves tropomyosin away from the myosin-binding sites on actin. Once these binding sites are free, myosin heads bind to them to form cross-bridges, and the contraction cycle begins. The events just described are referred to collectively as **excitation–contraction coupling**; they are the steps that connect excitation (a muscle action potential propagating along the sarcolemma and into the T tubules) to contraction (sliding of the filaments).

The membrane of the sarcoplasmic reticulum also contains **Ca^{2+} active transport pumps** that use ATP to move Ca^{2+} constantly from the sarcoplasm into the SR (Figure 10.8). While muscle action potentials continue to propagate through the T tubules, the Ca^{2+} release channels are open. Calcium ions flow into the sarcoplasm more rapidly than they are transported back by the pumps. After the last action potential has propagated throughout the T tubules, the Ca^{2+} release channels close. As the pumps move Ca^{2+} back into the SR, the concentration of calcium ions in the sarcoplasm decreases rapidly. Inside the SR, molecules of a calcium-binding protein, appropriately called **calsequestrin** (kal′-se-KWES-trin), bind to the Ca^{2+}, enabling even more Ca^{2+} to be sequestered or stored within the SR. As a result, the concentration of Ca^{2+} is 10,000 times higher in the SR than in the sarcoplasm of a relaxed muscle fiber. As the Ca^{2+} level in the sarcoplasm drops, tropomyosin covers the myosin-binding sites, and the muscle fiber relaxes.

Figure 10.8 The role of Ca^{2+} in the regulation of contraction by troponin and tropomyosin. (a) During relaxation, the level of Ca^{2+} in the sarcoplasm is low, only 0.1 μM (0.001 μM), because calcium ions are pumped into the sarcoplasmic reticulum by Ca^{2+} active transport pumps. (b) A muscle action potential propagating along a transverse tubule opens Ca^{2+} release channels in the sarcoplasmic reticulum, calcium ions flow into the sarcoplasm, and contraction begins.

🔑 An increase in the Ca^{2+} level in the sarcoplasm starts the sliding of thin filaments. When the level of Ca^{2+} in the sarcoplasm declines, sliding stops.

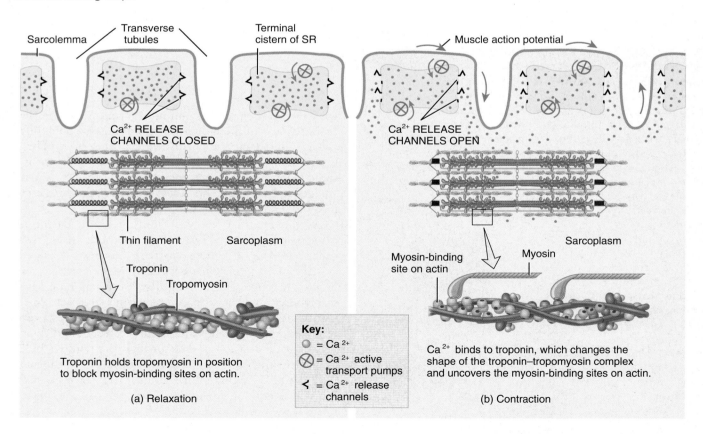

Key:
- ○ = Ca^{2+}
- ⊗ = Ca^{2+} active transport pumps
- < = Ca^{2+} release channels

(a) Relaxation

Troponin holds tropomyosin in position to block myosin-binding sites on actin.

(b) Contraction

Ca^{2+} binds to troponin, which changes the shape of the troponin–tropomyosin complex and uncovers the myosin-binding sites on actin.

❓ **What are the three functions of ATP in muscle contraction?**

After death, cellular membranes become leaky. Calcium ions leak out of the sarcoplasmic reticulum into the sarcoplasm and allow myosin heads to bind to actin. ATP synthesis ceases shortly after breathing stops, however, so the cross-bridges cannot detach from actin. The resulting condition, in which muscles are in a state of rigidity (cannot contract or stretch), is called **rigor mortis** (=rigidity of death). Rigor mortis begins 3–4 hours after death and lasts about 24 hours; it then disappears as proteolytic enzymes from lysosomes digest the cross-bridges. •

Muscle Tone

Even while at rest, a skeletal muscle exhibits **muscle tone** (*tonos-*=tension), a small amount of tautness or tension in the muscle due to weak, involuntary contractions of its motor units. Recall that skeletal muscle contracts only after it is activated by acetylcholine released by nerve impulses in its somatic motor neurons. Hence, muscle tone is established by neurons in the brain and spinal cord that excite the muscle's somatic motor neurons. To sustain muscle tone, small groups of motor units are alternately active and inactive in a constantly shifting pattern. Muscle tone keeps skeletal muscles firm, but it does not result in a force strong enough to produce movement. For example, when you are awake the muscles in the back of the neck are in normal tonic contraction. They keep the head upright and prevent it from slumping forward on the chest. Muscle tone also is important in smooth muscle tissues, such as those found in the gastrointestinal tract, where the walls of the digestive organs maintain a steady pressure on their contents. The tone of smooth muscle fibers in the walls of blood vessels plays a crucial role in maintaining blood pressure.

Hypotonia (*hypo-*=below) refers to decreased or lost muscle tone. Such muscles are said to be **flaccid** (FLAK-sid or FLAS-sid= flabby). Flaccid muscles are loose and appear flattened rather than rounded. Certain disorders of the nervous system and disruptions in the balance of electrolytes (especially sodium, calcium, and, to a lesser extent, magnesium) may result in flaccid paralysis, which is characterized by loss of muscle tone, loss or reduction of tendon reflexes, and atrophy (wasting away) and degeneration of muscles.

Hypertonia (*hyper-*=above) refers to increased muscle tone and is expressed in two ways: spasticity or rigidity. **Spasticity** (spas-TIS-i-tē) is characterized by increased muscle tone (stiffness) associated with an increase in tendon reflexes and pathological reflexes (such as the Babinski sign, in which the great toe extends with or without fanning of the other toes in response to stroking the outer margin of the sole). Certain disorders of the nervous system and electrolyte disturbances such as those previously noted may result in **spastic paralysis**, partial paralysis in which the muscles exhibit spasticity. **Rigidity** refers to increased muscle tone in which reflexes are not affected, as occurs in tetanus. Tetanus is a disease caused by a bacterium, *Clostridium tetani*, that enters the body through exposed wounds. It leads to muscle stiffness and spasms that can make breathing difficult and can become life-threatening as a result. The bacteria produce a toxin that interferes with the nerves controlling the muscles. The first signs are typically spasms and stiffness in the muscles of the face and jaws. •

Isotonic and Isometric Contractions

Muscle contractions may be either isotonic or isometric. In an **isotonic contraction** (i-sō-TON-ik; *iso-*=equal; *-tonic*=tension), the *tension* (force of contraction) in the muscle remains almost constant while the muscle changes its length. Isotonic contractions are used to produce body movements and for moving objects. The two types of isotonic contractions are concentric and eccentric. In a **concentric isotonic contraction** (kon-SEN-trik), if the tension generated is great enough to overcome the resistance of the object to be moved, the muscle shortens and pulls on its tendon to produce movement and to reduce the angle at a joint. Picking up a book from a table involves concentric isotonic contractions of the biceps brachii muscle in the arm. By contrast, as you lower the book to place it back on the table, the previously shortened biceps lengthens in a controlled manner while it continues to contract. When the length of a muscle increases during a contraction, the contraction is an **eccentric isotonic contraction** (ek-SEN-trik). During an eccentric contraction, the tension exerted by the myosin cross-bridges resists movement of a load (the book, in this case) and slows the lengthening process. For reasons that are not well understood, repeated eccentric isotonic contractions (for example, walking downhill) produce more muscle damage and more delayed-onset muscle soreness than do concentric isotonic contractions.

In an **isometric contraction** (ī-sō-MET-rik; *-metro*=measure or length), the tension generated is not enough to exceed the resistance of the object to be moved and the muscle does not change its length. An example would be holding a book steady using an outstretched arm. These contractions are important for maintaining posture and for supporting objects in a fixed position. Although isometric contractions do not result in body movement, energy is still expended, as you well know if you have ever tried to hold your anatomy book in an outstretched hand for any length of time. The book pulls the arm downward, stretching the shoulder and arm muscles. The isometric contraction of the shoulder and arm muscles counteracts the stretch. Isometric contractions are important because they stabilize some joints as others are moved. Most activities include both isotonic and isometric contractions.

 CHECKPOINT

4. Outline the types of connective tissue that cover skeletal muscles and explain how they are related.
5. Describe the nerve supply to a skeletal muscle fiber, and explain why a rich blood supply is so important to muscle contraction.
6. Describe the components of a sarcomere.
7. What is the sliding filament mechanism?
8. Explain the role of the neuromuscular junction in skeletal muscle contraction.
9. Outline the main events of the contraction cycle, including the role of calcium ions.
10. How are calcium ions involved in skeletal muscle contraction?
11. Why is muscle tone important?
12. Define each of the following terms: concentric isotonic contraction, eccentric isotonic contraction, and isometric contraction.

10.3 TYPES OF SKELETAL MUSCLE FIBERS

 OBJECTIVE

• Compare the structure and function of the three types of skeletal muscle fibers.

Skeletal muscle fibers vary in their content of myoglobin, the red protein that binds oxygen in muscle fibers. Those with a high myoglobin content are called **red muscle fibers**, while those that have a low myoglobin content are called **white muscle fibers**. Red muscle fibers also contain more mitochondria and are supplied by more blood capillaries than white muscle fibers.

Skeletal muscle fibers also contract and relax at different speeds, and can be categorized as either slow or fast depending on how rapidly the ATPase in their myosin heads hydrolyzes ATP. In addition, the metabolic reactions that skeletal muscle fibers use to generate ATP vary, as does the length of time it takes them to experience **fatigue** (inability of a muscle to maintain force of contraction after prolonged activity). All of these structural and functional characteristics are taken into account in classifying a skeletal muscle fiber as one of three main types: (1) slow oxidative fibers, (2) fast oxidative-glycolytic fibers, and (3) fast glycolytic fibers.

Slow Oxidative Fibers

Slow oxidative (SO) fibers, also called **type I fibers**, appear dark red because they contain large amounts of myoglobin and many blood capillaries. Because they have many large mitochondria, SO fibers generate ATP mainly by aerobic (oxygen-requiring) cellular respiration, which is why they are called oxidative fibers. These fibers are said to be "slow" because they use ATP at a slow rate. As a result, SO fibers have a slow speed of contraction. However, slow fibers are very resistant to fatigue and are capable of prolonged, sustained contractions for many hours. These fibers are adapted for maintaining posture and for aerobic, endurance-type activities such as running a marathon.

Fast Oxidative-Glycolytic Fibers

Fast oxidative-glycolytic (FOG) fibers or **type IIa fibers** are typically the largest fibers. Like slow oxidative fibers, they contain large amounts of myoglobin and many blood capillaries, giving them a dark red appearance. FOG fibers can generate considerable ATP by aerobic cellular respiration, which gives them a moderately high resistance to fatigue. Because their intracellular glycogen level is high, they also generate ATP by anaerobic (oxygen-free) glycolysis. FOG fibers are "fast" because they use ATP at a fast rate, which makes their speed of contraction faster than SO fibers. FOG fibers contribute to activities such as walking and sprinting.

Fast Glycolytic Fibers

Fast glycolytic (FG) fibers or **type IIb fibers** have low myoglobin content, relatively few blood capillaries and few mitochondria, and appear white in color. They contain large amounts of glycogen and generate ATP mainly by anaerobic (nonoxygen-requiring) cellular respiration (glycolysis). Due

The term **muscular dystrophy** (DIS-trō-fē; *dys-*=difficult; *-trophy*=nourishment) refers to a group of inherited muscle-destroying diseases that cause progressive degeneration of skeletal muscle fibers. The most common form of muscular dystrophy is *Duchenne muscular dystrophy* (doo-SHĀN) (*DMD*). Because the mutated gene is on the X chromosome, and because males have only one X chromosome, DMD strikes boys almost exclusively. Worldwide, about 1 in every 3500 male babies—about 21,000 in all—are born with DMD each year. The disorder usually becomes apparent between the ages of 2 and 5, when parents notice that their child falls often and has difficulty running, jumping, and hopping. By age 12 most boys with DMD are unable to walk. Respiratory or cardiac failure usually causes death between the ages of 20 and 30.

In DMD, the gene that codes for the protein dystrophin is mutated, and little or no dystrophin is present. Without the reinforcing effect of dystrophin, the sarcolemma tears easily during muscle contraction. Because their plasma membranes are damaged, muscle fibers slowly rupture and die. The dystrophin gene was discovered in 1987, and by 1990 the first attempts were made to treat DMD patients with gene therapy. The muscles of three boys with DMD were injected with myoblasts bearing functional dystrophin genes, but only a few muscle fibers gained the ability to produce dystrophin. Similar clinical trials with additional patients have also failed. An alternative approach to the problem is to find a way to induce muscle fibers to produce the protein utrophin, which is similar to dystrophin. Experiments with dystrophin-deficient mice suggest that this approach may work. •

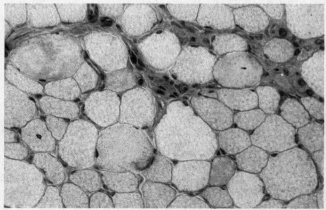

LM 25x

Variations in skeletal muscle fiber size and increased amounts of connective tissue (red) in Duchenne muscular dystrophy

to their ability to use ATP at a fast rate, FG fibers contract strongly and quickly. These fast-twitch fibers are adapted for intense anaerobic movements of short duration, such as weight lifting or throwing a ball, but they fatigue quickly. Strength training programs that engage a person in activities requiring great strength for short times produce increases in the size, strength, and glycogen content of fast glycolytic fibers. The FG fibers of a weight lifter may be 50 percent larger than those of a sedentary person or an endurance athlete because of increased synthesis of muscle proteins. The overall result is muscle enlargement due to hypertrophy of the FG fibers.

Most skeletal muscles are a mixture of all three types of skeletal muscle fibers. The proportions vary somewhat, depending on the action of the muscle, the person's training regimen, and genetic factors. For example, the continually active postural muscles of the neck, back, and legs have a high proportion of SO fibers. Muscles of the shoulders and arms, in contrast, are not constantly active but are used briefly now and then to produce large amounts of tension, such as in lifting and throwing. These muscles have a high proportion of FG fibers. Leg muscles, which not only support the body but are also used for walking and running, have large numbers of both SO and FOG fibers.

The skeletal muscle fibers of any given motor unit are all of the same type. However, the different motor units in a muscle are recruited in a specific order, depending on need. For example, if weak contractions are enough to perform a task, only SO motor units are activated. If more force is needed, the motor units of FOG fibers are also recruited. Finally, if maximal force is required, motor units of FG fibers are also called into action along with the other two types. Activation of various motor units is controlled by the brain and spinal cord.

Table 10.4 summarizes the characteristics of the three types of skeletal muscle fibers.

 CHECKPOINT

13. What is the basis for classifying skeletal muscle fibers into three types?

10.4 EXERCISE AND SKELETAL MUSCLE TISSUE

 OBJECTIVE

• Describe the effects of exercise on different types of skeletal muscle fibers.

The relative ratio of fast glycolytic (FG) and slow oxidative (SO) fibers in each muscle, which is genetically determined, helps account for individual differences in physical performance. For example, people with a higher proportion of FG fibers (see Table 10.4) often excel in activities that require periods of intense activity, such as weight lifting or sprinting. People with higher percentages of SO fibers are better at activities that require endurance, such as long-distance running.

Although the total number of skeletal muscle fibers usually does not increase, the characteristics of those present can change to some extent. Various types of exercises can induce changes in the fibers in a skeletal muscle. Endurance-type (aerobic) exercises, such as running or swimming, cause a gradual transformation of some FG fibers into fast oxidative-glycolytic (FOG) fibers. The transformed muscle fibers show slight increases in diameter, number of mitochondria, blood supply, and strength. Endurance exercises also result in cardiovascular and respiratory

---

TABLE 10.4

Characteristics of the Three Types of Skeletal Muscle Fibers

Slow oxidative fiber
Fast glycolytic fiber
Fast oxidative-glycolytic fiber

LM 440x

Transverse section of three types of skeletal muscle fibers

	SLOW OXIDATIVE (SO) OR TYPE I FIBERS	**FAST OXIDATIVE-GLYCOLYTIC (FOG) OR TYPE IIa FIBERS**	**FAST GLYCOLYTIC (FG) OR TYPE IIb FIBERS**
STRUCTURAL CHARACTERISTICS			
Myoglobin content	Large amount	Large amount	Small amount
Mitochondria	Many	Many	Few
Capillaries	Many	Many	Few
Color	Red	Red	White (pale)
FUNCTIONAL CHARACTERISTICS			
Capacity for generating ATP and method used	High capacity, by aerobic cellular respiration	Intermediate capacity, by both aerobic cellular respiration and anaerobic cellular respiration (glycolysis)	Low capacity, by anaerobic cellular respiration (glycolysis)
Rate of ATP use	Slow	Fast	Fast
Contraction velocity	Slow	Fast	Fast
Fatigue resistance	High	Intermediate	Low
Location where fibers are abundant	Postural muscles such as those of the neck	Lower limb muscles	Upper limb muscles
Primary functions of fibers	Maintaining posture and aerobic endurance activities	Walking, sprinting	Rapid, intense movements of short duration

changes that cause skeletal muscles to receive better supplies of oxygen and nutrients but do not increase muscle mass. By contrast, exercises that require great strength for short periods produce an increase in the size and strength of FG fibers because of an increased synthesis of thick and thin filaments. The overall result is muscle enlargement (hypertrophy), as evidenced by the bulging muscles of body builders.

A certain degree of elasticity is an important attribute of skeletal muscles and their connective tissue attachments. Greater elasticity contributes to a greater degree of flexibility, increasing the range of motion of a joint. When a relaxed muscle is physically stretched, its ability to lengthen is limited by connective tissue structures, such as fasciae. Regular stretching gradually elongates these structures, but the process occurs very slowly. To see an improvement in flexibility, stretching exercises must be performed regularly—daily, if possible—for many weeks.

Effective Stretching

Stretching cold muscles does not increase flexibility and may cause injury. Tissues stretch best when slow, gentle force is applied at elevated tissue temperatures. An external source of heat, such as hot packs or ultrasound, may be used, but 10 or more minutes of muscular contraction is also a good way to raise muscle temperature. Exercise heats muscle more deeply and thoroughly than external measures. That's where the term "warm-up" comes from. Many people stretch before they engage in exercise, but it's important to warm up (for example, walking, jogging, easy swimming or easy aerobics) *before* stretching to avoid injury.

Strength Training

Strength training refers to the process of exercising with progressively heavier resistance for the purpose of strengthening the musculoskeletal system. This activity results not only in stronger muscles, but in many other health benefits as well. Strength training also helps to increase bone strength by increasing the deposition of bone minerals in young adults and helping to prevent, or at least slow, their loss in later life. By increasing muscle mass, strength training raises resting metabolic rate, the amount of energy expended at rest, so a person can eat more food without gaining weight. Strength training helps to prevent back injury and other injuries from participation in sports and other physical activities. Psychological benefits include reductions in feelings of stress and fatigue. As repeated training builds exercise tolerance, it takes increasingly longer before lactic acid is produced in the muscle, resulting in a reduced probability of muscle spasms.

CLINICAL CONNECTION | Anabolic Steroids

The use of **anabolic steroids**, commonly called "roids," by athletes, has received widespread attention. The term "anabolic" means to build up proteins. These steroid hormones, similar to testosterone, are taken to increase muscle size and strength in order to enhance performance during athletic contests. However, the doses required to produce an effect are large and have damaging, sometimes even devastating side effects; these include liver cancer, kidney damage, increased risk of heart disease, stunted growth, wide mood swings, increased acne, and increased irritability and aggression. In addition, females who take anabolic steroids may also experience atrophy of the breasts and uterus, menstrual irregularities, sterility, facial hair growth, and deepening of the voice. Males may also experience diminished testosterone secretion, atrophy of the testes, sterility, and baldness. •

CHECKPOINT

14. Explain how the characteristics of skeletal muscle fibers are able to change with exercise.

10.5 CARDIAC MUSCLE TISSUE

OBJECTIVE

• Describe the main structural and functional characteristics of cardiac muscle tissue.

The principal tissue in the heart wall is **cardiac muscle tissue**. Although it is striated like skeletal muscle, its activity cannot be controlled voluntarily. Also, certain cardiac muscle fibers (as well as some smooth muscle fibers and nerve cells in the brain and spinal cord) display **autorhythmicity** (aw′-tō-rith-MIS-i-tē), the ability to repeatedly generate spontaneous action potentials. In the heart, these action potentials cause alternating contraction and relaxation of the heart muscle fibers. Compared with skeletal muscle fibers, cardiac muscle fibers are shorter in length and less circular in transverse section (Figure 10.9a). They also exhibit branching, which gives individual cardiac muscle fibers a "stair-step" appearance. A typical cardiac muscle fiber is 50–100 μm long and has a diameter of about 14 μm. Usually one centrally located nucleus is present, although an occasional cell may have two nuclei. The ends of cardiac muscle fibers connect to neighboring fibers by irregular transverse thickenings of the sarcolemma referred to as **intercalated discs** (in-TER-kā-lāt-ed; *intercalat-*=to insert between). The discs contain **desmosomes**, which hold the fibers together, and **gap junctions**, which allow muscle action potentials to spread from one muscle fiber to its neighbors. Cardiac muscle tissue has an endomysium, but lacks a perimysium and epimysium.

Mitochondria are larger and more numerous in cardiac muscle fibers than in skeletal muscle fibers. Cardiac muscle fibers have the same arrangement of actin and myosin, and the same bands, zones, and discs, as skeletal muscle fibers (Figure 10.9b). The transverse (T) tubules of cardiac muscle are wider but less abundant than those of skeletal muscle; there is one T tubule per sarcomere, located at the Z disc. The sarcoplasmic reticulum of cardiac muscle fibers is somewhat smaller than the SR of skeletal muscle fibers.

Under normal resting conditions, cardiac muscle tissue contracts and relaxes about 75 times per minute. This continuous, rhythmic activity is a major functional difference between cardiac and skeletal muscle tissue. Another difference is the source of stimulation. Skeletal muscle tissue contracts only when stimulated by acetylcholine released by an action potential in a somatic motor neuron. In contrast, cardiac muscle tissue can contract without extrinsic (outside) nervous or hormonal stimulation. Its source of stimulation is a conducting network of specialized cardiac muscle fibers within the heart. Stimulation from the body's nervous system or endocrine system merely causes the conducting fibers to increase or decrease their rate of discharge. Cardiac muscle tissue remains contracted 10 to 15 times longer than skeletal muscle tissue, allowing time for the chambers of the heart to relax and fill with blood between beats. This pattern permits the heart rate to increase significantly while preventing *tetanus* (sustained contraction), which would stop blood flow within the heart. Like skeletal muscle, cardiac muscle fibers can undergo hypertrophy in response to an increased workload. This is called a *physiological enlarged heart* and it is why many athletes have enlarged hearts. By contrast, a *pathological enlarged heart* is related to significant heart disease.

CHECKPOINT

15. How do cardiac and skeletal muscle tissue differ in structure and function?

Figure 10.9 Histology of cardiac muscle. A photomicrograph of cardiac muscle tissue is shown in Table 3.9B.

🔑 Cardiac muscle fibers display autorhythmicity, the ability to repeatedly generate spontaneous action potentials.

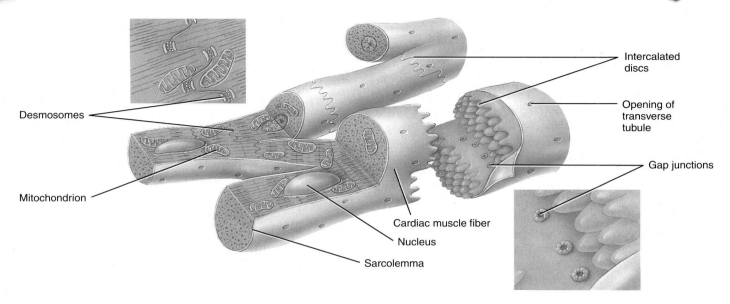

Intercalated discs

Opening of transverse tubule

Gap junctions

Desmosomes

Mitochondrion

Cardiac muscle fiber

Nucleus

Sarcolemma

(a) Cardiac muscle fibers

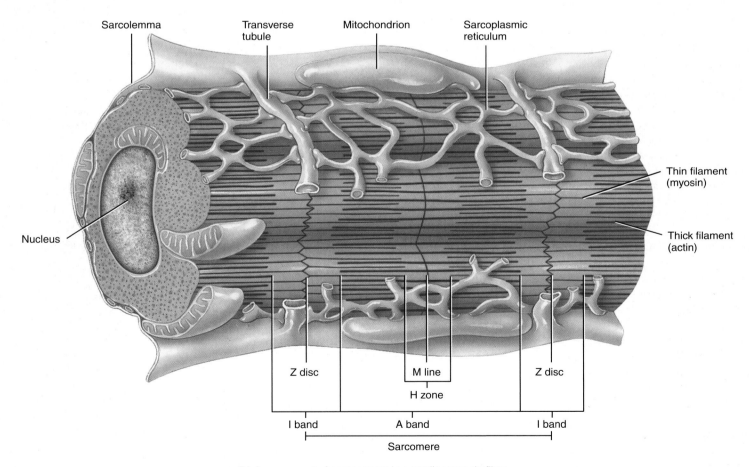

Sarcolemma

Transverse tubule

Mitochondrion

Sarcoplasmic reticulum

Thin filament (myosin)

Nucleus

Thick filament (actin)

Z disc

M line

Z disc

H zone

I band

A band

I band

Sarcomere

(b) Arrangement of components in a cardiac muscle fiber

❓ **What are the functions of intercalated discs in cardiac muscle fibers?**

10.6 SMOOTH MUSCLE TISSUE

● OBJECTIVE

• Describe the main structural and functional characteristics of smooth muscle tissue.

Like cardiac muscle tissue, **smooth muscle tissue** is usually activated involuntarily. There are two types of smooth muscle tissue: visceral (single-unit) smooth muscle tissue, and multiunit smooth muscle tissue. The more common type, **visceral (single-unit) smooth muscle tissue** (Figure 10.10a), is found in the skin, in wraparound sheets that form part of the walls of small arteries and veins, and in the walls of hollow viscera such as the stomach, intestines, uterus, and urinary bladder. Because the fibers connect to one another by gap junctions, muscle action potentials spread rapidly throughout the network. For example, when a neurotransmitter, hormone, or autorhythmic signal stimulates one fiber, the muscle action potential spreads to neighboring fibers, which then contract as a single unit.

The second kind of smooth muscle tissue, **multiunit smooth muscle tissue** (Figure 10.10b), consists of individual fibers, each of which has its own motor neuron terminals. There are few gap junctions between neighboring fibers. As you just learned, stimulation of one visceral muscle fiber causes contraction of many adjacent fibers; in contrast, stimulation of one multiunit smooth muscle fiber causes contraction of that fiber only. The walls of large arteries, the airways to the lungs, the arrector pili muscles that attach to hair follicles, the muscles of the iris that adjust pupil diameter, and the ciliary body that adjusts focus of the lens in the eye all contain multiunit smooth muscle tissue.

Smooth muscle fibers are considerably smaller than skeletal muscle fibers (Table 10.5). A single relaxed smooth muscle fiber is 30–200 μm long, thickest in the middle (3–8 μm), and tapered at each end (Figure 10.10c). Within each fiber is a single, oval, centrally located nucleus. The sarcoplasm of smooth muscle fibers contains both thick filaments and thin filaments, in ratios between about 1:10 and 1:15 respectively, but they are not arranged in orderly sarcomeres as in striated muscle. Smooth muscle fibers also contain **intermediate filaments**. These filaments, which contain the protein *desmin*, appear to have a structural rather than contractile role. Because the various filaments have no regular pattern of overlap, smooth muscle fibers do not exhibit striations—thus the name *smooth*. Smooth muscle fibers also lack transverse tubules and have little sarcoplasmic reticulum for storage of Ca^{2+}. Smooth muscle tissue has an endomysium, but lacks a perimysium and epimysium.

In smooth muscle fibers, intermediate filaments attach to structures called **dense bodies**, which are functionally similar to Z discs in striated muscle fibers. Some dense bodies are dispersed throughout the sarcoplasm; others are attached to the sarcolemma. Bundles of intermediate filaments stretch from one dense body to another (Figure 10.10c). During contraction, the tension generated by the thick and thin filaments in the sliding filament mechanism is transmitted to intermediate filaments. These in turn pull on the dense bodies attached to the sarcolemma, causing a shortening of the muscle fiber (Figure 10.10c). When a smooth muscle fiber contracts, it turns like a corkscrew; when it relaxes, it rotates in the opposite direction.

Although the principles of contraction are similar in all three types of muscle tissue, smooth muscle tissue exhibits some important physiological differences from skeletal and cardiac muscle tissue. Compared with contraction in a skeletal muscle fiber, contraction in a smooth muscle fiber starts more slowly and lasts much longer. In addition, smooth muscle can both shorten and stretch to a greater extent than other muscle types.

Figure 10.10 Histology of smooth muscle tissue. In (a), one autonomic motor neuron synapses with several visceral smooth muscle fibers, and action potentials spread to neighboring fibers through gap junctions. In (b), three autonomic motor neurons synapse with individual multiunit smooth muscle fibers. Stimulation of one multiunit fiber causes contraction of that fiber only. (c) Comparison between a relaxed and contracted smooth muscle fiber. A photomicrograph of smooth muscle tissue is shown in Table 3.9C.

🔑 Smooth muscle fibers have thick and thin filaments but no transverse tubules and little sarcoplasmic reticulum.

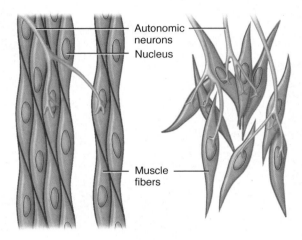

(a) Visceral (single-unit) smooth muscle tissue

(b) Multiunit smooth muscle tissue

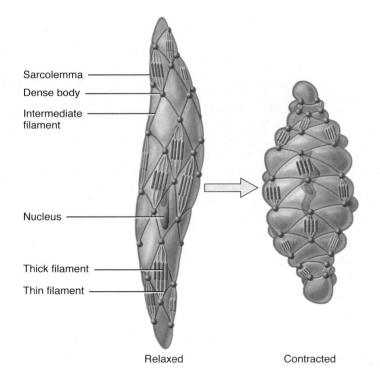

Relaxed   Contracted

(c) Microscopic anatomy of a relaxed and contracted smooth muscle fiber

 Which type of smooth muscle is more like cardiac muscle than skeletal muscle, with respect to both its structure and function?

As in striated muscle, an increase in the concentration of Ca^{2+} in the cytosol of smooth muscle initiates contraction. There is far less sarcoplasmic reticulum (the reservoir for Ca^{2+} in striated muscle) in smooth muscle than in skeletal muscle. Calcium ions flow into smooth muscle sarcoplasm from both the interstitial fluid and sarcoplasmic reticulum, but because there are no transverse tubules in smooth muscle fibers, it takes longer for Ca^{2+} to reach the filaments in the center of the fiber and trigger the contractile process. This accounts, in part, for the slow onset and prolonged contraction of smooth muscle.

Calcium ions also move out of the muscle fiber slowly, which delays relaxation. The prolonged presence of Ca^{2+} in the cytosol provides for **smooth muscle tone**, a state of continuous partial contraction. This long-term tone is important in the gastrointestinal tract, where the walls maintain a steady pressure on the contents of the tract, and in the walls of blood vessels called arterioles, which maintain a steady pressure on blood.

Most smooth muscle fibers contract or relax in response to action potentials from the autonomic nervous system. In addition, many smooth muscle fibers contract or relax in response to stretching, hormones, or local changes in pH, oxygen and carbon dioxide levels, temperature, and ion concentrations.

Unlike striated muscle fibers, smooth muscle fibers can stretch considerably and still maintain their contractile function. When smooth muscle fibers are stretched, they initially contract, developing increased tension. Within a minute or so, the tension decreases. This **stress–relaxation response** allows smooth muscle to undergo great changes in length while still retaining the ability to contract effectively. Even though smooth muscle in the walls of blood vessels and hollow organs can stretch, the pressure on their contents changes very little. After the organ empties, the smooth muscle rebounds, and the wall of the organ retains its firmness.

TABLE 10.5

Summary of the Major Features of the Three Types of Muscular Tissue

CHARACTERISTIC	SKELETAL MUSCLE	CARDIAC MUSCLE	SMOOTH MUSCLE
Microscopic appearance and features	Long cylindrical fiber with many peripherally located nuclei; unbranched; striated	Branched cylindrical fiber with one centrally located nucleus; intercalated discs join neighboring fibers; striated	Fiber is thickest in middle, tapered at each end, and has one centrally positioned nucleus; not striated
Location	Most commonly attached by tendons to bones	Heart	Walls of hollow viscera, airways, blood vessels, iris and ciliary body of eye, arrector pili muscles of hair follicles
Fiber diameter	Very large (10–100 μm)*	Large (10–20 μm)	Small (3–8 μm)
Connective tissue components	Endomysium, perimysium, and epimysium	Endomysium and perimysium	Endomysium
Fiber length	Very large (100 μm–30 cm = 12 inches)	Large (50–100 μm)	Intermediate (30–200 μm)
Contractile proteins organized into sarcomeres	Yes	Yes	No
Sarcoplasmic reticulum	Abundant	Some	Very little
Transverse tubules present	Yes, aligned with each A–I band junction	Yes, aligned with each Z disc	No
Junctions between fibers	None	Intercalated discs contain gap junctions and desmosomes	Gap junctions in visceral smooth muscle; none in multiunit smooth muscle
Autorhythmicity	No	Yes	Yes, in visceral smooth muscle
Source of Ca^{2+} for contraction	Sarcoplasmic reticulum	Sarcoplasmic reticulum and interstitial fluid	Sarcoplasmic reticulum and interstitial fluid
Speed of contraction	Fast	Moderate	Slow
Nervous control	Voluntary (somatic nervous system)	Involuntary (autonomic nervous system)	Involuntary (autonomic nervous system)
Capacity for regeneration	Limited, via satellite cells	Limited, under certain conditions	Considerable, via pericytes (compared with other muscular tissues, but limited compared with epithelium)

*1 micrometer (μm) = 1/25,000 of an inch.

Smooth muscle tissue, like skeletal and cardiac muscle tissue, can undergo hypertrophy. In addition, certain smooth muscle fibers retain their capacity for division and thus can grow by hyperplasia. Also, new smooth muscle fibers can arise from cells called *pericytes*, stem cells found in association with blood capillaries and small veins. Smooth muscle fibers can also proliferate in certain pathological conditions, such as atherosclerosis (see Section 13.7). Although smooth muscle tissue has considerably greater powers of regeneration than skeletal muscle or cardiac muscle, such powers are limited when compared with tissues such as epithelium.

Table 10.5 summarizes the major characteristics of the three types of muscular tissue.

✓ CHECKPOINT

16. How do visceral and multiunit smooth muscle differ?
17. Compare the properties of skeletal and smooth muscle.

10.7 DEVELOPMENT OF MUSCLES

OBJECTIVE

• Describe the development of muscles.

Except for the muscles of the iris of the eyes, all muscles of the body are derived from **mesoderm**. Recall from Chapter 4 that by about the seventeenth day after fertilization, the mesoderm adjacent to the notochord and neural tube forms paired longitudinal columns of paraxial mesoderm (see Figure 4.9b). The paraxial mesoderm soon undergoes segmentation. Initially the segments form **somitomeres** (sō-MIT-ō-merz), small bulges in the paraxial mesoderm. Posterior to the developing head, the somitomeres undergo further development into paired, cube-shaped structures called **somites** (SŌ-mīts) (Figure 10.11a). The first pair of somites appears on day 20 of embryonic development. By the end of week 5, 42 to 44 pairs of somites are formed. The number of somites can be correlated to the approximate age of the embryo. In the head region the somitomeres never become somites. Rather, the cells of the somitomeres migrate into the developing pharyngeal arches around the anterior end of the pharynx (throat).

As a result, all the skeletal muscles develop from the *paraxial mesoderm*. The somites give rise to all the muscles of the trunk wall and the limbs. The migrating somitomeres of the pharyngeal arches form the skeletal muscles of the head region.

As you learned in Chapter 4, the cells of a somite differentiate into three regions: (1) a **myotome** (MĪ-ō-tōm), which as its name suggests forms the skeletal muscles of the trunk and limbs; (2) a **dermatome** (DER-ma-tōm), which forms the connective tissues, including the dermis of the skin; and (3) a **sclerotome** (SKLE-rō-tōm), which gives rise to the vertebrae and ribs (Figure 10.11b).

Cardiac muscle develops from **mesodermal cells** that migrate to and envelop the developing heart while it is still in the form of endocardial tubes (see Figure 14.11b).

Figure 10.11 Location and structure of somites, key structures in the development of the muscular system.

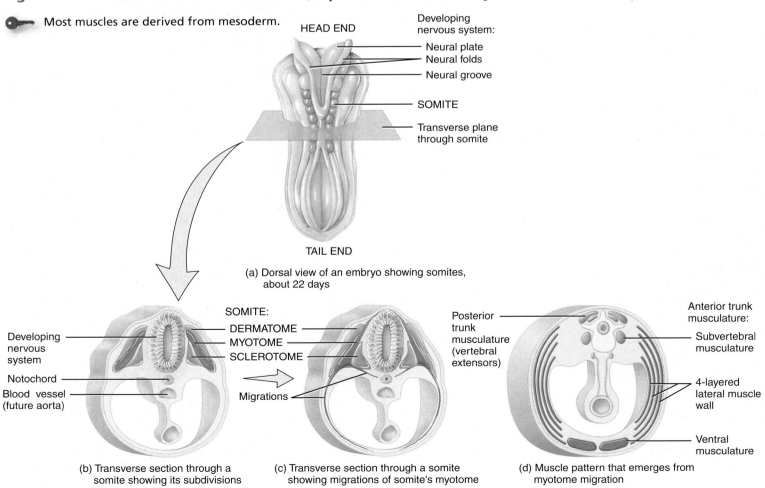

Most muscles are derived from mesoderm.

HEAD END

Developing nervous system:
— Neural plate
— Neural folds
— Neural groove

— SOMITE

— Transverse plane through somite

TAIL END

(a) Dorsal view of an embryo showing somites, about 22 days

SOMITE:
DERMATOME
MYOTOME
SCLEROTOME

Developing nervous system

Notochord

Blood vessel (future aorta)

Migrations

(b) Transverse section through a somite showing its subdivisions

(c) Transverse section through a somite showing migrations of somite's myotome

Posterior trunk musculature (vertebral extensors)

Anterior trunk musculature:
Subvertebral musculature

4-layered lateral muscle wall

Ventral musculature

(d) Muscle pattern that emerges from myotome migration

? **Which part of a somite differentiates into skeletal muscle?**

Most *smooth muscle* develops from lateral plate **mesodermal cells** that migrate to and envelop the developing gastrointestinal tract and viscera. Other smooth muscle cells that form in the walls of blood vessels develop from mesenchyme throughout the mesodermal regions of the embryo.

 CHECKPOINT

18. From which area of the somite does the quadriceps muscle of your thigh develop?

 ## 10.8 AGING AND MUSCULAR TISSUE

 OBJECTIVE

• Explain **the effects of aging on skeletal muscle.**

Between the ages of 30 and 50, humans undergo a slow, progressive loss of skeletal muscle mass as it is replaced largely by fibrous connective tissue and adipose tissue. An estimated 10 percent of muscle mass is lost during these years. In part, this decline may be due to decreased levels of physical activity.

Accompanying the loss of muscle mass is a decrease in maximal strength, a slowing of muscle reflexes, and a loss of flexibility. In some muscles, a selective loss of muscle fibers of a given type may occur. With aging, the relative number of SO fibers appears to increase. This could be due to either atrophy of the other fiber types or their conversion into SO fibers. Aerobic activities and strength training programs are effective in slowing or even reversing the age-associated decline in muscular performance. Another 40 percent of muscle is typically lost between ages 50 and 80. Loss of muscle strength is usually not perceived until the age of 60 to 65. Muscles of the lower limbs usually weaken before those of the upper limbs, limiting independence as it becomes difficult to climb stairs or get up from a seated position.

In the absence of a chronic medical condition, exercise has been shown to be effective at any age. Aerobic activities and strength training programs are just as effective in older people as in younger people and can slow or even reverse age-associated muscular decline.

CHECKPOINT

19. Why does muscle strength decrease with aging?

KEY MEDICAL TERMS ASSOCIATED WITH MUSCULAR TISSUE

Fasciculation (fa-sik-ū-LĀ-shun) An involuntary, brief twitch of an entire motor unit that is visible under the skin; it occurs irregularly and is not associated with movement of the affected muscle. Fasciculations may be seen in multiple sclerosis or in amyotrophic lateral sclerosis (Lou Gehrig's disease).

Fibrillation (fi-bri-LĀ-shun) A spontaneous contraction of a single muscle fiber that is not visible under the skin but can be recorded by electromyography. Fibrillations may signal destruction of motor neurons.

Myalgia (mī-AL-jē-a; *-algia*=painful condition) Pain in or associated with muscles.

Myoma (mī-Ō-ma; *-oma*=tumor) A tumor consisting of muscular tissue.

Myomalacia (mī-Ō-ma-LĀ-shē-a; *-malacia*=soft) Pathological softening of muscular tissue.

Myositis (mī′-ō-SĪ-tis; *-itis*=inflammation of) Inflammation of muscle fibers (cells).

Myotonia (mī′-ō-TŌ-nē-a; *-tonia*=tension) Increased muscular excitability and contractility, with decreased power of relaxation; tonic spasm of the muscle.

Volkmann contracture (FŌLK-man kon-TRAK-chur; *contra-*= against) Permanent shortening (contracture) of a muscle due to replacement of destroyed muscle fibers by fibrous connective tissue, which lacks extensibility. Typically occurs in forearm flexor muscles. Destruction of muscle fibers may occur from interference with circulation caused by a tight bandage, a piece of elastic, or a cast.

CHAPTER REVIEW AND RESOURCE SUMMARY

WileyPLUS

Review | **Resource**

Introduction

1. Muscles constitute 40–50 percent of total body weight.
2. The prime function of muscle is changing chemical energy into mechanical energy to perform work.

10.1 Overview of Muscular Tissue

Anatomy Overview - Muscular Tissue

1. The three types of muscular tissue are skeletal, cardiac, and smooth. Skeletal muscle tissue is primarily attached to bones; it is striated and under voluntary control. Cardiac muscle tissue forms the wall of the heart; it is striated and involuntary. Smooth muscle tissue is located primarily in internal organs; it is nonstriated (smooth) and involuntary.
2. Through contraction and relaxation, muscular tissue performs four important functions: producing body movements, stabilizing body positions, storing and moving substances within the body, and producing heat.
3. Four special properties of muscular tissues are electrical excitability, the property of responding to stimuli by producing action potentials; contractility, the ability to generate tension to do work; extensibility, the ability to be extended (stretched); and elasticity, the ability to return to original shape after contraction or extension.

Review	**Resource**

10.2 Skeletal Muscle Tissue

1. Connective tissues surrounding skeletal muscles are epimysium, covering the entire muscle; perimysium, covering fascicles; and endomysium, covering muscle fibers. Fascia covers all muscles of a region and separates muscle from the skin. Tendons and aponeuroses are extensions of connective tissue within the muscle belly beyond the muscle fibers that attach muscle to bone or to other muscle.

2. Each skeletal muscle fiber has 100 or more nuclei because it arises from fusion of many myoblasts. Satellite cells are myoblasts that persist after birth. The sarcolemma is a muscle fiber's plasma membrane; it surrounds the sarcoplasm. T tubules are invaginations of the sarcolemma.

3. Generally, an artery and one or two veins accompany each nerve that penetrates a skeletal muscle. Blood capillaries bring in oxygen and nutrients and remove heat and waste products of muscle metabolism.

4. Each fiber contains myofibrils, the contractile elements of skeletal muscle. Sarcoplasmic reticulum surrounds each myofibril. Within a myofibril are thin and thick filaments arranged in compartments called sarcomeres. The overlapping of thick and thin filaments produces striations; darker A bands alternate with lighter I bands (see Table 10.1).

5. Myofibrils are built from three types of proteins: contractile, regulatory, and structural. The contractile proteins are myosin (thick filament) and actin (thin filament). Regulatory proteins are tropomyosin and troponin (both are part of the thin filament). Structural proteins include titin (links Z disc to M line and stabilizes thick filament). See Table 10.2.

6. Muscle contraction occurs because cross-bridges attach to and "walk" along the thin filaments at both ends of a sarcomere, pulling the thin filaments toward the center of a sarcomere. As the thin filaments slide inward, the Z discs come closer together, and the sarcomere shortens.

7. The neuromuscular junction (NMJ) is the synapse between a somatic motor neuron and a skeletal muscle fiber. The NMJ includes the axon terminals and synaptic end bulbs of a motor neuron, plus the adjacent motor end plate of the muscle fiber sarcolemma. When a nerve impulse reaches the synaptic end bulbs of a somatic motor neuron, it triggers exocytosis of the synaptic vesicles, which releases acetylcholine (ACh). ACh diffuses across the synaptic cleft and binds to ACh receptors, initiating a muscle action potential. Acetylcholinesterase then quickly breaks down ACh into its component parts.

8. The contraction cycle is the repeating sequence of events that causes sliding of the filaments: (1) Myosin ATPase breaks down ATP and becomes energized; (2) the myosin head attaches to actin, forming a cross-bridge; (3) the cross-bridge generates force as it rotates toward the center of the sarcomere (power stroke); and (4) binding of ATP to the myosin head detaches it from actin. The myosin head again breaks down the ATP, returns to its original position, and binds to a new site on actin as the cycle continues.

9. An increase in Ca^{2+} concentration in the sarcoplasm starts filament sliding; a decrease turns off the sliding process. The muscle action potential propagating into the T tubule system causes opening of Ca^{2+} release channels in the SR membrane. Calcium ions diffuse from the SR into the sarcoplasm and combine with troponin. This binding causes tropomyosin to move away from the myosin-binding sites on actin. Ca^{2+} active transport pumps continually remove Ca^{2+} from the sarcoplasm into the SR. When the concentration of calcium ions in the sarcoplasm decreases, tropomyosin slides back over and blocks the myosin-binding sites, and the muscle fiber relaxes.

10. In a concentric isotonic contraction, the muscle shortens to produce movement and reduce the angle at a joint. During an eccentric isotonic contraction, the muscle lengthens. Isometric contractions, in which tension is generated without muscle changing its length, stabilize some joints as others are moved.

Anatomy Overview - Cross-section of Skeletal Muscle
Animation - Contraction and Movement
Animation - Neuromuscular Junctions
Animation - Contraction of a Sarcomere
Animation - Control of Muscle Tension
Animation - Muscle Metabolism
Figure 10.5 - Sliding Filament Mechanism of Muscle Contraction
Figure 10.7 - The Contraction Cycle
Exercise - Contraction Connections
Exercise - Muscle Car
Exercise - Increase Muscle Tension
Exercise - Fueling Contraction and Recovery

10.3 Types of Skeletal Muscle Fibers

1. On the basis of structure and function, skeletal muscle fibers are classified as slow oxidative (SO) or type I, fast oxidative-glycolytic (FOG) or type IIa, and fast glycolytic (FG) or type IIb fibers. Most skeletal muscles contain a mixture of all three. Their proportions vary with muscle action.

2. The motor units of a muscle are recruited in the following order: first SO fibers, then FOG fibers, and finally FG fibers.

3. Table 10.4 summarizes the three types of skeletal muscle fibers.

Anatomy Overview - Skeletal Muscle
Animation - Muscle Cell Structures

10.4 Exercise and Skeletal Muscle Tissue

1. Various types of exercises can induce changes in the fibers in a skeletal muscle. Endurance-type (aerobic) exercises cause a gradual transformation of some fast glycolytic (FG) fibers into fast oxidative-glycolytic (FOG) fibers.

2. Exercises that require great strength for short periods produce an increase in the size and strength of fast glycolytic (FG) fibers. The increase in size is due to increased synthesis of thick and thin filaments.

10.5 Cardiac Muscle Tissue

1. Cardiac muscle tissue, found only in the heart, is striated and involuntary. Cardiac muscle fibers are branching cylinders and usually contain a single centrally-located nucleus.

2. Compared to skeletal muscle tissue, cardiac muscle tissue has more mitochondria, smaller sarcoplasmic reticulum, and wider transverse tubules located at Z discs rather than at A–I band junctions.

3. Cardiac muscle fibers branch and are connected end-to-end via desmosomes.

4. Intercalated discs provide strength and aid in conduction of muscle action potentials by way of gap junctions located in the discs.

Anatomy Overview - Cardiac Muscle
Figure 10.9 - Histology of Cardiac Muscle

Review

5. Unlike skeletal muscle tissue, cardiac muscle tissue contracts continuously and rhythmically and can contract without extrinsic stimulation. It can remain contracted longer than skeletal muscle tissue.

10.6 Smooth Muscle Tissue

1. The fibers of smooth muscle tissue, which is nonstriated and involuntary, contain intermediate filaments and dense bodies that function as Z discs.
2. Visceral (single-unit) smooth muscle is found in the walls of viscera and small blood vessels. The fibers are arranged in a network.
3. Multiunit smooth muscle is found in large blood vessels, arrector pili muscles, and the iris of the eye. The fibers operate independently rather than in unison.
4. The duration of contraction and relaxation of smooth muscle is longer than in skeletal muscle.
5. Smooth muscle fibers contract in response to nerve impulses, hormones, and local factors, and can stretch considerably without developing tension.
6. Table 10.5 summarizes the principal characteristics of the three types of muscle tissue.

10.7 Development of Muscles

1. With few exceptions, muscles develop from mesoderm.
2. Skeletal muscles develop from the paraxial mesoderm, via somites and migrating somitomeres. Cardiac muscle and smooth muscle develop from mesodermal cells that migrate during the development process to the heart and to the gastrointestinal tract and viscera, respectively.

10.8 Aging and Muscular Tissue

1. With aging, there is a slow, progressive loss of skeletal muscle mass, which is replaced by fibrous connective tissue and fat.
2. Aging also results in a decrease in muscle strength, slower muscle reflexes, and loss of flexibility, which can be compensated for to an extent by increased physical activity.

Resource

Anatomy Overview - Smooth Muscle Tissue
Figure 10.10 - Histology of Smooth Muscle Tissue

CRITICAL THINKING QUESTIONS

1. A marathon runner and a weight-lifter ask you to help them understand how their fiber types differ. Explain to them how the fiber types in the lower limb muscles of the runner differ from those of the lower limb muscles of the power lifter.

2. Bill tore some ligaments in his knee while skiing. He was in a toe-to-thigh cast for 6 weeks. When the cast was removed, the newly healed leg was noticeably thinner than the uncasted leg. What happened to his leg?

3. The newspaper reported several cases of botulism poisoning following a potluck fund-raising dinner for the local clinic. The cause appeared to be a potato salad that contained the toxin of the soil bacteria *Clostridium botulinum*. This toxin blocks the release of acetylcholine. What would you expect the major effect of botulism poisoning to be, and why?

4. Research is under way to grow new cardiac muscle cells for ailing hearts. Skeletal muscle transplants have been tried, but they don't work as well as transplants of cardiac muscle. Both skeletal and cardiac muscle are striated, so why does cardiac muscle support rhythmic contraction while skeletal muscle does not?

5. Your study partner says that the contractility, extensibility, and elasticity of muscle tissue can be explained by the "stretchiness" of the contractile proteins in muscle cells. "After all," he says, "why else would they be called 'contractile' proteins?" How should you respond to your study partner? Explain your answer.

ANSWERS TO FIGURE QUESTIONS

10.1 Perimysium is the connective tissue layer that bundles groups of muscle fibers into fascicles.

10.2 The sarcoplasmic reticulum releases calcium ions to trigger muscle contraction.

10.3 Size, from smallest to largest: thick filament, myofibril, muscle fiber.

10.4 Actin and titin anchor into the Z disc. A bands contain myosin, actin, troponin, tropomyosin, and titin; I bands contain actin, troponin, tropomyosin, and titin.

10.5 During muscle contraction, the I bands and H zones of the sarcomere disappear. The lengths of the thin and thick filaments do not change.

10.6 The part of the sarcolemma that contains acetylcholine receptors is the motor end plate.

10.7 If ATP were not available, the cross-bridges would not be able to detach from actin. The muscles would remain in a state of rigidity, as occurs in rigor mortis.

10.8 Three functions of ATP in muscle contraction include the following: (1) Its hydrolysis by an ATPase activates the myosin head so it can bind to actin and rotate; (2) its binding to myosin causes detachment from actin after the power stroke; and (3) it powers the pumps that transport Ca^{2+} from the sarcoplasm back into the sarcoplasmic reticulum.

10.9 The intercalated discs contain desmosomes that hold the cardiac muscle fibers together and gap junctions that enable action potentials to be spread from one muscle fiber to another.

10.10 Visceral smooth muscle and cardiac muscle are similar in that both contain gap junctions, which allow action potentials to spread from one cell to its neighbors.

10.11 The myotome of a somite differentiates into skeletal muscle.

11 | THE MUSCULAR SYSTEM

INTRODUCTION Movements such as throwing a ball, biking, walking, and keyboarding require interactions among bones, joints, and skeletal muscles, which together form an integrated system called the *musculoskeletal system*. To better understand the movements produced by the musculoskeletal system, this chapter will introduce you to the names of specific skeletal muscles, how they attach to specific bones, the actions they produce, and their nerve supply.

Probably some of your first observations of movement involve walking, jogging, or running, activities that transport us from one location to another. This type of movement is easy to recognize, and has unquestionable survival value. But we also move in other ways. Think, for example, about grasping something with your hands or throwing something at your roommate to wake him up in time for class. These activities occur without moving from one location to another, yet they are movements nonetheless. Reflect for a moment on the wide variety of movements that your roommate makes when he finally stumbles out of bed to get dressed for the day. These range from the simple movements of putting on clothing, to the more intricate ones of buttoning his shirt and tying his shoelaces. A variety of intricate movements is also required to eat a meal, such as grasping, manipulating, cutting, chewing, and swallowing food. Communication also involves movement, whether it is writing, typing, smiling, or using your voice as well as your throwing arm to wake up your roommate.

Together, the voluntarily controlled muscles of your body comprise the **muscular system**. This chapter presents many of the major muscles in the body. For each muscle, we will identify the attachment sites, actions, and innervation—the nerve or nerves that stimulate a muscle to contract—of each muscle described. Developing a working knowledge of these key aspects of skeletal muscle anatomy will help you understand how normal movements occur. •

 Did you ever wonder why carpal tunnel syndrome occurs?

11.1 HOW SKELETAL MUSCLES PRODUCE MOVEMENTS

OBJECTIVES

- Describe the relationship between bones and skeletal muscles in producing body movements.
- Define lever and fulcrum, and compare the three types of levers based on location of the fulcrum, effort, and load.
- Identify the types of fascicle arrangements in a skeletal muscle, and relate the arrangements to strength of contraction and range of motion.
- Explain how the prime mover, antagonist, synergist, and fixator in a muscle group work together to produce movement.

Muscle Attachment Sites: Origin and Insertion

Those skeletal muscles that produce movements do so by exerting force on tendons, which in turn pull on bones or other structures (such as skin). Most muscles cross at least one joint and are usually attached to articulating bones that form the joint (Figure 11.1).

When a skeletal muscle contracts, it moves one of the articulating bones. The two articulating bones usually do not move equally in response to contraction. One bone remains stationary or near its original position, either because other muscles stabilize that bone by contracting and pulling it in the opposite direction or because its structure makes it less movable. Ordinarily, the attachment of a muscle's tendon to the stationary bone is called the **origin** (ŌR-i-jin); the attachment of the muscle's other tendon to the movable bone is called the **insertion** (in-SER-shun). A good analogy is a spring on a door. In this example, the part of the spring attached to the frame is the origin; the part attached to the door represents the insertion. A useful rule of thumb is that the origin is usually proximal and the insertion distal, especially in the limbs; the insertion is usually pulled toward the origin. The fleshy portion of the muscle between the tendons is called the **belly** (*body*) (the coiled middle portion of the spring in our example). The actions of a muscle are the main movements that occur when the muscle contracts. In our spring example, this would be the closing of the door. Certain muscles are also capable of **reverse muscle action (RMA)**, also called *closed kinetic chain (CKC) exercises*. During specific movements of the body the actions are reversed and therefore the positions of the origin and insertion of a specific muscle are switched.

Figure 11.1 Relationship of skeletal muscles to bones. Muscles are attached to bones by tendons at their origin and insertion.

🔑 In the limbs, the origin of a muscle is usually proximal and the insertion is usually distal.

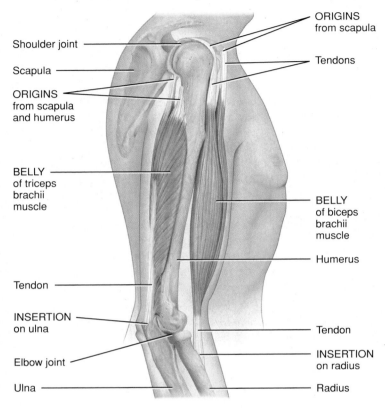

Origin and insertion of a skeletal muscle

 Where is the belly of the muscle that extends the forearm located?

Muscles that move a body part often do not cover the moving part. Figure 11.2a shows that although one of the functions of the biceps brachii muscle is to move the forearm, the belly of the muscle lies over the humerus, not over the forearm. You will also see that muscles that cross two joints, such as the rectus femoris and sartorius of the thigh, have more complex actions than muscles that cross only one joint.

Lever Systems

A *lever* is a rigid structure that can move around a fixed point called a **fulcrum**, symbolized by ⚠. A lever is acted on at two different points by two different forces: the **effort** (E), which causes movement, and the **load** 🅛 or **resistance**, which opposes movement. The effort is the force exerted by muscular contraction; the load is typically the weight of the body part that is moved or some resistance that the moving body part is trying to overcome (such as the weight of a book you might be picking up). Motion occurs when the effort applied to the bone at the insertion exceeds the load. Consider the biceps brachii flexing the forearm at the elbow as an object is lifted (Figure 11.2a). When the forearm is raised, the elbow is the fulcrum. The weight of the forearm plus the weight of the object in the hand is the load. The force of contraction of the biceps brachii pulling the forearm up is the effort.

The relative distance between the fulcrum and load and the point at which the effort is applied determine whether a given lever operates at a mechanical advantage or a mechanical disadvantage. For example, if the load is closer to the fulcrum and the effort farther from the fulcrum, then only a relatively small effort is required to move a large load over a small distance. This is called a **mechanical advantage**. If, instead, the load is farther from the fulcrum and the effort is applied closer to the fulcrum, then a relatively large effort is required to move a small load (but at greater speed). This is called a **mechanical disadvantage**. Compare chewing something hard (the load) with your front teeth and the teeth in the back of your mouth. It is much easier to crush the hard food item with the back teeth because they are closer to the fulcrum (the jaw or temporomandibular joint) than are the front teeth. Here is one more example you can try. Straighten out a paper clip. Now get a pair of scissors and try to cut the paper clip with the tip of the scissors (mechanical disadvantage) versus near the pivot point of the scissors (mechanical advantage).

Levers are categorized into three types according to the positions of the fulcrum, the effort, and the load:

1. The fulcrum is between the effort and the load in **first-class levers** (Figure 11.2b). (Think E*F*L.) Scissors and seesaws are examples of first-class levers. A first-class lever can produce either a mechanical advantage or mechanical disadvantage depending on whether the effort or the load is closer to the fulcrum. (Think of an adult and a child on a seesaw.) As we've seen in the preceding examples, if the effort (child) is farther from the fulcrum than the load (adult), a heavy load can be moved, but not very far or fast. If the effort is closer to the fulcrum than the load, only a lighter load can be moved, but it moves far and fast. There are few first-class levers in the body. One example is the lever formed by the head resting on the vertebral column (Figure 11.2b). When the head is raised, the contraction of the posterior neck muscles provides the effort (E), the joint between the atlas and the occipital bone (atlanto-occipital joint) forms the fulcrum ⚠, and the weight of the anterior portion of the skull is the load 🅛.

2. The load is between the fulcrum and the effort in **second-class levers** (Figure 11.2c). (Think E*L*F.) Second-class levers operate like a wheelbarrow. They always produce a mechanical advantage because the load is always closer to the fulcrum than the effort. This arrangement sacrifices speed and range of motion for force; this type of lever produces the most force. This class of lever is uncommon in the human body. An example is standing up on your toes (Figure 11.2c). The fulcrum ⚠ is the ball of the foot. The load 🅛 is the weight of the body. The effort (E) is the contraction of the muscles of the calf, which raise the heel off the ground.

3. The effort is between the fulcrum and the load in **third-class levers** (Figure 11.2d). (Think F*E*L.) These levers operate like a pair of forceps and are the most common levers in the body. Third-class levers always produce a mechanical disadvantage because the effort is always closer to the fulcrum than the load. In the body, this arrangement favors speed and range of motion over force. The elbow joint, the biceps brachii muscle, and the bones of the arm and forearm are one example of a third-class lever (Figure 11.2d). As we have seen, in flexing the forearm at the elbow, the elbow joint is the fulcrum ⚠, the contraction of the biceps brachii muscle provides the effort (E), and the weight of the hand and forearm is the load 🅛.

Figure 11.2 Lever structure and types of levers. Skeletal muscles produce movements by pulling on bones. Bones serve as levers, and joints act as fulcrums for the levers. Here the lever–fulcrum principle is illustrated by the movement of the forearm. Note where the load (resistance) and effort are applied in (a).

Levers are divided into three types based on the placement of the fulcrum, effort, and load (resistance).

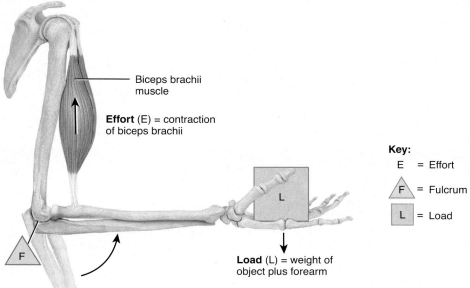

Biceps brachii
muscle

Effort (E) = contraction
of biceps brachii

Key:
E = Effort
F = Fulcrum
L = Load

Load (L) = weight of
object plus forearm

Fulcrum (F) = elbow joint

(a) Movement of the forearm lifting a weight

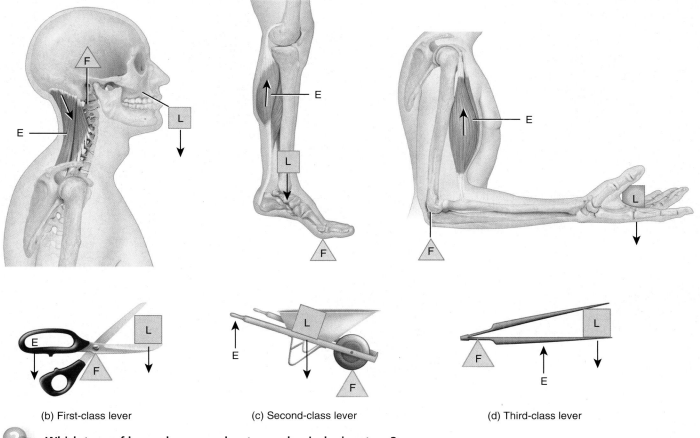

(b) First-class lever

(c) Second-class lever

(d) Third-class lever

Which type of lever always works at a mechanical advantage?

Effects of Fascicle Arrangement

Recall from Chapter 10 that the skeletal muscle fibers (cells) within a muscle are arranged in bundles known as **fascicles** (FAS-i-kuls). Within a fascicle, all muscle fibers are parallel to one another. The fascicles, however, may form one of five patterns with respect to the tendons: parallel, *fusiform* (spindle-shaped, narrow toward the ends and wide in the middle), circular, triangular, or *pennate* (shaped like a feather) (Table 11.1).

Fascicular arrangement affects a muscle's power and range of motion. As a muscle fiber contracts, it shortens to about 70 percent of its resting length. Thus, the longer the fibers in a mus-cle, the greater the range of motion it can produce. However, the power of a muscle depends not on length but on its total cross-sectional area. Therefore, because a short fiber can contract as forcefully as a long one, the more fibers per unit of cross-sectional area a muscle has, the more power it can produce. Fascicular arrangement often represents a compromise between power and range of motion. Pennate muscles, for instance, have a large number of short-fibered fascicles distributed over their tendons, giving them greater power but a smaller range of motion. In contrast, parallel muscles have comparatively fewer fascicles, but they have long fibers that extend the length of the muscle, so they have a greater range of motion but less power.

TABLE 11.1

Arrangement of Fascicles

PARALLEL

Fascicles parallel to longitudinal axis of muscle; terminate at either end in flat tendons.

Example: Sternohyoid muscle (see Figure 11.8a)

FUSIFORM

Fascicles nearly parallel to longitudinal axis of muscle; terminate in flat tendons; muscle tapers toward tendons, where diameter is less than at belly.

Example: Digastric muscle (see Figure 11.8a)

CIRCULAR

Fascicles in concentric circular arrangements form sphincter muscles that enclose an orifice (opening).

Example: Orbicularis oculi muscle (see Figure 11.4a)

TRIANGULAR

Fascicles spread over broad area converge at thick central tendon; gives muscle a triangular appearance.

Example: Pectoralis major muscle (see Figure 11.3a)

PENNATE

Short fascicles in relation to total muscle length; tendon extends nearly entire length of muscle.

Unipennate
Fascicles are arranged on only one side of tendon.

Bipennate
Fascicles are arranged on both sides of centrally positioned tendons.

Multipennate
Fascicles attach obliquely from many directions to several tendons.

Example: Extensor digitorum longus muscle (see Figure 11.24b)

Example: Rectus femoris muscle (see Figure 11.3a)

Example: Deltoid muscle (see Figure 11.17d)

Muscle Actions

In the study of muscles, it is common to describe the actions that individual muscles produce at their associated joints. It is important to recognize what movements muscles are capable of producing at the joints they cross, but it is equally important to realize that, in reality, muscles do not work in isolation. Sometimes the movements attributed to muscles are possible only for a certain range of the joint's movement, or occur in combination with the actions of other muscles. As you study the muscles in this chapter, we will introduce you to the primary actions they generate at the joints they cross. With selected muscles we will explore a broader view of their functional roles in the body.

In Chapter 10 we learned that the connective tissues surrounding the contractile components within the muscle belly emerge from either end of the muscle as the tendons, to blend with the periosteum and attach to the bone. When these contractile components work, they generate tension within the muscle as the sliding filaments of the sarcomere attempt to shorten. This results in two potential types of isotonic contraction—concentric and eccentric. Recall that an *isotonic contraction* is one in which enough muscle fibers are contracting to shorten the muscle against the load (see Section 10.2). This is in contrast to an *isometric contraction*, where the number of fibers contracting and generating a force are equal to the opposite force of the load, so the muscle does not change length. During a **concentric contraction** (kon-SEN-trik), the muscle shortens as it produces a constant tension and overcomes the load it is moving. In an **eccentric contraction** (ek-SEN-trik), the muscle produces a constant tension but lengthens as it gives in to the load it is moving. As a result, the same muscle is the active controller of two opposite movements at a joint. Imagine the following situation. You pick up a book from a table by flexing the elbow joint to lift the book. You then hold the book steady in front of you as you look at it. Next, you slowly lower the book back to the table by extending the elbow joint. The biceps brachii muscle controls this full range of activity. Isotonic concentric contraction of the biceps brachii overcomes the weight of the book and raises it off the table as the elbow joint is flexed. Isometric contraction of the biceps brachii holds the book steady in front of you with a flexed (bent) elbow as you look at it. Finally, isotonic eccentric contraction of the biceps brachii slowly gives way to the load and the book is lowered back down to the table as the elbow joint extends.

Throughout this chapter the actions described for muscles will be the actions produced by concentric (shortening) contractions of the muscles. However, realize that joint actions are not always simply the result of opposing muscles. As we just explained, a single muscle can control opposite joint movements through its concentric and eccentric contractions.

Coordination Among Muscles

Movements often are the result of several skeletal muscles acting as a group. Most skeletal muscles are arranged in opposing (antagonistic) pairs at joints—that is, flexors–extensors, abductors–adductors, and so on. Within opposing pairs, one mus-cle, called the **prime mover** or **agonist** (=leader), contracts to cause an action while the other muscle, the **antagonist** (*anti-*=against), stretches and yields to the effects of the prime mover. In the process of flexing the forearm at the elbow, for instance, the biceps brachii is the prime mover, and the triceps brachii is the antagonist (see Figure 11.1). The antagonist and prime mover are usually located on opposite sides of the bone or joint, as is the case in this example.

With an opposing pair of muscles, the roles of the prime mover and antagonist can switch for different movements. For example, while extending the forearm at the elbow against resistance (such as pushing against some resistance with the palm as you straighten the elbow), the triceps brachii becomes the prime mover, and the biceps brachii is the antagonist. If a prime mover and its antagonist contract at the same time with equal force, there will be no movement.

Sometimes a prime mover crosses other joints before it reaches the joint at which its primary action occurs. The biceps brachii, for example, spans both the shoulder and elbow joints, with primary action on the forearm. To prevent unwanted movements at intermediate joints or to otherwise aid the movement of the prime mover, muscles called **synergists** (SIN-er-jists; *syn-*=together; *-ergon*=work) contract and stabilize the intermediate joints. As an example, muscles that flex the fingers (prime movers) cross the intercarpal and radiocarpal joints (intermediate joints). If movement at these intermediate joints was unrestrained, you would not be able to flex your fingers without flexing the wrist at the same time. (Try to make a strong fist while also flexing the wrist. It is hard to do, isn't it?) Synergistic contraction of the wrist extensor muscles stabilizes the wrist joint and prevents unwanted movement, while the flexor muscles of the fingers contract to bring about the primary action, efficient flexion of the fingers. Synergists are usually located close to the prime mover.

Some muscles in a group also act as **fixators**, stabilizing the origin of the prime mover so that the prime mover can act more efficiently. Fixators steady the proximal end of a limb while movements occur at the distal end. For example, the scapula is a freely movable bone that serves as the origin for several muscles that move the arm. When the arm muscles contract, the scapula must be held steady. In abduction of the arm, the deltoid muscle serves as the prime mover, and fixators (pectoralis minor, trapezius, subclavius, serratus anterior muscles, and others) hold the scapula firmly against the back of the chest (see Figure 11.17a–e). The insertion of the deltoid muscle pulls on the humerus to abduct the arm. Under different conditions—that is, for different movements—and at different times, many muscles may act as prime movers, antagonists, synergists, or fixators.

Structure and Function of Muscle Groups

Muscles arise from common masses of muscle tissue in the developing embryo and fetus. The three regions of the body—the free limbs, the trunk, and the head—each have distinct patterns of muscle development.

As developing muscle tissue migrates into the embryonic free limbs, it forms two principal masses, an anterior mass of muscle and a posterior mass of muscle. These developing masses of muscle are separated by the developing bones and connective tissue of the free limb. As joints form between the developing bones of the free limb, the muscle masses differentiate into multiple muscles that are enveloped in fascia and separated by the bones, creating anterior and posterior compartments of muscle in the different regions of the free limb. Therefore, a limb muscle compartment is a group of skeletal muscles that arose from a common developmental origin. As the muscles of a compartment develop, the nerves and blood vessels develop along with them. Because of this, the muscles of a compartment share common blood and nerve supply. Also, because the muscles of a compartment are grouped on the same side of joints, the anterior compartment muscles are typically flexors of the joints they cross, and posterior compartment muscles are typically extensors of the joints they cross.

Muscles of the head also arise as functional groups from the embryonic pharyngeal arches and some of the cranial somites. For example, the muscles of mastication arise from the muscle tissue of the first pharyngeal arch. The muscles of facial expression arise from the second pharyngeal arch. Each arch is supplied by a unique cranial nerve; therefore, all the muscles of an arch, or functional group, are innervated by one nerve. For example, the trigeminal (V) nerve innervates all the muscles of mastication (first pharyngeal arch muscles) and the facial (VII) nerve innervates all the muscles of facial expression (second pharyngeal arch muscles).

 CHECKPOINT

1. Using the terms origin, insertion, and belly, describe how skeletal muscles produce body movements by pulling on bones.
2. List the three types of levers, and give an example of a first-, second-, and third-class lever found in the body.
3. Describe the various arrangements of fascicles.
4. Explain the difference between a concentric contraction and an eccentric contraction.
5. Define the roles of the prime mover (agonist), antagonist, synergist, and fixator in producing various movements of the free upper limb.

11.2 HOW SKELETAL MUSCLES ARE NAMED

 OBJECTIVE

• Explain **seven features used in naming skeletal muscles.**

The names of most of the skeletal muscles contain combinations of the word roots of their distinctive features. This works two ways. You can learn the names of muscles by remembering the terms that refer to muscle features, such as the pattern of the muscle's fascicles; the size, shape, action, number of origins, and location of the muscle; and the sites of origin and insertion of the muscle. Knowing the names of a muscle will then give you clues about its features. Study Table 11.2 to become familiar with the terms used in muscle names.

 CHECKPOINT

6. Select 10 muscles in Figure 11.3 and identify the features on which their names are based. (*Hint:* Use the prefix, suffix, and root of each muscle's name as a guide.)

11.3 PRINCIPAL SKELETAL MUSCLES

Exhibits 11.A–11.V (Figures 11.4–11.25) will assist you in learning the names of the principal skeletal muscles in various regions of the body. The muscles in the exhibits are divided into groups according to the part of the body on which they act. As you study groups of muscles in the exhibits, refer to Figure 11.3 to see how each group is related to the others.

The exhibits contain the following elements:

• **Objective.** This statement describes what you should learn from the exhibit.
• **Overview.** These paragraphs provide a general introduction to the muscles under consideration and emphasize how the muscles are organized within various regions. The discussion also highlights any distinguishing features of the muscles.
• **Muscle names.** The word roots indicate how the muscles are named. As noted previously, once you have mastered the naming of the muscles, you can more easily understand their actions.
• **Origins, insertions, and actions.** You are also given the origin, insertion, and actions of each muscle.
• **Innervation.** This section lists the nerve or nerves that cause contraction of each muscle. In general, cranial nerves, which arise from the lower parts of the brain, serve muscles in the head region. Spinal nerves, which arise from the spinal cord within the vertebral column, innervate muscles in the rest of the body. Cranial nerves are designated by both a name and a Roman numeral—for example, the facial (VII) nerve. Spinal nerves are numbered in groups according to the part of the spinal cord from which they arise: C=cervical (neck region), T=thoracic (chest region), L=lumbar (lower-back region), and S=sacral (buttocks region). An example is T1, the first thoracic spinal nerve.
• **Relating muscles to movements.** These exercises will help you organize the muscles in the body region under consideration according to the actions they produce.
• **Checkpoint questions.** These knowledge checkpoints relate specifically to information in each exhibit, and take the form of review, critical thinking, and/or application questions.
• **Clinical Connections.** Selected exhibits include clinical applications, which explore the clinical, professional, or everyday relevance of a particular muscle or its function through descriptions of disorders or clinical procedures.
• **Figures.** The figures in the exhibits may present superficial and deep, anterior and posterior, or medial and lateral views to show each muscle's position as clearly as possible. The muscle names in all capital letters are specifically referred to in the tabular part of the exhibit.

TABLE 11.2

Characteristics Used to Name Muscles

NAME	MEANING	EXAMPLE	FIGURE
DIRECTION: Orientation of muscle fascicles relative to the body's midline			
Rectus	Parallel to midline	Rectus abdominis	11.13c,e
Transverse	Perpendicular to midline	Transversus abdominis	11.13c,h
Oblique	Diagonal to midline	External oblique	11.13b,f
SIZE: Relative size of the muscle			
Maximus	Largest	Gluteus maximus	11.22b,d
Minimus	Smallest	Gluteus minimus	11.22c
Longus	Long	Adductor longus	11.23a,b
Brevis	Short	Adductor brevis	11.23c,d
Latissimus	Widest	Latissimus dorsi	11.18c,e
Longissimus	Longest	Longissimus thoracis	11.12a,b
Magnus	Large	Adductor magnus	11.23c,d
Major	Larger	Pectoralis major	11.3a
Minor	Smaller	Pectoralis minor	11.17a,c
Vastus	Huge	Vastus lateralis	11.23a,b,d
SHAPE: Relative shape of the muscle			
Deltoid	Triangular	Deltoid	11.3a
Trapezius	Trapezoid	Trapezius	11.17d,f
Serratus	Saw-toothed	Serratus anterior	11.17b
Rhomboid	Diamond-shaped	Rhomboid major	11.17e,f
Orbicularis	Circular	Orbicularis oculi	11.4a
Pectinate	Comblike	Pectineus	11.23a,b
Piriformis	Pear-shaped	Piriformis	11.22c,d
Platys	Flat	Platysma	11.4c
Quadratus	Square, four-sided	Quadratus femoris	11.22c
Gracilis	Slender	Gracilis	11.23a,b
ACTION: Principal action of the muscle			
Flexor	Decreases a joint angle	Flexor carpi radialis	11.20a,d
Extensor	Increases a joint angle	Extensor carpi ulnaris	11.20g
Abductor	Moves a bone away from the midline	Abductor pollicis longus	11.20h,j
Adductor	Moves a bone closer to the midline	Adductor longus	11.23a,b
Levator	Raises or elevates a body part	Levator scapulae	11.17b,e,f
Depressor	Lowers or depresses a body part	Depressor labii inferioris	11.4a
Supinator	Turns palm anteriorly	Supinator	11.20c,f
Pronator	Turns palm posteriorly	Pronator teres	11.20a
Sphincter	Decreases the size of an opening	External anal sphincter	11.15
Tensor	Makes a body part rigid	Tensor fasciae latae	11.22a
Rotator	Rotates a bone around its longitudinal axis	Rotatore	11.12a
NUMBER OF ORIGINS: Number of tendons of origin			
Biceps	Two origins	Biceps brachii	11.19a,c
Triceps	Three origins	Triceps brachii	11.19b,d
Quadriceps	Four origins	Quadriceps femoris	11.23a
LOCATION: Structure near which a muscle is found. Example: temporalis, a muscle near the temporal bone			11.4d
ORIGIN AND INSERTION: Sites where muscle originates and inserts. Example: sternocleidomastoid, originating on the sternum and clavicle and inserting on mastoid process of temporal bone			11.11d,e

Figure 11.3 **Principal superficial skeletal muscles.**

Most movements require several skeletal muscles acting in groups rather than individually.

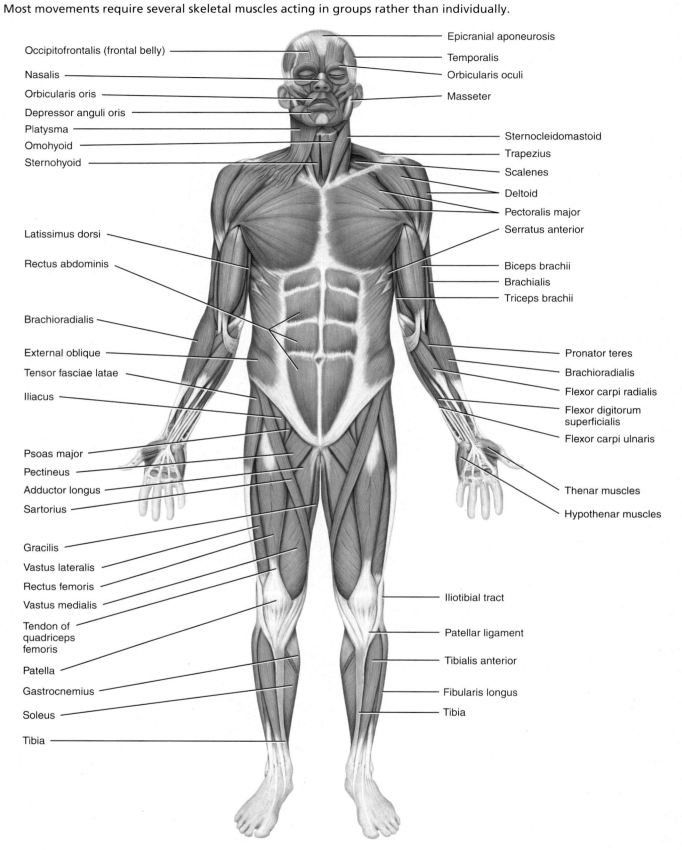

Occipitofrontalis (frontal belly)
Nasalis
Orbicularis oris
Depressor anguli oris
Platysma
Omohyoid
Sternohyoid

Epicranial aponeurosis
Temporalis
Orbicularis oculi
Masseter

Sternocleidomastoid
Trapezius
Scalenes
Deltoid
Pectoralis major
Serratus anterior

Latissimus dorsi
Rectus abdominis

Biceps brachii
Brachialis
Triceps brachii

Brachioradialis
External oblique
Tensor fasciae latae
Iliacus

Pronator teres
Brachioradialis
Flexor carpi radialis
Flexor digitorum superficialis
Flexor carpi ulnaris

Psoas major
Pectineus
Adductor longus
Sartorius

Thenar muscles
Hypothenar muscles

Gracilis
Vastus lateralis
Rectus femoris
Vastus medialis
Tendon of quadriceps femoris

Iliotibial tract

Patellar ligament

Patella
Gastrocnemius
Soleus
Tibia

Tibialis anterior
Fibularis longus
Tibia

(a) Anterior view

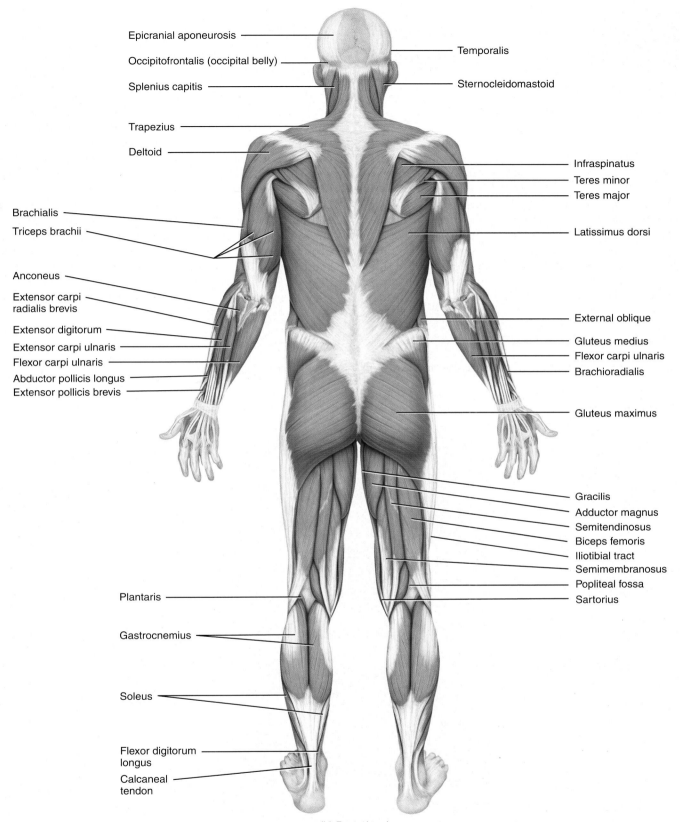

Epicranial aponeurosis

Occipitofrontalis (occipital belly)

Splenius capitis

Trapezius

Deltoid

Brachialis

Triceps brachii

Anconeus

Extensor carpi radialis brevis

Extensor digitorum

Extensor carpi ulnaris

Flexor carpi ulnaris

Abductor pollicis longus

Extensor pollicis brevis

Plantaris

Gastrocnemius

Soleus

Flexor digitorum longus

Calcaneal tendon

Temporalis

Sternocleidomastoid

Infraspinatus

Teres minor

Teres major

Latissimus dorsi

External oblique

Gluteus medius

Flexor carpi ulnaris

Brachioradialis

Gluteus maximus

Gracilis

Adductor magnus

Semitendinosus

Biceps femoris

Iliotibial tract

Semimembranosus

Popliteal fossa

Sartorius

(b) Posterior view

Give an example of a muscle named for each of the following characteristics: direction of fibers, shape, action, size, origin and insertion, location, and number of tendons of origin.

EXHIBIT 11.A

Muscles of the Head That Produce
Facial Expressions *(Figure 11.4)*

OBJECTIVE

• Describe the origin, insertion, action, and innervation of the muscles of facial expression.

The muscles of facial expression, which provide us with the ability to express a wide variety of emotions, lie within the subcutaneous layer (Figure 11.4). They usually originate from the fascia or bones of the skull and insert into the skin. Because of their insertions, the muscles of facial expression move the skin rather than a joint when they contract.

Among the noteworthy muscles in this group are those surrounding the orifices (openings) of the head such as the eyes, nose, and mouth. These muscles function as *sphincters* (SFINGK-ters), which close the orifices, and *dilators* (DĪ-lā-tors),

which dilate or open the orifices. For example, the **orbicularis oculi** muscle closes the eye, and the levator palpebrae superioris muscle (discussed in Exhibit 11.B) opens it. The **occipitofrontalis** is an unusual muscle in this group because it is made up of two parts: an anterior part called the **frontal belly (frontalis)**, which is superficial to the frontal bone, and a posterior part called the **occipital belly (occipitalis)**, which is superficial to the occipital bone. The two muscular portions are held together by a strong **aponeurosis** (sheetlike tendon), the **epicranial aponeurosis** (ep-i-KRĀ-nē-al ap′-ō-noo-RŌ-sis), also called the **galea aponeurotica** (GA-lē-a ap′-ō-noo′-RO-ti-ka), that covers the superior and lateral surfaces of the skull. The **buccinator** muscle forms the major muscular portion of the cheek. The duct of the parotid gland (a salivary gland) passes through the buccinator muscle to

Figure 11.4 Muscles of the head that produce facial expressions.

 When they contract, muscles of facial expression move the skin rather than a joint.

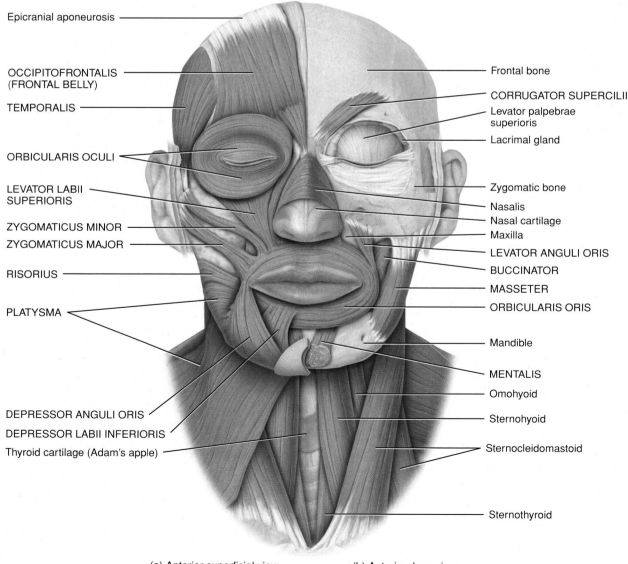

(a) Anterior superficial view   (b) Anterior deep view

FIGURE 11.4 CONTINUES ▶

EXHIBIT 11.A **321**

MUSCLE	ORIGIN	INSERTION	ACTION	INNERVATION
SCALP MUSCLES				
Occipitofrontalis (ok-sip'-i-tō-frun-TĀ-lis)				
Frontal belly (frontalis)	Epicranial aponeurosis	Skin superior to supraorbital margin	Draws scalp anteriorly, raises eyebrows, and wrinkles skin of forehead horizontally as in a look of surprise	Facial (VII) nerve
Occipital belly (occipitalis) (*occipit-*=back of the head)	Occipital bone and mastoid process of temporal bone	Epicranial aponeurosis	Draws scalp posteriorly	Facial (VII) nerve
MOUTH MUSCLES				
Orbicularis oris (or-bi'-kū-LAR-is OR-is; *orb-*=circular; *oris*=of the mouth)	Muscle fibers surrounding opening of mouth	Skin at corner of mouth	Closes and protrudes lips, as in kissing; compresses lips against teeth; and shapes lips during speech	Facial (VII) nerve
Zygomaticus major (zī-gō-MA-tī-kus; *zygomatic*=cheek bone; *major*=greater)	Zygomatic bone	Skin at angle of mouth and blends with fibers of orbicularis oris	Draws angle of mouth superiorly and laterally, as in smiling	Facial (VII) nerve
Zygomaticus minor (*minor*=lesser)	Zygomatic bone	Upper lip	Raises (elevates) upper lip, exposing maxillary (upper) teeth	Facial (VII) nerve
Levator labii superioris (le-VĀ-tor LĀ-bē-ī soo-per'-ē-OR-is; *levator*=raises or elevates; *labii*=lip; *superioris*=upper)	Maxilla superior to infraorbital foramen	Skin at angle of mouth and blends with fibers of orbicularis oris	Raises upper lip	Facial (VII) nerve
Depressor labii inferioris (de-PRE-sor LĀ-bē-ī; *depressor*=depresses or lowers; *inferioris*=lower)	Mandible	Skin of lower lip	Depresses (lowers) lower lip	Facial (VII) nerve
Depressor anguli oris (ANG-ū-lī; *angul*=angle or corner; *oris*=of the mouth)	Mandible	Angle of mouth	Draws angle of mouth laterally and inferiorly, as in opening mouth	Facial (VII) nerve
Levator anguli oris	Maxilla inferior to infraorbital foramen	Skin of lower lip	Draws angle of mouth laterally and superiorly	Facial (VII) nerve
Buccinator (BUK-si-nā'-tor; *bucc-*=cheek)	Alveolar processes of maxilla and mandible and pterygomandibular raphe	Blends with fibers of orbicularis oris	Presses cheeks against teeth and lips, as in whistling, blowing, and sucking; draws corner of mouth laterally	Facial (VII) nerve
Risorius (ri-ZOR-ē-us; *risor*=laughter)	Fascia over parotid (salivary) gland	Skin at the angle of mouth	Draws angle of mouth laterally, as in grimacing	Facial (VII) nerve
Mentalis (men-TĀ-lis; *ment-*=the chin)	Mandible	Skin of chin	Elevates and protrudes lower lip and pulls skin of chin up as in pouting	Facial (VII) nerve
Platysma (pla-TIZ-ma; *platys*=flat, broad)	Fascia over deltoid and pectoralis major muscles	Mandible, blends with muscles around angle of mouth, and skin of lower face	Draws outer part of lower lip inferiorly and posteriorly as in pouting; depresses mandible	Facial (VII) nerve
EYE REGION MUSCLES				
Orbicularis oculi (or-bi'-kū-LAR-is OK-ū-lī; *oculi*=of the eye)	Medial wall of orbit	Circular path around orbit	Closes eye	Facial (VII) nerve
Corrugator supercilii (KOR-a-gā'-tor soo'-per-SIL-ē-ī; *corrugat*=wrinkle; *supercilii*=of the eyebrow)	Medial end of superciliary arch of frontal bone	Skin of eyebrow	Draws eyebrows inferiorly and wrinkles skin of forehead vertically as in frowning	Facial (VII) nerve

EXHIBIT 11.A Muscles of the Head That Produce Facial Expressions *(Figure 11.4)* CONTINUED

reach the oral cavity. The buccinator muscle is so named because it compresses the cheeks (*bucc-*=cheek) during blowing—for example, when a musician plays a brass instrument such as a trumpet. It functions in whistling, blowing, and sucking and assists in chewing.

Relating Muscles to Movements

Arrange the muscles in this exhibit into two groups: (1) those that act on the mouth and (2) those that act on the eyes.

 CHECKPOINT

7. Why do the muscles of facial expression move the skin rather than a joint?

CLINICAL CONNECTION | *Bell's Palsy*

Bell's palsy, also known as **facial paralysis**, is a unilateral paralysis of the muscles of facial expression. It is due to damage or disease of the facial (VII) nerve. Possible causes include inflammation of the facial nerve due to an ear infection, ear surgery that damages the facial nerve, or infection by the herpes simplex virus. The paralysis causes the entire side of the face to droop in severe cases. The person cannot wrinkle the forehead, close the eye, or pucker the lips on the affected side. Drooling and difficulty in swallowing also occur. Eighty percent of patients recover completely within a few weeks to a few months. For others, paralysis is permanent. The symptoms of Bell's palsy mimic those of a stroke. •

FIGURE 11.4 CONTINUED

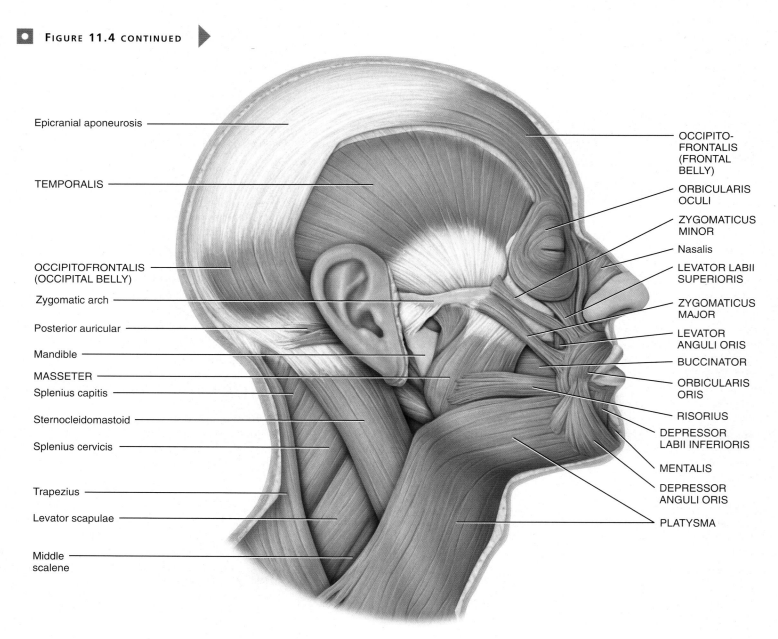

(c) Right lateral superficial view

EXHIBIT 11.A **323**

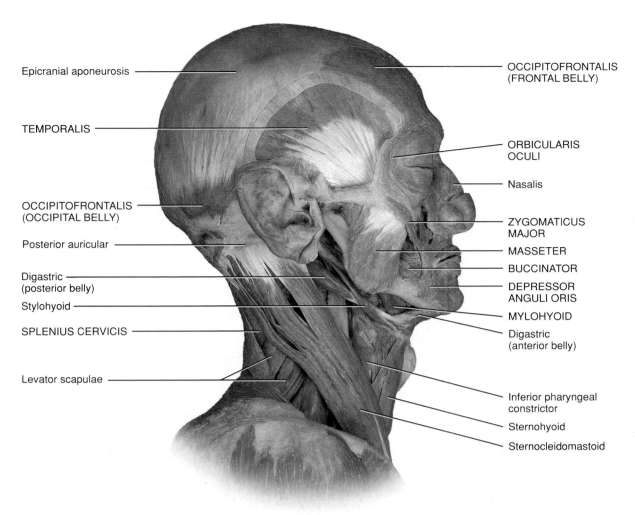

Epicranial aponeurosis

TEMPORALIS

OCCIPITOFRONTALIS
(OCCIPITAL BELLY)

Posterior auricular

Digastric
(posterior belly)

Stylohyoid

SPLENIUS CERVICIS

Levator scapulae

OCCIPITOFRONTALIS
(FRONTAL BELLY)

ORBICULARIS
OCULI

Nasalis

ZYGOMATICUS
MAJOR

MASSETER

BUCCINATOR

DEPRESSOR
ANGULI ORIS

MYLOHYOID

Digastric
(anterior belly)

Inferior pharyngeal
constrictor

Sternohyoid

Sternocleidomastoid

(d) Right lateral deep view

■ Origin

■ Insertion

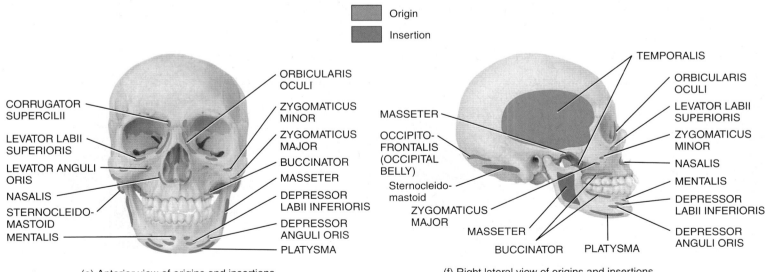

CORRUGATOR
SUPERCILII

LEVATOR LABII
SUPERIORIS

LEVATOR ANGULI
ORIS

NASALIS

STERNOCLEIDO-
MASTOID

MENTALIS

ORBICULARIS
OCULI

ZYGOMATICUS
MINOR

ZYGOMATICUS
MAJOR

BUCCINATOR

MASSETER

DEPRESSOR
LABII INFERIORIS

DEPRESSOR
ANGULI ORIS

PLATYSMA

(e) Anterior view of origins and insertions

MASSETER

OCCIPITO-
FRONTALIS
(OCCIPITAL
BELLY)

Sternocleido-
mastoid

ZYGOMATICUS
MAJOR

MASSETER

BUCCINATOR

PLATYSMA

TEMPORALIS

ORBICULARIS
OCULI

LEVATOR LABII
SUPERIORIS

ZYGOMATICUS
MINOR

NASALIS

MENTALIS

DEPRESSOR
LABII INFERIORIS

DEPRESSOR
ANGULI ORIS

(f) Right lateral view of origins and insertions

 Which muscles of facial expression cause frowning, smiling, pouting, and squinting?

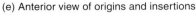
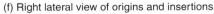

Muscles of the Head That Move the Eyeballs and Upper Eyelids *(Figure 11.5)*

EXHIBIT 11.B

 O B J E C T I V E

* Describe the origin, insertion, action, and innervation of the eye muscles that move the eyeballs and upper eyelids.

Muscles that move the eyeballs are called **extrinsic eye muscles** because they originate outside the eyeballs (in the orbit) and insert on the outer surface of the sclera ("white of the eye") (Figure 11.5). The extrinsic eye muscles are some of the fastest contracting and most precisely controlled skeletal muscles in the body.

Three pairs of extrinsic eye muscles control movements of the eyeballs: (1) superior and inferior recti, (2) lateral and medial recti, and (3) superior and inferior obliques. The four recti muscles (superior, inferior, lateral, and medial) arise from a tendinous ring in the posterior orbit and insert into the sclera of the eye. As their names imply, the **superior** and **inferior recti** move the eyeballs superiorly and inferiorly; the **lateral** and **medial recti** move the eyeballs laterally and medially, respectively.

The actions of the oblique muscles cannot be deduced from their names. The **superior oblique** muscle originates posteriorly near the tendinous ring, then passes anteriorly superior to the medial rectus, and ends in a round tendon. The tendon extends through a pulleylike loop of fibrocartilaginous tissue called the *trochlea* (=pulley) on the anterior and medial part of the roof of the orbit. Finally, the tendon turns and expands into a broad flat sheet that inserts on the posterolateral aspect of the eyeball. Accordingly, the superior oblique muscle moves the eyeballs inferiorly and laterally. The **inferior oblique** muscle originates on the maxilla at the anteromedial aspect of the floor of the orbit. It then passes posteriorly and laterally and inserts on the posterolateral aspect of the eyeball. Because of this arrangement, the inferior oblique muscle moves the eyeballs superiorly and laterally.

Developmentally related to the extrinsic eye muscles is the **levator palpebrae superioris**. This muscle splits off of the superior rectus during development. Unlike the recti and oblique muscles, it does not move the eyeballs as its tendon passes the eyeball to insert into the upper eyelid. Rather, it raises the upper eyelids, that is, opens the eyes. It is therefore an antagonist to the orbicularis oculi, which closes the eyes.

Relating Muscles to Movements

Arrange the muscles in this exhibit according to their actions on the eyeballs: (1) elevation, (2) depression, (3) abduction, (4) adduction, (5) medial rotation, and (6) lateral rotation. The same muscle may be mentioned more than once.

 CHECKPOINT

8. Which muscles that move the eyeballs contract and relax as you look to your left without moving your head?

CLINICAL CONNECTION | Strabismus

Strabismus (stra-BIZ-mus; *strabismos*=squinting) is a condition in which the two eyeballs are not properly aligned. This can be hereditary or it can be due to birth injuries, poor attachments of the muscles, problems with the brain's control center, or localized disease. Strabismus can be constant or intermittent. In strabismus, each eye sends an image to a different area of the brain, and because the brain usually ignores the messages sent by one of the eyes, the ignored eye becomes weaker, hence "lazy eye" or *amblyopia*, develops. *External strabismus* results when a lesion in the oculomotor (III) nerve causes the eyeball to move laterally when at rest, and results in an inability to move the eyeball medially and inferiorly. A lesion in the abducens (VI) nerve results in *internal strabismus*, a condition in which the eyeball moves medially when at rest and cannot move laterally.

Treatment options for strabismus depend on the specific type of problem and include surgery, visual therapy (retraining the brain's control center), and orthoptics (eye muscle training to straighten the eyes). •

Figure 11.5 **Muscles of the head that move the eyeballs (extrinsic eye muscles) and upper eyelid.**

🔑 The extrinsic muscles of the eyeball are among the fastest contracting and most precisely controlled skeletal muscles in the body.

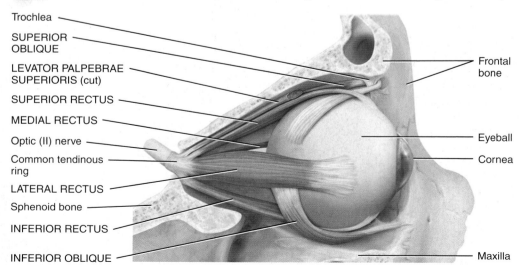

(a) Lateral view of right eyeball

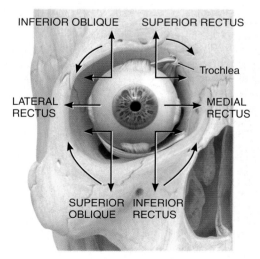

(b) Movements of right eyeball in response to contraction of extrinsic muscles

EXHIBIT 11.B **325**

MUSCLE	ORIGIN	INSERTION	ACTION	INNERVATION
Superior rectus (*rectus*=fascicles parallel to midline)	Common tendinous ring (attached to orbit around optic foramen)	Superior and central part of eyeballs	Moves eyeballs superiorly (elevation) and medially (adduction), and rotates them medially (intorsion)	Oculomotor (III) nerve
Inferior rectus	Same as above	Inferior and central part of eyeballs	Moves eyeballs inferiorly (depression) and medially (adduction), and rotates them laterally (extorsion)	Oculomotor (III) nerve
Lateral rectus	Same as above	Lateral side of eyeballs	Moves eyeballs laterally (abduction)	Abducens (VI) nerve
Medial rectus	Same as above	Medial side of eyeballs	Moves eyeballs medially (adduction)	Oculomotor (III) nerve
Superior oblique (*oblique*=fascicles diagonal to midline)	Sphenoid bone, superior and medial to the common tendinous ring in the orbit	Eyeball between superior and lateral recti. The muscle inserts into the superior and lateral surfaces of the eyeballs via a tendon that passes through the trochlea (a fibrous band on the supero-medial aspect of the orbit)	Moves eyeballs inferiorly (depression) and laterally (abduction), and rotates them medially (intorsion)	Trochlear (IV) nerve
Inferior oblique	Maxilla in floor of orbit	Eyeballs between inferior and lateral recti	Moves eyeballs superiorly (elevation) and laterally (abduction) and rotates them laterally (extorsion)	Oculomotor (III) nerve
Levator palpebrae superioris (le-VĀ-tor PAL-pebrē soo'-per'-ē-OR-is; *palpebrae*=eyelids)	Roof of orbit (lesser wing of sphenoid bone)	Skin and tarsal plate of upper eyelid	Elevates upper eyelids (opens eyes)	Oculomotor (III) nerve

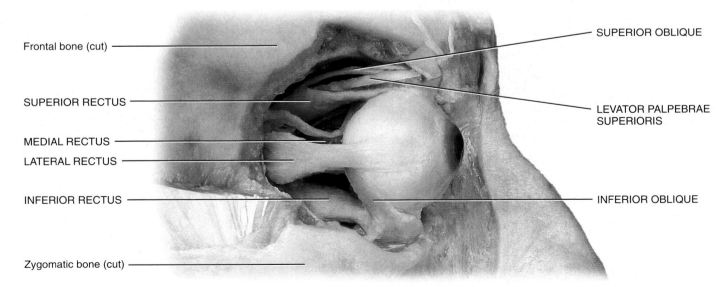

Frontal bone (cut)

SUPERIOR RECTUS

MEDIAL RECTUS

LATERAL RECTUS

INFERIOR RECTUS

Zygomatic bone (cut)

SUPERIOR OBLIQUE

LEVATOR PALPEBRAE SUPERIORIS

INFERIOR OBLIQUE

(c) Right lateral view

 How does the inferior oblique muscle move the eyeball superiorly and laterally?

| EXHIBIT 11.C | Muscles That Move the Mandible and Assist in Mastication and Speech *(Figure 11.6)* |

OBJECTIVE

• Describe the origin, insertion, action, and innervation of the muscles that move the mandible and assist in mastication and speech.

The muscles that move the mandible (lower jaw bone) at the *temporomandibular joint (TMJ)* are known as the muscles of *mastication* (chewing) (Figure 11.6). Of the four pairs of muscles involved in mastication, three are powerful closers of the jaw and account for the strength of the bite: **masseter**, **temporalis**, and **medial pterygoid**. Of these, the masseter is the strongest muscle of mastication. The medial and **lateral pterygoid** muscles assist in mastication by moving the mandible from side to side to help grind food. Additionally, the lateral pterygoids protract the mandible (thrust it forward). The masseter muscle has been removed in Figure 11.6 to illustrate the deeper pterygoid muscles; the masseter can be seen in Figure 11.4c–e. Note the enormous bulk of the temporalis and masseter muscles in Figure 11.4c, d compared to the smaller mass of the two pterygoid muscles.

CLINICAL CONNECTION | *Gravity and the Mandible*

As just noted, three of the four muscles of mastication close the mandible and only the lateral pterygoid opens the mouth. The force of **gravity on the mandible** offsets this imbalance. When the masseter, temporalis, and medial pterygoid muscles relax, the mandible drops. Now you know why the mouths of many persons, particularly the elderly, are open while they are asleep in a chair. In contrast, astronauts in zero gravity must work hard to open their mouths. •

Relating Muscles to Movements

Arrange the muscles in this exhibit according to their actions on the mandible: (1) elevation, (2) depression, (3) retraction, (4) protraction, and (5) side-to-side movement. The same muscle may be mentioned more than once.

CHECKPOINT

9. What would happen if you lost tone in the masseter and temporalis muscles?

Figure 11.6 Muscles that move the mandible and assist in mastication and speech.

🔑 The muscles that move the mandible are also known as muscles of mastication.

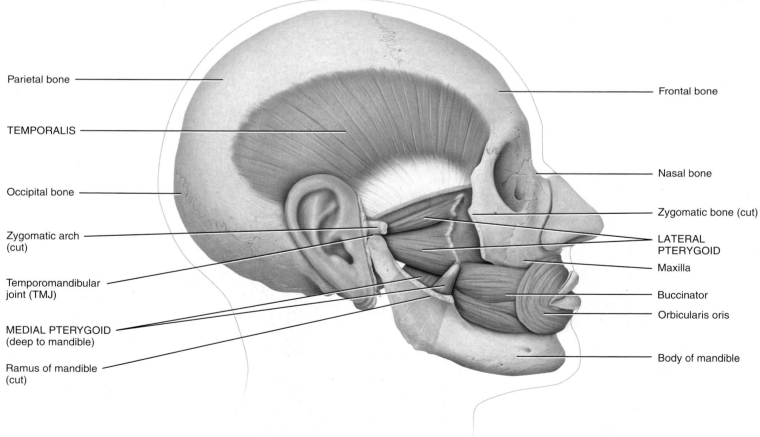

(a) Right lateral superficial view

EXHIBIT 11.C **327**

MUSCLE	ORIGIN	INSERTION	ACTION	INNERVATION
Masseter (MA-se-ter=chewer) (see Figure 11.4b, c)	Maxilla and zygomatic arch	Angle and ramus of mandible	Elevates mandible, as in closing mouth	Mandibular division of trigeminal (V) nerve
Temporalis (tem'-pō-RĀ-lis; *tempor-*=time or temples)	Temporal bone	Coronoid process and ramus of mandible	Elevates and retracts mandible	Mandibular division of trigeminal (V) nerve
Medial pterygoid (TER-i-goyd; *medial*= closer to midline; *pterygoid*=wing-like)	Medial surface of lateral portion of pterygoid process of sphenoid bone; maxilla	Angle and ramus of mandible	Elevates and protracts (protrudes) mandible and moves mandible from side to side	Mandibular division of trigeminal (V) nerve
Lateral pterygoid (TER-i-goyd; *lateral*= farther from midline)	Greater wing and lateral surface of lateral portion of pterygoid process of sphenoid bone	Condyle of mandible; temporomandibular joint (TMJ)	Protracts mandible, depresses mandible as in opening mouth, and moves mandible from side to side	Mandibular division of trigeminal (V) nerve

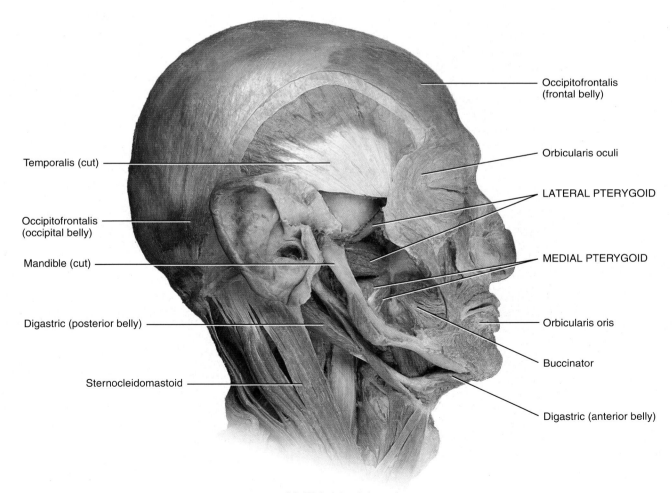

(b) Right lateral deep view

? **Which is the strongest muscle of mastication?**

EXHIBIT 11.D

Muscles of the Head That Move the Tongue and Assist in Mastication and Speech *(Figure 11.7)*

 OBJECTIVE

• Describe the origin, insertion, action, and innervation of the muscles that move the tongue and assist in mastication and speech.

The tongue is a highly mobile structure that is vital to digestive functions such as *mastication* (chewing), detection of taste, and *deglutition* (swallowing). It is also important in speech. The tongue's mobility is greatly aided by its attachment to the mandible, styloid process of the temporal bone, and hyoid bone.

The tongue is divided into lateral halves by a median fibrous septum. The septum extends throughout the length of the tongue. Inferiorly, the septum attaches to the hyoid bone. Muscles of the tongue are of two principal types: extrinsic and intrinsic. **Extrinsic tongue muscles** originate outside the tongue and insert into it (Figure 11.7). They move the entire tongue in various directions, such as anteriorly, posteriorly, and laterally. **Intrinsic tongue muscles** originate and insert within the tongue. These muscles alter the shape of the tongue rather than moving the entire tongue. The extrinsic and intrinsic muscles of the tongue insert into both lateral halves of the tongue.

When you study the extrinsic tongue muscles, you will notice that all of their names end in *glossus*, meaning tongue. You will also notice that the actions of the muscles are obvious, considering the positions of the mandible, styloid process, hyoid bone, and soft palate, which serve as origins for these muscles. For example, the **genioglossus** (origin: the mandible) pulls the tongue downward and forward, the **styloglossus** (origin: the styloid process) pulls the tongue upward and backward, the **hyoglossus** (origin: the hyoid bone) pulls the tongue downward and flattens it, and the **palatoglossus** (origin: the soft palate) raises the back portion of the tongue.

CLINICAL CONNECTION | *Intubation During Anesthesia*

When general anesthesia is administered during surgery, a total relaxation of the muscles results. Once the various types of drugs for anesthesia have been given (especially the paralytic agents), the patient's airway must be protected and the lungs ventilated because the muscles involved with respiration are among those paralyzed. Paralysis of the genioglossus muscle causes the tongue to fall posteriorly, which may obstruct the airway to the lungs. To avoid this, the mandible is either manually thrust forward and held in place (known as the "sniffing position"), or a tube is inserted from the lips through the laryngopharynx (inferior portion of the throat) into the trachea **(endotracheal intubation)**. People can also be intubated nasally (through the nose). •

Relating Muscles to Movements

Arrange the muscles in this exhibit according to the following actions on the tongue: (1) depression, (2) elevation, (3) protraction, and (4) retraction. The same muscle may be mentioned more than once.

 CHECKPOINT

10. When your physician says, "Open your mouth, stick out your tongue, and say *ahh*," to examine the inside of your mouth for possible signs of infection, which muscles do you contract?

MUSCLE	ORIGIN	INSERTION	ACTION	INNERVATION
Genioglossus (jē′-nē-ō-GLOS-us; *genio-*=chin; *glossus*=tongue)	Mandible	Undersurface of tongue and hyoid bone	Depresses tongue and thrusts it anteriorly (protraction)	Hypoglossal (XII) nerve
Styloglossus (stī′-lō-GLOS-us; *stylo*=stake or pole; styloid process of temporal bone)	Styloid process of temporal bone	Side and undersurface of tongue	Elevates tongue and draws it posteriorly (retraction)	Hypoglossal (XII) nerve
Hyoglossus (hī′-ō-GLOS-us; *hyo*=U-shaped)	Greater horn and body of hyoid bone	Side of tongue	Depresses tongue and draws down its sides	Hypoglossal (XII) nerve
Palatoglossus (pal′-a-tō-GLOS-us; *palato-*=roof of mouth or palate)	Anterior surface of soft palate	Side of tongue	Elevates posterior portion of tongue and draws soft palate down on tongue	Pharyngeal plexus, which contains axons from the vagus (X) nerve

EXHIBIT 11.D **329**

Figure 11.7 Muscles of the head that move the tongue and assist in mastication and speech—extrinsic tongue muscles.

The extrinsic and intrinsic muscles of the tongue are arranged in both lateral halves of the tongue.

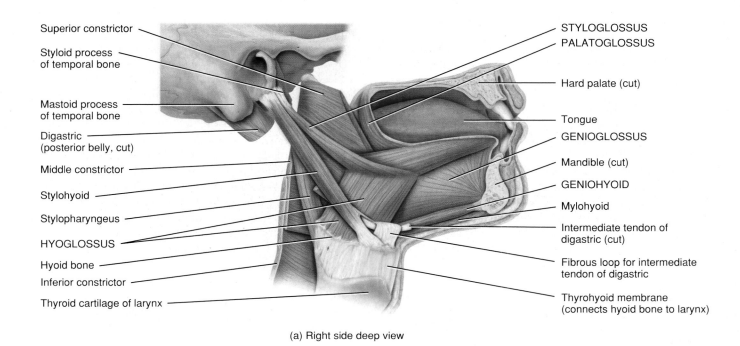

Superior constrictor

Styloid process
of temporal bone

Mastoid process
of temporal bone

Digastric
(posterior belly, cut)

Middle constrictor

Stylohyoid

Stylopharyngeus

HYOGLOSSUS

Hyoid bone

Inferior constrictor

Thyroid cartilage of larynx

STYLOGLOSSUS

PALATOGLOSSUS

Hard palate (cut)

Tongue

GENIOGLOSSUS

Mandible (cut)

GENIOHYOID

Mylohyoid

Intermediate tendon of
digastric (cut)

Fibrous loop for intermediate
tendon of digastric

Thyrohyoid membrane
(connects hyoid bone to larynx)

(a) Right side deep view

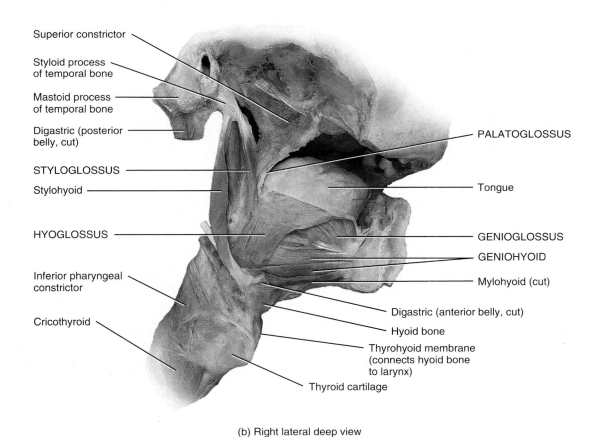

Superior constrictor

Styloid process
of temporal bone

Mastoid process
of temporal bone

Digastric (posterior
belly, cut)

STYLOGLOSSUS

Stylohyoid

HYOGLOSSUS

Inferior pharyngeal
constrictor

Cricothyroid

PALATOGLOSSUS

Tongue

GENIOGLOSSUS

GENIOHYOID

Mylohyoid (cut)

Digastric (anterior belly, cut)

Hyoid bone

Thyrohyoid membrane
(connects hyoid bone
to larynx)

Thyroid cartilage

(b) Right lateral deep view

What are the functions of the tongue?

EXHIBIT 11.E

Muscles of the Anterior Neck That Assist in Deglutition and Speech *(Figure 11.8)*

 OBJECTIVE

- Describe the origin, insertion, action, and innervation of the muscles of the anterior neck that assist in deglutition and speech.

Two groups of muscles are associated with the anterior aspect of the neck: (1) the **suprahyoid muscles,** so called because they are located superior to the hyoid bone, and (2) the **infrahyoid muscles,** named for their position inferior to the hyoid bone (Figure 11.8). Both groups of muscles stabilize the hyoid bone, allowing it to serve as a firm base on which the tongue can move.

As a group, the suprahyoid muscles elevate the hyoid bone, floor of the oral cavity, and tongue during *deglutition* (swallowing). As its name suggests, the **digastric** muscle has two bellies, anterior and posterior, united by an intermediate tendon that is held in position by a fibrous loop. Developmentally, each belly arises from a different pharyngeal arch, which accounts for its dual innervation. This muscle elevates the hyoid bone and larynx (voice box) during swallowing and speech. In a *reverse muscle action (RMA),* when the hyoid is stabilized, the digastric depresses the mandible and is therefore synergistic to the lateral pterygoid in the opening of the mandible. The **stylohyoid** muscle elevates and draws the hyoid bone posteriorly, thus elongating the floor of the oral cavity during swallowing. The **mylohyoid** muscle elevates the hyoid bone and helps press the tongue against the roof of the oral cavity during swallowing to move food from the oral cavity into the throat. The **geniohyoid** muscle (see Figure 11.7) elevates and draws the hyoid bone anteriorly to shorten the floor of the oral cavity and to widen the throat to receive food that is being swallowed. It also depresses the mandible.

MUSCLE	ORIGIN	INSERTION	ACTION	INNERVATION
SUPRAHYOID MUSCLES				
Digastric (dī'-GAS-trik; *di-*=two; *gastr-*=belly)	Anterior belly from inner side of inferior border of mandible; posterior belly from temporal bone	Body of hyoid bone via an intermediate tendon	Elevates hyoid bone RMA: Depresses mandible, as in opening the mouth	Anterior belly: mandibular division of trigeminal (V) nerve Posterior belly: facial (VII) nerve
Stylohyoid (stī'-lō-HĪ-oyd; *stylo-*=stake or pole, styloid process of temporal bone; *hyo-*=U-shaped, pertaining to hyoid bone)	Styloid process of temporal bone	Body of hyoid bone, posteriorly	Elevates hyoid bone and draws it posteriorly	Facial (VII) nerve
Mylohyoid (mī'-lō-HĪ-oyd); *mylo-*=mill)	Inner surface of mandible	Body of hyoid bone	Elevates hyoid bone and floor of mouth and depresses mandible	Mandibular division of trigeminal (V) nerve
Geniohyoid (jē'-nē-ō-HĪ-oyd; *genio-*=chin) (see Figure 11.7)	Inner surface of mandible	Body of hyoid bone	Elevates hyoid bone, draws hyoid bone and tongue anteriorly, and depresses mandible	First cervical spinal nerve (C1)
INFRAHYOID MUSCLES				
Omohyoid (ō-mō-HĪ-oyd; *omo-*=relationship to the shoulder)	Superior border of scapula and superior transverse ligament	Body of hyoid bone	Depresses hyoid bone	Branches of spinal nerves C1–C3
Sternohyoid (ster'-nō-HĪ-oyd; *sterno*=sternum)	Medial end of clavicle and manubrium of sternum	Body of hyoid bone	Depresses hyoid bone	Branches of spinal nerves C1–C3
Sternothyroid (ster'-nō-THĪ-royd; *thyro-*=thyroid gland)	Manubrium of sternum	Thyroid cartilage of larynx	Depresses thyroid cartilage of larynx	Branches of spinal nerves C1–C3
Thyrohyoid (thī'-rō-HĪ-oyd)	Thyroid cartilage of larynx	Greater horn of hyoid bone	Elevates thyroid cartilage RMA: Depresses hyoid bone	Branches of spinal nerves C1–C2 and descending hypoglossal (XII) nerve

EXHIBIT 11.E **331**

Figure 11.8 **Muscles of the anterior neck that assist in deglutition and speech.**

The suprahyoid muscles elevate the hyoid bone, the floor of the oral cavity, and the tongue during swallowing.

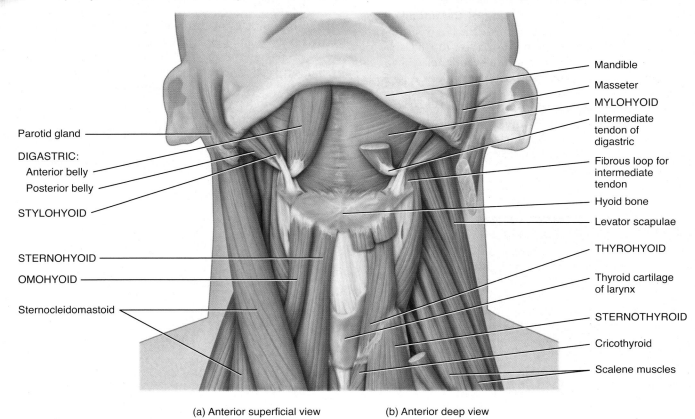

(a) Anterior superficial view   (b) Anterior deep view

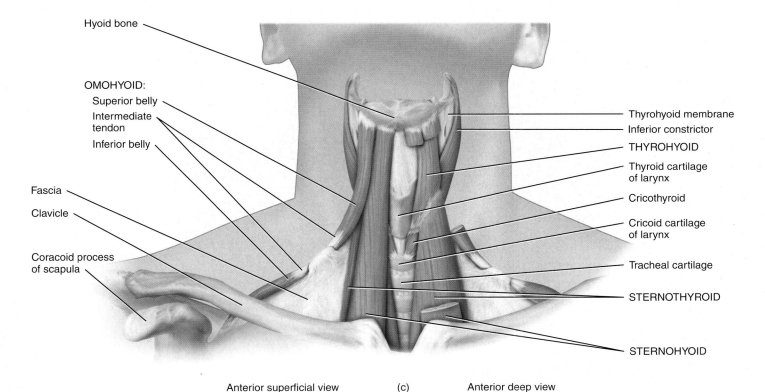

Anterior superficial view   (c)   Anterior deep view

FIGURE 11.8 CONTINUES

EXHIBIT 11.E Muscles of the Anterior Neck That Assist in Deglutition and Speech *(Figure 11.8)* CONTINUED

The infrahyoid muscles are sometimes called "strap" muscles because of their ribbonlike appearance. Most of the infrahyoid muscles depress the hyoid bone and some move the larynx during swallowing and speech. The **omohyoid** muscle, like the digastric muscle, is composed of two bellies connected by an intermediate tendon. In this case, however, the two bellies are referred to as *superior* and *inferior*, rather than anterior and posterior. Together, the omohyoid, **sternohyoid**, and **thyrohyoid** muscles depress the hyoid bone. In addition, the **sternothyroid** muscle depresses the thyroid cartilage (Adam's apple) of the larynx to produce low sounds; the RMA of the thyrohyoid muscle elevates the thyroid cartilage to produce high sounds.

Relating Muscles to Movements

Arrange the muscles in this exhibit according to the following actions on the hyoid bone: (1) elevating it, (2) drawing it anteriorly, (3) drawing it posteriorly, and (4) depressing it; and on the thyroid cartilage: (1) elevating it and (2) depressing it. The same muscle may be mentioned more than once.

✔ CHECKPOINT
11. Which tongue, facial, and mandibular muscles do you use for chewing?
12. Why do the two bellies of the digastric muscle have different innervations?

CLINICAL CONNECTION | *Dysphagia*

Dysphagia (dis-FĀ-jē-a; *dys-*=abnormal; *-phagia*=to eat) is a clinical term for difficulty in swallowing. Some individuals are unable to swallow while others have difficulty swallowing liquids, foods, or saliva. Causes include nervous system disorders that weaken or damage muscles of deglutition (stroke, Parkinson's disease, cerebral palsy); infections; cancer of the head, neck, or esophagus; and injuries to the head, neck, or chest. •

▣ **FIGURE 11.8 CONTINUED** ▶

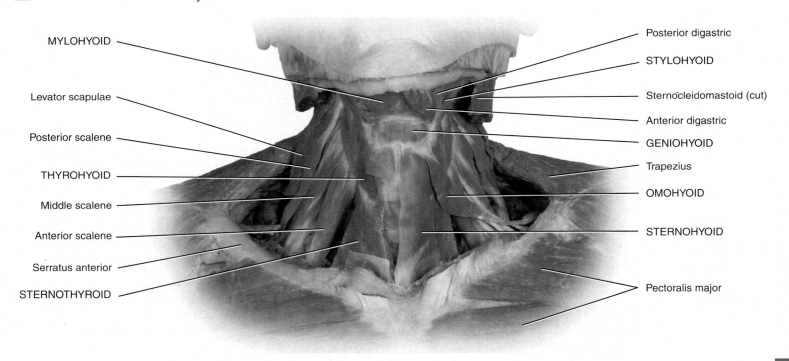

MYLOHYOID

Levator scapulae

Posterior scalene

THYROHYOID

Middle scalene

Anterior scalene

Serratus anterior

STERNOTHYROID

Posterior digastric

STYLOHYOID

Sternocleidomastoid (cut)

Anterior digastric

GENIOHYOID

Trapezius

OMOHYOID

STERNOHYOID

Pectoralis major

Anterior deep view   (d)   Anterior superficial view

 What is the combined action of the suprahyoid and infrahyoid muscles?

EXHIBIT 11.F **333**

EXHIBIT 11.F   Muscles of the Larynx That Assist in Speech   *(Figure 11.9)*

● **OBJECTIVE**
  • Describe the origin, insertion, action, and innervation of the muscles of the larynx that assist in speech.

The muscles of the larynx (voice box), like those of the eyeballs and tongue, are grouped into **extrinsic muscles of the larynx** and **intrinsic muscles of the larynx**. The extrinsic muscles of the larynx, which are associated with the anterior aspect of the neck, are called **infrahyoid muscles** because they lie inferior to the hyoid bone. Refer to Exhibit 11.E and Figures 11.8 and 11.9a and b for descriptions of these muscles. In this exhibit, we discuss the intrinsic muscles of the larynx (Figure 11.9c and d).

The extrinsic muscles move the larynx as a whole; the intrinsic muscles of the larynx move only parts of the larynx. Based on their actions, the intrinsic muscles may be grouped into three functional sets. The first set includes the **cricothyroid** and **thyroarytenoid** muscles, which regulate the tension of the vocal folds (true vocal cords). The second set varies the size of the rima glottidis (space between the vocal folds) which adjusts the tension of the vocal folds. Those muscles include the **lateral cricoarytenoid**, which brings the vocal folds together (adduction), thus closing the rima glottidis, and the **posterior cricoarytenoid**, which moves the vocal folds apart (abduction), thus opening the rima glottidis. The **transverse arytenoid** closes the posterior portion of the rima glottidis. The last intrinsic muscle functions as a sphincter to control the size of the inlet of the larynx, which is the opening anteriorly from the pharynx (throat) into the larynx. This muscle is the **oblique arytenoid**.

Relating Muscles to Movements

Arrange the intrinsic muscles of the larynx in this exhibit according to the following actions on the vocal cords: (1) increasing tension of the vocal cords; (2) moving the vocal cords apart, and (3) moving the vocal cords together. The same muscle may be mentioned more than once.

 CHECKPOINT
  13. How are the intrinsic muscles of the larynx grouped functionally?

MUSCLE	ORIGIN	INSERTION	ACTION	INNERVATION
EXTRINSIC MUSCLES OF THE LARYNX (INFRAHYOID MUSCLES)				
Omohyoid				
Sternohyoid	(See Exhibit 11.E for more details about these muscles)			
Sternothyroid				
Thyrohyoid				
INTRINSIC MUSCLES OF THE LARYNX				
Cricothyroid (kri-kō-THĪ -royd; *crico-*=cricoid cartilage of larynx)	Anterior and lateral portion of cricoid cartilage of larynx	Anterior border of thyroid cartilage of larynx and posterior part of inferior border of thyroid cartilage of larynx	Elongates and places tension on vocal folds	Recurrent laryngeal branch of vagus (X) nerve
Thyroarytenoid (thī-rō-ar'-i-TĒ-noyd; *-arytaina*=shaped like a jug)	Inferior portion of thyroid cartilage of larynx and middle of cricothyroid ligament	Base and anterior surface of arytenoid cartilage of larynx	Shortens and relaxes vocal folds	Recurrent laryngeal branch of vagus (X) nerve
Lateral cricoarytenoid (kri'-kō-ar'-i-TĒ-noyd)	Superior border of cricoid cartilage of larynx	Anterior surface of arytenoid cartilage of larynx	Brings vocal folds together (adduction), thus closing the rima glottidis	Recurrent laryngeal branch of vagus (X) nerve
Posterior cricoarytenoid	Posterior surface of cricoid cartilage of larynx	Posterior surface of arytenoid cartilage of larynx	Moves the vocal folds apart (abduction), thus opening the rima glottidis	Recurrent laryngeal branch of vagus (X) nerve
Transverse arytenoid (ar'-i-TĒ-noyd)	Posterior surface and lateral border of one arytenoid cartilage of larynx	Corresponding parts of opposite arytenoid cartilage of larynx	Closes the posterior portion of the rima glottidis	Recurrent laryngeal branch of vagus (X) nerve
Oblique arytenoid	Posterior surface and lateral border of arytenoid cartilage	Apex of opposite arytenoid cartilage	Regulates the size of the inlet of the larynx	Recurrent laryngeal branch of vagus (X) nerve

EXHIBIT 11.F Muscles of the Larynx That Assist in Speech *(Figure 11.9)* **CONTINUED**

Figure 11.9 Muscles of the larynx that assist in speech.

 Intrinsic muscles of the larynx adjust the tension of the vocal folds and open or close the rima glottidis.

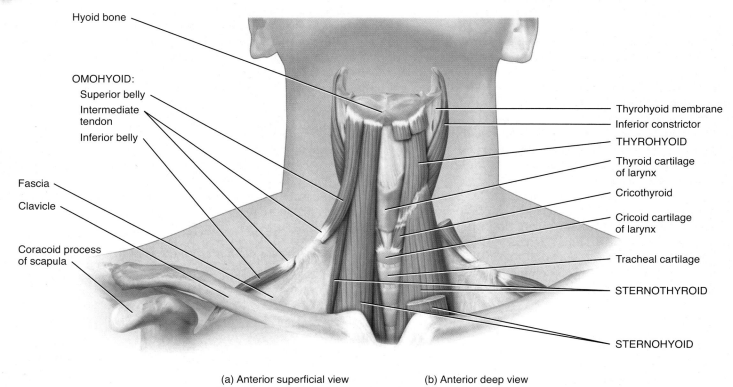

(a) Anterior superficial view   (b) Anterior deep view

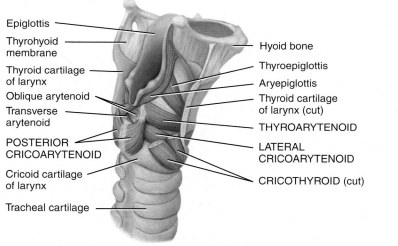

(c) Right posterolateral view

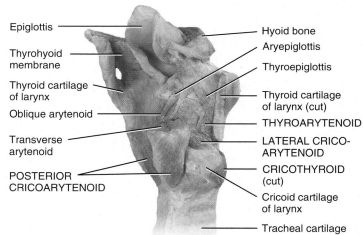

(d) Right posterolateral view

How do the extrinsic and intrinsic muscles of the larynx differ functionally?

EXHIBIT 11.G **335**

EXHIBIT 11.G

Muscles of the Pharynx That Assist in Deglutition and Speech *(Figure 11.10)*

● OBJECTIVE

• Describe the origin, insertion, action, and innervation of the muscles of the pharynx that assist in deglution and speech.

The pharynx (throat) is a somewhat funnel-shaped tube posterior to the nasal and oral cavities. It is a common chamber for the respiratory and digestive systems, opening anteriorly into the larynx and posteriorly into the esophagus.

The muscles of the pharynx are arranged in two layers, an outer circular layer and an inner longitudinal layer (Figure 11.10). The **circular layer** is composed of three constrictor muscles, each overlapping the muscle above it, an arrangement that resembles stacked flowerpots. Their names are the inferior, middle, and superior constrictor muscles. The **longitudinal layer** is composed of three muscles that descend from the styloid process of the temporal bone, auditory (eustachian) tube, and soft palate. Their names are the stylopharyngeus, salpingopharyngeus, and palatopharyngeus, respectively.

The **inferior constrictor** muscle is the thickest of the constrictor muscles. Its inferior fibers are continuous with the musculature of the esophagus, whereas its superior fibers overlap the middle constrictor. The **middle constrictor** is fan-shaped and smaller than the inferior constrictor, and it overlaps the superior constrictor. The **superior constrictor** is quadrilateral and thinner than the other constrictors. As a group, the constrictor muscles constrict the pharynx during deglutition (swallowing). The sequential contraction of these muscles moves food and drink from the mouth into the esophagus.

The **stylopharyngeus** muscle is a long, thin muscle that, as its name suggests, arises from the styloid process and enters the

pharynx between the middle and superior constrictors. Its fibers blend into the constrictors and some insert along with the palatopharyngeus on the thyroid cartilage. The **salpingopharyngeus** muscle is a thin muscle that descends in the lateral wall of the pharynx and also inserts along with the palatopharyngeus muscle. The **palatopharyngeus** muscle also descends in the lateral wall of the pharynx and inserts along with the preceding muscles on the posterior border of the thyroid cartilage. All three muscles of the longitudinal layer elevate the pharynx and larynx during deglutition and speech. Elevation of the pharynx widens it to receive food and liquids, and elevation of the larynx causes a structure called the epiglottis to close over the rima glottidis (space between the vocal folds) and seal the respiratory passageway. Food and drink are further kept out of the respiratory tract by the suprahyoid muscles of the larynx, which elevate the hyoid bone. Additionally, the respiratory passageway is sealed by the action of the intrinsic muscles of the larynx, which bring the vocal folds together to close off the rima glottidis. After deglutition, the infrahyoid muscles of the larynx depress the hyoid bone and larynx.

Relating Muscles to Movements

Arrange the muscles in this exhibit according to the following actions on the pharynx: (1) constriction and (2) elevation, and according to the following action on the larynx: elevation. The same muscle may be mentioned more than once.

 CHECKPOINT

14. What happens when the pharynx and larynx are elevated?

MUSCLE	ORIGIN	INSERTION	ACTION	INNERVATION
CIRCULAR LAYER				
Inferior constrictor (*inferior-*=below; *constrictor*=decreases diameter of a lumen)	Cricoid and thyroid cartilages of larynx	Posterior median raphe (slender band of collagen fibers) of pharynx	Constricts inferior portion of pharynx to propel food and drink into esophagus	Pharyngeal plexus, branches of vagus (X) nerve
Middle constrictor	Greater and lesser horns of hyoid bone and stylohyoid ligament	Posterior median raphe of pharynx	Constricts middle portion of pharynx to propel food and drink into esophagus	Pharyngeal plexus, branches of vagus (X) nerve
Superior constrictor (*superior*=above)	Pterygoid process of sphenoid, pterygomandibular raphe, and medial surface of mandible	Posterior median raphe of pharynx	Constricts superior portion of pharynx to propel food and drink into esophagus	Pharyngeal plexus, branches of vagus (X) nerve
LONGITUDINAL LAYER				
Stylopharyngeus (stī-lō-far-IN-jē-us; *stylo-*=stake or pole; styloid process of temporal bone; *pharyngo-*=pharynx)	Styloid process of temporal bone	Thyroid cartilage with the palatopharyngeus	Elevates larynx and pharynx	Glossopharyngeal (IX) nerve
Salpingopharyngeus (sal-pin'-gō-far-IN-jē-us; *salping-*=horn, trumpet)	Inferior portion of auditory (eustachian) tube	Thyroid cartilage with the palatopharyngeus	Elevates larynx and pharynx and opens orifice of auditory (eustachian) tube	Pharyngeal plexus, branches of vagus (X) nerve
Palatopharyngeus (pal'-a-tō-far-IN-jē-us; *palato-*=palate)	Soft palate	Thyroid cartilage with the stylopharyngeus	Elevates larynx and pharynx and helps close nasopharynx during swallowing	Pharyngeal plexus, branches of vagus (X) nerve

EXHIBIT 11.G Muscles of the Pharynx That Assist in Deglutition and Speech *(Figure 11.10)* **CONTINUED**

Figure 11.10 Muscles of the pharynx that assist in deglutition and speech.

🔑 The muscles of the pharynx assist in swallowing and in closing off the respiratory passageway.

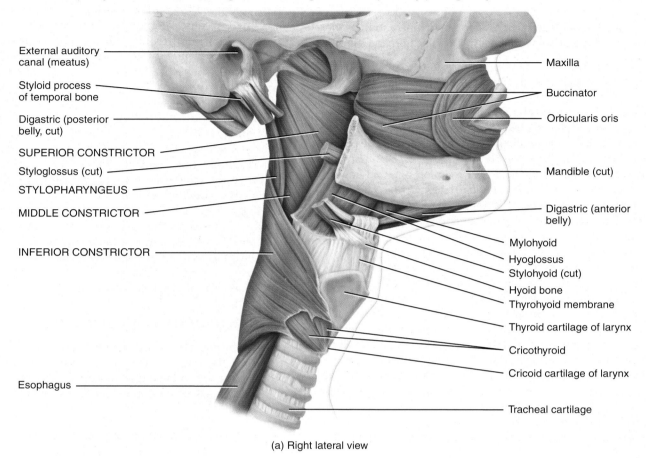

(a) Right lateral view

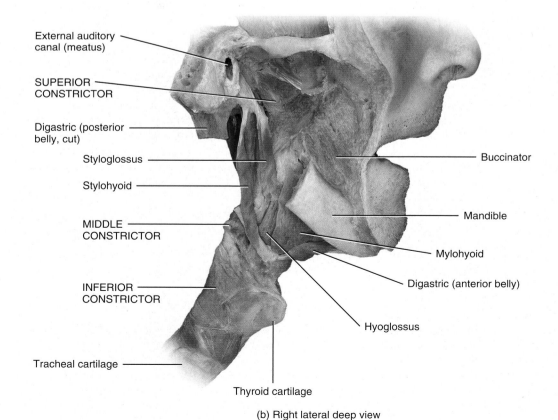

(b) Right lateral deep view

EXHIBIT 11.G **337**

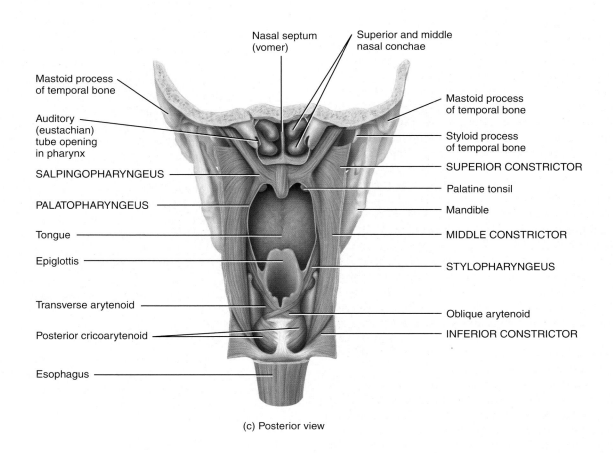

Nasal septum (vomer)

Superior and middle nasal conchae

Mastoid process of temporal bone

Mastoid process of temporal bone

Auditory (eustachian) tube opening in pharynx

Styloid process of temporal bone

SALPINGOPHARYNGEUS

SUPERIOR CONSTRICTOR

Palatine tonsil

PALATOPHARYNGEUS

Mandible

Tongue

MIDDLE CONSTRICTOR

Epiglottis

STYLOPHARYNGEUS

Transverse arytenoid

Oblique arytenoid

Posterior cricoarytenoid

INFERIOR CONSTRICTOR

Esophagus

(c) Posterior view

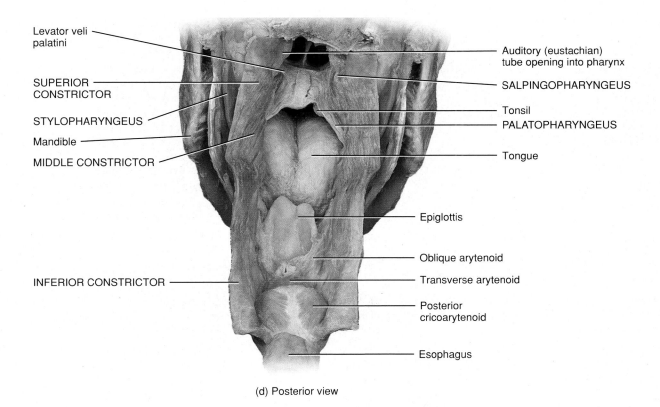

Levator veli palatini

Auditory (eustachian) tube opening into pharynx

SUPERIOR CONSTRICTOR

SALPINGOPHARYNGEUS

Tonsil

STYLOPHARYNGEUS

PALATOPHARYNGEUS

Mandible

Tongue

MIDDLE CONSTRICTOR

Epiglottis

Oblique arytenoid

INFERIOR CONSTRICTOR

Transverse arytenoid

Posterior cricoarytenoid

Esophagus

(d) Posterior view

What are the antagonistic roles of the longitudinal muscles of the pharynx and the infrahyoid muscles?

EXHIBIT 11.H   Muscles of the Neck That Move the Head *(Figure 11.11)*

OBJECTIVE

• Describe the origin, insertion action, and innervation of the muscles that move the head.

The head is attached to the vertebral column at the atlanto-occipital joints formed by the atlas and occipital bone. Balance and movement of the head on the vertebral column involves the action of several neck muscles. For example, acting together (bilaterally), contraction of the two **sternocleidomastoid (SCM)** muscles flexes the cervical portion of the vertebral column and flexes the head. Acting singly (unilaterally), each sternocleidomastoid muscle laterally flexes and rotates the head. Each SCM consists of two bellies (Figure 11.11d); they are more evident near the anterior attachments. The separation of the two bellies is variable and thus more evident in some persons than in others. The two bellies insert as the **sternal head** and the **clavicular head** of the SCM. The bellies also function differently; muscular spasm in the two bellies cause somewhat different symptoms. The large trapezius muscle, which has an important functional role in moving and stabilizing the scapula (see Exhibit 11.N), extends the head. It is assisted by the bilateral contraction of the **spinalis capitis, semispinalis capitis, splenius capitis,** and **longissimus capitis** muscles, which also extend the head (Figure 11.9a). However, when these same muscles contract unilaterally, their actions are quite different, involving primarily rotation of the head.

The sternocleidomastoid muscle is an important landmark that divides the neck into two major triangles: anterior and posterior (Figure 11.11d). The triangles are important anatomically and surgically because of the structures that lie within their boundaries.

The **anterior triangle** is bordered superiorly by the mandible, medially by the cervical midline, and laterally by the anterior border of the sternocleidomastoid muscle. It has its apex at the sternum (Figure 11.11d). The anterior triangle is subdivided into three paired triangles: *submandibular, carotid,* and *muscular.* An unpaired *submental triangle* is formed by the upper part of the combined right and left anterior triangles. The anterior triangle contains submental, submandibular, and deep cervical lymph nodes; the submandibular salivary gland and a portion of the parotid salivary gland; the facial artery and vein; carotid arteries and internal jugular vein; the thyroid gland and infrahyoid muscles; and the following cranial nerves: glossopharyngeal (IX), vagus (X), accessory (XI), and hypoglossal (XII).

The **posterior triangle** is bordered inferiorly by the clavicle, anteriorly by the posterior border of the sternocleidomastoid muscle, and posteriorly by the anterior border of the trapezius muscle (Figure 11.11d). The posterior triangle is subdivided into two triangles, *occipital* and *supraclavicular (omoclavicular),* by the inferior belly of the omohyoid muscle. The posterior triangle contains part of the subclavian artery, external jugular vein, cervical lymph nodes, brachial plexus, and the accessory (XI) nerve.

Relating Muscles to Movements

Arrange the muscles in this exhibit according to the following actions on the head: (1) flexion, (2) lateral flexion, (3) extension, (4) rotation to side opposite contracting muscle, and (5) rotation to same side as contracting muscle. The same muscle may be mentioned more than once.

 CHECKPOINT

15. What muscles do you contract to signify "yes" and "no"?

MUSCLE	ORIGIN	INSERTION	ACTION	INNERVATION
Trapezius	(see Exhibit 11.N for more details about this muscle)		Extends the head	
Sternocleidomastoid (ster'-nō-klī'-dō-MAS-toid; *sterno-*=breastbone; *cleido-*=clavicle; *mastoid*=mastoid process of temporal bone)	Sternal head: manubrium of sternum Clavicular head: medial third of clavicle	Mastoid process of temporal bone and lateral half of superior nuchal line of occipital bone	Acting together (bilaterally), flex cervical portion of vertebral column; flex head at atlanto-occipital joints; acting singly (unilaterally), laterally rotate and flex head to opposite side of contracting muscle. The posterior fibers of the muscle can assist in extension of head RMA: Elevate the sternum during forced inhalation	Motor supply: Accessory (XI) nerve Sensory supply: C2 and C3
Semispinalis capitis (se'-mē-spi-NĀ-lis KAP-i-tis; *semi-*=half; *spine*=spinous process; *capit-*=head)	Articular processes of C4–C6 and transverse processes of C7–T7	Occipital bone between superior and inferior nuchal lines	Acting together, extend head; acting singly, rotate head to side opposite contracting muscle	Cervical spinal nerves
Splenius capitis (SPLE-nē-us KAP-i-tis; *splenium-*=bandage)	Ligamentum nuchae and spinous processes of C7–T4	Occipital bone and mastoid process of temporal bone	Acting together, extend head; acting singly, rotate head to same side as contracting muscle	Cervical spinal nerves
Longissimus capitis (lon-JIS-i-mus KAP-i-tis; *longissimus*=longest)	Articular processes of C4–C7 and transverse processes of T1–T4	Mastoid process of temporal bone	Acting together, extend head; acting singly, laterally flex and rotate head to same side as contracting muscle	Cervical spinal nerves
Spinalis capitis (spi-NĀ-lis KAP-i-tis; *spinal*=vertebral column)	Often absent or very small; arises with semispinalis capitis	Occipital bone	Extends the head	Cervical spinal nerves

EXHIBIT 11.H **339**

Figure 11.11 **Muscles of the neck that move the head.**

The sternocleidomastoid muscle divides the neck into two principal triangles: anterior and posterior.

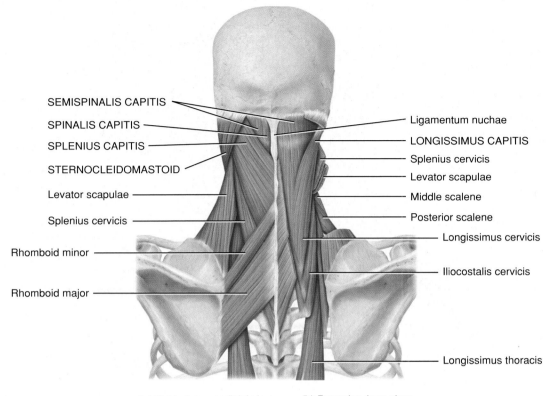

SEMISPINALIS CAPITIS
SPINALIS CAPITIS
SPLENIUS CAPITIS
STERNOCLEIDOMASTOID
Levator scapulae
Splenius cervicis
Rhomboid minor
Rhomboid major

Ligamentum nuchae
LONGISSIMUS CAPITIS
Splenius cervicis
Levator scapulae
Middle scalene
Posterior scalene
Longissimus cervicis
Iliocostalis cervicis
Longissimus thoracis

(a) Posterior superficial view   (b) Posterior deep view

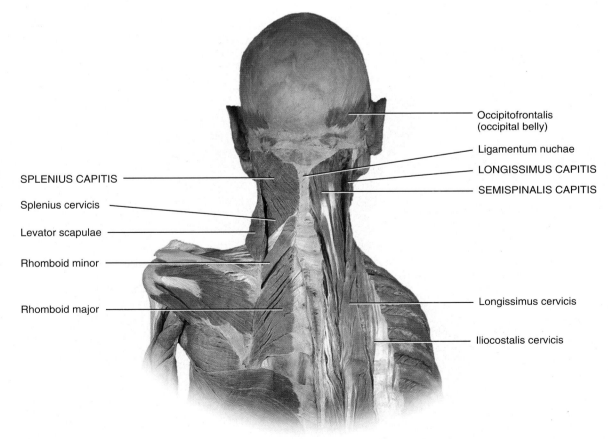

SPLENIUS CAPITIS
Splenius cervicis
Levator scapulae
Rhomboid minor
Rhomboid major

Occipitofrontalis
(occipital belly)
Ligamentum nuchae
LONGISSIMUS CAPITIS
SEMISPINALIS CAPITIS
Longissimus cervicis
Iliocostalis cervicis

Posterior superficial view   (c)   Posterior deep view

FIGURE 11.11 CONTINUES

EXHIBIT 11.H Muscles of the Neck That Move the Head *(Figure 11.11)* CONTINUED

FIGURE **11.11** CONTINUED ▶

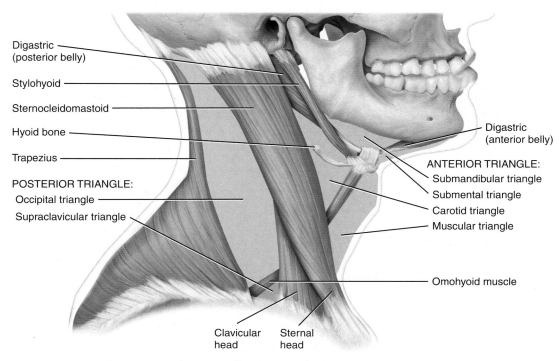

(d) Right lateral view of triangles of neck

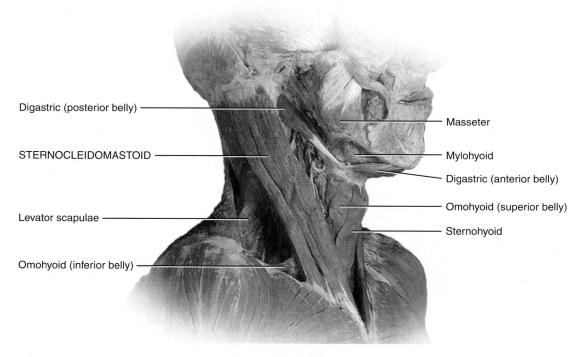

(e) Right lateral view of neck

? **Why are triangles of the neck important?**

EXHIBIT 11.I   **341**

EXHIBIT 11.I

Muscles of the Neck and Back That Move the Vertebral Column *(Figure 11.12)*

OBJECTIVE

• Describe the origin, insertion, action, and innervation of the muscles that move the vertebral column.

The muscles that move the vertebral column (backbone) are quite complex because they have multiple origins and insertions and there is considerable overlap among them. One way to group the muscles is on the basis of the general direction of the muscle bundles and their approximate lengths. For example, the splenius muscles arise from the midline and extend laterally and superiorly to their insertions (Figure 11.12a). The erector spinae muscle group (consisting of the iliocostalis, longissimus, and spinalis muscles) arises either from the midline or more laterally but usually runs almost longitudinally, with neither a significant lateral nor medial direction as it is traced superiorly. The muscles of the transversospinalis group (semispinalis, multifidus, rotatores) arise laterally but extend toward the midline as they are traced superiorly. Deep to these three muscle groups are small segmental muscles that extend between spinous processes or transverse processes of vertebrae. Note that the rectus abdominis, external oblique, internal oblique, and quadratus lumborum muscles also play a role in moving the vertebral column (see Exhibit 11.J).

The bandage-like **splenius** muscles are attached to the sides and back of the neck. The two muscles in this group are named on the basis of their superior attachments (insertions): *splenius capitis* (head region) and **splenius cervicis** (cervical region). They extend the head and neck and laterally flex the neck and rotate the head.

The **erector spinae** is the largest muscle mass of the back, forming a prominent bulge on either side of the vertebral column. It is the chief extensor of the vertebral column. It is also important in controlling flexion, lateral flexion, and rotation of the vertebral column and in maintaining the lumbar curve. As noted above, it consists of three groups: iliocostalis (laterally placed), longissimus (intermediately placed), and spinalis (medially placed). These groups in turn consist of a series of overlapping muscles, and the muscles within the groups are named according to the regions of the body with which they are associated. The **iliocostalis group** consists of three muscles: the **iliocostalis cervicis** (cervical region), **iliocostalis thoracis** (thoracic region), and **iliocostalis lumborum** (lumbar region). The **longissimus group** resembles a herringbone and consists of three muscles: the *longissimus capitis* (head region), **longissimus cervicis** (cervical region), and *longissimus thoracis* (thoracic region). The **spinalis group** also consists of three muscles: the **spinalis capitis**, **spinalis cervicis**, and **spinalis thoracis**.

The **transversospinales** are so named because their fibers run from the transverse processes to the spinous processes of the vertebrae. The semispinalis muscles in this group are also named according to the region of the body with which they are associated: *semispinalis capitis* (head region), **semispinalis cervicis** (cervical region), and **semispinalis thoracis** (thoracic region). These muscles extend the vertebral column and rotate the head. The **multifidus** muscle in this group, as its name implies, is segmented into several bundles. It extends and laterally flexes the vertebral column. This muscle is large and thick in the lumbar region and is important in maintaining the lumbar curve. The **rotatores** mus-

cles of this group are short and are found along the entire length of the vertebral column. These small muscles contribute little to vertebral movement, but play important roles in monitoring the position of the vertebral column and providing proprioceptive feedback to the stronger vertebral muscles.

Within the **segmental group** (Figure 11.12d), the **interspinales** and **intertransversarii** muscles unite the spinous and transverse processes of consecutive vertebrae. They function primarily in stabilizing the vertebral column during its movements, and providing proprioceptive feedback.

Within the **scalene group** (Figure 11.12e), the **anterior scalene** muscle is anterior to the middle scalene muscle, the **middle scalene** muscle is intermediate in placement and is the longest and largest of the scalene muscles, and the **posterior scalene** muscle is posterior to the middle scalene muscle and is the smallest of the scalene muscles. These muscles flex, laterally flex, and rotate the head and assist in deep inhalation.

CLINICAL CONNECTION | *Back Injuries and Heavy Lifting*

The four factors associated with increased risk of **back injury** are amount of force, repetition, posture, and stress applied to the backbone. Poor physical condition, poor posture, lack of exercise, and excessive body weight contribute to the number and severity of sprains and strains. Back pain caused by a muscle strain or ligament sprain will normally heal within a short time and may never cause further problems. However, if ligaments and muscles are weak, discs in the lower back can become weakened and may herniate (rupture) with excessive lifting or a sudden fall, causing considerable pain.

Full flexion at the waist, as in touching your toes, overstretches the erector spinae muscles. Muscles that are overstretched cannot contract effectively. Straightening up from such a position is therefore initiated by the hamstring muscles on the back of the thigh and the gluteus maximus muscles of the buttocks. The erector spinae muscles join in as the degree of flexion decreases. Improperly lifting a heavy weight, however, can strain the erector spinae muscles. The result can be painful muscle spasms, tearing of tendons and ligaments of the lower back, and herniating of intervertebral discs. The lumbar muscles are adapted for maintaining posture, not for lifting. This is why it is important to bend at the knees and use the powerful extensor muscles of the thighs and buttocks while lifting a heavy load. •

Relating Muscles to Movements

Arrange the muscles in this exhibit according to the following actions on the head at the atlanto-occipital and intervertebral joints: (1) extension, (2) lateral flexion, (3) rotation to same side as contracting muscle, and (4) rotation to opposite side as contracting muscle; and arrange the muscles according to the following actions on the vertebral column at the intervertebral joints: (1) flexion, (2) extension, (3) lateral flexion, (4) rotation, and (5) stabilization. The same muscle may be mentioned more than once.

CHECKPOINT

16. What is the largest muscle group of the back?

EXHIBIT 11.I Muscles of the Neck and Back That Move the Vertebral Column *(Figure 11.12)* CONTINUED

MUSCLE	ORIGIN	INSERTION	ACTION	INNERVATION
SPLENIUS				
Splenius capitis (SPLĒ -ne-us KAP-i-tis; *splenium*=bandage; *capit-*=head)	Ligamentum nuchae and spinous processes of C7–T4	Occipital bone and mastoid process of temporal bone	Acting together (bilaterally), extend head; acting singly (unilaterally), laterally flex and/or rotate head to same side as contracting muscle, and extend vertebral column	Posterior rami of middle cervical spinal nerves
Splenius cervicis (SPLĒ -ne-us SER-vi-sis; *cervic-*=neck)	Spinous processes of T3–T6	Transverse processes of C1–C2 or C1–C4	Acting together, extend vertebral column; acting singly, laterally flex and/or rotate vertebral column to same side as contracting muscle	Posterior rami of inferior cervical spinal nerves
ERECTOR SPINAE (e-REK-tor SPI-nē) Consists of Iliocostalis muscles (lateral), longissimus muscles (intermediate), and spinalis muscles (medial).				
Iliocostalis Group (Lateral)				
Iliocostalis cervicis (il'-ē-ō-kos-TĀL-is SER-vi-sis; *ilio-*=flank; *costa-*=rib)	Ribs 1–6	Transverse processes of C4–C6	Acting together, muscles of each region (cervical, thoracic, and lumbar) extend and maintain erect posture of vertebral column of their respective regions; acting singly, laterally flex vertebral column of their respective regions to the same side as the contracting muscle	Posterior rami of cervical and thoracic spinal nerves
Iliocostalis thoracis (il'-ē-ō-kos-TĀL-is thō-RĀ-sis; *thorac-*=chest)	Ribs 7–12	Ribs 1–6		Posterior rami of thoracic spinal nerves
Iliocostalis lumborum (il'-ē-ō-kos-TĀL-is lum-BOR-um)	Iliac crest	Ribs 7–12		Posterior rami of lumbar spinal nerves
LONGISSIMUS GROUP (Intermediate)				
Longissimus capitis (lon-JIS-i-mus KAP-i-tis; *longissimus*=longest)	Articular processes of C4–C7 and transverse processes of T1–T4	Mastoid process of temporal bone	Acting together, both longissimus capitis muscles extend head; acting singly, rotate head to same side as contracting muscle and extend vertebral column	Posterior rami of middle and inferior cervical spinal nerves
Longissimus cervicis (lon-JIS-i-mus SER-vi-sis)	Transverse processes of T4–T5	Transverse processes of C2–C6	Acting together, longissimus cervicis and both longissimus thoracis muscles extend vertebral column of their respective regions; acting singly, laterally flex vertebral column of their respective regions	Posterior rami of cervical and superior thoracic spinal nerves
Longissimus thoracis (lon-JIS-i-mus thō-RĀ-sis)	Transverse processes of lumbar vertebrae	Transverse processes of all thoracic and superior lumbar vertebrae and ribs 9 and 10		Posterior rami of thoracic and lumbar spinal nerves
SPINALIS GROUP (Medial)				
Spinalis capitis (spi-NĀ-lis KAP-i-tis; *spinal-*=vertebral column)	Often absent or very small; arises with semispinalis capitis	Occipital bone	Extend head	Posterior rami of cervical spinal nerves
Spinalis cervicis (spi-NĀ-lis SER-vi-sis)	Ligamentum nuchae and spinous process of C7	Spinous process of axis	Acting together, muscles of each region (cervical and thoracic) extend vertebral column of their respective regions	Posterior rami of inferior cervical and thoracic spinal nerves
Spinalis thoracis (spi-NĀ-lis thō-RĀ-sis)	Spinous processes of T10–L2	Spinous processes of superior thoracic vertebrae		Posterior rami of thoracic spinal nerves

EXHIBIT 11.I **343**

MUSCLE	ORIGIN	INSERTION	ACTION	INNERVATION
TRANSVERSOSPINALES (trans-ver-sō-spī-NĀ-lēz)				
Semispinalis capitis (sem'-ē-spi-NĀ-lis KAP-i-tis; *semi-*= partially or one half)	Articular processes of C4–C6 and transverse processes of C7–T7	Occipital bone	Acting together, extend head; acting singly, rotate head to side opposite contracting muscle and extend vertebral column	Posterior rami of cervical and thoracic spinal nerves
Semispinalis cervicis (sem'-ē-spi-NĀ-lis SER-vi-sis)	Transverse processes of T1–T5	Spinous processes of C1–C5	Acting together, both semispinalis cervicis and both semispinalis thoracis muscles extend vertebral column of their respective regions; acting singly, rotate vertebral column to side opposite contracting muscle	Posterior rami of cervical and thoracic spinal nerves
Semispinalis thoracis (sem'-e-spi-NĀ-lis thō-RĀ-sis)	Transverse processes of T6–T10	Spinous processes of C6–T4		Posterior rami of thoracic spinal nerves
Multifidus (mul-TIF-i-dus; *multi*=many; *fid-*=segmented)	Sacrum, ilium, transverse processes of L1–L5, T1–T12, and C4–C7	Spinous process of a more superior vertebra	Acting together, extend vertebral column; acting singly, weakly laterally flex vertebral column and weakly rotate vertebral column to side opposite contracting muscle	Posterior rami of cervical, thoracic, and lumbar spinal nerves
Rotatores (rō'-ta-TO-rez; singular is rotatore; *rotatore*=to rotate)	Transverse processes of all vertebrae	Spinous process of vertebra superior to the one of origin	Acting together, weakly extend vertebral column; acting singly, weakly rotate vertebral column to side opposite contracting muscle	Posterior rami of cervical, thoracic, and lumbar spinal nerves
SEGMENTAL (seg-MEN-tal)				
Interspinales (in-ter-spi-NĀ-lēz; *inter-*=between)	Superior surface of all spinous processes	Inferior surface of spinous process of vertebra superior to the one of origin	Acting together, weakly extend vertebral column; acting singly, stabilize vertebral column during movement	Posterior rami of cervical, thoracic, and lumbar spinal nerves
Intertransversarii (in'-ter-trans-vers-AR-ē-ī; singular is intertransversarius)	Transverse processes of all vertebrae	Transverse process of vertebra superior to the one of origin	Acting together, weakly extend vertebral column; acting singly, weakly laterally flex vertebral column and stabilize it during movements	Posterior rami of cervical, thoracic, and lumbar spinal nerves
SCALENES (SKĀ-lēnz)				
Anterior scalene (SKĀ-lēn; *anterior*=front; *scalene*=uneven)	Transverse processes of C3–C6	First rib	Acting together, right and left anterior scalene and middle scalene muscles elevate ribs during deep inhalation	Anterior rami of cervical spinal nerves
Middle scalene	Transverse processes of C2–C7	First rib	RMA: Flex cervical vertebrae; acting singly, laterally flex and slightly rotate cervical vertebrae	Anterior rami of cervical spinal nerves
Posterior scalene	Transverse processes of C4–C6	Second rib	Acting together, right and left posterior scalenes elevate second ribs during deep inhalation RMA: Flex cervical vertebrae and acting singly, laterally flex and slightly rotate cervical vertebrae	Anterior rami of cervical spinal nerves

EXHIBIT 11.I Muscles of the Neck and Back That Move the Vertebral Column *(Figure 11.12)* CONTINUED

Figure 11.12 **Muscles of the neck and back that move the vertebral column.** The trapezius and occipitofrontalis muscles have been removed.

🔑 The erector spinae group (iliocostalis, longissimus, and spinalis muscles) is the largest muscular mass of the body and is the chief extensor of the vertebral column.

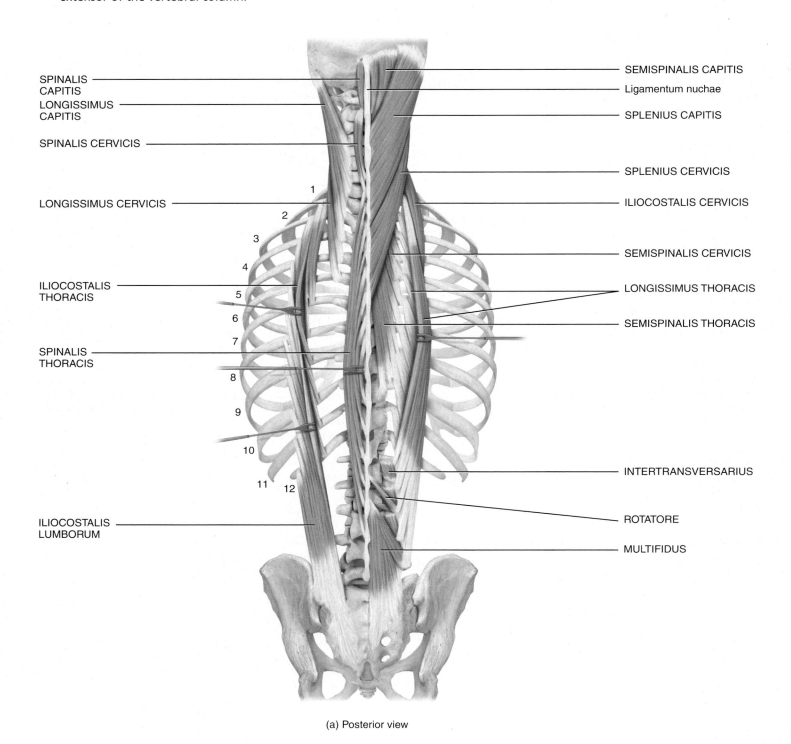

SPINALIS CAPITIS
LONGISSIMUS CAPITIS
SPINALIS CERVICIS
LONGISSIMUS CERVICIS
ILIOCOSTALIS THORACIS
SPINALIS THORACIS
ILIOCOSTALIS LUMBORUM

SEMISPINALIS CAPITIS
Ligamentum nuchae
SPLENIUS CAPITIS
SPLENIUS CERVICIS
ILIOCOSTALIS CERVICIS
SEMISPINALIS CERVICIS
LONGISSIMUS THORACIS
SEMISPINALIS THORACIS
INTERTRANSVERSARIUS
ROTATORE
MULTIFIDUS

(a) Posterior view

EXHIBIT 11.1 **345**

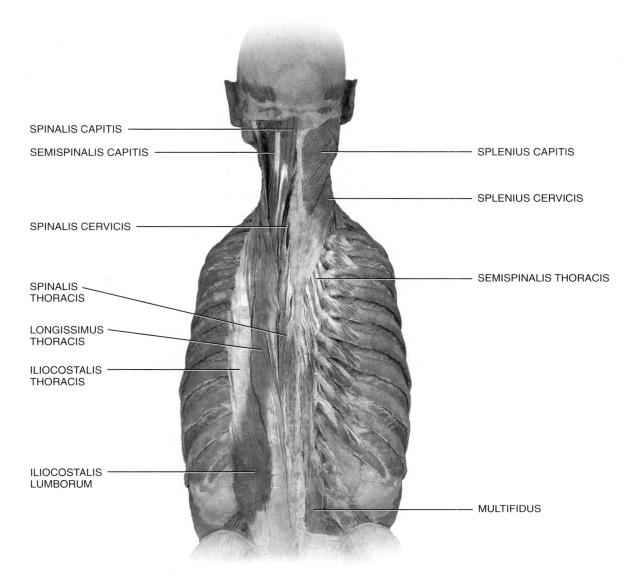

SPINALIS CAPITIS

SEMISPINALIS CAPITIS

SPLENIUS CAPITIS

SPLENIUS CERVICIS

SPINALIS CERVICIS

SEMISPINALIS THORACIS

SPINALIS THORACIS

LONGISSIMUS THORACIS

ILIOCOSTALIS THORACIS

ILIOCOSTALIS LUMBORUM

MULTIFIDUS

(b) Posterior view

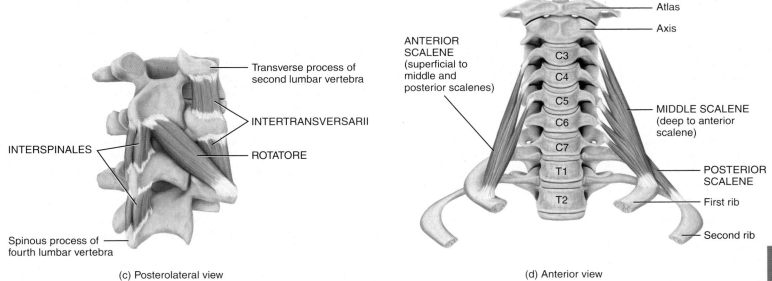

Transverse process of second lumbar vertebra

INTERTRANSVERSARII

INTERSPINALES

ROTATORE

Spinous process of fourth lumbar vertebra

(c) Posterolateral view

Atlas

Axis

ANTERIOR SCALENE (superficial to middle and posterior scalenes)

C3

C4

C5

C6

C7

T1

T2

MIDDLE SCALENE (deep to anterior scalene)

POSTERIOR SCALENE

First rib

Second rib

(d) Anterior view

Which muscles originate at the midline and extend laterally and superiorly to their insertions?

EXHIBIT 11.J

Muscles of the Abdomen That Protect Abdominal Viscera and Move the Vertebral Column *(Figure 11.13)*

 OBJECTIVE

• Describe the origin, insertion, action, and innervation of the muscles that protect abdominal viscera and move the vertebral column.

The anterolateral abdominal wall is composed of skin, fascia, and four pairs of muscles: the external oblique, internal oblique, transversus abdominis, and rectus abdominis (Figure 11.13). The first three muscles named are arranged from superficial to deep.

The **external oblique** is the superficial muscle. Its fascicles extend inferiorly and medially. The **internal oblique** is the intermediate flat muscle. Its fascicles extend at right angles to those of the external oblique. The **transversus abdominis** is the deep muscle, with most of its fascicles directed transversely around the abdominal wall. Together, the external oblique, internal oblique, and transversus abdominis form three layers of muscle around the abdomen. In each layer, the muscle fascicles extend in a different direction. This is a structural arrangement that affords considerable protection to the abdominal viscera, especially when the muscles have good tone.

The **rectus abdominis** muscle is a long muscle that extends the entire length of the anterior abdominal wall, originating at the pubic crest and pubic symphysis and inserting on the cartilages of ribs 5–7 and the xiphoid process of the sternum. The anterior surface of the muscle is interrupted by three transverse fibrous bands of tissue called **tendinous intersections**, believed to be remnants of septa that separated myotomes during embryological development (see Figure 10.11). There are usually three tendinous intersections, one at the level of the umbilicus, one near the xiphoid process, and one midway between the other two. A fourth intersection is sometimes found below the level of the umbilicus. These tendinous intersections are fused with the anterior wall of the rectus sheath but have no connections to the posterior wall of the sheath. Muscular persons may possess easily demonstrated intersections as the result of exercise and the ensuing hypertrophy of the rectus muscle. Hypertrophy of the muscle tissue, of course, has no effect on the connective tissue of the intersections. Body builders focus on the development of the **"six-pack"** effect of the abdomen. Small percentages of the population have a variant of the intersections and are able to develop an "eight-pack."

MUSCLE	ORIGIN	INSERTION	ACTION	INNERVATION
Rectus abdominis (REK-tus ab-DOM-in-is; *rectus-*=fascicles parallel to midline; *abdomin*=abdomen)	Pubic crest and pubic symphysis	Cartilage of ribs 5–7 and xiphoid process	Flexes vertebral column (especially lumbar portion), and compresses abdomen to aid in defecation, urination, forced exhalation, and childbirth RMA: Flexes pelvis on the vertebral column	Anterior rami of thoracic spinal nerves T7–T12
External oblique (ō-BLĒK; *external*=closer to surface; *oblique*=fascicles diagonal to midline)	Ribs 5–12	Iliac crest and linea alba	Acting together (bilaterally), compress abdomen and flex vertebral column; acting singly (unilaterally), laterally flex vertebral column, especially lumbar portion, and rotate vertebral column	Anterior rami of thoracic spinal nerves T7–T12 and the iliohypogastric nerve
Internal oblique (*internal*=farther from surface)	Iliac crest, inguinal ligament, and thoracolumbar fascia	Cartilage of ribs 7–10 and linea alba	Acting together, compress abdomen and flex vertebral column; acting singly, laterally flex vertebral column, especially lumbar portion, and rotate vertebral column	Anterior rami of thoracic spinal nerves T8–T12, iliohypogastric nerve, and ilioinguinal nerve
Transversus abdominis (tranz-VER-sus; *transverse*=fascicles perpendicular to midline)	Iliac crest, inguinal ligament, lumbar fascia, and cartilages of ribs 5–10	Xiphoid process, linea alba, and pubis	Compresses abdomen	Anterior rami of thoracic spinal nerves T8–T12, iliohypogastric nerve, and ilioinguinal nerve
Quadratus lumborum (kwod-RĀ-tus lum-BOR-um; *quad-*=four; *lumbo-*=lumbar region) (see Figure 11.14)	Iliac crest and iliolumbar ligament	Inferior border of rib 12 and L1–L4	Acting together, pull twelfth ribs interiorly during forced exhalation, fix twelfth ribs to prevent their elevation during deep inhalation, and help extend lumbar portion of vertebral column; acting singly, laterally flex vertebral column, especially lumbar portion RMA: Elevates hip bone, commonly on one side	Anterior rami of thoracic spinal nerve T12 and lumbar spinal nerves L1–L3 or L1–L4

EXHIBIT 11.J **347**

As a group, the muscles of the anterolateral abdominal wall help contain and protect the abdominal viscera; flex, laterally flex, and rotate the vertebral column (backbone) at the intervertebral joints; compress the abdomen during forced exhalation; and produce the force required for defecation, urination, and childbirth.

The aponeuroses (sheathlike tendons) of the external oblique, internal oblique, and transversus abdominis muscles form the **rectus sheaths**, which enclose the rectus abdominis muscles. The sheaths meet at the midline to form the **linea alba** (=white line), a tough, fibrous band that extends from the xiphoid process of the sternum to the pubic symphysis. In the latter stages of pregnancy, the linea alba stretches to increase the distance between the rectus abdominis muscles. The inferior free border of the external oblique aponeurosis forms the **inguinal ligament**, which runs from the anterior superior iliac spine to the pubic tubercle (Figure 11.23a, b, and c). Just superior to the medial end of the inguinal ligament is a triangular slit in the aponeurosis referred to as the **superficial inguinal ring**, the outer opening of the inguinal canal (see Figure 26.2). The **inguinal canal** contains the spermatic cord and ilioinguinal nerve in males, and the round ligament of the uterus and ilioinguinal nerve in females.

The posterior abdominal wall is formed by the lumbar vertebrae, parts of the ilia of the hip bones, psoas major and iliacus muscles (described in Exhibit 11.S), and quadratus lumborum muscle. The anterolateral abdominal wall can contract and distend; the posterior abdominal wall is bulky and stable by comparison.

CLINICAL CONNECTION | *Inguinal Hernia*

A **hernia** (HER-nē-a) is a protrusion of an organ through a structure that normally contains it, which creates a lump that can be seen or felt through the skin's surface. The inguinal region is a weak area in the abdominal wall. It is often the site of an **inguinal hernia**, a rupture or separation of a portion of the inguinal area of the abdominal wall resulting in the protrusion of a part of the small intestine. A hernia is much more common in males than in females because the inguinal canals in males are larger to accommodate the spermatic cord and ilioinguinal nerve. Treatment of hernias most often involves surgery. The organ that protrudes is "tucked" back into the abdominal cavity and the defect in the abdominal muscles is repaired. In addition, a mesh is often applied to reinforce the area of weakness. •

Relating Muscles to Movements

Arrange the muscles in this exhibit according to the following actions on the vertebral column: (1) flexion, (2) lateral flexion, (3) extension, and (4) rotation. The same muscle may be mentioned more than once.

✓ CHECKPOINT

17. Which muscles do you contract when you "suck in your gut," thereby compressing the anterior abdominal wall?

Figure 11.13 Muscles of the abdomen that protect abdominal viscera and move the vertebral column.

🔑 The anterolateral abdominal muscles protect the abdominal viscera, move the vertebral column, and assist in forced exhalation, defecation, urination, and childbirth.

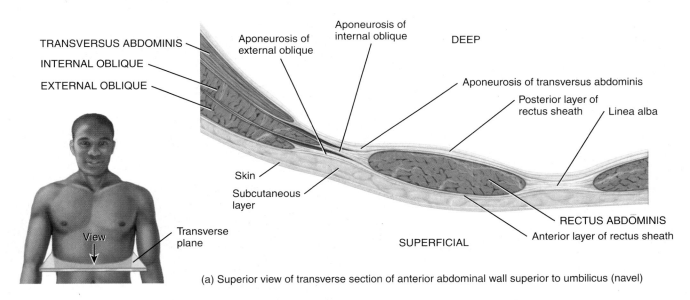

(a) Superior view of transverse section of anterior abdominal wall superior to umbilicus (navel)

FIGURE 11.13 CONTINUES ▶

EXHIBIT 11.J Muscles of the Abdomen That Protect Abdominal Viscera and Move the Vertebral Column *(Figure 11.13)* CONTINUED

FIGURE **11.13** CONTINUED

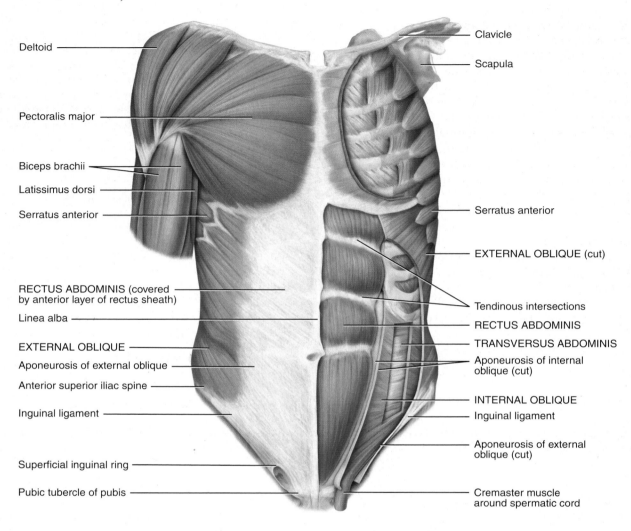

(b) Anterior superficial view   (c) Anterior deep view

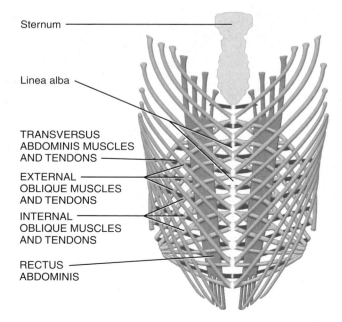

(d) Schematic of anterolateral abdominal wall

EXHIBIT 11.J **349**

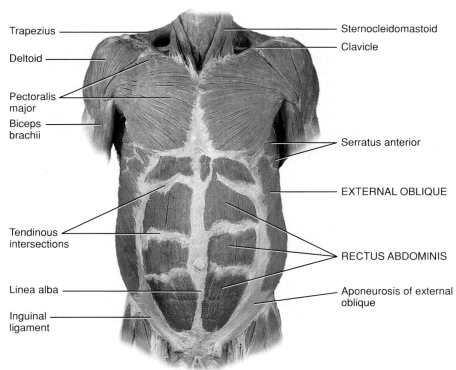

Trapezius

Deltoid

Pectoralis major

Biceps brachii

Tendinous intersections

Linea alba

Inguinal ligament

Sternocleidomastoid

Clavicle

Serratus anterior

EXTERNAL OBLIQUE

RECTUS ABDOMINIS

Aponeurosis of external oblique

(e) Anterior view

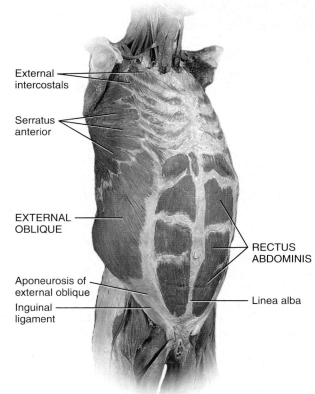

External intercostals

Serratus anterior

EXTERNAL OBLIQUE

Aponeurosis of external oblique

Inguinal ligament

RECTUS ABDOMINIS

Linea alba

(f) Right anterolateral superficial view

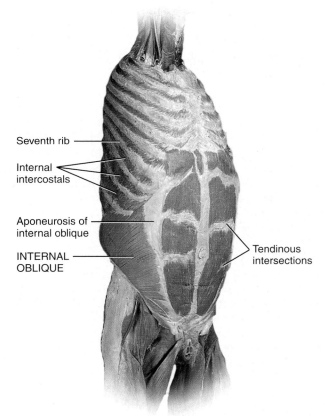

Seventh rib

Internal intercostals

Aponeurosis of internal oblique

INTERNAL OBLIQUE

Tendinous intersections

(g) Right anterolateral deep view

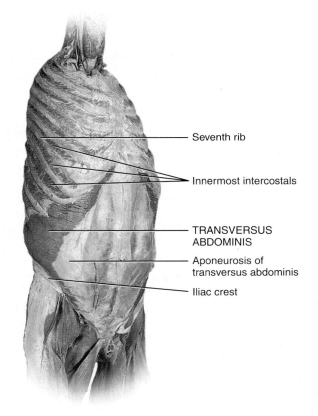

Seventh rib

Innermost intercostals

TRANSVERSUS ABDOMINIS

Aponeurosis of transversus abdominis

Iliac crest

(h) Right anterolateral deeper view

Which abdominal muscle aids in urination?

EXHIBIT 11.K | Muscles of the Thorax That Assist in Breathing *(Figure 11.14)*

 OBJECTIVE

• Describe the origin, insertion, action, and innervation of the muscles of the thorax that assist in breathing.

The muscles of the thorax (chest) alter the size of the thoracic cavity so that breathing can occur. Inhalation (breathing in) occurs when the thoracic cavity increases in size, and exhalation (breathing out) occurs when the thoracic cavity decreases in size.

The dome-shaped **diaphragm** is the most important muscle that powers breathing. It also separates the thoracic and abdominal cavities. The diaphragm has a convex superior surface that forms the floor of the thoracic cavity and concave, inferior surface that forms the roof of the abdominal cavity (Figure 11.14b). The **peripheral muscular portion** of the diaphragm originates on the xiphoid process of the sternum, the inferior six ribs and their costal cartilages, and the lumbar vertebrae and their intervertebral discs and the twelfth rib (Figure 11.14c, d). From their various origins, the fibers of the muscular portion converge and insert into the **central tendon,** a strong aponeurosis located near the center of the muscle (Figure 11.14c, d). The central tendon fuses with the inferior surface of the pericardium (covering of the heart) and the pleurae (coverings of the lungs).

The diaphragm has three major openings through which various structures pass between the thorax and abdomen. These structures include the aorta, along with the thoracic duct and azygos vein, which pass through the **aortic hiatus**; the esophagus with accompanying vagus (X) nerves, which pass through the **esophageal hiatus**; and the inferior vena cava, which passes through the **caval opening (foramen for the vena cava)**. In a condition called a hiatus hernia, the stomach protrudes superiorly through the esophageal hiatus.

Movements of the diaphragm also help return venous blood passing through abdominal veins to the heart. Together with the anterolateral abdominal muscles, the diaphragm helps to increase intra-abdominal pressure to evacuate the pelvic contents during defecation, urination, and childbirth. This mechanism is further assisted when you take a deep breath and close the rima glottidis (the space between the vocal folds). The trapped air in the respiratory system prevents the diaphragm from elevating. The increase in intra-abdominal pressure also helps support the vertebral column and helps prevent flexion during weightlifting. This greatly assists the back muscles in lifting a heavy weight.

Other muscles involved in breathing, called **intercostals**, span the intercostal spaces, the spaces between ribs (Figure 11.4a, b). These muscles are arranged in three layers. The 11 pairs of **external intercostals** occupy the superficial layer, and their fibers run in an oblique direction inferiorly and anteriorly from the rib above to the rib below. They elevate the ribs during inhalation to help expand the thoracic cavity. The 11 pairs of **internal intercostals** occupy the intermediate layer of the intercostal spaces. The fibers of these muscles run at right angles to the external intercostals, in an oblique direction inferiorly and posteriorly from the inferior border of the rib above to the superior border of the rib below. They draw adjacent ribs together during forced exhalation to help decrease the size of the thoracic cavity. The deepest muscle layer is made up of the paired **innermost intercostals**. These poorly developed muscles extend in the same direction as the internal intercostals and may have the same role.

Note: A mnemonic for the action of the intercostal muscles is singing "Old MacDonald had a farm, **E, I, E, I, O**" = **E**xternal **I**ntercostals **E**levate during **I**nhalation, **O**h!"

MUSCLE	ORIGIN	INSERTION	ACTION	INNERVATION
Diaphragm (DI-a-fram; *dia-*=across; *-praghm*=wall)	Xiphoid process of the sternum, costal cartilages and adjacent portions of inferior ribs 7–12, lumbar vertebrae and their intervertebral discs	Central tendon	Contraction of the diaphragm causes it to flatten and increases the vertical dimension of the thoracic cavity, resulting in inhalation; relaxation of the diaphragm causes it to move superiorly and decreases the vertical dimension of the thoracic cavity, resulting in exhalation	Phrenic nerve, which contains axons from cervical spinal nerves C3–C5
External intercostals (in'-ter-KOS-tals; *external*=closer to surface; *inter-*=between; *costa-*=rib)	Interior border of rib above	Superior border of rib below	Contraction elevates the ribs and increases the anteroposterior and lateral dimensions of the thoracic cavity, resulting in inhalation; relaxation depresses the ribs and decreases the anteroposterior and lateral dimensions of the thoracic cavity, resulting in exhalation	Anterior rami of thoracic spinal nerves T2–T12
Internal intercostals *internal*=farther from surface)	Superior border of rib below	Inferior border of rib above	Contraction draws adjacent ribs together to further decrease the anteroposterior and lateral dimensions of the thoracic cavity during forced exhalation	Anterior rami of thoracic spinal nerves T2–T12
Innermost intercostals	Superior border of rib below	Inferior border of rib above	Action is the same as for internal intercostals; formerly considered to be a deep layer of internal intercostals	Anterior rami of thoracic spinal nerves T2–T12

EXHIBIT 11.K **351**

As you will see in Chapter 24, the diaphragm and external intercostal muscles are used during quiet inhalation and exhalation. However, during deep, forceful inhalation (during exercise or playing a wind instrument), the sternocleidomastoid, scalene, and pectoralis minor muscles are also used; during deep, forceful exhalation, the external oblique, internal oblique, transversus abdominis, rectus abdominis, and internal intercostals are also used.

Relating Muscles to Movements

Arrange the muscles in this exhibit according to the following actions: (1) increase in vertical length, (2) increase in lateral and anteroposterior dimensions, and (3) decrease in lateral and anteroposterior dimensions of the thorax.

✅ CHECKPOINT

18. What are the names of the three openings in the diaphragm, and which structures pass through each?

Figure 11.14 Muscles of the thorax that assist in breathing.

🔑 Openings in the diaphragm permit the passage of the aorta, esophagus, and inferior vena cava.

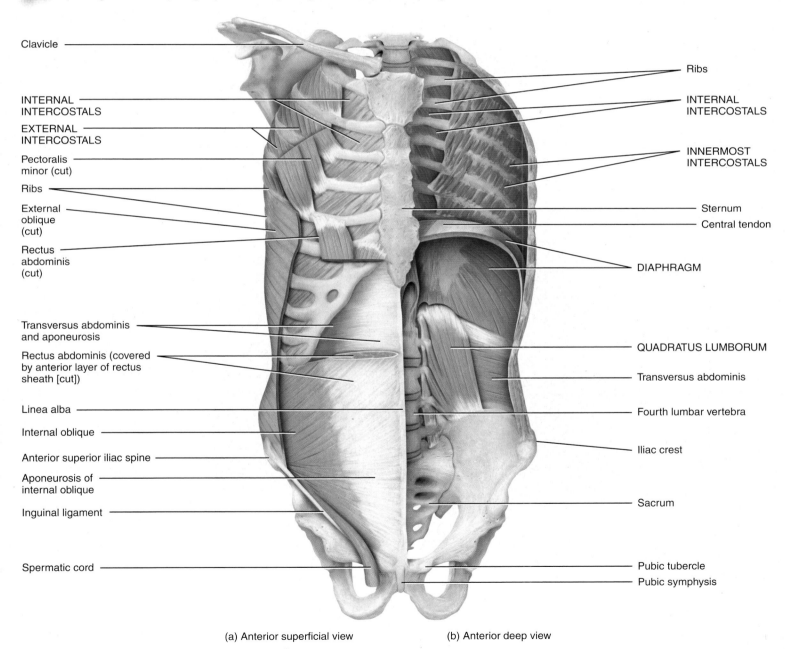

(a) Anterior superficial view (b) Anterior deep view

FIGURE 11.14 CONTINUES ▶

EXHIBIT 11.K Muscles of the Thorax That Assist in Breathing *(Figure 11.14)* CONTINUED

⬤ **FIGURE 11.14** CONTINUED ▶

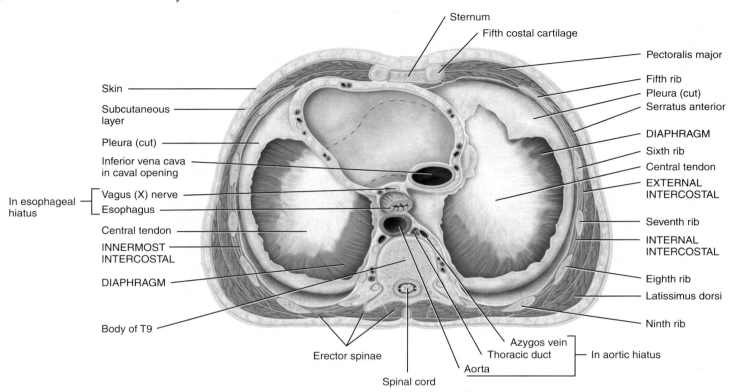

(c) Superior view of diaphragm

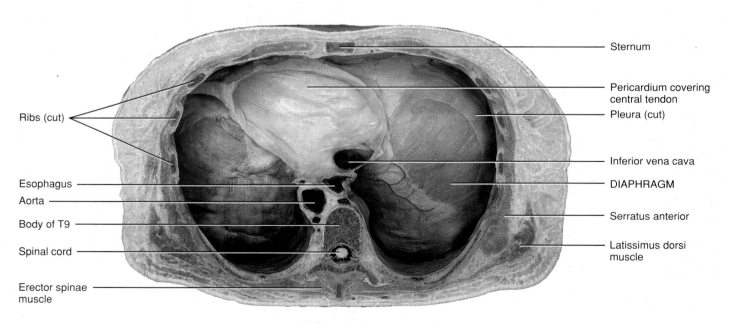

(d) Superior view

EXHIBIT 11.K **353**

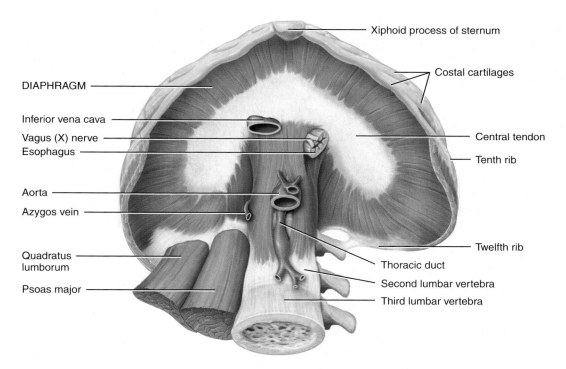

Xiphoid process of sternum

DIAPHRAGM

Costal cartilages

Inferior vena cava

Central tendon

Vagus (X) nerve

Esophagus

Tenth rib

Aorta

Azygos vein

Twelfth rib

Quadratus lumborum

Thoracic duct

Psoas major

Second lumbar vertebra

Third lumbar vertebra

(e) Inferior view of diaphragm

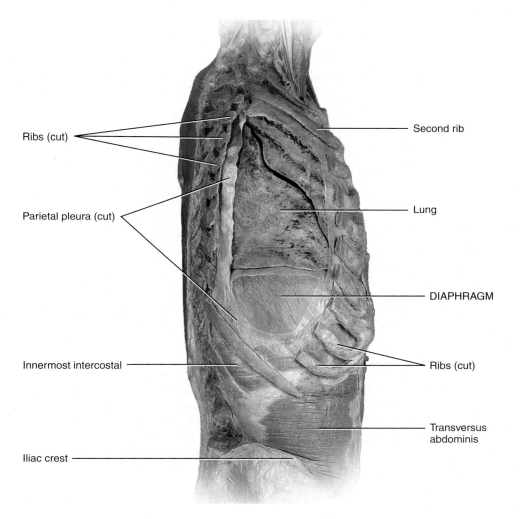

Ribs (cut)

Second rib

Parietal pleura (cut)

Lung

DIAPHRAGM

Innermost intercostal

Ribs (cut)

Transversus abdominis

Iliac crest

(f) Right lateral view

Which muscle associated with breathing is innervated by the phrenic nerve?

EXHIBIT 11.L

Muscles of the Pelvic Floor That Support the Pelvic Viscera and Function as Sphincters *(Figure 11.15)*

 OBJECTIVE

• Describe the origin, insertion, action, and innervation of the muscles of the pelvic floor that support the pelvic viscera and function as sphincters.

The muscles of the pelvic floor are the levator ani and ischiococcygeus. Along with the fascia covering their internal and external surfaces, these muscles are referred to as the **pelvic diaphragm**, which extends from the pubis anteriorly to the coccyx posteriorly, and from one lateral wall of the pelvis to the other. This arrangement gives the pelvic diaphragm the appearance of a funnel suspended from its attachments. The pelvic diaphragm separates the pelvic cavity above from the perineum below (see Exhibit 11.M). The anal canal and urethra pierce the pelvic diaphragm in both sexes, and the vagina also goes through it in females.

The three components of the **levator ani** muscle are the **pubococcygeus, puborectalis**, and **iliococcygeus**. Figure 11.15 shows these muscles in the female and Figure 11.16 in Exhibit 11.M illustrates them in the male. The levator ani is the largest and most important muscle of the pelvic floor. It supports the pelvic viscera and resists the inferior thrust that accompanies increases in intra-abdominal pressure during functions such as forced exhalation, coughing, vomiting, urination, and defecation. The muscle also functions as a sphincter at the anorectal junction, urethra, and vagina. In addition to assisting the levator ani, the **ischiococcygeus** pulls the coccyx anteriorly after it has been pushed posteriorly during defecation or childbirth.

Relating Muscles to Movements

Arrange the muscles in this exhibit according to the following actions: (1) supporting and maintaining the position of the pelvic viscera; (2) resisting an increase in intra-abdominal pressure; (3) constriction of the anus, urethra, and vagina. The same muscle may be mentioned more than once.

 CHECKPOINT

19. Which muscles are strengthened by Kegel exercises?

MUSCLE	ORIGIN	INSERTION	ACTION	INNERVATION
Levator ani (le-VĀ-tor Ā-nē; *levator*=raises; *ani*=anus)	This muscle is divisible into three parts: the pubococcygeus, puborectalis, and iliococcygeus muscles			
Pubococcygeus (pū'-bō-kok-SIJ-ē-us; *pubo-*=pubis; *-coccygeus*=coccyx)	Pubis and ischial spine	Coccyx, urethra, anal canal, perineal body of the perineum (a wedge-shaped mass of fibrous tissue in the center of the perineum), and anococcygeal ligament (narrow fibrous band that extends from anus to coccyx)	Supports and maintains position of pelvic viscera; resists increase in intra-abdominal pressure during forced exhalation, coughing, vomiting, urination, and defecation; constricts anus, urethra, and vagina	Anterior rami of sacral spinal nerves S2–S4
Puborectalis (pū-bō-rek-TĀ-lis; *rectal*=rectum)	Posterior surface of the pubic body	Decussates behind the anorectal junction as a thick muscular sling	Helps maintain fecal continence and assists in defecation	Anterior rami of sacral spinal nerves S2–S4
Iliococcygeus (il'-ē-ō-kok-SIJ-ē-us; *ilio-*=ilium)	Ischial spine	Coccyx	Supports and maintains position of pelvic viscera; resists increase in intra-abdominal pressure during forced exhalation, coughing, vomiting, urination, and defecation; constricts anus, urethra, and vagina	Anterior rami of sacral spinal nerves S2-S4
Ischiococcygeus (is'-kē-ō-kok-SIJ-ē-us; *ischio-*=hip)	Ischial spine	Lower sacrum and upper coccyx	Supports and maintains position of pelvic viscera; resists increase in intra-abdominal pressure during forced exhalation, coughing, vomiting, urination, and defecation; pulls coccyx anteriorly following defecation or childbirth	Anterior rami of sacral spinal nerves S4–S5

EXHIBIT 11.L **355**

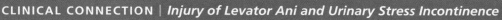

CLINICAL CONNECTION | *Injury of Levator Ani and Urinary Stress Incontinence*

During childbirth, the levator ani muscle supports the head of the fetus, and the muscle may be injured during a difficult childbirth or traumatized during an *episiotomy* (a cut made with surgical scissors to prevent or direct tearing of the perineum during the birth of a baby). The consequence of such injuries may be **urinary stress incontinence**, that is, the leakage of urine whenever intra-abdominal pressure is increased—for example, during coughing. One way to treat urinary stress incontinence is to strengthen and tighten the muscles that support the pelvic viscera. This is accomplished by *Kegel exercises*, the alternate contraction and relaxation of muscles of the pelvic floor. To find the correct muscles, the person imagines that she is urinating and then contracts the muscles as if stopping in midstream. The muscles should be held for a count of three, then relaxed for a count of three. This should be done 5–10 times each hour—sitting, standing, and lying down. Kegel exercises are also encouraged during pregnancy to strengthen the muscles for delivery. •

Figure 11.15 Muscles of the pelvic floor that support the pelvic viscera, assist in resisting intra-abdominal pressure, and function as sphincters.

 The pelvic diaphragm supports the pelvic viscera.

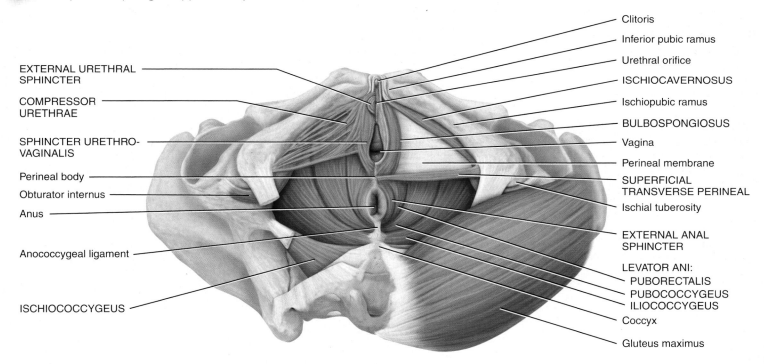

Inferior superficial view of a female perineum

❓ **What are the borders of the pelvic diaphragm?**

EXHIBIT 11.M Muscles of the Perineum *(Figure 11.16)*

OBJECTIVE

• Describe the origin, insertion, action, and innervation of the muscles of the perineum.

The **perineum** is the region of the trunk inferior to the pelvic diaphragm. It is a diamond-shaped area that extends from the pubic symphysis anteriorly, to the coccyx posteriorly, and to the ischial tuberosities laterally. The female and the male perineums may be compared in Figures 11.15 and 11.16, respectively. A transverse line drawn between the ischial tuberosities divides the perineum into an anterior **urogenital triangle** that contains the external genitals and a posterior **anal triangle** that contains the anus (see Figure 26.22). Several perineal muscles insert into the *perineal body* of the perineum, a muscular intersection anterior to the anus. Clinically, the perineum is very important to physicians who care for women during pregnancy and treat disorders related to the female genital tract, urogenital organs, and the anorectal region.

The muscles of the perineum are arranged in two layers; **superficial** and **deep**. The muscles of the superficial layer are the **superficial transverse perineal** muscle, the **bulbospongiosus**,

and the **ischiocavernosus** (Figures 11.15 and 11.16). The deep muscles of the male perineum are the **deep transverse perineal** muscle and **external urethral sphincter** (Figure 11.16). The deep muscles of the female perineum are the **compressor urethrae**, **sphincter urethrovaginalis**, and **external urethral sphincter** (see Figure 11.15). The deep muscles of the perineum assist in urination and ejaculation in males and urination and compression of the vagina in females. The **external anal sphincter** closely adheres to the skin around the margin of the anus and keeps the anal canal and anus closed except during defecation.

Relating Muscles to Movements

Arrange the muscles in this exhibit according to the following actions: (1) expulsion of urine and semen, (2) erection of the clitoris and penis, (3) closure of the anal orifice, and (4) constriction of the vaginal orifice. The same muscle may be mentioned more than once.

 CHECKPOINT

20. What are the borders and contents of the urogenital triangle and the anal triangle?

MUSCLE	ORIGIN	INSERTION	ACTION	INNERVATION
SUPERFICIAL PERINEAL MUSCLES				
Superficial transverse perineal (per-i-NĒ-al; *superficial*=closer to surface; *transverse*=across; *perineus*=perineum)	Ischial tuberosity	Perineal body of perineum	Stabilizes perineal body of perineum and supports pelvic floor	Perineal branch of the pudendal nerve of the sacral plexus
Bulbospongiosus (bul′-bō-spun′-jē-Ō-sus; *bulb-*=a bulb; *spongio-*=sponge)	Perineal body of perineum	Perineal membrane of deep muscles of perineum, corpus spongiosum of penis, and deep fascia on dorsum of penis in male; pubic arch and root and dorsum of clitoris in female	Helps expel urine during urination, helps propel semen along urethra, assists in erection of the penis in male; constricts vaginal orifice and assists in erection of clitoris in female	Perineal branch of the pudendal nerve of the sacral plexus
Ischiocavernosus (is′-kē-ō-ka′-ver-NŌ-sus; *ischio-*=the hip)	Ischial tuberosity and ischial and pubic rami	Corpora cavernosa of penis in male and clitoris in female, and pubic symphysis	Maintains erection of penis in male and clitoris in female by decreasing urine drainage	Perineal branch of the pudendal nerve of the sacral plexus
DEEP PERINEAL MUSCLES				
Deep transverse perineal (per-i-NĒ-al; *deep*=farther from surface)	Ischial ramus	Perineal body of perineum	Helps expel last drops of urine and semen in male	Perineal branch of the pudendal nerve of the sacral plexus
External urethral sphincter (ū-RĒ-thral SFINGK-ter)	Ischial and pubic rami	Median raphe in male and vaginal wall in female	Helps expel last drops of urine and semen in male and urine in female	Sacral spinal nerve S4 and the inferior rectal branch of the pudendal nerve
Compressor urethrae (ū-RĒ-thrē) (see Figure 11.15)	Ischiopubic ramus	Blends with same muscle of the opposite side anterior to urethra	Serves as an accessory sphincter of the urethra	Perineal branch of the pudendal nerve of the sacral plexus
Sphincter urethrovaginalis (ū-rē-thrō-vaj-i-NAL-is) (see Figure 11.15)	Perineal body	Blends with same muscle of the opposite side anterior to urethra	Serves as an accessory sphincter of the urethra and facilitates closing of the vagina	Perineal branch of the pudendal nerve of the sacral plexus
External anal sphincter (Ā-nal)	Anococcygeal ligament	Perineal body of perineum	Keeps anal canal and anus closed	Sacral spinal nerve S4 and the inferior rectal branch of the pudendal nerve

EXHIBIT 11.M **357**

Figure 11.16 Muscles of the perineum.

🗝 The urogenital diaphragm assists in urination in females and males, ejaculation in males, and helps strengthen the pelvic floor.

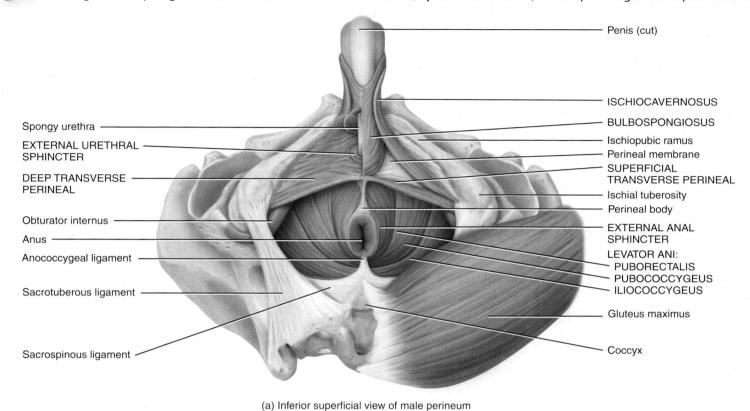

Penis (cut)

ISCHIOCAVERNOSUS

BULBOSPONGIOSUS

Spongy urethra

EXTERNAL URETHRAL SPHINCTER

Ischiopubic ramus

Perineal membrane

SUPERFICIAL TRANSVERSE PERINEAL

DEEP TRANSVERSE PERINEAL

Ischial tuberosity

Perineal body

Obturator internus

EXTERNAL ANAL SPHINCTER

Anus

Anococcygeal ligament

LEVATOR ANI:
 PUBORECTALIS
 PUBOCOCCYGEUS
 ILIOCOCCYGEUS

Sacrotuberous ligament

Gluteus maximus

Sacrospinous ligament

Coccyx

(a) Inferior superficial view of male perineum

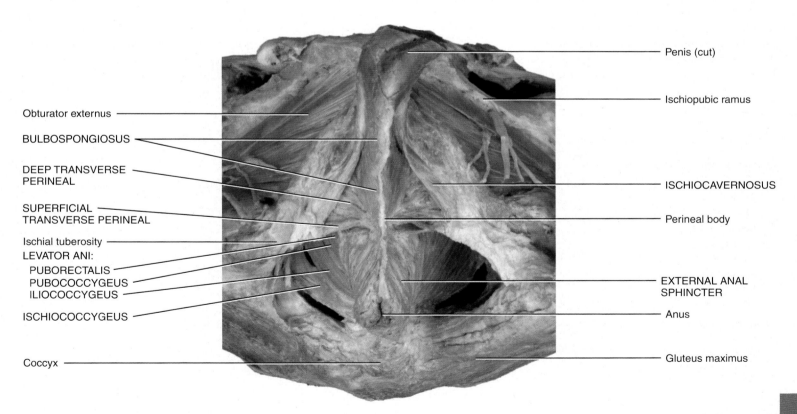

Penis (cut)

Ischiopubic ramus

Obturator externus

BULBOSPONGIOSUS

DEEP TRANSVERSE PERINEAL

ISCHIOCAVERNOSUS

SUPERFICIAL TRANSVERSE PERINEAL

Perineal body

Ischial tuberosity

LEVATOR ANI:
 PUBORECTALIS
 PUBOCOCCYGEUS
 ILIOCOCCYGEUS

EXTERNAL ANAL SPHINCTER

ISCHIOCOCCYGEUS

Anus

Coccyx

Gluteus maximus

(b) Inferior superficial view of male perineum

 What are the borders of the perineum?

EXHIBIT 11.N

Muscles of the Thorax That Move the Pectoral Girdle *(Figure 11.17)*

OBJECTIVE

• Describe the origin, insertion, action, and innervation of the muscles of the thorax that move the pectoral girdle.

Muscles of the proximal upper limb are arranged in diverse groups; muscles are superficial, deep, or very deep and grouped either as stabilizers of the pectoral or shoulder girdle (clavicle and scapula) or muscles that act on the shoulder joint (Figure 11.17). Four muscles, the pectoralis major, deltoid, trapezius, and latissimus dorsi muscles, are not only superficial but also have a large surface area and dominate the superficial musculature of the shoulder, chest, and upper back.

The main action of the muscles that move the pectoral girdle is to stabilize the scapula so it can function as a steady origin for most of the muscles that move the humerus. Because scapular movements usually accompany humeral movements in the same direction, the muscles also move the scapula to increase the range of motion of the humerus. For example, it would not be possible to raise the arm above the head if the scapula did not move with the humerus. During abduction, the scapula follows the humerus by rotating upward.

Muscles that move the pectoral girdle can be classified into two groups based on their location in the thorax: **anterior** and **posterior thoracic muscles**. The anterior thoracic muscles are the subclavius, pectoralis minor, and serratus anterior. The **subclavius** is a small, cylindrical muscle under the clavicle that extends from the clavicle to the first rib. It steadies the clavicle during movements of the pectoral girdle.

The **pectoralis minor** is a thin, flat, triangular muscle that is deep to the pectoralis major. This muscle causes, among other actions, abduction of the scapula. The scapulae of persons who spend a lot of time with their arms in front of them, such as pianists, factory workers, and those who use computers, may develop chronically contracted pectoralis minor muscles. Contracted muscles become shorter and wider and since the brachial plexus (the major nerve network to the upper limb) runs between the pectoralis minor and the rib cage, chronic contraction can compress nerves. The compressed nerves emulate symptoms of carpal tunnel syndrome.

The **serratus anterior** is a large, flat, fan-shaped muscle between the ribs and scapula. It is so named because of the saw-toothed appearance of its origins on the ribs. This muscle can be highly developed in body builders and athletes. It is an antagonist of the rhomboids and is responsible for abduction of the scapula. A large portion of the belly is deep to the anterior scapula. The muscle is thus riding over the rib cage. The lateral and inferior portion of the breast lies superficial to the serratus anterior muscle.

The posterior thoracic muscles are the trapezius, levator scapulae, rhomboid major, and rhomboid minor.

The **trapezius** is a large, flat, triangular sheet of muscle extending from the skull and vertebral column medially to the pectoral girdle laterally. It is the most superficial back muscle and covers the posterior neck region and superior portion of the trunk. The two trapezius muscles form a trapezoid (diamond-shaped quadrangle)—hence its name. The three sets of fibers (superior, middle, and inferior) enable this muscle to cause multiple actions. The superior fibers extend the head and neck. The weight of the head (about 12 pounds) is functionally doubled with each inch that the head flexes from the neutral position (directly over the atlas). For example, for the person who has a head-forward posture of

MUSCLE	ORIGIN	INSERTION	ACTION	INNERVATION
ANTERIOR THORACIC MUSCLES				
Subclavius (sub-KLĀ-vē-us, *sub-*=under; *clavius*=clavicle)	Rib 1	Clavicle	Depresses and moves clavicle anteriorly and helps stabilize pectoral girdle	Subclavian nerve
Pectoralis minor (pek'-tō-RĀ-lis; *pector-*=breast, chest, thorax; *minor*=lesser)	Ribs 2–5, 3–5, or 2–4	Coracoid process of scapula	Abducts scapula and rotates it downward / RMA: Elevates third through fifth ribs during forced inhalation when scapula is fixed	Medial pectoral nerve
Serratus anterior (ser-Ā-tus; *serratus*=saw-toothed; *anterior*=front)	Ribs 1–8, or 1–9	Vertebral border and inferior angle of scapula	Abducts scapula and rotates it upward / RMA: Elevates ribs when scapula is stabilized; known as "boxer's muscle" because it is important in horizontal arm movements such as punching and pushing	Long thoracic nerve

EXHIBIT 11.N **359**

2 in. from neutral, the trapezius and smaller muscles of the posterior neck need to contract as though the head weighed 36 pounds. The trapezius is thus overworked in such persons and becomes painful. The head-forward position is usually a habit that the patient can correct with practice.

The **levator scapulae** is a narrow, elongated muscle in the posterior portion of the neck. It is deep to the sternocleidomastoid and trapezius muscles. As its name suggests, one of its actions is to elevate the scapula. This muscle contains a twist in the belly. The twist inverts the superior and inferior fibers as they approach the insertion and ensures that the muscle will elevate the scapula rather than rotating it. Its reverse muscle action (RMA), when the origin and insertion are switched, is to extend the neck.

The **rhomboid major** and **rhomboid minor** lie deep to the trapezius and are not always distinct from each other. They appear as parallel bands that pass inferiorly and laterally from the vertebrae to the scapula. Their names are based on their shape—that is, a rhomboid (an oblique parallelogram). The rhomboid major is about two times wider than the rhomboid minor. The two muscles are often identified by their attachments. The muscles lie deep to the trapezius and superficial to the erector spinae. The rhomboids and the trapezius are functionally the only muscles holding the upper limb to the posterior axial skeleton. Both muscles are used when forcibly lowering the raised upper limbs, as in driving a stake with a sledgehammer.

Movements of the Scapula

To understand the actions of muscles that move the scapula, it is first helpful to review the various movements of the scapula:

- *Elevation:* superior movement of the scapula, such as shrugging the shoulders or lifting a weight over the head.
- *Depression:* inferior movement of the scapula, as in pulling down on a rope attached to a pulley.
- *Abduction (protraction):* movement of the scapula laterally and anteriorly, as in doing a push-up or punching.
- *Adduction (retraction):* movement of the scapula medially and posteriorly, as in pulling the oars in a rowboat.
- *Upward rotation:* movement of the inferior angle of the scapula laterally so that the glenoid cavity is moved upward. This movement is required to move the humerus past the horizontal, as in raising the arms in a "jumping jack."
- *Downward rotation:* movement of the inferior angle of the scapula medially so that the glenoid cavity is moved downward. This movement is seen when a gymnast on parallel bars supports the weight of the body on the hands.

Relating Muscles to Movements

Arrange the muscles in this exhibit according to the following actions on the scapula: (1) depression, (2) elevation, (3) abduction, (4) adduction, (5) upward rotation, and (6) downward rotation. The same muscle may be mentioned more than once.

 CHECKPOINT

21. What muscles in this exhibit are used to raise your shoulders, lower your shoulders, join your hands behind your back, and join your hands in front of your chest?

MUSCLE	ORIGIN	INSERTION	ACTION	INNERVATION
POSTERIOR THORACIC MUSCLES				
Trapezius (tra-PĒ-zē-us; *trapezi-*=trapezoid-shaped)	Superior nuchal line of occipital bone, ligamentum nuchae, and spines of C7–T12	Clavicle and acromion and spine of scapula	Superior fibers upward rotate scapula; middle fibers adduct scapula; inferior fibers depress and upward rotate scapula; superior and inferior fibers together rotate scapula upward; stabilizes scapula RMA: Superior fibers can help extend head	Accessory (XI) nerve and cervical spinal nerves C3–C5
Levator scapulae (le-VĀ-tor SKA-pu-lē; *levator*=raises; *scapulae*=of the scapula)	Transverse processes of C1–C4	Superior vertebral border of scapula	Elevates scapula and rotates it downward	Dorsal scapular nerve and cervical spinal nerves C3–C5
Rhomboid major (ROM-boyd; *rhomboid*=rhomboid- or diamond-shaped)	Spines of T2–T5	Vertebral border of scapula inferior to spine	Elevates and adducts scapula and rotates it downward; stabilizes scapula	Dorsal scapular nerve
Rhomboid minor	Spines of C7–T1	Vertebral border of scapula superior to spine	Elevates and adducts scapula and rotates it downward; stabilizes scapula	Dorsal scapular nerve

EXHIBIT 11.N Muscles of the Thorax That Move the Pectoral Girdle *(Figure 11.17)* **CONTINUED**

Figure 11.17 Muscles of the thorax that move the pectoral girdle.

Muscles that move the pectoral girdle originate on the axial skeleton and insert on the clavicle or scapula.

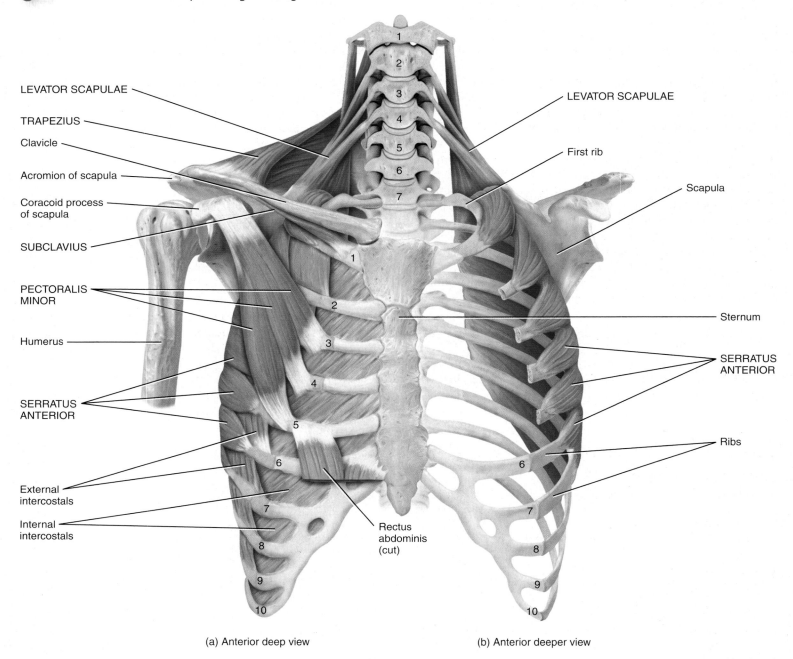

(a) Anterior deep view   (b) Anterior deeper view

EXHIBIT 11.N **361**

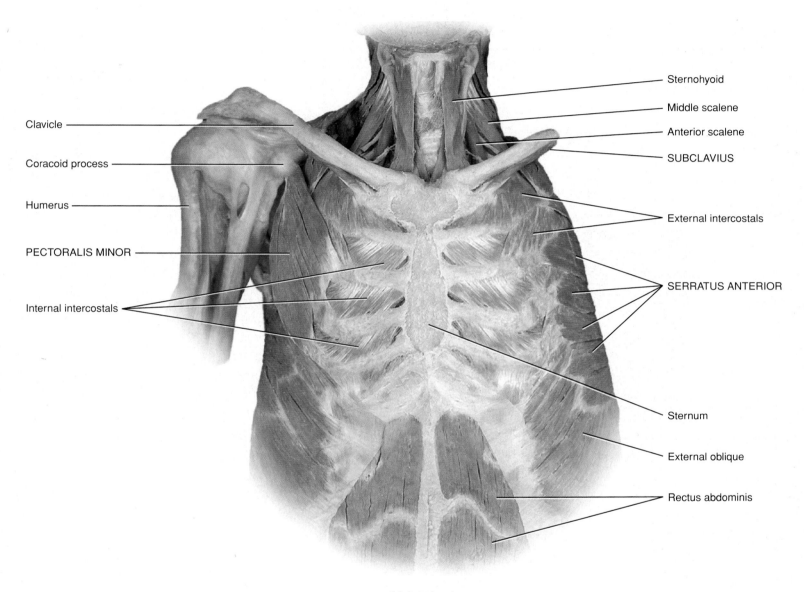

Clavicle

Coracoid process

Humerus

PECTORALIS MINOR

Internal intercostals

Sternohyoid

Middle scalene

Anterior scalene

SUBCLAVIUS

External intercostals

SERRATUS ANTERIOR

Sternum

External oblique

Rectus abdominis

(c) Anterior view

FIGURE 11.17 CONTINUES ▶

EXHIBIT 11.N Muscles of the Thorax That Move the Pectoral Girdle *(Figure 11.17)* CONTINUED

▣ **FIGURE 11.17** CONTINUED ▶

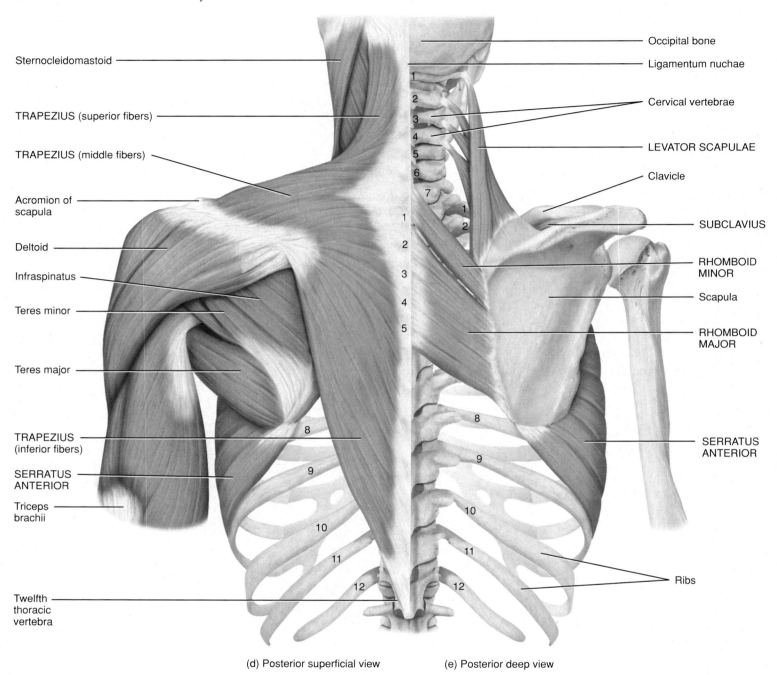

(d) Posterior superficial view   (e) Posterior deep view

EXHIBIT 11.N **363**

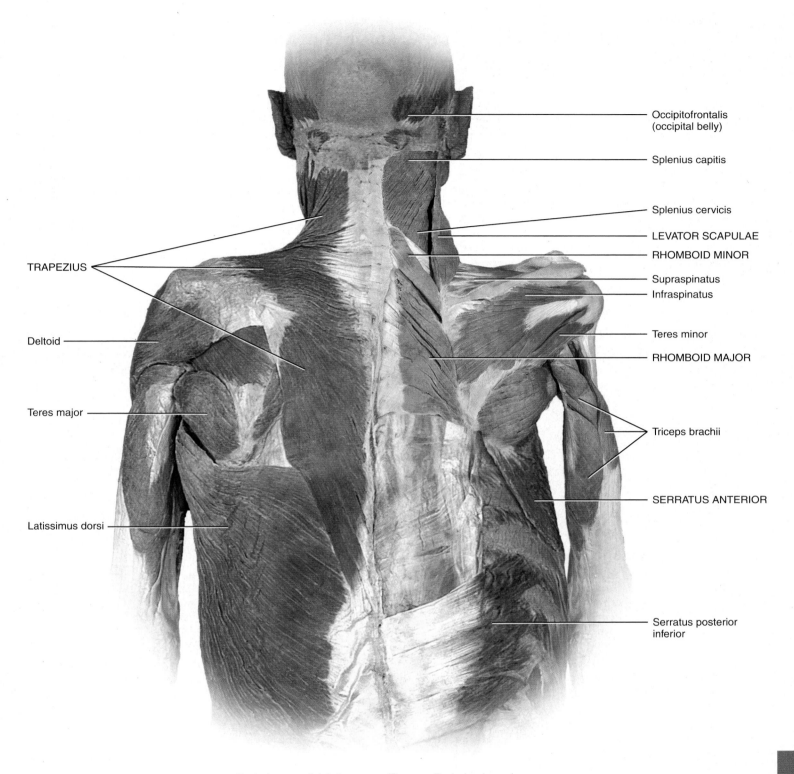

Occipitofrontalis
(occipital belly)

Splenius capitis

Splenius cervicis

LEVATOR SCAPULAE

RHOMBOID MINOR

Supraspinatus

Infraspinatus

Teres minor

RHOMBOID MAJOR

Triceps brachii

SERRATUS ANTERIOR

Serratus posterior
inferior

TRAPEZIUS

Deltoid

Teres major

Latissimus dorsi

Posterior superficial view (f) Posterior deep view

What is the main action of the muscles that move the pectoral girdle?

EXHIBIT 11.O

Muscles of the Thorax and Shoulder That Move the Humerus *(Figure 11.18)*

  OBJECTIVE

• Describe the origin, insertion, action, and innervation of the muscles of the thorax that move the humerus.

Of the nine muscles that cross the shoulder joint, all except the pectoralis major and latissimus dorsi originate on the scapula (shoulder blade). The pectoralis major and latissimus dorsi thus are called **axial muscles**, because they originate on the axial skeleton. The remaining seven muscles, the **scapular muscles**, arise from the scapula (Figure 11.18).

Of the two axial muscles that move the humerus (arm bone), the **pectoralis major** is a large, thick, fan-shaped muscle that

covers the superior part of the thorax and forms the anterior fold of the axilla. When this muscle and the latissimus dorsi are well developed, the axilla is deepened. It has two origins: a smaller clavicular head and a larger sternocostal head. The superior clavicular head attaches more distally on the humerus than its inferior counterpart, the sternal head, giving the tendon a twisted appearance when the arm is in the anatomical position. This improves the mechanical advantage of the muscle.

The **latissimus dorsi** is a broad, triangular muscle located on the inferior part of the back that forms most of the posterior wall of the axilla. The reverse muscle action (RMA) of the latissimus

MUSCLE	ORIGIN	INSERTION	ACTION	INNERVATION
AXIAL MUSCLES THAT MOVE THE HUMERUS				
Pectoralis major (pek'-tō-RĀ-lis; *pector-*=chest; *major*=larger) (see also Figure 11.13b)	Clavicle (clavicular head), sternum, and costal cartilages of ribs 2–6 and sometimes of ribs 1–7 (sternocostal head)	Greater tubercle and lateral lip of the intertubercular sulcus of humerus	As a whole, adducts and medially rotates arm at shoulder joint; clavicular head flexes arm, and sternocostal head extends the flexed arm to side of trunk	Medial and lateral pectoral nerves
Latissimus dorsi (la-TIS-i-mus DOR-sī; *latissimus*=widest; *dorsi*=of the back)	Spines of T7–L5, crests of sacrum and ilium, ribs 9–12	Intertubercular sulcus of humerus	Extends, adducts, and medially rotates arm at shoulder joint; draws arm inferiorly and posteriorly RMA: Elevates vertebral column and torso	Thoracodorsal nerve
SCAPULAR MUSCLES THAT MOVE THE HUMERUS				
Deltoid (DEL-toyd= triangularly shaped)	Acromial extremity of clavicle (anterior fibers), acromion of scapula (lateral fibers), and spine of scapula (posterior fibers)	Deltoid tuberosity of humerus	Lateral fibers abduct arm at shoulder joint; anterior fibers flex and medially rotate arm at shoulder joint; posterior fibers extend and laterally rotate arm at shoulder joint	Axillary nerve
Subscapularis (sub-scap'-ū-LĀ-ris; *sub-*=below; *scapularis*= scapula)	Subscapular fossa of scapula	Lesser tubercle of humerus	Medially rotates arm at shoulder joint	Upper and lower subscapular nerve
Supraspinatus (soo-pra-spī-NĀ-tus; *supra-*=above; *spina-* =spine [of the scapula])	Supraspinous fossa of scapula	Greater tubercle of humerus	Assists deltoid muscle in abducting arm at shoulder joint	Suprascapular nerve
Infraspinatus (in'-fra-spī-NĀ-tus; *infra-*=below)	Infraspinous fossa of scapula	Greater tubercle of humerus	Laterally rotates arm at shoulder joint	Suprascapular nerve
Teres major (TE-rēz; *teres*=long and round)	Inferior angle of scapula	Medial lip of intertubercular sulcus of humerus	Extends arm at shoulder joint and assists in adduction and medial rotation of arm at shoulder joint	Lower subscapular nerve
Teres minor	Inferior lateral border of scapula	Greater tubercle of humerus	Laterally rotates and extends arm at shoulder joint	Axillary nerve
Coracobrachialis (kor'-a-kō-brā-kē-Ā-lis; *coraco-*=coracoid process [of the scapula]; *brachi-*=arm)	Coracoid process of scapula	Middle of medial surface of shaft of humerus	Flexes and adducts arm at shoulder joint	Musculocutaneous nerve

EXHIBIT 11.O **365**

dorsi enables the vertebral column and torso to be elevated, as in doing a pullup. It is commonly called the "swimmer's muscle" because its many actions are used while swimming; consequently, many competitive swimmers have well-developed "lats." Like the pectoralis major and levator scapulae muscles, the latissimus dorsi has a twist in it near the insertion that increases its mechanical advantage. Sharing a similar insertion with the pectoralis major and latissimus dorsi is the teres major. The **teres major** is a thick, rounded muscle in cross-section (therefore its name) that also helps form part of the posterior wall of the axilla.

Among the scapular muscles, the **deltoid** is a thick, powerful shoulder muscle that covers the shoulder joint and forms the rounded contour of the shoulder. This muscle is a frequent site of intramuscular injections. As you study the deltoid, note that the muscle has three sets of fibers (anterior, lateral, and posterior) that enable it to function as three distinct muscles used in flexion, abduction, rotation, and extension of the humerus. The points of origin on the clavicle, acromion, and spine of the scapula are near the same three points as the insertions of the trapezius.

Four deep muscles of the shoulder—subscapularis, supraspinatus, infraspinatus, and teres minor—strengthen and stabilize the shoulder joint. These muscles join the scapula to the humerus. Their flat tendons fuse together to form the **rotator (musculotendinous) cuff**, a nearly complete circle of tendons around the

Figure 11.18 **Muscles of the thorax (chest) and shoulder that move the humerus.**

The strength and stability of the shoulder joint are provided by the tendons that form the rotator cuff.

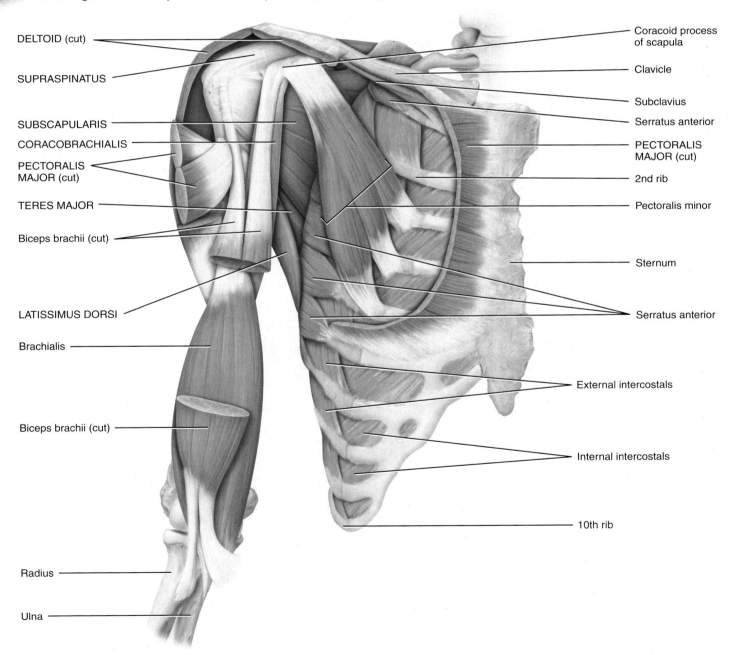

(a) Anterior deep view (the intact pectoralis major muscle is shown in Figure 11.3a)

FIGURE 11.18 CONTINUES

EXHIBIT 11.O Muscles of the Thorax and Shoulder That Move the Humerus *(Figure 11.18)* CONTINUED

shoulder joint, like the cuff on a shirtsleeve. These four muscles are often referred to as the "SITS" muscles and this could serve as a mnemonic for remembering their names. The **subscapularis** is a large triangular muscle that fills the subscapular fossa of the scapula and forms a small part in the apex of the posterior wall of the axilla. The **supraspinatus**, a rounded muscle named for its location in the supraspinous fossa of the scapula, lies deep to the trapezius. The tendon of insertion slides across the superior aspect of the shoulder joint beneath the acromion; inflammation in this tunnel-like region can cause swelling and accompanying

pain. The **infraspinatus** is a triangular muscle, also named for its location in the infraspinous fossa of the scapula. A portion of the muscle is superficial and other portions are deep to the trapezius and deltoid. The **teres minor** is a cylindrical, elongated muscle, located between the teres major and infraspinatus muscles. Its belly lies parallel to the inferior edge of the infraspinatus and is sometimes indistinguishable from it.

The **coracobrachialis** is an elongated, narrow muscle in the arm, located in the lateral wall of the axilla along with the biceps brachii.

🔘 **FIGURE 11.18** CONTINUED ▶

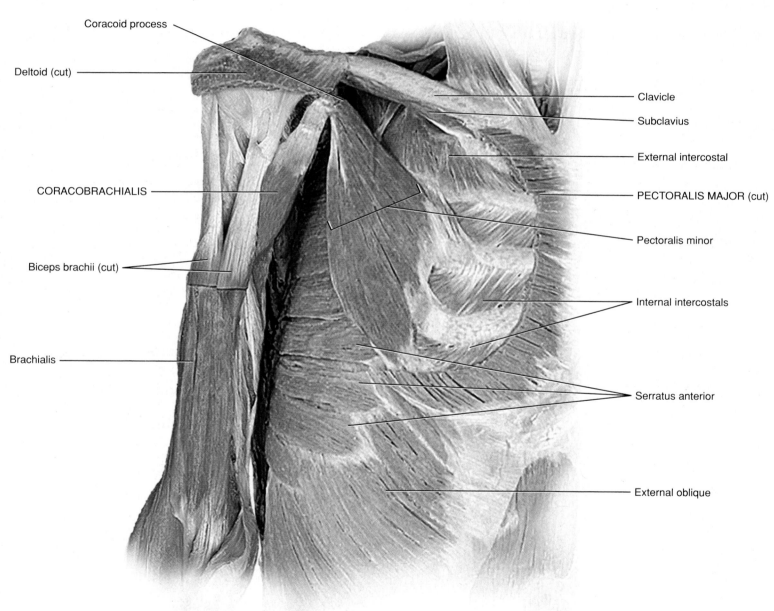

(b) Anterior deep view

EXHIBIT 11.O **367**

CLINICAL CONNECTION | *Rotator Cuff Injuries and Impingement Syndrome*

Rotator cuff injury is a strain or tear in the rotator cuff muscles and common among baseball pitchers, volleyball players, racket sports players, and swimmers due to shoulder movements that involve vigorous circumduction. It also occurs as a result of wear and tear, trauma, and repetitive motions in certain occupations, such as painters or those required to place items on a shelf above the head. Although muscle bellies may be damaged, the tendon of one or more of the four muscles is usually partially or completely torn. Most often, there is tearing of the supraspinatus muscle tendon or the rotator cuff. This tendon is especially predisposed to wear-and-tear because of its location between the head of the humerus and acromion of the scapula, which compresses the tendon during shoulder movements.

One of the most common causes of shoulder pain and dysfunction in athletes is known as **impingement syndrome**. The repetitive movement of the arm over the head may put athletes at risk. Impingement syndrome may also be caused by a direct blow or stretch injury. Continual pinching of the supraspinatus tendon as a result of overhead motions causes it to become inflamed and results in pain. If movement is continued despite the pain, the tendon may degenerate near the attachment to the humerus and ultimately may tear away from the bone (rotator cuff injury). Treatment consists of resting the injured tendons, strengthening the shoulder through exercise, massage therapy, and finally surgery if the injury is particularly severe. During surgery, an inflamed bursa may be removed, bone may be trimmed, and/or the coracoacromial ligament may be detached. Torn rotator cuff tendons may be trimmed and then reattached with sutures, anchors, or surgical tacks. These steps make more space, thus relieving pressure and allowing the arm to move freely. •

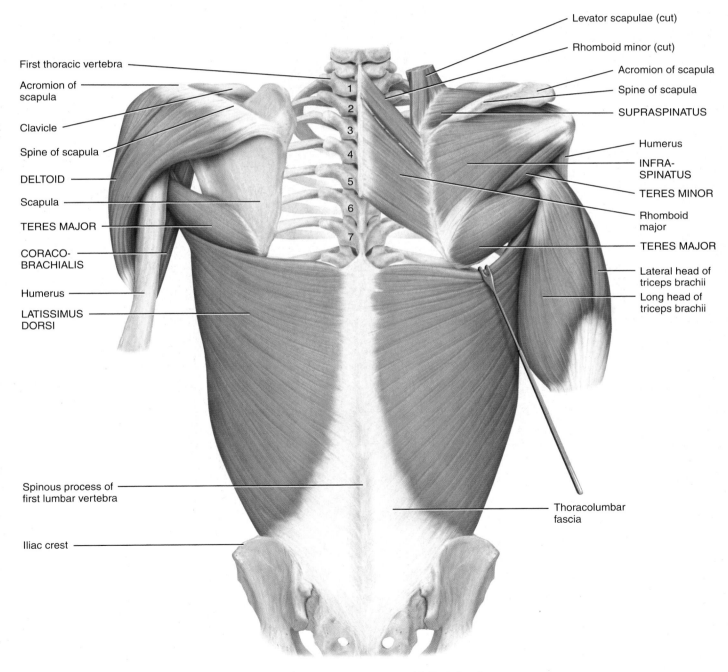

(c) Posterior view   (d) Posterior view

FIGURE 11.18 CONTINUES ▶

EXHIBIT 11.O Muscles of the Thorax and Shoulder That Move the Humerus *(Figure 11.18)* CONTINUED

Relating Muscles to Movements

Arrange the muscles in this exhibit according to the following actions on the humerus at the shoulder joint: (1) flexion, (2) extension, (3) abduction, (4) adduction, (5) medial rotation, and (6) lateral rotation. The same muscle may be mentioned more than once.

 CHECKPOINT

22. Why are the two muscles that cross the shoulder joint called axial muscles, and the seven others called scapular muscles?

◼ **FIGURE 11.18** CONTINUED ▶

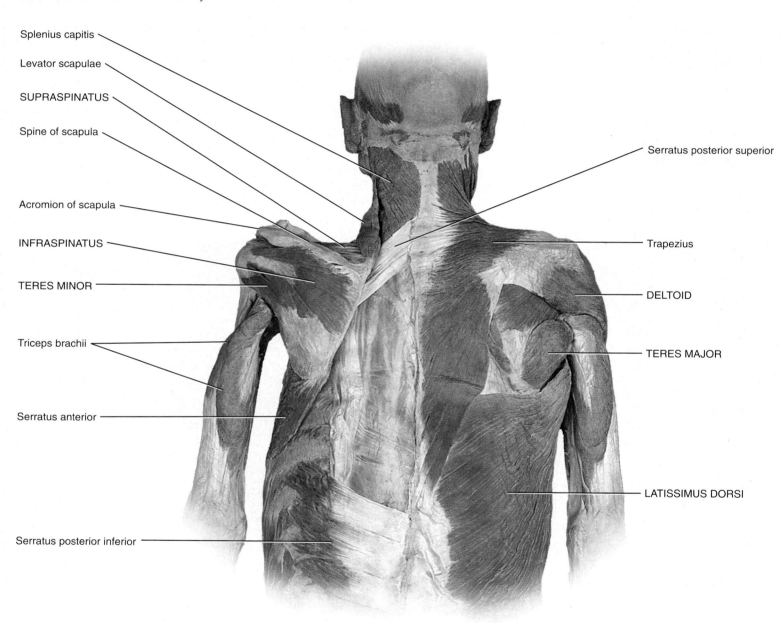

(e) Posterior view

EXHIBIT 11.P **369**

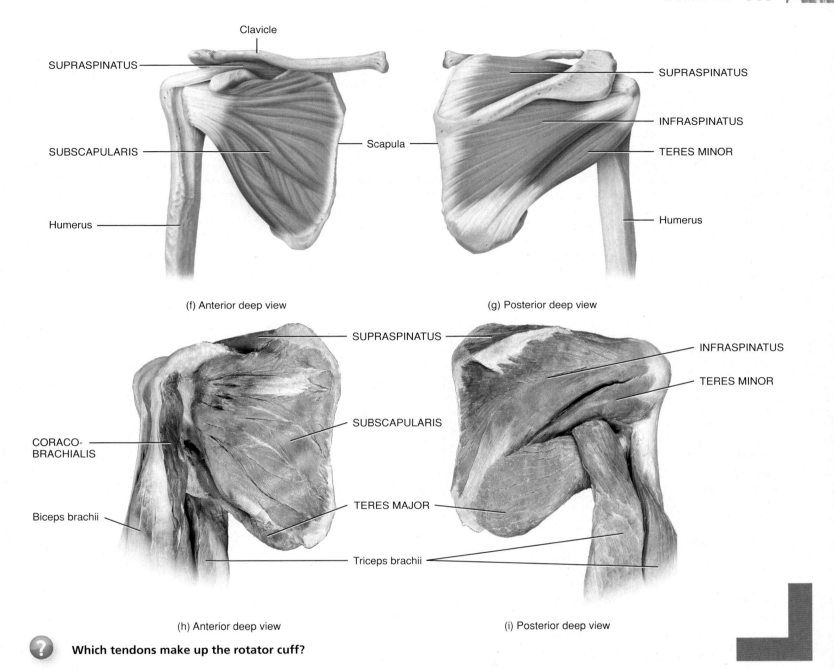

Clavicle

SUPRASPINATUS

SUBSCAPULARIS

Scapula

Humerus

SUPRASPINATUS

INFRASPINATUS

TERES MINOR

Humerus

(f) Anterior deep view

(g) Posterior deep view

SUPRASPINATUS

CORACO-
BRACHIALIS

Biceps brachii

SUBSCAPULARIS

TERES MAJOR

Triceps brachii

INFRASPINATUS

TERES MINOR

(h) Anterior deep view

(i) Posterior deep view

 Which tendons make up the rotator cuff?

EXHIBIT 11.P — Muscles of the Arm That Move the Radius and Ulna *(Figure 11.19)*

OBJECTIVE

• Describe the origin, insertion, action, and innervation of the muscles of the arm that move the radius and ulna.

Most of the muscles that move the radius and ulna (forearm bones) cause flexion and extension at the elbow, which is a hinge joint. The biceps brachii, brachialis, and brachioradialis muscles are the flexor muscles. The extensor muscles are the triceps brachii and the anconeus (Figure 11.19).

The **biceps brachii** is the large muscle located on the anterior surface of the arm. As indicated by its name, it has two heads of origin (long and short), both from the scapula. The muscle spans both the shoulder and elbow joints. In addition to its role in flex-

ing the forearm at the elbow joint, it also supinates the forearm at the radioulnar joints and flexes the arm at the shoulder joint. At its distal attachment in the forearm, a thin flat tendon, the bicipital aponeurosis, splits away from the rest of the tendon. This flat tendon descends medially across the brachial artery and vein and fuses with the fascia over the forearm flexor muscles. It helps protect the median nerve and brachial vessels.

The **brachialis** is deep to the biceps brachii muscle. It is the most powerful flexor of the forearm at the elbow joint. For this reason, it is called the "workhorse" of the elbow flexors. Its thick belly is wider than that of the biceps brachii. It is visible and easily palpated on the lateral aspect of the arm where it is sandwiched between the biceps brachii and triceps brachii muscles.

EXHIBIT 11.P Muscles of the Arm That Move the Radius and Ulna (*Figure 11.19*) CONTINUED

The **brachioradialis** flexes the forearm at the elbow joint, especially when a quick movement is required or when a weight is lifted slowly during flexion of the forearm.

The **triceps brachii** is the large muscle located on the posterior surface of the arm. It is the more powerful of the extensors of the forearm at the elbow joint. As its name implies, it has three heads of origin, one from the scapula (long head) and two from the humerus (lateral and medial heads). The long head crosses the shoulder joint; the other heads do not. The **anconeus** is a small muscle located on the lateral part of the posterior aspect of the elbow that assists the triceps brachii in extending the forearm at the elbow joint.

Some muscles that move the radius and ulna are involved in pronation and supination at the radioulnar joints. The pronators, as suggested by their names, are the **pronator teres** and **pronator quadratus** muscles. The supinator of the forearm is aptly named

the **supinator** muscle and it assists the biceps brachii in producing this action. You use the powerful action of the supinator when you twist a corkscrew or turn a screw with a screwdriver.

In the limbs, functionally related skeletal muscles and their associated blood vessels and nerves are grouped together by fascia into regions called **compartments**. In the arm, the biceps brachii, brachialis, and coracobrachialis muscles comprise the **anterior (flexor) compartment**. The triceps brachii muscle forms the **posterior (extensor) compartment**.

Relating Muscles to Movements

Arrange the muscles in this exhibit according to the following actions on the elbow joint: (1) flexion and (2) extension; the following actions on the forearm at the radioulnar joints: (1) supination

MUSCLE	ORIGIN	INSERTION	ACTION	INNERVATION
FOREARM FLEXORS				
Biceps brachii (BĪ-seps BRĀ-kē-ī; *biceps*=two heads of origin; *brachii*=arm)	Long head: tubercle above glenoid cavity of scapula (supraglenoid tubercle) Short head: coracoid process of scapula	Radial tuberosity of radius and bicipital aponeurosis	Flexes forearm at elbow joint, supinates forearm at radioulnar joints, and flexes arm at shoulder joint	Musculocutaneous nerve
Brachialis (brā-kē-Ā-lis)	Distal, anterior surface of humerus	Ulnar tuberosity and coronoid process of ulna	Flexes forearm at elbow joint	Musculocutaneous and radial nerves
Brachioradialis (brā'-kē-ō-rā-dē-Ā-lis: *radi*=radius)	Lateral border of distal end of humerus	Superior to styloid process of radius	Flexes forearm at elbow joint; supinates and pronates forearm at radioulnar joints to neutral position	Radial nerve
FOREARM EXTENSORS				
Triceps brachii (TRĪ-seps BRĀ-kē-ī; *triceps*=three heads of origin)	Long head: infraglenoid tubercle, a projection inferior to glenoid cavity of scapula Lateral head: lateral and posterior surface of humerus Medial head: entire posterior surface of humerus inferior to a groove for the radial nerve	Olecranon of ulna	Extends forearm at elbow joint and extends arm at shoulder joint	Radial nerve
Anconeus (an-KŌ-nē-us; *ancon*=the elbow)	Lateral epicondyle of humerus	Olecranon and superior portion of shaft of ulna	Extends forearm at elbow joint	Radial nerve
FOREARM PRONATORS				
Pronator teres (PRŌ-nā-tor TE-rēz; *pronator*=turns palm posteriorly; *teres*=round and long)	Medial epicondyle of humerus and coronoid process of ulna	Midlateral surface of radius	Pronates forearm at radioulnar joints and weakly flexes forearm at elbow joint	Median nerve
Pronator quadratus (PRŌ-nā-tor kwod-RĀ-tus =square, four-sided)	Distal portion of shaft of ulna	Distal portion of shaft of radius	Pronates forearm at radioulnar joint	Median nerve
FOREARM SUPINATOR				
Supinator (SOO-pi-nā-tor; *supinator*=turns palm anteriorly)	Lateral epicondyle of humerus and ridge near radial notch of ulna (supinator crest)	Lateral surface of proximal one-third of radius	Supinates forearm at radioulnar joints	Deep radial nerve

EXHIBIT 11.P **371**

Figure 11.19 Muscles of the arm that move the radius and ulna.

The anterior arm muscles flex the forearm, and the posterior arm muscles extend it.

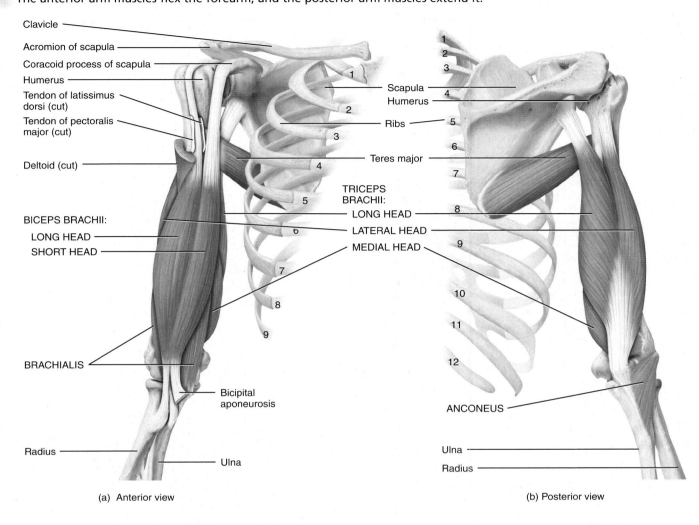

Clavicle
Acromion of scapula
Coracoid process of scapula
Humerus
Tendon of latissimus dorsi (cut)
Tendon of pectoralis major (cut)
Deltoid (cut)

BICEPS BRACHII:
LONG HEAD
SHORT HEAD

BRACHIALIS

Radius
Ulna

Scapula
Humerus
Ribs
Teres major
TRICEPS BRACHII:
LONG HEAD
LATERAL HEAD
MEDIAL HEAD

ANCONEUS

Ulna
Radius

Bicipital aponeurosis

(a) Anterior view

(b) Posterior view

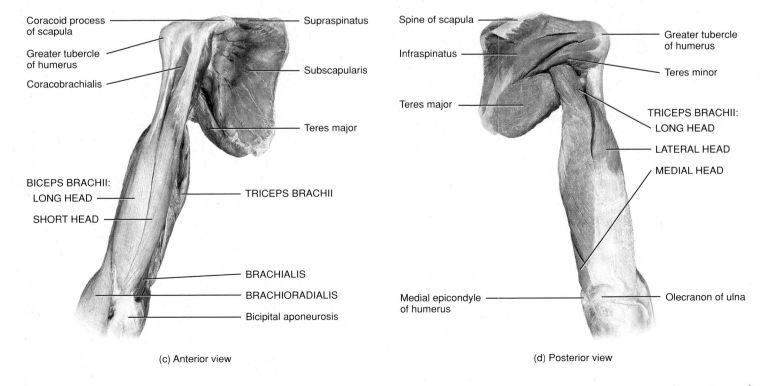

Coracoid process of scapula
Greater tubercle of humerus
Coracobrachialis

Supraspinatus
Subscapularis
Teres major

BICEPS BRACHII:
LONG HEAD
SHORT HEAD

TRICEPS BRACHII

BRACHIALIS
BRACHIORADIALIS
Bicipital aponeurosis

(c) Anterior view

Spine of scapula
Infraspinatus
Teres major

Greater tubercle of humerus
Teres minor

TRICEPS BRACHII:
LONG HEAD
LATERAL HEAD
MEDIAL HEAD

Medial epicondyle of humerus
Olecranon of ulna

(d) Posterior view

FIGURE 11.19 CONTINUES

EXHIBIT 11.P Muscles of the Arm That Move the Radius and Ulna *(Figure 11.19)* **CONTINUED**

and (2) pronation; and the following actions on the humerus at the shoulder joint: (1) flexion and (2) extension. The same muscle may be mentioned more than once.

✔ **CHECKPOINT**

23. Flex your forearm. Which group of muscles is contracting? Which group of muscles must relax so that you can flex your forearm?

⬤ **FIGURE 11.19 CONTINUED** ▶

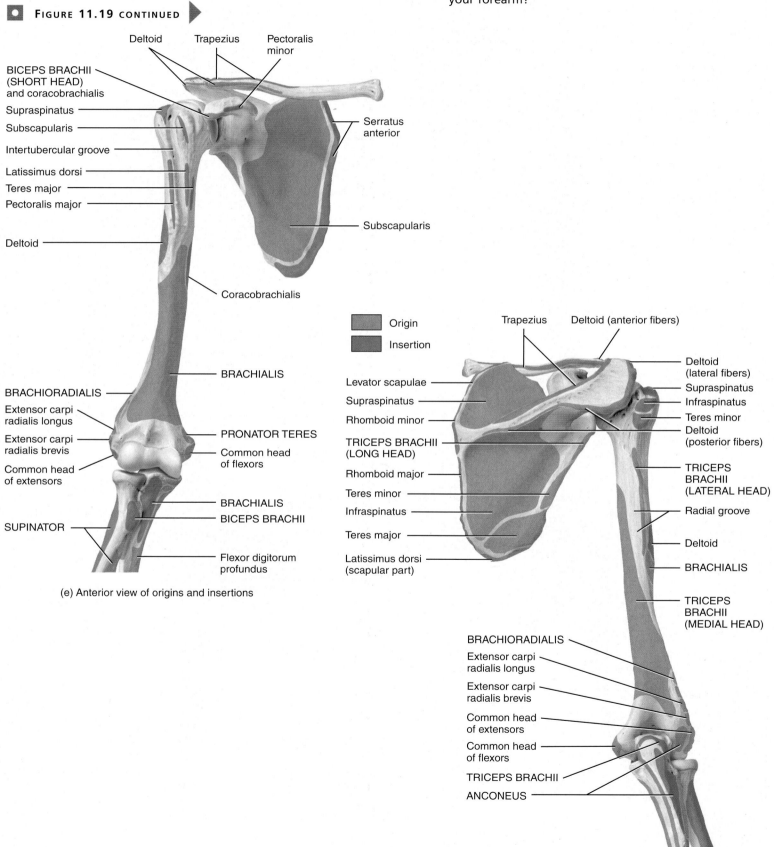

(e) Anterior view of origins and insertions

(f) Posterior view of origins and insertions

EXHIBIT 11.P **373**

View

Transverse plane

MEDIAL

POSTERIOR

LATERAL

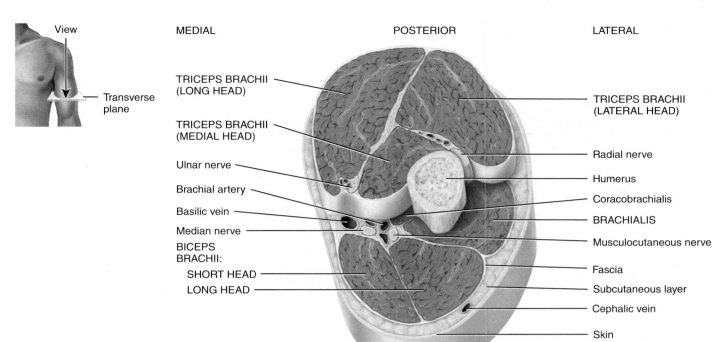

TRICEPS BRACHII
(LONG HEAD)

TRICEPS BRACHII
(MEDIAL HEAD)

Ulnar nerve

Brachial artery

Basilic vein

Median nerve

BICEPS
BRACHII:

SHORT HEAD

LONG HEAD

TRICEPS BRACHII
(LATERAL HEAD)

Radial nerve

Humerus

Coracobrachialis

BRACHIALIS

Musculocutaneous nerve

Fascia

Subcutaneous layer

Cephalic vein

Skin

ANTERIOR

(g) Superior view of transverse section of arm

POSTERIOR

MEDIAL

LATERAL

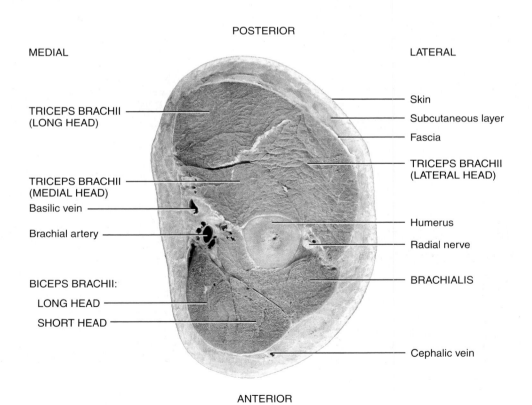

TRICEPS BRACHII
(LONG HEAD)

TRICEPS BRACHII
(MEDIAL HEAD)

Basilic vein

Brachial artery

BICEPS BRACHII:

LONG HEAD

SHORT HEAD

Skin

Subcutaneous layer

Fascia

TRICEPS BRACHII
(LATERAL HEAD)

Humerus

Radial nerve

BRACHIALIS

Cephalic vein

ANTERIOR

(h) Superior view of transverse section of arm

FIGURE **11.19** CONTINUES ▶

EXHIBIT 11.P Muscles of the Arm That Move the Radius and Ulna *(Figure 11.19)* CONTINUED

◼ FIGURE **11.19** CONTINUED ▶

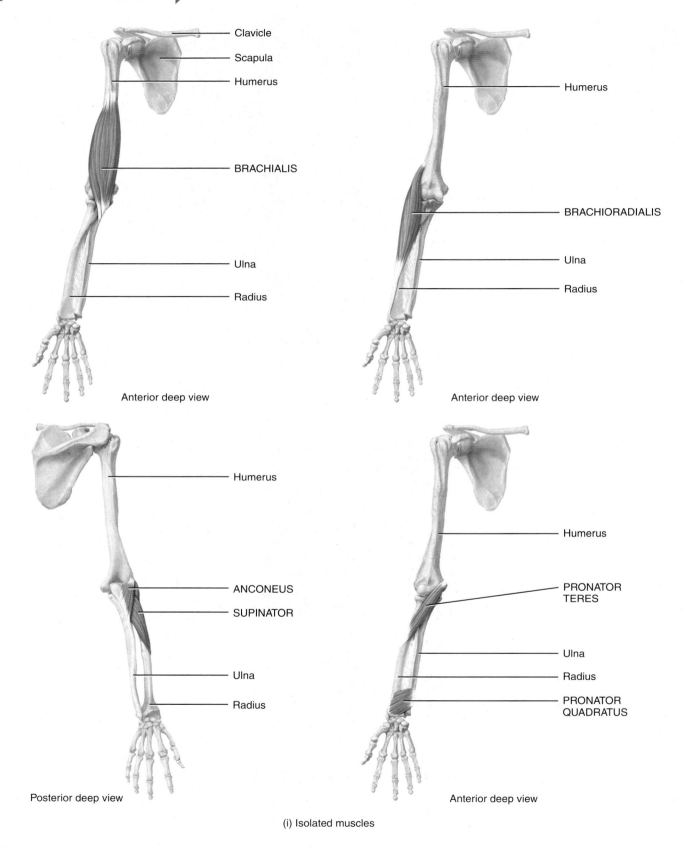

Anterior deep view

Anterior deep view

Posterior deep view

Anterior deep view

(i) Isolated muscles

❓ **Which muscles are the most powerful flexor and the most powerful extensor of the forearm?**

EXHIBIT 11.Q **375**

EXHIBIT 11.Q

Muscles of the Forearm That Move the Wrist, Hand, and Digits *(Figure 11.20)*

OBJECTIVE

• Describe the origin, insertion, action, and innervation of the muscles of the forearm that move the wrist, hand, and digits.

Muscles of the forearm that move the wrist, hand, and digits are many and varied (Figure 11.20). Those in this group that act on the digits are known as **extrinsic muscles of the hand** (*ex*=outside) because they originate *outside* the hand and insert within it. As you will see, the names for the muscles that move the wrist, hand, and digits give some indication of their origin, insertion, or action. Based on location and function, the muscles of the forearm are divided into two groups: (1) anterior compartment muscles and (2) posterior compartment muscles. The muscles of the **anterior**

(flexor) compartment of the forearm share a common origin on the medial epicondyle of the humerus (5 of the 8 muscles), typically insert on the carpals, metacarpals, and phalanges, and function primarily as flexors. The bellies of these muscles form the bulk of the forearm. One of the muscles in the superficial anterior compartment, the palmaris longus muscle, is missing in about 20 percent of individuals (usually in the left forearm) and is commonly used for tendon repair. The muscles of the **posterior (extensor) compartment of the forearm** share a common origin on the lateral epicondyle of the humerus (8 of the 12 muscles), insert on the metacarpals and phalanges, and function as extensors. Within each compartment, the muscles are grouped as superficial or deep.

MUSCLE	ORIGIN	INSERTION	ACTION	INNERVATION
SUPERFICIAL ANTERIOR (FLEXOR) COMPARTMENT OF THE FOREARM				
Flexor carpi radialis (FLEK-sor KAR-pē rā'-dē-Ā-lis: *flexor*=decreases angle at joint; *carpi*=of the wrist; *radi-*=radius)	Medial epicondyle of humerus	Second and third metacarpals	Flexes and abducts (radial deviation) hand at wrist joint	Median nerve
Palmaris longus (pal-MA-ris LON-gus; *palma*=palm; *longus*=long)	Medial epicondyle of humerus	Flexor retinaculum and palmar aponeurosis (fascia in center of palm)	Weakly flexes hand at wrist joint	Median nerve
Flexor carpi ulnaris (ūl-NAR-is; *ulnar-*=ulna)	Medial epicondyle of humerus and superior posterior border of ulna	Pisiform, hamate, and base of fifth metacarpal	Flexes and adducts (ulnar deviation) hand at wrist joint	Ulnar nerve
Flexor digitorum superficialis (di-ji-TOR-um soo'-per-fish'-ē-Ā-lis; *digit*=finger or toe; *superficialis*=closer to surface)	Medial epicondyle of humerus, coronoid process of ulna, and a ridge along lateral margin of anterior surface (anterior oblique line) of radius	Middle phalanx of each finger*	Flexes middle phalanx of each finger at proximal interphalangeal joint, proximal phalanx of each finger at metacarpophalangeal joint, and hand at wrist joint	Median nerve
DEEP ANTERIOR (FLEXOR) COMPARTMENT OF THE FOREARM				
Flexor pollicis longus (POL-li-sis; *pollic-*=thumb)	Anterior surface of radius and interosseous membrane (sheet of fibrous tissue that holds shafts of ulna and radius together)	Base of distal phalanx of thumb	Flexes distal phalanx of thumb at interphalangeal joint	Median nerve
Flexor digitorum profundus (prō-FUN-dus=deep)	Anterior medial surface of body of ulna	Base of distal phalanx of each finger	Flexes distal and middle phalanges of each finger at interphalangeal joints, proximal phalanx of each finger at metacarpophalangeal joint, and hand at wrist joint	Median and ulnar nerves
SUPERFICIAL POSTERIOR (EXTENSOR) COMPARTMENT OF THE FOREARM				
Extensor carpi radialis longus (eks-TEN-sor=increases angle at joint)	Lateral supracondylar ridge of humerus	Second metacarpal	Extends and abducts (radial deviation) hand at wrist joint	Radial nerve
Extensor carpi radialis brevis (BREV-is=short)	Lateral epicondyle of humerus	Third metacarpal	Extends and abducts (radial deviation) hand at wrist joint	Radial nerve

EXHIBIT 11.Q Muscles of the Forearm That Move the Wrist, Hand, and Digits *(Figure 11.20)* CONTINUED

MUSCLE	ORIGIN	INSERTION	ACTION	INNERVATION
Extensor digitorum	Lateral epicondyle of humerus	Distal and middle phalanges of each finger	Extends distal and middle phalanges of each finger at interphalangeal joints, proximal phalanx of each finger at metacarpophalangeal joint, and hand at wrist joint	Radial nerve
Extensor digiti minimi (DIJ-i-tē MIN-i-mē; *digit*=finger or toe; *minimi*=smallest)	Lateral epicondyle of humerus	Tendon of extensor digitorum on fifth phalanx	Extends proximal phalanx of little finger at metacarpophalangeal joint and hand at wrist joint	Deep radial nerve
Extensor carpi ulnaris	Lateral epicondyle of humerus and posterior border of ulna	Fifth metacarpal	Extends and adducts (ulnar deviation) hand at wrist joint	Deep radial nerve
DEEP POSTERIOR (EXTENSOR) COMPARTMENT OF THE FOREARM				
Abductor pollicis longus (ab-DUK-tor=moves part away from midline)	Posterior surface of middle of radius and ulna and interosseous membrane	First metacarpal	Abducts and extends thumb at carpometacarpal joint and abducts hand at wrist joint	Deep radial nerve
Extensor pollicis brevis	Posterior surface of middle of radius and interosseous membrane	Base of proximal phalanx of thumb	Extends proximal phalanx of thumb at metacarpophalangeal joint, first metacarpal of thumb at carpometacarpal joint, and hand at wrist joint	Deep radial nerve
Extensor pollicis longus	Posterior surface of middle of ulna and interosseous membrane	Base of distal phalanx of thumb	Extends distal phalanx of thumb at interphalangeal joint, first metacarpal of thumb at carpometacarpal joint, and abducts hand at wrist joint	Deep radial nerve
Extensor indicis (IN-di-sis; *indicis*=index)	Posterior surface of ulna	Tendon of extensor digitorum of index finger	Extends distal and middle phalanges of index finger at interphalangeal joints, proximal phalanx of index finger at metacarpophalangeal joint, and hand at wrist joint	Deep radial nerve

*Reminder: The thumb or pollex is the first digit and has two phalanges—proximal and distal. The remaining digits, the fingers, are numbered II–V (2–5), and each has three phalanges—proximal, middle, and distal.

The **superficial anterior compartment** muscles are arranged in the following order from lateral to medial: pronator teres (discussed in Exhibit 11.P), **flexor carpi radialis, palmaris longus**, and **flexor carpi ulnaris** (the ulnar nerve and artery are just lateral to the tendon of this muscle at the wrist). The **flexor digitorum superficialis** muscle is deep to the other three muscles and is the largest superficial muscle in the forearm. These muscles make up the fleshy mass that is deep to the hairless skin of the anterior forearm. The common origin is on the medial epicondyle. The palmaris longus muscle inserts into the thickened palmar aponeurosis and lies superficial to the flexor retinaculum.

The **deep anterior compartment** muscles are arranged in the following order from lateral to medial: **flexor pollicis longus** (the only flexor of the distal phalanx of the thumb) and **flexor digitorum profundus** (ends in four tendons that insert into the distal phalanges of the fingers). The tendons of the flexor digitorum profundus, along with the more superficial flexor digitorum superficialis, pass with the flexor pollicis longus through the carpal tunnel. Also in this group is the pronator quadratus at the distal ends of the radius and ulna (discussed in Exhibit 11.P).

The **superficial posterior compartment** muscles are arranged in the following order from lateral to medial: **brachioradialis** (discussed in Exhibit 11.P), **extensor carpi radialis longus, extensor carpi radialis brevis, extensor digitorum** (occupies most of the posterior surface of the forearm and divides into four tendons that insert into the middle and distal phalanges of the fingers), **extensor digiti minimi** (a slender muscle usually connected to the extensor digitorum), and the **extensor carpi ulnaris**.

The **deep posterior compartment** muscles are arranged in the following order from lateral to medial: **abductor pollicis longus, extensor pollicis longus, extensor pollicis brevis**, and **extensor indicis**.

The tendons of the muscles of the forearm that attach to the wrist or continue into the hand, along with blood vessels and nerves, are held close to bones by strong fasciae. The tendons are also surrounded by tendon sheaths. At the wrist, the deep fascia is thickened into fibrous bands called **retinacula** (*retinacul*=holdfast). The **flexor retinaculum** is located over the palmar surface of the carpal bones. The long flexor tendons of the digits and wrist and the median nerve pass deep to the flexor retinaculum.

EXHIBIT 11.Q **377**

Figure 11.20 Muscles of the forearm that move the wrist, hand, and digits.

🔑 The anterior compartment muscles function as flexors, and the posterior compartment muscles function as extensors.

- Biceps brachii
- Brachialis
- Medial epicondyle of humerus
- Lateral epicondyle of humerus
- Tendon of biceps brachii
- Bicipital aponeurosis
- PRONATOR TERES
- BRACHIORADIALIS
- PALMARIS LONGUS
- FLEXOR CARPI RADIALIS
- FLEXOR CARPI ULNARIS
- EXTENSOR CARPI RADIALIS LONGUS
- FLEXOR DIGITORUM SUPERFICIALIS
- FLEXOR POLLICIS LONGUS
- PRONATOR QUADRATUS
- Radius
- Tendon of abductor pollicis longus
- Flexor retinaculum
- Palmar aponeurosis
- Metacarpal
- Tendon of flexor pollicis longus
- Tendons of flexor digitorum superficialis (splitting)
- Tendons of flexor digitorum profundus

- Humerus
- SUPINATOR
- FLEXOR DIGITORUM PROFUNDUS
- FLEXOR POLLICIS LONGUS
- PRONATOR QUADRATUS
- Tendon of flexor pollicis longus
- Tendons of flexor digitorum profundus

(a) Anterior superficial view

(b) Anterior intermediate view

(c) Anterior deep view

FIGURE 11.20 CONTINUES ▶

EXHIBIT 11.Q Muscles of the Forearm That Move the Wrist, Hand, and Digits *(Figure 11.20)* CONTINUED

The **extensor retinaculum** is located over the dorsal surface of the carpal bones. The extensor tendons of the wrist and digits pass deep to it.

CLINICAL CONNECTION | *Golfer's Elbow*

Golfer's elbow is a condition that can be caused by strain of the flexor muscles, especially the flexor carpi radialis, as a result of repetitive movements such as swinging a golf club. Strain can, however, be caused by many actions. Pianists, violinists, movers, weight lifters, bikers, and those who use computers are among those who may develop pain near the medial epicondyle (*medial epicondylitis*). •

Relating Muscles to Movements

Arrange the muscles in this exhibit according to the following actions on the wrist joint: (1) flexion, (2) extension, (3) abduction (radial deviation), and (4) adduction (ulnar deviation); the following actions on the fingers at the metacarpophalangeal joints: (1) flexion and (2) extension; the following actions on the fingers at the interphalangeal joints: (1) flexion and (2) extension; the following actions on the thumb at the carpometacarpal, metacarpophalangeal, and interphalangeal joints: (1) extension and (2) abduction; and the following action on the thumb at the interphalangeal joint: flexion. The same muscle may be mentioned more than once.

✓ CHECKPOINT

24. Which muscles and actions of the wrist, hand, thumb, and fingers are used when writing?

■ FIGURE 11.20 CONTINUED ▶

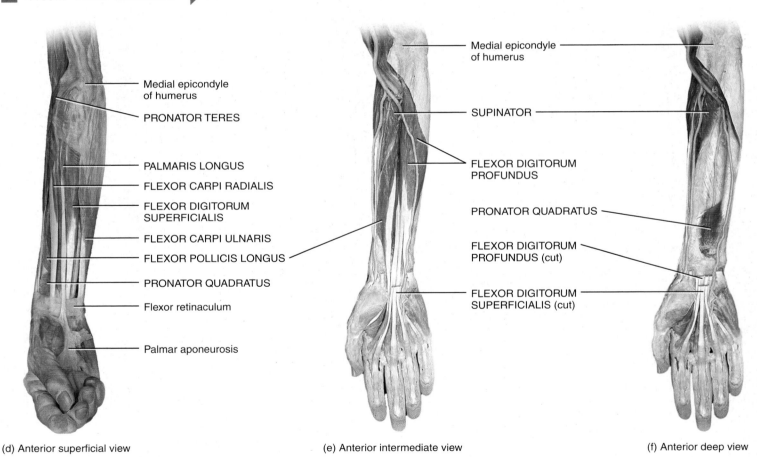

(d) Anterior superficial view

(e) Anterior intermediate view

(f) Anterior deep view

EXHIBIT 11.Q **379**

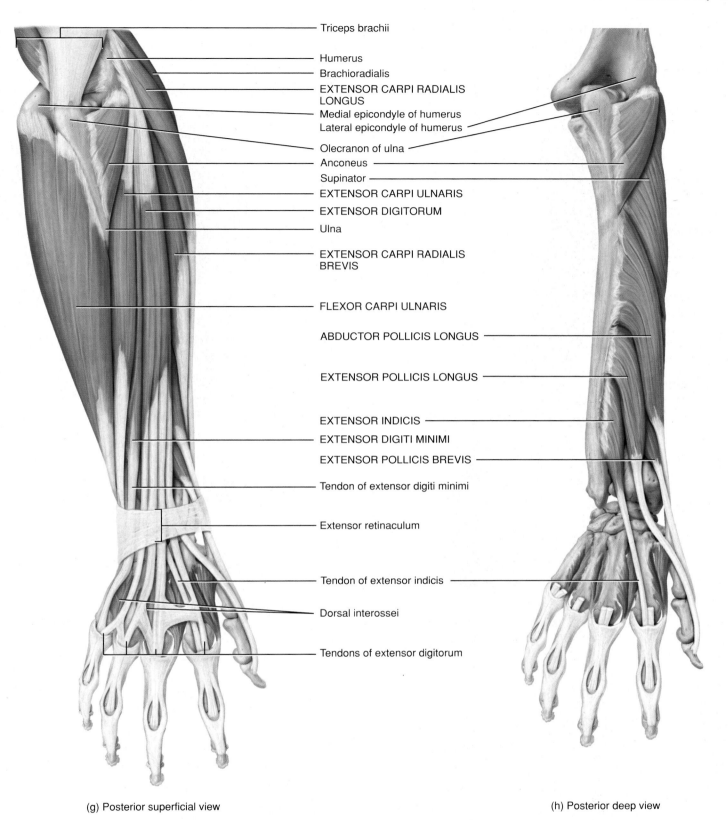

Triceps brachii

Humerus
Brachioradialis
EXTENSOR CARPI RADIALIS LONGUS
Medial epicondyle of humerus
Lateral epicondyle of humerus
Olecranon of ulna
Anconeus
Supinator
EXTENSOR CARPI ULNARIS
EXTENSOR DIGITORUM
Ulna
EXTENSOR CARPI RADIALIS BREVIS

FLEXOR CARPI ULNARIS

ABDUCTOR POLLICIS LONGUS

EXTENSOR POLLICIS LONGUS

EXTENSOR INDICIS
EXTENSOR DIGITI MINIMI
EXTENSOR POLLICIS BREVIS

Tendon of extensor digiti minimi

Extensor retinaculum

Tendon of extensor indicis

Dorsal interossei

Tendons of extensor digitorum

(g) Posterior superficial view

(h) Posterior deep view

FIGURE 11.20 CONTINUES ▶

EXHIBIT 11.Q Muscles of the Forearm That Move the Wrist, Hand, and Digits *(Figure 11.20)* **CONTINUED**

⬛ **FIGURE 11.20 CONTINUED** ▶

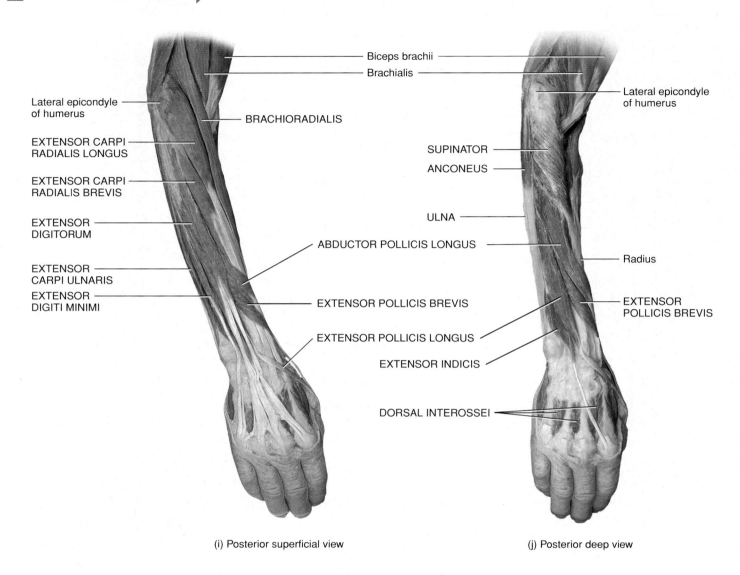

Biceps brachii

Brachialis

Lateral epicondyle of humerus

BRACHIORADIALIS

Lateral epicondyle of humerus

SUPINATOR

ANCONEUS

EXTENSOR CARPI RADIALIS LONGUS

EXTENSOR CARPI RADIALIS BREVIS

EXTENSOR DIGITORUM

ULNA

ABDUCTOR POLLICIS LONGUS

Radius

EXTENSOR CARPI ULNARIS

EXTENSOR DIGITI MINIMI

EXTENSOR POLLICIS BREVIS

EXTENSOR POLLICIS BREVIS

EXTENSOR POLLICIS LONGUS

EXTENSOR INDICIS

DORSAL INTEROSSEI

(i) Posterior superficial view

(j) Posterior deep view

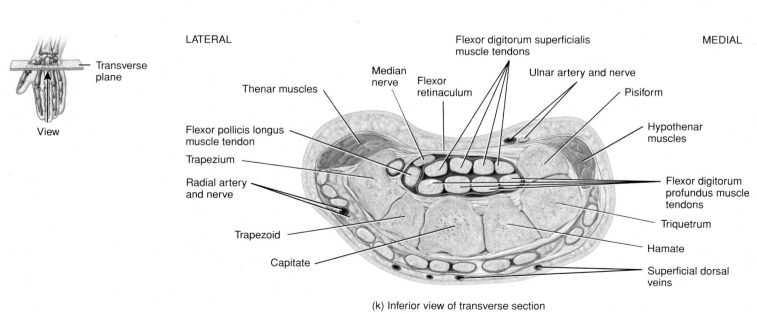

Transverse plane

View

LATERAL

MEDIAL

Flexor digitorum superficialis muscle tendons

Median nerve

Flexor retinaculum

Ulnar artery and nerve

Thenar muscles

Pisiform

Hypothenar muscles

Flexor pollicis longus muscle tendon

Trapezium

Radial artery and nerve

Flexor digitorum profundus muscle tendons

Triquetrum

Trapezoid

Hamate

Capitate

Superficial dorsal veins

(k) Inferior view of transverse section

EXHIBIT 11.Q **381**

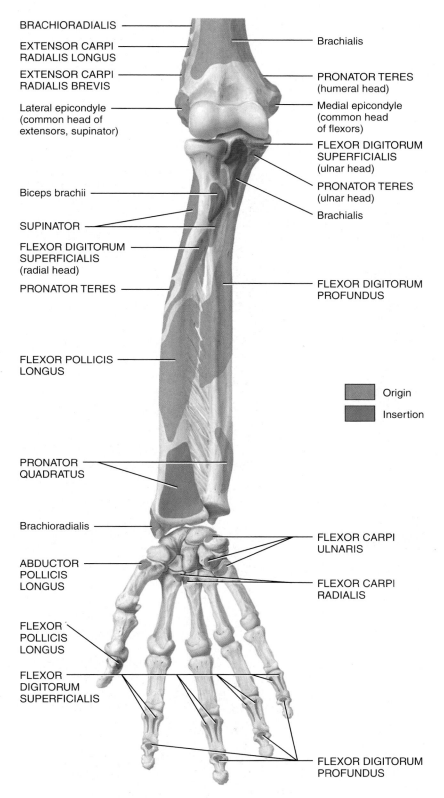

BRACHIORADIALIS

EXTENSOR CARPI
RADIALIS LONGUS

EXTENSOR CARPI
RADIALIS BREVIS

Lateral epicondyle
(common head of
extensors, supinator)

Biceps brachii

SUPINATOR

FLEXOR DIGITORUM
SUPERFICIALIS
(radial head)

PRONATOR TERES

FLEXOR POLLICIS
LONGUS

PRONATOR
QUADRATUS

Brachioradialis

ABDUCTOR
POLLICIS
LONGUS

FLEXOR
POLLICIS
LONGUS

FLEXOR
DIGITORUM
SUPERFICIALIS

Brachialis

PRONATOR TERES
(humeral head)

Medial epicondyle
(common head
of flexors)

FLEXOR DIGITORUM
SUPERFICIALIS
(ulnar head)

PRONATOR TERES
(ulnar head)

Brachialis

FLEXOR DIGITORUM
PROFUNDUS

Origin

Insertion

FLEXOR CARPI
ULNARIS

FLEXOR CARPI
RADIALIS

FLEXOR DIGITORUM
PROFUNDUS

(l) Anterior view of origins and insertions

FIGURE 11.20 CONTINUES ▶

EXHIBIT 11.Q Muscles of the Forearm That Move the Wrist, Hand, and Digits *(Figure 11.20)* **CONTINUED**

◼ **FIGURE 11.20 CONTINUED** ▶

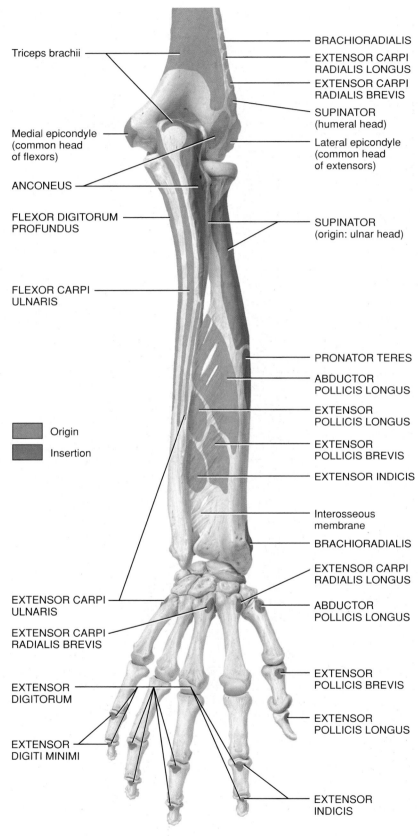

Triceps brachii

BRACHIORADIALIS

EXTENSOR CARPI
RADIALIS LONGUS

EXTENSOR CARPI
RADIALIS BREVIS

SUPINATOR
(humeral head)

Medial epicondyle
(common head
of flexors)

Lateral epicondyle
(common head
of extensors)

ANCONEUS

SUPINATOR
(origin: ulnar head)

FLEXOR DIGITORUM
PROFUNDUS

FLEXOR CARPI
ULNARIS

PRONATOR TERES

ABDUCTOR
POLLICIS LONGUS

EXTENSOR
POLLICIS LONGUS

Origin

Insertion

EXTENSOR
POLLICIS BREVIS

EXTENSOR INDICIS

Interosseous
membrane

BRACHIORADIALIS

EXTENSOR CARPI
RADIALIS LONGUS

EXTENSOR CARPI
ULNARIS

ABDUCTOR
POLLICIS LONGUS

EXTENSOR CARPI
RADIALIS BREVIS

EXTENSOR
POLLICIS BREVIS

EXTENSOR
DIGITORUM

EXTENSOR
POLLICIS LONGUS

EXTENSOR
DIGITI MINIMI

EXTENSOR
INDICIS

(m) Posterior view of origin and insertions

❓ **What structures pass through the flexor retinaculum?**

EXHIBIT 11.R **383**

EXHIBIT 11.R

Muscles of the Palm That Move the Digits—
Intrinsic Muscles of the Hand *(Figure 11.21)*

● OBJECTIVE

• Describe the origin, insertion, action, and innervation of the muscles of the palm that move the digits—the intrinsic muscles of the hand.

Several of the muscles discussed in Exhibit 11.P move the digits in various ways and are known as extrinsic muscles of the hand. They produce the powerful but crude movements of the digits. The **intrinsic muscles of the hand** in the palm produce the weak but intricate and precise movements of the digits that characterize the human hand (Figure 11.21a). The muscles in this group are so named because their origins and insertions are *within* the hand.

The intrinsic muscles of the hand are divided into three groups: (1) **thenar**, (2) **hypothenar**, and (3) **intermediate**. The thenar muscles include the abductor pollicis brevis, opponens pollicis, flexor pollicis brevis, and adductor pollicis (acts on the thumb but is not in the thenar eminence). The **abductor pollicis brevis** is a thin,

MUSCLE	ORIGIN	INSERTION	ACTION	INNERVATION
THENAR (LATERAL ASPECT OF PALM)				
Abductor pollicis brevis (ab-DUK-tor POL-li-sis BREV-is; *abductor*=moves part away from middle; *pollicis-*=the thumb; *brevis*=short)	Flexor retinaculum, scaphoid, and trapezium	Lateral side of proximal phalanx of thumb	Abducts thumb at carpometacarpal joint	Median nerve
Opponens pollicis (op-PŌ-nenz=opposes)	Flexor retinaculum and trapezium	Lateral side of metacarpal I (thumb)	Moves thumb across palm to meet any finger (opposition) at the carpometacarpal joint	Median nerve
Flexor pollicis brevis (FLEK-sor=decreases angle at joint)	Flexor retinaculum, trapezium, capitate, and trapezoid	Lateral side of proximal phalanx of thumb	Flexes thumb at carpometacarpal and metacarpophalangeal joints	Median and ulnar nerves
Adductor pollicis (ad-DUK-tor =moves part toward midline)	Oblique head: capitate and metacarpals II and III Transverse head: metacarpal III	Medial side of proximal phalanx of thumb by a tendon containing a sesamoid bone	Adducts thumb at carpometacarpal and metacarpophalangeal joints	Ulnar nerve
HYPOTHENAR (MEDIAL ASPECT OF PALM)				
Abductor digiti minimi (DIJ-i-tē MIN-i-mē; *digit*=finger or toe; *minimi*=smallest)	Pisiform and tendon of flexor carpi ulnaris	Medial side of proximal phalanx of little finger	Abducts and flexes little finger at metacarpophalangeal joint	Ulnar nerve
Flexor digiti minimi brevis	Flexor retinaculum and hamate	Medial side of proximal phalanx of little finger	Flexes little finger at carpometacarpal and metacarpophalangeal joints	Ulnar nerve
Opponens digiti minimi	Flexor retinaculum and hamate	Medial side of metacarpal V (little finger)	Moves little finger across palm to meet thumb (opposition) at the carpometacarpal joint	Ulnar nerve
INTERMEDIATE (MIDPALMAR)				
Lumbricals (LUM-bri-kals; *lumbric*=earthworm) (four muscles)	Lateral sides of tendons and flexor digitorum profundus of each finger	Lateral sides of tendons of extensor digitorum on proximal phalanges of each finger	Flex each finger at metacarpophalangeal joints and extend each finger at interphalangeal joints	Median and ulnar nerves
Palmar interossei (PAL-mar in′-ter-OS-ē-ī; *palmar*=palm; *inter-*=between; *ossei*=bones) (three muscles)	Sides of shafts of metacarpals of all digits (except the middle one)	Sides of bases of proximal phalanges of all digits (except the middle one)	Adduct and flex each finger except the middle finger at metacarpophalangeal joints and extend these fingers at interphalangeal joints	Ulnar nerve
Dorsal interossei (DOR-sal=back surface) (four muscles)	Adjacent sides of metacarpals	Proximal phalanx of each finger	Abduct fingers 2–4 at metacarpophalangeal joints; flex fingers 2–4 at metacarpophalangeal joints; and extend each finger at interphalangeal joints	Ulnar nerve

EXHIBIT 11.R Muscles of the Palm That Move the Digits—Intrinsic Muscles of the Hand *(Figure 11.21)* CONTINUED

short, relatively broad superficial muscle on the lateral side of the thenar eminence. The **flexor pollicis brevis** is a short, wide muscle that is medial to the abductor pollicis brevis muscle. The **opponens pollicis** is a small, triangular muscle that is deep to the flexor pollicis brevis and abductor pollicis brevis muscles. The three thenar muscles plus the adductor pollicis form the **thenar eminence**, the lateral rounded contour on the palm that is also called the *ball of the thumb*. The **adductor pollicis** also acts on the thumb. The muscle is fan-shaped and has two heads (oblique and transverse) separated by a gap through which the radial artery passes.

The three hypothenar muscles act on the little finger and form the **hypothenar eminence**, the medial rounded contour on the palm that is also called the ball of the little finger. The hypothenar muscles are the abductor digiti minimi, flexor digiti minimi brevis, and opponens digiti minimi. The **abductor digiti minimi** is a short, wide muscle and is the most superficial of the hypothenar muscles. It is a powerful muscle that plays an important role in grasping an object with outspread fingers. The **flexor digiti minimi brevis** muscle is also short and wide and is lateral to the abductor digiti minimi muscle. The **opponens digiti minimi** muscle is triangular and deep to the other two hypothenar muscles.

The 11 intermediate (midpalmar) muscles include the lumbricals, palmar interossei, and dorsal interossei. The **lumbricals**, as their name indicates, are worm-shaped. They originate from and insert into the tendons of other muscles (flexor digitorum profundus and extensor digitorum). The **palmar interossei** are the smallest and most anterior of the interossei muscles. The **dorsal interossei** are the most posterior of this series of muscles. Both sets of interossei muscles are located between the metacarpals and are important in abduction, adduction, flexion, and extension of the fingers, and in movements in skilled activities such as writing, typing, and playing a piano.

The functional importance of the hand is readily apparent when you consider that certain hand injuries can result in permanent disability. Most of the dexterity of the hand depends on movements of the thumb. The general activities of the hand are free motion, power grip (forcible movement of the fingers and thumb against the palm, as in squeezing), precision handling (a change in position of a handled object that requires exact control of finger and thumb positions as in winding a watch or threading a needle), and pinch (compression between the thumb and index finger or between the thumb and first two fingers).

Movements of the thumb are very important in the precise activities of the hand, and they are defined in different planes from comparable movements of other digits because the thumb is positioned at a right angle to the other digits. The five principal movements of the thumb are illustrated in Figure 11.21a and include *flexion* (movement of the thumb medially across the palm), *extension* (movement of the thumb laterally away from the palm), *abduction* (movement of the thumb in an anteroposterior plane away from the palm), *adduction* (movement of the thumb in an anteroposterior plane toward the palm), and *opposition* (movement of the thumb across the palm so that the tip of the thumb meets the tip of a finger). Opposition is the single most distinctive digital movement that gives humans and other primates the ability to grasp and manipulate objects precisely.

CLINICAL CONNECTION | *Carpal Tunnel Syndrome*

The **carpal tunnel** is a narrow passageway formed anteriorly by the flexor retinaculum and posteriorly by the carpal bones. Through this tunnel pass the median nerve, the most superficial structure, and the long flexor tendons for the digits (see Figure 11.20j). Structures within the carpal tunnel, especially the median nerve, are vulnerable to compression, and the resulting condition is called **carpal tunnel syndrome**. Compression of the median nerve leads to sensory changes over the lateral side of the hand and muscle weakness in the thenar eminence. This results in pain, numbness, and tingling of the fingers. The condition may be caused by inflammation of the digital tendon sheaths, fluid retention, excessive exercise, infection, trauma, and/or repetitive activities that involve flexion of the wrist, such as keyboarding, cutting hair, and playing a piano. Treatment may involve the use of nonsteroidal anti-inflammatory drugs (such as ibuprofen or aspirin), wearing a wrist splint, corticosteroid injections, or surgery to cut the flexor retinaculum and release pressure on the median nerve. •

Relating Muscles to Movements

Arrange the muscles in this exhibit according to the following actions on the thumb at the carpometacarpal and metacarpophalangeal joints: (1) abduction, (2) adduction, (3) flexion, and (4) opposition; and the following actions on the fingers at the metacarpophalangeal and interphalangeal joints: (1) abduction, (2) adduction, (3) flexion, and (4) extension. The same muscle may be mentioned more than once.

✓ CHECKPOINT

25. How do the actions of the extrinsic and intrinsic muscles of the hand differ?

Figure 11.21 Muscles of the palm that move the digits—intrinsic muscles of the hand.

🔑 The intrinsic muscles of the hand produce the intricate and precise movements of the digits that characterize the human hand.

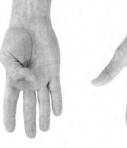   

| Flexion | Extension | Abduction | Adduction | Opposition |

(a) Movements of the thumb

EXHIBIT 11.R **385**

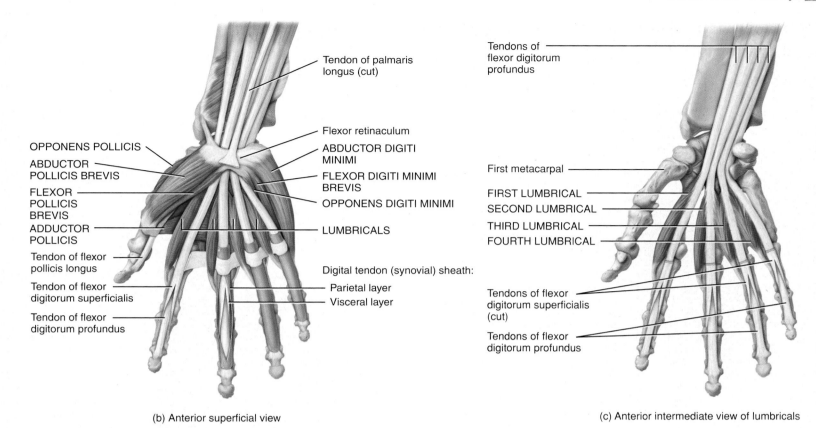

OPPONENS POLLICIS

ABDUCTOR POLLICIS BREVIS

FLEXOR POLLICIS BREVIS

ADDUCTOR POLLICIS

Tendon of flexor pollicis longus

Tendon of flexor digitorum superficialis

Tendon of flexor digitorum profundus

Tendon of palmaris longus (cut)

Flexor retinaculum

ABDUCTOR DIGITI MINIMI

FLEXOR DIGITI MINIMI BREVIS

OPPONENS DIGITI MINIMI

LUMBRICALS

Digital tendon (synovial) sheath:
Parietal layer
Visceral layer

(b) Anterior superficial view

Tendons of flexor digitorum profundus

First metacarpal

FIRST LUMBRICAL

SECOND LUMBRICAL

THIRD LUMBRICAL

FOURTH LUMBRICAL

Tendons of flexor digitorum superficialis (cut)

Tendons of flexor digitorum profundus

(c) Anterior intermediate view of lumbricals

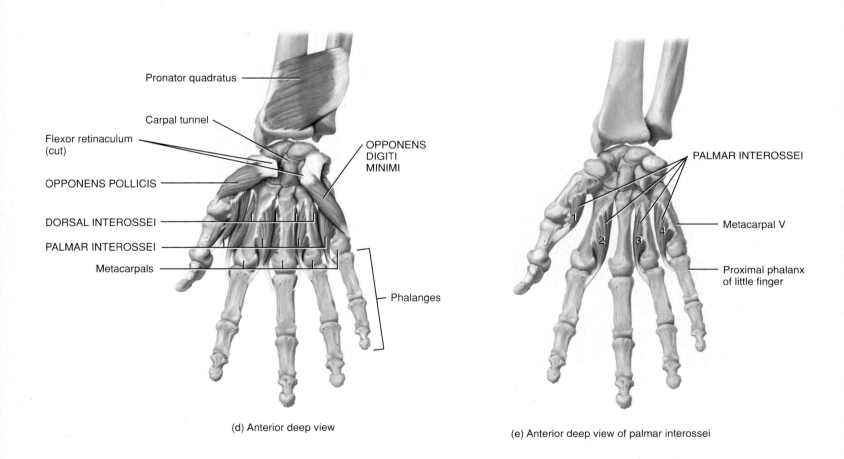

Pronator quadratus

Carpal tunnel

Flexor retinaculum (cut)

OPPONENS POLLICIS

DORSAL INTEROSSEI

PALMAR INTEROSSEI

Metacarpals

OPPONENS DIGITI MINIMI

Phalanges

(d) Anterior deep view

PALMAR INTEROSSEI

Metacarpal V

Proximal phalanx of little finger

(e) Anterior deep view of palmar interossei

FIGURE 11.21 CONTINUES ▶

EXHIBIT 11.R Muscles of the Palm That Move the Digits—Intrinsic Muscles of the Hand *(Figure 11.21)* CONTINUED

⬤ FIGURE **11.21** CONTINUED ▶

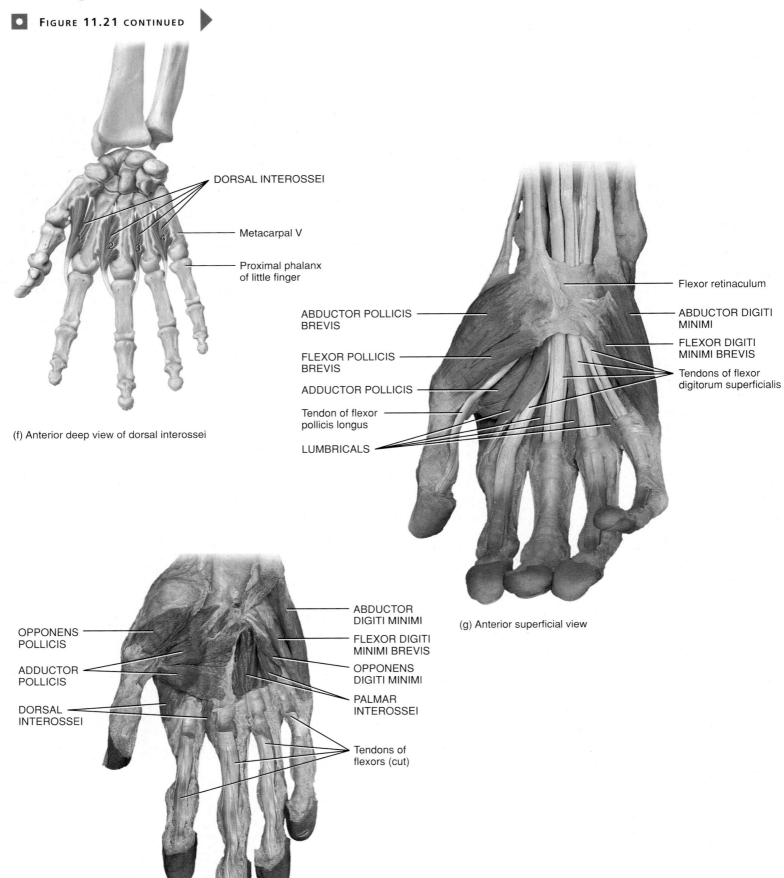

(f) Anterior deep view of dorsal interossei

(g) Anterior superficial view

(h) Anterior deep view

EXHIBIT 11.R **387**

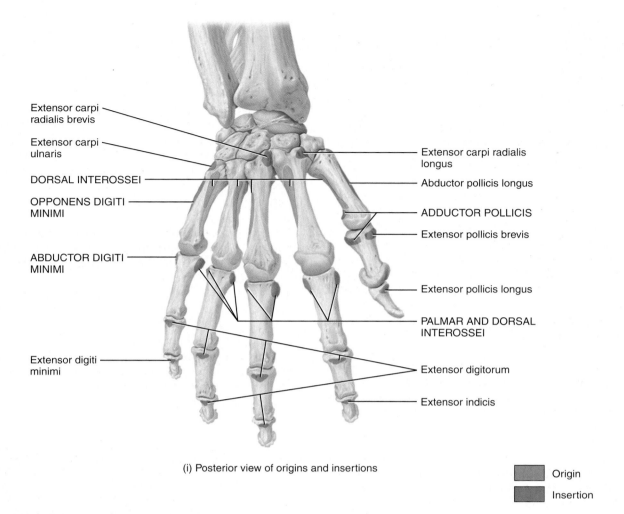

Extensor carpi radialis brevis
Extensor carpi ulnaris
DORSAL INTEROSSEI
OPPONENS DIGITI MINIMI
ABDUCTOR DIGITI MINIMI
Extensor digiti minimi

Extensor carpi radialis longus
Abductor pollicis longus
ADDUCTOR POLLICIS
Extensor pollicis brevis
Extensor pollicis longus
PALMAR AND DORSAL INTEROSSEI
Extensor digitorum
Extensor indicis

(i) Posterior view of origins and insertions

Origin
Insertion

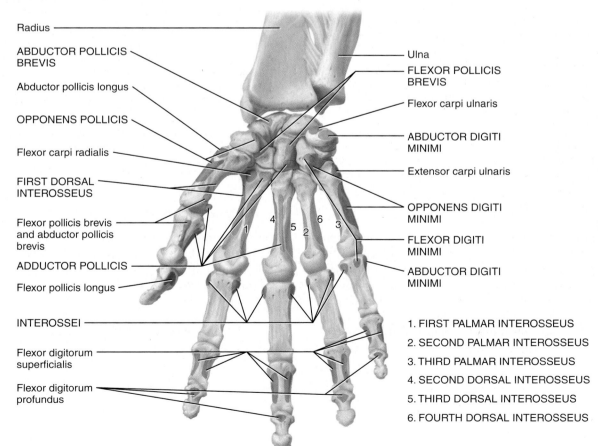

Radius
ABDUCTOR POLLICIS BREVIS
Abductor pollicis longus
OPPONENS POLLICIS
Flexor carpi radialis
FIRST DORSAL INTEROSSEUS
Flexor pollicis brevis and abductor pollicis brevis
ADDUCTOR POLLICIS
Flexor pollicis longus
INTEROSSEI
Flexor digitorum superficialis
Flexor digitorum profundus

Ulna
FLEXOR POLLICIS BREVIS
Flexor carpi ulnaris
ABDUCTOR DIGITI MINIMI
Extensor carpi ulnaris
OPPONENS DIGITI MINIMI
FLEXOR DIGITI MINIMI
ABDUCTOR DIGITI MINIMI

1. FIRST PALMAR INTEROSSEUS
2. SECOND PALMAR INTEROSSEUS
3. THIRD PALMAR INTEROSSEUS
4. SECOND DORSAL INTEROSSEUS
5. THIRD DORSAL INTEROSSEUS
6. FOURTH DORSAL INTEROSSEUS

Muscles of the thenar eminence act on which digit?

(j) Anterior view of origins and insertions

EXHIBIT 11.S

Muscles of the Gluteal Region That Move the Femur *(Figure 11.22)*

OBJECTIVE

• Describe the origin, insertion, action, and innervation of the muscles of the gluteal region that move the femur.

As you will see, muscles of the lower limbs are larger and more powerful than those of the upper limbs because of differences in function. While upper limb muscles are characterized by versatility of movement, lower limb muscles function in stability, locomotion, and maintenance of posture. In addition, muscles of the lower limbs often cross two joints and act equally on both.

The majority of muscles that move the femur (thigh bone) originate on the pelvic girdle and insert on the femur (Figure 11.22). The **psoas major** and **iliacus** muscles share a common insertion (lesser trochanter of femur) and are collectively known as the **iliopsoas** muscle. The iliopsoas is the most powerful flexor of the thigh and therefore is important for walking, running, and standing.

There are three gluteal muscles: gluteus maximus, gluteus medius, and gluteus minimus. The **gluteus maximus** is the largest and heaviest of the three muscles and is one of the largest muscles in the body. This large, superficial muscle of the buttock region is quadrilateral in form. It is coarsely fasciculated with the fascicles directed inferiorly and laterally. It is a powerful extensor of the thigh or, in its reverse muscle action (RMA), a powerful extensor of the torso at the hip joint. Its insertion into the iliotibial tract helps hold (lock) the knee in extension. This action is critical for standing and also for walking when the center of gravity is passing over the extended knee as each step is taken. The **gluteus medius** is mostly deep to the gluteus maximus and is a powerful abductor of the hip joint. It is a common site for an intramuscular injection. The **gluteus minimus** is the smallest of the gluteal muscles and lies deep to the gluteus medius. Its insertion, like that of the gluteus medius, is on the greater trochanter, and therefore the actions of the two muscles are essentially the same. Both muscles abduct and medially rotate the femur. This is very important in walking and especially when running. When one foot is raised off the ground, these two muscles abduct the hip joint of the supported limb (limb on the ground) and the RMA abducts the pelvic bone toward the greater trochanter so that it does not adduct under the gravitational load and collapse toward the unsupported side. At the same time, the rotary action helps swing the pelvis forward with each step.

The **tensor fasciae latae** muscle is a fusiform muscle located on the lateral surface of the thigh. The muscle lies beneath the *fascia lata*, which is the dense connective tissue fascia that encircles the entire thigh. The tensor fasciae latae muscle attaches to the underside of the fascia on the lateral side along with the gluteus maximus muscle. At this junction of these muscles the fascia thickens to form the **iliotibial tract (IT band)**, a strong band of connective tissue that spans the lateral aspect of the thigh from the ilium to the tibia. The tract inserts into the lateral condyle of the tibia. The iliotibial tract, approximately three fingers wide, has such strength that it commonly flattens the lateral side of the otherwise rounded thigh. The belly of the tensor fasciae latae lies between the gluteus medius and sartorius muscles. It keeps the iliotibial tract taut and thus helps maintain the extended knee in the erect posture.

The **piriformis, obturator internus, obturator externus, superior gemellus, inferior gemellus,** and **quadratus femoris** muscles are all deep to the gluteus maximus muscle and function as lateral rotators of the femur at the hip joint. These muscles share essentially the same insertion, on or near the greater trochanter of the femur, and they all have essentially the same actions—to laterally rotate the femur. The piriformis muscle is of particular interest because of its relationship to the sciatic nerve (the main nerve supply to the posterior thigh and entire leg musculature) (Figure 11.22c). If this muscle is injured or swells from overwork, it can put pressure on the sciatic nerve that can result in pain and muscle weakness.

CLINICAL CONNECTION | *Groin Pull*

The five major muscles of the inner thigh function to move the legs medially. This muscle group is important in activities such as sprinting, hurdling, and horseback riding. A rupture or tear of one or more of these muscles can cause a **groin pull**. Groin pulls most often occur during sprinting or twisting, or from kicking a solid, perhaps stationary object. Symptoms of a groin pull may be sudden, or may not surface until the day after the injury and include sharp pain in the inguinal region, swelling, bruising, or inability to contract the muscles. As with most strain injuries, treatment involves PRICE therapy, which stands for *Protection, Rest, Ice, Compression,* and *Elevation.* After the injured part is protected from further injury, ice should be applied immediately, and the injured part should be elevated and rested. An elastic bandage should be applied, if possible, to compress the injured tissue. •

Relating Muscles to Movements

Arrange the muscles in this exhibit according to the following actions on the thigh at the hip joint: (1) flexion, (2) extension, (3) abduction, (4) adduction, (5) medial rotation, and (6) lateral rotation. The same muscle may be mentioned more than once.

✔ CHECKPOINT

26. What is the origin of most muscles that move the femur?

EXHIBIT 11.S **389**

MUSCLE	ORIGIN	INSERTION	ACTION	INNERVATION
Iliopsoas (il-ē-ō-SŌ-as)				
Psoas major (SŌ-as; *psoa*=a muscle of the loin; *major*=larger)	Transverse processes and bodies of lumbar vertebrae	With iliacus into lesser trochanter of femur	Psoas major and iliacus muscles acting together flex thigh at hip joint, rotate thigh laterally, and flex trunk on the hip as in sitting up from the supine position	Lumbar spinal nerves L2–L3
Iliacus (il'-ē-A-cus; *iliac-*=ilium)	Iliac fossa and sacrum	With psoas major into lesser trochanter of femur		Femoral nerve
Gluteus maximus (GLOO-tē-us MAK-si-mus; *glute-*=rump or buttock; *maximus*=largest)	Iliac crest, sacrum, coccyx, and aponeurosis of sacrospinalis	Iliotibial tract of fascia lata and superior lateral part of linea aspera (gluteal tuberosity) under greater trochanter of femur	Extends thigh at hip joint and laterally rotates thigh	Inferior gluteal nerve
Gluteus medius (MĒ-dē-us; *medi-*=middle)	Ilium	Greater trochanter of femur	Abducts thigh at hip joint and medially rotates thigh	Superior gluteal nerve
Gluteus minimus (MIN-i-mus; *minimus*=smallest)	Ilium	Greater trochanter of femur	Abducts thigh at hip joint and medially rotates thigh	Superior gluteal nerve
Tensor fasciae latae (TEN-sor FA-shē-ē LĀ-tē; *tensor*=makes tense; *fasciae*=of the band; *lat-*=wide)	Iliac crest	Tibia by way of the iliotibial tract	Flexes and abducts thigh at hip joint	Superior gluteal nerve
Piriformis (pir-i-FOR-mis; *piri-*=pear; *form-*=shape)	Anterior sacrum	Superior border of greater trochanter of femur	Laterally rotates and abducts thigh at hip joint	Sacral spinal nerves S1 or S2, mainly S1
Obturator internus (OB-too-rā'-tor in-TER-nus; *obturator*=obturator foramen; *intern-*=inside)	Inner surface of obturator foramen, pubis, and ischium	Medial surface of greater trochanter of femur	Laterally rotates and abducts thigh at hip joint	Nerve to obturator internus
Obturator externus (ex-TER-nus; *extern-*=outside)	Outer surface of obturator membrane	Deep depression inferior to greater trochanter (trochanteric fossa) of femur	Laterally rotates and abducts thigh at hip joint	Obturator nerve
Superior gemellus (jem-EL-lus; *superior*=above; *gemell-*=twins)	Ischial spine	Medial surface of greater trochanter of femur	Laterally rotates and abducts thigh at hip joint	Nerve to obturator internus
Inferior gemellus (*inferior*=below)	Ischial tuberosity	Medial surface of greater trochanter of femur	Laterally rotates and abducts thigh at hip joint	Nerve to quadratus femoris
Quadratus femoris (kwod-RĀ-tus FEM-or-is; *quad-*=square, four-sided; *femoris*=femur)	Ischial tuberosity	Elevation superior to mid-portion of intertrochanteric crest (quadrate tubercle) on posterior femur	Laterally rotates and stabilizes hip joint	Nerve to quadratus femoris

EXHIBIT 11.S Muscles of the Gluteal Region That Move the Femur *(Figure 11.22)* CONTINUED

Figure 11.22 **Muscles of the gluteal region that move the femur (thigh bone).**

🔑 Most muscles that move the femur originate on the pelvic (hip) girdle and insert on the femur.

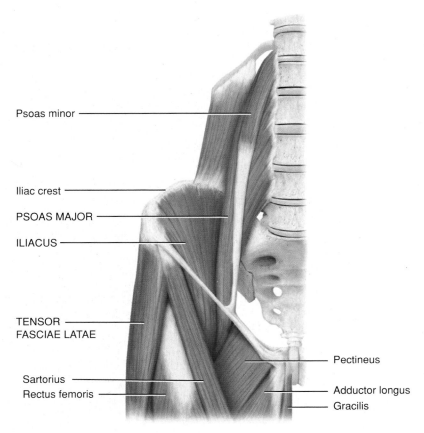

Psoas minor

Iliac crest

PSOAS MAJOR

ILIACUS

TENSOR FASCIAE LATAE

Sartorius

Rectus femoris

Pectineus

Adductor longus

Gracilis

(a) Anterior deep view

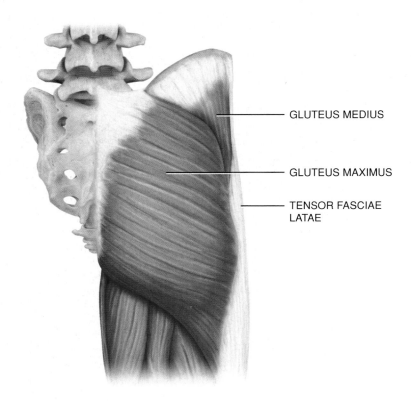

GLUTEUS MEDIUS

GLUTEUS MAXIMUS

TENSOR FASCIAE LATAE

(b) Posterior superficial view

EXHIBIT 11.S   **391**

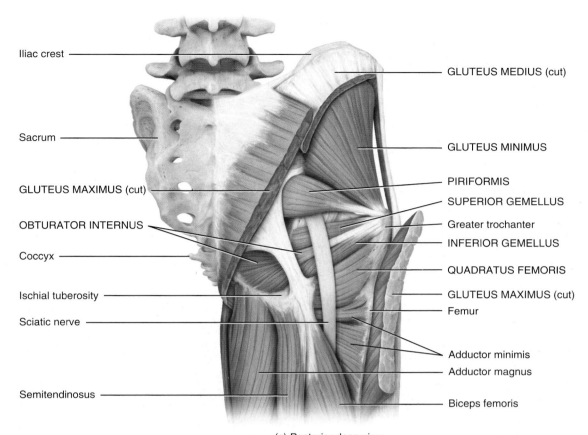

Iliac crest

Sacrum

GLUTEUS MAXIMUS (cut)

OBTURATOR INTERNUS

Coccyx

Ischial tuberosity

Sciatic nerve

Semitendinosus

GLUTEUS MEDIUS (cut)

GLUTEUS MINIMUS

PIRIFORMIS

SUPERIOR GEMELLUS

Greater trochanter

INFERIOR GEMELLUS

QUADRATUS FEMORIS

GLUTEUS MAXIMUS (cut)

Femur

Adductor minimis

Adductor magnus

Biceps femoris

(c) Posterior deep view

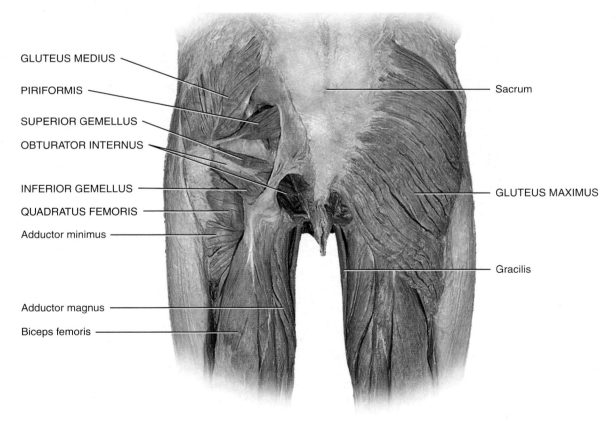

GLUTEUS MEDIUS

PIRIFORMIS

SUPERIOR GEMELLUS

OBTURATOR INTERNUS

INFERIOR GEMELLUS

QUADRATUS FEMORIS

Adductor minimus

Adductor magnus

Biceps femoris

Sacrum

GLUTEUS MAXIMUS

Gracilis

(d) Posterior view

FIGURE 11.22 CONTINUES ▶

EXHIBIT 11.S Muscles of the Gluteal Region That Move the Femur *(Figure 11.22)* CONTINUED

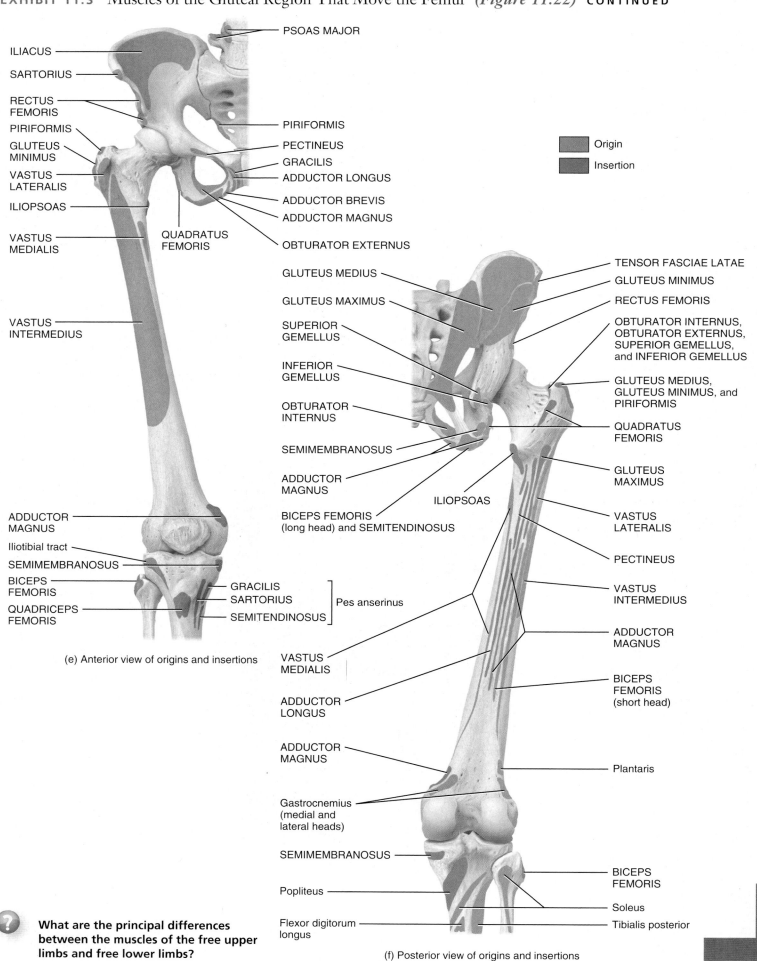

Origin

Insertion

ILIACUS

SARTORIUS

RECTUS FEMORIS

PIRIFORMIS

GLUTEUS MINIMUS

VASTUS LATERALIS

ILIOPSOAS

VASTUS MEDIALIS

QUADRATUS FEMORIS

VASTUS INTERMEDIUS

PSOAS MAJOR

PIRIFORMIS

PECTINEUS

GRACILIS

ADDUCTOR LONGUS

ADDUCTOR BREVIS

ADDUCTOR MAGNUS

OBTURATOR EXTERNUS

ADDUCTOR MAGNUS

Iliotibial tract

SEMIMEMBRANOSUS

BICEPS FEMORIS

QUADRICEPS FEMORIS

GRACILIS

SARTORIUS

SEMITENDINOSUS

Pes anserinus

(e) Anterior view of origins and insertions

GLUTEUS MEDIUS

GLUTEUS MAXIMUS

SUPERIOR GEMELLUS

INFERIOR GEMELLUS

OBTURATOR INTERNUS

SEMIMEMBRANOSUS

ADDUCTOR MAGNUS

BICEPS FEMORIS (long head) and SEMITENDINOSUS

VASTUS MEDIALIS

ADDUCTOR LONGUS

ADDUCTOR MAGNUS

Gastrocnemius (medial and lateral heads)

SEMIMEMBRANOSUS

Popliteus

Flexor digitorum longus

TENSOR FASCIAE LATAE

GLUTEUS MINIMUS

RECTUS FEMORIS

OBTURATOR INTERNUS, OBTURATOR EXTERNUS, SUPERIOR GEMELLUS, and INFERIOR GEMELLUS

GLUTEUS MEDIUS, GLUTEUS MINIMUS, and PIRIFORMIS

QUADRATUS FEMORIS

GLUTEUS MAXIMUS

ILIOPSOAS

VASTUS LATERALIS

PECTINEUS

VASTUS INTERMEDIUS

ADDUCTOR MAGNUS

BICEPS FEMORIS (short head)

Plantaris

BICEPS FEMORIS

Soleus

Tibialis posterior

(f) Posterior view of origins and insertions

? What are the principal differences between the muscles of the free upper limbs and free lower limbs?

EXHIBIT 11.T **393**

EXHIBIT 11.T

Muscles of the Thigh That Move the Femur, Tibia, and Fibula *(Figure 11.23)*

● OBJECTIVE

• Describe the origin, insertion, action, and innervation of the muscles that move the femur, tibia, and fibula.

Deep fascia separates the muscles of the thigh that act on the femur (thigh bone) and tibia and fibula (leg bones) into medial, anterior, and posterior compartments (Figure 11.23). Most of the muscles of the **medial (adductor) compartment of the thigh** have a similar orientation and adduct the femur at the hip joint. The **gracilis** is the exception; it is a long, straplike muscle on the medial aspect of the thigh and knee. It adducts the thigh at the hip joint, medially rotates the thigh, and flexes the leg at the knee joint.

The **adductor longus**, **adductor brevis**, and **adductor magnus** originate on the anterior aspect of the pubic and ischial bones and insert on the posterior aspect of the femur. These three muscles all adduct the thigh and are unique in their ability to both medially and laterally rotate the thigh. When the foot is on the ground these muscles medially rotate the thigh, but when the foot is off the ground, they are lateral rotators of the thigh. This results because of the oblique course of the muscle fibers from their anterior origin to their posterior insertion. In addition, the adductor longus flexes the thigh and the adductor magnus can both extend and flex the thigh because some of its fibers are anterior to the plane of the femoral head and assist in flexion, while other fibers are posterior to the plane of the femoral head and assist in extension. The distal tendons of the adductor magnus separate just superior to the medial epicondyle of the femur; the space formed by the separation is called the **adductor hiatus** (Figure 11.23c, d). As you will learn in Chapter 14, the femoral vessels pass through this space to become the popliteal vessels on the posterior side of the knee. The adductor longus and brevis muscles assist in the flexion of the thigh from the neutral (anatomical) position to nearly 80 degrees, but they assist in extension when the insertion on the linea aspera becomes superior to the origin on the pubis. Most activities, however, do not require flexion of more than 80 degrees of the thigh at the hip joint. The **pectineus** muscle also adducts and flexes the femur at the hip joint. This muscle lies between the iliopsoas and the adductor longus muscles.

At the junction between the trunk and lower limb is a space called the **femoral triangle**. The base is formed superiorly by the inguinal ligament, medially by the lateral border of the adductor longus muscle, and laterally by the medial border of the sartorius muscle. The apex is formed by the crossing of the adductor longus by the sartorius muscle (Figure 11.23a). The contents of the femoral triangle, from lateral to medial, are the femoral nerve and its branches, the femoral artery and several of its branches, the femoral vein and its proximal tributaries, and the deep inguinal lymph nodes. The femoral artery is easily accessible within the triangle and is the site for insertion of catheters that may extend into the aorta and ultimately into the coronary vessels of the heart. Such catheters are utilized during cardiac catheterization, coronary angiography, and other procedures involving the heart. Inguinal and femoral hernias frequently appear in this area.

The muscles of the **anterior (extensor) compartment of the thigh** extend the leg (and flex the thigh). This compartment contains the quadriceps femoris and sartorius muscles. The **quadriceps femoris** muscle is the largest muscle in the body, covering most of the anterior surface and sides of the thigh, and is the

great extensor of the knee joint. The muscle is actually a composite muscle, usually described as four separate muscles: (1) **rectus femoris**, on the anterior aspect of the thigh; (2) **vastus lateralis**, on the lateral aspect of the thigh; (3) **vastus medialis**, on the medial aspect of the thigh; and (4) **vastus intermedius**, located deep to the rectus femoris between the vastus lateralis and vastus medialis. The rectus femoris originates on the anterior inferior iliac spine (AIIS) and thus crosses two joints; it functions with the iliopsoas and pectineus muscles in their role as flexors of the femur at the hip. Chronic contraction or weakness of any part of the quadriceps femoris can cause the patella to track abnormally in the patellar groove, resulting in weakness and pain in the knee area. Physicians or physical therapists can prescribe specific exercises that selectively strengthen just one of the muscles to help correct problems at the knee. The common tendon for the four muscles, known as the **quadriceps tendon**, inserts into the patella. The tendon continues below the patella as the **patellar ligament**, which attaches to the tibial tuberosity. Any issue involving abnormal movement of the patella is known as patellofemoral dysfunction.

The **sartorius**, the longest single muscle in the body, is a narrow muscle that forms a band across the thigh from the ilium of the hip bone to the medial side of the tibia. The various movements it produces (flexion of the leg at the knee joint and flexion, abduction, and lateral rotation at the hip joint) help effect the cross-legged sitting position in which the heel of one limb is placed on the knee of the opposite limb. For this reason it is known as the tailor's muscle because tailors often assume a cross-legged sitting position. It originates on the anterior superior iliac spine (ASIS), crosses the proximal superficial thigh obliquely, descends almost vertically in the distal thigh, crosses the medial condyle of the femur posteriorly, and inserts into the tibia medial to the tibial tuberosity. The sartorius assists other muscles of the hip in its multiple actions at that joint.

The muscles of the **posterior (flexor) compartment of the thigh** flex the leg (and extend the thigh). This compartment is composed of three muscles collectively called the **hamstrings**: (1) biceps femoris, (2) semitendinosus, and (3) semimembranosus. The hamstrings are so named because their tendons are long and stringlike in the popliteal area. Because the hamstrings span two joints (hip and knee), they extend the hip or flex the knee or both. In walking, as the foot leaves the ground to take a step forward, the hamstrings initially contract and take the weight of the partially flexed leg. As soon as hip flexion begins, the hamstrings relax and permit knee extension in the advancing limb.

The **biceps femoris** has two heads; the long head originates with the tendon of the semitendinosus on the ischial tuberosity and also from the sacrotuberous ligament. The short head arises from the linea aspera. The two heads are innervated by two nerves: The tibial portion of the sciatic nerve often sends two branches into the long head, and the short head is innervated by a branch of the common fibular nerve.

The **semitendinosus**, as just stated, arises with the tendon of the long head of the biceps femoris from the ischial tuberosity. The rounded tendon of insertion forms the medial border of the **popliteal fossa**. The semitendinosus contains a tendinous intersection near the midpoint of its length that shortens the fiber length in the muscle.

EXHIBIT 11.T Muscles of the Thigh That Move the Femur, Tibia, and Fibula *(Figure 11.23)* CONTINUED

MUSCLE	ORIGIN	INSERTION	ACTION	INNERVATION
MEDIAL (ADDUCTOR) COMPARTMENT OF THE THIGH				
Adductor longus (ad-DUK-tor LONG-us; *adductor*=moves part closer to midline; *longus*=long)	Pubic crest and pubic symphysis	Linea aspera of femur	Adducts and flexes thigh at hip joint and rotates thigh* RMA: Extends thigh	Obturator nerve
Adductor brevis (BREV-is; *brevis*=short)	Inferior ramus of pubis	Superior half of linea aspera of femur	Adducts and flexes thigh at hip joint and rotates thigh* RMA: Extends thigh	Obturator nerve
Adductor magnus (MAG-nus; *magnus*=large)	Inferior ramus of pubis and ischium to ischial tuberosity	Linea aspera of femur	Adducts thigh at hip joint and rotates thigh; anterior part flexes thigh at hip joint, and posterior part extends thigh at hip joint*	Obturator and sciatic nerves
Pectineus (pek-TIN-ē-us; *pectin-*=a comb)	Superior ramus of pubis	Pectineal line of femur, between lesser trochanter and linea aspera	Flexes and adducts thigh at hip joint	Femoral nerve
Gracilis (GRAS-i-lis; *gracilis*=slender)	Body and inferior ramus of pubis	Medial surface of body of tibia	Adducts thigh at hip joint, medially rotates thigh, and flexes leg at knee joint	Obturator nerve
ANTERIOR (EXTENSOR) COMPARTMENT OF THE THIGH				
Quadriceps femoris (KWOD-ri-seps FEM-or-is: *quadriceps*=four heads [of origin]; *femoris*=femur)				
Rectus femoris (REK-tus=fascicles parallel to midline)	Anterior inferior iliac spine	Patella via quadriceps tendon and then tibial tuberosity via patellar ligament	All four heads extend leg at knee joint; rectus femoris muscle acting alone also flexes thigh at hip joint	Femoral nerve
Vastus lateralis (VAS-tus lat'-e-RA-lis; *vast*=huge; *lateralis*=lateral)	Greater trochanter and linea aspera of femur			
Vastus medialis (mē-dē-Ā-lis; *medialis*=medial)	Linea aspera of femur			
Vastus intermedius (in'-ter-MĒ-dē-us =middle)	Anterior and lateral surfaces of body of femur			
Sartorius (sar-TOR-ē-us; *sartor*=tailor; longest muscle in body)	Anterior superior iliac spine	Medial surface of body of tibia	Weakly flexes leg at knee joint; weakly flexes, abducts, and laterally rotates thigh at hip joint	Femoral nerve
POSTERIOR (FLEXOR) COMPARTMENT OF THE THIGH				
Hamstrings A collective designation for three separate muscles.				
Biceps femoris (BĪ-seps=two heads of origin)	Long head arises from ischial tuberosity; short head arises from linea aspera of femur	Head of fibula and lateral condyle of tibia	Flexes leg at knee joint and extends thigh at hip joint	Tibial and common fibular nerves from the sciatic nerve
Semitendinosus (sem'-ē-ten-di-NŌ-sus; *semi-*=half; *tendo* =tendon)	Ischial tuberosity	Proximal part of medial surface of shaft of tibia	Flexes leg at knee joint and extends thigh at hip joint	Tibial nerve from the sciatic nerve
Semimembranosus (sem'-ē-mem-bra-NŌ-sus; *membran-*=membrane)	Ischial tuberosity	Medial condyle of tibia	Flexes leg at knee joint and extends thigh at hip joint	Tibial nerve from the sciatic nerve

*All adductors are unique muscles that cross the thigh joint obliquely from an anterior origin to a posterior insertion. As a result they laterally rotate the hip joint when the foot is off the ground, but medially rotate the hip joint when the foot is on the ground.

EXHIBIT 11.T **395**

Figure 11.23 Muscles of the thigh that move the femur, tibia, and fibula. The femoral triangle is outlined by dashed lines in part a. The illustrations of the origins and insertions of the muscles discussed in this exhibit are shown in Figure 11.22e, f.

🔑 Muscles that act on the leg originate in the hip and thigh and are separated into compartments by deep fascia.

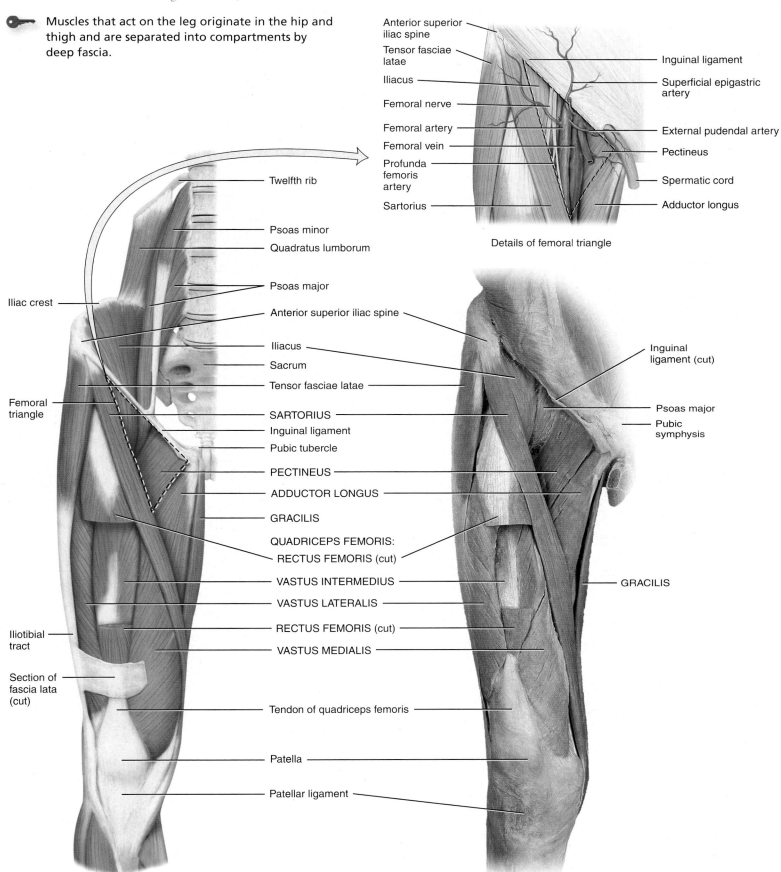

Details of femoral triangle

(a) Anterior superficial view (the femoral triangle is indicated by a dashed line)

(b) Anterior superficial view

FIGURE 11.23 CONTINUES ▶

EXHIBIT 11.T Muscles of the Thigh That Move the Femur, Tibia, and Fibula *(Figure 11.23)* CONTINUED

FIGURE 11.23 CONTINUED

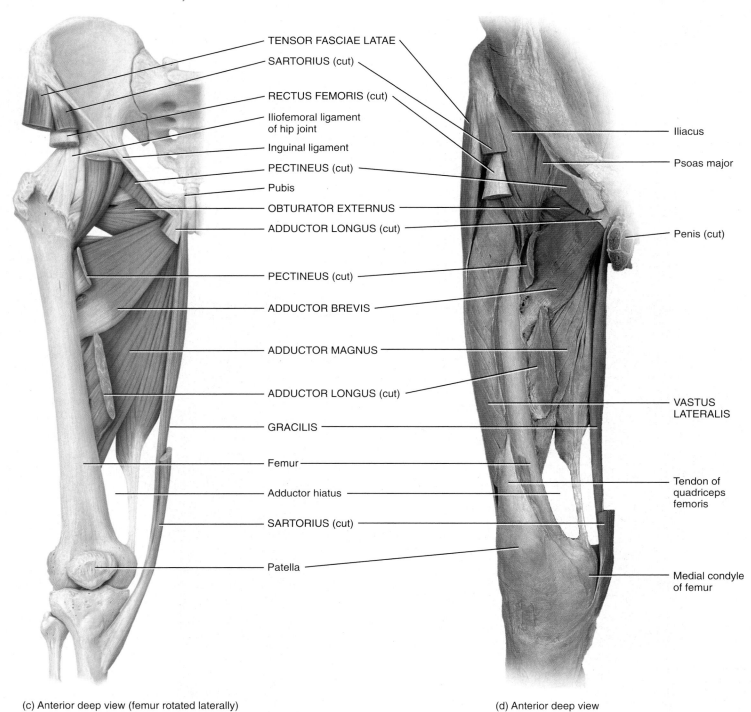

TENSOR FASCIAE LATAE
SARTORIUS (cut)
RECTUS FEMORIS (cut)
Iliofemoral ligament of hip joint
Inguinal ligament
PECTINEUS (cut)
Pubis
OBTURATOR EXTERNUS
ADDUCTOR LONGUS (cut)
PECTINEUS (cut)
ADDUCTOR BREVIS
ADDUCTOR MAGNUS
ADDUCTOR LONGUS (cut)
GRACILIS
Femur
Adductor hiatus
SARTORIUS (cut)
Patella

Iliacus
Psoas major
Penis (cut)
VASTUS LATERALIS
Tendon of quadriceps femoris
Medial condyle of femur

(c) Anterior deep view (femur rotated laterally)

(d) Anterior deep view

EXHIBIT 11.T **397**

The **semimembranosus** originates from the ischial tuberosity by a long flat tendon that lies deep to the proximal half of the muscle and adjacent to the tendon of the adductor magnus.

The **popliteal fossa** is a diamond-shaped space on the posterior aspect of the knee bordered laterally by the tendons of the biceps femoris muscle and medially by the tendons of the semitendinosus and semimembranosus muscles.

Relating Muscles to Movements

Arrange the muscles in this exhibit according to the following actions on the thigh at the hip joint: (1) abduction, (2) adduction,

(3) lateral rotation, (4) flexion, and (5) extension; and according to the following actions on the leg at the knee joint: (1) flexion and (2) extension. The same muscle may be mentioned more than once.

CHECKPOINT

27. Which muscles are part of the medial, anterior, and posterior compartments of the thigh?

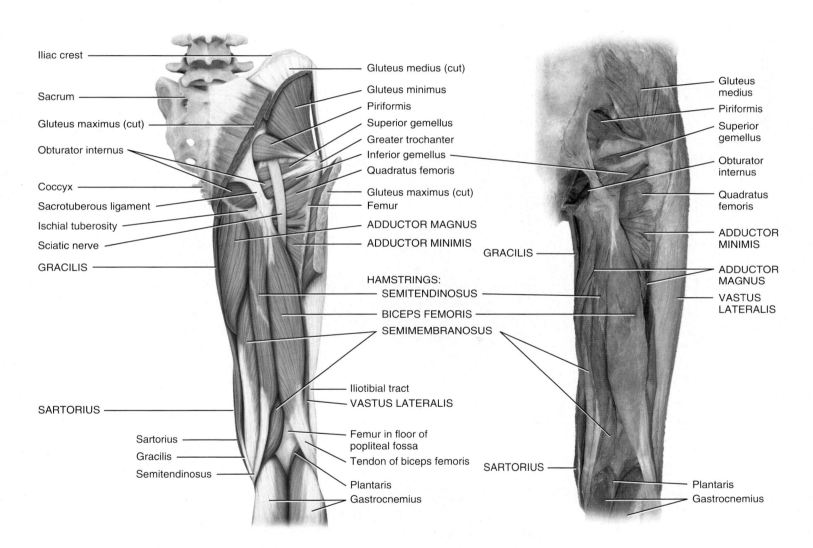

(e) Posterior superficial view of thigh and deep view of gluteal region

(f) Posterior superficial view of thigh and deep view of gluteal region

FIGURE 11.23 CONTINUES ▶

EXHIBIT 11.T Muscles of the Thigh That Move the Femur, Tibia, and Fibula *(Figure 11.23)* CONTINUED

> **CLINICAL CONNECTION | Pulled Hamstrings and Charley Horse**
>
> A strain or partial tear of the proximal hamstring muscles is referred to as **pulled hamstrings** or **hamstring strains**. Like pulled groins (see Exhibit 11.S), they are common sports injuries in individuals who run very hard and/or are required to perform quick starts and stops. Sometimes the violent muscular exertion required to perform a feat tears away a part of the tendinous origins of the hamstrings, especially the biceps femoris, from the ischial tuberosity. This is usually accompanied by a contusion (bruising), tearing of some of the muscle fibers, and rupture of blood vessels, producing a hematoma (collection of blood) and sharp pain. Adequate training with good balance between the quadriceps femoris and hamstrings and stretching exercises before running or competing are important in preventing this injury.
>
> The slang term **charley horse** (CHAR-lē HŌRS) is a popular name for a cramp or stiffness of muscles due to tearing of the muscle, followed by bleeding into the area. It is a common sports injury due to trauma or excessive activity and frequently occurs in the quadriceps femoris muscle, especially among football players. •

FIGURE 11.23 CONTINUED

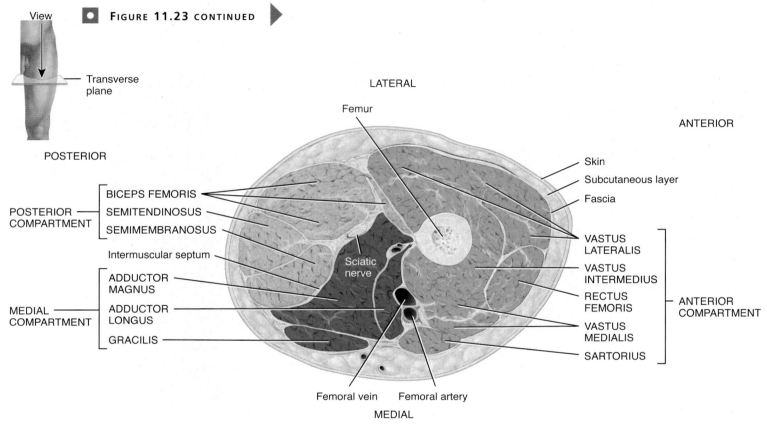

(g) Superior view of transverse section of thigh

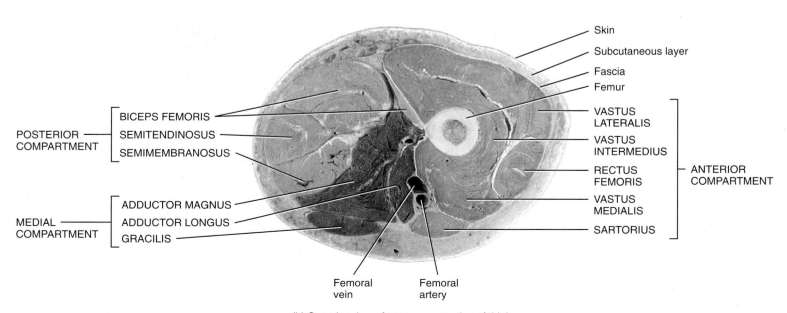

(h) Superior view of transverse section of thigh

EXHIBIT 11.T **399**

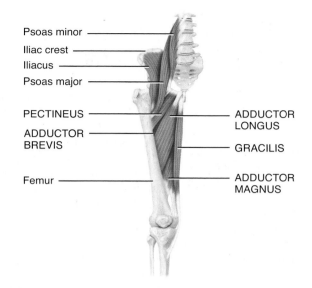

Anterior deep view

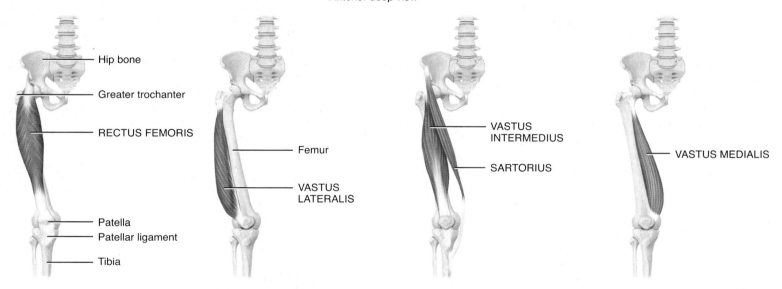

Anterior views

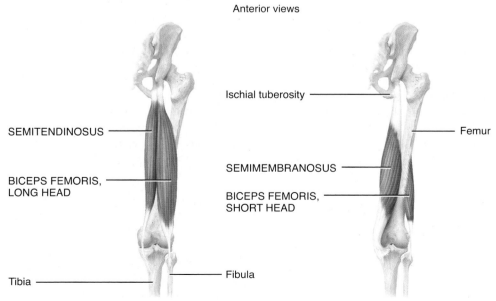

Posterior deep views

(i) Isolated muscles

Which muscles constitute the quadriceps femoris and hamstring muscles?

EXHIBIT 11.U Muscles of the Leg That Move the Foot and Toes *(Figure 11.24)*

OBJECTIVE
• Describe the origin, insertion, action, and innervation of the muscles of the leg that move the foot and toes.

Muscles that move the foot and toes are located in the leg (Figure 11.24). The muscles of the leg, like those of the thigh, are divided by deep fascia into three compartments: anterior, lateral, and posterior. The **anterior compartment of the leg** consists of muscles that dorsiflex the foot. In a situation analogous to the wrist, the tendons of the muscles of the anterior compartment are held firmly to the ankle by thickenings of deep fascia called the **superior extensor retinaculum** (*transverse ligament of the ankle*) and **inferior extensor retinaculum** (*cruciate ligament of the ankle*).

Within the anterior compartment, the **tibialis anterior** is a long, thick muscle against the lateral surface of the tibia, where it is easy to palpate (feel). The **extensor digitorum longus** is a featherlike muscle that is lateral to the tibialis anterior muscle, and it can also be palpated easily. The **extensor hallucis longus** is a thin muscle between and partly deep to the tibialis anterior and extensor digitorum longus muscles. The **fibularis (peroneus) tertius** muscle is part of the extensor digitorum longus, with which it shares a common origin.

The **lateral (fibular) compartment of the leg** contains two muscles that plantar flex and evert the foot: the **fibularis (peroneus) longus** and **fibularis (peroneus) brevis**.

The **posterior compartment of the leg** consists of muscles in superficial and deep groups. The superficial muscles share a common tendon of insertion, the **calcaneal (Achilles) tendon**, the strongest tendon of the body. It inserts into the calcaneal bone of the ankle. The superficial and most of the deep muscles plantar flex the foot at the ankle joint. The superficial muscles of the posterior compartment are the gastrocnemius, soleus, and plantaris—the so-called calf muscles. The large size of these muscles is directly related to the characteristic upright stance of humans. The **gastrocnemius** is the most superficial muscle and forms the prominence of the calf. The **soleus**, which lies deep to the gastrocnemius, is broad and flat. It derives its name from its resemblance to a flat fish (sole). The **plantaris** is a small muscle that may be absent; conversely, sometimes there are two of them in each leg. It runs obliquely between the gastrocnemius and soleus muscles and is often referred to as the freshman's nerve because first-year medical students sometimes confuse its long slender tendon with the tibial nerve during dissection.

The deep muscles of the posterior compartment are the popliteus, tibialis posterior, flexor digitorum longus, and flexor hallucis longus. The **popliteus** is a triangular muscle that forms the floor of the popliteal fossa. The **tibialis posterior** is the deepest muscle in the posterior compartment. It lies between the flexor digitorum longus and flexor hallucis longus muscles. The **flexor digitorum longus** is smaller than the **flexor hallucis longus**, even though the former flexes four toes, and the latter flexes only the great toe at the interphalangeal joint.

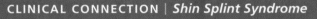

CLINICAL CONNECTION | *Shin Splint Syndrome*

Shin splint syndrome, or simply **shin splints**, refers to pain or soreness along the tibia, specifically the medial, distal two-thirds. It may be caused by tendonitis of the anterior compartment muscles, especially the tibialis anterior muscle, inflammation of the periosteum (periostitis) around the tibia, or stress fractures of the tibia. The tendonitis usually occurs when poorly conditioned runners run on hard or banked surfaces with poorly supportive running shoes. The condition may also occur with vigorous activity of the legs following a period of relative inactivity or running in cold weather without proper warmup. The muscles in the anterior compartment (mainly the tibialis anterior) can be strengthened to balance the stronger posterior compartment muscles. •

Relating Muscles to Movements

Arrange the muscles in this exhibit according to the following actions on the foot at the ankle joint: (1) dorsiflexion and (2) plantar flexion; according to the following actions on the foot at the intertarsal joints: (1) inversion and (2) eversion; and according to the following actions on the toes at the metatarsophalangeal and interphalangeal joints: (1) flexion and (2) extension. The same muscle may be mentioned more than once.

CHECKPOINT
28. What are the superior extensor retinaculum and inferior extensor retinaculum?

EXHIBIT 11.U **401**

MUSCLE	ORIGIN	INSERTION	ACTION	INNERVATION
ANTERIOR COMPARTMENT OF THE LEG				
Tibialis anterior (tib'-ē-Ā-lis=tibia; *anterior*=front)	Lateral condyle and body of tibia and interosseous membrane (sheet of fibrous tissue that holds shafts of tibia and fibula together)	Metatarsal I and first (medial) cuneiform	Dorsiflexes foot at ankle joint and inverts (supinates) foot at intertarsal joints	Deep fibular (peroneal) nerve
Extensor hallucis longus (HAL-ū-sis LON-gus; *extensor*=increases angle at joint; *hallucis-*=hallux or great toe; *longus*=long)	Anterior surface of fibula and interosseous membrane	Distal phalanx of great toe	Dorsiflexes foot at ankle joint and extends proximal phalanx of great toe at metatarsophalangeal joint	Deep fibular (peroneal) nerve
Extensor digitorum longus (di'-ji-TOR-um)	Lateral condyle of tibia, anterior surface of fibula, and interosseous membrane	Middle and distal phalanges of toes II–V*	Dorsiflexes foot at ankle joint and extends distal and middle phalanges of each toe at interphalangeal joints and proximal phalanx of each toe at metatarsophalangeal joint	Deep fibular (peroneal) nerve
Fibularis (Peroneus) tertius (fib-ū-LĀ-ris TER-shus; *peron-*=fibula; *tertius*=third)	Distal third of fibula and interosseous membrane	Base of metatarsal V	Dorsiflexes foot at ankle joint and everts (pronates) foot at intertarsal joints	Deep fibular (peroneal) nerve
LATERAL (FIBULAR) COMPARTMENT OF THE LEG				
Fibularis (Peroneus) longus	Head and body of fibula	Metatarsal I and first cuneiform	Plantar flexes foot at ankle joint and everts foot at intertarsal joints	Superficial fibular (peroneal) nerve
Fibularis (Peroneus) brevis (BREV-is=short)	Body of fibula	Base of metatarsal V	Plantar flexes foot at ankle joint and everts (pronates) foot at intertarsal joints	Superficial fibular (peroneal) nerve
SUPERFICIAL POSTERIOR COMPARTMENT OF THE LEG				
Gastrocnemius (gas'-trok-NĒ-mē-us; *gastro-*=belly; *cnem-*=leg)	Lateral and medial condyles of femur and capsule of knee	Calcaneus by way of calcaneal (Achilles) tendon	Plantar flexes foot at ankle joint and flexes leg at knee joint	Tibial nerve
Soleus (SŌ-lē-us; *sole*=a type of flat fish)	Head of fibula and medial border of tibia	Calcaneus by way of calcaneal (Achilles) tendon	Plantar flexes foot at ankle joint	Tibial nerve
Plantaris (plan-TĀR-is; *plantar-*=sole)	Femur superior to lateral condyle	Calcaneus by way of calcaneal (Achilles) tendon	Plantar flexes foot at ankle joint and flexes leg at knee joint	Tibial nerve
DEEP POSTERIOR COMPARTMENT OF THE LEG				
Popliteus (pop-LIT-ē-us; *poplit-*=the back of the knee)	Lateral condyle of femur	Proximal tibia	Flexes leg at knee joint and medially rotates tibia to unlock the extended knee	Tibial nerve
Tibialis posterior (tib'-ē-Ā-lis; *posterior*=back)	Tibia, fibula, and interosseous membrane	Metatarsals II–V; navicular; all three cuneiforms; and cuboid	Plantar flexes foot at ankle joint and inverts (supinates) foot at intertarsal joints	Tibial nerve
Flexor digitorum longus (di'-ji-TOR-um *digit*=finger or toe)	Posterior surface of tibia	Distal phalanges of toes II–V	Plantar flexes foot at ankle joint; flexes distal and middle phalanges of each toe at interphalangeal joints and proximal phalanx of each toe at metatarsophalangeal joint	Tibial nerve
Flexor hallucis longus (HAL-ū-sis; *flexor*=decreases angle at joint)	Inferior two-thirds of fibula	Distal phalanx of great toe	Plantar flexes foot at ankle joint; flexes distal phalanx of great toe at interphalangeal joint and proximal phalanx of great toe at metatarsophalangeal joint	Tibial nerve

*Reminder: The great toe or hallux is the first toe and has two phalanges—proximal and distal. The remaining toes are numbered II–V (2–5), and each has three phalanges—proximal, middle, and distal.

EXHIBIT 11.U Muscles of the Leg That Move the Foot and Toes *(Figure 11.24)* **CONTINUED**

Figure 11.24 **Muscles of the leg that move the foot and toes.**

🔑 The superficial muscles of the posterior compartment share a common tendon of insertion, the calcaneal (Achilles) tendon, which inserts into the calcaneal bone of the ankle.

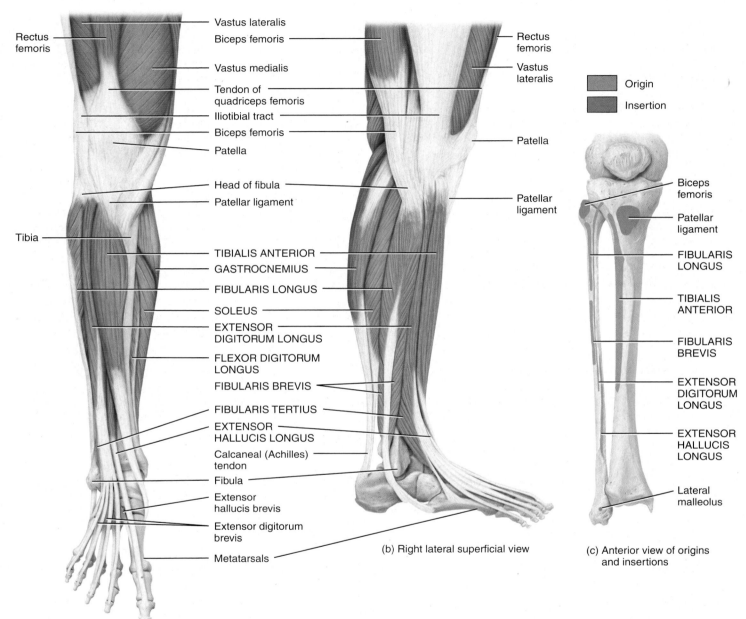

Rectus femoris
Vastus lateralis
Biceps femoris
Vastus medialis
Tendon of quadriceps femoris
Iliotibial tract
Biceps femoris
Patella
Head of fibula
Patellar ligament
Tibia
TIBIALIS ANTERIOR
GASTROCNEMIUS
FIBULARIS LONGUS
SOLEUS
EXTENSOR DIGITORUM LONGUS
FLEXOR DIGITORUM LONGUS
FIBULARIS BREVIS
FIBULARIS TERTIUS
EXTENSOR HALLUCIS LONGUS
Calcaneal (Achilles) tendon
Fibula
Extensor hallucis brevis
Extensor digitorum brevis
Metatarsals

(a) Anterior superficial view

Rectus femoris
Vastus lateralis
Patella
Patellar ligament

(b) Right lateral superficial view

Origin
Insertion

Biceps femoris
Patellar ligament
FIBULARIS LONGUS
TIBIALIS ANTERIOR
FIBULARIS BREVIS
EXTENSOR DIGITORUM LONGUS
EXTENSOR HALLUCIS LONGUS
Lateral malleolus

(c) Anterior view of origins and insertions

EXHIBIT 11.U **403**

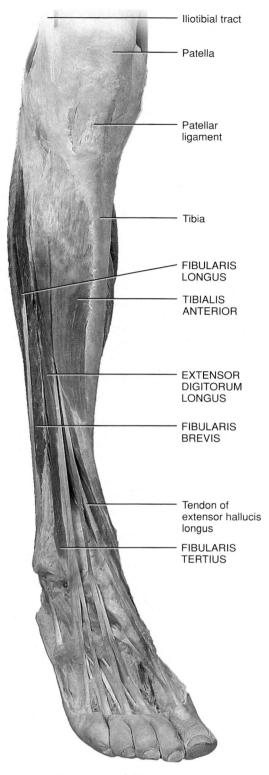

Iliotibial tract

Patella

Patellar
ligament

Tibia

FIBULARIS
LONGUS

TIBIALIS
ANTERIOR

EXTENSOR
DIGITORUM
LONGUS

FIBULARIS
BREVIS

Tendon of
extensor hallucis
longus

FIBULARIS
TERTIUS

(d) Anterior superficial view

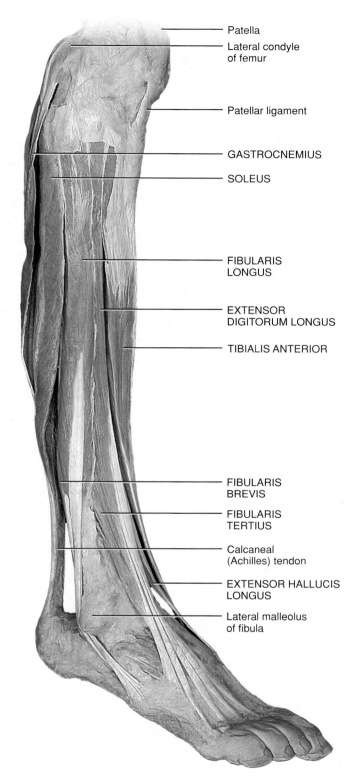

Patella

Lateral condyle
of femur

Patellar ligament

GASTROCNEMIUS

SOLEUS

FIBULARIS
LONGUS

EXTENSOR
DIGITORUM LONGUS

TIBIALIS ANTERIOR

FIBULARIS
BREVIS

FIBULARIS
TERTIUS

Calcaneal
(Achilles) tendon

EXTENSOR HALLUCIS
LONGUS

Lateral malleolus
of fibula

(e) Lateral superficial view

FIGURE 11.24 CONTINUES ▶

EXHIBIT 11.U Muscles of the Leg That Move the Foot and Toes *(Figure 11.24)* CONTINUED

⬛ FIGURE **11.24** CONTINUED ▶

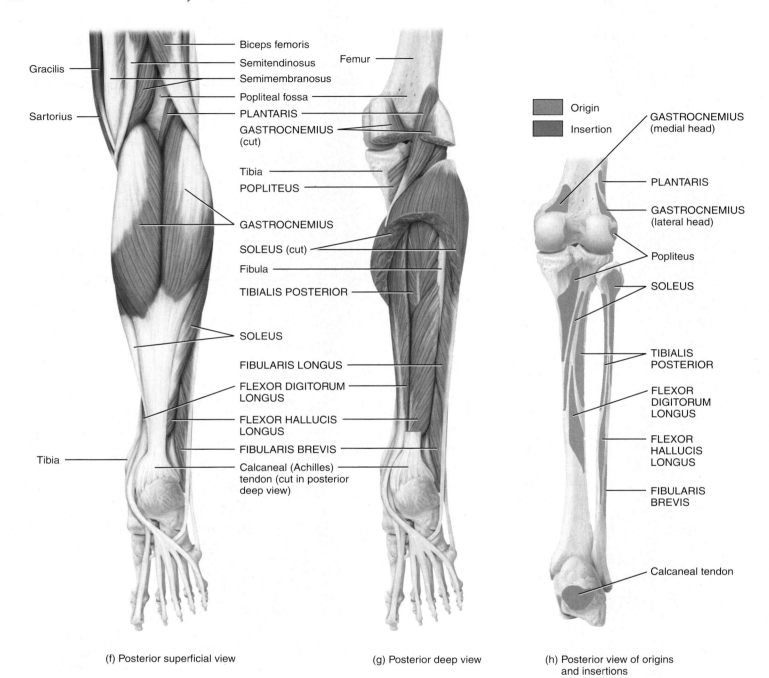

(f) Posterior superficial view

(g) Posterior deep view

(h) Posterior view of origins and insertions

EXHIBIT 11.U **405**

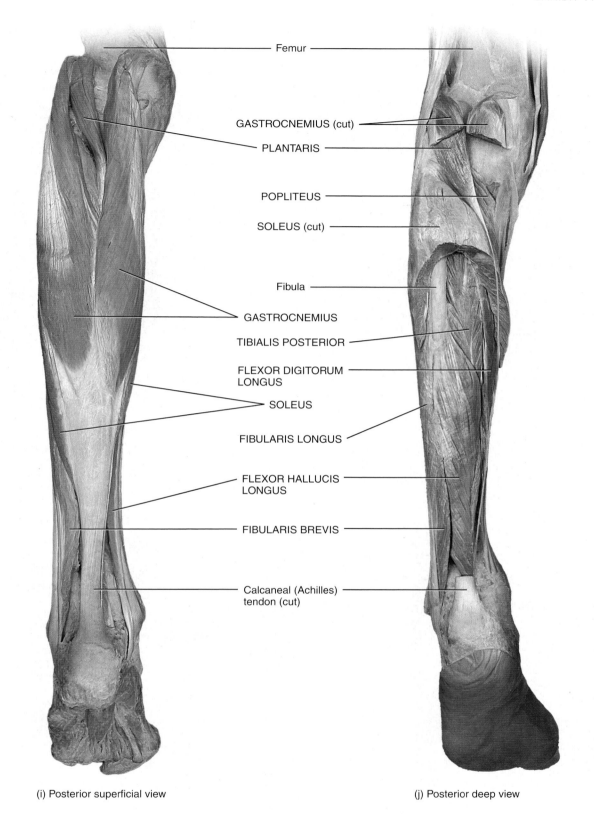

Femur

GASTROCNEMIUS (cut)

PLANTARIS

POPLITEUS

SOLEUS (cut)

Fibula

GASTROCNEMIUS

TIBIALIS POSTERIOR

FLEXOR DIGITORUM
LONGUS

SOLEUS

FIBULARIS LONGUS

FLEXOR HALLUCIS
LONGUS

FIBULARIS BREVIS

Calcaneal (Achilles)
tendon (cut)

(i) Posterior superficial view

(j) Posterior deep view

FIGURE 11.24 CONTINUES ▶

EXHIBIT 11.U Muscles of the Leg That Move the Foot and Toes *(Figure 11.24)* CONTINUED

⬜ FIGURE **11.24** CONTINUED ▶

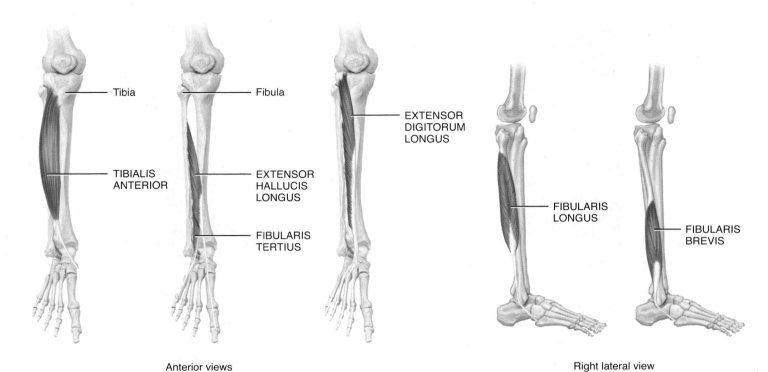

Anterior views

Right lateral view

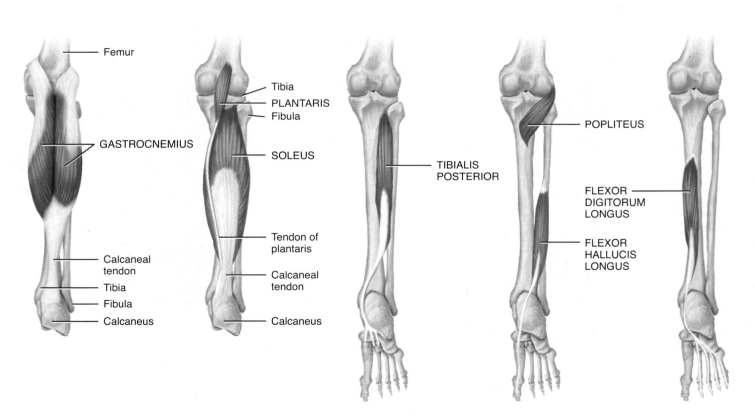

Posterior deep views

(k) Isolated muscles

 What structures firmly hold the tendons of the anterior compartment muscles to the ankle?

EXHIBIT 11.V **407**

EXHIBIT 11.V

Intrinsic Muscles of the Foot
That Move the Toes *(Figure 11.25)*

 OBJECTIVE

• Describe the origin, insertion, action, and innervation of the intrinsic muscles of the foot that move the toes.

The muscles in this exhibit are termed **intrinsic muscles of the foot** because they originate and insert *within* the foot. The muscles of the hand are specialized for precise and intricate movements, but those of the foot are limited to support and locomotion. The deep fascia of the foot forms the **plantar aponeurosis (fascia)** that extends from the calcaneus bone to the phalanges of the toes. The aponeurosis supports the longitudinal arch of the foot and encloses the flexor tendons of the foot.

The intrinsic muscles of the foot are divided into two groups: **dorsal muscles of the foot** and **plantar muscles of the foot** (Figure 11.25). There are two dorsal muscles, the **extensor hallucis brevis** and **extensor digitorum brevis**. The latter is a four-part muscle deep to the tendons of the extensor digitorum longus muscle, which extends toes II–V at the metatarsophalangeal joints.

The plantar muscles are arranged in four layers. The most superficial layer is called the first layer. Three muscles are in the first layer. The **abductor hallucis**, which lies along the medial border of the sole and is comparable to the abductor pollicis brevis in the hand, abducts the great toe at the metatarsophalangeal joint. The **flexor digitorum brevis**, which lies in the middle of the sole, flexes toes II–V at the interphalangeal and metatarsophalangeal joints. The **abductor digiti minimi**, which lies along the lateral border of the sole and is comparable to the same muscle in the hand, abducts the little toe.

The second layer consists of the **quadratus plantae**, a rectangular muscle that arises by two heads and flexes toes II–V at the metatarsophalangeal joints, and the **lumbricals**, four small muscles that are similar to the lumbricals in the hands. They flex the proximal phalanges and extend the distal phalanges of toes II–V.

Three muscles comprise the third layer. The **flexor hallucis brevis**, which lies adjacent to the plantar surface of the metatarsal of the great toe, is comparable to the flexor pollicis brevis muscle in the hand, and flexes the great toe. The **adductor hallucis**, which has an oblique and transverse head like the adductor pollicis in the hand, adducts the great toe. The **flexor digiti minimi brevis**, which lies superficial to the metatarsal of the little toe, is comparable to the same muscle in the hand, and flexes the little toe.

The fourth layer is the deepest and consists of two muscle groups. The **dorsal interossei** are four muscles that abduct toes II–IV, flex the proximal phalanges, and extend the distal phalanges. The three **plantar interossei** adduct toes III–V, flex the proximal phalanges, and extend the distal phalanges. The interossei of the feet are similar to those of the hand. However, their actions are relative to the midline of the second digit rather than the third digit as in the hand.

CLINICAL CONNECTION | *Plantar Fasciitis*

Plantar fasciitis (fas-ē-Ī-tis) or **painful heel syndrome** is an inflammatory reaction due to chronic irritation of the plantar aponeurosis (fascia) at its origin on the calcaneus (heel bone). The aponeurosis becomes less elastic with age. This condition is also related to weight-bearing activities (walking, jogging, lifting heavy objects), improperly constructed or fitting shoes, excess weight (puts pressure on the feet), and poor biomechanics (flat feet, high arches, and abnormalities in gait may cause uneven distribution of weight on the feet). Plantar fasciitis is the most common cause of heel pain in runners and arises in response to the repeated impact of running. Treatments include ice, deep heat, stretching exercises, weight loss, prosthetics (such as shoe inserts or heel lifts), steroid injections, and surgery. •

Relating Muscles to Movements

Arrange the muscles in this exhibit according to the following actions on the great toe at the metatarsophalangeal joint: (1) flexion, (2) extension, (3) abduction, and (4) adduction; and according to the following actions on toes II–V at the metatarsophalangeal and interphalangeal joints: (1) flexion, (2) extension, (3) abduction, and (4) adduction. The same muscle may be mentioned more than once.

 CHECKPOINT

29. How do the intrinsic muscles of the hand and foot differ in function?

EXHIBIT 11.V Intrinsic Muscles of the Foot That Move the Toes *(Figure 11.25)* CONTINUED

MUSCLE	ORIGIN	INSERTION	ACTION	INNERVATION
DORSAL				
Extensor hallucis brevis (eks-TEN-sor HAL-ū-sis BREV-is; *extensor*=increases angle at joint; *hallucis*= hallux or great toe; *brevis*= short) (see Figure 11.24a)	Calcaneus and inferior extensor retinaculum	Proximal phalanx of great toe	Extends great toe at metatarsophalangeal joint	Deep fibular (peroneal) nerve
Extensor digitorum brevis (*digit*=finger or toe) (see Figure 11.24a)	Calcaneus and inferior extensor retinaculum	Middle phalanges of toes II–IV	Extends toes II–IV at interphalangeal joints	Deep fibular (peroneal) nerve
PLANTAR				
First layer (most superficial)				
Abductor hallucis (*abductor*=moves part away from midline; *hallucis*=hallux or great toe)	Calcaneus, plantar aponeurosis, and flexor retinaculum	Medial side of proximal phalanx of great toe with the tendon of the flexor hallucis brevis	Abducts and flexes great toe at metatarsophalangeal joint	Medial plantar nerve
Flexor digitorum brevis (*flexor*=decreases angle at joint)	Calcaneus, plantar aponeurosis, and flexor retinaculum	Sides of middle phalanx of toes II–V	Flexes toes II–V at proximal interphalangeal and metatarsophalangeal joints	Medial plantar nerve
Abductor digiti minimi (*minimi*=little)	Calcaneus, plantar aponeurosis, and flexor retinaculum	Lateral side of proximal phalanx of little toe with the tendon of the flexor digiti minimi brevis	Abducts and flexes little toe at metatarsophalangeal joint	Lateral plantar nerve
Second layer				
Quadratus plantae (kwod-RĀ-tus PLAN-tē; *quad-*=square, four-sided; *planta*=the sole)	Calcaneus	Tendon of flexor digitorum longus	Assists flexor digitorum longus to only flex toes II–V at interphalangeal and metatarsophalangeal joints	Lateral plantar nerve
Lumbricals (LUM-bri-kals; *lumbric-*=earthworm)	Tendons of flexor digitorum longus	Tendons of extensor digitorum longus on proximal phalanges of toes II–V	Extend toes II–V at interphalangeal joints and flex toes II–V at metatarsophalangeal joints	Medial and lateral plantar nerves
Third layer				
Flexor hallucis brevis	Cuboid and third (lateral) cuneiform	Medial and lateral sides of proximal phalanx of great toe via a tendon containing a sesamoid bone	Flexes great toe at metatarsophalangeal joint	Medial plantar nerve
Adductor hallucis	Metatarsals II–IV, ligaments of III–V metatarsophalangeal joints, and tendon of fibularis (peroneus) longus	Lateral side of proximal phalanx of great toe	Adducts and flexes great toe at metatarsophalangeal joint	Lateral plantar nerve
Flexor digiti minimi brevis	Metatarsal V and tendon of fibularis (peroneus) longus	Lateral side of proximal phalanx of little toe	Flexes little toe at metatarsophalangeal joint	Lateral plantar nerve
Fourth layer (deepest)				
Dorsal interossei (in-ter-OS-ē-ī)	Adjacent side of all metatarsals	Proximal phalanges: both sides of toe II and lateral side of toes III–IV	Abduct and flex toes II–IV at metatarsophalangeal joints and extend toes at interphalangeal joints	Lateral plantar nerve
Plantar interossei	Metatarsals III–V	Medial side of proximal phalanges of toes III–V	Adduct and flex proximal metatarsophalangeal joints and extend toes at interphalangeal joints	Lateral plantar nerve

EXHIBIT 11.V **409**

Figure 11.25 Intrinsic muscles of the foot that move the toes.

The muscles of the hand are specialized for precise and intricate movements; those of the foot are limited to support and movement.

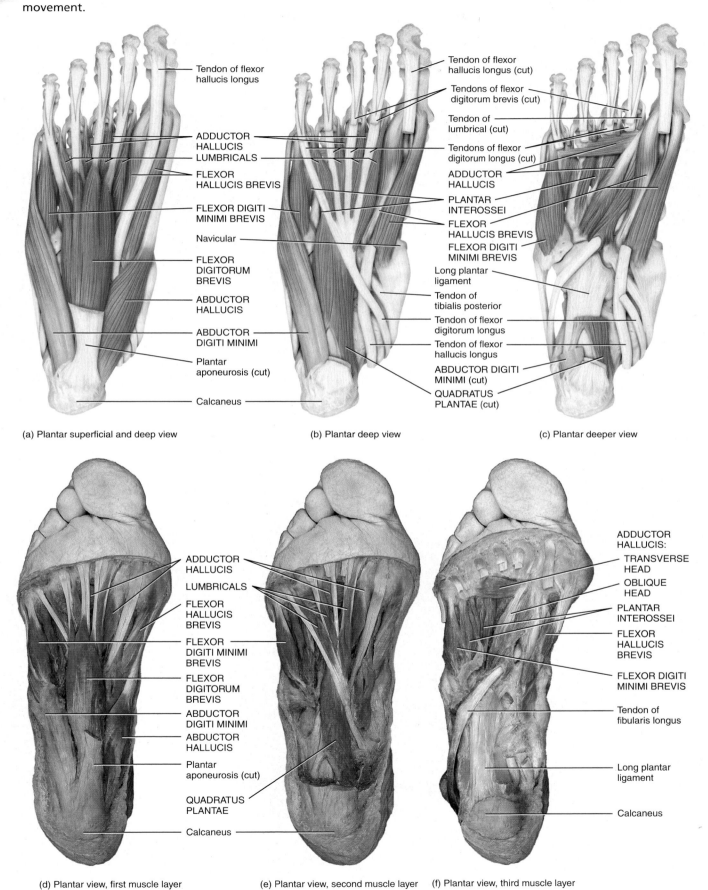

(a) Plantar superficial and deep view

(b) Plantar deep view

(c) Plantar deeper view

(d) Plantar view, first muscle layer

(e) Plantar view, second muscle layer

(f) Plantar view, third muscle layer

FIGURE 11.25 CONTINUES

EXHIBIT 11.V Intrinsic Muscles of the Foot That Move the Toes *(Figure 11.25)* CONTINUED

⬤ FIGURE **11.25** CONTINUED ▶

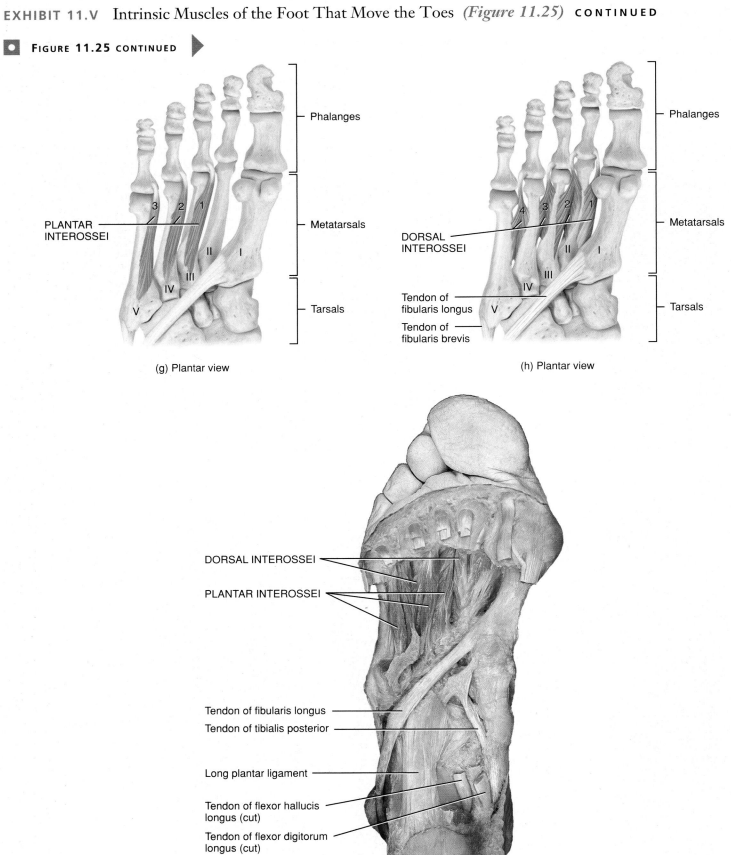

(g) Plantar view

(h) Plantar view

(i) Plantar view, fourth muscle layer

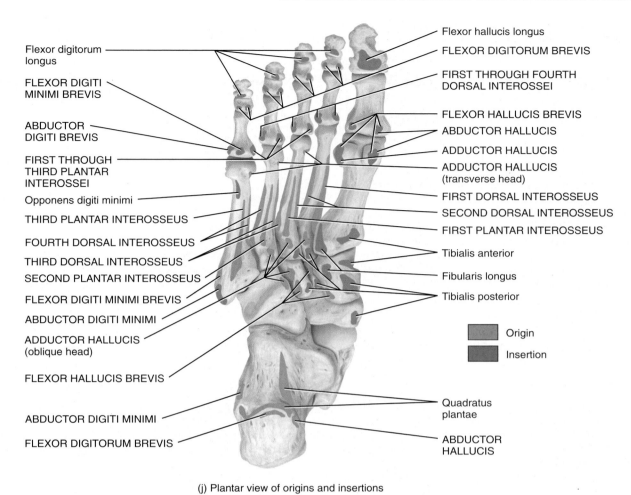

Flexor digitorum longus

FLEXOR DIGITI MINIMI BREVIS

ABDUCTOR DIGITI BREVIS

FIRST THROUGH THIRD PLANTAR INTEROSSEI

Opponens digiti minimi

THIRD PLANTAR INTEROSSEUS

FOURTH DORSAL INTEROSSEUS

THIRD DORSAL INTEROSSEUS

SECOND PLANTAR INTEROSSEUS

FLEXOR DIGITI MINIMI BREVIS

ABDUCTOR DIGITI MINIMI

ADDUCTOR HALLUCIS (oblique head)

FLEXOR HALLUCIS BREVIS

ABDUCTOR DIGITI MINIMI

FLEXOR DIGITORUM BREVIS

Flexor hallucis longus

FLEXOR DIGITORUM BREVIS

FIRST THROUGH FOURTH DORSAL INTEROSSEI

FLEXOR HALLUCIS BREVIS

ABDUCTOR HALLUCIS

ADDUCTOR HALLUCIS

ADDUCTOR HALLUCIS (transverse head)

FIRST DORSAL INTEROSSEUS

SECOND DORSAL INTEROSSEUS

FIRST PLANTAR INTEROSSEUS

Tibialis anterior

Fibularis longus

Tibialis posterior

Origin

Insertion

Quadratus plantae

ABDUCTOR HALLUCIS

(j) Plantar view of origins and insertions

 What structure supports the longitudinal arch and encloses the flexor tendons of the foot?

KEY MEDICAL TERMS ASSOCIATED WITH THE MUSCULAR SYSTEM

Cramp Painful spasmodic contraction caused by inadequate blood flow, overuse of a muscle, injury, holding a position for prolonged periods, and low levels of electrolytes (such as potassium).

Intramuscular injection (IM injection) The administration of a drug that penetrates the skin and subcutaneous layer to enter the muscle itself. The common sites for intramuscular injections include the buttock (gluteus medius muscle), the lateral side of the thigh (vastus lateralis muscle), and the deltoid region of the arm (deltoid muscle). Muscles in these areas are fairly thick and absorption is promoted by the extensive blood supply to such large muscles. To avoid injury, intramuscular injections are given deep within the muscle and away from major nerves and blood vessels.

Muscle strain Tearing of fibers in a skeletal muscle or its tendon that attaches the muscle to bone. The tearing can also damage small blood vessels, causing local bleeding (bruising) and pain (caused by irritation of nerve endings in the region). Muscle strains usually occur when a muscle is stretched beyond its limit, for example, in response to sudden, quick heavy lifting, during sports activities, or while performing work tasks. A similar injury occurs if there is a direct blow to a muscle. Factors that contribute to muscle strain include muscle tightness due to insufficient stretching, not working antagonistic muscles equally, poor conditioning, muscle fatigue, and insufficient warmup. The condition is treated by PRICE therapy: protection (P), rest (R), ice immediately after the injury (I), compression via a supportive wrap (C), and elevation of the limb (E). Also called **muscle pull** or **muscle tear**.

Paralysis (pa-RAL-i-sis; *para-*=departure from normal; *-lysis*=loosening) Loss of muscle function (voluntary movement) through injury, disease, or its nerve supply. Most paralysis is due to stroke or spinal cord injury.

Repetitive strain or motion injuries (RSIs) Include a large number of conditions resulting from overuse of equipment, poor posture, poor body mechanics, or activity that requires repeated movements; for example, various conditions of assembly line workers. Examples of overuse of equipment include overuse of a computer, hammer, guitar, or piano, to name a few.

Rhabdomyosarcoma (rab'-dō-mī'-ō-sar-KŌ-ma; *rhab-*=rod-shaped; *myo*=muscle; *sarc*=flesh; *-oma*=tumor) A tumor of skeletal muscle. Usually occurs in children and is highly malignant, with rapid metastasis.

Spasm (SPAZM) A sudden, involuntary contraction of one or more muscles.

Torticollis (tor-ti-KŌL-is; *tortus-*=twisted; *column*=neck) A contraction or shortening of the sternocleidomastoid muscle that causes the head to tilt toward the affected side and the chin to rotate toward the opposite side. It may be acquired or congenital. Also called **wryneck**.

Tic Spasmodic twitching made involuntarily by muscles that are usually under conscious control, for example, twitching of an eyelid.

CHAPTER REVIEW AND RESOURCE SUMMARY

WileyPLUS

Review

Resource

11.1 How Skeletal Muscles Produce Movements

1. Skeletal muscles that produce movement do so by pulling on bones.
2. The attachment to the more stationary bone is the origin; the attachment to the more movable bone is the insertion.
3. Bones serve as levers, and joints serve as fulcrums. Two different forces act on the lever: load (resistance) and effort.
4. Levers are categorized into three types—first-class, second-class, and third-class (most common)—according to the positions of the fulcrum, the effort, and the load on the lever.
5. Fascicular arrangements include parallel, fusiform, circular, triangular, and pennate. Fascicular arrangement affects a muscle's power and range of motion.
6. Muscles control joints by contracting both concentrically (shortening contractions) and eccentrically (lengthening contractions) to regulate fine muscle control at a joint.
7. A prime mover produces the desired action; an antagonist produces an opposite action. Synergists assist a prime mover by reducing unnecessary movement. Fixators stabilize the origin of a prime mover so that it can act more efficiently.

Anatomy Overview - Skeletal Muscle
Figure 11.1 - Relationship of Skeletal Muscle to Bones

11.2 How Skeletal Muscles Are Named

1. Distinctive features of different skeletal muscles include direction of muscle fascicles; size, shape, action, number of origins (or heads), and location of the muscle; and sites of origin and insertion of the muscle.
2. Most skeletal muscles are named based on combinations of these features.

11.3 Principal Skeletal Muscles

1. Muscles of the head that produce facial expression move the skin rather than a joint when they contract, and they permit us to express a wide variety of emotions (see Exhibit 11.A). The muscles of the head that move the eyeballs are among the fastest contracting and most precisely controlled skeletal muscles in the body. They permit us to elevate, depress, abduct, adduct, and medially and laterally rotate the eyeballs. The muscles that move the eyelids open the eyes (see Exhibit 11.B).
2. Muscles that move the mandible are also known as the muscles of mastication because they are involved in chewing and speech (see Exhibit 11.C). The muscles of the head that move the tongue are important in mastication and speech (see Exhibit 11.D).
3. Muscles of the anterior neck that assist in deglutition and speech, called suprahyoid muscles, are located above the hyoid bone (see Exhibit 11.E).
4. Most muscles of the larynx that assist in speech depress the hyoid bone and larynx and move the internal portion of the larynx (see Exhibit 11.F). Muscles of the pharynx are arranged into a circular layer, which functions in deglutition, and a longitudinal layer, which functions in deglutition and speech (see Exhibit 11.G).
5. Muscles of the neck that move the head alter its position and help balance it on the vertebral column (see Exhibit 11.H). Muscles of the neck that move the vertebral column are quite complex because they have multiple origins and insertions and because there is considerable overlap among them (see Exhibit 11.I).
6. Muscles of the abdomen help contain and protect the abdominal viscera, move the vertebral column, compress the abdomen, and produce the force required for defecation, urination, vomiting, and childbirth (see Exhibit 11.J). Muscles of the thorax used in breathing alter the size of the thoracic cavity so that inhalation and exhalation can occur and assist in venous return of blood to the heart (see Exhibit 11.K).
7. Muscles of the pelvic floor support the pelvic viscera, resist the thrust that accompanies increases in intra-abdominal pressure, and function as sphincters at the anorectal junction, urethra, and vagina (see Exhibit 11.L). Muscles of the perineum assist in urination, erection of the penis and clitoris, ejaculation, and defecation (see Exhibit 11.M).
8. Muscles of the thorax that move the pectoral (shoulder) girdle stabilize the scapula so it can function as a stable point of origin for most of the muscles that move the humerus (see Exhibit 11.N).
9. Muscles of the thorax that move the humerus originate for the most part on the scapula (scapular muscles); the remaining muscles originate on the axial skeleton (axial muscles) (see Exhibit 11.O). Muscles of the arm that move the radius and ulna are involved in flexion and extension at the elbow joint and are organized into flexor and extensor compartments (see Exhibit 11.P).
10. Muscles of the forearm that move the wrist, hand, thumb, and fingers are many and varied; those muscles that act on the digits are called extrinsic muscles (see Exhibit 11.Q). The muscles of the palm that move the digits (intrinsic muscles) are important in skilled activities and provide humans with the ability to grasp and manipulate objects precisely (see Exhibit 11.R).
11. Muscles of the gluteal region that move the femur originate for the most part on the pelvic girdle and insert on the femur; these muscles are larger and more powerful than comparable muscles in the upper limb (see Exhibit 11.S). Muscles of the thigh that move the femur and tibia and fibula are separated into medial (adductor), anterior (extensor), and posterior (flexor) compartments (see Exhibit 11.T).
12. Muscles of the leg that move the foot and toes are divided into anterior, lateral, and posterior compartments (see Exhibit 11.U). Muscles of the foot that move the toes (intrinsic muscles), unlike those of the hand, are limited to the functions of support and locomotion (see Exhibit 11.V).

Anatomy Overview - Muscles of Facial Expression
Anatomy Overview - Muscles Moving Eyeballs
Anatomy Overview - Muscles for Speech, Swallowing, and Chewing
Anatomy Overview - Muscles That Move the Head
Anatomy Overview - Muscles of the Torso
Anatomy Overview - Muscles for Breathing
Anatomy Overview - Muscles That Move the Pectoral Girdle
Anatomy Overview - Muscles That Move the Arm
Anatomy Overview - Muscles That Move the Forearm
Anatomy Overview - Muscles That Move the Hand
Anatomy Overview - Muscles That Move the Thigh
Anatomy Overview - Muscles of the Foot
Figure 11.3 - Principal Superficial Skeletal Muscles
Figure 11.14 - Muscles of the Thorax That Assist in Breathing
Figure 11.20 - Muscles of the Forearm That Move the Wrist, Hand, and Digits
Figure 11.23 - Muscles of the Thigh That Move the Femur, Tibia, and Fibula
Exercise - Shoulder Movements Target Practice
Exercise - Lower Limb Movements Target Practice
Exercise - Manage Those Muscles

CRITICAL THINKING QUESTIONS

1. Three-year-old Ming likes to form her tongue into a cylinder shape and use it as a straw when she drinks her milk. Name the muscles that Ming uses to protrude her lips and tongue and to suck up the milk.

2. Baby Eddie weighed 13 lb 4 oz at birth and he's been gaining ever since. His mom has noticed a sharp pain between her shoulder blades when she picks up Eddie to put him in his high chair (again). Which muscle may Eddie's mother have strained?

3. Anthropology graduate students are learning about "bite force" in Neanderthals. Since this is a significant feature of chewing, it is critical to understand muscle attachments onto the mandible, and specifically what actions these muscles produce. Distinguish the attachment sites and actions of the four muscles of mastication.

4. Perry was leading his fantasy baseball league until his best pitcher went on the disabled list with a shoulder injury. What is the most likely injury to this player? What specific muscles are likely to be involved?

5. Wyman has been doing a lot of heavy lifting lately. He has suddenly noticed a bulge on the anterior aspect of his torso down near his groin. What has probably happened to him? Do you think he needs to see a doctor? Why or why not?

? ANSWERS TO FIGURE QUESTIONS

11.1 The belly of the muscle that extends the forearm, the triceps brachii, is located posterior to the humerus.

11.2 Second-class levers produce the most force.

11.3 For muscles named after their various characteristics, here are possible correct responses (for others, see Table 11.2): direction of fibers: external oblique; shape: deltoid; action: extensor digitorum; size: gluteus maximus; origin and insertion: sternocleidomastoid; location: tibialis anterior; number of tendons of origin: biceps brachii.

11.4 The corrugator supercilii muscle is involved in frowning; the zygomaticus major muscle contracts when you smile; the mentalis muscle contributes to pouting; the orbicularis oculi muscle contributes to squinting.

11.5 The inferior oblique muscle moves the eyeball superiorly and laterally because it originates at the anteromedial aspect of the floor of the orbit and inserts on the posterolateral aspect of the eyeball.

11.6 The masseter is the strongest muscle of mastication.

11.7 Functions of the tongue include chewing, perception of taste, swallowing, and speech.

11.8 The suprahyoid and infrahyoid muscles stabilize the hyoid bone to assist in tongue movements.

11.9 The extrinsic muscles move the larynx as a whole, while the intrinsic muscles move parts of the larynx.

11.10 The longitudinal muscles elevate the larynx, and the infrahyoid muscles depress the larynx.

11.11 The triangles in the neck formed by the sternocleidomastoid muscles are important anatomically and surgically because of the structures that lie within their boundaries.

11.12 The splenius muscles arise from the midline and extend laterally and superiorly to their insertion.

11.13 The rectus abdominis muscle aids in urination.

11.14 The diaphragm is innervated by the phrenic nerve.

11.15 The borders of the pelvic diaphragm are the pubic symphysis anteriorly, the coccyx posteriorly, and the walls of the pelvis laterally.

11.16 The borders of the perineum are the pubic symphysis anteriorly, the coccyx posteriorly, and the ischial tuberosities laterally.

11.17 The main action of the muscles that move the pectoral girdle is to stabilize the scapula to assist in movements of the humerus.

11.18 The rotator cuff consists of the flat tendons of the subscapularis, supraspinatus, infraspinatus, and teres minor muscles that form a nearly complete circle around the shoulder joint.

11.19 The brachialis is the most powerful forearm flexor; the triceps brachii is the most powerful forearm extensor.

11.20 Flexor tendons of the digits and wrist and the median nerve pass beneath the flexor retinaculum.

11.21 Muscles of the thenar eminence act on the thumb (pollex).

11.22 Free upper limb muscles exhibit diversity of movement; free lower limb muscles function in stability, locomotion, and maintenance of posture. In addition, free lower limb muscles usually cross two joints and act equally on both.

11.23 The quadriceps femoris consists of the rectus femoris, vastus lateralis, vastus medialis, and vastus intermedius; the hamstrings consist of the biceps femoris, semitendinosus, and semimembranosus.

11.24 The superior and inferior extensor retinacula firmly hold the tendons of the anterior compartment muscles to the ankle.

11.25 The plantar aponeurosis (fascia) supports the longitudinal arch and encloses the flexor tendons of the foot.

12 THE CARDIOVASCULAR SYSTEM: BLOOD

INTRODUCTION The **cardiovascular system** (*cardio-*=heart; *vascular*=blood or blood vessels) consists of three interrelated components: blood, the heart, and blood vessels. The focus of this chapter is blood; the next two chapters will examine the heart and blood vessels, respectively. The branch of science concerned with the study of blood, blood-forming tissues, and the disorders associated with them is **hematology** (hēm-a-TOL-ō-jē; *hema-* or *hemato-* =blood; *-logy*=study of). Because of their common origins, the development of blood in the embryo and fetus is considered in Chapter 14 with the development of the blood vessels.

Most cells of a multicellular organism cannot move around to obtain oxygen and nutrients and get rid of carbon dioxide and other wastes. Instead, these needs are met by two fluids: blood and interstitial fluid. **Blood** is a liquid connective tissue that consists of cells surrounded by an extracellular matrix. The extracellular matrix is a liquid portion called plasma and the cellular portion consists of various cells and cell fragments. **Interstitial fluid** is the watery fluid that bathes body cells and is constantly renewed by the blood. Oxygen brought into the lungs and water and nutrients brought into the gastrointestinal tract are transported by the blood, diffuse from the blood into the interstitial fluid, and then diffuse into body cells. Carbon dioxide and other wastes move in the reverse direction, from the body cells into the interstitial fluid and then into the blood. Blood then transports the wastes to various organs—the lungs, kidneys, skin, and digestive system—for elimination from the body.

Blood transports various substances, helps regulate several life processes, and affords protection against disease. Health-care professionals routinely examine blood and analyze its differences through various blood tests when trying to determine the cause of different diseases. Despite these differences, blood is the most easily and widely shared of human tissues, saving many thousands of lives every year through blood transfusions. •

Did you ever wonder why anemia causes so many widespread symptoms?

12.1 FUNCTIONS OF BLOOD

OBJECTIVE
- Outline the functions of blood.

Blood is closely related to other body fluids. In fact, many of the extracellular body fluids (including interstitial fluid, lymph, cerebrospinal fluid, and aqueous humor) arise from the blood during development and are continually replenished by it. The extracellular fluids that nourish, protect, and exchange materials with every cell of the body are derived from the blood, renewed by the blood, and returned to the blood. Based on these relationships, blood has three general functions:

1. *Transportation.* Blood transports oxygen from the lungs to the cells of the body and carries carbon dioxide from the body cells to the lungs for exhalation. It carries nutrients from the gastrointestinal tract to body cells and hormones from endocrine glands to cells throughout the body. Blood also transports heat and waste products to the lungs, kidneys, and skin for elimination from the body.

2. *Regulation.* Circulating blood helps maintain homeostasis in all body fluids. Blood plays a role in the regulation of pH through buffers. (*Buffers* are chemicals that convert strong acids or bases into weak ones.) It also assists in the adjustment of body temperature; the heat-absorbing and coolant properties of the water in blood plasma and its variable rate of flow through the skin allow excess heat to be lost from the blood to the environment. Blood osmotic pressure also influences the water content of cells, mainly through interactions of dissolved ions and proteins.

3. *Protection.* Blood can clot (become gel-like), which protects against its excessive loss from the cardiovascular system after an injury. In addition, white blood cells protect against disease by carrying on phagocytosis. Several types of blood proteins, including antibodies, interferons, and complement, help protect against disease in a variety of ways (Chapter 15).

CHECKPOINT
1. What substances does blood transport?

12.2 PHYSICAL CHARACTERISTICS OF BLOOD

OBJECTIVE
- List the principal physical characteristics of blood.

Blood is denser and more viscous (thicker) than water, which is part of the reason it flows more slowly than water. The temperature of blood is about 38°C (100.4°F), which is slightly higher than normal body temperature, and it has a slightly alkaline pH ranging from 7.35 to 7.45. The color of blood varies. When saturated with oxygen it is bright red; when unsaturated with oxygen, the blood is dark red to purple. Blood constitutes about 8 percent of the total body weight. The blood volume is 5–6 liters (1.5 gal) in an average-sized adult male and 4–5 liters (1.2 gal) in an average-sized adult female. This gender difference in volume is due to the difference in average body size.

CHECKPOINT
2. How much does the blood of a 150-pound person weigh?

12.3 COMPONENTS OF BLOOD

OBJECTIVES
- Describe the principal components of blood.
- Explain the importance of hematocrit.

Whole blood is composed of two portions: (1) blood plasma, a watery liquid extracellular matrix that contains dissolved substances, and (2) formed elements, which are cells and cell fragments. If a sample of blood is centrifuged (spun) in a small glass tube, the cells (which are more dense) sink to the bottom of the tube and the plasma (which is less dense) forms a layer on top (Figure 12.1a). Blood is about 45 percent formed elements and 55 percent blood plasma. Normally, more than 99 percent of the formed elements are red-colored cells called red blood cells (RBCs). Pale or colorless white blood cells (WBCs) and platelets make up less than 1 percent of the formed elements. They form a very thin layer, called the *buffy coat*, between the packed RBCs and blood plasma in centrifuged blood. Figure 12.1b shows the composition of blood plasma and the numbers of the various types of formed elements in blood.

Blood Plasma

When the formed elements are removed from blood, a straw-colored liquid called **blood plasma** (or simply *plasma*) is left. Plasma is about 91.5 percent water and 8.5 percent solutes, most of which (7 percent by weight) are proteins. Some of the proteins in plasma are also found elsewhere in the body, but those confined to blood are called **plasma proteins**. Hepatocytes (liver cells) synthesize most of the plasma proteins, which include the **albumins** (al′-BŪ-mins) (54 percent of plasma proteins), **globulins** (GLOB-ū-lins) (38 percent), and **fibrinogens** (fī-BRIN-ō-jens) (7 percent). Some blood cells develop into plasma cells that produce globulins called **immunoglobulins** (im′-ū-nō-GLOB-ū-lins), also called *antibodies* because they are

415

Figure 12.1 Components of blood in a normal adult.

 Blood is a connective tissue that consists of blood plasma (liquid) plus formed elements (red blood cells, white blood cells, and platelets).

FUNCTIONS OF BLOOD

1. Transports oxygen, carbon dioxide, nutrients, hormones, heat, and wastes.
2. Regulates pH, body temperature, and water content of cells.
3. Protects against blood loss through clotting, and against disease through phagocytic white blood cells and proteins such as antibodies, interferons, and complement.

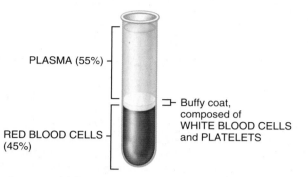

PLASMA (55%)

Buffy coat, composed of WHITE BLOOD CELLS and PLATELETS

RED BLOOD CELLS (45%)

(a) Appearance of centrifuged blood

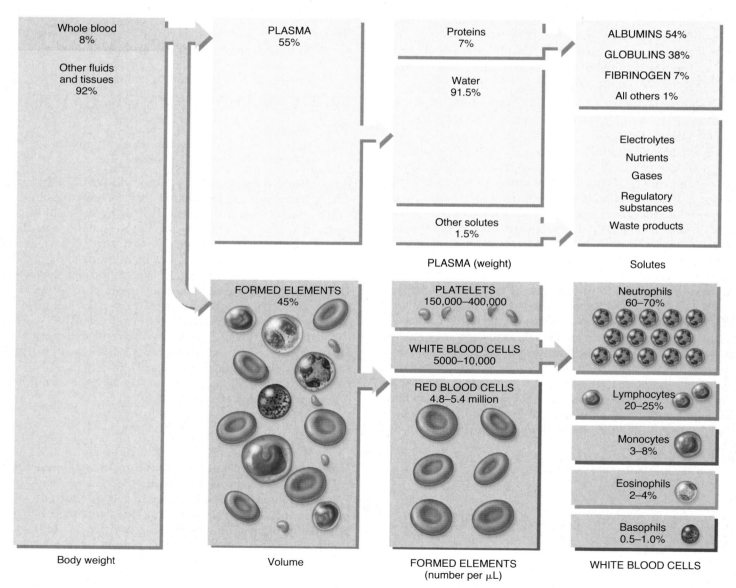

Whole blood 8%	**PLASMA** 55%	**Proteins** 7%	**ALBUMINS 54%**
Other fluids and tissues 92%		**Water** 91.5%	**GLOBULINS 38%**
			FIBRINOGEN 7%
			All others 1%

Electrolytes

Nutrients

Gases

Regulatory substances

Other solutes 1.5%

Waste products

PLASMA (weight)

Solutes

FORMED ELEMENTS 45%

PLATELETS 150,000–400,000

Neutrophils 60–70%

WHITE BLOOD CELLS 5000–10,000

RED BLOOD CELLS 4.8–5.4 million

Lymphocytes 20–25%

Monocytes 3–8%

Eosinophils 2–4%

Basophils 0.5–1.0%

Body weight

Volume

FORMED ELEMENTS (number per μL)

WHITE BLOOD CELLS

(b) Components of blood

 **What is the approximate volume of blood in your body?**

TABLE 12.1

Substances in Blood Plasma

CONSTITUENT	DESCRIPTION	FUNCTION
Water (91.5%)	Liquid portion of blood	Solvent and suspending medium; absorbs, transports, and releases heat
Plasma proteins (7%)	Most produced by liver	
Albumins	Smallest and most numerous of proteins	Responsible for colloid osmotic pressure; major contributors to blood viscosity; transport hormones (steroid), fatty acids, and calcium; help regulate blood pH
Globulins	Large proteins (plasma cells produce immunoglobulins)	Immunoglobulins help attack viruses and bacteria; alpha and beta globulins transport iron, lipids, and fat-soluble vitamins
Fibrinogen	Large protein	Plays essential role in blood clotting
Other solutes (1.5%)		
Electrolytes	Inorganic salts; positively charged ions (cations) Na^+, K^+, Ca^{2+}, Mg^{2+}; negatively charged ions (anions) Cl^-, HPO_4^{2-}, SO_4^{2-}, HCO_3^-	Help maintain osmotic pressure and play essential roles in cell functions
Nutrients	Products of digestion, such as amino acids, glucose, fatty acids, glycerol, vitamins, and minerals	Essential roles in cell functions, growth, and development
Gases	Oxygen (O_2)	Important in many cellular functions
	Carbon dioxide (CO_2)	Involved in the regulation of blood pH
	Nitrogen (N_2)	No known function
Regulatory substances	Enzymes	Catalyze chemical reactions
	Hormones	Regulate metabolism, growth, and development
	Vitamins	Cofactors for enzymatic reactions
Waste products	Urea, uric acid, creatine, creatinine, bilirubin, ammonia	Most are breakdown products of protein metabolism that are carried by the blood to organs of excretion

produced during certain immune responses (see Chapter 15). Foreign substances (antigens) such as bacteria and viruses stimulate production of antibodies. An antibody binds specifically to the antigen that stimulated its production and this disables the invading antigen.

Other solutes in plasma include electrolytes, nutrients, regulatory substances such as enzymes and hormones, gases, and waste products such as urea, uric acid, creatinine, ammonia, and bilirubin. Table 12.1 describes the chemical composition of blood plasma.

Formed Elements

The **formed elements** of the blood include **red blood cells (RBCs)**, **white blood cells (WBCs)**, and **platelets** (Figure 12.2). Unlike RBCs and platelets, which perform limited roles, WBCs have a number of more general functions. There are several distinct types of WBCs—neutrophils, lymphocytes, monocytes, eosinophils, and basophils—each having a unique microscopic appearance. The roles of each type of WBC are discussed later in this chapter.

Figure 12.2 The formed elements of blood.

🔑 The formed elements of blood include red blood cells (RBCs), white blood cells (WBCs), and platelets.

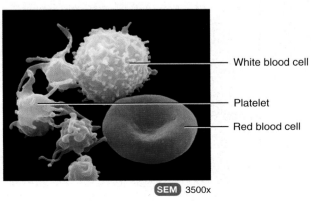

White blood cell

Platelet

Red blood cell

SEM 3500x

(a) Scanning electron micrograph

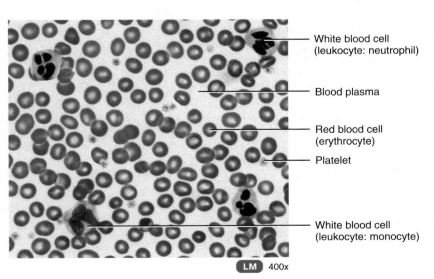

White blood cell (leukocyte: neutrophil)

Blood plasma

Red blood cell (erythrocyte)

Platelet

White blood cell (leukocyte: monocyte)

LM 400x

(b) Blood smear (thin film of blood spread on a glass slide)

 Which formed elements of the blood are cell fragments?

Following is the classification of the formed elements in blood:

I. Red blood cells (erythrocytes)
II. White blood cells (leukocytes)
   A. Granular leukocytes (contain conspicuous granules that are visible under light microscope after staining)
     1. Neutrophils
     2. Eosinophils
     3. Basophils
   B. Agranular leukocytes (no granules are visible under a light microscope after staining)
     1. T and B lymphocytes and natural killer (NK) cells
     2. Monocytes
III. Platelets (thrombocytes)

The percentage of total blood volume occupied by RBCs is called the **hematocrit** (hē-MAT-ō-krit). For example, a hematocrit of 40 means that 40 percent of the volume of blood is composed of RBCs. The normal range of hematocrit for adult females is about 38–46 percent (average = 42); for adult males it is about 40–54 percent (average = 47). The direct stimulus for the production of red blood cells is **hypoxia** (hī-POKS-ē-a) (cellular oxygen deficiency) in kidney cells. This in turn stimulates the synthesis of a hormone called **erythropoietin (EPO)** (e-rith′-rō-POY-e-tin) by the kidneys. EPO then stimulates the production of RBCs. Lower hematocrit values in women during their reproductive years may be due to excessive loss of blood during menstruation. A significant drop in hematocrit due to hemorrhage, red bone marrow failure, or excessive red blood cell destruction may be one cause of **anemia**, a condition in which the blood has a reduced oxygen-carrying capacity (described shortly). **Polycythemia** (pol′-ē-sī-THĒ-mē-a) is an increase in the percentage of RBCs in which the hematocrit is above 54 percent (the upper limit of normal). Polycythemia may be caused by conditions such as an unregulated increase in RBC production, tissue hypoxia, dehydration, or blood doping by athletes.

 CHECKPOINT

3. What are some functions of blood plasma proteins?
4. What is the significance of a hematocrit that is lower than normal? Higher than normal?

12.4 FORMATION OF BLOOD CELLS

 OBJECTIVE

• Explain the origin of blood cells.

Although some lymphocytes have a lifetime measured in years, most formed elements of the blood are continually dying and being replaced within hours, days, or weeks. Negative feedback systems regulate the total number of RBCs and platelets in circulation, and their numbers normally remain steady. However, the proportions of the different types of WBCs vary in response to challenges by invading pathogens and other foreign antigens.

The process by which the formed elements of blood develop is called **hemopoiesis** (hē-mō-poy-Ē-sis; *-poiesis*=making) or *hematopoiesis*. Hemopoiesis first occurs before birth in the yolk sac of an embryo and later in the liver, spleen, thymus, and lymph nodes of a fetus. In the last three months before birth, red bone marrow becomes the primary site of hemopoiesis and continues to be the source of blood cells after birth and throughout life.

Red bone marrow is a highly vascularized connective tissue located in the microscopic spaces between trabeculae of spongy bone tissue. It is present chiefly in bones of the axial skeleton, pectoral and pelvic girdles, and the proximal epiphyses of the humerus and femur. About 0.05–0.1 percent of red bone marrow cells are called **pluripotent stem cells** (ploo-RIP-ō-tent; *pluri-*=several), which are derived from mesenchyme (the tissue from which almost all connective tissues develop). Pluripotent stem cells are cells that have the capacity to develop into many different types of cells (Figure 12.3).

In newborns, all bone marrow is red and thus active in blood cell production. As an individual ages, the rate of blood cell formation decreases; the red bone marrow in the medullary (marrow) cavity of long bones becomes inactive and is replaced by yellow bone marrow, which consists largely of fat cells. Under certain conditions, such as severe bleeding, yellow bone marrow can revert to red bone marrow; this occurs

---

CLINICAL CONNECTION | *Bone Marrow Examination*

Sometimes a sample of red bone marrow must be obtained in order to diagnose certain blood disorders, such as leukemia and severe anemias. Bone marrow examination may involve **bone marrow aspiration** (withdrawal of a small amount of red bone marrow with a fine needle and syringe) or a **bone marrow biopsy** (removal of a core of red bone marrow with a larger needle). Both types of samples are usually taken from the iliac crest of the hip bone, although samples are sometimes aspirated from the sternum. In young children, bone marrow samples are taken from a vertebra or tibia (shin bone). The tissue or cell sample is then sent to a pathology lab for analysis. Specifically, the lab technicians look for signs of neoplastic (cancer) cells or other diseased cells to assist in diagnosis. •

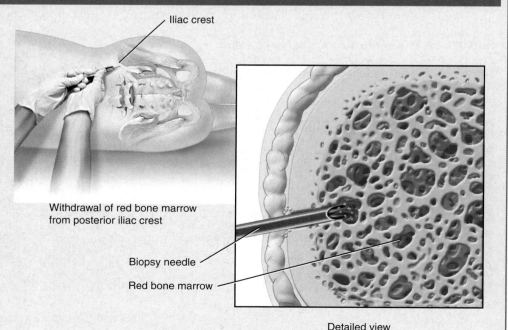

Iliac crest

Withdrawal of red bone marrow from posterior iliac crest

Biopsy needle

Red bone marrow

Detailed view

Figure 12.3 Origin, development, and structure of blood cells. A few of the generations of some cell lines have been omitted.

🔑 Blood cell production is called hemopoiesis and occurs mainly in red bone marrow after birth.

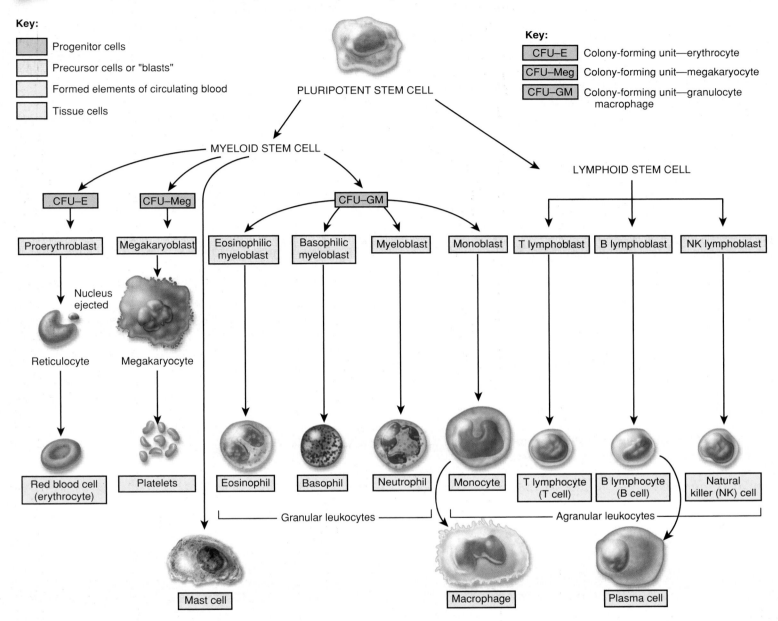

Key:

- Progenitor cells
- Precursor cells or "blasts"
- Formed elements of circulating blood
- Tissue cells

Key:

- CFU–E Colony-forming unit—erythrocyte
- CFU–Meg Colony-forming unit—megakaryocyte
- CFU–GM Colony-forming unit—granulocyte macrophage

PLURIPOTENT STEM CELL

MYELOID STEM CELL

LYMPHOID STEM CELL

CFU–E | CFU–Meg | CFU–GM

Proerythroblast | Megakaryoblast | Eosinophilic myeloblast | Basophilic myeloblast | Myeloblast | Monoblast | T lymphoblast | B lymphoblast | NK lymphoblast

Nucleus ejected

Reticulocyte | Megakaryocyte

Red blood cell (erythrocyte) | Platelets | Eosinophil | Basophil | Neutrophil | Monocyte | T lymphocyte (T cell) | B lymphocyte (B cell) | Natural killer (NK) cell

—— Granular leukocytes —— | —— Agranular leukocytes ——

Mast cell | Macrophage | Plasma cell

❓ **From which connective tissue do pluripotent stem cells develop?**

as blood-forming stem cells from the red bone marrow move into the yellow bone marrow, which is then repopulated by pluripotent stem cells.

Stem cells in red bone marrow reproduce themselves, proliferate, and differentiate into cells that give rise to blood cells, macrophages, reticular cells, mast cells, and adipocytes. Some stem cells can also form osteoblasts, chondroblasts, and muscle cells, and may be destined for use as a source of bone, cartilage, and muscular tissue for tissue and organ replacement. The reticular cells produce reticular fibers, which form the stroma (framework) that supports red bone marrow cells. Blood from nutrient and metaphyseal arteries (see Figure 6.4) enters a bone and passes into the enlarged and leaky capillaries, called *sinuses*, that surround red bone marrow cells and fibers. After blood cells form, they enter

the sinuses and other blood vessels and leave the bone through nutrient and periosteal veins (see Figure 6.4). With the exception of lymphocytes, formed elements do not divide once they leave red bone marrow.

In order to form blood cells, pluripotent stem cells produce two further types of stem cells, which have the capacity to develop into several kinds of cells. These stem cells are called **myeloid stem cells** and **lymphoid stem cells**. Myeloid stem cells begin and complete their development in red bone marrow and give rise to red blood cells, platelets, monocytes, neutrophils, eosinophils, basophils, and mast cells. Lymphoid stem cells, which give rise to lymphocytes, begin to develop in red bone marrow but complete development in lymphatic tissues. Lymphoid stem cells also give rise to natural killer (NK) cells. Although each type of stem cell has distinctive cell identity markers in its plasma membrane, they cannot be distinguished histologically and resemble lymphocytes.

During hemopoiesis, myeloid stem cells differentiate into **progenitor cells** (prō-JEN-i-tor). Progenitor cells are no longer capable of reproducing themselves and are committed to giving rise to more specific elements of blood. Some progenitor cells are known as *colony-forming units (CFUs)*. Following the CFU designation is an abbreviation that designates the mature elements in blood that they will produce: CFU—E ultimately produces erythrocytes (red blood cells), CFU—Meg produces megakaryocytes (the source of platelets), and CFU—GM ultimately produces granulocytes (specifically, neutrophils) and monocytes. Progenitor cells, like stem cells, resemble lymphocytes and cannot be distinguished by their microscopic appearance alone. Other myeloid stem cells develop directly into cells called precursor cells (described next). Lymphoid stem cells differentiate into T lymphoblasts and B lymphoblasts, which ultimately develop into T lymphocytes (T cells) and B lymphocytes (B cells), respectively.

In the next generation, the cells are called **precursor cells** or **blasts**. Over several cell divisions they develop into the actual formed elements of blood. For example, monoblasts develop into monocytes, eosinophilic myeloblasts develop into eosinophils, and so on. Each type of precursor cell has a recognizable microscopic appearance.

Several hormones called **hemopoietic growth factors** regulate the differentiation and proliferation of particular progenitor cells. Erythropoietin (EPO) increases the number of red blood cell precursors. The main producers of EPO are cells in the kidneys that lie between the kidney tubules (peritubular interstitial cells). If renal failure occurs, EPO release slows and RBC production is inadequate. This leads to a decreased hematocrit, which leads to a decreased ability to deliver oxygen to the body's cells. **Thrombopoietin (TPO)** (throm′-bō-POY-ē-tin) is a hormone produced by the liver that stimulates the formation of platelets (thrombocytes) from megakaryocytes. Several different **cytokines**—small glycoproteins produced by red bone marrow

cells, leukocytes, macrophages, fibroblasts, and endothelial cells, regulate development of different blood cell types. They typically act as local hormones (autocrines or paracrines). Cytokines stimulate proliferation of progenitor cells in red bone marrow and regulate the activities of cells involved in nonspecific defenses (such as phagocytes) and immune responses (such as B cells and T cells). Two important families of cytokines that stimulate white blood cell formation are **colony-stimulating factors (CSFs)** and **interleukins**.

 CHECKPOINT

5. Which hemopoietic growth factors regulate differentiation and proliferation of CFU–E and formation of platelets from megakaryocytes?

12.5 RED BLOOD CELLS

 OBJECTIVES

• Describe the structure and functions of red blood cells.
• Explain how red blood cells are produced.
• Describe the basis for the ABO and Rh grouping systems.

Red blood cells (RBCs) or **erythrocytes** (e-RITH-rō-sīts; *erythro-*=red; *-cyte*=cell) contain the oxygen-carrying protein **hemoglobin (Hb)**, which is a pigment that gives whole blood its red color. A healthy adult male has about 5.4 million red blood cells per microliter (μL) of blood,* and a healthy adult female has about 4.8 million. (One drop of blood is about 50 μL.) To maintain normal numbers of RBCs, new mature cells must enter the circulation at the astonishing rate of at least 2 million per second, a pace that balances the equally high rate of RBC destruction.

*1 μL = 1 mm^3 = 10^{-6} liter.

CLINICAL CONNECTION | *Anemia*

Anemia (a-NĒ-mē-a) is a condition in which the oxygen-carrying capacity of blood is reduced. All of the many types of anemia are characterized by reduced numbers of RBCs or a decreased amount of hemoglobin in the blood. The person feels fatigued and is intolerant of cold, both of which are related to lack of oxygen needed for ATP and heat production. Also, the skin appears pale, due to the low content of red-colored hemoglobin circulating in skin blood vessels. Among the most important causes and types of anemia are the following:

• *Inadequate absorption of iron, excessive loss of iron, increased iron requirement, or insufficient intake of iron* causes **iron-deficiency anemia**, the most common type of anemia. Women are at greater risk for iron-deficiency anemia due to menstrual blood losses and increased iron demands of the growing fetus during pregnancy. Gastrointestinal losses, such as those that occur with malignancy or ulceration, also contribute to this type of anemia.

• *Inadequate intake of vitamin B₁₂ or folic acid* causes **megaloblastic anemia** in which red bone marrow produces large, abnormal red blood cells (megaloblasts). It may also be caused by drugs that alter gastric secretion or are used to treat cancer.

• *Insufficient hemopoiesis* resulting from an inability of the stomach to produce intrinsic factor, which is needed for absorption of vitamin B₁₂ in the small intestine, causes **pernicious anemia**.

• *Excessive loss of RBCs* through bleeding resulting from large wounds, stomach ulcers, or especially heavy menstruation leads to **hemorrhagic anemia**.

• *RBC plasma membranes rupture prematurely* in **hemolytic anemia**. The released hemoglobin pours into the plasma and may damage the filtering units (glomeruli) in the kidneys. The condition may result from inherited defects such as abnormal red blood cell enzymes, or from outside agents such as parasites, toxins, or antibodies from incompatible transfused blood.

• *Deficient synthesis of hemoglobin* occurs in **thalassemia** (thal′-a-SĒ-mē-a), a group of hereditary hemolytic anemias. The RBCs are small (microcytic), pale (hypochromic), and short-lived. Thalassemia occurs primarily in populations from countries bordering the Mediterranean Sea.

• *Destruction of red bone marrow* results in **aplastic anemia**. It is caused by toxins, gamma radiation, and certain medications that inhibit enzymes needed for hemopoiesis. •

RBC Anatomy

RBCs are biconcave discs with a diameter of 7–8 μm (Figure 12.4a). (Recall that 1 μm = 1/25,000 of an inch or 1/10,000 of a centimeter, which is 1/1000 of a millimeter (mm).) Mature red blood cells have a simple structure. Their plasma membrane is both strong and flexible, which allows them to deform without rupturing as they squeeze through narrow blood capillaries. As you will see later, certain glycolipids in the plasma membrane of RBCs are antigens that account for the various blood groups such as the ABO and Rh groups. RBCs lack a nucleus and other organelles and can neither reproduce nor carry on extensive metabolic activities. The cytosol of RBCs contains hemoglobin molecules, which were synthesized before loss of the nucleus during RBC production and which constitute about 33 percent of the cell's weight. RBCs sometimes adhere to one another at their broad surfaces, and the arrangement, which resembles a pile of stacked coins, is called *rouleaux* (roo-LŌ).

RBC Functions

Red blood cells are highly specialized for their oxygen transport function. Because mature RBCs have no nucleus, all their internal space is available for oxygen transport. Because RBCs lack mitochondria and generate ATP anaerobically (without oxygen), they do not use up any of the oxygen they transport. Even the shape of an RBC facilitates its function. A biconcave disc has a

Figure 12.4 **The shapes of a red blood cell (RBC) and a hemoglobin molecule.** In (b), note that each of the four polypeptide chains of a hemoglobin molecule (blue) has one heme group (gold), which contains an iron ion (Fe^{2+}), shown in red.

 The iron portion of a heme group binds oxygen for transport by hemoglobin.

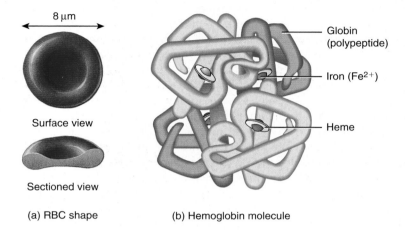

(a) RBC shape

8 μm

Surface view

Sectioned view

(b) Hemoglobin molecule

Globin (polypeptide)

Iron (Fe^{2+})

Heme

? **How many molecules of O_2 can one hemoglobin molecule transport?**

CLINICAL CONNECTION | *Sickle-cell Disease*

The RBCs of a person with **sickle-cell disease (SCD)** contain Hb-S, an abnormal kind of hemoglobin. When Hb-S gives up oxygen to the interstitial fluid, it forms long, stiff, rodlike structures that bend the erythrocyte into a sickle shape. The sickled cells rupture easily. Even though erythropoiesis is stimulated by the loss of the cells, it cannot keep pace with hemolysis. Signs and symptoms of SCD are caused by the sickling of red blood cells. When red blood cells sickle, they break down prematurely (sickled cells die in about 10 to 20 days). This leads to anemia, which can cause shortness of breath, fatigue, paleness, and delayed growth and development in children. The rapid breakdown and loss of blood cells may also cause *jaundice*, yellowing of the eyes and skin. Sickled cells do not move easily through blood vessels and they tend to stick together and form clumps that cause blockages in blood vessels. This deprives body organs of sufficient oxygen and causes pain, for example, in bones and the abdomen; serious infections; and organ damage, especially in the lungs, brain, spleen, and kidneys. Other symptoms of SCD include fever, rapid heart rate, swelling and inflammation of the hands and/or feet, leg ulcers, eye damage, excessive thirst, frequent urination, and painful and prolonged erections in males. Almost all individuals with SCD have painful episodes that can last from hours to days. Some people have one episode every few years; others have several episodes a year. The episodes may range from mild to those that require hospitalization. Any activity that reduces the amount of oxygen in the blood, such as vigorous exercise, may produce a **sickle-cell crisis** (worsening of the anemia, pain in the abdomen and long bones of the limbs, fever, and shortness of breath).

Sickle-cell disease is inherited. People with two sickle-cell genes have severe anemia; those with only one defective gene have the sickle-cell trait. Sickle-cell genes are found primarily among populations (or their descendants) that live in the malaria belt around the world, including parts of Mediterranean Europe, sub-Saharan Africa, and tropical Asia. The genes responsible for the tendency of the RBCs

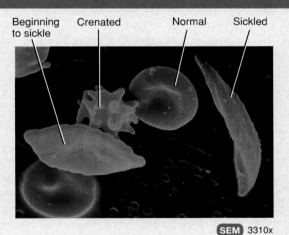

Beginning to sickle   Crenated   Normal   Sickled

SEM 3310x

Red blood cells

to sickle also alter the permeability of the plasma membranes of sickled cells, causing potassium ions to leak out. Low levels of potassium kill the malaria parasites that may infect sickled cells. Because of this effect, a person with one normal gene and one sickle-cell gene has higher-than-average resistance to malaria. The possession of a single sickle-cell gene thus confers a survival benefit. Treatment of SCD consists of administration of analgesics to relieve pain, fluid therapy to maintain hydration, oxygen to reduce oxygen deficiency, antibiotics to counter infections, and blood transfusions. People who suffer from SCD have normal fetal hemoglobin (Hb-F), a slightly different form of hemoglobin that predominates at birth and is present in small amounts after birth. In some patients with sickle-cell disease, a drug called hydroxyurea promotes transcription of the normal Hb-F gene, elevates the level of Hb-F, and reduces the chance that the RBCs will sickle. Unfortunately, this drug also has toxic effects on the bone marrow; thus, its safety for long-term use is questionable. •

much greater surface area for its volume than, say, a sphere or a cube. Its shape thus provides a large surface area for the diffusion of gas molecules into and out of the RBC; in addition, the thinness of the cell enhances the rapid diffusion of O_2 between the exterior and innermost regions of the cell.

Each RBC contains about 280 million hemoglobin molecules; each hemoglobin molecule can carry up to four oxygen molecules. A hemoglobin molecule consists of a protein called **globin**, composed of four polypeptide chains (two alpha and two beta chains), plus four nonprotein pigments called **hemes** (Figure 12.4b). One ringlike heme binds to each polypeptide chain. At the center of each heme ring is an iron ion (Fe^{2+}) that can combine reversibly with one oxygen molecule. The oxygen picked up in the lungs is transported bound to the iron of the heme group. As blood flows through tissue capillaries, the iron–oxygen reaction reverses. Hemoglobin releases oxygen, which diffuses first into the interstitial fluid and then into cells.

Hemoglobin also transports about 23 percent of the total carbon dioxide, a waste product of metabolism. (The remainder of the carbon dioxide is dissolved in plasma or carried as bicarbonate ions.) Blood flowing through tissue capillaries picks up carbon dioxide, some of which combines with amino acids in the globin part of the hemoglobin molecule. As blood flows through the lungs, the carbon dioxide is released from hemoglobin and then exhaled.

In addition to its key role in transporting oxygen and carbon dioxide, hemoglobin also plays a role in the regulation of blood flow and blood pressure. The gaseous hormone **nitric oxide (NO)**, produced by the endothelial cells that line blood vessels, binds to hemoglobin. Under some circumstances, hemoglobin releases NO. The released NO causes *vasodilation*, an increase in blood vessel diameter that occurs when the smooth muscle in the vessel wall relaxes. Vasodilation improves blood flow and enhances oxygen delivery to cells near the site of NO release.

RBC Life Cycle

Red blood cells live only about 120 days because of the wear and tear their plasma membranes undergo as they squeeze through blood capillaries. Without a nucleus and other organelles, RBCs cannot synthesize new components to replace damaged ones. The plasma membrane becomes more fragile with age, and the cells are more likely to burst, especially as they squeeze through narrow channels in the spleen. Ruptured red blood cells are removed from circulation and destroyed by fixed phagocytic macrophages in the spleen and liver, and the breakdown products are recycled and used in numerous metabolic processes, including the formation of new red blood cells.

Erythropoiesis: Production of RBCs

Erythropoiesis (e-rith-rō-poy-Ē-sis), the production of RBCs, starts in the red bone marrow with a precursor cell called a **proerythroblast** (prō-e-RITH-rō-blast) (see Figure 12.3). The proerythroblast divides several times, producing cells that begin to synthesize hemoglobin. Ultimately, a cell near the end of the developmental sequence ejects its nucleus and becomes a **reticulocyte** (re-TIK-ū-lō-sīt). Loss of the nucleus causes the center of the cell to indent, producing the distinctive biconcave shape. Reticulocytes, which are composed of about 34 percent hemoglobin and retain some mitochondria, ribosomes, and endoplasmic reticulum, pass from red bone marrow into the bloodstream by squeezing through holes in the plasma membrane

of the endothelial cells of blood capillaries. Reticulocytes usually develop into erythrocytes, or mature red blood cells, within 1–2 days after their release from red bone marrow.

Normally, erythropoiesis and red blood cell destruction proceed at roughly the same pace. If the oxygen-carrying capacity of the blood falls because erythropoiesis is not keeping up with RBC destruction, RBC production is increased. Hypoxia may occur if not enough oxygen enters the blood. For example, the reduced oxygen content of air at high altitudes results in a reduced level of oxygen in the blood. Oxygen delivery may also fall due to anemia, which has many causes. A diet deficient in iron, certain amino acids, and/or vitamin B_{12} are common causes of anemia. Circulatory problems that reduce blood flow to tissues may also reduce oxygen delivery. Whatever the cause, hypoxia stimulates the kidneys to step up the release of erythropoietin. This hormone circulates through the blood to the red bone marrow, where it speeds the development of proerythroblasts into reticulocytes. When the number of circulating RBCs increases, more oxygen can be delivered to body tissues.

Premature newborns often exhibit anemia, due in part to inadequate production of erythropoietin. During the first weeks after birth, it is the liver that produces most of the EPO, not the kidneys. Because the liver is less sensitive than the kidneys to hypoxia, newborns have a smaller EPO response to anemia than do adults. Also, after birth the loss of the more efficient oxygen-carrying fetal hemoglobin can compound the situation, making the anemia worse.

CLINICAL CONNECTION | *Blood Doping*

Delivery of oxygen to muscles is a limiting factor in muscular feats from weightlifting to running a marathon. As a result, increasing the oxygen-carrying capacity of the blood enhances athletic performance, especially in endurance events. Because RBCs transport oxygen, athletes have tried several means of increasing their RBC count, known as **blood doping** or **artificially induced polycythemia** (an abnormally high number of RBCs), to gain a competitive edge. Athletes have enhanced their RBC production by injecting Epoetin alfa (Procrit® or Epogen®), a drug that is used to treat anemia by stimulating the production of RBCs by red bone marrow. Practices that increase the number of RBCs are dangerous because they raise the viscosity of the blood, which increases the resistance to blood flow and makes the blood more difficult for the heart to pump. Increased viscosity also contributes to high blood pressure and increased risk of stroke. During the 1980s, at least 15 competitive cyclists died from heart attacks or strokes linked to suspected use of Epoetin alfa. Although the International Olympic Committee bans the use of Epoetin alfa, enforcement is difficult because the drug is identical to naturally occurring erythropoietin (EPO).

So-called **natural blood doping** is seemingly the key to the success of marathon runners from Kenya. The average altitude throughout Kenya's highlands is about 6600 feet (2000 meters) above sea level; other areas of Kenya are even higher. Altitude training greatly improves fitness, endurance, and performance. At these higher altitudes, the body increases the production of red blood cells, which means that exercise greatly oxygenates the blood. When these runners compete in Boston, for example, at an altitude just above sea level, the bodies of these runners contain more erythrocytes than do the bodies of competitors who trained in Boston. A number of training camps have been established in Kenya and now attract endurance athletes from all over the world. •

Blood Group Systems

More than 100 types of genetically determined antigens have been detected on the surface of red blood cells. Many of these antigens appear in characteristic patterns, a fact that enables scientists or health-care professionals to identify a person's blood as belonging to one or more blood groupings; there are at least 14 currently recognized blood group systems. Each system is characterized by the presence or absence of specific antigens on the surface of a red blood cell's plasma membrane. The two most commonly used categories are the ABO and Rh blood grouping systems.

The **ABO blood grouping system** is based on two antigens, symbolized as *A* and *B*. Individuals whose erythrocytes manufacture only antigen *A* are said to have blood type A. Those who manufacture only antigen *B* are type B. Individuals who manufacture both *A* and *B* are type AB. Those who manufacture neither are type O. Blood plasma usually contains **antibodies** that react with the *A* or *B* antigens if the two are mixed. These are the **anti-A antibody**, which reacts with antigen A, and the **anti-B antibody**, which reacts with antigen B. You do not have antibodies that react with the antigens of your own RBCs, but you do have antibodies for any antigens that your RBCs lack. For example, if your blood type is B, you have B antigens on your red blood cells, and you have *anti-A antibodies* in your blood plasma.

The **Rh blood grouping system** is so named because the Rh antigen, called *Rh factor*, was first found in the blood of the *Rh*esus monkey. Individuals whose erythrocytes have the Rh antigens (D antigens) are designated *Rh*⁺. Those who lack Rh antigens are designated *Rh*⁻.

CLINICAL CONNECTION | *Hemolytic Disease of the Newborn*

The most common problem with Rh incompatibility, **hemolytic disease of the newborn (HDN)**, arises during pregnancy. Normally, no direct contact occurs between maternal and fetal blood while a woman is pregnant. However, if a small amount of Rh⁺ blood leaks from the fetus through the placenta into the bloodstream of an Rh⁻ mother, the mother will start to make *anti-Rh antibodies*. Because the greatest possibility of fetal blood leakage into the maternal circulation occurs at delivery, the firstborn baby usually is not affected. If the mother becomes pregnant again, however, her anti-Rh antibodies can cross the placenta and enter the bloodstream of the fetus. If the fetus is Rh⁻, there is no problem, because Rh⁻ blood does not have the Rh antigen. If the fetus is Rh⁺, however, **agglutination** or clumping and **hemolysis** (hē-MOL-i-sis) or rupture of RBCs brought on by fetal–maternal incompatibility may occur in the fetal blood.

An injection of anti-Rh antibodies called anti-Rh gamma globulin (RhoGAM®) can be given to prevent HDN. Rh⁻ women receive RhoGAM® before delivery, and soon after every delivery, miscarriage, or abortion. These antibodies bind to and inactivate the fetal Rh antigens before the mother's immune system can respond to the foreign antigens by producing her own anti-Rh antibodies. •

As just noted, the presence or absence of certain antigens on red blood cells is the basis for classifying blood into several different groups. Such information is very important when a transfusion is given. A **transfusion** (trans-FŪ-zhun) is the transfer of whole blood or blood components (red blood cells or blood plasma, for example) into the bloodstream. A transfusion may be given to treat low blood volume, anemia, or a low platelet count. However, in an incompatible blood transfusion, antibodies in the recipient's plasma bind to the antigens on the donated RBCs,

which causes **agglutination** (a-gloo-ti-NĀ-shun) (clumping) of the RBCs. Agglutination is an antigen–antibody response in which RBCs become cross-linked to one another. When these antigen–antibody complexes form, they activate complement proteins (globulins), which make the plasma membrane of the donated RBCs leaky, causing hemolysis (rupture) of the RBCs and the release of hemoglobin into the blood plasma. The liberated hemoglobin may cause kidney damage. Although quite rare, it is possible for the viruses that cause AIDS and hepatitis B and C to be transmitted through transfusion of contaminated blood products.

✓ **CHECKPOINT**

6. Describe the size, microscopic appearance, and functions of RBCs.
7. Define erythropoiesis. Relate erythropoiesis to hematocrit. What factors accelerate and slow erythropoiesis?
8. What is the basis for the ABO and Rh blood grouping systems?

12.6 WHITE BLOOD CELLS

● **OBJECTIVES**

• Describe the structure and functions of white blood cells.
• Define differential white blood cell count.

WBC Anatomy and Types

Unlike red blood cells, **white blood cells**, or **leukocytes** (LOO-kō-sīts; *leuko-*=white), have a nucleus and a full complement of other organelles but they do not contain hemoglobin (Figure 12.5). WBCs are classified as either granular or agranular, depending on whether they contain conspicuous cytoplasmic granules that are made visible by staining when viewed through a light microscope. *Granular leukocytes* include neutrophils, eosinophils, and basophils; *agranular leukocytes* include lymphocytes and monocytes. As shown in Figure 12.3, monocytes and granular leukocytes develop from myeloid stem cells. In contrast, lymphocytes develop from lymphoid stem cells.

Granular Leukocytes

After staining, each of the three types of granular leukocytes displays conspicuous granules with distinctive coloration that can be recognized under a light microscope. Granular leukocytes can be distinguished as follows:

• **Neutrophil** (NOO-trō-fil). The granules of a neutrophil are smaller, evenly distributed, and pale lilac in color (Figure 12.5a). Because the granules do not strongly attract either the acidic (red) or basic (blue) stain, these WBCs are neutrophilic (=neutral loving). The nucleus has two to five lobes, connected by very thin strands of nuclear material. As the cells age, the number of nuclear lobes increases. Because older neutrophils thus have several differently shaped nuclear lobes, they are often called *polymorphonuclear leukocytes (PMNs)*, or "polys."

• **Eosinophil** (ē-ō-SIN-ō-fil). The large, uniform-sized granules within an eosinophil are *eosinophilic* (=eosin-loving)—they stain red-orange with acidic dyes (Figure 12.5b). The granules usually do not cover or obscure the nucleus, which most often has two or three lobes connected by either a thin strand or a thick strand of nuclear material.

Figure 12.5 Structure of white blood cells.

 White blood cells are distinguished from one another by the shape of their nuclei and the staining properties of their cytoplasmic granules.

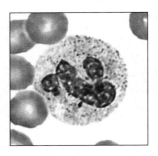 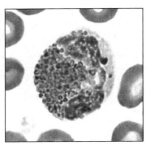 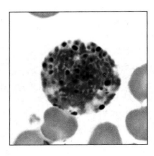 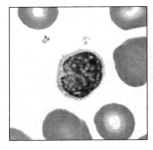 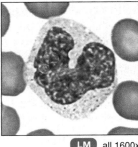

LM all 1600x

(a) Neutrophil   (b) Eosinophil   (c) Basophil   (d) Lymphocyte   (e) Monocyte

 Which WBCs are called granular leukocytes? Why?

- **Basophil** (BĀ-sō-fil). The round, variable-sized granules of a basophil are *basophilic* (=basic loving)—they stain blue-purple with basic dyes (Figure 12.5c). The granules commonly obscure the nucleus, which has two lobes.

Agranular Leukocytes

Even though so-called agranular leukocytes possess cytoplasmic granules, the granules are not visible under a light microscope because of their small size and poor staining qualities. Characteristics of agranular leukocytes are as follows:

- **Lymphocyte** (LIM-fō-sīt). The nucleus of a lymphocyte stains dark and is round or slightly indented (Figure 12.5d). The cytoplasm stains sky blue and forms a rim around the nucleus. The larger the cell, the more cytoplasm is visible. Lymphocytes are classified by cell diameter as large lymphocytes (10–14 μm) or small lymphocytes (6–9 μm). Although the functional significance of the size difference between small and large lymphocytes is unclear, the distinction is still clinically useful because an increase in the number of large lymphocytes has diagnostic significance in acute viral infections and in some immunodeficiency diseases.
- **Monocyte** (MON-ō-sīt). The nucleus of a monocyte is usually kidney-shaped or horseshoe-shaped, and the cytoplasm is blue-gray and has a foamy appearance (Figure 12.5e). The cytoplasm's color and appearance are due to very fine *azurophilic granules* (az′-ū-rō-FIL-ik; *azur*=blue; *philos*=loving), which are lysosomes. Blood is merely a conduit for monocytes, which migrate from the blood into the tissues, where they enlarge and differentiate into **macrophages** (MAK-rō-fā-jez=large eaters). Some become **fixed (tissue) macrophages,** which means they reside in a particular tissue; examples are alveolar macrophages in the lungs or macrophages in the spleen. Others become **wandering macrophages,** which roam the tissues and gather at sites of infection or inflammation.

White blood cells and other nucleated body cells have proteins, called *major histocompatibility (MHC) antigens,* protruding from their plasma membranes into the extracellular fluid. These cell identity markers are unique for each person (except identical twins). Although RBCs possess blood group antigens, they lack the MHC antigens.

WBC Functions

The skin and mucous membranes of the body are continuously exposed to microscopic organisms, such as bacteria, some of which are capable of invading deeper tissues and causing disease. Once pathogens enter the body, the general function of white blood cells is to combat them by phagocytosis or immune responses (see Chapter 15). To accomplish these tasks, many WBCs leave the bloodstream and collect at points of pathogen invasion or inflammation. Once granular leukocytes and monocytes leave the bloodstream to fight injury or infection, they never return to it. Lymphocytes, on the other hand, continually recirculate—from blood to interstitial spaces of tissues to lymphatic fluid and back to blood. Only 2 percent of the total lymphocyte population is circulating in the blood at any given time; the rest are in lymphatic fluid and organs such as skin, lungs, lymph nodes, and spleen.

RBCs are contained within the bloodstream, but WBCs are able to cross capillary walls by a process currently termed **emigration** (em′-i-GRĀ-shun; *e-*=out; *migra-*=wander); this process was formerly called *diapedesis*. During emigration, WBCs roll along the endothelium that forms capillary walls, stick to it, and then squeeze between endothelial cells. The precise signals that stimulate emigration through a particular blood vessel vary for the different types of WBCs.

Neutrophils and macrophages are active in **phagocytosis** (fāg′-ō-sī-TŌ-sis); they can ingest bacteria and dispose of dead matter. Several different chemicals released by microbes and inflamed tissues attract phagocytes, a phenomenon called chemotaxis (kē-mō-TAK-sis).

Among WBCs, neutrophils respond most quickly to tissue destruction by bacteria. After engulfing a pathogen during phagocytosis, a neutrophil unleashes several destructive chemicals to destroy the ingested pathogen. These chemicals include the enzyme **lysozyme** (LĪ-sō-zīm), which destroys certain bacteria, and **strong oxidants,** such as hydrogen peroxide (H_2O_2) and the hypochlorite anion (OCl^-), which is similar to household bleach. Neutrophils also contain **defensins,** proteins that exhibit a broad range of antibiotic activity against bacteria and fungi. Within a neutrophil, granules containing defensins form peptide "spears" that poke holes in microbe membranes; the resulting loss of cellular contents kills the invader.

Eosinophils leave the capillaries and enter tissue fluid. They are believed to release enzymes, such as histaminase, that combat

CLINICAL CONNECTION | *Leukemia*

The term **leukemia** (loo-KĒ-mē-a; *leuko-*=white) refers to a group of red bone marrow cancers in which abnormal white blood cells multiply uncontrollably. The accumulation of the cancerous white blood cells in red bone marrow interferes with the production of red blood cells, white blood cells, and platelets. As a result, the oxygen-carrying capacity of the blood is reduced, an individual is more susceptible to infection, and blood clotting is abnormal. In most leukemias, the cancerous white blood cells spread to the lymph nodes, liver, and spleen, causing them to enlarge. All leukemias produce the usual symptoms of anemia (fatigue, intolerance to cold, and pale skin). In addition, weight loss, fever, night sweats, excessive bleeding, and recurrent infections may occur.

In general, leukemias are classified as **acute** (symptoms develop rapidly) or **chronic** (symptoms may take years to develop). Leukemias are also classified on the basis of the type of white blood cell that becomes malignant. **Lymphoblastic leukemia** (lim-fō-BLAS-tik) involves cells derived from lymphoid stem cells (lymphoblasts) and/ or lymphocytes. **Myelogenous leukemia** (mī-e-LOJ-e-nus) involves

cells derived from myeloid stem cells (myeloblasts). Combining onset of symptoms and cells involved, there are four types of leukemia.

1. **Acute lymphoblastic leukemia (ALL)** is the most common leukemia in children, but adults can also get it.
2. **Acute myelogenous leukemia (AML)** affects both children and adults.
3. **Chronic lymphoblastic anemia (CLA)** is the most common leukemia in adults, usually those older than 55.
4. **Chronic myelogenous leukemia (CML)** occurs mostly in adults.

The cause of most types of leukemia is unknown. However, certain risk factors have been implicated. These include exposure to radiation or chemotherapy for other cancers, genetics (some genetic disorders such as Down syndrome), environmental factors (smoking and benzene), and microbes such as the human T cell leukemia-lymphoma virus-1 (HTLV-1) and the Epstein-Barr virus.

Treatment options include chemotherapy, radiation, stem cell transplantation, interferon, antibodies, and blood transfusion. •

---

the effects of histamine and other mediators of inflammation in allergic reactions. Eosinophils also phagocytize antigen–antibody complexes and are effective against certain parasitic worms. A high eosinophil count often indicates an allergic condition or a parasitic infection.

At sites of inflammation, basophils leave capillaries, enter tissues, and release heparin, histamine, and serotonin. These substances intensify the inflammatory reaction and are involved in hypersensitivity (allergic) reactions.

Lymphocytes are the major combatants in immune responses (described in detail in Chapter 15). During a life span of approximately 100 to 300 days, most lymphocytes continually move among the lymphoid tissues, lymph, and blood, spending only a few hours at a time in the blood. Therefore, only a small proportion of the total lymphocytes are present in the blood at any given moment. Three main types of lymphocytes are B cells, T cells, and natural killer (NK) cells. B cells develop into plasma cells, which produce antibodies that help destroy bacteria and inactivate their toxins. T cells attack viruses, fungi, transplanted cells, cancer cells, and some bacteria. Immune responses carried out by B cells and T cells help combat infection and provide protection against some diseases. T cells also are responsible for transfusion reactions, allergies, and the rejection of transplanted organs. Natural killer cells attack a wide variety of infectious microbes and certain spontaneously arising tumor cells.

Monocytes take longer to reach a site of infection than neutrophils, but they arrive in large numbers. Upon their arrival monocytes enlarge and differentiate into wandering macrophages, which phagocytize many more microbes than neutrophils can. They also clean up cellular debris following an infection.

Leukocytosis (loo′-kō-sī-TŌ-sis), an increase in the number of WBCs, is a normal, protective response to stresses such as invading microbes, strenuous exercise, anesthesia, and surgery. Leukocytosis usually indicates an inflammation or infection. Because each type of white blood cell plays a different role, determining the percentage of each type in the blood assists in diagnosing the condition. This test, called a **differential white blood cell count** or "**diff**," measures the number of each kind of white cell in a sample of 100 white blood cells (see Table 12.2). An abnormally low level of white blood cells (below 5000 cells/μL), called **leukopenia** (loo′-kō-PĒ-nē-a), is never beneficial; it may be caused by exposure to radiation, shock, and certain chemotherapeutic agents.

Table 12.2 lists the significance of both high and low WBC counts.

✓ CHECKPOINT

9. Explain the importance of emigration, chemotaxis, and phagocytosis in fighting bacterial invaders.
10. What effect does stress have on white blood cell count? (*Hint:* See Table 12.2.)
11. What functions are performed by B cells, T cells, and natural killer cells?

TABLE 12.2

Significance of *High* and *Low* White Blood Cell Counts

WBC TYPE	*HIGH* COUNT	*LOW* COUNT
Neutrophils	Bacterial infection, burns, stress, inflammation	Radiation exposure, drug toxicity, vitamin B_{12} deficiency, systemic lupus erythematosus (SLE)
Lymphocytes	Viral infections, some leukemias, infectious mononucleosis	Prolonged illness, HIV infection, immunosuppression, treatment with cortisol
Monocytes	Viral or fungal infections, tuberculosis, some leukemias, other chronic diseases	Bone marrow suppression, treatment with cortisol
Eosinophils	Allergic reactions, parasitic infections, autoimmune diseases	Drug toxicity, stress, acute allergic reaction
Basophils	Allergic reactions, leukemias, cancers, hypothyroidism	Pregnancy, ovulation, stress, hypothyroidism

12.7 PLATELETS

 OBJECTIVE

• Describe the structure, function, and origin of platelets.

In addition to becoming erythrocytes and leukocytes, hemopoietic stem cells also differentiate into cells that produce platelets. Under the influence of the hormone **thrombopoietin**, myeloid stem cells develop into megakaryocyte-colony-forming units; these in turn develop into precursor cells called *megakaryoblasts* (see Figure 12.3). Megakaryoblasts transform into *megakaryocytes*, huge cells that splinter into 2000–3000 fragments. Each fragment, enclosed by a piece of the cell membrane, is a **platelet** or **thrombocyte**. Platelets break off from the megakaryocytes in red bone marrow and then enter the blood circulation. Between 150,000 and 400,000 platelets are present in each μL of blood. They are irregularly disc-shaped, 2–4 μm in diameter, and exhibit many granules but no nucleus. Once released, the chemicals within their granules promote blood clotting. Platelets can initiate a series of chemical reactions that culminates in the formation of a network of insoluble protein threads called *fibrin*. A **blood clot** is a gel-like mass that consists of fibrin threads, platelets, and any blood cells trapped in the fibrin (Figure 12.6). The blood clot not only provides a seal that prevents blood loss from the damaged area of a blood vessel, but also pulls the edges of the vessel together to help repair the damage. Platelets come together to form a platelet plug that fills the gap in the blood vessel wall. Their life span is short, normally just 5–9 days. Aged and dead platelets are removed from the circulation by fixed macrophages in the spleen and liver.

Table 12.3 summarizes the quantities, characteristics, and functions of the formed elements in blood.

 CHECKPOINT

12. Compare RBCs, WBCs, and platelets with respect to size, number per mL, and life span.

12.8 STEM CELL TRANSPLANTS FROM BONE MARROW AND CORD-BLOOD

 OBJECTIVE

• Explain the importance of bone marrow transplants and stem cell transplants.

A **bone marrow transplant** is the replacement of cancerous or abnormal red bone marrow with healthy red bone marrow in order to establish normal blood cell counts. In patients with cancer or certain genetic diseases, the defective red bone marrow is destroyed by high doses of chemotherapy and whole body radiation just before the transplant takes place. These treatments kill the cancer cells and destroy the patient's immune system in order to decrease the chance of transplant rejection.

Healthy red bone marrow for transplanting may be supplied by a donor or by the patient when the underlying disease is inactive. The red bone marrow from a donor is usually removed from the iliac crest of the hip bone under general anesthesia with a syringe and is then injected into the recipient's vein, much like a blood transfusion. The injected marrow migrates to the recipient's red bone marrow cavities where the donor's stem cells multiply. If all goes well, the recipient's red bone marrow is replaced entirely by healthy, noncancerous cells.

Bone marrow transplants have been used to treat aplastic anemia, certain types of leukemia, severe combined immunodeficiency disease (SCID), Hodgkin disease, non-Hodgkin lymphoma, multiple myeloma, thalassemia, sickle-cell disease, breast cancer, ovarian

Figure 12.6 Scanning electron micrograph (SEM) of a portion of a blood clot. The SEM shows a platelet and red blood cells trapped by fibrin threads.

 A blood clot is a gel that contains formed elements of the blood entangled in fibrin threads.

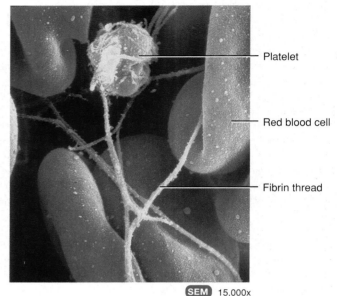

Platelet

Red blood cell

Fibrin thread

SEM 15,000x

 What are the two functions of a blood clot?

cancer, testicular cancer, and hemolytic anemia. However, there are some drawbacks. Since the recipient's white blood cells have been completely destroyed by chemotherapy and radiation, the patient is extremely vulnerable to infection. (It takes about 2–3 weeks for transplanted bone marrow to produce enough white blood cells to protect against infection.) Another problem is that transplanted red bone marrow may produce T cells that attack the recipient's tissues, a reaction called *graft-versus-host disease*. Moreover, any of the recipient's T cells that survived the chemotherapy and radiation can attack donor transplant cells. Another drawback is that patients must take immunosuppressive drugs for life. Because these drugs reduce the level of immune system activity, they increase the risk of infection. Immunosuppressive drugs also have side effects such as fever, muscle aches, headache, nausea, fatigue, depression, high blood pressure, and kidney and liver damage.

A more recent procedure used to obtain stem cells involves a **cord-blood transplant**. Recall from Chapter 4 that the connection between the placenta and embryo (and later the fetus) is the umbilical cord. The placenta contains stem cells, which may be obtained from the umbilical cord shortly after birth. The stem cells are removed from the cord with a syringe and then frozen. Stem cells from the cord have several advantages over those obtained from red bone marrow:

1. They are easily collected following permission of the newborn's parents.
2. They are more abundant than stem cells in red bone marrow.
3. They are less likely to cause graft-versus-host disease, so the match between donor and recipient does not have to be as close as in a bone marrow transplant. This provides a larger number of potential donors.
4. They are less likely to transmit infections.
5. They can be stored indefinitely in cord-blood banks.

 CHECKPOINT

13. What are the similarities between cord-blood transplants and bone marrow transplants? How do they differ?

TABLE 12.3

Summary of Formed Elements in Blood

NAME AND APPEARANCE	NUMBER	CHARACTERISTICS*	FUNCTIONS
Red Blood Cells (RBCs) or **Erythrocytes**	4.8 million/μL in females; 5.4 million/μL in males	7–8 μm diameter, biconcave discs, without nuclei; live for about 120 days	Hemoglobin within RBCs transports most of the oxygen and part of the carbon dioxide in the blood
White Blood Cells (WBCs) or **Leukocytes**	5,000–10,000 cells/μL	Most live for a few hours to a few days†	Combat pathogens and other foreign substances that enter the body
Granular leukocytes *Neutrophils*	60–70 percent of all WBCs	10–12 μm diameter; nucleus has 2–5 lobes connected by thin strands of chromatin; cytoplasm has very fine, pale lilac granules	Phagocytosis Destruction of bacteria with lysozyme
Eosinophils	2–4 percent of all WBCs	10–12 μm diameter; nucleus usually has 2 lobes connected by a thick strand of chromatin; large, red-orange granules fill the cytoplasm	Combat the effects of histamine in allergic reactions, phagocytize antigen–antibody complexes, and destroy certain parasitic worms
Basophils	0.5–1 percent of all WBCs	8–10 μm diameter; nucleus with 2 lobes; large cytoplasmic granules appear deep blue-purple	Liberate heparin, histamine, and serotonin in allergic reactions that intensify the overall inflammatory response
Agranular leukocytes *Lymphocytes (T cells, B cells, and natural killer cells)*	20–25 percent of all WBCs	Small, 6–9 μm in diameter; large, 10–14 μm in diameter; round or slightly indented nucleus; cytoplasm forms a rim around the nucleus that looks sky blue; the larger the cell, the more cytoplasm is visible	Mediate immune responses, including antigen–antibody reactions. B cells develop into plasma cells, which secrete antibodies. T cells attack invading viruses, cancer cells, and transplanted tissue cells. Natural killer cells attack a wide variety of infectious microbes and certain spontaneously arising tumor cells
Monocytes	3–8 percent of all WBCs	12–20 μm diameter; kidney-shaped or horseshoe-shaped nucleus; blue-gray cytoplasm with foamy appearance	Phagocytosis (after transforming into fixed or wandering macrophages)
Platelets (Thrombocytes)	150,000–400,000/μL	2–4 μm diameter cell fragments that live for 5–9 days; contain many granules but no nucleus	Form platelet plug in hemostasis; release chemicals that promote vascular spasm and blood clotting

*Colors are those seen when using Wright's stain.
† Some lymphocytes, called T and B memory cells, can live for many years.

KEY MEDICAL TERMS ASSOCIATED WITH BLOOD

Acute normovolemic hemodilution (nor-mō-vō-LĒ-mik hē-mō-dī-LOO-shun) Removal of blood immediately before surgery and its replacement with a cell-free solution to maintain sufficient blood volume for adequate circulation.

Anticoagulant (an'tē-kō-AG-ū-lant). A drug that delays, suppresses, or prevents blood clotting. Examples are heparin and warfarin.

Autologous preoperative transfusion (aw-TOL-o-gus trans-FŪ-zhun; *auto-*=self) Donating one's own blood; can be done up to 6 weeks before elective surgery. Also called predonation.

Blood bank A facility that collects and stores a supply of blood for future use by the donor or others. Because they have additional and diverse functions, they are more appropriately referred to as **centers of transfusion medicine**.

Cyanosis (sī-a-NŌ-sis; *cyano-*=blue) Slightly bluish/dark-purple skin discoloration, most easily seen in the nail beds and mucous membranes, due to an increased quantity of *methemoglobin*, hemoglobin not combined with oxygen in systemic blood.

Embolus (EM-bō-lus=plug) A blood clot, bubble of air, or particle of fat from broken bones, mass of bacteria, or other foreign material transported by the blood.

Gamma globulin (GLOB-ū-lin) Solution of immunoglobulins from blood consisting of antibodies that react with specific pathogens, such as viruses.

Hemochromatosis (hē-mō-krō-ma-TŌ-sis; *chroma*=color) Disorder of iron metabolism characterized by excessive absorption of ingested iron and excess deposits of iron in tissues (especially the liver, heart, pituitary gland, gonads, and pancreas) that result in bronze discoloration of the skin, cirrhosis, diabetes mellitus, and bone and joint abnormalities.

Hemophilia (hē-mō-FIL-ē-a; *-philia*=loving) An inherited deficiency of clotting in which bleeding may occur spontaneously or after only minor trauma. Hemophilia is characterized by spontaneous or traumatic subcutaneous and intramuscular hemorrhaging, nosebleeds, blood in the urine, and hemorrhages in joints that produce pain and tissue damage.

Hemorrhage (HEM-or-ij; *rhegnynai*=bursting forth) Loss of a large amount of blood; can be either internal (from blood vessels into tissues) or external (from blood vessels directly to the surface of the body).

Jaundice (*jaund-*=yellow) An abnormal yellowish discoloration of the sclerae (white of the eyes), skin, and mucous membranes due to excess bilirubin (yellow-orange pigment) in the blood.

Multiple myeloma (mī-e-LŌ-ma) Malignant disorder of plasma cells in red bone marrow; symptoms (pain, osteoporosis, hypercalcemia, thrombocytopenia, kidney damage) are caused by the growing tumor cell mass or antibodies produced by malignant cells.

Phlebotomist (fle-BOT-ō-mist; *phlebo-*=vein; *-tom*=cut) A technician who specializes in withdrawing blood.

Septicemia (sep′-ti-SĒ-mē-a; *septic-*=decay; *-emia*=condition of blood) Toxins or disease-causing bacteria in the blood. Also called "blood poisoning."

Thrombocytopenia (throm′-bō-sī′-tō-PĒ-nē-a; *-penia*=poverty) Very low platelet count that results in a tendency to bleed from capillaries.

Thrombus (THROM-bus=clot) A clot in the cardiovascular system formed in an unbroken blood vessel (usually a vein). The clot consists of a network of insoluble fibrin threads in which the formed elements of blood are trapped.

Transfusion (trans-FŪ-zhun) Transfer of whole blood, blood components (red blood cells only or plasma only), or red bone marrow directly into the bloodstream.

Venesection (vē-ne-SEK-shun; *ven*=vein) Opening of a vein for withdrawal of blood. Although **phlebotomy** (fle-BOT-ō-mē) is a synonym for venesection, in clinical practice phlebotomy refers to therapeutic bloodletting, such as the removal of some blood to lower its viscosity in a patient with polycythemia.

Whole blood Blood containing all formed elements, plasma, and plasma solutes in natural concentrations.

CHAPTER REVIEW AND RESOURCE SUMMARY   WileyPLUS

Review	Resource

Introduction

1. The cardiovascular system consists of the blood, heart, and blood vessels.
2. Blood is a connective tissue composed of blood plasma (liquid portion) and formed elements (cells and cell fragments).

12.1 Functions of Blood

1. Blood transports oxygen, carbon dioxide, nutrients, wastes, and hormones.
2. It helps regulate pH, body temperature, and water content of cells.
3. It provides protection through clotting and by combating toxins and microbes, a function of certain phagocytic white blood cells or specialized plasma proteins.

Anatomy Overview - Blood

12.2 Physical Characteristics of Blood

1. Physical characteristics of blood include a viscosity greater than that of water; a temperature of 38°C (100.4°F); and a pH of 7.35–7.45.
2. Blood constitutes about 8 percent of body weight, and its volume is 4–6 liters in adults.

Anatomy Overview - Blood

12.3 Components of Blood

1. Blood consists of about 55 percent blood plasma and 45 percent formed elements.
2. The hematocrit is the percentage of total blood volume occupied by red blood cells.
3. Blood plasma consists of 91.5 percent water and 8.5 percent solutes.
4. Principal solutes include proteins (albumins, globulins, fibrinogen), nutrients, vitamins, hormones, respiratory gases, electrolytes, and waste products.
5. The formed elements in blood include red blood cells (erythrocytes), white blood cells (leukocytes), and platelets.

Anatomy Overview - Blood
Concepts and Connections - Blood

12.4 Formation of Blood Cells

1. Hemopoiesis is the formation of blood cells from hemopoietic stem cells in red bone marrow.
2. Myeloid stem cells form RBCs, platelets, granulocytes, and monocytes. Lymphoid stem cells give rise to lymphocytes.
3. Several hemopoietic growth factors stimulate differentiation and proliferation of the various blood cells.

Anatomy Overview - Erythrocytes
Figure 12.3 - Origin, Development, and Structure of Blood Cells

12.5 Red Blood Cells

1. Mature RBCs are biconcave discs that lack nuclei and contain hemoglobin.
2. The function of the hemoglobin in red blood cells is to transport oxygen and some carbon dioxide.
3. RBCs live about 120 days. A healthy male has about 5.4 million RBCs/mL of blood; a healthy female, about 4.8 million/mL.

Anatomy Overview - Erythrocytes
Animation - Erythropoietin
Figure 12.4 - The Shapes of a Red Blood Cell and a Hemoglobin Molecule

Review   **Resource**

4. After phagocytosis of aged RBCs by macrophages, hemoglobin is recycled.

5. RBC formation, called erythropoiesis, occurs in adult red bone marrow of certain bones. It is stimulated by hypoxia, which stimulates the release of erythropoietin by the kidneys.

6. A reticulocyte count is a diagnostic test that indicates the rate of erythropoiesis.

12.6 White Blood Cells

Anatomy Overview - White Blood Cells

Figure 12.5 - Structure of White Blood Cells

1. WBCs are nucleated cells. The two principal types are granulocytes (neutrophils, eosinophils, and basophils) and agranulocytes (lymphocytes and monocytes).

2. The general function of WBCs is to combat inflammation and infection. Neutrophils and macrophages (which develop from monocytes) act by phagocytosis.

3. Eosinophils combat the effects of histamine in allergic reactions, phagocytize antigen–antibody complexes, and combat parasitic worms; basophils liberate heparin, histamine, and serotonin in allergic reactions that intensify the inflammatory response.

4. B lymphocytes, in response to the presence of foreign substances called antigens, differentiate into plasma cells that produce antibodies. Antibodies attach to the antigens and render them harmless. This antigen–antibody response combats infection and provides immunity. T lymphocytes destroy foreign invaders directly. Natural killer cells attack infectious microbes and tumor cells.

5. Except for lymphocytes, which may live for years, WBCs usually live for only a few hours or a few days. Normal blood contains 5000–10,000 WBCs/mL.

12.7 Platelets

Anatomy Overview - Platelets

1. Platelets (thrombocytes) are disc-shaped structures without nuclei.

2. They are fragments derived from megakaryocytes and are involved in clotting.

3. Normal blood contains 150,000–400,000 platelets/mL.

12.8 Stem Cell Transplants from Bone Marrow and Cord-Blood

1. Bone marrow transplants involve removal of marrow as a source of stem cells from the iliac crest.

2. In a cord-blood transplant, stem cells from the placenta are removed from the umbilical cord.

3. Cord-blood transplants have several advantages over bone marrow transplants.

CRITICAL THINKING QUESTIONS

1. A professional bicyclist finished second in a grueling cross-country bike race. Although he appeared to be in great physical condition, he suffered a heart attack a few hours after the race. The press speculated that his condition was caused by blood doping. Explain.

2. Maddy has been sniffling and sneezing her way through Human Anatomy class all semester. While checking a smear of her own blood, she noticed that her blood had a lot more of the bluish-black granular cells than her lab partner's blood. Maddy's eyes were watery and itchy so she asked her instructor to confirm her finding. What are these cells and why does Maddy have a higher number of them than usual?

3. Gus suffers from chronic renal (kidney) failure and undergoes dialysis regularly. One of his associated problems is chronic anemia,

which the doctor says is directly related to the renal failure. What is likely to be the connection between the two, and how might it be treated?

4. Millie was at her grandmother's house and found a very old bottle of a tonic that promised to relieve the symptoms of "iron-poor blood." What do you think the symptoms of "iron-poor blood" would be, and how would enriching the iron content of blood relieve these symptoms?

5. Raoul plans on having some elective surgery performed about a month after the semester is over. His doctor suggested that he donate some of his own blood now. Raoul's not sure about this; he figures he's going to need all the blood he has for the surgery. Why should he give up some of his blood now?

ANSWERS TO FIGURE QUESTIONS

12.1 Blood volume averages about 6 liters in males and 4–5 liters in females, representing about 8 percent of body weight.

12.2 Platelets are cell fragments.

12.3 Pluripotent stem cells develop from mesenchyme.

12.4 One hemoglobin molecule can transport four O_2 molecules—one bound to each heme group.

12.5 Neutrophils, eosinophils, and basophils are called granular leukocytes because all have cytoplasmic granules that are visible through a light microscope when stained.

12.6 The two functions of a blood clot are sealing the damaged blood vessel and pulling the edges of the vessel together to prevent blood loss.

13 THE CARDIOVASCULAR SYSTEM: THE HEART

INTRODUCTION In Chapter 12 we examined the composition and functions of blood. For blood to reach body cells and exchange materials with them, it must be pumped continuously by the heart through the body's blood vessels. The heart beats about 100,000 times every day, which adds up to about 35 million beats in a year, and approximately 2.5 billion beats in an average lifetime. The left side of the heart pumps blood through an estimated 120,000 km (75,000 mi) of blood vessels, which is equivalent to traveling around the earth's equator about 3 times. The right side of the heart pumps blood through the lungs, enabling blood to pick up oxygen and unload carbon dioxide. Even while you are sleeping, your heart pumps 30 times its own weight each minute, which amounts to about 5 liters (5.3 qt) to the lungs and the same volume to the rest of the body. At this rate, your heart pumps more than about 14,000 liters (3,600 gal) of blood in a day, or 5 million liters (1.3 million gal) in a year. You don't spend all your time sleeping, however, and your heart pumps more vigorously when you are active. Thus, the actual blood volume your heart pumps in a single day is much larger.

The scientific study of the normal heart and the diseases associated with it is **cardiology** (kar-dē-OL-ō-jē; *cardio-*=heart; *-logy*=study of). This chapter explores the structure of the heart and the unique properties that permit it to pump for a lifetime without rest. •

 Did you ever wonder about the difference between good cholesterol and bad cholesterol?

13.1 LOCATION AND SURFACE PROJECTION OF THE HEART

OBJECTIVES
- Describe the location of the heart.
- Trace the outline of the heart on the surface of the chest.

For all its might, the cone-shaped heart is relatively small, roughly the same size as a closed fist—about 12 cm (5 in.) long, 9 cm (3.5 in.) wide at its broadest point, and 6 cm (2.5 in.) thick. Its mass averages 250 g (8 oz) in adult females and 300 g (10 oz) in adult males. The heart rests on the diaphragm, near the midline of the thoracic cavity in the **mediastinum** (mē-dē-as-TĪ-num), an anatomical region that extends from the sternum to the vertebral column, the first rib to the diaphragm, and between the coverings (pleurae) of the lungs (Figure 13.1a, b). (Recall that the *midline* is an imaginary vertical line that divides the body into equal right and left sides.)

About two-thirds of the mass of the heart lies to the left of the body's midline. The position of the heart in the mediastinum is more readily appreciated by examining its ends, surfaces, and borders (Figure 13.1b). Visualize the heart as a cone lying on its side. The pointed end of the heart, the **apex**, is formed by the tip of the left ventricle (a lower chamber of the heart) and rests on the diaphragm. It is directed anteriorly,

inferiorly, and to the left. The **base** of the heart, its posterior aspect, is formed by the atria (upper chambers of the heart). The base is opposite the apex and is formed mostly by the left atrium, into which the four pulmonary veins open, and a portion of the right atrium, which receives the superior and inferior vena cavae and the coronary sinus (see Figure 13.3c, d). In addition to the apex and base, the heart has several surfaces that are useful in determining its surface projection (described shortly). The **anterior surface** of the heart is deep to the sternum and ribs. The **inferior surface** is the portion of the heart that rests mostly on the diaphragm and is found between the apex and right surface. The **right surface** faces the right lung and extends from the inferior surface to the base; the **left surface** faces the left lung and extends from the base to the apex.

Determining an organ's **surface projection** means outlining its dimensions with respect to landmarks on the surface of the body. This practice is useful when conducting diagnostic procedures (such as a lumbar puncture), auscultation (listening to heart and lung sounds), and anatomical studies. We can project the heart on the anterior surface of the chest by locating the following landmarks (Figure 13.1d): The **superior right point** is located at the superior border of the third right costal cartilage, about 3 cm (1 in.) to the right of the midline. The **superior left point** is located at the inferior border of the second

Figure 13.1 Position of the heart in the mediastinum. The positions of the heart and associated structures in the mediastinum are indicated by dashed outlines.

 The heart is located in the mediastinum; two-thirds of its mass is to the left of the midline.

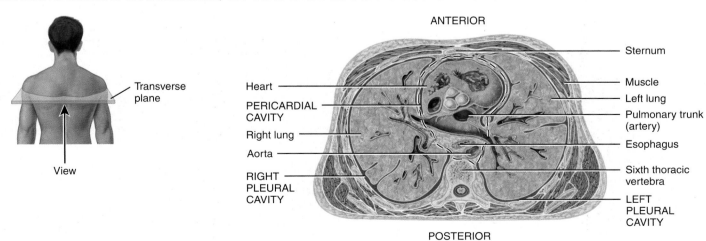

(a) Inferior view of transverse section of thoracic cavity showing the heart in the mediastinum

FIGURE 13.1 CONTINUES

431

FIGURE 13.1 CONTINUED ▶

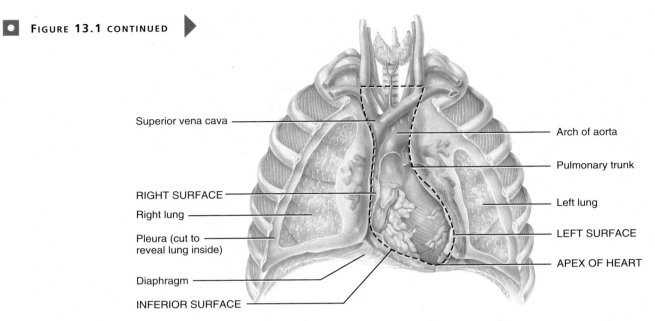

Superior vena cava

Arch of aorta

Pulmonary trunk

RIGHT SURFACE

Left lung

Right lung

Pleura (cut to reveal lung inside)

LEFT SURFACE

APEX OF HEART

Diaphragm

INFERIOR SURFACE

(b) Anterior view of the heart in the thoracic cavity

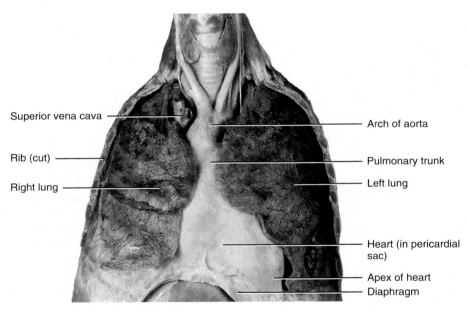

Superior vena cava

Arch of aorta

Rib (cut)

Pulmonary trunk

Right lung

Left lung

Heart (in pericardial sac)

Apex of heart

Diaphragm

(c) Anterior view

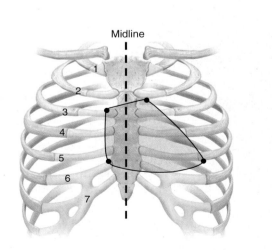

Midline

1
2
3
4
5
6
7

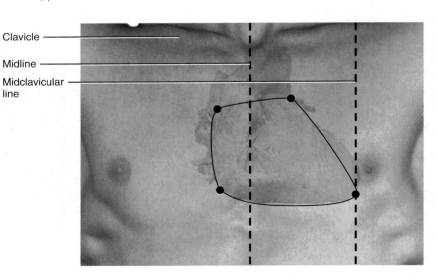

Clavicle

Midline

Midclavicular line

(d) Surface projection of the heart

What is the mediastinum?

CLINICAL CONNECTION | *Cardiopulmonary Resuscitation*

Cardiopulmonary resuscitation (CPR) (kar-dē-ō-PUL-mo-nar′-ē rē-sus-i-TĀ-shun) refers to an emergency procedure for establishing a normal heartbeat and rate of breathing. Standard CPR uses a combination of cardiac compression and artificial ventilation of the lungs via mouth-to-mouth respiration and, for many years, this combination was the sole method of CPR. Recently, however, hands-only CPR is usually the preferred method.

Because the heart lies between two rigid structures—the sternum and vertebral column—pressure on the chest (compression) can be used to force blood out of the heart and into the circulation. After calling 911, hands-only CPR should be administered. In the procedure, chest compressions should be given hard and fast at a rate of 100 per minute and two inches deep in adults. This should be continued until trained medical professionals arrive or an automated external defibrillator is available. Standard CPR is still recommended for infants and children, as well as anyone who suffers from lack of oxygen, for example, victims of near-drowning, drug overdose, or carbon monoxide poisoning.

It is estimated that hands-only CPR saves about 20 percent more lives than the standard method. Moreover, hands-only CPR boosts the survival rate from 18 percent to 34 percent compared to the traditional method or none at all. It is also easier for an emergency dispatcher to give instructions limited to hands-only CPR to frightened, nonmedical bystanders. Finally, as public fear of contracting contagious diseases such as HIV, hepatitis, and tuberculosis continues to rise, bystanders are much more likely to perform hands-only CPR than treatment involving the standard method. •

left costal cartilage, about 3 cm to the left of the midline. A line connecting these two points corresponds to the superior margin of the base of the heart. The **inferior left point** (superficial to the apex of the heart) is located in the fifth left intercostal space, about 9 cm (3.5 in.) to the left of the midline (at the midclavicular line). A line connecting the superior and inferior left points corresponds to the left border of the heart. The **inferior right point** is located at the superior border of the sixth right costal cartilage, about 3 cm to the right of the midline. A line

connecting the inferior left and right points corresponds to the inferior surface of the heart, and a line connecting the inferior and superior right points corresponds to the right border of the heart. When all four points are connected, they form an outline that roughly reveals the size and shape of the heart.

 CHECKPOINT

1. Describe the position of the heart in the mediastinum by defining its apex, base, anterior and posterior surfaces, and right and left borders.
2. Explain the location of the superior right point, superior left point, inferior left point, and inferior right point. Why are these points significant?

13.2 STRUCTURE AND FUNCTION OF THE HEART

 OBJECTIVES

• Describe the structure of the pericardium.
• Identify the layers of the heart wall.
• Discuss the external and internal anatomy of the chambers of the heart.
• Relate the thickness of the chambers of the heart to their functions.
• Describe the functions of the fibrous skeleton of the heart.
• List the structure and function of each valve of the heart.

Pericardium

The membrane that surrounds and protects the heart is the **pericardium** (per′-i-KAR-dē-um; *peri-*=around). It confines the heart to its position in the mediastinum while allowing sufficient freedom of movement for vigorous and rapid contraction. The pericardium consists of two principal portions: (1) the fibrous pericardium and (2) the serous pericardium (Figure 13.2a, c). (See also Figure 1.7e.) The superficial **fibrous pericardium** is composed of tough, inelastic, dense irregular connective tissue. It resembles a bag that rests on and attaches to the diaphragm; the

Figure 13.2 Pericardium and heart wall.

 The pericardium is a sac that surrounds and protects the heart.

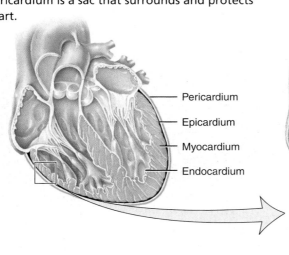

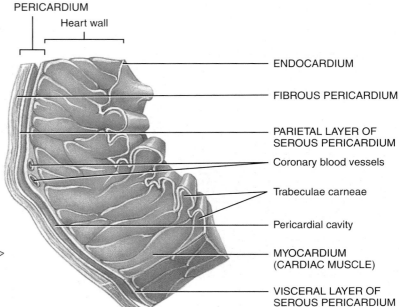

(a) Portion of pericardium and right ventricular heart wall showing the divisions of the pericardium and layers of the heart wall

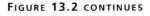**FIGURE 13.2 CONTINUES**

◻ **FIGURE 13.2 CONTINUED** ▶

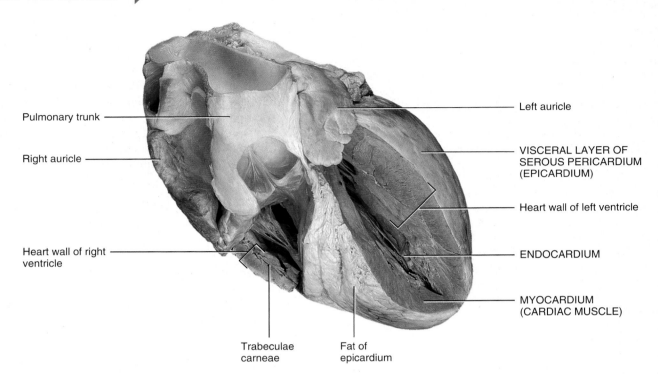

Pulmonary trunk

Right auricle

Heart wall of right ventricle

Left auricle

VISCERAL LAYER OF SEROUS PERICARDIUM (EPICARDIUM)

Heart wall of left ventricle

ENDOCARDIUM

MYOCARDIUM (CARDIAC MUSCLE)

Trabeculae carneae

Fat of epicardium

(b) Anterior view of layers of the heart wall

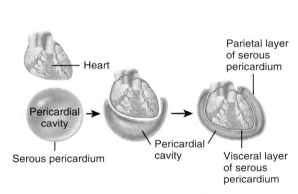

Heart

Pericardial cavity

Serous pericardium

Parietal layer of serous pericardium

Pericardial cavity

Visceral layer of serous pericardium

(c) Simplified relationship of the serous pericardium to the heart

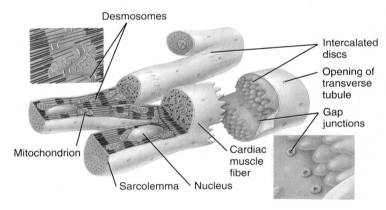

Desmosomes

Intercalated discs

Opening of transverse tubule

Gap junctions

Mitochondrion

Cardiac muscle fiber

Sarcolemma   Nucleus

(e) Cardiac muscle fibers

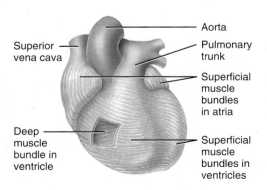

Aorta

Superior vena cava

Pulmonary trunk

Superficial muscle bundles in atria

Deep muscle bundle in ventricle

Superficial muscle bundles in ventricles

(d) Cardiac muscle bundles of the myocardium

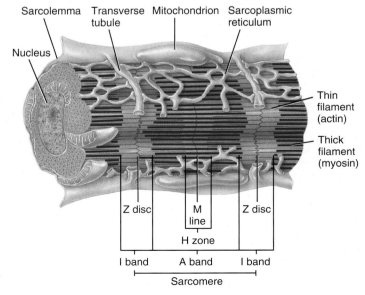

Sarcolemma   Transverse tubule   Mitochondrion   Sarcoplasmic reticulum

Nucleus

Thin filament (actin)

Thick filament (myosin)

Z disc   M line   Z disc

H zone

I band   A band   I band

Sarcomere

(f) Arrangement of components in a cardiac muscle fiber

Which layer is both a part of the pericardium and a part of the heart wall?

open end of the bag is fused to the connective tissues of the blood vessels entering and leaving the heart. The fibrous pericardium prevents overstretching of the heart, provides protection, and anchors the heart in the mediastinum. Because the fibrous pericardium near the apex of the heart is partially fused to the membrane that covers the diaphragm, movement of the diaphragm, as in deep breathing, facilitates the movement of blood by the heart.

The deeper **serous pericardium** is a thinner, more delicate membrane that forms a double layer around the heart (Figure 13.2a, c). The outer **parietal layer** of the serous pericardium is fused to the fibrous pericardium. The inner **visceral layer** of the serous pericardium, also called the epicardium when combined with the underlying areolar or adipose tissue, adheres tightly to the surface of the heart. Between the parietal and visceral layers of the serous pericardium is a thin film of lubricating fluid. This fluid, known as **pericardial fluid**, is a slippery secretion of the pericardial cells that reduces friction between the membranes as the heart moves. The space that contains the few milliliters of pericardial fluid is called the **pericardial cavity**.

CLINICAL CONNECTION | *Pericarditis*

Inflammation of the pericardium is called **pericarditis** (per-i-kar-DĪ-tis). The most common type, **acute pericarditis**, begins suddenly; in most cases it has no known cause, but it is sometimes linked to a viral infection. As a result of irritation to the pericardium, there is chest pain that may extend to the left shoulder and down the left arm (often mistaken for a heart attack), and *pericardial friction rub* (a scratchy or creaking sound heard through a stethoscope as the visceral layer of the serous pericardium rubs against the parietal layer of the serous pericardium). Acute pericarditis usually lasts for about one week and is treated with drugs that reduce inflammation and pain, such as ibuprofen or aspirin.

Chronic pericarditis begins gradually and is long-lasting. In one form of this condition, there is a buildup of pericardial fluid. If a great deal of fluid accumulates, this is a life-threatening condition because the fluid compresses the heart, a condition called *cardiac tamponade* (tam'-pon-ĀD). As a result of the compression, ventricular filling is decreased, cardiac output is reduced, venous return to the heart is diminished, blood pressure falls, and breathing is difficult. Most causes of chronic pericarditis involving cardiac tamponade are unknown, but it is sometimes caused by conditions such as cancer and tuberculosis. Treatment consists of draining the excess fluid through a needle passed into the pericardial cavity. •

Layers of the Heart Wall

The wall of the heart consists of three layers (Figure 13.2a, b): the epicardium (external layer), the myocardium (middle layer), and the endocardium (inner layer). The **epicardium** (ep'-i-KAR-dē-um; *epi-*=on top of) is composed of two tissue layers. The outermost, as you just learned, is also called the *visceral layer of the serous pericardium*. This thin, transparent outer layer of the heart wall is composed of mesothelium. Beneath the mesothelium is a variable layer of delicate fibroelastic tissue and adipose tissue. The adipose tissue predominates and becomes thickest over the ventricular surfaces, where it houses the major coronary and cardiac vessels of the heart. The amount of fat varies from person to person, corresponds to the general extent of body fat in an individual, and typically increases with age. The epicardium imparts a smooth, slippery texture to the outermost surface of the heart. The epicardium contains blood vessels, lymphatics, and nerves that supply the myocardium.

The middle **myocardium** (mī'-ō-KAR-dē-um; *myo-*=muscle) is responsible for the pumping action of the heart and is composed of cardiac muscle tissue. It makes up approximately 95 percent of the heart wall. The muscle fibers (cells), like those of striated skeletal muscle tissue, are wrapped and bundled with connective tissue sheaths composed of endomysium and perimysium. The muscle is organized in a pattern of struts and weaves (crisscrossing bundles of muscular beams) that generate the strong pumping actions of the heart (Figure 13.2d). Although it is striated like skeletal muscle, recall that cardiac muscle is involuntary like smooth muscle (see Section 10.5). Recall what you learned in Chapter 10 about cardiac muscle:

1. Cardiac muscle fibers are shorter in length and less circular in transverse section than skeletal muscle fibers (see Figure 13.2c and Table 3.9).
2. They exhibit branching, so that individual cardiac muscle fibers have a "stair-step" appearance (see Table 3.9).
3. The ends of cardiac muscle fibers connect to neighboring fibers by **intercalated discs** (in-TER-ka-lāt-ed) containing **desmosomes**, cell junctions that hold fibers together, and **gap junctions**, cell junctions that allow muscle action potentials to conduct from one muscle fiber to its neighbors (Figure 13.2e).
4. Mitochondria are larger and more numerous in cardiac muscle fibers than in skeletal muscle fibers (Figure 13.2f). In a cardiac muscle fiber, mitochondria take up 25 percent of the fiber. This is a reflection of the greater dependence on cellular respiration for ATP.
5. Recall that actin, myosin, A and I bands, H zones, and Z discs are present in skeletal muscle. The pattern of thick and thin filaments is similar to that found in skeletal muscle. The transverse tubules in skeletal muscle are located over the zone of overlap, but in cardiac muscle the transverse tubules are located over the Z discs and the sarcoplasmic reticulum does not contain terminal cisternae (Figure 13.2f). Because the sarcoplasmic reticulum of cardiac muscle fibers is smaller than the SR of skeletal muscle fibers, there is a smaller intracellular reserve of Ca^{2+} in cardiac muscle.

Myocarditis (mī-ō-kar-DĪ-tis) is an inflammation of the myocardium that usually occurs as a complication of a viral infection, rheumatic fever, or exposure to radiation or certain chemicals or medications.

The innermost **endocardium** (en'-dō-KAR-dē-um; *endo-*=within) is a thin layer of endothelium overlying a thin layer of connective tissue. It provides a smooth lining for the chambers of the heart and covers the valves of the heart. The smooth endothelial lining minimizes the surface friction as blood passes through the heart. The endocardium is continuous with the endothelial lining of the large blood vessels attached to the heart. **Endocarditis** (en'-dō-kar-DĪ-tis) refers to an inflammation of the endocardium and typically involves the heart valves. Most cases are caused by bacteria (bacterial endocarditis).

Chambers of the Heart

The heart is a dual pump that contains four chambers, two upper or receiving chambers called the **atria** (=entry halls or chambers) and the two lower or pumping chambers called the **ventricles** (=little bellies). The paired atria receive blood from blood vessels returning blood to the heart, called veins, while the ventricles eject the blood from the heart into blood vessels called arteries. The right pump, consisting of the right atrium and right ventricle, is the weaker **pulmonary pump**. The pulmonary pump moves deoxygenated blood through the blood vessels of the lungs. The left pump, comprised of the left atrium and left ventricle, is the stronger **systemic pump**. The systemic pump circulates oxygenated blood to all the systems of the body. On the anterior surface of each atrium is a wrinkled pouchlike structure called an **auricle** (OR-i-kul; *auri-*=ear), so named because of its resemblance to a dog's ear (Figure 13.3a, b). Each auricle increases the capacity of an atrium

slightly so that it can hold a greater volume of blood. Also on the surface of the heart are a series of grooves, called **sulci** (SUL-sī), which contain coronary blood vessels and a variable amount of fat (Figure 13.3a–c). Each sulcus (SUL-kus; singular form of *sulci*) marks the external boundary between two chambers of the heart. The deep **coronary sulcus** (*coron-*=resembling a crown) encircles most of the heart and marks the external boundary between the

Figure 13.3 Structure of the heart: surface features. Throughout this book, illustrations of blood vessels that carry oxygenated blood (which looks bright red) are colored red, and blood vessels that carry deoxygenated blood (which looks dark red) are colored blue.

🔑 Sulci are grooves that contain blood vessels and fat and mark the external boundaries between the various chambers.

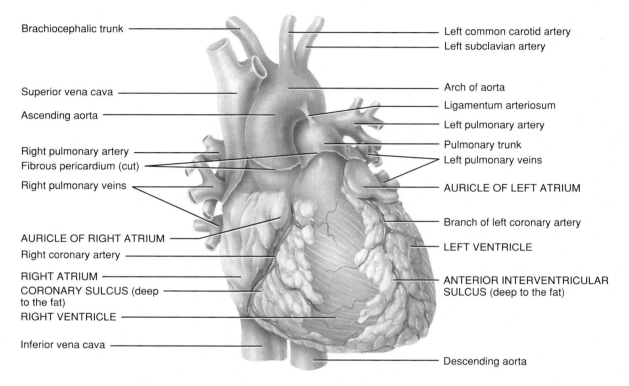

(a) Anterior external view showing surface features

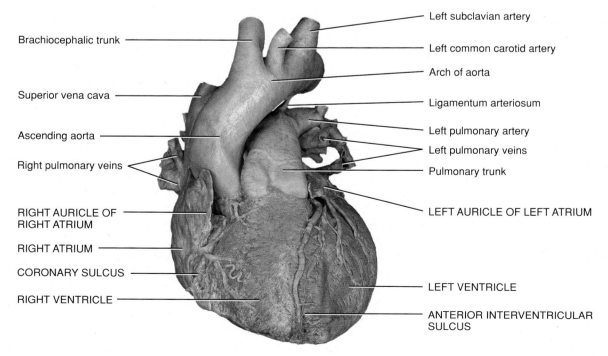

(b) Anterior external view

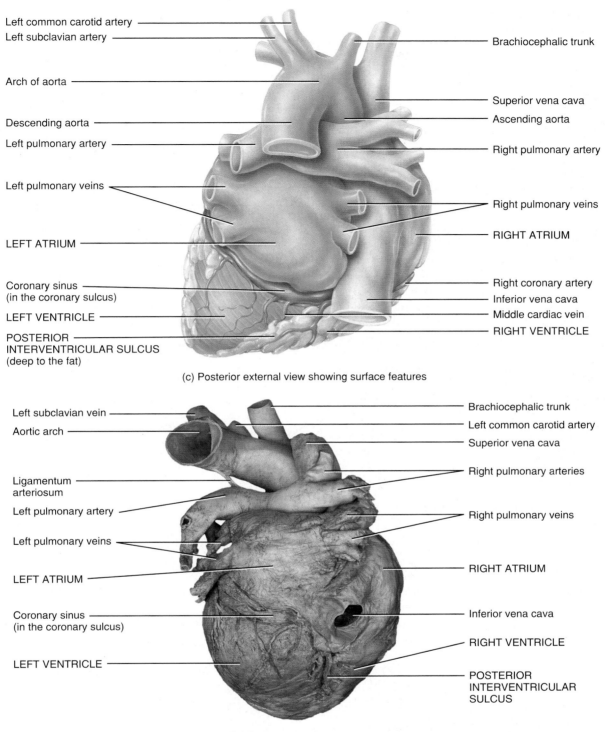

Left common carotid artery

Left subclavian artery

Arch of aorta

Descending aorta

Left pulmonary artery

Left pulmonary veins

LEFT ATRIUM

Coronary sinus
(in the coronary sulcus)

LEFT VENTRICLE

POSTERIOR
INTERVENTRICULAR SULCUS
(deep to the fat)

Brachiocephalic trunk

Superior vena cava

Ascending aorta

Right pulmonary artery

Right pulmonary veins

RIGHT ATRIUM

Right coronary artery

Inferior vena cava

Middle cardiac vein

RIGHT VENTRICLE

(c) Posterior external view showing surface features

Left subclavian vein

Aortic arch

Ligamentum
arteriosum

Left pulmonary artery

Left pulmonary veins

LEFT ATRIUM

Coronary sinus
(in the coronary sulcus)

LEFT VENTRICLE

Brachiocephalic trunk

Left common carotid artery

Superior vena cava

Right pulmonary arteries

Right pulmonary veins

RIGHT ATRIUM

Inferior vena cava

RIGHT VENTRICLE

POSTERIOR
INTERVENTRICULAR
SULCUS

(d) Posterior external view

 The coronary sulcus forms an external boundary between which chambers of the heart?

superior atria and inferior ventricles. The **anterior interventricular sulcus** (in′-ter-ven-TRIK-ū-lar) is a shallow groove on the anterior surface of the heart that marks the external boundary between the right and left ventricles. This sulcus continues around to the posterior surface of the heart as the **posterior interventricular sulcus,** which marks the external boundary between the ventricles on the posterior aspect of the heart.

Right Atrium

The **right atrium** forms the right border of the heart (see Figure 13.1b) and is about 2 to 3 mm (0.08 to 0.12 in.) in average

thickness. It receives blood from three veins: *superior vena cava, inferior vena cava,* and *coronary sinus* (Figure 13.4c). Veins always carry blood toward the heart. The anterior and posterior walls within the right atrium differ considerably. During development of the heart the atrial chambers enlarge by absorbing a significant portion of the associated entry veins into their walls. This posterior venous portion of the adult atria is characterized by a smooth internal wall. This contrasts with the parallel ridges, the **pectinate muscles** (PEK-ti-nāt; *pectin-* =comb), that line the remainder of the atrial wall. The ridged regions of the atria signify the original embryonic atrial

Figure 13.4 Structure of the heart: internal anatomy.

The thickness of the four chambers varies according to their functions.

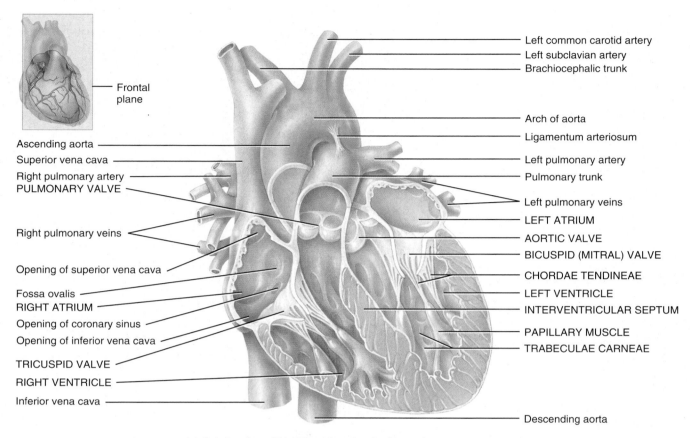

Frontal plane

Left common carotid artery
Left subclavian artery
Brachiocephalic trunk

Arch of aorta
Ligamentum arteriosum

Ascending aorta
Superior vena cava
Right pulmonary artery
PULMONARY VALVE

Left pulmonary artery
Pulmonary trunk
Left pulmonary veins
LEFT ATRIUM
AORTIC VALVE
BICUSPID (MITRAL) VALVE

Right pulmonary veins

Opening of superior vena cava

Fossa ovalis
RIGHT ATRIUM
Opening of coronary sinus
Opening of inferior vena cava

TRICUSPID VALVE

RIGHT VENTRICLE

Inferior vena cava

CHORDAE TENDINEAE
LEFT VENTRICLE
INTERVENTRICULAR SEPTUM
PAPILLARY MUSCLE
TRABECULAE CARNEAE

Descending aorta

(a) Anterior view of frontal section showing internal anatomy

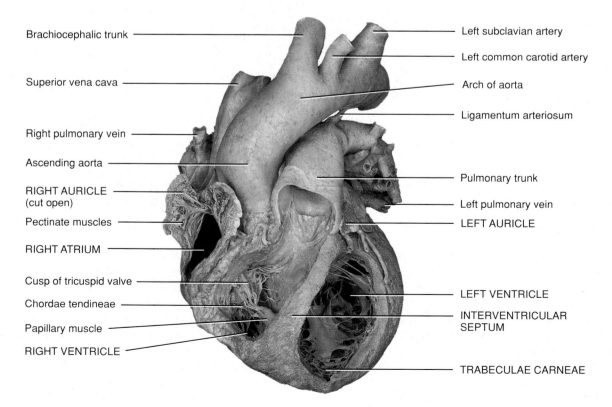

Brachiocephalic trunk

Superior vena cava

Right pulmonary vein

Ascending aorta

RIGHT AURICLE
(cut open)

Pectinate muscles

RIGHT ATRIUM

Cusp of tricuspid valve

Chordae tendineae

Papillary muscle

RIGHT VENTRICLE

Left subclavian artery

Left common carotid artery

Arch of aorta

Ligamentum arteriosum

Pulmonary trunk

Left pulmonary vein

LEFT AURICLE

LEFT VENTRICLE

INTERVENTRICULAR SEPTUM

TRABECULAE CARNEAE

(b) Anterior view of partially sectioned heart

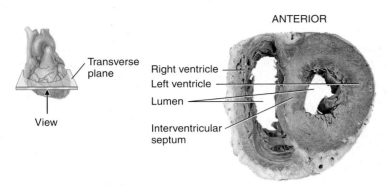

ANTERIOR

Transverse
plane

Right ventricle
Left ventricle
Lumen

Interventricular
septum

View

POSTERIOR

(c) Inferior view of transverse section showing differences
in thickness of ventricular walls

 Which chamber has the thickest wall?

chambers, the majority of which correspond to the muscular auricles (Figure 13.4b). Between the right atrium and left atrium is a thin partition called the **interatrial septum** (*inter-*=between; *septum*=a dividing wall or partition). A prominent feature of this septum is an oval depression called the **fossa ovalis**, which is the remnant of the *foramen ovale*, an opening in the interatrial septum of the fetal heart that directs blood from the right atrium to the left atrium in order to bypass the nonfunctioning fetal lungs. The foramen ovale normally closes soon after birth (see Figure 14.17). Blood passes from the right atrium into the right ventricle through a valve called the **tricuspid valve** (trī-KUS-pid; *tri-*=three; *cuspid*=point) because it consists of three folds or flaps called **cusps** or **leaflets** (Figures 13.4a and 13.5). It is also called the **right atrioventricular valve** (a′-trē-ō-ven-TRIK-ū-lar). The valves of the heart, which will be described in more detail later in the chapter, are composed of dense connective tissue covered by endocardium.

Right Ventricle

The **right ventricle**, about 4 to 5 mm (0.16 to 0.2 in.) in average thickness, forms most of the anterior surface of the heart (see Figure 13.3a). The inside of the right ventricle contains a series of ridges formed by raised bundles of cardiac muscle fibers called **trabeculae carneae** (tra-BEK-ū-lē KAR-nē-ē; *trabeculae*=little beams; *carneae*=fleshy) (Figure 13.4a–b). Some of the trabeculae carneae contain part of the conduction system of the heart (described in Section 13.4). The cusps of the tricuspid valve are connected to tendonlike cords, the **chordae tendineae** (KOR-dē ten-DIN-ē-ē; *chord-*=cord; *tend-*=tendon), which in turn are connected to cone-shaped trabeculae carneae called **papillary muscles** (*papill-*=nipple) (see Figure 13.6c). Internally, the right ventricle is separated from the left ventricle by a partition called the **interventricular septum**. Blood passes from the right ventricle through the **pulmonary valve** into a large vessel called the *pulmonary trunk*. The pulmonary trunk divides into right and left *pulmonary arteries*, which carry the blood to the lungs. Arteries always take blood *away* from the heart (a mnemonic to help you: artery=away).

Left Atrium

The **left atrium** forms most of the base of the heart (recall that this is the posterior aspect; see Figure 13.3c, d). It receives

blood from the lungs through four *pulmonary veins.* Like the right atrium, the inside of the left atrium has a smooth posterior wall where the pulmonary veins were absorbed into the heart during development. Because the ridged pectinate muscles are confined to the auricle of the left atrium, the anterior wall of the left atrium is smooth. Blood passes from the left atrium into the left ventricle through the **bicuspid (mitral) valve** (*bi-*=two), which has two cusps, as its name implies (see Figure 13.4a). The term *mitral* refers to the resemblance of the valve to a bishop's miter (hat), which is two-sided. It is also called the **left atrioventricular valve**.

Left Ventricle

The **left ventricle**, the thickest part of the heart (averaging 10 to 15 mm or 0.4 to 0.6 in. thick), forms the apex of the heart (see Figure 13.1b). Like the right ventricle, the left ventricle contains trabeculae carneae and has chordae tendineae that anchor the cusps of the bicuspid valve to papillary muscles. Blood passes from the left ventricle through the **aortic valve** into the largest artery of the body, the *ascending aorta* (*aorte*=to suspend, because the aorta once was believed to lift up the heart). Some of the blood in the aorta flows into the *coronary arteries*, which branch from the ascending aorta and carry blood to the heart wall; the remainder of the blood passes into the *arch of the aorta* and *descending aorta*. Branches of the arch of the aorta and descending aorta carry blood throughout the body.

During fetal life, a temporary blood vessel, called the *ductus arteriosus*, shunts blood from the pulmonary trunk into the aorta. Hence, only a small amount of blood enters the nonfunctioning fetal lungs (see Figure 14.17). The ductus arteriosus normally closes shortly after birth, leaving a remnant known as the **ligamentum arteriosum** (lig′-a-MEN-tum ar-ter-ē-Ō-sum), which connects the arch of the aorta and pulmonary trunk (Figure 13.4a, b).

Myocardial Thickness and Function

The thickness of the myocardium of the four chambers varies according to function. The atria are thin-walled because they deliver blood under less pressure into the adjacent ventricles; since the ventricles pump blood under higher pressure over greater distances, their walls are thicker (Figure 13.4a–b). Although the right and left ventricles act as two separate pumps that simultaneously eject equal volumes of blood, the right side has a much smaller workload. It pumps blood under lower pressure only a short distance to the lungs. The left ventricle pumps blood under higher pressure over great distances to all other parts of the body. Thus, the left ventricle works harder than the right ventricle to maintain the same rate of blood flow. The anatomy of the two ventricles confirms this functional difference: The muscular wall of the left ventricle is considerably thicker than that of the right ventricle (Figure 13.4c; see also 13.2b). Note also that the perimeter of the lumen (space) of the left ventricle is circular, in contrast to that of the right ventricle, which is crescent-shaped (see Figure 13.4c).

Fibrous Skeleton of the Heart

In addition to cardiac muscle tissue, the heart wall also contains dense connective tissue that forms the **fibrous skeleton of the heart**. Essentially, the fibrous skeleton consists of four dense connective tissue rings: the *pulmonary fibrous ring, aortic fibrous*

Figure 13.5 Fibrous skeleton of the heart.

 The fibrous skeleton provides a base for the attachment of heart valves, prevents overstretching of the valves, serves as a point of insertion for cardiac muscle bundles, and prevents the direct spread of action potentials from the atria to the ventricles.

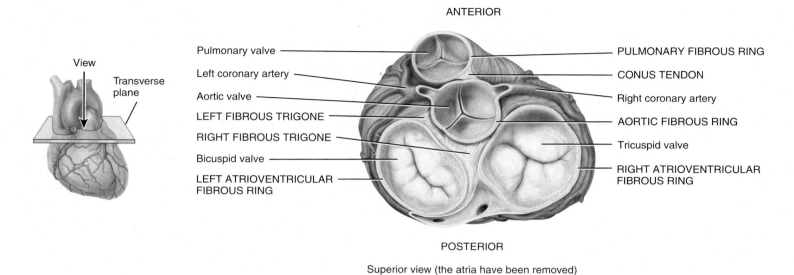

ANTERIOR

Pulmonary valve — — PULMONARY FIBROUS RING
Left coronary artery — — CONUS TENDON
Aortic valve — — Right coronary artery
LEFT FIBROUS TRIGONE — — AORTIC FIBROUS RING
RIGHT FIBROUS TRIGONE — — Tricuspid valve
Bicuspid valve — — RIGHT ATRIOVENTRICULAR FIBROUS RING
LEFT ATRIOVENTRICULAR FIBROUS RING

View
Transverse plane

POSTERIOR

Superior view (the atria have been removed)

 What is the composition of the fibrous skeleton of the heart?

ring, right atrioventricular fibrous ring, and *left atrioventricular fibrous ring* (Figure 13.5). The rings surround the valves of the heart, fuse with one another, and merge with the interventricular septum. As well as forming a structural foundation for the heart valves, the fibrous skeleton prevents overstretching of the valves as blood passes through them. It also serves as a point of insertion for bundles of cardiac muscle fibers and acts as an electrical insulator between the atria and ventricles.

Heart Valves

As each chamber of the heart contracts, it pushes a volume of blood into a ventricle or out of the heart into an artery. Valves open and close in response to pressure changes as the heart contracts and relaxes. Each of the four valves helps to ensure one-way flow of blood by opening to let blood through and then closing to prevent its backflow.

Atrioventricular Valves

Because they are located between an atrium and a ventricle, the tricuspid and bicuspid valves are termed **atrioventricular (AV) valves**. When an AV valve is open, the rounded ends of the cusps (leaflets) project into the ventricle (Figure 13.6a, d). Blood moves from the atria into the ventricles through open AV valves when atrial pressure is higher than ventricular pressure. At this time, the papillary muscles are relaxed, and the chordae

Figure 13.6 Valves of the heart.

 Heart valves prevent the backflow of blood.

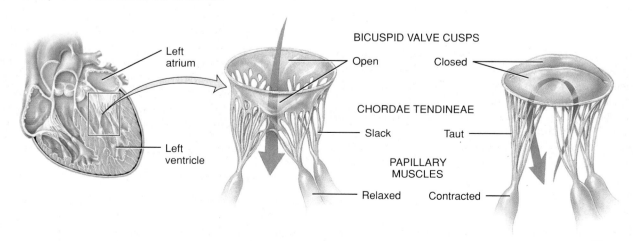

BICUSPID VALVE CUSPS

Left atrium
Left ventricle

Open
Closed

CHORDAE TENDINEAE
Slack
Taut

PAPILLARY MUSCLES
Relaxed
Contracted

(a) Bicuspid valve open   (b) Bicuspid valve closed

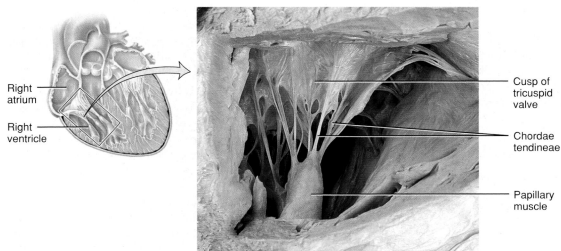

Right atrium

Right ventricle

Cusp of tricuspid valve

Chordae tendineae

Papillary muscle

(c) Tricuspid valve open

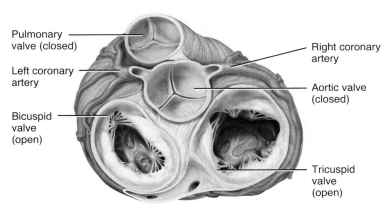

ANTERIOR

Pulmonary valve (closed)

Left coronary artery

Bicuspid valve (open)

Right coronary artery

Aortic valve (closed)

Tricuspid valve (open)

POSTERIOR

(d) Superior view with atria removed: pulmonary and aortic valves closed, bicuspid and tricuspid valves open

ANTERIOR

Pulmonary valve (open)

Bicuspid valve (closed)

Aortic valve (open)

Tricuspid valve (closed)

POSTERIOR

(e) Superior view with atria removed: pulmonary and aortic valves open, bicuspid and tricuspid valves closed

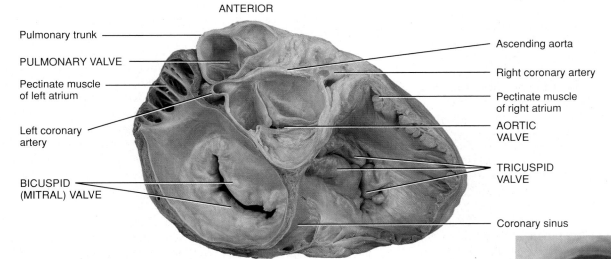

ANTERIOR

Pulmonary trunk

PULMONARY VALVE

Pectinate muscle of left atrium

Left coronary artery

BICUSPID (MITRAL) VALVE

Ascending aorta

Right coronary artery

Pectinate muscle of right atrium

AORTIC VALVE

TRICUSPID VALVE

Coronary sinus

POSTERIOR

(f) Superior view of atrioventricular and semilunar valves

Semilunar cusp of aortic valve

(g) Superior view of aortic valve

 How do papillary muscles prevent atrioventricular valve cusps from everting (swinging upward) into the atria?

tendineae are slack. When the ventricles contract, the pressure of the blood drives the cusps upward until their edges meet and close the opening (Figure 13.6b, e). At the same time, the papillary muscles are also contracting, which pulls on and tightens the chordae tendineae, preventing the valve cusps from everting (opening in the opposite direction into the atria due to the high ventricular pressure). If the AV valves or chordae tendineae are damaged, blood may regurgitate (flow back) into the atria when the ventricles contract.

Semilunar Valves

The aortic and pulmonary valves are known as the **semilunar (SL) valves** (sem′-ē-LOO-nar; *semi*=half; *lunar*=moon-shaped) because they are made up of three crescent moon–shaped cusps (Figure 13.6d–g). Each cusp attaches to the arterial wall by its convex outer margin. The SL valves allow ejection of blood from the heart into arteries, but prevent backflow of blood into the ventricles. The free borders of the cusps project into the lumen of the artery. When the ventricles contract, pressure builds up within the chambers. The semilunar valves open when pressure in the ventricles exceeds the pressure in the arteries, permitting ejection of blood from the ventricles into the pulmonary trunk and aorta (Figure 13.6e). As the ventricles relax, blood starts to flow back toward the heart. This back-flowing blood fills the valve cusps, which causes the free edges of semilunar valves to contact each other tightly and close the opening between the ventricle and artery (Figure 13.6d).

Surprisingly, perhaps, there are no valves guarding the junctions between the venae cavae and the right atrium or the pulmonary veins and the left atrium. As the atria contract, a small amount of blood does flow backward from the atria into these vessels. However, backflow is minimized by a different mechanism; as the atrial muscle contracts, it compresses and nearly collapses the weak walls of the venous entry points.

 CHECKPOINT

3. Define the following external features of the heart: auricle, coronary sulcus, anterior interventricular sulcus, and posterior interventricular sulcus.
4. Describe the location and functions of the different layers of the pericardium.
5. How do the functions of the layers of the heart wall differ?
6. What are the characteristic internal features of each chamber of the heart?
7. List the blood vessels that deliver blood to or receive ejected blood from each chamber of the heart, and name the valve that blood passes through on its way to the next heart chamber or blood vessel.
8. Describe the relationship between wall thickness and function for each heart chamber.
9. How does the fibrous skeleton of the heart assist the operation of heart valves?
10. What causes the heart valves to open and to close?

13.3 CIRCULATION OF BLOOD

 OBJECTIVES
• Outline the flow of blood through the chambers of the heart and through the systemic and pulmonary circulations.
• Describe the arteries and veins of the coronary circulation.

Systemic and Pulmonary Circulations

In postnatal (after birth) circulation, with each beat the heart pumps blood into two circuits—the **systemic circulation** and the **pulmonary circulation** (*pulmon-*=lung). As you will see later, the two circuits are arranged in series so that the output of one becomes the input of the other, as would happen if you attached two garden hoses together. The left side of the heart, which receives bright-red freshly *oxygenated* (oxygen-rich) *blood* from the lungs, is the pump for systemic circulation (Figure 13.7a). The left ventricle ejects blood into the *aorta*, which branches into progressively smaller *systemic arteries* that carry the blood to all organs throughout the body—except for the air sacs (alveoli) of the lungs, which are supplied by the pulmonary circulation. In systemic tissues, arteries give rise to smaller-diameter *arterioles*, which finally lead into extensive beds of *systemic capillaries*. Exchange of nutrients and gases occurs across the thin capillary walls: In the tissues, blood unloads O_2 (oxygen) and picks up CO_2 (carbon dioxide). In most cases, blood flows through only one capillary and then enters a *systemic venule*. Venules carry *deoxygenated* (oxygen-poor) *blood* away from tissues and merge to form larger *systemic veins*, and ultimately the blood flows back to the right atrium.

The right side of the heart is the pump for pulmonary circulation (Figure 13.7a); it receives all the dark-red deoxygenated blood returning from the systemic circulation. Blood ejected from the right ventricle flows into the *pulmonary trunk*, which branches into *pulmonary arteries* that carry blood to the right and left lungs. In pulmonary capillaries, blood unloads CO_2,

Figure 13.7 Systemic and pulmonary circulations.

 The left side of the heart pumps freshly oxygenated blood into the systemic circulation, which supplies all tissues of the body except the air sacs (alveoli) of the lungs; the right side of the heart pumps deoxygenated blood into the pulmonary circulation, which includes the air sacs (alveoli) of the lungs.

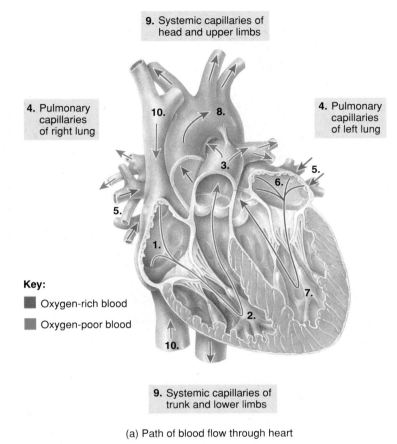

(a) Path of blood flow through heart

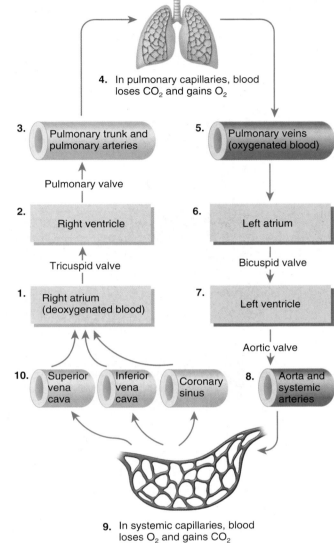

(b) Path of blood flow through systemic and pulmonary circulations

? **In part (b), which numbers represent pulmonary circulation? Which represent systemic circulation?**

which is exhaled, and picks up O_2 from inhaled air. The freshly oxygenated blood then flows into pulmonary veins and returns to the left atrium. The flowchart in Figure 13.7b shows the route of blood flow through the chambers and valves of the heart and the pulmonary and systemic circulations.

Coronary Circulation

Nutrients could not possibly diffuse from blood within the chambers of the heart through all the layers of cells that make up the very sturdy heart wall. For this reason, the wall of the heart has its own blood supply. The flow of blood through the many vessels that pierce the myocardium is called the **coronary (cardiac) circulation** because the arteries encircle the heart like a crown encircles the head (*corona*=crown). While it is contracting, the heart receives little oxygenated blood by way of the **coronary arteries**, which branch from the ascending aorta (Figure 13.8a, c, d). When the heart relaxes, however, the high

pressure of blood in the aorta propels blood through the coronary arteries, into capillaries, and then into **coronary veins** (Figure 13.8b, c, d).

Coronary Arteries

Two coronary arteries, the left and right coronary arteries, branch from the ascending aorta and supply oxygenated blood to the myocardium (Figure 13.8a, c). The **left coronary artery** passes inferior to the left auricle and divides into the anterior interventricular and circumflex branches. The **anterior interventricular branch** or *left anterior descending (LAD) artery* is in the anterior interventricular sulcus and supplies oxygenated blood to the walls of both ventricles. The **circumflex branch** (SER-kum-fleks) lies in the coronary sulcus and distributes oxygenated blood to the walls of the left ventricle and left atrium.

The **right coronary artery** supplies small branches (*atrial branches*) to the right atrium. It continues inferior to the right auricle and ultimately divides into the posterior interventricular and

Figure 13.8 Coronary (cardiac) circulation. The views of the heart are drawn as if the heart were transparent to reveal blood vessels on the posterior aspect.

🔑 The left and right coronary arteries deliver blood to the heart; the coronary veins drain blood from the heart into the coronary sinus.

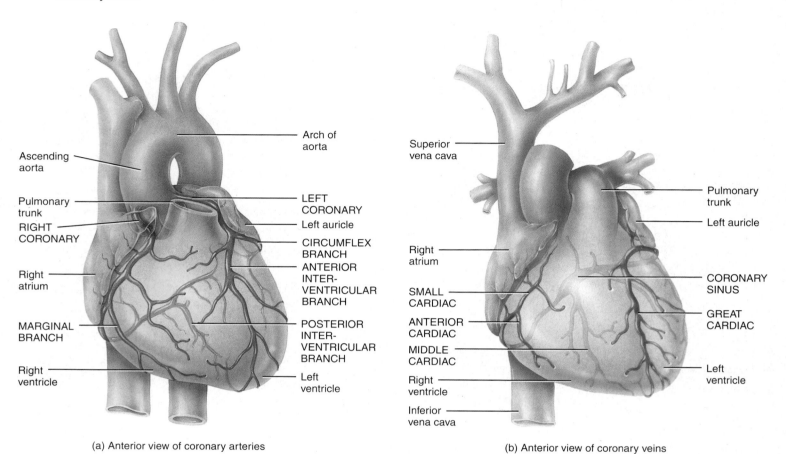

(a) Anterior view of coronary arteries

(b) Anterior view of coronary veins

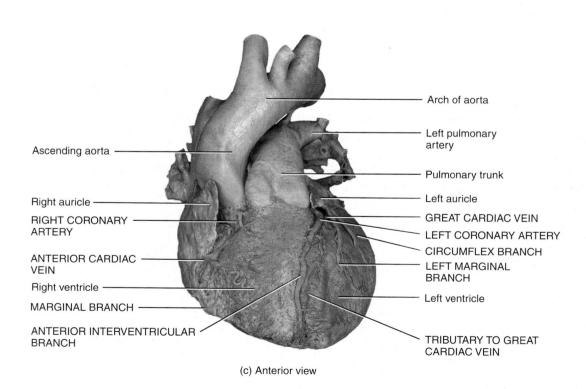

(c) Anterior view

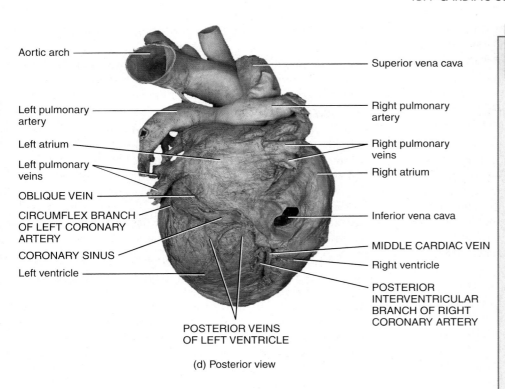

Aortic arch

Left pulmonary artery

Left atrium

Left pulmonary veins

OBLIQUE VEIN

CIRCUMFLEX BRANCH OF LEFT CORONARY ARTERY

CORONARY SINUS

Left ventricle

Superior vena cava

Right pulmonary artery

Right pulmonary veins

Right atrium

Inferior vena cava

MIDDLE CARDIAC VEIN

Right ventricle

POSTERIOR INTERVENTRICULAR BRANCH OF RIGHT CORONARY ARTERY

POSTERIOR VEINS OF LEFT VENTRICLE

(d) Posterior view

CLINICAL CONNECTION | *Myocardial Ischemia and Myocardial Infarction*

Partial obstruction of blood flow in the coronary arteries may cause **myocardial ischemia** (is-KĒ-mē-a; *ische-*=to obstruct; *-emia*=in the blood), a condition of reduced blood flow to the myocardium. Usually, ischemia causes **hypoxia** (hī-POKS-ē-a) (reduced oxygen supply), which may weaken cells without killing them. **Angina pectoris** (an-JĪ-na, or AN-ji-na, PEK-tō-ris), which literally means "strangled chest," is a severe pain that usually accompanies myocardial ischemia.

A complete obstruction to blood flow in a coronary artery may result in a **myocardial infarction (MI)** (in-FARK-shun), commonly called a heart attack. Infarction means the death of an area of tissue because of interrupted blood supply. Because the heart tissue distal to the obstruction dies and is replaced by noncontractile scar tissue, the heart muscle loses some of its strength. Depending on the size and location of the infarcted (dead) area, an infarction may disrupt the conduction system of the heart and cause sudden death by triggering ventricular fibrillation. •

 Which coronary blood vessel delivers oxygenated blood to the walls of the left atrium and left ventricle?

---

marginal branches. The **posterior interventricular branch** follows the posterior interventricular sulcus and supplies the walls of the two ventricles with oxygenated blood. The **marginal branch** beyond the coronary sulcus runs along the right margin of the heart and transports oxygenated blood to the wall of the right ventricle.

Most parts of the body receive blood from branches of more than one artery, and where two or more arteries supply the same region, they usually connect. These connections, called **anastomoses** (a-nas-tō-MŌ-sēs), provide alternate routes called **collateral circulation** for blood to reach a particular organ or tissue. The myocardium contains many anastomoses that connect different branches of the same coronary artery or extend between branches of different coronary arteries. They provide detours for arterial blood if a main route becomes obstructed. Thus, the heart muscle may receive sufficient oxygen even if one of its coronary arteries is partially blocked.

Coronary Veins

After blood passes through the arteries of the coronary circulation, it passes into capillaries where it delivers oxygen and nutrients to the heart muscle and collects carbon dioxide and wastes. From the capillaries the blood then enters veins. Most of the deoxygenated blood from the myocardium drains into a large vascular sinus. (A *vascular sinus* is a thin-walled vein that has no smooth muscle to alter its diameter.) The vascular sinus in the coronary sulcus on the posterior surface of the heart, called the **coronary sinus** (Figure 13.8b, d), empties into the right atrium. The principal tributaries carrying blood into the coronary sinus include the following:

- **Great cardiac vein** in the anterior interventricular sulcus, which drains the areas of the heart supplied by the left coronary artery (left and right ventricles and left atrium)
- **Middle cardiac vein** in the posterior interventricular sulcus, which drains the areas supplied by the posterior interven-

tricular branch of the right coronary artery (left and right ventricles)
- **Small cardiac vein** in the coronary sulcus, which drains the right atrium and right ventricle
- **Anterior cardiac veins**, which drain the right ventricle and open directly into the right atrium

 CHECKPOINT

11. In correct sequence, list the heart chambers, heart valves, and blood vessels encountered by a drop of blood as it flows from the right atrium to the aorta.

12. Which arteries deliver oxygenated blood to the myocardium of the left and right ventricles?

13.4 CARDIAC CONDUCTION SYSTEM AND INNERVATION

 OBJECTIVES

- Explain the structural and functional features of the cardiac conduction system.
- Define an electrocardiogram and indicate its diagnostic importance.
- Describe the innervation of the heart.

Cardiac Conduction System

During embryonic development, approximately 1 percent of the cardiac muscle fibers become **autorhythmic cells** (aw′-tō-RITH-mik; *auto-*=self), that is, cells that repeatedly and rhythmically generate action potentials. They continue to stimulate a heart to beat even after it is removed from the body—for example, to be transplanted into another person—and all of its nerves have been cut. The nerves regulate the heart rate, but do not determine it. Autorhythmic cells act as a natural **pacemaker**, setting the rhythm for the contraction of the entire heart, and they form the **cardiac conduction system**, the route that delivers action potentials throughout the heart muscle. The conduction system ensures that cardiac chambers are stimulated to contract in a coordinated manner, which makes the heart an effective pump. Cardiac action potentials propagate through the following components of the conduction system (Figure 13.9a):

❶ Normally, cardiac excitation begins in the **sinoatrial (SA) node**, located in the right atrial wall just inferior and lateral to the opening of the superior vena cava. Each action potential from the SA node conducts through conduction cells to the contractile cardiac muscle cells of both atria via gap junctions in the intercalated discs of these fibers. With the arrival of the action potential, the two atria contract at the same time.

❷ By conducting along modified atrial muscle fibers called *internodal fibers*, the action potential reaches the **atrioventricular (AV) node**, located in the interatrial septum, just anterior to the opening of the coronary sinus. At the AV node, the action potential slows considerably as a result of various differences in cell structure in the AV node. This delay provides time for the atria to empty their blood into the ventricles.

❸ From the AV node, the action potential enters the **atrioventricular (AV) bundle** (also known as the **bundle of His**, pronounced *hiz*), the only site where action potentials can conduct from the atria to the ventricles. (Elsewhere, the fibrous skeleton of the heart electrically insulates the atria from the ventricles.)

❹ After conducting through the AV bundle, the action potential then enters both the **right** and **left bundle branches**, which course through the interventricular septum toward the apex of the heart.

❺ Finally, the large-diameter **Purkinje fibers** (pur-KIN-jē) leave their insulating connective tissue sheaths near the apex of the heart and relay the action potential to the contractile cells of the ventricular myocardium. As the wave of ventricular contraction moves upward from the apex, the blood is pushed toward the semilunar valves. The ventricular contraction occurs about 0.20 sec (20 milliseconds) after atrial contraction.

The SA node initiates action potentials 90 to 100 times per minute, faster than any other region of the conducting system. Thus, the SA node sets the rhythm for contraction of the heart—it is the *natural pacemaker* of the heart. Various hormones and neurotransmitters can speed or slow pacing of the heart by SA node fibers. In a person at rest, for example, acetylcholine released by the parasympathetic division of the autonomic nervous system typically slows SA node pacing to about 75 action potentials per minute, causing 75 heartbeats per minute.

Transmission of action potentials through the conduction system generates an electric current that can be detected on the body's surface. A recording of the electrical changes that accompany the

Figure 13.9 The conduction system of the heart and a normal electrocardiogram. The route of action potentials through the numbered components of the conduction system in part (a) is described in the text.

🔑 The conduction system ensures that cardiac chambers contract in a coordinated manner.

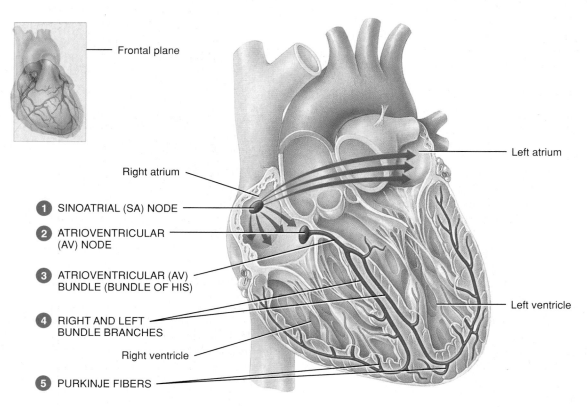

(a) Anterior view of frontal section

CLINICAL CONNECTION | *Artificial Pacemakers*

If the SA node becomes damaged or diseased, the slower AV node can pick up the pacemaking task. Its spontaneous pacing rate is 40 to 60 times per minute. If the activity of both nodes is suppressed, the heartbeat may still be maintained by autorhythmic fibers in the ventricles—the AV bundle, a bundle branch, or Purkinje fibers. However, the pacing rate is so slow (20–35 beats per minute) that blood flow to the brain is inadequate. When this condition occurs, normal heart rhythm can be restored and maintained by surgically implanting an **artificial pacemaker,** a device that sends out small electrical currents to stimulate the heart to contract. A pacemaker consists of a battery and impulse generator and is usually implanted beneath the skin just inferior to the clavicle. The pacemaker is connected to one or two flexible leads (wires) that are threaded through the superior vena cava and then passed into various chambers of the heart. Many of the newer pacemakers, referred to as activity-adjusted pacemakers, automatically speed up the heartbeat during exercise. •

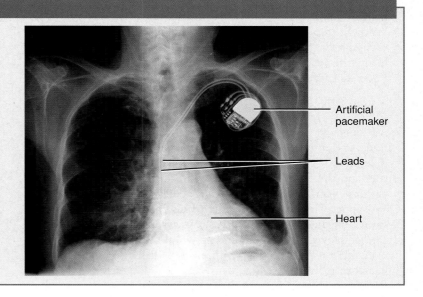

heartbeat is called an **electrocardiogram** (e-lek-trō-KAR-dē-ō-gram), which is abbreviated as either *EKG* or *ECG*. The procedure is referred to as *electrocardiography* (e-LEK-trō-kar-dē-OG-ra-fē). An EKG is a composite of all the action potentials generated by nodal and contractile cells.

Before discussing the different phases of an electrocardiogram, an understanding of the electrochemical properties of neurons will be helpful. The plasma membrane of excitable cells exhibits a **membrane potential**, that is, a difference in electrical potential (voltage) on either side of the membrane. This difference in voltage in an unstimulated cell, called the **resting** **membrane potential**, exists because of the buildup of negative ions in the cytosol along the inside surface of the membrane and an equal buildup of positive ions in the extracellular fluid along the outside surface of the membrane. Such a membrane is said to be **polarized**. The term **depolarization** refers to a reduction in the membrane potential: the inside of the membrane becomes less negative than the resting membrane potential. Initially, the membrane becomes less and less negative, reaches zero, and then becomes positive. The term **repolarization** refers to restoration of the resting membrane potential. An **action potential** or **impulse** is a sequence of rapidly occurring events that decrease and reverse the resting membrane potential and then restore it to the resting state.

Three clearly recognizable up-and-down waves normally accompany each cardiac cycle (Figure 13.9b). The first, called the **P wave**, is the spread of depolarization from the SA node through the two atria. A fraction of a second after the P wave begins, an action potential occurs and the atria contract. The second wave, called the **QRS wave**, is the spread of the depolarization through the ventricles. Shortly after the QRS wave begins, an action potential is reached and the ventricles contract. The third wave, the **T wave**, indicates ventricular repolarization, which corresponds to the relocation of ventricular muscle fibers. There is no wave to show atrial repolarization because the stronger QRS wave masks this event.

Variations in the size and duration of deflection waves of an ECG are useful in diagnosing abnormal cardiac rhythms, detecting an enlarged heart, determining whether certain regions of the heart are damaged, and identifying the cause of chest pain.

The usual rhythm of heartbeats, established by the SA node, is called **normal sinus rhythm**. The term **arrhythmia** (ā-RITH-mē-a) or **dysrhythmia** refers to an abnormal rhythm as a result of a defect in the conduction system of the heart. The heart may beat irregularly, too quickly, or too slowly. Symptoms include chest pain, shortness of breath, lightheadedness, dizziness, and fainting. Arrhythmias may be caused by factors that stimulate the heart such as stress, caffeine, or other stimulants. Arrhythmias may also be caused by a congenital defect, coronary artery disease, myocardial infarction, hypertension, defective heart valves, rheumatic heart disease, hyperthyroidism, and potassium deficiency.

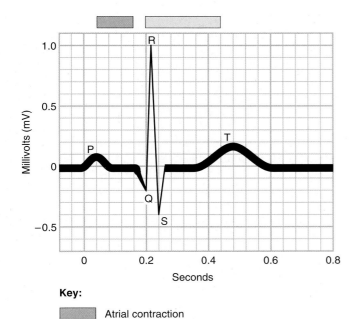

Key:

■ Atrial contraction

□ Ventricular contraction

(b) Waves associated with a normal electrocardiogram of a single heartbeat

Which component of the conduction system provides the only electrical connection between the atria and the ventricles?

Arrhythmias are categorized by their speed, rhythm, and origination of the problem. **Bradycardia** (brād-ē-KAR-dē-a; *brady-*=slow) refers to a slow heart rate (below 50 beats per minute); **tachycardia** (tak'-i-KAR-dē-a; *tachy-*=quick) refers to a rapid heart rate (over 100 beats per minute); and **fibrillation** (fib-bri-LĀ-shun) refers to rapid, uncoordinated heartbeats.

Cardiac Nerves

Although the initiation of the heartbeat originates in the SA node, it is influenced by nerves of the autonomic nervous system (Chapter 19), which form synaptic junctions with the nodal tissues and the coronary vessels. Cardiac branches of the parasympathetic vagus (X) nerve and cardiac branches from the cervical and upper thoracic sympathetic trunk unite around the heart to form a **cardiac plexus** of nerves. Nerves are distributed from the cardiac plexus to the heart and its blood vessels.

Sympathetic nerves relay impulses to the heart that cause the heartbeat to speed up and cause the coronary arteries to dilate. Parasympathetic nerves, on the other hand, slow the heartbeat and cause constriction of the coronary arteries.

 CHECKPOINT

13. What is the function of autorhythmic cells?
14. Trace an action potential through the conduction system of the heart.
15. What is an electrocardiogram? What is its diagnostic significance?
16. What effect do sympathetic and parasympathetic nerves have on the heartbeat?

13.5 CARDIAC CYCLE (HEARTBEAT)

 OBJECTIVE

• Describe the phases associated with a cardiac cycle.

A single **cardiac cycle** comprises all the events associated with one heartbeat. In a normal cardiac cycle, the two atria contract while the two ventricles relax. Then, while the two ventricles contract, the two atria relax. **Systole** (SIS-tō-lē=contraction) refers to the phase of contraction of a chamber of the heart; **diastole** (dī-AS-tō-lē=dilation or expansion) is the phase of relaxation. For the purposes of our discussion, we will divide the cardiac cycle into the following phases (Figure 13.10):

❶ *Relaxation period.* At the end of a cardiac cycle when the ventricles start to relax, all four chambers are in diastole. This is the beginning of the relaxation period. As the ventricles relax, pressure within the chambers drops, and blood starts to flow from the pulmonary trunk and aorta back toward the ventricles. As this blood becomes trapped in the semilunar cusps, the semilunar valves close. As the ventricles continue to relax, the space inside expands, and the pressure falls. When ventricular pressure drops below atrial pressure, the AV valves open and ventricular filling begins. The major part of ventricular filling (75 percent) occurs just after the AV valves open, *without atrial systole.*

❷ *Atrial systole (contraction).* Atrial systole marks the end of the relaxation period and accounts for the remaining 25 percent of the blood that fills the ventricles. Throughout the period of ventricular filling, the AV valves are still open and the semilunar valves are still closed.

Figure 13.10 The cardiac cycle (heartbeat).

 A cardiac cycle comprises all the events associated with a single heartbeat.

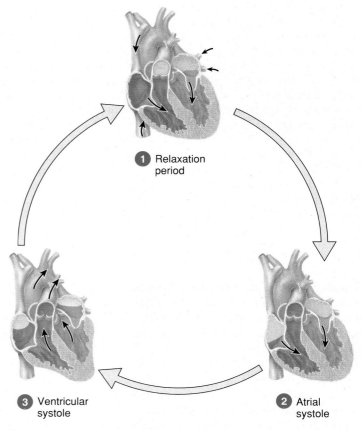

1 Relaxation period

3 Ventricular systole

2 Atrial systole

 What is the status of the four heart valves during ventricular filling?

❸ *Ventricular systole (contraction).* Ventricular contraction pushes blood up against the AV valves, forcing them shut. For a very brief period, all four valves are closed again. As ventricular contraction continues, pressure inside the chambers rises sharply. When left ventricular pressure rises above the pressure in the arteries, both semilunar valves open, and ejection of blood from the heart begins. This lasts until the ventricles start to relax. Then the semilunar valves close and another relaxation period begins.

 CHECKPOINT

17. What is a cardiac cycle?
18. Outline the major events of each of the three phases of the cardiac cycle.

13.6 HEART SOUNDS

 OBJECTIVE

• Describe how heart sounds are produced.

As you learned in Chapter 1, *auscultation* (aws-kul-TĀ-shun; *ausculta-*=listening) is the act of listening to sounds within the body, and it is usually done with a stethoscope. The sound of the heartbeat comes primarily from blood turbulence caused by the closing of the heart valves. Smoothly flowing blood is silent. Compare the sounds made by whitewater rapids or a waterfall to the silence of a smoothly flowing river. During each cardiac cycle, there are four **heart sounds**, but in a normal heart only the first and second heart sounds (S1 and S2) are loud enough to be heard through a stethoscope.

The *first sound (S1)*, which can be described as a lubb sound, is louder and a bit longer than the second sound. S1 is caused by blood turbulence associated with closure of the atrioventricular (AV) valves soon after ventricular systole begins. The *second sound (S2)*, which is shorter and not as loud as the first, can be described as a dupp sound. S2 is caused by blood turbulence associated with closure of the semilunar (SL) valves at the beginning of ventricular diastole. Although S1 and S2 are due to blood turbulence associated with the closure of valves, they are best heard at the surface of the chest in locations that are slightly different from the locations of the valves (Figure 13.11). This is because the sound is carried by the blood flow away from the valves. Normally not loud enough to be heard, S3 is due to blood turbulence during rapid ventricular filling, and S4 is due to blood turbulence during atrial systole.

CLINICAL CONNECTION | *Heart Murmur*

Heart sounds provide valuable information about the mechanical operation of the heart. A **heart murmur** is an abnormal sound consisting of a clicking, rushing, or gurgling noise that is heard before, between, or after the normal heart sounds, or that may mask the normal heart sounds. Heart murmurs in children are extremely common and usually do not represent a health condition. These types of heart murmurs often subside or disappear with growth. Although some heart murmurs in adults are innocent, most often an adult heart murmur indicates a valve disorder. •

 **CHECKPOINT**

19. What is the basis for heart sounds?
20. Where are the two heart sounds best heard on the surface of the chest?

13.7 EXERCISE AND THE HEART

OBJECTIVE

• Explain how the heart is affected by exercise.

A person's cardiovascular fitness can be improved at any age with regular exercise. Some types of exercise are more effective than others for improving the health of the cardiovascular system. **Aerobics**, any activity that works large body muscles for at least 20 minutes, elevates cardiac output and accelerates metabolic rate. Three to five such sessions a week are usually recommended for improving the health of the cardiovascular system. Brisk walking, running, bicycling, cross-country skiing, and swimming are examples of aerobic activities.

Sustained exercise increases the oxygen demand of the muscles. Whether the demand is met depends mainly on the adequacy of cardiac output and proper functioning of the respiratory system. After several weeks of training, a healthy person increases maximal cardiac output (the amount of blood

Figure 13.11 Location of valves (purple) and auscultation sites (red) for heart sounds.

Listening to sounds within the body is called auscultation; it is usually done with a stethoscope.

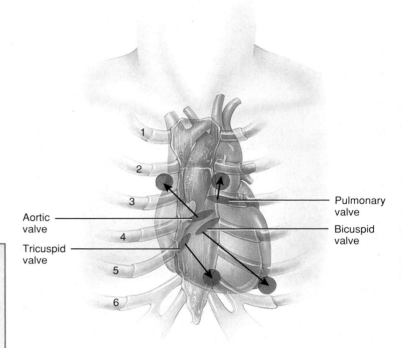

Aortic valve

Tricuspid valve

Pulmonary valve

Bicuspid valve

Anterior view of heart valve locations and auscultation sites

Which heart sound is related to blood turbulence associated with closure of the atrioventricular valves?

ejected from the ventricles into their respective arteries per minute), thereby increasing the maximal rate of oxygen delivery to the tissues. Oxygen delivery also rises because skeletal muscles develop more capillary networks in response to long-term training.

During strenuous activity, a well-trained athlete can achieve a cardiac output double that of a sedentary person, in part because training causes hypertrophy (enlargement) of the heart. This condition is referred to as **physiological cardiomegaly** (kar-dē-ō-MEG-a-lē; *mega*=large). A **pathological cardiomegaly** is related to significant heart disease. Even though the heart of a well-trained athlete is larger, *resting* cardiac output is about the same as in a healthy untrained person, because *stroke volume* (volume of blood pumped by each beat of a ventricle) is increased while heart rate is decreased. The resting heart rate of a trained athlete often is only 40–60 beats per minute (*resting bradycardia*). Regular exercise also helps reduce blood pressure, anxiety, and depression; control weight; and increase the body's ability to dissolve blood clots.

CHECKPOINT

21. What are some of the cardiovascular benefits of regular exercise?

CLINICAL CONNECTION | *Coronary Artery Disease*

Coronary artery disease (CAD) is a serious medical problem that affects about 7 million people annually. Responsible for nearly three-quarters of a million deaths in the United States each year, it is the leading cause of death for both men and women. CAD is defined as the effects of the accumulation of atherosclerotic plaques (described shortly) in coronary arteries that lead to a reduction in blood flow to the myocardium. Some individuals have no signs or symptoms; others experience angina pectoris (chest pain), and still others suffer heart attacks.

Risk Factors for CAD

People who possess combinations of certain *risk factors* (characteristics, symptoms, or signs present in a disease-free person that are statistically associated with a greater chance of developing a disease) are more likely to develop CAD. Risk factors include smoking, high blood pressure, diabetes, high cholesterol levels, obesity, "type A" personality, sedentary lifestyle, and family history of CAD. Most of these are modifiable; that is, they can be altered by changing diet and other habits or can be controlled by taking medications. However, other risk factors are unmodifiable (beyond our control), including genetic predisposition (family history of CAD at an early age), age, and gender. (For example, adult males are more likely than adult females to develop CAD.) Smoking is undoubtedly the number-one risk factor in all CAD-associated disease, doubling the risk of morbidity and mortality.

Development of Atherosclerotic Plaques

Thickening of the walls of arteries and loss of elasticity are the main characteristics of a group of diseases referred to as **arteriosclerosis** (ar-tē′-rē-ō-skle-RŌ-sis; *sclero-*=hardening). One form is **atherosclerosis** (ath′-er-ō-skle-RŌ-sis), a progressive disease characterized by the formation of lesions in the walls of large and medium-sized arteries called **atherosclerotic plaques** (ath′-er-ō-skle-RO-tik) (Figure A, right).

To understand how atherosclerotic plaques develop, you need to know something about molecules produced by the liver and small intestine called **lipoproteins**. These spherical particles consist of an inner core of triglycerides and other lipids and an outer shell of proteins, phospholipids, and cholesterol. Most lipids, including cholesterol, do not dissolve in water; they must be made water-soluble in order to be transported in the blood. This is accomplished by combining the lipids with lipoproteins. Two major classes of lipoproteins are **low-density lipoproteins (LDLs)** and **high-density lipoproteins (HDLs).** LDLs transport cholesterol from the liver to body cells for use in cell membrane repair and the production of steroid hormones and bile salts, but excessive amounts promote atherosclerosis, so the cholesterol in these particles is sometimes referred to as "bad cholesterol." HDLs remove excess cholesterol from body cells and transport it to the liver for elimination. Because HDLs decrease blood cholesterol level, the cholesterol in HDLs is referred to as "good cholesterol." When you get your blood test results, you want your LDL to be low and your HDL to be high.

Inflammation, a defensive response of the body to tissue damage, plays a key role in the development of atherosclerotic plaques. As a result of tissue damage, blood vessels dilate and increase their permeability, and phagocytes, including macrophages, appear in larger numbers. The formation of atherosclerotic plaques begins when excess LDLs from the blood accumulate in the inner layer of an artery wall (layer closest to the bloodstream). The lipids and proteins in the LDLs then undergo oxidation (which removes electrons from the molecules), and the proteins bind to sugars (glycation). Macrophages ingest and become so filled with the oxidized LDL particles that they have a foamy appearance when viewed microscopically (foam cells). T cells (lymphocytes) follow monocytes into the inner lining of an artery and there release chemicals that intensify the inflammatory response. Together, the foam cells, macrophages, and T cells form a **fatty streak**, the beginning of an atherosclerotic plaque. If the plaque is large enough, it can significantly decrease or stop the flow of blood and result in a heart attack.

In most inflammatory responses, macrophages release chemicals that promote healing following fatty streak formation. However, macrophages and endothelial cells also secrete chemicals that cause smooth muscle cells of the middle layer of an artery to migrate to the top of the atherosclerotic plaque, forming a cap over it and thus walling it off from the blood.

Because most atherosclerotic plaques expand away from the bloodstream rather than into it, blood can still flow through the affected artery with relative ease, often for decades. Only about 15 percent of heart attacks occur when plaque in a coronary artery expands into the bloodstream and restricts blood flow. Most heart attacks occur when the cap over the plaque breaks open in response to chemicals produced by foam cells. In addition, T cells induce foam cells to produce tissue factor (TF), a chemical that begins the cascade of reactions that results in blood clot formation.

A number of other risk factors (all modifiable) have also been identified as significant predictors of CAD when their levels are elevated. **C-reactive proteins (CRPs)** are proteins produced by the liver or present in blood in an inactive form that are converted to an active form during inflammation. CRPs may play a direct role in the development of atherosclerosis by promoting the uptake of LDL by macrophages. **Lipoprotein (a)**, an LDL-like particle that attaches to endothelial cells, macrophages, and blood platelets, inhibits blood clot breakdown and may promote smooth muscle fiber proliferation. **Fibrinogen**, a glycoprotein involved in blood clotting, may help regulate cellular proliferation, vasoconstriction, and platelet aggregation. **Homocysteine** (hō-mō-SIS-tēn) is an amino acid that may induce blood vessel damage by promoting platelet aggregation and smooth muscle fiber proliferation.

Diagnosis of CAD

Many procedures may be employed to diagnose CAD; the specific procedure depends on the affected individual's signs and symptoms.

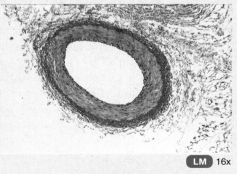

LM 16x

Normal artery

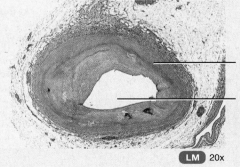

ATHEROSCLEROTIC PLAQUE

Partially obstructed lumen (space through which blood flows)

LM 20x

Obstructed artery

(A) Normal and obstructed arteries

A resting electrocardiogram (ECG or EKG) is the standard test employed to diagnose CAD. **Stress testing** can also be performed. In an *exercise stress test*, the functioning of the heart is monitored when placed under physical stress by exercising using a treadmill, an exercise bicycle, or arm exercises. During the procedure, ECG recordings are monitored continuously and blood pressure is monitored at intervals. A *nonexercise (pharmacologic) stress test* is used for individuals who cannot exercise due to conditions such as arthritis. A medication is injected that stresses the heart to mimic the effects of exercise. During both exercise and nonexercise stress testing, radionuclide imaging may be performed to evaluate blood flow through heart muscle. In **radionuclide imaging** (rā-dē-ō-NOO-klīd; -*radio*=x-ray or gamma ray), a radioactively labeled substance (radionuclide) is introduced into a vein, distributed throughout the body, and imaged to provide information about organ structure and function. (For representative images, see Table 1.3.)

Diagnosis of CAD may also involve **echocardiography** (ek'-ō-kar-dē-OG-ra-fē), a technique that uses ultrasound waves to image the interior of the heart. It allows the heart to be seen in motion and can be used to determine the size, shape, and functions of the chambers; the volume and velocity of blood pumped from the heart and the status of heart valves; the presence of birth defects; and abnormalities of the pericardium. A fairly recent technique is **electron beam computerized tomography (EBCT),** which detects calcium deposits in coronary arteries. These calcium deposits are indicators of atherosclerosis.

Coronary (cardiac) computed tomography angiography (CCTA) is a computer-assisted radiography procedure in which a contrast medium is injected into a vein and a beta-blocker is given to decrease heart rate. These X-ray beams trace an arc around the heart and ultimately produce an image called a CCTA scan. This procedure is used primarily to detect blockages such as atherosclerotic plaque or calcium (see Table 1.3).

Cardiac catheterization (kath'-e-ter-i-ZĀ-shun) is an invasive procedure used to visualize the heart's chambers, valves, and great vessels in order to diagnose and treat disease not related to abnormalities of the coronary arteries. It may also be used to measure pressure in the heart and great vessels; assess cardiac output; measure the flow of blood through the heart and great vessels; identify the location of septal and valvular defects; and take tissue and blood samples. The procedure involves inserting a long, flexible, radiopaque catheter (plastic tube) into a peripheral vein (for right heart catheterization) or a peripheral artery (for left heart catheterization) and guiding it under fluoroscopy (x-ray observation).

Coronary angiography (an'-jē-OG-ra-fē; *angio-*=blood vessel; -*grapho*=to write) is an invasive procedure used to obtain information about the coronary arteries. A catheter is inserted into an artery in the groin or wrist and threaded under fluoroscopy toward the heart and then into the coronary arteries. After the tip of the catheter is in place, a radiopaque contrast medium is injected into the coronary arteries. The radiographs of the arteries, called angiograms, appear in motion on a monitor, and the information is recorded on a videotape or computer disc. Coronary angiography may be used to visualize coronary arteries and to inject clot-dissolving drugs, such as streptokinase or tissue plasminogen activator (t-PA), into a coronary artery to dissolve an obstructing thrombus.

Treatment of CAD

Treatment options include drugs (antihypertensives, nitroglycerine, beta blockers, cholesterol-lowering drugs, and clot-dissolving agents) and various surgical and nonsurgical procedures designed to increase blood supply.

Coronary artery bypass grafting (CABG) is a surgical procedure in which a blood vessel from another part of the body is at-

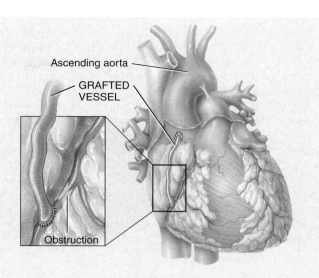

(B) Coronary artery bypass grafting (CABG)

tached (grafted) to a coronary artery to bypass an area of blockage (Figure B). A piece of the grafted blood vessel is sutured between the aorta and the unblocked portion of the coronary artery (Figure B).

A nonsurgical procedure is **percutaneous transluminal coronary angioplasty (PTCA)** (*percutaneous*=through the skin; *trans-*=across; *lumen*=an opening or channel in a tube; *angio-*=blood vessel; -*plasty*=to mold or to shape). A balloon catheter is inserted into an artery of an arm or a leg and gently guided into a coronary artery (Figure C). While dye is released, angiograms (x-rays of blood vessels) are taken to locate the plaques. Next, the catheter is advanced to the point of obstruction, and an intra-aortic balloon pump is inflated with air to squash the plaque against the blood vessel wall. Because 30 to 50 percent of PTCA-opened arteries fail due to restenosis (renarrowing) within six months after the procedure, a stent may be inserted via a catheter. A stent is a metallic, fine wire tube that is permanently placed in an artery to keep the artery *patent* (open) (Figures D, E). Restenosis may be due to damage from the procedure itself, for PTCA may damage the arterial

BALLOON Atherosclerotic Narrowed lumen Coronary
         plaque   of artery   artery

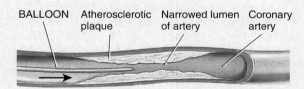

Balloon catheter with uninflated balloon is threaded to obstructed area in artery

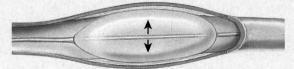

When balloon is inflated, it stretches arterial wall and squashes atherosclerotic plaque

After lumen is widened, balloon is deflated and catheter is withdrawn

(C) Percutaneous transluminal coronary angioplasty (PTCA)

wall, leading to platelet activation, proliferation of smooth muscle fibers, and plaque formation. Recently, drug-coated (drug-eluting) coronary stents have been used to prevent restenosis. The stents are coated with one of several antiproliferative drugs (drugs that inhibit the proliferation of smooth muscle fibers of the middle layer of an artery) and anti-inflammatory drugs. It has been shown that drug-coated stents reduce the rate of restenosis when compared to bare-metal (noncoated) stents.

One area of current research involves cooling the body's core temperature during procedures such as coronary artery bypass grafting (CABG). There have been some promising results from the application of cold therapy during a cerebral vascular accident (CVA) or stroke. This research stemmed from observations of people who had suffered a hypothermic incident (such as cold-water drowning) and recovered with relatively minimal neurologic deficits. •

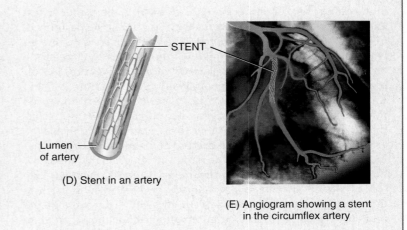

(D) Stent in an artery

(E) Angiogram showing a stent in the circumflex artery

CLINICAL CONNECTION | *Help for Failing Hearts*

As the heart fails, a person has decreasing ability to exercise or even to move around. A variety of surgical techniques and medical devices exist to aid a failing heart. For some patients, even a 10 percent increase in the volume of blood ejected from the ventricles can mean the difference between being bedridden and having limited mobility.

A **cardiac (heart) transplant** is the replacement of a severely damaged heart with a normal heart from a brain-dead or recently deceased donor. Cardiac transplants are performed on patients with end-stage heart failure or severe coronary artery disease. Once a suitable heart is located, the chest cavity is exposed through a midsternal cut. After the patient is placed on a heart–lung bypass machine, which oxygenates and circulates blood, the pericardium is cut to expose the heart. Next, the diseased heart is removed (usually except for the posterior wall of the left atrium) (Figure A) and the donor heart is trimmed and sutured into position (Figure B) so that the remaining left atrium and great vessels are connected to the donor heart (Figure C). The new heart is started as blood flows through it (an electrical shock may be used to correct an abnormal rhythm), the patient is weaned from the heart–lung bypass machine, and the chest is closed. The patient must remain on immunosuppressant drugs for a lifetime to prevent rejection. Since the vagus (X) nerve is severed during the surgery, the new heart will beat at about 100 times per minute (compared to a normal heart rate of about 75 times per minute).

Usually, a donor heart is perfused with a cold solution and then preserved in sterile ice. This can keep the heart viable for about 4–5 hours. In May 2007, surgeons in the United States performed the first beating-heart transplant. The donor heart was maintained at normal body temperature and hooked up to an organ-care system that allowed it to keep beating with warm, oxygenated blood flowing through it. This greatly prolongs the time between removal of the heart from the donor and transplantation into a recipient; it also decreases injury to the heart while being deprived of blood, which can lead to rejection. The safety and benefits of the oxygen-care system are still being evaluated.

Cardiac transplants are common today and produce good results, but the availability of donor hearts is very limited. Another approach is the use of cardiac assist devices and other surgical procedures that assist heart function without removing the heart, including the intra-aortic balloon pump and ventricular assist device (see Key Medical Terms section). •

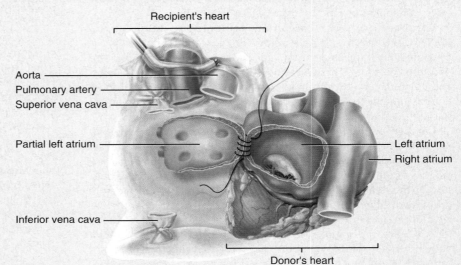

(A) The donor's left atrium is sutured to the recipient's left atrium

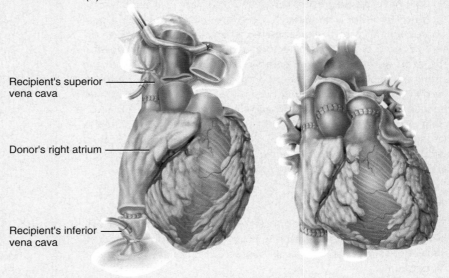

(B) The donor's right atrium is sutured to the recipient's superior and inferior venae cavae

(C) Transplanted heart with sutures

13.8 DEVELOPMENT OF THE HEART

OBJECTIVE
• Describe the development of the heart.

The cardiovascular system is the first system to form and function in an embryo, and the heart is the first functional organ. The heart must develop early on because the rapidly growing embryo needs an efficient way to obtain oxygen and nutrients and get rid of wastes. Recall that oxygen and nutrients in the mother's intervillous spaces diffuse into embryonic chorionic villi and wastes diffuse in the opposite direction as early as 21 days after fertilization (see Figure 4.10). Blood vessels in the chorionic villi connect to the embryonic heart by way of the umbilical arteries and umbilical vein (see Figure 4.10c).

As we trace the development of the heart, keep in mind that many congenital (present at birth) disorders of the heart develop during embryonic life. Such disorders are responsible for almost half of all deaths from birth defects.

The *heart* begins its development from **mesoderm** on day 18 or 19 following fertilization. In the head end of the embryo, the heart develops from a group of mesodermal cells called the **cardiogenic area** (kar-dē-Ō-JEN-ik; *cardio-*=heart; *-genic*= producing) (Figure 13.12a). In response to induction signals

from the underlying endoderm, the mesoderm in the cardiogenic area forms a pair of elongated strands called **cardiogenic cords**. Shortly after, these cords develop a hollow center and then become known as **endocardial tubes** (Figure 13.12b). With lateral folding of the embryo, the paired endocardial tubes approach each other and fuse into a single tube called the **primitive heart tube** on day 21 following fertilization (Figure 13.12c).

On the twenty-second day, the primitive heart tube develops into five distinct regions and begins to pump blood. From tail end to head end (also the direction of blood flow) they are the (1) **sinus venosus**, (2) **primitive atrium**, (3) **primitive ventricle**, (4) **bulbus cordis**, and (5) **truncus arteriosus** (Figure 13.12d). The sinus venosus initially receives blood from all the veins in the embryo; contractions of the heart begin in this region and follow sequentially in the other regions. Thus, at this stage, the heart consists of a series of unpaired regions. The fates of the regions are shown in Table 13.1.

On day 23, the primitive heart tube elongates. Because the bulbus cordis and ventricle grow more rapidly than other parts of the tube and because the atrial and venous ends of the tube are confined by the pericardium, the tube begins to loop and fold. At first, the primitive heart tube assumes a U-shape; later it becomes S-shaped (Figure 13.12e). As a result of these movements, which are completed by day 28, the atria and ventricles of the future heart are reoriented to assume their final adult

Figure 13.12 Development of the heart. Arrows within the structures indicate the direction of blood flow.

 The heart begins its development from mesoderm on day 18 or 19 following fertilization.

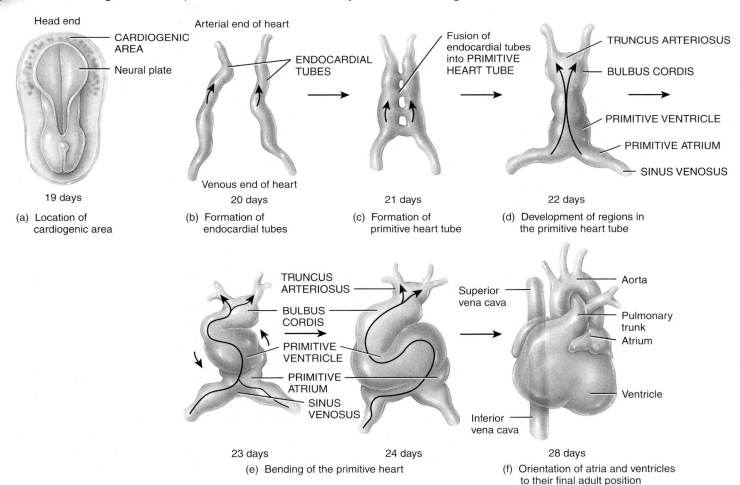

(a) Location of cardiogenic area — 19 days

(b) Formation of endocardial tubes — 20 days

(c) Formation of primitive heart tube — 21 days

(d) Development of regions in the primitive heart tube — 22 days

(e) Bending of the primitive heart — 23 days, 24 days, 28 days

(f) Orientation of atria and ventricles to their final adult position

 When during embryonic development does the primitive heart begin to contract?

TABLE 13.1

Development of the Heart and Great Vessels

EMBRYONIC STRUCTURE	ADULT DERIVATIVE
Sinus venosus	Part of right atrium (posterior wall), coronary sinus, and sinoatrial (SA) node
Primitive atrium	Part of right atrium (anterior wall), right auricle, part of left atrium (anterior wall), and left auricle
Primitive ventricle	Left ventricle
Bulbus cordis	Right ventricle
Truncus arteriosus	Ascending aorta and pulmonary trunk

Figure 13.13 Partitioning of the heart into four chambers.

 Partitioning of the heart begins on about the twenty-eighth day after fertilization.

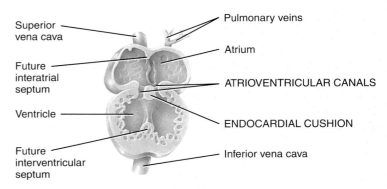

(a) Anterior view of frontal section at about 28 days

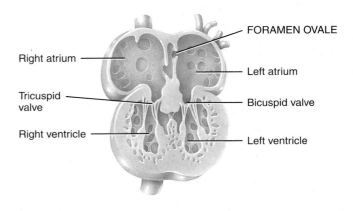

(b) Anterior view of frontal section at about 8 weeks

 When is partitioning of the heart complete?

positions (Figure 13.12f). The remainder of heart development consists of remodeling of the chambers and the formation of septa and valves to form a four-chambered heart.

On about day 28, thickenings of mesoderm of the inner lining of the heart wall appear. These **endocardial cushions** (Figure 13.13) grow toward each other, fuse, and divide the single **atrioventricular canal** (region between atria and ventricles) into smaller, separate left and right atrioventricular canals. Also, the *interatrial septum* begins its growth toward the fused endocardial cushions. Ultimately, the interatrial septum and endocardial cushions unite and an opening in the septum, called the **foramen ovale** (ō-VAL-ē), develops. The interatrial septum divides the atrial region into a *right atrium* and a *left atrium*. Before birth, the foramen ovale allows most blood entering the right atrium to pass into the left atrium. After birth, this opening normally closes so that the interatrial septum is a complete partition. The remnant of the foramen ovale is the fossa ovalis (see Figure 13.4a). Formation of the *interventricular septum* partitions the ventricular region into a *right ventricle* and a *left ventricle*. Partitioning of the atrioventricular canal, atrial region, and ventricular region is basically complete by the end of the fifth week. The *atrioventricular valves* form between the fifth and eighth weeks. The *semilunar valves* form between the fifth and ninth weeks.

 CHECKPOINT

22. Why is the cardiovascular system the first body system to develop?
23. From which primary germ layer does the heart develop?

KEY MEDICAL TERMS ASSOCIATED WITH THE HEART

Atrial premature contraction (APC) A heartbeat that occurs earlier than expected and briefly interrupts the normal heart rhythm. It often causes a sensation of a skipped heartbeat followed by a more forceful heartbeat. APCs originate in the atrial myocardium and are common in healthy individuals.

Cardiac arrest (KAR-dē-ak a-REST) Cessation of an effective heartbeat. The heart may be completely stopped or may be in *ventricular fibrillation* (see below).

Cardiac rehabilitation (rē-ha-bil-i-TĀ-shun) A supervised program of progressive exercise, psychological support, education, and training to enable a patient to resume normal activities following a myocardial infarction.

Cardiomyopathy (kar-dē-ō-mī-OP-a-thē; *myo*=muscle; *-pathos*=disease) A progressive disorder in which ventricular structure or function is impaired.

Congestive heart failure (CHF) A loss of pumping efficiency by the heart. Causes of CHF include coronary artery disease, congenital defects, long-term high blood pressure (which increases the afterload), myocardial infarctions, and valve disorders.

Echocardiography (ek′-ō-kar-dē-OG-ra-fē; *cardio*=heart; *-grapho*=to record) The use of ultrasound to study the structure of the heart and its motions and the presence of pericardial fluid.

Electrophysiological testing (e-lek′-trō-fiz′-ē-OL-ō-je-kal) A procedure in which a catheter with an electrode is passed through blood vessels and introduced into the heart to detect the exact locations of abnormal electrical conduction pathways.

Heart block An arrhythmia that occurs when the electrical pathways between the atria and ventricles are blocked, slowing the transmission of electrical signals throughout the contractile cardiac muscle cells. The most common site of blockage is the atrioventricular node, a condition called *atrioventricular (AV) block*.

Palpitation (pal-pi-TĀ-shun) A fluttering of the heart or an abnormal rate or rhythm of the heart.

Paroxysmal tachycardia (par-ok-SIZ-mal tak′-i-KAR-dē-a; *tachy-*=swift or fast) A period of rapid heartbeats that begins and ends suddenly.

Sick sinus syndrome An abnormally functioning SA node that initiates heartbeats too slowly or rapidly, pauses too long between heartbeats, or stops producing heartbeats. Symptoms include lightheadedness, shortness of breath, loss of consciousness, and palpitations.

Ventricular assist device (VAD) A mechanical pump that helps a weakened ventricle to pump blood throughout the body so that the heart does not have to work as hard.

Ventricular fibrillation (VF or **V-fib)** The most deadly arrhythmia, in which contractions of the ventricular fibers are completely asynchronous so that the ventricles quiver rather than contract in a coordinated way.

Ventricular premature contraction An arrhythmia that arises when an *ectopic focus* (ek-TŌP-ik), a region of the heart other than the conduction system, becomes more excitable than normal and causes an occasional abnormal action potential to occur.

Ventricular tachycardia (VT or **V-tach)** An arrhythmia that originates in the ventricles, is characterized by four or more ventricular premature contractions, and causes the ventricles to beat too fast (at least 120 beats/min).

CHAPTER REVIEW AND RESOURCE SUMMARY

WileyPLUS

Review   **Resource**

13.1 Location and Surface Projection of the Heart

1. The heart is located in the mediastinum; about two-thirds of its mass is to the left of the midline.
2. The heart is basically a cone lying on its side; it consists of an apex, a base, anterior and inferior surfaces, and right and left borders.
3. Four points are used to project the heart's location to the surface of the chest.

Anatomy Overview - The Cardiovascular System

13.2 Structure and Function of the Heart

1. The pericardium, the membrane that surrounds and protects the heart, consists of an outer fibrous layer and an inner serous pericardium, which is composed of a parietal layer and a visceral layer. Between the parietal and visceral layers of the serous pericardium is the pericardial cavity, a potential space filled with a few milliliters of lubricating pericardial fluid that reduces friction between the two membranes.
2. The wall of the heart has three layers: epicardium (visceral layer of the serous pericardium plus the underlying adipose tissue), myocardium, and endocardium. The epicardium consists of mesothelium and connective tissue, the myocardium is composed of cardiac muscle tissue, and the endocardium consists of endothelium and connective tissue.
3. Cardiac muscle fibers usually contain a single centrally located nucleus. Compared to skeletal muscle fibers, cardiac muscle fibers have more and larger mitochondria, slightly smaller sarcoplasmic reticulum, and wider transverse tubules, which are located at Z discs.
4. Cardiac muscle fibers are connected via end-to-end intercalated discs. Desmosomes in the discs provide strength, and gap junctions allow muscle action potentials to conduct from one muscle fiber to its neighbors.
5. The heart chambers include two superior chambers, the right and left atria, and two inferior chambers, the right and left ventricles. External features of the heart include the auricles (flaps of each atrium that increase their volume), the coronary sulcus between the atria and ventricles, and the anterior and posterior sulci between the ventricles on the anterior and posterior surfaces of the heart, respectively.
6. The right atrium, which receives blood from the superior vena cava, inferior vena cava, and coronary sinus, is separated internally from the left atrium by the interatrial septum, which contains the fossa ovalis. Blood exits the right atrium through the tricuspid valve. The right ventricle, which receives blood from the right atrium, is separated internally from the left ventricle by the interventricular septum and pumps blood to the lungs through the pulmonary valve and pulmonary trunk.
7. Oxygenated blood enters the left atrium from the pulmonary veins and exits through the bicuspid (mitral) valve. The left ventricle pumps oxygenated blood into the systemic circulation through the aortic valve and aorta.
8. The thickness of the myocardium of the four chambers varies according to the chamber's function. The left ventricle has the thickest wall because of its high workload.
9. The fibrous skeleton of the heart is dense connective tissue that surrounds and supports the valves of the heart.
10. Heart valves prevent the backflow of blood within the heart. The atrioventricular (AV) valves, which lie between atria and ventricles, are the tricuspid valve on the right side of the heart and the bicuspid (mitral) valve on the left. The chordae tendineae and papillary muscles stabilize the flaps of the AV valves and stop blood from backing up into the atria. Each of the two arteries that leaves the heart has a semilunar valve (aortic and pulmonary).

Anatomy Overview - Cardiac Muscle
Anatomy Overview - Heart Structures
Figure 13.3 - Structure of the Heart: Surface Features
Exercise - Paint the Heart

13.3 Circulation of Blood

1. The left side of the heart is the pump for systemic circulation, the circulation of blood throughout the body except for the air sacs of the lungs. The left ventricle ejects blood into the aorta, and blood then flows into systemic arteries, arterioles, capillaries, venules, and veins, which carry it back to the right atrium.
2. The right side of the heart is the pump for pulmonary circulation, the circulation of blood through the lungs. The right ventricle ejects blood into the pulmonary trunk, and blood then flows into pulmonary arteries, pulmonary capillaries, and pulmonary veins, which carry it back to the left atrium.

Anatomy Overview - Heart Structures
Anatomy Overview - Pulmonary Circulation
Figure 13.7 - Systemic and Pulmonary Circulations
Exercise - Drag and Drop Blood Flow

Review	Resource

3. The flow of blood through the vessels that supply the heart is called coronary (cardiac) circulation.
4. The principal arteries of coronary circulation are the left and right coronary arteries; the principal veins are the cardiac veins and the coronary sinus.

13.4 Cardiac Conduction System and Innervation

1. Autorhythmic cells form the cardiac conduction system; these are modified cardiac muscle fibers that spontaneously generate action potentials.
2. Components of the conduction system are the sinoatrial (SA) node (pacemaker), atrioventricular (AV) node, atrioventricular (AV) bundle (bundle of His), bundle branches, and Purkinje fibers.
3. The record of electrical changes during the course of cardiac cycles is called an electrocardiogram (ECG). A normal ECG consists of a P wave (atrial depolarization), a QRS wave (onset of ventricular depolarization), and a T wave (ventricular repolarization).

Anatomy Overview - Heart Structures
Animation - Cardiac Conduction
Animation - Cardiac Cycle and ECG
Figure 13.9 - The Conduction System of the Heart
Exercise - Sequence Cardiac Conduction

13.5 Cardiac Cycle (Heartbeat)

1. A cardiac cycle consists of the systole (contraction) and diastole (relaxation) of both atria, plus the systole and diastole of both ventricles.
2. The phases of the cardiac cycle are (a) relaxation period, (b) ventricular filling, and (c) ventricular systole.

Animation - Cardiac Cycle and ECG
Animation - Cardiac Cycle
Exercise - Cardiac Cycle
Exercise - ECG Jigsaw Puzzle
Concepts and Connections - Cardiac Cycle

13.6 Heart Sounds

1. S1, the first heart sound (lubb), is caused by blood turbulence associated with the closing of the atrioventricular valves.
2. S2, the second sound (dupp), is caused by blood turbulence associated with the closing of semilunar valves.

Animation - Cardiac Cycle

13.7 Exercise and the Heart

1. Sustained exercise increases oxygen demand on muscles.
2. Among the benefits of aerobic exercise are increased cardiac output, decreased blood pressure, weight control, increased fibrinolytic activity, and decreased resting heart rate.

Animation - Cardiac Output

13.8 Development of the Heart

1. The heart develops from mesoderm.
2. The endothelial tubes develop into the four-chambered heart and great vessels of the heart.

Figure 13.13 - Partitioning of the Heart into Four Chambers

CRITICAL THINKING QUESTIONS

1. Explain why the AV valves don't flip open backward when the ventricles contract.

2. Mr. Williams was diagnosed with blockages in his anterior interventricular branch or left anterior descending (LAD) and circumflex branch of the left coronary artery. What regions of his heart may be affected by these blockages?

3. The heart is constantly beating, which means it's constantly moving. Why doesn't the heart beat its way out of position? Why doesn't the muscle contraction pull the muscle fibers apart?

4. Aleesha's three-year-old daughter has a sore throat. Her pediatrician wants Aleesha to bring her daughter into the office so she can be tested for a streptococcal infection. Aleesha wonders why this is necessary, since children get so many sore throats and they usually recover quickly. What would you tell her?

5. Franz has been diagnosed with aortic valve stenosis. What does this mean, and how do you think it will affect his heart function?

⟨?⟩ ANSWERS TO FIGURE QUESTIONS

13.1 The mediastinum is the anatomical region that extends from the sternum to the vertebral column from the first rib to the diaphragm and between the pleurae of the lungs.

13.2 The visceral layer of the serous pericardium (epicardium) is both a part of the pericardium and a part of the heart wall.

13.3 The coronary sulcus forms an external boundary between the superior atria and inferior ventricles.

13.4 The left ventricle is the chamber of the heart that has the thickest wall.

13.5 The fibrous skeleton is composed of dense connective tissue.

13.6 The papillary muscles contract, which pulls on the chordae tendineae and prevents valve cusps of the atrioventricular valves from everting and letting blood flow back into the atria.

13.7 In part (b), numbers 2 (right ventricle) through 6 (left atrium) depict pulmonary circulation, and numbers 7 (left ventricle) through 10 and 1 (right atrium) depict systemic circulation.

13.8 The circumflex branch delivers oxygenated blood to the walls of the left atrium and left ventricle.

13.9 The only electrical connection between the atria and the ventricles is the atrioventricular bundle.

13.10 The atrioventricular valves are open and the semilunar valves are closed during ventricular filling.

13.11 The first heart sound (S1), or lubb, is associated with the closure of atrioventricular valves.

13.12 The heart begins to contract by the twenty-second day of gestation.

13.13 Partitioning of the heart is complete by the end of the fifth week.

14 THE CARDIOVASCULAR SYSTEM: BLOOD VESSELS

INTRODUCTION If you have ever planted a large garden, you probably know all about the importance of irrigation. At its simplest, an irrigation system is a network of channels or furrows that deliver needed water from one main source to the roots of all of a garden's plants. Similarly, the body's blood vessels form an extensive network of "irrigation channels" to deliver needed fluid—in this case the homeostatically maintained blood—to every cell in the body. In fact, this vascular network is part of one of the most phenomenal irrigation networks imaginable. Emanating from a muscular pump, the heart, these vessels form an extensive system of tubular roadways that carry nourishing blood away from the heart toward the tissues, then make a U-turn through small permeable vessels at the tissue level. Here, life-supporting exchanges occur between the blood and surrounding cells, including O_2 and nutrient delivery and waste pickup. The waste-laden fluid then flows back to the heart through a set of return vessels with routes that parallel those of the delivery vessels. This circular pattern of flow to and from the heart constitutes the vascular component of the cardiovascular system. This system of tubular roadways is so incredibly extensive that, if all of the individual vessels were placed end-to-end, they would extend about 75,000 miles, roughly three times the circumference of the earth. Furthermore, the small, permeable vessels that supply the tissues are so intimately distributed among the trillions of body cells that even the most minor tissue damage leads to the rupture of small vessels. This chapter focuses on the structure and functions of the various types of blood vessels and how they work together to form the major circulatory routes of the human body. •

Did you ever wonder why untreated hypertension has so many damaging effects?

CONTENTS AT A GLANCE

14.1 ANATOMY OF BLOOD VESSELS

OBJECTIVES

- Describe the basic structure of a blood vessel.
- Contrast the structure of arteries, arterioles, capillaries, venules, and veins.
- Compare the functions of arteries, arterioles, capillaries, venules, and veins.
- Distinguish between muscular and distributing arteries.
- Describe the types of capillaries and their functions.

The five main types of blood vessels are arteries, arterioles, capillaries, venules, and veins. **Arteries** (AR-ter-ēz; *ar-*=air; *ter-*=to carry) carry blood *away from the heart* to other organs. A learning trick to help you remember this is **a**rtery=**a**way. Large, elastic arteries leave the heart and divide into medium-sized, muscular arteries that branch out into the various regions of the body. Medium-sized arteries then divide into small arteries, which in turn divide into still smaller arteries called **arterioles** (ar-TER-ē-ōls). As the arterioles enter a tissue, they branch into numerous tiny vessels called **capillaries** (KAP-i-lar′-ēz=hairlike). The thin walls of capillaries allow exchange of substances between the blood and body tissues. Groups of capillaries within a tissue reunite to form small veins called **venules** (VEN-ūls=little veins). These in turn merge to form progressively larger blood vessels called veins. **Veins** (VĀNZ) are the blood vessels that convey blood from the tissues *back to the heart.*

Angiogenesis (an′-jē-ō-JEN-e-sis; *angio-*=blood vessel; *genesis*=production) refers to the growth of new blood vessels. It is an important process in embryonic and fetal development, and in postnatal life serves important functions such as wound healing, formation of a new uterine lining after menstruation, formation of the corpus luteum after ovulation, and development of blood vessels around obstructed arteries in the coronary circulation. Several proteins (peptides) are known to promote and inhibit angiogenesis.

CLINICAL CONNECTION | *Angiogenesis and Disease*

Clinically, angiogenesis is important because cells of a **malignant tumor** secrete proteins called *tumor angiogenesis factors* (*TAFs*) that stimulate blood vessel growth to provide nourishment for the tumor cells. Scientists are seeking chemicals that would inhibit angiogenesis and thus stop the growth of tumors. In **diabetic retinopathy** (ret′-i-NOP-a-thē), angiogenesis may be important in the development of blood vessels that actually cause blindness, so finding inhibitors of angiogenesis may also prevent the blindness associated with diabetes. •

Basic Structure of a Blood Vessel

The wall of a blood vessel consists of three layers, or tunics, of different tissues: an endothelial inner lining, a middle layer consisting of smooth muscle and elastic connective tissue, and a connective tissue outer covering. (Recall that the term *endothelium* refers to simple squamous epithelium that lines the cardiovascular and lymphatic systems.) From innermost to outermost, the three structural layers of a generalized blood vessel are the tunica interna (intima), tunica media, and tunica externa (adventitia) (Figure 14.1). Subtle modifications of this basic design account for the five different types of blood vessels and the structural and functional differences among the various vessel types. It will be easier to learn the structures of the various vessels if you remember that structural variations are correlated to differences in function throughout the cardiovascular system.

Tunica Interna (Intima)

The **tunica interna (intima)** (TOO-ni-ka; *tunic*=garment or coat; *interna* or *intima*=innermost) forms the inner lining of a blood vessel and is in direct contact with the blood as it flows through the *lumen* (LOO-men), or interior opening, of the vessel (Figure 14.1a, b). Although the tunica interna has multiple parts, its tissue components contribute minimally to the thickness of the vessel wall. Its innermost layer is called **endothelium**, which is continuous with the endocardial lining of the heart. The endothelium is a thin layer of flattened cells that lines the inner surface of the entire cardiovascular system (heart and blood vessels). Until recently, endothelial cells were regarded as little more than a passive barrier between the blood and the remainder of the vessel wall. It is now known that endothelial cells are active participants in a variety of vessel-related activities, including physical influences on blood flow, secretion of locally acting chemical mediators that influence the contractile state of the vessel's overlying smooth muscle, and assistance with capillary permeability. In addition, their smooth luminal surface facilitates efficient blood flow by reducing surface friction.

The second component of the tunica interna is a *basement membrane* deep to the endothelium. It provides a physical support base for the overlying epithelial layer. Its framework of collagen fibers affords the basement membrane significant tensile strength, yet its properties also provide resilience for stretching and recoil. The basal lamina anchors the endothelium to the underlying connective tissue while also regulating molecular movement. It appears to play an important role in guiding cell movements during tissue repair of blood vessel walls. The outermost part of the tunica

Figure 14.1 Comparative structure of blood vessels. The size of the capillary in (c) is enlarged relative to the artery (a) and vein (b).

Arteries carry blood from the heart to tissues; veins carry blood from tissues to the heart.

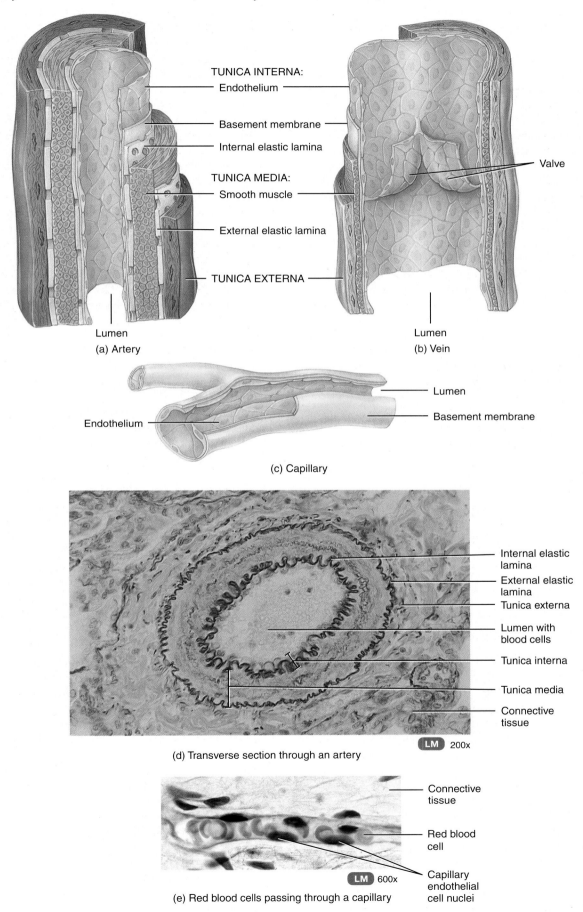

TUNICA INTERNA:
Endothelium
Basement membrane
Internal elastic lamina
TUNICA MEDIA:
Smooth muscle
External elastic lamina
TUNICA EXTERNA
Lumen
(a) Artery

Valve
Lumen
(b) Vein

Lumen
Endothelium
Basement membrane
(c) Capillary

Internal elastic lamina
External elastic lamina
Tunica externa
Lumen with blood cells
Tunica interna
Tunica media
Connective tissue
LM 200x
(d) Transverse section through an artery

Connective tissue
Red blood cell
Capillary endothelial cell nuclei
LM 600x
(e) Red blood cells passing through a capillary

Which vessel—the femoral artery or the femoral vein—has a thicker wall? Which has a wider lumen?

interna, which forms the boundary between the tunica interna and tunica media, is the **internal elastic lamina** (*lamina*=thin plate). The internal elastic lamina is a thin sheet of elastic fibers with a variable number of window-like openings that give it the look of Swiss cheese. These openings facilitate diffusion of materials through the tunica interna to the thicker tunica media.

Tunica Media

The **tunica media** (*media*=middle) is a layer composed of smooth muscle and connective tissue. This layer displays the greatest variation among the different vessel types (Figure 14.1a, b). In most vessels, it is a relatively thick layer comprised mainly of smooth muscle cells and substantial amounts of elastic fibers. The primary role of the smooth muscle cells, which extend circularly around the lumen like a ring encircles your finger, is to regulate the diameter of the lumen wall. As you will learn in more detail shortly, the rate of blood flow through different parts of the vascular network is regulated by the extent of smooth muscle contraction in the walls of particular vessels. Furthermore, the extent of smooth muscle contraction in particular vessel types is crucial in the regulation of blood pressure.

In addition to regulating blood flow and blood pressure, smooth muscle contracts when a small artery or arteriole is damaged (*vascular spasm*) to help limit loss of blood through the injured vessel. Smooth muscle cells also produce the elastic fibers within the tunica media that allow the vessels to stretch and recoil under the applied pressure of the blood. A less prominent network of elastic fibers, the **external elastic lamina**, forms the outer part of the tunica media and separates the tunica media from the outer tunica externa.

Sympathetic fibers of the autonomic nervous system innervate the smooth muscle of blood vessels. An increase in sympathetic stimulation typically stimulates the smooth muscle to contract, squeezing the vessel wall and narrowing the lumen. Such a decrease in the diameter of the lumen of a blood vessel is called **vasoconstriction** (vā′-sō-kon-STRIK-shun). In contrast, when sympathetic stimulation decreases, in the presence of certain chemicals (such as nitric oxide, H^+, and lactic acid), or in response to the pressure of blood, smooth muscle fibers relax. The resulting increase in lumen diameter is called **vasodilation** (vā′-sō-dī-LA-shun).

Tunica Externa

The outer covering of a blood vessel, the **tunica externa** (*externa*=outermost), consists of elastic and collagenous fibers (Figure 14.1a, b). It ranges in size from a thin connective tissue wrapping to the thickest layer of the blood vessel. The tunica externa contains numerous nerves and, especially in larger vessels, tiny blood vessels that supply the tissue of the vessel wall. These small vessels that supply blood to the tissues of the vessel are called **vasa vasorum** (VA-sa va-SŌ-rum; *vas*=vessel), or vessels to the vessels. They are easily seen on large vessels such as the aorta. In addition to the important role of supplying the vessel wall with nerves and vasa vasorum, the tunica externa helps anchor the vessels to surrounding tissues.

Arteries

Because arteries were found empty at death, in ancient times they were thought to contain only air. Like other blood vessels, the wall of an artery has three layers, but the tunica media may be thicker or more elastic as outlined in the following discussion (Figure 14.1a). Due to their plentiful elastic fibers, arteries normally have high

compliance, which means that their walls stretch easily or expand without tearing in response to a small increase in pressure.

Elastic Arteries

Elastic arteries are the largest arteries in the body, ranging from the garden hose-sized aorta and pulmonary trunk to the finger-sized branches of the aorta. They have the largest diameter among arteries, but their vessel walls (approximately one-tenth of the vessel's total diameter) are relatively thin compared to the overall size of the vessel. These vessels are characterized by well-defined internal and external elastic laminae, along with a thick tunica media that is dominated by elastic fibers, the **elastic lamellae** (la-MEL-ē=little plate). The elastic lamellae give the wall a yellowish tint. Elastic arteries include the two major trunks that exit the heart (the aorta and the pulmonary trunk), along with the aorta's major branches, including the brachiocephalic, subclavian, common carotid, and common iliac arteries (see Figure 14.6). Elastic arteries perform an important function: They help propel blood onward while the ventricles are relaxing. As blood is ejected from the heart into elastic arteries, their walls stretch, easily accommodating the surge of blood. As they stretch, the elastic fibers momentarily store mechanical energy, functioning as a **pressure reservoir** (REZ-er-vwar). Then, the elastic fibers recoil and convert the stored (potential) energy in the vessel into kinetic energy of the blood. Thus, blood continues to move through the arteries even while the ventricles are relaxed. Because they conduct blood from the heart to medium-sized, more muscular arteries, elastic arteries also are called *conducting arteries*.

Muscular Arteries

Medium-sized arteries are called **muscular arteries** because their tunica media contains more smooth muscle and fewer elastic fibers than elastic arteries. The large amount of smooth muscle, approximately three-quarters of the total mass, makes the walls of muscular arteries relatively thick. Thus, muscular arteries are capable of greater vasoconstriction and vasodilation to adjust the rate of blood flow. Muscular arteries have a well-defined internal elastic lamina, but a thin external elastic lamina. These two elastic laminae form the inner and outer boundaries of the muscular tunica media. In large arteries, the thick tunica media can have as many as 40 layers of circumferentially arranged smooth muscle cells; in smaller arteries there are as few as three layers.

Muscular arteries span a range of sizes from the pencil-sized femoral and axillary arteries to the string-sized arteries that enter organs, which measure as little as 0.5 mm ($\frac{1}{64}$ inch) in diameter. Compared to elastic arteries, the vessel wall of muscular arteries comprises a larger percentage (25 percent) of the total vessel diameter. Because the muscular arteries continue to branch and ultimately distribute blood to each of the various organs, they are called **distributing arteries**. Examples include the brachial artery in the arm and the radial artery in the forearm (see Figure 14.6).

The tunica externa is often thicker than the tunica media in muscular arteries. This outer layer contains fibroblasts, collagen fibers, and elastic fibers, all of which are oriented longitudinally. The loose structure of this layer permits changes in the diameter of the vessel to take place but also prevents shortening or retraction of the vessel when it is cut.

Because of the reduced amount of elastic tissue in the walls of muscular arteries, these vessels do not have the ability to recoil and help propel the blood like the elastic arteries. The ability of the muscle to contract and maintain a state of partial contraction is referred to as *vascular tone*. Vascular tone stiffens the vessel wall and is important in maintaining vessel pressure and efficient blood flow.

Anastomoses

Most tissues of the body receive blood from more than one artery. The union of the branches of two or more arteries supplying the same body region is called an **anastomosis** (a-nas-tō-MŌ-sis= connecting; plural is *anastomoses*) (see Figure 14.8c). Anastomoses between arteries provide alternative routes for blood to reach a tissue or organ. If blood flow stops momentarily when normal movements compress a vessel, or if a vessel is blocked by disease, injury, or surgery, then circulation to a part of the body can continue. The alternative route of blood flow to a body part through an anastomosis is known as **collateral circulation.** Anastomoses may also occur between veins and between arterioles and venules. Arteries that do not anastomose are known as **end arteries.** Obstruction of an end artery interrupts the blood supply to a whole segment of an organ, producing necrosis (death) of that segment. Alternative blood routes may also be provided by other, unconnected (non-anastomosing) vessels that supply the same region of the body.

Arterioles

Literally meaning "small arteries," arterioles are abundant microscopic vessels that regulate the flow of blood into the capillary networks of the body's tissues (Figure 14.2). The approximately 400 million arterioles have diameters that range in size from 15 μm to 30 μm. The wall thickness of arterioles is one-half of the total vessel diameter.

Arterioles have a thin tunica interna with a thin internal elastic lamina containing small pores that disappear at the terminal end. The tunica media consists of one to two layers of smooth muscle cells having a circular (rather than longitudinal) orientation in the vessel wall. The terminal end of the arteriole, the region called the **metarteriole** (met′-ar-TĒR-ē-ōl; *meta*=after), tapers toward the capillary junction. At the metarteriole–capillary junction, the most distal muscle cell forms the **precapillary sphincter** (SFINGH-ter=to bind tight), which monitors the blood flow into the capillary; the other muscle cells in the arteriole regulate *resistance* (opposition to blood flow) (Figure 14.2).

Since arterioles play a key role in regulating blood flow from arteries into capillaries by regulating resistance, they are known as **resistance vessels.** In a blood vessel, resistance is due mainly to friction between blood and the inner walls of blood vessels. When blood vessel diameter is smaller, the friction is greater, so there is more resistance. Contraction of arteriolar smooth muscle causes vasoconstriction, which further increases resistance and decreases blood flow into capillaries supplied by that arteriole. By contrast, relaxation of arteriolar smooth muscle causes vasodilation, which decreases resistance and increases blood flow into capillaries. A change in arteriole diameter can also affect blood pressure. Vasoconstriction of arterioles increases blood pressure, and vasodilation of arterioles decreases blood pressure.

The tunica externa of an arteriole consists of areolar connective tissue containing abundant unmyelinated sympathetic nerves. This sympathetic nerve supply, along with the actions of local chemical mediators, controls the diameter of arterioles and thus variations in the rate of blood flow and resistance through these vessels.

Capillaries

Capillaries, the smallest of blood vessels, have diameters of 5–10 μm, and form the "U-turns" that connect the arterial outflow to the venous return (Figure 14.2). Since red blood cells have a diameter of 8 μm, they must often fold upon themselves in order to pass single file through the lumens of these vessels (see

Figure 14.2 Arterioles, capillaries, and venules.

 Arterioles regulate blood flow into capillaries, where nutrients, gases, and wastes are exchanged between the blood and interstitial fluid.

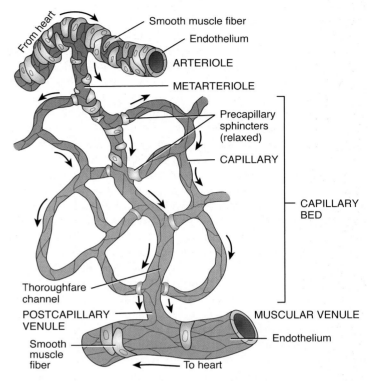

(a) Sphincters relaxed: blood flowing through capillaries

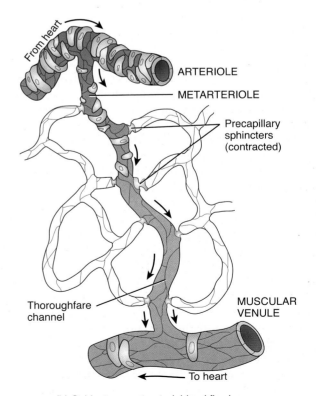

(b) Sphincters contracted: blood flowing through thoroughfare channel

Why do metabolically active tissues have extensive capillary networks?

Figure 14.1e). Approximately 20 billion in number, capillaries form an extensive network of short (hundreds of μm in length), branched, interconnecting vessels that course among and make contact with the individual cells of the body. The flow of blood from a metarteriole through capillaries and into a **postcapillary venule** (a venule that receives blood from a capillary) is called the **microcirculation** (*micro*=small) of the body.

Capillaries are found near almost every cell in the body, but their number varies with the metabolic activity of the tissue they serve. Body tissues with high metabolic requirements, such as muscles, the brain, the liver, the kidneys, and the nervous system, use more O_2 and nutrients and thus have extensive capillary networks. Tissues with lower metabolic requirements, such as tendons and ligaments, contain fewer capillaries. Capillaries are absent in a few tissues, such as all covering and lining epithelia, the cornea and lens of the eye, and cartilage.

Because the primary function of capillaries is the exchange of substances between the blood and interstitial fluid, these thin-walled vessels are referred to as **exchange vessels.** The structure of capillaries is well suited to this function because they lack both a tunica media and a tunica externa. Because capillary walls are composed of only a single layer of endothelial cells (see Figure 14.1e) and a basement membrane, a substance in the blood must pass through just one cell layer to reach the interstitial fluid and tissue cells. Exchange of materials occurs only through the walls of capillaries and the beginning of venules; the walls of arteries, arterioles, most venules, and veins present too thick a barrier. Capillaries form extensive branching networks that increase the surface area available for rapid exchange of materials. In most tissues, blood flows through only a small part of the capillary network when metabolic needs are low. However, when a tissue is active, such as contracting muscle, the entire capillary network fills with blood.

Throughout the body, capillaries function as part of a **capillary bed** (Figure 14.2a), a network of 10–100 capillaries that arises from a single metarteriole. In most parts of the body, blood can flow through a capillary network from an arteriole into a venule as follows:

1. ***Capillaries.*** In this route, blood flows from an arteriole into capillaries and then into postcapillary venules. As noted earlier, there are rings of smooth muscle fibers called precapillary sphincters at the junctions between the metarteriole and the capillaries that control the flow of blood through the capillaries. When the precapillary sphincters are relaxed (open), blood flows into the capillaries (Figure 14.2a); when precapillary sphincters contract (close or partially close), blood flow through the capillaries ceases or decreases (Figure 14.2b). Typically, blood flows intermittently through capillaries due to alternating contraction and relaxation of the smooth muscle of metarterioles and the precapillary sphincters. This intermittent contraction and relaxation, which may occur 5 to 10 times per minute, is called **vasomotion** (vā′-sō′-MŌ-shun). In part, vasomotion is due to chemicals released by the endothelial cells; nitric oxide is one example. At any given time, blood flows through only about 25 percent of the capillaries.

2. ***Thoroughfare channel.*** The proximal end of a metarteriole is surrounded by scattered smooth muscle fibers whose contraction and relaxation help regulate blood flow. The distal end of the vessel, which has no smooth muscle and resembles a capillary, is called a **thoroughfare channel** (Figure 14.2). Such a channel provides a direct route for blood from an arteriole to a venule, thus bypassing capillaries.

The body contains three different types of capillaries: continuous capillaries, fenestrated capillaries, and sinusoids (Figure 14.3). Most capillaries are **continuous capillaries,** in which the plasma membranes of endothelial cells form a continuous tube that is interrupted only by **intercellular clefts,** gaps between neighboring endothelial cells (Figure 14.3a). Continuous capillaries are found in the central nervous system, lungs, skin, skeletal and smooth muscle, and connective tissues.

Figure 14.3 Types of capillaries. The capillaries are shown in transverse sections.

 Capillaries are microscopic blood vessels that connect arterioles and venules.

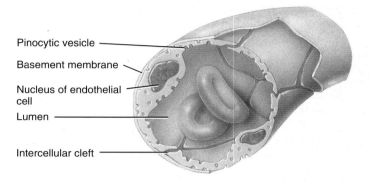

Pinocytic vesicle
Basement membrane
Nucleus of endothelial cell
Lumen
Intercellular cleft

(a) Continuous capillary formed by endothelial cells

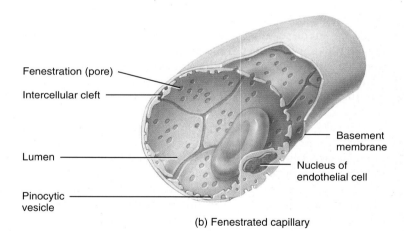

Fenestration (pore)
Intercellular cleft
Lumen
Pinocytic vesicle
Basement membrane
Nucleus of endothelial cell

(b) Fenestrated capillary

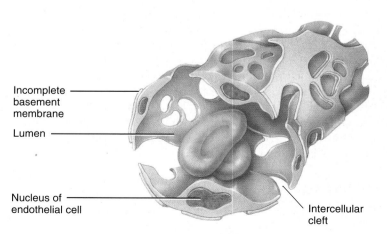

Incomplete basement membrane
Lumen
Nucleus of endothelial cell
Intercellular cleft

(c) Sinusoid

 How do materials cross capillary walls?

Other capillaries of the body are **fenestrated capillaries** (fen′-es-TRĀ-ted; *fenestr-*=window). The plasma membranes of the endothelial cells in these capillaries have many **fenestrations** (fen′-es-TRĀ-shuns), small pores (holes) ranging from 70 to 100 nm in diameter (Figure 14.3b). Fenestrated capillaries are found in the kidneys, villi of the small intestine, choroid plexuses of the ventricles in the brain, ciliary processes of the eyes, and most endocrine glands.

Sinusoids (SĪ-nū-soyds; *sinus*=curve) are wider and more winding than other capillaries. Their endothelial cells may have unusually large fenestrations. In addition to having an incomplete or absent basement membrane (Figure 14.3c), sinusoids have very large intercellular clefts that allow proteins and in some cases even blood cells to pass from a tissue into the bloodstream. For example, newly formed blood cells enter the bloodstream through the sinusoids of red bone marrow. In addition, sinusoids contain specialized lining cells that are adapted to the function of the tissue. For example, sinusoids in the liver contain phagocytic cells that remove bacteria and other debris from the blood. The spleen, anterior pituitary, parathyroid, and adrenal glands also contain sinusoids.

Usually, blood passes from the heart and then in sequence through arteries, arterioles, capillaries, venules, and veins and then back to the heart. In some parts of the body, however, blood passes from one capillary network into another through a vessel called a *portal vessel*. Such a circulation of blood is called a **portal system**. The name of the portal system gives the location of the second capillary network. As you will see later in the chapter (Section 14.2), there are portal systems associated with the liver (hepatic portal circulation), the pituitary gland (hypophyseal portal system), and the kidney (renal portal system).

Venules

Unlike their thick-walled arterial counterparts, venules and veins have thin walls that do not readily maintain their shape. Venules drain the capillary blood and begin the return flow of blood back toward the heart (see Figure 14.2). Because they carry blood toward the heart, veins are referred to as *afferent vessels* (*af-*=toward; *ferent*=carried).

As noted earlier, venules that initially receive blood from capillaries are called postcapillary venules. They are the smallest venules, measuring 10 μm to 50 μm in diameter, and have loosely organized intercellular junctions. Because they are the weakest endothelial contacts encountered along the entire vascular tree, venules are very porous. They function as significant sites of exchange of nutrients and wastes and white blood cell emigration, and for this reason form part of the microcirculatory exchange unit along with the capillaries.

As the postcapillary venules move away from the capillaries, they increase in size and acquire one or two layers of circularly arranged smooth muscle cells. These **muscular venules** (50 μm to 200 μm) have thicker walls across which exchanges with the interstitial fluid can no longer occur. The thin walls of the postcapillary and muscular venules are the most distensible elements of the vascular system; this allows them to expand and serve as excellent reservoirs for accumulating large volumes of blood. Blood volume increases of 360 percent have been measured in the postcapillary and muscular venules.

Veins

While veins do show structural changes as they increase in size from small to medium to large, the structural changes are not as distinct as they are in arteries. Veins, in general, have very thin walls relative to their total diameter (average thickness is less than one-tenth of the vessel diameter). They range in size from 0.5 mm in diameter for small veins to 3 cm in the large superior and inferior venae cavae entering the heart.

Although veins are composed of essentially the same three layers as arteries, the relative thicknesses of the layers are different. The tunica interna of veins is thinner than that of arteries; the tunica media of veins is much thinner than in arteries, with relatively little smooth muscle and elastic fibers. The tunica externa of a vein is its thickest layer and consists of collagen and elastic fibers. Veins lack the external or internal elastic laminae found in arteries (see Figure 14.1b). They are distensible enough to adapt to variations in the volume and pressure of blood passing through them, but are not designed to withstand high pressure. The lumen of a vein is larger than that of a comparable artery, and veins often appear collapsed (flattened) when sectioned.

The pumping action of the heart is a major factor in moving venous blood back to the heart. The contraction of skeletal muscles in the free lower limbs also helps boost venous return to the heart (Figure 14.4). The average blood pressure in veins is considerably lower than in arteries. Because of the difference in pressure, it is easy to tell whether a cut vessel is an artery or a vein. Blood leaves a cut vein in an even, slow flow but spurts rapidly from a cut artery. Most of the structural differences between arteries and veins reflect this pressure difference. For example, the walls of veins are not as strong as those of arteries.

Many veins, especially those in the limbs, also contain **valves**, thin folds of tunica interna that form flaplike cusps. The valve cusps project into the lumen, pointing toward the heart (see Figures 14.1b and 14.4a and b). The low blood pressure in veins allows the flow of blood returning to the heart to slow and even back up; the valves aid in venous return by preventing backflow.

A **vascular (venous) sinus** is a vein with a thin endothelial wall that has no smooth muscle to alter its diameter. In a vascular sinus, the supporting role played by the tunica media and tunica externa in other vessels is taken on by the dense connective tissue surrounding the sinus. For example, dural venous sinuses, which are supported by the dura mater, convey deoxygenated blood from the brain to the heart. Another example of a vascular sinus is the coronary sinus of the heart (see Figure 13.3c).

Veins are more numerous than arteries. Some veins are paired and accompany medium- to small-sized muscular arteries. These double sets of veins escort the arteries and connect with one another via venous channels called **anastomotic veins** (a-nas′-tō- MŌT-ik). The anastomotic veins cross the accompanying artery to form ladder-like rungs between the paired veins (see Figure 14.12c). The greatest number of paired veins occurs within the free limbs. The subcutaneous layer deep to the skin is another source of veins. These veins, called **superficial veins**, course through the subcutaneous layer unaccompanied by parallel arteries. Along their course, the superficial veins form small connections (anastomoses) with the **deep veins** that travel below the fascia between the skeletal muscles. These connections allow communication between the deep and superficial blood flows. The amount of blood flow through superficial veins varies from location to location within the body. In the free upper limb, the superficial veins are much larger than the deep veins and serve as the major pathways from the capillaries of the free upper limb back to the heart. In the free lower limb, the opposite is true; the deep veins serve as the principal return pathways. In fact, one-way valves in small anastomosing vessels allow blood to pass from the superficial veins to the deep veins, but prevent the blood from passing in the reverse direction. This design has important implications in the development of varicose veins.

In some individuals the superficial veins can be seen as blue-colored tubes passing under the skin. While the venous blood is a

Figure 14.4 **Role of skeletal muscle contractions and venous valves in returning blood to the heart.** (a) When skeletal muscles contract, the proximal valve opens, and blood is forced toward the heart. (b) Sections of a venous valve.

🔑 Venous return depends on the pumping action of the heart, skeletal muscle contractions, and valves in veins.

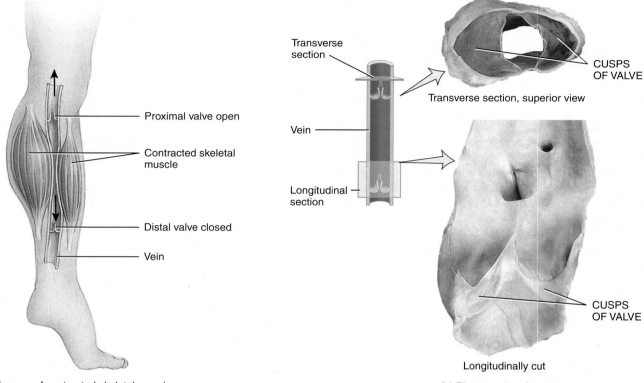

(a) Diagram of contracted skeletal muscles

(b) Photographs of a valve in a vein

 Why are valves more important in arm veins and leg veins than in neck veins?

CLINICAL CONNECTION | *Varicose Veins*

Leaky venous valves can cause veins to become dilated and twisted in appearance, a condition called **varicose veins** (VAR-i-kōs) or *varices* [VAR-i-sēz; *varic-*=a swollen vein; singular is *varix* (VAR-iks)]. The condition may occur in the veins of almost any body part, but it is most common in the esophagus, anal canal, and superficial veins of the lower limbs. Varicose veins in the lower limbs can range from cosmetic problems to serious medical conditions. The valvular defect may be congenital, or it may result from mechanical stress (pregnancy or prolonged standing) or aging. The leaking venous valves allow the backflow of blood from the deep veins to the less efficient superficial veins, where the blood pools. This creates pressure that distends the vein and allows fluid to leak into surrounding tissue. As a result, the affected vein and the tissue around it may become inflamed and painfully tender. Veins close to the surface of the legs, especially the saphenous vein (see Figure 14.14), are highly susceptible to varicosities; deeper veins are not as vulnerable because surrounding skeletal muscles prevent their walls from stretching excessively. Varicose veins in the anal canal are referred to as *hemorrhoids* (HEM-o-roids). Esophageal varices result from dilated veins in the walls of the lower part of the esophagus and sometimes the upper part of the stomach. Bleeding esophageal varices are life-threatening and are usually a result of chronic liver disease.

Several treatment options are available for varicose veins in the lower limbs. *Elastic stockings* (support hose) may be used for individuals with mild symptoms or for whom other options are not recommended. *Sclerotherapy* (skle-rō-THER-a-pē) involves injection of a solution into varicose veins that damages the tunica interna by producing a harmless superficial thrombophlebitis (inflammation

involving a blood clot). Healing of the damaged part leads to scar formation that occludes the vein. *Radiofrequency endovenous occlusion* involves the application of radiofrequency energy to heat up and close off varicose veins. *Laser occlusion* (ō-KLOO-zhun) uses laser therapy to shut down veins. In a surgical procedure called *stripping*, veins may be removed. In this more invasive procedure, a flexible wire is threaded through the vein and then pulled out to strip (remove) it from the body. •

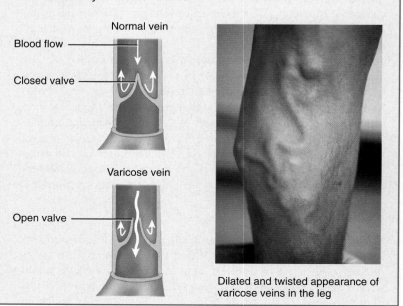

Dilated and twisted appearance of varicose veins in the leg

deep dark red, the veins appear blue because their thin walls and the tissues of the skin absorb the red-light wavelengths, allowing the blue light to pass through the surface to our eyes, where we see them as blue.

Blood Distribution

The largest portion of your blood volume at rest—about 64 percent—is in systemic veins and venules. Systemic arteries hold about 13 percent of the blood volume, systemic capillaries hold about 7 percent, pulmonary blood vessels hold about 9 percent, and the heart holds about 7 percent. Because systemic veins and venules contain a large percentage of the blood volume, they function as **blood reservoirs** from which blood can be diverted

quickly if the need arises. For example, when there is increased muscular activity, the cardiovascular center in the brain stem sends more sympathetic impulses to veins. The result is *venoconstriction* (vasoconstriction in veins), which reduces the volume of blood in reservoirs and allows a greater blood volume to flow to skeletal muscles, where it is needed most. A similar mechanism operates in cases of hemorrhage, when blood volume and pressure decrease; in this case, venoconstriction helps counteract the drop in blood pressure. Among the principal blood reservoirs are the veins of the abdominal organs (especially the liver and spleen) and the veins of the skin.

A summary of the distinguishing features of blood vessels is presented in Table 14.1.

TABLE 14.1

Distinguishing Features of Blood Vessels

BLOOD VESSEL	SIZE	TUNICA INTERNA	TUNICA MEDIA	TUNICA EXTERNA	FUNCTION
Elastic arteries	Largest arteries in the body	Well-defined internal elastic lamina	Thick and dominated by elastic fibers; well-defined external elastic lamina	Thinner than tunica media	Conduct blood from the heart to the muscular arteries
Muscular arteries	Medium-sized arteries	Well-defined internal elastic lamina	Thick and dominated by smooth muscle; thin external elastic lamina	Thicker than tunica media	Distribute blood to arterioles
Arterioles	Microscopic (15–30 μm in diameter)	Thin with a fenestrated internal elastic lamina that disappears distally	One or two layers of circularly oriented smooth muscle; distal-most smooth muscle cell forms a precapillary sphincter	Loose collagenous connective tissue and sympathetic nerves	Deliver blood to capillaries and help regulate blood flow to capillaries
Capillaries	Microscopic; smallest blood vessels (5–10 μm in diameter)	Endothelium and basement membrane	None	None	Permit exchange of nutrients and wastes between blood and interstitial fluid; distribute blood to postcapillary venules
Postcapillary venules	Microscopic (10–50 μm in diameter)	Endothelium and basement membrane	None	Sparse	Pass blood into muscular venules; permit exchange of nutrients and wastes between blood and interstitial fluid and function in white blood cell emigration
Muscular venules	Microscopic (50–200 μm in diameter)	Endothelium and basement membrane	One or two layers of circularly oriented smooth muscle	Sparse	Pass blood into veins; reservoirs for accumulating large volumes of blood (along with postcapillary venules)
Veins	Range from 0.5 μm to 3 cm in diameter	Endothelium and basement membrane; no internal elastic lamina; contain valves; lumen is much larger than in accompanying artery	Much thinner than in arteries; no external elastic lamina	Thickest of three layers	Return blood to the heart, facilitated by valves in veins in limbs

✔ CHECKPOINT

1. Describe the basic structure of a blood vessel.
2. Discuss the importance of elastic fibers and smooth muscle in the tunica media of arteries.
3. Distinguish between elastic and muscular arteries in terms of location, histology, and function.
4. Describe the relationship between anastomoses and collateral circulation.
5. Why are arterioles called resistance vessels?
6. Describe the structural features of capillaries that allow exchange of materials between blood and body cells.
7. What are the main structural and functional differences between arteries and veins?
8. Why are systemic veins and venules called blood reservoirs?

14.2 CIRCULATORY ROUTES

● OBJECTIVES

• Define the systemic circulation and explain its function.
• Describe the importance of the hepatic portal circulation.
• Explain why the pulmonary circulation is important.
• Describe the fates of the fetal structures once postnatal circulation begins.

Arteries, arterioles, capillaries, venules, and veins are organized into **circulatory routes** that deliver blood throughout the body. Now that you understand the structures of each of these vessel types, we can now look at the basic routes the blood takes as it is transported throughout the body.

Figure 14.5 shows the circulatory routes for blood flow. The routes are parallel; that is, in most cases a portion of the cardiac output flows separately to each tissue of the body. Thus, each organ receives its own supply of freshly oxygenated blood. The two basic postnatal (after birth) routes for blood flow are the systemic circulation and the pulmonary circulation. The **systemic circulation** includes all arteries and arterioles that carry oxygenated blood from the left ventricle to systemic capillaries, plus the veins and venules that return deoxygenated blood to the right atrium after flowing through capillaries in the body organs. Blood leaving the aorta and flowing through the systemic arteries is a bright red color. As it moves through capillaries, it loses some of its oxygen and picks up carbon dioxide, so that blood in systemic veins is dark red.

Some of the subdivisions of the systemic circulation include the **coronary (cardiac) circulation** (see Figure 13.8), which supplies the myocardium of the heart; **cerebral circulation**, which supplies the brain (see Figure 14.7c); and the **hepatic portal circulation** (he-PAT-ik; *hepat-*=liver), which extends from the gastrointestinal tract to the liver (see Figure 14.15). The nutrient arteries to the lungs, such as the bronchial arteries, are also part of the systemic circulation.

When blood returns to the heart from the systemic route, it is pumped out of the right ventricle through the **pulmonary circulation** (PUL-mo-ner′-ē; *pulmo*=lung) to the lungs (see Figure 14.16). In capillaries of the air sacs (alveoli) of the lungs, the blood loses some of its carbon dioxide and takes on oxygen. Bright red again, it returns to the left atrium of the heart and reenters the systemic circulation as it is pumped out by the left ventricle.

Another major route—the **fetal circulation**—exists only in the fetus and contains special structures that allow the developing fetus to exchange materials with its mother (see Figure 14.17).

Systemic Circulation

The systemic circulation carries oxygen and nutrients to body tissues and removes carbon dioxide and other wastes and heat from the tissues. All systemic arteries branch from the **aorta**. Deoxygenated blood returns to the heart through the systemic veins. All the veins of the systemic circulation drain into the **superior vena cava** (KĀ-va), **inferior vena cava**, or **coronary sinus**, which in turn empty into the right atrium.

The principal arteries and veins of the systemic circulation are described and illustrated in Exhibits 14.A through 14.L and Figures 14.6 through 14.14 to assist you in learning their names. The blood vessels are organized in the exhibits according to regions of the body. Figure 14.6a shows an overview of the major arteries, and Figure 14.10a shows an overview of the major veins. As you study the various blood vessels in the exhibits, refer to these two figures to see the relationships of the blood vessels under consideration to other regions of the body.

Each of the exhibits contains the following information:

• *An overview.* This information provides a general orientation to the blood vessels under consideration, with emphasis on how the blood vessels are organized into various regions as well as distinguishing and/or interesting features of the blood vessels.

• *Blood vessel names.* Students often have difficulty with the pronunciations and meanings of blood vessels' names. To learn them more easily, study the phonetic pronunciations and word origins that indicate how blood vessels get their names.

• *Region supplied or drained.* For each artery listed, there is a description of the parts of the body that receive blood from the vessel. For each vein listed, there is a description of the parts of the body that are drained by the vessel.

• *Illustrations and photographs.* The figures that accompany the exhibits contain several elements. Many include illustrations of the blood vessels under consideration and flow diagrams to indicate the patterns of blood distribution or drainage. Cadaver photographs are also included in selected exhibits to provide more realistic views of the blood vessels.

Figure 14.5 Schematic of circulatory routes. Long black arrows indicate the systemic circulation (detailed in Exhibits 14.A–14.L), short blue arrows identify the pulmonary circulation (detailed in Figure 14.16), and red arrows highlight the hepatic portal circulation (detailed in Figure 14.15). Refer to Figure 13.8 for details of the coronary circulation, and to Figure 14.17 for details of the fetal circulation.

Blood vessels are organized into various routes that deliver blood to tissues of the body.

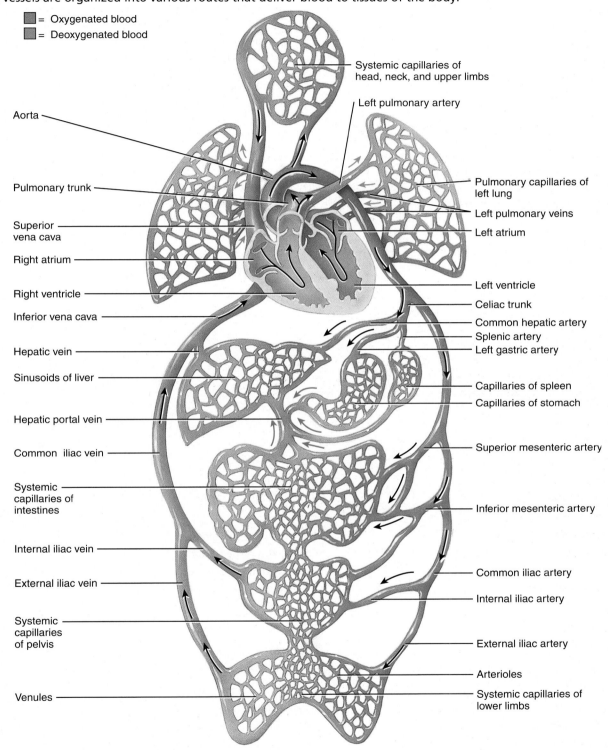

= Oxygenated blood
= Deoxygenated blood

Systemic capillaries of head, neck, and upper limbs

Left pulmonary artery

Aorta

Pulmonary trunk

Pulmonary capillaries of left lung

Left pulmonary veins

Left atrium

Superior vena cava

Right atrium

Left ventricle

Celiac trunk

Right ventricle

Common hepatic artery

Inferior vena cava

Splenic artery

Left gastric artery

Hepatic vein

Sinusoids of liver

Capillaries of spleen

Capillaries of stomach

Hepatic portal vein

Common iliac vein

Superior mesenteric artery

Systemic capillaries of intestines

Inferior mesenteric artery

Internal iliac vein

External iliac vein

Common iliac artery

Internal iliac artery

Systemic capillaries of pelvis

External iliac artery

Arterioles

Venules

Systemic capillaries of lower limbs

What are the body's two main circulatory routes?

EXHIBIT 14.A The Aorta and Its Branches *(Figure 11.6)*

● OBJECTIVE

• Identify the four principal divisions of the aorta, and locate the major arterial branches arising from each division.

The *aorta* (ā-OR-ta=to lift up) is the largest artery of the body, with a diameter of 2–3 cm (about 1 in.). Its four principal divisions are the ascending aorta, arch of the aorta, thoracic aorta, and abdominal aorta (Figure 14.6). The portion of the aorta that emerges from the left ventricle posterior to the pulmonary trunk is the **ascending aorta** (see Exhibit 14.B). The beginning of the aorta contains the aortic valve (see Figure 13.4a). The ascending aorta gives off two coronary arteries that supply the myocardium of the heart. Then the ascending aorta arches to the left, forming the **arch of the aorta** (see Exhibit 14.C), which descends and ends at the level of the intervertebral disc between the fourth and fifth thoracic vertebrae. As the aorta continues to descend, it lies close to the vertebral bodies, and is called the **thoracic aorta** (see Exhibit 14.D). When the thoracic aorta reaches the bottom of the thorax it passes through the aortic hiatus of the diaphragm to become the **abdominal aorta** (see Exhibit 14.E). The abdominal aorta descends to the level of the fourth lumbar vertebra where it divides into two **common iliac arteries** (see Exhibit 14.F), which carry blood to the pelvis and free lower limbs. Each division of the aorta gives off arteries that branch into distributing arteries that lead to various organs. Within the organs, the arteries divide into arterioles and then into capillaries that service the systemic tissues (all tissues except the alveoli of the lungs).

 CHECKPOINT

9. What general regions do each of the four principal divisions of the aorta supply?

DIVISION AND BRANCHES	REGION SUPPLIED
ASCENDING AORTA	
Right and left coronary arteries	Heart
ARCH OF THE AORTA	
Brachiocephalic trunk (brā'-kē-ō-se-FAL-ik)	
Right common carotid artery (ka-ROT-id)	Right side of head and neck
Right subclavian artery (sub-KLā-vē-an)	Right upper limb
Left common carotid artery	Left side of head and neck
Left subclavian artery	Left upper limb
THORACIC AORTA (*thorac-*=chest)	
Pericardial arteries (per-i-KAR-dē-al)	Pericardium
Bronchial arteries (BRONG-kē-al)	Bronchi of lungs
Esophageal arteries (e-sof'-a-JĒ-al)	Esophagus
Mediastinal arteries (mē'-dē-as-TĪ-nal)	Structures in mediastinum
Posterior intercostal arteries (in'-ter-KOS-tal)	Intercostal and chest muscles
Subcostal arteries (sub-KOS-tal)	Upper abdominal muscles
Superior phrenic arteries (FREN-ik)	Superior and posterior surfaces of diaphragm
ABDOMINAL AORTA	
Inferior phrenic arteries	Inferior surface of diaphragm
Lumbar arteries (LUM-bar)	Abdominal muscles
Celiac trunk (SĒ-lē-ak)	
Common hepatic artery (he-PAT-ik)	Liver, stomach, duodenum, pancreas
Left gastric artery (GAS-trik)	Stomach and esophagus
Splenic artery (SPLĒN-ik)	Spleen, pancreas, and stomach
Superior mesenteric artery (MES-en-ter'-ik)	Small intestine, cecum, ascending and transverse colons, and pancreas
Suprarenal arteries (soo-pra-RĒ-nal)	Adrenal (suprarenal) glands
Renal arteries (RĒ-nal)	Kidneys
Gonadal arteries (gō-NAD-al)	
Testicular arteries (tes-TIK-ū-lar)	Testes (male)
Ovarian arteries (ō-VAR-ē-an)	Ovaries (female)
Inferior mesenteric artery	Transverse, descending, and sigmoid colons; rectum
Common iliac arteries (IL-ē-ak)	
External iliac arteries	Lower limbs
Internal iliac arteries	Uterus (female), prostate (male), muscles of buttocks, and urinary bladder

EXHIBIT 14.A **469**

Figure 14.6 Aorta and its principal branches.

All systemic arteries branch from the aorta.

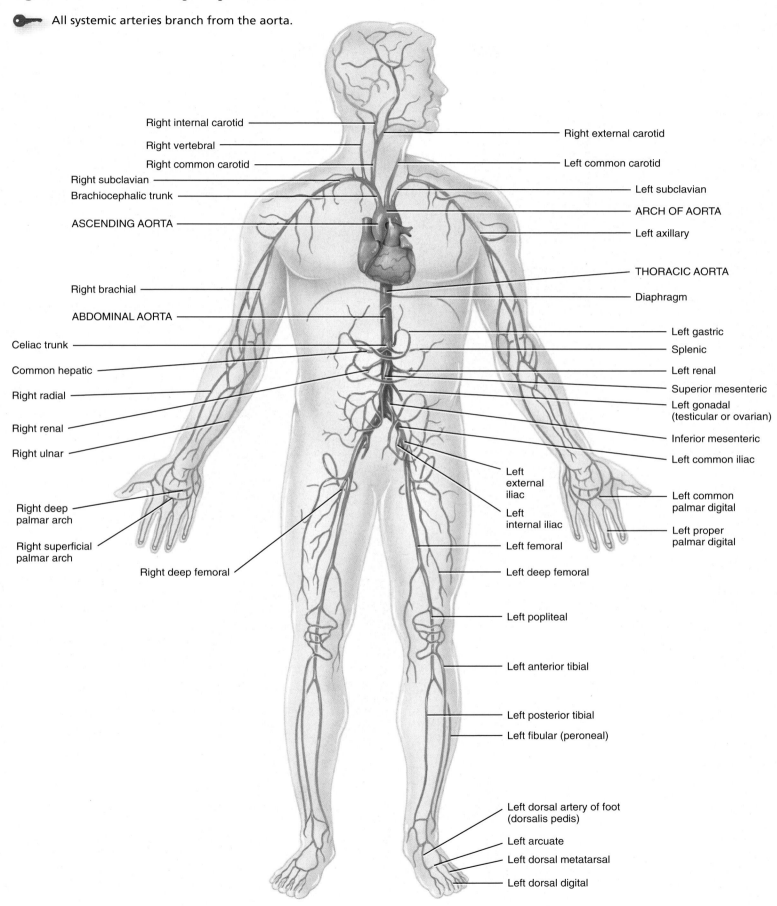

Right internal carotid
Right vertebral
Right common carotid
Right subclavian
Brachiocephalic trunk
ASCENDING AORTA
Right brachial
ABDOMINAL AORTA
Celiac trunk
Common hepatic
Right radial
Right renal
Right ulnar
Right deep palmar arch
Right superficial palmar arch
Right deep femoral

Right external carotid
Left common carotid
Left subclavian
ARCH OF AORTA
Left axillary
THORACIC AORTA
Diaphragm
Left gastric
Splenic
Left renal
Superior mesenteric
Left gonadal (testicular or ovarian)
Inferior mesenteric
Left common iliac
Left external iliac
Left internal iliac
Left femoral
Left common palmar digital
Left proper palmar digital
Left deep femoral
Left popliteal
Left anterior tibial
Left posterior tibial
Left fibular (peroneal)
Left dorsal artery of foot (dorsalis pedis)
Left arcuate
Left dorsal metatarsal
Left dorsal digital

(a) Overall anterior view of the principal branches of the aorta

FIGURE 14.6 CONTINUES ▶

EXHIBIT 14.A The Aorta and Its Branches *(Figure 14.6)* CONTINUED

▣ FIGURE **14.6** CONTINUED ▶

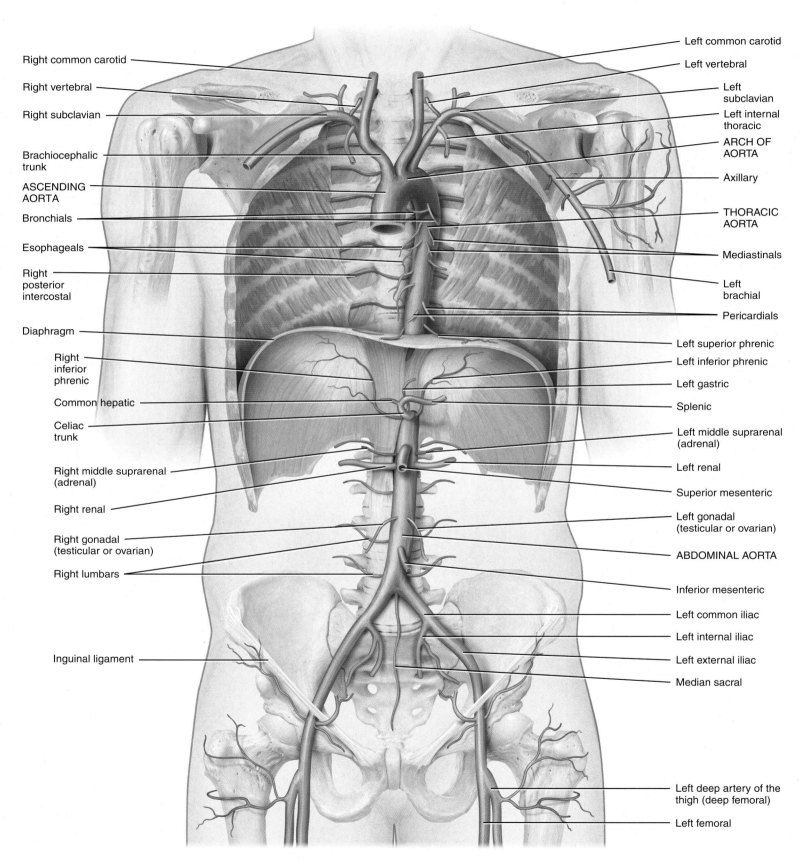

Right common carotid
Right vertebral
Right subclavian
Brachiocephalic trunk
ASCENDING AORTA
Bronchials
Esophageals
Right posterior intercostal
Diaphragm
Right inferior phrenic
Common hepatic
Celiac trunk
Right middle suprarenal (adrenal)
Right renal
Right gonadal (testicular or ovarian)
Right lumbars
Inguinal ligament

Left common carotid
Left vertebral
Left subclavian
Left internal thoracic
ARCH OF AORTA
Axillary
THORACIC AORTA
Mediastinals
Left brachial
Pericardials
Left superior phrenic
Left inferior phrenic
Left gastric
Splenic
Left middle suprarenal (adrenal)
Left renal
Superior mesenteric
Left gonadal (testicular or ovarian)
ABDOMINAL AORTA
Inferior mesenteric
Left common iliac
Left internal iliac
Left external iliac
Median sacral
Left deep artery of the thigh (deep femoral)
Left femoral

(b) Detailed anterior view of the principal branches of the aorta

EXHIBIT 14.A **471**

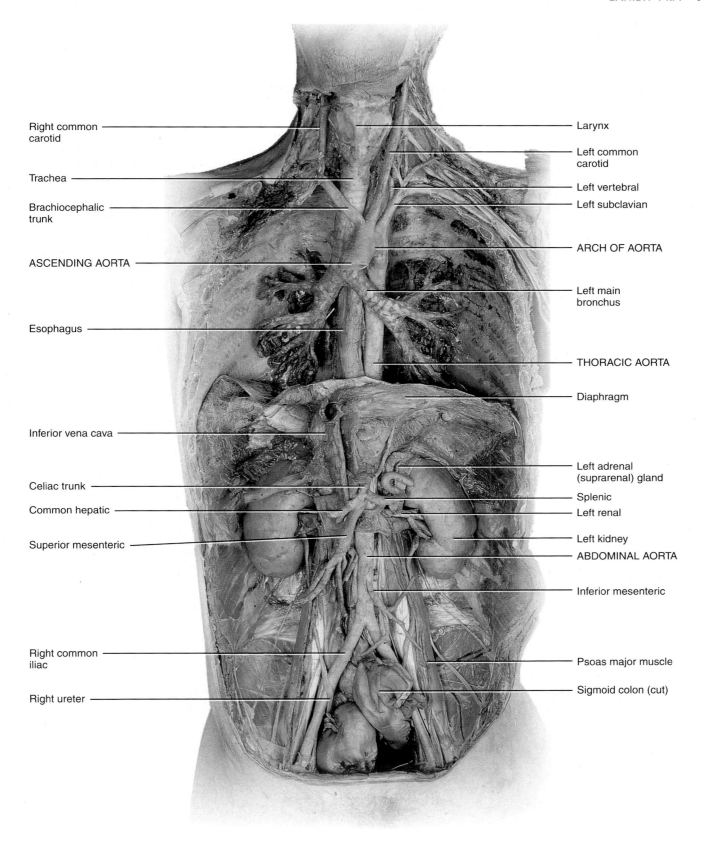

Right common carotid

Trachea

Brachiocephalic trunk

ASCENDING AORTA

Esophagus

Inferior vena cava

Celiac trunk

Common hepatic

Superior mesenteric

Right common iliac

Right ureter

Larynx

Left common carotid

Left vertebral

Left subclavian

ARCH OF AORTA

Left main bronchus

THORACIC AORTA

Diaphragm

Left adrenal (suprarenal) gland

Splenic

Left renal

Left kidney

ABDOMINAL AORTA

Inferior mesenteric

Psoas major muscle

Sigmoid colon (cut)

(c) Anterior view of the principal branches of the aorta

What are the four principal subdivisions of the aorta?

EXHIBIT 14.B Ascending Aorta

 OBJECTIVE

• Identify the two primary arterial branches of the ascending aorta.

The ascending aorta is about 5 cm (2 in.) in length and begins at the aortic valve. It is directed superiorly, slightly anteriorly, and to the right. It ends at the level of the sternal angle, where it becomes the arch of the aorta. The beginning of the ascending aorta is posterior to the pulmonary trunk and right auricle; the right pulmonary artery is posterior to it. At its origin, the ascending aorta contains three dilations called *aortic sinuses*. Two of these, the right and left sinuses, give rise to the right and left coronary arteries, respectively.

The right and left **coronary arteries** (*coron-*=crown) arise from the ascending aorta just superior to the aortic valve. They form a crownlike ring around the heart, giving off branches to the atrial and ventricular myocardium. The **posterior interventricular branch** (in-ter-ven-TRIK-ū-lar; *inter-*=between) of the right coronary artery supplies both ventricles, and the **marginal branch** supplies the right ventricle. The **anterior interventricular branch**, also known as the **left anterior descending (LAD) branch**, of the left coronary artery supplies both ventricles, and the **circumflex branch** (SER-kum-flex; *circum-*=around; *flex*=to bend) supplies the left atrium and left ventricle.

 CHECKPOINT

10. Which branches of the coronary arteries supply the left ventricle?

SCHEME OF DISTRIBUTION

Ascending aorta

Right coronary artery

Left coronary artery

Posterior interventricular branch

Marginal branch

Anterior interventricular branch

Circumflex branch

Arch of aorta

ASCENDING AORTA

Pulmonary trunk

Right atrium

POSTERIOR INTER-VENTRICULAR BRANCH

Right ventricle

LEFT CORONARY

RIGHT CORONARY

Left auricle

Right auricle

CIRCUMFLEX BRANCH

ANTERIOR INTERVENTRICULAR BRANCH

ANTERIOR CARDIAC VEIN

MARGINAL BRANCH

Left ventricle

Arch of aorta

Left pulmonary artery

ASCENDING AORTA

Pulmonary trunk

Left auricle

CIRCUMFLEX BRANCH

GREAT CARDIAC VEIN

LEFT MARGINAL BRANCH

TRIBUTARY TO GREAT CARDIAC VEIN

Left ventricle

Right ventricle

Anterior view of coronary arteries and their major branches

EXHIBIT 14.C **473**

EXHIBIT 14.C   The Arch of the Aorta *(Figure 14.7)*

OBJECTIVE

- Identify the three principal arteries that branch from the arch of the aorta.

The arch of the aorta is 4–5 cm (almost 2 in.) in length and is the continuation of the ascending aorta (Figure 14.7). It emerges from the pericardium posterior to the sternum at the level of the sternal angle (see Figure 14.3). The arch of the aorta is directed superiorly and posteriorly to the left and then inferiorly; it ends at the intervertebral disc between the fourth and fifth thoracic vertebrae, where it becomes the thoracic aorta. Three major arteries branch from the superior aspect of the arch of the aorta: the brachiocephalic trunk, the left common carotid, and the left subclavian. The first and largest branch from the arch of the aorta is the

BRANCH	DESCRIPTION AND BRANCHES	REGIONS SUPPLIED
Brachiocephalic	First branch of the arch of the aorta; divides to form the right subclavian artery and the right common carotid artery (Figure 14.7a)	Head, neck, upper limb, and thoracic wall
Right subclavian artery* (sub-KLĀ-vē-an)	Extends from the brachiocephalic artery to the inferior border of the first rib; gives rise to a number of branches at the base of the neck	Brain, spinal cord, neck, shoulder, thoracic muscle wall, and scapular muscles
Internal thoracic (mammary) artery (thor-AS-ik; *thorac-*=chest)	Arises from the first part of the subclavian artery and descends posterior to the costal cartilages of the superior six ribs just lateral to the sternum; terminates at the sixth intercostal space by *bifurcating* (branching into two arteries) and sends branches into the intercostal spaces	Anterior thoracic wall
	Clinical note: In coronary artery bypass grafting, if only a single vessel is obstructed, the internal thoracic (usually the left) is used to create the bypass. The upper end of the artery is left attached to the subclavian artery and the cut end is connected to the coronary artery at a point distal to the blockage. The lower end of the internal thoracic is tied off. Artery grafts are preferred over vein grafts because arteries can withstand the greater pressure of blood flowing through coronary arteries and are less likely to become obstructed over time.	
Vertebral artery (VER-te-bral)	Major branch of the right subclavian artery to the brain before it passes into the axilla (Figure 14.7b); ascends through the neck, passes through the transverse foramina of the cervical vertebrae, and enters the skull via the foramen magnum to reach the inferior surface of the brain; unites with the left vertebral artery to form the basilar (BĀS-i-lar) artery; the basilar artery passes along the midline of the anterior aspect of the brain stem and gives off several branches (posterior cerebral and cerebellar arteries)	Posterior portion of the cerebrum, the cerebellum, pons, and inner ear
Axillary artery* (AK-sil-ar-ē=armpit)	Continuation of the right subclavian artery into the axilla; begins where the subclavian artery passes the inferior border of the first rib and ends as it crosses the distal margin of the teres major muscle; gives rise to numerous branches in the axilla	Thoracic, shoulder, and scapular muscles and the humerus
Brachial artery* (BRĀ-kē-al=arm)	Continuation of the axillary artery into the arm; begins at the distal border of the teres major muscle and terminates by bifurcating into the radial and ulnar arteries just distal to the bend of the elbow; superficial and palpable along the medial side of the arm; as it descends toward the elbow it curves laterally and passes through the cubital fossa, a triangular depression anterior to the elbow, where you can easily detect the pulse of the brachial artery and listen to the various sounds when taking a person's blood pressure	Muscles of the arm, humerus, elbow joint
	Clinical note: Blood pressure is usually measured in the brachial artery. In order to control hemorrhage, the best place to compress the brachial artery is near the middle of the arm where it is superficial and easily pressed against the humerus.	
Radial artery (RĀ-dē-al=radius)	Smaller branch of the brachial bifurcation; a direct continuation of the brachial artery; passes along the lateral (radial) aspect of the forearm and enters the wrist where it bifurcates into superficial and deep branches that anastomose with the corresponding branches of the ulnar artery to form the palmar arches of the hand; makes contact with the distal end of the radius at the wrist, where it is covered only by fascia and skin	Major blood source to the muscles of the posterior compartment of the forearm
	Clinical note: Because of its superficial location at this point, the radial artery is a common site for measuring the radial pulse.	
Ulnar artery (UL-nar=ulna)	The larger branch of the brachial artery passes along the medial (ulnar) aspect of the forearm and then into the wrist, where it branches into superficial and deep branches that enter the hand; branches anastomose with the corresponding branches of the radial artery to form the palmar arches of the hand	Major blood source to the muscles of the anterior compartment of the forearm

*This is an example of the practice of giving the same vessel different names as it passes through different regions. See the subclavian, axillary, and brachial arteries.

EXHIBIT 14.C The Arch of the Aorta *(Figure 14.7)* CONTINUED

BRANCH	DESCRIPTION AND BRANCHES	REGIONS SUPPLIED
Superficial palmar arch (*palma*=palm)	Formed mainly by the superficial branch of the ulnar artery, with a contribution from the superficial branch of the radial artery; superficial to the long flexor tendons of the fingers and extends across the palm at the bases of the metacarpals; gives rise to common palmar digital arteries, each of which divides into proper palmar digital arteries	Muscles, bones, joints, and skin of the palm and fingers
Deep palmar arch	Arises mainly from the deep branch of the radial artery, but receives a contribution from the deep branch of the ulnar artery; deep to the long flexor tendons of the fingers and extends across the palm just distal to the bases of the metacarpals; gives rise to the palmar metacarpal arteries, which anastomose with the common palmar digital arteries from the superficial arch	Muscles, bones, and joints of the palm and fingers
Right common carotid	Begins at the bifurcation of the brachiocephalic trunk, posterior to the right sternoclavicular joint; passes superiorly into the neck to supply structures in the head (Figure 14.7b); divides into the right external and right internal carotid arteries at the superior border of the larynx (voice box) **Clinical note:** Pulse may be detected in the common carotid artery, just lateral to the larynx. It is convenient to detect a carotid pulse when exercising or when administering cardiopulmonary resuscitation.	Head and neck
External carotid artery	Begins at the superior border of the larynx and terminates near the temporomandibular joint of the parotid gland, where it divides into two branches: the superficial temporal and maxillary arteries **Clinical note:** The carotid pulse can be detected in the external carotid artery just anterior to the sternocleidomastoid muscle at the superior border of the larynx.	Major blood source to all structures of the head except the brain; supplies skin, connective tissues, muscles, bones, joints, dura and arachnoid mater in the head and supplies much of the neck anatomy
Internal carotid artery	Arises from the common carotid artery; enters the cranial cavity through the carotid foramen in the temporal bone and emerges in the cranial cavity near the base of the hypophyseal fossa of the sphenoid bone; gives rise to numerous branches inside the cranial cavity and terminates as the anterior cerebral and middle cerebral arteries. The anterior cerebral artery passes forward toward the frontal lobe of the cerebrum and the middle cerebral artery passes laterally between the temporal and parietal lobes of the cerebrum. Inside the cranium (Figure 14.7c), anastomoses of the left and right internal carotid arteries via the anterior communicating artery between the two anterior cerebral arteries, along with internal carotid–basilar artery anastomoses, form an arrangement of blood vessels at the base of the brain called the **cerebral arterial circle** (circle of Willis) (Figure 14.7c). The internal carotid–basilar anastomosis occurs where the posterior communicating arteries arising from the internal carotid artery anastomose with the posterior cerebral arteries from the basilar artery to link the internal carotid blood supply with the vertebral blood supply. The cerebral arterial circle equalizes blood pressure to the brain and provides alternate routes for blood flow to the brain, should the arteries become damaged.	Eyeball and other orbital structures, ear, and parts of nose and nasal cavity; frontal, temporal, parietal lobes of the cerebrum of brain; pituitary gland; and pia mater
Left common carotid artery	Arises as the second branch of the arch of the aorta and ascends through the mediastinum to enter the neck deep to the clavicle, then follows a similar path to the right common carotid artery	Distribution similar to the right common carotid artery
Left subclavian artery	Arises as the third and final branch of the arch of the aorta; passes superior and lateral through the mediastinum and deep to the clavicle at the base of the neck as it courses toward the upper limb; has a similar course to the right subclavian artery after leaving the mediastinum	Distribution similar to the right subclavian artery

brachiocephalic trunk (brā′-kē-ō-se-FAL-ik; *brachio-*=arm; *cephalic*=head). It extends superiorly, bending slightly to the right, and divides at the right sternoclavicular joint to form the right subclavian artery and right common carotid artery. The second branch from the arch of the aorta is the **left common carotid artery** (ka-ROT-id), which divides into the same branches with the same names as the right common carotid artery. The third branch from the arch of the aorta is the **left subclavian artery**

(sub-KLĀ-vē-an), which distributes blood to the left vertebral artery and vessels of the left upper limb. Arteries branching from the left subclavian artery are similar in distribution and name to those branching from the right subclavian artery.

 CHECKPOINT

11. What general regions do the arteries that arise from the arch of the aorta supply?

EXHIBIT 14.C **475**

SCHEME OF DISTRIBUTION

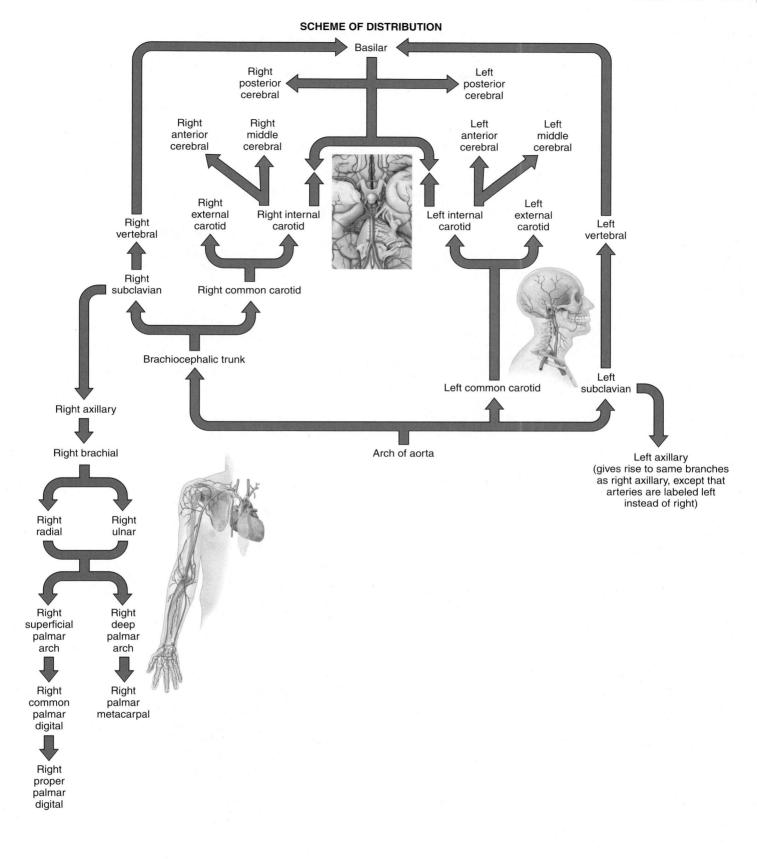

Basilar

Right posterior cerebral

Left posterior cerebral

Right anterior cerebral

Right middle cerebral

Left anterior cerebral

Left middle cerebral

Right vertebral

Right external carotid

Right internal carotid

Left internal carotid

Left external carotid

Left vertebral

Right subclavian

Right common carotid

Brachiocephalic trunk

Left common carotid

Left subclavian

Right axillary

Right brachial

Arch of aorta

Left axillary
(gives rise to same branches as right axillary, except that arteries are labeled left instead of right)

Right radial

Right ulnar

Right superficial palmar arch

Right deep palmar arch

Right common palmar digital

Right palmar metacarpal

Right proper palmar digital

EXHIBIT 14.C   The Arch of the Aorta   *(Figure 14.7)*   **CONTINUED**

Figure 14.7   Arch of the aorta and its branches. Note in (c) the arteries that constitute the cerebral arterial circle (circle of Willis).

🔑   The arch of the aorta ends at the level of the intervertebral disc between the fourth and fifth thoracic vertebrae.

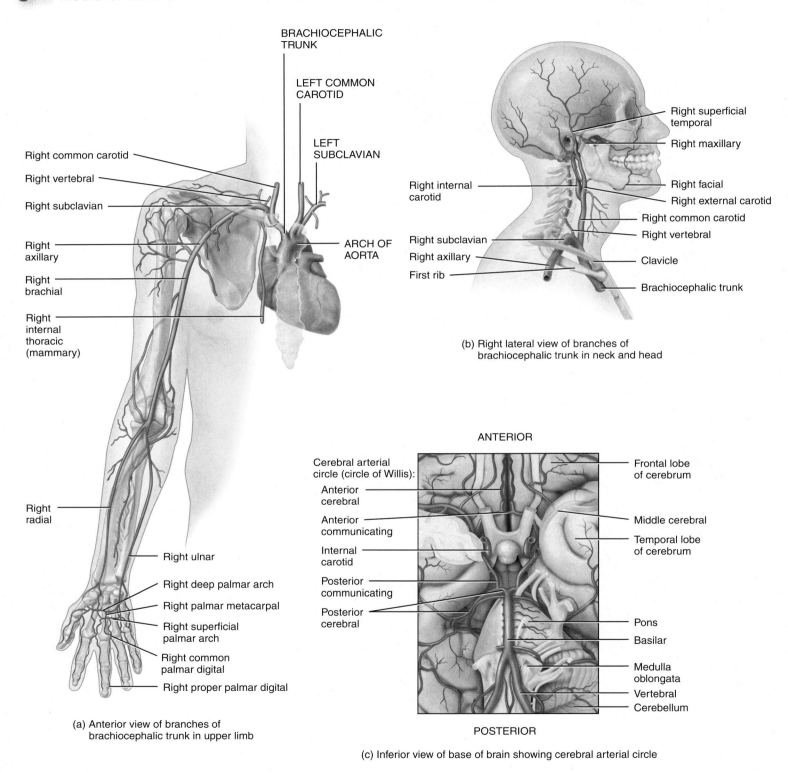

(a) Anterior view of branches of brachiocephalic trunk in upper limb

(b) Right lateral view of branches of brachiocephalic trunk in neck and head

(c) Inferior view of base of brain showing cerebral arterial circle

EXHIBIT 14.C **477**

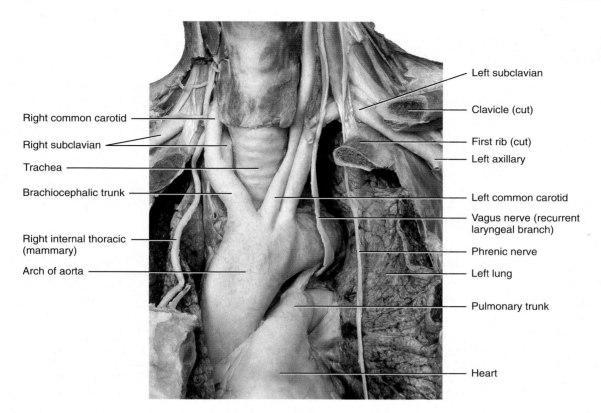

Right common carotid

Right subclavian

Trachea

Brachiocephalic trunk

Right internal thoracic (mammary)

Arch of aorta

Left subclavian

Clavicle (cut)

First rib (cut)

Left axillary

Left common carotid

Vagus nerve (recurrent laryngeal branch)

Phrenic nerve

Left lung

Pulmonary trunk

Heart

(d) Anterior view of branches of arch of aorta

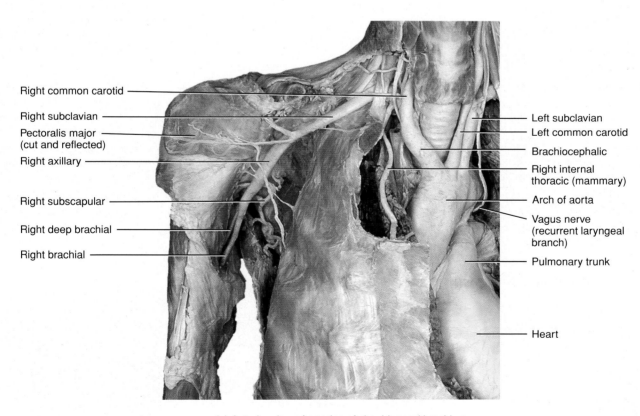

Right common carotid

Right subclavian

Pectoralis major (cut and reflected)

Right axillary

Right subscapular

Right deep brachial

Right brachial

Left subclavian

Left common carotid

Brachiocephalic

Right internal thoracic (mammary)

Arch of aorta

Vagus nerve (recurrent laryngeal branch)

Pulmonary trunk

Heart

(e) Anterior view of arteries of shoulder and brachium

What are the three major branches of the arch of the aorta, in order of their origination?

EXHIBIT 14.D ## Thoracic Aorta

 OBJECTIVE

• Identify the visceral and parietal branches of the thoracic aorta.

The thoracic aorta is about 20 cm (8 in.) long and is a continuation of the arch of the aorta (see Figure 14.6b). It begins at the level of the intervertebral disc between the fourth and fifth thoracic vertebrae, where it lies to the left of the vertebral column. As it descends, it moves closer to the midline and extends through an opening in the diaphragm (aortic hiatus), which is located an-

terior to the vertebral column at the level of the intervertebral disc between the twelfth thoracic and first lumbar vertebrae.

Along its course, the thoracic aorta sends off numerous small arteries, **visceral branches** (VIS-e-ral) to viscera, and **parietal branches** (pa-RĪ-e-tal) to body wall structures.

CHECKPOINT

12. What general regions do the visceral and parietal branches of the thoracic aorta supply?

BRANCH	DESCRIPTION AND BRANCHES	REGIONS SUPPLIED
VISCERAL BRANCHES		
Pericardial arteries (per′-i-KAR-dē-al; *peri-*= around; *cardia-*=heart)	Two to three small arteries that arise from variable levels of the thoracic aorta and pass forward to the pericardial sac surrounding the heart	Tissues of the pericardial sac
Bronchial arteries (BRONG-kē-al=windpipe)	Arise from the thoracic aorta or one of its branches; the right bronchial artery typically arises from the third posterior intercostal artery; the two left bronchial arteries arise from the upper end of the thoracic aorta; all follow the bronchial tree into the lungs	Supply the tissues of the bronchial tree and surrounding lung tissue down to the level of the alveolar ducts
Esophageal arteries (e-sof′-a-JĒ-al; *eso-*=to carry; *phage-*=food)	Four to five arteries that arise from the anterior surface of the thoracic aorta and pass forward to branch onto the esophagus	All the tissues of the esophagus
Mediastinal arteries (mē-dē-as-TĪ-nal)	Arise from various points on the thoracic aorta	Assorted tissues within the mediastinum, primarily connective tissue and lymph nodes
PARIETAL BRANCHES		
Posterior intercostal arteries (in′ter-KOS-tal; *inter-*= between; *costa*=rib)	Typically, nine pairs of arteries that arise from the posterolateral aspect on each side of the thoracic aorta; each passes laterally and then anteriorly through the intercostal space where they will eventually anastomose with anterior branches from the internal thoracic arteries	Skin, muscles, and ribs of the thoracic wall; thoracic vertebrae, meninges, and spinal cord; mammary glands
Subcostal arteries (sub-KOS-tal; *sub-*=under)	The lowest segmental branches of the thoracic aorta; one on each side passes into the thoracic body wall inferior to the twelfth rib and courses forward into the upper abdominal region of the body wall	Skin, muscles, and ribs; twelfth thoracic vertebra, meninges, and spinal cord
Superior phrenic arteries (FREN-ik=pertaining to the diaphragm)	Arise from the lower end of the thoracic aorta and pass onto the superior surface of the diaphragm	Diaphragm muscle and pleura covering the diaphragm

SCHEME OF DISTRIBUTION

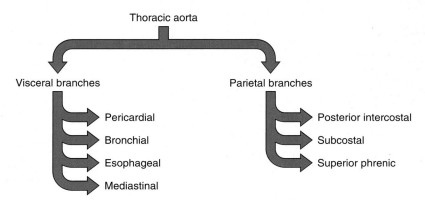

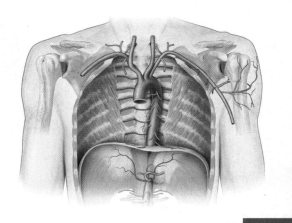

EXHIBIT 14.E **479**

EXHIBIT 14.E Abdominal Aorta *(Figure 14.8)*

● OBJECTIVE
• Identify the visceral and parietal branches of the abdominal aorta.

The abdominal aorta is the continuation of the thoracic aorta after it passes through the diaphragm (Figure 14.8). It begins at the aortic hiatus in the diaphragm and ends at about the level of the fourth lumbar vertebra, where it divides into the right and left common iliac arteries. The abdominal aorta lies anterior to the vertebral column.

As with the thoracic aorta, the abdominal aorta gives off visceral and parietal branches. The unpaired visceral branches arise from the anterior surface of the aorta and include the **celiac trunk** and the **superior mesenteric** and **inferior mesenteric arteries**

BRANCH	DESCRIPTION AND BRANCHES	REGIONS SUPPLIED
UNPAIRED VISCERAL BRANCHES		
Celiac trunk (artery) (SĒ-lē-ak)	First visceral branch of the aorta inferior to the diaphragm; arises from the abdominal aorta at the level of the twelfth thoracic vertebra as the aorta passes through the hiatus in the diaphragm; divides into three branches: the left gastric, splenic, and common hepatic arteries (Figure 14.8a)	Supplies all organs of the gastrointestinal tract from the abdominal part of the esophagus to the duodenum, and also the spleen
	1. The **left gastric artery** (GAS-trik=stomach), the smallest of the three celiac branches, arises superiorly to the left toward the esophagus and then turns to follow the lesser curvature of the stomach. On the lesser curvature of the stomach it anastomoses with the right gastric artery.	Abdominal part of esophagus, lesser curvature of stomach, and lesser omentum
	2. The **splenic artery** (SPLEN-ik=spleen), the largest branch of the celiac trunk, arises from the left side of the celiac trunk distal to the left gastric artery, and passes horizontally to the left along the pancreas. Before reaching the spleen, it gives rise to three named arteries:	Spleen, pancreas, fundus and greater curvature of stomach, and greater omentum
	• **Pancreatic arteries** (pan-krē-AT-ik), a series of small arteries that arise from the splenic and descend into the tissue of the pancreas.	Pancreas
	• **Left gastro-omental (gastro-epiploic) artery** (gas′-trō-ō-MEN-tal; *epiplo-*=omentum) arises from the terminal end of the splenic artery and passes from left to right along the greater curvature of the stomach.	Greater curvature of stomach and greater omentum
	• **Short gastric arteries** arise from the terminal end of the splenic artery and pass onto the fundus of the stomach.	Fundus of the stomach
	3. The **common hepatic artery** (he-PAT-ik=liver), intermediate in size between the left gastric and splenic arteries, arises from the right side of the celiac trunk and gives rise to three arteries:	Liver, gallbladder, lesser omentum, stomach, pancreas, and duodenum
	• **Proper hepatic artery** branches from the common hepatic artery and ascends along the bile ducts into the liver and gallbladder.	Liver, gallbladder, and lesser omentum
	• **Right gastric artery** arises from the common hepatic artery and curves back to the left along the lesser curvature of the stomach where it anastomoses with the left gastric artery.	Pyloric end of lesser curvature of stomach and lesser omentum
	• **Gastroduodenal artery** (gas′-trō-doo′-ō-DĒ-nal) passes inferiorly toward the stomach and duodenum and sends branches along the greater curvature of the stomach.	Pyloric end of stomach, duodenum, and p ancreas
Superior mesenteric artery (MES-en-ter′-ik; *meso-*=middle; *-enteric*=pertaining to the intestines)	Arises from the anterior surface of the abdominal aorta about 1 cm inferior to the celiac trunk at the level of the first lumbar vertebra (Figure 14.8b); extends inferiorly and anteriorly between the layers of the mesentery (the portion of the peritoneum that attaches the small intestine to the posterior abdominal wall); anastomoses extensively and has five branches:	Supplies all organs of the gastrointestinal tract from the duodenum to the transverse colon
	1. The **inferior pancreaticoduodenal artery** (pan′-krē-at′-i-kō-doo′-ō-DĒ-nal) passes superiorly and to the right toward the head of the pancreas and the duodenum.	Pancreas and duodenum
	2. The **jejunal** (je-JOO-nal) and **ileal arteries** (IL-ē-al) spread through the mesentery to pass to the loops of the jejunum and ileum (small intestine).	Jejunum and ileum
	3. The **ileocolic artery** (il′-ē-ō-KOL-ik) passes inferiorly and laterally toward the right side toward the terminal part of the ileum, cecum, appendix, and first part of ascending colon.	Terminal part of the ileum, cecum, appendix, and first part of ascending colon
	4. The **right colic artery** (KOL-ik) passes laterally to the right toward the ascending colon.	Ascending colon and first part of transverse colon
	5. The **middle colic artery** ascends slightly to the right toward the transverse colon.	Most of the transverse colon

EXHIBIT 14.E   Abdominal Aorta   *(Figure 14.8)*   CONTINUED

BRANCH	DESCRIPTION AND BRANCHES	REGIONS SUPPLIED
Inferior mesenteric artery	Arises from the anterior aspect of the abdominal aorta at the level of the third lumbar vertebra and then passes inferiorly to the left of the aorta (Figure 14.8c); anastomoses extensively and has three branches:	Supplies all organs of the gastrointestinal tract from the transverse colon to the rectum
	1. The **left colic artery** ascends laterally to the left toward the distal end of the transverse colon and the descending colon.	End of the transverse colon and descending colon
	2. The **sigmoid arteries** (SIG-moyd) descend laterally to the left toward the sigmoid colon.	Sigmoid colon
	3. The **superior rectal artery** (REK-tal) passes inferiorly to the superior part of the rectum.	Upper part of the rectum
PAIRED VISCERAL BRANCHES		
Suprarenal arteries (soo′-pra-RĒ-nal; *supra-*=above; *ren-*=kidney)	Typically three pairs (superior, middle, and inferior), but only the middle pair originates directly from the abdominal aorta (see Figure 14.6b). The middle suprarenal arteries arise from the abdominal aorta at the level of the first lumbar vertebra at or superior to the renal arteries. The superior suprarenal arteries arise from the inferior phrenic arteries, and the inferior suprarenal arteries originate from the renal arteries.	Suprarenal (adrenal) glands
Renal arteries (RĒ-nal; *ren*=kidney)	The right and left renal arteries usually arise from the lateral aspects of the abdominal aorta at the superior border of the second lumbar vertebra, about 1 cm inferior to the superior mesenteric artery (see Figure 15.6b). The right renal artery, which is longer than the left, arises slightly lower than the left and passes posterior to the right renal vein and inferior vena cava. The left renal artery is posterior to the left renal vein and is crossed by the inferior mesenteric vein.	All tissues of the kidneys
Gonadal (gō-NAD-al; *gon-*=seed) **[testicular** (test-TIK-ū-lar) or **ovarian arteries** (ō-VAR-ē-an)]	Arise from the anterior aspect of the abdominal aorta at the level of the second lumbar vertebra just inferior to the renal arteries (see Figure 14.13a). In males, the gonadal arteries are specifically referred to as the testicular arteries. They descend along the posterior abdominal wall to pass through the inguinal canal and descend into the scrotum. In females, the gonadal arteries are called the ovarian arteries. They are much shorter than the testicular arteries and remain within the abdominal cavity.	Males: testis, epididymis, ductus deferens, and ureters Females: ovaries, uterine (fallopian) tubes, and ureters
UNPAIRED PARIETAL BRANCH		
Median sacral artery (SĀ-kral=pertaining to the sacrum)	Arises from the posterior surface of the abdominal aorta about 1 cm superior to the *bifurcation* (division into two branches) of the aorta into the right and left common iliac arteries (see Figure 14.6b).	Sacrum, coccyx, sacral spinal nerves, and piriformis muscle
PAIRED PARIETAL BRANCH		
Inferior phrenic arteries (FREN-ik=pertaining to the diaphragm)	First paired branches of the abdominal aorta; arise immediately superior to the origin of the celiac trunk (see Figure 14.6b). (They may also arise from the renal arteries.)	Diaphragm and suprarenal (adrenal) glands
Lumbar arteries (LUM-bar=pertaining to the loin)	The four pairs arise from the posterolateral surface of the abdominal aorta similar to the pattern of the posterior intercostal arteries of the thorax (see Figure 14.6b); pass laterally into the abdominal muscle wall and curve toward the anterior aspect of the wall.	Lumbar vertebrae, spinal cord and meninges, skin and muscles of the posterior and lateral part of the abdominal wall

(see Figure 14.6b). The paired visceral branches arise from the lateral surfaces of the aorta and include the **suprarenal, renal,** and **gonadal arteries**. The lone unpaired parietal branch is the **median sacral artery**. The paired parietal branches arise from the posterolateral surfaces of the aorta and include the **inferior phrenic** and **lumbar arteries**.

 CHECKPOINT

13. Indicate the general regions supplied by the paired visceral and parietal branches and the unpaired visceral and parietal branches of the abdominal aorta.

EXHIBIT 14.E **481**

SCHEME OF DISTRIBUTION

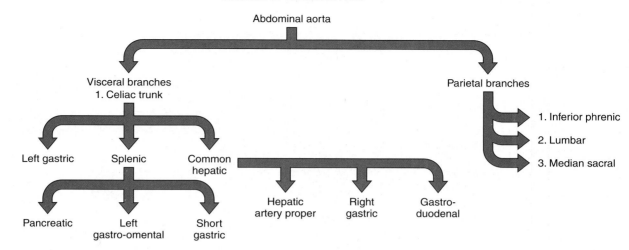

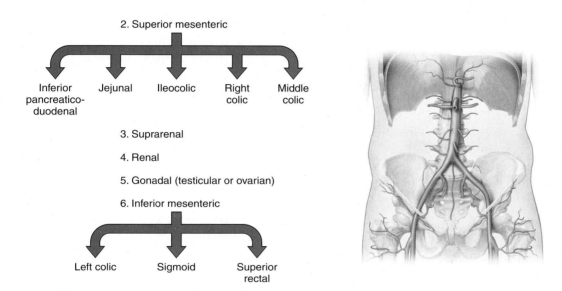

Figure 14.8 Abdominal aorta and principal branches.

The abdominal aorta is the continuation of the thoracic aorta.

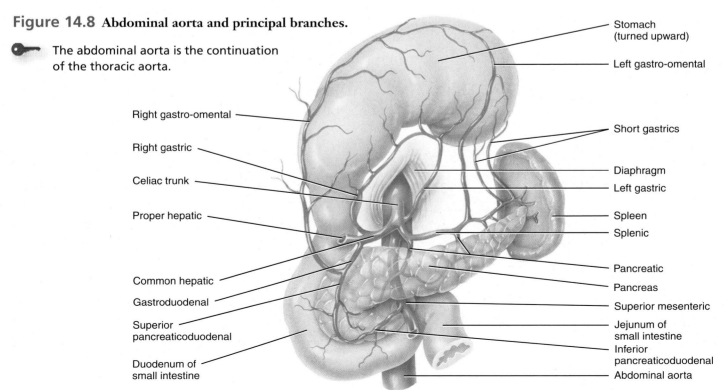

(a) Anterior view of celiac trunk and its branches

FIGURE 14.8 CONTINUES ▶

EXHIBIT 14.E *Abdominal Aorta* *(Figure 14.8)* CONTINUED

FIGURE 14.8 CONTINUED

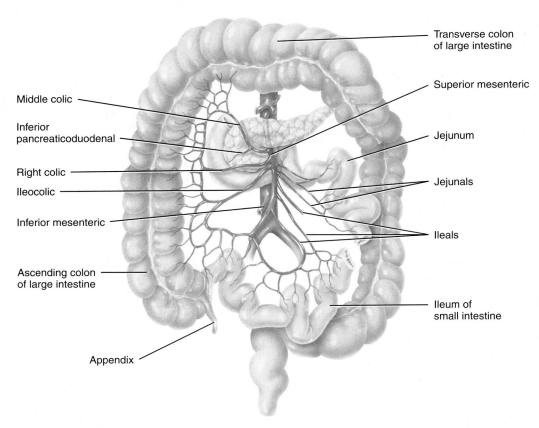

(b) Anterior view of superior mesenteric artery and its branches

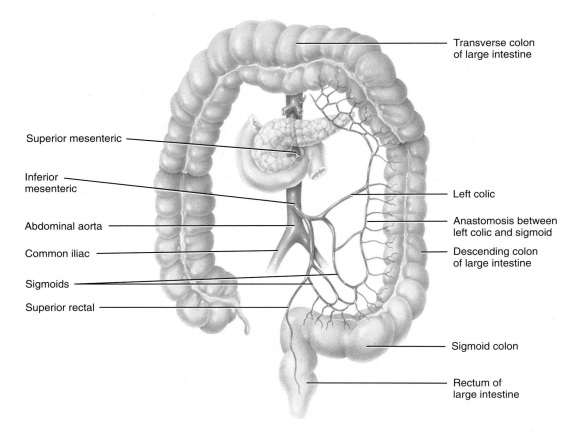

(c) Anterior view of inferior mesenteric artery and its branches

EXHIBIT 14.E **483**

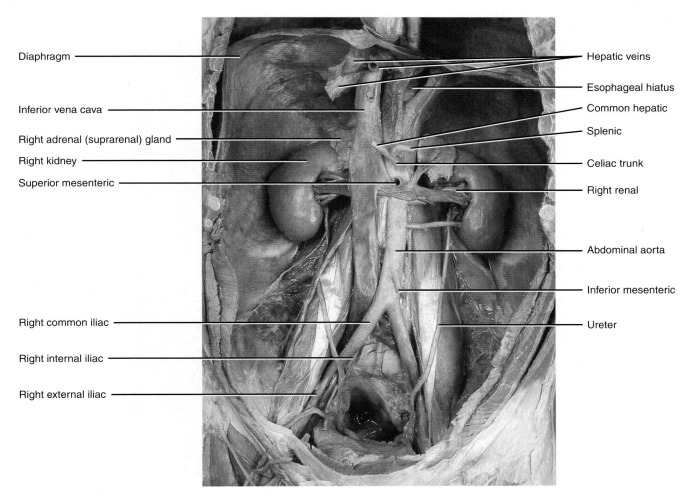

Diaphragm

Inferior vena cava

Right adrenal (suprarenal) gland

Right kidney

Superior mesenteric

Right common iliac

Right internal iliac

Right external iliac

Hepatic veins

Esophageal hiatus

Common hepatic

Splenic

Celiac trunk

Right renal

Abdominal aorta

Inferior mesenteric

Ureter

(d) Anterior view of arteries of abdomen and pelvis

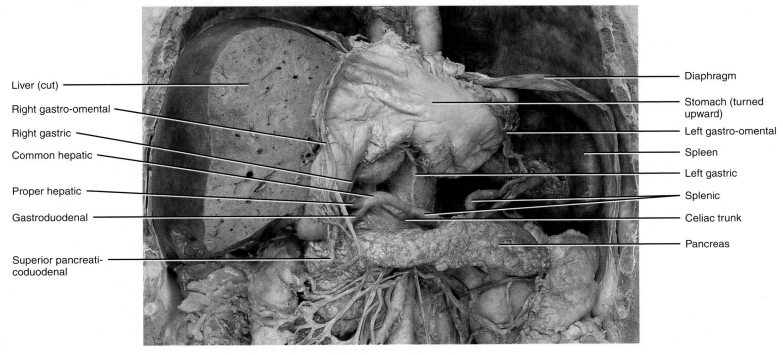

Liver (cut)

Right gastro-omental

Right gastric

Common hepatic

Proper hepatic

Gastroduodenal

Superior pancreaticoduodenal

Diaphragm

Stomach (turned upward)

Left gastro-omental

Spleen

Left gastric

Splenic

Celiac trunk

Pancreas

(e) Anterior view of celiac trunk and its branches

FIGURE 14.8 CONTINUES

EXHIBIT 14.E Abdominal Aorta *(Figure 14.8)* CONTINUED

FIGURE 14.8 CONTINUED ▶

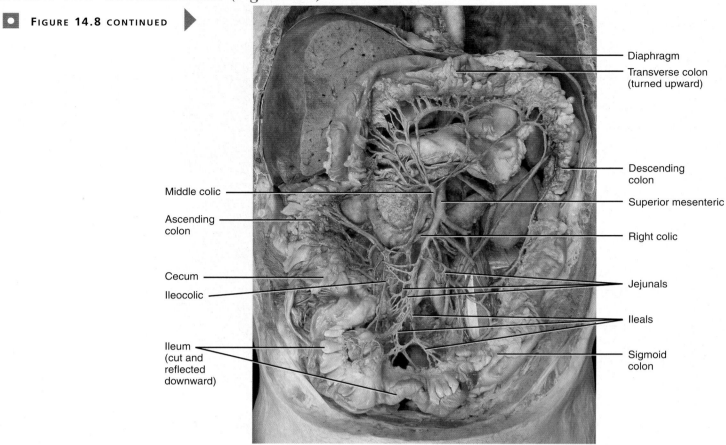

Diaphragm
Transverse colon (turned upward)

Descending colon

Middle colic
Superior mesenteric

Ascending colon
Right colic

Cecum
Jejunals
Ileocolic

Ileals

Ileum (cut and reflected downward)
Sigmoid colon

(f) Anterior view of superior mesenteric artery and its branches

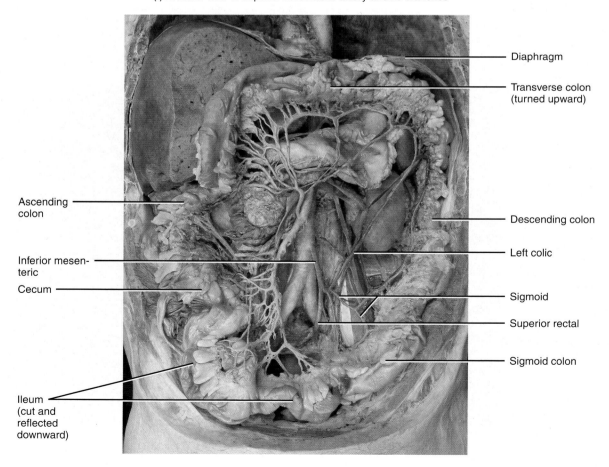

Diaphragm

Transverse colon (turned upward)

Ascending colon

Descending colon

Inferior mesenteric
Left colic

Cecum
Sigmoid

Superior rectal

Sigmoid colon

Ileum (cut and reflected downward)

(g) Anterior view of inferior mesenteric artery and its branches

Where does the abdominal aorta begin?

EXHIBIT 14.F **485**

EXHIBIT 14.F Arteries of the Pelvis and Lower Limbs *(Figure 14.9)*

 OBJECTIVE

- Identify the two major branches of the common iliac arteries.

The abdominal aorta ends by dividing into the right and left common iliac arteries (Figure 14.9). Each common iliac artery in turn divides into an **internal iliac artery** and an **external iliac artery**. Just as the subclavian arteries take on regional names as they pass

into the upper limbs, the external iliac arteries, in sequence, become the **femoral arteries** in the thighs and the **popliteal arteries** posterior to the knee before terminating by bifurcating into the **anterior** and **posterior tibial arteries** in the legs.

CHECKPOINT

14. What general regions do the internal and external iliac arteries supply?

SCHEME OF DISTRIBUTION

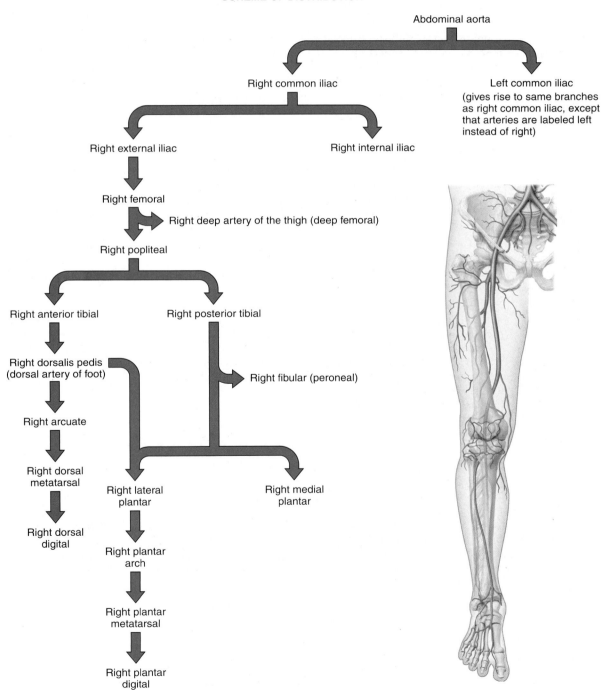

EXHIBIT 14.F Arteries of the Pelvis and Lower Limbs *(Figure 14.9)* CONTINUED

BRANCH	DESCRIPTION AND BRANCHES	REGIONS SUPPLIED
Common iliac arteries (IL-ē-ak=pertaining to the ilium)	Arise from the abdominal aorta at about the level of the fourth lumbar vertebra; each passes inferiorly and slightly laterally for about 5 cm (2 in.) and gives rise to two branches: internal and external iliac arteries	Pelvic muscle wall, pelvic organs, external genitals, and lower limbs
Internal iliac arteries	Primary arteries of the pelvis; begin at the *bifurcation* (division into two branches) of the common iliac arteries anterior to the sacroiliac joint at the level of the lumbosacral intervertebral disc; pass posteriorly as they descend into the pelvis and divide into anterior and posterior divisions	Pelvic muscle wall, pelvic organs, buttocks, external genitals, and medial muscles of the thigh
External iliac arteries	Larger than the internal iliac arteries and begin at the bifurcation of the common iliac arteries; descend along the medial border of the psoas major muscles following the pelvic brim, pass posterior to the midportion of the inguinal ligaments, and become the femoral arteries as they pass beneath the inguinal ligament and enter the thigh	Lower abdominal wall, cremaster muscle in males and round ligament of uterus in females, and the lower limb
Femoral arteries (FĒM-o-ral=pertaining to the thigh)	Continuations of the external iliac arteries as they enter the thigh; in the *femoral triangle* of the upper thighs they are superficial along with the femoral vein and nerve and deep inguinal lymph nodes (see Figure 11.22a); pass beneath the sartorius muscle as they descend along the anteromedial aspects of the thighs and follow its course to the distal end of the thigh where they pass through an opening in the tendon of the adductor magnus muscle to end at the posterior aspect of the knee, where they become the popliteal arteries	Muscles of the thigh—quadriceps, adductors, and hamstrings—femur, and ligaments and tendons around the knee joint
	Clinical note: In cardiac catheterization, a catheter is inserted through a blood vessel and advanced into the major vessels to access a heart chamber. A catheter often contains a measuring instrument or other device at its tip. To reach the left side of the heart, the catheter is inserted into the femoral artery and passed into the aorta to the coronary arteries or heart chamber.	
Popliteal arteries (pop′li-TĒ-al=posterior surface of the knee)	Originate as the femoral arteries; pass through the hiatus in the adductor magnus muscle and continue through the popliteal fossa (space behind the knee); descend to the inferior border of the popliteus muscles, where they divide into the anterior and posterior tibial arteries	Muscles of the distal thigh, skin of the knee region, muscles of the proximal leg, knee joint, femur, patella, tibia, and fibula
Anterior tibial arteries (TIB-ē-al=pertaining to the shin)	Descend from the bifurcation of the popliteal arteries at the distal border of the popliteus muscles; smaller than the posterior tibial arteries; pass over the interosseous membrane of the tibia and fibula to descend through the anterior muscle compartment of the leg; become the dorsalis pedis arteries (dorsal arteries of the foot) at the ankles. On the dorsum of the feet, the dorsal arteries of the foot give off a transverse branch at the first medial cuneiform bone called the arcuate arteries (*arcuat-*=bowed) that run laterally over the bases of the metatarsals. From the arcuate arteries branch the dorsal metatarsal arteries, which course along the metatarsal bones; the dorsal metatarsal arteries terminate by dividing into the dorsal digital arteries, which pass into the toes.	Tibia, fibula, anterior muscles of the leg, dorsal muscles of the foot, tarsal bones, metatarsal bones, and phalanges
Posterior tibial arteries	Direct continuations of the popliteal arteries; descend from the bifurcation of the popliteal arteries; pass down the posterior muscular compartment of the legs deep to the soleus muscles; pass posterior to the medial malleolus at the distal end of the leg and curve forward toward the plantar surface of the feet; pass deep to the flexor retinaculum on the medial side of the feet and terminate by branching into medial and lateral plantar arteries; give rise to the fibular (peroneal) arteries in the upper third of the leg, which course laterally as they descend into the lateral compartment of the leg. The smaller medial plantar arteries (PLAN-tar=sole) pass along the medial side of the sole and the larger lateral plantar arteries angle toward the lateral side of the sole and unite with a branch of the dorsalis pedis arteries of the foot to form the plantar arch; the arch begins at the base of the fifth metatarsal and extends medially across the metatarsals. As the arch crosses the foot, it gives off plantar metatarsal arteries, which course along the plantar surface of the metatarsal bones; these arteries terminate by dividing into plantar digital arteries that pass into the toes.	Posterior and lateral muscle compartments of the leg, plantar muscles of the foot, tibia, fibula, tarsal, metatarsal, and phalangeal bones

EXHIBIT 14.F **487**

Figure 14.9 Arteries of the pelvis and right free lower limb.

🔑 The internal iliac arteries carry most of the blood supply to the pelvic viscera and wall.

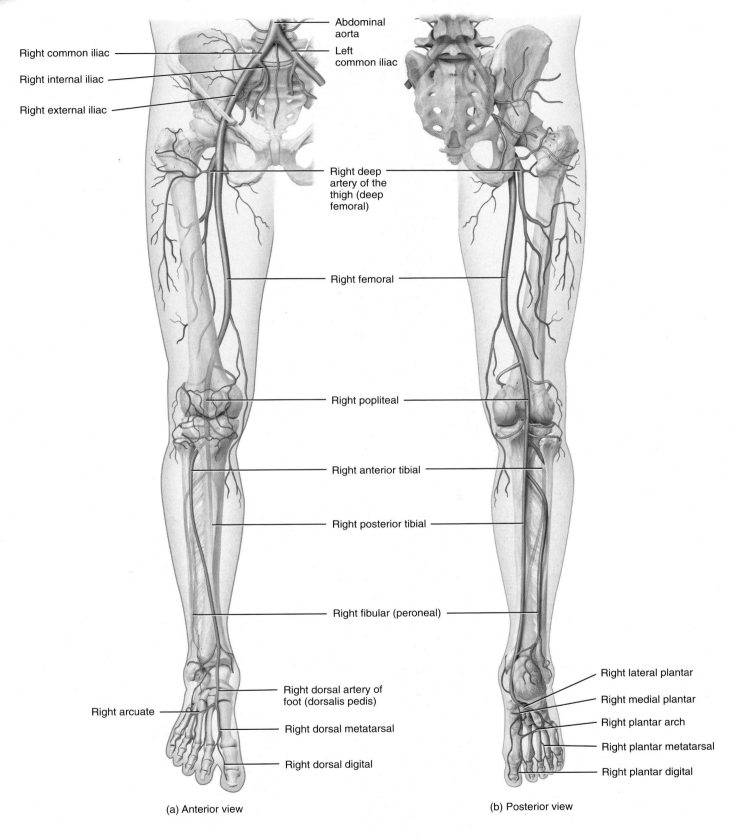

Right common iliac

Right internal iliac

Right external iliac

Abdominal aorta

Left common iliac

Right deep artery of the thigh (deep femoral)

Right femoral

Right popliteal

Right anterior tibial

Right posterior tibial

Right fibular (peroneal)

Right arcuate

Right dorsal artery of foot (dorsalis pedis)

Right dorsal metatarsal

Right dorsal digital

Right lateral plantar

Right medial plantar

Right plantar arch

Right plantar metatarsal

Right plantar digital

(a) Anterior view

(b) Posterior view

FIGURE 14.9 CONTINUES ▶

EXHIBIT 14.F Arteries of the Pelvis and Lower Limbs *(Figure 14.9)* CONTINUED

FIGURE 14.9 CONTINUED ▶

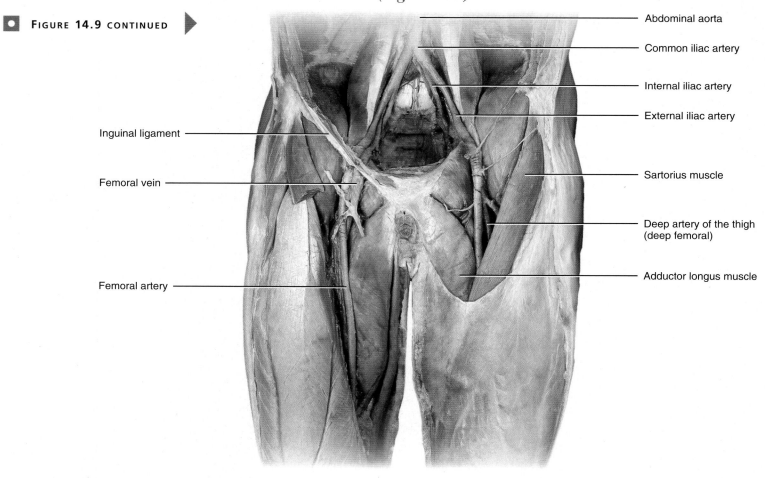

Inguinal ligament

Femoral vein

Femoral artery

Abdominal aorta

Common iliac artery

Internal iliac artery

External iliac artery

Sartorius muscle

Deep artery of the thigh (deep femoral)

Adductor longus muscle

(c) Anterior view of arteries of pelvis and thigh

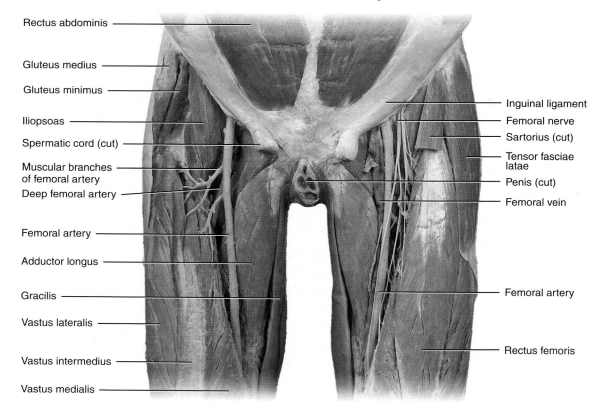

Rectus abdominis

Gluteus medius

Gluteus minimus

Iliopsoas

Spermatic cord (cut)

Muscular branches of femoral artery

Deep femoral artery

Femoral artery

Adductor longus

Gracilis

Vastus lateralis

Vastus intermedius

Vastus medialis

Inguinal ligament

Femoral nerve

Sartorius (cut)

Tensor fasciae latae

Penis (cut)

Femoral vein

Femoral artery

Rectus femoris

(d) Anterior view of arteries of thigh with right side deep and left side more superficial

At what point does the abdominal aorta divide into the common iliac arteries?

EXHIBIT 14.G **489**

EXHIBIT 14.G Veins of the Systemic Circulation *(Figure 14.10)*

O B J E C T I V E
- Identify the three systemic veins that return deoxygenated blood to the heart.

As you have already learned, arteries distribute blood from the heart to various parts of the body and veins drain blood away from the various parts and return the blood to the heart. For the

Figure 14.10 Principal veins.

Deoxygenated blood returns to the heart via the superior and inferior venae cavae and the coronary sinus.

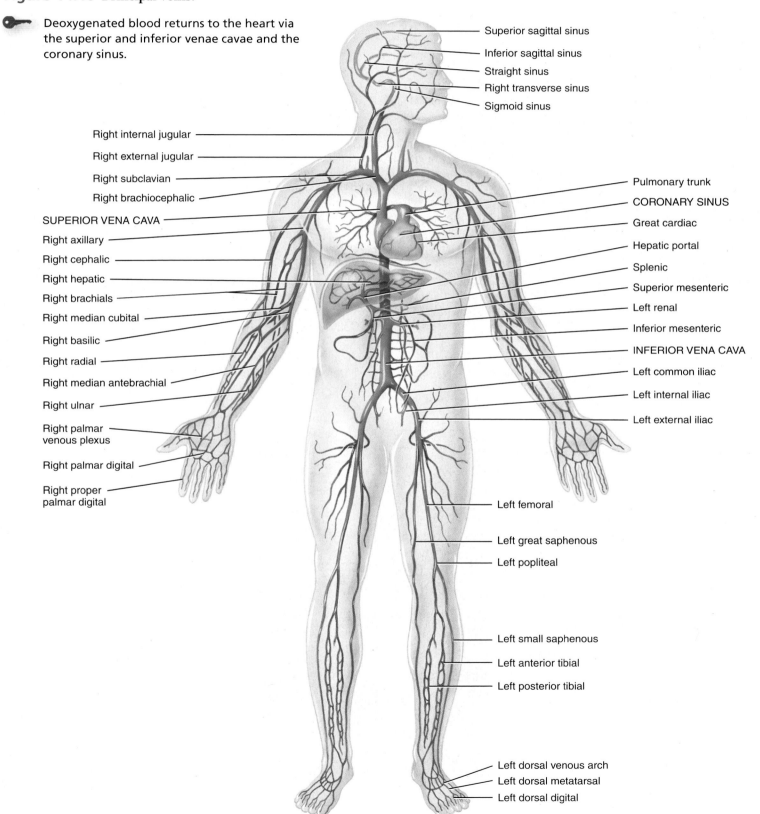

(a) Overall anterior view of the principal veins

most part, arteries are deep; veins may be superficial or deep. Superficial veins are located just beneath the skin and can be seen easily. Because there are no large superficial arteries, the names of superficial veins do not correspond to those of arteries. Superficial veins are clinically important as sites for withdrawing blood or giving injections. Deep veins generally travel alongside arteries and usually bear the same name. Arteries usually follow definite pathways; veins are more difficult to follow because they connect in irregular networks in which many tributaries merge to form a large vein. Although only one systemic artery, the aorta, takes oxygenated blood away from the heart (left ventricle), three systemic veins, the coronary sinus, superior vena cava, and inferior vena cava, return deoxygenated blood to the heart (right atrium). The coronary sinus receives blood from the cardiac veins that drain the capillaries of the heart; with a few exceptions, the superior vena cava receives blood from veins that drain capillaries from tissues superior to the diaphragm (except the air sacs (alveoli) of the lungs); the inferior vena cava receives blood from veins that drain capillaries from tissues inferior to the diaphragm.

 CHECKPOINT

15. What are the three tributaries of the coronary sinus?

VEINS	DESCRIPTION AND TRIBUTARIES	REGIONS DRAINED
Coronary sinus (KOR-ō-nar-ē; *corona*=crown)	The main vein of the heart; receives almost all venous blood from the myocardium; located in the coronary sulcus (see Figure 13.3c) on the posterior aspect of the heart and opens into the right atrium between the orifice of the inferior vena cava and the tricuspid valve; a wide venous channel into which three veins drain; receives the great cardiac vein (from the anterior interventricular sulcus) into its left end, and the middle cardiac vein (from the posterior interventricular sulcus) and the small cardiac vein into its right end; several anterior cardiac veins drain directly into the right atrium	All tissues of the heart
Superior vena cava (SVC) (VĒ-na CĀ-va; *vena*=vein; *cava*=cavelike)	About 7.5 cm (3 in.) long and 2 cm (1 in.) in diameter; empties its blood into the superior part of the right atrium; begins posterior to the right first costal cartilage by the union of the right and left brachiocephalic veins and ends at the level of the right third costal cartilage, where it enters the right atrium	Head, neck, upper limbs, and thorax
Inferior vena cava (IVC)	The largest vein in the body, about 3.5 cm (1.4 in.) in diameter; begins anterior to the fifth lumbar vertebra by the union of the common iliac veins, ascends behind the peritoneum to the right of the midline, pierces the caval opening of the diaphragm at the level of the eighth thoracic vertebra, and enters the inferior part of the right atrium	Abdomen, pelvis, and lower limbs
	Clinical note: The inferior vena cava is commonly compressed during the later stages of pregnancy by the enlarging uterus, producing edema of the ankles and feet and temporary varicose veins.	

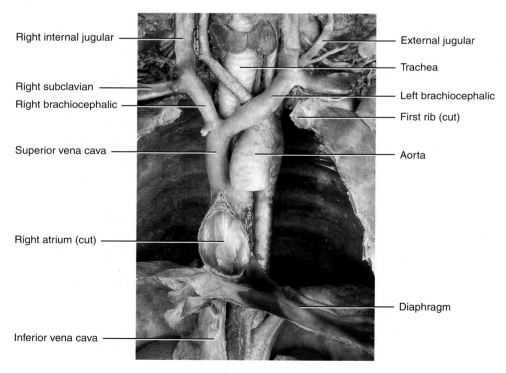

(b) Anterior view of superior vena cava and its tributaries

Right internal jugular — External jugular — Trachea — Right subclavian — Left brachiocephalic — Right brachiocephalic — First rib (cut) — Superior vena cava — Aorta — Right atrium (cut) — Diaphragm — Inferior vena cava

 Which general regions of the body are drained by the superior vena cava and the inferior vena cava?

EXHIBIT 14.H **491**

EXHIBIT 14.H Veins of the Head and Neck *(Figure 14.11)*

OBJECTIVE
- Identify the three major veins that drain blood from the head.

Most blood draining from the head passes into three pairs of veins: the **internal jugular** (JUG-ū-lar); **external jugular**, and **vertebral veins** (Figure 14.11). Within the cranial cavity, all veins drain into dural venous sinuses and then into the internal jugular veins. **Dural venous sinuses** are endothelial-lined venous channels between layers of the cranial dura mater.

CHECKPOINT
16. Which vein of the head is the major drainage pathway from the brain?

VEINS	DESCRIPTION AND TRIBUTARIES	REGIONS DRAINED
Brachiocephalic veins	(See Exhibit 14.C)	
Internal jugular veins (JŪG-ū-lar=throat)	Begin at the base of the cranium as the sigmoid sinus and inferior petrosal sinus converge at the opening of the jugular foramen; descend within the carotid sheath lateral to the internal and common carotid arteries, deep to the sternocleidomastoid muscles. Receive numerous tributaries from the face and neck. The internal jugular veins anastomose with the subclavian veins to form the brachiocephalic veins (brā'-kē-ō-se-FAL-ik; *brachi-*=arm; *cephal-*=head) deep and slightly lateral to the sternoclavicular joints; the major dural venous sinuses that contribute to the internal jugular vein are as follows:	Brain, meninges, bones of the cranium, muscles and tissues of the face and neck
	1. The **superior sagittal sinus** (SAJ-i-tal=arrow) begins at the frontal bone, where it receives a vein from the nasal cavity; and passes posteriorly to the occipital bone along the midline of the skull deep to the sagittal suture. It usually angles to the right and drains into the right transverse sinus.	Nasal cavity; superior, lateral, medial aspects of cerebrum; skull bones; meninges
	2. The **inferior sagittal sinus** is much smaller than the superior sagittal sinus. It begins posterior to the attachment of the falx cerebri and receives the great cerebral vein to become the straight sinus.	Medial aspects of the cerebrum and diencephalon
	3. The **straight sinus** runs in the tentorium cerebelli and is formed by the union of the inferior sagittal sinus and the great cerebral vein. It typically drains into the left transverse sinus.	Medial and inferior aspects of the cerebrum and the cerebellum
	4. The **sigmoid sinuses** (SIG-moyd=S-shaped) are located along the posterior aspect of the petrous temporal bone. They begin where the transverse sinuses and superior petrosal sinuses anastomose and terminate in the internal jugular vein at the jugular foramen.	Lateral and posterior aspect of the cerebrum and the cerebellum
	5. The **cavernous sinuses** (KAV-er-nus=cave-like) are located on both sides of the body of the sphenoid bone. The ophthalmic veins from the orbits and the cerebral veins from the cerebral hemispheres, along with other small sinuses, empty into the cavernous sinuses. They drain posteriorly to the petrosal sinuses to eventually return to the internal jugular veins. The cavernous sinuses are unique because they have major blood vessels and nerves passing through them on their way to the orbit and face. The oculomotor (III) nerve, trochlear (IV) nerve, ophthalmic and maxillary branches of the trigeminal (V) nerve, abducens (VI) nerve, and internal carotid arteries pass through the cavernous sinuses.	Orbits, nasal cavity, frontal regions of the cerebrum, and the superior aspect of the brain stem
Subclavian veins	(See Exhibit 14.I)	
External jugular veins	Begin in the parotid glands near the angle of the mandible; descend through the neck across the sternocleidomastoid muscles; terminate at a point opposite the middle of the clavicles, where they empty into the subclavian veins; become very prominent along the side of the neck when venous pressure rises, for example, during heavy coughing or straining or in cases of heart failure	Scalp and skin of the head and neck, muscles of the face and neck, and oral cavity and pharynx
Vertebral veins (VER-te-bral; *vertebra*=vertebrae)	The right and left vertebral veins originate inferior to the occipital condyles; they descend through successive transverse foramina of the first six cervical vertebrae and emerge from the foramina of the sixth cervical vertebra to enter the brachiocephalic veins in the root of the neck	Cervical vertebrae, cervical spinal cord and meninges, and some deep muscles in the neck

EXHIBIT 14.H Veins of the Head and Neck *(Figure 14.11)* CONTINUED

SCHEME OF DRAINAGE

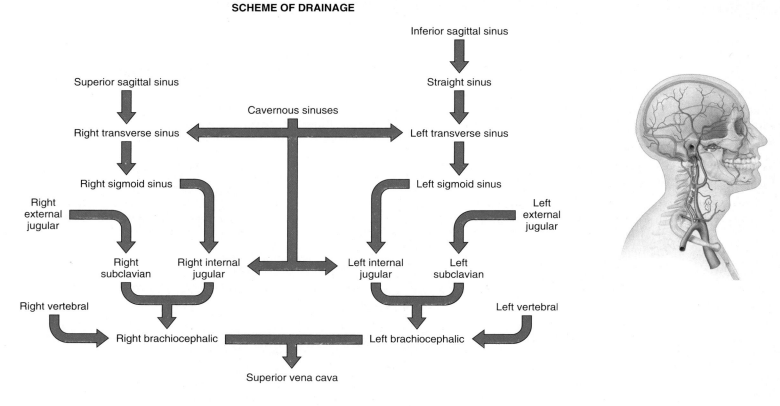

Figure 14.11 Principal veins of the head and neck.

🔑 Blood draining from the head passes into the internal jugular, external jugular, and vertebral veins.

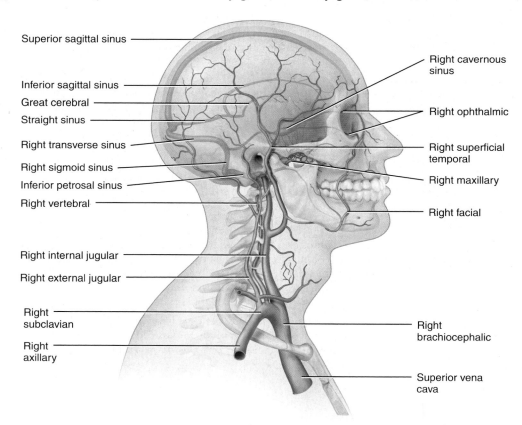

(a) Right lateral view

EXHIBIT 14.H **493**

SUPERIOR

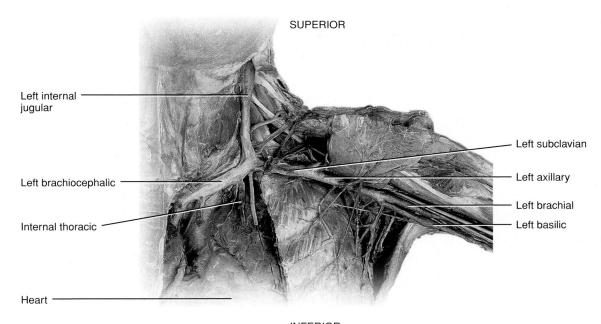

Left internal jugular

Left brachiocephalic

Internal thoracic

Heart

Left subclavian

Left axillary

Left brachial

Left basilic

INFERIOR

(b) Anterior view

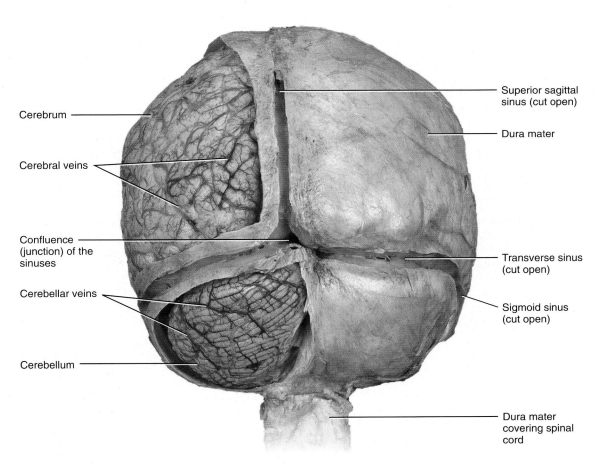

Cerebrum

Cerebral veins

Confluence (junction) of the sinuses

Cerebellar veins

Cerebellum

Superior sagittal sinus (cut open)

Dura mater

Transverse sinus (cut open)

Sigmoid sinus (cut open)

Dura mater covering spinal cord

(c) Posterior view of brain and spinal cord with meningeal coverings dissected to show dural sinuses and veins on brain

Into which veins in the neck does all venous blood in the brain drain?

EXHIBIT 14.I Veins of the Upper Limbs *(Figure 14.12)*

 OBJECTIVE

• Identify the principal veins that drain the upper limbs.

Both superficial and deep veins return blood from the upper limbs to the heart (Figure 14.12). Superficial veins are located just deep to the skin and are often visible. They anastomose extensively with one another and with deep veins, and they do not accompany arteries. Superficial veins are larger than deep veins and return most of the blood from the upper limbs. The principal superficial veins of the upper limbs are the cephalic and basilic veins. Deep veins are located deep in the upper limbs. They usually accompany arteries and have the same names as the corresponding arteries.

VEINS	DESCRIPTION AND TRIBUTARIES	REGIONS DRAINED
DEEP VEINS		
Brachiocephalic veins	(See Exhibit 14.J)	
Subclavian veins (sub-KLĀ-vē-an; *sub-* =under; *clavian*=pertaining to the clavicle)	Continuations of the axillary veins; pass over the first rib deep to the clavicle to terminate at the sternal end of the clavicle, where they unite with the internal jugular veins to form the brachiocephalic veins; the thoracic duct of the lymphatic system delivers lymph into the junction between the left subclavian and left internal jugular veins; the right lymphatic duct delivers lymph into the junction between the right sub-clavian and right internal jugular veins (see Figure 15.3) **Clinical note:** In a procedure called central line placement, the right subclavian vein is frequently used to administer nutrients and medication and measure venous pressure.	Skin, muscles, bones of the arms, shoulders, neck, and superior thoracic wall
Axillary veins (AK-sil-ar-ē; *axilla*=armpit)	Arise as the brachial veins and basilic veins unite near the base of the axilla (armpit); ascend to outer borders of first ribs, where they become the subclavian veins; receive numerous tributaries in the axilla that correspond to the branches of the axillary arteries	Skin, muscles, bones of the arm, axilla, shoulder, and superolateral chest wall
Brachial veins (BRĀ-kē-al; *brachi-*=arm)	Accompany the brachial arteries; begin in the anterior aspect of the elbow region where the radial and ulnar veins join one another; as they ascend through the arm, the basilic veins join them to form the axillary vein near the distal border of the teres major muscle	Muscles and bones of the elbow and brachial regions
Ulnar veins (UL-nar=pertaining to the ulna)	Begin at the superficial palmar venous arches, which drain the common palmar digital veins and the proper palmar digital veins in the fingers; course along the medial aspect of the forearms, pass alongside the ulnar arteries, and join with the radial veins to form the brachial veins	Muscles, bones, and skin of the hand, and muscles of the medial aspect of the forearm
Radial veins (RĀ-dē-al= pertaining to the radius)	Begin at the deep palmar venous arches (Figure 14.12c), which drain the palmar meta-carpal veins in the palms; drain the lateral aspects of the forearms and pass alongside the radial arteries; unite with the ulnar veins to form the brachial veins just inferior to the elbow joint	Muscles and bones of the lateral hand and forearm
SUPERFICIAL VEINS		
Cephalic veins (se-FAL-ik= pertaining to the head)	Begin on the lateral aspect of the dorsal venous networks of the hands (dorsal venous arches), networks of veins on the dorsum of the hands formed by the dorsal metacarpal veins (Figure 14.12a); these in turn drain the dorsal digital veins, which pass along the sides of the fingers; arch around the radial side of the forearms to the anterior surface and ascend through the entire limbs along the anterolateral surface; end where they join the axillary veins, just inferior to the clavicles; accessory cephalic veins originate either from a venous plexus on the dorsum of the forearms or from the medial aspects of the dorsal venous networks of the hands, and unite with the cephalic veins just inferior to the elbow	Integument and superficial muscles of the lateral aspect of the upper limb
Basilic veins (ba-SIL-ik=royal, of prime importance)	Begin on the medial aspects of the dorsal venous networks of the hands and ascend along the posteromedial surface of the forearm and anteromedial surface of the arm (Figure 14.12b); connected to the cephalic veins anterior to the elbow by the median cubital veins (*cubital*=pertaining to the elbow); after receiving the median cubital veins, the basilic veins continue ascending until they reach the middle of the arm. There they penetrate the tissues deeply and run alongside the brachial arteries until they join with the brachial veins to form the axillary veins. **Clinical note:** If veins must be punctured for an injection, transfusion, or removal of a blood sample, the median cubital veins are preferred.	Integument and superficial muscles of the medial aspect of the upper limb
Median antebrachial veins (median veins of the forearm) (an'-tē-BRĀ-kē-al; *ante-*=before, in front of)	Begin in the palmar venous plexuses, networks of veins on the palms; drain the palmar digital veins in the fingers; ascend anteriorly in the forearms to join the basilic or me-dian cubital veins, sometimes both	Integument and superficial muscles of the palm and anterior aspect of the upper limb

EXHIBIT 14.I **495**

In the free upper limb each artery is accompanied by a network of surrounding veins called a venae comitans (VE-nē KOM-i-tans). For example, two to three ulnar veins on each side surround the corresponding ulnar artery. The two to three ulnar veins form numerous anastomoses with one another as they ascend through the forearm. Both superficial and deep veins have valves, but valves are more numerous in the deep veins.

 CHECKPOINT

17. Where do the cephalic, basilic, median antebrachial, radial, and ulnar veins originate?

SCHEME OF DRAINAGE

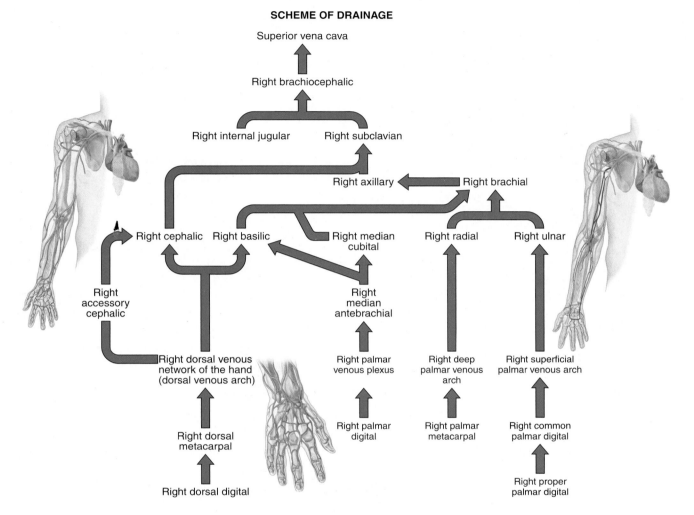

Figure 14.12 Principal veins of the right upper limb.

Deep veins usually accompany arteries that have similar names.

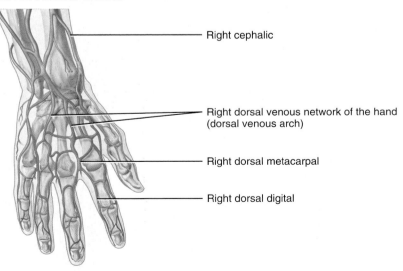

(a) Posterior view of superficial veins of the hand

FIGURE **14.12** CONTINUES ▶

EXHIBIT 14.I Veins of the Upper Limbs *(Figure 14.12)* CONTINUED

FIGURE **14.12** CONTINUED

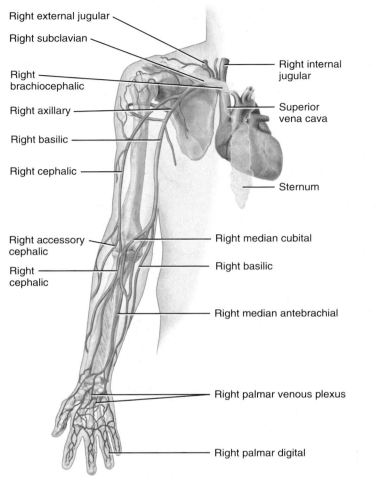

Right external jugular
Right subclavian
Right brachiocephalic
Right axillary
Right basilic
Right cephalic
Right internal jugular
Superior vena cava
Sternum
Right accessory cephalic
Right cephalic
Right median cubital
Right basilic
Right median antebrachial
Right palmar venous plexus
Right palmar digital

(b) Anterior view of superficial veins

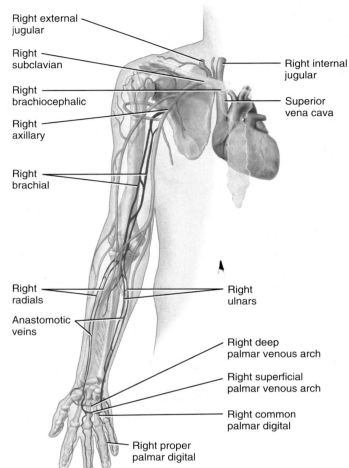

Right external jugular
Right subclavian
Right brachiocephalic
Right axillary
Right brachial
Right internal jugular
Superior vena cava
Right radials
Right ulnars
Anastomotic veins
Right deep palmar venous arch
Right superficial palmar venous arch
Right common palmar digital
Right proper palmar digital

(c) Anterior view of deep veins

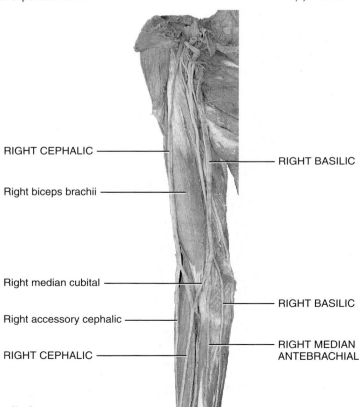

RIGHT CEPHALIC
RIGHT BASILIC
Right biceps brachii
Right median cubital
RIGHT BASILIC
Right accessory cephalic
RIGHT CEPHALIC
RIGHT MEDIAN ANTEBRACHIAL

(d) Anterior view of superficial veins of arm and forearm

From which vein in the upper limb is a blood sample often taken?

EXHIBIT 14.J **497**

EXHIBIT 14.J Veins of the Thorax *(Figure 14.13)*

OBJECTIVE

• Identify the components of the azygos system of veins.

Although the brachiocephalic veins drain some portions of the thorax, most thoracic structures are drained by a network of veins, called the **azygos system** (az-Ī-gus or a-ZĪ-gos)**,** which runs on either side of the vertebral column (Figure 14.13). The system consists of three veins—the **azygos, hemiazygos,** and **accessory hemiazygos veins**—that show considerable variation in origin, course, tributaries, anastomoses, and termination. Ultimately they empty into the superior vena cava. The azygos system, besides collecting blood from the thorax and abdominal wall, may

serve as a bypass for the inferior vena cava, which drains blood from the lower body. Several small veins directly link the azygos system with the inferior vena cava. Larger veins that drain the lower limbs and abdomen also connect into the azygos system. If the inferior vena cava or hepatic portal vein becomes obstructed, blood that typically passes through the inferior vena cava can detour into the azygos system to return blood from the lower body to the superior vena cava.

 CHECKPOINT

18. How is the azygos system related to the inferior vena cava?

VEINS	DESCRIPTION AND TRIBUTARIES	REGIONS DRAINED
Brachiocephalic veins (brā′-kē-ō-se-FAL-ik; *brachio-*=arm; *cephalic*=pertaining to the head)	Form by the union of the subclavian and internal jugular veins and the two brachiocephalic veins, then unite to form the superior vena cava; because the superior vena cava is to the right of the body's midline, the left brachiocephalic vein is longer than the right; the right brachiocephalic vein is anterior and to the right of the brachiocephalic trunk and follows a more vertical course; the left brachiocephalic vein is anterior to the brachiocephalic trunk, the left common carotid and left subclavian arteries, the trachea, the left vagus (X) nerve, and the phrenic nerve; it approaches a more horizontal position as it passes from left to right	Head, neck, upper limbs, mammary glands, and superior thorax
Azygos vein (az-Ī-gus=unpaired)	An unpaired vein that is anterior to the vertebral column, slightly to the right of the midline; usually begins at the junction of the right ascending lumbar and right subcostal veins near the diaphragm; arches over the root of the right lung at the level of the fourth thoracic vertebra to end in the superior vena cava; receives the following tributaries: right posterior intercostal, hemiazygos, accessory hemiazygos, esophageal, mediastinal, pericardial, and bronchial veins	Right side of thoracic wall, thoracic viscera, and posterior abdominal wall
Hemiazygos vin (hem′-ē-az-Ī-gus; *hemi-*=half)	Anterior to the vertebral column and slightly to the left of the midline; often begins at the junction of the left ascending lumbar and left subcostal veins; terminates by joining the azygos vein at about the level of the ninth thoracic vertebra; receives the following tributaries: ninth through eleventh left posterior intercostal, esophageal, mediastinal, and sometimes the accessory hemiazygos veins	Left side of the lower thoracic wall, thoracic viscera, and left posterior abdominal wall
Accessory hemiazygos vein	Anterior to the vertebral column and to the left of the midline; begins at the fourth or fifth intercostal space and descends from the fifth to the eighth thoracic vertebra or ends in the hemiazygos vein; terminates by joining the azygos vein at about the level of the eighth thoracic vertebra; receives the following tributaries: fourth through eighth left posterior intercostal veins (the first through third posterior intercostal veins drain into the left brachiocephalic vein), left bronchial, and mediastinal veins	Left side of upper thoracic wall and thoracic viscera

SCHEME OF DRAINAGE

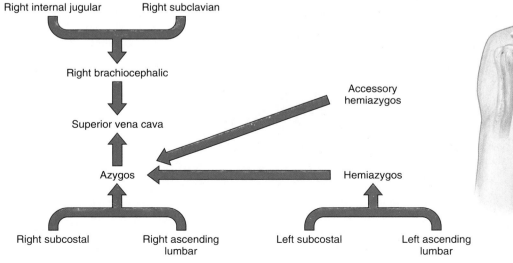

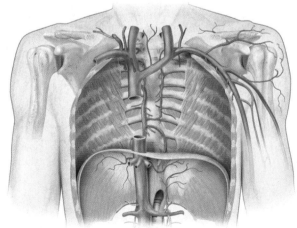

EXHIBIT 14.J Veins of the Thorax *(Figure 14.13)* CONTINUED

Figure 14.13 **Principal veins of the thorax, abdomen, and pelvis.**

🔑 Most thoracic structures are drained by the azygos system of veins.

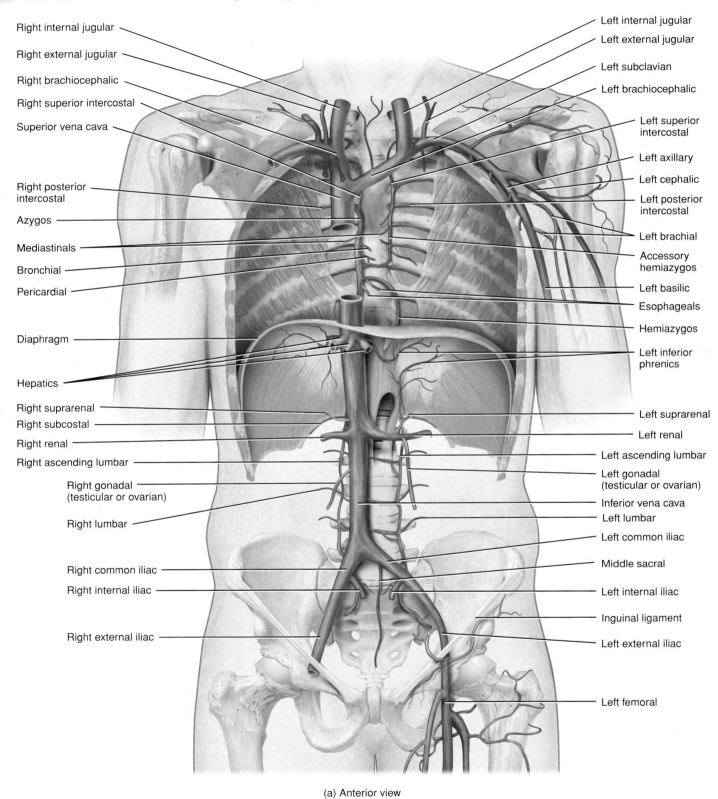

(a) Anterior view

EXHIBIT 14.J **499**

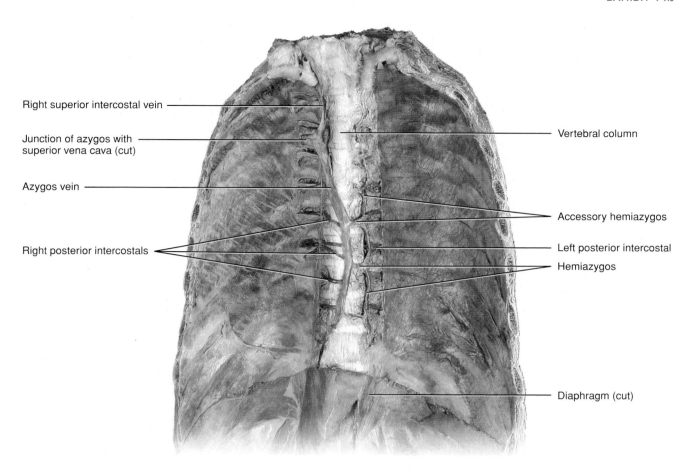

Right superior intercostal vein

Junction of azygos with superior vena cava (cut)

Azygos vein

Right posterior intercostals

Vertebral column

Accessory hemiazygos

Left posterior intercostal

Hemiazygos

Diaphragm (cut)

(b) Anterior view of posterior thoracic wall

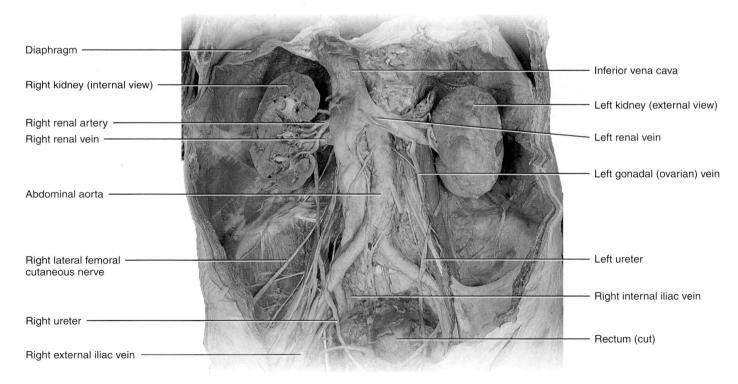

Diaphragm

Right kidney (internal view)

Right renal artery

Right renal vein

Abdominal aorta

Right lateral femoral cutaneous nerve

Right ureter

Right external iliac vein

Inferior vena cava

Left kidney (external view)

Left renal vein

Left gonadal (ovarian) vein

Left ureter

Right internal iliac vein

Rectum (cut)

(c) Anterior view

Which vein returns blood from the abdominopelvic viscera to the heart?

EXHIBIT 14.K Veins of the Abdomen and Pelvis

OBJECTIVE

• Identify the principal veins that drain the abdomen and pelvis.

Blood from the abdominal and pelvic viscera and lower half of the abdominal wall returns to the heart via the inferior vena cava (see Figure 14.13a). Many small veins enter the inferior vena cava. Most carry return flow from parietal branches of the abdominal aorta, and their names correspond to the names of the arteries.

The inferior vena cava does not receive veins directly from the gastrointestinal tract, spleen, pancreas, and gallbladder.

These organs pass their blood into a common vein, the **hepatic portal vein**, which delivers the blood to the liver. The superior mesenteric and splenic veins unite to form the hepatic portal vein (see Figure 14.16). This special flow of venous blood, called the hepatic portal circulation, is described shortly. After passing through the liver for processing, blood drains into the hepatic veins, which then empty into the inferior vena cava.

CHECKPOINT

19. What structures do the lumbar, gonadal, renal, suprarenal, inferior phrenic, and hepatic veins drain?

SCHEME OF DRAINAGE

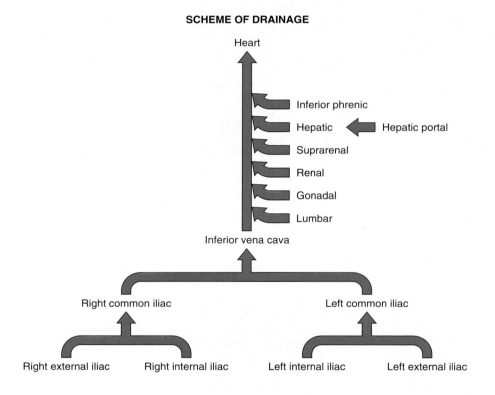

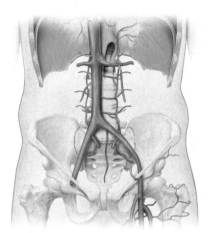

EXHIBIT 14.K **501**

VEINS	DESCRIPTION AND TRIBUTARIES	REGIONS DRAINED
Inferior vena cava	(See Exhibit 14.G)	
Inferior phrenic veins (FREN-ik=pertaining to the diaphragm)	Arise on the inferior surface of the diaphragm; the left inferior phrenic vein usually sends one tributary to the left suprarenal vein, which empties into the left renal vein, and another tributary into the inferior vena cava; the right inferior phrenic vein empties into the inferior vena cava	Inferior surface of the diaphragm and adjoining peritoneal tissues
Hepatic veins (he-PAT-ik= pertaining to the liver)	Typically two or three in number, **hepatic veins** drain the sinusoidal capillaries of the liver; the capillaries of the liver receive venous blood from the capillaries of the gastrointestinal organs via the hepatic portal vein, which receives the following tributaries from the gastrointestinal organs:	
	1. The left gastric vein arises from the left side of the lesser curvature of the stomach and joins the left side of the hepatic portal vein in the lesser omentum.	Terminal esophagus, stomach, liver, gallbladder, spleen, pancreas, small intestine, and large intestine
	2. The right gastric vein arises from the right aspect of the lesser curvature of the stomach and joins the hepatic portal vein on its anterior surface within the lesser omentum.	Lesser curvature of stomach and abdominal portion of the esophagus; stomach and duodenum
	3. The splenic vein arises in the spleen and crosses the abdomen transversely posterior to the stomach to anastomose with the superior mesenteric vein to form the hepatic portal vein. Near its junction with the hepatic portal vein, it receives the inferior mesenteric vein, which receives tributaries from the second half of the large intestine.	Spleen, fundus and greater curvature of the stomach, pancreas, greater omentum, descending colon, sigmoid colon, and rectum
	4. The superior mesenteric vein arises from numerous tributaries from most of the small intestine and the first half of the large intestine and ascends to join the splenic vein to form the hepatic portal vein.	Duodenum, jejunum, ileum, cecum, appendix, ascending colon, and transverse colon
Lumbar veins (LUM-bar= pertaining to the loin)	Usually four on each side; course horizontally through the posterior abdominal wall with the lumbar arteries; connect at right angles with the right and left ascending lumbar veins, which form the origin of the corresponding azygos or hemiazygos vein; join the ascending lumbar veins and then connect from the ascending lumbar veins to the inferior vena cava	Posterior and lateral abdominal muscle wall, lumbar vertebrae, spinal cord and spinal nerves (cauda equina) within the vertebral canal, and meninges
Suprarenal veins (soo'-pra-RĒ =nal; supra-=above)	Pass medially from the adrenal (suprarenal) glands (the left suprarenal vein joins the left renal vein, and the right suprarenal vein joins the inferior vena cava)	Adrenal (suprarenal) glands
Renal veins (RĒ-nal; ren-=kidney)	Pass anterior to the renal arteries; the left renal vein is longer than the right renal vein and passes anterior to the abdominal aorta; it receives the left testicular (or ovarian), left inferior phrenic, and usually left suprarenal veins; the right renal vein empties into the inferior vena cava posterior to the duodenum	Kidneys
Gonadal veins (gō-NAD-al; gono=seed) [**testicular** (tes-TIK-ū-lar) or **ovarian** (ō-VAR-ē-an)]	Ascend with the gonadal arteries along the posterior abdominal wall. Called testicular veins in the male; the testicular veins drain the testes; the left testicular vein joins the left renal vein, and the right testicular vein joins the inferior vena cava. Called ovarian veins in the female; the ovarian veins drain the ovaries; the left ovarian vein joins the left renal vein, and the right ovarian vein joins the inferior vena cava	Testes, epididymis, ductus deferens, ovaries, and ureters
Common iliac veins (IL-ē-ak=pertaining to the ilium)	Formed by the union of the internal and external iliac veins anterior to the sacroiliac joint and anastomose anterior to the fifth lumbar vertebra to form the inferior vena cava; the right common iliac is much shorter than the left and is also more vertical, as the inferior vena cava sits to the right of the midline	Pelvis, external genitals, and the lower limbs
Internal iliac veins	Begin near the superior portion of the greater sciatic notch and are medial to their corresponding arteries	Muscles of the pelvic wall and gluteal region, pelvic viscera, and external genitals
External iliac veins	Companions of the external iliac arteries; begin at the inguinal ligaments as the continuations of the femoral veins; end anterior to the sacroiliac joints where they join with the internal iliac veins to form the common iliac veins	The lower abdominal wall anteriorly, cremaster muscle in males, and the external genitals and lower limb

EXHIBIT 14.L Veins of the Lower Limbs *(Figure 14.14)*

 OBJECTIVE

• Identify the principal superficial and deep veins that drain the lower limbs.

As with the upper limbs, blood from the lower limbs is drained by both superficial and deep veins (Figure 14.14). The superficial veins often anastomose with one another and with deep veins

along their length. For the most part, deep veins have the same names as corresponding arteries. All veins of the lower limbs have valves, which are more numerous than in veins of the upper limbs.

CHECKPOINT

20. What is the clinical importance of the great saphenous veins?

SCHEME OF DRAINAGE

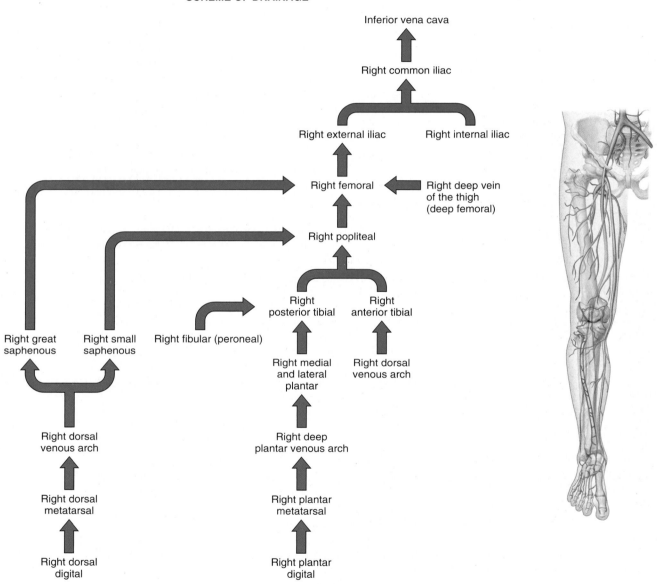

EXHIBIT 14.L **503**

VEINS	DESCRIPTION AND TRIBUTARIES	REGIONS DRAINED
DEEP VEINS		
Common iliac veins	(See Exhibit 14.K)	
External iliac veins	(See Exhibit 14.K)	
Femoral veins (FĒM-o-ral)	Accompany the femoral arteries and are the continuations of the popliteal veins just superior to the knee where the veins pass through an opening in the adductor magnus muscle; ascend deep to the sartorius muscle and emerge from beneath the muscle in the femoral triangle at the proximal end of the thigh; receive the deep veins of the thigh (deep femoral veins) and the great saphenous veins just before penetrating the abdominal wall; pass below the inguinal ligament and enter the abdominopelvic region to become the external iliac veins	Skin, lymph nodes, muscles, and bones of the thigh, and the external genitals
	Clinical note: In order to take blood samples or pressure recordings from the right side of the heart, a catheter is inserted into the femoral vein as it passes through the femoral triangle. The catheter passes through the external and common iliac veins, then into the inferior vena cava, and finally into the right atrium.	
Popliteal veins (pop'-li-TĒ-al= pertaining to the hollow behind the knee)	Formed by the union of the anterior and posterior tibial veins at the proximal end of the leg; ascend through the popliteal fossa with the popliteal arteries and the tibial nerve; terminate where they pass through a window in the adductor magnus muscle and pass to the front of the knee to become the femoral veins; also receive blood from the small saphenous veins and tributaries that correspond to branches of the popliteal artery	Knee joint and the skin, muscles, and bones around the knee joint
Posterior tibial veins (TIB-ē-al)	Begin posterior to the medial malleolus at the union of the medial and lateral plantar veins from the plantar surface of the foot; ascend through the leg with the posterior tibial artery and the tibial nerve deep to the soleus muscle; join the posterior tibial veins about two-thirds of the way up the leg; join the anterior tibial veins near the top of the interosseous membrane to form the popliteal veins; on the plantar surface of the foot the plantar digital veins unite to form the plantar metatarsal veins, which parallel the metatarsals; they in turn unite to form the deep plantar venous arches; the medial and lateral plantar veins emerge from the deep plantar venous arches	Skin, muscles, and bones on the plantar surface of the foot, and skin, muscles, and bones from the posterior and lateral aspects of the leg
Anterior tibial veins	Arise in the dorsal venous arch and accompany the anterior tibial artery; ascend deep to the tibialis anterior muscle on the anterior surface of the interosseous membrane; pass through an opening at the superior end of the interosseous membrane to join the posterior tibial veins to form the popliteal veins	Dorsal surface of the foot, ankle joint, anterior aspect of the leg, knee joint, and tibiofibular joint
SUPERFICIAL VEINS		
Great (long) saphenous veins (sa-FĒ-nus; *saphen-*=clearly visible)	The longest veins in the body; ascend from the foot to the groin in the subcutaneous layer; begin at the medial end of the dorsal venous arches of the foot; the dorsal venous arches (VĒ-nus) are networks of veins on the dorsum of the foot formed by the dorsal digital veins, which collect blood from the toes, and then unite in pairs to form the dorsal metatarsal veins, which parallel the metatarsals; as the dorsal metatarsal veins approach the foot, they combine to form the dorsal venous arches, pass anterior to the medial malleolus of the tibia and then superiorly along the medial aspect of the leg and thigh approximately four finger breadths posterior to the patella and just deep to the skin; receive tributaries from superficial tissues and connect with the deep veins as well; empty into the femoral veins at the groin; have from 10 to 20 valves along their length, with more located in the leg than the thigh	The integumentary tissues and superficial muscles of the lower limbs, groin, and lower abdominal wall
	Clinical note: These veins are more likely to be subject to varicosities than other veins in the lower limbs because they must support a long column of blood and are not well supported by skeletal muscles. The great saphenous veins are often used for prolonged administration of intravenous fluids. This is particularly important in very young children and in patients of any age who are in shock and whose veins are collapsed. In coronary artery bypass grafting, if multiple blood vessels need to be grafted, sections of the great saphenous vein are used along with at least one artery as a graft (see **Clinical note** in Exhibit 14.C). After the great saphenous vein is removed and divided into sections, the sections are used to bypass the blockages. The vein grafts are reversed so that the valves do not obstruct the flow of blood.	
Small saphenous veins	Begin at the lateral aspect of the dorsal venous arches of the foot; pass posterior to the lateral malleolus of the fibula and ascend deep to the skin along the posterior aspect of the leg; empty into the popliteal veins in the popliteal fossa, posterior to the knee; have from 9 to 12 valves; may communicate with the great saphenous veins in the proximal leg	The integumentary tissues and superficial muscles of the foot and posterior aspect of the leg

EXHIBIT 14.L Veins of the Lower Limbs *(Figure 14.14)* CONTINUED

Figure 14.14 **Principal veins of the pelvis and free lower limbs.**

Deep veins usually bear the names of their companion arteries.

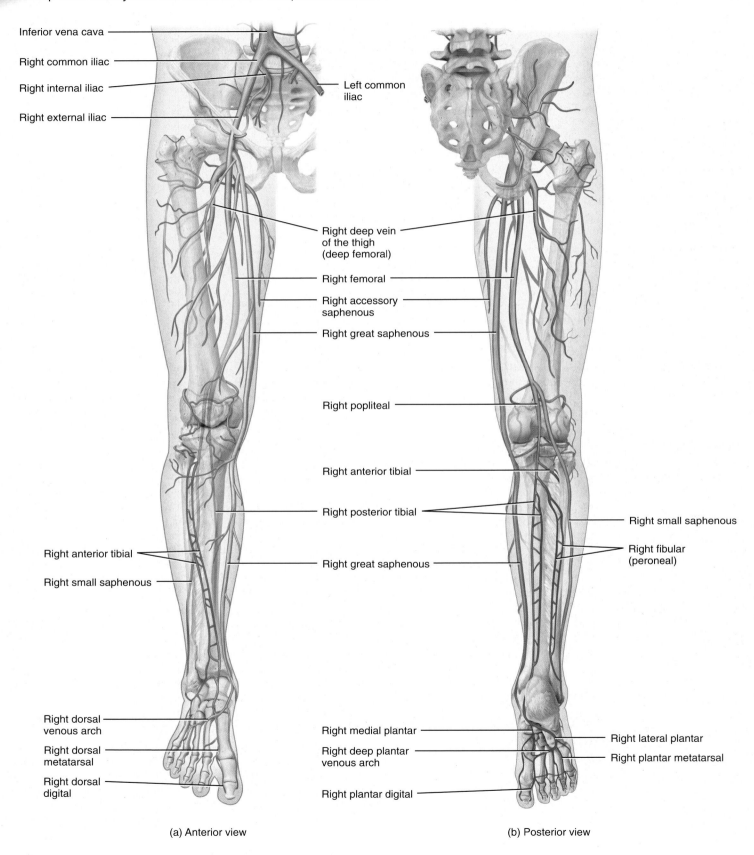

Inferior vena cava

Right common iliac

Right internal iliac

Right external iliac

Left common iliac

Right deep vein of the thigh (deep femoral)

Right femoral

Right accessory saphenous

Right great saphenous

Right popliteal

Right anterior tibial

Right posterior tibial

Right small saphenous

Right anterior tibial

Right great saphenous

Right fibular (peroneal)

Right small saphenous

Right dorsal venous arch

Right dorsal metatarsal

Right dorsal digital

Right medial plantar

Right deep plantar venous arch

Right plantar digital

Right lateral plantar

Right plantar metatarsal

(a) Anterior view

(b) Posterior view

EXHIBIT 14.L **505**

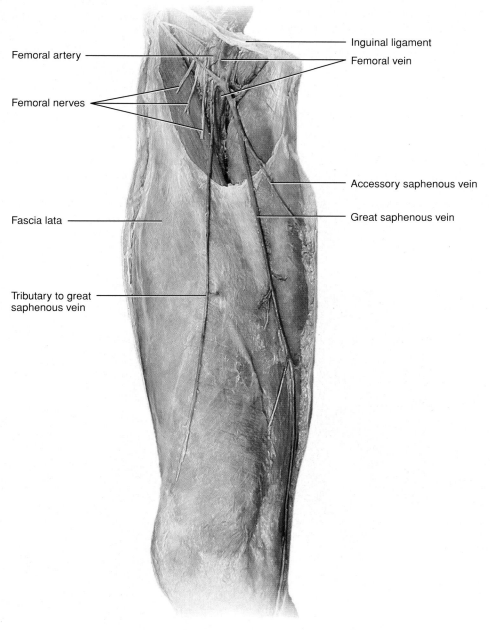

Femoral artery

Femoral nerves

Fascia lata

Tributary to great
saphenous vein

Inguinal ligament

Femoral vein

Accessory saphenous vein

Great saphenous vein

(c) Anteromedial view of thigh

Which veins of the lower limb are superficial?

The Hepatic Portal Circulation

The hepatic portal circulation carries venous blood from the gastrointestinal organs and spleen to the liver. A vein that carries blood from one capillary network to another is called a **portal vein**. The hepatic portal vein receives blood from capillaries of gastrointestinal organs and the spleen and delivers it to the sinusoids of the liver (Figure 14.15). After a meal, hepatic portal blood is rich in nutrients absorbed from the gastrointestinal tract. The liver stores some of them and modifies others before they pass into the general circulation. For example, the liver converts glucose into glycogen for storage, reducing blood glucose level shortly after a meal. The liver also detoxifies harmful substances, such as alcohol, that have been absorbed from the gastrointestinal tract, and destroys bacteria by phagocytosis.

The superior mesenteric and splenic veins unite to form the hepatic portal vein. The **superior mesenteric vein** (MES-en-ter′-ik) drains blood from the small intestine and portions of the large intestine, stomach, and pancreas through the *jejunal, ileal, ileocolic* (il′-ē-ō-KOL-ik), *right colic, middle colic, pancreaticoduodenal* (pan′-krē-at-i-kō-doo-ō-DĒ-nal), and *right gastro-omental veins*. The **splenic vein** drains blood from the stomach, pancreas, and portions of the large intestine through the *short gastric, left gastro-omental, pancreatic,* and *inferior mesenteric veins*. The **inferior mesenteric vein**, which passes into the splenic vein, drains portions of the large intestine through the *superior rectal, sigmoidal,* and *left colic veins*. The *right* and *left gastric veins*, which open directly into the hepatic portal vein, drain the stomach. The *cystic vein*, which also opens into the hepatic portal vein, drains the gallbladder.

Figure 14.15 **Hepatic portal circulation.** A schematic diagram of blood flow through the liver, including arterial circulation, is shown in (b); deoxygenated blood is indicated in blue, oxygenated blood in red.

🔑 The hepatic portal circulation delivers venous blood from the organs of the gastrointestinal tract and spleen to the liver.

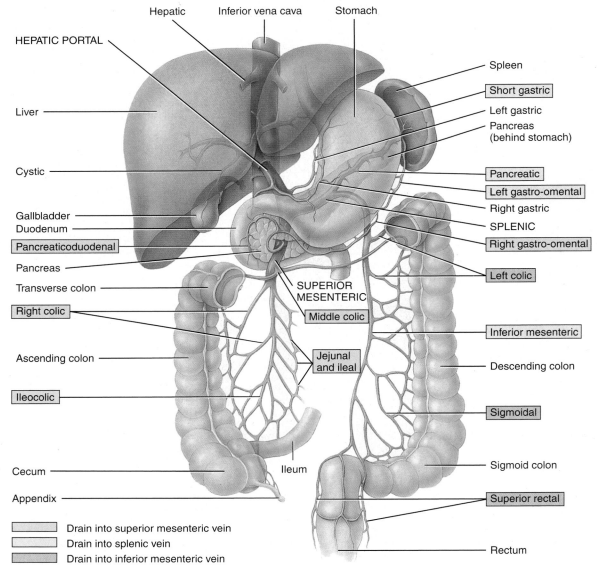

Drain into superior mesenteric vein
Drain into splenic vein
Drain into inferior mesenteric vein

(a) Anterior view of veins draining into the hepatic portal vein

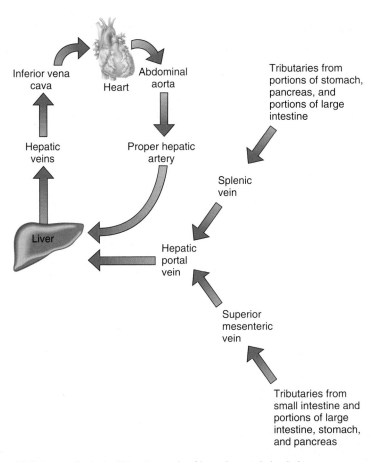

 Which veins carry blood away from the liver?

(b) Scheme of principal blood vessels of hepatic portal circulation and arterial supply and venous drainage of liver

At the same time the liver is receiving nutrient-rich but deoxygenated blood via the hepatic portal vein, it also is receiving oxygenated blood via the hepatic artery, a branch of the celiac trunk. The oxygenated blood mixes with the deoxygenated blood in sinusoids. Eventually, blood leaves the sinusoids of the liver through the **hepatic veins,** which drain into the inferior vena cava.

The Pulmonary Circulation

The pulmonary circulation carries deoxygenated blood from the right ventricle to the air sacs (alveoli) within the lungs and returns oxygenated blood from the air sacs to the left atrium (Figure 14.16). The **pulmonary trunk** emerges from the right ventricle and passes superiorly, posteriorly, and to the left. It then divides into two branches: the **right pulmonary artery** to the right lung and the **left pulmonary artery** to the left lung. After birth, the pulmonary arteries are the only arteries that carry deoxygenated blood. On entering the lungs, the branches divide and subdivide until finally they form capillaries around the air sacs (alveoli) within the lungs. CO_2 passes from the blood into the air sacs and is exhaled. Inhaled O_2 passes from the air within the lungs into the blood. The pulmonary capillaries unite to form venules and eventually **pulmonary veins,** which exit the lungs and carry the oxygenated blood to the left atrium. Two left and two right pulmonary veins enter the left atrium. After birth, the pulmonary veins are the only veins that carry oxygenated blood. Contractions of the left ventricle then eject the oxygenated blood into the systemic circulation.

Figure 14.16 Pulmonary circulation.

The pulmonary circulation brings deoxygenated blood from the right ventricle to the lungs and returns oxygenated blood from the lungs to the left atrium.

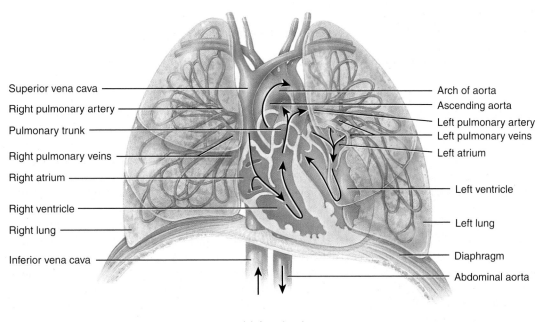

(a) Anterior view

FIGURE 14.16 CONTINUES ▶

FIGURE **14.16** CONTINUED

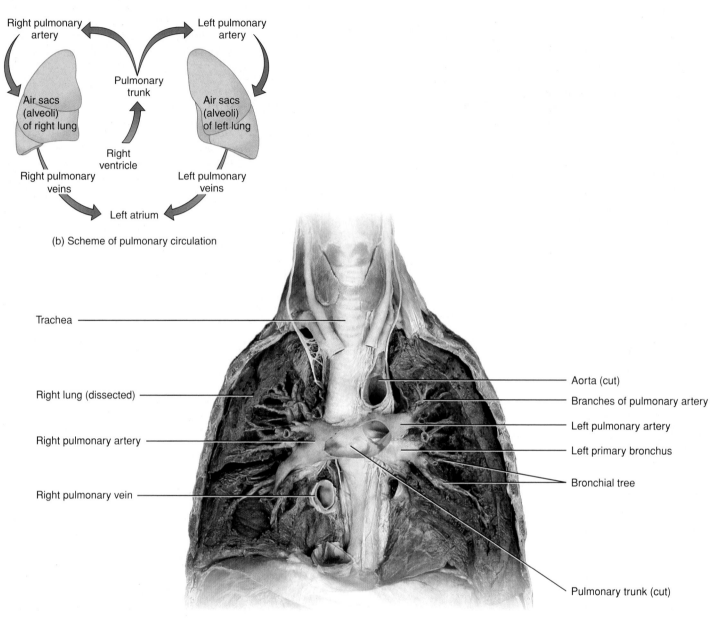

(b) Scheme of pulmonary circulation

(c) Pulmonary circulation into the lungs

After birth, which are the only arteries that carry deoxygenated blood?

The Fetal Circulation

The circulatory system of a fetus, called the fetal circulation, exists only in the fetus and contains special structures that allow the developing fetus to exchange materials with its mother (Figure 14.17). It differs from the postnatal (after birth) circulation because the lungs, kidneys, and gastrointestinal organs do not begin to function until birth. The fetus obtains O_2 and nutrients from the maternal blood and eliminates CO_2 and other wastes into it.

The exchange of materials between fetal and maternal circulations occurs through the **placenta** (pla-SEN-ta), which forms inside the mother's uterus and attaches to the umbilicus (navel) of the fetus by the **umbilical cord** (um-BIL-i-kal). The placenta communicates with the mother's cardiovascular system through many small blood vessels that emerge from the uterine wall. The umbilical cord contains blood vessels that branch into capillaries in the placenta. Wastes from the fetal blood diffuse out of

the capillaries, into spaces containing maternal blood (intervillous spaces) in the placenta, and finally into the mother's uterine veins. Nutrients travel the opposite route—from the maternal blood vessels to the intervillous spaces to the fetal capillaries. Normally, there is no direct mixing of maternal and fetal blood because all exchanges occur by diffusion through capillary walls.

Blood passes from the fetus to the placenta via two **umbilical arteries** (Figure 14.17) in the umbilical cord. These branches of the internal iliac (hypogastric) arteries are within the umbilical cord. At the placenta, fetal blood picks up O_2 and nutrients and eliminates CO_2 and wastes. The oxygenated blood returns from the placenta via a single **umbilical vein** in the umbilical cord. This vein ascends to the liver of the fetus, where it divides into two branches. Some blood flows through the branch that joins the hepatic portal vein and enters the liver, but most of the blood flows into the second branch, the **ductus venosus** (DUK-tus ve-NŌ-sus), which drains into the inferior vena cava.

Figure 14.17 Fetal circulation and changes at birth. The gold boxes between parts (a) and (b) describe the fate of certain fetal structures once postnatal circulation is established.

🔑 The lungs and gastrointestinal organs do not begin to function until birth.

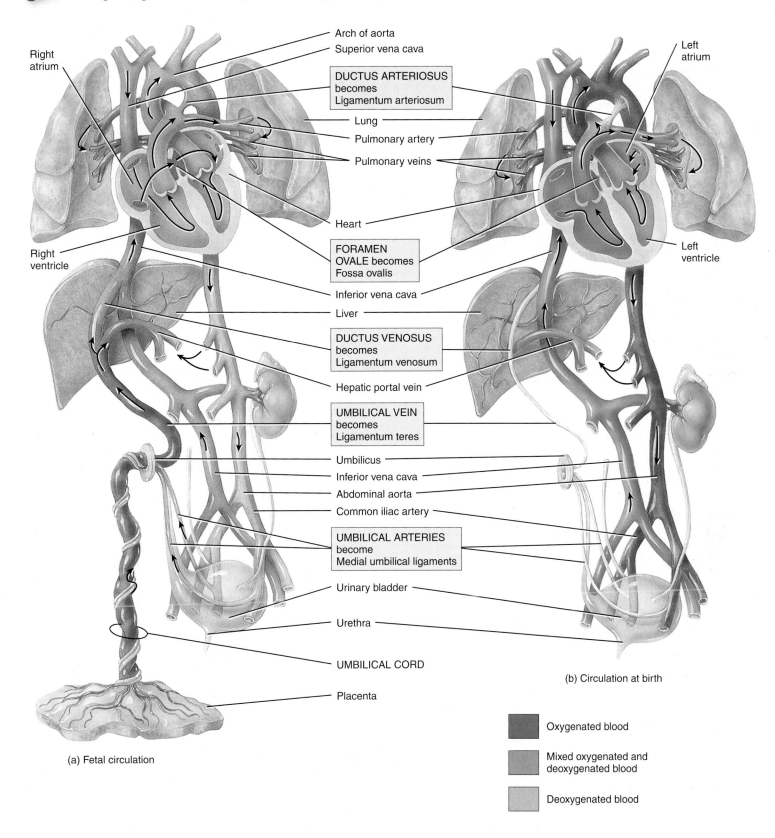

(a) Fetal circulation

(b) Circulation at birth

■ Oxygenated blood

■ Mixed oxygenated and deoxygenated blood

■ Deoxygenated blood

FIGURE 14.17 CONTINUES ▶

FIGURE **14.17** CONTINUED

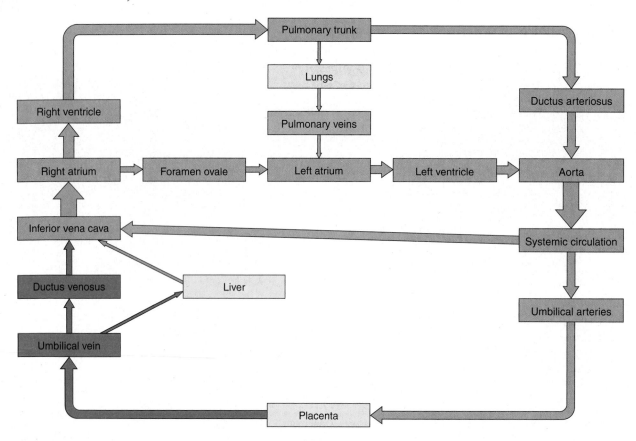

(c) Scheme of fetal circulation

Through which structure does the exchange of materials between mother and fetus occur?

Deoxygenated blood returning from lower body regions of the fetus mingles with oxygenated blood from the ductus venosus in the inferior vena cava. This mixed blood then enters the right atrium. Deoxygenated blood returning from upper body regions of the fetus enters the superior vena cava and also passes into the right atrium.

Most of the fetal blood does not pass from the right ventricle to the lungs, as it does in postnatal circulation, because an opening called the **foramen ovale** (fō-RA-men ō-VAL-ē) exists in the septum between the right and left atria. Most of the blood that enters the right atrium passes through the foramen ovale into the left atrium and joins the systemic circulation. The blood that does pass into the right ventricle is pumped into the pulmonary trunk, but little of this blood reaches the nonfunctioning fetal lungs. Instead, most is sent through the **ductus arteriosus** (ar-tē-rē-Ō-sus), a vessel that connects the pulmonary trunk with the aorta. The blood in the aorta is carried to all fetal tissues through the systemic circulation. When the common iliac arteries branch into the external and internal iliacs, part of the blood flows into the internal iliacs, into the umbilical arteries, and back to the placenta for another exchange of materials.

After birth, when pulmonary (lung), renal (kidney), and digestive functions begin, the following vascular changes occur (Figure 14.17b):

1. When the umbilical cord is tied off, blood no longer flows through the umbilical arteries, they fill with connective tissue, and the distal portions of the umbilical arteries become fibrous cords called the **medial umbilical ligaments**. Although the arteries are closed functionally only a few minutes after birth, complete obliteration of the lumens may take 2 to 3 months.

2. The umbilical vein collapses but remains as the **ligamentum teres (round ligament)** (TE-rēz), a structure that attaches the umbilicus to the liver.

3. The ductus venosus collapses but remains as the **ligamentum venosum** (ve-NŌ-sum), a fibrous cord on the inferior surface of the liver.

4. The placenta is expelled as the **afterbirth**.

5. The foramen ovale normally closes shortly after birth to become the **fossa ovalis**, a depression in the interatrial septum. When an infant takes its first breath, the lungs expand and blood flow to the lungs increases. Blood returning from the lungs to the heart increases pressure in the left atrium. This closes the foramen ovale by pushing the valve that guards it against the interatrial septum. Permanent closure occurs in about a year.

6. The ductus arteriosus closes by vasoconstriction almost immediately after birth and becomes the **ligamentum arteriosum** (ar-tēr-ē-Ō-sum). Complete anatomical obliteration of the lumen takes 1 to 3 months.

✓ CHECKPOINT

21. What is the purpose of the systemic circulation?

22. Diagram the hepatic portal circulation. Why is this route important?

23. Outline the route of the pulmonary circulation.

24. Discuss the anatomy and physiology of the fetal circulation, including the functions of the umbilical arteries, umbilical vein, ductus venosus, foramen ovale, and ductus arteriosus.

14.3 DEVELOPMENT OF BLOOD VESSELS AND BLOOD

OBJECTIVE

• Describe the development of blood vessels and blood.

The development of blood cells and the formation of blood vessels begins outside the embryo as early as the beginning of the third week in the **mesoderm** of the wall of the yolk sac, chorion, and connecting stalk (see Figure 4.10c). About 2 days later, blood vessels form within the embryo. The early formation of the cardiovascular system is linked to the small amount of yolk in the ovum and yolk sac. As the embryo develops rapidly during the third week, there is a greater need to develop a cardiovascular system to supply sufficient nutrients to the embryo and remove wastes from it.

Blood vessels and blood cells develop from the same precursor cell, called a **hemangioblast** (hē-MAN-jē-ō-blast; *hema-*=blood; *blast*=immature stage). Once mesenchyme develops into hemangioblasts, they can give rise to cells that produce blood vessels or cells that produce blood cells.

Blood vessels develop from **angioblasts** (AN-jē-ō-blasts), which are derived from hemangioblasts (Figure 14.18). Angioblasts aggregate to form isolated masses and cords throughout the embryonic disc called **blood islands** (Figure 14.18). Spaces soon appear in the islands and become the lumens of the blood vessels. Some of the angioblasts immediately around the spaces give rise to the *endothelial lining of the blood vessels*. Angioblasts around the endothelium form the *tunics* (interna, media, and externa) of the larger blood vessels. Growth and fusion of blood islands form an extensive network of blood vessels throughout the embryo. By continuous branching, blood vessels outside the embryo connect with those inside the embryo, linking the embryo with the placenta.

Blood cells develop from **pluripotent stem cells** (ploo-RIP-ō-tent), also derived from hemangioblasts. This development occurs in the walls of blood vessels in the yolk sac, chorion, and allantois at about the end of the third week after fertilization. Blood formation in the embryo itself begins at about the fifth week in the liver and the twelfth week in the spleen, red bone marrow, and thymus.

CHECKPOINT

25. What are the sites of blood cell production outside the embryo? Within the embryo?

14.4 AGING AND THE CARDIOVASCULAR SYSTEM

OBJECTIVE

• Explain the effects of aging on the cardiovascular system.

General changes in the cardiovascular system associated with aging include decreased compliance of the aorta, reduction in cardiac muscle fiber size, progressive loss of cardiac muscular strength, reduced cardiac output, a decline in maximum heart

Figure 14.18 Development of blood vessels and blood cells from blood islands.

 Blood vessel development begins in the embryo on about day 17 or 18.

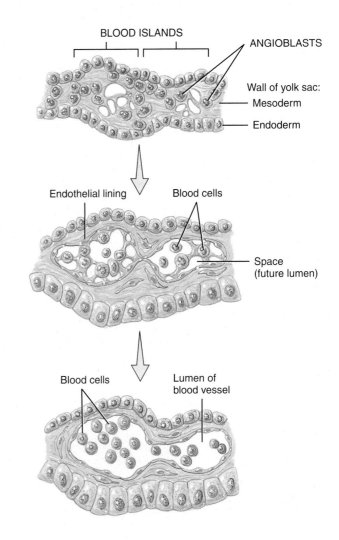

 From which germ cell layer are blood vessels and blood derived?

rate, and an increase in systolic blood pressure. Total blood cholesterol tends to increase with age, as does low-density lipoprotein (LDL); high-density lipoprotein (HDL) tends to decrease. There is an increase in the incidence of coronary artery disease (CAD), the major cause of heart disease and death in older Americans (see Clinical Connection in Section 13.7). Congestive heart failure (CHF), a set of symptoms associated with impaired pumping of the heart, is also prevalent in older individuals. Changes in blood vessels that serve brain tissue—such as atherosclerosis—reduce nourishment to the brain and result in the malfunction or death of brain cells. By age 80, cerebral blood flow is 20 percent less and renal blood flow is 50 percent less than in the same person at age 30 because of the effects of aging on the vessels.

CHECKPOINT

26. How does aging affect the blood vessels?

CLINICAL CONNECTION | *Hypertension*

About 50 million Americans have **hypertension** (hī′-per-TEN-shun), or persistently high blood pressure. It is the most common disorder affecting the heart and blood vessels and is the major cause of heart failure, kidney disease, and stroke. In May 2003, the Joint National Committee on Prevention, Detection, Evaluation, and Treatment of High Blood Pressure published new guidelines for hypertension:

Category	Systolic (mmHg)	Diastolic (mmHg)
Normal	Less than 120 *and*	Less than 80
Prehypertension	120–139 *or*	80–89
Stage 1 hypertension	140–159 *or*	90–99
Stage 2 hypertension	Greater than 160 *or*	Greater than 100

Types and Causes of Hypertension

Between 90 percent and 95 percent of all cases of hypertension are **primary hypertension**, a persistently elevated blood pressure that cannot be attributed to any identifiable cause. The remaining 5–10 percent are cases of **secondary hypertension**, which has an identifiable underlying cause. Several disorders cause secondary hypertension:

- *Obstruction of renal blood flow* or disorders that damage renal tissue may cause the kidneys to release excessive amounts of an enzyme called renin into the blood. Renin converts angiotensinogen (an′-jē-ō-ten-SIN-ō-jen) (a plasma protein produced by the liver) into the hormone angiotensin I. In the lungs, angiotensin I is converted into the hormone angiotensin II. The resulting high level of angiotensin II causes vasoconstriction, thus increasing systemic vascular resistance.
- *Hypersecretion of aldosterone* resulting, for instance, from a tumor of the adrenal cortex stimulates excess reabsorption of salt and water by the kidneys, which increases the volume of body fluids.
- *Hypersecretion of epinephrine and norepinephrine* by a **pheochromocytoma** (fē-ō-krō′-mō-sī-TŌ-ma), a tumor of the adrenal medulla. Epinephrine and norepinephrine increase heart rate and contractility and increase systemic vascular resistance.

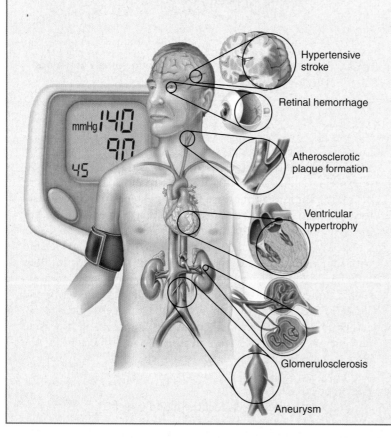

Hypertensive stroke

Retinal hemorrhage

Atherosclerotic plaque formation

Ventricular hypertrophy

Glomerulosclerosis

Aneurysm

Damaging Effects of Untreated Hypertension

High blood pressure is known as the "silent killer" because it can cause considerable damage to the blood vessels, heart, brain, and kidneys before it causes pain or other noticeable symptoms. It is a major risk factor for the number-one and number-three causes of death in the United States, which are heart disease and stroke, respectively. In blood vessels, hypertension causes thickening of the tunica media, accelerates development of atherosclerosis and coronary artery disease, and increases systemic vascular resistance. In the heart, hypertension increases and forces the ventricles to work harder to eject blood.

The normal response to the increased workload caused by vigorous and regular exercise is hypertrophy of the myocardium, especially in the wall of the left ventricle. This is a positive effect that makes the heart a more efficient pump. An increased afterload, however, leads to myocardial hypertrophy, accompanied by muscle damage and fibrosis (a buildup of collagen fibers between the muscle fibers). As a result, the left ventricle enlarges, weakens, and dilates. Because arteries in the brain are usually less protected by surrounding tissues than are the major arteries in other parts of the body, prolonged hypertension can eventually cause them to rupture, resulting in a stroke. Hypertension also damages kidney arterioles, causing them to thicken, which narrows the lumen; because the blood supply to the kidneys is thereby reduced, the kidneys secrete more renin, which elevates blood pressure even more.

Lifestyle Changes to Reduce Hypertension

The following lifestyle changes are effective in managing hypertension:

- *Lose weight.* Loss of even a few pounds helps reduce blood pressure in overweight hypertensive individuals.
- *Limit alcohol intake.* Drinking in moderation may lower the risk of coronary heart disease, mainly among males over 45 and females over 55. Moderation is defined as no more than one 12-oz beer per day for females and no more than two 12-oz beers per day for males.
- *Exercise.* Engaging in moderate activity several times a week for 30 to 45 minutes can lower systolic blood pressure by about 10 mmHg.
- *Reduce intake of sodium (salt).* Roughly half the people with hypertension are "salt sensitive." For them, a high-salt diet appears to promote hypertension, and a low-salt diet can lower their blood pressure.
- *Maintain recommended dietary intake of potassium, calcium, and magnesium.* Higher levels are associated with a lower risk of hypertension.
- *Don't smoke* or *quit smoking.* Smoking can augment the damaging effects of high blood pressure by promoting vasoconstriction.
- *Manage stress.* Meditation and biofeedback techniques help some people reduce high blood pressure by decreasing the daily release of epinephrine and norepinephrine by the adrenal medulla.

Drug Treatment of Hypertension

Drugs with several different mechanisms of action are effective in lowering blood pressure. Many people are successfully treated with *diuretics* (dī-ū-RET-iks), agents that decrease blood pressure by decreasing blood volume through the increased elimination of water and salt in the urine. *ACE (angiotensin converting enzyme) inhibitors* block formation of angiotensin II and thereby promote vasodilation and decrease the secretion of aldosterone. *Beta blockers* (BĀ-ta) inhibit the secretion of renin and decrease heart rate and contractility. *Vasodilators* relax the smooth muscle in arterial walls, causing vasodilation and lowering systemic vascular resistance. The *calcium channel blockers* are vasodilators that slow the inflow of Ca^{2+} into vascular smooth muscle cells, reducing the heart's workload by slowing Ca^{2+} entry into pacemaker cells and regular myocardial fibers, thereby decreasing heart rate and the force of myocardial contraction.

KEY MEDICAL TERMS ASSOCIATED WITH BLOOD VESSELS

Aneurysm (AN-ū-rizm) A thin, weakened section of the wall of an artery or a vein that bulges outward, forming a balloonlike sac.

Angiography (an′-jē-OG-ra-fē; *angio-*=vessel; *graphos*=to write) A diagnostic procedure in which a radiopaque dye is injected through a catheter that has been introduced into a blood vessel and guided to the blood vessel to be examined. The dye flows into the appropriate blood vessel, making abnormalities such as blockages visible on x-rays.

Arteritis (ar′-te-RĪ-tis; *itis*=inflammation of) Inflammation of an artery, probably due to an autoimmune response.

Carotid endarterectomy (ka-ROT-id end′-ar-ter-EK-tō-mē) The removal of atherosclerotic plaque from the carotid artery to restore greater blood flow to the brain.

Claudication (klaw′-di-KĀ-shun) Pain and lameness or limping caused by defective circulation of the blood in the vessels of the limbs.

Deep vein thrombosis The presence of a thrombus (blood clot) in a deep vein of the lower limbs. It may lead to (1) pulmonary embolism, if the thrombus dislodges and then lodges within the pulmonary arterial blood flow, and (2) postphlebitic syndrome, which consists of edema, pain, and skin changes due to destruction of venous valves.

Doppler ultrasound scanning Imaging technique commonly used to measure blood flow. A transducer is placed on the skin and an image is displayed on a monitor that provides the exact position and severity of a blockage.

Hypotension (hī-pō-TEN-shun) Low blood pressure; most commonly used to describe an acute drop in blood pressure, as occurs during excessive blood loss.

Normotensive (nor′-mō-TEN-siv) Normal blood pressure.

Occlusion (ō-KLOO-zhun) The closure or obstruction of the lumen of a structure such as a blood vessel.

Orthostatic hypotension (or′-thō-STAT-ik; *ortho-*=straight; *static-*=causing to stand) An excessive lowering of systemic blood pressure when a person assumes an erect or semierect posture.

Phlebitis (fle-BĪ-tis; *phleb-*=vein) Inflammation of a vein, often in a leg.

Shock Failure of the cardiovascular system to deliver enough oxygen and nutrients to meet cellular metabolic needs. It is caused by loss of body fluids; symptoms include systolic blood pressure less than 90 mmHg, rapid heart rate and pulse, cool and pale skin, sweating, decreased urine formation, thirst, nausea, and altered mental state.

Syncope (SIN-kō-pē=cutting short) Fainting; a sudden, temporary loss of consciousness followed by spontaneous recovery, usually due to decreased blood flow to the brain.

Thrombectomy (throm-BEK-tō-mē; *thrombo-*=clot) An operation to remove a blood clot from a blood vessel.

Thrombophlebitis (throm′-bō-fle-BĪ-tis) Inflammation of a vein involving clot formation.

Venipuncture (VEN-i-punk-chur; *vena-*=vein) The puncture of a vein, usually to withdraw blood for analysis or introduce a solution such as, for example, an antibiotic.

White coat (office) hypertension A stress-induced syndrome found in patients who have elevated blood pressure when being examined by health-care personnel, but otherwise have normal blood pressure.

CHAPTER REVIEW AND RESOURCE SUMMARY

WileyPLUS

Review

14.1 Anatomy of Blood Vessels

1. Arteries carry blood away from the heart. The wall of an artery consists of a tunica interna, a tunica media (which maintains elasticity and contractility), and a tunica externa. Large arteries are termed elastic (conducting) arteries, and medium-sized arteries are called muscular (distributing) arteries.

2. Many arteries anastomose (the distal ends of two or more vessels unite). An alternate blood route from an anastomosis is called collateral circulation. Arteries that do not anastomose are called end arteries.

3. Arterioles are small arteries that deliver blood to capillaries. Through constriction and dilation, they regulate blood flow from arteries into capillaries and alter arterial blood pressure.

4. Capillaries are microscopic blood vessels through which materials are exchanged between blood and tissue cells. Some capillaries are continuous; others are fenestrated.

5. Capillaries branch to form an extensive network throughout a tissue. This network increases the surface area available for the exchange of materials between the blood and the body tissue; it also allows rapid exchange of large quantities of materials. Precapillary sphincters regulate blood flow through capillaries.

6. Microscopic blood vessels in the liver are called sinusoids.

7. Venules are small vessels that form from the merging capillaries; venules merge to form veins. Veins consist of the same three tunics as arteries but have a thinner tunica interna and tunica media. The lumen of a vein is also larger than that of a comparable artery.

8. Veins contain valves to prevent backflow of blood. Weak valves can lead to varicose veins. Vascular (venous) sinuses are veins with very thin walls.

9. Systemic veins are collectively called blood reservoirs because they hold a large volume of blood. If the need arises, this blood can be shifted into other blood vessels through vasoconstriction. The principal blood reservoirs are the veins of the abdominal organs (liver and spleen) and skin.

Resource

Anatomy Overview - Arteries and Arterioles
Anatomy Overview - Capillaries
Animation - Capillary Exchange
Figure 14.1 - Comparative Structure of Blood Vessels
Exercise - Capillary Exchange Pick 'em
Exercise - Vein Archery
Concepts and Connections - Capillary Exchange

Review

14.2 Circulatory Routes

1. The two basic postnatal circulatory routes are the systemic and pulmonary circulations. Among the subdivisions of the systemic circulation are the coronary (cardiac) circulation and the hepatic portal circulation. Fetal circulation exists only in the fetus.

2. The systemic circulation carries oxygenated blood from the left ventricle through the aorta to all parts of the body and returns the deoxygenated blood to the right atrium.

3. The aorta is divided into the ascending aorta, the arch of the aorta, and the descending aorta. Each section gives off arteries that branch to supply the whole body.

4. Blood returns to the heart through the systemic veins. All veins of the systemic circulation drain into the superior or inferior venae cavae or the coronary sinus; these in turn empty into the right atrium.

5. The principal blood vessels of the systemic circulation are presented in Exhibits 14.A–14.L.

6. The hepatic portal circulation detours venous blood from the gastrointestinal organs and spleen and directs it into the hepatic portal vein of the liver before it is returned to the heart. It enables the liver to utilize nutrients and detoxify harmful substances in the blood.

7. The pulmonary circulation takes deoxygenated blood from the right ventricle to the alveoli within the lungs and returns oxygenated blood from the alveoli to the left atrium. It allows blood to be oxygenated for systemic circulation.

8. The fetal circulation involves the exchange of materials between fetus and mother. The fetus derives O_2 and nutrients and eliminates CO_2 and wastes through the maternal blood supply via the placenta. At birth, when pulmonary (lung), digestive, and liver functions begin to operate, the special structures of fetal circulation are no longer needed.

Anatomy Overview - Comparison of Circulatory Routes
Anatomy Overview - Pulmonary Circulation
Animation - MABP
Animation - Vascular Regulation
Animation - Lymph Flow
Figure 14.15 - Hepatic Portal Circulation
Figure 14.17 - Fetal Circulation
Concepts and Connections - Blood Flow

14.3 Development of Blood Vessels and Blood

1. Blood vessels develop from mesenchyme in mesoderm called blood islands.

2. Blood cells also develop from mesenchyme (hemangioblasts → pluripotent stem cells).

14.4 Aging and the Cardiovascular System

1. General changes in the cardiovascular system associated with aging include reduced elasticity of blood vessels, reduction in cardiac muscle size, reduced cardiac output, and increased systolic blood pressure.

2. The incidences of coronary artery disease (CAD), congestive heart failure (CHF), and atherosclerosis increase with age.

Animation - Regulating Blood Pressure

CRITICAL THINKING QUESTIONS

1. Which structures present in the fetal circulation are absent from the adult circulatory system? Why do these changes occur?

2. The word *sinus* has more than one anatomical meaning. In the skeletal system, what are "sinuses"? Name them. (*Hint:* See Section 7.2.) In the circulatory system, what does this same term mean? Identify the path of blood taken from the superior sagittal sinus to the superior vena cava.

3. You've read about varicose veins. Why aren't there varicose arteries?

4. Enrique is scheduled for a medical procedure called a *cardiac catheterization*. During this procedure, a tube will be inserted into

Enrique's femoral artery and snaked through the blood vessels to reach the heart. If the ultimate target is the pulmonary arteries, what vessels will the tube need to pass through?

5. Gina is 45 years old and works as an executive secretary in the college's biology department. She is 5'3" tall, weighs 163 pounds, and has smoked since she was 15 years old. Her doctor measured her blood pressure at "132 over 86." What do you think the doctor will be advising Gina to do, and why?

ANSWERS TO FIGURE QUESTIONS

14.1 The femoral artery has the thicker wall; the femoral vein has the wider lumen.

14.2 Metabolically active tissues have extensive capillary networks because they use O_2 and produce wastes more rapidly than less active tissues.

14.3 Materials cross capillary walls through intercellular clefts and fenestrations, via transcytosis in pinocytic vesicles, and through the plasma membranes of endothelial cells.

14.4 Valves are more important in arm veins and leg veins than in neck veins because gravity tends to cause pooling of blood in the veins of the free limbs when you are standing. The valves prevent backflow into the free limbs so the blood proceeds toward the right atrium after each heartbeat. When you are erect, gravity aids the flow of blood in neck veins back toward the heart.

14.5 The two main circulatory routes in the human body are the systemic circulation and the pulmonary circulation.

14.6 The four principal subdivisions of the aorta are the ascending aorta, arch of the aorta, thoracic aorta, and abdominal aorta.

14.7 The three major branches of the arch of the aorta, in order of their origination, are the brachiocephalic trunk, left common carotid artery, and left subclavian artery.

14.8 The abdominal aorta begins at the aortic hiatus in the diaphragm.

14.9 The abdominal aorta divides into the common iliac arteries at about the level of L4.

14.10 The superior vena cava drains regions above the diaphragm, and the inferior vena cava drains regions below the diaphragm.

14.11 All venous blood in the brain drains into the internal jugular veins.

14.12 The median cubital vein of the free upper limb is often used for withdrawing blood.

14.13 The inferior vena cava returns blood from abdominopelvic viscera to the heart.

14.14 The superficial veins of the free lower limbs are the dorsal venous arch and the great saphenous and small saphenous veins.

14.15 The hepatic veins carry blood away from the liver.

14.16 After birth, the only arteries that carry deoxygenated blood are the pulmonary arteries.

14.17 Exchange of materials between mother and fetus occurs across the placenta.

14.18 Blood vessels and blood cells are derived from mesoderm.

15 THE LYMPHATIC SYSTEM AND IMMUNITY

INTRODUCTION Maintaining physical health requires continuous combat against harmful agents in our internal and external environments. Despite constant exposure to a variety of **pathogens** (PATH-ō-jens), disease-producing microbes such as bacteria and viruses, most people remain healthy. The body's surface also responds to cuts and bumps, exposure to ultraviolet rays in sunlight, chemical toxins, and minor burns with an array of defensive ploys. **Immunity** (i-MŪ-ni-tē) or **resistance** is the ability to ward off damage or disease. Lack of resistance is termed **susceptibility**.

Immunity is divided into two general types: (1) innate (nonspecific) immunity and (2) adaptive (specific) immunity. **Innate (nonspecific) immunity** refers to defenses that are present from birth and are always available to protect us against disease. Innate immunity does not involve specific recognition of a microbe; it acts against all microbes in the same way. In addition, innate immunity does not have a memory component, that is, it cannot recall a previous contact with a foreign molecule. Among the components of innate immunity are the first line of defense (skin and mucous membranes) and the second line of defense (natural killer cells and phagocytes, inflammation, fever, and antimicrobial substances). Innate immune responses can be considered the body's early-warning system; they are designed to prevent microbes from gaining access into the body and to help eliminate those that do gain access.

Adaptive (specific) immunity refers to defenses that involve specific recognition of a microbe once it has breached the defenses of innate immunity. Adaptive immunity is based on a specific response to a specific microbe; it adapts or adjusts to handle a single type of invader. Adaptive immunity is slower to respond than innate immunity, but it does have a memory component; it involves lymphocytes called T lymphocytes (T cells) and B lymphocytes (B cells). •

Did you ever wonder how cancer can spread from one part of the body to another?

15.1 LYMPHATIC SYSTEM STRUCTURE AND FUNCTIONS

 OBJECTIVES

- Describe the components and major functions of the lymphatic system.
- Outline the organization of lymphatic vessels.
- Explain the formation and flow of lymph.
- Compare the structure and functions of the primary and secondary lymphatic organs and tissues.

The **lymphatic system** (lim-FAT-ik) consists of four elements: a fluid called lymph, lymphatic vessels that transport the lymph, a number of structures and organs containing lymphocytes within a filtering tissue (lymphatic tissue), and red bone marrow (Figure 15.1). The lymphatic system assists in circulating body fluids and helps defend the body against disease-causing agents.

Most components of blood plasma filter through blood capillary walls to form interstitial fluid. After interstitial fluid passes into lymphatic vessels, it is referred to as **lymph** (LIMF=clear fluid). So, interstitial fluid and lymph are very similar; the major difference between them is location. Interstitial fluid is found between cells, but lymph is located within lymphatic vessels and lymphatic tissue.

Lymphatic tissue is a specialized form of reticular connective tissue (see Table 3.4C) that contains large numbers of lymphocytes. Recall from Chapter 12 that lymphocytes are agranular white blood cells. Two types of lymphocytes participate in adaptive immune responses: B cells and T cells (described shortly).

Functions of the Lymphatic System

The lymphatic system has three primary functions:

1. **Drains excess interstitial fluid.** Lymphatic vessels drain excess interstitial fluid from tissue spaces and return it to the blood. This function closely allies it with the cardiovascular system. In fact, without this function the maintenance of circulating blood volume would not be possible.

2. **Transports dietary lipids.** Lymphatic vessels transport the lipids and lipid-soluble vitamins (A, D, E, and K), absorbed by the gastrointestinal tract, to the blood.

3. **Carries out immune responses.** Lymphatic tissue initiates highly specific responses directed against particular microbes or abnormal cells. With the assistance of macrophages, T and B cells recognize foreign cells, microbes, toxins, and cancer cells. These chemical substances that are recognized as foreign by the immune system are called **antigens** and provoke an immune response. T and B cells respond to antigens in several ways. **B cells** make up about 15–30 percent of the lymphocytes in the body. Most of the B cells differentiate into **plasma cells** that protect us against disease by producing **antibodies**, proteins that combine with and destroy specific foreign substances (antigens). Actually, the term *antigen* is so named because it is an *anti*body-*gen*erator. Some B cells become long-lived **memory B cells**, which can mount an even stronger immune response if the same antigen attacks the body at a later date.

   T cells, which make up 70–85 percent of the lymphocytes in the body, have several roles in the immune response. The four major types of T cells are helper T cells, cyototoxic T cells, regulatory T cells, and memory T cells. **Helper T cells** cooperate with B cells to amplify antibody production by plasma cells. Following activation by helper T cells, **cytotoxic T cells** destroy target cells on contact by causing them to rupture or by releasing cytotoxic (cell-killing) substances. **Regulatory T cells** (formerly called *suppressor T cells*) can turn off the immune response by suppressing T cells; this is important in combating an *autoimmune disease* (a disease caused by a reaction against the body's own cells). Regulatory T cells also protect beneficial intestinal bacteria, which aid digestion and produce some B vitamins and vitamin K. **Memory T cells** "remember" an antigen and mount a more vigorous response if the same antigen attacks the body in the future.

517

Figure 15.1 Components of the lymphatic system.

The lymphatic system consists of lymph, lymphatic vessels, lymphatic tissues, and red bone marrow.

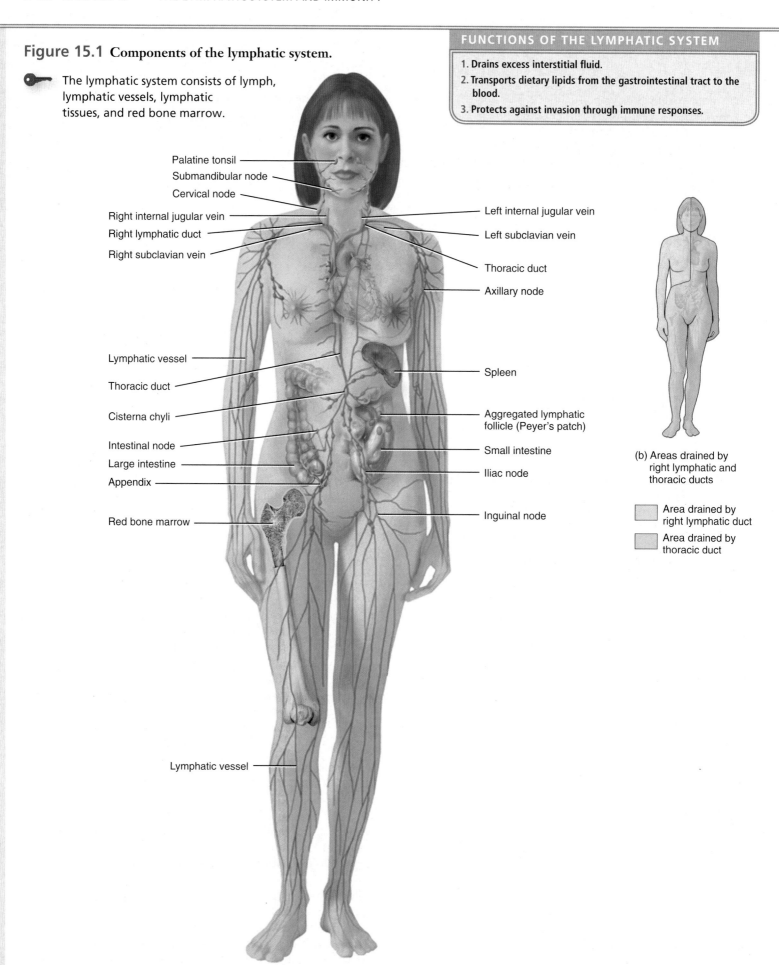

Palatine tonsil
Submandibular node
Cervical node
Right internal jugular vein
Right lymphatic duct
Right subclavian vein

Left internal jugular vein
Left subclavian vein
Thoracic duct
Axillary node

Lymphatic vessel
Thoracic duct
Cisterna chyli
Intestinal node
Large intestine
Appendix
Red bone marrow

Spleen
Aggregated lymphatic follicle (Peyer's patch)
Small intestine
Iliac node
Inguinal node

Lymphatic vessel

(b) Areas drained by right lymphatic and thoracic ducts

Area drained by right lymphatic duct
Area drained by thoracic duct

(a) Anterior view of principal components of lymphatic system

What tissue contains stem cells that develop into lymphocytes?

CLINICAL CONNECTION | *Allergic Reactions*

A person who is overly reactive to an antigen that is tolerated by most other people is said to be **allergic (hypersensitive).** Whenever an allergic reation takes place, some tissue injury occurs. The antigens that induce an allergic reaction are called **allergens** (AL-er-jens). Common allergens include certain foods (milk, peanuts, shellfish, eggs), antibiotics (penicillin, tetracycline), vaccines (pertussis, typhoid), venoms (honeybee, wasp, snake), cosmetics, chemicals in plants such as poison ivy, pollens, dust molds, iodine-containing dyes used in certain x-ray procedures, and even microbes.

Type I (anaphylactic) reactions (AN-a-fil-lak'-tik) are the most common type of allergic reaction and occur within a few minutes after a person previously exposed to an allergen is reexposed to it. In response to the first exposure to certain allergens, some people produce antibodies that bind to the surface of mast cells and basophils. The next time the same allergen enters the body, it attaches to the antibodies already present. In response, the mast cells and basophils release chemicals such as histamine. These mediators cause vasodilation, increased blood capillary permeability, increased smooth muscle contraction in the airways of the lungs, and increased mucus secretion. As a result, a person may experience inflammatory responses, difficulty in breathing through the constricted airways, and a runny nose from excess mucus secretion. In **anaphylactic shock,** which may occur in a susceptible individual who has just received a triggering drug or been stung by a wasp, wheezing and shortness of breath as airways constrict are usually accompanied by shock due to vasodilation and fluid loss from blood. This life-threatening emergency is usually treated by injecting epinephrine to dilate the airways and strengthen the heartbeat. •

Lymphatic Vessels and Lymph Circulation

Lymphatic vessels begin as **lymphatic capillaries**. These capillaries are closed at one end and located in the spaces between cells (Figure 15.2). Just as blood capillaries converge to form venules and then veins, lymphatic capillaries unite to form larger **lymphatic vessels** (see Figure 15.1), which resemble small veins in structure but have thinner walls and more valves. At intervals along the lymphatic vessels, lymph flows through **lymph nodes**, encapsulated masses of B cells and T cells. In the skin, lymphatic vessels lie in the subcutaneous tissue and generally follow the same route as veins; lymphatic vessels of the viscera generally follow arteries, forming plexuses (networks) around them. Tissues that lack lymphatic capillaries include avascular tissues (such as cartilage, the epidermis, and the cornea of the eye), the central nervous system, portions of the spleen, and red bone marrow.

Lymphatic Capillaries

Lymphatic capillaries have greater permeability than blood capillaries and thus can absorb large molecules such as proteins and lipids. Lymphatic capillaries are also slightly larger in diameter than blood capillaries and have a unique structure that permits interstitial fluid to flow into them but not out. The ends of endothelial cells that make up the wall of a lymphatic capillary overlap (Figure 15.2b). When pressure is greater in the interstitial fluid than in lymph, the cells separate slightly, like the opening of a one-way swinging door, and interstitial fluid enters the lymphatic capillary. When pressure is greater inside the lymphatic capillary, the cells adhere more closely, and lymph cannot escape back into interstitial fluid. Attached to the lymphatic capillaries

Figure 15.2 Lymphatic capillaries.

 Lymphatic capillaries are found throughout the body except in avascular tissues, the central nervous system, portions of the spleen, and red bone marrow.

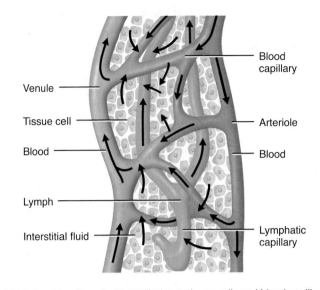

(a) Relationship of lymphatic capillaries to tissue cells and blood capillaries

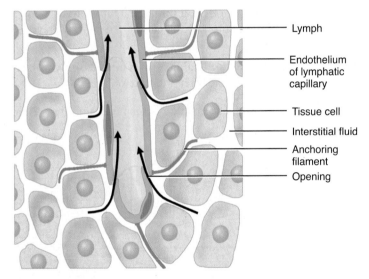

(b) Details of a lymphatic capillary

? Is lymph more similar to blood plasma or to interstitial fluid? Explain your answer.

are *anchoring filaments*, which contain elastic fibers. The anchoring filaments extend out from the lymphatic capillary, attaching (anchoring) lymphatic endothelial cells to surrounding tissues. When excess interstitial fluid accumulates and causes tissue swelling, the anchoring filaments are pulled, making the openings between cells even larger so that more fluid can flow into the lymphatic capillaries.

In the small intestine, specialized lymphatic capillaries called **lacteals** (LAK-tē-als; *lact-*=milky) carry dietary lipids into lymphatic vessels and ultimately into the blood (see Figure 24.17). The presence of these lipids causes the lymph draining from the small intestine to appear creamy white; such lymph is referred to as **chyle** (KĪL=juice). Elsewhere in the body, lymph is a clear, pale-yellow fluid.

Lymph Trunks and Ducts

Lymph passes from lymphatic capillaries into lymphatic vessels and then through lymph nodes. As lymphatic vessels exit lymph nodes in a particular region of the body, they unite to form **lymph trunks**. The principal lymph trunks are the lumbar, intestinal, bronchomediastinal, subclavian, and jugular trunks (Figure 15.3). The **lumbar trunks** drain lymph from the free lower limbs, the wall and viscera of the pelvis, the kidneys, the adrenal glands, and the abdominal wall. The **intestinal trunk** drains lymph from the stomach, intestines, pancreas, spleen, and part of the liver. The **bronchomediastinal trunks** (brong-kō-mē′-dē-as-TĪ-nal) drain lymph from the thoracic wall, lung, and heart. The **subclavian trunks** drain the free upper limbs. The **jugular trunks** drain the head and neck.

Lymph passes from lymph trunks into two main channels, the thoracic duct and the right lymphatic duct, and then drains into venous blood. The **thoracic (left lymphatic) duct**, the main duct for return of lymph to blood, is about 38–45 cm (15–18 in.) long. It begins as a dilation called the **cisterna chyli** (sis-TER-na KĪ-lē; *cisterna*=cavity or reservoir) anterior to the second lumbar vertebra. The cisterna chyli receives lymph from the right and left lumbar trunks and from the intestinal trunk. In the neck, the thoracic duct also receives lymph from the left jugular, left subclavian, and left bronchomediastinal trunks. So the thoracic duct receives lymph from the left side of the head, neck, and chest, the left free upper limb, and the entire body inferior to the ribs. The thoracic duct in turn drains lymph into venous blood at the junction of the left internal jugular and left subclavian veins.

The **right lymphatic duct** (Figure 15.3) is about 1.2 cm (0.5 in.) long and receives lymph from the right jugular, right subclavian, and right bronchomediastinal trunks. Thus, the right lymphatic duct receives lymph from the upper-right side of the body. From the right lymphatic duct, lymph drains into venous blood at the junction of the right internal jugular and right subclavian veins.

Formation and Flow of Lymph

Most components of blood plasma, such as nutrients, gases, and hormones, filter freely through the capillary walls to form interstitial fluid. More fluid filters out of blood capillaries, however, than returns to them by reabsorption. The excess filtered fluid—about 3 liters per day—drains into lymphatic vessels and becomes lymph. Because most blood plasma proteins are too large to leave blood vessels, interstitial fluid contains only a small amount of protein. Proteins that do leave blood plasma cannot return to the blood directly by diffusion because the concentration gradient (high level of proteins inside blood capillaries, low level outside) opposes such movement. The proteins can, however, move readily through the more permeable lymphatic capillaries into lymph. Thus, an important function of lymphatic vessels is to return lost blood plasma proteins and plasma to the bloodstream. Without this return of lymph (plasma lost from the blood) to the blood, the blood volume would drop precipitously and the cardiovascular system would cease to function. Therefore, the lymphatic vessels are a key part of the cardiovascular pathways in the body.

Figure 15.3 **Routes for drainage of lymph from lymph trunks into the thoracic and right lymphatic ducts.** The arrows indicate the direction of lymph flow.

🔑 All lymph returns to the bloodstream through the thoracic (left) lymphatic duct and right lymphatic duct.

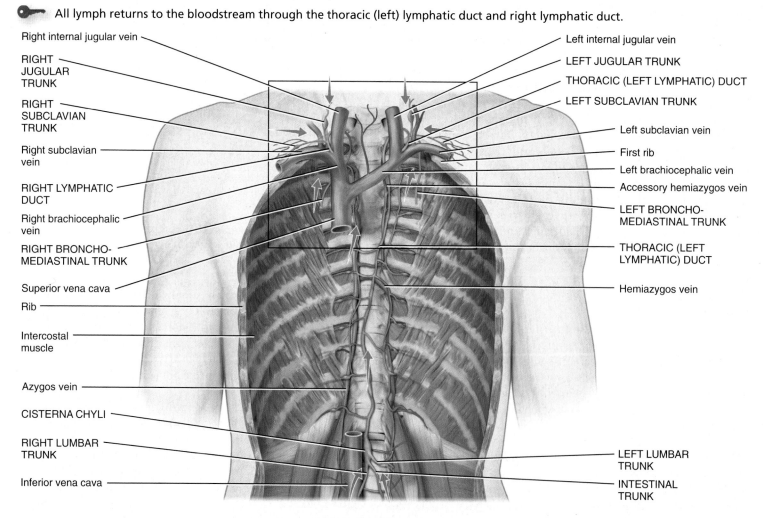

(a) Overall anterior view

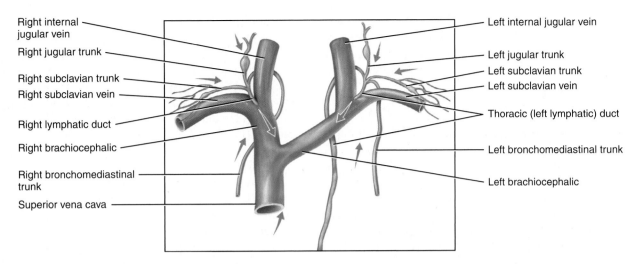

(b) Detailed anterior view of thoracic and right lymphatic duct

Which lymphatic vessels empty into the cisterna chyli, and which duct receives lymph from the cisterna chyli?

Like veins, lymphatic vessels contain valves, which ensure the one-way movement of lymph. As noted previously, lymph drains into venous blood through the right lymphatic duct and the thoracic duct at the junction of the internal jugular and subclavian veins (Figure 15.3). Thus, the sequence of fluid flow is blood capillaries (blood) → interstitial spaces (interstitial fluid) → lymphatic capillaries (lymph) → lymphatic vessels (lymph) → lymphatic ducts (lymph) → junction of the internal jugular and subclavian veins (blood). Figure 15.4 illustrates this sequence, along with the relationship of the lymphatic and cardiovascular systems. In essence, the lymphatic and cardiovascular systems form a very efficient circulatory system.

Two "pumps" that aid the return of venous blood to the heart maintain the flow of lymph.

1. **Skeletal muscle pump.** The "milking action" of skeletal muscle contractions compresses lymphatic vessels (as well as veins) and forces lymph toward the junction of the internal jugular and subclavian veins (see Figure 14.4a).
2. **Respiratory pump.** Lymph flow is also maintained by pressure changes that occur during inhalation (breathing in). Lymph flows from the abdominal region, where the pressure is higher, toward the thoracic region, where it is lower. When the pressures

Figure 15.4 The relationship of the lymphatic system to the cardiovascular system. Arrows show direction of flow of lymph and blood.

The sequence of fluid flow is blood capillaries (blood) → interstitial spaces (interstitial fluid) → lymphatic capillaries (lymph) → lymphatic vessels (lymph) → lymphatic ducts (lymph) → junction of internal jugular and subclavian veins (blood).

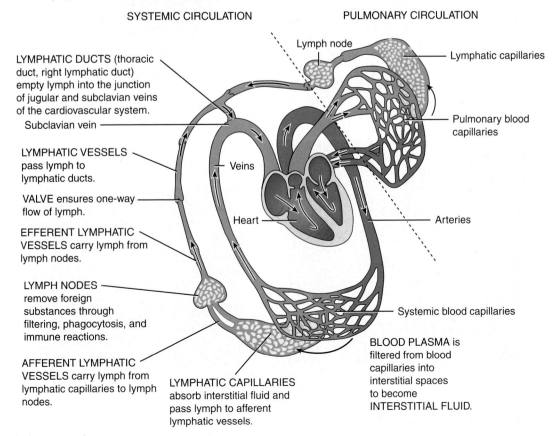

Does inhalation promote or hinder the flow of lymph?

reverse during exhalation (breathing out), the valves in lymphatic vessels prevent backflow of lymph. In addition, when a lymphatic vessel distends, the smooth muscle in its wall contracts, which helps move lymph from one segment of the vessel to the next.

Lymphatic Organs and Tissues

Lymphatic organs and tissues, which are widely distributed throughout the body, are classified into two groups based on their functions. **Primary lymphatic organs** are the sites where stem cells divide and become *immunocompetent* (im′-ū-nō-KOM-pe-tent), that is, capable of mounting an immune response. The primary lymphatic organs are the red bone marrow (in flat bones and the epiphyses of some long bones of adults) and the thymus (see Figure 15.1). Pluripotent stem cells in red bone marrow give rise to mature,

immunocompetent B cells and to pre-T cells (immature T cells), which migrate to and become immunocompetent T cells in the thymus. The **secondary lymphatic organs** and **tissues** are the sites where most immune responses occur. They include lymph nodes, the spleen, and lymphatic nodules (follicles) (see Figure 15.1). The thymus, lymph nodes, and spleen are considered organs because each is surrounded by a connective tissue capsule; lymphatic nodules, in contrast, are not organs because they lack such a capsule.

Thymus

The **thymus** is a bilobed organ located in the mediastinum between the sternum and the aorta. It extends from the top of the sternum or the inferior cervical region to the level of the fourth costal cartilages, anterior to the top of the heart and its great vessels (Figure 15.5a). An enveloping layer of connective tissue holds

Figure 15.5 Thymus.

 The bilobed thymus is largest at puberty and then the functioning portion atrophies with age.

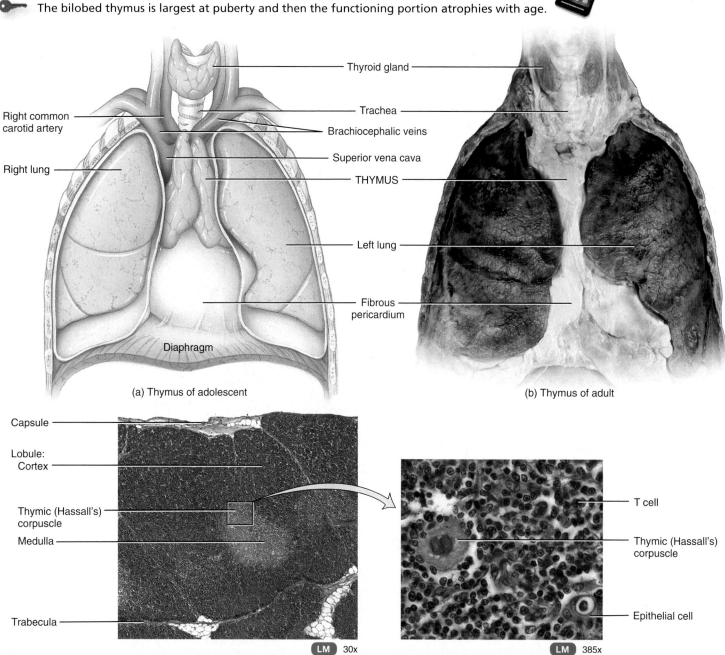

(a) Thymus of adolescent

(b) Thymus of adult

(c) Thymic lobules   LM 30x

(d) Details of the thymic medulla   LM 385x

? **Which types of lymphocytes mature in the thymus?**

the two lobes closely together, but a connective tissue **capsule** encloses each lobe separately. Extensions of the capsule, called **trabeculae** (tra-BEK-ū-lē=little beams), penetrate inward and divide each lobe into **lobules** (Figure 15.5c).

Each thymic lobule consists of a deeply staining outer cortex and a lighter-staining central medulla (Figure 15.5c). The **cortex** is composed of large numbers of T cells and scattered dendritic cells, epithelial cells, and macrophages. Immature T cells migrate from red bone marrow to the cortex of the thymus, where they proliferate and begin to mature. **Dendritic cells** (den-DRIT-ik; *dendr-*=a tree), so named because they have long, branched projections that resemble the dendrites of a neuron, assist the maturation process. As you will see shortly, dendritic cells in other parts of the body, such as lymph nodes, play another key role in immune responses. Each of the specialized **epithelial cells** in the cortex has several long processes that surround and serve as a framework for as many as 50 T cells. These epithelial cells help pre-T cells mature into T cells. Additionally, the epithelial cells produce thymic hormones that are thought to aid in the maturation of T cells. Only about 2 percent of developing T cells survive in the cortex. The remaining cells die via **apoptosis** (programmed cell death). Thymic macrophages help clear out the debris of dead and dying cells. The surviving T cells enter the medulla.

The **medulla** consists of widely scattered, more mature T cells, epithelial cells, dendritic cells, and macrophages (Figure 15.5d). Some of the epithelial cells become arranged into concentric layers of flat cells that degenerate and become filled with keratohyalin granules and keratin. These clusters are called **thymic corpuscles** or *Hassall's corpuscles*. Although their role is uncertain, they may serve as sites of T cell death in the medulla. T cells that leave the thymus via the blood are carried to lymph nodes, the spleen, and other lymphatic tissues where they colonize parts of these organs and tissues.

Because of its high content of lymphoid tissue and a rich blood supply, the thymus has a reddish appearance in a living body. With age, however, fatty infiltrations replace the lymphoid tissue and the thymus takes on more of the yellowish color of the invading fat, giving the false impression of reduced size. However, the actual size of the thymus, defined by its connective tissue capsule, does not change. In infants, the thymus has a mass of about 70 g (2.3 oz). It is after puberty that adipose and areolar connective tissues begin to replace the thymic tissue. By the time a person reaches maturity, the functional portion of the gland is reduced considerably, and in old age the

CLINICAL CONNECTION | *Edema*

If filtration greatly exceeds reabsorption, the result is **edema** (=swelling), an abnormal increase in interstitial fluid volume. Edema is not usually detectable in tissues until interstitial fluid volume has risen to 30 percent above normal. Edema can result from either excess filtration or inadequate reabsorption. For example, increased capillary blood pressure causes more fluid to be filtered from capillaries. Also, increased permeability of capillaries raises interstitial fluid pressure by allowing some plasma proteins to escape. Such leakiness may be caused by the destructive effects of chemical, bacterial, thermal, or mechanical agents on capillary walls. On the other hand, decreased concentration of plasma proteins causes inadequate reabsorption. Inadequate synthesis or dietary intake or loss of plasma proteins is associated with liver disease, burns, malnutrition, and kidney disease. •

functional portion may weigh only 3 g (0.1 oz). Before the thymus atrophies, it populates the secondary lymphatic organs and tissues with T cells. Some T cells do continue to proliferate in the thymus throughout an individual's lifetime, but this number decreases with age.

Lymph Nodes

Located along lymphatic vessels are about 600 bean-shaped lymph nodes. They are scattered throughout the body, both superficially and deep, and usually occur in groups (see Figure 15.1). Large groups of lymph nodes are present near the mammary glands and in the axillae and groin. Later in the chapter, the principal groups of lymph nodes in various regions of the body will be presented in a series of exhibits (see Section 15.2).

Lymph nodes are 1–25 mm (0.04–1 in.) long and, like the thymus, are covered by a capsule of dense connective tissue that extends into the node (Figure 15.6). The capsular extensions, called trabeculae, divide the node into compartments, give support, and provide a route for blood vessels into the interior of a node. Internal to the capsule is a supporting network of reticular fibers and fibroblasts. The capsule, trabeculae, reticular fibers, and fibroblasts constitute the *stroma* (framework tissue) of a lymph node.

The *parenchyma* (functional tissue) of a lymph node is divided into a superficial cortex and a deep medulla. Within the **outer cortex** are egg-shaped aggregates of B cells called **lymphatic nodules (follicles)**. A lymphatic nodule consisting chiefly of B cells is called a *primary lymphatic nodule*. Most lymphatic nodules in the outer cortex are *secondary lymphatic nodules* (Figure 15.6), which form in response to an antigen (a foreign substance) and are sites of plasma cell and memory B cell formation. After B cells in a primary lymphatic nodule recognize an antigen, the primary lymphatic nodule develops into a secondary lymphatic nodule. The center of a secondary lymphatic nodule contains a region of light-staining cells called a *germinal center*. In the germinal center are B cells, follicular dendritic cells (a special type of dendritic cell), and macrophages. When follicular dendritic cells "present" an antigen, B cells proliferate and develop into antibody-producing plasma cells or memory B cells. Memory B cells persist after an immune response and remember having encountered a specific antigen. B cells that do not develop properly undergo apoptosis and are destroyed by macrophages. In a secondary lymphatic nodule, the region surrounding the germinal center is composed of dense accumulations of B cells that have migrated away from their sites of origin within the nodule.

The **inner cortex**, also called the *paracortex*, does not contain lymphatic nodules. It consists mainly of T cells and dendritic cells that enter a lymph node from other tissues. The dendritic cells present antigens to T cells, causing the T cells to proliferate. The newly formed T cells then migrate from the lymph node to areas of the body where there is antigenic activity.

The medulla of a lymph node contains B cells, antibody-producing plasma cells that have migrated out of the cortex into the medulla, and macrophages. The various cells are embedded in a network of reticular fibers and reticular cells.

As you have already learned, lymph flows through a lymph node in one direction only (the left side of Figure 15.6a). It enters through several **afferent lymphatic vessels** (AF-er-ent=to carry toward), which penetrate the convex surface of the node at several points. The afferent vessels contain valves that open toward the center of the node so that the lymph is directed *inward*.

Figure 15.6 Structure of a lymph node. Arrows indicate the direction of lymph flow through a lymph node.

🔑 Lymph nodes are present throughout the body, usually clustered in groups.

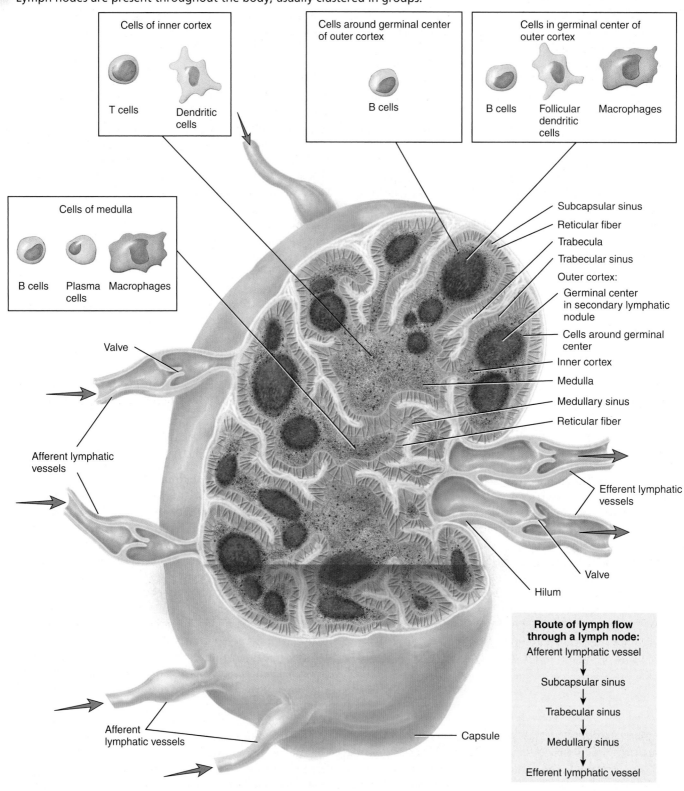

(a) Partially sectioned lymph node

Route of lymph flow through a lymph node:

Afferent lymphatic vessel

↓

Subcapsular sinus

↓

Trabecular sinus

↓

Medullary sinus

↓

Efferent lymphatic vessel

Within the node, lymph enters **sinuses**, a series of irregular channels that contain branching reticular fibers, lymphocytes, and macrophages. From the afferent lymphatic vessels, lymph flows into the **subcapsular sinus** (sub-KAP-soo-lar) immediately beneath the capsule. From there the lymph flows through **trabecular sinuses** (tra-BEK-ū-lar), which extend through the cortex parallel to the trabeculae, and into **medullary sinuses**, which extend through the medulla. The medullary sinuses drain into one or two **efferent lymphatic vessels** (EF-er-ent=to carry away), which are wider than afferent vessels and fewer in number. They contain valves that open away from the center of the lymph node to convey lymph, antibodies secreted by plasma cells, and activated T cells *out* of the lymph node. Efferent lymphatic vessels emerge from one side of the lymph node at a slight depression called a **hilum** (HĪ-lum). Blood vessels also enter and leave the node at the hilum.

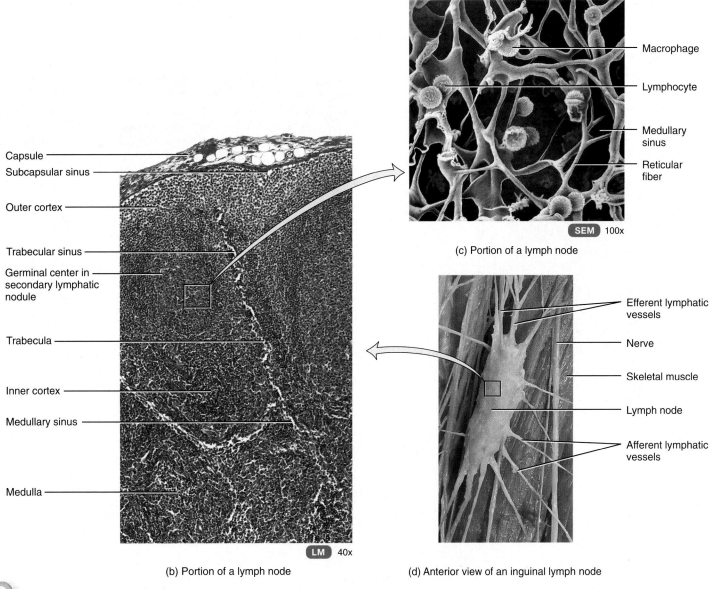

Capsule
Subcapsular sinus
Outer cortex
Trabecular sinus
Germinal center in secondary lymphatic nodule
Trabecula
Inner cortex
Medullary sinus
Medulla

LM 40x

(b) Portion of a lymph node

Macrophage
Lymphocyte
Medullary sinus
Reticular fiber

SEM 100x

(c) Portion of a lymph node

Efferent lymphatic vessels
Nerve
Skeletal muscle
Lymph node
Afferent lymphatic vessels

(d) Anterior view of an inguinal lymph node

? What happens to foreign substances in the lymph when they enter a lymph node?

Lymph nodes function as a type of filter. As lymph enters one end of a lymph node, foreign substances are trapped by the reticular fibers within the sinuses of the node. Then macrophages destroy some foreign substances by phagocytosis, while lymphocytes destroy others by immune responses. The filtered lymph then leaves the other end of the lymph node. Since there are many afferent lymphatic vessels that bring lymph into a lymph node and only one or two efferent lymphatic vessels that transport lymph out of a lymph node, the slow flow of lymph within the lymph nodes allows additional time for lymph to be filtered. Additionally, all lymph flows through multiple lymph nodes on its path through the lymph vessels. This exposes the lymph to multiple filtering events before it is returned to the blood.

Spleen

The oval **spleen** is the largest single mass of lymphatic tissue in the body. It is a soft, encapsulated organ of variable size, but on average, it fits in a person's open hand and measures about 12 cm (5 in.) in length (Figure 15.7a). It is located in the left hypochondriac region between the stomach and diaphragm. The superior surface of the spleen is smooth and convex and conforms to the concave surface of the diaphragm. Neighboring organs make indentations in the visceral surface of the spleen—the *gastric impression* (stomach), the *renal impression* (left kidney), and the *colic impression* (left colic flexure of the large intestine). Like lymph nodes, the spleen has a hilum. Through it pass the large, tortuous splenic artery and splenic vein, along with efferent lymphatic vessels and sympathetic nerves that regulate blood flow in the vessels.

A capsule of dense connective tissue surrounds the spleen and is covered in turn by a serous membrane, the visceral peritoneum. Trabeculae extend inward from the capsule. The capsule plus trabeculae, reticular fibers, and fibroblasts constitute the stroma of the spleen; the parenchyma of the spleen consists of two different kinds of tissue called white pulp and red pulp (Figure 15.7c, d). **White pulp** is lymphatic tissue, consisting mostly of lymphocytes and macrophages arranged around branches of the splenic artery called **central arteries**. The **red pulp** consists of blood-filled **venous sinuses** and cords of splenic tissue called **splenic cords** or *Billroth's cords*. Splenic cords consist of red blood cells, macrophages, lymphocytes, plasma cells, and granulocytes. Veins are closely associated with the red pulp.

Figure 15.7 Structure of the spleen.

The spleen is the largest single mass of lymphatic tissue in the body.

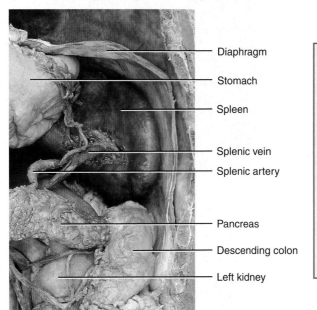

Diaphragm

Stomach

Spleen

Splenic vein

Splenic artery

Pancreas

Descending colon

Left kidney

(a) Anterior view of portion of abdominal cavity

CLINICAL CONNECTION | *Ruptured Spleen*

The spleen is the organ most often damaged in cases of abdominal trauma. Severe blows over the inferior left chest or superior abdomen can fracture the protecting ribs. Such crushing injury may result in a **ruptured spleen,** which causes significant hemorrhage and shock. Prompt removal of the spleen, called a **splenectomy** (splē-NEK-tō-mē), is needed to prevent death due to bleeding. Other structures, particularly red bone marrow and the liver, can take over some functions normally carried out by the spleen. Immune functions, however, decrease in the absence of a spleen. The spleen's absence also places the patient at higher risk for **sepsis** (a blood infection) due to loss of the filtering and phagocytic functions of the spleen. To reduce the risk of sepsis, patients who have undergone a splenectomy take prophylactic (preventive) antibiotics before any invasive procedures. •

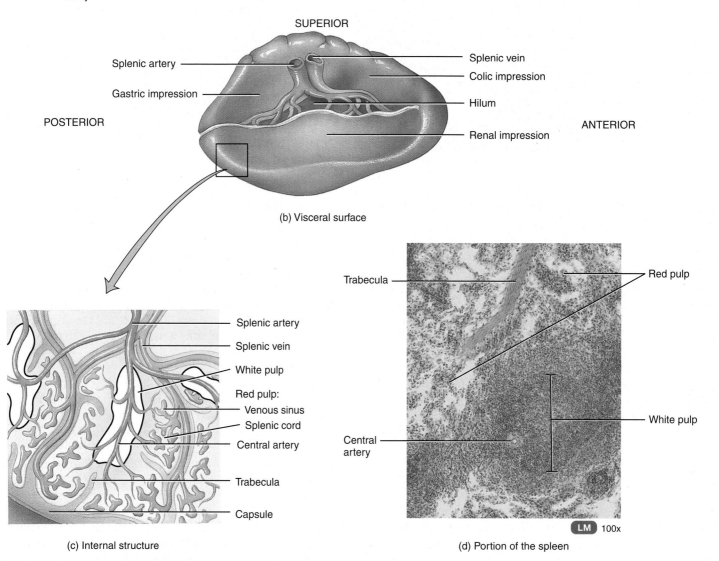

SUPERIOR

Splenic artery

Gastric impression

POSTERIOR

Splenic vein

Colic impression

Hilum

Renal impression

ANTERIOR

(b) Visceral surface

Splenic artery

Splenic vein

White pulp

Red pulp:
Venous sinus
Splenic cord

Central artery

Trabecula

Capsule

(c) Internal structure

Trabecula

Central artery

Red pulp

White pulp

LM 100x

(d) Portion of the spleen

After birth, what are the main functions of the spleen?

Blood flowing into the spleen through the splenic artery enters the central arteries of the white pulp. Within the white pulp, B cells and T cells carry out immune functions, similar to lymph nodes, while spleen macrophages destroy blood-borne pathogens by phagocytosis. Within the red pulp, the spleen performs three functions related to blood cells: (1) removal by macrophages of ruptured, worn out, or defective blood cells and platelets; (2) storage of platelets, up to one-third of the body's supply; and (3) production of blood cells (hemopoiesis) during fetal life (Chapter 12).

Lymphatic Nodules

Lymphatic nodules are egg-shaped masses of lymphatic tissue; unlike lymph nodes, they are not surrounded by a capsule. Because they are scattered throughout the lamina propria (connective tissue) of mucous membranes lining the gastrointestinal, urinary, and reproductive tracts, and the respiratory airways, lymphatic nodules in these areas are also referred to as **mucosa-associated lymphatic tissue (MALT)**.

Although many lymphatic nodules are small and solitary, some occur in multiple large aggregations in specific parts of the body. Among these are the tonsils in the pharyngeal region and the **aggregated lymphatic follicles** (*Peyer's patches*) in the ileum of the small intestine. Aggregations of lymphatic nodules also occur in the appendix.

Usually there are five **tonsils**, which form a tonsillar (Waldeyer's) ring at the junction of the oral cavity and oropharynx and at the junction of the nasal cavity and nasopharynx (see Figure 23.2b). The single **pharyngeal tonsil** (fa-RIN-jē-al) or **adenoid** is embedded in the posterior wall of the nasopharynx. This pyramid-shaped mass of lymphoid tissue is covered with a mucous membrane. The two **palatine tonsils** (PAL-a-tīn) lie in the lateral wall of the oropharynx in the **tonsilar fossa** just inferior to the soft palate; these are the tonsils commonly removed in a tonsillectomy. The almond-shaped palatine tonsils have numerous branched crypts that form a surface area of approximately 300 cm² (118 in.²). The paired **lingual tonsils** (LIN-gwal), located at the base of the tongue, may also require removal during a tonsillectomy. The tonsils are strategically positioned to participate in immune responses against inhaled or ingested foreign substances. Tonsils are masses of lymphoid tissue covered with mucosal epithelium. The epithelium forms narrow invaginations, called **crypts**, into the lymphoid tissue below. The crypts greatly increase the mucosal surface associated with the lymphoid tissue. In the crypts the mucosa becomes very thin, forming patches of **reticulated epithelium**. This specialized epithelium is well-designed for the transfer of antigen from the environment of the oral cavity and pharynx to the lymphoid cells of the tonsils.

CLINICAL CONNECTION | Tonsillitis

Tonsillitis is an infection or inflammation of the tonsils. Most often, it is caused by a virus, but it may also be caused by the same bacteria that cause strep throat. The principal symptom of tonsillitis is a sore throat. Additionally, fever, swollen lymph nodes, nasal congestion, difficulty in swallowing, and headache may also occur. Tonsillitis of viral origin usually resolves on its own. Bacterial tonsillitis is typically treated with antibiotics. **Tonsillectomy** (ton-si-LEK-tō-mē; *ectomy*=incision), the removal of a tonsil, may be indicated for individuals who do not respond to other treatments. Such individuals usually have tonsillitis lasting for more than three months (despite medication), obstructed air pathways, and difficulty in swallowing and talking. It appears that tonsillectomy does not interfere with a person's response to subsequent infections. •

CHECKPOINT

1. How are interstitial fluid and lymph similar, and how do they differ?
2. How do lymphatic vessels differ in structure from veins?
3. Construct a diagram of the route of lymph circulation.
4. What is the role of the thymus in immunity?
5. What functions do lymph nodes serve?
6. Describe the locations and functions of the spleen and tonsils.

15.2 PRINCIPAL GROUPS OF LYMPH NODES

OBJECTIVES

• Identify the locations and drainage regions of the principal groups of lymph nodes.

Lymph nodes are a part of your body that you don't usually notice until they are working really hard to fight an infection. The technical term for enlarged, sometimes tender lymph nodes is **lymphadenopathy** (lim-fad′-e-NOP-a-thē; *lymph-* = clear fluid; *pathy* = disease), commonly referred to as "swollen glands." The lymph nodes on either side of the neck below the mandible or those behind the ears may swell as a result of an infection such as a cold, sore throat, or ear infection, or an injury such as a bite or cut near the affected node. Enlarged lymph nodes above the clavicles may result from infections or tumors in the lungs, breasts, neck, or abdomen. In the axillary area, lymphadenopathy may be caused by an infection in the upper limbs or metastasis of a tumor of the breasts (see the Clinical Connection in Exhibit 15.C). Lymph nodes in the groin may swell as a result of infection or injury in the groin, genitals, or lower limbs.

Swelling of lymph nodes in two or more areas of the body is referred to as *generalized lymphadenopathy*. One cause is a viral illness such as AIDS, mononucleosis, measles, rubella, chickenpox, or mumps. Others include syphilis, strep throat, Lyme disease, cancer, cat scratch disease, rheumatoid arthritis, and lupus.

In some cases a lymph node may become so overwhelmed by the infection it is trying to combat that the skin overlying the swollen lymph nodes may also become red and tender, a condition called **lymphadenitis** (lim′-fad′-e-NĪ-tis; *-itis* = inflammation). Lymphadenitis may be treated with warm compresses, antibiotics, and over-the-counter pain relievers. Any enlarged lymph node that does not return to its normal size within about a month should be examined by a physician, who may order blood tests, radiographs, or a biopsy.

With this background in mind, refer to Exhibits 15.A–15.E, which describe the principal groups of lymph nodes by region and by the general areas they drain (Figures 15.8–15.12).

CHECKPOINT

7. Why do you think that lymph nodes are grouped more densely in some regions of the body than others?

EXHIBIT 15.A	Principal Lymph Nodes of the Head and Neck *(Figure 15.8)*

 OBJECTIVE
• Identify the principal lymph nodes of the head and neck.

 CHECKPOINT
8. Which lymph nodes drain the chin?

LYMPH NODES OF THE HEAD	LOCATION	DRAINAGE
Occipital nodes (ok-SIP-i-tal)	Near trapezius and semispinalis capitis muscles	Occipital portion of scalp and upper neck
Retroauricular nodes (re′-trō-aw-RIK-ū-lar=behind auricle)	Posterior to ear	Skin of ear and posterior parietal region of scalp
Pre-auricular nodes (*pre-*=before)	Anterior to ear	Auricle of ear and temporal region of scalp
Parotid nodes (pa-ROT-id; *para-*=beside; *ot-*=ear)	Embedded in and inferior to parotid gland	Root of nose, eyelids, anterior temporal region, external auditory meatus, tympanic cavity, nasopharynx, and posterior portions of nasal cavity
Facial nodes	Consist of three groups: infraorbital, buccal, and mandibular	
Infraorbital nodes (in′-fra-OR-bi-tal; *infra-*=below; *-orbital*=orbit)	Inferior to the orbit	Eyelids and conjunctiva
Buccal nodes (BUK-al; *bucca-*=cheek)	At angle of mouth	Skin and mucous membrane of nose and cheek
Mandibular nodes (man-DIB-ū-lar; *mand-*=to chew)	Over mandible	Skin and mucous membrane of nose and cheek
LYMPH NODES OF THE NECK	LOCATION	DRAINAGE
Submandibular nodes	Along inferior border of mandible	Chin, lips, nose, nasal cavity, cheeks, gums, inferior surface of palate, and anterior portion of tongue
Submental nodes (sub-MEN-tal; *sub-*=beneath; *-ment*=chin)	Between digastric muscles	Chin, lower lip, cheeks, tip of tongue, and floor of mouth
Superficial cervical nodes (SER-vi-kul; *cervic-*=neck)	Along external jugular vein	Inferior part of ear and parotid region
Deep cervical nodes	Largest group of nodes in neck, consisting of numerous large nodes forming a chain extending from base of skull to root of neck; arbitrarily divided into superior deep cervical nodes and inferior deep cervical nodes	
Superior deep cervical nodes	Deep to sternocleidomastoid muscle	Posterior head and neck, auricle, tongue, larynx, esophagus, thyroid gland, nasopharynx, nasal cavity, palate, and tonsils
Inferior deep cervical nodes	Near subclavian vein	Posterior scalp and neck, superficial pectoral region, and part of arm

EXHIBIT 15.A **529**

Figure 15.8 **Principal lymph nodes of the head and neck.**

 The facial lymph nodes include the infraorbital, buccal, and mandibular lymph nodes.

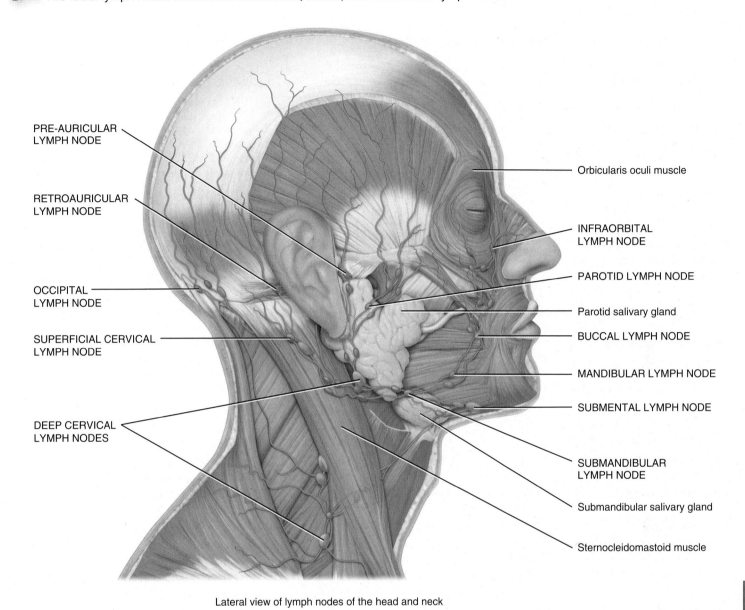

PRE-AURICULAR
LYMPH NODE

RETROAURICULAR
LYMPH NODE

OCCIPITAL
LYMPH NODE

SUPERFICIAL CERVICAL
LYMPH NODE

DEEP CERVICAL
LYMPH NODES

Orbicularis oculi muscle

INFRAORBITAL
LYMPH NODE

PAROTID LYMPH NODE

Parotid salivary gland

BUCCAL LYMPH NODE

MANDIBULAR LYMPH NODE

SUBMENTAL LYMPH NODE

SUBMANDIBULAR
LYMPH NODE

Submandibular salivary gland

Sternocleidomastoid muscle

Lateral view of lymph nodes of the head and neck

Which is the largest group of lymph nodes in the neck?

EXHIBIT 15.B   Principal Lymph Nodes of the Thorax   *(Figure 15.9)*

 OBJECTIVE
• Identify the principal lymph nodes of the thorax.

 CHECKPOINT
9. Which lymph nodes drain the bronchi?

PARIETAL LYMPH NODES	LOCATION	DRAINAGE
PARIETAL NODES DRAIN THE WALL OF THE THORAX.		
Sternal (parasternal) nodes	Alongside internal thoracic artery	Medial part of mammary gland, deeper structures of anterior abdominal wall superior to umbilicus, diaphragmatic surface of liver, and deeper parts of anterior portion of thoracic wall
Intercostal nodes (in'-ter-KOS-tal; *inter-*=between; *costa*=rib)	Near heads of ribs at posterior parts of intercostal spaces	Posterolateral aspect of thoracic wall
Phrenic (diaphragmatic) nodes (FREN-ik=diaphragm)	On thoracic aspect of diaphragm and divisible into three sets called anterior phrenic, middle phrenic, and posterior phrenic	
Anterior phrenic nodes	Posterior to base of xiphoid process	Convex surface of liver, diaphragm, and anterior abdominal wall
Middle phrenic nodes	Close to phrenic nerves where they pierce diaphragm	Medial part of diaphragm and convex surface of liver
Posterior phrenic nodes	Posterior surface of diaphragm near aorta	Posterior part of diaphragm
VISCERAL LYMPH NODES	**LOCATION**	**DRAINAGE**
VISCERAL NODES DRAIN THE VISCERA IN THE THORAX.		
Anterior mediastinal nodes (mē'-dē-as-TĪ-nal; *media-*=middle; *-stinum*=partition)	Anterior part of superior mediastinum, anterior to arch of aorta	Thymus and pericardium
Posterior mediastinal nodes	Posterior to pericardium	Esophagus, posterior aspect of the pericardium, diaphragm, and convex surface of liver
Tracheobronchial nodes	Are divided into five groups: tracheal, superior and inferior tracheobronchial, bronchopulmonary, and pulmonary nodes	
Tracheal nodes	Either side of trachea	Trachea and upper esophagus
Superior tracheobronchial nodes (trā'kē-ō-BRONG-kē-al)	Between trachea and bronchi	Trachea and bronchi
Inferior tracheobronchial nodes	Between bronchi	Trachea and bronchi
Bronchopulmonary nodes (brong-kō-PUL-mō-nar-ē)	In hilus of each lung	Lungs and bronchi
Pulmonary nodes	Within lungs on larger bronchial tube branches	Lungs and bronchi

EXHIBIT 15.B **531**

Figure 15.9 Principal lymph nodes of the thorax.

 The parietal lymph nodes drain the thoracic wall, while the visceral lymph nodes drain the viscera of the thorax.

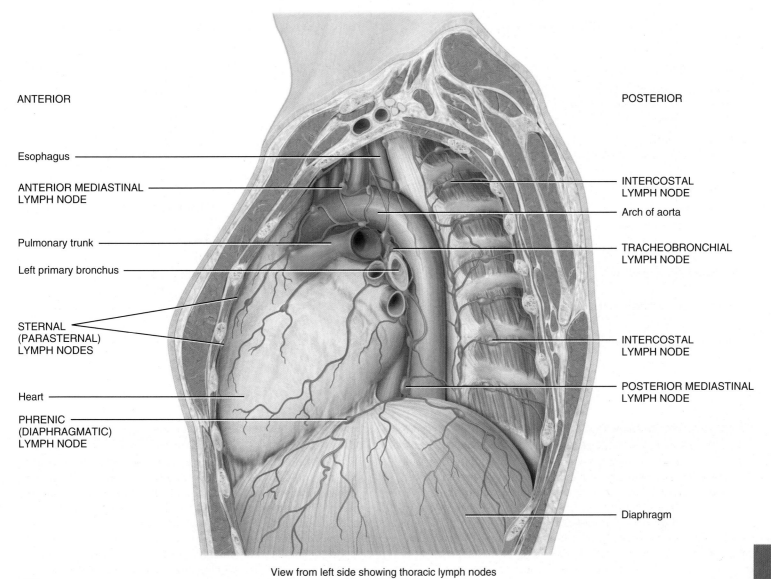

ANTERIOR

POSTERIOR

Esophagus

ANTERIOR MEDIASTINAL
LYMPH NODE

Pulmonary trunk

Left primary bronchus

STERNAL
(PARASTERNAL)
LYMPH NODES

Heart

PHRENIC
(DIAPHRAGMATIC)
LYMPH NODE

INTERCOSTAL
LYMPH NODE

Arch of aorta

TRACHEOBRONCHIAL
LYMPH NODE

INTERCOSTAL
LYMPH NODE

POSTERIOR MEDIASTINAL
LYMPH NODE

Diaphragm

View from left side showing thoracic lymph nodes

? **Which group of lymph nodes drains most of the diaphragm?**

EXHIBIT 15.C	Principal Lymph Nodes of the Upper Limbs *(Figure 15.10)*

 OBJECTIVE
• Identify the principal lymph nodes of the upper limbs.

 CHECKPOINT
10. Which lymph nodes drain the mammary glands?

LYMPH NODES	LOCATION	DRAINAGE
Supratrochlear nodes (soo-pra-TROK-lē-ar; *supra-*=above; *-trochlea*=pulley)	Superior to medial epicondyle of humerus	Medial fingers, palm, and forearm
Deltopectoral nodes (del'tō-PEK-tō-ral; *-pectus*=breast)	Inferior to clavicle	Lymphatic vessels on radial side of upper limb
Axillary nodes (AK-sil-ār-ē; *axil-*=armpit)	Include most deep lymph nodes of the upper limbs; large in size	
Lateral nodes	Medial and posterior aspects of axillary artery	Most of entire upper limb
Pectoral (anterior) nodes	Along inferior border of the pectoralis minor muscle	Skin and muscles of anterior and lateral thoracic walls and central and lateral portions of mammary gland
Subscapular (posterior) nodes	Along subscapular artery	Skin and muscles of posterior part of neck and thoracic wall
Central (intermediate) nodes	Base of axilla embedded in adipose tissue	Lateral, pectoral (anterior), and subscapular (posterior) nodes
Subclavicular (apical) nodes	Posterior and superior to pectoralis minor muscle	Deltopectoral nodes

CLINICAL CONNECTION | *Breast Cancer and Metastasis*

In Chapter 26, we will consider the pathology, detection, and treatment of breast cancer in detail. Very simply, **breast cancer** is the development of a malignant tumor within the breast. At this point we will concentrate on how breast cancer may spread to other parts of the body via the lymphatic system.

An understanding of the lymphatic drainage of the breasts is clinically important because knowledge of the direction of lymph flow can help predict the spread of breast cancer to other sites in the body. When considering the lymphatic drainage of the breasts, it is convenient to divide the breasts into quadrants: upper lateral, lower lateral, upper medial, and lower medial. About 75% of the lymph of the breasts drains into lymphatics located in the lateral breast quadrants. These lymphatics drain into the axillary lymph nodes. Most of the remainder of the lymph drains into lymphatics from the medial breast quadrants. These lymphatics drain into the sternal (parasternal) lymph nodes. The majority of breast cancers occur in the upper lateral quadrant, and the lymphatic vessels from these quadrants provide routes for the cancer to spread to the axillary lymph nodes, specifically the pectoral (anterior) nodes. From here, the cancer may spread to other axillary lymph nodes. The spread of cancer from the organ of origin to another part of the body is called **metastasis** (me-TAS-ta-sis; *meta-*=beyond; *stasi-*=to stand), and when it occurs through lymphatic vessels it is referred to as *lymphogenic metastasis*. Abundant communications among lymphatic vessels and among axillary, cervical, and sternal lymph nodes may also cause metastasis from the breast to develop in the opposite breast and abdomen. If a breast cancer spreads beyond the axillary nodes, it is called *distant metastasis*. The most common sites include the lungs, liver, and bones. In general, cancerous lymph nodes feel enlarged, firm, nontender, and fixed to underlying structures. By contrast, most lymph nodes that are enlarged due to an infection are softer, tender, and movable. It is ironic that the role of the lymphatic system in filtering lymph and returning it to the cardiovascular system is also unfortunately the pathway for metastasis. •

EXHIBIT 15.C **533**

Figure 15.10 Principal lymph nodes of the upper limbs.

🔑 Most of the lymph drainage of the breast is to the pectoral group of axillary lymph nodes.

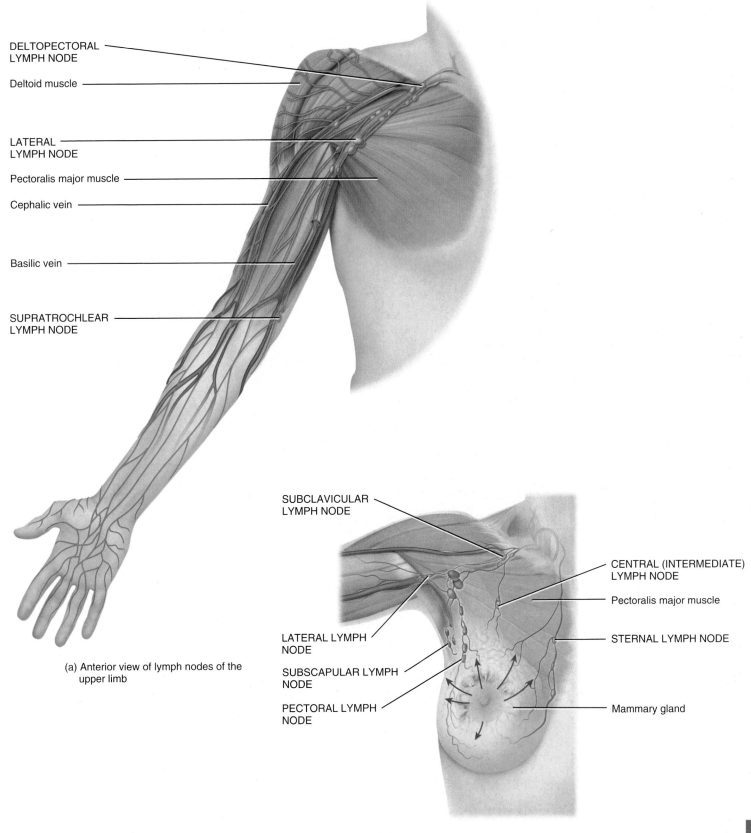

DELTOPECTORAL LYMPH NODE

Deltoid muscle

LATERAL LYMPH NODE

Pectoralis major muscle

Cephalic vein

Basilic vein

SUPRATROCHLEAR LYMPH NODE

(a) Anterior view of lymph nodes of the upper limb

SUBCLAVICULAR LYMPH NODE

CENTRAL (INTERMEDIATE) LYMPH NODE

Pectoralis major muscle

STERNAL LYMPH NODE

LATERAL LYMPH NODE

SUBSCAPULAR LYMPH NODE

PECTORAL LYMPH NODE

Mammary gland

(b) Anterior view of mostly axillary lymph nodes. Arrows indicate the direction of drainage.

 Which lymph nodes drain most of the upper limb?

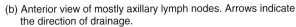

EXHIBIT 15.D

Principal Lymph Nodes of the Abdomen and Pelvis *(Figure 15.11)*

 OBJECTIVE
• Identify the principal lymph nodes of the abdomen and pelvis.

CHECKPOINT
11. Distinguish between parietal and visceral lymph nodes.

Figure 15.11 **Principal lymph nodes of the abdomen and pelvis.**

The parietal lymph nodes are retroperitoneal and in close association with larger blood vessels.

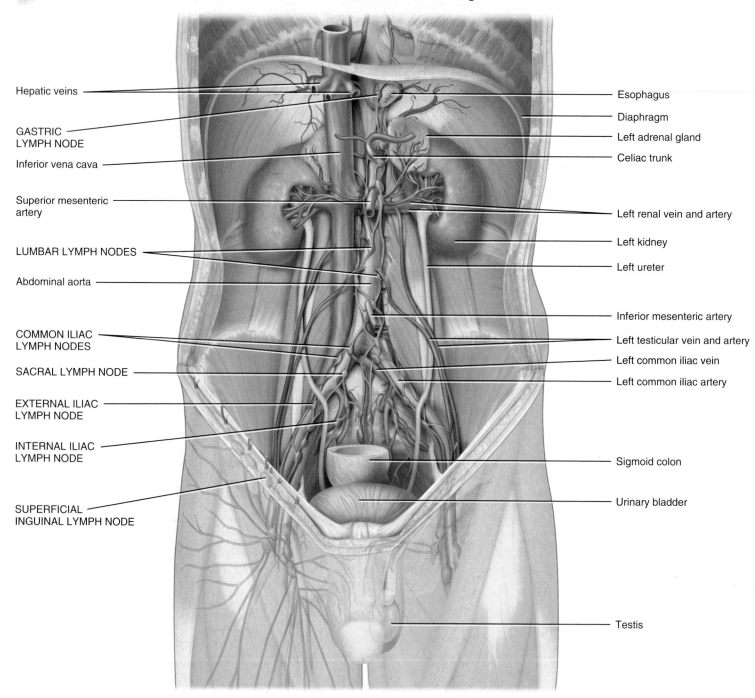

Hepatic veins

GASTRIC
LYMPH NODE

Inferior vena cava

Superior mesenteric
artery

LUMBAR LYMPH NODES

Abdominal aorta

COMMON ILIAC
LYMPH NODES

SACRAL LYMPH NODE

EXTERNAL ILIAC
LYMPH NODE

INTERNAL ILIAC
LYMPH NODE

SUPERFICIAL
INGUINAL LYMPH NODE

Esophagus

Diaphragm

Left adrenal gland

Celiac trunk

Left renal vein and artery

Left kidney

Left ureter

Inferior mesenteric artery

Left testicular vein and artery

Left common iliac vein

Left common iliac artery

Sigmoid colon

Urinary bladder

Testis

(a) Anterior view of abdominal and pelvic lymph nodes

EXHIBIT 15.D **535**

PARIETAL LYMPH NODES	LOCATION	DRAINAGE
PARIETAL NODES ARE LOCATED RETROPERITONEALLY (BEHIND THE PARIETAL PERITONEUM) AND IN CLOSE ASSOCIATION WITH LARGER BLOOD VESSELS.		
External iliac nodes (IL-ē-ak=ilium)	Along external iliac vessels	Deep lymphatics of abdominal wall inferior to umbilicus, adductor region of thigh, urinary bladder, prostate, ductus (vas) deferens, seminal vesicles, prostatic and membranous urethra, uterine (fallopian) tubes, uterus, and vagina
Common iliac nodes	Along course of common iliac vessels	Pelvic viscera
Internal iliac nodes	Near internal iliac artery	Pelvic viscera, perineum, gluteal region, and posterior surface of thigh
Sacral nodes (SĀ-krul=holy bone)	In hollow of sacrum	Rectum, prostate, and posterior pelvic wall
Lumbar nodes (LUM-bar; *lumb-*=loin)	From aortic bifurcation to diaphragm; arranged around aorta and designated as *right lateral aortic nodes, left lateral aortic nodes, preaortic nodes,* and *retroaortic nodes*	Efferents from testes, ovaries, uterine (fallopian) tubes, uterus, kidneys, adrenal (suprarenal) glands, abdominal surface of diaphragm, and lateral abdominal wall

VISCERAL LYMPH NODES	LOCATION	DRAINAGE
VISCERAL NODES ARE FOUND IN ASSOCIATION WITH VISCERAL ARTERIES.		
Celiac nodes (SE-lē-ak; *koilia-*=belly)	Consist of three groups: gastric, hepatic, and pancreaticosplenic	
Gastric nodes	Along lesser curvature of stomach	Lesser curvature of stomach; inferior, anterior, and posterior aspects of stomach; esophagus
Hepatic nodes	Along hepatic artery (not shown)	Stomach, duodenum, liver, gallbladder, and pancreas
Pancreaticosplenic nodes (pan-krē-at'-i-kō-SPLĒN-ik)	Along splenic artery (not shown)	Stomach, spleen, and pancreas
Superior mesenteric nodes (MES-en-ter'-ik; *meso-*=middle; *-enteric*=intestines)	Consist of three groups: mesenteric, ileocolic, and transverse mesocolic	
Mesenteric nodes	Along superior mesenteric artery	Jejunum and all parts of ileum, except for terminal portion
Ileocolic nodes (il'-ē-ō-KŌL-ik)	Along ileocolic artery	Terminal portion of ileum, appendix, cecum, and ascending colon
Transverse mesocolic nodes	Between layers of transverse mesocolon	Descending and sigmoid parts of colon
Inferior mesenteric nodes	Near left colic, sigmoid, and superior rectal arteries	Descending and sigmoid parts of colon; superior part of rectum; superior anal canal

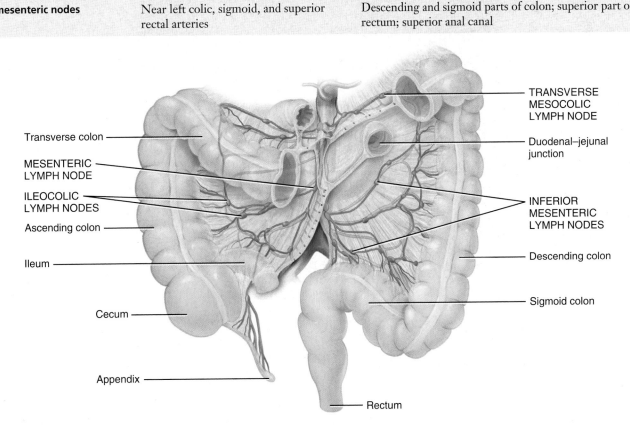

(b) Anterior view of superior and inferior mesenteric lymph nodes

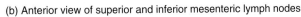 **What are the three groups of celiac lymph nodes?**

EXHIBIT 15.E Principal Lymph Nodes of the Lower Limbs *(Figure 15.12)*

OBJECTIVE

• Identify the principal lymph nodes of the lower limbs.

CHECKPOINT

12. What lymph nodes are located in the popliteal fossa?

LYMPH NODES	LOCATION	DRAINAGE
Popliteal nodes (pop-LIT-ē-al; *poples-*=ham of the knee)	In adipose tissue in popliteal fossa (not shown)	Knee and portions of leg and foot, especially heel
Superficial inguinal nodes (ING-gwi-nal; *inguen-*=groin)	Parallel to saphenous vein	Anterior and lateral abdominal wall to level of umbilicus, gluteal region, external genitals, perineal region, and entire superficial lymphatics of lower limb
Deep inguinal nodes	Medial to femoral vein	Deep lymphatics of lower limb, penis, and clitoris

Figure 15.12 **Principal lymph nodes of the lower limbs.**

The inguinal lymph nodes drain the lymphatic vessels of the lower limbs.

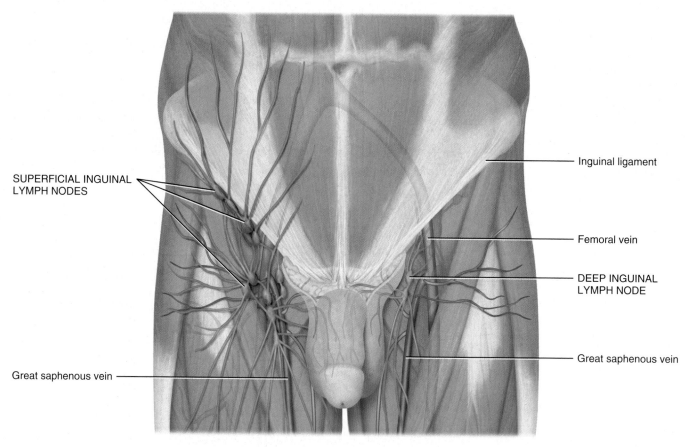

(a) Anterior view of inguinal lymph nodes

EXHIBIT 15.E **537**

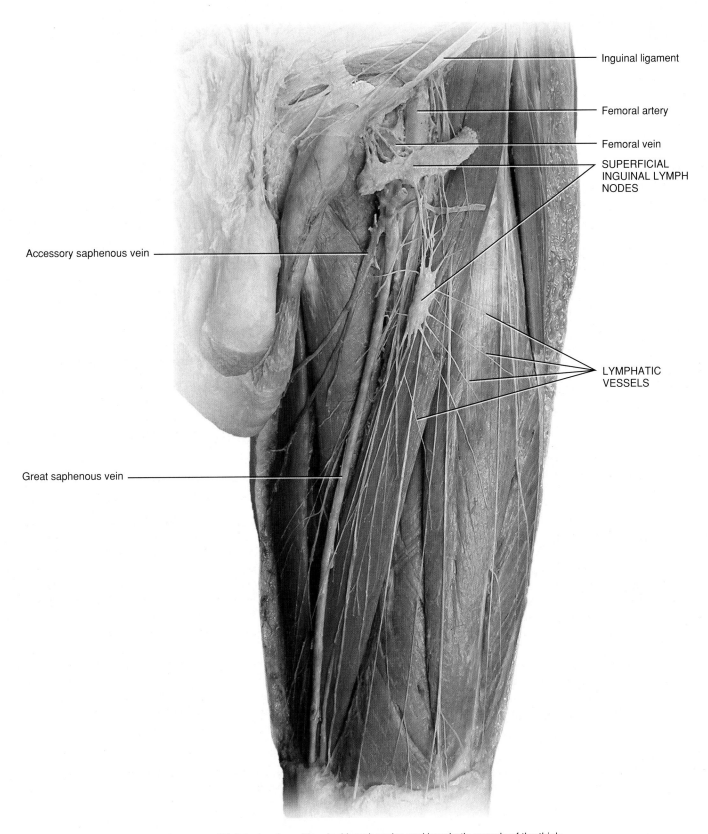

Inguinal ligament

Femoral artery

Femoral vein

SUPERFICIAL INGUINAL LYMPH NODES

Accessory saphenous vein

LYMPHATIC VESSELS

Great saphenous vein

(b) Anterior view of inguinal lymph nodes and lymphatic vessels of the thigh

Which lymph nodes are parallel to the saphenous vein?

15.3 DEVELOPMENT OF LYMPHATIC TISSUES

OBJECTIVE

• Outline the development of lymphatic tissues.

Lymphatic tissues begin to develop by the end of the fifth week of embryonic life. *Lymphatic vessels* develop from **lymph sacs**; lymph sacs arise from developing veins, which are derived from **mesoderm**.

The first lymph sacs to appear are the paired **jugular lymph sacs** at the junction of the internal jugular and subclavian veins (Figure 15.13). From the jugular lymph sacs, lymphatic capillary plexuses spread to the thorax, upper limbs, neck, and head. Some of the plexuses enlarge and form lymphatic vessels in their respective regions. Each jugular lymph sac retains at least one connection with its jugular vein; the left one develops into the superior portion of the thoracic duct (left lymphatic duct).

The next lymph sac to appear is the unpaired **retroperitoneal lymph sac** (ret′-rō-per′-i-tō-NĒ-al) at the root of the mesentery of the intestine. It develops from the primitive vena cava and mesonephric (primitive kidney) veins. Capillary plexuses and lymphatic vessels spread from the retroperitoneal lymph sac to the abdominal viscera and diaphragm. The sac establishes connections with the cisterna chyli but loses its connections with neighboring veins.

At about the time the retroperitoneal lymph sac is developing, another lymph sac, the cisterna chyli, develops inferior to the diaphragm on the posterior abdominal wall. It gives rise to the inferior portion of the *thoracic duct* and the *cisterna chyli* of the thoracic duct. Like the retroperitoneal lymph sac, the cisterna chyli also loses its connections with surrounding veins.

The last of the lymph sacs, the paired **posterior lymph sacs**, develop from the iliac veins. The posterior lymph sacs produce lymphatic capillary plexuses and lymphatic vessels of the abdominal wall, pelvic region, and lower limbs. The posterior lymph sacs join the cisterna chyli and lose their connections with adjacent veins.

With the exception of the anterior part of the sac from which the cisterna chyli develops, all lymph sacs are invaded during development by **mesenchymal cells** (me-SENG-ki-mal) and are converted into groups of *lymph nodes* with their specialized filtering tissue.

The *spleen* develops from mesenchymal cells between layers of the dorsal mesentery of the stomach. The *thymus* arises as an outgrowth of the **third pharyngeal pouch** (fa-RIN-jē-al) (see Figure 22.8).

✔ CHECKPOINT

13. Name the four lymph sacs from which lymphatic vessels develop.

Figure 15.13 Development of lymphatic tissue.

🔑 The lymphatic system is derived from mesoderm.

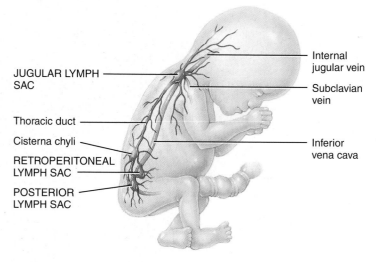

JUGULAR LYMPH SAC

Thoracic duct

Cisterna chyli

RETROPERITONEAL LYMPH SAC

POSTERIOR LYMPH SAC

Internal jugular vein

Subclavian vein

Inferior vena cava

? When do lymphatic tissues begin to develop?

15.4 AGING AND THE LYMPHATIC SYSTEM

OBJECTIVE

• Describe the effects of aging on the lymphatic system and on the immune response.

With advancing age, elderly individuals become more susceptible to all types of infections and malignancies. Their response to vaccines is decreased, and they tend to produce more autoantibodies (antibodies against their body's own molecules). In addition, the immune system exhibits lowered levels of function. For example, T cells become less responsive to antigens (foreign substances), and fewer T cells respond to infections because of age-related atrophy of the thymus or decreased production of thymic hormones. Because the T cell population decreases with age, B cells are also less responsive. Consequently, antibody levels do not increase as rapidly in response to a challenge by an antigen, resulting in increased susceptibility to various infections. It is for this key reason that elderly individuals are encouraged to get influenza (flu) vaccinations each year.

✔ CHECKPOINT

14. What changes occur in the lymphatic system with advancing age?

CLINICAL CONNECTION | *AIDS: Acquired Immunodeficiency Syndrome*

Acquired immunodeficiency syndrome (AIDS) is a condition in which a person experiences a telltale assortment of infections due to the progressive destruction of immune system cells by the human immunodeficiency virus (HIV). AIDS represents the end stage of infection by HIV. A person who is infected with HIV may be symptom-free for many years, even while the virus is actively attacking the immune system. In the two decades after the first five cases were reported in 1981, 22 million people died of AIDS. Worldwide, 35 to 40 million people are currently infected with HIV.

HIV Transmission

Because HIV is present in the blood and some body fluids, it is most effectively transmitted (spread from one person to another) by actions or practices that involve the exchange of blood or body fluids.

HIV is transmitted in semen or vaginal fluid during unprotected (without a condom) anal, vaginal, or oral sex. HIV also is transmitted by direct blood-to-blood contact, such as occurs among intravenous drug users who share hypodermic needles or health-care professionals who may be accidentally stuck by HIV-contaminated hypodermic needles. In addition, HIV can be transmitted from an HIV-infected mother to her baby at birth or during breast-feeding.

The chance of transmitting or of being infected by HIV during vaginal or anal intercourse can be greatly reduced—although not entirely eliminated—by the use of latex condoms. Public health programs aimed at encouraging drug users not to share needles have proven effective at checking the increase in new HIV infections in this population. Also, giving certain drugs to pregnant HIV-infected women greatly reduces the risk of transmission of the virus to their babies.

HIV is a very fragile virus; it cannot survive for long outside the human body. The virus is not transmitted by insect bites. One cannot become infected by casual physical contact with an HIV-infected person, such as by hugging or sharing household items. The virus can be eliminated from personal care items and medical equipment by exposing them to heat (135°F for 10 minutes) or by cleaning them with common disinfectants such as hydrogen peroxide, rubbing alcohol, household bleach, or germicidal cleansers such as Betadine® or Hibiclens®. Standard dishwashing and clothes washing also kills HIV.

HIV: Structure and Infection

HIV consists of an inner core of ribonucleic acid (RNA) covered by a protein coat (capsid) surrounded in turn by an outer layer called the envelope. The envelope is composed of a lipid bilayer penetrated by glycoproteins (see figure). HIV is classified as a *retrovirus* (RET-rō-vī-rus) meaning that its genetic information is carried in RNA instead of DNA. Glycoproteins assist both the binding of HIV to a host cell and its entry into the cell. Outside a living host cell, a virus is unable to replicate. However, when the virus infects and enters a host cell, its RNA uses the resources of the host cell to make thousands of copies of itself. The new viruses eventually leave the cell and then infect other cells.

HIV mainly damages T cells. Over 10 billion viral copies may be made each day. The viruses bud so rapidly from an infected T cell's plasma membrane that the cell ruptures and dies. In most HIV-infected people, T cells are initially replaced as fast as they are destroyed. After several years, however, the body's ability to replace T cells is slowly exhausted, and the number of T cells in circulation gradually declines.

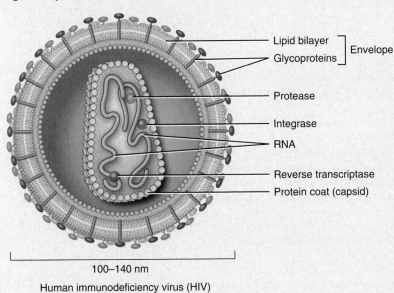

Lipid bilayer ⎤
Glycoproteins ⎦ Envelope

Protease

Integrase

RNA

Reverse transcriptase

Protein coat (capsid)

100–140 nm

Human immunodeficiency virus (HIV)

Signs, Symptoms, and Diagnosis of HIV Infection

Soon after being infected with HIV, most people experience a brief flu-like illness. Common signs and symptoms are fever, fatigue, rash, headache, joint pain, sore throat, and swollen lymph nodes. About 50 percent of infected people also experience night sweats. As early as three to four weeks after HIV infection, plasma cells begin secreting antibodies against HIV. These antibodies are detectable in blood plasma and form the basis for some of the screening tests for HIV. When people test "HIV-positive," it usually means they have antibodies to HIV antigens in their bloodstream.

Progression to AIDS

After a period of 2 to 10 years, the virus destroys enough helper T cells that most infected people begin to experience symptoms of immunodeficiency. HIV-infected people commonly have enlarged lymph nodes and experience persistent fatigue, involuntary weight loss, night sweats, skin rashes, diarrhea, and various lesions of the mouth and gums. In addition, the virus may begin to infect neurons in the brain, affecting the person's memory and producing visual disturbances.

As the immune system slowly collapses, an HIV-infected person becomes susceptible to a host of *opportunistic infections*. These are diseases caused by microorganisms that are normally held in check but now proliferate because of the defective immune system. AIDS is diagnosed when the helper T cell count drops below 200 cells per microliter (5 cubic millimeters) of blood or when opportunistic infections arise, whichever occurs first. In time, opportunistic infections usually are the cause of death.

Treatment of HIV Infection

At present, infection with HIV cannot be cured. Vaccines designed to block new HIV infections and to reduce the viral load (the number of copies of HIV RNA in a microliter of blood plasma) in those who are already infected are in clinical trials. Meanwhile, two categories of drugs have proved successful in extending the life of many of those infected with HIV:

1. *Reverse transcriptase inhibitors* interfere with the action of reverse transcriptase, the enzyme that the virus uses to convert its RNA into a DNA copy. Among the drugs in this category are zidovudine (ZDV, previously called AZT), didanosine (ddI), and stavudine (d4T). Trizivir®, approved in 2000 for treatment of HIV infection, combines three reverse transcriptase inhibitors in one pill.
2. *Protease inhibitors* interfere with the action of protease, a viral enzyme that cuts proteins into pieces to assemble the coat of newly produced HIV particles. Drugs in this category include nelfinavir, saquinavir, ritonavir, and indinavir.

In 1996, many physicians treating HIV-infected patients adopted *highly active antiretroviral therapy (HAART)*—a combination of two differently acting reverse transcriptase inhibitors and one protease inhibitor. Most HIV-infected individuals receiving HAART experience a drastic reduction in viral load and an increase in the number of T cells in their blood. Not only does HAART delay the progression of HIV infection to AIDS, but many people with AIDS have seen the remission or disappearance of opportunistic infections and an apparent return to health. Unfortunately, HAART is very costly (exceeding $10,000 per year), the dosing schedule is grueling, and not all people can tolerate the toxic side effects of these drugs. Although HIV may virtually disappear from the blood with drug treatment (and thus a blood test may be "negative" for HIV), the virus typically still lurks in various lymphatic tissues. In such cases, the infected person can still transmit the virus to another person. •

KEY MEDICAL TERMS ASSOCIATED WITH THE LYMPHATIC SYSTEM AND IMMUNITY

Adenitis (ad-e-NĪ-tis; *aden-*=gland; *itis*=inflammation of) Enlarged, tender, and inflamed lymph nodes resulting from an infection.

Autoimmune disease (aw-tō-i-MŪN) A disease in which the immune system fails to recognize self-antigens and attacks a person's own cells. Examples are rheumatoid arthritis (RA), systemic lupus erythematosus (SLE), rheumatic fever, hemolytic and pernicious anemias, Addison's disease, Graves disease, insulin-dependent diabetes mellitus, myasthenia gravis, multiple sclerosis (MS), and ulcerative colitis. Also called *autoimmunity*.

Chronic fatigue syndrome (CFS) A disorder, usually occurring in young adults and primarily in females, characterized by (1) extreme fatigue that impairs normal activities for at least 6 months and (2) the absence of other known diseases (cancer, infections, drug abuse, toxicity, or psychiatric disorders) that might produce similar symptoms.

Gamma globulin (GLOB-ū-lin) Suspension of immunoglobulins from blood consisting of antibodies that react with a specific pathogen. It is prepared by injecting the pathogen into animals, removing blood from the animals after antibodies have been produced, isolating the antibodies, and injecting them into a human to provide short-term immunity.

Graft Any tissue or organ used for transplantation; also refers to the transplant of such structures.

Hypersplenism (hī′-per-SPLĒN-izm; *hyper-*=over) Abnormal splenic activity due to splenic enlargement; associated with an increased rate of destruction of normal blood cells.

Lymphedema (lim′-fe-DĒ-ma; *edema*=swelling) Accumulation of lymph in lymphatic vessels, causing painless swelling of a limb.

Lymphomas (lim-FŌ-mas; *lymph-*=clear water; *-oma*=tumor) Cancers of the lymphatic organs, especially the lymph nodes; most have no

known cause. The two main types of lymphomas are Hodgkin disease (HOJ-kin) and non-Hodgkin lymphoma.

Severe combined immunodeficiency disease (SCID) A rare inherited disorder in which both B cells and T cells are missing or inactive; in some cases, an infusion of red bone marrow cells from a sibling having very similar MHC (HLA) antigens can provide normal stem cells that give rise to normal B and T cells.

Splenomegaly (splē′-nō-MEG-a-lē; *mega-*=large) Enlarged spleen.

Systemic lupus erythematosus (SLE) (er-e′-thĕm-a-TŌ-sus) or *lupus* (*lupus*=wolf) An autoimmune, noncontagious, inflammatory disease of connective tissue, occurring mostly in young women, in which damage to blood vessel walls results in the release of chemicals that mediate inflammation; symptoms include joint pain, slight fever, fatigue, oral ulcers, weight loss, enlarged lymph nodes and spleen, photosensitivity, rapid loss of large amounts of scalp hair, and sometimes an eruption across the bridge of the nose and cheeks called a "butterfly rash."

Transplantation (tranz-plan-TĀ-shun; *trans-*=across; *planto*=to plant) The transfer of living cells, tissues, or organs from a donor to a recipient or from one part of the body to another part of the same body in order to restore a lost function; success depends on *major histocompatibility (MHC) antigens* on the surfaces of white blood cells and other body cells (except for red blood cells), which are unique for each person—the better the match between donor and recipient, the greater the probability that graft rejection will be avoided.

Xenograft (zen-ō-graft; *xeno-*=strange or foreign) A transplant between animals of different species, including xenografts from porcine (pig) or bovine (cow) tissue, which may be used as a physiological dressing for severe burns; other xenografts include pig heart valves and baboon hearts.

CHAPTER REVIEW AND RESOURCE SUMMARY

WileyPLUS

Review

Introduction

1. The ability to ward off disease is called immunity or resistance. Lack of resistance is called susceptibility.
2. Innate immunity refers to defenses that are present at birth; they are always present and provide immediate or general protection against a wide variety of pathogens. Adaptive immunity refers to defenses that respond to a particular invader; it involves activation of specific lymphocytes that can combat a specific invader.

15.1 Lymphatic System Structure and Functions

1. The lymphatic system carries out immune responses and consists of lymph, lymphatic vessels, and structures and organs that contain lymphatic tissue (specialized reticular tissue containing many lymphocytes).
2. The lymphatic system drains interstitial fluid, transports dietary lipids, and protects against pathogens through immune responses.
3. Lymphatic vessels begin as lymph capillaries with one closed end in tissue spaces between cells. Interstitial fluid drains into lymphatic capillaries, thus forming lymph.
4. Lymph capillaries merge to form larger vessels, called lymphatic vessels, which convey lymph into and out of structures called lymph nodes. The route of lymph flow is from lymph capillaries to lymphatic vessels to lymph trunks to the thoracic duct or right lymphatic duct to the subclavian veins.
5. Lymph flows as a result of skeletal muscle contractions and respiratory movements. It is also aided by valves in lymphatic vessels.
6. The primary lymphatic organs are red bone marrow and the thymus. Secondary lymphatic organs are lymph nodes, spleen, and lymphatic nodules. The thymus, which lies between the sternum and the large blood vessels above the heart, is the site of T cell maturation.
7. Lymph nodes are encapsulated, oval structures located along lymphatic vessels. Lymph enters nodes through afferent lymphatic vessels, is filtered, and exits through efferent lymphatic vessels. Lymph nodes are the site of proliferation of plasma cells and T cells.

Resource

Anatomy Overview - The Lymphatic System and Disease Resistance
Anatomy Overview - The Integument and Disease Resistance
Animation - Introduction to Disease Resistance
Animation - Nonspecific Disease Resistance
Exercise - Integument vs. Disease

Anatomy Overview - Lymphatic Vessels
Anatomy Overview - The Thymus
Anatomy Overview - The Spleen and Lymph Nodes
Animation - Lymph Formation and Flow
Animation - Lymphatic System Functions
Figure 15.5 - Thymus
Figure 15.6 - Lymph Node
Exercise - Lymphatic Highway

Review   Resource

8. The spleen is the largest single mass of lymphatic tissue in the body. It is a site of B cell proliferation into plasma cells and phagocytosis of bacteria and worn-out red blood cells.
9. Lymphatic nodules are scattered throughout the mucosa of the gastrointestinal, respiratory, urinary, and reproductive tracts. This lymphatic tissue is termed mucosa-associated lymphoid tissue (MALT).

15.2 Principal Groups of Lymph Nodes

1. Lymph nodes are scattered throughout the body in superficial and deep groups.
2. The principal groups of lymph nodes are found in the head and neck, thorax, upper limbs, abdomen and pelvis, and lower limbs. See Exhibits 15.A–15.E.

Anatomy Overview - The Lymphatic System and Disease Resistance
Anatomy Overview - The Spleen and Lymph Nodes

15.3 Development of Lymphatic Tissues

1. Lymphatic vessels develop from lymph sacs, which arise from developing veins. Thus, they are derived from mesoderm.
2. Lymph nodes develop from lymph sacs that become invaded by mesenchymal cells to form the specialized filtering tissue.

15.4 Aging and the Lymphatic System

1. With advancing age, individuals become more susceptible to infections and malignancies, respond less well to vaccines, and produce more autoantibodies.
2. Immune responses also diminish with age.

CRITICAL THINKING QUESTIONS

1. Jamal got a splinter in his right heel while he was playing beach volleyball. He neglected to clean it properly and it became infected. The first aid station warned him to have the wound taken care of before the infection spreads to his blood. How can an infection travel from his foot to his cardiovascular system?

2. Four-year-old Kelsey had a history of repeated throat infections, averaging 10 per year. She breathed loudly through her mouth and even snored. Following the pediatrician's recommendation, she had an operation to remove the troublesome organs. When she returned to preschool, she told the other kids about all the ice cream she got to eat following her "tonsil-X-me." Where are these tonsils located, and what is their function?

3. After several years of struggling with a chronic autoimmune disease, Kelly's doctor recommended a splenectomy. Since then, every time she goes to the dentist, she is advised to take antibiotics for a period of time prior to her appointment. Why would the doctor

have recommended removal of such an important organ, and why would the dentist recommend the antibiotics?

4. Nan was ordering her dinner at an Italian restaurant. She informed the waiter that she was "highly allergic to eggplant" and that it was extremely important that her meal be eggplant-free. Dinner arrived, and shortly after she started eating her grilled vegetables, she started having trouble breathing and talking. There were no visible pieces of eggplant on her plate, so she announced to her dinner companion that the vegetables had been cooked on the same grill as the eggplant for other diners. How would her difficulty breathing lead her to this conclusion? How does being allergic to eggplant relate to difficulty breathing? If the difficulty breathing continues to worsen, what treatment might be required? Explain why such treatment would help.

5. An infection with a tropical parasite may block a lymphatic vessel. What would be the effect of blockage of the left subclavian trunk?

ANSWERS TO FIGURE QUESTIONS

15.1 Red bone marrow contains stem cells that develop into lymphocytes.

15.2 Lymph is more similar to interstitial fluid than to blood plasma because the protein content of lymph is low.

15.3 The left and right lumbar trunks and the intestinal trunk empty into the cisterna chyli, which then drains into the thoracic duct.

15.4 Inhalation promotes the movement of lymph from abdominal lymphatic vessels toward the thoracic region.

15.5 T cells mature in the thymus.

15.6 When foreign substances enter a lymph node, they may be phagocytized by macrophages or attacked by lymphocytes that mount immune responses.

15.7 After birth, white pulp of the spleen functions in immunity, and red pulp of the spleen removes worn-out blood cells and stores platelets.

15.8 The deep cervical nodes are the largest group of lymph nodes in the neck.

15.9 The phrenic lymph nodes drain most of the diaphragm.

15.10 The lateral lymph nodes drain most of the upper limb.

15.11 The three groups of celiac lymph nodes are the gastric, hepatic, and pancreaticosplenic nodes.

15.12 The superficial inguinal lymph nodes are parallel to the saphenous vein.

15.13 Lymphatic tissues begin to develop by the end of the fifth week of embryonic life.

16 | NERVOUS TISSUE

INTRODUCTION As it did throughout the 1980s and 1990s, the computer continues to revolutionize our world today. In the late 1970s, the first desktop computers operated with a total RAM of 16 KB. Today it is not uncommon to have a desktop or even a notebook with 1 gig of RAM, increasing capacity by *one million times* over the past 30 years.

However, even the most advanced supercomputers pale in comparison with the machine that created them—the human nervous system. In this chapter we will introduce the basic organization of this human computer and study the fundamental components that function as its wires and circuitry.

Because the nervous system is quite complex, we will consider different aspects of its structure and function in several related chapters. This chapter focuses on the organization of the nervous system and the properties of the cells that make up nervous tissue—neurons (nerve cells) and neuroglia (cells that support the activities of neurons). In chapters that follow, we will examine the structure and functions of the spinal cord and spinal nerves (Chapter 17), and of the brain and cranial nerves (Chapter 18). Then we will discuss the autonomic nervous system, the part of the nervous system that operates without voluntary control (Chapter 19). Next, we examine the somatic senses—touch, pressure, warmth, cold, pain, and others—and the sensory and motor pathways to understand how nerve impulses pass into the spinal cord and brain or from the spinal cord and brain to muscles and glands (Chapter 20). Our exploration of this complex yet fascinating system concludes with a discussion of the special senses: smell, taste, vision, hearing, and equilibrium (Chapter 21). The branch of medical science that deals with the normal functioning and disorders of the nervous system is **neurology** (noo-ROL-ō-jē; *neuro-*=nerve or nervous system; *logy*=study of). A **neurologist** (noo-ROL-ō-jist) is a physician who specializes in the diagnosis and treatment of disorders of the nervous system. •

Did you ever wonder how local anesthetics work?

16.1 OVERVIEW OF THE NERVOUS SYSTEM

OBJECTIVES

• List the structures and basic functions of the nervous system.
• Describe the organization of the nervous system.
• Explain the functional organization of the peripheral nervous system.

Structures of the Nervous System

With a mass of only 2 kg (4.5 lb), about 3 percent of total body weight, the **nervous system** is one of the smallest and yet the most complex of the 11 body systems. The nervous system is a highly organized network of two types of cells; it contains billions of neurons and even more neuroglia. The structures that make up the nervous system include the brain, cranial nerves and their branches, the spinal cord, spinal nerves and their branches, ganglia, enteric plexuses, and sensory receptors (Figure 16.1).

Organization of the Nervous System

Although we have only one nervous system, it is convenient to organize it into various components to make it easier to study. Broadly speaking, the nervous system can be organized both anatomically and functionally.

Figure 16.1 Components of the nervous system and anatomical organization of the nervous system.

 The major structures of the nervous system are the brain and cranial nerves, spinal cord and spinal nerves, ganglia, enteric plexuses, and sensory receptors.

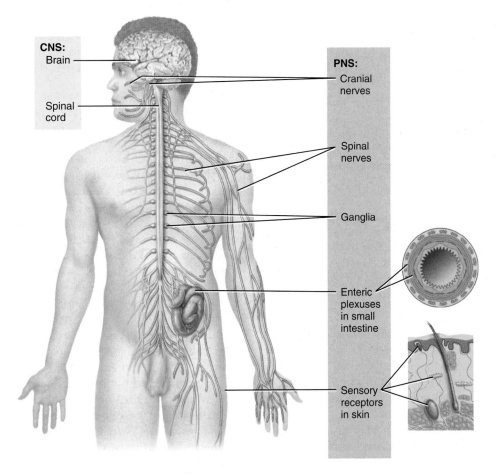

 How is the nervous system organized anatomically?

543

Anatomical Organization

The nervous system differs from other body systems because many of its cells are extremely long. Thus it is difficult to divide the nervous system into discrete organs with their own unique cell populations as we do for the other body systems. Anatomically the nervous system consists of two intimately interconnected divisions: (1) the central nervous system and (2) the peripheral nervous system (Figure 16.1).

CENTRAL NERVOUS SYSTEM The **central nervous system (CNS)** is composed of the brain and spinal cord. The **brain** is enclosed and protected by the skull in the cranial cavity and contains about 85 billion neurons. The **spinal cord** is enclosed and protected by the bones of the vertebral column in the vertebral canal and contains about 100 million neurons. The brain and spinal cord are continuous with one another through the foramen magnum of the occipital bone. The CNS processes many different kinds of incoming sensory information. It is also the source of thoughts, emotions, and memories. Most nerve impulses that stimulate muscles to contract and glands to secrete originate in the CNS.

PERIPHERAL NERVOUS SYSTEM The **peripheral nervous system (PNS)** (pe-RIF-e-ral) is composed of all nervous structures outside the CNS, such as cranial nerves and their branches, spinal nerves and their branches, ganglia, enteric plexuses, and sensory receptors. These structures link all parts of the body to the CNS. Twelve pairs (right and left) of **cranial nerves**, numbered I through XII, emerge from the base of the brain. A **nerve** is a bundle of hundreds to thousands of axons (nerve cell fibers) plus associated connective tissue and blood vessels that lies outside the brain and spinal cord. Each nerve follows a defined path and serves a specific region of the body. For example, the median nerve carries signals for motor output and sensory input to and from the muscles and skin of the upper limb. Thirty-one pairs of **spinal nerves** emerge from the spinal cord, each serving a specific region on the right or left side of the body. **Ganglia** (GANG-lē-a=swelling or knot; singular is *ganglion*) are small masses of nervous tissue, consisting primarily of neuron cell bodies, that are located outside the brain and spinal cord. Ganglia are closely associated with cranial and spinal nerves. In the walls of organs of the gastrointestinal tract, extensive networks of neurons, called **enteric plexuses** (PLEK-sus-ēz), help regulate the digestive system. **Sensory receptors** are structures that monitor changes in the internal and external environment, such as receptors in the skin that detect touch sensations.

Functional Organization

The nervous system carries out a complex array of tasks. It allows you to sense various smells, produce speech, and remember past events; it also provides signals that control body movements, and regulates the operation of internal organs. These diverse activities can be grouped into three basic functions: sensory (input), integrative (control), and motor (output).

1. *Sensory function.* Sensory receptors *detect* internal stimuli, such as an increase in blood pressure, and external stimuli, such as a raindrop landing on your arm. Neurons called **sensory** or **afferent neurons** (AF-er-ent; *af-*=toward; *ferrent*=carried) carry this sensory information into the brain and spinal cord through cranial and spinal nerves.

2. *Integrative function.* The nervous system processes sensory information by analyzing and storing some of it and by making decisions for appropriate responses—an activity known as **integration.** An important integrative function

is **perception**, the conscious awareness of sensory stimuli. Perception occurs in the brain. Many of the neurons that participate in integration are **interneurons** (neurons that interconnect with other neurons), with axons that extend for only a short distance and contact nearby neurons in the brain or spinal cord to set up the complex "circuit boards" of the central nervous system. The vast majority of neurons in the body are interneurons. These neurons make up the majority of the central nervous system.

3. *Motor function.* Once sensory information is integrated, the nervous system may elicit an appropriate motor response, such as muscular contraction or glandular secretion. The neurons that serve this function are called **motor** or **efferent neurons** (EF-er-ent; *ef-*=away from). Motor neurons carry information from the brain toward the spinal cord or out of the brain and spinal cord to **effectors** (muscles and glands) through cranial and spinal nerves. Stimulation of the effectors by motor neurons causes muscles to contract and glands to secrete.

Let's now take a closer look at the sensory and motor functions of the peripheral nervous system. The peripheral nervous system is divided into three functional components: (1) somatic nervous system, (2) autonomic nervous system, and (3) enteric nervous system (Figure 16.2).

SOMATIC NERVOUS SYSTEM The **somatic nervous system (SNS)** (sō-MAT-ik; *somat-*=body) of the PNS consists of sensory neurons, called *somatic sensory neurons*, that convey information to the CNS from sensory receptors in the skin, skeletal muscles, and joints, and from the receptors for the special senses (vision, hearing, equilibrium, taste, and smell). These *somatic sensory pathways* are involved in the *input* of information to the CNS for integration (processing). The SNS also consists of motor neurons, called *somatic motor neurons*, that convey information from the CNS to *skeletal muscles only*. These *somatic motor pathways* are involved in the *output* of information from the CNS that results in a muscular contraction. Because these motor responses can be consciously controlled, the actions of these parts of the SNS are *voluntary*.

AUTONOMIC NERVOUS SYSTEM The **autonomic nervous system (ANS)** (aw'-tō-NOM-ik; *auto-*=self; *nomic*=law) of the PNS also has sensory and motor components. Sensory neurons, called *autonomic (visceral) sensory neurons*, convey information to the CNS from autonomic sensory receptors, located primarily in the visceral organs (smooth muscle organs in the thorax, abdomen, and pelvis). *Autonomic motor neurons* convey information from the CNS to *smooth muscle, cardiac muscle,* and *glands* and cause the muscles to contract and the glands to secrete. Because its motor responses are not normally under conscious control, the action of the ANS is involuntary. The motor part of the ANS consists of two branches, the **sympathetic division** and the **parasympathetic division**. With a few exceptions, effectors receive nerves from both divisions, and usually the two divisions have opposing actions. For example, sympathetic neurons increase heart rate, and parasympathetic neurons slow it down. In general, the sympathetic division helps support exercise or emergency actions, so-called "fight-or-flight" responses, and the parasympathetic division controls "rest-and-digest" activities (Chapter 19). Because the sympathetic division is the major regulator of the smooth muscle of the cardiovascular system, it has a wider distribution, as blood vessels are located everywhere in the body. The parasympathetic division is the major regulator of the smooth muscle of the digestive and respiratory systems, which are derived from the embryonic gut tube.

ENTERIC NERVOUS SYSTEM The **enteric nervous system (ENS)** (en-TER-ik; *enteron*=intestines) of the PNS is called the "brain of the gut" and consists of over 100 million neurons that occur throughout most of the length of the gastrointestinal (GI) tract. The ENS also has both sensory and motor components and can operate independently from the CNS. *Sensory neurons* of the ENS monitor chemical changes within the GI tract as well as the stretching of its walls. *Motor neurons* of the ENS govern contraction of GI tract smooth muscle to propel food through the GI tract. These neurons also control secretions of the GI tract organs such as acid from the stomach, and endocrine cells, which secrete hormones. Like the ANS, the ENS is involuntary.

 CHECKPOINT

1. What are the structural components of the CNS and PNS?
2. What types of problems would result from damage to each of the following: sensory neurons, interneurons, and motor neurons?
3. Describe the structural components and functions of the SNS, ANS, and ENS, and indicate which of these subdivisions of the PNS control voluntary actions and which regulate involuntary actions.

16.2 HISTOLOGY OF NERVOUS TISSUE

 OBJECTIVES

• Compare the histological characteristics and functions of neurons and neuroglia.
• Distinguish between gray matter and white matter.
• Define a neuromuscular junction.
• Distinguish between electrical and chemical synapses.
• Define a neurotransmitter and provide several examples.

As noted earlier in the chapter, nervous tissue is comprised of two types of cells—neurons and neuroglia. These cells combine in a variety of ways in different regions of the nervous system. In addition to forming complex processing networks within the brain and spinal cord, neurons comprise the circuitry that connects all regions of the body to this central processing unit. As highly specialized cells capable of reaching great lengths and making extremely intricate connections with other cells, neurons provide most of the unique functions of the nervous system, such as sensing, thinking, remembering, controlling muscle activity, and regulating glandular secretions. As a result of their specialization, most neurons have lost the ability to undergo mitotic divisions. Neuroglia are smaller but outnumber neurons by as much as 25 times, according to some estimates. Neuroglia support, nourish, and protect neurons, and maintain the interstitial fluid that bathes them. Unlike neurons, neuroglia continue to divide throughout an individual's lifetime. The structures of both neurons and neuroglia differ depending on whether they are located in the central nervous system or the peripheral nervous system. These differences in structure correlate with differences in function in these two branches of the nervous system.

Neurons

Neurons (NOO-rons) or **nerve cells** possess **electrical excitability** (ek-sīt′-a-BIL-i-tē), the ability to respond to a stimulus and convert it into a nerve impulse. A *stimulus* is any change in the environment that is strong enough to initiate a nerve impulse. Very simply, a **nerve impulse (action potential)** is an electrical signal that propagates (travels) along the surface of the membrane of a neuron. It begins and travels due to the movement of ions (such as sodium and potassium) between interstitial fluid and the inside of a neuron through specific ion channels in the neuron's plasma membrane. Once begun, a nerve impulse travels rapidly and at a constant strength.

Some neurons are tiny and propagate impulses over a short distance (less than 1 mm) within the CNS. Others are the longest cells in your body. Motor neurons that cause muscles to wiggle your toes, for example, extend from the lumbar region of your

Figure 16.2 Functional organization of the nervous system.

 Sensory pathways are involved in the input of information to the CNS; motor pathways are involved in the output of information from the CNS.

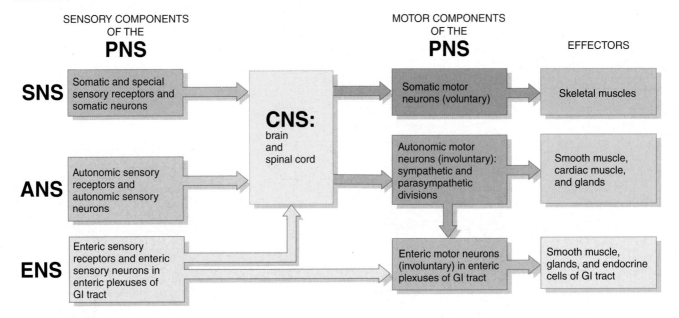

 Which motor component of the peripheral nervous system is voluntary?

spinal cord (just above waist level) to your foot. Some sensory neurons are even longer. Those that allow you to feel sensations in your toes stretch all the way from your foot to the lower portion of your brain. Nerve impulses travel these great distances at speeds ranging from 0.5 to 130 meters per second (1 to 280 mi/hr).

Parts of a Neuron

While neurons come in a wide variety of shapes and sizes, a general pattern of design is shared by all of them (Figure 16.3). A neuron typically consists of two basic parts: (1) the cell body and (2) a variable number of processes called nerve fibers. The nerve fibers exhibit great variation in length and size and are classified, based on distinct structural and functional differences, as dendrites or an axon.

Cell Body

The **cell body** (*perikaryon*) (per′-i-KAR-ē-on) contains a nucleus surrounded by cytoplasm. Its cytoplasm includes typical cellular organelles such as lysosomes, mitochondria, and a Golgi complex.

Figure 16.3 **Structure of a typical neuron and a synapse between neurons.** Arrows in (a) indicate the direction of information flow: dendrites → cell body → axon → axon terminals. The break indicates that the axon is actually much longer than shown.

🔑 The basic parts of a neuron are several dendrites, a cell body, and an axon.

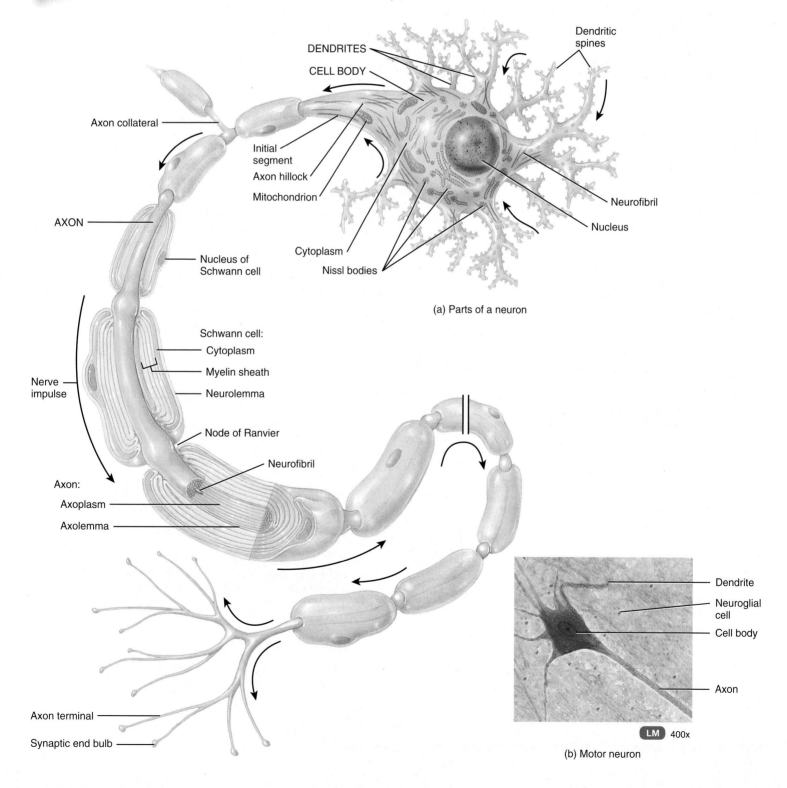

(a) Parts of a neuron

(b) Motor neuron

Neuronal cell bodies also contain prominent clusters of rough endoplasmic reticulum, termed **Nissl bodies** (NIS-el). The Nissl bodies are responsible for high levels of protein synthesis; proteins play important roles in maintenance and repair, transmission of nerve impulses, and reception of stimuli. The cytoskeleton includes both **neurofibrils** (noo-rō-FĪ-brils), composed of bundles of intermediate filaments that provide the cell shape and support, and **microtubules** (mī′-krō-TOO-būls), which assist in moving materials between the cell body and axon. Aging neurons also contain **lipofuscin** (līp-ō-FYŪS-in), a pigment that occurs as clumps of yellowish brown granules in the cytoplasm. Lipofuscin is a product of neuronal lysosomes that accumulates in a neuron, but does not seem to cause harm to the neuron as it ages. The plasma membrane of the neuronal cell body ranges from smooth to very bumpy; these bumps are caused by many small projections called **somatic gemmules** (JEM-yūls) or **somatic spines**, which increase the surface area available for interactions with other nerve cells.

Nerve Fibers

A **nerve fiber** is a general term for any neuronal process (extension) that emerges from the cell body of a neuron. Processes are termed dendrites and axons. Most neurons have multiple dendrites and one axon. **Dendrites** (DEN-drīts=little trees) are the receiving or input portions of a neuron. They usually are short, tapering, and highly branched: The greater the branching, the greater the surface area of the neuron for receiving synaptic communication from other neurons. In many neurons the dendrites form a tree-shaped array of processes extending from the cell body. While there are many different dendritic branching patterns, each type of neuron has a similar branching pattern.

Their cytoplasm contains Nissl bodies, mitochondria, and other organelles. The plasma membranes of the dendrites, like the cell body, contain numerous receptor sites for binding chemical messengers from other cells. In fact, the dendrites typically have many more sites than the cell body. The number of these receptor sites on the plasma membrane of the dendrites of many neurons is increased by small bumps or projections of the plasma membrane, similar to the less numerous projections on the cell body, called **dendritic gemmules** or **dendritic spines**.

The **axon** (=axis) is the other type of nerve fiber, and it varies significantly from the dendrites. Axons vary in length from less than a millimeter in neurons that communicate only with neighboring cells, to longer than a meter in neurons that communicate with distant parts of the nervous system or with peripheral organs. The single axon of a neuron carries nerve impulses toward another neuron, a muscle fiber, or a gland cell. An axon is a long, thin, cylindrical projection that often joins the cell body at a cone-shaped elevation called the **axon hillock** (HIL-lok=small hill). The part of the axon closest to the axon hillock is the **initial segment.** In most neurons, impulses arise at the junction of the axon hillock and the initial segment, an area called the **trigger zone**, and then travel along the axon. The trigger zone is free of Nissl bodies and has numerous voltage-sensitive channels in the plasma membrane. An axon contains mitochondria, microtubules, and neurofibrils. The cytoplasm of an axon, called **axoplasm**, is surrounded by a plasma membrane known as the **axolemma** (*lemma*=sheath or husk). Along the length of an axon, side branches called **axon collaterals** may branch off, typically at a right angle to the axon. The axon and its collaterals end by dividing into many fine processes called **axon terminal arborizations** (*telodendria*) (tēl′-ō-DEN-drē-a).

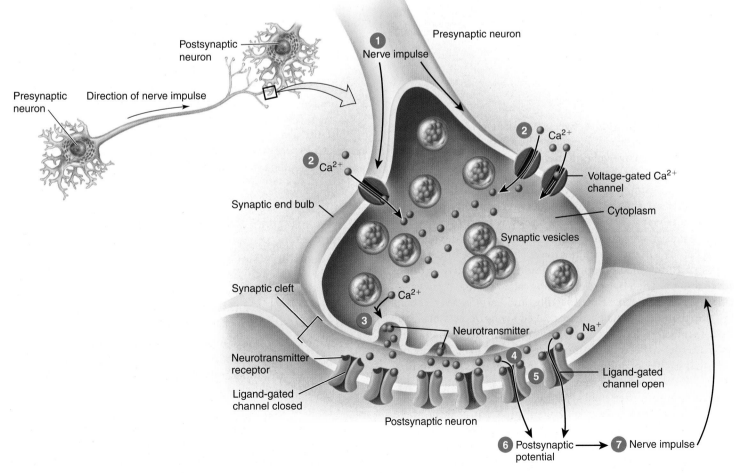

(c) Synapse between neurons

 What roles do the dendrites, cell body, and axon play in communication of nerve impulses?

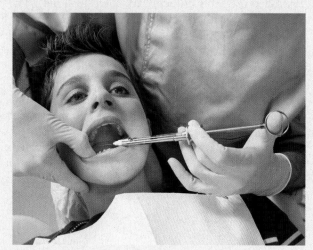

Synapses

The site of communication between two neurons or between a neuron and an effector cell is called a **synapse** (SIN-aps) (Figure 16.3c). The term **presynaptic neuron** (*pre-*=before) refers to a nerve cell that carries a nerve impulse toward a synapse. It is the cell that sends a signal. A **postsynaptic cell** (*post-*=after) is the cell that receives a signal. It may be a nerve cell called a **postsynaptic neuron** that carries a nerve impulse away from a synapse, or an *effector* (muscle or glandular cell) that responds to the impulse at the synapse. The tips of some axon terminals swell into bulb-shaped structures called **synaptic end bulbs (terminal boutons)**; others exhibit a string of swollen bumps called **varicosities** (var′-i-KOS-i-tēz). Both synaptic end bulbs and varicosities contain many tiny membrane-enclosed sacs called **synaptic vesicles** that store a chemical called a neurotransmitter. A **neurotransmitter** (noo′-rō-trans′-MIT-ter) is a molecule released from a synaptic vesicle that excites or inhibits postsynaptic neurons, muscle fibers, or gland cells. Neurons were long thought to liberate just one type of neurotransmitter at all synaptic end bulbs. We now know that many neurons contain two or even three neurotransmitters and that the release of the second or third neurotransmitter from a presynaptic neuron may depend on the frequency of activation of the neuron, as well as other factors. Neurotransmitters will be described in more detail shortly.

Neuromuscular Junction

If any of this sounds familiar, it's because back in Chapter 10 you learned about the synapse between a motor neuron and a muscle fiber, called a **neuromuscular junction** (see Figure 10.6). (The synapse between a neuron and a glandular cell is called a **neuroglandular junction.**) Recall from Chapter 10 that most synapses contain a small gap between cells called a **synaptic cleft** (see Figure 10.6c). A nerve impulse cannot jump the gap to excite the next cell, so it must rely on neurotransmitters. At a neuromuscular junction synaptic vesicles release the neurotransmitter acetylcholine (ACh) (see Figure 10.6c). The ACh is picked up by ACh receptors, proteins embedded in the cell membranes of muscle fibers, triggering an action potential. The action potential causes the muscle fibers to contract. Once ACh has accomplished its task, it is broken down rapidly by an enzyme in the synaptic cleft called *acetylcholinesterase (AChE)* (a-sē′-til-kō′-lin-ES-ter-ās).

Synapses Between Neurons

For synapses between neurons, the neuron sending the signal is called the **presynaptic neuron**, and the neuron receiving the message is referred to as the **postsynaptic neuron** (Figure 16.3c). Most synapses between neurons are **axodendritic** (ak′-sō-den-DRIT-ik), from presynaptic axon to postsynaptic dendrite, while others are **axosomatic** (ak′-sō-sō-MAT-ik), from presynaptic axon to postsynaptic cell body (soma), or **axoaxonic** (ak′-sō-ak-SON-ik), from presynaptic axon to postsynaptic axon.

Synapses between neurons may be electrical or chemical. In an **electrical synapse**, the plasma membranes of the presynaptic and postsynaptic neurons are tightly bound by gap junctions that contain connexons (see Figure 3.2e). As ions flow from one cell to another through the connexons, a nerve impulse is generated and passes from one cell to another. Although electrical synapses are not as common as chemical synapses in the brain, they are quite common in visceral smooth muscles, cardiac muscle tissue, and the developing embryo. Electrical synapses permit very rapid communication and uniform, coordinated movements, such as those required to make the heart beat.

Chemical synapses, which involve the release of a neurotransmitter from a presynaptic neuron, occur between most neurons and between all neurons and effectors (muscle cells and glandular cells).

A typical chemical synapse operates as follows (Figure 16.3c):

❶ A nerve impulse arrives at a synaptic end bulb of a presynaptic axon.

❷ The nerve impulse opens voltage-gated Ca^{2+} channels, present in the membrane of synaptic end bulbs, which allows Ca^{2+} to flow into the synaptic end bulb. Recall from Chapter 10 that voltage-gated channels are integral membrane proteins that open in response to changes in membrane potential (voltage).

❸ An increase in the concentration of Ca^{2+} inside the synaptic end bulb triggers exocytosis of some of the synaptic vesicles, which releases thousands of neurotransmitter molecules into the synaptic cleft.

❹ The neurotransmitter molecules diffuse across the synaptic cleft and bind to **neurotransmitter receptors** in the postsynaptic neuron's plasma membrane.

❺ Binding of neurotransmitter molecules opens ion channels, which allows certain ions to flow across the membrane.

6 As ions flow through the opened channels, the voltage across the membrane changes. Depending on which ions the channels admit, the voltage change may result in the generation of a nerve impulse if sodium (Na^+) channels open (*excitatory*) or inhibition of a nerve impulse if chloride (Cl^-) or potassium (K^+) channels open (*inhibitory*). A typical neuron in the CNS receives input from 1000 to 10,000 synapses. Some of this input is excitatory and some is inhibitory. The sum of all the excitatory and inhibitory effects at any given time determines whether one or more impulses will occur in the postsynaptic neuron.

At most synapses, only one-way information transfer can occur—from a presynaptic neuron to either a postsynaptic neuron or an effector such as a muscle fiber or gland cell. For example, synaptic transmission at a neuromuscular junction (NMJ) proceeds from a somatic motor neuron to a skeletal muscle fiber (but not in the opposite direction). Only synaptic end bulbs of presynaptic neurons can release neurotransmitters, and only the postsynaptic neuron's membrane has the correct receptor proteins to recognize and bind that neurotransmitter.

A neurotransmitter affects the postsynaptic neuron, muscle fiber, or gland cell as long as it remains bound to its receptors. Thus, removal of the neurotransmitter is essential for normal synaptic function. Neurotransmitter is removed in three ways: (1) Some of the released neurotransmitter molecules diffuse away from the synaptic cleft. Once a neurotransmitter molecule is out of reach of its receptors, it can no longer exert an effect; (2) some neurotransmitters are destroyed by enzymes; (3) many neurotransmitters are actively transported back into the neuron that released them (reuptake); others are transported into neighboring neuroglia (uptake).

Neurotransmitters

About 100 substances are either known or suspected to be neurotransmitters. **Acetylcholine (ACh)** (a-sē′-til-KŌ-lēn) is released by many PNS neurons and by some CNS neurons. ACh is an excitatory neurotransmitter at the neuromuscular junction. It is also known to be an inhibitory neurotransmitter at other synapses. For example, parasympathetic neurons slow heart rate by releasing ACh at inhibitory synapses.

Several amino acids are neurotransmitters in the CNS. **Glutamate** (gloo-TA-māt) and **aspartate** (as-PAR-tāt) have powerful excitatory effects. Other amino acids, such as **gamma aminobutyric acid (GABA)** (GAM-ma am-ē-nō-bū-TER-ik), are important inhibitory neurotransmitters. Antianxiety drugs such as diazepam (Valium) enhance the action of GABA.

Some neurotransmitters are modified amino acids, including norepinephrine, dopamine, and serotonin. **Norepinephrine (NE)** (nōr′-ep-i-NEF-rin) plays roles in arousal (awakening from deep sleep), dreaming, and regulating mood. Brain neurons containing the neurotransmitter **dopamine (DA)** (DŌ-pa-mēn) are active during emotional responses, addictive behaviors, and pleasurable experiences. In addition, dopamine-releasing neurons help regulate upper motor neurons that ultimately affect skeletal muscle tone and some aspects of movement due to contraction of skeletal muscles. One form of schizophrenia is due to accumulation of excess dopamine. **Serotonin** (ser′-ō-TŌ-nin) is thought to be involved in sensory perception, temperature regulation, control of mood, appetite, and the onset of sleep.

CLINICAL CONNECTION | Depression

Depression is a disorder that affects over 18 million people each year in the United States. People who are depressed feel sad and helpless, have a lack of interest in activities that they once enjoyed, and experience suicidal thoughts. There are several types of depression. A person with **major depression** experiences symptoms of depression that last for more than two weeks. A person with **dysthymia** (dis-THĪ-mē-a) experiences episodes of depression that alternate with periods of feeling normal. A person with **bipolar disorder,** or *manic-depressive illness*, experiences recurrent episodes of depression and extreme elation (mania). A person with **seasonal affective disorder (SAD)** experiences depression during the winter months, when daylength is short (see Section 22.9). Although the exact cause of depression is unknown, research suggests that depression is linked to an imbalance of the neurotransmitters serotonin, norepinephrine, and dopamine in the brain. Factors that may contribute to depression include heredity, stress, chronic illnesses, certain personality traits (such as low self-esteem), and hormonal changes. Medication is the most common treatment for depression. For example, *selective serotonin reuptake inhibitors (SSRIs)* are drugs that provide relief from some forms of depression. By inhibiting reuptake of serotonin by serotonin transporters, SSRIs prolong the activity of this neurotransmitter at synapses in the brain. SSRIs include fluoxetine (Prozac®), paroxetine (Paxil®), and sertraline (Zoloft®). •

Neurotransmitters consisting of amino acids linked by peptide bonds are called **neuropeptides** (noor-ō-PEP-tīds). The **endorphins** (en-DOR-fins) are neuropeptides that serve as the body's natural painkillers. Acupuncture may produce analgesia (loss of pain sensation) by increasing the release of endorphins. Endorphins have also been linked to improved memory and learning and to feelings of pleasure or euphoria.

Another neurotransmitter is the simple gas **nitric oxide (NO)**, which is different from all previously known neurotransmitters because it is not synthesized in advance and packaged into synaptic vesicles. Rather, it is formed on demand, diffuses out of cells that produce it and into neighboring cells, does not react with receptors, and acts immediately. Additionally, NO reacts with so many different molecules that it is quickly consumed close to where it is synthesized and acts only on cells near its point of synthesis. NO is an excitatory neurotransmitter secreted in the brain, spinal cord, adrenal glands, and nerves to the penis. Some research suggests that NO plays a role in learning and memory. NO also functions in blood vessel dilation and immune response.

Carbon monoxide (CO), like NO, is not produced in advance or packaged into synaptic vesicles. It too is formed as needed and diffuses out of cells that produce it into adjacent cells and does not react with receptors. CO is an excitatory neurotransmitter produced in the brain and in response to some neuromuscular and neuroglandular functions. CO may be related to dilation of blood vessels, memory, olfaction (sense of smell), vision, thermoregulation, insulin release, and anti-inflammatory activity.

Structural Diversity in Neurons

Neurons display great diversity in size and shape. Their cell bodies range in diameter from 5 micrometers (μm) (slightly smaller than a red blood cell) up to 135 mm (barely large enough to see with the unaided eye). The pattern of dendritic branching is varied and distinctive for neurons in different parts of the nervous

Figure 16.4 Structural classification of neurons. Breaks indicate that axons are longer than shown.

 A multipolar neuron has many processes extending from the cell body; a bipolar neuron has two and a unipolar neuron has one.

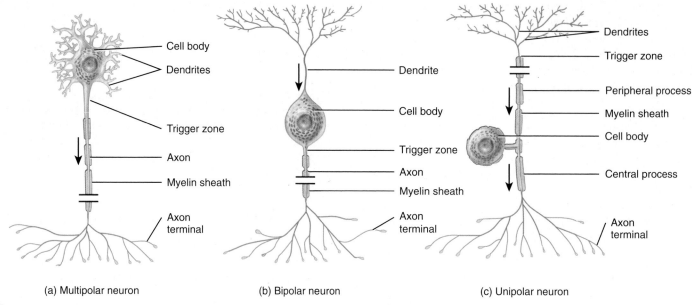

(a) Multipolar neuron   (b) Bipolar neuron   (c) Unipolar neuron

 What occurs at a trigger zone?

system (Figure 16.4). A few small neurons lack an axon, and many others have very short axons. As noted earlier in the chapter, your longest axons are almost as long as you are tall, extending from your toes to the lowest part of your brain.

Both structural and functional features are used to classify the various neurons in the body. Structurally, neurons are classified according to the number of processes extending from the cell body (Figure 16.4).

- **Multipolar neurons** usually have several dendrites and one axon (Figure 16.4a; see also Figure 16.3). Most neurons in the brain and spinal cord (interneurons) are multipolar, as are all motor (efferent) neurons.
- **Bipolar neurons** have one main dendrite and one axon (Figure 16.4b). They are found in the retina of the eye, in the inner ear, and in the olfactory area of the brain.
- **Unipolar** or **pseudounipolar neurons** (soo′-dō-ū′-ni-PŌ-lar) are sensory neurons that begin in the embryo as bipolar neurons. As a result of differential growth of the plasma membrane of the cell body between the axon and dendrite, the two processes are brought together and fuse into a single process that divides into two branches a short distance from the cell body. Both branches have the characteristic structure and function of an axon. They are long, cylindrical processes that propagate action potentials. However, the axon branch that extends into the periphery, called the **peripheral process**, has dendritic branches at its distal tip, whereas the axon branch that extends into the CNS, called the **central process**, ends in synaptic end bulbs. The dendrites monitor a sensory stimulus such as touch or stretching. The trigger zone for nerve impulses in a unipolar neuron is at the junction of the dendrites and axon (Figure 16.4c). The impulses then propagate toward the synaptic end bulbs.

Some neurons are named for the histologist who first described them or for an aspect of their shape or appearance; examples include **Purkinje cells** (pur-KIN-jē) in the cerebellum (Figure 16.5a) and **pyramidal cells** (pi-RAM-i-dal), found in the cerebral cortex of the brain, which have pyramid-shaped cell bodies (Figure 16.5b). Often, a distinctive pattern of dendritic

branching allows identification of a particular type of neuron in the CNS.

Neuroglia

Neuroglia (noo-ROG-lē-a; *glia*=glue), **glia** (GLĒ-a), or **glial cells** constitute about half the volume of the CNS. Their name derives from the idea of early histologists that they were the "glue" that held nervous tissue together. We now know that neuroglia are not merely passive bystanders but rather active participants in nervous tissue function. Generally, neuroglia are smaller than neurons, and

Figure 16.5 Two examples of CNS neurons. Arrows indicate the direction of information flow.

 The dendritic branching pattern often is distinctive for a particular type of neuron.

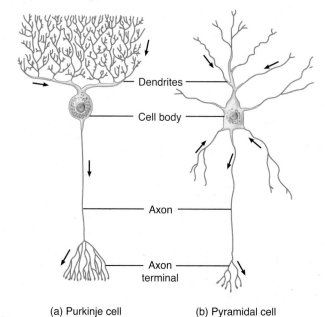

(a) Purkinje cell   (b) Pyramidal cell

How did pyramidal cells get their name?

as stated previously they are much more numerous. In contrast to neurons, glia do not generate or propagate nerve impulses, and they have the ability to multiply and divide in the mature nervous system. In cases of injury or disease, neuroglia multiply to fill in the spaces formerly occupied by neurons. Brain tumors derived from glia, called **gliomas** (glē-Ō-mas), tend to be highly malignant and rapidly growing. Of the six types of neuroglia, four—astrocytes, oligodendrocytes, microglia, and ependymal cells—are found only in the CNS. The remaining two types—Schwann cells (neurolemmocytes) and satellite cells—are present in the PNS.

Neuroglia of the CNS

Neuroglia of the CNS can be distinguished on the basis of size, cytoplasmic processes, and intracellular organization into four types: astrocytes, oligodendrocytes, microglia, and ependymal cells (Figure 16.6).

ASTROCYTES (AS-trō-sīts; *astro-*=star; *cyte*=cell) **Astrocytes** are the largest and most numerous of the neuroglia. They are star-shaped cells that have many armlike processes. There are two types of astrocytes. *Protoplasmic astrocytes* have many short branching processes and are found in gray matter (described shortly). *Fibrous astrocytes* have many long unbranched processes and are located mainly in white matter (also described shortly). The processes of astrocytes make contact with blood capillaries, neurons, and the pia mater (a thin membrane around the brain and spinal cord).

Astrocytes have the following functions: (1) They contain microfilaments that provide them with considerable strength, which enables them to support neurons. (2) Since neurons of the CNS must be isolated from various potentially harmful substances in blood, the endothelial cells of CNS blood capillaries have very selective permeability characteristics. Processes of astrocytes wrapped around blood capillaries secrete chemicals that maintain the unique permeability characteristics of the endothelial cells. In effect, the endothelial cells create a *blood–brain barrier*, which restricts the movement of substances between the blood and interstitial fluid of the CNS. Details of the blood–brain barrier are discussed in Chapter 18. (3) In the embryo, astrocytes secrete chemicals that appear to regulate the growth, migration, and interconnections among neurons in the brain. (4) Astrocytes help to maintain the appropriate chemical environment for the generation of nerve impulses. For example, they regulate the concentration of important ions such as K^+; take up excess neurotransmitters; and serve as a conduit for the passage of nutrients and other substances between blood capillaries and neurons. (5) Astrocytes may also play a role in learning and memory by influencing the formation of neural synapses.

Figure 16.6 Neuroglia of the central nervous system (CNS).

 Neuroglia of the CNS are distinguished on the basis of size, cytoplasmic processes, and intracellular organization.

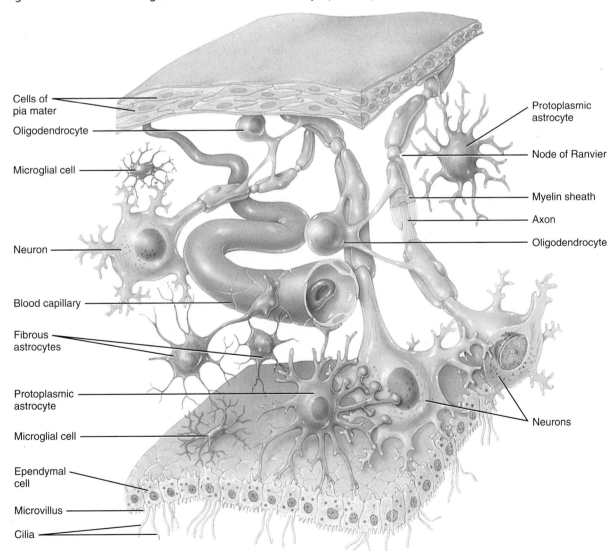

Ventricle

 Which CNS neuroglia function as phagocytes?

OLIGODENDROCYTES (OL-i-gō-den′-drō-sīts; *oligo-*=few; *dendro-*=tree) **Oligodendrocytes** resemble astrocytes, but are smaller and contain fewer processes. Oligodendrocyte processes are responsible for forming and maintaining the protective covering around CNS axons. As you will see shortly, the myelin sheath is a lipid and protein covering around some axons that insulates the axon and increases the speed of nerve impulse conduction.

MICROGLIAL CELLS OR MICROGLIA (mī-KRO-glē-a; *micro-*=small) **Microglial cells** or **microglia** are small cells with slender processes that give off numerous spinelike projections. Unlike other neuroglial cells, which develop from the neural tube, microglial cells originate in red bone marrow and migrate into the CNS as it develops. Microglial cells function as phagocytes. Like tissue macrophages, they remove cellular debris formed during normal development of the nervous system and phagocytize microbes and damaged nervous tissue.

EPENDYMAL CELLS (ep-EN-de-mal; *epen-*=above; *dym-*=garment) **Ependymal cells** are cuboidal to columnar cells arranged in a single layer that possess microvilli and cilia. These cells line the ventricles of the brain and central canal of the spinal cord (spaces filled with cerebrospinal fluid). Functionally, ependymal cells produce (possibly), monitor, and assist in the circulation of cerebrospinal fluid. They also form the blood–cerebrospinal fluid barrier, which is discussed in Chapter 18.

Neuroglia of the PNS

Neuroglia of the PNS completely surround axons and cell bodies. The two types of glial cells in the PNS are Schwann cells (neurolemmocytes) and satellite cells (Figure 16.7).

SCHWANN CELLS (SCHVON or SCHWON) **Schwann cells**, also called **neurolemmocytes** (noo′-rō-LEM-mō-sīts), are flat cells that encircle PNS axons. Like the oligodendrocytes of the CNS, they form the myelin sheath around axons. A single oligodendrocyte myelinates several axons (Figure 16.6), but each Schwann cell myelinates a single axon (Figure 16.7a; see also Figure 16.8a, c). A single Schwann cell can also enclose as many as 20 or more unmyelinated axons (axons that lack a myelin sheath) (Figure 16.7b). Schwann cells participate in axon regeneration, which is more easily accomplished in the PNS than in the CNS.

SATELLITE CELLS (SAT-i-līt) **Satellite cells** are flat cells that surround the cell bodies of neurons of PNS ganglia (Figure 16.7c). (Recall that *ganglia* are collections of neuronal cell bodies outside the CNS.) Besides providing structural support, satellite cells regulate the exchange of materials between neuronal cell bodies and interstitial fluid.

Myelination

Axons that are surrounded by a multilayered lipid and protein covering, called the **myelin sheath**, are said to be **myelinated** (MĪ′-e-li-nāt′-ed) (Figure 16.8a). The sheath electrically insulates the axon of a neuron and increases the speed of nerve impulse conduction. Axons without such a covering are said to be **unmyelinated** (Figure 16.8b).

Two types of neuroglia produce myelin sheaths: Schwann cells (in the PNS) and oligodendrocytes (in the CNS). In the PNS, Schwann cells begin to form myelin sheaths around axons during fetal development. Each Schwann cell wraps about 1 millimeter (1 mm = 0.04 in.) of a single axon's length by spiraling many times around the axon (Figure 16.8a). Eventually, multiple layers of glial plasma membrane surround the axon, with the Schwann cell's cytoplasm and nucleus forming the outermost layer. The inner portion, consisting of up to 100 layers of Schwann cell membrane, is the myelin sheath. The outer nucleated cytoplasmic layer of the Schwann cell, which encloses myelinated or unmyelinated axons, is the **neurolemma (sheath of Schwann)**. A neurolemma is found only around axons in the PNS. When an axon is injured, the neurolemma aids regeneration by forming

Figure 16.7 Neuroglia of the peripheral nervous system (PNS).

🔑 Neuroglia of the PNS completely surround axons and cell bodies of neurons.

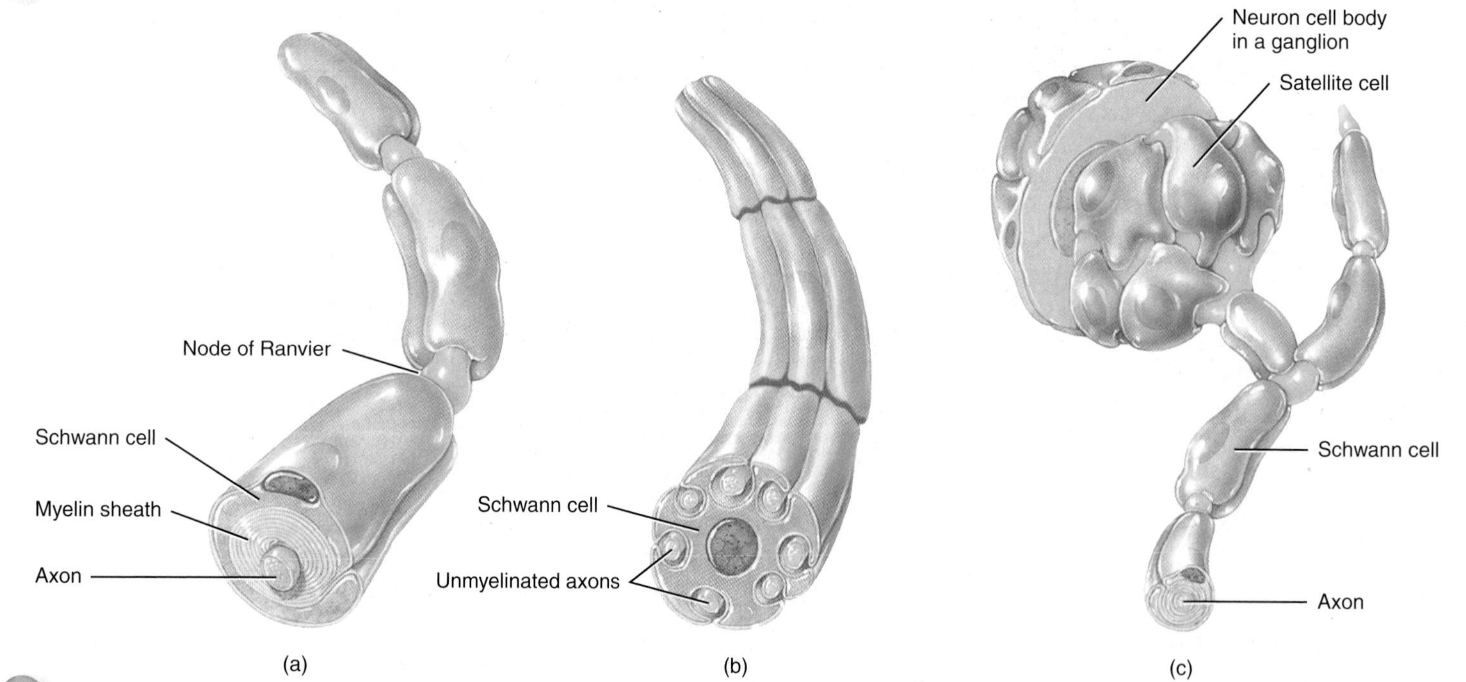

(a) (b) (c)

❓ **How do Schwann cells and oligodendrocytes differ with respect to number of axons myelinated?**

Figure 16.8 Myelinated and unmyelinated axons.

🔑 Axons surrounded by a myelin sheath produced either by Schwann cells or by oligodendrocytes are said to be myelinated.

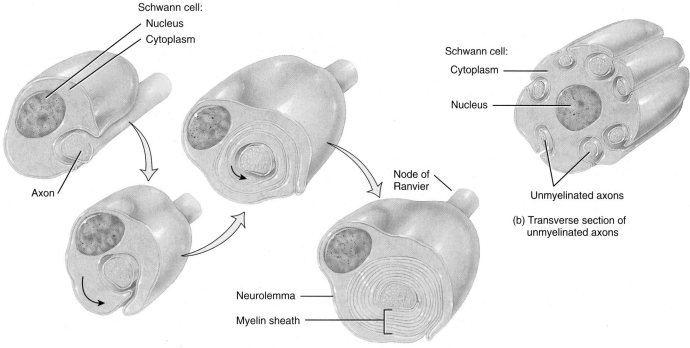

(a) Transverse sections of stages in the formation of a myelin sheath

(b) Transverse section of unmyelinated axons

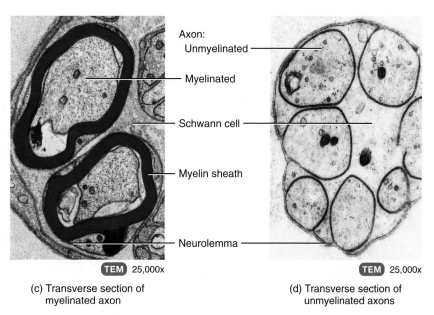

(c) Transverse section of myelinated axon

(d) Transverse section of unmyelinated axons

❓ **What is the functional advantage of myelination?**

a regeneration tube that guides and stimulates regrowth of the axon. Gaps in the myelin sheath, called **nodes of Ranvier** (RON-vē-ā) (*myelin sheath gaps*), appear at intervals along the axon (see Figures 16.3 and 16.7a). Each Schwann cell wraps one axon segment between two nodes. Nerve impulses are conducted more rapidly in myelinated axons because the impulses are formed more quickly at nodes of Ranvier and appear to "leap" from node to node as opposed to being conducted more slowly through every part of the membrane in unmyelinated axons.

In the CNS, an oligodendrocyte myelinates parts of several axons. Each oligodendrocyte puts forth about 15 broad, flat processes that spiral around CNS axons, forming a myelin sheath (see Figure 16.6). A neurolemma is not present, however, because the oligodendrocyte cell body and nucleus do not envelop the axon. Nodes of Ranvier

are present, but they are fewer in number. Axons in the CNS display little regrowth after injury. This is thought to be due, in part, to the absence of a neurolemma, and in part to an inhibitory influence on axon regrowth exerted by the oligodendrocytes.

The amount of myelin increases from birth to maturity, and its presence greatly increases the speed of nerve impulse conduction. An infant's responses to stimuli are neither as rapid nor as coordinated as those of an older child or a young adult, in part because myelination is still in progress during infancy. **Demyelination** (dē-mī-e-li-NĀ-shun) refers to the loss or destruction of myelin sheaths around axons. It may result from disorders such as multiple sclerosis or Tay-Sachs disease, or from medical treatments such as radiation therapy and chemotherapy. Any single episode of demyelination may cause deterioration of affected nerves.

Gray and White Matter

In a freshly dissected section of the brain or spinal cord, some regions look white and glistening, and others appear gray (Figure 16.9). The **white matter** is aggregations of myelinated and unmyelinated axons of many neurons. The whitish color of myelin gives white matter its name. The **gray matter** of the nervous system contains neuronal cell bodies, dendrites, unmyelinated axons, axon terminals, and neuroglia. It appears grayish, rather than white, because the Nissl bodies impart a gray color and there is little or no myelin in these areas. Blood vessels are present in both white and gray matter.

In the spinal cord, the white matter surrounds an inner core of gray matter shaped like a butterfly or the letter H (in transverse section); in the brain, a thin shell of gray matter covers the surface of the largest portions of the brain, the cerebrum and cerebellum (Figure 16.9). When used to describe nervous tissue, a **nucleus** is a cluster of neuronal cell bodies within the CNS. (Recall that the term *ganglion* refers to a similar arrangement within the PNS.) Many nuclei of gray matter also lie deep within the brain. Much of the CNS white matter consists of **tracts**, bundles of axons in the CNS that extend for some distance up or down the spinal cord or connect parts of the brain with each other and with the spinal cord (Table 16.1). The arrangements of gray and white matter in the spinal cord and brain are discussed more extensively in Chapters 17 and 18, respectively.

TABLE 16.1

Summary of Terminology

	COLLECTION OF NERVE CELL BODIES	COLLECTION OF NERVE FIBERS
Central nervous system	**Nucleus** *Example:* red nucleus (see Figure 18.8b)	**Tract** *Example:* posterior column tract (see Table 20.3)
Peripheral nervous system	**Ganglion** *Example:* vestibular ganglion (see Figure 21.13b)	**Nerve** *Example:* vagus (X) nerve (see Figure 18.25)

✓ CHECKPOINT

4. Describe the parts of a neuron and the functions of each.
5. What are the differences between electrical and chemical synapses?
6. What is a neuromuscular junction?
7. What role do neurotransmitters play in the transmission of impulses at a chemical synapse?
8. How do bipolar and unipolar neurons differ?
9. What is a neurolemma, and why is it important?
10. With reference to the central nervous system, what is a nucleus?
11. What is the difference between gray matter and white matter?

Figure 16.9 Distribution of gray matter and white matter in the spinal cord and brain.

White matter consists of myelinated and unmyelinated axons of many neurons. Gray matter consists of neuron cell bodies, dendrites, axon terminals, bundles of unmyelinated axons, and neuroglia.

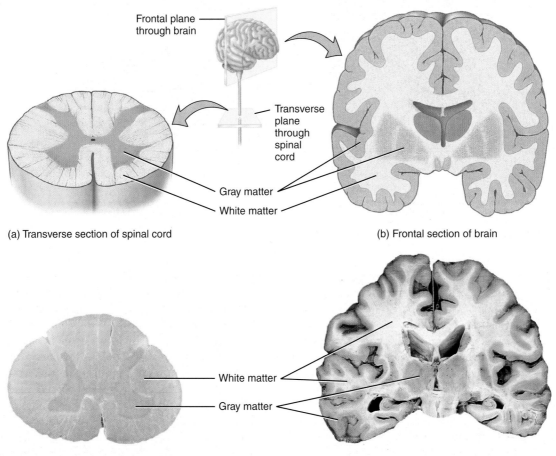

(a) Transverse section of spinal cord

(b) Frontal section of brain

(c) Transverse section of spinal cord

(d) Frontal section of brain

 What is responsible for the white appearance of white matter?

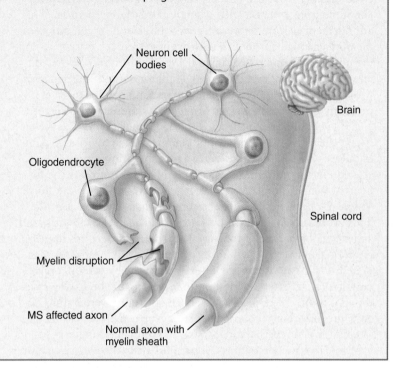

CLINICAL CONNECTION | *Multiple Sclerosis*

Multiple sclerosis (MS) is a disease that causes a progressive destruction of myelin sheaths surrounding neurons in the CNS. It afflicts about 350,000 people in the United States and 2 million people worldwide. It usually appears between the ages of 20 and 40, affecting females twice as often as males. MS is most common in whites, less common in blacks, and rare in Asians. MS is an autoimmune disease—the body's own immune system spearheads the attack. The condition's name describes the anatomical pathology: In *multiple* regions the myelin sheaths deteriorate to *scleroses*, which are hardened scars or plaques. Magnetic resonance imaging (MRI) studies reveal numerous plaques in the white matter of the brain and spinal cord. The destruction of myelin sheaths slows and then short-circuits propagation of nerve impulses.

The most common form of the condition is *relapsing-remitting MS*, which usually appears in early adulthood. The first symptoms may include a feeling of heaviness or weakness in the muscles, abnormal sensations, or double vision. An attack is followed by a period of remission during which the symptoms temporarily disappear. One attack follows another over the years, usually every year or two. The result is a progressive loss of function interspersed with remission periods, during which symptoms abate.

Although the cause of MS is unclear, both genetic susceptibility and exposure to some environmental factor (perhaps a herpes virus) appear to contribute. Many patients with relapsing-remitting MS are treated with injections of beta interferon. This treatment lengthens the time between relapses, decreases the severity of relapses, and slows formation of new lesions in some cases. Unfortunately, not all MS patients can tolerate beta interferon, and therapy becomes less effective as the disease progresses. •

Neuron cell bodies

Brain

Oligodendrocyte

Spinal cord

Myelin disruption

MS affected axon

Normal axon with myelin sheath

16.3 NEURAL CIRCUITS

OBJECTIVE

- Identify the structure and function of the various types of neural circuits in the nervous system.

The CNS contains billions of neurons organized into complicated networks called **neural circuits**, each of which is a functional group of neurons that processes a specific kind of information. In a **simple series circuit**, a presynaptic neuron stimulates a single postsynaptic neuron. The second neuron then stimulates another, and so on. However, most neural circuits are more complex.

A single presynaptic neuron may synapse with several postsynaptic neurons. Such an arrangement, called **divergence**, permits one presynaptic neuron to influence several postsynaptic neurons (or several muscle fibers or gland cells) at the same time. In a **diverging circuit**, the nerve impulse from a single presynaptic neuron causes the stimulation of increasing numbers of cells along the circuit (Figure 16.10a). For example, a small number of neurons in the brain that govern a particular body movement

Figure 16.10 Examples of neural circuits.

 A neural circuit is a functional group of neurons that processes a specific kind of information.

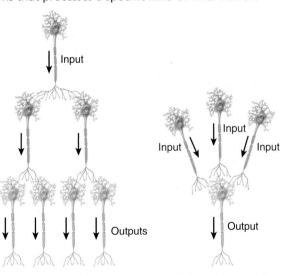

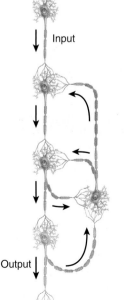

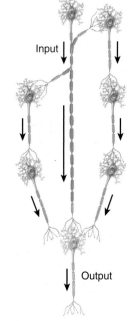

(a) Diverging circuit   (b) Converging circuit   (c) Reverberating circuit   (d) Parallel after-discharge circuit

? A motor neuron in the spinal cord typically receives input from neurons that originate in several different regions of the brain. Is this an example of convergence or divergence?

stimulate a much larger number of neurons in the spinal cord. Sensory signals also spread into diverging circuits and are often relayed to several regions of the brain. This arrangement has the effect of amplifying the signal.

In another arrangement, called **convergence**, several presynaptic neurons synapse with a single postsynaptic neuron. This arrangement permits more effective stimulation or inhibition of the postsynaptic neuron. In a **converging circuit** (Figure 16.10b), the postsynaptic neuron receives nerve impulses from several different sources. For example, a single motor neuron that synapses with skeletal muscle fibers at neuromuscular junctions receives input from several pathways that originate in different brain regions.

Some circuits are constructed so that stimulation of the presynaptic cell causes the postsynaptic cell to transmit a series of nerve impulses. One such circuit is called a **reverberating circuit** (Figure 16.10c). In this pattern, the incoming impulse stimulates the first neuron, which stimulates the second, which stimulates the third, and so on. Branches from later neurons synapse with earlier ones. This arrangement sends impulses back through the circuit again and again. The output signal may last from a few seconds to many hours, depending on the number of synapses and the arrangement of neurons in the circuit. Inhibitory neurons may turn off a reverberating circuit after a period of time. Among the body responses thought to be the result of output signals from reverberating circuits are breathing, coordinated muscular activities, waking up, sleeping (when reverberation stops), and short-term memory.

A fourth type of circuit is the **parallel after-discharge circuit** (Figure 16.10d). In this circuit, a single presynaptic cell stimulates a group of neurons, each of which synapses with a common postsynaptic cell. Parallel after-discharge circuits may be involved in precise activities such as mathematical calculations.

 **CHECKPOINT**

12. What is a neural circuit?
13. What are the functions of diverging, converging, reverberating, and parallel after-discharge circuits?

16.4 REGENERATION AND NEUROGENESIS

 OBJECTIVE

• Define regeneration and explain the concept of neurogenesis.

Throughout your life, your nervous system exhibits **plasticity** (plas-TIS-i-tē), the capability to constantly change, grow, and remap itself over the course of your lifetime. At the level of individual neurons, the changes that can occur include the sprouting of new dendrites, synthesis of new proteins, and changes in synaptic contacts with other neurons. Undoubtedly, both chemical and electrical signals drive the changes that occur. Despite this plasticity, mammalian neurons have very limited powers of **regeneration**, the capability to replicate or repair themselves. In the PNS, damage to dendrites and myelinated axons may be repaired if the cell body remains intact and if the Schwann cells that produce myelination remain active. In the CNS, little or no repair of damage to neurons occurs. Even when the cell body remains intact, a severed axon cannot be repaired or regrown.

Neurogenesis (noo′-rō-JEN-e-sis)—the birth of new neurons from undifferentiated stem cells—occurs regularly in some animals. For example, new neurons appear and disappear every year in some songbirds. Until relatively recently, the dogma was "no new neurons" in the adult brains of humans and other primates. Then, in 1992, Canadian researchers published their unexpected finding that **epidermal growth factor (EGF)** stimulated cells taken from the brains of adult mice to proliferate into both neurons and astrocytes. Previously, EGF was known to trigger mitosis in a variety of nonneuronal cells and to promote wound healing and tissue regeneration. In 1998, scientists discovered that significant numbers of new neurons do arise in the adult human hippocampus, an area of the brain that is crucial for learning. More recently, it has been demonstrated that neurogenesis occurs in the olfactory bulb, caudate nucleus, and cerebellum of some mammals.

The nearly complete lack of neurogenesis in other regions of the brain and spinal cord seems to result from two factors: (1) inhibitory influences from neuroglia, particularly oligodendrocytes, and (2) absence of growth-stimulating cues that were present during fetal development. Axons in the CNS are myelinated by oligodendrocytes rather than Schwann cells, and this CNS myelin is one of the factors inhibiting regeneration of neurons. It is thought that this same mechanism may stop axonal growth once a target region has been reached during development. Also, after axonal damage, nearby astrocytes proliferate rapidly, forming a type of scar tissue that acts as a physical barrier to regeneration. Thus, injury of the brain or spinal cord usually is permanent. Ongoing research seeks ways to improve the environment for existing spinal cord axons to bridge the injury gap. Scientists also are trying to find ways to stimulate dormant stem cells to replace neurons lost through damage or disease and to develop tissue-cultured neurons that can be used for transplantation purposes.

 CHECKPOINT

14. What factors contribute to the absence of neurogenesis in most parts of the brain?

KEY MEDICAL TERMS ASSOCIATED WITH NERVOUS TISSUE

Acupuncture (ak-ū-PUNK-chur) The use of fine needles (lasers, ultrasound, or electricity) inserted into specific exterior body locations (acupoints) and manipulated to relieve pain and provide therapy for various conditions. The placement of needles may cause the release of neurotransmitters such as endorphins.

Excitotoxicity The destruction of neurons through prolonged activation of excitatory synaptic transmission caused by a high level of glutamate in the interstitial fluid of the CNS. The most common cause is oxygen deprivation of the brain due to *ischemia* (inadequate blood flow), as happens during a stroke. Lack of oxygen causes glutamate transporter failure, and glutamate accumulates in the interstitial spaces between neurons and glia, stimulating the neurons to death.

Guillain-Barré Syndrome (GBS) (GHĒ-an ba-RĀ) An acute demyelinating disorder in which macrophages strip myelin from axons in the PNS. It is the most common cause of acute paralysis in North America and Europe.

Neuroblastoma (noor-ō-blas-TŌ-ma) A malignant tumor that consists of immature nerve cells (neuroblasts); occurs most commonly in the abdomen and most frequently in the adrenal glands.

Neuropathy (noo-ROP-a-thē; *neuro-*=a nerve; *pathy*=disease) Any disorder that affects the nervous system, but particularly a disorder of a cranial or spinal nerve. An example is *facial neuropathy* (Bell's palsy), a disorder of the facial (VII) nerve (See Exhibit 18.E).

Rabies (RĀ-bēz; *rabi-*=mad, raving) A fatal disease caused by a virus that reaches the CNS via an axon, usually transmitted by the bite of a dog or other meat-eating animal. The symptoms are excitement, aggressiveness, and madness, followed by paralysis and death.

Review   **Resource**

16.1 Overview of the Nervous System

1. Structures that make up the nervous system are the brain, 12 pairs of cranial nerves and their branches, the spinal cord, 31 pairs of spinal nerves and their branches, ganglia, enteric plexuses, and sensory receptors.
2. Anatomically, the nervous system consists of two divisions. (1) The central nervous system (CNS) consists of the brain and spinal cord. (2) The peripheral nervous system (PNS) consists of all nervous tissue outside the CNS.
3. Functionally, the nervous system integrates all body activities by sensing changes (sensory function), interpreting them (integrative function), and reacting to them (motor function).
4. Sensory (afferent) neurons carry sensory information from cranial and spinal nerves into the brain and spinal cord or from a lower to a higher level in the spinal cord and brain. Interneurons have short axons that contact nearby neurons in the brain, spinal cord, or a ganglion. Motor (efferent) neurons carry information from the brain toward the spinal cord or out of the brain and spinal cord into cranial or spinal nerves.
5. Components of the PNS include the somatic nervous system (SNS), autonomic nervous system (ANS), and enteric nervous system (ENS).
6. The SNS consists of somatic sensory neurons that conduct impulses from somatic and special sense receptors to the CNS (input) and somatic motor neurons from the CNS to skeletal muscles (output).
7. The ANS contains autonomic sensory neurons from visceral organs (input) and autonomic motor neurons that convey impulses from the CNS to smooth muscle tissue, cardiac muscle tissue, and glands (output).
8. The ENS consists of neurons in enteric plexuses in the gastrointestinal (GI) tract that function independently of the ANS and CNS to some extent. Sensory neurons of the ENS monitor chemical changes and stretching of the GI tract (input) and motor neurons of the ENS generate contractions and secretions of the GI tract (output).

Anatomy Overview - Nervous System Overview
Animation - Structure and Function of the Nervous System - System Organization

16.2 Histology of Nervous Tissue

1. Nervous tissue consists of neurons (nerve cells) and neuroglia. Neurons have the property of electrical excitability and are responsible for most unique functions of the nervous system.
2. Most neurons have three parts: dendrites, cell body, and axon. The dendrites are the main receiving or input region. Integration occurs in the cell body, which includes typical cellular organelles. The output part typically is a single axon, which propagates nerve impulses toward another neuron, a muscle fiber, or a gland cell.
3. Synapses are the sites of communication between two neurons or neurons and effectors. The synapse between a neuron and muscle fiber is called a neuromuscular junction.
4. Synapses between neurons may be electrical or chemical. In electrical synapses, impulses pass through connexons. Chemical synapses involve the release of a neurotransmitter by a presynaptic neuron.
5. On the basis of their structure, neurons are classified as multipolar, bipolar, or unipolar.
6. Neuroglia support, nurture, and protect neurons and maintain the interstitial fluid that bathes them. Neuroglia in the CNS include astrocytes, oligodendrocytes, microglial cells, and ependymal cells. Neuroglia in the PNS include Schwann cells and satellite cells.
7. Two types of neuroglia produce myelin sheaths: oligodendrocytes myelinate axons in the CNS, and Schwann cells myelinate axons in the PNS.
8. White matter consists of aggregates of myelinated processes; gray matter contains cell bodies, dendrites, and axon terminals of neurons, unmyelinated axons, and neuroglia.
9. In the spinal cord, gray matter forms an H-shaped inner core that is surrounded by white matter. In the brain, a thin, superficial shell of gray matter covers the cerebral and cerebellar hemispheres.

Anatomy Overview - Nervous Tissue - Neuron
Anatomy Overview - Nerve
Anatomy Overview - Nervous Tissue - Neuroglia
Animation - Neuron Structure and Function
Animation - Factors That Affect Conduction Rates - Myelination
Animation - Events at the Synapse
Figure 16.3 - Structure of a Typical Neuron and Synapse Between Neurons
Figure 16.6 - Neuroglia of the Central Nervous System
Exercise - Paint a Neuron
Exercise - Are Your Synapses Working?

16.3 Neural Circuits

1. Neurons in the central nervous system are organized into networks called neural circuits.
2. Neural circuits include simple series, diverging, converging, reverberating, and parallel after-discharge circuits.

16.4 Regeneration and Neurogenesis

1. The nervous system exhibits plasticity (the capability to change based on experience), but it has very limited powers of regeneration (the capability to replicate or repair damaged neurons).
2. Neurogenesis, the birth of new neurons from undifferentiated stem cells, is normally very limited in adult humans. Repair of damaged axons does not occur in most regions of the CNS.

Animation - Introduction to Membrane Potentials
Figure 16.10 - Neural Circuits

CRITICAL THINKING QUESTIONS

1. Varudhini was reviewing her anatomy notes when she heard a catchy advertising jingle on the radio. Hours later, on her way to class, the jingle was still running through her mind. Varudhini was suddenly reminded of a neuronal circuit from her anatomy notes. Name the type of circuit and explain its uses in the body.

2. Following a brain injury in her early thirties, Jaqueline, an artist, could no longer draw pictures with her hands, but within a few years was able to produce beautiful artwork using her feet "as her hands." What characteristic of the nervous system made this possible?

3. Long-distance runners often describe a "runner's high." What neurotransmitter is responsible for producing this "high," and how does it work?

4. The buzzing of the alarm clock woke Mohamed. He yawned, stretched, and started to salivate as he smelled the coffee brewing.

He could feel his stomach rumble. List the divisions of the nervous system involved in each of these actions.

5. Althea is in research and development for a large pharmaceutical company. She has just been pleasantly surprised by the results of her recent experiments on a new drug to treat a viral infection of the brain. Her secretary exclaims, "Great! With my shares of stock in the company, I'll be rich!" Althea told her secretary, "Don't buy that new Porsche just yet. What works in the Petri dish in the lab may not work in a real person. First of all, the drug has to be able to reach the infection." The secretary was perplexed, and said, "Well, why not? Just inject it into a vein, and it will get there." What would your response be?

? ANSWERS TO FIGURE QUESTIONS

16.1 The anatomical subdivisions of the nervous system are the CNS (brain and spinal cord) and PNS (all nervous structures outside the CNS).

16.2 The somatic motor component of the peripheral nervous system is voluntary.

16.3 Dendrites receive (motor neurons or interneurons) or generate (sensory neurons) inputs; the cell body also receives input signals; the axon conducts nerve impulses (action potentials) and transmits the message to another neuron or effector cell by releasing a neurotransmitter at its synaptic end bulbs.

16.4 Nerve impulses arise at the trigger zone.

16.5 The cell body of a pyramidal cell is shaped like a pyramid.

16.6 Microglia function as phagocytes in the CNS.

16.7 One Schwann cell myelinates a single axon; one oligodendrocyte myelinates several axons.

16.8 Myelination increases the speed of nerve impulse conduction.

16.9 Myelin makes white matter look shiny and white.

16.10 A motor neuron receiving input from several other neurons is an example of convergence.

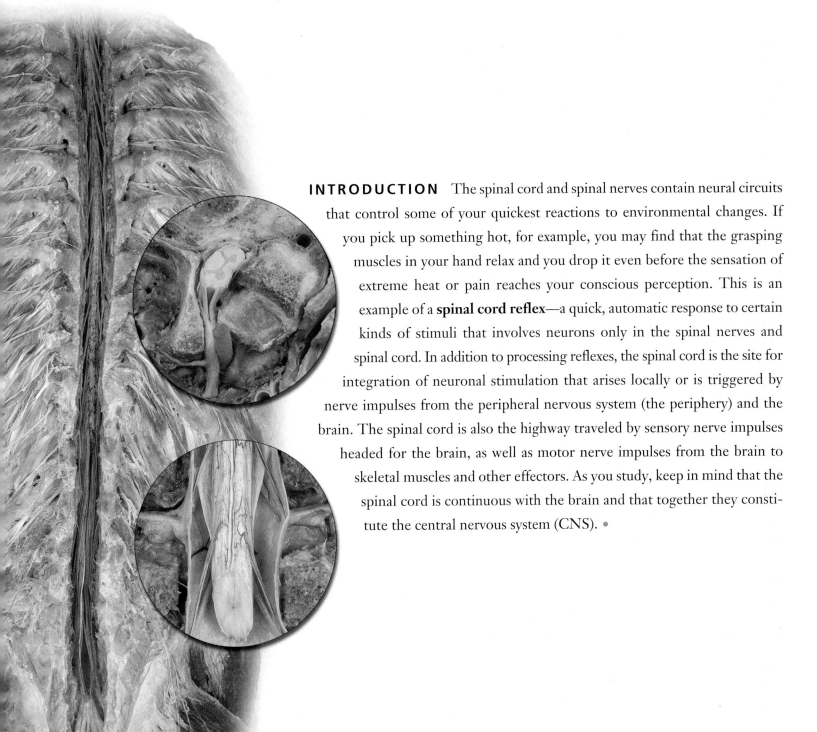

17 | THE SPINAL CORD AND THE SPINAL NERVES

INTRODUCTION The spinal cord and spinal nerves contain neural circuits that control some of your quickest reactions to environmental changes. If you pick up something hot, for example, you may find that the grasping muscles in your hand relax and you drop it even before the sensation of extreme heat or pain reaches your conscious perception. This is an example of a **spinal cord reflex**—a quick, automatic response to certain kinds of stimuli that involves neurons only in the spinal nerves and spinal cord. In addition to processing reflexes, the spinal cord is the site for integration of neuronal stimulation that arises locally or is triggered by nerve impulses from the peripheral nervous system (the periphery) and the brain. The spinal cord is also the highway traveled by sensory nerve impulses headed for the brain, as well as motor nerve impulses from the brain to skeletal muscles and other effectors. As you study, keep in mind that the spinal cord is continuous with the brain and that together they constitute the central nervous system (CNS). •

Did you ever wonder why spinal cord injuries can have such widespread effects on the body?

17.1 SPINAL CORD ANATOMY

OBJECTIVES

- Describe the protective structures of the spinal cord.
- Explain the external features of the spinal cord.
- Describe the external anatomy of the spinal cord.

Protective Structures

Recall from the previous chapter that the nervous tissue of the central nervous system is very delicate and does not respond well to injury or damage. Accordingly, nervous tissue requires considerable protection. The first layer of protection for the central nervous system is the hard bony skull and vertebral column. The skull encases the brain and the vertebral column surrounds the spinal cord, providing strong protective defenses against damaging blows or bumps. The second protective layer is the meninges, three membranes that lie between the bony encasement and the nervous tissue in both the brain and spinal cord. Finally, a space between two of the meningeal membranes contains cerebrospinal fluid, a buoyant liquid that suspends the central nervous tissue in a weightless environment while surrounding it with a shock-absorbing, hydraulic cushion.

Vertebral Column

The spinal cord is located within the vertebral canal of the vertebral column. The vertebral foramina of all the vertebrae, stacked one on top of the other, form the vertebral canal. The surrounding vertebrae provide a sturdy shelter for the enclosed spinal cord (see Figure 17.1b). The vertebral ligaments provide additional protection.

Meninges

The **meninges** (me-NIN-jēz; singular is *meninx* (ME-ninks)) are three protective, connective tissue coverings that encircle the spinal cord and brain. From superficial to deep they are the: (1) dura mater; (2) arachnoid mater; and (3) pia mater. The arachnoid mater and pia mater are often referred to collectively as the **leptomeninges**. Between these two meninges is a space, the **subarachnoid space**, which contains the shock-absorbing cerebrospinal fluid (see Section 18.2). The **spinal meninges** surround the spinal cord (Figure 17.1a) and are continuous with the **cranial meninges**, which encircle the brain (shown in Figure 18.5b). All three spinal meninges cover the spinal nerves up to the point where they exit the spinal column through the intervertebral foramina. The spinal cord is also protected by a cushion of fat and connective tissue located in the **epidural**

space (ep'-i-DOO-ral), a space between the dura mater and the wall of the vertebral canal (Figure 17.1b). Following is a description of each meningeal layer.

1. **Dura mater** (DOO-ra MĀ-ter=tough mother). The most superficial of the three spinal meninges is a thick strong layer composed of dense irregular connective tissue called the dura mater. *Mater* meaning mother was chosen because early anatomists erroneously believed that all of the tissues of the body arose from the dura mater and other meningeal layers. The dura mater forms a sac from the level of the foramen magnum in the occipital bone, where it is continuous with the meningeal dura mater of the brain, to the second sacral vertebra. The dura mater is also continuous with the epineurium, the outer covering of spinal and cranial nerves.

2. **Arachnoid mater** (a-RAK-noyd MĀ-ter; *arachn-*=spider; *-oid*=similar to). This layer, the middle of the meningeal membranes, is a thin, avascular covering comprised of cells and thin, loose arrays of collagen. It is called the arachnoid mater because of its spider's web arrangement of delicate collagen fibers and some elastic fibers. It is deep to the dura mater and is continuous through the foramen magnum with the arachnoid mater of the brain. Between the dura mater and the arachnoid mater is a thin **subdural space**, which contains interstitial fluid.

3. **Pia mater** (PĒ-a MĀ-ter; *pia*=delicate). This is the innermost meninx, a thin transparent connective tissue layer that adheres to the surface of the spinal cord and brain. It consists of thin squamous to cuboidal cells within interlacing bundles of collagen fibers and some fine elastic fibers. Within the pia mater are many blood vessels that supply oxygen and nutrients to the spinal cord. Triangular-shaped membranous extensions of the pia mater suspend the spinal cord in the middle of its dural sheath. These extensions, called **denticulate ligaments** (den-TIK-ū-lāt=small tooth), are thickenings of the pia mater. They project laterally and fuse with the arachnoid mater and inner surface of the dura mater between the anterior and posterior nerve roots of spinal nerves on either side (Figure 17.1a, b). Extending along the entire length of the spinal cord, the denticulate ligaments protect the spinal cord against sudden displacement that could result in shock.

External Anatomy of the Spinal Cord

The **spinal cord** is roughly oval in shape, being flattened slightly in the anterior–posterior axis. In adults, it extends from the medulla oblongata (the most inferior part of the brain) to the superior border of the second lumbar vertebra. In newborn

Figure 17.1 Gross anatomy of the spinal cord. The spinal meninges are evident in both views.

 Meninges are connective tissue coverings that surround the spinal cord and brain.

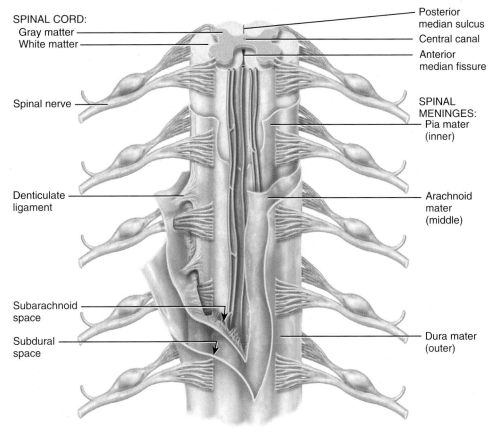

SPINAL CORD:
Gray matter
White matter

Spinal nerve

Denticulate
ligament

Subarachnoid
space

Subdural
space

Posterior
median sulcus

Central canal

Anterior
median fissure

SPINAL
MENINGES:
Pia mater
(inner)

Arachnoid
mater
(middle)

Dura mater
(outer)

(a) Anterior view and transverse section through spinal cord

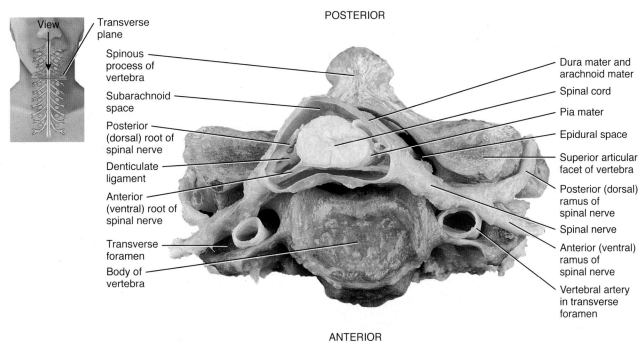

POSTERIOR

View
Transverse
plane

Spinous
process of
vertebra

Subarachnoid
space

Posterior
(dorsal) root of
spinal nerve

Denticulate
ligament

Anterior
(ventral) root of
spinal nerve

Transverse
foramen

Body of
vertebra

Dura mater and
arachnoid mater

Spinal cord

Pia mater

Epidural space

Superior articular
facet of vertebra

Posterior (dorsal)
ramus of
spinal nerve

Spinal nerve

Anterior (ventral)
ramus of
spinal nerve

Vertebral artery
in transverse
foramen

ANTERIOR

(b) Transverse section of the spinal cord within a cervical vertebra

CLINICAL CONNECTION | *Spinal Cord Compression*

Although the spinal cord is normally protected by the vertebral column, certain disorders may put pressure on it and disrupt its normal functions. **Spinal cord compression** may result from fractured vertebrae, herniated intervertebral discs, tumors, osteoporosis, or infections. If the source of the compression is determined before neural tissue is destroyed, spinal cord function usually returns to normal. Depending on the location and degree of compression, symptoms include pain, weakness or paralysis, and either decreased or complete loss of sensation below the level of the injury. •

What are the superior and inferior boundaries of the spinal dura mater?

infants, the spinal cord extends to the third or fourth lumbar vertebra. During early childhood, both the spinal cord and the vertebral column grow longer as part of overall body growth. Elongation of the spinal cord stops around age 4 or 5, but growth of the vertebral column continues, which explains why the spinal cord does not extend the entire length of the verte-bral column. The length of the adult spinal cord ranges from 42 to 45 cm (16–18 in.). Its maximum diameter is approximately 1.5 cm (0.6 in.) in the lower cervical region and is smaller in the thoracic region and at its inferior tip.

When the spinal cord is viewed externally, two conspicuous en-largements can be seen (Figure 17.2a). The superior enlargement,

Figure 17.2 External anatomy of the spinal cord and the spinal nerves.

The spinal cord extends from the medulla oblongata of the brain to the superior border of the second lumbar vertebra.

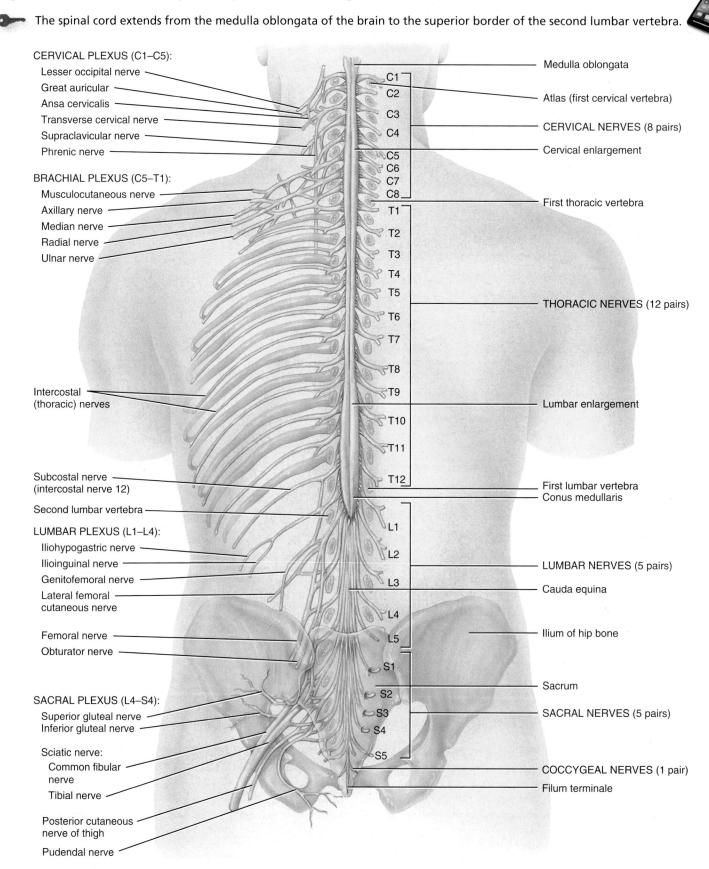

(a) Posterior view of entire spinal cord and portions of spinal nerves

In a **spinal tap (lumbar puncture),** a local anesthetic is given, and a long hollow needle is inserted into the subarachnoid space to withdraw cerebrospinal fluid (CSF) for diagnostic purposes; to introduce antibiotics, contrast media for myelography, or anesthetics; to administer chemotherapy; to measure CSF pressure; and/or to evaluate the effects of treatment for diseases such as meningitis. During this procedure, the patient lays on his or her side with the vertebral column flexed. Flexion of the vertebral column increases the distance between the spinous processes of the vertebrae, which allows easy access to the subarachnoid space. The spinal cord ends around the second lumbar vertebra (L2); however, the spinal meninges and circulating cerebrospinal fluid extend to the second sacral vertebra (S2). Between vertebrae L2 and S2 the spinal meninges are present, but the spinal cord is absent. Consequently, a spinal tap is normally performed in adults between the L3 and L4 or L4 and L5 lumbar vertebrae because this region provides safe access to the subarachnoid space without the risk of damaging the spinal cord. (A line drawn across the highest points of the iliac crests, called the *supracristal line,* passes through the spinous process of the fourth lumbar vertebra, and is used as a landmark for administering a spinal tap.) •

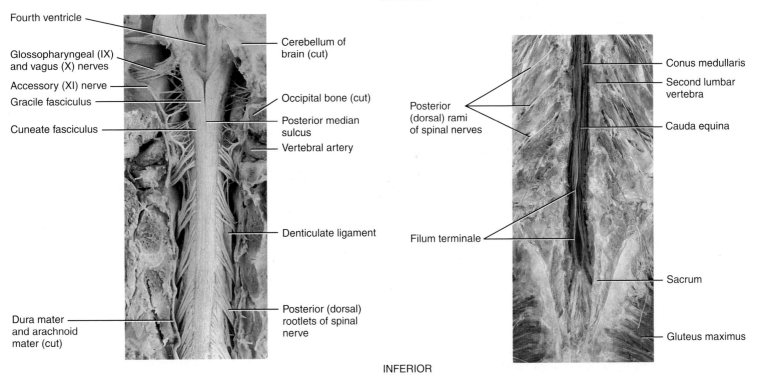

the **cervical enlargement,** extends from the fourth cervical vertebra (C4) to the first thoracic vertebra (T1). Nerves to and from the upper limbs arise from the cervical enlargement. The inferior enlargement, called the **lumbar enlargement,** extends from the ninth to the twelfth thoracic vertebra (T9–T12). Nerves to and from the lower limbs arise from the lumbar enlargement.

Inferior to the lumbar enlargement, the spinal cord terminates as a tapering, conical structure called the **conus medullaris** (KŌ-nus med-ū-LAR-is; *conus*=cone). The conus medullaris ends at the level of the intervertebral disc between the first and second lumbar vertebrae (L1–L2) in adults (Figure 17.2a, c). Arising from the conus medullaris is the **filum terminale**

SUPERIOR

(b) Posterior view of cervical region of spinal cord

(c) Posterior view of inferior portion of spinal cord

INFERIOR

What portion of the spinal cord connects with sensory and motor nerves of the upper limbs?

(FĪ-lum ter-mi-NAL-ē=terminal filament), an extension of the pia mater that extends inferiorly and fuses with the arachnoid mater and dura mater to anchor the spinal cord to the coccyx.

As spinal nerves branch from the spinal cord, they pass laterally to exit the spinal canal through the intervertebral foramina between adjacent vertebrae. However, because the spinal cord is shorter than the vertebral column, nerves that arise from the lumbar, sacral, and coccygeal regions of the spinal cord do not leave the vertebral column at the same level they exit the cord. The roots of these lower spinal nerves angle inferiorly alongside the filum terminale in the vertebral canal like wisps of hair. Accordingly,

the roots of these nerves are collectively named the **cauda equina** (KAW-da ē-KWĪ-na), meaning "horse's tail" (Figure 17.2a, c).

Internal Anatomy of the Spinal Cord

A transverse section of the spinal cord reveals its internal structure, which at all levels is characterized by a central letter H or butterfly-shaped gray matter region that is surrounded by white matter. Two grooves penetrate the white matter of the spinal cord and divide it into right and left sides (Figure 17.3). The **anterior median fissure** is a wide groove on the anterior (ventral)

Figure 17.3 Internal anatomy of the spinal cord: the organization of gray matter and white matter. For simplicity, dendrites are not shown in this and several other illustrations of transverse sections of the spinal cord. Blue and red arrows in (a) indicate the direction of nerve impulse propagation.

FUNCTIONS OF THE SPINAL CORD

1. White matter tracts propagate sensory impulses from receptors to the brain and motor impulses from the brain to effectors.
2. Gray matter receives and integrates incoming and outgoing information.

🔑 In the spinal cord, white matter surrounds the gray matter.

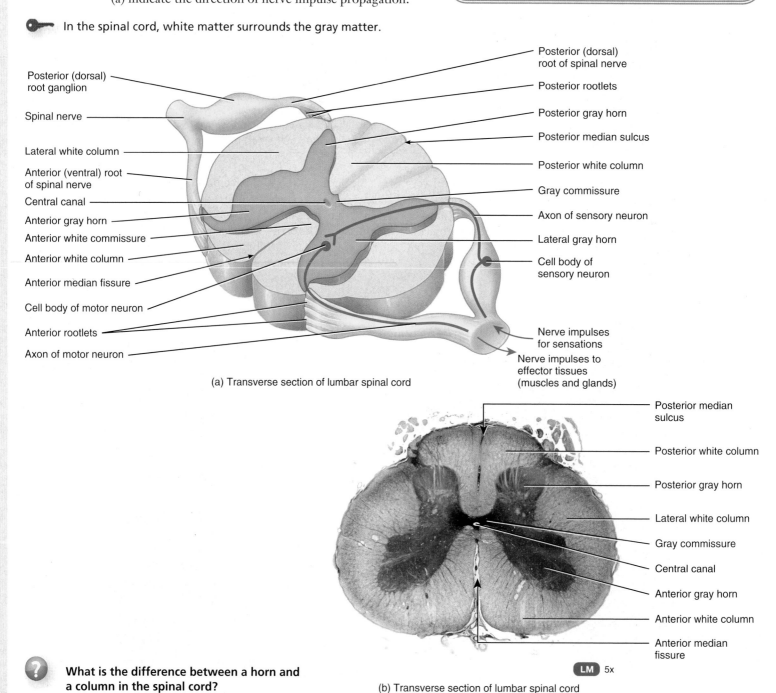

(a) Transverse section of lumbar spinal cord

LM 5x

(b) Transverse section of lumbar spinal cord

❓ **What is the difference between a horn and a column in the spinal cord?**

side. The **posterior median sulcus** is a narrow groove on the posterior (dorsal) side. The gray matter consists primarily of the cell bodies of neurons, neuroglia, unmyelinated axons, and the dendrites of interneurons and motor neurons. The white matter consists of bundles of myelinated axons of sensory neurons, interneurons, and motor neurons. The **gray commissure** (KOM-mi-shur) forms the crossbar of the H (or body of the butterfly, depending on how imaginative you are). In the center of the gray commissure is a small space called the **central canal**; it extends the entire length of the spinal cord and contains cerebrospinal fluid. At its superior end, the central canal is continuous with the fourth ventricle (a space that also contains cerebrospinal fluid) in the medulla oblongata of the brain. Anterior to the gray commissure is the **anterior (ventral) white commissure**, which connects the white matter of the right and left sides of the spinal cord.

In the gray matter of the spinal cord and brain, clusters of neuronal cell bodies form functional groups called **nuclei**. *Sensory nuclei* receive input from receptors via sensory neurons, and *motor nuclei* provide output to effector tissues via motor neurons. The gray matter on each side of the spinal cord is subdivided into regions called **horns**. The **anterior (ventral) gray horns** contain *somatic motor nuclei*, which are clusters of cell bodies of somatic motor neurons that provide nerve impulses for contraction of skeletal muscles. The **posterior (dorsal) gray horns** contain cell bodies and axons of interneurons as well as axons of incoming sensory neurons. Between the anterior and posterior gray horns are the **lateral gray horns**, which are present only in the thoracic, upper lumbar, and sacral segments of the spinal cord. The lateral horns contain the cell bodies of *autonomic motor nuclei* that regulate activity of smooth muscle, cardiac muscle, and glands (see Section 19.2).

The white matter, like the gray matter, is organized into regions. The anterior and posterior gray horns divide the white matter on each side into three broad areas called **columns**: (1) **anterior (ventral) white columns**, (2) **posterior (dorsal) white columns**, and (3) **lateral white columns**. Each column in turn contains distinct bundles of axons having a common origin or destination and carrying similar information. These bundles, which may extend long distances up or down the spinal cord, are called **tracts**. **Sensory (ascending) tracts** consist of axons that conduct nerve impulses from the spinal cord toward the brain. Tracts consisting of axons that carry nerve impulses away from the brain down the spinal cord are called **motor (descending) tracts**. Sensory and motor tracts of the spinal cord are continuous with sensory and motor tracts in the brain.

The white matter and the gray matter exhibit differences depending on the region of the spinal cord (Table 17.1). There is more white matter at the cranial end of the spinal cord than at the caudal end. Because all ascending tracts are headed toward the brain, the ascending tracts become thicker as they progress from caudal to cranial. When you think about it, it makes sense that the descending tracts are thickest in the cranial region as well. The gray matter, especially the anterior horn, which is the location of motor neurons to skeletal muscles, is the largest at lower cervical levels, and at lower and lumbar upper sacral levels. These levels correspond to upper and lower limb anatomy respectively, where large amounts of skeletal muscle tissue in the limbs require motor innervation.

CHECKPOINT

1. Where are the spinal meninges, epidural space, subdural space, and arachnoid space located?

TABLE 17.1

Comparison of Various Spinal Cord Segments

SEGMENT	DISTINGUISHING CHARACTERISTICS
Cervical	Relatively large diameter, relatively large amounts of white matter, oval in shape; in upper cervical segments (C1–C4), posterior gray horn is large, but anterior gray horn is relatively small; in lower cervical segments (C5 and below), posterior gray horns are enlarged and anterior gray horns are well-developed
Thoracic	Small diameter is due to relatively small amounts of gray matter; except for first thoracic segment, anterior and posterior gray horns are relatively small; a small lateral gray horn is present
Lumbar	Nearly circular; very large anterior and posterior gray horns; relatively less white matter than cervical segments
Sacral	Relatively small, but with relatively large amounts of gray matter; relatively small amounts of white matter; anterior and posterior gray horns are large and thick
Coccygeal	Resemble lower sacral spinal segments, but much smaller

2. What are the cervical and lumbar enlargements?
3. Define conus medullaris, filum terminale, and cauda equina.
4. How is the spinal cord partially divided into right and left sides?
5. Define each of the following terms: gray commissure, central canal, anterior gray horn, lateral gray horn, posterior gray horn, anterior white column, lateral white column, posterior white column, ascending tract, descending tract.

17.2 SPINAL NERVES

 OBJECTIVES
- Explain the basic structure of a nerve.
- Identify the components, connective tissue coverings, and branching of a spinal nerve.
- Describe the distribution of nerves of the cervical, brachial, lumbar, and sacral plexuses.
- Compare the clinical significance of dermatomes and cutaneous fields.

Spinal nerves are nerves associated with the spinal cord and, like all nerves of the peripheral nervous system (PNS), are parallel bundles of axons and their associated neuroglial cells wrapped in several layers of connective tissue. Spinal nerves connect the CNS to sensory receptors, muscles, and glands in all parts of the body. There are 31 pairs of spinal nerves. The spinal cord appears to be segmented because the 31 pairs of spinal nerves emerge at regular intervals from the spinal cord through intervertebral

CLINICAL CONNECTION | *Spinal Cord Injury*

Most **spinal cord injuries** are due to trauma as a result of factors such as automobile accidents, falls, contact sports, diving, and acts of violence. The effects of the injury depend on the extent of direct trauma to the spinal cord or compression of the cord by fractured or displaced vertebrae or blood clots. Although any segment of the spinal cord may be involved, the most common sites of injury are in the cervical, lower thoracic, and upper lumbar regions. Depending on the location and extent of spinal cord damage, paralysis may occur. **Monoplegia** (mon'-ō-PLĒ-jē-a; *mono-*=one; *-plegia*=blow or strike) is paralysis of one limb only. **Diplegia** (*di-*=two) is paralysis of both upper limbs or both lower limbs. **Paraplegia** (*para-*=beyond) is paralysis of both lower limbs. **Hemiplegia** (*hemi-*=half) is paralysis of the upper limb, trunk, and lower limb on one side of the body, and **quadriplegia** (*quad-*=four) is paralysis of all four limbs.

Complete transection (tran-SEK-shun; *trans-*=across; *-section*=a cut) of the spinal cord means that the cord is severed from one side to the other, thus cutting all sensory and motor tracts. It results in a loss of all sensations and voluntary movement below the level of the transection. A person will have permanent loss of all sensations in dermatomes below the injury because ascending nerve impulses cannot propagate past the transection to reach the brain. A **dermatome** (DER-ma-tōm; *derma*=skin, *-tome*=thin segment) is an area of skin that provides sensory input to the CNS via a pair of spinal nerves. At the same time, all voluntary muscle contractions will be lost below the transection because nerve impulses descending from the brain also cannot pass. The extent of paralysis of skeletal muscles depends on the level of injury. The closer the injury is to the head, the greater the area of the body that may be affected. The following list outlines which muscle functions may be *retained* at progressively lower levels of spinal cord transection. (These are spinal cord levels and not vertebral column levels. Recall that spinal cord levels differ from vertebral column levels because of the differential growth of the cord versus the column, especially as you progress inferiorly.)

- C1–C3: no function maintained from the neck down; ventilator needed for breathing; electric wheelchair with breath, head, or shoulder-controlled device required (see Figure A)
- C4–C5: diaphragm, which allows breathing
- C6–C7: some arm and chest muscles, which allows feeding, some dressing, and manual wheelchair required (see Figure B)
- T1–T3: intact arm function
- T4–T9: control of trunk above the umbilicus
- T10–L1: most thigh muscles, which allows walking with long leg braces (see Figure C)
- L1–L2: most leg muscles, which allows walking with short leg braces (see Figure D)

Hemisection is a partial transection of the cord on either the right or the left side. After hemisection, three main symptoms, known together as *Brown-Séquard syndrome* (sē-KAR), occur below the level of the injury: (1) Damage of the posterior column (sensory tracts) causes loss of proprioception and fine touch sensations on the *ipsilateral* (same) side as the injury. (2) Damage of the lateral corticospinal tract (motor tract) causes ipsilateral paralysis. (3) Damage of the spinothalamic tracts (sensory tracts) causes loss of pain and temperature sensations on the *contralateral* (opposite) side. These tracts are discussed in more detail in Chapter 20.

Following complete transection, and to varying degrees after hemisection, spinal shock occurs. **Spinal shock** is an immediate response to spinal cord injury characterized by temporary **areflexia** (a'-re-FLEK-sē-a), loss of reflex function. The areflexia occurs in parts of the body served by spinal nerves below the level of the injury. Signs of acute spinal shock include slow heart rate, low blood pressure, flaccid paralysis of skeletal muscles, loss of somatic sensations, and urinary bladder dysfunction. Spinal shock may begin within 1 hour after injury and may last from several minutes to several months, after which reflex activity gradually returns.

In many cases of traumatic injury of the spinal cord, the patient may have an improved outcome if an anti-inflammatory corticosteroid drug called methylprednisolone is given within 8 hours of the injury. This is because the degree of neurologic deficit is greatest immediately following traumatic injury as a result of *edema* (collection of fluid within tissues) as the immune system responds to the injury. •

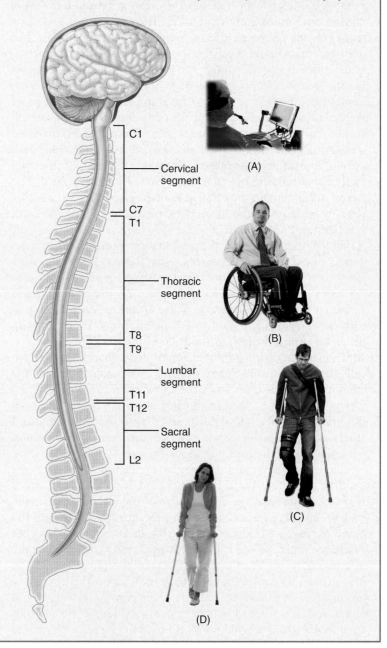

foramina (see Figure 17.2a). Indeed, each pair of spinal nerves is said to arise from a *spinal segment*. Within the spinal cord there is no obvious segmentation but, for convenience, the naming of spinal nerves is based on the segment in which they are located.

There are 8 pairs of *cervical nerves* (represented in Figure 17.2a as C1–C8), 12 pairs of *thoracic nerves* (T1–T12), 5 pairs of *lumbar nerves* (L1–L5), 5 pairs of *sacral nerves* (S1–S5), and 1 pair of *coccygeal nerves* (Co1). The first cervical pair emerges between the atlas (first cervical vertebra) and the occipital bone. All other spinal nerves emerge from the vertebral column through the intervertebral foramina between adjoining vertebrae.

Not all spinal cord segments are aligned with their corresponding vertebrae because of the differential growth between the spinal cord and vertebral column. Recall that the spinal cord ends near the level of the superior border of the second lumbar vertebra, and that the roots of the lumbar, sacral, and coccygeal nerves descend at an angle to reach their respective foramina before emerging from the vertebral column. This arrangement constitutes the cauda equina (see Figure 17.2a, c).

Structure of a Single Nerve

Spinal nerves arise from both the brain and spinal cord. Before examining the spinal nerves in more detail, it is important to understand the design of a single nerve and the difference between neurons and nerves. Recall that *neurons* are the conductive cells of the nervous tissue. **Nerves** are bundles of axons and their associated neuroglial cells wrapped in layers of connective tissue. Nerves, like skeletal muscles, consist of long cells.

While the function of nerves and muscles is quite different, their structures are similar in many ways. In a skeletal muscle, the long muscle cells have a loose connective tissue surrounding the endomysium, which distributes capillaries throughout the muscle tissue. Thicker sheets of connective tissue, the perimysium, bundle groups of muscle fibers together into units within the muscle called fasciculi. Many fasciculi are held together into a larger bundle of fibers, the individual muscle, by the thicker connective tissue epimysium.

Nerves have an identical design, except the connective tissue wraps have the root word *neurium* (meaning *nerve*) instead of *mysium*, (which means *muscle*). Figure 17.4 illustrates the structure of a nerve. Within the nerve are many axons of neurons with their surrounding neurilemma and myelin sheaths. The axon and its associated glial cells form the **nerve fiber**. Each nerve fiber sits in a loose connective tissue covering, the **endoneurium** (en′-dō-NOO-rē-um; *endo-*=within or inner), which consists of a mesh of collagen fibers, fibroblasts, and macrophages surrounded by endoneurial fluid (extracellular fluid) derived from the capillaries. This fluid nourishes the neuron and provides the necessary environment for its function of propagating action potentials.

A thicker sheath of connective tissue, the **perineurium** (per′-i-NOO-rē-um; *peri-*=around), holds many nerve fibers together into bundles called **fasciculi** (singular=*fasciculus*). The perineurium consists of collagenous sheaths with up to 15 layers of fibroblasts distributed in a meshwork of collagen. This important sheath functions as a diffusion barrier that, along with the tight junctions in the capillaries (blood–nerve barrier), maintains the osmotic environment and fluid pressure within the endoneurium. The perineurium terminates by blending with the connective tissue capsules of various types of nerve endings and muscle junctions.

Completing the structure of the nerve is an outer connective tissue sheath, the **epineurium** (ep′-i-NOO-rē-um; *epi-*=over), which bundles all the fasciculi together to form a single nerve. The epineurium is continuous with the dura mater and consists of fibroblasts and thick collagen strands that primarily parallel the long axis of the nerve. Extensions of the epineurium fill the spaces between fascicles. The epineurium makes up, on average, 50 percent of the cross-sectional area of a nerve. This important sheath gives the nerve the necessary tensile strength to resist the forces that can so easily damage the delicate nervous tissue. Within any nerve in the body there is a greater amount of collagenous connective tissue than nervous tissue, which makes it possible for the nervous tissue to leave the protection of the bony skull and vertebral column and course throughout the peripheral tissues of the body. The strong collagenous sheaths protect the neurons from being torn apart by the strong tensile forces generated by muscular activities and movements of the body. The epineurium also contains the small blood vessels and lymphatic vessels for the nerve.

Figure 17.4 Organization and connective tissue coverings of a spinal nerve.

Three layers of connective tissue wrappings protect axons: Endoneurium surrounds individual axons, perineurium surrounds bundles of axons (fascicles), and epineurium surrounds an entire nerve.

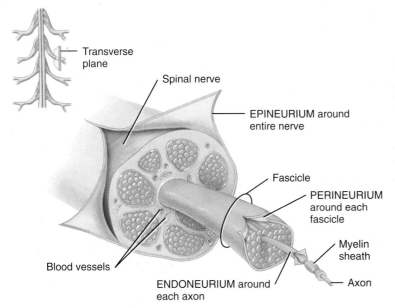

Transverse plane

Spinal nerve

EPINEURIUM around entire nerve

Fascicle

PERINEURIUM around each fascicle

Myelin sheath

Axon

Blood vessels

ENDONEURIUM around each axon

(a) Transverse section showing the coverings of a spinal nerve

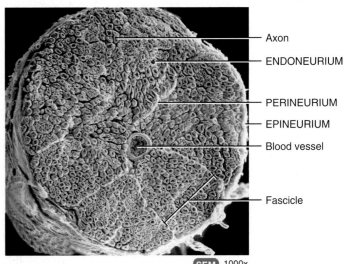

Axon

ENDONEURIUM

PERINEURIUM

EPINEURIUM

Blood vessel

Fascicle

SEM 1000x

(b) Transverse section of several nerve fascicles

 Why are all spinal nerves classified as mixed nerves?

Organization of Spinal Nerves

The design of a spinal nerve is like that of a tree. Beneath the ground, tiny rootlets converge to form large tree roots. These large roots come together at the surface to form the trunk, which then divides into numerous large branches that fork into smaller and smaller branches. Spinal nerves follow the same pattern and use much of the same terminology (see Figures 17.3 and 17.5).

The spinal nerves arise from the spinal cord as a series of small **rootlets**. The two types of rootlets are *anterior (ventral) rootlets* and *posterior (dorsal) rootlets*. From the anterolateral aspect of the cord, anterior rootlets emerge in two or three irregular rows. The anterior rootlets contain the axons of multipolar motor neurons arising from cell bodies in the anterior regions of the spinal cord gray matter. These axons transmit action potentials to the muscles and glands of the body. Projecting from the *posterolateral sulcus* of the spinal cord is another series of rootlets, the posterior rootlets, which contain the central processes of the sensory unipolar neurons. These neurons transmit action potentials from peripheral receptor organs to the central nervous system.

Each series of anterior rootlets converges to form larger **anterior (ventral) roots**. Likewise, each series of posterior rootlets converges to form larger **posterior (dorsal) roots**. Each posterior root has a swelling, the **posterior (dorsal) root ganglion**, which contains the cell bodies of sensory neurons. The anterior and posterior roots on each side of the spinal cord correspond to one developmental segment or level of the body. As the sensory posterior root and motor anterior root project laterally from the spinal cord, they converge to form a mixed nerve called the **spinal nerve trunk**. (A *mixed nerve* contains both motor and sensory axons.) The spinal nerve trunk runs for a short distance before branching into two large branches and a variable series of smaller branches (Figure 17.5).

Branches of Spinal Nerves

Each large spinal nerve branch, named a **ramus** (=branch; plural is *rami* [RĀ-mī]), follows a specific course to different peripheral regions. The two largest branches, the anterior (ventral) ramus and posterior (dorsal) ramus, are somatic branches that run in the musculoskeletal wall of the body. The **posterior (dorsal) ramus** serves the deep muscles and skin of the posterior surface of the trunk. The **anterior (ventral) ramus** serves the muscles and structures of the upper and lower limbs and the muscles and skin of the lateral and anterior regions of the trunk. Smaller visceral branches, such as the meningeal branch and the communicating rami, form the autonomic pathways to smooth muscle and glandular tissue. The **meningeal branch** (me-NIN-jē-al) reenters the vertebral canal through the intervertebral foramen and supplies the vertebrae, vertebral ligaments, blood vessels of the spinal cord, and meninges. The communicating rami or **rami communicantes** (ko-mū-ni-KAN-tēz) are components of the autonomic nervous system and will be discussed in Chapter 19.

Figure 17.5 Branches of a typical spinal nerve. (See also Figure 17.1b.)

🔑 The principal branches of a spinal nerve are the posterior (dorsal) ramus, the anterior (ventral) ramus, the meningeal branch, and the rami communicantes.

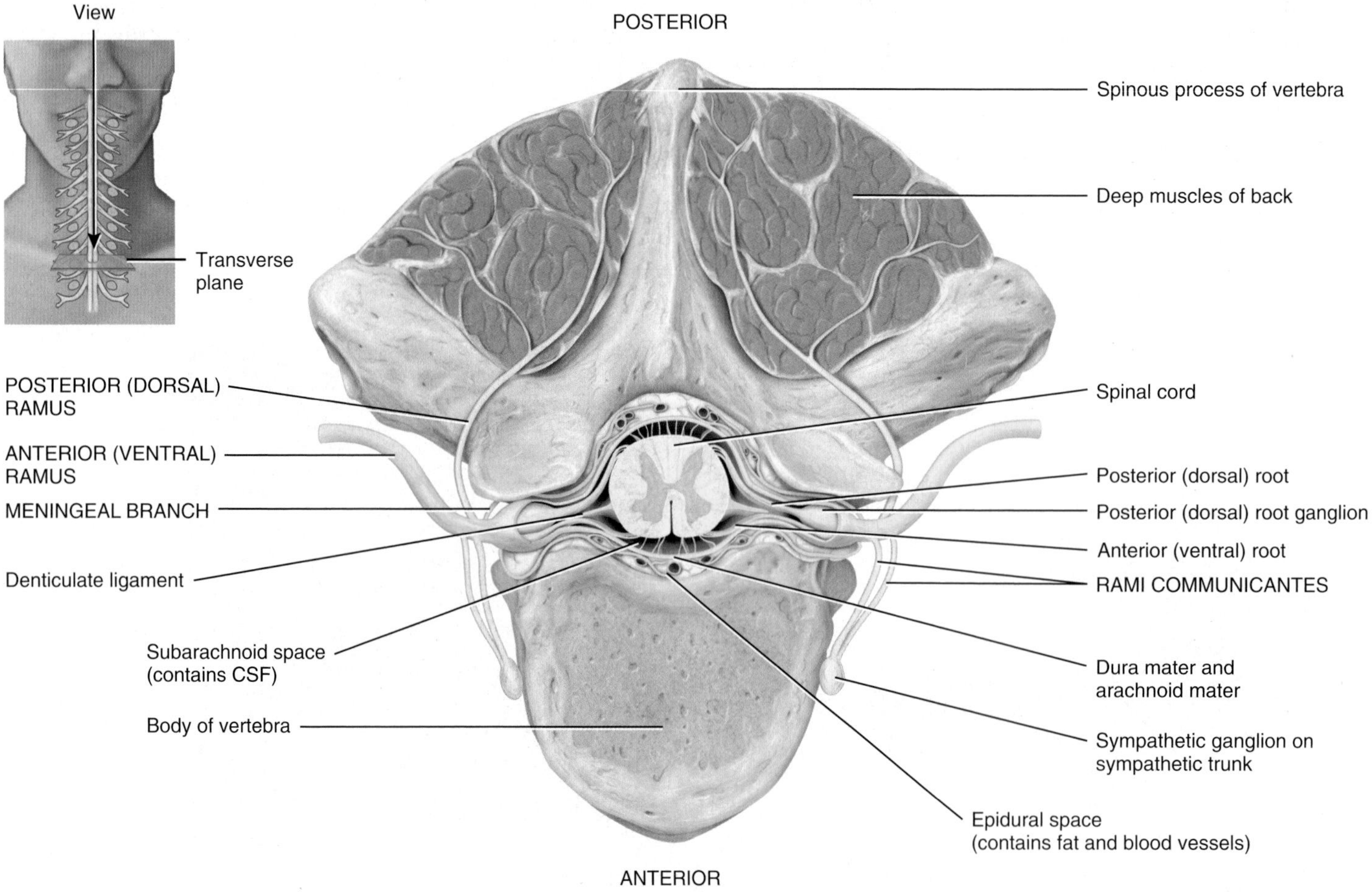

View

POSTERIOR

Transverse plane

Spinous process of vertebra

Deep muscles of back

POSTERIOR (DORSAL) RAMUS

ANTERIOR (VENTRAL) RAMUS

MENINGEAL BRANCH

Denticulate ligament

Subarachnoid space (contains CSF)

Body of vertebra

Spinal cord

Posterior (dorsal) root

Posterior (dorsal) root ganglion

Anterior (ventral) root

RAMI COMMUNICANTES

Dura mater and arachnoid mater

Sympathetic ganglion on sympathetic trunk

Epidural space (contains fat and blood vessels)

ANTERIOR

(a) Superior view

Plexuses

Axons from the anterior rami of spinal nerves, except for thoracic nerves T2–T12, do not go directly to the body structures they supply. Instead, they form networks on both the left and right sides of the body by joining with various numbers of axons from anterior rami of adjacent nerves. Such a network of axons is called a **plexus** (PLEK-sus=braid or network). The principal spinal nerve plexuses are the **cervical plexus**, **brachial plexus**, **lumbar plexus**, and **sacral plexus**. A smaller **coccygeal plexus** is also present. Refer to Figure 17.2 to see the relationships of the plexuses to one another. Emerging from the plexuses are nerves bearing names that often describe the general regions they serve or the course they take. Each of these nerves may in turn have several branches named for the specific structures they innervate.

Exhibits 17.A–17.D (Figures 17.6–17.9) summarize the principal plexuses.

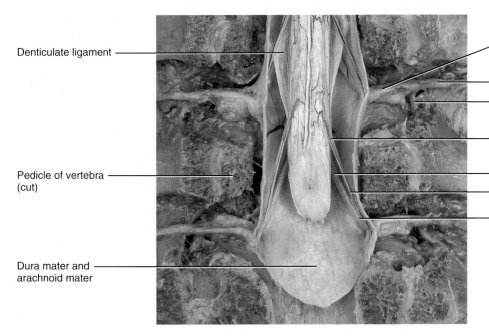

Denticulate ligament

Pedicle of vertebra (cut)

Dura mater and arachnoid mater

Spinal nerve

ANTERIOR (VENTRAL) RAMUS

POSTERIOR (DORSAL) RAMUS

Anterior (ventral) rootlets

Posterior (dorsal) rootlets

Anterior (ventral) root

Posterior (dorsal) root

(b) Anterior view and oblique section of spinal cord

CLINICAL CONNECTION | *Spinal Nerve Root Damage*

The most common cause of **spinal nerve root damage** is a herniated intervertebral disc. Damage to vertebrae as a result of osteoporosis, osteoarthritis, cancer, or trauma can also damage spinal nerve roots. Symptoms of spinal nerve root damage include pain, muscle weakness, and loss of feeling. Rest, manual therapy, pain medications, and epidural injections are the most widely used conservative treatments. It is recommended that 6 to 12 weeks of conservative therapy be attempted first. If the pain continues, is intense, or is impairing normal functioning, surgery is often the next step. •

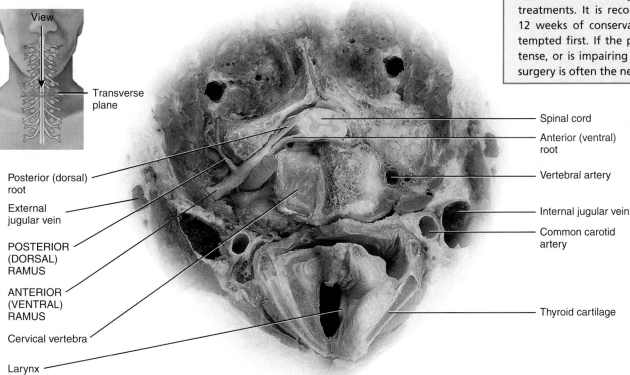

View

Transverse plane

Posterior (dorsal) root

External jugular vein

POSTERIOR (DORSAL) RAMUS

ANTERIOR (VENTRAL) RAMUS

Cervical vertebra

Larynx

Spinal cord

Anterior (ventral) root

Vertebral artery

Internal jugular vein

Common carotid artery

Thyroid cartilage

(c) Superolateral view of transverse section of neck

? **Which spinal nerve branches serve the upper and lower limbs?**

EXHIBIT 17.A   Cervical Plexus   *(Figure 17.6)*

OBJECTIVE

• Describe the origin and distribution of the cervical plexus.

The **cervical plexus** (SER-vi-kul) is formed by the roots (anterior rami) of the first four cervical nerves (C1–C4), with contributions from C5 (Figure 17.6). There is one on each side of the neck alongside the first four cervical vertebrae.

The cervical plexus supplies the skin and muscles of the head, neck, and superior part of the shoulders and chest. The phrenic nerves arise from the cervical plexuses and supply motor fibers to the diaphragm. Branches of the cervical plexus also run parallel to two cranial nerves, the accessory (XI) nerve and hypoglossal (XII) nerve.

NERVE	ORIGIN	DISTRIBUTION
SUPERFICIAL (SENSORY) BRANCHES		
Lesser occipital	C2	Skin of scalp posterior and superior to ear
Great auricular (aw-RIK-ū-lar)	C2–C3	Skin anterior, inferior, and over ear, and over parotid glands
Transverse cervical	C2–C3	Skin over anterior and lateral aspect of neck
Supraclavicular	C3–C4	Skin over superior portion of chest and shoulder
DEEP (LARGELY MOTOR) BRANCHES		
Ansa cervicalis (AN-sa ser-vi-KAL-is)		This nerve divides into superior and inferior roots
Superior root	C1	Infrahyoid and geniohyoid muscles of neck
Inferior root	C2–C3	Infrahyoid muscles of neck
Phrenic (FREN-ik)	C3–C5	Diaphragm
Segmental branches	C1–C5	Prevertebral (deep) muscles of neck, levator scapulae, and middle scalene muscles

Figure 17.6 Cervical plexus in anterior view.

The cervical plexus supplies the skin and muscles of the head, neck, superior portion of the shoulders and chest, and diaphragm.

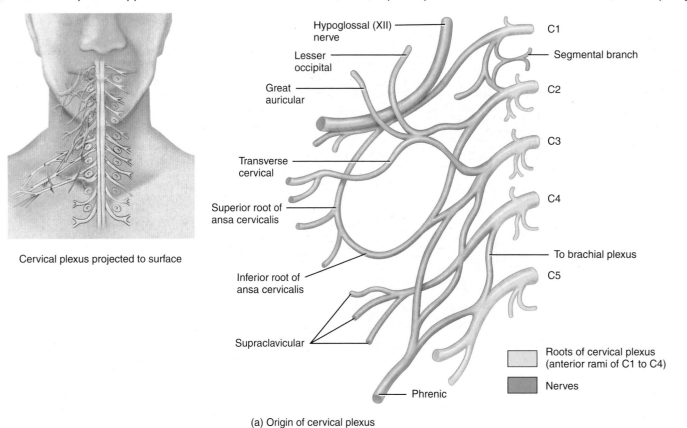

(a) Origin of cervical plexus

EXHIBIT 17.A **571**

CLINICAL CONNECTION | *Injuries to the Phrenic Nerves*

The phrenic nerves originate from C3, C4, and C5 and supply the diaphragm. Complete severing of the spinal cord above the origin of the phrenic nerves causes respiratory arrest. Breathing stops because the phrenic nerves no longer send impulses to the diaphragm. The phrenic nerves may also be damaged due to pressure from malignant tumors in the mediastinum, such as tracheal and esophageal tumors. •

CHECKPOINT

6. Which nerve from the cervical plexus causes contraction of the diaphragm?

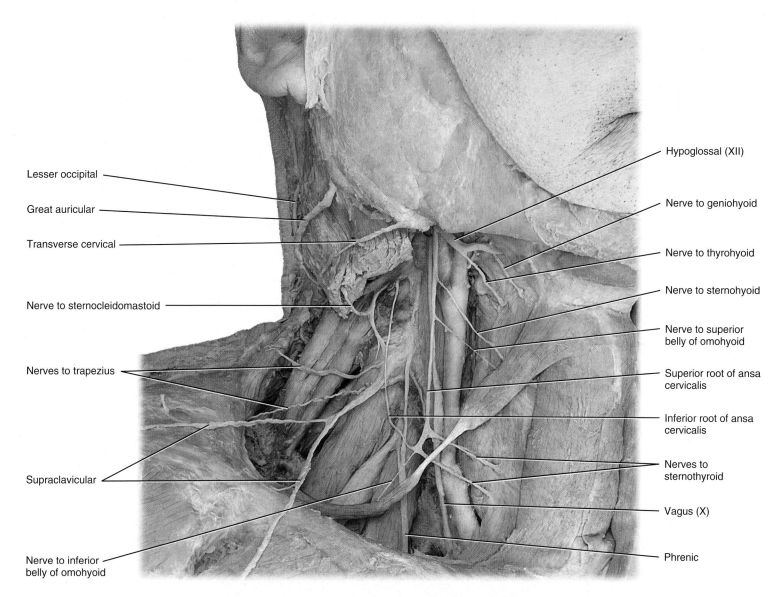

Lesser occipital

Great auricular

Transverse cervical

Nerve to sternocleidomastoid

Nerves to trapezius

Supraclavicular

Nerve to inferior belly of omohyoid

Hypoglossal (XII)

Nerve to geniohyoid

Nerve to thyrohyoid

Nerve to sternohyoid

Nerve to superior belly of omohyoid

Superior root of ansa cervicalis

Inferior root of ansa cervicalis

Nerves to sternothyroid

Vagus (X)

Phrenic

(b) Anterior view of cervical plexus

 Why does complete severing of the spinal cord at level C2 cause respiratory arrest?

| EXHIBIT 17.B | Brachial Plexus *(Figure 17.7)* |

OBJECTIVE

• Describe the origin, distribution, and effects of damage to the brachial plexus.

The anterior rami of spinal nerves C5–C8 and T1 form the roots of the **brachial plexus** (BRĀ-kē-al), which extends inferiorly and laterally on either side of the last four cervical and first thoracic vertebrae (Figure 17.7a). It passes above the first rib posterior to the clavicle and then enters the axilla.

Since the brachial plexus is so complex, an explanation of its various parts is helpful. As with the cervical and other plexuses,

the **roots** of the brachial plexus are the anterior rami of the spinal nerves. The roots of several spinal nerves unite to form **trunks** in the inferior part of the neck. These are the *superior, middle,* and *inferior trunks.* Posterior to the clavicles, the trunks divide into **divisions** called the *anterior* and *posterior divisions.* In the axillae, the divisions unite to form **cords** called the *lateral, medial,* and *posterior cords.* The cords are named for their relationship to the axillary artery, a large artery that supplies blood to the upper limb. The **branches** of the brachial plexus form the principal nerves of the brachial plexus.

Figure 17.7 Brachial plexus in anterior view.

🔑 The brachial plexus supplies the shoulders and upper limbs.

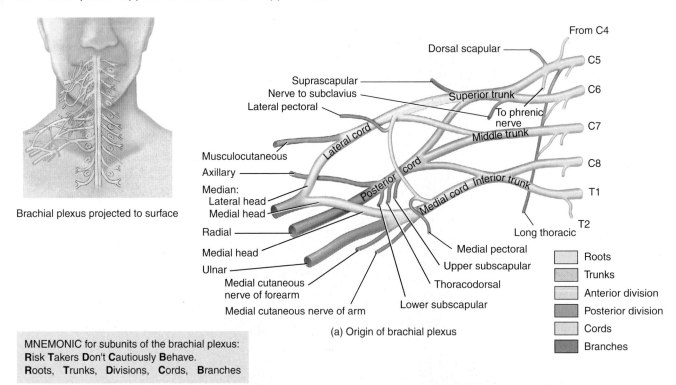

(a) Origin of brachial plexus

MNEMONIC for subunits of the brachial plexus:
Risk **T**akers **D**on't **C**autiously **B**ehave.
Roots, **T**runks, **D**ivisions, **C**ords, **B**ranches

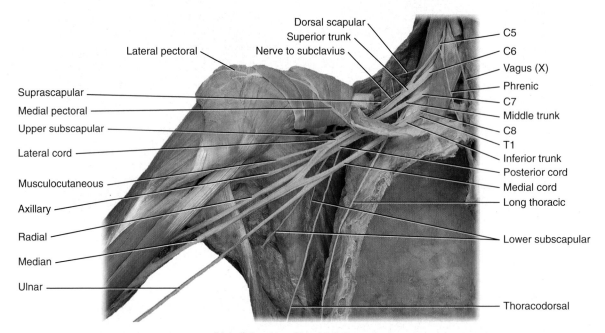

(b) Anterior view of brachial plexus

EXHIBIT 17.B **573**

The brachial plexus provides almost the entire nerve supply of the shoulders and upper limbs (Figure 17.7b). Five large terminal branches arise from the brachial plexus: (1) The **axillary nerve** supplies the deltoid and teres minor muscles. (2) The **musculocutaneous nerve** supplies the anterior muscles of the arm. (3) The **radial nerve** supplies the muscles on the posterior aspect of the arm and forearm. (4) The **median nerve** supplies most of the muscles of the anterior forearm (6 1/2 out of 8) and some of the muscles of the hand. (5) The **ulnar nerve** supplies the anteromedial muscles of the forearm (the other 1 1/2) and most of the muscles of the hand.

 CHECKPOINT

7. Injury of which nerve could cause paralysis of the serratus anterior muscle?

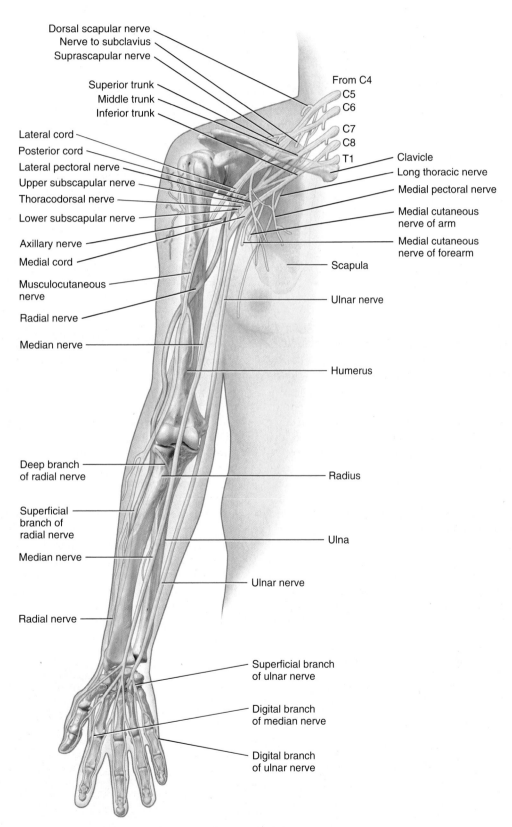

(c) Distribution of nerves from brachial plexus

What five important nerves arise from the brachial plexus?

EXHIBIT 17.B Brachial Plexus *(Figure 17.7)* CONTINUED

NERVE	ORIGIN	DISTRIBUTION
Dorsal scapular (SKAP-ū-lar)	C5	Levator scapulae, rhomboid major, and rhomboid minor muscles
Long thoracic (thor-RAS-ik)	C5–C7	Serratus anterior muscle
Nerve to subclavius (sub-KLĀ-vē-us)	C5–C6	Subclavius muscle
Suprascapular	C5–C6	Supraspinatus and infraspinatus muscles
Musculocutaneous (mus'-kū-lō-kū-TĀN-ē-us)	C5–C7	Coracobrachialis, biceps brachii, and brachialis muscles
Lateral pectoral (PEK-to-ral)	C5–C7	Pectoralis major muscle
Upper subscapular	C5–C6	Subscapularis muscle
Thoracodorsal (tho-RĀ-kō-dor-sal)	C6–C8	Latissimus dorsi muscle
Lower subscapular	C5–C6	Subscapularis and teres major muscles
Axillary (AK-si-lar-ē)	C5–C6	Deltoid and teres minor muscles; skin over deltoid and superior posterior aspect of arm
Median	C5–T1	Flexors of forearm, except flexor carpi ulnaris: the ulnar half of the flexor digitorum profundus, and some muscles of the hand (lateral palm); skin of lateral two-thirds of palm of hand and fingers
Radial	C5–T1	Triceps brachii, anconeus, and extensor muscles of forearm; skin of posterior arm and forearm, lateral two-thirds of dorsum of hand, and fingers over proximal and middle phalanges
Medial pectoral	C8–T1	Pectoralis major and pectoralis minor muscles
Medial cutaneous nerve of arm (kū'-TĀ-nē-us)	C8–T1	Skin of medial and posterior aspects of distal third of arm
Medial cutaneous nerve of forearm	C8–T1	Skin of medial and posterior aspects of forearm
Ulnar	C8–T1	Flexor carpi ulnaris, ulnar half of the flexor digitorum profundus, and most muscles of the hand; skin of medial side of hand, little finger, and medial half of ring finger

---

CLINICAL CONNECTION | *Injuries to Nerves Emerging from the Brachial Plexus*

Injury to the superior roots of the brachial plexus (C5–C6) may result from forceful pulling away of the head from the shoulder, as might occur from a heavy fall on the shoulder or excessive stretching of an infant's neck during childbirth. The presentation of this injury is characterized by an upper limb in which the shoulder is adducted, the arm is medially rotated, the elbow is extended, the forearm is pronated, and the wrist is flexed (Figure 17.7d). This condition is called **Erb-Duchenne palsy** or **waiter's tip position**. There is loss of sensation along the lateral side of the arm.

Radial (and axillary) **nerve injury** can be caused by improperly administered intramuscular injections into the deltoid muscle. The radial nerve may also be injured when a cast is applied too tightly around the mid-humerus. Radial nerve injury is indicated by **wrist drop**, the inability to extend the wrist and fingers (Figure 17.7d). Sensory loss is minimal due to the overlap of sensory innervation by adjacent nerves.

Median nerve injury may result in **median nerve palsy**, which is indicated by numbness, tingling, and pain in the palm and fingers. There is also inability to pronate the forearm and flex the proximal interphalangeal joints of all digits and the distal interphalangeal joints of the second and third digits (Figure 17.7d). In addition, wrist flexion is weak and is accompanied by adduction, and thumb movements are weak.

Ulnar nerve injury may result in **ulnar nerve palsy**, which is indicated by an inability to abduct or adduct the fingers, atrophy of the interosseus muscles of the hand, hyperextension of the metacarpophalangeal joints, and flexion of the interphalangeal joints, a condition called **clawhand** (Figure 17.7d). There is also loss of sensation over the little finger.

Long thoracic nerve injury results in paralysis of the serratus anterior muscle. The medial border of the scapula protrudes, giving it the appearance of a wing. When the arm is raised, the vertebral border and inferior angle of the scapula pull away from the thoracic wall and protrude outward, causing the medial border of the scapula to protrude; because the scapula looks like a wing, this condition is called **winged scapula** (Figure 17.7d). The arm cannot be abducted beyond the horizontal position. •

Erb-Duchenne palsy (waiter's tip)

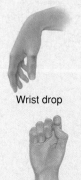

Wrist drop

Median nerve palsy

Ulnar nerve palsy

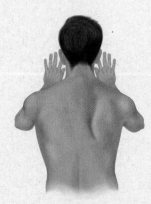

Winging of right scapula

Injuries to nerves of brachial plexus

EXHIBIT 17.C **575**

 EXHIBIT 17.C Lumbar Plexus *(Figure 17.8)*

 OBJECTIVE

• Describe the origin and distribution of the lumbar plexus.

The anterior rami of spinal nerves L1–L4 form the roots of the **lumbar plexus** (LUM-bar) (Figure 17.8). Unlike the brachial plexus, the intricate intermingling of fibers is minimal in the lumbar plexus. On either side of the first four lumbar vertebrae, the lumbar plexus passes obliquely outward, between the superficial and deep heads of the psoas major muscle and anterior to the quadratus lumborum muscle. Between the heads of the psoas major, the roots of the plexus split into anterior and posterior divisions, which then give rise to the peripheral branches of the plexus.

The lumbar plexus supplies the anterolateral abdominal wall, external genitals, and part of the lower limbs.

CHECKPOINT

8. What structures are supplied by the lumbar plexus?

NERVE	ORIGIN	DISTRIBUTION
Iliohypogastric (il′-ē-ō-hī-pō-GAS-trik)	L1	Muscles of anterolateral abdominal wall; skin of inferior abdomen and buttock
Ilioinguinal (il′-ē-ō-ING-gwi-nal)	L1	Muscles of anterolateral abdominal wall; skin of superior and medial aspect of thigh, root of penis and scrotum in male, and labia majora and mons pubis in female
Genitofemoral (jen′-i-tō-FEM-or-al)	L1–L2	Cremaster muscle; skin over middle anterior surface of thigh, scrotum in male, and labia majora in female
Lateral cutaneous nerve of thigh	L2–L3	Skin over lateral, anterior, and posterior aspects of thigh
Femoral	L2–L4	Largest nerve arising from the lumbar plexus, distributed to the flexor muscles of hip joint and extensor muscles of knee joint, and to skin over anterior and medial aspect of thigh and medial side of leg and foot
Obturator (OB-too-rā-tor)	L2–L4	Adductor muscles of hip joint; skin over medial aspect of thigh

Figure 17.8 Lumbar plexus in anterior view.

🔑 The lumbar plexus supplies the anterolateral abdominal wall, external genitals, and part of the lower limbs.

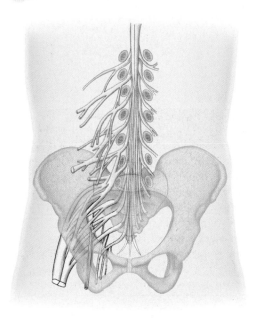

Lumbar plexus projected to surface

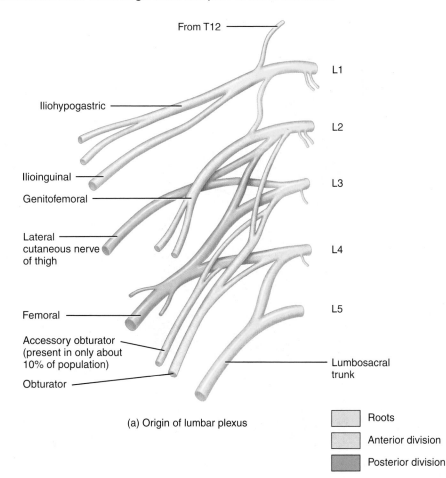

(a) Origin of lumbar plexus

Roots

Anterior division

Posterior division

FIGURE 17.8 CONTINUES ▶

EXHIBIT 17.C Lumbar Plexus *(Figure 17.8)* CONTINUED

 FIGURE **17.8** CONTINUED

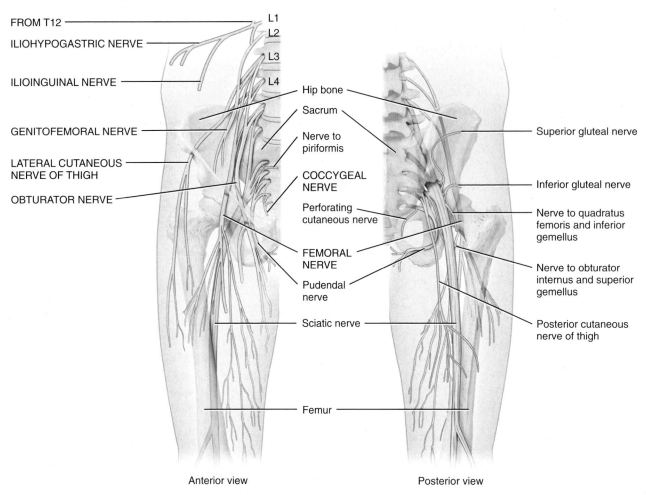

FROM T12

ILIOHYPOGASTRIC NERVE

ILIOINGUINAL NERVE

GENITOFEMORAL NERVE

LATERAL CUTANEOUS
NERVE OF THIGH

OBTURATOR NERVE

L1
L2
L3
L4

Hip bone

Sacrum

Nerve to
piriformis

COCCYGEAL
NERVE

Perforating
cutaneous nerve

FEMORAL
NERVE

Pudendal
nerve

Sciatic nerve

Femur

Superior gluteal nerve

Inferior gluteal nerve

Nerve to quadratus
femoris and inferior
gemellus

Nerve to obturator
internus and superior
gemellus

Posterior cutaneous
nerve of thigh

Anterior view

Posterior view

(b) Distribution of nerves from lumbar plexus

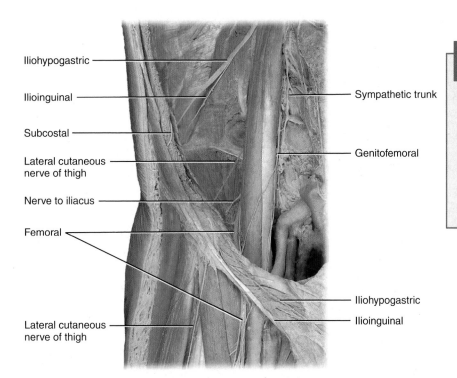

Iliohypogastric

Ilioinguinal

Subcostal

Lateral cutaneous
nerve of thigh

Nerve to iliacus

Femoral

Lateral cutaneous
nerve of thigh

Sympathetic trunk

Genitofemoral

Iliohypogastric

Ilioinguinal

(c) Anterior view of lumbar plexus in right pelvic region

? **What is the origin of the lumbar plexus?**

CLINICAL CONNECTION | *Injuries to the Lumbar Plexus*

The largest nerve arising from the lumbar plexus is
the femoral nerve. **Femoral nerve injury**, which
can occur in stab or gunshot wounds, is indicated by an
inability to extend the leg and by loss of sensation in the
skin over the anteromedial aspect of the thigh.

Obturator nerve injury results in paralysis of the
adductor muscles of the thigh and loss of sensation over
the medial aspect of the thigh. It may result from pressure on the nerve by the fetal head during pregnancy. •

EXHIBIT 17.D **577**

EXHIBIT 17.D Sacral and Coccygeal Plexuses *(Figure 17.9)*

 OBJECTIVE

• Describe the origin and distribution of the sacral and coccygeal plexuses.

The anterior rami of spinal nerves L4–L5 and S1–S4 form the roots of the **sacral plexus** (SĀ-kral) (Figure 17.9). This plexus is situated largely anterior to the sacrum. The sacral plexus supplies the buttocks, perineum, and lower limbs. The largest nerve in the body—the sciatic nerve—arises from the sacral plexus.

The roots (anterior rami) of spinal nerves S4–S5 and the coccygeal nerves form a small **coccygeal plexus** (kok-SIJ-ē-al). From this plexus arises the anococcygeal nerves (Figure 17.9b), which supply a small area of skin in the coccygeal region.

 CHECKPOINT

9. Injury of which branch of the sciatic nerve causes foot drop?

Figure 17.9 Sacral and coccygeal plexuses in anterior view.

 The sacral plexus supplies the buttocks, perineum, and lower limbs.

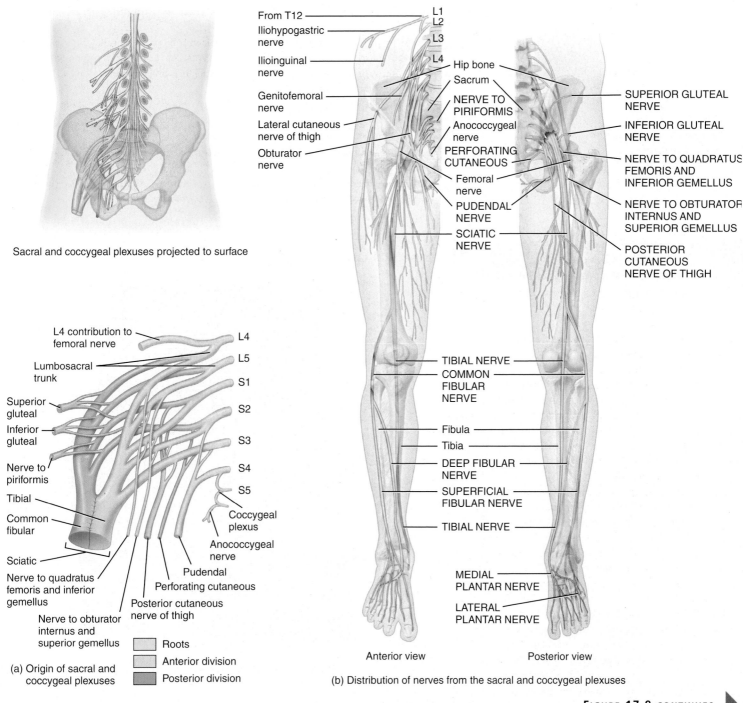

Sacral and coccygeal plexuses projected to surface

From T12
Iliohypogastric nerve
Ilioinguinal nerve
Genitofemoral nerve
Lateral cutaneous nerve of thigh
Obturator nerve

L1
L2
L3
L4

Hip bone
Sacrum
NERVE TO PIRIFORMIS
Anococcygeal nerve
PERFORATING CUTANEOUS
Femoral nerve
PUDENDAL NERVE
SCIATIC NERVE

SUPERIOR GLUTEAL NERVE
INFERIOR GLUTEAL NERVE
NERVE TO QUADRATUS FEMORIS AND INFERIOR GEMELLUS
NERVE TO OBTURATOR INTERNUS AND SUPERIOR GEMELLUS
POSTERIOR CUTANEOUS NERVE OF THIGH

L4 contribution to femoral nerve
Lumbosacral trunk
Superior gluteal
Inferior gluteal
Nerve to piriformis
Tibial
Common fibular
Sciatic
Nerve to quadratus femoris and inferior gemellus
Nerve to obturator internus and superior gemellus

L4
L5
S1
S2
S3
S4
S5

Coccygeal plexus
Anococcygeal nerve
Pudendal
Perforating cutaneous
Posterior cutaneous nerve of thigh

 Roots
 Anterior division
 Posterior division

(a) Origin of sacral and coccygeal plexuses

TIBIAL NERVE
COMMON FIBULAR NERVE
Fibula
Tibia
DEEP FIBULAR NERVE
SUPERFICIAL FIBULAR NERVE
TIBIAL NERVE
MEDIAL PLANTAR NERVE
LATERAL PLANTAR NERVE

Anterior view   Posterior view

(b) Distribution of nerves from the sacral and coccygeal plexuses

FIGURE 17.9 CONTINUES ▶

EXHIBIT 17.D Sacral and Coccygeal Plexuses *(Figure 17.9)* CONTINUED

NERVE	ORIGIN	DISTRIBUTION
Superior gluteal (GLOO-tē-al)	L4–L5 and S1	Gluteus minimus, gluteus medius, and tensor fasciae latae muscles
Inferior gluteal	L5–S2	Gluteus maximus muscle
Nerve to piriformis (pir-i-FORM-is)	S1–S2	Piriformis muscle
Nerve to quadratus femoris (quad-RĀ-tus FEM-or-is) **and inferior gemellus** (jem-EL-us)	L4–L5 and S1	Quadratus femoris and inferior gemellus muscles
Nerve to obturator internus (OB-too-rā′-tor in-TER-nus) **and superior gemellus**	L5–S2	Obturator internus and superior gemellus muscles
Perforating cutaneous (kū′-TĀ-ne-us)	S2–S3	Skin over inferior medial aspect of buttock
Posterior cutaneous nerve of thigh	S1–S3	Skin over anal region, inferior lateral aspect of buttock, superior posterior aspect of thigh, superior part of calf, scrotum in male, and labia majora in female
Pudendal (pū-DEN-dal)	S2–S4	Muscles of perineum; skin of penis and scrotum in male and clitoris, labia majora, labia minora, and vagina in female
Sciatic (sī-AT-ik)	L4–S3	Actually two nerves—tibial and common fibular—bound together by common sheath of connective tissue; it splits into its two divisions, usually at the knee (see below for distributions); as it descends through the thigh, sends branches to hamstring muscles and the adductor magnus
Tibial (TIB-ē-al)	L4–S3	Gastrocnemius, plantaris, soleus, popliteus, tibialis posterior, flexor digitorum longus, and flexor hallucis longus muscles; branches in foot are medial plantar nerve and lateral plantar nerve
Medial plantar (PLAN-tar) (see Figure 13.10b)		Abductor hallucis, flexor digitorum brevis, and flexor hallucis brevis muscles; skin over medial two-thirds of plantar surface of foot
Lateral plantar (see Figure 13.10b)		Remaining muscles of foot not supplied by medial plantar nerve; skin over lateral third of plantar surface of foot
Common fibular (FIB-ū-lar)	L4–S2	Divides into a superficial fibular and a deep fibular branch
Superficial fibular		Fibularis longus and fibularis brevis muscles; skin over distal third of anterior aspect of leg and dorsum of foot
Deep fibular		Tibialis anterior, extensor hallucis longus, fibularis tertius, and extensor digitorum longus and extensor digitorum brevis muscles; skin on adjacent sides of great and second toes

CLINICAL CONNECTION | *Injury to the Sciatic Nerve*

The most common form of back pain is caused by compression or irritation of the sciatic nerve, the longest nerve in the human body. **Injury to the sciatic nerve** results in **sciatica** (sī-AT-i-ka), pain that may extend from the buttock down the posterior and lateral aspect of the leg and the lateral aspect of the foot. The nerve may be injured because of a herniated (slipped) disc, dislocated hip, osteoarthritis of the lumbosacral spine, pathological shortening of the lateral rotator muscles of the thigh, pressure from the uterus during pregnancy, inflammation, irritation, or an improperly administered gluteal intramuscular injection.

In many sciatic nerve injuries, the common fibular portion is the most affected, frequently from fractures of the fibula or by pressure from casts or splints over the thigh or leg. Damage to the common fibular nerve causes the foot to be plantar flexed, a condition called **foot drop**, and inverted, a condition called **equinovarus** (e-KWī-nō-va-rus). There is also loss of function along the anterolateral apects of the leg and dorsum of the foot and toes. Injury to the tibial portion of the sciatic nerve results in dorsiflexion of the foot plus eversion, a condition called **calcaneovalgus** (kal-KĀ-nē-ō-val′-gus). Loss of sensation on the sole also occurs. Treatments for sciatica are similar to those for a herniated (slipped) disc—rest, pain medications, exercises, ice or heat, and massage. •

FIGURE 17.9 CONTINUED

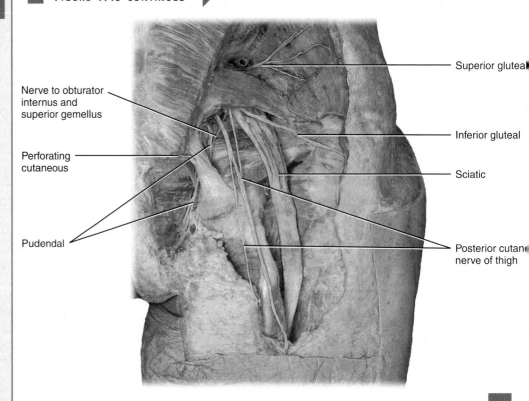

(c) Posterior view of sacral plexus in right gluteal region

 What is the origin of the sacral plexus?

Intercostal Nerves

The anterior rami of spinal nerves T2–T12 do not enter into the formation of plexuses and are known as **intercostal** or **thoracic nerves** (see Figure 17.2a). Because these nerves connect directly to the structures they supply in the intercostal spaces and are mainly distributed to a single body segment, they are referred to as segmental nerves. After leaving its intervertebral foramen, the anterior ramus of nerve T2 innervates the intercostal muscles of the second intercostal space and supplies the skin of the axilla and posteromedial aspect of the arm. Nerves T3–T6 extend along the costal grooves of the ribs and then to the intercostal muscles and skin of the anterior and lateral chest wall. Nerves T7–T12 supply the intercostal muscles, the abdominal muscles, and the overlying skin. The posterior rami of the intercostal nerves supply the deep back muscles and skin of the posterior aspect of the thorax.

Dermatomes versus Cutaneous Fields

The skin over the entire body is supplied by somatic sensory neurons that carry nerve impulses from the skin into the spinal cord and brain stem. Likewise, somatic motor neurons that carry impulses out of the spinal cord innervate the underlying skeletal muscles. Each spinal nerve contains sensory neurons that serve a specific, predictable segment of the body. The trigeminal (V) nerve serves most of the skin of the face and scalp. Recall that the area of the skin that provides sensory input to the CNS via one pair of spinal nerves or the trigeminal (V) nerve is called a **dermatome** (Figure 17.10a). Knowing which spinal cord segments supply each dermatome makes it possible to locate damaged regions of the spinal cord. **Cutaneous fields**, on the other hand, are regions of skin supplied by a specific nerve arising from a plexus (Figure 17.10b). For example, the median nerve from the

Figure 17.10 Distribution of dermatomes and cutaneous fields.

A dermatome is an area of skin that provides sensory input via the posterior roots of one pair of spinal nerves or via the trigeminal (V) nerve.

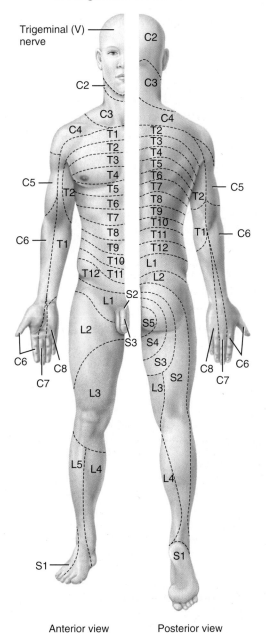

(a) Distribution of dermatomes

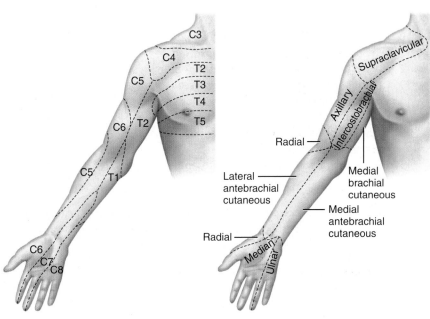

(b) Comparison between distributions of dermatomes (left) and cutaneous fields (right)

CLINICAL CONNECTION | Shingles

Shingles is an acute infection of the peripheral nervous system caused by *herpes zoster* (HER-pēz ZOS-ter), the virus that also causes chickenpox. After a person recovers from chickenpox, the virus retreats to a posterior root ganglion. If the virus is reactivated, the immune system usually prevents it from spreading. From time to time, however, the reactivated virus overcomes a weakened immune system, leaves the ganglion, and travels down sensory neurons of the skin. The result is pain, discoloration of the skin, and a characteristic line of skin blisters. The line of blisters marks the distribution (dermatome) of the particular cutaneous sensory nerve belonging to the infected posterior root ganglion. •

 Which is the only spinal nerve that does not have a corresponding dermatome?

brachial plexus has a distinct cutaneous field; this cutaneous field overlaps multiple dermatomes because the median nerve contains neurons from multiple spinal nerve levels. Because nerves arising from a plexus can contain neurons from more than one spinal nerve level, damage within a cutaneous field typically alerts a clinician to peripheral nerve damage rather than spinal root or spinal cord damage. If the skin in a particular region is stimulated but the sensation is not perceived, it is important to assess whether the loss of sensation is within a dermatome or a cutaneous field in order to properly diagnose the site of the injury. There is a variable amount of overlap between neighboring dermatomes. In regions where the overlap is considerable, little loss of sensation may result if only one of the nerves supplying the dermatome is damaged. Information about the innervation patterns of spinal nerves can also be used therapeutically. Cutting posterior roots or infusing local anesthetics can block pain either permanently or transiently. Because dermatomes overlap, deliberate production of a region of complete anesthesia may require that at least three adjacent spinal nerves be cut or blocked by an anesthetic drug.

CHECKPOINT

10. How are the spinal nerves named and numbered?
11. Why are spinal nerves classified as mixed nerves?
12. How is a spinal nerve connected to the spinal cord?
13. Describe the coverings of a spinal nerve.
14. What are the branches and innervations of a typical spinal nerve?
15. What is a plexus? Name the major plexuses and the regions they supply.
16. Distinguish between dermatomes and cutaneous fields.

17.3 SPINAL CORD FUNCTIONS

OBJECTIVES

• Describe the functions of the major sensory and motor tracts of the spinal cord.
• Explain the functional components of a reflex arc and the ways reflexes maintain homeostasis.

The spinal cord has two principal functions in maintaining homeostasis: nerve impulse propagation and integration of information. The *white matter tracts* in the spinal cord are highways for nerve impulse propagation. Along these tracts, sensory impulses from receptors flow toward the brain, and motor impulses flow from the brain toward skeletal muscles and other effector tissues. The *gray matter* of the spinal cord receives and integrates incoming and outgoing information.

Sensory and Motor Tracts

The first spinal cord function that promotes homeostasis is the conduction of nerve impulses along tracts. The name of a tract often indicates its position in the white matter of the spinal cord as well as where it begins and ends. For example, the anterior spinothalamic tract is located in the *anterior* white column; it begins in the *spinal cord* and ends in the *thalamus* (a region of the brain). Notice that the location of the axon terminals comes last in the name. This regularity in naming allows you to determine the direction of information flow along any tract identified according to this convention. Thus, because the anterior spinothalamic tract conveys nerve impulses from the spinal cord toward the brain, it is a sensory (ascending) tract. Figure 17.11 highlights the major sensory and motor tracts in the spinal cord. These tracts are described in detail in Chapter 20 and summarized in Tables 20.3 and 20.4.

Figure 17.11 The locations of selected sensory and motor tracts, shown in a transverse section of the spinal cord. Sensory tracts are indicated on one half and motor tracts on the other half of the cord, but in fact all tracts are present on both sides.

🔑 The name of a tract often indicates its location in the white matter and where it begins and ends.

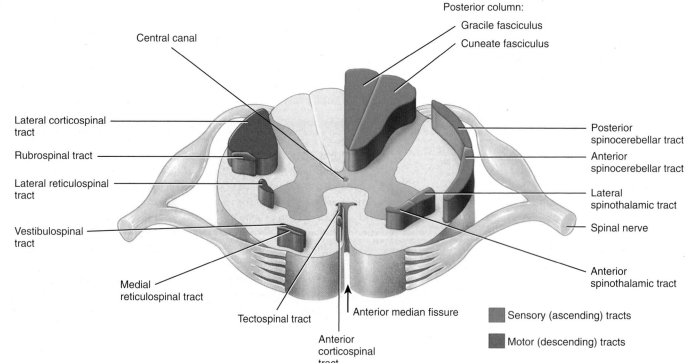

 Based on its name, what are the position in the spinal cord, origin, and destination of the anterior corticospinal tract? Is this a sensory or a motor tract?

Nerve impulses from sensory receptors propagate up the spinal cord to the brain along two main routes on each side: the spinothalamic tracts and the posterior columns. The **lateral** and **anterior spinothalamic tracts** (spī′-nō-tha-LAM-ik) convey nerve impulses for sensing pain, warmth, coolness, itching, tickling, and deep pressure, and a crude, poorly localized sense of touch. The right and left **posterior columns** carry nerve impulses for several types of sensations. These include (1) *proprioception*, awareness of the positions and movements of muscles, tendons, and joints; (2) *discriminative touch*, the ability to feel exactly what part of the body is touched; (3) *two-point discrimination*, the ability to distinguish the touching of two different points on the skin, even though they are close together; (4) *light pressure sensations*; and (5) *vibration sensations*.

The sensory systems keep the CNS informed of changes in the external and internal environments. Responses to this information are brought about by motor systems, which allow you to move about and change your physical relationship to the world around you. As sensory information is conveyed to the CNS, it becomes part of a large pool of sensory input. Each piece of incoming information is integrated with all of the other information arriving from activated sensory neurons.

Through the activity of interneurons, integration occurs in several regions of the spinal cord and brain. As a result, motor impulses to make a muscle contract or a gland secrete can be initiated at several levels. Most regulation of involuntary activities of smooth muscle, cardiac muscle, and glands by the autonomic nervous system (ANS) originates in the brain stem (the lower part of the brain that is continuous with the spinal cord) and in a nearby brain region called the hypothalamus.

The cerebral cortex (superficial gray matter of the cerebrum) plays a major role in controlling precise, voluntary muscular movements. Other brain regions integrate automatic movements, such as arm swinging during walking. Motor output to skeletal muscles travels down the spinal cord in two types of descending pathways: direct and indirect. The **direct pathways** include the **lateral corticospinal tract** (kor′-ti-kō-SPĪ-nal), the **anterior corticospinal tract**, and the **corticobulbar tract** (kor′-ti-kō-BUL-bar). Each of these tracts conveys nerve impulses that originate in the cerebral cortex and are destined to cause precise, *voluntary* movements of skeletal muscles. **Indirect pathways** include the **rubrospinal tract** (ROO-brō-spī-nal), **tectospinal tract** (TEK-tō-spī-nal), **vestibulospinal tract** (ves-TIB-ū-lō-spī-nal), **lateral reticulospinal tract** (re-TIK-ū-lō-spī-nal), and **medial reticulospinal tract**. These tracts convey nerve impulses from the brain stem and other parts of the brain that govern *automatic movements* and help coordinate body movements with visual stimuli. Indirect pathways also maintain skeletal muscle tone, sustain contraction of postural muscles, and play a major role in equilibrium by regulating muscle tone in response to movements of the head.

Reflexes and Reflex Arcs

The spinal cord also promotes homeostasis by serving as an integrating center for some reflexes. A **reflex** is a fast, involuntary, unplanned sequence of actions that occurs in response to a particular stimulus. Some reflexes are inborn, such as pulling your hand away from a hot surface before you even feel the heat. Other reflexes are learned or acquired. For instance, you learn many reflexes while acquiring driving expertise. Slamming on the brakes in an emergency is one example. When integration takes place in the spinal cord gray matter, the reflex is a **spinal reflex**. An example is the familiar patellar reflex (knee jerk). By contrast, if integration occurs in the brain stem rather than the

spinal cord, the reflex is a **cranial reflex**. An example is the tracking movements of your eyes as you read this sentence. You are probably most aware of **somatic reflexes**, which involve contraction of skeletal muscles. Equally important, however, are the **autonomic (visceral) reflexes**, which generally are not consciously perceived. These reflexes involve responses of smooth muscle, cardiac muscle, and glands. As you will see in Chapter 19, body functions such as heart rate, digestion, urination, and defecation are controlled by the autonomic nervous system through autonomic reflexes.

Nerve impulses propagating into, through, and out of the CNS follow specific pathways, depending on the type of information, its origin, and its destination. The pathway followed by nerve impulses that produce a reflex is a **reflex arc** (*reflex circuit*). Using the **patellar reflex** (knee jerk) as an example, the basic components of a reflex arc are as follows (Figure 17.12):

❶ **Sensory receptor.** The distal end of a sensory neuron (dendrite) or an associated sensory structure serves as a sensory receptor. Sensory receptors respond to a specific type of *stimulus* (a change in the internal or external environment) by generating one or more nerve impulses. In the patellar reflex, sensory receptors known as *muscle spindles* detect slight stretching of the quadriceps femoris muscle (anterior thigh) when the patellar (knee cap) ligament is tapped with a reflex hammer.

❷ **Sensory neuron.** The nerve impulses conduct from the sensory receptor along the axon of a sensory neuron to its axon terminals, which are located in the CNS gray matter. From here, relay neurons send nerve impulses to the area of the brain that allows conscious awareness that the reflex has occurred.

❸ **Integrating center.** One or more regions of gray matter in the CNS act as an integrating center. In the simplest type of reflex, such as the patellar reflex in our example, the integrating center is a single synapse between a sensory neuron and a motor neuron in the spinal cord. A reflex pathway in the CNS that involves one synapse is called a *monosynaptic reflex arc* (mon′-ō-sī=NAP-tik; *mono-*=one). In other types of reflexes, the integrating center includes one or more interneurons and thus more than one synapse. These reflex pathways are referred to as *polysynaptic reflex arcs* (*poly-*=many).

❹ **Motor neuron.** Impulses triggered by the integrating center pass out of the spinal cord (or brain stem, in the case of a cranial reflex) along a motor neuron to the part of the body that will respond. In the patellar reflex, the axon of the motor neuron extends to the quadriceps femoris muscle. While the quadriceps femoris muscle is contracting, the antagonist hamstring muscles are relaxed.

❺ **Effector.** The part of the body that responds to the motor nerve impulse, such as a muscle or gland, is the effector. The patellar reflex is a *somatic reflex* because its effector is a skeletal muscle, the quadriceps femoris muscle, which contracts and thereby relieves the stretching that initiated the reflex. In sum, the patellar reflex causes extension of the knee by contraction of the quadriceps femoris muscle in response to tapping the patellar ligament. If the effector is smooth muscle, cardiac muscle, or a gland, the reflex is an *autonomic (visceral) reflex*.

✔ CHECKPOINT

17. What are the functions of the anterior spinothalamic tract and the posterior columns?

18. Describe the components of the patellar reflex arc.

19. Why are reflexes important clinically?

Figure 17.12 Patellar reflex showing the general components of a reflex arc. The arrows show the direction of nerve impulse propagation.

🔑 Reflexes are fast, involuntary responses to particular stimuli.

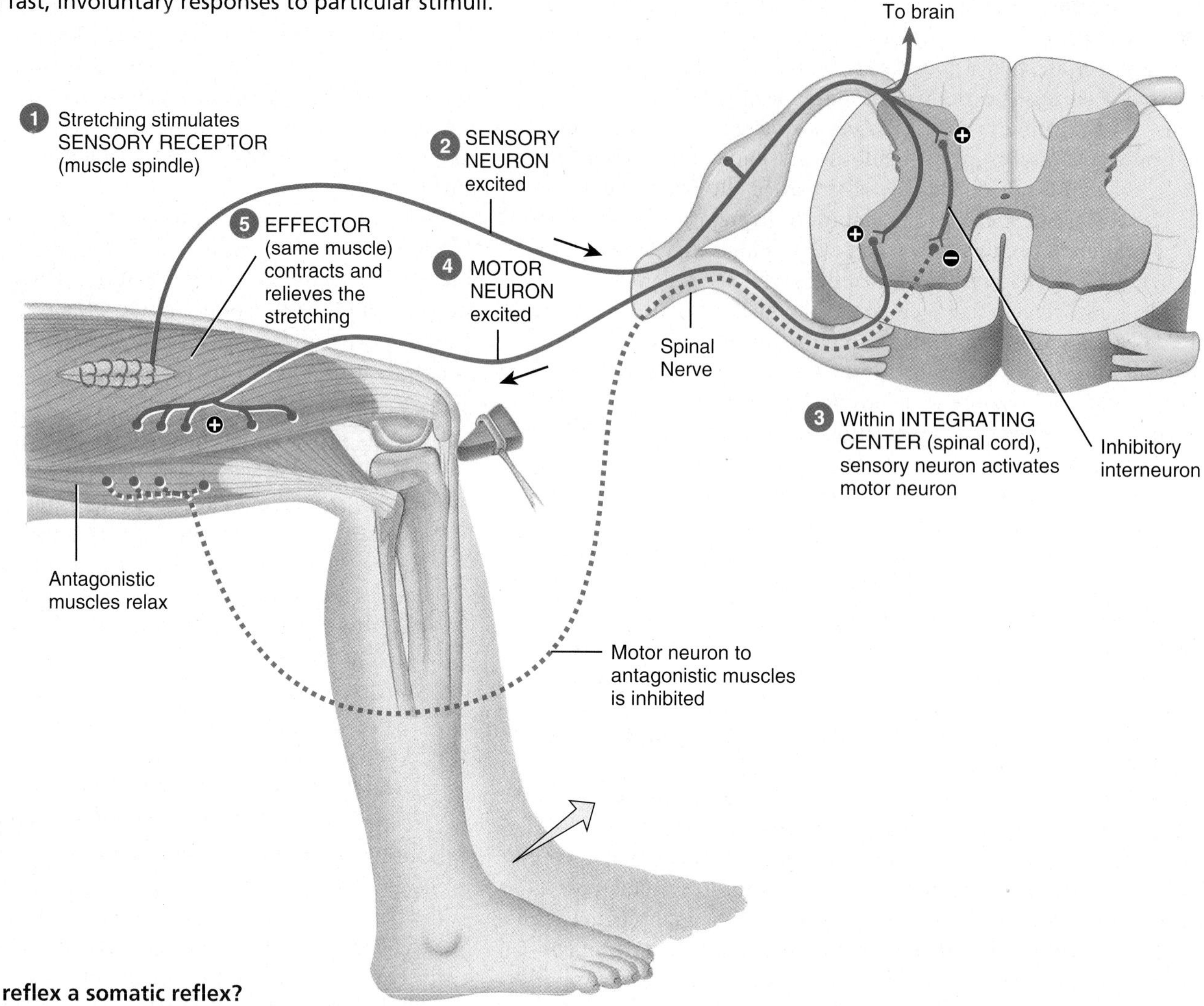

1 Stretching stimulates
SENSORY RECEPTOR
(muscle spindle)

2 SENSORY
NEURON
excited

5 EFFECTOR
(same muscle)
contracts and
relieves the
stretching

4 MOTOR
NEURON
excited

Spinal
Nerve

3 Within INTEGRATING
CENTER (spinal cord),
sensory neuron activates
motor neuron

Inhibitory
interneuron

To brain

Antagonistic
muscles relax

Motor neuron to
antagonistic muscles
is inhibited

❓ **Why is this reflex a somatic reflex?**

KEY MEDICAL TERMS ASSOCIATED WITH THE SPINAL CORD AND THE SPINAL NERVES

Epidural block (ep'-i-DOO-ral) Injection of an anesthetic drug into the epidural space, the space between the dura mater and the vertebral column, in order to cause a temporary loss of sensation. Such injections in the lower lumbar region are used to control pain during childbirth.

Meningitis (men-in-JĪ-tis; *-itis*=inflammation) Inflammation of the meninges due to an infection, usually caused by a bacterium or virus. Symptoms include fever, headache, stiff neck, vomiting, confusion, lethargy, and drowsiness. Bacterial meningitis is much more serious and is treated with antibiotics. Viral meningitis has no specific treatment. Bacterial meningitis may be fatal if not treated promptly; viral meningitis usually resolves on its own in 1–2 weeks. A vaccine is available to help protect against some types of bacterial meningitis.

Myelitis (mī-ē-LĪ-tis; *myel-*=spinal cord) Inflammation of the spinal cord.
Nerve block Loss of sensation in a region due to injection of a local anesthetic; an example is local dental anesthesia.
Neuralgia (noo-RAL-jē-a; *neur-*=nerve; *-algia*=pain) Attacks of pain along the entire course or a branch of a sensory nerve.
Neuritis (*-itis*=inflammation) Inflammation of one or several nerves that may result from irritation to the nerve produced by direct blows, bone fractures, contusions, or penetrating injuries. Additional causes include infections, vitamin deficiency (usually thiamine), and poisons such as carbon monoxide, carbon tetrachloride, heavy metals, and some drugs.
Paresthesia (par-es-THĒ-zē-a; *par-*=departure from normal; *-esthesia*=sensation) An abnormal sensation such as burning, pricking, tickling, or tingling resulting from a disorder of a sensory nerve.

CHAPTER REVIEW AND RESOURCE SUMMARY

WileyPLUS

Review

Resource

17.1 Spinal Cord Anatomy

1. The spinal cord is protected by the vertebral column, the meninges, cerebrospinal fluid, and denticulate ligaments.

Anatomy Overview - The
Spinal Cord
Figure 17.1 - Gross Anatomy
of the Spinal Cord

Review	**Resource**

2. The three meninges are coverings that run continuously around the spinal cord and brain. They are the dura mater, arachnoid mater, and pia mater.

3. The spinal cord begins as a continuation of the medulla oblongata and ends at about the second lumbar vertebra in an adult. It contains cervical and lumbar enlargements that serve as points of origin for nerves to the limbs.

4. The tapered inferior portion of the spinal cord is the conus medullaris, from which arise the filum terminale and cauda equina.

5. Spinal nerves connect to each segment of the spinal cord by two roots. The posterior or dorsal root contains sensory axons, and the anterior or ventral root contains motor neuron axons.

6. The anterior median fissure and the posterior median sulcus partially divide the spinal cord into right and left sides.

7. The gray matter in the spinal cord is divided into horns, and the white matter into columns. In the center of the spinal cord is the central canal, which runs the length of the spinal cord and is filled with cerebrospinal fluid.

8. Parts of the spinal cord observed in transverse section are the gray commissure; central canal; anterior, posterior, and lateral gray horns; and anterior, posterior, and lateral white columns, which contain ascending and descending tracts. Each part has specific functions.

9. The spinal cord conveys sensory and motor information by way of ascending and descending tracts, respectively.

Resource: Figure 17.2 - External Anatomy of the Spinal Cord and Spinal Nerves

17.2 Spinal Nerves

1. The 31 pairs of spinal nerves are named and numbered according to the region and level of the spinal cord from which they emerge. There are 8 pairs of cervical, 12 pairs of thoracic, 5 pairs of lumbar, 5 pairs of sacral, and 1 pair of coccygeal nerves.

2. Three connective tissue coverings associated with spinal nerves are the endoneurium, perineurium, and epineurium. Spinal nerves typically are connected with the spinal cord by a posterior root and an anterior root. All spinal nerves contain both sensory and motor axons (are mixed nerves).

3. Branches of a spinal nerve include the posterior ramus, anterior ramus, meningeal branch, and rami communicantes. The anterior rami of spinal nerves, except for T2–T12, form networks of nerves called plexuses. Emerging from the plexuses are nerves bearing names that typically describe the general regions they supply or the route they follow.

4. Nerves of the cervical plexus supply the skin and muscles of the head, neck, and upper part of the shoulders; they connect with some cranial nerves and innervate the diaphragm.

5. Nerves of the brachial plexus supply the upper limbs and several neck and shoulder muscles.

6. Nerves of the lumbar plexus supply the anterolateral abdominal wall, external genitals, and part of the lower limbs.

7. Nerves of the sacral plexus supply the buttocks, perineum, and part of the lower limbs.

8. Nerves of the coccygeal plexus supply the skin of the coccygeal region.

9. Anterior rami of nerves T2–T12 do not form plexuses and are called intercostal (thoracic) nerves. They are distributed directly to the structures they supply in intercostal spaces.

10. Sensory neurons within spinal nerves and the trigeminal (V) nerve serve specific, constant segments of the skin called dermatomes. Knowledge of dermatomes helps a physician determine which segment of the spinal cord or which spinal nerve is damaged. Cutaneous fields are regions of skin supplied by a specific nerve arising from a plexus.

Resource: Anatomy Overview - Spinal Nerves

17.3 Spinal Cord Functions

1. The white matter tracts in the spinal cord are highways for nerve impulse propagation. Along these tracts, sensory input travels toward the brain, and motor output travels from the brain toward skeletal muscles and other effector tissues.

2. Sensory input travels along two main routes in the white matter of the spinal cord: the posterior columns and the spinothalamic tracts.

3. Motor output travels along two main routes in the white matter of the spinal cord: direct pathways and indirect pathways.

4. A second major function of the spinal cord is to serve as an integrating center for spinal reflexes. This integration occurs in the gray matter.

5. A reflex is a fast, predictable sequence of involuntary actions, such as muscle contractions or glandular secretions, which occurs in response to certain changes in the environment.

6. Reflexes may be spinal or cranial and somatic or autonomic (visceral).

7. The components of a reflex arc are sensory receptor, sensory neuron, integrating center, motor neuron, and effector.

Resource: Animation - Somatic Sensory and Motor Pathways
Animation - Reflex Arcs
Animation - Reflexes
Figure 17.11 - The Locations of Selected Sensory and Motor Tracts
Exercise - Assemble an Arc
Exercise - Stretch Reflex
Figure 17.12 - Patellar Reflex Showing the General Components of a Reflex Arc

CRITICAL THINKING QUESTIONS

1. A high school senior dove headfirst into a murky pond. Tragically, his head hit a submerged log and he is now paralyzed from the neck down (quadriplegia). Can you deduce the location of the injury? What is the likelihood of recovery from his injury?

2. The spinal cord is covered in protective layers and enclosed by the vertebral column. How is it able to send and receive messages from the periphery of the body?

3. Why doesn't the spinal cord "creep" up toward the head every time you bend over? Why doesn't it get all twisted out of position when you exercise?

4. Nischal slipped on the ice and broke his "tailbone." However, he did no damage to his spinal cord. How is it possible that he broke part of his vertebral column without damaging the spinal cord?

5. Not long after Nischal fell on the ice, he began experiencing pain in his lower back and down his leg, as well as some numbness in his foot. When he walks, he has trouble controlling the flexion of his foot. He told his friend that "the doctor said something about compression of the lumbar vertebrae." What has probably happened to Nischal?

? ANSWERS TO FIGURE QUESTIONS

17.1 The superior boundary of the spinal dura mater is the foramen magnum of the occipital bone. The inferior boundary is the second sacral vertebra.

17.2 The cervical enlargement of the spinal cord connects with sensory and motor nerves of the upper limbs.

17.3 A horn is an area of gray matter, and a column is a region of white matter in the spinal cord.

17.4 All spinal nerves are mixed because they contain a posterior root with sensory axons and an anterior root with motor axons.

17.5 The anterior rami serve the upper and lower limbs.

17.6 Severing the spinal cord at level C2 causes respiratory arrest because it prevents descending nerve impulses from reaching the phrenic nerve, which stimulates contraction of the diaphragm, the main muscle needed for breathing.

17.7 The axillary, musculocutaneous, radial, median, and ulnar nerves are five important nerves that arise from the brachial plexus.

17.8 The lumbar plexus originates from the roots of spinal nerves L1–L4.

17.9 The origin of the sacral plexus is the anterior rami of spinal nerves L4–L5 and S1–S4.

17.10 The only spinal nerve without a corresponding dermatome is C1.

17.11 The anterior corticospinal tract is located on the anterior side of the spinal cord, originates in the cortex of the cerebrum, and ends in the spinal cord. It contains descending axons and thus is a motor tract.

17.12 It is a somatic reflex because the effector is a skeletal muscle.

18 THE BRAIN AND THE CRANIAL NERVES

INTRODUCTION Solving an equation, feeling hungry, laughing—the neural processes needed for each of these activities occur in different regions of the **brain**, that portion of the central nervous system contained within the cranium. About 85 billion neurons and 10–50 trillion neuroglia make up the brain, which has a mass of about 1300 grams (almost 3 lb) in adults. On average, each neuron forms 1000 synapses with other neurons. Thus, the total number of synapses in each human brain, about a thousand trillion (10^{15}), is larger than the number of stars in the galaxy.

The brain is the center for registering sensations, correlating them with one another and with stored information, making decisions, and taking actions. It is also the center for the intellect, emotions, behavior, and memory. But this fascinating organ encompasses an even larger domain: It directs our behavior toward others. With ideas that excite, artistry that dazzles, or rhetoric that mesmerizes, one person's thoughts and actions may influence and shape the lives of many others. As you will see shortly, different regions of the brain are specialized for different functions, but can also work together to accomplish certain shared tasks. This chapter explores how the brain is protected and nourished, what functions occur in the major regions of the brain, and how the spinal cord and the 12 pairs of cranial nerves connect with the brain to form the control center of the human body. •

Did you ever wonder how cerebrovascular accidents (strokes) occur and how they are treated?

18.1 DEVELOPMENT AND GENERAL STRUCTURE OF THE BRAIN

OBJECTIVES

- Describe how the brain develops and relates to the different parts of the postnatal brain.
- Identify the major parts of the brain.

Brain Development

In order to understand the terminology used for the principal parts of the adult brain, it will be helpful to know how the brain develops. Recall from Chapter 4 that the brain and spinal cord develop from the ectodermal **neural tube** (see Figure 4.9). The anterior part of the neural tube expands, along with the associated neural crest tissue. Constrictions in this expanded tube soon appear, creating three regions called **primary brain vesicles:** *prosencephalon, mesencephalon,* and *rhombencephalon* (Figure 18.1). Both the prosencephalon and rhombencephalon subdivide further, forming **secondary brain vesicles.** The *prosencephalon* (PRŌS-en-sef′-a-lon), or forebrain, gives rise to the telencephalon and diencephalon, and the *rhombencephalon* (ROM-ben-sef′-a-lon), or hindbrain, develops into the metencephalon and myelen-

cephalon. The various brain vesicles give rise to the following adult structures:

- The **telencephalon** develops into the *cerebrum* and *lateral ventricles.*
- The **diencephalon** (di-en-SEF-a-lon; *di-*=through; *-encephalon*=brain) forms the *thalamus, hypothalamus, epithalamus,* and the *third ventricle.*
- The **mesencephalon** (MEZ-en-sef′-a-lon) or midbrain gives rise to the *midbrain* and *aqueduct of the midbrain (cerebral aqueduct).*
- The **metencephalon** becomes the *pons, cerebellum,* and *upper part of the fourth ventricle.*
- The **myelencephalon** forms the *medulla oblongata* and *lower part of the fourth ventricle.*

The various parts of the brain will be described shortly. The walls of these brain regions develop into the nervous tissue of the brain, while the hollow interior of the tube is transformed into the various ventricles (fluid-filled spaces) of the brain. The expanded neural crest tissue becomes prominent in head development. Most of the protective structures of the brain, that is, most of the bones of the skull, associated connective tissues, and meningeal membranes arise from this expanded neural crest tissue.

These relationships are summarized in Table 18.1.

Figure 18.1   Development of the brain and spinal cord.

 The various parts of the brain develop from the primary brain vesicles.

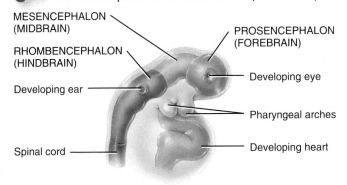

MESENCEPHALON (MIDBRAIN)

RHOMBENCEPHALON (HINDBRAIN)

Developing ear

Spinal cord

PROSENCEPHALON (FOREBRAIN)

Developing eye

Pharyngeal arches

Developing heart

Lateral view of right side

(a) Three–four week embryo showing primary brain vesicles

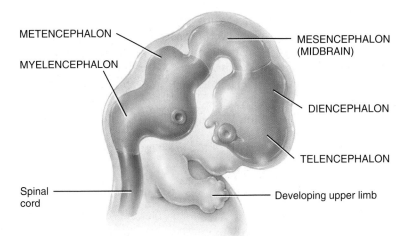

METENCEPHALON

MYELENCEPHALON

MESENCEPHALON (MIDBRAIN)

DIENCEPHALON

TELENCEPHALON

Spinal cord

Developing upper limb

(b) Seven-week embryo showing secondary brain vesicles

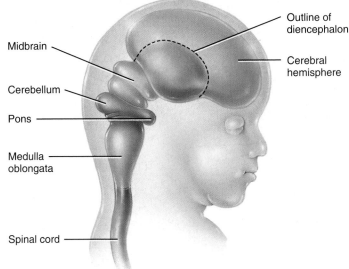

Midbrain

Cerebellum

Pons

Medulla oblongata

Spinal cord

Outline of diencephalon

Cerebral hemisphere

(c) Eleven-week fetus showing expanding cerebral hemispheres overgrowing the diencephalon

Cerebral hemisphere

Diencephalon

Cerebellum

Brainstem: (covered by cerebrum) Midbrain Pons

Medulla oblongata

Spinal cord

(d) Brain at birth (the diencephalon and superior portion of the brain stem have been projected to the surface)

?   Which primary brain vesicle does not develop into a secondary brain vesicle?

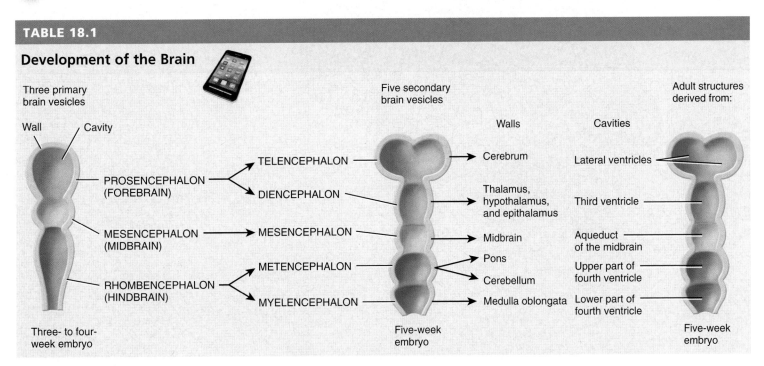

TABLE 18.1

Development of the Brain

Three primary brain vesicles	Five secondary brain vesicles		Adult structures derived from:
		Walls	Cavities
PROSENCEPHALON (FOREBRAIN)	TELENCEPHALON	Cerebrum	Lateral ventricles
	DIENCEPHALON	Thalamus, hypothalamus, and epithalamus	Third ventricle
MESENCEPHALON (MIDBRAIN)	MESENCEPHALON	Midbrain	Aqueduct of the midbrain
RHOMBENCEPHALON (HINDBRAIN)	METENCEPHALON	Pons Cerebellum	Upper part of fourth ventricle
	MYELENCEPHALON	Medulla oblongata	Lower part of fourth ventricle

Wall   Cavity

Three- to four-week embryo

Five-week embryo

Five-week embryo

Major Parts of the Brain

The adult brain consists of four major parts: brain stem, cerebellum, diencephalon, and cerebrum (Figure 18.2). The **brain stem** is continuous with the spinal cord and consists of the medulla oblongata, pons, and midbrain. Posterior to the brain stem is the **cerebellum** (ser'-e-BEL-um=little brain). Superior to the brain stem is the **diencephalon**, which as noted previously consists of the thalamus, hypothalamus, and epithalamus. Supported on the diencephalon and brain stem is the **cerebrum** (se-RĒ-brum=brain), the largest part of the brain.

✔ CHECKPOINT

1. What parts of the brain develop from each primary brain vesicle?
2. Compare the sizes and locations of the cerebrum and cerebellum.

Figure 18.2 The brain. The pituitary gland is discussed with the endocrine system in Chapter 22.

 The four principal parts of the brain are the brain stem, cerebellum, diencephalon, and cerebrum.

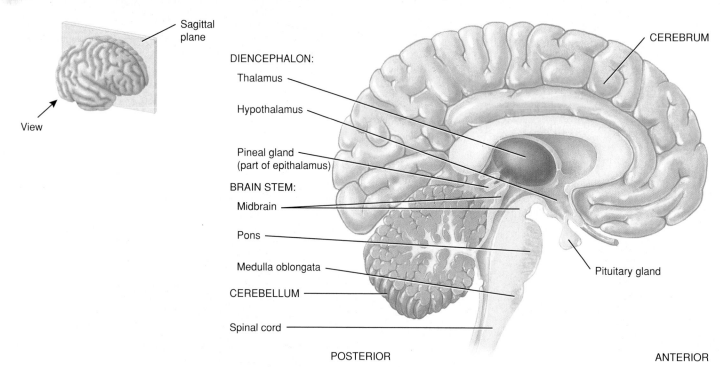

(a) Sagittal section, medial view

(b) Sagittal section, medial view

❓ **Which part of the brain is the largest?**

18.2 PROTECTION AND BLOOD SUPPLY

OBJECTIVES

• Explain how the brain is protected.
• Describe the formation and circulation of cerebrospinal fluid.
• Outline the blood supply of the brain.

Protective Coverings of the Brain

The cranium and the cranial meninges surround and protect the brain. The **cranial meninges** (me-NIN-jēz) are continuous with the spinal meninges you learned about in the last chapter.

They have the same basic structure, and bear the same names: the outer **dura mater** (DOO-ra-MĀ-ter), the middle **arachnoid mater** (a-RAK-noyd), and the inner **pia mater** (PĪ-a or PĒ-a) (Figure 18.3). Note that the cranial dura mater has two layers, compared to the single layer of the spinal dura mater. The external layer is called the *periosteal layer* and the internal layer is called the *meningeal layer*. The two dural layers around the brain are fused together except where they separate to enclose the dural venous sinuses (endothelial-lined venous channels). These sinuses drain venous blood from the brain and deliver it into the internal jugular veins. The *epidural space* is a potential space between the periosteal layer of the dura mater and skull bones. Blood vessels that enter brain tissue pass along the surface of the brain; as they penetrate inward, the vessels become sheathed by a loose-fitting sleeve of pia mater. Three extensions of the dura mater separate parts of the brain. (1) The **falx cerebri** (FALKS SER-e-brī;

Figure 18.3 The protective coverings of the brain.

🔑 Cranial bones and the cranial meninges protect the brain.

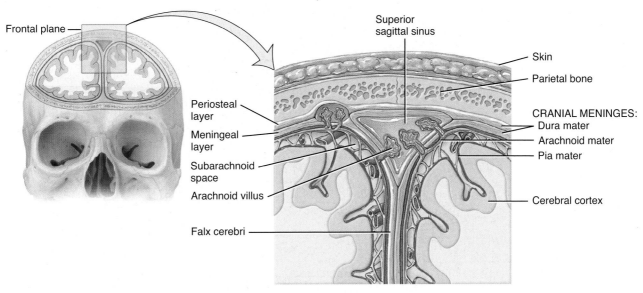

(a) Anterior view of frontal section through skull showing the cranial meninges

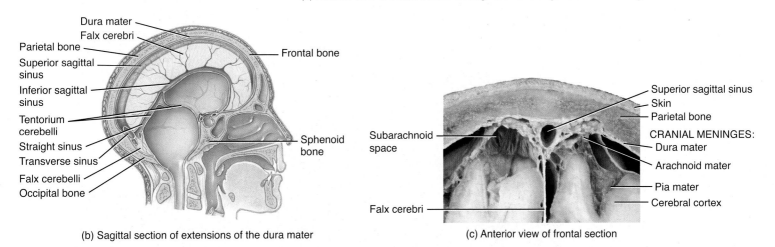

(b) Sagittal section of extensions of the dura mater

(c) Anterior view of frontal section

CLINICAL CONNECTION | *Meningitis*

Meningitis (men-in-JĪ-tis; *-itis*=inflammation) is an inflammation of the meninges due to an infection, usually caused by a bacterium or virus. Symptoms include fever, headache, stiff neck, vomiting, confusion, lethargy, and drowsiness. Bacterial meningitis is much more serious and is treated with antibiotics. Bacterial meningitis may be fatal if not treated promptly; viral meningitis usually resolves on its own in 1–2 weeks. A vaccine is available to help protect against some types of bacterial meningitis. Viral meningitis has no specific treatment. •

❓ What are the three layers of the cranial meninges, from superficial to deep?

falx=sickle-shaped) separates the two hemispheres (sides) of the cerebrum. (2) The **falx cerebelli** (ser-e-BEL-ī) separates the two hemispheres of the cerebellum. (3) The **tentorium cerebelli** (ten-TŌ-rē-um=tent) separates the cerebrum from the cerebellum.

Cerebrospinal Fluid

Cerebrospinal fluid (CSF) is a clear, colorless liquid comprised primarily of water that protects the brain and spinal cord against chemical and physical injuries. It also carries small amounts of oxygen, glucose, and other needed chemicals from the blood to neurons and neuroglia. CSF circulates slowly and continuously through cavities in the brain and spinal cord and around the brain and spinal cord in the *subarachnoid space* (space between the arachnoid mater and pia mater). The total volume of CSF is 80 to 150 mL (3 to 5 oz) in an adult. CSF also contains small amounts of proteins, lactic acid, urea, cations (Na^+, K^+, Ca^{2+}, Mg^{2+}), anions (Cl^- and HCO_3^-), and some white blood cells.

Formation of CSF in the Ventricles

Figure 18.4 shows the four CSF-filled cavities within the brain, which are called **ventricles** (VEN-tri-kuls=little cavities). There

Figure 18.4 Locations of ventricles within a "transparent" brain. The lateral ventricles connect by interventricular foramina to the third ventricle, and the aqueduct of the midbrain connects the third ventricle to the fourth ventricle.

 Ventricles are cavities within the brain that are filled with cerebrospinal fluid.

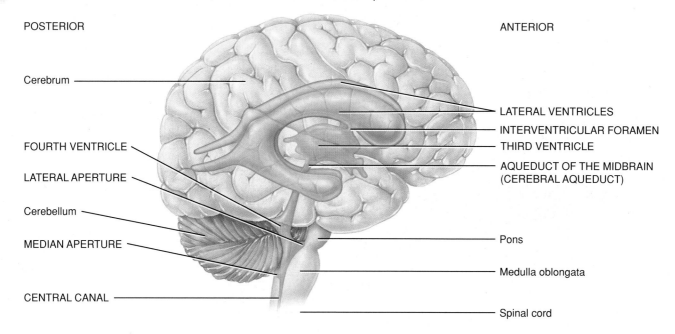

(a) Right lateral view of brain

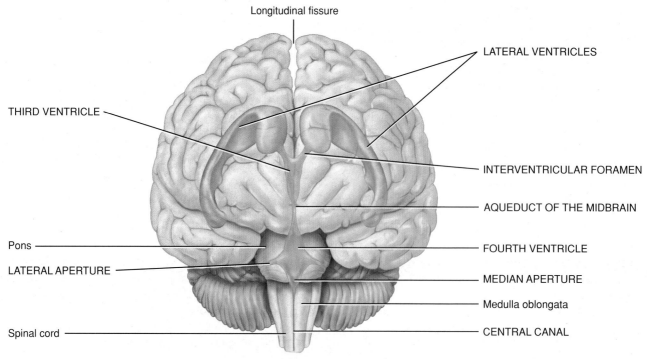

(b) Anterior view of brain

Which brain region is anterior to the fourth ventricle? Which is posterior to it?

is one **lateral ventricle** located in each hemisphere of the cerebrum (think of them as ventricles 1 and 2). Anteriorly, the lateral ventricles are separated by a thin membrane, the **septum pellucidum** (SEP-tum pe-LOO-si-dum; *pellucid*=transparent; see Figure 18.5a). The **third ventricle** is a narrow, slit-like cavity along the midline superior to the hypothalamus and between the right and left halves of the thalamus. The **fourth ventricle** lies between the brain stem and the cerebellum.

The majority of CSF production is from the **choroid plexuses** (KŌ-royd=membrane-like), networks of modified blood capillaries in the walls of the ventricles (Figure 18.5a). Ependymal cells joined by tight junctions cover the capillaries of the choroid plexuses. Selected substances from the blood plasma (primarily water) are filtered from the capillaries and secreted by the ependymal cells to produce CSF. Because of the tight junctions between ependymal cells, materials entering CSF from choroid capillaries cannot leak between these cells; instead, they must pass through the ependymal cells. This **blood–cerebrospinal fluid barrier** permits certain substances to enter the CSF but excludes others, protecting the brain and spinal cord from potentially harmful blood-borne substances. In contrast, the blood–brain barrier (described shortly) is formed mainly by tight junctions of brain capillary endothelial cells rather than ependymal cells.

Functions of CSF

The CSF functions in three main ways:

1. ***Mechanical protection.*** The primary function of the CSF is to serve as a shock-absorbing medium. It protects the delicate tissues of the brain and spinal cord from jolts that would otherwise cause them to hit the bony walls of the cranial cavity and vertebral canal. This important fluid also buoys the brain so that it "floats" in the cranial cavity and reduces its weight within the skull to approximately 50 grams (0.1 lb).

2. ***Chemical protection.*** CSF provides an optimal chemical environment for efficient neuronal signaling. Even slight changes in the ionic composition of CSF within the brain can seriously disrupt production of action potentials.

3. ***Circulation.*** CSF is a medium for the minor exchange of nutrients and waste products between the blood and adjacent nervous tissue. The subarachnoid space through which CSF flows is continuous with the *perivascular spaces* (spaces around the blood vessels that penetrate the brain tissue); together, the CSF and these spaces provide a lymphatic function for the tissue of the brain. See the following section for more on the circulation of CSF.

Circulation of CSF

The CSF formed in the choroid plexuses of each lateral ventricle flows into the third ventricle through two narrow, oval openings, the **interventricular foramina** (in′-ter-ven-TRIK-ū-lar) (Figure 18.5b). More CSF is added by the choroid plexus in the roof of the third ventricle. The fluid then flows through the **aqueduct of the midbrain** (*cerebral aqueduct*) (AK-we-dukt), which passes through the midbrain, into the fourth ventricle. The choroid plexus of the fourth ventricle contributes more fluid. From here, a small amount of CSF passes downward into the central canal of the spinal cord, while the majority of the CSF enters the subarachnoid space through three openings in

Figure 18.5 Pathways of circulating cerebrospinal fluid.

🔑 CSF is formed from blood plasma by ependymal cells that cover the choroid plexuses of the ventricles.

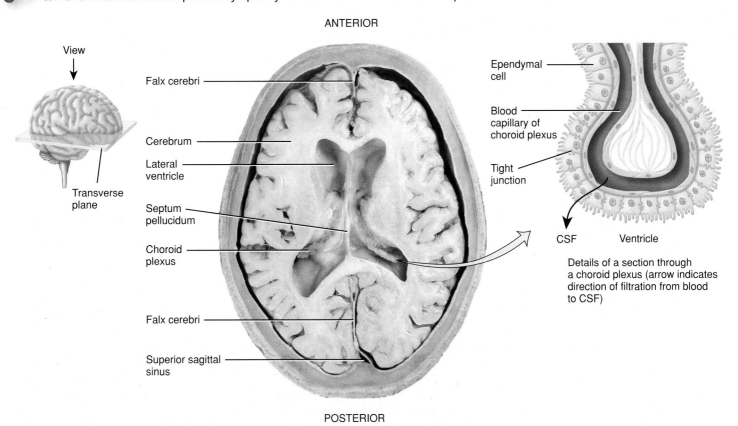

(a) Superior view of transverse section of brain showing choroid plexuses

FIGURE 18.5 CONTINUES ▶

FIGURE 18.5 CONTINUED

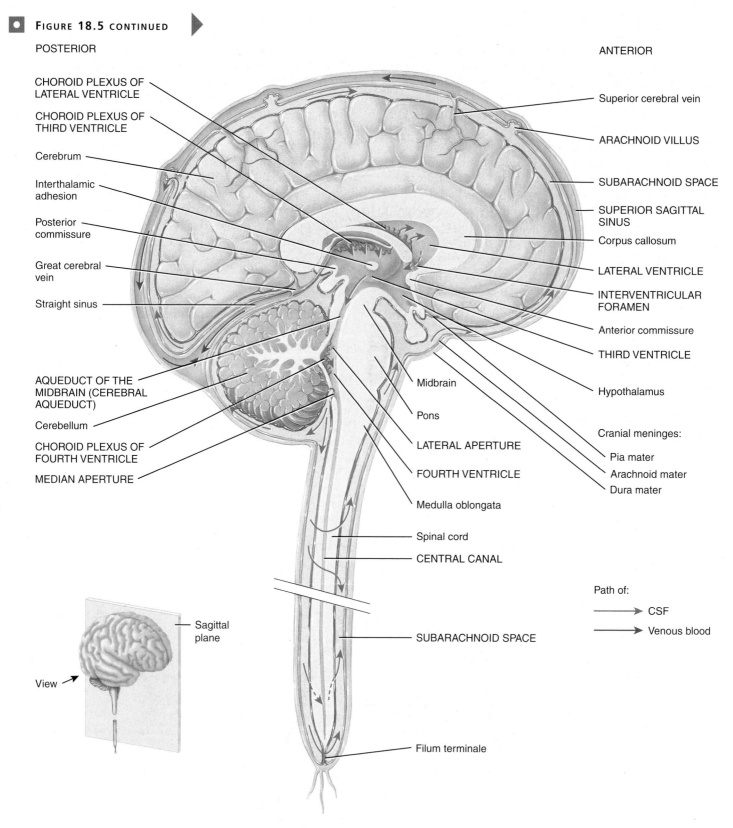

(b) Sagittal section of brain and spinal cord

the roof of the fourth ventricle: a single **median aperture** (AP-er-chur), which exits posteriorly, and paired **lateral apertures**, one on each side. The CSF then circulates in the subarachnoid space around the surface of the brain and spinal cord.

CSF is gradually reabsorbed into the blood through **arachnoid villi**, fingerlike extensions of the arachnoid that project into the dural venous sinuses, especially the **superior sagittal sinus** (see also Figure 18.3). (A *dural venous sinus* is an endothelial-lined

space in the thick-walled dura mater.) Unlike most veins, these cannot collapse because the endothelium adheres to the rigid wall of the dura mater. This ensures more efficient venous drainage of the brain tissue. Normally, CSF is reabsorbed as rapidly as it is formed by the choroid plexuses, at a rate of about 20 mL/hr (480 mL/day). Because the rates of formation and reabsorption are the same, the pressure of CSF is normally constant, and the volume of CSF remains constant.

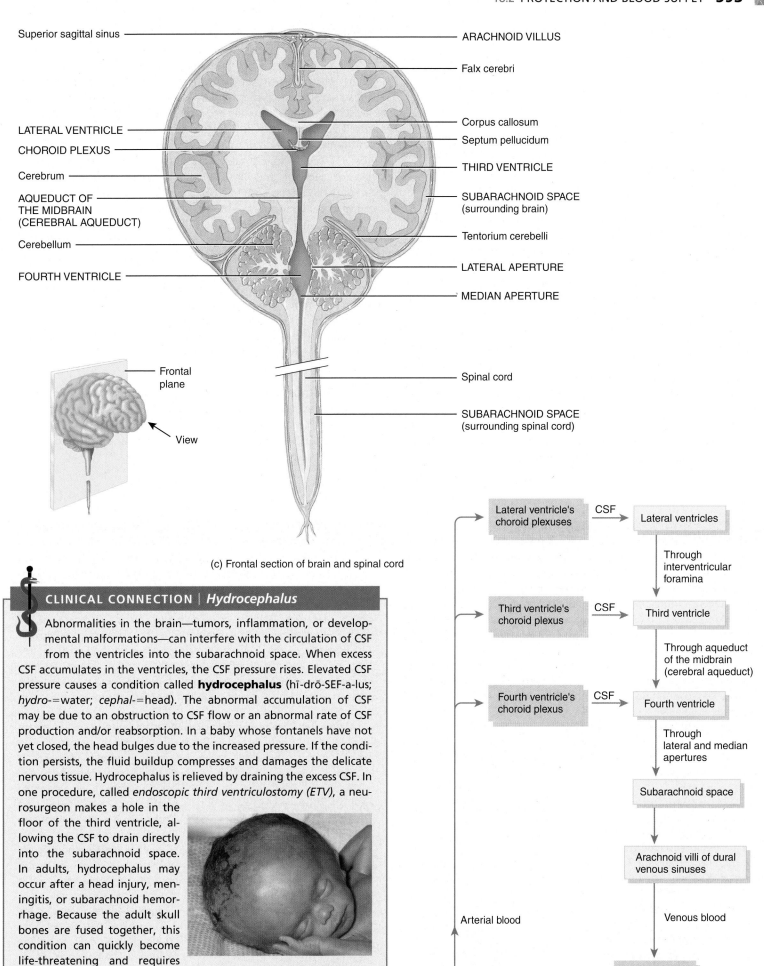

Superior sagittal sinus

ARACHNOID VILLUS

Falx cerebri

LATERAL VENTRICLE

CHOROID PLEXUS

Corpus callosum

Septum pellucidum

THIRD VENTRICLE

Cerebrum

AQUEDUCT OF
THE MIDBRAIN
(CEREBRAL AQUEDUCT)

SUBARACHNOID SPACE
(surrounding brain)

Tentorium cerebelli

Cerebellum

FOURTH VENTRICLE

LATERAL APERTURE

MEDIAN APERTURE

Frontal
plane

View

Spinal cord

SUBARACHNOID SPACE
(surrounding spinal cord)

(c) Frontal section of brain and spinal cord

CLINICAL CONNECTION | *Hydrocephalus*

Abnormalities in the brain—tumors, inflammation, or developmental malformations—can interfere with the circulation of CSF from the ventricles into the subarachnoid space. When excess CSF accumulates in the ventricles, the CSF pressure rises. Elevated CSF pressure causes a condition called **hydrocephalus** (hī-drō-SEF-a-lus; *hydro-*=water; *cephal-*=head). The abnormal accumulation of CSF may be due to an obstruction to CSF flow or an abnormal rate of CSF production and/or reabsorption. In a baby whose fontanels have not yet closed, the head bulges due to the increased pressure. If the condition persists, the fluid buildup compresses and damages the delicate nervous tissue. Hydrocephalus is relieved by draining the excess CSF. In one procedure, called *endoscopic third ventriculostomy (ETV)*, a neurosurgeon makes a hole in the floor of the third ventricle, allowing the CSF to drain directly into the subarachnoid space. In adults, hydrocephalus may occur after a head injury, meningitis, or subarachnoid hemorrhage. Because the adult skull bones are fused together, this condition can quickly become life-threatening and requires immediate intervention. •

Hydrocephalus in a newborn

 Where is CSF reabsorbed?

Lateral ventricle's choroid plexuses → CSF → Lateral ventricles

Through interventricular foramina

Third ventricle's choroid plexus → CSF → Third ventricle

Through aqueduct of the midbrain (cerebral aqueduct)

Fourth ventricle's choroid plexus → CSF → Fourth ventricle

Through lateral and median apertures

Subarachnoid space

Arachnoid villi of dural venous sinuses

Arterial blood

Venous blood

Heart and lungs

(d) Summary of the formation, circulation, and absorption of cerebrospinal fluid (CSF)

Brain Blood Flow and the Blood–Brain Barrier

Blood flows to the brain mainly via the internal carotid and vertebral arteries (see Figure 14.7b, e); the dural venous sinuses drain into the internal jugular veins to return blood from the head to the heart (see Figure 14.11a).

In an adult, the brain represents only 2 percent of total body weight, but consumes about 20 percent of the oxygen and glucose used by the body, even when you are resting. Neurons synthesize ATP almost exclusively from glucose via reactions that use oxygen. When the activity of neurons and neuroglia increases in a particular region of the brain, blood flow to that area also increases. Even a brief slowing of brain blood flow may cause disorientation or a loss of consciousness, such as when you stand up too quickly after sitting for a long period of time. Typically, an interruption in blood flow for 1 or 2 minutes impairs neuronal function, and total deprivation of oxygen for about 4 minutes causes permanent injury. Because virtually no glucose is stored in the brain, the supply of glucose also must be continuous. If blood entering the brain has a low level of glucose, mental confusion, dizziness, convulsions, and loss of consciousness may occur. People with diabetes must be vigilant about their blood sugar levels because these levels can drop quickly, leading to diabetic shock, which is characterized by seizure, coma, and possibly death.

The **blood–brain barrier (BBB)** consists mainly of tight junctions that seal together the endothelial cells of brain blood capillaries and a thick basement membrane that surrounds the capillaries. As you learned in Chapter 16, astrocytes are one type of neuroglia; the processes of many astrocytes press up against the capillaries and secrete chemicals that maintain the permeability characteristics of the tight junctions. A few water-soluble substances, such as glucose, cross the BBB by active transport. Other substances, such as creatinine, urea, and most ions, cross the BBB very slowly. Still other substances—proteins and most antibiotic drugs—do not pass at all from the blood into brain tissue. However, lipid-soluble substances, such as oxygen, carbon dioxide, alcohol, and most anesthetic agents, are able to access brain tissue freely. Trauma, certain toxins, and inflammation can cause a breakdown of the blood–brain barrier.

Because it is so effective, the BBB prevents the passage of helpful substances as well as those that are potentially harmful. Researchers are exploring ways to move drugs that could be therapeutic for brain cancer or other CNS disorders past the BBB. In one method, the drug is injected in a concentrated sugar solution. The high osmotic pressure of the sugar solution causes the endothelial cells of the capillaries to shrink, which opens gaps between their tight junctions, making the BBB more leaky and allowing the drug to enter the brain tissue.

 CHECKPOINT

3. Describe the locations of the cranial meninges.
4. What structures are the sites of CSF production, and where are they located?
5. Explain the blood supply to the brain and the importance of the blood–brain barrier.

CLINICAL CONNECTION | *Cerebrovascular Accident and Transient Ischemic Attack*

The most common brain disorder is a **cerebrovascular accident (CVA)**, also called a **stroke** or **brain attack.** CVAs affect 500,000 people each year in the United States and represent the third leading cause of death, behind heart attacks and cancer. A CVA is characterized by abrupt onset of persistent neurological symptoms, such as paralysis or loss of sensation, that arise from destruction of brain tissue. Common causes of CVAs are intracerebral hemorrhage (bleeding from a blood vessel in the pia mater or brain), emboli (blood clots), and atherosclerosis (formation of cholesterol-containing plaques that block blood flow) of the cerebral arteries.

Among the risk factors implicated in CVAs are high blood pressure, high blood cholesterol, heart disease, narrowed carotid arteries, transient ischemic attacks (TIAs; discussed next), diabetes, smoking, obesity, and excessive alcohol intake.

A clot-dissolving drug called *tissue plasminogen activator (t-PA)* is now being used to open up blocked blood vessels in the brain. The drug is most effective when administered within three hours of the onset of the CVA, however, and is helpful only for CVAs due to a blood clot (*ischemic CVA*). Use of t-PA can decrease the permanent disability associated with these types of CVAs by 50 percent. T-PA should not be administered to individuals

with strokes caused by hemorrhaging (*hemorrhagic CVA*) since it can cause further injury or even death. The distinction between the types of CVA is made on the basis of a CT scan.

New studies show that "cold therapy" might be successful in limiting the amount of residual damage from a CVA. States of hypothermia, such as those experienced by cold-water drowning victims, seem to trigger a survival response in which the body requires less oxygen; application of this principle to stroke victims has been showing promise. Some commercial companies now provide "CVA survival kits," which include cooling blankets that can be kept in the home.

A **transient ischemic attack (TIA)** is an episode of temporary cerebral dysfunction caused by impaired blood flow to part of the brain. Symptoms include dizziness, weakness, numbness, or paralysis in a limb or on one side of the body; drooping of one side of the face; headache; slurred speech or difficulty understanding speech; and/or a partial loss of vision or double vision. Sometimes nausea or vomiting also occur. The onset of symptoms is sudden and reaches maximum intensity almost immediately. A TIA usually persists for 5 to 10 minutes and only rarely lasts as long as 24 hours. It leaves no persistent neurological deficits. The causes of TIAs include blood clots, atherosclerosis, and certain blood disorders. ·

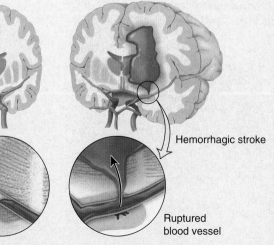

Ischemic stroke

Atherosclerotic blood vessel (or blood clot)

Hemorrhagic stroke

Ruptured blood vessel

18.3 THE BRAIN STEM AND RETICULAR FORMATION

 OBJECTIVE

• Describe the structure and functions of the medulla oblongata, pons, midbrain, and reticular formation.

The brain stem is the part of the brain between the spinal cord and the diencephalon. It consists of three structures: (1) medulla oblongata, (2) pons, and (3) midbrain. Extending throughout the brain stem is the reticular formation, a netlike region of interspersed gray and white matter.

Medulla Oblongata

The **medulla oblongata** (me-DOOL-la ob'-long-GA-ta), or more simply the **medulla**, is a continuation of the superior part of the spinal cord; it forms the inferior part of the brain stem (Figure 18.6; see also Figure 18.2). The medulla begins at the foramen magnum and extends to the inferior border of the pons, a distance of about 3 cm (1.2 in.). This short region of the central nervous system resembles the spinal cord in many ways. Like the spinal cord, it gives rise to many nerve roots; however, these are the roots of cranial nerves rather than spinal nerves. Six of the 12 pairs of cranial nerves arise from this region. While its surface features are similar to those of the spinal cord, its internal anatomy shows significant differences in the arrangement of the gray and white matter. Key external landmarks that distinguish the medulla oblongata from the spinal cord are the pyramids and olives at its slightly expanded cranial end; these structures will be discussed shortly.

Within the medulla's white matter are all the sensory (ascending) and motor (descending) tracts extending between the spinal cord and other parts of the brain. Some of the white matter forms

Figure 18.6 Medulla oblongata in relation to the rest of the brain stem.

🔑 The brain stem consists of the medulla oblongata, pons, and midbrain.

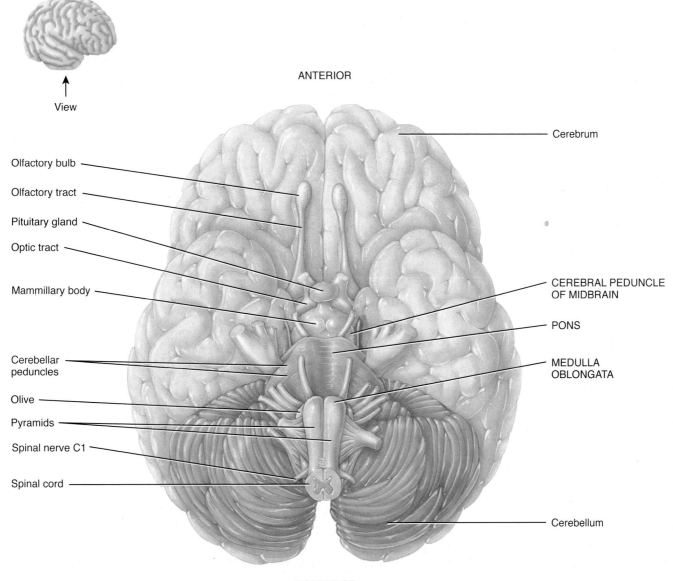

Inferior aspect of brain

 What part of the brain stem contains the pyramids? The cerebral peduncles? Literally means "bridge"?

bulges on the anterior aspect of the medulla. These protrusions are the **pyramids** (Figure 18.7; see also Figure 18.6a), formed by the largest motor tracts that pass from the cerebrum to the spinal cord (see Section 17.3). Just superior to the junction of the medulla with the spinal cord, 90 percent of the axons in the left pyramid cross to the right side, and 90 percent of the axons in the right pyramid cross to the left side. This crossing is called the **decussation of pyramids** (dē′-ka-SĀ-shun; *decuss-*=crossing) and explains how each side of the brain controls movements on the opposite side of the body.

The medulla also contains several **nuclei**, masses of gray matter where neurons form synapses with one another. Several of these nuclei control vital body functions. The **cardiovascular center** regulates the rate and force of the heartbeat and the diameter of blood vessels. The **medullary respiratory center** adjusts the basic rhythm of breathing (see Figure 23.13a). Other nuclei in the medulla control reflexes for vomiting, coughing, and sneezing.

Just lateral to each pyramid is an oval-shaped swelling called an **olive** (see Figures 18.6 and 18.7). Within the olive is the **inferior olivary nucleus**. Neurons here relay impulses from *proprioceptors* (receptors that monitor joint and muscle positions) to the cerebellum.

Nuclei associated with sensations of touch, conscious proprioception, pressure, and vibration are located in the posterior part of the medulla. These nuclei, the right and left **gracile nucleus** (GRAS-il=slender) and **cuneate nucleus** (KŪ-nē-āt=wedge), receive neurons from the gracile and cuneate fasciculi (see Section 20.3). Many ascending sensory axons form synapses in these nuclei, and postsynaptic neurons then relay the sensory information to the thalamus on the opposite side of the brain (see Figure 20.4a). The axons ascend to the thalamus in a band of white matter called the **medial lemniscus** (lem-NIS-kus=ribbon), which extends through the medulla, pons, and midbrain (see Figure 18.8b).

Finally, the medulla contains nuclei associated with the following five pairs of cranial nerves (see Figure 18.17a):

1. *Vestibulocochlear (VIII) nerves.* Several cochlear nuclei in the medulla receive sensory input from and provide motor output to the cochlea of the internal ear via the cochlear branches of the vestibulocochlear nerves. These nerves convey impulses related to hearing. See Exhibit 18.F.

2. *Glossopharyngeal (IX) nerves.* Nuclei in the medulla relay sensory and motor impulses related to taste, swallowing, and salivation via the glossopharyngeal nerves. See Exhibit 18.G.

3. *Vagus (X) nerves.* Nuclei in the medulla receive sensory impulses from and provide motor impulses to the pharynx and larynx and many thoracic and abdominal viscera via the vagus nerves. See Exhibit 18.H.

4. *Accessory (XI) nerves (cranial portion).* These fibers are actually part of the vagus (X) nerves. Nuclei in the medulla are the origin for nerve impulses that control swallowing via the vagus nerves (cranial portion of the accessory nerves). See Exhibit 18.I.

5. *Hypoglossal (XII) nerves.* Hypoglossal nuclei in the medulla are the origin for nerve impulses that control tongue movements during speech and swallowing via the hypoglossal nerves. See Exhibit 18.J.

Figure 18.7 Internal anatomy of the medulla oblongata.

🔑 The pyramids of the medulla contain the largest motor tracts that run from the cerebrum to the spinal cord.

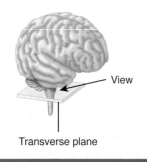
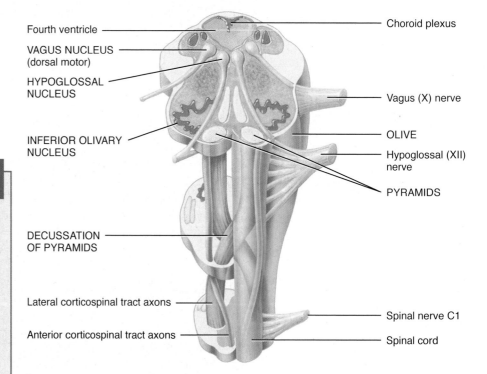

Transverse section and anterior surface of medulla oblongata

CLINICAL CONNECTION |
Injury to the Medulla

Given the many vital activities controlled by the medulla, it is not surprising that **injury to the medulla** from a hard blow to the back of the head or upper neck can be fatal. The medulla can also be damaged, even fatally damaged, by a blow such as an uppercut from a boxer when the skull is moved violently on the vertebral column and the dens of the axis impinges on the medulla. Damage to the medullary respiratiory center is particularly serious and can rapidly lead to death. Symptoms of nonfatal injury to the medulla may include paralysis and loss of sensation on the opposite side of the body, and irregularities in breathing or heart rhythm. Alcohol overdose also suppresses the medullary rhythmicitiy area and may result in death. •

❓ **What does decussation mean? What is the functional consequence of decussation of the pyramids?**

Pons

The **pons** (=bridge) lies directly superior to the medulla and anterior to the cerebellum and is about 2.5 cm (1 in.) long (see Figures 18.2 and 18.6). Like the medulla, the pons consists of both nuclei and tracts. As its name implies, the pons is a bridge that connects parts of the brain with one another. These connections are provided by bundles of axons. Some axons of the pons connect the right and left sides of the cerebellum. Others are part of ascending sensory tracts and descending motor tracts.

The pons has two major structural components: a ventral region and a dorsal region. The ventral region of the pons forms a large synaptic relay station consisting of scattered gray centers called the **pontine nuclei** (PON-tīn), shown in Figure 23.13. Entering and exiting these nuclei are numerous white matter tracts, each of which provides a connection between the cortex (outer layer) of a cerebral hemisphere and that of the opposite hemisphere of the cerebellum. This complex circuitry plays an essential role in coordinating and maximizing the efficiency of voluntary motor output throughout the body. The dorsal region of the pons is more like the other regions of the brainstem, the medulla and midbrain. It contains ascending and descending tracts along with the nuclei of cranial nerves.

Also within the pons is the **pontine respiratory group**, shown in Figure 23.13a. Together, the medullary respiratory center and the pontine respiratory group help control breathing.

The pons also contains nuclei associated with the following four pairs of cranial nerves (see Figure 18.17a):

1. *Trigeminal (V) nerves.* Nuclei in the pons receive sensory impulses for somatic sensations from the head and face and provide motor impulses that govern chewing via the trigeminal nerves. See Exhibit 18.D.

2. *Abducens (VI) nerves.* Abducens nuclei in the pons provide motor impulses that control eyeball movement via the abducens nerves. See Exhibit 18.C.

3. *Facial (VII) nerves.* Nuclei in the pons receive sensory impulses for taste and provide motor impulses to regulate secretion of saliva and tears and contraction of muscles of facial expression via the facial nerves. See Exhibit 18.E.

4. *Vestibulocochlear (VIII) nerves.* Vestibular nuclei in the pons receive sensory impulses from and provide motor impulses to the vestibular apparatus via the vestibular branches of the vestibulocochlear nerves. These nerves convey impulses related to balance and equilibrium. See Exhibit 18.F.

Midbrain

This short segment of the brainstem sits just superior to the pons where it is obscured by the large, overlapping cerebral hemispheres. The **midbrain** (mesencephalon) extends from the pons to the diencephalon (see Figures 18.2 and 18.6) and is about 2.5 cm (1 in.) long. The aqueduct of the midbrain (cerebral aqueduct) passes through the midbrain, connecting the third ventricle above with the fourth ventricle below. Like the medulla and the pons, the midbrain contains both tracts and nuclei.

The anterior part of the midbrain contains a pair of tracts called **cerebral peduncles** (pe-DUNG-kuls or PĒ-dung-kuls=little feet; see Figures 18.6 and 18.8a, c). They contain axons of corticospinal, corticobulbar, and corticopontine motor neurons, which conduct nerve impulses from the cerebrum to the spinal cord, medulla, and pons, respectively. The cerebral peduncles also contain axons of sensory neurons that extend from the medulla to the thalamus.

The posterior part of the midbrain, called the **tectum** (TEK-tum=roof), contains four rounded elevations (Figure 18.8a). The

Figure 18.8 Midbrain and reticular activating system (RAS).

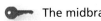

The midbrain connects the pons to the diencephalon.

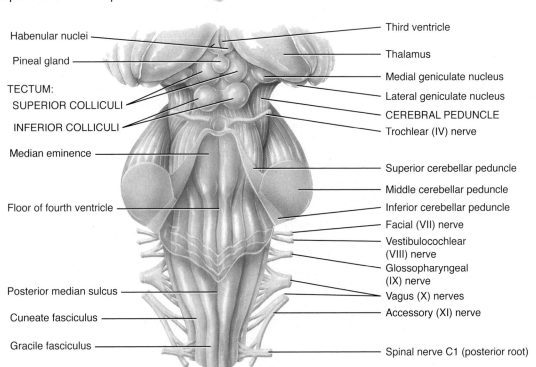

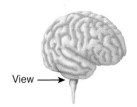

View →

Habenular nuclei

Pineal gland

TECTUM:
  SUPERIOR COLLICULI
  INFERIOR COLLICULI

Median eminence

Floor of fourth ventricle

Posterior median sulcus

Cuneate fasciculus

Gracile fasciculus

Third ventricle

Thalamus

Medial geniculate nucleus

Lateral geniculate nucleus

CEREBRAL PEDUNCLE

Trochlear (IV) nerve

Superior cerebellar peduncle

Middle cerebellar peduncle

Inferior cerebellar peduncle

Facial (VII) nerve

Vestibulocochlear (VIII) nerve

Glossopharyngeal (IX) nerve

Vagus (X) nerves

Accessory (XI) nerve

Spinal nerve C1 (posterior root)

(a) Posterior view of midbrain in relation to brain stem

FIGURE 18.8 CONTINUES ▶

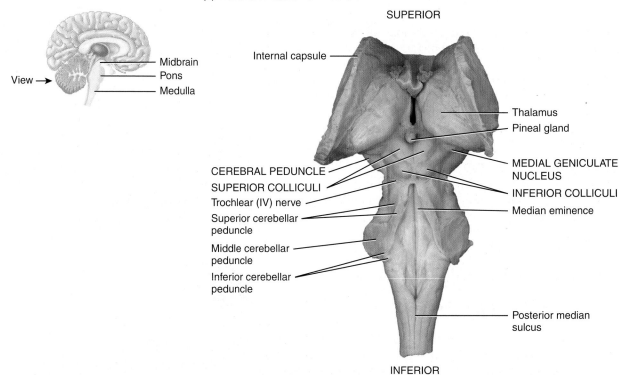

FIGURE 18.8 CONTINUED

(b) Transverse section of midbrain

(c) Posterior view of midbrain in relation to brain stem (much of the brain has been cut away)

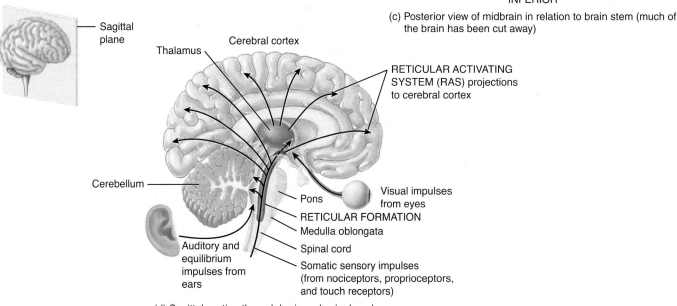

(d) Sagittal section through brain and spinal cord showing the reticular formation

 What is the importance of the cerebral peduncles?

two superior elevations are known as the **superior colliculi** (kō-LIK-ū-lī=little hills; singular is *colliculus*). These nuclei serve as reflex centers for certain visual activities. Through neural circuits from the retina of the eye to the superior colliculi to the extrinsic eye muscles, visual stimuli elicit eye movements for tracking moving images (such as a moving car) and scanning stationary images (as you are doing to read this sentence). Other superior colliculi reflexes are the accommodation reflex that adjusts the shape of the lens for close versus far vision and reflexes that govern movements of the eyes, head, and neck in response to visual stimuli. The two inferior elevations, the **inferior colliculi**, are part of the auditory pathway, relaying impulses from the receptors for hearing in the ear to the thalamus. These two nuclei also are reflex centers for the *startle reflex*, sudden movements of the head and body that occur when you are surprised by a loud noise such as a gunshot.

The midbrain contains several nuclei, including the left and right **substantia nigra** (sub-STAN-shē-a=substance; NĪ-gra= black), which are large, darkly pigmented nuclei (Figure 18.8b). Neurons that release dopamine extend from the substantia nigra to the basal nuclei and help control subconscious muscle activities. Loss of these neurons is associated with Parkinson's disease (see Section 20.4). Also present are the left and right **red nuclei** (Figure 18.8b), which look reddish due to their rich blood supply and an iron-containing pigment in their neuronal cell bodies. Axons from the cerebellum and cerebral cortex form synapses in the red nuclei, which function with the cerebellum to coordinate muscular movements.

Situated in the grey matter that surrounds the aqueduct of the midbrain is a unique nucleus, the **mesencephalic nucleus.** This is the only nucleus in the central nervous system that is not a synaptic relay station between neurons. Instead, this nucleus contains cell bodies of sensory (unipolar) neurons carrying proprioceptive signals (the feeling or sense of muscle position and tension) from skeletal muscles of the head. Other than this unique exception, all other cell bodies of sensory neurons reside in the ganglia of cranial or spinal nerves outside the central nervous system. The sensory neurons in the mesencephalic nucleus relay the proprioceptive sense to the various motor nuclei of the brainstem to influence motor control to the skeletal muscles of the head.

Finally, nuclei in the midbrain are associated with two pairs of cranial nerves (see Figure 18.17a):

1. *Oculomotor (III) nerves.* Oculomotor nuclei in the midbrain provide motor impulses that control movements of the eyeball, while accessory oculomotor nuclei provide motor control to the smooth muscles that regulate constriction of the pupil and changes in shape of the lens via the oculomotor nerves. See Exhibit 18.C.
2. *Trochlear (IV) nerves.* Trochlear nuclei in the midbrain provide motor impulses that control movements of the eyeball via the trochlear nerves. See Exhibit 18.C.

Reticular Formation

In addition to the well-defined nuclei already described, much of the brain stem consists of small clusters of neuronal cell bodies

(gray matter) interspersed among small bundles of myelinated axons (white matter). This broad region where white matter and gray matter exhibit a netlike arrangement is known as the **reticular formation** (re-TIK-ū-lar; *ret-*=net; Figure 18.8d). It extends from the superior part of the spinal cord, throughout the brain stem, and into the inferior part of the diencephalon. Neurons within the reticular formation have both ascending (sensory) and descending (motor) functions.

The ascending portion of the reticular formation is called the **reticular activating system (RAS)**, which consists of sensory axons that project to the cerebral cortex, both directly and through the thalamus. Many sensory stimuli can activate the ascending portion of the RAS. Among these are visual and auditory stimuli; mental activities; stimuli from pain, touch, pressure receptors; and receptors in our limbs and head that keep us aware of the position of our body parts. Perhaps the most important function of the RAS is **consciousness**, a state of wakefulness in which an individual is fully alert, aware, and oriented. Visual and auditory stimuli and mental activities can stimulate the RAS to help maintain consciousness. The RAS is also active during **arousal** or awakening from sleep. Another function of the RAS is to help maintain **attention** and *alertness*. The RAS also prevents sensory overload by filtering out insignificant information so that it does not reach consciousness. For example, while waiting in the hallway for your anatomy class to begin, you may be unaware of all the noise around you while reviewing your notes for class. Inactivation of the RAS produces **sleep**, a state of partial consciousness from which an individual can be aroused. Damage to the RAS, on the other hand, results in **coma**, a state of unconsciousness from which an individual cannot be aroused. In the lightest stages of coma, brain stem and spinal cord reflexes persist, but in the deepest stages even these reflexes are lost, and if respiratory and cardiovascular controls are lost, the patient dies. Drugs such as melatonin affect the RAS by helping to induce sleep, and general anesthetics turn off consciousness via the RAS. The descending portion of the RAS has connections to the cerebellum and spinal cord and helps regulate **muscle tone**, the slight degree of involuntary contraction in normal resting skeletal muscles. This portion of the RAS also assists in the regulation of heart rate, blood pressure, and respiratory rate.

Even though the RAS receives input from the eyes, ears, and other sensory receptors, there is no input from receptors for the sense of smell; even strong odors may fail to cause arousal. People who die in house fires usually succumb to smoke inhalation without awakening. For this reason, all sleeping areas should have a nearby smoke detector that emits a loud alarm. A vibrating pillow or flashing light can serve the same purpose for those who are hearing impaired.

The functions of the brain stem are summarized in Table 18.2.

 CHECKPOINT

6. Define decussation of pyramids. Why is it important?
7. What body functions are governed by nuclei in the pons?
8. What are the functions of the superior and inferior colliculi?
9. Describe several functions of the reticular formation.

18.4 THE CEREBELLUM

OBJECTIVES

• Describe the structure and functions of the cerebellum.
• Explain the location and importance of the cerebellar peduncles.

The **cerebellum**, the second-largest part of the brain, occupies the inferior and posterior aspects of the cranial cavity. Like the cerebrum, it has a highly folded surface that greatly increases the surface area of its outer gray matter cortex, allowing for a greater number of neurons. The cerebellum accounts for about a tenth of the brain mass yet contains nearly half of the neurons in the brain. The cerebellum is posterior to the medulla and pons and inferior to the posterior portion of the cerebrum (see Figure 18.2). A deep groove between the cerebrum and cerebellum known as the **transverse fissure** (see Figure 18.12b) is occupied by the **tentorium cerebelli**, which supports the posterior part of the cerebrum and separates it from the cerebellum (see Figure 18.3b). As mentioned previously, the tentorium cerebelli is a tentlike fold of the dura mater that is attached to the temporal and occipital bones.

In superior or inferior views, the shape of the cerebellum resembles a butterfly. The central constricted area is the **vermis** (=worm), and the lateral "wings" or lobes are the **cerebellar hemispheres** (Figure 18.9a, b). Each hemisphere consists of lobes separated by deep and distinct fissures. The **anterior lobe** and **posterior lobe** govern subconscious aspects of skeletal muscle movements. The **flocculonodular lobe** (flok-ū-lō-NOD-ū-lar; *flocculo-*=wool-like tuft) on the inferior surface contributes to equilibrium and balance.

The superficial layer of the cerebellum, called the **cerebellar cortex**, consists of gray matter in a series of slender, parallel ridges called **folia** (=leaves) (Figure 18.9c, d). Deep to the gray matter are tracts of white matter called **arbor vitae** (AR-bor VĪ-tē=tree of life) that resemble branches of a tree. Even deeper, within the white matter, are the **cerebellar nuclei**, regions of gray matter that give rise to axons carrying impulses from the cerebellum to other brain centers and to the spinal cord.

Three paired **cerebellar peduncles** (pe-DUNG-kuls) attach the cerebellum to the brain stem (see also Figure 18.8c). These bundles of white matter consist of axons that conduct impulses between the cerebellum and other parts of the brain. The **inferior cerebellar peduncles** carry sensory information from the vestibular apparatus of the inner ear and from proprioceptors throughout the body into the cerebellum; their axons extend from the inferior olivary nucleus of the medulla and from the spinocerebellar tracts of the spinal cord into the cerebellum. The **middle cerebellar peduncles** are the largest peduncles; their axons carry impulses for voluntary movements (those that originate in motor areas of the cerebral cortex) to nuclei in the pons. From the nuclei they pass into the cerebellum. The **superior cerebellar peduncles** contain axons that extend from the cerebellum to the red nuclei of the midbrain and to several nuclei of the thalamus.

The primary function of the cerebellum is to evaluate how well movements initiated by motor areas in the cerebrum are actually being carried out. When movements initiated by the cerebral motor areas are not being carried out correctly, the cerebellum detects the discrepancies. It then sends inhibitory feedback signals to motor areas of the cerebral cortex, via its connections to the red nucleus and thalamus. The feedback signals help correct the errors, smooth the movements, and coordinate complex sequences of skeletal muscle contractions. The cerebellum is also the main brain region that regulates posture and balance. These aspects of cerebellar function make possible all skilled muscular activities, from catching a baseball to dancing to speaking. The presence of reciprocal connections between the cerebellum and association areas of the cerebral cortex (see Section 18.7) suggest that the cerebellum may also have nonmotor functions such as cognition (acquisition of knowledge) and language processing. This view is supported by imaging studies like MRI and PET. Studies also suggest that the cerebellum may play a role in processing sensory information.

CLINICAL CONNECTION | *Ataxia*

Damage to the cerebellum can result in a loss of ability to coordinate muscular movements, a condition called **ataxia** (a-TAK-sē-a; *a-*=without; *-taxia*=order). Blindfolded people with ataxia cannot touch the tip of their nose with a finger because they cannot coordinate movement with their sense of where a body part is located. Another sign of ataxia is a changed speech pattern due to uncoordinated speech muscles. Cerebellar damage may also result in staggering or abnormal walking movements. People who consume too much alcohol show signs of ataxia because alcohol inhibits activity of the cerebellum. Such individuals have difficulty in passing sobriety tests. Ataxia can also occur as a result of degenerative diseases (multiple sclerosis and Parkinson's disease), trauma, brain tumors, genetic factors, and as a side effect of medications prescribed for bipolar disorder. •

The functions of the cerebellum are summarized in Table 18.2.

 CHECKPOINT

10. Describe the location and principal parts of the cerebellum.
11. Where do the axons of each of the three pairs of cerebellar peduncles begin and end? What are their functions?

18.5 THE DIENCEPHALON

OBJECTIVES

• Describe the components and functions of the thalamus, hypothalamus, and epithalamus.
• Define circumventricular organs and their functions.

The **diencephalon** forms a central core of brain tissue just superior to the midbrain. It is almost completely surrounded by the cerebral hemispheres and contains numerous nuclei involved in a wide variety of sensory and motor processing between higher and lower brain centers. The diencephalon extends from the brain stem to the cerebrum and surrounds the third ventricle; it includes the thalamus, hypothalamus, and epithalamus. Projecting from the hypothalamus is the hypophysis, or pituitary gland. Portions of the diencephalon in the wall of the third ventricle are called circumventricular organs and will be discussed shortly. The optic tracts carrying neurons from the retina enter the diencephalon.

Figure 18.9 Cerebellum.

The cerebellum coordinates skilled movements and regulates posture and balance.

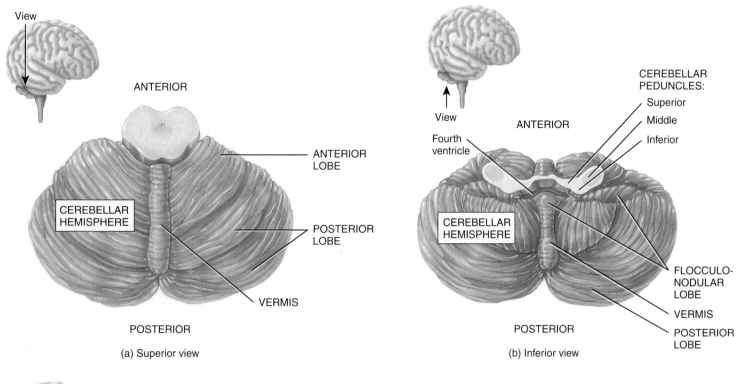

View

ANTERIOR

ANTERIOR LOBE

CEREBELLAR HEMISPHERE

POSTERIOR LOBE

VERMIS

POSTERIOR

(a) Superior view

View

ANTERIOR

CEREBELLAR PEDUNCLES:
Superior
Middle
Inferior

Fourth ventricle

CEREBELLAR HEMISPHERE

FLOCCULO-NODULAR LOBE

VERMIS

POSTERIOR

POSTERIOR LOBE

(b) Inferior view

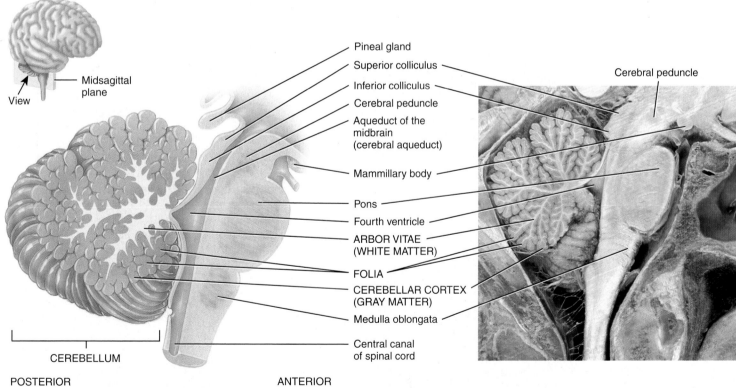

Midsagittal plane

View

Pineal gland

Superior colliculus

Inferior colliculus

Cerebral peduncle

Aqueduct of the midbrain (cerebral aqueduct)

Mammillary body

Pons

Fourth ventricle

ARBOR VITAE (WHITE MATTER)

FOLIA

CEREBELLAR CORTEX (GRAY MATTER)

Medulla oblongata

Central canal of spinal cord

Cerebral peduncle

CEREBELLUM

POSTERIOR

ANTERIOR

(c) Midsagittal section of cerebellum and brain stem

(d) Midsagittal section

 Which fiber tracts carry information into and out of the cerebellum?

Thalamus

The **thalamus** (THAL-a-mus=inner chamber), which measures about 3 cm (1.2 in.) in length, makes up 80 percent of the diencephalon. Thin layers of white matter partially outline the thalamus, which consists of paired oval masses of gray matter organized into nuclei with interspersed tracts of white matter (Figure 18.10). A bridge of gray matter called the **interthalamic adhesion (intermediate mass)** joins the right and left halves of the thalamus in about 70 percent of human brains. The interthalamic adhesion forms during development as the medial surfaces of the thalami enlarge and push against one another across the third ventricle. In some individuals, the ependymal cells covering this aspect of the thalamus fuse and some of the surface neurons grow into this fused region to form the interthalamic adhesion. It has no functional significance.

The thalamus is the major relay station for sensory impulses (except smell) that reach the primary sensory areas of the cerebral cortex from the spinal cord, brain stem, and midbrain. Although crude perception of painful, thermal, and pressure sensations arises at the level of the thalamus, precise localization of these sensations depends on nerve impulses arriving at the cerebral cortex.

The thalamus contributes to motor functions by transmitting information from the cerebellum and basal nuclei to the primary motor area of the cerebral cortex. It also relays nerve impulses between different areas of the cerebrum and plays a role in the regulation of autonomic activities and the maintenance of consciousness. Axons that connect the thalamus and cerebral cortex pass through the **internal capsule**, a thick band of white matter lateral to the thalamus (see Figure 18.14b, c).

A vertical Y-shaped sheet of white matter called the **internal medullary lamina** divides the gray matter of the right and left sides of the thalamus (Figure 18.10c). It consists of myelinated axons that enter and leave the various thalamic nuclei.

Based on their positions and functions, there are seven major groups of nuclei on each side of the thalamus (Figure 18.10c, d).

1. The **anterior nucleus** connects to the hypothalamus and limbic system (the latter is described in Section 18.6). It functions in emotions, regulation of alertness, and memory.
2. The **medial nuclei** connect to the cerebral cortex, limbic system, and basal nuclei. They function in emotions, learning, memory, awareness, and cognition (thinking and knowing).
3. Nuclei in the **lateral group** connect to the superior colliculi, limbic system, and cortex in all lobes of the cerebrum.

Figure 18.10 Thalamus. Note the position of the thalamus in (a), the lateral view, and (b), the medial view. Various thalamic nuclei in (c) and (d) are correlated by color to the cortical regions to which they project in (a) and (b).

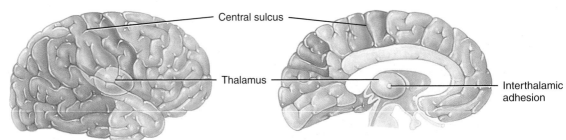

The thalamus is the principal relay station for sensory impulses that reach the cerebral cortex from other parts of the brain and the spinal cord.

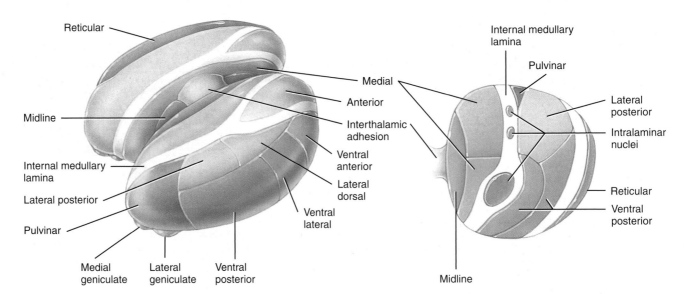

(a) Lateral view of right cerebral hemisphere

(b) Medial view of left cerebral hemisphere

(c) Superolateral view of thalamus showing locations of thalamic nuclei (reticular nucleus is shown on the left side only; all other nuclei are shown on the right side)

(d) Transverse section of right side of thalamus showing locations of thalamic nuclei

What structure connects the right and left halves of the thalamus?

The **lateral dorsal nucleus** functions in the expression of emotions. The **lateral posterior nucleus** and **pulvinar nucleus** help integrate sensory information.

4. Five nuclei are part of the **ventral group**. The **ventral anterior nucleus** contributes to motor functions, and possibly movement planning. The **ventral lateral nucleus** connects to the cerebellum and motor parts of the cerebral cortex. Its neurons are active during movements on the opposite side of the body. The **ventral posterior nucleus** relays impulses for somatic sensations such as touch, pressure, proprioception, vibration, heat, cold, and pain from the face and body to the cerebral cortex. The **lateral geniculate nucleus** (je-NIK-ū-lāt=bent like a knee) relays visual impulses for sight from the retina to the primary visual area of the cerebral cortex. The **medial geniculate nucleus** relays auditory impulses for hearing from the ear to the primary auditory area of the cerebral cortex.

5. **Intralaminar nuclei** (in′-tra-LA-mi-nar) lie within the internal medullary lamina and make connections with the reticular formation, cerebellum, basal nuclei, and wide areas of the cerebral cortex. They function in pain perception, integration of sensory and motor information, and arousal (activation of the cerebral cortex from the brain stem reticular formation).

6. The **midline nucleus** forms a thin band adjacent to the third ventricle and has a presumed function in memory and olfaction.

7. The **reticular nucleus** surrounds the lateral aspect of the thalamus, next to the internal capsule. This nucleus monitors, filters, and integrates activities of other thalamic nuclei.

Hypothalamus

The **hypothalamus** (hī′-pō-THAL-a-mus; *hypo-*=under) is a small part of the diencephalon located inferior to the thalamus.

Weighing only about 4 grams (0.14 oz), this tiny area of the brain is much more important than its size suggests. It is composed of a dozen or so nuclei in four major regions:

1. The **mammillary region** (MAM-i-lar′-ē; *mammill-*=nipple-shaped), adjacent to the midbrain, is the most posterior part of the hypothalamus. It includes the mammillary bodies and posterior hypothalamic nuclei (Figure 18.11). The **mammillary bodies** are two small, rounded projections that serve as relay stations for reflexes related to the sense of smell (see also Figures 18.6 and 18.9c, d).

2. The **tuberal region** (TOO-ber-al), the widest part of the hypothalamus, includes the *dorsomedial, ventromedial,* and *arcuate nuclei* (AR-kū-at), plus the stalklike **infundibulum** (in-fun-DIB-ū-lum=funnel), which connects the pituitary gland to the hypothalamus (Figure 18.11). The **median eminence** is a slightly raised region that encircles the infundibulum.

3. The **supraoptic region** (*supra-*=above; *-optic*=eye) lies superior to the optic chiasm (point of crossing of optic nerves) and contains the *paraventricular nucleus, supraoptic nucleus, anterior hypothalamic nucleus,* and *suprachiasmatic nucleus* (soo′-pra-kī′-az-MA-tik) (Figure 18.11). Axons from the paraventricular and supraoptic nuclei form the hypothalamohypophyseal tract (hī′-pō-thal′-a-mō-hī-pō-FIZ-ē-al), which extends through the infundibulum to the posterior lobe of the pituitary.

4. The **preoptic region** anterior to the supraoptic region is usually considered part of the hypothalamus because it participates with the hypothalamus in regulating certain autonomic activities. The preoptic region contains the medial and lateral preoptic nuclei (Figure 18.11).

The hypothalamus controls many body activities and is one of the major regulators of homeostasis. Sensory impulses related to both somatic and visceral senses arrive at the hypothalamus,

Figure 18.11 Hypothalamus. Selected portions of the hypothalamus and a three-dimensional representation of hypothalamic nuclei are shown (after Netter).

The hypothalamus controls many body activities and is an important regulator of homeostasis.

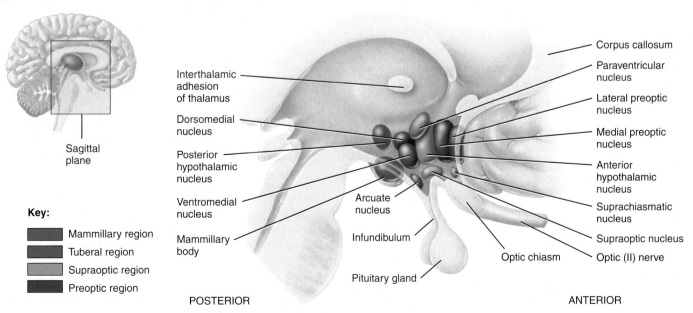

Sagittal section of brain showing hypothalamic nuclei

 What are the four major regions of the hypothalamus, from posterior to anterior?

as do impulses from receptors for vision, taste, and smell. Other receptors within the hypothalamus itself continually monitor osmotic pressure, blood glucose level, certain hormone concentrations, and the temperature of blood. The hypothalamus has several very important connections with the pituitary gland and produces a variety of hormones, which are described in more detail in Chapter 22. Some functions can be attributed to specific hypothalamic nuclei, but others are not so precisely localized. Important functions of the hypothalamus include the following:

1. **Control of the ANS.** The hypothalamus controls and integrates activities of the autonomic nervous system (ANS) (Chapter 19), which regulates contraction of smooth and cardiac muscle and the secretions of many glands. Axons extend from the hypothalamus to sympathetic and parasympathetic nuclei in the brain stem and spinal cord. Through the ANS, the hypothalamus is a major regulator of visceral activities, including heart rate, movement of food through the gastrointestinal tract, and contraction of the urinary bladder.

2. **Production of hormones.** The hypothalamus produces several hormones and has two important connections with the pituitary gland, an endocrine gland located inferior to the hypothalamus (see Figure 18.2a). First, hypothalamic hormones are released into capillary networks in the median eminence. The bloodstream carries these hormones directly to the anterior lobe of the pituitary, where they stimulate or inhibit secretion of anterior pituitary hormones. Second, axons extend from the paraventricular and supraoptic nuclei through the infundibulum into the posterior lobe of the pituitary. The cell bodies of these neurons make one of two hormones (*oxytocin* or *antidiuretic hormone*). Their axons transport the hormones to the posterior pituitary, where they are released.

3. **Regulation of emotional and behavioral patterns.** Together with the limbic system, the hypothalamus participates in expressions of rage, aggression, pain, and pleasure, and the behavioral patterns related to sexual arousal.

4. **Regulation of eating and drinking.** The hypothalamus regulates food intake through the arcuate and paraventricular nuclei. It also contains a thirst center. When certain cells in the hypothalamus are stimulated by rising osmotic pressure of the extracellular fluid, they cause the sensation of thirst. The intake of water by drinking restores the osmotic pressure to normal, removing the stimulation and relieving the thirst.

5. **Control of body temperature.** If the temperature of blood flowing through the hypothalamus is above normal, the hypothalamus directs the autonomic nervous system to stimulate activities that promote heat loss. When blood temperature is below normal, the hypothalamus generates impulses that promote heat production and retention.

6. **Regulation of circadian rhythms and states of consciousness.** The suprachiasmatic nucleus establishes patterns of awakening and sleep that occur on a *circadian (daily) rhythm* (ser-KĀ-dē-an). This nucleus receives input from the eyes (retina) and sends output to other hypothalamic nuclei, the reticular formation, and the pineal gland.

Epithalamus

The **epithalamus** (ep′-i-THAL-a-mus; *epi-*=above), a small region superior and posterior to the thalamus, consists of the pineal gland and habenular nuclei. The **pineal gland** (PĪN-ē-al=pinecone-like) is about the size of a small pea and protrudes from the posterior midline of the third ventricle (see Figures 18.2a and 18.8a). The pineal gland is part of the endocrine system because it secretes the hormone **melatonin**. As more melatonin is liberated during darkness than in light, this hormone is thought to promote sleepiness. When taken orally, melatonin also appears to contribute to the setting of the body's biological clock by inducing sleep and helping the body to adjust to jet lag. The **habenular nuclei** (ha-BEN-ū-lar), shown in Figure 18.8a, are involved in olfaction, especially emotional responses to odors such as a loved one's cologne or Mom's chocolate chip cookies baking in the oven.

The functions of the three parts of the diencephalon are summarized in Table 18.2.

Circumventricular Organs

Parts of the diencephalon, called **circumventricular organs (CVOs)** (ser′-kum-ven-TRIK-ū-lar) because they lie in the wall of the third **ventricle**, can monitor chemical changes in the blood because they lack a blood–brain barrier. CVOs include part of the hypothalamus, the pineal gland, the pituitary gland, and a few other nearby structures. Functionally, these regions coordinate homeostatic activities of the endocrine and nervous systems, such as the regulation of blood pressure, fluid balance, hunger, and thirst. CVOs are also thought to be the sites of entry into the brain of HIV, the virus that causes AIDS. Once in the brain, HIV may cause dementia (irreversible deterioration of mental state) and other neurological disorders.

 CHECKPOINT

12. Why is the thalamus considered a "relay station" in the brain?
13. In what respect is the hypothalamus part of both the nervous system and the endocrine system?
14. What are the functions of the epithalamus?
15. Define circumventricular organ and explain its function.

18.6 THE CEREBRUM

 OBJECTIVES

• Describe the cortex, convolutions, fissures, and sulci of the cerebrum.
• Outline the lobes of the cerebrum and indicate their locations.
• Describe the tracts that comprise the cerebral white matter.
• Identify the nuclei that comprise the basal nuclei.
• List the structures of the limbic system and describe their functions.

The **cerebrum** is the "seat of intelligence." It provides us with the ability to read, write, and speak; to make calculations and compose music; and to remember the past, plan for the future, and imagine things that have never existed before.

Structure of the Cerebrum

By far the largest portion of the human brain, the cerebrum consists of the cerebral hemispheres and the basal nuclei. The right and left halves of the cerebrum, called **cerebral hemispheres**, are separated by a deep groove called the longitudinal fissure that is occupied by the falx cerebri (see Section 18.2). The hemispheres consist of an outer rim of gray matter, an internal region of cerebral white matter, and gray matter nuclei deep within the white matter. The outer rim of gray matter is the **cerebral cortex** (*cortex*=rind or bark) (Figure 18.12a). Although only 2–4 mm (0.08–0.16 in.) thick, the cerebral cortex contains bil-lions of neurons arranged in layers. Deep to the cerebral cortex lies the cerebral white matter.

During embryonic development, when brain size increases rapidly, the gray matter of the cortex enlarges much faster than the deeper white matter. As a result, the cortical region rolls and folds upon itself. The folds are called **gyri** (JĪ-rī=circles; singular is *gyrus*) or **convolutions** (kon´-vō-LOO-shuns) (Figure 18.12a, b). The deepest grooves between folds are known as **fissures**; the shallower grooves between folds are termed **sulci** (SUL-sī=grooves; singular is *sulcus* [SUL-kus]). As mentioned previously, the most prominent fissure, the **longitudinal fissure**, separates the cerebrum into right and left halves called **cerebral hemispheres**. The

Figure 18.12 Cerebrum. Because the insula cannot be seen externally, it has been projected to the surface in (b).

🔑 The cerebrum is the "seat of intelligence"; it provides us with the ability to read, write, and speak; to make calculations and compose music; to remember the past and plan for the future; and to create.

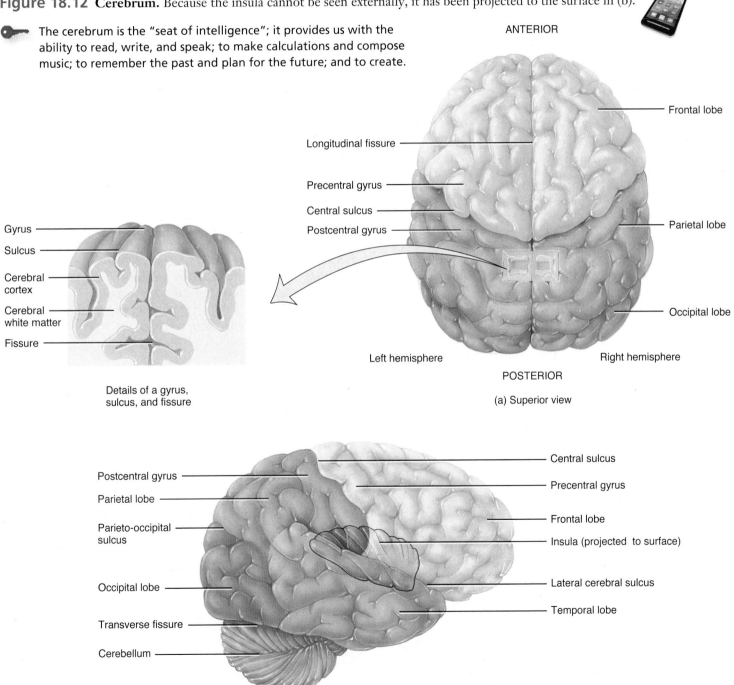

(a) Superior view

(b) Right lateral view

FIGURE **18.12** CONTINUES ▶

FIGURE **18.12** CONTINUED

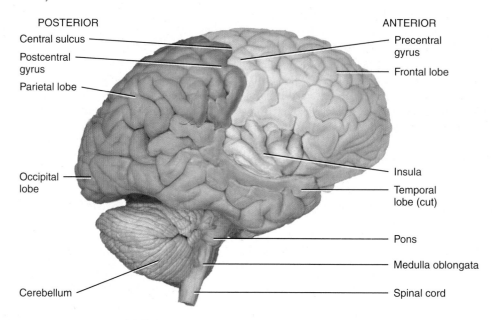

(c) Right lateral view with temporal lobe cut away

During development, does the gray matter or the white matter enlarge more rapidly? What are the brain folds, shallow grooves, and deep grooves that develop in the cortical region called?

hemispheres are connected internally by the **corpus callosum** (kal-LŌ-sum; *corpus*=body; *callosum*=hard), a broad band of white matter containing axons that extend between the hemispheres at the floor of the longitudinal fissure (Figure 18.13).

Each cerebral hemisphere can be further subdivided into several lobes. The lobes are named after the bones that cover them: frontal, parietal, temporal, and occipital lobes (Figure 18.12a–c). The **central sulcus** separates the **frontal lobe** from the **parietal lobe**. A major gyrus, the **precentral gyrus**—located immediately anterior to the central sulcus—contains

the primary motor area of the cerebral cortex. Another major gyrus, the **postcentral gyrus**, which is located immediately posterior to the central sulcus, contains the primary somatosensory area of the cerebral cortex. The **lateral cerebral sulcus (fissure)** separates the **frontal lobe** from the **temporal lobe**. The **parieto-occipital sulcus** separates the **parietal lobe** from the **occipital lobe**. A fifth part of the cerebrum, the **insula**, cannot be seen at the surface of the brain because it lies within the lateral cerebral sulcus, deep to the parietal, frontal, and temporal lobes (Figure 18.12b, c).

Figure 18.13 Organization of fibers into white matter tracts of the left cerebral hemisphere.

Association tracts, commissural tracts, and projection tracts form white matter tracts in the cerebral hemispheres.

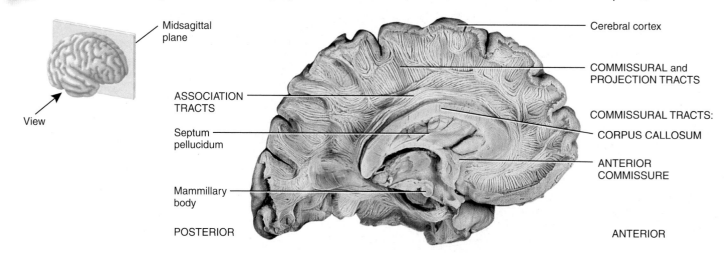

Medial view of tracts revealed by removing gray matter from a midsagittal section

 Which fibers carry impulses between gyri of the same hemisphere? Between gyri in opposite hemispheres? Between the cerebrum and thalamus, brain stem, and spinal cord?

Cerebral White Matter

The **cerebral white matter** consists primarily of myelinated axons in three types of tracts (Figure 18.13):

1. **Association tracts** contain axons that conduct nerve impulses between gyri in the same hemisphere.
2. **Commissural tracts** (kom'-i-SYŪR-al) contain axons that conduct nerve impulses from gyri in one cerebral hemisphere to corresponding gyri in the other cerebral hemisphere. Three important groups of commissural tracts are the **corpus callosum** (the largest fiber bundle in the brain, containing about 300 million fibers), **anterior commissure**, and **posterior commissure**.
3. **Projection tracts** contain axons that conduct nerve impulses from the cerebrum to lower parts of the CNS (thalamus, brain stem, or spinal cord) or from lower parts of the CNS to the cerebrum. An example is the **internal capsule**, a thick band of white matter that contains both ascending and descending axons (see Figure 18.14b).

Basal Nuclei

Deep within each cerebral hemisphere are three nuclei (masses of gray matter) that are collectively termed the **basal nuclei** (Figure 18.14). (Historically, these nuclei have been called the *basal ganglia*. However, this is a misnomer because a *ganglion* is an aggregate of neuronal cell bodies in the peripheral nervous system. While both terms still appear in the literature, we use *nuclei* as this is the correct term as determined by the *Terminologia Anatomica*, the final say on correct anatomical terminology.)

Two of the basal nuclei are side-by-side, just lateral to the thalamus. They are the **globus pallidus** (GLŌ-bus PAL-i-dus; *globus*=ball; *pallidus*=pale), which is closer to the thalamus, and the **putamen** (pu-TA-men=shell), which is closer to the cerebral cortex. Together, the globus pallidus and putamen are referred to as the **lentiform nucleus** (LEN-ti-form=shaped like a lens). The third of the basal nuclei is the **caudate nucleus** (KAW-dāt; *caud-*=tail), which has a large "head" connected to a smaller "tail" by a long comma-shaped "body." Together, the lentiform and caudate nuclei are known as the **corpus striatum** (strī-Ā-tum; *corpus*=body; *striatum*=striated). The term corpus striatum refers to the striated (striped) appearance of the internal capsule as it passes among the basal nuclei. Nearby structures that are functionally linked to the basal nuclei are the *substantia nigra* of the midbrain (see Figure 18.8b) and the *subthalamic nuclei* of the diencephalon (Figure 18.14b). Axons from the substantia nigra terminate in the caudate nucleus and putamen. The subthalamic nuclei interconnect with the globus pallidus.

The **claustrum** (KLAWS-trum) is a thin sheet of gray matter situated lateral to the putamen. It is considered by some to be a subdivision of the basal nuclei. The function of the claustrum in humans has not been clearly defined but it may be involved in visual attention.

The basal nuclei receive input from the cerebral cortex and provide output to motor parts of the cortex via the medial nuclei and ventral group nuclei of the thalamus. In addition, the basal nuclei have extensive connections with one another. A major function of the basal nuclei is to help regulate initiation and termination of movements. Activity of neurons in the putamen precedes or anticipates body movements; activity of neurons in the caudate nucleus occurs prior to eye movements. The globus pallidus helps regulate the muscle tone required for specific body movements. The basal nuclei also control subconscious contractions of skeletal muscles. Examples include automatic arm swings while walking and true laughter in response to a joke (not the kind you consciously initiate to humor your anatomy instructor).

In addition to influencing motor functions, the basal nuclei have other roles. They help initiate and terminate some cognitive processes, such as attention, memory, and planning, and may act with the limbic system to regulate emotional behaviors.

Disorders of the basal nuclei can affect body movements, cognition, and behavior. Uncontrollable shaking (tremor) and muscle rigidity (stiffness) are hallmark signs of **Parkinson's disease (PD)**. In this disorder, dopamine-releasing neurons that extend from the substantia nigra to the putamen and caudate nucleus degenerate.

Figure 18.14 Basal nuclei. In (a) and (b), the basal nuclei have been projected to the surface and are shown in purple.

🔑 The basal nuclei control automatic movements of skeletal muscles and muscle tone.

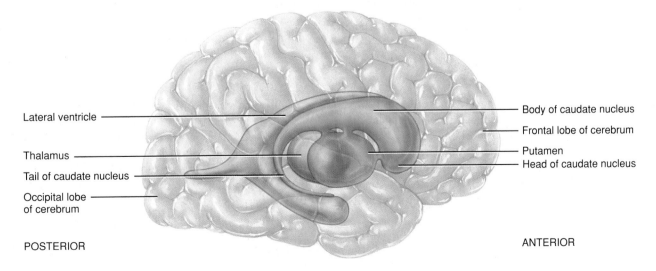

(a) Lateral view of right side of brain

POSTERIOR

ANTERIOR

Lateral ventricle

Thalamus

Tail of caudate nucleus

Occipital lobe of cerebrum

Body of caudate nucleus

Frontal lobe of cerebrum

Putamen

Head of caudate nucleus

FIGURE 18.14 CONTINUES ▶

FIGURE 18.14 CONTINUED

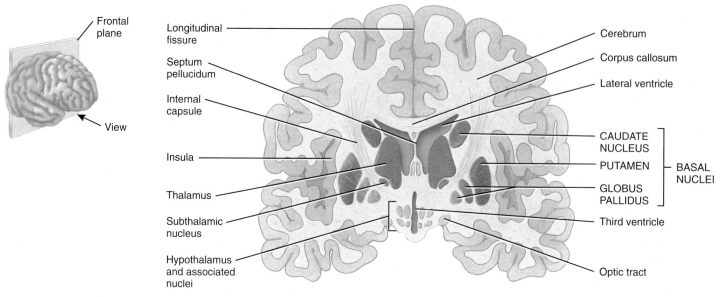

(b) Anterior view of frontal section

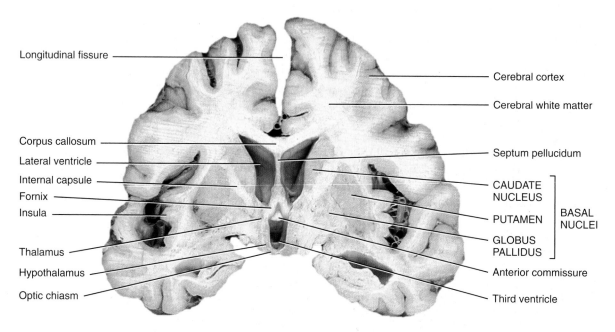

(c) Anterior view of frontal section

Where are the basal nuclei located relative to the thalamus?

Huntington disease (HD) is an inherited disorder in which the caudate nucleus and putamen degenerate, with loss of neurons that normally release GABA or acetylcholine. A key sign of HD is **chorea** (KŌ-rē-a=a dance), in which rapid, jerky movements occur involuntarily and without purpose. Progressive mental deterioration also occurs. Symptoms of HD often do not appear until age 30 or 40. Death occurs 10 to 20 years after symptoms first appear.

Tourette syndrome is a disorder that is characterized by involuntary body movements (motor tics) and the use of inappropriate or unnecessary sounds or words (vocal tics). Although the cause is unknown, research suggests that this disorder involves a dysfunction of the cognitive neural circuits between the basal nuclei and the prefrontal cortex.

Some psychiatric disorders, such as schizophrenia and obsessive-compulsive disorder, are thought to involve dysfunction of the behavioral neural circuits between the basal nuclei and the limbic system. In **schizophrenia**, excess dopamine activity in the brain causes a person to experience delusions, distortions of reality, paranoia, and hallucinations. People who have **obsessive-compulsive disorder (OCD)** experience repetitive thoughts (obsessions) that cause repetitive behaviors (compulsions) that they feel obligated to perform. For example, a person with OCD might have repetitive thoughts about someone breaking into the house; these thoughts might drive that person to check the doors of the house over and over again (for minutes or hours at a time) to make sure that they are locked.

The Limbic System

Encircling the upper part of the brain stem and the corpus callosum is a ring of structures on the inner border of the cerebrum and floor of the diencephalon that constitutes the **limbic system** (*limbic*=border). The main components of the limbic system are as follows (Figure 18.15):

1. The so-called **limbic lobe** is a rim of cerebral cortex on the medial surface of each hemisphere. It includes the **cingulate gyrus** (SIN-gū-lāt; *cingul-*=belt), which lies above the corpus callosum, and the **parahippocampal gyrus** (par′-a-hip-ō-KAM-pal), which is in the temporal lobe below. The **hippocampus** (hip′-ō-KAM-pus=seahorse) is a portion of the parahippocampal gyrus that extends into the floor of the lateral ventricle.
2. The **dentate gyrus** (*dentate*=toothed) lies between the hippocampus and parahippocampal gyrus.
3. The **amygdala** (a-MIG-da-la; *amygda-*=almond-shaped) is composed of several groups of neurons located close to the tail of the caudate nucleus.
4. The **septal nuclei** are located within the septal area formed by the regions under the corpus callosum and the paraterminal gyrus (a cerebral gyrus).
5. The **mammillary bodies** of the hypothalamus are two round masses close to the midline near the cerebral peduncles.
6. Two nuclei of the thalamus, the **anterior nucleus** and the **medial nucleus**, participate in limbic circuits.
7. The **olfactory bulbs** are flattened bodies of the olfactory pathway that rest on the cribriform plate.

8. The **fornix, stria terminalis, stria medullaris, medial forebrain bundle**, and **mammillothalamic tract** (mam-i-lō-tha-LAM-ik) are linked by bundles of interconnecting myelinated axons.

The limbic system is sometimes called the "emotional brain" because it plays a primary role in a range of emotions, including pain, pleasure, docility, affection, and anger. It also is involved in olfaction (smell) and memory. Experiments have shown that when different areas of animals' limbic systems are stimulated, the animals' reactions indicate that they are experiencing intense pain or extreme pleasure. Stimulation of other limbic system areas in animals produces tameness and signs of affection. Stimulation of a cat's amygdala or certain nuclei of the hypothalamus produces a behavioral pattern called rage—the cat extends its claws, raises its tail, opens its eyes wide, hisses, and spits. By contrast, removal of the amygdala produces an animal that lacks fear and aggression. Likewise, a person whose amygdala is damaged fails to recognize fearful expressions in others or to express fear in situations where this emotion would normally be appropriate, for example, while being attacked by an animal.

Together with parts of the cerebrum, the limbic system also functions in memory; damage to the limbic system causes memory impairment. One portion of the limbic system, the hippocampus, is seemingly unique among structures of the central nervous system—it has cells reported to be capable of mitosis. Thus, the portion of the brain that is responsible for some aspects of memory may develop new neurons, even in the elderly.

Figure 18.15 Components of the limbic system (shaded green) and surrounding structures.

🔑 The limbic system governs emotional aspects of behavior.

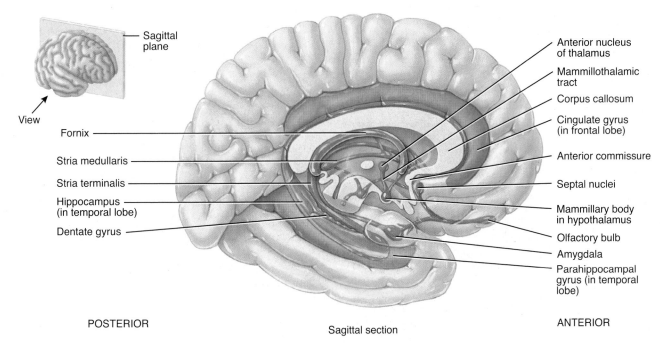

POSTERIOR   Sagittal section   ANTERIOR

❓ **Which part of the limbic system functions with the cerebrum in memory?**

  CHECKPOINT

16. List and locate the lobes of the cerebrum. How are they separated from one another? What is the insula?
17. Describe the organization of cerebral white matter and indicate the function of each major group of fibers.

18. Give the name and function of each of the nuclei that form basal nuclei, and describe the effects of basal nuclei damage.
19. Define the limbic system and list several of its functions.

18.7 FUNCTIONAL ORGANIZATION OF THE CEREBRAL CORTEX

OBJECTIVES

• Outline the locations and functions of the sensory, association, and motor areas of the cerebral cortex.
• Explain what is meant by hemispheric lateralization.
• Describe the importance of brain waves.

Specific types of sensory, motor, and integrative signals are processed in certain cerebral regions (Figure 18.16). Generally, **sensory areas** receive and interpret sensory impulses, **motor areas** initiate movements, and **association areas** deal with more complex integrative functions such as memory, emotions, reasoning, will, judgment, personality traits, and intelligence. In this section we will also discuss hemispheric lateralization, memory, and brain waves.

Sensory Areas

Sensory impulses arrive mainly in the posterior half of both cerebral hemispheres, in regions behind the central sulci. In the cerebral cortex, primary sensory areas receive sensory information that has been relayed through lower regions of the brain from peripheral sensory receptors.

Secondary sensory areas and sensory association areas often are adjacent to the primary areas. They usually receive input both from the primary areas and from other brain regions. Secondary sensory areas and sensory association areas integrate sensory experiences to generate meaningful patterns of recognition and awareness. A person with damage in the *primary* visual area would be blind in at least part of his visual field, but a person with damage to a visual *association* area might see normally yet be unable to recognize a friend.

The following are some important sensory areas (the area numbers referenced below are the Brodmann numbers discussed in Figure 18.16):

• The **primary somatosensory area** (areas 1, 2, and 3) is located directly posterior to the central sulcus of each cerebral hemisphere in the postcentral gyrus of each parietal lobe. It extends from the lateral cerebral sulcus, along the lateral surface of the parietal lobe to the longitudinal fissure, and then along the medial surface of the parietal lobe within the longitudinal fissure.

The primary somatosensory area receives nerve impulses for touch, proprioception (joint and muscle position), pain, itch, tickle, and thermal sensations. A "map" of the entire body is present in the primary somatosensory area: Each point within the area receives impulses from a specific part of the body (see Figure 20.5a). The size of the cortical area receiving impulses from a particular part of the body depends on the number of receptors present there rather than on the size of the body part. For example, a larger region of the somatosensory area receives

Figure 18.16 Functional areas of the cerebrum. Broca's speech area and Wernicke's area are in the left cerebral hemisphere of most people; they are shown here to indicate their relative locations. The numbers, still used today, are from K. Brodmann's map of the cerebral cortex, first published in 1909.

🔑 Particular areas of the cerebral cortex process sensory, motor, and integrative signals.

Central sulcus
PRIMARY SOMATOSENSORY AREA (POSTCENTRAL GYRUS)
SOMATOSENSORY ASSOCIATION AREA
Parietal lobe
COMMON INTEGRATIVE AREA
WERNICKE'S (POSTERIOR LANGUAGE) AREA
VISUAL ASSOCIATION AREA
PRIMARY VISUAL AREA
Occipital lobe
Temporal lobe

PRIMARY MOTOR AREA (PRECENTRAL GYRUS)
PREMOTOR AREA
PRIMARY GUSTATORY AREA
FRONTAL EYE FIELD AREA
Frontal lobe
BROCA'S SPEECH AREA
PREFRONTAL CORTEX
Lateral cerebral sulcus
PRIMARY AUDITORY AREA
AUDITORY ASSOCIATION AREA

POSTERIOR
ANTERIOR

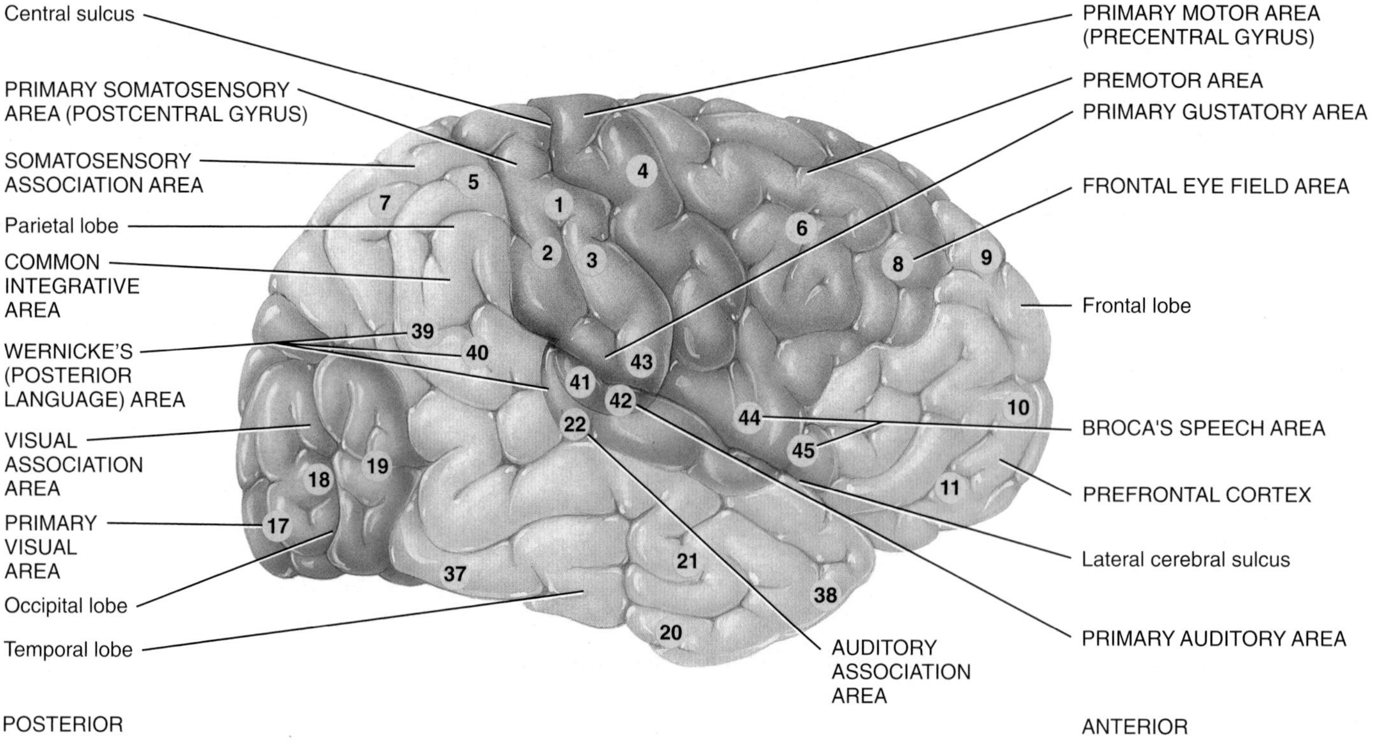

Lateral view of right cerebral hemisphere

 What area(s) of the cerebrum integrate(s) interpretation of visual, auditory, and somatic sensations? Translate(s) thoughts into speech? Control(s) skilled muscular movements? Interpret(s) sensations related to taste? Interpret(s) pitch and rhythm? Interpret(s) shape, color, and movement of objects? Control(s) voluntary scanning movements of the eyes?

impulses from the lips and fingertips than from the thorax or hip. The major function of the primary somatosensory area is to pinpoint the areas where sensations originate, so that you know exactly where on your body to swat at that mosquito.

- The **primary visual area** (area 17), located at the posterior tip of the occipital lobe mainly on the medial surface (next to the longitudinal fissure), receives visual information and is involved in visual perception.
- The **primary auditory area** (areas 41 and 42), located in the superior part of the temporal lobe near the lateral cerebral sulcus, receives information from auditory receptors and is involved in auditory perception.
- The **primary gustatory area** (area 43), located at the base of the postcentral gyrus superior to the lateral cerebral sulcus in the parietal cortex, receives impulses for taste and is involved in gustatory perception.
- The **primary olfactory area** (area 28), located on the medial aspect of the temporal lobe (and thus not visible in Figure 18.16), receives impulses for smell and is involved in olfactory perception.

Motor Areas

Motor output from the cerebral cortex flows mainly from the anterior part of each hemisphere. Among the most important motor areas are the following (Figure 18.16):

- The **primary motor area** (area 4) is located in the precentral gyrus of the frontal lobe. Each region in the primary motor area controls voluntary contractions of specific muscles or groups of muscles (see Figure 20.5b). Electrical stimulation of any point in the primary motor area causes contraction of specific skeletal muscle fibers on the opposite side of the body. As is true for the primary somatosensory area, body parts do not "map" to the primary motor area in proportion to their size. More cortical area is devoted to those muscles involved in skilled, complex, or delicate movements. For instance, the cortical region devoted to muscles that move the fingers is much larger than the region for muscles that move the toes.
- **Broca's speech area** (BRŌ-kaz) (areas 44 and 45) is located in the frontal lobe close to the lateral cerebral sulcus. Speaking and understanding language are complex activities that involve several sensory, association, and motor areas of the cortex. In about 97 percent of the population, these language areas are localized in the *left* hemisphere. The planning and production of speech occur in the *left* frontal lobe in most people. From Broca's speech area, nerve impulses pass to the premotor regions that control the muscles of the larynx, pharynx, and mouth. The impulses from the premotor area result in specific, coordinated muscle contractions. Simultaneously, impulses propagate from Broca's speech area to the primary motor area. From here, impulses also control the breathing muscles to regulate the proper flow of air past the vocal cords. The coordinated contractions of your speech and breathing muscles enable you to speak your thoughts. People who suffer a cerebrovascular accident (CVA) or stroke in this area can still have clear thoughts, but are unable to form words, a phenomenon referred to as *nonfluent aphasia* (described shortly).

Cerebral palsy (CP) is a motor disorder that results in the loss of muscle control and coordination; it is caused by damage of the motor areas of the brain during fetal life, birth, or infancy. Radiation during fetal life, temporary lack of oxygen during birth, and hydrocephalus during infancy may also cause cerebral palsy.

Association Areas

The association areas of the cerebrum consist of some motor and sensory areas, plus large areas on the lateral surfaces of the occipital, parietal, and temporal lobes and on the frontal lobes anterior to the motor areas. Association areas are connected with one another by **association tracts** and include the following (Figure 18.16):

- The **somatosensory association area** (areas 5 and 7) is just posterior to and receives input from the primary somatosensory area, as well as from the thalamus and other parts of the brain. This area permits you to determine the exact shape and texture of an object without looking at it, to determine the orientation of one object with respect to another as they are felt, and to sense the relationship of one body part to another. Another role of the somatosensory association area is the storage of memories of past sensory experiences, enabling you to compare current sensations with previous experiences. For example, the somatosensory association area allows you to recognize objects such as a pencil and a paperclip simply by touching them.
- The **prefrontal cortex (frontal association area)** is an extensive area in the anterior portion of the frontal lobe that is well-developed in primates, especially humans (areas 9, 10, 11, and 12; area 12 is not illustrated since it can be seen only in a medial view). This area has numerous connections with other areas of the cerebral cortex, thalamus, hypothalamus, limbic system, and cerebellum. The prefrontal cortex is concerned with the makeup of a person's personality, intellect, complex learning abilities, recall of information, initiative, judgment, foresight, reasoning, conscience, intuition, mood, planning for the future, and development of abstract ideas. A person with bilateral damage to the prefrontal cortices typically becomes rude, inconsiderate, incapable of accepting advice, moody, inattentive, less creative, unable to plan for the future, and incapable of anticipating the consequences of rash or reckless words or behavior.
- The **visual association area** (areas 18 and 19), located in the occipital lobe, receives sensory impulses from the primary visual area and the thalamus. It relates present and past visual experiences and is essential for recognizing and evaluating what is seen. For example, the visual association area allows you to recognize an object such as a spoon simply by looking at it.
- The **auditory association area** (area 22), located inferior and posterior to the primary auditory area in the temporal cortex, allows you to recognize a particular sound as speech, music, or noise.
- **Wernicke's (posterior language) area** (VER-ni-kēz) (area 22, and possibly areas 39 and 40), a broad region in the *left* temporal and parietal lobes, interprets the meaning of speech by recognizing spoken words. It is active as you translate words into thoughts. The regions in the *right* hemisphere that correspond to Broca's and Wernicke's areas also contribute to verbal communication by adding emotional content, such as anger or joy, to spoken words. Unlike those who have CVAs in Broca's area, people who suffer strokes in Wernicke's area can still speak, but cannot arrange words in a coherent fashion.
- The **common integrative area** (areas 5, 7, 39, and 40) is bordered by somatosensory, visual, and auditory association areas. It receives nerve impulses from these areas and from the primary gustatory area, primary olfactory area, the thalamus, and parts of the brain stem. This area integrates sensory information from the association areas and impulses from other areas, allowing the formation of thoughts based on a variety of sensory inputs. It then transmits signals to other parts of the brain for the appropriate response to the sensory signals it has interpreted.

- The **premotor area** (area 6) is a motor association area that is immediately anterior to the primary motor area. Neurons in this area communicate with the primary motor cortex, the sensory association areas in the parietal lobe, the basal nuclei, and the thalamus. The premotor area deals with learned motor activities of a complex and sequential nature. It generates nerve impulses that cause specific groups of muscles to contract in a specific sequence, as when you write your name. The premotor area also serves as a memory bank for such movements.
- The **frontal eye field area** (area 8) in the frontal cortex is sometimes included in the premotor area. It controls voluntary scanning movements of the eyes—like those you just used in reading this sentence.

The functions of the cerebrum are summarized in Table 18.2.

CLINICAL CONNECTION | *Aphasia*

Injury to language areas of the cerebral cortex results in **aphasia** (a-FĀ-zē-a; *a-*=without; *-phasia*=speech), an inability to use or comprehend words. Damage to Broca's speech area results in *nonfluent aphasia*, an inability to properly form words. People with nonfluent aphasia know what they wish to say but cannot properly speak the words. Damage to Wernicke's (posterior language) area, the common integrative area, or the auditory association area results in *fluent aphasia*, characterized by faulty understanding of spoken or written words. A person experiencing this type of aphasia may produce strings of words that have no meaning ("word salad"). For example, someone with fluent aphasia might say, "I rang car porch dinner light river pencil." •

TABLE 18.2

Summary of Functions of Principal Parts of Brain

PART	FUNCTION	PART	FUNCTION
Cerebrum 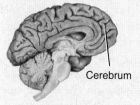	Sensory areas are involved in the perception of sensory information; motor areas control muscular movement; and association areas deal with more complex integrative functions such as memory, personality traits, and intelligence. Basal nuclei coordinate gross, automatic muscle movements and regulate muscle tone. Limbic system functions in emotional aspects of behavior related to survival.	**Brain stem** 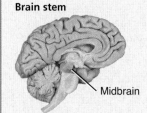	*Midbrain:* Relays motor output from the cerebral cortex to the pons and sensory input from the spinal cord to the thalamus. Superior colliculi coordinate movements of the eyeballs in response to visual and other stimuli, and the inferior colliculi coordinate movements of the head and trunk in response to auditory stimuli. Most of substantia nigra and red nucleus contribute to control of movement. Contains nuclei of origin for oculomotor (III) and trochlear (IV) nerves.
Diencephalon Epithalamus Thalamus 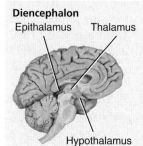 Hypothalamus	*Thalamus:* Relays almost all sensory input to the cerebral cortex. Provides crude perception of touch, pressure, pain, and temperature. Includes nuclei involved in movement planning and control. *Hypothalamus:* Controls and integrates activities of the autonomic nervous system and pituitary gland. Regulates emotional and behavioral patterns and circadian rhythms. Controls body temperature and regulates eating and drinking behavior. Helps maintain the waking state and establishes patterns of sleep. Produces the hormones oxytocin and antidiuretic hormone (ADH). *Epithalamus.* Consists of pineal gland, which secretes melatonin, and the habenular nuclei.	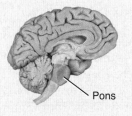	*Pons:* Relays impulses from one side of the cerebellum to the other and between the medulla and midbrain. Pontine respiratory group, together with the medulla, helps control breathing. Contains nuclei of origin for trigeminal (V), abducens (VI), facial (VII), and vestibulocochlear (VIII) nerves.
Cerebellum 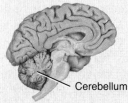	Compares intended movements with what is actually happening to smooth and coordinate complex, skilled movements. Regulates posture and balance. May have a role in cognition and language processing.	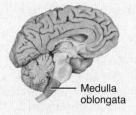	*Medulla oblongata:* Relays sensory input and motor output between other parts of the brain and the spinal cord. Reticular formation (also in pons, midbrain, and diencephalon) functions in consciousness and arousal. Vital centers regulate heartbeat and blood vessel diameter (cardiovascular center) and breathing (medullary respiratory center), together with the pons. Other centers coordinate swallowing, vomiting, coughing, sneezing, and hiccuping. Contains nuclei of origin for vestibulocochlear (VIII), glossopharyngeal (IX), vagus (X), accessory (XI), and hypoglossal (XII) nerves.

Hemispheric Lateralization

Although the brain is almost symmetrical on its right and left sides, subtle anatomical differences between the two hemispheres exist. For example, in about two-thirds of the population, the planum temporale, a region of the temporal lobe that includes Wernicke's (posterior language) area, is 50 percent larger on the left side than on the right side. This asymmetry appears in the human fetus at about 30 weeks of gestation. Physiological differences also exist; although the two hemispheres share performance of many functions, each hemisphere also specializes in performing certain unique functions. This functional asymmetry is termed **hemispheric lateralization** (Table 18.3).

Despite some dramatic differences in functions of the two hemispheres, there is considerable variation from one person to another. Also, lateralization seems less pronounced in females than in males, both for language (left hemisphere) and for visual and spatial skills (right hemisphere). For instance, females are less likely than males to suffer aphasia after damage to the left hemisphere. A possibly related observation is that the anterior commissure is 12 percent larger and the corpus callosum has a broader posterior portion in females. Recall that both the anterior commissure and the corpus callosum are commissural tracts that provide communication between the two hemispheres.

Table 18.3 summarizes some of the functional differences between the two cerebral hemispheres.

Memory

Memory is the process by which information acquired through learning is stored and retrieved. Without memory, we would repeat mistakes and be unable to learn. Similarly, we would not be able to repeat our successes or accomplishments, except by chance. For an experience to become part of memory, it must produce structural and functional changes in the brain. The parts of the brain known to be involved with memory include the association areas of the frontal, parietal, occipital, and temporal lobes; and parts of the limbic system, including the hippocampus, which is the one area of the brain that is seemingly able to produce new neurons. Memories for motor skills, such as how to serve a tennis ball, are stored in the basal nuclei and cerebellum as well as in the cerebral cortex.

Brain Waves

At any instant, brain neurons are generating millions of action potentials (nerve impulses). Taken together, these electrical signals are called **brain waves**. Brain waves generated by neurons close to the brain surface, mainly neurons in the cerebral cortex, can be detected by sensors called *electrodes* placed on the forehead and scalp. A record of such waves is called an **electroencephalogram (EEG)** (e-lek'-trō-en-SEF-a-lō-gram; *electro-* =electricity; *-gram* =recording). Electroencephalograms are useful in the study of normal brain functions, such as changes that occur during sleep, and in diagnosis of a variety of brain disorders, such as epilepsy, tumors, metabolic abnormalities, sites of trauma, and degenerative diseases. The EEG is also utilized to determine whether "life" is present, that is, to establish or confirm that brain death has occurred.

✓ CHECKPOINT

20. Compare the functions of the sensory, motor, and association areas of the cerebral cortex.
21. What is hemispheric lateralization?
22. Define memory.
23. What is the diagnostic value of an EEG?

TABLE 18.3

Functional Differences Between Right and Left Cerebral Hemispheres

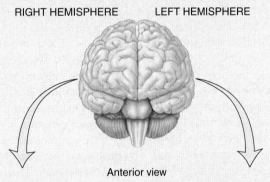

RIGHT HEMISPHERE   LEFT HEMISPHERE

Anterior view

RIGHT HEMISPHERE FUNCTIONS

Receives somatic sensory signals from and controls muscles on left side of body

Musical and artistic awareness

Space and pattern perception

Recognition of faces and emotional content of facial expressions

Generating emotional content of language

Generating mental images to compare spatial relationships

Identifying and discriminating among odors

Patients with damage in right hemisphere regions that correspond to Broca's and Wernicke's areas in the left hemisphere speak in a monotonous voice, having lost the ability to impart emotional inflection to what they say.

LEFT HEMISPHERE FUNCTIONS

Receives somatic sensory signals from and controls muscles on right side of body

Reasoning

Numerical and scientific skills

Ability to use and understand sign language

Spoken and written language

Persons with damage in the left hemisphere often exhibit aphasia.

Alzheimer's disease (AD) (ALTZ-hī-merz) is a disabling senile dementia, the loss of reasoning and ability to care for oneself, that afflicts about 11 percent of the population over age 65. In the United States, about 4 million people suffer from AD. Claiming over 100,000 lives a year, AD is the fourth leading cause of death among the elderly, after heart disease, cancer, and stroke. The cause of most AD cases is still unknown, but evidence suggests it is due to a combination of genetic factors, environmental or lifestyle factors, and the aging process. Mutations in three different genes (coding for presenilin-1, presenilin-2, and amyloid precursor protein) lead to early-onset forms of AD in afflicted families but account for less than 1 percent of all cases. An environmental risk factor for developing AD is a history of head injury. A similar dementia occurs in boxers, probably caused by repeated blows to the head.

Individuals with AD initially have trouble remembering recent events. They then become confused and forgetful, often repeating questions or getting lost while traveling to familiar places. Disorientation grows, memories of past events disappear, and episodes of paranoia, hallucination, or violent changes in mood may occur. As their minds continue to deteriorate, they lose their ability to read, write, talk, eat, or walk. The disease culminates in dementia. A person with AD usually dies of some complication that afflicts bedridden patients, such as pneumonia.

At autopsy, brains of AD victims show three distinct structural abnormalities:

1. **Loss of neurons that liberate acetylcholine.** A major center of neurons that liberate ACh is the *nucleus basalis*, a group of large cells below the globus pallidus. Axons of these neurons project widely throughout the cerebral cortex and limbic system. Their destruction is a hallmark of Alzheimer's disease.

2. **Beta-amyloid plaques** (bā′-ta-AM-i-loyd). Clusters of abnormal proteins deposited outside neurons.

3. **Neurofibrillary tangles** (noo′-rō-FI-bril-ler-ē). Abnormal bundles of filaments inside neurons in affected brain regions. These filaments consist of a protein called *tau* (TAW) that has been *hyperphosphorylated* (hī-per-fos-FOR-i-lā-ted), meaning too many phosphate groups have been added to it.

Drugs that inhibit acetylcholinesterase (AChE), the enzyme that inactivates ACh, improve alertness and behavior in about 5 percent of AD patients. Tacrine®, the first anticholinesterase inhibitor approved for treatment of AD in the United States, has significant side effects and requires dosing four times a day. Donepezil®, approved in 1998, is less toxic to the liver and has the advantage of once-a-day dosing. Some evidence suggests that vitamin E (an antioxidant), estrogen, ibuprofen, and ginkgo biloba extract may have slight beneficial effects in AD patients. In addition, researchers are currently exploring ways to develop drugs that will prevent beta-amyloid plaque formation by inhibiting the enzymes involved in beta-amyloid synthesis and by increasing the activity of the enzymes involved in beta-amyloid degradation. Researchers are also trying to develop drugs that will reduce the formation of neurofibrillary tangles by inhibiting the enzymes that hyperphosphorylate tau. •

18.8 AGING AND THE NERVOUS SYSTEM

 OBJECTIVE

• Describe the effects of aging on the nervous system.

The brain grows rapidly during the first few years of life. Growth is due mainly to an increase in the size of neurons already present, the proliferation and growth of neuroglia, the development of dendritic branches and synaptic contacts, and continuing myelination of axons. From early adulthood onward, brain mass declines. By the time a person reaches age 80, the brain weighs about 7 percent less than it did in young adulthood. Although the number of neurons present does not decrease very much, the number of synaptic contacts declines. Associated with the decrease in brain mass is a decreased capacity for sending nerve impulses to and from the brain. As a result, processing of information diminishes, conduction velocity decreases, voluntary motor movements slow down, and reflex times increase.

 CHECKPOINT

24. How is brain mass related to age?

18.9 CRANIAL NERVES

OBJECTIVE

• Identify the cranial nerves by name, number, and type, and give the function of each.

The 12 pairs of **cranial nerves** are so named because they pass through various foramina in the bones of the cranium and arise from the brain inside the cranial cavity. Like the 31 pairs of spinal nerves, they are part of the peripheral nervous system (PNS). Each cranial nerve has both a number, designated by a roman numeral, and a name (Figure 18.17). The numbers indicate the order, from carnial to caudal, in which the nerves arise from the brain. The names designate a nerve's distribution or function.

Three cranial nerves (I, II, and VIII) carry axons of sensory neurons into the brain and thus are called **special sensory nerves**. These nerves are unique to the head and are associated with the special senses of smelling, seeing, and hearing. The cell bodies of most sensory neurons are located in ganglia outside the brain. (One exception is the proprioceptive cell bodies of the mesencephalic nucleus of the trigeminal (V) nerve, which has pseudounipolar cell bodies typical of a sensory ganglion, but is located in the central nervous system, classifying it as a nucleus and not a ganglion.)

Five cranial nerves (III, IV, VI, XI, and XII) are classified as **motor nerves** because they contain only axons of motor neurons as they leave the brain stem. The cell bodies of motor neurons lie in nuclei within the brain. Motor axons that innervate skeletal muscles are of two types:

1. *Branchial motor axons* innervate skeletal muscles that develop from the pharyngeal (branchial) arches (see Figure 4.13). These neurons leave the brain through the mixed cranial nerves and the accessory nerve.

2. *Somatic motor axons* innervate skeletal muscles that develop from head somites (eye muscles and tongue muscles). These neurons exit the brain through four motor cranial nerves (III, IV, VI, and XII).

Motor axons that innervate smooth muscle, cardiac muscle, and glands are called *autonomic motor axons* and are part of the parasympathetic division.

The remaining four cranial nerves (V, VII, IX, and X) are **mixed nerves**—they contain axons of both sensory neurons entering the brain stem and motor neurons leaving the brain stem.

Figure 18.17 Cranial nerves.

🔑 Cranial nerves are so named because they pass through various foramina in the cranial bones and they originate from the brain inside the cranial cavity.

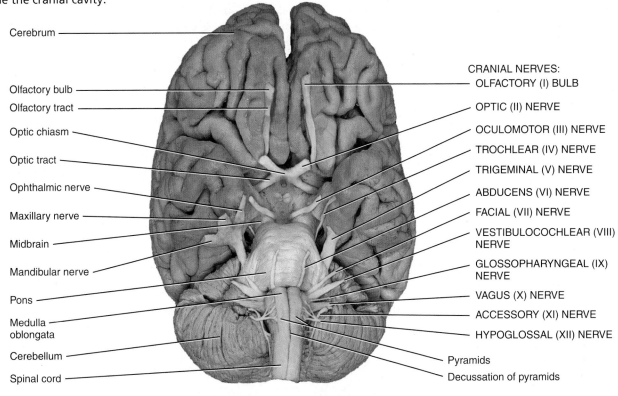

Cerebrum

Olfactory bulb
Olfactory tract
Optic chiasm
Optic tract
Ophthalmic nerve
Maxillary nerve
Midbrain
Mandibular nerve
Pons
Medulla oblongata
Cerebellum
Spinal cord

CRANIAL NERVES:
OLFACTORY (I) BULB
OPTIC (II) NERVE
OCULOMOTOR (III) NERVE
TROCHLEAR (IV) NERVE
TRIGEMINAL (V) NERVE
ABDUCENS (VI) NERVE
FACIAL (VII) NERVE
VESTIBULOCOCHLEAR (VIII) NERVE
GLOSSOPHARYNGEAL (IX) NERVE
VAGUS (X) NERVE
ACCESSORY (XI) NERVE
HYPOGLOSSAL (XII) NERVE
Pyramids
Decussation of pyramids

(a) Inferior aspect of brain

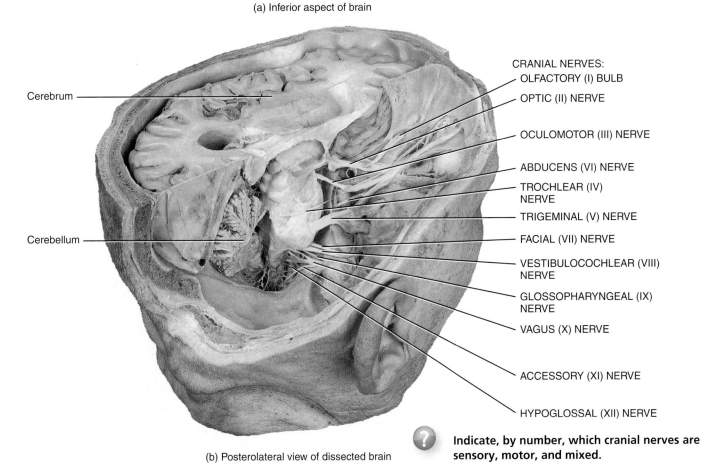

Cerebrum

Cerebellum

CRANIAL NERVES:
OLFACTORY (I) BULB
OPTIC (II) NERVE
OCULOMOTOR (III) NERVE
ABDUCENS (VI) NERVE
TROCHLEAR (IV) NERVE
TRIGEMINAL (V) NERVE
FACIAL (VII) NERVE
VESTIBULOCOCHLEAR (VIII) NERVE
GLOSSOPHARYNGEAL (IX) NERVE
VAGUS (X) NERVE
ACCESSORY (XI) NERVE
HYPOGLOSSAL (XII) NERVE

(b) Posterolateral view of dissected brain

 **Indicate, by number, which cranial nerves are sensory, motor, and mixed.**

Each cranial nerve is covered in detail in Exhibits 18.A–18.J. Although the cranial nerves are mentioned singly in Exhibits with regard to their type, location, and function, remember that they are paired structures.

Following the exhibits, Table 18.4 presents a summary of the components and principal functions of the cranial nerves.

✔ CHECKPOINT
25. How are cranial nerves named and numbered?
26. What is the difference between a mixed cranial nerve and a sensory cranial nerve?

EXHIBIT 18.A Olfactory (I) Nerve *(Figure 18.18)*

OBJECTIVE

• Identify the termination of the olfactory (I) nerve in the brain, the foramen through which it passes, and its function.

The **olfactory (I) nerve** (ol-FAK-tō-rē; *olfact-*=to smell) is entirely sensory; it contains axons that conduct nerve impulses for olfaction, the sense of smell (Figure 18.18). The olfactory epithelium occupies the superior part of the nasal cavity, covering the inferior surface of the cribriform plate and extending down along the superior nasal concha. The olfactory receptors within the olfactory epithelium are bipolar neurons. Each has a single odor-sensitive, knob-shaped dendrite projecting from one side of the cell body and an unmyelinated axon extending from the other side. Bundles of axons of olfactory receptors extend through about 20 olfactory foramina in the cribriform plate of the ethmoid bone on each side of the nose. These 40 or so bundles of axons collectively form the right and left olfactory nerves.

Olfactory nerves join the brain in paired masses of gray matter called the **olfactory bulbs**, two extensions of the brain that rest on the cribriform plate. Within the olfactory bulbs, the axon terminals of olfactory receptors form synapses with the dendrites and cell bodies of the next neurons in the olfactory pathway. The axons of these neurons make up the **olfactory tracts**, which extend posteriorly from the olfactory bulbs (see Figures 18.17 and 18.18). Axons in the olfactory tracts end in the primary olfactory area in the temporal lobe of the cerebral cortex.

✓ CHECKPOINT

27. Where is the olfactory epithelium located?

Figure 18.18 Olfactory (I) nerve.

🔑 The olfactory epithelium is located on the inferior surface of the cribriform plate and superior nasal conchae.

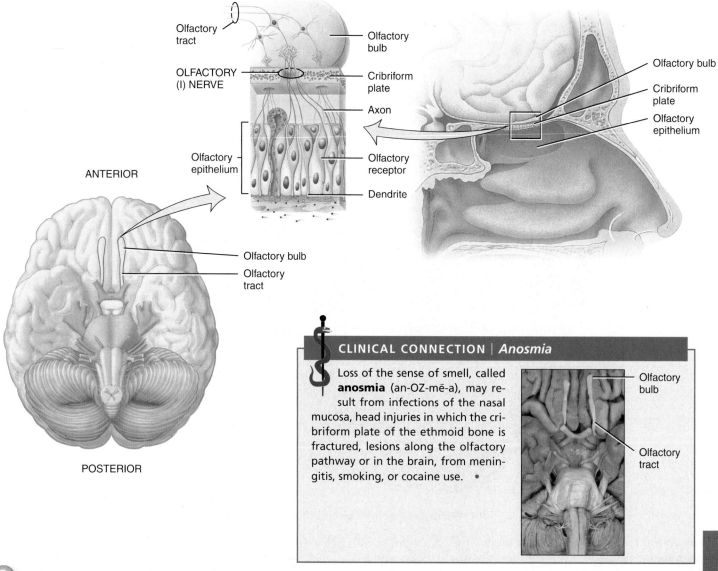

CLINICAL CONNECTION | *Anosmia*

Loss of the sense of smell, called **anosmia** (an-OZ-mē-a), may result from infections of the nasal mucosa, head injuries in which the cribriform plate of the ethmoid bone is fractured, lesions along the olfactory pathway or in the brain, from meningitis, smoking, or cocaine use. •

❓ Where do axons in the olfactory tracts terminate?

EXHIBIT 18.B **617**

EXHIBIT 18.B Optic (II) Nerve *(Figure 18.19)*

OBJECTIVE

• Identify **termination of the optic (II) nerve in the brain, the foramen through which it exits the skull, and its function.**

The **optic (II) nerve** (OP-tik; *opti-*=the eye, vision) is entirely sensory; it contains axons that conduct nerve impulses for vision (Figure 18.19). In the retina, rods and cones initiate visual signals and relay them to bipolar cells, which transmit the signals to ganglion cells. Axons of all ganglion cells in the retina of each eye join to form an optic nerve, which passes through the optic foramen. About 10 mm (0.4 in.) posterior to the eyeball, the two optic nerves merge to form the **optic chiasm** (KĪ-azm=a crossover, as in the letter X). Within the chiasm, axons from the medial half of each eye cross to the opposite side; axons from the lateral half

remain on the same side. Posterior to the chiasm, the regrouped axons, some from each eye, form the **optic tracts**. Most axons in the optic tracts end in the lateral geniculate nucleus of the thalamus. There they synapse with neurons whose axons extend to the primary visual area in the occipital lobe of the cerebral cortex (area 17 in Figure 18.16). A few axons pass through the lateral geniculate nucleus and then extend to the superior colliculi of the midbrain and to motor nuclei of the brainstem where they synapse with motor neurons that control the extrinsic and intrinsic eye muscles.

CHECKPOINT

28. Trace the sequence of nerve cells that process visual impulses within the retina.

Figure 18.19 Optic (II) nerve.

In sequence, visual signals are relayed from rods and cones to bipolar cells to ganglion cells.

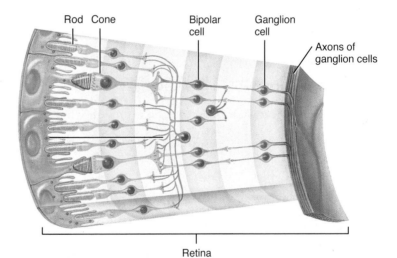
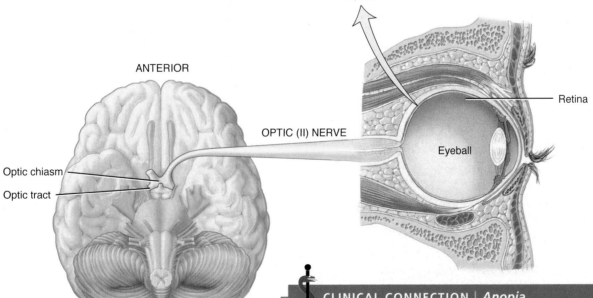

Where do most axons in the optic tracts terminate?

CLINICAL CONNECTION | *Anopia*

Fractures in the orbit, brain lesions, damage along the visual pathway, diseases of the nervous system (such as multiple sclerosis), pituitary gland tumors, or cerebral aneurysms (enlargements of blood vessels due to weakening of their walls) may result in visual field defects and loss of visual acuity. Blindness due to a defect in or loss of one or both eyes is called **anopia** (an-Ō-pē-a). •

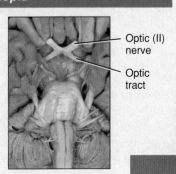

EXHIBIT 18.C

Oculomotor (III), Trochlear (IV), and Abducens (VI) Nerves *(Figure 18.20)*

 OBJECTIVE

• Identify the origins of the oculomotor (III), trochlear (IV), and abducens (VI) nerves in the brain, the foramen through which each exits the skull, and their functions.

The oculomotor, trochlear, and abducens nerves are the cranial nerves that control the muscles that move the eyeballs. They are all motor nerves that contain only motor axons as they exit the brain stem. Sensory axons from the extrinsic eyeball muscles begin their course toward the brain in each of these nerves, but eventually these sensory axons leave these nerves to join the ophthalmic branch of the trigeminal nerve. The sensory axons *do not* return to the brain in the oculomotor, trochlear, or abducens nerves. The cell bodies of the unipolar sensory neurons reside in the mesencephalic nucleus and they enter the midbrain

via the trigeminal nerve. These axons convey nerve impulses from the extrinsic eyeball muscles for *proprioception*, the perception of the movements and position of the body independent of vision.

The **oculomotor (III) nerve** (ok′-ū-lō-MŌ-tor; *oculo-*=eye; *-motor*=a mover) has its motor nucleus in the anterior part of the midbrain. The oculomotor nerve extends anteriorly and divides into superior and inferior branches, both of which pass through the superior orbital fissure into the orbit (Figure 18.20a). Axons in the superior branch innervate the superior rectus (an extrinsic eyeball muscle) and the levator palpebrae superioris (the muscle of the upper eyelid). Axons in the inferior branch supply the medial rectus, inferior rectus, and inferior oblique muscles—all extrinsic eyeball muscles. These somatic motor neurons control movements of the eyeball and upper eyelid.

Figure 18.20 Oculomotor (III), trochlear (IV), and abducens (VI) nerves.

The oculomotor (III) nerve has the widest distribution among extrinsic eye muscles.

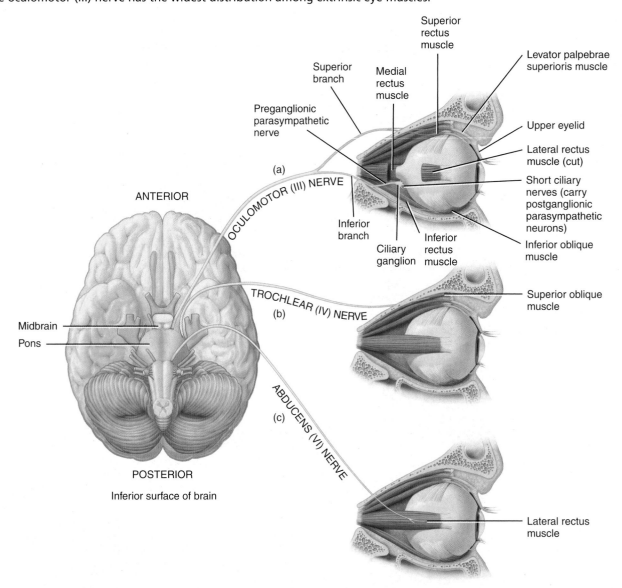

 Which branch of the oculomotor (III) nerve is distributed to the superior rectus muscle? Which is the smallest cranial nerve?

EXHIBIT 18.C **619**

The inferior branch of the oculomotor nerve also supplies parasympathetic motor axons to intrinsic eyeball muscles, which consist of smooth muscle. They include the ciliary muscle of the eyeball and the circular muscles (sphincter pupillae) of the iris. Parasympathetic impulses propagate from the accessory oculomotor nucleus in the midbrain to the **ciliary ganglion**, a synaptic relay center for the two motor neurons of the parasympathetic nervous system. From the ciliary ganglion, parasympathetic motor axons extend to the ciliary muscle, which adjusts the lens for near vision (*accommodation*). Other parasympathetic motor axons stimulate the circular muscles of the iris to contract when bright light stimulates the eye, causing a decrease in the size of the pupil (*constriction*).

The **trochlear (IV) nerve** (TRŌK-lē-ar; *trochle-*=a pulley) is the smallest of the 12 cranial nerves and is the only one that arises from the posterior aspect of the brain stem. The somatic motor neurons originate in the trochlear nucleus in the midbrain, and axons from the nucleus cross to the opposite side as they exit the brain on its posterior aspect. The nerve then wraps around the pons and exits through the superior orbital fissure into the orbit. These somatic motor axons innervate the superior oblique muscle of the eyeball, another extrinsic eyeball muscle that controls movement of the eyeball (Figure 18.20b).

Neurons of the **abducens (VI) nerve** (ab-DOO-senz; *ab-*=away; *-ducens*=to lead) originate from the abducens nucleus in the pons. Somatic motor axons extend from the nucleus to the lateral rectus muscle of the eyeball, an extrinsic eyeball muscle, through the superior orbital fissure into the orbit (Figure 18.20c). The abducens nerve is so named because nerve impulses cause abduction (lateral rotation) of the eyeball.

✔ CHECKPOINT

29. How are the oculomotor (III), trochlear (IV), and abducens (VI) nerves related functionally?

CLINICAL CONNECTION | *Strabismus, Ptosis, and Diplopia*

Damage to the oculomotor (III) nerve causes **strabismus** (stra-BIZ-mus) (a condition in which both eyes do not fix on the same object since one or both eyes may turn inward or outward), **ptosis** (TŌ-sis) (drooping) of the upper eyelid, dilation of the pupil, movement of the eyeball downward and outward on the damaged side, loss of accommodation for near vision, and **diplopia** (di-PLŌ-pē-a) (double vision).

Trochlear (IV) nerve damage can also result in strabismus and diplopia.

With damage to the abducens (VI) nerve, the affected eyeball cannot move laterally beyond the midpoint, and the eyeball is usually directed medially. This leads to strabismus and diplopia.

Causes of damage to the oculomotor, trochlear, and abducens nerves include trauma to the skull or brain, compression resulting from aneurysms, and lesions of the superior orbital fissure. Individuals with damage to these nerves are forced to tilt their heads in various directions to help bring the affected eyeball into the correct frontal plane. •

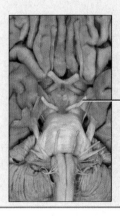
Oculomotor (III) nerve

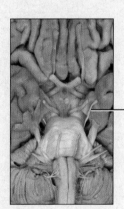
Trochlear (IV) nerve

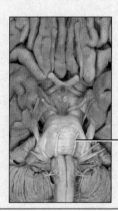
Abducens (VI) nerve

EXHIBIT 18.D Trigeminal (V) Nerve *(Figure 18.21)*

OBJECTIVE

• Identify the origin of the trigeminal (V) nerve from the brain, describe the foramina through which each of its three major branches exits the skull, and explain the function of each branch.

The **trigeminal (V) nerve** (trī-JEM-i-nal=triple, for its three branches) is a mixed cranial nerve and the largest of the cranial nerves. The trigeminal nerve emerges from two roots on the anterolateral surface of the pons. The large sensory root has a swelling called the **trigeminal (semilunar) ganglion**, which is located in a fossa on the inner surface of the petrous portion of the temporal bone. The ganglion contains cell bodies of most of the primary sensory neurons. Neurons of the smaller motor root originate in a nucleus in the pons.

As indicated by its name, the trigeminal nerve has three branches: ophthalmic, maxillary, and mandibular (Figure 18.21). The **ophthalmic nerve** (of-THAL-mik; *ophthalm-*=the eye), the smallest branch, passes into the orbit via the superior orbital fissure. The **maxillary nerve** (*maxilla*=upper jaw bone) is intermediate in size between the ophthalmic and mandibular nerves and passes through the foramen rotundum. The **mandibular nerve** (*mandibula*=lower jaw bone), the largest branch, passes through the foramen ovale.

Sensory axons in the trigeminal nerve carry nerve impulses for touch, pain, and thermal sensations (heat and cold). The ophthalmic nerve contains sensory axons from the skin over the upper eyelid, cornea, lacrimal glands, upper part of the nasal cavity, side of the nose, forehead, and anterior half of the scalp. The maxillary nerve includes sensory axons from the mucosa of the nose,

Figure 18.21 Trigeminal (V) nerve.

The three branches of the trigeminal (V) nerve leave the cranium through the superior orbital fissure, foramen rotundum, and foramen ovale.

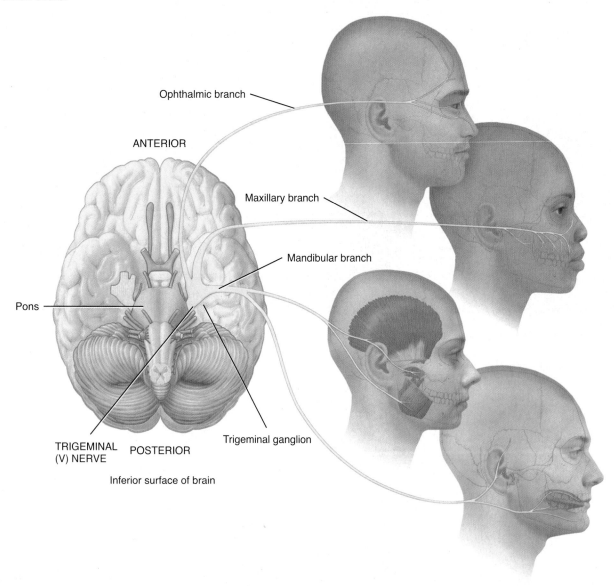

How does the trigeminal (V) nerve compare in size to the other cranial nerves?

EXHIBIT 18.D **621**

palate, part of the pharynx, upper teeth, upper lip, and lower eyelid. The mandibular nerve contains sensory axons from the anterior two-thirds of the tongue (not taste), cheek and mucosa deep to it, lower teeth, skin over the mandible and side of the head anterior to the ear, and mucosa of the floor of the mouth. The sensory axons from the three branches enter the trigeminal ganglion, where their cell bodies are located, and terminate in nuclei in the pons. The trigeminal nerve also contains sensory axons from proprioceptors (receptors that provide information regarding body position and movements) located in the muscles of mastication and extrinsic muscles of the eyeball, but the cell bodies of these neurons are located in the mesencephalic nucleus.

Branchial motor neurons of the trigeminal nerve are part of the mandibular nerve and supply muscles of mastication (masseter, temporalis, medial pterygoid, lateral pterygoid, anterior belly of digastric, and mylohyoid muscles, as well as the tensor veli palatini muscle in the soft palate and tensor tympani muscle in the middle ear). These motor neurons mainly control chewing movements.

● CHECKPOINT

30. What are the three branches of the trigeminal (V) nerve, and which branch is the largest?

CLINICAL CONNECTION | *Trigeminal Neuralgia*

Neuralgia (pain) relayed via one or more branches of the trigeminal (V) nerve, caused by conditions such as inflammation or lesions, is called **trigeminal neuralgia (tic douloureux)**. This is a sharp cutting or tearing pain that lasts for a few seconds to a minute and is caused by anything that presses on the trigeminal nerve or its branches. It occurs almost exclusively in people over 60 and can be the first sign of a disease, such as multiple sclerosis or diabetes, or lack of vitamin B_{12}, which damages the nerves. Injury of the mandibular nerve may cause paralysis of the chewing muscles and a loss of the sensations of touch, temperature, and proprioception in the lower part of the face. Dentists apply anesthetic drugs to branches of the maxillary nerve for anesthesia of upper teeth and to branches of the mandibular nerve for anesthesia of lower teeth. •

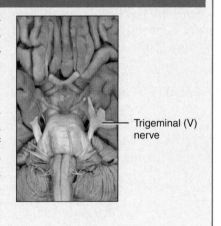

Trigeminal (V) nerve

EXHIBIT 18.E Facial (VII) Nerve *(Figure 18.22)*

 OBJECTIVE

• Identify the origins of the facial (VII) nerve in the brain, the foramen through which it exits the skull, and its function.

The **facial (VII) nerve** (FĀ-shal=face) is a mixed cranial nerve. Its sensory axons extend from the taste buds of the anterior two-thirds of the tongue and enter the temporal bone to join the facial nerve. From here the sensory axons pass to the **geniculate ganglion** (je-NIK-ū-lāt), a cluster of cell bodies of sensory neurons of the facial nerve within the temporal bone, and end in the pons. From the pons, axons extend to the thalamus, and then to the gustatory areas of the cerebral cortex (Figure 18.22). The sensory portion of the facial nerve also contains axons from skin in the ear canal that relay touch, pain, and thermal sensations. Additionally, proprioceptors from muscles of the face and scalp relay information through their cell bodies in the mesencephalic nucleus.

Axons of branchial motor neurons arise from a nucleus in the pons and exit the stylomastoid foramen to innervate middle ear, facial, scalp, and neck muscles. Nerve impulses propagating along these axons cause contraction of the muscles of facial expression plus the stylohyoid muscle, the posterior belly of the digastric muscle, and the stapedius muscle. The facial nerve innervates more named muscles than any other nerve in the body.

Axons of the parasympathetic motor neurons run in branches of the facial nerve and end in two ganglia: the **pterygopalatine ganglion** (ter′-i-gō-PAL-a-tīn) and the **submandibular ganglion**, respectively. From synaptic relays in the two ganglia, postganglionic parasympathetic motor axons extend to lacrimal glands (which secrete tears), nasal glands, palatine glands, and saliva-producing sublingual and submandibular glands.

 CHECKPOINT

31. Why is the facial (VII) nerve considered the major motor nerve of the head?

Figure 18.22 Facial (VII) nerve.

🔑 The facial (VII) nerve causes contraction of the muscles of facial expression.

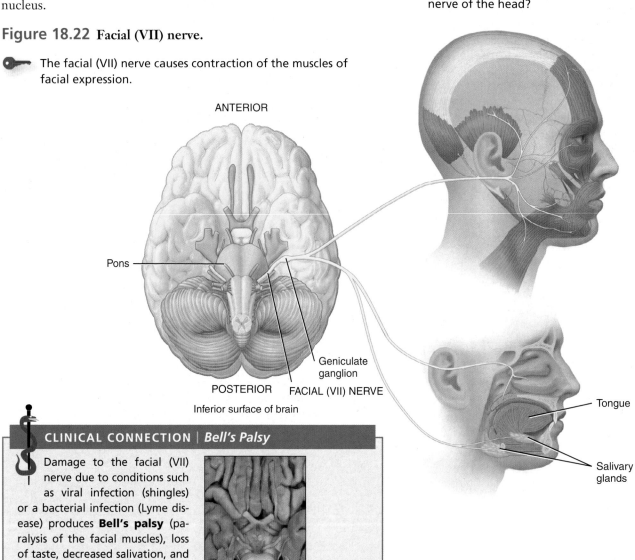

ANTERIOR

Pons

Geniculate ganglion

POSTERIOR   FACIAL (VII) NERVE

Inferior surface of brain

Tongue

Salivary glands

CLINICAL CONNECTION | *Bell's Palsy*

 Damage to the facial (VII) nerve due to conditions such as viral infection (shingles) or a bacterial infection (Lyme disease) produces **Bell's palsy** (paralysis of the facial muscles), loss of taste, decreased salivation, and loss of ability to close the eyes, even during sleep. The nerve can also be damaged by trauma, tumors, and stroke. •

Facial (VII) nerve

❓ **Where do the motor fibers for the facial (VII) nerve originate?**

EXHIBIT 18.F **623**

EXHIBIT 18.F Vestibulocochlear (VIII) Nerve *(Figure 18.23)*

 OBJECTIVE

- Identify the origin of the vestibulocochlear (VIII) nerve in the brain, the foramen through which it exits the skull, and the functions of each of its branches.

The **vestibulocochlear (VIII) nerve** (vest-tib-ū-lō-KOK-lē-ar; *vestibulo-*=small cavity; *-cochlear*=a spiral, snail-like) was formerly known as the *acoustic* or *auditory nerve*. It is a sensory cranial nerve and has two branches, the vestibular branch and the cochlear branch (Figure 18.23). The **vestibular branch** carries impulses for equilibrium and the **cochlear branch** carries impulses for hearing.

Sensory axons in the vestibular branch extend from the semicircular canals, the saccule, and the utricle of the inner ear to the **vestibular ganglion**, where the cell bodies of the neurons are located (see Figure 21.13b), and end in vestibular nuclei in the pons and cerebellum. Some sensory axons also enter the cerebellum via the inferior cerebellar peduncle.

Sensory axons in the cochlear branch arise in the spiral organ (organ of Corti) in the cochlea of the internal ear. The cell bodies of cochlear branch sensory neurons are located in the **spiral ganglion** of the cochlea (see Figure 21.13b). From there, axons extend to cochlear nuclei in the medulla oblongata and end in the thalamus.

The nerve contains some motor fibers but they do not innervate muscle tissue. Instead, they modulate the hair cells in the inner ear.

 CHECKPOINT

32. What are the functions of each of the two branches of the vestibulocochlear (VIII) nerve?

Figure 18.23 Vestibulocochlear (VIII) nerve.

 The vestibular branch of the vestibulocochlear (VIII) nerve carries impulses for equilibrium, while the cochlear branch carries impulses for hearing.

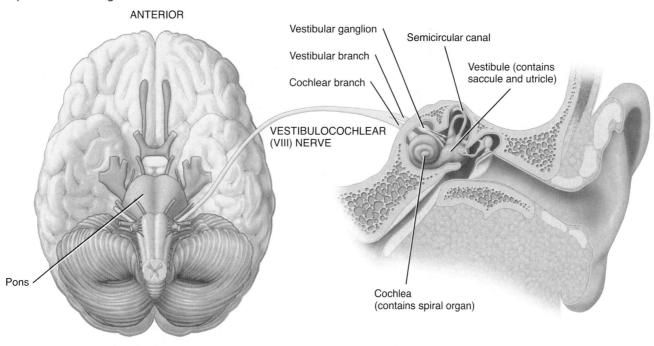

ANTERIOR

Vestibular ganglion

Semicircular canal

Vestibular branch

Vestibule (contains saccule and utricle)

Cochlear branch

VESTIBULOCOCHLEAR (VIII) NERVE

Pons

Cochlea (contains spiral organ)

POSTERIOR

CLINICAL CONNECTION | *Vertigo, Ataxia, and Nystagmus*

Injury to the vestibular branch of the vestibulocochlear (VIII) nerve may cause **vertigo** (ver-TĪ-gō), a subjective feeling that one's own body or the environment is rotating, **ataxia** (a-TAK-sē-a) (muscular incoordination), and **nystagmus** (nis-TAG-mus) (involuntary rapid movement of the eyeball). Injury to the cochlear branch may cause **tinnitus** (ringing in the ears) or deafness. The vestibulocochlear nerve may be injured as a result of conditions such as trauma, lesions, or middle ear infections. •

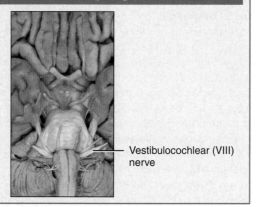

Vestibulocochlear (VIII) nerve

 **Where do sensory axons of the cochlear branch originate?**

EXHIBIT 18.G   Glossopharyngeal (IX) Nerve *(Figure 18.24)*

OBJECTIVE

• Identify the origin of the glossopharyngeal (IX) nerve in the brain, the foramen through which it exits the skull, and its function.

The **glossopharyngeal (IX) nerve** (glos'-Ō-fa-RIN-jē-al; *glosso-*=tongue; *-pharyngeal*=throat) is a mixed cranial nerve (Figure 18.24). Sensory axons of the glossopharyngeal nerve arise from (1) taste buds on the posterior one-third of the tongue, (2) proprioceptors from some swallowing muscles supplied by the motor portion, (3) baroreceptors (pressure-monitoring receptors) in the carotid sinus that monitor blood pressure, (4) chemoreceptors (receptors that monitor blood levels of oxygen and carbon dioxide) in the carotid bodies near the carotid arteries (see Figure 23.14) and aortic bodies near the arch of the aorta (see Figure 23.14), and

(5) the external ear to convey touch, pain, and thermal (heat and cold) sensations. The cell bodies of these sensory neurons are located in the **superior** and **inferior ganglia**. From the ganglia, sensory axons pass through the jugular foramen and end in the medulla.

Axons of motor neurons in the glossopharyngeal nerve arise in nuclei of the medulla and exit the skull through the jugular foramen. Branchial motor neurons innervate the stylopharyngeus muscle, which assists in swallowing, and axons of parasympathetic motor neurons stimulate the parotid gland to secrete saliva. The postganglionic cell bodies of parasympathetic motor neurons are located in the **otic ganglion**.

 CHECKPOINT

33. Which other cranial nerves are also distributed to the tongue?

Figure 18.24 Glossopharyngeal (IX) nerve.

 Sensory fibers of the glossopharyngeal (IX) nerve supply the taste buds.

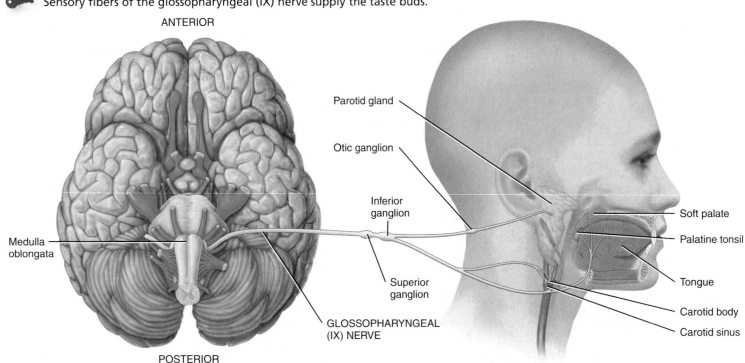

Inferior surface of brain

CLINICAL CONNECTION | *Dysphagia, Aptyalia, and Ageusia*

Injury to the glossopharyngeal (IX) nerve causes **dysphagia** (dis-FĀ-jē-a) or difficulty in swallowing, **aptyalia** (ap'-tē-A-lē-a) or reduced secretion of saliva, loss of sensation in the throat, and **ageusia** (a-GOO-sē-a) or loss of taste sensation. The glossopharyngeal nerve may be injured as a result of conditions such as trauma or lesions.

The **pharyngeal (gag) reflex** is a rapid and intense contraction of the pharyngeal muscles. Except for normal swallowing, the pharyngeal reflex is designed to prevent choking by not allowing objects to enter the throat. The reflex is initiated by contact of an object with the roof of the mouth, back of the tongue, area around the tonsils, and back of the throat. Stimulation of receptors in these areas sends sensory information to the brain via the glossopharyngeal (IX) and vagus (X) nerves. Returning motor information via the same nerves results in contraction of the pharyngeal muscles. People with a hyperactive pharyngeal reflex have difficulty swallowing pills and are very sensitive to various medical and dental procedures. •

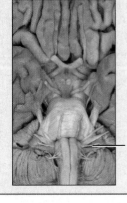

Glossopharyngeal (IX) nerve

 Through which foramen does the glossopharyngeal (IX) nerve exit the skull?

EXHIBIT 18.H **625**

EXHIBIT 18.H Vagus (X) Nerve *(Figure 18.25)*

 OBJECTIVE

- Identify the origin of the vagus (X) nerve in the brain, the foramen through which it exits the skull, and its function.

The **vagus (X) nerve** (VĀ-gus=vagrant or wandering) is a mixed cranial nerve that is distributed from the head and neck into the thorax and abdomen (Figure 18.25). The nerve derives its name from its wide distribution. In the neck, it lies medial and posterior to the internal jugular vein and common carotid artery.

Sensory axons in the vagus nerve arise from the skin of the external ear for touch, pain, and thermal sensations; a few taste buds in the epiglottis and pharynx; and proprioceptors in muscles of the neck and throat. Also, sensory axons come from baroreceptors in the carotid sinus and chemoreceptors in the carotid and aortic bodies. The majority of sensory neurons come from visceral sensory receptors in most organs of the thoracic and abdominal cavities that convey sensations (such as hunger, fullness, and discomfort) from these organs. The sensory neurons have cell bodies in the **superior** and **inferior ganglia** and then pass through the jugular foramen to end in the medulla and pons.

The branchial motor neurons, which run briefly with the accessory nerve, arise from nuclei in the medulla oblongata and supply muscles of the pharynx, larynx, and soft palate that are used in swallowing, vocalization, and coughing. Historically these motor neurons have been called the cranial accessory nerve, but these fibers actually belong to the vagus (X) nerve.

Axons of parasympathetic motor neurons in the vagus nerve originate in nuclei of the medulla and almost all of the organs that arise from the embryonic gut tube. These axons supply the lungs, heart, glands of the gastrointestinal (GI) tract and smooth muscle of the respiratory passageways, esophagus, stomach, gallbladder, small intestine, and most of the large intestine (see Figure 19.3). Parasympathetic motor axons initiate smooth muscle contractions in the gastrointestinal tract to aid motility and stimulate secretion by digestive glands; activate smooth muscle to constrict respiratory passageways; and decrease heart rate.

✓ CHECKPOINT

34. On what basis is the vagus (X) nerve named?

Figure 18.25 Vagus (X) nerve.

🔑 The vagus (X) nerve is widely distributed in the head, neck, thorax, and abdomen.

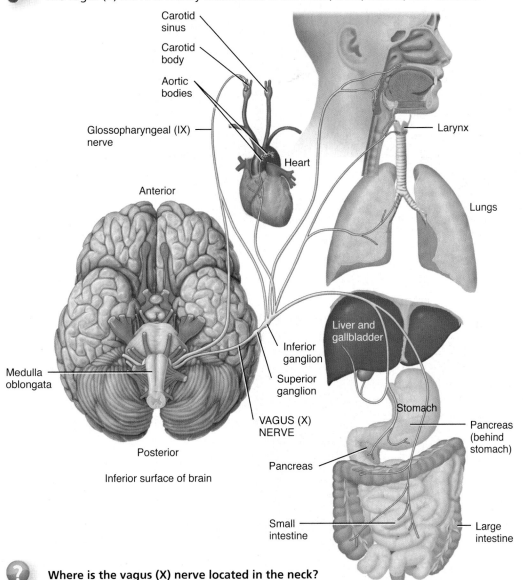

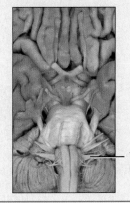

CLINICAL CONNECTION | *Vagal Neuropathy, Dysphagia, and Tachycardia*

Injury to the vagus (X) nerve due to conditions such as trauma or lesions causes **vagal neuropathy** or interruptions of sensations from many organs in the thoracic and abdominal cavities, **dysphagia** (dis-FĀ-jē-a) or difficulty in swallowing, paralysis of the vocal cords, and **tachycardia** (tak′-i-KAR-dē-a) or increased heart rate. •

❓ **Where is the vagus (X) nerve located in the neck?**

EXHIBIT 18.I   Accessory (XI) Nerve *(Figure 18.26)*

OBJECTIVE

• Identify the origin of the accessory (XI) nerve in the spinal cord, the foramina through which it first enters and then exits the skull, and its function.

The **accessory (XI) nerve** (ak-SES-ō-rē=assisting) is a branchial motor cranial nerve (Figure 18.26). Historically it has been divided into two parts, a cranial accessory nerve and a spinal accessory nerve. The cranial accessory nerve actually is part of the vagus (X) nerve (see Exhibit 18.H). The "old" spinal accessory nerve is the accessory nerve we discuss in this Exhibit. Its motor axons arise in the anterior gray horn of the first five segments of the cervical portion of the spinal cord. The axons from the segments exit the spinal cord laterally and come together, ascend through the foramen magnum, and then exit through the jugular foramen

along with the vagus and glossopharyngeal nerves. The accessory nerve conveys motor impulses to the sternocleidomastoid and trapezius muscles to coordinate head movements. Sensory axons in the accessory nerve, which originate from proprioceptors in the sternocleidomastoid and trapezius muscles, begin their course toward the brain in the accessory nerve, but eventually leave the nerve to join nerves of the cervical plexus. From the cervical plexus they enter the spinal cord via the posterior roots of cervical spinal nerves, where their cell bodies are located in the posterior root ganglia of those nerves. In the spinal cord the axons ascend to nuclei in the medulla oblongata.

CHECKPOINT

35. Where do the motor axons of the accessory (XI) nerve originate?

Figure 18.26 Accessory (XI) nerve.

 The accessory (XI) nerve exits the cranium through the jugular foramen.

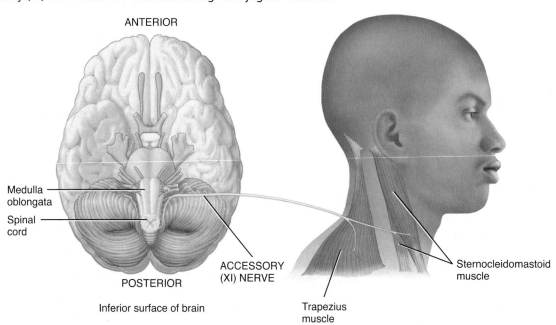

Medulla oblongata

Spinal cord

ANTERIOR

POSTERIOR

Inferior surface of brain

ACCESSORY (XI) NERVE

Trapezius muscle

Sternocleidomastoid muscle

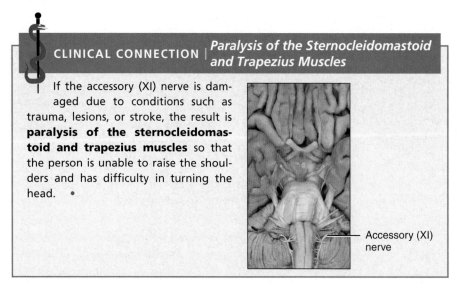

CLINICAL CONNECTION | *Paralysis of the Sternocleidomastoid and Trapezius Muscles*

If the accessory (XI) nerve is damaged due to conditions such as trauma, lesions, or stroke, the result is **paralysis of the sternocleidomastoid and trapezius muscles** so that the person is unable to raise the shoulders and has difficulty in turning the head. •

Accessory (XI) nerve

?   How does the accessory (XI) nerve differ from the other cranial nerves?

EXHIBIT 18.J **627**

EXHIBIT 18.J   Hypoglossal (XII) Nerve *(Figure 18.27)*

 OBJECTIVE

• Identify the origin of the hypoglossal (XII) nerve in the brain, the foramen through which it exits the skull, and its function.

The **hypoglossal (XII) nerve** (hī-pō-GLOS-al; *hypo-*=below; *-glossal*=tongue) is a motor cranial nerve. The somatic motor axons originate in the hypoglossal nucleus in the medulla oblongata, exit the medulla on its anterior surface, and pass through the hypoglossal canal to supply the muscles of the tongue. These axons conduct nerve impulses for speech and swallowing. The sensory axons do not return to the brain in the hypoglossal nerve. Instead, sensory axons that originate from proprioceptors in the tongue muscles begin their course toward the brain in the hypoglossal nerve but leave the nerve to join cervical spinal nerves and end in the medulla oblongata, again entering the central nervous system via posterior roots of cervical spinal nerves (Figure 18.27).

 CHECKPOINT

36. In what portion of the brain does the hypoglossal nucleus originate?

Figure 18.27 Hypoglossal (XII) nerve.

🔑 The hypoglossal (XII) nerve exits the cranium through the hypoglossal canal.

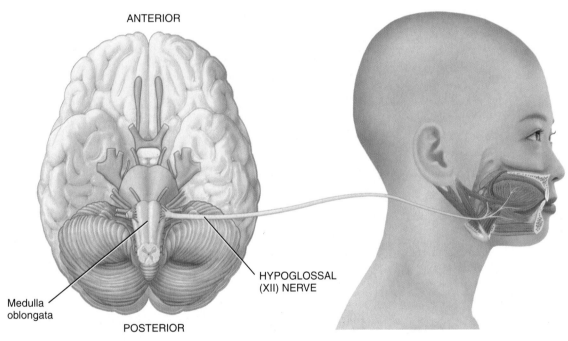

ANTERIOR

HYPOGLOSSAL (XII) NERVE

Medulla oblongata

POSTERIOR

Inferior surface of brain

CLINICAL CONNECTION | *Dysarthria and Dysphagia*

Injury to the hypoglossal (XII) nerve results in difficulty in chewing, **dysarthria** (dis-AR-thrē-a) or difficulty in speaking, and **dysphagia** (dis-FĀ-jē-a) or difficulty in swallowing. The tongue, when protruded, curls toward the affected side, and the affected side atrophies. The hypoglossal nerve may be injured as a result of conditions such as trauma, lesions, stroke, amyotropic lateral sclerosis (Lou Gehrig's disease), or infections in the brain stem. •

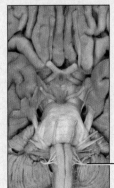

Hypoglossal (XII) nerve

 What important motor functions are related to the hypoglossal (XII) nerve?

TABLE 18.4

SUMMARY OF CRANIAL NERVES*

CRANIAL NERVE	COMPONENTS	PRINCIPAL FUNCTIONS
Olfactory (I)	*Special sensory*	Olfaction (smell)
Optic (II)	*Special sensory*	Vision (sight)
Oculomotor (III)	*Motor*	
	• Somatic	Movement of eyeballs and upper eyelid
	• Motor (autonomic)	Adjusts lens for near vision (accommodation)
		Constriction of the pupil
Trochlear (IV)	*Motor*	
	• Somatic	Movement of eyeballs
Trigeminal (V)	*Mixed*	
	• Sensory	Touch, pain, and thermal sensations from the scalp, face, and oral cavity (including the teeth and anterior 2/3 of the tongue)
	• Motor (branchial)	Chewing and control of middle ear muscle
Abducens (VI)	*Motor*	
	• Somatic	Movement of eyeballs
Facial (VII)	*Mixed*	
	• Sensory	Taste from anterior 2/3 of tongue
		Touch, pain, and thermal sensations from skin in external ear canal
	• Motor (branchial)	Control of muscles of facial expression and middle ear muscle
	• Motor (autonomic)	Secretion of tears and saliva
Vestibulocochlear (VIII)	*Special sensory*	Hearing and equilibrium
Glossopharyngeal (IX)	*Mixed*	
	• Sensory	Taste from posterior 1/3 of tongue
		Proprioception in some swallowing muscles
		Monitoring of blood pressure and oxygen and carbon dioxide levels in blood
		Touch, pain, and thermal sensations from skin of external ear and upper pharynx
	• Motor (branchial)	Assists in swallowing
	• Motor (autonomic)	Secretion of saliva
Vagus (X)	*Mixed*	
	• Sensory	Taste from epiglottis
		Proprioception from throat and voice box muscles
		Monitoring of blood pressure and oxygen and carbon dioxide levels in blood
		Touch, pain, and thermal sensations from skin of external ear
		Sensations from thoracic and abdominal organs
	• Motor (branchial)	Swallowing, vocalization, and coughing
	• Motor (autonomic)	Motility and secretion of gastrointestinal organs
		Constriction of respiratory passageways
		Decrease in heart rate
Accessory (XI)	*Motor*	
	• Branchial	Movement of head and pectoral girdle
Hypoglossal (XII)	*Motor*	
	• Somatic	Speech, manipulation of food, and swallowing

*MNEMONIC FOR CRANIAL NERVES:

Oh	Oh	Oh	To	Touch	And	Feel	Very	Green	Vegetables		AH!
Olfactory	Optic	Oculomotor	Trochlear	Trigeminal	Abducens	Facial	Vestibulocochlear	Glossopharyngeal	Vagus	Accessory	Hypoglossal

KEY MEDICAL TERMS ASSOCIATED WITH THE BRAIN AND THE CRANIAL NERVES

Apraxia (a-PRAK-sē-a; *-praxia*=coordinated) Inability to carry out purposeful movements in the absence of paralysis.

Brain injuries Injuries commonly associated with head trauma and hypoxia (cellular oxygen deficiency). Result in part from displacement and distortion of neuronal tissue at the moment of impact. Additional tissue damage occurs when normal blood flow is restored after a period of ischemia (reduced blood flow).

Concussion (kon-KUSH-un) An injury characterized by an abrupt, but temporary, loss of consciousness (from seconds to hours), disturbances of vision, and problems with equilibrium. It is caused by a blow to the head or the sudden stopping of a moving head and is the most common brain injury.

Contusion (kon-TOO-zhun) Bruising of the brain due to trauma, and includes the leakage of blood from microscopic vessels. It is usually associated with a concussion. In a contusion, the pia mater may be torn, allowing blood to enter the subarachnoid space. The area most commonly affected is the frontal lobe.

Epilepsy (ep′-i-LEP-sē) A disorder characterized by short, recurrent attacks (epileptic seizures) involving motor, sensory, or psychological malfunction; it almost never affects intelligence. The attacks are initiated by abnormal, synchronous electrical discharges from millions of neurons in the brain. Epilepsy has many causes, including brain damage at birth; metabolic disturbances; infections; toxins; vascular disturbances; head injuries; and tumors and abscesses of the brain.

Insomnia (in-SOM-nē-a; *in-*=not; *-somnia*=sleep) Difficulty in falling asleep and staying asleep.

Laceration (las-er-Ā-shun) A tear of the brain, usually from a skull fracture or a gunshot wound. A laceration results in rupture of large blood vessels, with bleeding into the brain and subarachnoid space. Consequences include cerebral hematoma (localized pool of blood, usually clotted, that swells against the brain tissue), edema, and increased intracranial pressure.

Microcephaly (mī-krō-SEF-a-lē; *micro-*=small; *-cephal*=head) A congenital condition that involves the development of a small brain and skull and frequently results in mental retardation.

Reye's syndrome (RĪZ) A disorder characterized by vomiting and brain dysfunction (disorientation, lethargy, and personality changes) that may progress to coma and death; occurs after a viral infection, particularly chickenpox or influenza, most often in children or teens who have taken aspirin.

Stupor (STOO-por) Unresponsiveness from which a patient can be aroused only briefly and only by vigorous and repeated stimulation.

CHAPTER REVIEW AND RESOURCE SUMMARY

WileyPLUS

Review

Resource

18.1 Development and General Structure of the Brain

1. During embryological development, primary brain vesicles form from the neural tube and serve as forerunners of various parts of the brain.
2. The telencephalon forms the cerebrum, the diencephalon develops into the thalamus and hypothalamus, the mesencephalon develops into the midbrain, the metencephalon develops into the pons and cerebellum, and the myelencephalon forms the medulla.
3. The major parts of the brain are the brain stem, cerebellum, diencephalon, and cerebrum.

Anatomy Overview - The Brain
Table 18.1 - Development of the Brain
Cadaver Video - The Nervous System - Brain

18.2 Protection and Blood Supply

1. The brain is protected by cranial bones and the cranial meninges.
2. The cranial meninges are continuous with the spinal meninges. From superficial to deep the cranial meninges are the dura mater, arachnoid mater, and pia mater.
3. Cerebrospinal fluid (CSF) is formed in the choroid plexuses and circulates through the lateral ventricles, third ventricle, fourth ventricle, subarachnoid space, and central canal. Most of the fluid is absorbed into the blood across the arachnoid villi of the superior sagittal blood sinus.
4. Cerebrospinal fluid provides mechanical protection, chemical protection, and circulation of nutrients.
5. Blood flow to the brain is mainly via the internal carotid and vertebral arteries.
6. Any interruption of the oxygen or glucose supply to the brain can result in weakening of, permanent damage to, or death of brain cells.
7. The blood–brain barrier (BBB) causes different substances to move between the blood and the brain tissue at different rates and prevents the movement of some substances from blood into the brain.

Anatomy Overview - Ventricles

18.3 The Brain Stem and Reticular Formation

1. The medulla oblongata is continuous with the superior part of the spinal cord and contains both motor and sensory tracts. It contains nuclei that are reflex centers for regulation of heart rate and blood vessel diameter (cardiovascular center), respiratory rate (medullary respiratory center), vasoconstriction, swallowing, coughing, vomiting, and sneezing. It also contains nuclei associated with the vestibulocochlear (VIII), glossopharyngeal (IX), vagus (X), accessory (XI), and hypoglossal (XII) nerves.
2. The pons is superior to the medulla. It connects the spinal cord with the brain and links parts of the brain with one another by way of tracts. Pontine nuclei relay nerve impulses related to voluntary skeletal movements from the cerebral cortex to the cerebellum. The pons contains the pontine respiratory group, which helps control breathing. It contains nuclei associated with the trigeminal (V), abducens (VI), and facial (VII) nerves and the vestibular branch of the vestibulocochlear (VIII) nerve.
3. The midbrain, which connects the pons and diencephalon and surrounds the cerebral aqueduct, conveys motor impulses from the cerebrum to the cerebellum and spinal cord, sends sensory impulses from the spinal cord to the thalamus, and regulates auditory and visual reflexes. It also contains nuclei associated with the oculomotor (III) and trochlear (IV) nerves.

Anatomy Overview - The Brainstem
Figure 18.8 - Midbrain and Reticular Activating System (RAS)

Review	**Resource**

4. A large part of the brain stem consists of small areas of gray matter and white matter called the reticular formation, which helps maintain consciousness, causes awakening from sleep, and contributes to regulating muscle tone.

18.4 The Cerebellum

1. The cerebellum occupies the inferior and posterior aspects of the cranial cavity. It consists of two lateral hemispheres and a medial, constricted vermis.
2. It connects to the brain stem by three pairs of cerebellar peduncles.
3. The cerebellum coordinates contractions of skeletal muscles and maintains normal muscle tone, posture, and balance.

Anatomy Overview - The Cerebellum

18.5 The Diencephalon

1. The diencephalon thalamus, hypothalamus, and epithalamus surrounds the third ventricle.
2. The thalamus is superior to the midbrain and contains nuclei that serve as relay stations for all sensory impulses to the cerebral cortex. It also allows crude appreciation of pain, temperature, and pressure and mediates some motor activities.
3. The hypothalamus is inferior to the thalamus. It controls and integrates the autonomic nervous system, connects the nervous and endocrine systems, functions in rage and aggression, controls body temperature, regulates food and fluid intake, and establishes circadian rhythms.
4. The epithalamus consists of the pineal gland and the habenular nuclei. The pineal gland secretes melatonin, which is thought to promote sleep and to help set the body's biological clock.
5. Circumventricular organs (CVOs) can monitor chemical changes in the blood because they lack the blood–brain barrier.

Anatomy Overview - The Diencephalon
Figure 18.11 - Hypothalamus

18.6 The Cerebrum

1. The cerebrum is the largest part of the brain. Its cortex contains gyri (convolutions), fissures, and sulci.
2. The cerebral hemispheres are divided into lobes: frontal, parietal, temporal, occipital, and insula.
3. The white matter of the cerebrum is deep to the cortex and consists of myelinated and unmyelinated axons extending to other regions as association, commissural, and projection fibers.
4. The basal nuclei, several groups of nuclei in each cerebral hemisphere, help control large, automatic movements of skeletal muscles and help regulate muscle tone.
5. The limbic system encircles the upper part of the brain stem and the corpus callosum. It functions in emotional aspects of behavior and memory.
6. Table 18.2 summarizes the functions of various parts of the brain.

Anatomy Overview - The Cerebrum
Figure 18.12 - Cerebrum

18.7 Functional Organization of the Cerebral Cortex

1. The sensory areas of the cerebral cortex allow perception of sensory impulses. The motor areas govern muscular movement. The association areas are concerned with more complex integrative functions.
2. The primary somatosensory area receives nerve impulses from somatic sensory receptors for touch, proprioception, pain, and temperature. Each point within the area receives impulses from a specific part of the face or body.
3. The primary visual area receives impulses that convey visual information. The primary auditory area interprets the basic characteristics of sound such as pitch and rhythm. The primary gustatory area receives impulses for taste. The primary olfactory area receives impulses for smell.
4. Motor areas include the primary motor area, which controls voluntary contractions of specific muscles or groups of muscles, and Broca's speech area, which controls production of speech.
5. The somatosensory association area permits you to determine the exact shape and texture of an object without looking at it and to sense the relationship of one body part to another. The visual association area relates present to past visual experiences and is essential for recognizing and evaluating what is seen. The auditory association area deals with the meanings of sounds.
6. Wernicke's (posterior language) area interprets the meaning of speech by translating words into thoughts. The common integrative area integrates sensory interpretations from the association areas and impulses from other areas, allowing thoughts based on sensory inputs.
7. The premotor area generates nerve impulses that cause specific groups of muscles to contract in specific sequences. The frontal eye field area controls voluntary scanning movements of the eyes.
8. Subtle anatomical differences exist between the two hemispheres, and each has unique functions. Each hemisphere receives sensory signals from and controls the opposite side of the body. The left hemisphere is more important for language, numerical and scientific skills, and reasoning. The right hemisphere is more important for musical and artistic awareness, spatial and pattern perception, recognition of faces, emotional content of language, identifying odors, and generating mental images of sight, sound, touch, taste, and smell.
9. Memory, the ability to store and recall thoughts, involves persistent changes in the brain.
10. Brain waves generated by the cerebral cortex are recorded from the surface of the head in an electroencephalogram (EEG). The EEG may be used to diagnose epilepsy, infections, and tumors.

Exercise - Paint the Functional Areas of the Cerebral Cortex

18.8 Aging and the Nervous System

1. The brain grows rapidly during the first few years of life.
2. Age-related effects involve loss of brain mass and decreased capacity for sending nerve impulses.

Review

18.9 Cranial Nerves

1. Twelve pairs of cranial nerves originate from the nose, eyes, inner ear, brain stem, and spinal cord.
2. The cranial nerves are named primarily based on their distribution and are numbered I–XII in order of attachment to the brain. Each one is covered in detail in Exhibits 18.A–18.J. Table 18.4 summarizes the components and principal functions of the cranial nerves.

Resource

Animation - Cranial Somatic Sensory Pathways - Cranial Nerves
Animation - Direct Somatic Motor Pathways - Corticobulbar Tracts
Exercise - Cranial Nerve Target Practice - Sensory Functions
Exercise - Cranial Nerve Target Practice - Motor Functions

CRITICAL THINKING QUESTIONS

1. An elderly relative suffered a stroke and now has difficulty moving her right arm and also has speech problems. What areas of the brain were damaged by the stroke?

2. Wolfgang partied a little too hard one night and passed out drunk in his bathroom at home. He awoke with a lump on his head from hitting the sink, but he thought he was okay. However, when Tony came over a day later, he said, "Man, are you still drunk? You're not making any sense, and you look like you're going to pass out again!" Before Wolfgang could answer, he vomited. What do you think—is Wolfgang still drunk? Explain your answer.

3. Dr. M. D. Hatter has developed a drug that inhibits the activity of the amygdala of the limbic system. Do you think this is a good thing or a bad thing? Explain your answer.

4. One of the postmortem findings in the brains of Alzheimer's patients is exaggerated widening of cerebral sulci as well as narrowing of cerebral gyri. Can you explain the relationship between this anatomical change and the symptoms, which include loss of reasoning and ability to care for oneself?

5. Dwayne's first trip to the dentist after a 10-year absence resulted in extensive dental work. He received numbing injections of anesthetic in several locations during the session. While having lunch right after the appointment, soup dribbled out of Dwayne's mouth because he had no feeling in his left upper lip, right lower lip, and tip of his tongue. What happened to Dwayne?

ANSWERS TO FIGURE QUESTIONS

18.1 The mesencephalon does not develop into a secondary brain vesicle.

18.2 The largest part of the brain is the cerebrum.

18.3 From superficial to deep, the three cranial meninges are the dura mater, arachnoid mater, and pia mater.

18.4 The brain stem is anterior to the fourth ventricle, and the cerebellum is posterior to it.

18.5 Cerebrospinal fluid is reabsorbed into the blood by the arachnoid villi that project into the dural venous sinuses.

18.6 The medulla oblongata contains the pyramids; the midbrain contains the cerebral peduncles; pons means "bridge."

18.7 Decussation means crossing to the opposite side. The functional consequence of decussation of the pyramids is that each side of the cerebrum controls muscles on the opposite side of the body.

18.8 The cerebral peduncles conduct motor impulses from the cerebrum to the pons, medulla, and spinal cord and sensory impulses from the medulla to the thalamus.

18.9 The cerebellar peduncles carry information into and out of the cerebellum.

18.10 In about 70 percent of human brains, the interthalamic adhesion (intermediate mass) connects the right and left halves of the thalamus.

18.11 Posterior to anterior, the four major hypothalamic regions are the mammillary, tuberal, supraoptic, and preoptic regions.

18.12 The gray matter enlarges more rapidly during development, in the process producing convolutions or gyri (folds), sulci (shallow grooves), and fissures (deep grooves).

18.13 Association tracts connect gyri of the same hemisphere; commissural tracts connect gyri in opposite hemispheres; projection tracts connect the cerebrum with the thalamus, brain stem, and spinal cord.

18.14 The basal nuclei are lateral, superior, and inferior to the thalamus.

18.15 The hippocampus is the part of the limbic system that functions with the cerebrum in memory.

18.16 Common integrative area integrates interpretation of visual, auditory, and somatic sensations; motor speech area translates thoughts into speech; premotor area controls skilled muscular movements; gustatory areas interpret sensations related to taste; auditory areas interpret pitch and rhythm; visual areas interpret shape, color, and movement of objects; frontal eye field area controls voluntary scanning movements of the eyes.

18.17 Cranial nerves I, II, and VII are sensory; II, IV, VI, XI, and XII are motor; and V, VII, IX, and X are mixed.

18.18 Axons in the olfactory tracts terminate in the primary olfactory area in the temporal lobe of the cerebral cortex.

18.19 Most axons in the optic tracts terminate in the primary visual area in the occipital lobe of the cerebral cortex.

18.20 The superior branch of the oculomotor nerve is distributed to the superior rectus muscle; the trochlear nerve is the smallest cranial nerve.

18.21 The trigeminal nerve is the largest cranial nerve.

18.22 Motor axons of the facial nerve originate in the pons.

18.23 Sensory axons of the cochlear branch originate in the spiral organ, the organ of hearing.

18.24 The glossopharyngeal nerve exits the skull through the jugular foramen.

18.25 The vagus nerve is located medial and posterior to the internal jugular vein and common carotid artery in the neck.

18.26 The accessory nerve is the only cranial nerve that originates from both the brain and spinal cord.

18.27 Two important motor functions of the hypoglossal nerve are speech and swallowing.

19 THE AUTONOMIC NERVOUS SYSTEM

INTRODUCTION It is the end of the semester, you have studied diligently for your anatomy final, and now it is time to take the exam. As you enter the crowded lecture hall and take a seat, you sense tension in the room as the other students nervously chatter about last-minute details they think will be important to know for the test. Suddenly you feel your heart race with excitement—or is that apprehension? You notice that your mouth becomes somewhat dry, and you break out in a cold sweat. You also notice that your breathing is a little bit faster and deeper. As you wait for the professor to pass out the test, these symptoms become more and more pronounced. Finally the test arrives at your desk. As you slowly flip through the exam to get a feel for the questions being asked, you recognize that you can answer them all with confidence. What a relief! Your symptoms begin to disappear as you focus on transferring your knowledge from your brain to the paper.

Most of the effects just described fall under the control of the autonomic nervous system. As you learned in Chapter 16, the **autonomic nervous system (ANS)** (aw′-tō-NOM-ik; *auto-*=self; *-nomic*=law) is a division of the nervous system that regulates cardiac and smooth muscle and glandular tissue. The word *autonomic* is based on the Latin words for *self* and *law* because the ANS was originally believed to be self-governing. The ANS consists of (1) *autonomic sensory neurons* in visceral organs and in blood vessels that convey information to (2) *integrating centers* in the central nervous system (CNS), (3) *autonomic motor neurons* that propagate from the CNS to various effector tissues to regulate the activity of smooth muscle, cardiac muscle, and many glands, and (4) the *enteric division*, a specialized network of nerves and ganglia forming an independent nerve network within the wall of the gastrointestinal (GI) tract. Functionally, the ANS usually operates without conscious control. However, centers in the hypothalamus and brain stem do regulate ANS reflexes, so it is not completely self-governing.

In this chapter, we compare structural and functional features of the autonomic nervous system with those of the somatic nervous system, which was introduced in Chapter 16. Then we discuss the anatomy of the ANS and compare the organization and actions of its two major parts, the sympathetic and parasympathetic divisions. •

Did you ever wonder how some blood pressure medications exert their effects through the autonomic nervous system?

19.1 COMPARISON OF SOMATIC AND AUTONOMIC NERVOUS SYSTEMS

 OBJECTIVE
- Compare the structures and functions of the somatic and autonomic nervous systems.

Somatic Nervous System

As you learned in Chapter 16, the somatic nervous system includes both sensory and motor neurons. Sensory neurons convey input from receptors for somatic senses (pain, thermal, tactile, and proprioceptive sensations; see Chapter 20) and from receptors for the special senses (vision, audition, gustation, olfaction, and equilibrium; see Chapter 21). Normally, all of these sensations are consciously perceived. Somatic motor neurons innervate skeletal muscles—the effectors of the somatic nervous system—and produce both reflexive and voluntary movements of the musculoskeletal system. When a somatic motor neuron stimulates a skeletal muscle, the muscle contracts; the effect always is excitation. If somatic motor neurons cease to stimulate a muscle, the result is a paralyzed, limp muscle that has no muscle tone. In addition, even though we are generally not conscious of breathing, the muscles that generate respiratory movements are skeletal muscles controlled by somatic motor neurons. If the respiratory motor neurons become inactive, breathing stops. A few skeletal muscles, such as those in the middle ear, are controlled by reflexes and cannot be contracted voluntarily.

Autonomic Nervous System

The main input to the ANS comes from **autonomic (visceral) sensory neurons**. Mostly, these neurons are associated with **interoceptors** (receptors inside the body), such as chemoreceptors that monitor blood CO_2 level, and mechanoreceptors that detect the degree of stretch in the walls of organs or blood vessels. These sensory signals are not consciously perceived most of the time, although intense activation may produce conscious sensations. Two examples of perceived visceral sensations are sensations of pain from damaged viscera and angina pectoris (chest pain) from inadequate blood flow to the heart. Some sensations monitored by somatic sensory (Chapter 20) and special sensory neurons (Chapter 21) also influence the ANS. For example, pain can produce dramatic changes in some autonomic activities.

Autonomic motor neurons regulate visceral activities by either increasing (exciting) or decreasing (inhibiting) activities in their effector tissues, which are cardiac muscle, smooth muscle, and glands. Changes in the diameter of the pupils, dilation and constriction of blood vessels, and adjustment of the rate and force of the heartbeat are examples of autonomic motor responses. Unlike skeletal muscle, tissues innervated by the ANS often function to some extent even if their nerve supply is damaged. For example, the heart continues to beat when it is removed for transplantation. Single-unit smooth muscle, like that found in the lining of the gastrointestinal tract, contracts rhythmically on its own, and glands produce some secretions in the absence of ANS control.

Most autonomic responses cannot be consciously altered or suppressed to any great degree. You probably cannot voluntarily slow your heartbeat to half its normal rate. For this reason, some autonomic responses are the basis for polygraph ("lie detector") tests. Nevertheless, practitioners of yoga or other techniques of meditation may learn how to regulate at least some of their autonomic activities through long practice. (*Biofeedback*, in which monitoring devices display information about a body function such as heart rate or blood pressure, enhances the ability to learn such conscious control.) Signals from the general somatic and special senses, acting via the limbic system, also influence responses of autonomic motor neurons. For example, seeing a bike about to hit you, hearing the squealing brakes of a nearby car as you cross the street, or being grabbed from behind by an attacker would increase the rate and force of your heartbeat.

Comparison of Somatic and Autonomic Motor Neurons

Recall from Chapter 10 that the axon of a single, myelinated somatic motor neuron extends from the central nervous system (CNS) all the way to the skeletal muscle fibers in its motor unit

633

(Figure 19.1a). By contrast, most autonomic motor pathways consist of two motor neurons *in series*, one following the other (Figure 19.1b). The first neuron (the **preganglionic neuron**) has its cell body in the CNS; its myelinated axon extends from the CNS to an **autonomic ganglion**. (Recall that a *ganglion* is a collection of neuronal cell bodies outside the CNS.) The cell body of the second neuron (the **postganglionic neuron**) is in that autonomic ganglion; its unmyelinated axon extends directly from the ganglion to the effector (smooth muscle, cardiac muscle, or a gland). In some autonomic pathways, the preganglionic neuron extends to specialized cells of the adrenal

medullae (inner portions of the adrenal glands) called *chromaffin cells*, which secrete epinephrine and norepinephrine. These cells develop from the same embryonic cells that give rise to the autonomic ganglia. All somatic motor neurons release only acetylcholine (ACh) as their neurotransmitter; autonomic motor neurons release either ACh or norepinephrine (NE).

Unlike somatic output (motor), the output part of the ANS has two divisions: the **sympathetic division** and the **parasympathetic division**. Most organs have **dual innervation**, that is, they receive impulses from both sympathetic and parasympathetic neurons. In some organs, nerve impulses from one division of

Figure 19.1 Motor neuron pathways in the (a) somatic nervous system and (b) autonomic nervous system (ANS). Note that autonomic motor neurons release either acetylcholine (ACh) or norepinephrine (NE); somatic motor neurons release only ACh.

🔑 Somatic nervous system stimulation always excites its effectors (skeletal muscle fibers); stimulation by the autonomic nervous system either excites or inhibits visceral effectors.

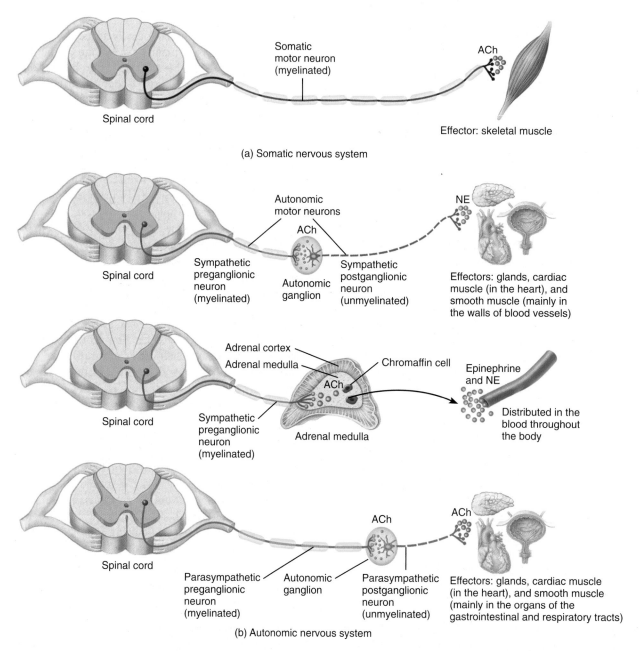

(a) Somatic nervous system

(b) Autonomic nervous system

❓ **What does dual innervation mean?**

TABLE 19.1

Comparison of the Somatic and Autonomic Nervous Systems

	SOMATIC NERVOUS SYSTEM	AUTONOMIC NERVOUS SYSTEM
Sensory input	Special senses and somatic senses	Mainly from interoceptors; some from special senses and somatic senses
Control of motor output	Voluntary control from cerebral cortex, with contributions from basal nuclei, cerebellum, spinal cord	Involuntary control from limbic system, hypothalamus, brain stem, and spinal cord; limited control from brain stem and cerebral cortex
Motor neuron pathway	One-neuron pathway: Somatic motor neurons extending from CNS synapse directly with effector	Usually two-neuron pathway: Preganglionic CNS neurons → synapse in an autonomic ganglion → postganglionic neurons → a visceral effector OR Preganglionic CNS neurons → chromaffin cells of adrenal medullae → via blood vessels → effector
Neurotransmitters and hormones	All somatic motor neurons release ACh	All preganglionic axons release acetylcholine (ACh); most sympathetic postganglionic neurons release norepinephrine (NE); those to most sweat glands release ACh; all parasympathetic postganglionic neurons release ACh; adrenal medullae release epinephrine and norepinephrine
Effectors	Skeletal muscle	Smooth muscle, cardiac muscle, and glands
Responses	Contraction of skeletal muscle	Contraction or relaxation of smooth muscle; increased or decreased rate and force of contraction of cardiac muscle; increased or decreased secretions of glands

the ANS stimulate the organ to increase its activity (excitation), and impulses from the other division decrease the organ's activity (inhibition). For example, an increased rate of nerve impulses from the *sympathetic* division increases heart rate; an increased rate of nerve impulses from the *parasympathetic* division decreases heart rate. The majority of the output of the sympathetic division, often called the *fight-or-flight division*, is directed at the smooth muscle of blood vessels. Sympathetic activities result in increased alertness and enhanced metabolic activities in order to prepare the body for an emergency situation. Responses to such situations, which may occur during physical activity or emotional stress, include a rapid heart rate, faster breathing rate, dilation of the pupils, dry mouth, sweaty but cool skin, dilation of blood vessels to organs involved in combating stress (such as the heart and skeletal muscles), constriction of blood vessels to organs not involved in combating stress (for example, the gastrointestinal tract and kidneys), and the release of glucose from the liver.

The parasympathetic division is often referred to as the *rest-and-digest division* because its activities conserve and restore body energy during times of rest or digesting a meal; the majority of its output is directed to the smooth muscle and glandular tissue of the gastrointestinal and respiratory tracts. The parasympathetic division conserves energy and replenishes nutrient stores. Although both the sympathetic and parasympathetic divisions are concerned with maintaining homeostasis, they do so in dramatically different ways.

Table 19.1 summarizes the comparisons of the somatic and autonomic nervous systems presented in this section.

 CHECKPOINT

1. Why is the autonomic nervous system so named?
2. What are the main input and output components of the autonomic nervous system?

19.2 ANATOMY OF AUTONOMIC MOTOR PATHWAYS

 OBJECTIVES

- Outline the development of the autonomic motor pathways and explain how it relates to their adult anatomy.
- Compare preganglionic and postganglionic neurons of the autonomic nervous system.
- Describe the anatomy of the autonomic ganglia.

Understanding Autonomic Motor Pathways

When studying the autonomic motor pathways, the student of anatomy is confronted with many questions:

1. Why are there two distinct divisions—sympathetic and parasympathetic?
2. Why is there a series of two motor neurons from the CNS to an effector organ?
3. Why is the sympathetic system more widely distributed in the body than the parasympathetic system?
4. Why does the parasympathetic system originate in the cranial and sacral regions of the CNS, while the sympathetic system originates from the thoracic and lumbar regions of the CNS?
5. Why is there no autonomic output from the cervical region and from lower lumbar and upper sacral regions?

The answers to these questions are seldom clarified and yet they are the true keys to understanding the anatomy of the autonomic pathways. To help answer these questions we must briefly review the development of the nervous system and enhance it with a few important details.

Figure 19.2 Development of autonomic motor pathways. (a) Neurulation and (b) migration of the neural crest cells.

🔑 The nervous system begins developing in the third week from a thickening of ectoderm called the neural plate.

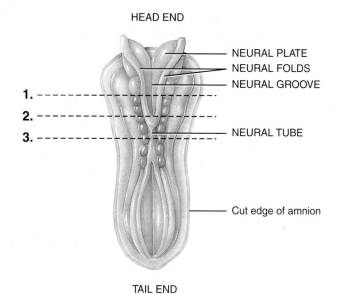

HEAD END

NEURAL PLATE
NEURAL FOLDS
NEURAL GROOVE

1.
2.
3.
NEURAL TUBE

Cut edge of amnion

TAIL END

(a) Dorsal view

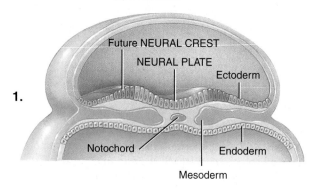

1.

Future NEURAL CREST
NEURAL PLATE
Ectoderm

Notochord
Endoderm

Mesoderm

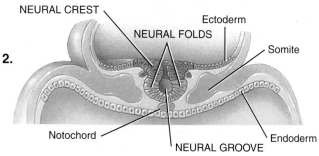

NEURAL CREST
Ectoderm
NEURAL FOLDS
Somite

2.

Notochord
Endoderm
NEURAL GROOVE

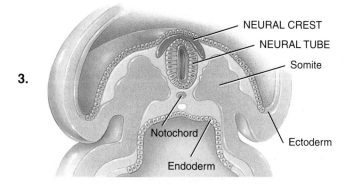

NEURAL CREST
NEURAL TUBE
Somite

3.

Notochord
Ectoderm
Endoderm

(b) Transverse sections

As you learned in Chapter 4, development of the nervous system begins in the third week of gestation with a thickening of the ectoderm called the **neural plate** (Figure 19.2). During the process of neurulation the plate folds inward and forms a longitudinal groove, the **neural groove**. The raised edges of the neural plate are called **neural folds**. As development continues, the neural folds increase in height and meet to form a tube called the **neural tube**. This tube becomes the central nervous system. During this folding process, a mass of tissue from the edge of the fold, the **neural crest**, migrates between the neural tube and the skin ectoderm (Figure 19.2b). The neural crest tissue plays a prominent role in the formation of the peripheral nervous system and is a key player in the formation of the autonomic motor neuron pathways.

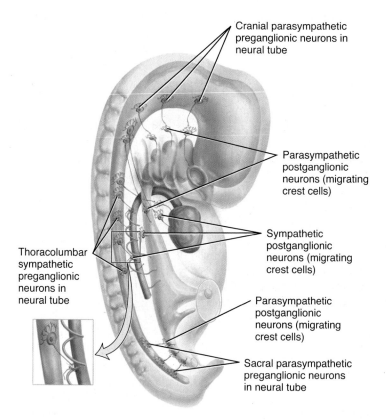

Cranial parasympathetic preganglionic neurons in neural tube

Parasympathetic postganglionic neurons (migrating crest cells)

Sympathetic postganglionic neurons (migrating crest cells)

Thoracolumbar sympathetic preganglionic neurons in neural tube

Parasympathetic postganglionic neurons (migrating crest cells)

Sacral parasympathetic preganglionic neurons in neural tube

(c) Relationship of neural tube neurons and neural crest neurons

 What is the origin of the neural crest?

Migration of the Neural Crest Tissue

Some of the neural crest tissue cells give rise to the cell bodies of the dorsal root and cranial ganglia of all somatic and visceral sensory neurons in the body. However, other neural crest cells migrate toward the developing smooth muscle of blood vessels and the gut tube. These *migrating crest cells* will develop into postganglionic neurons, the *second* of the two motor neurons of the autonomic motor pathway. As these cells migrate they are followed by axons that grow from cells in the ventrolateral part of the neural tube. These ventrolateral tube cells become preganglionic neurons, the *first* motor neurons in the autonomic motor pathway, and eventually form synapses with the migrating crest cells (Figure 19.2c).

The migrating crest cells form two distinct populations of developing neurons. One population migrates into the cranial and caudal ends of the gut tube, as these are the initial developing regions of the gut tube, and the second takes up positions near the yolk sac in the central region of the embryo around developing blood vessels. Because the gut tube develops more rapidly at its two ends (the pharynx and the cloaca), the initial neuronal relationship between the neural tube cell (first motor neuron), the migrating crest cell (second motor neuron), and smooth muscle and gland cells in the developing gut wall (autonomic effectors) arise in the cranial and sacral regions of the body and developing central nervous system. As the gut tube completes its development from the two ends, this neuronal relationship is carried toward the midgut. This neuronal pattern establishes the *craniosacral outflow* to the gut tube and establishes the parasympathetic division of the autonomic nervous system.

As major blood vessels begin to emerge around the developing yolk sac in the central region of the embryo, the migrating neural crest cells in the thoracic and upper lumbar regions establish connections with the developing smooth muscle cells in the walls of the blood vessels. From this central region of the embryo the initial neuronal relationship between the neural tube cell (first motor neuron), the migrating crest cell (second motor neuron), and smooth muscle cells in the developing blood vessels (autonomic effectors) arise in the thoracic and lumbar regions of the body and developing central nervous system. This neuronal pattern establishes the *thoracolumbar outflow* to the smooth muscle of the cardiovascular system and becomes the anatomy of the sympathetic division of the autonomic system.

With this knowledge of embryonic development we can answer the question about the division of the ANS into sympathetic and parasympathetic divisions. These two distinct divisions emerged to control the two distinct populations of developing smooth muscle—gut tube smooth muscle (parasympathetic) and cardiovascular smooth muscle (sympathetic). The migration of the crest cells, which were pursued by neural tube axons, explains the two-neuron pathway in motor control. The reason that the sympathetic division is more widespread than the parasympathetic division is that the sympathetic division controls blood vessels, and blood vessels develop throughout the entire body, and the parasympathetic distribution is limited to the gut tube and its derivatives. The initial migration of the two separate populations of crest cells clarifies why sympathetic control comes from the thoracolumbar regions of the CNS and parasympathetic control comes from the craniosacral regions. The final question

raised is why there are gaps in the autonomic output. The answer is limb development. The lack of autonomic output from the cervical and lower lumbar–upper sacral regions is a result of the massive dominance of skeletal muscle innervation arising in the developing limbs, which displaces the autonomic output from these regions of the CNS. In other words, the somatic motor neurons "muscle" the autonomic neurons out of the way as the limbs develop.

Shared Anatomical Components of an Autonomic Motor Pathway

As a result of their development, both the sympathetic and parasympathetic divisions share certain features, while other aspects of their anatomy are unique. We first describe the features common to both divisions, and then explore each of the two divisions in greater detail.

Motor Neurons and Autonomic Ganglia

As you learned in Section 19.1, each ANS pathway has two motor neurons (see Figure 19.1b). The cell body of the preganglionic neuron (the first neuron in the pathway) is in the brain or spinal cord, and its axon exits the CNS as part of a cranial or spinal nerve. The axon of a preganglionic neuron is a small-diameter, myelinated fiber that usually extends to an autonomic ganglion, an aggregate of migrated neural crest cells. The autonomic ganglia may be divided into three general groups: Two groups are components of the sympathetic division (the **sympathetic ganglia**), and one is a component of the parasympathetic division (the **parasympathetic ganglia**). The preganglionic neuron synapses with a postganglionic neuron (the second neuron in the pathway) within the autonomic ganglion (see Figure 19.1b). Notice that, because of its origin from migrating neural crest tissue, the postganglionic neuron lies entirely outside the CNS in the PNS. Its cell body and dendrites are located in the autonomic ganglion, and its axon is a small-diameter, unmyelinated type C fiber that terminates in a visceral effector. Unlike other neurons, postganglionic autonomic fibers do not end in a single terminal swelling like a synaptic knob or end plate. The terminal branches of autonomic fibers contain numerous swellings, called **varicosities**, which simultaneously release neurotransmitter over a large area of the innervated organ. This extensive release of neurotransmitter and the greater number of postganglionic neurons means that entire organs, rather than discrete cells, are typically influenced by autonomic activity.

Autonomic Plexuses

In the thorax, abdomen, and pelvis, axons of preganglionic neurons of both the sympathetic and parasympathetic divisions form tangled networks called **autonomic plexuses**, many of which lie along major arteries. The autonomic plexuses also may contain *autonomic ganglia* (groups of cell bodies for the postganglionic neurons in the plexuses) and axons of autonomic sensory neurons. The major plexuses in the thorax are the **cardiac plexus**, which supplies the heart, and the **pulmonary plexus**, which supplies the bronchial tree (see Figure 19.3; see also Figure 19.4).

Figure 19.3 Autonomic plexuses in the thorax, abdomen, and pelvis.

An autonomic plexus is a network of sympathetic and parasympathetic axons that sometimes also includes autonomic sensory axons and sympathetic ganglia.

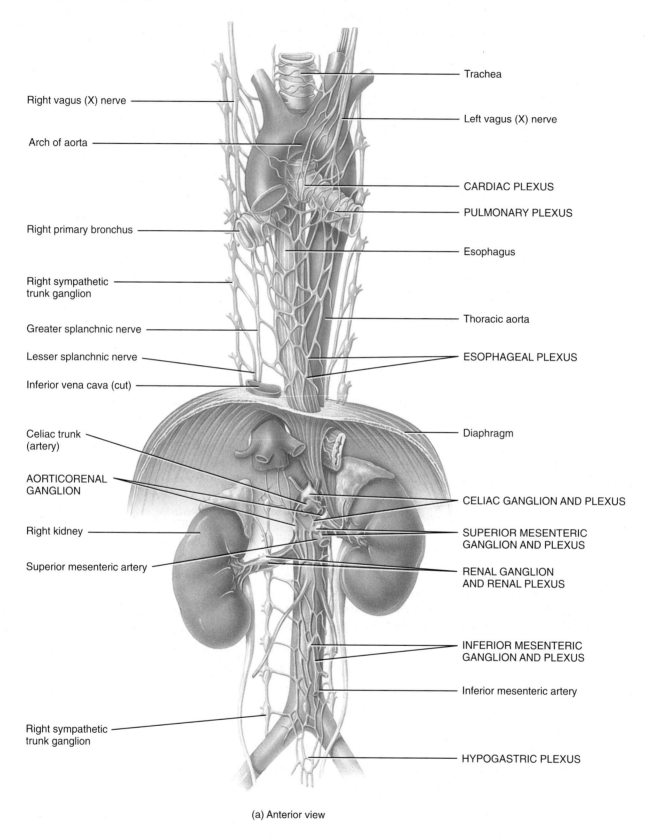

Right vagus (X) nerve

Arch of aorta

Right primary bronchus

Right sympathetic trunk ganglion

Greater splanchnic nerve

Lesser splanchnic nerve

Inferior vena cava (cut)

Celiac trunk (artery)

AORTICORENAL GANGLION

Right kidney

Superior mesenteric artery

Right sympathetic trunk ganglion

Trachea

Left vagus (X) nerve

CARDIAC PLEXUS

PULMONARY PLEXUS

Esophagus

Thoracic aorta

ESOPHAGEAL PLEXUS

Diaphragm

CELIAC GANGLION AND PLEXUS

SUPERIOR MESENTERIC GANGLION AND PLEXUS

RENAL GANGLION AND RENAL PLEXUS

INFERIOR MESENTERIC GANGLION AND PLEXUS

Inferior mesenteric artery

HYPOGASTRIC PLEXUS

(a) Anterior view

The abdomen and pelvis also contain major autonomic plexuses that often are named after the artery along which they are distributed (Figure 19.3). The **celiac (solar) plexus**, the largest autonomic plexus, surrounds the celiac and superior mesenteric arteries. It contains two large celiac ganglia, two aorticorenal ganglia, and a dense network of autonomic axons and is distributed to the liver, gallbladder, stomach, pancreas, spleen, kidneys, medullae (inner regions) of the adrenal glands, testes, and ovaries. The **superior mesenteric plexus** contains the superior mesenteric ganglion and supplies the small and large intestine. The **inferior mesenteric plexus** contains the inferior mesenteric ganglion,

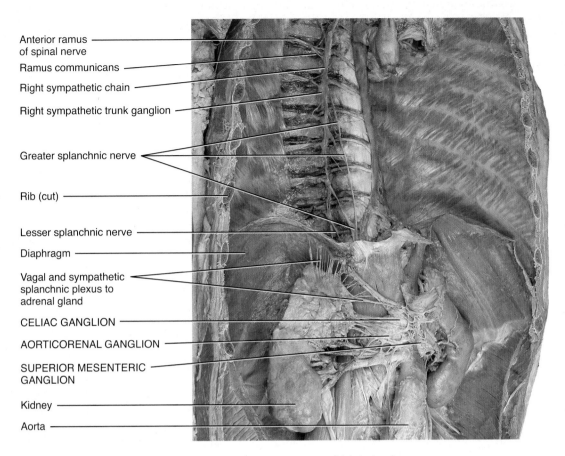

Anterior ramus
of spinal nerve

Ramus communicans

Right sympathetic chain

Right sympathetic trunk ganglion

Greater splanchnic nerve

Rib (cut)

Lesser splanchnic nerve

Diaphragm

Vagal and sympathetic
splanchnic plexus to
adrenal gland

CELIAC GANGLION

AORTICORENAL GANGLION

SUPERIOR MESENTERIC
GANGLION

Kidney

Aorta

(b) Anterior view

 Which is the largest autonomic plexus?

which innervates the large intestine. The **hypogastric plexus** supplies pelvic viscera. The **renal plexuses** contain the renal ganglion and supply the renal arteries within the kidneys and the ureters.

With this understanding of the development and the basic anatomical features shared by the sympathetic and parasympathetic divisions of the autonomic motor pathways, we are now ready to explore the two divisions in greater detail.

 CHECKPOINT
3. Describe the migrations of neural crest tissue in the early development of the autonomic motor pathways.
4. Describe the shared anatomical features of the autonomic motor pathways.

19.3 STRUCTURE OF THE SYMPATHETIC DIVISION

 OBJECTIVES

• Explain the central nervous system origin of the sympathetic division.
• Describe the location of the sympathetic ganglia.
• List the synapses between the preganglionic and postganglionic motor neurons of the sympathetic division and the different pathways of the postganglionic neurons to their effector organs.

Sympathetic Preganglionic Neurons

In the sympathetic division, the preganglionic neurons have their cell bodies in the lateral horns of the gray matter in the 12 thoracic segments and the first two or three lumbar segments of the spinal cord (Figure 19.4). For this reason, the sympathetic division is also called the **thoracolumbar division** (thor′-a-kō-LUM-bar), and the axons of the sympathetic preganglionic neurons are known as the **thoracolumbar outflow**.

The preganglionic axons leave the spinal cord along with the somatic motor neurons via the anterior rootlets of the spinal nerve. After exiting via the spinal nerve trunk through the intervertebral foramina, the myelinated preganglionic sympathetic axons pass into the anterior ramus of a spinal nerve and enter a short pathway called a **white ramus** (RA̅-mus) before passing to the nearest sympathetic trunk ganglion on the same side. Collectively, the white rami are called the **white rami communicantes** (kō-mū-ni-KAN-tēz; singular is *ramus communicans*). The "white" in their name indicates that they contain myelinated axons. Only the thoracic and first two or three lumbar nerves have white rami communicantes, because these thoracolumbar output levels are the only levels from which sympathetic preganglionic motor neurons (the myelinated neurons of the autonomic motor pathway) leave the spinal cord (as a result of the development pattern discussed previously). The white rami communicantes connect the anterior ramus of the spinal nerve with the ganglia of the sympathetic trunk.

Figure 19.4 **The sympathetic division of the autonomic nervous system.** Solid lines represent preganglionic axons; dashed lines represent postganglionic axons. Although the innervated structures are shown for only one side of the body for diagrammatic purposes, the sympathetic division actually innervates tissues and organs on both sides.

Cell bodies of sympathetic preganglionic neurons are located in the lateral horns of gray matter in the 12 thoracic and first two lumbar segments of the spinal cord.

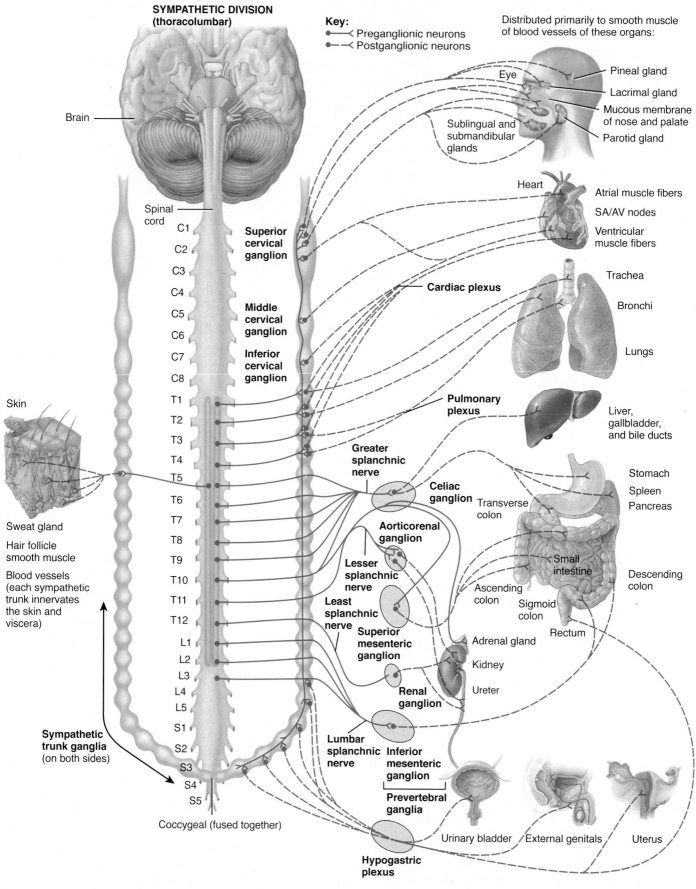

Which division, sympathetic or parasympathetic, has longer preganglionic axons? Why?

640

As preganglionic axons extend from a white ramus communicans into the sympathetic trunk ganglion, they give off several axon collaterals (branches). These collaterals terminate and synapse in several ways (Figure 19.5):

❶ Some synapse in the first ganglion at the level of entry.

❷ Others pass up or down the sympathetic trunk for a variable distance to form the **sympathetic chains**, the fibers on which the ganglia are strung.

❸ Some preganglionic axons pass through the sympathetic trunk without terminating in it. Beyond the trunk, they form nerves known as **splanchnic nerves** (SPLANK-nik; see Figure 19.4), which extend to and terminate in the outlying prevertebral ganglia. These ganglia, formed by neural crest cells that migrated toward the major blood vessels, supply the organs that arise from the abdominal portion of the gut tube.

A single sympathetic preganglionic fiber has many axon collaterals (branches) and may synapse with 20 or more postganglionic neurons. This pattern of projection is an example of divergence (see Chapter 16) and helps explain why many sympathetic responses affect almost the entire body simultaneously.

Sympathetic Ganglia and Postganglionic Neurons

The sympathetic ganglia are the sites of synapses between sympathetic preganglionic and postganglionic neurons and contain the postganglionic neuron cell bodies. There are two groups of sympathetic ganglia—the sympathetic trunk ganglia and prevertebral ganglia.

Sympathetic Trunk Ganglia

Sympathetic trunk ganglia (also called *vertebral chain ganglia* or *paravertebral ganglia*) lie in a vertical row on either side of the vertebral column. The position of the sympathetic trunk ganglia is established in the embryo as blood vessels branch from the aorta into each segment of the developing embryonic trunk. (Recall that the sympathetic pathways are following blood vessels during development and establish positions along these branches of the aorta.) These ganglia extend from the base of the skull to the coccyx (Figure 19.4).

The paired sympathetic trunk ganglia are arranged anterior and lateral to the vertebral column, one on either side.

Figure 19.5   Connections between ganglia and postganglionic neurons in the sympathetic division of the ANS. Also illustrated are the gray and white rami communicantes. See also Figure 17.5a.

 Sympathetic ganglia lie in two chains on either side of the vertebral column (sympathetic trunk ganglia) and near large abdominal arteries anterior to the vertebral column (prevertebral ganglia).

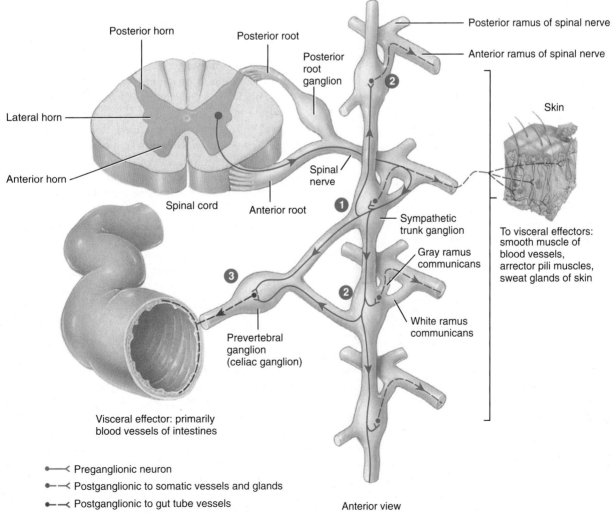

●——< Preganglionic neuron
●——< Postganglionic to somatic vessels and glands
●——< Postganglionic to gut tube vessels

Anterior view

? What substance gives the white rami their white appearance?

Typically, there are 3 cervical, 11 or 12 thoracic, 4 or 5 lumbar, and 4 or 5 sacral sympathetic trunk ganglia, and 1 coccygeal ganglion. The right and left coccygeal ganglia are fused together and usually lie at the midline. The sympathetic trunk ganglia extend inferiorly from the neck, chest, and abdomen to the coccyx (recall that these were sites of migration of neural crest cells to locations near segmental vessels arising from the embryonic aorta); however, they receive preganglionic axons only from the thoracic and lumbar segments of the spinal cord, as this region was the initial sight of blood vessel formation (see Figure 19.4).

Postganglionic neurons arising from the sympathetic trunk ganglia do one of the following (see Figure 19.4):

1. From all the ganglia of the sympathetic chain they return via gray communicating rami to the anterior ramus of a spinal nerve where they are distributed to blood vessels, sweat glands, and arrector pili muscles in the body wall.
2. From cervical sympathetic chain ganglia they exit into nerve branches that supply the heart or that follow blood vessels into the head, neck, and shoulder region.
3. From upper thoracic, lower abdominal, and pelvic sympathetic trunk ganglia they exit the trunk in nerves that enter plexuses that follow blood vessels of those regions.

The cervical portion of each sympathetic trunk ganglion is located in the neck and is subdivided into superior, middle, and inferior ganglia (see Figure 19.4). Postganglionic neurons leaving the **superior cervical ganglion** serve the head and heart. They are distributed primarily to blood vessels in the head, but also innervate sweat glands, smooth muscle of the eye, lacrimal glands, nasal mucosa, salivary glands (submandibular, sublingual, and parotid), and the heart. Gray rami communicantes (described shortly) from the superior cervical ganglion also pass to the upper two to four cervical spinal nerves, through which they supply blood vessels, sweat glands, and arrector pili muscles in the occipital region of the head and in the neck. Postganglionic neurons leaving the **middle cervical ganglion** and **inferior cervical ganglion** innervate the heart and blood vessels of the neck, shoulder, and upper limb.

The thoracic portion of each sympathetic trunk ganglion lies anterior to the necks of the corresponding ribs. This region of the sympathetic trunk receives most of the sympathetic preganglionic axons, and its postganglionic neurons innervate the thoracic blood vessels, heart, lungs, and bronchial tree. In the skin, these neurons also innervate blood vessels, sweat glands, and arrector pili muscles of hair follicles.

The lumbar portion of each sympathetic trunk ganglion lies lateral to the corresponding lumbar vertebrae. The sacral region of the sympathetic trunk ganglion lies in the pelvic cavity on the medial side of the anterior sacral foramina. Unmyelinated postganglionic axons from the lumbar and sacral sympathetic trunk ganglia enter a short pathway called a **gray ramus** and then merge with a spinal nerve to supply somatic blood vessels and glands, or they exit the ganglia as direct visceral nerves and join the hypogastric plexus.

The **gray rami communicantes** are structures containing the postganglionic axons that connect the ganglia of the various portions of the sympathetic trunk ganglion to spinal nerves (Figure 19.5). The axons of postganglionic neurons in the gray rami are unmyelinated. Gray rami communicantes outnumber the white rami because there is a gray ramus leading to each of the 31 pairs of spinal nerves that carries sympathetic output

to the smooth muscle and glands of the body wall and limbs, primarily the smooth muscle of blood vessels.

Prevertebral Ganglia

As noted earlier, some preganglionic neurons pass through the sympathetic trunk ganglia and chain and exit ventrally as the splanchnic nerves. These nerves join the second group of sympathetic ganglia, the **prevertebral (collateral) ganglia**, which lie anterior to the vertebral column and close to the large abdominal arteries that supply the derivatives of the embryonic gut. Postganglionic axons leaving the prevertebral ganglia follow the course of various arteries to abdominal and pelvic visceral effectors.

There are four major prevertebral ganglia (Figure 19.4; see also Figure 19.3a): (1) The **celiac ganglion** (SĒ-lē-ak) is on either side of the celiac artery just inferior to the diaphragm; (2) the **superior mesenteric ganglion** (MEZ-en-ter′-ik) is near the beginning of the superior mesenteric artery in the upper abdomen; (3) the **inferior mesenteric ganglion** is near the beginning of the inferior mesenteric artery in the middle of the abdomen; (4) the **aorticorenal ganglion** (ā-or′-ti-kō-RĒ-nal) is near the renal artery as it branches from the aorta.

Splanchnic nerves from the thoracic area form synapses with postganglionic cell bodies in the celiac ganglion. Preganglionic axons from the fifth through ninth or tenth thoracic ganglia (T5–T9 or T10) form the **greater splanchnic nerve**, which pierces the diaphragm and enters the celiac ganglion of the celiac plexus. From there, postganglionic neurons follow and innervate blood vessels to the stomach, spleen, liver, kidneys, and small intestine. Preganglionic axons from the tenth and eleventh thoracic ganglia (T10–T11) form the **lesser splanchnic nerve**, which pierces the diaphragm and passes through the celiac plexus to enter the aorticorenal ganglion and superior mesenteric ganglion of the superior mesenteric plexus. Postganglionic neurons from the superior mesenteric ganglion follow and innervate blood vessels of the small intestine and proximal colon. The **least** or **lowest splanchnic nerve**, which is not always present, is formed by preganglionic axons from the twelfth thoracic ganglia (T12) or a branch of the lesser splanchnic nerve. It passes through the diaphragm and enters the renal plexus near the kidney. Postganglionic neurons from the renal plexus supply kidney arterioles and the ureter.

Preganglionic axons that form the **lumbar splanchnic nerves** from the first through fourth lumbar ganglia (L1–L4) enter the inferior mesenteric plexus and terminate in the inferior mesenteric ganglion, where they synapse with postganglionic neurons. Axons of postganglionic neurons extend through the hypogastric plexus and principally supply blood vessels of the distal colon and rectum, urinary bladder, and genital organs.

Sympathetic preganglionic neurons also extend to the **adrenal medullae** (me-DUL-ē). Developmentally, the adrenal medullae and sympathetic ganglia are derived from the same tissue, the neural crest (see Figure 19.2). The adrenal medullae arise from migrating neural crest cells that develop into *chromaffin cells*, which are developmentally similar to sympathetic postganglionic neurons. Rather than extending to another organ, however, these cells release hormones into the blood. Upon stimulation by sympathetic preganglionic neurons, the adrenal medullae release a mixture of hormones—about 80 percent **epinephrine**, 20 percent **norepinephrine**, and a trace amount of **dopamine**. These hormones circulate throughout the body and intensify responses elicited by sympathetic postganglionic neurons.

 CHECKPOINT

5. Why is the sympathetic division called the thoracolumbar division even though its ganglia extend from the cervical region to the sacral region?
6. List the organs served by each sympathetic and parasympathetic ganglion.
7. Where are sympathetic trunk ganglia and prevertebral ganglia located?

19.4 STRUCTURE OF THE PARASYMPATHETIC DIVISION

 OBJECTIVES

• Explain the central nervous system origin of the parasympathetic division.
• Describe the location of the sympathetic ganglia.

Parasympathetic Preganglionic Neurons

Cell bodies of preganglionic neurons of the parasympathetic division are located in the nuclei of four cranial nerves in the brain stem—oculomotor (III), facial (VII), glossopharyngeal (IX), and vagus (X) and in the lateral gray horns of the second through fourth sacral segments of the spinal cord (Figure 19.6). (This results from the development we discussed previously.) Hence, the parasympathetic division is also known as the **craniosacral division** (krā′-nē-ō-SĀ-kral), and the axons of the parasympathetic preganglionic neurons are referred to as the **craniosacral outflow**. Their axons emerge as part of a cranial nerve or as part of the anterior root of a sacral spinal nerve. The **cranial parasympathetic outflow** consists of preganglionic axons that extend from the brain stem in four cranial nerves. The **sacral parasympathetic outflow** consists of preganglionic axons in anterior roots of the second through fourth sacral nerves. The preganglionic axons of both the cranial and sacral outflows end in terminal ganglia, where they synapse with postganglionic neurons.

The cranial outflow has five components: four pairs of ganglia and the plexuses associated with the vagus (X) nerve. The four pairs of cranial parasympathetic ganglia innervate structures in the head and are located close to the organs they innervate (Figure 19.6). Preganglionic axons that leave the brain as part of the vagus (X) nerves carry nearly 80 percent of the total craniosacral outflow. Vagal axons extend to many terminal ganglia in the thorax and abdomen. As the vagus nerve passes through the thorax, it sends axons to the heart and to the airways of the lungs. In the abdomen, it supplies the liver, gallbladder, stomach, pancreas, small intestine, and part of the large intestine.

The sacral parasympathetic outflow consists of preganglionic axons from the anterior roots of the second through fourth sacral nerves (S2–S4), which form the **pelvic splanchnic nerves** (Figure 19.6). These nerves synapse with parasympathetic postganglionic neurons located in terminal ganglia in the walls of the innervated viscera. From the ganglia, parasympathetic postganglionic axons innervate smooth muscle and glands in the walls of the colon, ureters, urinary bladder, and reproductive organs. Because the axons of parasympathetic preganglionic neurons extend from the CNS to a terminal ganglion in an innervated organ, they are longer than most of the axons of sympathetic preganglionic neurons.

Parasympathetic Ganglia and Postganglionic Neurons

The parasympathetic ganglia are the sites of synapses between parasympathetic preganglionic and postganglionic neurons, and contain the postganglionic neuron cell bodies. Parasympathetic ganglia are often referred to as **terminal ganglia** (neural crest cells that migrated into the developing gut wall) because most of these ganglia are located close to or actually within the wall of a visceral organ (the preganglionic neurons terminate at the organ). Most terminal ganglia do not have individual names. Only the terminal ganglia in the head have specific names (Figure 19.6):

1. The **ciliary ganglia** lie lateral to each optic (II) nerve near the posterior aspect of the orbit. Preganglionic axons pass with the oculomotor (III) nerves to the ciliary ganglia. Postganglionic axons from the ciliary ganglia innervate smooth muscle fibers in the eyeball.
2. The **pterygopalatine ganglia** (ter′-i-gō-PAL-a-tīn) are located lateral to the sphenopalatine foramen in the pterygopalatine fossa, between the sphenoid and palatine bones. Each ganglion receives preganglionic axons from a branch of the facial (VII) nerve and sends postganglionic axons to the nasal mucosa, palate, pharynx, and lacrimal glands.
3. The **submandibular ganglia** are found near the ducts of the submandibular salivary glands. Each ganglion receives preganglionic axons from a branch of the facial (VII) nerve and sends postganglionic axons to the submandibular and sublingual salivary glands.
4. The **otic ganglia** are situated just inferior to each foramen ovale. Each ganglion receives preganglionic axons from a branch of the glossopharyngeal (IX) nerve and sends postganglionic axons to the parotid salivary glands.

In a parasympathetic ganglion, the presynaptic neuron usually synapses with only four or five postsynaptic neurons, all of which supply a single visceral effector. Thus, parasympathetic responses can be localized to a single effector. Because the terminal ganglia are close to or in the walls of their visceral effectors, postganglionic parasympathetic axons are very short.

 CHECKPOINT

8. Name the organs served by each parasympathetic ganglion.
9. Where are the pterygopalatine ganglia located, and what type of ganglia are they?

19.5 STRUCTURE OF THE ENTERIC DIVISION

 OBJECTIVES

• Describe the relationship of the enteric division to the sympathetic and parasympathetic divisions of the autonomic nervous system.
• Explain how the enteric division of the autonomic nervous system is different from other parts of the peripheral nervous system.

It is important to realize that the gastrointestinal tract, like the surface of the body, forms an extensive area of contact with the environment. Although this environment is inside the body, it is still considered part of the external environment. Just as the surface of the body must respond to important environmental stimuli in order to function properly, the surface of the gastrointestinal tract must

Figure 19.6 The parasympathetic division of the autonomic nervous system. Solid lines represent preganglionic axons; dashed lines represent postganglionic axons. Although the innervated structures are shown for only one side of the body for diagrammatic purposes, the parasympathetic division actually innervates tissues and organs on both sides.

Cell bodies of parasympathetic preganglionic neurons are located in brain stem nuclei and in the lateral horns of gray matter in the second through fourth sacral segments of the spinal cord.

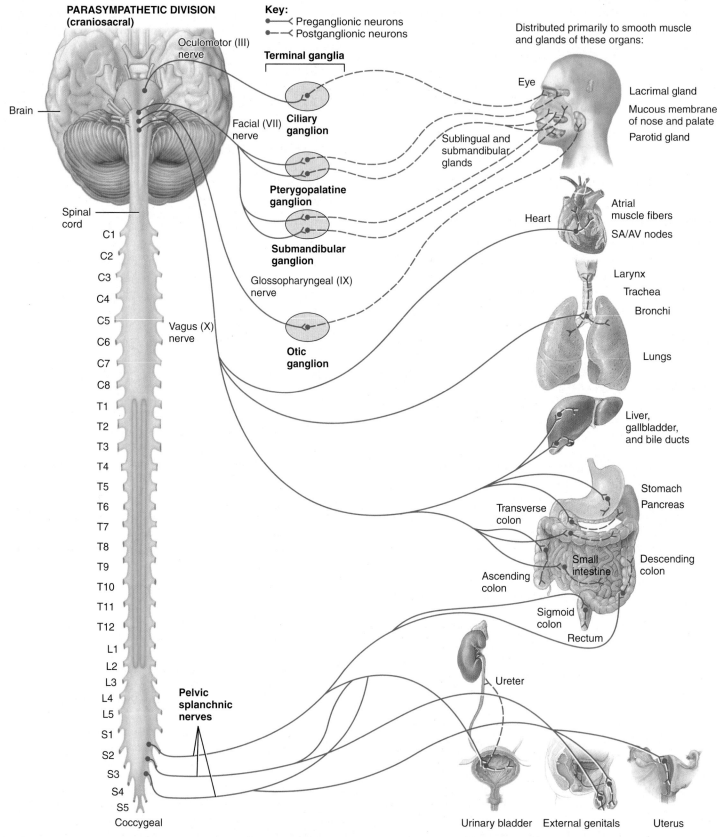

Which ganglia are associated with the parasympathetic division? Sympathetic division?

respond to surrounding stimuli to generate proper homeostatic controls. In fact, these responses and controls are so important that the gastrointestinal tract has its own nervous system with intrinsic input, processing, and output. This division can and does function independently of central nervous system activity, but can also receive controlling input from the central nervous system.

The **enteric division** (en-TER-ik) of the autonomic nervous system is the specialized network of nerves and ganglia forming a complex, integrated neuronal network within the wall of the gastrointestinal tract, pancreas, and gallbladder. This incredible nerve network contains in the neighborhood of 100 million neurons, approximately the same number as the spinal cord, and is capable of continued function without input from the central nervous system. The enteric network of nerves and ganglia contains sensory neurons capable of monitoring tension in the intestinal wall and assessing the composition of the intestinal contents. These sensory neurons relay their input signals to interneurons within the enteric ganglia. The interneurons establish an integrative network that processes the incoming signals and generates regulatory output signals to motor neurons throughout plexuses within the wall of the digestive organs. The motor neurons carry the output signals to the smooth muscle and glands of the gastrointestinal tract, as well as the smooth muscle of blood vessels, to exert control over its motility, secretory activities, and blood supply.

Most of the nerve fibers that innervate the digestive organs arise from two plexuses within the enteric system. The largest, the **myenteric plexus** (mī-en-TER-ik), is positioned between the outer longitudinal and circular muscle layers from the upper esophagus to the anus. The myenteric plexus communicates extensively with a somewhat smaller plexus, the **submucosal plexus**, which occupies the gut wall between the circular muscle layer and the muscularis mucosae (see Section 24.2) and runs from the stomach to the anus. Neurons emerge from the ganglia of these two plexuses to form smaller plexuses around blood vessels and within the muscle layers and mucosa of the gut wall. It is this system of nerves that makes possible the normal motility and secretory functions of the gastrointestinal tract.

 CHECKPOINT

10. How does the enteric division differ from the sympathetic and parasympathetic divisions of the autonomic nervous system?

19.6 ANS NEUROTRANSMITTERS AND RECEPTORS

 OBJECTIVE

• Describe the neurotransmitters and receptors involved in autonomic responses.

Autonomic neurons are classified based on the neurotransmitter they produce and release. The receptors for the neurotransmitters are integral membrane proteins located in the plasma membrane of the postsynaptic neuron or effector cell.

Cholinergic Neurons and Receptors

Cholinergic neurons (kō′-lin-ER-jik) release the neurotransmitter **acetylcholine (Ach)** (as′-ē-til-KŌ-lēn). (Remember: acetyl*choline*=*choline*rgic.) In the ANS, the cholinergic neurons include (1) all sympathetic and parasympathetic preganglionic neurons, (2) sympathetic postganglionic neurons that innervate most sweat glands, and (3) all parasympathetic postganglionic neurons (Figure 19.7).

ACh is stored in synaptic vesicles and released by exocytosis. It then diffuses across the synaptic cleft and binds with specific **cholinergic receptors**, integral membrane proteins in the *postsynaptic* plasma membrane. The two types of cholinergic receptors, both of which bind ACh, are nicotinic receptors and muscarinic receptors. **Nicotinic receptors** (nik-ō-TIN-ik) are present in the plasma membranes of dendrites and cell bodies of sympathetic and parasympathetic postganglionic neurons (Figure 19.7a, b), and in the motor end plate at the neuromuscular junction. They are so named because nicotine mimics the action of ACh by binding to

Figure 19.7 Cholinergic neurons and adrenergic neurons in the sympathetic and parasympathetic divisions. Cholinergic neurons release acetylcholine; adrenergic neurons release norepinephrine. Cholinergic and adrenergic receptors are integral membrane proteins located in the plasma membrane of a postsynaptic neuron or an effector cell.

Most sympathetic postganglionic neurons are adrenergic; other autonomic neurons are cholinergic.

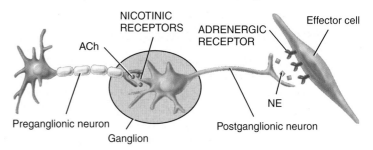

(a) Sympathetic division–innervation to most effector tissues

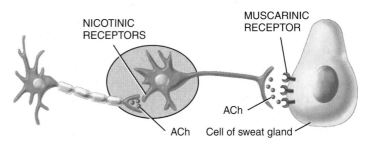

(b) Sympathetic division–innervation to most sweat glands

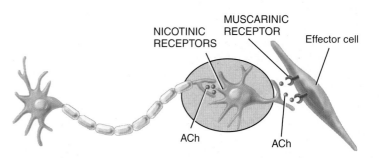

(c) Parasympathetic division

 Which neurons are cholinergic and possess nicotinic ACh receptors? What type of ACh receptors do their effector tissues possess?

these receptors. (Nicotine, a natural substance in tobacco leaves, is not normally present in the bodies of nonsmokers.) **Muscarinic receptors** (mus′-ka-RIN-ik) are present in the plasma membranes of all effectors innervated by parasympathetic postganglionic axons (smooth muscle, cardiac muscle, and glands). Most sweat glands, which receive their innervation from cholinergic sympathetic postganglionic neurons, possess muscarinic receptors (see Figure 19.7). These receptors are also named for a substance that does not naturally occur in the human body; a mushroom poison called muscarine mimics the actions of ACh by binding to muscarinic receptors.

Activation of nicotinic receptors by ACh always causes depolarization and thus excitation of the postsynaptic cell, which can be a postganglionic neuron, an autonomic effector, or a skeletal muscle fiber. Activation of muscarinic receptors by ACh sometimes causes depolarization (excitation) and sometimes causes hyperpolarization (inhibition), depending on which particular cell bears the muscarinic receptors. For example, binding of ACh to muscarinic receptors inhibits (relaxes) smooth muscle sphincters in the gastrointestinal tract. By contrast, ACh excites smooth muscle fibers in the circular muscles of the iris of the eye, causing them to contract. Because acetylcholine is quickly inactivated by the enzyme **acetylcholinesterase (AChE)**, effects triggered by cholinergic neurons are brief.

Adrenergic Neurons and Receptors

In the ANS, **adrenergic neurons** (ad′-ren-ER-jik) release **norepinephrine (NE)** (nor′-ep-i-NEF-rin), also known as **noradrenalin** (Figure 19.7a). (Remember: *adren*ergic= nor*adren*alin.) Most sympathetic postganglionic neurons are adrenergic. Like ACh, NE is synthesized and stored in synaptic vesicles and released by exocytosis. Molecules of NE diffuse across the synaptic cleft and bind to specific adrenergic receptors on the postsynaptic membrane, causing either excitation or inhibition of the effector cell.

Adrenergic receptors bind both NE and epinephrine, a hormone with actions similar to NE. As noted previously, NE is released as a neurotransmitter by sympathetic postganglionic neurons. In addition, both epinephrine and NE are released as hormones into the blood by the chromaffin cells of the adrenal medullae. The two main types of adrenergic receptors are **alpha (α) receptors** and **beta (β) receptors**, which are found on visceral effectors innervated by most sympathetic postganglionic axons. These receptors are further classified into subtypes—α_1, α_2, β_1, β_2, and β_3—based on the specific responses they elicit and by their selective binding of drugs that activate or block them. Although there are some exceptions, activation of α_1 and β_1 receptors generally produces excitation, in contrast to activation of α_2 and β_2 receptors, which causes inhibition of effector tissues. β_3 receptors are present only on cells of brown adipose tissue, where their activation causes *thermogenesis* (heat production). Cells of most effectors contain either α or β receptors; some visceral effector cells contain both. NE stimulates alpha receptors more strongly than beta receptors; epinephrine is a potent stimulator of both alpha and beta receptors.

The activity of NE at a synapse is terminated when (1) the NE is taken up by the axon that released it or (2) when NE is enzymatically inactivated by either *catechol-O-methyltransferase*

(COMT) (kat-e-kōl′-ō-meth-il-TRANS-fer-ās) or *monoamine oxidase (MAO)* (mon-ō-AM-ēn OK-si-dās′). NE lingers in the synaptic cleft for a longer time than ACh. Thus, effects triggered by adrenergic neurons typically are longer lasting than those triggered by cholinergic neurons.

 CHECKPOINT

11. Why are cholinergic and adrenergic neurons so named?
12. What substances bind to adrenergic receptors?

19.7 FUNCTIONS OF THE ANS

 OBJECTIVES

• Describe the major responses of the body to stimulation by the sympathetic division of the ANS.
• Explain the reactions of the body to stimulation by the parasympathetic division.

As noted earlier in the chapter, most body organs are innervated by both divisions of the ANS, which typically work in opposition to one another. The balance between sympathetic and parasympathetic activity is regulated by the hypothalamus. The hypothalamus typically increases sympathetic activity at the same time it decreases parasympathetic activity, and vice versa. As you learned in Section 19.6, the two divisions affect body organs differently because of the different neurotransmitters released by their postganglionic neurons and the different adrenergic and cholinergic receptors on the cells of their effector organs. A few structures receive only sympathetic innervation—sweat glands, arrector pili muscles attached to hair follicles in the skin, the kidneys, the spleen, most blood vessels, and the adrenal medullae (see Figure 19.4). In these structures there is no opposition from the parasympathetic division; increases and decreases in sympathetic activity are responsible for the changes.

CLINICAL CONNECTION | *Autonomic Dysreflexia*

Autonomic dysreflexia (dis-rē-FLEKS-sē-a) is an exaggerated response of the sympathetic division of the ANS that occurs in about 85 percent of individuals with spinal cord injury at or above the level of T6. The condition occurs due to interruption of the control of ANS neurons by higher centers. When certain sensory impulses are unable to ascend the spinal cord, such as those resulting from stretching of a full urinary bladder, mass stimulation of the sympathetic nerves below the level of injury occurs. Among the effects of increased sympathetic activity is severe vasoconstriction, which elevates blood pressure. In response, the cardiovascular center in the medulla oblongata (1) increases parasympathetic output via the vagus nerve, which decreases heart rate, and (2) decreases sympathetic output, which causes dilation of blood vessels above the level of the injury. Autonomic dysreflexia is characterized by a pounding headache; severe high blood pressure (hypertension); flushed, warm skin with profuse sweating above the injury level; pale, cold, and dry skin below the injury level; and anxiety. It is an emergency condition that requires immediate intervention. If untreated, autonomic dysreflexia can cause seizures, stroke, or heart attack. •

Sympathetic Responses

During physical or emotional stress, the sympathetic division dominates the parasympathetic division. High sympathetic activity favors body functions that can support vigorous physical activity and rapid production of ATP. At the same time, the sympathetic division decreases body functions that favor the storage of energy. Physical exertion and a variety of emotions—such as fear, embarrassment, or rage—stimulate the sympathetic division. Visualizing body changes that occur during "E situations" such as exercise, emergency, excitement, and embarrassment, will help you remember most of the sympathetic responses. Activation of the sympathetic division and release of hormones by the adrenal medullae set in motion a series of physiological responses collectively called the **fight-or-flight response**, which includes the following effects (many of which were experienced by the student in the introduction to this chapter):

1. The pupils of the eyes dilate.
2. Heart rate, force of heart contraction, and blood pressure increase.
3. The airways dilate, allowing faster movement of air into and out of the lungs.
4. Blood vessels that supply organs involved in exercise or fighting off danger—skeletal muscles, cardiac muscle, liver, and adipose tissue—dilate, allowing greater blood flow through these tissues.
5. Liver cells perform glycogenolysis (breakdown of glycogen to glucose), and adipose tissue cells perform lipolysis (breakdown of triglycerides to fatty acids and glycerol).
6. Release of glucose by the liver increases blood glucose level.
7. Processes that are not essential for meeting the stressful situation are inhibited. For example, the blood vessels that supply the kidneys and gastrointestinal tract constrict, which decreases blood flow through these tissues. The result is a slowing of urine formation and digestive activities.

The effects of sympathetic stimulation are longer lasting and more widespread than the effects of parasympathetic stimulation for three reasons: (1) Sympathetic postganglionic axons diverge more extensively; as a result, many tissues are activated simultaneously. (2) AChE quickly inactivates ACh, but NE lingers in the synaptic cleft for a longer period. (3) The secretion of epinephrine and NE into the blood from the adrenal medulla (as hormones) intensifies and prolongs the responses caused by NE released as a neurotransmitter from sympathetic postganglionic axons. These blood-borne hormones circulate throughout the body, affecting all tissues that have α and β receptors. In time, blood-borne NE and epinephrine are destroyed by enzymes in the liver.

Parasympathetic Responses

In contrast to the fight-or-flight activities of the sympathetic division, the parasympathetic division enhances **rest-and-digest** activities. Parasympathetic responses support body functions that conserve and restore body energy during times of rest and recovery. In the quiet intervals between periods of exercise, parasympathetic impulses to the digestive glands and the smooth muscle of the gastrointestinal tract predominate over sympathetic impulses. This allows energy-supplying food to be digested and absorbed. At the same time, parasympathetic responses decrease body functions that support physical activity.

CLINICAL CONNECTION | *Drugs and Receptor Selectivity*

A large variety of drugs and natural products can selectively activate or block specific cholinergic or adrenergic receptors. An **agonist** (*agon*=a contest) is a substance that binds to and activates a receptor, in the process mimicking the effect of a natural neurotransmitter or hormone. Phenylephrine, an adrenergic agonist at α_1 receptors, is a common ingredient in cold and sinus medications. Because it constricts blood vessels in the nasal mucosa, phenylephrine reduces production of mucus, thus relieving nasal congestion. An **antagonist** (*anti-*=against) is a substance that binds to and blocks a receptor, thereby preventing a natural neurotransmitter or hormone from exerting its effect. For example, atropine, which blocks muscarinic ACh receptors, dilates the pupils, reduces glandular secretions, and relaxes smooth muscle in the gastrointestinal tract. It is used to dilate the pupils during eye examinations, in the treatment of smooth muscle disorders such as iritis and intestinal hypermotility, and as an antidote for chemical warfare agents that inactivate AChE.

Propranolol (Inderal®) often is prescribed for patients with hypertension (high blood pressure). It is a nonselective beta blocker, meaning it binds to all types of beta receptors and prevents their activation by epinephrine and norepinephrine. The desired effects of propranolol are due to its blockade of β_1 receptors—namely, decreased heart rate and force of contraction and a consequent decrease in blood pressure. Undesired effects due to blockade of β_2 receptors may include hypoglycemia (low blood glucose), resulting from decreased glycogen breakdown and decreased gluconeogenesis (the conversion of a noncarbohydrate into glucose in the liver), and mild bronchoconstriction (narrowing of the airways). If these side effects pose a threat to the patient, a selective β_1 blocker that binds only to specific beta receptors, such as metoprolol (Lopressor®), can be prescribed. •

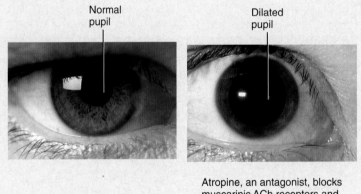

Normal pupil   Dilated pupil

Atropine, an antagonist, blocks muscarinic ACh receptors and dilates the pupils

---

The acronym *SLUDD* can be helpful in remembering five parasympathetic responses: salivation (S), lacrimation (L), urination (U), digestion (D), and defecation (D). All of these activities are stimulated mainly by the parasympathetic division. Other important parasympathetic responses are "three decreases": decreased heart rate, decreased diameter of airways (bronchoconstriction), and decreased diameter (constriction) of the pupils.

Table 19.2 compares the structural and functional features of the sympathetic and parasympathetic divisions of the ANS. Table 19.3 lists the responses of glands, cardiac muscle, and smooth muscle to stimulation by the sympathetic and parasympathetic divisions of the ANS.

TABLE 19.2

Structure and Function of Sympathetic and Parasympathetic Divisions of the ANS

	SYMPATHETIC (THORACOLUMBAR)	PARASYMPATHETIC (CRANIOSACRAL)
Distribution	Wide regions of the body: skin, sweat glands, arrector pili muscles of hair follicles, adipose tissue, smooth muscle of blood vessels	Limited mainly to head and to gut tube derived viscera of thorax, abdomen, and pelvis; some blood vessels
Location of preganglionic neuron cell bodies and site of outflow	Lateral gray horns of spinal cord segments T1–L2; axons of preganglionic neurons constitute thoracolumbar outflow	Nuclei of oculomotor (III), facial (VII), glossopharyngeal (IX), and vagus (X) nerves and the lateral gray matter of spinal cord segments S2–S4; axons of preganglionic neurons constitute craniosacral outflow
Associated ganglia	Two types: sympathetic trunk ganglia and prevertebral ganglia	One type: terminal ganglia
Ganglia locations	Closer to CNS and more distant from visceral effectors	Typically near or within wall of visceral effectors
Axon length and divergence	Generally, preganglionic neurons with shorter axons synapse with many postganglionic neurons with longer axons that pass to many visceral effectors	Preganglionic neurons with longer axons usually synapse with four to five postganglionic neurons with shorter axons that pass to a single visceral effector
White and gray rami communicantes	Both present; white rami communicantes contain myelinated preganglionic axons, and gray rami communicantes contain unmyelinated postganglionic axons	Neither present
Neurotransmitters	Preganglionic neurons release acetylcholine (ACh), which is excitatory and stimulates postganglionic neurons; most postganglionic neurons release norepinephrine (NE); postganglionic neurons that innervate most sweat glands and some blood vessels in skeletal muscle release ACh	Preganglionic neurons release acetylcholine (ACh), which is excitatory and stimulates postganglionic neurons; postganglionic neurons release ACh
Physiological effects	Fight-or-flight responses	Rest-and-digest activities

TABLE 19.3

Effects of Sympathetic and Parasympathetic Divisions of the ANS

VISCERAL EFFECTOR	EFFECT OF SYMPATHETIC STIMULATION (α OR β ADRENERGIC RECEPTORS, EXCEPT AS NOTED)*	EFFECT OF PARASYMPATHETIC STIMULATION (MUSCARINIC ACh RECEPTORS)
GLANDS		
Adrenal medullae	Secretion of epinephrine and norepinephrine (nicotinic ACh receptors)	No known effect
Lacrimal (tear)	Slight secretion of tears (α)	Secretion of tears
Pancreas	Inhibits secretion of digestive enzymes and the hormone insulin (α_2); promotes secretion of the hormone glucagon (β_2)	Secretion of digestive enzymes and the hormone insulin
Posterior pituitary	Secretion of antidiuretic hormone (ADH) (β_1)	No known effect
Pineal	Increases synthesis and release of melatonin (β)	No known effect
Sweat	Increases sweating in most body regions (muscarinic ACh receptors); sweating on palms and soles (α_1)	No known effect
Adipose tissue†	Lipolysis (breakdown of triglycerides into fatty acids and glycerol) (β_1); release of fatty acids into blood (β_1 and β_3)	No known effect
Liver†	Glycogenolysis (conversion of glycogen into glucose); gluconeogenesis (conversion of noncarbohydrates into glucose); decreased bile secretion (α and β_2)	Glycogen synthesis; increased bile secretion
Kidney, juxtaglomerular cells†	Secretion of renin (β_1)	No known effect
CARDIAC (HEART) MUSCLE		
	Increased heart rate and force of atrial and ventricular contractions (β_1)	Decreased heart rate; decreased force of atrial contraction

TABLE 19.3 CONTINUED

Effects of Sympathetic and Parasympathetic Divisions of the ANS

VISCERAL EFFECTOR	EFFECT OF SYMPATHETIC STIMULATION (α OR β ADRENERGIC RECEPTORS, EXCEPT AS NOTED)*	EFFECT OF PARASYMPATHETIC STIMULATION (MUSCARINIC ACh RECEPTORS)
SMOOTH MUSCLE		
Iris, radial muscle	Contraction → dilation of pupil (α_1)	No innervation
Iris, circular muscle	No known effect	Contraction → constriction of pupil
Ciliary muscle of eye	Relaxation to adjust shape of lens for distant vision (β_2)	Contraction for close vision
Lungs, bronchial muscle	Relaxation → airway dilation (β_2)	Contraction → airway constriction
Gallbladder and ducts	Relaxation to facilitate storage of bile in the gallbladder (β_2)	Contraction → release of bile into small intestine
Stomach and intestines	Decreased motility and tone (α_1, α_2, β_2); contraction of sphincters (α_1)	Increased motility and tone; relaxation of sphincters
Spleen	Contraction and discharge of stored blood into general circulation (α_1)	No innervation
Ureter	Increases motility (α_1)	Increases motility (?)
Urinary bladder	Relaxation of muscular wall (β_2); contraction of internal urethral sphincter (α_1)	Contraction of muscular wall; relaxation of internal urethral sphincter
Uterus	Inhibits contraction in nonpregnant women (β_2); promotes contraction in pregnant women (α_1)	Minimal effect
Sex organs	In males: contraction of smooth muscle of ductus (vas) deferens, prostate, and seminal vesicle → ejaculation of semen (α_1)	Vasodilation; erection of clitoris (females) and penis (males)
Hair follicles, arrector pili muscle	Contraction → erection of hairs resulting in goosebumps (α_1)	No innervation
VASCULAR SMOOTH MUSCLE		
Salivary gland arterioles	Vasoconstriction, which decreases secretion of saliva (α_1)	Vasodilation, which increases secretion of saliva
Gastric gland arterioles	Vasoconstriction, which inhibits secretion (α_1)	Secretion of gastric juice
Intestinal gland arterioles	Vasoconstriction, which inhibits secretion (α_1)	Secretion of intestinal juice
Coronary (heart) arterioles	Relaxation → vasodilation (β_2); contraction → vasoconstriction (α_1, α_2); contraction → vasoconstriction (muscarinic ACh receptors)	Contraction → vasoconstriction
Skin and mucosal arterioles	Contraction → vasoconstriction (α_1)	No innervation
Skeletal muscle arterioles	Contraction → vasoconstriction (α_1); relaxation → vasodilation (β_2); relaxation → vasodilation (muscarinic ACh receptors)	No innervation
Abdominal viscera arterioles	Contraction → vasoconstriction (α_1, β_2)	No innervation
Brain arterioles	Slight contraction → vasoconstriction (α_1)	No innervation
Kidney arterioles	Constriction of blood vessels → decreased urine volume (α_1)	No innervation
Systemic veins	Contraction → constriction (α_1); relaxation → dilation (β_2)	No innervation

*Subcategories of α and β receptors are listed if known.
†Grouped with glands because they release substances into the blood.

CLINICAL CONNECTION | *Raynaud Phenomenon*

In **Raynaud phenomenon** (rā-NŌ), the fingers and toes become ischemic (lack blood) after exposure to cold or with emotional stress. The condition is due to excessive sympathetic stimulation of smooth muscle in the arterioles of the fingers and toes. When the arterioles constrict in response to sympathetic stimulation, blood flow is greatly diminished. Symptoms are colorful—red, white, and blue. Fingers and toes may look white due to blockage of blood flow or look blue (cyanotic) due to deoxygenated blood in capillaries. With rewarming after cold exposure, the arterioles may dilate, causing the fingers and toes to look red. The disorder is most common in young women and occurs more often in cold climates. Drugs used to treat Raynaud phenomenon include nifedipine, a calcium channel blocker that relaxes smooth muscle by blocking alpha receptors. Smoking and the use of alcohol or illicit drugs can exacerbate the symptoms of this condition. •

CHECKPOINT

13. What are some antagonistic effects of the sympathetic and parasympathetic divisions of the autonomic nervous system?

14. What happens during the fight-or-flight response?

15. Why is the parasympathetic division of the ANS sometimes called an energy conservation/restoration system?

16. Using Table 19.3 as a reference, describe the sympathetic response in a frightening situation for each of the following body parts: hair follicles, iris of eye, lungs, spleen, adrenal medullae, urinary bladder, stomach, intestines, gallbladder, liver, heart, arterioles of the abdominal viscera, and arterioles of skeletal muscles.

19.8 INTEGRATION AND CONTROL OF AUTONOMIC FUNCTIONS

 OBJECTIVES

• Describe the components of an autonomic reflex.
• Explain the relationship of the hypothalamus to the ANS.

Autonomic Reflexes

Autonomic reflexes are responses that occur when nerve impulses travel over an autonomic reflex arc. Recall from Chapter 17 that a **reflex arc** is a neural pathway that elicits a reflex and contains a receptor, sensory neuron, integrating center, motor neuron, and effector. These reflexes play a key role in regulating controlled conditions in the body, such as *blood pressure*, by adjusting heart rate, force of ventricular contraction, and blood vessel diameter; *digestion*, by adjusting the motility (movement) and muscle tone of the gastrointestinal tract; and *defecation* and *urination*, by regulating the opening and closing of sphincters.

The components of an autonomic reflex arc are as follows:

1. **Receptor.** Like the receptor in a somatic reflex arc (see Figure 17.12), the receptor in an autonomic reflex arc is the distal end of a sensory neuron, which responds to a stimulus and produces a change that will ultimately trigger nerve impulses. Autonomic sensory receptors are usually associated with interoceptors.

2. **Sensory neuron.** Conducts nerve impulses from receptors to the CNS.

3. **Integrating center.** Interneurons within the CNS relay signals from sensory neurons to motor neurons. The main integrating centers for most autonomic reflexes are located in the hypothalamus and brain stem. Some autonomic reflexes, such as those for urination and defecation, have integrating centers in the spinal cord.

4. **Motor neurons.** Nerve impulses triggered by the integrating center propagate out of the CNS along motor neurons to an effector. In an autonomic reflex arc, two motor neurons connect the CNS to an effector: The preganglionic neuron conducts motor impulses from the CNS to an autonomic ganglion, and the postganglionic neuron conducts motor impulses from an autonomic ganglion to an effector (see Figure 19.1).

5. **Effector.** In an autonomic reflex arc, the effectors are smooth muscle, cardiac muscle, and glands, and the reflex is called an autonomic reflex.

Autonomic Control by Higher Centers

Normally, you are not aware of muscular contractions of your digestive organs, your heartbeat, changes in the diameter of your blood vessels, and pupil dilation and constriction because the integrating centers for these autonomic responses are in your spinal cord or the lower regions of your brain. Somatic or autonomic sensory neurons deliver input to these centers, and autonomic motor neurons provide output that adjusts activity in the visceral effector, usually without conscious perception.

The hypothalamus is the major control and integration center of the ANS. The hypothalamus receives sensory input related to visceral functions, olfaction (smell), and gustation (taste), along with input related to changes in temperature and levels of various substances in blood. In addition, the hypothalamus receives input relating to emotions from the limbic system. Output from the hypothalamus influences autonomic centers in the brain stem (such as the cardiovascular, salivation, swallowing, and vomiting centers) and in the spinal cord (such as the defecation and urination reflex centers in the sacral spinal cord).

Anatomically, the hypothalamus is connected to both the sympathetic and parasympathetic divisions of the ANS by axons of neurons whose dendrites and cell bodies are in various hypothalamic nuclei. The axons form tracts from the hypothalamus to sympathetic and parasympathetic nuclei in the brain stem and spinal cord through relays in the reticular formation. The posterior and lateral parts of the hypothalamus control the sympathetic division. As you might expect, experimental stimulation of the posterior or lateral hypothalamus produces an increase in heart rate and force of contraction, a rise in blood pressure due to constriction of blood vessels, an increase in body temperature, dilation of the pupils, and inhibition of the gastrointestinal tract. In contrast, stimulation of the anterior and medial parts of the hypothalamus, which control the parasympathetic division, results in a decrease in heart rate, lowering of blood pressure, constriction of the pupils, and increased secretion and motility of the gastrointestinal tract.

 CHECKPOINT

17. Give three examples of activities in the body that are controlled by autonomic reflexes.

18. How does an autonomic reflex arc differ from a somatic reflex arc?

KEY MEDICAL TERMS ASSOCIATED WITH THE AUTONOMIC NERVOUS SYSTEM

Autonomic nerve neuropathy (noo-ROP-a-thē) A *neuropathy* (disorder of a cranial or spinal nerve) that affects one or more autonomic nerves, with multiple effects on the autonomic nervous system, including constipation, urinary incontinence, impotence, and fainting and low blood pressure when standing (*orthostatic hypotension*) due to decreased sympathetic control of the cardiovascular system. Often caused by long-term diabetes mellitus (*diabetic neuropathy*).

Biofeedback A technique in which an individual is provided with information regarding an autonomic response such as heart rate, blood pressure, or skin temperature via various electronic monitoring devices. By concentrating on positive thoughts, individuals learn to alter autonomic responses. For example, biofeedback has been used to decrease heart rate and blood pressure and increase skin temperature in order to decrease the severity of migraine headaches.

Dysautonomia (dis-aw-tō-NŌ-mē-a; *dys-*=difficult; *autonomia*=self-governing) An inherited disorder in which the autonomic nervous system functions abnormally, resulting in reduced tear gland secretions, poor vasomotor control, motor incoordination, skin blotching, absence of pain sensation, difficulty in swallowing, hyporeflexia, excessive vomiting, and emotional instability.

Horner syndrome A condition in which the sympathetic innervation to one side of the face is lost due to an inherited mutation, an injury, or a disease that affects sympathetic outflow through the superior cervical ganglion. Symptoms occur on the affected side and include ptosis (drooping of the upper eyelid), miosis (constricted pupil), and anhidrosis (lack of sweating).

Hyperhidrosis (hī′-per-hī-DRŌ-sis; *hyper-*=above or too much; *hidro*=sweat; *-osis*=condition) Excessive or profuse sweating due to intense stimulation of sweat glands.

Mass reflex In cases of severe spinal cord injury above the level of the sixth thoracic vertebra, stimulation of the skin or overfilling of a visceral organ (such as the urinary bladder or colon) below the level of the injury results in intense activation of autonomic and somatic output from the spinal cord as reflex activity returns. The exaggerated response occurs because there is no inhibitory input from the brain. The mass reflex consists of flexor spasms of the lower limbs, evacuation of the urinary bladder and colon, and profuse sweating below the level of the lesion.

Megacolon (*mega-*=big) An abnormally large colon. In congenital megacolon, parasympathetic nerves to the distal segment of the colon do not develop properly. Loss of motor function in the segment causes massive dilation of the normal proximal colon. The condition results in extreme constipation, abdominal distension, and, occasionally, vomiting.

Reflex sympathetic dystrophy (RSD) A syndrome that includes spontaneous pain, painful hypersensitivity to stimuli such as light touch, and excessive coldness and sweating in the involved body part. The disorder frequently involves the forearms, hands, knees, and feet. It appears to involve activation of the sympathetic division of the ANS due to traumatized nociceptors as a result of trauma or surgery on bones or joints. Also called **complex regional pain syndrome type 1**.

Vagotomy (vā-GOT-ō-mē; *-tome*=incision) Cutting the vagus (X) nerve. This procedure is frequently used to decrease the production of hydrochloric acid in persons with ulcers.

CHAPTER REVIEW AND RESOURCE SUMMARY

WileyPLUS

Review	Resource

19.1 Comparison of Somatic and Autonomic Nervous Systems

1. The somatic nervous system operates under conscious control; the ANS usually operates without conscious control.
2. Sensory input to the somatic nervous system is mainly from the special senses and somatic senses; sensory input to the ANS is primarily from interoceptors, with some contributions from the special senses and somatic senses.
3. The axons of somatic motor neurons extend from the CNS and synapse directly with an effector. Autonomic motor pathways consist of two motor neurons in series. The axon of the first motor neuron extends from the CNS and synapses in a ganglion with the second motor neuron; the second neuron synapses with an effector.
4. The output (motor) portion of the ANS has two divisions: sympathetic and parasympathetic. Most body organs receive dual innervation; usually one ANS division causes excitation and the other causes inhibition.
5. Somatic nervous system effectors are skeletal muscles; ANS effectors include cardiac muscle, smooth muscle, and glands.
6. Table 19.1 compares the somatic and autonomic nervous systems.

Resource:
Anatomy Overview - The Nervous System: Overview
Anatomy Overview - Organization of the ANS
Figure 19.1 - Motor Neuron Pathways in the Somatic Nervous System and Autonomic Nervous System (ANS)
Exercise - Assemble the Structure of the ANS

19.2 Anatomy of Autonomic Motor Pathways

1. Neural crest cells migrate to positions near developing smooth muscle in the gut tube and near the major blood vessels to become the postganglionic neurons of the autonomic motor pathways.
2. Neurons from the ventrolateral part of the developing neural tube grow axons toward the migrating crest cells to become the preganglionic neurons of the autonomic motor pathways.
3. Preganglionic neurons are myelinated; postganglionic neurons are unmyelinated.
4. Preganglionic neurons and postganglionic neurons synapse in autonomic ganglia.

Resource:
Animation - ANS: Motor Pathways
Anatomy Overview - Visceral Receptors
Anatomy Overview - Visceral Effectors

19.3 Structure of the Sympathetic Division

1. The cell bodies of sympathetic preganglionic neurons are in the lateral gray horns of the 12 thoracic and the first two or three lumbar segments of the spinal cord.
2. Sympathetic ganglia are classified as sympathetic trunk ganglia (on both sides of the vertebral column) or prevertebral ganglia (anterior to the vertebral column).
3. Sympathetic preganglionic neurons synapse with postganglionic neurons in sympathetic trunk ganglia or prevertebral ganglia.

Resource:
Figure 19.4 - The Sympathetic Division of the Autonomic Nervous System

19.4 Structure of the Parasympathetic Division

1. The cell bodies of parasympathetic preganglionic neurons are in four cranial nerve nuclei—oculomotor (III), facial (VII), glossopharyngeal (IX), and vagus (X)—in the brain stem and lateral gray horns of the second through fourth sacral segments of the spinal cord.
2. Parasympathetic ganglia are referred to as terminal ganglia because they are located near or inside visceral effectors.
3. Parasympathetic preganglionic neurons synapse with postganglionic neurons in terminal ganglia.

Resource:
Figure 19.6 - The Parasympathetic Division of the Autonomic Nervous System

Review	Resource

19.5 Structure of the Enteric Division

1. The enteric nervous system consists of nerves and ganglia in the wall of the gastrointestinal tract, pancreas, and gallbladder.

19.6 ANS Neurotransmitters and Receptors

1. Cholinergic neurons release acetylcholine, which binds to nicotinic or muscarinic cholinergic receptors.
2. In the ANS, the cholinergic neurons include all sympathetic and parasympathetic preganglionic neurons, all parasympathetic postganglionic neurons, and sympathetic postganglionic neurons that innervate most sweat glands.
3. In the ANS, adrenergic neurons release norepinephrine (noradrenalin). Both epinephrine and norepinephrine bind to alpha and beta adrenergic receptors.
4. Most sympathetic postganglionic neurons are adrenergic.
5. An agonist is a substance that binds to and activates a receptor, mimicking the effect of a natural neurotransmitter or hormone. An antagonist is a substance that binds to and blocks a receptor, thereby preventing a natural neurotransmitter or hormone from exerting its effect.

Anatomy Overview - Neurotransmitters
Animation - The ANS: Types of Neurotransmitters and Neurons
Figure 19.7 - Cholinergic Neurons and Adrenergic Neurons in the Sympathetic and Parasympathetic Divisions

19.7 Functions of the ANS

1. The sympathetic division favors body functions that can support vigorous physical activity and rapid production of ATP in a series of physiological responses called the fight-or-flight response; the parasympathetic division regulates activities that conserve and restore body energy.
2. The effects of sympathetic stimulation are longer-lasting and more widespread than the effects of parasympathetic stimulation.
3. Table 19.2 compares structural and functional features of the sympathetic and parasympathetic divisions.
4. Table 19.3 lists the effects of sympathetic and parasympathetic stimulation on effectors throughout the body.

Anatomy Overview - Effectors
Animation - Physiological Effects of the ANS
Animation - The Alarm Reaction
Exercise - Sort ANS Functions
Exercise - What Is Your ANS Status?

19.8 Integration and Control of Autonomic Functions

1. An autonomic reflex adjusts the activities of smooth muscle, cardiac muscle, and glands.
2. An autonomic reflex arc consists of a receptor, a sensory neuron, an integrating center, two autonomic motor neurons, and a visceral effector.
3. The hypothalamus is the major control and integration center of the ANS. It is connected to both the sympathetic division and the parasympathetic division.

Anatomy Overview - The ANS Control Centers

CRITICAL THINKING QUESTIONS

1. Skydiving, hang gliding, and bungee jumping can all give you a great rush (or get you killed). How do these activities cause this rush?

2. "The quickest way to a man's heart is through his stomach," Sophia's grandmother is fond of saying. Trace the pathway that an impulse would follow from a full stomach to a happy heart.

3. Mando has been living in a refugee camp for two years. The guards are hostile to the refugees; they regularly fire off rounds from their machine guns at night and frequently make threatening remarks and gestures to the women in the camp. Every time a guard comes near, Mando feels her heart race and her mouth go dry. Physiologically, what is happening to Mando, and what do you think some of the long-term physiological effects might be?

4. The autonomic and the enteric divisions of the nervous system both control the digestive system. How do they compare?

5. While reviewing the chapter on the brain, your study partner says, "The hypothalamus can't be very important. It's so small." After reading this chapter, would you agree with your friend? Why or why not?

? ANSWERS TO FIGURE QUESTIONS

19.1 Dual innervation means that a body organ receives nerve impulses from both sympathetic and parasympathetic neurons of the ANS.

19.2 The neural crest arises from ectoderm in the process of neurulation.

19.3 The celiac (solar) plexus is the largest autonomic plexus.

19.4 Most parasympathetic preganglionic axons are longer than most sympathetic preganglionic axons because most parasympathetic ganglia are in the walls of visceral organs, and most sympathetic ganglia are close to the spinal cord in the sympathetic trunk.

19.5 White rami look white due to the presence of myelin.

19.6 Terminal ganglia are associated with the parasympathetic division; sympathetic trunk ganglia and prevertebral ganglia are associated with the sympathetic division.

19.7 Cholinergic neurons with nicotinic ACh receptors include sympathetic postganglionic neurons innervating sweat glands and all parasympathetic postganglionic neurons. The effectors innervated by these cholinergic neurons possess muscarinic receptors.

20 SOMATIC SENSES AND MOTOR CONTROL

INTRODUCTION Imagine a camping trip to a beautiful rocky coastline cradling a sandy patch of beach. As you rouse from your night of slumber on the packed sand, you slowly stretch your stiffened joints and cautiously climb out of your sleeping bag to greet the crisp morning air. You rub the sleep from your eyes and see the fog rolling in off the white crests of the slapping waves. As you walk toward the ocean you take a deep breath, smell the salty scent of the tide, and feel individual grains of sand between your wriggling toes. Suddenly you stop to rub your exposed arms vigorously, as the cool, crisp air sends a chill through your still-sleepy body. You see and hear noisy gulls gliding through the air overhead, and listen to a distant boat sounding its horn. As you walk toward the water line, where the sounds of the water are playing their tune against the rocks, you glance down into the tide pools left behind by the receding waves, and notice a colorful array of intertidal life—sea stars, mussels, anemones, and scurrying crabs. Bending down to take a closer look, your face is splashed by an incoming wave, giving you a taste of the salty sea. You think for a minute about the beauty you have *sensed* in the past few minutes. Your mind is flooded with what you have *seen*, what you have *felt*, what you have *smelled*, what you have *heard*, and what you have *tasted*.

The previous four chapters described the organization of the nervous system. Now we will see how certain parts cooperate to carry out its three basic functions: (1) receiving sensory input; (2) integrating the sensory input—that is, processing and interpreting it, deciding a course of action, and storing information; and (3) transmitting motor impulses that result in a response (muscular contraction or glandular secretion). In this chapter we will explore the nature and types of sensations, the pathways that convey the sensory input from the body to the brain, and the pathways that carry motor commands from the brain to effectors. Chapter 21 deals with the special senses of smell, taste, vision, hearing, and equilibrium. •

Did you ever wonder why patients who have had a limb amputated still experience sensations as if the limb were still there?

20.1 OVERVIEW OF SENSATIONS

 OBJECTIVES

- Define **sensation**.
- Describe the conditions necessary for a sensation to occur.
- Explain the different ways sensory receptors can be classified.

Definition of Sensations

A **sensation** is the conscious or subconscious awareness of changes in the external or internal conditions of the body. For a sensation to occur, four conditions must be satisfied:

1. A *stimulus*, or change in the environment, capable of activating certain sensory neurons, must occur.
2. A *sensory receptor* must convert the stimulus to nerve impulses.
3. The nerve impulses must be *conducted* along a neural pathway from the sensory receptor to the brain.
4. A region of the brain must receive and *integrate* the nerve impulses, producing a sensation.

A stimulus may be in the form of light, heat, pressure, mechanical energy, or chemical energy. A sensory receptor responds to a stimulus by altering its membrane's permeability to small ions. In most types of sensory receptors, the resulting flow of ions across the membrane produces a change that triggers one or more nerve impulses. The impulses are then conducted along the sensory neuron toward the CNS.

Characteristics of Sensations

A **perception** is the conscious interpretation of a sensation. Perceptions are integrated in the cerebral cortex. You seem to see with your eyes, hear with your ears, and feel pain in an injured part of your body. This is because sensory impulses from each part of the body arrive in a specific region of the cerebral cortex, which interprets the sensation as coming from the stimulated sensory receptors.

Each unique type of sensation, such as touch, pain, vision, or hearing, is called a **sensory modality** (mō-DAL-i-tē). Based on the receptor stimulated, a sensory neuron carries information for one sensory modality only. For example, neurons relaying impulses for touch do not transmit impulses for pain. The specialization of sensory neurons allows nerve impulses from the eyes to

be perceived as sight, and those from the ears to be perceived as sounds.

A characteristic of most sensory receptors is **adaptation**, a decrease in sensation during a prolonged stimulus. Adaptation is caused in part by a decrease in the responsiveness of sensory receptors. As a result of adaptation, the perception of a sensation decreases or disappears even though the stimulus persists. For example, when you first step into a hot shower, the water may feel very hot, but soon the sensation decreases to one of comfortable warmth even though the stimulus (the high temperature of the water) does not change. Receptors vary in how quickly they adapt.

Classification of Sensations

The senses can be grouped into two classes: general senses and special senses.

1. The **general senses** refer to two types of senses: somatic senses and visceral senses. **Somatic senses** (*somat-*=of the body) are tactile sensations (touch, pressure, vibration, itch, and tickle); thermal sensations (warm and cold); pain sensations; and proprioceptive sensations, which allow perception of both the static (nonmoving) positions of limbs and body parts (joint and muscle position sense) and movements of the limbs and head. **Visceral senses** provide information about conditions within internal organs, such as pressure, stretch, chemicals, nausea, hunger, and temperature.
2. The **special senses** include smell, taste, vision, hearing, and equilibrium (balance).

In addition to introducing the mechanism of sensation, this chapter deals with the somatic senses; the special senses will be covered in Chapter 21.

Types of Sensory Receptors

Several structural and functional characteristics of sensory receptors can be used to group them into different classes. On a microscopic level, sensory receptors may be defined by the *structures of their sensory receptors* (Figure 20.1), which include the following:

- **Free nerve endings** of sensory neurons
- **Encapsulated nerve endings** of sensory neurons
- **Separate cells** that synapse with sensory neurons

Figure 20.1 Sensory receptors in the skin and subcutaneous layer.

 The somatic sensations of touch, pressure, vibration, warmth, cold, and pain arise from sensory receptors in the skin, subcutaneous layer, and mucous membranes.

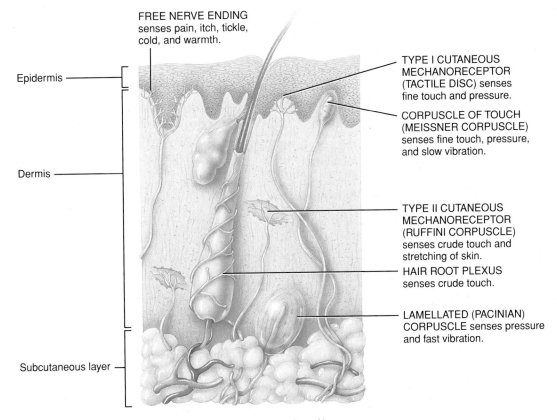

FREE NERVE ENDING
senses pain, itch, tickle, cold, and warmth.

Epidermis

TYPE I CUTANEOUS MECHANORECEPTOR (TACTILE DISC) senses fine touch and pressure.

CORPUSCLE OF TOUCH (MEISSNER CORPUSCLE) senses fine touch, pressure, and slow vibration.

Dermis

TYPE II CUTANEOUS MECHANORECEPTOR (RUFFINI CORPUSCLE) senses crude touch and stretching of skin.

HAIR ROOT PLEXUS senses crude touch.

LAMELLATED (PACINIAN) CORPUSCLE senses pressure and fast vibration.

Subcutaneous layer

 Which sensations can arise when free nerve endings are stimulated?

A second method of categorizing sensory receptors is to group them by the *type of stimulus they detect*:

- **Photoreceptors** detect light that strikes the retina.
- **Mechanoreceptors** are sensitive to mechanical stimuli, such as touch or pressure.
- **Thermoreceptors** detect changes in temperature.
- **Osmoreceptors** detect the osmotic pressure of the body fluids.
- **Chemoreceptors** detect chemicals.
- **Nociceptors** (nō′-sē-SEP-tors; *noci*=harmful) (pain receptors) respond to painful stimuli resulting from physical or chemical damage to tissue.

Finally, the *location of the receptors* and the *origin of the stimuli* that activate them can be used to classify sensory receptors:

- **Exteroceptors** (EKS-ter-ō-sep′-tors) are located at or near the external surface of the body.
- **Interoceptors** (IN-ter-ō-sep′-tors) or *visceroceptors* are located in blood vessels, visceral organs, and muscles.
- **Proprioceptors** (PRŌ-prē-ō-sep′-tors; *proprio*=one's own) are located in muscles, tendons, joints, and the inner ear.

Table 20.1 summarizes the classification of sensory receptors and presents more detail regarding their functions.

 CHECKPOINT
1. Distinguish between sensation and perception.
2. Define sensory modality and adaptation.
3. What events are necessary for a sensation to occur?
4. Use the three classification schemes to categorize a receptor that detects the painful sensation of sand in the eye. (*Hint:* Refer to Table 20.1.)

20.2 SOMATIC SENSATIONS

 OBJECTIVES
- Describe the location and function of the receptors for tactile, thermal, and pain sensations.
- Identify the receptors for proprioception and describe their functions.

Somatic sensations arise from stimulation of sensory receptors embedded in the skin or subcutaneous layer; in mucous membranes of the mouth, vagina, and anus; in muscles, tendons, and joints; and in the inner ear. The sensory receptors for somatic sensations are distributed unevenly—some parts of the body surface are densely populated with receptors, but other parts contain only a few. Areas with the highest density of somatic sensory receptors are the tip of the tongue, the lips, and the fingertips. Somatic sensations that arise from stimulating the skin surface are referred to as **cutaneous sensations** (kū-TĀ-nē-us; *cutane-*= skin). Somatic sensations are of four sensory modalities: tactile, thermal, pain, and proprioceptive.

Tactile Sensations

The **tactile sensations** (TAK-tīl; *tact-*=touch) include touch, pressure, vibration, itch, and tickle. Several types of encapsulated mechanoreceptors mediate sensations of touch, pressure, and vibration. Other touch sensations, as well as itch and tickle sensations, are detected by free nerve endings. Tactile receptors in the skin or subcutaneous layer include corpuscles of touch, hair root plexuses, type I and II cutaneous mechanoreceptors, lamellated corpuscles, and free nerve endings (Figure 20.1).

TABLE 20.1

Classification of Sensory Receptors

BASIS OF CLASSIFICATION	DESCRIPTION
MICROSCOPIC FEATURES	
Free nerve endings	Bare dendrites associated with pain, thermal, tickle, itch, and some touch sensations
Encapsulated nerve endings	Dendrites enclosed in a connective tissue capsule for pressure, vibration, and some touch sensations
Separate cells	Receptor cells synapse with first-order sensory neurons; located in the retina of the eye (photoreceptors), inner ear (hair cells), and taste buds of the tongue (gustatory receptor cells)
TYPE OF STIMULUS DETECTED	
Photoreceptors	Detect light that strikes the retina of the eye
Mechanoreceptors	Detect mechanical stimuli; provide sensations of touch, pressure, vibration, proprioception, and hearing and equilibrium; also monitor stretching of blood vessels and internal organs
Thermoreceptors	Detect changes in temperature
Osmoreceptors	Sense the osmotic pressure of body fluids
Chemoreceptors	Detect chemicals in mouth (taste), nose (smell), and body fluids
Nociceptors	Respond to painful stimuli resulting from physical or chemical damage to tissue
RECEPTOR LOCATION AND ACTIVATING STIMULI	
Exteroceptors	Located at or near body surface; sensitive to stimuli originating outside body; provide information about *external* environment; convey visual, smell, taste, touch, pressure, vibration, thermal, and pain sensations
Interoceptors	Located in blood vessels, visceral organs, and nervous system; provide information about *internal* environment; impulses produced usually are not consciously perceived but occasionally may be felt as pain or pressure
Proprioceptors	Located in muscles, tendons, joints, and inner ear; provide information about body position, muscle length and tension, position and motion of joints, and equilibrium (balance)

Touch

Sensations of **touch** generally result from stimulation of tactile receptors in the skin or subcutaneous layer. **Fine touch** provides specific information about a touch sensation, such as exactly what point on the body is touched plus the shape, size, and texture of the source of stimulation. **Crude touch** is the ability to perceive that something has contacted the skin, even though its exact location, shape, size, or texture cannot be determined.

There are two types of rapidly adapting touch receptors, one for fine touch and one for crude touch. **Corpuscles of touch**, or *Meissner corpuscles* (MĪS-ner), are receptors for fine touch that are located in the dermal papillae of hairless skin. Each corpuscle is a mass of dendrites enclosed by an egg-shaped capsule of connective tissue. Because corpuscles of touch are rapidly adapting receptors, they generate nerve impulses mainly at the onset of a touch. They are abundant in the hands, eyelids, tip of the tongue, lips, nipples, soles, clitoris, and tip of the penis. **Hair root plexuses** are rapidly adapting crude touch receptors found in hairy skin; they consist of free nerve endings wrapped around hair follicles. Hair root plexuses detect movements on the surface of the skin that disturb hairs. For example, an insect landing on a hair causes movement of the hair shaft that stimulates the free nerve endings.

There also are two types of slowly adapting touch receptors, again one for fine touch and one for crude touch. **Type I cutaneous mechanoreceptors**, also known as *tactile discs*,

function in fine touch. Type I cutaneous mechanoreceptors are saucer-shaped, flattened free nerve endings that make contact with tactile epithelial cells (*Merkel cells*) of the stratum basale (see Figure 5.2d). These mechanoreceptors are plentiful in the fingertips, hands, lips, and external genitalia. **Type II cutaneous mechanoreceptors**, or *Ruffini corpuscles*, are elongated, encapsulated receptors for crude touch located deep in the dermis, and in ligaments and tendons. Present in the hands and abundant on the soles, they are most sensitive to stretching that occurs as digits or free limbs are moved.

Pressure and Vibration

Pressure, such as that felt by the weight of your heavily laden backpack, is a sustained sensation that is felt over a larger area and occurs in deeper tissues than those supplied by the receptors for touch. Pressure occurs with deformation of deeper tissues. Receptors that contribute to sensations of pressure include corpuscles of touch, type I mechanoreceptors, and lamellated corpuscles. **Lamellated corpuscles**, or *pacinian corpuscles* (pa-SIN-ē-an), are large oval structures composed of a multilayered connective tissue capsule that encloses a dendrite. Like corpuscles of touch, lamellated corpuscles adapt rapidly. They are widely distributed in the body: in the dermis and subcutaneous layer; in submucosal tissues that underlie mucous and serous membranes; around joints, tendons, and muscles; in the periosteum; and in the mammary glands, external genitalia, and certain viscera, such as the pancreas and urinary bladder.

Sensations of **vibration**, such as using an electric knife to carve a turkey, result from rapidly repetitive sensory signals from tactile receptors. The receptors for vibratory sensations are corpuscles of touch and lamellated corpuscles. Corpuscles of touch can detect lower-frequency vibrations; lamellated corpuscles detect higher-frequency vibrations.

Itch and Tickle

The **itch** sensation results from stimulation of free nerve endings by certain chemicals, such as antigens in mosquito saliva injected from a bite, often because of a local inflammatory response. Free nerve endings are thought to mediate the **tickle** sensation. This intriguing sensation typically arises only when someone else touches you, not when you touch yourself. The explanation of this puzzling phenomenon seems to lie in the impulses that conduct to and from the cerebellum when you are moving your fingers and touching yourself that do not occur when someone else is tickling you.

Thermal Sensations

Thermoreceptors are free nerve endings. Two distinct **thermal sensations**—coldness and warmth—are detected by different receptors. **Cold receptors** are located in the stratum basale of the epidermis. Temperatures between 10° and 40°C (50–105°F) activate cold receptors. **Warm receptors** are located in the dermis and are activated by temperatures between 32° and 48°C (90–118°F). Cold and warm receptors both adapt rapidly at the onset of a stimulus but continue to generate impulses at a lower frequency throughout a prolonged stimulus. Temperatures below 10°C and above 48°C stimulate mainly pain receptors rather than thermoreceptors, producing painful sensations to protect the body from damage.

Pain Sensations

Pain is indispensable for survival. It serves a protective function by signaling the presence of noxious, tissue-damaging conditions. From a medical standpoint, the subjective description of pain, along with an indication of where it occurs, may help pinpoint the underlying cause of disease.

Nociceptors, the receptors for pain, are free nerve endings found in every tissue of the body except the brain (Figure 20.1). Intense thermal, mechanical, or chemical stimuli can activate nociceptors. Tissue irritation or injury releases chemicals such as prostaglandins, kinins, and potassium ions (K^+) that stimulate nociceptors. Pain may persist even after a pain-producing stimulus is removed because pain-mediating chemicals linger, and because nociceptors exhibit very little adaptation. Conditions that elicit pain include excessive distension (stretching) of a structure, prolonged muscular contractions, muscle spasms, or ischemia (inadequate blood flow to an organ).

Types of Pain

There are two types of pain: fast and slow. The perception of **fast pain** occurs very rapidly, usually within 0.1 second after a stimulus is applied. This type of pain is also known as acute, sharp, or pricking pain. The sensations resulting from a needle puncture or knife cut to the skin are examples of fast pain. Fast pain is not felt in deeper tissues of the body. The perception of

slow pain, by contrast, begins a second or more after a stimulus is applied. It then gradually increases in intensity over a period of several seconds or minutes. This type of pain, which may be excruciating, is also referred to as *chronic, burning, aching,* or *throbbing* pain. Slow pain can occur both in the skin and in deeper tissues or internal organs. An example is the pain associated with a toothache.

Pain that arises from stimulation of receptors in the skin is called **superficial somatic pain**. Stimulation of receptors in skeletal muscles, joints, tendons, and fascia causes **deep somatic pain**. **Visceral pain** results from stimulation of nociceptors in visceral organs.

Localization of Pain

Fast pain is very precisely localized to the stimulated area. For example, if someone pricks you with a pin, you know exactly which part of your body was stimulated. Slow somatic pain also is well localized but more diffuse (involves large areas); it usually appears to come from a larger area of the skin. In some instances of visceral slow pain, the pain is felt directly in the affected area. If the pleural membranes around the lungs are inflamed, for example, you experience chest pain.

In many instances of visceral pain, the pain is felt in or just deep to the skin that overlies the stimulated organ, or in a surface area far from the stimulated organ. This phenomenon is called

Figure 20.2 Distribution of referred pain. The colored portions of the diagrams indicate skin areas to which visceral pain is referred.

 Nociceptors are present in almost every tissue of the body.

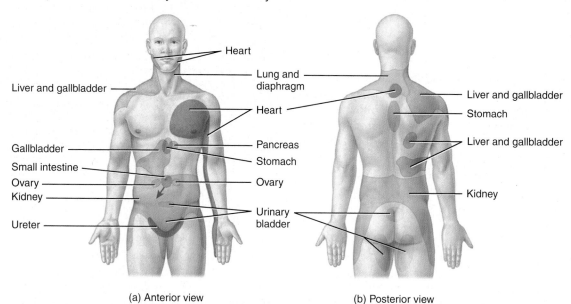

(a) Anterior view   (b) Posterior view

 Which visceral organs have the broadest area for referred pain?

referred pain. Figure 20.2 shows skin regions to which visceral pain may be referred. In general, the visceral organ involved and the area to which the pain is referred are served by the same segment of the spinal cord. For example, sensory fibers from the heart, the skin superficial to the heart, and the skin along the medial aspect of the left arm enter spinal cord segments T1 to T5. Thus, the pain of a heart attack typically is felt in the skin over the heart and along the left arm.

Proprioceptive Sensations

Proprioceptive sensations are also called *proprioception* (prō'-prē-ō-SEP-shun; *proprius* = self or one's own). These sensations allow us to recognize that body parts belong to us (self). They also allow us to know where our head and free limbs are located and how they are moving even if we are not looking at them, so that we can walk, type, or put on a shirt without using our eyes. **Kinesthesia** (kin'-es-THĒ-zē-a; *kin-*=motion; *esthesia*= perception) is the perception of body movements. Proprioceptive sensations arise in receptors called **proprioceptors**, which are embedded in muscles (especially postural muscles) and tendons to inform us of the degree to which muscles are contracted, the amount of tension on tendons, and the positions of joints. Hair cells of the vestibular apparatus in the inner ear monitor the orientation of the head relative to the ground and head position during movements, allowing maintenance of balance and equilibrium (described in Chapter 21). Because proprioceptors adapt slowly and only slightly, the brain continually receives nerve impulses related to the position of different body parts and makes adjustments to ensure coordination.

Proprioceptors also allow **weight discrimination**, the ability to assess the weight of an object. This helps you determine the muscular effort necessary to perform a task. For example, as you pick up a shopping bag, you quickly realize whether it contains books or feathers; you then exert the correct amount of effort needed to lift it.

Here we discuss three types of proprioceptors: muscle spindles within skeletal muscles, tendon organs within tendons, and joint kinesthetic receptors within synovial joint capsules.

Muscle Spindles

Muscle spindles are the proprioceptors in skeletal muscles that monitor changes in the length of skeletal muscles and participate in stretch reflexes. By adjusting how vigorously a muscle spindle responds to stretching of a skeletal muscle, the brain sets an overall level of **muscle tone**, the small degree of contraction that is present while the muscle is at rest.

Each muscle spindle consists of several slowly adapting sensory nerve endings that wrap around 3 to 10 specialized muscle fibers, called **intrafusal muscle fibers** (in'-tra-FŪ-sal; *intrafusal-*=within a spindle) (Figure 20.3). A connective tissue capsule encloses the sensory nerve endings and intrafusal fibers and anchors the spindle to endomysium and perimysium. Muscle spindles are interspersed among ordinary skeletal muscle fibers and are aligned parallel to them. In muscles that produce finely controlled movements, such as movements of the fingers and eyes as you read music and play a musical instrument, muscle spindles are plentiful. Muscles involved in coarser but more forceful movements, such as the quadriceps femoris and hamstring muscles of the thigh, have fewer muscle spindles. The only skeletal muscles that lack spindles are the tiny muscles of the middle ear.

The main function of muscle spindles is to measure *muscle length*—how much a muscle is being stretched. Either sudden or prolonged stretching of the central areas of the intrafusal muscle fibers stimulates the sensory nerve endings. The resulting nerve impulses propagate into the CNS. Information from muscle spindles arrives quickly at the somatic sensory areas of the cerebral cortex, which allows conscious perception of free limb positions and movements. At the same time, impulses from muscle spindles also pass to the cerebellum, where the input is used to coordinate muscle contractions.

In addition to their sensory nerve endings near the middle of intrafusal fibers, muscle spindles contain motor neurons called **gamma motor neurons**. The motor neurons terminate near both ends of the intrafusal fibers and adjust the tension in a muscle spindle to variations in the length of the muscle. For example, when your biceps muscle shortens in response to lifting a weight,

Figure 20.3 **Two types of proprioceptors: a muscle spindle and a tendon organ.** In muscle spindles, which monitor changes in skeletal muscle length, sensory nerve endings wrap around the central portion of intrafusal muscle fibers. In tendon organs, which monitor the force of muscle contraction, sensory nerve endings are activated by increasing tension on a tendon.

 Proprioceptors provide information about body position and movement.

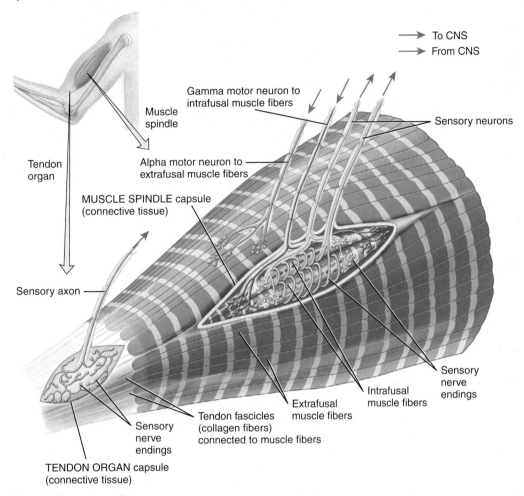

 How is a muscle spindle activated?

gamma motor neurons stimulate the ends of the intrafusal fibers to contract slightly. This keeps the intrafusal fibers taut even though the contracting muscle fibers surrounding the spindle are reducing spindle tension. This maintains the sensitivity of the muscle spindle to stretching of the muscle. As the frequency of impulses in its gamma motor neuron increases, a muscle spindle becomes more sensitive to stretching of its midregion.

Surrounding the muscle spindles are ordinary skeletal muscle fibers, called **extrafusal muscle fibers** (*extrafusal*=outside a spindle), which are innervated by **alpha motor neurons**. The cell bodies of both gamma and alpha motor neurons are located in the anterior gray horn of the spinal cord (or in the brain stem for muscles in the head). During stretching of a muscle, such as the knee-jerk reflex when a physician strikes the patellar tendon, impulses in muscle spindle sensory axons propagate into the spinal cord and brain stem and activate alpha motor neurons that innervate extrafusal muscle fibers in the same muscle. In this way, activation of its muscle spindles causes contraction of an entire skeletal muscle, which relieves the stretching.

Tendon Organs

When a muscle contracts, it exerts a force that tends to pull the points of attachment at either end toward each other; this force

is the *muscle tension.* **Tendon organs** are located at the junction of a tendon and a muscle and protect tendons and their associated muscles from damage due to excessive muscle tension. Each tendon organ consists of a thin capsule of connective tissue that encloses a few tendon fascicles (bundles of collagen fibers) (see Figure 20.3). One or more sensory nerve endings penetrate the capsule, winding among the collagen fibers. When tension is applied to a muscle, the tendon organs generate nerve impulses that propagate into the CNS, providing information about changes in muscle tension. The resulting tendon reflexes decrease muscle tension by causing relaxation.

Joint Kinesthetic Receptors

Several types of **joint kinesthetic receptors** (kin-es-THET-ik) are present within and around the articular capsules of synovial joints (see Section 9.4). Free nerve endings and type II cutaneous mechanoreceptors (Ruffini corpuscles) in the capsules of joints respond to pressure. Small lamellated (pacinian) corpuscles in the connective tissue outside articular capsules respond to acceleration and deceleration in the movement of joints. Joint ligaments contain receptors similar to tendon organs that adjust the adjacent muscles when excessive strain is placed on the joint.

TABLE 20.2

Summary of Receptors for Somatic Sensations

RECEPTOR TYPE	RECEPTOR STRUCTURE AND LOCATION	SENSATIONS	ADAPTATION RATE
TACTILE RECEPTORS			
Corpuscles of touch (Meissner corpuscles)	Capsule surrounds mass of dendrites in dermal papillae of hairless skin	Fine touch, pressure, and slow vibrations	Rapid
Hair root plexuses	Free nerve endings wrapped around hair follicles in skin	Crude touch	Rapid
Type I cutaneous mechanoreceptors (tactile discs)	Saucer-shaped free nerve endings make contact with tactile epithelial cells in epidermis	Fine touch and pressure	Slow
Type II cutaneous mechanoreceptors (Ruffini corpuscles)	Elongated capsule surrounds dendrites deep in dermis and in ligaments and tendons	Crude touch and stretching of skin	Slow
Lamellated (pacinian) corpuscles	Oval, layered capsule surrounds dendrites; present in dermis and subcutaneous layer, submucosal tissues, joints, periosteum, and some viscera	Pressure and fast vibrations	Rapid
Itch and tickle receptors	Free nerve endings in skin and mucous membranes	Itching and tickling	Both slow and rapid
THERMORECEPTORS			
Warm receptors and cold receptors	Free nerve endings in skin and mucous membranes of mouth, vagina, and anus	Warmth or cold	Initially rapid, then slow
PAIN RECEPTORS			
Nociceptors	Free nerve endings in every tissue of the body except the brain	Pain	Slow
PROPRIOCEPTORS			
Muscle spindles	Sensory nerve endings wrap around central area of encapsulated intrafusal muscle fibers within most skeletal muscles	Muscle length	Slow
Tendon organs	Capsule encloses collagen fibers and sensory nerve endings at junction of tendon and muscle	Muscle tension	Slow
Joint kinesthetic receptors	Lamellated (pacinian) corpuscles, type II cutaneous mechanoreceptors (Ruffini corpuscles), tendon organs, and free nerve endings	Joint position and movement	Rapid

Table 20.2 summarizes the somatic sensory receptors and the sensations they convey.

 CHECKPOINT

5. Which somatic sensory receptors are encapsulated?
6. Why do some somatic sensory receptors adapt slowly, and others adapt rapidly? Why do others not adapt at all?
7. What is the difference between fine touch and crude touch? Which somatic sensory receptors mediate each type of sensation?
8. How does fast pain differ from slow pain?
9. What is referred pain, and how is it useful in diagnosing internal disorders?
10. How do muscle spindles and tendon organs monitor muscle function?

20.3 SOMATIC SENSORY PATHWAYS

 OBJECTIVES

• Describe the neuronal components and functions of the posterior column–medial lemniscus pathway.

• Outline the structures and functions of the anterolateral pathway.
• Explain the activities controlled by the spinocerebellar pathway and the structures involved.

On reaching the spinal cord, sensory information has two possible destinations: (1) It may become part of a reflex arc, bringing about an appropriate effector response, or (2) it may be relayed upward to the brain via ascending pathways for further processing and possible conscious awareness. **Somatic sensory pathways** relay information from the somatic sensory receptors just described to two locations: (1) the primary somatosensory area in the cerebral cortex and (2) the cerebellum. The pathways to the cerebral cortex consist of thousands of sets of three neurons: first-order neurons, second-order neurons, and third-order neurons.

1. **First-order neurons** conduct impulses from somatic receptors into the brain stem or spinal cord. From the face, mouth, teeth, and eyes, somatic sensory impulses propagate along *cranial nerves* into the brain stem. From the neck, trunk, limbs, and posterior (dorsal) aspect of the head, somatic sensory impulses propagate along *spinal nerves* into the spinal cord.
2. **Second-order neurons** conduct impulses from the brain stem and spinal cord to the thalamus or cerebellum. Axons

of second-order neurons *decussate* (cross over to the opposite side) in the brain stem or spinal cord before ascending to the ventral posterior nucleus of the thalamus. Thus, all somatic sensory information from one side of the body reaches the thalamus on the opposite side.

3. **Third-order neurons** conduct impulses from the thalamus to the primary somatosensory area of the cortex on the same side.

Regions within the CNS where neurons synapse with other neurons that are a part of a particular sensory or motor pathway are known as **relay stations** because neural signals are being relayed from one region of the CNS to another. For example, the neurons of many sensory pathways synapse with neurons in the thalamus; therefore the thalamus functions as a major relay station. Many other regions of the CNS, including the spinal cord and brain stem, can also function as relay stations.

Somatic sensory impulses entering the spinal cord ascend to the cerebral cortex via two general pathways: (1) the posterior column–medial lemniscus pathway and (2) the anterolateral (spinothalamic) pathways. Somatic sensory impulses entering the spinal cord reach the cerebellum via the spinocerebellar tracts.

Posterior Column–Medial Lemniscus Pathway to the Cortex

Nerve impulses for conscious proprioception and most tactile sensations ascend to the cerebral cortex along the **posterior column–medial lemniscus pathway** (lem-NIS-kus) (Figure 20.4a). The name of the pathway comes from the names of two white-matter tracts that convey the impulses: the posterior column of the spinal cord and the medial lemniscus of the brain stem.

Figure 20.4 Somatic sensory pathways.

 Nerve impulses are conducted along sets of first-order, second-order, and third-order neurons to the primary somatosensory area (postcentral gyrus) of the cerebral cortex.

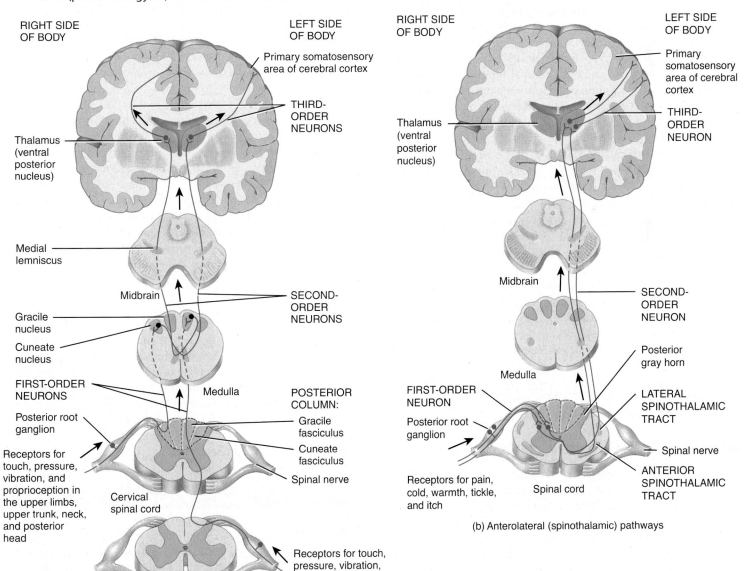

(a) Posterior column–medial lemniscus pathway

(b) Anterolateral (spinothalamic) pathways

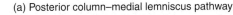

What types of sensory deficits could be produced by damage to the right lateral spinothalamic tract?

First-order neurons in the posterior column–medial lemniscus pathway extend from sensory receptors in the trunk and limbs into the spinal cord and ascend to the medulla oblongata on the same side of the body. These are the longest neurons in the body. The cell bodies of these first-order neurons are in the posterior (dorsal) root ganglia of spinal nerves. In the spinal cord, their axons form the **posterior (dorsal) columns**, which consist of two parts: the **gracile fasciculus** (GRAS-il fa-SIK-ū-lus) and the **cuneate fasciculus** (KŪ-nē-at). (See Table 20.3.) After ascending through their respective fascicles, the axons synapse with *dendrites of* second-order neurons whose cell bodies are located in the gracile nucleus or the cuneate nucleus of the medulla oblongata. Impulses from the posterior head, neck, free upper limbs, and upper thorax (above T6) propagate along axons in the cuneate fasciculus and arrive at the cuneate nucleus. Impulses from the lower thorax (below T6), abdomen, pelvis, and lower limbs propagate along axons in the gracile fasciculus and arrive at the gracile nucleus. (The exception to this rule is the proprioceptive neurons from the lower limbs. They relay in the nucleus dorsalis or thoracic nucleus of Clarke in the thoracic spinal cord and then travel to the brainstem in the posterior cerebellar tract.) Impulses for somatic sensations from the head arrive in the brain stem via sensory axons that are part of the trigeminal (V), facial (VII), glossopharyngeal (IX), and vagus (X) nerves. Most somatic sensory axons carrying impulses from the face are part of the trigeminal (V) nerve.

The axons of the second-order neurons cross to the opposite side of the medulla and enter the **medial lemniscus** (lem-NIS-kus=ribbon), a thin ribbonlike projection tract that extends from the medulla to the ventral posterior nucleus of the thalamus (see Figure 20.4a). In the thalamus, the axon terminals of second-order neurons synapse with third-order neurons, which project their axons to the primary somatosensory area of the cerebral cortex.

Impulses conducted along the posterior column–medial lemniscus pathway give rise to several highly evolved and refined sensations:

• Fine touch is the ability to recognize specific information about a touch sensation, such as what point on the body is touched plus the shape, size, and texture of the source of stimulation (see Section 20.2).
• **Stereognosis** (ster'-ē-og-NŌ-sis) is the ability to recognize the size, shape, and texture of an object by feeling it. Examples are reading Braille or identifying a paperclip by feeling it.
• Proprioception is the awareness of the precise position of body parts, and kinesthesia is the awareness of directions of movement. Proprioceptors also allow weight discrimination, the ability to assess the weight of an object. (See Section 20.2.)
• Vibratory sensations arise when rapidly fluctuating touch stimuli are present (see Section 20.2).

Anterolateral Pathways to the Cortex

Like the posterior column–medial lemniscus pathway just described, the **anterolateral** or **spinothalamic pathways** (spī-nō-tha-LAM-ik) are composed of three-neuron sets (Figure 20.4b). The first-order neurons connect a receptor of the neck, trunk, or free limbs with the spinal cord. The cell bodies of the first-order neurons are in the posterior root ganglion. The axon terminals of the first-order neurons synapse with the dendrites of second-order neurons, whose cell bodies are located in the posterior gray horn of the spinal cord.

The axons of the second-order neurons cross to the opposite side of the spinal cord. Then, they pass upward to the brain stem in either the **lateral spinothalamic tract** or the **anterior**

spinothalamic tract. The lateral spinothalamic tract conveys sensory impulses for pain and temperature; the anterior spinothalamic tract conveys impulses for tickle, itch, crude touch, pressure, and vibrations. The axons of the second-order neurons end in the ventral posterior nucleus of the thalamus, where they synapse with the third-order neurons. The axons of the third-order neurons project to the primary somatosensory area on the same side of the cerebral cortex as the thalamus.

Mapping the Primary Somatosensory Area

Specific areas of the cerebral cortex receive somatic sensory input from particular parts of the body, and other areas of the cerebral cortex provide output in the form of instructions for movement of particular parts of the body. The *somatic sensory map* and the *somatic motor map* relate body parts to these cortical areas.

Precise localization of somatic sensations occurs when nerve impulses arrive at the **primary somatosensory area** (areas 1, 2, and 3 in Figure 18.16), which occupies the postcentral gyri of the parietal lobes of the cerebral cortex. Each region in this area receives sensory input from a different part of the body. Figure 20.5a, which appears in the next section, maps the destination of somatic sensory signals from different parts of the left side of the body in the somatosensory area of the right cerebral hemisphere. The left cerebral hemisphere has a similar primary somatosensory area that receives sensory input from the right side of the body.

Note that some parts of the body—chiefly the lips, face, tongue, and hand—provide input to large regions in the somatosensory area. Other parts of the body, such as the trunk and lower limbs, project to much smaller cortical regions. The relative sizes of these regions in the somatosensory area are proportional to the number of specialized sensory receptors within the corresponding part of the body. For example, there are many sensory receptors in the skin of the lips but few in the skin of the trunk. The size of the cortical region that represents a body part may expand or shrink somewhat, depending on the quantity of sensory impulses received from that body part. For example, people who learn to read Braille eventually develop a larger cortical region in the somatosensory area representing the fingertips. The representation of the concept of the body within the brain is referred to as the *cortical homunculus*.

Somatic Sensory Pathways to the Cerebellum

Two tracts in the spinal cord—the **posterior spinocerebellar tract** (spī-nō-ser-e-BEL-ar) and the **anterior (ventral) spinocerebellar tract**—are the major routes proprioceptive impulses take to reach the cerebellum. Although not consciously perceived, sensory impulses conveyed to the cerebellum along these two pathways are critical for posture, balance, and coordination of skilled movements. Table 20.3 summarizes the major somatic sensory tracts in the spinal cord and pathways in the brain.

CLINICAL CONNECTION | *Syphilis*

Syphilis is a sexually transmitted disease caused by the bacterium *Treponema pallidum*. Because it is a bacterial infection, it can be treated with antibiotics. However, if the infection is not treated, the third stage of syphilis typically causes debilitating neurological symptoms. A common outcome is progressive degeneration of the posterior portions of the spinal cord, including the posterior columns, posterior spinocerebellar tracts, and posterior roots. Somatic sensations are lost, and the person's gait becomes uncoordinated and jerky because proprioceptive impulses fail to reach the cerebellum. •

TABLE 20.3

Major Somatic Sensory Tract Pathways

TRACTS AND LOCATIONS	PATHWAY FUNCTIONS
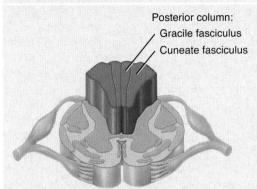 Posterior column: Gracile fasciculus Cuneate fasciculus 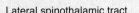 Spinal cord	**Posterior column–medial lemniscus pathway** **Cuneate fasciculus:** Conveys nerve impulses for touch, pressure, vibration, and conscious proprioception from the upper limbs, upper trunk, neck, and posterior head **Gracile fasciculus:** Conveys nerve impulses for touch, pressure, and vibration from the lower limbs and lower trunk
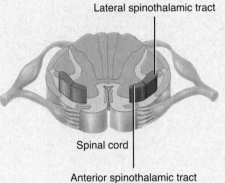 Lateral spinothalamic tract 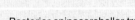 Spinal cord Anterior spinothalamic tract	**Anterolateral Pathway** **Lateral spinothalamic tract:** Conveys nerve impulses for pain and temperature sensations **Anterior spinothalamic tract:** Conveys nerve impulses for itch, tickle, pressure, vibrations, and crude, poorly localized touch sensations
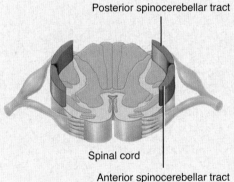 Posterior spinocerebellar tract  Spinal cord Anterior spinocerebellar tract	**Spinocerebellar Pathway** **Anterior and posterior spinocerebellar tracts:** Convey nerve impulses from proprioceptors in the trunk and lower limb of one side of the body to the same side of the cerebellum

 CHECKPOINT

11. What are three differences between the posterior column–medial lemniscus pathway and the anterolateral pathways?
12. Which body parts have the largest representation in the primary somatosensory area?
13. What type of sensory information is carried in the spinocerebellar tracts, and what is its function?

20.4 SOMATIC MOTOR PATHWAYS

OBJECTIVES

- Identify the locations and functions of the different types of neurons in the somatic motor pathways.
- Compare the locations and functions of the direct and indirect motor pathways.

- Explain how the basal nuclei and cerebellum contribute to movements.

Neural circuits in the brain and spinal cord orchestrate all voluntary and involuntary movements. Ultimately, all excitatory and inhibitory signals that control movement converge on the alpha motor neurons that extend out of the brain stem and spinal cord to innervate skeletal muscles in the head and body. These neurons, also known as **lower motor neurons (LMNs)**, have their cell bodies in the brain stem and spinal cord. Their axons extend from the motor nuclei of cranial nerves to skeletal muscles of the face and head and from the anterior gray horns at all levels of the spinal cord to skeletal muscles of the limbs and trunk. Only lower motor neurons provide output from the CNS to skeletal muscle fibers. For this reason, lower motor neurons are also referred to collectively as the *final common pathway*.

Four groups of neurons participate in control of movement by providing input to lower motor neurons.

Figure 20.5 Somatic sensory and somatic motor maps in the cerebral cortex. Shown here is (a) a sensory homunculus and (b) a motor homunculus. A **homunculus** (hō-MUNGK-ū-lus=diminutive man) is a distorted figure of a human placed on the surface of the brain. It illustrates those proportions of the body supplied by various sensory and motor regions. (a) Primary somatosensory area (postcentral gyrus) and (b) primary motor area (precentral gyrus) of the right cerebral hemisphere. The left hemisphere has similar representation. (After Penfield and Rasmussen.)

🔑 Each point on the body surface maps to a specific region in both the primary somatosensory area and the primary motor area.

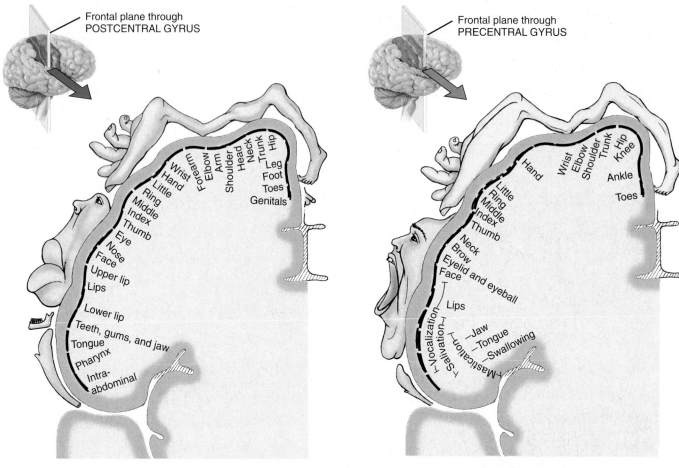

(a) Frontal section of primary somatosensory area in right cerebral hemisphere

(b) Frontal section of primary motor area in right cerebral hemisphere

 How do the somatic sensory and somatic motor representations compare for the hand, and what does this difference imply?

1. **Local circuit neurons.** Input arrives at lower motor neurons from nearby interneurons called *local circuit neurons.* These neurons are located close to the lower motor neuron cell bodies in the brain stem and spinal cord. Local circuit neurons receive input from somatic sensory receptors, such as nociceptors and muscle spindles, as well as from higher centers in the brain. They help coordinate rhythmic activity in specific muscle groups, such as alternating flexion and extension of the free lower limbs during walking.

2. **Upper motor neurons.** Both local circuit neurons and lower motor neurons receive input from *upper motor neurons (UMNs).* Most upper motor neurons synapse with local circuit neurons, which in turn synapse with lower motor neurons. (A few upper motor neurons synapse directly with lower motor neurons.) UMNs from the cerebral cortex are essential for planning, initiating, and directing sequences of voluntary movements. Other UMNs originate in motor centers of the brain stem: the red nucleus, the vestibular nucleus, the superior colliculus, and the

reticular formation. UMNs from the brain stem regulate muscle tone, control postural muscles, and help maintain balance and orientation of the head and body. Both the basal nuclei and cerebellum exert an influence on upper motor neurons.

3. **Basal nuclei neurons.** *Basal nuclei neurons* assist movement by providing input through the thalamus to upper motor neurons. Neural circuits (see Section 16.3) interconnect the basal nuclei with motor areas of the cerebral cortex, thalamus, subthalamic nucleus, and substantia nigra. These circuits help initiate and terminate movements, suppress unwanted movements, and establish a normal level of muscle tone.

4. **Cerebellar neurons.** *Cerebellar neurons* also aid movement by controlling the activity of upper motor neurons through the thalamus. Neural circuits interconnect the cerebellum with motor areas of the cerebral cortex and the brain stem. A prime function of the cerebellum is to monitor differences between intended movements and movements actually performed. It then issues commands to upper motor neurons to reduce

errors in movement. The cerebellum thus coordinates body movements and helps maintain normal posture and balance.

The axons of upper motor neurons extend from the brain to lower motor neurons via two types of somatic motor pathways—direct and indirect. **Direct motor pathways** provide input to lower motor neurons via axons that extend directly from the cerebral cortex. **Indirect motor pathways** provide input to lower motor neurons from motor centers in the brain stem. These brain stem centers in turn receive signals from neurons in the basal nuclei, cerebellum, and cerebral cortex. Direct and indirect pathways both govern generation of nerve impulses in the lower motor neurons, the neurons that stimulate contraction of skeletal muscles.

Before we examine these pathways we consider the role of the motor cortex in voluntary movement.

CLINICAL CONNECTION | *Paralysis*

Damage or disease of *lower* motor neurons produces **flaccid paralysis** (FLAK-sid or FLAS-id) of muscles on the same side of the body. There is neither voluntary nor reflex action of the innervated muscle fibers, muscle tone is decreased or lost, and the muscle remains limp or flaccid. Injury or disease of *upper* motor neurons in the cerebral cortex causes **spastic paralysis** of muscles on the opposite side of the body. In this condition muscle tone is increased, reflexes are exaggerated, and pathological reflexes such as the Babinski sign appear (extension of the great toe, with or without fanning of the other toes; see Clinical Connection in Section 17.3). •

Mapping the Motor Areas

Control of body movements occurs via neural circuits in several regions of the brain. The **primary motor area** (area 4 in Figure 18.16), located in the precentral gyrus of the frontal lobe of the cerebral cortex (Figure 20.5b), is a major control region for planning and initiating voluntary movements. The adjacent **premotor area** (area 6 in Figure 18.16) and even the primary somatosensory area in the postcentral gyrus also contribute axons to the descending motor pathways. Like somatic sensory representation in the somatosensory area, different muscles are represented unequally in the primary motor area. The cortical area devoted to a muscle is proportional to the relative number of motor units in that muscle. Muscles in the thumb, fingers, lips, tongue, and vocal cords have large representations, in contrast to the much smaller representation of the trunk. By comparing Figures 20.5a and b, you can see that somatosensory and somatic motor representations are similar but not identical for most parts of the body.

Direct Motor Pathways

Nerve impulses for voluntary movements propagate from the cerebral cortex to lower motor neurons via the direct motor pathways (Figure 20.6), also known as the *pyramidal pathways*. Areas of the cerebral cortex that contain large, pyramid-shaped cell bodies of upper motor neurons include not only the primary motor area in the precentral gyrus (area 4 in Figure 18.16) but also the premotor area (area 6) and even the primary somatosensory area in the postcentral gyrus (areas 1, 2, and 3). Axons of these cortical UMNs descend through the internal capsule of the cerebrum. In the medulla oblongata, the axon bundles form the ventral bulges known as the *pyramids*.

Figure 20.6 Direct motor pathways. Signals initiated by the primary motor area in the right hemisphere control skeletal muscles on the left side of the body. Spinal cord tracts carrying impulses of direct motor pathways are the lateral corticospinal tract and anterior corticospinal tract.

 Direct pathways convey impulses that result in precise, voluntary movements.

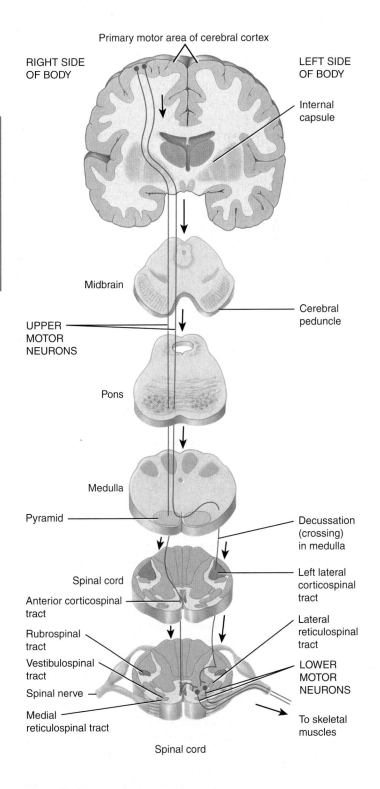

What other tracts (not shown in the figure) convey impulses that result in precise, voluntary movements?

CLINICAL CONNECTION | *Amyotrophic Lateral Sclerosis (ALS)*

Amyotrophic lateral sclerosis (ā'-mī-ō-TROF-ik; *a-*=without; *myo-*=muscle; *trophic*=nourishment) is a progressive degenerative disease that attacks motor areas of the cerebral cortex, axons of upper motor neurons in the lateral white columns (corticospinal and rubrospinal tracts), and lower motor neuron cell bodies. It causes progressive muscle weakness and atrophy. ALS often begins in sections of the spinal cord that serve the hands and arms but rapidly spreads to involve the whole body and face, without affecting intellect or sensations. Death typically occurs in 2 to 5 years. ALS is commonly known as Lou Gehrig's disease after the New York Yankees baseball player who died from it at age 37 in 1941.

Inherited mutations account for about 15 percent of all cases of ALS (familial ALS). Noninherited (sporadic) cases of ALS appear to have several implicating factors. According to one theory there is a buildup in the synaptic cleft of the neurotransmitter glutamate released by motor neurons due to a mutation of the protein that normally deactivates and recycles the neurotransmitter. The excess glutamate causes motor neurons to malfunction and eventually die. The drug riluzole, which is used to treat ALS, reduces damage to motor neurons by decreasing the release of glutamate. Other factors may include damage to motor neurons by free radicals, autoimmune responses, viral infections, deficiency of nerve growth factor, apoptosis (programmed cell death), environmental toxins, and trauma.

In addition to riluzole, ALS is treated with drugs that relieve symptoms such as fatigue, muscle pain and spasticity, excessive saliva, and difficulty sleeping. The only other treatment is supportive care provided by physical, occupational, and speech therapists; nutritionists; social workers; and home care and hospice nurses. •

About 90 percent of the axons of upper motor neurons *decussate* (cross over) to the *contralateral* (opposite) side in the medulla oblongata. The 10 percent that remain on the *ipsilateral* (same) side eventually decussate at the spinal cord levels where they synapse with an interneuron or lower motor neuron. Thus, the right cerebral cortex controls most of the muscles on the left side of the body, and the left cerebral cortex controls most of the muscles on the right side of the body.

Three tracts contain axons of upper motor neurons belonging to the direct motor pathways:

1. **Lateral corticospinal tracts.** Axons of UMNs that decussate in the medulla form the **lateral corticospinal tracts** (kor'-ti-kō-SPĪ-nal) in the right and left lateral white columns of the spinal cord (Figure 20.6 and Table 20.4). These motor neurons control muscles located in distal parts of the free limbs. The distal muscles are responsible for precise, agile, and highly skilled movements of the hands and feet. Examples include the movements needed to button a shirt or play the piano.

2. **Anterior corticospinal tracts.** Axons of cortical UMNs that do not decussate in the medulla form the **anterior corticospinal tracts** in the right and left anterior white columns (Figure 20.6 and Table 20.4). At each spinal cord level, some of these axons decussate via the anterior white commissure. Then they synapse with interneurons or lower motor neurons in the anterior gray horn. Axons of these lower motor neurons exit the cervical and upper thoracic segments of the cord in the anterior roots of spinal nerves. They terminate in skeletal muscles that control movements of the trunk, thus coordinating movements of the axial skeleton, and also control muscles in the proximal parts of the free limbs.

3. **Corticobulbar tracts.** Some axons of upper motor neurons that conduct impulses for the control of skeletal muscles in the head extend through the internal capsule to the midbrain, where they join the **corticobulbar tracts** (kor'-ti-kō-BUL-bar) in the right and left cerebral peduncles (see Table 20.4). Some of the axons in the corticobulbar tracts have decussated; others have not. The axons terminate in the motor nuclei of nine pairs of cranial nerves in the pons and medulla oblongata: the oculomotor (III), trochlear (IV), trigeminal (V), abducens (VI), facial (VII), glossopharyngeal (IX), vagus (X), accessory (XI), and hypoglossal (XII). The lower motor neurons of cranial nerves convey impulses that control precise, voluntary movements of the eyes, tongue, and neck, plus chewing, facial expression, speech, and swallowing.

Table 20.4 summarizes the functions and pathways of the tracts in the direct motor pathways.

Indirect Motor Pathways

The **indirect motor pathways** or *extrapyramidal pathways* include all somatic motor tracts other than the corticospinal and corticobulbar tracts. Nerve impulses conducted along the indirect pathways follow complex, polysynaptic circuits that involve the motor cortex, basal nuclei, thalamus, cerebellum, reticular formation, and nuclei in the brain stem. Axons of upper motor neurons descend from various nuclei of the brain stem into five major tracts of the spinal cord and terminate on local circuit neurons or lower motor neurons. These tracts are the **rubrospinal tract** (ROO-brō-spī-nal), **tectospinal tract** (TEK-tō-spī-nal), **vestibulospinal tract** (ves-TIB-ū-lō-spī-nal), **lateral reticulospinal tract** (re-TIK-ū-lō-spī-nal), and **medial reticulospinal tract**.

Table 20.4 summarizes the functions and pathways of the tracts in the indirect motor pathways.

Roles of the Basal Nuclei

As previously noted, the basal nuclei and cerebellum influence movement through their effects on upper motor neurons. The functions of the basal nuclei include the following:

1. **Initiation and termination of movements.** Two parts of the basal nuclei, the caudate nucleus and the putamen, receive input from sensory, association, and motor areas of the cerebral cortex and from the substantia nigra. Output from the basal nuclei comes from the globus pallidus and substantia nigra, which send feedback signals to the upper motor neurons in the motor cortex by way of the thalamus. (Figure 18.14b shows these parts of the basal nuclei.) This circuit—from cortex to basal nuclei to thalamus to cortex—appears to function in the initiation and termination of movements. Neurons in the putamen generate impulses just before body movements occur, and neurons in the caudate nucleus generate impulses just before eye movements occur.

2. **Suppression of unwanted movements.** Unwanted movements are suppressed by the inhibitory effects of the basal nuclei on the thalamus and superior colliculus.

3. **Control of muscle tone.** The globus pallidus sends impulses into the reticular formation that reduce muscle tone. Damage or destruction of some basal nuclei connections causes a generalized increase in muscle tone.

TABLE 20.4	
Major Somatic Motor Tract Pathways	
TRACTS AND LOCATIONS	PATHWAY FUNCTIONS

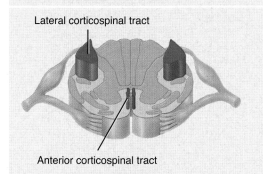

Lateral corticospinal tract

Anterior corticospinal tract

Spinal cord

Direct (Pyramidal) Pathways

Lateral corticospinal: Conveys nerve impulses from motor cortex of cerebrum to skeletal muscles on opposite side of body for precise, voluntary movements of distal parts of free limbs

Anterior corticospinal: Conveys nerve impulses from motor cortex to skeletal muscles on opposite side of body for movements of trunk and proximal parts of free limbs

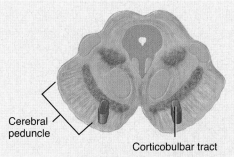

Cerebral peduncle

Corticobulbar tract

Midbrain of brain stem

Corticobulbar: Conveys nerve impulses from motor cortex to skeletal muscles of head and neck to coordinate precise, voluntary movements

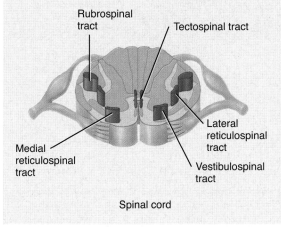

Rubrospinal tract

Tectospinal tract

Lateral reticulospinal tract

Medial reticulospinal tract

Vestibulospinal tract

Spinal cord

Indirect (Extrapyramidal) Pathways

Rubrospinal: Conveys nerve impulses from the red nucleus (which receives input from the cerebral cortex and cerebellum) to contralateral skeletal muscles that govern precise, voluntary movements of the distal parts of the free upper limbs

Tectospinal: Conveys nerve impulses from the superior colliculus to contralateral skeletal muscles that reflexively move the head, eyes, and trunk in response to visual or auditory stimuli

Vestibulospinal: Conveys nerve impulses from the vestibular nucleus (which receives input about head movements from the inner ear) to ipsilateral skeletal muscles of the trunk and proximal parts of the free limbs for maintaining posture and balance in response to head movements

Medial and lateral reticulospinal: Convey nerve impulses from reticular formation to ipsilateral skeletal muscles of the trunk and proximal parts of the free limbs for maintaining posture and regulating muscle tone in response to ongoing body movements

4. ***Influence on cortical function.*** The basal nuclei influence sensory, limbic, cognitive, and linguistic functions of the cerebral cortex. For example, the basal nuclei help initiate and terminate some cognitive processes, such as attention, memory, and planning. In addition, the basal nuclei may act with the limbic system to regulate emotional behaviors.

Roles of the Cerebellum

In addition to maintaining proper posture and balance, the cerebellum is active in learning and performing rapid, coordinated, highly skilled movements such as hitting a golf ball, speaking, and swimming. Cerebellar function involves four activities:

1. ***Monitoring intentions for movement.*** The cerebellum receives impulses from the motor cortex and basal nuclei via the pontine nuclei in the pons regarding what movements are planned.

2. ***Monitoring actual movement.*** The cerebellum receives input from proprioceptors in joints and muscles that reveals what actually is happening. These nerve impulses travel in the anterior and posterior spinocerebellar tracts. Nerve impulses from the vestibular (equilibrium-sensing) apparatus in the inner ear and from the eyes also enter the cerebellum.

3. ***Comparing command signals with sensory information.*** The cerebellum compares intentions for movement with the actual movement performed.

4. ***Sending out corrective feedback.*** If there is a discrepancy between intended and actual movement, the cerebellum sends feedback to upper motor neurons. This information travels via the thalamus to UMNs in the cerebral cortex but goes directly to UMNs in brain stem motor centers. As movements occur, the cerebellum continuously provides error corrections to upper motor neurons, which decreases errors and smoothes the motion. It also contributes over longer periods to the learning of new motor skills.

CLINICAL CONNECTION | *Parkinson's Disease*

Parkinson's disease (PD) is a progressive disorder of the CNS that typically affects its victims around age 60. It involves the degeneration of neurons that extend from the substantia nigra to the putamen and caudate nucleus, where they normally release the neurotransmitter dopamine (DA). In the caudate nucleus of the basal nuclei are neurons that liberate the neurotransmitter acetylcholine (ACh) (see figure). Although the level of ACh does not change as the level of DA declines, the imbalance of neurotransmitter activity—too little DA and too much ACh—is thought to cause most of the symptoms.

In PD patients, involuntary skeletal muscle contractions often interfere with voluntary movement. For instance, the muscles of the upper limb may alternately contract and relax, causing the hand to shake. This shaking, called **tremor**, is the most common symptom of PD. Also, muscle tone may increase greatly, causing rigidity of the involved body part. Rigidity of the facial muscles gives the face a mask-like appearance. The expression is characterized by a wide-eyed, un-blinking stare and a slightly open mouth with uncontrolled drooling.

Motor performance is also impaired by **bradykinesia** (*brady-* =slow), slowness of movements. Activities such as shaving, cutting food, and buttoning a shirt take longer and become increasingly more difficult as the disease progresses. Muscular movements also exhibit **hypokinesia** (*hypo-*=under), decreasing range of motion. For example, words are written smaller, letters are poorly formed, and eventually handwriting becomes illegible. Often, walking is impaired; steps become shorter and shuffling, and arm swing diminishes. Even speech may be affected.

The cause of PD is unknown, but toxic environmental chemicals, such as pesticides, herbicides, and carbon monoxide, are suspected contributing agents. Only 5 percent of PD patients have a family history of the disease.

Treatment of PD is directed toward increasing levels of DA and decreasing levels of ACh. Although people with PD do not manufacture enough dopamine, taking it orally is useless because DA cannot cross the blood–brain barrier. Even though symptoms are partially relieved by a drug developed in the 1960s called levodopa (L-dopa), a precursor of DA, the drug does not slow the progression of the disease. As more and more affected brain cells die, the drug becomes useless. Another drug called selegiline (Deprenyl®) is used to inhibit monoamine oxidase, an enzyme that degrades dopamine. This drug slows progression of PD and may be used with levodopa. Anticholinergic drugs (such as benzotropine and trihexyphenidyl) can also be used to block the effects of ACh at some of the synapses between basal nuclei neurons, helping restore the balance between ACh and DA. Anticholinergic drugs effectively reduce symptomatic tremor, rigidity, and drooling.

For more than a decade, surgeons have sought to reverse the effects of Parkinson's disease by transplanting dopamine-rich fetal nervous tissue into the basal nuclei (usually the putamen) of patients with severe PD. Only a few postsurgical patients have shown any degree of improvement, such as less rigidity and improved quickness of motion. Another surgical technique that has produced improvement for some patients is *pallidotomy*, in which a part of the globus pallidus that generates tremors and produces muscle rigidity is destroyed. In addition, some patients are being treated with a surgical procedure called *deep-brain stimulation (DBS)*, which involves the implantation of electrodes into the subthalamic nucleus. The electrical currents released by the implanted electrodes reduce many of the symptoms of PD. •

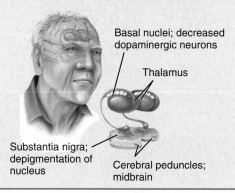

Basal nuclei; decreased dopaminergic neurons

Thalamus

Substantia nigra; depigmentation of nucleus

Cerebral peduncles; midbrain

Skilled activities such as tennis or volleyball provide good examples of the contribution of the cerebellum to movement. To make a good serve or to block a spike, you must bring your racket or arms forward just far enough to make solid contact. How do you stop at exactly the right point? Before you even hit the ball, the cerebellum has sent nerve impulses to the cerebral cortex and basal nuclei informing them where your arm or racket swing must stop. In response to impulses from the cerebellum, the cortex and basal nuclei transmit motor impulses to opposing body muscles to stop the swing.

 CHECKPOINT

14. Which parts of the body have the largest representation in the motor cortex? Which have the smallest?

15. Why are the two main somatic motor pathways referred to as "direct" and "indirect"?

16. Compare the effects of disease of the basal nuclei with damage to the cerebellum.

20.5 INTEGRATION OF SENSORY INPUT AND MOTOR OUTPUT

 OBJECTIVE

• Explain how sensory input and motor output are integrated in the central nervous system.

Sensory pathways provide the input that keeps the central nervous system informed of changes in the external and internal environment. Output from the CNS is then conveyed through motor pathways, which enable movement and glandular secretions to occur. As sensory information reaches the CNS, it becomes part of a large pool of sensory information. However, the CNS does not necessarily respond to every impulse. Rather, the incoming sensory information is **integrated**; that is, it is processed and interpreted and a course of action is taken.

The integration process occurs not just once but at many places along the pathways of the CNS and at both conscious and subconscious levels. It occurs within the spinal cord, brain stem, cerebellum, basal nuclei, and cerebral cortex. As a result, the output descending along a motor pathway that makes a muscle contract or a gland secrete can be modified and responded to at any of these levels. Motor portions of the cerebral cortex play the major role for initiating and controlling precise movements of muscles. The basal nuclei largely integrate semivoluntary movements such as walking, swimming, and laughing. The cerebellum assists the motor cortex and basal nuclei by making body movements smooth and coordinated and by contributing significantly to the maintenance of normal posture and balance.

 CHECKPOINT

17. Where does integration of the movements required to perform the backstroke occur?

KEY MEDICAL TERMS ASSOCIATED WITH SOMATIC SENSES AND MOTOR CONTROL

Acupuncture (ak-ū-PUNK-chur) The use of fine needles (lasers, ultrasound, or electricity) inserted into specific exterior body locations (*acupoints*) and manipulated to relieve pain and provide therapy for various conditions. The placement of needles may cause the release of neurotransmitters such as endorphins, painkillers that may inhibit pain pathways.

Cerebral palsy (CP) A motor disorder that results in the loss of muscle control and coordination; caused by damage of the motor areas of the brain during fetal life, birth, or infancy. Radiation during fetal life, temporary lack of oxygen during birth, and hydrocephalus during infancy may also cause cerebral palsy.

Coma (KŌ-ma) A state of unconsciousness in which a person's responses to stimuli are reduced or absent. In a *light coma*, an individual may respond to certain stimuli, such as sound, touch, or light, and move his or her eyes, cough, and even murmur. In a *deep coma*, a person does not respond to any stimuli and does not make any movements. Causes of coma include head injuries, cardiac arrest, stroke, brain tumors, infections (encephalitis and meningitis), seizures, alcoholic intoxica-

tion, drug overdose, severe lung disorders (chronic obstructive pulmonary disease, pulmonary edema, pulmonary embolism), inhalation of large amounts of carbon monoxide, liver or kidney failure, low or high blood sugar or sodium levels, and low or high body temperature.

Dysarthria (dis-AR-thrē-a; *dys-*=difficult; *arthro*=to articulate) Difficult or imperfect speech due to loss of muscular control as a result of disorders of motor pathways.

Pain threshold The smallest intensity of a painful stimulus at which a person perceives pain. All individuals have the same pain threshold.

Pain tolerance The greatest intensity of painful stimulation that a person is able to tolerate. Individuals vary in their tolerance to pain.

Synesthesia (sin-es-THĒ-zē-a; *syn-*=together; *aisthesis*=sensation) A condition in which sensations of two or more modalities accompany one another. In some cases, a stimulus for one sensation is perceived as a stimulus for another; for example, a sound produces a sensation of color. In other cases, a stimulus from one part of the body is experienced as coming from a different part.

CHAPTER REVIEW AND RESOURCE SUMMARY

WileyPLUS

Review

Resource

20.1 Overview of Sensations

1. Sensation is the awareness of changes in the external and internal conditions of the body.
2. For a sensation to occur, a stimulus must reach a sensory receptor, the stimulus must be converted to a nerve impulse, and the impulse conducted to the brain; finally, the impulse must be integrated by a region of the brain.
3. When stimulated, most sensory receptors ultimately produce one or more nerve impulses.
4. Sensory impulses from each part of the body arrive in specific regions of the cerebral cortex.
5. Modality is the distinct quality that makes one sensation different from others.
6. Adaptation is a decrease in sensation during a prolonged stimulus.
7. Two classes of senses are general senses, which include somatic senses and visceral senses, and special senses, which include the modalities of smell, taste, vision, hearing, and equilibrium (balance).
8. Sensory receptors can be classified on the basis of microscopic structure, location and origin of the stimuli that activate them, and type of stimulus they detect.
9. Table 20.1 summarizes the classification of sensory receptors.

Anatomy Overview - Chemoreceptors
Anatomy Overview - Baroreceptors
Anatomy Overview - Proprioceptors

20.2 Somatic Sensations

1. Somatic sensations include tactile sensations (touch, pressure, vibration, itch, and tickle), thermal sensations (warmth and cold), pain, and proprioception.
2. Receptors for tactile, thermal, and pain sensations are located in the skin, subcutaneous layer, and mucous membranes of the mouth, vagina, and anus.
3. Receptors for proprioceptive sensations (position and movement of body parts) are located in muscles, tendons, joints, and the inner ear.
4. Receptors for touch are (a) hair root plexuses and corpuscles of touch (Meissner corpuscles), which are rapidly adapting, and (b) slowly adapting type I cutaneous mechanoreceptors (tactile discs). Type II cutaneous mechanoreceptors (Ruffini corpuscles), which are slowly adapting, are sensitive to stretching. Receptors for pressure include corpuscles of touch, type I mechanoreceptors, and lamellated (pacinian) corpuscles. Receptors for vibration are corpuscles of touch and lamellated corpuscles. Itch receptors are free nerve endings. Both free nerve endings and lamellated corpuscles mediate the tickle sensation.
5. Thermoreceptors are free nerve endings. Cold receptors are located in the stratum basale of the epidermis; warm receptors are located in the dermis.
6. Pain receptors (nociceptors) are free nerve endings that are located in nearly every body tissue.
7. Proprioceptors include muscle spindles, tendon organs, joint kinesthetic receptors, and hair cells of the inner ear.
8. Table 20.2 summarizes the somatic sensory receptors and the sensations they convey.

Anatomy Overview - Sensory
Receptors of the Skin
Anatomy Overview - Proprioceptors
Figure 20.2 - Referred Pain
Figure 20.3 - Proprioception

20.3 Somatic Sensory Pathways

1. Somatic sensory pathways from receptors to the cerebral cortex involve three-neuron sets: first-order, second-order, and third-order.
2. Axon collaterals (branches) of somatic sensory neurons simultaneously carry signals into the cerebellum and the reticular formation of the brain stem.

Anatomy Overview - Cerebrum
Animation - Somatic Sensory
Pathways
Figure 20.4 - Somatic Sensory
Pathways

Review

3. Impulses propagating along the posterior column–medial lemniscus pathway relay fine touch, stereognosis, proprioception, and vibratory sensations.
4. The neural pathway for pain and thermal sensations is the lateral spinothalamic tract.
5. The neural pathway for tickle, itch, crude touch, and pressure sensations is the anterior spinothalamic pathway.
6. The pathways to the cerebellum are the anterior and posterior spinocerebellar tracts, which transmit impulses for subconscious muscle and joint position sense from the trunk and free lower limbs.
7. Table 20.3 summarizes the major somatic sensory pathways.
8. Specific regions of the primary somatosensory area (postcentral gyrus) of the cerebral cortex receive somatic sensory input from different parts of the body.

20.4 Somatic Motor Pathways

Animation - Somatic Motor Pathways
Figure 20.5 - Somatic Sensory and Somatic Motor Maps in the Cerebral Cortex
Figure 20.6 - Direct Motor Pathways

1. All excitatory and inhibitory signals that control movement converge on the alpha motor neurons, also known as lower motor neurons (LMNs) or the final common pathway.
2. Several groups of neurons participate in control of movement by providing input to lower motor neurons: local circuit neurons, upper motor neurons, basal nuclei neurons, and cerebellar neurons.
3. The primary motor area (precentral gyrus) of the cortex is the major control region for planning and initiating voluntary movements.
4. The axons of upper motor neurons extend from the brain to lower motor neurons via direct and indirect motor pathways. The direct (pyramidal) pathways include the lateral and anterior corticospinal tracts and corticobulbar tracts. Indirect (extrapyramidal) pathways extend from several motor centers of the brain stem into the spinal cord.
5. Neurons of the basal nuclei assist movement by providing input to the upper motor neurons. They help initiate and terminate movements, suppress unwanted movements, and establish a normal level of muscle tone.
6. The cerebellum is active in learning and performing rapid, coordinated, highly skilled movements. It also contributes to the maintenance of balance and posture.
7. Table 20.4 summarizes the major somatic motor pathways.

20.5 Integration of Sensory Input and Motor Output

Animation - Somatic Sensory Pathways
Animation - Somatic Motor Pathways

1. Sensory input keeps the CNS informed of changes in the environment.
2. Incoming sensory information is integrated at many stations along the CNS at both conscious and subconscious levels.
3. A motor response makes a muscle contract or a gland secrete.

CRITICAL THINKING QUESTIONS

1. When Sau Lan held the cup of hot cocoa in her hands, at first it felt comfortably warm and within minutes, she didn't notice the temperature at all. Absently she took a big gulp and almost choked when she felt the hot cocoa burn her mouth and throat. What happened—did the hot cocoa get hotter?

2. Jenny was scheduled for surgery. As a part of her preparation, she was to take a course of antibiotic treatment for 10 days prior to the surgery. Unfortunately, in a rare reaction to the drugs, the antibiotics destroyed her muscle spindles. How will this affect Jenny's life?

3. Stella was visiting a long-term-care facility as a part of her observational experiences in her allied health program, where she saw a patient who was having great difficulty controlling his body

movements. She was not allowed to look at the patient's charts, so she was trying to figure out all the possible nervous system problems he could have. Damage to what specific parts of the brain might result in such a loss of motor control?

4. Very young children are usually given only spoons (no forks or knives) and cups with lids at the dinner table. Why?

5. Jon used to predict the weather by how much the bunion (abnormally swollen joint on the big toe) on his left foot was bothering him. Last year Jon's left foot was amputated due to complications from diabetes, but sometimes he still thinks he feels that bunion. Explain Jon's weather toe sensations.

? ANSWERS TO FIGURE QUESTIONS

20.1 Pain, thermal sensations, tickle, and itch involve activation of different free nerve endings.

20.2 The kidneys have the broadest area for referred pain.

20.3 Muscle spindles are activated when the central areas of their intrafusal fibers are stretched.

20.4 Damage to the right lateral spinothalamic tract could result in loss of pain and thermal sensations on the left side of the body.

20.5 The hand has a larger representation in the motor area than in the somatosensory area, which implies greater precision in the hand's movement control than discriminative ability in its sensation.

20.6 The corticobulbar and rubrospinal tracts (see Table 20.4) convey impulses that result in precise, voluntary movements.

21 | SPECIAL SENSES

INTRODUCTION Recall from Chapter 20 that the *general senses* include somatic senses (tactile, thermal, pain, and proprioceptive) and visceral sensations. Receptors for the general senses are scattered throughout the body and are relatively simple in structure. They range from modified dendrites of sensory neurons to specialized structures associated with the ends of dendrites. Receptors for the *special senses*—smell, taste, vision, hearing, and equilibrium—are anatomically distinct from one another and are concentrated in specific locations in the head. They are usually embedded in the epithelial tissue within complex sensory organs such as the eyes and ears. Neural pathways for the special senses are more complex than those for the general senses.

In this chapter we examine the structure and function of the special sense organs, and the pathways involved in conveying their information to the central nervous system. **Ophthalmology** (of-thal-MOL-ō-jē; *ophthalmo-*=eye; *logy*=study of) is the science that deals with the eyes and their disorders. The other special senses are, in large part, the concern of **otorhinolaryngology** (ō′-tō-rī′-nō-lar-in-GOL-ō-jē; *oto-*=ear; *rhino-*=nose; *laryngo-*=larynx), the science that deals with the ears, nose, pharynx (throat), and larynx (voice box) and their disorders. •

Did you ever wonder how LASIK is performed?

CONTENTS AT A GLANCE

21.1 OLFACTION: SENSE OF SMELL

OBJECTIVES
• Describe the structure of the olfactory receptors and other cells involved in olfaction.
• Outline the neural pathway for olfaction.

Last night as you were studying anatomy in the lounge, all of a sudden you were surrounded by the smell of freshly baked brownies. When you followed your nose and begged for one, biting into the moist, flavorful treat transported you back 10 years into your mother's kitchen. Both smell and taste are chemical senses; the sensations arise from the interaction of molecules with smell or taste receptors. To be detected by either sense, the stimulating molecules must be dissolved. Because impulses for smell and taste propagate to the limbic system (and to higher cortical areas as well), certain odors and tastes can evoke strong emotional responses or a flood of memories.

Figure 21.1 Olfactory receptors and olfactory pathway. (a) Location of olfactory epithelium in nasal cavity. (b) Anatomy of olfactory receptor cells. (c) Histology of the olfactory epithelium. (d) Olfactory pathway.

The olfactory epithelium consists of olfactory receptors, supporting cells, and basal cells.

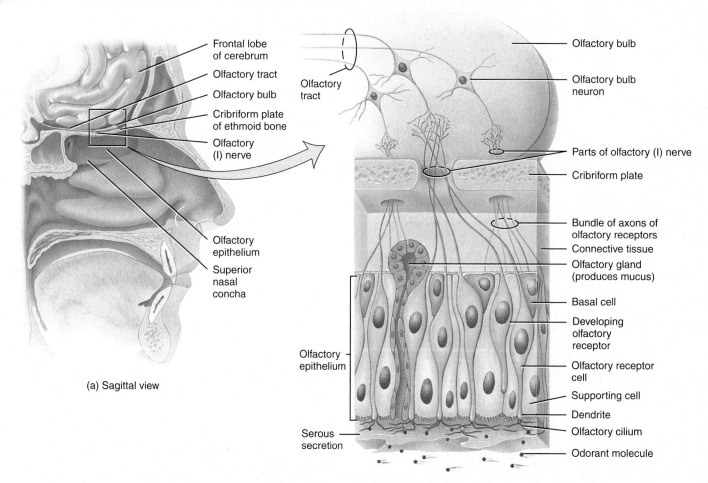

(a) Sagittal view

(b) Enlarged aspect of olfactory receptors

Anatomy of Olfactory Receptors

It is estimated that humans can recognize more than 10,000 different odors. To make this possible, the nose contains 10–100 million receptors for the sense of smell or **olfaction** (ol-FAK-shun; *olfact-*=smell), contained within an area called the **olfactory epithelium** (ol-FAK-tō-rē). With a total area of 5 cm^2 (a little less than 1 in.^{2}), the olfactory epithelium occupies the superior part of the nasal cavity, covering the inferior surface of the cribriform plate and extending along the superior nasal concha (Figure 21.1a). The olfactory epithelium consists of three kinds of cells: olfactory receptors, supporting cells, and basal cells (Figure 21.1b, c).

Olfactory receptor cells are the first-order neurons of the olfactory pathway. Each olfactory receptor cell is a bipolar neuron with a knob-shaped dendrite and an axon. Extending from the dendrite are several long, thin, nonmotile **olfactory cilia**. Within the plasma membrane of the cilia are **olfactory receptors** that interact with **odorants**, chemicals that have an oder. This interaction is the beginning of a process that results in the generation of an action potential and the olfactory response. The axons of olfactory receptor cells are grouped into bundles that pass through the cribriform plate of the ethmoid bone. The collections of axons from the olfactory receptor cells form the olfactory (I) nerves.

Supporting cells are columnar epithelial cells of the mucous membrane lining the nose. They provide physical support, nourishment, and electrical insulation for the olfactory receptor cells. Additionally, they produce odorant-binding proteins that transport odorants to the olfactory receptors to initiate an action potential and help detoxify chemicals that come in contact with the olfactory epithelium.

Basal cells are stem cells located between the bases of the supporting cells. They continually undergo cell division to produce new olfactory receptor cells, which live for only a month or so before being replaced. This process is remarkable considering that olfactory receptor cells are neurons, and as you have already learned, mature neurons are generally not replaced.

Within the connective tissue that supports the olfactory epithelium are **olfactory glands** or *Bowman's glands*, which produce mucus that is carried to the surface of the epithelium by ducts. The secretion moistens the surface of the olfactory epithelium and dissolves odorants so that transduction can occur. Both supporting cells of the nasal epithelium and olfactory glands are innervated by parasympathetic neurons within branches of the facial (VII) nerve, which can be stimulated by certain chemicals. Impulses in these nerves in turn stimulate the lacrimal glands in the eyes and nasal mucous glands. The result is tears and a runny nose after inhaling substances such as pepper or the vapors of household ammonia.

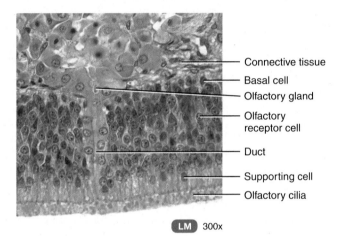

Connective tissue
Basal cell
Olfactory gland
Olfactory receptor cell
Duct
Supporting cell
Olfactory cilia

LM 300x

(c) Histology of olfactory epithelium

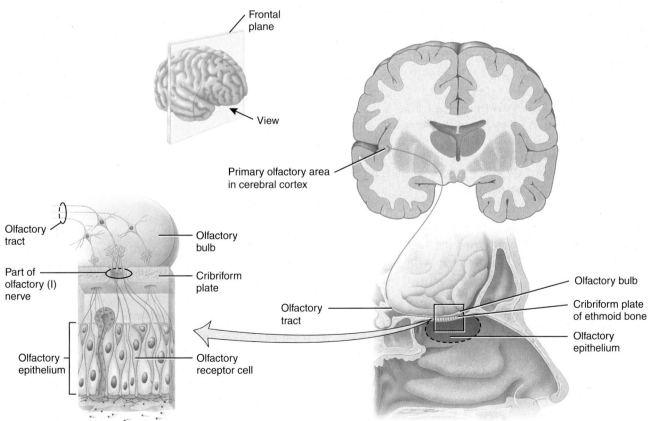

Frontal plane

View

Primary olfactory area in cerebral cortex

Olfactory tract

Olfactory bulb

Part of olfactory (I) nerve

Cribriform plate

Olfactory tract

Olfactory epithelium

Olfactory receptor cell

Olfactory bulb

Cribriform plate of ethmoid bone

Olfactory epithelium

(d) Olfactory pathway

 Which part of an olfactory receptor detects an odorant molecule?

The Olfactory Pathway

On each side of the nose, about 40 bundles of the slender, unmyelinated axons of olfactory receptor cells extend through about 20 olfactory foramina in the cribriform plate of the ethmoid bone (Figure 21.1b). These bundles of axons collectively form the right and left **olfactory (I) nerves**. The olfactory nerves terminate in the brain in paired masses of gray matter called the **olfactory bulbs**, which are located below the frontal lobes of the cerebrum and lateral to the crista galli of the ethmoid bone. Within the olfactory bulbs, the axon terminals of olfactory receptor cells form synapses with the dendrites and cell bodies of neurons in the olfactory pathway.

Axons of olfactory bulb neurons extend posteriorly and form the **olfactory tract** (Figure 21.1a, b). Olfactory sensations are the only sensations that reach the cerebral cortex without first synapsing in the thalamus. Some of the axons of the olfactory tract project to the **primary olfactory area**; located at the inferior and medial surface of the temporal lobe, the primary olfactory area (area 28 in Figure 18.16) is where conscious awareness of smell begins (Figure 21.1d). Other axons of the olfactory tract project to the limbic system and hypothalamus; these connections account for the aforementioned emotional and memory-evoked responses to odors. Examples include sexual excitement upon smelling a certain perfume, nausea upon smelling a food that once made you violently ill, or an odor-evoked memory of a childhood experience.

From the primary olfactory area, pathways also extend to the **frontal lobe**. An important region for odor identification and discrimination is the **orbitofrontal area** (area 11 in Figure 18.16). People who suffer damage in this area have difficulty identifying different odors. Positron emission tomography (PET) studies suggest some degree of hemispheric lateralization: The orbitofrontal area of the *right* hemisphere exhibits greater activity during olfactory processing.

CHECKPOINT

1. How do basal cells contribute to olfaction?
2. What is the sequence of events from the binding of an odorant molecule to an olfactory cilium to the arrival of a nerve impulse in an olfactory bulb?

21.2 GUSTATION: SENSE OF TASTE

OBJECTIVES

• Describe the gustatory receptors and the neural pathway for gustation.

Like olfaction, taste or **gustation** (gus-TĀ-shun; *gust*=taste) is a chemical sense. However, it is much simpler than olfaction in that only five primary tastes can be distinguished: *sour, sweet, bitter, salty,* and *umami* (ū-MAM-ē). The umami taste, first reported by Japanese scientists, is described as "meaty" or "savory." Umami is believed to arise from taste receptors that are stimulated by L-glutamates and nucleotides, substances present in many foods. Monosodium glutamate (MSG) added to foods as a flavor enhancer confers the umami taste to foods. All other flavors, such as chocolate, pepper, and coffee, are combinations of the five primary tastes, plus accompanying olfactory and tactile (touch) sensations. Odors from food can pass upward from the mouth into the nasal cavity, where they stimulate olfactory receptors. Because olfaction is much more sensitive than taste, a given concentration of a food substance may stimulate the olfactory system thousands of times more strongly than it stimulates the gustatory system. When you have a cold or are suffering from allergies and cannot taste your food, it is actually olfaction that is blocked, not taste.

Anatomy of Gustatory Receptors

The receptors for taste are located in the taste buds (Figure 21.2). The nearly 10,000 taste buds of a young adult are mainly on the tongue, but they are also found on the soft palate (posterior portion of roof of mouth), pharynx (throat), and epiglottis (cartilage lid over the voice box). The number of taste buds declines with age. Each **taste bud** is an oval body consisting of three kinds of epithelial cells: supporting cells, gustatory receptor cells, and basal cells (Figure 21.2c, d). The **supporting cells** contain microvilli and surround a group of about 50 **gustatory receptor cells** (GUS-ta-tōr-ē). **Gustatory microvilli** (*gustatory hairs*) project from gustatory receptor cells and pass to the external surface through the **taste pore**, an opening in the taste bud. **Basal cells**, stem cells found at the periphery of the taste bud near the connective tissue layer, produce supporting cells. The supporting cells then develop into gustatory receptor cells, which have a life span of about 10 days. This is why it doesn't take your taste senses or the tongue too long to recover from being burned by that too-hot cup of coffee or cocoa. At their base, the gustatory receptor cells synapse with dendrites of the first-order sensory neurons that form the first part of the gustatory pathway. Each first-order neuron has many dendrites, which receive input from many gustatory receptor cells in several taste buds.

Taste buds are found in elevations on the tongue called **papillae** (pa-PIL-ē), which increase the surface area and provide a rough texture to the upper surface of the tongue (Figure 21.2a, b). Three types of papillae contain taste buds:

1. About 12 very large, circular **vallate (circumvallate) papillae** (VAL-āt=wall-like) form an inverted V-shaped row at the back of the tongue. Each of these papillae houses 100–300 taste buds.
2. **Fungiform papillae** (FUN-ji-form=mushroomlike) are mushroom-shaped elevations scattered over the entire surface of the tongue; each one contain about five taste buds.
3. **Foliate papillae** (FŌ-lē-āt=leaflike) are located in small trenches on the lateral margins of the tongue, but most of their taste buds degenerate in early childhood.

In addition, the entire surface of the tongue has **filiform papillae** (FIL-i-form=threadlike). These pointed, threadlike structures contain tactile receptors but no taste buds. They increase friction between the tongue and food, making it easier for the tongue to move food around in the oral cavity.

Chemicals that stimulate gustatory receptor cells are known as **tastants**. Once a tastant is dissolved in saliva, it can make contact with the gustatory microvilli, ultimately triggering nerve impulses in the first-order sensory neurons that synapse with gustatory receptor cells.

Figure 21.2 Gustatory receptor cells and gustatory pathway.

Gustatory receptor cells are located in taste buds.

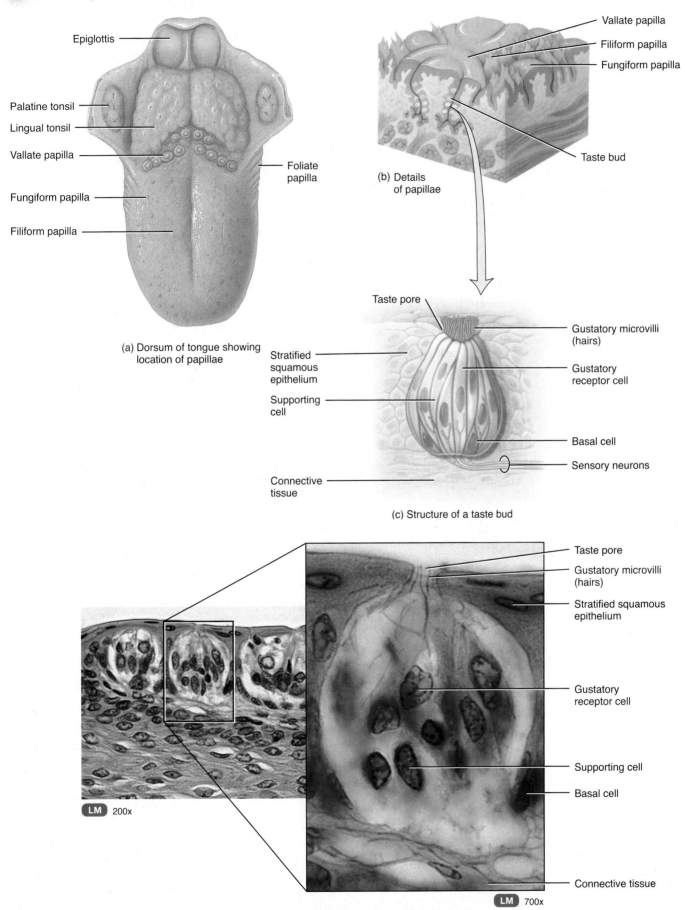

Epiglottis

Palatine tonsil

Lingual tonsil

Vallate papilla

Fungiform papilla

Filiform papilla

Foliate papilla

(a) Dorsum of tongue showing location of papillae

Vallate papilla

Filiform papilla

Fungiform papilla

Taste bud

(b) Details of papillae

Taste pore

Stratified squamous epithelium

Supporting cell

Connective tissue

Gustatory microvilli (hairs)

Gustatory receptor cell

Basal cell

Sensory neurons

(c) Structure of a taste bud

LM 200x

Taste pore

Gustatory microvilli (hairs)

Stratified squamous epithelium

Gustatory receptor cell

Supporting cell

Basal cell

Connective tissue

LM 700x

(d) Histology of a taste bud from a vallate papilla

FIGURE 21.2 CONTINUES ▶

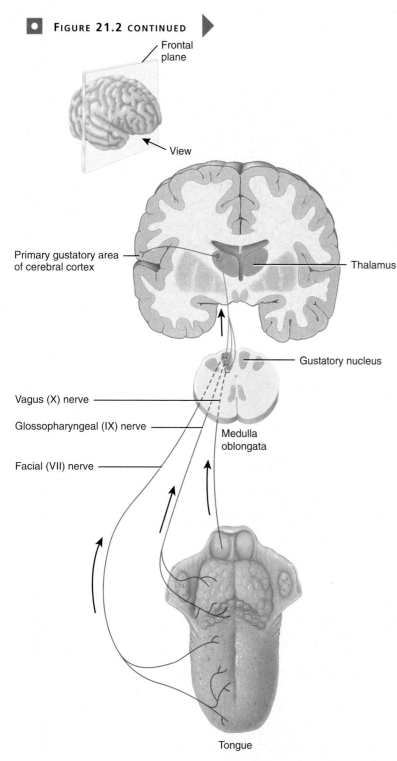

Frontal plane

View

Primary gustatory area of cerebral cortex

Thalamus

Gustatory nucleus

Vagus (X) nerve

Glossopharyngeal (IX) nerve

Medulla oblongata

Facial (VII) nerve

Tongue

(e) Gustatory pathway

Beginning at the gustatory receptor cells, what structures form the gustatory pathway?

CLINICAL CONNECTION | Taste Aversion

Probably because of taste projections to the hypothalamus and limbic system, there is a strong link between taste and pleasant or unpleasant emotions. Sweet foods evoke reactions of pleasure, while bitter ones cause expressions of disgust, even in newborn babies. This phenomenon is the basis for **taste aversion**, in which people and animals quickly learn to avoid a food if it upsets the digestive system. The advantage of avoiding foods that cause such illness is longer survival. However, the drugs and radiation treatments used to combat cancer often cause nausea and gastrointestinal upset regardless of what foods are consumed. Thus, cancer patients may lose their appetite because they develop taste aversions for most foods. •

The Gustatory Pathway

Three cranial nerves contain axons of sensory neurons from **taste buds** (Figure 21.2e). The **facial (VII) nerve** serves the anterior two-thirds of the tongue, the **glossopharyngeal (IX) nerve** serves the posterior one-third of the tongue, and the **vagus (X) nerve** serves the throat and epiglottis (see Figures 18.22, 18.24, and 18.25). From taste buds, impulses propagate along these cranial nerves to the **gustatory nucleus** in the medulla oblongata. From the medulla, some axons carrying taste signals project to the **limbic system** and the **hypothalamus**; others project to the **thalamus**. Taste signals that project from the thalamus to the **primary gustatory area** in the parietal lobe of the cerebral cortex (see area 43 in Figure 18.16) give rise to the conscious perception of taste.

✓ CHECKPOINT

3. How do olfactory receptor cells and gustatory receptor cells differ in structure and function?
4. Compare the olfactory and gustatory pathways.

21.3 VISION

● **OBJECTIVES**
• List and describe the accessory structures of the eye.
• Describe the structural components of the eyeball.
• Outline the neural pathway for vision.

Vision, or the act of seeing, is extremely important to human survival. More than half the sensory receptors in the human body are located in the eyes, and a large part of the cerebral cortex is devoted to processing visual information. In this section of the chapter, we examine the accessory structures of the eye, the eyeball itself, the formation of visual images, and the visual pathway from the eye to the brain.

Accessory Structures of the Eye

The **accessory structures of the eye** include the eyelids, eyelashes, eyebrows, lacrimal (tear-producing) apparatus, and extrinsic eye muscles.

Eyelids

The upper and lower **eyelids**, or **palpebrae** (PAL-pe-brē; singular is *palpebra*), shade the eyes during sleep, protect the eyes from excessive light and foreign objects, and spread lubricating secretions over the eyeballs (Figure 21.3; see also Figure 27.3). The upper eyelid is more movable than the lower and contains in its superior region the **levator palpebrae superioris muscle**. Sometimes a person may experience an annoying *twitch* in an eyelid, an involuntary quivering similar to muscle twitches in the hand, forearm, leg, or foot. Twitches are almost always harmless and usually last for only a few seconds. They are often associated with stress and fatigue. The space between the upper and lower eyelids that exposes the eyeball is the **palpebral fissure** (PAL-pe-bral). Its angles are known as the **lateral commissure** (KOM-i-shur), which is narrower and closer to the temporal bone, and the **medial commissure**, which is broader and nearer the nasal bone. In the medial commissure is a small, reddish elevation, the **lacrimal caruncle** (KAR-ung-kul), which contains sebaceous (oil) glands and sudoriferous (sweat) glands. The whitish material that sometimes collects in the medial commissure comes from these glands.

Figure 21.3 Accessory structures of the eye.

🔑 Accessory structures of the eye include the eyelids, eyelashes, eyebrows, lacrimal apparatus, and extrinsic eye muscles.

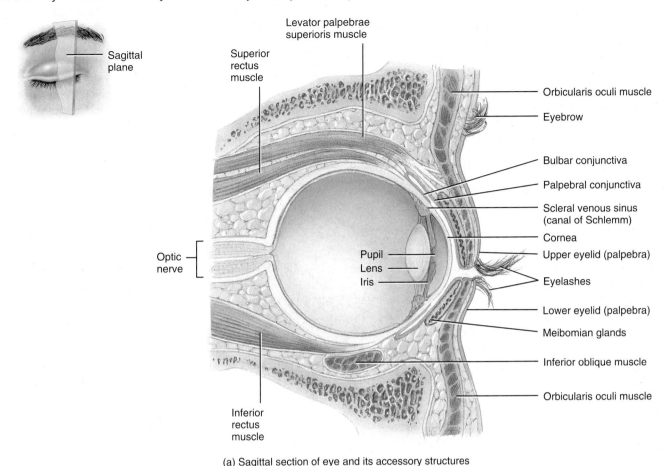

(a) Sagittal section of eye and its accessory structures

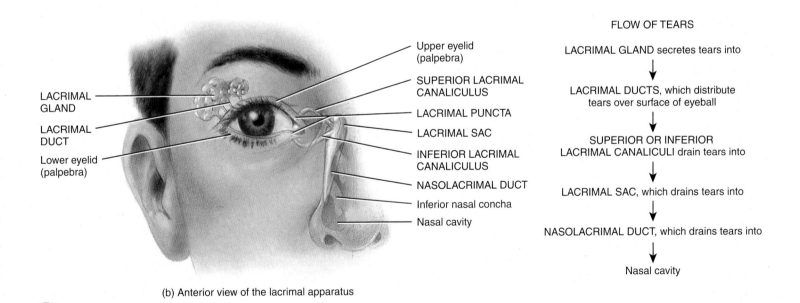

(b) Anterior view of the lacrimal apparatus

❓ **What is lacrimal fluid, and what are its functions?**

From superficial to deep, each eyelid consists of epidermis, dermis, subcutaneous tissue, fibers of the orbicularis oculi muscle, a tarsal plate, tarsal glands, and conjunctiva (Figure 21.3). The **tarsal plate** is a thick fold of connective tissue that gives form and support to the eyelids. Embedded in each tarsal plate is a row of elongated modified sebaceous glands, known as **tarsal glands** or *Meibomian glands* (mī-BŌ-mē-an), that secrete a fluid that helps keep the eyelids from adhering to each other. Infection

of the tarsal glands produces a tumor or cyst on the eyelid called a **chalazion** (ka-LĀ-zē-on=small bump). The **conjunctiva** (kon'-junk-TĪ-va) is a thin, protective mucous membrane composed of nonkeratinized stratified squamous epithelium with numerous goblet cells that is supported by areolar connective tissue. The **palpebral conjunctiva** lines the inner aspect of the eyelids, and the **bulbar conjunctiva** passes from the eyelids onto the surface of the eyeball. Over the sclera (the "white" of the eye) the conjunctiva is vascular, but over the cornea (a transparent region that forms the outer anterior surface of the eyeball) it loses its connective tissue component and consists of clear avascular epithelium. Both the sclera and the cornea will be discussed in more detail shortly. Dilation and congestion of the blood vessels of the bulbar conjunctiva due to local irritation or infection are the causes of **bloodshot eyes**.

Eyelashes and Eyebrows

The **eyelashes**, which project from the border of each eyelid, and the **eyebrows**, which arch transversely above the upper eyelids, help protect the eyeballs from foreign objects, perspiration, and the direct rays of the sun. Sebaceous glands at the base of the hair follicles of the eyelashes, called **sebaceous ciliary glands**, release a lubricating fluid into the follicles. Infection of these glands, usually by bacteria, causes a painful, pus-filled swelling called a **sty**.

The Lacrimal Apparatus

The **lacrimal apparatus** (LAK-ri-mal; *lacrim-*=tears) is a group of structures that produces and drains **lacrimal fluid** or **tears**. The **lacrimal glands** are each about the size and shape of an almond. They are supplied by parasympathetic fibers of the facial (VII) nerves and secrete lacrimal fluid, which drains into 6–12 **lacrimal ducts** that empty tears onto the surface of the conjunctiva of the upper lid (Figure 21.3b). From here the tears pass medially over the anterior surface of the eyeball to enter two small openings called **lacrimal puncta**. Tears then pass into two ducts, the superior and inferior **lacrimal canaliculi**, which lead into the **lacrimal sac** (within the lacrimal fossa) and then into the **nasolacrimal duct**. This duct carries the lacrimal fluid into the nasal cavity just inferior to the inferior nasal concha where it mixes with mucus. An infection of the lacrimal sacs is called **dacryocystitis** (dak′-rē-ō-sis-TĪ-tis; *dacryo-*=lacrimal sac; *itis*=inflammation of). It is usually caused by a bacterial infection and results in blockage of the nasolacrimal ducts.

Lacrimal fluid is a watery solution containing salts, some mucus, and **lysozyme** (LĪ-sō-zīm), a protective bactericidal enzyme. The fluid protects, cleans, lubricates, and moistens the eyeball. After being secreted, lacrimal fluid is spread medially over the surface of the eyeball by the blinking of the eyelids. Each gland produces about 1 mL of lacrimal fluid per day.

Normally, tears are cleared away as fast as they are produced, either by evaporation or by passing into the lacrimal canals and then into the nasal cavity. If an irritating substance makes contact with the conjunctiva, however, the lacrimal glands are stimulated to oversecrete, and tears accumulate (*watery eyes*). This is a protective mechanism, as the tears dilute and wash away the irritating substance. Watery eyes also occur when an inflammation of the nasal mucosa obstructs the nasolacrimal ducts and blocks drainage of tears, such as occurs with a cold. Humans are unique in expressing emotions, both happiness and sadness, by **crying**. In response to parasympathetic stimulation, the lacrimal glands produce excessive lacrimal fluid that may spill over the edges of the eyelids and even fill the nasal cavity with fluid. Thus, crying often produces a runny nose.

Extrinsic Eye Muscles

Positioned at the front of the face, the eyes sit in the bony depressions of the skull called the *orbits*. The orbits help protect the eyes, stabilize them in three-dimensional space, and anchor them to the muscles that produce their essential movements. The extrinsic eye muscles extend from the walls of the orbit to the sclera (white) of the eye and are surrounded in the orbit by a

significant quantity of **periorbital fat** (per′-ē-OR-bi-tal). These muscles are capable of moving the eye in almost any direction. Six extrinsic eye muscles move each eye: the **superior rectus**, **inferior rectus**, **lateral rectus**, **medial rectus**, **superior oblique**, and **inferior oblique** (see Figures 21.3a and 21.4). They receive innervation from the oculomotor (III), trochlear (IV), and abducens (VI) nerves. In general, the motor units in these muscles are small. Some motor neurons serve only two or three muscle fibers—fewer than in any other part of the body except the larynx (voice box). Such small motor units permit smooth, precise, and rapid movement of the eyes. As indicated in Exhibit 11.2, the extrinsic eye muscles move the eyeball laterally, medially, superiorly, and inferiorly. For example, looking to the right requires simultaneous *contraction* of the lateral rectus of the right eyeball and medial rectus of the left eyeball and *relaxation* of the lateral rectus of the left eyeball and medial rectus of the right eyeball. The oblique muscles preserve rotational stability of the eyeball. Circuits in the brain stem and cerebellum coordinate and synchronize the movements of the eyes.

The surface anatomy of the accessory structures of the eye and the eyeball may be reviewed in Figure 27.3.

Anatomy of the Eyeball

The adult **eyeball** measures about 2.5 cm (1 in.) in diameter. Of its total surface area, only the anterior one-sixth is exposed; the remainder is recessed and protected by the orbit, into which it fits. Anatomically, the wall of the eyeball consists of three layers: (1) fibrous tunic, (2) vascular tunic, and (3) retina (inner tunic).

Fibrous Tunic

The **fibrous tunic** (TOO-nik), the outer layer of the eyeball, is a strong, dense collagenous connective tissue layer (Figure 21.4). The greater part of this tunic forms a tough outer covering, the **sclera** (SKLE-ra; *sclera*=hard), which forms the visible white part of the eye. The sclera provides protection for the more delicate internal structures of the eye, while helping maintain the shape of the eyeball against the pressure in the eye. It also serves as an important site of muscle attachment for the extrinsic muscles of the eye, allowing the eye to undergo the critical movements necessary for proper vision. At its anterior surface, the fibrous tunic becomes transparent and has a more exaggerated curvature. This dome-like region, called the **cornea** (KOR-nē-a), is superficial to the colored iris and allows light to enter the interior of the eye. Its curved surface is the major refractory structure for incoming light. The outer surface of the cornea is covered by and continuous with the same nonkeratinized stratified squamous epithelium of the bulbar conjunctiva. While both the sclera and cornea consist of dense collagenous connective tissue, the sclera is opaque and the cornea is transparent because the corneal tissue is avascular and has regularly spaced collagen fibers that are smaller than the wavelength of light. At the corneoscleral junction a hollow channel forms within the tissue. This channel, the **scleral venous sinus** (*canal of Schlemm*), connects with the aqueous fluid in the anterior aspect of the eye and veins within the sclera. It plays an important role in the drainage of aqueous fluid, which will be described later in this section (see Figure 21.8).

Vascular Tunic

The **vascular tunic** or *uvea* (ū-vē-a) is the middle layer of the eyeball. It is composed of three parts: choroid, ciliary body, and iris (Figure 21.4). The posterior portion of the vascular tunic is the highly vascularized **choroid** (KŌ-royd), which lines most of the

Figure 21.4 Anatomy of the eyeball.

The wall of the eyeball consists of three layers: the fibrous tunic, the vascular tunic, and the retina.

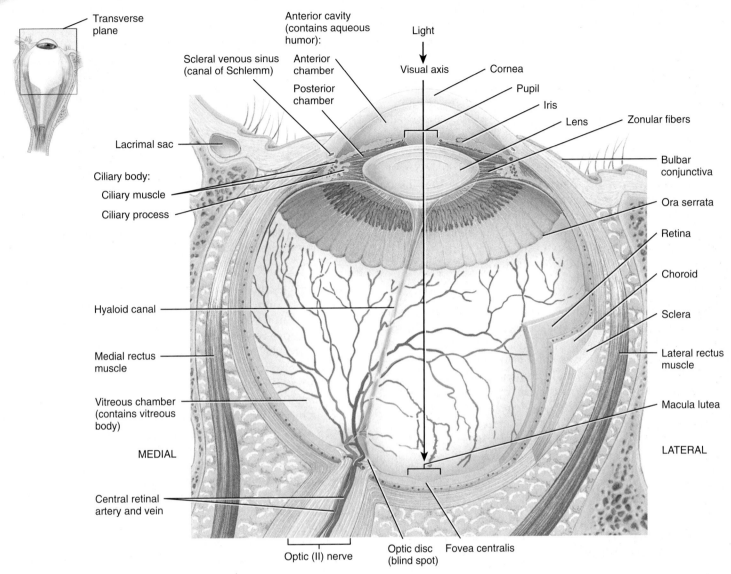

Superior view of transverse section of right eyeball

What are the components of the fibrous tunic and vascular tunic?

internal surface of the sclera. Its numerous blood vessels provide nutrients to the posterior surface of the retina. The choroid also contains melanocytes that produce the pigment melanin, which causes this layer to appear dark brown in color. Melanin in the choroid absorbs stray light rays, which prevents reflection and scattering of light within the eyeball. As a result, the image cast on the retina by the cornea and lens remains sharp and clear. Albinos lack melanin in all parts of the body, including the eye. They often need to wear sunglasses, even indoors, because even moderately bright light is perceived as bright glare due to light scattering.

In the anterior portion of the vascular tunic, the choroid becomes the **ciliary body** (SIL-ē-ar′-ē). It extends from the **ora serrata** (Ō-ra ser-RĀ-ta), the jagged anterior margin of the retina, to a point just posterior to the junction of the sclera and cornea. Like the choroid, the ciliary body appears dark brown in color because it contains melanin-producing melanocytes. In addition, the ciliary body consists of ciliary processes and ciliary muscle. The **ciliary processes** are protrusions or folds on the internal surface of the ciliary body. They contain blood capillaries that secrete aqueous humor. Extending from the ciliary process are **zonular**

fibers (*suspensory ligaments*) that attach to the lens. The zonular fibers are composed of thin, hollow fibrils that resemble elastic connective tissue fibers. The smooth muscle of the ciliary body is called the **ciliary muscle**. With its complex arrangement of longitudinal, oblique, and circularly oriented smooth muscle fibers, the ciliary muscle produces a narrowing or sphincter action of the ringlike ciliary body. Contraction or relaxation of the ciliary muscle changes the tightness of the zonular fibers, which alters the shape of the lens, adapting it for near or far vision.

The **iris** (=rainbow), the colored portion of the eyeball, is shaped like a flattened donut and the opening in its center (the donut hole) is called the **pupil** (*pupil*=little person; because this is where you see a reflection of yourself when looking into someone's eyes). The iris is suspended between the cornea and the lens as an anterior projection of the ciliary body. It consists of melanocytes and circular and radial smooth muscle fibers. The amount of melanin in the iris determines eye color. The eyes appear brown to black when the iris contains a large amount of melanin, blue when melanin concentration is very low, and green when melanin concentration is moderate.

The two distinct fiber arrangements of the smooth muscle form the **sphincter pupillae muscle** (pū-PIL-ē), which is a flat, thin band of circularly oriented muscle fibers at the pupillary boundary of the iris and the **dilator pupillae muscle**, which attaches to the outer circumference of the sphincter pupillae and projects like the spokes of a wheel toward the base of the iris. A principal function of the iris is to regulate the amount of light entering the eyeball through the pupil. The pupil appears black because, as you look through the lens, you see the heavily pigmented back of the eye (choroid and retina). However, if bright light is directed into the pupil, the reflected light is red because of the blood vessels on the surface of the retina. It is for this reason that a person's eyes appear red in a photograph ("red eye") when the flash is directed into the pupil. Autonomic reflexes regulate pupil diameter in response to light levels (Figure 21.5). When bright light stimulates the eye, parasympathetic fibers of the oculomotor (III) nerve stimulate the sphincter pupillae muscle of the iris to contract, causing a decrease in the size of the pupil (constriction). In dim light, sympathetic neurons stimulate the dilator pupillae muscle of the iris to contract, causing an increase in the pupil's size (dilation).

Retina

The third and inner coat of the eyeball, the **retina**, lines the posterior three-quarters of the eyeball and is the beginning of the visual pathway (see Figure 21.4). The anatomy of this layer can be viewed with an *ophthalmoscope* (of-THAL-mō-skōp; *ophthalmos-* =eye; *skopeo*=to examine), an instrument that shines light into the eye and allows an observer to peer through the pupil, providing a magnified image of the retina and its blood vessels as well as the optic (II) nerve (Figure 21.6). The surface of the retina is the only place in the body where blood vessels can be viewed directly and examined for pathological changes, such as those that occur with hypertension, diabetes mellitus, cataracts, and age-related macular disease. Several landmarks are visible through an ophthalmoscope. The **optic disc** is the site where the optic (II) nerve exits the eyeball. Bundled together with the optic nerve are the **central retinal artery**, a branch of the ophthalmic artery, and the **central retinal vein** (see Figure 21.4). Branches of the central retinal artery fan out to nourish the anterior surface of the retina;

Figure 21.5 Responses of the pupil to light of varying brightness.

 Contraction of the circular muscles causes constriction of the pupil; contraction of the radial muscles causes dilation of the pupil.

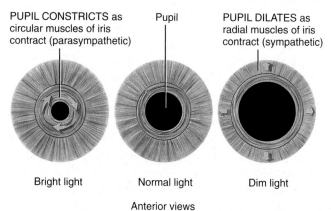

PUPIL CONSTRICTS as circular muscles of iris contract (parasympathetic)

Pupil

PUPIL DILATES as radial muscles of iris contract (sympathetic)

Bright light   Normal light   Dim light

Anterior views

 Which division of the autonomic nervous system causes pupillary constriction? Which causes pupillary dilation?

Figure 21.6 A normal retina, as seen through an ophthalmoscope.

 Blood vessels in the retina can be viewed directly and examined for pathological changes.

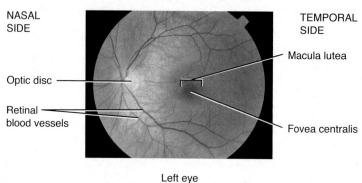

NASAL SIDE

TEMPORAL SIDE

Macula lutea

Optic disc

Retinal blood vessels

Fovea centralis

Left eye

 Evidence of what diseases may be seen through an ophthalmoscope?

the central retinal vein drains blood from the retina through the optic disc. Also visible are the macula lutea and fovea centralis, which are described shortly.

The retina consists of a pigmented layer and a neural layer. The **pigmented layer** is a sheet of melanin-containing epithelial cells located between the choroid and the neural part of the retina. The melanin in the pigmented layer of the retina, as in the choroid, also helps to absorb stray light rays. The **neural (sensory) layer** of the retina is a multilayered outgrowth of the brain that processes visual data extensively before sending nerve impulses into axons that form the optic (II) nerve. Three distinct layers of retinal neurons—the **photoreceptor layer**, the **bipolar cell layer**, and the **ganglion cell layer**—are separated by two zones, the *outer* and *inner synaptic layers*, where synaptic contacts are made (Figure 21.7). Note that light passes through the ganglion and bipolar cell layers and both synaptic layers before it reaches the photoreceptor layer. Two other types of cells present in the bipolar cell layer of the retina are called **horizontal cells** and **amacrine cells** (AM-a-krin). These cells form laterally directed neural circuits that modify the signals being transmitted along the pathway from photoreceptors to bipolar cells to ganglion cells.

CLINICAL CONNECTION | *Detached Retina*

A **detached retina** may occur due to trauma, such as a blow to the head, in various eye disorders, or as a result of age-related degeneration. The detachment occurs between the neural portion of the retina and the pigment epithelium. Fluid accumulates between these layers, forcing the thin, pliable retina to billow outward. The result is distorted vision and blindness in the corresponding field of vision. The retina may be reattached by laser surgery or cryosurgery (localized application of extreme cold), and reattachment must be accomplished quickly to avoid permanent damage to the retina. •

Photoreceptors are specialized cells in the photoreceptor layer that begin the process by which light rays are ultimately converted to nerve impulses. There are two types of photoreceptors: rods and cones. Each retina has about 6 million cones and 120 million rods. **Rods** allow us to see in dim light, such as moonlight. Because they do not provide color vision, in dim light we see only black, white, and all shades of gray in between. Bright lights stimulate the **cones**, which produce color vision. Most of our visual experiences

Figure 21.7 Microscopic structure of the retina. Eventually, nerve impulses arise in ganglion cells and propagate along their axons, which make up the optic (II) nerve.

🔑 In the retina, visual signals pass from photoreceptors to bipolar cells to ganglion cells.

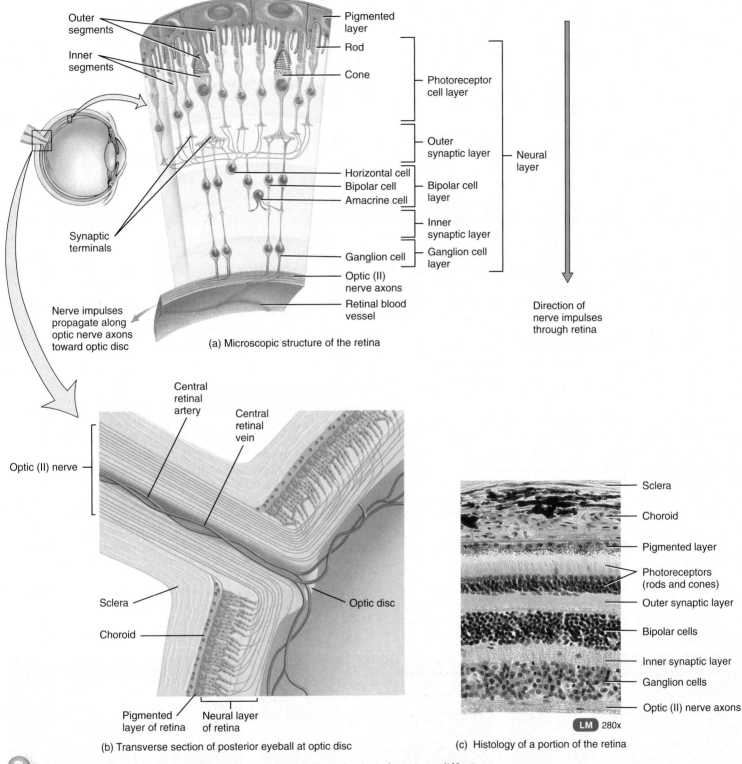

(a) Microscopic structure of the retina

(b) Transverse section of posterior eyeball at optic disc

(c) Histology of a portion of the retina

❓ **What are the two types of photoreceptors, and how do their functions differ?**

are mediated by the cone system, the loss of which produces legal blindness. A person who loses rod vision mainly has difficulty seeing in dim light and thus should not drive at night. The photoreceptive rods and cones are situated next to the outer pigment layer and their photoreceptive ends face away from the incoming light. The rods and cones consist of three parts (Figure 21.7a): (1) an outer segment, which detects the light stimulus; (2) an inner seg-

ment forming the midregion of the cell, which contains the metabolic machinery; and (3) a synaptic terminal, which lies closest to the eye's interior, facing the bipolar neurons. The outer segment, which is rod-shaped in rods and cone-shaped in cones, is composed of stacked, flattened, membranous discs containing an abundance of **photopigment** molecules. Over a billion of these molecules may be packed into the outer segment of each photoreceptor.

The **macula lutea** (MAK-ū-la LOO-tē-a; *macula*=a small, flat spot; *lute-*=yellowish) is in the exact center of the posterior portion of the retina, at the visual axis of the eye (see Figure 21.6). The **fovea centralis** (FŌ-vē-a) (see Figure 21.4), a small depression in the center of the macula lutea, contains only cones. In addition, the layers of bipolar and ganglion cells, which scatter light to some extent, do not cover the cones here; these layers are displaced to the periphery of the fovea centralis. As a result, the fovea centralis is the area of highest **visual acuity** (a-KŪ-i-tē) or *resolution* (sharpness of vision). A main reason that you move your head and eyes while looking at something is to place images of interest on your fovea centralis—as you do to read the words in this sentence! Rods are absent in the fovea centralis and are more plentiful toward the periphery of the retina. Because rod vision is more sensitive than cone vision, you can see a faint object (such as a dim star) better if you gaze slightly to one side rather than looking directly at it.

From photoreceptors, information flows through the outer synaptic layer to bipolar cells and then from bipolar cells through the inner synaptic layer to ganglion cells. The axons of ganglion cells extend posteriorly to the optic disc and exit the eyeball as the optic (II) nerve. The optic disc is also called the **blind spot**. Because it contains no rods or cones, we cannot see an image that strikes the blind spot. Normally, you are not aware of having a blind spot, but you can easily demonstrate its presence. Hold this page about 15 in. from your face with the cross shown below directly in front of your right eye. You should be able to see the cross and the square when you close your left eye. Now, keeping your left eye closed, slowly bring the page closer to your face while keeping your right eye focused on the cross. At a certain distance, the square will disappear from your field of vision because its image falls on the blind spot.

CLINICAL CONNECTION | *Major Causes of Blindness*

Cataracts

A common cause of blindness is a loss of transparency of the lens known as a **cataract** (CAT-a-rakt=waterfall) (Figure A). The lens becomes cloudy (less transparent) due to changes in the structure of the lens proteins. Cataracts often occur with aging but may also be caused by injury, excessive exposure to ultraviolet rays, certain medications (such as long-term use of steroids), or complications of other diseases (such as diabetes). People who smoke also have increased risk of developing cataracts. Fortunately, sight can usually be restored by surgical removal of the old lens and implantation of a new artificial one.

Glaucoma

Glaucoma (glaw-KŌ-ma) is the most common cause of blindness in the United States, afflicting about 2 percent of the population over age 40. In many cases, glaucoma is due to an abnormally high intraocular pressure as a result of a buildup of aqueous humor within the anterior cavity. The fluid compresses the lens into the vitreous body and puts pressure on the neurons of the retina. Persistent pressure results in a progression from mild visual impairment to irreversible destruction of neurons of the retina, damage to the optic nerve, and blindness (Figure B). Glaucoma is painless, and the other eye compensates largely, so a person may experience considerable retinal damage and loss of vision before the condition is diagnosed. Because glaucoma occurs more often with advancing age, regular measurement of intraocular pressure is an increasingly important part of an eye exam as people grow older. Risk factors include race (blacks are more susceptible), increasing age, family history, and past eye injuries and disorders.

Some individuals have another form of glaucoma called **normal-tension (low-tension) glaucoma**. In this condition, there is damage to the optic nerve with a corresponding loss of vision, even though intraocular pressure is normal. Although the cause is unknown, it appears to be related to a fragile optic nerve, vasospasm of blood vessels around the optic nerve, and ischemia due to narrowed or obstructed blood vessels around the optic nerve. The incidence of normal-tension glaucoma is higher among Japanese and Koreans and females.

Age-Related Macular Disease

Age-related macular disease (AMD), also known as *macular degeneration,* is a degenerative disorder of the retina and pigmented layer in persons 50 years of age and older. In AMD, abnormalities occur in the region of the macula lutea, which is ordinarily the area of most acute vision. Victims of advanced AMD retain their peripheral vision but lose the ability to see straight ahead. For instance, they cannot see facial features to identify a person in front of them (Figure C). AMD is the leading cause of blindness in those over age 75, afflicting 13 million Americans, and is 2.5 times more common in pack-a-day smokers than in nonsmokers. Initially, a person may experience blurring and distortion at the center of the visual field. In "dry" AMD, central vision gradually diminishes because the pigmented layer atrophies and degenerates. There is no effective treatment. In about 10 percent of cases, dry AMD progresses to "wet" AMD, in which new blood vessels form in the choroid and leak plasma or blood under the retina. Vision loss can be slowed by using laser surgery to destroy leaking blood vessels. •

(A) A scene as it might be viewed by a person with a cataract.

(B) A scene as it might be viewed by a person with glaucoma.

(C) A scene as it might be viewed by a person with age-related macular degeneration.

Lens

Posterior to the pupil and iris, within the cavity of the eyeball, is the **lens** (see Figure 21.4). Within the cells of the lens, proteins called **crystallins** (KRIS-ta-lins), arranged like the layers of an onion, make up the refractive media of the lens, which normally is perfectly transparent and lacks blood vessels. It is enclosed by a clear connective tissue capsule and held in position by encircling zonular fibers, which attach to the ciliary processes. The lens helps focus images on the retina to facilitate clear vision.

CLINICAL CONNECTION | *Presbyopia*

With aging, the lens loses elasticity and thus its ability to curve to focus on objects that are close. Therefore, older people cannot read print at the same close range as can younger people. This condition is called **presbyopia** (prez-bē-Ō-pē-a; *presby-*=old; *-opia*= pertaining to the eye or vision). By age 40 the near point of vision may have increased to 20 cm (8 in.), and at age 60 it may be as much as 80 cm (31 in.). Presbyopia usually begins in the mid-forties. At about that age, people who have not previously worn glasses begin to need them for reading. Those who already wear glasses typically start to need bifocals, lenses that can correct both distant and close vision. •

Interior of the Eyeball

In addition to the cornea and the lens, other features of the eye transmit and refract the light to the retina. The interior of the eye consists of two fluid-filled cavities, separated by the lens, each of which is transparent to permit light to pass through the eye from the cornea to the retina. The **anterior cavity**, between the cornea and lens, contains a clear watery fluid, the **aqueous humor** (ĀK-wē-us HŪ-mor; *aqua*=water). The anterior cavity consists of two chambers. The **anterior chamber** lies between the cornea and the iris, and the **posterior chamber** lies behind the iris and in front of the zonular fibers and lens (Figure 21.8). Aqueous humor continually filters out of blood capillaries in the ciliary processes of the ciliary body and enters the posterior chamber. It then flows forward between the iris and the lens, through the pupil, and into the anterior chamber. From the anterior chamber, aqueous humor drains into the scleral venous sinus (canal of Schlemm) and then into the blood. Normally, aqueous humor is completely replaced about every 90 minutes.

The larger posterior cavity of the eyeball is the **vitreous chamber** (VIT-rē-us), which lies between the lens and the retina. Within the vitreous chamber is the **vitreous body**, a jellylike substance that contributes to intraocular pressure. The vitreous body, which occupies approximately four-fifths of the eyeball and plays an important role in maintaining the spherical shape of the eye, is an oval structure with an anterior depression that houses the lens and is firmly affixed to all neighboring structures. Formed during embryonic life, this colorless body consists of approximately 99 percent water interspersed with collagen fibers and hyaluronic acid. Structurally, the vitreous body has a gel-like periphery surrounding a fluid center. Around the

---

Figure 21.8 The anterior and posterior chambers of the eye. The section shown is through the anterior portion of the eyeball at the junction of the cornea and sclera. Arrows indicate the direction of flow of aqueous humor.

Whereas the iris separates the anterior and posterior chambers of the anterior cavity, the lens separates the posterior chamber of the anterior cavity from the vitreous chamber.

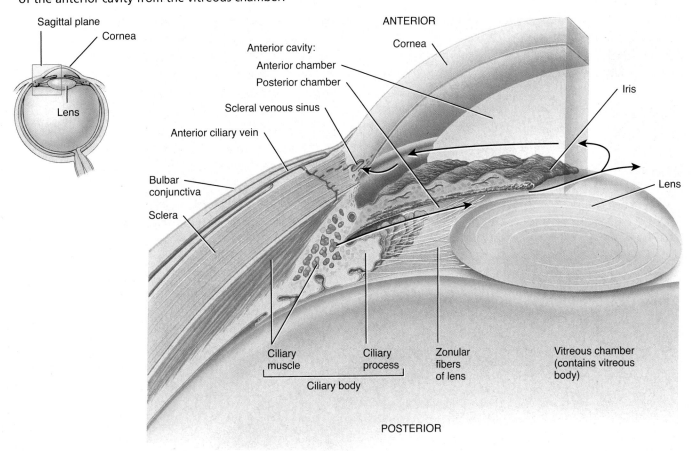

Where is aqueous humor produced, what is its circulation path, and where does it drain from the eyeball?

perimeter of the lens, the outer layer of the vitreous body forms the fibrous suspension network of the lens, the zonular fibers (see Figure 21.4). The pressure of the vitreous body holds the retina flush against the choroid, so that the retina provides an even surface for the reception of clear images. The vitreous body also contains phagocytic cells that remove debris, keeping this part of the eye clear for unobstructed vision. Occasionally, collections of debris may cast a shadow on the retina and create the appearance of specks, hairs, and fine strings that dart in and out of the field of vision. These **vitreal floaters** result from changes in the vitreous body related to aging and from certain conditions such as nearsightedness and inflammation. Vitreal floaters are usually harmless and do not require treatment. The **hyaloid canal** (HĪ-a-loyd) (see Figure 21.4) is a narrow channel that is inconspicuous in adults and runs through the vitreous body from the optic disc to the posterior aspect of the lens. In the fetus, this channel is occupied by the hyaloid artery (see Figure 21.19d).

The pressure in the eye, called **intraocular pressure**, is produced mainly by the aqueous humor and partly by the vitreous body; normally it is about 16 mmHg (millimeters of mercury). The intraocular pressure maintains the shape of the eyeball and prevents the eyeball from collapsing. Puncture wounds to the eyeball may lead to the loss of aqueous humor and fluid from the vitreous body. This can cause a decrease in intraocular pressure, a detached retina, and even blindness.

Table 21.1 summarizes the structures associated with the eyeball.

CLINICAL CONNECTION | *LASIK*

An increasingly popular alternative to wearing glasses or contact lenses is refractive surgery to correct the curvature of the cornea for conditions such as farsightedness, nearsightedness, and astigmatism. The most common type of refractive surgery is **LASIK** (laser-assisted in-situ keratomileusis). After anesthetic drops are placed in the eye, a circular flap of tissue is cut from the center of the cornea. The flap is folded out of the way, and the underlying layer of cornea is reshaped with a laser, one microscopic layer at a time. A computer assists the physician in removing very precise layers of the cornea. After the sculpting is complete, the corneal flap is repositioned over the treated area. A patch is placed over the eye overnight and the flap quickly reattaches to the rest of the cornea. •

The Visual Pathway

After considerable processing of visual signals at synapses among the various types of neurons in the neural layer of the retina, the axons of retinal ganglion cells provide output from the retina to the brain. They exit the eyeball as the optic (II) nerve (see Figure 21.7).

TABLE 21.1

Summary of the Structures of the Eyeball

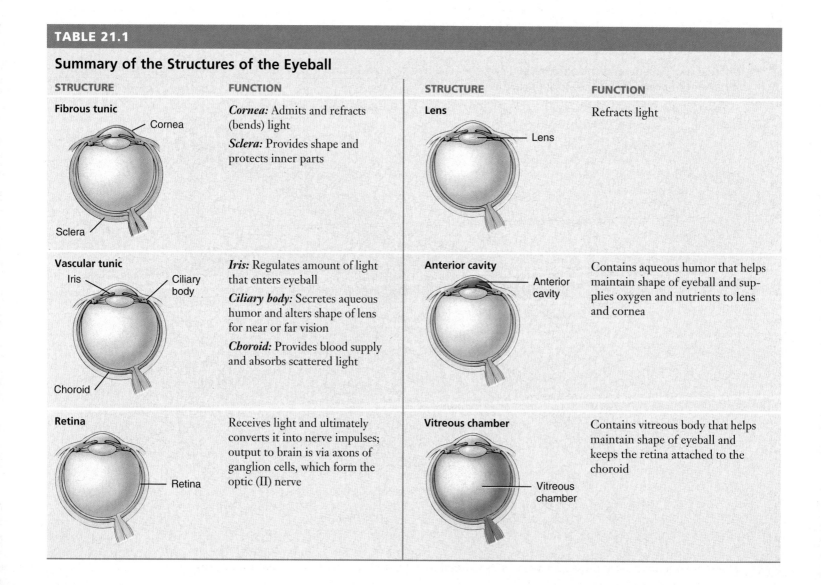

STRUCTURE	FUNCTION	STRUCTURE	FUNCTION
Fibrous tunic (Cornea, Sclera)	*Cornea:* Admits and refracts (bends) light *Sclera:* Provides shape and protects inner parts	**Lens**	Refracts light
Vascular tunic (Iris, Ciliary body, Choroid)	*Iris:* Regulates amount of light that enters eyeball *Ciliary body:* Secretes aqueous humor and alters shape of lens for near or far vision *Choroid:* Provides blood supply and absorbs scattered light	**Anterior cavity**	Contains aqueous humor that helps maintain shape of eyeball and supplies oxygen and nutrients to lens and cornea
Retina	Receives light and ultimately converts it into nerve impulses; output to brain is via axons of ganglion cells, which form the optic (II) nerve	**Vitreous chamber**	Contains vitreous body that helps maintain shape of eyeball and keeps the retina attached to the choroid

Processing of Visual Input in the Retina

Within the neural layer of the retina, certain features of visual input are enhanced while other features may be discarded. Input from several cells may either converge upon a smaller number of postsynaptic neurons or diverge to a larger number. On the whole, convergence predominates: 1 million ganglion cells receive input from about 126 million photoreceptor cells.

Chemicals (neurotransmitters) released by rods and cones induce changes in both **bipolar cells** and horizontal cells that lead to the generation of nerve impulses (see Figure 21.7). Amacrine cells synapse with ganglion cells and also transmit information to them. When bipolar, horizontal, or amacrine cells transmit signals to ganglion cells, the ganglion cells initiate nerve impulses.

Pathway in the Brain

The axons of the **optic (II) nerve** pass through the **optic chiasm** (KĪ-azm=a crossover, as in the letter X), a crossing point of the optic nerves (Figure 21.9). Some fibers cross to the opposite side; others remain uncrossed. After passing through the optic chiasm, the fibers, now part of the **optic tract**, enter the brain and most of them terminate in the **lateral geniculate nucleus** of the thalamus. Here they synapse with neurons whose axons form the **optic radiations**, which project to the **primary visual areas** in the occipital lobes of the cerebral cortex (area 17 in Figure 18.16). Some of the fibers in the optic tracts terminate in the **superior colliculi**, which control the extrinsic eye muscles, and the **pretectal nuclei**, which control pupillary and accommodation reflexes.

✓ CHECKPOINT

5. What features of the eyelids, eyelashes, and eyebrows help them accomplish their function?
6. What does the lacrimal apparatus do?
7. How are the structures of the fibrous tunic, vascular tunic, and retina related to their functions?
8. Describe the histology of the neural portion of the retina.
9. Outline the pathway of a light stimulus from where it enters the eye to where it ends in the brain.

Figure 21.9 The visual pathway. Partial dissection of the brain in part (a) reveals the optic radiations (axons extending from the thalamus to the occipital lobe), along with the rest of the visual pathway, which is diagrammed in part (b).

🔑 The optic chiasm is the crossing point of the optic nerves.

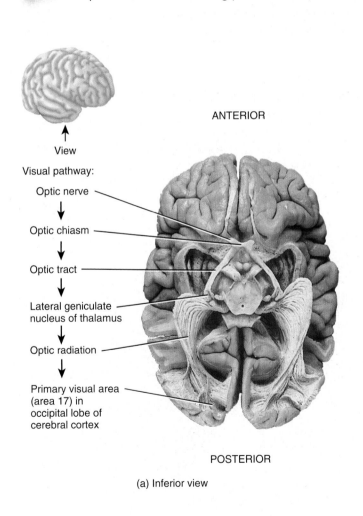

(a) Inferior view

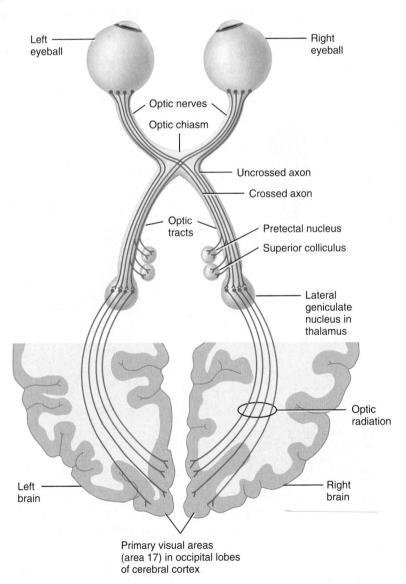

(b) Superior view of transverse section through eyeballs and brain

 Where does the optic tract terminate?

21.4 HEARING AND EQUILIBRIUM

OBJECTIVES

• Describe the anatomy of the structures in the three principal regions of the ear.
• Explain the principal events involved in hearing.
• Outline the auditory pathway.
• List the receptor organs for equilibrium, and describe how they function.
• Identify the equilibrium pathway.

Hearing is the ability to perceive sounds. The ear is an engineering marvel. Its sensory receptors can transduce sound vibrations with amplitudes as small as the diameter of an atom of gold (0.3 nm) into electrical signals 1000 times faster than photoreceptors can respond to light. The ear also contains receptors for **equilibrium** (ē-kwi-LIB-rē-um), the sense that helps you maintain your balance and be aware of your orientation in space.

Anatomy of the Ear

The **ear** is divided into three main regions: (1) the external ear, which collects sound waves and channels them inward; (2) the middle ear, which conveys sound vibrations to the oval window; and (3) the internal ear, which houses the receptors for hearing and equilibrium.

External (Outer) Ear

The **external (outer) ear** consists of the auricle, external auditory canal, and tympanic membrane (eardrum) (Figure 21.10). The **auricle** (AW-rik-kul) or **pinna** is a flap of elastic cartilage shaped like the flared end of a trumpet and covered by skin. The rim of the auricle is the **helix**; the inferior portion is the **lobule**. Ligaments and muscles attach the auricle to the head. The **external auditory canal** (*audit-*=hearing) is a curved tube about 2.5 cm (1 in.) long that lies in the cartilage of the auricle and the temporal bone and leads to the tympanic membrane. The **tympanic membrane** (tim-PAN-ik; *tympan-*=a drum) or *eardrum* is a thin, semitransparent partition between the external auditory canal and middle ear. It attaches by a fibrocartilaginous ring to the temporal bone at the base of the external acoustic meatus and stretches across the opening as a three-layered window. From outside to inside, the three layers are a covering of epidermis; a middle layer of dense connective tissue containing collagen, elastic fibers, and fibroblasts; and a lining of simple cuboidal epithelium. The tympanic membrane is somewhat convex toward the middle ear cavity, with the apex forming the **umbo** (UM-bō). Attached along the superior inner surface of the tympanic membrane, to the point of the umbo, is the first middle ear ossicle, the malleus. The handle of this bone can be easily viewed through an **otoscope** (Ō-tō-skōp; *oto-*=ear; *skopeo*=to view), a viewing instrument that illuminates and magnifies the external ear and tympanic membrane. Tearing

Figure 21.10 Anatomy of the ear.

🔑 The ear has three principal regions: the external (outer) ear, the middle ear, and the internal (inner) ear. (See key below.)

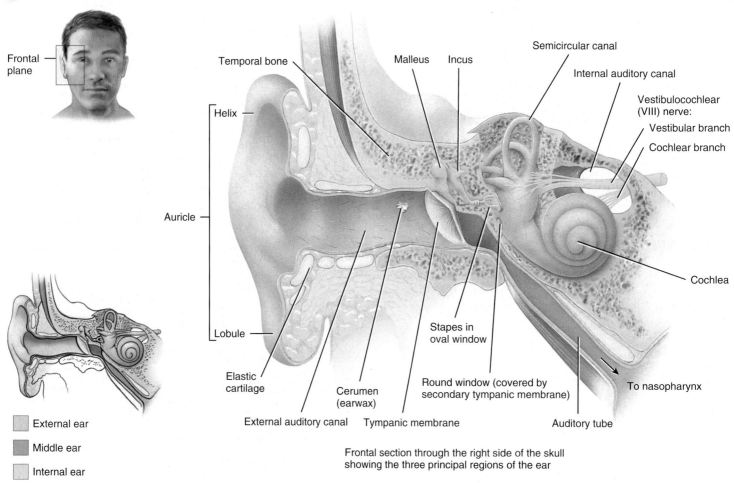

Frontal plane

Temporal bone
Malleus
Incus
Semicircular canal
Internal auditory canal
Vestibulocochlear (VIII) nerve:
Vestibular branch
Cochlear branch
Helix
Auricle
Cochlea
Lobule
Stapes in oval window
Elastic cartilage
Cerumen (earwax)
Round window (covered by secondary tympanic membrane)
To nasopharynx
External auditory canal
Tympanic membrane
Auditory tube

☐ External ear
☐ Middle ear
☐ Internal ear

Frontal section through the right side of the skull showing the three principal regions of the ear

❓ **To which structure of the external ear does the malleus of the middle ear attach?**

of the tympanic membrane is called a **perforated eardrum**. It may be due to pressure from a cotton swab, trauma, or a middle ear infection, and usually heals within a month.

Near the exterior, the external auditory canal contains a few hairs and specialized sebaceous (oil) glands called **ceruminous glands** (se-ROO-mi-nus) that secrete earwax or **cerumen** (se-ROO-men). The combination of hairs and cerumen helps prevent dust and foreign objects from entering the ear. Cerumen also prevents damage by water and adventurous insects to the delicate skin of the external auditory canal. Cerumen usually dries up and falls out of the ear canal. Some people, however, produce a large amount of cerumen, which can become impacted and can muffle incoming sounds. The treatment for **impacted cerumen** is usually periodic ear irrigation or removal of wax with a blunt instrument; the latter should be attempted only by trained medical personnel.

Middle Ear

The **middle ear** is a small, air-filled cavity in the petrous portion of the temporal bone that is lined by epithelium (Figure 21.11a). It is separated from the external ear by the tympanic membrane and from the internal ear by a thin bony partition that contains two small openings: the oval window and the round window. Extending across the middle ear and attached to it by ligaments are the three smallest bones in the body, the **auditory ossicles** (OS-si-kuls), which are connected to one another by synovial joints. The bones, named for their shapes, are the malleus, incus, and stapes—commonly called the hammer, anvil, and stirrup, respectively. The "handle" of the **malleus** (MAL-ē-us) attaches to the internal surface of the tympanic membrane. The head of the malleus articulates with the body

of the incus. The **incus** (ING-kus), the middle bone in the series, articulates with the head of the stapes. The base or footplate of the **stapes** (STĀ-pēz) fits into the **oval window** (*fenestra vestibuli*). Directly below the oval window is another opening, the **round window** (*fenestra cochlea*), which is enclosed by a membrane called the **secondary tympanic membrane**.

Two tiny skeletal muscles also attach to the ossicles (Figure 21.11a, c). The **tensor tympani muscle** (TIM-pan-ē), which is innervated by the mandibular branch of the trigeminal (V) nerve, limits movement and increases tension on the eardrum to prevent damage to the inner ear from loud noises. The **stapedius muscle** (sta-PĒ-dē-us), which is innervated by the facial (VII) nerve, is the smallest of all skeletal muscles. By dampening large vibrations of the stapes due to loud noises, it protects the oval window, but it also decreases the sensitivity of hearing. For this reason, paralysis of the stapedius muscle is associated with **hyperacusia** (hī-per-a-KŪ-sē-a), abnormally sensitive hearing. Because it takes a fraction of a second for the tensor tympani and stapedius muscles to contract, they can protect the inner ear from prolonged loud noises such as thunder, but not from brief ones such as a gunshot.

The anterior wall of the middle ear contains an opening that leads directly into the **auditory tube** (or *pharyngotympanic tube*), commonly known as the *eustachian tube* (ū′-STĀ-kē-an or ū-STĀ-shun). The auditory tube, which consists of both bone and elastic cartilage, connects the middle ear with the nasopharynx (superior portion of the throat). It is normally closed at its medial (pharyngeal) end. During swallowing and yawning, it opens, allowing air to enter or leave the middle ear until the pressure in the middle ear equals the atmospheric pressure. When the pressures are balanced, the tympanic

Figure 21.11 The right middle ear and the auditory ossicles.

🔑 Common names for the malleus, incus, and stapes are the hammer, anvil, and stirrup, respectively.

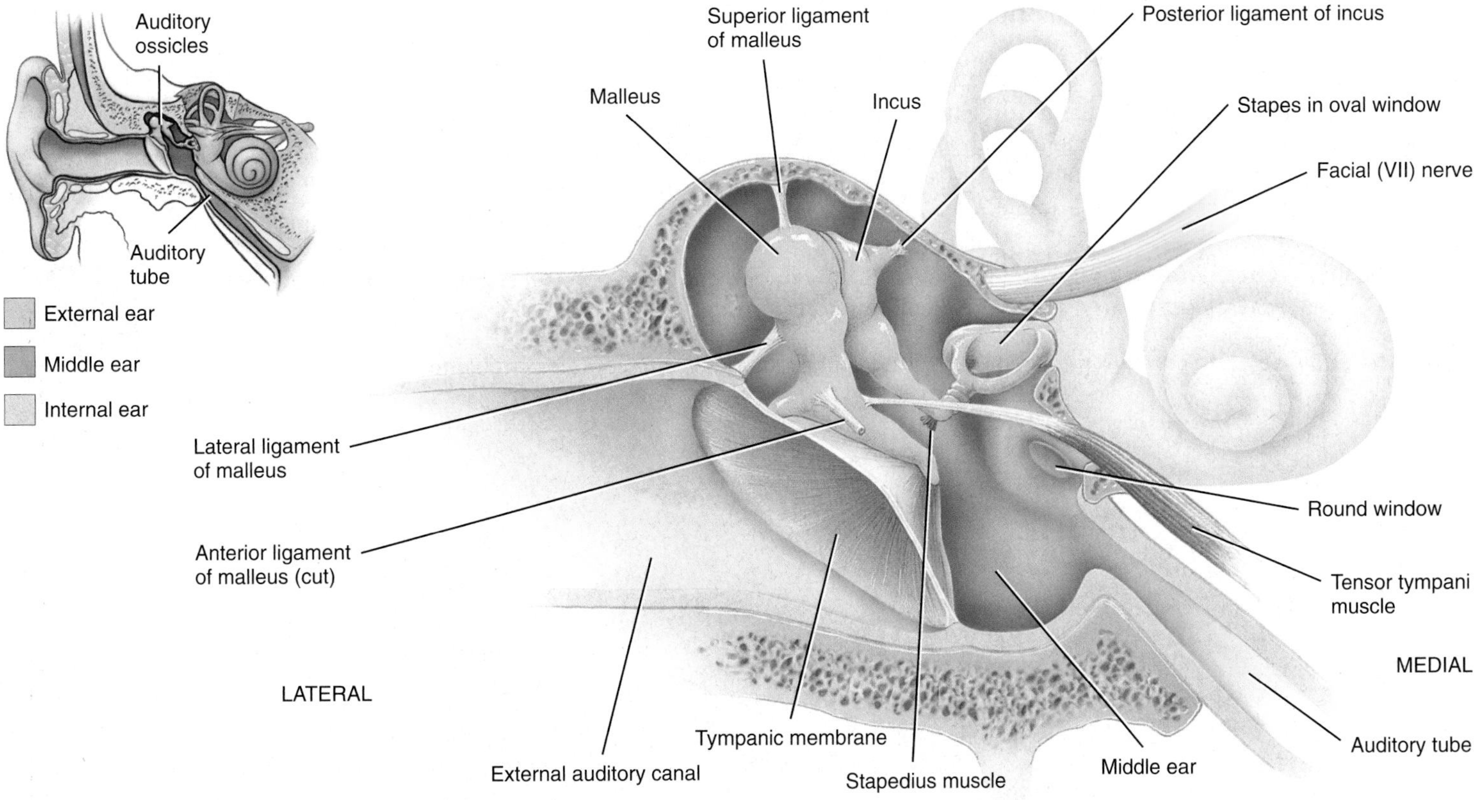

(a) Frontal section showing location of auditory ossicles in the middle ear

FIGURE 21.11 CONTINUES ▶

FIGURE 21.11 CONTINUED ▶

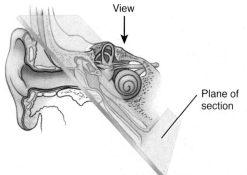

View

Plane of
section

CLINICAL CONNECTION | *Otitis Media*

Otitis media (ō-TĪ-tis MĒ-dē-a) is an acute infection of the middle ear caused mainly by bacteria and associated with infections of the nose and throat. Symptoms include pain, malaise, fever, and a reddening and outward bulging of the eardrum, which may rupture unless prompt treatment is received. (This may involve draining pus from the middle ear.) Bacteria passing into the auditory tube from the nasopharynx are the primary cause of middle ear infections. Children are more susceptible than adults to middle ear infections because their auditory tubes are almost horizontal, which decreases drainage. If otitis media occurs frequently, a surgical procedure called **tympanotomy** (tim′-pa-NOT-ō-mē; *tympano-*=drum; *-tome*=incision) is often employed. This consists of the insertion of a small tube into the eardrum to provide a pathway for the drainage of fluid from the middle ear •

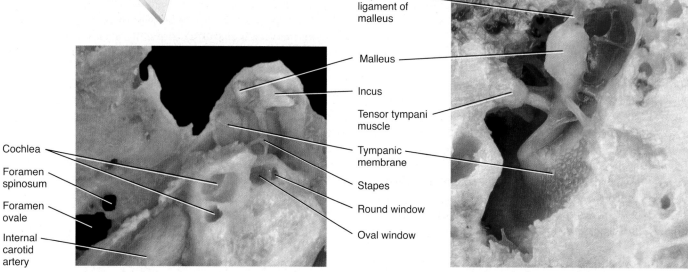

Superior
ligament of
malleus

Malleus

Incus

Tensor tympani
muscle

Tympanic
membrane

Stapes

Round window

Oval window

Cochlea

Foramen
spinosum

Foramen
ovale

Internal
carotid
artery

(b) Right middle and inner ear viewed from medial side

(c) Right middle ear viewed from posterosuperior
with incus and stapes removed

? **What structures separate the middle ear from the internal ear?**

membrane vibrates freely as sound waves strike it. If the pressure is not equalized, intense pain, hearing impairment, ringing in the ears, and vertigo could develop. The auditory tube also is a route by which pathogens may travel from the nose and throat to the middle ear, causing the most common type of ear infection.

Internal (Inner) Ear

The **internal (inner) ear** is also called the **labyrinth** (LAB-i-rinth) because of its complicated series of canals (Figure 21.12). Structurally, it consists of two main divisions: an outer bony labyrinth that encloses an inner membranous labyrinth. It is like long

Figure 21.12 **The right internal ear.** The outer, cream-colored area [shown in (b)] is part of the bony labyrinth; the inner, pink-colored area is the membranous labyrinth.

🔑 **The bony labyrinth contains perilymph, and the membranous labyrinth contains endolymph.**

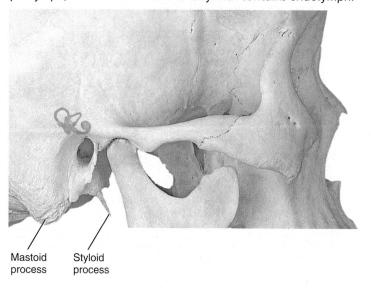

Mastoid
process

Styloid
process

(a) Position of semicircular canals and cochlea (blue) projected to surface

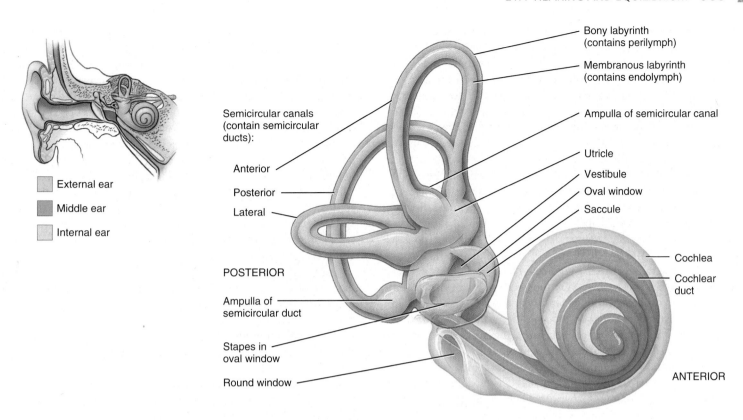

Bony labyrinth
(contains perilymph)

Membranous labyrinth
(contains endolymph)

Ampulla of semicircular canal

Utricle

Vestibule

Oval window

Saccule

Cochlea

Cochlear
duct

External ear

Middle ear

Internal ear

Semicircular canals
(contain semicircular
ducts):

Anterior

Posterior

Lateral

POSTERIOR

Ampulla of
semicircular duct

Stapes in
oval window

Round window

ANTERIOR

(b) Components of the right internal ear

Incus

Malleus

Stapes

Semicircular
canals

Cochlea

Utricle

(c) Auditory ossicles (left) and cast of internal ear (right) compared to
a dime for size

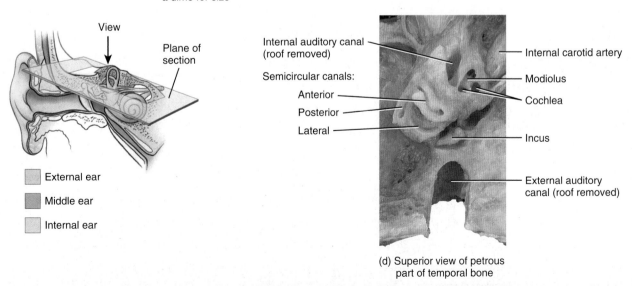

View

Plane of
section

Internal auditory canal
(roof removed)

Semicircular canals:

Anterior

Posterior

Lateral

Internal carotid artery

Modiolus

Cochlea

Incus

External auditory
canal (roof removed)

External ear

Middle ear

Internal ear

(d) Superior view of petrous
part of temporal bone

CLINICAL CONNECTION | *Ménière's Disease*

Ménière's disease (men'-ē-ARZ) results from an increased amount of endolymph that enlarges the membranous labyrinth. Among the symptoms are fluctuating hearing loss (caused by distortion of the basilar membrane of the cochlea) and roaring tinni- tus (ringing). Spinning or whirling vertigo (dizziness) is characteristic of Ménière's disease. Almost total destruction of hearing may occur over a period of years. •

?   **What are the names of the two sacs that lie in the vestibule?**

balloons put inside a rigid tube. The **bony labyrinth** is a series of cavities in the petrous part of the temporal bone divided into three areas: (1) the semicircular canals, (2) the vestibule, and (3) the cochlea. The bony labyrinth is lined with periosteum and contains **perilymph**. This fluid, which is chemically similar to cerebrospinal fluid, surrounds the **membranous labyrinth**, a series of epithelial sacs and tubes inside the bony labyrinth that have the same general form as the bony labyrinth and house the receptors for equilibrium and hearing. The epithelial membranous labyrinth contains **endolymph**. The level of potassium ions (K^+) in endolymph is unusually high for an interstitial fluid, and potassium ions play a role in the generation of auditory signals (described shortly).

The **vestibule** (VES-ti-būl) is the oval central portion of the bony labyrinth. The membranous labyrinth in the vestibule consists of two sacs called the **utricle** (Ū-tri-kul=little bag) and the **saccule** (SAK-ūl=little sac), which are connected by a small duct. Projecting superiorly and posteriorly from the vestibule are the three bony **semicircular canals**, each of which lies at approximately right angles to the other two. Based on their positions, they are named the anterior, posterior, and lateral semicircular canals. The anterior and posterior semicircular canals are vertically oriented; the lateral one is horizontally oriented. At one end of each canal is a swollen enlargement called the **ampulla** (am-PŪL-la=saclike duct). The portions of the membranous labyrinth that lie inside the bony semicircular canals are called the **semicircular ducts**. These structures communicate with the utricle of the vestibule.

The vestibular (ves-TIB-ū-lar) branch of the vestibulocochlear (VIII) nerve consists of *ampullary*, *utricular*, and *saccular nerves*. These nerves contain both first-order sensory neurons and motor neurons that synapse with receptors for equilibrium. The first-order sensory neurons carry sensory information from the receptors, and the motor neurons carry feedback signals to the receptors,

apparently to modify their sensitivity. Cell bodies of the sensory neurons are located in the **vestibular ganglia** (see Figure 21.13b).

Anterior to the vestibule is the **cochlea** (KŌK-lē-a=snail-shaped), a bony spiral canal (Figure 21.13a) that resembles a snail's shell and makes almost three turns around a central bony core called the **modiolus** (mō-DĪ-ō′-lus; Figure 21.13b). The cochlea is divided into three channels: cochlear duct, scala vestibuli, and scala tympani (Figure 21.13a–c). The **cochlear duct** or *scala media* is a continuation of the membranous labyrinth and is filled with endolymph. The channel above the cochlear duct is the **scala vestibuli**, which ends at the oval window. The channel below is the **scala tympani**, which ends at the round window. Because both the scala vestibuli and scala tympani are part of the bony labyrinth of the cochlea, these chambers are filled with perilymph. The scala vestibuli and scala tympani are completely separated by the cochlear duct, except for an opening at the apex of the cochlea, the **helicotrema** (hel-i-kō-TRĒ-ma; see Figure 21.14). The cochlea adjoins the wall of the vestibule, into which the scala vestibuli opens. The perilymph in the vestibule is continuous with that of the scala vestibuli.

The **vestibular membrane** separates the cochlear duct from the scala vestibuli, and the **basilar membrane** (BĀS-i-lar) separates the cochlear duct from the scala tympani. Resting on the basilar membrane is the **spiral organ** or *organ of Corti* (KOR-tē) (Figure 21.13c, d). The spiral organ is a coiled sheet of epithelial cells, including supporting cells and about 16,000 hair cells, which are the receptors for hearing. There are two groups of hair cells: The *inner hair cells* are arranged in a single row and the *outer hair cells* are arranged in three rows. At the apical tip of each hair cell is a hair bundle, consisting of 40–80 **stereocilia** and a **kinocilium** (cilium) that extend into the endolymph of the cochlear duct. Despite their name, stereocilia are not cilia but long, hairlike microvilli arranged in several rows of graded height.

Figure 21.13 Semicircular canals, vestibule, and cochlea of the right ear. Note that the cochlea makes almost three complete turns.

🔑 The three channels in the cochlea are the scala vestibuli, the scala tympani, and the cochlear duct.

External ear
Middle ear
Internal ear

POSTERIOR

Utricle
Stapes in oval window
Saccule
Scala vestibuli

ANTERIOR

Cochlea
Scala tympani
Cochlear duct
Scala vestibuli

Vestibular membrane
Cochlear duct
Basilar membrane
Secondary tympanic membrane in round window
Scala tympani

Transmission of sound waves from scala vestibuli to scala tympani by way of helicotrema

(a) Sections through the cochlea

A **cochlear implant** (KŌK-lē-ar) is a device that translates sounds into electrical signals that can be interpreted by the brain. Such a device is useful for people with deafness that is caused by damage to hair cells in the cochlea. The artificially induced electrical signals propagate over their normal pathways to the brain. The perceived sounds are crude compared to normal hearing, but they provide a sense of rhythm and loudness; information about certain noises, such as those made by telephones and automobiles; and the pitch and cadence of speech. Some patients hear well enough with a cochlear implant to use the telephone. •

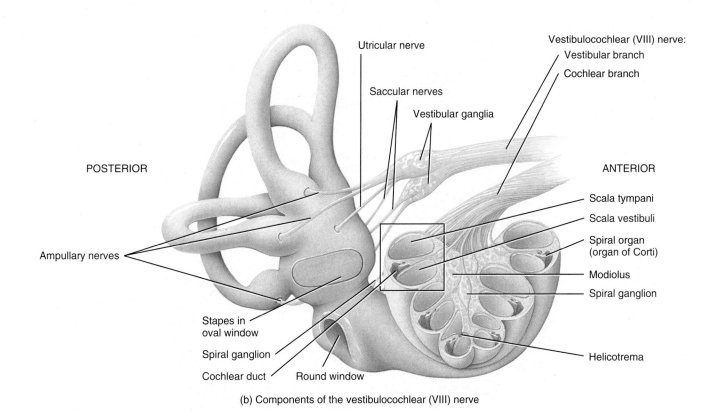

(b) Components of the vestibulocochlear (VIII) nerve

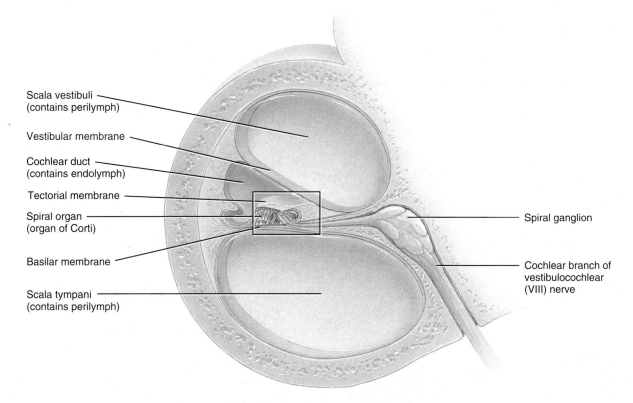

(c) Section through one turn of the cochlea

FIGURE 21.13 CONTINUES ▶

Figure 21.13 continued ▶

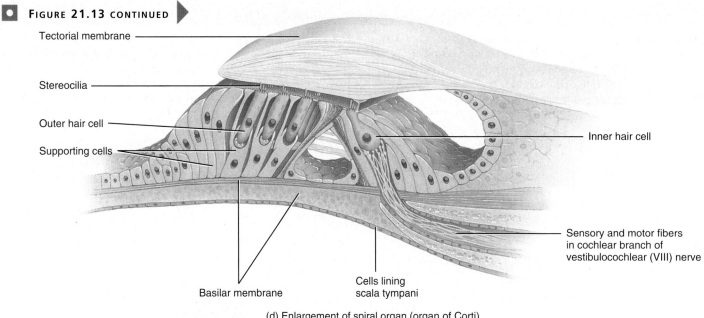

Tectorial membrane

Stereocilia

Outer hair cell

Supporting cells

Inner hair cell

Sensory and motor fibers
in cochlear branch of
vestibulocochlear (VIII) nerve

Basilar membrane

Cells lining
scala tympani

(d) Enlargement of spiral organ (organ of Corti)

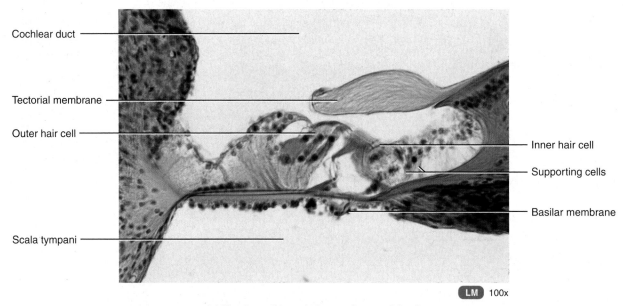

Cochlear duct

Tectorial membrane

Outer hair cell

Inner hair cell

Supporting cells

Basilar membrane

Scala tympani

LM 100x

(e) Histology of the spiral organ (organ of Corti)

 What are the three subdivisions of the bony labyrinth?

At their basal ends, inner and outer hair cells synapse both with first-order sensory neurons and with motor neurons from the cochlear branch of the vestibulocochlear (VIII) nerve. Cell bodies of the sensory neurons are located in the **spiral ganglion** (Figure 21.13b, c). Although outer hair cells outnumber them by 3 to 1, the inner hair cells synapse with 90–95 percent of the first-order sensory neurons in the cochlear nerve, which relay auditory information to the brain. By contrast, 90 percent of the motor neurons in the cochlear nerve synapse with outer hair cells. The **tectorial membrane** (tek-TŌ-rē-al; *tector-*=covering), a flexible gelatinous membrane, overlies the hair cells of the spiral organ (Figure 21.13d). The ends of the stereocilia of the hair cells are embedded in the tectorial membrane, while the bodies of the hair cells rest on the basilar membrane.

Mechanism of Hearing

Sound waves are a series of alternating high- and low-pressure regions traveling in the same direction through some medium (such as air). They originate from a vibrating object in much the same way that ripples arise and travel over the surface of a pond when you toss a stone into it.

The following events are involved in hearing (Figure 21.14):

❶ The auricle directs sound waves into the external auditory canal.

❷ When sound waves strike the tympanic membrane, the alternating waves of high and low pressure in the air cause the tympanic membrane to vibrate back and forth. The tympanic membrane vibrates slowly in response to low-frequency (low-pitched) sounds and rapidly in response to high-frequency (high-pitched) sounds.

❸ The central area of the tympanic membrane connects to the malleus, which vibrates along with the tympanic membrane. This vibration is transmitted from the malleus to the incus and then to the stapes.

❹ As the stapes moves back and forth, its oval-shaped footplate, which is attached via a ligament to the circumference of the oval window, vibrates in the oval window. The vibrations at the oval window are about 20 times more vigorous than those at the tympanic membrane because the auditory ossicles efficiently transmit small vibrations spread over a large surface area (the tympanic membrane) into larger vibrations at a smaller surface (the oval window).

Figure 21.14 Stimulation of auditory receptors in the right ear. The cochlea has been uncoiled to more easily visualize the transmission of sound waves and their distortion of the vestibular and basilar membranes of the cochlear duct.

🔑 The function of hair cells of the spiral organ (organ of Corti) is to ultimately convert a mechanical vibration (stimulus) into an electrical signal (nerve impulse).

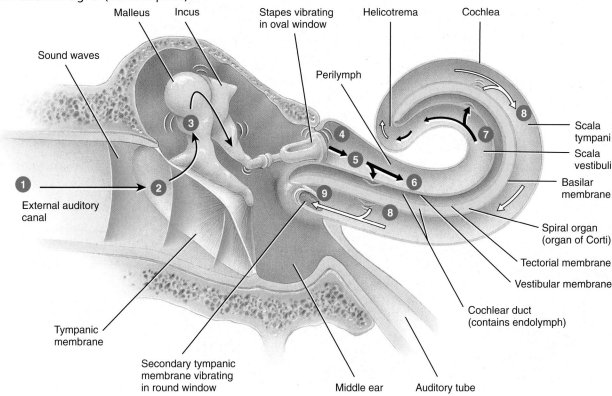

❓ **What structure vibrates and sets up pressure waves in the perilymph?**

⑤ The movement of the stapes at the oval window sets up fluid pressure waves in the perilymph of the cochlea. As the oval window bulges inward, it pushes on the perilymph of the scala vestibuli.

⑥ Pressure waves are transmitted from the scala vestibuli to the scala tympani and eventually to the round window, causing it to bulge outward into the middle ear. (See ⑨ in the figure.)

⑦ The pressure waves travel through the perilymph of the scala vestibuli, then the vestibular membrane, and then move into the endolymph inside the cochlear duct.

⑧ The pressure waves in the endolymph cause the basilar membrane to vibrate, which moves the hair cells of the spiral organ against the tectorial membrane. This leads to bending of the stereocilia and ultimately to the generation of nerve impulses in first-order neurons in cochlear nerve fibers.

⑨ Sound waves of various frequencies cause certain regions of the basilar membrane to vibrate more intensely than other regions. Each segment of the basilar membrane is "tuned" for a particular pitch. Because the membrane is narrower and stiffer at the base of the cochlea (closer to the oval window), high-frequency (high-pitched) sounds induce maximal vibrations in this region. Toward the apex of the cochlea, the basilar membrane is wider and more flexible; low-frequency (low-pitched) sounds cause maximal vibration of the basilar membrane there. Loudness is determined by the intensity of sound waves. High-intensity sound waves cause larger vibrations of the basilar membrane, which leads to a higher frequency of nerve impulses reaching the brain. Louder sounds also may stimulate a larger number of hair cells.

Besides its role in detecting sounds, the cochlea has the surprising ability to produce sounds. These usually inaudible sounds, called *otoacoustic emissions*, can be picked up by placing a sensitive microphone next to the eardrum. They are caused by vibrations of the outer hair cells that occur in response to sound waves and to signals from motor neurons. This vibratory behavior appears to change the stiffness of the tectorial membrane and is thought to enhance the movement of the basilar membrane, which amplifies the responses of the inner hair cells. At the same time, the outer hair cell vibrations set up a traveling wave that goes back toward the stapes and leaves the ear as an otoacoustic emission. Detection of these inner ear–produced sounds is a fast, inexpensive, and noninvasive way to screen newborns for hearing defects. In deaf babies, otoacoustic emissions are not produced or are greatly reduced in size.

The Auditory Pathway

Bending of the stereocilia of the hair cells of the spiral organ causes the release of a neurotransmitter (probably glutamate), which generates nerve impulses in the sensory neurons attached to the hair cells. The cell bodies of the sensory neurons are located in the **spiral ganglia**. Nerve impulses pass along the axons of the neurons, which form the cochlear branch of the vestibulocochlear (VIII) nerve (Figure 21.15). These axons synapse with neurons in the **cochlear nuclei** in the medulla oblongata on the same side. Some of the axons from the cochlear nuclei decussate (cross over) in the medulla, ascend in a tract called the **lateral lemniscus** on the opposite side, and terminate in the **inferior colliculus** in the midbrain. Other axons from the cochlear nuclei end in the **superior olivary nucleus** in the pons on each side. Slight differences in the timing of nerve impulses arriving from the two ears at the superior olivary

Figure 21.15 The auditory pathway.

 Since many auditory axons decussate in the medulla while others stay on the same side, the right and left primary auditory areas receive nerve impulses from both ears.

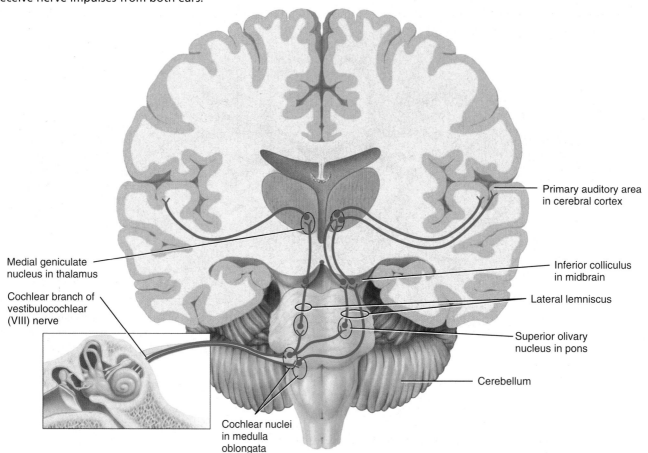

❓ **Where are the cell bodies of the first sensory neurons in the auditory pathway located?**

nuclei allow you to locate the source of a sound. Axons from the superior olivary nuclei also ascend in the lateral lemniscus tracts on both sides and end in the inferior colliculi. From each inferior colliculus, nerve impulses are conveyed to the **medial geniculate nucleus** in the thalamus and finally to the **primary auditory area** of the cerebral cortex in the temporal lobe of the cerebrum (see areas 41 and 42 in Figure 18.16). Because many auditory axons decussate in the medulla while others remain on the same side, the right and left primary auditory areas receive nerve impulses from both ears.

Mechanism of Equilibrium

There are two types of equilibrium or balance. **Static equilibrium** refers to the maintenance of the position of the body (mainly the head) relative to the force of gravity. Body movements that stimulate the receptors for static equilibrium include tilting the head and linear acceleration or deceleration, such as when the body is being moved in an elevator or in a car that is speeding up or slowing down. **Dynamic equilibrium** is the maintenance of body position (mainly the head) in response to sudden movements such as rotational acceleration or deceleration. The receptor organs for equilibrium include the saccule, utricle, and semicircular ducts, and are collectively referred to as the **vestibular apparatus** (ves-TIB-ū-lar).

Otolithic Organs: Saccule and Utricle

The saccule and utricle are referred to as *otolithic organs* for reasons that will become clear shortly. The walls of both the utricle and the saccule contain a small, thickened region called a **macula** (MAK-ū-la; Figure 21.16). The maculae of the utricle and saccule are the sense organs of static equilibrium. The two *maculae* (MAK-ū-lē; plural), which are perpendicular to one another, are the receptors for static equilibrium, and contribute to some aspects of dynamic equilibrium as well. Their role in static equilibrium is to provide sensory information on the position of the head in space, essential to maintaining appropriate posture and balance. The maculae detect linear acceleration and deceleration—for example, the sensations you feel while in an elevator or a car that is speeding up or slowing down.

The maculae consist of two kinds of cells: **hair cells**, which are the sensory receptors, and **supporting cells**. On the surface of each hair cell there are 40–80 *stereocilia* of graded height, plus one *kinocilium*, a conventional cilium anchored firmly to its basal body and extending beyond the longest stereocilium. (Recall that stereocilia are actually microvilli.) Collectively, the stereocilia and kinocilium are called a **hair bundle**. Scattered among the hair cells are columnar supporting cells; these supporting cells secrete the thick, gelatinous, glycoprotein layer, called the **otolithic membrane** (ō-tō-LITH-ik), that rests on the hair cells. A layer of dense calcium carbonate crystals, called **otoliths** (Ō-tō-liths; *oto*=ear; *liths*=stones), extends over the entire surface of the otolithic membrane.

Because the otolithic membrane sits on top of the macula, when you tilt your head forward, the otolithic membrane and otoliths are pulled by gravity and slide downhill over the hair cells in the direction of the tilt, bending the hair bundles. The movement of the hair bundles initiates responses that ultimately lead to the generation of nerve impulses. The hair cells synapse with first-order sensory neurons in the vestibular branch of the vestibulocochlear (VIII) nerve.

Figure 21.16 Receptors in the maculae of the right ear. Both sensory neurons (blue) and motor neurons (red) synapse with the hair cells.

🔑 The movement of stereocilia initiates responses that ultimately lead to the generation of nerve impulses.

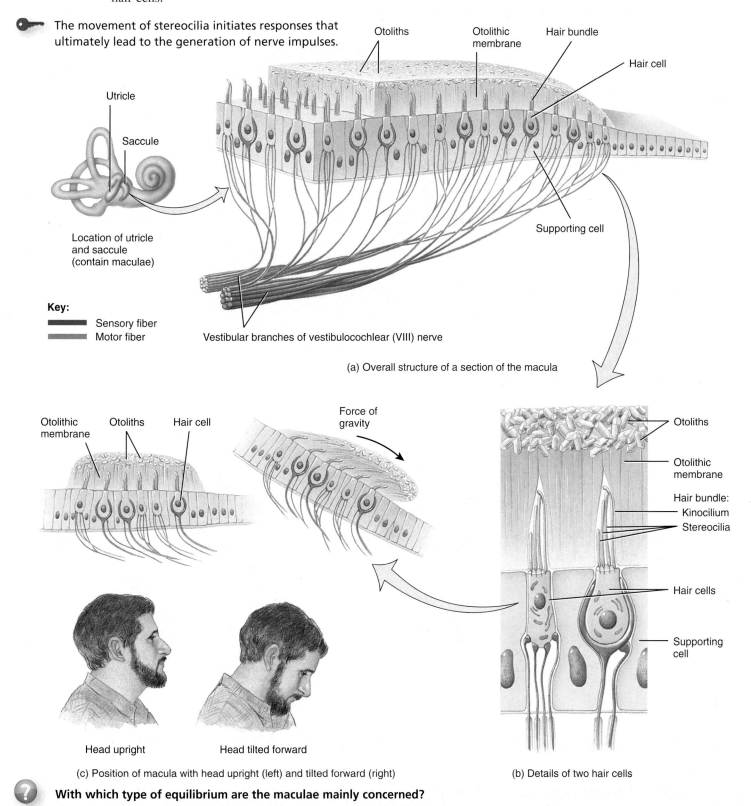

(a) Overall structure of a section of the macula

(c) Position of macula with head upright (left) and tilted forward (right)

(b) Details of two hair cells

❓ **With which type of equilibrium are the maculae mainly concerned?**

Semicircular Ducts

The three semicircular ducts function in dynamic equilibrium. The ducts lie at right angles to one another in three planes. The two vertical ducts are the anterior and posterior semicircular ducts, and the horizontal one is the lateral semicircular duct (see Figure 21.12). This positioning permits detection of rotational acceleration or deceleration. In the **ampulla**, the dilated portion of each duct, is a small elevation called the **crista** (KRIS-ta=crest; plural is cristae; KRIS-tē) (Figure 21.17). Each crista contains a group of hair cells and supporting cells covered by a mass of gelatinous material called the **cupula** (KŪ-pū-la). When the head moves, the attached semicircular ducts and hair cells move with it. However, the endolymph within the ampulla is not attached and lags behind due to inertia. As the moving hair cells drag along the stationary fluid, the hair bundles bend (Figure 21.17b). Bending of the hair bundles produces responses that lead to nerve impulses that pass along the ampullary nerve, a branch of the vestibular division of the vestibulocochlear (VIII) nerve.

Figure 21.17 Semicircular ducts of the right ear. Both sensory neurons (blue) and motor neurons (red) synapse with the hair cells.

🔑 The ampullary nerves are branches of the vestibular division of the vestibulocochlear (VIII) nerve. The positions of the semicircular ducts permit detection of rotational movements.

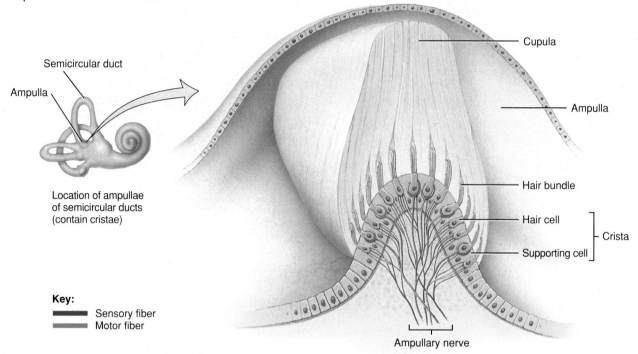

Semicircular duct

Ampulla

Location of ampullae
of semicircular ducts
(contain cristae)

Cupula

Ampulla

Hair bundle

Hair cell

Supporting cell

⎤
⎦ Crista

Key:

━━━ Sensory fiber
━━━ Motor fiber

Ampullary nerve

(a) Details of a crista

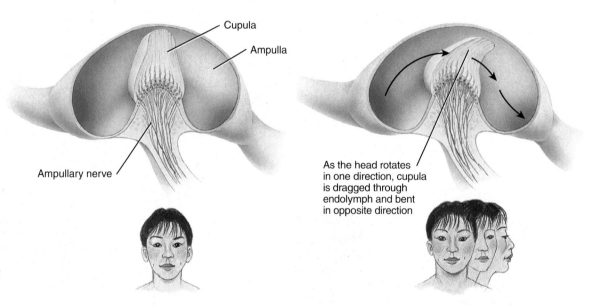

Cupula

Ampulla

Ampullary nerve

As the head rotates
in one direction, cupula
is dragged through
endolymph and bent
in opposite direction

Head in still position

Head rotating

❓ **Are the semicircular ducts associated with static equilibrium or dynamic equilibrium?**

(b) Position of a cupula with the head in the still position (left)
and when the head rotates (right)

CLINICAL CONNECTION | *Motion Sickness*

Motion sickness is a condition that results when there is a conflict among the senses with regard to motion. For example, the vestibular apparatus senses angular and vertical motion, while the eyes and proprioceptors in muscles and joints determine the position of the body in space. If you are in the cabin of a moving ship, your vestibular apparatus informs the brain that there is movement from waves. But your eyes don't see any movement. This leads to the conflict among the senses. Motion sickness can also be experienced in other situations that involve movement, for example, in a car or airplane or on a train or amusement park ride.

Symptoms of motion sickness include paleness, restlessness, excess salivation, nausea, dizziness, cold sweats, headache, and malaise that may progress to vomiting. Once the motion is stopped, the symptoms disappear. If it is not possible to stop the motion, you might try sitting in the front seat of a car, the forward car of a train, the upper deck on a boat, or the wing seats in a plane. Looking at the horizon and not reading also help. Medications for motion sickness are usually taken in advance of travel and include scopolamine in time-release patches or tablets, dimenhydrinate (Dramamine®), and meclizine (Bonine®). •

Equilibrium Pathways

Bending of hair bundles of the hair cells in the semicircular canals, utricle, or saccule causes the release of a neurotransmitter (probably glutamate), which generates nerve impulses in the sensory neurons attached to hair cells. The cell bodies of the hair cells are located in the **vestibular ganglia**. Nerve impulses pass along the axons of the neurons, which form the **vestibular branch of the vestibulocochlear (VIII) nerve** (Figure 21.18). Most of these axons synapse with sensory neurons in **vestibular nuclei**, the major integrating centers for equilibrium, in the medulla oblongata and pons. The vestibular nuclei also receive input from the eyes and proprioceptors, especially proprioceptors in the neck and limb muscles that indicate the positions of the head and limbs. The remaining axons enter the cerebellum through the **inferior cerebellar peduncles** (see Figure 18.8). Bidirectional pathways connect the cerebellum and vestibular nuclei.

The vestibular nuclei integrate information from vestibular, visual, and proprioceptors and then send commands to the following areas:

1. *Nuclei of the oculomotor (III), trochlear (IV), and abducens (VI) nerves.* These cranial nerves control coupled movements of the eyes with those of the head to help maintain focus on the visual field.
2. *Nuclei of the accessory (XI) nerves.* The accessory (XI) nerves help control head and neck movements and assist in maintaining equilibrium.
3. *Vestibulospinal tract.* The **vestibulospinal tract** conveys impulses down the spinal cord to maintain muscle tone in skeletal muscles to help maintain equilibrium.

4. *Ventral posterior nucleus in the thalamus → the vestibular area in the parietal lobe of the cerebral cortex.* This part of the primary somatosensory area (see areas 1, 2, and 3 in Figure 18.16) provides us with the conscious awareness of the position and movements of the head and limbs.

Various pathways among the vestibular nuclei, cerebellum, and cerebrum enable the cerebellum to play a key role in maintaining static and dynamic equilibrium. The cerebellum continuously receives updated sensory information from the utricle and saccule. It monitors this information and makes corrective adjustments. Essentially, in response to input from the utricle, saccule, and semicircular ducts, the cerebellum continuously sends nerve impulses to the motor areas of the cerebrum. This feedback allows correction of signals from the motor cortex to specific skeletal muscles to smooth movements and coordinate complex sequences of muscle contractions to help maintain equilibrium.

Table 21.2 summarizes the structures of the ear related to hearing and equilibrium.

CHECKPOINT

10. List the components of the external, middle, and internal ear and their functions.
11. Explain the mechanism of hearing.
12. Describe the auditory pathway.
13. Compare the function of the maculae in maintaining static equilibrium with the role of the cristae in maintaining dynamic equilibrium.
14. Outline the equilibrium pathways.
15. Describe the role of vestibular input to the cerebellum in the maintenance of equilibrium.

Figure 21.18 The equilibrium pathway.

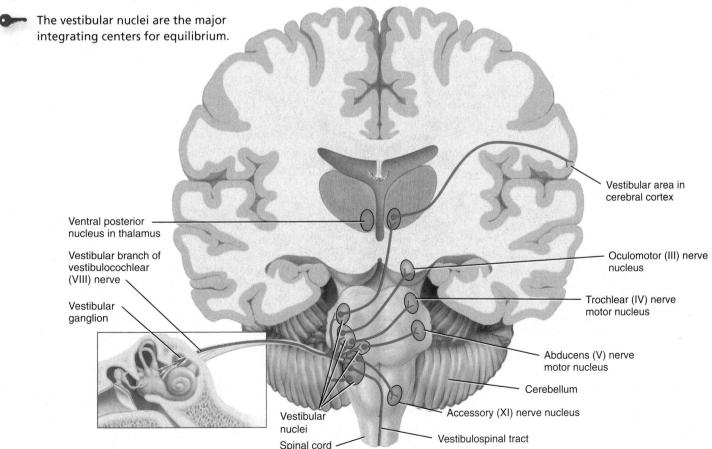

The vestibular nuclei are the major integrating centers for equilibrium.

What is the function of the vestibulospinal tract?

CLINICAL CONNECTION | Deafness

Deafness is significant or total hearing loss. **Sensorineural deafness** (sen'-sō-rē-NOO-ral) is caused by either impairment of hair cells in the cochlea or damage of the cochlear branch of the vestibulocochlear (VIII) nerve. This type of deafness may be caused by atherosclerosis, which reduces blood supply to the ears; by certain drugs such as aspirin and streptomycin; and/or by repeated exposure to loud noise, which destroys hair cells of the spiral organ. Because prolonged noise exposure causes hearing loss, employers in the United States must require workers to use hearing protectors when occupational noise levels are high. Continued exposure to high-intensity sounds can cause a significant or total hearing loss. The louder the sounds, the more rapid the hearing loss. Deafness usually begins with loss of sensitivity for high-pitched sounds. If you are listening to music through ear buds and bystanders can hear it, the noise level is in the damaging range. Most people fail to notice their progressive hearing

loss until destruction is extensive and they begin having difficulty understanding speech. Wearing earplugs while engaging in noisy activities can protect the sensitivity of your ears.

Conduction deafness is caused by impairment of the external and middle ear mechanisms for transmitting sounds to the cochlea. Causes of conduction deafness include otosclerosis, the deposition of new bone around the oval window; impacted cerumen; injury to the eardrum; and aging, which often results in thickening of the eardrum and stiffening of the joints of the auditory ossicles.

A hearing test called *Weber's test* is used to distinguish between sensorineural and conduction deafness. In the test, the stem of a vibrating fork is held to the forehead. In people with normal hearing, the sound is heard equally in both ears. If the sound is heard best in the affected ear, the deafness is probably of the conduction type; if the sound is heard best in the normal ear, it is probably of the sensorineural type. •

TABLE 21.2

Summary of Structures of the Ear

REGIONS OF THE EAR AND KEY STRUCTURES	FUNCTION
External (outer) ear 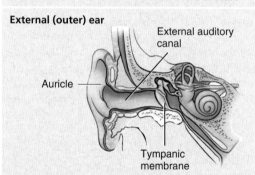	*Auricle (pinna):* Collects sound waves *External auditory canal (meatus):* Directs sound waves to the eardrum *Tympanic membrane (eardrum):* Sound waves cause it to vibrate, which in turn causes the malleus to vibrate
Middle ear 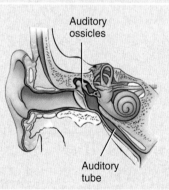	*Auditory ossicles:* Transmit and amplify vibrations from tympanic membrane to oval window *Auditory (eustachian) tube:* Equalizes air pressure on both sides of the tympanic membrane
Internal (inner) ear 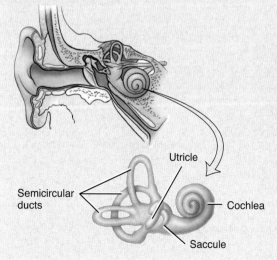	*Cochlea:* Contains a series of fluids, channels, and membranes that transmit vibrations to the spiral organ (organ of Corti), the organ of hearing; hair cells in the spiral organ produce receptor potentials, which elicit nerve impulses in the cochlear branch of the vestibulocochlear (VIII) nerve *Vestibular apparatus:* Includes semicircular ducts, utricle, and saccule, which generate nerve impulses that propagate along the vestibular branch of the vestibulocochlear (VIII) nerve *Semicircular ducts:* Contain cristae, site of hair cells for dynamic equilibrium (maintenance of body position, mainly the head, in response to rotation, acceleration, and deceleration movements) *Utricle:* Contains macula, site of hair cells for static equilibrium (maintenance of body position, mainly the head, relative to the force of gravity) *Saccule:* Contains macula, site of hair cells for static equilibrium (maintenance of body position, mainly the head, relative to the force of gravity)

21.5 DEVELOPMENT OF THE EYES AND EARS

OBJECTIVE

• Describe the development of the eyes and the ears.

Development of the Eyes

The eyes begin to develop about 22 days after fertilization when the **ectoderm** of the lateral walls of the prosencephalon (forebrain) bulges out to form a pair of shallow grooves called the **optic grooves** (Figure 21.19a). Within a few days, as the neural tube is closing, the optic grooves enlarge and grow toward the surface ectoderm and become known as the **optic** vesicles (Figure 21.19b). When the optic vesicles reach the surface ectoderm, the surface ectoderm thickens to form the **lens placodes** (PLAK-ōds). In addition, the distal portions of the optic vesicles invaginate (Figure 21.19c), forming the **optic cups**; they remain attached to the prosencephalon by narrow, hollow proximal structures called **optic stalks** (Figure 21.19d).

The lens placodes also invaginate and develop into **lens vesicles** that sit in the optic cups. The lens vesicles eventually develop into the *lenses* of the eye. Blood is supplied to the developing lenses (and retina) by the hyaloid arteries. These arteries gain access to the developing eyes through a groove on the inferior surface of the optic cup and optic stalk called the **choroid fissure**. As the lenses mature, part of the hyaloid arteries that pass through the vitreous chamber degenerate; the remaining portions of the hyaloid arteries become the *central retinal arteries.*

Figure 21.19 Development of the eyes.

The eyes begin to develop about 22 days after fertilization from ectoderm of the prosencephalon.

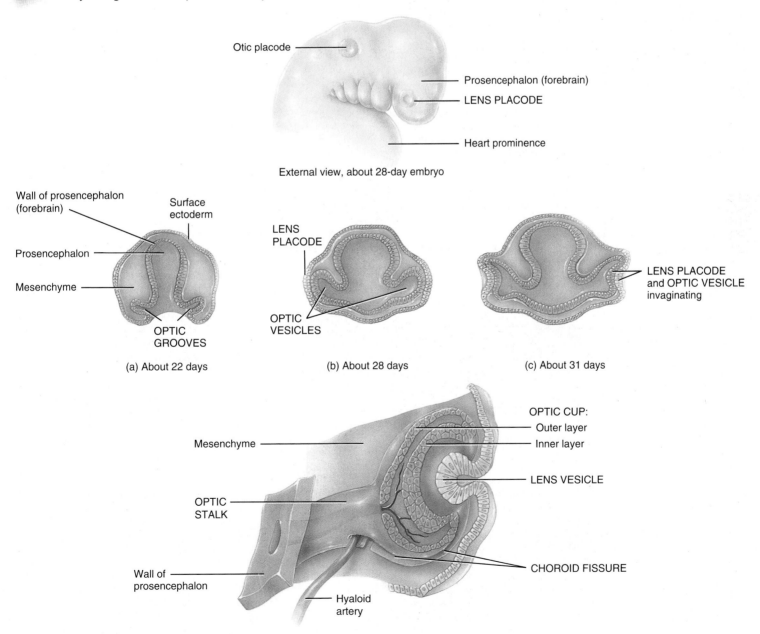

External view, about 28-day embryo

(a) About 22 days

(b) About 28 days

(c) About 31 days

(d) About 32 days

 Which structure gives rise to the neural and pigmented layers of the retina?

The inner wall of the optic cup forms the *neural layer* of the retina, while the outer layer forms the *pigmented layer* of the retina. Axons from the neural layer grow through the optic stalk to the brain, converting the optic stalk to the *optic (II) nerve*. Although myelination of the optic nerves begins late in fetal life, it is not completed until the tenth week after birth.

The anterior portion of the optic cup forms the epithelium of the *ciliary body, iris,* and *circular and radial muscles* of the iris. The connective tissue of the ciliary body, *ciliary muscle,* and *zonular fibers* of the lens develop from **mesenchyme** around the anterior portion of the optic cup.

Mesenchyme surrounding the optic cup and optic stalk differentiates into an inner layer that gives rise to the *choroid* and an outer layer that develops into the *sclera* and part of the *cornea*. The remainder of the cornea is derived from surface ectoderm.

The *anterior chamber* develops from a cavity that forms in the mesenchyme between the iris and cornea; the *posterior chamber* develops from a cavity that forms in the mesenchyme between the iris and lens.

Some mesenchyme around the developing eye enters the optic cup through the choroid fissure. This mesenchyme occupies the space between the lens and retina and differentiates into a delicate network of fibers. Later, the spaces between the fibers fill with a jellylike substance, thus forming the *vitreous body* in the vitreous chamber.

The *eyelids* form from surface ectoderm and mesenchyme. The upper and lower eyelids meet and fuse at about eight weeks of development and remain closed until about 26 weeks of development.

Development of the Ears

The first portion of the ear to develop is the *internal ear*. It begins to form about 22 days after fertilization as a thickening of the surface ectoderm, called **otic placodes** (Figure 21.20a), which appear on either side of the rhombencephalon (hindbrain). The otic placodes invaginate quickly (Figure 21.20b) to form the **otic pits** (Figure 21.20c). Next, the otic pits pinch off from the sur-

Figure 21.20 Development of the ears.

🔑 The first part of the ears to develop are the internal ears, which begin to form about 22 days after fertilization as thickenings of surface ectoderm.

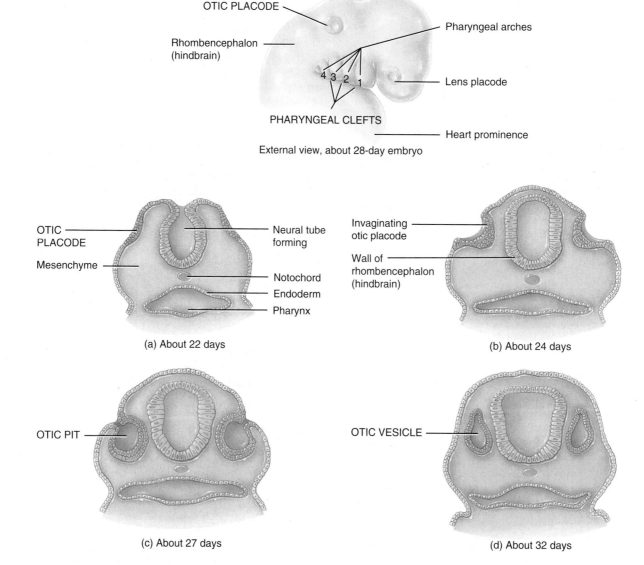

External view, about 28-day embryo

(a) About 22 days

(b) About 24 days

(c) About 27 days

(d) About 32 days

❓ **How do the three parts of the ear differ in origin?**

face ectoderm to form the **otic vesicles** within the mesenchyme of the head (Figure 21.20d). During later development, the otic vesicles will form the structures associated with the *membranous labyrinth* of the internal ear. Mesenchyme around the otic vesicles produces cartilage that later ossifies to form the bone associated with the *bony labyrinth* of the internal ear.

The *middle ear* develops from a structure called the first **pharyngeal (branchial) pouch**, an **endoderm**-lined outgrowth of the primitive pharynx (see the Figure 22.8a). The pharyngeal pouches were discussed in detail in Section 4.1. The *auditory ossicles* develop from the first and second pharyngeal arches.

The *external ear* develops from the first **pharyngeal cleft**, an endoderm-lined groove between the first and second pharyngeal arches (see the inset in Figure 21.20). The pharyngeal clefts were discussed in detail in Section 4.1.

 CHECKPOINT

16. How do the origins of the eyes and ears differ?

21.6 AGING AND THE SPECIAL SENSES

 OBJECTIVE

- Describe the age-related changes that occur in the eyes and ears.

Most people do not experience any problems with the senses of smell and taste until about age 50. This is due to a gradual loss of olfactory receptor cells and gustatory receptor cells coupled with their slower rate of replacement as we age.

Several age-related changes occur in the eyes. As noted earlier, the lens loses some of its elasticity and thus cannot change shape as easily, resulting in presbyopia. Cataracts (loss of transparency of the lenses) also occur with aging. In old age, the sclera ("white" of the eye) becomes thick and rigid and develops a yellowish or brownish coloration due to many years of exposure to ultraviolet light, wind, and dust. The sclera may also develop random splotches of pigment, especially in people with dark complexions. The iris fades or develops irregular pigment. The muscles that regulate the size of the pupil weaken with age and the pupils become smaller, react more slowly to light, and dilate more slowly in the dark. For these reasons, elderly people find that objects are not as bright, their eyes may adjust more slowly when going outdoors, and they have problems going from brightly lit to darkly lit places. Some diseases of the retina are more likely to occur in old age, including age-related macular disease and detached retina. A disorder called glaucoma develops in the eyes of aging people as a result of the buildup of aqueous humor. Tear production and the number of mucous cells in the conjunctiva may decrease with age, resulting in dry eyes. The eyelids lose their elasticity, becoming baggy and wrinkled. The amount of fat around the orbits may decrease, causing the eyeballs to sink into the orbits. Finally, as we age the sharpness of vision decreases, color and depth perception are reduced, and the number of "vitreal floaters" increases.

By about age 60, around 25 percent of individuals experience a noticeable hearing loss, especially for higher-pitched sounds. The age-associated progressive loss of hearing in both ears is called **presbycusis** (pres′-bi-KŪ-sis; *presby-*=old; *acou*=hearing; *sis*=condition). It may be related to damaged and lost hair cells in the spiral organ or degeneration of the nerve pathway for hearing. **Tinnitus** (ti-NĪ-tus; a ringing, roaring, or clicking in the ears) and vestibular imbalance also occur more frequently in the elderly.

 CHECKPOINT

17. How do the eyes change during the aging process?

KEY MEDICAL TERMS ASSOCIATED WITH SPECIAL SENSES

Ageusia (a-GOO-sē-a; *a-*=without; *geusis*=taste) Loss of the sense of taste.

Amblyopia (am′-blē-Ō-pē-a; *ambly-*=dull or dim) Term used to describe the loss of vision in an otherwise normal eye that, because of muscle imbalance, cannot focus in synchrony with the other eye. Sometimes called "wandering eyeball" or a "lazy eye."

Anosmia (an-OZ-mē-a; *a-*=without; *osmi*=smell, odor) Total lack of the sense of smell.

Barotrauma (bar′-ō-TRAW-ma; *baros-*=weight) Damage or pain, mainly affecting the middle ear, as a result of pressure changes. It occurs when pressure on the outer side of the tympanic membrane is higher than on the inner side, for example, when flying in an airplane or driving. Swallowing or holding your nose and exhaling with your mouth closed usually opens the auditory tubes, allowing air into the middle ear to equalize the pressure.

Blepharitis (blef-a-RĪ-tis; *blephar-*=eyelid; *itis*=inflammation of) An inflammation of the eyelid.

Color blindness An inherited inability to distinguish between certain colors, resulting from the absence or deficiency of one of the three types of cones. The most common type is *red–green color blindness*, in which red cones or green cones are missing. As a result, the person cannot distinguish between red and green.

Conjunctivitis (pinkeye) An inflammation of the conjunctiva; the type caused by bacteria such as pneumococci, staphylococci, or *Haemophilus influenzae*, is very contagious and more common in children. Conjunctivitis may also be caused by irritants, such as dust, smoke, or pollutants in the air, in which case it is not contagious.

Corneal abrasion (KOR-nē-al a-BRĀ-zhun) A scratch on the surface of the cornea, for example, from a speck of dirt or damaged contact lenses. Symptoms include pain, redness, watering, blurry vision, sensitivity to bright light, and frequent blinking.

Corneal transplant A procedure in which a defective cornea is removed and a donor cornea of similar diameter is sewn in. It is the most common and most successful transplant operation. Since the cornea is avascular, antibodies in the blood that might cause rejection do not enter the transplanted tissue, and rejection rarely occurs.

Diabetic retinopathy (ret-i-NOP-a-thē; *retino-*=retina; *pathos*=suffering) Degenerative disease of the retina due to diabetes mellitus, in which blood vessels in the retina are damaged or new ones grow and interfere with vision.

Exotropia (ek′-sō-TRŌ-pē-a; *ex-*=out; *tropia*=turning) Turning outward of the eyes.

Keratitis (ker′-a-TĪ-tis; *kerat-*=cornea) An inflammation or infection of the cornea.

Miosis (mī-Ō-sis) Constriction of the pupil.

Mydriasis (mī-DRĪ-a-sis) Dilation of the pupil.

Nystagmus (nis-TAG-mus; *nystagm-*=nodding or drowsy) A rapid involuntary movement of the eyeballs, possibly caused by a disease of the central nervous system. It is associated with conditions that cause vertigo.

Otalgia (ō-TAL-jē-a; *oto-*=ear; *algia*=pain) Earache.

Photophobia (fō′-tō-FŌ-bē-a; *photo-*=light; *phobia*=fear) Abnormal visual intolerance to light.

Ptosis (TŌ-sis=fall) Falling or drooping of the eyelid (or slippage of any organ below its normal position).

Retinoblastoma (ret-i-nō-blas-TŌ-ma; *oma*=tumor) A highly malignant tumor arising from immature retinal cells; it accounts for 2 percent of childhood cancers.

Scotoma (skō-TŌ-ma=darkness) An area of reduced or lost vision in the visual field.

Strabismus (stra-BIZ-mus; *strabismus*=squinting) Misalignment of the eyeballs so that the eyes do not move in unison when viewing an object; the affected eye turns either medially or laterally with respect to the normal eye and the result is double vision (*diplopia*). It may be caused by physical trauma, vascular injuries, or tumors of the extrinsic eye muscle or the oculomotor (III), trochlear (IV), or abducens (VI) nerves.

Trachoma (tra-KŌ-ma) A serious form of conjunctivitis and the greatest single cause of blindness in the world. It is caused by the bacterium *Chlamydia trachomatis*. The disease produces an excessive growth of subconjunctival tissue and invasion of blood vessels into the cornea, which progresses until the entire cornea is opaque.

Vertigo (VER-ti-gō=dizziness) A sensation of spinning or movement in which the world seems to revolve or the person seems to revolve in space, often associated with nausea and, in some cases, vomiting. It may be caused by arthritis of the neck or an infection of the vestibular apparatus.

CHAPTER REVIEW AND RESOURCE SUMMARY

WileyPLUS

Review	Resource
21.1 Olfaction: Sense of Smell	Anatomy Overview - The Special Senses
1. The olfactory receptor cells, which are bipolar neurons, are in the nasal epithelium.	Anatomy Overview - Chemoreceptors
2. Axons of olfactory receptor cells form the olfactory (I) nerves, which convey nerve impulses to the olfactory bulbs, olfactory tracts, limbic system, and cerebral cortex (temporal and frontal lobes).	Anatomy Overview - Olfactory (I) Nerve

21.2   Gustation: Sense of Taste

1. The receptors for gustation, the gustatory receptor cells, are located in taste buds.
2. Dissolved chemicals, called tastants, stimulate gustatory receptor cells.
3. Gustatory receptor cells trigger nerve impulses in the facial (VII), glossopharyngeal (IX), and vagus (X) nerves. Taste signals then pass to the medulla oblongata, thalamus, and cerebral cortex (parietal lobe).

Anatomy Overview - Chemoreceptors
Anatomy Overview - Facial (VII) Nerve
Anatomy Overview - Glossopharyngeal (IX) Nerve
Anatomy Overview - Vagus (X) Nerve

21.3   Vision

1. Accessory structures of the eyes include the eyebrows, eyelids, eyelashes, lacrimal apparatus, and extrinsic eye muscles.
2. The lacrimal apparatus consists of structures that produce and drain tears.
3. The eye is constructed of three layers: (a) fibrous tunic (sclera and cornea), (b) vascular tunic (choroid, ciliary body, and iris), and (c) retina.
4. The retina consists of a pigmented layer and a neural layer that includes a photoreceptor layer, bipolar cell layer, ganglion cell layer, horizontal cells, and amacrine cells.
5. The anterior cavity contains aqueous humor; the vitreous chamber contains the vitreous body.
6. Chemicals (neurotransmitters) released by rods and cones induce changes in bipolar cells and horizontal cells that ultimately lead to the generation of nerve impulses.
7. Impulses from ganglion cells are conveyed into the optic (II) nerve, through the optic chiasm and optic tract, to the thalamus. From the thalamus, impulses for vision propagate to the cerebral cortex (occipital lobe). Axon collaterals of retinal ganglion cells extend to the midbrain and hypothalamus.

Anatomy Overview - Oculomotor (III) Nerve
Anatomy Overview - Trochlear (IV) Nerve
Anatomy Overview - Abducens (VI) Nerve
Anatomy Overview - Optic (II) Nerve
Figure 21.4 - Anatomy of the Eyeball

21.4   Hearing and Equilibrium

1. The external (outer) ear consists of the auricle, external auditory canal, and tympanic membrane (eardrum).
2. The middle ear consists of the auditory (eustachian) tube, ossicles, oval window, and round window.
3. The internal (inner) ear consists of the bony labyrinth and membranous labyrinth. The internal ear contains the spiral organ (organ of Corti), the organ of hearing.
4. Sound waves enter the external auditory canal and strike the tympanic membrane, causing it to vibrate. The vibrations pass through the ossicles, strike the oval window, set up waves in the perilymph, strike the vestibular membrane and scala tympani, increase pressure in the endolymph, vibrate the basilar membrane, and stimulate hair bundles on the spiral organ (organ of Corti).
5. Bending of stereocilia ultimately leads to generation of nerve impulses in sensory neurons.
6. Impulses pass into the cochlear branch of the vestibulocochlear (VIII) nerve and then to the cochlear nuclei in the medulla. Most fibers decussate in the medulla, ascend in the lateral lemniscus, and terminate in the inferor colliculus in the midbrain. Some axons from the cochlear nuclei end in the superior olivary nuclei in the pons on both sides. Axons from the superior olivary nuclei ascend and end in the inferior colliculi. Nerve impulses from the inferior colliculi are passed to the medial geniculate nuclei in the thalamus and auditory areas in the temporal lobes of the cerebrum.
7. Static equilibrium is the orientation of the body relative to the pull of gravity. The maculae of the utricle and saccule are the sense organs of static equilibrium.
8. Dynamic equilibrium is the maintenance of body position in response to rotational acceleration or deceleration. The cristae in the semicircular ducts are the principal sense organs of dynamic equilibrium.
9. Impulses pass into the vestibular branch of the vestibulocochlear (VIII) nerve. Most of the axons synapse with neurons in vestibular nuclei in the medulla and pons. The remaining axons enter the cerebellum.

Anatomy Overview - Vestibulocochlear (VIII) Nerve
Figure 21.11 - The Right Middle Ear and Auditory Ossicles
Figure 21.12 - The Right Internal Ear
Figure 21.14 - Stimulation of Auditory Receptors in the Right Ear

Review

Resource

The vestibular nuclei send impulses to cranial nerve nuclei that control eye, head, and neck movements, muscle tone in skeletal muscles, and the posterior nuclei in the thalamus and vestibular areas in the cerebral cortex.

21.5 Development of the Eyes and Ears

1. The eyes begin their development about 22 days after fertilization from ectoderm of the lateral walls of the prosencephalon (forebrain).
2. The ears begin their development about 22 days after fertilization from a thickening of ectoderm on either side of the rhombencephalon (hindbrain). The sequence of development of the ear is internal ear, middle ear, and external ear.

21.6 Aging and the Special Senses

1. Most people do not experience problems with the senses of smell and taste until about age 50.
2. Among the age-related changes to the eyes are presbyopia, cataracts, difficulty adjusting to light, macular disease, glaucoma, dry eyes, and decreased sharpness of vision.
3. With age there is a progressive loss of hearing, and tinnitus and vestibular imbalance occur more frequently.

CRITICAL THINKING QUESTIONS

1. Suestia was diagnosed with a brain tumor. Early symptoms had included false sensations of smell. How/why might this have occurred?

2. Three-year-old Atticus developed sinusitis from his original rhinitis. This eventually developed into a severe pharyngitis and laryngitis. Can you explain how an otitis would finally arise? What would eventually happen if his otitis continued untreated?

3. Lucas noticed that the colored ring in one of his mother's eyes was a different shape than the other eye. His mother explained that she had gotten poked in the eye with a stick when she was a little girl about his age and that had caused the damage to her eye. What structure of the eye was injured by the stick? What effect would the damage have on her vision?

4. Reuben was on his first cruise and he felt miserable! The ship's doctor mentioned something about messages from his eyes confusing his ears (or the other way around) but Reuben was too sick to listen. He took his medication and went to bed. Explain the cause of Reuben's seasickness.

5. Why do the eyes in some flash photographs appear to be red?

? ANSWERS TO FIGURE QUESTIONS

21.1 The olfactory cilia detect odorant molecules.

21.2 The gustatory pathway is as follows: gustatory receptor cells → facial (VII), glossopharyngeal (IX), or vagus (X) nerve → medulla oblongata → either (1) limbic system and hypothalamus, or (2) thalamus → primary gustatory area in the parietal lobe of the cerebral cortex.

21.3 Lacrimal fluid, or tears, is a watery solution containing salts, some mucus, and lysozyme that protects, cleans, lubricates, and moistens the eyeball.

21.4 The fibrous tunic consists of the cornea and sclera; the vascular tunic consists of the choroid, ciliary body, and iris.

21.5 The parasympathetic division of the ANS causes pupillary constriction, and the sympathetic division causes pupillary dilation.

21.6 An ophthalmoscopic examination of the blood vessels of the eye can reveal evidence of hypertension, diabetes mellitus, cataracts, and age-related macular disease.

21.7 The two types of photoreceptors are rods and cones. Rods provide black-and-white vision in dim light; cones provide high visual acuity and color vision in bright light.

21.8 After its secretion by the ciliary process, aqueous humor flows into the posterior chamber, around the iris, into the anterior chamber, and out of the eyeball through the scleral venous sinus.

21.9 The optic tract terminates in the lateral geniculate nucleus of the thalamus.

21.10 The malleus of the middle ear is attached to the tympanic membrane, which is considered part of the external ear.

21.11 The oval and round windows separate the middle ear from the internal ear.

21.12 The two sacs that lie in the vestibule are the utricle and saccule.

21.13 The three subdivisions of the bony labyrinth are the semicircular canals, vestibule, and cochlea.

21.14 The oval window vibrates and sets up pressure waves in the perilymph.

21.15 The cell bodies of the first sensory neurons in the auditory pathway are located in the spiral ganglia.

21.16 The maculae are associated primarily with static equilibrium. They provide sensory information on the position of the head in space.

21.17 The semicircular ducts are associated with dynamic equilibrium.

21.18 The vestibulospinal tract conveys impulses down the spinal cord to maintain muscle tone in skeletal muscles and to help maintain equilibrium.

21.19 The optic cup forms the neural and pigmented layers of the optic part of the retina.

21.20 The internal ear develops from surface ectoderm, the middle ear develops from pharyngeal pouches, and the external ear develops from a pharyngeal cleft.

22 THE ENDOCRINE SYSTEM

INTRODUCTION Together, the nervous and endocrine systems coordinate functions of all body systems. As you learned in the past several chapters, the nervous system exerts its control through nerve impulses conducted along axons of neurons. At synapses, nerve impulses trigger the release of mediator (messenger) molecules called *neurotransmitters*. In contrast, the endocrine system releases regulating molecules called **hormones** (*hormon*=to excite or get moving) into interstitial fluid and then the bloodstream. The circulating blood delivers the hormones, to virtually all of the cells of the body; cells recognize a particular hormone, then respond. The science that deals with the structure and function of the endocrine glands and the diagnosis and treatment of disorders of the endocrine system is **endocrinology** (en′-dō-kri-NOL-ō-jē; *endo-*=within; *crino*=to secrete; *logy*=study of).

The nervous and endocrine systems are coordinated as an interlocking supersystem called the *neuroendocrine system*. Certain parts of the nervous system stimulate or inhibit the release of hormones, which in turn may promote or inhibit the generation of nerve impulses. The nervous system causes muscles to contract and glands to secrete either more or less of their products. The endocrine system not only helps regulate the activity of smooth muscle, cardiac muscle, and some glands, it affects virtually all other tissues as well. Hormones alter metabolism, regulate growth and development, and influence reproductive processes.

The nervous and endocrine systems respond to stimuli at different rates. Nerve impulses most often produce an effect within a few milliseconds; some hormones can act within seconds, but others can take several hours or more to cause a response. The effects of nervous system activation are generally more brief than the effects produced by the endocrine system. Finally, the nervous system produces more localized responses, in contrast to the more widespread effects of the endocrine system. In other words, the effects of nervous system stimulation are fast and direct, while those of the endocrine system are slow and indirect. Table 22.1 compares the characteristics of the nervous and endocrine systems.

In this chapter we examine the principal endocrine glands and hormone-producing tissues, along with their roles in coordinating body activities. •

Did you ever wonder why thyroid gland disorders affect all major body systems?

22.1 ENDOCRINE GLANDS DEFINED

OBJECTIVE
- Distinguish between exocrine glands and endocrine glands.

Recall from Chapter 3 that the body contains two kinds of glands: exocrine glands and endocrine glands. **Exocrine glands** (EK-sō-krin; *exo-*=outside) secrete their products (sweat, oil, mucus, and digestive juices) into ducts that carry the secretions into body cavities, into the lumen of an organ, or to the outer surface of the body. Exocrine glands include sudoriferous (sweat) glands, sebaceous (oil) glands, mucous glands, and digestive glands. **Endocrine glands** (EN-dō-krin; *endo-*=within) by contrast, secrete their products (hormones) into the interstitial fluid surrounding the secretory cells, rather than into ducts. From the interstitial fluid, hormones move into blood capillaries and are carried by the blood throughout the body. Because of their dependence on the circulatory system to distribute their products, endocrine glands are some of the most vascular tissues in the body. Very small amounts of most hormones are required in most cases to produce a response, so circulating levels typically are low. The endocrine glands include the pituitary, thyroid, parathyroid, adrenal, and pineal glands (Figure 22.1). In addition, there are several other organs and tissues that do not function exclusively as endocrine glands, but contain cells that secrete hormones. These include the hypothalamus, thymus, pancreas, ovaries, testes, kidneys, stomach, liver, small intestine, skin, heart, adipose tissue, and placenta. Taken together, all endocrine glands and hormone-secreting cells constitute the **endocrine system**.

CHECKPOINT
1. Which of the following are exclusively exocrine glands and which are exclusively endocrine glands: thymus, pancreas, placenta, pituitary, kidneys, sudoriferous glands, mucous glands, testes, ovaries?

22.2 HORMONES

OBJECTIVE
- Describe how hormones interact with target-cell receptors.

Hormones have powerful effects, even when present in very low concentrations. As a rule, most of the body's 50 or so hormones affect only a few types of cells. The reason that some cells respond to a particular hormone and others do not depends on whether the cells have hormone receptors.

Although a given hormone travels throughout the body in the blood, it affects only certain **target cells**. Hormones, like neurotransmitters, influence their target cells by chemically binding to specific protein **receptors**. Only the target cells for a given hormone have receptors (hormone receptors) that bind and recognize that hormone. For example, thyroid-stimulating hormone (TSH) binds to receptors on cells of the thyroid gland, but it does not bind to cells of the ovaries because ovarian cells do not have TSH receptors.

Receptors, like other cellular proteins, are constantly being synthesized and broken down. Generally, a target cell has 2000–100,000 receptors for a particular hormone. When a hormone is present in excess, the number of target-cell receptors may decrease (down regulation). This decreases the responsiveness of target cells to the hormone. In contrast, when a hormone (or neurotransmitter) is deficient, the number of receptors may increase to make the target tissue more sensitive (up regulation).

CHECKPOINT
2. Explain the relationship between target-cell receptors and hormones.

TABLE 22.1

Comparison of Control by the Nervous and Endocrine Systems

CHARACTERISTIC	NERVOUS SYSTEM	ENDOCRINE SYSTEM
Mediator molecules	Neurotransmitters released locally in response to nerve impulses	Hormones delivered to tissues throughout the body by the blood
Site of mediator action	Close to site of release, at a synapse; binds to receptors in postsynaptic membrane	Far from site of release (usually); binds to receptors on or in target cells
Types of target cells	Muscle (smooth, cardiac, and skeletal) cells, gland cells, other neurons	Cells throughout the body
Time to onset of action	Typically within milliseconds (thousandths of a second)	Seconds to hours or days
Duration of action	Generally short term (milliseconds)	Generally longer (seconds to days)

705

Figure 22.1 Location of many endocrine glands. Also shown are other organs that contain endocrine tissue, and associated structures.

🔑 Endocrine glands secrete hormones, which circulating blood delivers to target tissues.

> **FUNCTIONS OF HORMONES**
>
> 1. **Help regulate:**
>   - Chemical composition and volume of internal environment (interstitial fluid)
>   - Metabolism and energy balance
>   - Contraction of smooth and cardiac muscle fibers
>   - Glandular secretions
>   - Some immune system activities
> 2. **Control growth and development.**
> 3. **Regulate operation of reproductive systems.**
> 4. **Help establish circadian rhythms.**

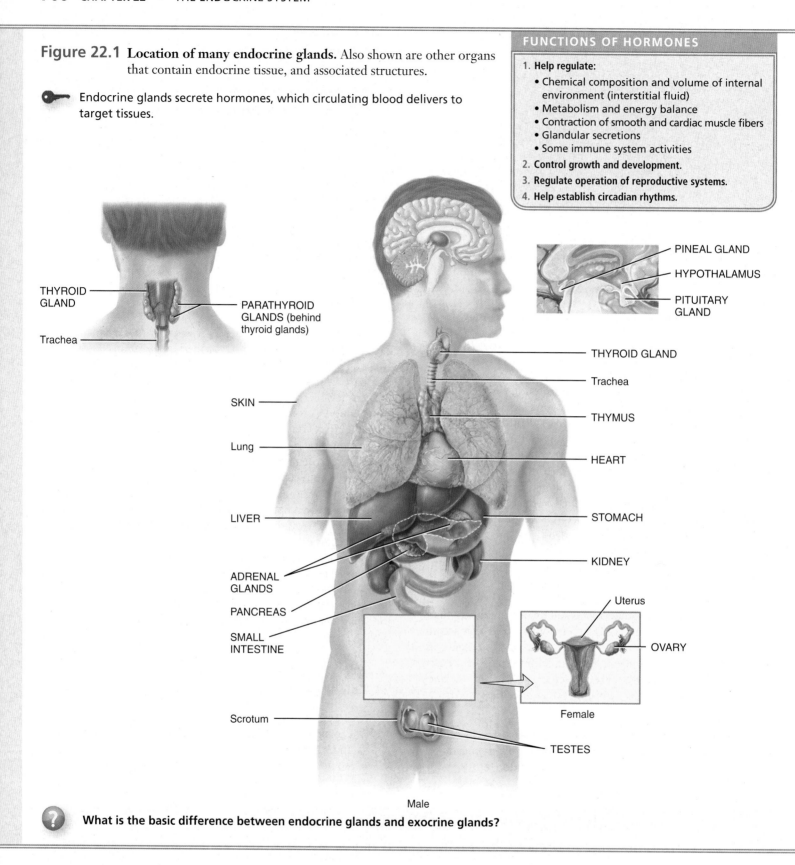

THYROID GLAND

Trachea

PARATHYROID GLANDS (behind thyroid glands)

PINEAL GLAND

HYPOTHALAMUS

PITUITARY GLAND

THYROID GLAND

Trachea

SKIN

THYMUS

Lung

HEART

LIVER

STOMACH

KIDNEY

ADRENAL GLANDS

PANCREAS

Uterus

SMALL INTESTINE

OVARY

Scrotum

Female

TESTES

Male

❓ **What is the basic difference between endocrine glands and exocrine glands?**

CLINICAL CONNECTION | *Blocking Hormone Receptors*

Synthetic hormones that **block the receptors** for certain naturally occurring hormones are available as drugs. For example, RU486 (mifepristone), which is used to induce abortion, binds to the receptors for progesterone (a female sex hormone) and prevents progesterone from exerting its normal effects. When RU486 is given to a pregnant woman, the uterine conditions needed for nurturing an embryo are not maintained, embryonic development stops, and the embryo is sloughed off along with the uterine lining. RU486 is often used in conjunction with one or two other drugs, methotrexate and misoprostol. This example illustrates an important endocrine principle: If a hormone is prevented from interacting with its receptors, the hormone cannot perform its normal functions. •

22.3 HYPOTHALAMUS AND PITUITARY GLAND

● OBJECTIVES

- Explain why the hypothalamus is classified as an endocrine gland.
- Describe the location, histology, hormones, and functions of the anterior pituitary.
- Describe the location, histology, hormones, and functions of the posterior pituitary.

For many years the **pituitary gland** (pi-TOO-i-tar-ē) or **hypophysis** (hī-POF-i-sis) was called the "master" endocrine gland because it secretes several hormones that control other endocrine glands. We now know that the pituitary gland itself has a master—the **hypothalamus** (Figure 22.2). This small region of the brain, inferior to the thalamus, is the major integrating link between the nervous and endocrine systems. It receives input from several other regions of the brain, including the limbic system, cerebral cortex, thalamus, and reticular activating system. It also receives sensory signals from internal organs and from the retina.

Painful, stressful, and emotional experiences all cause changes in hypothalamic activity. In turn, the hypothalamus controls the autonomic nervous system, regulating body temperature, thirst, hunger, sexual behavior, and defensive reactions such as fear and rage. Thus, not only is the hypothalamus an important regulatory center in the nervous system, it is also a crucial endocrine gland. Hormones secreted by the hypothalamus (described shortly) and the pituitary

Figure 22.2 Hypothalamus and pituitary gland, and their blood supply. As shown in part (b), releasing and inhibiting hormones synthesized by hypothalamic neurons are transported within axons and released from the axon terminals. The hormones diffuse into capillaries of the primary plexus of the hypophyseal portal system and are carried by the hypophyseal portal veins to the secondary plexus of the hypophyseal portal system for distribution to target cells in the anterior pituitary.

🔑 Hypothalamic hormones are an important link between the nervous and endocrine systems.

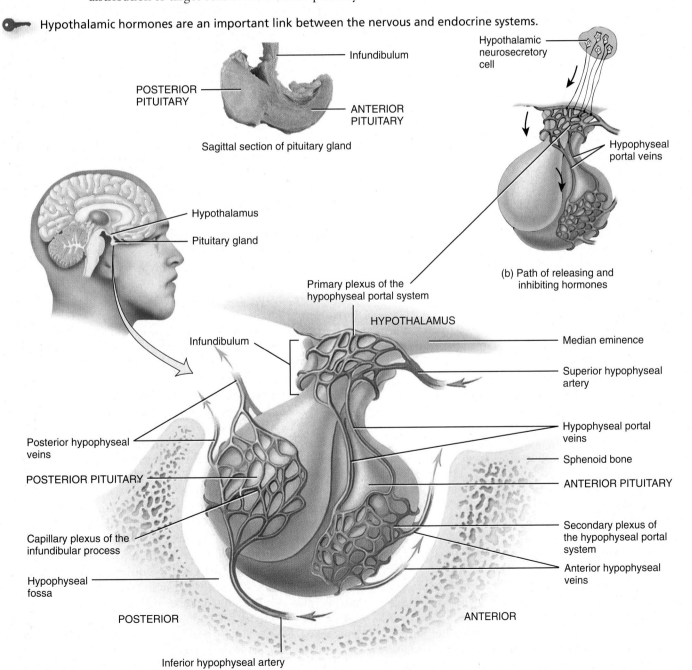

Sagittal section of pituitary gland

(b) Path of releasing and inhibiting hormones

(a) Relationship of the hypothalamus to the pituitary gland

FIGURE 22.2 CONTINUES ▶

⊙ **FIGURE 22.2 CONTINUED** ▶

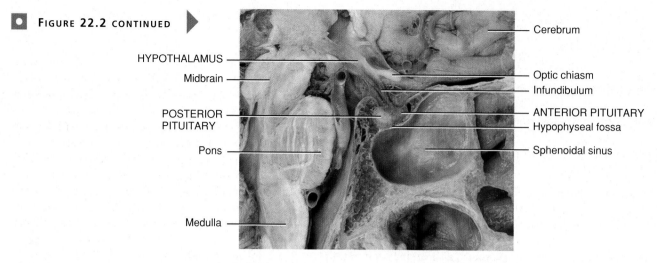

(c) Sagittal section of hypothalamus and pituitary gland of adult

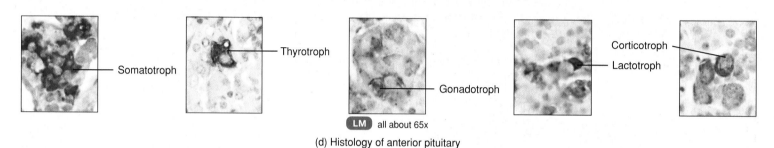

LM all about 65x

(d) Histology of anterior pituitary

❓ **What is the functional importance of the hypophyseal portal veins?**

gland play important roles in the regulation of virtually all aspects of growth, development, metabolism, and homeostasis.

The pituitary gland is a pea-shaped structure measuring about 1–1.5 cm (0.5 in.) in diameter that lies in the hypophyseal fossa of the sphenoid bone and attaches to the hypothalamus by a stalk, the **infundibulum** (in′-fun-DIB-ū-lum=a funnel; Figure 22.2a). The pituitary gland has two anatomically and functionally separate portions, the anterior pituitary and posterior pituitary. The *anterior pituitary (anterior lobe)* accounts for about 75 percent of the total weight of the gland and is composed of epithelial tissue. It develops from an outgrowth of ectoderm called the *hypophyseal (Rathke's) pouch* in the roof of the mouth (see Figure 22.8b). The anterior pituitary consists of two parts in an adult: The **pars distalis** is the larger bulbar portion, and the **pars tuberalis** (PARS too-be-RAL-is) forms a sheath around the infundibulum. The *posterior pituitary (posterior lobe)* also develops from an ectodermal outgrowth; this one, called the *neurohypophyseal bud*, is an outgrowth of the neural tube (see Figure 22.8b). The posterior pituitary is composed of neural tissue and also consists of two parts: the **pars nervosa** (ner-VŌ-sa), the larger bulbar portion, and the infundibulum. The posterior pituitary contains axons and axon terminals of more than 10,000 neurosecretory cells whose cell bodies are located in the supraoptic and paraventricular nuclei of the hypothalamus (see Figure 18.11). The axon terminals in the posterior pituitary gland are associated with specialized neuroglia called **pituicytes** (pi-TOO-i-sītz). These cells have a supporting role similar to that of astrocytes (see Chapter 16).

A third region, called the **pars intermedia**, atrophies during fetal development and ceases to exist as a separate lobe in adults (see Figure 22.8b). However, some of its cells migrate into adjacent parts of the anterior pituitary, where they persist.

Anterior Pituitary

The **anterior pituitary**, or **adenohypophysis** (ad′-ē-nō-hī-POF-i-sis; *adeno-*=gland; *hypophysis*=undergrowth), secretes hormones that regulate a wide range of bodily activities, from growth to reproduction. Release of anterior pituitary hormones is stimulated by **releasing hormones** and suppressed by **inhibiting hormones** from the hypothalamus (Figure 22.2b).

Hypothalamic hormones that release or inhibit anterior pituitary hormones reach the anterior pituitary through a **portal system**. As you learned in Chapter 14, blood usually passes from the heart through an artery to a capillary to a vein and back to the heart. In a portal system, blood flows from one capillary network into a portal vein, and then into a second capillary network before returning to the heart. The name of the portal system gives the location of the *second* capillary network. In the **hypophyseal portal system** (hī′-pō-FIZ-ē-al), also referred to as the hypothalamic-hypophyseal portal system, blood flows from capillaries in the hypothalamus into portal veins that carry blood to capillaries of the anterior pituitary.

The **superior hypophyseal arteries**, branches of the internal carotid arteries, bring blood into the hypothalamus (Figure 22.2a). At the juncture of the median eminence of the hypothalamus and the infundibulum, these arteries divide into a capillary network called the **primary plexus of the hypophyseal portal system**. From the primary plexus, blood drains into the **hypophyseal portal veins** that pass down the surface of the infundibulum. In the anterior pituitary, the hypophyseal portal veins divide again and form another capillary network called the **secondary plexus of the hypophyseal portal system**.

Above the optic chiasm are clusters of specialized neurons, called **neurosecretory cells**. They synthesize the hypothalamic releasing and inhibiting hormones in their cell bodies and package the hormones inside vesicles, which reach the axon terminals by axonal transport. When nerve impulses reach the axon terminals, they stimulate the vesicles to undergo exocytosis. The hormones then diffuse into the primary plexus of the hypophyseal portal system.

Quickly, the hypothalamic hormones flow with the blood through the portal veins and into the secondary plexus. This direct route permits hypothalamic hormones to act immediately on anterior pituitary cells, before the hormones are diluted or destroyed in the general circulation. Hormones secreted by anterior pituitary cells pass into the secondary plexus capillaries, which drain into the anterior hypophyseal veins. Now in the general circulation, anterior pituitary hormones travel to target tissues throughout the body.

The following list describes the major hormones secreted by five types of anterior pituitary cells (Figure 22.2c):

1. **Human growth hormone (hGH)**, or **somatotropin** (sō′-ma-tō-TRŌ-pin; *somato-*=body; *tropin*=change), is secreted by cells called **somatotrophs** (sō-MAT-ō-trofs). Human growth hormone in turn stimulates several tissues to secrete **insulinlike growth factors (IGFs)**, hormones that stimulate general body growth and regulate various aspects of metabolism.
2. **Thyroid-stimulating hormone (TSH)**, or **thyrotropin** (thī-rō-TRŌ-pin; *thyro-*=shield), which controls the secretions and other activities of the thyroid gland, is secreted by cells called **thyrotrophs** (THĪ-rō-trofs).
3. **Follicle-stimulating hormone (FSH)** and **luteinizing hormone (LH)** (LOO-tē-in′-īz-ing) are secreted by cells called **gonadotrophs** (gō-NAD-ō-trofs; *gonado-*=seed). FSH and LH both act on the gonads: They stimulate secretion of estrogens and progesterone and the maturation of oocytes in the ovaries, and they stimulate secretion of testosterone and sperm production in the testes.

4. **Prolactin (PRL)**, which initiates milk production in the mammary glands, is released by cells called **lactotrophs** (LAK-tō-trofs; *lacto-*=milk).
5. **Adrenocorticotropic hormone (ACTH)** (a-drē-nō-kor-ti-kō-TRŌ-pik), or **corticotropin** (*cortico-*=rind or bark), which stimulates the adrenal cortex to secrete glucocorticoids, is synthesized by cells called **corticotrophs** (KOR-ti-kō-trofs). Some corticotrophs also secrete **melanocyte-stimulating hormone (MSH)**.

The five different types of cells of the anterior pituitary (somatotrophs, thyrotrophs, gonadotrophs, lactotrophs, and corticotrophs) may be grouped according to their reaction to chemical staining (Figure 22.2d):

- **Basophils** (BĀ-sō-fils) (thyrotrophs, gonadotrophs, and corticotrophs) make up about 10 percent of anterior pituitary cells, stain blue with basic dyes, and contain secretory granules. These cells should not be confused with white blood cells of the same name.
- **Acidophils** (a-SID-ō-fils) (somatotrophs and lactotrophs) make up about 40 percent of anterior pituitary cells, stain red with acidic dyes, and also contain secretory granules.
- **Chromophobes** (KRŌ-mō-fōbs) make up about 50 percent of anterior pituitary cells and have little affinity for basic or acidic dyes. They have few or no secretory granules and do not secrete hormones. They are believed to be basophils or acidophils that have already released their granules.

Four anterior pituitary hormones influence the activity of another endocrine gland and are called **tropic hormones** (TRŌ-pik), or **tropins**. These include TSH, ACTH, FSH, and LH. The two **gonadotropins** (gō-nad-ō-TRŌ-pins), FSH and LH, are tropic hormones that specifically regulate the functions of the gonads (ovaries and testes).

The principal actions of the hormones of the anterior pituitary gland are summarized in Table 22.2.

TABLE 22.2

Summary of the Principal Actions of Anterior Pituitary Hormones

HORMONE AND TARGET TISSUES	PRINCIPAL ACTIONS	HORMONE AND TARGET TISSUES	PRINCIPAL ACTIONS
Human growth hormone (hGH) or somatotropin Liver (and other tissues)	Stimulates liver, muscle, cartilage, bone, and other tissues to synthesize and secrete insulinlike growth factors (IGFs); IGFs promote growth of body cells, protein synthesis, tissue repair, lipolysis, and elevation of blood glucose concentration	**Luteinizing hormone (LH)** Ovaries Testes	In females, stimulates secretion of estrogens and progesterone, ovulation, and formation of corpus luteum; in males, stimulates testes to produce testosterone
Thyroid-stimulating hormone (TSH) or thyrotropin Thyroid gland	Stimulates the synthesis and secretion of thyroid hormones by the thyroid gland	**Prolactin (PRL)** Mammary glands	Together with other hormones, promotes milk secretion by the mammary glands
Follicle-stimulating hormone (FSH) Ovaries Testes	In females, initiates development of oocytes and induces ovarian secretion of estrogens; in males, stimulates testes to produce sperm	**Adrenocorticotropic hormone (ACTH) or corticotropin** Adrenal cortex	Stimulates secretion of glucocorticoids (mainly cortisol) by the adrenal cortex
		Melanocyte-stimulating hormone (MSH) Brain	Exact role in humans is unknown but may influence brain activity; when present in excess, can cause darkening of skin

Disorders of the endocrine system often involve either **hyposecretion** (*hypo-*=too little or under), inadequate release of a hormone, or **hypersecretion** (*hyper-*=too much or above), excessive release of a hormone. Several disorders of the anterior pituitary involve human growth hormone (hGH). Hyposecretion of hGH during the growth years results in **pituitary dwarfism** (see Section 6.11). Hypersecretion during the growth years causes **giantism**, an abnormal increase in the length of long bones. The person grows to be very tall, but body proportions are about normal. Hypersecretion of hGH during adulthood is called **acromegaly** (ak'-rō-MEG-a-lē). Although hGH cannot produce further lengthening of the long bones because the epiphyseal plates are already closed, the bones of the hands, feet, cheeks, and jaws thicken and other tissues enlarge. In addition, the eyelids, lips, tongue, and nose enlarge, and the skin thickens and develops furrows, especially on the forehead and soles. •

Posterior Pituitary

Although the **posterior pituitary**, or **neurohypophysis** (noo'-rō-hī-POF-i-sis; *neuro-*=nerve; *-hypophysis*=undergrowth), does not *synthesize* hormones, it does *store* and *release* two hormones. As noted earlier, it consists of pituicytes and axon terminals of hypothalamic neurosecretory cells. The cell bodies of the neurosecretory cells are in the **paraventricular** and **supraoptic nuclei** of the hypothalamus; their axons form the **hypothalamohypophyseal tract** (hī'-pō-thal'-a-mō-hī-pa-FIZ-ē-al), which begins

in the hypothalamus and ends near blood capillaries in the posterior pituitary gland (Figure 22.3). Different neurosecretory cells produce two hormones: **oxytocin (OT)** (ok'-sē-TŌ-sin; *oxytoc-*=quick birth) and **antidiuretic hormone (ADH)** (an-tī-dī-ū-RET-ik; *anti-*=against; *diuretic*=increases urine production), also called **vasopressin** (vā-sō-PRES-in; *vaso-*=blood vessel; *pressus*=to press).

After their production in the cell bodies of neurosecretory cells, oxytocin and antidiuretic hormone are packed into vesicles and transported to the axon terminals in the posterior pituitary gland. Nerve impulses that propagate along the axon and reach the axon terminals trigger exocytosis of these secretory vesicles.

During and after delivery of a baby, oxytocin has two target tissues: the mother's uterus and breasts. During delivery, stretching of the cervix of the uterus stimulates the release of oxytocin, which in turn enhances contraction of smooth muscle cells in the wall of the uterus; after delivery, it stimulates milk ejection ("letdown") from the mammary glands in response to the mechanical stimulus provided by a suckling infant (a neuroendocrine reflex). The function of oxytocin in males and in nonpregnant females is unclear. Experiments with animals have suggested that its actions within the brain foster parental caretaking behavior toward young offspring. It may also be responsible, in part, for the feelings of sexual pleasure during and after intercourse.

Years before oxytocin was discovered, midwives commonly let a firstborn twin nurse at the mother's breast to speed the birth of the second child. Now we know why this practice is helpful—it stimulates release of oxytocin. Even after a single birth, nursing promotes expulsion of the placenta (afterbirth) and helps the

Figure 22.3 The hypothalamohypophyseal tract. Hormone molecules synthesized in the neurosecretory cell body are packaged into secretory vesicles that move down to the axon terminals, where nerve impulses trigger exocytosis of the vesicles and release of the hormone.

Oxytocin and antidiuretic hormone are synthesized in the hypothalamus and released into capillaries of the posterior pituitary.

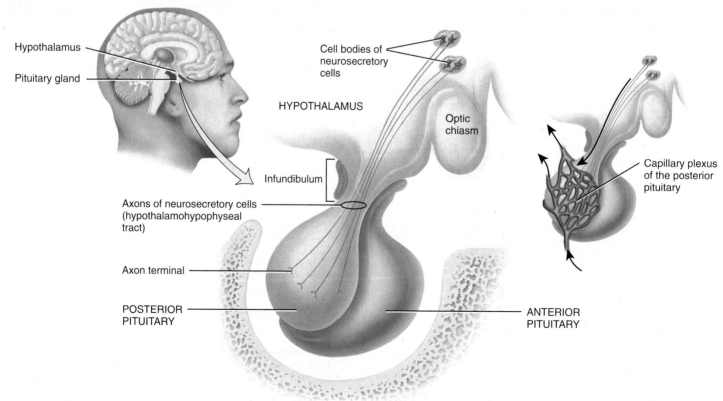

Functionally, how are the hypothalamohypophyseal tract and the hypophyseal portal veins similar? Structurally, how are they different?

TABLE 22.3

Summary of Posterior Pituitary Hormones

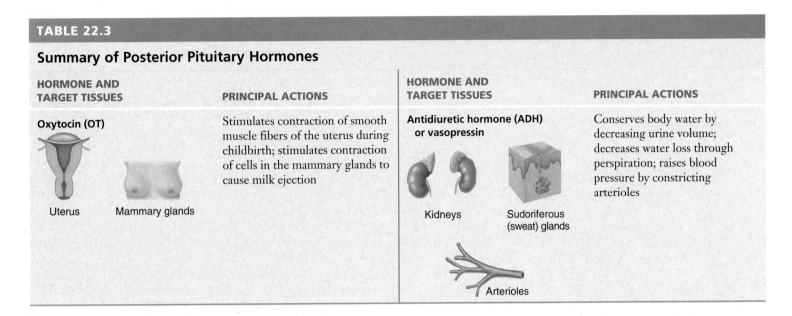

HORMONE AND TARGET TISSUES	PRINCIPAL ACTIONS	HORMONE AND TARGET TISSUES	PRINCIPAL ACTIONS
Oxytocin (OT) Uterus Mammary glands	Stimulates contraction of smooth muscle fibers of the uterus during childbirth; stimulates contraction of cells in the mammary glands to cause milk ejection	**Antidiuretic hormone (ADH) or vasopressin** Kidneys Sudoriferous (sweat) glands Arterioles	Conserves body water by decreasing urine volume; decreases water loss through perspiration; raises blood pressure by constricting arterioles

uterus regain its smaller size. Synthetic oxytocin (Pitocin®) often is given to induce labor or to increase uterine tone and control hemorrhage just after giving birth.

An **antidiuretic** is a substance that decreases urine production. ADH causes the kidneys to return more water to the blood, thus decreasing urine volume. In the absence of ADH, urine output increases more than tenfold, from the normal 1–2 liters to about 20 liters a day. Drinking alcohol often causes frequent and copious urination because the alcohol inhibits secretion of ADH. ADH also decreases the water lost through sweating and causes constriction of arterioles, which increases blood pressure. The other name for this hormone, *vasopressin*, reflects its effect on blood pressure.

Blood is supplied to the posterior pituitary by the **inferior hypophyseal arteries** (see Figure 22.2a), which branch from the internal carotid arteries. In the posterior pituitary, the inferior hypophyseal arteries drain into the **capillary plexus of the infundibular process**, a capillary network that receives oxytocin and antidiuretic hormone secreted from the neurosecretory cells of the hypothalamus (see Figure 22.2a). From this plexus, hormones pass into the **posterior hypophyseal veins** for distribution to target cells in other tissues. Table 22.3 lists the posterior pituitary hormones and their principal actions.

 CHECKPOINT

3. In what respect is the pituitary gland actually two glands?
4. What are the functions of the anterior pituitary hormones?
5. Describe the structure and importance of the hypothalamohypophyseal tract.
6. List the posterior pituitary hormones and their functions.

22.4 PINEAL GLAND AND THYMUS

 OBJECTIVES

• Describe the location, histology, hormones, and functions of the pineal gland.
• Describe the role of the thymus in immunity.

The **pineal gland** (PIN-ē-al=pinecone shape) is a small endocrine gland attached to the roof of the third ventricle of the brain at the midline (see Figure 22.1). Part of the epithalamus, it is positioned between the two superior colliculi and weighs 0.1–0.2 g. The pineal gland consists of masses of neuroglia and secretory cells called

pinealocytes (pin-ē-AL-ō-sīts), which are covered by a capsule formed by the pia mater. Sympathetic postganglionic fibers from the superior cervical ganglion terminate in the pineal gland.

Although many anatomical features of the pineal gland have been known for years, its physiological role is still unclear. One hormone secreted by the pineal gland is **melatonin**. Melatonin contributes to the setting of the body's biological clock, which is controlled from the suprachiasmatic nucleus of the hypothalamus. During sleep, levels of melatonin in the bloodstream increase tenfold and then decline to a low level again before awakening. Small doses of melatonin given orally can induce sleep and reset daily or circadian rhythms, which might benefit workers whose shifts alternate between daylight and nighttime hours. Melatonin also is a potent antioxidant that may provide some protection against damaging oxygen free radicals. In animals with specific breeding seasons, melatonin inhibits reproductive functions outside the breeding season. What effect, if any, melatonin exerts on human reproductive function is still unclear.

The posterior cerebral artery supplies the pineal gland with blood, and the great cerebral vein drains it.

Seasonal affective disorder (SAD) is a type of depression that afflicts some people during the winter months, when day length is short. It is thought to be due, in part, to overproduction of melatonin. Full-spectrum bright-light therapy—repeated doses of several hours of exposure to artificial light that is as bright as sunlight—provides relief for some people. Three to six hours of exposure to bright light also appears to speed recovery from jet lag, the fatigue suffered by travelers who quickly cross several time zones.

Because of its role in immunity, the details of the structure and functions of the **thymus** are discussed in Chapter 15, which examines the lymphatic system and immunity. In this chapter, only its hormonal role in immunity will be discussed. As you learned in Chapter 12, lymphocytes are one type of white blood cell. There are two types of lymphocytes, called T cells and B cells, based on their specific roles in immunity. Hormones produced by the thymus (thymic hormones), called **thymosin, thymic humoral factor (THF), thymic factor (TF),** and **thymopoietin** (thī-mō-poy-Ē-tin), promote the proliferation and maturation of T cells, which destroy foreign substances and microbes. There is also some evidence that thymic hormones may retard the aging process.

 CHECKPOINT

7. What is the relationship between melatonin and sleep?
8. Which thymic hormones play a role in immunity?

22.5 THYROID GLAND AND PARATHYROID GLANDS

 OBJECTIVES

• Describe the location, histology, hormones, and functions of the thyroid gland.
• Describe the location, histology, hormones, and functions of the parathyroid glands.

The butterfly-shaped **thyroid gland** is located just inferior to the larynx (voice box); the right and left **lateral lobes** lie on either side of the trachea (Figure 22.4a, c, d). Connecting the lobes is a mass of tissue called an **isthmus** (IS-mus) that lies anterior to the trachea. About 50 percent of thyroid glands have a small third lobe called the *pyramidal lobe*. It extends superiorly from the isthmus. The gland usually weighs about 30 g (1 oz) and has a rich blood supply, receiving 80–120 mL of blood per minute.

Microscopic spherical sacs called **thyroid follicles** (Figure 22.4b) make up most of the thyroid gland. The wall of each follicle consists primarily of cells called **follicular cells** (fō-LIK-ū-lar), most of which extend to the lumen (internal space) of the follicle. When the follicular cells are inactive, their shape is low cuboidal to squamous, but under the influence of TSH they become cuboidal or low columnar and actively secrete hormones. The follicular cells produce two hormones: **thyroxine** (thī-ROK-sen), which is also called **tetraiodothyronine (T₄)** (tet-ra-ī-ō-dō-THĪ-rō-nēn), because it contains four atoms of iodine, and **triiodothyronine (T₃)** (trī-ī′-ō-dō-THĪ-rō-nēn), which

contains three atoms of iodine. T₃ and T₄ are also collectively referred to as **thyroid hormones**. The thyroid hormones regulate (1) oxygen use and basal metabolic rate, (2) cellular metabolism, and (3) growth and development. A few cells called **parafollicular cells** (par′-a-fō-LIK-ū-lar), or **C cells**, may be embedded within a follicle or lie between follicles. They produce the hormone **calcitonin** (kal-si-TŌ-nin), which helps regulate calcium homeostasis.

Miacalcin®, a calcitonin extract from salmon, is an effective treatment for osteoporosis, a disorder in which the pace of bone breakdown exceeds the pace of bone rebuilding. Miacalcin inhibits breakdown of bone and accelerates uptake of calcium and phosphates into bone.

The main blood supply of the thyroid gland is from the superior thyroid artery (a branch of the external carotid artery) and the inferior thyroid artery (a branch of the thyrocervical trunk from the subclavian artery). The thyroid gland is drained by the superior and middle thyroid veins, which pass into the internal jugular veins, and the inferior thyroid veins, which join either the brachiocephalic veins or the internal jugular veins (Figure 22.4a).

The nerve supply of the thyroid gland consists of postganglionic fibers from the superior and middle cervical sympathetic ganglia. Preganglionic fibers to these ganglia are derived from the second through seventh thoracic segments of the spinal cord.

A summary of thyroid gland hormones and their actions is presented in Table 22.4.

Partially embedded in the posterior surface of the lateral lobes of the thyroid gland are several small, round masses of tissue

TABLE 22.4

Summary of Thyroid Gland and Parathyroid Gland Hormones

HORMONE AND SOURCE	PRINCIPAL ACTIONS
THYROID GLAND	
T₃ (triiodothyronine) and T₄ (thyroxine), called thyroid hormones, from follicular cells 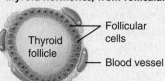	Increase basal metabolic rate, stimulate synthesis of proteins, increase use of glucose and fatty acids for ATP production, increase lipolysis, enhance cholesterol excretion, accelerate body growth, and contribute to development of the nervous system
Calcitonin (CT) from parafollicular cells 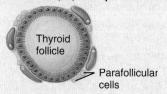	Lowers blood levels of Ca^{2+} and HPO_4^{2-} by inhibiting bone resorption by osteoclasts and by accelerating uptake of calcium and phosphates into bone extracellular matrix
PARATHYROID GLAND HORMONE	
Parathyroid hormone (PTH) from chief cells 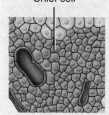	Increases blood Ca^{2+} and Mg^{2+} levels and decreases blood HPO_4^{2-} level; increases bone resorption by osteoclasts; increases Ca^{2+} reabsorption and HPO_4^{2-} excretion by kidneys; and promotes formation of calcitriol (active form of vitamin D)

Figure 22.4 The thyroid gland.

Thyroid hormones regulate (1) oxygen use and basal metabolic rate, (2) cellular metabolism, and (3) growth and development.

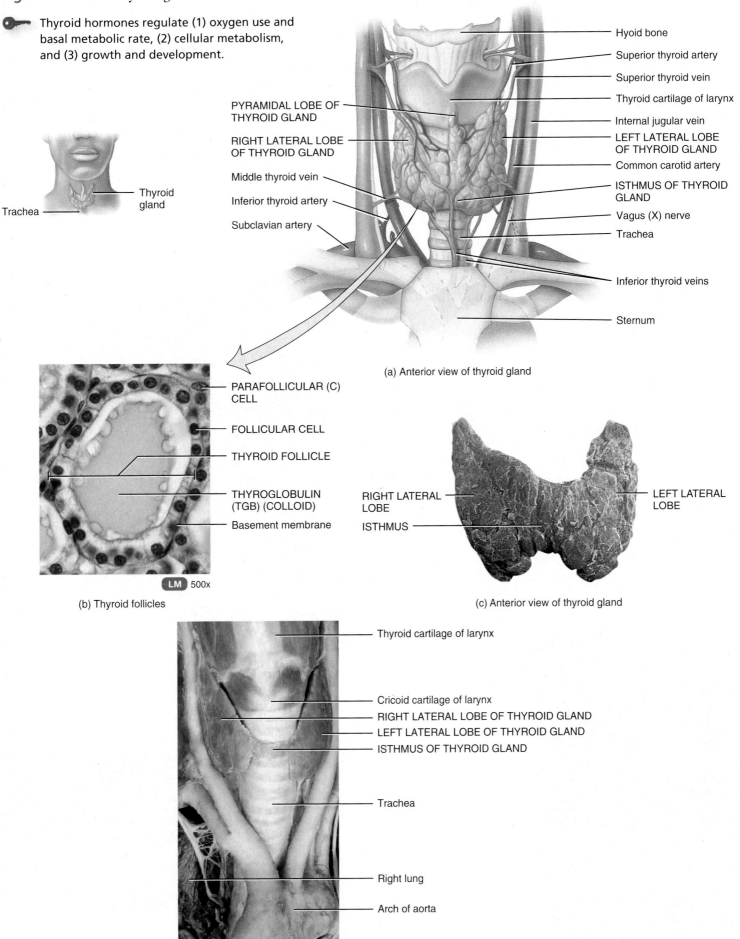

Thyroid gland

Trachea

Hyoid bone

Superior thyroid artery

Superior thyroid vein

Thyroid cartilage of larynx

Internal jugular vein

PYRAMIDAL LOBE OF THYROID GLAND

RIGHT LATERAL LOBE OF THYROID GLAND

LEFT LATERAL LOBE OF THYROID GLAND

Common carotid artery

Middle thyroid vein

Inferior thyroid artery

Subclavian artery

ISTHMUS OF THYROID GLAND

Vagus (X) nerve

Trachea

Inferior thyroid veins

Sternum

(a) Anterior view of thyroid gland

PARAFOLLICULAR (C) CELL

FOLLICULAR CELL

THYROID FOLLICLE

THYROGLOBULIN (TGB) (COLLOID)

Basement membrane

LM 500x

(b) Thyroid follicles

RIGHT LATERAL LOBE

ISTHMUS

LEFT LATERAL LOBE

(c) Anterior view of thyroid gland

Thyroid cartilage of larynx

Cricoid cartilage of larynx

RIGHT LATERAL LOBE OF THYROID GLAND

LEFT LATERAL LOBE OF THYROID GLAND

ISTHMUS OF THYROID GLAND

Trachea

Right lung

Arch of aorta

(d) Anterior view

Which cells secrete T_3 and T_4? Which secrete calcitonin? Which of these hormones are also called thyroid hormones?

called the **parathyroid glands** (*para-*=beside). Each parathyroid gland has a mass of about 40 mg (0.04 g). Usually, one superior and one inferior parathyroid gland are attached to each lateral thyroid lobe (Figure 22.5a, d).

Microscopically, the parathyroid glands contain two kinds of epithelial cells (Figure 22.5b). The more numerous cells are called **chief (principal) cells** and produce **parathyroid hormone (PTH)**, or **parathormone**. The function of the other type of parathyroid epithelial cell, called an **oxyphil cell**, is unknown in a normal parathyroid gland. However, its presence clearly helps to identify the parathyroid gland histologically due to its unique staining characteristics. Furthermore, in a cancer of the parathyroid glands, oxyphil cells secrete excess PTH.

PTH decreases blood HPO_4^{2-} level and increases blood Ca^{2+} and Mg^{2+} levels. With respect to blood Ca^{2+} level, PTH and calcitonin are *antagonists*; that is, they have opposite actions. A third effect of PTH on the kidneys is to promote formation of the hormone **calcitriol** (kal'-si-TRĪ-ol), the active form of vitamin D.

The parathyroid glands are abundantly supplied with blood from branches of the superior and inferior thyroid arteries. Blood is drained by the superior, middle, and inferior thyroid veins. The nerve supply of the parathyroid glands is derived from the thyroid branches of cervical sympathetic ganglia.

A summary of parathyroid hormone and its actions is presented in Table 22.4.

Figure 22.5 The parathyroid glands.

The parathyroid glands, normally four in number, are embedded in the posterior surface of the thyroid gland.

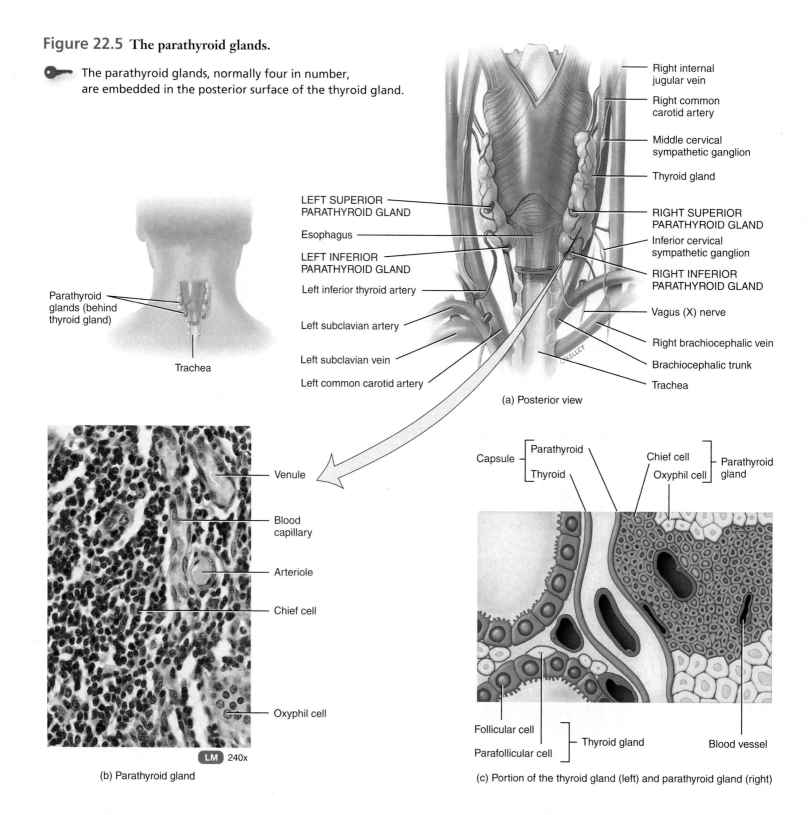

(a) Posterior view

(b) Parathyroid gland

(c) Portion of the thyroid gland (left) and parathyroid gland (right)

CLINICAL CONNECTION | *Thyroid Gland Disorders and Parathyroid Gland Disorders*

Thyroid gland disorders affect all major body systems and are among the most common endocrine disorders. **Congenital hypothyroidism**, hyposecretion of thyroid hormones that is present at birth, causes severe mental retardation and stunted bone growth. At birth, the baby typically is normal because lipid-soluble maternal thyroid hormones crossed the placenta during pregnancy and allowed normal development. If congenital hypothyroidism exists, oral thyroid hormone treatment must be started soon after birth and continued for life.

Hypothyroidism during the adult years produces **myxedema** (mix-e-DĒ-ma), which occurs about five times more often in females than in males. A hallmark of this disorder is *edema* (accumulation of interstitial fluid) that causes the facial tissues to swell and look puffy. A person with myxedema has a slow heart rate, low body temperature, sensitivity to cold, dry hair and skin, muscular weakness, general lethargy, and a tendency to gain weight easily. Because the brain has already reached maturity, mental retardation does not occur, but the person may be less alert. Oral thyroid hormones reduce the symptoms.

The most common form of hyperthyroidism is **Graves disease**, which also occurs seven to ten times more often in females than in males, usually before age 40. Graves disease is an autoimmune disorder in which the person produces antibodies that mimic the action of thyroid-stimulating hormone (TSH). The antibodies continually stimulate the thyroid gland to grow and produce thyroid hormones. A primary sign is an enlarged thyroid, which may be two to three times the normal size. Graves patients often have a peculiar

edema behind the eyes, called **exophthalmos** (ek'-sof-THAL-mos), which causes the eyes to protrude. Treatment may include surgical removal of part or all of the thyroid gland (thyroidectomy), the use of radioactive iodine (^{131}I) to selectively destroy thyroid tissue, and the use of antithyroid drugs to block synthesis of thyroid hormones.

A **goiter** (GOY-ter; *guttur*=throat), an enlarged thyroid gland, may be associated with hyperthyroidism, hypothyroidism, or **euthyroidism** (ū-THĪ-royd-izm; *eu*=good); the latter is normal secretion of thyroid hormone. In some areas, dietary iodine intake is inadequate; the resultant low level of thyroid hormone in the blood stimulates secretion of TSH, which causes thyroid gland enlargement.

Hypoparathyroidism (hī-pō-par'-a-THĪ-royd-izm)—too little parathyroid hormone—leads to a deficiency of blood Ca^{2+}, which causes neurons and muscle fibers to depolarize and produce action potentials spontaneously. This leads to twitches, spasms, and *tetany* (maintained contraction) of skeletal muscle. The leading cause of hypoparathyroidism is accidental damage to the parathyroid glands or to their blood supply during thyroidectomy surgery.

Hyperparathyroidism, an elevated level of parathyroid hormone, most often is due to a tumor of one of the parathyroid glands. An elevated level of PTH causes excessive resorption of bone matrix, raising the blood levels of calcium and phosphate ions and causing bones to become soft and easily fractured. High blood calcium level promotes formation of kidney stones. Fatigue, personality changes, and lethargy are also seen in patients with hyperparathyroidism. •

 CHECKPOINT

9. Name the hormones produced by follicular and parafollicular cells and describe their actions.
10. Compare the effects of PTH and calcitonin on blood Ca^{2+} level.

22.6 ADRENAL GLANDS

OBJECTIVES

• Describe the location, histology, hormones, and functions of the adrenal cortex.
• Describe the location, histology, hormones, and functions of the adrenal medulla.

The paired **adrenal** or **suprarenal glands** (*supra-*=above; *renal*=kidney), one of which lies superior to each kidney in the retroperitoneal space (Figure 22.6a, c), have a flattened pyramidal

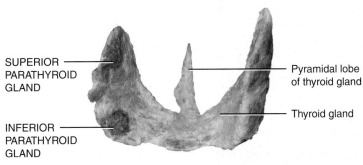

SUPERIOR PARATHYROID GLAND

INFERIOR PARATHYROID GLAND

Pyramidal lobe of thyroid gland

Thyroid gland

(d) Posterior view of parathyroid glands

 What are the secretory products of (1) parafollicular cells of the thyroid gland and (2) chief cells of the parathyroid glands?

shape. In an adult, each adrenal gland is 3–5 cm in height, 2–3 cm in width, and a little less than 1 cm thick; it weighs 3.5–5 g, only half its weight at birth. During embryonic development, the adrenal glands differentiate into two structurally and functionally distinct regions: A large, peripherally located **adrenal cortex**, representing 80–90 percent of the gland by weight, develops from mesoderm; a small, centrally located **adrenal medulla** develops from ectoderm (Figure 22.6b). The adrenal cortex produces hormones that are essential for life. Complete loss of adrenocortical hormones leads to death in a few days to a week due to dehydration and electrolyte imbalances, unless hormone replacement therapy begins promptly. The adrenal medulla produces two hormones: norepinephrine and epinephrine. A connective tissue capsule covers the gland.

Adrenal Cortex

The adrenal cortex is subdivided into three zones, each of which secretes different hormones (Figure 22.6b, d). The outer zone, just deep to the connective tissue capsule, is called the **zona glomerulosa** (glō-mer'-ū-LŌ-sa; *zona*=belt; *glomerul-*=little ball). Its cells, which are closely packed and arranged in spherical clusters and arched columns, secrete hormones called **mineralocorticoids** (min'-er-al-ō-KOR-ti-koyds) because they affect metabolism of the minerals sodium and potassium. **Aldosterone** is a mineralocorticoid. The middle zone, or **zona fasciculata** (fa-sik'-ū-LA-ta; *fascicul-*=little bundle), is the widest of the three zones and consists of cells arranged in long, straight cords. The cells of the zona fasciculata secrete mainly **glucocorticoids** (gloo'-kō-KOR-ti-koyds), primarily cortisol. The glucocorticoids are so named because they affect glucose metabolism. The cells of the inner zone, the **zona reticularis** (re-tik'-ū-LAR-is; *reticul-*=network), are arranged in branching cords. They synthesize small amounts of weak **androgens** (*andro-*=a man), hormones that have masculinizing effects.

Figure 22.6 The adrenal (suprarenal) glands.

🔑 The adrenal cortex secretes hormones that are essential for life; the adrenal medulla secretes norepinephrine and epinephrine, which mimic the response of the sympathetic division of the ANS.

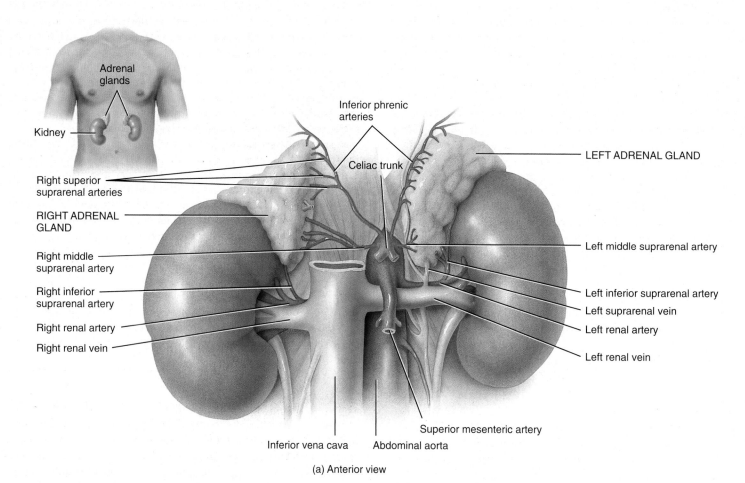

Adrenal glands

Kidney

Inferior phrenic arteries

Celiac trunk

Right superior suprarenal arteries

RIGHT ADRENAL GLAND

LEFT ADRENAL GLAND

Right middle suprarenal artery

Left middle suprarenal artery

Right inferior suprarenal artery

Left inferior suprarenal artery

Left suprarenal vein

Right renal artery

Left renal artery

Right renal vein

Left renal vein

Inferior vena cava   Abdominal aorta

Superior mesenteric artery

(a) Anterior view

Capsule

ADRENAL CORTEX

ADRENAL MEDULLA

(b) Section through left adrenal gland

Capsule

ADRENAL CORTEX:

ZONA GLOMERULOSA secretes mineralocorticoids, mainly aldosterone

ZONA FASCICULATA secretes glucocorticoids, mainly cortisol

ADRENAL GLAND

ZONA RETICULARIS secretes androgens

Kidney

ADRENAL MEDULLA chromaffin cells secrete epinephrine and norepinephrine (NE)

LM 50x

(c) Anterior view of adrenal gland and kidney

(d) Subdivisions of the adrenal gland

❓ **What is the position of the adrenal glands relative to the kidneys?**

TABLE 22.5

Summary of Adrenal Gland Hormones

HORMONES AND SOURCE	PRINCIPAL ACTIONS
ADRENAL CORTEX HORMONES	
Mineralocorticoids (min'-er-al-ō-KOR-ti-koyds) **(mainly aldosterone)** from zona glomerulosa cells	Increase blood levels of Na$^+$ and water and decrease blood level of K$^+$
Glucocorticoids (gloo'-kō-KOR-ti-koyds) **(mainly cortisol) from zona** fasciculata cells	Increase protein breakdown (except in liver), stimulate gluconeogenesis and lipolysis, provide resistance to stress, decrease inflammation, and depress immune responses
Androgens (mainly dehydroepiandrosterone or DHEA) from zona reticularis cells 	Assist in early growth of axillary and pubic hair in both sexes; in females, contribute to libido and are source of estrogens after menopause
ADRENAL MEDULLA HORMONES	
Epinephrine and norepinephrine from chromaffin cells 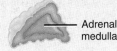	Produce effects that mimic those of the sympathetic division of the autonomic nervous system (ANS) during stress

Adrenal Medulla

The inner region of the adrenal gland, the **adrenal medulla** (Figure 22.6b, d), is a modified sympathetic ganglion of the autonomic nervous system (ANS). It develops from the same embryonic tissue as all other sympathetic ganglia (embryonic neural crest tissue), but its cells lack axons and form clusters around large blood vessels. Rather than releasing a neurotransmitter, the cells of the adrenal medulla secrete hormones. The hormone-producing cells, called **chromaffin cells** (KRŌ-maf-in; *chrom-*=color; *affin*=affinity for; Figure 22.6d), are innervated by sympathetic preganglionic neurons of the greater splanchnic nerve from lower thoracic spinal levels. Because the ANS controls the chromaffin cells directly, hormone release can occur very quickly.

The two principal hormones synthesized by the adrenal medulla are **epinephrine** (ep'-i-NEF-rin) and **norepinephrine (NE)**, also called adrenaline and noradrenaline, respectively. Epinephrine constitutes about 80 percent of the total secretion of the gland. Both hormones are **sympathomimetic** (sim'-pa-thō-mi-MET-ik)—

their effects mimic those brought about by the sympathetic division of the ANS. To a large extent, they are responsible for the fight-or-flight response. Like the glucocorticoids of the adrenal cortex, these hormones help resist stress. Unlike the hormones of the adrenal cortex, however, the medullary hormones are not essential for life.

The main arteries that supply the adrenal glands are the several superior suprarenal arteries arising from the inferior phrenic artery, the right and left middle suprarenal arteries from the aorta, and the inferior suprarenal arteries from the renal arteries. The suprarenal vein of the right adrenal gland drains directly into the inferior vena cava; the suprarenal vein of the left adrenal gland empties into the left renal vein (see Figure 22.6a).

The principal nerve supply to the adrenal glands is from preganglionic fibers from the thoracic splanchnic nerves, which pass through the celiac and associated sympathetic plexuses. These myelinated fibers innervate the secretory cells of the gland found in a region of the medulla.

A summary of adrenal gland hormones and their actions is presented in Table 22.5.

CLINICAL CONNECTION | *Adrenal Gland Disorders*

Congenital adrenal hyperplasia (CAH) (hī-per-PLĀ-zē-a) is a genetic disorder in which one or more enzymes needed for synthesis of cortisol are absent. Because the cortisol level is low, secretion of ACTH by the anterior pituitary is high due to lack of negative feedback inhibition. ACTH in turn stimulates growth and secretory activity of the adrenal cortex. As a result, both adrenal glands are enlarged. Precursor molecules accumulate, and some of these are weak androgens that can undergo conversion to testosterone. The result is **virilism** (VIR-i-lizm), or masculinization. In a female, virile characteristics include growth of a beard, development of a much deeper voice and a masculine distribution of body hair, growth of the clitoris so it may resemble a penis, atrophy of the breasts, and increased muscularity that produces a masculine physique. In prepubertal males, the syndrome causes the same characteristics as in females, plus rapid development of the male sexual organs and emergence of male sexual desires.

Hypersecretion of cortisol by the adrenal cortex produces **Cushing's syndrome**. Causes include a tumor of the adrenal gland that secretes cortisol, or a tumor elsewhere that secretes adrenocorticotropic hormone (ACTH), which in turn stimulates excessive secretion of cortisol. The condition is characterized by breakdown of muscle proteins and redistribution of body fat, resulting in spindly arms and legs accompanied by a rounded "moon face," "buffalo hump" on the back, and pendulous (hanging) abdomen. Facial skin is flushed,

and the skin covering the abdomen develops stretch marks. The person also bruises easily, and wound healing is poor. The elevated level of cortisol causes hyperglycemia, osteoporosis, weakness, hypertension, increased susceptibility to infection, decreased resistance to stress, and mood swings.

Hyposecretion of glucocorticoids and aldosterone causes **Addison's disease** (*chronic adrenocortical insufficiency*). The majority of cases are autoimmune disorders in which antibodies cause adrenal cortex destruction or block binding of ACTH to its receptors. Symptoms include mental lethargy, anorexia, nausea and vomiting, weight loss, hypoglycemia, and muscular weakness. Loss of aldosterone leads to elevated potassium and decreased sodium in the blood, low blood pressure, dehydration, decreased cardiac output, arrhythmias, and even cardiac arrest. Treatment consists of replacing glucocorticoids and mineralocorticoids and increasing sodium in the diet.

Usually benign tumors of the chromaffin cells of the adrenal medulla, called **pheochromocytomas** (fē-ō-krō'-mō-sī-TŌ-mas; *pheo-*=dusky; *chromo-*=color; *cyto-*=cell), cause hypersecretion of epinephrine and norepinephrine. The result is a prolonged version of the fight-or-flight response: rapid heart rate, high blood pressure, high levels of glucose in blood and urine, an elevated basal metabolic rate (BMR), flushed face, nervousness, sweating, and decreased gastrointestinal motility. Treatment is surgical removal of the tumor. •

CHECKPOINT

11. Compare the location and histology of the adrenal cortex and the adrenal medulla.
12. Describe the relationship of the adrenal medulla to the autonomic nervous system.

22.7 PANCREAS

 OBJECTIVE

• Describe the location, histology, hormones, and functions of the pancreas.

The **pancreas** (*pan-*=all; *creas*=flesh) is both an endocrine gland and an exocrine gland. Its exocrine functions are discussed in Chapter 24. The pancreas is a flattened organ that measures about 12.5–15 cm (5–6 in.) in length. It is located posterior and slightly inferior to the stomach and consists of a head, a body, and a tail (Figure 22.7a, d). Roughly 99 percent of the pancreatic exocrine cells are arranged in clusters called **acini** (AS-i-nī; singular is *acinus*); these cells produce digestive enzymes, which flow into the gastrointestinal tract through a network of ducts (see Section 24.8). Scattered among the exocrine acini are 1–2 million tiny clusters of endocrine cells called **pancreatic islets** (Ī-lets) or **islets of Langerhans** (LAHNG-er-hanz; Figure 22.7b, c). Abundant capillaries serve both the exocrine and endocrine portions of the pancreas.

Each pancreatic islet contains four types of hormone-secreting cells (see also Table 22.6):

1. **Alpha cells** constitute about 15 percent of pancreatic islet cells and secrete **glucagon** (GLOO-ka-gon).

Figure 22.7 The pancreas.

 Pancreatic hormones regulate blood glucose level.

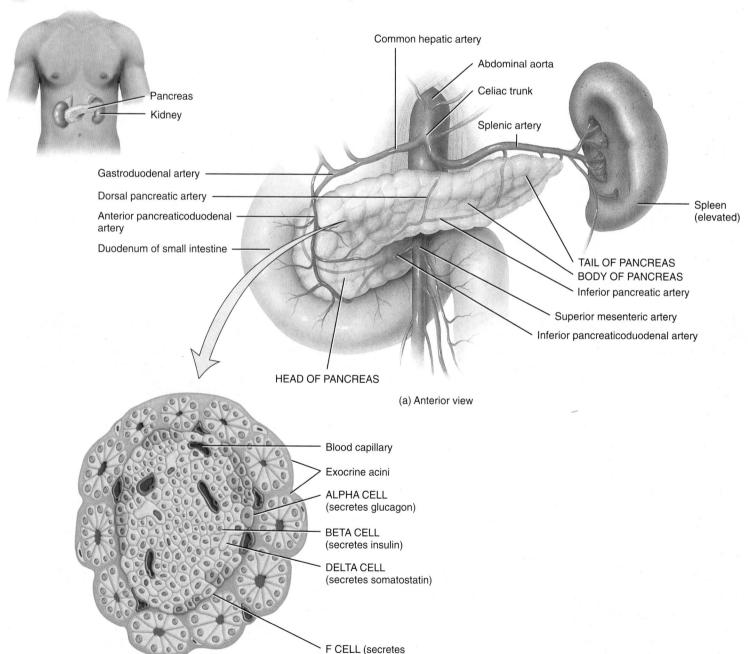

(a) Anterior view

(b) Pancreatic islet and surrounding acini

Exocrine
acinus

PANCREATIC
ISLET

BETA
CELL

ALPHA
CELL

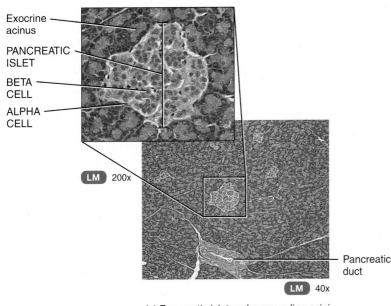

LM 200x

LM 40x

Pancreatic
duct

(c) Pancreatic islet and surrounding acini

Pancreas

Pancreatic
duct

Duodenum
(cut open)

(d) Anterior view of pancreas dissected
to reveal pancreatic duct

? **Is the pancreas an exocrine gland or an endocrine gland?**

TABLE 22.6

Summary of Pancreatic Islet Hormones

HORMONE AND SOURCE	PRINCIPAL ACTIONS
Glucagon from alpha cells of pancreatic islets 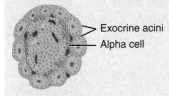 Exocrine acini — Alpha cell	Raises blood glucose level by accelerating breakdown of glycogen into glucose in liver (glycogenolysis), converting other nutrients into glucose in liver (gluconeogenesis), and releasing glucose into the blood
Insulin from beta cells of pancreatic islets 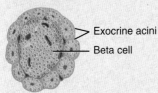 Exocrine acini — Beta cell	Lowers blood glucose level by accelerating transport of glucose into cells, converting glucose into glycogen (glycogenesis), and decreasing glycogenolysis and gluconeogenesis; also increases lipogenesis and stimulates protein synthesis
Somatostatin from delta cells of pancreatic islets 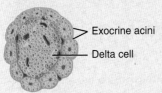 Exocrine acini — Delta cell	Inhibits secretion of insulin and glucagon and slows absorption of nutrients from the gastrointestinal tract
Pancreatic polypeptide from F cells of pancreatic islets 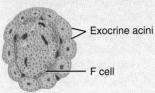 Exocrine acini — F cell	Inhibits somatostatin secretion, gallbladder contraction, and secretion of pancreatic digestive enzymes

CLINICAL CONNECTION | *Diabetes Mellitus*

The most common endocrine disorder is **diabetes mellitus** (MEL-i-tus; *melli-*=honey sweetened), caused by an inability to produce or use insulin. Diabetes mellitus is the fourth leading cause of death by disease in the United States, primarily because of its damage to the cardiovascular system. Because insulin is unavailable to aid transport of glucose into body cells, blood glucose level is high and glucose "spills" into the urine (glucosuria). Hallmarks of diabetes mellitus are the three "polys": *polyuria,* excessive urine production due to an inability of the kidneys to reabsorb water; *polydipsia,* excessive thirst; and *polyphagia,* excessive eating.

Type 1 diabetes, previously known as **insulin-dependent diabetes mellitus (IDDM)**, occurs because the person's immune system destroys the pancreatic beta cells. As a result, the pancreas produces little or no insulin. Type 1 diabetes usually develops in people younger than age 20, and it persists throughout life. By the time symptoms of type 1 diabetes arise, 80–90 percent of the islet beta cells have been destroyed. In the United States, type 1 diabetes is 1.5–2.0 times more common in whites than in African American or Asian populations.

The cellular metabolism of an untreated type 1 diabetic is similar to that of a starving person. Because insulin is not present to aid the entry of glucose into body cells, most cells use fatty acids to produce ATP. The byproducts of fatty acid breakdown—organic acids called *ketones* or *ketone bodies*—accumulate. Buildup of ketones causes blood pH to fall, a condition known as **ketoacidosis** (kē′-tō-as-i-DŌ-sis). Unless treated quickly, ketoacidosis can cause coma and death.

The breakdown of stored triglycerides also causes weight loss. As lipids are transported by the blood from storage depots to cells, lipid particles are deposited on the walls of blood vessels, leading to atherosclerosis and a multitude of cardiovascular problems, including cerebrovascular insufficiency, ischemic heart disease, peripheral vascular disease, and gangrene. A major complication of diabetes is loss of vision due either to cataracts (attachment of excess glucose to lens proteins, causing cloudiness) or damage to blood vessels of the retina. Severe kidney problems also may result from damage to renal blood vessels.

Type 1 diabetes is treated through self-monitoring of blood glucose level (up to 7 times daily), a diet of regular meals containing 45–50 percent carbohydrates and less than 30 percent fats, exercise, and periodic insulin injections (up to 3 times a day). Several implantable pumps are available to provide insulin without the need for repeated injections. Because they lack a reliable glucose sensor, however, the person must self-monitor blood glucose level to determine insulin doses. Pancreas transplant has also proved successful, but immunosuppressive drugs must then be taken for life. Most pancreatic transplants are done in people who also need a kidney transplant because of renal failure.

Type 2 diabetes, also called **non-insulin-dependent diabetes mellitus (NIDDM)**, is much more common than type 1, representing more than 90 percent of all cases. Type 2 diabetes most often occurs in obese people over the age of 35. Clinical symptoms are mild, and the high glucose levels in the blood often can be controlled by diet, exercise, and weight loss. Sometimes, drugs such as *glyburide* (DiaBeta) and *metformin* (Fortamet) are used to stimulate secretion of insulin by pancreatic beta cells. Although some type 2 diabetics need insulin, many have a sufficient amount (or even a surplus) of insulin in the blood. In such cases, diabetes arises not from a shortage of insulin but because target cells become less sensitive to it due to down-regulation of insulin receptors.

Hyperinsulinism most often results when a diabetic injects too much insulin. The main symptom is **hypoglycemia**, decreased blood glucose level, which occurs because the excess insulin stimulates too much uptake of glucose by body cells. The resulting hypoglycemia stimulates the secretion of epinephrine, glucagon, and human growth hormone. As a consequence, anxiety, sweating, tremor, increased heart rate, hunger, and weakness occur. When blood glucose falls, brain cells are deprived of the steady supply of glucose they need to function effectively. Severe hypoglycemia leads to mental disorientation, convulsions, unconsciousness, and shock. Shock due to an insulin overdose is termed **insulin shock**. Death can occur quickly unless blood glucose level is restored to normal levels. From a clinical standpoint, a diabetic suffering from either a hyperglycemia or a hypoglycemia crisis can have very similar symptoms—mental changes, insulin-induced coma, seizures, and so on. It is important to quickly and correctly identify the cause of the underlying symptoms and treat them appropriately. •

2. **Beta cells** constitute about 80 percent of pancreatic islet cells and secrete **insulin** (IN-soo-lin).
3. **Delta cells** constitute about 5 percent of pancreatic islet cells and secrete **somatostatin** (sō-ma-tō-STAT-in), which is identical to the growth hormone–inhibiting hormone secreted by the hypothalamus.
4. **F cells** constitute the remainder of pancreatic islet cells and secrete **pancreatic polypeptide**.

The superior and inferior pancreaticoduodenal arteries and the splenic and superior mesenteric arteries supply blood to the pancreas (Figure 22.7a). The veins, in general, correspond to the arteries. Venous blood reaches the hepatic portal vein by means of the splenic and superior mesenteric veins.

The nerves to the pancreas are autonomic nerves derived from the celiac and superior mesenteric plexuses. They include preganglionic vagal fibers, postganglionic sympathetic fibers, and sensory fibers. Parasympathetic vagal fibers are said to terminate at both acinar (exocrine) and islet (endocrine) cells. Although it is assumed that nervous innervation influences enzyme formation, pancreatic secretion is controlled largely by the hormones secre-

tin and cholecystokinin (CCK) released by the small intestine. The sympathetic fibers that enter the islets are vasomotor (they innervate blood vessels) and are accompanied by sensory fibers that transmit impulses, especially for pain.

 CHECKPOINT

13. Identify the cells in a pancreatic islet and the secretions of each.

22.8 OVARIES AND TESTES

 OBJECTIVE

• Describe the location, hormones, and endocrine functions of the male and female gonads.

The female gonads, called the **ovaries**, are paired oval bodies located in the pelvic cavity. The ovaries produce female sex hormones called **estrogens** and **progesterone**. Along with the gonadotropic hormones of the pituitary gland, the sex hormones regulate the female reproductive cycle, maintain pregnancy, and

TABLE 22.7	
Summary of Hormones of the Ovaries and Testes	
HORMONE	**PRINCIPAL ACTIONS**
OVARIAN HORMONES	
Estrogens and progesterone Ovaries	Together with gonadotropic hormones of the anterior pituitary, regulate the female reproductive cycle, maintain pregnancy, prepare the mammary glands for lactation, and promote development and maintenance of female secondary sex characteristics
Relaxin	Increases flexibility of pubic symphysis during pregnancy and helps dilate uterine cervix during labor and delivery
Inhibin	Inhibits secretion of FSH from the anterior pituitary
TESTICULAR HORMONES	
Testosterone Testes	Stimulates descent of the testes before birth, regulates production of sperm, and promotes development and maintenance of male secondary sex characteristics
Inhibin	Inhibits secretion of FSH from the anterior pituitary

prepare the mammary glands for lactation. These hormones are also responsible for the development and maintenance of female secondary sexual characteristics. In addition, the ovaries produce **inhibin**, a hormone that inhibits secretion of follicle-stimulating hormone (FSH) from the anterior pituitary. During pregnancy, the ovaries and placenta produce a hormone called **relaxin (RLX)**, which increases the flexibility of the pubic symphysis during pregnancy and helps dilate the uterine cervix during labor and delivery. These actions help ease the baby's passage by enlarging the birth canal.

The male has two oval gonads, called **testes**, that produce **testosterone**, the primary androgen. Testosterone stimulates descent of the testes before birth, regulates production of sperm, and stimulates the development and maintenance of male secondary sexual characteristics such as beard growth. The testes also produce inhibin, which inhibits secretion of FSH. The specific roles of gonadotropic hormones and sex hormones are discussed in Chapter 26.

Table 22.7 summarizes the hormones produced by the ovaries and testes and their principal actions.

✔ CHECKPOINT

14. Explain why the ovaries and testes are considered endocrine glands.

22.9 OTHER ENDOCRINE TISSUES

● OBJECTIVE

• List the hormones secreted by cells in other tissues and organs, and describe their functions.

As you learned at the beginning of this chapter, cells in organs other than those usually classified as endocrine glands have an endocrine function and secrete hormones. You learned about

several of these in this chapter, including the hypothalamus, thymus, pancreas, ovaries, and testes. Other tissues with endocrine functions are summarized in Table 22.8.

CLINICAL CONNECTION | *Stress, Hormones, and Disease*

Although the exact role of stress in human diseases is not known, it is clear that stress can lead to particular diseases by temporarily inhibiting certain components of the immune system. Stress-related disorders include gastritis, ulcerative colitis, irritable bowel syndrome, hypertension, asthma, rheumatoid arthritis (RA), migraine headaches, anxiety, and depression. People under stress are at a greater risk of developing chronic disease or dying prematurely.

Interleukin-1, a substance secreted by macrophages of the immune system, is an important link between stress and immunity. One action of interleukin-1 is to stimulate secretion of ACTH, which in turn stimulates the production of cortisol. Not only does cortisol provide resistance to stress and inflammation, but it also suppresses further production of interleukin-1. Thus, the immune system turns on the stress response, and the resulting cortisol then turns off one immune system mediator. This negative feedback system keeps the immune response in check once it has accomplished its goal. Because of this activity, cortisol and other glucocorticoids are used as immunosuppressive drugs for organ transplant recipients. •

✔ CHECKPOINT

15. List the hormones secreted by the gastrointestinal tract, placenta, kidneys, skin, adipose tissue, and heart, and indicate their functions.

TABLE 22.8

Summary of Hormones Produced by Other Organs and Tissues That Contain Endocrine Cells

HORMONE	PRINCIPAL ACTIONS
GASTROINTESTINAL TRACT	
Gastrin	Promotes secretion of gastric juice and increases movements of the stomach
Glucose-dependent insulinotropic peptide (GIP)	Stimulates release of insulin by pancreatic beta cells
Secretin	Stimulates secretion of pancreatic juice and bile
Cholecystokinin (CCK)	Stimulates secretion of pancreatic juice, regulates release of bile from the gallbladder, and brings about a feeling of fullness after eating
PLACENTA	
Human chorionic gonadotropin (hCG)	Stimulates the corpus luteum in the ovary to continue the production of estrogens and progesterone to maintain pregnancy
Estrogens and progesterone	Maintain pregnancy and help prepare mammary glands to secrete milk
Human chorionic somatomammotropin (hCS)	Stimulates the development of the mammary glands for lactation
KIDNEYS	
Renin	Part of a sequence of reactions that raises blood pressure by bringing about vasoconstriction and secretion of aldosterone
Erythropoietin (EPO)	Increases rate of red blood cell formation
Calcitriol (active form of vitamin D)*	Aids in the absorption of dietary calcium and phosphorus
HEART	
Atrial natriuretic peptide (ANP)	Decreases blood pressure
ADIPOSE TISSUE	
Leptin	Suppresses appetite and may increase the activity of FSH and LH

*Synthesis begins in the skin, continues in the liver, and ends in the kidneys.

22.10 DEVELOPMENT OF THE ENDOCRINE SYSTEM

OBJECTIVE

• Describe the development of endocrine glands.

The development of the endocrine system is not as localized as the development of other systems because endocrine organs develop in widely separated parts of the embryo. About three weeks after fertilization, the *pituitary gland (hypophysis)* begins to develop from two different regions of the **ectoderm**. The *posterior pituitary (neurohypophysis)* is derived from an outgrowth of ectoderm called the **neurohypophyseal bud** (noo′-rō-hī-pō-FIZ-ē-al), located on the floor of the hypothalamus (Figure 22.8). The *infundibulum,* also an outgrowth of the neurohypophyseal bud, connects the posterior pituitary to the hypothalamus. The *anterior pituitary (adenohypophysis)* is derived from an outgrowth of ectoderm from the roof of the mouth called the **hypophyseal pouch** or *Rathke's pouch.* The pouch grows toward the neurohypophyseal bud and eventually loses its connection with the roof of the mouth.

The *thyroid gland* develops during the fourth week as a midventral outgrowth of **endoderm**, called the **thyroid diverticulum** (dī-ver-TIK-ū-lum), from the floor of the pharynx at the level of the second pair of pharyngeal pouches. The outgrowth projects inferiorly and differentiates into the right and left lateral lobes and the isthmus of the gland.

The *parathyroid glands* develop during the fourth week from endoderm as outgrowths from the third and fourth **pharyngeal pouches** (fa-RIN-jē-al), which help to form structures of the head and neck.

The adrenal cortex and adrenal medulla develop during the fifth week and have completely different embryological origins. The *adrenal cortex* is derived from intermediate **mesoderm** from the same region that produces the gonads. Endocrine tissues that secrete steroid hormones all are derived from mesoderm. The *adrenal medulla* is derived from ectoderm from **neural crest** cells that migrate to the superior portion of the kidney. Recall that neural crest cells also give rise to sympathetic ganglia and other structures of the nervous system (see Figure 18.26).

The *pancreas* develops during the fifth through seventh weeks from two outgrowths of endoderm from the part of the **foregut** that later becomes the duodenum (see Figure 4.12c). The two outgrowths eventually fuse to form the pancreas. The origin of the ovaries and testes is discussed in the section on the reproductive system.

The *pineal gland* arises during the seventh week as an outgrowth between the thalamus and colliculi of the midbrain from ectoderm associated with the **diencephalon** (see Figure 18.27).

The *thymus* arises during the fifth week from endoderm of the third pharyngeal pouches.

 CHECKPOINT

16. Which endocrine gland is derived from both mesoderm and ectoderm?

Figure 22.8 Development of the endocrine system.

Glands of the endocrine system develop from all three primary germ layers.

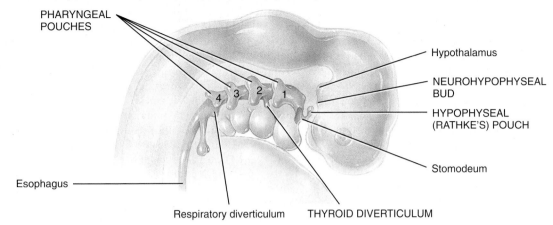

(a) Location of the neurohypophyseal bud, hypophyseal (Rathke's) pouch, thyroid diverticulum, and pharyngeal pouches in a 28-day embryo

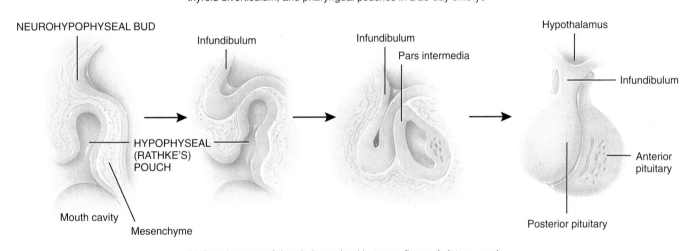

(b) Development of the pituitary gland between five and sixteen weeks

Which endocrine glands develop from tissues with two different embryological origins?

22.11 AGING AND THE ENDOCRINE SYSTEM

OBJECTIVE

• Describe the effects of aging on the endocrine system.

Although some endocrine glands shrink as we get older, their performance may or may not be compromised. Production of human growth hormone by the anterior pituitary decreases, which is one cause of muscle atrophy as aging proceeds. The thyroid gland often decreases its output of thyroid hormones with age, causing a decrease in metabolic rate, an increase in body fat, and hypothyroidism, which is why this disorder is seen more often in older people. Due to less negative feedback (lower levels of thyroid hormones), thyroid-stimulating hormone level increases with age.

The blood level of PTH rises with age, perhaps due to inadequate dietary intake of calcium. In a study of older women who took 2400 mg/day of supplemental calcium, blood levels of PTH were as low as those of younger women. Both calcitriol and calcitonin levels are lower in older persons. Together, the rise in PTH and the fall in calcitonin level heighten the age-related decrease in bone mass that leads to osteoporosis (see Section 6.7) and increased risk of fractures (see Section 6.10).

The amount of fibrous tissue in the adrenal glands increases, decreasing the production of cortisol and aldosterone with advancing age. However, production of epinephrine and norepinephrine remains normal. The pancreas releases insulin more slowly with age, and receptor sensitivity to glucose declines. As a result, blood glucose levels in older people increase faster and return to normal more slowly than in younger individuals.

The thymus is largest in infancy. After puberty, its size begins to decrease, and thymic tissue is replaced by adipose and areolar connective tissue. In older adults, the thymus has atrophied significantly. However, it still produces new T cells for immune responses.

The ovaries decrease in size with age, and they no longer respond to gonadotropins. As a result, the output of estrogens decreases, leading to conditions such as osteoporosis, high blood cholesterol, and atherosclerosis. FSH and LH levels are high due to less negative feedback inhibition of estrogens. Although testosterone production by the testes decreases with age, the effects are not usually apparent until very old age, and many elderly males can still produce active sperm in normal numbers, but there are higher numbers of morphologically abnormal sperm, and sperm motility is decreased.

 CHECKPOINT

17. Which hormone is related to the muscle atrophy associated with aging?

KEY MEDICAL TERMS ASSOCIATED WITH THE ENDOCRINE SYSTEM

Diabetes insipidus (DI) (dī-a-BĒ-tēs in-SIP-i-dus; *diabetes*=overflow; *insipidus*=tasteless). The most common abnormality associated with dysfunction of the posterior pituitary is due to defects in antidiuretic hormone (ADH) receptors or an inability to secrete ADH. *Neurogenic diabetes insipidus* results from hyposecretion of ADH, usually caused by a brain tumor, head trauma, or brain surgery that damages the posterior pituitary or the hypothalamus. In *nephrogenic diabetes insipidus*, the kidneys do not respond to ADH. The ADH receptors may be nonfunctional, or the kidneys may be damaged. A common symptom of both forms of DI is excretion of large volumes of urine, with resulting dehydration and thirst. Bed-wetting is common in afflicted children. Because so much water is lost in the urine, a person with DI may die of dehydration if deprived of water for only a day or so.

Gynecomastia (gī-ne′-kō-MAS-tē-a; *gyneco-*=woman; *mast-*=breast) Excessive development of mammary glands in a male, sometimes caused by a tumor of the adrenal gland.

Hirsutism (HER-soo-tizm; *hirsut-*=shaggy) Presence of excessive bodily and facial hair in a male pattern, especially in women; may be caused by excess androgen production due to tumors or drugs.

Thyroid crisis (storm) A severe, potentially life-threatening, state of hyperthyroidism characterized by high body temperature, rapid heart rate, high blood pressure, gastrointestinal symptoms (abdominal pain, vomiting, diarrhea), agitation, tremors, confusion, seizures, and possibly coma.

Virilizing adenoma (*aden*=gland; *oma*=tumor) Tumor of the adrenal gland that liberates excessive androgens, causing virilism (masculinization) in females. Occasionally, adrenal tumor cells liberate estrogens to the extent that a male patient develops gynecomastia. Such a tumor is called a **feminizing adenoma**.

CHAPTER REVIEW AND RESOURCE SUMMARY

WileyPLUS

Review	Resource
Introduction 1. The nervous system controls homeostasis through nerve impulses; the endocrine system uses hormones. 2. The nervous system causes muscles to contract and glands to secrete; the endocrine system affects virtually all body tissues. 3. Table 22.1 compares the characteristics of the nervous and endocrine systems.	Anatomy Overview - The Endocrine System
22.1 Endocrine Glands Defined 1. Exocrine glands (sudoriferous, sebaceous, and digestive) secrete their products through ducts into body cavities or onto body surfaces. 2. Endocrine glands secrete hormones into the blood. 3. The endocrine system consists of endocrine glands and several organs that contain endocrine tissue. 4. Hormones regulate the internal environment, metabolism, and energy balance, and help regulate muscular contraction, glandular secretion, and certain immune responses. 5. Hormones affect growth, development, and reproduction.	Anatomy Overview - The Endocrine System
22.2 Hormones 1. The amount of hormone released is determined by the body's need for the hormone. 2. Cells that respond to the effects of hormones are called target cells. 3. The combination of hormone and receptor activates a chain of events in a target cell that produce the physiological effects of the hormone.	Anatomy Overview - Hormones Animation - Introduction to Hormonal Regulation, Secretion, and Concentration Exercise - Hormone Actions Exercise - Produce That Hormone
22.3 Hypothalamus and Pituitary Gland 1. The hypothalamus is the major integrating link between the nervous and endocrine systems. 2. The hypothalamus and pituitary gland regulate virtually all aspects of growth, development, and metabolism, and they also affect other body activities. 3. The pituitary gland is located in the hypophyseal fossa and is divided into the anterior pituitary (glandular portion), the posterior pituitary (nervous portion), and pars intermedia. 4. Hormones of the anterior pituitary are controlled by releasing or inhibiting hormones produced by the hypothalamus. 5. The blood supply to the anterior pituitary is from the superior hypophyseal arteries. It carries releasing and inhibiting hormones from the hypothalamus. 6. Histologically, the anterior pituitary consists of somatotrophs that produce human growth hormone (hGH); lactotrophs that produce prolactin (PRL); corticotrophs that secrete adrenocorticotropic hormone (ACTH) and melanocyte-stimulating hormone (MSH); thyrotrophs that secrete thyroid-stimulating hormone (TSH); and gonadotrophs that synthesize follicle-stimulating hormone (FSH) and luteinizing hormone (LH). 7. hGH stimulates body growth. TSH regulates thyroid gland activities. FSH and LH both regulate the activities of the ovaries and testes. PRL helps initiate milk secretion. MSH increases skin pigmentation. ACTH regulates the activities of the adrenal cortex. 8. The neural connection between the hypothalamus and posterior pituitary is via the hypothalamohypophyseal tract.	Anatomy Overview - The Hypothalamus and Pituitary Gland Anatomy Overview - Hormones of the Anterior Pituitary Gland Anatomy Overview - The Hypothalamus Anatomy Overview - Hormones of the Hypothalamus Anatomy Overview - Hypothalamic Reproductive Hormones Animation - hGH - Growth and Development Animation - GHRH / hGH Animation - ACTH/Cortisol-Glycogenolysis Animation - TRH/TSH - Production Animation - hGH - Glycogenolysis and Lipolysis Animation - Antidiuretic Hormone Figure 22.2 - Hypothalamus and Pituitary Gland and Their Blood Supply

Review

9. Hormones made by the hypothalamus and stored in the posterior pituitary are oxytocin (OT), which stimulates contraction of the uterus and ejection of milk, and antidiuretic hormone (ADH), which stimulates water reabsorption by the kidneys and arteriole constriction.

10. The hormones of the anterior pituitary are summarized in Table 22.2, and the posterior pituitary hormones are summarized in Table 22.3.

22.4 Pineal Gland and Thymus

Anatomy Overview - The Pineal Gland

1. The pineal gland is attached to the roof of the third ventricle.
2. Histologically, the pineal gland consists of secretory cells called pinealocytes, neuroglial cells, and scattered postganglionic sympathetic fibers.
3. The pineal gland secretes melatonin, which contributes to setting the body's biological clock, a process controlled by the suprachiasmatic nucleus. During sleep, levels of melatonin in the bloodstream increase tenfold and then decline to a low level again before awakening.
4. The thymus secretes several hormones related to immunity.
5. Thymosin, thymic humoral factor (THF), thymic factor (TF), and thymopoietin promote the maturation of T cells.

22.5 Thyroid Gland and Parathyroid Glands

Anatomy Overview - The Thyroid
Anatomy Overview - Parathyroid Glands
Anatomy Overview - Hormones of the Parathyroid
Animation - Thyroid Hormones and Glucose and Lipid Catabolism
Animation - TRH/TSH
Animation - Calcitonin
Animation - Parathyroid Hormone

1. The thyroid gland is located inferior to the larynx.
2. Histologically, the thyroid gland consists of thyroid follicles composed of follicular cells, which secrete the thyroid hormones thyroxine (T_4) and triiodothyronine (T_3), and parafollicular cells, which secrete calcitonin (CT).
3. Thyroid hormones regulate the rate of metabolism, growth and development, and the reactivity of the nervous system. Calcitonin (CT) lowers the blood level of calcium. A summary of thyroid gland hormones and their actions is presented in Table 22.4.
4. The parathyroid glands are embedded on the posterior surfaces of the lateral lobes of the thyroid gland.
5. The parathyroid glands consist of chief cells and oxyphil cells.
6. Parathyroid hormone (PTH) increases blood calcium level and decreases blood phosphate level. A summary of parathyroid hormone and its actions is presented in Table 22.4.

22.6 Adrenal Glands

Anatomy Overview - Adrenal Glands
Anatomy Overview - Hormones of the Adrenal Glands
Animation - Epinephrine/NE
Figure 22.6 - The Adrenal Glands

1. The adrenal glands are located superior to the kidneys. They consist of an outer adrenal cortex and an inner adrenal medulla.
2. Histologically, the adrenal cortex is divided into a zona glomerulosa, zona fasciculata, and zona reticularis; the adrenal medulla consists of chromaffin cells and large blood vessels.
3. Cortical secretions include mineralocorticoids, glucocorticoids, and androgens. The medullary secretions epinephrine and norepinephrine (NE) produce effects similar to sympathetic responses and are released during stress. A summary of adrenal gland hormones and their actions is presented in Table 22.5.

22.7 Pancreas

Anatomy Overview - The Pancreas
Anatomy Overview - Hormones of the Pancreas
Animation - Glucagon
Animation - Insulin
Figure 22.7 - The Pancreas

1. The pancreas is posterior and slightly inferior to the stomach.
2. Histologically, the pancreas consists of pancreatic islets (endocrine cells), and clusters of enzyme-producing cells (acini) (exocrine cells). The four types of cells in the endocrine portion are alpha, beta, delta, and F cells.
3. Alpha cells secrete glucagon, beta cells secrete insulin, delta cells secrete somatostatin, and F cells secrete pancreatic polypeptide.
4. Glucagon increases blood sugar level. Insulin decreases blood sugar level. A summary of pancreatic hormones and their actions is presented in Table 22.6.

22.8 Ovaries and Testes

Anatomy Overviews: The Ovaries; Ovarian Hormones; Testes; Testicular Hormones
Animations: Hormonal Control of Male Reproductive Function; Hormonal Regulation of Female Reproductive System
Exercises: Match Female Hormones; Match Male Hormones

1. The ovaries are located in the pelvic cavity and produce sex hormones that function in the development and maintenance of female secondary sexual characteristics, the reproductive cycle, pregnancy, lactation, and normal reproductive functions.
2. The testes lie inside the scrotum and produce sex hormones that function in the development and maintenance of male secondary sexual characteristics and normal reproductive functions.
3. Table 22.7 summarizes the hormones produced by the ovaries and testes and their principal actions.

22.9 Other Endocrine Tissues

Animation - Hormonal Control of Digestive Activities
Animation - Hormonal Regulation of Pregnancy

1. The gastrointestinal tract synthesizes several hormones, including gastrin, gastric inhibitory peptide (GIP), secretin, and cholecystokinin (CCK).
2. The placenta produces human chorionic gonadotropin (hCG), estrogens, progesterone, and human chorionic somatomammotropin (hCS).
3. The kidneys release erythropoietin.
4. The skin begins the synthesis of vitamin D.

Review	Resource

5. The atria of the heart produce atrial natriuretic peptide (ANP).
6. Adipose tissue produces leptin.
7. A summary of hormones secreted by other endocrine tissues is included in Table 22.8.

22.10 Development of the Endocrine System

1. The development of the endocrine system is not as localized as in other systems because endocrine organs develop in widely separated parts of the embryo.
2. The pituitary gland, adrenal medulla, and pineal gland develop from ectoderm; the adrenal cortex develops from mesoderm; and the thyroid gland, parathyroid glands, pancreas, and thymus develop from endoderm.

22.11 Aging and the Endocrine System

1. Although some endocrine glands shrink as we get older, their performance may or may not be compromised.
2. Production of human growth hormone, thyroid hormones, cortisol, aldosterone, and estrogens decrease with advancing age.
3. With aging, the blood levels of TSH, LH, FSH, and PTH rise.
4. The pancreas releases insulin more slowly with age, and receptor sensitivity to glucose declines.
5. After puberty, thymus size begins to decrease, and thymic tissue is replaced by adipose and areolar connective tissue.

CRITICAL THINKING QUESTIONS

1. You've won a trip to beautiful Tropicanaland, a 12-hour time difference from where you live. Your co-workers have given you a bottle of melatonin, a bottle of melanocyte-stimulating hormone, and a very bright flashlight as a bon voyage present. You'll be arriving at 8 P.M. Tropicanaland time, which is 8 A.M. your time. How can you adjust to Tropicanaland time most quickly?

2. Amadu, who has just arrived in the United States from Africa, has what appears to be a tumor in his neck. His doctor says it is not a tumor but a goiter. After some blood work, the doctor determines that a diet rich in seafoods and iodized salt should be sufficient to solve the problem. What is a goiter, and how is a change of diet going to help Amadu?

3. Lester's son isn't growing as tall as he had hoped, although he is not abnormally short. "He's not going to be any taller than me," said Lester. "He'll never make it in the NBA and be able to support me in my old age!" Lester thinks that maybe he should find a doctor who will treat his son with human growth hormone. Besides stimulating growth of tissues, human growth hormone raises blood glucose. Do you think this treatment is a good idea (physiologically speaking) in the long-term? Explain your answer.

4. For several years, military pilots were given radiation treatments in their nasal cavities to reduce sinus problems that interfered with flying. Years later, some of these former pilots began to exhibit problems with their pituitary gland hormones. Can you propose an explanation for this relationship?

5. A patient was found to have markedly elevated blood sugar. Tests showed that his insulin level was actually a bit elevated. How can someone with elevated insulin have elevated blood sugar?

? ANSWERS TO FIGURE QUESTIONS

22.1 Secretions of endocrine glands diffuse into interstitial fluid and then into the blood; exocrine secretions flow into ducts that lead into body cavities or to the body surface.

22.2 The hypophyseal portal veins carry blood from the median eminence of the hypothalamus (where hypothalamic releasing and inhibiting hormones are secreted) to the anterior pituitary (where these hormones act).

22.3 Functionally, both the hypothalamohypophyseal tract and the hypophyseal portal veins carry hypothalamic hormones to the pituitary gland. Structurally, the tract is composed of axons of neurons that extend from the hypothalamus to the posterior pituitary; the portal veins are blood vessels that extend to the anterior pituitary.

22.4 Follicular cells secrete T_3 and T_4, also known as thyroid hormones. Parafollicular cells secrete calcitonin.

22.5 Parafollicular cells of the thyroid gland secrete calcitonin; chief cells of the parathyroid gland secrete PTH.

22.6 The adrenal glands are superior to the kidneys in the retroperitoneal space.

22.7 The pancreas is both an endocrine and an exocrine gland.

22.8 The adrenal cortex of the adrenal gland is derived from mesoderm, while the adrenal medulla is derived from ectoderm.

23 THE RESPIRATORY SYSTEM

INTRODUCTION Have you ever swallowed something and had it go down the wrong tube, making you cough or choke uncontrollably? This uncomfortable (and sometimes embarrassing) situation occurs because the respiratory and digestive systems both arise from the embryonic gut tube and share the nose, mouth, and throat as a common initial pathway. While the majority of the gut tube gives rise to the digestive system (Chapter 24), the tube that will become the respiratory system forms a highly branched network of respiratory airways that terminate in the lungs. The respiratory tubes have the basic design features shared by all tubular anatomy: an epithelial inner lining, a muscular and connective tissue middle layer, and an outer covering layer of connective tissue. Adaptations of this basic structural plan account for the principal functions associated with the respiratory system—gas transport and gas exchange.

Cells continually use oxygen (O_2) for the metabolic reactions that release energy from nutrient molecules to produce ATP. At the same time, these reactions release carbon dioxide (CO_2). Because an excessive amount of the CO_2 that is produced can be toxic to cells, excess CO_2 must be eliminated quickly and efficiently by the cardiovascular and respiratory systems. The **respiratory system** is responsible for gas exchange—intake of O_2 and elimination of CO_2—and the cardiovascular system transports blood containing the gases between the lungs and body cells. The respiratory system also participates in regulating blood pH, contains receptors for the sense of smell, filters inhaled air, produces sounds, and rids the body of small amounts of water and heat in exhaled air.

The branch of medicine that deals with the diagnosis and treatment of disease of the ears, nose, and throat (ENT) is called **otorhinolaryngology** (o′-tō-rī-nō-lar-in-GOL-ō-jē; *oto-*=ear; *rhino-*=nose; *laryngo-*=voice box, *-logy*=study of). A **pulmonologist** (pul-mō-NOL-ō-gist; *pulmo-*=lung) is a specialist in the diagnosis and treatment of diseases of the lungs. •

Did you ever wonder how smoking affects the respiratory system?

727

23.1 RESPIRATORY SYSTEM ANATOMY

OBJECTIVES

- Describe the anatomy and histology of the nose.
- Outline the structure and function of the pharynx.
- Identify the features and purpose of the larynx.
- List the structures of voice production.
- Describe the anatomy and histology of the trachea.
- Identify the functions of each bronchial structure.
- Explain how the anatomy of the lungs makes breathing possible.

The **respiratory system** consists of the nose, nasal cavity, pharynx (throat), larynx (voice box), trachea (windpipe), bronchi, and lungs (Figure 23.1). *Structurally*, the respiratory system consists of two parts: (1) the **upper respiratory system** includes the nose, nasal cavity, pharynx, and associated structures; (2) the **lower respiratory system** includes the larynx, trachea, bronchi, and lungs.

Functionally, the respiratory system also consists of two parts. (1) The **conducting zone** consists of a series of interconnecting cavities and tubes both outside and within the lungs. These passageways include the nose, nasal cavity, pharynx, larynx, trachea, bronchi, bronchioles, and terminal bronchioles; their function is to filter, warm, and moisten air and conduct it into the lungs. (2) The **respiratory zone** consists of tubes and tissues within the lungs where gas exchange occurs. These tubes and tissues include the respiratory bronchioles, alveolar ducts, alveolar sacs, and alveoli, and are the main sites of gas exchange between air and blood.

Nose

The head has two openings through which substances such as air and food can enter the body—the nose and the mouth. While air can enter through either of these passageways, it is the nose that forms the primary entryway for inhaled air. The nose consists of much more than what you see on someone's face. In fact, the visible part of the nose makes up only about one-fourth of the entire nasal region. The **nose** is a special organ at the entrance to the respiratory system that is divided into a visible external portion and an internal portion inside the skull called the nasal cavity. The **external nose**, the skin and muscle-covered portion of the nose visible on the face, is an extension of bone and cartilage with an internal dividing wall and two entryways (the nostrils). The nasal bones project anteriorly to form the upper bony framework or "bridge" of the external nose on which a pair of glasses rest. The cartilaginous framework of the external nose is made up of several pieces of hyaline cartilage connected to each other and to the bones by tough fibrous connective tissue (Figure 23.2a). The

unpaired **septal nasal cartilage** forms the anterior portion of the **nasal septum**, a partition that divides the external and internal nose into right and left chambers (see Figure 7.9). The septal nasal cartilage is connected to the perpendicular plate of the ethmoid and vomer to form the remainder of the nasal septum. It is also connected to the nasal bones and the lateral nasal cartilages. The paired **lateral nasal cartilages** form the sides of the midportion of the external nose. They are connected to the nasal bones, maxillae, septal nasal cartilage, and major alar cartilages. The paired **major alar cartilages** (Ā-lar) form the sides of the inferior portion of the external nose. They are connected to the lateral nasal cartilages and septal nasal cartilage. The major alar cartilages form the medial and lateral borders of the nostrils. When the muscles of the nose contract and relax, the major alar cartilages dilate and constrict the nostrils. Finally, there are three or four small pieces of cartilage posterior to the major alar cartilages called the **minor alar cartilages**. Because it consists of pliable hyaline cartilage, the cartilaginous framework of the external nose is somewhat flexible. The surface anatomy of the nose is shown in Figure 27.5.

CLINICAL CONNECTION | Rhinoplasty

Rhinoplasty (RĪ-nō-plas'-tē; *rhin*=nose; *-plasty*=to mold or to shape), commonly called a "nose job," is a surgical procedure to alter the shape of the external nose. Although rhinoplasty is often done for cosmetic reasons, it is sometimes performed to repair a fractured nose or a deviated nasal septum. With anesthesia, instruments inserted through the nostrils are used to reshape the nasal cartilage and fracture and reposition the nasal bones to achieve the desired shape. An internal packing and splint keep the nose in the desired position while it heals. •

The openings into the external nose are the **external nares** (NĀ-rēz; singular is *naris*) or **nostrils**, which lead into cavities about the size of a finger tip called the **nasal vestibules**. (This is the area that is occupied by a finger that is placed in the nose.) The lower half of each nasal vestibule is lined with skin continuous with the skin of the face. This skin has numerous hairs, with sebaceous and sweat glands that secrete onto its surface. The upper lining of each nasal vestibule transitions into a mucous membrane that continues deeper into the nasal cavity.

Deeper into the skull, beyond the region of the nasal vestibules, is the **internal nose**, also called the **nasal cavity**. It is a large space in the anterior aspect of the skull that lies inferior to the nasal bone and superior to the oral cavity and forms the majority of the nose. The bony and cartilaginous framework of the nose help to keep the vestibule and nasal cavity *patent*, that is, open or unobstructed. Anteriorly, the nasal cavity merges with the external nose, and posteriorly the internal nose communicates with the pharynx through

Figure 23.1 Structures of the respiratory system.

 The upper respiratory system includes the nose, pharynx, and associated structures; the lower respiratory system includes the larynx, trachea, bronchi, and lungs.

FUNCTIONS
1. Provides for gas exchange—intake of O_2 for delivery to body cells and removal of CO_2 produced by body cells. 2. Helps regulate blood pH.	3. Contains receptors for the sense of smell, filters inspired air, produces sounds (phonation), and excretes small amounts of water and heat.

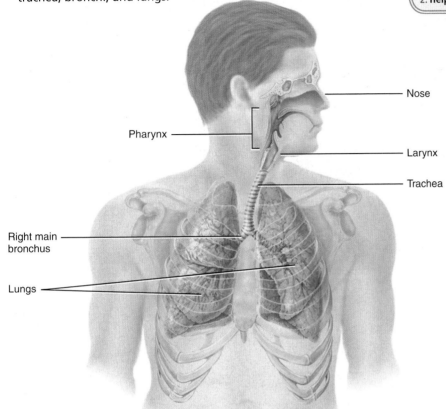

Nose

Pharynx

Larynx

Trachea

Right main bronchus

Lungs

(a) Anterior view showing organs of respiration

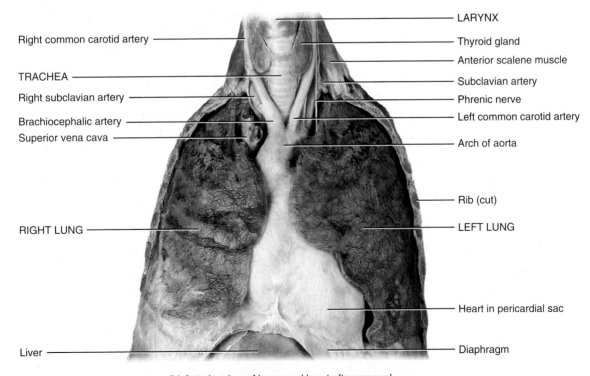

LARYNX

Right common carotid artery

Thyroid gland

Anterior scalene muscle

TRACHEA

Subclavian artery

Right subclavian artery

Phrenic nerve

Brachiocephalic artery

Left common carotid artery

Superior vena cava

Arch of aorta

Rib (cut)

RIGHT LUNG

LEFT LUNG

Heart in pericardial sac

Liver

Diaphragm

(b) Anterior view of lungs and heart after removal of the anterolateral thoracic wall and pleura

? **Which structures are part of the conducting portion of the respiratory system?**

Figure 23.2 Respiratory structures in the head and neck.

🔑 As air passes through the nose, it is warmed, filtered, and moistened, and olfaction occurs.

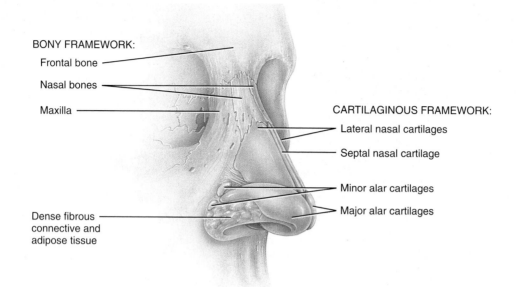

BONY FRAMEWORK:
Frontal bone
Nasal bones
Maxilla

Dense fibrous connective and adipose tissue

CARTILAGINOUS FRAMEWORK:
Lateral nasal cartilages
Septal nasal cartilage
Minor alar cartilages
Major alar cartilages

(a) Anterolateral view of nose showing cartilaginous and bony frameworks

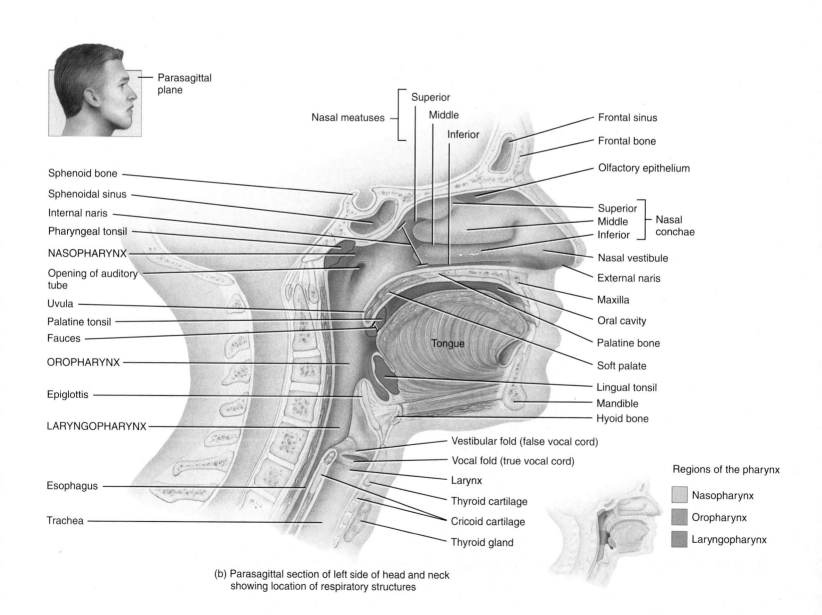

Parasagittal plane

Nasal meatuses
Superior
Middle
Inferior

Sphenoid bone
Sphenoidal sinus
Internal naris
Pharyngeal tonsil
NASOPHARYNX
Opening of auditory tube
Uvula
Palatine tonsil
Fauces
OROPHARYNX
Epiglottis
LARYNGOPHARYNX
Esophagus
Trachea

Frontal sinus
Frontal bone
Olfactory epithelium
Superior
Middle
Inferior
Nasal conchae
Nasal vestibule
External naris
Maxilla
Oral cavity
Palatine bone
Soft palate
Lingual tonsil
Mandible
Hyoid bone

Tongue

Vestibular fold (false vocal cord)
Vocal fold (true vocal cord)
Larynx
Thyroid cartilage
Cricoid cartilage
Thyroid gland

Regions of the pharynx
Nasopharynx
Oropharynx
Laryngopharynx

(b) Parasagittal section of left side of head and neck showing location of respiratory structures

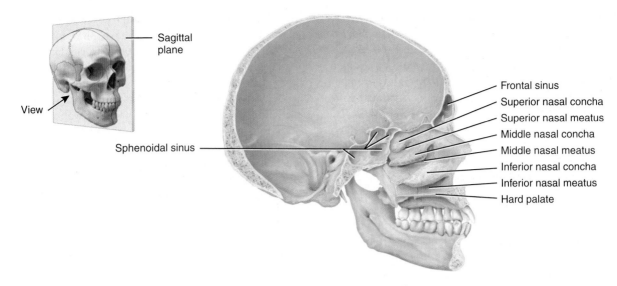

Sagittal plane

View

Frontal sinus
Superior nasal concha
Superior nasal meatus
Middle nasal concha
Middle nasal meatus
Inferior nasal concha
Inferior nasal meatus
Hard palate

Sphenoidal sinus

(c) Medial view of sagittal section

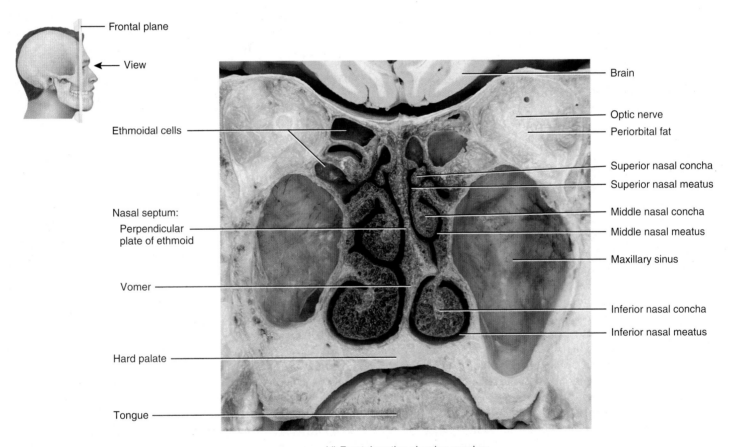

Frontal plane

View

Ethmoidal cells

Nasal septum:
  Perpendicular plate of ethmoid

Vomer

Hard palate

Tongue

Brain

Optic nerve
Periorbital fat

Superior nasal concha
Superior nasal meatus

Middle nasal concha
Middle nasal meatus

Maxillary sinus

Inferior nasal concha
Inferior nasal meatus

(d) Frontal section showing conchae

CLINICAL CONNECTION | *Tonsillectomy*

Tonsillectomy (ton-si-LEK-tō-mē; *-ektome*=excision or to cut out) is surgical removal of the tonsils. The procedure is usually performed under general anesthesia on an outpatient basis. Tonsillectomies are performed in individuals who have frequent *tonsillitis* (ton'-si-LĪ-tis), inflammation of the tonsils; those who have tonsils that develop an abcess or tumor; or when the tonsils obstruct breathing during sleep.

 What is the path taken by air molecules into and through the nose?

two openings called the **internal nares** or **choanae** (kō-A-nē) (Figure 23.2b). Ducts from the *paranasal sinuses* (frontal, sphenoidal, maxillary, and ethmoidal paranasal sinuses) and the *nasolacrimal ducts*, which drain tears from the lacrimal glands, also open into the nasal cavity (see Figure 7.13). The lateral walls of the nasal cavity are formed by the ethmoid, maxillae, lacrimal, palatine, and inferior nasal conchae bones (see Figure 7.7); the ethmoid also forms the roof. The horizontal plates of the palatine bones and palatine processes of the maxillae, which together constitute the hard palate, form the floor of the nasal cavity. The nasal cavity is divided into two regions—the large inferior *respiratory region* and the small superior *olfactory region*. The nasal cavity, like the nasal vestibules, is divided by an intermediate **nasal septum** into right and left halves. A strong impact to the nasal region can break the delicate nasal septal bones or separate the cartilage portion of the nasal septum from the bony portion. During the healing process the bones and cartilage can become displaced to one side, resulting in a **deviated septum**. This displaced septum can lead to a narrowing of one side of the nasal cavity. This makes it more difficult to breathe through that side of the nose.

Three shelves called **conchae**, formed by projections of the **superior, middle,** and **inferior nasal conchae,** extend out of each lateral wall of the nasal cavity. The conchae, almost reaching the bony nasal septum, subdivide each side of the nasal cavity into a series of groove-like passageways—the **superior, middle,** and **inferior nasal meatuses** (mē-Ā-tus-ēz=openings or passages). A mucous membrane lines the nasal cavity and its shelves. The arrangement of conchae and meatuses increases surface area in the nasal cavity and prevents dehydration by acting as a baffle that traps water droplets during exhalation.

The olfactory receptor cells, supporting cells, and basal cells lie in the olfactory region, the membrane lining the superior nasal conchae, and adjacent nasal septum. This region is called the **olfactory epithelium** (see Figure 21.1). It contains cilia but no goblet cells. Inferior to the olfactory epithelium, the mucous membrane contains capillaries and pseudostratified ciliated columnar epithelium with many goblet cells; this epithelium in the respiratory region is called the **respiratory epithelium.** As inhaled air whirls around the conchae and meatuses, it is warmed by blood circulating in the abundant capillaries. Mucus secreted by the goblet cells moistens the air and traps dust particles. Drainage from the nasolacrimal ducts and perhaps secretions from the paranasal sinuses also help moisten the air. The cilia move the mucus and trapped dust particles toward the pharynx, at which point they can be removed (i.e., swallowed or spit out) from the respiratory tract.

In summary, the interior structures of the nose have three functions: (1) warming, moistening, and filtering incoming air; (2) detecting olfactory (smell) stimuli; and (3) modifying speech vibrations as they pass through the large, hollow resonating chambers. *Resonance* refers to prolonging, amplifying, or modifying a sound by vibration.

 CHECKPOINT

1. What functions do the respiratory and cardiovascular systems have in common?
2. How do the the structures and functions of the upper and lower respiratory systems differ?
3. What is the difference between the structures and functions of the external nose and the internal nose?

Pharynx

The **pharynx** (FAR-inks), or throat, is a funnel-shaped tube about 13 cm (5 in.) long that starts at the internal nares and extends to the level of the cricoid cartilage, the most inferior cartilage of the larynx (voice box) (Figure 23.2). The pharynx lies just posterior to the nasal and oral cavities, superior to the larynx and esophagus, and just anterior to the cervical vertebrae. Its wall is composed of skeletal muscles and is lined with a mucous membrane. Relaxed skeletal muscles help keep the pharynx patent. Contraction of the skeletal muscles assists in deglutition (swallowing). The pharynx functions as a passageway for air and food, provides a resonating chamber for speech sounds, and houses the tonsils, which participate in immunological reactions against foreign invaders. The pharynx can be divided into three anatomical regions: (1) nasopharynx, (2) oropharynx, and (3) laryngopharynx. (See the lower orientation diagram in Figure 23.2b.)

The superior portion of the pharynx, called the **nasopharynx**, lies posterior to the nasal cavity and extends to the plane of the soft palate. The *soft palate*, which forms the posterior portion of the roof of the mouth, is an arch-shaped muscular partition between the nasopharynx and oropharynx that is covered by mucous membrane. There are five openings in the wall of the nasopharynx: two internal nares, two openings that lead into the *auditory (pharyngotympanic) tubes* (commonly known as the *eustachian tubes*), and the single opening into the oropharynx. The posterior wall also contains the **pharyngeal tonsil** (fa-RIN-jē-al) or **adenoid.** Through the internal nares, the nasopharynx receives air from the nasal cavity and receives packages of dust-laden mucus. The nasopharynx is lined with pseudostratified ciliated columnar epithelium, and the cilia move the mucus down toward the most inferior part of the pharynx. The nasopharynx also exchanges small amounts of air with the auditory tubes to equalize air pressure between the middle ear and the atmosphere.

The intermediate portion of the pharynx, the **oropharynx,** lies posterior to the oral cavity and extends from the soft palate inferiorly to the level of the hyoid bone. In addition to communicating upward with the nasopharynx and downward with the laryngopharynx, it has an anterior opening, the **fauces** (FAW-sēz = throat), the opening from the mouth. This portion of the pharynx has both respiratory and digestive functions because it is a common passageway for air, food, and drink. Because the oropharynx is subject to abrasion by food particles, it is lined with nonkeratinized stratified squamous epithelium. Two pairs of tonsils, the **palatine tonsils** and **lingual tonsils,** are found in the oropharynx.

The inferior portion of the pharynx, the **laryngopharynx** (la-rin′-gō-FAR-inks), or **hypopharynx,** begins at the level of the hyoid bone. At its inferior end, it opens into the esophagus (food tube) posteriorly and the larynx (voice box) anteriorly. Like the oropharynx, the laryngopharynx is both a respiratory and a digestive pathway and is lined by nonkeratinized stratified squamous epithelium.

The arterial supply of the pharynx includes the ascending pharyngeal artery, the ascending palatine branch of the facial artery, the descending palatine and pharyngeal branches of the maxillary artery, and the muscular branches of the superior thyroid artery. The veins of the pharynx are similar in name to the arteries and drain into the pterygoid plexus and the internal jugular veins.

Most of the muscles of the pharynx are innervated by nerve branches from the pharyngeal plexus supplied by the glossopharyngeal (IX) and vagus (X) nerves.

Larynx

The **larynx** (LAIR-inks), or voice box, is a short passageway that connects the laryngopharynx with the trachea. It lies in the

Hundreds of viruses can cause **coryza** (kō-RĪ-za) or the *common cold*, but a group of viruses called *rhinoviruses* (RĪ-nō-vī-rus-es) is responsible for about 40 percent of all colds in adults. Typical symptoms include sneezing, excessive nasal secretion, dry cough, and congestion. The uncomplicated common cold is not usually accompanied by a fever. Complications include sinusitis, asthma, bronchitis, ear infections, and laryngitis. Recent investigations suggest an association between emotional stress and the common cold. The higher the stress level, the greater the frequency and duration of colds.

Seasonal influenza (flu) is also caused by a virus. Its symptoms include chills, fever (usually higher than 101°F = 39°C), headache, and muscular aches. Seasonal influenza can become life-threatening and may develop into pneumonia. It is important to recognize that influenza is a respiratory disease, not a gastrointestinal (GI) disease. Many people mistakenly report having seasonal flu when they are suffering from a GI illness.

H1N1 influenza (flu), also known as *swine flu*, is a type of influenza caused by a virus called *influenza H1N1*. The term swine flu originated because early laboratory testing indicated that many of the genes in the new virus were similar to ones found in pigs (swine) in North America. However, subsequent testing revealed that the new virus is very different from the one that circulates in North American pigs.

H1N1 flu is a respiratory disorder first detected in the United States in April 2009. In June 2009, the World Health Organization declared H1N1 flu to be a *global pandemic disease* (a disease that affects large numbers of individuals in a short period of time and that occurs worldwide). The virus is spread in the same way that seasonal flu spreads: from person-to-person through coughing or sneezing or by touching infected objects and then touching one's mouth or nose. Most individuals infected with the virus have mild disease and recover without medical treatment, but some people have severe disease, and some have even died. The symptoms of H1N1 flu include fever, cough, runny or stuffy nose, headache, body aches, chills, and fatigue. Some people also have vomiting and diarrhea. Most people who have been hospitalized for H1N1 flu have had one or more preexisting medical conditions such as diabetes, heart disease, asthma, kidney disease, or pregnancy. People infected with the virus can infect others from one day before symptoms occur to 5–7 days or more after they occur.

Treatment of H1N1 flu involves taking antiviral drugs, such as Tamiflu and Relenza. A vaccine is also available. But the H1N1 flu vaccine is not a substitute for seasonal flu vaccines. In order to prevent infection, the Centers for Disease Control and Prevention (CDC) recommends washing your hands often with soap and water or with an alcohol-based hand cleaner; covering your mouth and nose with a tissue when coughing or sneezing and disposing of the tissue; avoiding touching your mouth, nose, or eyes; avoiding close contact (within six feet) with people who have flu-like symptoms; and staying home for seven days after symptoms begin or after being symptom-free for 24 hours, whichever is longer. •

midline of the neck anterior to the fourth through sixth cervical vertebrae (C4–C6).

The wall of the larynx is composed of nine pieces of cartilage (Figure 23.3). Three occur singly (thyroid cartilage, epiglottis, and cricoid cartilage), and three occur in pairs (arytenoid, cuneiform, and corniculate cartilages). Of the paired cartilages, the arytenoid cartilages are the most important because they influence the posi-

tions and tensions of the vocal folds (true vocal cords). The extrinsic muscles of the larynx connect the cartilages to other structures in the throat; the intrinsic muscles connect the cartilages to each other (see Figures 11.9 and 11.10). The **cavity of the larynx** is the space that extends from the laryngeal entrance to the inferior border of the cricoid cartilage. The portion of the cavity of the larynx above the vestibular folds is called the **laryngeal vestibule**. The portion of the cavity of the larynx below the vocal folds is called the **infraglottic cavity** (*infra*=below).

Figure 23.3 Larynx.

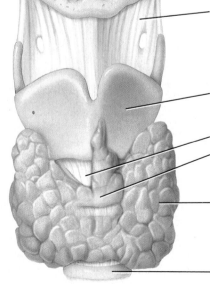

🔑 The larynx is composed of nine pieces of cartilage.

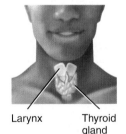

Larynx   Thyroid gland

Epiglottis
Hyoid bone
Thyrohyoid membrane
Epiglottis:
   Leaf
   Stem
Corniculate cartilage
Thyroid cartilage (Adam's apple)
Arytenoid cartilage
Cricothyroid ligament
Cricoid cartilage
Cricotracheal ligament
Thyroid gland
Parathyroid glands (4)
Tracheal cartilage

(a) Anterior view   (b) Posterior view

FIGURE 23.3 CONTINUES ▶

FIGURE 23.3 CONTINUED ▶

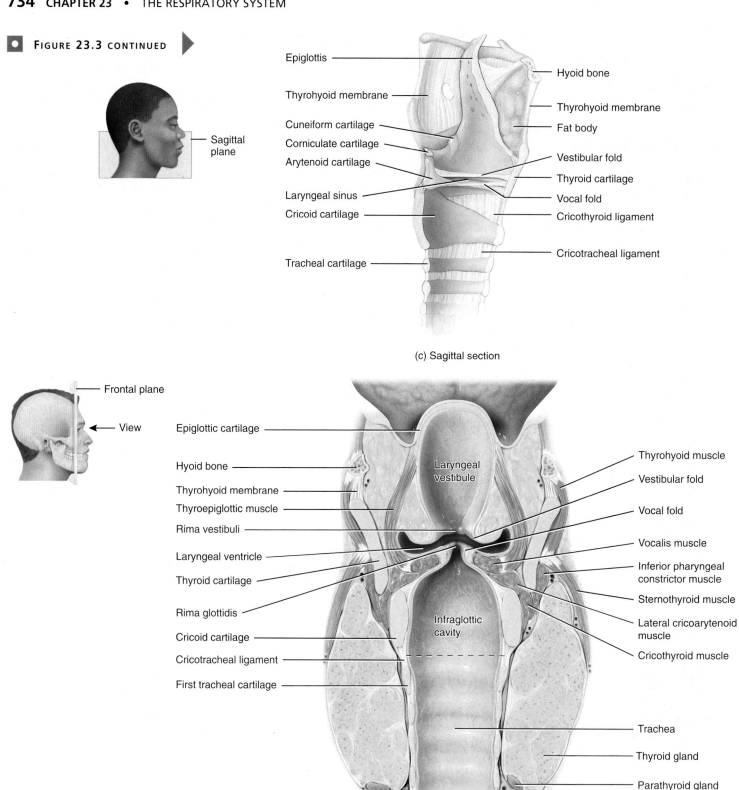

Sagittal plane

Epiglottis

Thyrohyoid membrane

Cuneiform cartilage

Corniculate cartilage

Arytenoid cartilage

Laryngeal sinus

Cricoid cartilage

Tracheal cartilage

Hyoid bone

Thyrohyoid membrane

Fat body

Vestibular fold

Thyroid cartilage

Vocal fold

Cricothyroid ligament

Cricotracheal ligament

(c) Sagittal section

Frontal plane

View

Epiglottic cartilage

Hyoid bone

Thyrohyoid membrane

Thyroepiglottic muscle

Rima vestibuli

Laryngeal ventricle

Thyroid cartilage

Rima glottidis

Cricoid cartilage

Cricotracheal ligament

First tracheal cartilage

Laryngeal vestibule

Infraglottic cavity

Thyrohyoid muscle

Vestibular fold

Vocal fold

Vocalis muscle

Inferior pharyngeal constrictor muscle

Sternothyroid muscle

Lateral cricoarytenoid muscle

Cricothyroid muscle

Trachea

Thyroid gland

Parathyroid gland

? **How does the epiglottis prevent aspiration of foods and liquids?**

(d) Frontal section

The **thyroid cartilage**, the largest cartilage of the larynx, consists of two fused plates of hyaline cartilage that form the upper anterior and lateral walls of the larynx and give it a triangular shape. The anterior junction of the two plates forms the laryngeal prominence (*Adam's apple*). It is usually larger in males than in females due to the influence of male sex hormones on its growth during puberty. Above the prominence is a V-shaped notch that can be palpated with your fingertip. The ligament that connects the thyroid cartilage to the hyoid bone just superior to it is called the **thyrohyoid membrane**.

The **epiglottis** (*epi-*=over; *glottis*=tongue) is a large, leaf-shaped piece of elastic cartilage that is covered with epithelium (Figure 23.3b, c, h). The "stem" of the epiglottis is the tapered inferior portion that is attached to the anterior rim of the thyroid cartilage. The broad superior "leaf" portion of the epiglottis is unattached and is free to move up and down like a trap door. During swallowing, the pharynx and larynx rise. Elevation of the pharynx widens it to receive food or drink; elevation of the larynx causes the epiglottis to move down and form a lid over the opening into the larynx, closing it off. The narrowed passageway through the larynx is called the

glottis. The glottis consists of a pair of folds of mucous membrane, the vocal folds in the larynx, and the space between them called the **rima glottidis** (RĪ-ma GLOT-ti-dis; Figure 23.3h). The closing of the larynx during swallowing routes liquids and foods into the esophagus and keeps them out of the larynx and airways. When small particles of dust, smoke, food, or liquids pass into the larynx, a cough reflex occurs, usually expelling the material.

The **cricoid cartilage** (KRĪ-koyd=ringlike) is a ring of hyaline cartilage that forms the inferior wall of the larynx. It is attached to the first ring of cartilage of the trachea by the **cricotracheal ligament** (krī′-kō-TRĀ-kē-al). The thyroid cartilage is connected to the cricoid cartilage by the **cricothyroid ligament**. The cricoid cartilage is the landmark for making an emergency airway called a tracheotomy.

The paired **arytenoid cartilages** (ar′-i-TĒ-noyd=ladle-like) are triangular pieces of mostly hyaline cartilage located at the posterior, superior border of the cricoid cartilage. They form synovial joints with the cricoid cartilage and have a wide range of mobility.

The paired **corniculate cartilages** (kor-NIK-ū-lāt=shaped like a small horn), horn-shaped pieces of elastic cartilage, are located at the apex of each arytenoid cartilage. The paired **cuneiform cartilages** (KŪ-nē-i-form=wedge-shaped) are club-shaped elastic cartilages anterior to the corniculate cartilages at the lateral aspect of the epiglottis.

The lining of the larynx superior to the vocal folds is non-keratinized stratified squamous epithelium. The lining of the larynx inferior to the vocal folds is pseudostratified ciliated columnar epithelium consisting of ciliated columnar cells, goblet cells, and basal cells. The mucus secreted by these cells helps trap dust not removed in the upper passages. In contrast to the action of the cilia in the upper respiratory tract, which move mucus and trapped particles *down* toward the pharynx, the cilia in the lower respiratory tract move the mucus *up* toward the pharynx.

Substances in cigarette smoke inhibit movement of cilia. If the cilia are paralyzed, only coughing can remove mucus–dust packages from the airways. This is why smokers cough so much and are more prone to respiratory infections.

The Structures of Voice Production

The mucous membrane of the larynx forms two pairs of folds (Figure 23.3c): a superior pair called the **vestibular folds (false vocal cords)** and an inferior pair called simply the **vocal folds (true vocal cords)**. The space between the vestibular folds is known as the **rima vestibuli**. The **laryngeal ventricle (sinus)** is a lateral expansion of the middle portion of the laryngeal cavity; it is bordered superiorly by the vestibular folds and inferiorly by the vocal folds. While the vestibular folds do not function in voice production, they do have other important functional roles. When the vestibular folds are brought together, they function in holding the breath against pressure in the thoracic cavity, such as might occur when a person strains to lift a heavy object.

The vocal folds are the principal structures of voice production (Figure 23.4). Deep to the mucous membrane of the vocal folds, which is nonkeratinized stratified squamous epithelium, are bands

---

Figure 23.4 Movement of the vocal folds.

 The glottis consists of a pair of folds of mucous membrane, the vocal folds in the larynx, and the space between them (the rima glottidis).

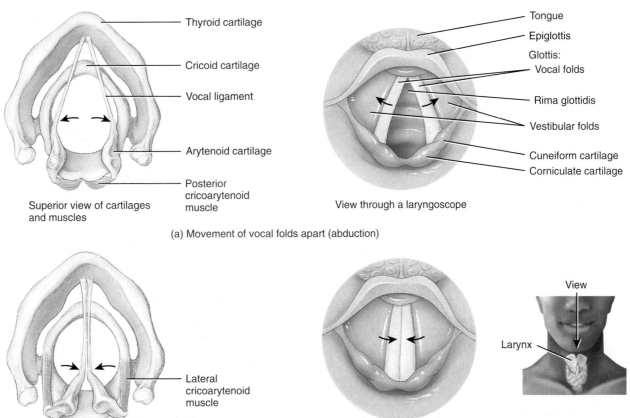

(a) Movement of vocal folds apart (abduction)

(b) Movement of vocal folds together (adduction)

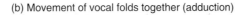 What is the main function of the vocal folds?

of elastic ligaments stretched between the rigid cartilages of the larynx like the strings on a guitar. Intrinsic laryngeal muscles attach to both the rigid cartilages and the vocal folds. When the muscles contract they move the cartilages, which pulls the elastic ligaments tight; this stretches the vocal folds out into the airways, narrowing the rima glottidis. Contracting and relaxing the muscles varies the tension in the vocal folds, much like loosening or tightening a guitar string. Air passing through the larynx vibrates the folds and produces sound (phonation) by setting up sound waves in the column of air in the pharynx, nose, and mouth. The variation in the pitch of the sound is related to the tension in the vocal folds. The greater the pressure of air, the louder the sound produced by the vibrating vocal folds.

When the intrinsic muscles of the larynx contract, they pull on the arytenoid cartilages, which causes the cartilages to pivot and slide. Contraction of the posterior cricoarytenoid muscles, for example, moves the vocal folds apart (abduction), opening the rima glottidis (Figure 23.4a). By contrast, contraction of the lateral cricoarytenoid muscles moves the vocal folds together (adduction), closing the rima glottidis (Figure 23.4b). Other intrinsic muscles can elongate (and place tension on) or shorten (and relax) the vocal folds.

Pitch is controlled by the tension on the vocal folds. If they are pulled taut by the muscles, they vibrate more rapidly, and a higher vocal pitch results. Decreasing the muscular tension on the vocal folds produces lower-pitch sounds. Due to the influence of androgens (male sex hormones), the vocal folds are usually thicker and longer in males than in females, and therefore they vibrate more slowly. Thus, men's voices generally have a lower range of pitch than women's.

Sound originates from the vibration of the vocal folds, but other structures are necessary for converting the sound into recognizable speech. The pharynx, mouth, nasal cavity, and paranasal sinuses all act as resonating chambers that give the voice its human and individual quality. We produce the vowel sounds by constricting and relaxing the muscles in the wall of the pharynx. Muscles of the face, tongue, and lips help us enunciate words.

Whispering is accomplished by closing all but the posterior portion of the rima glottidis. Because the vocal folds do not vibrate during whispering, there is no pitch to this form of speech. However, we can still produce intelligible speech while whispering by changing the shape of the oral cavity as we enunciate. As the size of the oral cavity changes, its resonance qualities change, which imparts a vowel-like pitch to the air as it rushes toward the lips.

The arteries of the larynx are the superior and inferior laryngeal arteries. The superior and inferior laryngeal veins accompany the arteries. The superior laryngeal vein empties into the superior thyroid vein, and the inferior laryngeal vein empties into the inferior thyroid vein.

The nerves of the larynx are both branches of the vagus (X) nerve. The superior laryngeal nerve enters the larynx from above, and the recurrent laryngeal nerve ascends through the base of the neck to enter the larynx from below.

CLINICAL CONNECTION | *Laryngitis and Cancer of the Larynx*

Laryngitis is an inflammation of the larynx that is most often caused by a respiratory infection or irritants such as cigarette smoke. Inflammation of the vocal folds causes hoarseness or loss of voice by interfering with the contraction of the folds or by causing them to swell to the point where they cannot vibrate freely. Many long-term smokers acquire a permanent hoarseness from the damage done by chronic inflammation. **Cancer of the larynx** is found almost exclusively in individuals who smoke. The condition is characterized by hoarseness, pain on swallowing, or pain radiating to an ear. Treatment consists of radiation therapy and/or surgery. •

Trachea

The **trachea** (TRĀ-kē-a=sturdy), or windpipe, is a tubular passageway for air that is about 12 cm (5 in.) long and 2.5 cm (1 in.) in diameter. It is located anterior to the esophagus (Figure 23.5) and

Figure 23.5 Location of the trachea in relation to the esophagus.

The trachea is anterior to the esophagus and extends from the larynx to the superior border of the fifth thoracic vertebra.

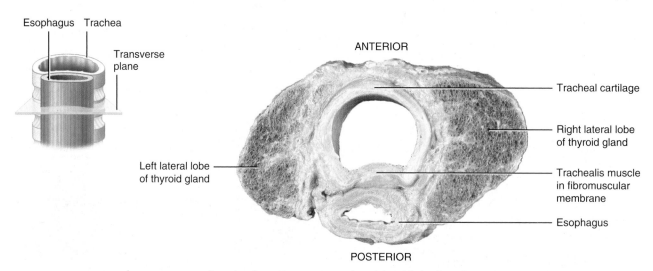

Superior view of transverse section of thyroid gland, trachea, and esophagus

 What is the benefit of not having complete rings of tracheal cartilage between the trachea and the esophagus?

extends from the larynx to the superior border of the fifth thoracic vertebra (T5), where it divides into the right and left main bronchi (see Figure 23.6). The layers of the tracheal wall, from deep to superficial, are (1) the mucosa, (2) the submucosa, (3) the fibromusculocartilaginous layer, and (4) the adventitia. The **mucosa** of the trachea consists of an epithelial layer of pseudostratified ciliated columnar epithelium and an underlying layer of lamina propria that contains elastic and reticular fibers. The epithelium consists of ciliated columnar cells and goblet cells that reach the luminal surface, plus basal cells that do not (see Table 3.1E). The epithelium provides the same protection against dust as the membrane lining the nasal cavity and larynx. The **submucosa** consists of areolar connective tissue that contains seromucous glands and their ducts.

In the fibromusculocartilaginous layer, the 16–20 incomplete horizontal rings of hyaline cartilage resemble the letter C. The rings are stacked one above another and are joined together by dense connective tissue. They may be felt through the skin inferior to the larynx. The open part of each C-shaped cartilage ring faces posteriorly toward the esophagus (Figure 23.5) and is spanned by a **fibromuscular membrane**. Within this membrane are transverse smooth muscle fibers, called the **trachealis muscle** (trā-kē-A-lis), and elastic connective tissue that allows the diameter of the trachea to change subtly during inhalation and exhalation, which is important in maintaining an efficient flow of air. The solid C-shaped cartilage rings provide a semi-rigid support to maintain patency so that the tracheal wall does not collapse inward (especially during inhalation) and obstruct the air passageway. The most superficial layer of the trachea, the **adventitia**, consists of areolar connective tissue that joins the trachea to surrounding tissues.

CLINICAL CONNECTION | *Tracheotomy and Intubation*

Several conditions may block airflow by obstructing the trachea. The tracheal cartilage may be accidentally crushed, the mucous membrane may become inflamed and swell so much that it closes off the passageway, excess mucus secreted by inflamed membranes may clog the lower respiratory passages, a large object may be aspirated (breathed in), or a cancerous tumor may protrude into the airway. Two methods are used to reestablish airflow past a tracheal obstruction. If the obstruction is above the level of the larynx, a **tracheotomy** (tra-kē-O-tō-me) may be performed. In this procedure, also called a *tracheostomy*, a skin incision is followed by a short longitudinal incision into the trachea below the cricoid cartilage. A tracheal tube is then inserted to create an emergency air passageway. The second method is **intubation** (in'-too-BĀ-shun), in which a tube is inserted into the mouth or nose and passed inferiorly through the larynx and trachea. The firm wall of the tube pushes aside any flexible obstruction, and the lumen of the tube provides a passageway for air; any mucus clogging the trachea can be suctioned out through the tube. •

The arteries of the trachea are branches of the inferior thyroid, internal thoracic, and bronchial arteries. The veins of the trachea terminate in the inferior thyroid veins.

The smooth muscle and glands of the trachea are innervated parasympathetically by branches of the vagus (X) nerves. Sympathetic innervation is through branches from the sympathetic trunk and its ganglia.

Bronchi

At the superior border of the fifth thoracic vertebra, the trachea divides into a **right main (primary) bronchus** (BRON-kus=windpipe), which goes into the right lung, and a **left main (primary) bronchus**, which goes into the left lung (Figure 23.6). The right main bronchus is more vertical, shorter, and wider than the left. As a result, an aspirated object is more likely to enter and lodge in the right main bronchus than the left. Like the trachea, the main bronchi (BRON-kī) contain incomplete rings of cartilage and are lined by pseudostratified ciliated columnar epithelium.

At the point where the trachea divides into right and left main bronchi is an internal ridge called the **carina** (ka-RĪ-na=keel of a boat). It is formed by a posterior and somewhat inferior projection of the last tracheal cartilage. The mucous membrane of the carina is one of the most sensitive areas of the entire larynx and trachea for triggering a cough reflex. Widening and distortion of the carina is a serious sign because it usually indicates a carcinoma of the lymph nodes around the region where the trachea divides.

On entering the lungs, the main bronchi divide to form smaller bronchi—the **lobar (secondary) bronchi**, one for each lobe of the lung. (The right lung has three lobes; the left lung has two.) The lobar bronchi continue to branch, forming still smaller bronchi, called **segmental (tertiary) bronchi** (TER-shē-e-rē) that supply the specific bronchopulmonary segments within the lobes. The segmental bronchi then divide into **bronchioles**. Bronchioles in turn branch repeatedly, and the smallest ones branch into even smaller tubes called **terminal bronchioles**. These bronchioles contain *Clara cells*, columnar, nonciliated cells interspersed among the epithelial cells. Clara cells may protect against harmful effects of inhaled toxins and carcinogens, produce surfactant (discussed shortly), and function as stem cells (reserve cells), which can give rise to the various cells of the epithelium. The terminal bronchioles represent the end of the conducting zone of the respiratory system. Because this extensive branching from the trachea through the terminal bronchioles resembles an inverted tree, it is commonly referred to collectively as the **bronchial tree**.

As the branching in the bronchial tree becomes more extensive, several structural changes may be noted.

1. The mucous membrane in the bronchial tree changes from pseudostratified ciliated columnar epithelium in the main bronchi, lobar bronchi, and segmental bronchi to ciliated simple columnar epithelium with some goblet cells in larger bronchioles, to mostly ciliated simple cuboidal epithelium with no goblet cells in smaller bronchioles, to mostly nonciliated simple cuboidal epithelium in terminal bronchioles. Recall that ciliated epithelium of the respiratory membrane removes inhaled particles in two ways. Mucus produced by goblet cells traps the particles and the cilia move the mucus and trapped particles toward the pharynx for removal. In regions where nonciliated simple cuboidal epithelium is present, inhaled particles are removed by macrophages.

2. Plates of cartilage gradually replace the incomplete rings of cartilage in main bronchi and finally disappear in the distal bronchioles.

3. As the amount of cartilage decreases, the amount of smooth muscle increases. Smooth muscle encircles the lumen in spiral bands and helps maintain patency. However, because there is no supporting cartilage, muscle spasms such as those that occur during an asthma attack can close off the airways, a potentially life-threatening situation.

Figure 23.6 **Branching of airways from the trachea: the bronchial tree.**

The bronchial tree consists of macroscopic airways that begin at the trachea and continue through the terminal bronchioles.

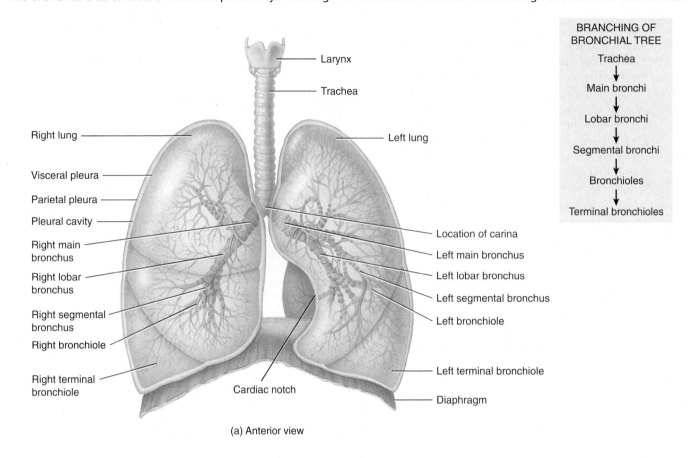

BRANCHING OF
BRONCHIAL TREE

Trachea
↓
Main bronchi
↓
Lobar bronchi
↓
Segmental bronchi
↓
Bronchioles
↓
Terminal bronchioles

(a) Anterior view

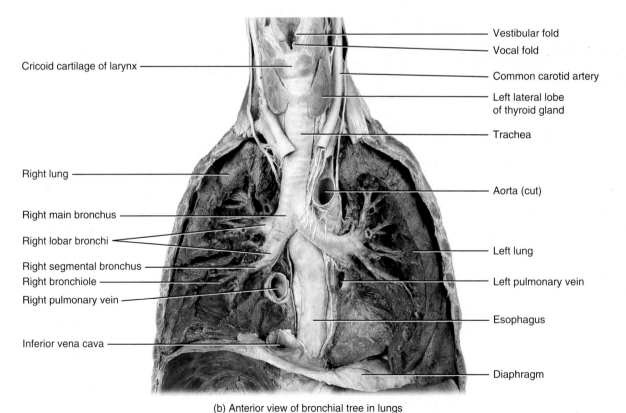

(b) Anterior view of bronchial tree in lungs

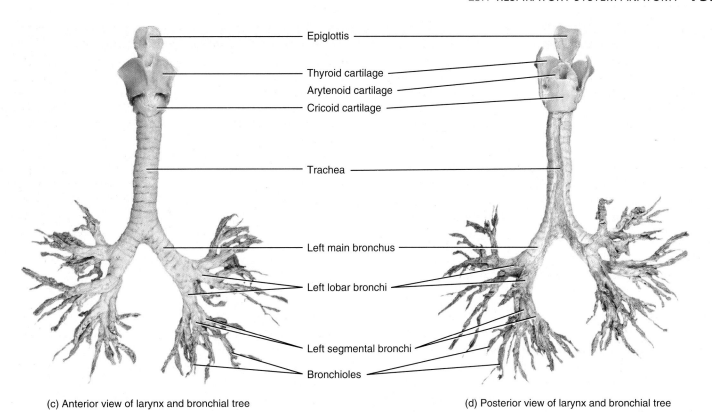

Epiglottis

Thyroid cartilage

Arytenoid cartilage

Cricoid cartilage

Trachea

Left main bronchus

Left lobar bronchi

Left segmental bronchi

Bronchioles

(c) Anterior view of larynx and bronchial tree

(d) Posterior view of larynx and bronchial tree

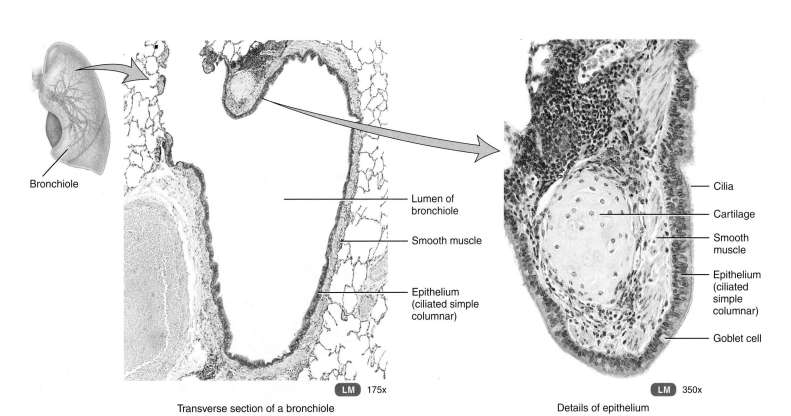

Bronchiole

Lumen of bronchiole

Smooth muscle

Epithelium (ciliated simple columnar)

LM 175x

Transverse section of a bronchiole

Cilia

Cartilage

Smooth muscle

Epithelium (ciliated simple columnar)

Goblet cell

LM 350x

Details of epithelium

(e) Histology of a bronchiole

 How many lobes and lobar bronchi are present in each lung?

CLINICAL CONNECTION | *Asthma and Chronic Bronchitis*

During an **asthma** (AZ-ma) attack, bronchiolar smooth muscle goes into spasm. Because there is no supporting cartilage, the spasms can reduce the lumen or even close off the air passageways. Movement of air through constricted bronchioles causes breathing to be more labored. The parasympathetic division of the ANS and mediators of allergic reactions such as histamine also cause narrowing of bronchioles (bronchoconstriction) due to contraction of bronchiolar smooth muscle. Because air moving through a restricted lumen causes a noise, the breathing of a true asthmatic can often be heard across the room. The principle is similar to that of a vacuum cleaner: It is so noisy because a large volume of air is moving through a small or restricted tube.

Asthmatics typically react to low concentrations of stimuli that do not normally cause symptoms in people without asthma. Sometimes the trigger is an allergen such as pollen, dust mites, molds, or a particular food. Other common triggers include emotional upset, aspirin, sulfating agents (used in wine and beer and to keep greens fresh in salad bars), exercise, and breathing cold air or cigarette smoke. Symptoms include difficult breathing, coughing, wheezing, chest tightness, tachycardia, fatigue, moist skin, and anxiety.

Chronic **bronchitis** (brong-KĪ-tis) is a disorder characterized by excessive secretion of bronchial mucus accompanied by a cough. Inhaled irritants lead to chronic inflammation with an increase in the size and number of mucous glands and goblet cells in the airway epithelium. The thickened and excessive mucus narrows the airway and impairs the action of cilia. Thus, inhaled pathogens become embedded in airway secretions and multiply rapidly. Besides a cough, symptoms of chronic bronchitis are shortness of breath, wheezing, cyanosis, and pulmonary hypertension. •

During exercise, activity in the sympathetic division of the autonomic nervous system (ANS) increases and causes the adrenal medullae to release the hormones epinephrine and norepinephrine; these hormones cause relaxation of smooth muscle in the bronchioles, which dilates the airways. The result is improved lung ventilation because air reaches the alveoli more quickly. The parasympathetic division of the ANS and mediators of allergic reactions such as histamine cause contraction of bronchiolar smooth muscle, resulting in constriction of distal bronchioles.

The blood supply to the bronchi is via the left and right bronchial arteries. The veins that drain the bronchi are the right bronchial vein, which enters the azygos vein, and the left bronchial vein, which empties into the accessory hemiazygos vein or the left superior intercostal vein.

CHECKPOINT

4. What are the roles of the three anatomical regions of the pharynx in respiration?
5. How does the larynx function in respiration and voice production?
6. Describe the location, structure, and function of the trachea.
7. What is the structure and function of the bronchial tree?

Lungs

The **lungs** (=lightweights, because they float) (Figure 23.7) are paired cone-shaped organs in the thoracic cavity. The lungs are separated from each other by the heart and other structures of the

Figure 23.7 Relationship of the pleural membranes to the lungs.

The parietal pleura lines the thoracic cavity; the visceral pleura covers the lungs.

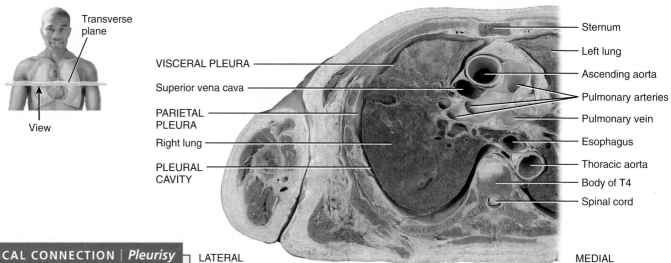

Inferior view of a transverse section through the thoracic cavity showing the pleural cavity and pleural membranes

---

CLINICAL CONNECTION | *Pleurisy*

Inflammation of the pleural membrane, called **pleurisy** or **pleuritis**, may in its early stages cause pain due to friction between the parietal and visceral layers of the pleura. If the inflammation persists, excess fluid accumulates in the pleural space, a condition known as **pleural effusion**. •

? **What type of membrane is the pleural membrane?**

mediastinum, which separates the thoracic cavity into two anatomically distinct chambers. Because of this separation, if trauma causes one lung to collapse, the other may remain expanded. Each lung is surrounded by a protective, double-layered serous membrane called the **pleural membrane** (PLOOR-al; *pleur-*=side). This membrane is easily visualized with the following analogy: Imagine that the lung is your fist and you push your fist into a balloon. The two layers of the balloon wrap around your fist separated

by the space inside the balloon (see Figure 1.7e). This is similar to the design of the pleural membrane. The superficial layer of the pleural membrane lining the wall of the thoracic cavity is called the **parietal pleura** (the part of the balloon not touching your fist); the deep layer, the **visceral pleura**, adheres to the lungs (the part of the balloon in contact with your fist). The two layers are continuous with one another where the bronchi enter the lung (at your wrist, where the balloon folds off of your fist). Between the visceral and parietal pleurae is a small space, the **pleural cavity** (the inside of the balloon), which contains a small amount of lubricating fluid secreted by the two layers. This fluid reduces friction between the membranes, allowing them to slide easily over one another during breathing. Pleural fluid also causes the pleurae to adhere to one another just as a film of water causes two glass microscope slides to stick together, a phenomenon called *surface tension*. Separate pleural cavities surround the left and right lungs.

The lungs extend from the diaphragm to just slightly superior to the clavicles and lie against the ribs anteriorly and posteriorly (Figure 23.8a). The broad inferior portion of the lung, the **base**, is concave and fits over the convex area of the diaphragm. The narrow superior portion of the lung is the **apex**. The surface of the lung lying against the ribs, the **costal surface**, matches the rounded curvature of the ribs. The **mediastinal (medial) surface** of each lung contains a region, the **hilum**, through which bronchi, pulmonary blood vessels, lymphatic vessels, and nerves enter and exit (Figure 23.8e, g). These structures are held together by the pleura and connective tissue and constitute the **root** of the lung. Medially, the left lung also contains a concavity, the **cardiac notch**, into which the apex of the heart projects. Due to the space occupied by the heart, the left lung is about 10 percent smaller than the right lung. The right lung is thicker and broader,

Figure 23.8 Surface anatomy of the lungs.

🔑 The oblique fissure divides the left lung into two lobes. The oblique and horizontal fissures divide the right lung into three lobes.

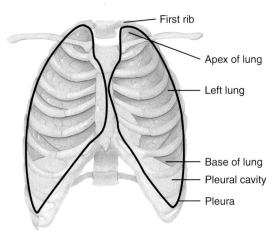

(a) Anterior view of lungs and pleurae in thorax

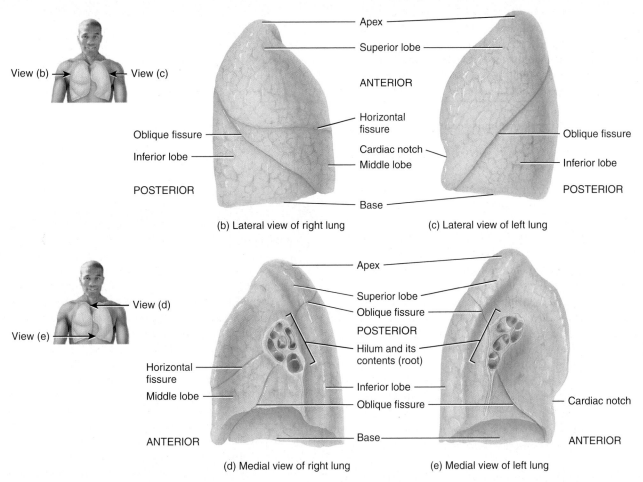

(b) Lateral view of right lung

(c) Lateral view of left lung

(d) Medial view of right lung

(e) Medial view of left lung

FIGURE 23.8 CONTINUES ▶

FIGURE 23.8 CONTINUED ▶

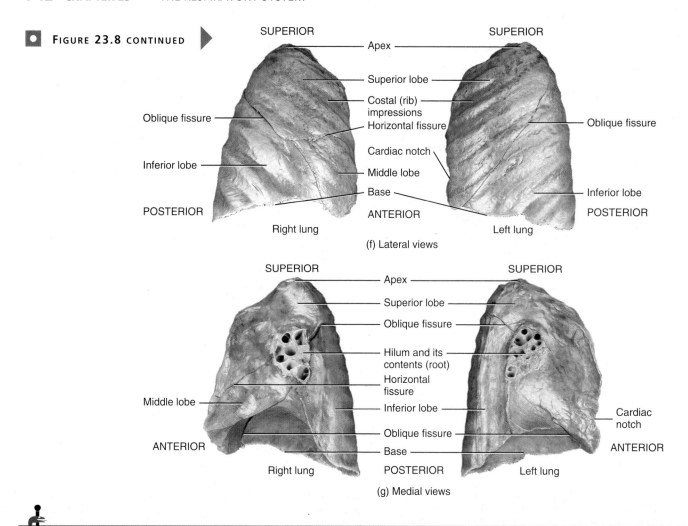

(f) Lateral views

(g) Medial views

 CLINICAL CONNECTION | *Pneumothorax and Hemothorax*

In certain conditions, the pleural cavities may fill with air (**pneumothorax**; noo′-mō-THOR-aks; *pneumo-*=air or breath), blood (**hemothorax**), or pus. Air in the pleural cavities, most commonly introduced in a surgical opening of the chest or as a result of a stab or gunshot wound, may cause the lungs to collapse. This collapse of a part of a lung, or rarely an entire lung, is called **atelectasis** (at′-e-LEK-ta-sis; *ateles-*=incomplete; *-ectasis*=expansion). The goal of treatment is the evacuation of air (or blood) from the pleural space, which allows the lung to reinflate. A small pneumothorax may resolve on its own, but it is often necessary to insert a chest tube to assist in evacuation. •

? **Why are the right and left lungs slightly different in size and shape?**

---

but it is also somewhat shorter than the left lung because the diaphragm is higher on the right side to accommodate the liver, which lies inferior to it.

The lungs almost fill the thorax (Figure 23.8a). The apex of the lungs lies superior to the medial third of the clavicles and this is the only area that can be palpated. The anterior, lateral, and posterior surfaces of the lungs lie against the ribs. The base of the lungs extends from the sixth costal cartilage anteriorly to the spinous process of the tenth thoracic vertebra posteriorly. The pleura extends about 5 cm (2 in.) below the base from the sixth costal cartilage anteriorly to the twelfth rib posteriorly. Thus, the lungs do not completely fill the pleural cavity in this area (see Figure 23.8a).

Removal of excessive fluid in the pleural cavity can be accomplished without injuring lung tissue by inserting a needle anteriorly through the seventh intercostal space, a procedure termed **thoracentesis** (thor′-a-sen-TĒ-sis; *-centesis*=puncture). The needle is passed along the superior border of the eighth rib to avoid damage to the intercostal nerves and blood vessels. Inferior to the seventh intercostal space there is danger of penetrating the diaphragm.

CLINICAL CONNECTION | *Malignant Mesothelioma*

Malignant mesothelioma (mē-zō-thē-lē-Ō-ma) is a rare cancer that affects the mesothelium of a serous membrane. The most common form, about 75 percent of all cases, affects the pleurae of the lungs (*pleural mesothelioma*). About 2000–3000 cases are diagnosed each year in the United States, accounting for about 3 percent of all cancers. The disease is almost entirely caused by asbestos, which has been widely used in insulation, textiles, cement, brake linings, gaskets, roof shingles, and floor products.

The signs and symptoms of pleural mesothelioma, which may not appear until 20–50 years or more after asbestos exposure, include chest pain, shortness of breath, pleural effusion (fluid surrounding the lungs), fatigue, anemia, blood in the sputum (fluid) coughed up, wheezing, hoarseness, and unexplained weight loss. Diagnosis is based on medical history, physical examination, radiographs, CT scans, and biopsy.

There is usually no cure, and the prognosis is poor. Chemotherapy, radiation therapy, immunotherapy, and multimodality therapy may be used to help decrease symptoms. •

Lobes, Fissures, and Lobules

Fissures divide each lung into lobes (Figure 23.8b–g). Both lungs have an **oblique fissure**, which extends inferiorly and anteriorly; the right lung also has a **horizontal fissure**. The oblique fissure in the left lung separates the **superior lobe** from the **inferior lobe**. In the right lung, the superior part of the oblique fissure separates the superior lobe from the inferior lobe; the inferior part of the oblique fissure separates the inferior lobe from the **middle lobe**, which is bordered superiorly by the horizontal fissure.

Each lobe receives its own lobar bronchus. Thus, the right main bronchus gives rise to three lobar bronchi called the **superior**, **middle**, and **inferior lobar bronchi**, and the left main bronchus

gives rise to superior and inferior lobar bronchi. Within the lung, the lobar bronchi give rise to the **segmental bronchi**; there are 10 segmental bronchi in each lung. The segment of lung tissue that each segmental bronchus supplies is called a **broncho-pulmonary segment** (brong′-kō-PUL-mō-nar′-ē) (Figure 23.9). Bronchial and pulmonary disorders (such as tumors or abscesses) that are localized in a particular bronchopulmonary segment may be surgically removed without seriously disrupting the surrounding lung tissue.

Each bronchopulmonary segment of the lungs has many small compartments called **lobules**; each lobule is wrapped in elastic connective tissue and contains a lymphatic vessel, an arteriole, a

Figure 23.9 Bronchopulmonary segments of the lungs. The bronchial branches are shown in the center of the figure. The bronchopulmonary segments within the lungs are numbered and named for convenience.

There are 10 segmental bronchi in each lung; each is composed of smaller compartments called lobules.

Superior lobe:
1 Apical
2 Posterior
3 Anterior

Middle lobe:
4 Lateral
5 Medial

Inferior lobe:
6 Superior
7 Medial basal*
8 Anterior basal
9 Lateral basal
10 Posterior basal
*Cannot be seen from this view.

POSTERIOR

Trachea

Lateral view of right lung

Right main bronchus

Left main bronchus

Lateral view of left lung

POSTERIOR

Superior lobe:
1 Apical
2 Posterior
3 Anterior

Middle lobe:
4 Lateral*
5 Medial

Inferior lobe:
6 Superior
7 Medial basal
8 Anterior basal
9 Lateral basal
10 Posterior basal
*Cannot be seen from this view.

Lobar bronchi

Right

Segmental bronchi

Left

Superior lobe:
1 Apical
2 Posterior
3 Anterior
4 Superior
5 Inferior

Inferior lobe:
6 Superior
7 Medial basal*
8 Anterior basal
9 Lateral basal
10 Posterior basal
*Cannot be seen from this view.

ANTERIOR

Superior lobe:
1 Apical
2 Posterior
3 Anterior
4 Superior
5 Inferior

Inferior lobe:
6 Superior
7 Medial basal
8 Anterior basal
9 Lateral basal
10 Posterior basal

Hilum

ANTERIOR

Medial and basal views of right lung

Medial and basal views of left lung

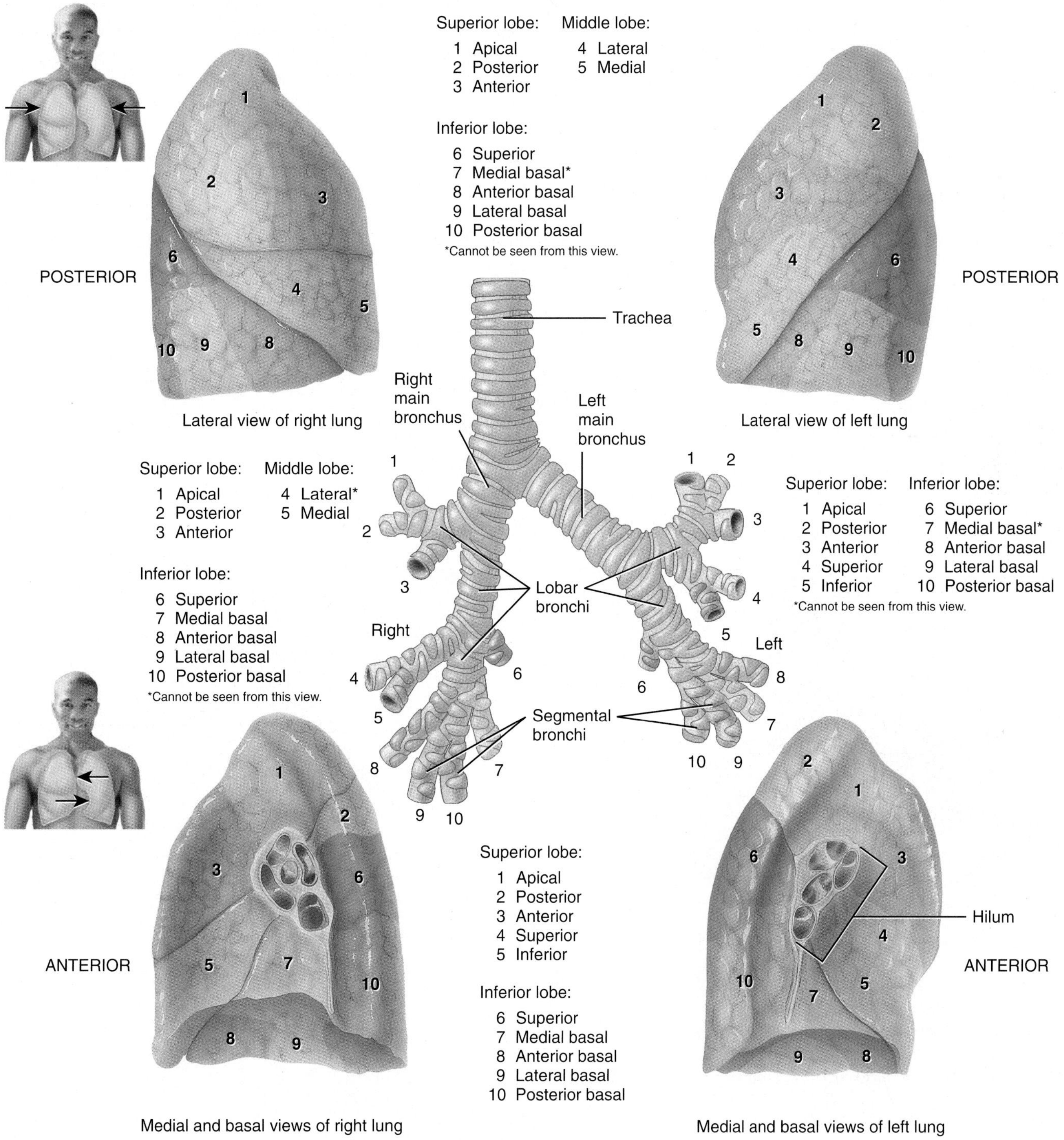

Which bronchi supply a bronchopulmonary segment?

venule, and a branch from a terminal bronchiole (Figure 23.10a). Terminal bronchioles subdivide into microscopic branches called **respiratory bronchioles** (Figure 23.10b). They also have alveoli (described shortly) budding from their walls. Because alveoli participate in gas exchange, respiratory bronchioles are the first structure in the respiratory zone of the respiratory system. As the respiratory bronchioles penetrate more deeply into the lungs, the epithelial lining changes from simple cuboidal to simple squamous. Respiratory bronchioles in turn subdivide into several (2–11) **alveolar ducts** (al-VĒ-ō-lar), which consist of simple squamous epithelium. The respiratory passages from the trachea to the alveolar ducts contain about 25 orders of branching; that is, branching occurs about 25 times—from the trachea into main bronchi (first-order branching) into lobar bronchi (second-order branching) and so on down to the alveolar ducts.

Alveoli

Around the circumference of the alveolar ducts are numerous alveoli and alveolar sacs. An **alveolus** (al-VĒ-ō-lus) is a cup-shaped outpouching lined by simple squamous epithelium and

Figure 23.10 Microscopic anatomy of a lobule of the lungs.

Alveolar sacs consist of two or more alveoli that share a common opening.

MICROSCOPIC AIRWAYS

Respiratory bronchioles
↓
Alveolar ducts
↓
Alveolar sacs
↓
Alveoli

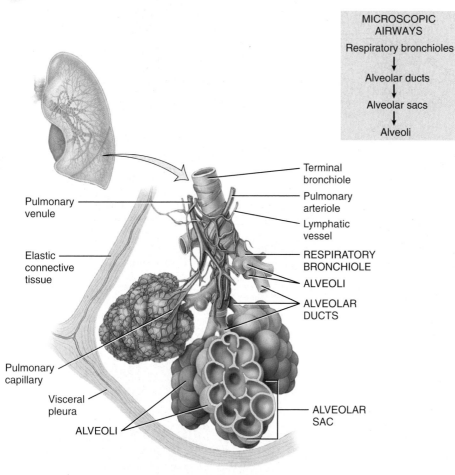

(a) Diagram of a portion of a lobule of the lung

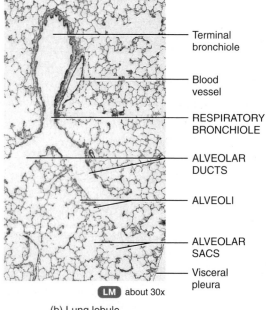

LM about 30x

(b) Lung lobule

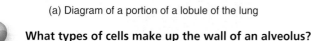

What types of cells make up the wall of an alveolus?

Figure 23.11 **Structural components of an alveolus.** The respiratory membrane consists of a layer of type I and type II alveolar cells, an epithelial basement membrane, a capillary basement membrane, and the capillary endothelium.

🔑 The exchange of respiratory gases occurs by diffusion across the respiratory membrane.

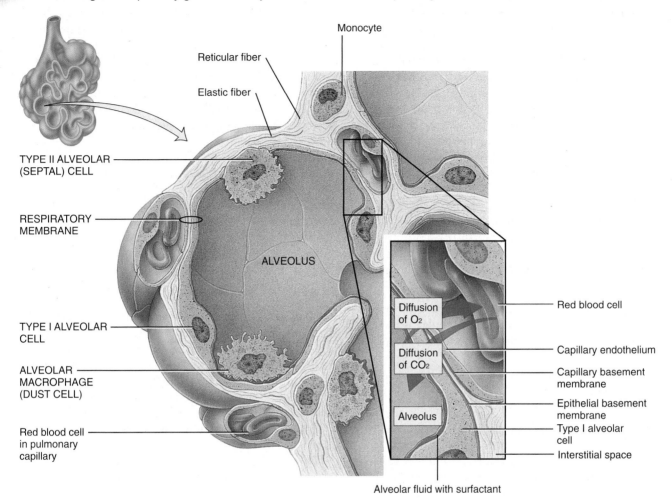

(a) Section through an alveolus showing its cellular components

(b) Details of respiratory membrane

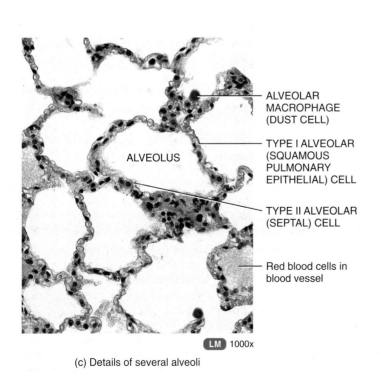

LM 1000x

(c) Details of several alveoli

❓ **How thick is the respiratory membrane?**

CLINICAL CONNECTION | *Emphysema*

Emphysema (em'-fi-SĒ-ma=blown up or full of air) is a disorder characterized by destruction of the walls of the alveoli, which produces abnormally large air spaces that remain filled with air during exhalation. With less surface area for gas exchange, O_2 diffusion across the respiratory membrane is reduced. Blood O_2 level is somewhat lowered, and any mild exercise that raises the O_2 requirements of the cells leaves the patient breathless. As increasing numbers of alveolar walls are damaged, lung elastic recoil decreases due to loss of elastic fibers, and an increasing amount of air becomes trapped in the lungs at the end of exhalation. Over several years, added respiratory exertion increases the size of the chest cage, resulting in a "barrel chest." Emphysema is a common precursor to the development of lung cancer. •

supported by a thin elastic basement membrane; an **alveolar sac** consists of two or more alveoli that share a common opening (see Figure 23.10a, b). The walls of alveoli consist of two types of alveolar epithelial cells (Figure 23.11). **Type I alveolar cells**, the predominant cells, are simple squamous epithelial cells that form a nearly continuous lining of the alveolar wall. **Type II alveolar cells**, also called **septal cells**, are fewer in number and are found between type I alveolar cells. The thin type I alveolar cells are the main sites of gas exchange. Type II alveolar cells, which are rounded or cuboidal epithelial cells with free surfaces containing microvilli, secrete alveolar fluid. This fluid keeps the surface between the cells and the air moist. Included in the alveolar fluid is surfactant (sur-FAK-tant), a complex mixture of phospholipids and lipoproteins. Surfactant lowers the surface tension of alveolar fluid, which reduces the tendency of alveoli to collapse and thus maintains their patency (described later).

Associated with the alveolar wall are **alveolar macrophages** (dust cells), wandering phagocytes that remove fine dust particles and other debris in the alveolar spaces. Also present are fibroblasts that produce reticular and elastic fibers. Underlying the layer of type I alveolar cells is an elastic basement membrane. On the outer surface of the alveoli, the pulmonary arterioles and venules disperse into a network of blood capillaries (see Figure 23.10a) that consist of a single layer of endothelial cells and basement membrane.

The exchange of O_2 and CO_2 between the air spaces in the lungs and the blood takes place by diffusion across the alveolar and capillary walls, which together form the **respiratory membrane**. Extending from the alveolar air space to blood plasma, the respiratory membrane consists of four layers (see Figure 23.11b):

1. A layer of type I and type II alveolar cells and associated alveolar macrophages that constitutes the **alveolar wall**
2. An **epithelial basement membrane** underlying the alveolar wall
3. A **capillary basement membrane** that is often fused to the epithelial basement membrane
4. The **endothelial cells** of the capillary wall

Despite having several layers, the respiratory membrane is very thin—only 0.5 μm thick, about one-sixteenth the diameter of a red blood cell. This thinness allows rapid diffusion of gases. It has been estimated that the lungs contain 300 million alveoli, providing an immense surface area of 70 m^2 (750 ft^2)—about the size of a handball court—for the exchange of gases.

A summary of the epithelial linings and special features of the organs of the respiratory system is presented in Table 23.1.

Blood Supply to the Lungs

The lungs receive blood via two sets of arteries: pulmonary arteries and bronchial arteries. Deoxygenated blood passes through the pulmonary trunk, which divides into a left pulmonary artery that enters the left lung and a right pulmonary artery that enters the right lung. Return of the oxygenated blood to the heart occurs by way of the four pulmonary veins, which drain into the left atrium (see Figure 14.16). A unique feature of pulmonary

blood vessels is their constriction in response to localized hypoxia (low O_2 level). In all other body tissues, hypoxia causes dilation of blood vessels, which serves to increase blood flow. In the lungs, however, blood vessels constrict in response to hypoxia, diverting deoxygenated blood from poorly ventilated areas to well-ventilated regions of the lungs for more efficient gas exchange.

Bronchial arteries, which branch from the aorta, deliver oxygenated blood to the lungs. This blood mainly passes to the muscular walls of the bronchi and bronchioles. Connections do exist between branches of the bronchial arteries and branches of the pulmonary arteries; most blood returns to the heart via pulmonary veins. Some blood, however, drains into bronchial veins, which are tributaries of the azygos system (see Exhibit 14.J), and returns to the heart via the superior vena cava.

The nerve supply of the lungs is derived from the pulmonary plexus, located anterior and posterior to the roots of the lungs. The pulmonary plexus is formed by branches of the vagus (X) nerves and sympathetic trunks. Motor parasympathetic fibers arise from the dorsal nucleus of the vagus (X) nerve, and motor sympathetic fibers are postganglionic fibers of the second to fifth thoracic paravertebral ganglia of the sympathetic trunk.

Patency of the Respiratory System

Throughout the discussion of the respiratory organs we have included examples of structures or secretions that help to maintain patency of the system so that air passageways are kept free of obstruction. These include the bony and cartilaginous frameworks of the nose, skeletal muscles of the pharynx, cartilages of the larynx, C-shaped rings of cartilage in the trachea and bronchi, smooth muscle in the bronchioles, and surfactant in the alveoli.

Unfortunately, numerous factors can compromise patency. These include crushing injuries to bone and cartilage, a deviated nasal septum, nasal polyps, inflammation of mucous membranes, spasms of smooth muscle, and a deficiency of surfactant.

✓ **CHECKPOINT**

8. Where are the lungs located in relation to the clavicles, ribs, and heart? Where can the lungs be palpated?
9. Distinguish the parietal pleura from the visceral pleura.
10. Define each of the following parts of a lung: base, apex, costal surface, medial surface, hilum, root, cardiac notch, lobe, and lobule.
11. What is a bronchopulmonary segment?
12. Describe the histology and function of the respiratory membrane.
13. Give several examples of structures that maintain the patency of the respiratory system.

CLINICAL CONNECTION | *Lung Cancer*

In the United States, **lung cancer** is the leading cause of cancer death in both males and females, accounting for 160,000 deaths annually. At the time of diagnosis, lung cancer is usually well advanced, with distant metastases present in about 55 percent of patients, and regional lymph node involvement in an additional 25 percent. Most people with lung cancer die within a year of the initial diagnosis; the overall survival rate is only 10–15 percent. Cigarette smoke is the most common cause of lung cancer. Roughly 85 percent of lung cancer cases are related to smoking, and the disease is 10 to 30 times more common in smokers than nonsmokers.

Exposure to secondhand smoke is also associated with lung cancer and heart disease. In the United States, secondhand smoke causes an estimated 4000 deaths a year from lung cancer, and nearly 40,000 deaths a year from heart disease. Other causes of lung cancer are ionizing radiation and inhaled irritants, such as asbestos and radon gas. Emphysema is a common precursor to the development of lung cancer.

The most common type of lung cancer, **bronchogenic carcinoma** (brong′-kō-JEN-ik), starts in the epithelium of the bronchial tubes. Bronchogenic tumors are named based on where they arise. For example, *adenocarcinomas* (ad-ēn-ō-kar-si-NŌ-mas; *adeno-*=gland)

TABLE 23.1

Summary of the Respiratory System

STRUCTURE	EPITHELIUM	CILIA	GOBLET CELLS	SPECIAL FEATURES
NOSE				
Vestibule	Nonkeratinized stratified squamous	No	No	Contains numerous hairs
Respiratory region	Pseudostratified ciliated columnar	Yes	Yes	Contains conchae and meatuses
Olfactory region	Olfactory epithelium (olfactory receptors)	Yes	No	Functions in olfaction
PHARYNX				
Nasopharynx	Pseudostratified ciliated columnar	Yes	Yes	Passageway for air; contains internal nares, openings for auditory tubes, and pharyngeal tonsil
Oropharynx	Nonkeratinized stratified squamous	No	No	Passageway for both air and food and drink; contains opening from mouth (fauces)
Laryngopharynx	Nonkeratinized stratified squamous	No	No	Passageway for both air and food and drink
LARYNX	Nonkeratinized stratified squamous above the vocal folds; pseudostratified ciliated columnar below the vocal folds	No above folds; Yes below folds	No above folds; Yes below folds	Passageway for air; contains vocal folds for voice production
TRACHEA	Pseudostratified ciliated columnar	Yes	Yes	Passageway for air; contains C-shaped rings of cartilage to keep trachea open
BRONCHI				
Main bronchi	Pseudostratified ciliated columnar	Yes	Yes	Passageway for air; contain C-shaped rings of cartilage to maintain patency
Lobar bronchi	Pseudostratified ciliated columnar	Yes	Yes	Passageway for air; contain plates of cartilage to maintain patency
Segmental bronchi	Pseudostratified ciliated columnar	Yes	Yes	Passageway for air; contain plates of cartilage to maintain patency
Larger bronchioles	Ciliated simple columnar	Yes	Yes	Passageway for air; contain more smooth muscle than in the bronchi
Smaller bronchioles	Ciliated simple columnar	Yes	No	Passageway for air; contain more smooth muscle than in the larger bronchioles
Terminal bronchioles	Nonciliated simple columnar	No	No	Passageway for air; contain more smooth muscle than in the smaller bronchioles
LUNGS				
Respiratory bronchioles	Simple cuboidal to simple squamous	No	No	Passageway for air; gas exchange
Alveolar ducts	Simple squamous	No	No	Passageway for air; gas exchange; produce surfactant
Alveoli	Simple squamous	No	No	Passageway for air; gas exchange; produce surfactant to maintain patency

☐ Conducting structures   ☐ Respiratory structures

develop in peripheral areas of the lungs from bronchial glands and alveolar cells, *squamous cell carcinomas* develop from the squamous cells in the epithelium of larger bronchial tubes, and *small (oat) cell carcinomas* develop from epithelial cells in main bronchi near the hilum of the lungs. They get their name due to their flat cell shape with little cytoplasm, and they tend to involve the mediastinum early on. Depending on the type, bronchogenic tumors may be aggressive, locally invasive, and undergo widespread metastasis. The tumors begin as epithelial lesions that grow to form masses that obstruct the bronchial tubes or invade adjacent lung tissue. Bron-

chogenic carcinomas metastasize to lymph nodes, the brain, bones, liver, and other organs.

Symptoms of lung cancer are related to the location of the tumor. These may include a chronic cough, spitting blood from the respiratory tract, wheezing, shortness of breath, chest pain, hoarseness, difficulty swallowing, weight loss, anorexia, fatigue, bone pain, confusion, problems with balance, headache, anemia, thrombocytopenia, and jaundice.

Treatment consists of partial or complete surgical removal of a diseased lung (pulmonectomy), radiation therapy, and chemotherapy. •

23.2 MECHANICS OF PULMONARY VENTILATION (BREATHING)

OBJECTIVES

• **Distinguish** among pulmonary ventilation, external respiration, and internal respiration.
• **Describe** how inhalation and exhalation occur.

Respiration is the exchange of gases between the atmosphere, blood, and body cells. It takes place in three basic steps:

1. **Pulmonary ventilation.** The first process, pulmonary ventilation (*pulmo*=lung), or **breathing**, consists of inhalation (inflow) and exhalation (outflow) of air and is the exchange of air between the atmosphere and the air spaces of the lungs.
2. **External (pulmonary) respiration.** This is the exchange of gases between the air spaces of the lungs and blood in pulmonary capillaries across the respiratory membrane. The blood gains O_2 and loses CO_2.
3. **Internal (tissue) respiration.** This is the exchange of gases between systemic capillary blood and tissue cells. The blood loses O_2 and gains CO_2.

The flow of air between the atmosphere and lungs occurs for the same reason that blood flows through the body: A pressure gradient (difference) exists. Air moves into the lungs when the pressure inside the lungs is less than the air pressure in the atmosphere. Air moves out of the lungs when the pressure inside the lungs is greater than the pressure in the atmosphere.

Inhalation

Breathing in is called **inhalation (inspiration)**. Just before each inhalation, the air pressure inside the lungs is equal to the pressure of the atmosphere, which at sea level is about 760 millimeters of mercury (mmHg), or 1 atmosphere (atm). For air to flow into the lungs, the pressure inside the alveoli must become lower than the atmospheric pressure. This condition is achieved by increasing the size of the lungs.

For inhalation to occur, the lungs must expand. This increases lung volume and thus decreases the pressure in the lungs below atmospheric pressure. The first step in expanding the alveoli of the lungs during normal quiet breathing involves contraction of the principal muscle of inhalation—the diaphragm, with assistance from the external intercostals (Figure 23.12).

The diaphragm, the most important muscle of inhalation, is a dome-shaped skeletal muscle that forms the floor of the thoracic cavity. It is innervated by fibers of the phrenic nerves, which emerge from both sides of the spinal cord at cervical levels 3, 4, and 5. Contraction of the diaphragm causes it to flatten, lowering its dome. This increases the vertical diameter of the thoracic cavity and accounts for the movement of about 75 percent of the air that enters the lungs during normal quiet inhalation. The distance the diaphragm moves during inspiration ranges from 1 cm (0.4 in.) during normal quiet breathing up to about 10 cm (4 in.) during strenuous exercise. Advanced pregnancy, excessive obesity, or confining abdominal clothing can prevent a complete descent of the diaphragm. At the same time the diaphragm is contracting, the external intercostals are at their most active stage (these muscles contract during all phases of breathing).

These skeletal muscles run obliquely downward and forward between adjacent ribs, and when these muscles contract, the ribs are pulled superiorly and the sternum is pushed anteriorly. This increases the anteroposterior and lateral diameters of the thoracic cavity. The primary role of all intercostal muscles is to keep the intercostal spaces from collapsing inward during the descent of the diaphragm, as this would reduce the thoracic volume and increase pressure.

As the diaphragm and external intercostals contract and the overall size of the thoracic cavity increases, the walls of the lungs are pulled outward. The parietal and visceral pleurae normally adhere strongly to each other because of the below-atmospheric pressure between them and because of the surface tension created by their moist adjoining surfaces. As the thoracic cavity expands, the parietal pleura lining the cavity is pulled outward in all directions, and the visceral pleura and lungs are pulled along with it, increasing the volume of the lungs.

When the volume of the lungs increases, alveolar pressure decreases from 760 to 758 mmHg. A pressure gradient is thus established between the atmosphere and the alveoli. Air moves from the atmosphere into the lungs due to a gas pressure difference, and inhalation takes place. Air continues to move into the lungs as long as the pressure difference exists.

During deep, forceful inhalation, accessory muscles of inhalation also participate in increasing the size of the thoracic cavity (Figure 23.12a). The muscles are so named because they make little, if any, contribution during normal quiet inhalation, but during exercise or forced inhalation they may contract vigorously. The accessory muscles of inhalation include the sternocleidomastoid muscles, which elevate the sternum; the scalene muscles, which elevate the first two ribs; and the pectoralis minor muscles, which elevate the third through fifth ribs.

Exhalation

Breathing out, called **exhalation (expiration)**, is also achieved by a pressure gradient, but in this case the gradient is reversed: The pressure in the lungs is greater than the pressure of the atmosphere. Normal exhalation during quiet breathing depends on two factors: (1) the recoil of elastic fibers that were stretched during inhalation and (2) the inward pull of surface tension due to the film of alveolar fluid.

Exhalation starts when the muscles of inhalation relax. As the diaphragm relaxes and the external intercostals become less active, the ribs move inferiorly; the diaphragm dome moves superiorly owing to its elasticity. These movements decrease the vertical, anteroposterior, and lateral diameters of the thoracic cavity. Also, surface tension exerts an inward pull between the parietal and visceral pleurae, and the elastic basement membranes of the alveoli and elastic fibers in bronchioles and alveolar ducts recoil. As a result, lung volume decreases and the alveolar pressure increases to 762 mmHg. Air then flows from the area of higher pressure in the alveoli to the area of lower pressure in the atmosphere.

During labored breathing and when air movement out of the lungs is impeded, muscles of exhalation—abdominal and internal intercostals—contract. Contraction of the abdominal muscles moves the inferior ribs downward and compresses the abdominal viscera, thus forcing the diaphragm superiorly. Contraction of the internal intercostals, which extend inferiorly and posteriorly between adjacent ribs, also pulls the ribs downward.

Figure 23.12 Muscles of inhalation and exhalation and their actions. The pectoralis minor muscle, a muscle of deep inhalation (not shown here), is illustrated in Figure 11.13a.

 During deep, labored breathing, accessory muscles of inhalation (sternocleidomastoid, scalene, and pectoralis minor muscles) and accessory muscles of exhalation (internal intercostal and abdominal muscles) participate.

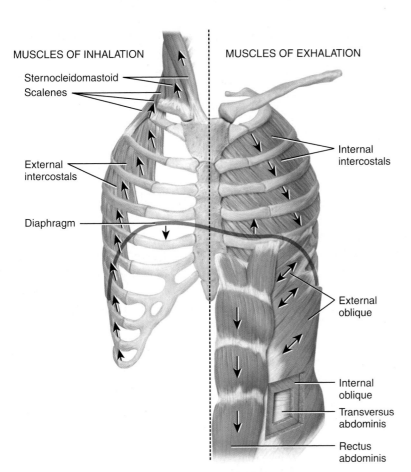

MUSCLES OF INHALATION

MUSCLES OF EXHALATION

Sternocleidomastoid
Scalenes

External intercostals

Diaphragm

Internal intercostals

External oblique

Internal oblique

Transversus abdominis

Rectus abdominis

(a) Muscles of inhalation (left); muscles of exhalation (right); arrows indicate the direction of muscle contraction

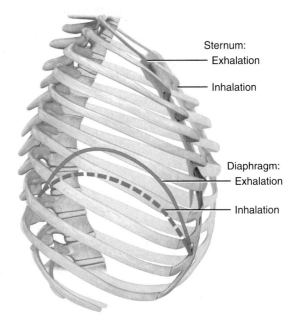

Sternum:
Exhalation
Inhalation

Diaphragm:
Exhalation
Inhalation

(b) Changes in size of thoracic cavity during inhalation and exhalation

(c) During inhalation, the lower ribs (7–10) move upward and outward like the handle on a bucket

 What is the main muscle that powers quiet breathing?

Breathing also provides humans with several methods for expressing emotions such as laughing, sighing, and sobbing. Moreover, the air involved in breathing can be used to expel foreign matter from the lower air passages through actions such as sneezing and coughing. Breathing movements can also be modified and controlled when you talk or sing. Some of the modified breathing movements that express emotion or clear the airways are listed in Table 23.2. All of these movements are reflexes, but some of them can be initiated voluntarily.

 CHECKPOINT

14. What are the basic differences among pulmonary ventilation, external respiration, and internal respiration?
15. Compare the events of quiet breathing and forceful breathing.
16. What causes hiccups?

23.3 REGULATION OF BREATHING

OBJECTIVES
• Explain the role of the respiratory center in breathing.
• Describe the various factors that regulate the rate and depth of breathing.

Although breathing can be controlled voluntarily for short periods, the nervous system usually controls breathing automatically to meet the body's demand without conscious effort.

Role of the Respiratory Center

As you have already learned, the size of the thorax is altered by the action of the respiratory muscles, which contract and relax as a result of nerve impulses transmitted to them from centers in the

TABLE 23.2

Modified Breathing Movements

MOVEMENT	DESCRIPTION
Coughing	A long-drawn and deep inhalation followed by a complete closure of the rima glottidis, which results in a strong exhalation that suddenly pushes the rima glottidis open and sends a blast of air through the upper respiratory passages. Stimulus for this reflex act may be a foreign body lodged in the larynx, trachea, or epiglottis.
Sneezing	Spasmodic contraction of muscles of exhalation that forcefully expels air through the nose and mouth. Stimulus may be an irritation of the nasal mucosa.
Sighing	A long-drawn and deep inhalation immediately followed by a shorter but forceful exhalation.
Yawning	A deep inhalation through the widely opened mouth producing an exaggerated depression of the mandible. It may be stimulated by drowsiness, or someone else's yawning, but the precise cause is unknown.
Sobbing	A series of convulsive inhalations followed by a single prolonged exhalation. The rima glottidis closes earlier than normal after each inhalation so only a little air enters the lungs with each inhalation.
Crying	An inhalation followed by many short convulsive exhalations, during which the rima glottidis remains open and the vocal folds vibrate; accompanied by characteristic facial expressions and tears.
Laughing	The same basic movements as crying, but the rhythm of the movements and the facial expressions usually differ from those of crying. Laughing and crying are sometimes indistinguishable.
Hiccupping	Spasmodic contraction of the diaphragm followed by a spasmodic closure of the rima glottidis, which produces a sharp sound on inhalation. Stimulus is usually irritation of the sensory nerve endings of the gastrointestinal tract.
Valsalva (val-SAL-va) maneuver	Forced exhalation against a closed rima glottidis as may occur during periods of straining while defecating.
Pressurizing the middle ear	The nose and mouth are held closed and air from the lungs is forced through the auditory tube into the middle ear. Employed by those snorkeling or scuba diving during descent to equalize the pressure of the middle ear with that of the external environment.

brain. The area from which nerve impulses are sent to breathing muscles consists of clusters of neurons located bilaterally in the medulla oblongata and pons of the brain stem. This area, called the **respiratory center**, can be divided into two principal areas on the basis of location and function: (1) the medullary respiratory center in the medulla oblongata and (2) the pontine respiratory group in the pons (Figure 23.13a).

Medullary Respiratory Center

The **medullary respiratory center** is made up of two collections of neurons called the **dorsal respiratory group (DRG)** formerly known as the *inspiratory area*, and the **ventral respiratory group (VRG)**, formerly called the *expiratory area* (Figure 23.13a).

During *normal quiet breathing*, neurons of the DRG generate impulses to the diaphragm via the phrenic nerves and the external intercostal muscles via the intercostal nerves (Figure 23.13a, b and Figure 23.14a). These impulses are released in bursts, which begin weakly, increase in strength for about two seconds, and then stop altogether. When the nerve impulses reach the diaphragm and external intercostal muscles, the diaphragm contracts, the external intercostal muscles contract during their most active phase, and inhalation occurs. When the DRG becomes inactive after two seconds, the diaphragm relaxes and the external intercostal muscles become less active and relax for about three seconds, allowing passive recoil of the lungs and thoracic wall. This results in exhalation. Then the cycle repeats itself. Even when all incoming nerve impulses to the DRG are cut or blocked, neurons in this area still rhythmically discharge impulses that cause inhalation. However, traumatic injury to both phrenic nerves causes paralysis of the diaphragm and cessation of breathing.

The neurons of the VRG do not participate in normal quiet inhalation. However, when forceful breathing is required, such as during exercise, playing a wind instrument, or at high altitudes, the VRG becomes activated as follows. During *forceful inhalation* (Figure 23.14b), nerve impulses from the DRG not only stimulate the diaphragm and external intercostal muscles to contract, they also activate neurons of the VRG involved in forceful inhalation to send nerve impulses to the accessory muscles of inhalation (sternocleidomastoid, scalene, and pectoralis minor muscles). Contraction of these muscles results in forceful inhalation.

During *forceful exhalation* (Figure 23.14b), the DRG is inactive along with the neurons of the VRG that result in forceful exhalation, but neurons of the VRG involved in forceful exhalation send nerve impulses to the accessory muscles of exhalation (internal intercostal, external oblique, internal oblique, transversus abdominis, and rectus abdominis muscles). Contraction of these muscles results in forceful exhalation.

Also located in the VRG is a cluster of neurons called the *pre-Bötzinger complex* (BOT-zin-ger) that is believed to be important in the generation of the rhythm of breathing (Figure 23.13a). This rhythm generator, analogous to the cardiac conduction system of the heart, is composed of pacemaker cells that set the basic rhythm of breathing. The exact mechanism of the pacemaker cells is unknown and is the topic of much ongoing research. However, it is thought that the pacemaker cells provide input to the DRG, driving the rate at which DRG neurons fire nerve impulses.

Pontine Respiratory Group

The **pontine respiratory group (PRG)** (PON-tēn), formerly called the *pneumotaxic center*, is a collection of neurons in the pons (see Figure 23.13a). The neurons in the PRG are active during inhalation and exhalation. The PRG transmits nerve impulses to the DRG in the medulla. The PRG may play a role in both inhalation and exhalation by modifying the basic rhythm of breathing by the DRG, for example, when exercising, speaking, or sleeping.

Figure 23.13 Location of areas of the respiratory center.

🔑 The respiratory center is composed of neurons in the medullary respiratory center in the medulla oblongata plus the pontine respiratory group in the pons.

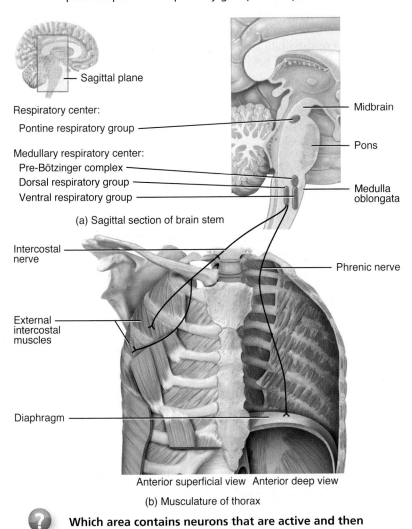

(a) Sagittal section of brain stem

Sagittal plane

Respiratory center:
Pontine respiratory group

Medullary respiratory center:
Pre-Bötzinger complex
Dorsal respiratory group
Ventral respiratory group

Midbrain

Pons

Medulla oblongata

Intercostal nerve

Phrenic nerve

External intercostal muscles

Diaphragm

Anterior superficial view Anterior deep view

(b) Musculature of thorax

❓ **Which area contains neurons that are active and then inactive in a repeating cycle?**

Regulation of the Respiratory Center

Although the basic rhythm of breathing is set and coordinated by the DRG of the medullary respiratory center, the rhythm can be modified in response to inputs from other brain regions, receptors in the peripheral nervous system, and other factors.

Cortical Influences on Breathing

Because the cerebral cortex has connections with the respiratory center, you can voluntarily alter your pattern of breathing. You can even refuse to breathe at all for a short time. Voluntary control is protective because it enables us to prevent water or irritating gases from entering the lungs. However, the ability to not breathe is limited by the buildup of CO_2 and H^+ in the body. When CO_2 and H^+ concentrations increase to a certain level, the DRG neurons are strongly stimulated, nerve impulses are sent along the phrenic and intercostal nerves to inspiratory muscles, and breathing resumes, whether you want it to or not. It is impossible for people to kill themselves by voluntarily holding their breath. Even if you hold your breath for so long that you faint, breathing resumes when consciousness is lost. Nerve impulses from the hypothalamus and limbic system also stimulate the respiratory center, allowing emotional stimuli to alter breathing as, for example, when you laugh or cry.

Figure 23.14 Medullary respiratory center. Roles of the medullary respiratory center in controlling (a) the basic rhythm of breathing and (b) forceful breathing.

🔑 During normal quiet breathing, the ventral respiratory group is inactive; during forceful breathing the dorsal respiratory group activates the ventral respiratory group.

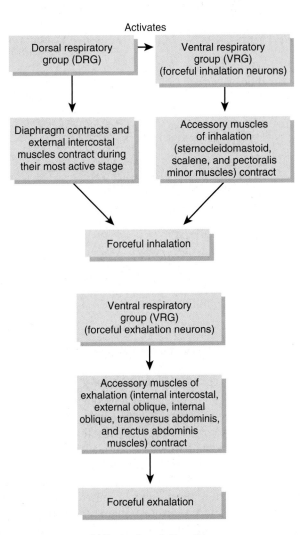

Dorsal respiratory group (DRG)

Active — 2 seconds → Diaphragm contracts and external intercostal muscles contract during their most active phase → Normal quiet inhalation

Inactive — 3 seconds → Diaphragm relaxes and external intercostal muscles become less active and relax, followed by elastic recoil of lungs → Normal quiet exhalation

(a) During normal quiet breathing

Activates

Dorsal respiratory group (DRG) → Ventral respiratory group (VRG) (forceful inhalation neurons)

Diaphragm contracts and external intercostal muscles contract during their most active stage

Accessory muscles of inhalation (sternocleidomastoid, scalene, and pectoralis minor muscles) contract

Forceful inhalation

Ventral respiratory group (VRG) (forceful exhalation neurons)

Accessory muscles of exhalation (internal intercostal, external oblique, internal oblique, transversus abdominis, and rectus abdominis muscles) contract

Forceful exhalation

(b) During forceful breathing

❓ **Which nerves convey impulses from the respiratory center to the diaphragm?**

Figure 23.15 Locations of peripheral chemoreceptors.

 Chemoreceptors are sensory neurons that respond to changes in the levels of certain chemicals in the body.

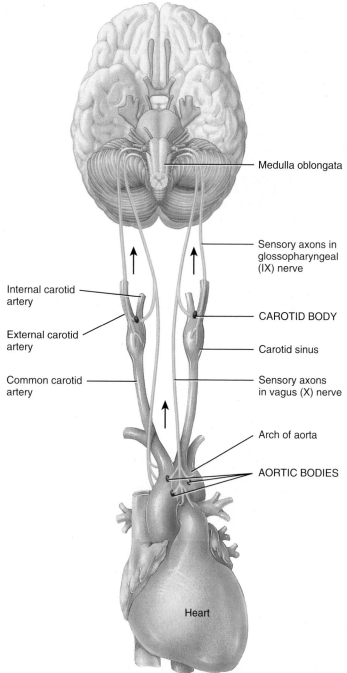

- Medulla oblongata
- Sensory axons in glossopharyngeal (IX) nerve
- Internal carotid artery
- CAROTID BODY
- External carotid artery
- Carotid sinus
- Common carotid artery
- Sensory axons in vagus (X) nerve
- Arch of aorta
- AORTIC BODIES
- Heart

 Which chemicals stimulate peripheral chemoreceptors?

Chemoreceptor Regulation of Breathing

Certain chemical stimuli determine how quickly and how deeply we breathe. The respiratory system functions to maintain proper levels of CO_2 and O_2 and is very responsive to changes in the levels of either in body fluids. Sensory neurons that are responsive to chemicals are termed **chemoreceptors**. Chemoreceptors in two locations of the respiratory system monitor levels of CO_2, H^+, and O_2 and provide input to the respiratory center. **Central chemoreceptors** are located in the medulla oblongata in the *central* nervous system. They respond to changes in H^+ or CO_2 concentration, or both, in cerebrospinal fluid. **Peripheral chemoreceptors** are located in the **aortic bodies**, clusters of chemoreceptors located in the wall of the arch of the aorta, and in the **carotid bodies**, which are oval nodules in the wall of the left and right common carotid arteries where they divide into the internal and external carotid arteries.

These chemoreceptors are part of the *peripheral* nervous system and are sensitive to changes in O_2, H^+, and CO_2 in the blood. Axons of sensory neurons from the aortic bodies are part of the vagus (X) nerves, and those from the carotid bodies are part of the right and left glossopharyngeal (IX) nerves and vagus (X) nerves (Figure 23.15).

If there is even a slight increase in CO_2, central and peripheral chemoreceptors are stimulated. The chemoreceptors send nerve impulses to the brain that cause the DRG to become highly active, and the rate of breathing increases. This allows the body to expel more CO_2 until the CO_2 is lowered to normal. If arterial CO_2 is lower than normal, the chemoreceptors are not stimulated, and stimulatory impulses are not sent to the DRG. Consequently, the rate of breathing decreases until CO_2 accumulates and the CO_2 level rises to normal.

Role of Lung Inflation in Stimulation of Breathing

In addition to all of the previously mentioned factors, receptors in the musculature of the bronchi and bronchioles throughout the lungs themselves can also modify breathing. Within these air passageways are stretch-sensitive receptors called **baroreceptors** (bar'-ō-re-SEP-tors) or **stretch receptors.** When these receptors become stretched during overinflation of the lungs, nerve impulses are sent along the vagus (X) nerves to the dorsal respiratory group (DRG) of neurons in the medullary respiratory center. In response, the DRG is inhibited, the diaphragm relaxes and the external intercostals become less active, and further inhalation is stopped. As a result, exhalation begins. As air leaves the lungs during exhalation, the lungs deflate and the stretch receptors are no longer stimulated. Thus, the DRG is no longer inhibited, and a new inhalation begins. This reflex is referred to as the **inflation (Hering-Breuer) reflex** (HER-ing BROY-er). In infants, the reflex appears to function in normal breathing. In adults, however, the reflex is not activated until tidal volume (normally 500 mL) reaches more than 1500 mL. Therefore, the reflex in adults is a protective mechanism that prevents excessive inflation of the lungs (as during exercise), rather than a key component in the normal control of breathing.

 CHECKPOINT

17. How does the medullary respiratory center function in regulating breathing?
18. How is the pontine respiratory group related to the control of breathing?
19. Explain how each of the following modifies breathing: cerebral cortex, inflation reflex, CO_2 levels, and O_2 levels.

23.4 EXERCISE AND THE RESPIRATORY SYSTEM

OBJECTIVE
- Describe the effects of exercise on the respiratory system.

The respiratory and cardiovascular systems make adjustments in response to both intensity and duration of exercise. The effects of exercise on the heart are discussed in Chapter 13. Here we focus on how exercise affects the respiratory system.

The heart pumps the same amount of blood to the lungs as to all the rest of the body. Thus, as cardiac output rises, the rate of blood flow through the lungs also increases. As blood flows more rapidly through the lungs, it picks up more O_2. In addition, the rate at which O_2 diffuses from alveolar air into the blood increases during maximal exercise because blood flows through a larger percentage of the pulmonary capillaries, providing a greater surface area for the diffusion of O_2 into blood.

Pneumonia or **pneumonitis** (nū'-mō-NĪ-tis) is an acute infection or inflammation of the alveoli. It is the most common infectious cause of death in the United States, where an estimated 4 million cases occur annually. When certain microbes enter the lungs of susceptible individuals, they release harmful toxins, stimulating inflammation and immune responses that have damaging side effects. The toxins and immune response damage alveoli and bronchial mucous membranes; inflammation and edema cause the alveoli to fill with fluid, interfering with ventilation and gas exchange.

The most common cause of pneumonia is the pneumococcal bacterium *Streptococcus pneumoniae* (see figure), but other microbes may also cause pneumonia. Those who are most susceptible to pneumonia are the elderly, infants, immunocompromised individuals, cigarette smokers, and individuals with an obstructive lung disease. Most cases of pneumonia are preceded by an upper respiratory infection that often is viral. Individuals then develop fever, chills, productive or dry cough, malaise, chest pain, and sometimes dyspnea (difficult breathing) and hemoptysis (spitting blood).

Treatment may involve antibiotics, bronchodilators, oxygen therapy, increased fluid intake, and chest physiotherapy (percussion, vibration, and postural drainage). •

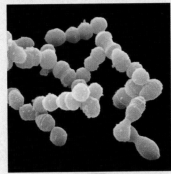

SEM about 10,000x

Streptococcus pneumoniae, the most common cause of pneumonia

When muscles contract during exercise, they consume large amounts of O_2 and produce large amounts of CO_2. During vigorous exercise, O_2 consumption and breathing both increase dramatically. At the onset of exercise, an abrupt increase in breathing is due to neural changes that send excitatory impulses to the DRG in the medulla oblongata. The more gradual increase in breathing during moderate exercise is due to chemical and physical changes in the bloodstream. During moderate exercise, the depth of breathing changes more than the breathing rate. When exercise is more strenuous, the frequency of breathing also increases.

At the end of an exercise session, an abrupt decrease in breathing is followed by a more gradual decline to the resting level. The initial decrease is due mainly to changes in neural factors when movement stops or slows; the more gradual phase reflects the slower return of blood chemistry levels and temperature to the resting state.

 CHECKPOINT

20. How does exercise affect the dorsal respiratory group?
21. Describe the changes in breathing caused by a brisk walk in the park (considered moderate exercise).

23.5 DEVELOPMENT OF THE RESPIRATORY SYSTEM

 OBJECTIVE

• Describe the development of the respiratory system.

The development of the mouth and pharynx are discussed in Chapter 24. Here we consider the development of the remainder of the respiratory system.

At about four weeks of development, the respiratory system begins as an outgrowth of the foregut just inferior to the pharynx. This outgrowth is called the **respiratory diverticulum** (dī-ver-TIK-ū-lum) or **lung bud** (Figure 23.16; see also Figure 22.8a). The **endoderm** lining the respiratory diverticulum gives rise to the epithelium and glands of the trachea, bronchi, and alveoli. **Splanchnic mesoderm** and **neural crest tissue** (see Figure 4.9d) surrounding the respiratory diverticulum gives rise to the connective tissue, cartilage, and smooth muscle of these structures.

The epithelial lining of the *larynx* develops from the endoderm of the respiratory diverticulum; the cartilages and muscles originate from the **fourth** and **sixth pharyngeal arches** (see Figure 4.13).

Figure 23.16 Development of the bronchial tubes and lungs.

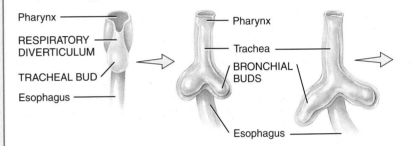 The respiratory system develops from endoderm and mesoderm.

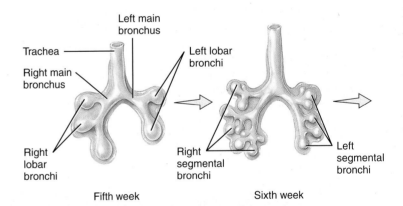

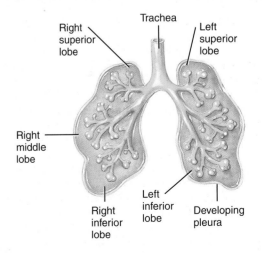

When does the respiratory system begin to develop in an embryo?

As the respiratory diverticulum elongates, its distal end enlarges to form a globular **tracheal bud**, which gives rise to the *trachea*. Soon after, the tracheal bud divides into **bronchial buds**, which branch repeatedly and develop with the *bronchi*. By 24 weeks, 17 orders of branches have formed and *respiratory bronchioles* have developed.

During weeks 6 to 16, all major elements of the *lungs* have formed, except for those involved in gaseous exchange (respiratory bronchioles, alveolar ducts, and alveoli). Since breathing is not possible at this stage, fetuses born during this time cannot survive.

During weeks 16 to 26, lung tissue becomes highly vascular and respiratory bronchioles, alveolar ducts, and some primitive alveoli develop. Although it is possible for a fetus born near the end of this period to survive if given intensive care, death frequently occurs due to the immaturity of the respiratory and other systems.

From 26 weeks to birth, many more primitive alveoli develop; they consist of type I alveolar cells (main sites of gaseous exchange) and type II surfactant-producing cells. Blood capillaries also establish close contact with the primitive alveoli. Recall that surfactant is necessary to lower surface tension of alveolar fluid and thus reduce the tendency of alveoli to collapse on exhalation. Although surfactant production begins by 20 weeks, it is present in only small quantities. Amounts sufficient to permit survival of a premature (preterm) infant are not produced until 26 to 28 weeks' gestation. Infants born before 26–28 weeks are severely at risk of **respiratory distress syndrome (RDS)**, in which the alveoli collapse during exhalation and must be reinflated during inhalation. The condition is treated by employing respirators that force air into the lungs and by administering surfactant.

At about 30 weeks, mature alveoli develop. However, it is estimated that only about one-sixth of the full complement of alveoli develop before birth; the remainder develop after birth during the first eight years.

As the lungs develop, they acquire their *pleural sacs*. The *visceral pleura* develops from splanchnic mesoderm and the *parietal pleura* develops from **somatic mesoderm** (see Figure 4.9d).

During development, breathing movements of the fetus cause the aspiration of fluid into the lungs. The fluid is a mixture of amniotic fluid, mucus from the bronchial glands, and surfactant. At birth, the lungs are about half-filled with fluid. When breathing begins at birth, most of the fluid is rapidly reabsorbed by blood and lymph capillaries; a small amount is expelled through the nose and mouth during delivery.

 CHECKPOINT

22. What structures develop from the respiratory diverticulum?
23. Which respiratory structures develop from endoderm? From mesoderm?
24. How many weeks old must a fetus be for it to survive as a preterm infant? Why?

 ## 23.6 AGING AND THE RESPIRATORY SYSTEM

OBJECTIVE

• Describe the effects of aging on the respiratory system.

With advancing age, the airways and tissues of the respiratory tract, including the alveoli, become less elastic and more rigid; the chest wall becomes more rigid as well. The result is a decrease in lung capacity. In fact, vital capacity (the maximum amount of air that can be exhaled after maximal inhalation) can decrease as much as 35 percent by age 70. A decrease in blood level of O_2, decreased activity of alveolar macrophages, and diminished ciliary action of the epithelium lining the respiratory tract also occur. Because of all these age-related factors, older people are more susceptible to pneumonia, bronchitis, emphysema, and other pulmonary disorders. Age-related changes in the structure and functions of the lung can also contribute to an older person's reduced ability to perform vigorous exercises, such as running.

 CHECKPOINT

25. What accounts for the decrease in lung capacity with aging?

KEY MEDICAL TERMS ASSOCIATED WITH THE RESPIRATORY SYSTEM

Abdominal thrust (Heimlich) maneuver (HĪM-lik ma-NOO-ver) First-aid procedure designed to clear the airways of obstructing objects. It is performed by applying a quick upward thrust between the navel and costal margin that causes sudden elevation of the diaphragm and forceful, rapid expulsion of air from the lungs; this action forces air out of the trachea to eject the obstructing object. The abdominal thrust maneuver is also used to expel water from the lungs of near-drowning victims before resuscitation is begun.

Apnea (AP-nē-a; *a*=without; *pnoia*=air or breath) Absence of breathing movements.

Carbon monoxide (CO) poisoning Elevated level of carbon monoxide in the body, which can cause the lips and oral mucosa to appear bright, cherry-red (the color of hemoglobin with carbon monoxide bound to it). Without prompt treatment, carbon monoxide poisoning is fatal. It is possible to rescue a victim of CO poisoning by administering pure oxygen, which speeds up the separation of carbon monoxide from hemoglobin.

Chronic obstructive pulmonary disease (COPD) A type of respiratory disorder characterized by chronic and recurrent obstruction of airflow, which increases airway resistance. COPD affects about 30 million Americans and is the fourth leading cause of death behind heart disease, cancer, and cerebrovascular disease. The principal types of COPD are emphysema and chronic bronchitis.

Hypoxia (hī-POK-sē-a; *hypo-*=below or under) A deficiency of O_2 at the tissue level.

Pulmonary edema An abnormal accumulation of fluid in the interstitial spaces and alveoli of the lungs. The edema may arise from increased permeability of the pulmonary capillaries (pulmonary origin) or increased pressure in the pulmonary capillaries (cardiac origin); the latter cause may coincide with congestive heart failure. The most common symptom is dyspnea. Others include wheezing, tachypnea (rapid breathing rate), restlessness, a feeling of suffocation, cyanosis, pallor (paleness), diaphoresis (excessive perspiration), and pulmonary hypertension.

Pulmonary embolism (EM-bō-lizm) The blockage of a pulmonary artery or its branches by a blood clot that travels to the lungs, usually from a vein in a leg or the pelvis.

Respirator (RES-pi-rā′-tor) An apparatus fitted to a mask over the nose and mouth, or hooked directly to an endotracheal or tracheotomy tube, that is used to assist or support breathing or to provide nebulized medication to the air passages.

Sudden infant death syndrome (SIDS) Death of infants between the ages of 1 week and 12 months thought to be due to hypoxia while sleeping in a prone position (on the stomach) and the rebreathing of exhaled air trapped in a depression of the mattress. It is now recommended that normal newborns be placed on their backs for sleeping: "back to sleep."

CHAPTER REVIEW AND RESOURCE SUMMARY

WileyPLUS

Review

Resource

23.1 Respiratory System Anatomy

1. The respiratory system consists of the nose, pharynx, larynx, trachea, bronchi, and lungs. The respiratory system acts with the cardiovascular system to supply oxygen (O_2) to and remove carbon dioxide (CO_2) from the blood.
2. The external nose is made of cartilage and skin and is lined with a mucous membrane. Openings to the exterior are the external nares. The internal nose communicates with the paranasal sinuses and nasopharynx through the internal nares. The nasal cavity is divided by a nasal septum. The anterior portion of the cavity is called the nasal vestibule. The nose warms, moistens, and filters air and functions in olfaction and speech.
3. The pharynx (throat) is a muscular tube lined by a mucous membrane. The anatomical regions of the pharynx are the nasopharynx, oropharynx, and laryngopharynx. The nasopharynx functions in respiration. The oropharynx and laryngopharynx function both in respiration and digestion.
4. The larynx (voice box) is a passageway that connects the pharynx with the trachea. It contains the thyroid cartilage (Adam's apple); the epiglottis, which prevents food from entering the larynx; the cricoid cartilage, which connects the larynx and trachea; and the paired arytenoid, corniculate, and cuneiform cartilages. The vocal folds of the larynx produce sound as they vibrate; taut folds produce high pitches, and relaxed ones produce low pitches.
5. The trachea (windpipe) extends from the larynx to the main bronchi. It is composed of C-shaped rings of cartilage and smooth muscle and is lined with pseudostratified ciliated columnar epithelium.
6. The bronchial tree consists of the trachea, main bronchi, lobar bronchi, segmental bronchi, bronchioles, and terminal bronchioles. Walls of bronchi contain rings of cartilage; walls of bronchioles contain increasingly smaller plates of cartilage and increasing amounts of smooth muscle.
7. Lungs are paired organs in the thoracic cavity enclosed by the pleural membrane. The parietal pleura is the superficial layer that lines the thoracic cavity; the visceral pleura is the deep layer that covers the lungs. The right lung has three lobes separated by two fissures; the left lung has two lobes separated by one fissure, along with a depression, the cardiac notch.
8. Lobar bronchi give rise to branches called segmental bronchi, which supply segments of lung tissue called bronchopulmonary segments. Each bronchopulmonary segment consists of lobules, which contain lymphatics, arterioles, venules, terminal bronchioles, respiratory bronchioles, alveolar ducts, alveolar sacs, and alveoli.
9. Alveolar walls consist of type I alveolar cells, type II alveolar cells, and associated alveolar macrophages.
10. Gas exchange occurs across the respiratory membranes.

Anatomy Overview - Overview of Respiratory Organs
Anatomy Overview - Nasal Cavity and Pharynx
Anatomy Overview - Respiratory Tissues
Anatomy Overview - Respiratory Epithelium
Figure 23.3 - Larynx
Figure 23.7 - Relationship of the Pleural Membrane to the Lung
Figure 23.11 - Structure of an Alveolus
Exercise - Build an Airway
Exercise - Concentrate on Respiratory Structures
Concepts and Connections - Functional Anatomy of the Respiratory System

23.2 Mechanics of Pulmonary Ventilation (Breathing)

1. Pulmonary ventilation, or breathing, consists of inhalation (inspiration) and exhalation (expiration).
2. Inhalation occurs when alveolar pressure falls below atmospheric pressure. Contraction of the diaphragm and external intercostals increases the size of the thorax, decreasing the intrapleural pressure so that the lungs expand. Expansion of the lungs decreases alveolar pressure so that air moves down a pressure gradient from the atmosphere into the lungs.
3. During forceful inhalation, accessory muscles of inhalation (sternocleidomastoids, scalenes, and pectoralis minors) are also used.
4. Exhalation occurs when alveolar pressure is higher than atmospheric pressure. Relaxation of the diaphragm and external intercostals results in elastic recoil of the chest wall and lungs, which increases intrapleural pressure; lung volume decreases and alveolar pressure increases, so air moves from the lungs to the atmosphere.
5. Forceful exhalation involves contraction of the internal intercostal and abdominal muscles.

Animation - Pulmonary Ventilation
Animation - Gas Exchange Introduction
Animation - Gas Exchange - Internal and External Respiration
Animation - Gas Transport
Exercise - Carbon Dioxide Transport Try-out
Interactive Exercise - Gas Exchange Match-up
Exercise - Concentrate on Respiration
Concepts and Connections - Carbon Dioxide Transport
Concepts and Connections - Ventilation

23.3 Regulation of Breathing

1. The respiratory center consists of a medullary respiratory center in the medulla oblongata and the pontine respiratory group in the pons.
2. The medullary respiratory center is made up of a dorsal respiratory group (DRG), which controls normal quiet breathing, and a ventral respiratory group (VRG), which is used during forceful breathing and controls the rhythm of breathing.
3. The pontine respiratory group may modify the rhythm of breathing during exercise, speaking, and sleep.
4. Breathing may be modified by a number of factors, including cortical influences; the inflation reflex; and chemical stimuli, such as O_2, CO_2, and H^+ levels.

Anatomy Overview - Structures That Control Respiration
Animation - Regulation of Ventilation

Review   **Resource**

23.4 Exercise and the Respiratory System

1. The rate and depth of breathing change in response to both the intensity and duration of exercise.
2. An increase in pulmonary blood flow and O_2-diffusing capacity occurs during exercise.
3. The abrupt increase in breathing at the start of exercise is due to neural changes that send excitatory impulses to the DRG in the medulla oblongata. The more gradual increase in breathing during moderate exercise is due to chemical and physical changes in the bloodstream.

23.5 Development of the Respiratory System

1. The respiratory system begins as an outgrowth of endoderm called the respiratory diverticulum.
2. Smooth muscle, cartilage, and connective tissue of the bronchial tubes and pleural sacs develop from mesoderm.

23.6 Aging and the Respiratory System

1. Aging results in decreased vital capacity, decreased blood level of O_2, and diminished alveolar macrophage activity.
2. Older people are more susceptible to pneumonia, emphysema, bronchitis, and other pulmonary disorders.

CRITICAL THINKING QUESTIONS

1. Your friend Hedge wants to pierce his nose to go along with the 6 earrings in his ear. He thinks a ring through the center would be awesome but wonders if there's a difference between piercing the center versus the side of the nose. Is there?

2. Suzanne is traveling high into the Andes Mountains of South America to visit some ancient Incan ruins. Though she is in excellent physical condition, she finds herself breathing rapidly. What is happening to Suzanne and why?

3. Upon placement of an endotracheal tube (ET) into an anesthetized patient, the anesthesiology resident realized that air sounds were coming from the epigastric region rather than from the lungs. What went wrong?

4. Gretchen, who is nine years old, is upset because after finally persuading her daddy to play soccer with her, he had to sit down and rest after only 15 minutes. She indignantly berated him, saying, "You know, Daddy, if you'd stop smoking, you could play longer." Outline the specific breathing difficulties that make it so hard for Gretchen's father to catch his breath.

5. Latasha is losing her patience with her little sister, LaTonya. "I'm going to hold my breath 'til I turn blue and die and then you're gonna get it!" screams LaTonya. Latasha is not too worried. Why not?

? ANSWERS TO FIGURE QUESTIONS

23.1 The conducting portion of the respiratory system includes the nose, nasal cavity, pharynx, larynx, trachea, bronchi, and bronchioles (except the respiratory bronchioles).

23.2 The path of air is external nares → nasal vestibule → nasal cavity → internal nares.

23.3 During swallowing, the epiglottis closes over the rima glottidis, the entrance to the trachea, to prevent aspiration of food and liquids into the lungs.

23.4 The main function of the vocal folds is voice production.

23.5 Because the tissues between the esophagus and trachea are soft, the esophagus can bulge and press against the trachea during swallowing.

23.6 The left lung has two lobes and two lobar bronchi; the right lung has three of each.

23.7 The pleural membrane is a serous membrane.

23.8 Because two-thirds of the heart lies to the left of the midline, the left lung contains a cardiac notch to accommodate the position of the heart. The right lung is shorter than the left because the diaphragm is higher on the right side to accommodate the liver.

23.9 Segmental bronchi supply a bronchopulmonary segment.

23.10 The wall of an alveolus is made up of type I alveolar cells, type II alveolar cells, and associated alveolar macrophages.

23.11 The respiratory membrane averages 0.5 mm in thickness.

23.12 The diaphragm is responsible for about 75 percent of each inhalation during quiet breathing.

23.13 The DRG contains neurons that have cycles of activity/inactivity.

23.14 The phrenic nerves innervate the diaphragm.

23.15 Peripheral chemoreceptors are responsive to changes in the partial pressures of oxygen and carbon dioxide, and concentrations of H^+ in the blood.

23.16 The respiratory system begins to develop about 4 weeks after fertilization.

24 | THE DIGESTIVE SYSTEM

INTRODUCTION Food contains a variety of *nutrients*, the molecules needed to build new body tissues, repair damaged tissues, and sustain needed chemical reactions. Food is also vital to life—it is the energy source that drives the chemical reactions occurring in every cell of the body. Despite the many jokes to the contrary, you can't apply that piece of chocolate cake directly to your stomach or hips; most food cannot be used as a source of cellular energy. It must first be broken down into molecules small enough to cross the plasma membranes of the cells of the digestive tract. This breakdown process is known as **digestion** (*dis=* apart; *gerere=*to carry). The passage of these smaller molecules through cells into the blood and lymph is termed **absorption** (ab-SORP-shun). The organs that perform these functions—collectively called the **digestive system**—are the focus of this chapter. The digestive system forms an extensive surface area in contact with the external environment and is closely associated with the cardiovascular system. The combination of extensive environmental exposure and close association with blood vessels is essential for processing the food that we eat.

The medical specialty that deals with the structure, function, diagnosis, and treatment of diseases of the stomach and intestines is called **gastroenterology** (gas'-trō-en'-ter-OL-ō-jē; *gastro-=*stomach; *enter-=*intestines; *-ology=*study of). The medical specialty that deals with the diagnosis and treatment of disorders of the rectum and anus is called **proctology** (prok-TOL-ō-jē; *proct-=*rectum). •

Did you ever wonder why some people are sensitive to dairy products?

24.1 OVERVIEW OF THE DIGESTIVE SYSTEM

OBJECTIVES
- Identify the organs of the digestive system.
- Describe the basic processes performed by the digestive system.

The digestive system (Figure 24.1) is composed of two groups of organs: the gastrointestinal (GI) tract and the accessory digestive organs. The **gastrointestinal (GI) tract**, or **alimentary canal** (*alimentary*=nourishment), is a continuous tube that extends from the mouth to the anus through the thoracic and abdominopelvic cavities. Organs of the gastrointestinal tract include the mouth, pharynx, esophagus, stomach, small intestine, and large intestine. The length of the GI tract is variable. It is about 5–7 meters (16.5–23 ft) in a living person when the muscles along the walls of GI tract organs are in a state of *tonus* (sustained contraction). It is longer in a cadaver (about 7–9 meters or 23–29.5 ft) because of the loss of muscle tone after death. The **accessory digestive organs** include the teeth, tongue, salivary glands, liver, gallbladder, and pancreas. Teeth aid in the physical breakdown of food, and the tongue assists in chewing and swallowing. The other accessory digestive organs never come into direct contact with food. They produce or store secretions that flow into the GI tract through ducts and aid in the chemical breakdown of food.

The GI tract contains food and its byproducts from the time it is eaten until it is digested and absorbed or eliminated. Muscular contractions in the wall of the GI tract physically break down the food by churning it, and propel the food along the tract, from the esophagus to the anus. The contractions also help to dissolve foods by mixing them with fluids secreted into the tract. Enzymes secreted by accessory digestive organs and cells that line the tract break down the food chemically.

Overall, the digestive system performs six basic functions:

1. *Ingestion.* **Ingestion** involves taking foods and liquids into the mouth (eating).
2. *Secretion.* Each day, cells within the walls of the GI tract and accessory digestive organs secrete a total of about 7 liters of water, acid, buffers, and enzymes into the lumen (interior space) of the tract; this process is called **secretion**.
3. *Mixing and propulsion.* Alternating contraction and relaxation of smooth muscle in the walls of the GI tract mix food and secretions and move them toward the anus. This capability of the GI tract to mix and move material along its length is termed **motility** (mō-TIL-i-tē).
4. *Digestion.* Mechanical and chemical processes break down ingested food into small molecules. In **mechanical digestion** the teeth cut and grind food before it is swallowed, and then smooth muscles of the stomach and small intestine churn the food to further assist the process. As a result, food molecules become dissolved and thoroughly mixed with digestive enzymes. **Chemical digestion** is the breakdown of the large carbohydrate, lipid, protein, and nucleic acid molecules present in food into smaller molecules that can be absorbed (see next step). Digestive enzymes produced by the salivary glands, tongue, stomach, pancreas, and small intestine speed up these breakdown reactions. A few substances in food can be absorbed without chemical digestion, including amino acids, cholesterol, glucose, vitamins, minerals, and water.
5. *Absorption.* The entrance of ingested and secreted fluids, ions, and the products of digestion into the epithelial cells lining the lumen of the GI tract is called **absorption**. The absorbed substances pass into the blood or lymph and circulate to cells throughout the body.
6. *Defecation.* Wastes, indigestible substances, bacteria, cells sloughed from the lining of the GI tract, and digested materials that were not absorbed leave the body through the anus in

Figure 24.1 Organs of the digestive system. Accessory digestive organs, including the teeth, tongue, salivary glands, liver, gallbladder, and pancreas, are indicated in red.

The organs of the gastrointestinal (GI) tract include the mouth, pharynx, esophagus, stomach, small intestine, and large intestine.

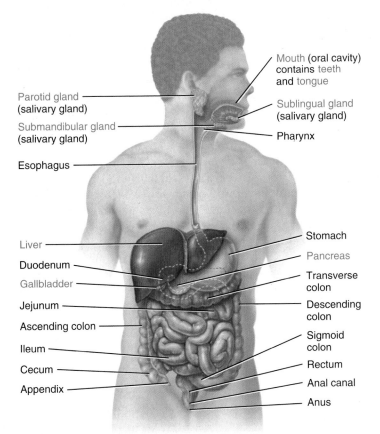

(a) Right lateral view of head and neck and anterior view of trunk

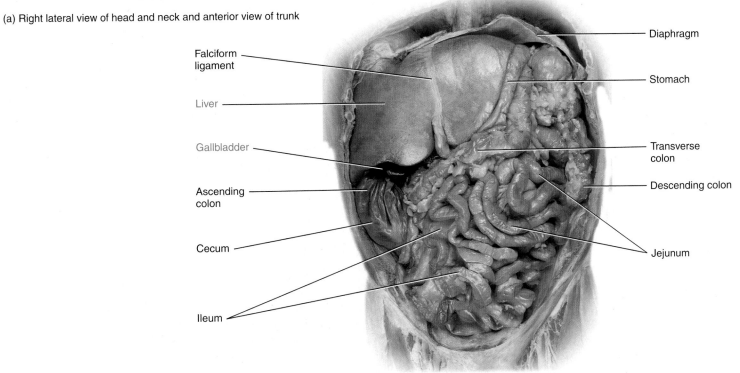

(b) Anterior view

Which structures of the digestive system secrete digestive enzymes?

a process called **defecation** (def'-e-KĀ-shun). The eliminated material is called **feces** (FĒ-sēz) or **stool**.

CHECKPOINT

1. Which components of the digestive system are GI tract organs and which are accessory digestive organs?
2. Which organs of the digestive system come in contact with food and help to break it down?

24.2 LAYERS OF THE GI TRACT

OBJECTIVE

• Describe the layers that form the wall of the gastrointestinal tract.

Like other tubular systems in the body, the wall of the GI tract, from the esophagus to the anal canal, has a basic arrangement of four layers. The four layers of the tract, from deep to superficial, are (1) the mucosa, (2) the submucosa, (3) the muscularis, and (4) the serosa/adventitia (Figure 24.2).

Mucosa

The **mucosa**, or inner lining of the GI tract, is a mucous membrane. It is composed of (1) a layer of epithelium in direct contact with the contents of the GI tract, (2) areolar connective tissue, and (3) a thin layer of smooth muscle (muscularis mucosae).

1. The **epithelium** in the mouth, pharynx, esophagus, and anal canal is mainly nonkeratinized stratified squamous epithelium that serves a protective function. Simple columnar epithelium, which functions in secretion and absorption, lines the stomach and intestines. Neighboring simple columnar epithelial cells are firmly sealed to one another by tight junctions that restrict leakage between the cells. The rate of renewal of GI tract epithelial cells is rapid: Every 5–7 days they slough off and are replaced by new cells. Located among the absorptive epithelial cells are exocrine cells that secrete mucus and fluid into the lumen of the tract, and several types of endocrine cells, collectively called **enteroendocrine cells** (en'-ter-ō-EN-dō-krin), which secrete hormones.

2. The **lamina propria** (*lamina*=thin, flat plate; *propria*=one's own) is an areolar connective tissue layer containing many blood and lymphatic vessels that carry the nutrients absorbed by the GI tract back to the heart. This layer supports the epithelium and binds it to the muscularis mucosae (discussed next). The lamina propria also contains most of the cells of the **mucosa-associated lymphatic tissue (MALT)**. These prominent lymphatic nodules contain immune system cells that protect against disease. MALT is present all along the GI tract, especially in the tonsils, small intestine, appendix, and large intestine, and it contains about as many immune cells as are present in all the rest of the body. The lymphocytes and macrophages in MALT mount immune responses against microbes, such as bacteria, that may penetrate the epithelium.

Figure 24.2 Layers of the gastrointestinal tract.

🔑 The four layers of the gastrointestinal tract, from deep to superficial, are the mucosa, submucosa, muscularis, and serosa.

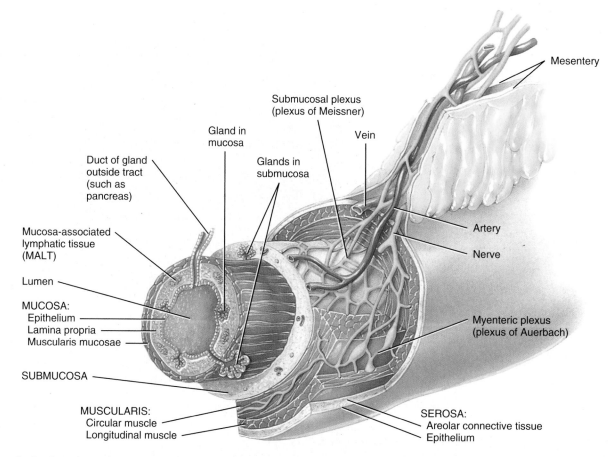

 What is the function of the enteric plexuses in the wall of the gastrointestinal tract?

3. A thin layer of smooth muscle fibers called the **muscularis mucosae** (mū-KŌ-sē) causes the mucous membrane of the stomach and small intestine to form many small folds, increasing the surface area for digestion and absorption. Movements of the muscularis mucosae ensure that all absorptive cells are fully exposed to the contents of the GI tract. The muscularis mucosae of the small intestine also surround the lacteal lymph vessels. Contractions of these smooth muscle cells help move the lymph along these vessels.

Submucosa

The **submucosa** is a thin meshwork of collagenous fibers, nerves, and blood vessels. It consists of areolar connective tissue that binds the mucosa to the middle layer, the muscularis. The submucosa is highly vascular and contains the **submucosal plexus**, or *plexus of Meissner* (MĪS-ner), a portion of the autonomic nervous system (ANS) called the **enteric nervous system (ENS)** (see Section 19.5). The ENS is the "brain of the gut" and consists of approximately 100 million neurons in two main enteric plexuses that extend the entire length of the GI tract. The submucosal plexus (plexus of Meissner) contains sensory and motor enteric neurons, plus parasympathetic and sympathetic postganglionic fibers that innervate the mucosa and submucosa. It regulates movements of the mucosa and vasoconstriction of blood vessels. Because it also innervates secretory cells of mucosal glands, the submucosal plexus is important in controlling secretions of the GI tract. The submucosa may also contain glands and lymphatic tissue.

Muscularis

The **muscularis** of the mouth, pharynx, and superior and middle parts of the esophagus contains *skeletal muscle* that produces voluntary swallowing. Skeletal muscle also forms the external anal sphincter, which permits voluntary control of defecation. Throughout the rest of the tract, the muscularis consists of *smooth muscle* that is generally found in two sheets: an inner sheet of *circular fibers* and an outer sheet of *longitudinal fibers*. Involuntary contractions of the smooth muscles assist in the mechanical breakdown of food, mix it with digestive secretions, and propel it along the tract. The muscularis also contains the second major plexus of the enteric nervous system—the **myenteric plexus** (*my-*=muscle), or *plexus of Auerbach* (OW-er-bak), which contains enteric neurons, parasympathetic ganglia and postganglionic fibers, and sympathetic postganglionic fibers that are vasomotor to the blood vessels of this layer. This plexus mostly controls GI tract motility (movement), in particular the frequency and strength of the contractions of the muscularis.

Serosa

Those portions of the GI tract that are suspended in the abdominal cavity have a superficial layer called the **serosa**. As its name implies, the serosa is a serous membrane composed of areolar connective tissue and simple squamous epithelium (mesothelium). The epithelial portion of the serosa is also called the *visceral peritoneum* because it forms the portion of the peritoneum that surrounds the organs suspended in the peritoneal cavity, which we will examine in detail shortly. The name *serous membrane* arises from the fact that the epithelium has a lubricating coat of *serous fluid*, a watery solution of electrolytes and other solutes derived from interstitial fluid of the adjacent tissues, along with blood plasma from local capillaries. The serous fluid also

contains various white blood cells. When the highly coiled digestive tube rubs against neighboring areas of the tube or against the inside of the body wall, the serous membrane protects the outer wall of the gut from abrasion. The esophagus and lower aspect of the rectum are the only organs of the GI tract that completely lack a serosa; instead, only a single layer of areolar connective tissue called the *adventitia* surrounds them.

 CHECKPOINT

3. Where along the GI tract is the muscularis composed of skeletal muscle? Is control of this skeletal muscle voluntary or involuntary?

4. What two plexuses form the enteric nervous system, and where are they located?

24.3 PERITONEUM

 OBJECTIVE

• Describe the peritoneum and its folds.

The **peritoneum** (per'-i-tō-NĒ-um; *peri-*=around) is the largest serous membrane of the body; it consists of a layer of simple squamous epithelium (mesothelium) with an underlying supporting layer of connective tissue. The peritoneum is divided into the **parietal peritoneum**, which lines the wall of the abdominal cavity, and the **visceral peritoneum** or serosa, which as you just learned covers some of the organs in the cavity (Figure 24.3a).

The slim space between the parietal and visceral portions of the peritoneum, called the **peritoneal cavity**, contains lubricating serous fluid. In certain diseases, the peritoneal cavity may become distended by the accumulation of several liters of fluid, a condition called **ascites** (a-SĪ-tēz).

As we will see, some organs lie against the posterior abdominal wall and do not project into the peritoneum. These organs are called **retroperitoneal organs** (*retro*=behind). Some, such as the ascending and descending colon, duodenum, and pancreas, are covered by peritoneum only on their anterior surfaces. Other retroperitoneal organs are separated from the peritoneum by fat and have no peritoneum on them at all, including the kidneys and adrenal glands.

Unlike the pericardium and pleurae, which smoothly cover the heart and lungs, the peritoneum contains large fat-filled folds where the mesothelium *reflects* (folds back) from the parietal peritoneum to the visceral peritoneum or from visceral peritoneum of one organ to that of another organ. The folds bind the organs to one another and to the walls of the abdominal cavity. They also contain blood vessels, lymphatic vessels, and nerves that supply the abdominal organs. There are six major peritoneal folds: the greater omentum, falciform ligament, lesser omentum, mesentery, transverse mesocolon, and sigmoid mesocolon.

The **greater omentum** (ō-MEN-tum=fat skin), a large peritoneal fold, drapes over the transverse colon and coils of the small intestine like a "fatty apron" (Figure 24.3). The greater omentum is a double sheet that folds back upon itself, giving it a total of four layers. From attachments along the greater curvature of the stomach and initial part of the duodenum, the greater omentum extends downward anterior to the small intestine, then turns and extends upward and attaches to the transverse colon. The greater omentum normally contains a considerable amount of adipose tissue. Its adipose tissue content can greatly expand with weight gain, contributing to the characteristic "beer belly" seen in some overweight individuals. The many lymph nodes of the greater omentum contribute macrophages and antibody-producing plasma cells that help combat and contain infections of the GI tract.

The **falciform ligament** (FAL-si-form; *falc-*=sickle-shaped) attaches the liver to the anterior abdominal wall and diaphragm (Figure 24.3b, e). This remnant of the ventral mesentery of the embryo was the path of the umbilical vein from the umbilical cord to the inferior vena cava in the fetus. The liver is the only digestive organ that is attached to the anterior abdominal wall.

The **lesser omentum** arises as an anterior fold of the serosa of the stomach and duodenum. It connects the stomach and duodenum to the liver (Figure 24.3a, c). It is the pathway of blood vessels entering the liver and contains the hepatic portal vein, the common hepatic artery, and the common bile duct, along with some lymph nodes.

Another fold of the peritoneum, called the **mesentery** (MEZ-en-ter'-ē; *mes-*=middle), is fan-shaped and binds the jejunum and ileum of the small intestine to the posterior abdominal wall (Figure 24.3a, d). This is the largest peritoneal fold and is typically laden with fat, which contributes extensively to a large abdomen in obese individuals. It extends from the posterior abdominal wall to wrap around almost the entire length of the small intestine and then returns to its origin, forming a double-layered structure. Between the two layers are blood vessels (branches and tributaries of the superior mesenteric artery and vein), lymphatic vessels, and lymph nodes associated with the jejunum and ileum. In order to fit the entire small intestine with its large mesentery into the abdominal cavity, the intestine and mesentery are folded together like a fan to make them more compact.

Two separate folds of peritoneum called the **mesocolon** (mez'-ō-KŌ-lon) bind the transverse colon (*transverse mesocolon*) and sigmoid colon (*sigmoid mesocolon*) of the large intestine to the posterior abdominal wall (Figure 24.3a); these folds carry blood vessels (superior and inferior mesenteric vessels) and lymphatic vessels to the intestines. The mesentery and mesocolon hold the intestines loosely in place, allowing for a great amount of movement as muscular contractions mix and move the intestinal contents along the GI tract.

CLINICAL CONNECTION | *Peritonitis*

A common cause of **peritonitis** (per'-i-tō-NĪ-tis), an acute inflammation of the peritoneum, is contamination of the peritoneum by infectious microbes, which can result from accidental or surgical wounds in the abdominal wall, or from perforation or rupture of abdominal organs. For example, if bacteria gain access to the peritoneal cavity through an intestinal perforation or rupture of the appendix, they can produce an acute, life-threatening form of peritonitis. A less serious (but still painful) form of peritonitis can result from the rubbing together of inflamed peritoneal surfaces. The increased risk of peritonitis is of particular concern to those who rely on peritoneal dialysis, a procedure in which the peritoneum is used to filter the blood when the kidneys do not function properly. •

Figure 24.3 Relationship of the peritoneal folds to each other and to organs of the digestive system. The size of the peritoneal cavity in (a) is exaggerated for emphasis.

🔑 The peritoneum is the largest serous membrane in the body.

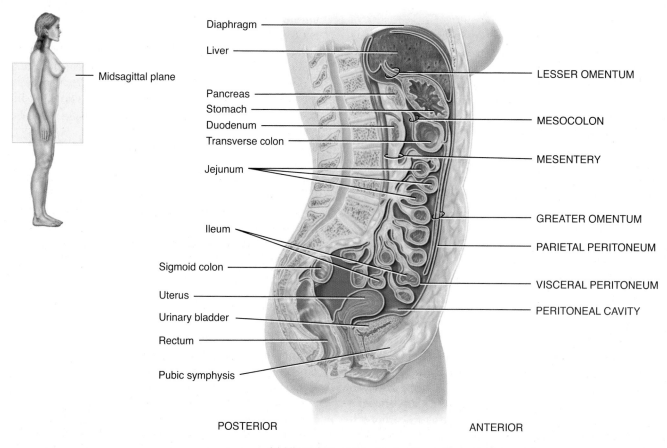

(a) Midsagittal section showing the peritoneal folds

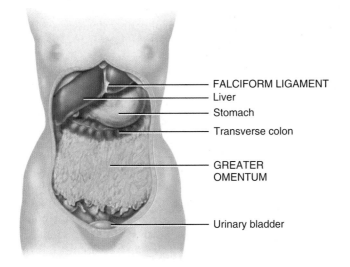

FALCIFORM LIGAMENT
Liver
Stomach
Transverse colon
GREATER OMENTUM
Urinary bladder

(b) Anterior view

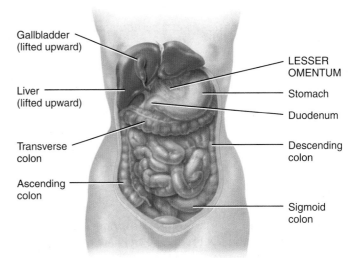

Gallbladder (lifted upward)
Liver (lifted upward)
Transverse colon
Ascending colon
LESSER OMENTUM
Stomach
Duodenum
Descending colon
Sigmoid colon

(c) Lesser omentum, anterior view
(liver and gallbladder lifted)

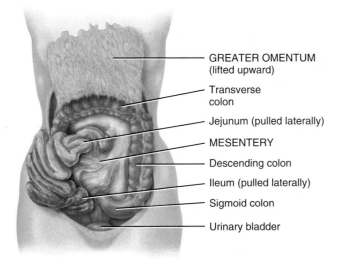

GREATER OMENTUM (lifted upward)
Transverse colon
Jejunum (pulled laterally)
MESENTERY
Descending colon
Ileum (pulled laterally)
Sigmoid colon
Urinary bladder

(d) Anterior view (greater omentum lifted and small intestine moved to right side)

SUPERIOR

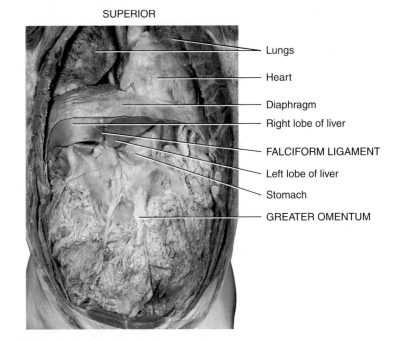

Lungs
Heart
Diaphragm
Right lobe of liver
FALCIFORM LIGAMENT
Left lobe of liver
Stomach
GREATER OMENTUM

(e) Anterior view

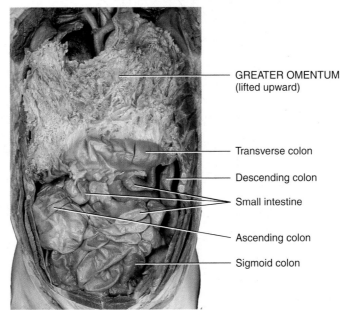

GREATER OMENTUM (lifted upward)
Transverse colon
Descending colon
Small intestine
Ascending colon
Sigmoid colon

 Which peritoneal fold binds the small intestine to the posterior abdominal wall?

(f) Anterior view

CHECKPOINT

5. Describe the locations of the visceral peritoneum and parietal peritoneum.
6. Describe the attachment sites and functions of the mesentery, mesocolon, falciform ligament, lesser omentum, and greater omentum.

24.4 MOUTH

 OBJECTIVES

• Identify the locations of the salivary glands, and describe the functions of their secretions.
• Describe the structure and functions of the tongue.
• Identify the parts of a typical tooth, and compare the deciduous and permanent dentitions.

The **mouth**, also referred to as the **oral** or **buccal cavity** (BUK-al; *bucca*=cheeks), is formed by the cheeks, hard and soft palates, and tongue (Figure 24.4). The **cheeks** form the lateral walls of the oral cavity. They are covered externally by skin and internally by a mucous membrane, which consists of nonkeratinized stratified squamous epithelium. Buccinator muscles and connective tissue lie between the skin and mucous membranes of the cheeks. The anterior portions of the cheeks end at the lips.

The **lips** or **labia** (=fleshy borders) are fleshy folds surrounding the opening of the mouth. They contain the orbicularis oris muscle and are covered externally by skin and internally by a mucous membrane. The inner surface of each lip is attached to its corresponding gum by a midline fold of mucous membrane called the **labial frenulum** (LĀ-bē-al FREN-ū-lum; *frenulum*=small bridle). During chewing, contraction of the buccinator muscles in the cheeks and orbicularis oris muscle in the lips helps keep food between the upper and lower teeth. These muscles also assist in speech.

The cheeks and lips play important roles in procuring food and keeping it in the mouth. It would be impossible to suckle milk without these structures. Look at a newborn puppy: It has a short snout, with a reduced mouth opening and definite cheeks, much different from an adult dog. This arrangement allows the puppy to suckle. As the dog grows, the snout and mouth opening increase in size, while the cheeks are greatly reduced. Without the cheeks and lips it is impossible to create a tight seal to generate suction. The cheeks and lips also serve important nondigestive functions. For example, the creation of many sounds we make during speech depends on certain lip formations. Blowing air from your mouth to whistle or extinguish a candle is also a function of your cheeks and lips.

The **oral vestibule** (*vestibule*=entrance to a canal) of the oral cavity is the space bounded externally by the cheeks and lips and internally by the gums and teeth. The **oral cavity proper** is the space that extends from the gums and teeth to the **fauces** (FAW-sēs=passages), the opening between the oral cavity and the oropharynx.

The **palate** is a wall or septum that separates the oral cavity from the nasal cavity, and forms the roof of the mouth. This important structure makes it possible to chew and breathe at the same time. The **hard palate**—the anterior two-thirds of the palate—is formed by the maxillae and palatine bones and is covered by a mucous membrane. The **soft palate**, which forms the posterior

Figure 24.4 Structures of the mouth (oral cavity).

 The mouth is formed by the cheeks, hard and soft palates, and tongue.

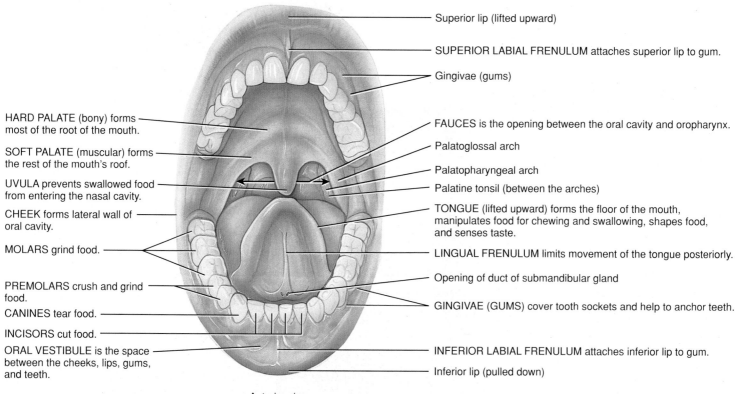

Anterior view

 What is the function of the uvula?

portion of the roof of the mouth, is an arch-shaped muscular partition between the oropharynx and nasopharynx that is lined with mucous membrane.

Hanging from the free border of the soft palate is a finger-like muscular structure called the **uvula** (Ū-vū-la=little grape). During swallowing, the soft palate and uvula are drawn superiorly, closing off the nasopharynx and preventing swallowed foods and liquids from entering the nasal cavity. Lateral to the base of the uvula are two muscular folds that run down the lateral sides of the soft palate: (1) anteriorly, the **palatoglossal arch** (pal-a-tō-GLOS-al) extends to the side of the base of the tongue; (2) posteriorly, the **palatopharyngeal arch** (pal-a-tō-fa-RIN-jē-al) extends to the side of the pharynx. The palatine tonsils are situated between the arches, and the lingual tonsils are situated at the base of the tongue. At the posterior border of the soft palate, the mouth opens into the oropharynx through the fauces (see Figure 24.4).

Salivary Glands

A **salivary gland** (SAL-i-ver-ē) is a gland that releases a secretion called saliva into the oral cavity. Ordinarily, just enough saliva is secreted to keep the mucous membranes of the mouth and pharynx moist and to cleanse the mouth and teeth. When food enters the mouth, however, secretion of saliva increases to lubricate, dissolve, and begin the chemical breakdown of the food.

The mucous membrane of the mouth and tongue contains many small salivary glands that open directly into the oral cavity, or indirectly via short ducts. These glands, which include *labial*, *buccal*, and *palatal glands* in the lips, cheeks, and palate, respectively, and *lingual glands* in the tongue, all make a small contribution to saliva.

However, most saliva is secreted by the **major salivary glands**, which lie beyond the oral mucosa. Their secretions empty into ducts that lead to the oral cavity. The three pairs of major salivary glands are the parotid, submandibular, and sublingual glands (Figure 24.5a). The **parotid glands** (pa-ROT-id; *par-*=near; *ot-*=ear) are located inferior and anterior to the ears, between the skin and the masseter muscle. Each secretes saliva into the oral cavity via a **parotid duct** or *Stensen's duct* that pierces the buccinator muscle to open into the vestibule opposite the second maxillary (upper) molar tooth. The **submandibular glands** (sub'-man-DIB-ū-lar), found in the floor of the oral cavity beneath the base of the tongue, are medial and partly inferior to the mandible. Their ducts, the **submandibular ducts** (*Wharton's ducts*), run under the mucosa on either side of the midline of the floor of the mouth and enter the oral cavity proper lateral to the lingual frenulum. The **sublingual glands** (sub'-LING-gwal) are superior to the submandibular glands. Their ducts, the **lesser sublingual ducts** (*Rivinus' ducts*), open into the floor of the mouth in the oral cavity proper.

Figure 24.5 The three major salivary glands—parotid, sublingual, and submandibular. The submandibular glands are magnified in the light micrograph in (c).

⚷ Saliva lubricates and dissolves foods and begins the chemical breakdown of carbohydrates and lipids.

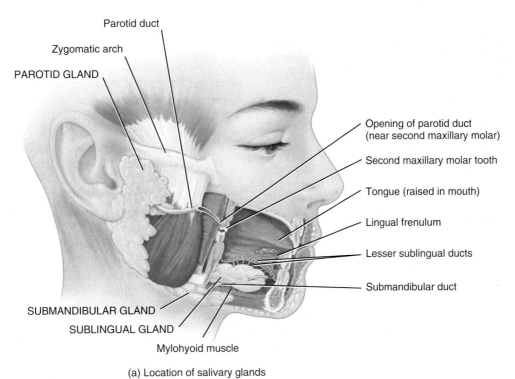

(a) Location of salivary glands

FIGURE 24.5 CONTINUES ▶

FIGURE 24.5 CONTINUED ▶

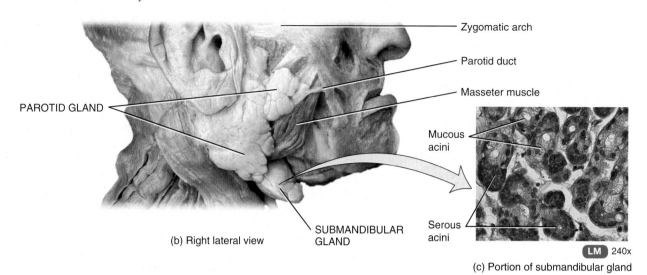

Zygomatic arch

Parotid duct

Masseter muscle

PAROTID GLAND

Mucous acini

Serous acini

(b) Right lateral view

SUBMANDIBULAR GLAND

LM 240x

(c) Portion of submandibular gland

? **The ducts of which salivary glands empty on either side of the lingual frenulum?**

The submandibular glands consist mostly of serous acini (serous-fluid-secreting portions of gland) and a few mucous acini (mucus-secreting portions of gland). The parotid glands consist of serous acini only. The sublingual glands consist of mostly mucous acini and a few serous acini.

The parotid gland receives its blood supply from branches of the external carotid artery and is drained by tributaries of the external jugular vein. The submandibular gland is supplied by branches of the facial artery and drained by tributaries of the facial vein. The sublingual gland is supplied by the sublingual branch of the lingual artery and the submental branch of the facial artery and is drained by tributaries of the sublingual and submental veins.

CLINICAL CONNECTION | *Mumps*

Although any of the salivary glands may be the target of a nasopharyngeal infection, the mumps virus (*paramyxovirus*) typically attacks the parotid glands. **Mumps** is an inflammation and enlargement of the parotid glands accompanied by moderate fever, malaise (general discomfort), and extreme pain in the throat, especially when swallowing sour foods or acidic juices. Swelling occurs on one or both sides of the face, just anterior to the ramus of the mandible. In about 30 percent of males past puberty, the testes may also become inflamed; sterility rarely occurs because testicular involvement is usually unilateral (one testis only). Since a vaccine became available for mumps in 1967, the incidence of the disease has declined dramatically. •

The salivary glands receive both sympathetic and parasympathetic innervation. The sympathetic fibers form plexuses on the blood vessels that supply the glands and initiate vasoconstriction, which decreases the production of saliva. The parotid gland receives sympathetic fibers from the plexus on the external carotid artery; the submandibular and sublingual glands receive sympathetic fibers that contribute to the sympathetic plexus and accompany the facial artery to the glands. The parasympathetic fibers of the glands produce vasodilation and stimulate the glandular cells of the glands, thus increasing the production of saliva.

The fluids secreted by the buccal glands, minor salivary glands, and the three pairs of major salivary glands constitute **saliva**. The amount of saliva that is secreted daily varies considerably but ranges from 1000 to 1500 mL (1 to 1.6 qt). Chemically, saliva is 99.5 percent water and 0.5 percent solutes and has a slightly acidic pH (6.35 to 6.85). The solute portion includes mucus, an enzyme that destroys bacteria (*lysozyme*), the digestive enzymes salivary amylase and lingual lipase, and traces of salts, proteins, and other organic compounds. **Salivary amylase** (AM-i-lās) plays a minor role in the breakdown of starch in the mouth into maltose, maltotriose, and α_1-dextrins.

The secretion of saliva, called **salivation** (sal-i-VĀ-shun), is controlled by the nervous system. Normally, parasympathetic stimulation promotes continuous secretion of a moderate amount of saliva, which keeps the mucous membranes moist and lubricates the movements of the tongue and lips during speech. The saliva is then swallowed and helps moisten the esophagus. Eventually, most components of saliva are reabsorbed, which prevents fluid loss. Sympathetic stimulation dominates during stress, resulting in dryness of the mouth. Dehydration also causes dryness of the mouth. The salivary glands stop secreting saliva to conserve water, and the resulting dryness contributes to the sensation of thirst. Drinking will then not only restore the homeostasis of body water but also moisten the mouth.

The feel and taste of food also are potent stimulators of salivary gland secretions. Chemicals in the food stimulate receptors in taste buds on the tongue, and impulses are conveyed from the taste buds to two salivary nuclei in the brain stem (**superior** and **inferior salivatory nuclei**). Returning parasympathetic impulses in fibers from these nuclei pass through the facial (VII) nerve to the sublingual and submandibular glands and via the glossopharyngeal (IX) nerve to the parotid gland to stimulate the secretion of saliva. Saliva continues to be heavily secreted for some time after food is swallowed; this flow of saliva washes out the mouth and dilutes and buffers the remnants of irritating chemicals. The smell, sight, sound, or thought of food may also stimulate secretion of saliva, hence the description "mouth-watering."

Tongue

The **tongue** is an accessory digestive organ composed of skeletal muscle covered with mucous membrane. Together with its associated muscles, it forms the floor of the oral cavity. The tongue is divided into symmetrical lateral halves by a median septum that extends its entire length, and it is attached to the hyoid bone, styloid process of the temporal bone, and mandible. Each half of the tongue consists of an identical complement of extrinsic and intrinsic muscles.

The **extrinsic muscles of the tongue**, which originate outside the tongue (attach to bones in the area) and insert into connective tissues in the tongue, include the *hyoglossus, genioglossus*, and *styloglossus* muscles (see Figure 11.7). The extrinsic muscles move the tongue from side to side and in and out to maneuver food for chewing, shape the food into a rounded mass, and force the food to the back of the mouth for swallowing. They also form the floor of the mouth, hold the tongue in position, and assist in speech. The **intrinsic muscles of the tongue** originate in and insert into connective tissue within the tongue and alter the shape and size of the tongue for speech and swallowing. The intrinsic muscles include the *longitudinalis superior, longitudinalis inferior, transversus linguae*, and *verticalis linguae* muscles. The **lingual frenulum** (*lingua*=the tongue), a fold of mucous membrane in the midline of the undersurface of the tongue, is attached to the floor of the mouth and aids in limiting the movement of the tongue posteriorly (see Figures 24.4 and 24.5).

If a lingual frenulum is abnormally short or rigid—a condition called **ankyloglossia** (ang′-kē-lō-GLOSS-ē-a)—the person is said to be "tongue-tied" because of the resulting speech impairment. It can be corrected surgically.

The dorsum and lateral surfaces of the tongue are covered with **papillae** (pa-PIL-ē=nipple-shaped), projections of the lamina propria covered with stratified squamous epithelium (see Figure 21.2). Many papillae contain taste buds, the receptors for gustation (taste; see Section 21.2). As their name implies, **fungiform papillae** (FUN-ji-form=mushroomlike) are mushroomlike elevations distributed among the more numerous filiform papillae. They are scattered over the dorsum of the tongue, but are concentrated mainly around the margins of the tongue. They appear as red dots on the surface of the tongue, and most of them contain taste buds. They are red because of a high density of capillaries beneath a thin, non keratinized epithelium. **Vallate papillae** (VAL-āt=wall-like) or *circumvallate papillae* are arranged in an inverted V-shape on the posterior surface of the tongue; all of them contain taste buds. **Foliate papillae** (FŌ-lē-āt=leaflike) are located in small trenches on the lateral margins of the tongue, but most of their taste buds degenerate in early childhood. **Filiform papillae** (FIL-i-form=threadlike) are pointed, threadlike projections distributed in parallel rows over the anterior two-thirds of the tongue. Although filiform papillae lack taste buds, they contain receptors for touch and increase friction between the tongue and food, making it easier for the tongue to move food in the oral cavity. The mottled white color of the tongue results from the dead keratinized cells on the ends of these plentiful filiform papillae.

Lingual lipase and mucus are secreted by **lingual glands** on the dorsum (upper surface) of the tongue. This enzyme, which is active in the stomach, can digest as much as 30 percent of dietary triglycerides (fats and oils) into simpler fatty acids and diglycerides.

Teeth

The **teeth**, or **dentes** (Figure 24.6), are accessory digestive organs located in sockets of the alveolar processes of the mandible and maxillae. The alveolar processes are covered by the **gingivae** (jin-JI-vē), or gums, which extend slightly into each socket to form the **gingival sulcus**. The sockets are lined by the **periodontal ligament** or **membrane** (per′-ē-ō-DON-tal; *odont-*=tooth), which consists of dense fibrous connective tissue and is attached to the socket walls and outer covering (cementum) of the roots of the teeth. The periodontal ligament anchors the teeth in position and acts as a shock absorber during chewing.

Figure 24.6 A typical tooth and surrounding structures.

 Teeth are anchored in sockets of the alveolar processes of the mandible and maxillae.

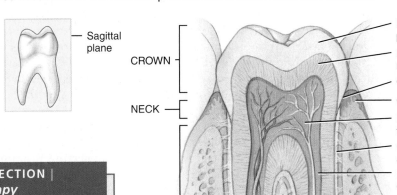

Sagittal plane

CROWN

NECK

ROOT

ENAMEL (made of calcium salts) protects the tooth from wear and tear.

DENTIN (calcified connective tissue) makes up the majority of the tooth.

Gingival sulcus

Gingiva (gum)

PULP CAVITY contains pulp (connective tissue containing nerves and blood vessels).

CEMENTUM is a bone-like substance that attaches the root to the periodontal ligament.

ROOT CANAL is an extension of the pulp cavity that contains nerves and blood vessels.

Alveolar bone

PERIODONTAL LIGAMENT helps anchor the tooth to the underlying bone.

APICAL FORAMEN is an opening at the base of a root canal through which blood vessels, lymphatic vessels, and nerves enter a tooth.

Nerve

Blood supply

Sagittal section of a mandibular (lower) molar

 **What type of tissue is the main component of teeth?**

Periodontal disease is a collective term for a variety of conditions characterized by inflammation and degeneration of the gingivae (gums), alveolar bone, periodontal ligament, and cementum. In one such condition, called **pyorrhea**, initial symptoms include enlargement and inflammation of the soft tissue and bleeding of the gums. Without treatment, the soft tissue may deteriorate and the alveolar bone may be resorbed, causing loosening of the teeth and recession of the gums. Periodontal diseases are often caused by poor oral hygiene; by local irritants, such as bacteria, impacted food, and cigarette smoke; or by a poor "bite."

Dental caries (KAR-ēz), or tooth decay, involves a gradual demineralization (softening) of the enamel and dentin. If untreated, microorganisms may invade the pulp, causing inflammation and infection, with subsequent death of the pulp and abscess of the alveolar bone surrounding the root's apex, requiring root canal therapy.

Dental caries begin when bacteria, acting on sugars, produce acids that demineralize the enamel. **Dextran**, a sticky polysaccharide produced from sucrose, causes the bacteria to stick to the teeth. Masses of bacterial cells, dextran, and other debris adhering to teeth constitute **dental plaque** (PLAK). Saliva cannot reach the tooth surface to buffer the acid because the plaque covers the teeth. Brushing the teeth after eating removes the plaque from flat surfaces before the bacteria can produce acids. Dentists also recommend that the plaque between the teeth be removed every 24 hours with dental floss. •

A typical tooth consists of three principal regions. The **crown** is the visible portion above the level of the gums. One to three **roots** are embedded in each socket. The **neck** is the constricted junction of the crown and root near the gum line.

Internally, **dentin** forms the majority of the tooth. Dentin consists of a calcified connective tissue that gives the tooth its basic shape and rigidity. It is harder than bone because of its higher content of hydroxyapatite (70 percent of dry weight). It also differs from bone in that it lacks blood vessels, gets deposited incrementally (in successive waves a little at a time), and does not remodel. Dentin contains *dentinal tubules*, parallel microscopic tubules radiating through the dentin from the pulp cavity. Cells that produce the dentin, odontoblasts, line the pulp cavity and send cytoplasmic processes into the dentinal tubules.

The dentin of the crown is covered by **enamel**, which consists primarily of calcium phosphate and calcium carbonate. Enamel is also harder than bone because of its even higher content of calcium salts (about 95 percent of dry weight). In fact, enamel is the hardest substance in the body. It is thickest over the cusps and tapers to a thin edge at the neck of the tooth, where it terminates. Enamel arises as a cellular secretion during development. After development is complete the body cannot replace or repair it, which is why it is so important to take good care of your teeth. Enamel protects the tooth from the wear and tear of chewing and from acids that could easily dissolve the softer underlying dentin. The dentin of the root is covered by **cementum**, another bonelike substance, which attaches the root to the periodontal ligament.

Within the dentin of a tooth is an enclosed space. The enlarged part of the space, the **pulp cavity**, lies within the crown and is filled with **pulp**, a connective tissue containing blood vessels, nerves, and lymphatic vessels. Narrow extensions of the pulp cavity, called **root canals**, run through the root of the tooth. Each root canal has an opening at its base, the **apical foramen**, through which blood vessels, lymphatic vessels, and nerves enter a tooth. The vessels bring nourishment, lymphatic vessels offer protection, and nerves provide sensation.

The arteries that supply blood to the teeth are distributed to the pulp cavity and surrounding periodontal ligament. These include the superior alveolar branches of the maxillary artery (anterior and posterior) and the incisive and dental branches of the inferior alveolar artery.

The teeth receive sensory fibers from branches of the maxillary and mandibular divisions of the trigeminal (V) nerve. The maxillary teeth receive sensory fibers from branches of the maxillary division, and the mandibular teeth receive theirs from branches of the mandibular division.

The branch of dentistry that is concerned with the prevention, diagnosis, and treatment of diseases that affect the pulp, root, periodontal ligament, and alveolar bone is known as **endodontics** (en'-dō-DON-tiks; *endo-*=within). **Orthodontics** (or'-thō-DON-tiks; *ortho-*=straight) is a branch of dentistry that is concerned with the prevention and correction of abnormally aligned teeth, and **periodontics** (per'-ē-ō-DON-tiks) is a branch of dentistry concerned with the treatment of abnormal conditions of the tissues immediately surrounding the teeth.

Humans have two sets of teeth, or **dentitions**: deciduous and permanent. The first set is the **deciduous teeth** (*decidu-*= falling out), also called *primary teeth*, *milk teeth*, or *baby teeth*. They begin to erupt (emerge) at about 6 months of age, and one pair of teeth appears at about each month thereafter, until all 20 are present (Figure 24.7a, c). The incisors, which are closest to the midline, are chisel-shaped and adapted for cutting into food. They are referred to as either **central incisors** or **lateral incisors** on the basis of their position. Next to the incisors, moving posteriorly, are the **canines**, which have a pointed surface called a cusp. Canines are used to tear and shred food. Incisors and canines have only one root. Posterior to them are the **first** and **second deciduous molars**, which have four cusps. *Maxillary (upper) molars* have three roots; *mandibular (lower) molars* have two roots. The molars crush and grind food to prepare for swallowing.

All deciduous teeth are lost—generally between the ages of 6 and 12 years—and are replaced by the second set of teeth, the **permanent (secondary) teeth** (Figure 24.7b, d). The permanent dentition contains 32 teeth that erupt between age 6 and adulthood. The pattern resembles the deciduous dentition, with the following exceptions. The deciduous molars are replaced by the **first** and **second premolars (bicuspids)**, which have two cusps and one root and are used for crushing and grinding. The permanent molars, which erupt into the mouth posterior to the premolars, do not replace any deciduous teeth and erupt as the jaw grows to accommodate them—the **first permanent molars** at age 6, the **second permanent molars** at age 12, and the **third permanent molars (wisdom teeth)** after age 17 or not at all.

Often the human jaw does not have enough room posterior to the second molars to accommodate the eruption of the third molars (wisdom teeth). In this case, the third molars remain embedded in the alveolar bone and are said to be **impacted**. They often cause pressure and pain and must be removed surgically.

Figure 24.7 Dentitions and times of eruptions. A designated letter (deciduous teeth) or number (permanent teeth) uniquely identifies each tooth, and the time of eruption is indicated in parentheses in parts (a) and (b).

There are 20 teeth in a complete deciduous set and 32 teeth in a complete permanent set.

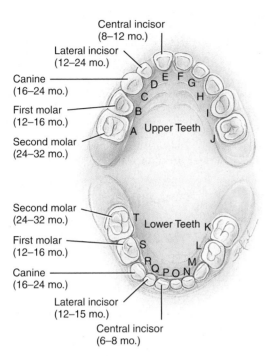

Central incisor (8–12 mo.)
Lateral incisor (12–24 mo.)
Canine (16–24 mo.)
First molar (12–16 mo.)
Second molar (24–32 mo.)
Upper Teeth

Second molar (24–32 mo.)
First molar (12–16 mo.)
Canine (16–24 mo.)
Lateral incisor (12–15 mo.)
Central incisor (6–8 mo.)
Lower Teeth

(a) Deciduous (primary) dentition; teeth designated by letters (with times of eruption)

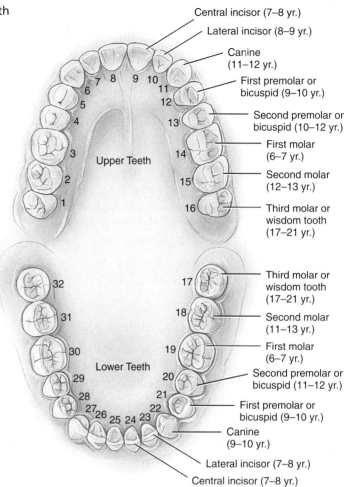

Central incisor (7–8 yr.)
Lateral incisor (8–9 yr.)
Canine (11–12 yr.)
First premolar or bicuspid (9–10 yr.)
Second premolar or bicuspid (10–12 yr.)
First molar (6–7 yr.)
Second molar (12–13 yr.)
Third molar or wisdom tooth (17–21 yr.)
Upper Teeth

Third molar or wisdom tooth (17–21 yr.)
Second molar (11–13 yr.)
First molar (6–7 yr.)
Second premolar or bicuspid (11–12 yr.)
First premolar or bicuspid (9–10 yr.)
Canine (9–10 yr.)
Lateral incisor (7–8 yr.)
Central incisor (7–8 yr.)
Lower Teeth

(b) Permanent (secondary) dentition; teeth designated by numbers (with times of eruption)

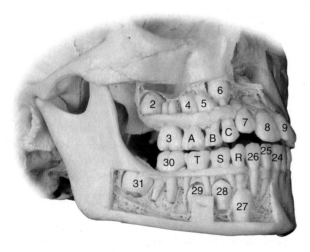

Right anterolateral view

DECIDUOUS TEETH:
A - First molar (maxillary)
B - Second molar (maxillary)
C - Canine (maxillary)
R - Canine (mandibular)
S - First molar (mandibular)
T - Second molar (mandibular)

PERMANENT TEETH:
8, 9, 24, 25 - Central incisor
7, 26 - Lateral incisor
6, 27 - Canine
5, 28 - First premolar
4, 29 - Second premolar
3, 30 - First molar
2, 31 - Second molar

(c) Mandible and maxilla of an eight-year old child showing erupted deciduous teeth and unerupted permanent teeth

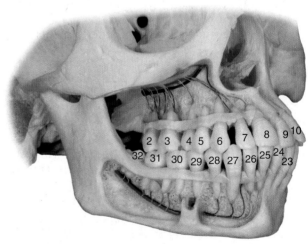

Right anterolateral view

8, 9, 24, 25 - Central incisor
7, 10, 23, 26 - Lateral incisor
6, 27 - Canine
5, 28 - First premolar

4, 29 - Second premolar
3, 30 - First molar
2, 31 - Second molar
32 - Third molar

(d) Mandible and maxilla showing permanent teeth and blood and nerve supply to them

 Which permanent teeth do not replace any deciduous teeth?

TABLE 24.1

Summary of Digestive Activities in the Mouth

STRUCTURE	ACTIVITY	RESULT
Cheeks and lips	Keep food between teeth	Foods uniformly chewed during mastication
Salivary glands	Secrete saliva	Lining of mouth and pharynx moistened and lubricated; saliva softens, moistens, and dissolves food and cleanses mouth and teeth; salivary amylase splits starch into smaller fragments (maltose, maltotriose, and α-dextrins)
Tongue		
Extrinsic tongue muscles	Move tongue from side to side and in and out	Food maneuvered for mastication, shaped into bolus, and maneuvered for swallowing
Intrinsic tongue muscles	Alter shape of tongue	Swallowing and speech
Taste buds	Serve as receptors for gustation (taste) and presence of food in mouth	Secretion of saliva stimulated by nerve impulses from taste buds to salivary nuclei in brain stem to salivary glands
Lingual glands	Secrete lingual lipase	Triglycerides broken down into fatty acids and diglycerides
Teeth	Cut, tear, and pulverize food	Solid foods reduced to smaller particles for swallowing

Table 24.1 summarizes the digestive activities in the mouth.

 CHECKPOINT

7. What structures form the mouth?
8. How are the major salivary glands distinguished on the basis of location and structure?
9. How do the extrinsic and intrinsic muscles of the tongue differ in function?
10. Contrast the functions of incisors, cuspids, premolars, and molars.

24.5 PHARYNX

 OBJECTIVE

• Describe the structure and function of the pharynx.

Through chewing, or **mastication** (mas′-ti-KĀ-shun; *masticare*= to chew), the tongue manipulates food, the teeth grind it, and the food is mixed with saliva. As a result, the food is reduced to a soft, flexible mass called a **bolus** (*bolos*=lump) that is easily swallowed. When food is first swallowed, it passes from the mouth into the pharynx.

The **pharynx** (FAR-inks; *pharynx*=throat) is a funnel-shaped tube that extends from the internal nares to the esophagus posteriorly and the larynx anteriorly (see Figure 23.2b). The pharynx is composed of skeletal muscle and lined by mucous membrane. The nasopharynx functions only in respiration, but the oropharynx and laryngopharynx have both digestive and respiratory functions. **Swallowing**, or **deglutition** (de-gloo-TISH-un), is a mechanism that moves food from the mouth to the stomach. It is helped by saliva and mucus and involves the mouth, pharynx, and esophagus. Food that is swallowed passes from the mouth into the oropharynx and laryngopharynx before passing into the esophagus. Muscular contractions of the oropharynx and laryngopharynx help propel food into the esophagus and then into the stomach.

 CHECKPOINT

11. What is a bolus? How is it formed?
12. Where does the pharynx begin and end?

24.6 ESOPHAGUS

 OBJECTIVES

• Describe the location and histology of the esophagus.
• Explain the function of the esophagus in the digestive process.

The **esophagus** (e-SOF-a-gus=eating gullet) is a 25-cm (10-in.) long collapsible muscular tube that lies posterior to the trachea. It begins at the inferior end of the laryngopharynx, passes through the inferior aspect of the neck, enters the mediastinum and descends anterior to the vertebral column, pierces the diaphragm through an opening called the **esophageal hiatus** (e-sof-a-JĒ-al hī-Ā-tus), and ends in the superior portion of the stomach (see Figure 24.1). Sometimes, a portion of the stomach protrudes above the diaphragm through the esophageal hiatus. This condition, called a *hiatal hernia* (hī-Ā-tal HER-nē-a), is described in the Key Medical Terms list at the end of this chapter.

The arteries of the esophagus, the esophageal arteries, arise from the following arteries along its length: inferior thyroid, thoracic aorta, intercostal arteries, phrenic, and left gastric arteries. The esophagus is drained by the adjacent esophageal veins, which primarily drain into the various azygos veins. The esophagus is innervated by recurrent laryngeal branches of the vagus nerves, other branches of the vagus (X) nerves, and the cervical sympathetic chain.

Histology of the Esophagus

The **mucosa** of the esophagus consists of nonkeratinized stratified squamous epithelium, lamina propria (areolar connective tissue), and a muscularis mucosae (smooth muscle) (Figure 24.8a). Near the stomach, the mucosa of the esophagus also contains mucous glands. The stratified squamous epithelium associated with the lips, mouth, tongue, oropharynx, laryngopharynx, and esophagus affords considerable protection against abrasion and wear-and-tear from food particles that are chewed, mixed with secretions, and swallowed. The **submucosa** contains areolar connective tissue with blood vessels and mucous glands and numerous elastic fibers that assist in closing the distended tube. The **muscularis** of the superior third of the esophagus is skeletal muscle, the intermediate third is skeletal and smooth muscle, and the inferior third is

Figure 24.8 **Histology of the esophagus.** A higher magnification of nonkeratinized, stratified squamous epithelium is shown in Table 3.1F.

🔑 The esophagus secretes mucus and transports food to the stomach.

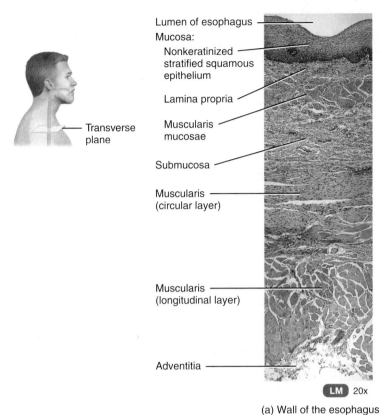

Lumen of esophagus

Mucosa:
  Nonkeratinized stratified squamous epithelium
  Lamina propria
  Muscularis mucosae

Submucosa

Muscularis (circular layer)

Muscularis (longitudinal layer)

Adventitia

Transverse plane

LM 20x

(a) Wall of the esophagus

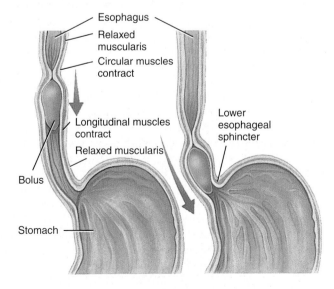

Mucosa

Submucosa

Muscularis (circular layer)

Muscularis (longitudinal layer)

Vagus (X) nerve

(b) Anterosuperior view of step dissection of esophagus

❓ **In which layers of the esophagus are the glands that secrete lubricating mucus located?**

smooth muscle. The superficial layer of the esophagus is known as the **adventitia** (ad-ven-TISH-a). Unlike the serosa of the stomach and intestines, the areolar connective tissue of this layer is not covered by mesothelium (the esophagus does not push into a serous-lined cavity). The connective tissue of the adventitia merges with the connective tissue of surrounding structures of the mediastinum through which it passes, supporting the esophagus and supplying it with blood vessels.

Functions of the Esophagus

The esophagus secretes mucus and transports food into the stomach. It does not produce digestive enzymes, and it does not carry on absorption. The passage of food from the laryngopharynx into the esophagus is regulated at the entrance to the esophagus by a sphincter (a circular band or ring of muscle that is normally contracted) called the **upper esophageal sphincter (UES)** (e-sof-a-JĒ-al) or *pharyngoesophageal sphincter* (fa′-ring′-gō-e-sof-a-JĒ-al). It consists of skeletal muscle (cricopharyngeus muscle) attached to the cricoid cartilage. The elevation of the larynx causes the sphincter to relax, allowing the bolus to enter the esophagus. This sphincter also relaxes during exhalation.

Food is pushed through the esophagus by a progression of involuntary coordinated contractions and relaxations of the circular and longitudinal layers of the muscularis called **peristalsis** (per′-i-STAL-sis; *stalsis*=constriction). Peristalsis occurs in other tubular structures, including other portions of the GI tract, the ureters, bile ducts, and uterine tubes; in the esophagus it is controlled by the medulla oblongata. The steps of peristalsis are as follows (Figure 24.9):

1. The circular muscle fibers in the section of the esophagus above the bolus contract, constricting the wall of the esophagus and squeezing the bolus downward.
2. Longitudinal muscle fibers around the bottom of the bolus contract, shortening the section below the bolus and pushing its walls outward.

Figure 24.9 **Peristalsis during deglutition (swallowing).**

🔑 Peristalsis consists of progressive, wavelike contractions of the muscularis.

Esophagus

Relaxed muscularis

Circular muscles contract

Longitudinal muscles contract

Relaxed muscularis

Bolus

Lower esophageal sphincter

Stomach

Anterior view of frontal sections showing peristalsis in esophagus

❓ **Does peristalsis "push" or "pull" food along the gastrointestinal tract?**

TABLE 24.2

Summary of Digestive Activities in the Pharynx and Esophagus

STRUCTURE	ACTIVITY	RESULT
Pharynx	Pharyngeal stage of deglutition	Moves bolus from oropharynx to laryngopharynx and into esophagus; closes air passageways
Esophagus	Relaxation of upper esophageal sphincter	Permits entry of bolus from laryngopharynx into esophagus
	Esophageal stage of deglutition (peristalsis)	Pushes bolus down esophagus
	Relaxation of lower esophageal sphincter	Permits entry of bolus into stomach
	Secretion of mucus	Lubricates esophagus for smooth passage of bolus

3. After the bolus moves into the new section of the esophagus, the circular muscles above it contract, and the cycle repeats. The contractions move the bolus down the esophagus toward the stomach. As the bolus approaches the end of the esophagus, the lower esophageal sphincter relaxes and the bolus moves into the stomach.

Mucus secreted by esophageal glands lubricates the bolus and reduces friction.

Just superior to the level of the diaphragm, the esophagus narrows slightly. This narrowing is a physiological sphincter in the inferior part of the esophagus composed of smooth muscle, known as the **lower esophageal sphincter (LES)** or *gastroesophageal sphincter* (gas′-trō-e-sof-a-JĒ-al). It is also called the *cardiac sphincter* because of its proximity to the heart. (A *physiological sphincter* is a section of a tubular structure, in this case the esophagus, which functions like a sphincter even though no sphincter muscle is actually present.) The lower esophageal sphincter relaxes during swallowing and thus allows the bolus to pass from the esophagus into the stomach.

Table 24.2 summarizes the digestive activities in the pharynx and esophagus.

CLINICAL CONNECTION | *Gastroesophageal Reflux Disease*

If the lower esophageal sphincter fails to close adequately after food has entered the stomach, the stomach contents can *reflux* (back up) into the inferior portion of the esophagus. This condition is known as **gastroesophageal reflux disease (GERD)** (gas′-trō-e-sof-a-JĒ-al). Hydrochloric acid (HCl) from the stomach contents can irritate the esophageal wall, resulting in a burning sensation that is called **heartburn** because it is experienced in a region very near the heart; it is unrelated to any cardiac problem. Drinking alcohol and smoking can cause the sphincter to relax, worsening the problem. The symptoms of GERD often can be controlled by avoiding foods that strongly stimulate stomach acid secretion (coffee, chocolate, tomatoes, fatty foods, orange juice, peppermint, spearmint, and onions). Other acid-reducing strategies include taking over-the-counter histamine-2 (H₂) blockers such as Tagamet HB® or Pepcid AC® 30 to 60 minutes before eating to block acid secretion, and neutralizing acid that has already been secreted with antacids such as Tums® or Maalox®. Symptoms are less likely to occur if food is eaten in smaller amounts and if the person does not lie down immediately after a meal. GERD may be associated with cancer of the esophagus. •

CHECKPOINT

13. Where is the esophagus located? What type of tissue does it contain that allows it to expand during swallowing?

14. What is the role of the esophagus in digestion?
15. How do the upper and lower esophageal sphincters work? Which one is a physiological sphincter?

24.7 STOMACH

OBJECTIVE

• Describe the location, anatomy, histology, and functions of the stomach.

The **stomach** is typically a J-shaped enlargement of the GI tract directly inferior to the diaphragm in the epigastric, umbilical, and left hypochondriac regions of the abdomen (see Figure 1.8). The stomach connects the esophagus to the duodenum, the first part of the small intestine (Figure 24.10). Because a meal can be eaten much more quickly than the intestines can digest and absorb it, the stomach functions as a mixing area and holding reservoir. At appropriate intervals after food is ingested, the stomach forces a small quantity of material into the first portion of the small intestine. The position and size of the stomach vary continually; the diaphragm pushes it inferiorly with each inhalation and pulls it superiorly with each exhalation. The stomach is the most distensible portion of the GI tract and can accommodate a large quantity of food, up to 6.4 liters (6 qt.). In the stomach, the digestion of starch and triglycerides that began in the mouth continues, digestion of proteins begins, the semisolid bolus is converted to a liquid, and certain substances are absorbed.

Anatomy of the Stomach

The stomach has four main regions: the cardia, fundus, body, and pyloric part (see Figure 24.10). The **cardia** (KAR-dē-a) surrounds the opening of the esophagus into the stomach. The rounded portion superior and to the left of the cardia is the **fundus** (FUN-dus). Inferior to the fundus is the large central portion of the stomach, called the **body**. The **pyloric part** is divisible into three regions. The first region, the **pyloric antrum**, connects to the body of the stomach. The next region, the **pyloric canal**, leads to the third region, the **pylorus** (pī-LOR-us; *pyl-* =gate; *orus*=guard), which in turn connects to the duodenum. When the stomach is empty, the mucosa lies in large folds, called **rugae** (ROO-gē=wrinkles), which can be seen with the unaided eye. The pylorus communicates with the duodenum of the small intestine via a smooth muscle sphincter called the **pyloric sphincter (valve)**. The concave medial border of the stomach is called the **lesser curvature**, and the convex lateral border is called the **greater curvature**.

Figure 24.10 External and internal anatomy of the stomach.

The four regions of the stomach are the cardia, fundus, body, and pylorus.

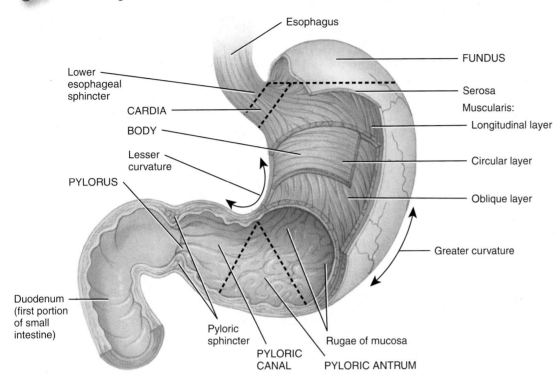

(a) Anterior view of regions of stomach

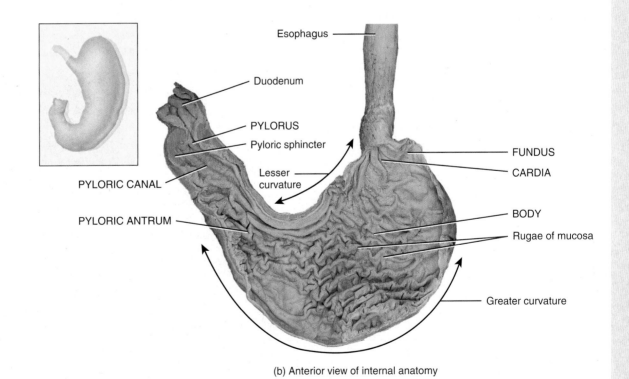

(b) Anterior view of internal anatomy

After a very large meal, does your stomach have rugae?

The arterial supply of the stomach arises from the celiac trunk. The right and left gastric arteries form an anastomosing arch along the lesser curvature, and the right and left gastro-omental arteries form a similar arch on the greater curvature. Short gastric arteries supply the fundus. The veins of the same name accompany the arteries and drain, directly or indirectly, into the hepatic portal vein.

The vagus (X) nerves convey parasympathetic fibers to the stomach. These fibers form synapses within the submucosal plexus in the submucosa and the myenteric plexus in the muscularis. The sympathetic nerves arise from the celiac ganglia, and the nerves reach the stomach along the branches of the celiac artery.

Histology of the Stomach

The stomach wall is composed of the same basic layers as the rest of the GI tract, with certain modifications (Figure 24.11a). The surface of the **mucosa** is a layer of nonciliated simple columnar epithelial cells called **surface mucous cells**. The mucosa contains a **lamina propria** (areolar connective tissue) and a **muscularis mucosae** (smooth muscle). Inward folds of epithelial cells extend down into the lamina propria, where they form columns of secretory cells called **gastric glands**. Several gastric glands open into the bottom of narrow channels called **gastric pits**. Secretions from several gastric glands flow into each gastric pit and then into the lumen of the stomach.

The gastric glands contain three types of *exocrine gland cells* that secrete their products into the stomach lumen: mucous neck cells, chief cells, and parietal cells. Both mucous surface cells and **mucous neck cells** secrete mucus (Figure 24.11b). **Parietal cells** produce intrinsic factor (needed for absorption of vitamin B_{12}) and hydrochloric acid. The **chief (zymogenic) cells** secrete pepsinogen and gastric lipase. The secretions of the mucous, parietal, and chief cells form **gastric juice**, about 2000–3000 mL (roughly 2–3 qt) per day. In addition, gastric glands include a type of enteroendocrine cell, the **G cell**, which is located mainly in the pyloric antrum and secretes the hormone gastrin into the bloodstream. Gastrin stimulates growth of the gastric glands and secretion of large amounts of gastric juice. It also strengthens contraction of the lower esophageal sphincter, increases motility of the stomach, and relaxes the pyloric and ileocecal sphincters (described later).

Figure 24.11 Histology of the stomach.

🔑 The muscularis of the stomach has three layers of smooth muscle tissue.

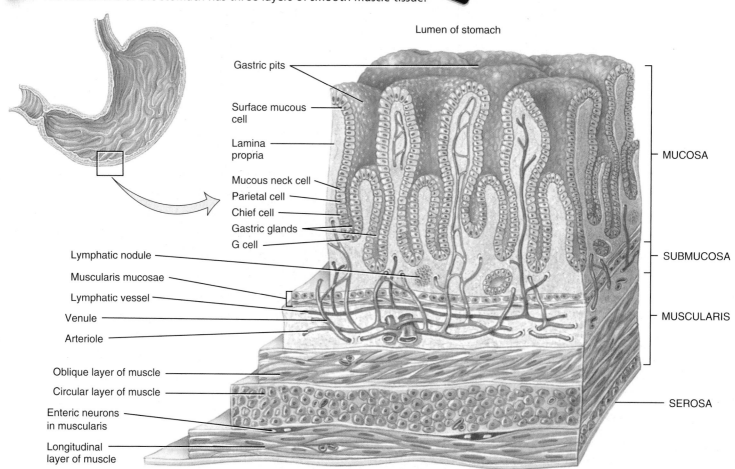

Lumen of stomach

Gastric pits
Surface mucous cell
Lamina propria
Mucous neck cell
Parietal cell
Chief cell
Gastric glands
G cell
Lymphatic nodule
Muscularis mucosae
Lymphatic vessel
Venule
Arteriole
Oblique layer of muscle
Circular layer of muscle
Enteric neurons in muscularis
Longitudinal layer of muscle

MUCOSA
SUBMUCOSA
MUSCULARIS
SEROSA

(a) Three-dimensional view of layers of stomach

The **submucosa** of the stomach is composed of areolar connective tissue, as it is throughout the entire GI tract. The **muscularis** has three layers of smooth muscle (rather than the two found in the lower esophagus and small and large intestines): an outer longitudinal layer, a middle circular layer, and an inner oblique layer. The oblique layer is limited primarily to the body of the stomach. This arrangement of muscle allows the stomach to more effectively churn and mix the food. The **serosa** is composed of simple squamous epithelium (mesothelium), which is also called the visceral peritoneum, and an underlying areolar connective tissue. At the lesser curvature, the stomach's mesothelium bends back from the stomach and extends superiorly to connect to the liver as the lesser omentum. At the greater curvature, the stomach's mesothelium forms a large inferior fold, the greater omentum, which drapes over the intestines and doubles back on itself to attach to the transverse colon.

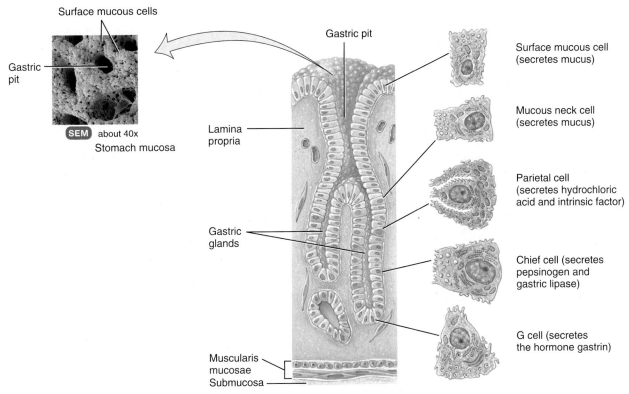

(b) Sectional view of the stomach mucosa showing gastric glands and cell types

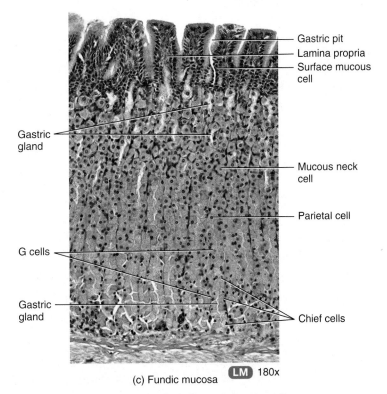

(c) Fundic mucosa   **LM** 180x

?   **What types of cells are found in gastric glands, and what does each secrete?**

Functions of the Stomach

Several minutes after food enters the stomach, waves of peristalsis pass over the stomach every 15 to 25 seconds. Few peristaltic waves are observed in the fundus, which primarily has a storage function. Instead, most waves begin at the body of the stomach and intensify as they reach the antrum. Each peristaltic wave moves gastric contents from the body of the stomach down into the antrum, a process known as **propulsion**. The pyloric sphincter normally remains almost, but not completely, closed. Because most food particles in the stomach initially are too large to fit through the narrow pyloric sphincter, they are forced back into the body of the stomach, a process referred to as **retropulsion**. Another round of propulsion then occurs, moving the food particles back down into the antrum. If the food particles are still too large to pass through the pyloric sphincter, retropulsion occurs again as the particles are squeezed back into the body of the stomach. Then yet another round of propulsion occurs, and the cycle continues to repeat. The net result of these movements is that gastric contents are mixed with gastric juice, eventually becoming reduced to a soupy liquid called **chyme** (KĪM=juice). Once the food particles in chyme are small enough, they can pass through the pyloric sphincter, a phenomenon known as **gastric emptying**. Gastric emptying is a slow process: only about 3 mL of chyme move through the pyloric sphincter at this time.

The enzymatic digestion of proteins begins in the stomach. In the adult, this is achieved mainly through the enzyme **pepsin**, secreted by chief cells in an inactive form called *pepsinogen*. Pepsin breaks certain peptide bonds between the amino acids making up proteins. Thus, a protein chain of many amino acids is broken down into smaller fragments called **peptides**. Pepsin also brings about the clumping and digestion of milk proteins. Another enzyme of the stomach is **gastric lipase**. Gastric lipase splits triglycerides (fats and oils) in fat molecules (such as those found in milk) into fatty acids and monoglycerides (a glyceride molecule attached to one fatty acid molecule). This enzyme has a limited role in the adult stomach. To digest fats and oils, adults rely almost exclusively on the **lingual lipase** secreted by lingual glands in the tongue, in the acid environment of the stomach, and **pancreatic lipase**, an enzyme secreted by the pancreas into the small intestine.

Within 2–4 hours after eating a meal, the stomach has emptied its contents into the duodenum. Foods rich in carbohydrate spend the least time in the stomach; high-protein foods remain somewhat longer; and emptying is slowest after a fat-laden meal containing large amounts of triglycerides.

The stomach wall is impermeable to the passage of most materials into the blood; most substances are not absorbed until they reach the small intestine. However, the stomach does participate in the absorption of some water, electrolytes, certain drugs (especially aspirin), and alcohol.

Table 24.3 summarizes the digestive activities in the stomach.

CLINICAL CONNECTION | *Vomiting*

Vomiting, or *emesis*, is the forcible expulsion of the contents of the upper GI tract (stomach and sometimes duodenum) through the mouth. The strongest stimuli for vomiting are irritation and distension of the stomach; other stimuli include unpleasant sights, general anesthesia, dizziness, and certain drugs, such as morphine and derivatives of digitalis. Nerve impulses are transmitted to the vomiting center in the medulla oblongata, and returning impulses travel to the upper GI tract organs, diaphragm, and abdominal muscles. Vomiting basically involves squeezing the stomach between the diaphragm and abdominal muscles and expelling the contents through open esophageal sphincters. Prolonged vomiting, especially in infants and elderly people, can be serious because the loss of acidic gastric juice can lead to alkalosis (higher than normal blood pH), dehydration, and damage to the esophagus and teeth. •

✓ CHECKPOINT

16. Describe the location and anatomical features of the stomach.
17. Compare the epithelium of the esophagus with that of the stomach. How is each adapted to the function of the organ?
18. Describe the importance of rugae, surface mucous cells, mucous neck cells, chief cells, parietal cells, and G cells in the stomach.

TABLE 24.3

Summary of Digestive Activities in the Stomach

STRUCTURE	ACTIVITY	RESULT
MUCOSA		
Surface mucous cells and mucous neck cells	Secrete mucus	Forms a protective barrier that prevents digestion of stomach wall
	Absorption	Small quantity of water, ions, short-chain fatty acids, and some drugs enter the bloodstream
Parietal cells	Secrete hydrochloric acid	Kills microbes in food; denatures proteins; converts pepsinogen into pepsin
	Secrete intrinsic factor	Needed for absorption of vitamin B_{12}, which is used in red blood cell formation (erythropoiesis)
Chief cells	Secrete pepsinogen	Pepsin, the activated form, breaks down proteins into peptides
	Secrete gastric lipase	Splits triglycerides into fatty acids and monoglycerides
G cells	Secrete gastrin	Stimulates parietal cells to secrete HCl and chief cells to secrete pepsinogen; contracts lower esophageal sphincter, increases motility of the stomach, and relaxes pyloric sphincter
MUSCULARIS	Mixing waves (gentle peristaltic movements)	Churn and physically break down food and mix it with gastric juice, forming chyme; force chyme through pyloric sphincter
PYLORIC SPHINCTER	Opens to permit passage of chyme into duodenum	Regulates passage of chyme from stomach to duodenum; prevents backflow of chyme from duodenum to stomach

19. How does mechanical digestion occur in the stomach?
20. What are the functions of gastric lipase and lingual lipase in the stomach?
21. How does the stomach help absorb nutrients from food?

24.8 PANCREAS

OBJECTIVES

• Describe the location and structure of the pancreas.
• Explain the role of the pancreas in digestion.

From the stomach, chyme passes into the small intestine. Because chemical digestion in the small intestine depends on activities of the pancreas, liver, and gallbladder, we first consider the activities of these accessory digestive organs and their contributions to digestion in the small intestine.

Anatomy of the Pancreas

The **pancreas** (*pan-*=all; *-creas*=flesh), a retroperitoneal gland that is about 12–15 cm (5–6 in.) long and 2.5 cm (1 in.) thick, lies posterior to the greater curvature of the stomach (see Figure 24.1). Along with the liver and gallbladder, it develops as an embryonic epithelial outgrowth of the duodenum. The pancreas consists of a head, a body, and a tail and is usually connected to the duodenum by two ducts (Figure 24.12a, c, d). The **head** is the expanded portion of the organ near the curve of the duodenum. Projecting from the lower portion of the head is the hooklike **uncinate process** that arches behind the superior mesenteric artery and vein, encircling them with pancreatic tissue. Superior to and to the left of the head are the central **body** and the tapering **tail**.

Pancreatic secretions pass from the secreting cells into small ducts that ultimately unite to form two larger ducts that convey the secretions into the duodenum of the small intestine. The larger of the two ducts is called the **pancreatic duct** (*duct of Wirsung* [VER-sung]). In most people, the pancreatic duct joins the common bile duct from the liver and gallbladder and enters the duodenum as a dilated common duct called the **hepatopancreatic ampulla** (hep'-a-tō-pan-krē-A-tik) (*ampulla of Vater* [FAH-ter]). The ampulla opens onto an elevation of the duodenal mucosa, the **major duodenal papilla**, that lies about 10 cm (4 in.) inferior to the pyloric sphincter of the stomach. The smaller of the two ducts, the **accessory duct** (*duct of Santorini*), leads from the pancreas and empties into the duodenum about 2.5 cm (1 in.) superior to the hepatopancreatic ampulla.

The arterial supply of the pancreas is from the superior and inferior pancreaticoduodenal arteries and from the splenic and superior mesenteric arteries. The veins, in general, correspond to the arteries. Venous blood reaches the hepatic portal vein by means of the splenic and superior mesenteric veins.

The nerves to the pancreas are autonomic nerves that branch from the celiac and superior mesenteric plexuses. Included are preganglionic vagal, postganglionic sympathetic, and sensory fibers. Parasympathetic vagal fibers are said to terminate at both acinar (exocrine) and islet (endocrine) cells. Although the innervation is presumed to influence enzyme formation, pancreatic secretion is controlled largely by the hormones secretin and cholecystokinin (CCK) released by the small intestine. The sympathetic fibers enter the islets and also end on blood vessels; these fibers are vasomotor and accompanied by sensory fibers, especially for pain.

Figure 24.12 Relationship of the pancreas to the liver, gallbladder, and duodenum. The inset shows details of the common bile duct and pancreatic duct forming the hepatopancreatic ampulla and emptying into the duodenum.

Pancreatic enzymes digest starches (polysaccharides), proteins, triglycerides, and nucleic acids.

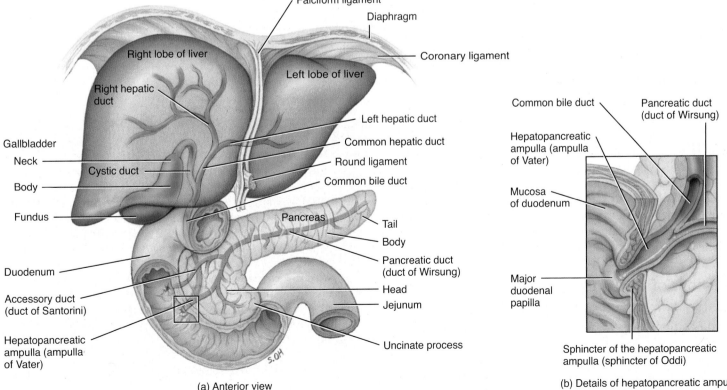

(a) Anterior view

(b) Details of hepatopancreatic ampulla

FIGURE 24.12 CONTINUES ▶

FIGURE 24.12 CONTINUED ▶

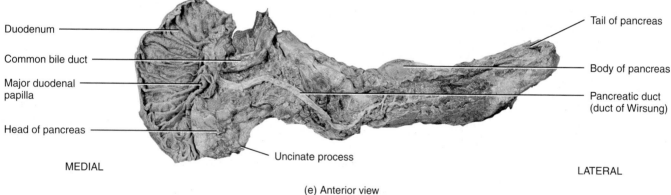

Right hepatic duct → Left hepatic duct

Common hepatic duct from liver

Cystic duct from gallbladder

Common bile duct

Pancreatic duct from pancreas

Key:
- Liver
- Gallbladder
- Pancreas

Sphincter

Duodenum

(c) Ducts carrying bile from liver and gallbladder and pancreatic juice from pancreas to the duodenum

Falciform ligament
Liver
Hepatic duct
Cystic duct
Gallbladder
Common bile duct
Major duodenal papilla
Duodenum

Diaphragm
Spleen
Tail of pancreas
Pancreatic duct (duct of Wirsung)
Body of pancreas
Head of pancreas

(d) Anterior view

Duodenum
Common bile duct
Major duodenal papilla
Head of pancreas

Tail of pancreas
Body of pancreas
Pancreatic duct (duct of Wirsung)

Uncinate process

MEDIAL

LATERAL

(e) Anterior view

 What type of fluid is found in the pancreatic duct? The common bile duct? The hepatopancreatic ampulla?

Histology of the Pancreas

The pancreas is made up of small clusters of glandular epithelial cells, about 99 percent of which are arranged in clusters called **acini** (AS-i-nī) and constitute the *exocrine* portion of the organ (see Figure 22.7b, c). The cells within acini secrete a mixture of fluid and digestive enzymes called **pancreatic juice**. The remaining 1 percent of the cells are organized into clusters called **pancreatic islets (islets of Langerhans)** (Ī-lets), the *endocrine* portion of the pancreas. These cells secrete the hormones glucagon, insulin, somatostatin, and pancreatic polypeptide. The functions of these hormones are discussed in Table 22.6.

Functions of the Pancreas

Each day the pancreas produces 1200–1500 mL (about 1.2–1.5 qt) of **pancreatic juice**, a clear, colorless liquid consisting mostly of water, some salts, sodium bicarbonate, and several enzymes. The sodium bicarbonate gives pancreatic juice a slightly alkaline pH (7.1–8.2) that buffers acidic gastric juice in chyme, stops the action of pepsin from the stomach, and creates the

CLINICAL CONNECTION | *Pancreatitis and Pancreatic Cancer*

Inflammation of the pancreas, as may occur in association with alcohol abuse or chronic gallstones, is called **pancreatitis** (pan'-krē-a-TĪ-tis). In a more severe condition known as **acute pancreatitis**, which is associated with heavy alcohol intake or biliary tract obstruction, the pancreatic cells may release an enzyme (trypsin) instead of trypsinogen or insufficient amounts of trypsin inhibitor, and the trypsin begins to digest the pancreatic cells.

Pancreatic cancer usually affects people over 50 years of age and occurs more frequently in males. Typically, there are few symptoms until the disorder reaches an advanced stage and often not until it has metastasized to other parts of the body such as the lymph nodes, liver, or lungs. The disease is nearly always fatal and is the fourth most common cause of death from cancer in the United States. Pancreatic cancer has been linked to fatty foods, high alcohol consumption, genetic factors, smoking, and chronic pancreatitis. •

proper pH for the action of digestive enzymes in the small intestine. The enzymes in pancreatic juice include a starch-digesting enzyme called **pancreatic amylase**; several enzymes that digest proteins into peptides called **trypsin** (TRIP-sin), **chymotrypsin** (kī′-mō-TRIP-sin), **carboxypeptidase** (kar-bok′-sē-PEP-ti-dās), and **elastase** (ē-LAS-tās); the principal triglyceride (fat and oil)-digesting enzyme in adults, called **pancreatic lipase**; and enzymes called **ribonuclease** (rī′-bō-NOO-klē-ās) and **deoxyribonuclease** (dē-oks-ē-rī′-bō-NOO-klē-ās) that digest ribonucleic acid (RNA) and deoxyribonucleic acid (DNA) into nucleotides.

 CHECKPOINT

22. Describe the duct system that connects the pancreas to the duodenum.
23. What are pancreatic acini? How do their functions compare with those of the pancreatic islets (islets of Langerhans)?
24. Describe the composition and functions of pancreatic juice.

24.9 LIVER AND GALLBLADDER

 OBJECTIVES

• Describe the location and structure of the liver and gallbladder.
• Explain the roles of the liver and gallbladder in the digestive process.

The **liver** is the largest internal organ and heaviest gland of the body, weighing about 1.4 kg (about 3 lb) in an average adult. Of the organs of the body, it is second in size only to the skin. The liver is inferior to the diaphragm and occupies most of the right hypochondriac and part of the epigastric regions of the abdominopelvic cavity (see Figure 1.8).

The **gallbladder** (*gall-*=bile) is a pear-shaped sac that is located on the inferior surface of the liver. It is 7–10 cm (3–4 in.) long and part of it typically hangs below the anterior inferior margin of the liver (Figure 24.12a).

Anatomy of the Liver and Gallbladder

The liver is almost completely covered by visceral peritoneum and *is* completely covered by a capsule composed of dense irregular connective tissue that lies deep to the peritoneum. The liver is divided into two principal lobes—a large **right lobe** and a smaller **left lobe**—by the **falciform ligament**, a mesenteric fold from the parietal peritoneum of the diaphragm and anterior abdominal wall to the visceral peritoneum of the liver (Figures 24.12 and 24.13). The right lobe is considered by many anatomists to include an inferior **quadrate lobe** (KWA-drāt) and a posterior **caudate lobe** (KAW-dāt). However, on the basis of internal morphology (primarily the distribution of blood vessels), the quadrate and caudate lobes more appropriately belong to the left lobe. The falciform ligament extends from the undersurface of the diaphragm between the two principal lobes of the liver to the superior surface of the liver, helping to suspend the liver in the

Figure 24.13 External anatomy of the liver. The anterior view is illustrated in Figure 24.12a.

The two principal lobes of the liver, the right and left lobes, are separated by the falciform ligament.

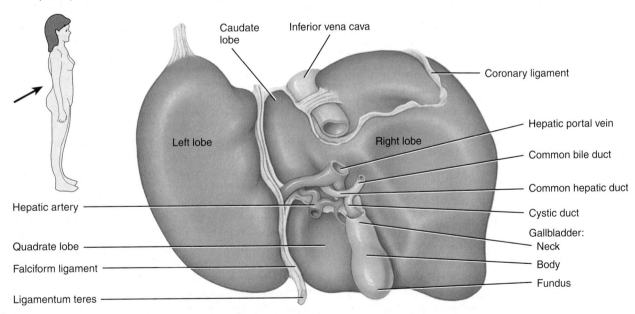

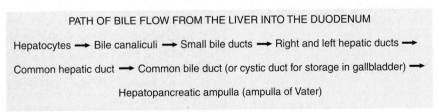

(a) Posteroinferior surface of liver

FIGURE 24.13 CONTINUES ▶

FIGURE 24.13 CONTINUED

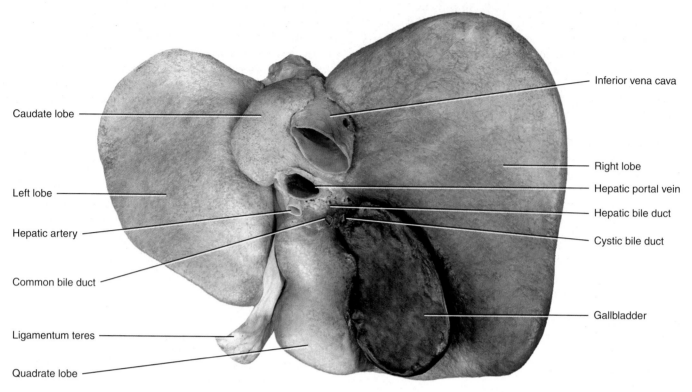

Caudate lobe

Left lobe

Hepatic artery

Common bile duct

Ligamentum teres

Quadrate lobe

Inferior vena cava

Right lobe

Hepatic portal vein

Hepatic bile duct

Cystic bile duct

Gallbladder

(b) Posteroinferior surface of liver

 Within which abdominopelvic region (see Figure 1.8a) could you palpate (feel) most of the liver to decide if it is enlarged?

abdominal cavity. The free border of the falciform ligament is the **ligamentum teres (round ligament)**, a fibrous cord that is a remnant of the umbilical vein of the fetus; it extends from the liver to the umbilicus. The right and left **coronary ligaments** are narrow extensions of the parietal peritoneum that suspend the liver from the diaphragm.

The parts of the gallbladder are the broad **fundus**, which projects downward beyond the inferior border of the liver; the central portion, called the **body**; and a tapered portion called the **neck**. The body and neck project superiorly (Figures 24.12a and 24.13a).

Histology of the Liver and Gallbladder

Histologically, the liver is composed of several components (Figure 24.14a–c):

1. **Hepatocytes** (*hepat-*=liver; *-cytes*=cells). Hepatocytes are the major functional cells of the liver and perform a wide array of metabolic, secretory, and endocrine functions. These are specialized epithelial cells with 5 to 12 sides that make up about 80 percent of the volume of the liver. Hepatocytes are arranged in rows called **hepatic laminae** (LAM-i-nē). The hepatic laminae are plates of hepatocytes one cell thick bordered on either side by endothelial-lined vascular spaces called hepatic sinusoids. The hepatic laminae are highly branched, irregular structures. Grooves in the cell membranes between neighboring hepatocytes provide spaces for canaliculi (described next), into which the hepatocytes secrete bile. Bile, a yellow, brownish, or olive-green liquid secreted by hepatocytes, serves as both an excretory product and a digestive secretion.

Figure 24.14 Histology of the liver.

Histologically, the lobule is composed of hepatocytes, bile canaliculi, and hepatic sinusoids.

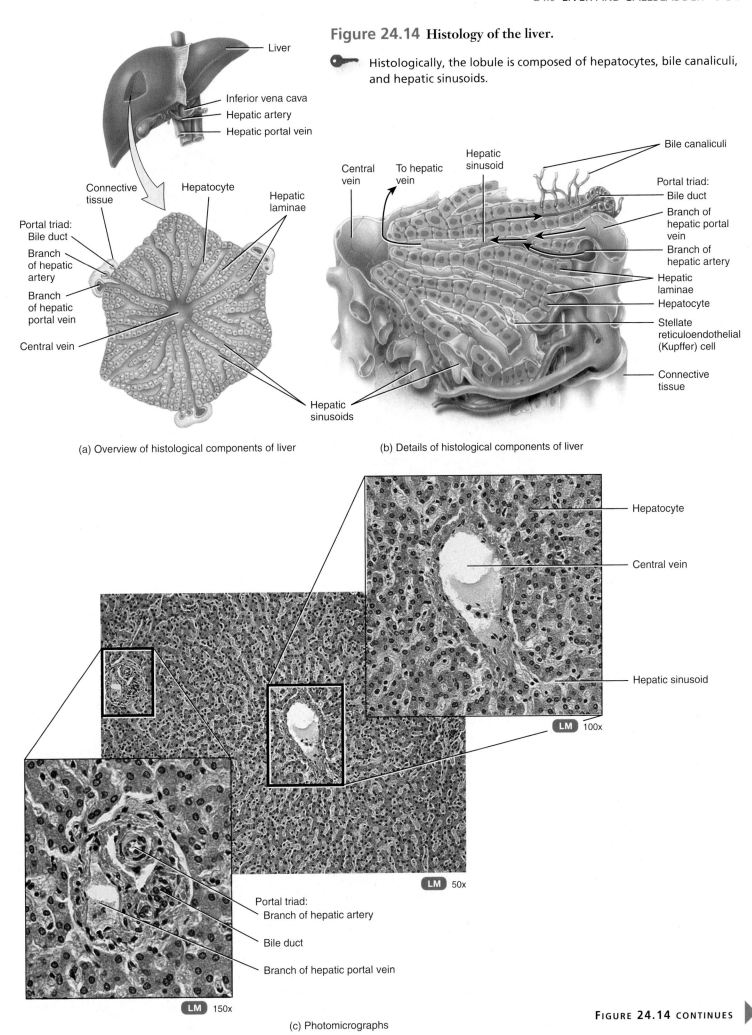

Liver

Inferior vena cava
Hepatic artery
Hepatic portal vein

Connective tissue

Hepatocyte

Hepatic laminae

Portal triad:
Bile duct

Branch of hepatic artery

Branch of hepatic portal vein

Central vein

Hepatic sinusoids

(a) Overview of histological components of liver

Central vein

To hepatic vein

Hepatic sinusoid

Bile canaliculi

Portal triad:
Bile duct

Branch of hepatic portal vein

Branch of hepatic artery

Hepatic laminae

Hepatocyte

Stellate reticuloendothelial (Kupffer) cell

Connective tissue

(b) Details of histological components of liver

Hepatocyte

Central vein

Hepatic sinusoid

LM 100x

LM 50x

Portal triad:
Branch of hepatic artery

Bile duct

Branch of hepatic portal vein

LM 150x

(c) Photomicrographs

FIGURE 24.14 CONTINUES

🔲 **FIGURE 24.14 CONTINUED** ▶

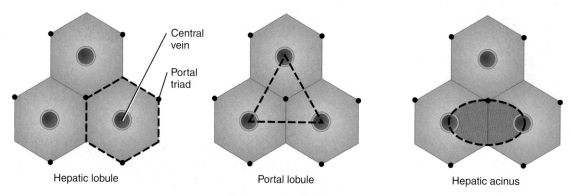

(d) Comparison of three units of liver structure and function

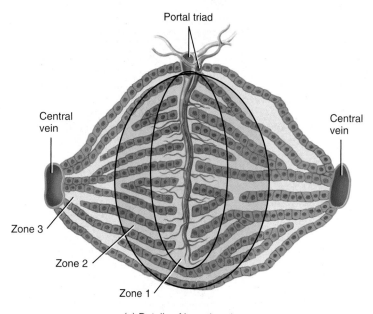

(e) Details of hepatic acinus

 Which type of liver cell is phagocytic?

2. **Bile canaliculi** (kan-a-LIK-ū-lī=small canals). These are small ducts between hepatocytes that collect bile produced by the hepatocytes. From bile canaliculi, bile passes into **bile ductules** and then **bile ducts**. The bile ducts merge and eventually form the larger **right** and **left hepatic ducts**, which unite and exit the liver as the **common hepatic duct** (see Figure 24.12a–b). The common hepatic duct joins the **cystic duct** (*cystic*=bladder) from the gallbladder to form the **common bile duct**. From here, bile enters the duodenum of the small intestine to participate in digestion. When the small intestine is empty, the sphincter around the common bile duct at the entrance to the duodenum closes, and bile backs up into the cystic duct to the gallbladder for storage.

3. **Hepatic sinusoids.** These highly permeable blood capillaries between hepatic laminae receive oxygenated blood from branches of the hepatic artery and nutrient-rich deoxygenated blood from branches of the hepatic portal vein. Recall that the hepatic portal vein brings venous blood from the gastrointestinal organs and spleen into the liver. Hepatic sinusoids converge and deliver blood into a **central vein**. From central veins the blood flows into the **hepatic veins**, which drain into the inferior vena cava (see Figure 14.5). Whereas blood flows

toward a central vein, bile flows in the opposite direction. Also present in the hepatic sinusoids are a modified macrophage, fixed phagocytes called **stellate reticuloendothelial cells** (STEL-āt re-tik′-ū-lō-en′-dō-THĒ-lē-al), *Kupffer cells* (KUP-fer), or *hepatic macrophages*, which destroy worn-out white and red blood cells, bacteria, and other foreign matter in the venous blood draining from the gastrointestinal tract.

Together, a bile duct, branch of the hepatic artery, and branch of the hepatic vein are referred to as a **portal triad** (*tri*=three).

The hepatocytes, bile duct system, and hepatic sinusoids can be organized into anatomical and functional units in three different ways:

1. *Hepatic lobule.* For years, anatomists described the **hepatic lobule** as the functional unit of the liver. According to this model, each hepatic lobule is shaped like a hexagon (six-sided structure) (Figure 24.14d, left). At its center is the central vein and radiating out from it are rows of hepatocytes and hepatic sinusoids. Located at three corners of the hexagon is a portal triad. This model is based on a description of the liver of adult pigs. In the human liver it is difficult to find such well-defined hepatic lobules surrounded by thick layers of connective tissue.

2. *Portal lobule.* This model emphasizes the exocrine function of the liver, that is, bile secretion. Accordingly, the bile duct of a portal triad is taken as the center of the portal lobule. The triangular shape of the portal lobule is defined by three imaginary straight lines that connect three central veins closest to the portal triad (Figure 24.14d, center). This model has not gained widespread acceptance.

3. *Hepatic acinus.* In recent years, the preferred structural and functional unit of the liver has become the hepatic acinus (AS-i-nus). Each hepatic acinus is an approximately oval mass that includes portions of two neighboring hepatic lobules. The short axis of the hepatic acinus is defined by branches of the portal triad—branches of the hepatic artery, vein, and bile ducts—that run along the border of the hepatic lobules. The long axis of the acinus is defined by two imaginary curved lines, which connect the two central veins closest to the short axis (Figure 24.14d, right). Hepatocytes in the hepatic acinus are arranged in three zones around the short axis, with no sharp boundaries between them (Figure 24.14e). Cells in zone 1 are closest to the branches of the portal triad and the first to receive incoming oxygen, nutrients, and toxins from incoming blood. These cells are the first ones to take up glucose, store it as glycogen after a meal, and break down glycogen to glucose during fasting. They are also the first to show morphological changes following bile duct obstruction or exposure to toxic substances. Zone 1 cells are the last ones to die if circulation is impaired and the first ones to regenerate. Cells in zone 3 are farthest from branches of the portal triad and are the last to show the effects of bile obstruction or exposure to toxins, the last to regenerate, and the first ones to show the effects of impaired circulation. Zone 3 cells also are the first to show evidence of fat accumulation. Cells in zone 2 have structural and functional characteristics intermediate between the cells in zones 1 and 3.

The hepatic acinus is the smallest structural and functional unit of the liver. Its popularity and appeal are based on the fact that it provides a logical description and interpretation of (1) patterns of glycogen storage and release and (2) toxic effects, degeneration, and regeneration relative to the proximity of the acinar zones to branches of the portal triad.

The mucosa of the gallbladder consists of simple columnar epithelium arranged in rugae resembling those of the stomach. The wall of the gallbladder lacks a submucosa. The middle, muscular coat consists of smooth muscle fibers; the contraction of these fibers ejects the contents of the gallbladder into the cystic duct. The gallbladder's outer coat is the visceral peritoneum. The functions of the gallbladder are to store bile and concentrate it (up to tenfold) until the bile is needed in the small intestine.

Blood and Nerve Supply of the Liver and Gallbladder

The liver receives blood from two sources (Figure 24.15). From the hepatic artery it obtains oxygenated blood, and from the hepatic portal vein it receives deoxygenated blood containing newly absorbed nutrients, drugs, and possibly microbes and toxins from the gastrointestinal tract. Branches of both the hepatic artery and the hepatic portal vein carry blood into hepatic sinusoids, where oxygen, most of the nutrients, and certain toxic substances are taken up by the hepatocytes. Products manufactured by the

Figure 24.15 Hepatic blood flow: sources, path through the liver, and return to the heart.

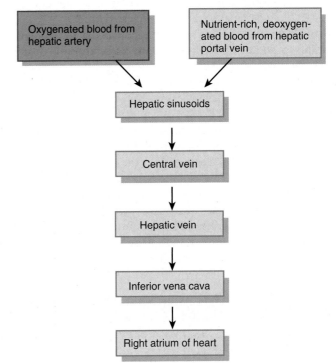

The liver receives oxygenated blood via the hepatic artery and nutrient-rich deoxygenated blood via the hepatic portal vein.

 Why is the liver often a site for metastasis of cancer that originates in the gastrointestinal tract?

hepatocytes and nutrients needed by other cells are secreted back into the blood, which then drains into the central vein and eventually passes into a hepatic vein. Because blood from the gastrointestinal tract passes through the liver as part of the hepatic portal circulation, the liver is often a site for metastasis of cancer that originates in the GI tract.

The nerve supply to the liver consists of parasympathetic innervation from the vagus (X) nerves and sympathetic innervation from the greater splanchnic nerves through the celiac ganglia.

The gallbladder is supplied by the cystic artery, which usually arises from the right hepatic artery. The cystic veins drain the gallbladder. The nerves to the gallbladder include branches from the celiac plexus and the vagus (X) nerve.

CLINICAL CONNECTION | *Liver Function Tests*

Liver function tests are blood tests designed to determine the presence of certain chemicals released by liver cells. These include albumin globulinase, alanine aminotransferase (ALT), aspartate aminotransferase (AST), alkaline phosphatase (ALP), gamma-glutamyl-transpeptidase (GGT), and bilirubin. The tests are used to evaluate and monitor liver disease or damage. Common causes of elevated liver enzymes include nonsteroidal anti-inflammatory drugs, cholesterol-lowering medications, some antibiotics, alcohol, diabetes, infections (viral hepatitis and mononucleosis), gallstones, tumors of the liver, and excessive use of herbal supplements such as kava, comfrey, pennyroyal, dandelion root, skullcap, and ephedra. •

Functions of the Liver and Gallbladder

Hepatocytes continuously secrete 800–1000 mL (about 1 qt) of bile per day. Bile salts, which are sodium salts and potassium salts of bile acids (mostly cholic acid and chenodeoxycholic acid), play roles in (1) **emulsification** (ē-mul-si-fi-KĀ-shun), the breakdown of large lipid globules into a suspension of droplets about 1 μm in diameter, and (2) the **absorption** of digested lipids.

Between meals, bile flows into the gallbladder for storage because the **sphincter of the hepatopancreatic ampulla** or *sphincter of Oddi* (OD-ē) (see Figure 24.12b) closes off the entrance to the duodenum. The sphincter surrounds the hepatopancreatic ampulla. After a meal, several neural and hormonal stimuli promote the production and release of bile. Parasympathetic impulses along the vagus (X) nerve fibers can stimulate the liver to increase bile production to more than twice the baseline rate. Fatty acids and amino acids in chyme entering the duodenum stimulate some duodenal enteroendocrine cells to secrete the hormone cholecystokinin (CCK) into the blood. CCK causes contraction of the walls of the gallbladder, which squeezes stored bile out of the gallbladder into the cystic duct and through the common bile duct. CCK also causes relaxation of the sphincter of the hepatopancreatic ampulla, which allows bile to flow into the duodenum.

CLINICAL CONNECTION | Gallstones

If bile contains either insufficient bile salts or lecithin or excessive cholesterol, the cholesterol may crystallize to form **gallstones**. As they grow in size and number, gallstones may cause minimal, intermittent, or complete obstruction to the flow of bile from the gallbladder into the duodenum. Treatment consists of using gallstone-dissolving drugs, lithotripsy (shock-wave therapy), or surgery. For people with a history of gallstones or for whom drugs or lithotripsy are not options, **cholecystectomy** (kō′-le-sis-TEK-tō-me)—the removal of the gallbladder and its contents—is necessary. More than half a million cholecystectomies are performed each year in the United States. To prevent side effects resulting from a loss of the gallbladder, patients should make lifestyle and dietary changes, including the following: (1) limiting the intake of saturated fat; (2) avoiding the consumption of alcoholic beverages; (3) eating smaller amounts of food during a meal and eating five to six smaller meals per day instead of two to three larger meals; and (4) taking vitamin and mineral supplements. •

In addition to secreting bile, the liver performs many other vital functions:

- **Carbohydrate metabolism.** The liver is especially important in maintaining a normal blood glucose level. When blood glucose is low, the liver can break down glycogen to glucose and release glucose into the bloodstream. The liver can also convert certain amino acids and lactic acid to glucose, and it can convert other sugars, such as fructose and galactose, into glucose. When blood glucose is high, as occurs just after eating a meal, the liver converts glucose to glycogen and triglycerides for storage.
- **Lipid metabolism.** Hepatocytes store some triglycerides; break down fatty acids to generate ATP; synthesize lipoproteins (HDLs, LDLs, VLDLs), which transport fatty acids, triglycerides, and cholesterol to and from body cells; synthesize cholesterol; and use cholesterol to make bile salts.
- **Protein metabolism.** Hepatocytes *deaminate* [remove the amino group (2NH₂) from] amino acids so that the amino acids can be

used for ATP production or converted to carbohydrates or fats. The resulting toxic ammonia (NH_3) is then converted into the much less toxic urea, which is excreted in urine. Hepatocytes also synthesize most plasma proteins, such as alpha and beta globulins, albumin, prothrombin, and fibrinogen.

- **Processing of drugs and hormones.** The liver can detoxify substances such as alcohol or secrete drugs such as penicillin, erythromycin, and sulfonamides into bile. It can also inactivate hormones such as thyroid hormones, estrogens, and aldosterone.
- **Excretion of bilirubin.** Bilirubin, derived from the heme of aged red blood cells, is absorbed by the liver from the blood and secreted into bile. Most of the bilirubin in bile is metabolized in the small intestine by bacteria and eliminated in feces.
- **Synthesis of bile salts.** Bile salts are used in the small intestine for the emulsification and absorption of lipids, cholesterol, phospholipids, and lipoproteins.
- **Storage.** In addition to glycogen, the liver is a prime storage site for certain vitamins (A, B₁₂, D, E, and K) and minerals (iron and copper), which are released from the liver when needed elsewhere in the body.
- **Phagocytosis.** The stellate reticuloendothelial (Kupffer) cells of the liver phagocytize aged red blood cells and white blood cells and some bacteria.
- **Activation of vitamin D.** The skin, liver, and kidneys participate in synthesizing the active form of vitamin D.

 CHECKPOINT

25. Draw and label a diagram of the cells/cell zones of a hepatic acinus.
26. Describe the pathways of blood flow into, through, and out of the liver.
27. How are the liver and gallbladder connected to the duodenum?
28. Describe the roles of the liver and gallbladder in the digestion of fats.

24.10 SMALL INTESTINE

 OBJECTIVES

- Describe the location and structure of the small intestine.
- Identify the functions of the small intestine.

Most digestion and absorption of nutrients occur in a long tube called the **small intestine**. Because of this, its structure is specially adapted for this function. Its length alone provides a large surface area for digestion and absorption, and that area is further increased by circular folds, villi, and microvilli. The small intestine begins at the pyloric sphincter of the stomach, coils through the central and inferior part of the abdominal cavity, and eventually opens into the large intestine. It averages 2.5 cm (1 in.) in diameter; its length is about 3 m (10 ft) in a living person and about 6.5 m (21 ft) in a cadaver due to the loss of smooth muscle tone after death.

Anatomy of the Small Intestine

The small intestine is divided into three regions (Figure 24.16). The first part of the small intestine, the **duodenum** (doo′-ō′-DĒ-num), is the shortest region, and is retroperitoneal. *Duodenum* means "12"; it is so named because it is about as long as the width of 12 fingers. It starts at the pyloric sphincter of the stomach and is in the form of a C-shaped tube that extends about 25 cm (10 in.) until it merges with the next section, called the

Figure 24.16 Regions of the small intestine. See also Figure 24.1b.

🔑 Most digestion and absorption occur in the small intestine.

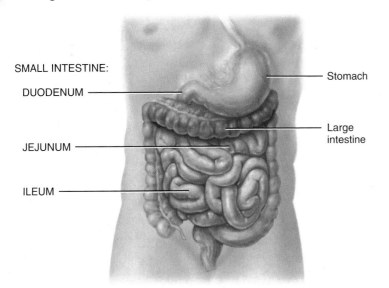

SMALL INTESTINE:

DUODENUM

JEJUNUM

ILEUM

Stomach

Large
intestine

Anterior view of external anatomy

❓ **Which portion of the small intestine is the longest?**

FUNCTIONS OF THE SMALL INTESTINE

1. Segmentations mix chyme with digestive juices and bring food into contact with the mucosa for absorption; peristalsis propels chyme through the small intestine.
2. Completes the digestion of carbohydrates (starches), proteins, and lipids; begins and completes the digestion of nucleic acids.
3. Absorbs about 90 percent of nutrients and water.

jejunum. The **jejunum** (jē-JOO-num), the next portion, is about 1 m (3 ft) long, and extends to the ileum. *Jejunum* means "empty," which is how it is found at death. The jejunum is mostly in the left upper quadrant (LUQ). The final and longest region of the small intestine, the **ileum** (IL-ē-um=twisted), measures about 2 m (6 ft) and joins the large intestine at a smooth muscle sphincter called the **ileocecal sphincter (valve)** (il'-ē-ō-SĒ-kal). The ileum is mostly in the right lower quadrant (RLQ).

The arterial blood supply of the small intestine is from the superior mesenteric artery and the gastroduodenal artery, which arises from the hepatic artery of the celiac trunk. Blood is returned by way of the superior mesenteric vein, which anastomoses with the splenic vein to form the hepatic portal vein.

The nerves to the small intestine are supplied by the superior mesenteric plexus. The branches of the plexus contain postganglionic sympathetic fibers, preganglionic parasympathetic fibers, and sensory fibers. The sensory fibers are components of the vagus (X) nerves and spinal nerves via the sympathetic pathways. In the wall of the small intestine are two autonomic plexuses: the myenteric plexus between the muscular layers and the submucosal plexus in the submucosa. The nerve fibers to the smooth muscle of the blood vessels arise chiefly from the sympathetic division of the autonomic nervous system, while the nerve fibers to the smooth muscle of the intestinal wall originate from the vagus (X) nerves.

Histology of the Small Intestine

The wall of the small intestine is composed of the same four layers that make up most of the GI tract: mucosa, submucosa, muscularis, and serosa (Figure 24.17b). The mucosa is composed of a layer of epithelium, lamina propria, and muscularis mucosae. The epithelial layer of the small intestinal mucosa consists of simple columnar epithelium that contains several types of cells: absorptive, goblet, enteroendocrine, and paneth (Figure 24.17c). **Absorptive cells** of the epithelium release enzymes that digest

food and contain microvilli that absorb nutrients in the small intestine. **Goblet cells** secrete mucus. The small intestinal mucosa contains many deep crevices lined with glandular epithelium. Cells lining the crevices form the **intestinal glands** or *crypts of Lieberkühn* (LĒ-ber-kēn) and secrete intestinal juice (to be discussed shortly). In addition to absorptive cells and goblet cells, the intestinal glands also contain enteroendocrine cells and paneth cells. Three types of **enteroendocrine cells** (cells that secrete hormones) are found in the intestinal glands of the small intestine: **S cells**, **CCK cells**, and **K cells**, which secrete the hormones **secretin** (se-KRĒ-tin), **cholecystokinin (CCK)** (kō-le-sis'-tō-KĪ N-in), and **glucose-dependent insulinotropic peptide (GIP)** (in-soo-lin-ō-TRŌ-pik), respectively. **Paneth cells** secrete *lysozyme*, a bactericidal enzyme, and are capable of phagocytosis. Paneth cells may have a role in regulating the microbial population in the small intestine.

The lamina propria of the small intestinal mucosa contains areolar connective tissue and has an abundance of mucosa-associated lymphoid tissue (MALT). **Solitary lymphatic nodules** are most numerous in the distal part of the ileum (see Figure 24.18c). Groups of lymphatic nodules referred to as **aggregated lymphatic follicles** or *Peyer's patches* (PĪ-ers) are also present in the ileum. As it does throughout the digestive tract, the muscularis mucosae consists of smooth muscle.

⚕ **CLINICAL CONNECTION | *Gastroenteritis***

Gastroenteritis (gas'-trō-en-ter-Ī-tis; *gastro-*=stomach; *enteron*= intestine; *-itis*=inflammation) is the inflammation of the lining of the stomach and intestine (especially the small intestine). It is usually caused by a viral or bacterial infection that may be acquired by contaminated food or water or by people in close contact. Symptoms include diarrhea, vomiting, fever, loss of appetite, cramps, and abdominal discomfort. •

Figure 24.17 Histology of the small intestine.

🔑 Circular folds, villi, and microvilli increase the surface area of the small intestine for digestion and absorption.

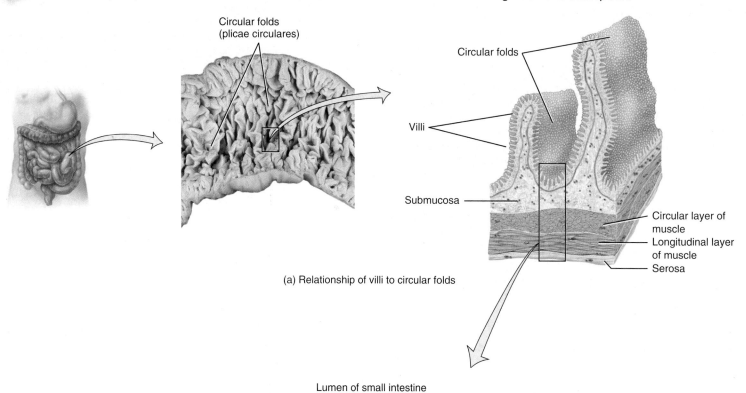

(a) Relationship of villi to circular folds

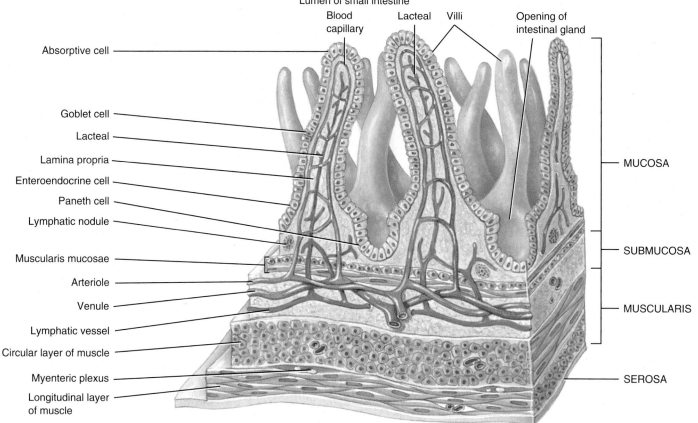

(b) Three-dimensional view of layers of the small intestine showing villi

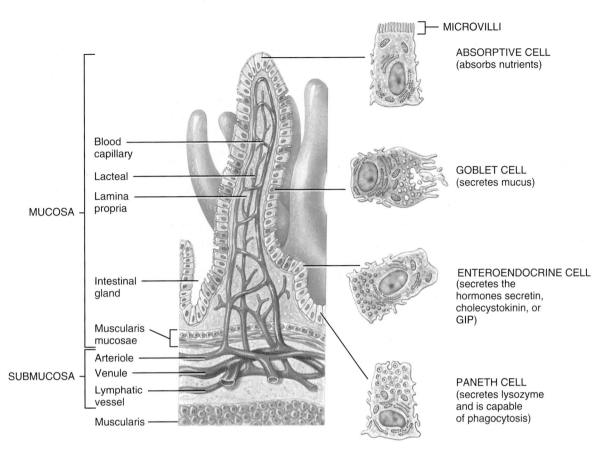

— MICROVILLI

ABSORPTIVE CELL
(absorbs nutrients)

MUCOSA
- Blood capillary
- Lacteal
- Lamina propria
- Intestinal gland
- Muscularis mucosae

GOBLET CELL
(secretes mucus)

ENTEROENDOCRINE CELL
(secretes the hormones secretin, cholecystokinin, or GIP)

SUBMUCOSA
- Arteriole
- Venule
- Lymphatic vessel
- Muscularis

PANETH CELL
(secretes lysozyme and is capable of phagocytosis)

(c) Enlarged villus showing lacteal, capillaries, intestinal glands, and cell types

 What is the functional significance of the blood capillary network and lacteal in the center of each villus?

The **submucosa** of the duodenum contains **duodenal glands** or *Brunner's glands* (BROO-ners) (Figure 24.18a), which secrete an alkaline mucus that helps neutralize gastric acid in the chyme. Sometimes the lymphatic tissue of the lamina propria extends through the muscularis mucosae into the submucosa.

The **muscularis** of the small intestine consists of two layers of smooth muscle. The outer, thinner layer contains longitudinal fibers; the inner, thicker layer contains circular fibers. Except for a major portion of the duodenum, which is retroperitoneal, the **serosa** (or visceral peritoneum) completely surrounds the small intestine.

Even though the wall of the small intestine is composed of the same four basic layers as the rest of the GI tract, special structural features of the small intestine facilitate the processes of digestion and absorption. These structural features include circular folds, villi, and microvilli. **Circular folds** or *plicae circulares* (PLĪ-sē ser-kyū-LA-res) are folds of the mucosa and submucosa (Figure 24.17a). These permanent ridges, which are about 10 mm (0.4 in.) long, begin near the proximal portion of the duodenum and end at about the midportion of the ileum. Some extend all the way around the circumference of the intestine; others extend only part of the way around. Circular folds enhance absorption by increasing surface area and causing the chyme to spiral, rather than move in a straight line, as it passes through the small intestine.

Also present in the small intestine are **villi** (=tufts of hair), fingerlike projections of the mucosa that are 0.5–1 mm long (Figure 24.18). The large number of villi (20–40 per square millimeter)

vastly increases the surface area of the epithelium available for absorption and digestion and gives the intestinal mucosa a velvety appearance. Each *villus* (singular form) is covered by epithelium and has a core of lamina propria; embedded in the connective tissue of the lamina propria are an arteriole, a venule, a blood capillary network, and a **lacteal** (LAK-tē-al=milky), which is a lymphatic capillary (see Figure 24.17c). Nutrients absorbed by the epithelial cells covering the villus pass through the wall of a capillary or a lacteal to enter blood or lymph, respectively.

Besides circular folds and villi, the small intestine also has **microvilli** (mī-krō-VIL-ī; *micro-*=small), which are projections of the apical (free) membrane of the absorptive cells. Each microvillus is a 1-μm-long projection of the cell membrane that contains a bundle of 20–30 actin filaments. When viewed through a light microscope, the microvilli are too small to be seen individually; instead they form a fuzzy line, called the **brush border**, extending into the lumen of the small intestine (Figure 24.18d). There are an estimated 200 million microvilli per square millimeter of small intestine. Because the microvilli greatly increase the surface area of the plasma membrane, larger amounts of digested nutrients can diffuse into absorptive cells in a given period. The brush border also contains several brush-border enzymes that have digestive functions (discussed shortly).

Functions of the Small Intestine

Chyme entering the small intestine contains partially digested carbohydrates, proteins, and lipids (mostly triglycerides). The

Figure 24.18 Histology of the duodenum and ileum.

🔑 Microvilli greatly increase the surface area of the small intestine for digestion and absorption.

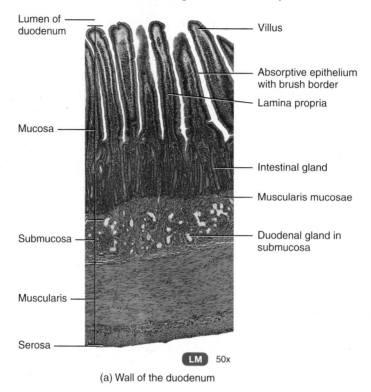

Lumen of duodenum — Villus

Absorptive epithelium with brush border

Lamina propria

Mucosa —

Intestinal gland

Muscularis mucosae

Submucosa — Duodenal gland in submucosa

Muscularis —

Serosa —

LM 50x

(a) Wall of the duodenum

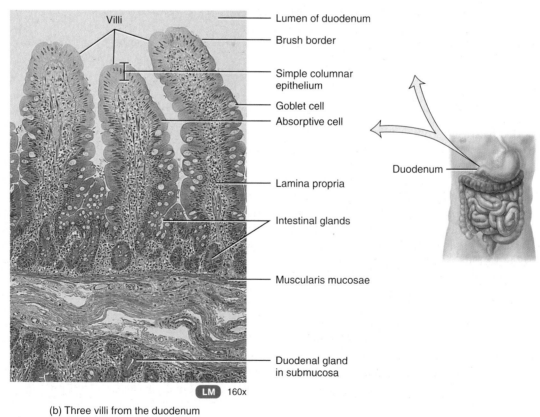

Villi — Lumen of duodenum

Brush border

Simple columnar epithelium

Goblet cell

Absorptive cell

Duodenum —

Lamina propria

Intestinal glands

Muscularis mucosae

Duodenal gland in submucosa

LM 160x

(b) Three villi from the duodenum

completion of the digestion of carbohydrates, proteins, and lipids is the result of the collective action of pancreatic juice, bile, and intestinal juice in the small intestine.

Intestinal juice is a clear yellow fluid secreted in amounts of 1 to 2 liters (about 1 to 2 quarts) each day. It has a pH of 7.6, which is slightly alkaline (due to its high concentration of bicarbonate ions), and contains water and mucus. Together, pancreatic juice

and intestinal juice provide a vehicle for the absorption of substances from chyme as they come in contact with the villi.

The absorptive epithelial cells synthesize several digestive enzymes, called **brush-border enzymes**, and insert them in the plasma membrane of the microvilli. Thus, some enzymatic digestion occurs at the surface of the epithelial cells that line the villi; in other parts of the GI tract, enzymatic digestion

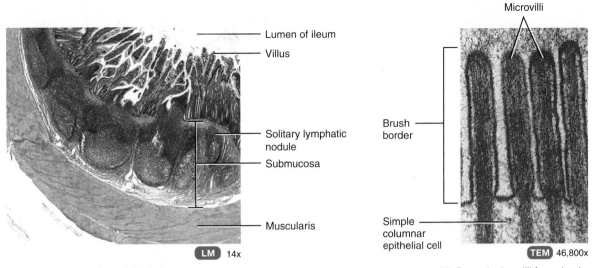

(c) Lymphatic nodules in ileum

(d) Several microvilli from duodenum

 What is the function of the fluid secreted by duodenal (Brunner's) glands?

occurs in the lumen exclusively. Among the brush-border enzymes are four carbohydrate-digesting enzymes called α-**dextrinase**, **maltase**, **sucrase**, and **lactase**; protein-digesting enzymes called **peptidases** (**aminopeptidase** and **dipeptidase**); and two types of nucleotide-digesting enzymes, **nucleosidases** and **phosphatases**. Also, as cells slough off into the lumen of the small intestine, they break apart and release enzymes that help digest nutrients in the chyme.

CLINICAL CONNECTION | *Lactose Intolerance*

In some people, the absorptive cells of the small intestine fail to produce enough of the enzyme lactase, which is essential for the digestion of lactose. This results in a condition called **lactose intolerance**, in which undigested lactose in chyme causes fluid to be retained in the feces; bacterial fermentation of the undigested lactose results in the production of gases. Symptoms of lactose intolerance include diarrhea, gas, bloating, and abdominal cramps after consumption of milk and other dairy products. The symptoms can be relatively minor or serious enough to require medical attention. The hydrogen breath test is often used to aid in diagnosis of lactose intolerance. Very little hydrogen can be detected in the breath of a normal person, but hydrogen is among the gases produced when undigested lactose in the colon is fermented by bacteria. The hydrogen is absorbed from the intestines and carried through the bloodstream to the lungs, where it is exhaled. Persons with lactose intolerance should select a diet that restricts lactose (but not calcium) and take dietary supplements to aid in the digestion of lactose. •

The two types of movements of the small intestine—segmentations and a type of peristalsis called migrating motility complexes—are governed mainly by the myenteric plexus of the enteric nervous system. **Segmentations** are localized, mixing contractions that occur in portions of the intestine distended by a large volume of chyme. Segmentations mix chyme with the digestive juices and bring the particles of food into contact with the mucosa for absorption; they do not push the intestinal contents

along the tract. A segmentation starts with the contractions of circular muscle fibers in a portion of the small intestine, an action that constricts the intestine into segments. Next, muscle fibers that encircle the middle of each segment also contract, dividing each segment again. Finally, the fibers that first contracted relax, and each small segment unites with an adjoining small segment so that large segments are formed again. As this sequence of events repeats, the chyme sloshes back and forth. Segmentations occur most rapidly in the duodenum, about 12 times per minute, and progressively decrease to about 8 times per minute in the ileum. This movement is similar to alternately squeezing the middle and then the ends of a capped tube of toothpaste.

After most of a meal has been absorbed, which lessens distension of the wall of the small intestine, segmentation stops and peristalsis begins. The type of peristalsis that occurs in the small intestine, termed a **migrating motility complex (MMC)**, begins in the lower portion of the stomach and pushes chyme forward along a short stretch of small intestine before dying out. The MMC slowly migrates down the small intestine, reaching the end of the ileum in 90–120 minutes. Then another MMC begins in the stomach. Altogether, chyme remains in the small intestine for 3–5 hours.

All the chemical and mechanical phases of digestion from the mouth through the small intestine are directed toward changing food into forms that can pass through the epithelial cells lining the mucosa into the underlying blood and lymphatic vessels. These forms are monosaccharides (glucose, fructose, and galactose) from carbohydrates; single amino acids, dipeptides, and tripeptides from proteins; fatty acids, glycerol, and monoglycerides from lipids; and pentoses and nitrogenous bases from nucleic acids. Passage of these digested nutrients from the gastrointestinal tract into the blood or lymph is called **absorption**. Absorption occurs by diffusion, facilitated diffusion, osmosis, and active transport.

About 90 percent of all absorption of nutrients takes place in the small intestine. The other 10 percent occurs in the stomach and large intestine. Any undigested or unabsorbed material left in the small intestine passes on to the large intestine.

CLINICAL CONNECTION | *Peptic Ulcer Disease*

In the United States, 5–10 percent of the population develops **peptic ulcer disease (PUD)**. An **ulcer** is a craterlike lesion in a membrane; ulcers that develop in areas of the GI tract exposed to acidic gastric juice are called **peptic ulcers**. The most common complication of peptic ulcers is bleeding, which can lead to anemia if enough blood is lost. In acute cases, peptic ulcers can lead to shock and death. Three distinct causes of PUD are recognized: (1) the bacterium *Helicobacter pylori* (hēl-i-kō-BAK-ter pī-Lō-rē); (2) nonsteroidal anti-inflammatory drugs (NSAIDs) such as aspirin; and (3) hypersecretion of HCl, as occurs in Zollinger-Ellison syndrome (ZOL-in-jer EL-i-son), which involves a gastrin-producing tumor, usually of the pancreas.

Helicobacter pylori (previously named *Campylobacter pylori*) is the most frequent cause of PUD. The bacterium produces an enzyme called *urease*, which splits urea into ammonia and carbon dioxide. While shielding the bacterium from the acidity of the stomach, the ammonia also damages the protective mucous layer of the stomach and the underlying gastric cells. *H. pylori* also produces catalase, an enzyme that may protect the microbe from phagocytosis by neutrophils, plus several adhesion proteins that allow the bacterium to attach itself to gastric cells.

Several therapeutic approaches are helpful in the treatment of PUD. Because cigarette smoke, alcohol, caffeine, and NSAIDs can impair mucosal defensive mechanisms, in the process increasing mucosal susceptibility to the damaging effects of HCl, these substances should

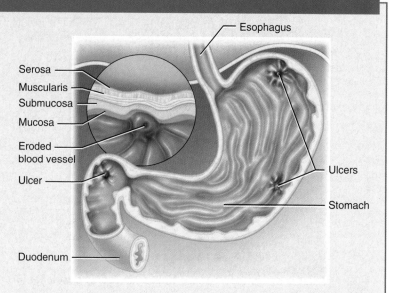

be avoided. In cases associated with *H. pylori*, treatment with an antibiotic drug often resolves the problem. Oral antacids such as Tums® or Maalox® can help temporarily by buffering gastric acid. When hypersecretion of HCl is the cause of PUD, H₂ blockers (such as Tagamet®) or proton pump inhibitors such as omeprazole (Prilosec®) may be used to block secretion of H⁺ from parietal cells. •

Table 24.4 summarizes the digestive activities in the pancreas, liver, gallbladder, and small intestine.

Table 24.5 summarizes the digestive enzymes and their functions for the digestive system.

TABLE 24.4

Summary of Digestive Activities in the Pancreas, Liver, Gallbladder, and Small Intestine

STRUCTURE	ACTIVITY
Pancreas	Delivers pancreatic juice into the duodenum via the pancreatic duct to assist absorption (see Table 24.5 for pancreatic enzymes and their functions)
Liver	Produces bile (bile salts) necessary for emulsification and absorption of lipids
Gallbladder	Stores, concentrates, and delivers bile into the duodenum via the common bile duct
Small intestine	Major site of digestion and absorption of nutrients and water in the gastrointestinal tract
Mucosa/submucosa	
Intestinal glands	Secrete intestinal juice to assist absorption
Absorptive cells	Digest and absorb nutrients
Goblet cells	Secrete mucus
Enteroendocrine cells (S, CCK, G)	Secrete secretin, cholecystokinin, and glucose-dependent insulinotropic peptide
Paneth cells	Secrete lysozyme (a bactericidal enzyme) and are capable of phagocytosis
Duodenal (Brunner's) glands	Secrete alkaline mucus to buffer stomach acids
Circular folds	Folds of mucosa and submucosa that increase the surface area for digestion and absorption
Villi	Fingerlike projections of mucosa that are the sites of absorption of digested food and increase the surface area for digestion and absorption
Microvilli	Microscopic, membrane-covered projections of absorptive epithelial cells that contain brush-border enzymes (listed in Table 24.5) and that increase the surface area for digestion and absorption
Muscularis	
Segmentation	A type of peristalsis consisting of alternating contractions of circular smooth muscle fibers that produce segmentation and resegmentation of sections of the small intestine; mixes chyme with digestive juices and brings food into contact with the mucosa for absorption
Migrating motility complex (MMC)	A type of peristalsis consisting of waves of contraction and relaxation of circular and longitudinal smooth muscle fibers passing the length of the small intestine; moves chyme toward ileocecal sphincter

TABLE 24.5

Summary of Digestive Enzymes

ENZYME	SOURCE	SUBSTRATES	PRODUCTS
SALIVA			
Salivary amylase	Salivary glands	Starches (polysaccharides)	Maltose (disaccharide), maltotriose (trisaccharide), and α-dextrins
Lingual lipase	Lingual glands in the tongue	Triglycerides (fats and oils) and other lipids	Fatty acids and diglycerides
GASTRIC JUICE			
Pepsin (activated from pepsinogen by pepsin and hydrochloric acid)	Stomach chief cells	Proteins	Peptides
Gastric lipase	Stomach chief cells	Triglycerides (fats and oils)	Fatty acids and monoglycerides
PANCREATIC JUICE			
Pancreatic amylase	Pancreatic acinar cells	Starches (polysaccharides)	Maltose (disaccharide), maltotriose (trisaccharide), and α-dextrins
Trypsin	Pancreatic acinar cells	Proteins	Peptides
Chymotrypsin	Pancreatic acinar cells	Proteins	Peptides
Elastase	Pancreatic acinar cells	Proteins	Peptides
Carboxypeptidase	Pancreatic acinar cells	Amino acid at carboxyl end of peptides	Amino acids and peptides
Pancreatic lipase	Pancreatic acinar cells	Triglycerides (fats and oils) that have been emulsified by bile salts	Fatty acids and monoglycerides
Ribonuclease	Pancreatic acinar cells	Ribonucleic acid	Nucleotides
Deoxyribonuclease	Pancreatic acinar cells	Deoxyribonucleic acid	Nucleotides
BRUSH BORDER ENZYMES IN MICROVILLI PLASMA MEMBRANE			
α-Dextrinase	Small intestine	α-Dextrins	Glucose
Maltase	Small intestine	Maltose	Glucose
Sucrase	Small intestine	Sucrose	Glucose and fructose
Lactase	Small intestine	Lactose	Glucose and galactose
Peptidases Aminopeptidase	Small intestine	Amino acid at amino end of peptides	Amino acids and peptides
Dipeptidase	Small intestine	Dipeptides	Amino acids
Nucleosidases and phosphatases	Small intestine	Nucleotides	Nitrogenous bases, pentoses, and phosphates

CLINICAL CONNECTION | *Bariatric Surgery*

Bariatric surgery (bar'-ē-AT-rik; *baros*=weight; *atreia*=medical treatment) is a surgical procedure that limits the amount of food that can be ingested and absorbed to bring about a significant weight loss in obese individuals. The most commonly performed type is called *gastric bypass surgery*. In one variation of this procedure, a small pouch about the size of a walnut is created at the top of the stomach. The pouch, which is only 5–10 percent of the stomach volume, is sealed off using surgical staples or a plastic band. The pouch is connected to the jejunum of the small intestine, thus bypassing the rest of the stomach and the duodenum. The result is that smaller amounts of food are ingested and fewer nutrients are absorbed in the small intestine. This leads to weight loss.   •

✓ CHECKPOINT

29. What are the characteristics of the different regions of the small intestine?

30. In what ways are the mucosa and submucosa of the small intestine adapted for digestion and absorption?
31. Describe the types of movement in the small intestine.
32. In what form are the products of carbohydrate, protein, and lipid digestion absorbed?

24.11 LARGE INTESTINE

 OBJECTIVES

• Describe the anatomy and structure of the large intestine.
• Explain the functions of the large intestine.

The large intestine is the terminal portion of the GI tract and is divided into four principal regions. As chyme moves through the large intestine, bacteria act on it and water, ions, and vitamins are absorbed. As a result, feces are formed and then eliminated from the body.

Anatomy of the Large Intestine

The **large intestine**, which is about 1.5 m (5 ft) long and 6.5 cm (2.5 in.) in diameter in living humans and cadavers, extends from the ileum to the anus (Figure 24.19). The ascending colon and descending colon are retroperitoneal, while the remaining parts of the colon and cecum are attached to the posterior abdominal wall by their **mesocolon** (mez'-ō-KŌ-lon), a double layer of peritoneum

Figure 24.19 Anatomy of the large intestine.

The regions of the large intestine are the cecum, colon, rectum, and anal canal.

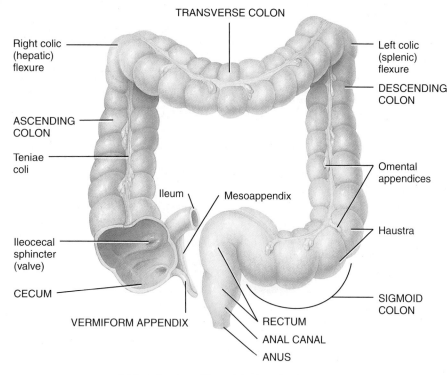

(a) Anterior view of large intestine showing major regions

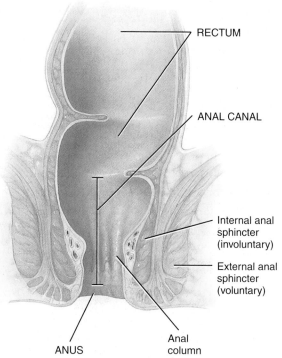

(b) Frontal section of anal canal

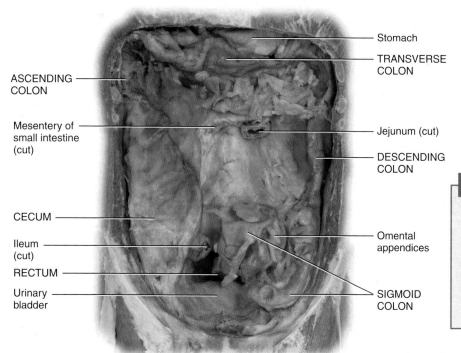

(c) Anterior view

 Which portions of the colon are retroperitoneal?

connecting the parietal peritoneum to the visceral peritoneum that contains the vascular and nervous supply to the organs (see Figure 24.3a). Structurally, the four principal regions of the large intestine are the cecum, colon, rectum, and anal canal (Figure 24.19a).

The opening from the ileum into the large intestine is guarded by a fold of mucous membrane called the **ileocecal sphincter** or **valve** (il'-ē-ō-SĒ-kal), which allows materials from the small intestine to pass into the large intestine. Hanging inferior to the ileocecal valve is the **cecum**, a small pouch about 6 cm (2.4 in.) long. Attached to the cecum is a twisted, coiled tube, measuring about 8 cm (3 in.) in length, called the **appendix** or **vermiform appendix** (VER-mi-form=worm-shaped; *appendix*=appendage). The mesentery of the appendix, called the **mesoappendix** (mes'-ō-a-PEN-diks), attaches the appendix to the inferior part of the mesentery of the ileum. The appendix has a high concentration of lymphatic nodules, which control the bacteria entering the large intestine by immune responses.

The open end of the cecum merges with a long tube called the **colon** (=food passage), which is divided into ascending, transverse, descending, and sigmoid portions. Both the ascending and descending colon are retroperitoneal; the transverse and sigmoid colon are not. The **ascending colon** ascends on the right side of the abdomen, reaches the inferior surface of the liver, and turns abruptly to the left to form the **right colic (hepatic) flexure**. The colon continues across the abdomen to the left side as the **transverse colon**. It curves beneath the inferior end of the spleen on the left side as the **left colic (splenic) flexure** and passes inferiorly to the level of the left iliac crest as the **descending colon**. The **sigmoid colon** (*sigm-*=S-shaped) begins near the left iliac crest, projects medially to the midline, and terminates as the rectum at about the level of the third sacral vertebra.

The **rectum**, approximately 15 cm (6 in.) in length, lies anterior to the sacrum and coccyx. The terminal 2–3 cm (1 in.) of the large intestine is called the **anal canal** (Figure 24.19b). The mucous membrane of the anal canal is arranged in longitudinal folds called **anal columns** that contain a network of arteries and veins. The opening of the anal canal to the exterior, called the **anus**, is guarded by an **internal anal sphincter** of smooth muscle (involuntary) and an **external anal sphincter** of skeletal muscle (voluntary). Normally the anus is closed except during the elimination of feces.

The arterial supply of the cecum and colon is derived from branches of the superior mesenteric and inferior mesenteric arteries. The distal end of the transverse colon, near the left colic flexure, is the transition zone between superior and inferior mesenteric blood supply and drainage. Within this zone the two vessels form numerous collateral circuits. The venous return is by way of the superior and inferior mesenteric veins ultimately to the hepatic portal vein and into the liver. The arterial supply of the rectum and anal canal branches from the superior, middle, and inferior rectal arteries. The rectal veins correspond to the rectal arteries.

The nerves to the large intestine consist of sympathetic, parasympathetic, and sensory components. The sympathetic innervation arises from the celiac, superior, and inferior mesenteric ganglia and superior and inferior mesenteric plexuses. The fibers reach the viscera by way of the thoracic and lumbar splanchnic nerves. The parasympathetic innervation is derived from the vagus (X) and pelvic splanchnic nerves. Similar to its function in the vascular supply, the left colic flexure serves as the transition zone between vagal and pelvic splanchnic innervation.

Histology of the Large Intestine

The wall of the large intestine contains the typical four layers found in the rest of the GI tract: mucosa, submucosa, muscularis, and serosa. The **mucosa** consists of simple columnar epithelium, lamina propria (areolar connective tissue), and muscularis mucosae (smooth muscle) (Figure 24.20a). The

Figure 24.20 Histology of the large intestine.

Intestinal glands formed by simple columnar epithelial cells and goblet cells extend the full thickness of the mucosa.

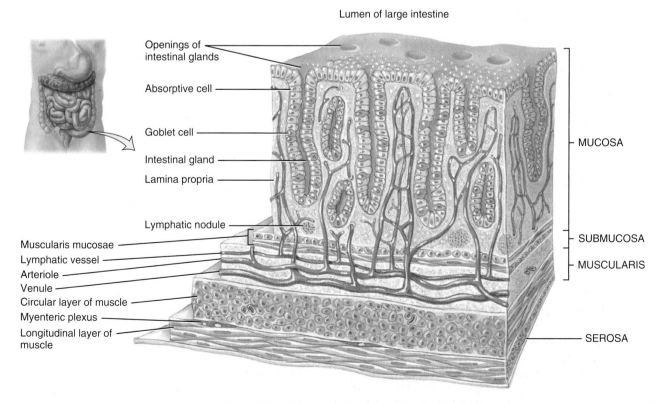

(a) Three-dimensional view of layers of the large intestine

FIGURE 24.20 CONTINUES ▶

FIGURE 24.20 CONTINUED

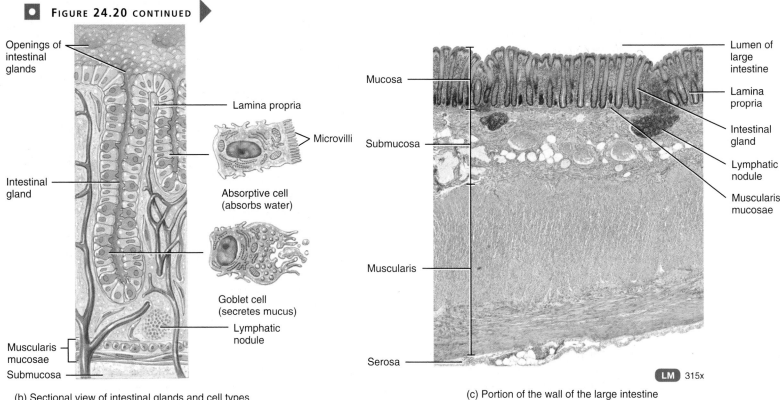

Openings of intestinal glands

Lamina propria

Microvilli

Absorptive cell (absorbs water)

Intestinal gland

Goblet cell (secretes mucus)

Lymphatic nodule

Muscularis mucosae

Submucosa

(b) Sectional view of intestinal glands and cell types

Mucosa

Lumen of large intestine

Lamina propria

Submucosa

Intestinal gland

Lymphatic nodule

Muscularis

Muscularis mucosae

Serosa

LM 315x

(c) Portion of the wall of the large intestine

Opening of intestinal gland

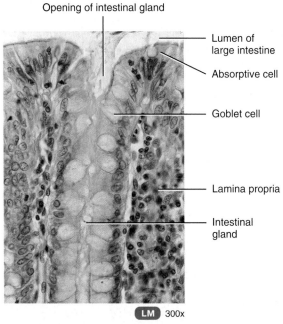

Lumen of large intestine

Absorptive cell

Goblet cell

Lamina propria

Intestinal gland

LM 300x

(d) Details of mucosa of large intestine

 What is the function of the goblet cells of the large intestine?

epithelium contains mostly absorptive and goblet cells (Figure 24.20b and c). The absorptive cells function primarily in water absorption; the goblet cells secrete mucus that lubricates the passage of the colonic contents. Both absorptive and goblet cells are located in long, straight, tubular **intestinal glands** or *crypts of Lieberkühn* that extend the full thickness of the mucosa. Solitary lymphatic nodules are also found in the lamina propria of the mucosa and may extend through the muscularis mucosae into the submucosa. Compared to the small intestine, the mucosa of the large intestine does not have as many structural adaptations that increase surface area. There are no

circular folds or villi; however, microvilli are present on the absorptive cells. Consequently, much more absorption occurs in the small intestine than in the large intestine.

The **submucosa** of the large intestine consists of areolar connective tissue. The **muscularis** consists of an external layer of longitudinal smooth muscle and an internal layer of circular smooth muscle. Unlike other parts of the GI tract, portions of the longitudinal muscles are condensed and thickened, forming three conspicuous bands called the **teniae coli** (TE-nē-ē KŌ-lī; *teniae*=flat bands) that run most of the length of the large intestine (see Figure 24.19a). The portions of the wall between the teniae coli have little or no longitudinal muscle. Tonic contractions of the bands gather the colon into a series of pouches called **haustra** (HAWS-tra=shaped like pouches; singular is *haustrum*), which give the colon a puckered appearance. A single layer of circular smooth muscle lies deep to the teniae coli. The **serosa** of the large intestine is part of the visceral peritoneum. Small pouches of visceral peritoneum filled with fat are attached to teniae coli and are called **omental (fatty) appendices** or *epiploic appendices* (see Figure 24.19a).

Functions of the Large Intestine

The passage of chyme from the ileum into the cecum is regulated by the action of the ileocecal sphincter. While this has been called a valve, and there are valve-like folds of tissue from the wall of the cecum that meet at the opening of the ileum, the mechanism of closure of the opening lies in the muscle wall of the terminal ileum and is not affected by the "valve." Normally, the sphincter remains partially closed, and the passage of chyme into the cecum is a slow process. Immediately after a meal, ileal peristalsis intensifies, the sphincter relaxes, and chyme is forced from the ileum into the cecum. As food passes through the ileocecal sphincter, it fills the cecum and accumulates in the ascending colon, and movements of the colon begin.

There are three types of movement characteristic of the large intestine:

1. In **haustral churning**, the haustra remain relaxed and distended while they fill up. When the distension reaches a certain point, the wall contracts and squeezes the contents into the next haustrum.
2. **Peristalsis** also occurs, although at a slower rate (3 to 12 contractions per minute) than in other portions of the GI tract.
3. In **mass peristalsis**, a strong peristaltic wave begins at about the middle of the transverse colon and quickly drives the colonic contents into the rectum. Mass peristalsis usually takes place three or four times a day, during or immediately after a meal.

The final stage of digestion occurs in the colon through the activity of bacteria that inhabit the lumen. Mucus is secreted by the glands of the large intestine, but no enzymes are secreted. Chyme is prepared for elimination by the action of bacteria, which ferment any remaining carbohydrates and release hydrogen, carbon dioxide, and methane gases. These gases contribute to flatus (gas) in the colon, termed *flatulence* when it is excessive. Bacteria also convert any remaining proteins to amino acids and break down the amino acids into simpler substances: indole, skatole, hydrogen sulfide, and fatty acids. Some of the indole and skatole is eliminated in the feces and contributes to their odor; the rest is absorbed and transported to the liver, where these compounds are converted to less toxic compounds and excreted in the urine. Bacteria also decompose bilirubin to simpler pigments, including stercobilin, which gives feces their brown color. Several vitamins needed for normal metabolism, including some B vitamins and vitamin K, are bacterial products that are absorbed in the colon.

By the time chyme has remained in the large intestine 3–10 hours, it has become solid or semisolid as a result of water absorption and is now called **feces**. Chemically, feces consist of water, inorganic salts, sloughed-off epithelial cells from the mucosa of the gastrointestinal tract, bacteria, products of bacterial decomposition, unabsorbed digested materials, and indigestible parts of food.

Although 90 percent of all water absorption occurs in the small intestine, the large intestine absorbs enough to make it an important organ in maintaining the body's water balance. Of the 0.5–1.0 liter of water that enters the large intestine, all but about 100–200 mL is absorbed via osmosis. The large intestine also absorbs electrolytes, including sodium and chloride, and some vitamins.

Mass peristaltic movements push fecal material from the sigmoid colon into the rectum. The resulting distension of the rectal wall stimulates stretch receptors, which initiates **defecation**, the elimination of feces from the rectum through the anus. The **defecation reflex** occurs as follows: In response to distension of the rectal wall, the receptors send sensory nerve impulses to the sacral spinal cord. Motor impulses from the cord travel along parasympathetic nerves back to the descending colon, sigmoid colon, rectum, and anus. The resulting contraction of the longitudinal rectal muscles shortens the rectum, thereby increasing the pressure within it. This pressure, along with voluntary contractions of the diaphragm and abdominal muscles and parasympathetic stimulation, opens the internal sphincter.

The external sphincter is voluntarily controlled. If it is voluntarily relaxed, defecation occurs and the feces are expelled through the anus; if it is voluntarily constricted, defecation can be postponed. Voluntary contractions of the diaphragm and abdominal muscles aid defecation by increasing the pressure within the abdomen, which pushes the walls of the sigmoid colon and rectum inward. If defecation does not occur, the feces back up into the sigmoid colon until the next wave of mass peristalsis again stimulates the stretch receptors, further creating the urge to defecate. In infants, the defecation reflex causes automatic emptying of the rectum because voluntary control of the external anal sphincter has not yet developed.

Table 24.6 summarizes the digestive activities of the large intestine.

Dietary fiber consists of indigestible plant carbohydrates—such as cellulose, lignin, and pectin—found in fruits, vegetables, grains, and beans. **Insoluble fiber**, which does not dissolve in water, includes the woody or structural parts of plants such as the skins of fruits and vegetables and the bran coating around wheat and corn kernels. Insoluble fiber passes through the GI tract largely unchanged and speeds up the passage of material through the tract. **Soluble fiber** dissolves in water and forms a gel, which slows the passage of material through the tract; it is found in abundance in beans, oats, barley, broccoli, prunes, apples, and citrus fruits.

People who choose a fiber-rich diet may reduce their risk of developing obesity, diabetes, atherosclerosis, gallstones, hemorrhoids, diverticulitis, appendicitis, and colorectal cancer. Soluble fiber also may help lower blood cholesterol. The liver normally

---

TABLE 24.6

Summary of Digestive Activities in the Large Intestine

STRUCTURE	ACTIVITY	FUNCTION(S)
Lumen	Bacterial activity	Breaks down undigested carbohydrates, proteins, and amino acids into products that can be expelled in feces or absorbed and detoxified by liver; synthesizes certain B vitamins and vitamin K
Mucosa	Secretes mucus	Lubricates colon and protects mucosa
	Absorption	Water absorption solidifies feces and contributes to the body's water balance; solutes absorbed include ions and some vitamins
Muscularis	Haustral churning	Moves contents from haustrum to haustrum by muscular contractions
	Peristalsis	Moves contents along length of colon by contractions of circular and longitudinal muscles
	Mass peristalsis	Forces contents into sigmoid colon and rectum
	Defecation reflex	Eliminates feces by contractions in sigmoid colon and rectum

converts cholesterol to bile salts, which are released into the small intestine to help fat digestion. Having accomplished their task, the bile salts are reabsorbed by the small intestine and recycled back to the liver. Since soluble fiber binds to bile salts to prevent their reabsorption, the liver makes more bile salts to replace those lost in feces. Thus, the liver uses more cholesterol to make more bile salts and blood cholesterol level is lowered.

A summary of the digestive organs and their functions is presented in Table 24.7.

CLINICAL CONNECTION | Diarrhea and Constipation

Diarrhea (dī-a-RĒ-a; *dia-*=through; *rrhea*=flow) is an increase in the frequency, volume, and fluid content of the feces caused by increased motility of and decreased absorption by the intestines. When chyme passes too quickly through the small intestine and feces pass too quickly through the large intestine, there is not enough time for absorption. Frequent diarrhea can result in dehydration and electrolyte imbalances. Excessive motility may be caused by lactose intolerance, stress, or microbes that irritate the gastrointestinal mucosa.

Constipation (kon-sti-PĀ-shun; *con-*=together; *stip-*=to press) refers to infrequent or difficult defecation caused by decreased motility of the intestines. Because the feces remain in the colon for prolonged periods of time, excessive water absorption occurs, and the feces become dry and hard. Constipation may be caused by poor habits (delaying defecation), spasms of the colon, insufficient fiber in the diet, inadequate fluid intake, lack of exercise, emotional stress, and certain drugs. A common treatment is a mild laxative, such as milk of magnesia, which induces defecation. However, many physicians maintain that laxatives are habit-forming, and that adding fiber to the diet, increasing the amount of exercise, and increasing fluid intake are safer ways of controlling this common problem. •

CHECKPOINT

33. What are the principal regions of the large intestine?
34. How does the muscularis of the large intestine differ from that of the rest of the gastrointestinal tract? What are haustra?
35. Describe the mechanical movements that occur in the large intestine.
36. What is defecation, and how does it occur?
37. Explain the activities of the large intestine that change its contents into feces.

24.12 DEVELOPMENT OF THE DIGESTIVE SYSTEM

OBJECTIVE

• Describe the development of the digestive system.

During the fourth week of development, the cells of the **endoderm** form a cavity called the **primitive gut**, the forerunner of the gastrointestinal tract (see Figure 4.12b). Soon afterward the mesoderm forms and splits into two layers (somatic and splanchnic), as shown in Figure 4.9d. The splanchnic mesoderm associates with the endoderm of the primitive gut; as a result, the primitive gut has a double-layered wall. The **endodermal layer** gives rise to the *epithelial lining* and *glands* of most of the gastrointestinal tract; the **mesodermal layer** produces the *smooth muscle*, *blood vessels*, and *connective tissue* of the tract.

The primitive gut elongates and differentiates into an anterior **foregut**, an intermediate **midgut**, and a posterior **hindgut** (see Figure 4.12c). Until the fifth week of development, the midgut opens into the yolk sac; after that time, the yolk sac constricts and detaches from the midgut, and the midgut seals. In the region of

TABLE 24.7

Summary of Organs of the Digestive System and Their Functions

ORGANS	FUNCTIONS
Tongue	Maneuvers food for mastication, shapes food into a bolus, maneuvers food for deglutition, detects sensations for taste, and initiates digestion of triglycerides
Salivary glands	Saliva produced by these glands softens, moistens, and dissolves foods; cleanses mouth and teeth; initiates the digestion of starch
Teeth	Cut, tear, and pulverize food to reduce solids to smaller particles for swallowing
Pancreas	Pancreatic juice buffers acidic gastric juice in chyme, stops the action of pepsin from the stomach, creates the proper pH for digestion in the small intestine, and participates in the digestion of carbohydrates, proteins, triglycerides, and nucleic acids
Liver	Produces bile, which is required for the emulsification and absorption of lipids in the small intestine
Gallbladder	Stores and concentrates bile and releases it into the small intestine
Mouth	See the functions of the tongue, salivary glands, and teeth, all of which are in the mouth. Additionally, the lips and cheeks keep food between the teeth during mastication, and buccal glands lining the mouth produce saliva
Pharynx	Receives a bolus from the oral cavity and passes it into the esophagus
Esophagus	Receives a bolus from the pharynx and moves it into the stomach; this requires relaxation of the upper esophageal sphincter and secretion of mucus
Stomach	Mixing waves combine saliva, food, and gastric juice, which activates pepsin, initiates protein digestion, kills microbes in food, helps absorb vitamin B_{12}, contracts the lower esophageal sphincter, increases stomach motility, relaxes the pyloric sphincter, and moves chyme into the small intestine
Small intestine	Segmentation mixes chyme with digestive juices; peristalsis propels chyme toward the ileocecal sphincter; digestive secretions from the small intestine, pancreas, and liver complete the digestion of carbohydrates, proteins, lipids, and nucleic acids; circular folds, villi, and microvilli help absorb about 90 percent of digested nutrients
Large intestine	Haustral churning, peristalsis, and mass peristalsis drive the colonic contents into the rectum; bacteria produce some B vitamins and vitamin K; absorption of some water, ions, and vitamins occurs; defecation

the foregut, a depression consisting of ectoderm, the **stomodeum** (stō-mō-DĒ-um), appears (see Figure 4.12d). This develops into the *oral cavity*. The **oropharyngeal membrane** (ōr'-ō-fa-RIN-jē-al) is a depression of fused ectoderm and endoderm on the surface of the embryo that separates the foregut from the stomodeum. The membrane ruptures during the fourth week of development, so that the foregut is continuous with the outside of the embryo through the oral cavity. Another depression consisting of ectoderm, the **proctodeum** (prok-tō-DĒ-um), forms in the hindgut and goes on to develop into the *anus*. (See Figure 4.12d.) The **cloacal membrane** (klō-Ā-kul) is a fused membrane of ectoderm and endoderm that separates the hindgut from the proctodeum. After the cloacal membrane ruptures during the seventh week, the hindgut forms a continuous tube from mouth to anus.

The foregut develops into the *pharynx, esophagus, stomach,* and *part of the duodenum*. The midgut is transformed into the *remainder of the duodenum*, the *jejunum*, the *ileum*, and *portions of the large intestine* (cecum, appendix, ascending colon, and most of the transverse colon). The hindgut develops into the *remainder of the large intestine*, except for a portion of the anal canal that is derived from the proctodeum.

As development progresses, the endoderm at various places along the foregut develops into hollow buds that grow into the mesoderm. These buds will develop into the *salivary glands, liver, gallbladder,* and *pancreas*. Each of these organs retains a connection with the gastrointestinal tract through ducts.

 CHECKPOINT

38. What structures develop from the foregut, midgut, and hindgut?

  ## 24.13 AGING AND THE DIGESTIVE SYSTEM

 OBJECTIVE

• Describe the effects of aging on the digestive system.

Overall changes of the digestive system associated with aging include decreased secretory mechanisms, decreased motility of the digestive organs, loss of strength and tone of the muscular tissue

and its supporting structures, changes in neurosensory feedback regarding enzyme and hormone release, and diminished response to pain and internal sensations. In the upper portion of the GI tract, common changes include reduced sensitivity to mouth irritations and sores, loss of taste, periodontal disease, difficulty in swallowing, hiatal hernia, gastritis, and peptic ulcer disease. Changes in the small intestine may include duodenal ulcers, appendicitis, malabsorption, and maldigestion. Other pathologies that increase in incidence with age are gallbladder problems, jaundice, cirrhosis, and acute pancreatitis. Large intestinal changes such as constipation, hemorrhoids, and diverticular disease may also occur. Cancer of the colon or rectum is quite common, as are bowel obstructions and impactions.

CLINICAL CONNECTION | *Colorectal Cancer*

Colorectal cancer is among the deadliest of malignancies, ranking second to lung cancer in males and third after lung cancer and breast cancer in females. Genetics plays a very important role; an inherited predisposition contributes to more than half of all cases of colorectal cancer. Intake of alcohol and diets high in animal fat and protein are associated with increased risk of colorectal cancer; dietary fiber, aspirin, calcium, and selenium may be protective. Signs and symptoms of colorectal cancer include diarrhea, constipation, cramping, abdominal pain, and rectal bleeding, either visible or occult. Precancerous growths on the mucosal surface, called **polyps**, also increase the risk of developing colorectal cancer. Screening includes testing for blood in the feces, digital rectal examination, sigmoidoscopy, colonoscopy, and barium enema. Tumors may be removed endoscopically or surgically. •

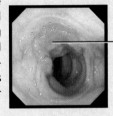

Malignant tumor in colon

 CHECKPOINT

39. What overall changes in the digestive system appear with age?

KEY MEDICAL TERMS ASSOCIATED WITH THE DIGESTIVE SYSTEM

Anorexia nervosa (an'-ō-REK-sē-a ner-VŌ-sa) A chronic disorder characterized by self-induced weight loss, negative perception of body image, and physiological changes that result from nutritional depletion. Patients have a fixation on weight control and often abuse laxatives, which worsens their fluid and electrolyte imbalances and nutrient deficiencies.

Appendicitis Inflammation of the appendix preceded by obstruction of the lumen of the appendix by chyme, inflammation, a foreign body, a carcinoma of the cecum, stenosis, or kinking of the organ. Characterized by high fever and elevated white blood cell count, especially of neutrophils, which increase to a count of higher than 75 percent.

Barrett's esophagus A pathological change in the epithelium of the esophagus from nonkeratinized stratified squamous epithelium to columnar epithelium so that the lining resembles that of the stomach or small intestine due to long-term exposure of the esophagus to stomach acid; increases the risk of developing cancer of the esophagus.

Bulimia (bū-LĒM-ē-a; *bu-*=ox; *limia*=hunger or **binge–purge syndrome**) A disorder that typically affects young, single, middle-class white females, characterized by overeating at least twice a week followed by purging by self-induced vomiting, strict dieting or fasting, vigorous exercise, or use of laxatives or diuretics; it occurs in response to fears of being overweight or to stress, depression, and physiological disorders such as hypothalamic tumors.

Canker sore (KANG-ker) Painful ulcer on the mucous membrane of the mouth that affects females more often than males, usually between ages 10 and 40; may be an autoimmune reaction or a food allergy.

Celiac disease (SĒ-lē-ak; *koilia*=hollow) Common malabsorption disorder caused by sensitivity to gluten, a protein found in grains such as wheat, rye, barley, and oats. The gluten damages intestinal villi and reduces the length of the microvilli; symptoms include foul-smelling diarrhea, bloating, weight loss, abdominal pain, and anemia.

Cirrhosis (si-RŌ-sis) Distorted or scarred liver as a result of chronic inflammation due to hepatitis, chemicals that destroy hepatocytes, parasites that infect the liver, or alcoholism. Symptoms include jaundice, edema in the legs, uncontrolled bleeding, and increased sensitivity to drugs.

Colitis (kō-LĪ-tis) Inflammation of the mucosa of the colon and rectum in which absorption of water and salts is reduced, producing watery, bloody feces and, in severe cases, dehydration and salt depletion.

Colostomy (kō-LŌS-tō-mē; *-stomy*=provide an opening) Diversion of feces through an opening in the colon; a surgical "stoma" (artificial opening) is made in the exterior of the abdominal wall that serves as a substitute anus; feces are eliminated into a bag.

Diverticulitis (dī'-ver-tik'-ū-LĪ-tis) Inflammation of diverticula; may be characterized by pain, either constipation or increased frequency of defecation, nausea, vomiting, and low-grade fever. Patients who change to high-fiber diets show marked relief of symptoms.

Flatus (FLA-tus) Air (gas) in the stomach or intestine, usually expelled through the anus. If the gas is expelled through the mouth, it is called **eructation** or **belching** (burping). Flatus may result from gas released during the breakdown of foods in the stomach or from swallowing air- or gas-containing substances such as carbonated drinks.

Food poisoning A sudden illness caused by ingesting food or drink contaminated by an infectious microbe or a toxin. The most common cause is the toxin produced by the bacterium *Staphylococcus aureus.* Most types cause diarrhea and/or vomiting, often associated with abdominal pain.

Heartburn A burning sensation in a region near the heart due to irritation of the mucosa of the esophagus from hydrochloric acid in stomach contents. Caused by failure of the lower esophageal sphincter to close properly, so that the stomach contents enter the inferior esophagus.

Hemorrhoids (HEM-ō-royds; *hemo*=blood; *rhoia*=flow) Varicosed superior rectal veins in the upper rectum that develop when the veins are put under pressure and become engorged with blood. Bleeding or itching is usually the first symptom. Hemorrhoids may be caused by constipation, which may be brought on by low-fiber diets. Also called **piles**.

Hernia (HER-nē-a) Protrusion of all or part of an organ through a membrane or cavity wall, usually the abdominal cavity. *Hiatal hernia* is the protrusion of a part of the stomach into the thoracic cavity through the esophageal hiatus. *Inguinal hernia* is the protrusion of the hernial sac into the inguinal opening; it may contain a portion of the bowel in an advanced stage and may extend into the scrotal compartment in males, causing strangulation of the herniated part.

Indigestion (in-di-JES-chun) A nonspecific term used to describe many symptoms associated with abdominal distress, especially after eating. Symptoms include discomfort or a feeling of fullness in the upper abdomen, nausea, and a sensation of bloating, often relieved by belching. Also called **dyspepsia**.

Inflammatory bowel disease (in-FLAM-a-tō′-rē BOW-el) Inflammation of the gastrointestinal tract that exists in two forms. (1) **Crohn's disease** is an inflammation of any part of the gastrointestinal tract in which the inflammation extends from the mucosa through the submucosa, muscularis, and serosa. (2) **Ulcerative colitis** is an inflammation of the mucosa of the colon and rectum, usually accompanied by rectal bleeding.

Irritable bowel syndrome (IBS) Disease of the entire gastrointestinal tract in which a person reacts to stress by developing symptoms (such as cramping and abdominal pain) associated with alternating patterns of diarrhea and constipation. Also known as **irritable colon** or **spastic colitis**.

Jaundice (JAWN-dis=yellowed) A yellowish coloration of the sclerae (whites of the eyes), skin, and mucous membranes due to buildup of a yellow compound called bilirubin.

Nausea (NAW-sē-a; *nausia*=seasickness) Discomfort characterized by a loss of appetite and the sensation of impending vomiting. Its causes include local irritation of the gastrointestinal tract, a systemic disease, brain disease or injury, overexertion, or the effects of medication or drug overdose.

Occult blood (*occult*=hidden) Blood not detectable by the human eye. The main diagnostic value of occult blood testing is to screen for colorectal cancer. Two substances often examined are feces and urine.

Traveler's diarrhea Infectious disease of the gastrointestinal tract that results in loose, urgent bowel movements, cramping, abdominal pain, malaise, nausea, and occasionally fever and dehydration. Acquired through ingestion of food or water contaminated with fecal material typically containing bacteria.

CHAPTER REVIEW AND RESOURCE SUMMARY

WileyPLUS

Review

Resource

Introduction

1. The breakdown of larger food molecules into smaller molecules is called digestion.
2. The passage of these smaller molecules into blood and lymph is termed absorption.

24.1 Overview of the Digestive System

1. The organs that collectively perform digestion and absorption constitute the digestive system and are usually divided into two main groups: the gastrointestinal (GI) tract and accessory digestive organs.
2. The GI tract is a continuous tube extending from the mouth to the anus.
3. The accessory digestive organs include the teeth, tongue, salivary glands, liver, gallbladder, and pancreas.
4. Digestion includes six basic processes: ingestion, secretion, mixing and propulsion, mechanical and chemical digestion, absorption, and defecation.
5. Mechanical digestion consists of mastication and GI tract movements that aid chemical digestion.
6. Chemical digestion is a series of hydrolysis reactions that break down large carbohydrates, lipids, proteins, and nucleic acids in foods into smaller molecules that are usable by body cells.

Anatomy Overview - The Digestive System
Animation - Chemical Digestion - Enzymes
Exercise - Concentrate on Digestion

24.2 Layers of the GI Tract

1. The basic arrangement of layers in most of the gastrointestinal tract, from deep to superficial, is mucosa, submucosa, muscularis, and serosa.
2. Associated with the lamina propria of mucosa are extensive patches of lymphatic tissue called mucosa-associated lymphoid tissue (MALT).

Anatomy Overview - GI Tract Histology
Figure 24.2 - Layers of the Gastrointestinal Tract
Exercise - Paint the Gastrointestinal Tract

24.3 Peritoneum

1. The peritoneum is the largest serous membrane of the body; it lines the wall of the abdominal cavity and covers some abdominal organs.
2. Folds of the peritoneum include the mesentery, mesocolon, falciform ligament, lesser omentum, and greater omentum.

24.4 Mouth

1. The mouth is formed by the cheeks, hard and soft palates, lips, and tongue. The vestibule is the space bounded externally by the cheeks and lips and internally by the teeth and gums. The oral cavity proper extends from the vestibule to the fauces.
2. The tongue, together with its associated muscles, forms the floor of the oral cavity. It is composed of skeletal muscle covered with mucous membrane. The upper surface and sides of the tongue are covered with papillae, some of which contain taste buds.

Anatomy Overview - Oral Cavity
Anatomy Overview - Salivary Glands
Animation - Mastication
Animation - Neural Regulation of Mechanical Digestion
Figure 24.7 - Dentitions

Review

3. The major portion of saliva is secreted by the salivary glands, which lie outside the mouth and pour their contents into ducts that empty into the oral cavity. There are three pairs of salivary glands: parotid, submandibular (submaxillary), and sublingual glands.

4. Saliva lubricates food and starts the chemical digestion of carbohydrates. Salivation is controlled by the nervous system.

5. The teeth (dentes) project into the mouth and are adapted for mechanical digestion. A typical tooth consists of three principal regions: crown, root, and neck.

6. Teeth are composed primarily of dentin and are covered by enamel, the hardest substance in the body. There are two dentitions: deciduous and permanent.

7. Through mastication, food is mixed with saliva and shaped into a soft, flexible mass called a bolus. Salivary amylase begins the digestion of starches, and lingual lipase acts on triglycerides.

24.5 Pharynx

1. Deglutition, or swallowing, moves a bolus from the mouth to the stomach.
2. Muscular contractions of the oropharynx and laryngopharynx propel a bolus into the esophagus.

Anatomy Overview - Pharynx and Esophagus
Animation - Deglutition

24.6 Esophagus

1. The esophagus is a collapsible, muscular tube that connects the pharynx to the stomach.
2. It passes a bolus into the stomach by peristalsis.
3. It contains an upper and a lower esophageal sphincter.

Anatomy Overview - Pharynx and Esophagus
Anatomy Overview - Esophagus Histology

24.7 Stomach

1. The stomach connects the esophagus to the duodenum. The principal anatomical regions of the stomach are the cardia, fundus, body, and pylorus.
2. Adaptations of the stomach for digestion include rugae; glands that produce mucus, hydrochloric acid, pepsin, gastric lipase, and intrinsic factor; and a three-layered muscularis.
3. Mechanical digestion consists of mixing waves. Chemical digestion consists mostly of the conversion of proteins into peptides by pepsin.
4. The stomach wall is impermeable to most substances. Among the substances the stomach can absorb are water, certain ions, drugs, and alcohol.

Anatomy Overviews:
  The Stomach; Stomach Histology
Animations: Stomach Peristalsis;
  Protein Digestion in the
  Stomach; Lipid Digestion
  in the Stomach; Chemical
  Digestion - Gastric Acid
Figure 24.11 - Histology of the
  Stomach

24.8 Pancreas

1. The pancreas consists of a head, a body, and a tail and is connected to the duodenum via the pancreatic duct and accessory duct.
2. Endocrine pancreatic islets secrete hormones, and exocrine acini secrete pancreatic juice.
3. Pancreatic juice contains enzymes that digest starch (pancreatic amylase), proteins (trypsin, chymotrypsin, carboxypeptidase, and elastase), triglycerides (pancreatic lipase), and nucleic acids (ribonuclease and deoxyribonuclease).

Anatomy Overviews:
Pancreas; Pancreas Histology
Animations: Carb. Digestion -
  Pancreas; Protein Digestion -
  Pancreatic Juice; Lipid Digestion -
  Bile Salts and Pancreatic Lipase
Figure 24.12 - Relationship
  of the Pancreas to the Liver,
  Gallbladder, and Duodenum

24.9 Liver and Gallbladder

1. The liver has left and right lobes; the left lobe includes a quadrate lobe and a caudate lobe. The gallbladder is a sac located in a depression on the posterior surface of the liver that stores and concentrates bile.
2. The lobes of the liver are made up of lobules that contain hepatocytes (liver cells), sinusoids, stellate reticuloendothelial (Kupffer) cells, a vein, and bile ducts.
3. Hepatocytes produce bile that is carried by a duct system to the gallbladder for concentration and temporary storage. Cholecystokinin (CCK) stimulates ejection of bile into the common bile duct.
4. Bile's contribution to digestion is the emulsification of dietary lipids.
5. The liver also functions in carbohydrate, lipid, and protein metabolism; processing of drugs and hormones; excretion of bilirubin; synthesis of bile salts; storage of vitamins and minerals; phagocytosis; and activation of vitamin D.
6. Bile secretion is regulated by neural and hormonal mechanisms.

Anatomy Overviews:
  Liver and Gallbladder; Liver Histology
Animations:
  Chemical Digestion - Bile; Lipid
  Digestion - Bile Salts and
  Pancreatic Lipase; Carbohydrate
  Metabolism; Lipid Metabolism;
  Protein Metabolism

24.10 Small Intestine

1. The small intestine extends from the pyloric sphincter to the ileocecal sphincter. It is divided into three parts: duodenum, jejunum, and ileum.
2. Its glands secrete fluid and mucus, and the circular folds, villi, and microvilli of its wall provide a large surface area for digestion and absorption.
3. Brush-border enzymes digest α-dextrins, maltose, sucrose, lactose, peptides, and nucleotides at the surface of mucosal epithelial cells.
4. Pancreatic and intestinal brush-border enzymes break down carbohydrates, proteins, and nucleic acids.
5. Mechanical digestion in the small intestine involves segmentation and migrating motility complexes.
6. Absorption is the passage of digested nutrients from the gastrointestinal tract into blood or lymph. Absorbed nutrients include monosaccharides, amino acids, fatty acids, monoglycerides, pentoses, and nitrogenous bases.

Anatomy Overview - Small Intestine
Anatomy Overview - Small Intestine
  Histology
Animation - Segmentation
Animation - Carbohydrate Digestion
  in the Small Intestine
Animation - Protein Digestion in the
  Small Intestine
Animation - Nucleic Acid Digestion in
  the Small Intestine
Animation - Carbohydrate, Protein,
  Nucleic Acid and Lipid Absorption
  in the Small Intestine

Review

24.11 Large Intestine

1. The large intestine extends from the ileocecal sphincter to the anus. Its regions include the cecum, colon, rectum, and anal canal.
2. The mucosa contains many goblet cells, and the muscularis consists of teniae coli and haustra.
3. Mechanical movements of the large intestine include haustral churning, peristalsis, and mass peristalsis.
4. The last stages of chemical digestion occur in the large intestine through bacterial action. Substances are further broken down, and some vitamins are synthesized.
5. The large intestine absorbs water, electrolytes, and vitamins.
6. Feces consist of water, inorganic salts, epithelial cells, bacteria, and undigested foods.
7. Defecation, the elimination of feces from the rectum, is a reflex action aided by voluntary contractions of the diaphragm and abdominal muscles and relaxation of the external anal sphincter.

Anatomy Overview - The Large Intestine
Anatomy Overview - Large Intestine Histology
Animation - Mechanical Digestion in the Large Intestine
Exercise - Match the Movement

24.12 Development of the Digestive System

1. The endoderm of the primitive gut forms the epithelium and glands of most of the gastrointestinal tract.
2. The mesoderm of the primitive gut forms the smooth muscle and connective tissue of the GI tract.

24.13 Aging and the Digestive System

1. General changes include decreased secretory mechanisms, decreased motility, and loss of tone.
2. Specific changes may include loss of taste, pyorrhea, hernias, peptic ulcer disease, constipation, hemorrhoids, and diverticular disease.

CRITICAL THINKING QUESTIONS

1. If you could leave French fried potatoes in your mouth long enough after chewing them, they would start to taste sweeter and maybe even a little tart. Why?

2. When Zelda turned 50, her doctor said, "Well, it's time to schedule your first colonoscopy!" What is a colonoscopy? Describe the specific anatomical characteristics the doctor would be examining with the fiber optic scope.

3. The small intestine is the primary site of digestion and absorption in the gastrointestinal tract. What structural modifications are unique to the small intestine and what are their functions?

4. Krystal and her friend were giggling uncontrollably over their milk and fries at her birthday party. In mid-giggle, milk started coming out of Krystal's nose. How did this happen?

5. Four-year-old Billy was lying down with his head resting on his mother's abdomen. He started laughing and said, "Mommy, your tummy sure is making a lot of funny noises!" What (specifically) is Billy hearing?

? ANSWERS TO FIGURE QUESTIONS

24.1 Digestive enzymes are secreted by the salivary glands, tongue, stomach, pancreas, and small intestine.

24.2 The enteric plexuses help regulate secretions and motility of the gastrointestinal tract.

24.3 Mesentery binds the small intestine to the posterior abdominal wall.

24.4 The uvula helps prevent foods and liquids from entering the nasal cavity during swallowing.

24.5 The ducts of the submandibular glands empty on either side of the lingual frenulum.

24.6 The main component of teeth is a type of connective tissue called dentin.

24.7 The first, second, and third molars do not replace any deciduous teeth.

24.8 The esophageal mucosa and submucosa contain mucus-secreting glands that lubricate the gastrointestinal tract to allow food to pass through more easily.

24.9 Food is pushed along the gastrointestinal tract by contraction of smooth muscle behind the bolus and relaxation of smooth muscle in front of it.

24.10 After a large meal, the rugae stretch and disappear as the stomach fills; as the stomach empties, the rugae reappear.

24.11 In the gastric glands of the stomach, surface mucous cells and mucous neck cells secrete mucus; chief cells secrete pepsinogen and gastric lipase; parietal cells secrete HCl and intrinsic factor; and G cells secrete gastrin.

24.12 The pancreatic duct contains pancreatic juice (fluid and digestive enzymes); the common bile duct contains bile; the hepatopancreatic ampulla contains pancreatic juice and bile.

24.13 The epigastric region, which contains most of the liver, can be palpated in a clinical examination to check for enlargement.

24.14 The phagocytic cell in the liver is the stellate reticuloendothelial (Kupffer) cell.

24.15 The liver is often a site for metastasis of cancer that originates in the gastrointestinal tract because blood from the tract passes through the liver as part of the hepatic portal circulation.

24.16 The ileum is the longest part of the small intestine.

24.17 Nutrients being absorbed enter the blood via the capillaries or the lymph via the lacteals in the center of each villus.

24.18 The fluid secreted by duodenal glands—alkaline mucus—neutralizes gastric acid and protects the mucosal lining of the duodenum.

24.19 The ascending and descending portions of the colon are retroperitoneal.

24.20 The goblet cells of the large intestine secrete mucus to lubricate the colonic contents.

25 THE URINARY SYSTEM

INTRODUCTION As body cells carry out their metabolic functions, they consume oxygen and nutrients and produce substances, such as carbon dioxide, that have no useful functions and need to be eliminated from the body. While the respiratory system rids the body of carbon dioxide, the urinary system disposes of most other unneeded substances. As you will learn in this chapter, however, the urinary system is not merely concerned with waste disposal; it carries out a number of other important functions as well.

The kidneys allow you to conserve water by secreting hormones to reabsorb water that is normally lost to the urine. When you exercise, you remove toxins in your body through increased sweating and therefore lose more water than when you are at rest. Not only do you lose water, but important ions such as sodium and potassium (important to electrical transmission in your body) are also lost. This is why athletes often drink sports drinks that contain water as well as sodium, potassium, and glucose (for quick energy supply). Most people think that the first sign of dehydration is thirst and are unaware that muscle fatigue and soreness occur first and thirst later. In order to ensure that you are getting enough fluids you should really never let yourself get to the point of thirst. When you participate in exercise or activities that increase your metabolism, it is even more important to drink fluids or suffer the consequences of lactic acid buildup and muscle fatigue to the point of muscle spasm.

Nephrology (nef-ROL-ō-jē; *nephr-*=kidney; *-logy*=study of) is the scientific study of the anatomy, physiology, and pathology of the kidneys. The branch of medicine that deals with the male and female urinary systems and the male reproductive system is called **urology** (ū-ROL-ō-jē; *uro-*=urine). A physician who specializes in this branch of medicine is called a **urologist** (ū-ROL-ō-jist). •

 Did you ever wonder how diuretics work and why they are used?

25.1 OVERVIEW OF THE URINARY SYSTEM

  OBJECTIVE

- Describe the major structures of the urinary system and the functions they perform.

The **urinary system** consists of two kidneys, two ureters, one urinary bladder, and one urethra (Figure 25.1). Like the respiratory and digestive systems, the urinary system forms an extensive area of contact with the cardiovascular system. After the kidneys filter blood plasma, they return most of the water and solutes to the bloodstream. The remaining water and solutes constitute **urine**, which passes through the ureters and is stored in the urinary bladder until it is removed from the body through the urethra.

The kidneys do the major work of the urinary system. The other parts of the system are mainly passageways and storage areas. Functions of the kidneys include the following:

- ***Regulation of blood ionic composition.*** The kidneys help regulate the blood levels of several ions, most importantly sodium

Figure 25.1 Organs of the urinary system. The organs are shown in relation to the surrounding structures in a female.

 Urine formed by the kidneys passes first into the ureters, then to the urinary bladder for storage, and finally through the urethra for elimination from the body.

FUNCTIONS OF THE URINARY SYSTEM

1. The kidneys regulate blood volume and composition; help regulate blood pressure, pH, and glucose levels; produce two hormones (calcitriol and erythropoietin); and excrete wastes in the urine.
2. The ureters transport urine from the kidneys to the urinary bladder.
3. The urinary bladder stores urine and expels it into the urethra.
4. The urethra discharges urine from the body.

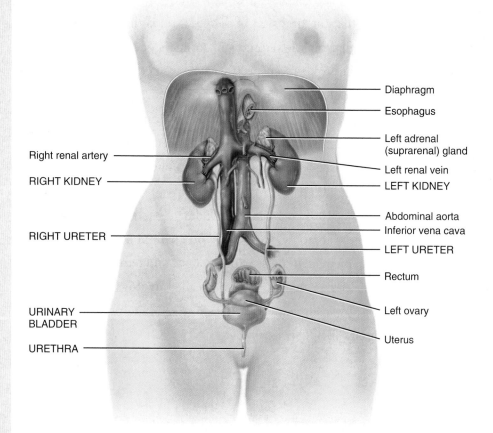

Right renal artery

RIGHT KIDNEY

RIGHT URETER

URINARY BLADDER

URETHRA

Diaphragm

Esophagus

Left adrenal (suprarenal) gland

Left renal vein

LEFT KIDNEY

Abdominal aorta

Inferior vena cava

LEFT URETER

Rectum

Left ovary

Uterus

(a) Anterior view of urinary system

ions (Na^+), potassium ions (K^+), calcium ions (Ca^{2+}), chloride ions (Cl^-), and phosphate ions (HPO_4^{2-}).

- *Regulation of blood pH.* The kidneys excrete a variable amount of hydrogen ions (H^+) into the urine and conserve bicarbonate ions (HCO_3^-). Both of these activities help regulate blood pH.
- *Regulation of blood volume.* The kidneys adjust blood volume by conserving or eliminating water in the urine. An increase in blood volume increases blood pressure; a decrease in blood volume decreases blood pressure.
- *Enzymatic regulation of blood pressure.* The kidneys *also* help regulate blood pressure by secreting the enzyme renin, which indirectly causes an increase in blood pressure.
- *Maintenance of blood osmolarity.* By separately regulating loss of water and loss of solutes in the urine, the kidneys maintain a relatively constant blood osmolarity. The **osmolarity** of a solution is a measure of the total number of dissolved particles per liter of solution.
- *Production of hormones.* The kidneys produce two hormones. *Calcitriol,* the active form of vitamin D, helps regulate calcium

homeostasis, and *erythropoietin* stimulates the production of red blood cells.

- *Regulation of blood glucose level.* Like the liver, the kidneys can use the amino acid glutamine in *gluconeogenesis,* the synthesis of new glucose molecules. They can then release glucose into the blood to help maintain a normal blood glucose level.
- *Excretion of wastes and foreign substances.* By forming urine, the kidneys help excrete **wastes**—substances that have no useful function in the body. Some wastes excreted in urine result from metabolic reactions in the body. These include ammonia and urea from the deamination of amino acids; bilirubin from the catabolism of hemoglobin; creatinine from the breakdown of creatine phosphate in muscle fibers; and uric acid from the catabolism of nucleic acids. Other wastes excreted in urine are foreign substances from the diet, such as drugs and environmental toxins.

 CHECKPOINT

1. Which organ does most of the work of the urinary system? Explain.

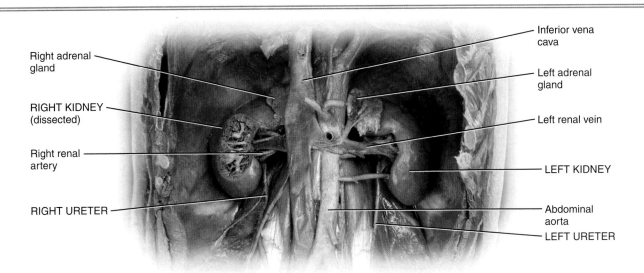

(b) Anterior view of urinary system

Right adrenal gland
RIGHT KIDNEY (dissected)
Right renal artery
RIGHT URETER

Inferior vena cava
Left adrenal gland
Left renal vein
LEFT KIDNEY
Abdominal aorta
LEFT URETER

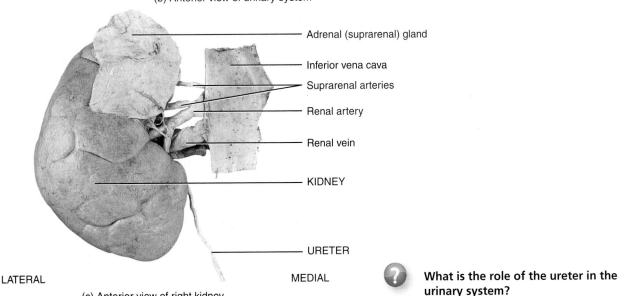

Adrenal (suprarenal) gland
Inferior vena cava
Suprarenal arteries
Renal artery
Renal vein
KIDNEY
URETER

LATERAL

MEDIAL

(c) Anterior view of right kidney

What is the role of the ureter in the urinary system?

25.2 ANATOMY AND HISTOLOGY OF THE KIDNEYS

OBJECTIVES

• Describe the external gross anatomical features of the kidneys.
• Explain the internal gross anatomical features of the kidneys.
• Trace the path of blood flow through the kidneys.
• Describe the structure of renal corpuscles and renal tubules.

The paired **kidneys** are reddish, kidney-bean-shaped organs located just above the waist between the peritoneum and the posterior wall of the abdomen. Because their position is posterior to the peritoneum of the abdominal cavity, they are said to be **retroperitoneal** (re′-trō-per-i-tō-NĒ-al; *retro-*=behind) organs (Figure 25.2). The kidneys are located between the levels of the last thoracic and third lumbar vertebrae, a position where they are partially protected by ribs 11 and 12. This is a double-edged sword, however, because if these lower ribs are fractured they can slice into the kidneys and create significant, even life-threatening, damage. The right kidney is slightly lower than the left (see Figure 25.1) because the liver occupies considerable space on the right side superior to the kidney.

External Anatomy of the Kidneys

A typical kidney in an adult is 10–12 cm (4–5 in.) long, 5–7 cm (2–3 in.) wide, and 3 cm (1 in.) thick—about the size of a bar of bath soap—and has a mass of 125–170 g (4.5–5 oz). The concave medial border of each kidney faces the vertebral column (see Figure 25.1). Near the center of the concave border is an indentation called the **renal hilum** (RĒ-nal; *ren-*=kidney) (see Figure 25.3), through which the ureter emerges from the kidney along with blood vessels, lymphatic vessels, and nerves.

Three layers of tissue surround each kidney (see Figure 25.2). The deep layer, the **renal capsule**, is a smooth, transparent sheet of dense irregular connective tissue that is continuous with the outer coat of the ureter. It serves as a barrier against trauma and helps maintain the shape of the kidney. The middle layer, the **adipose capsule**, is a mass of fatty tissue surrounding the renal capsule. It also protects the kidney from trauma and holds it firmly in place within the abdominal cavity. The superficial layer, the **renal fascia** (FASH-ē-a), is another thin layer of dense irregular connective tissue that anchors the kidney to the surrounding structures and to the abdominal wall. On the anterior surface of the kidneys, the renal fascia is deep to the peritoneum.

CLINICAL CONNECTION | *Nephroptosis*

Nephroptosis (nef′-ro-TŌ-sis; *ptosis*=falling), or *floating kidney*, is an inferior displacement or dropping of the kidney. It occurs when the kidney slips from its normal position because it is not securely held in place by adjacent organs or its covering of fat. Nephroptosis develops most often in very thin people whose adipose capsule or renal fascia is deficient. It is dangerous because the ureter may kink and block urine flow. The resulting backup of urine puts pressure on the kidney, which damages the tissue. Twisting of the ureter also causes pain. Nephroptosis is very common; about one in four people has some degree of weakening of the fibrous bands that hold the kidney in place. It is 10 times more common in females than males. Because it happens during life, it is very easy to distinguish from congenital anomalies. •

Internal Anatomy of the Kidneys

A frontal section through the kidney reveals two distinct regions: a superficial, light red region called the **renal cortex** (*cortex*=rind

Figure 25.2 Position and coverings of the kidneys.

🔑 The kidneys are surrounded by a renal capsule, adipose capsule, and renal fascia.

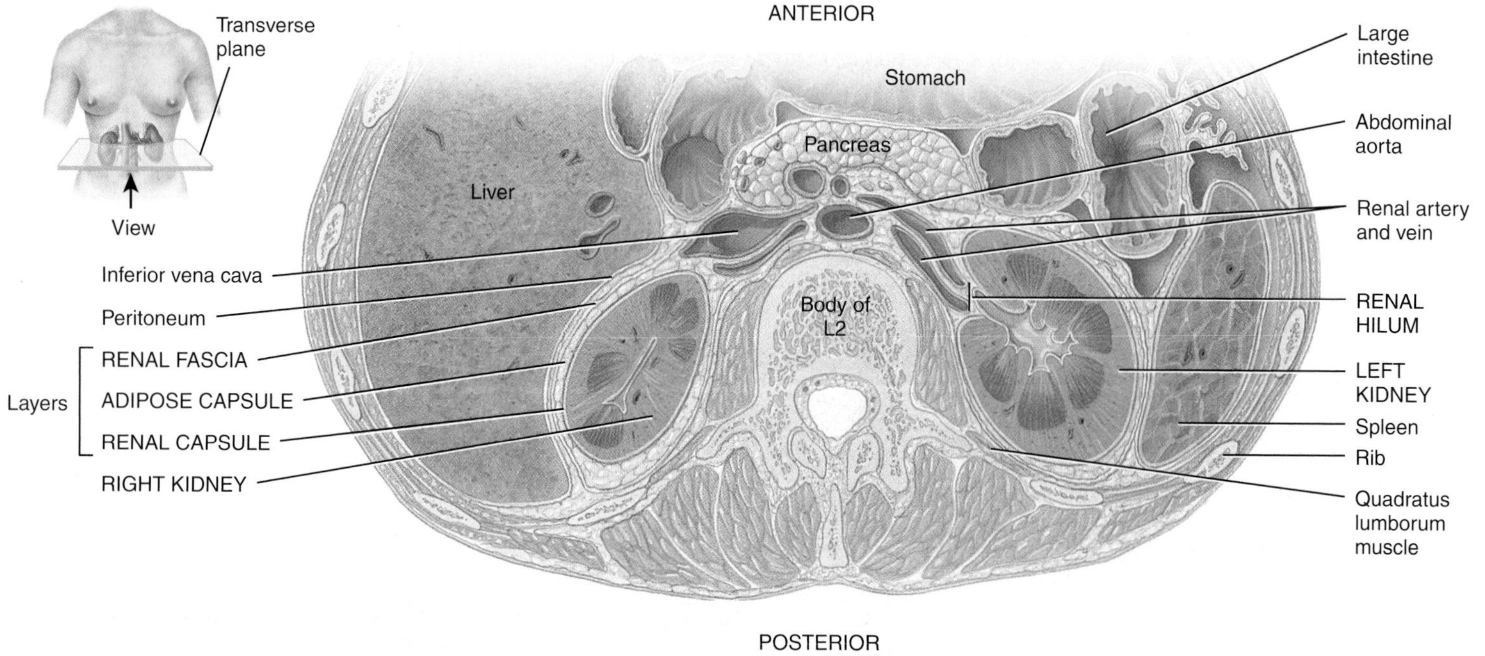

Transverse plane

ANTERIOR

Stomach

Pancreas

Large intestine

Liver

View

Abdominal aorta

Inferior vena cava

Renal artery and vein

Peritoneum

Body of L2

RENAL HILUM

Layers {

RENAL FASCIA

ADIPOSE CAPSULE

RENAL CAPSULE

RIGHT KIDNEY

LEFT KIDNEY

Spleen

Rib

Quadratus lumborum muscle

POSTERIOR

(a) Inferior view of transverse section of abdomen (L2)

or bark) and a deep, darker red-brown region called the **renal medulla** (*medulla*=inner portion) (Figure 25.3). The renal medulla consists of several cone-shaped **renal pyramids**. The base (wider end) of each pyramid faces the renal cortex, and its apex (narrower end), called a **renal papilla**, points toward the renal hilum. The smooth-textured renal cortex extends from the renal capsule to the bases of the renal pyramids and into the spaces between them.

It is divided into an outer *cortical zone* and an inner *juxtamedullary zone* (juks′-ta-MED-ū-lar′-ē). Those portions of the renal cortex that extend between renal pyramids are called **renal columns**. A **renal lobe** consists of a renal pyramid, its overlying area of renal cortex, and one-half of each adjacent renal column.

Together, the renal cortex and renal pyramids of the renal medulla constitute the **parenchyma** (pa-RENG-ki-ma) or

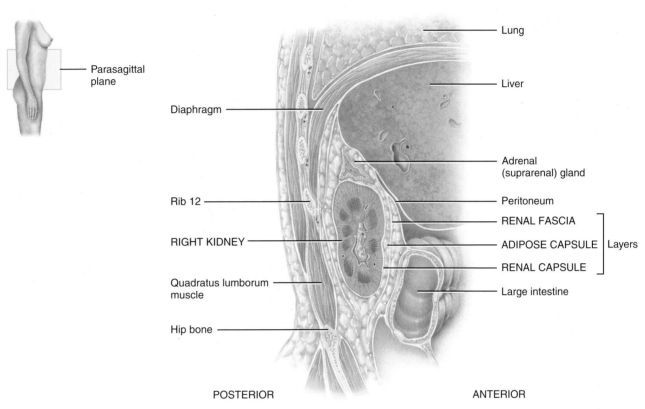

SUPERIOR

Parasagittal plane

Diaphragm

Rib 12

RIGHT KIDNEY

Quadratus lumborum muscle

Hip bone

Lung

Liver

Adrenal (suprarenal) gland

Peritoneum

RENAL FASCIA

ADIPOSE CAPSULE ⎱ Layers

RENAL CAPSULE

Large intestine

POSTERIOR   ANTERIOR

(b) Sagittal section through the right kidney

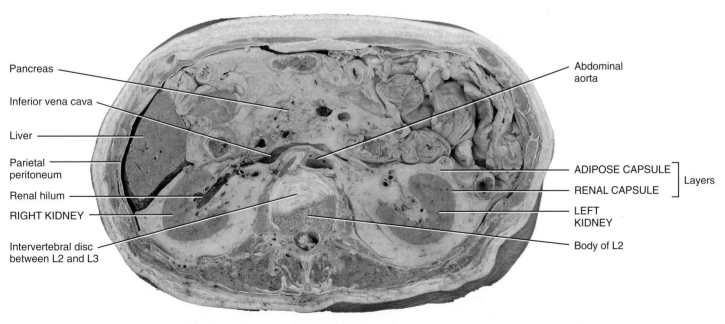

ANTERIOR

Pancreas

Inferior vena cava

Liver

Parietal peritoneum

Renal hilum

RIGHT KIDNEY

Intervertebral disc between L2 and L3

Abdominal aorta

ADIPOSE CAPSULE ⎱ Layers

RENAL CAPSULE

LEFT KIDNEY

Body of L2

POSTERIOR

(c) Inferior view of transverse section of abdomen between L2 and L3

 Why are the kidneys said to be retroperitoneal?

Figure 25.3 Internal anatomy of the kidneys.

 The two main regions of the kidney are the superficial light red region called the renal cortex and the deep dark red region called the renal medulla.

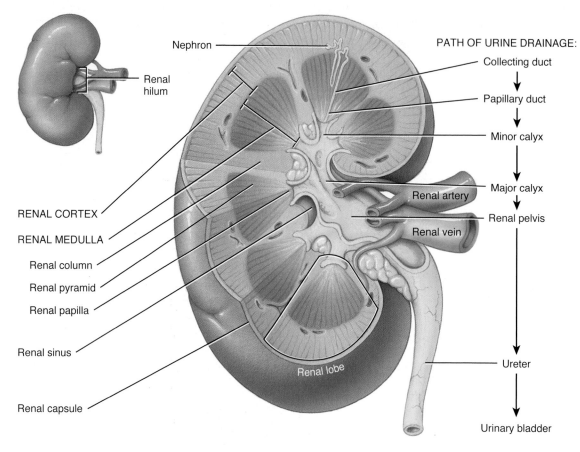

Nephron

Renal hilum

RENAL CORTEX

RENAL MEDULLA

Renal column

Renal pyramid

Renal papilla

Renal sinus

Renal lobe

Renal capsule

PATH OF URINE DRAINAGE:

Collecting duct
↓
Papillary duct
↓
Minor calyx
↓
Major calyx
↓
Renal pelvis
↓
Ureter
↓
Urinary bladder

Renal artery

Renal vein

(a) Anterior view of dissection of right kidney

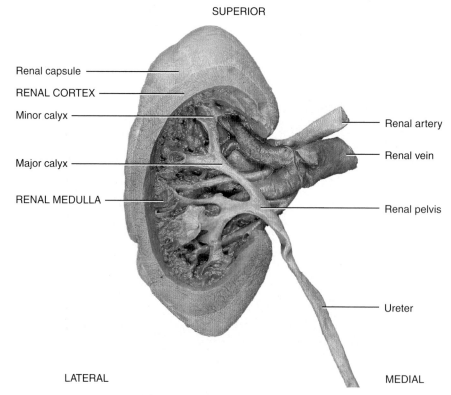

SUPERIOR

Renal capsule

RENAL CORTEX

Minor calyx

Major calyx

RENAL MEDULLA

Renal artery

Renal vein

Renal pelvis

Ureter

LATERAL

MEDIAL

(b) Posterior view of dissection of left kidney

 What structures pass through the renal hilum?

functional portion of the kidney. Within the parenchyma are the functional units of the kidney—about 1 million microscopic structures called **nephrons** (NEF-rons). Filtrate (filtered fluid) formed by the nephrons drains into large **papillary ducts** (PAP-i-lar′-ē), which extend through the renal papillae of the pyramids. The papillary ducts drain into cuplike structures called **minor** and **major calyces** (KĀ-li-sēz=cups; singular is *calyx* [KĀ-liks]). Each kidney has 8 to 18 minor calyces and 2 to 3 major calyces. A minor calyx receives filtrate from the papillary ducts of one renal papilla and delivers it to a major calyx. Once the filtrate enters the calyces it becomes urine because no further reabsorption can occur. The reason for this is that the simple epithelium of the nephron and ducts becomes transitional epithelium in the calyces. From the major calyces, urine drains into a single large cavity called the **renal pelvis** (*pelv-*=basin) and then out through the ureter to the urinary bladder.

The hilum expands into a cavity within the kidney called the **renal sinus**, which contains part of the renal pelvis, the calyces,

and branches of the renal blood vessels and nerves. Adipose tissue helps stabilize the position of these structures in the renal sinus.

Blood and Nerve Supply of the Kidneys

Because the kidneys remove wastes from the blood and regulate its volume and ionic composition, it is not surprising that they are abundantly supplied with blood vessels. Although the kidneys constitute less than 0.5 percent of total body mass, they receive 20–25 percent of the resting cardiac output via the right and left **renal arteries** (Figure 25.4). In adults, **renal blood flow**, the blood flow through the kidneys, is about 1200 mL per minute.

Within the kidney, the renal artery divides into several **segmental arteries** (seg-MENT-al), which supply different segments (areas) of the kidney. Each segmental artery gives off several branches that enter the parenchyma and pass through the renal columns between the renal lobes of the kidneys as the **interlobar arteries** (in′-ter-LŌ-bar). A **renal lobe** consists of

Figure 25.4 Blood supply of the kidneys. The arteries are red, the veins are blue, and the urine-draining structures are yellow.

🔑 The renal arteries deliver 20–25 percent of the resting cardiac output to the kidneys.

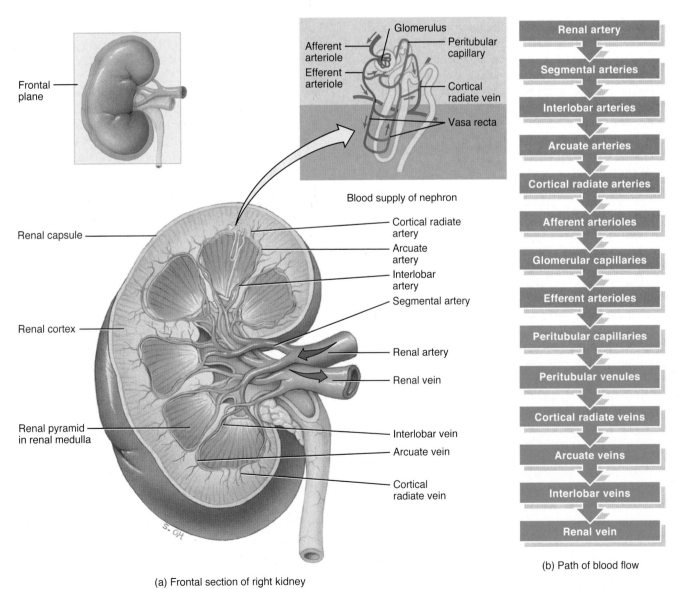

(a) Frontal section of right kidney

(b) Path of blood flow

FIGURE 25.4 CONTINUES ▶

FIGURE 25.4 CONTINUED ▶

SUPERIOR

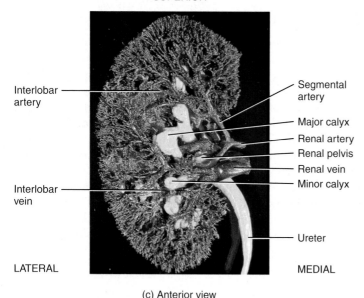

Interlobar artery

Interlobar vein

LATERAL

Segmental artery

Major calyx

Renal artery

Renal pelvis

Renal vein

Minor calyx

Ureter

MEDIAL

(c) Anterior view

What volume of blood enters the renal arteries per minute?

a renal pyramid, some of the renal column on either side of the renal pyramid, and the renal cortex at the base of the renal pyramid (see Figure 25.5a). At the bases of the renal pyramids, the interlobar arteries arch between the renal medulla and cortex. Here they are known as the **arcuate arteries** (AR-kū-āt=shaped like a bow), because they arch over the bases of the renal pyramids. Branches of the arcuate arteries produce a series of **cortical radiate** (KOR-ti-kal RĀ-dē-āt) or *interlobular arteries*. These arteries radiate outward and enter the renal cortex where they give off branches called **afferent arterioles** (AF-er-ent; *af-*=toward; *ferrent*=to carry).

Each nephron receives one afferent arteriole, which divides into a tangled, ball-shaped capillary network called the glomerulus (glō-MER-ū-lus=little ball; plural is *glomeruli*). The capillaries of the glomerulus then reunite to form an **efferent arteriole** (EF-er-ent; *ef-*=out), which has a smaller diameter than the afferent arteriole. The efferent arteriole carries blood out of the glomerulus toward a second capillary plexus (the renal portal system). Capillaries of the glomerulus are unique among capillaries in the body because they are positioned between two arterioles, rather than between an arteriole and a venule. Because the efferent arteriole is smaller in diameter than the afferent arteriole, resistance to the outflow of blood from the glomerulus is high. As a result, blood pressure in capillaries of the glomerulus is considerably higher than in capillaries elsewhere in the body. This is important in urine formation.

The efferent arterioles divide to form the **peritubular capillaries** (per-i-TOOB-ū-lar; *peri-*=around), which surround tubular parts of the nephron in the renal cortex. Extending from some efferent arterioles are long loop-shaped capillaries called **vasa recta** (VĀ-sa REK-ta; *vasa*=vessels; *recta*=straight) that supply tubular portions of the nephron in the renal medulla (see Figure 25.5b).

The peritubular capillaries eventually reunite to form **peritubular venules** and then **cortical radiate** (*interlobular*) **veins**, which also receive blood from the vasa recta. Then the blood drains through the **arcuate veins** to the **interlobar veins** running between the renal pyramids. Blood leaves the kidney through a

single **renal vein** that exits at the renal hilum and carries venous blood to the inferior vena cava.

Most renal nerves originate in the *celiac* and *aorticorenal ganglia* as postganglionic neurons of the sympathetic division of the autonomic nervous system (see Figure 19.3a) and pass through the *renal plexus* into the kidneys along with the renal arteries. Because renal nerves are part of the sympathetic division of the autonomic nervous system (see Section 19.3), most are vasomotor nerves that regulate the flow of blood through the kidney by causing vasodilation or vasoconstriction of renal arterioles.

The Nephron

Parts of a Nephron

Nephrons are the functional units of the kidneys. Each nephron (Figure 25.5) consists of two parts: a **renal corpuscle** (KOR-pus-el=tiny body), where blood plasma is filtered, and a **renal tubule** into which the filtered fluid (glomerular filtrate) passes. Closely associated with a nephron is its blood supply, which was just described. The two components of a renal corpuscle are the glomerulus (capillary network) and the **glomerular capsule** (*Bowmen's capsule*), a dead-end tube that surrounds the glomerular capillaries. The glomerular capsule is a double-walled cup with a cavity between the two layers that receives filtrate from the capillaries. In the order that fluid (filtrate) passes through them, the renal tubule consists of a (1) **proximal convoluted tubule (PCT)** (kon'-vō-LOOT-ed), (2) **nephron loop** (*loop of Henle*), and (3) **distal convoluted tubule (DCT)**. *Proximal* denotes the part of the tubule attached to the glomerular capsule, and *distal* denotes the part that is farther away. *Convoluted* means the tubule is tightly coiled rather than straight. The renal corpuscle and both convoluted tubules lie within the renal cortex; some nephron loops remain in the renal cortex and some extend into the renal medulla, make a hairpin turn, and then return to the renal cortex.

The distal convoluted tubules of several nephrons empty into a single **collecting duct**. Collecting ducts then unite and converge until eventually there are only several hundred large **papillary ducts**, which drain into the minor calyces. The collecting ducts and papillary ducts extend from the renal cortex through the renal medulla to the renal pelvis. Although one kidney has about 1 million nephrons, it has a much smaller number of collecting ducts and even fewer papillary ducts.

In a nephron, the nephron loop connects the proximal and distal convoluted tubules. The first part of the **nephron loop** begins at the point where the proximal convoluted tubule takes its final turn downward. It begins in the renal cortex and extends downward into the renal medulla, where it is called the **descending limb of the nephron loop** (Figure 25.5). It then makes that hairpin turn and returns to the renal cortex where it terminates at the distal convoluted tubule and is known as the **ascending limb of the nephron loop**. About 80–85 percent of the nephrons are **cortical nephrons** (KOR-ti-kal). Their renal corpuscles lie in the outer portion of the renal cortex, and they have *short* nephron loops that lie mainly in the cortex and penetrate only into the outer region of the renal medulla (Figure 25.5a). The short nephron loops receive their blood supply from peritubular capillaries that arise from efferent arterioles. The other 15–20 percent of the nephrons are **juxtamedullary nephrons** (juks'-ta-MED-ū-lar'-ē; *juxta-*=near to). Their renal corpuscles lie deep in the cortex, close to the medulla, and they have *long* nephron loops that extend into the deepest region of the medulla (Figure 25.5b). Long nephron loops receive their blood supply from peritubular capillaries and from the vasa recta that arise from efferent arterioles. In addition, in juxtamedullary nephrons the ascending limb of the nephron loop consists of two portions: a **thin ascending limb** followed by a **thick ascending limb** (Figure 25.5b). The lumen diameter of the thin ascending limb is the same as in other areas of the renal tubule; it is only the epithelium that is thinner. Nephrons with long nephron loops enable the kidneys to excrete very concentrated urine.

Histology of the Nephron and Collecting Duct

A single layer of epithelial cells forms the entire wall of the glomerular capsule, renal tubule, and ducts. Each part, however, has distinctive histological features that reflect its particular functions. In the order that fluid flows through them, the parts are the glomerular capsule, the renal tubule, and the collecting duct.

Figure 25.5 The structure of nephrons and associated blood vessels. Note that the collecting duct and papillary duct are not part of the nephron. (a) A cortical nephron. (b) A juxtamedullary nephron.

Nephrons are the functional units of the kidneys.

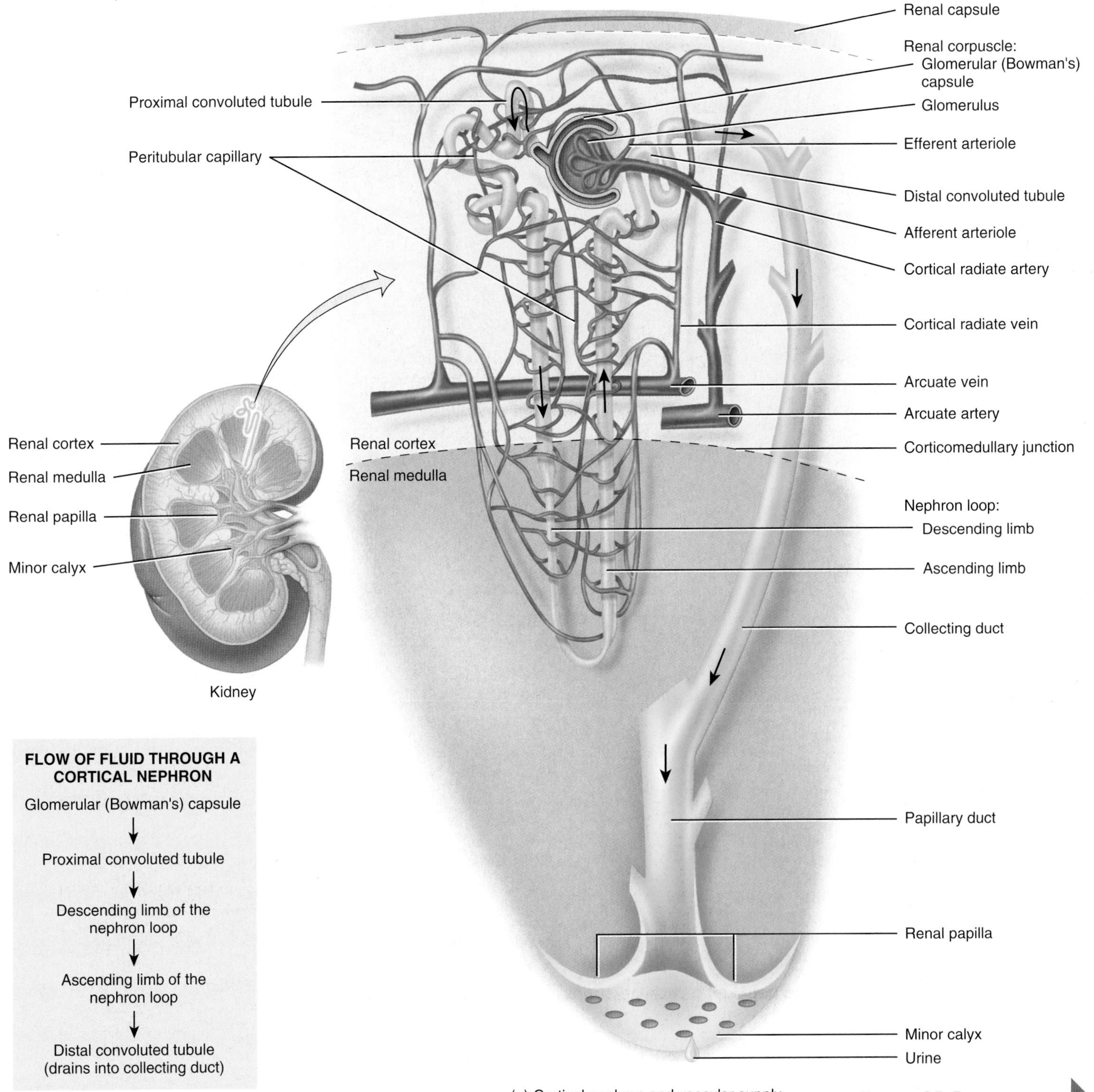

FLOW OF FLUID THROUGH A CORTICAL NEPHRON

Glomerular (Bowman's) capsule
↓
Proximal convoluted tubule
↓
Descending limb of the nephron loop
↓
Ascending limb of the nephron loop
↓
Distal convoluted tubule (drains into collecting duct)

(a) Cortical nephron and vascular supply

FIGURE 25.5 CONTINUES

FIGURE 25.5 CONTINUED

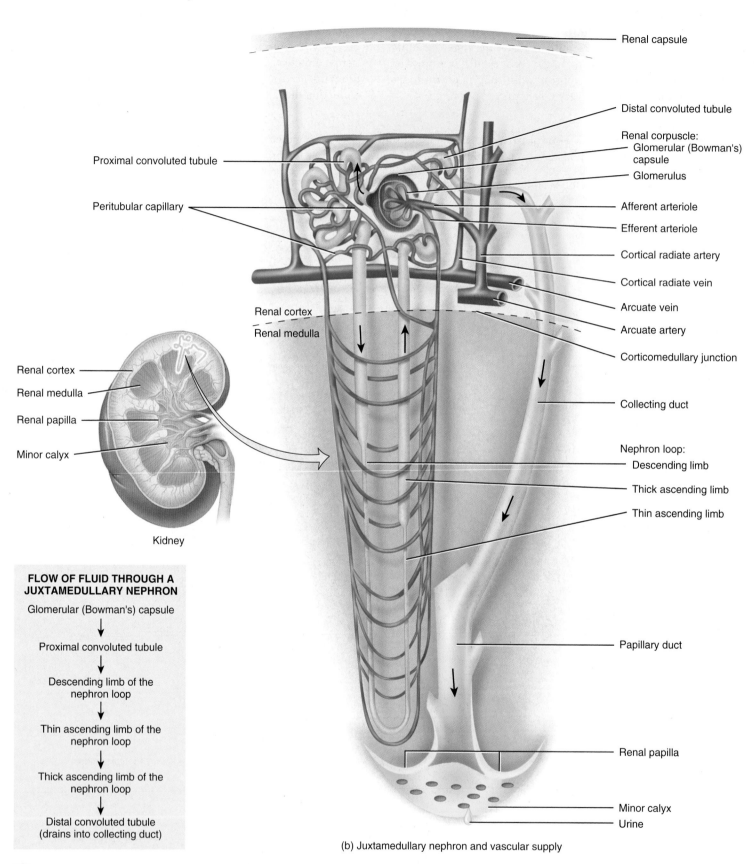

FLOW OF FLUID THROUGH A
JUXTAMEDULLARY NEPHRON

Glomerular (Bowman's) capsule
↓
Proximal convoluted tubule
↓
Descending limb of the
nephron loop
↓
Thin ascending limb of the
nephron loop
↓
Thick ascending limb of the
nephron loop
↓
Distal convoluted tubule
(drains into collecting duct)

(b) Juxtamedullary nephron and vascular supply

 What are the basic differences between cortical nephrons and juxtamedullary nephrons?

GLOMERULAR CAPSULE The **glomerular capsule** or *Bowman's capsule* consists of visceral and parietal layers (Figure 25.6a). The visceral layer consists of modified simple squamous epithelial cells called **podocytes** (PŌ-dō-sīts; *podo-*=foot; *cytes*=cells). The many footlike projections of these cells (pedicels) wrap around the single endothelial cell layer of the glomerular capillaries and form the

CLINICAL CONNECTION | *Kidney Transplant*

A **kidney transplant** is the transfer of a kidney from a living donor or a cadaver to a recipient whose kidney(s) no longer function. In the procedure, the donor kidney is placed in the pelvis of the recipient through an abdominal incision. The renal artery and vein of the transplanted kidney are attached to a nearby artery and vein in the pelvis, and the ureter is then attached to the urinary bladder. During a kidney transplant, the patient receives only one donor kidney, since only one kidney is needed to maintain sufficient renal function. The diseased kidneys are usually left in place. As with all organ transplants, kidney transplant patients must be ever vigilant for signs of infection or organ rejection. The transplant patient will take immunosuppressive drugs for the rest of his or her life to avoid rejection of the "foreign" organ. •

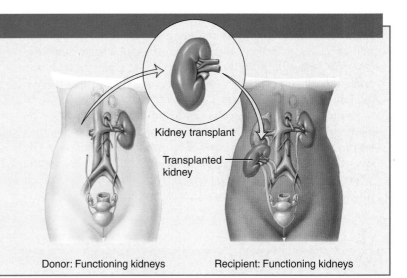

Kidney transplant

Transplanted kidney

Donor: Functioning kidneys

Recipient: Functioning kidneys

Figure 25.6 Histology of a renal corpuscle.

A renal corpuscle consists of a glomerular (Bowman's) capsule and a glomerulus.

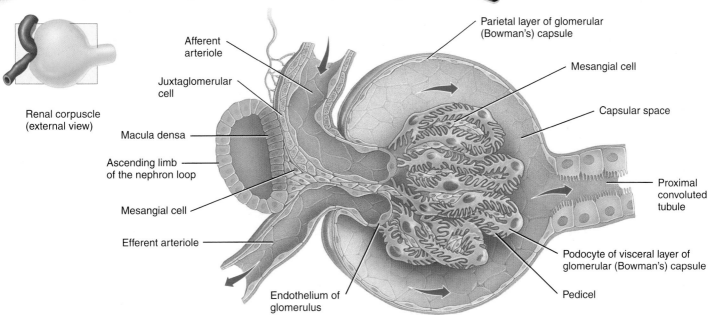

Renal corpuscle (external view)

Afferent arteriole

Juxtaglomerular cell

Macula densa

Ascending limb of the nephron loop

Mesangial cell

Efferent arteriole

Endothelium of glomerulus

Parietal layer of glomerular (Bowman's) capsule

Mesangial cell

Capsular space

Proximal convoluted tubule

Podocyte of visceral layer of glomerular (Bowman's) capsule

Pedicel

(a) Renal corpuscle (internal view)

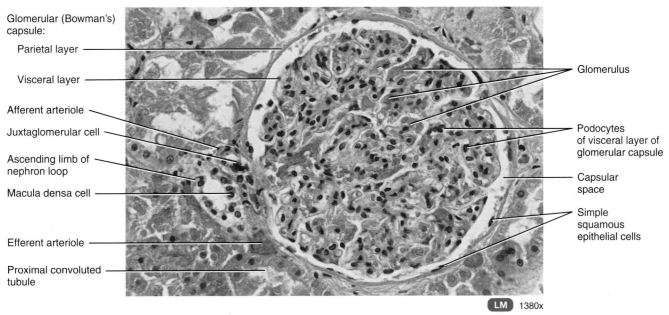

Glomerular (Bowman's) capsule:

Parietal layer

Visceral layer

Afferent arteriole

Juxtaglomerular cell

Ascending limb of nephron loop

Macula densa cell

Efferent arteriole

Proximal convoluted tubule

Glomerulus

Podocytes of visceral layer of glomerular capsule

Capsular space

Simple squamous epithelial cells

LM 1380x

(b) Renal corpuscle

Is the photomicrograph in (b) from a section through the renal cortex or renal medulla? How can you tell?

inner wall of the capsule. The parietal layer of the glomerular capsule consists of simple squamous epithelium and forms the outer wall of the capsule. Fluid filtered from the glomerular capillaries enters the space between the two layers of the glomerular capsule called the **capsular space** or *Bowman's space*. The capsular space is the lumen of the urinary tube. Think of the relationship between the glomerulus and glomerular capsule in the following way. The glomerulus is like a fist punched into a limp balloon (the glomerular capsule) until the fist is covered by two layers of balloon. The layer of the balloon touching the fist is the visceral layer, and the layer not against the fist is the parietal layer. The space in between (the inside of the balloon) is the capsular space.

CLINICAL CONNECTION | *Glomerulonephritis*

Glomerulonephritis (glō-mer'-ū-lō-ne-FRĪ-tis) is an inflammation of the kidney that involves the glomeruli. One of the most common causes is an allergic reaction to the toxins produced by streptococcal bacteria that have recently infected another part of the body, especially the throat. The glomeruli become so inflamed, swollen, and engorged with blood that the filtration membranes allow blood cells and plasma proteins to enter the filtrate. As a result, the urine contains many erythrocytes (hematuria) and a lot of protein. The glomeruli may be permanently damaged, leading to chronic renal failure. •

RENAL TUBULE AND COLLECTING DUCT Table 25.1 illustrates the histology of the cells that form the renal tubule and collecting duct. In the proximal convoluted tubule, the cells are simple cuboidal epithelial cells with a prominent brush border of microvilli on their apical surface (surface facing the lumen). These microvilli, like those of the small intestine, increase the surface area for reabsorption and secretion. The descending limb of the nephron loop and the first part of the ascending limb of the nephron loop (the thin ascending limb) are composed of simple squamous epithelium. (Recall that cortical or short-loop nephrons lack the thin ascending limb.) The thick ascending limb of the nephron loop is composed of simple cuboidal to low columnar epithelium.

In each nephron, the final part of the ascending limb of the nephron loop makes contact with the afferent arteriole serving that renal corpuscle (Figure 25.6a). Because the columnar tubule cells in this region are crowded together, they are known as the **macula densa** (MAK-ū-la DEN-sa; *macula*=spot; *densa*=dense). Alongside the macula densa, the wall of the afferent arteriole (and sometimes the efferent arteriole) contains modified smooth muscle fibers called **juxtaglomerular (JG) cells** (juks'-ta-glō-MER-ū-lar). Together with the macula densa, they constitute the **juxtaglomerular apparatus (JGA)**. The JGA helps regulate blood pressure within the kidneys. The distal convoluted tubule (DCT) begins a short distance past the macula densa. In the last part of the DCT and continuing into the collecting ducts, two different types of cells are

TABLE 25.1

Histological Features of the Renal Tubule and Collecting Duct

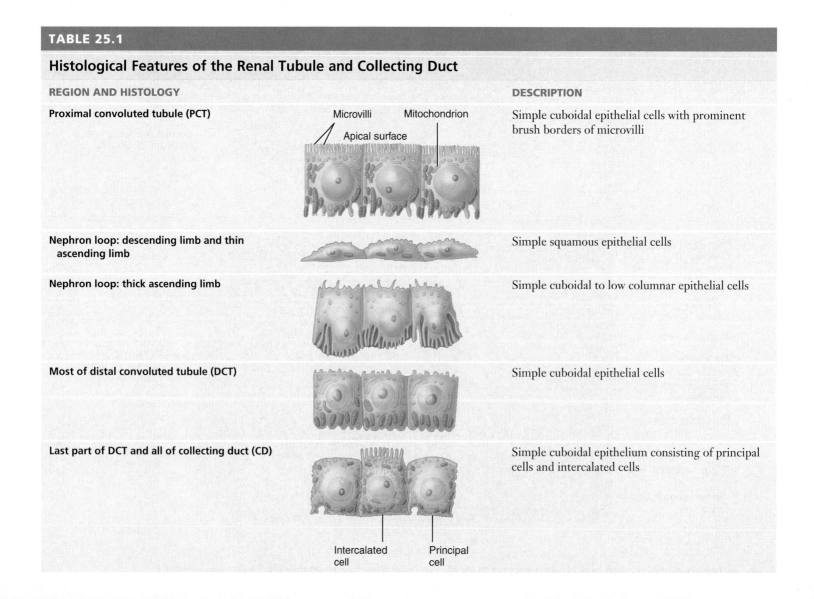

REGION AND HISTOLOGY	DESCRIPTION
Proximal convoluted tubule (PCT) Microvilli Mitochondrion Apical surface	Simple cuboidal epithelial cells with prominent brush borders of microvilli
Nephron loop: descending limb and thin ascending limb	Simple squamous epithelial cells
Nephron loop: thick ascending limb	Simple cuboidal to low columnar epithelial cells
Most of distal convoluted tubule (DCT)	Simple cuboidal epithelial cells
Last part of DCT and all of collecting duct (CD) Intercalated cell Principal cell	Simple cuboidal epithelium consisting of principal cells and intercalated cells

present. Most are **principal cells**, which have receptors for both antidiuretic hormone (ADH) and aldosterone, two hormones that regulate their functions. A smaller number are **intercalated cells** (in-TER-ka-lā-ted), which play a role in the homeostasis of blood pH. The collecting ducts drain into large papillary ducts, which are lined by simple columnar epithelium.

The number of nephrons is constant from birth. Any increase in kidney size is due solely to the growth of individual nephrons. If nephrons are injured or become diseased, new ones do not form. Signs of kidney dysfunction usually do not become apparent until function declines to less than 25 percent of normal because the remaining functional nephrons adapt to handle a larger-than-normal load. Surgical removal of one kidney, for example, stimulates hypertrophy (enlargement) of the remaining kidney, which eventually is able to filter blood at 80 percent of the rate of two normal kidneys.

✔ **CHECKPOINT**

2. Describe the location of the kidneys. Why are they said to be retroperitoneal?
3. Identify the three layers that surround the kidney from internal to external.
4. Describe the components of the renal cortex and renal medulla.
5. Trace a drop of blood into a renal artery, through the kidney, and out a renal vein.
6. Which branch of the autonomic nervous system innervates renal blood vessels?
7. How do cortical nephrons and juxtamedullary nephrons differ structurally?
8. Describe the histology of the various portions of a nephron and collecting duct.
9. Describe the structure of the juxtaglomerular apparatus (JGA).

25.3 FUNCTIONS OF NEPHRONS

 OBJECTIVES

• Explain the role of the nephrons and collecting ducts in glomerular filtration.
• Outline the steps of tubular reabsorption and explain where it occurs.
• Identify the events of tubular secretion and where they occur.
• Describe the filtration membrane.

To produce urine, nephrons and collecting ducts perform three basic processes—glomerular filtration, tubular reabsorption, and tubular secretion (Figure 25.7):

❶ **Glomerular filtration.** In the first step of urine production, water and most solutes in blood plasma move across the wall of capillaries in the glomeruli where they are filtered and move into the glomerular capsule and then into the renal tubule.

❷ **Tubular reabsorption.** As filtered fluid flows through the renal tubules and the collecting ducts, tubule cells reabsorb about 99 percent of the filtered water and many useful solutes. The water and solutes return to the blood as it flows through the peritubular capillaries and vasa recta. Note that the term *reabsorption* refers to the return of filtered water and solutes to the bloodstream. The term *absorption*, by contrast, means entry of new substances into the body, as occurs in the gastrointestinal tract.

❸ **Tubular secretion.** As fluid flows through the renal tubules and collecting ducts, the renal tubule and duct cells secrete other materials, such as wastes, drugs, and excess ions, into the fluid. Notice that tubular secretion *removes a substance from* the blood.

Figure 25.7 **Relationship of a nephron's structure to its three basic functions: glomerular filtration, tubular reabsorption, and tubular secretion.** Secreted substances remain in the urine and are subsequently excreted by the body.

🔑 Glomerular filtration occurs in the renal corpuscle; tubular reabsorption and tubular secretion occur all along the renal tubule and collecting duct.

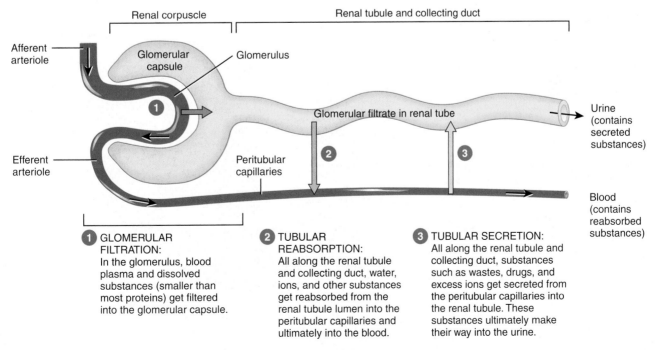

 When cells of the renal tubules secrete the drug penicillin, is the drug being added to or removed from the bloodstream?

Solutes and the fluid that drain into the minor and major calyces and renal pelvis constitute urine and are excreted. The rate of urinary excretion of any solute is equal to its rate of glomerular filtration, plus its rate of secretion, minus its rate of reabsorption (excretion = filtration + secretion − reabsorption).

By filtering, reabsorbing, and secreting, nephrons help maintain homeostasis of blood volume and blood composition. The situation is somewhat analogous to a recycling center: Garbage trucks dump garbage into an input hopper, where the smaller garbage passes onto a conveyor belt (filtration of plasma by the glomerulus). As the conveyor belt carries the garbage along, workers remove useful items, such as aluminum cans, plastics, and glass containers (reabsorption). Other workers place additional garbage left at the center and larger items onto the conveyor belt (secretion). At the end of the belt, all remaining garbage falls into a truck for transport to the landfill (excretion of wastes in urine).

CLINICAL CONNECTION | *Diuretics*

Diuretics (dī′-ū-RET-iks) are substances that slow renal reabsorption of water and thereby cause *diuresis*, an elevated urine flow rate, which in turn reduces blood volume. Diuretic drugs often are prescribed to treat *hypertension* (high blood pressure) because lowering blood volume usually reduces blood pressure. Naturally occurring diuretics include *caffeine* in coffee, tea, and sodas, and *alcohol* in beer, wine, and mixed drinks. •

Glomerular Filtration

The fluid that enters the capsular space is called the **glomerular filtrate** because it is filtered by the glomerulus. On average, the daily volume of glomerular filtrate in adults is 150 liters in females and 180 liters in males, a volume that represents about 65 times the entire blood plasma volume. Because more than 99 percent of the glomerular filtrate returns to the bloodstream via tubular reabsorption, only 1–2 liters (about 1–2 qt) are excreted as urine.

Together, glomerular capillaries and podocytes, which completely encircle the glomerulus, form a leaky barrier referred to as the **filtration membrane** or **endothelial–capsular membrane**. This sandwich-like assembly permits filtration of water and small solutes but prevents filtration of most plasma proteins, blood cells, and platelets. Substances move from the bloodstream through three filtration barriers—endothelial cells of the glomerulus, the basal lamina, and a filtration slit formed by a podocyte (Figure 25.8):

❶ Endothelial cells of the glomerulus are quite leaky because they have large **fenestrations** (fen′-es-TRĀ-shuns) (pores) that are 0.07–0.1 μm in diameter. This size permits all solutes in blood plasma to exit the glomerulus but prevents filtration of blood cells and platelets. Located among the glomerular capillaries and in the cleft between afferent and efferent arterioles are **mesangial cells** (mes-AN-jē-al; *mes-*=in the middle; *angi*=blood vessel), contractile cells that help regulate glomerular filtration (see Figure 25.6a).

Figure 25.8 The filtration (endothelial–capsular) membrane. The size of the endothelial fenestrations and filtration slits have been exaggerated for emphasis.

🔑 During glomerular filtration, water and solutes pass from blood plasma into the capsular space.

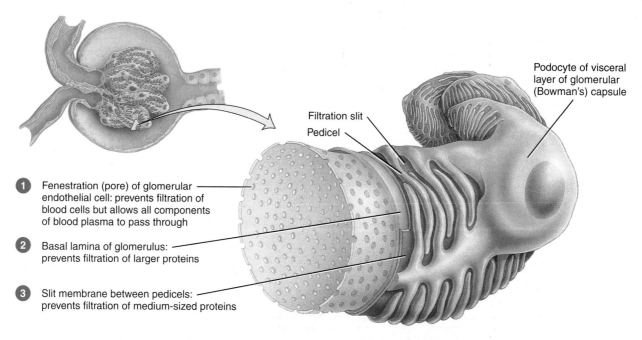

❶ Fenestration (pore) of glomerular endothelial cell: prevents filtration of blood cells but allows all components of blood plasma to pass through

❷ Basal lamina of glomerulus: prevents filtration of larger proteins

❸ Slit membrane between pedicels: prevents filtration of medium-sized proteins

Filtration slit
Pedicel
Podocyte of visceral layer of glomerular (Bowman's) capsule

Details of filtration membrane

❓ **Which part of the filtration membrane prevents red blood cells from entering the capsular space?**

❷ The **basal lamina**, a layer of material between the endothelium of the glomerulus and the podocytes, consists of minute fibers in a glycoprotein matrix; negative charges within this matrix prevent filtration of larger negatively charged plasma proteins.

❸ Extending from each podocyte are thousands of footlike processes termed **pedicels** (PED-i-sels=little feet) that wrap around a glomerulus. The spaces between pedicels are the **filtration slits**. A thin membrane, the **slit membrane**, extends across each filtration slit; it permits the passage of molecules having a diameter smaller than 0.006–0.007 μm, including water, glucose, vitamins, amino acids, very small plasma proteins, ammonia, urea, and ions. Because the most plentiful plasma protein—albumin—has a diameter of 0.007 μm, less than 1 percent of it passes the slit membrane.

The principle of *filtration*—the use of pressure to force fluids and solutes through a membrane—is the same in capillaries of the glomerulus as in capillaries elsewhere in the body. However, the volume of fluid filtered by the renal corpuscle is much larger than in other blood capillaries of the body for three reasons:

1. Glomeruli present a large surface area for filtration because they are long and extensive. The mesangial cells regulate how much of this surface area is available for filtration. When mesangial cells are relaxed, surface area is maximal, and glomerular filtration is very high. Contraction of mesangial cells reduces the available surface area, and glomerular filtration decreases.

2. The filtration membrane is thin and porous. Despite having several layers, the thickness of the filtration membrane is only 0.1 μm. Capillaries of the glomeruli also are about 50 times leakier than blood capillaries in most other tissues, mainly because of their large fenestrations.

3. Blood pressure in glomeruli is high. Because the efferent arteriole is smaller in diameter than the afferent arteriole, resistance to the outflow of blood from the glomerulus is high. As a result, blood pressure in capillaries of the glomeruli is considerably higher than in blood capillaries elsewhere in the body, and a higher pressure produces more filtrate.

CLINICAL CONNECTION | Renal Failure

Renal failure is a decrease or cessation of glomerular filtration. In **acute renal failure (ARF)**, the kidneys abruptly stop working entirely (or almost entirely). The main feature of ARF is the suppression of urine flow, usually characterized either by **oliguria** (ol-i-GŪ-rē-a) (daily urine output between 50 mL and 250 mL), or **anuria** (an-Ū-rē-a) (daily urine output less than 50 mL). Causes include low blood volume (for example, due to hemorrhage), decreased cardiac output, damaged renal tubules, kidney stones, reactions to the dyes used to visualize blood vessels in angiograms, nonsteroidal anti-inflammatory drugs, some antibiotic drugs, and traumatic injury.

Chronic renal failure (CRF) refers to a progressive and usually irreversible decline in glomerular filtration rate (GFR). CRF may result from chronic glomerulonephritis, pyelonephritis, polycystic kidney disease, or traumatic loss of kidney tissue. The final stage, called **end-stage renal failure**, occurs when about 90 percent of the nephrons have been lost. At this stage, GFR diminishes to 10–15 percent of normal, oliguria is present, and blood levels of nitrogen-containing wastes and creatinine increase further. People with end-stage renal failure need dialysis therapy and are possible candidates for a kidney transplant operation. •

Tubular Reabsorption

The normal rate of glomerular filtration is so high that the volume of fluid entering the proximal convoluted tubules in half an hour is greater than the total volume of blood plasma. Fortunately, most of the contents of the renal tubules and collecting ducts are reclaimed and pass into the blood by reabsorption, the second function of the renal tubules and collecting ducts. Although all epithelial cells of the renal tubules and collecting ducts carry out reabsorption, the proximal convoluted tubules make the largest contribution. Reabsorption occurs in two ways. In *transcellular reabsorption* (trans′-SEL-ū-lar), reabsorbed substances move *through* the epithelial cells and then into peritubular capillaries. In *paracellular reabsorption* (par′-a-SEL-ū-lar), reabsorbed substances move *between* renal tubule epithelial cells. Even though the epithelial cells are connected by tight junctions, the tight junctions between cells in the proximal convoluted tubules are "leaky" and permit some reabsorbed substances to pass between cells into peritubular capillaries. Solutes that are reabsorbed by both active and passive processes include glucose, amino acids, urea, and ions such as Na^+ (sodium), K^+ (potassium), Ca^{2+} (calcium), Cl^- (chloride), HCO_3^- (bicarbonate), and HPO_4^{2-} (phosphate). Cells located more distally fine-tune the reabsorption processes to maintain the appropriate concentrations of water and selected ions. Most small proteins and peptides that pass through the filter also are reabsorbed, usually via bulk-phase endocytosis.

When the blood concentration of glucose is above 200 mg/mL, the proximal convoluted tubule cannot work fast enough to reabsorb all the glucose that enters the glomerular filtrate. As a result, some glucose remains in the urine, a condition called **glucosuria** (gloo′-kō-SOO-rē-a). The most common cause is diabetes mellitus, in which the blood glucose level may rise far above normal because insulin activity is deficient. Excessive glucose in the glomerular filtrate inhibits water reabsorption by kidney tubules. This leads to increased urinary output (polyuria), decreased blood volume, and dehydration.

Tubular Secretion

The third function of nephrons and collecting ducts is tubular secretion, the transfer of materials from the blood and tubule cells into the glomerular filtrate. Secreted substances include H^+, K^+, ammonium ions (NH_4^+), creatinine, and certain drugs such as penicillin. Tubular secretion has two important outcomes: The secretion of H^+ helps control blood pH, and the secretion of other substances helps eliminate them from the body.

 CHECKPOINT

10. What is the function of glomerular filtration?
11. Explain the composition of the filtration membrane.
12. Why is the volume of fluid filtered by a renal corpuscle much larger than in other blood capillaries in the body?
13. What happens during tubular reabsorption?
14. What substances are secreted during tubular secretion?
15. Describe the factors that allow a considerably greater filtration through capillaries in glomeruli than through capillaries elsewhere in the body.

If a person's kidneys are so impaired by disease or injury that they are unable to function adequately, then blood must be cleansed artificially by **dialysis** (dī-AL-i-sis), which utilizes the same methods as kidney filtration: the separation of large solutes from smaller ones through use of a selectively permeable membrane. The leading cause of renal failure is diabetes. One method of dialysis is the artificial kidney machine, which performs **hemodialysis** (hē′-mō-dī-AL-i-sis; *hemo-*=blood); it directly filters the patient's blood. As blood flows through tubing made of selectively permeable dialysis membrane, waste products diffuse from the blood into a dialysis solution surrounding the membrane. The dialysis solution is continuously replaced to maintain favorable concentration gradients for diffusion of solutes into and out of the blood. After passing through the dialy-

sis tubing, the cleansed blood flows back into the body. As a general rule, most affected people require 6–12 hours on dialysis each week (roughly every other day).

Continuous ambulatory peritoneal dialysis (CAPD) uses the peritoneal lining of the abdominal cavity as the dialysis membrane to filter the blood. The tip of a catheter is surgically placed in the patient's peritoneal cavity and connected to a sterile dialysis solution. The dialysis solution flows into the peritoneal cavity from a plastic container by gravity. The solution remains in the cavity until metabolic waste products, excess electrolytes, and extracellular fluid diffuse into the dialysis solution. The solution is then drained from the cavity by gravity into a sterile bag that is discarded. The procedure is repeated several times each day. •

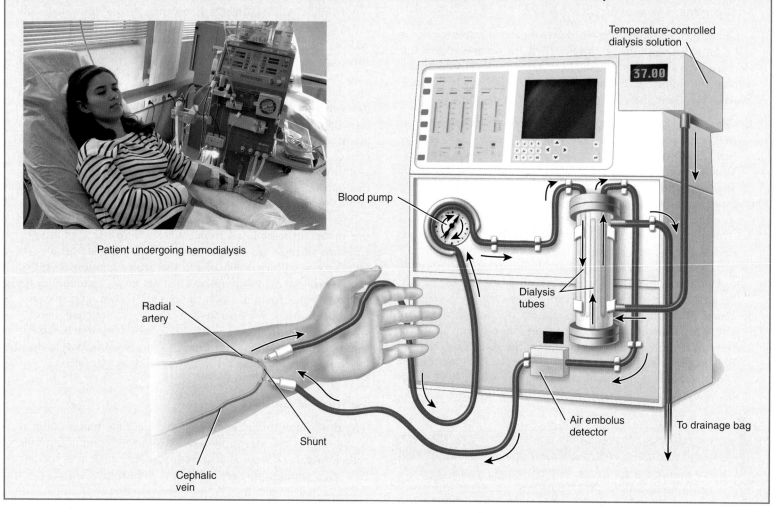

Patient undergoing hemodialysis

25.4 URINE TRANSPORTATION, STORAGE, AND ELIMINATION

🔴 **OBJECTIVES**

• Describe the structure and function of the ureters.
• Explain the histology of the urinary bladder and its role in micturition.
• Indicate the differences in structure in the urethras of males and females, and explain how they affect the organ's function.

Urine drains through the minor calyces, which join to become major calyces that unite to form the renal pelvis (see Figure 25.3). From the renal pelvis, urine first drains into the ureters and then

into the urinary bladder; urine is then discharged from the body through the single urethra (see Figure 25.1).

Ureters

Each of the two **ureters** (Ū-rē-ters) transports urine from the renal pelvis of one kidney to the urinary bladder. Peristaltic contractions of the muscular walls of the ureters push urine toward the urinary bladder, but hydrostatic pressure and gravity also contribute. Peristaltic waves that pass from the renal pelvis to the urinary bladder vary in frequency from one to five per minute, depending on how fast urine is being formed.

The ureters are 25–30 cm (10–12 in.) long and are thick-walled, narrow tubes that vary in diameter from 1 mm to 10 mm along their course between the renal pelvis and the urinary bladder. Like the kidneys, the ureters are retroperitoneal. At the base of the urinary bladder the ureters curve medially and pass obliquely through the wall of the posterior aspect of the urinary bladder (Figure 25.9).

Even though there is no anatomical valve at the opening of each ureter into the urinary bladder, there is a physiological one that is quite effective. As the urinary bladder fills with urine, pressure within it compresses the oblique openings into the ureters and prevents the backflow of urine. When this physiological valve is not operating properly, it is possible for microbes to travel up the ureters from the urinary bladder to infect one or both kidneys.

CLINICAL CONNECTION | *Cystoscopy*

Cystoscopy (sis-TOS-kō-pē; *cysto-*=bladder; *-skopy*=to examine) is a very important procedure for direct examination of the mucosa of the urethra and urinary bladder and prostate in males. In the procedure, a *cystoscope* (a flexible narrow tube with a light) is inserted into the urethra to examine the structures through which it passes. With special attachments, tissue samples can be removed for examination (biopsy) and small stones can be removed. Cystoscopy is useful for evaluating urinary bladder problems such as cancer and infections. It can also evaluate the degree of obstruction resulting from an enlarged prostate. •

Figure 25.9 Ureters, urinary bladder, and urethra in a female.

Urine is stored in the urinary bladder before being expelled by micturition.

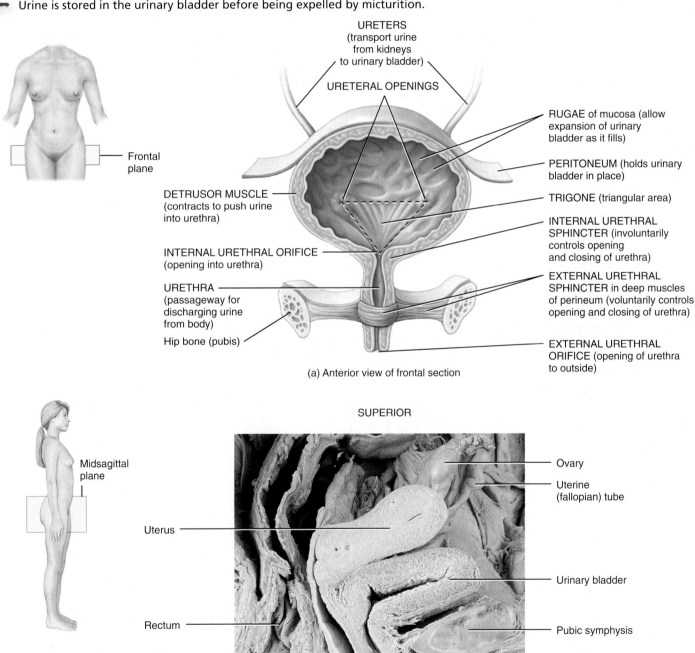

(a) Anterior view of frontal section

(b) Midsagittal section

 Why is transitional epithelium important in the ureters and urinary bladder?

Similar to other tubular structures in the body, the ureters are formed of three coats or layers (Figure 25.10). The deepest coat, or **mucosa**, is a mucous membrane with **transitional epithelium** (see Table 3.1) and an underlying **lamina propria** of areolar connective tissue with considerable collagen, elastic fibers, and lymphatic tissue. Transitional epithelium is able to stretch—a marked advantage for any organ that must accommodate a variable volume of fluid. Mucus secreted by the mucosa prevents the cells of the walls of the ureter from coming in contact with urine, which is important because the solute concentration and pH of urine may differ drastically from those of the cytosol of the cells of the ureter walls.

Throughout most of the length of the ureters, the intermediate coat, the **muscularis**, is composed of inner longitudinal and outer circular layers of smooth muscle fibers, an arrangement opposite that of the gastrointestinal tract, which contains inner circular and outer longitudinal layers. The muscularis of the distal third of the ureters also contains a third, outer layer of longitudinal muscle fibers. Peristalsis is the major function of the muscularis.

The superficial coat of the ureters is the **adventitia**, a layer of areolar connective tissue containing blood vessels, lymphatic vessels, and nerves that serve the muscularis and mucosa. The adventitia blends in with surrounding connective tissue and anchors the ureters in place.

The arterial supply of the ureters is from the renal, testicular or ovarian, common iliac, and inferior vesical arteries (arising from the internal iliac artery, a trunk with the internal pudendal and superior gluteal arteries, or a branch of the internal pudendal artery). The veins have names corresponding to those of the arteries and eventually terminate in the inferior vena cava.

The ureters are innervated by the renal plexuses, which are supplied by sympathetic and parasympathetic fibers from the lesser and lowest splanchnic nerves.

Urinary Bladder

The **urinary bladder** is a hollow, distensible muscular organ situated in the pelvic cavity posterior to the pubic symphysis. In males, it is directly anterior to the rectum; in females it is anterior to the vagina and inferior to the uterus (see Figure 25.12). It is held in position by folds of the peritoneum. The shape of the urinary bladder depends on how much urine it contains. Empty, it is collapsed; when slightly distended it becomes spherical; as urine volume increases it becomes pear-shaped and rises into the abdominal cavity. Urinary bladder capacity averages 700–800 mL; it is smaller in females because the uterus occupies the space just superior to the urinary bladder.

Figure 25.10 Histology of the ureter.

🔑 Three coats of tissue form the wall of the ureters: mucosa, muscularis, and adventitia.

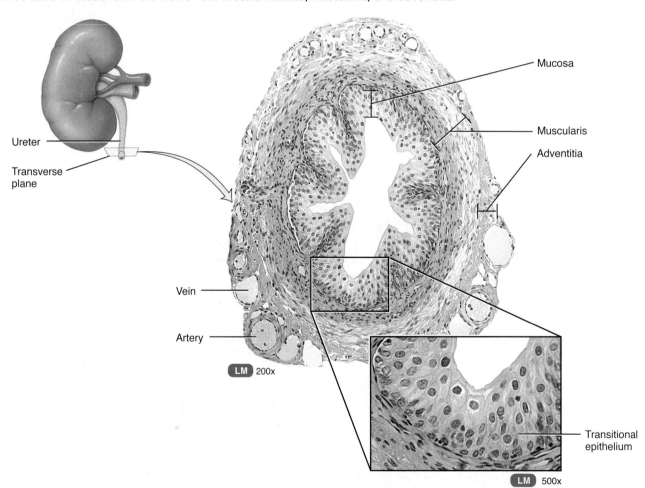

Transverse section of ureter

  How does the muscularis of most of the ureter differ from that of the gastrointestinal tract?

In the floor of the urinary bladder is a small triangular area called the **trigone** (TRĪ-gōn=triangle). The two posterior corners of the trigone contain the two ureteral openings; the opening into the urethra, the **internal urethral orifice** (OR-i-fis), lies in the anterior corner (see Figure 25.9a). Because its mucosa is firmly bound to the muscularis, the trigone has a smooth appearance.

Three coats make up the wall of the urinary bladder (Figure 25.11). The deepest is the **mucosa**, a mucous membrane composed of **transitional epithelium** and an underlying **lamina propria** similar to that of the ureters. The transitional epithelium permits stretching. **Rugae** (folds in the mucosa) are also present. Surrounding the mucosa is the intermediate **muscularis**, also called the **detrusor muscle** (de-TROO-ser=to push down),

Figure 25.11 Histology of the urinary bladder.

Discharge of urine from the urinary bladder is a combination of voluntary and involuntary muscular contractions called micturition.

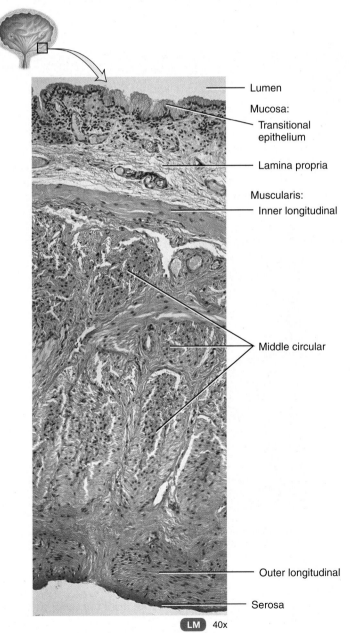

Lumen

Mucosa:
Transitional epithelium

Lamina propria

Muscularis:
Inner longitudinal

Middle circular

Outer longitudinal

Serosa

LM 40x

Transverse section of urinary bladder

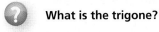

What is the trigone?

which consists of three layers of smooth muscle fibers: the inner longitudinal, middle circular, and outer longitudinal layers. Around the opening to the urethra the circular fibers form an **internal urethral sphincter** (see Figure 25.9a); inferior to it is the **external urethral sphincter**, which is composed of skeletal muscle. The most superficial coat of the urinary bladder on the posterior and inferior surfaces is the **adventitia**, a layer of areolar connective tissue that is continuous with that of the ureters. Over the superior surface of the urinary bladder is the **serosa**, a layer of peritoneum.

Discharge of urine from the urinary bladder, called **micturition** (mik′-too-RISH-un; *mictur-*=urinate), is also known as *urination* or *voiding*. Micturition occurs via a combination of involuntary and voluntary muscle contractions. When the volume of urine in the urinary bladder exceeds 200–400 mL, pressure within the urinary bladder increases considerably, and stretch receptors in its wall transmit nerve impulses into the spinal cord. These impulses propagate to the **micturition center** in sacral spinal cord segments S2 and S3 and trigger a spinal reflex called the **micturition reflex**. In this reflex arc, parasympathetic impulses from the micturition center propagate to the urinary bladder wall and internal urethral sphincter. The nerve impulses cause *contraction* of the detrusor muscle and *relaxation* of the internal urethral sphincter muscle. Simultaneously, the micturition center inhibits somatic motor neurons that innervate skeletal muscle in the external urethral sphincter. Upon contraction of the urinary bladder wall and relaxation of the sphincters, urination takes place. Urinary bladder filling causes a sensation of fullness that initiates a conscious desire to urinate before the micturition reflex actually occurs. Although

CLINICAL CONNECTION | *Incontinence*

A lack of voluntary control over micturition is called **urinary incontinence** (in-KON-ti-nens). In infants and children under 2–3 years old, incontinence is normal because neurons to the external urethral sphincter muscle are not completely developed; voiding occurs whenever the urinary bladder is sufficiently distended to stimulate the micturition reflex. Urinary incontinence also occurs in adults. There are four types of urinary incontinence—stress, urge, overflow, and functional. **Stress incontinence** is the most common type of incontinence in young and middle-aged females, and results from weakness of the deep muscles of the pelvic floor. As a result, any physical stress that increases abdominal pressure, such as coughing, sneezing, laughing, exercising, straining, lifting heavy objects, and pregnancy, causes leakage of urine from the urinary bladder. **Urge incontinence** is most common in older people and is characterized by an abrupt and intense urge to urinate followed by an involuntary loss of urine. It may be caused by irritation of the urinary bladder wall by infection or kidney stones, stroke, multiple sclerosis, spinal cord injury, or anxiety. **Overflow incontinence** refers to the involuntary leakage of small amounts of urine caused by some type of blockage or weak contractions of the musculature of the urinary bladder. When urine flow is blocked (for example, from an enlarged prostate or kidney stones) or the urinary bladder muscles can no longer contract, the urinary bladder becomes overfilled and the pressure inside increases until small amounts of urine dribble out. **Functional incontinence** is urine loss resulting from the inability to get to a toilet facility in time as a result of conditions such as stroke, severe arthritis, and Alzheimer's disease. Choosing the right treatment option depends on correct diagnosis of the type of incontinence. Treatments include Kegel exercises, urinary bladder training, medication, and possibly even surgery. •

emptying of the urinary bladder is a reflex, in early childhood we learn to initiate it and stop it voluntarily. Through learned control of the external urethral sphincter muscle and certain muscles of the pelvic floor, the cerebral cortex can initiate micturition or delay its occurrence for a limited period of time.

The arteries of the urinary bladder are the superior vesical (arises from the umbilical artery), the middle vesical (arises from the umbilical artery or a branch of the superior vesical), and the inferior vesical (arises from the internal iliac artery, a trunk with the internal pudendal and superior gluteal arteries, or a branch of the internal pudendal artery). The veins from the urinary bladder pass to the internal iliac vein.

The nerves supplying the urinary bladder arise partly from the hypogastric sympathetic plexus and partly from the second and third sacral nerves (pelvic splanchnic nerve).

Urethra

The **urethra** (ū-RĒ-thra) is a small tube leading from the internal urethral orifice in the floor of the urinary bladder to the exterior of the body (see Figure 25.9a). In both males and females, the urethra is the terminal portion of the urinary system and the passageway for discharging urine from the body; in males it discharges semen as well.

In males, the urethra also extends from the internal urethral orifice to the exterior, but its length and passage through the body are considerably different than in females (Figure 25.12a). The male urethra first passes through the prostate, then through the deep perineal muscles, and finally through the penis, a distance of about 20 cm (8 in.).

The male urethra, which also consists of a deep **mucosa** and a superficial **muscularis**, is subdivided into three anatomical regions: (1) The **prostatic urethra** passes through the prostate; (2) the **intermediate (membranous) urethra**, the shortest portion, passes through the deep perineal muscles; and (3) the **spongy urethra**, the longest portion, passes through the penis. The mucosa of the prostatic urethra is continuous with that of the urinary bladder and consists of transitional epithelium that becomes stratified columnar or pseudostratified columnar epithelium more distally. The mucosa of the intermediate urethra contains stratified columnar or pseudostratified columnar epithelium. The epithelium of the spongy urethra is stratified columnar or pseudostratified columnar epithelium, except near the external urethral orifice, which is nonkeratinized stratified squamous epithelium. The **lamina propria** of the male urethra is areolar connective tissue with elastic fibers and a plexus of veins.

The muscularis of the prostatic urethra is composed of wisps of mostly circular smooth muscle fibers superficial to the lamina propria; these circular fibers help form the internal urethral sphincter of the urinary bladder. The muscularis of the intermediate urethra consists of circularly arranged skeletal muscle fibers of the urogenital diaphragm that help form the external urethral sphincter.

Several glands and other structures associated with reproduction deliver their contents into the male urethra (see Figure 26.9). The prostatic urethra contains the openings of (1) ducts that transport secretions from the **prostate** and (2) the **seminal vesicles** that provide secretions that both neutralize the acidity of the female reproductive tract and contribute to sperm motility and viability, and the **ductus (vas) deferens**, which deliver sperm into the urethra. The openings of the ducts of the **bulbourethral glands** (bul'-bō-u-RĒ-thral) or *Cowper's glands* empty into the spongy urethra. They deliver an alkaline substance prior to ejaculation

that neutralizes the acidity of the urethra. The glands also secrete mucus, which lubricates the end of the penis during sexual arousal. Throughout the urethra, but especially in the spongy urethra, the openings of the ducts of **urethral glands** or *Littré glands* (LĒ-trē) discharge mucus during sexual arousal and ejaculation.

In females, the urethra lies directly posterior to the pubic symphysis, is directed obliquely inferiorly and anteriorly, and has a length of 4 cm (1.5 in.) (Figure 25.12b). The opening of the urethra to the exterior, the **external urethral orifice**, is located between the clitoris and the vaginal opening. The wall of the female urethra consists of a deep **mucosa** and a superficial **muscularis**. The mucosa is a mucous membrane composed of **epithelium** and **lamina propria** (areolar connective tissue with elastic fibers and a plexus of veins). The muscularis consists of circularly arranged smooth muscle fibers and is continuous with that of the urinary bladder. Near the urinary bladder, the mucosa contains transitional epithelium that is continuous with that of the urinary bladder; near the external urethral orifice the epithelium is nonkeratinized stratified squamous epithelium. Between these areas, the mucosa contains stratified columnar or pseudostratified columnar epithelium.

A summary of the organs of the urinary system is presented in Table 25.2.

CLINICAL CONNECTION | *Urinalysis*

An analysis of the volume and physical, chemical, and microscopic properties of urine, called a **urinalysis** (ū-ri-NAL-i-sis), reveals much information about the health of the body. Of the 1–2 liters (about 1–2 quarts) of urine eliminated per day by a normal adult, about 95 percent is water. The remaining 5 percent consists of solutes, among them urea, sodium, potassium, phosphate, and sulfate ions; creatinine; and uric acid. In addition, much smaller amounts of calcium, magnesium, and bicarbonate ions are also found in urine. If disease alters body metabolism or kidney function, traces of substances not normally present may appear in the urine, or normal constituents may appear in abnormal amounts. •

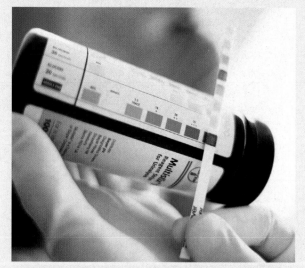

Test stick previously placed in urine is compared to a chart to detect abnormal constituents in urine.

✔ CHECKPOINT

16. What forces help propel urine from the renal pelvis to the urinary bladder?

17. What is micturition? Describe the micturition reflex.

18. Why are the lengths of the urethra so different in males and females?

Figure 25.12 Comparison between female and male urethras.

 The urethra carries urine from the urinary bladder to the exterior.

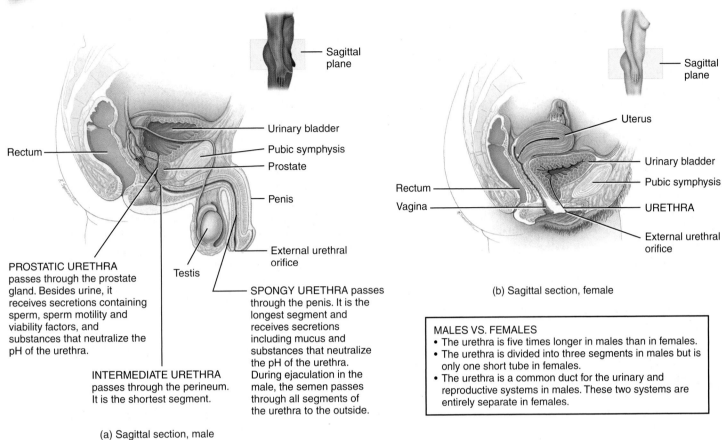

Sagittal plane

Rectum

Urinary bladder
Pubic symphysis
Prostate
Penis

PROSTATIC URETHRA passes through the prostate gland. Besides urine, it receives secretions containing sperm, sperm motility and viability factors, and substances that neutralize the pH of the urethra.

Testis

External urethral orifice

SPONGY URETHRA passes through the penis. It is the longest segment and receives secretions including mucus and substances that neutralize the pH of the urethra. During ejaculation in the male, the semen passes through all segments of the urethra to the outside.

INTERMEDIATE URETHRA passes through the perineum. It is the shortest segment.

(a) Sagittal section, male

Sagittal plane

Uterus

Rectum
Vagina

Urinary bladder
Pubic symphysis
URETHRA
External urethral orifice

(b) Sagittal section, female

MALES VS. FEMALES
• The urethra is five times longer in males than in females.
• The urethra is divided into three segments in males but is only one short tube in females.
• The urethra is a common duct for the urinary and reproductive systems in males. These two systems are entirely separate in females.

? **Through which three main structures does the male urethra pass?**

TABLE 25.2

Summary of the Urinary System Organs

STRUCTURE	LOCATION	DESCRIPTION	FUNCTION
Kidneys	Posterior abdomen between last thoracic and third lumbar vertebrae posterior to peritoneum (retroperitoneal); lie against eleventh and twelfth ribs	Solid, reddish bean-shaped organs; internal structure consists of three tubular systems—arteries, veins, and urinary tubes—that form an intimate interface	Regulate blood volume and composition, help regulate blood pressure, synthesize glucose, release erythropoietin, participate in vitamin D synthesis, and excrete wastes in urine
Ureters	Posterior to the peritoneum (retroperitoneal); descend from kidney to urinary bladder along anterior surface of psoas major muscle and cross back of pelvis to reach inferoposterior surface of urinary bladder anterior to the sacrum	Thick, muscular walled tubes with three structural layers—mucosa of transitional epithelium, muscularis with circular and longitudinal layers of smooth muscle, and adventitia of areolar connective tissue	Transport tubes that move urine from kidneys to urinary bladder
Urinary bladder	Located in pelvic cavity anterior to sacrum and rectum in males and sacrum, rectum, and vagina in females and posterior to the pubis in both sexes; in males, superior surface is covered with parietal peritoneum; in females, uterus covers superior aspect	Hollow, distensible, muscular organ with variable shape depending on amount of urine; three basic layers—inner mucosa of transitional epithelium, middle smooth muscle coat called detrusor muscle, and outer adventitia or serosa over superior aspect in males	Storage organ that temporarily stores urine until convenient to discharge from the body
Urethra	Exits urinary bladder in both sexes; in females, runs through perineal floor of pelvis to exit between labia minora; in males, passes through prostate, then perineal floor of pelvis, and then through penis (in male) to exit at tip	Thin-walled tubes with three structural layers—inner mucosa (transitional, stratified columnar, and stratified squamous epithelium), thin middle layer of circular smooth muscle, and thin connective tissue exterior	Drainage tube that transports stored urine from body

25.5 DEVELOPMENT OF THE URINARY SYSTEM

 ● OBJECTIVE

• Describe the development of the urinary system.

Starting in the third week of fetal development, a portion of the mesoderm along the posterior aspect of the embryo, the **intermediate mesoderm**, differentiates into the kidneys. The intermediate mesoderm is located in paired elevations called **urogenital ridges** (ū-rō-JEN-i-tal). Three pairs of kidneys form within the intermediate mesoderm in succession: the pronephros, the mesonephros, and the metanephros (Figure 25.13). Only the last pair remains as the functional kidneys of the newborn.

The first kidney to form, the **pronephros** (prō-NEF-rōs; *pro-* = before; *nephros* = kidney), is the most superior of the three and has an associated **pronephric duct**. This duct empties into the **cloaca** (klō-Ā-ka), the expanded terminal part of the hindgut, which functions as a common outlet for the urinary, digestive, and reproductive ducts. The pronephros begins to degenerate during the fourth week and is completely gone by the sixth week.

The second kidney, the **mesonephros** (mez′-ō-NEF-rōs; *meso-* = middle), replaces the pronephros. The retained portion of the pronephric duct, which connects to the mesonephros, develops into the **mesonephric duct**. The mesonephros begins to degenerate by the sixth week and is almost gone by the eighth week.

At about the fifth week, a mesodermal outgrowth, called a **ureteric bud** (ū-re-TER-ik), develops from the distal portion of the mesonephric duct near the cloaca. The **metanephros** (met-a-NEF-rōs; *meta-* = after), or ultimate kidney, develops from the ureteric bud and metanephric mesoderm. The ureteric bud forms the *collecting ducts, calyces, renal pelvis,* and *ureter.* The **metanephric mesoderm** (met′-a-NEF-rik) forms the *nephrons* of the kidneys. By the third month, the fetal kidneys begin excreting urine into the surrounding amniotic fluid; indeed, fetal urine makes up most of the amniotic fluid.

During development, the cloaca divides into a **urogenital sinus**, into which urinary and genital ducts empty, and a *rectum* that discharges into the anal canal. The *urinary bladder* develops from the urogenital sinus. In females, the *urethra* develops as a result of lengthening of the short duct that extends from the urinary bladder to the urogenital sinus. In males, the urethra is considerably longer and more complicated, but it is also derived from the urogenital sinus.

Although the metanephric kidneys form in the pelvis, they ascend to their ultimate destination in the abdomen. As they do so, they receive renal blood vessels. Although the inferior blood vessels usually degenerate as superior ones appear, sometimes the inferior vessels do not degenerate. Consequently, some individuals (about 30 percent) develop multiple renal vessels.

Figure 25.13 Development of the urinary system.

 Three pairs of kidneys form within intermediate mesoderm in successive time periods: pronephros, mesonephros, and metanephros.

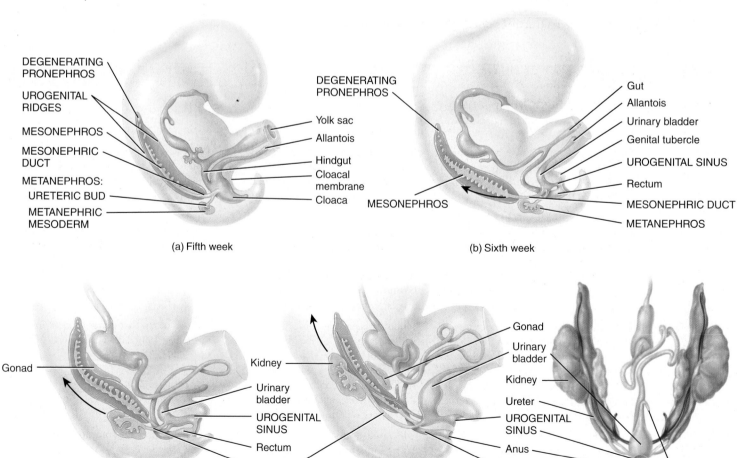

(a) Fifth week

(b) Sixth week

(c) Seventh week

(d) Eighth week

(e) Anterior view 8-week embryo

? **When do the kidneys begin to develop?**

In a condition called **unilateral renal agenesis** (ā-JEN-e-sis; *a-*=without; *genesis*=production; *unilateral*=one side), only one kidney develops (usually the right) due to the absence of a ureteric bud. The condition occurs once in every 1000 newborn infants and usually affects males more than females. Other kidney abnormalities that occur during development are **malrotated kidneys** (the hilum faces anteriorly, posteriorly, or laterally instead of medially); **ectopic kidney** (abnormal position of one or both kidneys, usually inferior); and **horseshoe kidney** (the fusion of the two kidneys, usually inferiorly, into a single U-shaped kidney).

 CHECKPOINT

19. Which type of embryonic tissue develops into nephrons?
20. Which embryonic tissue gives rise to collecting ducts, calyces, renal pelves, and ureters?

25.6 AGING AND THE URINARY SYSTEM

OBJECTIVE

• Outline the effects of aging on the urinary system.

With aging, the kidneys shrink in size, have a decreased blood flow, and filter less blood. The mass of the two kidneys decreases from an average of nearly 300 g in 20-year-olds to less than 200 g by age 80, a decrease of about one-third. Likewise, renal blood flow and filtration rates decline by 50 percent between ages 40 and 70. By age 80, about 40 percent of glomeruli are not functioning and thus filtration, reabsorption, and secretion decrease. Kidney diseases that become more common with age include acute and chronic kidney inflammations and renal calculi (kidney stones). Because the sensation of thirst diminishes with age, older individuals also are susceptible to dehydration. Urinary bladder changes that occur with aging include a reduction in size and capacity and weakening of the muscles. Urinary tract infections are more common among the elderly, as are **polyuria** (pol'-ē-Ū-rē-a; *poly-*=too much) (excessive urine production), **nocturia** (excessive urination at night), increased frequency of urination, **dysuria** (dis-Ū-rē-a; *dys*=painful; *uria*=urine) (painful urination), urinary retention or incontinence, and hematuria (blood in the urine).

 CHECKPOINT

21. By how much do kidney mass and filtration rate decrease with age?

KEY MEDICAL TERMS ASSOCIATED WITH THE URINARY SYSTEM

Azotemia (az-ō-TĒ-mē-a; *azot-*=nitrogen; *emia*=condition of blood) Presence of urea or other nitrogen-containing substances in blood.

Cystocele (SIS-tō-sēl; *cysto-*=bladder; *cele*=hernia or rupture) Hernia of the urinary bladder.

Diabetic kidney disease A disorder caused by diabetes mellitus in which glomeruli are damaged. Results in leakage of proteins into urine and reduction in ability of kidney to remove water and waste.

Enuresis (en'-ū-RĒ-sis=to void urine) Involuntary voiding of urine after the age at which voluntary control has typically been attained.

Hydronephrosis (hī'-drō-ne-FRŌ-sis; *hydro-*=water; *nephros*=kidney; *osis*=condition) Swelling of the kidney due to dilation of the renal pelvis and calyces as a result of an obstruction to the flow of urine.

Intravenous pyelogram (IVP) (in'-tra-VĒ-nus PĪ-e-lō-gram'; *intra-*=within; *veno-*=vein; *pyelo-*=pelvis of kidney; *gram*=record) Radiograph (x-ray) of the kidneys, ureters, and urinary bladder after venous injection of a radiopaque contrast medium.

Nephropathy (ne-FROP-a-thē; *neph-*=kidney; *pathos*=suffering) Any disease of the kidneys. Types include *analgesic nephropathy* (from long-term and excessive use of drugs such as ibuprofen), *lead nephropathy* (from ingestion of lead-based paint), and *solvent nephropathy* (from carbon tetrachloride and other solvents).

Nephrotic syndrome (nef-ROT-ik) A condition characterized by *proteinuria* (prō-ten-OO-rē-a), protein in the urine, and *hyperlipidemia* (hī'-per-lip-i-DĒ-mē-a), high blood levels of cholesterol, phospholipids, and triglycerides.

Nocturnal enuresis (nok-TUR-nal en'-ū-RĒ-sis) Discharge of urine during sleep, resulting in bed-wetting; occurs in about 15 percent of 5-year-old children and generally resolves spontaneously, afflicting only about 1 percent of adults. Possible causes include smaller-than-normal urinary bladder capacity, failure to awaken in response to a full urinary bladder, and above-normal production of urine at night.

Polycystic kidney disease (PKD) (pol'-ē-sis-tik) Common inherited disorder in which kidney tubules become riddled with hundreds or thousands of cysts (fluid-filled cavities). Inappropriate apoptosis (programmed cell death) of cells in noncystic tubules leads to progressive impairment of renal function and eventually end-stage renal failure.

Renal calculi (KAL-kū-lī=pebbles) Kidney stones formed from crystals of salts present in urine; commonly contain calcium oxalate, uric acid, or calcium phosphate. Conditions leading to calculus formation include ingestion of excessive calcium, low water intake, abnormally alkaline or acidic urine, and overactivity of parathyroid glands.

Shock-wave lithotripsy (LITH-ō-trip'-sē; *litho*=stone) Procedure that uses high-energy shock waves to disintegrate kidney stones as an alternative to surgical removal.

Uremia (ū-RĒ-mē-a; *emia*=condition of blood) Toxic levels of urea in the blood resulting from severe malfunction of the kidneys.

Urinary retention A failure to completely or normally void urine; may be due to an obstruction in the urethra or neck of the urinary bladder, to nervous contraction of the urethra, or to lack of urge to urinate.

Urinary tract infection (UTI) An infection of a part of the urinary system or the presence of large numbers of microbes in urine; more common in females due to the shorter length of the urethra. Symptoms include painful or burning urination, urgent and frequent urination, low back pain, and bed-wetting.

CHAPTER REVIEW AND RESOURCE SUMMARY

WileyPLUS

Review	Resource
25.1 Overview of the Urinary System 1. The organs of the urinary system are the kidneys, ureters, urinary bladder, and urethra. 2. After the kidneys filter blood and return most water and many solutes to the bloodstream, the remaining water and solutes constitute urine.	Anatomy Overview - The Urinary System Overview Exercise - Assemble the Urinary Tract Exercise - Concentrate on the Urinary System

Review

3. The kidneys regulate ionic composition, osmolarity, volume, and pH of blood, and blood pressure.
4. The kidneys also release calcitriol and erythropoietin, and excrete wastes and foreign substances.

25.2 Anatomy and Histology of the Kidneys

1. The kidneys are retroperitoneal organs attached to the posterior abdominal wall.
2. Three layers of tissue surround the kidneys: the renal capsule, adipose capsule, and renal fascia.
3. Internally, the kidneys consist of a renal cortex, a renal medulla, renal pyramids, renal papillae, renal columns, major and minor calyces, and a renal pelvis.
4. Blood flows into the kidney through the renal artery and successively into segmental, interlobar, arcuate, and cortical radiate arteries; afferent arterioles; glomeruli; efferent arterioles; peritubular capillaries and vasa recta; and cortical radiate, arcuate, and interlobar veins and flows out through the renal vein.
5. Vasomotor nerves from the sympathetic division of the autonomic nervous system supply kidney blood vessels; they help regulate blood flow through the kidney.
6. The nephron is the functional unit of the kidneys. A nephron consists of a renal corpuscle (glomerulus and glomerular or Bowman's capsule) and a renal tubule.
7. A renal tubule consists of a proximal convoluted tubule, a nephron loop, and a distal convoluted tubule, which drains into a collecting duct. The nephron loop consists of descending and ascending limbs.
8. A cortical nephron has a short loop that dips only into the superficial region of the renal medulla; a juxtamedullary nephron has a long loop that stretches through the medulla almost to the renal papilla.
9. The wall of the entire glomerular capsule, renal tubule, and ducts consists of a single layer of epithelial cells. The epithelium has distinctive histological features in different parts of the tubule. Table 25.1 summarizes the histological features of the renal tubule and collecting duct.
10. The juxtaglomerular apparatus (JGA) consists of the juxtaglomerular cells of an afferent arteriole and the macula densa of the final portion of the ascending limb of the nephron loop.

Anatomy Overview - Kidney Overview
Anatomy Overview - Blood Supply to the Kidney
Anatomy Overview - Types of Nephrons
Anatomy Overview - Fluid Flow Through a Nephron
Figure 25.3 - Internal Anatomy of the Kidney
Figure 25.6 - Histology of a Renal Corpuscle
Exercise - Paint the Nephron

25.3 Functions of Nephrons

1. Fluid that is filtered by glomeruli and enters the capsular space is called glomerular filtrate.
2. The filtration (endothelial–capsular) membrane consists of the endothelium of the glomeruli, basal lamina, and filtration slits between pedicels of podocytes.
3. Most substances in plasma pass easily through the glomeruli. However, blood cells and most proteins normally are not filtered.
4. Glomerular filtrate amounts to up to 180 liters per day. This is because the filter is porous and thin, glomerular capillaries are long, and capillary blood pressure is high.
5. Tubular reabsorption is a selective process that reclaims materials from tubular fluid and returns them to the bloodstream. Reabsorbed substances include water, glucose, amino acids, urea, and ions, such as sodium, chloride, potassium, bicarbonate, and phosphate.
6. Some substances not needed by the body are removed from the blood and discharged into the urine via tubular secretion. Included are ions (K^+, H^+, and NH_4^+), urea, creatinine, and certain drugs.

Anatomy Overview - Overview of Fluids
Animation - Renal Filtration
Animation - Renal Reabsorption and Secretion
Animation - Hormonal Control of Blood Volume and Pressure
Animation - Regulation of pH Filtration Finale
Exercise - Magical Renal Ride

25.4 Urine Transportation, Storage, and Elimination

1. The ureters are retroperitoneal and consist of a mucosa, muscularis, and adventitia. They transport urine from the renal pelvis to the urinary bladder, primarily via peristalsis.
2. The urinary bladder is located in the pelvic cavity posterior to the pubic symphysis; its function is to store urine prior to micturition.
3. The urinary bladder consists of a mucosa with rugae, a muscularis (detrusor muscle), and an adventitia (serosa over the superior surface).
4. The micturition reflex discharges urine from the urinary bladder via parasympathetic impulses that cause contraction of the detrusor muscle and relaxation of the internal urethral sphincter muscle, and via inhibition of impulses in somatic motor neurons to the external urethral sphincter.
5. The urethra is a tube leading from the floor of the urinary bladder to the exterior. In both sexes the urethra functions to discharge urine from the body; in males it discharges semen as well.

Anatomy Overview - Ureters, Urinary Bladder, and Urethra
Figure 25.12 - Comparison Between Male and Female Urethras

25.5 Development of the Urinary System

1. The kidneys develop from intermediate mesoderm.
2. The kidneys develop in the following sequence: pronephros, mesonephros, and metanephros. Only the metanephros remains and develops into a functional kidney.

25.6 Aging and the Urinary System

1. With aging, the kidneys shrink in size, have a decreased blood flow, and filter less blood.
2. Common problems related to aging include urinary tract infections, increased frequency of urination, urinary retention or incontinence, and renal calculi.

CRITICAL THINKING QUESTIONS

1. Imagine that a new superbug has emerged from a nuclear waste dump. It produces a toxin that blocks renal tubule function but leaves the glomerulus unaffected. Predict the effects of this toxin.

2. Whenever 35-year-old Barbara jumps on the trampoline with her two young sons, she notices that she slightly "wets her pants," even though she just went to the bathroom before playing. Describe the muscles and nerves involved in the micturition reflex that may be causing Barbara's problem.

3. Although urinary catheters come in one length only, the number of centimeters that must be inserted to release the urine differs significantly in males and females. Why? Why is volume of urine released from a full bladder different in males and females?

4. As Juan focused his microscope on his own urine sample during his Human Anatomy lab, he was concerned to see many cells in the field of view; his urine had appeared to be fluid. There's no evidence of blood or infection. What are these cells and where do they come from?

5. Mr. Chase has suffered from *polycystic kidney disease* for his entire life, but it has only been in the past 10 years that the progression of the disease has destroyed a significant amount of kidney parenchyma (nephrons). In order to compensate for the destroyed nephron's functions, Mr. Chase undergoes *hemodialysis* four times per week. What are the functions of the nephron that are now being carried out by the dialysis machine?

? ANSWERS TO FIGURE QUESTIONS

25.1 The kidneys are the primary organs of the urinary system; the other structures serve as storage areas and passageways.

25.2 The kidneys are retroperitoneal because they are posterior to the peritoneum of the abdominal cavity.

25.3 Blood vessels, lymphatic vessels, nerves, and a ureter pass through the renal hilum.

25.4 About 1200 mL of blood enters the renal arteries each minute.

25.5 Cortical nephrons have glomeruli in the superficial renal cortex, and their short nephron loops penetrate only into the superficial renal medulla; juxtamedullary nephrons have glomeruli deep in the renal cortex, and their long nephron loops extend through the renal medulla nearly to the renal papilla.

25.6 The section shown must be part of the renal cortex because there are no renal corpuscles in the renal medulla.

25.7 Penicillin secreted by the cells of the renal tubule is being removed from the bloodstream.

25.8 Endothelial fenestrations (pores) in glomerular capillaries prevent red blood cells from entering the capsular space because they are too small for red blood cells to pass through.

25.9 Transitional epithelium is important in the ureters and urinary bladder because it permits the organs to stretch to accommodate a variable amount of fluid.

25.10 The muscularis of most of the ureter consists of an inner longitudinal layer and an outer circular layer, an arrangement opposite that of the gastrointestinal tract.

25.11 The trigone is a triangular area in the urinary bladder formed by the ureteral openings (posterior corners) and the internal urethral orifice (anterior corner).

25.12 The male urethra passes through the prostate, urogenital diaphragm, and penis.

25.13 The kidneys start to develop during the third week of gestation.

26 THE REPRODUCTIVE SYSTEMS

INTRODUCTION **Sexual reproduction** is the process by which organisms produce offspring through the union of germ cells called **gametes** (GAM-ēts= spouses). After the male gamete (sperm cell) unites with the female gamete (secondary oocyte)—an event called **fertilization** (fer-til-i-ZĀ-shun)—the resulting cell contains one set of chromosomes from each parent. Males and females have anatomically distinct reproductive organs that are adapted for producing gametes, facilitating fertilization, and, in females, sustaining the growth of the embryo and fetus.

The male and female reproductive organs can be grouped by function. The **gonads**—testes in males and ovaries in females—produce gametes and secrete sex hormones. Various **ducts** then store and transport the gametes, and **accessory sex glands** produce substances that protect the gametes and facilitate their movement. Finally, **supporting structures**, such as the penis in males and the vagina in females, assist the delivery of gametes, and the uterus in females assists in the growth of the embryo and fetus during pregnancy.

Gynecology (gī-ne-KOL-ō-jē; *gynec-*=woman; *logy*=study of) is the specialized branch of medicine concerned with the diagnosis and treatment of diseases of the female reproductive system. As noted in Chapter 25, **urology** (ū-ROL-ō-jē) is the study of the urinary system. Urologists also diagnose and treat diseases and disorders of the male reproductive system. The branch of medicine that deals with male disorders, especially infertility and sexual dysfunction, is called **andrology** (an-DROL-ō-jē; *andro-*=masculine). •

Did you ever wonder how breast augmentation and breast reduction are performed?

26.1 MALE REPRODUCTIVE SYSTEM

OBJECTIVES

- Describe the location, structure, and functions of the scrotum.
- Explain where the testes are located and outline their functions.
- Discuss the process of spermatogenesis in the testes.
- Describe the structure and functions of each part of a mature sperm.

The organs of the **male reproductive system** include the testes (male gonads), a system of ducts (including the epididymis, ductus deferens, ejaculatory ducts, and urethra), accessory sex glands (seminal vesicles, prostate, and bulbourethral glands), and several supporting structures, including the scrotum and the penis (Figure 26.1). The testes produce sperm and secrete hormones. The duct system transports and stores sperm, assists in their maturation, and conveys them to the exterior. Semen contains sperm plus the secretions provided by the accessory sex glands. The supporting structures have various functions. The penis delivers sperm into the female reproductive tract and the scrotum supports the testes.

Scrotum

The **scrotum** (SKRŌ-tum=bag), the supporting structure for the testes, consists of loose skin and an underlying subcutaneous layer that hangs from the root (attached portion) of the penis (Figure 26.1a). Externally, the scrotum looks like a single pouch of skin separated into lateral portions by a median ridge called the **raphe** (RĀ-fē=seam) (Figure 26.2). Internally, the **scrotal septum** divides the scrotum into two sacs, each containing a single testis. The septum is made up of a subcutaneous layer and muscle tissue called the **dartos muscle** (DAR-tōs=skinned), which is composed of bundles of smooth muscle fibers. The dartos muscle is also found in the subcutaneous layer of the scrotum. Associated

with each testis in the scrotum is the **cremaster muscle** (kre-MAS-ter=suspender). The cremaster muscle consists of a series of small bands of skeletal muscle that descend, as an extension of the internal oblique muscle, through the spermatic cord to surround the testis.

The location of the scrotum and the contraction of its muscle fibers regulate the temperature of the testes. Normal sperm production requires a temperature about 2–3°C below core body temperature. This lowered temperature is maintained within the scrotum because it is outside the pelvic cavity. In response to cold temperatures, the cremaster and dartos muscles contract. Contraction of the cremaster muscles moves the testes closer to the body, where they can absorb body heat. Contraction of the dartos muscle causes the scrotum to become tight (wrinkled in appearance), which reduces heat loss. Exposure to warmth reverses these actions.

The blood supply of the scrotum arises from the internal pudendal branch of the internal iliac artery, the cremasteric branch of the inferior epigastric artery, and the external pudendal artery from the femoral artery. The scrotal veins follow the arteries.

The scrotal nerves arise from the pudendal nerve, posterior cutaneous nerve of the thigh, and ilioinguinal nerves.

CLINICAL CONNECTION | Cryptorchidism

The condition in which the testes do not descend into the scrotum is called **cryptorchidism** (krip-TOR-ki-dizm; *crypt-*=hidden; *orchid*=testis). It occurs in about 3 percent of full-term infants and about 30 percent of premature infants. Untreated bilateral cryptorchidism causes sterility due to the higher temperature of the pelvic cavity. The chance of testicular cancer is 30 to 50 times greater in cryptorchid testes, possibly due to abnormal division of germ cells caused by the higher temperature of the pelvic cavity. The testes of about 80 percent of boys with cryptorchidism will descend spontaneously during the first year of life. When the testes remain undescended, the condition can be corrected surgically, ideally before 18 months of age. •

Figure 26.1 Male reproductive organs and surrounding structures.

 Reproductive organs are adapted for producing new individuals and passing on genetic material from one generation to the next.

FUNCTIONS OF THE MALE REPRODUCTIVE SYSTEM

1. The testes produce sperm and the male sex hormone testosterone.
2. The ducts transport, store, and assist in maturation of sperm.
3. The accessory sex glands secrete most of the liquid portion of semen.
4. The penis contains the urethra, a passageway for ejaculation of semen and excretion of urine.

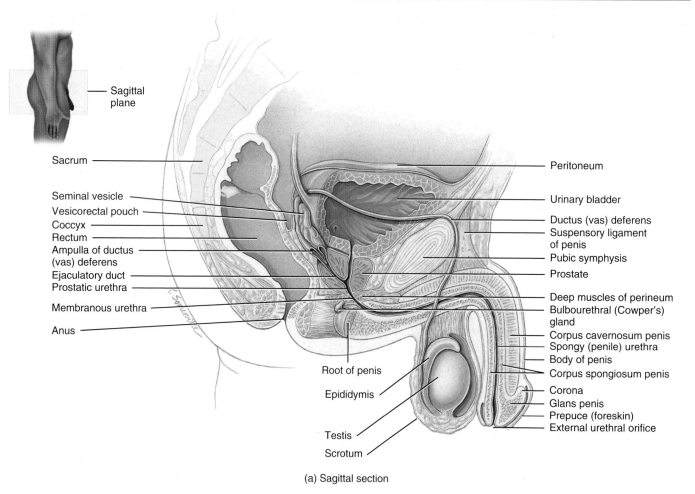

Sagittal plane

Sacrum

Seminal vesicle
Vesicorectal pouch
Coccyx
Rectum
Ampulla of ductus (vas) deferens
Ejaculatory duct
Prostatic urethra
Membranous urethra
Anus

Root of penis
Epididymis
Testis
Scrotum

Peritoneum
Urinary bladder
Ductus (vas) deferens
Suspensory ligament of penis
Pubic symphysis
Prostate
Deep muscles of perineum
Bulbourethral (Cowper's) gland
Corpus cavernosum penis
Spongy (penile) urethra
Body of penis
Corpus spongiosum penis
Corona
Glans penis
Prepuce (foreskin)
External urethral orifice

(a) Sagittal section

SUPERIOR

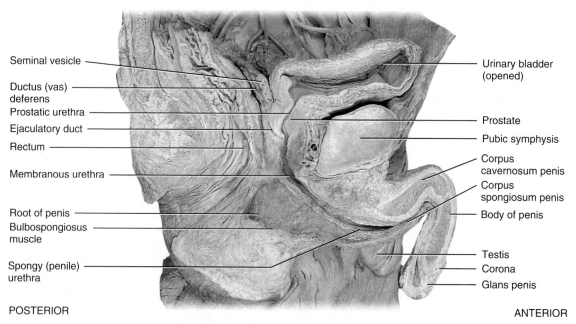

Seminal vesicle
Ductus (vas) deferens
Prostatic urethra
Ejaculatory duct
Rectum
Membranous urethra
Root of penis
Bulbospongiosus muscle
Spongy (penile) urethra

Urinary bladder (opened)
Prostate
Pubic symphysis
Corpus cavernosum penis
Corpus spongiosum penis
Body of penis
Testis
Corona
Glans penis

POSTERIOR

ANTERIOR

(b) Sagittal section

What are the groups of reproductive organs in males, and what are the functions of each group?

Figure 26.2 The scrotum, the supporting structure for the testes.

 The scrotum consists of loose skin and superficial fascia and supports the testes.

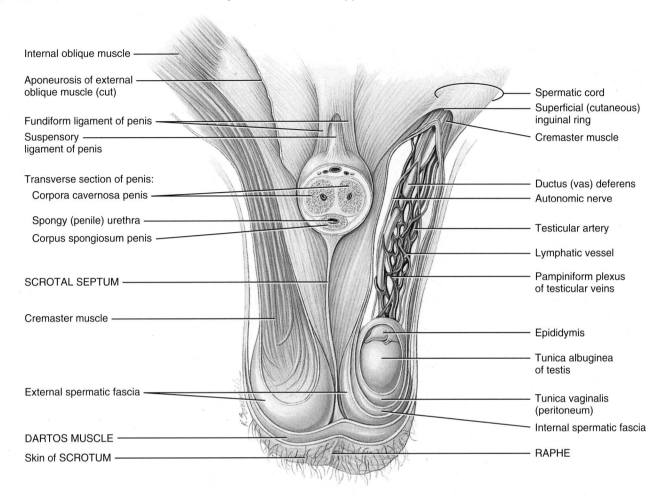

Internal oblique muscle

Aponeurosis of external oblique muscle (cut)

Fundiform ligament of penis

Suspensory ligament of penis

Transverse section of penis:

Corpora cavernosa penis

Spongy (penile) urethra

Corpus spongiosum penis

SCROTAL SEPTUM

Cremaster muscle

External spermatic fascia

DARTOS MUSCLE

Skin of SCROTUM

Spermatic cord

Superficial (cutaneous) inguinal ring

Cremaster muscle

Ductus (vas) deferens

Autonomic nerve

Testicular artery

Lymphatic vessel

Pampiniform plexus of testicular veins

Epididymis

Tunica albuginea of testis

Tunica vaginalis (peritoneum)

Internal spermatic fascia

RAPHE

Anterior view of scrotum and testes and transverse section of penis

 **Which muscles help regulate the temperature of the testes?**

Testes

The **testes** (TES-tēz=witness), or **testicles**, are paired oval glands in the scrotum measuring about 5 cm (2 in.) long and 2.5 cm (1 in.) in diameter (Figure 26.3). Each testis (singular) weighs 10–15 grams. The testes develop near the kidneys, in the posterior portion of the abdomen, and they usually begin their descent into the scrotum through the inguinal canals (passageways in the lower anterior abdominal wall; see Figure 26.2) during the latter half of the seventh month of fetal development.

The testes are partially covered by a serous membrane called the **tunica vaginalis** (TOO-nik-a vaj-i-NAL-is; *tunica*=sheath), which is derived from the peritoneum and forms during the descent of the testes. Like other serous membranes, it has a visceral layer and a parietal layer and forms a fist-in-balloon relationship with the testis. A collection of serous fluid in the cavity of the tunica vaginalis is called a **hydrocele** (HĪ-drō-sēl; *hydro-*=water; *kele*=hernia). It may be caused by injury to the testes or inflammation of the epididymis. Usually, no treatment is required. Internal to the visceral layer of the tunica vaginalis, the testis is surrounded by a white fibrous capsule composed of dense irregular connective tissue, the **tunica albuginea** (al′-bū-JIN-ē-a; *albu-*=white); it extends inward, forming septa that divide each testis into a series of

internal compartments called **lobules**. Each of the 200–300 lobules contains one to three tightly coiled tubules, the **seminiferous tubules** (sem′-i-NIF-er-us; *semin-*=seed; *fer-*=to carry), where sperm are produced (see Figure 26.4). The process by which the seminiferous tubules of the testes produce sperm is called **spermatogenesis** (sper′-ma-tō-JEN-e-sis; *genesis*=to be born).

CLINICAL CONNECTION | *Vasectomy*

The principal method for sterilization of males is a **vasectomy** (vas-EK-tō-mē; *-ectomy*=cut out), in which a portion of each ductus deferens is removed. An incision is made on either side of the scrotum, the ducts are located and cut, each is tied (ligated) in two places with stitches, and the portion between the ties is removed. Although sperm production continues in the testes, sperm can no longer reach the exterior. The sperm degenerate and are destroyed by phagocytosis. Because the blood vessels are not cut, testosterone levels in the blood remain normal, so vasectomy has no effect on sexual desire, performance, and ejaculation. If done correctly, vasectomy is close to 100 percent effective. The procedure can be reversed, but the chance of regaining fertility is only 30–40 percent. •

Figure 26.3 Internal and external anatomy of a testis.

The testes are the male gonads, which produce haploid sperm.

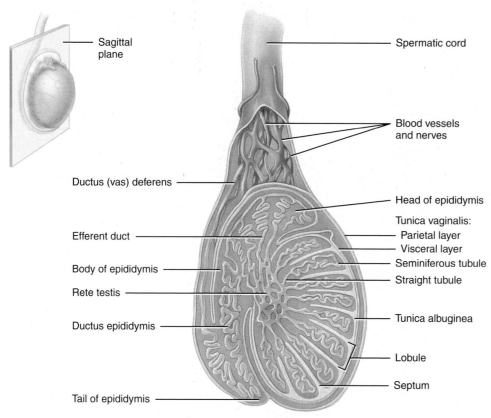

Sagittal plane

Spermatic cord

Blood vessels and nerves

Ductus (vas) deferens

Head of epididymis

Tunica vaginalis:
— Parietal layer
— Visceral layer
Seminiferous tubule
Straight tubule

Efferent duct

Body of epididymis

Rete testis

Tunica albuginea

Ductus epididymis

Lobule

Septum

Tail of epididymis

(a) Sagittal section of a testis showing seminiferous tubules

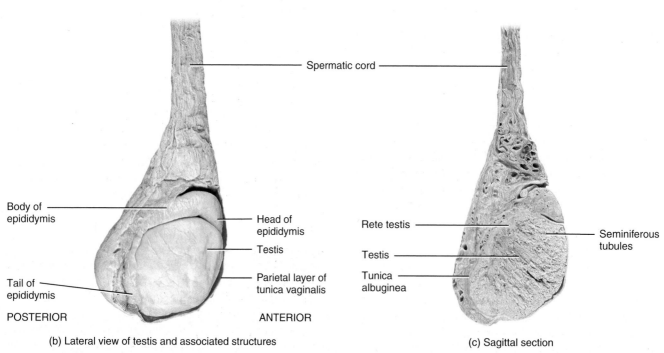

SUPERIOR

Spermatic cord

Body of epididymis

Head of epididymis

Rete testis

Seminiferous tubules

Testis

Testis

Tail of epididymis

Parietal layer of tunica vaginalis

Tunica albuginea

POSTERIOR

ANTERIOR

(b) Lateral view of testis and associated structures

(c) Sagittal section

What tissue layers cover and protect the testes?

The walls of the seminiferous tubules contain two types of cells: **spermatogenic cells** (sper'-ma-tō-JEN-ik), the sperm-forming cells, and **sustentacular cells** (sus'-ten-TAK-ū-lar) or *Sertoli cells* (ser-TŌ-lē), which have several functions in supporting spermatogenesis (Figure 26.4). Starting at puberty, sperm production begins at the periphery of the seminiferous tubules in stem cells called **spermatogonia** (sper′-ma-tō-GŌ-nē-a; *gonia*=offspring; singular is *spermatogonium*). These cells develop from **primordial germ cells** (prī-MOR-dē-al; *primordi-*=primitive or early form) that arise from the yolk sac endoderm and enter the testes during the fifth week of development. In the embryonic testes, the primordial germ cells differentiate into spermatogonia, which remain dormant during childhood and become active at puberty. Toward the lumen of the tubule are layers of progressively more mature cells. In order of advancing maturity, these

are primary spermatocytes, secondary spermatocytes, spermatids, and sperm cells. After a **sperm cell** (*sperma*=seed), or **spermatozoon** (sper′-ma-tō-ZŌ-on; *zoon*=life), has formed, it is released into the lumen of the seminiferous tubule. (The plural terms are *sperm* and *spermatozoa*.)

Embedded among the spermatogenic cells in the tubules, the large sustentacular cells extend from the basement membrane to the lumen of the tubule. Internal to the basement membrane and spermatogonia, tight junctions join neighboring sustentacular cells to

Figure 26.4 **The seminiferous tubules and stages of sperm production (spermatogenesis).** Arrows in (b) indicate the progression of spermatogenic cells from least mature to most mature. The (*n*) and (2*n*) refer to haploid and diploid chromosome number, respectively.

🔑 Spermatogenesis occurs in the seminiferous tubules of the testes.

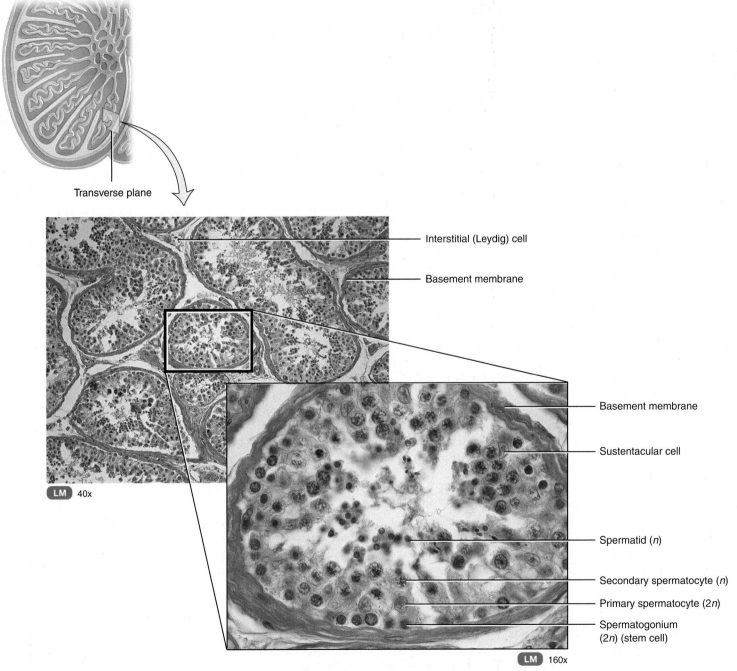

(a) Transverse section of several seminiferous tubules

FIGURE 26.4 CONTINUES ▶

FIGURE 26.4 CONTINUED

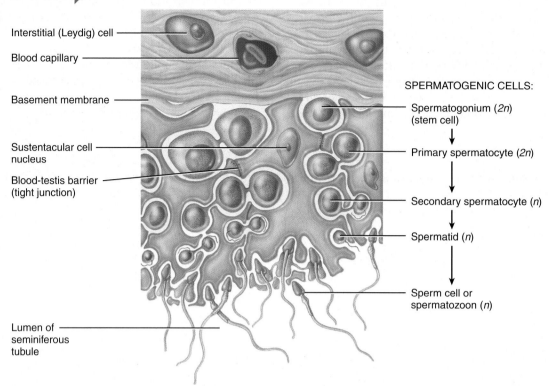

Interstitial (Leydig) cell

Blood capillary

Basement membrane

Sustentacular cell
nucleus

Blood-testis barrier
(tight junction)

Lumen of
seminiferous
tubule

SPERMATOGENIC CELLS:

Spermatogonium (*2n*)
(stem cell)

Primary spermatocyte (*2n*)

Secondary spermatocyte (*n*)

Spermatid (*n*)

Sperm cell or
spermatozoon (*n*)

(b) Transverse section of a part of a seminiferous tubule

Which cells produce testosterone?

one another. These junctions form an obstruction known as the **blood–testis barrier** because substances must pass through the sustentacular cells before they can reach the developing sperm. By isolating the developing gametes from the blood, the blood–testis barrier prevents an immune response against the spermatogenic cell's surface antigens, which are recognized as "foreign" by the immune system. The blood–testis barrier does not include spermatogonia.

Sustentacular cells support and protect developing spermatogenic cells in several ways. They nourish spermatocytes, spermatids, and sperm; phagocytize excess spermatid cytoplasm as development proceeds; and control movements of spermatogenic cells and the release of sperm into the lumen of the seminiferous tubule. They also produce fluid for sperm transport, secrete the hormone inhibin, which decreases the rate of spermatogenesis, and regulate the effects of testosterone and FSH (follicle-stimulating hormone).

In the spaces between adjacent seminiferous tubules are clusters of cells called **interstitial cells** or *Leydig cells* (LĪ-dig) (Figure 26.4). These cells secrete testosterone, the most important androgen. An **androgen** (AN-drō-jen) is a hormone that promotes the development of masculine characteristics. Testosterone also promotes a man's *libido* (sex drive).

Before you read this section, please review the topic of reproductive cell division in Section 2.5. Also, pay particular attention to Figures 2.20 and 2.21 in Section 2.5.

In humans, spermatogenesis takes about 65–75 days. It begins in the spermatogonia, which contain the diploid (*2n*) chromosome number (Figure 26.5). Spermatogonia are a type of *stem cells*; when they undergo mitosis, some cells remain near the basement membrane of the seminiferous tubule in an undifferentiated state to serve as a reservoir of cells for future cell division and sub-

sequent sperm production. The rest of the cells lose contact with the basement membrane, squeeze through the tight junctions of the blood–testis barrier, undergo developmental changes, and differentiate into **primary spermatocytes** (SPER-ma-tō-sītz′). Primary spermatocytes, like spermatogonia, are diploid (*2n*); that is, they have 46 chromosomes.

Shortly after it forms, each primary spermatocyte replicates its DNA, and then meiosis begins (Figure 26.5). In meiosis I, homologous pairs of chromosomes line up at the metaphase plate, and crossing-over occurs. Then the meiotic spindle pulls one (duplicated) chromosome of each pair to an opposite pole of the dividing cell. The two cells formed by meiosis I are called **secondary spermatocytes**. Each secondary spermatocyte has 23 chromosomes, the haploid number (*n*). Each chromosome within a secondary spermatocyte, however, is made up of two chromatids (two copies of the DNA) still attached by a centromere. No replication of DNA occurs in the secondary spermatocytes.

In meiosis II, no replication of DNA occurs. The chromosomes line up in single file along the metaphase plate, and the two chromatids of each chromosome separate. The four haploid cells resulting from meiosis II are called **spermatids** (SPER-ma-tids). A single primary spermatocyte therefore produces four spermatids through two rounds of cell division (meiosis I and meiosis II).

A unique process occurs during spermatogenesis. As the sperm cells proliferate following their production by spermatogonia, they fail to complete cytoplasmic separation (cytokinesis). The cells remain in contact by cytoplasmic bridges through their entire development (Figure 26.5). This pattern of development most likely accounts for the synchronized production of sperm in any given area of a seminiferous tubule. It may have survival value in that half of the sperm contain an X chromosome and half contain a Y chromosome. The larger X chromosome may carry genes needed for spermatogenesis that are lacking on the smaller Y chromosome.

The final stage of spermatogenesis, **spermiogenesis** (sper'-mē-ō-JEN-e-sis), is the maturation of haploid spermatids into sperm. Because no cell division occurs in spermiogenesis, each spermatid develops into a single **sperm cell**. During this process, spherical spermatids transform into elongated, slender sperm. An acrosome (described shortly) forms atop the condensing and elongating nucleus, a flagellum develops, and mitochondria multiply. Sustentacular cells dispose of the excess cytoplasm that is sloughed off during this process. Finally, sperm are released from their connections to sustentacular cells, an event known as **spermiation** (sper'-mē-Ā-shun). Sperm then enter the lumen of the seminiferous tubule. Fluid secreted by sustentacular cells pushes sperm along their way, toward the ducts of the testes. At this point, sperm are not yet able to swim.

Sperm

Spermatogenesis produces about 300 million sperm per day. A sperm is about 60 μm long and contains several structures

Figure 26.5 Spermatogenesis. Diploid cells (2*n*) have 46 chromosomes; haploid cells (*n*) have 23 chromosomes.

 Spermiogenesis involves the maturation of spermatids into sperm.

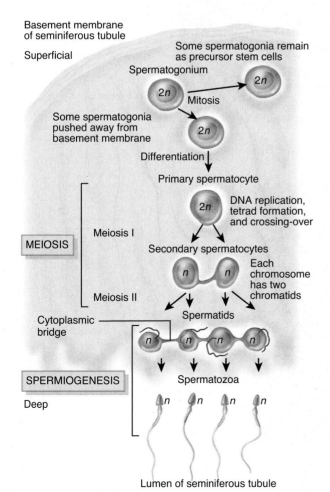

 What is the outcome of meiosis I?

Figure 26.6 Parts of sperm cell (spermatozoon).

 About 300 million sperm mature each day.

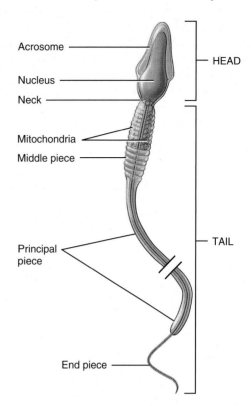

 What are the functions of each part of a sperm cell?

that are highly adapted for reaching and penetrating a secondary oocyte (Figure 26.6). The major parts of a sperm are the head and the tail. The flattened, pointed **head** of the sperm is about 4–5 μm long. It contains a **nucleus** that has highly condensed haploid chromosomes (23). Covering the anterior two-thirds of the nucleus is the **acrosome** (AK-rō-sōm; *acro-*=atop; *some*=body), a caplike vesicle filled with enzymes that help a sperm to penetrate a secondary oocyte to bring about fertilization. Among the enzymes are hyaluronidase and proteases. The **tail** of a sperm is subdivided into four parts: neck, middle piece, principal piece, and end piece. The **neck** is the constricted region just behind the head that contains centrioles. The centrioles form the microtubules that comprise the remainder of the tail. The **middle piece** contains mitochondria arranged in a spiral, which provide the energy (ATP) for locomotion of sperm to the site of fertilization and for sperm metabolism. The **principal piece** is the longest portion of the tail and the **end piece** is the terminal, tapering portion of the tail. Once ejaculated, most sperm do not survive more than 48 hours within the female reproductive tract.

CHECKPOINT

1. What is the function of the scrotum in protecting the testes from temperature fluctuations?
2. Describe the internal structure of a testis. Where in the testes are sperm cells produced? What are the functions of sustentacular cells and interstitial cells?
3. What are the principal events of spermatogenesis?
4. How does the middle piece of a sperm cell contribute to successful fertilization?

Reproductive System Ducts in Males

Ducts of the Testis

Pressure generated by the fluid secreted by sustentacular cells pushes sperm and fluid along the lumen of seminiferous tubules and then into a series of very short ducts called **straight tubules**. The straight tubules lead to a network of ducts in the testis called the **rete testis** (Rē-tē=network) (see Figure 26.3a). From the rete testis, sperm move into a series of coiled **efferent ducts** (EF-e-rent) in the epididymis that empty into a single tube called the **ductus epididymis**.

Epididymis

The **epididymis** (ep′-i-DID-i-mis; *epi-*=above or over; *-didymis*= testis; plural is *epididymides* [ep′-i-did-ĪM-i-dēs]) is a comma-shaped organ about 4 cm (1.5 in.) long that lies along the posterior border of each testis (see Figure 26.3b). Each epididymis consists mostly of the tightly coiled **ductus epididymis**. The efferent ducts from the testis join the ductus epididymis at the larger, superior portion of the epididymis called the **head**. The **body** is the narrow midportion of the epididymis, and the **tail** is the smaller, inferior portion. At its distal end, the tail of the epididymis continues as the ductus (vas) deferens (discussed shortly).

The ductus epididymis would measure about 6 m (20 ft) in length if it were straightened out. It is lined with pseudostratified columnar epithelium and encircled by a layer of smooth muscle (Figure 26.7). The free surfaces of the columnar cells contain

Figure 26.7 Histology of the ductus epididymis.

 Stereocilia increase the surface area for the reabsorption of degenerated sperm.

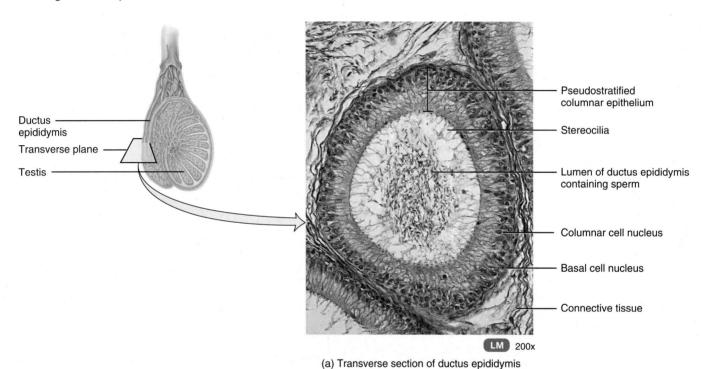

(a) Transverse section of ductus epididymis

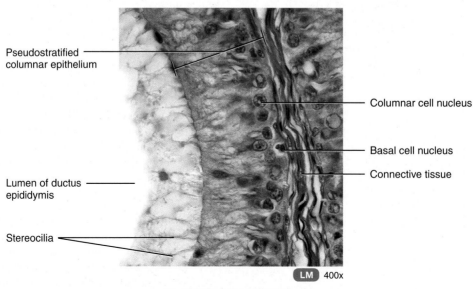

(b) Details of the epithelium

? **What are the functions of the ductus epididymis?**

Figure 26.8 Histology of the ductus (vas) deferens.

🔑 The ductus (vas) deferens enters the pelvic cavity through the inguinal canal.

Transverse plane

Ductus deferens

Testis

Testicular blood vessels with red blood cells

Muscularis:
Outer longitudinal
Middle circular
Inner longitudinal

Lumen of ductus deferens
Pseudostratified columnar epithelium

LM 30x

(a) Transverse section of the ductus deferens

Lumen of ductus deferens

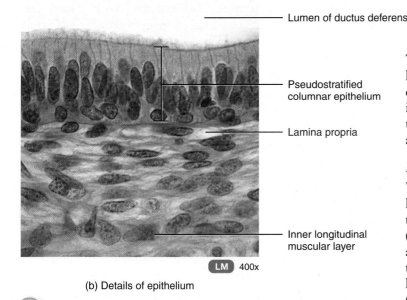

Pseudostratified columnar epithelium

Lamina propria

Inner longitudinal muscular layer

LM 400x

(b) Details of epithelium

❓ **What is the function of the ductus (vas) deferens?**

stereocilia (ster′-ē-ō-SIL-ē-a), long, branching microvilli that increase surface area for the reabsorption of degenerated sperm. Connective tissue around the muscle layer attaches the loops of the ductus epididymis to one another and carries blood vessels and nerves.

Functionally, the epididymis is the site of **sperm maturation,** the process by which sperm acquire motility and the ability to fertilize an ovum. This occurs over a period of about 14 days.

The epididymis also stores sperm and during sexual arousal helps propel sperm into the ductus (vas) deferens by peristaltic contraction of its smooth muscle. Sperm may remain in storage in the epididymis for up to several months. Any stored sperm that are not ejaculated by that time are eventually phagocytized and reabsorbed.

Ductus Deferens

Within the tail of the epididymis, the ductus epididymis becomes less convoluted, and its diameter increases. Beyond this point, the duct is referred to as the **ductus deferens** or **vas deferens** (DEF-er-enz) (see Figure 26.3a). The ductus deferens, which is about 45 cm (18 in.) long, ascends along the posterior border of the epididymis, through the spermatic cord to the point in the lower abdominal wall where it passes through the inguinal canal (see Figure 26.2) to enter the pelvic cavity; there it loops over the ureter and passes over the side and down the posterior surface of the urinary bladder (see Figure 26.1a). The dilated terminal portion of the ductus deferens is known as the **ampulla** (am-POOL-la=little jar) (see Figure 26.9). The mucosa of the ductus deferens consists of pseudostratified columnar epithelium and lamina propria (made up of areolar connective tissue). The muscularis is composed of three layers; the inner and outer layers are longitudinal, and the middle layer is circular (Figure 26.8). Functionally, the ductus deferens conveys sperm during sexual arousal from the epididymis toward the urethra by peristaltic contractions of its muscular coat. Like the epididymis, the ductus deferens also can store sperm for several months. Any stored sperm that are not ejaculated by that time are eventually reabsorbed.

Ejaculatory Ducts

Each **ejaculatory duct** (e-JAK-ū-la-tō'-rē; *ejacul-*=to expel) is about 2 cm (1 in.) long and is formed by the union of the duct from the seminal vesicle and the ampulla of the ductus deferens (see Figure 26.9). The short ejaculatory ducts form just superior to the base (superior portion) of the prostate and pass inferiorly and anteriorly through the prostate. They terminate in the prostatic urethra, where they eject sperm and seminal vesicle secretions just before the release of semen from the urethra to the exterior.

Urethra

In males, the **urethra** (ū-RĒ-thra) is the shared terminal duct of the reproductive and urinary systems; it serves as a passageway for both semen and urine. About 20 cm (8 in.) long, it passes through the prostate, the deep muscles of the perineum, and the penis, and is subdivided into three parts (see Figures 26.1a and 26.9). The **prostatic urethra** (pros-TAT-ik) is 2–3 cm (1 in.) long and passes through the prostate. As this duct continues inferiorly, it passes through the deep muscles of the perineum (see Figure 11.15), where it is known as the **intermediate (membranous) urethra** (MEM-bra-nus). The intermediate urethra is about 1 cm (0.5 in.) in length. As this duct passes through the corpus spongiosum of the penis, it is known as the **spongy (penile) urethra**, which is about 15–20 cm (6–8 in.) long. The spongy urethra ends at the **external urethral orifice**. The histology of the male urethra may be reviewed in Section 25.3.

Spermatic Cord

The **spermatic cord** is a supporting structure of the male reproductive system that ascends out of the scrotum (see Figure 26.2). It consists of the ductus deferens as it ascends through the scrotum, the testicular artery, both somatic and autonomic nerves, veins that drain the testes and carry testosterone into the circulation (the *pampiniform plexus*), lymphatic vessels, the cremaster muscle and fascial covering. The spermatic cord and ilioinguinal nerve pass through the **inguinal canal** (ING-gwin-al=groin), an oblique passageway in the anterior abdominal wall just superior and parallel to the medial half of the inguinal ligament. The canal, which is about 4–5 cm (about 2 in.) long, originates at the **deep (abdominal) inguinal ring**, a slitlike opening in the aponeurosis of the transversus abdominis muscle; the canal ends at the **superficial (subcutaneous) inguinal ring** (see Figure 26.2), a somewhat triangular opening in the aponeurosis of the external oblique muscle. In females, the round ligament of the uterus and ilioinguinal nerve pass through the inguinal canal.

The term **varicocele** (VAR-i-kō-sēl; *varico-*=varicose; *kele*=hernia) refers to a swelling in the scrotum due to varicosities in the veins that drain the testes. It usually disappears when the person lies down, and typically does not require treatment.

✅ CHECKPOINT

5. Which ducts transport sperm within the testes?
6. Compare the functions of the ductus epididymis, ductus (vas) deferens, and ejaculatory duct.
7. Where are the three subdivisions of the male urethra located?
8. What route do sperm take through the system of ducts from the seminiferous tubules to the urethra?
9. What is the spermatic cord, and what structures does it contain?

Accessory Sex Glands in Males

The ducts of the male reproductive system store and transport sperm cells, but the **accessory sex glands** secrete most of the liquid portion of semen. The accessory sex glands include the seminal vesicles, the prostate, and the bulbourethral glands.

Seminal Vesicles

The paired **seminal vesicles** (VES-i-kuls) or *seminal glands* are convoluted pouchlike structures, about 5 cm (2 in.) in length, lying posterior to and at the base of the urinary bladder anterior to the rectum (Figure 26.9). Through the seminal vesicle ducts they secrete an alkaline, viscous fluid that contains fructose (a monosaccharide sugar), prostaglandins, and clotting proteins (discussed shortly) unlike those found in blood. The alkaline nature of the fluid helps to neutralize the acidic environment of the male urethra and female reproductive tract that otherwise would inactivate and kill sperm. The fructose is used for the production of ATP by sperm. Prostaglandins contribute to sperm motility and viability and may also stimulate muscular contractions within the female reproductive tract. The clotting proteins help semen coagulate after ejaculation. Fluid secreted by the seminal vesicles normally constitutes about 60 percent of the volume of semen.

Prostate

The **prostate** (PROS-tāt; *prostata*=one who stands before) is a single, doughnut-shaped gland about the size of a ping-pong ball. It measures about 4 cm (1.6 in.) from side to side, about 3 cm (1.2 in.) from top to bottom, and about 2 cm (0.8 in.) from front to back. It is inferior to the urinary bladder and surrounds the prostatic urethra (Figure 26.9). The prostate slowly increases in size from birth to puberty, and then expands rapidly. The size attained by age 30 typically remains stable until about age 45, when further enlargement may occur.

The prostate secretes a milky, slightly acidic fluid (pH about 6.5) that contains several substances:

1. *Citric acid* in prostatic fluid is used by sperm for ATP production.
2. Several *proteolytic enzymes*, such as *prostate-specific antigen (PSA)*, pepsinogen, lysozyme, amylase, and hyaluronidase, eventually break down the clotting proteins from the seminal vesicles.
3. The function of the *acid phosphatase* secreted by the prostate is unknown.
4. *Seminalplasmin* in prostatic fluid is an antibiotic that can destroy bacteria. Seminalplasmin may help decrease the number of naturally occurring bacteria in semen and in the lower female reproductive tract.

Secretions of the prostate enter the prostatic urethra through many prostatic ducts. Prostatic secretions make up about 25 percent of the volume of semen and contribute to sperm motility and viability.

Bulbourethral Glands

The paired **bulbourethral glands** (bul'-bō-ū-RĒ-thral), or *Cowper's glands* (KOW-pers), each about the size of a pea, lie inferior to the prostate on either side of the membranous urethra within the deep muscles of the perineum; their ducts open into the spongy urethra (Figure 26.9). During sexual arousal, the bulbourethral glands secrete an alkaline substance that protects the passing sperm by neutralizing acids from urine in the urethra. At the same time, they secrete mucus that lubricates the end of

Figure 26.9 Locations of several accessory reproductive glands in males. The prostate, urethra, and penis have been sectioned to show internal details.

🔑 The male urethra has three subdivisions: the prostatic, membranous (intermediate), and spongy (penile) urethra.

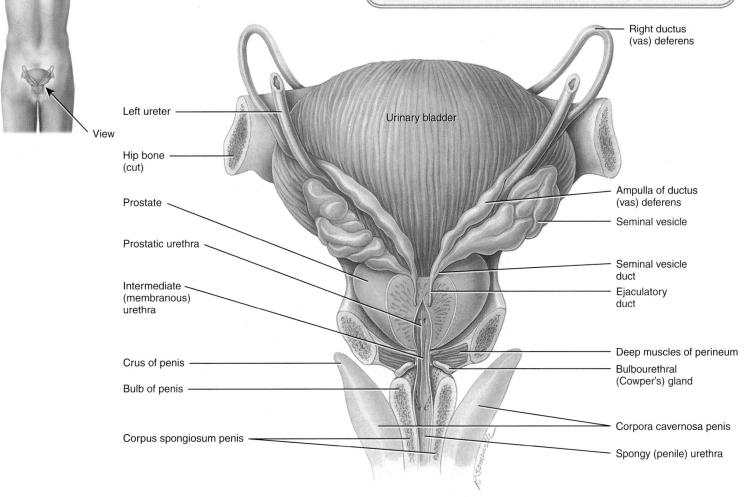

(a) Posterior view of male accessory organs of reproduction

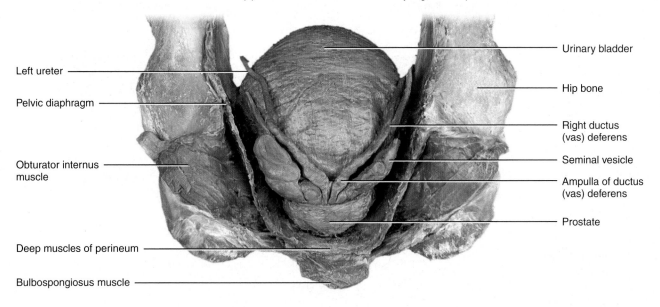

(b) Posterior view of male accessory organs of reproduction

❓ **What accessory sex gland contributes the majority of the seminal fluid?**

the penis and the lining of the urethra, thereby decreasing the number of sperm damaged during ejaculation. Some males release a drop or two of this mucus upon sexual arousal and erection. The fluid does not contain sperm cells.

Semen

Semen (=seed) is a mixture of sperm and **seminal fluid**, a liquid that consists of the secretions of the seminiferous tubules, seminal vesicles, prostate, and bulbourethral glands. The volume of semen in a typical ejaculation is 2.5–5 mL, with a sperm count (concentration) of 50–150 million sperm/mL. A male whose sperm count falls below 20 million/mL is likely to be infertile. The very large number is required for successful fertilization because only a tiny fraction ever reach the secondary oocyte.

Despite the slight acidity of prostatic fluid, semen has a slightly alkaline pH of 7.2–7.7 due to the higher pH and larger volume of fluid from the seminal vesicles. The prostatic secretion gives semen a milky appearance, and fluids from the seminal vesicles and bulbourethral glands give it a sticky consistency. Seminal fluid provides sperm with a transportation medium, nutrients, and protection from the hostile acidic environment of the male's urethra and the female's vagina.

Once ejaculated, liquid semen coagulates within 5 minutes due to the presence of clotting proteins from the seminal vesicles. The functional role of semen coagulation is not known, but the proteins involved are different from those that cause blood coagulation. After about 10–20 minutes, semen reliquefies because prostate-specific antigen (PSA) and other proteolytic enzymes produced by the prostate break down the clot. Abnormal or delayed liquefaction of clotted semen may cause complete or partial immobilization of sperm, thereby inhibiting their movement through the cervix of the uterus. After passing through the uterus to the uterine tube, the sperm are affected by secretions of the uterine tube in a process called *capacitation* (see Section 26.2).

The presence of blood in semen is called **hemospermia** (hē-mō-SPER-mē-a; *hemo-*=blood; *sperma*=seed). In most cases, it is caused by inflammation of the blood vessels lining the seminal vesicles; it is usually treated with antibiotics.

Penis

The **penis** (*penis*=tail) is a supporting structure of the male reproductive system that contains the urethra and is a passageway for the ejaculation of semen and the excretion of urine (Figure 26.10). It is cylindrical in shape and consists of a body, glans penis, and a root (see Figure 26.1). The **body of the penis** is composed of three cylindrical masses of tissue, each surrounded by fibrous tissue called the **tunica albuginea** (al-bū-JIN-ē-a; Figure 26.10). The two dorsolateral masses of the body of the penis are called the **corpora cavernosa penis** (*corpora*=main bodies; *cavernosa*=hollow; singular is *corpus cavernosum penis*).

CLINICAL CONNECTION | *Prostatitis and Prostate Cancer*

Because the prostate surrounds part of the urethra, any infection, enlargement, or tumor can obstruct the flow of urine. Acute and chronic infections of the prostate are common in postpubescent males, often in association with inflammation of the urethra. Symptoms may include fever, chills, urinary frequency, frequent urination at night, difficulty in urinating, burning or painful urination, low back pain, joint and muscle pain, blood in the urine, or painful ejaculation. However, often there are no symptoms. Antibiotics are used to treat most cases that result from a bacterial infection. In **acute prostatitis**, the prostate becomes swollen and tender. **Chronic prostatitis** (pros'-ta-TĪ-tis) is one of the most common chronic infections in men of the middle and later years. On examination, the prostate feels enlarged, soft, and very tender, and its surface outline is irregular.

Prostate cancer is the leading cause of death from cancer in men in the United States, having surpassed lung cancer in 1991. Each year it is diagnosed in almost 200,000 U.S. men and causes nearly 40,000 deaths. The amount of PSA (prostate-specific antigen), which is produced only by prostate epithelial cells, increases with enlargement of the prostate (circled area in Figure B showing enlarged prostate narrowing the prostatic urethra) and may indicate infection, benign enlargement, or prostate cancer. A blood test can measure the level of PSA in the blood. Males over the age of 40 should have an annual examination of the prostate gland. In a **digital rectal exam**, a physician palpates the gland through the rectum with the fingers (digits). Many physicians also recommend an annual PSA test for males over age 50. Treatment for prostate cancer may involve surgery, cryotherapy, radiation, hormonal therapy, and chemotherapy. Because many prostate cancers grow very slowly, some urologists recommend "watchful waiting" before treating small tumors in men over age 70. •

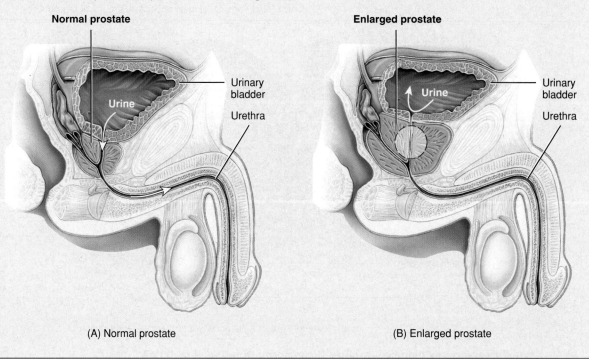

Normal prostate — Urinary bladder — Urine — Urethra

Enlarged prostate — Urinary bladder — Urine — Urethra

(A) Normal prostate

(B) Enlarged prostate

The smaller midventral mass, the **corpus spongiosum penis**, contains the spongy urethra and keeps it open during ejaculation. Skin and a subcutaneous layer enclose all three masses, which consist of erectile tissue. *Erectile tissue* is composed of numerous blood sinuses (vascular spaces) lined by endothelial cells and surrounded by smooth muscle and elastic connective tissue.

The distal end of the corpus spongiosum penis is a slightly enlarged, acorn-shaped region called the **glans penis**; its margin is the **corona** (kō-RŌ-na). The distal urethra enlarges within the glans penis and forms a terminal slitlike opening, the **external urethral orifice**. Covering the glans in an uncircumcised penis is the loosely fitting **prepuce** (PRĒ-poos), or **foreskin**.

Figure 26.10 Internal structure of the penis. The inset in (b) shows details of the skin and fasciae.

🔑 The penis contains the urethra, a common pathway for semen and urine.

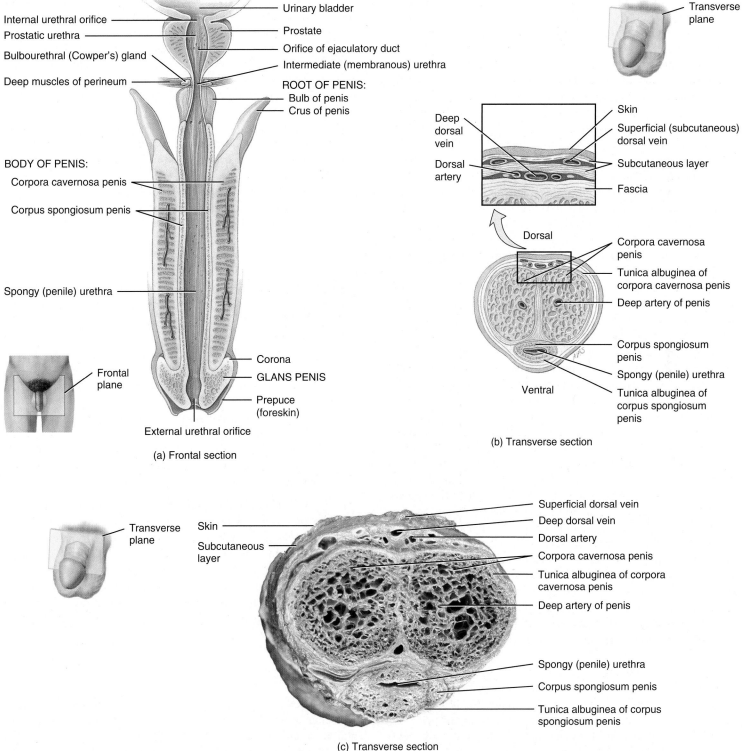

(a) Frontal section

(b) Transverse section

(c) Transverse section

❓ **Which tissue masses form the erectile tissue in the penis, and why do they become rigid during sexual arousal?**

The **root of the penis** is the attached portion (proximal portion). It consists of the **bulb of the penis**, the expanded posterior continuation of the base of the corpus spongiosum penis, and the **crura of the penis** (KROO-ra; singular is *crus*= resembling a leg), the two separated and tapered continuations of the corpora cavernosa penis. The bulb of the penis is attached to the inferior surface of the deep muscles of the perineum and is enclosed by the bulbospongiosus muscle. Contraction of the bulbospongiosus muscle aids ejaculation. Each crus of the penis bends laterally away from the bulb of the penis to attach to the ischial and inferior pubic rami and is surrounded by the ischiocavernosus muscle (see Figure 11.15). The weight of the penis is supported by two ligaments that are continuous with the fascia of the penis: (1) The **fundiform ligament** (FUN-di-form) arises from the inferior part of the linea alba; (2) the **suspensory ligament of the penis** arises from the pubic symphysis.

Upon sexual stimulation (visual, tactile, auditory, olfactory, or imagined), parasympathetic fibers from the sacral portion of the spinal cord initiate and maintain an **erection**, the enlargement and stiffening of the penis. The parasympathetic fibers produce and release and cause local production of nitric oxide (NO). The NO causes smooth muscle in the walls of arterioles supplying erectile tissue to relax, which allows these blood vessels to dilate. This in turn causes large amounts of blood to enter the erectile tissue of the penis. NO also causes the smooth muscle within the erectile tissue to relax, resulting in widening of the blood sinuses. The combination of increased blood flow and widening of the blood sinuses results in an erection. Expansion of the blood sinuses also compresses the veins that drain the penis; the slowing of blood outflow helps to maintain the erection.

The term **priapism** (PRĪ-a-pizm) refers to a persistent and usually painful erection of the penis that does not involve sexual desire or excitement. The condition may last up to several hours and is accompanied by pain and tenderness. It results from abnormalities of blood vessels and nerves, usually in response to medication used to produce erections in males who otherwise cannot attain them. Other causes include a spinal cord disorder, leukemia, sickle-cell disease, or a pelvic tumor.

Ejaculation (ē-jak-ū-LĀ-shun; *ejectus*=to throw out), the powerful release of semen from the urethra to the exterior, is a sympathetic reflex coordinated by the lumbar portion of the spinal cord. As part of the reflex, the smooth muscle sphincter at the base of the urinary bladder closes. Therefore, urine is not expelled during ejaculation, and semen does not normally enter the urinary bladder. Even before ejaculation occurs, peristaltic contractions in the ampulla of the ductus deferens, seminal vesicles, ejaculatory ducts, and prostate propel semen into the penile portion of the urethra (spongy urethra). Typically, this leads to **emission** (ē-MISH-un), the discharge of a small volume of semen before ejaculation. Emission may also occur during sleep (*nocturnal emission*). The musculature of the penis (bulbospongiosus, ischiocavernosus, and superficial transverse perineal muscles), which is supplied by the pudendal nerve, also contracts at ejaculation (see Figure 11.15).

Once sexual stimulation of the penis has ended, the arterioles supplying the erectile tissue of the penis constrict and the smooth muscle within erectile tissue contracts, making the blood sinuses smaller. This relieves pressure on the veins draining the penis and allows the blood to drain through them. Consequently, the penis returns to its flaccid (relaxed) state.

A **premature ejaculation** is ejaculation that occurs too early, for example, during foreplay or upon or shortly after penetration. It is usually caused by anxiety, other psychological causes, or an unusually sensitive foreskin or glans penis. For most males, premature ejaculation can be overcome by various mechanical techniques (such as squeezing the penis between the glans penis and shaft as ejaculation approaches), behavioral therapy, or medication.

The penis has a very rich blood supply from the internal pudendal artery and the femoral artery. The capillaries drain via veins that parallel the arteries and have corresponding names.

The sensory nerves to the penis are branches from the pudendal and ilioinguinal nerves. The corpora have a parasympathetic and a sympathetic supply. As noted previously, parasympathetic stimulation causes the blood vessels to dilate, increasing the flow of blood into the erectile tissue and trapping blood within the penis to maintain the erection. At ejaculation, sympathetic stimulation causes the smooth muscle located in the walls of the ducts and accessory sex glands of the reproductive tract to contract and propel the semen along its course.

 CHECKPOINT

10. What are the locations and functions of the seminal vesicles, the prostate, and the bulbourethral (Cowper's) glands?
11. What is semen? What is its function?
12. Explain the physiological processes involved in erection and ejaculation.

26.2 FEMALE REPRODUCTIVE SYSTEM

 OBJECTIVES

• Describe the location and functions of the ovaries.
• Discuss the process of oogenesis in the ovaries.
• Describe the location and functions of the uterine tubes.
• Describe the location and functions of the uterus.
• Explain the functions of the vagina.
• List and describe the components of the vulva.
• Describe the location and functions of the mammary glands.

The organs of the **female reproductive system** (Figure 26.11) include the ovaries, which produce secondary oocytes and hormones such as progesterone and estrogens (the female sex hormones), inhibin, and relaxin; the uterine (fallopian) tubes, or oviducts, which transport secondary oocytes and fertilized ova to the uterus; the uterus, in which embryonic and fetal development occur; the vagina; and external organs that constitute the vulva, or pudendum. The mammary glands also are considered part of the female reproductive system.

Ovaries

The **ovaries** (=egg receptacles) are paired glands that resemble unshelled almonds in size and shape; they are the female gonads and are homologous to the testes. (Here *homologous* means that two organs have the same embryonic origin.) The ovaries, one

Figure 26.11 Female organs of reproduction and surrounding structures.

🔑 The organs of reproduction in females include the ovaries, uterine (fallopian) tubes, uterus, vagina, vulva, and mammary glands.

FUNCTIONS OF THE FEMALE REPRODUCTIVE SYSTEM

1. The ovaries produce secondary oocytes and hormones, including progesterone and estrogens (female sex hormones), inhibin, and relaxin.
2. The uterine tubes transport a secondary oocyte to the uterus and normally are the sites where fertilization occurs.
3. The uterus is the site of implantation of a fertilized ovum, development of the fetus during pregnancy, and labor.
4. The vagina receives the penis during sexual intercourse and is a passageway for childbirth.
5. The mammary glands synthesize, secrete, and eject milk for nourishment of the newborn.

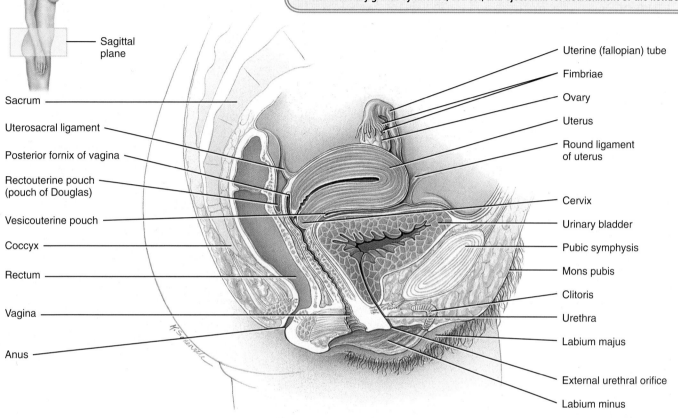

(a) Sagittal section

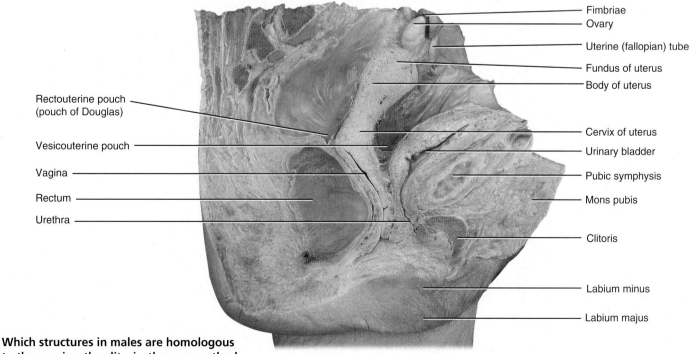

(b) Sagittal section

❓ **Which structures in males are homologous to the ovaries, the clitoris, the paraurethral glands, and the greater vestibular glands?**

on either side of the uterus, descend to the brim of the superior portion of the pelvic cavity during the third month of development. A series of ligaments holds them in position (Figure 26.12). The **broad ligament** of the uterus, which is a fold of the peritoneum, attaches to the ovaries by a subset of this peritoneal fold called the **mesovarium** (mez′-ō-VA-rē-um). The **ovarian ligament** anchors the ovaries to the uterus, and the **suspensory ligament** attaches them to the pelvic wall. Each ovary contains a **hilum** (HĪ-lum), the point of entrance and exit for blood vessels and nerves along which the mesovarium is attached (see Figure 26.13).

Figure 26.12 Relative positions of the ovaries, the uterus, and the ligaments that support them.

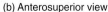 Ligaments holding the ovaries in position include the mesovarium, the ovarian ligament, and the suspensory ligament.

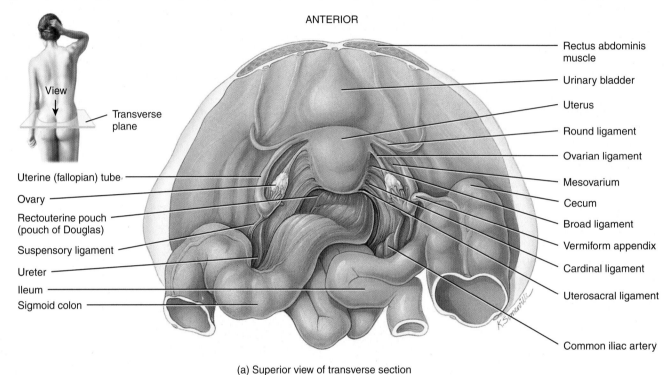

(a) Superior view of transverse section

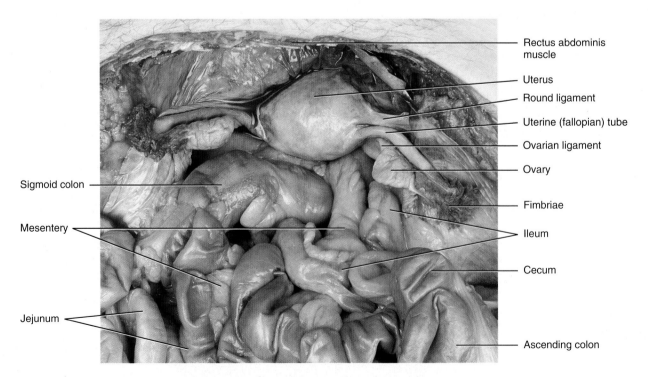

(b) Anterosuperior view

 To which structures do the mesovarium, ovarian ligament, and suspensory ligament anchor the ovary?

Histology of the Ovaries

Each ovary consists of the following parts (Figure 26.13):

- The **germinal epithelium** (*germen*=sprout or bud) is a layer of simple epithelium (low cuboidal or squamous) that covers the surface of the ovary. It is continuous with the mesothelium of the mesovarium and peritoneum. We now know that the term germinal epithelium is inappropriate because the cells of this tissue do not give rise to ova, but at one time people believed that it did. We have since learned that the cells that produce ova arise from the endoderm of the yolk sac and migrate to the ovaries during embryonic development.
- The **tunica albuginea** (al-bū-JIN-ē-a) is a whitish capsule of dense, irregular connective tissue immediately deep to the germinal epithelium.
- The **ovarian cortex** is a region just deep to the tunica albuginea. It consists of ovarian follicles (described shortly) surrounded by dense irregular connective tissue that contains collagen fibers and fibroblast-like cells called stromal cells.
- The **ovarian medulla** is deep to the ovarian cortex. The border between the cortex and medulla is indistinct, but the medulla consists of more loosely arranged connective tissue and contains blood vessels, lymphatic vessels, and nerves.
- **Ovarian follicles** (*folliculus*=little bag) are located in the ovarian cortex and consist of **oocytes** (Ō-ō-sīts) in various stages of development, plus the cells surrounding them. When the surrounding cells form a single layer, they are called **follicular cells** (fō-LIK-ū-lar); later in development, when they form

several layers, they are referred to as **granulosa cells** (gran-ū-LŌ-sa). The surrounding cells nourish the developing oocyte and begin to secrete estrogens as the follicle grows larger.

- A **mature** (*graafian*) **follicle** (GRĀ-fē-an) is a large, fluid-filled follicle that is ready to rupture and expel its secondary oocyte, a process known as **ovulation** (ov-ū-LĀ-shun).
- A **corpus luteum** (=yellow body) contains the remnants of a mature follicle after ovulation. The corpus luteum produces progesterone, estrogens, relaxin, and inhibin until it degenerates into fibrous scar tissue called the **corpus albicans** (AL-bi-kanz=white body).

The ovarian blood supply is furnished by the ovarian arteries, which anastomose with branches of the uterine arteries. The ovaries are drained by the ovarian veins. On the right side they drain into the inferior vena cava, and on the left side they drain into the renal vein. Sympathetic and parasympathetic nerve fibers to the ovaries terminate on the blood vessels and enter the ovaries.

An **ovarian cyst** is a fluid-filled sac in or on an ovary. Such cysts are relatively common, usually noncancerous, and frequently disappear on their own. Cancerous cysts are more likely to occur in women over 40. Ovarian cysts may cause pain, pressure, a dull ache, or fullness in the abdomen; pain during sexual intercourse; delayed, painful, or irregular menstrual periods; abrupt onset of sharp pain in the lower abdomen; and/or vaginal bleeding. Most ovarian cysts require no treatment, but larger ones (more than 5 cm or 2 in.) may be removed surgically.

Figure 26.13 Histology of the ovary. The arrows indicate the sequence of developmental stages that occur as part of the maturation of an ovum during the ovarian cycle.

 The ovaries are the female gonads; they produce haploid oocytes.

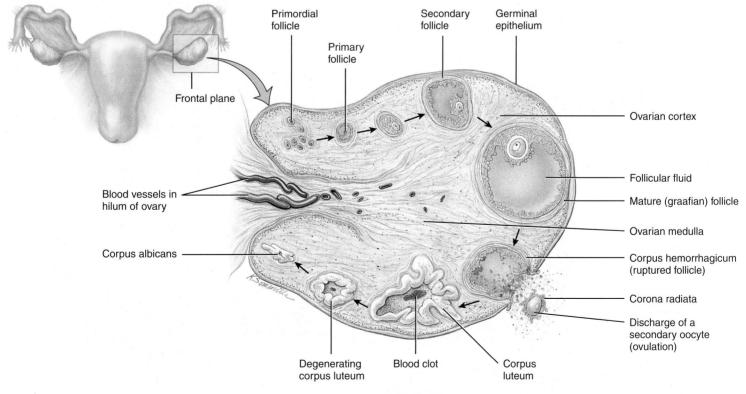

Frontal section

 What structures in the ovary contain endocrine tissue, and what hormones do they secrete?

Oogenesis and Follicular Development

The formation of gametes in the ovaries is termed **oogenesis** (ō-ō-JEN-e-sis; *oo-*=egg). Unlike spermatogenesis, which begins in males at puberty, oogenesis begins in females before they are born. As in spermatogenesis, meiosis (see Section 2.5) takes place and the resulting germ cells undergo maturation.

During early fetal development, primordial (primitive) germ cells migrate from the yolk sac to the ovaries. There, germ cells differentiate within the ovaries into **oogonia** (ō'-ō-GŌ-nē-a; singular is *oogonium*). Oogonia are diploid (*2n*) stem cells that divide mitotically to produce millions of germ cells. Even before birth, most of these germ cells degenerate in a process known as **atresia** (a-TRĒ-zē-a). However, a few develop into larger cells called **primary oocytes** (Ō-ō-sītz) that enter prophase of meiosis I during fetal development but do not complete that phase until after puberty. During this arrested stage of development, each primary oocyte is surrounded by a single layer of flat follicular cells, and the entire structure is called a **primordial follicle** (prī-MOR-dē-al) (Figure 26.14a). The ovarian cortex surrounding the

Figure 26.14 Development of ovarian follicles.

🔑 As an ovarian follicle enlarges, follicular fluid accumulates in a cavity called the antrum.

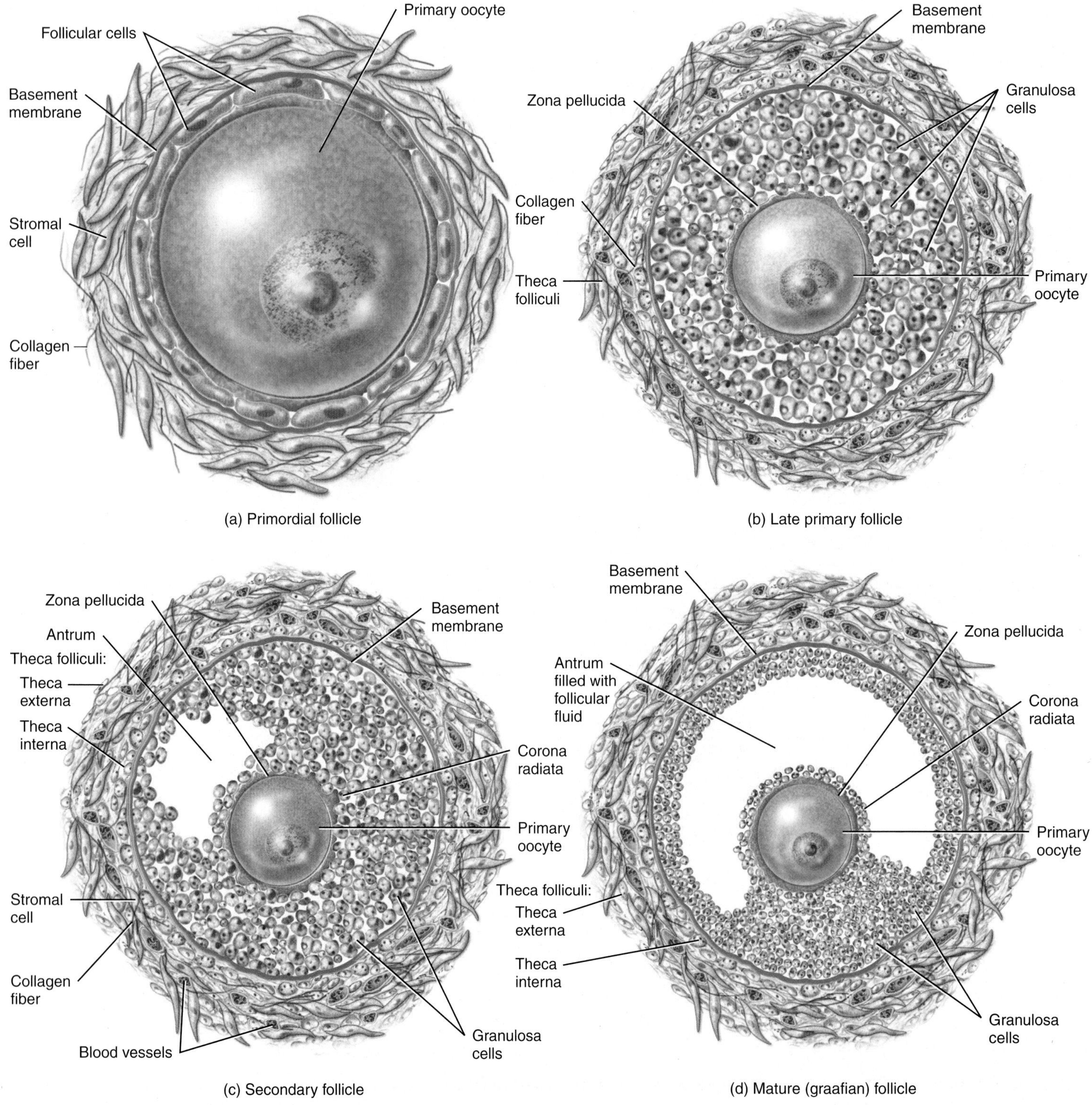

(a) Primordial follicle

(b) Late primary follicle

(c) Secondary follicle

(d) Mature (graafian) follicle

primordial follicles consists of collagen fibers and fibroblast-like **stromal cells**. At birth, approximately 200,000 to 2,000,000 primary oocytes remain in each ovary. Of these, about 40,000 are still present at puberty, and around 400 will mature and ovulate during a woman's reproductive lifetime. The remainder of the primary oocytes undergo atresia.

Each month after puberty until menopause, gonadotropins (FSH and LH) secreted by the anterior pituitary further stimu-

late the development of several primordial follicles, although only one will typically reach the maturity needed for ovulation. A few primordial follicles start to grow, developing into **primary follicles** (Figure 26.14b). Each primary follicle consists of a primary oocyte that is surrounded in a later stage of development by several layers of cuboidal and low-columnar cells called **granulosa cells** (gran-ū-LŌ-sa). The outermost granulosa cells rest on a basement membrane. As the primary follicle grows, it forms

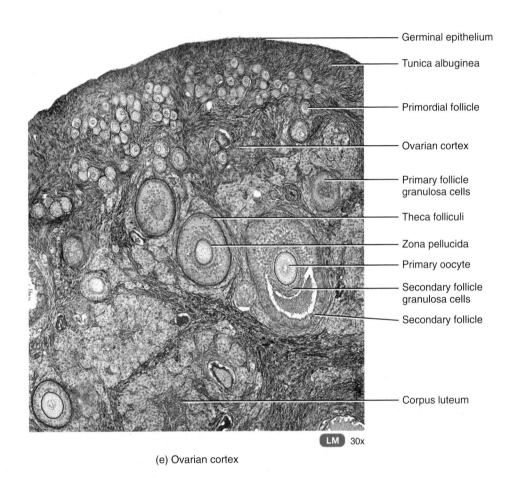

Germinal epithelium
Tunica albuginea
Primordial follicle
Ovarian cortex
Primary follicle granulosa cells
Theca folliculi
Zona pellucida
Primary oocyte
Secondary follicle granulosa cells
Secondary follicle
Corpus luteum

LM 30x

(e) Ovarian cortex

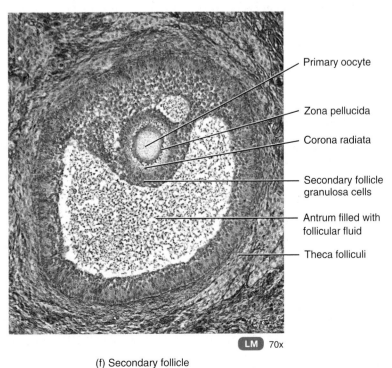

Primary oocyte
Zona pellucida
Corona radiata
Secondary follicle granulosa cells
Antrum filled with follicular fluid
Theca folliculi

LM 70x

(f) Secondary follicle

 What happens to most ovarian follicles?

a clear glycoprotein layer called the **zona pellucida** (pe-LOO-si-da) between the primary oocyte and the granulosa cells. In addition, stromal cells surrounding the basement membrane begin to form an organized layer called the **theca folliculi** (THĒ-ka fo-LIK-ū-lī).

With continuing maturation, a primary follicle develops into a secondary follicle (Figure 26.14c, f). In a **secondary follicle**, the theca differentiates into two layers: (1) the **theca interna**, a highly vascularized internal layer of cuboidal secretory cells that secrete estrogens, and (2) the **theca externa**, an outer layer of stromal cells and collagen fibers. In addition, the granulosa cells begin to secrete follicular fluid, which builds up in a cavity called the **antrum** in the center of the secondary follicle. The innermost layer of granulosa cells becomes firmly attached to the zona pellucida and is now called the **corona radiata** (*corona*=crown; *radiata*=radiation) (Figure 26.14c, f).

The secondary follicle eventually becomes larger, turning into a **mature** (*graafian*) **follicle** (Figure 26.14d). While in this follicle, and just before ovulation, the diploid primary oocyte completes meiosis I, producing two haploid (*n*) cells of unequal size—each with 23 chromosomes (Figure 26.15). The smaller cell produced by meiosis I, called the **first polar body**, is essentially a packet of discarded nuclear material. The larger cell, known as the **secondary oocyte**, receives most of the cytoplasm. Once a secondary oocyte is formed, it begins meiosis II but then stops in metaphase. The mature (graafian) follicle soon ruptures and releases its secondary oocyte, a process known as **ovulation**.

At ovulation, the secondary oocyte is expelled into the peritoneal cavity together with the first polar body and corona radiata. Normally these cells are swept into the uterine tube. If fertilization does not occur, the cells degenerate. However, if sperm are present in the uterine tube and one penetrates the secondary oocyte, meiosis II resumes. The secondary oocyte splits into two haploid cells, again of unequal size. The larger cell is the **ovum**, or mature egg; the smaller one is the **second polar body**. The nuclei of the sperm cell and the ovum then unite, forming a diploid **zygote**. If the first polar body undergoes another division to produce two polar bodies, then the primary oocyte ultimately gives rise to three haploid polar bodies, which all degenerate, and a single haploid ovum. Thus, one primary oocyte gives rise to a single gamete (an ovum). By contrast, recall that in males one primary spermatocyte produces four gametes (sperm).

Table 26.1 summarizes the events of oogenesis and follicular development.

CHECKPOINT

13. What is the male homolog to the ovary?
14. What is the microscopic structure of an ovary, and what are its functions?
15. How does oogenesis transform a germ cell into a mature follicle?

Figure 26.15 Oogenesis. Diploid cells (*2n*) have 46 chromosomes; haploid cells (*n*) have 23 chromosomes.

 In an oocyte, meiosis II is completed only if fertilization occurs.

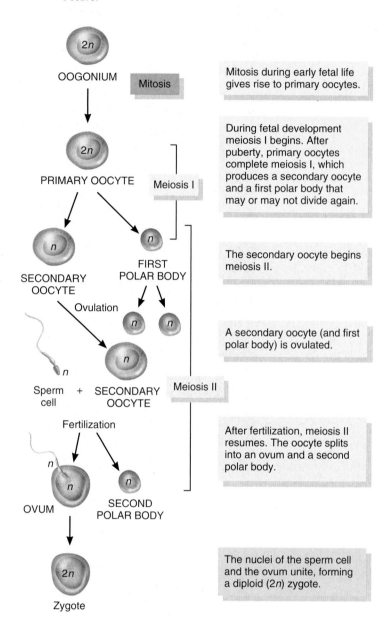

Mitosis during early fetal life gives rise to primary oocytes.

During fetal development meiosis I begins. After puberty, primary oocytes complete meiosis I, which produces a secondary oocyte and a first polar body that may or may not divide again.

The secondary oocyte begins meiosis II.

A secondary oocyte (and first polar body) is ovulated.

After fertilization, meiosis II resumes. The oocyte splits into an ovum and a second polar body.

The nuclei of the sperm cell and the ovum unite, forming a diploid (*2n*) zygote.

How does the age of a primary oocyte in a female compare with the age of a primary spermatocyte in a male?

TABLE 26.1

Summary of Oogenesis and Follicular Development

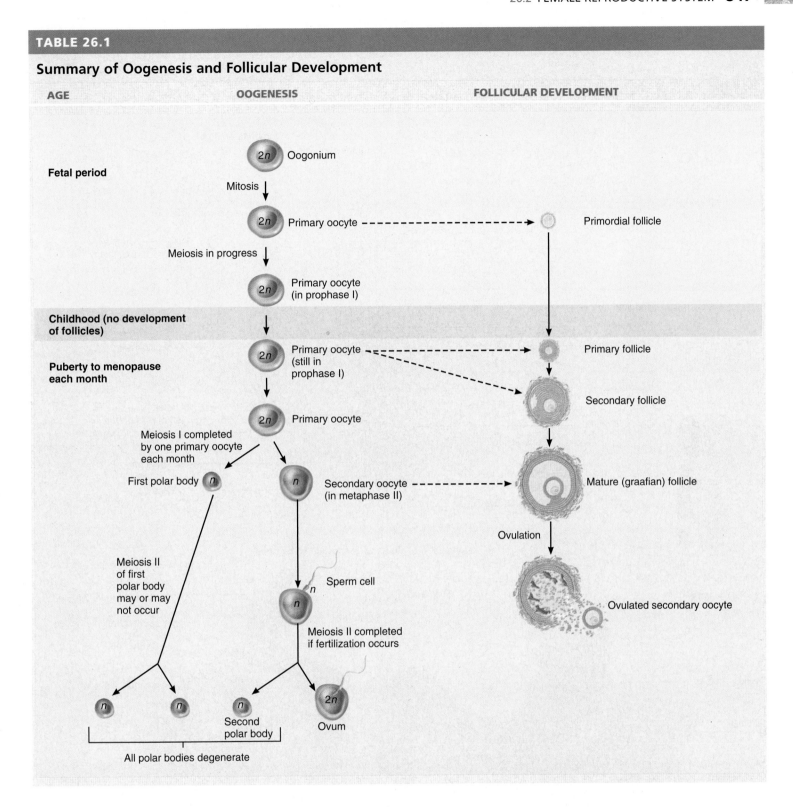

Uterine Tubes

Females have two **uterine tubes**, also called *fallopian tubes* or *oviducts*, that extend laterally from the uterus (Figure 26.16). The tubes, which measure about 10 cm (4 in.) long and lie within the folds of the broad ligaments of the uterus, transport secondary oocytes and fertilized ova from the ovaries to the uterus. (This portion of the broad ligament is called the *mesosalpinx* [mez′-ō-SAL-pinks].) The funnel-shaped portion of each uterine tube, called the **infundibulum** (in′-fun-DIB-ū-

lum), is close to the ovary but is open to the peritoneal cavity. It ends in a fringe of fingerlike projections called **fimbriae** (FIM-brē-ē = fringe), one of which is attached to the lateral end of the ovary. From the infundibulum, the uterine tube extends medially and eventually inferiorly and attaches to the superior lateral angle of the uterus. The **ampulla** of the uterine tube is the widest, longest portion, making up about the lateral two-thirds of its length. The **isthmus** (IS-mus) of the uterine tube is the more medial, short, narrow, thick-walled portion that joins the uterus.

Figure 26.16 **Relationship of the uterine (fallopian) tubes to the ovaries, uterus, and associated structures.** In the left side of the drawing the uterine tube and uterus have been sectioned to show internal structures.

🔑 After ovulation, a secondary oocyte and its corona radiata move from the pelvic cavity into the infundibulum of the uterine tube. The uterus is the site of menstruation, implantation of a fertilized ovum, development of the fetus, and labor.

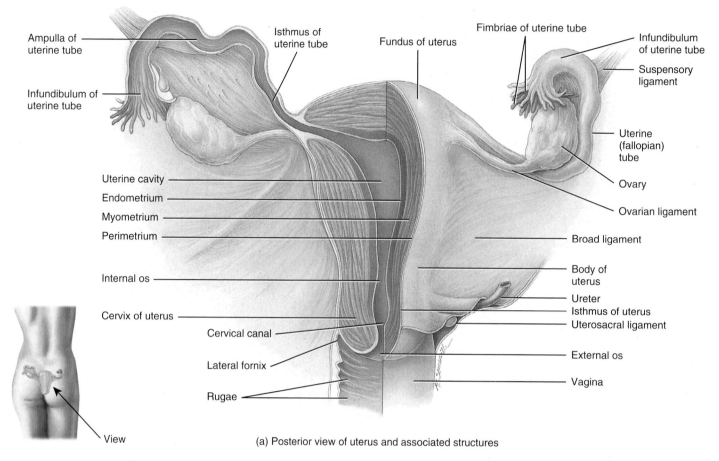

Ampulla of uterine tube

Isthmus of uterine tube

Fundus of uterus

Fimbriae of uterine tube

Infundibulum of uterine tube

Suspensory ligament

Infundibulum of uterine tube

Uterine (fallopian) tube

Ovary

Uterine cavity

Endometrium

Myometrium

Perimetrium

Ovarian ligament

Broad ligament

Body of uterus

Internal os

Ureter

Isthmus of uterus

Uterosacral ligament

Cervix of uterus

Cervical canal

External os

Lateral fornix

Vagina

Rugae

View

(a) Posterior view of uterus and associated structures

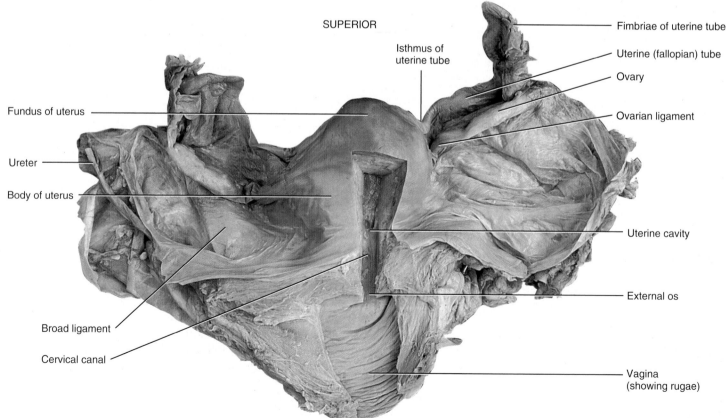

SUPERIOR

Isthmus of uterine tube

Fimbriae of uterine tube

Uterine (fallopian) tube

Ovary

Fundus of uterus

Ovarian ligament

Ureter

Body of uterus

Uterine cavity

External os

Broad ligament

Cervical canal

Vagina (showing rugae)

INFERIOR

(b) Posterior view of uterus and associated structures

❓ **Where does fertilization usually occur?**

CLINICAL CONNECTION | *Uterine Prolapse*

A condition called **uterine prolapse** (PRŌ-laps; *prolapse*=falling down or downward displacement) may result from weakening of supporting ligaments and pelvic musculature associated with age or disease, traumatic vaginal delivery, chronic straining from coughing or difficult bowel movements, or pelvic tumors. The prolapse may be characterized as first degree (mild), in which the cervix remains within the vagina; second degree (marked), in which the cervix protrudes through the vagina to the exterior; or third degree (complete), in which the entire uterus is outside the vagina. Depending on the degree of prolapse, treatment may involve pelvic exercises, dieting if a patient is overweight, a stool softener to minimize straining during defecation, pessary therapy (placement of a rubber device around the uterine cervix that helps prop up the uterus), or surgery. •

Histologically, the uterine tubes are composed of three layers: mucosa, muscularis, and serosa (Figure 26.17a). The mucosa consists of epithelium and lamina propria (areolar connective tissue). The epithelium contains ciliated simple columnar cells, which function as a ciliary "conveyor belt" to help move a fertilized ovum (or secondary oocyte) within the tube toward the uterus, and nonciliated cells called **peg cells**, which have microvilli and secrete a fluid that provides nutrition for the ovum (Figure 26.17b). The middle layer, the muscularis, is composed of an inner, thick, circular ring of smooth muscle and an outer, thin region of longitudinal smooth muscle. Peristaltic contractions of the muscularis and the ciliary action of the mucosa help move the oocyte or fertilized ovum toward the uterus. The outer layer of the uterine tubes is a serous membrane, the serosa.

After ovulation, local currents are produced by movements of the fimbriae, which surround the surface of the mature follicle

Figure 26.17 Histology of the uterine tube.

Peristaltic contractions of the muscularis and ciliary action of the mucosa of the uterine tube help move the oocyte or fertilized ovum toward the uterus.

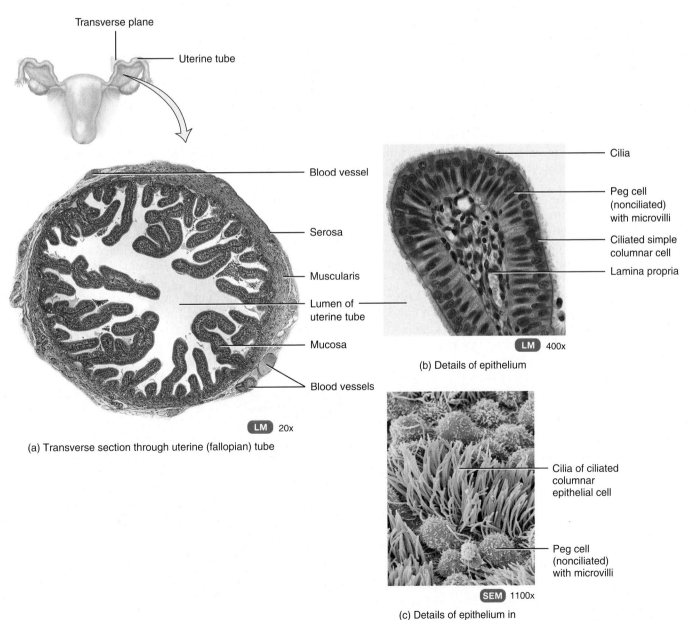

Transverse plane

Uterine tube

Blood vessel

Serosa

Muscularis

Lumen of uterine tube

Mucosa

Blood vessels

LM 20x

(a) Transverse section through uterine (fallopian) tube

Cilia

Peg cell (nonciliated) with microvilli

Ciliated simple columnar cell

Lamina propria

LM 400x

(b) Details of epithelium

Cilia of ciliated columnar epithelial cell

Peg cell (nonciliated) with microvilli

SEM 1100x

(c) Details of epithelium in surface view

What types of cells line the uterine tubes?

just before ovulation occurs. These currents sweep the ovulated secondary oocyte from the peritoneal cavity into the uterine tube. A sperm cell usually encounters and fertilizes a secondary oocyte in the ampulla of the uterine tube, although fertilization in the peritoneal cavity is not uncommon. Fertilization can occur up to about 24 hours after ovulation. Some hours after fertilization, the nuclear materials of the haploid ovum and sperm unite. The diploid fertilized ovum is now called a **zygote** and begins to undergo cell divisions while moving toward the uterus. It arrives in the uterus 6 to 7 days after ovulation. Unfertilized secondary oocytes disintegrate.

The uterine tubes are supplied by branches of the uterine arteries (see Figure 26.19) and ovarian arteries (see Figure 14.6b). Venous return is via the uterine veins.

The uterine tubes are supplied with sympathetic and parasympathetic nerve fibers from the hypogastric plexus and the pelvic splanchnic nerves. The fibers are distributed to the muscular coat of the tubes and their blood vessels.

Uterus

The **uterus** (womb) serves as part of the pathway for sperm deposited in the vagina to reach the uterine tubes (see Figure 26.16). It is also the site of implantation of a fertilized ovum, development of the fetus during pregnancy, and labor. During reproductive cycles when implantation does not occur, the uterus is the source of menstrual flow.

Situated between the urinary bladder and the rectum, the uterus is the size and shape of an inverted pear (see Figure 26.11). In females who have never been pregnant, it is about 7.5 cm (3 in.) long, 5 cm (2 in.) wide, and 2.5 cm (1 in.) thick. The uterus is larger in females who have recently been pregnant, and smaller (atrophied) when sex hormone levels are low, as occurs after menopause.

Anatomical subdivisions of the uterus include the following (see Figure 26.16): (1) a dome-shaped portion superior to the uterine tubes called the **fundus**, (2) a tapering central portion called the **body**, and (3) an inferior narrow portion called the **cervix** that opens into the vagina. Between the body of the uterus and the cervix is the **isthmus**, a constricted region about 1 cm (0.5 in.) long. The interior of the body of the uterus is called the **uterine cavity**, and the interior of the narrow cervix is called the **cervical canal**. The cervical canal opens into the uterine cavity at the **internal os** (os=mouthlike opening) and into the vagina at the **external os**.

Normally, the body of the uterus projects anteriorly and superiorly over the urinary bladder in a position called **anteflexion** (an'-tē-FLEK-shun; ante=before). The cervix projects inferiorly and posteriorly and enters the anterior wall of the vagina at nearly a right angle (see Figure 26.11). Several ligaments that are either extensions of the parietal peritoneum or fibromuscular cords maintain the position of the uterus (see Figure 26.12). The paired **broad ligaments** are double folds of peritoneum attaching the uterus to either side of the pelvic cavity. The paired **uterosacral ligaments** (ū-ter-ō-SĀ-kral), also peritoneal extensions, lie on either side of the rectum and connect the uterus to the sacrum. The **cardinal (lateral cervical) ligaments** are located inferior to the bases of the broad ligaments and extend from the pelvic wall to the cervix and vagina. The **round ligaments** are bands of fibrous connective tissue between the layers of the broad ligament; they extend from a point on the uterus just inferior to the uterine tubes to a

portion of the labia majora of the external genitalia. Although the ligaments normally maintain the anteflexed position of the uterus, they also allow the uterine body enough movement that the uterus may become malpositioned.

A posterior tilting of the uterus, called **retroflexion** (ret-rō-FLEK-shun; retro-=backward or behind), is a harmless variation of the normal position of the uterus. There is often no cause for the condition, but it may occur after childbirth or because of an ovarian cyst.

Histologically, the uterus consists of three layers of tissue: the perimetrium, myometrium, and endometrium (Figure 26.18). The outer layer—the **perimetrium** (per'-i-MĒ-trē-um; peri-= around; metrium=uterus) or serosa—is part of the peritoneum; it is composed of simple squamous epithelium and a thin layer of areolar connective tissue. Laterally, the peritoneum becomes the broad ligament. Anteriorly, the peritoneum covers the urinary bladder and forms a shallow pouch, the **vesicouterine pouch** (ves'-i-kō-Ū-ter-in; vesico-=bladder; see Figure 26.11) between the urinary bladder and the uterus. Posteriorly, the peritoneum covers the rectum and forms a deep pouch between the uterus and the rectum, the **rectouterine pouch** (rek-tō-Ū-ter-in; recto-=rectum) or pouch of Douglas—the most inferior point in the peritoneal cavity.

The middle layer of the uterus, the **myometrium** (myo-= muscle), consists of three layers of smooth muscle fibers that are thickest in the fundus and thinnest in the cervix. The thicker middle layer is circular; the inner and outer layers are longitudinal or oblique. During labor and childbirth, coordinated contractions of the myometrium in response to stimulation by the hormone oxytocin from the posterior pituitary help expel the fetus from the uterus.

The inner layer of the uterus, the **endometrium** (endo-= within), is highly vascularized and has three components: (1) An innermost layer composed of simple columnar epithelium (ciliated and secretory cells) lines the lumen. (2) An underlying endometrial stroma is a very thick region of lamina propria (areolar connective tissue). (3) Endometrial (uterine) glands develop as invaginations of the luminal epithelium and extend almost to the myometrium. The endometrium is divided into two layers. The **stratum functionalis** (=functional layer) lines the uterine cavity and sloughs off during menstruation as a result of declining levels of progesterone from the ovaries. The deeper layer, the **stratum basalis** (=basal layer), is permanent and gives rise to a new stratum functionalis after each menstruation.

CLINICAL CONNECTION | *Endometriosis*

Endometriosis (en-dō-mē-trē-Ō-sis; endo-=within; metri-= uterus; osis=condition) is characterized by the growth of endometrial tissue outside the uterus. The tissue enters the pelvic cavity via the open uterine tubes and may be found in any of several sites—on the ovaries, the rectouterine pouch, the outer surface of the uterus, the sigmoid colon, pelvic and abdominal lymph nodes, the cervix, the abdominal wall, the kidneys, and the urinary bladder. Endometrial tissue responds to hormonal fluctuations, whether it is inside or outside the uterus. With each reproductive cycle, the tissue proliferates and then breaks down and bleeds. When this occurs outside the uterus, it can cause inflammation, pain, scarring, and infertility. Symptoms include premenstrual pain or unusually severe menstrual pain. •

Figure 26.18 Histology of the uterus.

🔑 The three layers of the uterus from superficial to deep are the perimetrium (serosa), the myometrium, and the endometrium.

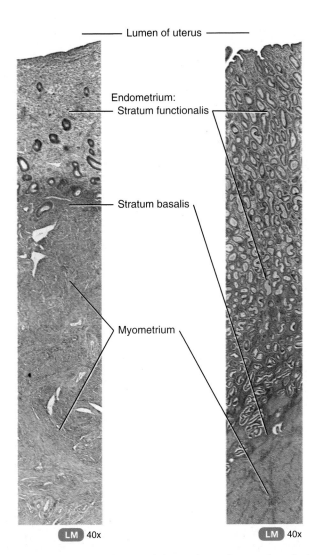

Lumen of uterus

Endometrium:
— Stratum functionalis

— Stratum basalis

Myometrium

LM 40x   LM 40x

(a) Transverse section through the uterine wall: second week of menstrual cycle (left) and third week of menstrual cycle (right)

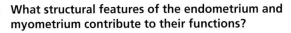

❓ **What structural features of the endometrium and myometrium contribute to their functions?**

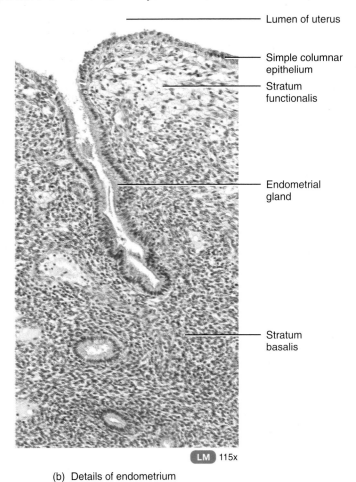

Lumen of uterus

Simple columnar epithelium

Stratum functionalis

Endometrial gland

Stratum basalis

LM 115x

(b) Details of endometrium

Branches of the internal iliac artery called **uterine arteries** (Figure 26.19) supply blood to the uterus. Uterine arteries give off branches called **arcuate arteries** (AR-kū-āt=shaped like a bow) that are arranged in a circular fashion in the myometrium. These arteries branch into **radial arteries** that penetrate deeply into the myometrium. Just before the branches enter the endometrium, they divide into two kinds of arterioles: **Straight arterioles** supply the stratum basalis with the materials needed to regenerate the stratum functionalis; **spiral arterioles** supply the stratum functionalis and change markedly during the menstrual cycle. Blood leaving the uterus is drained by the **uterine veins** into the internal iliac veins. The extensive blood supply of the uterus is essential to support regrowth of a new stratum functionalis after menstruation, implantation of a fertilized ovum, and development of the placenta.

The secretory cells of the mucosa of the cervix produce a secretion called **cervical mucus**, a mixture of water, glycoproteins, lipids, enzymes, and inorganic salts. During their reproductive years, females secrete 20–60 mL of cervical mucus per day.

Cervical mucus is more hospitable to sperm at or near the time of ovulation because it is then less viscous and more alkaline (pH 8.5). At other times, a more viscous mucus forms a cervical plug that physically impedes sperm penetration. Cervical mucus supplements the energy needs of sperm, and both the cervix and cervical mucus protect sperm from phagocytes and the hostile environment of the vagina and uterus. They may also play a role in **capacitation** (ka-pas'-i-TĀ'-shun)—a functional change that sperm undergo in the female reproductive tract before they are able to fertilize a secondary oocyte. Capacitation causes a sperm cell's tail to beat even more vigorously and it prepares the sperm cell's plasma membrane to fuse with the oocyte's plasma membrane.

CLINICAL CONNECTION | *Cervical Cancer*

Cervical cancer, carcinoma of the cervix of the uterus, starts with cervical dysplasia (dis-PLĀ-sē-a), a change in the shape, growth, and number of cervical cells. The cells may either return to normal or progress to cancer. In most cases, cervical cancer may be detected in its earliest stages by a Pap test (see Clinical Connection in Section 3.4). Some evidence links cervical cancer to the virus that causes genital warts, human papillomavirus (HPV). Increased risk is associated with having a large number of sexual partners, having first intercourse at a young age, and smoking cigarettes. •

Figure 26.19 Blood supply of the uterus. The inset shows histological details of the blood vessels of the endometrium.

 Straight arterioles supply the materials needed for regeneration of the stratum functionalis after menstruation.

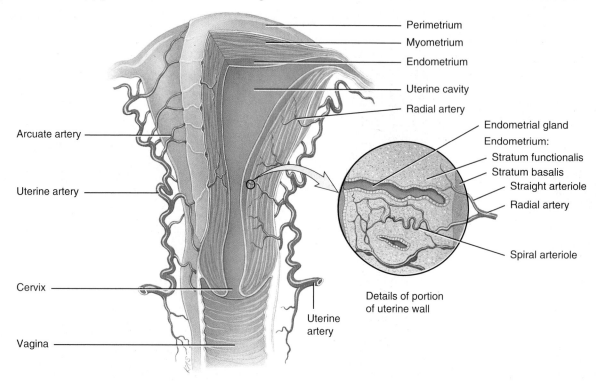

Anterior view with left side of uterus partially sectioned

 **What is the functional significance of the stratum basalis layer of the endometrium?**

CHECKPOINT

16. Where are the uterine tubes located and what is their function?
17. What are the principal parts of the uterus? Where are they located in relation to one another?
18. How do ligaments hold the uterus in its normal position?
19. What is the microscopic structure of the uterus?
20. Why is an abundant blood supply important to the uterus?

Vagina

The **vagina** (=sheath) is a tubular, fibromuscular canal lined with mucous membrane that extends from the exterior of the body to the uterine cervix (see Figures 26.11 and 26.16). It is about 10 cm (4 in.) long, and serves as the receptacle for the penis during sexual intercourse, the outlet for menstrual flow, and the passageway for childbirth. Situated posterior to the urinary bladder and urethra and anterior to the rectum, the vagina is directed superiorly and posteriorly, to attach to the cervix of the uterus. A recess called the **fornix** (=arch or vault) surrounds the vaginal attachment to the cervix (see Figure 26.16a). When properly inserted, a contraceptive diaphragm rests in the fornix, where it is held in place as it covers the cervix.

The mucosa of the vagina is continuous with that of the uterus. Histologically, it consists of nonkeratinized stratified squamous epithelium and lamina propria (areolar connective tissue)

(Figure 26.20b) that lies in a series of transverse folds called **rugae** (ROO-gē; see Figure 26.16). Dendritic cells in the mucosa are antigen-presenting cells. Unfortunately, they also participate in the transmission of viruses—for example, HIV (the virus that causes AIDS)—to a female during intercourse with an infected male. The mucosa of the vagina contains large stores of glycogen, the decomposition of which produces organic acids. The resulting acidic environment retards microbial growth, but it is also harmful to sperm. Alkaline components of semen, mainly from the seminal vesicles, raise the pH of fluid in the vagina and increase viability of the sperm.

The muscularis is composed of an outer longitudinal layer and an inner circular layer of smooth muscle (Figure 26.20a) that can stretch considerably to accommodate the penis during sexual intercourse and an infant during birth.

The adventitia, the superficial layer of the vagina, consists of areolar connective tissue (Figure 26.20a). It anchors the vagina to adjacent organs such as the urethra and urinary bladder anteriorly, and the rectum and anal canal posteriorly.

A thin fold of vascularized mucous membrane, called the **hymen** (=membrane), forms a border around and partially closes the inferior end of the vaginal opening to the exterior, the **vaginal orifice** (see Figure 26.21). After its rupture, usually following the first sexual intercourse, only remnants of the hymen remain.

Sometimes the hymen completely covers the orifice, a condition called **imperforate hymen** (im-PER-fō-rāt). Surgery may be needed to open the orifice and permit the discharge of menstrual flow.

Figure 26.20 Histology of the vagina.

🔑 The muscularis of the vagina can stretch considerably to accommodate the penis during sexual intercourse and an infant during birth.

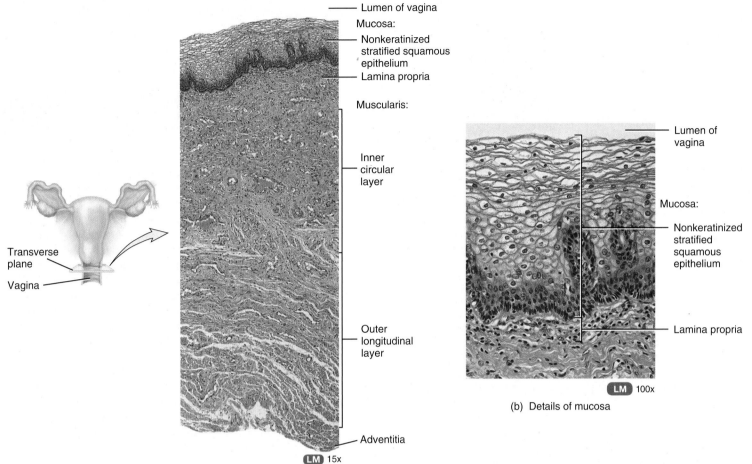

(a) Transverse section through the vaginal wall

(b) Details of mucosa

❓ **What are the functions of the vagina?**

Vulva

The term **vulva** (VUL-va = to wrap around), or **pudendum** (pū-DEN-dum), refers to the external genitals of the female (Figure 26.21). The following are the components of the vulva:

- Anterior to the vaginal and urethral openings is the **mons pubis** (MONZ PŪ-bis; *mons* = mountain), an elevation of adipose tissue covered by skin and coarse pubic hair that cushions the pubic symphysis.
- From the mons pubis, two longitudinal folds of skin, the **labia majora** (LĀ-bē-a ma-JŌ-ra; *labia* = lips; *majora* = larger; singular is *labium majus*), extend inferiorly and posteriorly. The labia majora are covered by pubic hair and contain an abundance of adipose tissue, sebaceous (oil) glands, and apocrine sudoriferous (sweat) glands. They are homologous to the scrotum.
- Medial to the labia majora are two smaller folds of skin called the **labia minora** (mī-NŌ-ra; *minora* = smaller; singular is *labium minus*). Unlike the labia majora, the labia minora are devoid of pubic hair and fat and have few sudoriferous glands, but they do contain many sebaceous glands. The labia minora are homologous to the spongy (penile) urethra.
- The **clitoris** (KLI-to-ris) is a small cylindrical mass of erectile tissue and nerves located at the anterior junction of the labia minora. A layer of skin called the **prepuce of the clitoris** (PRĒ-pus) is formed at the point where the labia minora unite and covers

the body of the clitoris. The body of the clitoris has two bodies of erectile tissue, the corpora cavernosa. Like the similar bodies in the male penis, the corpora cavernosa bend posteriorly to attach to the rami of the pubis and ischium as the crura of the clitoris. The exposed portion of the clitoris is the **glans clitoris**. The glans clitoris is homologous to the glans penis in males. Like the male structure, the clitoris is capable of enlargement upon tactile stimulation and has a role in sexual excitement in the female.

- The region between the labia minora is the **vestibule**. Within the vestibule are the hymen (if still present), the vaginal orifice, the external urethral orifice, and the openings of the ducts of several glands. The vestibule is homologous to the intermediate urethra of males. The **vaginal orifice**, the opening of the vagina to the exterior, occupies the greater portion of the vestibule and is bordered by the hymen. Anterior to the vaginal orifice and posterior to the clitoris is the **external urethral orifice**, the opening of the urethra to the exterior. On either side of the external urethral orifice are the openings of the ducts of the **paraurethral glands** (par'-a-ū-RĒ-thral) or *Skene's glands* (SKENS). These mucus-secreting glands are embedded in the wall of the urethra. The paraurethral glands are homologous to the prostate. On either side of the vaginal orifice itself are the **greater vestibular glands** or *Bartholin's glands* (BAR-tō-lins) (see Figure 26.22), which open by ducts into a groove between the hymen and labia minora. They produce a small quantity of mucus during sexual arousal and intercourse that adds to cervical mucus and provides lubrication. The greater vestibular glands are homologous to the bulbourethral glands in males. Several **lesser vestibular glands** also open into the vestibule.

Figure 26.21 Vulva (pudendum).

 The vulva refers to the external genitals of the female.

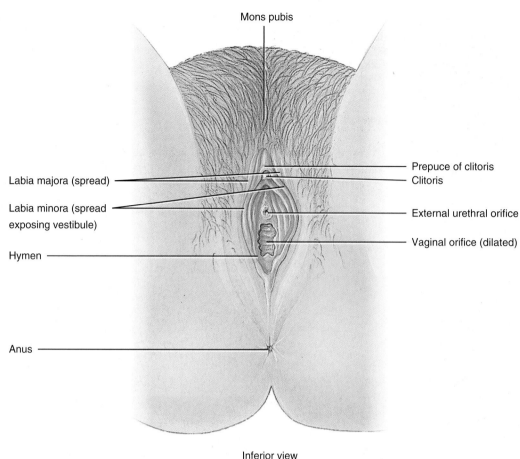

Inferior view

CLINICAL CONNECTION | *Episiotomy*

During childbirth, the emerging fetus normally stretches the perineal region. However, if it appears that the stretching could be excessive, a physician may elect to perform an **episiotomy** (e-piz-ē-OT-ō-mē; *episi-* =vulva or pubic region; *-otomy*= incision), a perineal cut between the vagina and anus made with surgical scissors to widen the birth canal. The cut is made along the midline or at about a 45° angle to the midline. Reasons for an episiotomy include a very large fetus, breech presentation (buttocks or lower limbs coming first), fetal distress (such as an abnormal heart rate), forceps delivery, or a short perineum. Following delivery, the incision is closed in layers with sutures that are absorbed within a few weeks.

? **What surface structures are anterior to the vaginal opening? Lateral to it?**

• The **bulb of the vestibule** (see Figure 26.22) consists of two elongated masses of erectile tissue just deep to the labia on either side of the vaginal orifice. The bulb of the vestibule becomes engorged with blood during sexual arousal, narrowing the vaginal orifice and placing pressure on the penis during intercourse. The bulb of the vestibule is homologous to the corpus spongiosum and bulb of the penis in males.

Table 26.2 summarizes the homologous structures of the female and male reproductive systems.

Perineum

The **perineum** (per'-i-NĒ-um) is the diamond-shaped area medial to the thighs and buttocks of both males and females (Figure 26.22). It contains the external genitals and anus. The perineum is bounded anteriorly by the pubic symphysis, laterally by the ischial tuberosities, and posteriorly by the coccyx. A transverse line drawn between the ischial tuberosities divides the perineum into an anterior **urogenital triangle** (ū'-rō-JEN-i-tal) that encloses the urethral (*uro-*) and vaginal (*-genital*) orifices and a posterior **anal triangle** that contains the anus.

TABLE 26.2

Summary of Homologous Structures of Female and Male Reproductive Systems

EMBRYONIC STRUCTURE (See Figures 26.26 and 26.27)	FEMALE STRUCTURES	MALE STRUCTURES
Gonadal ridge	Ovaries	Testes
Yolk sac endoderm	Ovum	Sperm cell
Labioscrotal swellings	Labia majora	Scrotum
Urethral folds	Labia minora	Spongy (penile) urethra
Urogenital sinus	Vestibule Paraurethral glands Greater vestibular glands	Intermediate (membranous) urethra Prostate Bulbourethral (Cowper's) glands
Genital tubercle	Bulb of vestibule Clitoris	Corpus spongiosum penis and bulb of penis Glans penis and corpora cavernosa

Figure 26.22 Perineum of a female. (Figure 11.15 shows the perineum of a male.)

🔑 The perineum is a diamond-shaped area medial to the thighs and buttocks that contains the external genitals and anus.

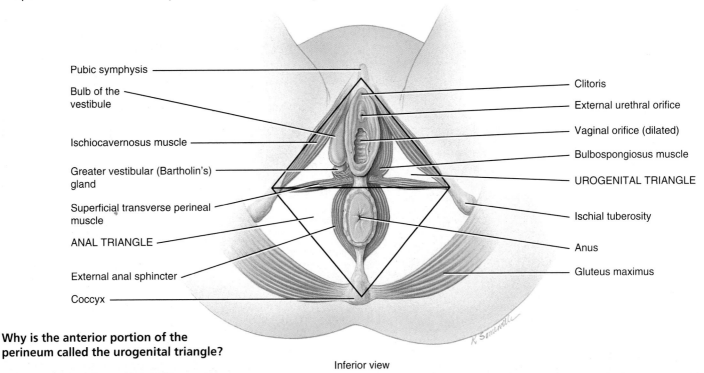

Pubic symphysis

Bulb of the vestibule

Ischiocavernosus muscle

Greater vestibular (Bartholin's) gland

Superficial transverse perineal muscle

ANAL TRIANGLE

External anal sphincter

Coccyx

Clitoris

External urethral orifice

Vaginal orifice (dilated)

Bulbospongiosus muscle

UROGENITAL TRIANGLE

Ischial tuberosity

Anus

Gluteus maximus

Inferior view

❓ **Why is the anterior portion of the perineum called the urogenital triangle?**

CLINICAL CONNECTION | *Breast Cancer*

One in eight women in the United States faces the prospect of **breast cancer**, the second-leading cause of death from cancer in U.S. women. An estimated 5 percent of the 180,000 cases diagnosed each year in the United States stem from inherited genetic mutations (changes in the DNA). Researchers have identified two genes that increase susceptibility to breast cancer: *BRCA1 (breast cancer 1)* and *BRCA2*. In addition, mutations of the *p53* gene increase the risk of breast cancer in both males and females, and mutations of the androgen receptor gene are associated with the occurrence of breast cancer in some males. Because breast cancer generally is not painful until it becomes quite advanced, any lump, no matter how small, should be reported to a physician at once. Early detection—by breast self-examination and mammograms—is the best way to increase the chance of survival.

The most effective technique for detecting tumors less than 1 cm (0.4 in.) in diameter is **mammography** (mam-OG-ra-fē; *graphy*=to record), a type of radiography using very sensitive x-ray film. The image of the breast, called a **mammogram** (see Table 1.3), is best obtained by compressing the breasts, one at a time, using flat plates. A supplementary procedure for evaluating breast abnormalities is **ultrasound**. Although ultrasound cannot detect tumors smaller than 1 cm in diameter, it can help determine whether a lump is a benign, fluid-filled cyst or a solid (and therefore possibly malignant) tumor.

Among the factors that increase the risk of developing breast cancer are (1) a family history of breast cancer, especially in a mother or sister; (2) nulliparity (never having borne a child) or having a first child after age 35; (3) previous cancer in one breast; (4) exposure to ionizing radiation, such as x-rays; (5) excessive alcohol intake; and (6) cigarette smoking.

The American Cancer Society recommends the following step guidelines:

- All women over age 20 should develop the habit of monthly breast self-examination.
- Breasts should be examined by a physician every 3 years between ages 20 and 40, and every year after age 40.
- A baseline mammogram should be taken in women between ages 35 and 39.

- Women with no symptoms should have a mammogram every year after age 40.
- Women of any age with a history of breast cancer, a strong family history of the disease, or other risk factors should consult a physician to determine a schedule for mammography.

In November 2009, the United States Preventive Services Task Force (USPSTF) issued a series of recommendations relative to breast cancer screening for females at normal risk for breast cancer:

- Women aged 50–74 should have a mammogram every two years.
- Women over 75 should not have mammograms.
- Breast self-examination is not required.

Treatment may involve hormone therapy, chemotherapy, radiation therapy, **lumpectomy** (lump-EK-tō-mē) (removal of the tumor and the immediate surrounding tissue), a modified or radical mastectomy, or a combination of these approaches. A **radical mastectomy** (mas-TEK-tō-mē; *mast*-=breast) involves removal of the affected breast along with the underlying pectoral muscles and the axillary lymph nodes. (Lymph nodes are removed because metastasis of cancerous cells usually occurs through lymphatic or blood vessels.) Radiation treatment and chemotherapy may follow surgery to ensure destruction of any stray cancer cells.

Several types of chemotherapeutic drugs are used to decrease the risk of relapse or disease progression. Tamoxifen (*Nolvadex*®) is an antagonist to estrogens that binds to and blocks receptors for estrogens, thus decreasing the stimulating effect of estrogens on breast cancer cells. Tamoxifen has been used for 20 years, and greatly reduces the risk of cancer recurrence. *Herceptin*,® a monoclonal antibody drug, targets an antigen on the surface of breast cancer cells. It is effective in causing regression of tumors and retarding progression of the disease. The early data from clinical trials of two new drugs, *Femara*® and *Amimidex*,® show relapse rates that are lower than those for tamoxifen. These drugs are inhibitors of aromatase, the enzyme needed for the final step in synthesis of estrogens. Finally, two drugs—tamoxifen and *Evista (raloxifene)*—are being marketed for breast cancer *prevention*. •

Figure 26.23 Mammary glands.

The mammary glands function in the synthesis, secretion, and ejection of milk (lactation).

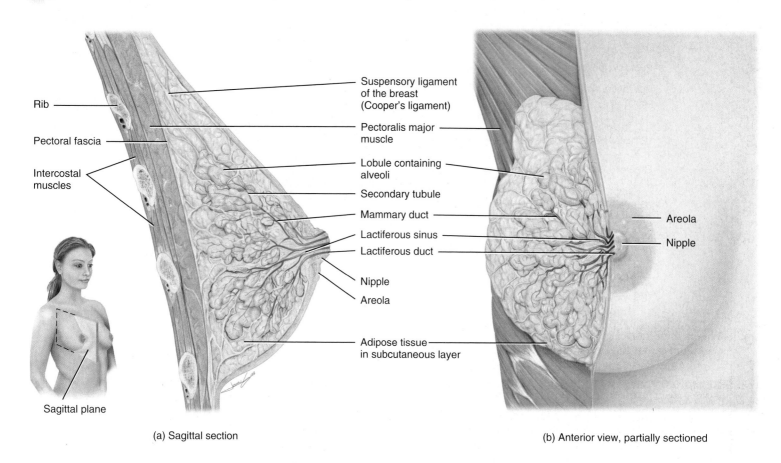

(a) Sagittal section

(b) Anterior view, partially sectioned

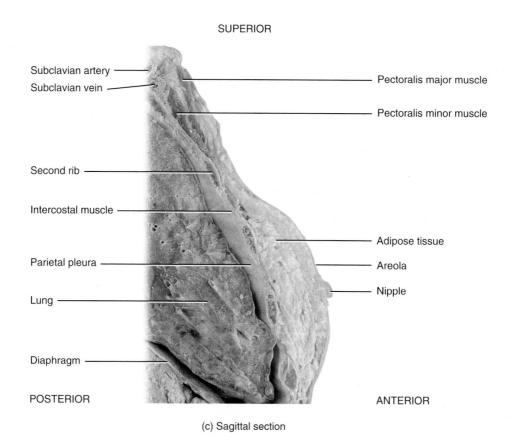

(c) Sagittal section

What hormones regulate the synthesis and ejection of milk?

Mammary Glands

Each **breast** is a hemispheric projection of variable size anterior to the pectoralis major and serratus anterior muscles and attached to them by a layer of fascia composed of dense irregular connective tissue (Figure 26.23).

Each breast has one pigmented projection, the **nipple**, which has a series of closely spaced openings of ducts called **lactiferous ducts** (lak-TIF-er-us), where milk emerges. The circular pigmented area of skin surrounding the nipple is called the **areola** (a-RĒ-ō-la=small space); it appears rough because it contains modified sebaceous (oil) glands. Strands of connective tissue called the **suspensory ligaments of the breast** (*Cooper's ligaments*) run between the skin and fascia and support the breast. These ligaments become looser with age or with excessive strain, as occurs in long-term jogging or high-impact aerobics. Wearing a supportive bra can slow this process and help maintain the integrity of the suspensory ligaments.

Within each breast is a **mammary gland**, a modified sudoriferous (sweat) gland that produces milk (Figure 26.23). A mammary gland consists of 15 to 20 **lobes**, or compartments, separated by a variable amount of adipose tissue. In each lobe are several smaller compartments called **lobules**, composed of grapelike clusters of milk-secreting glands termed **alveoli** (al-VĒ-ō-lī=small cavities) embedded in connective tissue (Figure 26.24). Contraction of **myoepithelial cells** (mī'-ō-ep'-i-THĒ-lē-al) surrounding the alveoli helps propel milk toward the nipples. When milk is being produced, it passes from the alveoli into a series of **secondary tubules** and then into the **mammary ducts**. Near the nipple, the mammary ducts expand slightly to form sinuses called **lactiferous sinuses** (*lact-*=milk), where some milk may be stored before draining into a **lactiferous duct**. Each lactiferous duct typically carries milk from one of the lobes to the exterior.

The functions of the mammary glands are the synthesis, secretion, and ejection of milk; these functions, called **lactation** (lak-TĀ-shun), are associated with pregnancy and childbirth. Milk production is stimulated largely by the hormone prolactin from the anterior pituitary, with contributions from progesterone and estrogens. The ejection of milk is stimulated by oxytocin, which is released from the posterior pituitary in response to the sucking of an infant on the mother's nipple (suckling).

 CHECKPOINT

21. What is the function of the vagina?
22. How does the histology of the vagina differ from the histology of the uterus?
23. What is the function of each of the parts of the vulva?
24. Define the perineum.
25. Describe the structure of the mammary glands and explain how they are supported.
26. What is the route of milk from the alveoli of the mammary gland to the nipple?

26.3 FEMALE REPRODUCTIVE CYCLE

● OBJECTIVES
• Define the female reproductive cycle.
• Describe the uterine events during menstruation.
• Compare the ovarian and uterine events during the preovulatory and postovulatory phases.
• Define ovulation.

During their reproductive years, nonpregnant females normally exhibit cyclical changes in the ovaries and uterus. Each cycle takes about a month and involves both oogenesis and preparation of the uterus to receive a fertilized ovum. Hormones secreted by the hypothalamus, anterior pituitary, and ovaries control the main events. The **ovarian cycle** is a series of events in the ovaries that occur during and after the maturation of an oocyte. Steroid hormones released by the ovaries control the **uterine (menstrual) cycle**, a concurrent series of changes in the endometrium

Figure 26.24 Histology of the mammary glands.

🔑 Contraction of myoepithelial cells helps propel milk toward the nipples.

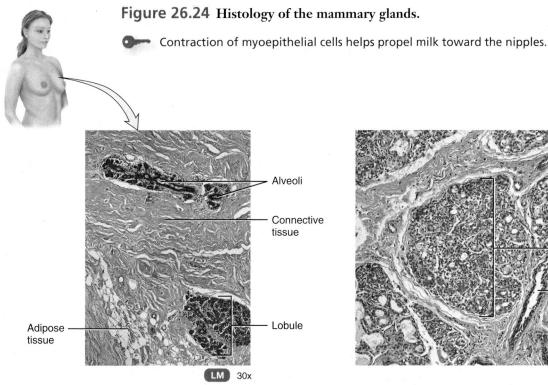

(a) Section of a nonlactating (inactive) mammary gland

Alveoli
Connective tissue
Adipose tissue
Lobule
LM 30x

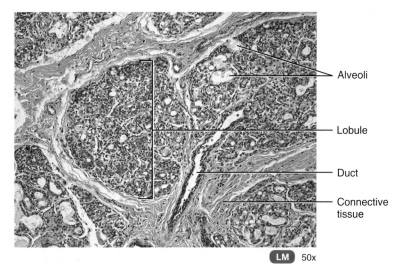

(b) Section of a lactating (active) mammary gland

Alveoli
Lobule
Duct
Connective tissue
LM 50x

 In which portion of the mammary gland are alveoli located?

of the uterus to prepare it for the arrival and development of a fertilized ovum. If fertilization does not occur, levels of ovarian hormones decrease, which causes the stratum functionalis of the endometrium to slough off. The general term **female reproductive cycle** encompasses the ovarian and uterine cycles, the hormonal changes that regulate them, and the related cyclical changes in the breasts and cervix.

Gonadotropin-releasing hormone (GnRH) secreted by the hypothalamus controls the events of the female reproductive cycle (Figure 26.25). GnRH stimulates the release of **follicle-stimulating hormone (FSH)** and **luteinizing hormone (LH)** from the anterior pituitary. FSH in turn initiates follicular growth and the secretion of estrogens by the growing ovarian follicles. LH stimulates the further development of follicles and their full secretion of estrogens. At midcycle, LH triggers ovulation and then promotes formation of the corpus luteum (the reason for the name luteinizing hormone). Stimulated by LH, the corpus luteum produces and secretes estrogens, progesterone, relaxin, and inhibin.

The duration of the female reproductive cycle typically ranges from 24 to 35 days. For this discussion, we assume a duration of 28 days and divide it into four phases: the menstrual phase, the preovulatory phase, ovulation, and the postovulatory phase (see Figure 26.25).

Menstrual Phase

The **menstrual phase** (MEN-stroo-al), also called **menstruation** (men'-stroo-Ā-shun) or **menses** (=month), lasts for roughly the first 5 days of the cycle. (By convention, the first day of menstruation is day one of a new cycle.)

Events in the Ovaries

Under the influence of FSH, several primordial follicles develop into primary follicles and then into secondary follicles. Follicular fluid, secreted by the granulosa cells and filtered from blood in the capillaries of the theca folliculi, accumulates in the enlarging antrum (space that forms within the follicle) while the oocyte remains near the edge of the follicle (see Figure 26.14b).

Events in the Uterus

Menstrual flow from the uterus consists of 50–150 mL of blood, tissue fluid, mucus, and epithelial cells shed from the endometrium. This discharge occurs because the declining level of

Figure 26.25 The female reproductive cycle. Events in the ovarian and uterine cycles and the release of anterior pituitary gland hormones are correlated with the sequence of the cycle's four phases. In the cycle shown, fertilization and implantation have not occurred.

The length of the female reproductive cycle typically is 24–36 days; the preovulatory phase is more variable in length than the other phases.

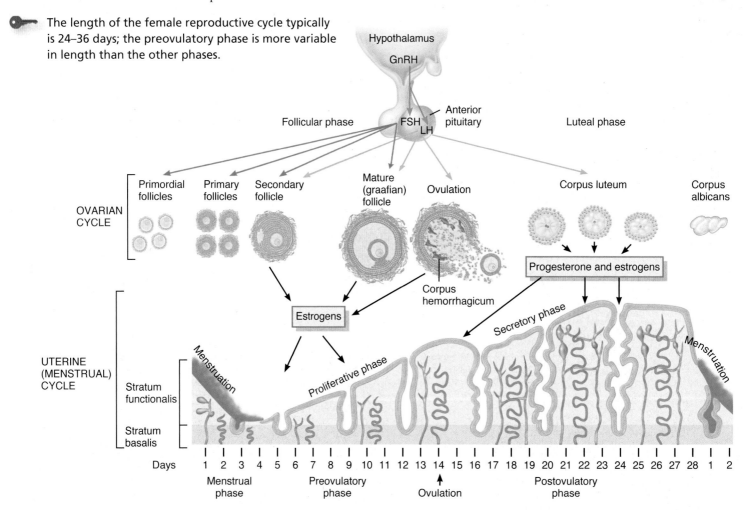

Hormonal regulation of changes in the ovary and uterus

 Which hormones are responsible for the proliferative phase of endometrial growth, for ovulation, for growth of the corpus luteum, and for the surge of LH at midcycle?

ovarian progesterone and estrogens stimulates release of prostaglandins that cause the uterine spiral arterioles to constrict. As a result, the cells they supply become oxygen-deprived and start to die. Eventually, the entire stratum functionalis sloughs off. At this time the endometrium is very thin, about 2–5 mm, because only the stratum basalis remains. The menstrual flow passes from the uterine cavity through the cervix and vagina to the exterior.

Preovulatory Phase

The **preovulatory phase** (prē′-OV-ū-la-tō-rē) is the time between the end of menstruation and ovulation. The preovulatory phase of the cycle is more variable in length than the other phases and accounts for most of the difference when cycles are shorter or longer than 28 days. It lasts from days 6 to 13 in a 28-day cycle.

Events in the Ovaries

Some of the secondary follicles in the ovaries begin to secrete estrogens and inhibin. By about day 6, a single secondary follicle in one of the two ovaries has outgrown all the others to become the **dominant follicle**. Estrogens and inhibin secreted by the dominant follicle decrease the secretion of FSH, which causes other, less well-developed follicles to stop growing and undergo atresia. Fraternal (nonidentical) twins or triplets result when two or three secondary follicles become codominant and later are ovulated and fertilized at about the same time.

Normally, the one dominant secondary follicle becomes the **mature (graafian) follicle**, which continues to enlarge until it is more than 20 mm in diameter and ready for ovulation (see Figure 26.13a). This follicle forms a blisterlike bulge due to the swelling antrum on the surface of the ovary. During the final maturation process, the mature follicle continues to increase its production of estrogens.

With reference to the ovarian cycle, the menstrual and preovulatory phases together are termed the **follicular phase** (fo-LIK-ū-lar) because ovarian follicles are growing and developing.

Events in the Uterus

Estrogens liberated into the blood by growing ovarian follicles stimulate the repair of the endometrium; cells of the stratum basalis undergo mitosis and produce a new stratum functionalis. As the endometrium thickens, the short, straight endometrial glands develop, and the arterioles coil and lengthen as they penetrate the stratum functionalis. The thickness of the endometrium approximately doubles, to about 4–10 mm. The preovulatory phase is also referred to as the **proliferative phase** (prō-LIF-er-ā-tiv) of the uterine cycle because the endometrium is proliferating.

Ovulation

Ovulation, the rupture of the mature (graafian) follicle and the overlying germinal epithelium of the ovary and the release of the secondary oocyte into the peritoneal cavity, usually occurs on day 14 in a 28-day cycle. During ovulation, the secondary oocyte remains surrounded by its zona pellucida and corona radiata. Development of a secondary follicle into a fully mature follicle generally takes a total of about 20 days (spanning the last 6 days of the previous cycle and the first 14 days of the current cycle). During this time the primary oocyte completes meiosis I to become a secondary oocyte; the secondary oocyte then begins meiosis II but halts in metaphase until it is fertilized. From time to time, an oocyte is lost into the peritoneal cavity, where it later disintegrates. The small amount of blood that sometimes leaks into the peritoneal

cavity from the ruptured follicle can cause pain, known as **mittelschmerz** (MIT-el-shmarts=pain in the middle), at the time of ovulation. An over-the-counter home test that detects a rising level of LH can be used to predict ovulation a day in advance.

Postovulatory Phase

The **postovulatory phase** of the female reproductive cycle is the time between ovulation and onset of the next menses. In duration, it is the most constant part of the female reproductive cycle. It lasts for 14 days in a 28-day cycle, from day 15 to day 28 (see Figure 26.25).

The female reproductive cycle can be disrupted by many factors, including weight loss, low body weight, disordered eating, and vigorous physical activity. The observation that three conditions—disordered eating, amenorrhea, and osteoporosis—often occur together in female athletes led researchers to coin the term **female athlete triad**.

Many athletes experience intense pressure from coaches, parents, peers, and themselves to lose weight to improve performance. Hence, they may develop disordered eating behaviors and engage in other harmful weight-loss practices in a struggle to maintain a very low body weight. **Amenorrhea** (ā-men′-ō-RĒ-a; *a-*=without; *men-*=month; *rrhea*=a flow) is the absence of menstruation. The most common causes of amenorrhea are pregnancy and menopause. In female athletes, amenorrhea results from reduced secretion of gonadotropin-releasing hormone, which decreases the release of LH and FSH. As a result, ovarian follicles fail to develop, ovulation does not occur, synthesis of estrogens and progesterone wanes, and monthly menstrual bleeding ceases. Most cases of the female athlete triad occur in young women with very low amounts of body fat. Low levels of the hormone leptin, secreted by adipose cells, may be a contributing factor.

Because estrogens help bones retain calcium and other minerals, chronically low levels of estrogens are associated with loss of bone mineral density. The female athlete triad causes "old bones in young women." In one study, amenorrheic runners in their twenties had low bone mineral densities, similar to those of postmenopausal women 50 to 70 years old! Short periods of amenorrhea in young athletes may cause no lasting harm. However, long-term cessation of the reproductive cycle may be accompanied by a loss of bone mass, and adolescent athletes may fail to achieve an adequate bone mass; both of these situations can lead to premature osteoporosis and irreversible bone damage.

Events in the Ovaries

After ovulation, the mature follicle collapses, and the basement membrane between the granulosa cells and theca interna breaks down. Once a blood clot forms from minor bleeding of the ruptured follicle, the follicle becomes the **corpus hemorrhagicum** (hem′-ō-RĀJ-i-kum; *hemo-*=blood; *rrhagic-*=bursting forth) (see Figure 26.13a). Theca interna cells mix with the granulosa cells as they all become transformed into **corpus luteum** cells under the influence of LH. Stimulated by LH, the corpus luteum secretes progesterone, estrogens, relaxin, and inhibin. The luteal cells also absorb the blood clot. Because of this activity, the postovulatory phase is also referred to as the *luteal phase* (LOO-tē-al) of the ovarian cycle.

Later events in an ovary that has ovulated an oocyte depend on whether the oocyte is fertilized. If the oocyte is *not fertilized*, the corpus luteum has a life span of only 2 weeks. At the end of this time period, its secretory activity declines, and it degenerates into a corpus albicans (see Figure 26.13a). As the levels of

CLINICAL CONNECTION | *Sexually Transmitted Diseases*

A **sexually transmitted disease (STD)** is a disease that is spread by sexual contact. In most developed countries of the world, such as those of Western Europe, Japan, Australia, and New Zealand, the incidence of STDs has declined markedly during the past 25 years. In the United States, by contrast, STDs have been rising to near-epidemic proportions; they currently affect more than 65 million people. AIDS and hepatitis B, which are sexually transmitted diseases that also may be contracted in other ways, are discussed in Chapters 15 and 24, respectively.

Chlamydia

Chlamydia (kla-MID-ē-a) is a sexually transmitted disease caused by the bacterium *Chlamydia trachomatis* (*chlamy-*=cloak) (Figure A). This unusual bacterium cannot reproduce outside body cells; it "cloaks" itself inside cells, where it divides. At present, chlamydia is the most prevalent sexually transmitted disease in the United States. In most cases, the initial infection is asymptomatic and thus difficult to recognize clinically. In males, urethritis is the principal result, causing a clear discharge, burning on urination, frequent urination, and painful urination. Without treatment, the epididymides may also become inflamed, leading to sterility. In 70 percent of females with chlamydia, symptoms are absent, but chlamydia is the leading cause of pelvic inflammatory disease. Moreover, the uterine tubes may also

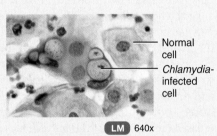

Normal cell

Chlamydia-infected cell

LM 640x

(A) Cervical smear showing normal and *Chlamydia*-infected cells

become inflamed, which increases the risk of ectopic pregnancy (implantation of a fertilized ovum outside the uterus) and infertility due to the formation of scar tissue in the tubes.

Gonorrhea

Gonorrhea (gon-ō-RĒ-a) (Figure B) or "the clap" is caused by the bacterium *Neisseria gonorrhoeae*. In the United States, 1–2 million new cases of gonorrhea appear each year, mostly among individuals aged 15–29 years. Discharges from infected mucous membranes are the source of transmission of the bacteria either during sexual contact or during the passage of a newborn through the birth canal. The infection site can be in the mouth and throat after oral–genital contact, in the vagina and penis after genital intercourse, or in the rectum after recto–genital contact.

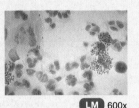

LM 600x

(B) *Neisseria gonorrhoeae* bacteria (tiny spheres) in a vaginal smear

Males usually experience urethritis with profuse pus drainage and painful urination. The prostate and epididymis may also become infected. In females, infection typically occurs in the vagina, often with a discharge of pus. Both infected males and females may harbor the disease without any symptoms, however, until it has progressed to a more advanced stage; about 5–10 percent of males and 50 percent of females are asymptomatic. In females, the infection and consequent inflammation can proceed from the vagina into the uterus, uterine tubes, and pelvic cavity. An estimated 50,000 to 80,000 women in the United States are made infertile by gonorrhea every year as a result of scar tissue forma-

tion that closes the uterine tubes. If bacteria in the birth canal are transmitted to the eyes of a newborn, blindness can result. Administration of a 1 percent silver nitrate solution in the infant's eyes prevents infection.

Syphilis

Syphilis, caused by the bacterium *Treponema pallidum* (trep-ō-NĒ-ma PAL-i-dum) (Figure C), is transmitted through sexual contact or exchange of blood, or through the placenta to a fetus. The disease progresses through several stages. During the *primary stage,* the chief sign is a painless open sore, called a **chancre** (SHANG-ker), at the point of contact. The chancre heals within 1 to 5 weeks. From 6 to 24 weeks later, signs and symptoms such as a skin rash, fever, and aches in the joints and muscles usher in the *secondary stage,* which is systemic—the infection spreads to all major body systems. When signs of organ degeneration appear, the disease is said to be in the *tertiary stage*. If the nervous system is involved, the tertiary stage is called **neurosyphilis**. As motor areas become extensively damaged, victims may be

SEM 4000x

(C) *Treponema pallidum* bacteria (thread-like cells) on the heads of two sperm cells

unable to control urine and bowel movements. Eventually they may become bedridden and unable even to feed themselves. In addition, damage to the cerebral cortex produces memory loss and personality changes that range from irritability to hallucinations.

Genital Herpes

Genital herpes is an incurable STD. Type II herpes simplex virus (HSV-2) causes genital infections (Figure D), producing painful blisters on the prepuce, glans penis, and penile shaft in males and on the vulva or sometimes high up in the vagina in females. The blisters disappear and reappear in most patients, but the virus itself remains in the body. A related virus, type I herpes simplex virus (HSV-1), causes cold sores on the mouth and lips and is not considered a sexually transmitted disease. Infected individuals typically experience recurrences of symptoms several times a year.

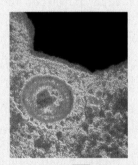

TEM 75,000x

(D) Type II herpes simplex virus (red-orange sphere) in the cytoplasm (blue) of an infected cell

Genital Warts

Warts are an infectious disease caused by viruses. *Human papillomavirus (HPV)* causes **genital warts**, which can be transmitted sexually (Figure E). Nearly one million people a year develop genital warts in the United States. Patients with a history of genital warts may be at increased risk for cancers of the cervix, vagina, anus, vulva, and penis. There is no cure for genital warts. A vaccine (Gardasil®) against certain types of HPV that cause cervical

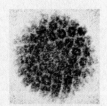

TEM 350,000x

(E) Human papillomavirus (HPV), which causes genital warts

cancer and genital warts is available and recommended for 11- and 12-year-old girls and boys. •

progesterone, estrogens, and inhibin decrease, release of GnRH, FSH, and LH rises due to loss of negative feedback suppression by the ovarian hormones. Follicular growth resumes, and a new ovarian cycle begins.

If the secondary oocyte *is fertilized* and begins to divide, the corpus luteum persists past its normal 2-week life span. It is "rescued" from degeneration by **human chorionic gonadotropin (hCG)** (kō-rē-ON-ik). This hormone is produced by the chorion of the embryo beginning about 8 days after fertilization. Like LH, hCG stimulates the secretory activity of the corpus luteum. The presence of hCG in maternal blood or urine is an indicator of pregnancy and is the hormone detected by **home pregnancy tests**.

Events in the Uterus

Progesterone and estrogens produced by the corpus luteum promote growth and coiling of the endometrial glands, vascularization of the superficial endometrium, and thickening of the endometrium to 12–18 mm (0.48–0.72 in.). Because of the secretory activity of the endometrial glands, which begin to secrete glycogen, this period is called the **secretory phase** of the uterine cycle. These preparatory changes peak about one week after ovulation, at the time a fertilized ovum might arrive in the uterus. If fertilization does not occur, the levels of progesterone and estrogens decline due to degeneration of the corpus luteum. Withdrawal of progesterone and estrogens causes menstruation.

 CHECKPOINT

27. What is the effect of GnRH on FSH and LH during the female reproductive cycle?
28. What is menstruation?
29. What are the major ovarian and uterine events of the preovulatory and postovulatory phases?
30. What is ovulation?

26.4 BIRTH CONTROL METHODS AND ABORTION

 OBJECTIVES

- Explain the differences among the various types of birth control methods and compare their effectiveness.
- Distinguish between spontaneous and induced abortion.

Birth control refers to restricting the number of children by various methods designed to control fertility and prevent conception. No single, ideal method of birth control exists. The only method of preventing pregnancy that is 100 percent reliable is complete **abstinence**, the avoidance of sexual intercourse. Several other methods are available, including surgical sterilization, hormonal methods, intrauterine devices, spermicides, barrier methods, and periodic abstinence. Each has its advantages and disadvantages. Table 26.3 provides the failure rates for various methods of birth control. Although it is not a form of birth control, in this section we will also discuss abortion, the premature expulsion of the products of conception from the uterus.

Birth Control Methods

Surgical Sterilization

Sterilization is a procedure that renders an individual incapable of further reproduction. The principal method for sterilization of males is a **vasectomy** (vas-EK-tō-mē; *-ectomy*=cut out),

TABLE 26.3

Failure Rates for Several Birth Control Methods

METHOD	FAILURE RATES*	
	PERFECT USE†	TYPICAL USE
Complete abstinence	0%	0%
Surgical sterilization		
Vasectomy	0.10%	0.15%
Tubal ligation	0.5%	0.5%
Non-incisional sterilization (Essure®)	0.2%	0.2%
Hormonal methods		
Oral contraceptives		
Combined pill	0.3%	1–2%
Extended cycle birth control pill	0.3%	1–2%
Minipill	0.5%	2%
Non-oral contraceptives		
Contraceptive skin patch	0.1%	1–2%
Vaginal contraceptive ring	0.1%	1–2%
Emergency contraception	25%	25%
Hormone injections	0.3%	1–2%
Intrauterine devices (Copper T 380A®)	0.6%	0.8%
Spermicides (alone)	15%	29%
Barrier methods		
Male condom	2%	15%
Vaginal pouch	5%	21%
Diaphragm (with spermicide)	6%	16%
Cervical cap (with spermicide)	9%	16%
Periodic abstinence		
Rhythm method	9%	25%
Sympto-thermal method (STM)	2%	20%
No method	85%	85%

*Defined as percentage of women having an unintended pregnancy during the first year of use.
†Failure rate when the method is used correctly and consistently.

in which a portion of each ductus deferens is removed. In order to gain access to the ductus deferens, an incision is made with a scalpel (conventional procedure) or a puncture is made with special forceps (non-scalpel vasectomy). Next the ducts are located and cut, each is tied (ligated) in two places with stitches, and the portion between the ties is removed. Although sperm production continues in the testes, sperm can no longer reach the exterior. The sperm degenerate and are destroyed by phagocytosis. Because the blood vessels are not cut, testosterone levels in the blood remain normal, so vasectomy has no effect on sexual desire and performance. If done correctly, it is close to 100 percent effective. The procedure can be reversed, but the chance of regaining fertility is only 30–40 percent. Sterilization in females most often is achieved by performing a **tubal ligation** (lī-GĀ-shun) in which both uterine tubes are tied closed and then cut. This can be achieved in a few different ways. "Clips" or "clamps" can be placed on the uterine tubes, the tubes can be tied and/or cut, and sometimes the tubes are cauterized. In any case, the result is that the secondary oocyte cannot pass through the uterine tubes, and sperm cannot reach the oocyte.

Non-incisional Sterilization

Non-incisional sterilization (Essure®) is an alternative to tubal ligation. In the procedure, a soft coil made of polyester fibers and metals (nickel-titanium and stainless steel) is inserted with a catheter into the vagina, through the uterus, and into each uterine tube. Over a three-month period, the insert stimulates the growth of scar tissue in and around itself, blocking the uterine tubes. As with tubal ligation, the secondary oocyte cannot pass through the uterine tubes, and sperm cannot reach the oocyte. Unlike tubal ligation, non-incisional sterilazation does not require general anesthesia.

Hormonal Methods

Aside from complete abstinence or surgical sterilization, hormonal methods are the most effective means of birth control. **Oral contraceptives** (the pill) contain hormones designed to prevent pregnancy. Some, called *combined oral contraceptives (COCs)*, contain both progestin (hormone with actions similar to progesterone) and estrogens. The primary action of COCs is to inhibit ovulation by suppressing the gonadotropins FSH and LH. The low levels of FSH and LH usually prevent the development of a dominant follicle in the ovary. As a result, levels of estrogens do not rise, the midcycle LH surge does not occur, and ovulation does not take place. Even if ovulation does occur, as it does in some cases, COCs may also block implantation in the uterus and inhibit the transport of ova and sperm in the uterine tubes.

Progestins thicken cervical mucus and make it more difficult for sperm to enter the uterus. *Progestin-only pills* thicken cervical mucus and may block implantation in the uterus but they do not consistently inhibit ovulation.

Among the noncontraceptive benefits of oral contraceptives are regulation of the length of the menstrual cycle and decreased menstrual flow (which decreases the risk of anemia). The pill also provides protection against endometrial and ovarian cancers and reduces the risk of endometriosis. However, oral contraceptives may not be advised for women with a history of blood clotting disorders, cerebral blood vessel damage, migraine headaches, hypertension, liver malfunction, or heart disease. Women who take the pill and smoke face far higher odds of having a heart attack or stroke than do nonsmoking pill users. Smokers should quit smoking or use an alternative method of birth control.

Oral hormonal methods of contraception include the following:

- *Combined pill.* Contains both progestin and estrogens and is typically taken once a day for 3 weeks to prevent pregnancy and regulate the menstrual cycle. The pills taken during the fourth week are inactive (do not contain hormones) and permit menstruation to occur.
- *Extended cycle birth control pill.* Contains both progestin and estrogens and is taken once a day in 3-month cycles of 12 weeks of hormone-containing pills followed by 1 week of inactive pills. Menstruation occurs during the thirteenth week.
- *Minipill.* Contains progestin only and is taken every day of the month.

A number of *non-oral* hormonal methods of contraception are also available:

- *Contraceptive skin patch (Ortho Evra®).* Contains both progestin and estrogens delivered in a skin patch placed on the skin (upper outer arm, back, lower abdomen, or buttocks) once a week for 3 weeks. After 3 weeks, the patch is removed from one location and then a new one is placed elsewhere. During the fourth week no patch is used.
- *Vaginal contraceptive ring (NuvaRing®).* A flexible doughnut-shaped ring about 5 cm (2 in.) in diameter that contains estrogens and progesterone and is inserted by the user into the vagina. It is left in the vagina for 3 weeks to prevent conception and then removed for 1 week to permit menstruation.

- *Emergency contraception (EC) (morning-after pill).* Consists of progestin and estrogens or progestin alone to prevent pregnancy following unprotected sexual intercourse. The relatively high levels of progestin and estrogens in EC pills provide inhibition of FSH and LH secretion. Loss of the stimulating effects of these gonadotropic hormones causes the ovaries to cease secretion of their own estrogens and progesterone. In turn, declining levels of estrogens and progesterone induce shedding of the uterine lining, thereby blocking implantation. One pill is taken as soon as possible but within 72 hours of unprotected sexual intercourse. The second pill must be taken 12 hours after the first. The pills work in the same way as regular birth control pills.
- *Hormone injections (Depo-provera®).* An injectable progestin given intramuscularly by a health-care practitioner once every 3 months.

Intrauterine Devices

An **intrauterine device (IUD)** is a small object made of plastic, copper, or stainless steel that is inserted by a health-care professional into the cavity of the uterus. IUDs prevent fertilization from taking place by blocking sperm from entering the uterine tubes. The IUD most commonly used in the United States today is the Copper T 380A®, which is approved for up to 10 years of use and has long-term effectiveness comparable to that of tubal ligation. Some women cannot use IUDs because of expulsion, bleeding, or discomfort.

Spermicides

Various foams, creams, jellies, suppositories, and douches that contain sperm-killing agents, or **spermicides** (SPER-mi-sīds), make the vagina and cervix unfavorable for sperm survival and are available without prescription. They are placed in the vagina before sexual intercourse. The most widely used spermicide is *nonoxynol-9*, which kills sperm by disrupting their plasma membranes. A spermicide is more effective when used with a barrier method such as a male condom, vaginal pouch, diaphragm, or cervical cap.

Barrier Methods

As the name implies, **barrier methods** use a physical barrier to prevent sperm from gaining access to the uterine cavity and uterine tubes. In addition to preventing pregnancy, certain barrier methods (male condom and vaginal pouch) may also provide some protection against sexually transmitted diseases (STDs) such as AIDS. In contrast, oral contraceptives and IUDs confer no such protection. Among the barrier methods are the male condom, vaginal pouch, diaphragm, and cervical cap.

A **male condom** is a nonporous, latex covering placed over the penis that prevents deposition of sperm in the female reproductive tract. A **vaginal pouch**, sometimes called a **female condom**, is designed to prevent sperm from entering the uterus. It is made of two flexible rings connected by a polyurethane sheath. One ring lies inside the sheath and is inserted to fit over the cervix; the other ring remains outside the vagina and covers the female external genitals. A **diaphragm** is a rubber, dome-shaped structure that fits over the cervix and is used in conjunction with a spermicide. It can be inserted by the female up to 6 hours before intercourse. The diaphragm stops most sperm from passing into the cervix, and the spermicide kills most sperm that do get around it. Although diaphragm use does decrease the risk of some STDs, it does not fully protect against HIV infection because the vagina is still exposed. A

cervical cap resembles a diaphragm but is smaller and more rigid. It fits snugly over the cervix and must be fitted by a health-care professional. Spermicides should be used with the cervical cap.

Periodic Abstinence

A couple can use their knowledge of the physiological changes that occur during the female reproductive cycle to decide either to abstain from intercourse on those days when pregnancy is a likely result, or to plan intercourse on those days if they wish to conceive a child. In females with normal and regular menstrual cycles, these physiological events help to predict the day on which ovulation is likely to occur.

The first physiologically based method, developed in the 1930s, is known as the **rhythm method**. It involves abstaining from sexual activity on the days that ovulation is likely to occur in each reproductive cycle. During this time (3 days before ovulation, the day of ovulation, and 3 days after ovulation) the couple abstains from intercourse. The effectiveness of the rhythm method for birth control is poor in many women due to the irregularity of the female reproductive cycle.

Another system is the **sympto-thermal method** (STM), a natural, fertility-awareness–based method of family planning that is used to either avoid or achieve pregnancy. STM utilizes normally fluctuating physiological markers to determine ovulation such as increased basal body temperature and the production of abundant, clear, stretchy cervical mucus that resembles uncooked egg white. These indicators, reflecting the hormonal changes that govern female fertility, provide a double-check system by which a female knows when she is or is not fertile. Sexual intercourse is avoided during the fertile time to avoid pregnancy. STM users observe and chart these changes and interpret them according to precise rules.

Abortion

Abortion refers to the premature expulsion of the products of conception from the uterus, usually before the twentieth week of pregnancy. An abortion may be *spontaneous* (naturally occurring; also called a *miscarriage*) or *induced* (intentionally performed).

There are several types of induced abortions. One involves **mifepristone** (MIF-pris-tōn), also known as **RU 486**. It is a hormone approved only for pregnancies 9 weeks or less when taken with a misopostol (a prostaglandin). Mifepristone is an antiprogestin; it blocks the action of progesterone by binding to and blocking progesterone receptors. Progesterone prepares the uterine endometrium for implantation and then maintains the uterine lining after implantation. If the level of progesterone falls during pregnancy or if the action of the hormone is blocked, menstruation occurs, and the embryo sloughs off along with the uterine lining. Within 12 hours after taking mifepristone, the endometrium starts to degenerate, and within 72 hours it begins to slough off. Misoprostol stimulates uterine contractions, and is given after mifepristone to aid in expulsion of the endometrium.

Another type of induced abortion is called **vacuum aspiration (suction)** and can be performed up to the sixteenth week of pregnancy. A small, flexible tube attached to a vacuum source is inserted into the uterus through the vagina. The embryo or fetus, placenta, and lining of the uterus are then removed by suction. For pregnancies between 13 and 16 weeks, a technique called **dilation and evacuation** is commonly used. After the cervix is dilated, suction and forceps are used to remove the fetus, placenta, and uterine lining. From the 16th to 24th week, a **late-stage abortion** may be induced using surgical methods similar to dilation and evacuation or nonsurgical methods utiliz-

ing a saline solution or medications. Labor may be induced by using vaginal suppositories, intravenous infusion, or injections into the amniotic fluid through the uterus.

 CHECKPOINT

31. How do oral contraceptives work to reduce the likelihood of pregnancy?
32. Which methods of birth control protect against sexually transmitted diseases, and how do they do so?
33. Distinguish between spontaneous and induced abortions.

26.5 DEVELOPMENT OF THE REPRODUCTIVE SYSTEMS

OBJECTIVE

• Describe the development of the male and female reproductive systems.

The male and female *gonads* develop from **gonadal ridges** that arise from growth of the **intermediate mesoderm**. During the fifth week of development, the gonadal ridges appear as bulges just medial to the mesonephros (intermediate kidney) (Figure 26.26). Adjacent to the gonads are the **mesonephric ducts** (mez'-ō-NEF-rik) or *Wolffian ducts* (WULF-ē-an), which eventually develop into structures of the reproductive system in males. A second pair of ducts, the **paramesonephric ducts** (par'-a-mes'-ō-NEF-rik) or *Müllerian ducts* (mil-E-rē-an), develop lateral to the mesonephric ducts and eventually form structures of the reproductive system in females. Both sets of ducts empty into the urogenital sinus. An early embryo has the potential to follow either the male or the female pattern of development because it contains both sets of ducts and genital ridges that can differentiate into either testes or ovaries.

Cells of a male embryo have one X chromosome and one Y chromosome. The male pattern of development is initiated by a Y chromosome "master switch" gene named *SRY*, which stands for *Sex-determining Region of the Y* chromosome. When the *SRY* gene is expressed during development, its protein product causes the primitive sustentacular cells to begin to differentiate in the testes during the seventh week. The developing sustentacular cells secrete a hormone called **Müllerian-inhibiting substance (MIS)**, which causes apoptosis of cells within the paramesonephric (Müllerian) ducts. As a result, those cells do not contribute any functional structures to the male reproductive system. Stimulated by human chorionic gonadotropin (hCG), primitive interstitial (Leydig) cells in the testes begin to secrete the androgen **testosterone** during the eighth week. Testosterone then stimulates development of the mesonephric duct on each side into the *epididymis, ductus (vas) deferens, ejaculatory duct*, and *seminal vesicle*. The *testes* connect to the mesonephric duct through a series of tubules that eventually become the *seminiferous tubules*. The *prostate* and *bulbourethral glands* are **endodermal** outgrowths of the urethra.

Cells of a female embryo have two X chromosomes and no Y chromosome. Because *SRY* is absent, the genital ridges develop into *ovaries*, and because MIS is not produced, the paramesonephric ducts flourish. The distal ends of the paramesonephric ducts fuse to form the *uterus* and *vagina*, and the unfused proximal portions become the *uterine (fallopian) tubes*. The mesonephric ducts degenerate without contributing any functional structures to the female reproductive system because testosterone is absent. The *greater* and *lesser vestibular glands* develop from **endodermal** outgrowths of the vestibule.

Figure 26.26 Development of the internal reproductive systems.

The gonads develop from intermediate mesoderm.

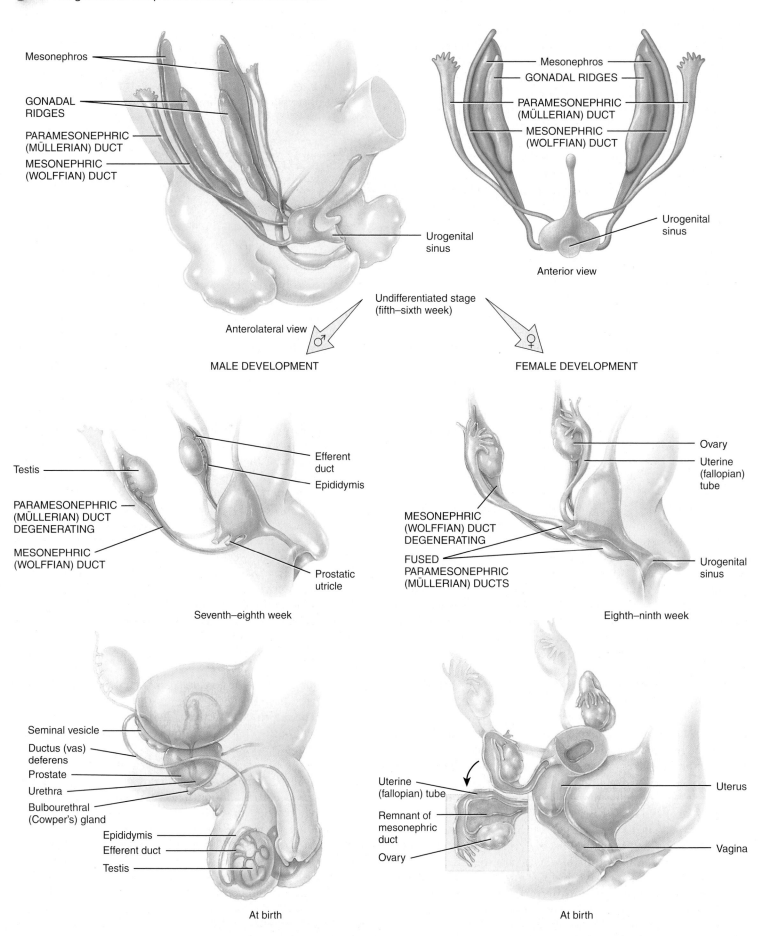

Mesonephros

GONADAL RIDGES

PARAMESONEPHRIC (MÜLLERIAN) DUCT

MESONEPHRIC (WOLFFIAN) DUCT

Urogenital sinus

Anterolateral view

Mesonephros
GONADAL RIDGES
PARAMESONEPHRIC (MÜLLERIAN) DUCT
MESONEPHRIC (WOLFFIAN) DUCT

Urogenital sinus

Anterior view

Undifferentiated stage (fifth–sixth week)

♂ MALE DEVELOPMENT

♀ FEMALE DEVELOPMENT

Testis

Efferent duct

Epididymis

PARAMESONEPHRIC (MÜLLERIAN) DUCT DEGENERATING

MESONEPHRIC (WOLFFIAN) DUCT

Prostatic utricle

Seventh–eighth week

Ovary

Uterine (fallopian) tube

MESONEPHRIC (WOLFFIAN) DUCT DEGENERATING

FUSED PARAMESONEPHRIC (MÜLLERIAN) DUCTS

Urogenital sinus

Eighth–ninth week

Seminal vesicle

Ductus (vas) deferens

Prostate

Urethra

Bulbourethral (Cowper's) gland

Epididymis

Efferent duct

Testis

At birth

Uterine (fallopian) tube

Remnant of mesonephric duct

Ovary

Uterus

Vagina

At birth

Which gene is responsible for the development of the gonads into testes?

The *external genitals* of both male and female embryos (penis and scrotum in males and clitoris, labia, and vaginal orifice in females) also remain undifferentiated until about the eighth week. Before differentiation, all embryos have the following external structures (Figure 26.27):

1. *Urethral (urogenital) folds.* These paired structures develop from mesoderm in the cloacal region (see Figure 25.13).
2. *Urethral groove.* An indentation between the urethral folds, which is the opening into the urogenital sinus.
3. *Genital tubercle.* A rounded elevation just anterior to the urethral folds.
4. *Labioscrotal swelling* (lā-bē-ō-SKRŌ-tal). Paired, elevated structures lateral to the urethral folds.

Figure 26.27 Development of the external genitals.

 The external genitals of male and female embryos remain undifferentiated until about the eighth week.

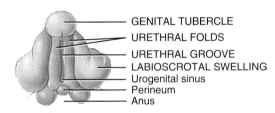
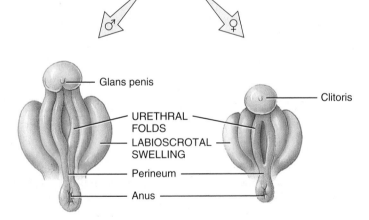
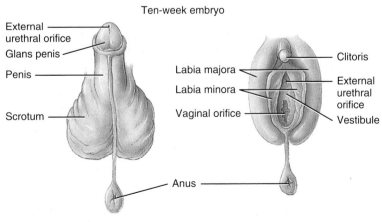

MALE DEVELOPMENT   FEMALE DEVELOPMENT

 Which hormone is responsible for the differentiation of the external genitals?

In male embryos, some testosterone is converted to a second androgen called **dihydrotestosterone (DHT)** (dī-hī-drō-test-TOS-ter-ōn). DHT stimulates development of the urethra, prostate, and external genitals (scrotum and penis). Part of the genital tubercle elongates and develops into a penis. Fusion of the urethral folds forms the *spongy (penile) urethra* and leaves an opening to the exterior only at the distal end of the penis, the *external urethral orifice*. The labioscrotal swellings develop into the *scrotum*. In the absence of DHT, the genital tubercle gives rise to the *clitoris* in female embryos. The urethral folds remain open as the *labia minora*, and the labioscrotal swellings become the *labia majora*. The urethral groove becomes the *vestibule*. After birth, androgen levels decline because hCG is no longer present to stimulate secretion of testosterone.

CHECKPOINT

34. Describe the role of hormones in differentiation of the gonads, the mesonephric ducts, the paramesonephric ducts, and the external genitals.

26.6 AGING AND THE REPRODUCTIVE SYSTEMS

OBJECTIVE

• Describe the effects of aging on the reproductive systems.

During the first decade of life, the reproductive system is in a juvenile state. At about age 10, hormone-directed changes start to occur in both sexes. **Puberty** (PŪ-ber-tē=a ripe age) is the period when secondary sexual characteristics begin to develop and the potential for sexual reproduction is reached. The onset of puberty is marked by pulses or bursts of LH and FSH secretion, each triggered by a pulse of GnRH. Most pulses occur during sleep. As puberty advances, the hormone pulses occur during the day as well as at night. The pulses increase in frequency during a 3- to 4-year period until the adult pattern is established. The stimuli that cause the GnRH pulses are still unclear, but a role for the hormone leptin is starting to unfold. Just before puberty, leptin levels rise in proportion to adipose tissue mass. Interestingly, leptin receptors are present in both the hypothalamus and anterior pituitary. Mice that lack a functional leptin gene from birth are sterile and remain in a prepubertal state. Giving leptin to such mice elicits secretion of gonadotropins, and they become fertile. Leptin may signal the hypothalamus that long-term energy stores (triglycerides in adipose tissue) are adequate for reproductive functions to begin.

In females, the reproductive cycle normally occurs once each month from **menarche** (me-NAR-kē), the first menses, to **menopause**, the permanent cessation of menses. Thus, the female reproductive system has a time-limited span of fertility between menarche and menopause. For the first 1 to 2 years after menarche, ovulation occurs in only about 10 percent of the cycles and the luteal phase is short. Gradually, the percentage of ovulatory cycles increases, and the luteal phase reaches its normal duration of 14 days. With age, fertility declines. Between the ages of 40 and 50 the pool of remaining ovarian follicles becomes exhausted. As a result, the ovaries become less responsive to hormonal stimulation. The production of estrogens declines, despite copious secretion of FSH and LH by the anterior pituitary. Many women experience hot flashes and heavy sweating, which coincide with bursts of GnRH release. Other symptoms of menopause are headache, hair loss, muscular pains, vaginal dryness, insomnia, depression, weight gain, and mood swings. Some atrophy of the ovaries, uterine tubes, uterus, vagina, external genitalia, and breasts occurs in postmenopausal women. Due to loss of estrogens, most women experience a decline in bone mineral density after menopause. Sexual desire (libido) does not show a parallel decline; it may be maintained by adrenal sex steroids. The risk of having uterine cancer peaks at about 65 years of age, but cervical cancer is more common in younger women.

In males, declining reproductive function is much more subtle than in females. Healthy men often retain reproductive capacity into their eighties or nineties. At about age 55 a decline in testosterone synthesis leads to reduced muscle strength, fewer viable sperm, and decreased sexual desire. Although sperm production decreases 50–70 percent between ages 60 and 80, abundant sperm may still be present even in old age.

Enlargement of the prostate to two to four times its normal size occurs in approximately one-third of all males over age 60. This condition, called **benign prostatic hyperplasia (BPH)** (hī-per-PLĀ-zē-a), decreases the size of the prostatic urethra and is characterized by frequent urination, nocturia (bed-wetting), hesitancy in urination, decreased force of urinary stream, postvoiding leakage, and a sensation of incomplete emptying.

 CHECKPOINT

35. What changes occur in males and females at puberty?
36. What do the terms menarche and menopause mean?

KEY MEDICAL TERMS ASSOCIATED WITH THE REPRODUCTIVE SYSTEMS

Breast augmentation (awg-men-TĀ-shun) Technically called augmentation mammaplasty (MAM-a-plas-tē), this surgical procedure is used to increase breast size and shape to enhance breast size, restore breast volume due to weight loss or following pregnancy, improve the shape of sagging breasts, or improve breast appearance following surgery, trauma, or congenital abnormalities.

Breast reduction Technically called reduction mammaplasty, this surgical procedure involves reducing breast size by removing fat, skin, and glandular tissue and is performed for several reasons, including chronic back, neck, and shoulder pain; circulation or breathing problems; restricted levels of activity; self-esteem problems; and difficulty wearing or fitting into certain clothing.

Castration (kas-TRĀ-shun=to prune) Removal, inactivation, or destruction of the gonads; commonly used in reference to removal of the testes only.

Circumcision (=to cut around) A surgical procedure in which part of or the entire prepuce is removed. It is usually performed just after delivery, several days after birth, and is done for cultural, religious, or (more rarely) medical reasons. Although most health-care professionals find no medical justification for the procedure, some feel that it has benefits, such as a lower risk of urinary tract infections, protection against penile cancer, and possibly a lower risk for sexually transmitted diseases.

Colposcopy (kol-POS-kō-pē; *colpo-*=vagina; *scopy*=to view) Visual inspection of the vagina and cervix of the uterus using a culposcope, an instrument that has a magnifying lens (between 5 and 50×) and a light.

Culdoscopy (kul-DOS-kō-pē; *cul-*=cul-de-sac; *scopy*=to view) A procedure in which a culdoscope (endoscope) is inserted through the posterior wall of the vagina to view the rectouterine pouch in the pelvic cavity.

Dysmenorrhea (dis′-men-or-Ē-a; *dys-*=difficult or painful) Pain associated with menstruation; the term is usually reserved to describe menstrual symptoms that are severe enough to prevent a woman from functioning normally for one or more days each month.

Dyspareunia (dis-pa-ROO-nē-a; *dys-*=difficult; *para*=beside; *enue*=bed) Pain during sexual intercourse. It may occur in the genital area or in the pelvic cavity, and may be due to inadequate lubrication, inflammation, infection, an improperly fitting diaphragm or cervical cap, endometriosis, pelvic inflammatory disease, pelvic tumors, or weakened uterine ligaments.

Endocervical curettage (ku-re-TAHZH; *curette*=scraper) A procedure in which the cervix is dilated and the endometrium of the uterus is scraped with a spoon-shaped instrument called a curette; commonly called a *D and C* (dilation and curettage).

Erectile dysfunction (ED) The consistent inability of an adult male to ejaculate or to attain or hold an erection long enough for sexual intercourse; previously termed impotence. Many cases are caused by insufficient release of nitric oxide (NO), which relaxes the smooth muscle of the penile arterioles and erectile tissue. Other causes include diabetes mellitus, physical abnormalities of the penis, systemic disorders such as syphilis, vascular disturbances, neurological disorders, surgery, testosterone deficiency, and drugs. Psychological factors include anxiety or depression, fear of causing pregnancy, fear

of sexually transmitted diseases, religious inhibitions, and emotional immaturity.

Fibrocystic disease (fī′-bro-SIS-tik) The most common cause of breast lumps in females, one or more cysts (fluid-filled sacs) and thickenings of alveoli develop. Occurs mainly in females between the ages of 30 and 50, and is probably due to a relative excess of estrogens or a deficiency of progesterone in the postovulatory (luteal) phase of the reproductive cycle. Usually causes one or both breasts to become lumpy, swollen, and tender a week or so before menstruation begins.

Fibroids (FĪ-broyds; *fibro-*=fiber; *eidos*=resemblance) Noncancerous tumors in the myometrium of the uterus composed of muscular and fibrous tissue. Their growth appears to be related to high levels of estrogens. They do not occur before puberty and usually stop growing after menopause. Symptoms include abnormal menstrual bleeding, and pain or pressure in the pelvic area.

Hermaphroditism (her-MAF-rō-dīt-izm) The presence of both ovarian and testicular tissue in one individual.

Hypospadias (hī′-pō-SPĀ-dē-as; *hypo-*=below) A common congenital abnormality in which the urethral opening is displaced. In males, the displaced opening may be on the underside of the penis, at the penoscrotal junction, between the scrotal folds, or in the perineum; in females, the urethra opens into the vagina. The problem can be corrected surgically.

Leukorrhea (loo′-kō-RĒ-a; *leuko-*=white) A whitish (nonbloody) vaginal discharge containing mucus and pus cells that may occur at any age and affects most women at some time.

Menorrhagia (men-ō-RĀ-jē-a; *meno-*=menstruation; *rhage*=to burst forth) Excessively prolonged or profuse menstrual period. May be due to a disturbance in hormonal regulation of the menstrual cycle, pelvic infection, medications (anticoagulants), fibroids, endometriosis, or intrauterine devices.

Oophorectomy (ō′-of-ō-REK-tō-me; *oophor-*=bearing eggs) Removal of the ovaries.

Orchitis (or-KĪ-tis; *orchi-*=testes; *itis*=inflammation) Inflammation of the testes, for example, as a result of the mumps virus or a bacterial infection.

Pelvic inflammatory disease (PID) A collective term for any extensive bacterial infection of the pelvic organs, especially the uterus, uterine tubes, or ovaries, which is characterized by pelvic soreness, lower back pain, abdominal pain, and urethritis.

Polycystic ovary syndrome (PCOS) (pol-ē-SIS-tik) A disorder that usually develops during puberty and is characterized by enlarged ovaries with many fluid-filled sacs (cysts) and a tendency to have high levels of male hormones (androgens). Symptoms include failure to menstruate, unpredictable menstrual periods, an unusual amount of facial or body hair, high blood sugar, obesity, and increased risk of cardiovascular disease.

Premenstrual syndrome (PMS) Cyclical disorder of severe physical and emotional distress. It appears during the postovulatory phase of the female reproductive cycle and dramatically disappears when menstruation begins. The signs and symptoms may include edema, weight gain, breast swelling and tenderness, abdominal distension, backache, joint pain, constipation, skin eruptions, fatigue and lethargy, greater need for sleep, depression or anxiety, irritability, mood swings, headache, poor coordination and clumsiness, and/or cravings for sweet or salty food. The cause is unknown.

Premenstrual dysphoric disorder (PMDD) (dis-FOR-ik) A severe syndrome in which PMS-like signs and symptoms do not resolve after the onset of menstruation. May be caused by abnormal responses to normal levels of ovarian hormones.

Salpingectomy (sal′-pin-JEK-tō-mē; *salpingo*=tube) Removal of a uterine (fallopian) tube.

Smegma (SMEG-ma) A secretion found chiefly around the external genitalia and especially under the foreskin of the male, consisting principally of desquamated epithelial cells.

Testicular cancer Cancer of the testes, the most common cancer in males between the ages of 20 and 35. More than 95 percent of cases arise from spermatogenic cells within the seminiferous tubules. An early sign is a mass in the testis, often associated with a sensation of testicular heaviness or a dull ache in the lower abdomen; pain usually does not occur.

Vulvovaginal candidiasis (vul-vō-VAJ-i-nal can-di-DĪ-a-sis) The most common form of vaginitis (vaj-i-NĪ-tis), inflammation of the vagina. Candidiasis is characterized by severe itching; a thick, yellow, cheesy discharge; a yeasty odor; and pain. The disorder, experienced at least once by about 75 percent of females, is usually a result of proliferation of a yeast-like fungus (*Candida albicans*) following antibiotic therapy for another condition.

CHAPTER REVIEW AND RESOURCE SUMMARY

WileyPLUS

Review	Resource

Introduction

1. Reproduction is the process by which new individuals of a species are produced and the genetic material is passed from generation to generation.
2. The organs of reproduction are grouped as gonads (produce gametes), ducts (transport and store gametes), accessory sex glands (produce materials that support gametes), and supporting structures (have various roles in reproduction).

26.1 Male Reproductive System

1. The male structures of reproduction include the testes, ductus epididymis, ductus (vas) deferens, ejaculatory duct, urethra, seminal vesicles, prostate, bulbourethral (Cowper's) glands, and penis.
2. The scrotum is a sac that hangs from the root of the penis and consists of loose skin and superficial fascia; it supports the testes. The testes are paired oval glands (gonads) in the scrotum containing seminiferous tubules, in which sperm cells are made; sustentacular cells, which nourish sperm cells and secrete inhibin; and interstitial (Leydig) cells, which produce the male sex hormone testosterone. The testes descend into the scrotum through the inguinal canals during the seventh month of fetal development.
3. Secondary oocytes and sperm, both of which are called gametes, are produced in the gonads. Spermatogenesis, which occurs in the testes, is the development of immature spermatogonia into mature sperm. The spermatogenesis sequence (meiosis I, meiosis II, and spermiogenesis) results in the formation of four haploid sperm (spermatozoa) from each primary spermatocyte.
4. The principal parts of a mature sperm are a head and a tail. The function of sperm is to fertilize a secondary oocyte.
5. The duct system of the testes includes the seminiferous tubules, straight tubules, and rete testis. The ductus epididymis is the site of sperm maturation and storage. The ductus (vas) deferens stores sperm and propels them toward the urethra during ejaculation. Each ejaculatory duct, formed by the union of the duct from the seminal vesicle and ductus (vas) deferens, is the passageway for ejection of sperm and secretions of the seminal vesicles. The seminal vesicles secrete an alkaline, viscous fluid that constitutes about 60 percent of the volume of semen and contributes to sperm viability.
6. The urethra in males is subdivided into three portions: the prostatic, intermediate, and spongy (penile) urethra. The prostatic urethra or prostate secretes a slightly acidic fluid that constitutes about 25 percent of the volume of semen and contributes to sperm motility. The bulbourethral (Cowper's) glands secrete mucus for lubrication and an alkaline substance that neutralizes acid.
7. Semen is a mixture of sperm and seminal fluid; it provides the fluid in which sperm are transported, supplies nutrients, and neutralizes the acidity of the male urethra and the vagina.
8. The penis consists of a root, a body, and a glans penis. Engorgement of the penile blood sinuses under the influence of sexual excitation is called erection.

26.2 Female Reproductive System

1. Female organs of reproduction include ovaries (gonads), uterine (fallopian) tubes or oviducts, uterus, vagina, and vulva. Mammary glands are considered part of the reproductive system in females.
2. The ovaries, the female gonads, are located in the superior portion of the pelvic cavity, lateral to the uterus. Ovaries produce secondary oocytes, discharge secondary oocytes (ovulation), and secrete estrogens, progesterone, relaxin, and inhibin.
3. Oogenesis (production of haploid secondary oocytes) begins in the ovaries. The oogenesis sequence includes meiosis I and meiosis II, which goes to completion only after an ovulated secondary oocyte is fertilized by a sperm cell.
4. The uterine (fallopian) tubes transport secondary oocytes from the ovaries to the uterus and are the normal sites of fertilization. Ciliated cells and peristaltic contractions help move a secondary oocyte or fertilized ovum toward the uterus.
5. The uterus is an organ the size and shape of an inverted pear that functions in menstruation, implantation of a fertilized ovum, development of a fetus during pregnancy, and labor. It is also part of the pathway for sperm to reach the uterine tubes to fertilize a secondary oocyte. Normally, the uterus is

Review

held in position by a series of ligaments. Histologically, the layers of the uterus are an outer perimetrium (serosa), a middle myometrium, and an inner endometrium.

6. The vagina is a passageway for sperm and the menstrual flow, the receptacle of the penis during sexual intercourse, and the inferior portion of the birth canal.

7. The vulva, a collective term for the external genitals of the female, consists of the mons pubis, labia majora, labia minora, clitoris, vestibule, vaginal and urethral orifices, hymen, bulb of the vestibule, and the paraurethral (Skene's), greater vestibular (Bartholin's), and lesser vestibular glands.

8. The perineum is a diamond-shaped area at the inferior end of the trunk medial to the thighs and buttocks.

9. The mammary glands are modified sweat glands lying superficial to the pectoralis major muscles. Their function is to synthesize, secrete, and eject milk (lactation). Mammary gland development depends on estrogens and progesterone. Milk production is stimulated by prolactin, estrogens, and progesterone; milk ejection is stimulated by oxytocin.

26.3 Female Reproductive Cycle

1. The function of the ovarian cycle is to develop a secondary oocyte. The function of the uterine (menstrual) cycle is to prepare the endometrium each month to receive a fertilized egg. The female reproductive cycle includes both the ovarian and uterine cycles.

2. The female reproductive cycle is controlled by GnRH from the hypothalamus, which stimulates the release of FSH and LH by the anterior pituitary gland.

3. FSH stimulates development of secondary follicles and initiates secretion of estrogens by the follicles. LH stimulates further development of the follicles, secretion of estrogens by follicular cells, ovulation, formation of the corpus luteum, and the secretion of progesterone and estrogens by the corpus luteum.

4. During the menstrual phase, the stratum functionalis of the endometrium is shed, discharging blood, tissue fluid, mucus, and epithelial cells.

5. During the preovulatory phase, follicles in the ovaries begin to undergo final maturation. One follicle outgrows the others and becomes dominant while the others degenerate. At the same time, endometrial repair occurs in the uterus. Estrogens are the dominant ovarian hormones during this phase.

6. Ovulation is the rupture of the dominant mature (graafian) follicle and the release of a secondary oocyte into the pelvic cavity, brought about by a surge of LH. Signs and symptoms include increased basal body temperature; clear, stretchy cervical mucus; changes in the uterine cervix; and ovarian pain.

7. During the postovulatory phase, both progesterone and estrogens are secreted in large quantity by the corpus luteum of the ovary, and the uterine endometrium thickens in readiness for implantation.

8. If fertilization and implantation do not occur, the corpus luteum degenerates; the resulting low progesterone level allows discharge of the endometrium followed by initiation of another reproductive cycle.

9. If fertilization and implantation occur, the corpus luteum is maintained by placental hCG, and the corpus luteum and later the placenta secrete progesterone and estrogens to support pregnancy and breast development for lactation.

Anatomy Overview - Ovarian Hormones
Animation - Hormonal Control of Female Reproduction
Anatomy Overview - Hypothalamic Reproductive Hormones
Animation - Oogenesis
Figure 26.25 - The Female Reproductive Cycle
Exercise - Assemble the Cycle
Exercise - Match the Female Hormones
Exercise - Organize Oogenesis
Exercise - Assemble the Cycle
Concepts and Connections - Regulation of Female Reproduction

26.4 Birth Control Methods and Abortion

1. Birth control methods include complete abstinence, surgical sterilization, non-incisional sterilization, hormonal methods, intrauterine devices, spermicides, barrier methods, and periodic abstinence. Table 26.3 provides failure rates for the various types of birth control.

2. Contraceptive pills of the combination type contain progestin and estrogens in concentrations that decrease the secretion of FSH and LH, inhibiting development of ovarian follicles and ovulation, inhibiting transport of ova and sperm in the uterine tubes, and blocking implantation in the uterus.

3. An abortion is the premature expulsion from the uterus of the products of conception; it may be spontaneous or induced.

26.5 Development of the Reproductive Systems

1. The gonads develop from intermediate mesoderm. In the presence of the *SRY* gene, the gonads begin to differentiate into testes during the seventh week. The gonads differentiate into ovaries when the *SRY* gene is absent.

2. In males, testosterone stimulates development of each mesonephric duct into an epididymis, ductus (vas) deferens, ejaculatory duct, and seminal vesicle, and Müllerian-inhibiting substance (MIS) causes the paramesonephric duct cells to die. In females, testosterone and MIS are absent; the paramesonephric ducts develop into the uterine tubes, uterus, and vagina and the mesonephric ducts degenerate.

3. The external genitals develop from the genital tubercle and are stimulated to develop into typical male structures by the hormone dihydrotestosterone (DHT). The external genitals develop into female structures when DHT is not produced, the normal situation in female embryos.

26.6 Aging and the Reproductive Systems

1. Puberty is the period when secondary sex characteristics begin to develop and the potential for sexual reproduction is reached.

2. Onset of puberty is marked by pulses or bursts of LH and FSH secretion, each triggered by a pulse of GnRH. The hormone leptin, released by adipose tissue, may signal the hypothalamus that long-term energy stores (triglycerides in adipose tissue) are adequate for reproductive functions to begin.

Review   **Resource**

3. In females, the reproductive cycle normally occurs once each month from menarche, the first menses, to menopause, the permanent cessation of menses.

4. Between the ages of 40 and 50, the pool of remaining ovarian follicles becomes exhausted and levels of progesterone and estrogens decline. Most women experience a decline in bone mineral density after menopause, together with some atrophy of the ovaries, uterine tubes, uterus, vagina, external genitalia, and breasts. Uterine and breast cancers increase in incidence with age.

5. In older males, decreased levels of testosterone are associated with decreased muscle strength, waning sexual desire, and fewer viable sperm; prostate disorders are common.

CRITICAL THINKING QUESTIONS

1. Esther nearly died from peritonitis (inflammation of the peritoneum) that her doctor said had spread from an infection in her reproductive tract. How was this possible?

2. Occasionally, someone feels that he or she has been born the wrong gender and undergoes a "sex-change" or "gender-reassignment" process involving hormone treatment and surgery. However, a born male can never truly become a fully biological female or a born female become a fully biological male. Why not?

3. Thirty-nine-year-old Meg has been advised to have a hysterectomy due to medical problems. She is worried that the procedure will cause menopause. Is this a valid concern?

4. Darby has been doing peritoneal dialysis for kidney failure at home for several years. However, the weight of the dialysis fluid in his peritoneal cavity has caused him to develop inguinal hernias repeatedly. Darby is tired of this, and in consultation with his doctor, decides to have the inguinal canal surgically closed off. As a result, Darby will also have to have his testes removed. Why?

5. Phil has promised his wife that he will get a vasectomy after the birth of their next child. However, he's concerned about possible effects on his virility. How would you respond to Phil's concerns?

ANSWERS TO FIGURE QUESTIONS

26.1 The *testes* produce gametes (sperm) and hormones; the *ducts* transport, store, and receive gametes; the *accessory sex glands* secrete materials that help transport and protect gametes; and *supporting structures*, such as the penis, convey semen to the exterior.

26.2 The cremaster and dartos muscles help regulate the temperature of the testes.

26.3 The tunica vaginalis and tunica albuginea are tissue layers that cover and protect the testes.

26.4 The interstitial (Leydig) cells of the testes secrete testosterone.

26.5 As a result of meiosis I, the number of chromosomes in each cell is reduced by half.

26.6 The sperm head contains the nucleus with highly condensed haploid chromosomes and an acrosome that contains enzymes for penetration of a secondary oocyte; the neck contains centrioles that produce microtubules for the rest of the tail; the midpiece contains mitochondria for ATP production for locomotion and metabolism; the principal and end pieces of the tail provide motility.

26.7 The functions of the ductus epididymis include sperm maturation, sperm storage, and propulsion of sperm into the ductus (vas) deferens.

26.8 The ductus deferens stores sperm and conveys sperm toward the urethra.

26.9 Seminal vesicles are the accessory sex glands that contribute the largest volume to seminal fluid.

26.10 Two corpora cavernosa penis and one corpus spongiosum penis contain blood sinuses that fill with blood that cannot flow out of the penis as quickly as it flows in. The trapped blood engorges and stiffens the tissue, producing an erection. The corpus spongiosum penis keeps the spongy urethra open so that ejaculation can occur.

26.11 The testes are homologous to the ovaries; the glans penis is homologous to the clitoris; the prostate is homologous to the paraurethral glands; and the bulbourethral gland is homologous to the greater vestibular glands.

26.12 The mesovarium anchors the ovary to the broad ligament of the uterus and the uterine tube; the ovarian ligament anchors it to the uterus; the suspensory ligament anchors it to the pelvic wall.

26.13 Ovarian follicles secrete estrogens; the corpus luteum secretes progesterone, estrogens, relaxin, and inhibin.

26.14 Most ovarian follicles undergo atresia (degeneration).

26.15 Primary oocytes are present in the ovary at birth, so they are as old as the woman. In males, primary spermatocytes are continually being formed from stem cells (spermatogonia) and thus are only a few days old.

26.16 Fertilization most often occurs in the ampulla of the uterine tube.

26.17 Ciliated simple columnar epithelial cells and nonciliated (peg) cells with microvilli line the uterine tubes.

26.18 Endometrium is a highly vascular, secretory epithelium that provides the oxygen and nutrients needed to sustain a fertilized egg. Myometrium is a thick smooth muscle layer that supports the uterine wall during pregnancy and contracts to expel the fetus at birth.

26.19 The stratum basalis of the endometrium provides cells to replace those that shed (the stratum functionalis) during each menstruation.

26.20 The vagina receives the penis during sexual intercourse, serves as the outlet for menstrual flow, and is the passageway for childbirth.

26.21 Anterior to the vaginal opening are the mons pubis, clitoris, prepuce, and external urethral orifice. Lateral to the vaginal opening are the labia minora and labia majora.

26.22 The anterior portion of the perineum is called the urogenital triangle because its borders form a triangle that encloses the urethral (*uro-*) and vaginal (*-genital*) orifices.

26.23 Prolactin, estrogens, and progesterone regulate the synthesis of milk. Oxytocin regulates the ejection of milk.

26.24 Alveoli are located in lobules of the mammary glands.

26.25 The hormones responsible for the proliferative phase of endometrial growth are estrogens; for ovulation, LH; for growth of the corpus luteum, LH; and for the midcycle surge of LH, estrogens.

26.26 The *SRY* gene on the Y chromosome is responsible for the development of the gonads into testes.

26.27 The presence of dihydrotestosterone (DHT) stimulates differentiation of the external genitals in males; its absence allows differentiation of the external genitals in females.

27 SURFACE ANATOMY

INTRODUCTION In Chapter 1 we introduced several branches of anatomy and noted their relationship to our understanding of the body's structure. Now that you have learned about all of the systems of the body, in this final chapter we will take a closer look at surface anatomy. Recall that **surface anatomy** is the study of the anatomical landmarks on the exterior of the body. A knowledge of surface anatomy will not only help you to identify structures on the body's exterior, but it will also assist you in locating the positions of various internal structures. In fact, this is the real value of surface anatomy, especially in a clinical setting—to visualize the locations of anatomy that cannot be seen by mapping it on the surface.

The study of surface anatomy involves two related but distinct activities: visualization and palpation. **Visualization** involves looking in a very selective and purposeful manner at a specific part of the body. **Palpation** (pal-PĀ-shun) means using the sense of touch to determine the location of an internal part of the body through the skin. Like visualization, palpation is performed selectively and purposefully, and it supplements information already gained through other methods, including visualization. Because of variations in depth from the surface and gender differences in the thickness of the dermis and subcutaneous layer over different parts of the body, palpations may range from light, to moderate, to deep. •

Did you ever wonder why health-care professionals use their knowledge of surface anatomy for conducting physical examinations and performing certain diagnostic tests?

27.1 OVERVIEW OF SURFACE ANATOMY

OBJECTIVE

- Explain how surface anatomy can be used in a physical examination.

Knowledge of surface anatomy has many applications, both anatomically and clinically. From an anatomical viewpoint, knowledge of surface anatomy can provide valuable information regarding the location of structures such as bones, muscles, blood vessels, nerves, lymph nodes, and internal organs. Clinically, expertise in surface anatomy is essential for conducting a physical examination and performing certain diagnostic tests. Healthcare professionals use their understanding of surface anatomy to learn where to take the pulse, measure blood pressure, draw blood, stop bleeding, insert needles and tubes, make surgical incisions, *reduce* (straighten) fractured bones, and listen to sounds made by the heart, lungs, and intestines. Clinicians also draw on their knowledge of surface anatomy to assess the status of lymph nodes and identify the presence of tumors or other unusual masses in the body.

Recall that there are five principal regions of the body: (1) **head**, (2) **neck**, (3) **trunk**, (4) **upper limbs**, and (5) **lower limbs**. These are discussed in this chapter in Exhibits 27.A through 27.E. Using these regions as our guide, we will begin our study of the surface anatomy with the head and then move to the other regions, concluding with the lower limbs. Because living bodies are best suited for studying surface anatomy, you will find it helpful to visualize and/or palpate the various structures described on your own body, or on a study partner, as you study each region. Following a brief introduction to a region of the body, you will be presented with a list of the prominent structures to locate. This directed approach will help to organize your learning efforts. Labeled photographs illustrate most of the structures listed in each region.

Also, in some of the exhibits we will examine how key surface structures allow you to visualize *transections* (transverse-sectional cuts) of the body to better understand the location of the body's internal anatomy. This will give you the ability to look at the surface of a body and understand where internal organs are located. Recall that throughout the book sections were used in many of the orientation diagrams to help you better visualize the anatomy illustrated in the figures.

Once you have studied each exhibit in this chapter, you will see how knowledge of the surface features of the principal regions of the body can help you locate the positions of many internal structures—such as bones, joints, muscles, vessels, nerves, and organs—in the chest, abdomen, and pelvis. Visualize and palpate as many of these structures as possible on your own body to gain a better appreciation of the application of surface anatomy to your anatomical and clinical studies.

CHECKPOINT

1. What is surface anatomy? Why are visualization and palpation important in learning surface anatomy?
2. What are some of the applications of knowledge of surface anatomy?

EXHIBIT 27.A   Surface Anatomy of the Head *(Figures 27.1–27.5)*

OBJECTIVE
• Describe the surface features of the head.

The **head** (*cephalic region*, or *caput*) contains the brain and sense organs (eyes, ears, nose, and tongue) and is divided into the cranium and face. The **cranium** (*skull*, or *brain case*) is that portion of the head that surrounds and protects the brain; the **face** is the anterior portion of the head.

Regions of the Cranium and Face

The cranium and face are divided into several regions, which are described here and illustrated in Figure 27.1.

• **Frontal region:** forms the front of the skull and includes the frontal bone.

• **Parietal region:** forms the crown of the skull and includes the parietal bones.
• **Temporal region:** forms the sides of the skull and includes the temporal bones.
• **Occipital region:** forms the base of the skull and includes the occipital bone.
• **Orbital** (*ocular*) **region:** includes the eyeball, eyebrow, and eyelid.
• **Infraorbital region:** the region inferior to the orbit.
• **Zygomatic region:** the region inferolateral to the orbit that includes the zygomatic (cheek) bone.
• **Nasal region:** region of the nose.
• **Oral region:** region of the mouth.
• **Mental region:** anterior portion of the mandible, or region of the chin.

Figure 27.1 Principal regions of the cranium and face.

🔑 The head consists of an anterior face and a cranium that surrounds the brain.

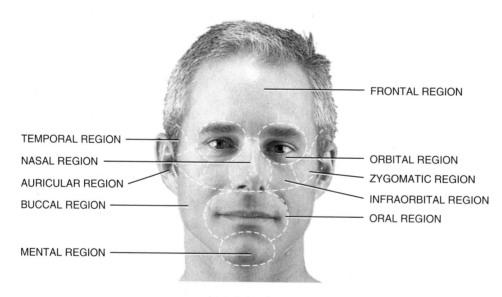

(a) Anterior view

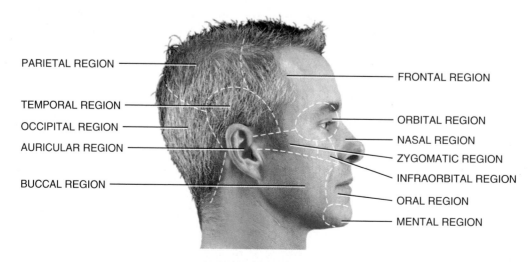

(b) Right lateral view

 Which regions of the head are named for bones that are deep to them?

EXHIBIT 27.A **873**

- **Buccal region** (BUK-al): region of the cheek.
- **Auricular region** (aw-RIK-ū-lar): region of the external ear.

In addition to locating the various regions of the head, it is also possible to visualize and/or palpate certain structures within each region. We will examine various bone and muscle structures of the head and then look at specific areas such as the eyes, ears, and nose.

Many of the bony landmarks of the head and muscles of facial expression are labeled in Figure 27.2. As you read about each feature, refer to the figure, and try to visualize and/or palpate each feature on your own head or on the head of a study partner.

Bony Landmarks of the Head

Several bony structures of the skull can be detected through palpation, including the following:

- **Sagittal suture.** Move the fingers from side to side over the superior aspect of the scalp in order to palpate this suture (see Figure 7.2a).
- **Coronal** and **lambdoid sutures.** Move the fingers from anterior to posterior to palpate these structures located on the frontal and occipital regions of the skull (see Figure 7.3a).

Figure 27.2 Surface anatomy of the head.

🔑 Several muscles of facial expression can be palpated while they are contracting.

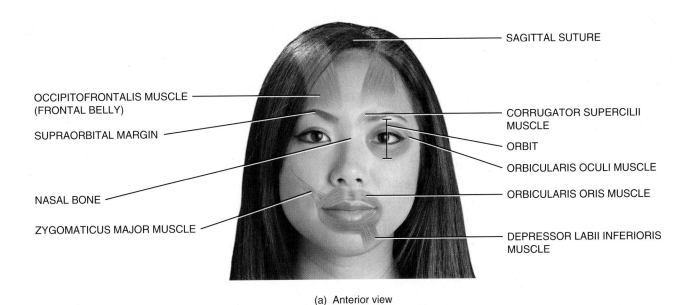

(a) Anterior view

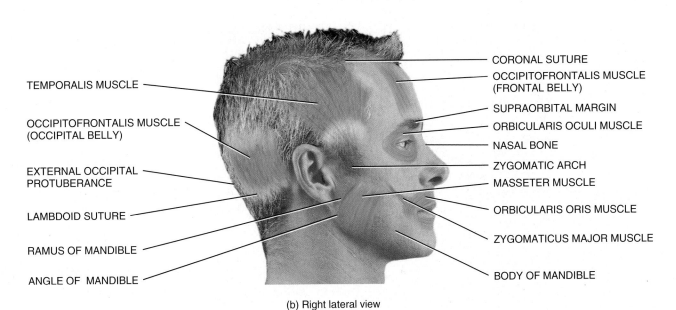

(b) Right lateral view

❓ **When palpating the sagittal suture, what sheetlike tendon is also palpated deep to the skin?**

EXHIBIT 27.A Surface Anatomy of the Head *(Figures 27.1–27.5)* CONTINUED

- **External occipital protuberance.** This is the most prominent bony landmark on the occipital region of the skull (see Figure 7.3a).
- **Orbit.** The entire circumference of the orbit can be palpated. Deep to the eyebrow, the superior aspect of the orbit, you can palpate the **supraorbital margin** of the frontal bone.
- **Nasal bones.** These can be palpated in the nasal region between the orbits on either side of the midline. If you wear glasses, the bridge of the glasses sits on these bones.
- **Mandible.** The **ramus** (vertical portion), **body** (horizontal portion), and **angle** (area where the ramus meets the body) of the mandible can be easily palpated at the mental and buccal regions of the head.
- **Zygomatic arch** and **zygomatic bones.** These can be palpated in the zygomatic region (see Figure 7.3c).
- **Mastoid process.** This is the prominent bony landmark situated posterior to the ear that is easily palpated in the auricular region (see Figure 7.3a).

- **Corrugator supercilii.** By frowning and drawing the eyebrows toward each other, this muscle can be felt above the nose near the medial end of the eyebrow.
- **Zygomaticus major.** By smiling, this muscle can be palpated between the corner of the mouth and zygomatic bone.
- **Depressor labii inferioris.** This muscle can be palpated in the mental region between the lower lip and chin when moving the lower lip inferiorly to expose the lower teeth.
- **Orbicularis oris.** By closing the lips tightly, this muscle can be palpated in the oral region around the margin of the lips.

Muscles of Mastication

- **Temporalis.** By alternately clenching the teeth and then opening the mouth, it is possible to palpate this muscle in the temporal region just superior to the zygomatic arch.
- **Masseter.** Again, by alternately clenching the teeth and then opening the mouth, this muscle can be palpated over the ramus of the mandible.

Muscles of Facial Expression

Several muscles of facial expression can be palpated while they are contracting:

- **Occipitofrontalis.** By raising and lowering the eyebrows, it is possible to palpate the frontal belly and the occipital belly in the frontal and occipital regions, respectively, as they alternately contract and the scalp moves forward and backward.
- **Orbicularis oculi.** By closing the eyes and placing the fingers on the eyelids, this muscle can be palpated in the orbital region by tightly squeezing the eyes shut.

Surface Features of the Eyes

Now we will examine the surface features of the eyes, illustrated in Figure 27.3.

- **Iris** (=rainbow). Circular, pigmented muscular structure behind the cornea (transparent covering over the iris).
- **Pupil.** Opening in center of iris through which light travels.
- **Sclera** *(skleros*=hard). "White" of the eye; a coat of fibrous tissue that covers the entire eyeball and is continuous with the transparent cornea anteriorly.
- **Conjunctiva** (kon-junk-TĪ-va). Membrane that covers the exposed surface of the eyeball and lines the eyelids.

Figure 27.3 **Surface anatomy of the right eye.**

🔑 The eyeball is protected by the eyebrow, eyelashes, and eyelids.

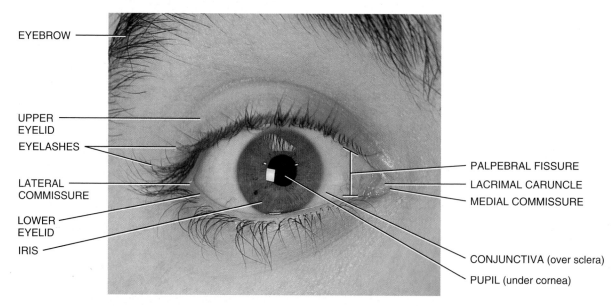

Anterior view

 Through which part of the eye does light enter?

EXHIBIT 27.A **875**

- **Eyelids** *(palpebrae)* (PAL-pe-brē). Folds of skin and muscle lined on their inner aspect by conjunctiva.
- **Palpebral fissure.** Space between the eyelids when they are open; when the eye is closed, it lies just inferior to the level of the pupil.
- **Medial commissure** (KOM-i-shur). Medial site of the union of the upper and lower eyelids.
- **Lateral commissure.** Lateral site of the union of the upper and lower eyelids.
- **Lacrimal caruncle** *(lacrim=*tears). Fleshy, yellowish projection of the medial commissure that contains modified sudoriferous (sweat) and sebaceous (oil) glands.
- **Eyelashes.** Hairs on the margin of the eyelids, usually arranged in two or three rows.
- **Eyebrows** *(supercilia).* Several rows of hairs superior to the upper eyelids.

Surface Features of the Ears

Next, locate the surface features of the ears in Figure 27.4.

- **Auricle** (AW-ri-kul). Shell-shaped portion of the external ear; also called the *pinna*. It funnels sound waves into the external auditory canal and plays an important role in helping to *localize* sound (determine its location). Just posterior to the auricle, the mastoid process of the temporal bone can be palpated.

- **Tragus** (TRĀ-gus; *tragos*=goat; when hair grows on the tragus it is thought to resemble the beard on a goat's chin). Cartilaginous projection anterior to the external auditory canal. Just anterior to the tragus and posterior to the neck of the mandible you can palpate the **superficial temporal artery** and feel the pulse in the vessel. Inferior to this point, it is possible to feel the **temporomandibular joint (TMJ).** As you open and close your jaw, you can feel the movement of the mandibular condyle.
- **Antitragus.** Cartilaginous projection opposite the tragus.
- **Concha** (KON-ka). Hollow of the auricle.
- **Helix.** Superior and posterior free margin of the auricle.
- **Antihelix.** Semicircular ridge superior and posterior to the tragus.
- **Triangular fossa.** Depression in the superior portion of the antihelix.
- **Lobule.** Inferior portion of the auricle; it does not contain cartilage. Commonly referred to as the earlobe.
- **External auditory canal (meatus).** Canal about 3 cm (1 in.) long extending from the external ear to the eardrum (tympanic membrane). It contains ceruminous glands that secrete cerumen (earwax). The **condylar process** of the mandible can be palpated by placing your little finger in the canal and opening and closing your mouth (see Figure 9.11e).

Figure 27.4 Surface anatomy of the right ear.

 The most conspicuous surface feature of the ear is the auricle.

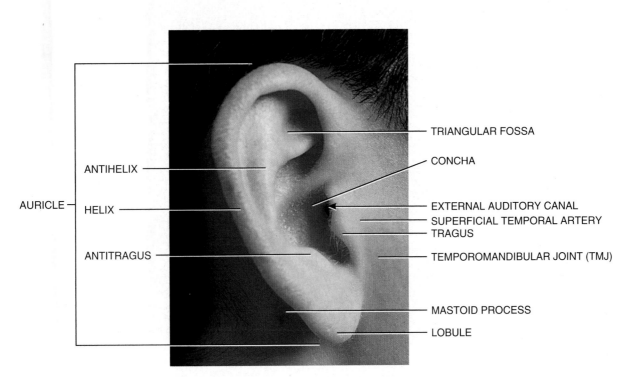

Right lateral view

Which part of the temporal bone can be palpated just posterior to the auricle?

EXHIBIT 27.A Surface Anatomy of the Head *(Figures 27.1–27.5)* CONTINUED

Surface Features of the Nose and Mouth

To complete the discussion of the surface anatomy of the head, locate the following features of the nose and mouth (Figure 27.5):

- **Root.** Superior attachment of the nose at the forehead, between the eyes.
- **Apex.** Tip of the nose.
- **Dorsum nasi.** Rounded anterior border connecting the root and apex; in profile, it may be straight, convex, concave, or wavy.
- **External naris.** External opening into the nose.
- **Ala.** Convex flared portion of the inferior lateral surface of the nose.
- **Bridge.** Superior portion of the dorsum nasi, superficial to and formed by the nasal bones.
- **Philtrum** (FIL-trum=love charm or potion). The vertical groove on the upper lip that extends along the midline to the inferior portion of the nose.
- **Lips** *(labia)*. Superior and inferior fleshy borders of the oral cavity.

✓ CHECKPOINT

3. Outline the regions of the head and name one structure in each.
4. List and define five surface features of the eyes, ears, and nose.
5. How would you locate the superficial temporal artery, temporomandibular joint, zygomatic arch, mastoid process, and condylar process of the mandible?

Figure 27.5 **Surface anatomy of the nose and lips.**

🔑 The external nares permit air to move into and out of the nose during breathing.

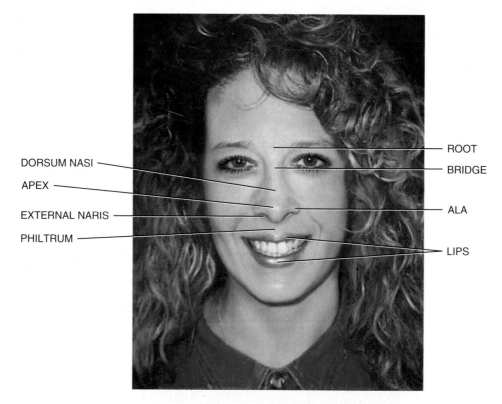

DORSUM NASI
APEX
EXTERNAL NARIS
PHILTRUM

ROOT
BRIDGE
ALA
LIPS

Anterior view

 What is the name of the anterior border of the nose that connects the root and apex?

EXHIBIT 27.B **877**

EXHIBIT 27.B Surface Anatomy of the Neck *(Figures 27.6–27.7)*

O B J E C T I V E
• Describe the surface features of the neck.

The neck (*cervical*) is the superior portion of the trunk that connects the head to the thorax. It is divided into an *anterior cervical region*, two *lateral cervical regions*, and a *posterior cervical region (nucha)*.

The following are the major surface features of the neck, most of which are illustrated in Figure 27.6:

• **Thyroid cartilage.** The largest of the cartilages that compose the larynx (voice box). It is the most prominent structure in the midline of the anterior cervical region. The anterior junction of the two plates of cartilage of the thyroid cartilage forms the **laryngeal prominence** or *Adam's apple*. The common carotid artery *bifurcates* (divides) at the level of the superior border of the thyroid cartilage to form the internal and external carotid arteries.

• **Hyoid bone.** Located just superior to the thyroid cartilage. It is the first structure palpated in the midline inferior to the chin. It is easily palpated laterally as you move posterior from its midline body onto the greater cornu (see Figure 7.14).

• **Cricoid cartilage.** A laryngeal cartilage located just inferior to the thyroid cartilage, it attaches the larynx to the trachea (windpipe). After you pass your finger over the cricoid cartilage moving inferiorly, your fingertip sinks in. The cricoid cartilage is used as a landmark for performing a tracheotomy.

• **Thyroid gland.** Two-lobed gland just inferior to the larynx with one lobe on either side.

• **Sternocleidomastoid.** Muscle that forms the major portion of the lateral aspect of the neck. If you rotate your head to either side, you can palpate the muscle from its origin on the sternum and clavicle to its insertion on the mastoid process of the temporal bone. Recall that the sternocleidomastoid

muscle divides the neck into anterior and posterior triangles (see Figure 11.9).

• **Common carotid artery.** Lies just deep to the sternocleidomastoid muscle along its anterior border.

• **Internal jugular vein.** Located lateral to the common carotid artery.

• **Subclavian artery.** Located just lateral to the inferior portion of the sternocleidomastoid muscle. Pressure on this artery can stop bleeding in the upper limb because this vessel supplies blood to the entire limb.

• **External carotid artery.** Located superior to the larynx, just anterior to the sternocleidomastoid muscle, this artery is the site of the carotid (neck) pulse.

• **External jugular vein.** Located superficial to the sternocleidomastoid muscle, this vessel is readily seen if you are angry or if your collar is too tight.

• **Trapezius.** Muscle that extends inferiorly and laterally from the base of the skull and occupies a portion of the lateral cervical region. A "stiff neck" is frequently associated with inflammation of this muscle.

• **Vertebral spines.** The spinous processes of the cervical vertebrae may be felt along the midline of the posterior region of the neck. Especially prominent at the base of the neck is the spinous process of the seventh cervical vertebra, called the vertebra prominens, which is followed by the prominent spinous process of the first thoracic vertebra (see Figure 27.9). The sternocleidomastoid muscle is not only a landmark for several arteries and veins, it is also the muscle that divides the neck into two major triangles: anterior and posterior (see Figure 11.9). The triangles are important because of the structures that lie within their boundaries (see Exhibit 11.H).

Figure 27.6 Surface anatomy of the neck.

🔑 The anatomical subdivisions of the neck are the anterior cervical region, lateral cervical regions, and posterior cervical region.

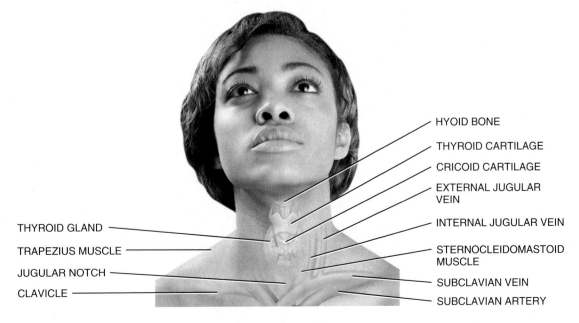

 Which muscle divides the neck into anterior and posterior triangles?

EXHIBIT 27.B Surface Anatomy of the Neck *(Figures 27.6–27.7)* CONTINUED

Cross Sections and Surface Relationships

The superior aspect of the neck is defined by the atlas, or first cervical vertebra. If a transverse section is projected forward in the horizontal plane from this vertebra, it intersects the uvula of the soft palate and the superior aspect of the tongue as it passes through the upper teeth. Figure 27.7 illustrates this section through the upper neck and head. Study this section and realize that the next time you look at someone's clenched teeth, you are looking at the uppermost level of their neck. This is the level of the joint between the atlas and axis that allows you to rotate your head as you indicate "no." Notice that this is also the level of the palatine tonsils at the junction of the oral cavity and the pharynx.

The anatomy of the base of the neck can be visualized readily by imagining a transverse section passing forward from the spinous process of the prominent seventh cervical vertebra (vertebra prominens) (Figure 27.7). Notice that the horizontal section through this vertebral level corresponds to a level through the upper part of the trachea and thyroid gland just inferior to the cricoid cartilage of the larynx. The common carotid artery has not yet split into the internal and external carotid arteries. The respiratory passageway (the trachea) and the digestive passageway

(the esophagus) have separated from the laryngopharynx. What muscles would be sectioned at this level? The following landmarks define other key levels in the neck:

- The fourth cervical vertebra and hyoid bone reside at the same transverse-sectional level.
- The superior aspect of the fifth cervical vertebra is at the same level as the top of the thyroid cartilage, which you will recall is also where the common carotid artery bifurcates into the internal and external carotid arteries.
- The thyroid cartilage spans the levels of the fifth and sixth cervical vertebrae.
- The cricoid cartilage is at the level of the sixth and seventh cervical vertebrae.

 **CHECKPOINT**

6. What is the neck?
7. How would you palpate the hyoid bone, cricoid cartilage, thyroid gland, common carotid artery, external carotid artery, and external jugular vein?
8. What muscles would be sectioned at the level of the first and seventh cervical vertebrae?

Figure 27.7 Transverse-sectional planes of the neck.

The superior and inferior boundaries of the neck are defined by transverse-sectional planes.

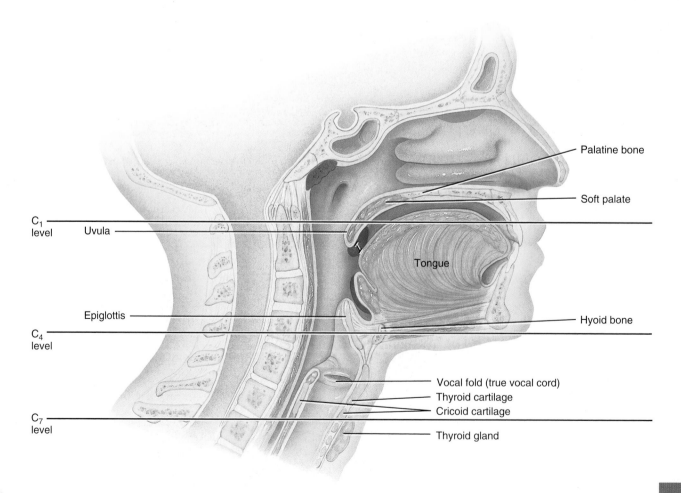

C$_1$ level
Uvula

Palatine bone

Soft palate

Tongue

Epiglottis

C$_4$ level

Hyoid bone

Vocal fold (true vocal cord)
Thyroid cartilage
Cricoid cartilage

C$_7$ level

Thyroid gland

Which synovial joint is located at the superior plane of the neck?

EXHIBIT 27.C **879**

EXHIBIT 27.C Surface Anatomy of the Trunk *(Figures 27.8–27.11)*

OBJECTIVE
• Describe the surface anatomy of the various regions of the trunk—back, chest, abdomen, and pelvis.

The **trunk** is divided into the neck, chest, abdomen, and pelvis. We will consider the lower three trunk regions separately here, along with their posterior surface, the back.

Surface Features of the Back

Among the prominent features of the **back**, or *dorsum*, are several superficial bones and muscles (Figure 27.8).

• **Vertebral spines.** The spinous processes of vertebrae, especially the thoracic and lumbar vertebrae, are quite prominent when the vertebral column is flexed.

• **Scapulae.** These easily identifiable surface landmarks on the back lie between ribs 2 and 7. In fact, it is also possible to palpate some ribs on the back. Depending on how lean a person is, it might be possible to palpate various parts of the scapula, such as the **vertebral border**, **axillary border**, **inferior angle**, **spine**, and **acromion**. The **spinous process of T3** is at about the same level as the spine of the scapula, and the **spinous process of T7** is approximately opposite the inferior angle of the scapula.

• **Latissimus dorsi.** A broad, flat, triangular muscle of the lumbar region that extends superiorly to the axilla and,

when well developed, gives a V-shape to the trunk of the body.

• **Erector spinae** *(sacrospinalis)*. Muscle located on either side of the vertebral column between the skull and iliac crests.

• **Infraspinatus.** Muscle located inferior to the spine of the scapula.

• **Trapezius.** Muscle that extends from the cervical and thoracic vertebrae to the spine and acromion of the scapula and the lateral end of the clavicle. It also occupies a portion of the lateral region of the neck and forms the posterior border of the posterior triangle of the neck.

• **Teres major.** Muscle located inferior to the infraspinatus muscle; together with the latissimus dorsi muscle it forms the inferior border of the posterior axillary fold (posterior wall of the axilla or armpit region).

• **Posterior axillary fold.** Formed by the latissimus dorsi and teres major muscles, the posterior axillary fold can be palpated between the fingers and thumb at the posterior aspect of the axilla (armpit region); forms the posterior wall of the axilla.

• **Triangle of auscultation** (aw-skul-TĀ-shun; *ausculto-=* listening). A triangular region of the back just medial to the inferior part of the scapula, where the rib cage is not covered by superficial muscles. It is bounded by the latissimus dorsi and trapezius muscles and vertebral border of the scapula. The triangle of auscultation is a landmark of clinical significance

Figure 27.8 Surface anatomy of the back.

🗝 The posterior boundary of the axilla—the posterior axillary fold—is formed mainly by the latissimus dorsi and teres major muscles.

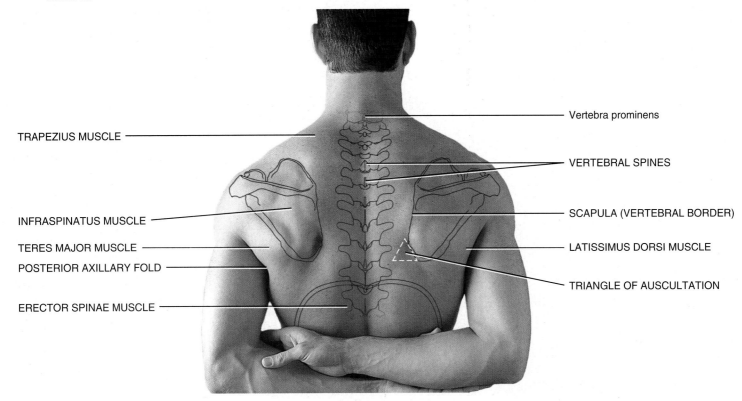

TRAPEZIUS MUSCLE
INFRASPINATUS MUSCLE
TERES MAJOR MUSCLE
POSTERIOR AXILLARY FOLD
ERECTOR SPINAE MUSCLE

Vertebra prominens
VERTEBRAL SPINES
SCAPULA (VERTEBRAL BORDER)
LATISSIMUS DORSI MUSCLE
TRIANGLE OF AUSCULTATION

Posterior view

 What is the clinical significance of the triangle of auscultation?

EXHIBIT 27.C Surface Anatomy of the Trunk *(Figures 27.8–27.11)* CONTINUED

because in this area respiratory sounds can be heard clearly through a stethoscope pressed against the skin. If a patient folds the arms across the chest and bends forward, the lung sounds can be heard clearly in the intercostal space between ribs 6 and 7.

Surface Features of the Chest

The surface anatomy of the **chest**, or anterior aspect of the *thorax*, includes several distinguishing surface landmarks (Figure 27.9), as well as some surface markings used to identify the location of the heart and lungs within the chest.

- **Clavicle.** Visible at the junction of the neck and thorax. Inferior to the clavicle, especially where it articulates with the manubrium (superior portion) of the sternum, the **first rib** can be palpated. In a depression superior to the medial end of the clavicle, just lateral to the sternocleidomastoid muscle, the **trunks of the brachial plexus** and axillary artery can be palpated (see Figure 17.7a).
- **Sternoclavicular joint.** Formed where the medial end of the clavicle articulates with the manubrium of the sternum. Move your upper limb around and feel the subtle movements that occur at this joint. Just posterior to these joints, the internal jugular and subclavian veins unite to form the brachiocephalic veins.
- **Suprasternal notch.** A depression on the superior border of the manubrium of the sternum between the medial ends of the clavicles; the **trachea** (windpipe) can be palpated posterior to the notch. A horizontal line through this landmark is at the level of the second thoracic vertebra. Also called the *jugular notch* of sternum.
- **Manubrium of sternum.** Superior portion of the sternum at the same levels as the bodies of the third and fourth thoracic vertebrae and anterior to the arch of the aorta.

- **Sternal angle.** Formed at the junction between the manubrium and body of the sternum, located about 4 cm (1.5 in.) inferior to the suprasternal notch. It is palpable deep to the skin and locates the costal cartilage of the second rib. It is the most reliable landmark of the chest and the starting point from which ribs are counted. At or inferior to the sternal angle and slightly to the right, the trachea divides into right and left primary bronchi. It also represents the top of the heart, the branching of the pulmonary trunk, and the level where the azygos vein joins the superior vena cava. A horizontal line through this landmark is at the level of the intervertebral disc between the fourth and fifth thoracic vertebrae.
- **Body of sternum.** Midportion of the sternum anterior to the heart and the vertebral bodies of T5–T8.
- **Xiphoid process of sternum.** Inferior portion of the sternum, medial to the seventh costal cartilages. The joint between the xiphoid process and body of the sternum is called the **xiphisternal joint**. The heart lies on the diaphragm deep to this joint. A horizontal line through this joint is at the level of the ninth thoracic vertebra.
- **Costal margin.** Inferior edges of the costal cartilages of ribs 7 through 10. The first costal cartilage lies inferior to the medial end of the clavicle; the seventh costal cartilage is the most inferior costal cartilage to articulate directly with the sternum; the tenth costal cartilage forms the most inferior part of the costal margin when viewed anteriorly. At the superior end of the costal margin is the xiphisternal joint.
- **Serratus anterior.** Muscle inferior and lateral to the pectoralis major muscle.
- **Ribs.** Twelve pairs of ribs help to form the bony cage of the thoracic cavity. Depending on body leanness, they may or may not be visible. One site for listening to the heartbeat in adults is the left fifth intercostal space, just medial to the left midclavicular line. This point marks the apex of the heart.

Figure 27.9 Surface anatomy of the chest.

 Thoracic viscera are protected by the sternum, ribs, and thoracic vertebrae, which form the skeleton of the thorax.

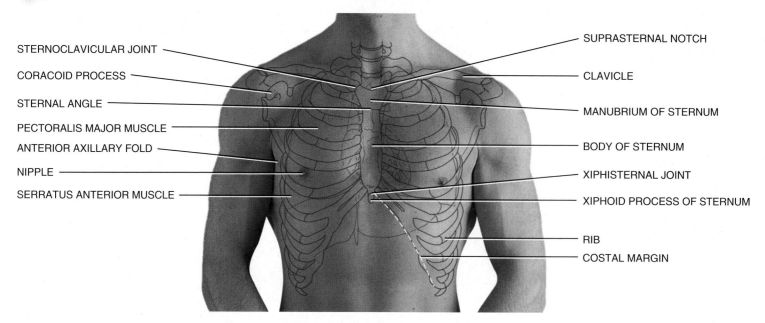

Anterior view of surface features of the chest

What is the most reliable landmark of the chest?

EXHIBIT 27.C **881**

- **Mammary glands.** Accessory organs of the female reproductive system located inside the breasts. These glands overlie the pectoralis major muscle (two-thirds) and serratus anterior muscle (one-third). After puberty, the mammary glands enlarge to their hemispherical shape, and in young adult females they extend from the second through the sixth ribs and from the lateral margin of the sternum to the midaxillary line. The **midaxillary line** is a vertical line that extends downward on the lateral thoracic wall from the center of the axilla.
- **Nipples.** Superficial to the fourth intercostal space or fifth rib, about 10 cm (4 in.) from the midline in males and most females. The position of the nipples in females is variable depending on the size and pendulousness of the breasts. The right dome of the diaphragm is just inferior to the right nipple, the left dome is about 2–3 cm (1 in.) inferior to the left nipple, and the central tendon is at the level of the xiphisternal joint.
- **Anterior axillary fold.** Formed by the lateral border of the pectoralis major muscle; can be palpated between the fingers and thumb; forms the anterior wall of the axilla (armpit region).
- **Pectoralis major.** Principal upper chest muscle; in the male, the inferior border of the muscle forms a curved line leading to the anterior wall of the axilla and serves as a guide to the fifth rib. In the female, the inferior border is mostly covered by the breast.

Although structures within the chest are almost totally hidden by the sternum, ribs, and thoracic vertebrae, it is important to indicate the surface markings of the heart and lungs. The projection (shape of an organ on the surface of the body) of the heart on the anterior surface of the chest is indicated by four points as follows (see Figure 13.1c). The *inferior left point* is the apex (inferior, pointed end) of the heart, which projects downward and to the left and can be palpated in the fifth intercostal space, about 9 cm (3.5 in) to the left of the midline, or in the fifth intercostal space at the midclavicular line (a vertical line extended down from the middle of the clavicle). The *inferior right point* is at the lower border of the costal cartilage of the right sixth rib, about 3 cm to the right of the midline (one finger breadth to the right of the lateral edge of the sternum). The *superior right point* is located at the superior border of the costal cartilage of the right third rib, about 3 cm to the right of the midline (one finger breadth to the right of the lateral edge of the sternum). The *superior left point* is located at the inferior border of the costal cartilage of the left second rib, about 3 cm to the left of the midline (one finger breadth to the left of the lateral edge of the sternum). If you connect the four points, you can determine the heart's location and size (about the size of a closed fist).

The lungs almost totally fill the thorax. The *apex* (superior, pointed end) of the lungs lies just superior to the medial third of the clavicles and is the only area that can be palpated (see Figure 23.9a). The anterior, lateral, and posterior surfaces of the lungs lie against the ribs. The *base* (inferior, broad end) of the lungs is concave and fits over the convex area of the diaphragm. It extends from the sixth costal cartilage anteriorly to the spinous process of the tenth thoracic vertebra posteriorly. (The base of the right lung is usually slightly higher than that of the left due to the position of the liver below it.) The covering around the lungs is called the *pleura*. At the base of each lung, the pleura extends about 5 cm below the base from the sixth costal cartilage anteriorly to the twelfth rib posteriorly. Thus, the lungs do not completely fill the pleural cavity in this area. A needle is typically inserted into this space in the pleural cavity to withdraw fluid from the chest.

Surface Features of the Abdomen and Pelvis

Following are some of the prominent surface anatomy features of the **abdomen** and **pelvis** (Figure 27.10):

- **Umbilicus.** Also called the *navel*; it marks the site of attachment of the umbilical cord to the fetus. It is level with the intervertebral disc between the bodies of vertebrae L3 and L4. The **abdominal aorta**, which branches into the right and left common iliac arteries anterior to the body of vertebra L4, can be palpated with deep pressure through the upper part of the anterior abdominal wall just to the left of the midline. The **inferior vena cava** lies to the right of the abdominal aorta and is wider; it arises anterior to the body of vertebra L5.
- **External oblique.** Muscle located inferior to the serratus anterior muscle. The aponeurosis of the muscle on its inferior border is the **inguinal ligament**, a structure along which hernias frequently occur.
- **Rectus abdominis.** Muscles located just lateral to the midline of the abdomen. They can be seen by raising the shoulders off the ground while in the supine position without using the arms (doing an abdominal crunch exercise).
- **Linea alba.** Flat, tendinous *raphe* (intersection of muscle tendons) forming a furrow along the midline between the rectus abdominis muscles. The furrow extends from the xiphoid process to the pubic symphysis. It is broad superior to the umbilicus and narrow inferior to it. The linea alba is a frequently selected site for abdominal surgery because an incision through it severs no muscles and only a few blood vessels and nerves.
- **Tendinous intersections of rectus abdominis.** Fibrous bands that run transversely or obliquely across the rectus abdominis muscles. On average, one intersection is at the level of the umbilicus, one is at the level of the xiphoid process, and one is midway between the two. The rectus abdominis muscles thus contain four prominent bulges and three tendinous intersections.
- **Linea semilunaris.** The lateral edge of the rectus abdominis muscle can be seen as a curved line that extends from the costal margin at the top of the ninth costal cartilage to the pubic tubercle.
- **McBurney's point.** A clinically important site (see next section) located two-thirds of the way down an imaginary line drawn between the umbilicus and anterior superior iliac spine. McBurney's point is an important landmark related to the appendix. Pressure of the finger on McBurney's point produces pain in acute **appendicitis**, inflammation of the appendix, aiding in diagnosis.
- **Iliac crest.** Superior margin of the ilium of the hip bone. It forms the outline of the superior border of the buttock. When you rest your hands on your hips, they rest on the iliac crests. A horizontal line drawn across the highest point of each iliac crest is called the **supracristal line**, which intersects the spinous process of the fourth lumbar vertebra. This vertebra is a landmark for performing a spinal tap.
- **Anterior superior iliac spine.** The anterior end of the iliac crest that lies at the upper lateral end of the fold of the groin.
- **Posterior superior iliac spine.** The posterior end of the iliac crest, indicated by a dimple in the skin that coincides with the middle of the sacroiliac joint, where the hip bone attaches to the sacrum.

EXHIBIT 27.C Surface Anatomy of the Trunk *(Figures 27.8–27.11)* **CONTINUED**

Figure 27.10 **Surface anatomy of the abdomen and pelvis.**

🔑 The linea alba is a frequent site for an abdominal incision because cutting through it severs no muscles and only a few blood vessels and nerves.

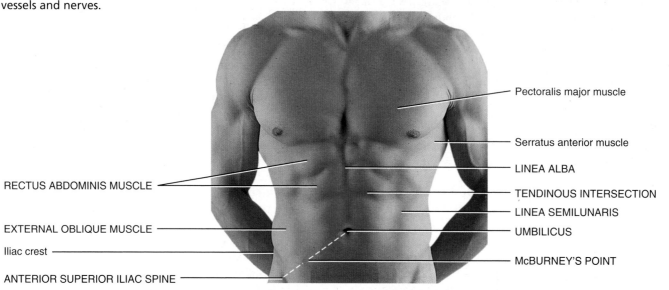

(a) Anterior view of abdomen

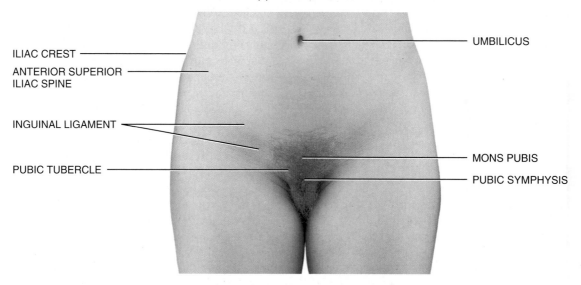

(b) Anterior view of pelvis

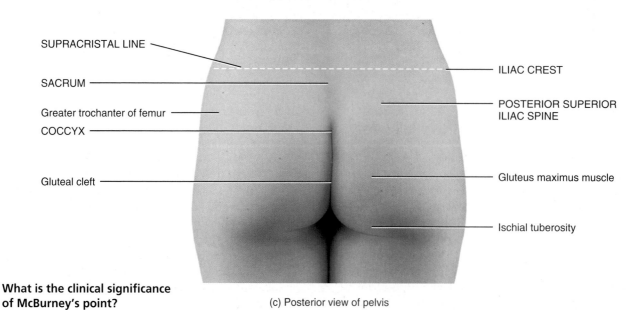

❓ **What is the clinical significance of McBurney's point?**

(c) Posterior view of pelvis

EXHIBIT 27.C **883**

- **Pubic tubercle.** Projection on the superior border of the pubis of the hip bone. Attached to it is the medial end of the **inguinal ligament**, the inferior free edge of the aponeurosis of the external oblique muscle that forms the **inguinal canal**. The lateral end of the ligament is attached to the anterior superior iliac spine. The spermatic cord in males and the round ligament of the uterus in females pass through the inguinal canal.
- **Pubic symphysis.** Anterior joint of the hip bones; palpated as a firm resistance in the midline at the inferior portion of the anterior abdominal wall.
- **Mons pubis.** An elevation of adipose tissue covered by skin and pubic hair that is anterior to the pubic symphysis.
- **Sacrum.** The fused spinous processes of the sacrum, called the **median sacral crest**, can be palpated beneath the skin superior to the **gluteal cleft** or *natal cleft*, a depression along the midline that separates the buttocks (and is part of the discussion on the buttocks in Exhibit 27.E).
- **Coccyx.** The inferior surface of the tip of the coccyx can be palpated in the gluteal cleft, about 2.5 cm (1 in.) superior to the anus.

Transverse Sections and Surface Relationships

The locations of the internal organs of the trunk are generally not well understood by most people. Organs such as the heart, lungs, stomach, and liver typically are thought to occur in much lower positions within the trunk than they are actually located. Understanding a few key anterior bony landmarks and correlating horizontal sections from these landmarks to vertebral levels can enhance your understanding of the position of the internal organs of the trunk. Study each of the sections that follow (Figure 27.11) and visualize their relationship to key surface features.

❶ *Suprasternal notch to second thoracic intervertebral disc.* A transverse section through the suprasternal notch corresponds to the intervertebral disc between the second and third thoracic vertebrae. Note the apex of the lungs project above this level into the inferior aspect of the neck. Most people do not realize that the lungs occupy the lower portion of the neck and are susceptible to injury here. At this level the major blood vessels of the upper limbs sit just deep to the sternal end of the clavicles. The brachiocephalic veins anastomose to form the superior vena cava about two finger-breadths to the left and inferior to the suprasternal notch.

❷ *Sternal angle to fourth thoracic intervertebral disc.* A transverse section through the sternal angle corresponds to the level of the intervertebral disc between the fourth and fifth thoracic vertebrae. This is one of the busiest areas in the thorax, marking the intersection between the upper border of the heart and the root of the lungs. This superior level on your chest wall marks the following: the beginning and end of the aortic arch; the bifurcation of the trachea; the level of the pulmonary arteries branching from the pulmonary trunk; the level where the azygos vein arches over the principal bronchus and pulmonary artery to enter the superior vena cava.

❸ *Fifth costosternal joint to eighth thoracic intervertebral disc.* A transverse section from the junction of the fifth costal cartilage with the sternum intersects the intervertebral

Figure 27.11 Transverse-sectional planes of the trunk. Transverse-sectional planes of the trunk illustrate how the deep anatomy of the internal viscera is projected to key surface landmarks on the trunk wall.

The sternal angle marks the top of the heart and the busy intersection between the heart and lungs.

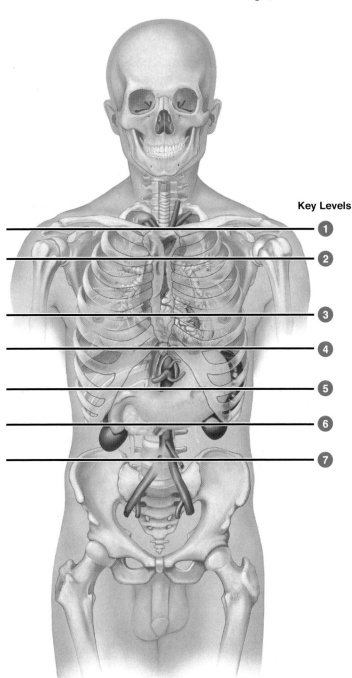

Key Levels

 Which vertebral level is marked by the plane of the umbilicus?

disc between the eighth and ninth thoracic vertebrae. This level marks the superiormost aspect of the dome of the diaphragm and the level where the inferior vena cava passes through the diaphragm to enter the right atrium. This level also marks the top of the liver.

EXHIBIT 27.C Surface Anatomy of the Trunk *(Figures 27.8–27.11)* CONTINUED

❹ ***Xiphoid process to tenth thoracic vertebra.*** The transverse section that passes through the middle of the xiphoid process corresponds to the level of the body of the tenth thoracic vertebra. This level marks the esophageal hiatus, where the esophagus passes through the diaphragm to enter the abdomen and stomach. The branches of the vagus (X) nerve follow the esophagus through this opening. The fundus of the stomach is situated at this level.

❺ ***Transpyloric plane.*** The transpyloric line is formed by the transverse section that passes between the junction of the eighth and ninth costal cartilages and the intervertebral disc between the twelfth thoracic and first lumbar vertebrae. Just superior to this line, anterior to the body of the twelfth thoracic vertebra, is the entry point through the diaphragm of the aorta into the abdomen. The transpyloric plane corresponds to the level of the pylorus of the stomach, which marks the end of the stomach and the beginning of the duodenum. The plane intersects the body of the stomach, so that part of the stomach is above it and part of the stomach is below it. The pancreas and the hilum of the kidneys are also at this level.

❻ ***Subcostal plane.*** The transverse section that passes through the lowest points of the rib cage and the intervertebral disc between the second and third lumbar vertebrae is called the subcostal plane. This marks the lowest aspect of the liver. Note that the anterior projection of the liver onto the trunk wall spans from the lower aspect of the sternum to the bottom of the rib cage; however, it follows the contour of the ribs and is almost entirely hidden beneath the rib cage and sternum. This section intersects the inferior part of the duodenum and the transverse colon. The majority of the small intestine is situated below this level.

❼ ***Supracristal plane.*** This plane is defined as the intersection of the tubercles at the top of the iliac crests and the body of

the fourth lumbar vertebra. It is approximately at the level of the umbilicus. The ascending and descending portions of the colon are present laterally, and the coils of the jejunum and ileum dominate the central part of this section. Posteriorly, the common iliac vessels separate from the aorta and inferior vena cava at this level.

Chapter 1 discussed the division of the abdomen and pelvis into nine regions (see Figure 1.8). The nine-region designation subdivided the abdomen and pelvis by drawing two vertical lines (left and right clavicular), an upper horizontal line (subcostal), and a lower horizontal line (transtubercular). Take time to examine Figure 1.8 carefully, so that you can see which organs or parts of organs lie within each region.

In Chapter 11, as part of your study of skeletal muscles, you were introduced to the **perineum**, the diamond-shaped region medial to the thigh and buttocks. It is bounded by the pubic symphysis anteriorly, the ischial tuberosities laterally, and the coccyx posteriorly. A transverse line drawn between the ischial tuberosities divides the perineum into an anterior *urogenital triangle* that contains the external genitals, and a posterior *anal triangle* that contains the anus. Details concerning the functions of the perineum are provided in Chapter 11. At this point keep in mind that the musculature of the perineum constitutes the inferior wall of the pelvic cavity.

✅ **CHECKPOINT**

9. What is the significance of the xiphisternal joint as a landmark?
10. Why is the midaxillary line an important anatomical landmark?
11. Why do the lungs not completely fill the pleural cavity? Why is this important clinically?
12. Why are the linea alba, linea semilunaris, and supracristal line important landmarks?
13. What is the inguinal ligament, and why is it important?

EXHIBIT 27.D **885**

EXHIBIT 27.D Surface Anatomy of the Upper Limb *(Figures 27.12–27.15)*

OBJECTIVE

• Describe the surface anatomy of the various regions of the upper limb.

The **upper limb** consists of the pectoral (shoulder) girdle and free upper limb. The free upper limb can in turn be divided into the arm, forearm, and hand. We will consider each of these regions separately.

Surface Features of the Shoulder

The **shoulder**, or *acromial region*, is located at the lateral aspect of the clavicle, where the clavicle joins the scapula and the scapula joins the humerus. The region presents several conspicuous surface features (Figure 27.12).

• **Acromioclavicular joint.** A slight elevation at the lateral end of the clavicle. It is the joint between the acromion of the scapula and the clavicle.
• **Acromion.** The expanded lateral end of the spine of the scapula that forms the top of the shoulder. It can be palpated about 2.5 cm (1 in.) distal to the acromioclavicular joint.
• **Humerus.** The most laterally palpable bony structure. The **greater tubercle** of the humerus may be palpated on the superior aspect of the shoulder.

• **Deltoid.** Triangular muscle that forms the rounded prominence of the shoulder. The deltoid muscle is a frequent site for an intramuscular injection. To avoid injury to major blood vessels and nerves, the injection is given in the midportion of the muscle about 2–3 finger-widths inferior to the acromion of the scapula and lateral to the axilla.
• **Coracoid process.** Anterior projection of the scapula that can be palpated at the medial border of the deltoid muscle just inferior to the clavicle.

Surface Features of the Armpit

The armpit region, or **axilla**, is a pyramid-shaped area at the junction of the arm and the chest that enables blood vessels and nerves to pass between the neck and the free upper limbs (Figure 27.13a, b).

• **Apex.** The apex of the axilla is surrounded by the clavicle, scapula, and first rib.
• **Base.** The base of the axilla is formed by the concave skin and fascia that extends from the arm to the chest wall. It contains hair. Deep to the base the axillary lymph nodes can be palpated.
• **Anterior wall.** The anterior wall of the axilla is composed mainly of the pectoralis major muscle (anterior axillary fold; see also Figure 27.9).

Figure 27.12 Surface anatomy of the shoulder.

 The deltoid muscle gives the shoulder its rounded prominence.

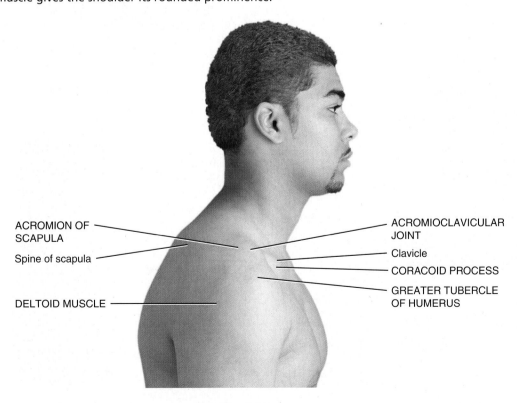

ACROMION OF SCAPULA

Spine of scapula

DELTOID MUSCLE

ACROMIOCLAVICULAR JOINT

Clavicle

CORACOID PROCESS

GREATER TUBERCLE OF HUMERUS

Right lateral view

 Which structure forms the top of the shoulder?

EXHIBIT 27.D Surface Anatomy of the Upper Limb *(Figures 27.12–27.15)* **CONTINUED**

- **Posterior wall.** The posterior wall of the axilla is formed mainly by the teres major and latissimus dorsi muscles (posterior axillary fold; see also Figure 27.8).
- **Medial wall.** The medial wall of the axilla is formed by ribs 1–4 and their corresponding intercostal muscles, plus the overlying serratus anterior muscle.
- **Lateral wall.** Finally, the lateral wall of the axilla is formed by the triceps brachii, coracobrachialis, and biceps brachii muscles and the superior portion of the shaft of the humerus. Passing through the axilla are the axillary artery and vein, branches of the brachial plexus, and axillary lymph nodes. All of these structures are surrounded by a considerable amount of axillary fat.

Surface Features of the Arm and Elbow

The **arm**, or *brachium*, is the region between the shoulder and elbow. The **elbow**, or *cubitus*, is the region where the arm and forearm join. The arm and elbow present several surface anatomy features (Figure 27.13b–d).

- **Humerus.** This arm bone may be palpated along its entire length, especially near the elbow (see descriptions of the medial and lateral epicondyles that follow).
- **Biceps brachii.** Muscle that forms the bulk of the anterior surface of the arm. On the medial side of the muscle is a groove that contains the **brachial artery**.

Figure 27.13 Surface anatomy of the axilla, arm, and elbow. The location of the muscles that form the walls of the axilla are shown in Figure 11.17a.

🔑 The biceps brachii and triceps brachii muscles form the bulk of the musculature of the arm.

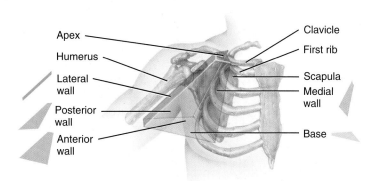

(a) Location and parts of the axilla

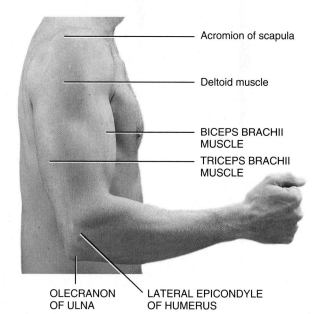

(c) Right lateral view of arm

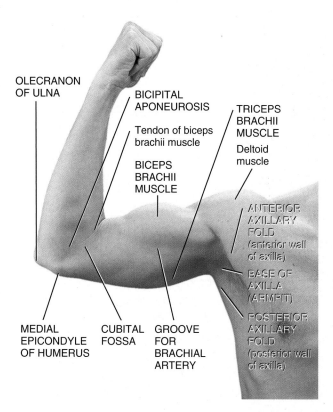

(b) Medial view of arm

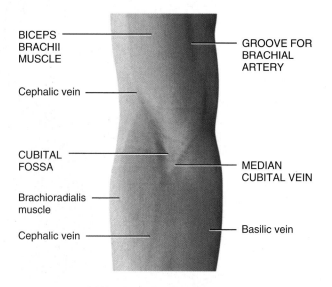

(d) Anterior view of cubital fossa

 Which blood vessel in the cubital fossa is frequently used to withdraw blood?

EXHIBIT 27.D **887**

- **Triceps brachii.** Muscle that forms the bulk of the posterior surface of the arm.
- **Medial epicondyle.** Medial projection of the humerus near the elbow.
- **Lateral epicondyle.** Lateral projection of the humerus near the elbow.
- **Olecranon.** Projection of the proximal end of the ulna between and slightly superior to the epicondyles when the forearm is extended; it forms the elbow.
- **Ulnar nerve.** Can be palpated in a groove posterior to the medial epicondyle. The "funny bone" is the region where the ulnar nerve rests against the medial epicondyle. Hitting the nerve at this point produces a sharp pain along the medial side of the forearm that almost everyone would agree isn't very funny.
- **Cubital fossa.** Triangular space in the anterior region of the elbow bounded proximally by an imaginary line between the humeral epicondyles, laterally by the medial border of the brachioradialis muscle, and medially by the lateral border of the pronator teres muscle; contains the tendon of the biceps brachii muscle, the median cubital vein, brachial artery and its terminal branches (radial and ulnar arteries), and parts of the median and radial nerves.
- **Median cubital vein.** Crosses the cubital fossa obliquely and connects the laterally positioned cephalic vein with the medially positioned basilic vein. The median cubital vein is frequently used to withdraw blood from a vein for diagnostic purposes or to introduce substances into blood, such as medications, contrast media for radiographic procedures, nutrients, and blood cells and/or plasma for transfusion.
- **Brachial artery.** Continuation of the axillary artery that passes posterior to the coracobrachialis muscle and then medial to the biceps brachii muscle. It enters the middle of the cubital fossa and passes deep to the bicipital aponeurosis, which separates it from the median cubital vein. **Blood pressure** is usually measured in the brachial artery, when the cuff of a *sphygmomanometer* (blood pressure instrument) is wrapped around the arm and a stethoscope is placed over the brachial artery in the cubital fossa. Pulse can also be detected in the artery in the cubital fossa. However, blood pressure can be measured at any artery where you can obstruct blood flow. This becomes important in case the brachial artery cannot be utilized. In such situations, the radial or popliteal arteries might be used to obtain a blood pressure reading.
- **Bicipital aponeurosis.** An aponeurosis that inserts the biceps brachii muscle into the deep fascia in the medial aspect of the forearm (see Figure 11.19a, c). It can be felt when the muscle contracts.

Surface Features of the Forearm and Wrist

The **forearm**, or *antebrachium*, is the region between the elbow and wrist. The **wrist**, or *carpus*, is between the forearm and palm. Following are some prominent surface anatomy features of the forearm and wrist (Figure 27.14).

- **Ulna.** The medial bone of the forearm. It can be palpated along its entire length from the olecranon (see Figure 12.13a, b) to the **styloid process**, a projection on the distal end of the bone at the medial (little finger) side of the wrist. The **head of the ulna** is a conspicuous enlargement just proximal to the styloid process.
- **Radius.** The distal half of the radius can be palpated proximal to the thumb side of the hand. The proximal half is covered by muscles. The **styloid process** of the radius is a projection on the distal end of the bone at the lateral (thumb) side of the wrist.
- **Muscles.** Because of their close proximity it is difficult to identify individual muscles of the forearm. Instead, it is much easier to identify tendons as they approach the wrist and then trace them proximally to the following muscles:
  - **Brachioradialis.** Muscle located at the superior and lateral aspect of the forearm.
  - **Flexor carpi radialis.** The tendon of this muscle is on the lateral side of the forearm about 1 cm medial to the styloid process of the radius.

Figure 27.14 Surface anatomy of the forearm and wrist.

🔑 Muscles of the forearm are most easily identified by locating their tendons near the wrist and tracing them proximally.

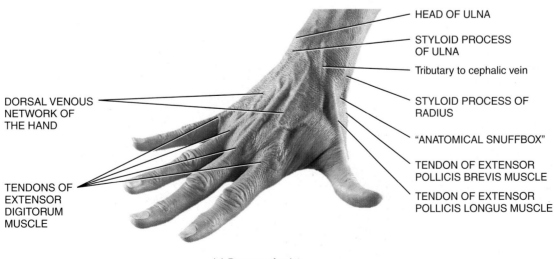

HEAD OF ULNA

STYLOID PROCESS OF ULNA

Tributary to cephalic vein

STYLOID PROCESS OF RADIUS

"ANATOMICAL SNUFFBOX"

TENDON OF EXTENSOR POLLICIS BREVIS MUSCLE

TENDON OF EXTENSOR POLLICIS LONGUS MUSCLE

DORSAL VENOUS NETWORK OF THE HAND

TENDONS OF EXTENSOR DIGITORUM MUSCLE

(a) Dorsum of wrist

FIGURE 27.14 CONTINUES ▶

EXHIBIT 27.D Surface Anatomy of the Upper Limb *(Figures 27.12–27.15)* CONTINUED

- **Palmaris longus.** The tendon of this muscle is medial to the flexor carpi radialis tendon and can be seen quite prominently if the wrist is slightly flexed and the base of the thumb and little finger are drawn together. About 15 to 20 percent of individuals do not have this muscle in at least one arm.
- **Flexor digitorum superficialis.** The tendon of this muscle is medial to the palmaris longus tendon and can be palpated by flexing the fingers at the metacarpophalangeal and proximal interphalangeal joints.
- **Flexor carpi ulnaris.** The tendon of this muscle is on the medial aspect of the forearm.
- **Radial artery.** Located on the lateral aspect of the wrist between the flexor carpi radialis tendon and styloid process of the radius. It is frequently used to take a pulse.
- **Pisiform bone.** Medial bone of the proximal row of carpals that can be palpated as a projection distal and anterior to the styloid process of the ulna.
- **"Anatomical snuffbox."** A triangular depression between tendons of extensor pollicis brevis and extensor pollicis longus

muscles. It derives its name from a habit in previous centuries of taking a pinch of snuff (powdered tobacco or scented powder) and placing it in the depression before sniffing it into the nose. The styloid process of the radius, the base of the first metacarpal, trapezium, scaphoid, and deep branch of the radial artery can all be palpated in this depression.
- **Wrist creases.** Three more-or-less constant lines on the anterior aspect of the wrist (named *proximal*, *middle*, and *distal*) where the skin is firmly attached to underlying deep fascia.

Surface Features of the Hand

The **hand**, or *manus*, is the region from the wrist to the termination of the upper limb; it has several conspicuous surface features (Figure 27.15).

- **Knuckles.** Commonly refers to the dorsal aspect of the heads of metacarpals II–V (or 2–5), but also includes the dorsal aspects of the metacarpophalangeal and interphalangeal joints.

■ **FIGURE 27.14 CONTINUED** ▶

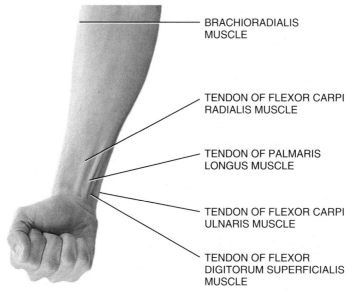

(b) Anterior aspect of forearm and wrist

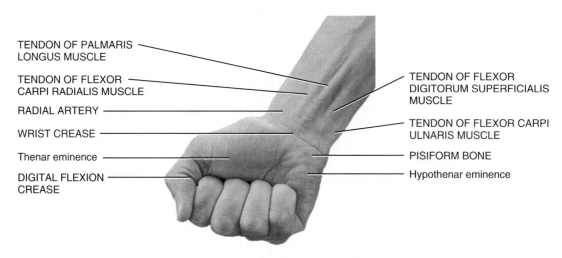

(c) Anterior aspect of wrist

 Which blood vessel is frequently used to take a pulse?

EXHIBIT 27.D **889**

- **Dorsal venous network of the hand** (*dorsal venous arch*). Superficial veins on the dorsum of the hand that drain blood into the cephalic vein. It can be displayed by compressing the blood vessels at the wrist for a few moments as the hand is opened and closed.
- **Extensor digiti minimi.** The tendon of this muscle can be seen on the dorsum of the hand in line with the phalanx of the little finger.
- **Extensor digitorum.** The tendons of this muscle can be seen on the dorsum of the hand in line with the phalanges of the ring, middle, and index fingers.
- **Extensor pollicis brevis.** This tendon (described previously) is in line with the phalanx of the thumb (see Figure 27.14a).
- **Thenar eminence.** Larger, rounded contour on the lateral aspect of the palm formed by muscles that move the thumb. Also called the ball of the thumb.
- **Hypothenar eminence.** Smaller, rounded contour on the medial aspect of the palm formed by muscles that move the little finger. Also called the ball of the little finger.

- **Palmar flexion creases.** Skin creases on the palm.
- **Digital flexion creases.** Skin creases on the anterior surface of the fingers.

✔ **CHECKPOINT**

14. What is the clinical significance of the deltoid muscle?
15. In which arteries of the upper limb can a pulse be detected?
16. What is the "funny bone"?
17. What is the clinical importance of the median cubital vein?
18. What artery is normally used to measure blood pressure?
19. Explain the easiest way to identify the muscles of the forearm.
20. What are the knuckles?
21. Why are the thenar and hypothenar eminences important?

Figure 27.15 Surface anatomy of the hand.

🔑 Several tendons on the dorsum of the hand can be identified by their alignment with the phalanges of the digits.

KNUCKLES (proximal interphalangeal joints)

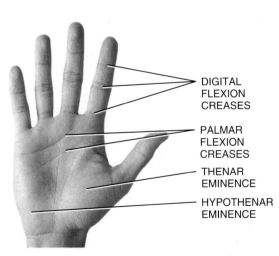

THENAR EMINENCE

HYPOTHENAR EMINENCE

(a) Palmar and dorsal view

KNUCKLES (metacarpophalangeal joints)

(b) Dorsal view

DIGITAL FLEXION CREASES

PALMAR FLEXION CREASES

THENAR EMINENCE

HYPOTHENAR EMINENCE

(c) Palmar view

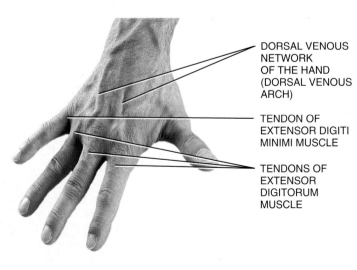

DORSAL VENOUS NETWORK OF THE HAND (DORSAL VENOUS ARCH)

TENDON OF EXTENSOR DIGITI MINIMI MUSCLE

TENDONS OF EXTENSOR DIGITORUM MUSCLE

(d) Dorsum

 Muscles that form the thenar eminence move which digit?

EXHIBIT 27.E Surface Anatomy of the Lower Limb *(Figures 27.16–27.18)*

OBJECTIVE
• Describe the surface features of the various regions of the lower limb.

The **lower limb** consists of the pelvic (hip) girdle, covered by the gluteal muscles forming the buttocks, and free lower limb. The free lower limb in turn is divided into four regions: the thigh, knee, leg, and ankle/foot. We will consider each of these regions separately.

Surface Features of the Buttock

The **buttock**, or *gluteal region*, is formed mainly by the gluteus maximus muscle. The outline of the superior border of the buttock is formed by the iliac crests (see Figure 27.10c). Following are some surface features of the buttock (Figure 27.16).

• **Gluteus maximus.** Muscle that forms the major portion of the prominence of the buttocks. The sciatic nerve is deep to this muscle.
• **Gluteus medius.** Muscle that is superior, anterior, and lateral to the gluteus maximus muscle. Another common site for an intramuscular injection is the gluteus medius muscle. In order to give this injection, the buttock is divided into quadrants and the upper outer quadrant is used as an injection site. The iliac crest serves as the landmark for this quadrant. This site is chosen because the gluteus medius muscle in this area is quite thick, and there is less chance of injury to the sciatic nerve or major blood vessels.
• **Gluteal cleft.** Depression along the midline that separates the left and right buttocks. Also called the *natal cleft*.
• **Gluteal fold.** Inferior limit of the buttock that roughly corresponds to the inferior margin of the gluteus maximus muscle.
• **Ischial tuberosity.** Just superior to the medial side of the gluteal fold, the ischial tuberosity bears the weight of the body when a person is seated.

• **Greater trochanter.** A projection of the proximal end of the femur on the lateral side of the thigh. It is about 20 cm (8 in.) inferior to the highest point of the iliac crest.

Surface Features of the Thigh and Knee

The **thigh**, or *femoral region*, is the region from the hip to the knee. The **knee**, or *genu*, is the region where the thigh and leg join. Several muscles are clearly visible in the thigh. Following are several surface features of the thigh and knee (Figure 27.17).

• **Sartorius.** Superficial anterior muscle that can be traced from the lateral aspect of the thigh to the medial aspect of the knee.
• **Quadriceps femoris.** Three of the four components of the quadriceps femoris muscle can be seen: **rectus femoris** at the midpoint of the anterior aspect of the thigh; **vastus medialis** at the anteromedial aspect of the thigh; and **vastus lateralis** at the anterolateral aspect of the thigh. The fourth component, the vastus intermedius, is deep to the rectus femoris (see Figure 11.22a). The vastus lateralis muscle of the quadriceps femoris group is another site that may be used for an intramuscular injection. The injection site is located at a point midway between the greater trochanter (see Figure 27.16) of the femur and the patella (knee cap). Injection in this area reduces the chance of injury to major blood vessels and nerves.
• **Adductor longus.** Muscle located at the superior aspect of the medial thigh. It is the most anterior of the three adductor muscles (adductor magnus, adductor brevis, and adductor longus; see Figure 11.22b).
• **Femoral triangle.** A space at the proximal end of the thigh formed by the inguinal ligament superiorly, the sartorius muscle laterally, and the adductor longus muscle medially. The triangle contains, from lateral to medial, the femoral nerve,

Figure 27.16 Surface anatomy of the buttock.

🔑 The gluteus medius muscle is a frequent site for an intramuscular injection.

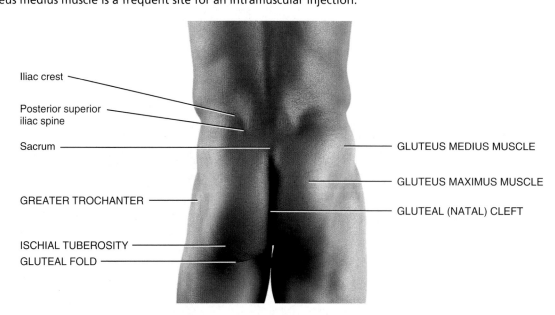

Iliac crest

Posterior superior iliac spine

Sacrum

GREATER TROCHANTER

ISCHIAL TUBEROSITY
GLUTEAL FOLD

GLUTEUS MEDIUS MUSCLE

GLUTEUS MAXIMUS MUSCLE

GLUTEAL (NATAL) CLEFT

Posterior view

 Which muscle forms the bulk of the buttock?

EXHIBIT 27.E **891**

Figure 27.17 Surface anatomy of the thigh and knee.

🔑 The quadriceps femoris and hamstrings form the bulk of the musculature of the thigh.

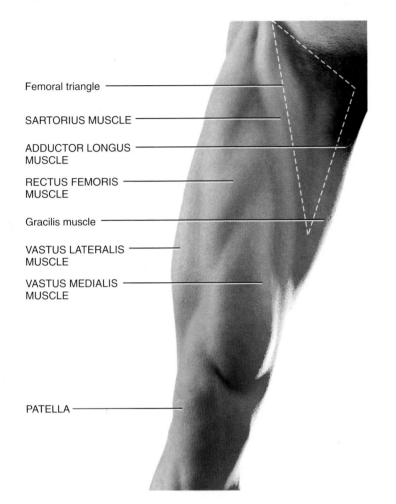

Femoral triangle

SARTORIUS MUSCLE

ADDUCTOR LONGUS MUSCLE

RECTUS FEMORIS MUSCLE

Gracilis muscle

VASTUS LATERALIS MUSCLE

VASTUS MEDIALIS MUSCLE

PATELLA

(a) Anterior view of thigh

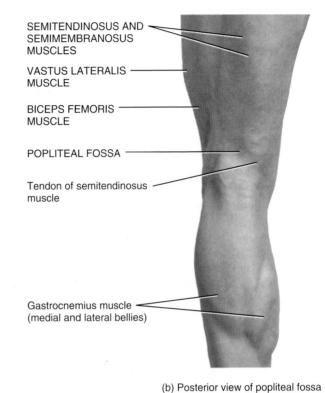

SEMITENDINOSUS AND SEMIMEMBRANOSUS MUSCLES

VASTUS LATERALIS MUSCLE

BICEPS FEMORIS MUSCLE

POPLITEAL FOSSA

Tendon of semitendinosus muscle

Gastrocnemius muscle (medial and lateral bellies)

(b) Posterior view of popliteal fossa

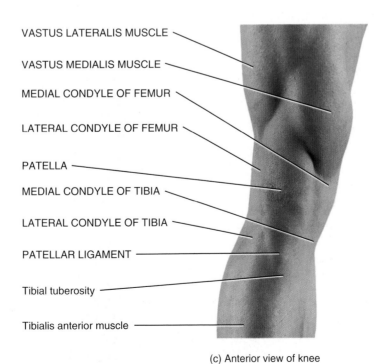

VASTUS LATERALIS MUSCLE

VASTUS MEDIALIS MUSCLE

MEDIAL CONDYLE OF FEMUR

LATERAL CONDYLE OF FEMUR

PATELLA

MEDIAL CONDYLE OF TIBIA

LATERAL CONDYLE OF TIBIA

PATELLAR LIGAMENT

Tibial tuberosity

Tibialis anterior muscle

(c) Anterior view of knee

❓ **Which muscles form the borders of the popliteal fossa?**

EXHIBIT 27.E Surface Anatomy of the Lower Limb *(Figures 27.16–27.18)* CONTINUED

artery, vein, and deep inguinal lymph nodes. The triangle is an important arterial pressure point in cases of severe hemorrhage of the lower limb. Hernias frequently occur in this area as they follow the vessels through the abdominal wall into the thigh.

• **Hamstrings.** Superficial, posterior thigh muscles located below the gluteal folds. They are the **biceps femoris**, which lies more laterally as it passes inferiorly to the knee, and the **semitendinosus** and **semimembranosus**, which lie medially as they pass inferiorly to the knee. The tendons of the hamstring muscles can be palpated laterally and medially on the posterior aspect of the knee.

The knee presents several distinguishing surface features.

• **Patella.** Also called the *kneecap*, this large sesamoid bone is located within the tendon of the quadriceps femoris muscle on the anterior surface of the knee along the midline.
• **Patellar ligament.** Continuation of the quadriceps femoris tendon inferior to the patella. Infrapatellar fat pads cushion the patellar ligament on both sides.
• **Medial condyle of femur.** Medial projection on the distal end of the femur.
• **Medial condyle of tibia.** Medial projection on the proximal end of the tibia.
• **Lateral condyle of femur.** Lateral projection on the distal end of the femur.
• **Lateral condyle of tibia.** Lateral projection on the proximal end of the tibia. All four condyles can be palpated just inferior to the patella on either side of the patellar ligament.

• **Popliteal fossa** (pop-LIT-ē-al). A diamond-shaped area on the posterior aspect of the knee that is clearly visible when the knee is flexed. The fossa is bordered superolaterally by the biceps femoris muscle, superomedially by the semimembranosus and semitendinosus muscles, and inferolaterally and inferomedially by the lateral and medial heads of the gastrocnemius muscle, respectively. The **head of the fibula** can be easily palpated on the lateral side of the popliteal fossa. The fossa also contains the popliteal artery and vein. It is sometimes possible to detect a pulse in the popliteal artery.

Surface Features of the Leg, Ankle, and Foot

The **leg**, or *crus*, is the region between the knee and ankle. The **ankle**, or *tarsus*, is between the leg and foot. The **foot** is the region from the ankle to the termination of the free lower limb. Following are several surface anatomy features of the leg, ankle, and foot (Figure 27.18).

• **Tibial tuberosity.** Bony prominence on the superior, anterior surface of the tibia into which the patellar ligament inserts.
• **Tibialis anterior.** Muscle that lies against the lateral surface of the tibia, where it is easy to palpate, particularly when the foot is dorsiflexed. The distal tendon of the muscle can be traced to its insertion on the medial cuneiform and base of the first metatarsal.
• **Tibia.** The medial surface and anterior border (shin) of the tibia are subcutaneous and can be palpated throughout the length of the bone.

Figure 27.18 Surface anatomy of the leg, ankle, and foot.

🔑 The calcaneal tendon is a common tendon for the gastrocnemius and soleus (calf muscles) and inserts into the calcaneus (heel bone).

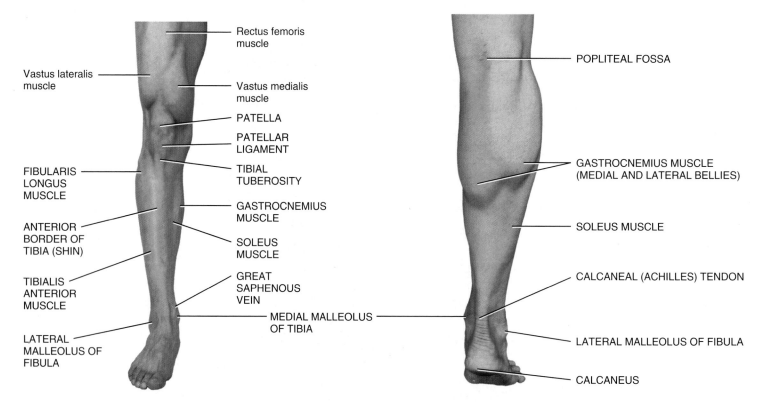

(a) Anterior view of leg, ankle, and foot

(b) Posterior view of leg and ankle

- **Fibularis longus.** A superficial lateral muscle that overlies the fibula. Also called the *peroneus longus*.
- **Gastrocnemius.** Muscle that forms the bulk of the midportion and superior portion of the posterior aspect of the leg. The medial and lateral bellies can be seen clearly in a person standing on tiptoe.
- **Soleus.** Muscle located deep to the gastrocnemius muscle; the soleus and the gastrocnemius are collectively referred to as the calf muscles.
- **Calcaneal tendon.** Prominent tendon of the gastrocnemius and soleus muscles on the posterior aspect of the ankle; inserts into the **calcaneus** (heel bone) of the foot. Also called the *Achilles tendon*.
- **Lateral malleolus of fibula.** Projection of the distal end of the fibula that forms the lateral prominence of the ankle. The head of the fibula, at the proximal end of the bone, lies at the same level as the tibial tuberosity.
- **Medial malleolus of tibia.** Projection of the distal end of the tibia that forms the medial prominence of the ankle.
- **Dorsal venous arch.** Superficial veins on the dorsum of the foot that unite to form the small and great saphenous veins.

The great saphenous vein is the longest vein of the body, extending from the foot to the groin, where it joins the femoral vein in the femoral triangle.
- **Extensor digitorum longus.** Tendons of this muscle are visible in line with phalanges II–V (2–5).
- **Extensor hallucis longus.** Tendon of this muscle is visible in line with phalanx I (great toe). In most people, pulsations in the dorsalis pedis artery may be felt just lateral to this tendon where the blood vessel passes over the navicular and cuneiform bones of the tarsus.

CHECKPOINT
22. Why is the greater trochanter of the femur important?
23. Why is the femoral triangle important? What structures does it contain?
24. What are the borders of the popliteal fossa? What structures pass through it?
25. Define the buttock. What is the gluteal cleft? Gluteal fold?
26. Which veins are formed by the dorsal venous arch?

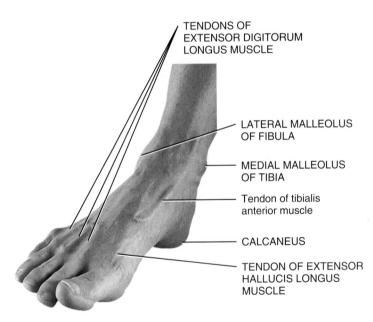

(c) Dorsum of foot

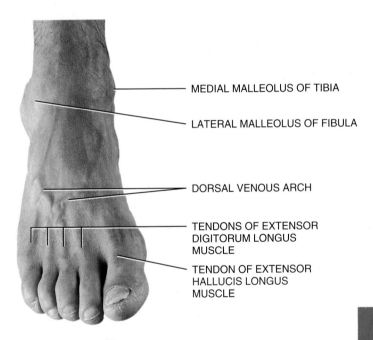

(d) Dorsum of foot

Which artery is just lateral to the tendon of the extensor digitorum muscle and is a site where a pulse may be detected?

CHAPTER REVIEW AND RESOURCE SUMMARY

WileyPLUS

Review

27.1 Overview of Surface Anatomy

1. Surface anatomy is the study of anatomical landmarks on the exterior of the body. Surface features may be noted by visualization and/or palpation.
2. The regions of the body include the head, neck, trunk, upper limbs, and lower limbs.
3. The head is divided into a cranium and face, each of which is divided into distinct regions; surface anatomy of the head is discussed in Exhibit 27.A.
4. Many muscles of facial expression and skull bones are easily palpated. The eyes, ears, and nose present numerous readily identifiable surface features (see Exhibit 27.A).
5. The neck connects the head to the trunk; surface anatomy of the neck is discussed in Exhibit 27.B.
6. The neck contains several important arteries and veins, and is divisible into several triangles by specific muscles and bones. Transverse planes through the neck help relate internal anatomy to key surface landmarks (see Exhibit 27.B).

Resource

Figure 27.1 - Principal Regions of the Cranium and Face
Figure 27.10a - Surface Anatomy of the Abdomen
Figure 27.16 - Surface Anatomy of the Buttock

Review	Resource

7. The trunk is divided into the back, chest, abdomen, and pelvis; see Exhibit 27.C.
8. The back presents several landmarks. The triangle of auscultation is a region of the back not covered by superficial muscles where respiratory sounds can be heard clearly (see Exhibit 27.C).
9. The skeleton of the chest protects the organs within and provides several surface landmarks for identifying the position of the heart and lungs (see Exhibit 27.C).
10. The abdomen and pelvis contain several important landmarks, including the linea alba, linea semilunaris, McBurney's point, and supracristal line. The perineum forms the floor of the pelvis (see Exhibit 27.C).
11. Transverse planes through the trunk help map internal organs to key landmarks on the surface of the trunk (see Exhibit 27.C).
12. The upper limb consists of the pectoral (shoulder) girdle and free upper limb; see Exhibit 27.D.
13. The prominence of the shoulder is formed by the deltoid muscle. The axilla, a pyramid-shaped area at the junction of the arm and chest, contains blood vessels and nerves that pass between the neck and upper limbs. The arm contains several blood vessels used for withdrawing blood, introducing fluids, taking the pulse, and measuring blood pressure (see Exhibit 27.D).
14. Several muscles of the forearm are more easily identifiable by their tendons. The radial artery is a principal blood vessel for detecting a pulse. The hand contains the origin of the cephalic vein and extensor tendons on the dorsum, and the thenar and hypothenar eminences on the palm (see Exhibit 27.D).
15. The lower limb consists of the pelvic (hip) girdle and free lower limb; see Exhibit 27.E.
16. The buttock is formed mainly by the gluteus maximus muscle. An important landmark in the thigh is the femoral triangle. The popliteal fossa is a diamond-shaped area on the posterior aspect of the knee (see Exhibit 27.E).
17. The gastrocnemius muscle forms the bulk of the midportion of the posterior aspect of the leg and, together with the soleus muscle, forms the calf of the leg. The foot contains the origin of the great saphenous vein, the largest vein in the body (see Exhibit 27.E).

CRITICAL THINKING QUESTIONS

1. Fred, the football fanatic, ran into the goal post and landed on his back. He told the coach that he felt all right but when he reached behind his head, he felt a bump on the back of his skull and another at the base of his neck. The team physician examined Fred and told him that the bumps were nothing to worry about; that they had always been there. What were these bumps?

2. Great Aunt Edith heard you're bringing your new boyfriend, the one with an anatomically correct heart tattooed on his bicep (he can even make it beat) and "I brake for ambulances" bumper sticker, to Thanksgiving dinner. She told your mother that she's having "palpations." Your boyfriend, a first-year medical student, said she probably meant "palpitations." Is there a difference? Explain your answer.

3. After Great Aunt Edith meets your new boyfriend, she decides he's not so bad after all, and says to him, "Why did you tattoo that heart on your arm? Why didn't you put it on your chest where it belongs?" Your boyfriend thinks this is a great idea. Describe the projection points on the surface of the chest that the tattoo artist should use to be anatomically correct.

4. Randy has been told that he needs to receive a large dose of an antibiotic to treat his infection. When he hears that this will be administered by a gluteal intramuscular injection, he says he isn't worried because, "They're injecting me there because it's so padded with fat, I won't feel it." If the nurse injected him in the fatty part of his buttocks, would the injection be administered correctly? What muscle should the nurse be aiming for, and what is the major landmark for locating the muscle? What are the risks of not injecting the correct muscle?

5. An ad for a new telephoto camera appears in a medical journal. It states that "you'll see the pores on your neighbor's ala, the sweat on his philtrum, the hairs in his external nares, the studs in his helix, the ring in his supercilia, and the color of his sclera." From this description, do you think this is a great camera? Explain your answer.

❓ ANSWERS TO FIGURE QUESTIONS

27.1 The frontal, parietal, temporal, occipital, zygomatic, and nasal regions are all named for bones that are deep to them.

27.2 When palpating the sagittal suture, the sheetlike epicranial aponeurosis (galea aponeurotica) is also palpated deep to the skin.

27.3 Light enters the eye through the pupil.

27.4 The mastoid process of the temporal bone can be palpated just posterior to the auricle.

27.5 The anterior border of the nose that connects the root and apex is the dorsum nasi.

27.6 The sternocleidomastoid muscle divides the neck into anterior and posterior triangles.

27.7 The atlanto-axial joint is located at the superior plane of the neck.

27.8 With the aid of a stethoscope, respiratory sounds can be heard clearly in the triangle of auscultation.

27.9 The sternal angle is the most reliable landmark of the chest.

27.10 Pain produced by pressure of the finger on McBurney's point indicates acute appendicitis.

27.11 The fourth lumbar vertebra is marked by the plane of the umbilicus.

27.12 The acromion of the scapula forms the top of the shoulder.

27.13 Blood is usually withdrawn from the median cubital vein located in the cubital fossa; this vein is also used to administer medication.

27.14 A pulse is frequently taken by placing pressure on the radial artery in the wrist.

27.15 Muscles that form the thenar eminence move the thumb.

27.16 The gluteus maximus muscle forms the bulk of the buttock.

27.17 The popliteal fossa is bordered superolaterally by the biceps femoris, superomedially by the semimembranosus and semitendinosus, and inferolaterally and inferomedially by the lateral and medial bellies of the gastrocnemius.

27.18 A pulse may be felt in the dorsalis pedis artery, located just lateral to the tendon of the extensor digitorum muscle.

MEASUREMENTS

U.S. Customary System

PARAMETER	UNIT	RELATION TO OTHER U.S. UNITS	SI (METRIC) EQUIVALENT
Length	inch	1/12 foot	2.54 centimeters
	foot	12 inches	0.305 meter
	yard	36 inches	0.914 meters
	mile	5,280 feet	1.609 kilometers
Mass	grain	1/1000 pound	64.799 milligrams
	dram	1/16 ounce	1.772 grams
	ounce	16 drams	28.350 grams
	pound	16 ounces	453.6 grams
	ton	2,000 pounds	907.18 kilograms
Volume (Liquid)	ounce	1/16 pint	29.574 milliliters
	pint	16 ounces	0.473 liter
	quart	2 pints	0.946 liter
	gallon	4 quarts	3.785 liters
Volume (Dry)	pint	1/2 quart	0.551 liter
	quart	2 pints	1.101 liters
	peck	8 quarts	8.810 liters
	bushel	4 pecks	35.239 liters

International System (SI)

BASE UNITS

UNIT	QUANTITY	SYMBOL
meter	length	m
kilogram	mass	kg
second	time	s
liter	volume	L
mole	amount of matter	mol

PREFIXES

PREFIX	MULTIPLIER	SYMBOL
tera-	$10^{12} = 1,000,000,000,000$	T
giga-	$10^{9} = 1,000,000,000$	G
mega-	$10^{6} = 1,000,000$	M
kilo-	$10^{3} = 1,000$	k
hecto-	$10^{2} = 100$	h
deca-	$10^{1} = 10$	da
deci-	$10^{-1} = 0.1$	d
centi-	$10^{-2} = 0.01$	c
milli-	$10^{-3} = 0.001$	m
micro-	$10^{-6} = 0.000,001$	μ
nano-	$10^{-9} = 0.000,000,001$	n
pico-	$10^{-12} = 0.000,000,000,001$	p

ANSWERS

Answers to Critical Thinking Questions

Chapter 1

1. On the x-ray, the bone (humerus) of the arm (brachium) would be located most proximal. Two bones of the forearm (antebrachium), the ulna and radius, would be located distal to the humerus with the ulna situated medial to the radius. In the hand, 15 phalanges would comprise the distal fingers, while the 5 metacarpals would be found in the more proximal palm. Distal to the ulna and radius, and proximal to the metacarpals, the wrist would be comprised of 8 carpal bones.

2. The alien would have 2 tails, 4 arms, 2 legs, and a mouth where the navel is usually located.

3. Lung, diaphragm, stomach, large intestine, small intestine; possibly part of pancreas, ovary or uterine tube, kidney.

4. The peritoneum, the largest serous membrane in the body, covers most organs in the abdominal cavity. Therefore, an infection in this structure can spread to any or all organs in the cavity.

Chapter 2

1. Maternal inheritance is due to the DNA found in extranuclear organelles such as the mitochondria. The sperm contributes only nuclear DNA. Mitochondria are inherited only from the mother.

2. Water would move out of the red blood cells, where the concentration of water is higher, causing them to shrink.

3. Arsenic will poison the mitochondrial enzymes involved in cellular respiration. This will halt production of ATP by the mitochondria. Cells with a high metabolic rate such as muscle cells will be particularly affected.

4. Disruption of microtubules would halt cell division. Microtubules are also involved in cell motility, transportation, and movement of cilia, all of which would be affected.

5. The cell imports the virus via receptor-mediated endocytosis. A potential obstacle is having a healthy immune system react to this "foreign" virus and thus try to eliminate it along with the healthy gene.

Chapter 3

1. The drug will make it easier for the cilia on Mike's epithelial cells to move the mucus, in which microbes are trapped, away from the lungs. Coughing up the thinned mucus should also be easier.

2. The cells are nonkeratinized stratified squamous epithelial cells, which, when separated and spread over a glass microscope slide, would have the appearance of "tiles" (i.e., thin and flat). In cervical dysplasia the cells appear altered in size, shape, and organization.

3. Janelle's kidneys may drop out of position from lack of supporting fat. There would also be less fat for padding in joints and buttocks, and her eyes may appear sunken.

4. The tissue that Jonathan pierced is epithelium, which constitutes the most superficial layer of skin. Epithelium is avascular, and therefore, there was no visible bleeding.

5. Bacteria would need a mechanism to pass through or between keratin-filled cells of the epidermis and to combat any phagocytic cells present. They would also need to survive any enzymes produced by cells the bacteria might encounter.

Chapter 4

1. Alcohol and other substances capable of producing birth defects are able to pass freely through the placenta.

2. All tissues and organs form from the embryo's three primary germ layers. The ectoderm, the outermost germ layer, develops into the nervous system and the epithelial layer of the skin. This early embryological connection may sometimes result in disorders showing signs in both tissues.

3. Josefina is in false labor. In true labor, the contractions are regular and the pain may localize in the back. The pain may be intensified by walking. True labor is indicated by the "show" of bloody mucus and cervical dilation.

4. Certain microorganisms can infect the placenta itself to infect the embryo/fetus (transplacental transmission). Risk is greatest during the first trimester, a crucial time in the differentiation of cells.

5. Pregnancy increases appetite to accommodate the increase in nutritional demands. Increasing pressure on the urinary bladder increases frequency of urination. As the uterus increases in size, abdominal contents are pushed upward, and the stomach and its contents may push up into the esophagus (heartburn). Elena may need to prop herself up higher at night to help prevent heartburn and make breathing easier.

Chapter 5

1. Chronic exposure to UV light causes damage to elastin and collagen, promoting wrinkles. The suspicious growth may be skin cancer caused by UV photodamage to skin cells.

2. Dilation of the blood vessels in the skin increases blood flow, which means more body heat will reach the surface of the skin and be radiated away. Therefore, although they will feel warmer superficially, body temperature will decrease. They should wear their coats!

3. In part, the increased risk of melanoma is due to depletion of the ozone layer, which absorbs some UV light high in the atmosphere. But the main reason for the increase is that more people are spending more time in the sun. Malignant melanomas metastasize rapidly and can kill a person within months of diagnosis.

4. The surface layer of the skin is keratinized stratified squamous epithelium. The keratinocytes are dead and filled with the protective protein keratin. Lipids from lamellar granules and sebum also function in waterproofing.

5. The higher the humidity, the less evaporation can occur, which means the less cooling can occur. Therefore people perceive the same temperature as hotter at a higher humidity.

Chapter 6

1. Lynda's diet lacks several nutrients necessary for bone health, including essential vitamins (such as A, D), minerals (such as calcium), and proteins. Her lack of exercise will weaken bone strength. Her age and smoking habit may result in a lack of estrogens, causing demineralization of bone.

2. In bone tissue, there is a balance between the actions of osteoblasts and osteoclasts. There also exists a balance between the inorganic (mineral) and organic (collagen) components of bone matrix. When these balances are altered, the consistency of the bone matrix will change accordingly. If osteoblast activity outpaces osteoclast activity, then too much new tissue will be formed, causing the bones to become abnormally thick and heavy. In this particular case, Scott noticed the spurs that were produced by too much mineral material being deposited in the bone. The surplus formed thick bumps, called spurs, on the bones that most likely interfered with movement at joints.

3. At Aunt Edith's age, the production of several hormones necessary for bone remodeling (such as estrogens and human growth hormone) would be decreased. She may have loss of bone mass, brittleness, and possibly osteoporosis. Increased susceptibility to fractures results in damage to the vertebrae and loss of height.

4. Exercise causes mechanical stress on bones but since there is no gravity in space, the pull of gravity on bones is missing. The lack of stress results in demineralization of bone and weakness.

5. Because babies have fontanels (soft spots), the cranial bones tend to be more movable than after the sutures close. As the baby passed through the birth canal (and possibly because of the use of forceps to assist the delivery), the bones could move into a more "cone-shaped" position. Likewise, laying the baby on its back all the time flattens the occipital bone. In addition, infant bones tend to be more flexible than older bones, which lose moisture and become more brittle over time.

Chapter 7

1. Fontanels, the soft spots between cranial bones, are fibrous connective tissue membrane–filled spaces. They allow the infant's head to be molded during its passage through the birth canal and allow for brain and skull growth in the infant.

2. The pituitary gland lies under the brain as it sits in the sella turcica of the sphenoid bone. Thus, the gland is heavily protected and difficult to reach from any direction. In addition, there are important nerves, such as the optic nerve, and blood vessels passing nearby.

3. Infants are born with a single concave curve. Adults have four curves in their vertebral column at the cervical, thoracic, lumbar, and sacral regions.

4. The occipital bone protects key areas of the cerebrum, but also the brain stem, which joins the spinal cord at the foramen magnum. Damage to the spinal cord and/or the vital centers in the medulla oblongata controlling breathing, heart activity, and blood pressure could be fatal.

5. The x-rays were taken to view the paranasal sinuses—the "fuzzy-looking holes." John probably had sinusitis due to infected paranasal sinuses.

Chapter 8

1. There are 14 phalanges in each hand: 2 bones in the thumb and 3 in each of the other fingers. Farmer Ramsey has lost 5 phalanges on his left hand, so he has 9 remaining on his left and 14 remaining on his right for a total of 23.

2. The talus receives all of the body's weight. Half that weight is normally transmitted to the calcaneus but the weight distribution will be shifted on the hallux by dancing. Problems may appear in the talocrural (ankle) joint and hallux. Rose also has a deviated hallux or bunion that may have been caused by tight ballet shoes. The dancing did strengthen the foot muscles and correct the flat feet.

3. Bone growth occurs in response to mechanical stress. Therefore, any activity that causes the deltoid muscle to pull on the deltoid tuberosity would increase its growth. Such activity might include paddling the large kayaks frequently over time and/or against turbulent waters.

4. Grandmother Amelia has probably fractured the neck of her femur. This is a common fracture in the elderly and is often called a "broken hip." The sacrum does indeed articulate with the ilium, and Grandpa Jeremiah could have osteoarthritis in that joint, which could be quite painful. There are, however, other lower back issues (e.g., herniated discs, sciatica, etc.) that could be equally painful.

5. Derrick has patellofemoral syndrome (runner's knee). Running daily in the same direction on a banked track has stressed the downhill knee. The patella is tracking laterally to its proper position, resulting in pain after exercise.

Chapter 9

1. Structurally the hip joint is a simple ball-and-socket synovial joint composed of the head of the femur (ball) and the acetabulum of the hip bone (socket). The knee is really three joints: patellofemoral, lateral tibiofemoral, and medial tibiofemoral. It is a combination gliding and hinge joint.

2. Flexion at knee, extension at hip, extension of vertebral column except for hyperextension at neck, abduction at shoulder, flexion of fingers, adduction of thumb, flexion/extension at elbow, and depression of mandible.

3. No more body surfing for Lars. He has a dislocated shoulder. The head of the humerus was displaced from the glenoid cavity, causing tearing of the supporting ligaments and tendons (rotator cuff) of the shoulder joint.

4. Most likely the cartilage of the menisci were torn in the accident. Pieces of cartilage (and possibly bone) may be in the joint cavity. In addition, it is possible that the articular cartilage of the tibia (and possibly the femur) was damaged. Damage to the menisci and ligaments associated with the knee would contribute to stability problems.

5. There are many possibilities in Chuck's situation. The problems could be all unrelated, and the back pain could be completely related to vertebral column issues. On the opposite end of the spectrum, it is possible that a long-term ankle problem caused Chuck to walk differently in order to favor the damaged ankle, thus causing the bowing of the legs as the muscles pull the leg in a slightly different direction. From there, the change in gait could cause the muscles that attach to the pelvis to pull more forcefully on points that they normally would not, which could pull the pelvic bones slightly out of alignment in relation to the vertebral column, thus causing pain at the sacroiliac joint.

Chapter 10

1. The runner's leg muscles will contain a higher amount of slow fibers—rich in myoglobin, mitochondria, and blood. The weightlifter's arm muscles will contain abundant fast fibers—high in glycogen, with less myoglobin and blood capillaries than the slow fibers. The slow fibers appear red while the fast fibers are white in color.

2. Bill's muscles in the casted leg were not being used, so they decreased in size due to loss of myofibrils. Bill's leg shows disuse atrophy.

3. Acetylcholine (ACh) is the neurotransmitter used to "bridge the gap" at the neuromuscular junction. If ACh release is blocked, the neuron cannot send a signal to the muscle and the muscle will not contract.

4. Some cardiac muscle cells are autorhythmic and contraction is started intrinsically. Gap junctions connect the cardiac cells, spreading the impulse through the fiber network so it contracts as a unit. Cardiac muscle has a long refractory period so it will not exhibit tetanus like skeletal muscle does.

5. First, it is the term *extensibility* (not *contractility*) that refers to stretching of a muscle fiber. In addition, the contractile proteins never change length during muscle contraction or during stretching of a muscle fiber—they change position only relative to each other.

Chapter 11

1. The orbicularis oris protrudes the lips. The genioglossus protracts the tongue. The buccinator aids in sucking.

2. One likely possibility is the rhomboid major muscle.

3. Refer to Exhibit 11.C, "Muscles that Move the Mandible." You should have included temporalis (retract, elevate, i.e., closing the mouth); masseter (retract, elevate, i.e., closing the mouth); medial pterygoid (protract, elevate, i.e., closing the mouth); and lateral pterygoid (protract, depress, i.e., opening the mouth). Note that it is a suprahyoid group of muscles that aids the lateral pterygoid in depressing the mandible to open the mouth (see Exhibit 11.E) but not with as much force as afforded by the temporalis and masseter muscles in closing the mouth; closing the mouth is a much more powerful movement.

4. The most likely injury is a rotator cuff injury. The rotator cuff includes tendons of the subscapularis, supraspinatus, infraspinatus, and teres minor muscles. The most common injury is to the supraspinatus muscle and/or tendon (impingement syndrome).

5. Wyman probably has an inguinal hernia. He should see a doctor, since constriction of the intestines pushing out through the opening can cause serious damage.

Chapter 12

1. Blood doping is used by some athletes to improve performance but the practice strains the heart. A blood test would reveal polycythemia, an increased number of red blood cells.

2. Basophils appear as bluish-black granular cells in a stained blood smear. An elevated number of basophils may suggest an allergic response such as "hay fever."

3. As the kidney fails, it secretes less erythropoietin, which results in a lower rate of red blood cell production. Recombinant erythropoietin can be administered to boost red blood cell production.

4. Iron is an important element found in the heme portion of hemoglobin. If levels of iron are low, then hemoglobin production is low. Low levels of hemoglobin reduce the oxygen-carrying capacity of blood, and a person with low hemoglobin levels would demonstrate symptoms of anemia. Adding iron to the diet increases one of the raw materials needed for adequate hemoglobin production.

5. Raoul's doctor suggested autologous preoperative transfusion (predonation). If Raoul does need a transfusion during surgery it is safer to use his own blood. Possible problems with matching blood types and blood-borne disease are eliminated with predonation.

Chapter 13

1. The cusps of the valves "catch" the blood like the fabric of the parachute "catches" the air. The cusps are anchored to the papillary muscles of the ventricle by the chordae tendineae, just as the fabric of the parachute is anchored by the lines, so the cusps do not flip open backward.

2. The anterior interventricular branch is located in the anterior interventricular sulcus and supplies both ventricles. The circumflex branch is located in the coronary sulcus and supplies the left ventricle and the left atrium.

3. The pericardium surrounds the heart. The fibrous pericardium anchors the heart to the diaphragm, sternum, and major blood vessels. The intercalated discs contain desmosomes, which hold the cardiac muscle fibers together.

4. Some people develop an immune response to streptococcal infections; this can result in rheumatic fever, which can damage the valves of the heart, especially the mitral valve. It would be important to pinpoint the cause of the sore throat to ensure that antibiotic treatment is started if the infection is streptococcal to reduce the possibility of heart damage.

5. Aortic valve stenosis is a narrowing and/or stiffening of the valve, which makes the valve harder to open. Therefore, the heart must pump harder to open the valve sufficiently to pump out an amount of blood equal to that of a normal heart. If the heart is unable to pump strongly enough, more blood can remain in the left ventricle following ventricular systole than is normal, and the system backs up from there.

Chapter 14

1. The foramen ovale and ductus arteriosus close to establish the separate pulmonary and systemic circulations. The umbilical artery and veins close since the placenta is no longer functioning. The ductus venosus closes so that the liver is no longer bypassed.

2. In the skeletal system "sinuses" are cavities within certain cranial and facial bones near the nasal cavity. They are lined with mucous membranes that are continuous with the lining of the nasal cavity. Besides producing mucus, the paranasal sinuses serve as resonating chambers for sounds as we speak or sing. The four sinuses of the skull are the frontal, sphenoid, ethmoid, and maxillary. In the circulatory system, "sinuses" are endothelial-lined venous channels between layers of the cranial dura mater that circulate blood in the internal skull. The pathway of blood from the superior sagittal sinus goes to the right transverse sinus, which communicates with the left transverse sinus, which was formed from the inferior sagittal sinus becoming the straight sinus. The caverous sinuses also flow into the transverse sinuses. The transverse sinuses flow into the sigmoid sinuses, which lead out of the skull through the jugular foramen. Outside of the skull, the sigmoid sinus is termed the *internal jugular vein*. Each internal jugular vein unites with its respective subclavian vein from the upper limb to form a brachiocephalic vein. The right brachiocephalic vein unites with the left brachiocephalic vein to form the superior vena cava.

3. Varicose veins are caused by faulty valves. The weak valves allow pooling of blood in the veins. Valves are found only in veins and not in arteries.

4. The catheter would snake from the femoral artery to the external iliac artery to the common iliac artery to the abdominal aorta to the arch of the aorta to the left ventricle.

5. Gina's doctor will either say her blood pressure is too high or is "borderline high." The doctor should encourage Gina to stop smoking and to lose some weight, since both smoking and being overweight are known risk factors for high blood pressure. The doctor should also suggest regularly monitoring her blood pressure to make sure that this one reading was not unusual for Gina. If Gina's blood pressure stays high and she is making no progress with the weight loss and smoking cessation, she may require medication to control her blood pressure.

Chapter 15

1. The route is from lymph capillaries to lymphatic vessels to the popliteal nodes to the superficial inguinal nodes to the right lumbar trunk to the cisterna chyli to the thoracic duct to the junction of the left internal jugular and left subclavian veins.

2. The two palatine tonsils are in the lateral, posterior oral cavity. The two lingual tonsils are at the base of the tongue. The single pharyngeal tonsil is in the posterior nasopharynx. The five tonsils function in the immune response to inhaled or ingested foreign invaders.

3. The spleen is an important secondary lymphatic organ. By removing the organ, the doctor hopes to reduce the level of the immune response causing Kelly's symptoms. However, because the immune response would be reduced, taking antibiotics prior to invasive procedures, such as dental work, reduce the risk of serious infection.

4. Nan recognizes the difficulty breathing as a part of her anaphylactic allergic reaction to the eggplant. Release of histamine from mast cells causes her airways to constrict, thus reducing the flow of air. In very serious reactions, the airway can narrow so much that an affected individual cannot take in sufficient oxygen. In such cases, epinephrine should be administered; this will dilate the airways.

5. The left subclavian trunk drains lymph from the left upper limb. Blockage would cause a buildup of lymph and interstitial fluid in the upper limb, causing edema (swelling).

Chapter 16

1. The jingle that won't stop is similar to a reverberating circuit in that a signal will be repeated over and over again. Reverberating circuits function in breathing, memory, waking, and coordinated muscle activity.

2. The nervous system exhibits plasticity, "the capability to change based on experience." New pathways and synapses needed for foot manipulation

replaced damaged pathways and synapses that were once needed for hand movements.

3. The endorphins are neuropeptides that function as the body's natural painkillers. They have been linked to feelings of pleasure or euphoria.

4. The somatic, afferent, peripheral division would detect sound and smell. The somatic, efferent, peripheral division sends messages to the skeletal muscles for stretching and yawning. Stomach rumbling and salivation are controlled by the autonomic, efferent, peripheral division. The enteric, efferent, peripheral division also controls GI tract organs.

5. The drug may not be able to pass the blood–brain barrier to reach the affected nervous tissue.

Chapter 17

1. The senior damaged the cervical region of his spinal cord. Because the upper limbs are paralyzed, the injury has to be superior to C5. This is a high cervical spinal cord injury. Recovery is unlikely.

2. The spinal cord is connected to the periphery by the spinal nerves. The nerves exit through the intervertebral foramina. Spaces are maintained between the vertebrae by the intervertebral discs. The dura mater of the spinal cord fuses with the epineurium. The nerves are connected to the cord by the posterior and anterior roots.

3. The spinal cord is anchored in place by the filum terminale and denticulate ligaments.

4. The spinal cord extends only to the level of the superior border of the second lumbar vertebra—well above the level of the coccygeal vertebrae.

5. The compression of the vertebrae may cause herniation of intervertebral discs, which then can affect spinal nerves. Nischal may have damage to one or more spinal nerves. In this case, damage to the sciatic nerve is a likely possibility.

Chapter 18

1. Movement of the right upper limb is controlled by the left hemisphere's primary motor area, located in the postcentral gyrus. Speech is controlled by Broca's area in the left hemisphere's frontal lobe just superior to the lateral cerebral sulcus.

2. While it is conceivable that he could still be drunk, the signs suggest Wolfgang may have suffered a subdural hemorrhage, which is very dangerous.

3. The amygdala is the center for fear, rage, and aggression. While it might seem useful to be "fearless" or to be nonaggressive, this could prove dangerous, if not fatal, in some circumstances in which fear might prevent someone from doing risky things or aggression might be an appropriate response.

4. There is such an enormous *loss of neurons* within the cerebrum that the gyri shrink and the sulci enlarge. While the lost neurons cannot be replaced, treatments for the disease center around increasing the availability of the neurotransmitter that is lost along with the neurons.

5. The dentist has injected anesthetic into the inferior alveolar nerve, a branch of the mandibular nerve that numbs the lower teeth and the lower lip. The tongue is numbed by blocking the lingual nerve. The upper teeth and lip are numbed by injecting the superior alveolar nerve, a branch of the maxillary nerve.

Chapter 19

1. The exciting and potentially dangerous activities activate the sympathetic nervous system, resulting in a "fight-or-flight" reaction. The sympathetic nervous system stimulates the release of epinephrine (adrenalin) and norepinephrine from the adrenal medulla. These hormones prolong the fight-or-flight response.

2. The autonomic sensory neurons detect stretch in the stomach and send impulses to the brain. The hypothalamus sends impulses through the parasympathetic division, along the vagus nerve, to the heart, resulting in decreased heart rate and force of contraction.

3. Mando's fight-or-flight response is continually being stimulated, which means, among other effects, that her heart rate is often high, her blood glucose may become chronically high, and her adrenal glands are being continually stimulated. Such chronically altered physiological conditions can lead to other systemic problems. (See Chapter 22 The Endocrine System, for possibilities.)

4. The autonomic division controls the digestive system and many other organs, including the heart, lungs, and eyes. Most organs are controlled by dual innervation, with the sympathetic and parasympathetic divisions causing opposite effects. Stimulation of the sympathetic division results in widespread effects throughout the body, while the effects of parasympathetic stimulation are more localized. The enteric system controls only the digestive system and may function independently of the ANS.

5. The hypothalamus controls many aspects of behavior and physiology, including those regulated by the autonomic nervous system. Though it is a physically small area of the brain, its effects are wide-reaching.

Chapter 20

1. The warm (thermal) receptors in her hands were at first activated by the warmth of the cup, then they adapted to the stimulus as she continued to hold the cup. The high temperature of the hot cocoa stimulated the thermal receptors in her mouth as well as pain receptors.

2. Muscle spindles monitor the change in length of skeletal muscles, sending proprioceptive information (information about body position) to the brain. Without this input, Jenny would find it difficult to know where her body parts are without actually seeing them. If the spindle function does not return, Jenny may need to learn to move her arms, legs, facial muscles, and so on, all over again in a different way.

3. Many brain areas are involved in controlling skeletal muscle activity, including the basal ganglia of the cerebrum, the premotor and primary motor areas of the cerebrum, the cerebellum, and so on. In addition, sensory spinal and brain tracts are necessary for providing input and motor tracts for transmitting the "orders" for movement.

4. In an infant or very young child, the myelination of the CNS is not complete. The corticospinal tracts, which control fine, voluntary motor movement, are not fully myelinated until the child is about 2 years old. An infant would not be able to manipulate a knife or fork safely due to lack of complete motor control.

5. Jon's perception of feeling in his amputated foot is called *phantom limb sensation*. Impulses from the remaining proximal section of the sensory neuron are perceived by the brain as still coming from the amputated foot.

Chapter 21

1. The pressure from the tumor may be stimulating Suestia's olfactory areas in the temporal lobe, which would make her think she is smelling something that is not there.

2. Otitis media (an inner ear infection) would arise from the other regions of inflammation directly from the nasopharynx via the eustacian (auditory) tube. Left untreated, otitis media has the potential to spread into the mastoid air cells (mastoiditis) and even into the skull, producing inflammation of the meninges (meningitis).

3. The stick damaged the iris (the colored ring), which is part of the vascular tunic. The iris controls the diameter of the pupil but does not directly affect the focusing of light rays on the retina. If permanent damage was done to the cornea or ciliary muscles, then focusing may be affected.

4. The endolymph in the saccule, utricle, and membranous semicircular ducts moves in response to the up-and-down motion of the ship on the ocean. The vestibular apparatus is stimulated and sends impulses to the medulla, pons, and cerebellum. The eyes also send feedback to the cerebellum to help maintain balance. Excessive stimulation and conflicting messages (eyes say the ship is still, ears say it's moving) cause motion sickness.

5. Pigments in the iris and choroid are usually sufficient to absorb light entering the eye. A flash photograph captures the picture before the iris has time to constrict in response to the high intensity of light. The picture captures the reflection of the light back through the pupil (and iris if it is lightly pigmented). The eyes appear red due to the vascular tunic. Albinos have no pigment in their eyes so the "red eye" effect is more pronounced.

Chapter 22

1. Melatonin could be taken to help induce sleep during the evening at Tropicanaland. For the quickest adjustment, exposure to very bright light in the morning would help reset the body's clock. Sunlight would be better for this than a bright flashlight. The MSH may help with getting a tan but won't help reset the body's clock.

2. Amadu has a goiter, which is an increase in the size of the thyroid gland. In this case the thyroid gland has increased in size because Amadu's diet lacked sufficient iodine to produce normal levels of thyroid hormones. Because levels of the thyroid hormones were low, TSH levels increased in an attempt to stimulate the thyroid gland to produce more thyroid hormones, but without the iodine, this was not possible. Adding iodine back into the diet should provide the necessary raw materials, and the gland will not be stimulated to grow more.

3. Because blood glucose might be chronically elevated by such treatment, this could make the pancreas work harder to produce insulin to lower the levels, which could have an adverse effect on the pancreas.

4. The pituitary gland is located in the hypophyseal fossa in the base of the sphenoid bone. The sphenoid is posterior to the ethmoid bone, which makes up a major portion of the walls and roof of the nasal cavity. Due to its proximity to the nasal cavity, the pituitary gland could have been affected by the radiation.

5. All hormones, including insulin, need an adequate number of functioning receptors to perform their function. The problem in this patient is a lack of functioning receptors, not a problem with insulin level.

Chapter 23

1. The septum is composed of hyaline cartilage covered with mucous membrane. The lateral wall of the nose is skin and muscle, lined with mucous membrane. The upper ear is skin covering elastic cartilage.

2. Lower oxygen levels at the higher altitudes means lower oxygen levels in Suzanne's blood and cerebrospinal fluid. Central and peripheral chemoreceptors will detect changes in blood gas levels and proportions and signal the medullary rhythmicity center to increase the rate of breathing.

3. Once the endotracheal tube was introduced either nasally or orally, the progression of the tube did not enter the laryngopharynx to proceed to the larynx and trachea. Instead the tube remained in the pharynx and entered the esophagus, which eventually leads to the stomach. The sounds that the resident heard in the epigastric region were therefore coming from the stomach rather than the lungs.

4. Several factors decrease respiratory efficiency in smokers: (1) Nicotine constricts terminal bronchioles, which decreases airflow into and out of the lungs. (2) Carbon monoxide in smoke binds to hemoglobin and reduces its oxygen-carrying capability. (3) Irritants in smoke cause increased mucus secretion by the mucosa of the bronchial tree and swelling of the mucosal lining, both of which impede airflow into and out of the lungs. (4) Irritants in smoke also inhibit the movement of cilia and destroy cilia in the lining of the respiratory system. Thus, excess mucus and foreign debris are not easily removed, which further adds to the difficulty in breathing. (5) With time, smoking leads to destruction of elastic fibers in the lungs and is the prime cause of emphysema. These changes cause collapse of small bronchioles and trapping of air in alveoli at the end of exhalation. The result is less efficient gas exchange.

5. LaTonya's cerebral cortex can allow her to voluntarily hold her breath for a short time. Increasing levels of carbon dioxide and H$^+$ will stimulate the inspiratory area and normal breathing will resume despite LaTonya's desire to get her sister in trouble.

Chapter 24

1. Saliva contains salivary amylase, which will break down the starch of the potatoes into sugar (sweet), and lingual lipase, which will break down the fat from the frying into fatty acids (tart) and monoglycerides.

2. The doctor will be looking at the mucosa of the regions of the large intestine to check for smoothness (vs. polyp growth), normal-appearing simple columnar epithelium, normal-appearing haustra, and so forth.

3. Modifications to increase surface area include overall length, circular folds, villi, and microvilli (brush border). The circular folds also enhance absorption by imparting a spiraling motion to the chyme.

4. During swallowing, the soft palate and uvula close off the nasopharynx. If the nervous system sends conflicting signals (such as breathe, swallow, and giggle), the palate may be in the wrong position to block food from entering the nasal cavity.

5. Billy is hearing the normal bowel sounds produced as a result of the movement of intestinal contents via peristalsis, haustral churning, and mass peristalsis. In addition, gas is being produced as bacteria digest undigested materials, which contributes to the sounds as the bubbles move through the intestinal contents.

Chapter 25

1. The glomerulus would continue to filter the blood, producing filtrate containing water, glucose, ions, and other compounds. The toxin would block the normal reabsorption of 99% of the filtrate by the renal tubules. The infected person would rapidly become dehydrated from lost water and would also lose ions, glucose, and essential vitamins and nutrients. The infected person would most likely soon die.

2. The micturition reflex is primarily a parasympathetic reflex with input from the sympathetic system in "fight-flight-or fright" situations.

Parasympathetic pathways traveling from the micturition center and through S2, S3, and S4 reach the inferior hypogastric (pelvic) plexus to be distributed to the smooth muscle of the urinary bladder and internal urethral sphincter. This will cause a contraction of the detrusor muscle and relaxation of the sphincter. Concurrently, relaxation of the external urethral sphincter is produced by inhibiting somatic efferent impulses to the pelvic floor (levator ani) musculature. The synergistic functions of contracting the bladder and relaxing the sphincters cause urination.

3. In females the urethra is about 4 cm long. In males, the urethra is about 15–20 cm long, including its passage through the penis, the urogenital diaphragm, and prostate. The female urinary bladder holds less urine than the male bladder due to the presence of the uterus in females.

4. Urine samples contain an abundance of epithelial cells that are shed from the mucosal lining of the urinary system.

5. Functions normally performed by the healthy nephrons include filtering the blood (glomerular filtration); *keeping* necessary substances for the body (tubular reabsorption); and *removing* unnecessary substances from the body (tubular secretion). Unnecessary substances (waste) are then excreted as a liquid (urine). Because of the large number of damaged nephrons resulting from the disease, these functions are now being carried out by a hemodialysis machine with the use of semipermeable membranes to separate the necessary substances from the wastes.

Chapter 26

1. The infection could pass through the vagina to the uterus through the uterine tubes, which are open to the pelvic cavity.

2. The presence or absence of the SRY gene on the Y chromosome determines the development of the urogenital structures before birth. The presence of SRY and secretion of testosterone by the fetal testes results in a male. Absence of SRY results in a female. Structures can be altered by surgery and hormones somewhat, but the female will still lack receptors and organs necessary for sperm production and the male will lack receptors and organs necessary for the production of ova. The chromosome composition cannot be changed.

3. A hysterectomy will not cause menopause. A hysterectomy is the removal of the uterus, which does not produce hormones. The ovaries are left intact and continue to produce estrogens and progesterone.

4. The testicular artery and veins that drain the testes pass through the inguinal ring; if their blood supply is cut off by the closure of the inguinal canal, the testes would have no source of nourishment or waste removal.

5. A vasectomy cuts only the vas (ductus) deferens and leaves the testes untouched. Male secondary sex characteristics and libido (sex drive) are maintained by androgens including testosterone. The secretion of these hormones by the testes is not interrupted by the vasectomy.

Chapter 27

1. The spine of C7 forms a prominence along the midline at the base of the neck. The spine of T1 forms a second prominence slightly inferior to C7.

2. Palpation is touching or examining using the hands. Palpitation is a forceful beating of the heart that can often be felt by the patient.

3. The projection (shape of an organ on the surface of the body) of the heart on the anterior surface of the chest is indicated by four points as follows: The *inferior left point* is the apex (inferior, pointed end) of the heart, which projects downward and to the left and can be palpated in the fifth intercostal space, about 9 cm (3.5 in.) to the left of the midline. The *inferior right point* is at the lower border of the costal cartilage of the right sixth rib, about 3 cm to the right of the midline. The *superior right point* is located at the superior border of the costal cartilage of the right third rib, about 3 cm to the right of the midline. The *superior left point* is located at the inferior border of the costal cartilage of the left second rib, about 3 cm to the left of the midline. If you connect the four points, you can determine the heart's location and size (about the size of a closed fist).

4. The *gluteus medius muscle* is the site for an intramuscular injection, not the gluteus maximus, which Randy seems to think. In order to give this injection, the buttock is divided into quadrants and the upper outer quadrant is used as an injection site, where there is less chance of injury to the sciatic nerve or major blood vessels.

5. The nostrils (ala), upper lip (philtrum), nose (external nares) hairs, edge of the ear (helix), eyebrow (supercilia) ring, and the sclera (whites of the eyes) should all be visible using even the least sophisticated camera, so this is probably not a particularly great camera.

GLOSSARY

Pronunciation Key

1. The most strongly accented syllable appears in capital letters, for example, bilateral (bī-LAT-er-al) and diagnosis (dī-ag-NŌ-sis).

2. If there is a secondary accent, it is noted by a prime ('), for example, constitution (kon'-sti-TOO-shun) and physiology (fiz'-ē-OL-o¯-jē). Any additional secondary accents are also noted by a prime, for example, decarboxylation (dē'-kar-bok'-si-LĀ-shun).

3. Vowels marked by a line above the letter are pronounced with the long sound, as in the following common words:

ā as in māke	ō as in pōle
ē as in bē	ū as in cūte
ī as in īvy	

4. Vowels not marked by a line above the letter are pronounced with the short sound, as in the following words:

a as in above or at	o as in not
e as in bet	u as in bud
i as in sip	

5. Other vowel sounds are indicated as follows:

oy as in oil
oo as in root

6. Consonant sounds are pronounced as in the following words:

b as in bat	m as in mother
ch as in chair	n as in no
d as in dog	p as in pick
f as in father	r as in rib
g as in get	s as in so
h as in hat	t as in tea
j as in jump	v as in very
k as in can	w as in welcome
ks as in tax	z as in zero
kw as in quit	zh as in lesion
l as in let	

A

Abdomen The area between the diaphragm and pelvis.

Abdominal cavity Superior portion of the abdominopelvic cavity that contains the stomach, spleen, liver, gallbladder, most of the small intestine, and part of the large intestine.

Abdominopelvic cavity A body cavity that is subdivided into a superior abdominal cavity and an inferior pelvic cavity.

Abduction Movement away from the midline of the body.

Abortion (a-BOR-shun) The premature loss (spontaneous) or removal (induced) of the embryo or nonviable fetus; miscarriage due to a failure in the normal process of developing or maturing.

Abscess (AB-ses) A localized collection of pus and liquefied tissue in a cavity.

Absorption (ab-SORP-shun) Intake of fluids or other substances by cells of the skin or mucous membranes; the passage of digested foods from the gastrointestinal tract into blood or lymph.

Accessory duct A duct of the pancreas that empties into the duodenum about 2.5 cm (1 in.) superior to the hepatopancreatic ampulla (ampulla of Vater). Also called the duct of Santorini (san'-tō-RĒ-nē).

Accessory sex glands Glands that produce substances that protect the gametes and facilitate their movements.

Acetylcholine (as'ē-til-KŌ-lēn) **(ACh)** A neurotransmitter liberated by many peripheral nervous system neurons and some central nervous system neurons. It is excitatory at neuromuscular junctions but inhibitory at some other synapses (for example, it slows heart rate).

Acini (AS-i-nē) Groups of cells in the pancreas that secrete digestive enzymes. Also, functional units within the liver lobule. *Singular* is **acinus**.

Acoustic Pertaining to sound or the sense of hearing.

Acquired immunodeficiency syndrome (AIDS) A disease caused by the human immunodeficiency virus (HIV). Characterized by a positive HIV-antibody test, low helper T cell count, and certain indicator diseases (for example Kaposi's sarcoma, pneumocystis carinii pneumonia, tuberculosis, fungal diseases). Other symptoms include fever or night sweats, coughing, sore throat, fatigue, body aches, weight loss, and enlarged lymph nodes.

Acrosome (AK-rō-sōm) A lysosomelike organelle in the head of a sperm cell containing enzymes that facilitate the penetration of a sperm cell into a secondary oocyte.

Actin A contractile protein that is part of thin filaments in muscle fibers.

Action potentials (impulses) Electrical signals that propagate along the membrane of a neuron or muscle fiber (cell); rapid changes in membrane potential that involve a depolarization followed by a repolarization. Also called nerve action potentials or nerve impulses as they relate to a neuron, and muscle action potentials as they relate to a muscle fiber.

Activation (ak´-ti-VĀ-shun) **energy** The minimum amount of energy required for a chemical reaction to occur.

Active transport The movement of substances across cell membranes against a concentration gradient, requiring the expenditure of cellular energy (ATP).

Acute (a-KŪT) Having rapid onset, severe symptoms, and a short course; not chronic.

Adaptation The adjustment of the pupil of the eye to changes in light intensity. The property by which a sensory neuron relays a decreased frequency of action potentials from a receptor, even though the strength of the stimulus remains constant; the decrease in perception of a sensation over time while the stimulus is still present.

Adduction Movement toward the midline of the body.

Adenoid Structure on the posterior wall of the nasopharynx that filters contaminants. Also called the pharyngeal tonsil.

Adenosine triphosphate (ATP) Molecule used to generate the energy needed by the cell.

Adipocyte Fat cell, derived from a fibroblast.

Adipose tissue Tissue composed of adipocytes specialized for triglyceride storage and present in the form of soft pads between various organs for support, protection, and insulation.

Adrenal cortex The outer portion of an adrenal gland, divided into three zones; the zona glomerulosa secretes mineralocorticoids, the zona fasciculata secretes glucocorticoids, and the zona reticularis secretes androgens.

Adrenal glands Two glands located superior to each kidney. Also called the suprarenal (soo´-pra-RE-nal) glands.

Adrenal medulla The inner part of an adrenal gland, consisting of cells that secrete epinephrine, norepinephrine, and a small amount of dopamine in response to stimulation by sympathetic preganglionic neurons.

Adrenergic neuron A neuron that releases epinephrine (adrenaline) or norepinephrine (noradrenaline) as its neurotransmitter.

Adrenocorticotropic hormone (ACTH) A hormone produced by the anterior pituitary that influences the production and secretion of certain hormones of the adrenal cortex. Also called corticotropin.

Adventitia The outermost covering of a structure or organ.

Aerobics Any activity that works large body muscles for at least 20 minutes; elevates cardiac output and accelerates metabolic rate.

Afferent arteriole A blood vessel of a kidney that divides into the capillary network called a glomerulus; there is one for each glomerulus.

Afferent neurons Neurons that carry sensory information from cranial and spinal nerves into the brain and spinal cord or from a lower to a higher level in the spinal cord and brain. Also called sensory neurons.

Agglutination (a-gloo′-ti-NĀ-shun) Clumping of microorganisms or blood cells, typically due to an antigen–antibody reaction.

Aggregated lymphatic follicles Clusters of lymph nodules that are most numerous in the ileum. Also called Peyer's (PĪ-erz) patches.

Aging A progressive alteration of the body's homeostatic adaptive responses that produces observable changes in structure and function.

Albinism Abnormal, nonpathological, partial, or total absence of pigment in skin, hair, and eyes.

Aldosterone A mineralocorticoid produced by the adrenal cortex that promotes sodium and water reabsorption by the kidneys and potassium excretion in urine.

All-or-none principle If a stimulus depolarizes a neuron to threshold, the neuron fires at its maximum voltage (all); if threshold is not reached, the neuron does not fire at all (none). Given above threshold, stronger stimuli do not produce stronger action potentials.

Allantois A small, vascularized outpouching of the yolk sac that serves as an early site for blood formation and development of the urinary bladder.

Alleles (a-LĒLZ) Alternate forms of a single gene that control the same inherited trait (such as type A blood) and are located at the same position on homologous chromosomes.

Allergen (AL-er-jen) An antigen that evokes a hypersensitivity reaction.

Alopecia The partial or complete lack of hair as a result of factors such as genetics, aging, endocrine disorders, chemotherapy, and skin diseases.

Alpha cell A type of cell in the pancreatic islets (islets of Langerhans) in the pancreas that secretes the hormone glucagon. Also termed an A cell.

Alpha (α) receptor A type of receptor for norepinephrine and epinephrine; present on visceral effectors innervated by sympathetic postganglionic neurons.

Alveolar duct Branch of a respiratory bronchiole around which alveoli and alveolar sacs are arranged.

Alveolar macrophage Highly phagocytic cell found in the alveolar walls of the lungs. Also called a dust cell.

Alveolar sac A cluster of alveoli that share a common opening.

Alveolus A small hollow or cavity; an air sac in the lungs; milk-secreting portion of a mammary gland.

Alzheimer's (ALTZ-hī-merz) **disease (AD)** Disabling neurological disorder characterized by dysfunction and death of specific cerebral neurons, resulting in widespread intellectual impairment, personality changes, and fluctuations in alertness.

Amenorrhea (ā-men′-ō-RĒ-a) Absence of menstruation.

Amnion (AM-nē-on) A thin, protective fetal membrane that develops from the epiblast; holds the fetus suspended in amniotic fluid.

Amniotic (am′-nē-OT-ik) **fluid** Fluid in the amniotic cavity, the space between the developing embryo (or fetus) and amnion; the fluid is initially produced as a filtrate from maternal blood and later includes fetal urine. It functions as a shock absorber, helps regulate fetal body temperature, and helps prevent desiccation.

Amphiarthrosis (am′-fē-ar-THRŌ-sis) A slightly movable joint, in which the articulating bony surfaces are separated by fibrous connective tissue or fibrocartilage to which both are attached; types are syndesmosis and symphysis.

Ampulla A saclike dilation of a canal or duct, such as the ampulla at one end of each semicircular canal in the internal ear. Dilated terminal portion of the ductus deferens. Widest, longest portion of the uterine tube.

Anabolism (a-NAB-ō-lizm) Synthetic, energy-requiring reactions whereby small molecules are built up into larger ones.

Anaerobic (an-ar-Ō-bik) Not requiring oxygen.

Anal canal The last 2 or 3 cm (1 in.) of the rectum; opens to the exterior through the anus.

Anal column A longitudinal fold in the mucous membrane of the anal canal that contains a network of arteries and veins.

Anal triangle The subdivision of the female or male perineum that contains the anus.

Anaphase The third stage of mitosis in which the chromatids that have separated at the centromeres move to opposite poles of the cell.

Anastomosis An end-to-end union or joining of blood vessels, lymphatic vessels, or nerves.

Anatomic dead space Spaces of the nose, pharynx, larynx, trachea, bronchi, and bronchioles totaling about 150 mL of the 500 mL in a quiet breath (tidal volume); air in the anatomic dead space does not reach the alveoli to participate in gas exchange.

Anatomical position A position of the body universally used in anatomical descriptions in which the body is erect, the head is level, the eyes face forward, the upper limbs are at the sides, the palms face forward, and the feet are flat on the floor.

Anatomical snuffbox Triangular depression between tendons of extensor pollicis brevis and extensor pollicis longus muscles; radial styloid process, base of first metacarpal, trapezium, scaphoid, and deep branch of the radial artery can be palpated in this depression.

Anatomy The structure or study of structure of the body and the relationships of its parts to each other.

Androgens Masculinizing sex hormones produced by the testes in males and the adrenal cortex in both sexes; responsible for libido (sexual desire); the two main androgens are testosterone and dihydrotestosterone.

Anemia (a-NĒ-mē-a) Condition of the blood in which the number of functional red blood cells or their hemoglobin content is below normal.

Angiogenesis The formation of blood vessels; occurs in the extraembryonic mesoderm of the

yolk sac, connecting stalk and chorion at the beginning of the third week of development, and throughout the body after birth.

Antagonist A muscle that has an action opposite that of the prime mover (agonist) and yields to the movement of the prime mover.

Anterior Nearer to or at the front of the body. Equivalent to ventral in bipeds.

Anterior (ventral) roots The structure composed of axons of motor (efferent) neurons that emerges from the anterior aspect of the spinal cord and extends laterally to join a posterior root, forming a spinal nerve. Also called a ventral root.

Anterior cavity Area of the eye between the cornea and lens.

Anterior chamber Chamber of the anterior cavity that lies between the cornea and the iris.

Anterior pituitary Anterior lobe of the pituitary gland. Also called the adenohypophysis (ad′-e-no-hi-POF-i-sis).

Anterolateral pathway Sensory pathway that conveys information related to pain, temperature, crude touch, pressure, tickle, and itch. Also called the spinaothalamic pathway.

Aneurysm (AN-ū-rizm) A saclike enlargement of a blood vessel caused by a weakening of its wall.

Antibody (AN-ti-bod′-ē) A protein produced by plasma cells in response to a specific antigen; combines with that antigen to neutralize, inhibit, or destroy it. Also called an **immunoglobulin** (im-ū-nō-GLOB-ū-lin) or Ig.

Anticoagulant (an-tī-cō-AG-ū-lant) A substance that can delay, suppress, or prevent the clotting of blood.

Antidiuretic Substance that inhibits urine formation.

Antidiuretic hormone (ADH) Hormone produced by neurosecretory cells in the paraventricular and supraoptic nuclei of the hypothalamus that stimulates water reabsorption from kidney tubule cells into the blood and vasoconstriction of arterioles. Also called vasopressin (vaz-o-PRES-in).

Antigen (AN-ti-jen) **(Ag)** A substance that has immunogenicity (the ability to provoke an immune response) and reactivity (the ability to react with the antibodies or cells that result from the immune response); term is a contraction of antibody generator.

Antigen-presenting cell (APC) Special class of migratory cell that processes and presents antigens to T cells during an immune response; APCs include macrophages, B cells, and dendritic cells, which are present in the skin, mucous membranes, and lymph nodes.

Antioxidant A substance that inactivates oxygen-derived free radicals. Examples are selenium, zinc, beta-carotene, and vitamins C and E.

Antrum Any nearly closed cavity or chamber, especially one within a bone, such as a sinus. Cavity in the center of a secondary follicle.

Anus The distal end and outlet of the rectum.

Aorta (ā-OR-ta) The main systemic trunk of the arterial system of the body that emerges from the left ventricle.

Aortic body Cluster of chemoreceptors on or near the arch of the aorta that respond to changes in blood levels of oxygen, carbon dioxide, and hydrogen ions (H^+).

Aortic reflex A reflex that helps maintain normal systemic blood pressure; initiated by baroreceptors in the wall of the ascending aorta and arch of the aorta. Nerve impulses from aortic baroreceptors reach the cardiovascular center via sensory axons of the vagus (X) nerves.

Apex (Ā-peks) The pointed end of a conical structure, such as the apex of the lungs or heart, or the tip of the nose, or the apex of the axilla.

Apnea (AP-nē-a) Temporary cessation of breathing.

Apocrine gland A type of gland in which the secretory products gather at the free end of the secreting cell and are pinched off, along with some of the cytoplasm, to become the secretion, as in mammary glands.

Aponeurosis (ap´-ō-noo-RŌ-sis) A sheetlike tendon joining one muscle with another or with bone.

Apoptosis Programmed cell death; a normal type of cell death that removes unneeded cells during embryological development, regulates the number of cells in tissues, and eliminates many potentially dangerous cells such as cancer cells.

Appositional (a-pō-ZISH-o-nal) **growth** Growth due to surface deposition of material, as in the growth in diameter of cartilage and bone. Also called exogenous (eks-OJ-e-nus) growth.

Aqueduct of the midbrain (cerebral aqueduct) A channel through the midbrain connecting the third and fourth ventricles and containing cerebrospinal fluid.

Aqueous humor The watery fluid, similar in composition to cerebrospinal fluid, that fills the anterior cavity of the eye.

Arachnoid mater The middle of the three meninges (coverings) of the brain and spinal cord.

Arachnoid villus Berrylike tuft of the arachnoid mater that protrudes into the superior sagittal sinus and through which cerebrospinal fluid is reabsorbed into the bloodstream.

Arbor vitae (AR-bor VĪ-tē) The white matter tracts of the cerebellum, which have a treelike appearance when seen in midsagittal section.

Arch of the aorta The most superior portion of the aorta, lying between the ascending and descending segments of the aorta.

Areola Any tiny space in a tissue. The pigmented ring around the nipple of the breast.

Arm The part of the free upper limb from the shoulder to the elbow. Also called the brachium.

Arousal (a-ROW-zal) Awakening from sleep, a response due to stimulation of the reticular activating system (RAS).

Arrector pili (a-REK-tor PĪ-lē) Smooth muscles attached to hairs; contraction pulls the hairs into a vertical position, resulting in "goose bumps."

Arrhythmia (a-RITH-mē-a) An irregular heart rhythm. Also called a dysrhythmia.

Arteriole (ar-TĒ-rē-ōl) A small, almost microscopic, artery that delivers blood to a capillary.

Arteriosclerosis (ar-tē-rē-ō-skle-RŌ-sis) Group of diseases characterized by thickening of the walls of arteries and loss of elasticity.

Artery (AR-ter-ē) A blood vessel that carries blood away from the heart.

Arthritis (ar-THRĪ-tis) Inflammation of a joint.

Arthrology (ar-THROL-ō-jē) The study or description of joints.

Arthroplasty (AR-thrō-plas´-tē) Surgical replacement of joints, for example, the hip and knee joints.

Arthroscopy (ar-THROS-kō-pē) A procedure for examining the interior of a joint, usually the knee, by inserting an arthroscope into a small incision; used to determine extent of damage, remove torn cartilage, repair cruciate ligaments, and obtain samples for analysis.

Arthrosis (ar-THRŌ-sis) A joint or articulation.

Articular (ar-TIK-ū-lar) **capsule** Sleevelike structure around a synovial joint composed of a fibrous membrane and a synovial membrane.

Articular cartilage Hyaline cartilage attached to articular bone surfaces.

Articular disc Fibrocartilage pad between articular surfaces of bones of some synovial joints. Also called a meniscus (men-IS-kus).

Articulation A joint; a point of contact between bones, cartilage and bones, or teeth and bones.

Arytenoid cartilages A pair of small, pyramidal cartilages of the larynx that attach to the vocal folds and intrinsic pharyngeal muscles and can move the vocal folds.

Ascending colon The part of the large intestine that passes superiorly from the cecum to the inferior border of the liver, where it bends at the right colic (hepatic) flexure to become the transverse colon.

Ascites Abnormal accumulation of serous fluid in the peritoneal cavity.

Association areas Large cortical regions on the lateral surfaces of the occipital, parietal, and temporal lobes and on the frontal lobes anterior to the motor areas connected by many motor and sensory axons to other parts of the cortex.

Association tracts One of three types of tracts in the cerebral white matter; conduct nerve impulses between gyri in the same hemisphere.

Asthma Usually allergic reaction characterized by smooth muscle spasms in bronchi resulting in wheezing and difficult breathing. Also called bronchial asthma.

Atherosclerosis (ath-er-ō-skle-RŌ-sis) A progressive disease characterized by the formation in the walls of large and medium-sized arteries of lesions called atherosclerotic plaques.

Atherosclerotic plaque A lesion that results from accumulated cholesterol and smooth muscle fibers (cells) of the tunica media of an artery; may become obstructive.

Atom Unit of matter that makes up a chemical element; consists of a nucleus (containing positively charged protons and uncharged neutrons) and negatively charged electrons that orbit the nucleus.

Atresia (a-TRĒ-zē-a) Degeneration and reabsorption of an ovarian follicle before it fully matures and ruptures; abnormal closure of a passage, or absence of a normal body opening.

Atria Superior chambers of the heart. *Singular* is **atrium**.

Atrial natriuretic (nā´-trē-ū-RET-ik) **peptide (ANP)** Peptide hormone, produced by the atria of the heart in response to stretching, that inhibits aldosterone production and thus lowers blood pressure; causes natriuresis, increased urinary excretion of sodium.

Atrioventricular (AV) bundle The part of the conduction system of the heart that begins at the atrioventricular (AV) node, passes through the cardiac skeleton separating the atria and the ventricles, then extends a short distance down the interventricular septum before splitting into right and left bundle branches. Also called the bundle of His (HISS).

Atrioventricular (AV) node The part of the conduction system of the heart made up of a compact mass of conducting cells located in the septum between the two atria.

Atrioventricular (AV) valve A heart valve made up of membranous flaps or cusps that allows blood to flow in one direction only, from an atrium into a ventricle.

Atrophy (AT-rō-fē) Wasting away or decrease in size of a part, due to a failure, abnormality of nutrition, or lack of use.

Auditory ossicle One of the three small bones of the middle ear called the malleus, incus, and stapes.

Auditory tube The tube that connects the middle ear with the nose and nasopharynx region of the throat. Also called the eustachian (ū-STĀ-shun or ū-STĀ-kē-an) tube or pharyngotympanic tube.

Auricle (OR-i-kul) The projecting part of the external ear composed of elastic cartilage and covered by skin and shaped like the flared end of a trumpet. Also called the **pinna**. Wrinkled pouchlike structure on the anterior surface of each atrium that increases the capacity of an atrium slightly so that it can hold a greater volume of blood.

Auscultation Examination by listening to sounds in the body.

Autoimmunity An immunological response against a person's own tissues.

Autolysis Self-destruction of cells by their own lysosomal digestive enzymes after death or in a pathological process.

Autonomic ganglion (aw´-tō-NOM-ik GANG-lē-on) A cluster of cell bodies of sympathetic or parasympathetic neurons located outside the central nervous system.

Autonomic nervous system (ANS) Visceral sensory (afferent) and visceral motor (efferent) neurons. Autonomic motor neurons, both sympathetic and parasympathetic, conduct nerve impulses from the central nervous system to smooth muscle, cardiac muscle, and glands. So named because this part of the nervous system was thought to be self-governing or spontaneous.

Autonomic plexus A network of sympathetic and parasympathetic axons; examples are the cardiac, celiac, and pelvic plexuses, which are located in the thorax, abdomen, and pelvis, respectively.

Autophagy Process by which worn-out organelles are digested within lysosomes.

Autopsy The examination of the body after death.

Autorhythmicity (aw´-tō-rith-MIS-i-tē) The ability to repeatedly generate spontaneous action potentials.

Autosome (AW-tō-sōm) Any chromosome other than the X and Y chromosome (sex chromosomes).

Avascular (ā-VAS-kū-lar) Without a blood supply.

Axilla The small hollow beneath the arm where it joins the body at the shoulders. Also called the armpit.

Axon The usually single, long process of a nerve cell that propagates a nerve impulse toward the axon terminals.

Axon terminal Terminal branch of an axon where synaptic vesicles undergo exocytosis to release neurotransmitter molecules.

B

B cell A lymphocyte that can develop into a clone of antibody-producing plasma cells or memory cells when properly stimulated by a specific antigen.

Back The posterior part of the body; the dorsum.

Ball-and-socket joint A synovial joint in which the rounded surface of one bone moves within a cup-shaped depression or socket of another bone, as in the shoulder or hip joint. Also called a spheroid (SFE-royd) joint.

Baroreceptor Neuron capable of responding to changes in blood, air, or fluid pressure. Also called a **stretch receptor**.

Basal cells Stem cells located between the bases of supporting cells that produce new olfactory receptor stem cells; found at the periphery of the taste bud that produce supporting cells.

Basal nuclei Paired clusters of gray matter deep in each cerebral hemisphere including the globus pallidus, putamen, and caudate nucleus.

Basement membrane Thin, extracellular layer between epithelium and connective tissue consisting of a basal lamina and a reticular lamina.

Basilar membrane A membrane in the cochlea of the internal ear that separates the cochlear duct from the scala tympani and on which the spiral organ (organ of Corti) rests.

Basophil (BA-sō-fil) A type of white blood cell characterized by a pale nucleus and large vesicles that stain blue-purple with basic dyes.

Belly The abdomen. The prominent, fleshy part of a skeletal muscle.

Beta cell A type of cell in the pancreatic islets (islets of Langerhans) in the pancreas that secretes the hormone insulin.

Beta (β) receptor A type of adrenergic receptor for epinephrine and norepinephrine; found on visceral effectors innervated by sympathetic postganglionic neurons.

Bicuspid valve Atrioventricular (AV) valve on the left side of the heart. Also called the mitral valve or left atrioventricular valve.

Bilateral (bī-LAT-er-al) Pertaining to two sides of the body.

Bile (BĪL) A secretion of the liver consisting of water, bile salts, bile pigments, cholesterol, lecithin, and several ions; it emulsifies lipids prior to their digestion.

Bile canaliculi Small ducts between hepatocytes that collect bile produced by hepatocytes.

Bile duct Passageway that transports bile from the liver and gall bladder to the duodenum.

Biopsy (BĪ-op-sē) The removal of a sample of living tissue to help diagnose a disorder, for example, cancer.

Birth control Various methods designed to control fertility and prevent conception.

Blastocyst In the development of an embryo, a hollow ball of cells that consists of a blastocyst cavity, trophoblast (outer cells), and inner cell mass or embryoblast.

Blastocyst cavity The fluid-filled cavity within the blastocyst.

Blastomere One of the cells resulting from the cleavage of a fertilized ovum.

Blind spot Area in the retina at the end of the optic (II) nerve in which there are no photoreceptors. Also called the **optic disc**.

Blood The fluid that circulates through the heart, arteries, capillaries, and veins and that constitutes the chief means of transport within the body.

Blood clot Gel-like mass consisting of fibrin threads, platelets, and any blood cells trapped in the fibrin.

Blood island Isolated mass of mesoderm derived from angioblasts and from which blood vessels develop.

Blood plasma The extracellular fluid found in blood vessels; blood minus the formed elements; also called plasma.

Blood pressure Force exerted by blood against the walls of blood vessels due to contraction of the heart and influenced by the elasticity of the vessel walls; clinically, a measure of the pressure in the brachial artery during ventricular systole and ventricular diastole.

Blood reservoir Systemic veins that contain large amounts of blood that can be moved quickly to parts of the body requiring the blood.

Blood–brain barrier (BBB) A barrier consisting of specialized brain capillaries and astrocytes that prevents the passage of materials from the blood to the cerebrospinal fluid and brain.

Blood–testis barrier A barrier formed by sustentacular cells that prevents an immune response against antigens produced by spermatogenic cells by isolating the cells from the blood.

Body cavity A space within the body that contains various internal organs.

Bolus A soft, rounded mass, usually food, that is swallowed.

Bone remodeling Ongoing replacement of old bone tissue by new bone tissue.

Bone tissue A connective tissue that consists of abundant extracellular matrix (hydroxyapatite, collagen fibers, and water) that surrounds widely scattered cells (osteogenic cells, osteoblasts, osteocytes, and osteoclasts). Also called osseous tissue (OS-ē-us).

Bony labyrinth A series of cavities within the petrous portion of the temporal bone forming the vestibule, cochlea, and semicircular canals of the inner ear.

Brachial plexus A network of nerve axons of the ventral rami of spinal nerves C5, C6, C7, C8, and T1. The nerves that emerge from the brachial plexus supply the upper limb.

Bradycardia (brād´-i-KAR-dē-a) A slow resting heart or pulse rate (under 50 beats per minute).

Brain The part of the central nervous system contained within the cranial cavity.

Brain stem The portion of the brain immediately superior to the spinal cord, made up of the medulla oblongata, pons, and midbrain.

Brain waves Electrical signals that can be recorded from the skin of the head due to electrical activity of brain neurons.

Broad ligament A double fold of parietal peritoneum attaching the uterus to the side of the pelvic cavity.

Broca's speech area Motor area of the brain in the frontal lobe that translates thoughts into speech. Also called the motor speech area.

Bronchi (BRON-kī) Branches of the respiratory passageway including main (primary) bronchi (the two divisions of the trachea), lobar or secondary bronchi (divisions of the main bronchi that are distributed to the lobes of the lung), and segmental or tertiary bronchi (divisions of the lobar bronchi that are distributed to bronchopulmonary segments of the lung). *Singular* is **bronchus**.

Bronchial tree The trachea, bronchi, and their branching structures up to and including the terminal bronchioles.

Bronchioles Branches of segmental bronchi further dividing into terminal bronchioles (distributed to lobules of the lung), which divide into respiratory bronchioles (distributed to alveolar sacs).

Bronchitis Inflammation of the mucous membrane of the bronchial tree; characterized by hypertrophy and hyperplasia of seromucous glands and goblet cells that line the bronchi which results in a productive cough.

Bronchopulmonary segment One of the smaller divisions of a lobe of a lung supplied by its own segmental bronchi.

Buccal (BUK-al) Pertaining to the cheek or mouth.

Bulb of penis Expanded portion of the base of the corpus spongiosum penis.

Bulbourethral gland One of a pair of glands located inferior to the prostate on either side of the urethra that secretes an alkaline fluid into the cavernous urethra. Also called Cowper's (KOW-perz) gland.

Bulimia (boo-LIM-ē-a *or* boo-LĒ-mē-a) A disorder characterized by overeating at least twice a week followed by purging by self-induced vomiting, strict dieting or fasting, vigorous exercise, or use of laxatives or diuretics. Also called binge–purge syndrome.

Bulk-phase endocytosis A process by which most body cells can ingest membrane-surrounded droplets of interstitial fluid. Also called pinocytosis.

Bursae Sacs or pouches of synovial fluid located at friction points, especially around joints. *Singular* is **bursa**.

Bursitis Inflammation of a bursa.

Buttocks The two fleshy masses on the posterior aspect of the inferior trunk, formed by the gluteal muscles. Also called the gluteal region.

C

Calcaneal (kal-KĀ-nē-al) **tendon** The tendon of the soleus, gastrocnemius, and plantaris muscles at the back of the heel. Also called the Achilles (a-KIL-ēz) tendon.

Calcification Deposition of calcium salts, primarily hydroxyapatite, in a framework formed by collagen fibers in which the tissue hardens. Also called mineralization (min´-e-ral-i-ZĀ-shun).

Calcitonin A hormone produced by the parafollicular cells of the thyroid gland that can lower the amount of blood calcium and phosphates by inhibiting bone resorption (breakdown of bone matrix) and by accelerating uptake of calcium and phosphates into bone matrix.

Calculus (KAL-kū-lus) A stone, or insoluble mass of crystallized salts or other material, formed within the body, as in the gallbladder, kidney, or urinary bladder.

Callus An abnormal thickening of the stratum corneum because of constant exposure of skin to friction. An acquired, localized thickening.

Canaliculi Small channels or canals, as in bones, where they connect lacunae. *Singular* is **canaliculus** (kan′-a-LIK-ū-lus).

Cancer A group of diseases characterized by uncontrolled or abnormal cell division.

Capacitation The functional changes that sperm undergo in the female reproductive tract that allow them to fertilize a secondary oocyte.

Capillary A microscopic blood vessel located between an arteriole and venule through which materials are exchanged between blood and interstitial fluid.

Carbohydrate Organic compound consisting of carbon, oxygen, and hydrogen; the ratio of hydrogen to oxygen atoms is usually 2:1. Examples include sugars, glycogen, starches, and glucose.

Carcinogen A chemical substance or radiation that causes cancer.

Cardiac (KAR-dē-ak) **arrest** Cessation of an effective heartbeat in which the heart is completely stopped or in ventricular fibrillation.

Cardiac conduction system A group of autorhythmic cardiac muscle fibers that generates and distributes electrical impulses to stimulate coordinated contraction of the heart chambers; includes the sinoatrial (SA) node, the atrioventricular (AV) node, the atrioventricular (AV) bundle, the right and left bundle branches, and the Purkinje fibers.

Cardiac cycle A complete heartbeat consisting of systole (contraction) and diastole (relaxation) of both atria plus systole and diastole of both ventricles.

Cardiac muscle tissue Muscle tissue that forms most of the heart; striated and involuntary.

Cardiac notch An angular notch in the anterior border of the left lung into which part of the heart fits.

Cardinal (lateral cervical) ligament A ligament of the uterus, extending laterally from the cervix and vagina as a continuation of the broad ligament.

Cardiogenic area A group of mesodermal cells in the head end of an embryo that gives rise to the heart.

Cardiology The study of the heart and diseases associated with it.

Cardiovascular center Groups of neurons scattered within the medulla oblongata that regulate heart rate, force of contraction, and blood vessel diameter.

Cardiovascular system Body system that consists of blood, the heart, and blood vessels.

Carotene Antioxidant precursor of vitamin A, which is needed for synthesis of photopigments; yellow-orange pigment present in the stratum corneum of the epidermis. Accounts for the yellowish coloration of skin.

Carotid body Cluster of chemoreceptors on or near the carotid sinus that respond to changes in blood levels of oxygen, carbon dioxide, and hydrogen ions.

Carotid sinus A dilated region of the internal carotid artery just superior to where it branches from the common carotid artery; it contains baroreceptors that monitor blood pressure.

Carpals The eight bones of the wrist.

Carpus A collective term for the eight bones of the wrist.

Cartilages A dense network of collagen fibers and elastic fibers firmly embedded in chondroitin sulfate, a gel-like component of the ground substance.

Cartilaginous (kar′-ti-LAJ-i-nus) **joint** A joint without a synovial (joint) cavity where the articulating bones are held tightly together by cartilage, allowing little or no movement.

Catabolism (ka-TAB-ō-lizm) Chemical reactions that break down complex organic compounds into simple ones, with the net release of energy.

Cataract (KAT-a-rakt) Loss of transparency of the lens of the eye or its capsule or both.

Cauda equina (KAW-da ē-KWĪ-na) A tail-like array of roots of spinal nerves at the inferior end of the spinal cord.

Caudal (KAW-dal) Pertaining to any tail-like structure; inferior in position.

Cecum A blind pouch at the proximal end of the large intestine that attaches to the ileum.

Celiac plexus A large mass of autonomic ganglia and axons located at the level of the superior part of the first lumbar vertebra. Also called the solar plexus.

Cell The basic structural and functional unit of all organisms; the smallest structure capable of performing all the activities vital to life.

Cell biology The study of cellular structure and function. Also called cytology.

Cell cycle Growth and division of a single cell into two identical cells; consists of interphase and cell division.

Cell division Process by which a cell reproduces itself that consists of a nuclear division (mitosis) and a cytoplasmic division (cytokinesis); types include somatic and reproductive cell division.

Cell junction Point of contact between plasma membrane of tissue cells.

Cellular respiration The oxidation of glucose to produce ATP that involves glycolysis, acetyl coenzyme A formation, the Krebs cycle, and the electron transport chain.

Cementum Calcified tissue covering the root of a tooth.

Central canal A microscopic tube running the length of the spinal cord in the gray commissure. A circular channel running longitudinally in the center of an osteon (haversian system) of mature compact bone, containing blood and lymphatic vessels and nerves. Also called a haversian (ha-VER-shan) canal.

Central nervous system That portion of the nervous system that consists of the brain and spinal cord.

Centrioles (SEN-trē-ōlz) Paired, cylindrical structures of a centrosome, each consisting of a ring of microtubules and arranged at right angles to each other.

Centromere (SEN-trō-mēr) The constricted portion of a chromosome where the two chromatids are joined; serves as the point of attachment for the microtubules that pull chromatids during anaphase of cell division.

Centrosome A dense network of small protein fibers near the nucleus of a cell, containing a pair of centrioles and pericentriolar material.

Cerebellar peduncle (ser-e-BEL-ar pe-DUNG-kul) A bundle of nerve axons connecting the cerebellum with the brain stem.

Cerebellum The part of the brain lying posterior to the medulla oblongata and pons; governs balance and coordinates skilled movements.

Cerebral aqueduct (SER-ē-bral AK-we-dukt) A channel through the midbrain connecting the third and fourth ventricles and containing cerebrospinal fluid. Also termed the aqueduct of the midbrain.

Cerebral arterial circle A ring of arteries forming an anastomosis at the base of the brain between the internal carotid and basilar arteries and arteries supplying the cerebral cortex. Also called the circle of Willis.

Cerebral cortex The surface of the cerebral hemispheres, 2–4 mm thick, consisting of gray matter; arranged in six layers of neuronal cell bodies in most areas.

Cerebral hemispheres Right and left halves of the cerebrum separated by the longitudinal fissure and connected internally by the corpus callosum.

Cerebral peduncle One of a pair of nerve axon bundles located on the anterior surface of the midbrain, conducting nerve impulses between the pons and the cerebral hemispheres.

Cerebrospinal fluid (CSF) A fluid produced by ependymal cells that cover choroid plexuses in the ventricles of the brain; the fluid circulates in the ventricles, the central canal, and the subarachnoid space around the brain and spinal cord.

Cerebrovascular accident Destruction of brain tissue (infarction) resulting from obstruction or rupture of blood vessels that supply the brain. Also called a stroke or brain attack.

Cerebrum The two hemispheres of the forebrain (derived from the telencephalon), making up the largest part of the brain.

Cerumen Waxlike secretion produced by ceruminous glands in the external auditory meatus (ear canal). Also termed ear wax.

Ceruminous gland A modified sudoriferous (sweat) gland in the external auditory meatus that secretes cerumen (ear wax).

Cervical enlargement Superior enlargement of the spinal cord that extends from the fourth cervical vertebra to the first thoracic vertebra.

Cervical ganglion (SER-vi-kul GANG-glē-on) A cluster of cell bodies of postganglionic sympathetic neurons located in the neck, near the vertebral column.

Cervical plexus A network formed by nerve axons from the ventral rami of the first four cervical nerves and receiving gray rami communicantes from the superior cervical ganglion.

Cervix Neck; any constricted portion of an organ, such as the inferior cylindrical part of the uterus.

Chemical synapses Synapses involving the release of a neurotransmitter from a presynaptic neuron; occur between most neurons and all neurons and effectors.

Chemoreceptor Sensory receptor that detects the presence of a specific chemical.

Chief cell The secreting cell of a gastric gland that produces pepsinogen, the precursor of the enzyme pepsin, and the enzyme gastric lipase. Also called a zymogenic (zi′-mo-JEN-ik) cell. Cell in the parathyroid glands that secretes parathyroid hormone (PTH). Also called a principal cell.

Cholecystectomy (kō′-lē-sis-TEK-tō-mē) Surgical removal of the gallbladder.

Cholecystitis (kō′-lē-sis-TĪ-tis) Inflammation of the gallbladder.

Cholesterol (kō-LES-te-rol) Classified as a lipid, the most abundant steroid in animal tissues; located in cell membranes and used for the synthesis of steroid hormones and bile salts.

Cholinergic neuron A neuron that liberates acetylcholine as its neurotransmitter.

Chondrocyte Cell of mature cartilage.

Chondroitin (kon-DROY-tin) **sulfate** An amorphous extracellular matrix material found outside connective tissue cells.

Chordae tendineae Tendonlike, fibrous cords that connect atrioventricular valves of the heart with papillary muscles.

Chorion (KŌ–rē-on) The most superficial fetal membrane that becomes the principal embryonic portion of the placenta; serves a protective and nutritive function.

Chorionic villi Fingerlike projections of the chorion that grow into the decidua basalis of the endometrium and contain fetal blood vessels.

Chorionic villi sampling (CVS) The removal of a sample of chorionic villus tissue by means of a catheter to analyze the tissue for prenatal genetic defects.

Choroid One of the vascular coats of the eyeball.

Choroid plexus A network of capillaries located in the roof of each of the four ventricles of the brain; ependymal cells around these produce cerebrospinal fluid.

Chromaffin cell Cell that has an affinity for chromium salts, due in part to the presence of the precursors of the neurotransmitter epinephrine; found, among other places, in the adrenal medulla.

Chromatid One of a pair of identical connected nucleoprotein strands that are joined at the centromere and separate during cell division, each becoming a chromosome of one of the two daughter cells.

Chromatin The threadlike mass of genetic material, consisting of DNA and histone proteins, that is present in the nucleus of a nondividing or interphase cell.

Chromosome One of the small, threadlike structures in the nucleus of a cell, normally 46 in a human diploid cell, that bears the genetic material; composed of DNA and proteins (histones) that form a delicate chromatin thread during interphase; becomes packaged into compact rodlike structures that are visible under the light microscope during cell division.

Chronic (KRON-ik) Long term or frequently recurring; applied to a disease that is not acute.

Chronic obstructive pulmonary disease (COPD) A disease, such as bronchitis or emphysema, in which there is some degree of obstruction of airways and consequent increase in airway resistance.

Chyle (KĪL) The milky-appearing fluid found in the lacteals of the small intestine after absorption of lipids in food.

Chyme (KĪM) The semifluid mixture of partly digested food and digestive secretions found in the stomach and small intestine during digestion of a meal.

Cilia Hairs or hairlike process projecting from a cell that may be used to move the entire cell or to move substances along the surface of the cell. *Singular* is **cilium**.

Ciliary body One of the three parts of the vascular tunic of the eyeball, the others being the choroid and the iris; includes the ciliary muscle and the ciliary processes.

Ciliary ganglion A very small parasympathetic ganglion whose preganglionic axons come from the oculomotor (III) nerve and whose postganglionic axons carry nerve impulses to the ciliary muscle and the sphincter muscle of the iris.

Ciliary muscle Smooth muscle of the ciliary body that produces a narrowing or sphincter action of the ringlike ciliary body.

Circadian (ser-KĀ-dē-an) **rhythm** The pattern of biological activity on a 24-hour cycle, such as the sleep-wake cycle.

Circular folds Permanent, deep, transverse folds in the mucosa and submucosa of the small intestine that increase the surface area for absorption. Also called plicae circulares (PLĪ-kē SER-kū-lar-ēs).

Circulation time The time required for a drop of blood to pass from the right atrium, through pulmonary circulation, back to the left atrium, through systemic circulation down to the foot, and back again to the right atrium.

Circulatory routes Organizations of arteries, arterioles, capillaries, venules, and veins that deliver blood to specific areas of the body.

Circumduction A movement at a synovial joint in which the distal end of a bone moves in a circle while the proximal end remains relatively stable.

Cirrhosis (si-RŌ-sis) A liver disorder in which the parenchymal cells are destroyed and replaced by connective tissue.

Cisterna chyli The origin of the thoracic duct.

Cleavage The rapid mitotic divisions following the fertilization of a secondary oocyte, resulting in an increased number of progressively smaller cells, called blastomeres.

Clitoris (KLI-to-ris) An erectile organ of the female, located at the anterior junction of the labia minora, that is homologous to the male penis.

Clone (KLŌN) A population of identical cells.

Coarctation (kō-ark-TĀ-shun) **of the aorta** A congenital heart defect in which a segment of the aorta is too narrow. As a result, the flow of oxygenated blood to the body is reduced, the left ventricle is forced to pump harder, and high blood pressure develops.

Coccyx The fused bones at the inferior end of the vertebral column.

Cochlea A winding, cone-shaped tube forming a portion of the inner ear and containing the spiral organ (organ of Corti).

Cochlear duct The membranous cochlea consisting of a spirally arranged tube enclosed in the bony cochlea and lying along its outer wall. Also called the scala media (SCA-la MĒ-dē-a).

Collagen (KOL-a-jen) A protein that is the main organic constituent of connective tissue.

Collateral circulation Alternate routes for the blood to reach a particular organ or tissue.

Colliculus A small elevation. *Plural* is **colliculi**.

Colon The portion of the large intestine consisting of ascending, transverse, descending, and sigmoid portions.

Colony-stimulating factors (CSFs) One of a group of molecules that stimulates development of white blood cells. Examples are macrophage CSF and granulocyte CSF.

Colostrum (kō-LOS-trum) A thin, cloudy fluid secreted by the mammary glands a few days prior to or after delivery before true milk is produced.

Column Group of white matter tracts in the spinal cord.

Commissural tracts (kom′-i-SYŪR-al) Within cerebral white matter, contain axons that conduct nerve impulses from gyri in the other cerebral hemisphere. Includes corpus callosum, anterior commissure, and posterior commissure.

Common bile duct A tube formed by the union of the common hepatic duct and the cystic duct that empties bile into the duodenum at the hepatopancreatic ampulla (ampulla of Vater).

Common hepatic duct Duct emerging from the liver that joins the cystic duct from the gallbladder to form the common bile duct.

Compact bone Type of bone tissue consisting of an osteon with four parts--lamellae, lacunae, canaliculi, and a central-haversian) canal.

Compact bone tissue Bone tissue that contains few spaces between osteons (haversian systems); forms the external portion of all bones and the bulk of the diaphysis (shaft) of long bones; is found immediately deep to the periosteum and external to spongy bone. Also called cortical bone.

Compartments Groupings of functionally related skeletal muscles and their associated blood vessels and nerves.

Concentric contraction A contraction in which a muscle shortens as it produces a constant tension and overcomes the load it is moving.

Concentric lamellae Circular plates of mineralized extracellular matrix that form the rings of bony material in the osteon.

Concha A scroll-like bone found in the skull. The hollow of the auricle.

Concussion Traumatic injury to the brain that produces no visible bruising but may result in abrupt, temporary loss of consciousness.

Conduction system A group of autorhythmic cardiac muscle fibers that generates and distributes electrical impulses to stimulate coordinated contraction of the heart chambers; includes the sinoatrial (SA) node, the atrioventricular (AV) node, the atrioventricular (AV) bundle, the right and left bundle branches, and the Purkinje fibers.

Condyloid joint A synovial joint structured so that an oval-shaped condyle of one bone fits into an elliptical cavity of another bone, permitting side-to-side and back-and-forth movements, such as the joint at the wrist between the radius and carpals. Also called an ellipsoidal (ē-lip-SOYD-al) joint.

Cone The type of photoreceptor in the retina that is specialized for highly acute color vision in bright light.

Congenital (kon-JEN-i-tal) Present at the time of birth.

Conjunctiva The delicate membrane covering the eyeball and lining the eyes.

Connective tissue One of the most abundant of the four basic tissue types in the body, performing the functions of binding and supporting; consists of relatively few cells in a gelatinous matrix (the ground substance and fibers between the cells).

Consciousness (KON-shus-nes) A state of wakefulness in which an individual is fully alert, aware, and oriented, partly as a result of feedback between the cerebral cortex and reticular activating system.

Contraception (kon´-tra-SEP-shun) The prevention of fertilization or impregnation without destroying fertility.

Contractility The ability of cells or parts of cells to actively generate force to undergo shortening for movements.

Conus medullaris (KŌ-nus med-ū-LAR-is) The tapered portion of the spinal cord inferior to the lumbar enlargement.

Convergence A synaptic arrangement in which the synaptic end bulbs of several presynaptic neurons terminate on one postsynaptic neuron. The medial movement of the two eyeballs so that both are directed toward a near object being viewed in order to produce a single image.

Cornea The nonvascular, transparent fibrous coat through which the iris of the eye can be seen.

Corneocytes Thin, flat, plasma-membrane enclosed keratin cells that form the stratum corneum of the epidermis. Also called squames.

Corona Margin of the glans penis.

Corona radiata The innermost layer of granulosa cells that is firmly attached to the zona pellucida around a secondary oocyte.

Coronary (cardiac) circulation The pathway followed by the blood from the ascending aorta through the blood vessels supplying the heart and returning to the right atrium. Also called cardiac circulation.

Coronary artery disease A condition such as atherosclerosis that causes narrowing of coronary arteries so that blood flow to the heart is reduced. The result is coronary heart disease (CHD), in which the heart muscle receives inadequate blood flow due to an interruption of its blood supply.

Coronary sinus A wide venous channel on the posterior surface of the heart that collects the blood from the coronary circulation and returns it to the right atrium.

Corpus albicans (KOR-pus AL-bi-kanz) A white fibrous patch in the ovary that forms after the corpus luteum regresses.

Corpus callosum The great commissure of the brain between the cerebral hemispheres.

Corpus luteum A yellowish body in the ovary formed when a follicle has discharged its secondary oocyte; secretes estrogens, progesterone, relaxin, and inhibin.

Corpus striatum An area in the interior of each cerebral hemisphere composed of the caudate and lentiform nuclei of the basal nuclei and white matter of the internal capsule, arranged in a striated manner.

Corpuscle of touch The sensory receptor for the sensation of touch; found in the dermal papillae, especially in palms and soles. Also called a **Meissner corpuscle** (MĪZ-ner).

Cortex An outer layer of an organ. The convoluted layer of gray matter covering each cerebral hemisphere.

Cranial cavity A subdivision of the dorsal body cavity formed by the cranial bones and containing the brain.

Cranial nerve One of 12 pairs of nerves that leave the brain; pass through foramina in the skull; and supply sensory and motor neurons to the head, neck, part of the trunk, and viscera of the thorax and abdomen. Each is designated by a Roman numeral and a name.

Craniosacral outflow The axons of parasympathetic preganglionic neurons, which have their cell bodies located in nuclei in the brain stem and in the lateral gray matter of the sacral portion of the spinal cord.

Cranium The skeleton of the skull that protects the brain and the organs of sight, hearing, and balance; includes the frontal, parietal, temporal, occipital, sphenoid, and ethmoid bones.

Crista A crest or ridged structure. A small elevation in the ampulla of each semicircular duct that contains receptors for dynamic equilibrium.

Crossing-over The exchange of a portion of one chromatid with another during meiosis. It permits an exchange of genes among chromatids and is one factor that results in genetic variation of progeny.

Crura of penis Separated, tapered portion of the corpora cavernosa penis. *Singular* is **crus of penis**.

Cubital fossa (KŪ-bi-tal FOS-a) Triangular space in the anterior region of the elbow bounded proximally by an imaginary line between the humeral epicondyles, laterally by the medial border of the brachiradialis, and medially by the lateral border of the pronator teres.

Cuneate nucleus A group of neurons in the inferior part of the medulla oblongata in which axons of the cuneate fasciculus terminate.

Cupula A mass of gelatinous material covering the hair cells of a crista.

Cushing's syndrome Condition caused by a hypersecretion of glucocorticoids characterized by spindly legs, "moon face," "buffalo hump," pendulous abdomen, flushed facial skin, poor wound healing, hyperglycemia, osteoporosis, hypertension, and increased susceptibility to disease.

Cutaneous membrane (kū-TĀ-nē-us) The external covering of the body (skin) that consists of a superficial, thinner epidermis (epithelial tissue) and a deep, thicker dermis (connective tissue) that is anchored to the subcutaneous layer.

Cyst (SIST) A sac with a distinct connective tissue wall, containing a fluid or other material.

Cystic duct The duct that carries bile from the gallbladder to the common bile duct.

Cytokinesis (sī´-tō-ki-NĒ-sis) Distribution of the cytoplasm into two separate cells during cell division; coordinated with nuclear division (mitosis).

Cytolysis (sī-TOL-i-sis) The rupture of living cells in which the contents leak out.

Cytoplasm (SĪ-tō-plasm) Cytosol plus all organelles except the nucleus.

Cytoskeleton Complex internal structure of cytoplasm consisting of microfilaments, microtubules, and intermediate filaments.

Cytosol Semifluid portion of cytoplasm in which organelles and inclusions are suspended and solutes are dissolved. Also called **intracellular fluid.**

D

Decidua That portion of the endometrium of the uterus (all but the deepest layer) that is modified during pregnancy and shed after childbirth.

Deciduous (dē-SID-ū-us) Falling off or being shed seasonally or at a particular stage of development. In the body, referring to the first set of teeth.

Decussation (dē´-ku-SĀ-shun) A crossing-over to the opposite (contralateral) side; an example is the crossing of 90% of the axons in the large motor tracts to opposite sides in the medullary pyramids.

Deep Away from the surface of the body or an organ.

Deep (abdominal) inguinal ring A slitlike opening in the aponeurosis of the transversus abdominis muscle that represents the origin of the inguinal canal.

Deep-venous thrombosis (DVT) The presence of a thrombus in a vein, usually a deep vein of the lower limbs.

Defecation (def-e-KĀ-shun) The discharge of feces from the rectum.

Deglutition (dē-gloo-TISH-un) The act of moving food from the mouth to the stomach; also called swallowing.

Dehydration (dē-hī-DRĀ-shun) Excessive loss of water from the body or its parts.

Delta cell A cell in the pancreatic islets (islets of Langerhans) in the pancreas that secretes somatostatin. Also termed a D cell.

Demineralization Loss of calcium and phosphorus from bones.

Dendrite A neuronal process that carries electrical signals, usually graded potentials, toward the cell body.

Dendritic cell One type of antigen-presenting cell with long branchlike projections that commonly is present in mucosal linings such as the vagina, in the skin (intraepidermal macrophage cells in the epidermis), and in lymph nodes (follicular dendritic cells).

Dental caries (KA-rēz) Gradual demineralization of the enamel and dentin of a tooth that may invade the pulp and alveolar bone. Also called tooth decay.

Denticulate (den-TIK-ū-lat) **ligaments** Triangular-shaped membranous extensions of the pia mater that suspend the spinal cord in the middle of its dural sheath.

Dentin The bony tissues of a tooth enclosing the pulp cavity.

Dentition The eruption of teeth. The number, shape, and arrangement of teeth.

Deoxyribonucleic (dē-ok´-sē-rī-bō-nū-KLĒ-ik) **acid (DNA)** A nucleic acid constructed of nucleotides consisting of one of four bases (adenine, cytosine, guanine, or thymine), deoxyribose, and a phosphate group; encoded in the nucleotides as genetic information.

Depression Movement in which a part of the body moves inferiorly.

Dermal papilla Fingerlike projection of the papillary region of the dermis that may contain blood capillaries or corpuscles of touch (Meissner corpuscles).

Dermatology The medical specialty dealing with diseases of the integumentary system.

Dermatome The cutaneous area developed from one embryonic spinal cord segment and receiving most of its sensory innervation from one spinal nerve. An instrument for incising the skin or cutting thin transplants of skin.

Dermis A layer of dense irregular connective tissue lying deep to the epidermis.

Descending colon The part of the large intestine descending from the left colic (splenic) flexure to the level of the left iliac crest.

Detrusor (de-TROO-ser) **muscle** Smooth muscle that forms the wall of the urinary bladder.

Developmental biology The study of development from the fertilized egg to the adult form.

Deviated septum Displacement of the bones and cartilage of the nose by an injury that results in a narrowing of one side of the nasal cavity, making it difficult to breathe through that side of the nose.

Diabetes mellitus (dī-a-BĒ-tēz MEL-i-tus) An endocrine disorder caused by an inability to produce or use insulin. It is characterized by the three "polys": polyuria (excessive urine production), polydipsia (excessive thirst), and polyphagia (excess eating).

Diagnosis Distinguishing one disease from another or determining the nature of a disease from signs and symptoms by inspection, palpation, laboratory tests, and other means.

Dialysis (dī-AL-i-sis) The removal of waste products from blood by diffusion through a selectively permeable membrane.

Diaphragm Any partition that separates one area from another, especially the dome-shaped skeletal muscle between the thoracic and abdominal cavities. Also a dome-shaped device that is placed over the cervix, usually with a spermicide, to prevent conception.

Diaphysis The shaft of a long bone.

Diarrhea (dī-a-RĒ-a) Frequent defecation of liquid feces caused by increased motility of the intestines.

Diarthrosis (dī-ar-THRŌ-sis) A freely movable joint; types are gliding, hinge, pivot, condyloid, saddle, and ball-and-socket.

Diastole In the cardiac cycle, the phase of relaxation or dilation of the heart muscle, especially of the ventricles.

Diastolic (dī-as-TOL-ik) **blood pressure (DBP)** The force exerted by blood on arterial walls during ventricular relaxation; the lowest blood pressure measured in the large arteries, normally about 80 mmHg in a young adult.

Diencephalon A part of the brain consisting of the thalamus, hypothalamus, epithalamus, and subthalamus; ectodermal structure from which the pineal gland develops.

Differentiation The process unspecialized cells go through to become specialized cells; one of the six life processes.

Diffusion A passive process in which there is a net or greater movement of molecules or ions from a region of high concentration to a region of low concentration until equilibrium is reached.

Digestion The mechanical and chemical breakdown of food to simple molecules that can be absorbed and used by body cells.

Digestive system A system that consists of the gastrointestinal tract (mouth, pharynx, esophagus, stomach, small intestine, and large intestine) and accessory digestive organs (teeth, tongue, salivary glands, liver, gallbladder, and pancreas). Its function is to break down foods into small molecules that can be used by body cells.

Dilate (DĪ-lāt) To expand or swell.

Diploë The spongy bone sandwiched between the outer plates of compact bone in flat bones.

Diploid cell (2n) A cell with the number of chromosomes characteristically found in the somatic cells of an organism, two haploid sets of chromosomes, one each from the mother and father.

Direct motor pathways Collections of upper motor neurons with cell bodies in the motor cortex that project axons into the spinal cord, where they synapse with lower motor neurons or interneurons in the anterior horns. Also called the pyramidal pathways.

Disease Any change from a state of health.

Dislocation Displacement of a bone from a joint with tearing of ligaments, tendons, and articular capsules. Also called luxation (luks-Ā-shun).

Dissection Separation of tissues and parts of a cadaver or an organ for anatomical study.

Distal (DIS-tal) Farther from the attachment of a limb to the trunk; farther from the point of origin or attachment.

Diuretic (dī´-ū-RET-ik) A chemical that increases urine volume by decreasing reabsorption of water, usually by inhibiting sodium reabsorption.

Divergence A synaptic arrangement in which the synaptic end bulbs of one presynaptic neuron terminate on several postsynaptic neurons.

Diverticulum (dī-ver-TIK-ū-lum) A sac or pouch in the wall of a canal or organ, especially in the colon.

Dorsal ramus (RĀ-mus) A branch of a spinal nerve containing motor and sensory axons supplying the muscles, skin, and bones of the posterior part of the head, neck, and trunk.

Dorsal respiratory group (DRG) Collection of neurons in the medullary respiratory center that controls normal breathing; formerly called the inspiratory area.

Dorsiflexion Bending the foot in the direction of the dorsum (upper surface).

Ductus arteriosus A small vessel connecting the pulmonary trunk with the aorta; found only in the fetus.

Ductus deferens The duct that carries sperm from the epididymis to the ejaculatory duct. Also called the vas deferens or seminal duct.

Ductus epididymis A tightly coiled tube inside the epididymis, distinguished into a head, body, and tail, in which sperm undergo maturation.

Ductus venosus A small vessel in the fetus that helps the circulation bypass the liver.

Duodenal (doo-ō-DĒ-nal) **gland** Gland in the submucosa of the duodenum that secretes an alkaline mucus to protect the lining of the small intestine from the action of enzymes and to help neutralize the acid in chyme. Also called Brunner's (BRUN-erz) gland.

Duodenal papilla (pa-PIL-a) An elevation on the duodenal mucosa that receives the hepatopancreatic ampulla (ampulla of Vater).

Duodenum (doo´-ō-DĒ-num or doo-OD-e-num) The first 25 cm (10 in.) of the small intestine, which connects the stomach and the ileum.

Dura mater The outermost of the three meninges (coverings) of the brain and spinal cord.

Dynamic equilibrium The maintenance of body position, mainly the head, in response to sudden movements such as rotation.

Dysmenorrhea (dis´-men-ō-RĒ-a) Painful menstruation.

Dysplasia (dis-PLĀ-zē-a) Change in the size, shape, and organization of cells due to chronic irritation or inflammation; may either revert to normal if stress is removed or progress to neoplasia.

Dyspnea (DISP-nē-a) Shortness of breath; painful or labored breathing.

E

Eccentric contraction Contraction in which a muscle lengthens as it produces a constant tension and overcomes the load it is moving.

Eccrine glands Gland made up of secretory cells that remain intact throughout the process of formation and discharge of the secretory product, as in the salivary and pancreatic glands. Also called **merocrine glands**.

Ectoderm (EK-tō-derm) The primary germ layer that gives rise to the nervous system and the epidermis of skin and its derivatives.

Ectopic (ek-TOP-ik) Out of the normal location, as in ectopic pregnancy.

Edema (e-DĒ-ma) An abnormal accumulation of interstitial fluid.

Effector An organ of the body, either a muscle or a gland, that is innervated by somatic or autonomic motor neurons.

Efferent arteriole A vessel of the renal vascular system that carries blood from a glomerulus to a peritubular capillary.

Efferent ducts A series of coiled tubes that transport sperm from the rete testis to the epididymis.

Efferent neurons Neurons that conduct impulses from the brain toward the spinal cord or out of the brain and spinal cord into cranial or spinal nerves to effectors that may be either muscles or glands. Also called **motor neurons**.

Ejaculation The reflex ejection or expulsion of semen from the penis.

Ejaculatory duct A tube that transports sperm from the ductus (vas) deferens to the prostatic urethra.

Elasticity The ability of tissue to return to its original shape after contraction or extension.

Electrical excitability (ek-sīt´-a-BIL-i-tē) Ability of muscle and nerve cells to respond to certain stimuli by producing electrical signals.

Electrical synapse Synapse in which the plasma membranes of the presynaptic and postsynaptic neurons are tightly bound by gap junctions that contain connexons; ion flow through the gap junctions generates a nerve impulse.

Electrocardiogram A recording of the electrical changes accompanying the cardiac cycle that can be detected at the surface of the body; may be resting, stress, or ambulatory.

Elevation Movement in which a part of the body moves superiorly.

Embolus (EM-bō-lus) A blood clot, bubble of air or fat from broken bones, mass of bacteria, or other debris or foreign material transported by the blood.

Embryo The young of any organism in an early stage of development; in humans, the developing organism from fertilization to the end of the eighth week of development.

Embryoblast A region of cells of a blastocyst that differentiates into the three primary germ layers — ectoderm, mesoderm, and endoderm — from which all tissues and organs develop. Also called an inner cell mass.

Embryology The study of development from the fertilized egg to the end of the eighth week of development.

Embryonic connective tissue Connective tissue present primarily in the embryo.

Emesis (EM-e-sis) Vomiting.

Emigration Process whereby white blood cells (WBCs) leave the bloodstream by rolling along the endothelium, sticking to it, and squeezing between the endothelial cells. Adhesion molecules help WBCs stick to the endothelium. Formerly known as diapedesis.

Emission Propulsion of sperm into the urethra due to peristaltic contractions of the ducts of the testes, epididymides, and ductus (vas) deferens as a result of sympathetic stimulation.

Emphysema (em-fi-SĒ-ma) A lung disorder in which alveolar walls disintegrate, producing abnormally large air spaces and loss of elasticity in the lungs; typically caused by exposure to cigarette smoke.

Emulsification (ē-mul-si-fi-KĀ-shun) The dispersion of large lipid globules into smaller, uniformly distributed particles in the presence of bile.

Enamel The hard, white substance covering the crown of a tooth.

Endocardium The layer of the heart wall, composed of endothelium and smooth muscle, that lines the inside of the heart and covers the valves and tendons that hold the valves open.

Endochondral ossification The replacement of cartilage by bone. Also called intracartilaginous (in´-tra-kar´-ti-LAJ-i-nus) ossification.

Endocrine gland A gland composed of glandular epithelium that secretes hormones into interstitial fluid and then the blood; a ductless gland.

Endocrine system The body system composed of all endocrine glands and hormone-secreting cells.

Endocrinology The science concerned with the structure and functions of endocrine glands and the diagnosis and treatment of disorders of the endocrine system.

Endocytosis The uptake into a cell of large molecules and particles in which a segment of plasma membrane surrounds the substance, encloses it, and brings it in.

Endoderm A primary germ layer of the developing embryo; gives rise to the gastrointestinal tract, urinary bladder, urethra, and respiratory tract.

Endodontics The branch of dentistry concerned with the prevention, diagnosis, and treatment of diseases that affect the pulp, root, periodontal ligament, and alveolar bone.

Endolymph The fluid within the membranous labyrinth of the internal ear.

Endometriosis (en´-dō-me´-trē-Ō-sis) The growth of endometrial tissue outside the uterus.

Endometrium The mucous membrane lining the uterus.

Endomysium (en´-dō-MĪZ-ē-um) Invagination of the perimysium separating each individual muscle fiber (cell).

Endoneurium Connective tissue wrapping around individual nerve axons (cells).

Endoplasmic reticulum (en´-dō-PLAS-mik re-TIK-ū-lum) **(ER)** A network of channels running through the cytoplasm of a cell that serves in intracellular transportation, support, storage, synthesis, and packaging of molecules.

Endosteum The membrane that lines the medullary (marrow) cavity of bones, consisting of osteogenic cells and scattered osteoclasts.

Endothelial-capsular membrane A filtration membrane in a nephron of a kidney consisting of the endothelium and basement membrane of the glomerulus and the epithelium of the visceral layer of the glomerular (Bowman's) capsule. Also called the **filtration membrane**.

Endothelium The layer of simple squamous epithelium that lines the cavities of the heart, blood vessels, and lymphatic vessels.

Enteric nervous system (ENS) The part of the nervous system that is embedded in the submucosa and muscularis of the gastrointestinal (GI) tract; governs motility and secretions of the GI tract.

Enteroendocrine cell A cell of the mucosa of the gastrointestinal tract that secretes a hormone that governs function of the GI tract; hormones secreted include gastrin, cholecystokinin, glucose-dependent insulinotropic peptide (GIP), and secretin.

Enzyme A substance that accelerates chemical reactions; an organic catalyst, usually a protein.

Eosinophil A type of white blood cell characterized by vesicles that stain red or pink with acid dyes.

Ependymal (ep-EN-de-mal) **cells** Neuroglial cells that cover choroid plexuses and produce cerebrospinal fluid (CSF); they also line the ventricles of the brain and probably assist in the circulation of CSF.

Epicardium The thin outer layer of the heart wall, composed of adipose tissue and mesothelium. The mesothelium alone is called the visceral pericardium.

Epicranial aponeurosis (ep-i-KRĀ-nē-al ap´-ō-noo-RO-sis) Sheetlike tendon that joins the frontal and occipital bellies of the occipitofrontalis muscle. Also called the galea aponeurotica.

Epidemiology Study of the occurrence and distribution of diseases and disorders in human populations.

Epidermis The superficial, thinner layer of skin, composed of keratinized stratified squamous epithelium.

Epididymis A comma-shaped organ that lies along the posterior border of the testis and contains the ductus epididymis, in which sperm undergo maturation. *Plural* is **epididymides**.

Epidural space A space between the spinal dura mater and the vertebral canal, containing areolar connective tissue and a plexus of veins.

Epiglottis A large, leaf-shaped piece of cartilage lying on top of the larynx, attached to the thyroid cartilage; its unattached portion is free to move up and down to cover the glottis (vocal folds and rima glottidis) during swallowing.

Epimysium (ep-i-MĪZ-ē-um) Fibrous connective tissue around muscles.

Epinephrine Hormone secreted by the adrenal medulla that produces actions similar to those that result from sympathetic stimulation. Also called adrenaline (a-DREN-a-lin).

Epineurium (ep´-i-NOO-rē-um) The superficial connective tissue covering around an entire nerve.

Epiphyseal line (ep´-i-FIZ-ē-al) The remnant of the epiphyseal plate in the metaphysis of a long bone.

Epiphyseal plate The hyaline cartilage plate in the metaphysis of a long bone; site of lengthwise growth of long bones. Also called the growth plate.

Epiphysis The end of a long bone, usually larger in diameter than the shaft (diaphysis).

Epiphysis cerebri (se-RĒ-brē) Pineal gland.

Episiotomy (e-piz´-ē-OT-ō-mē) A cut made with surgical scissors to avoid tearing of the perineum at the end of the second stage of labor.

Epistaxis (ep´-i-STAK-sis) Loss of blood from the nose due to trauma, infection, allergy, neoplasm, and bleeding disorders. Also called nosebleed.

Epithalamus Part of the diencephalon superior and posterior to the thalamus, comprising the pineal gland and associated structures.

Epithelial tissue The tissue that forms innermost and outermost surfaces of body structures and forms glands. Also called epithelium.

Eponychium Narrow band of stratum corneum at the proximal border of a nail that extends from the margin of the nail wall. Also called the cuticle.

Equilibrium (ē-kwi-LIB-rē-um) The state of being balanced.

Erectile dysfunction (ED) Failure to maintain an erection long enough for sexual intercourse. Previously known as impotence (IM-pō-tens).

Erection The enlarged and stiff state of the penis or clitoris resulting from the engorgement of the spongy erectile tissue with blood.

Eructation (e-ruk′-TĀ-shun) The forceful expulsion of gas from the stomach. Also called belching.

Erythema (er-e-THĒ-ma) Skin redness usually caused by dilation of the capillaries.

Erythrocyte A mature red blood cell.

Erythropoietin (EPO) A hormone released by the juxtaglomerular cells of the kidneys that stimulates red blood cell production.

Esophagus The hollow muscular tube that connects the pharynx and the stomach.

Estrogens Feminizing sex hormones produced by the ovaries; govern development of oocytes, maintenance of female reproductive structures, and appearance of secondary sex characteristics; also affect fluid and electrolyte balance, and protein anabolism. Examples are b-estradiol, estrone, and estriol.

Eupnea (ŪP-nē-a) Normal quiet breathing.

Eversion The movement of the sole laterally at the ankle joint or of an atrioventricular valve into an atrium during ventricular contraction.

Excitability (ek-sīt′-a-BIL-i-tē) The ability of muscle fibers to receive and respond to stimuli; the ability of neurons to respond to stimuli and generate nerve impulses.

Excretion (eks-KRĒ-shun) The process of eliminating waste products from the body; also the products excreted.

Exhalation (eks-ha-LĀ-shun) Breathing out; expelling air from the lungs into the atmosphere. Also called expiration.

Exocrine gland A gland composed of glandular epithelium that secretes its products into ducts that carry the secretions into body cavities, into the lumen of an organ, or to the outer surface of the body.

Exocytosis (ek-sō-sī-TŌ-sis) A process in which membrane-enclosed secretory vesicles form inside the cell, fuse with the plasma membrane, and release their contents into the interstitial fluid; achieves secretion of materials from a cell.

Extensibility The ability of tissue to stretch when it is pulled.

Extension An increase in the angle between two bones; restoring a body part to its anatomical position after flexion.

External Located on or near the surface.

External (outer) ear The outer ear, consisting of the pinna, external auditory canal, and tympanic membrane (eardrum).

External auditory canal A curved tube in the temporal bone that leads to the middle ear.

External nares The openings into the nasal cavity on the exterior of the body. Also called the nostrils.

External respiration The exchange of respiratory gases between the lungs and blood. Also called pulmonary respiration.

Exteroceptor A sensory receptor adapted for the reception of stimuli from outside the body.

Extracellular fluid (ECF) Fluid outside body cells.

Extracellular matrix The ground substance and fibers between cells in a connective tissue.

Eyebrow The hairy ridge superior to the eye.

Eyelids Supporting structures of the eye that shade the eyes during sleep protect them from excessive light and foreign objects, and spread lubricating secretions over the eyeballs. Also called the palpebrae.

F

F cell A cell in the pancreatic islets (islets of Langerhans) that secretes pancreatic polypeptide.

Face The anterior aspect of the head.

Facilitated diffusion (fa- SIL-i-tā′-ted di-FŪ-zhun) Transport process in which an integral membrane protein assists a specific substance across the plasma membrane.

Falciform ligament (FAL-si-form LIG-a-ment) A sheet of parietal peritoneum between the two principal lobes of the liver. The ligamentum teres, or remnant of the umbilical vein, lies within its fold.

Falx cerebelli (FALKS ser-e-BEL-lī) A small triangular process of the dura mater attached to the occipital bone in the posterior cranial fossa and projecting inward between the two cerebellar hemispheres.

Falx cerebri (FALKS SER-e-brē) A fold of the dura mater extending deep into the longitudinal fissure between the two cerebral hemispheres.

Fascia A fibrous membrane covering, supporting, and separating muscles.

Fascicle A small bundle or cluster, especially of nerve or muscle fibers (cells). Also called a fasciculus (fa-SIK-ū-lus) for *singular* and fasciculi for *plural*.

Fasciculation (fa-sik-ū-LĀ-shun) Abnormal, spontaneous twitch of all skeletal muscle fibers in one motor unit that is visible at the skin surface; not associated with movement of the affected muscle; present in progressive diseases of motor neurons, for example, poliomyelitis.

Fast glycolytic (FG) fibers Skeletal muscle fibers that have a low myoglobin content, relatively few blood capillaries and mitochondria, and appear white in color; they contract strongly and quickly and are adapted for intense anaerobic movements of short duration, for example, weight lifting. Also called type IIb fibers.

Fast oxidative-glycolytic (FOG) fibers Skeletal muscle fibers that have a high hemoglobin content and many blood capillaries and appear dark red; they contribute to activities such as running and walking. Also called type IIa fibers.

Fat A triglyceride that is a solid at room temperature.

Fatty acid A simple lipid that consists of a carboxyl group and a hydrocarbon chain; used to synthesize triglyceride and phospholipids.

Fauces The opening from the mouth into the pharynx.

Feces Material discharged from the rectum and made up of bacteria, excretions, and food residue. Also called stool.

Female reproductive cycle General term for the ovarian and uterine cycles, the hormonal changes that accompany them, and cyclic changes in the breasts and cervix; includes changes in the endometrium of a nonpregnant female that prepares the lining of the uterus to receive a fertilized ovum. Less correctly termed the menstrual cycle.

Femoral triangle (FEM-ō-ral) Anatomical region at the proximal end of the thigh formed by the inguinal ligament superiorly, sartorius muscle laterally, and adductor longus muscle medially; contains the femoral nerve, artery, vein, and deep inguinal lymph nodes.

Fertilization Penetration of a secondary oocyte by a sperm cell, meiotic division of secondary oocyte to form an ovum, and subsequent union of the nuclei of the gametes.

Fetal circulation The cardiovascular system of the fetus, including the placenta and special blood vessels involved in the exchange of materials between fetus and mother.

Fetus In humans, the developing organism in utero from the beginning of the third month to birth.

Fever An elevation in body temperature above the normal temperature of 37°C (98.6°F) due to a resetting of the hypothalamic thermostat.

Fibroblast A large, flat cell that secretes most of the extracellular matrix of areolar and dense connective tissues.

Fibrosis The process by which fibroblasts synthesize collagen fibers and other extracellular matrix materials that aggregate to form scar tissue.

Fibrous (FĪ-brus) **joint** A synarthrosis in which the connective tissue core is dense collagenous connective tissue; allows little or no movement; examples include a suture or a syndesmosis.

Fibrous pericardium (FĪ-brus per-i-KAR-dē-um) Superficial layer of the pericardium composed of tough, inelastic, dense irregular connective tissue.

Fibrous tunic The superficial coat of the eyeball, made up of the posterior sclera and the anterior cornea.

Fight-or-flight response The effects produced upon stimulation of the sympathetic division of the autonomic nervous system.

Filiform papilla One of the conical projections that are distributed in parallel rows over the anterior two-thirds of the tongue and lack taste buds.

Filtration membrane A leaky barrier that completely surrounds the glomerulus, formed from glomerular capillaries and podocytes. Also called the **endothelial-capsular membrane**.

Filum terminale Non-nervous fibrous tissue of the spinal cord that extends inferiorly from the conus medullaris to the coccyx.

Fimbriae (FIM-brē-ē) Fingerlike structures, especially the lateral ends of the uterine (fallopian) tubes.

Fissure (FISH-ur) A groove, fold, or slit that may be normal or abnormal.

Fixator A muscle that stabilizes the origin of the prime mover so that the prime mover can act more efficiently.

Fixed (tissue) macrophage Stationary phagocytic cell found in the liver, lungs, brain, spleen, lymph nodes, subcutaneous tissue, and red bone marrow. Also called a histiocyte (HIS-tē-ō-sīt).

Flaccid Relaxed, flabby, or soft; lacking muscle tone.

Flagella Hairlike, motile processes on the extremities of bacterium, protozoa, or sperm cells.

Flat bone Bone that is generally thin and composed of two nearly parallel plates of compact bone enclosing a layer of spongy bone.

Flatus (FLĀ-tus) Gas in the stomach or intestines; commonly used to denote expulsion of gas through the anus.

Flexion Movement in which there is a decrease in the angle between two bones.

Follicle (FOL-i-kul) A small secretory sac or cavity; the group of cells that contains a developing oocyte in the ovaries.

Follicle-stimulating hormone (FSH) Hormone secreted by the anterior pituitary; it initiates development of ova and stimulates the ovaries to secrete estrogens in females, and initiates sperm production in males.

Fontanel A fibrous connective tissue membrane-filled space where bone formation is not yet complete, especially between the cranial bones of an infant's skull.

Foot The terminal part of the lower limb, from the ankle to the toes.

Foramen (fō-RĀ-men) A passage or opening; a communication between two cavities of an organ, or a hole in a bone for passage of vessels or nerves. *Plural is* **foramina**.

Foramen ovale (fō-RĀ-men ō-VAL-ē) An opening in the fetal heart in the septum between the right and left atria. A hole in the greater wing of the sphenoid bone that transmits the mandibular branch of the trigeminal (V) nerve.

Forearm The part of the free upper limb between the elbow and the wrist.

Fornix An arch or fold; a tract in the brain made up of association fibers, connecting the hippocampus with the mammillary bodies; a recess around the cervix of the uterus where it protrudes into the vagina.

Fossa A furrow or shallow depression.

Fossa ovalis (FOS-a ō-VAL-is) Oval depression in the interatrial septum that is a remnant of the foramen ovale, an opening between the fetal atria that allows blood to bypass the nonfunctioning fetal lungs.

Fourth ventricle A cavity filled with cerebrospinal fluid within the brain lying between the cerebellum and the medulla oblongata and pons.

Fovea (FŌ-vē-a) **centralis** A depression in the center of the macula lutea of the retina, containing cones only and lacking blood vessels; the area of highest visual acuity (sharpness of vision).

Fracture Any break in a bone.

Free lower limb (extremity) The appendage attached at the pelvic (hip) girdle, consisting of the thigh, knee, leg, ankle, foot, and toes.

Free nerve endings Dendrites without structural specialization that initiate signals that give rise to sensations of warmth, coolness, pain, tickling, and itching.

Free radical An atom or group of atoms with an unpaired electron in the outermost shell.

Free upper limb (extremity) The appendage attached at the shoulder girdle, consisting of the arm, forearm, wrist, hand, and fingers.

Frontal plane A plane at a right angle to a midsagittal plane that divides the body or organs into anterior and posterior portions. Also called a coronal (kō-RŌ-nal) plane.

Fundus (FUN-dus) The part of a hollow organ farthest from the opening; the rounded portion of the stomach superior and to the left of the cardia; the broad portion of the gallbladder that projects downard beyond the inferior border of the liver.

Fungiform papilla (FUN-ji-form pa-PIL-a) A mushroomlike elevation on the upper surface of the tongue appearing as a red dot; most contain taste buds.

Furuncle (FŪ-rung-kul) A boil; painful nodule caused by bacterial infection and inflammation of a hair follicle or sebaceous (oil) gland.

G

Gallbladder A small pouch, located inferior to the liver, that stores bile and empties by means of the cystic duct.

Gallstone A solid mass, usually containing cholesterol, in the gallbladder or a bile-containing duct; formed anywhere between bile canaliculi in the liver and the hepatopancreatic ampulla (ampulla of Vater), where bile enters the duodenum. Also called a biliary calculus.

Gamete (GAM-ēt) A male or female reproductive cell; a sperm cell or secondary oocyte.

Ganglion (GANG-glē-on) A group of neuronal cell bodies lying outside the central nervous system (CNS). *Plural is* **ganglia**.

Gap junctions Cell junctions that allow muscle action potentials to spread from one cardiac muscle fiber to its neighbors.

Gastric glands Glands in the mucosa of the stomach composed of cells that empty their secretions into narrow channels called gastric pits.

Gastroenterology The medical specialty that deals with the structure, function, diagnosis, and treatment of diseases of the stomach and intestines.

Gastrointestinal (GI) tract A continuous tube running through the ventral body cavity extending from the mouth to the anus. Also called the alimentary (al′-i-MEN-tar-ē) canal.

Gastrulation The migration of groups of cells from the epiblast that transforms a bilaminar embryonic disc into a trilaminar embryonic disc with three primary germ layers; transformation of the blastula into the gastrula.

Gene Biological unit of heredity; a segment of DNA located in a definite position on a particular chromosome; a sequence of DNA that codes for a particular mRNA, rRNA, or tRNA.

Genetic engineering The manufacture and manipulation of genetic material.

Genetics The study of genes and heredity.

Genome The complete set of genes of an organism.

Geriatrics The branch of medicine devoted to the medical problems and care of elderly persons.

Gestation (jes-TĀ-shun) The period of development from fertilization to birth.

Gingivae Gums. They cover the alveolar processes of the mandible and maxilla and extend slightly into each socket.

Gland One of the paired mucus-secreting glands with ducts that open on either side of the urethral orifice in the vestibule of the female.

Glans clitoris Exposed portion of the erectile organ of the female, located at the anterior junction of the labia minora.

Glans penis The slightly enlarged region at the distal end of the penis.

Glaucoma (glaw-KŌ-ma) An eye disorder in which there is increased intraocular pressure due to an excess of aqueous humor.

Gliding A simple movement in which nearly flat bone surfaces move black-and-forth and from side-to-side with respect to one another.

Glomerular capsule A double-walled globe at the proximal end of a nephron that encloses the glomerular capillaries. Also called Bowman's (BŌ-manz) capsule.

Glomerular filtrate The fluid produced when blood is filtered by the filtration membrane in the glomeruli of the kidneys.

Glomerular filtration The first step in urine formation in which substances in blood pass through the filtration membrane and the filtrate enters the proximal convoluted tubule of a nephron.

Glomerular filtration rate The amount of filtrate formed in all renal corpuscles per minute. It averages 125 mL/min in males and 105 mL/min in females.

Glomerulus (glō-MER-ū-lus) A rounded mass of nerves or blood vessels, especially the microscopic tuft of capillaries that is surrounded by the glomerular (Bowman's) capsule of each kidney tubule. *Plural is* **glomeruli**.

Glottis The vocal folds (true vocal cords) in the larynx plus the space between them (rima glottidis).

Glucagon A hormone produced by the alpha cells of the pancreatic islets (islets of Langerhans) that increases blood glucose level.

Glucocorticoids Hormones secreted by the cortex of the adrenal gland, especially cortisol, that influence glucose metabolism.

Glucose (GLOO-kōs) A hexose (six-carbon sugar), $C_6H_{12}O_6$, that is a major energy source for the production of ATP by body cells.

Glucosuria (gloo′-kō-SOO-rē-a) The presence of glucose in the urine; may be temporary or pathological. Also called glycosuria.

Glycogen (GLĪ-kō-jen) A highly branched polymer of glucose containing thousands of subunits; functions as a compact store of glucose molecules in liver and muscle fibers (cells).

Goblet cell A goblet-shaped unicellular gland that secretes mucus; present in epithelium of the airways and intestines.

Goiter (GOY-ter) An enlarged thyroid gland.

Golgi (GOL-jē) **complex** An organelle in the cytoplasm of cells consisting of four to six flattened sacs (cisternae), stacked on one another, with expanded areas at their ends; functions in processing, sorting, packaging, and delivering proteins and lipids to the plasma membrane, lysosomes, and secretory vesicles.

Gomphosis A fibrous joint in which a cone-shaped peg fits into a socket.

Gonad A gland that produces gametes and hormones; the ovary in the female and the testis in the male.

Gonadotropic hormone Anterior pituitary hormone that affects the gonads.

Gout Hereditary condition associated with excessive uric acid in the blood; the acid crystallizes and deposits in joints, kidneys, and soft tissue.

Gracile nucleus A group of nerve cells in the inferior part of the medulla oblongata in which axons of the gracile fasciculus terminate.

Gray commissure A narrow strip of gray matter connecting the two lateral gray masses within the spinal cord.

Gray matter Areas in the central nervous system and ganglia containing neuronal cell bodies, dendrites, unmyelinated axons, axon terminals, and neuroglia; Nissl bodies impart a gray color and there is little or no myelin in gray matter.

Gray ramus communicans (RĀ-mus kō-MŪ-ni-kans) A short nerve containing axons of sympathetic postganglionic neurons; the cell bodies of the neurons are in a sympathetic chain ganglion, and the unmyelinated axons extend via the gray ramus to a spinal nerve and then to the periphery to supply smooth muscle in blood vessels, arrector pili muscles, and sweat glands. *Plural* is **gray rami communicantes**.

Greater omentum A large fold in the serosa of the stomach that hangs down like an apron anterior to the intestines.

Greater vestibular glands A pair of glands on either side of the vaginal orifice that open by a duct into the space between the hymen and the labia minora. Also called Bartholin's (BAR-to-linz) glands.

Groin Area on the front surface of the body marked by a crease on each side where the trunk attaches to the thighs.

Gross anatomy The branch of anatomy that deals with structures that can be studied without using a microscope. Also called macroscopic anatomy.

Ground substance The fluid, semifluid, gelatinous, or calcified component of connective tissue between the cells and fibers.

Growth An increase in body size; one of the six life processes.

Gustation (gus-TĀ-shun). The sense of taste.

Gustatory (GUS-ta-tō′-rē) Pertaining to taste.

Gustatory microvilli Microvilli extending from a gustatory receptor cell through a taste pore.

Gustatory receptor cells Receptors for taste containing gustatory microvilli through taste pores.

Gynecology The branch of medicine dealing with the study and treatment of disorders of the female reproductive system.

Gynecomastia (gīn′-e-kō-MAS-tē-a) Excessive growth (benign) of the male mammary glands due to secretion of estrogens by an adrenal gland tumor (feminizing adenoma).

Gyri (JĪ-rī) Folds of the cerebral cortex of the brain. *Singular* is **gyrus**. Also called convolutions.

H

Hair A threadlike structure produced by hair follicles that develops in the dermis. Also called a pilus (PĪ-lus). *Plural* is **pili**.

Hair follicle Structure, composed of epithelium and surrounding the root of a hair, from which hair develops.

Hair root plexus A network of dendrites arranged around the root of a hair as free or naked nerve endings that are stimulated when a hair shaft is moved.

Hand The terminal portion of a free upper limb, including the carpus, metacarpus, and phalanges. Also called the manus.

Haploid (*n*) cell Having half the number of chromosomes characteristically found in the somatic cells of an organism; characteristic of mature gametes.

Hard palate The anterior portion of the roof of the mouth, formed by the maxillae and palatine bones and lined by mucous membrane.

Haustra (HAWS-tra) A series of pouches that characterize the colon; caused by tonic contractions of the teniae coli. *Singular* is **haustrum**.

Head The superior part of a human, cephalic to the neck. The superior or proximal part of a structure, such as a sperm cell or proximal portion of the epididymis or the head of the pancreas.

Heart A hollow muscular organ lying slightly to the left of the midline of the chest that pumps the blood through the cardiovascular system.

Heart block An arrhythmia (dysrhythmia) of the heart in which the atria and ventricles contract independently because of a blocking of electrical impulses through the heart at some point in the conduction system.

Heart murmur (MER-mer) An abnormal sound that consists of a flow noise that is heard before, between, or after the normal heart sounds, or that may mask normal heart sounds.

Hemangioblast (hē-MAN-jē-ō-blast) A precursor mesodermal cell that develops into blood and blood vessels.

Hematocrit (he-MAT-ō-krit) **(Hct)** The percentage of blood made up of red blood cells. Usually measured by centrifuging a blood sample in a graduated tube and then reading the volume of red blood cells and dividing it by the total volume of blood in the sample.

Hematology (hēm-a-TOL-ō-jē) The study of blood.

Hematoma (hē′-ma-TŌ-ma) A tumor or swelling filled with blood.

Hemiplegia (hem-i-PLĒ-jē-a) Paralysis of the upper limb, trunk, and lower limb on one side of the body.

Hemodialysis (hē-mō-dī-AL-i-sis) Direct filtration of blood by removing wastes and excess electrolytes and fluid and then returning the cleansed blood.

Hemoglobin (hē′-mō-GLŌ-bin) **(Hb)** A substance in red blood cells consisting of the protein globin and the iron-containing red pigment heme that transports most of the oxygen and some carbon dioxide in blood.

Hemolysis (hē-MOL-i-sis) The escape of hemoglobin from the interior of a red blood cell into the surrounding medium; results from disruption of the cell membrane by toxins or drugs, freezing or thawing, or hypotonic solutions.

Hemolytic disease of the newborn (HDN) A hemolytic anemia of a newborn child that results from the destruction of the infant's erythrocytes (red blood cells) by antibodies produced by the mother; usually the antibodies are due to an Rh blood type incompatibility. Also called erythroblastosis fetalis (e-rith′-rō-blas-TŌ-sis fe-TAL-is).

Hemophilia (hē-mō-FIL-ē-a) A hereditary blood disorder in which there is a deficient production of certain factors involved in blood clotting, resulting in excessive bleeding into joints, deep tissues, and elsewhere.

Hemopoiesis Blood cell production, which occurs in red bone marrow after birth. Also called hematopoiesis (hem′-a-to-poy-Ē-sis).

Hemorrhage (HEM-o-rij) Bleeding; the escape of blood from blood vessels, especially when the loss is profuse.

Hemorrhoids (HEM-ō-royds) Dilated or varicosed blood vessels (usually veins) in the anal region. Also called piles.

Hepatic (he-PAT-ik) Refers to the liver.

Hepatic acinus The structural and functional unit of the liver. It is an oval mass of hepatocytes with a short axis defined by the branches of the portal triad.

Hepatic duct A duct that receives bile from the bile capillaries. Small hepatic ducts merge to form the larger right and left hepatic ducts that unite to leave the liver as the common hepatic duct.

Hepatic laminae Plates of hepatocytes one cell thick bordered on either side by endothelial lined vascular spaces called hepatic sinusoids.

Hepatic lobule Functional model of the liver; consists of a central vein with rows of hepatocytes and hepatic sinusoids radiating outward from it.

Hepatic portal circulation The flow of blood from the gastrointestinal organs to the liver before returning to the heart.

Hepatic sinusoid Highly permeable blood capillary between rows of hepatocytes.

Hepatocyte (he-PAT-ō-cyte) A liver cell.

Hepatopancreatic (hep′-a-tō-pan′-krē-A-tik) **ampulla** A small, raised area in the duodenum where the combined common bile duct and main pancreatic duct empty into the duodenum. Also called the ampulla of Vater (VA-ter).

Herniated (HER-nē-ā′-ted) **disc** A rupture of an intervertebral disc so that the nucleus pulposus protrudes into the vertebral cavity. Also called a slipped disc.

Hiatus (hī-Ā-tus) An opening; a foramen.

Hilum An area, depression, or pit where lymphatic vessels, blood vessels, and nerves enter or leave an organ.

Hinge joint A synovial joint in which a convex surface of one bone fits into a concave surface of another bone, such as the elbow, knee, ankle, and interphalangeal joints. Also called a ginglymus (JIN-gli-mus) joint.

Hirsutism (HER-soo-tizm) An excessive growth of hair in females and children, with a distribution similar to that in adult males, due to the conversion of vellus hairs into large terminal hairs in response to higher-than-normal levels of androgens.

Histamine (HISS-ta-mēn) Substance found in many cells, especially mast cells, basophils, and platelets, that is released when the cells are injured; results in vasodilation, increased permeability of blood vessels, and constriction of bronchioles.

Histology Microscopic study of the structure of tissues.

Holocrine gland A type of gland in which entire secretory cells, along with their accumulated secretions, make up the secretory product of the gland, as in the sebaceous (oil) glands.

Homeostasis (hō-mē-ō-STĀ-sis) The condition in which the body´s internal environment remains relatively constant within physiological limits.

Homologous chromosomes Two chromosomes that belong to a pair. Also called homologs.

Hormone A secretion of endocrine cells that alters the physiological activity of target cells of the body.

Horn An area of gray matter (anterior, lateral, or posterior) in the spinal cord.

Human chorionic gonadotropin (hCG) A hormone produced by the developing placenta that maintains the corpus luteum.

Human chorionic somatomammotropin (hCS) Hormone produced by the chorion of the placenta that stimulates breast tissue for lactation, enhances body growth, and regulates metabolism. Also called human placental lactogen (hPL).

Human growth hormone (hGH) Hormone secreted by the anterior pituitary that stimulates growth of body tissues, especially skeletal and muscular tissues. Also known as somatotropin.

Hyaluronic acid (hī´-a-loo-RON-ik) A viscous, amorphous extracellular material that binds cells together, lubricates joints, and maintains the shape of the eyeballs.

Hymen A thin fold of vascularized mucous membrane at the vaginal orifice.

Hyperextension Continuation of extension beyond the normal range of motion.

Hyperplasia (hī-per-PLĀ-zē-a) An abnormal increase in the number of normal cells in a tissue or organ, increasing its size.

Hypersecretion (hī-per-se-KRĒ-shun) Overactivity of glands resulting in excessive secretion.

Hypertension High blood pressure.

Hyperthermia (hī´-per-THERM-ē -a) An elevated body temperature.

Hypertonia (hī´-per-TŌ-nē-a) Increased muscle tone that is expressed as spasticity or rigidity.

Hypertonic (hī´-per-TON-ik) Solution that causes cells to shrink due to loss of water by osmosis.

Hypertrophy An excessive enlargement or overgrowth of tissue without cell division.

Hyperventilation (hī ´-per-ven-ti-LĀ -shun) A rate of inhalation and exhalation higher than that required to maintain a normal partial pressure of carbon dioxide in the blood.

Hyponychium Portion of the nail beneath the free edge composed of a thickened region of stratum corneum.

Hypophyseal fossa (hī´-pō-FIZ-ē-al FOS-a) A depression on the superior surface of the sphenoid bone that houses the pituitary gland.

Hypophyseal pouch An outgrowth of ectoderm from the roof of the mouth from which the anterior pituitary develops. Also called Rathke's pouch.

Hypophysis (hī-POF-i-sis) A small endocrine gland occupying the hypophyseal fossa of the sphenoid bone and attached to the hypothalamus by the infundibulum. Also called the **pituitary gland**.

Hyposecretion (hī´-pō-se-KRĒ-shun) Underactivity of glands resulting in diminished secretion.

Hypothalamohypophyseal (hī´-pō-thal´-a-mō-hī-pō-FIZ-ē-al) **tract** A bundle of axons containing secretory vesicles filled with oxytocin or antidiuretic hormone that extend from the hypothalamus to the posterior pituitary.

Hypothalamus A portion of the diencephalon, lying beneath the thalamus and forming the floor and part of the wall of the third ventricle.

Hypothermia (hī´-pō-THER-mē-a) Lowering of body temperature below 35°C (95°F); in surgical procedures, it refers to deliberate cooling of the body to slow down metabolism and reduce oxygen needs of tissues.

Hypotonia (hī´-pō-TŌ-nē-a) Decreased or lost muscle tone in which muscles appear flaccid.

Hypotonic (hī´-pō-TON-ik) Solution that causes cells to swell and perhaps rupture due to gain of water by osmosis.

Hypoventilation (hī´-pō-ven-ti-LĀ-shun) A rate of inhalation and exhalation lower than that required to maintain a normal partial pressure of carbon dioxide in plasma.

Hypoxia Lack of adequate oxygen at the tissue level.

Hysterectomy (hiss-te-REK-tō-mē) The surgical removal of the uterus.

I

Ileocecal (il-ē-ō-SĒ-kal) **sphincter** A fold of mucous membrane that guards the opening from the ileum into the large intestine. Also called the ileocecal valve.

Ileum (IL-ē-um) The terminal part of the small intestine.

Immunity The state of being resistant to injury, particularly by poisons, foreign proteins, and invading pathogens. Also called resistance.

Immunoglobulin An antibody synthesized by plasma cells derived from B lymphocytes in response to the introduction of an antigen. Immunoglobulins are divided into five kinds (IgG, IgM, IgA, IgD, IgE).

Immunology (im´-ū-NOL-ō-jē) The study of the responses of the body when challenged by antigens.

Imperforate hymen Condition in which the hymen completely covers the vaginal orifice; may require surgery.

Implantation The insertion of a tissue or a part into the body. The attachment of the blastocyst to the stratum basalis of the endometrium about 6 days after fertilization.

Incontinence (in-KON-ti-nens) Inability to retain urine, semen, or feces through loss of sphincter control.

Indirect motor pathways Motor tracts that convey information from the brain down the spinal cord for automatic movements, coordination of body movements with visual stimuli, skeletal muscle tone and posture, and balance. Also known as extrapyramidal pathways.

Induction The process by which one tissue (inducting tissue) stimulates the development of an adjacent unspecialized tissue (responding tissue) into a specialized one.

Infarction (in-FARK-shun) A localized area of necrotic tissue, produced by inadequate oxygenation of the tissue.

Infection (in-FEK-shun) Invasion and multiplication of microorganisms in body tissues, which may be inapparent or characterized by cellular injury.

Inferior (in-FĒR-ē-or) Away from the head or toward the lower part of a structure. Also called caudal (KAW-dal).

Inferior vena cava Large vein that collects blood from parts of the body inferior to the heart and returns it to the right atrium.

Infertility Inability to conceive or to cause conception. Also called sterility in males.

Inflammation (in´-fla-MĀ-shun) Localized, protective response to tissue injury designed to destroy, dilute, or wall off the infecting agent or injured tissue; characterized by redness, pain, heat, swelling, and sometimes loss of function.

Infundibulum (in-fun-DIB-ū-lum) The stalklike structure that attaches the pituitary gland to the hypothalamus of the brain. The funnel-shaped, open, distal end of the uterine (fallopian) tube.

Ingestion (in-JES-chun) The taking in of food, liquids, or drugs, by mouth.

Inguinal (ING-gwin-al) Pertaining to the groin.

Inguinal canal An oblique passageway in the anterior abdominal wall just superior and parallel to the medial half of the inguinal ligament that transmits the spermatic cord and ilioinguinal nerve in the male and round ligament of the uterus and ilioinguinal nerve in the female.

Inguinal ligament (ING-gwin-al) Ligament formed from the inferior free border of the external oblique aponeurosis; runs from the anterior superior iliac spine to the pubic tubercle.

Inhalation The act of drawing air into the lungs. Also termed inspiration.

Inheritance The acquisition of body traits by transmission of genetic information from parents to offspring.

Inhibin A hormone secreted by the gonads that inhibits release of follicle-stimulating hormone (FSH) by the anterior pituitary.

Inhibiting hormone Hormone secreted by the hypothalamus that can suppress secretion of hormones by the anterior pituitary.

Insertion Inner layer of muscle of the pharynx composed of three muscles that elevate the larynx and pharynx during deglutition and speech.

Insula (IN-soo-la) A triangular area of the cerebral cortex that lies deep within the lateral cerebral fissure, under the parietal, frontal, and temporal lobes.

Insulin (IN-soo-lin) A hormone produced by the beta cells of a pancreatic islet (islet of Langerhans) that decreases the blood glucose level.

Insulinlike growth factors (IGFs) Hormones that stimulate general body growth and regulate various aspects of metabolism.

Integrins (IN-te-grinz) A family of transmembrane glycoproteins in plasma membranes that functions in cell adhesion; they are present in hemidesmosomes, which anchor cells to a basement membrane, and they mediate adhesion of neutrophils to endothelial cells during emigration.

Integumentary (in-teg′-ū-MEN-tar-ē) Relating to the skin.

Integumentary system Body system composed of the skin, hair, oil and sweat glands, nails, and sensory receptors.

Intercalated (in-TER-ka-lāt-ed) **disc** An irregular transverse thickening of sarcolemma that contains desmosomes, which hold cardiac muscle fibers (cells) together, and gap junctions, which aid in conduction of muscle action potentials from one fiber to the next.

Intercostal nerve A nerve supplying a muscle located between the ribs. Also called a thoracic nerve.

Intermediate Between two structures, one of which is medial and one of which is lateral. One of three groups of intrinsic muscles of the hand; includes the lumbricals, palmar interossei, and dorsal interossei.

Intermediate filament Protein filament, ranging from 8 to 12 nm in diameter, that may provide structural reinforcement, hold organelles in place, and give shape to a cell.

Internal Away from the surface of the body.

Internal capsule A large tract of projection fibers lateral to the thalamus that is the major connection between the cerebral cortex and the brain stem and spinal cord; contains axons of sensory neurons carrying auditory, visual, and somatic sensory signals to the cerebral cortex plus axons of motor neurons descending from the cerebral cortex to the thalamus, subthalamus, brain stem, and spinal cord.

Internal (inner) ear The inner ear or labyrinth, lying inside the temporal bone, containing the organs of hearing and balance.

Internal (tissue) respiration The exchange of respiratory gases between blood and body cells.

Internal nares The two openings posterior to the nasal cavities opening into the nasopharynx. Also called the choanae (kō-A-nē).

Internal respiration The exchange of respiratory gases between blood and body cells. Also called tissue respiration or systemic gas exchange.

Interneurons Neurons whose axons extend only for a short distance and contact nearby neurons in the brain, spinal cord, or a ganglion; comprise the vast majority of neurons in the body.

Interoceptor (IN-ter-ō-sep′-tor) Sensory receptor located in blood vessels and viscera that provides information about the body's internal environment.

Interosseous (in′-ter-OS-ē-es) **membrane** A substantial sheet of dense irregular connective tissue that binds neighboring long bones and permits slight movement (amphiarthrosis).

Interphase The period of the cell cycle between cell divisions, consisting of the G1 (gap or growth) phase, when the cell is engaged in growth, metabolism, and production of substances required for division; S (synthesis) phase, during which chromosomes are replicated; and G2 phase.

Interstitial cell A type of cell that secretes testosterone; located in the connective tissue

between seminiferous tubules in a mature testis. Also called Leydig cell.

Interstitial fluid (in′-ter-STISH-al) The portion of extracellular fluid that fills the microscopic spaces between the cells of tissues; the internal environment of the body. Also called **intercellular** or **tissue fluid**.

Interstitial growth Growth from within, as in the growth of cartilage. Also called endogenous (en-DOJ-e-nus) growth.

Interthalamic adhesion Bridge of gray matter that joins the right and left halves of the thalamus in about 70 percent of human brains. Also called the intermediate mass.

Interventricular foramen A narrow, oval opening through which the lateral ventricles of the brain communicate with the third ventricle.

Intervertebral disc A pad of fibrocartilage located between the bodies of two vertebrae.

Intervillous space Space within the placenta containing fetal capillaries bathed with maternal blood that leaves the uterine arteries.

Intestinal gland A gland that opens onto the surface of the intestinal mucosa and secretes digestive enzymes. Also called a crypt of Lieberkühn (LĒ-ber-kēn).

Intracellular fluid (ICF) Fluid located within cells.

Intraepidermal macrophage cell Epidermal dendritic cell that functions as an antigen-presenting cell (APC) during an immune response. Also called an intraepidermal macrophage.

Intrafusal muscle fibers Three to ten specialized muscle fibers (cells), partially enclosed in a spindle-shaped connective tissue capsule, that make up a muscle spindle.

Intramembranous (in′-tra-MEM-bra-nus) **ossification** The method of bone formation in which the bone is formed directly in membranous tissue.

Intramuscular injection An injection that penetrates the skin and subcutaneous layer to enter a skeletal muscle. Common sites are the deltoid, gluteus medius, and vastus lateralis muscles.

Intraocular pressure Pressure in the eyeball, produced mainly by aqueous humor.

Intrinsic factor (IF) A glycoprotein, synthesized and secreted by the parietal cells of the gastric mucosa, that facilitates vitamin B_{12} absorption in the small intestine.

Invagination (in-vaj′-i-NĀ-shun) The pushing of the wall of a cavity into the cavity itself; infolding of the ectoderm neural plate cells into the underlying mesoderm.

Inversion The movement of the sole medially at the ankle joint.

In vitro (VĒ-trō) Literally, in glass; outside the living body and in an artificial environment such as a laboratory test tube.

Ipsilateral (ip-si-LAT-er-al) On the same side, affecting the same side of the body.

Iris The colored portion of the vascular tunic of the eyeball seen through the cornea that contains circular and radial smooth muscle; the hole in the center is the pupil.

Irregular bone Bone that has a complex shape and cannot be placed into the other bone shape categories.

Irritable bowel syndrome (IBS) Disease of the entire gastrointestinal tract in which a person reacts to stress by developing symptoms (such as cramping and abdominal pain) associated with alternating patterns of diarrhea and constipation. Excessive amounts of mucus may appear in feces, and other symptoms include flatulence, nausea, and loss of appetite. Also known as irritable colon or spastic colitis.

Ischemia (is-KE′-mē-a) A lack of sufficient blood to a body part due to obstruction or constriction of a blood vessel.

Islet of Langerhans A cluster of endocrine gland cells in the pancreas that secretes insulin, glucagon, somatostatin, and pancreatic polypeptide. Also called a **pancreatic islet**.

Isotonic (ī′-sō-TON-ik) Having equal tension or tone. A solution having the same concentration of impermeable solutes as cytosol.

Isthmus (IS-mus) A narrow strip of tissue or narrow passage connecting two larger parts. The medial, short, narrow, thick-walled portion of the uterine tube that joins the uterus. Constricted region of the uterus between the body and cervix. The mass of tissue connecting the right and left lateral lobes of the thyroid gland.

J

Jaundice (JON-dis) A condition characterized by yellowness of the skin, the white of the eyes, mucous membranes, and body fluids because of a buildup of bilirubin.

Jejunum (je-JOO-num) The middle part of the small intestine.

Joint A point of contact between two bones, between bone and cartilage, or between bone and teeth. Also called an articulation or arthrosis.

Joint kinesthetic receptor A proprioceptive receptor located in a joint, stimulated by joint movement.

Junctional folds Deep grooves in the motor end plate that provide a large surface area for ACh.

Juxtaglomerular (juks-ta-glō-MER-ū-lar) **apparatus (JGA)** Consists of the macula densa (cells of the distal convoluted tubule adjacent to the afferent and efferent arteriole) and juxtaglomerular cells (modified cells of the afferent and sometimes efferent arteriole); secretes renin when blood pressure starts to fall.

K

Keratin An insoluble protein found in the hair, nails, and other keratinized tissues of the epidermis.

Keratinocyte The most numerous of the epidermal cells; produces keratin.

Kidney One of the paired reddish organs located in the lumbar region that regulates the composition, volume, and pressure of blood and produces urine.

Kidney stone A solid mass, usually consisting of calcium oxalate, uric acid, or calcium phosphate crystals, that may form in any portion of the urinary tract. Also called a **renal calculus.**

Kinesiology (ki-nē-sē-OL-ō-jē) The study of the movement of body parts.

Kinesthesia (kin′-es-THĒ-zē-a) The perception of the extent and direction of movement of body parts; this sense is possible due to nerve impulses generated by proprioceptors.

Kinetochore (ki-NET-ō-kor) Protein complex attached to the outside of a centromere to which kinetochore microtubules attach.

Kyphosis (kī-FŌ-sis) An exaggeration of the thoracic curve of the vertebral column, resulting in a "round-shouldered" appearance. Also called hunchback.

L

Labia Fleshy folds surrounding the opening of the mouth; contain the orbicularis oris muscle and are covered externally by skin and internally by a mucous membrane. *Singular is* **labium**. Also called lips.

Labia majora (LĀ-bē-a ma-JŌ-ra) Two longitudinal folds of skin extending downward and backward from the mons pubis of the female. *Singular is* **labium majus**.

Labia minora (min-OR-a) Two small folds of mucous membrane lying medial to the labia majora of the female. *Singular is* **labium minus**.

Labial frenulum (LĀ-bē-al FREN-ū-lum) A medial fold of mucous membrane between the inner surface of the lip and the gums.

Labor The process of giving birth in which a fetus is expelled from the uterus through the vagina.

Labrum A fibrocartilaginous lip that extends from the edge of a ball-and-socket joint to deepen the socket.

Labyrinth The inner ear, lying inside the temporal bone, that contains the organs of hearing and balance.

Lacrimal canaliculus A duct, one on each eyelid, beginning at the punctum at the medial margin of an eyelid and conveying tears medially into the nasolacrimal sac. *Plural is* **lacrimal canaliculi**.

Lacrimal gland Secretory cells, located at the superior anterolateral portion of each orbit, that secrete tears into excretory ducts that open onto the surface of the conjunctiva.

Lacrimal sac The superior expanded portion of the nasolacrimal duct located within the lacrimal fossa that receives the tears from a lacrimal canal.

Lactation The secretion and ejection of milk by the mammary glands.

Lacteal One of many lymphatic vessels in villi of the intestines that absorb triglycerides and other lipids from digested food.

Lacunae (la-KOO-nē) Small, hollow spaces, such as those found in bones in which the osteocytes lie. Small hollow space in endometrium and yolk sac during embryonic development. *Singular is* **lacuna** (la-KOO-na).

Lambdoid (LAM-doyd) **suture** The joint in the skull between the parietal bones and the occipital bone; sometimes contains sutural (Wormian) bones. Can be palpated on the occipital region of the skull.

Lamellae (la-MEL-ē) Concentric rings of hard, calcified extracellular matrix found in compact bone.

Lamellated corpuscle Oval-shaped pressure receptor located in the dermis or subcutaneous tissue and consisting of concentric layers of connective tissue wrapped around the dendrites of a sensory neuron. Also called a pacinian (pa-SIN-ē-an) corpuscle.

Lamina propria The connective tissue layer of a mucosa.

Lanugo Fine downy hairs that cover the fetus.

Large intestine The portion of the gastrointestinal tract extending from the ileum of the small intestine to the anus, divided structurally into the cecum, colon, rectum, and anal canal.

Laryngeal ventricle (sinus) Lateral expansion of the middle portion of the laryngeal cavity bordered superiorly by the vestibular folds and inferiorly by the vocal folds.

Laryngopharynx The inferior portion of the pharynx, extending downward from the level of the hyoid bone, that divides posteriorly into the esophagus and anteriorly into the larynx. Also called the hypopharynx.

Larynx The voice box, a short passageway that connects the pharynx with the trachea.

Lateral Farther from the midline of the body or a structure.

Lateral ventricle A cavity within a cerebral hemisphere that communicates with the lateral ventricle in the other cerebral hemisphere and with the third ventricle by way of the interventricular foramen.

Left bundle branch Left one of the two branches of the atrioventricular (AV) bundle made up of specialized muscle fibers (cells) that transmit electrical impulses to the ventricles.

Leg The part of the lower limb between the knee and the ankle.

Lens A transparent organ constructed of proteins (crystallins) lying posterior to the pupil and iris of the eyeball and anterior to the vitreous body.

Leptomeninges The inner two meningeal membranes consisting of the pia mater and the arachnoid mater.

Lesion Anatomical sign of disease.

Lesser omentum A fold of the peritoneum that extends from the liver to the lesser curvature of the stomach and the first part of the duodenum.

Lesser vestibular gland One of the paired mucus-secreting glands with ducts that open on either side of the urethral orifice in the vestibule of the female.

Leukemia (loo-KĒ-mē-a) A malignant disease of the blood-forming tissues characterized by either uncontrolled production and accumulation of immature leukocytes in which many cells fail to reach maturity (acute) or an accumulation of mature leukocytes in the blood because they do not die at the end of their normal life span (chronic).

Leukocyte (LOO-kō-sīt) A white blood cell.

Ligament Dense regular connective tissue that attaches bone to bone.

Ligamentum teres (round ligament) A band of fibrous connective tissue enclosed between the folds of the broad ligament of the uterus, emerging from the uterus just inferior to the uterine tube, extending laterally along the pelvic wall and through the deep inguinal ring to end in the labia majora.

Ligand (LĪ-gand) A chemical substance that binds to a specific receptor.

Limbic system A part of the forebrain, sometimes termed the visceral brain, concerned with various aspects of emotion and behavior; includes the limbic lobe, dentate gyrus, amygdala, septal nuclei, mammillary bodies, anterior thalamic nucleus, olfactory bulbs, and bundles of myelinated axons.

Lingual frenulum (LIN-gwal FREN-ū-lum) A fold of mucous membrane that connects the tongue to the floor of the mouth.

Lipase An enzyme that splits fatty acids from triglycerides and phospholipids.

Lipid (LIP-id) An organic compound composed of carbon, hydrogen, and oxygen that is usually insoluble in water but soluble in alcohol, ether, and chloroform; examples include triglycerides (fats and oils), phospholipids, steroids, and eicosanoids.

Lipid bilayer Arrangement of phospholipid, glycolipid, and cholesterol molecules in two parallel sheets in which the hydrophilic "heads" face outward and the hydrophobic "tails" face inward; found in cellular membranes.

Lipoprotein (lip´-ō-PRŌ-tēn) One of several types of particles containing lipids (cholesterol and triglycerides) and proteins that make it water soluble for transport in the blood; high levels of low-density lipoproteins (LDLs) are associated with increased risk of atherosclerosis, and high levels of high-density lipoproteins (HDLs) are associated with decreased risk of atherosclerosis.

Liver Large organ under the diaphragm that occupies most of the right hypochondriac region and part of the epigastric region. Functionally, it produces bile and synthesizes most plasma proteins; interconverts nutrients; detoxifies substances; stores glycogen, iron, and vitamins; carries on phagocytosis of worn-out blood cells and bacteria; and helps synthesize the active form of vitamin D.

Lobar bronchi Branches of main bronchi that enter the lobes of the lungs. Also called secondary bronchi.

Long bone Bone that has a greater length than width and consists of a shaft and a variable number of extremities (ends).

Lordosis An exaggeration of the lumbar curve of the vertebral column. Also called swayback.

Lower limbs The appendages attached at the pelvic (hip) girdle, consisting of the thigh, knee, leg, ankle, foot, and toes. Also called the lower extremities.

Lumbar enlargement Inferior enlargement of the spinal cord that extends from the ninth to the twelfth thoracic vertebra.

Lumbar plexus A network formed by the anterior (ventral) branches of spinal nerves L1 through L4.

Lumen (LOO-men) The space within an artery, vein, intestine, renal tubule, or other tubular structure.

Lungs Main organs of respiration that lie on either side of the heart in the thoracic cavity.

Lunula The moon-shaped white area at the base of a nail.

Luteinizing (LOO-tē-in´-īz-ing) **hormone (LH)** A hormone secreted by the anterior pituitary that stimulates ovulation, stimulates progesterone secretion by the corpus luteum, and readies the mammary glands for milk secretion in females; stimulates testosterone secretion by the testes in males.

Lymph (LIMF) Fluid confined in lymphatic vessels and flowing through the lymphatic system until it is returned to the blood.

Lymph node An oval or bean-shaped structure located along lymphatic vessels.

Lymph trunks Vessels of the lymphatic system formed from the union of lymphatic vessels.

Lymphatic capillary (lim-FAT-ik) Closed-ended microscopic lymphatic vessel that begins in spaces between cells and converges with others to form lymphatic vessels.

Lymphatic system Body system consisting of lymph, lymphatic vessels, lymphatic tissues, and red bone marrow; responsible for immune responses.

Lymphatic tissue A specialized form of reticular tissue that contains large numbers of lymphocytes.

Lymphatic vessel A large vessel that collects lymph from lymphatic capillaries and converges with other lymphatic vessels to form the thoracic and right lymphatic ducts.

Lymphocyte A type of white blood cell that helps carry out cell-mediated and antibody-mediated immune responses; found in blood and in lymphatic tissues.

Lysosome An organelle in the cytoplasm of a cell, enclosed by a single membrane and containing powerful digestive enzymes.

Lysozyme One of the chemicals released by neutrophils to destroy bacteria. A bactericidal enzyme found in tears, saliva, and perspiration.

M

Macrophage (MAK-rō-fāj) Phagocytic cell derived from a monocyte; may be fixed or wandering.

Macula (MAK-ū-la) A discolored spot or a colored area. A small, thickened region on the wall of the utricle and saccule that contains receptors for static equilibrium.

Macula lutea The yellow spot in the center of the retina.

Main bronchus (BRON-kus) One of two divisions of the trachea that enter each lung. Also called a primary bronchus.

Major histocompatibility (MHC) antigens Surface proteins on white blood cells and other nucleated cells that are unique for each person (except for identical siblings); used to type tissues and help prevent rejection of transplanted tissues. Also known as human leukocyte antigens (HLA).

Malignant (ma-LIG-nant) Referring to diseases that tend to become worse and cause death.

Mammary gland Accessory organ of the female reproductive system located inside the breasts that is a modified sudoriferous (sweat) gland of the female that produces milk for the nourishment of the young.

Mammillary bodies Two small rounded bodies on the inferior aspect of the hypothalamus that are involved in reflexes related to the sense of smell; also part of the limbic system.

Marrow (MAR-ō) Soft, spongelike material in the cavities of bone. *Red bone marrow* produces blood cells; *yellow bone marrow* contains adipose tissue that stores triglycerides.

Mast cell A cell found in areolar connective tissue that releases histamine, a dilator of small blood vessels, during inflammation.

Mastication (mas´-ti-KA-shun) Chewing.

Mature ovarian follicle A large, fluid-filled follicle containing a secondary oocyte and surrounding granulosa cells that secrete estrogens. Also called a Graafian follicle.

McBurney's point Important landmark related to the appendix located two-thirds of the way down an imaginary line between the umbilicus and anterior superior iliac spine.

Meatus (mē-A-tus) A passage or opening, especially the external portion of a canal.

Mechanoreceptor (MEK-an-ō-rē-sep´-tor) Sensory receptor that detects mechanical deformation of the receptor itself or adjacent cells; stimuli so detected include those related to touch, pressure, vibration, proprioception, hearing, equilibrium, and blood pressure.

Medial lemniscus A white matter tract that originates in the gracile and cuneate nuclei of the medulla oblongata and extends to the thalamus on the same side; sensory axons in this tract conduct nerve impulses for the sensations of proprioception, fine touch, vibration, hearing, and equilibrium.

Median aperture One of the three openings in the roof of the fourth ventricle through which cerebrospinal fluid enters the subarachnoid space of the brain and cord.

Median Situated in the middle.

Median plane A vertical plane dividing the body into right and left halves.

Mediastinum (mē´-dē-as-TI-num) The broad, median partition between the pleurae of the lungs that extends from the sternum to the vertebral column in the thoracic cavity.

Medulla (me-DOOL-la) An inner layer of an organ, such as the medulla of the thymic lobules, lymph nodes, or kidneys. Alternate name for the **medulla oblongata**.

Medulla oblongata (me-DOOL-la ob´-long-GA-ta) The most inferior part of the brain stem. Also called the **medulla**.

Medullary (MED-ū-lar´-ē) **cavity** The space within the diaphysis of a bone that contains yellow bone marrow. Also called the marrow cavity.

Medullary respiratory center Neurons of the respiratory center consisting of the dorsal respiratory group (DRG) and ventral respiratory group (VRG). Formerly called the medullary rhythmicity area.

Meiosis A type of cell division that occurs during production of gametes, involving two successive nuclear divisions that result in cells with the haploid (n) number of chromosomes.

Meissner corpuscle See **Corpuscle of touch**.

Melanin A dark black, brown, or yellow pigment found in some parts of the body such as the skin, hair, and pigmented layer of the retina.

Melanocyte (MEL-a-nō-sīt´) A pigmented cell, located between or beneath cells of the deepest layer of the epidermis, that synthesizes melanin.

Melanocyte-stimulating hormone (MSH) A hormone secreted by the anterior pituitary that stimulates the dispersion of melanin granules in melanocytes in amphibians; continued administration produces darkening of skin in humans.

Melatonin A hormone secreted by the pineal gland that helps set the timing of the body's biological clock.

Membrane A thin, flexible sheet of tissue composed of an epithelial layer and an underlying connective tissue layer.

Membrane proteins Proteins associated with the plasma membrane (integral and peripheral) that function as ion channels, carriers, receptors, linkers, or cell-identity markers.

Membranous labyrinth The part of the labyrinth of the internal ear that is located inside the bony labyrinth and separated from it by the perilymph; made up of the semicircular ducts, the saccule and utricle, and the cochlear duct.

Memory The ability to recall thoughts; commonly classifed as short-term (activated) and long-term.

Menarche (me-NAR-kē) The first menses (menstrual flow) and beginning of ovarian and uterine cycles.

Meninges (me-NIN-jēz) Three membranes covering the brain and spinal cord, called the dura mater, arachnoid mater, and pia mater. *Singular* is **meninx** (MEN-inks).

Menopause The termination of the menstrual cycles.

Menstruation (men´-stroo-A-shun) Periodic discharge of blood, tissue fluid, mucus, and epithelial cells that usually lasts for 5 days; caused by a sudden reduction in estrogens and progesterone. Also called the menstrual phase or menses.

Merocrine glands Glands made up of secretory cells that remain intact throughout the process of formation and discharge of the secretory product, as in the salivary and pancreatic glands. Also called **eccrine glands**.

Mesenchyme (MEZ-en-kīm) An embryonic connective tissue from which all other connective tissues arise.

Mesentery (MEZ-en-ter´-ē) A fold of peritoneum attaching the small intestine to the posterior abdominal wall.

Mesocolon A fold of peritoneum attaching the colon to the posterior abdominal wall.

Mesoderm (MES-ō-derm) The middle primary germ layer that gives rise to lymphatic vessels, connective tissues, blood and blood vessels, and muscles.

Mesothelium (mez´-ō-THE-lē-um) The layer of simple squamous epithelium that lines serous membranes.

Mesovarium A short fold of peritoneum that attaches an ovary to the broad ligament of the uterus.

Metabolism The sum of all chemical processes that occur in the body; one of the six life processes.

Metacarpus A collective term for the five bones that make up the palm.

Metaphase (MET-a-fāz) The second stage of mitosis, in which chromatid pairs line up on the metaphase plate of the cell.

Metaphysis (me-TAF-i-sis) Region of a long bone between the diaphysis and epiphysis that contains the epiphyseal plate in a growing bone.

Metarteriole A blood vessel that emerges from an arteriole.

Metastasis The spread of cancer to surrounding tissues (local) or to other body sites (distant).

Metatarsus A collective term for the five bones located in the foot between the tarsals and the phalanges.

Microcirculation The flow of blood from a metarteriole through capillaries and into a postcapillary venule.

Microfilaments Thinnest elements of the cytoskeleton composed of the proteins actin and myosin.

Microglia (mī-KROG-lē-a) Neuroglial cells that carry on phagocytosis.

Microtubule (mī-krō-TOO-būl) The largest component of the cytoskeleton composed primarily of the protein tubulin.

Microvilli (mī-krō-VIL-ī) Microscopic, fingerlike projections of the plasma membranes of cells that increase surface area for absorption, especially in the small intestine and proximal convoluted tubules of the kidneys.

Micturition (mik-choo-RISH-un) The act of expelling urine from the urinary bladder. Also called urination (ū-ri-NĀ-shun).

Midline An imaginary vertical line that divides the body into equal left and right sides.

Midbrain The part of the brain between the pons and the diencephalon. Also called the mesencephalon (mes′-en-SEF-a-lon).

Middle ear A small, epithelial-lined cavity hollowed out of the temporal bone, separated from the external ear by the eardrum and from the internal ear by a thin bony partition containing the oval and round windows; extending across the middle ear are the three auditory ossicles. Also called the tympanic (tim-PAN-ik) cavity.

Midsagittal plane A vertical plane through the midline of the body that divides the body or organs into equal right and left sides. Also called a **median plane**.

Mineralocorticoids (min′-er-al-ō-KOR-ti-koyds) A group of hormones of the adrenal cortex that help regulate sodium and potassium balance.

Mitochondria (mī-tō-KON-drē-a) Double-membraned organelles that play a central role in the production of ATP; known as the "powerhouses" of the cell. *Singular* is **mitochondrion**.

Mitosis (mī-TŌ-sis) The orderly division of the nucleus of a cell that ensures that each new nucleus has the same number and kind of chromosomes as the original nucleus. The process includes the replication of chromosomes and the distribution of the two sets of chromosomes into two separate and equal nuclei.

Mitotic (M) phase Phase of the cell cycle that consists of a nuclear division (mitosis) and a cytoplasmic division (cytokinesis) to form two identical cells.

Mitotic spindle Collective term for a football-shaped assembly of microtubules (nonkinetochore, kinetochore, and aster) that is responsible for the movement of chromosomes during cell division.

Modality (mō-DAL-i-tē) Any of the specific sensory entities, such as vision, smell, taste, or touch.

Modiolus The central pillar or column of the cochlea.

Molecule Two or more atoms joined together.

Monocyte The largest type of white blood cell, characterized by agranular cytoplasm.

Mons pubis (MONZ PŪ-bis) The rounded, fatty prominence over the pubic symphysis, covered by coarse pubic hair.

Monounsaturated fat A fatty acid that contains one double covalent bond between its carbon atoms; it is not completely saturated with hydrogen atoms. Plentiful in triglycerides of olive and peanut oils.

Morula (MOR-ū-la) A solid sphere of cells produced by successive cleavages of a fertilized ovum about four days after fertilization.

Motor area The region of the cerebral cortex that governs muscular movement, particularly the precentral gyrus of the frontal lobe.

Motor end plate Region of the sarcolemma of a muscle fiber (cell) that includes acetylcholine (ACh) receptors, which bind ACh released by synaptic end bulbs of somatic motor neurons.

Motor neurons Neurons that conduct impulses from the brain toward the spinal cord or out of the brain and spinal cord into cranial or spinal nerves to effectors that may be either muscles or glands. Also called **efferent neurons**.

Motor unit A motor neuron together with the muscle fibers (cells) it stimulates.

Mouth Portion of the digestive system formed by the cheeks, hard and soft palates, and tongue. Also called the oral cavity or buccal cavity.

Mucosa-associated lymphatic tissue (MALT) Lymphatic nodules scattered throughout the lamina propria (connective tissue) of mucous membranes lining the gastrointestinal tract, respiratory airways, urinary tract, and reproductive tract.

Mucous connective tissue An embryonic tissue found in the umbilical cord of the fetus. Also called Wharton's jelly.

Mucous membrane A membrane that lines a body cavity that opens to the exterior. Also called a mucosa.

Mucus The thick fluid secretion of goblet cells, mucous cells, mucous glands, and mucous membranes.

Multiunit smooth muscle tissue Less common type of smooth muscle tissue that consists of individual fibers, each of which has its own motor neuron terminals; found in walls of larger arteries, airways to the lungs, arrector pili muscle of hair follicles, muscles that control pupil diameter, and ciliary body that focuses the lens of the eye.

Muscarinic (mus′-ka-RIN-ik) **receptor** Receptor for the neurotransmitter acetylcholine found on all effectors innervated by parasympathetic postganglionic axons and on sweat glands innervated by cholinergic sympathetic postganglionic axons; so named because muscarine activates these receptors but does not activate nicotinic receptors for acetylcholine.

Muscle An organ composed of one of three types of muscle tissue (skeletal, cardiac, or smooth), specialized for contraction to produce voluntary or involuntary movement of parts of the body.

Muscle action potential A stimulating impulse that propagates along the sarcolemma and transverse tubules; in skeletal muscle, it is generated by acetylcholine, which increases the permeability of the sarcolemma to cations, especially sodium ions (Na^+).

Muscle fatigue (fa-TĒG) Inability of a muscle to maintain its strength of contraction or tension; may be related to insufficient oxygen, depletion of glycogen, and/or lactic acid buildup.

Muscle fibers Skeletal muscle cells; have an elongated shape.

Muscle spindle An encapsulated proprioceptor in a skeletal muscle, consisting of specialized intrafusal muscle fibers and nerve endings; stimulated by changes in length or tension of muscle fibers.

Muscle strain Tearing of skeletal muscle fibers or tendon. Also called a muscle pull or muscle tear.

Muscle tone A sustained, partial contraction of portions of a skeletal or smooth muscle in response to activation of stretch receptors or a baseline level of action potentials in the innervating motor neurons.

Muscular dystrophies (DIS-trō-fēz) Inherited muscle-destroying diseases, characterized by degeneration of muscle fibers (cells), which causes progressive atrophy of the skeletal muscle.

Muscular system Usually refers to the voluntarily controlled muscles of the body that are composed of skeletal muscle tissue.

Muscular tissue A tissue specialized to produce motion in response to muscle action potentials by its qualities of contractility, extensibility, elasticity, and excitability; types include skeletal, cardiac, and smooth.

Muscularis (MUS-kū-la-ris) A muscular layer (coat or tunic) of an organ.

Muscularis mucosae A thin layer of smooth muscle fibers that underlie the lamina propria of the mucosa of the gastrointestinal tract.

Musculoskeletal (mus′-kū-lō-SKEL-e-tal) **system** An integrated system formed by the bones, muscles, and joints.

Mutation (mū-TĀ-shun) Permanent changes in the DNA base sequence of a gene.

Myasthenia (mī-as-THĒ-nē-a) **gravis** Weakness and fatigue of skeletal muscles caused by antibodies directed against acetylcholine receptors.

Myelin sheath Multilayered lipid and protein covering, formed by Schwann cells and oligodendrocytes, around axons of many peripheral and central nervous system neurons.

Myenteric plexus A network of autonomic axons and postganglionic cell bodies located in the muscularis of the gastrointestinal tract.

Myenteric plexus A network of autonomic axons and postganglionic cell bodies located in the muscularis of the gastrointestinal tract. Also called the plexus of Auerbach (OW-er-bak).

Myocardial infarction (mī-ō-KAR-dē-al in-FARK-shun) **(MI)** Gross necrosis of myocardial tissue due to interrupted blood supply. Also called a heart attack.

Myocardium The middle layer of the heart wall, made up of cardiac muscle tissue, lying between the epicardium and the endocardium and constituting the bulk of the heart.

Myofibril A threadlike structure, extending longitudinally through a muscle fiber (cell) consisting mainly of thick filaments (myosin) and thin filaments (actin, troponin, and tropomyosin).

Myoglobin The oxygen-binding, iron-containing protein present in the sarcoplasm of muscle fibers (cells); contributes the red color to muscle.

Myogram (MĪ-ō-gram) The record or tracing produced by a myograph, an apparatus that measures and records the force of muscular contractions.

Myology The study of muscles.

Myometrium (mī-ō-MĒ-trē-um) The smooth muscle layer of the uterus.

Myopathy (mī-OP-a-thē) Any abnormal condition or disease of muscle tissue.

Myopia (mī-Ō-pē-a) Defect in vision in which objects can be seen distinctly only when very close to the eyes; nearsightedness.

Myosin The contractile protein that makes up the thick filaments of muscle fibers.

Myotome (MĪ-ō-tōm) A group of muscles innervated by the motor neurons of a single spinal segment. In an embryo, the portion of a somite that develops into some skeletal muscles.

N

Nail A hard plate, composed largely of keratin, that develops from the epidermis of the skin to form a protective covering on the dorsal surface of the distal phalanges of the fingers and toes.

Nail matrix The portion of the epithelium proximal to the nail root.

Nasal cavity A mucosa-lined cavity on either side of the nasal septum that opens onto the face at the external nares and into the nasopharynx at the internal nares.

Nasal septum A vertical partition composed of bone (perpendicular plate of ethmoid and vomer) and cartilage, covered with a mucous membrane, separating the nasal cavity into left and right sides.

Nasolacrimal (nā-zō-LAK-ri-mal) **duct** A canal that transports the lacrimal secretion (tears) from the nasolacrimal sac into the nose.

Nasopharynx (nā-zō-FAR-inks) The superior portion of the pharynx, lying posterior to the nose and extending inferiorly to the soft palate.

Neck The part of the body connecting the head and the trunk. A constricted portion of an organ, such as the neck of the femur or uterus. The constricted junction of the crown and root of the tooth near the gum line, or the tapered portion of the gallbladder. The part of the tail of a sperm most proximal to the sperm head; contains centrioles.

Necrosis (ne-KRŌ-sis) A pathological type of cell death that results from disease, injury, or lack of blood supply in which many adjacent cells swell, burst, and spill their contents into the interstitial fluid, triggering an inflammatory response.

Neonatal (nē-ō-NĀ-tal) **period** The first four weeks after birth.

Neoplasm (NĒ-ō-plazm) A new growth that may be benign or malignant.

Nephron (NEF-ron) The functional unit of the kidney.

Nerve A cordlike bundle of neuronal axons and/or dendrites and associated connective tissue coursing together outside the central nervous system.

Nerve action potentials (nerve impulses) An electrical signal that propagates along the membrane of a neuron or muscle fiber (cell); a rapid change in membrane potential that involves a depolarization followed by a repolarization. Also called an action potential, and a muscle action potential as it relates to a muscle fiber.

Nerve fiber General term for any process (axon or dendrite) projecting from the cell body of a neuron.

Nerve impulse A wave of depolarization and repolarization that self-propagates along the plasma membrane of a neuron. Also called a nerve action potential.

Nervous system Body system composed of a network of billions of neurons and neuroglia divided into two main subdivisions, the central nervous system and the peripheral nervous system.

Nervous tissue Tissue containing neurons that initiate and conduct nerve impulses to coordinate homeostasis, and neuroglia that provide support and nourishment to neurons.

Neural plate A thickening of ectoderm, induced by the notochord, that forms early in the third week of development and represents the beginning of the development of the nervous system.

Neuralgia (noo-RAL-jē-a) Attacks of pain along the entire course or branch of a peripheral sensory nerve.

Neuroglia Cells of the nervous system that perform various supportive functions. In the central nervous system include the astrocytes, oligodendrocytes, microglia, and ependymal cells; in the peripheral nervous system include Schwann cells and satellite cells. Also called glial (GLĒ-al) cells.

Neurohypophyseal (noo-rō-hī-pō-FIZ-ē-al) **bud** An outgrowth of ectoderm located on the floor of the hypothalamus that gives rise to the posterior pituitary.

Neurolemma (noo-rō-LEM-ma) **(sheath of Schwann)** The peripheral, nucleated cytoplasmic layer of the Schwann cell.

Neurology The study of the normal functioning and disorders of the nervous system.

Neuromuscular junction A synapse between the axon terminals of a motor neuron and the sarcolemma of a muscle fiber (cell).

Neuron A nerve cell, consisting of a cell body, dendrites, and an axon.

Neurosecretory (noo-rō-SĒK-re-tō-rē) **cell** A neuron that secretes a hypothalamic releasing hormone or inhibiting hormone into blood capillaries of the hypothalamus; a neuron that secretes oxytocin or antidiuretic hormone into blood capillaries of the posterior pituitary.

Neurotransmitter One of a variety of molecules within axon terminals that are released into the synaptic cleft in response to a nerve impulse, and that change the membrane potential of the postsynaptic neuron.

Neurovascular (noo-rō-VAS-kū-lar) **bundle** The nerves and blood vessels innervating a muscle.

Neurulation The process by which the neural plate, neural folds, and neural tube develop.

Neutrophil A type of white blood cell characterized by vesicles that stain pale lilac with a combination of acidic and basic dyes.

Nicotinic (nik-ō-TIN-ik) **receptor** Receptor for the neurotransmitter acetylcholine found on both sympathetic and parasympathetic postganglionic neurons and on skeletal muscle in the motor end plate.

Nipple A pigmented, wrinkled projection on the surface of the breast that in the female is the location of the openings of the lactiferous ducts for milk release.

Nociceptor (nō-sē-SEP-tor) A free (naked) nerve ending that detects painful stimuli.

Node of Ranvier (RON-vē-ā) A space, along a myelinated axon, between the individual Schwann cells that form the myelin sheath and the neurolemma.

Norepinephrine (NE) A hormone secreted by the adrenal medulla that produces actions similar to those that result from sympathetic stimulation. Also called noradrenalin.

Notochord A flexible rod of mesodermal tissue that lies where the future vertebral column will develop and plays a role in induction.

Nucleic (noo-KLĒ-ik) **acid** An organic compound that is a long polymer of nucleotides, with each nucleotide containing a pentose sugar, a phosphate group, and one of four possible nitrogenous bases (adenine, cytosine, guanine, and thymine or uracil).

Nucleoli Spherical bodies within a cell nucleus composed of protein, DNA, and RNA that are the site of the assembly of small and large ribosomal subunits. *Singular* is **nucleolus**.

Nucleosome (NOO-klē-ō-sōm) Structural subunit of a chromosome consisting of histones and DNA.

Nucleus (NOO-klē-us) A spherical or oval organelle of a cell that contains the hereditary factors of the cell, called genes. A cluster of unmyelinated nerve cell bodies in the central nervous system. The central part of an atom made up of protons and neutrons.

Nucleus pulposus (pul-PŌ-sus) A soft, pulpy, highly elastic substance in the center of an intervertebral disc; a remnant of the notochord.

O

Obesity (ō-BĒS-i-tē) Body weight more than 20% above a desirable standard due to excessive accumulation of fat.

Oblique (ō-BLĒK) **plane** A plane that passes through the body or an organ at an angle between the transverse plane and either the midsagittal, parasagittal, or frontal plane.

Obstetrics The specialized branch of medicine that deals with pregnancy, labor, and the period of time immediately after delivery (about 6 weeks).

Olfaction (ōl-FAK-shun) The sense of smell.

Olfactory (ōl-FAK-tō-rē) Pertaining to smell.

Olfactory bulb A mass of gray matter containing cell bodies of neurons that form synapses with neurons of the olfactory (I) nerve, lying inferior to the frontal lobe of the cerebrum on either side of the crista galli of the ethmoid bone; also part of the limbic system.

Olfactory cilia Cilia projecting from the dendrites of olfactory receptors that respond to inhaled chemicals.

Olfactory receptor cell A bipolar neuron with its cell body lying between supporting cells located in the mucous membrane lining the superior portion of each nasal cavity; transduces odors into neural signals.

Olfactory tract A bundle of axons that extends from the olfactory bulb posteriorly to olfactory regions of the cerebral cortex.

Oligodendrocyte (OL-i-gō-den´-drō-sīt) A neuroglial cell that supports neurons and produces a myelin sheath around axons of neurons of the central nervous system.

Oliguria (ol´-i-GŪ-rē-a) Daily urinary output usually less than 250 mL.

Olive A prominent oval mass on each lateral surface of the superior part of the medulla oblongata.

Oncology The study of tumors.

Oogenesis Formation and development of female gametes (oocytes).

Oophorectomy (ō-of-ō-REK-tō-me) Surgical removal of the ovaries.

Ophthalmic (of-THAL-mik) Pertaining to the eye.

Ophthalmologist (of´-thal-MOL-ō-jist) A physician who specializes in the diagnosis and treatment of eye disorders using drugs, surgery, and corrective lenses.

Ophthalmology The study of the structure, function, and diseases of the eye.

Optic (OP-tik) Refers to the eye, vision, or properties of light.

Optic chiasm A crossing point of the two branches of the optic (II) nerve, anterior to the pituitary gland.

Optic disc A small area of the retina containing openings through which the axons of the ganglion cells emerge as the optic (II) nerve. Also called the **blind spot**.

Optic tract A bundle of axons that carry nerve impulses from the retina of the eye between the optic chiasm and the thalamus.

Ora serrata (Ō-ra ser-RĀ-ta) The irregular margin of the retina lying internal and slightly posterior to the junction of the choroid and ciliary body.

Orbit The bony, pyramidal-shaped cavity of the skull that holds the eyeball.

Organ A structure composed of two or more different kinds of tissues with a specific function and usually a recognizable shape.

Organ of Corti The organ of hearing, consisting of supporting cells and hair cells that rest on the basilar membrane and extend into the endolymph of the cochlear duct. Also called the **spiral organ**.

Organelle A permanent structure within a cell with characteristic morphology that is specialized to serve a specific function in cellular activities.

Organism A total living form; one individual.

Organogenesis (or´-ga-nō-JEN-e-sis) The formation of body organs and systems. By the end of the eighth week of development, all major body systems have begun to develop.

Orifice (OR-i-fis) Any aperture or opening.

Origin The attachment of a muscle tendon to a stationary bone or the end opposite the insertion.

Oropharynx (or´-ō-FAR-inks) The intermediate portion of the pharynx, lying posterior to the mouth and extending from the soft palate to the hyoid bone.

Orthopedics (or´-thō-PĒ-diks) The branch of medicine that deals with the preservation and restoration of the skeletal system, articulations, and associated structures.

Osmoreceptor Receptor in the hypothalamus that is sensitive to changes in blood osmolarity and, in response to high osmolarity (low water concentration), stimulates synthesis and release of antidiuretic hormone (ADH).

Osmosis The net movement of water molecules through a selectively permeable membrane from an area of higher water concentration to an area of lower water concentration until equilibrium is reached.

Osseous (OS-ē-us) Bony.

Ossicle (OS-si-kul) One of the small bones of the middle ear (malleus, incus, stapes).

Ossification The production of a special mineralized extracellular matrix by bone cells.

Ossification center An area in the cartilage model of a future bone where the cartilage cells hypertrophy, secrete enzymes that calcify their extracellular matrix, and die, and the area they occupied is invaded by osteoblasts that then lay down bone.

Osteoblast (OS-tē-ō-blast) Cell formed from an osteogenic cell that participates in bone formation by secreting some organic components and inorganic salts.

Osteoclast (OS-tē-ō-klast) A large, multinuclear cell that resorbs (destroys) bone matrix.

Osteocyte (OS-tē-ō-sīt´) A mature bone cell that maintains the daily activities of bone tissue.

Osteogenic cell (os´-tē-ō-JEN-ik) Stem cell derived from mesenchyme that has mitotic potential and the ability to differentiate into an osteoblast.

Osteogenic layer The inner layer of the periosteum that contains cells responsible for forming new bone during growth and repair.

Osteology The study of bones.

Osteon The basic unit of structure in adult compact bone, consisting of a central (haversian) canal with its concentrically arranged lamellae, lacunae, osteocytes, and canaliculi. Also called a haversian (ha-VER-shan) system.

Osteoporosis Age-related disorder characterized by decreased bone mass and increased susceptibility to fractures, often as a result of decreased levels of estrogens.

Osteoprogenitor cell Stem cell derived from mesenchyme that has mitotic potential and the ability to differentiate into an osteoblast.

Otolith A particle of calcium carbonate embedded in the otolithic membrane that functions in maintaining static equilibrium.

Otolithic membrane (ō-tō-LITH-ik) Thick, gelatinous, glycoprotein layer located directly over hair cells of the macula in the saccule and utricle of the internal ear.

Otorhinolaryngology (ō-tō-rī´-nō-lar-in-GOL-ō-jē) The branch of medicine that deals with the diagnosis and treatment of diseases of the ears, nose, and throat.

Oval window A small, membrane-covered opening between the middle ear and inner ear into which the footplate of the stapes fits.

Ovarian cycle A monthly series of events in the ovary associated with the maturation of a secondary oocyte.

Ovarian follicle A general name for oocytes (immature ova) in any stage of development, along with its surrounding epithelial cells.

Ovarian ligament A rounded cord of connective tissue that attaches the ovary to the uterus.

Ovary Female gonad that produces oocytes and hormones including the estrogens, progesterone, inhibin, and relaxin.

Ovulation The rupture of a mature ovarian (graafian) follicle with discharge of a secondary oocyte into the pelvic cavity.

Ovum The female reproductive or germ cell; an egg cell; arises through completion of meiosis in a secondary oocyte after penetration by a sperm. *Plural* is **ova**.

Oxytocin (ok´-sē-TŌ-sin) **(OT)** A hormone secreted by neurosecretory cells in the paraventricular and supraoptic nuclei of the hypothalamus that stimulates contraction of smooth muscle in the pregnant uterus and myoepithelial cells around the ducts of mammary glands.

P

P wave The deflection wave of an electrocardiogram that signifies atrial depolarization.

Palate The horizontal structure separating the oral and the nasal cavities; the roof of the mouth.

Palpate (PAL-pāt) To examine by touch; to feel.

Pancreas (PAN-krē-as) A soft, oblong organ lying along the greater curvature of the stomach and connected by a duct to the duodenum. It is both an exocrine gland (secreting pancreatic juice) and an endocrine gland (secreting insulin, glucagon, somatostatin, and pancreatic polypeptide).

Pancreatic (pan´-krē-AT-ik) **duct** A single large tube that unites with the common bile duct from the liver and gallbladder and drains pancreatic juice into the duodenum at the hepatopancreatic ampulla (ampulla of Vater). Also called the duct of Wirsung.

Pancreatic islet (Ī-let) A cluster of endocrine gland cells in the pancreas that secretes insulin, glucagon, somatostatin, and pancreatic polypeptide. Also called an **islet of Langerhans** (LANG-er-hanz).

Papanicolaou (pa-pa-NI-kō-lō) **test** A cytological staining test for the detection and diagnosis of premalignant and malignant conditions of the female genital tract. Cells scraped from the epithelium of the cervix of the uterus are examined microscopically. Also called a Pap test or Pap smear.

Papilla of the hair Nipple-shaped indentation of the hair bulb that contains areolar connective tissue and blood vessels.

Paralysis (pa-RAL-a-sis) Loss or impairment of motor function due to a lesion of nervous or muscular origin.

Paranasal sinus A mucus-lined air cavity in a skull bone that communicates with the nasal cavity. Paranasal sinuses are located in the frontal, maxillary, ethmoid, and sphenoid bones.

Paraplegia Paralysis of both lower limbs.

Parasagittal plane A vertical plane that does not pass through the midline and that divides the body or organs into unequal left and right portions.

Parasympathetic division One of the two subdivisions of the autonomic nervous system, having cell bodies of preganglionic neurons in nuclei in the brain stem and in the lateral gray horn of the sacral portion of the spinal cord; primarily concerned with activities that conserve and restore body energy. Also known as the craniosacral division.

Parathyroid gland One of usually four small endocrine glands embedded in the posterior surfaces of the lateral lobes of the thyroid gland.

Parathyroid hormone (PTH) A hormone secreted by the chief (principal) cells of the parathyroid glands that increases blood calcium level and decreases blood phosphate level. Also called parathormone.

Paraurethral glands Glands embedded in the wall of the urethra with ducts opening on either side of the urethral orifice and secreting mucus. Also called Skene's (SKĒNZ) glands.

Parenchyma (pa-RENG-ki-ma) The functional parts of any organ, as opposed to tissue that forms its stroma or framework.

Parietal cell A type of secretory cell in gastric glands that produces hydrochloric acid and intrinsic factor.

Parietal pleura The outer layer of the serous pleural membrane that encloses and protects the lungs; the layer that is attached to the wall of the pleural cavity. *Plural* is **pleurae**.

Parkinson's disease Progressive degeneration of the basal ganglia and substantia nigra of the cerebrum resulting in decreased production of dopamine (DA) that leads to tremor, slowing of voluntary movements, and muscle weakness.

Parotid (pa-ROT-id) **gland** One of the paired salivary glands located inferior and anterior to the ears and connected to the oral cavity via a parotid duct that opens into the inside of the cheek opposite the maxillary (upper) second molar tooth.

Pars intermedia A small avascular zone between the anterior and posterior pituitary glands.

Parturition (par-toor-ISH-un) Act of giving birth to young; childbirth; delivery.

Patellar ligament Extension of the quadriceps tendon that attaches to the tibial tuberosity.

Patent (PĀ-tent) **ductus arteriosus (PDA)** A congenital heart defect in which the ductus arteriosus remains open. As a result, aortic blood flows into the lower-pressure pulmonary trunk, increasing pulmonary trunk pressure and overworking both ventricles.

Pathogen (PATH-ō-jen) A disease-producing microbe.

Pathological anatomy (path´-ō-LOJ-i-kal) The study of structural changes caused by disease.

Pectinate muscles Projecting muscle bundles of the anterior atrial walls and the lining of the auricles.

Pectoral Pertaining to the chest or breast.

Pedicel Footlike structure, as on podocytes of a glomerulus.

Pelvic cavity Inferior portion of the abdominopelvic cavity that contains the urinary bladder, sigmoid colon, rectum, and internal female and male reproductive structures.

Pelvic splanchnic nerves Consist of preganglionic parasympathetic axons from the levels of S2, S3, and S4 that supply the urinary bladder, reproductive organs, and the descending and sigmoid colon and rectum.

Pelvis The basinlike structure formed by the two hip bones, the sacrum, and the coccyx. The expanded, proximal portion of the ureter, lying within the kidney and into which the major calyces open.

Penis The organ of urination and copulation in males; used to deposit semen into the female vagina.

Pepsin Protein-digesting enzyme secreted by chief cells of the stomach in the inactive form pepsinogen, which is converted to active pepsin by hydrochloric acid.

Peptic ulcer An ulcer that develops in areas of the gastrointestinal tract exposed to hydrochloric acid; classified as a gastric ulcer if in the lesser curvature of the stomach and as a duodenal ulcer if in the first part of the duodenum.

Percussion (pur-KUSH-un) The act of striking (percussing) an underlying part of the body with short, sharp blows as an aid in diagnosing the part by the quality of the sound produced.

Perforating canal A minute passageway by means of which blood vessels and nerves from the periosteum penetrate into compact bone. Also called Volkmann's (FŌLK-manz) canal.

Perforating fibers Thick bundles of collagen that extend from the periosteum into the bone extracellular matrix to attach the periosteum to the underlying bone. Also called Sharpey's fibers.

Pericardial (per´-i-KAR-dē-al) **cavity** Small potential space between the visceral and parietal layers of the serous pericardium that contains pericardial fluid.

Pericardium A loose-fitting membrane between the visceral and parietal layers of the serous pericardium that encloses the heart, consisting of a superficial fibrous layer and a deep serous layer, and that contains pericardial fluid.

Perichondrium The membrane that covers cartilage.

Perilymph The fluid contained between the bony and membranous labyrinths of the inner ear.

Perimetrium (pe´-i-MĒ-trē-um) The serosa of the uterus.

Perimysium (per-i-MĪZ-ē-um) Invagination of the epimysium that divides muscles into bundles.

Perineum (per´-i-NĒ-um) The portion of the pelvic floor beneath the pelvic diaphragm; the space between the anus and the scrotum in the male and between the anus and the vulva in the female.

Perineurium (per´-i-NOO-rē-um) Connective tissue wrapping around fascicles in a nerve.

Periocentriolar (per´-ē-sen´-trē-Ō-lar) **material** Ring-shaped complexes composed of the protein tubulin that surround the centrioles in a centrosome.

Periodontal ligament The periosteum lining the alveoli (sockets) for the teeth in the alveolar processes of the mandible and maxillae. Also called the periodontal membrane.

Periosteum (per´-ē-OS-tē-um) The covering of a bone that consists of connective tissue, osteogenic cells, and osteoblasts; essential for bone growth, repair, and nutrition.

Peripheral (pe-RIF-er-al) Located on the outer part or a surface of the body.

Peripheral nervous system (PNS) The part of the nervous system that lies outside the central nervous system, consisting of nerves and ganglia.

Peristalsis (per´-i-STAL-sis) Successive muscular contractions along the wall of a hollow muscular structure.

Peritoneum (per´-i-tō-NĒ-um) The largest serous membrane of the body that lines the abdominal cavity and covers the viscera.

Peritonitis (per´-i-tō-NĪ-tis) Inflammation of the peritoneum.

Peroxisome Organelle similar in structure to a lysosome containing enzymes that use molecular oxygen to oxidize various organic compounds; such reactions produce hydrogen peroxide; abundant in liver cells.

Perspiration Sweat; produced by sudoriferous (sweat) glands and containing water, salts, urea, uric acid, amino acids, ammonia, sugar, lactic acid, and ascorbic acid. Helps maintain body temperature and eliminate wastes.

pH A measure of the concentration of hydrogen ions (H^+) in a solution. The pH scale extends from 0 to 14, with a value of 7 expressing neutrality, values lower than 7 expressing increasing acidity, and values higher than 7 expressing increasing alkalinity.

Phagocytosis (fag´-ō-sī-TŌ-sis) The process by which phagocytes ingest and destroy microbes, cell debris, and other foreign matter.

Phalanges (fa-LAN-gēs) The bones of the fingers or toes. *Singular* is **phalanx.**

Pharmacology (far´-ma-KOL-ō-jē) The science of the effects and uses of drugs in the treatment of disease.

Pharynx (FAR-inks) The throat; a tube that starts at the internal nares and runs partway down the neck, where it opens into the esophagus posteriorly and the larynx anteriorly.

Phlebitis (fle-BĪ-tis) Inflammation of a vein, usually in a lower limb.

Photopigment A substance that can absorb light and undergo structural changes that can lead to the development of a receptor potential. An example is rhodopsin.

Photoreceptor Receptor that detects light shining on the retina of the eye.

Physiology (fiz´-ē-OL-ō-jē) Science that deals with the functions of an organism or its parts.

Pia mater (PĪ-a MĀ-ter *or* PĒ-a MA-ter) The innermost of the three meninges (coverings) of the brain and spinal cord.

Pineal gland A cone-shaped gland located in the roof of the third ventricle that secretes melatonin.

Pinealocyte (pin-ē-AL-ō-sīt) Secretory cell of the pineal gland that releases melatonin.

Pinna The projecting part of the external ear composed of elastic cartilage and covered by skin and shaped like the flared end of a trumpet. Also called the **auricle** (OR-i-kul).

Pituicyte Supporting cell of the posterior pituitary.

Pituitary gland A small endocrine gland occupying the hypophyseal fossa of the sphenoid bone and attached to the hypothalamus by the infundibulum. Also called the **hypophysis** (hi-POF-i-sis).

Pivot joint A synovial joint in which a rounded, pointed, or conical surface of one bone articulates with a ring formed partly by another bone and partly by a ligament, as in the joint between the atlas and axis and between the proximal ends of the radius and ulna. Also called a trochoid (TRŌ-koyd) joint.

Placenta The special structure through which the exchange of materials between fetal and maternal circulations occurs. Called the afterbirth following birth.

Plane joint Joint in which the articulating surfaces are flat or slightly curved; permits back-and-forth and side-to-side movements and rotation between the flat surfaces.

Plantar flexion Bending the foot in the direction of the plantar surface (sole).

Plasma The extracellular fluid found in blood vessels; blood minus the formed elements.

Plasma cell Cell that develops from a B cell (lymphocyte) and produces antibodies.

Plasma membrane Outer, limiting membrane that separates the cell's internal parts from extracellular fluid or the external environment.

Platelet A fragment of cytoplasm enclosed in a cell membrane and lacking a nucleus; found in the circulating blood; plays a role in hemostasis. Also called a thrombocyte (THROM-bō-sīt).

Platelet plug Aggregation of platelets (thrombocytes) at a site where a blood vessel is damaged that helps stop or slow blood loss.

Pleura The serous membrane that covers the lungs and lines the walls of the chest and the diaphragm.

Pleural cavity Small potential space between the visceral and parietal pleurae.

Plexus A network of nerves, veins, or lymphatic vessels.

Pluripotent stem cell Immature stem cell in red bone marrow that gives rise to precursors of all of the different mature blood cells.

Pneumotaxic (noo-mō-TAK-sik) **area** A part of the respiratory center in the pons that continually sends inhibitory nerve impulses to the inspiratory area, limiting inhalation and facilitating exhalation.

Polycythemia (pol-ē-sī-THĒ-mē-a) Disorder characterized by an above-normal hematocrit (above 55%) in which hypertension, thrombosis, and hemorrhage can occur.

Polyunsaturated fat A fatty acid that contains more than one double covalent bond between its carbon atoms; abundant in triglycerides of corn oil, safflower oil, and cottonseed oil.

Polyuria (pol-ē-Ū-rē-a) An excessive production of urine.

Pons The part of the brain stem that forms a "bridge" between the medulla oblongata and the midbrain, anterior to the cerebellum.

Pontine respiratory group (PRG) Neurons of the respiratory center in the pons that may modify the rhythm of inhalation and exhalation.

Popliteal fossa (pop-LIT-ē-al) Diamond-shaped space on the posterior aspect of the knee bordered laterally by the tendons of the biceps femoris muscle and medially by the tendons of the semitendinosus and semimembranosus muscles.

Portal system The circulation of blood from one capillary network into another through a vein.

Portal triad A microscopic association of a bile duct, hepatic artery, and hepatic vein within the liver.

Postcentral gyrus Gyrus of cerebral cortex located immediately posterior to the central sulcus; contains the primary somatosensory area.

Posterior (pos-TĒR-ē-or) Nearer to or at the back of the body. Equivalent to dorsal in bipeds.

Posterior (dorsal) ramus A branch of a spinal nerve containing motor and sensory axons supplying the muscles, skin, and bones of the posterior part of the head, neck, and trunk.

Posterior root ganglion A group of cell bodies of sensory neurons and their supporting cells located along the posterior root of a spinal nerve. Also called a dorsal (sensory) root ganglion.

Posterior (dorsal) roots The structure composed of sensory axons lying between a spinal nerve and the dorsolateral aspect of the spinal cord.

Posterior column–medial lemniscus pathways Sensory pathways that carry information related to proprioception, fine touch, two-point discrimination, pressure, and vibration via first-order neurons, second-order neurons, and third-order neurons.

Posterior pituitary Posterior lobe of the pituitary gland. Also called the neurohypophysis (noo-ro-hi-POF-i-sis).

Postganglionic neuron The second autonomic motor neuron in an autonomic pathway, having its cell body and dendrites located in an autonomic ganglion and its unmyelinated axon ending at cardiac muscle, smooth muscle, or a gland.

Postsynaptic (pōst'-sin-AP-tik) **neuron** The nerve cell that is activated by the release of a neurotransmitter from another neuron and carries nerve impulses away from the synapse.

Precapillary sphincter A ring of smooth muscle fibers (cells) at the metarteriole-capillary junction that regulate blood flow into capillaries.

Precentral gyrus Gyrus of cerebral cortex located immediately anterior to the central sulcus; contains the primary motor area.

Preganglionic neuron The first autonomic motor neuron in an autonomic pathway, with its cell body and dendrites in the brain or spinal cord and its myelinated axon ending at an autonomic ganglion, where it synapses with a postganglionic neuron.

Pregnancy Sequence of events that normally includes fertilization, implantation, embryonic growth, and fetal growth and terminates in birth.

Premenstrual syndrome (PMS) Severe physical and emotional stress occurring late in the postovulatory phase of the menstrual cycle and sometimes overlapping with menstruation.

Prepuce The loose-fitting skin covering the glans of the penis. Also called the foreskin.

Presbyopia (prez-bē-Ō-pē-a) A loss of elasticity of the lens of the eye due to advancing age with resulting inability to focus clearly on near objects.

Presynaptic neuron A neuron that propagates nerve impulses toward a synapse.

Prevertebral ganglion A cluster of cell bodies of postganglionic sympathetic neurons anterior to the spinal column and close to large abdominal arteries. Also called a collateral ganglion.

Primary germ layer One of three layers of embryonic tissue, called ectoderm, mesoderm, and endoderm, that give rise to all tissues and organs of the body.

Primary motor area A region of the cerebral cortex in the precentral gyrus of the frontal lobe of the cerebrum that controls specific muscles or groups of muscles.

Primary somatosensory area A region of the cerebral cortex posterior to the central sulcus in the postcentral gyrus of the parietal lobe of the cerebrum that localizes exactly the points on the body where somatic sensations originate.

Prime mover The muscle directly responsible for producing a desired motion. Also called an agonist (AG-ō-nist).

Primitive gut Embryonic structure formed from the dorsal part of the yolk sac that gives rise to most of the gastrointestinal tract.

Primordial (prī-MŌR-dē-al) Existing first; especially primordial egg cells in the ovary.

Principal cell Cell type in the distal convoluted tubules and collecting ducts of the kidneys that is stimulated by aldosterone and antidiuretic hormone.

Proctology The branch of medicine concerned with the rectum and its disorders.

Progenitor cells Cells that arise from myeloid stem cells during hemopoiesis; different types produce erythrocytes, megakaryocytes, granulocytes, and monocytes.

Progeny (PROJ-e-nē) Offspring or descendants.

Progesterone (prō-JES-te-rōn) A female sex hormone produced by the ovaries that helps prepare the endometrium of the uterus for implantation of a fertilized ovum and the mammary glands for milk secretion.

Projection tracts Cerebral white matter tract containing axons that conduct nerve impulses from the cerebrum to lower parts of the CNS or from lower parts of the CNS to the cerebrum.

Prolactin (PRL) A hormone secreted by the anterior pituitary that initiates and maintains milk secretion by the mammary glands.

Prolapse (PRŌ-laps) A dropping or falling down of an organ, especially the uterus or rectum.

Proliferation (prō-lif'-er-Ā-shun) Rapid and repeated reproduction of new parts, especially cells.

Pronation A movement of the forearm in which the palm is turned posteriorly.

Prophase The first stage of mitosis during which chromatid pairs are formed and aggregate around the metaphase plate of the cell.

Proprioceptive sensations The perception of the position of body parts, especially the limbs, independent of vision; this sense is possible due to nerve impulses generated by proprioceptors. Also called proprioception.

Proprioceptor (PRŌ-prē-ō-sep'-tor) A receptor located in muscles, tendons, joints, or the internal ear (muscle spindles, tendon organs, joint kinesthetic receptors, and hair cells of the vestibular apparatus) that provides information about body position and movements.

Prostaglandin (pros'-ta-GLAN-din) **(PG)** A membrane-associated lipid; released in small quantities and acts as a local hormone.

Prostate A doughnut-shaped gland inferior to the urinary bladder that surrounds the superior portion of the male urethra and secretes a slightly acidic solution that contributes to sperm motility and viability.

Proteasome Tiny barrel-shaped cellular organelle in cytosol and nucleus containing proteases that destroy unneeded, damaged, or faulty proteins.

Protein An organic compound consisting of carbon, hydrogen, oxygen, nitrogen, and sometimes sulfur and phosphorus; synthesized on ribosomes and made up of amino acids linked by peptide bonds.

Proto-oncogene Gene responsible for some aspect of normal growth and development; it may transform into an oncogene, a gene capable of causing cancer.

Protraction The movement of the mandible or shoulder girdle forward on a plane parallel with the ground.

Proximal (PROK-si-mal) Nearer the attachment of a limb to the trunk; nearer to the point of origin or attachment.

Pseudopods Temporary protrusions of the leading edge of a migrating cell; cellular projections that surround a particle undergoing phagocytosis.

Pterygopalatine ganglion A cluster of cell bodies of parasympathetic postganglionic neurons ending at the lacrimal and nasal glands.

Ptosis (TŌ-sis) Drooping, as of the eyelid or the kidney.

Puberty The time of life during which the secondary sex characteristics begin to appear and the capability for sexual reproduction is possible; usually occurs between the ages of 10 and 17.

Pubic symphysis A slightly movable cartilaginous joint between the anterior surfaces of the hip bones.

Puerperium (pū-er-PER-ē-um) The period immediately after childbirth, usually 4–6 weeks.

Pulmonary (PUL-mo-ner´-ē) Concerning or affected by the lungs.

Pulmonary circulation The flow of deoxygenated blood from the right ventricle to the lungs and the return of oxygenated blood from the lungs to the left atrium.

Pulmonary edema (e-DĒ-ma) An abnormal accumulation of interstitial fluid in the tissue spaces and alveoli of the lungs due to increased pulmonary capillary permeability or increased pulmonary capillary pressure.

Pulmonary embolism (EM-bō-lizm) **(PE)** The presence of a blood clot or a foreign substance in a pulmonary arterial blood vessel that obstructs circulation to lung tissue.

Pulmonary ventilation The inflow (inhalation) and outflow (exhalation) of air between the atmosphere and the lungs. Also called breathing.

Pulp cavity A cavity within the crown and neck of a tooth, which is filled with pulp, a connective tissue containing blood vessels, nerves, and lymphatic vessels.

Pulse (PULS) The rhythmic expansion and elastic recoil of a systemic artery after each contraction of the left ventricle.

Pupil The hole in the center of the iris, the area through which light enters the posterior cavity of the eyeball.

Purkinje cell Neuron in the cerebellum with extensive branching of dendrites.

Purkinje fiber Muscle fiber (cell) in the ventricular tissue of the heart specialized for conducting an action potential to the myocardium; part of the conduction system of the heart.

Pus The liquid product of inflammation containing leukocytes or their remains and debris of dead cells.

Pyloric sphincter A thickened ring of smooth muscle through which the pylorus of the stomach communicates with the duodenum. Also called the pyloric valve.

Pyorrhea (pī-ō-RĒ-a) A discharge or flow of pus, especially in the alveoli (sockets) and the tissues of the gums.

Pyramid A pointed or cone-shaped structure. One of two roughly triangular structures on the anterior aspect of the medulla oblongata composed of the largest motor tracts that run from the cerebral cortex to the spinal cord. A triangular structure in the renal medulla.

Q

QRS wave The deflection waves of an electrocardiogram that represent onset of ventricular depolarization.

Quadrant One of four parts.

Quadriplegia (kwod´-ri-PLĒ-jē-a) Paralysis of four limbs: two upper and two lower.

R

Radiographic (rā´-dē-ō-GRAF-ik) **anatomy** Diagnostic branch of anatomy that includes the use of x rays.

Rami communicantes (RĀ-mē kō-mū-ni-KAN-tēz) Branches of a spinal nerve that are components of the autonomic nervous system *Singular* is **ramus communicans.**

Ramus (RĀ-mus) Branch of a spinal nerve. *Plural* is **rami.**

Range of motion (ROM) The range through which the bones of a join can be moved, measured in degrees of a circle.

Receptor A specialized cell or a distal portion of a neuron that responds to a specific sensory modality, such as touch, pressure, cold, light, or sound, and converts it to an electrical signal (generator or receptor potential). A specific molecule or cluster of molecules that recognizes and binds a particular ligand.

Receptor-mediated endocytosis A highly selective process whereby cells take up specific ligands, which usually are large molecules or particles, by enveloping them within a sac of plasma membrane.

Recombinant DNA Synthetic DNA, formed by joining a fragment of DNA from one source to a portion of DNA from another.

Rectouterine (rek-tō-Ū-ter-in) **pouch** A pocket formed by the parietal peritoneum as it moves posteriorly from the surface of the uterus and is reflected onto the rectum; the most inferior point in the pelvic cavity. Also called the pouch of Douglas.

Rectum The last 20 cm (8 in.) of the gastrointestinal tract, from the sigmoid colon to the anus.

Recumbent (re-KUM-bent) Lying down.

Red bone marrow A highly vascularized connective tissue located in microscopic spaces between trabeculae of spongy bone tissue.

Red nucleus A cluster of cell bodies in the midbrain, occupying a large part of the tectum from which axons extend into the rubroreticular and rubrospinal tracts.

Red pulp That portion of the spleen that consists of venous sinuses filled with blood and thin plates of splenic tissue called splenic (Billroth's) cords.

Referred pain Pain that is felt at a site remote from the place of origin.

Reflex Fast response to a change (stimulus) in the internal or external environment that attempts to restore homeostasis.

Reflex arc The most basic conduction pathway through the nervous system, connecting a receptor and an effector and consisting of a receptor, a sensory neuron, an integrating center in the central nervous system, a motor neuron, and an effector.

Regional anatomy The division of anatomy dealing with a specific region of the body, such as the head, neck, chest, or abdomen.

Regurgitation (rē-gur´-ji-TĀ-shun) Return of solids or fluids to the mouth from the stomach; backward flow of blood through incompletely closed heart valves.

Relaxin (RLX) A female hormone produced by the ovaries and placenta that increases flexibility of the pubic symphysis and helps dilate the uterine cervix to ease delivery of a baby.

Releasing hormone Hormone secreted by the hypothalamus that can stimulate secretion of hormones of the anterior pituitary.

Renal (RĒ-nal) Pertaining to the kidneys.

Renal corpuscle A glomerular (Bowman's) capsule and its enclosed glomerulus.

Renal lobe Collective name for a renal pyramid, its overlying area of renal cortex, and one-half of each adjacent renal column.

Renal pelvis A cavity in the center of the kidney formed by the expanded, proximal portion of the ureter, lying within the kidney, and into which the major calyces open.

Renal pyramid A triangular structure in the renal medulla containing the straight segments of renal tubules and the vasa recta.

Reproduction (rē-prō-DUK-shun) The formation of new cells for growth, repair, or replacement, or the production of a new individual.

Reproductive cell division Type of cell division in which gametes (sperm and oocytes) are produced; consists of meiosis and cytokinesis.

Respiration (res-pi-RĀ-shun) Overall exchange of gases between the atmosphere, blood, and body cells.

Respiratory center Neurons in the pons and medulla oblongata of the brain stem that regulate the rate and depth of breathing.

Respiratory membrane Four-layered membrane through which respiratory gases (O_2 and CO_2) diffuse; formed from the alveolar wall, epithelial basement membrane, capillary basement membrane, and endothelial cells of the capillary walls.

Respiratory (RES-pir-a-to´-rē) **physiology** Study of the functions of the air passageways and lungs.

Respiratory system Body system consisting of the nose, nasal cavity, pharynx, larynx, trachea, bronchi, and lungs.

Rete (RĒ-tē) **testis** The network of ducts in the testes.

Retention (rē-TEN-shun) A failure to void urine due to obstruction, nervous contraction of the urethra, or absence of sensation of desire to urinate.

Reticular activating system (RAS) A portion of the reticular formation that has many ascending connections with the cerebral cortex; when this area of the brain stem is active, nerve impulses pass to the thalamus and widespread areas of the cerebral cortex, resulting in generalized alertness or arousal from sleep.

Reticular formation A network of small groups of neuronal cell bodies scattered among bundles of axons (mixed gray and white matter) beginning in the medulla oblongata and extending superiorly through the central part of the brain stem.

Reticulocyte (re-TIK-ū-lō-sīt) An immature red blood cell.

Retina The deep coat of the posterior portion of the eyeball consisting of nervous tissue (where the process of vision begins) and a pigmented layer of epithelial cells that contact the choroid.

Retinacula (ret-i-NAK-ū-la) Thickenings of deep fascia that hold structures in place, for example, the superior and inferior retinacula of the ankle. *Singular* is **retinaculum**.

Retraction The movement of a protracted part of the body posteriorly on a plane parallel to the ground, as in pulling the lower jaw back in line with the upper jaw.

Retroperitoneal Posterior to the peritoneal lining of the abdominal cavity.

Rh factor An inherited antigen on the surface of red blood cells in Rh⁺ individuals; not present in Rh⁻ individuals.

Rhinology (rī-NOL-ō-jē) The study of the nose and its disorders.

Ribonucleic (rī-bō-noo-KLĒ-ik) **acid (RNA)** A single-stranded nucleic acid made up of nucleotides, each consisting of a nitrogenous base (adenine, cytosine, guanine, or uracil), ribose, and a phosphate group; major types are messenger RNA (mRNA), transfer RNA (tRNA), and ribosomal RNA (rRNA), each of which has a specific role during protein synthesis.

Ribosome (RĪ-bō-sōm) An organelle in the cytoplasm of cells, composed of a small subunit and a large subunit that contain ribosomal RNA and ribosomal proteins; the site of protein synthesis.

Right lymphatic duct A vessel of the lymphatic system that drains lymph from the upper right side of the body and empties it into the right subclavian vein.

Rigidity (ri-JID-i-tē) Hypertonia characterized by increased muscle tone, that does not affect reflexes.

Rigor mortis State of partial contraction of muscles after death due to lack of ATP; myosin heads (cross-bridges) remain attached to actin, thus preventing relaxation.

Rod One of two types of photoreceptor in the retina of the eye; specialized for vision in dim light.

Root canal A narrow extension of the pulp cavity lying within the root of a tooth.

Root of penis Attached portion of penis that consists of the bulb and crura.

Rotation Moving a bone around its own axis, with no other movement.

Rotator cuff Refers to the tendons of four deep shoulder muscles (subscapularis, supraspinatus, infraspinatus, and teres minor) that form a complete circle around the shoulder; they strengthen and stabilize the shoulder joint.

Round ligament A band of fibrous connective tissue enclosed between the folds of the broad ligament of the uterus, emerging from the uterus just inferior to the uterine tube, extending laterally along the pelvic wall and through the deep inguinal ring to end in the labia majora.

Round window A small opening between the middle and internal ear, directly inferior to the oval window, covered by the secondary tympanic membrane.

Rugae Large folds in the mucosa of an empty hollow organ, such as the stomach and vagina.

S

Saccule (SAK-ūl) The inferior and smaller of the two chambers in the membranous labyrinth inside the vestibule of the internal ear containing a receptor organ for static equilibrium.

Sacral plexus (SĀ-kral PLEK-sus) A network formed by the ventral branches of spinal nerves L4 through S3.

Sacral promontory (PROM-on-tor´-ē) The superior surface of the body of the first sacral vertebra that projects anteriorly into the pelvic cavity; a line from the sacral promontory to the superior border of the pubic symphysis divides the abdominal and pelvic cavities.

Sacrum (SĀ-krum) Vertebral bone consisting of five fused sacral vertebrae.

Saddle joint A synovial joint in which the articular surface of one bone is saddle-shaped and the articular surface of the other bone is shaped like the legs of the rider sitting in the saddle, as in the joint between the trapezium and the metacarpal of the thumb. Also called a sellar joint.

Sagittal (SAJ-i-tal) **plane** A plane that divides the body or organs into left and right portions. Such a plane may be midsagittal (median), in which the divisions are equal, or parasagittal, in which the divisions are unequal.

Saliva (sa-LĪ-va) A clear, alkaline, somewhat viscous secretion produced mostly by the three pairs of salivary glands; contains various salts, mucin, lysozyme, salivary amylase, and lingual lipase (produced by glands in the tongue).

Salivary amylase (SAL-i-ver-ē AM-i-lās) An enzyme in saliva that initiates the chemical breakdown of starch.

Salivary gland One of three pairs of glands that lie external to the mouth and pour their secretory product (saliva) into ducts that empty into the oral cavity; the parotid, submandibular, and sublingual glands.

Satellite (SAT-i-līt) **cell** Flat neuroglial cells that surround cell bodies of peripheral nervous system ganglia to provide structural support and regulate the exchange of material between a neuronal cell body and interstitial fluid.

Sarcolemma (sar´-kō-LEM-ma) The cell membrane of a muscle fiber (cell), especially of a skeletal muscle fiber.

Sarcomere (SAR-kō-mēr) A contractile unit in a striated muscle fiber (cell) extending from one Z disc to the next Z disc.

Sarcoplasm (SAR-kō-plazm) The cytoplasm of a muscle fiber (cell).

Sarcoplasmic reticulum (SR) A network of saccules and tubes surrounding myofibrils of a muscle fiber (cell), comparable to endoplasmic reticulum; functions to reabsorb calcium ions during relaxation and release them to cause contraction.

Saturated fat A fatty acid that contains only single bonds (no double bonds) between its carbon atoms; all carbon atoms are bonded to the maximum number of hydrogen atoms; prevalent in triglycerides of animal products such as meat, milk, milk products, and eggs.

Scala tympani (SKA-la TIM-pan-ē) The inferior spiral-shaped channel of the bony cochlea, filled with perilymph.

Scala vestibuli The superior spiral-shaped channel of the bony cochlea, filled with perilymph.

Schwann (SCHVON *or* SCHWON) **cell** A neuroglial cell of the peripheral nervous system that forms the myelin sheath and neurolemma around a nerve axon by wrapping around the axon in a jellyroll fashion.

Sciatica (sī-AT-i-ka) Inflammation and pain along the sciatic nerve; felt along the posterior aspect of the thigh extending down the inside of the leg.

Sclera (SKLE-ra) The white coat of fibrous tissue that forms the superficial protective covering over the eyeball except in the most anterior portion; the posterior portion of the fibrous tunic.

Scleral venous sinus A circular venous sinus located at the junction of the sclera and the cornea through which aqueous humor drains from the anterior chamber of the eyeball into the blood. Also called the canal of Schlemm (SHLEM).

Sclerosis (skle-RŌ-sis) A hardening with loss of elasticity of tissues.

Scoliosis (skō-lē-Ō-sis) An abnormal lateral curvature from the normal vertical line of the backbone.

Scrotum (SKRŌ-tum) A skin-covered pouch that contains the testes and their accessory structures.

Sebaceous ciliary glands Sebaceous glands at the base of the hair follicles of the eyelashes that release a lubricating fluid into the follicles.

Sebaceous (se-BĀ-shus) **gland** An exocrine gland in the dermis of the skin, almost always associated with a hair follicle, that secretes sebum. Also called an oil gland.

Sebum Secretion of sebaceous (oil) glands.

Secondary sex characteristic A characteristic of the male or female body that develops at puberty under the influence of sex hormones but is not directly involved in sexual reproduction; examples are distribution of body hair, voice pitch, body shape, and muscle development.

Secretion (se-KRĒ-shun) Production and release from a cell or a gland of a physiologically active substance.

Segmental bronchi Branches of lobar bronchi that enter bronchopulmonary segments. Also called tertiary bronchi.

Selective permeability (per´-mē-a-BIL-i-tē) The property of a membrane by which it permits the passage of certain substances but restricts the passage of others.

Semen (SĒ-men) A fluid discharged at ejaculation by a male that consists of a mixture of sperm and the secretions of the seminiferous tubules, seminal vesicles, prostate, and bulbourethral (Cowper's) glands.

Semicircular canals Three bony channels (anterior, posterior, lateral), filled with perilymph, in which lie the membranous semicircular canals filled with endolymph. They contain receptors for equilibrium.

Semicircular ducts The membranous semicircular canals filled with endolymph and floating in the perilymph of the bony semicircular canals; they contain cristae that are concerned with dynamic equilibrium.

Semilunar (SL) valve A valve between the aorta or the pulmonary trunk and a ventricle of the heart.

Seminal vesicle (SEM-i-nal VES-i-kul) One of a pair of convoluted, pouchlike structures, lying posterior and inferior to the urinary bladder and anterior to the rectum, that secrete a component of semen into the ejaculatory ducts. Also called a seminal gland.

Seminiferous tubule (sem´-i-NI-fer-us TOO-būl) A tightly coiled duct, located in the testis, where sperm are produced.

Sensation A state of awareness of external or internal conditions of the body.

Sensory area A region of the cerebral cortex concerned with the interpretation of sensory impulses.

Sensory neurons Neurons that carry sensory information from cranial and spinal nerves into the brain and spinal cord or from a lower to a higher level in the spinal cord and brain. Also called afferent neurons.

Septal defect An opening in the atrial septum (atrial septal defect) because the foramen ovale fails to close, or in the ventricular septum (ventricular septal defect) due to incomplete development of the ventricular septum.

Septum (SEP-tum) A wall dividing two cavities.

Serous membrane A membrane that lines a body cavity that does not open to the exterior. The external layer of an organ formed by a serous membrane. The membrane that lines the pleural, pericardial, and peritoneal cavities. Also called a serosa (se-RŌ-sa).

Serum Blood plasma minus its clotting proteins.

Sesamoid bones Small bones usually found in tendons.

Sex chromosomes The twenty-third pair of chromosomes, designated X and Y, which determine the genetic sex of an individual; in males, the pair is XY; in females, XX.

Sexual intercourse The insertion of the erect penis of a male into the vagina of a female. Also called coitus (KŌ-i-tus).

Shock Failure of the cardiovascular system to deliver adequate amounts of oxygen and nutrients to meet the metabolic needs of the body due to inadequate cardiac output. It is characterized by hypotension; clammy, cool, and pale skin; sweating; reduced urine formation; altered mental state; acidosis; tachycardia; weak, rapid pulse; and thirst. Types include hypovolemic, cardiogenic, vascular, and obstructive.

Short bone Bone that is somewhat cube-shaped and nearly equal in length and width.

Shoulder joint A synovial joint where the humerus articulates with the scapula.

Sigmoid colon (SIG-moyd KŌ-lon) The S-shaped part of the large intestine that begins at the level of the left iliac crest, projects medially, and terminates at the rectum at about the level of the third sacral vertebra.

Sign Any objective evidence of disease that can be observed or measured such as a lesion, swelling, or fever.

Sinoatrial (si-nō-Ā-trē-al) **(SA) node** A small mass of cardiac muscle fibers (cells) located in the right atrium inferior to the opening of the superior vena cava that spontaneously depolarize and generate a cardiac action potential about 100 times per minute. Also called the pacemaker.

Sinus (SĪ-nus) A hollow in a bone (paranasal) or other tissue; a channel for lymph (lymphatic) or blood (vascular); any cavity having a narrow opening.

Sinusoid (SĪ-nū-soyd) A large, thin-walled, and leaky type of capillary, having large intercellular clefts that may allow proteins and blood cells to pass from a tissue into the bloodstream; present in the liver, spleen, anterior pituitary, parathyroid glands, and red bone marrow.

Skeletal muscle An organ specialized for contraction, composed of striated muscle fibers (cells), supported by connective tissue, attached to a bone by a tendon or an aponeurosis, and stimulated by somatic motor neurons.

Skeletal muscle tissue Muscle tissue that moves the bones of the skeleton; striated and voluntary.

Skeletal system The entire framework of bones and their cartilages.

Skin The external covering of the body that consists of a superficial, thinner epidermis (epithelial tissue) and a deep, thicker dermis (connective tissue) that is anchored to the subcutaneous layer.

Skin graft The transfer of a patch of healthy skin taken from a donor site to cover a wound.

Skull The skeleton of the head consisting of the cranial and facial bones.

Sleep A state of partial unconsciousness from which a person can be aroused; associated with a low level of activity in the reticular activating system.

Sliding filament mechanism A model that describes muscle contraction in which thin filaments slide past thick ones so that the filaments overlap, causing shortening of a sarcomere and thus shortening of muscle fibers, and ultimately shortening of the entire muscle.

Slow oxidative (SO) fibers Skeletal muscle fibers that contain a high myoglobin content, many blood capillaries, and appear red in color; they are adapted for maintaining posture and endurance-type activities such as running a marathon; also called type I fibers.

Small intestine A long tube of the gastrointestinal tract that begins at the pyloric sphincter of the stomach, coils through the central and inferior part of the abdominal cavity, and ends at the large intestine; divided

into three segments: duodenum, jejunum, and ileum.

Smooth muscle tissue A tissue specialized for contraction, composed of smooth muscle fibers (cells), located in the walls of hollow internal organs, and innervated by autonomic motor neurons; nonstriated and involuntary.

Soft palate The posterior portion of the roof of the mouth, extending from the palatine bones to the uvula. It is a muscular partition lined with mucous membrane.

Somatic cell division Type of cell division in which a single starting cell duplicates itself to produce two identical cells; consists of mitosis and cytokinesis.

Somatic motor neurons Neurons that stimulate skeletal muscle fibers to contract.

Somatic motor pathway Pathway that carries information from the cerebral cortex, basal nuclei, and cerebellum to stimulate contraction of skeletal muscles.

Somatic nervous system (SNS) The portion of the peripheral nervous system consisting of somatic sensory (afferent) neurons and somatic motor (efferent) neurons.

Somatic sensory pathway Pathway that carries information from somatic sensory receptor to the primary somatosensory area in the cerebral cortex and cerebellum.

Somite (SŌ-mīt) Block of mesodermal cells in a developing embryo that is distinguished into a myotome (which forms most of the skeletal muscles), dermatome (which forms connective tissues), and sclerotome (which forms the vertebrae).

Spasm (SPAZM) A sudden, involuntary contraction of large groups of muscles.

Spasticity (spas-TIS-i-tē) Hypertonia characterized by increased muscle tone, increased tendon reflexes, and pathological reflexes (Babinski sign).

Sperm cell A mature male gamete. Also termed spermatozoon (sper´-ma-tō-ZŌ-on).

Spermatic cord A supporting structure of the male reproductive system, extending from a testis to the deep inguinal ring, that includes the ductus (vas) deferens, arteries, veins, lymphatic vessels, nerves, cremaster muscle, and connective tissue.

Spermatogenesis (sper´-ma-tō-JEN-e-sis) The formation and development of sperm in the seminiferous tubules of the testes.

Spermiogenesis (sper´-mē-ō-JEN-e-sis) The maturation of spermatids into sperm.

Sphincter of the hepatopancreatic ampulla A circular muscle at the opening of the common bile and main pancreatic ducts in the duodenum. Also called the sphincter of Oddi (OD-ē).

Spinal cord A mass of nerve tissue located in the vertebral canal from which 31 pairs of spinal nerves originate.

Spinal nerve One of the 31 pairs of nerves that originate on the spinal cord from posterior and anterior roots.

Spinal shock A period from several days to several weeks following transection of the spinal cord that is characterized by the abolition of all reflex activity.

Spinothalamic (spī-nō-tha-LAM-ik) **tract** Sensory (ascending) tract that conveys information up the spinal cord to the thalamus for sensations of pain, temperature, itch, and tickle.

Spinous (SPĪ-nus) **process** A sharp or thornlike process or projection. Also called a spine. A sharp ridge running diagonally across the posterior surface of the scapula.

Spiral organ The organ of hearing, consisting of supporting cells and hair cells that rest on the basilar membrane and extend into the endolymph of the cochlear duct. Also called the **organ of Corti** (KOR-tē).

Splanchnic (SPLANK-nik) Pertaining to the viscera.

Spleen (SPLĒN) Large mass of lymphatic tissue between the fundus of the stomach and the diaphragm that functions in formation of blood cells during early fetal development, phagocytosis of ruptured blood cells, and proliferation of B cells during immune responses.

Spongy bone Type of bone tissue that has trabeculae instead of osteons.

Spongy bone tissue Bone tissue that consists of an irregular latticework of thin plates of bone called trabeculae; spaces between trabeculae of some bones are filled with red bone marrow; found inside short, flat, and irregular bones and in the epiphyses (ends) of long bones.

Sprain Forcible wrenching or twisting of a joint with partial rupture or other injury to its attachments without dislocation.

Starvation (star-VĀ-shun) The loss of energy stores in the form of glycogen, triglycerides, and proteins due to inadequate intake of nutrients or inability to digest, absorb, or metabolize ingested nutrients.

Static equilibrium The maintenance of posture in response to changes in the orientation of the body, mainly the head, relative to the ground.

Stellate reticuloendothelial (STEL-āt retik´-ū-lō-en´-dō-THĒ-lē-al) **cell** Phagocytic cell bordering a sinusoid of the liver. Also called a Kupffer (KOOP-fer) cell or hepatic macrophage.

Stem cell An unspecialized cell that has the ability to divide for indefinite periods and give rise to a specialized cell.

Stenosis An abnormal narrowing or constriction of a duct or opening.

Stereocilia Groups of extremely long, slender, nonmotile microvilli projecting from epithelial cells lining the epididymis.

Sterile (STE-ril) Free from any living microorganisms. Unable to conceive or produce offspring.

Sterilization Elimination of all living microorganisms. Any procedure that renders an individual incapable of reproduction (for example, castration, vasectomy, hysterectomy, or oophorectomy).

Stimulus Any stress that changes a controlled condition; any change in the internal or external environment that excites a sensory receptor, a neuron, or a muscle fiber.

Stomach The J-shaped enlargement of the gastrointestinal tract directly inferior to the diaphragm in the epigastric, umbilical, and left hypochondriac regions of the abdomen, between the esophagus and small intestine.

Straight tubule A duct in a testis leading from a convoluted seminiferous tubule to the rete testis.

Stratum (STRĀ-tum) A layer.

Stratum basale The deepest layer of the epidermis. Also called the stratum germinativum.

Stratum basalis The layer of the endometrium next to the myometrium that is maintained during menstruation and gestation and produces a new stratum functionalis following menstruation or parturition.

Stratum functionalis The layer of the endometrium next to the uterine cavity that is shed during menstruation and that forms the maternal portion of the placenta during gestation.

Stretch receptor Receptor in the walls of blood vessels, airways, or organs that monitors the amount of stretching. Also termed **baroreceptor**.

Striae (STRĪ-ē) Internal scarring due to overstretching of the skin in which collagen fibers and blood vessels in the dermis are damaged. Also called stretch marks.

Stroma The tissue that forms the ground substance, foundation, or framework of an organ, as opposed to its functional parts (parenchyma).

Subarachnoid (sub´-a-RAK-noyd) **space** A space between the arachnoid mater and the pia mater that surrounds the brain and spinal cord and through which cerebrospinal fluid circulates.

Subcutaneous Beneath the skin. Also called hypodermic (hi-pō-DER-mik).

Subcutaneous (subQ) layer A continuous sheet of areolar connective tissue and adipose tissue between the dermis of the skin and the deep fascia of the muscles. Also called the hypodermis.

Subdural space A space between the dura mater and the arachnoid mater of the brain and spinal cord that contains a small amount of fluid.

Sublingual gland One of a pair of salivary glands situated in the floor of the mouth deep to the mucous membrane and to the side of the lingual frenulum, with a lesser sublingual duct that opens into the floor of the mouth.

Submandibular gland One of a pair of salivary glands found inferior to the base of the tongue deep to the mucous membrane in the posterior part of the floor of the mouth, posterior to the sublingual glands, with a submandibular duct situated to the side of the lingual frenulum.

Submucosa A layer of connective tissue located deep to a mucous membrane, as in the gastrointestinal tract or the urinary bladder; the submucosa connects the mucosa to the muscularis layer.

Submucosal plexus A network of autonomic nerve fibers located in the superficial part of the submucous layer of the small intestine. Also called the plexus of Meissner (MĪZ-ner).

Sudoriferous gland An apocrine or eccrine exocrine gland in the dermis or subcutaneous layer that produces perspiration. Also called a sweat gland.

Subthalamus (sub-THAL-a-mus) Part of the diencephalon inferior to the thalamus; the substantia nigra and red nucleus extend from the midbrain into the subthalamus.

Sulcus (SUL-kus) A groove or depression between parts, especially on the surface of the heart or between the convolutions of the brain. *Plural* is **sulci**.

Superficial Located on or near the surface of the body or an organ.

Superficial inguinal ring A triangular opening in the aponeurosis of the external oblique muscle that represents the termination of the inguinal canal. Also called subcutaneous inguinal ring.

Superior Toward the head or upper part of a structure. Also called cephalad (SEF-a-lad) or craniad.

Superior vena cava Large vein that collects blood from parts of the body superior to the heart and returns it to the right atrium.

Supination (soo-pi-NĀ-shun) A movement of the forearm in which the palm is turned anteriorly.

Surface anatomy The study of the structures that can be identified from the outside of the body.

Surfactant (sur-FAK-tant) Complex mixture of phospholipids and lipoproteins, produced by type II alveolar (septal) cells in the lungs, that decreases surface tension.

Susceptibility Lack of resistance to the damaging effects of an agent such as a pathogen.

Suspensory ligament (sus-PEN-so-rē LIG-a-ment) A fold of peritoneum extending laterally from the surface of the ovary to the pelvic wall.

Sustentacular cell A supporting cell in the seminiferous tubules that secretes fluid for supplying nutrients to sperm and the hormone inhibin, removes excess cytoplasm from spermatogenic cells, and mediates the effects of FSH and testosterone on spermatogenesis.

Sutural (SOO-chur-al) **bone** A small bone located within a suture between certain cranial bones. Also called wormian (WER-mē-an) bone.

Suture (SOO-cher) An immovable fibrous joint that joins skull bones.

Sympathetic division One of the two subdivisions of the autonomic nervous system, having cell bodies of preganglionic neurons in the lateral gray columns of the thoracic segment and the first two or three lumbar segments of the spinal cord; primarily concerned with processes involving the expenditure of energy. Also referred to as the thoracolumbar division.

Sympathetic trunk ganglion A cluster of cell bodies of sympathetic postganglionic neurons lateral to the vertebral column, close to the body of a vertebra. These ganglia extend inferiorly through the neck, thorax, and abdomen to the coccyx on both sides of the vertebral column and are connected to one another to form a chain on each side of the vertebral column. Also called sympathetic chain or vertebral chain ganglia.

Symphysis (SIM-fi-sis) A line of union. A slightly movable cartilaginous joint such as the pubic symphysis.

Symptom A subjective change in body function not apparent to an observer, such as pain or nausea, that indicates the presence of a disease or disorder of the body.

Synapse The functional junction between two neurons or between a neuron and an effector, such as a muscle or gland; may be electrical or chemical.

Synapsis The pairing of homologous chromosomes during prophase I of meiosis.

Synaptic cleft The narrow gap at a chemical synapse that separates the axon terminal of one neuron from another neuron or muscle fiber (cell) and across which a neurotransmitter diffuses to affect the postsynaptic cell.

Synaptic end bulb Expanded distal end of an axon terminal that contains synaptic vesicles. Also called a terminal bouton.

Synaptic vesicle Membrane-enclosed sac in a synaptic end bulb that stores neurotransmitters.

Synarthrosis (sin´-ar-THRŌ-sis) An immovable joint such as a suture, gomphosis, or synchondrosis.

Synchondrosis (sin´-kon-DRŌ-sis) A cartilaginous joint in which the connecting material is hyaline cartilage.

Syndesmosis (sin´-dez-MŌ-sis) A slightly movable joint in which articulating bones are united by fibrous connective tissue.

Synergist A muscle that assists the prime mover by reducing undesired action or unnecessary movement.

Synergistic (syn-er-JIS-tik) **effect** A hormonal interaction in which the effects of two or more hormones acting together is greater or more extensive than the effect of each hormone acting alone.

Synostosis (sin´-os-TŌ-sis) A joint in which the dense fibrous connective tissue that unites bones at a suture has been replaced by bone, resulting in a complete fusion across the suture line.

Synovial cavity The space between the articulating bones of a synovial joint, filled with synovial fluid. Also called a joint cavity.

Synovial fluid Secretion of synovial membranes that lubricates joints and nourishes articular cartilage.

Synovial joint A fully movable or diarthrotic joint in which a synovial (joint) cavity is present between the two articulating bones.

Synovial membrane The deeper of the two layers of the articular capsule of a synovial joint, composed of areolar connective tissue that secretes synovial fluid into the synovial (joint) cavity.

System An association of organs that have a common function.

Systemic (sis-TEM-ik) Affecting the whole body; generalized.

Systemic anatomy The anatomical study of particular systems of the body, such as the skeletal, muscular, nervous, cardiovascular, or urinary systems.

Systemic circulation The routes through which oxygenated blood flows from the left ventricle through the aorta to all the organs of the body and deoxygenated blood returns to the right atrium.

Systole (SIS-tō-lē) In the cardiac cycle, the phase of contraction of the heart muscle, especially of the ventricles.

Systolic (sis-TOL-ik) **blood pressure (SBP)** The force exerted by blood on arterial walls during ventricular contraction; the highest pressure measured in the large arteries, about 120 mmHg under normal conditions for a young adult.

T

T cell A lymphocyte that becomes immunocompetent in the thymus and can differentiate into a helper T cell or a cytotoxic T cell, both of which function in cell-mediated immunity.

T wave The deflection wave of an electrocardiogram that represents ventricular repolarization.

Tachycardia (tak´-i-KAR-dē-a) An abnormally rapid resting heartbeat or pulse rate (over 100 beats per minute).

Tactile (TAK-tīl) Pertaining to the sense of touch.

Tactile disc Modified epidermal cell in the stratum basale of hairless skin that functions as a cutaneous receptor for discriminative touch. Also called a Merkel (MER-kel) cell.

Tactile epithelial cell Type of cell in the epidermis of hairless skin that makes contact with a tactile disc, which functions in touch.

Target cell A cell whose activity is affected by a particular hormone.

Tarsal bones The seven bones of the ankle.

Tarsal gland Sebaceous (oil) gland that opens on the edge of each eyelid. Also called a Meibomian (mī-BŌ-mē-an) gland.

Tarsal plate A thin, elongated sheet of connective tissue, one in each eyelid, giving the eyelid form and support. The aponeurosis of the levator palpebrae superioris is attached to the tarsal plate of the superior eyelid.

Tarsus (TAR-sus) A collective term for the seven bones of the ankle.

Tectorial membrane A gelatinous membrane projecting over and in contact with the hair cells of the spiral organ (organ of Corti) in the cochlear duct.

Teeth Accessory structures of digestion, composed of calcified connective tissue and embedded in bony sockets of the mandible and maxilla, that cut, shred, crush, and grind food. Also called dentes (DEN-tēz).

Telophase (TEL-ō-fāz) The final stage of mitosis.

Tendon A white fibrous cord of dense regular connective tissue that attaches muscle to bone.

Tendon (synovial) sheath Tubelike bursa that wraps around a tendon to protect it from friction as it runs through a tunnel of connective tissue and bone.

Tendon organ A proprioceptive receptor, sensitive to changes in muscle tension and force of contraction, found chiefly near the junctions of tendons and muscles.

Teniae coli (TĒ-nē-ē KŌ-lī) The three flat bands of thickened, longitudinal smooth muscle running the length of the large intestine, except in the rectum.

Tension lines (lines of cleavage) Orientation of dermal collagen fibers because of natural tension resulting from bony projections, orientation of muscles, and movements at joints; important in surgical repair.

Tentorium cerebelli (ten-TŌ-rē-um ser´-e-BEL-ī) A transverse shelf of dura mater that forms a partition between the occipital lobe of the cerebral hemispheres and the cerebellum and that covers the cerebellum.

Teratogen (TER-a-tō-jen) Any agent or factor that causes physical defects in a developing embryo.

Terminal ganglion A cluster of cell bodies of parasympathetic postganglionic neurons either lying very close to the visceral effectors or located within the walls of the visceral effectors supplied by the postganglionic neurons.

Testis Male gonad that produces sperm and the hormones testosterone and inhibin. *Plural* is **testes**. Also called a testicle.

Testosterone (tes-TOS-te-rōn) A male sex hormone (androgen) secreted by interstitial cells (Leydig cells) of a mature testis; needed for development of sperm; together with a second androgen termed dihydrotestosterone (DHT), controls the growth and development of male reproductive organs, secondary sex characteristics, and body growth.

Thalamus (THAL-a-mus) A large, oval structure located bilaterally on either side of the third ventricle, consisting of two masses of gray matter organized into nuclei; main relay center for sensory impulses ascending to the cerebral cortex.

Thermoreceptor Sensory receptor that detects changes in temperature.

Thigh The portion of the free lower limb between the hip and the knee. Also called the femoral region.

Third ventricle A slitlike cavity between the right and left halves of the thalamus and between the lateral ventricles of the brain.

Thoracic (left lymphatic) duct A lymphatic vessel that begins as a dilation called the cisterna chyli, receives lymph from the left side of the head, neck, and chest, left arm, and the entire body below the ribs, and empties into the junction between the internal jugular and left subclavian veins.

Thoracic (thor-AS-ik) **cavity** Superior portion of the ventral body cavity that contains two pleural cavities, the mediastinum, and the pericardial cavity.

Thoracolumbar (thōr´-a-kō-LUM-bar) **outflow** The axons of sympathetic preganglionic neurons, which have their cell bodies in the lateral gray columns of the thoracic segments and first two or three lumbar segments of the spinal cord.

Thorax (THŌ-raks) The entire chest region.

Thrombosis (THROM-BŌ-sis) The formation of a clot in an unbroken blood vessel, usually a vein.

Thrombus (THROM-bus) A stationary clot formed in an unbroken blood vessel, usually a vein.

Thymus (THĪ-mus) A bilobed organ, located in the superior mediastinum posterior to the sternum and between the lungs, in which T cells develop immunocompetence.

Thyroid cartilage The largest single cartilage of the larynx, consisting of two fused plates that form the anterior wall of the larynx.

Thyroid follicle Spherical sac that forms the parenchyma of the thyroid gland and consists of follicular cells that produce thyroxine (T4) and triiodothyronine (T3).

Thyroid gland An endocrine gland with right and left lateral lobes on either side of the trachea connected by an isthmus; located anterior to the trachea just inferior to the cricoid cartilage; secretes thyroxine (T4), triiodothyronine (T3), and calcitonin.

Thyroid-stimulating hormone (TSH) A hormone secreted by the anterior pituitary that stimulates the synthesis and secretion of thyroxine (T4) and triiodothyronine (T3). Also called thyrotropin.

Thyroxine (thī-ROK-sēn) A hormone secreted by the thyroid gland that regulates metabolism, growth and development, and the activity of the nervous system. Also called tetraiodothyronine.

Tic Spasmodic, involuntary twitching of muscles that are normally under voluntary control.

Tissue A group of similar cells and their intercellular substance joined together to perform a specific function.

Tissue rejection Phenomenon by which the body recognizes the protein (HLA antigens) in transplanted tissues or organs as foreign and produces antibodies against them.

Tongue A large skeletal muscle covered by a mucous membrane located on the floor of the oral cavity.

Tonsil An aggregation of large lymphatic nodules embedded in the mucous membrane of the throat.

Torn cartilage A tearing of an articular disc (meniscus) in the knee.

Trabeculae (tra-BEK-ū-lē) Irregular latticework of thin plates of spongy bone tissue. Fibrous cord of connective tissue serving as supporting fiber by forming a septum extending into an organ from its wall or capsule. *Singular* is **trabecula** (tra-BEK-ū-la).

Trabeculae carneae Ridges and folds of the myocardium in the ventricles.

Trachea (TRĀ-kē-a) Tubular air passageway extending from the larynx to the fifth thoracic vertebra. Also called the windpipe. Can be palpated posterior to the suprasternal notch of the sternum.

Tract A bundle of nerve axons in the central nervous system.

Transcytosis (tranz´-sī-TŌ-sis) Transport in vesicles that moves a substance into, across, and out of a cell.

Transplantation (tranz-plan-TĀ-shun) The transfer of living cells, tissues, or organs from a donor to a recipient or from one part of the body to another in order to restore a lost function.

Transverse colon The portion of the large intestine extending across the abdomen from the right colic (hepatic) flexure to the left colic (splenic) flexure.

Transverse fissure The deep cleft that separates the cerebrum from the cerebellum.

Transverse plane A plane that divides the body or organs into superior and inferior portions. Also called a horizontal plane.

Transverse section A flat, two-dimensional surface of a three-dimensional structure produced by passing a transverse plane through it. Also called a cross section.

Transverse tubules (T tubules) Small, cylindrical invaginations of the sarcolemma of striated muscle fibers (cells) that conduct muscle action potentials toward the center of the muscle fiber.

Tremor (TREM-or) Rhythmic, involuntary, purposeless contraction of opposing muscle groups.

Triad (TRĪ-ad) A complex of three units in a muscle fiber composed of a transverse tubule and the sarcoplasmic reticulum terminal cisterns on both sides of it.

Triangle of auscultation (aus´-cul-TĀ-shun) Triangular region of the back just medial to the inferior part of the scapula, where the rib cage is not covered by superficial muscles; bounded by the latissimus dorsi and trapezius muscles and the vertebral border of the scapula.

Tricuspid valve Atrioventricular (AV) valve on the right side of the heart; also called right atrioventricular valve.

Triglyceride (trī-GLI-ser-īd) A lipid formed from one molecule of glycerol and three molecules of fatty acids that may be either solid (fat) or liquid (oil) at room temperature; the body´s most highly concentrated source of chemical potential energy. Found mainly within adipocytes.

Trigone (TRĪ-gōn) A triangular region at the base of the urinary bladder.

Triiodothyronine (trī-ī-ō-dō-THĪ-rō-nēn) **(T3)** A hormone produced by the thyroid gland that regulates metabolism, growth and development, and the activity of the nervous system.

Trophoblast The superficial covering of cells of the blastocyst.

Tropic hormone A hormone whose target is another endocrine gland. Also called tropins.

Trunk The part of the body to which the upper and lower limbs are attached. Union of several spinal nerves in the inferior part of the neck.

Tubal ligation A sterilization procedure in which the uterine (fallopian) tubes are tied and cut.

Tubular reabsorption The movement of filtrate from renal tubules back into blood in response to the body's specific needs.

Tubular secretion The movement of substances in blood into renal tubular fluid in response to the body's specific needs.

Tumor suppressor gene Gene that produces proteins that normally inhibit cell division; damage to this gene causes some types of cancer.

Tunica albuginea (TOO-ni-ka al´-bū-JIN-ē-a) A dense white fibrous capsule covering a testis or deep to the surface of an ovary.

Tunica externa The superficial coat of an artery or vein, composed mostly of elastic and collagen fibers.

Tunica interna (intima) The deep coat of an artery or vein, consisting of a lining of endothelium, basement membrane, and internal elastic lamina.

Tunica media The intermediate coat of an artery or vein, composed of smooth muscle and elastic fibers.

Tympanic membrane A thin, semitransparent partition of fibrous connective tissue between the external auditory meatus and the middle ear. Also called the eardrum.

Type I cutaneous mechanoreceptor Slowly adapting touch receptor for fine touch; also called a tactile disc or Merkel cell.

Type II cutaneous mechanoreceptor A sensory receptor embedded deeply in the dermis and deeper tissues that detects stretching of skin. Also called a Ruffini corpuscle.

U

Umbilical cord The long, ropelike structure containing the umbilical arteries and vein that connect the fetus to the placenta.

Umbilicus (um-BIL-i-kus *or* um-bi-LĪ-kus) A small scar on the abdomen that marks the former attachment of the umbilical cord to the fetus. Also called the navel.

Umbo The apex of the tympanic membrane that projects into the tympanic cavity.

Upper limb The appendage attached at the shoulder girdle, consisting of the arm, forearm, wrist, hand, and fingers. Also called the free upper extremity.

Uremia (ū-RĒ-mē-a) Accumulation of toxic levels of urea and other nitrogenous waste products in the blood, usually resulting from severe kidney malfunction.

Ureter (Ū-rē-ter) One of two tubes that connect the kidney with the urinary bladder.

Urethra (ū-RĒ-thra) The duct from the urinary bladder to the exterior of the body that conveys urine in females and urine and semen in males.

Urinalysis (ū-ri-NAL-i-sis) An analysis of the volume and physical, chemical, and microscopic properties of urine.

Urinary bladder A hollow, muscular organ situated in the pelvic cavity posterior to the pubic symphysis; receives urine via two ureters and stores urine until it is excreted through the urethra.

Urinary system The body system consisting of the kidneys, ureters, urinary bladder, and urethra.

Urine The fluid produced by the kidneys that contains wastes and excess materials; excreted from the body through the urethra.

Urogenital triangle The region of the pelvic floor inferior to the pubic symphysis, bounded by the pubic symphysis and the ischial tuberosities, and containing the external genitalia.

Urology (ū-ROL-ō-jē) The specialized branch of medicine that deals with the structure, function, and diseases of the male and female urinary systems and the male reproductive system.

Uterine cycle A series of changes in the endometrium of a nonpregnant female that prepares the lining of the uterus to receive a fertilized ovum. Also called the menstrual cycle.

Uterine tube Duct that transports ova from the ovary to the uterus. Also called the fallopian (fal-LŌ-pē-an) tube or oviduct.

Uterosacral ligament (ū-ter-ō-SĀ-kral LIG-a-ment) A fibrous band of tissue extending from the cervix of the uterus laterally to the sacrum.

Uterus (Ū-te-rus) The hollow, muscular organ in females that is the site of menstruation, implantation, development of the fetus, and labor. Also called the womb.

Utricle (Ū-tri-kul) The larger of the two divisions of the membranous labyrinth located inside the vestibule of the inner ear, containing a receptor organ for static equilibrium.

Uvea (Ū-vē-a) The three structures that together make up the **vascular tunic** of the eye.

Uvula (Ū-vū-la) A soft, fleshy mass, especially the V-shaped pendant part, descending from the soft palate.

V

Vagina (va-JĪ-na) A muscular, tubular organ that leads from the uterus to the vestibule, situated between the urinary bladder and the rectum of the female.

Vallate papilla (VAL-āt pa-PIL-a) One of the circular projections that is arranged in an inverted V-shaped row at the back of the tongue; the largest of the elevations on the upper surface of the tongue containing taste buds. Also called circumvallate papilla.

Varicocele (VAR-i-kō-sēl) A twisted vein; especially, the accumulation of blood in the veins of the spermatic cord.

Varicose (VAR-i-kōs) Pertaining to an unnatural swelling, as in the case of a vein.

Vasa recta (VĀ-sa REK-ta) Extensions of the efferent arteriole of a juxtamedullary nephron that run alongside the nephron loop (loop of Henle) in the medullary region of the kidney.

Vasa vasorum (va-SŌ-rum) Blood vessels that supply nutrients to the larger arteries and veins.

Vascular (venous) sinus A vein with a thin endothelial wall that lacks a tunica media and externa and is supported by surrounding tissue.

Vascular tunic The middle layer of the eyeball, composed of the choroid, ciliary body, and iris. Also called the **uvea** (Ū-ve-a).

Vasectomy (va-SEK-tō-mē) A means of sterilization of males in which a portion of each ductus (vas) deferens is removed.

Vasoconstriction A decrease in the size of the lumen of a blood vessel caused by contraction of the smooth muscle in the wall of the vessel.

Vasodilation An increase in the size of the lumen of a blood vessel caused by relaxation of the smooth muscle in the wall of the vessel.

Vein A blood vessel that conveys blood from tissues back to the heart.

Vena cava (VĒ-na KĀ-va) One of two large veins that open into the right atrium, returning to the heart all of the deoxygenated blood from the systemic circulation except from the coronary circulation.

Ventral Pertaining to the anterior or front side of the body; opposite of dorsal.

Ventral ramus (RĀ-mus) The anterior branch of a spinal nerve, containing sensory and motor fibers to the muscles and skin of the anterior surface of the head, neck, trunk, and the limbs.

Ventral respiratory group (VRG) Group of neurons in the medullary respiratory center that controls forceful breathing; formerly called the expiratory area.

Ventricle A cavity in the brain filled with cerebrospinal fluid. An inferior chamber of the heart.

Ventricular fibrillation (ven-TRIK-ū-lar fib-ri-LĀ-shun) **(VF or V-fib)** Asynchronous ventricular contractions; unless reversed by defibrillation, results in heart failure.

Venule A small vein that collects blood from capillaries and delivers it to a vein.

Vermiform appendix A twisted, coiled tube attached to the cecum. Also called the appendix.

Vermis The central constricted area of the cerebellum that separates the two cerebellar hemispheres.

Vertebrae (VER-te-brē) Bones that make up the vertebral column.

Vertebral canal A cavity within the vertebral column formed by the vertebral foramina of all the vertebrae and containing the spinal cord. Also called the spinal canal.

Vertebral column The 26 vertebrae of an adult and 33 vertebrae of a child; encloses and protects the spinal cord and serves as a point of attachment for the ribs and back muscles. Also called the backbone, spine, or spinal column.

Vesicle A small bladder or sac containing liquid.

Vesicouterine (ves´-ik-ō-Ū-ter-in) **pouch** A shallow pouch formed by the reflection of the peritoneum from the anterior surface of the uterus, at the junction of the cervix and the body, to the posterior surface of the urinary bladder.

Vestibular apparatus Collective term for the organs of equilibrium, which includes the saccule, utricle, and semicircular ducts.

Vestibular membrane The membrane that separates the cochlear duct from the scala vestibuli.

Vestibule Oval central portion of the bony labyrinth. A small space or cavity at the beginning of a canal, especially the inner ear, larynx, mouth, nose, and vagina.

Villi (VIL-lī) Projections of the intestinal mucosal cells containing connective tissue, blood vessels, and a lymphatic vessel; functions in the absorption of the end products of digestion. *Singular* is **villus**.

Viscera The organs inside the ventral body cavity.

Visceral (VIS-er-al) Pertaining to the organs or to the covering of an organ.

Visceral effectors (e-FEK-torz) Organs of the ventral body cavity that respond to neural stimulation, including cardiac muscle, smooth muscle, and glands.

Visceral smooth muscle tissue More common type of smooth muscle tissue found in skin, walls of small arteries and veins, and walls of hollow viscera. Also called single-unit smooth muscle tissue.

Vision The act of seeing.

Vitamin An organic molecule necessary in trace amounts that acts as a catalyst in normal metabolic processes in the body.

Vitreous (VIT-rē-us) **body** A soft, jellylike substance that fills the vitreous chamber of the eyeball, lying between the lens and the retina.

Vocal folds (true vocal cords) Pair of mucous membrane folds below the vestibular folds that function in voice production.

Vulva (VUL-va) Collective designation for the external genitalia of the female. Also called the pudendum (poo-DEN-dum).

W

Wallerian (wal-LE-rē-an) **degeneration** Degeneration of the portion of the axon and myelin sheath of a neuron distal to the site of injury.

Wandering macrophages Phagocytic cells that roam the tissues and gather at the sites of infection or inflammation.

White matter Aggregations or bundles of myelinated and unmyelinated axons located in the brain and spinal cord.

White pulp The regions of the spleen composed of lymphatic tissue, mostly B lymphocytes.

White ramus communicans The portion of a preganglionic sympathetic axon that branches from the anterior ramus of a spinal nerve to enter the nearest sympathetic trunk ganglion. *Plural* is **white rami communicantes**.

X

Xiphoid (ZĪ-foyd) Sword-shaped.

Xiphoid process The inferior portion of the sternum.

Y

Yolk sac An extraembryonic membrane composed of the exocoelomic membrane and hypoblast. It transfers nutrients to the embryo, is a source of blood cells, contains primordial germ cells that migrate into the gonads to form primitive germ cells, forms part of the gut, and helps prevent desiccation of the embryo.

Z

Zona fasciculata (ZŌ-na fa-sik´-ū-LA-ta) The middle zone of the adrenal cortex consisting of cells arranged in long, straight cords that secrete glucocorticoid hormones, mainly cortisol.

Zona glomerulosa (glo-mer´-ū-LŌ-sa) The outer zone of the adrenal cortex, directly under the connective tissue covering, consisting of cells arranged in arched loops or round balls that secrete mineralocorticoid hormones, mainly aldosterone.

Zona pellucida (pe-LOO-si-da) Clear glycoprotein layer between a secondary oocyte and the surrounding granulosa cells of the corona radiata.

Zona reticularis (ret-ik´-ū-LAR-is) The inner zone of the adrenal cortex, consisting of cords of branching cells that secrete sex hormones, chiefly androgens.

Zygote (ZĪ-gōt) The single cell resulting from the union of male and female gametes; the fertilized ovum.

CREDITS

Illustration Credits

Chapter 1   1.1, 1.5 1.7: Kevin Somerville/
Imagineering. 1.8, 1.9: Kevin Somerville. 1.2: Molly
Borman. 1.3, Table 1.2: DNA Illustrations. 1.4, 1.6:
Imagineering.

Chapter 2   2.1–2.2, 2.7, 2.8, 2.10, 2.11,
2.13–2.15, Table 2.2: Tomo Narashima. 2.3–2.6, 2.9,
2.12, 2.16–2.22: Imagineering.

Chapter 3   3.1, 3.5, 3.6: Kevin Somerville. 3.2
Hilda Muinos. 3.4, 3.7, 3.8, PAP Test: Imagineering.
3.9: Tables 3.1–3.6: Kevin Somerville/Imagineering.

Chapter 4   4.1: Kevin Somerville/Imagineering.
4.2–4.13, 4.15, Premature Infant: Kevin Somerville.

Chapter 5   5.1–5.3, 5.5–5.7, 5.8: Kevin
Somerville. 5.4: Imagineering. Burns: Kevin
Somerville/Imagineering.

Chapter 6   6.1, 6.2, 6.5, 6.7a, Table 6.2, 6.10:
John Gibb. 6.3: Lauren Keswick. 6.4, 6.6, 6.9a: Kevin
Somerville. 6.9b: John Gibb/Imagineering.

Chapter 7   7.1–7.24, Table 7.1, Table 7.2,
Table 7.5, Herniated Disc: John Gibb. Abnormal
Curves: Imagineering.

Chapter 8   8.1–8.12, Table 8.1: John Gibb. 8.13:
Kevin Somerville.

Chapter 9   9.1–9.4, 9.10–9.16: John
Gibb. 9.15e: Torn Cartilage and Arthroscopy:
Imagineering. Arthroplasty: John Gibb/
Imagineering.

Chapter 10   10.1, 10.2, 10.9, 10.11, Table 10.3:
Kevin Somerville/Imagineering. 10.3–10.8, 10.10,
Table 10.5, Electromyography: Imagineering.

Chapter 11   11.1–11.25: John Gibb.

Chapter 12   12.1, 12.3, 12.4, Table 12.2 and
12.3, Bone Marrow Examination: Imagineering.

Chapter 13   13.1a,b, Coronary Artery Disease
part B: Kevin Somerville. 13.2: Kevin Somerville/
Imagineering. 13.3, 13.4, 13.6a, 13.7a, 13.8–13.13:
Kevin Somerville. 13.5. 13.6d–e: John Gibb. 13.1d,
13.7b, Coronary Artery Disease part C and D, Help
for Failing Hearts: Imagineering.

Chapter 14   14.1, 14.3, 14.5, 14.6a, 14.8, 14.10,
14.15–14.18: Kevin Somerville. 14.2, Varicose Veins,
Hypertension: Imagineering. 14.4: Kevin Somerville/
Imagineering. 14.6b, 14.7, 14.9, 14.11–14.14, John
Gibb.

Chapter 15   15.1: Richard Combs/Kevin
Somerville/Imagineering. 15.2, 15.4, HIV:
Imagineering. 15.3, 15.7a, 15.8–15.12: John Gibb.
15.6, 15.13: Kevin Somerville. 15.5, 15.7bc: Steve
Oh.

Chapter 16   16.1, 16.3a, 16.6–16.9: Kevin
Somerville. 16.2, 16.3c, 16.4–16.5, 16.10, Multiple
Sclerosis: Imagineering.

Chapter 17   17.1–17.4: Kevin Somerville. 17.5,
17.7c, 17.8b, 17.9b: John Gibb. 17.6, 17.7a, 17.8a,

17.9a: Steve Oh. Injuries to Nerves Emerging from
the Brachial Plexus: DNA Illustrations. 17.10, Spinal
Tap, Spinal Cord Injury: Imagineering. 17.11: Kevin
Somerville/Imagineering. 17.12: Leonard Dank.

Chapter 18   18.1–18.4, 18.5b,c, 18.6–18.16:
Kevin Somerville. 18.18–18.27: Imagineering/Richard
Coombs. 18.5a,d, Table 18.3, Cerebrovascular
Accident: Imagineering.

Chapter 19   19.1–19.3a, 19.5: Kevin
Somerville/Imagineering. 19.4, 19.6, 19.7:
Imagineering.

Chapter 20   20.1, 20.2, 20.4, 20.6, Table:
Kevin Somerville. 20.3, 20.5, Parkinson's Disease:
Imagineering. Table 20.3, Table 20.4: Kevin
Somerville/Imagineering.

Chapter 21   21.1, 21.4, 21.7, 21.8, 21.10–21.14,
21.16a, b, 21.17a: Tomo Narashima. 21.2: Molly
Borman. 21.3, 21.16c, 21.17b: Sharon Ellis. 21.9,
21.15, 21.18, Table 21.1, Table 21.2: Imagineering.
21.19, 21.20: Kevin Somerville.

Chapter 22   22.1, 22.8: Kevin Somerville.
22.2–22.7: Lynn O'Kelley/Imagineering. Tables 22.2 –
22.8: Imagineering.

Chapter 23   23.1, 23.2a, 23.10, 23.11, 23.15,
23.16: Kevin Somerville. 23.2b, 23.3, 23.4, 23.6,
23.8b–d: Molly Borman. 23.8a, 23.9, 23.13, 23.14:
Imagineering. 23.12: John Gibb/Imagineering.

Chapter 24   24.1, 24.2, 24.11, 24.14ab, 24.16,
24.17, 24.20: Kevin Somerville. 24.3, 24,14d–e,
24.15, Peptic Ulcer Disease: Imagineering. 24.4,
24.6, 24.7, 24.9: Nadine Sokol. 24.5: DNA
Illustrations. 24.10a, 24.12, 24.13: Steve Oh. 24.19:
Molly Borman.

Chapter 25   25.1, 25.2, 25.6, 25.8, 25.12,
25.13: Kevin Somerville. 25.3, 25.4, 25.9: Steve Oh/
Imagineering. 25.5, 25.7, Table 25.1, Dialysis:
Imagineering. Kidney transplant: Kevin Somerville/
Imagineering.

Chapter 26   26.1–26.3, 26.6, 26.9–26.14, 26.16,
26.19, 26.21, 26.22, 26.26, 26.27: Kevin Somerville.
26.4, 26.5, 26.15, Table 26.1, Prostatitis and Prostate
Cancer, 26.25: Imagineering. 26.23: John Gibb.

Chapter 27   27.7: Molly Borman. 27.11: DNA
Illustrations. 27.13: Imagineering.

Photo Credits

**All photos in *Principles of Human
Anatomy*, 13th edition, are by
Mark Nielsen with the following
exceptions:**

Chapter 1   Clinical Connection, p. 3 (left):
Gary Conner/Phototake; Clinical Connection, p. 3
(center left): ©La/Bc.Aigo/Phototake; Clinical
Connection, p. 3 (center right): Jose Luis Pelaez/
Getty Images, Inc.; Clinical Connection, p. 3 (right):

Malvina Mendil/Photo Researchers Inc.; Figure 1.4a,
b, c: Dissection Shawn Miller; Photograph Mark
Nielsen; Figure 1.6c, d: Dissection Shawn Miller;
Photograph Mark Nielsen; Figure 1.7d: Dissection
Shawn Miller; Photograph Mark Nielsen;
Figure 1.8a: Andy Washnik; Table 1.3a: Warwick G./
Science Source/Photo Researchers, Inc.; Table 1.3b:
Breast Cancer Unit, Kings College Hospital,
London/Photo Researchers, Inc.; Table 1.3c: Zephyr
Photo Researchers, Inc.; Table 1.3d: Cardiothoracic
Centre, Freeman Hospital, Newcastle upon Tyne/
Science Source; Table 1.3e: CNRI/Science Photo
Library/Photo Researchers, Inc.; Table 1.3f: Science
Photo Library/Photo Researchers, Inc.; Table 1.3g:
Scott Camazine/Photo Researchers, Inc.; Table 1.3h:
Living Art Enterprises/Science Source/Photo
Researchers, Inc.; Table 1.3i: ISM/Phototake;
Table 1.3j: Andrew Joseph Tortora and Damaris
Soler; Table 1.3k: Department of Nuclear Medicine,
Charing Cross Hospital /Photo Researchers, Inc.;
Table 1.3l: Publiphoto/Photo Researchers, Inc.;
Table 1.3m: Dept. of Nuclear Medicine, Charing
Cross Hospital/Photo Researchers, Inc.; Table 1.3n:
Camal/Phototake.

Chapter 2   Opener: Montage created
by Mark Nielsen; Opener 2.1: Don W.
Fawcett/Photo Researchers, Inc.; Opener 2.2:
p. Motta/Photo Researchers Inc.; Opener
2.3: Courtesy Michael Ross, University of
Florida; Figure 2.5b, c: Omikron/Photo
Researchers, Inc.; Figure 2.6a, b: Albert Tousson/
Phototake; Figure 2.6c: Alexey Khodjakov/
Photo Researchers; Figure 2.7c: Don W.
Fawcett/Photo Researchers, Inc.; Figure 2.8b:
P. Motta/Photo Researchers, Inc.; Figure 2.8c:
Don W. Fawcett/Photo Researchers, Inc.;
Figure 2.10b: D. W. Fawcett/Photo Researchers,
Inc.; Figure 2.11b: Biophoto Associates/Photo
Researchers, Inc.; Figure 2.13b: Dr. Gopal Murti/
Phototake; Figure 2.14b: Don W. Fawcett/Photo
Researchers, Inc.; Figure 2.15c: D. W. Fawcett/
Photo Researchers, Inc.; Figure 2.18a, b, c, d, e,
f: Courtesy Michael Ross, University of Florida;
Clinical Connection, p. 43: U.S. Department of
Energy Genome Programs; Clinical Connection,
p. 52: Steve Gschmeissner/Science Photo Library.

Chapter 3   Table 3.9a: Courtesy Michael Ross,
University of Florida; Clinical Connection, p. 77:
Pascal Goetgheluck/Photo Researchers, Inc.

Chapter 4   Opener: Micrograph photos by
Mark Nielsen. Embryo provided courtesy of Gary C.
Schoenwolf, University of Utah School of Medicine;
Figure 4.2b: Don W. Fawcett/Photo Researchers,
Inc.; Figure 4.2c: Myriam Wharman/Phototake;
Figure 4.11b: Preparation and photography Mark
Nielsen; Figure 4.14a, g, h: Photo provided courtesy
of Kohei Shiota, Congenital Anomaly Research
Center, Kyoto University, Graduate School of
Medicine; Figure 4.14b, c, d, e: Courtesy National
Museum of Health and Medicine, Armed Forces
Institute of Pathology; Figure 4.14f: Lennart
Nilsson/Scanpix Sweden AB; Clinical Connection,
p. 97: Steve Gschmeissner/Photo Researchers, Inc.;
Clinical Connection, p. 98: Dominic Lipinski/PA
Photos/Landov.

Chapter 5 Figure 5.1b: Courtesy Michael Ross, University of Florida; Figure 5.1c: David Becker/Photo Researchers, Inc.; Figure 5.1d: Andrew J. Kuntzman; Figure 5.5b: VVG/Science Photo Library/Photo Researchers, Inc.; Clinical Connection, p. 124 (right): Biophoto Associates/Photo Researchers, Inc.; Clinical Connection, p. 124 (left): Alain Dex/Photo Researchers, Inc.; Clinical Connection, p. 133 (left): Sheila Terry/Science Photo Library/Photo Researchers, Inc.; Clinical Connection, p. 133 (center, right): St. Stephen's Hospital/Science Photo Library/Photo Researchers, Inc.

Chapter 6 Figure 6.3 (center): SPL/Science Source/Photo Researchers, Inc.; Figure 6.3 (left, right): Steve Gschmeissner/Science Source Images; Figure 6.7b: Scott Camazine/Photo Researchers, Inc.; Figure 6.8a: The Bergman Collection; Table 6.2a, c, d, e: Courtesy Dr. Brent Layton; Table 6.2b: Courtesy Per Amundson, M.D; Table 6.2f: Watney Collection/Phototake; Clinical Connection, p. 148: CNRI/Photo Researchers, Inc.; Clinical Connection, p. 155: P. Motta/Photo Researchers, Inc.

Chapter 7 Figure 7.13d: Dissection Shawn Miller, Photograph Mark Nielsen; Figure 7.19b: Mark Nielsen and Shawn Miller; Clinical Connection, p. 195 (left): Princess Margaret Rose Orthopaedic Hospital/Photo Researchers, Inc.; Clinical Connection, p. 195 (center): Dr. P. Marazzi/Photo Researchers, Inc.; Clinical Connection, p. 195 (right): Custom Medical Stock Photo, Inc.

Chapter 8 Clinical Connection p. 232: Neil Borden/Living Art Enterprises, LLC/Photo Researchers, Inc.

Chapter 9 Figure 9.1d: Dissection Shawn Miller; Photograph Mark Nielsen; Figure 9.11d: Dissection Shawn Miller, Photograph Mark Nielsen; Figure 9.12d: Dissection Shawn Miller, Photograph Mark Nielsen; Figure 9.14e: Dissection Shawn Miller, Photograph Mark Nielsen; Figure 9.15d: Dissection Shawn Miller, Photograph Mark Nielsen; Figure 9.15g: Dissection Shawn Miller, Photograph Mark Nielsen; Figure 9.15h: Dissection Shawn Miller, Photograph Mark Nielsen; Clinical Connection, p. 262 (left): CNRI/Photo Researchers, Inc.; Clinical Connection, p. 262 (right): Barts Medical Library/Phototake; Clinical Connection, p. 280 (top center): SIU BioMed/Custom Medical Stock Photo, Inc.; Clinical Connection, p. 280 (top right): ISM/Phototake; Clinical Connection, p. 280 (bottom right): Scott Camazine/Phototake

Chapter 10 Table 10.1: Courtesy Denah Appelt and Clara Franzini-Armstrong; Figure 10.5: Courtesy Hiroyouki Sasaki, Yale E. Goldman and Clara Franzini-Armstrong; Figure 10.6d: Don Fawcett/Photo Researchers, Inc.; Table 10.4: Biophoto Associates/Photo Researchers; Clinical Connection, p. 300: Custom Medical Stock Photo, Inc.

Chapter 11 Figure 11.6b: Dissection Shawn Miller, Photograph Mark Nielsen; Figure 11.7b: Dissection Shawn Miller, Photograph Mark Nielsen; Figure 11.9d: Dissection Shawn Miller, Photograph Mark Nielsen; Figure 11.10b, d: Dissection Shawn Miller, Photograph Mark Nielsen; Figure 11.12b: Dissection Nathan Mortensen and Shawn Miller; Photograph Mark Nielsen; Figure 11.13e, f, g, h: Dissection Nathan Mortensen and Shawn Miller; Photograph Mark Nielsen; Figure 11.14d: Dissection Shawn Miller, Photograph Mark Nielsen; Figure 11.14f: Dissection Shawn Miller and Nathan Mortensen; Photograph Mark Nielsen; Figure 11.16b: Dissection Nathan Mortensen; Photograph Mark Nielsen; Figure 11.17c: Dissection Nathan Mortensen and Shawn Miller; Photograph Mark Nielsen; Figure 11.19c, h: Dissection Shawn Miller, Photograph Mark Nielsen; Figure 11.19d: Dissection Nathan Mortensen and Shawn Miller; Photograph Mark Nielsen; Figure 11.21a, b, c, d, e: Andy Washnik; Figure 11.21h: Dissection Shawn Miller, Photograph Mark Nielsen; Figure 11.23b, d, f, h: Dissection Shawn Miller, Photograph Mark Nielsen; Figure 11.24d, e, i, j: Dissection Shawn Miller, Photograph Mark Nielsen; Figure 11.25d, e, f, i: Dissection Nathan Mortensen; Photograph Mark Nielsen.

Chapter 12 Figure 12.2a: Juergen Berger/Photo Researchers, Inc.; Figure 12.5a, b, c, d, e: Courtesy Michael Ross, University of Florida; Figure 12.6: © Dennis Kunkel Microscopy, Inc./Phototake; Clinical Connection, p. 421: Lewin/Royal Free Hospital/Photo Researchers, Inc.

Chapter 13 Figure 13.1c: Dissection Shawn Miller, Photograph Mark Nielsen; Figure 13.2b: Dissection Shawn Miller, Photograph Mark Nielsen; Figure 13.3b, d: Dissection Shawn Miller, Photograph Mark Nielsen; Figure 13.4b: Dissection Shawn Miller, Photograph Mark Nielsen; Figure 13.6c, f, g: Dissection Shawn Miller, Photograph Mark Nielsen; Figure 13.8c, d: Dissection Shawn Miller, Photograph Mark Nielsen; Clinical Connection, p. 447: Scott Camazine/Phototake; Clinical Connection, p. 450 (right): Carolina Biological Supply/Phototake; Clinical Connection, p 450 (left): Chuck Brown/Photo Researchers, Inc.; Clinical Connection, p.452: © ISM/Phototake.

Chapter 14 Clinical Connection, p. 464: Scott Camazine/Phototake; Figure 14.1d: Dennis Strete; Figure 14.1e: Courtesy Michael Ross, University of Florida; Figure 14.4b1, b2: Dissection Shawn Miller, Photograph Mark Nielsen; Figure 14.6c1, c2: Dissection Shawn Miller, Photograph Mark Nielsen; Figure 14.7d, e: Dissection Shawn Miller, Photograph Mark Nielsen; Figure 14.8d: Dissection Shawn Miller, Photograph Mark Nielsen; Figure 14.8e, f, g: Dissection Tory Meyer; Photograph Mark Nielsen; Figure 14.9c: Dissection Shawn Miller, Photograph Mark Nielsen; Figure 14.9d: Dissection Nathan Mortensen; Photograph Mark Nielsen; Figure 14.10b: Dissection Shawn Miller, Photograph Mark Nielsen; Figure 14.11c: Dissection Shawn Miller, Photograph Mark Nielsen; Figure 14.12d: Dissection Shawn Miller, Photograph Mark Nielsen; Figure 14.13b, c: Dissection Shawn Miller, Photograph Mark Nielsen; Figure 14.14c: Dissection Shawn Miller, Photograph Mark Nielsen; Figure 14.16c: Dissection Shawn Miller, Photograph Mark Nielsen.

Chapter 15 Figure 15.5c, d: Courtesy Michael Ross, University of Florida; Figure 15.6c: Lester Bergman & Associates; Figure 15.6d: Dissection Shawn Miller, Photograph Mark Nielsen; Figure 15.7a: Dissection Shawn Miller, Photograph Mark Nielsen; Figure 15.12b: Dissection Shawn Miller, Photograph Mark Nielsen.

Chapter 16 Clinical Connection, p. 548: Image Source/Getty Images, Inc.; Figure 16.8c, d: David M. Phillips/Photo Researchers, Inc.

Chapter 17 Figure 17.1b: Dissection Shawn Miller, Photograph Mark Nielsen; Table 17.1a, b, c, d: Dissection Shawn Miller, Photograph Mark Nielsen; Figure 17.2b: Dissection Chris Roach; Photograph Mark Nielsen; Figure 17.2c: Dissection Shawn Miller, Photograph Mark Nielsen; Figure 17.3b: Courtesy Michael Ross, University of Florida; Figure 17.4b: Thomas Deerinck, NCMIR/Photo Researchers, Inc.; Figure 17.5b: Dissection Shawn Miller, Photograph Mark Nielsen; Figure 17.6b: Dissection Shawn Miller, Photograph Mark Nielsen; Figure 17.7b: Dissection Richard Homer; Photograph Mark Nielsen; Figure 17.8c: Dissection Shawn Miller, Photograph Mark Nielsen; Figure 17.9c: Dissection Shawn Miller, Photograph Mark Nielsen; Clinical Connection, p. 566 (top): Bold Stock/Age Fotostock America, Inc.; Clinical Connection, p. 566 (center top, bottom): Masterfile; Clinical Connection, p. 566 (center bottom): Andersen Ross/Getty Images, Inc.

Chapter 18 Figure 18.2b: Dissection Shawn Miller, Photograph Mark Nielsen; Figure 18.3c: Dissection Shawn Miller, Photograph Mark Nielsen; Figure 18.5a: Dissection Shawn Miller, Photograph Mark Nielsen; Figure 18.8c: Dissection Shawn Miller, Photograph Mark Nielsen; Figure 18.9d: Dissection Shawn Miller, Photograph Mark Nielsen; Figure 18.13: From N. Gluhbegovic and T.H. Williams, *The Human Brain: A Photographic Guide*, Harper and Row, Publishers, Inc. Hagerstown, MD, 1980. Reproduced with permission; Figure 18.14c: Dissection Shawn Miller, Photograph Mark Nielsen; Figure 18.17a, b: Dissection Shawn Miller, Photograph Mark Nielsen; Table 18.2a, b, c, d, e, f, g, h: Dissection Shawn Miller, Photograph Mark Nielsen; Clinical Connection, p. 593: Steve Allen/Photo Researchers, Inc.; Clinical Connection, p. 616: Dissection Shawn Miller, Photograph Mark Nielsen; Clinical Connection, p. 617: Dissection Shawn Miller, Photograph Mark Nielsen; Clinical Connection, p. 619: Dissection Shawn Miller, Photograph Mark Nielsen; Clinical Connection, p. 621: Dissection Shawn Miller, Photograph Mark Nielsen; Clinical Connection, p. 622: Dissection Shawn Miller, Photograph Mark Nielsen; Clinical Connection, p. 623: Dissection Shawn Miller, Photograph Mark Nielsen; Clinical Connection, p. 624: Dissection Shawn Miller, Photograph Mark Nielsen; Clinical Connection, p. 625: Dissection Shawn Miller, Photograph Mark Nielsen; Clinical Connection, p. 626: Dissection Shawn Miller, Photograph Mark Nielsen; Clinical Connection, p. 627: Dissection Shawn Miller, Photograph Mark Nielsen.

Chapter 19 Figure 19.3b: Dissection Shawn Miller, Photograph Mark Nielsen; Clinical Connection, p. 647 (left): EH Stock/iStockphoto; Clinical Connection, p. 647 (right): P. Marazzi/Photo Researchers, Inc.

Chapter 21 Figure 21.1c: Courtesy Michael Ross, University of Florida; Figure 21.6: Paul Parker/Photo Researchers, Inc.; Figure 21.9: N. Gluhbegovic and T. H. Williams, *The Human Brain: A Photographic Guide*, Harper and Row, Publishers, Inc., Hagerstown, MD, 1980. Reproduced with permission; Figure 21.11b, c: Dissection Mark Nielsen; Photograph Mark Nielsen; Figure 21.12d: Dissection Mark Nielsen; Photograph Mark Nielsen; Figure 21.13e: Carolina Biological Supply Company/Phototake, Inc.; Clinical Connection, p. 682: Cordelia Molloy/Photo Researchers, Inc.

Chapter 22   Figure 22.2d1, d2, d3, d4, d5: Courtesy James Lowe, University of Nottingham, Nottingham, United Kingdom; Figure 22.4c, d: Dissection Shawn Miller, Photograph Mark Nielsen; Figure 22.6c: Dissection Shawn Miller, Photograph Mark Nielsen.

Chapter 23   Figure 23.1b: Dissection Shawn Miller, Photograph Mark Nielsen; Figure 23.2d: Dissection Shawn Miller, Photograph Mark Nielsen; Figure 23.5: Dissection Shawn Miller, Photograph Mark Nielsen; Figure 23.6b: Dissection Shawn Miller, Photograph Mark Nielsen; Figure 23.6c, d: Dissection Shawn Miller, Photograph Mark Nielsen; Figure 23.6e: Courtesy Michael Ross, University of Florida; Figure 23.7: Dissection Shawn Miller, Photograph Mark Nielsen; Figure 23.8f, g: Dissection Shawn Miller, Photograph Mark Nielsen; Figure 23.10b: Biophoto Associates/Science Source; Clinical Connection, p. 753: Juergen Berger/Photo Researchers, Inc.

Chapter 24   Figure 24.1b: Dissection Shawn Miller, Photograph Mark Nielsen; Figure 24.3e, f: Dissection Shawn Miller, Photograph Mark Nielsen; Figure 24.5c: Dissection Shawn Miller, Photograph Mark Nielsen; Figure 24.8b: Dissection Shawn Miller, Photograph Mark Nielsen; Figure 24.10b: Dissection Shawn Miller, Photograph Mark Nielsen;

Figure 24.11b (inset): Steve Gschmeissner/Photo Researchers, Inc.; Figure 24.12d, e: Dissection Shawn Miller, Photograph Mark Nielsen; Figure 24.13b: Dissection Shawn Miller, Photograph Mark Nielsen; Figure 24.17: Dissection Shawn Miller, Photograph Mark Nielsen; Figure 24.18c, d: Courtesy Michael Ross, University of Florida; Figure 24.19c: Dissection Shawn Miller, Photograph Mark Nielsen; Figure 24.20c, d: Courtesy Michael Ross, University of Florida; Clinical Connection, p. 797: David M. Martin/Photo Researchers, Inc.

Chapter 25   Opener: Dissection Shawn Miller, Photograph Mark Nielsen; Figure 25.1b, c: Dissection Shawn Miller, Photograph Mark Nielsen; Figure 25.2c: Dissection Shawn Miller, Photograph Mark Nielsen; Figure 25.3b: Mark Nielsen and Shawn Miller; Figure 25.6b: Dennis Strete; Figure 25.9b: Dissection Shawn Miller, Photograph Mark Nielsen; Clinical Connection, p. 816: BSIP/Phototake; Clinical Connection, p. 820: Ian Hooton/Photo Researchers, Inc.

Chapter 26   Figure 26.1b: Dissection Shawn Miller, Photograph Mark Nielsen; Figure 26.3b: Dissection Shawn Miller, Photograph Mark Nielsen; Figure 26.3c: Dissection Shawn Miller, Photograph Mark Nielsen; Figure 26.9b: Dissection Shawn Miller, Photograph Mark

Nielsen; Figure 26.10c: Dissection Shawn Miller, Photograph Mark Nielsen; Figure 26.11b: Dissection Shawn Miller, Photograph Mark Nielsen; Figure 26.12b: Dissection Shawn Miller, Photograph Mark Nielsen; Figure 26.16b: Dissection Shawn Miller, Photograph Mark Nielsen; Figure 26.17c: Steve Gschmeissner/Science Source/Photo Researchers, Inc.; Figure 26.23c: Dissection Shawn Miller, Photograph Mark Nielsen; Clinical Connection, p. 860 (center left): Photo Researchers, Inc.; Clinical Connection, p. 860 (bottom left): Biophoto Associates/Photo Researchers, Inc.; Clinical Connection, p. 860 (top right): Dennis Kunkel Microscopy, Inc./Phototake; Clinical Connection, p. 860 (center right): Institut Pasteur/Photo Researchers, Inc.; Clinical Connection, p. 860 (bottom right): Phototake.

Chapter 27   Figure 27.3: Geirge Diebold/Getty Images, Inc.; Figure 27.5: Reprinted with permission of John Wiley & Sons, Inc.

Icons   Question Mark Icon: Kristina Velickovic/iStockphoto; Check Mark Icon: Kutsal Lenger/iStockphoto; Life Cycle Icon: DonDesigns/iStockphoto; Key Icon: Marco Grassilli/iStockphoto; Clinical Connection Icon: JPL Design/Shutterstock; Smartphone photo icon: Oleksiy Mark/iStockphoto.

INDEX

The letter *t* following a page number denotes a table; *f* denotes a figure; *e* denotes an exhibit.

A

A band, 291, 291*t*
Abdomen:
    autonomic plexuses in, 638*f*–639*f*
    lymph nodes of, 534*e*–535*e*
    muscles of, 346*e*–349*e*
    surface features of, 881*e*–883*e*
    veins of, 500*e*–501*e*
Abdominal aorta, 468*e*, 479*e*–484*e*, 881*e*
Abdominal cavity, 14
Abdominal inguinal ring, 836
Abdominal thrust maneuver, 754
Abdominal viscera arterioles, 649*t*
Abdominopelvic cavity, 14–16, 14*f*
Abdominopelvic quadrants, 19, 19*f*
Abdominopelvic regions, 18–19, 18*f*
Abducens (VI) nerve, 597, 618*e*, 619*e*, 628*t*, 697
Abduction, 256, 256*f*, 258*t*, 359*e*, 384*e*
Abductor digiti minimi muscle, 383*e*, 384*e*, 407*e*, 408*e*
Abductor hallucis muscle, 407*e*, 408*e*
Abductor muscles, 317*t*
Abductor pollicis brevis muscle, 383*e*, 384*e*
Abductor pollicis longus muscle, 376*e*
Abnormal curves of vertebral column, 195
ABO blood grouping system, 423
Abortion, 863
Abrasion, 136
Abscess, 161, 629, 743, 768
Absorption:
    defined, 62, 813
    and digestion, 757, 758
    and skin, 132, 133
    in small intestine, 789
Absorptive cells, 785, 790*t*
Abstinence, 861, 861*t*
Accessory digestive organs, 758
Accessory duct, 777
Accessory hemiazygos vein, 497*e*
Accessory ligaments, 253
Accessory (XI) nerve, 596, 626*e*, 628*t*, 697
Accessory reproductive glands, 836–838, 837*f*
Accessory sex glands, 826
Accessory structures of eye, 676–678, 677*f*
Accomodation, 619*e*
ACE (angiotensin converting enzyme), 512
Acetabular labrum, 274*e*
Acetabular notch, 232*e*
Acetabulum, 231*e*, 232*e*, 235*t*, 274*e*
Acetylcholine (ACh), 294, 549, 614, 645
Acetylcholine receptors, 294, 296
Acetylcholinesterase (AChE), 296, 548, 646
ACh, *see* Acetylcholine
Achilles tendon, 400*e*, 893*e*
Aching pain, 657

Achondroplasia, 160
Achondroplastic dwarfism, 160
Acidophils, 709
Acid phosphatase, in prostate, 836
Acinar glands, 69
Acini, 718, 778
ACL (anterior cruciate ligament), 275*e*
Acne, 129
Acoustic nerve, 623*e*
Acquired immunodeficiency syndrome (AIDS), 538–539
Acromegaly, 160, 710
Acromial end of clavicle, 216
Acromial region, 885*e*. *See also* Shoulder
Acromioclavicular joints, 216*e*, 218*e*, 264*t*, 885*e*
Acromion, 218*e*, 879*e*, 885*e*
Acrosomal reaction, 94
Acrosome (sperm), 94, 833
ACTH, *see* Adrenocorticotropic hormone
Actin, 33, 292, 292*t*, 296
Actions, muscle, 317*t*
Action potentials, 87, 285–286, 296, 447, 545
Active processes, 30
Active transport, 30, 33*t*
Acupoints, 669
Acupuncture, 556, 669
Acute (term), 425
Acute leukemia, 425
Acute lymphoblastic leukemia (ALL), 425
Acute myelogenuous leukemia (AML), 425
Acute normovolemic hemodilution, 427
Acute pancreatitis, 778
Acute pericarditis, 435
Acute prostatitis, 838
Acute renal failure (ARF), 815
AD (Alzheimer's disease), 614
Adam's apple, 734, 877*e*
Adaptation, sensory receptors, 654
Adaptive (specific) immunity, 517
Addison's disease, 717
Adduction, 256, 256*f*, 258*t*, 359*e*, 384*e*
Adductor brevis muscle, 393*e*, 394*e*
Adductor compartment, thigh, 393*e*, 394*e*
Adductor hallucis muscle, 407*e*, 408*e*
Adductor hiatus, 393*e*
Adductor longus muscle, 393*e*, 394*e*, 890*e*
Adductor magnus muscle, 393*e*, 394*e*
Adductor muscles, 317*t*
Adductor pollicis muscle, 383*e*, 384*e*
Adductor tubercle, 237*e*
Adenitis, 540
Adenocarcinomas, 747, 748
Adenohypophysis (anterior pituitary gland), 708–710, 709*t*, 722
Adenoids, 527, 732
Adenomas, 53
Adenosine triphosphate (ATP), 30, 296
ADH, *see* Antidiuretic hormone
Adherens junctions, 60–61

Adhesions, 88
Adhesion belts, 61
Adhesion proteins, 73
Adhesiotomy, 88
Adipocytes, 72, 76*t*
Adipose capsule, 804
Adipose tissue, 72, 76*t*, 648*t*, 722*t*
Adolescents, bone growth in, 152–154, 153*f*, 154*f*
Adrenal cortex, 715, 717*t*, 722
Adrenal glands, 715–718, 716*f*, 717*t*
Adrenal medulla, 642, 648*t*, 715, 717, 717*t*, 722
Adrenergic neurons, 645*f*, 646
Adrenergic receptors, 645*f*, 646
Adrenocortical insufficiency, 717
Adrenocorticotropic hormone (ACTH), 709, 709*t*
Adult oligopotent stem cells, 97
Adult rickets, 147
Adventitia, 737, 761, 771, 818, 819
Aerobic exercise, 449
Afferent arterioles, 808
Afferent lymphatic vessels, 523
Afferent (sensory) neurons, 544, 650
Afferent vessels, 463
Afterbirth, 106, 510
Age:
    fertilization, 113
    gestational, 113
Age-related macular disease (AMD), 682
Age spots, 123
Ageusia, 624*e*, 701
Agglutination, 423
Aggregated lymphatic follicles (Peyer's patches), 527, 785
Aging:
    and blood vessels, 511–512
    and bone tissue, 159, 160*t*
    and cells, 52–54
    and digestive system, 797
    and endocrine system, 723
    and integumentary system, 136
    and joints, 279–280
    and lymphatic system, 538
    and muscular tissue, 307
    and nervous system/brain, 614
    and reproductive systems, 865–866
    and respiratory system, 754
    and skin cancer, 124
    and skull, 192–193
    and special senses, 701
    and tissues, 87, 88
    and urinary system, 823
    and vertebral column, 199
Agonists (neurotransmitters), 647
Agonist muscles, 315
Agranular leukocytes, 423, 424, 427*t*
AIDS (acquired immunodeficiency syndrome), 538–539
AIIS (anterior inferior iliac spine), 231*e*
Ala, 231*e*, 876*e*
Albinism, 123
Albinos, 123
Albumins, 415, 417*t*
Alcohol, 814

Aldosterone, 512, 715
Alertness, 599
Alimentary canal, 758. *See also* Gastrointestinal (GI) tract
ALL (acute lymphoblastic leukemia), 425
Allantois, 102
Allergens, 519
Allergic reactions, 519
Alopecia, 127
Alpha cells, 718
α-Dextrinase, 789, 791*t*
Alpha motor neurons, 659
Alpha (α) receptors, 646
ALS (amytrophic lateral sclerosis), 666
Alveolar ducts, 744, 747*t*
Alveolar glands, 69
Alveolar macrophages, 746
Alveolar process, 184*e*, 185*e*
Alveolar sac, 745
Alveolar wall, 746
Alveoli:
    of lungs, 744–746, 745*f*, 747*t*
    of mammary glands, 857
    for teeth, 184*e*, 185*e*
Alzheimer's disease (AD), 614
Amacrine cells, 680
Amblyopia, 324*e*, 701
AMD (age-related macular disease), 682
Amenorrhea, 859
Amimidex®, 855
Aminopeptidase, 789, 791*t*
AML (acute myelogenuous leukemia), 425
Amniocentesis, 98, 113
Amnion, development of, 98
Amniotic cavity, 98
Amniotic fluid, 98, 113, 198, 822
Amphiarthrosis, 249
Ampulla:
    of ductus deferens, 835
    of ear, 690, 695
    hepatopancreatic, 777, 784
    of uterine tube, 847
    of Vater, 777
Ampullary nerves, 690
Amygdala, 609
Amytrophic lateral sclerosis (ALS), 666
Anabolic steroids, 302
Anabolism, 9
Anaerobic (term), 300, 421
Anal canal, 793
Anal columns, 793
Analgesia, 657
Analgesic nephropathy, 823
Anal sphincters, 356*e*, 793
Anal triangle, 356*e*, 854, 884*e*
Anaphase (mitosis), 47, 47*t*
Anaphylactic reactions, 519
Anaphylactic shock, 519
Anaplasia, 54
Anastomoses, 445, 461
Anastomotic veins, 463
Anatomical neck of humerus, 220*e*
Anatomical position, 9, 10*f*
Anatomical snuffbox, 888*e*

EPONYMS USED
IN THIS TEXT

An **eponym** is a term that includes reference to a person's name; for example, you may be more familiar with *Achilles tendon* than you are with the technical, but correct, term *calcaneal tendon.* Because eponyms remain in frequent use, this glossary has been prepared to indicate which current terms have been used to replace eponyms in this book. In the body of the text, eponyms are cited in parentheses or immediately following the current terms where they are used for the first time in a chapter or later in the book. In addition, although eponyms are included in the index, they have been cross-referenced to their current terminology.

EPONYM	CURRENT TERMINOLOGY	EPONYM	CURRENT TERMINOLOGY
Achilles tendon	calcaneal tendon	loop of Henle (HEN-lē)	nephron loop
Adam's apple	thyroid cartilage	Meibomian (mī-BŌ-mē-an) gland	tarsal gland
ampulla of Vater (VA-ter)	hepatopancreatic ampulla	Meissner (MĪS-ner) corpuscle	corpuscle of touch
		Merkel (MER-kel) disc	type I cutaneous mechanoreceptor
Bartholin's (BAR-tō-linz) gland	greater vestibular gland		
Billroth's (BIL-rōtz) cord	splenic cord	Müllerian (mil-E-rē-an) duct	paramesonephric duct
Bowman's (BŌ-manz) capsule	glomerular capsule		
Bowman's (BŌ-manz) gland	olfactory gland	Nissl (NIS-l) bodies	chromatophilic substances
Broca's (BRŌ-kaz) area	motor speech area		
Brunner's (BRUN-erz) gland	duodenal gland	organ of Corti (KOR-tē)	spiral organ
bundle of His (HISS)	atrioventricular (AV) bundle		
		Pacinian (pa-SIN-ē-an) corpuscle	lamellated corpuscle
canal of Schlemm (SHLEM)	scleral venous sinus	Peyer's (PĪ-erz) patch	aggregated lymphatic follicle
circle of Willis (WIL-is)	cerebral arterial circle	plexus of Auerbach (OW-er-bak)	myenteric plexus
Cooper's (KOO-perz) ligament	suspensory ligament of the breast	plexus of Meissner (MĪS-ner)	submucosal plexus
		pouch of Douglas	rectouterine pouch
Cowper's (KOW-perz) gland	bulbourethral gland		
crypt of Lieberkühn (LĒ-ber-kūn)	intestinal gland	Rathke's (rath-KĒZ) pouch	hypophyseal pouch
		Ruffini corpuscle (roo-FĒ-nē)	type II cutaneous mechanoreceptor
duct of Rivinus (re-VĒ-nus)	lesser sublingual duct		
duct of Santorini (san'-tō-RĒ-nē)	accessory duct	Sertoli (ser-TŌ-lē) cell	sustentacular cell
duct of Wirsung (VĒR-sung)	pancreatic duct	Sharpey's (SHAR-pēz) fiber	perforating fiber
		Sheath of Schwann (SCHVON)	neurolemma
Eustachian (ū-STĀ-kē-an) tube	auditory tube	Skene's (SKĒNZ) gland	paraurethral gland
		sphincter of Oddi (OD-dē)	sphincter of the hepatopancreatic ampulla
Fallopian (fal-LŌ-pē-an) tube	uterine tube		
		Stensen's (STEN-senz) duct	parotid duct
gland of Littré (LĒ-trā)	urethral gland		
Golgi (GOL-jē) tendon organ	tendon organ	Volkmann's (FŌLK-manz) canal	perforating canal
Graafian (GRAF-ē-an) follicle	mature ovarian follicle		
		Wernicke's (VER-ni-kēz) area	auditory association area
Hassall's (HAS-alz) corpuscle	thymic corpuscle	Wharton's (HWAR-tunz) duct	submandibular duct
Haversian (ha-VĒR-shun) canal	central canal	Wharton's (HWAR-tunz) jelly	mucous connective tissue
Haversian (ha-VĒR-shun) system	osteon	Wormian (WER-mē-an) bone	sutural bone
Heimlich (HĪM-lik) maneuver	abdominal thrust maneuver		
interstitial cell of Leydig (LĪ-dig)	interstitial cell		
islet of Langerhans (LANG-er-hanz)	pancreatic islet		
Kupffer's (KOOP-ferz) cell	stellate reticuloendothelial cell		